The Hymenoptera of Costa Rica

The Natural History Museum, London

The Hymenoptera of Costa Rica

Edited by

PAUL E. HANSON

University of Costa Rica

and

IAN D. GAULD

The Natural History Museum, London

OXFORD UNIVERSITY PRESS

This book has been printed digitally and produced in a standard specification in order to ensure its continuing availability

OXFORD
UNIVERSITY PRESS

Great Clarendon Street, Oxford OX2 6DP

Oxford University Press is a department of the University of Oxford.
It furthers the UniversityÕs objective of excellence in research, scholarship, and education by publishing worldwide in

Oxford New York

Auckland Bangkok Buenos Aires Cape Town Chennai
Dar es Salaam Delhi Hong Kong Istanbul Karachi Kolkata
Kuala Lumpur Madrid Melbourne Mexico City Mumbai Nairobi
S‹o Paulo Shanghai Taipei Tokyo Toronto

Oxford is a registered trade mark of Oxford University Press in the UK and in certain other countries

Published in the United States by Oxford University Press Inc., New York

Reprinted 2003

ISBN 0-19-854905-9

Printed in Great Britain by
Antony Rowe Ltd., Eastbourne

*This work is dedicated to Dr Alvaro Wille,
pioneer in the study of Costa Rican
Hymenoptera and first Director of the
Museo de Insectos, Universidad de Costa Rica.*

Foreword

This book is a continuation of the tradition of collaboration between Costa Rica and the United Kingdom established by the publication of *Biología Centrali Americana* in 1883–1900. Its preparation involves not only personnel from the Universidad de Costa Rica and The Natural History Museum, but draws on a formidable battery of leading international experts from eight countries. Such collaboration is a prerequisite for the study of one of the largest and most diverse orders of insects in Costa Rica, a country which for its size, has a faunal diversity unsurpassed elsewhere on Earth. Given the unparalleled international interest in tropical biodiversity, this work could not have appeared at a more opportune time. From the perspective of the Universidad de Costa Rica, this book is a further valued contribution in a series of studies of the Costa Rican biota that includes volumes about mammals, birds, fishes and orchids. For The Natural History Museum *The Hymenoptera of Costa Rica* is further fruit of our policy of international collaboration, using the historical wealth of our collections and the global expertise of our staff to produce timely and authoritative studies. We warmly recommend this work.

Presentación

Este libro es un tributo a la tradición cooperativa entre Costa Rica y el Reino Unido, establecido por la publicación de *Biología Centrali Americana* en 1883–1900. Involucra no solo al personal de la Universidad de Costa Rica y el Museo de Historia Natural (Museo Británico), sino también a un equipo formidable de expertos internacionales de ocho países. Tal aporte es un prerequisito para el estudio de uno de los órdenes de insectos más grande y más diverso en Costa Rica, país que para su tamaño, tiene una variedad no sobrepasada en otro lugar en la Tierra. Dado el gran interés internacional en la biodiversidad tropical, este trabajo no podrá aparecer en un momento más propicio. Desde la perpectiva de la Universidad de Costa Rica, este libro es otra contribución valorada en una serie de estudios de la biota de Costa Rica que incluye volúmenes sobre mamiferos, pájaros, peces y orquídeas. Para el Museo de Historia Natural *The Hymenoptera of Costa Rica* es otro fruto de nuestra política de colaboración internacional, utilizando la riqueza histórica de nuestras colecciones y expertos mundiales para producir estudios de autoridad y oportunos. Nosotros recomendamos con gran satisfacción este trabajo.

Carlos Quesada Mateo
Vicerrector de Investigación
Universidad de Costa Rica
San José, Costa Rica

Neil R. Chalmers
Director
The Natural History Museum
London, U.K.

Preface

The conservation and sustainable utilization of biodiversity are major social and political issues of our time. Yet, as many authors have pointed out, we know woefully little about the organisms with which we share our planet. We have only the crudest estimates of actually how many species inhabit the Earth and we know even less of their interactions and their role in maintaining the delicate balance of life on which ultimately all species, including our own, depend. We do know, however, that the majority of terrestrial species inhabit the lush forests in tropical latitudes. Such forests are currently among the most threatened habitats on Earth and in many regions they are disappearing at an unprecedented rate. Consequently species of plants and animals that inhabit tropical forests are probably becoming extinct at a rate unparalleled in our planet's history. Most are lost before their role in the natural ecosystem, or their potential benefit to human society, has ever been assessed.

Most developing nations, beset as they are by economic and demographic problems, simply cannot afford to conserve tropical forests and the diversity of species they contain, unless these wilderness areas can be seen to be of financial benefit to society. But before biodiversity can be used in some way it must be catalogued and explored. Paradoxically, it is these tropical forests, the richest of terrestrial habitats, about which least is known. The information that is available is generally scattered throughout scientific journals—many of which are inaccessible to scientists in tropical countries—and the lacunae in our knowledge remain unexposed. For the most part, the basic faunistic overviews and syntheses have never been written. The principal aim of this book is to provide such an overview for one of the major orders of insects in one small tropical country. Our objectives have been to compile a book which will both provide a basis for further scientific investigation by workers throughout the region, and promote a greater understanding and appreciation of biodiversity.

Our chosen subject immediately poses two questions—why the Hymenoptera, and why Costa Rica? To address the former question first. Understanding the diversity of Hymenoptera is important for many reasons, and this topic will be discussed in considerable detail in the introductory chapters of this book, but here it is perhaps sufficient to say that these insects are of critical importance to humankind in two main ways—as allies and as adversaries. In the former role they are the single most important group of plant pollinators, and predators/parasitoids controlling the populations of other insects. As adversaries, leaf cutter ants are one of the most destructive groups of agricultural pests in the New World tropics and, as patio pests in our urban environment, ants, bees and wasps traumatize more people worldwide than do any other group of insects.

Costa Rica was chosen as the geographic focus for this work for one simple reason—it is the only tropical country for which a representative sample of the hymenopteran diversity is available to biologists for study. Building on a foundation laid by native Costa Ricans and others who have studied the hymenopterous fauna for a century or more, we undertook a Malaise trap survey of the entire country. Over 150 trap-years of samples have been processed from about 40 sites throughout the country, ranging from mangrove swamps at sea-level to paramo-like scrub at over 3000 metres on the highest mountains. Many of the specimens obtained were sent to contributing authors in eight countries, making this book an international collaborative effort. Synoptic collections of these specimens are being returned to Costa Rica and will form a reference base for future research by Costa Rican and other biologists. Augmenting our own collections are many others, including extensive material assembled by the efforts of the parataxonomists at INBio and kindly made available to us by the INBio staff.

Although this work is based on study of the Costa Rican fauna, the information is relevant to all of Mesoamerica from tropical Mexico to northern Colombia. The families discussed extend throughout all of tropical and subtropical America so that much of the text will be useful to workers from the southern United States south to Argentina and Chile.

Acknowledgements

We thank the following persons for looking after Malaise traps that have provided the material foundation for this book: Sr Abarca, Rafael Alberto Calderón, Carlos Chávez, William Eberhard, Roberto Espinoza, Luis Fournier, Miguel Gamboa, Francisco González, Virginia Herrero, Daniel Janzen, Jorge Arturo Lizano, Roberto Miranda, Edgar Quirós, Moises Pérez, María Salablanca, Doyle Santamaría, Federico Serrano Chávez, Francisco Serrano Chinchilla. We are also grateful to the assistants, students and miscellaneous persons at the Escuela de Biología (Universidad de Costa Rica) who helped separate the samples: Roberto Céspedes, Carolina Godoy, Gabriel Aguilar, Gerardo Rojas and others too numerous to mention. Humberto Lezama of the Museo de Insectos (Universidad de Costa Rica) helped organize the separation of other samples. In the UK Sondra Ward assisted in many ways, taking stereoscan photographs, handling coordination of the figures and generally doing the thousand and one things necessary to produce a book from a series of manuscripts. Pam Mitchell coordinated references, cross checked them with the text and checked them to source.

We thank all the contributing authors for the time and energy they have invested in their chapters, and the artists who did most of the whole insect drawings—Eugenia Mesén, Ana Rita Arguedas and David Morgan. Many unpublished observations were kindly made available by our colleagues and they are acknowledged in the text. Many of the chapters were reviewed by other people, to whom we are grateful: Ricardo Ayala (18), Barry Bolton (16), Brian Brown (2.7), Zdenek Bouček (11.14, 11.18), Gabriela Chavarría (18), Howard Evans (13.2), Don Feener (16), Gordon Frankie (18), Eric Grissell (11.18), Jeff Halstead (11.3), Tom Huddleston (12.2), Mohammad Hayat (11.2), Frank Kurczewski (17), John LaSalle (11), Jorge Lobo (2.2, 18), Bob Matthews (17), John Noyes (11.2) Frank Parker (17), Andy Polaszek (11.2), Scott Shaw (2.4), Roy Snelling (16), Dick Vane-Wright (2.7), Brad Vinson (2.4 and 2.5) and Jim Wetterer (16). We wish to extend special thanks to Bill Eberhard who read the entire manuscript and offered many helpful suggestions.

P.E.H. wishes to acknowledge financial support given by the Vicerectoría de Investigacion (Universidad de Costa Rica), Fulbright Foundation, the Royal Society, London, and British Airways. Thanks are also due to the Instituto Nacional de Biodiversidad de Costa Rica (INBio), The Natural History Museum (London), the U.S. National

Museum (Washington, D.C.) and Escuela Agrícola Panamericana (Zamorano, Honduras) for access to their collections.

I.D.G. wishes to acknowledge the facilities provided by the Escuela de Biología (Universidad de Costa Rica), and to thank Rodrigo Gámez, Jorge Jímenez and the staff of the Instituto Nacional de Biodiversidad de Costa Rica (INBio) for unfettered access to their collections. Access to other material in the American Entomological Institute, CATIE, the Museo Nacional de Costa Rica, and the Philadelphia Academy of Sciences is gratefully acknowledged, and colleagues at the NHM are thanked for their advice about so many matters.

And finally we would like to thank our wives, Carolina Godoy and Pam Mitchell, for still being our wives, despite the fact that the preparation of this book has consumed almost every waking moment (and some of our dreams) for the last two years.

London I.D.G
San Pedro de Montes de Oca P.E.H

Contents

Contributors

Denis J. Brothers Department of Zoology and Entomology, University of Natal, Pietermaritzburg 3200, SOUTH AFRICA

David Carmean Department of Entomology, University of California, Davis, California 95616, U.S.A.

James M. Carpenter American Museum of Natural History, Central Park West at 79th Street, New York, New York 10024-5192, U.S.A.

Lonny D. Coote Department of Entomology, Royal Ontario Museum, 100 Queen's Park, Toronto, Ontario M5S 2C6, CANADA

D. Christopher Darling Department of Entomology, Royal Ontario Museum, 100 Queen's Park, Toronto, Ontario M5S 2C6, CANADA

Gérard Delvare Centre de Cooperation Internationale en Recherche Agronomique pour le Developpement, Avenue du Val de Montferrand, BP 5035-34032 Montpellier cedex 1, FRANCE

Paul Dessart Institut Royal des Sciences Naturelles de Belgique, 31 Rue Vautier, B-1040 Bruxelles, BELGIUM

Albert Finnamore Provincial Museum of Alberta, 12845 102nd Avenue, Edmonton, Alberta T5N 0M6, CANADA

Nigel D.M. Fergusson Department of Entomology, The Natural History Museum, Cromwell Road, London SW7 5BD, UNITED KINGDOM

Kevin J. Gaston Department of Entomology, The Natural History Museum, Cromwell Road, London SW7 5BD, UNITED KINGDOM

Ian D. Gauld Biodiversity Division, The National History Museum, Cromwell Road, London SW 75BD, UNITED KINGDOM

Gary A.P. Gibson Biosystematic Research Institute, KW Neatby Building, CEF, Agriculture Canada, Research Branch, Ottawa, Ontario K1A 0C6, CANADA

Terry L. Griswold USDA, Bee Biology & Systematics Lab, Utah State University, Logan, Utah 84322-5310, U.S.A.

Paul E. Hanson Escuela de Biologia, Universidad de Costa Rica, Ciudad Universitaria, San Pedro de Montes de Oca, COSTA RICA

John M. Heraty Systematic Entomology Laboratory, National Museum of Natural History Smithsonian Institution, NHB-165 Washington D.C, 20560, U.S.A

Mike Huben c/o Museum of Comparative Zoology, Harvard University, Cambridge, Massachussetts, U.S.A.

John T. Huber Biosystematic Research Institute, KW Neatby Building, CEF, Agriculture Canada, Research Branch, Ottawa, Ontario K1A 0C6, CANADA

Lynn S. Kimsey R.M. Bohart Museum of Entomology, Department of Entomology, University of California, Davis, California 95616, U.S.A.

John LaSalle CABI International Institute of Entomology, 56 Queen's Gate, London 5W7 5JR, UNITED KINGDOM

John T. Longino The Evergreen State College, Olympia, Washington 98505, U.S.A.

Lubomir Masner Biosystematic Research Institute, KW Neatby Building, CEF, Agriculture Canada, Research Branch, Ottawa, Ontario K1A 0C6, CANADA

Arnold Menke Systematic Entomology Laboratory, PSIARS, USDA, c/o National Museum of Natural History, Washington, D.C. 20560, U.S.A.

John S. Noyes Department of Entomology, The Natural History Museum, Cromwell Road, London SW7 5BD, UNITED KINGDOM

Massimo Olmi Dipartimento di Protezione delle Piante, Universita degli Studi della Tuscia, Via S. Camillo de Lellis, 01100 Viterbo, ITALY

Frank D. Parker c/o Bee Biology of Systematics Lab, Utah State University, Logan, Utah 84322–5310, U.S.A

John D. Pinto Department of Entomology–41, University of California, Riverside, California 92521-0314, U.S.A.

William Ramirez B. Escuela de Fitotecnia, Universidad de Costa Rica, San Pedro de Montes de Oca, San José, COSTA RICA

Michael E. Schauff Systematic Entomology Laboratory PSI, ARS, USDA, c/o National Museum of Natural History, MRC-168, Washington D.C. 20560, U.S.A.

Scott R. Shaw Department of Plant, Soil and Insect Sciences, College of Agriculture, P.O. Box 3354, University of Wyoming, Laramie, Wyoming 82071-3354, U.S.A.

David R. Smith Systematic Entomology Laboratory, PSI, ARS U.S. Department of Agriculture c/o National Museum of Natural History, MRC-168, Washington, D.C. 20560, U.S.A.

Marius S. Wasbauer P.O. Box 6820, Brookings, Oregon 97415, U.S.A

Mary Jane West-Eberhard Smithsonian Tropical Research Institute, Apdo 2072, Balboa, PANAMA

1 Introduction

Ian D. Gauld

With more than 115,000 described species, ranging from diminutive 'fairy-flies' which are less than 0.2 mm in length and develop within the single egg of a leafhopper, to colossal tarantula-hawks—spider-hunting wasps with a wingspan of more than ten centimetres—the Hymenoptera is one of the largest and most diverse of all insect orders. However, the variety and composition of the group often remains unappreciated, a fact that is perhaps best reflected by their lack of a common name in either the English or Spanish languages. The Coleoptera are beetles, the Diptera flies and the Lepidoptera butterflies and moths, yet the great variety of predatory, parasitic and phytophagous hymenopterous insects that comprise the other hyperdiverse insect order share no common name. Most are generally overlooked, although some social species—certain bees, ants and wasps—are perhaps the most familiar of all insects, and have long fascinated humankind. For millenia diverse human cultures have marvelled at the complex societies of these insects and incorporated their attributes into folklore and fable. More than two and a half thousand years ago, in Greece, Aesop applauded the industriousness of the ant and contrasted its virtues with the indolence of grasshoppers, whilst in pre-Columbian Mexico red ants were involved in mystic beliefs of the Teotihuacan culture and the Mayans worshipped a bee god, *Ah Mucan Cab*. The ferocity of some social Hymenoptera has impressed people for generations. Ancient Greeks used images of wasps on their shields (Davies & Kathirithamby, 1986), and they have had martial connotations in Europe ever since, where as recently as the second world war, a '*Wespe*' (German for wasp) was an armoured fighting vehicle. Wasps are similarly regarded in tropical America, where the Yukpa indians use them in rituals following male births because bravery is believed to be transmitted from wasp to child (Ruddle, 1973). In a similar manner Brazilian indians deliberately provoked stings from the ferocious bullet ant (*Paraponera clavata*) as part of the ceremonies for initiating young boys into manhood (Bequaert, 1926). Amongst the Bribri indians in Costa Rica an extensive folk taxonomy exists for wasps and ants, and several species feature in myths. To encourage industriousness, hunting abilities and ferocity in defence of territory, elders bound a wasp with cotton and tied it to the back of a pregnant woman for a night. The Bribri also believed that the aggressive and armoured Spanish conquerors shared a common ancestry with wasps (Starr & Bozzoli de Wille, 1990). An unusual variation of the theme of 'wasp characteristic acquisition' occurs in Colombia where Sierra indians sting their dogs with digger-wasps (sphecids) in the belief this will help the dogs run prey into their dens (W. Eberhard, pers. comm.).

Not all human interactions with hymenopterous insects have been cultural. Many have been, and still are of a commercial or economic nature as Hymenoptera are both adversaries and allies of humankind. Perhaps nowhere is this adversarial interaction more noticeable than in the New World tropics. The ubiquitous leaf-cutting ants are a dominant group of herbivores that have devastated crops from pre-Columbian times to the present day. These insects remove more green plant tissue than any other group of comparable diversity, including both mammals or caterpillars (Cherrett *in* Lofgren & Vander Meer, 1986), and such is their impact on people that the early Portuguese settlers dubbed Brazil 'the kingdom of the ants' (Hölldobler & Wilson, 1990*a*). Several other ant species are commonly found in association with human dwellings, and although most are a relatively minor nuisance, today in the electronic age some cause serious problems. For example, in Costa Rica a species of *Camponotus* has recently damaged hundreds of FAX machines by constructing nests inside the machine casings.

Most people are aware that the social hymenopterans can inflict a painful sting, but few realize that the familiar honey-bee is, at least statistically, one of the most dangerous animals they are likely to encounter. Although bee stings usually only cause minor discomfort, in a very few instances they may provoke a serious, even fatal response, and in the United States more people are killed annually by honey-bee stings than are killed by all other venomous animals, rattlesnakes included! It is not surprising therefore, when a new race of the honey-bee, the notorious Africanized Bee, spread from Brazil northwards throughout Central America, its advent was greeted with widespread alarm by the media who dubbed it the 'killer bee'. It should be noted, however, that the venom of the 'killer bee' is no more potent than that of the normal honey-bee (Schumacher *et al.*, 1989); the difference is the greater propensity of the Africanized variety to sting.

Whilst it is most commonly the social Hymenoptera that are adversarial some solitary species are also pests. The phytophagous sawflies can be plant pests as their larvae consume foliage, though there are no tropical examples that reach the scale of devastation these insects have caused in temperate ecosystems, where today some species defoliate entire conifer plantations, and in the seventeenth century others caused widespread famine (Gauld *et al.*, 1990). Diprionids, however, do cause damage in Central American pine forests (Wilkinson & Drooz, 1979) and in Brazil the larvae of a pergid sawfly, which can be common in pastures, are sometimes consumed by and poison cattle (Camargo, 1956). Although it is possible to cite many other adverse interactions between Hymenoptera and human beings (see Chapter 3), it is usually not adversarial interactions that the order is best known for. No other order includes so many beneficial species, and it is this feature of the group that has probably stimulated most human investigation and study, from prehistoric times up to the present day.

Wherever they occur, in the Old or the New World, social bees have long been exploited for the provisions they store in their nests. A nine thousand year old cave painting in Spain depicts the gathering of wild honey, and native stingless-bees were kept in horizontal log hives by the Mayans in Belize as early as 300 BC (Crane & Graham, 1985). In addition to its use as food, stingless-bee honey has been used by indigenous American peoples to treat wounds and ulcers (Schwarz, 1948; Chittampalli & Mulcahy, 1990). In the Old World, honey-bee honey had similar uses, but another honey-bee product, beeswax, became a major commodity. It was widely used in mediaeval Europe for candle making and it was possibly Europe's almost insatiable demand for beeswax that prompted the introduction of the honey-bee into the New World in the sixteenth century. Although honey from stingless-bees is still harvested in some parts of Central America, during the last two hundred years the introduced honey-bee has dominated apiculture in the region. By 1800 Spain imported more than a hundred thousand dollars worth of beeswax annually from Cuba (Gauld *et al.*, 1990) and honey-bee keeping has become a major rural industry throughout much of Latin America. In Mexico and Central America the average hive density is 1.3 per square kilometre, higher than any other area except Europe, and in the last decade Mexico was the world's largest exporter of honey-bee commodities (Crane, 1990).

In addition to their use for the production of commodities, bee populations have been manipulated for agricultural purposes, because they are vitally important as plant pollinators—transferring pollen from stamens to stigmas and thus facilitating fertilization. A large number of human food crops, such as beans, tomatoes, melons, citrus, macadamia and coffee, are pollinated by bees (Free, 1970) and although some of these plants may self-pollinate to some extent, bees are necessary to produce economic yields in many agroecosystems (Corbet, 1987).

Bees are not the only social hymenopterous insects whose nests have been moved by humans for agricultural purposes. For millenia, farmers have been aware that ants are major predators of many arthropod pests, and more than 3000 years ago the Chinese transported predaceous ant's nests from woodlands into citrus groves (Groff & Howard, 1924; Liu, 1939; Huang & Yang, 1987). Similarly, in the western hemisphere the Kayapó Indians of Brazil regularly introduce nests of *Azteca* ants to protect their mixed crops against herbivorous insects (Overal & Posey, 1990). In some states in modern Europe the nests of wood ants, predators of many forest pests, enjoy legal protection. However, in the twentieth century the attention of agricultural entomologists has focused more upon the solitary Hymenoptera, and their role as the enemies of insect pests. Several parasitic wasps have been moved from one country to another, and throughout the world hundreds of millions of dollars have been, and are being saved by these insects destroying crop pests. A recent example is the control of the cassava mealybug. This minor South American pest was accidentally introduced into Africa where, during the 1970s and 1980s, it devastated cassava crops, the primary source of nutrition for 200 million people (Herren & Neuenschwander, 1991). Research in South America revealed that, in its natural habitat, populations of this mealybug were controlled by a tiny parasitic wasp. This wasp was introduced to Africa and promptly reduced mealybug infestation to a very low level where it caused no significant crop damage. The total savings from this biological control project are estimated to be at least $250 million per annum (Norgaard, 1988).

The above account has mentioned just a few of the ways hymenopterous insects are familiar to and interact with our own species, but this anthropocentric view should not be allowed to obscure the fact that many members of the order are important in the functioning of natural ecosystems. Predatory and parasitic hymenopterans are linked in foodwebs to more species of insects than are any other order of terrestrial arthropods (Schoenly, 1990). Hymenopterans are a food source for many vertebrates, from anteaters to woodpeckers; marauding columns of army ants are so effective in flushing out insects that some birds, such as the antbirds, make a living by following them (Willis & Oniki, 1978). In the lowland tropics ants generally move more soil than do earthworms (Abe, 1982) and in the forest canopy ants may be major providers of epiphyte substrate and initiators of epiphyte growth (Kleinfeldt, 1978; Longino, 1986). Many tropical trees are defended by ants and their presence may be necessary for the survival of young trees (Janzen, 1966*a*; Schupp, 1986). As in the temperate zone, bees are probably the single most important group of pollinators of tropical plants despite the greater diversity of animal pollinators present in equatorial areas (Schatz, 1990). For example, of 143 tree species examined at the La Selva Biological Station in northeastern Costa Rica, 41.5 percent were pollinated by bees, compared with 15.9 percent by moths, 7.3 percent by beetles, 4.3 percent by hummingbirds, and 3.0 percent by bats (Bawa *et al.*, 1985). Another critically important group of pollinators in tropical ecosystems are fig-wasps (Agaonidae); most of the 800 or so species of *Ficus* (Moraceae) are completely dependent upon these insects to effect their pollination (see Chapter 11.1).

Thus the Hymenoptera play a major role in maintaining the 'balance of nature' in tropical wildlands where many are keystone species (LaSalle & Gauld, 1991, 1993). Understanding their role in such habitats is critical if humans are ever going to manage tropical ecosystems sustainably, yet their role may be subtle and the interactions complex. Vinson *et al.* (1993) show that removal of dead wood by fire or for fuel can lead to loss of nesting sites for bees. If that bee is a keystone mutualist necessary for the survival of a plant then the plant may become endangered or extinct and the variety of other species it supports will be lost. Tiny agaonid wasps depend on finding, during their short adult life span, a particular species of fig tree at precisely the right stage in its development so that they can oviposit and carry out pollination. Reduction in the density of fig trees in a forest could break this chain and lead to the local extinction of the wasp, which will in turn cause the ultimate extinction of the fig tree, with catastrophic consequences for the charismatic vertebrate fauna that subsists on its fruits. Removal of large forest trees and the consequent loss of epiphytic orchids can disrupt the mating of orchid bees—leading to crop failure by Brazil-nut trees (*Bertholletia excelsa*: Lecythidaceae), which they pollinate (Prance, 1976). The clearing of herbaceous vegetation, and the consequent absence of flowers may cause the loss of parasitoids that feed from the flowers, which in turn may allow their host, a previously innocuous herbivorous insect, to assume pest status (Pruett, cited in Gauld & Mitchell, 1981).

Understanding these interactions can only be gained from a knowledge of the biology of the Hymenoptera involved, yet no overview of hymenopterous biology exists for the tropics. The results of previous studies are widely dispersed throughout the literature, as the exceptionally large bibliography in this volume attests. This book attempts to draw together much of this biology and relate it to an extensive survey of the hymenopterous fauna of one tropical country. In the following sections we outline more of the diversity and composition of the order (Chapter 1.2), then, to set the scene, present a brief overview of the physical and climatological features of Costa Rica (Chapter 1.3). This is followed by an account of the intensive sampling programme which has generated most of the specimens on which the faunistic part of the present study is based (Chapter 1.4). The concluding section (Chapter 1.5) utilizes the data generated in the course of the preparation of this book to examine the Costa Rican Hymenoptera fauna from a global perspective, comparing and contrasting its composition to that of other regions, and relating its size to estimates of the global fauna as a whole. But first (Chapter 1.1) I outline a few of the terms and conventions adopted in the text.

1.1 TERMINOLOGY AND CONVENTIONS

Ian D. Gauld

The morphological terminology used in this book is defined in Chapter 4 and a variety of other biological terms are defined in the Glossary, but a few general conventions are worth pointing out here. The classification of the Hymenoptera that we have adopted is explained in Chapter 5. The classification of other insects and spiders follows '*An Introduction to the Study of Insects*', by D.J. Borror, C.A. Triplehorn and N.F. Johnson (6th Edition, 1989—Saunders College Publishing) because this is one of the most widely used general textbooks in the New World. The plant classification and terminology adopted follows '*Flowering Plants of the World*' by V.H. Heywood (1985—Croom Helm) as this book, and its Spanish translation, are widely available in the Americas.

In the diagnoses of the various families certain features that are particularly characteristic of the group in Mesoamerica are italicized. Throughout this work there are many keys and synopses of the Costa Rican taxa. In some cases, where their ranges of distribution suggests they may also occur in Costa Rica, taxa found in neighbouring areas have also been mentioned. Such taxa, that have *not* been recorded from Costa Rica but are expected to occur here, are denoted by the suffix^{-EX}. Taxa thought unlikely to occur in the country but included for completeness in some keys are indicated by an asterisk. When it is known that a group is newly recorded from Costa Rica we have denoted this with the suffix^{-CR}, but in many cases it has been difficult to establish this for certain, so the suffix is omitted.

A number of geographical terms relate to America south of the United States. They are not exactly synonymous, and to avoid confusion they are here defined as they are used in this book.

America south of the United States: all mainland American states from Mexico south to and including Argentina/Chile, plus all the Caribbean Islands.

Neotropics: all of tropical America, including the extreme south of Florida, tropical Mexico, Central America, the Caribbean and tropical South America. Northern Mexico and the southern parts of Argentina and Chile are excluded.

The Caribbean: all Caribbean islands.

Mesoamerica: Central America plus the tropical part of Mexico.

Central America: Panama, Costa Rica, Nicaragua, Honduras, El Salvador, Guatemala and Belize.

South America: all mainland countries south of Panama.

1.2 THE ORDER HYMENOPTERA

Ian D. Gauld and Paul E. Hanson

The Hymenoptera is one of the nine or so orders of holometabolous insects—that is to say it belongs to the group of insects that undergo complete metamorphosis with distinct egg, larval, pupal and adult stages. Adult hymenopterans generally have two pairs of membranous wings, the hind pair of which are slightly smaller than the fore wings and coupled to them in flight by small hooks on their anterior margin; many adult Hymenoptera are, however, secondarily wingless. In most Hymenoptera the mouth parts are of a rather generalized mandibulate form, though the maxillae and labium may be modified into a tongue-like structure. The antennae are quite long, and in most usually with eight or more segments. Females of most species have a more or less well-developed ovipositor (that structurally is rather like that of some Orthoptera), although in some more specialized members of the group the ovipositor has lost its egg laying function and is used as a sting, and in a few (some ants, chrysidids and stingless bees) it is more or less lost. All Hymenoptera are haplodiploid, with males having n chromosomes and females with the normal $2n$ complement, and in the vast majority of cases reproduction is arrhenotokous with males developing from unfertilized and females from fertilized eggs.

The Hymenoptera is one of the four hyperdiverse orders of insects. It almost certainly exceeds both the Lepidoptera (butterflies and moths) and Diptera (flies) in the sheer numbers of species it includes, and it rivals the Coleoptera (beetles) in species-richness. Indeed, there is evidence to suggest that, the Creator's inordinate fondness not withstanding, the Hymenoptera is globally the most speciose of all insect orders (Gaston, 1991*b*). The actual species-richness of the group can only really be estimated crudely since the myriads of small, mainly tropical species are extremely poorly known, but certainly the order's total size exceeds 250,000 species and it may well comprise half a million or more species (Gaston, 1991*b*). Numbers of such magnitude are hard to visualize, but the size of the order can perhaps be appreciated when one considers that it contains about 80 families and just one of these, the Ichneumonidae, comprises more species than all the vertebrate groups combined.

Currently the Hymenoptera includes slightly more than 115,000 described species, but these are far from being a representative sample of the group. Descriptions are biased in favour of the larger species, and those from the north temperate regions (Gaston, 1991*a*). The higher aculeate families, and perhaps to a lesser extent, the symphytans, are reasonably well-known in the Neotropics, but amongst the parasitoid groups dozens of new genera and hundreds of new species are discovered whenever a new tropical area is sampled extensively (Noyes, 1989).

The Hymenoptera is classified in 18 superfamilies (see Table 1.1), but above the level of superfamily the present classification is unsatisfactory. This is because one of the two traditional suborders and one of the infraorders are paraphyletic assemblages, that is groups of taxa that are united only by the possession of shared ancestral features, not shared derived characters (see Chapter 5.3). In the following discussion these paraphyletic assemblages are denoted by the use of quotation marks ("). Traditionally the order has been divided into two suborders, the "SYMPHYTA" (sawflies and woodwasps) and the APOCRITA (wasp-waisted hymenopterans). The former comprises the superfamilies Xyeloidea, Megalodontoidea, Tenthredinoidea, Cephoidea and Siricoidea whilst the latter includes all other superfamilies; the Orussoidea may be placed in either suborder depending upon the classificatory criteria adopted. The "Symphyta" and Apocrita differ in the form of their larvae, the developmental stage that does most of the feeding; those of the former group are caterpillar-like (eruciform) and feed on plants, whilst apocritan larvae are legless and maggot-like (hymenopteriform) and generally feed on other arthropods. However, several groups of Apocrita (such as fig-wasps, leaf-cutter ants, and bees) are secondarily phytophagous or mycophagous (see Chapter 2.3) and orussoids, which are frequently classified in the "Symphyta", have carnivorous larvae (see Chapter 7).

The Apocrita have traditionally been subdivided into two groups—the "PARASITICA" (parasitic wasps) and the ACULEATA (the superfamilies Chrysidoidea, Vespoidea and Apoidea). The former is simply a heterogeneous assemblage of nine apocritan superfamlies that are not aculeates (the Stephanoidea, Trigonalyoidea, Megalyroidea, Evanioidea, Ceraphronoidea, Proctotrupoidea, Cynipoidea, Chalcidoidea and Ichneumonoidea) and in general the term is avoided in subsequent chapters of this book. In the earlier literature it was sometimes suggested that a biological difference existed between the two groups, with the "Parasitica" being parasitoids and the Aculeata being predators. However, this difference is not tenable because many aculeates, such as dryinids, rhopalosomatids, bethylids, pompilids and even some sphecids develop as parasitoids—feeding on a single food organism—rather than as predators, whilst a few "Parasitica" are predators, consuming several prey individuals. In practice, the division between predator and parasitoid is arbitrary since a continuum of biology exists between them (see Chapter 2.4).

It is a rather general feature of both phytophagous and carnivorous Hymenoptera that the larva consumes a concentrated and highly nutritious diet, be it insect tissue, gall tissue, fungal tissue or pollen, and it is principally only the tenthredinoid and megalodontoid sawfly lineages that have adapted to become consumers of green plant material. Leaf-cutting ants do not feed on green plant material but rather on the fungus that they cultivate on cut pieces of leaves. A striking behavioural feature of the order is that food is almost always provided for the hatching larva by the adult female parent. Like many other insects the adult female locates the larval food source, but in the Hymenoptera the female often does much more. For example, woodwasps inoculate the larval substrate (wood) with fungus, whilst parasitoids may paralyse a host or inject substances that disrupt the host's immune system prior to egg-laying (see Chapter 2.2). Like some parasitoids many predatory wasps paralyse their arthropod prey before oviposition, and the most specialized construct a nest in which the larval food source is subsequently concealed. In a few groups that construct nests EUSOCIAL behaviour has evolved, with females co-operating in the feeding and raising of young (see Chapter 2.6).

1.3 COSTA RICA: ITS GEOGRAPHY, TOPOGRAPHY, CLIMATE AND LIFE ZONES

Paul E. Hanson

Costa Rica is a small country that occupies approximately 51,000 square kilometres—about the size of West Virginia (Boza & Mendoza, 1981). Yet, for several groups of plants and animals, Costa Rica has more species than all of America north of Mexico and on a per area basis it also has a more diverse hymenopterous fauna (see Chapter 1.5). The reasons for this high diversity are: its equatorial position (between 8°00′ and 11°09′ N), extensive altitudinal range (sea level to 3820 metres on Cerro Chirripó), varied precipitation (ranging from 1.5 metres per year in the northwest to more than six metres on the Cordillera de Talamanca) and its geographic position between two major continents. Before discussing the topography, climate and life zones of the country, it is helpful to consider Costa Rica in its broader geographical context, that is, Mesoamerica.

GEOLOGICAL AND BIOTIC HISTORY OF MESOAMERICA

Mesoamerica is here defined as the area situated south of the Isthmus of Tehuantepec in southern Mexico (Chiapas and the Yucatan) south to the Panama/Colombia border. Geologically, this region comprises three distinct parts or terranes: the Maya Terrane (southern Mexico, Belize, north/central Guatemala), the Chortis Terrane (southern Guatemala to northern Nicaragua) and the Chortega-Chocó Terrane (Costa Rica to northwestern Colombia) (Donnelly, 1992). At the break up of the supercontinent Pangea during the Jurassic (180 million years ago), when the Laurasian (North American/Eurasian) and the Gondwanic (South American /African/Australian) tectonic plates were separating, the Maya Terrane was attached to southern Mexico, more or less as it is today. The Chortis Terrane, on the other hand, appears to have detached from western Mexico, moved southward and then eastward, colliding with the southern margin of the Maya Terrane in the late Cretaceous (65 mya). At about this time the eastward moving plate to the west created a subduction zone along the Pacific margin of the Chortega-Chocó Terrane, which then consisted of an island arc south of the Chortis Terrane. During most of the last 65 million years the northern part of Mesoamerica (southern Mexico to northern Nicaragua) existed as a peninsula of North America, while the land that would become Costa Rica and Panama existed only as a chain of volcanic islands. However, from the Oligocene onwards there was a steady migration northwards into southern Mexico of plants of Gondwanic origin (Gentry, 1982).

About three million years ago, in the Pliocene, orogenic events established a continuous land bridge connecting North and South America. The creation of this bridge allowed for an accelerated interchange of organisms between the two continents (Stehli & Webb, 1985). In the plants, northward migrating taxa of Gondwanic origin seem to have overwhelmed the southward moving Laurasian elements, especially in the lowlands. Much of this invasion was so recent that there has been little differentiation even at species-level and amongst some groups, for example of Bignoniaceae, virtually all the species that reach northern Central America are indistinguishable from South American taxa (Gentry, 1982). Although there is rather little evidence for a southern migration of Laurasian plants in the lowlands of Central and South America, a different situation prevails in the uplands where Laurasian elements clearly predominate ecologically, with families like the Pinaceae, Fagaceae, Juglandaceae, Magnoliaceae, Theaceae and Ulmaceae being especially important as canopy members of the montane forests (Gentry, 1982). These northern taxa gradually decrease southward with families such as the Pinaceae and Hamamelidaceae not extending south of the northern highlands in Nicaragua, the Garryaceae and many Juglandaceae reaching only to central Panama, and *Quercus* extending only to Colombia. However, even in the northern Andes Laurasian elements tend to prevail in montane floras. A similar interchange has occurred in many insects, thus providing the contemporary fauna of Mesoamerica with a diverse mixture of taxa, including many that originated in North or South America.

In addition to this biotic interchange, the closing of the land bridge probably brought climatic changes to the region, since the warm westward flowing sub-equatorial ocean current was now diverted northward, possibly resulting in a more humid climate on the Caribbean coast of Mesoamerica (Donnelly, 1992).

One other vegetational pattern present in Central America is the xerophytic region of dry tropical lowland forest that extends from Sinaloa in Mexico, south along the Pacific coast to northwestern Costa Rica, disappearing on the southern Pacific coast of Costa Rica and reappearing on the Pacific coast of western Panama. This region is characterized by the presence of many Gondwanic generic elements, such as *Agave* and *Opuntia*, but there is extensive endemism in southern Mexico. In northwestern Costa Rica *Quercus oleoides* (Fagaceae), the only lowland oak in the country, reaches the southern limit of its distribution.

The topography of Costa Rica

Between the Pacific and Atlantic (Caribbean) coasts of Costa Rica, which are separated in most places by less than 150 kilometres of land, mountains rise to over 3500 metres. Although frosts are common at such elevations, none of the peaks has a snowcap. The major mountain block, the Cordillera de Talamanca, occupies much of the southern part of the country. Here, the coastal plains are fairly narrow and the highlands, rising above 2500 metres, extend unbroken into northern Panama. Approximately in the centre of Costa Rica, at an altitude of about 1000 metres, is a more or less east-west orientated valley system that contains the principal cities of San José, Alajuela, Cartago and Heredia. This Central Valley, or Meseta Central, is bordered to the north and east by a series of volcanoes (including Poas, Barva and Irazú) that rise to between 2700 and 3400 metres. From this area of highlands a chain of mountains and volcanoes (Cordillerras de Tilaran and Guanacaste) stretches northwestwards, almost to the Nicaraguan border. These ranges become progressively narrower and lower, until in northwestern Costa Rica (Guanacaste Province) the more isolated peaks barely reach above 1500 metres. Here the mountain chain, which forms the continental divide, is very close to the west coast, so the Pacific plain is narrow, but the Atlantic lowland plain is very wide. To the north Lake Nicaragua forms a natural boundary between the geologically younger mountains of Costa Rica and the relatively older mountains of northern Nicaragua.

The climate of Costa Rica

The climate of Costa Rica has been summarized by Coen (*in* Janzen, 1983), the source of most of the following information, and treated in considerable detail by Herrera (1986). Average yearly rainfall varies from 1.5 metres on the northwestern Pacific coast to about six metres on the northern, Caribbean-facing slopes of Cerro Chirripó. As in much of Central America, a few rainy days (usually less than 15) account for 70 percent or more of the total yearly rainfall. On the Caribbean side of the country, which receives three to five metres of rain each year, the rainy season usually begins in mid to late April and continues through mid-December to late January, with the wettest months usually being July and November. There may be a brief interruption in the rains in August or September. In the Central Valley and the northern Pacific lowlands the rainy season usually begins in May and continues to mid or late November, with a short interruption in late June and July. In most of the southern Pacific region rains are heavier (average annual rainfall is three to five metres) and the dry season is shorter. Around the Golfo Dulce there is a drier period between December and March, though even during these months 100 mm or more rainfall can occur per month (Boinski & Fowler, 1989).

The yearly rainfall regimes described above are generated by the movement of the intertropical convergence zone, where northeast and southeast trade winds meet. This convergence zone extends farthest northward (to about 12°N) in August and recedes to near the equator in February—thus following the path where the sun is directly overhead but with a lag of about two months. Rainfall is also affected by the polar trough, which penetrates the air mass over the Caribbean Sea to as low as 10°N, generating the intense November–January rains on the Caribbean side of the country. The extremely varied physical topography of Costa Rica results in considerable variation in rainfall between sites, but in general the middle altitude, windward sides of mountains, receive the heaviest rains.

In Costa Rica the average temperature of the warmest month (March, April or May) does not exceed the average temperature of the coolest month (November, December or January) by more than five degrees centigrade at any given site. This is usually less than the diurnal fluctuation in temperature. The

Pacific side generally has less cloud cover than the Caribbean side and thus the average temperature of the warmest month is about 32.6°C at sea level in contrast to 29.9°C on the Caribbean coast. Mean annual temperature decreases with elevation at a rate of 6.5°C per 1000 metres. On the highest peak, Cerro Chirripó (3819 m), the maximum and minimum average annual temperatures are about 7.2°C and 2.2°C respectively. Somewhat anomalous temperatures may occur in some valleys where cold air descends from the mountains.

Life zones and vegetation patterns

Holdridge's classification system of Life Zones has been extensively used in Latin America in general, and Costa Rica in particular (Holdridge, 1967; Tosi, 1969; Holdridge *et al.*, 1971; Hartshorn *in* Janzen, 1983). This system is based on temperature and rainfall, and the seasonal variation and distribution of these two climatic parameters. Thus, Holdridge's system avoids the botanical bias of other classifications (e.g. that used by Gómez, 1986), although each Life Zone has a distinctive vegetational physiognomy and structure.

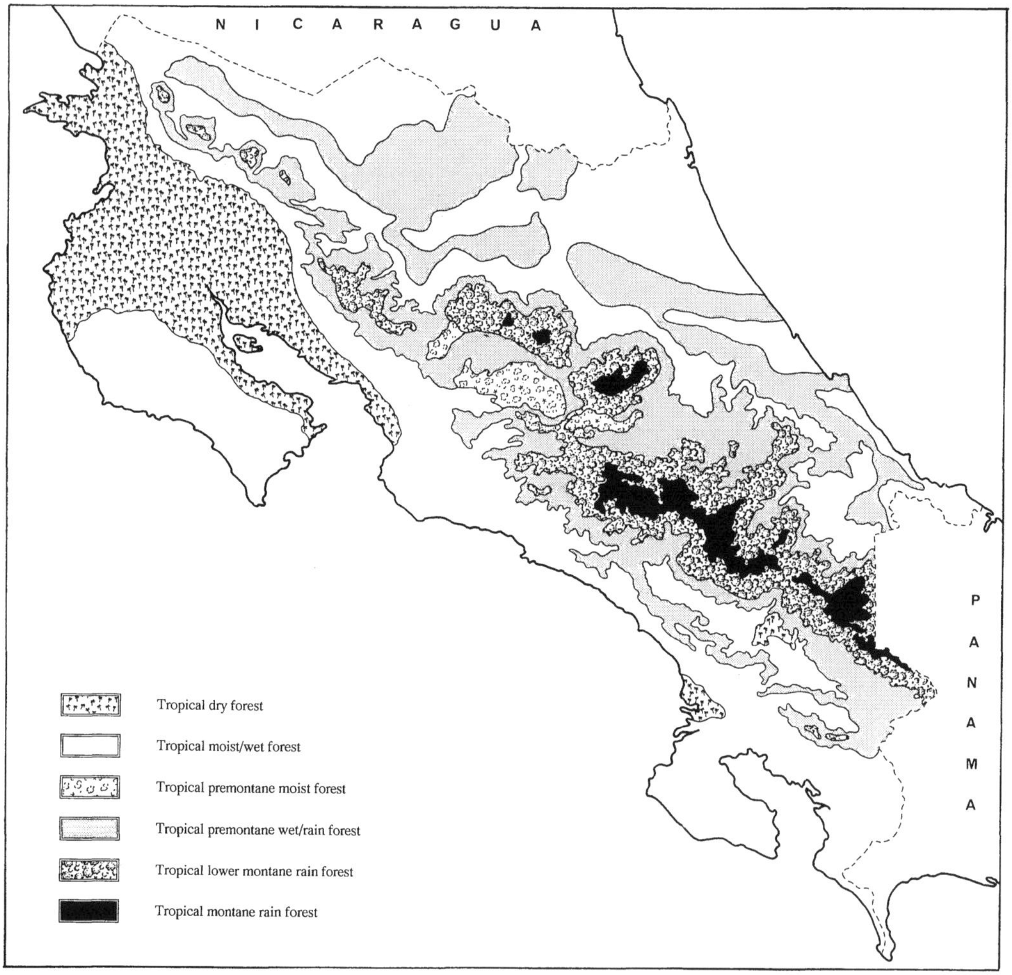

Fig. 1.01. Costa Rica, showing the approximate limits of the major Holdridge Life Zones discussed in the text.

Of the approximately 116 Life Zones on earth, 12 occur in Costa Rica, but only the most extensive ones (Fig. 1.01) are discussed below.

Tropical Dry Forest (TDF). This Life Zone is mainly found in the lowlands of Guanacaste Province in northwestern Costa Rica (10.3 percent of the land area). It is a semideciduous forest having a pronounced dry season with virtually no rain from December to late May. TDF has two strata, with canopy trees standing 20 to 30 metres tall. The most conspicuous components of the canopy are bipinnately-leaved legumes. The understorey contains more evergreen species than the canopy; woody vines are common but epiphytes are scarce (bromeliads being the most conspicuous).

Tropical Moist and Tropical Wet Forests (TMF, TWF). These are the two most extensive Life Zones in Costa Rica, together occupying 46.8 percent of the land area. They occur principally on the Caribbean lowlands and in the lowlands of southwestern Costa Rica surrounding the Golfo Dulce. TMF and TWF are evergreen forests with many strata. The canopy trees are 40 to 55 metres tall, and they often have thin, smooth, light-coloured bark and high buttresses. Those in the TMF are generally smaller in diameter (mostly less than 1 metre diameter at breast height) than those of the TWF (attaining 1 to 2 metres dbh). Woody lianas and epiphytic shrubs are more common in the former, but palms are common in the subcanopy and shrub layers of both Life Zones. TWF contains more species of plants than any other Life Zone in Costa Rica.

Tropical Premontane Moist Forest (TPMF). This is the most extensively altered Life Zone in Costa Rica, having once extended through the Central Valley and occupied 4.7 percent of the land area. No significant areas of primary forest remain in this zone.

Tropical Premontane Wet and Tropical Premontane Rain Forests (TPWF, TPRF). These zones occur at intermediate elevations and occupy 24.4 percent of the land area. They are characterized by evergreen forests with two or three strata. Canopy trees are mostly 30 to 40 metres tall and have moderately thick, dark-coloured, flaky or fissured bark; buttresses are common but small. The trees are covered with moss, but epiphytic angiosperms and woody vines are more conspicuous in TPRF. The ground layer is generally bare with scattered ferns in TPWF, but the fern covering is almost complete in TPRF.

Tropical Lower Montane Rain Forest (TLMRF). This occurs on the windward flanks of the Central Cordillera, both flanks of the Talamanca Cordillera, the top of the Tilarán Cordillera and around the volcanic summits in the Guanacaste Cordillera and occupies 7.4 percent of the land area. These are evergreen forests having two strata, with canopy trees reaching 25 to 30 metres tall (although oaks may reach 50 metres); trees have dark, rough bark, and buttresses are uncommon. Epiphytes such as Araceae, Bromeliaceae, Orchidaceae, Ericaceae, Gesneriaceae and Melastomataceae, are common.

Tropical Montane Rain Forest (TMRF). This occurs in the high Talamancas, with small outliers on the summits of the higher volcanoes in central Costa Rica where it occupies 5.4 percent of the land area. The canopy trees are quite similar to those of the TLMRF Zone (25 to 30 metres tall) but the forests differ in that large epiphytes are less common (restricted to a few species of bromeliads) and the understorey is more open.

Although this system of life zones is designed to avoid a botanical bias, in practice, the discussion centres around plants—for the obvious reason that animals tend to move between zones more readily. In considering the Hymenoptera of Costa Rica, it is helpful to consult two recent monographs on animals—one on butterflies of Costa Rica (DeVries, 1987) and the other on birds of Costa Rica (Stiles & Skutch, 1989). The latter provides a pictorial catalogue of avian habitats and a discussion of the affinities of the Costa Rican avifauna, while the former includes a discussion of butterfly zones of endemism, boundary zones and species pockets (Figure 15 in DeVries, 1987).

1.4 SAMPLING COSTA RICAN HYMENOPTERA

Paul E. Hanson

The rate at which tropical species are going extinct places a particular urgency on efficient means of sampling these species, not merely to preserve them in museums, but more importantly to catalogue and characterize them, making them accessible to humanity and facilitate bringing the value of biodiversity to public attention. Hopefully, in this way the habitat in which they live will be preserved (Wilson, 1988). The importance of conserving tropical Hymenoptera can be argued from a variety of viewpoints: the potential use of undiscovered parasitoids in biological control; the likely presence of useful chemical substances in hymenopteran venoms; the importance of pollinating bees in maintaining plant diversity; and so forth (LaSalle & Gauld, 1993; see also Chapter 3). However, the value of tropical Hymenoptera will remain illusory so long as the majority of species are unknown to science and can neither be recognized nor discriminated from their relatives. We estimate that roughly 70 percent of the Costa Rican hymenopteran species are presently undescribed. Therefore the first step in evaluating tropical biodiversity is getting the specimens from the forest to the taxonomist as efficiently as possible.

SAMPLING METHODS USED IN THE PRESENT STUDY

The specimens upon which this book is based come primarily from Malaise traps that we have been operating throughout Costa Rica over the last nine years. The present survey commenced in 1984 when a series of traps were placed in Santa Rosa National Park, northwestern Costa Rica, and operated continuously until 1987. These were managed by Dan Janzen and Ian Gauld, and the collections were sorted by Pam Mitchell. Shortly after this programme was begun more extensive collecting was undertaken by Dan Janzen, Winnie Hallwachs and colleagues throughout what is now Guanacaste National Park. Then from 1988 until the present Paul Hanson and Ian Gauld extended the survey throughout the country by operating traps, for varying periods of time, at more than 30 sites in all of the major life zones. Samples from the latter traps were separated at the University of Costa Rica and the specimens were then sent directly to the contributing authors. The extent of the survey was further broadened by examination of Malaise trap samples made in some National Parks by colleagues at INBio. The locations of the sites from which we have seen more than three continuous months of samples are shown in Fig. 1.02. We conservatively estimate that we have examined more than 150 Malaise trap-years of catches from Costa Rica, that is around 4.5 million specimens.

Malaise traps appear to be one of the most efficient methods of sampling Hymenoptera, as well as several other groups of insects such as Diptera and many Homoptera. These traps are tent-like structures, which usually have black vertical walls and a white roof-shaped top that is higher at one end (Townes, 1972; Darling & Packer, 1988). Flying insects collide with the walls of the trap, accumulate at the highest point, and drop into a removable bottle of alcohol (usually 75 to 80 percent). The trap is made of a fine mesh netting, of a type that is relatively resistant to the effects of sunlight and epiphyte growth. The traps we used were made from Dacron by Marris House Nets (54 Richmond Park Ave, Bournemouth BH8 9DR, U.K.) and their longevity in the field averaged about one year at low altitudes and up to two years at higher altitudes. The position in which a trap is sited greatly affects the size of the catch. In general we found that the greatest diversity of Hymenoptera was taken in traps placed at forest edges, especially in clearings that were uphill from the forest. Apparently this placement increases the capture of species that normally fly in the canopy. Since Malaise traps are highly esteemed as mosquito netting by local people, avoiding theft is another important consideration in their siting. We avoided such losses by placing traps on small farms, and paying the residents to look after them and change the bottle at the end of every month. This strategy proved to be very cost-effective, since it reduced the journeys around the country to collect samples to three or four times per year.

Malaise traps can yield almost astronomical quantities of specimens and therefore efficient methods of processing the samples are mandatory. Because of the number of traps we had in the field, and the relatively extended period of time we operated these traps, solutions to the processing problem were forced upon us. The samples were first separated into a large and small fraction by passing them through a one centimetre mesh wire screen, gently shaking the latter in a wide pan of alcohol. Samples from lowland sites,

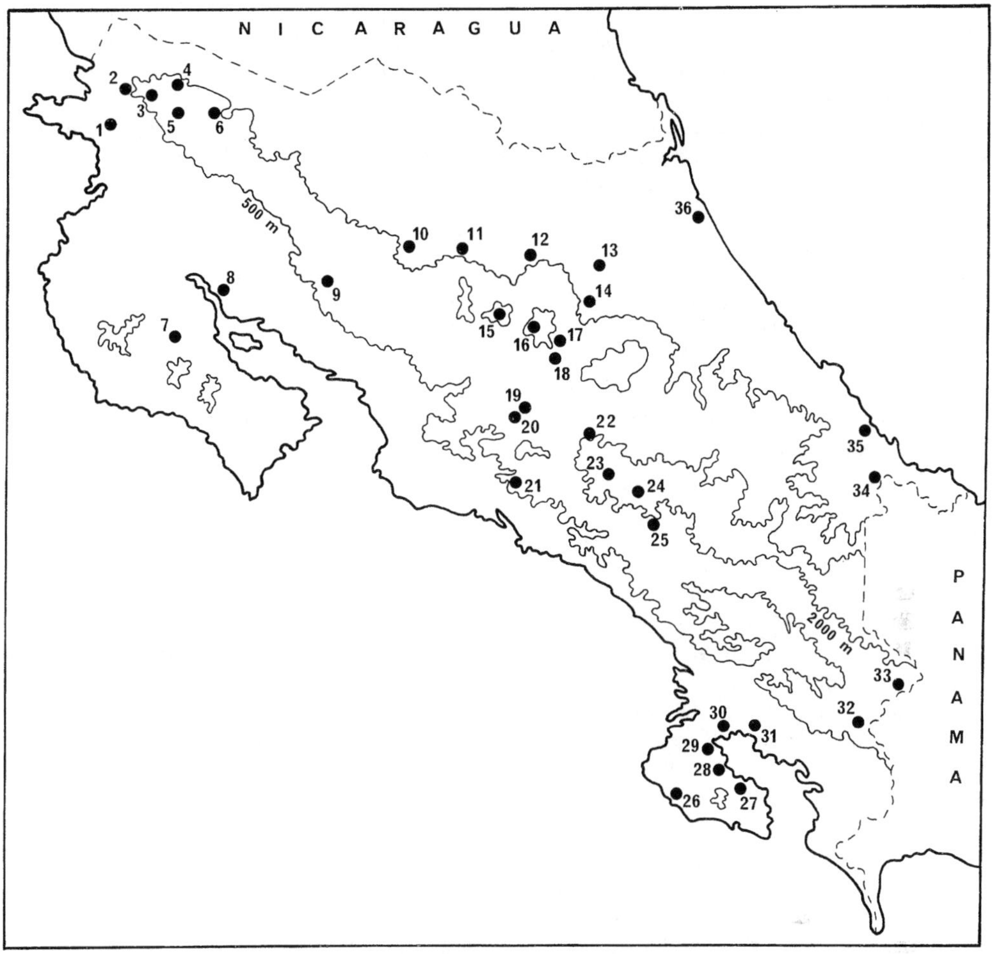

Fig. 1.02. Costa Rica, showing the collecting localities from which we have examined more than three consecutive months of Malaise trap catches. These are: **1**, Santa Rosa National Park; **2**, Cerro el Hacha; **3**, Estación Maritza; **4**, Estación Pitilla; **5**, Estación Cacao; **6**, Finca San Gabriel; **7**, Barro Honda National Park; **8**, Palo Verde N.P.; **9**, Monteverde; **10**, Jabillos; **11**, Aguas Zarcas; **12**, La Selva Biological Station; **13**, Río Frio; **14**, 16 km W. Guapiles; **15**, Volcán Poás; **16**, Vara Blanca; **17**, El Tunel; **18**, Zurquí de Moravia; **19**, San Antonio de Escazú; **20**, Río Agres; **21**, Carara National Park; **22**, La Cangreja; **23**, Cerro de la Muerte; **24**, Villa Mills; **25**, 26 km N San Isidro; **26**, Corcovado National Park; **27**, Puerto Jimenez; **28**, La Palma; **29**, Rincon de Osa; **30**, Golfo Dulce; **31**, 5 km W Peñas Blancas; **32**, San Vito, Las Cruces Botanic Garden; **33**, Las Alturas; **34**, Bribri; **35**, Cahuita National Park; **36**, Tortuguero National Park.

where very small hymenopterans (Aphelinidae and Trichogrammatidae) are often abundant, were sometimes further separated by passing the small fraction through an ordinary kitchen strainer. A fine mesh aquarium net is of course essential for retrieving the specimens from the alcohol. These sieving operations greatly facilitate the separation process.

Although our interest was in the Hymenoptera, we also wanted to make other taxa available. Thus we began by separating Hymenoptera, Coleoptera, Homoptera and a few other taxa. The bulk of the Diptera were left in the sample and these residues are being made available to dipterists. Following this initial separation the Hymenoptera were then separated to appropriate taxonomic units and sent in alcohol directly to the taxonomists. Because of the extremely large samples it is impractical, if not impossible, to mount all the material. A single, moderately productive Malaise trap will collect enough Hymenoptera to keep at least one full-time technician employed doing nothing but mounting and labelling. Even if such labour were available, the resulting collection would comprise tens of thousands of examples of a few common species, and relatively low numbers of many others. Samples in alcohol, initially sorted to family (or other, appropriate level), need to be scanned by a competent taxonomist who can then select material for mounting. This is particularly important for groups that require labour-intensive slide-mounting (e.g. Trichogrammatidae), where only a few tens of specimens can be prepared per day, but many hundreds may be collected.

No single sampling method will obtain all the hymenopteran species in an area and therefore it has been necessary to complement Malaise trapping with other collecting techniques, such as hand netting (Townes, 1973), sweeping with a net that has a screen over the opening (Noyes, 1982; Hanson, 1991), yellow pan trapping (Noyes, 1982), light trapping, and using flight interception traps other than Malaise traps (Masner & Goulet, 1981). All of these methods have been used by us or our colleagues to augment the hymenopterous material available from Costa Rica. In some cases collecting has been very intensive indeed; we have, for example, examined light trap collections made continuously over two or three years from some sites. Although these collecting techniques required more effort per insect captured they added species that are rarely captured in Malaise traps.

Ants and other wingless hymenopterans require a different set of collecting methods, particularly hand picking from rotten wood, and extraction from litter or soil using Berlese funnels or Winkler bags (Gauld & Bolton, 1988). Jack Longino, author of the ant chapter, has been using these specialist techniques to sample the Costa Rican ant fauna for more than a decade.

One very labour-intensive method of sampling involves collecting living potential hosts and rearing their hymenopteran parasitoids (Smith, 1974; Gauld & Bolton, 1988). This method not only provides critical biological information that can be used for ecological, biological control and other studies, it occasionally produces species not collected by other techniques (e.g. Gauld & Janzen, 1994). Considerable material has been reared during the course of the work and we have further benefited by having had access to the extensive material reared by Dan Janzen and Winnie Hallwachs in their survey of the Lepidoptera of Santa Rosa National Park.

Other sampling techniques

In tropical regions many insect species probably spend most of their time in the canopy and thus special attention must be given to collecting in this relatively inaccessible part of the forest. As mentioned above, a well-placed Malaise trap may capture many of these canopy dwellers, but until the same area is sampled by other methods we cannot know for certain what proportion of the canopy fauna is not being collected. Canopy fogging is one method that has been used, especially for relatively heavy-bodied, weak-flying insects such as beetles (Erwin, 1983; Paarmann & Stork, 1987), but we still do not know how effective this method is for small hymenopterans, which may stick to the nearly perpetually wet foliage and thus never drop to the ground, or large fast-flying hymenopterans which may escape before dropping. One interesting possibility is to place Malaise traps on a platform in the canopy but experience with Malaise traps suspended in the canopy in northwestern Costa Rica suggested that a modified design may be required. Climbing techniques (Perry & Williams, 1981) or other methods that allow access to the canopy are necessary for placing traps, and for retrieving epiphytes and aerial leaf litter samples from which hymenopterans can be extracted in the laboratory (e.g. Berlese funnel).

Preparation techniques

Hymenopterans that are to be mounted should be divided into two groups: the physically large (symphytans, evanioids, most ichneumonoids and aculeates, and a few proctotrupoids and chalcidoids) and the physically small (most proctotrupoids, cynipoids, chalcidoids and ceraphronoids, and some Braconidae). Before mounting the specimens should be washed in clean alcohol, taking care to remove plant fibres, moth wing scales and other detritus. The large hymenopterans can be taken directly out of alcohol and dried on filter paper, although hairy specimens (e.g. most bees) are best transferred to ethyl acetate for at least 24 hours before drying. The best method for drying small hymenopterans is with a critical point dryer (Gordh & Hall, 1979), although a chloroform technique (Sharkey, 1988) is said to give adequate results if a critical point dryer is unavailable. Large hymenopterans are mounted directly on stainless steel insect pins of the appropriate size, and the small ones are generally mounted on stiff card rectangles, although ant taxonomists and a few others prefer point mounting on triangles. Some hymenopterans that are too small for direct pin-mounting, but too large for cards, can be glued directly to the side of the pin using reduced white shellac. More details on mounting can be found in Noyes (1982) and Gauld and Bolton (1988).

1.5 THE COSTA RICAN HYMENOPTERA FAUNA

Ian D. Gauld and Kevin J. Gaston

In the preparation for this book very large samples of Hymenoptera were examined from all major habitat types in the country (see 1.4 above). Whilst these samples undoubtedly still under-represent the true fauna, they far surpass any previous inventory of tropical Hymenoptera. Thus they provide our best available estimate of a complete tropical hymenopterous biota, and as such a basis for analysis of the composition and species richness of a tropical fauna and comparison with other faunas. Here we examine these data from several perspectives. First, we look at the family-level composition of the fauna. Second, we look at the magnitude in terms of species richness of the Costa Rican fauna, and examine the implications of this for global species richness. Third, we look at some biological aspects of this tropical fauna and make preliminary comparisons with well-known temperate hymenopterous faunas. Finally we look at the fauna from a regional biogeographic perspective.

The family-level composition of the Costa Rican fauna

Sixty-one of the 79 currently recognized families of Hymenoptera have been collected in Costa Rica (Table 1.1). The only family that has a natural distribution that could include Costa Rica, but has not yet been collected in the country, is the Mymarommatidae, a small family of minute chalcidoids that are rarely collected though they may be locally common.

The largest group of families not represented in Costa Rica belong to the Symphyta. The Xyelidae, Pamphiliidae, Diprionidae, Cephidae and Siricidae all extend southwards into Mexico, and the Diprionidae and Siricidae reach northern Central America. Three of these families, the Siricidae, Diprionidae and Xyelidae, together with the cephalciine pamphiliids, are primarily associated with northern conifers, trees that are not native in Costa Rica. Caribbean pine (*Pinus caribbea*) however, is extensively planted in Costa Rica and the potential exists for species of one or more of these symphytan families to establish themselves in the country. The hartigiine Cephidae, which have been recorded as far south as Chiapas, Mexico (Smith, 1988), and the exclusively north temperate pamphiliine Pamphiliidae, feed as larvae on northern woody angiosperms, whilst the almost exclusively north temperate Cephini feed in the stems of grasses. The other symphytan families not represented in Costa Rica are the Megalodontidae and Blasticotomidae, which are restricted to the Old World, the monobasic Anaxyelidae (= Syntexidae), which is restricted to western North America, and the Cimbicidae, which is holarctic with the exception of the subfamily Pachylostictinae which is south temperate, and occurs in Brazil, Argentina and Paraguay (Smith, 1988).

Another set of families that are absent from Costa Rica are the relictual and often mainly, southern hemisphere apocritan groups: Megalyridae, Austroniidae, Peradeniidae, Rotoitidae and Plumariidae. With the exception of the Megalyridae (Shaw, 1990), all are very small groups comprising only a few species. The only other families absent from Costa Rica are the small relictual holarctic proctotrupoid taxa Roproniidae, Renyxidae and Vanhorniidae.

The magnitude of the Costa Rican hymenopterous fauna

Summing the estimates of species-richness provided by each of the contributing authors allows us to give what undoubtedly is the best estimate to date of the number of hymenopteran species present in a tropical region. An overall estimate of the species richness of the Hymenoptera of Costa Rica, summed across families, is approximately 20,000 species. This can be contrasted with estimated totals for several other regions (Table 1.2). The estimate for Costa Rica is similar to that of Canada and about half that for North America as a whole. The numbers of species per unit area of

Megalodontoidea	
Megalodontidae	(0)
Pamphiliidae	(0)
Tenthredinoidea	
Argidae	(50)
Blasticotomidae	(0)
Cimbicidae	(0)
Diprionidae	(0)
Pergidae	(40)
Tenthredinidae	(50)
Stephanoidea	
Stephanidae	(6)
Trigonalyoidea	
Trigonalyidae	(10)
Megalyroidea	
Megalyridae	(0)
Evanioidea	
Aulacidae	(2)
Evaniidae	(30)
Gasteruptiidae	(6)
Ceraphronoidea	
Ceraphronidae	(25)
Megaspilidae	(20)
Proctotrupoidea	
Austroniidae	(0)
Diapriidae	(1000)
Heloridae	(2)
Monomachidae	(1)
Pelecinidae	(1)
Peradeniidae	(0)
Platygastridae	(650)
Proctotrupidae	(20)
Renyxidae	(0)
Roproniidae	(0)
Scelionidae	(2000)
Vanhorniidae	(0)
Chrysidoidea	
Bethylidae	(150)
Chrysididae	(80)
Dryinidae	(90)
Embolemidae	(4)
Plumariidae	(0)
Sclerogibbidae	(2)
Scolebythidae	(2)
Apoidea	
Apidae	(700)
Sphecidae	(400)
Xyeloidea	
Xyelidae	(0)
Cephoidea	
Cephidae	(0)
Siricoidea	
Anaxyelidae	(0)
Siricidae	(0)
Xiphydriidae	(1)
Orussoidea	
Orussidae	(5)
Cynipoidea	
Cynipidae	(100)
Figitidae	(500)
Ibaliidae	(3)
Chalcidoidea	
Agaonidae	(100)
Aphelinidae	(100)
Chalcididae	(450)
Elasmidae	(20)
Encyrtidae	(1000)
Eucharitidae	(22)
Eulophidae	(1000)
Eupelmidae	(100)
Eurytomidae	(150)
Leucospidae	(10)
Mymaridae	(500)
Mymarommatidae	(0)
Ormyridae	(2)
Perilampidae	(25)
Pteromalidae	(1000)
Rotoitidae	(0)
Signiphoridae	(30)
Tanaostigmatidae	(25)
Tetracampidae	(1)
Torymidae	(100)
Trichogrammatidae	(70)
Ichneumonoidea	
Braconidae	(2000)
Ichneumonidae	(3000)
Vespoidea	
Bradynobaenidae	(3)
Formicidae	(620)
Mutillidae	(300)
Pompilidae	(250)
Rhopalosomatidae	(8)
Sapygidae	(2)
Scoliidae	(15)
Sierolomorphidae	(2)
Tiphiidae	(30)
Vespidae	(180)
TOTAL	**17,065**

Table 1.1. The families of Hymenoptera of the world, with the numbers of species present in Costa Rica indicated in brackets.

Country	Area in 1000 km^2	Estimated number of species	Species$^+$/ 1000 km^2
British Is.	313.7	6500	1544.6
Switzerland	412.9	9000	1996.6
Canada	9922.4	16665	1669.7
N. America	19285.5	36000	3054.9
Costa Rica	51.0	20000	7484.1

Table 1.2. The numbers of species estimated to occur in the hymenopteran faunas of different regions of the world, and the numbers per unit area. See text for a description of how species densities were calculated.

these regions provides a more informative basis for comparison. Let us assume that for all regions species-area relationships take the form $s = x/a^{0.25}$, where s = number of species per unit area, x = the number of species in the country or region, and a = area of the country or region (MacArthur & Wilson, 1967; Dixon *et al.*, 1987). This gives the estimates of the numbers of species per unit area in Table 1.2. These suggest that the numbers per unit area in Costa Rica are more than twice that of the temperate regions for which data are available.

Although we expect the total of the number of species of Hymenoptera to increase, there is substantial semi-quantitative evidence that the figure given above is not unduly low. We have, for example, found that collecting during the past two years has added very few species that we have not seen before. We have also collected quite large numbers of taxa previously considered to be exceptionally rare—such as bradynobaenids, scolebythids and sierolomorphids—and found Costa Rican representatives of poorly-known groups, such as the brachycistidine tiphiids, that have extended their ranges by several thousand kilometres. Furthermore, the additional sorting by colleagues of large samples of some groups known to be exceptionally species-rich, such as the microgastrine Braconidae (see Table 12.9), has yielded species totals that are within the estimates we have earlier made. Although we have not extensively undertaken rearing, aerial Malaise trapping and insecticidal fogging sampling, collections that we have seen, suggest there is not a large 'unknown' hymenopterous canopy fauna. For example, preliminary examination of samples from canopy fogging in La Selva Biological Station (part of the Arthropods of La Selva project directed by R. Colwell and J. Longino) suggest that in only a few families, primarily in the Chalcidoidea, will canopy samples yield more than isolated species not collected by Malaise traps. Thus we feel quite confident estimating the total hymenopterous fauna of Costa Rica to be within the range of 20,000 to 40,000 species.

What are the implications of this estimate for the global species-richness of the order? For comparatively well-studied groups, such as flowering plants, butterflies, amphibians, reptiles, birds and mammals, the Costa Rican fauna comprises between 3.3 and 9.4 percent of the global species total (Groombridge, 1992). Rather than be conservative, for present purposes we assume that the hymenopteran fauna of Costa Rica is at the lower end of this range, or a little less, say between two and six percent. Then, estimates of the size of the hymenopteran fauna of Costa Rica of between 20,000 and 40,000 species yield estimates of the global species-richness for the order of 0.3 to 2.0 million. These figures are not greatly at odds with previous suggestions that the Hymenoptera number less than a million species (Gaston, 1991*b*). If this is approximately correct then, unless the Hymenoptera comprises a vastly smaller proportion of the Insecta than data from temperate regions suggest, global estimates of insect species-richness of 30 to 80 million are untenable (Table 1.3).

The composition of the hymenopterous fauna compared to that of temperate regions

The estimate of 17,000 species of Hymenoptera present in Costa Rica (Table 1.1), ignoring taxa with fewer than 100 species, is composed as follows: Ichneumonoidea (5000), Chalcidoidea (4705), Proctotrupoidea (3674), Vespoidea (1410), Apoidea (1100), Cynipoidea (603), Chrysidoidea (328) and Symphyta (141). The proportions that the superfamilies comprise of the whole fauna are compared with data for Britain in Fig. 1.03. In both Britain and Costa Rica the Ichneumonoidea is the most species rich superfamily, although in Costa Rica it is of similar size to the Chalcidoidea, whereas in Britain it is more than twice as large as the Chalcidoidea. In all countries however, these two groups are extremely speciose and

Estimated number of Hymenopteran species in Costa Rica		20,000	40,000
(A) World total Hymenoptera if Costa Rican fauna is:			
	2%	1.0M	2.0M
	4%	0.5M	1.0M
	6%	0.3M	0.7M
(B) Range of world total insect species richness (assuming Costa Rican fauna is 2–6 percent of total) if the hymenopteran component of total is:			
	10%	3.0–10.0M	7.0–20.0M
	20%	1.5–5.0M	3.5–10.0M

Table 1.3. Estimates of global species richness of (A) the Hymenoptera and (B) the Insecta under various assumptions.

in both Costa Rica and Britain together they comprise more than 55 percent of the hymenopterous fauna. A striking difference in species richness in the Proctotrupoidea can be seen between Britain and Costa Rica. In the former country (as is also the case for North America) this group comprises about 10 percent of the fauna, but it accounts for nearly 25 percent of the Costa Rican one. Similarly the Vespoidea is comparatively more speciose in Costa Rica than Britain. Rather surprisingly the proportion of the fauna the Apoidea comprises is rather similar in Britain and Costa Rica (ca 6 percent) but it is probably higher in the subtropical latitudes (Michener, 1979). Finally the proportion of the Symphyta shows a decrease from more than 7 percent of the fauna in a temperate northern country to less than 1 percent of a tropical fauna.

A striking difference between the hymenopterous faunas of Costa Rica and non-tropical areas is that eusocial taxa are much more species-rich in the former than in the latter regions. For example, in Costa Rica approximately 4.6 percent of the hymenopterous fauna (781 species) are eusocial, whereas in Britain the figure is about 1.4 percent (80 species). This disproportionately high tropical species-richness of eusocial groups is particularly apparent when one compares very closely related eusocial and solitary groups in tropical and extratropical regions. For example, in the Vespidae the solitary Masarinae and Eumeninae are represented by more than 280 species in North America, but only 80 in Costa Rica, whilst the eusocial groups—the Polistinae and Vespinae—are represented by only 40 species in North America, but about 100 in Costa Rica. Although there are fewer species of bees in tropical regions than there are in subtropical dry areas (see Chapter 18), there are more eusocial species, notably the stingless bees, in tropical habitats. In Costa Rica there are 48 species of stingless-bees, whilst in all of Mexico there are 29, and none occur in the United States. A clear latitudinal gradient in increasing species-richness from temperate to tropical latitudes can also be seen in the Formicidae (all members of which are eusocial). For example, 43 different species of ants have been found in a single tree in Peru, a figure that is about equal to the total number of species present throughout the British Isles (Wilson, 1987).

Not only are the social Hymenoptera very species rich in tropical ecosystems, many are also extremely abundant. In lowland rainforests the ants outweigh all the vertebrates combined and in the Amazonian rainforest about one third of the entire animal biomass is composed of ants and termites, the latter being the only major group of non-hymenopteran social insects. Each hectare of soil in the Amazon contains in excess of eight million individual ants and one million termite individuals (Beck, 1971; Fittkau & Klinge, 1973). Rates of ant predation, measured in terms of time taken to exploit a food item,

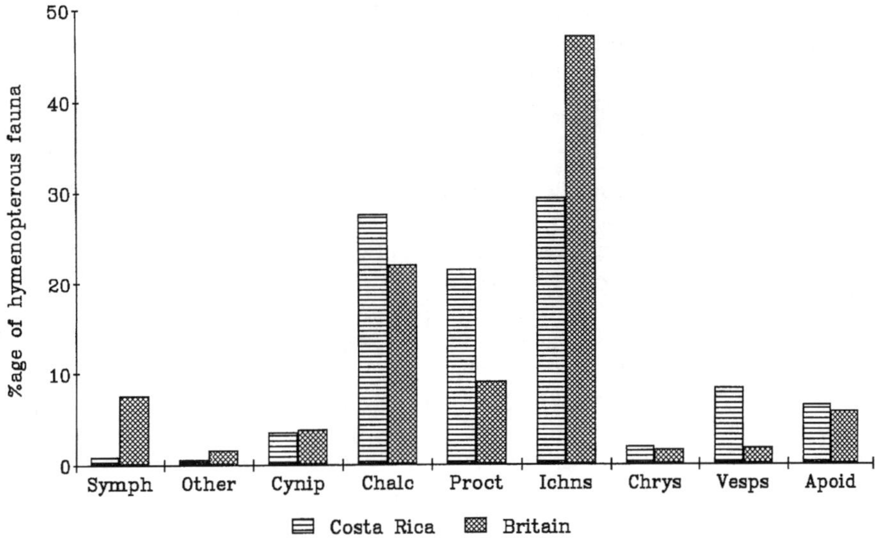

Fig. 1.03. The proportions of the major superfamilies in Costa Rican and British hymenopteran faunas. Apoid = Apoidea; Chalc = Chalcidoidea; Chrys = Chrysidoidea; Cynip = Cynipoidea; Ichns = Ichneumonoidea; Proct = Proctotrupoidea; Symph = Symphyta; Vesps = Vespoidea; Other = other superfamilies.

have been shown to be significantly higher in tropical regions than they are in temperate ones (Jeanne, 1979*a*).

The latitudinal distribution of many parasitoid groups is less clearly understood than that of the eusocial groups mentioned above and has been the subject of some controversy (Owen & Owen, 1974; Janzen & Pond, 1975; Hespenheide, 1979; Morrison *et al.*, 1979; Gauld, 1986, 1991). Interest in latitudinal gradients of parasitoid species-richness has been stimulated primarily by the observation that some taxa, notably the Ichneumonidae, fail to increase in species-richness in the tropics (Janzen, 1981; Noyes, 1989; Askew, 1990). This is surprising because some of the principal hosts of ichneumonids, such as Lepidoptera, are usually appreciably more species-rich in tropical latitudes than they are in temperate regions. Thus in Canada there are approximately 1.3 species of Lepidoptera per ichneumonid species that parasitizes them, whereas in Costa Rica the ratio is about six to one. Several possible explanations have been advanced to account for these observations. According to one suggestion, the 'resource fragmentation hypothesis', some lepidopterous species may be too scarce in the tropics to serve as hosts for specialist parasitoids (Janzen, 1981). Although low host population density is probably an important factor it does not explain all the observations (Gauld, 1986, 1987, 1991) and recently a separate but complementary mechanism, the 'nasty host hypothesis', has been proposed (Gauld *et al.*, 1992; Gauld & Gaston, 1994). This hypothesis is based on the observation that tropical trees are, in general, richer in toxic compounds than their extratropical counterparts (Levin & York, 1978). Therefore caterpillars that feed on tropical trees are likely to be chemically better defended against parasitoids than those in temperate regions since secondary plant compounds in the diet of the host are known to have deleterious effects on parasitoids (Flanders, 1942; Duffey *et al.*, 1986; and see review in Gauld & Gaston, 1994). A corollary of this hypothesis is that parasitoids which attack host stages that are chemically relatively undefended (such as insect eggs or pupae), or hosts that feed on relatively non-toxic plant substances (such as Homoptera that feed on phloem or xylem sap) should be more species rich in tropical habitats than they are in extratropical habitats.

Considerable evidence corroborates the predictions of the nasty host hypothesis. Families of parasitoids that attack chemically relatively unprotected hosts, such as the Encyrtidae (mostly parasitoids of Homoptera) appear to reach their peak of species richness in tropical forests (Noyes, 1989; Askew, 1990), as do groups of ichneumonids that attack vespids and insects boring in wood (Gauld, 1991). Similarly, the present data indicate that egg parasitoids comprise a much larger proportion of this tropical fauna than they do in a temperate area. Approximately 16.5 percent of the Costa Rican hymenopteran fauna (2810 species) are egg parasitoids compared with only 5.2 percent of the British fauna (305 species). On the other hand, groups of diurnally active ichneumonoids that attack leaf-feeding caterpillars are somewhat less speciose in tropical regions than they are in temperate areas. For example, there are approximately 250 species each of Campopleginae and Microgastrinae in Britain compared with 150 campoplegines and 200 microgastrines in Costa Rica (Chapter 12).

Supergeneric taxa:	
Acaenitinae	(Ichneumonidae)
Brachycistidinae	(Tiphiidae)
Exenterini	(Ichneumonidae: Tryphoninae)
Genera:	
Ametastegia	(Tenthredinidae: Allantinae)
Anthidiellum	(Apidae: Megachilinae)
Chalybion	(Sphecidae: Sphecinae)
Crabro	(Sphecidae: Crabroninae)
Dolichostelis	(Apidae: Megachilinae)
Enizemum	(Ichneumonidae: Diplazontinae)
Eucerceris	(Sphecidae: Philanthinae)
Glenosema	(Bethylidae: Epyrinae)
Hedychridium	(Chrysididae: Chrysidinae)
Lestiphorus	(Sphecidae: Nyssoninae)
Lycogaster	(Trigonalyidae)
Mimumesa	(Sphecidae: Pemphredoninae)
Paratiphia	(Tiphiidae: Tiphiinae)
Podalonia	(Sphecidae: Sphecinae)
Polyblastus	(Ichneumonidae: Tryphoninae)
Psammaletes	(Sphecidae: Nyssoninae)
Psenulus	(Sphecidae: Pemphredoninae)
Pseudometagea	(Eucharitidae)
Pseudoplisus	(Sphecidae: Nyssoninae)
Psilochalcis	(Chalcididae: Halticellinae)
Tanyoprymnus	(Sphecidae: Nyssoninae)
Xenoglossa	(Apidae: Anthophorinae)
Xylaspis	(Figitidae: Anacharitinae)

Table 1.4. Some superspecific taxa present in Costa Rica at the southern limits of their geographical range.

Biogeographic patterns in the hymenopterous fauna of Costa Rica

The mix of Hymenoptera encountered in Costa Rica probably reflects the biotic interchange that has occurred between North and South America in the last three million years. Many groups have affinities with the South American fauna. Like plants (Gentry, 1982), these often occur in the lowlands below 700 metres (e.g. pergid sawflies, scolebythids, labenine ichneumonids and most chalcidids), in humid forests at 700–1600 metres (e.g. monomachids, thyniine tiphiids), or in both (e.g. most ants, orchid and stingless-bees and epiponine vespids). Others show affinities with the North American fauna (e.g. acaenitine and exenterine ichneumonids, bumblebees, the sphecid genus *Podalonia*) and, again like plants, many of these occur at moderate to high altitudes. Many superspecific taxa in a wide range of families reach either the southern (Table 1.4) or northern (Table 1.5) limit of their distribution in Costa Rica.

A word of caution, however. Using contemporary distribution patterns to infer geographic origins is fraught with difficulties, a fact that can be best appreciated by considering the situation in certain groups of mammals, which unlike Hymenoptera, have left a very good fossil record. Without a fossil record tapirs and llamas would appear to be South American groups, when in fact they came down from North America and then went extinct in their previous homeland. The cricetid rodents, which came from North America and still occur there, spread to South America and have undergone very extensive diversification in the last three million years, resulting in 40 South American genera (Webb *in* Stehli & Webb, 1985).

In addition to containing elements from North America and South America, Costa Rica also contains a few apparently endemic higher hymenopterous taxa (such as the aberrant poemeniine genus *Rodrigama*). Although data at present are sparse, at the species-level most endemic taxa occur at higher altitudes, above 1500 metres, and seem to be restricted primarily to the Talamanca massif (which extends into Panama) and the volcanoes to the north (Gauld, 1991). This area seems to be an important zone of faunal endemicity in Mesoamerica.

Supergeneric taxa:	
Derecyrtinae	(Xiphydriidae)
Thynninae	(Tiphiidae)
Genera:	
Acrogymnia	(Argidae: Sterictiphorinae)
Agapostemonoides	(Apidae: Halictinae)
Aglaomelissa	(Apidae: Anthophorinae)
Aspirrhina	(Chalcididae: Haltichellinae)
Bohartella	(Sphecidae: Crabroninae)
Caenomymar	(Mymaridae)
Ceratogramma	(Trichogrammatidae)
Chartergellus	(Vespidae: Eumeninae)
Ctenioschelus	(Apidae: Anthophorinae)
Didymia	(Argidae: Sterictiphorinae)
Entrichopteris	(Mymaridae)
Foenatopus	(Stephanidae)
Hadromymar	(Mymaridae)
Incalia	(Pergidae: Syzygoniinae)
Leipomeles	(Vespidae: Polistinae)
Megommation	(Apidae: Halictinae)
Microstigmus	(Sphecidae: Pemphredoninae)
Nectarinella	(Vespidae: Polistinae)
Neogorytes	(Sphecidae: Nyssoninae)
Nomadina	(Trigonalyidae)
Parataruma	(Sphecidae: Crabroninae)
Platyfrons	(Mymaridae)
Prethophion	(Ichneumonidae: Ophioninae)
Protosiris	(Apidae: Anthophorinae)
Quexua	(Sphecidae: Crabroninae)
Restisoma	(Mymaridae)
Scapheutes	(Sphecidae: Crabroninae)
Sicophion	(Ichneumonidae: Ophioninae)
Trochophora	(Argidae: Sterictiphorinae)

Table 1.5. Some superspecific taxa present in Costa Rica at the northern limits of their geographical range.

Also present in the country are a few anomalous taxa that have their apparent closest relatives in the Old World. Examples include the ichneumonid *Leptopimpla hansoni* and *Syrphidepulo* sp., which belong to southeast Asian genera (Gauld, 1991), and *Erythrodolius* sp. which belongs to an Afrotropical ichneumonid genus (see Chapter 12.1). Such anomalies may merely be the result of under-collecting throughout much of highland tropical America.

Finally, several groups have been recorded to the north and south, but not in Central America. Although some of these possibly do occur in Central America and simply have not been collected yet, others probably have a disjunct distribution, or are distributed throughout the Caribbean but not Central America. Those with an apparent disjunct distribution include taxa that are mainly restricted to arid areas in the southwestern United States, Mexico and Brazil/Argentina (such as masarine vespids, chyphotine bradynobaenids and the ichneumonid genus *Compsocryptus*), and temperate species present in the United States and southern Brazil (e.g. Cimbicidae, the ichneumonid *Enicospilus americanus* and the pteromalid genus *Dryadochalcis*). Other taxa (such as the ichneumonid *Enicospilus flavus*) are apparently distributed in the southern United States and Mexico, throughout the Caribbean and into South America, but not Central America (Porter, 1986, 1989; Gauld, 1988*b*).

2 The biology of Hymenoptera

2.1 NATURAL HISTORY

Paul E. Hanson and Ian D. Gauld

In Costa Rica hymenopterans have been collected in all major terrestrial habitats, from the coastal mangrove swamps to the peaks of the highest mountains. Although none has yet actually been collected in freshwater, a species of *Hydrophylita*, a trichogrammatid genus known to attack submerged eggs of dragonflies, has recently been captured in Malaise traps in the country (see Chapter 11.19). The different life zones referred to in Chapter 1 support rather different hymenopterous communities. In the hot, seasonally dry and rather open tropical deciduous forests of the northwest a variety of aculeates are relatively common, including representatives of small and biologically little-studied groups such as the Sierolomorphidae and brachycistidine Tiphiidae. In these dry forests several species of the larger Hymenoptera are yellow and brown or brown and blackish banded. Some ichneumonid species are dichromic, with a yellow ochre and brown patterned individuals in dry forests and a black and lemon yellow coloured individuals in wet forests or in dry forests during the wet season (Gauld, 1991). As one moves from the deciduous forest to the lowland moist or wet forests the number of social insect species apparently increases (e.g. see Janzen, 1983). Species of stingless-bees and ants abound, as do the less noticeable non-social, but numerous, yellow and black patterned chalcidids, whose hovering flight resembles that of stingless-bees, and the tiny orange and black scelionids that parasitize katydid eggs. Reddish braconids with dark patterned wings may be seen among the undergrowth while black and white banded ichneumonids are occasionally glimpsed, almost invisible as they fly through shafts of sunlight. However, the family Ichneumonidae seems to reach its peak of species-richness between 700 and 1500 metres in altitude, in the premontane moist and wet forests. In a study of one large ichneumonid subfamily Gauld (1991) found between 48 and 68 species (mean 56.75) at well-collected mid altitude sites compared to 25 to 39 species (mean 30.5) in lowland sites. In premontane forests slender black pelecinids are also frequently encountered on the wet understorey vegetation. In these habitats occur representatives of small and predominantly southern hymenopterous groups, such as thynnine tiphiids and monomachids. At higher altitudes, in the lower montane forests above 1800 metres, the hymenopterous fauna is more depauperate—chalcidids, ants and stingless-bees become increasingly scarce. This zone is rich in local endemic species that seem to be restricted to the Talamanca massif and the volcanoes immediately to its north (Gauld, 1991). Amongst ichneumonids occurring in these forests are species with an obvious northern affinity, such as the tryphonine *Polyblastus*, the diplazontine *Enizemum* and the exenterine *Eridolius* (Gauld, in prep). In the cool and humid high altitude montane oak forests, small drab-coloured parasitoids of Diptera abound—diapriids, alysiine braconids and orthocentrine ichneumonids. Wingless or brachypterous ceraphronoids, diapriids, scelionids, and ichneumonids are also fairly common at high altitudes, whereas in the lowlands such wing reduction is primarily restricted to a few scelionids and diparine pteromalids.

The majority of hymenopterans are most active on bright sunny days and remain concealed, or motionless on the vegetation during inclement weather or at night. Some species of Tiphiidae, Vespidae, Sphecidae and Apidae form 'sleeping' aggregations on plants (e.g. Eickwort & Eickwort, 1971, 1973*a*), while the females of some other species remain in their nests at night. Although there are fewer field observations of microhymenopterans, monospecific aggregations of *Conura* (Chalcididae) have been observed during the dry season at night on the undersides of leaves (W. Eberhard, pers. comm.). *Neorileya* (Eurytomidae) and *Herbertia* (Pteromalidae) form aggregations in similar places during the dry season, but they remain together day and night (W. Eberhard, pers. comm.). Two types of resting behaviour have been distinguished in the scelionid *Gryon* (Vogt & Nechols, 1991): 'motionless' (temporary resting during the afternoon) and 'tucking' (aggregated individuals of both sexes at night).

Not all hymenopterans are inactive at night, and in the lowlands particularly a diverse set of species is nocturnal. Often these are characterized by being pale in colour and having very large ocelli (Kerfoot, 1967; Huddleston & Gauld, 1988). The vespid *Apoica* and the halictine bee *Megalopta* fly at night, as do species of *Rhopalosoma*. There are also nocturnal ants, braconids, mutillids and diapriids. However, the most conspicuous of the night-flying Hymenoptera are the orange-brown ichneumonids, members of the species-rich genera *Enicospilus* and *Netelia*, which search for their caterpillar hosts in the dark. A few hymenopterans are crepuscular, flying only at dawn or dusk. For example, *Ptiloglossa*, a large bee, often can be heard in deciduous forests flying just before and at dawn (Janzen, 1968*b*), though at other times of day they are very rarely encountered.

Seasonal change in abundance of adult Hymenoptera in the tropics is less well-known than in the temperate zone, and although much has been published about some eusocial species, almost nothing is recorded about solitary forms. From the Malaise trap samples we have monitored in diverse habitats in Costa Rica it is possible to make some preliminary observations on the patterns of adult activity of a few parasitoid species. Not surprisingly, in the seasonally dry forests of northwestern Costa Rica, where almost no rain falls between the end of December and mid-May, most species show pronounced peaks of seasonal abundance. In many cases these correspond with the onset of the wet season when trees flush with new growth and herbivorous larvae abound. Samples of ichneumonids that oviposit into young caterpillars frequently peak during May and June (Fig. 2.01). Other species that attack pupae appear to reach their peak of abundance later, towards the end of the wet season (Fig. 2.02) and these often remain active well into the dry season. A few species that search for concealed or protected immobile hosts (such as cocooned egg masses or pupae), only appear as adults during the dry season (Fig. 2.03) and in Santa Rosa National Park the Chalcidoidea, as a whole, appear to be both more numerous and diverse during the dry season. Even in seasonally dry habitats some species appear to be present throughout the year, although they apparently move from the rather open areas they frequent at the start of the wet season, to more shady patches of mature forest at the height of the dry season (Fig. 2.04).

In the perpetually wet premontane and montane forests one perhaps would expect Hymenoptera to show less seasonality, and in fact adults of many ichneumonids, such as *Ophion uraniae* and *Zagryphus* sp. 1 are present throughout the year (Gauld, 1988*b*; unpubl.). In some cases (e.g. *Enicospilus bozai*, Gauld, 1988*b*), species which show pronounced peaks of adult activity in seasonal habitats appear to be present as adults throughout the year in more uniformly wet sites. However, by no means all wet forest species are aseasonal (Denlinger, 1986; Boinski & Fowler, 1989). For example, adults of the thynnine tiphiid, *Aleurus septentrionalis*, have only been collected from March through July (Hanson, unpubl.).

Adult feeding

All adult hymenopterans possess mandibles yet few apparently bite off and consume large pieces of solid food, although there are isolated records of sawflies, ichneumonids and dryinids devouring arthropod tissue (Hobby, 1932; Leius, 1961*b*; Waloff, 1974), and a few records of parasitoids consuming plant trichomes and epidermis (see review by Jervis *et al.*, 1993). Like the non-mandibulate adults of Diptera and Lepidoptera most adult Hymenoptera probably consume a more or less fluid diet. The commonest liquid foods are nectar (sometimes fortified with pollen), honeydew secretions of homopterous insects, liquid oozing from a host or prey organism, and in the case of eusocial Hymenoptera, larval saliva and crop regurgitate from conspecific adults (Hunt, 1991).

Most adult hymenopterans feed on nectar to obtain their basic metabolic carbohydrates and possibly other nutrients (Vinson & Barbosa, 1987), but vespids, bees, and perhaps others also consume pollen (Hunt *et al.*, 1991). Thus flowers generally attract a wide diversity of Apocrita, and particularly large numbers of species can often be found on the blossoms of plants that flower during the dry season. Many of the larger hymenopterans with unspecialized mouthparts are restricted to feeding on flowers that have relatively accessible nectaries (e.g. Umbelliferae, see Jervis *et al.*, 1993) but some species have elongate mouthparts

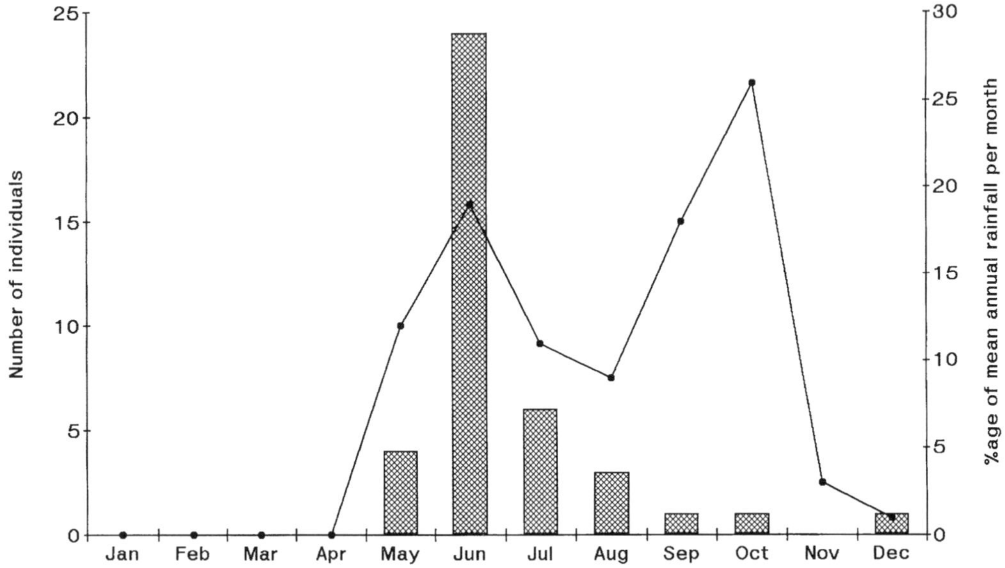

Fig. 2.01. Histogram showing the numbers of individuals of *Enicospilus lebophagus* (Ichneumonidae, Ophioninae) and collection dates in Santa Rosa National Park, Costa Rica; the solid line shows the percentage of the mean annual rainfall occurring per month. This ichneumonid is a koinobiont endoparasitoid that attacks the young larvae of *Rothschildia lebeau* (Saturniidae).

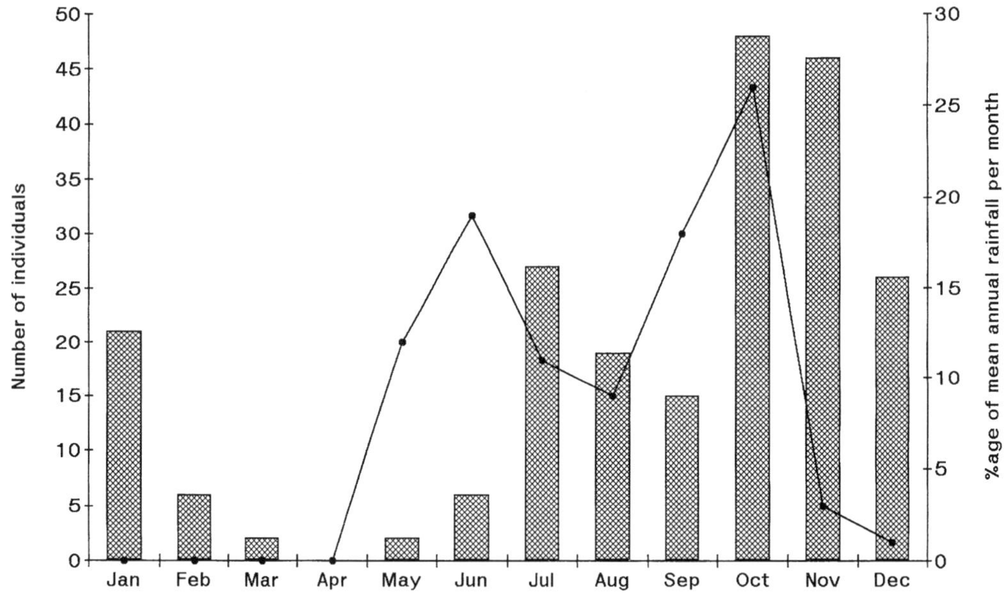

Fig. 2.02. Histogram showing the numbers of individuals of *Pimpla croceiventris* (Ichneumonidae, Pimplinae) and collection dates in Santa Rosa National Park, Costa Rica; the solid line shows the percentage of the mean annual rainfall occurring per month. This species is an idiobiont parasitoid that attacks prepupae and pupae of Lepidoptera.

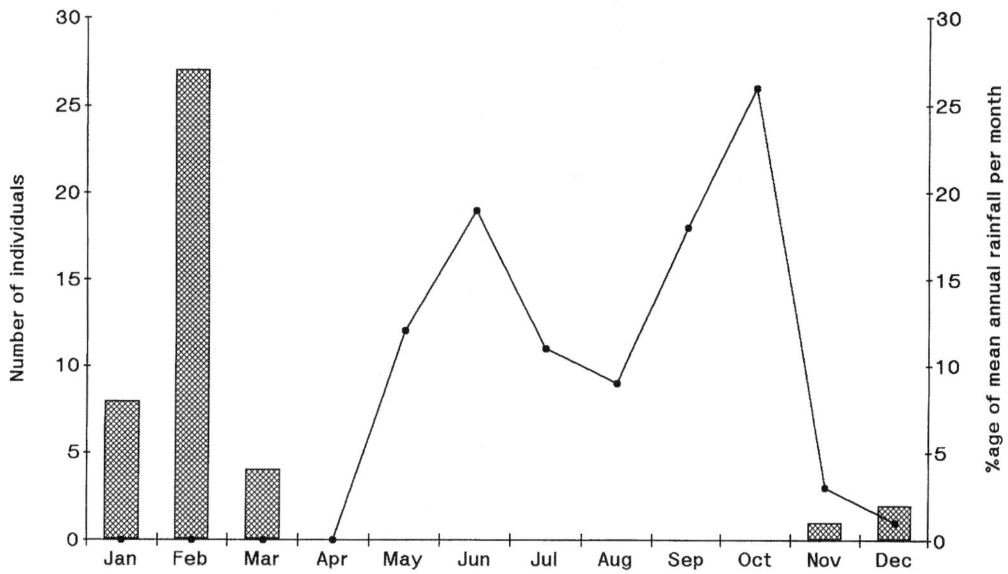

Fig. 2.03. Histogram showing the numbers of individuals of *Clistopyga calixtoi* (Ichneumonidae) and collection dates in Santa Rosa National Park, Costa Rica; the solid line shows the percentage of the mean annual rainfall occurring per month. This species is a dry season specialist that attacks either spider egg sacs or 'diapausing' spiders guarding egg sacs.

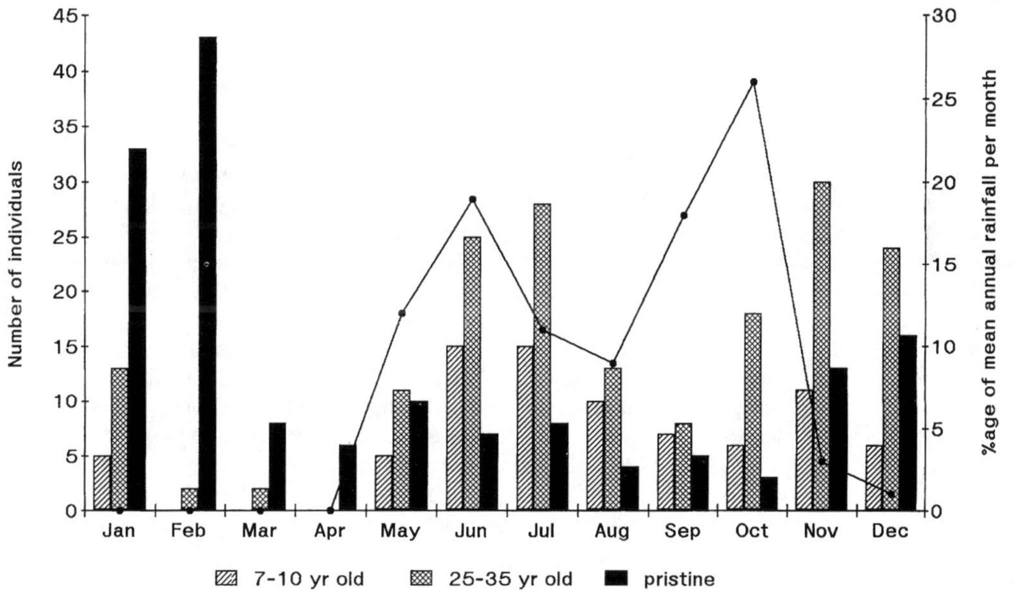

Fig. 2.04. Histogram showing the numbers of individuals of *Neotheronia mellosa* (Ichneumonidae) and collection dates in forest habitats of different ages in Santa Rosa National Park, Costa Rica; the solid line shows the percentage of the mean annual rainfall occurring per month. In drier months a larger proportion of the ichneumonid sample occurs in denser, closed-canopy forest.

which enables them to reach nectaries concealed deep within tubular-shaped flowers. Such adaptations are most common in bees, but they also occur in members of other families including the Leucospidae, agathidine Braconidae and tersilochine Ichneumonidae. One of the most extreme examples can be observed in orchid bees, which often have mouthparts that are longer than the body (Fig. 18.16).

Extra-floral nectaries are another important source of sustenance for a wide range of hymenopterans (Bentley, 1983). These are nectar producing organs found elsewhere on the plant, often on the floral receptacle or on the leaves. They are less common than floral nectaries, occurring on about one percent of the plants in the seasonally dry forests of north-western Costa Rica (Bentley, 1976). Forty families of insects have been found visiting extra-floral nectaries on *Ipomoea carnea* (Convolvulaceae), including ants, vespids, stingless-bees and chalcidoids (Keeler, 1978). In wet lowland forests of northeastern Costa Rica the extrafloral nectaries of *Byttneria aculeata* (Sterculiaceae) attract a variety of parasitic Hymenoptera (Hespenheide, 1985). Certain plants growing in the understorey of oil palm plantations are particularly attractive to Chalcididae (Mexzón & Chinchilla, 1991; Delvare & Genty, 1992). These include *Baltimora* (Compositae), *Cassia* (Leguminosae), *Melanthera* (Compositae), *Scleria* (Cyperaceae), *Solanum* (Solanaceae) and *Urena* (Malvaceae). However, not all of these plants have extrafloral nectaries. In *Scleria* the parasitoids are attracted by pollen and in *Solanum* by glandular trichomes. The latter attracts larger Chalcididae and Ichneumonidae whereas *Urena* (with extrafloral nectaries) attracts smaller Chalcididae, Eurytomidae, Eulophidae and Braconidae.

Plants infested with phloem-feeding homopteran insects often accumulate patches of honeydew that attract numerous hymenopterans. It is not necessarily the sugar in the honeydew that is the main attractant, but rather traces of nitrogenous substances, especially chemicals released by the hydrolysis of the amino acid tryptophan (Hagen, 1986). While many hymenopterans feed on the honeydew that has accumulated on foliage, only ants, polistine vespids, and some stingless-bees are known to obtain honeydew directly from the homopteran insects. These hymenopterans, especially ants, often defend the homopterans against predators (Way, 1963; Buckley, 1987; Letourneau & Choe, 1987).

Not all adult hymenopterans are limited to a sugary diet. Some sawflies, at least in temperate habitats, kill and consume smaller insects (Hobby, 1932), while adults of *Netelia* have been observed to attack other insects, sever their heads, and imbibe the fluids issuing from these wounds. The females of a large number of parasitoid species feed from the exudate of wounds inflicted by the ovipositor on larvae or pupae that can serve as hosts for their young (Jervis & Kidd, 1986). The females of predatory wasps and ants, in addition to capturing prey to bring to their nests, often feed themselves on the prey, but only after thoroughly chewing it. A few bees, ants and wasps feed on and provision their nests with carrion, although saprophagy in general is not very common in the order (see review by O'Donnell, 1995).

Mating behaviour

Mating behaviour in Hymenoptera, as in other groups of insects, varies with the diverse ecological conditions confronted by different species (Thornhill & Alcock, 1983). Monogamy, whereby a male mates with only one female during his lifetime, probably occurs in some social Hymenoptera such as honey-bees, where the queen is frequently inseminated by several males during her nuptial flight. The majority of Hymenoptera, however, are polygynous—that is, the male potentially mates with more than one female. Monandry (female mates with only one male) is presumed to be more common than polyandry (female mates with more than one male) in most species of Hymenoptera, although both are known among social hymenopterans (Page, 1986).

Where females or resources utilized by females are concentrated, males often defend territories containing such concentrations of females. For example, in *Podagrion*, a torymid parasitoid of mantid eggs, a male that successfully defends the mantid egg case will be able to mate with numerous emerging females (Grissell & Goodpasture, 1981). Similarly, males of some rhyssine ichneumonids establish territories on tree trunks from which females are emerging (Eggleton, 1990). In several aculeate wasps and bees the males maintain territories near nesting aggregations or plants frequently visited by the females. Sphecids of the subfamily Philanthinae and certain groups of bees mark their territories with chemical secretions. For example, males of some *Centris* species

(Apidae: Anthophorinae) maintain territories in the crowns of flowering trees and utilize scents produced by glands in their hind legs to mark their territories (Williams *et al.*, 1984).

When females are more dispersed, males sometimes establish 'symbolic' mating territories, or leks. Some male orchid bees for example defend perch sites, such as smooth barked trees in treefall gaps in lowland rainforests (Kimsey, 1980*b*). Male *Centris adani* bees establish territories in open, second growth forests in northwestern Costa Rica (Frankie *et al.*, 1980) and mark their territories with chemical scents secreted by their mandibular glands (Vinson *et al.*, 1982). In some solitary wasps and bees the nesting aggregations are so dense that it is not feasible for males to maintain territories; instead, they acquire mating opportunities through scramble competition. In such circumstances males are sometimes morphologically specialized to allow them to gain access to females before competitors (Nuttall, 1973). For example, in some parasitoids of wood-boring insects, such as the Costa Rican ichneumonid species *Dolichomitus irritator*, males are extremely elongate and are able to insert their metasoma into the emergence hole of the female and copulate with her before she has fully escaped from the log (Gauld, 1991). The wingless, blind males of fig-wasps (Agaonidae: Agaoninae), which never leave the fig, chew open seeds containing females, insert their metasoma, and copulate.

Females of many hymenopterous species release pheromones to attract mates, and sex pheromones have been found in a variety of superfamilies including the Tenthredinoidea, Proctotrupoidea, Ichneumonoidea, Chalcidoidea, Vespoidea and Apoidea (Cardé & Baker, 1984; Eller *et al.*, 1984). Wingless thynnine tiphiid females, for example, crawl up on vegetation and release a pheromone attractive to males. When a patrolling male locates her, he carries her off, copulating in flight (Alcock, 1981). Similarly in some primitive species of ants, the female releases sex pheromones while on the ground or on low vegetation (Haskins, 1978). Male chelonine braconids actively search leaves upon which females of the same species have rested, and a hexane extract from the female body elicited antennation by the male (Kamano *et al.*, 1989). A pheromone in this extract has recently been identified as (Z)-9-hexadecenal (Kainoh *et al.*, 1991). A rather similar pheromone is produced by a female banchine ichneumonid, *Exetastes*, and large numbers of males are attracted to its synthetic analogues, 8-dodecenyl and 11-tetradecenyl acetates (Hrdy & Sedivy, 1979).

Although swarming seems to be less common among Hymenoptera than it is in the Diptera, it has been reported in species of a variety of groups, including ants (Eberhard, 1978), braconids (S.R. Shaw, 1991), ichneumonids (Rotheray, 1981*a*), pteromalids and encyrtids (Nadel, 1987). This last author, working in California, observed that males of species of *Bothriothorax*, *Copidosoma* and *Pachyneura* began to form swarms in the early morning over boulders on a small ridge. Uninseminated females arrived at the ridge apparently to mate rather than to feed or search for hosts, and after mating the females seemed to show no further interest in the swarms. In the Monteverde cloud forest in Costa Rica, males of the ant *Acromyrmex cornatus* have been observed swarming at 10:00 a.m. above an oak on a ridge, 25 metres above the ground (J. Longino, pers. comm.). In the same location and at about the same time of day, males of a species of *Bracon* (Braconidae) were observed swarming one to two metres above the ground in sun-lit patches in the forest (P. Hanson, unpubl. obs.).

Courtship behaviour in parasitoids usually includes wing vibration, antennation, head movements and leg tapping prior to copulation (Matthews, 1975; Assem, 1986). In many cases courtship involves the production of sounds not audible to the human ear (Assem & Putters, 1980; Sivinski & Webb, 1989). Courtship behaviour is often fairly characteristic of a particular taxon and some investigators have used these behavioural characters to help reconstruct phylogenies (Gordh & DeBach, 1978; Bosch & Assem, 1986).

In non-social Hymenoptera it is often presumed that females mate just once with a single male, and this may be the case in many species (Abe, 1991), but in at least some species of parasitoids, females copulate with several males during their lifetime (Al-Maliky & Al-Izzi, 1990). In some chalcidoids and cynipoids, the female may, after stimulation by a male, expose her genitalia in readiness for copulation. Once she has exposed herself in this way she never repeats the process and so never again mates, even if the first exposure failed to result in a copulation (Barras, 1965). Such unmated, sexually unreceptive females have been called PSEUDO-MATED FEMALES (Yoshida, 1977) and their occurrence has been documented in both the Pteromalidae and Cynipoidea (Yoshida, 1977; Abe, 1991).

Life cycle

The Hymenoptera are holometabolous insects—that is, they have a life cycle progressing from egg to larva to pupa to adult. Unlike many species of Diptera, all Hymenoptera are oviparous, although a few tryphonine ichneumonids retain the egg under the subgenital plate or beneath the ovipositor until it is almost ready to hatch. Parental care for the young is common and extremely well documented in species that build nests (e.g. Malyshev, 1968; Evans & West Eberhard, 1970; Iwata, 1976). However, direct parental care occurs in other groups of Hymenoptera as well. Some bethylids guard their offspring until they pupate (Hardy & Blackburn, 1991), and among sawflies the adult females of some species of argids, pergids and pamphiliids have been observed to stand guard over eggs and larvae (Dias, 1975, 1976; Naumann, 1984; Kudô *et al.*, 1992).

Compared with coleopteran and dipteran parasitoids, the larvae of hymenopteran parasitoids are relatively inactive since the adult female usually places her egg on or very near the larval food. There are, however, a few exceptions to this generalization. Species of Eucharitidae, Perilampidae (Heraty & Darling, 1984) and eucerotine ichneumonids (Tripp, 1961) lay eggs that produce active, planidial first instar larvae, which seek out their hosts. Trigonalyids lay numerous tiny eggs on leaves and these may be swallowed by caterpillars. In the caterpillar's gut the eggs hatch and the minute larvae migrate to the body cavity. If the caterpillar is then attacked by another parasitoid or taken as prey by a vespid wasp, the trigonalyid larvae then feed on the parasitoid or predator; they cannot complete their development in the caterpillar (Clausen, 1940*b*; Weinstein & Austin, 1991).

The more primitive Hymenoptera, the Symphyta, have up to eight larval instars, with females often having one more than the male (Benson, 1950). Primitively, the Apocrita probably have five larval instars (Evans & West Eberhard, 1970), but the number is reduced in many endoparasitoids. However, in such species it is extremely difficult to ascertain the exact number of larval instars (Rojas-Rousse & Benoit, 1977) and accurate observations have yet to be made on the majority of species. The larvae of the exophytic (leaf-feeding) Symphyta resemble lepidopterous caterpillars, whereas those of the Apocrita, and to a lesser extent endophytic Symphyta, are generally legless maggots with few distinguishing features (Evans, 1987*a*). Most hymenopteran larvae consume a very nutritious diet of animal, plant, or fungal origin, and apparently only two lineages of symphytans have become adapted to feeding on a relatively less nutritious diet of green leaves.

Hymenoptera pupate in a variety of situations. Species of many families construct pupal chambers within the soil or concealed within plant tissue, while the higher aculeates pupate within the shelters (nests) provided for them by the female parent. The final instar larva of most hymenopterans spins a cocoon with silk produced by the modified salivary glands, although cocoons are not formed by most Proctotrupoidea (except some Platygastridae), Cynipoidea and Chalcidoidea. One of the few groups of chalcidoids that spin cocoons are species of *Euplectrus* (Eulophidae), which secrete silk from the Malpighian tubules. Although they do not form true cocoons, some chalcidoids such as *Scutellista* (Pteromalidae) utilize silk in binding together host remains (Clausen, 1940*b*). Cocoon formation is closely correlated with the type of pupation site, and the cocoon has been independently reduced in several braconids (Alysiinae, Opiinae, and most Rogadinae and Aphidiinae) and ichneumonids (some Anomaloninae, Pimplinae and Ichneumoninae) that pupate within the host. Among ants only Ponerinae and some Formicinae form cocoons. In some pergids it is the penultimate larva that spins an initial cocoon, and the final (prepupal) instar completes cocoon formation from within, spinning a separate chamber for the cast penultimate larval exuvium (Macdonald & Ohmart, 1993).

Hymenopterans that pupate in exposed situations often form thick, elaborately constructed cocoons. For example, some species of the ichneumonid genus *Enicospilus* that remain within their cocoons during the long dry season in northwestern Costa Rica, make double cocoons with a thick fibrous outer wall and an inner shiny layer (Gauld, 1988*a*). A number of parasitoid species that spin cocoons in exposed situations camouflage them in various ways. The ichneumonid *Casinaria* that commonly parasitizes *Sibene* (Limacodidae) in Costa Rican oil-palm plantations, for example, spins a black and white cocoon that resembles bird droppings, whilst some polysphinctine ich-

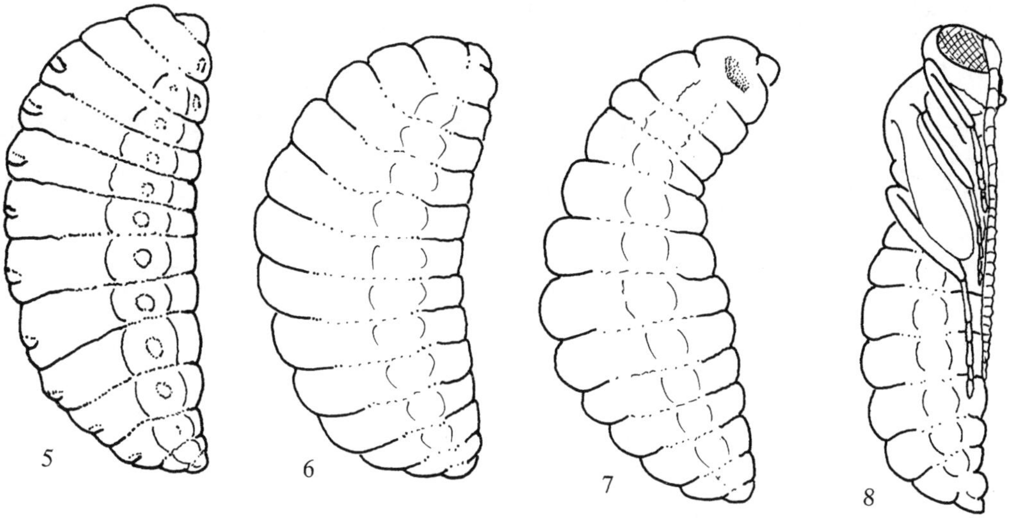

Figs 2.05–2.08. Late developmental stages of a tryphonine ichneumonid; 2.05, final instar larva prior to cocoon spinning; 2.06, eonymph—as contained in cocoon; 2.07, pronymph; 2.08, pupa.

neumonids that spin cocoons in spider webs disguise their cocoon as a dead insect wrapped in silk (Gauld, 1991). Other ichneumonoids, such as *Charops* and *Meteorus*, spin cocoons that hang from vegetation by a long silken thread, and thus perhaps escape the attention of crawling predators. In many gregarious species, such as some symphytans (Dias, 1976), pimpline ichneumonids and microgastrine braconids, the larvae collaborate to produce communal cocoon masses—structures that perhaps offer more protection against cocoon parasitoids than solitary cocoons do. In Costa Rica adults of both an ichneumonid (*Zonopimpla*) and its hyperparasitoid (*Theronia*) have been reared from an aggregated cocoon mass of a *Zonopimpla* species, showing that the hyperparasitoid female failed to successfully attack all potential hosts in the cocoon mass (Gauld, 1991).

In many Hymenoptera much of the period spent within the cocoon is actually passed as a virtually immobile last instar larva. In at least some sawflies there is a distinct non-feeding prepupal instar which follows ecdysis of the final-instar feeding-larva, but precedes the pupal ecdysis (Schedl, 1991). During this prepupal instar morphologically distinctive eonymphal and pronymphal phases occur (Eliescu, 1932). In some apocritans a 'prepupal phase' is also recognized, but this is a post-feeding phase of the final instar larva, and the two phases are not separated by an ecdysis. During this apocritan 'prepupal phase', eonymphal and pronymphal phases are also distinguishable (Morris, 1937; Slovák, 1984). In the eonymphal phase, the larva (which is now within its cocoon) becomes almost incapable of movement and is rather shorter and stouter than an active larva (Fig. 2.06). The pronymphal phase, a pharate pupa (Fig. 2.07), is narrower anteriorly and apparently incapable of motion (Morris, 1937). Hymenoptera that diapause for prolonged periods often do so as an eonymph. The pronymphal phase is of very short duration, and once a change from eonymph to pronymph has occurred pupation and generally adult emergence follow rapidly.

Some hymenopterans may spend a very long period diapausing as a prepupa within the cocoon. In the temperate region one sawfly is known to be able to diapause for up to six years (Prebble, 1941*a–c*), while in the seasonally dry parts of Costa Rica the ichneumonid *Enicospilus lebophagus* may remain in its cocoon for two years. In such cases it is almost always only a small fraction of each generation that enters a prolonged diapause. This behaviour is found in a number of insects living in climatically unpredictable habitats where inopportune periods of inclement weather could eliminate the entire population. Although hymenopterans often spend long periods within the cocoon, the duration of the actual pupal instar seldom lasts more than a few weeks. Some adults emerge from the pupa but remain within the cocoon or pupation chamber until weather conditions are favourable.

2.2 IMPORTANT BIOLOGICAL FEATURES IN THE EVOLUTION OF THE ORDER

Ian D. Gauld and Paul E. Hanson

Four biological features that are present in most hymenopteran species, including the most primitive, have probably played an important role in the evolution of the order. These are: the ovipositional mechanism, parental provision for the larva, the larval diet, and the sex determination mechanism. Only the sex determination mechanism is unquestionably a synapomorphy for the order; the ovipositional mechanism and possibly the larval diet were inherited from the ancestors of Hymenoptera, whereas parental provisioning for the larva may have arisen independently on numerous occasions within the order (perhaps as a consequence of the ovipositional mechanism).

Ovipositional mechanism

Amongst the holometabolous insect orders only the Hymenoptera retains the primitive, lepismatid form of ovipositor (Scudder, 1961; Smith, 1970), a structure whose original (ancestral) function presumably was to place an egg precisely in a more or less concealed location (as in many Orthoptera). Xyelid sawflies and siricoid woodwasps use their ovipositor to penetrate plant material and place an egg in concealment. The ovipositor is used in an essentially similar manner by many apocritans, allowing them to penetrate barriers and gain ovipositional access to hosts concealed in plant tissue, silken sacs or mud nests (Gauld, 1988*c*). In several groups of parasitoids the entire metasoma of the female wasp is modified to allow her to manipulate an extremely long ovipositor in reaching otherwise inaccessible hosts (see Chapter 4.3). In the more specialized endophagous parasitoids the ovipositor permits the placement of an egg within a host (Vinson, 1990), sometimes in an extremely precise location such as a nerve ganglion or salivary gland (Silvestri, 1921; Veen, 1981), or in the phytophagous agaonids, in a fig ovule (Verkerke, 1989). Dipterous parasitoids, which lack a well-developed ovipositor, are largely excluded from exploiting concealed hosts (Eggleton & Belshaw, 1992, 1993; Memmott *et al.*, 1994), and from developing as egg parasitoids.

Associated with the ovipositor and female reproductive tract are accessory glands that produce fluid secretions. Originally this fluid probably lubricated the valves of the ovipositor in order to facilitate the passage of the egg (Robertson, 1968). At some early stage in the evolution of the Hymenoptera these glands became modified to produce secretions that affect the ovipositional substrate, rendering it more favourable for larval development. Nematine tenthredinids produce gall-forming secretions that are injected into a plant during oviposition (McCalla *et al.*, 1962) and many siricids inject a mucus secretion that promotes the growth of a mutualistic fungus in wood where the larva feeds (Spradbery & Kirk, 1978). The ancestors of the Apocrita probably used the ovipositor in much the same way, but accessory gland secretions were probably modified in the carnivorous hymenopterans to venoms that incapacitate an arthropod host. This hypothesis is supported by the fact that the mucus gland and reservoir of siricids is homologous with the venom gland and reservoir of Apocrita. Many apocritan venoms, which are injected into the host prior to oviposition, paralyse the victim without killing it. Consequently the host remains alive and fresh until consumed by the hymenopteran larva. In many other lineages of parasitoids the venoms no longer cause paralysis in the host, but instead, subtly modify the physiology and development of the host (Vinson & Iwantsch, 1980*b*; M.R. Shaw, 1981, 1983; Bühler *et al.*, 1985; Coudron *et al.*, 1990).

The dual functions of the ovipositor—venom injection and egg laying—are operative in almost all Apocrita except the aculeate families. In most aculeate hymenopterans the ovipositor is no longer used for egg-laying; instead it has become modified into a sting that can penetrate and inject venom into animal tissue. In these families the venom glands are often well-developed and the venom is particularly effective for subduing large and active arthropods. In many groups, such as pompilids and scoliids the ovipositor is used primarily for subduing the prey, but in social aculeates, such as honey-bees and many vespids, its major use is defence (Hermann & Blum, 1981). It has been argued that the sting has been a major preadaptation facilitating the evolution of eusociality (Starr, 1985, 1989), although this view has been contested (Kukuk *et al.*, 1989). It is interesting to note that bradykinins, peptides present in the venoms of some solitary and eusocial vespoids, both block synaptic transmission in insects, and thus may be responsible for causing paralysis, and are important pain-producers in vertebrates (Piek, 1992).

Parental provision for the larva

Ancestral Hymenoptera almost certainly had phytophagous larvae, and these insects probably used their ovipositor to place the egg very precisely within nutritious plant tissue in much the same way that some of the modern representatives of the most primitive hymenopteran lineage, the Xyelidae, place their eggs in the developing sporophyll of gymnosperms. An important evolutionary trait of the Apocrita has been the continuation of this ancestral behaviour. In many apocritans, the adult female places the egg adjacent to an immobilized animal food source for the young. This has, of course, involved the female in locating and immobilizing a food resource for her offspring. Most other carnivorous insects have a more mobile larva that is obliged to seek out its own food. In ancestral apocritans the female probably oviposited onto a concealed, immobilized arthropod, and the hatching larva derived its sustenance feeding on this food resource, which it ultimately killed and more or less completely devoured.

The least specialized type of hymenopterous parasitoid is thus an ectoparasitoid developing externally on a concealed host that was immobilized by its female parent. Many groups show extraordinary specializations enabling them to utilize such hosts. An additional evolutionary step for apocritans has been the development of ways of using exposed hosts—hosts which cannot simply be paralysed and oviposited upon *in situ* lest they, and the apocritan egg, be devoured by a passing scavenger. This has been achieved in one of two main ways:

- By a reduction in the paralysing effect of the venom, so that the host can be stung, partially subdued and oviposited upon, then recover and develop further while the parasitoid larva slowly develops on or, more usually, within it.
- By paralysing the host and then concealing it prior to oviposition.

The former strategy has evolved repeatedly in a variety of parasitoid groups including the Proctotrupoidea, Cynipoidea, Chalcidoidea, and Ichneumonoidea, whereas the latter is characteristic of the nest-building Aculeata. In general the more primitive members of both evolutionary pathways still utilize relatively large hosts, but both strategies have offered a means of exploiting smaller, numerically more common and often less dangerous hosts (prey). Parasitoids may oviposit on or in a very young host larva (or even an egg) and delay development until the host is more or less fully grown, while many aculeates provision their nest with multiple prey organisms.

Diet of the larva

The most primitive Hymenoptera, such as Xyelidae, have larvae that are phytophagous, but many of these show a tendency to feed on a more nutritious diet than normal plant mesophyll, such as developing sporophyll tissue (Burdick, 1961). Gross leaf feeding, as practised by many sawflies, probably represents a later specialization in the Symphyta whereas the apocritan lineage has probably retained the ancestral habit of feeding on a relatively nutritious food source (Rasnitsyn, 1969, 1980).

The larvae of apocritans are adapted for consuming a very nourishing, semi-fluid diet, which means that only a relatively small amount of indigestible matter passes into the gut. This has allowed a simple structural modification to occur—failure of the proctodeal invagination in the embryo to join with the midgut, so that the gut remains closed towards its hind end. It remains closed for most of the larval life and faecal matter accumulates at the end of the midgut until being voided (as a MECONIUM) immediately prior to pupation, when intestinal development is completed. Within the apocritan Hymenoptera this feature appears to be present in all families except the Gasteruptiidae and many bees, but outside the apocritan Hymenoptera it occurs only rarely (e.g in Neuroptera, and some Cecidomyiidae). By postponing defecation until they have completed feeding apocritan larvae avoid fouling their immediate environment, which may be an important preadaptation for endoparasitism. In some bees early instar larvae defecate (e.g. megachilines, *Xylocopa*, *Ceratina*), in which case the faeces are solid and are often attached by silk threads to the cell wall to prevent them from contaminating the food of the larva (Roubik, 1989). In other bees most larval growth does not occur until the last larval instar, after connection of the midgut to the hindgut.

Comparing the diet of Hymenoptera with that of the other speciose orders of insects it is obvious that the success (in terms of speciation) of the Hymenoptera has been in its carnivorous species, particularly the parasitoids. In part, this species richness has perhaps resulted from their use of another arthropod, not only as a food source, but also as a place to live. In the case of endoparasitoids this requires physiological specializations to allow them to overcome the host's immunodefensive systems, yet not kill the host. Such specializations are often almost species-specific as, in many cases, closely related groups of endoparasitoid species differ in their ability to develop in particular host species (see Chapter 2.5). Another food resource that hymenopterans have managed to exploit more successfully than any other group of insects is pollen, which is collected and concentrated for larval consumption by bees and masarine vespids. In general, however, groups of hymenopteran herbivores are not very speciose; the tenthredinoid sawflies, for example, are not nearly as species-rich as the somewhat similar macrolepidoptera. Fungivory is not as prevalent in Hymenoptera (see Chapter 2.3) as it is in Coleoptera and Diptera, although the leaf-cutting ants have used fungivory to become (in terms of impact on the ecosystem) very successful herbivores. In contrast to the Diptera and Coleoptera, saprophagy, necrophagy and coprophagy are very uncommon within the Hymenoptera. These habits only occur in some ants and social vespids that scavenge from overripe fruit, corpses etc., and in one group of neotropical bees, the *Trigona hypogea* species-complex, which are obligate necrophages (Camargo & Roubik, 1991).

Sex determination mechanisms

In 1845 the bee-keeper Johannes Dzierzon suggested that drones (male bees) developed from unfertilized eggs, while workers and queens (females) came from fertilized eggs. Dzierzon's theory initially aroused considerable scepticism, but it is now known that this method of sex determination is a characteristic of the order Hymenoptera (Kerr, 1962*a*). Generally, males are haploid, having an unpaired set of chromosomes, and females are diploid, having a paired set of chromosomes (haploid chromosome numbers in Hymenoptera range from three to 42; Crozier, 1975). The process whereby unfertilized (haploid) eggs develop to become males is termed ARRHENOTOKY or arrhenotokous parthenogenesis. Oogenesis is similar to that in other insects but spermiogenesis in haploid males is ameiotic. As a result of arrhenotoky Hymenoptera males receive genes only from their mother and they transmit genes only to their daughters. Outside the Hymenoptera arrhenotoky is known to occur in micromalthine beetles, iceryine scale-insects, thrips, tetranychid mites and rotifers; in most other animals sex determination occurs via male or female heterogamety, that is, by a balance between autosomal and sex chromosomes (Bull, 1981).

Two general mechanisms have been proposed to explain the association between sex and ploidy in Hymenoptera (summarized in Luck *et al.*, 1993): the allelic diversity model (Crozier, 1977) and the genetic balance model (Kerr & Nielsen, 1967). According to the first model there are one or more sex-determining loci, each with multiple alleles (an estimated 5 to 25 alleles). Diploid offspring are female if they are heterozygous at one or more of these loci. Diploid offspring that are homozygous at all sex-determining loci are diploid males. Haploid offspring are always males because they are hemizygous. On the other hand, the genetic balance model postulates that the sex of the offspring is determined by the balance between the effects of female-determining loci and the male-determining loci. In haploid progeny the strength of the male-determining loci outweighs that of the female-determining loci.

The two models can be tested by continuous inbreeding and testing for the presence of diploid males in the progeny. The allelic diversity model predicts that inbreeding will yield diploid males whereas the genetic balance model does not. Moreover, because diploid males are non-functional, the allelic diversity model predicts that field populations should show minimal inbreeding. Laboratory inbreeding experiments have produced diploid males in sawflies, ichneumonids, braconids and bees (Luck *et al.*, 1993); indirect evidence suggests their presence in ants as well (Stouthamer *et al.*, 1992). However, inbreeding (sib-mating) appears to characterize field populations of other groups of Hymenoptera, such as many Scelionidae, Chalcidoidea and Bethylidae, yet diploid males have not been found in the field or in the laboratory. Thus the allelic diversity model (or more specifically, the single locus multiple allele model) appears to be operative in some, but not all Hymenoptera. In hymenopterans showing extensive inbreeding, but lacking diploid males, the genic balance model may be operative, although this mechanism of sex determination has still not been demonstrated unambiguously in any species (Crozier, 1977; Luck *et al.*, 1993).

Arrhenotoky has various consequences for the biology of Hymenoptera, as well as for human attempts to manipulate hymenopterans in biological control and conservation of endangered species (Unruh & Messing, 1993).

- Because the act of fertilization determines the sex of the zygote, the spermatheca (sperm storage organ) takes on a special significance in the Hymenoptera. By controlling the release of sperm from the spermatheca, inseminated females can select whether they lay a male or a female egg (Flanders, 1956). This allows parasitoids to place female eggs in larger hosts and male eggs in smaller hosts. Also, progeny sex ratio can be adjusted, to conditions of inbreeding for example (see Chapter 2.5).
- The fact that all females (sisters) derived from a singly-mated mother share an average of three-fourths of their genetic complement may have been one of the mechanisms facilitating the evolution of eusociality in Hymenoptera (see Chapter 2.6).
- Hymenoptera have roughly one-third the genetic variation (single locus polymorphism) observed in diploid insects. Various hypotheses have been proposed to explain this observation. For example, low genetic variability could result from a lower effective population size, especially in eusocial Hymenoptera where there are fewer breeding individuals (Graur, 1985).
- Haplodiploidy may provide a mechanism whereby deleterious recessive genes can be rapidly eliminated from a population through natural selection operating on hemizygous males, as the genes carried by the males are necessarily expressed (Havron *et al.*, 1987).
- In taxa that utilize the allelic diversity mechanism, the sex determining alleles are effectively recessive lethals in diploid individuals. For a single locus system this genetic load would exceed the mutational load. In addition, these taxa are probably more vulnerable to inbreeding depression (Werren, 1993). This may explain why, in general, Chalcidoidea have been more successful in biological control programmes than Ichneumonoidea (Stouthamer *et al.*, 1992).

Although arrhenotokous hymenopterans sometimes have extremely skewed sex ratios, males are still necessary to inseminate the female, so that she can lay fertilized, female-producing eggs. However, many species of Hymenoptera exhibit a different form of parthenogenesis, THELYTOKY, where unmated diploid females produce diploid female progeny by a variety of asexual means (White, 1973). Thelytokous reproduction is known to occur in at least 270 species belonging to eight superfamilies (the Tenthredinoidea, Proctotrupoidea, Cynipoidea, Chalcidoidea, Ichneumonoidea, Chrysidoidea, Vespoidea and Apoidea) as well as cyclically in more than 2000 species of cynipids (Luck *et al.*, 1993). At least two principal causes of thelytoky have been reported in the Hymenoptera: interspecific hybridization and infection with extrachromosomal elements. In the latter case, treatment with heat or antibiotics results in permanent reversion to arrhenotokous reproduction (Stouthamer *et al.*, 1990*a*).

There are two possible benefits for thelytokous females when compared with their arrhenotokous counterparts:

- unmated females can produce daughters, whereas arrhenotokous females cannot (Gerritsen, 1980);
- they may manifest a higher intrinsic rate of increase (Suomalainen *et al.*, 1987).

To date, there is little evidence for the latter (Luck *et al.*, 1993), but some evidence exists for the former—many thelytokous species do seem to be geographically widespread in comparison to their arrhenotokous relatives. For example, the thelytokous ichneumonid, *Diplazon laetatorius*, a larval parasitoid of cosmopolitan aphidophagous syrphids, has been found in all parts of the world including remote oceanic islands; no other diplazontine has a comparable range. A substantial number of parasitoid species that are used in classical biological control programmes are known to reproduce arrhenotokously in their original geographic range, but thelytokously in their new environment (Aeschlimann, 1990). In some cases this probably reflects the existence of arrhenotokous and thelytokous races (Stouthamer & Luck, 1991).

In several thelytokous species normal meiosis occurs but diploidy is restored by gamete duplication or terminal fusion, which results in increasing homozygosity (Suomalainen *et al.*, 1987). The allelic diversity model discussed above predicts the occurrence of diploid males under these circumstances, but such males have not been reported. Because many thelytokous taxa belong to groups in which inbreeding is common, there may be an association between sex determination (genetic balance model), mating structure (sib-mating) and the occurrence of thelytoky (Luck *et al.*, 1993). It has been assumed that thelytokous individuals cannot mate and are therefore reproductively isolated. However, in some cases males from arrhenotokous subpopulations can inseminate thelytokous females who incorporate the genes in some of their daughters (Rössler & DeBach, 1973; Legner, 1987). The resulting genetic recombination should create new thelytokous lineages with varying levels of genetic load, including some with loads less than their maternal lineage (Werren, 1993).

Traditionally, a third type of parthenogenesis—DEUTEROTOKY—has been distinguished. This is the production of male and female offspring from unfertilized eggs. Some species that were previously thought to be thelytokous have been shown to produce occasional males from unfertilized eggs, and in some species at least, these males mate and pass on their genes to subsequent generations which then reproduce via thelytoky. Consequently, the distinction between deuterotoky and thelytoky is now being questioned, except perhaps in the case of Cynipidae (Luck *et al.*, 1993).

The complex life cycles of cynipids ('gall wasps') entail alternating sexual and asexual (agamic) generations, and the mode of reproduction can be either arrhenotoky–thelytoky or thelytoky–deuterotoky (see Chapter 10.1). In some species females of the asexual generation produce males and females (the sexual generation) by deuterotoky. In other species fertilized (sexual) females produce offspring (asexuals) that are either androphores (females that produce only males) or gynephores (females that produce only females). In yet other species asexual androphores appear to produce two types of males that form spermatozoa with differing sexual potential (Folliot, 1964; Askew, 1984). Thus the evolutionary trend in gall wasps appears to be progressively earlier determination of the sex of individuals of the sexual generation. A somewhat similar, though less complex situation that may involve deuterotokous reproduction exists in the ichneumonid *Sphecophaga*. Inseminated females may produce brachypterous females which do not mate, but are capable of laying eggs that give rise to both males and females (Donovan, 1991).

2.3 PHYTOPHAGY IN THE LARVAL HYMENOPTERA

Paul E. Hanson

Hymenoptera with phytophagous larvae can be divided into two groups, which more or less coincide with the traditional division of the order into two suborders, Symphyta (sawflies and woodwasps) and Apocrita (fig wasps, gall wasps, bees, etc.). The former are primary phytophages, where feeding on plants represents a retention of the ancestral habit of the order. Although both the more primitive xyeline xyelids and the lineage from which the apocritans have been derived seem to specialize in feeding on highly nutritious parts of plants (such as developing male sporophylls and meristematic tissue), other symphytans have radiated to exploit a wide range of plant material. Most feed externally on leaves, like macrolepidopteran caterpillars, but a few are leaf miners (e.g. some Heterarthrinae), and one or two cause galls (e.g. some Nematinae). Most authors regard the plant-feeding apocritans as being secondarily phytophagous in that they are apparently descendants of carnivorous ancestors (Rasnitsyn, 1980; Königsmann, 1978*a* & *b*; but see Malyshev, 1968, for an alternative view). None of the secondarily phytophagous Apocrita are folivores. All feed endophytically or enclosed within a nest, and nearly all feed on very nutritious plant parts such as gall tissue, seeds, pollen or fungal tissue.

Exophytic leaf feeders

The larvae of primitive symphytans, such as xyelids, generally feed endophytically on fairly nutritious plant material (Burdick, 1961), but many tenthredinoid larvae feed externally on green plant tissue. Thus the latter are confronted with problems similar to those encountered by exophytic lepidopteran larvae (*cf.* Strong *et al.*, 1984), and not surprisingly the two groups have evolved similar adaptations. For example, both must maintain a hold on the plant, and exophytic sawfly larvae like those of the Lepidoptera, have abdominal prolegs. In sawfly larvae, however, the prolegs lack crochets, the minute hooks characteristic of lepidopterous caterpillars. Exophytic larvae must also be capable of surviving attacks by both parasitoids and predators. Both lepidopteran and sawfly larvae utilize a variety of similar defences against predators, such as cryptic coloration, cuticular spines, or toxic chemicals combined with aposematic coloration (Scoble, 1992). In sawflies, however, maternal guarding of the offspring is more common than in the Lepidoptera.

A major obstacle confronting exophytic plant feeders is obtaining adequate nutrition. Not only is green plant tissue generally less nutritious than the diet utilized by primitive sawflies, so more food must be consumed, it is also often richer in toxic secondary compounds. Folivores either have to avoid feeding on these leaves, or they have had to evolve detoxification mechanisms. In the north temperate region most sawfly species have apparently taken the first option. They feed as larvae on plants notably lacking in toxic alkaloids and complex terpenoids, such as Salicaceae, Rosaceae and Betulaceae (Gauld & Bolton, 1988). In the tropics, however, woody plants are generally rich in these toxins (Levin, 1976; Levin & York, 1978). Some symphytan lineages, such as Argidae, have overcome tropical plant defences and diversified to some extent, but other lineages (e.g. Pamphiliinae, Nematinae and Tenthredininae) are notably absent or very scarce, and in total there are far fewer sawfly species in tropical habitats than there are in comparable temperate ones. Moreover, tropical symphytans seem to be more uniformly distributed across woody plant taxa (see Chapter 6.1) than are their temperate zone relatives. Although more data are needed to substantiate this suggestion, we have found no indication of 'species-flocks' of tropical sawflies associated with a few plant species, such as occur in some sawfly groups on a few species of Salicaceae in northern habitats.

Gall-formers

Galls are 'abnormal' plant growths induced by another organism. They are abnormal only in the sense that morphogenesis is directed by the gall-former and not by the plant. However, the process of gall formation does not involve any unique substances or physiological processes. Unlike a tumour, a gall has a definite structure characteristic of the species that induced it (Mani, 1964; Shorthouse & Rohfritsch, 1992). Gall-inducing organisms include some fungi, a few nematodes, one group of mites, and several groups of insects. Among neotropical gall-forming insects the Hymenoptera probably rank second to Diptera in species-richness (if the 'seed parasitoids' are included—see below). Gall-forming Hymenoptera include both Symphyta and Apocrita, but only species of the latter form galls in the Neotropics. These include members of three superfamilies: Cynipoidea, Chalcidoidea and Ichneumonoidea.

The actual mechanism of gall formation is poorly understood, but undoubtedly involves plant hormones (Staden & Davey, 1978), and possibly the injection of a plasmid or virus into the plant (Cornell, 1983). Complicating our understanding of gall formation is the likelihood that the process may differ between groups of gall-formers. In Hymenoptera for example, sawfly galls are induced, at least initially, by the adult female during oviposition (McCalla *et al.*, 1962; Narendran, 1984). In Cynipidae on the other hand, the stimulus appears to come from the larva (Askew, 1984). Gall-formers are usually very host specific and in many cases the gall is formed on a specific part of the plant. Gall structure also varies between species of gall-formers, but all have an inner layer of very nutritious tissue on which the gall-forming organism feeds (Bronner, 1977). Many gall-formers require meristematic tissue for gall formation, although a few appear to be capable of causing developed tissue to revert to meristematic tissue.

In the past it has sometimes been stated that galls represent an adaptive response by the plant—that galls represent a plant's way of isolating a parasite (Mani, 1964). However, the fact that gall-forming insects have been successfully used in biological control of weeds (Dennill, 1988) demonstrates their potential negative impact on the plant. Thus gall formation is more appropriately viewed as a highly specialized form of phytophagy, with the gall-former inducing the host plant to channel nutrients to its protective chamber. Because gall-formers feed exclusively on the nutritive layer within the gall, secondary plant compounds probably offer little defence against gall-formers. Indeed cynipids appear to concentrate oak tannins in the outer part of the gall for their own defence against fungi (Taper & Case, 1987). Thus the principal defences of the plant against gall-formers are probably an early hypersensitive (necrotic) reaction at the tissue level, or later abortion (abcission) of heavily galled organs.

Not all Hymenoptera that live in galls are gall-formers. Many are parasitoids of the gall-formers, although some of these also develop, at least partially, on gall tissue. Organisms that occur in galls induced by other organisms, and subsist principally on gall tissue, are called INQUILINES. The presence of an inquiline may or may not result in the death of the original gall-former. Although they are not capable of initiating their own gall, some inquiline species induce modifications in the structure of the gall (Askew, 1984). In some cases careful study is required to determine whether a given species is a gall-former, an inquiline, or a parasitoid. The genus *Eurytoma* (Eurytomidae), for example, includes species of each.

Among Hymenoptera, the family Cynipidae consists entirely of gall-formers or obligate inquilines of galls, and the majority of species are associated with oaks (*Quercus*) and other Fagaceae. Virtually all species of Tanaostigmatidae are associated with galls, especially on plants of the family Leguminosae, but in most cases we do not know what the larva actually does within the gall. Many Eurytomidae are associated with galls, as parasitoids, inquilines, or as gall-formers. For example, species of *Tetramesa* cause inconspicuous galls on grasses (Claridge, 1961). Many, if not all Chalcidoidea that feed in seeds are probably gall-formers (see below) and, in other parts of the world, a few species of Eulophidae and Pteromalidae cause galls on other parts of plants, although thus far no examples are known from Costa Rica.

Until recently gall-formers were unknown in the superfamily Ichneumonoidea. However, there is now good evidence that the Brazilian braconid, *Allorhogas dyspistus*, induces galls in the seeds of *Pithecellobium tortum* (Leguminosae) (Macêdo & Monteiro, 1989; Marsh, 1991). In Costa Rica I have found an undescribed species of *Allorhogas* inducing galls in the seeds of *Pithecellobium macradenium* and it is possible that many of the species in this genus that were previously presumed to be parasitoids of bruchid beetle larvae (Whitehead, 1975), are in fact gall-formers. A possible second genus of gall-forming Braconidae is *Monitoriella*, which appears to be the causal agent of leaf galls on *Philodendron radiatum* (Araceae) (Hanson *et al.*, in prep.).

Although the ancestral biology of Cynipidae is uncertain, most of the other apocritan gall-formers appear to have originated from ectoparasitoids of hosts feeding in concealment in plant tissue. It is possible that many have evolved from parasitoids (or inquilines) of other gall-formers. In contrast, most non-apocritan gall-forming insects have probably evolved from phytophagous (or mycophagous) ancestors, and for these insects there are three hypothesized advantages to gall-formation, compared with ordinary phytophagy (Price *et al.*, 1987): enhanced nutrition, protection from hygrothermal stress, and reduced attack by natural enemies. However, among apocritan gall-formers, the first two hypotheses are probably untenable, since they are derived from groups that fed on a nutritious diet (meat), protected from hygrothermal stress within the plant. Thus, of the three hypotheses on the adaptive significance of insect gall formation, only the natural enemy hypothesis is tenable for apocrite Hymenoptera. There are, however, other possible advantages entailed in the evolution of gall-formers from gall associates—enhanced possibilities in locating a host and reduced competition.

SEED FEEDERS

Hymenopterous larvae feed on seeds in two very different ways (Table 2.1): as SEED PARASITOIDS or as SEED PREDATORS. Although such seed feeding is often injurious to the plant, and can hamper husbandry activities (see Chapter 3), in a few cases the plants have turned the relationship into a mutualism by employing the seed parasitoids as pollinators and the seed predatory ants as seed dispersers.

Seed parasitoids. Female seed parasitoids oviposit into developing seeds and their larvae feed within the

Characteristic	Seed parasitoids	Seed predators
Taxa	mostly Chalcidoidea	Formicidae
Ancestry	parasitoids	predators
Age of seed utilized	developing seed	mature seed
Manner of feeding	endophagous 'koinobiont'*	ectophagous 'idiobiont'*
Type of mutualism (when present)	pollination	seed dispersal

Table 2.1. Comparison of hymenopterous seed parasitoids and seed predators. *For an explanation of idiobiont versus koinobiont, see Chapter 2.3.

seed. A wide variety of holometabolous insects are seed parasitoids, but among the Hymenoptera, most seed parasitoids are probably also gall-formers—although in many cases gall formation is difficult to detect without microscopic examination. Hymenopterous seed parasitoids therefore encompass the same apocritan taxa discussed above: Cynipidae, Chalcidoidea and Braconidae. In general, seed parasitoids are less common in Cynipidae (a few species that form galls in acorns) and Braconidae (*Allorhogas*), and most of these develop as gregarious parasitoids. In Chalcidoidea, however, there are many species of seed parasitoids, and most of these develop solitarily. Among Chalcidoidea, seed parasitoids ('seed chalcids') occur primarily in the families Agaonidae, Eurytomidae and Torymidae, but also include a few species of Eulophidae (in the Neotropics, *Lisseurytomella flava*) and Pteromalidae.

All, or nearly all, Agaonidae are seed parasitoids of figs (Moraceae: *Ficus* species), but the larvae of different subfamilies may feed on different tissues within the seed. For example, Agaoninae feed within the embryo sac (Verkerke, 1989) whereas at least one species of Sycophaginae is known to feed on proliferating nucellar cells (Galil *et al.*, 1970). Besides being seed parasitoids of figs, agaonines are also their obligate pollinators, and for each of the 750 species of *Ficus* there is a corresponding species of fig wasp. Non-pollinating fig wasps (the other subfamilies of Agaonidae) are also seed parasitoids. A few are capable of developing in the absence of the pollinating fig wasps, but most appear to require the presence of pollinating Agaoninae since unpollinated flower receptacles are aborted by the plant.

Seed-feeding appears to have evolved more often in the family Eurytomidae than in any other family of Hymenoptera (see Chapter 11.9). Some *Bruchophagus* species develop in the seeds of Leguminosae from eggs placed in the endosperm of developing seeds. The larva lies between the seed coat and the developing plant embryo, at first feeding only on the growing cotyledons, but eventually consuming the entire contents of the seed (Batiste, 1967). Seeds infested by some species may be distorted and larger than uninfested seeds (Batiste, 1967), but in other cases infested seeds may be smaller than uninfested ones (Hedlin, 1956).

Among Torymidae the larvae of many species of *Megastigmus* feed in the seeds of conifers, Rosaceae, Aquifoliaceae and a few other plants (see Chapter 11.18). In rose seeds in North America, *M. nigrovariegatus* larvae feed on the developing cotyledons but do not consume the growing basal part of the embryo and the endosperm until later larval development (Milliron, 1949). An undescribed species of *Megastigmus* in Costa Rica feeds in the seeds of *Symplocos* (Symplocaceae) (Hanson, in prep.). The young larva feeds on the walls of the locule, before the ovules have grown to fill the space; presumably the ovules are consumed later.

Seed predators. Seed harvesting has arisen many times within three subfamilies of ants, the Myrmicinae, Ponerinae and Formicinae, although it is disproportionately concentrated in the first of these. The commitment of these harvesting ants to a seed diet varies from occasional to almost total. Harvesting ants are dominant elements in the deserts and drier grasslands in warm temperate and tropical regions (Hölldobler & Wilson, 1990*a*). Although less prevalent in Costa Rica, seed harvesting does occur to some extent in *Solenopsis geminata* and some species of *Pheidole*. Harvesting ants carry seeds to their nests where they are utilized as food for the larvae, but they sometimes end up dispersing seeds when they are forgotten or lost in their granaries.

A more dependable and specialized form of seed dispersal occurs in plants having a special appendage or outer covering, the ELAIOSOME, surrounding the seed (Beattie, 1985; Davidson & Epstein, 1989). Such seeds are highly attractive to many different ants, which carry them away. Only the elaiosome is consumed, the seed itself being discarded. Ants that harvest these seeds are therefore more comparable to frugivores than to seed predators.

Pollen and nectar feeders

Among holometabolous insects, bees and masarine vespids are the only ones to successfully exploit pollen and nectar as a larval food resource. Their ancestral habits of nest building and foraging for scattered prey items probably allowed them to exploit this very dispersed but nutritious food resource. In the process of gathering pollen and nectar for their larvae, these hymenopterans often pollinate plants. Plants that have come to depend upon bees for pollination often produce even more nutritious pollen than plants pollinated by other means. For example, plants that release pollen through apical pores in the anthers are usually bee pollinated (by a vibrational 'buzz pollination', see Chapter 18) and may contain the highest crude protein content of any pollen, from 55 to 65 percent (Buchmann, 1983). The protein content of pollen from other plants utilized by bees is usually in the range of 20 to 35 percent, which is still much higher than that of wind-pollinated plants which is under 10 percent (Stanley & Linskens, 1974). In addition to protein, pollen provides lipids (usually less than 5 percent), sterols, sugars, starches, vitamins and minerals.

In the best studied example, the honey-bee, foragers bring the pollen back to the nest where it is treated by other worker bees. Enzymes produced by the hypopharyngeal glands, mandibular glands, salivary glands or the gut are added to the pollen in order to begin the digestion process and to prepare it for long term storage (as 'bee bread'). Bacterial action may be responsible for the increase in histamine and vitamin K content (Winston, 1987). Similar processing of pollen appears to occur in other eusocial bees, such as stingless-bees (Delage-Darchen & Darchen, 1982). In solitary bees an individual female must do all the processing herself.

Floral nectar consists largely of sugars (5 to 80 percent) but also contains small quantities of nitrogen compounds (less than 0.2 percent protein), minerals, organic acids, vitamins (mostly vitamin C), lipids, pigments, and aromatic substances. Nectar from flowers visited by long-tongued bees appears to be generally rich in sucrose, whereas nectar from flowers visited by short-tongued bees is often rich in hexose (Baker & Baker, 1983), although the reason for this difference is not clear. Certain sugars such as mannose, galactose and rhamnose may be less suitable, or even toxic. In the case of the honey-bee, nectar is carried in the honey stomach to the nest, where it is transferred to nest workers for processing (Winston, 1987). Within the crop of the nest worker, enzymes from the hypopharyngeal glands are added to the nectar, which break down the sugars to simple 'inverted' forms. It is then exposed to evaporation on the worker's tongue and placed into a cell for further evaporation by wing fanning until the water content is less than 18 percent. With the water content thus reduced the honey is more resistant to yeast infection and ready for storage. Honey is likewise stored by stingless-bees and bumblebees, and its chemical composition is often influenced by the cells in which the honey is stored, for example by resins in the case of stingless-bees (Roubik, 1989).

Other, non-floral, sources of sugar occasionally used by bees and vespids as food for larvae include nectar from extrafloral nectaries and honeydew produced by Homoptera. Outside the bees and vespids, only ants utilize these sugar sources as a component in the larval diet. Although ants commonly harvest nectar from extra-floral nectaries and honeydew from Homoptera, they are often notably uncommon at flowers. They appear to be largely excluded by mechanical barriers and/or chemical deterrents in floral tissue, through which they would have to chew in order to reach the nectar (Guerrant & Fiedler, 1981).

In addition to the primary collectors of pollen and nectar, some hymenopterous cleptoparasites that attack bee nests, such as gasteruptiids and groteine ichneumonids, feed as larvae partially or entirely upon their host's pollen reserves.

Fungivores

Fungal tissue is qualitatively very different from vascular plant tissue (Martin, 1979). In general fungal tissue is relatively more nutritious than most green plant tissue and is certainly more nutritious than wood. Fungivory is less common in the Hymenoptera than it is in the three other major orders of endopterygote insects. It occurs only in a very few, specialized lineages, most notably the siricoids, the perreyine pergids and the attine ants. These three groups exemplify the three general categories of fungivory defined for insects in general (Kukor & Martin, 1986):

- The ingestion of relatively small quantities of fungal tissues along with larger amounts of the substrate on which the fungus is growing (generally wood), in order to assist in the digestion of the substrate;
- The ingestion of the fruiting bodies of larger macrofungi;
- Selective grazing on the vegetative portions and/or on the microscopic reproductive structures of fungi.

Fungus-assisted digestion of wood. To date there has been no compelling demonstration that any insect is capable of secreting the complete set of enzymes (cellulases) necessary for digesting cellulose. Nonetheless several groups of insects utilize cellulose as the principal component in their diet. They do so by harbouring endosymbionts, or by the ingestion of fungal cellulases (Martin, 1987). Among Hymenoptera, the larvae of Siricoidea utilize cellulases produced by fungi in order to feed on wood. Technically they are not true fungivores, since most of their food is derived from wood. However, the digestive enzymes necessary for processing the wood fragments are acquired from the relatively small quantity of fungus ingested, and without the fungus the woodwasp larva cannot survive beyond the first instar.

Adult female siricoids inoculate wood with a symbiotic fungus, carried in intersegmental pouches located at the base of the ovipositor (Francke-Grosmann, 1967; Stillwell, 1967; Spradbery, 1977; Madden & Coutts, 1979). The fungi involved are species of *Amylostereum* in the case of conifers, and *Daedalea unicolor* in hardwood trees (Stillwell, 1964). These fungi can occur as free-living species, but woodwasps are probably their major agents of dispersal in many areas. Following inoculation by the woodwasp, the fungus grows along the larval galleries and in the surrounding wood. The female larvae have a pair of deep folds (hypopleural organs) between the first and second abdominal segments that are filled with fungal fragments embedded in a clear secretion that hardens to a wax-like consistency before pupation. Upon eclosion these hardened fungal fragments attach to the ovipositor and are eventually taken up into the intersegmental pouches of the adult female.

Ingestion of fruiting bodies of macrofungi. Recently, the first record of any hymenopteran feeding on the fruiting body of a macrofungus was obtained in Costa Rica. About ten larvae of *Decameria rufiventris* (Pergidae) were found at about 1600 metres in the Monteverde cloud forest feeding on a jelly fungus (Basidiomycetes, *Auricularia* sp.), which was growing on rotten wood (D. Olson & F. Joyce, pers. comm.). Two were reared to adulthood. Some other pergids belonging to the same subfamily (Perreyinae) may have similar habits, as there are a few tantalizing records of them occurring in atypical situations. The wingless female of a New Guinea species of *Cladomacra*, has been found guarding its eggs under a log (Naumann, 1984)—a typical situation where jelly fungi grow. In the New World, masses of migrating larvae of a species of *Perreyia* have been observed crawling on the ground in Costa Rica (1200–1500 metres altitude), Mexico and South America. These aggregations of fifty or more larvae may be up to 30 cm long, 10 cm broad, and 5 cm high in the middle, and they progress as a single unit. These migrating masses of larvae have been observed only during the rainy season and are avoided by domestic chickens (C. Flores and INBio parataxonomists, pers. comm.). Recently these larvae have been reared through to adults on a diet of mouldy leaf litter (J. Ugalde and E. Quirós, pers. comm.)

Ingestion of microfungi. The only hymenopterans that are known to graze on microfungi are the fungus-growing ants (tribe Attini), which live in an obligate mutualistic association with a fungus that is found only in the nests of these ants. The fungus, the exact identity of which is still being debated, has lost its capacity for sexual reproduction; dispersal occurs by means of the queen ant carrying fungal hyphae to a new nest site. The ants in turn are dependent upon the specialized swollen hyphae of the fungus ('gongylidia') as food for their larvae (Weber, 1972, 1982; Lofgren & Vander Meer, 1986; Vander Meer *et al.*, 1990; Hölldobler & Wilson, 1990*a*).

In the more primitive attine ants, the adults cultivate the fungus on a substrate of insect excrement or fallen leaves. Species of *Atta* and *Acromyrmex*, which are known as the leaf-cutting ants, cut pieces of living leaves from plants and carry them to the nest, where smaller ants process the leaf material for incorporation into their fungal gardens (see Chapter 16). The mutualism between attine ants and their fungal

Host/prey group	**Families of Hymenoptera attacking**
Protura, Thysanura and Microcoryphyia	–
Collembola	For, Sph.
Diplura	For.
Ephemeroptera	**Sph**.
Odonata	*Eul*; *Mym*; *Sce*; **Sph**; *Tri*.
Grylloblattaria	–
Phasmida	*Chr*.
Orthoptera	*Aph*; *Enc*; *Eup*; *Eur*; *Mym*; Rho; *Sce*; Sph; Tip; *Tri*.
Mantodea	*Eul*; *Eup*; *Sce*; Sph; *Tor*.
Blattaria	*Eul*; *Eup*; *Eva*; *Mut*; *Pte*; Sph.
Isoptera	Bra, For.
Dermaptera	Pte.
Embiidina	Bra; *Sce*; Scl.
Plecoptera	–
Zoraptera	–
Psocoptera	Bra; *Mym*; Sph.
Phthiraptera	–
Hemiptera	*Aph*; Bra; Enc; *Eul*; *Eup*; *Eur*; *Mym*; *Pte*; *Sce*; Sph; *Tri*.
Homoptera-Auchenorrhyncha	*Aph*; Dry; Emb; Enc; Eul; *Eup*; *Mym*; *Pla*; Sph; *Tri*.
Homoptera-Sternorrhyncha	Aph; Bra; Cer; Enc; Eul; Eup; *Mym* (?); Pla; Pte; Sig; Sph.
Thysanoptera	Cer; Eul; *Mym*; Sph; *Tri*.
Neuroptera	Bra; Cer; Cha; Enc; Eup; Fig; Hel; Ich; Meg; Per; Pte; *Sce*; **Sph**; *Tri*.
Mecoptera	Bra, Ich, Meg; **Sph**.
Coleoptera	Aul; Bet; Bra; Cha; Dia; Ela; Enc; Eul; Eup; Eur; For, Iba; Ich; Meg; Mut; *Mym*; Oru; Pel; Per; *Pla*; Pro; Pte; *Sce*; Sco; Scy (?); **Sph**; Ste; *Tet*; Tip; Tor; *Tri*; Ves.
Strepsiptera	**Cha**; Pte.
Trichoptera	Ich; **Sph**.
Lepidoptera	*Aph*; Bet; Bra; Cha; Ela; Enc; Eul; Eup; Eur; For, Ich; Meg (?); Mut; *Mym*; Pte; *Sce*; Sph; Tor; *Tri*; Ves.

Table 2.2. continued opposite.

Host/prey group	Families of Hymenoptera attacking
Diptera	Aph; Bra; Cer; Cha; Dia; Enc; Eul; Eup; Eur; Fig; Ich; Meg; Mon; Mut; Orm (?); Pla; Pro; Pte; *Sce*; Sig; **Sph**; Tet; Tor; Trg; *Tri*.
Siphonaptera	Pte.
Hymenoptera-Symphyta	Aul; Bra; Cha; Chr; Eul; Eup; Eur; Iba; Ich; *Mym*; Oru; Per; Pte; Sph; Ste; Tor; Trg; *Tet*; *Tri*; Ves.
Hymenoptera-Apocrita	Aph; Bet; Bra; Cer; Cha; Chr; Dia; Ela; Enc; Euc; Eul; Eup; Eur; Fig; For, Gas; Ich; Leu; Meg; Mut; *Mym* (?); Orm; Per; Pte; Sap; Sig; **Sph**; Ste (?); Tor; Trg; *Tri*.
Araneae	*Eul*; *Eup*; *Eur*; *For*, Ich; Pom; *Pte*; *Sce*; Sph.
Solifugae	Bry.
Acariformes	Eul, For.
Parasitiformes (Ixodidae)	Enc.
Chilopoda	For, Pro (?).
Diplopoda	For.
Isopoda	For.
Oligochaeta	For.
Nematoda	Eul.

Table 2.2. The groups of terrestrial arthropods (plus Annelida and Nematoda) and the families of Hymenoptera that attack them as hosts or prey. Most Formicidae (excluding the specialized predators listed by Hölldobler and Wilson, 1990*a*, Table 15.1) and social Vespidae, which when considered as a group are extremely polyphagous, are excluded as are the very few hymenopteran scavengers and necrophages.
Aul = Aulacidae; Aph = Aphelinidae; Bet = Bethylidae; Bra = Braconidae; Bry = Bradynobaenidae; Cer = Ceraphronidae; Cha = Chalcididae; Chr = Chrysididae; Dia = Diapriidae; Dry = Dryinidae; Ela = Elasmidae; Emb = Embolemidae; Enc = Encyrtidae; Euc = Eucharitidae; Eul = Eulophidae; Eup = Eupelmidae; Eur = Eurytomidae; Eva = Evaniidae; Fig = Figitidae; For = Formicidae; Gas = Gasteruptiidae; Hel = Heloridae; Iba = Ibaliidae; Ich = Ichneumonidae; Leu = Leucospidae; Meg = Megaspilidae; Mon = Monomachidae; Mut = Mutillidae; Mym = Mymaridae; Orm = Ormyridae; Oru = Orussidae; Pel = Pelecinidae; Per = Perilampidae; Pla = Platygastridae; Pom = Pompilidae; Pro = Proctotrupidae; Pte = Pteromalidae; Rho = Rhopalosomatidae; Sap = Sapygidae; Sce = Scelionidae; Scl = Sclerogibbidae; Sco = Scoliidae; Scy = Scolebythidae; Sig = Signiphoridae; Sph = Sphecidae; Ste = Stephanidae; Tet = Tetracampidae; Tip = Tiphiidae; Tor = Torymidae; Trg = Trigonalyidae; Tri = Trichogrammatidae; Ves = Vespidae (Eumeninae). Abbreviations in *italics* mean that the group concerned only attacks the eggs of the particular order, and in **bold** the group concerned only attacks the adults of the particular order.

partner has allowed the leaf-cutting ants to overcome many of the nutritional limitations imposed on other phytophagous insects. This indirect means of utilizing green plants apparently permits greater polyphagy. The fact that some leaf-cutting ants have achieved the largest colony size known among insects, with as many as seven million individuals in a single colony, is evidence for the success of this partnership. The quantity of vegetation needed to sustain colonies of such size (up to 6000 kg of leaves during the life of a colony; Weber, 1966) makes these ants the single most important group of defoliators in the Neotropics (see Chapter 3.1).

2.4 CARNIVORY IN THE LARVAL HYMENOPTERA

Ian D. Gauld and Paul E. Hanson

The larvae of most apocritan Hymenoptera and the Orussidae are carnivorous. Those exceptions that devour plant material, the Apidae, Agaonidae, many Cynipidae and a few smaller groups that were mentioned above in Chapter 2.3, are probably secondarily phytophagous, and many share biological features with their carnivorous relatives. Although larval carnivory is a widespread feature in many insect orders, larval hymenopteran carnivores have three rather unusual characteristics.

- They are relatively immobile, poorly armoured and virtually never free-living. This contrasts with Coleoptera, whose predatory larvae are generally well armoured and highly mobile, and even some Diptera, such as Syrphidae, that have free-living, active, predacious larvae.
- In the vast majority of cases the larva feeds on (or within) a food organism provided for it by its female parent. Thus it is the adult female hymenopteran rather than the larva that catches and subdues the larval 'prey'. This too differs from most predatory Coleoptera and some Diptera whose active larvae seek out and subdue their own prey-organisms.
- The range of organisms used as larval food is more or less entirely restricted to terrestrial arthropods—insects, spiders and less commonly other arachnids (Table 2.2). Except for a eulophid that attacks nematodes, and a very few predatory social aculeates that use other invertebrates, other phyla with common terrestrial representatives (such as molluscs, annelids, nematodes or vertebrates) are not attacked. This contrasts strikingly with the Diptera, whose carnivorous taxa have radiated to use a very broad spectrum of food organisms (Eggleton & Belshaw, 1992).

Traditionally hymenopterans with carnivorous larvae have been subdivided into predatory and parasitic groups along taxonomic boundaries. Aculeates (i.e. members of the Chrysidoidea, Vespoidea and Apoidea) with carnivorous larvae have been described as 'predators' and the arthropod provided by the parent as food for the larva has been termed 'prey'. The non-aculeate carnivores (the so-called 'Parasitica') have been called 'parasitoids' and the larval food organism has been referred to as a 'host'. However, the implied differences between predators and parasitoids are a matter of degree, not of kind. A parasitoid larva develops on a single host individual whereas the larva of a predator utilizes more than one individual as food (Eggleton & Gaston, 1990). The distinction is completely arbitrary in a few species of Hymenoptera where the larva may develop on just one arthropod, or on two or three, depending on the size of the arthropod (e.g. a few species of Ichneumonidae and Sphecidae). Moreover, the predator/parasitoid dichotomy has tended to obscure the fact that development in the less advanced members of both groups is fundamentally the same—they develop by consuming an immobilized arthropod provided for them by the female parent (Gauld & Bolton, 1988).

The term parasitoid has been used to describe those insects whose larvae develop on an arthropod host, but unlike parasites, the host is nearly always killed. However, in terms of their effects on the population of the host, parasitoids are similar to parasites that castrate (and thus reproductively 'kill') their host (Kuris, 1974). In practice the term parasitoid has been taxonomically limited to insects, and more specifically to 56 families of Hymenoptera, 21 families of Diptera, 11 families of Coleoptera, two families of Lepidoptera and one family of Neuroptera (Eggleton & Belshaw, 1992). However, several groups of non-insects, such as mermithid nematodes and some fungi, also fit the definition of parasitoid: 'an organism which develops on or in another, single ("host") organism, extracts nourishment from it, and kills it as a direct or indirect result of that development' (Eggleton and Gaston,

1990). We generally accept this definition of a parasitoid, which it should be noted includes many solitary aculeates that provide a single arthropod as food for their larva. For convenience, however, we also include as parasitoids gregarious species, where a number of individual larvae develop to maturity on a single host, as well as the very few hymenopterans which, perhaps exceptionally, do not actually kill the host. A few hymenopterans that are evolutionarily derived from typical parasitoids feed on several hosts in close proximity, and are thus, strictly speaking, predators. Examples include evaniids, which feed on several eggs in a cockroach ootheca, *Tromatobia* species (Ichneumonidae), which feed on a series of eggs in a spider egg sac (Gauld, 1991), and some pteromalid larvae that feed on several cecidomyiid larvae in a seed pod (Parnell, 1963). Such species have been referred to as 'pseudoparasitoids' (Fitton *et al.*, 1988), but this term can be confusing and here we simply refer to them as predators.

The most primitive type of larval carnivory is exemplified by the family Orussidae, where adult females of at least some species immobilize an arthropod larva concealed in wood and oviposit upon it (Nuttall, 1980). The larva feeds externally on the immobilized arthropod provided for it by its mother. Essentially the same habits are shown by such primitive families as Stephanidae and Megalyridae, as well as the basal groups of most other Hymenoptera (e.g. Braconidae, Ichneumonidae and Chalcidoidea) (Fig. 2.09). From this ancestral habit all other higher hymenopteran lifeways have been derived (Gauld & Bolton, 1988).

More meaningful and useful than the parasitoid/predator dichotomy is the differentiation between idiobionts and koinobionts (see Table 2.3). IDIOBIONTS feed as larvae on a host whose development has been arrested by the female parent prior to oviposition (Askew & Shaw, 1986). Their larvae thus develop by consuming a defenceless, immobilized food resource. Consequently the host utilized is generally an arthropod living in a concealed or protected situation, such as the wood-boring larvae or the cocooned pupae or prepupae of endopterygote insects. The host's own concealment or thick cuticle offers it, and its idiobiont parasitoid, protection from scavengers after it is immobilized. KOINOBIONTS, on the other hand, feed on or within a host that continues to develop after being parasitized. As the host remains active, exposed or weakly concealed hosts can be utilized, and the host's death is often postponed until after it has gained the protection of its pupation retreat. Some biological features and consequences of these two modes of development are outlined in Table 2.3.

Idiobiont	**Koinobiont**
Host development curtailed at time of oviposition	Host allowed to continue development after oviposition
Adult often long-lived	Adult often rather short lived
Adult generally host-feeds	Adult often does not host-feed
Synovigenic; egg large	Often pro-ovigenic; egg often small
Host range often broad, sometimes several orders attacked	Host range often narrow, rarely more than one order attacked
Hosts concealed or protected by thick cuticle	Hosts often exposed or weakly concealed, or well-concealed in ephemeral substrate
Major differences between species tend to be in morphological features that enable adult to gain ovipositional access to host	Major differences between species tend to be in physiological features that allow larva to circumvent the host's immunodefences

Table 2.3. Biological differences between idiobiont and koinobiont hymenopterous parasitoids.

Mills (1992) has pointed out a problem with the idiobiont/koinobiont dichotomy—namely that an endoparasitoid showing continuous development would be considered to be a koinobiont if it attacked larvae, but an idiobiont if it attacked pupae. He advocated a more complex system of classification of parasitoid guilds, based on the stage of the host attacked, the mode of parasitism (ecto- versus endoparasitism) and the form of parasitoid development (continuous versus delayed). However, while we accept the existence of the anomaly Mills outlined, we suggest that the idiobiont/koinobiont division is a very practical classification for discussing parasitoid biology in general terms.

The idiobiont strategy

As idiobionts generally utilize hosts in concealment or some form of protection, one of the major problems they have had to overcome is how to gain access to such hosts. Often this has been achieved by the parasitoid inserting its usually long ovipositor through the intervening substrate, stinging the host *in situ*, then laying an egg onto this immobilized host. Consequently many such parasitoids show a wide range of morphological specializations associated both with manipulating the ovipositor and penetrating a particular substrate (e.g. Gauld, 1987; see also Chapter 4.3), and with making an exit from their concealed place of development when they are adults. Many ichneumonoid parasitoids of wood-boring beetle larvae, for example, have long ovipositors with characteristic patterns of apical teeth, complex ovipositor guides on the coxae or metasomal tergites, and strong mandibles with powerful muscles housed within broad genal cavities (Gauld, 1991).

Not all idiobionts attacking concealed hosts bore through the substrate with their ovipositor; some may burrow or chew their way to a prospective host. For example, ichneumonine ichneumonids push their way into leaf-rolls or tunnel in the soil to reach pupae (Lloyd, 1940; Hinz, 1983). This type of behaviour is best known within many groups of lower aculeates, which cannot use the ovipositor to pass an egg through a substrate. Scoliids and tiphiids attack beetle larvae in their burrows or earthen cells (Clausen *et al.*, 1932, 1933), while some chrysidids bite a hole in an insect's cocoon and insert their telescopic metasoma in order to sting and oviposit onto the host (Kimsey, 1992*b*). Bethylids frequently attack the concealed larvae of beetles and microlepidoptera by forcing their way into the host's place of concealment (Gordh, 1976*a*; Gordh & Hawkins, 1981).

The larvae of parasitoids that attack hosts in concealment with their long ovipositors are generally fairly safe from attack by other parasitoids and scavengers, but species that push or chew their way to a prospective host could leave it accessible to these competitors. They adopt various strategies for protecting the host. For example, cleptine chrysidids, which bite a hole in a sawfly cocoon, prevent other carnivores from gaining access to their offspring's host by sealing the hole with a mucilaginous oral secretion (Kimsey & Bohart, 1990). Other parasitoids, such as the tiphiid *Pterombrus*, bury a host after oviposition (Palmer, 1976), while some bethylid species drag the paralysed host into concealment prior to oviposition (e.g. Richards, 1932; Krombein, 1955; Yamada, 1955). This feature, actively hiding a paralysed host prior to oviposition, is most developed in the higher aculeate lineages, where the adult female often constructs a place of concealment (i.e. nest) before she searches for a host (see Chapter 2.6). Other idiobiont parasitoids remain with their host after oviposition, defending their offspring. Such behaviour is uncommon among the non-aculeates, but is known in species of the Old World braconid genus *Cedria* and the cosmopolitan eulophid genus *Melittobia* (Doutt, 1973). *Cedria paradoxus*, for example, remains with the host caterpillar after oviposition, and guards her brood against chalcid hyperparasitoids (Beeson & Chatterjee, 1935). In the aculeate lineage, however, brood care is more widespread, even in the more basal lineages. For example, some bethylids attacking hosts in weak concealment remain with the parasitized host until their offspring have pupated (Hardy & Blackburn, 1991).

Most idiobiont larvae that consume concealed hosts are ectophagous (ectoparasitic), that is they develop externally on the host which is consumed through an integumentary lesion. An important feature of such species is the fact that the venom of the female parent, injected prior to oviposition, has effectively immobilized the host. Such venom may cause the host's death, but more usually it paralyses the host (e.g. Beard, 1978; Bocchino & Sullivan, 1981), preventing it from dislodging or damaging the rather fragile parasitoid egg or young larva. Paralysed hosts may remain alive, and fresh, for several weeks (Askew, 1971).

Although many idiobionts only attack hosts in concealment, several idiobionts attack more exposed hosts. The nest-building aculeates are able to exploit a wide range of free-living or relatively exposed, active, hosts (e.g. nymphs and adults of exopterygote insects and the adults of endopterygotes—Table 2.2) because they subsequently conceal the immobilized host prior to oviposition. Some other idiobionts attack relatively exposed, immobile hosts and do not remove them to a concealed site. For example, scelionids and trichogrammatids attack insect eggs on leaves, while pimpline ichneumonids and some pteromalids attack naked lepidopterous pupae hanging from the vegetation (Gauld, 1991). Such parasitoids are generally endophagous (endoparasitic), feeding within the host and using the cuticle or chorion of the host for their own protection.

Endoparasitic idiobionts of larvae and pupae are likely to be subjected to the actions of the host immunodefensive system (see Chapter 2.5). Endoparasitic idiobionts prevent or restrict host reactions against their larvae in several ways, such as the injection of venoms that paralyse the host (Vinson & Iwantsch, 1980*a*) and physical destruction of the brain by the young larva (Führer & Kilincer, 1972). The larvae of some endophagous idiobionts that develop in prepupae or pupae may be exposed to detrimental effects resulting from damage to the host's internal organs. These include melanization from tissues injured by the parasitoid, and septicaemia resulting from midgut bacteria entering the haemocoel if the gut wall is damaged. The larva of one endophagous idiobiont ichneumonid is known to secrete a substance that inhibits melanization and has antibiotic properties against bacteria and fungi (Führer & Willers, 1986). This substance is produced by two greatly enlarged Malpighian tubules and is secreted via the anus. Similar substances will probably be found in other endoparasitoids with a similar biology, especially in some species of chalcidoids that devour only part of the host pupa but somehow appear to prevent decay from occurring in the unconsumed parts.

Endoparasitic idiobionts of eggs do not have to contend with the host's immunodefensive system since this is lacking in eggs (Salt, 1970). The most serious problem facing egg parasitoids is that, unless the female parasitoid oviposits almost immediately after the host has laid the egg, the host egg may hatch before that of the parasitoid (Strand, 1986). This problem is especially acute in the case of holometabolous hosts (i.e. Coleoptera, Lepidoptera and Diptera), which often hatch in three days or less. In such cases the female parasitoid may inject a venom that contains an arrestment factor, which prevents host embryonic development (Strand *et al.*, 1986). Another adaptation of some egg parasitoids is PHORESY, whereby the female wasp attaches to the female host, allowing the parasitoid immediate access to freshly laid eggs (Clausen, 1976).

Since the idiobiont larva usually feeds on a host that is little more than a defenceless piece of meat, many have the potential to successfully develop on a wide range of hosts. Species that attack deeply concealed hosts, and only encounter the host with the tip of the ovipositor, may have a very broad host range spanning several orders of insects (Fitton *et al.*, 1988; Gauld, 1991). Even those that are restricted in nature to a very narrow host range can often be cultured artificially on a variety of surrogate hosts. Idiobionts that have more intimate contact with the host than just reaching it with the ovipositor, are often restricted to hosts belonging to a single order of insects; if they do attack hosts in other orders then these are often other parasitoids associated with the host. In part, this narrower host range may reflect a need for more specific cues in host recognition, as is probably the case in the Ichneumoninae where host range is very narrow (Hinz, 1983; Mills, 1992). However, in the case of the endophagous idiobionts that attack exposed hosts, a narrow host range may result from such hosts being chemically better defended than deeply concealed hosts, and therefore only exploitable by relatively specialist parasitoids capable of avoiding toxins or rendering them harmless.

Except for egg parasitoids, the hosts of idiobionts are often concealed, quite mature, and tend to be infrequently encountered; consequently, female idiobionts are often long-lived. Such species are SYNOVIGENIC, that is, throughout their life they successively develop a number of eggs to maturity, and at any one time the female will only have a few mature eggs in the oviduct. Since these eggs are deposited on a defenceless host it is advantageous if they develop rapidly so that the parasitoid larva can consume the host before the host is attacked by scavengers or micro-organisms. Consequently synovigenic species generally produce relatively large, yolky eggs (Price, 1974). These eggs are enriched on their passage down the mother's ovariole and thus their production is physiologically 'expensive' for the parent. Adult females usually feed on prospective hosts to provide extra proteins for egg production (see Chapter 2.5, host-feeding).

The koinobiont strategy

As stated above, koinobionts allow their host to continue development after parasitization. In the case of egg-larval parasitoids the adult female oviposits into the host egg and her offspring emerges from the host larva; similarly, koinobionts include egg-pupal, larval-larval, larval-pupal, and nymphal-adult parasitoids. Many koinobionts often spend a protracted period as an egg or an early instar larva postponing destroying the host until after it has concealed itself for pupation. In the case of egg-pupal parasitoids they are thus dormant for their host's entire larval life. Since developing koinobionts often remain associated with their hosts for a long period of time, both the parasitoid and its host are vulnerable to mortality factors for a longer period than idiobionts. Nonetheless, the fact that koinobiosis has evolved repeatedly and independently in numerous lineages of Hymenoptera (Whitfield, 1992) suggests that there are certain advantages to this strategy.

Koinobiosis may be regarded as an adaptation to allow parasitoids to exploit hosts living in exposed situations, such as leaf-feeding caterpillars. By postponing permanent immobilization until after the host has constructed a pupation retreat, egg-pupal and larval-pupal koinobionts are able to complete their development in a relatively secure situation. They thus have the selective advantages of being able to attack relatively easily discoverable hosts, yet for the period they actually incapacitate the host, benefit from the concealed site in which the host larva pupates. Unlike the host of an idiobiont, the host of a koinobiont usually continues to grow, and therefore the female koinobiont can afford to lay her egg in a young host that is initially too small to sustain the growth of her offspring. Young hosts are more numerous and therefore generally easier to locate than older hosts (Price, 1972), and young host larvae usually offer less resistance to attack (e.g. smaller defensive droplets, Johansson, 1951; softer cuticle, Beckage & Riddiford, 1978). Attacking early-instar larvae may also provide a competitive advantage since the first parasitoid to become established in a host is often able to kill subsequent competitors.

Comparatively few koinobionts are ectoparasitoids. Examples include tryphonine ichneumonids, polysphinctine ichneumonids (Fincke *et al.*, 1990), some rogadine braconids (M.R. Shaw, 1983), most dryinids and a few pompilids (e.g. *Notocyphus*, Simons, 1989). In these parasitoids egg placement is generally very important. Many sting and temporarily paralyse the host prior to oviposition, then position the egg in a place where it cannot easily be dislodged, such as just behind the head (Baltensweiler & Moreau, 1957) or under overlapping sclerites as in some dryinids. Generally the egg is cemented in place, but tryphonines have eggs that are furnished with a barb that anchors them to the host's integument (Mason, 1967; Kasparayan, 1981). All koinobiont ectoparasitoids have to contend with the possibility of being sloughed off during moulting. Many tryphonines avoid this by attacking only final instar larvae, but the eulophid *Euplectrus* injects a venom which prevents its host from moulting (M.R. Shaw, 1981; Coudron *et al.*, 1990).

The majority of koinobionts are endoparasitoids. Although a few of these do temporarily paralyse the host prior to oviposition (e.g. see Wharton, 1984), most do not. Many koinobiont endoparasitoids insert their ovipositor into the host only once, but there is clear evidence they inject a venom along with the egg (e.g. Veen, 1981). This venom may have subtle and diverse effects on the physiology and development of the host (see review by Vinson & Iwantsch, 1980*b*). Endoparasitic koinobionts undergo a long developmental period within an active host, and therefore, compared to idiobionts and ectoparasitic koinobionts, they have to cope with a physiologically much more demanding existence. They must not seriously debilitate their host lest it (and thus also they) be consumed by a predator, but at the same time they must suppress or avoid the immunological defences of the host (see Chapter 2.5). Many koinobiont endoparasitoids pass the period between oviposition and host cocoon formation as a protracted first instar larva. This perhaps serves either to avoid debilitating the host by their feeding activity, and/or it enables them to fight off competing parasitoids, as first instar larvae generally have better developed defensive mandibles than other instars. Some koinobiont endoparasitoids cause the host to pupate prematurely. This is sometimes mediated by the endoparasitoid larva itself (Varley & Butler, 1933; Bühler *et al.*, 1985), or it results from venom injected at oviposition by the adult female (M.R. Shaw, 1981, 1983; Jones, 1987).

Not all koinobiont endoparasitoids delay their own development until after the host pupates. Some, especially species that oviposit in very young larvae, destroy their host prematurely in an exposed situation. For example, some campoplegine ichneumonids inject substances that halt development of the host by inhibiting ecdysteroid production and causing prothoracic gland degeneration (Dover *et al.*, 1987, 1989; Dover & Vinson, 1990). Such species have evolved a variety of means to protect themselves during pupation, including camouflaging the cocoon as a bird dropping or dead spider prey (Gauld, 1991), pupating under or in the remains of a spiny host larva or hard host (Balduf, 1926; Gauld & Bolton, 1988) or within a 'mummified' host, spinning a cocoon that hangs from the vegetation by a long thread or constructing complex thick walled cocoons (Gauld & Bolton, 1988).

Although some ectoparasitic koinobionts are synovigenic, endoparasitic koinobionts are predominantly PRO-OVIGENIC; that is, the adult completes ovigenesis soon after emergence from the pupa. Most pro-ovigenic koinobionts have moderately small eggs (Iwata, 1958) and some, such as certain braconids (Jackson, 1928; Parker, 1931*a*; Tremblay, 1966), cynipoids (Haviland, 1921; Chrystal, 1930), ichneumonids (Tothill, 1922) and platygastrids (M.C. Day, pers. comm.), lay very tiny, alecithal (yolkless) eggs. In physiological terms such eggs are relatively 'inexpensive' for the female to produce because they require little in the way of nutritive substances from her. The production of small, low-cost eggs is advantageous for koinobionts ovipositing in early stage hosts because such hosts are more likely to suffer high rates of mortality (Price, 1974). Egg enlargement and enrichment are actually postponed until after oviposition and, for successful embryological development to occur, the egg must draw nutritive substances from the host haemolymph. During this development an alecithal egg may increase in size dramatically (Oglobin, 1913; Jackson, 1928). Many, if not all, are characterized by the rapid development of an outer nutritive layer, which may or may not be composed of discrete cells. This nutritive layer, the TROPHAMNION (= serosa), seems to have an analogous function to the trophocytes associated with a developing lecithal egg in the ovariole (see Chapman, 1969; Pennacchio *et al.*, in press). Thus the alecithal egg itself is parasitic upon the host, and its trophamnionic 'feeding' prior to eclosion may have a noticeable effect upon the fitness of the host (Polaszek, 1986). This type of embryological development has sometimes given rise to polyembryony (see Chapter 2.5).

Upon eclosion of the egg the trophamnion may fragment and the resultant cellular parts, the TERATOCYTES ('giant cells'), are liberated into the host haemocoel (Vinson, 1970; Dahlman, 1990, 1991; Kitano *et al.*, 1990; Tanaka & Wago, 1990; Tawfik, 1991). After dissociation the teratocytes do not divide, but they absorb nutrients from the host and may increase in size ten to twenty times (Slüss, 1968). Except for Scelionidae, all of which are idiobionts, teratocytes are known only from certain koinobionts, namely Platygastridae and several evolutionary lineages of Braconidae (Vinson & Iwantsch, 1980*b*; Bühler *et al.*, 1985). A single teratocyte has been reported in association with a trichogrammatid (Voegele *et al.*, 1974), but this may be a polar body (Strand, 1986). These cells have otherwise not been found in the hosts of Chalcidoidea, nor have they been observed in hosts of Ichneumonidae. The number produced by a scelionid or a platygastrid is usually less than 40, but between one and eight hundred are produced by braconids (Vinson & Iwantsch, 1980*b*).

Teratocytes appear to have a variety of trophic, secretory and immunosuppressive functions (Dahlman, 1991). In some parasitoid species the developing larva feeds on teratocytes (Polaszek, 1986), but in other species this does not appear to be the case (Vinson & Lewis, 1973). Young teratocytes are often secretory, have an extensive endoplasmic reticulum, and bear microvilli, while older ones have an extensively ramified nucleus, lipid and autophagic vacuoles, and appear to be degenerating (Slüss & Leutenegger, 1968; Vinson, 1970; Gerling & Orion, 1973; Vinson & Scott, 1974). The substances secreted by teratocytes have been observed to have a variety of effects on their hosts (Table 2.4).

2.5 THE PARASITOID AND ITS INTERACTIONS WITH ITS HOST

Ian D. Gauld and Paul E. Hanson

The two basic strategies of carnivorous Hymenoptera, idiobiosis and koinobiosis, were outlined above. In this section host-parasitoid interactions are

Parasitoid species	Reference	Teratocyte secretion
Apanteles kariyai (Braconidae: Microgastrinae)	Tanaka & Wago, 1990	A substance that inhibits encapsulation of parasitoid by host
Cardiochiles nigriceps (Braconidae: Cardiochilinae)	Vinson, 1970	A substance that delayed host larval development and resulted in malformed host pupae
	Joiner *et al.*, 1973	Juvenile hormone
	Pennacchio *et al.*, 1992	Substance causing complete pupation failure by host
Chelonus sp. (Braconidae: Cheloninae)	Bühler *et al.*, 1985	A substance causing developmental disruption in the noctuid host
Cotesia glomerata (Braconidae: Microgastrinae)	Führer & Elsufty, 1979	A fungistatic agent
	Kitano *et al.*, 1990	A substance that suppresses phenyloxidase activity in pierid host
Microplitis croceipes (Braconidae: Microgastrinae)	Hegazi & Führer, 1985	A substance that helps suppress physiological development of competitors
	Dalman, 1991	Substances causing delayed host larval development and inhibition of juvenile hormone esterase production
	Zhang *et al.*, 1992	A substance that inhibits juvenile hormone esterase release
Telenomus heliothidis (Scelionidae: Telenominae)	Strand, 1986	Secretion of digestive enzymes

Table 2.4. Effects of substances produced by teratocytes of various parasitoids.

reviewed and, wherever possible, discussed from a phylogenetic perspective. The account presented here is intended as an introduction to parasitoid biology and as an entry into the vast and burgeoning literature on the subject. Godfray (1994) has presented an excellent and comprehensive review of this subject area.

Host-feeding

The majority of adult hymenopterous parasitoids require an intake of carbohydrates in the form of honeydew, nectar or other plant secretions (Vinson & Barbosa, 1987). In the laboratory, the adult longevity of many species can be greatly increased by feeding a honey-water mixture, ten percent sucrose solution, aphid honeydew, or pared ripe fruit (Laing & Caltagirone, 1969; Ridgway & Mahr, 1990). Many female parasitoids also feed extensively on the body fluids of prospective hosts, which exude from wounds inflicted by the parasitoid with its ovipositor (Leius, 1960; Bartlett, 1964; Askew & Shaw, 1979), and some ichneumonids and dryinids consume host tissue in addition to fluids (Leius, 1961*b*; Waloff, 1974). Such HOST-FEEDING, as it is now generally called, has been observed in members of 17 families in the Ichneumonoidea, Chalcidoidea, Proctotrupoidea, Chrysidoidea and Vespoidea (Jervis & Kidd, 1986), and is probably common in most parasitoid groups. An interesting parallel with host-feeding by carnivorous parasitoids is the observation that a species of phytophagous eurytomid feeds on the exudation from the wound it makes in a seed 'host' while ovipositing (Hedlin, 1956).

The occurrence of host-feeding seems, to some extent, to be correlated with egg production. Synovigenic species that continue to produce large, fully-yolked (lecithal) eggs throughout their lives need frequent protein meals in order to sustain egg production. Although many such species (e.g. *Bracon hebetor*) are capable of laying a few eggs without first host-feeding (Lum, 1977), host-feeding is necessary to maximize egg production. Other species, however, must host-feed before they can begin ovipositing (Leius, 1961*a*).

The female sometimes uses the same organism both for feeding and oviposition (concurrent host-feeding). In these cases the female parasitoid is usually small in relation to the host and feeding by the parasitoid is probably not detrimental to the development of her offspring. However, most parasitoids feed on prospective host individuals that are not used subsequently for oviposition (non-concurrent host-feeding). In such cases the female usually feeds on an earlier host stage, and oviposits on a later stage (Kidd & Jervis, 1991), as occurs for example in *Aphytis lingnanensis* (Rosenheim & Rosen, 1992). Often the individual on which the female has fed is severely debilitated or killed. Host-feeding adult parasitoids thus cause additional mortality in a host population over and above that caused by actual parasitism, and this sometimes exceeds the mortality caused by parasitism (Campbell, 1963). In Costa Rica several eulophid parasitoids of leaf-miners, such as *Liriomyza* and the coffee leaf-miner (*Leucoptera coffeella*, Lyonetiidae), appear to cause significant host mortality by host-feeding (P. Hanson, unpubl.).

Some synovigenic species are able to resorb mature eggs or oocytes when deprived of hosts on which to feed and/or oviposit (King, 1962, 1963; King & Richards, 1968; Sugimoto *et al.*, 1983). Through oösorption materials contained in the eggs are used both for maintenance of the adult parasitoid, and for sustaining oögenesis until host-feeding can be resumed (Jervis & Kidd, 1986). However, observations on *Coccophagus bartletti* (Aphelinidae) indicate that the oögenesis–oösorption cycle is possibly not a mechanism evolved to optimize nutrient use to the extent often envisaged (Walter, 1988). The usual low level of resorption that does occur in some species may simply be a consequence of old eggs sequestering juvenile hormone less competitively than younger ones (Chapman, 1969) or it may indicate a shortage of certain amino acids. No special function in terms of a constantly operating production line that economically maintains reproductive potential need be assigned (Walter, 1988).

Solitary and gregarious parasitoids

In the discussion of the idiobiont and koinobiont strategies above most of the examples given were SOLITARY parasitoids, where a single larva utilizes an entire host to complete its development. However, many parasitoids attack a host that is large enough to support the development of more than one larva. When this occurs the parasitoids are termed GREGARIOUS. Such parasitoids may be ectophagous or endophagous, idiobiont or koinobiont. Gregariousness is most commonly encountered in the Diapriidae, Platygastridae, Proctotrupidae, Encyrtidae, Eulophidae, Pteromalidae, Braconidae (especially in Microgastrinae) and Bethylidae but is very rare in the Ichneumonidae and, except for some Trichogrammatidae, is also rare in egg parasitoids (Scelionidae and Mymaridae).

Gregariousness is usually characteristic of a particular species, and facultative gregariousness in normally solitary species rarely occurs, even when the host resource appears to be sufficient to support multiple individuals. For example, some solitary microgastrine braconids develop in hosts large enough to support several similar-sized individuals of congeneric gregarious species, yet they are apparently never facultatively gregarious (LeMasurier, 1987). When facultative gregariousness does occur in a species, it is mostly in idiobionts. For example, occasionally multiple individuals of a normally solitary *Pimpla* species (Ichneumonidae) develop on a single host (Fuester *et al.*, 1989). In the cosmopolitan pteromalid *Muscidifurax raptorellus* both solitary and gregarious populations are known. In this case the solitary habit appears to be dominant and the gregarious habit is inherited polygenically (Legner, 1991).

In the Microgastrinae, which are koinobiont endoparasitoid braconids, there is a bimodal frequency distribution of brood sizes, with large numbers of species being solitary, large numbers having brood sizes between twelve and twenty-six, and relatively few with brood sizes between two and eleven (LeMasurier, 1987). This may be because there is a minimum clutch size for gregarious species below which no larvae survive (Kitano, 1986), probably because they are incapable of overwhelming the host's immunodefensive system. There is no such bimodality in many gregarious idiobionts. For example, trichogrammatids adjust the number of eggs they lay in a single host egg from one to sixty, depending upon host size (Klomp & Teerink, 1962; Pallewatta cited by LeMasurier, 1987). Similarly, a gregarious ectoparasitic braconid was observed to lay between one and about twelve eggs on its lepidopterous host depending upon the instar attacked (Knight & Croft, 1989).

Although gregarious parasitism is usually the result of multiple eggs being laid on or in one host individual, gregariousness may result from repeated division of a single egg, a phenomenon called POLYEMBRYONY. Polyembryony has evolved in several taxonomically distinct groups of pro-ovigenic endoparasitoids that have an alecithal, 'feeding' egg: Braconidae (*Macrocentrus*) (Daniel, 1932), Encyrtidae (e.g. *Copidosoma* and related genera) (Patterson, 1921; Leiby, 1922), Platygastridae (*Platygaster*) (Leiby & Hill, 1924), and Dryinidae (*Crovettia*) (Kornhauser, 1919). The most extreme form of polyembryonic development occurs in some species of *Copidosoma*, where one egg may give rise to several thousand individuals. The oviposition of at least two eggs is necessary to produce broods that contain both sexes (Hardy *et al.*, 1993). *C. floridanum* oviposits in the host egg, where its own egg cleaves to give rise to a morula. This morula proliferates to form secondary morulae during the first four larval instars of the host. By the second day of the host's fourth instar proliferation ceases and the secondary morulae increase in size. On the first day of the host's fifth instar morphogenesis begins (Baehrecke & Strand, 1990). The production of numerous genetically identical siblings in encyrtids has facilitated the evolution of larval castes to occur—an analogous situation to the development of adult castes in ants (see Chapter 16). In *Copidosomopsis* two different larval castes occur (Cruz, 1986*a*): a normal feeding morph that eventually gives rise to an adult encyrtid, and a precocious 'guard' morph that protects its siblings by attacking any other parasitoid larvae present, but fails to mature (Cruz, 1981, 1986*b*).

The evolution of multiple egg-laying gregariousness from solitary parasitism may involve both anatomical and behavioural changes. First, adult females of gregarious species need to lay more eggs in a short period than do those of solitary species. While this may not be a problem for pro-ovigenic species, it could be problematical for synovigenic species that generally have few ovarioles and few mature eggs at any one time. Not surprisingly, gregarious, synovigenic parasitoids often have a much greater number of ovarioles than similar sized, closely related solitary species (see Iwata, 1960). Second, multiple egg-laying requires a change in the ovipositional behaviour of the female, both in actually laying several eggs and, in at least some species (e.g. many trichogrammatids), being able to assess the size of the host and adjust the number of eggs laid accordingly (Klomp & Teerink, 1962; Schmidt & Smith, 1985*a*, 1987). Third, gregariousness requires a change in larval behaviour. Most solitary parasitoid larvae will destroy, by physical or chemical means, other parasitoids in the same host (see p. 60), whereas the larvae of gregarious species must tolerate their siblings. Indeed larvae of gregarious species rarely show the weaponry or inclination to be aggressive (Waage, 1986).

From a theoretical standpoint, the evolution of gregarious parasitism may be restricted by a parent–offspring conflict—that is, selection acts on the adult female to optimize her clutch size whereas selection acts on larvae to maximize their own adult lifetime fitness, which may include intolerance of competing siblings (Waage, 1986; Godfray, 1987). A gene for tolerance may spread in a population of aggressive, solitary parasitoids when fitness per larva in a clutch of one is less than that in larger clutch sizes. This can occur, for example, when several parasitoid larvae are required to overcome a large host's immunodefensive system. Another factor that may increase the likelihood that larval tolerance will evolve is increased relatedness of larvae in the host, the extreme case being that of polyembryonic species where the larvae are genetically identical.

HYPERPARASITOIDS

Many hymenopterous parasitoids, both idiobionts and koinobionts, develop by feeding on other parasitoids. These are called HYPERPARASITOIDS and hyperparasitism is a common phenomenon in many hymenopterous groups, most notably the Trigonalyidae, Ceraphronoidea, Figitidae (Charipini), Chalcidoidea and Ichneumonidae. Elsewhere in the Insecta hyperparasitism is an extremely uncommon strategy (Gordh, 1981; Sullivan, 1987). Hyperparasitoids may be FACULTATIVE (that is they may develop either as a primary parasitoid or as a hyperparasitoid depending upon circumstance), or

OBLIGATE; that is, they may develop only as a hyperparasitoid. Perhaps because of their physiologically rather unspecialized relationship with their host (see above) many, perhaps even most, idiobiont ectoparasitoids are facultative hyperparasitoids. However, some idiobionts, such as megaspilids, are obligate hyperparasitoids, as are probably all koinobiont hyperparasitoids. The majority of hyperparasitoids are secondary parasitoids, that is they attack a primary parasitoid, but tertiary and even quaternary parasitism also occur facultatively (Haviland, 1920; Sullivan, 1972; Levine & Sullivan, 1983; Matejko & Sullivan, 1984).

The term hyperparasitism describes a trophic interaction, but this interaction may, as stated above, be anything from an almost serendipitous event through to very highly specialized associations between species. Gauld and Bolton (1988) suggested that hyperparasitism may have evolved in four broad ways depending upon the association of the ancestor of the hyperparasitoid. In the following discussion the term primary host is used for the host of the parasitoid (definitive host) attacked by the hyperparasitoid. Gauld and Bolton's evolutionary types are:

A) hyperparasitism resulting from an association with the primary host;
B) hyperparasitism resulting from an association with the definitive host (primary parasitoid);
C) hyperparasitism resulting from an association with predators of the primary host;
D) hyperparasitism that has evolved without any close association with either the primary host or the primary host's guild of parasitoids or predators.

Type A hyperparasitism is perhaps the least specialized and most common form. Many idiobionts are facultative type A hyperparasitoids, paralysing and ovipositing on either a concealed primary host, or its parasitoid developing in the host's place of concealment. Gauld (1991) cites several examples from pimpline ichneumonid genera represented in Costa Rica. Most type A hyperparasitoids locate hosts by using cues derived from the presence of the primary host, so it is serendipitous whether it consumes the primary host, or a parasitoid of such a host. Given the rather generalized trophic requirements of many idiobiont ectoparasitoids it probably does not make much difference what the host is, so the type A relationship is most usually facultative. However, some type A hyperparasitoids do seem to be obligate hyperparasitoids, but these are usually endoparasitoids. For example, *Prochiloneurus* species are obligate hyperparasitoids of encyrtid parasitoids of coccids (Rosen, 1981). Many related genera of encyrtids are either primary or primary and facultative secondary parasitoids of coccoids, suggesting the obligate hyperparasitism exhibited by *Prochiloneurus* evolved in a type A manner—that is via an association with the primary host.

Type B hyperparasitoids include some of the most biologically specialized of all the hyperparasitic species, including trigonalyids and mesochorine ichneumonids. A common feature of this group is that they are all obligate hyperparasitoids and are apparently unable to develop on the primary host. All seem to have tracked their definitive hosts or its close phylogenetic relatives, through evolutionary time. For example, mesochorine ichneumonids are obligate koinobiont endoparasitoids of the larvae of ichneumonoid and tachinid endoparasitoids. Most attack hosts in endopterygote larvae, but some attack euphorine braconid larvae in a mirid primary host; no other group of Ichneumonidae is known to oviposit into Hemiptera. Presumably mesochorines have somehow tracked euphorines, possibly by locating an ovipositional marker or similar parasitoid-derived chemical, from the euphorines' ancestral endopterygote hosts (S.R. Shaw, 1988*a*) to a quite different host group. A similar example is found in the Chalcididae where *Brachymeria moerens* attacks parasitic fly larvae within a grasshopper (Léonide & Léonide, 1969); other species of *Brachymeria* are parasitoids of dipterous larvae, but no chalcidid is known to attack orthopterans.

Neither the primary nor the definitive hosts of type C hyperparasitoids are attacked by the parasitoid's relatives, but they do have relatives that are primary parasitoids of predators of the primary host. For example, some *Dendrocerus* species (Megaspilidae), develop as idiobiont hyperparasitoids of aphids through aphidiine braconids (Matejko & Sullivan, 1984; Sullivan, 1987), and although no close relatives of these hyperparasitoids are known to be either primary parasitoids of aphids or hyperparasitoids of other braconids, many are known to be primary parasitoids of a range of aphid-associated predators, including the larvae of Coccinellidae, Chrysopidae, Hemerobiidae and

Syrphidae. If one assumes that the hyperparasitic condition evolved from the primary parasitic condition exemplified by living relatives, then a host switch must have occurred. Other examples include *Syrphophagus* (Encyrtidae) and *Pachyneuron* (Pteromalidae), species of which are either aphid hyperparasitoids or syrphid primary parasitoids, and the Charipini which are hyperparasitoids of Psylloidea or Aphidoidea, while their relatives in the Figitinae include parasitoids of dipterous larvae that prey on Homoptera. This and other examples of such hyperparasitism seem to involve sternorrhynchous primary hosts, a type of host that is comparatively easily located by its honeydew (Hôgvar & Hofsvang, 1989; Cloutier & Baudin, 1990). Presumably the ancestral primary parasitoids located their dipterous or other hosts by responding to aphid-derived semiochemicals. The same chemical cues are apparently still used by the hyperparasitic species (Budenberg, 1990).

Type D hyperparasitoids attack mature larvae or pupae of primary parasitoids in their own more or less exposed cocoons. Such species, like members of the ichneumonid genus *Gelis*, are invariably idiobionts and most attack a range of cocoon-like objects constructed by members of various arthropod groups (Muesebeck & Dohanian, 1927). However, although most type D hyperparasitoids attack a range of cocoons, members of certain genera (e.g. in the ichneumonid subtribe Acrolytina) may be specialized to attack only certain types of ichneumonoid cocoons (Blunck & Kerrich, 1956) and thus may be considered obligate hyperparasitoids. Because type D hyperparasitoids attack their definitive hosts after these primary parasitoids have completed the destruction of their own (primary) hosts, Shaw and Askew (1976) refer to them as PSEUDOHYPERPARASITOIDS, in order to distinguish them biologically from true hyperparasitoids—i.e. those that attack their definitive hosts while these parasitoids are actually parasitizing other species.

A particularly bizarre form of hyperparasitism is found in some members of the family Aphelinidae, where the females develop as primary parasitoids of Homoptera but males develop as hyperparasitoids of female larvae. The male aphelinid larva feeds on a female larva of another species or of its own species (Chapter 11.2). Such a habit is called HETERONOMOUS HYPERPARASITISM ('adelphoparasitism') and has been discussed in detail by Flanders (1967) and Walter (1983) (but see Polaszek, 1991).

Superparasitism

Unlike hyperparasitism, where the hyperparasitoid develops by consuming a primary parasitoid (which in turn has consumed the host resource), superparasitism involves one parasitoid competing with another (usually to the death) for the host resource. SUPERPARASITISM results from the allocation of one or more eggs to previously parasitized hosts. This may come about because:

a) the same individual attacks a host more than once (self superparasitism *sensu* Waage, 1986);
b) two or more individual parasitoids of the same species oviposit on or in a single host (conspecific superparasitism *sensu* Waage, 1986);
c) one host is attacked by two different species of parasitoids (interspecific superparasitism *sensu* Alphen & Visser, 1990, heterospecific superparasitism *sensu* Mackauer, 1990, or MULTIPARASITISM *sensu* Smith, 1929).

Ovipositing females of many parasitoid species are able to discriminate between parasitized and unparasitized potential hosts (see below). They tend to avoid those hosts that have already been parasitized since, at least in solitary parasitoids, normally only one larva can survive (e.g. Wylie, 1972; Jalali *et al.*, 1988). Although superparasitism in solitary parasitoids has been regarded as a maladaptive mistake (e.g. by Huffaker & Matsumoto, 1982), various authors have considered conditions whereby superparasitism may be adaptive (e.g. Alphen & Nell, 1982; Charnov & Skinner, 1984; Waage & Godfray, 1985; Waage, 1986; Speirs *et al.*, 1991; Visser *et al.*, 1992). Self superparasitism may be advantageous if repeated ovipositions increase the probability of successful parasitism (Alphen & Visser, 1990) by, for example, overwhelming a host's physiological defences, as is the case with the ichneumonid *Bathyplectes curculionis* (Puttler, 1974) and at least one strain of the encyrtid *Comperiella bifasciata* (Blumberg & Luck, 1990). Conspecific superparasitism may increase a female's fitness if unparasitized hosts are very scarce and if the younger larva has a greater than zero chance of survival (Chow & Mackauer, 1986; Dijken & Waage, 1987; Takasu & Hirose, 1988; Bai & Mackauer, 1990, 1991).

Interspecific superparasitism (multiparasitism) may be advantageous to the second parasitoid in a variety of ways:

- by allowing it to locate a host more easily;
- by allowing it to attack a host that it is otherwise incapable of gaining access to;
- by allowing it to successfully develop in a host in which it is otherwise incapable of doing so.

An example of the first is the interaction between a species of *Temelucha* (Ichneumonidae) and *Orgilus* (Braconidae) attacking a lepidopterous larva (Arthur *et al.*, 1964). *Temelucha* may develop normally as a solitary endoparasitoid of the moth larva, but it can more easily locate larvae that have been 'marked' pheromonally by ovipositing *Orgilus*, than it can unparasitized larvae. Thus *Temelucha* oviposits preferentially in hosts that are already parasitized by the braconid, and its larva is usually competitively superior to the braconid larva (Arthur *et al.*, 1964), i.e. the *Temelucha* larva kills the braconid larva and then feeds on the lepidopterous larva.

An example of the second type of advantageous interspecific superparasitism is the association of a species of *Pseudorhyssa* (Ichneumonidae: Poemeniinae) with the ichneumonid *Rhyssa*, a parasitoid of siricid woodwasps. The *Pseudorhyssa* can only gain ovipositional access to the deeply concealed woodwasp larva via the ovipositional boring of the *Rhyssa*. The *Pseudorhyssa* larva is well adapted to fighting and destroys the rhyssine larva before consuming the siricid (Spradbery, 1969). In Costa Rica there is circumstantial evidence to suggest that another poemeniine, *Ganodes matai*, is doing something similar. It possesses a long but poorly sclerotized ovipositor, not unlike that of *Pseudorhyssa*, and it has always been found in company with species adapted for drilling (Gauld, 1991). Similarly, in the Chalcidoidea, an Asian species of *Eurytoma* can only attack a cocooned lepidopterous larva if the cocoon is first breached by a chrysidid (Piel, 1933).

The third way in which interspecific superparasitism can be advantageous is where a second species cannot develop in a particular host unless that host has previously been parasitized by another species (e.g. Pemberton & Willard, 1918*b*). For example, drosophilid larvae attacked by one species of eucoiline can successfully be parasitized by another congeneric species that normally cannot develop on this host (Streams & Greenberg, 1969). The eggs of *Hyposoter* (Ichneumonidae) are normally encapsulated and destroyed in the moth larva *Orgyia leucostigma*, but when the host larva has been previously parasitized by a species of microgastrine braconid the ichneumonid develops to maturity (Guzo & Stoltz, 1985). Appropriate injection experiments have demonstrated that both the virus (see below) and venom of the braconid are necessary for successful parasitism by *Hyposoter*.

All of the above examples of interspecific superparasitism result in only one species of parasitoid completing development on the host. The emergence of two different parasitoid species from a single host individual is very rare, but may occur in situations where the parasitoids develop gregariously and do not consume the entire host (Miller, 1982).

Host searching behaviour and oviposition

The adult female parasitoid searches for a host by responding to a series of cues in the environment (Vinson, 1978, 1985). When she succeeds in locating a potential host (or its immediate covering), she examines it more closely, often using her antennae and/or the ovipositor tip, in order to 'decide' whether or not it is acceptable for oviposition. In some species this stage may also involve choosing to feed on the host (host-feeding) instead of, or in addition to, ovipositing. Finally, if the host is accepted for oviposition, the female must decide how best to exploit it—how many eggs to lay per host (clutch size) and what sex ratio to produce among the eggs laid. These processes of host-finding, host acceptance and host exploitation have been extensively studied in temperate regions (e.g. Schmidt, 1992; Vinson, 1992), but virtually nothing is known about these aspects of parasitoid behaviour in the tropics. Although the processes involved in the tropics are likely to be very similar, additional complexities will probably be encountered. Tropical forests are very diverse habitats both in the species-richness of the plants and herbivores living there, and in the range of potentially toxic secondary plant chemicals present (Levin & York, 1978). Many tropical herbivores are very patchily distributed, or present at a lower average density than are herbivores in temperate forests. Low host density in the tropics could well select for parasitoids with a broader host range than temperate species, or for parasitoids that are extremely good at locating scarce hosts (Janzen, 1981). However, the diversity of toxic substances sequestered by many insect herbivores may well preclude some generalist parasitoids (Gauld *et al.*, 1992; Gauld & Gaston, in press) and favour specialists that are both extremely efficient at finding scarce hosts, and physiologically adapted to cope with particular allelochemicals.

Host finding

Searching (or foraging) behaviour is influenced by both extrinsic and intrinsic factors (Bell, 1991). Extrinsic factors include climatic conditions, the distribution of resources (hosts) and the risks entailed in obtaining them. Intrinsic factors can vary both between and within species. Species of parasitoids differ in their sensory and locomotory abilities, as well as in the time spent handling the resource (e.g. drilling through a substrate with the ovipositor). Intra-specific variation in searching behaviour has three sources (Lewis *et al.*, 1990):

- genetically fixed differences among individuals of a species;
- phenotypic plasticity allowing individuals to modify their behaviour through learning;
- physiological state of the parasitoid (e.g. egg load—Rosenheim & Rosen, 1991; food and mating status).

The female parasitoid searches for hosts by responding to a sequence of cues by which she approaches increasingly closer to the host. It is useful to distinguish two stages in this searching sequence: HOST HABITAT LOCATION—responding to a series of environmental cues culminating in searching for a 'patch', a microhabitat where hosts could potentially occur, and HOST LOCATION—searching within a patch for a host. When searching within a patch the female parasitoid generally responds to stimuli that result from the host's actual presence.

Host habitat location. In general a parasitoid species searches for hosts in only a small sector of the total environment, to which the female orients regardless of whether hosts are actually present (Vinson, 1975). Some hosts that are readily accepted in the laboratory may not be attacked in the field simply because the host does not normally occur in that part of the environment where the parasitoid searches.

The area searched by a particular parasitoid species can be defined at different spatial scales. At the largest or most inclusive scale, female parasitoids forage in a broadly defined climatic or altitudinal zone, as is the case for Costa Rican species of the ichneumonid genus *Pimpla*, where three different species are encountered flying among forest understorey vegetation as one progresses from sea-level to the mountain tops (Gauld, 1991). One species, *Pimpla croceiventris*, is commonly encountered in seasonally dry lowland forests and regenerating woodlands below about 800 metres, *P. croceipes* occurs in more humid primary forests between about 800 and 1600 metres, while *P. sedula* occurs at the highest altitudes (above 2000 m), though occasional individuals may occur down to about 1600 metres.

At a finer scale parasitoids usually restrict their searching in other ways, for example, to particular habitats, heights above the ground, types of plants, etc. Different species of *Trichogramma* are known to search in different habitats—open fields versus forests versus marshy areas (Flanders, 1937). In North America rates of parasitism of a corn borer (*Ostrinia nubilalis*) by a campoplegine ichneumonid have been shown to be higher near the wooded edges of fields than they are in the centres of fields (Landis & Haas, 1992), and in Costa Rica striking differences have been observed in the parasitoid faunas of young regenerating and mature dry forested areas (Gauld & Janzen, unpubl.). In some cases habitat preferences may vary seasonally, as is the case for the ichneumonid *Neotheronia mellosa* in northwestern Costa Rica. In the wet season this ichneumonid mostly flies in open forests, but in the dry season it inhabits denser shaded woodlands (Gauld, 1991 and Fig. 2.04). The amount of sunlight a potential host's microhabitat receives has in several instances been shown to affect parasitism rates. For example, in North America eggs of the gypsy moth are more frequently parasitized by a species of *Ooencyrtus* (Encyrtidae) when they occur on trees in clearings than in dense forest (Weseloh, 1972). The braconid *Cardiochiles nigriceps* readily searches tobacco plants in sunlit places, but rarely visits these plants in heavily shaded situations (Vinson, 1975), and another braconid, a species of *Cotesia*, is almost incapable of finding its pierid host's foodplant if this plant is shaded by overgrowing weeds (Sato & Ohsaki, 1987). The presence of nearby plants of other species may enhance (Read *et al.*, 1970) or reduce attraction to a specific plant. These patterns have implications for the use of monoculture versus intercropping (Kaas *et al.*, 1992).

Parasitoid species may search for hosts at different heights from the ground. For example, in Santa Rosa National Park *Cryptophion moragai* (Ichneumonidae) attacks the larvae of the saturniid *Syssphinx molina* in the tops of large *Pithecellobium* trees; this ichneumonid has never been collected (in four continuous years of sampling) in Malaise traps under these trees (Gauld & Janzen, 1994). In North America gypsy moth pupae located in tree tops are more frequently attacked by *Brachymeria intermedia* (Chalcididae) than pupae closer to the ground (Weseloh, 1972). Parasitoids of fly larvae in the leaf litter (e.g. diapriids, alysiine braconids, orthocentrine ichneumonids, etc.) appear to search primarily at ground level, at least in north temperate forests. Tropical forests often have a great deal of leaf litter in canopy epiphytes, and therefore such parasitoids probably show more variation in heights at which they search.

Many parasitoids are attracted by patch-specific cues such as volatile chemicals emanating from the host's food source (Vinson, 1984*b*) irrespective of whether or not a host is present in the vicinity. For example, *Venturia canescens*, a cosmopolitan ichneumonid parasitoid of the Mediterranean flour moth (*Ephestia kuehniella*: Pyralidae), is attracted to oatmeal even if moth larvae are not present (Thorpe & Jones, 1937). Diplazontine ichneumonids, which parasitize aphidophagous syrphid larvae, are strongly attracted by volatile substances originating from the honeydew produced by aphid colonies (Rotheray, 1981*b*). Another ichneumonid, *Campoletis sonorensis*, is not strongly attracted by the host's food-plant if the plant is in pristine condition, but is attracted if the plant is very slightly damaged by the feeding activities of non-host organisms such as thrips (Vinson & Williams, 1991). Some parasitoids that attack drosophilid larvae, such as the eucoiline *Leptopilina* and various alysiine braconids, are extremely specific in their response to patch-specific odours, such as fruit in a particular stage of decay (Vet *et al.*, 1984; Vet & Opzeeland, 1985; Vet, 1985).

In some cases parasitoids are known to develop in a host species feeding on one plant, but not in the same host feeding on a different plant species. In Costa Rica, for example, the ichneumonid *Cryptophion inaequalipes* is a host specific parasitoid of *Xylophanes turbata* (Sphingidae), but it has only been reared from host caterpillars feeding on *Hamelia patens* (Rubiaceae), despite the fact that this caterpillar also commonly feeds sympatrically on another rubiaceous understorey tree, *Psycotria microdon* (Janzen, 1984; Gauld & Janzen, in press).

At the finest level, patch location often involves the use of a combination of cues. *Campoletis sonorensis*, for example, uses both odour and vision in locating the host's food-plant, as significantly more females flew to plants that were visible than plants which were obscured from view (McAuslane *et al.*, 1991). Sometimes the detection of one cue, such as odour, 'turns on' another, specific perceptual ability, such as vision. For example, *Diaeretiella rapae* (Braconidae: Aphidiinae), a parasitoid of the cabbage aphid, is attracted initially by the odour of brassicas (Read *et al.*, 1970). If the female then encounters a leaf contaminated with honeydew she flies upward to an inflorescence; if the inflorescence contains no aphids she walks down a stem and up another until honeydew is again encountered (Ayal, 1987).

Some parasitoids, especially synovigenic species, seem to have a pre-ovipositional period when they are either not receptive to volatile substances from the host's food-plant (Nishida, 1956), or when they are actually repelled by such chemicals (Thorpe & Caudle, 1938). Their response changes several days or even weeks after emergence when they are ready to commence oviposition. Many parasitoids are able to change their behavioural responses to odours by associative learning. For example, the eucoiline parasitoid, *Leptopilina heterotoma*, is attracted by odours emanating from different substrates in which their drosophilid hosts feed. After encountering and ovipositing into hosts in one particular substrate, odour from this substrate is preferred over that from other substrates in choice experiments, and these behavioural changes may last for several days (Vet, 1988; Vet & Schoonman, 1988; Papaj & Vet, 1990; Vet *et al.*, 1990). Similar findings have been reported for other parasitoid species (e.g. Dmoch *et al.*, 1985; Lewis & Tumlinson, 1988; Kaas *et al.*, 1990; Vinson & Williams, 1991), and the general subject of learning in insects has recently been treated at length (Papaj & Lewis, 1992).

In some species there is evidence for pre-imaginal conditioning in parasitoids, whereby an individual parasitoid responds more readily to the odours of the plant species on which it developed (Vet & Groenewold, 1990). For example, the aphid parasitoid *Diaeretiella rapae* showed a greater positive response to the odour of potato when it had been reared from aphids feeding upon this host plant (Sheehan & Shelton, 1989). *Microplitis demolitor* (Braconidae) reared from *Helicoverpa zea* larvae on cowpea plants responded strongly to the odour of cowpeas, whereas females reared from hosts fed on an artificial diet did not (Hérard *et al.*, 1988). An interesting recent discovery is that certain species of parasitoids seem to be able to ascertain, from a distance, that another parasitoid species is searching a particular type of microhabitat, and they then avoid searching this microhabitat (Janssen *et al.*, 1992). Little is yet known of the source of the substance that the parasitoid is responding to, but it seems to be something produced by the host-searching parasitoid already in the microhabitat.

Host location. After a female parasitoid has found a suitable microhabitat she must locate a potentially acceptable host, and marked changes in behaviour sometimes take place between these two stages in host finding. Many koinobiont ichneumonoid parasitoids of leaf-feeding larvae fly rapidly horizontally in a zigzag pattern until they locate a plant on which the host may occur; they then change to an up-and-down 'scanning' flight as they begin searching for hosts (Vinson, 1975; Gauld, 1976*b*). Many idiobionts on the other hand cease flying and begin running over the substrate, sweeping with their antennae. Parasitoids in general often begin an even more intensive local search after perceiving cues associated with the host or after the first oviposition.

Parasitoids may respond to a variety of stimuli in locating hosts within a patch (Schmidt, 1992; Vinson, 1992). Physical stimuli, which in general are probably only detected at close range include:

- visual stimuli, for example seeing the host or its immediate covering (e.g. Sugimoto *et al.*, 1988*a*; Henriquez & Spence, 1993);
- detecting the host's movement through the substrate on which it feeds (O.J. Smith, 1952; Askew, 1961*b*; Richerson & DeLoach, 1972; Sokolowski & Turlings, 1987);
- the detection of sound or vibration from the host's feeding activities (Assem & Keunen, 1958; Ryan & Rudinsky, 1962; Soper *et al.*, 1976; Sugimoto *et al.*, 1988*b*).

Other possible physical methods of host location may include infra-red detection by a doryctine braconid parasitoid of bark-beetles (Richerson & Borden, 1972), and a form of acoustic probing (echo-location) by pimpline ichneumonid parasitoids of cocooned pupae (Henaut & Guerdoux, 1982). This latter method, which involves the parasitoid generating sound that makes hosts resonate, is apparently used to inspect prospective hosts prior to probing with the ovipositor (Henaut, 1990).

Chemical stimuli, resulting from the presence of the host and detectable over long or short range, are probably the most important factors enabling parasitoids to detect their hosts (Weseloh, 1981; Vinson & Williams, 1991; Vinson, 1992). The chemical may be the same as the plant volatile substances that attract parasitoids to the general habitat (Vinson & Williams, 1991), or it may be a substance that is released by the plant only in response to particular types of damage. Such attractants, the so-called green leaf volatiles, are released from the food-plant by the feeding of particular herbivores. Corn seedlings, for example, release large amounts of terpenoid volatiles only after they have been fed upon by caterpillars, and not when they have been artificially damaged (Turlings *et al.*, 1990). These volatiles, which attract the braconid *Cotesia marginiventris*, are also released by undamaged leaves on the same seedlings (Turlings & Tumlinson, 1992). Some parasitoids are capable of detecting extremely small quantities of green leaf volatiles (Whitman & Eller, 1992) and it has been suggested that these substances may have evolved to serve a tritrophic communicative function, i.e. plants release natural enemy attractants in direct response to herbivore attack (Whitman, 1988; Dicke *et al.*, 1990; Turlings *et al.*, 1990).

Frequently the chemical attractant may be a pheromone produced by the host (Noldus *et al.*, 1990), which acts as a KAIROMONE for the parasitoid—i.e. a substance produced by one organism that evokes a response in another species that is adaptively favourable to the receiver but not the transmitter. A variety of host pheromones are known to attract parasitoids: sex pheromones (Sternlicht, 1973; Noldus, 1989; Noldus *et al.*, 1991*a* & *b*), marking pheromones deposited by the host during oviposition

(Vinson, 1975; Prokopy & Webster, 1978; Roitberg & Lalonde, 1991) and aggregation pheromones (Rice, 1969; Kennedy, 1979). Non-pheromonal secretions by the host can also serve as kairomones. *Venturia canescens* (Ichneumonidae), a common parasitoid in warehouses, uses a chemical substance present in mandibular gland secretions of the Mediterranean flour moth as a kairomone (Corbet, 1971). *Aprostocetus hagenowi* (Eulophidae), a common parasitoid of cockroach eggs in houses, utilizes calcium oxalate (from the adult cockroach colleterial gland) and mucopolysaccharides (from the salivary gland) in host location (Vinson & Piper, 1986). In other cases the kairomone may be a chemical present in the cuticular wax of the host (Takabayashi & Takahashi, 1986), a substance associated with the host's silk (Weseloh, 1976), a chemical present in the frass of the host (e.g. Thibout *et al.*, 1988; Auger *et al.*, 1989; Fukushima *et al.*, 1989; Lewis *et al.*, 1991), or in the case of aphid parasitoids, in honeydew (Budenberg, 1990). Vinson (1984*b*) gives a list of kairomones known to be used as attractants by parasitoids.

As is the case in habitat location, parasitoids may learn to associate certain kairomones with a host and preferentially investigate such chemical stimuli (Turlings *et al.*, 1990; Lewis *et al.*, 1991), although learning may be less important in host location than in host habitat location (Lewis *et al.*, 1990).

After detecting cues associated with the presence of the host, the female parasitoid usually remains in the patch for a period of time. There are at least four ways of restricting the search to a profitable patch (Bell, 1991):

- looping/spiralling or zigzag motor patterns, as in *Trichogramma pretiosum* (Beevers *et al.*, 1981);
- turning around at the edge of the patch, as in *Venturia canescens* (Waage, 1979) and *Cardiochiles nigriceps* (Stand & Vinson, 1982);
- variable move lengths;
- change in arrival-departure direction.

Patches differ intrinsically in quality. Moreover, host availability within a patch declines as a result of parasitoid foraging (oviposition or host-feeding). Various models have been proposed to explain how parasitoids, using only innate responses, 'decide' to leave a patch when the profitability of remaining declines. In general, the time spent within a patch usually increases in response to contact chemical stimuli and in response to oviposition, but diminishes over time as the parasitoid becomes habituated to host-associated odours. Oviposition sometimes serves to 'reset' responsiveness to odours (Waage, 1979).

Host acceptance

Once a female parasitoid has located a potential host the next step is for her to choose whether or not to oviposit. This 'decision' may be based on the potential host's identity, size, developmental stage, period of time elapsed since its last moult, its defensive behaviour, etc.—all of which can be considered as HOST IDENTIFICATION (Vinson, 1975). Also affecting the female's decision is whether or not the host has already been parasitized. The ability to discern previously parasitized hosts is known as HOST DISCRIMINATION. It should be noted that host acceptance is occasionally affected by additional factors, such as how long it has been since the female parasitoid last oviposited, which may account for the occasional acceptance of 'abnormal' hosts.

Host identification. When a female wasp encounters a host or host construction she generally investigates it initially with her antennae. Important factors in eliciting an oviposition response include size (Duodu & Davis, 1974; Kouamé & Mackauer, 1992), shape (Wilson *et al.*, 1974; Strand & Vinson, 1983) and surface texture (Weseloh, 1974; Vinson, 1976) of the prospective host and, in trichogrammatids, host density (Renzik & Umarova, 1991). Movement of a host can also be an important cue, either for eliciting oviposition by the parasitoid, or for preventing host acceptance (Arthur, 1981). For example, the ichneumonid *Diadegma* does not attempt to insert its ovipositor into its lepidopterous host until the caterpillar moves (Lloyd, 1940), and the braconid *Perilitus* can be induced to oviposit on a movable model of its coccinellid host (Richerson & DeLoach, 1972).

Endoparasitoids generally also probe the prospective host with their ovipositor before deciding whether or not to lay an egg, and in some cases an egg is only laid if sense organs on the ovipositor apex are stimulated by appropriate chemicals (Arthur *et al.*, 1969; Rotheray, 1984). Substances have been identified in the haemolymph of the wax moth (*Galleria mellonella*) that elicit oviposition by its parasitoid, the ichneumonid *Itoplectis* (Arthur *et al.*, 1972; Hegdekar &

Arthur, 1973). Similarly, substances from the haemolymph of *Helicoverpa zea* elicit oviposition by *Microplitis croceipes* (Tilden & Ferkovitch, 1988).

Host discrimination. Many endoparasitoids are capable of discerning whether or not a prospective host has already been parasitized and they tend to avoid those that are (Wylie, 1965; Lenteren, 1981; Bai & Mackauer, 1990; Artigues *et al.*, 1992), although under certain conditions superparasitism occurs (see above). Although host discrimination is widespread a few species seem to lack this ability (Gardner *et al.*, 1984). Repeated observations of other species have shown that females without experience of ovipositing in unparasitized hosts (i.e. inexperienced) will superparasitize more readily than females that have oviposited in unparasitized hosts (experienced) (Lenteren, 1976). This has been interpreted as evidence for a lack of innate discriminatory ability, and it has been suggested that such an ability has to be learned by a female encountering both parasitized and unparasitized hosts (Lenteren & Bakker, 1975). However, Alphen *et al.* (1987) have shown that certain of these parasitoids are able to discriminate, even when they have never oviposited in an unparasitized host. These authors proposed an explanation of why inexperienced females superparasitize more readily. They suggested that an experienced parasitoid may reject parasitized hosts and leave an exploited patch because her experience in a previous patch supplied her with the information that less exploited patches exist. Inexperienced females lack such information and thus stay and superparasitize (Visser *et al.*, 1992).

In the process of host discrimination female parasitoids are known to respond to two types of stimuli (Lenteren, 1981; Hofsvang, 1990; Mackauer, 1990):

- an external chemical marker on the host or its surroundings left by themselves or by another female, usually of the same species. These markers are apparently detected by antennal examination;
- internal differences between parasitized and unparasitized hosts, detected by sense organs on the ovipositor.

External host-marking usually involves the application of a pheromone by the attacking parasitoid to the outside of the host, either before or after the actual act of oviposition (Rabb & Bradley, 1970; Wylie, 1970, 1971*b*; Eberhard, 1975; Völkl & Mackauer, 1990). Although little is known of the sources of these pheromones, they seem generally to be produced by glands or bodies associated with the female reproductive tract, but in different taxa they apparently have different origins. Among the Scelionidae, the host-marking pheromone of a species of *Telenomus* is produced in the female's accessory gland, and after ovipositing in a host egg it is applied by the scelionid wiping her extruded ovipositor across the surface of the host egg (Strand, 1986; Cave *et al.*, 1987). Some braconids and ichneumonids are known to secrete a marking pheromone from the Dufour's gland (Guillot & Vinson, 1972; Vinson & Guillot, 1972; Harrison *et al.*, 1985), while in the Ceraphronoidea a substance with a similar function seems to be produced by nurse cells in the oviducts (Höller *et al.*, 1993). Sometimes not only the host, but the surrounding substrate is also marked. For example, the braconid *Opius pallipes*, an endoparasitoid of an agromyzid leaf-miner, marks both the host and the leaf in which the host occurs. Other females of the same species devote less time investigating marked leaves than they spend examining unmarked leaves (Lenteren, 1981), and similar results have been obtained with another braconid parasitoid of leaf-miners (Sugimoto *et al.*, 1986).

Host-marking pheromones probably evolved to facilitate an individual female's foraging efficiency, allowing her to quickly detect areas that she has already searched and rapidly discriminate hosts in which she has previously oviposited (Strand, 1986). Although many such pheromones are species-specific (e.g. Alphen & Visser, 1990), some may be individually recognizable (Harrison *et al.*, 1985; Hubbard *et al.*, 1987; Völkl & Mackauer, 1990). A species of *Dendrocerus* (Megaspilidae) uses a two-component external marking pheromone to discriminate between hosts parasitized by themselves and hosts parasitized by conspecifics (Höller *et al.*, 1991). In a few cases marking pheromones are recognized by other sympatric species of parasitoids (e.g. Pijls *et al.*, 1992).

External markers are active immediately following deposition, and last from 12 hours to three days (Bosque & Rabinovich, 1979; Klomp *et al.*, 1980). Host discrimination that occurs after the external marker has faded results from detection of internal changes in a parasitized host, which are not species specific (Fisher & Ganesalingam, 1970; Takasu & Hirose, 1988), such as necrosis in the case of many egg parasitoids (Strand, 1986). A species of mymarid that attacks water beetle eggs under water depends entirely on internal stimuli for host discrimination, although the source of these stimuli is uncertain (D.J. Jackson, 1969).

Host discrimination usually relies on the detection of chemical stimuli, whether internal or external, but it is possible that non-chemical cues are used in a few cases. For example the protruding egg stalk of some encyrtids apparently functions as a marker that deters oviposition by other experienced conspecific females (Takasu & Hirose, 1988).

Host exploitation

After having accepted a host for oviposition the female parasitoid must 'decide' how best to exploit it—how many eggs to lay per host (clutch size) and what sex ratio to produce. Parasitoid species vary in the ways in which they optimize clutch size (progeny allocation) and sex ratio (sex allocation) (Waage, 1986).

Clutch size. For any given host there is an optimum number of individuals of a particular parasitoid that can develop on it. In many cases a parasitoid requires the entire host to successfully complete its development and superparasitism results in mortal combat between larvae (e.g. Escalante & Rabinovich, 1979). In gregarious endoparasitoids fitness, measured as probability of survival, rises with increasing clutch size to an optimum (see gregarious parasitism above). Beyond this optimum the fitness of each individual declines as clutch size increases (Skinner, 1985; Waage, 1986). Increasing clutch size generally results in smaller sized offspring, which have reduced fecundity, longevity and searching rates, as shown in studies of *Trichogramma* species (Pak & Oatman, 1982; Waage & Ng, 1984).

Optimization of clutch size by parasitoids assumes they are capable of perceiving the 'carrying capacity' of a particular host, and of responding to differences in host density. For idiobionts, one indication of the host's carrying capacity is its physical size, and numerous studies (see Schmidt, 1992) have demonstrated parasitoids' abilities to assess this and adjust the number of eggs laid accordingly. For example, in *Trichogramma* the number of eggs laid by the wasp in a host egg varies with the host egg's volume. The wasp determines this volume by the duration of its initial transit across the egg surface during host examination (Schmidt & Smith, 1987). In koinobionts there is some evidence that host instar, and not host size, is the stimulus for changing clutch size (Sato & Tanaka, 1984). Some gregarious endoparasitoids may be capable of assessing how many eggs have previously been laid in a host and adjusting their clutch size (and sex ratio) accordingly (Holmes, 1972; Werren, 1980; Dijken & Waage, 1987). Information on host abundance, at least in *Trichogramma*, may be provided by measurements of time between ovipositions (Waage & Ng, 1984; Waage & Godfray, 1985), or by host distribution; hosts packed in groups have less exposed surface and hence are allocated fewer eggs (Schmidt & Smith, 1985*a* & *b*).

Sex ratio. In diploid organisms sex is usually determined by Mendelian assortment of sex chromosomes, which may restrict the potential for sex ratios to evolve away from 50 percent females and 50 percent males. In Hymenoptera however, which are haplodiploid (see Chapter 2.2), there may be more variation in sex ratio since, in principle, selection may act on the control that females have over insemination—that is, regulating the release of sperm from the spermatheca (Flanders, 1956).

In solitary parasitoids, especially but not exclusively in idiobiotic species, the adult female tends to lay female eggs on large hosts and male eggs on smaller hosts (Arthur & Wylie, 1959; Kishi, 1970; Tillman & Cate, 1993), thus the female progeny are significantly larger than the male progeny (Sandlan, 1979; Gauld & Fitton, 1987; Cloutier *et al.*, 1991; see Gauld, 1991 for Costa Rican examples). Since egg synthesis generally requires more nourishment than that of sperm, females gain relatively more from larger hosts than do males, thus explaining the observed patterns of sex allocation (Charnov, 1982). Large females produce more mature eggs and often live longer than small females (Wylie, 1966; Pitcairn & Gutierrez, 1992; Tillman & Cate, 1993). However, if males compete with each other for females there could be a selective advantage for producing larger males.

During oviposition the female parasitoid probably assesses relative rather than absolute host size and adjusts the sex ratio of her progeny on the basis of both short- and long-term experience (Jones, 1982; Assem *et al.*, 1984; Avilla & Albajes, 1984; Luck & Podoler, 1985). Some parasitoids allocate male eggs preferentially to hosts at a less suitable stage for development (Alphen & Thunnissen, 1983).

If a patch of hosts is discovered by just one female parasitoid, she should produce only as many sons as necessary to mate all her daughters, at least in cases where offspring mate immediately upon emergence. This is the LOCAL MATE COMPETITION theory (Hamilton, 1967, 1979). In many gregarious parasitoids males emerge from the host first and mate with females as they emerge. As predicted by the theory, such species tend to have female-biased sex ratios. In polyembryonic encyrtids mixed broods (containing both sexes) are laid in the first host encountered; all female broods are produced if the host encounter rate is high (Hardy *et al.*, 1993). Exceptions to a female bias occur in those gregarious species where females disperse before mating (Tagawa & Kitano, 1981) or among species where females prefer to mate with non-sibling males (Grant *et al.*, 1980). Local mate competition theory also predicts that relatively more males should be produced if the female detects that other females have visited the patch, and this prediction has been supported by both controlled and natural experiments (Wylie, 1966, 1979; Shiga & Nakanishi, 1968; Werren, 1983; Waage & Lane, 1984; Herre, 1987). The theory thus helps explain why crowded laboratory cultures of parasitoids often yield mostly males.

There are various proximate mechanisms of sex allocation. The simplest is for the female to release a fixed rate of sperm from her spermatheca resulting in random fertilization (Owen, 1983). If the rate of oviposition is very high relatively more unfertilized (male) eggs may be laid than at a lower rate of oviposition (Kainoh, 1988). This is particularly the case in groups such as chalcidoids and braconids where many species have small spermathecal glands, which secrete substances that activate sperm (Flanders, 1956).

The above discussion has assumed nuclear genetic control over sex ratio. Autosomal nuclear genes are transmitted via both sexes and therefore selection favours a 1:1 sex ratio, except where populations are not panmictic (e.g. local mate competition). Cytoplasmic genes on the other hand are usually transmitted by just one sex, which creates strong selective pressure for cytoplasmic genes that distort the sex ratio toward production of the sex that transmits the gene (Bull, 1983). *Nasonia vitripennis* (Pteromalidae) is known to have three such extrachromosomal sex ratio distorters (Werren *et al.*, 1980; Huger *et al.*, 1985; Skinner 1985; Werren & Assem, 1986): a maternal sex ratio distorter (maternally transmitted, produces about 95 percent female sex ratio), a 'son killer' (a maternally transmitted bacterium that causes death of 50 to 80 percent of the males), and a paternal sex ratio distorter (paternally transmitted to all fertilized eggs, which remain haploid). The causative agent of the son-killer trait has recently been described as *Arsenophonus nasoniae*, and appears to belong to the Enterobacteriaceae (Gherna *et al.*, 1991). Since *N. vitripennis* is one of the few parasitoids that has been extensively studied it is probable that extrachromosal sex ratio distorters will eventually be found in other hymenopteran parasitoids.

The suitability of the host for parasitoid development

After the female parasitoid has accepted a host and oviposited, the immature stages must be able to survive and develop on or within the host. Different host species that are equally acceptable to the female are not necessarily equally suitable for her larvae (Mueller, 1983). HOST SUITABILITY can be measured by survival of larvae to adulthood and the size or fecundity of the emerging adults. Factors affecting host suitability (Vinson and Iwantsch, 1980*a*) include:

- nutrition;
- environmental factors;
- other organisms (competitors, hyperparasitoids and predators);
- in the case of endoparasitoids, the immunological defences of the host.

In many endoparasitic koinobionts host suitability can presumably be enhanced via HOST REGULATION, that is, the ability of the parasitoid larva or its female parent to alter the host's development, physiology or behaviour for its own ends (Vinson and Iwantsch, 1980*b*). Host regulation often involves alterations in the host's hormonal balance (Beckage, 1985; Lawrence, 1986*a*, 1990). This is induced by venom or virus particles injected by the female (Stoltz & Vinson, 1979; Krell, 1991; Tanaka & Vinson, 1991; Fleming, 1992), by the teratocytes liberated at the hatching of the parasitoid's egg (Pennacchio *et al.*, 1992), or by the parasitoid larva itself (Beckage & Riddiford, 1982).

Nutrition

Most parasitoid larvae feed by oral ingestion. Ectoparasitoids pierce the host's cuticle with their mandibles before feeding, whereas many endoparasitic larvae ingest host haemolymph directly. Some endoparasitoid larvae are reported to absorb nutrients directly through the cuticle (Doutt, 1952; Fisher, 1971), and in those that have an anal vesicle (see Chapter 4.4) membranal absorption may be localized in this organ (Edson & Vinson, 1977). In some endoparasitic koinobionts nutrient absorption may begin in the egg stage via a specialized egg membrane, the trophamnion. Upon eclosion of the egg this trophamnion may fragment into teratocytes (see Chapter 2.4). These often absorb nutrients from the host, and in some cases the parasitoid larva appears to feed on the teratocytes, thus avoiding damaging the host's organs early on in its development (Polaszek, 1986; Tawfik, 1991). Feeding on teratocytes could conceivably also be a way for a parasitoid to avoid toxins sequestered or manufactured by the host, and present in its haemolymph or fatty tissue.

The basic qualitative nutritional requirements of parasitoid larvae are quite similar to those of nonparasitic insects, although the form and balance in which nutrients need to be supplied may differ (Thompson, 1986). Endoparasitoid larvae exhibit high assimilation efficiencies, presumably because of the high nutritional quality of their food (haemolymph) and possibly because of selection for efficient utilization of a limited food supply (Slansky & Scriber, 1985). However, net growth efficiencies (i.e. the conversion of assimilated food into parasitoid biomass) are mostly moderate, possibly because of selection for rapid rather than efficient growth (Slansky, 1986).

Host suitability is affected by the nutritional status of the host, the types and quantities of toxins present (e.g. secondary plant compounds sequestered by the host; see Gauld & Gaston, 1993 for summary), and the complex ways in which nutrients and toxins interact (Duffey *et al.*, 1986). Not surprisingly, a given phytophagous host may differ in its suitability, depending upon the plant on which it is feeding. For example, California red scale (Diaspididae) reared on different cultivars of citrus differ in their suitability for the aphelinid ectoparasitoid, *Aphytis melinus* (Hare & Luck, 1991). Survivorship of *Microplitis croceipes* (Braconidae), an endoparasitoid of *Heliothis* spp. (Noctuidae), is higher in host larvae reared on cotton than in hosts reared on bean or tomato (Mueller, 1983). In both examples the plant that was most suitable for growth of the host was also the most suitable for the parasitoid. For egg parasitoids the nutritional suitability of the host egg can vary with the diet upon which the previous generation of the host was fed (Magrini & Botelho, 1991).

Many endoparasitic koinobionts appear to enhance the suitability of their host by altering the host's normal feeding behaviour. Hosts attacked by gregarious koinobionts frequently increase feeding (Hunter & Stoner, 1975; Führer & Keja, 1976; Beckage & Riddiford, 1983; Sato *et al.*, 1986), whereas those attacked by solitary parasitoids may show reduced feeding and growth (Vinson & Barras, 1970; Ashley, 1983; Slansky, 1986; Kumar & Ballal, 1992). Endoparasitic koinobionts also alter the nutrient composition of the host's haemolymph and fat body and the metabolic rate of the host (Vinson & Iwantsch, 1980*b*; Thompson, 1983). In most cases it is unclear which of these alterations are indirect effects of stress and which represent host regulation for the benefit of the parasitoid.

Environmental factors

Environmental factors such as temperature and humidity can affect the parasitoid directly, or indirectly via effects on host suitability. Suitability of lepidopteran eggs for *Trichogramma* can vary between species of Lepidoptera due to differences in host egg chorion, which in turn affect vulnerability to desiccation (Pak *et al.*, 1990). Presumably koinobionts are more susceptible to the indirect effects of environmental factors on host suitability than are idiobionts. Environmental factors also affect the synchronization of the parasitoid's development with that of its host. Where there is pronounced seasonality in the climate, as in the seasonally dry forests of northwestern Costa Rica, the life cycles of many parasitoids are seasonally synchronized with those of their hosts, as is the case of *Enicospilus lebophagus* with its saturniid host, *Rothschildia lebeau* (Gauld, 1988*b*). In the case of some idiobionts synchronization may be achieved by prolonged life spans of adult females and relative inactivity during periods when hosts are scarce (possibly many scelionids). In other cases synchronization is achieved by diapause in an immature stage, and this diapause seems to be regulated by one of two means, or by a combination of both (Tauber *et al.*, 1986):

- External factors that also affect the host. These may be abiotic factors, such as temperature, or possibly humidity, or external biotic factors such as food availability or parasitoid density (Eskafi & Legner, 1974*b*; Wylie, 1980; Freeman & Ittyeipe, 1982). In temperate regions photoperiod is the most common abiotic stimulus known to affect diapause (e.g. Fabres & Reymonet, 1991), but this is probably less important in the tropics where variation in daylength is very slight (Denlinger, 1986).
- The host's physiological status. Endoparasitic koinobionts commonly have a very protracted first larval instar and delay their development until the host reaches a specific stage in its metamorphosis (Doutt *et al.*, 1976). The duration of development of a parasitoid species may vary depending upon the growth rate of the particular species of host parasitized (Broodryk, 1969). Some endoparasitic koinobionts are known to use host hormones as cues for their own development. Thus, reduced levels of ecdysone may play a role in diapause regulation in some endoparasitoids (Claret *et al.*, 1978), and the larva of an opiine braconid apparently needs exposure to 20-hydroxyecdysone from its host in order to moult (Lawrence, 1986*b*). A polyembryonic encyrtid begins morphogenesis after a large drop in the host juvenile hormone titre in the penultimate host larval stadium (Strand *et al.*, 1991). In other cases the parasitoid may manipulate the hormonal balance of the host (see review by Beckage, 1985).

Competitors, hyperparasitoids and predators

The larva of a solitary parasitoid usually requires the entire host for its development. If it is to survive to maturity, or at least achieve an optimum adult size, it must have the capability of eliminating competing parasitoids. Usually it is the larval stage that must contend with competitors but in a few cases the adult female injects a substance that kills parasitoid eggs present in the host (Hågvar, 1988; Mackauer, 1990). A rather unusual means of avoiding competitors has been reported in an aphidiine braconid. Aphids subjected to a stabbing attack by the female wasp are stimulated to adopt a 'bucking' defensive behaviour when subjected to further attacks and such behaviour appears to reduce the likelihood of superparasitism (Gardner *et al.*, 1984).

The first instar larvae of parasitoids from a wide range of families are equipped with large curved mandibles (see Chapter 4.3) that are used in physical combat to destroy competitors (Salt, 1961; Eijsackers & Bakker, 1971; Fisher, 1971; Flanders, 1971; Wylie, 1971*a*). An aphidiine braconid has been observed using its bifurcate caudal appendage as a grappling hook during combat (Chow & Mackauer, 1986). Some first instar parasitoid larvae have small, more or less non-functional mandibles that appear quite unsuitable for fighting (Gauld, 1976*a*) and in some cases these species appear to eliminate competitors by physiological means. Young aphidiine braconid larvae, for example, possibly secrete a cytolytic enzyme that suppresses the development of embryos of competitors (Johnson, 1959). In most parasitoids later instar larvae lack the physical weaponry to eliminate competitors and instead they probably rely exclusively on physiological suppression (e.g. Vinson & Ables, 1980; Artigues *et al.*, 1992; Tillman & Powell, 1992). Reducing the oxygen content of the host's haemolymph offers a means for older parasitoid larvae to eliminate younger competitors, as there is evidence that a low oxygen content is more debilitating to eggs and young parasitoid larvae than it is to older larvae (Fisher, 1963). Suppression of younger larvae by older larvae may also occur via nutrient depletion (Tremblay, 1966).

The outcome of competition between two larvae of the same species often depends upon the relative age of the two competitors (Wylie, 1972; Mackauer, 1990). If both larvae are in the first instar the oldest larva often wins. On the other hand, if a second instar larva has not succeeded in physiologically suppressing a competing first instar it may be killed by the latter, which still has its large mandibles. A well-studied case is that of the solitary egg parasitoid, *Telenomus heliothidis* (Scelionidae) (Strand, 1986). If ovipositions occur within two hours of one another both eggs hatch and one of the larvae is eliminated by physical attack. If the interval between ovipositions is more than two hours, the second female's egg develops normally until the first female's egg hatches, whereupon the latter's teratocytes secrete a substance that causes necrosis in both the host and any unhatched eggs.

Since interspecific host discrimination by ovipositing females is less common than intraspecific discrimination, interspecific competition (multiparasitism) may be more important than intraspecific competition

(Force, 1975). Because of their paralyzing venoms idiobionts probably have an intrinsic superiority over koinobionts (Flanders, 1971), although there are exceptions (Sullivan, 1972). In competitive interactions between two idiobiont species the second attacking species sometimes facultatively hyperparasitizes the first (Haviland, 1920). The winner of competition between *Telenomus heliothidis* and *Trichogramma pretiosum* is usually the species whose egg hatches first (Strand, 1986), although when *Telenomus* reaches the third instar it is vulnerable to hyperparasitism by *Trichogramma* (Strand & Vinson, 1984). The outcome of competitive interactions between two species of koinobionts may also depend on the timing of the two attacks (e.g. Tillman & Powell, 1992), or one species may be intrinsically superior to the other. For example, the diplazontine ichneumonid *Syrphophilus tricinctorius* is competitively superior to *Diplazon pectoratorius* in a syrphid host (Rotheray, 1984), the aphidiine braconid *Aphidius ervi* generally wins in competition with *Aphelinus asychis* (Aphelinidae) in aphid hosts (Bai & Mackauer, 1991), and the encyrtid *Epidinocarsis lopezi* outcompetes its congener *E. diversicornis* in multiparasitized hosts (Pijls *et al.*, 1992). On the other hand, equal numbers of aphidiine braconids *Aphidius colemani* and *Lysiphlebus testaceipes* emerge from multiply parasitized hosts regardless of the attack sequence (Völkl & Stadler, 1991).

Besides other parasitoids, competitors also include microbes. Koinobionts sometimes have to compete with pathogenic infections of the host. In general female parasitoids appear to avoid ovipositing in hosts showing advanced stages of microbial infection but often they cannot apparently detect recently infected hosts. Oviposition into hosts recently infected with *Bacillus thuringiensis* often results in death of the parasitoid, although sublethal infections sometimes favour the parasitoid (Weseloh & Andreadis, 1982). The outcome of competition between parasitoid and microbe sometimes depends upon the timing of attack. For example, the braconid *Microplitis croceipes* is adversely affected by nuclear polyhedrosis virus in its noctuid host if the time between oviposition and ingestion of the virus is less than three days (Brown & Phillips, 1991). In koinobionts, defence against microbial competitors probably consists mostly in maintaining the host's immunological defences intact, although the teratocytes of one microgastrine braconid are known to secrete a fungistatic metabolite (Führer & Elsufty, 1979). In contrast to koinobionts, idiobionts principally have to compete with decomposers for the host food resource. Many inject a venom that not only paralyses the host, but also preserves it from decomposing (Askew, 1971) and, in most cases, the host is consumed very rapidly. In some endoparasitic idiobionts the larva itself secretes antimicrobial compounds (Führer & Willers, 1986).

To avoid hyperparasitoids the primary parasitoid larva may induce an alteration in host behaviour. For example, larvae of the butterfly *Euphydryas* parasitized by a species of microgastrine braconid crawl up out of dense vegetation to feed in more exposed places (Stamp, 1981). Similarly the aphid *Macrosiphum euphorbiae*, when parasitized by a species of *Aphidius*, migrates to the upper surface of the leaf where it appears to escape detection by hyperparasitoids (Brodeur & McNeil, 1992), and parasitized mealybugs often leave the foliage to hide in the bark (Bartlett, 1978). After successful parasitization by microgastrine braconids, the still living larva of *Pieris brassicae* (Lepidoptera: Pieridae) spins a web over its parasitoid's cocoons, and remains with them, apparently aggressively defending them from possible enemies (Brodeur, 1992).

The host's immunological defences

The most important defence against endoparasitoids by many hosts is the ENCAPSULATION of the parasitoid's eggs or larvae, a process that is mediated by haemocytes (see below). Endoparasitoids have evolved various methods of circumventing this defence (Salt, 1968; Vinson, 1990*a*). In his excellent review of this subject Vinson (1990*a*) enumerated five ways in which parasitoids may deal with the host's immune system: a) avoidance; b) evasion; c) destruction; d) suppression; and e) subversion.

Avoidance. Endoparasitoids may avoid encapsulation by depositing their eggs in sites within the host away from the host's haemocytes, such as within the salivary glands or nerve ganglia (Silvestri, 1921; Veen, 1981). A rather unusual method of avoidance is shown by a species of *Eretmocerus* (Aphelinidae) that attacks whiteflies. The female aphelinid oviposits externally, laying an egg underneath a whitefly nymph. The first instar larva commences feeding as an ectoparasitoid, but secretions it produces (possibly from enlarged salivary glands) cause proliferation of the host's epidermal cells which grow to form a dome-shaped structure within which the parasitoid larva continues to develop (Gerling *et al.*, 1991).

Another method of avoiding encapsulation seems to be ovipositing into eggs or very young hosts, in which the immune system is not fully developed. For example, a braconid parasitoid of a mirid preferentially accepts mostly young nymphs as hosts, since these have a lower rate of encapsulating its progeny (Debolt, 1991). Similarly the encapsulation of an encyrtid by a scale insect has been observed to be age-related, with significantly more parasitoids succumbing to encapsulation in old hosts than young ones (Blumberg, 1988). Similar observations have been made for campoplegine ichneumonids where progeny develop successfully from eggs deposited in young larvae, but are encapsulated if eggs are deposited in more mature hosts (Puttler, 1961; Bosch, 1984; Beckage & Templeton, 1985).

Evasion. The surface of the egg of some parasitoids is coated with a fine fibrous coat, which prevents it from being encapsulated (Davies & Vinson, 1986); if the fibrous coating is removed the egg is attacked by the host's haemocytes. In the ichneumonid *Venturia canescens*, virus-like particles produced in cell nuclei of the ovarian calyx share antigenic determinants with one or more host proteins. Thus the parasitoid egg, which is coated with these particles, is not recognized as foreign, thereby evading detection by the host immune system (Rotheram, 1967, 1973; Salt, 1968, 1970; Feddersen *et al.*, 1986; Schmidt & Schuchmann-Feddersen, 1989). However, the exact mechanism is not fully understood, and it is possible that the protein suppresses the haemocytic defence (Schmidt *et al.*, 1990). A second method of evading encapsulation is simply to grow very rapidly, like, for example, the ichneumonine *Phaeogenes nigridens*, which may complete three larval instars in 24 hours (Smith, 1932).

Destruction. Complete destruction of the host's immune system could have serious consequences for a koinobiont endoparasitoid, which depends upon the host's health for its own survival. However, idiobiont endoparasitoids, which do not need to keep the host alive, could conceivably destroy their host's immune system and yet protect their food supply by secreting antimicrobial agents (Vinson, 1990*a*).

Suppression. Koinobiont endoparasitoids tend to suppress their host's immune system in specific ways rather than destroy it generally. This is evidenced by the fact that, in cases of superparasitism where larvae engage in combat, vanquished 'losers' are often encapsulated (Vinson & Sroka, 1978; Vinson & Stoltz, 1986; Meloche & Guppy, 1990). During oviposition certain ichneumonids (primarily Campopleginae, but also Banchinae) and braconids (Microgastrinae, Cheloninae and Cardiochilinae) inject a unique type of virus, polydnavirus, into their host (Krell, 1991; Fleming, 1992; Stoltz & Whitfield, 1992). The polydnaviruses found in ichneumonids (ichnoviruses) differ from those present in braconids (bracoviruses), and each parasitoid species carries a different virus. Within parasitoid populations these viruses are transmitted chromosomally in the form of a provirus (Stoltz, 1990), but replicate only in the ovarian calyx (region between the ovaries and oviduct). The virus appears to be 'designed' for export into the host, where they are immunosuppressive but do not replicate. Immunosuppression apparently results, at least in part, from changes in haemocyte number, behaviour, and/or viability, thus protecting the egg or young larva from encapsulation (Edson *et al.*, 1981; Stoltz & Guzo, 1986; Davies *et al.*, 1987; Guzo & Stoltz, 1987; Wago & Tanaka, 1989; Prévost *et al.*, 1990). Melanization, which follows encapsulation as part of the immunodefensive reaction, is also suppressed by the polydnavirus in a species of *Cotesia* (Beckage *et al.*, 1990), by the inhibition of monophenoloxidase activity.

Other types of viruses or virus-like particles have been reported from various ichneumonids (Campopleginae and *Mesoleius*) and braconids (Microgastrinae, *Diachasmimorpha*, *Meteorus*) (Stoltz & Whitfield, 1992). However, in many of these species it is not known whether the viruses play a role in successful parasitization. Outside the Braconidae and Ichneumonidae, the only known example of viruses or virus-like particles playing a role in suppressing the host's immune system is in *Leptopilina heterotoma* (Cynipoidea: Eucoilini), where virus-like particles in the venom cause the lamellocytes (a type of blood cell) of its drosophilid host to lose their adhesiveness (Rizki & Rizki, 1984, 1990).

In parasitoids lacking viruses or virus-like particles, other factors must be responsible for preventing encapsulation. Thus, the venom of a pimpline ichneumonid is known to reduce the ability of haemocytes to form pseudopodia (Osman & Führer, 1979). Substances secreted by teratocytes may also be important in suppressing the host's immune system (Kitano *et al.*,

1990; Tanaka & Wago, 1990). Even in braconids that carry polydnaviruses, it is probable that a combination of venoms, teratocyte secretions and polydnaviruses are necessary for successful parasitism (Kitano, 1986; Tanaka, 1986, 1987*a*; Vinson, 1990*a*).

Subversion. Certain dipterous parasitoids subvert the immune system of the host by allowing the host haemocytes to form a sheath around them. They avoid asphyxiation by maintaining contact with the atmosphere through a respiratory funnel. Some encyrtids may similarly subvert the host immune system by maintaining contact with the exterior via the protruding eggshell (Askew, 1971), or by developing in a sheath that has anastomozed with the tracheal system of the host (Alam, 1957, 1959). The larvae of many aphelopine dryinids are also more or less ensheathed in a layer of tissue produced by the host, which may serve a similar function (see Chapter 13.3). The aphelinid, *Eretmocerus*, passes its first two larval instars within a capsule formed from the host's epidermal cells (Gerling *et al.*, 1990, 1991).

THE DEFENCES OF THE HOST

Insect parasitoids are capable of inflicting severe mortality on populations of their hosts; thus features that reduce the chance of an individual being successfully parasitized are likely to be selectively favoured. A wide variety of such mechanisms have evolved and these have been grouped (Gross, 1993) into three functional classes:

1. Characteristics reducing the probability of a prospective host being discovered or encountered by a parasitoid—here called environmental factors;
2. Morphological or behavioural features of a prospective host that reduce the probability of it being oviposited upon after it has been located by a parasitoid—here called proximate factors;
3. Physiological mechanisms that operate after oviposition, but prevent successful parasitoid development.

These are discussed separately below.

Environmental factors

Not all stages of all insects are equally attacked by hymenopterous parasitoids (see Table 2.2). Certain groups seem largely to escape attack, possibly because either they are highly mobile, or they live in aquatic or semiaquatic habitats that render them particularly difficult to locate. The former category, highly mobile insects, includes the adults of most insect orders, the nymphs of some exopterygotes (such as the orthopteroid orders, Hemiptera, Homoptera Auchenorrhyncha) and mobile, active larvae of adephagan beetles. Some of these are exploited by various aculeate idiobionts (such as sphecids and vespids) which are highly specialized for attacking very active insects, but they largely escape attack by koinobionts and the traditional 'Parasitica'. Perhaps the most striking exception to this generalization occurs in the Auchenorrhyncha which are exploited as hosts by the diverse family Dryinidae (see Chapter 13.6). Other exceptions are the few orthopterans attacked by rhopalosomatids (Chapter 14.6), a very few adult insects parasitized by the eulophid genus *Phymastichus* (Feldhege, 1992) and members of the braconid subfamily Euphorinae (S.R. Shaw, 1988*a*), and some adephagan beetle larvae attacked by proctotrupids (see Chapter 9.4). However, it is generally only the eggs of these groups that are extensively exploited by parasitoid Hymenoptera.

The second category of potential hosts largely escaping parasitoids includes mud-inhabiting larvae and most aquatic insects. Syrphids dwelling in semifluid substrates, for example, almost entirely escape the attentions of parasitoids, but their predaceous relatives living in association with easily-discoverable homopteran colonies are attacked by a wide range of different parasitoid species (Shaw & Askew, 1979). Except for their eggs, which are attacked by a few trichogrammatids, scelionids and mymarids, aquatic insects in general are also largely free of parasitoids. The larvae of rice water weevils (*Lissorhoptrus* spp.), for example, which feed on the submerged roots of aquatic grasses, appear to totally escape parasitoids (P. Hanson, unpubl. obs.).

It is insect eggs, the relatively immobile sap-sucking homopterans, and the larval and pupal stages of holometabolous insects associated with plants or living in discrete, and probably semiochemical-rich habitats, such as dung, rotting fruit or corpses, that are the principal hosts of hymenopteran parasitoids. At the population level, such insects may evolve ways of escaping, in space or time, high levels of parasitism. At the individual level a potential host may attempt to eliminate cues that lead parasitoids to discover it.

Spatial avoidance of parasitoids may occur both at macro- or microhabitat levels. An example of the former is the migratory habit of many species of Lepidoptera. This may help them escape from areas with high parasitoid populations. Several species of sphingids have a single generation in the seasonally dry forests of northwestern Costa Rica, then the adults migrate to wet forests on the Caribbean coast where they presumably have further generations (Janzen, 1987). In the dry forests some of these species are attacked by several species of *Thyreodon*, ichneumonids which are not known to occur in lowland wet forests (Gauld, 1988*b*). However, in some parts of the world, e.g. Australia, ichneumonids are known to migrate, perhaps to follow their migratory hosts (Common, 1954). In Japan one species of pierid, which is not able to encapsulate a common microgastrine braconid, apparently avoids these parasitoids by constantly colonizing new habitats. Other less vagile congeneric pierids are immunodefensively resistant to these braconids (Ohsaki & Sato, 1990).

At the microhabitat level one form of spatial escape is to feed in some type of concealment. Borers deep in plant tissue or root feeders tend to have smaller parasitoid assemblages than more exposed hosts (Hawkins, 1990). Price and Clancy (1986), working with gall-causing sawflies, demonstrated that mortality due to parasitoids is inversely correlated with gall size. Parasitism rates of bark beetles have been shown to be negatively correlated with bark thickness (Ball & Dahlsten, 1973). However, phytophages living in 'weak' concealment on plants, such as leaf-rollers, web-spinners and leaf-miners are attacked by larger assemblages of parasitoids, especially idiobiont ectoparasitoids, than are exposed hosts (Hawkins & Lawton, 1987; Hawkins, 1990).

Temporal escape may be effected in several ways. In the North Temperate region large numbers of Lepidoptera overwinter as pupae. Consequently the larvae of many species feed synchronously and are pupating at a similar time in mid to late summer—when idiobiont pupal parasitoids are particularly abundant. A few species, such as *Euproctis chrysorrhea* (Lymantriidae), largely escape attack by these idiobionts since they overwinter as larvae, complete larval development very early in spring and pupate when most other species are still early or mid instar larvae. Prospective hosts can also reduce parasitism by exposing their vulnerable stages at times when fewer parasitoids are likely to encounter them. A number of endopterygote larvae (such as many species of noctuids) conceal themselves during the day then emerge and feed at night. Although nocturnal feeding probably allows them to escape the attentions of diurnally active parasitoids, they are attacked by a smaller assemblage of nocturnal parasitoids. Another method of temporally avoiding parasitoids is to spend long periods in a developmental stage that is not attacked, such as the adult stage. Many hispine leaf-miners in the seasonally dry forests of northwestern Costa Rica have a single generation per year, then emerge and spend up to eleven months as an adult. Manipulations creating an artificial second generation have demonstrated that such a generation is likely to be subjected to rates of parasitism approaching 100 percent (J. Memmott, pers. comm.). Many moths in this same area also have only a single generation and spend the dry season as reproductively inactive adults (Janzen, 1987).

Many insects have behavioural features that may help them avoid leaving cues that could enable parasitoids to discover them. A Costa Rican dry forest caterpillar repeatedly pushed its faecal pellet off the leaf it rested on (D.H. Janzen, pers. comm.), and many nest-building aculeates make considerable efforts to remove spoilage from the area around their burrows.

Phytophagous larvae feeding on relatively unusual food-plants may escape parasitism because their potential parasitoids search for hosts by initially locating the host's usual food plant. For example, a species of *Cardiochiles* (Braconidae) commonly parasitizes the larvae of *Heliothis virescens* (Noctuidae) when they occur on a variety of cultivated plants, but not when they occur on peanuts (Vinson, 1981). Although a potential host may avoid its normal parasitoids by such behaviour, it may be attacked by a different set of parasitoids.

Proximate factors

Defences of prospective hosts that either physically prevent oviposition, or that deter attack by the parasitoids can be grouped into four overlapping classes: a) mechanical or structural features; b) behavioural characteristics; c) chemical defences; and d) ant-mediated defences. Insect behavioural and morphological defences against parasitoids were recently reviewed by Gross (1993), who gives many additional examples to those cited below.

Mechanical defences. The immotile forms of many insects, such as coccoid scales, and the eggs and pupae of holometabolous insects, are often defended by some form of 'armour'—a thick chorion in the case of eggs, a hard carapace in diaspidid coccoids and a heavily chitinized pupal exoskeleton. All of these features offer some form of protection from parasitoids. For example, the eggs of some insect species have a chorion thick enough to prevent oviposition by a trichogrammatid entirely (Salt, 1938). Thicker scale covers on later instars of coccids significantly increase handling time by an encyrtid (Noda *et al.*, 1982). Many butterflies, which pupate in exposed situations, have hard smooth pupal cuticle, which makes it difficult for a parasitoid to grip the pupa and insert its ovipositor (Cole, 1959*a*). This difficulty is compounded by the fact that many such pupae are suspended on threads, making it difficult for the parasitoid to gain the foothold necessary for getting enough leverage to pierce the cuticle. Some pupae wriggle violently making oviposition even more difficult.

In other cases relatively immotile stages may be defended by hardened cases or dense silken masses. For eggs, such defences are provided by the egg-laying adult. For example, *Microrhopala*, a chrysomelid beetle, covers its vertical stacks of eggs with a layer of liquid faeces which hardens; *Chrysonotomyia* species are only able to attack the bottom eggs where the covering is incomplete (Damman & Cappuccino, 1991). Some mantids lay their eggs in a frothy mass of secreted foam that hardens rapidly. However, one group of scelionids have circumvented this defence by being phoretic upon the mantid adult, so they can gain ovipositional access to the eggs before the protective foam hardens (Rabaud, 1922). Various other coverings, such as layers of scales and frass, may also protect eggs from parasitoids. For example, scelionid parasitoids of an embiid require three times as long to reach eggs concealed with silk, faeces and substrate material, than they do to attack eggs from which the covers have been experimentally removed (Edgerly, 1987). Spiders that enclose their eggs in dense silken sacs may escape baeine scelionid egg parasitoids (Austin, 1985); however, these sacs resemble insect cocoons and thus attract a suite of ichneumonid parasitoids that mostly attack cocooned insect hosts (Gauld, 1988*c*).

Pupae or prepupae may also be defended in a similar manner, by hard coverings or dense silken cocoons, but in such cases these are secreted by the prospective host itself. For example, the prepupae of many limacodid moths are concealed in cocoons that are extremely hard and smooth, and difficult for a parasitoid to penetrate. Certain ichneumonids, most notably species of baryceratine cryptines, have a specially modified ovipositor with a threaded tip which they screw into the limacodid cocoon (Momoi & Okamoto, 1965), but oviposition may take over one hour. Other hard pupae are equipped with spines between their abdominal segments (Hinton, 1955), which may serve to pinch prospective parasitoids and drive them away (Askew, 1971). Many Lepidoptera spin dense woolly cocoons that may entangle parasitoids and deter them from ovipositing, or it may prevent them from actually reaching the pupa with their ovipositor (Cole, 1967; Cronin & Gill, 1989).

Larvae, like pupae or prepupae may gain some measure of protection from parasitoids by living in silken webs (e.g. Stamp, 1982), though silk-associated semiochemicals may also serve to attract parasitoids (Weseloh, 1976). The integument of the larvae of many holometabolous insects is furnished with spines or dense pubescence which, in some cases, offers protection against parasitoids (see Gross, 1993). However, although these features may protect caterpillars from attacks by certain parasitoids, such caterpillars do have large parasitoid assemblages (Sheehan, 1991).

Behavioural defences. Behavioural responses elicited from a prospective host by a parasitoid may be grouped into two broad classes—evasive and aggressive actions (Gross, 1993). The larvae of many holometabolous insects, and to some extent the pupae, attempt to evade parasitism by violently wriggling or twitching when attacked (Beeson & Chatterjee, 1935; Cole, 1959*a*). In the curculionid, *Hypera postica*, the larva of one form rapidly rotates about its longitudinal axis when attacked by a eulophid, whereas the other form does not. In the laboratory only 33 percent of parasitoid encounters with the rotating form resulted in successful oviposition, compared to a 76 percent success rate against the stationary form (Volker & Simpson, 1975).

Thrashing about or rolling may result in the larva falling off the feeding substrate, or out of its leaf roll. In other cases the larva may simply let go and fall off its leaf, but in both instances the larva often remains attached by a long silken thread (e.g. see Yeargan & Braman, 1986, 1989). Such behaviour can substantially increase host-handling time by the parasitoid. Active host-escape behaviours may be especially problematic for idiobionts which have to sting and immobilize a prospective host prior to oviposition. Evidence for this was provided by a study of gelechiid leaf-miners (Gross & Price, 1988). The authors observed that less than two percent of the total parasitism of a vagile species, with an opening to the outside, was due to idiobionts, whereas idiobionts accounted for more than 40 percent of the total parasitism of a sympatric congeneric species that lacked such an escape route.

Many parasitoids that attack active exopterygote nymphs or various adult insects attempt to hold down the host during oviposition (Gross, 1993). Species in a few groups have obvious morphological adaptations to do this—such as the chelate fore tarsus of many dryinids or the flattened hirsute tarsi of some rhopalosomatids.

Although many prospective hosts adopt evasive behavioural tactics, several holometabolous larvae and aphids respond to a parasitoid's attention aggressively. Many caterpillars jerk their heads violently when approached, and some may produce sticky oral secretions (Gross, 1993). Several noctuids attempt, and sometimes succeed in biting prospective parasitoids, and certain gregarious symphytan larvae form defensive circles and attempt to bite parasitoids (Weinstein, 1989). Aphids may kick approaching parasitoids or strike them with their wings or antennae (e.g. Gardner *et al.*, 1984; Hofsvang & Hâgvar, 1986). Loomans *et al.* (1992) observed that *Ceranisus* was more successful at parasitizing smaller thrips larvae than it was at attacking larger ones, apparently because the larger larvae were capable of putting up a more vigorous defence preventing ovipositor insertion by the parasitoid.

A final category of behavioural defence is that of guarding by adults. Among the non-social insects parental care in the form of egg-guarding is practised sporadically in the Hemiptera, Coleoptera, Hymenoptera and very rarely in the Lepidoptera (Eickwort, 1981). Although such behaviour may protect the egg mass from parasitoids (Ralston, 1977) the presence of the adult may serve to attract egg parasitoids, and experimental removal of the guarding adult may not affect, or may even reduce percentage parasitism (Eberhard, 1975; Nafus & Schreiner, 1988).

Chemical defences. Many ALLOMONES (chemical substances that benefit the emitter, cf. kairomone) are protective against a wide array of natural enemies, but their effectiveness has more often been tested with predators than parasitoids (Gross, 1993). Indeed, Gross (*op. cit.*) has suggested that parasitoids may be less susceptible to host chemical defences than invertebrate predators because they are often more specialized, and oviposition may require less contact with a host than is necessary for a small predator to subdue and consume its prey. However, certain secretions do deter or impede parasitoids. Some mole-crickets produce a sticky exudate that entangles an attacking sphecid parasitoid (Castner, 1984), and aphids may smear attacking aphidiine braconids with a waxy secretion which hardens rapidly, causing the wasp to stop and clean itself (Calvert, 1973). Parasitoids often locate their prospective hosts by responding to extremely low concentrations of volatile chemicals, so a very effective form of defence for a prospective host is the ability to emit or eject such a substance at a high concentration to inundate the adult parasitoid's sensory system (Vinson, 1984*a*; Whitman *et al.*, 1990). However, some specialist endoparasitoids may approach and even oviposit in hosts without triggering a defensive reaction. For example, attack by the specialist papilionid parasitoid *Trogus* (Ichneumonidae) fails to cause a *Eurytides* larva to evert its osmeteria (Damman, 1986).

Ant-mediated defences. Honeydew-secreting Homoptera and the larvae of some lycaenids and riodinids are often tended by ants, which may protect them against attacks by parasitoids (Buckley, 1987; Gross, 1993), though the effectiveness of ant attendance as a defence varies depending upon the species involved. For example, parasitism of lycaenid prepupae by a species of chalcidid increases from zero to nearly 90 percent when ants are excluded, but ant attendance fails to provide any protection against attacks on young larvae by a microgastrine braconid (Pierce *et al.*, 1987). Although ants offer protection to aphids against a variety of predators (Way, 1963), there is little evidence that they protect the colonies against attack by parasitoids (see Gross, 1993). However, in some cases parasitized aphids, 'mummies', are removed and destroyed by attendant ants (Vinson & Scarborough, 1991). A different type of apparent ant-mediated defence occurs in at least one species of phasmid in northwestern Costa Rica. The female stick-insect flicks her new-laid eggs to the ground. These eggs have an operculum that is attractive to certain ants, which pick up the eggs and disperse them widely thereby reducing their immediate density on the ground below their parent. This may reduce levels of parasitization by amisegine chrysidids (Windsor, pers. comm.).

Physiological defences

Although the immune system of insects and other arthropods lacks antibodies, certain haemocytes (blood cells) are capable of detecting foreign objects (such as the eggs or young larvae of endoparasitoids) and encapsulating them (Bronskill, 1960; Salt, 1964; Brehélin, 1986; Gupta, 1986). In the process of encapsulation haemocytes surround and encase the parasitoid; the inner flattened layers of haemocytes then often melanize, isolating and killing the parasitoid by asphyxiation (for a more detailed description see Vinson, 1990*a*). A second defence system is called CUTICULAR ENCYSTMENT, which involves the formation of a bubble-like cyst between the cuticle and epidermis. The parasitoid larva moves to this cyst (in some way not yet fully understood) and is sloughed off with the exuviae at the next moult (Arthur & Ewen, 1975; Ewen & Arthur, 1976).

2.6 NESTING BEHAVIOUR AND THE EVOLUTION OF SOCIALITY

Mary Jane West-Eberhard and Paul E. Hanson

The distinguishing attribute of the aculeate Hymenoptera is, as the name implies, the modification of the ovipositor into a sting (Latin: *aculeus*). Most aculeates (with the exception of Dryinidae) do not use the ovipositor as a means of egg placement, but instead use it solely as a weapon, for hunting or for defence. The egg is extruded from the oviduct prior to entering the ovipositor valves and is manipulated and placed by the terminal abdominal segments. This mechanism probably evolved among wasps that entered a substrate in which the host lived, confronted the host directly and oviposited on the host's body surface, rather than using the ovipositor to penetrate the intervening substrate. It facilitated the paralysing of active or aggressive host individuals, either free-living or concealed. The oviposition site, externally on the host's body, and the subsequent mode of development of the larva, remain essentially the same as those of idiobiont parasitoids. The very numerous modifications and specializations of acquiring food for the larvae in aculeate families, the radical change of diet from animal- to plant-based larval food in the bees, and the repeated emergence of truly social (eusocial) colonies in several distinct groups, all occurred in groups that stem from the parasitoid lower aculeate families.

The diversification of aculeate biologies is, to some extent, reflected by two different modifications of the wings. Many groups (such as some bethylids and tiphiids, and all mutillids and worker ants) have apterous females. This condition is associated with host- or prey-finding in confined spaces and similar habitats and, in many insects, diversion of energetic resources to ovarian development (Matsuda, 1987). Other aculeates (Vespidae, Sphecidae and Apidae) have developed their load-carrying capabilities while flying in order to enhance transport of prey to concealed locations or specially constructed nests.

Another feature of many of the more specialized aculeates is the construction of a place of concealment (a nest) for caching larval food provision (see below). This construction may involve extensive habitat modification, and has apparently evolved independently in at least four lineages—Pompilidae, Vespidae, Formicidae and the Sphecidae/Apidae.

The evolution of provisioning behaviour

There is no clear-cut distinction between 'parasitoids' and 'predators' (see Chapter 2.4). The former has been defined as having a larva that develops by feeding on a single host individual (Eggleton & Gaston, 1990), whereas a predatory larva consumes more than one individual as food. In this respect a few species of Hymenoptera can develop either in the parasitoid or the predator mode. For example, the ichneumonid *Aritranis* develops on a single bee larva in one cell, or it consumes several larvae in adjacent cells (Daly, 1983). Most sphecids are predators, but a few (e.g. *Larra*) are parasitoids, completing development on just one host/prey individual. Thus this distinction between parasitoid and predator is not particularly useful in considering the evolution of the aculeate

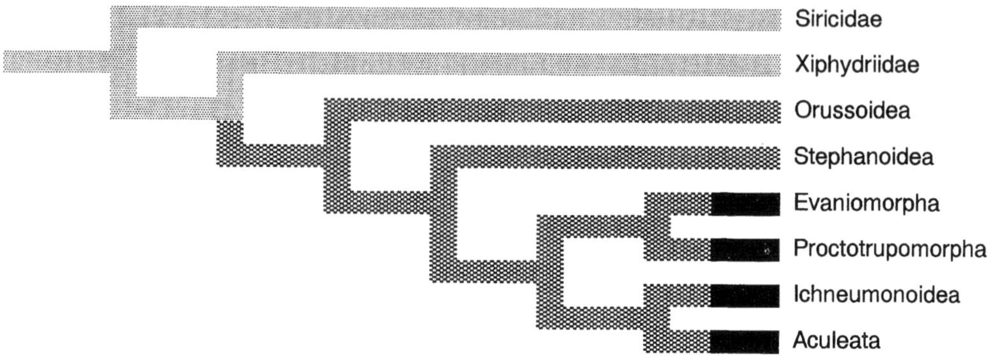

Fig. 2.09. Cladogram showing postulated phylogenetic inter-relationships of the higher Hymenoptera (after Rasnitsyn, 1988). Species belonging to the more primitive lineages (fine stippling) develop as fungivorous/xylophagous larvae in wood, from eggs inserted into wood by the adult female. In at least some cases the wood is 'envenomated' by the ovipositing female. The larvae of the more derived taxa (coarse stippling) develop from eggs placed on an insect wood-borer, which has been immobilized by an injection of venom from the adult female wasp. Primitive members of the most derived groups (solid shading) develop similarly, but members of the more derived groups have radiated to exploit a range of hosts in other situations. More details of phylogeny are given in Chapter 5.

Hymenoptera. On the other hand, modifications in provisioning behaviour can be grouped into an ascending series of informal grades (Malyshev, 1968; Evans & West-Eberhard, 1970), each successive grade being behaviourally more complex, though not necessarily phylogenetically more derived.

The simplest grade includes species that construct no nest and opportunistically occupy the victim's lair. Female Tiphiidae and Scoliidae locate and attack large subterranean beetle larvae in their burrows or earthen cells. The victim is stung to immobility and an egg laid on it. The female may arrange the substrate about the host, or close the victim's burrow, but otherwise departs to leave her offspring to develop without further parental attention. Some pompilids attack and paralyse spiders in their own retreats, oviposit, and leave the offspring to benefit from the host spider's concealment. Some of these pompilid biologies may be recent simplifications of more complex ancestral behaviour, but the tiphiid and scoliid biologies represent the retention of primitive (plesiomorphic) patterns (see Chapter 14).

Several aculeate groups have specialized to attack hosts in the friable substrate that are better protected than beetle larvae by, for example, having a thick covering. Since they are unable to use their ovipositor to pass the egg through the covering, they bite a hole through which the tip of the abdomen is inserted. The sting may or may not be used to immobilize the host. Mutillids, for example, bite through the host cell and lay an egg on the mature larvae or pupae of aculeate hosts. The terminal portion of the abdomen of chrysidids is a series of telescoping segments that are employed as an analogue of the ovipositor, facilitating placement of the egg (Kimsey, 1992*b*), but the female remains dependant upon use of her mandibles to penetrate the covering to gain access to the host.

Somewhat more complex behaviour is exhibited by some Bethylidae and many Pompilidae and Sphecidae, in which the female captures and paralyses her prey, and then drags or flies with it, searching for some cavity (beetle boring, hollow stem, etc.) in which to hide it. Once a concealed niche has been located, the female enters with her prey, lays an egg, and often seals the cavity before leaving. The cavity constitutes a primitive developmental chamber, but there is no chamber construction by such females.

The above examples all concern species that utilize a single prey individual for the development of each larva. However, some Bethylidae attack relatively enormous larval Coleoptera or Lepidoptera, which are dragged into concealment or sought out and paralysed in a cryptic location. The adult female bethylid may feed on the haemolymph of the victim, and lay one or several eggs onto its body. The larvae that emerge develop through to adulthood in the presence of their mother, and subsequent generations may co-exist on the corpse of a large prey. Although no nest is constructed, the continued presence of the protective mother represents an overlap of generations, which is sometimes considered important in the evolution of eusociality (see below).

The next grade is the excavation or construction of a nest by the female wasp. The primitive condition is probably the building of a nest after the prey is captured. For example, in several pompilids and a few sphecids (e.g. *Podalonia*) the female captures and paralyses the prey which she then drags to an appropriate location. The prey is temporarily concealed while she excavates a simple burrow with a single apical cell, after which the prey is retrieved, inserted into the nest, and an egg laid; the nest is then sealed and the wasp departs. Similar behaviour occurs in several Pompilidae, and in some of these the female, immediately before digging a burrow, caches the spider prey high on a plant stem to keep it out of reach of scavengers.

In other aculeates the female wasp prepares the nest before hunting for prey. Associated with this behaviour is orientation capability by the female, allowing her to find the nest again when she returns with prey. The ability to relocate the nest also permits the female to provision the cell with multiple small prey individuals, rather than one large prey. Each brood cell is provisioned with one or more prey items and an egg is laid in it, after which the female seals the cell. Some Pompilidae and most Sphecidae illustrate this general behavioural sequence, as do bees, which provision the nest with pollen and nectar instead of arthropod prey. Between hunting forays the female may temporarily close the nest entrance which presumably functions to hinder intruders, and thus she must reopen the nest when she returns with prey. Primitively the female carries the prey in her mandibles, which necessitates a temporary release of the prey while she uses her mandibles to open the nest. In the more derived condition the prey is carried with the middle or hind pair of legs, or the sting, thus allowing the female to open the nest without releasing the prey and therefore possibly reducing exposure to cleptoparasites (Evans, 1966*b*).

When multiple prey are placed in each cell the egg may be laid on the last prey item, or on the first. In the case of most eumenine vespids, the egg is laid in the empty cell before prey are introduced. Among those hymenopterans that provision each cell with several prey, the primitive condition is for the female to fully provision the cell and seal it before the larva begins to develop. This is called MASS PROVISIONING because the initial food that is amassed must be sufficient to feed the larva during its entire development. In grades beyond this level the practice of PROGRESSIVE PROVISIONING is initiated and becomes gradually more refined. Under this regime the larva is given prey progressively during its development; only after larval development is advanced or complete is the cell sealed. Some Sphecidae (e.g. *Trigonopsis*) and eumenine Vespidae (e.g. *Zethus miniatus*) show a stage that can be considered intermediate between mass and progressive provisioning. Here an egg is laid after the first prey item has been added to the cell, then after the egg hatches the larva is fed progressively for several days. More prey are rapidly stored and then the nest is sealed.

The evolution of nest architecture

More species of Hymenoptera build nests than any other order of insects; outside the Hymenoptera elaborate nest-building occurs principally in the termites, all of which are social. In Hymenoptera nests are constructed by pompilids, ants, vespids, sphecids and bees; most species in these groups are solitary, the nest being built by an individual female, but in parasocial and eusocial species (see below) the nest is built by several (up to a maximum of probably 100,000) females. Nests of solitary Hymenoptera are generally small and inconspicuous whereas those of eusocial species range from the centimetre long, delicate sac of the sphecid *Microstigmus* to the gigantic, resinous nest of the South American stingless bee, *Trigona amazonensis*, which is cemented to a tree trunk, may be six metres long by one metre across and can weigh several hundred kilograms. Other large nests include those of leaf-cutting ants (e.g. *Atta cephalotes*, see Chapter 16) and those of the Brazilian vespid, *Agelaia vicina*, which contains an estimated 33 square metres of comb and 1.3 million adults (Simoes unpublished, cited by Jeanne, 1991).

The nests of most ants are, in comparison with the other Hymenoptera, relatively unstructured, lacking permanent cells and consisting merely of interconnecting cavities. This lack of fixed cells may actually facilitate the ability of ants to move the offspring as a means of avoiding disturbance, pathogens, predators, or diurnal fluctuations in temperature. However, except for ants, hymenopteran nests are generally made of cells, each brood cell containing a cavity in which one (rarely more than one) larva is reared. In the nests of pompilids, eumenine vespids and sphecids the cell is usually a simple unlined cavity, whereas bees generally line the cavity with a secretion (Michener, 1964).

Burrows excavated in the ground are thought to be the ancestral type of nest in most nest-building Hymenoptera. Species in these groups that excavate nests generally show certain morphological adaptations for digging, removing soil, or compacting the soil in the tunnel. In the Sphecidae four types of digging behaviour can be recognized: 'rakers', 'pullers', 'carriers' and 'pushers' (see Chapter 17). Raking appears to be a primitive mode of digging in the Sphecidae, and also in the Pompilidae, and species showing this type of behaviour generally burrow in friable soil (e.g. sand). Many masarine and eumenine vespids, on the other hand, burrow in hard ground and carry water to the nest site in order to soften the soil. Most ants are 'carriers' whereas many bees that burrow in the ground are 'pushers', digging a vertical burrow by pushing soil with the pygidial plate as they back out of the burrow.

The most primitive type of nest consists simply of a cell excavated from the end of an existing burrow in the ground or in rotten wood (e.g. some pompilids and sphecids). A slight development from this pattern is shown by those hymenopterans that excavate both the burrow and the terminal cell (e.g. many pompilids, some sphecids and bees). Although unicellular nests are generally considered primitive, in some sphecids a unicellular nest may represent a derived condition, perhaps resulting from selection to avoid natural enemies (e.g. Vinson *et al.*, 1987). Multicellular nests comprise two or more cells, the simplest form being a branching tunnel with each branch terminating in a cell. A more advanced condition is the grouping of cells, either in a cluster, or in a linear sequence with the cells arranged end to end. The latter arrangement is also used by eumenines, sphecids, and bees that nest in hollow or pithy plant stems. Whether in the ground or in a stem, these tube-nesters show similar adaptations. Nesting in the confines of a tubular substrate selects for a larva that orients with its head toward the exit before spinning a cocoon, since there is often insufficient space available for the adult to turn around. It has been shown that stem-nesting Hymenoptera generally make cell partitions with the concave surface facing the exit and the convex surface facing inward, thus allowing the larva to orient with respect to the exit end of the stem (Cooper, 1957). Another problem confronting tube-nesting hymenopterans is that the innermost cell is the oldest and therefore if developmental time is the same in all cells the innermost individual may eclose first and destroy the outer cells in the process of emerging from the nest. The evolved solution to this problem is for individuals in the inner cells to wait and all emerge simultaneously (Krombein, 1967). In addition, many tube-nesters place female eggs in the inner cells and male eggs in the outer cells, since males generally require less time to develop. The result is more highly synchronized emergence.

Several hymenopterans do not utilize or excavate cavities but require instead a surface on which to build an exposed nest entirely from material gathered by the female. One of the most common construction materials is mud. This is employed by some lineages of pompilids (e.g. *Auplopus*), vespids (e.g. *Eumenes*), sphecids (e.g. *Sceliphron*, some Trypoxylonini), and bees (e.g. *Eulaema*). Another common construction material is masticated wood chips that are formed into a 'paper' or 'carton' nest (many *Azteca* ants and polistine Vespidae). Some orchid bees (Euglossinae) and stingless bees (Meliponinae) build exposed nests, utilizing mostly resin as a construction material.

The evolution of social behaviour

Many wasps and bees nest in aggregations. Factors other than shortage of nesting space (such as defence against parasites) may have led to aggregated nesting (Lin & Michener, 1972; Rosenheim, 1990). Aggregated nesting may lead to communal nesting, whereby several females use the same nest entrance but continue to maintain their own individual cells. The individual females in a communal nest may utilize either mass provisioning or progressive provisioning of their cells.

The apparently large gulf between the solitary and highly social Hymenoptera is spanned in the aculeate families by a finely stepped sequence of behavioural modifications that link the extremes. These modifications can be grouped into a number of categories, from the behaviourally least complex to the most complex. The following terms are commonly used for the different types of intermediate groups existing between solitary and eusocial (Wilson, 1971; Michener, 1974*a*; Eickwort, 1981; Cowan, 1991).

Solitary. Females nest alone, mass provision their nests, and do not interact with their developing offspring.

Primitively social (= presocial). Females exhibit social behaviour beyond sexual interactions, yet show no reproductive division of labour.

Subsocial: females nest alone but interact with their offspring, usually by progressive provisioning.

Parasocial: females of the same generation interact on the same nest. [It should be noted that different types of parasocial behaviour, which are listed below, may be present in the same species at different times.]

Communal: each female builds, oviposits in, and provisions her own cells.

Quasisocial: all females co-operate in building and provisioning brood cells, all females oviposit.

Semisocial. Some females lay most or all of the eggs and some of the same generation lay none.

Eusocial. Females co-operate in nesting, exhibit reproductive division of labour (as in semisocial), and show an overlap of generations.

Primitively eusocial: morphological differences between reproductive and worker females do not occur.

Highly eusocial: reproductive and worker castes are morphologically distinct.

Two lines of evolution towards eusociality are sometimes distinguished. Each route involves the gradual acquisition of the three characteristics of eusocial organization, but the characteristics are developed in a different order in the two routes. The PARASOCIAL ROUTE involves the transition from communal nesting to the quasisocial condition, whereby several reproductive females of the same generation co-operate in caring for each others' brood, although all the females are reproductives and all partake in construction and maintenance of the communal nest. In the next grade, the semisocial condition, not only do the females co-operate in brood care but there is also a reproductive division of labour. The females differentiate into functional reproductives (queens) and assistants which do not lay or which cease to lay eggs, these latter females being the first sign of the emergence of a sterile worker caste. In some cases the dominant female will destroy any eggs laid by subordinate females. If this semisocial condition is extended over two or more generations by extension of the reproductive's life, so that the generations overlap and offspring assist parents, then the group is termed eusocial. Specialization within the eusocial grade then may widen the gap between reproductive and non-reproductive females until they are strongly differentiated into morphologically distinct queen and worker castes.

Examples of stages in the parasocial route among neotropical species include the orchid bee *Euglossa imperialis*, which has reached the communal and quasisocial stages (Roberts & Dodson, 1967), and at least three halictine bees that have reached the semisocial stage—*Caenaugochlora costaricensis*, *Pseudaugochloropsis graminea*, and *P. sordicutis* (Michener & Kerfoot, 1967; Michener, 1974*a*).

Unlike the parasocial route the first increase in complexity shown in the SUBSOCIAL ROUTE is the development of an overlap of generations in forms that otherwise do not show cooperative brood care, nor reproductive division of labour. This is achieved by progenitor females living long enough to see the development of their brood to adulthood and to undertake more reproductive activity in company with their offspring. The next stage shows, in addition to overlapping generations, the development of cooperative brood care. The differentiation of a reproductive female caste and non-reproductive workers will again satisfy the criteria of eusociality. A possible example of the subsocial route is the halictine bee, *Augochlora nominata*, in which the first brood remains with the nest and helps with the second brood (Eickwort & Eickwort, 1972*a*).

The dichotomization of these two 'routes' to sociality in fact probably exaggerates the clarity of the distinction between 'parasocial' and 'subsocial' groups, and the importance of overlap of generations in the evolution of eusociality, where a reproductive division of labour is the more important biological innovation. Some primitively social (nest-sharing, workerless) wasp species, for example, use the same nest for several generations (up to four years of constant occupation in the tropical eumenine *Zethus miniatus*) and although all group members are descendants of a single foundress female, generational overlap sometimes occurs and sometimes does not (female longevity varies). Such a colony is simultaneously 'subsocial' and 'parasocial'. For this and other reasons West-Eberhard (1978*a*) considered the POLYGYNOUS FAMILY route a more accurate and general description of a hypothetical condition for eusociality in wasps and bees.

Amongst the Hymenoptera eusociality occurs in all ants, some vespids, one genus of Sphecidae, and some bees. In some eusocial Hymenoptera the colony contains just one queen (MONOGYNOUS colonies) while in others the colony contains more than one queen (POLYGYNOUS colonies). The evolution of queen number is a highly labile character in ants but is more conservative in bees (Ross & Carpenter, 1991). Among the highly eusocial bees, those having morphological worker-queen differences (Meliponinae and Apinae), monogyny appears to be ancestral and there has been little subsequent evolution in queen number. In the Polistinae (Vespidae) monogyny is ancestral, and is retained in the tribes Polistini and Mischocyttarini, but polygyny has arisen in the large tribe Epiponini (i.e. those that initiate new colonies in swarms; see Chapter 15). In this last group alternating polygyny and monogyny characterizes the relatively small colonies that have been studied; permanent polygyny probably occurs in large colonies, though it has never actually been demonstrated.

In some highly eusocial Hymenoptera there exists not only a reproductive division of labour (i.e. queen vs worker) but also a further division of labour among workers, based on differences in the probabilities of performing different tasks within the colony (Wilson, 1971; Oster & Wilson, 1978). This division of labour may be based on worker age polyethism, worker polymorphism (most prominent in some ants), and individual differences in task specialization (Jeanne, 1988). Despite the existence of increased worker specialization in highly eusocial Hymenoptera, the workers retain some of the behavioural flexibility associated with more primitive species, thus allowing the individuals and colonies to respond to changing conditions. Since there are no colony 'leaders' directing the activities, a major thrust in research on insect sociobiology is directed toward understanding how division of labour is regulated, i.e. how individual workers respond to fragmentary information with actions appropriate to the state of the whole colony (Robinson, 1992).

The origin of eusociality in ants remains problematical for comparative study since all extant species are fully social. Sociality in ants may have evolved from something resembling the communal parasitoid activities exhibited by some Bethylidae (Malyshev, 1968), or through a wasp-like progressive provisioning regime (regarded as more likely on the basis of current evidence; Hölldobler & Wilson, 1990*a*). In communally parasitoid bethylids the first brood may develop to adulthood without the original female leaving, and several generations may exist together around the corpse of a large prey. Such agglomerations of numerous individuals in a single spot, with the members of more than one generation interacting, may form a behavioural base from which social organization could develop. For instance, if the female parent were to develop dominance over the others, such that their fecundity was suppressed, or if they acted in concert to raise the offspring, say by transporting larvae to fresh prey (or vice versa), then a primitive form of sociality would have been achieved (Gauld & Bolton, 1988).

The larvae of the most primitive groups of ants are mobile and highly predaceous, and are fed whole or disarticulated prey by their worker nest-mates. For her first brood, the queen of these primitive forms goes out actively foraging for prey. This type of colony foundation, where the foundress is an active predator and not permanently enclosed in a cell, is exhibited by several primitive taxa of ants. In one such group, the amblyoponines (Ponerinae), there is a marked tendency for the larval brood to be transported to very large immobilized prey, rather than the prey being brought to the brood. In other groups the prey is captured by the foraging female and brought back to the incipient nest. This latter form of behaviour may represent a remnant of the progressive provisioning seen in various families of wasps, but it may also be secondary to the amblyoponine type of organization in that, rather than expose the larvae to danger by moving them to their food, the prey is brought to them while they remain in a relatively secure place. Conversely it may imply that sociality has arisen twice in the Formicidae, once by each of these techniques (see Hölldobler & Wilson, 1990*a*, p. 27 for a phylogenetic discussion leaving both possibilities open).

Among non-hymenopterous insects eusociality has arisen only in the termites. Thus considerable attention has been directed to the question of why eusociality has arisen so often in the order Hymenoptera and so infrequently in other insects. W.D. Hamilton (1964) postulated, and later elaborated (W.D. Hamilton, 1972, 1974) the novel hypothesis that the haplo-diploid genetic constitution of Hymenoptera may be the mechanism facilitating the evolution of eusociality. The reasoning advanced for this HAPLODIPLOIDY THEORY was that, as males are haploid and females diploid throughout the order Hymenoptera, all males (brothers) derived from a singly-mated female are genetically identical, whereas all females (sisters) thus derived will share an average of three-quarters of their genetic complement. Set against this three-quarters genetic similarity of sisters is the fact that the quantity of shared genetic material between mother and daughter is only one-half, and between sister and brother only one-quarter.

Assuming that the interest of the female should be most strongly directed towards those offspring that are genetically closest to her (i.e. those that share the highest proportion of common genetic material), then in cases where the mother lives long enough to see the development of her female offspring to maturity, those female offspring would maximize their inclusive fitness by forsaking their own reproductive potential for the sake of rearing their mother's later female offspring, provided that the increased reproductive rate of the mother more than offsets the lost potential brought about by her daughters' lack of reproduction. This 'altruistic' suppression of reproduction in the daughters would thus lead them to function as a worker caste.

The three-quarters relatedness thesis is very attractive, reducing as it does the concept of the evolution of hymenopterous eusociality to the simpler concept of genetic relatedness. Reviews of the theory and its ramifications are given by Starr (1979) and Andersson (1984). Besides these reviews both W.D. Hamilton (1964) and Wilson (1971) have discussed the pros and cons of the theory and mentioned factors that act against it, such as multiple mating by outbreeding females (Page, 1986; Sherman *et al.*, 1988) and the occurrence of functionally polygynous colonies (West-Eberhard, 1978*a*). However, genetic relatedness alone does not answer all questions regarding the taxonomic distribution of eusociality; factors other than relatedness must also be involved and in fact are contemplated by the cost-benefit side of Hamilton's theory (West-Eberhard, 1975). In the Hymenoptera eusociality is a trait peculiar to a very few families of Aculeata, and even among these few families there is no uniformity of eusocial development. On the basis of relatedness alone it is not easy to explain why eusociality has never developed in the Bethylidae, why there are relatively so few occurrences of eusociality in the Aculeata, and why a similar eusociality has never arisen in other haplodiploid insect groups such as the thrips, micromalthine beetles, or iceryine scale-insects, all of which form close aggregations of adults and offspring.

Alexander (1974), Evans (1977) and West-Eberhard (1987) discuss factors other than relatedness that may have contributed to the evolution of sterile workers, including extrinsic factors such as predation and parasitism, maternal manipulation of offspring roles, and the consequences of obligatory group life for intragroup dominance relations and reproductive suppression of some individuals, given characteristics widespread in hymenopteran females.

It is important in considering relatedness to remember that while doubts remain concerning the importance of haplodiploidy for social evolution, there is no doubt about the importance of KIN SELECTION. All worker-containing groups are composed of kin, if not matrifilial associations, at least extended families. Recent studies using electrophoresis have confirmed what was earlier evident from a few long-term studies of female genealogies (e.g. West-Eberhard, 1969, 1978*a*; Noonan, 1981): kinship among colony members is well above values for the population at large, and especially if the cost–benefit term of Hamilton's formulation is taken into account, the levels of relatedness observed in all studied social insects indicate a ubiquitous role of kin selection (the general, rather than special-case haplodiploidy, version of the theory). This even applies in one of its severest tests—the polygynous tropical social wasps, where if a multiple queen condition persists with outbreeding, relatedness is expected to decline to relatively low levels, approaching those of the population as a whole (West-Eberhard, 1973). So far, all such species studied using protein-electrophoresis show worker relatedness to be above relatedness of the population as a whole, ranging from 0.11 in one population of *Parachartergus colobopterus* (Queller *et al.*, 1988) to 0.39 in *Protopolybia exigua* (Gastreich *et al.*, in press) [in a later study by Strassmann *et al.*, 1991, *P. colobopterus* showed intracolony relatedness of 0.31, due to high kinship of queens ($r = 0.67$)]. In the Costa Rican species *Agelaia* *multipicta* average worker relatedness was 0.184, with that of queens 0.69 (West-Eberhard, Lobo & Azofeifa, unpublished; see West-Eberhard, 1990). For a review of relatedness values in polygynous social wasps see Gastreich *et al.* (1993). Whenever relatedness is above the population average, kin selection can be presumed to play a role in social evolution (Maynard Smith, 1982). In honey-bees a fine degree of kin discrimination has been demonstrated, with workers attacking half sisters more than full sisters in the same colony (Evers & Seeley, 1986).

Another appealing hypothesis that presents a selective context to account for the evolution of eusociality in the Hymenoptera is that postulated by Alexander (1974) and Michener and Brothers (1974). This is the PARENTAL MANIPULATION THEORY (reviewed in Brian, 1980; and Brockmann, 1984) and, like the kin selection theory outlined above, it becomes feasible only in taxa having a protracted mother–daughter relationship and an overlapping of generations. Unlike the kin selection theory, this system envisages the mother manipulating her offspring (either genetically, behaviourally, or physiologically; Fletcher & Ross, 1985) so that she dominates and controls her daughters in such a way that her own fitness is enhanced. Her reproductive potential is increased at the expense of her offspring, which therefore function as a worker caste, not by altruistic relinquishment of their reproductive potential, but through parental dominance and control. However, kin selection on daughters would lower the amount of mother–daughter conflict over social roles (Trivers, 1974; West-Eberhard, 1975).

Both the haplodiploidy theory and the parental manipulation theory are contradicted by observations of primitively social species (e.g. see West-Eberhard, 1978*a*) and supported by observations of eusocial ones (Alexander *et al.*, 1991). Thus haplodiploidy may have been more important in the maintenance and elaboration of eusociality than in the transition to worker production. It may eventually prove the case that both theories are partially correct, each accounting for some aspect of eusocial development, or acting synergistically to produce co-operation and maintain it.

Communication among members of a colony

The spectacular achievements of social insects are mass phenomena that emerge from the integration of simpler individual behaviour by means of communication (Wilson, 1971). Intraspecific communication in social Hymenoptera involves primarily, although not exclusively, the use of chemical signals, i.e. the use of pheromones. In ants more than ten organs have been implicated in the production of communicative chemicals. The worker ant is, as Hölldobler & Wilson (1990*a*) put it, 'a walking battery of exocrine glands, developed to a degree well beyond that typifying nonsocial hymenopterans'. Reviews of the exocrine glands are available for ants (Hölldobler & Wilson, 1990*a*), social wasps (Downing, 1991), and social bees (Free, 1987). Besides pheromones these exocrine glands produce venoms, substances used in nest building, etc. Communication among members of a colony of social Hymenoptera involves about a dozen categories of responses, of which the following have received the most attention.

Recognition

Members of a social colony recognize other members of the same colony and distinguish castes (e.g. queen recognition). 'Kin recognition' refers to the process by which individuals assess genetic relatedness of conspecifics to themselves. The ability to distinguish between nestmates and non-nestmates appears to be learned shortly after eclosing to the adult stage, and usually involves odours. Polistine vespids appear to learn recognition odours from the nest (Gamboa *et al.*, 1986), and in honey-bees, comb wax serves as a source and medium for transmission of recognition cues (Breed & Stiller, 1992). Callow ants, whose odours differ from the ambient odours of the colony, may be protected by a brood-masking substance while they acquire labels and learn 'templates' (Hölldobler & Wilson, 1990*a*).

In all social Hymenoptera queen recognition and queen control over workers are probably inseparable (Fletcher & Ross, 1985). In the honey-bee there are pheromonal stimuli that inform workers of the presence of a queen in the colony and at close quarters workers are attracted to the queen. A weak attractant is produced by Kischevnikov's gland (located in the sting chamber) which produces the 'queen substance' (trans-9-keto-2-decenoic acid), and a highly active attractant is produced by the mandibular glands. Stingless bees also show queen recognition as evidenced by the fact that they tend to avoid the queen. One of the most dramatic examples of attraction is that elicited by army ant queens, which are usually so covered with workers that they are hidden from view.

Alarm

A sting from a honey-bee or social wasp often attracts further attack by other members of the colony. In bees the signal provoking the mass assault includes a mixture of compounds, including isoamyl acetate, which has a banana-like odour and is produced by the lining of the sting pouch (Free, 1987). Honey-bees also secrete an alarm pheromone (2-heptanone) from their mandibular glands, although its function may be primarily to repel potential robber bees coming from other colonies. Stingless bees emit various compounds from their mandibular glands upon biting (Smith & Roubik, 1983; Johnson *et al.*, 1985), which may serve as alarm pheromones and possibly as allomones allowing them to construct nests in active ant or termite nests (Roubik, 1989, p. 218). In ants alarm pheromones are produced by the mandibular glands (many Myrmicinae, some Formicinae), Dufour's gland (Formicinae), and the pygidial gland (Dolichoderinae). In social wasps alarm pheromones are produced by the poison gland in Vespinae but in the Polistinae alarm pheromones *per se* appear to be absent, although the venom itself may elicit alarm in nestmates (Jeanne, 1981, 1982). An extreme alarm response in the tropical wasp *Synoeca surinama* elicited an allergic response in a nearby, hypersensitive human observer, suggesting release of an airborne substance (West-Eberhard, 1981).

The alarm pheromones vary between taxa but tend to be small, simple molecules that spread and fade quickly (Hölldobler & Wilson, 1990*a*). When the alarm pheromone is detected by another member of the colony the type of response elicited depends upon the species—in aggressive species alarm pheromones may lower the threshold for attack and help orient individuals toward the source of danger (in conjunction with visual cues), whereas in more docile species alarm pheromone may elicit escape behaviour. Moreover, closely related species may recognize each others' alarm pheromones; thus *Trigona fulviventris* responds (flight and defensive posture) to mandibular gland components of other stingless bees (Johnson, 1980).

Although *Polistes* (Vespidae) apparently lack alarm pheromones, they use tactile alarm signals transmitted via the paper nest when aroused by the presence of an ichneumonid parasitoid (West-Eberhard, 1969). Alarmed *Synoeca* wasps show a synchronized pecking of the nest envelope with the mandibles (West-Eberhard, pers. obs.), which may serve as both an alarm signal to nestmates, and a warning to predators. Ants belonging to the subfamilies Ponerinae, Pseudomyrmecinae and Myrmicinae stridulate, which may serve as an acoustical underground alarm signal, at least in some cases. In leaf-cutter ants (*Atta*) stridulation commonly occurs when part of the colony is buried by a cave-in of the nest and elicits a rescue response in nestmates (Markl, 1985). Some tropical *Camponotus* ants (Formicinae) make a tapping sound when disturbed (West-Eberhard, pers. obs.).

Recruitment

Recruitment is defined as communication that brings nestmates to some point in space where nestmates are active, whether it be in food retrieval, nest construction, nest defence, or migration (Wilson, 1971). The simplest types of recruitment result as a by-product of alarm and attraction pheromones. Stingless bees often mark a food source with odours that are recognized by other members of the species, including members of other colonies (Villa & Weiss, 1990). Another primitive form of recruitment utilizes the odour of food brought to the nest accompanied by erratic, 'excited' running movements. In honey-bees a successful forager can communicate the floral source simply by the odours adhering to her body and the scent of regurgitated nectar.

Some ants utilize a simple form of chemical recruitment known as 'tandem running'. If a worker ant encounters a food particle too large to carry, it returns to the nest, contacts another worker, and leads it from the nest. The leader runs a short distance and stops; the follower, who is now excited by secretions released by the leader, runs swiftly behind or circles until it contacts the leader. This process continues until they arrive at the food source.

The most elaborate form of chemical communication in ants is the odour trail system, which in some ants has probably evolved from tandem running. After discovering a food source a worker ant returns to the nest, stopping at intervals to mark the route with pheromones. Other workers which detect the airborne molecules emanating from the substrate, follow the freshly laid trail out from the nest and are capable of discriminating an increasing gradient of molecular concentration. The odour trail eventually fades away unless subsequent workers find the food source sufficiently rewarding to add more trail pheromones.

These pheromones are used not only in recruitment to a food source but also in moving to another nest site. Odour trail communication has evidently originated several times in ants since the glandular source of the trail pheromone varies between ant taxa: poison gland in Myrmicinae, sternal gland in Dolichoderinae, and hindgut in Formicinae. 'Graded recruitment', with more ants responding to rich food sources, has been demonstrated in the Costa Rican ponerine ant, *Paraponera clavata* (Breed *et al*., 1987), one of the few species in which this has been demonstrated. Both tandem running and odour trails may be used in exploration. In army ants for example trail pheromones from the pygidial gland are deposited almost continuously by exploring workers.

In stingless bees that lay odour trails (not all species do so), the source of the trail pheromone is the mandibular gland and larger quantities are secreted than in trail-laying ants; indeed they are detectable by humans. After returning to the nest a successful forager arouses her nestmates by producing a buzzing sound. They then leave to follow a trail of odour droplets deposited on the vegetation (e.g. Hubbell & Johnson, 1978). In social wasps recruitment to a food source is rare, but the odour trails laid down by glands on the fifth and/or sixth sternite are used in swarming to a new nest location (Downing, 1991).

The most famous insect recruitment system is the 'waggle dance' of the honey-bee (Frisch, 1967; Lindauer, 1967; Kirchner & Sommer, 1992; Michelsen *et al*., 1992; Seeley & Towne, 1992). The waggle dance is unique in that it represents an unusually complete message that guides a complex behavioural response. The message is constructed from a ritualized and miniaturized imitation of the journey that the signalling bee has taken in the past and upon which some of its sister bees are about to embark (Wilson, 1971). The signalling bee runs straight up the side of a vertical comb, turns to the left, circles back, repeats the straight run up the comb, turns to the right, circles back, etc.—alternately turning left and right after each straight run. During each straight run she buzzes her wings and wags her abdomen. The direction and duration of the straight run up the comb are closely correlated with the direction and distance, respectively, of the food source. For example, if the food source is located 20 degrees to the right of the sun, the straight run of the dance is directed at an angle of 20 degrees to the right of vertical. The duration of the straight run increases with length of the journey.

Trophallaxis

The exchange of liquid food among members of the same colony, known as TROPHALLAXIS, plays a prominent role in the social interactions of most social insects (Wilson, 1971). The transfer of regurgitated liquid may occur between adults, or between adults and larvae. In ants both types of trophallaxis occur and tend to follow a phylogenetic pattern, occurring sporadically in primitive groups (e.g. absent in army ants) but occurring commonly in more advanced groups. In general liquid food exchange is most common in ants which naturally include a large amount of liquid in the diet and in those which do not regularly feed the larvae and queen with trophic eggs. In some social wasps trophallaxis occurs between adults, and between adults and larvae, the latter representing a biochemical division of labour, i.e. only larvae possess certain enzymes necessary for converting proteins to carbohydrates (Ikan *et al*., 1968). Among bees trophallaxis appears to be restricted to the Old World allodapine bees (adult–larva trophallaxis only), and stingless bees and honey bees (adult–adult trophallaxis only). Unlike honey bees, stingless bee workers do not lick the queen and seldom feed her.

In some species each worker extensively exchanges liquid food with numerous nestmates of various castes. In those species where trophallaxis is common the combined crops of the adult population represent a communal stomach, which could function to keep workers informed on the nutritional status of the colony as a whole (Wilson, 1971). Hunt (1991) discusses the role of trophallaxis in the social organization and evolution of wasp colonies.

2.7 INTERSPECIFIC INTERACTIONS OF NESTING HYMENOPTERA

Paul E. Hanson, Mary Jane West-Eberhard and Ian D. Gauld

Hymenopteran nests, especially those containing large colonies, offer a concentrated food resource for a wide diversity of organisms ranging from large vertebrates (such as primates, bears and anteaters) to micro-organisms (e.g. fungi). Outside the Hymenoptera only the termites offer such a tempting target and it is perhaps not surprising that many groups of organisms have become facultative or obligatory symbionts in or near hymenopteran nests. Much has been written elsewhere about the associations between Hymenoptera and other organisms and no attempt has been made here to summarize most of this work. However, in this section we present a summary of the interspecific interactions, or symbioses (in the original, broad sense of the term), of nest-building Hymenoptera with other arthropods. For convenience, bearing in mind this is a book about the Hymenoptera, we have divided these associations into two classes: symbioses with other Hymenoptera; and symbioses with arthropods other than Hymenoptera.

The types of symbioses in both classes range from mutualism to commensalism to parasitism. Parasites, in the broad sense, include cleptoparasites, parasites *sensu strictu*, parasitoids, and predators. In general, eumenine vespids, sphecids, and solitary bees have a similar set of symbionts (Krombein, 1967) whereas the symbionts associated with ants are taxonomically quite distinct and, at least among the non-Hymenoptera, generally more diverse (Hölldobler & Wilson, 1990*a*).

Symbioses with other Hymenoptera

The hymenopterous symbionts of solitary and primitively social nest-builders are usually parasites in the broadest sense, that is they are either parasitoids which feed on the brood, cleptoparasites which feed on the provisions, or nest usurpers which take over the nesting site. The nests of eusocial Hymenoptera provide not only food and a nesting site, but also a labour force that may be exploited by other social hymenopterans. Eusocial species enter into an impressively diverse array of interspecific symbioses, encompassing almost every conceivable category of commensalism and parasitism. Thus, in addition to hosting parasites, eusocial hymenopterans may interact in compound nests and mixed nests (social parasitism), as explained below. Interspecific mutualism between Hymenoptera is very rare, perhaps limited to cases where two aggressive species nest together, thereby enhancing their mutual defence. For example, the 'ant trees' of the South American tropics, where a single large tree contains a set of carton nests built by *Azteca* ants, may be inhabited by numerous social wasps and stingless bees (see Richards, 1978, p. 28).

Parasitoids

Like the immature stages of most holometabolous insects, the larvae and pupae of nest-building Hymenoptera are subject to attack by a variety of hymenopteran parasitoids (Table 2.5). Very occasionally adults too are parasitized, and in the case of some parasitoids of ants it is not clear whether immature or adult ants are attacked.

There are at least three different evolutionary pathways whereby parasitoids have adapted to exploiting various nest-building Hymenoptera. First, the association of a parasitoid with solitary nest-building hymenopteran hosts has often arisen among groups of idiobiont pupal/prepupal parasitoids that primitively attack hosts tunnelling in stems or wood. In these cases parasitoids initially probably have exploited aculeates nesting in these tunnels, and a few have radiated to exploit aculeates nesting in other situations (Gauld, 1983, 1988*c*). Examples of this pathway include all labenine and most cryptine and pimpline Ichneumonidae, Megalyridae, Torymidae and perhaps also doryctine Braconidae, Chalcididae, Eurytomidae and Leucospidae.

A second evolutionary route has primarily developed with nest-building species that practice progressive provisioning. This involves parasitoids, such as Trigonalyidae, Eucharitidae and Perilampidae, that produce large numbers of minute eggs typically deposited on vegetation. These eggs eclose to produce larvae which are, by some means, transported back to the host's nest where they attack the aculeate larvae.

A third possible route has involved eusocial species being attacked by parasitoids that were perhaps already adapted to entering nests to attack non-hymenopterous symbionts. In this case it is envisaged that a host shift has occurred, from the parasitoid attacking this commensal in the nest, to attacking the Hymenoptera themselves. Such an evolutionary route may well have occurred in the Diapriidae, which probably primitively attacked the larvae and pupae of commensal Diptera. Some of the smaller chalcidoids may also have developed an association with nests in a similar way.

Very little is known about the biology of parasitoids in tropical hymenopterous nests, but in the case of species attacking very large nests there might well be

Parasitoid	Host	Reference
BETHYLIDAE		
Pristocerinae		
Pseudisobrachium?	Formicidae [L]	(Mann, 1915)
Bethylinae		
Parasierola?	Apidae[L, P]	(Rau, 1922)
BRACONIDAE		
Doryctinae		
Heterospilus microstigmi	*Microstigmus* (Sphecidae) [L, P]	(Matthews, 1991)
Euphorinae		
Syntretus sp.	*Bombus* spp. (Apidae) [A]	(Alford, 1968).
Neoneurinae	Formicidae [A]	(Hölldobler & Wilson, 1990*a*)
DIAPRIIDAE		
Diapriinae		
Gymnopria sp.	*Acromyrmex* (Formicidae) [L]	(Loiácono, 1987)
Caecopria, Labidopria, Mimopria, Xanthopria spp.	Formicidae [L, P]	(Huggert & Masner, 1983; Loiácono, 1981; Kistner, 1982)
CHALCIDIDAE		
Chalcidinae		
Brachymeria spp.	Pompilidae [P]	(Bouček, 1992)
	Trypoxylon spp. (Sphecidae) [P]	(Coville & Griswold, 1984; Bouček, 1992)
	eusocial Vespidae [P]	(Bouček, 1992)
CHRYSIDIDAE		
Chrysidinae	solitary Vespidae, Apidae [L, P]	(see Chapter 13.2)
ELASMIDAE		
Elasmus spp.	eusocial Vespidae [P]	(Reed & Vinson, 1979)
ENCYRTIDAE		
Coelopencyrtus spp.	solitary Apidae [L]	(Daly *et al.*, 1967)
EUCHARITIDAE[4]	Formicidae [L]	(see Chapter 11.6)
EULOPHIDAE		
Entodontinae		
Alachua spp.	*Camponotus* (Formicidae) [P]	(see chapter 11.7)
Pediobius sp.	solitary Apidae	(Krombein *et al.*, 1979)
Tetrastichinae		
Baryscapus spp.	solitary Apidae [L]	(Daly *et al.*, 1967 as *Aprostocetus*)
Chaenotetrastichus spp.	Pompilidae	(J. LaSalle, pers. comm.)
Chytotetrastichus sp.	Pompilidae	(J. LaSalle, pers. comm.)
Melittobia spp.	Apidae; eusocial Vespidae [L, P]	(see Chapter 11.7)
Tetrastichus spp.	solitary Apidae; Pompilidae	(J. LaSalle, pers. comm.)
EUPELMIDAE		
Eupelmus sp.	solitary Apidae	(Krombein *et al.*, 1979)
EURYTOMIDAE		
Axima spp.	*Ceratina* (Apidae) [P]	(Daly *et al.*, 1967)
Conaxima sp.	*Azteca* spp. (Formicidae) [A]	(Brues, 1922)
Eurytoma spp.	solitary Apoidea, Vespidae [L, P]	(Daly *et al.*, 1967; Parker & Bohart, 1967)
LEUCOSPIDAE	solitary Apidae, Vespidae [P]	(see Chapter 11.10)

Table 2.5. continued opposite.

Parasitoid	Host	Reference
ICHNEUMONIDAE		
Cryptinae		
Aritranis[2] spp.	stem-nesting aculeates [P]	(Townes, 1969*a*)
Pachysomoides spp.	*Polistes* spp. (Vespidae) [P]	(West-Eberhard, 1969; Richards, 1978)
Photocryptus sp.	*Trigonopsis* sp. (Sphecidae) [P]	(Eberhard, 1974)
+ other Osprynchotina	mud-nesting solitary Vespidae, Sphecidae, Pompilidae [P]	(Townes, 1969*a*)
Sphecophagina[1]	Vespinae [P]	(Townes, 1969*a*; Donovan, 1991)
Toechorychus spp.	*Mischocyttarus* (Vespidae) [P]	(Richards, 1978)
Trachysphyrus sp.	eumenine Vespidae [P]	(Porter, 1967)
Labeninae		
Grotea spp.	solitary Apidae [P]	(Slobodchikoff, 1967)
Neorhacodinae	*Spilomena* spp. (Sphecidae) [?L]	(Danks, 1971*b*)
Paxylommatinae		
Hybrizon spp.	Formicidae [?A]	(Hölldobler & Wilson, 1990*a*)
Pimplinae		
Camptotypus genus-group[3]	eusocial Vespidae [P]	(Keeping & Crewe, 1983; Brooks & Wahl, 1987)
Ephialtes genus-group	solitary Apidae, Sphecidae [P]	(Jussila & Käpylä, 1975)
Neotheronia sp.	solitary Apidae [P]	(Gauld, pers. comm.)
Nomosphecia spp.	social Vespidae [P]	(Gupta, 1962)
Perithous spp.	stem-nesting Sphecidae [P]	(Krombein *et al.*, 1979)
Poemeniinae		
Poemenia spp.	stem-nesting Sphecidae [P]	(Fitton *et al.*, 1988)
MUTILLIDAE	solitary Apoidea, Vespidae [P]	(see Chapter 14.2)
PERILAMPIDAE[5]		
Perilampus sp.	solitary Apidae/Sphecidae [L]	(Krombein, 1960)
PTEROMALIDAE		
Cleonyminae		
Cleonymus sp.	solitary Apidae [?P]	(Krombein *et al.*, 1979 as *Ptinobius*)
Epistenia spp.	stem-nesting Apoidea, Vespidae [P]	(Krombein, 1960; Parker & Bohart, 1967)
Pteromalinae		
Dibrachys spp.	solitary Apidae; Sphecidae [P]	(Grissell, 1974)
Habitrys spp.	stem-nesting Apoidea [?P]	(Krombein *et al.*, 1979)
Trichokaleva sp.	*Microstigmus* spp. (Sphecidae)	(Bouček, 1972*b*)
TORYMIDAE		
Toryminae		
Diomorus	stem-nesting Apoidea [?P]	(Parker & Bohart, 1967)
Monodontomerinae	solitary Apoidea and Vespidae [P]	(Torchio, 1974; Krombein *et al.*, 1979)
TRIGONALYIDAE		
Various spp.	eusocial Vespidae [L]	(see Chapter 8.1)

[1] Although this group has not been recorded from the New World one undescribed South American species may belong here.
[2] Includes *Hoplocryptus* and probably also *Kaltenbachia* records.
[3] No New World species of this group have yet been recorded from vespid nests, but the hosts of most neotropical species are not known.
[4] Some Old World groups tentatively placed in this family parasitize solitary bees (Michener, 1969).
[5] One Old World genus is a primary parasitoid of eumenine Vespidae.

Table 2.5. New World hymenopteran parasitoids of nest-building Hymenoptera. A = parasitoids of adults; L = larval parasitoids; P = pupal (or prepupal) parasitoids. It should be noted that some of the parasitoids found associated with nests may not be attacking the immature stages of the nest-builder, but rather other insects that are associated with the nest; these are not included in the above examples. Records suffixed by a interrogation mark require confirmation.

ample opportunities for several generations of a parasitoid to develop in a single nest. Such behaviour is known to occur in the ichneumonid *Sphecophaga*, a parasitoid of vespine vespids in temperate regions. A winged female ichneumonid enters a vespid nest early in the season and oviposits through recently capped cells onto vespid prepupae or pupae. The ectoparasitic larvae develop and then spin one of three types of cocoon—white cocoons giving rise to brachypterous females; thin yellow cocoons giving rise to winged females in the same season; thick yellow cocoons giving rise to winged males and females in the following or subsequent seasons. Whilst the winged forms eventually disperse, the brachypterous females remain within the nest attacking other vespid prepupae or pupae. There may be several generations of such brachypters in the nest during a season (Donovan, 1991), and these seem to be deuterotokous as they are capable of producing offspring making all three cocoon types, including males in thick yellow cocoons. Such reproductive behaviour is reminiscent of the cyclical parthenogenesis seen in many aphids (Godfray, 1994) and is presumably an adaptation allowing the parasitoid to capitalize on the abundant resources available.

Cleptoparasites

The term CLEPTOPARASITISM refers to the oviposition by a species that does not construct a nest of its own, into a provisioned nest built by another (host) species. They are commonly called 'cuckoo wasps' and 'cuckoo bees' because of their behavioural similarity to the European cuckoo (*Cuculus canorus*), a bird famed for laying its eggs in the nests of other birds. Rather rarely the host egg is destroyed by the adult cleptoparasite when she is ovipositing, but more usually, following eclosion from the egg, the larval cleptoparasite seeks out and destroys the host egg or young larva. After killing the host, the cleptoparasitic larva goes on to consume the food stored as provision for the host larva. Thus the provision initially stored by the host adult is the essential food supply for the cleptoparasitic larva. In the Hymenoptera, the term cleptoparasite is usually restricted to species that exploit non-eusocial nest-builders.

It is helpful to distinguish two possible evolutionary origins of cleptoparasitism (Gauld & Bolton, 1988). Taxa that were derived from ancestors that were solitary nest-builders, whose larvae originally fed on nest provisions, are PRIMARY CLEPTOPARASITES whereas taxa derived from ancestors that were parasitoids of the host larva in a nest are SECONDARY CLEPTOPARASITES. Appreciating these differing evolutionary origins help one understand differences in biology between various cleptoparasites.

Primary cleptoparasites. These cleptoparasites have abandoned the construction of their own nests to appropriate the completed and provisioned nests of another species. The nest and provision of the host, rather than the host egg or larva, has always been the primary objective of these cleptoparasites. The cleptoparasite's ancestral behaviour was essentially similar to that of the host and the elimination of the host offspring is a novel but incidental behaviour. The evolutionary pathway to primary cleptoparasitism may have involved initially intraspecific cleptoparasitism (see below). Primary cleptoparasites usually attack the nests of closely related species, always members of the same family and often members of the same subfamily. This type of cleptoparasitism has arisen several times in the Pompilidae where, for example, *Irenangelus* oviposits in the nests of *Auplopus* (see Chapter 14.3); in the Sphecidae where, for example, *Zanysson* attacks *Tachytes* (see Chapter 17); and in the Apidae where, for example, species of *Exarete* are cleptoparasites of *Eulaema* (see Chapter 18).

Secondary cleptoparasites. Here the provision of the host is a derived secondary objective and the consumption of the host offspring is ancestral behaviour. For example, in the ichneumonid subfamily Labeninae the ancestral behaviour is attacking endopterygote pupae or prepupae, but members of one Australian genus, *Labium*, develop as typical cleptoparasites of bees (Gauld, 1983, 1984). Secondary cleptoparasites attack nests of unrelated species, nearly always members of a different family. Examples of secondary cleptoparasites include some Chrysididae (in nests of eumenine vespids, sphecids, and solitary bees), all Sapygidae (in nests of solitary bees), and all Gasteruptiidae (in nests of solitary bees). Some of these parasitoids initially feed on the larval hymenopteran and complete their development on the provisions in the nest, thus illustrating how the transition from parasitoid to cleptoparasite may have occurred.

Cleptoparasitic interactions not only occur between different species, they often occur between individuals of a single species. *Intraspecific food cleptoparasitism* is common in both solitary wasps nesting in aggregations and in primitively social (communal and parasocial) wasps. Indeed, it appears to occur in most such species that have been carefully observed for extended periods including: *Auplopus* (Wcislo *et al.*, 1988); *Bembix* (Evans, 1966*b*); *Clypeadon* (Alexander, 1986); *Trigonopsis* (Eberhard, 1972); *Zethus* (West-Eberhard, 1987) and various solitary wasps (Yoshikawa, 1962). The 'joint provisioning' of the same cell in some species, such as *Cerceris cribosa* (Elliott & Elliott, 1987), *Sphex ichneumonius* (Brockmann, 1985) and *Trypoxylon politum* (Brockmann, 1980), may be regarded as a kind of cleptoparasitic race between conspecifics.

Although the concept of cleptoparasitism is generally applied only to nest-building aculeate Hymenoptera a similar type of interspecific association exists in some non-aculeates. Parasitoids that require the presence of another parasitoid before they can successfully utilize a host can also be considered cleptoparasites (see Chapter 2.5). Such a form of cleptoparasitism is, however, generally termed interspecific superparasitism. But all cleptoparasitism can be considered to be interspecific superparasitism if the nest provision is seen to be analogous to a parasitoid's 'host'. Furthermore, inquilines in galls can also be considered to be cleptoparasites, as they cannot elicit gall formation on their own and are dependent upon the gall tissue (= host or provision) provided by another species (see Chapter 2.3). The siricid *Xeris spectrum*, feeds as a larva on wood inoculated with a fungus, but unlike other siricines it does not itself transmit this fungus (Viitasaari, 1984). Thus it too may be considered a cleptoparasite as it can only develop in timber inoculated by other siricines which have previously oviposited in the wood. Here we suggest that diverse interspecific interactions between hymenopterans can be considered to be cleptoparasitism or a form of interspecific superparasitism. The evolutionary distinction outlined above between primary and secondary cleptoparasitism is equally applicable in all these examples. In all cases of cleptoparasitism in this broad sense, mixed broods may occur (i.e. the hymenopteran providing the resource survives), which is commensalism rather than parasitism. This is far more common in galls and siricines, where the food resource continues to grow.

Nest usurpation and supersedure

Nest-building Hymenoptera are often limited by availability of nest sites, which often results in intra- and interspecific competition (Krombein, 1967). In nest USURPATION a female hymenopteran forcefully evicts another female hymenopteran (of the same or different species), and the usurping female then uses the nest to rear her own offspring. In hollow twigs it is not uncommon to find a series of cells that have been built and provisioned by different hymenopteran species, which is termed SUPERSEDURE and may or may not involve competition (the original female may have died). For example, in the Caribbean lowlands of Costa Rica, artificial nests were sometimes concurrently occupied by a leaf-cutter bee, the sphecid *Trypoxylon superbum*, and the sphecid *Podium rufipes* (Coville & Griswold, 1984).

Compound nests

The diverse interspecific associations known to exist between species of eusocial Hymenoptera can be divided into two broad categories. In COMPOUND NESTS two or more species live very close to each other, but keep their broods separated. In MIXED NESTS the broods of the two species are mingled and reared together. In both cases one of the component species usually gains at the expense of the other. The diversity and complexity of interspecific associations is especially pronounced in ants (Hölldobler & Wilson, 1990*a*) and some of these are difficult to categorize. For example, in Panama, *Zacryptocerus maculatus* uses pheromone trails of *Azteca maculatus* but nests separately (Adams, 1990).

Compound nests are very common in ants and range from an accidental association through to total parasitism. The following are the types of associations that are commonly recognized:

PLESIOBIOSIS. A rudimentary association in which two species merely nest very close together, but engage in little or no communication.

CLEPTOBIOSIS. An association between a small (cleptobiotic) species and a much larger species in which the former feeds on the food stores or refuse of the larger, or steals food from foragers of the larger species as they return home. In Nicaragua, *Ectatomma ruidum* has been observed using foraging trails of *Pheidole radoszkowskii* and other ants, and robbing ants that were returning to their nests with provisions (Perfecto & Vandermeer, 1993). A slightly different type of cleptobiosis occurs in stingless bees belonging to the genus *Lestrimelitta* (Sakagami, Roubik & Zucchi, in press). *L. limao* robs food and nest material by raiding nests of other stingless bees. Occasionally this species usurps the nest for its own.

LESTOBIOSIS. An association between a smaller and a larger species, in which the smaller (lestobiotic) species nests next to or within the walls of the nest of the larger species, and enters that nest to prey on the brood therein.

PARABIOSIS. Two species utilize the same nest (but with separate brood chambers) and sometimes the same odour trails, although foraging is usually independent in each species and acquired food is not generally shared between the species. The co-occurrence of two ant species is quite common in Costa Rica, but in most cases the exact nature of the relationship remains unknown, and is often tentatively categorized as parabiosis. Examples from the Neotropics include species of the *Camponotus abdominalis* complex, which nest with species of the *Azteca instabilis* complex, and species of the *Crematogaster parabiotica* complex, which nest with various ant species (e.g. *Dolichoderus*, *Odontomachus panamensis*, etc.) (J. Longino, pers. comm.). In Brazilian rainforest canopies a species of the *Crematogaster parabiotica* complex nests with *Camponotus femoratus*. Although *Camponotus* may drive *Crematogaster* away from a food source, it may provide *Crematogaster* with added protection at the nest (Swain, 1980; as *Crematogaster limitata*).

XENOBIOSIS. One species nests within the walls or actually in the nest chambers of another species, and moves freely about the host nest. The xenobiont, which is often smaller than the host, not only shares the nest and trails of its host but also solicits, and receives, liquid food from the crop of the host species. In Panama *Megalomyrmex symmetochus* live in the fungus gardens of *Sericomyrmex amabilis*, but since liquid food exchange is rare among fungus-growing ants this xenobiont apparently subsists exclusively on the fungus (Wheeler, 1925).

Mixed nests and social parasitism

The queens of some eusocial species in the families Formicidae, Vespidae, and Apidae have secondarily abandoned the usual nest founding techniques seen in eusocial forms and instead begin their colonies by taking over the already functioning nest of another, usually closely related, eusocial species. The aim of such SOCIAL PARASITISM is not the acquisition of the host's brood as food but the utilization of the host's nest and available workers to rear at least the first generation of the parasite's brood. There is enormous variation in the biologies of socially parasitic species but two informal grades of social parasitism may be defined, depending upon whether the parasite has a worker caste or not.

In TEMPORARY SOCIAL PARASITISM the parasite has a worker caste. Here the socially parasitic female actively invades, or is passively accepted into, an established nest of the host species. Once inside the nest the social parasite destroys the foundress host queen or relegates her to subsidiary status; in some cases the workers turn on their own queen and destroy her. The social parasite begins to lay eggs, which are reared by the host workers, and these eggs eventually develop into workers of the socially parasitic species. These take over the running of the nest as the host workers gradually die off from natural causes. In time the nest comes to consist entirely of the social parasite's offspring, and continues to function as a normal colony.

PERMANENT SOCIAL PARASITISM or INQUILINISM is the final, degenerate stage in the evolution of social parasitism. As in temporary social parasitism a reproductive female of the parasite gains entrance to a host nest, usually but not always destroying or causing the death of the foundress host queen. In permanent social parasitism, however, the eggs laid by the parasite (the inquiline) give rise only to males and new reproductive females; the worker caste has been lost.

Social parasitism appears to be less common in the tropics than in the temperate zone, which may be due to the lack of seasonality, and hence lack of host synchrony, in the tropics (Wcislo, 1987). Among neotropical ants permanent social parasitism has been reported in *Crematogaster atitlanica* (*C. sumichrasti* as the host) in Guatemala (Wheeler, 1936) and

Monomorium inquilinum (*M. minimum* as the host) in Mexico (DuBois, 1986). Social parasitism is so far unknown in tropical social wasps. *Vespula squamosa* (Vespidae, Vespinae), which occurs from eastern North America to Guatemala, is an obligate social parasite in the northern part of its range but is facultatively free-living in the southern United States (MacDonald & Matthews, 1984). Among bumblebees species of *Psithyrus* are social parasites but the genus extends only as far south as Guatemala.

In a few ant species the workers of an established nest raid the nest of a closely related species and steal their pupae. The stolen pupae are taken back to the raider's nest where they eventually eclose and function thereafter as workers or 'slaves' within the raider's nest. This phenomenon is called slave-making, or DULOSIS. Freshly mated females of most slave-making species start a new nest by temporary social parasitism on a species belonging to the group which is later raided for slaves. Although most slave-making species begin new colonies by temporary social parasitism it is currently accepted that the phenomenon of dulosis has arisen from ancestral mass-provisioning behaviour in which the nests of other ants were the targets, or from territoriality, and is an evolutionary development separate from social parasitism (Gauld & Bolton, 1988).

Symbioses with other arthropods

There are relatively few summaries treating the symbioses between nest-building Hymenoptera in general and other arthropods (e.g. Krombein, 1967; Wilson, 1971). This contrasts with the voluminous literature treating the very diverse set of symbionts ('myrmecophiles') associated with ants (Kistner, 1979, 1982; Wojcik, 1989; Hölldobler & Wilson, 1990*a*). The latter are usually classified by how they are treated by the host ants: SYNECHTHRANS (treated in a hostile manner), SYNOEKETES (ignored), or SYMPHILES (accepted by the host ants as if they were members of the colony). Symphiles become integrated into the ant society by breaking into the ants' communication and recognition code (Kistner, 1979). This may involve behavioural imitation of solicitation signals (trophallaxis), tactile mimicry, acquisition of the colony odour via grooming the host ants, or production of pseudopheromones that mimic the 'brood pheromones' of their hosts. The visual resemblance between many myrmecophiles and ants may serve to deceive predators rather than the host ants (e.g. the staphylinids associated with army ants). The numerous groups of non-myrmecophiles that mimic ants (visually) have recently been reviewed by McIver and Stonedahl (1993) and are not included here.

Mutualists

The only arthropods showing mutualistic associations with solitary wasps and bees appear to be certain mites that feed on potentially damaging fungi in the nest (Cooper, 1955; Flechtmann & Camargo, 1979). The adult hymenopteran may possess special structures for carrying the mites, as in some eumenine vespids (Krombein, 1961) and halictine bees (Eickwort, 1979; McGinley, 1986). For example, all species of the eumenine genus *Parancistrocerus* have a complex symbiotic relationship with winterschmidtiid mites, each wasp having a host-specific mite. The hypopial stage of the mite congregates in large numbers in the acarinarium, a chamber at the base of the second metasomal tergite which is covered by the apex of the first tergite (Krombein, 1967).

Mutualism with other arthropods is more common in the eusocial Hymenoptera, the best known examples being ants that protect insects producing attractive secretions. The most frequently observed of such interactions involve honeydew-secreting Homoptera (Way, 1963; Buckley, 1987). The ants usually attend these insects (known as TROPHOBIONTS) outside the nest, but in a few instances they may be brought into the nest. For example, a species of *Myrmelachista* maintains *Dysmicoccus* (Pseudococcidae) inside the hollowed-out stems of *Ocotea* (Lauraceae) (Stout, 1979). Homoptera are tended not only by ants but also by stingless bees (Salt, 1929) and polistine vespids (Letourneau & Choe, 1987).

The caterpillars of a variety of lycaenoid butterflies (Lycaenidae and Riodinidae) have a somewhat similar mutualistic relationship with certain ants (Hinton, 1951; Pierce, 1987; DeVries, 1991*a*, *b*; Fiedler, 1991) although in a few cases this relationship is more one-sided as the butterfly larva completes development by consuming ant larvae. Myrmecophilous caterpillars mediate their interactions with ants by the possession of specialized ant organs. These are often glands producing chemical secretions, substances that are either nutritious and reward ants, or are semiochemicals that in some way modify the ant's behaviour. In some lycaenid and riodinid larvae structures are present which are capable of producing acoustic signals, and are presumably used to 'call' to ants (DeVries, 1990; DeVries *et al.*, 1993). Lycaenoid-ant interactions are often facultative and typically involve those species of ants (such as *Pheidole*, *Crematogaster*, *Myrmica*, *Camponotus*, *Iridomyrmex*, *Tapinoma* and *Azteca*) that are attracted to extrafloral nectaries and honeydew (DeVries, 1991*a*; Fiedler, 1991). Frequently one species of butterfly seems to be attended by different species of ants in different places (DeVries, 1991*b*), although in some species obligate and species-specific interactions occur (Fiedler, 1991). In the New World tropics the vast majority of the lycaenids belong to the subtribe Eumaeiti, and myrmecophily is less well-developed in this group than in other large subtribes. Probably only about 28 percent of the species will be found to be myrmecophilous, and obligate myrmecophily is expected to be very uncommon (Fiedler, 1991). Riodinids, on the other hand, are predominantly neotropical in distribution. Myrmecophily is known only among tropical American taxa, but even so it probably occurs in only about 25 percent of the species (DeVries, 1991*b*).

Commensals

Mites are the most abundant arthropod symbionts present in hymenopteran nests in general. The majority of these acarines are probably commensal scavengers and many are phoretic on the adult hymenopterans. Mites are particularly abundant in colonies of army ants (Rettenmeyer, 1960) but several taxa have also been reported from nests of bees (Eickwort, 1979; Roubik, 1989). Other common commensals associated with hymenopteran nests are flies of the family Phoridae. In Costa Rica, species of *Megaselia*, *Dohrniphora*, *Puliciphora* and *Chonocephalus* are generalized scavengers that occur in any decaying organic material, including that found in hymenopteran colonies (B.V. Brown, pers. comm.). Several beetles are also generalized scavengers (e.g. Ptiliidae), but two genera of Leiodidae (*Parabystus* and *Scotocryptus*) are apparently specialized scavengers in nests of stingless bees (Salt, 1929), and one species of Lathridiidae (*Coluocera maderae*) is a highly specialized scavenger in nests of the cosmopolitan 'crazy ant', *Paratrechina longicornis* (Kistner, 1982).

Ants with large colonies, such as army ants and leaf-cutter ants, provide large quantities of debris for commensal scavengers. Among phorid flies, species of *Acontistoptera*, *Adelopteromyia*, *Apterophora*, *Ecitomyia*, *Ecitophora*, and *Ecituncula* are apparently restricted to army ant colonies (Rettenmeyer & Akre, 1968). Females of these obligate associates are wingless or brachypterous, whereas males are fully winged. Millipedes of the family Pyrgodesmidae (e.g. *Calymnodesmus* and *Rettenmeyeria*) are scavengers that follow army ant trails, and are tolerated, and sometimes even carried by the ants (Rettenmeyer, 1962; Akre & Rettenmeyer, 1968). Several thysanurans of the family Nicoletiidae feed on the fluids oozing from booty captured by army ants (Rettenmeyer, 1963*b*). Although they are capable of following army ant trails (Torgerson & Akre, 1969), they are not integrated into the colony and may be attacked by the army ants.

Scavengers specifically associated with leaf-cutter ants include several groups of beetles, such as Cerylonidae (*Lapethus*), Histeridae (various taxa), and Scarabaeidae (Aphodiinae: *Euparixia* and *Lomanoxia*). Larvae of various flies are also probably scavengers of detritus in nests of leaf-cutters, for example Milichiidae (*Pholeomyia*), Fanniidae, and Mydidae (*Mydas*). Nine species of Collembola (family Cyphoderidae) were found in an *Atta* nest in Guatemala (Ellis, 1967) while other Central American members of this family are scavengers in nests of stingless bees (Mari Mutt, 1977).

In addition to scavengers, commensals also include organisms that occupy nests of aggressive hymenopterans in order to gain protection. For example, the katydid, *Ancistrocercus inficitus*, roosts near the nests of at least five polistine vespid nests, which apparently affords protection against predators (Downhower & Wilson, 1973).

Cleptoparasites

Cleptoparasites of solitary Hymenoptera include flies of the families Phoridae and Sarcophagidae, and beetles of the family Meloidae. *Phalacrotophora halictorum* (Phoridae) is found throughout the New World, and in Costa Rica the larvae are reported to be cleptoparasitic in the nests of the halictine bees *Agapostemon nasutus* (Eickwort & Eickwort, 1969) and *Dialictus umbripennis* (Eickwort & Eickwort, 1971), although other authors record this phorid species as a parasitoid. The larvae of *Melittophora salti* are found within sealed pupal cells of stingless bees and apparently share the food provided for the bee brood, with no deleterious effects on the host (Salt, 1929). In the Sarcophagidae most species of Miltogrammini are cleptoparasites of solitary, ground-nesting Hymenoptera, and the presence of these flies usually results in the death of the host larva. In many cases the female fly larviposits in the burrow entrance but some ('satellite flies') may larviposit directly on the provision while the female wasp is entering the nest. Several genera of Meloidae attack solitary bees, the female beetle ovipositing near the nest entrance, or on flowers (in which case the larvae attach to adult bees in order to gain access to the nest). Examples from Central America include *Cissites macultata* on *Xylocopa frontalis* (Hurd, 1978*b*), *Meleotyphlus* spp. on orchid bees (Dodson, 1966), and *Tetraonyx cyanipennis* on *Crawfordapis luctuosa* (Roubik & Michener, 1984).

Cleptoparasites of ants include various taxa that steal food being exchanged between workers. Carabid beetles of the genus *Helluomorphoides* run about in raiding columns of army ants, causing the latter to drop their prey (Plsek *et al.*, 1969; Topoff, 1969). Cleptoparasites of leaf-cutter ants feed on the fungal gardens or the leaf material brought into the nest as substrate for the fungus. These include larvae of Alleculidae (*Lobopoda*), Scarabaeidae (Dynastinae: *Coelosis*), Tineidae (*Atticonviva*), and apparently several taxa of cockroaches (Attaphilidae, Atticolidae and Nothoblattidae). Cockroaches of the genus *Attaphila* have highly developed arolia on their tarsi that enable females to remain attached to queen ants during swarming (Kistner, 1982).

Fly larvae of the family Braulidae (introduced into Costa Rica) feed on pollen in nests of honey-bees (Knutson, 1978); the highly modified adult flies cling to adult bees and feed on liquids entering the mouth of the bee. A few cleptoparasites may actually feed on the nest itself. The greater wax moth, *Galleria mellonella* (Pyralidae), is a cosmopolitan pest of honey-bee hives (Williams, 1978) whose larvae feed on wax and pollen. *Tinea carniella* (Tineidae) feeds on the paper nests of *Polistes* (Vespidae) (Nelson, 1968).

Parasites

Arthropod parasites of aculeate Hymenoptera include both ecto- and endoparasites. Certain mites belonging to the families Machrochelidae, Neoparasitidae, and Pyemotidae are ectoparasites on adult and larval army ants (*Eciton* spp.) (Rettenmeyer, 1960, 1962). *Varroa jacobsoni* (Varroidae) is an ectoparasite that feeds on the blood of honey-bee larvae, pupae and sometimes adults, and is a serious pest of honey-bees in many parts of the world, including Mexico and South America, but not yet in Costa Rica (DeJong *et al.*, 1982; W. Ramirez, pers. comm.). Roubik (1989) provides a table of probable mite parasites of bees in general. Very few mites have adapted to an endoparasitic habitat in invertebrate hosts, one of the few examples being *Acarapis woodi* (Tarsonemidae), which invades the tracheal system of honey-bees and may eventually cause suffocation. This honey-bee pest occurs throughout the Neotropics, and it may have arrived in Costa Rica with the Africanized honey-bee (Otis *et al.*, 1988).

The only other arthropod endoparasites of aculeate Hymenoptera are Strepsiptera. All members of the suborder occurring in the Neotropics, the Stylopidia, are extraordinarily sexually dimorphic (Kathirithamby, 1989, 1992). The male is winged, with antennae and well-developed eyes, whereas the female is larviform and never completely emerges from the host. Males mate with a female that has partially extruded between the sclerites of its host. The mated female releases thousands of minute larval triungulins into the host's haemocoel and these are eventually expelled. They wait on vegetation or in nests until a new host is encountered, which they penetrate. Once in the host the triungulin moults to an apodous second instar larva, which feeds by filtering blood from its host. The fully grown larva of the male partially extrudes and then pupates within a tanned puparium. It emerges from this and flies off, leaving a gaping orifice in the protruding puparium. In Costa Rica it is not uncommon to find vespids with several protruding empty puparia, and occasionally extruding females. Two of the nine strepsipteran families, Myrmecolacidae and Stylopidae, include species that attack Hymenoptera (Kathirithamby, 1992). Myrmecolacids have different host groups for each sex; males parasitize ants whereas females parasitize Mantodea and Orthoptera. In Stylopidae both sexes parasitize Hymenoptera. Known generic host records of neotropical strepsipterans are given in Table 2.5.

Specific host records for species occurring in Central America include *Pseudoxenos huastecae* from *Montezumia huasteca* (Vespidae: Eumeninae) and *Xenos araujoi* from *Apoica pallens* (Vespidae: Polistinae) (Kathirithamby, 1992). In the seasonally dry forests in Santa Rosa National Park, where vespids have been collected systematically for several years, patterns of parasitism are apparent. Towards the end of the wet

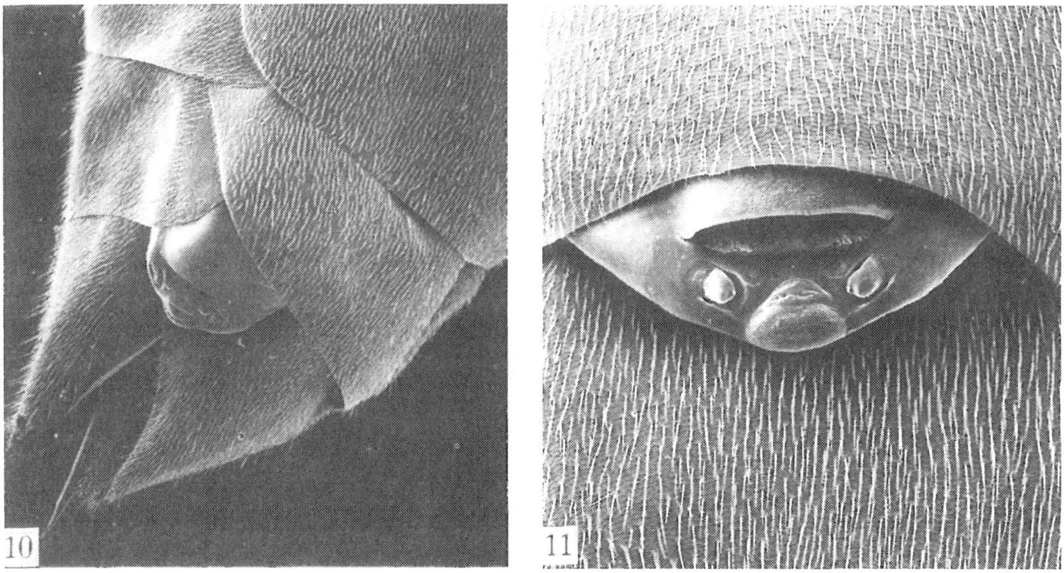

Figs 2.10–2.11. Scanning electron photomicrographs of female strepsipterans (*Xenos* spp.) extruding between metasomal tergites of: 2.10, *Polistes canadensis*; 2.11, *Mischocyttarus angulatus*.

season, large numbers of *Mischocyttarus angulatus* workers have been found to be parasitized by a species of *Xenos* (Fig. 2.11), and parasitism rates can reach almost 100 percent in some localities (Gauld, unpubl.). *Xenos* species have also been found, at a very much lower frequency, on *Polistes canadensis* (Fig. 2.10) and *P. pacificus*. Other very common large vespids at the site, notably *Agelaia myrmecophila*, *A. areata* and *Synoeca septentrionalis*, have not been found to be parasitized.

Parasitoids

At least one family of Coleoptera and two families of Diptera include species that are parasitoids of larvae and/or pupae of nest-building Hymenoptera. Beetles of the family Rhipiphoridae are koinobiont parasitoids of larval wasps (Sphecidae and Vespidae) and bees. They begin their development as endoparasitoids but terminate as ectoparasitoids (Clausen, 1940*b*). Examples from the Neotropics include *Macrosiagon lineare*, reared from the sphecid *Trigonopsis cameronii* (Eberhard, 1974), *Macrosiagon* sp. reared from the bee *Exomalopsis* (Raw, 1977) and *Rhipiphorus* sp. from the bee *Augochlora nominata* (Eickwort & Eickwort, 1972*a*).

Flies of the family Bombyliidae include a few cleptoparasites but most are probably ectoparasitoids (apparently koinobionts) of larvae and pupae of solitary Hymenoptera, including even Tiphiidae. Several bombyliids oviposit into the nest entrance of their host while in flight. In Costa Rica *Anthrax cintalapa* and *Mesoplia rufipes* have been reared from the anthophorine bee *Centris inermis* (Coville *et al.*, 1983). The phorid *Phalacrotophora halictorum*, although sometimes recorded as a cleptoparasite, is also reported to be a parasitoid of the larvae of various solitary bees (Wcislo, 1990). Other species of *Phalacrotophora* are known to parasitize the larvae of Sphecidae (Coville & Griswold, 1983, 1984). The widespread phorid, *Pseudohypocera kerteszi*, is frequently found in nests of honey bees and stingless bees, where its larvae begin by feeding on pollen, but later attack the developing brood (Salt, 1929; Chaud-Netto, 1980). Larvae of a species of *Megaselia* were reported to feed on the brood of a polistine wasp in Colombia (Litte, 1981).

Certain flies, while not strictly nest associates, are endoparasitoids in adult Hymenoptera, often ovipositing while the host is in flight. Several Conopidae attack wasps and bees, both solitary and eusocial. In

Strepsipteran	**Host**
Myrmecolacidae	
Caenocholax	Formicidae: *Pheidole* spp.
Myrmecolax	Formicidae: *Eciton* sp.
Stichotrema	Formicidae: *Camponotus* spp.; *Pseudomyrmex* sp.; *Solenopsis* sp.
Stylopidae: Paraxeninae	
Paraxenos	Sphecidae: *Ammophila* sp.; *Isodontia* sp.; *Prionyx* spp.; *Sceliphron* sp.; *Sphex* spp.
Stylopidae: Stylopinae	
Halictoxenos	Apidae: *Augochloropsis* spp.
Melittostylops	Apidae: *Hesperapis* spp.
Stylopidae: Xeninae	
Pseudoxenos	Vespidae: *Eumenes* sp.; *Hypodynerus* spp.; *Monobia* sp.; *Montezumia* spp.; *Pachodynerus* sp.
Xenos	Vespidae: *Apoica* sp.; *Polistes* spp.; *Mischocyttarus* spp.; *Polybia* spp.

Table 2.6. The known hymenopterous hosts, in the Neotropics, of Strepsiptera (after Kathirithamby, 1992).

Costa Rica *Zodium americanum* has been reported from halictine bees (Eickwort & Eickwort, 1969, 1971). Among Phoridae species of *Melaloncha* have been reported parasitizing adult bees: Anthophorinae in Guatemala (Batra & Schuster, 1977), *Bombus mexicanus* in Costa Rica (Ramirez, 1982), Meliponinae in Brazil (Simoes *et al.*, 1980), and *Apis mellifera* (the honey-bee) in various countries (e.g. Ramirez, 1984).

A large number of phorids are endoparasitoids of adult ants. *Apocephalus paraponerae* attacks *Paraponera clavata* and is especially attracted by injured individuals (Brown & Feener, 1991*a*, *b*). Similarly, *Rhyncophoromyia maculineura* are attracted to injured *Camponotus sericiventris* and the phorid larvae develop in the abdomen of the ant (Brown & Feener, 1993). Leaf-cutting ants transporting pieces of leaves can often be seen with smaller ants ('hitchhikers') riding on top of the leaf—apparently in defence against a species of *Apocephalus* (Feener & Moss, 1990). *Pseudaceteon* spp. can cause a reduction in foraging by the tropical fire ant, *Solenopsis geminata* (Disney, 1991; Feener & Brown, 1992). Species of *Menozziola* are specialized parasitoids of alate female *Camponotus* species (Brown *et al.*, 1991). Other phorid parasitoids of adult ants include species of *Acanthophorides*, *Cremersia*, *Diocophora*, *Myrmosicarius* and *Neodohrniphora* (B.V. Brown, pers. comm.). Species of *Borgmeieriphora* have structural modifications associated with parasitism and are found associated with ants, but have not been shown to parasitize the ants themselves (Brown, 1993).

Predators

Certain beetles associated with ant nests are predators of ant larvae, and in some cases these beetles are tolerated by the host ants or even integrated into the colony. Examples include Carabidae (the subfamily Paussinae, which is completely myrmecophilous), Staphylinidae (especially in nests of leafcutter ants), Pselaphidae (all species of Clavigerinae appear to be myrmecophilous; Park, 1964), Histeridae, and perhaps a few Scarabaeidae (the genus *Canthon* may be predatory in *Atta* colonies). In Histeridae the neotropical tribe Hetaeriomorphini (subfamily Hetaeriinae) is primarily associated with army ants, feeding on their brood and booty; *Euxenister caroli* for example is a specific associate of the army ant *Eciton burchelli* (Akre, 1968; Torgerson & Akre, 1970). Syrphid flies of the subfamily Microdoninae have slug-like larvae that may feed on the detritus in ant nests or on ant larvae (Hölldobler & Wilson, 1990*a*); in Costa Rica the Microdoninae include four genera (C. Thompson, pers. comm.): *Argentinomyia*, *Microdon*, *Mixogaster* and *Paramicrodon*. Crickets belonging to the cosmopolitan genus *Myrmecophila* (Gryllidae: Myrmecophilinae) are predators of ant eggs and snatch food being exchanged between ants (Kistner, 1982).

Larvae of Mantispidae (subfamily Symphrasinae) have occasionally been reported as brood predators of polistine vespids, bees and sphecids (Richards, 1978). Pyralid caterpillars (*Chalcoela* and *Dicymolomia*) are predators of *Polistes* larvae (Nelson, 1971) and are so effective that they frustrated attempts to establish *P. annularis* on some Caribbean islands as a biological control agent (Ballou, 1934). An undescribed genus of Tineidae has been reported to prey upon pupae of *Polistes canadensis* in Brazil (Jeanne, 1979*b*). Among beetles, clerid larvae (*Cymatodera*; Linsley *et al.*, 1980) and dermestid larvae (*Trogoderma*; Raw, 1984) are recorded as predators of solitary bee larvae.

Other predators capture adult hymenopterans at the nest entrance or on ant trails. Among assassin bugs (Reduviidae), for example, *Acanthaspis concinnula* preys on the ant *Solenopsis geminata*, camouflaging itself by placing dead ants on its back (Kistner, 1982), and *Apiomerus pictipes* preys on stingless bees (Johnson, 1983). The aleocharine staphylinid, *Tetradonia marginalis*, follows army ants (*Eciton* spp.) and preys on stragglers at the end of the column (Akre & Rettenmeyer, 1966; Hölldobler & Wilson, 1990*a*).

3 Economic importance of Hymenoptera

Paul E. Hanson

INTRODUCTION

From the human perspective, no other group of insects contains so many beneficial species as the order Hymenoptera. The majority of hymenopterans have larval stages that are carnivorous on other insects and numerous species are vital in controlling the populations of pest insects. Among the non-carnivorous members of the order, bees constitute the single most important group of plant pollinators. One species of bee, the honey-bee, is the most prevalent domesticated insect and provides us with honey, wax, and other products. On the other hand, the order Hymenoptera also contains some very serious pest species. Leaf-cutting ants are among the most damaging agricultural pests in the Neotropics. Many ants, wasps, and bees are capable of inflicting painful stings, and in persons who develop a hypersensitive reaction to the venom, a sting can be fatal.

The economically important Hymenoptera of the New World tropics include both native and introduced species. The latter category includes species that were deliberately introduced (the honey-bee and several parasitoids) as well as some that arrived unintentionally (several household ants and a few parasitoids). Perhaps the most dramatic recent introduction to the Neotropics is a new, African race of the honey-bee, the 'killer bee'. Whether the Africanized honey-bee is a menacing pest or a better bee for the tropics is still being debated, and the effect it will have on the native flora and fauna is only beginning to be investigated. Some introductions of other hymenopterans are known to have a devastating impact on the native fauna, such as the accidental introduction of the ant, *Wasmannia auropunctata*, to the Galapagos (Clark *et al.*, 1982; Lubin, 1984). On the other hand, the introduction of certain parasitoids for the biological control of pest insects has resulted in savings of millions of dollars (Jutsum, 1988). Whether these introduced parasitoids can occasionally have unintended, negative impacts has received relatively little attention until recently (Howarth, 1991; Bennett, 1993).

Just as the addition of exotic species can have unforeseen consequences, so too the subtraction of native species through extinction may result in incalculable economic losses. The Brazil nut, for example, is pollinated by native orchid bees, which depend upon other pollen sources in the forest for the eleven months of the year when the Brazil nut tree is not in flower; the male bees also require epiphytic orchids as a source of sexual attractants. Deforestation removes alternate pollen sources and orchids, resulting in the loss of the pollinator, which in turn results in failure of fruit production in plantation-grown Brazil nuts (Prance, 1986). Conservation of native bee species therefore requires more attention in the future (Frankie *et al.*, 1988*b*). In addition to loss of pollinators, deforestation may result in the extinction of potential biological control agents, the value of which are nearly impossible to calculate. The parasitic Hymenoptera may be essential in the functioning of natural ecosystems, yet an undetermined but probably significant proportion remain unknown to science (LaSalle & Gauld, 1991, 1993). We do not know at what rate we are losing species, much less the consequences these extinctions will have.

3.1 PEST HYMENOPTERA

Although a relatively small proportion of hymenopteran species are pests, some of these can cause substantial economic losses. Pestiferous species of Hymenoptera are treated here in four categories: crop pests, pests of medical importance, killer bees and household pests. Hymenopterans can also cause problems in unanticipated ways that are difficult to categorize. For example, a few parasitoids (e.g. Trichogrammatidae) are pests in butterfly farms by interfering with efforts to rear butterflies. Winged *Camponotus* ants are sometimes attracted to lights on ships being loaded with bananas, thus infesting the cargo and causing shipments of this valuable export commodity to be rejected by overseas quarantine officials. Similarly, the presence of *Solenopsis geminata* in candy exports from Costa Rica has resulted in the shipment being rejected in quarantine.

CROP PESTS

Some hymenopterans cause direct damage to cultivated plants while others are indirectly detrimental. The direct pests include stem and wood-borers (sawflies), ordinary leaf consumers (sawflies), leaf harvesters that use the leaves as substrate for fungal gardens (leaf-cutting ants), leaf-scrapers (stingless bees), gall-formers (Eurytomidae), and seed-feeders (Eurytomidae and Torymidae). Indirect pests include predators and hyperparasitoids of beneficial insects, ants that protect honeydew-secreting Homoptera, stinging hymenopterans that interfere with farm labour and, occasionally, bees that transmit plant pathogens while visiting flowers.

The most serious hymenopteran pests of crops in the Neotropics are ants, particularly the leaf-cutting ants (*Atta* and *Acromyrmex*). Leaf-cutting ants probably had profound effects on pre-Columbian agriculture, including that of the Mayan civilization, since several species are known to increase their populations when forests are cleared for crops. When the Spanish settlers arrived and began to plant their introduced crops, leaf-cutting ants were among the first pests that they encountered. For example, in 1557 Bartolomé de las Casas complained that ants had destroyed his cassia, oranges and pomegranate trees in Hispaniola (cited in Weber, 1972). Mariconi (1970) provides many similar disparaging quotations from early Portuguese settlers in Brazil. In his fascinating account of life in nineteenth century Nicaragua, Thomas Belt (1874) states, 'again and again have I been told . . . , when inquiring why no fruit trees were grown at particular places—it is no use planting them; the ants eat them up'.

Leaf-cutting ants take leaves from a wide variety of crop plants and are particularly fond of citrus, cocoa, coffee, cotton, maize and cassava (manioc). In many areas of South America citrus cultivation would be impossible in the absence of control measures and *Atta cephalotes* is known to kill virtually all trees in a citrus plantation by repeated defoliation within seven years of control being abandoned (Cherrett, 1989). In the state of Sao Paulo, Brazil, leaf-cutters reduce the carrying capacity of pastures by the equivalent of a million head of cattle, and throughout the country they devastate plantations of eucalyptus, pine and teak (Cherrett, 1986). Their huge nests may also cause structural damage to roads (Nogueira & Martinho, 1983). This scale of economic damage is in many respects comparable to that caused by migratory locusts in Africa, and in many Latin American countries laws have been enacted requiring the extermination of leaf-cutting ants. However, not all species of leaf-cutting ants cause economic damage, and indeed a few are in danger of extinction, partially as a result of indiscriminate applications of insecticides (Vander Meer *et al.*, 1990).

One of the most effective control measures used against leaf-cutting ants is insecticidal bait (Lofgren & Vander Meer, 1986; Vander Meer *et al.*, 1990). However, the preferred insecticide, Mirex, has been banned in many countries because of its persistence in the environment, and alternatives are being sought. There is also a need for baits that can be made cheaply and simply from locally available materials, for local development and distribution (Robinson & Cherrett, 1978). Another, less substantiated, control measure that has been advocated is to place the leaves of plants containing fungicidal compounds (such as jackbean, *Canavalia ensiformis*) over the leaf-cutter nest (Mullenax, 1979). It has also been suggested that there is a possibility of biological control, using scarab beetles, cockroaches, phorid flies or mites (Waller & Moser, 1990). *Azteca* ants are capable of defending older citrus trees against attack by leaf-cutters, but may create other problems by protecting Homoptera and molesting farm labourers (Jutsum *et al.*, 1981; Cherrett & Jutsum, 1983).

The ubiquitous Central American stingless bees (Meliponinae) occasionally cause direct damage to crops such as mango, citrus, palms, rambutan and macadamia, by scraping young leaves and shoots with their mandibles in order to obtain resins that they use in nest construction (Wille, 1965; Camacho, 1966). Eucalyptus plantations, for example, are an attractive resin source for stingless bees (Haueisen Freire & Gara, 1970). In some crops control of stingless bees is complicated by the fact that the bees are needed for pollination.

Other direct hymenopteran crop pests in Central America include a few Chalcidoidea. Species of

Bephratelloides (Eurytomidae), which feed on the seeds of *Annona* ('guanabana'), provoke premature fruit drop. In addition, the emergence holes of the adult through the skin of the fruit provides an entrance for micro-organisms, and damaged fruit are readily colonized by fungus, lowering their market value (Korytkowski & Peña, 1966; Nadel & Peña, 1991). Another eurytomid has been found attacking the seeds of vanilla, while eucharitids have sometimes been reported causing scars on bananas when they oviposit in the skin. Although these eucharitids do not actually damage the fruit, the scarred skin makes the banana less attractive to the consumer.

Sawflies cause considerable damage in the north temperate region, especially on conifers (Baker, 1972; Gauld *et al.*, 1990), but in the Neotropics these hymenopterans are much less conspicuous and rarely cause economic damage. The natural distribution of northern conifers (Pinaceae) reaches as far south as the highlands in northern Nicaragua and certain sawflies (e.g. Diprionidae) may occasionally cause damage in Belize and other parts of northern Central America (Wilkinson & Drooz, 1979). Since pines are planted in Costa Rica and elsewhere in the Neotropics there exists the very real possibility that a sawfly pest could accidentally be introduced in the future. Freed from their natural enemies, they could have devastating consequences, as has occurred when European sawfly species were inadvertently introduced into North America, Australasia, and most recently Chile (Baker, 1972; Neumann & Minko, 1981).

Hymenopterans may also have indirect effects on agriculture by:

- killing natural enemies of crop pests;
- protecting homopteran pests;
- stinging and traumatizing farm labourers;
- transmitting plant pathogens.

Those that kill natural enemies include various hyperparasitoids (parasitoids of parasitoids, see Chapter 2.5), parasitoids of useful predators (such as spiders, lacewings, ladybird beetles, syrphid fly larvae, etc.), and predators of predators (e.g. pompilids and sphecids that prey on spiders). The effects of hyperparasitoids on the performance of biological control agents have been documented (Bennett, 1981; Luck *et al.*, 1981), but the impact of hymenopterans that kill beneficial predators has received less attention. Syrphid predators of Homoptera in particular are heavily parasitized by encyrtids and pteromalids in Central America (pers. obs.), and it is possible that these parasitoids limit the effectiveness of these predators.

The other three categories of indirect pests include primarily the social Hymenoptera: ants, polistine vespids and stingless bees. Ants are especially well known for their mutualistic associations with Homoptera, whereby the ant receives honeydew, and in return, protects the aphid, mealybug, scale-insect, or other homopteran from predators and parasitoids (Way, 1963; Buckley, 1987). Although ants may occasionally devour some of their homopteran associates, the population of the homopterans usually increases significantly under the protection of ants. The same species of ant that protects homopterous pests may also sting farm labourers. For example, *Wasmannia auropunctata* protects Homoptera on a range of crops (Ulloa-Chacon & Cherix, 1990) and, because of its potent sting, it occasionally interferes with the harvesting of bananas and other crops (J. Longino, pers. comm.).

Another group of ants that protect Homoptera and sting people are the fire ants (*Solenopsis*, subgenus *Solenopsis*), so called because of the potent alkaloid-containing venom that they inject with their sting. Most of the species are confined to South America, but *S. invicta* was accidentally introduced into the United States where it is a serious pest (Lofgren & Vander Meer, 1986). The only fire ant present in Central America is *S. geminata*, which may harvest seeds, nip off young shoots, and protect Homoptera. In Costa Rica this fire ant, along with *Pheidole* species, protects a mealybug that is the vector for a serious virus disease of pineapple. In Nicaragua its nests interfere with farm machinery in cotton fields and its painful stings are a nuisance to farm workers (J.-M. Maes, pers. comm.).

Although polistine wasps and stingless bees also protect homopterans, their economic impact is probably less than that of ants. Polistines can of course interfere with farm labour, but because of their importance as predators, wanton destruction of their nests should be discouraged. Stingless bees occasionally serve as vectors of plant pathogens, such as bacterial wilt (*Pseudomonas solanacearum*) of bananas and plantain (Buddenhagen & Elsasser, 1962).

Pests of medical importance

Although hymenopterans are less important medically than certain other insects (flies, fleas, and lice), psychologically they may be more important. Most people's experience with wasps is limited to a sting from a vespid wasp, and the resulting attitude toward wasps is generally one of fear and loathing (Olkowski & Olkowski, 1976). Vespid wasps are a particular nuisance in the urban environment (Davis, 1978), and since the human populations of many countries are becoming increasingly urbanized, vespids will become increasingly important pests. It should be remembered of course that vespids are valuable predators.

Although a few non-aculeate hymenopterans are capable of inflicting a painful sting (e.g. some Ichneumonidae such as *Joppa*), aculeates are responsible for the majority of stinging incidents. With their modified ovipositor, the sting, many aculeates are capable of penetrating mammalian skin and injecting toxic and/or pain-causing venoms (Hermann & Blum, 1981; Piek, 1986; Schmidt, 1990). This facility is especially well-developed in the eusocial Hymenoptera and the fact that a sting from a small insect can cause a large mammal to retreat, demonstrates the effectiveness of this unique defensive weapon.

The venoms of vespid wasps, ants, and honey-bees are chemically similar in some respects. For example, all contain hyaluronidase and phospholipase, enzymes that destroy the adhesive substance between cells and digest the cell membrane, respectively. On the other hand there are notable chemical differences between the venoms of vespids, ants, and honey-bees. Pain-producing peptides (kinins) are known only from vespids, while ant venoms are distinctive in being dominated by proteinaceous constituents or alkaloids (Escoubas & Blum, 1990), and honey-bee venom contains a protein called melittin, which is a powerful cytolytic agent (Dotimas & Hider, 1987).

The venoms of some non-social wasps, such as Pompilidae and Mutillidae, may cause a great deal of pain, yet they result in very little tissue damage. On the other hand, the venoms from some eusocial hymenopterans may cause both pain and tissue damage such as swelling, haemorrhagic damage, gross metabolic dysfunction, etc. Such effects are often quantified by measuring the ability of the venom to lyse red blood cells, and according to these measurements, the most haemolytic venoms are usually those of vespids, certain ants (e.g. *Platythyrea*, *Pseudomyrmex*) and the honey-bee (Schmidt, 1990). In the Neotropics one of the most painful stings is that from the large 'bullet ant', *Paraponera clavata*. In Venezuela it is called the '24-hour ant' from the belief that its sting will kill a person within one day. Fortunately this is usually not the case, although the sting can result in the most excruciating 24 hours of one's life. Certain indigenous tribes in Brazil deliberately provoked stings from the bullet ant as part of their ceremonies initiating young boys into manhood (Bequaert, 1926). The venom from this notorious ant is proteinaceous, and has both haemolytic and neuromuscular effects (Hermann & Blum, 1966; Sevcik & Hernandez, 1990).

In a few cases a hymenopteran sting can cause death, not from tissue damage, but rather as a result of an extreme allergic reaction to the high molecular weight components of venom (Barr, 1971; Riches, 1988). In the mildest cases an allergic reaction produces a slight swelling and itching for a few days around the site of the sting. Further stings may produce no reaction in persons who become immune (desensitized). However, in other cases a person may become more and more allergic (hypersensitive) with successive stings until the stage is reached when the person is so allergic that a single sting could result in death through acute anaphylaxis. Symptoms of a severe allergic response may include nausea, difficulties in breathing, weakness, and confusion. Highly susceptible persons can undergo desensitization treatment, which consists of repeated subcutaneous injections of venom, administered under medical supervision. In the United States it is estimated that up to two percent of the population are hypersensitive to hymenopteran stings (not all life threatening), and treatment costs range from $300 to $1000 per person per year (Schmidt, 1986).

Some hymenopterans have lost their sting, notably dolichoderine and formicine ants, and stingless bees. Despite their inability to sting some members of these groups are nonetheless capable of inflicting caustic bites. For example, bites from stingless bees belonging to the genus *Oxytrigona* can result in long-lasting blisters.

'Killer bees'

Few insects have attracted as much public concern as the so-called 'killer bees'. Among researchers there is controversy surrounding the mechanism of the bee's rapid dispersion, the impact it is going to have, and what actions (if any) should be taken. Not surprisingly, this interesting and controversial bee has been the subject of many investigations (some of the more comprehensive works include: Michener, 1975; Needham *et al.*, 1988; Spivak *et al.*, 1991; Winston, 1992*a*, *b*).

The 'killer bee' is an African subspecies (*Apis mellifera scutellata*) of the honey-bee that, in appearance, looks just like an ordinary honey-bee. The African honey-bee was deliberately introduced into Brazil in 1956 in the hope of obtaining a honey-bee better adapted to tropical conditions (Kerr, 1957). Since then it has spread across the continent at the impressive rate of 300–500 kilometres per year. By 1979 it had reached Colombia, by 1981 Costa Rica, by 1986 Mexico and in 1990 it entered southern Texas. Nearly all honey-bees one now sees in Latin America are 'killer bees'. Whereas the European subspecies existed mostly in managed colonies the new subspecies occurs in both apiaries and as feral populations. The phenomenal success of the African honey-bee suggests that it is indeed better adapted to the tropics than the European subspecies that was previously present.

Although the African honey-bee is similar in appearance to the European honey-bee it behaves quite differently. Its nests are generally smaller and frequently more exposed; within the nest the worker population grows more rapidly. Reproductive swarms and abandonment of the nest site (absconding) occur more frequently. The best known characteristic of the African honey-bee is its more rapid arousal and the greater intensity with which it defends its nest; although it more readily stings, its venom is no different from that of the European honey-bee (Schumacher *et al.*, 1989). All these characteristics are probably adaptive in tropical climates where a long dormant season is absent, where the nectar resources are more unpredictable, and where predators are more prevalent. Although the African honey-bee tends to collect more honey under conditions of poor, dispersed honeyflows, it is more difficult to manage.

As the African honey-bee spread it built up large feral populations and very quickly 'Africanized' managed colonies; in other words the direction of gene flow has been almost exclusively from the feral Africanized bees to the managed European bees, and not vice versa. Thus, over a very short period of time the normally docile honey-bee has become more aggressive, increasing the number of human and livestock fatalities due to bee stings. The fatalities are usually due not to a hypersensitive reaction, but rather a systemic reaction to a massive exposure of venom—in most people this can occur when they receive more than ten consecutive stings (although there is considerable variation from person to person). Between 1975 and 1988 in Venezuela there were 2.1 human fatalities per million people per year, which compares to 0.08 fatalities per million people per year in the United States due to hypersensitive reactions to honey-bees (Winston, 1992*b*). However, the fatality rate due to Africanized honey-bees has usually diminished over time, as the local human population learned to be more cautious when approaching honey-bees.

Despite the medical importance of the Africanized honey-bee, the greatest economic impact of this insect is probably in the practice of bee keeping (see below). Moreover, the fact that it can survive in feral populations means that it will probably have an impact on native bee populations (Roubik *et al.*, 1986; Ramalho *et al.*, 1990; Spivak *et al.*, 1991).

Household pests

Certain hymenopterans, most notably ants, can be a nuisance in houses. In Costa Rica the most common species living in human habitations are the pantropical 'tramp' species: *Monomorium pharaonis*, *Paratrechina longicornis*, *Tapinoma melanocephala*, *Tetramorium bicarinatum*, and *Wasmannia auropunctata* (see Chapter 16). Only the last species is probably native to the region (J. Longino, pers. comm.). These omnipresent occupants of houses are annoying, but usually do not cause serious problems, except in hospitals where they may pose health hazards by contaminating sterile dressings and equipment (Edwards, 1980; Vander Meer *et al.*, 1990). The minute *T. melanocephala* is known as the 'hormiga loca' or 'crazy ant' because of its very rapid, erratic movement. Although this sugar-loving species is an irritation in houses, it is occasionally beneficial as a predator of eggs of houseflies (Pimental, 1955) and *Rhodnius prolixus* (Gómez-Nuñez, 1971), the latter being a vector of Chagas' disease.

In the north temperate regions carpenter ants (*Camponotus*, subgenus *Camponotus*) can cause serious structural damage to wood. Carpenter ants *sensu stricto* do not occur in the Neotropics, but members of a closely related subgenus (*Tanymyrmex*) are present, and some of these enter human habitations (J. Longino, pers. comm.). In Costa Rica an unidentified species belonging to this subgenus is common in buildings and, when they build their nest inside the casing of electronic equipment, serious damage may result. The nest consists of an accumulation of hair, pieces of dead cockroaches, paper clips and other debris that they scavenge from the building. In 1990 this species destroyed over 300 facsimile machines in Costa Rica, causing damage totalling hundreds of thousands of dollars. It is apparently the same species of ant that has been found periodically nesting in television sets, video cassette recorders and other domestic electronic apparatus. In the United States the imported fire ant (*Solenopsis invicta*) is known to damage outdoor electrical equipment (Vinson & MacKay, 1990).

Other Central American ants that are less common inside houses are nonetheless very common in the vicinity of buildings, and may be referred to as 'patio pests'. These include *Pheidole punctiticima* and *Solenopsis geminata* (the native fire ant) (J. Longino, pers. comm.). The latter can be particularly annoying because of its painful sting. It often protects honeydew-secreting homopterans on ornamental plants but it may also be a beneficial predator of other insect pests. Besides ants, patio pests also include various species of vespid wasps, which sometimes nest on or near buildings and traumatize the human occupants (see Chapter 15).

3.2 BENEFICIAL HYMENOPTERA

In order to conserve tropical rainforests we need to convince the public and the decision makers of the economic benefits in doing so. Most of the attention thus far has been directed toward the potential pharmaceutical value of tropical plants. The myriads of hymenopteran species inhabiting these forests provide a substantial additional reason for preserving biodiversity. Like chemical compounds in plants, the complex venoms of Hymenoptera represent a vast and unexplored source of potentially valuable natural products. Tropical hymenopterans could also provide extremely valuable biological control agents and pollinators.

Biological control

The control of pest insects in crops is big business, worth five billion dollars in 1985, with most of the money going to chemical control (Jutsum, 1988). Insecticides, however, pose risks to human health, environmental contamination, secondary outbreaks of previously innocuous insects, and the rapid development of insecticide-resistance in the pest. Since most Latin American countries have to import pesticides, their use contributes to the national debt, and when produce for export is rejected because of pesticide residues these countries incur yet another economic burden.

One of the most promising alternatives to pesticides is BIOLOGICAL CONTROL—the use of 'natural enemies' (predators, parasitoids and microbial pathogens) to lower the population of the pest (DeBach, 1964, 1974; Huffaker & Messenger, 1976). Since the order Hymenoptera includes the majority of parasitoid species as well as many important predators, it is arguably the single most important group of biological control agents for use against pest insects. The practice of biological control entails the manipulation of natural enemies in order to enhance their effectiveness in controlling the population of the pest. There are three ways of accomplishing this: importation of natural enemies ('classical biological control'), mass-rearing for periodic liberations in crops and conservation of natural enemies through environmental modification.

Hymenopterous parasitoids

The majority of successes in classical biological control have been with importations of Aphelinidae, Encyrtidae, and Braconidae (Clausen, 1978; Luck, 1981; Cock, 1985; Greathead, 1986; Altieri *et al.*, 1989). Aphelinidae and Encyrtidae have been used principally for the control of Homoptera—aphelinids for hard scale (Diaspididae) and whitefly (Aleyrodidae), and encyrtids for mealybugs (Pseudococcidae) and soft scale (Coccidae). Braconids have been most successfully employed against lepidopterous pests, but also against certain dipterous pests such as fruit flies (Tephritidae). Other parasitoids that have been used successfully in classical biological control include Eulophidae, Ichneumonidae, Pteromalidae, Mymaridae, Scelionidae, and Platygastridae.

The first successful classical biological control project in Central America was against the citrus blackfly, *Aleurocanthus woglumi* (Aleyrodidae), a native of Asia (DeBach, 1974; Quezada, 1986). Successful biological control was achieved in 1932–1933 from the offspring of 42 female *Eretmocerus serius* (Aphelinidae) that were sent from Asia to Cuba, and subsequently distributed in Panama and Costa Rica. The citrus blackfly invaded El Salvador around 1965 but was brought under control by the introduction of *Encarsia opulenta* (Aphelinidae) in 1971 (Quezada, 1974). The latter parasitoid apparently arrived unintentionally in Nicaragua, where it is now an important factor in reducing citrus blackfly populations (Swezey & Cano Vasquez, 1991). Such 'fortuitous biological control' appears to have occurred in the case of several scale-insect pests in Central America, whereby exotic aphelinid and encyrtid species either dispersed into the region from neighbouring countries or arrived unnoticed with their hosts (see Chapters 11.2 and 11.5). Alternatively, some of these may have been deliberately introduced but the introduction was never reported.

The Mediterranean fruit fly, *Ceratitis capitata* (Tephritidae), was first reported from Costa Rica in 1955 and shortly thereafter eight species of parasitoids were introduced from Hawaii, including five opiine braconids, one eucoiline cynipoid, one chalcidid and one eulophid. Only two braconids, *Diachasmimorpha longicaudata* and *Biosteres arisanus*, and the eulophid, *Aceratoneuromyia indica*, became established (Wharton *et al.*, 1981; Jirón & Mexzón, 1989). Because these parasitoids were not very effective in controlling Mediterranean fruit fly populations, further introductions were made from Hawaii and West Africa in 1981–1982 (Steck *et al.*, 1986; Gilstrap & Hart, 1987), but establishment has not been verified. Although biological control of tephritids in general has been fraught with difficulties (Wharton, 1989*a*), the increasing importance of export fruits (especially citrus and mango) in the economies of Latin American countries will require more work with tephritid parasitoids.

Since the overall success rate of classical biological control is about 16 percent (Hall *et al.*, 1980), there has been considerable discussion about which host attributes and which parasitoid attributes are most likely to yield success (Myers *et al.*, 1989; Stiling, 1990; Gross, 1991). In general, there have been more successes against pests that feed in relatively exposed locations than against concealed feeders. There has also been a greater success rate of parasitoids from temperate climates establishing in tropical regions than of tropical parasitoids establishing in temperate climates. Thus there seems to be a biological justification for reversing the historical flow of parasitoids (and other resources) from developing countries to industrialized countries (Altieri, 1991*b*).

The second type of biological control, mass-rearing for periodic liberations (Ridgway & Vinson, 1977), is currently being used in Costa Rica against the sugarcane borer *Diatraea saccharalis* (Lepidoptera: Pyralidae), using *Cotesia flavipes* (Braconidae). The egg parasitoid, *Trichogramma pretiosum* (Trichogrammatidae), is being reared in Nicaragua and elsewhere for release against a variety of lepidopterous pests. The bethylid, *Cephalonomia stephanoderis*, is being reared in various countries in Latin America for control of the coffee berry borer, *Hypothenemus hampei* (Scolytidae) (Murphy & Moore, 1990).

The third type of biological control, conservation of existing parasitoids (see Table 3.1), has received less attention despite the fact that it offers many unexplored possibilities (Hagen, 1986; Powell, 1986). The effectiveness of existing parasitoids may be enhanced by eliminating unfavourable factors, such as insecticides and dust. In Costa Rica, at present, natural biological control appears to be occurring in banana (Stephens, 1984), coffee (Hanson, 1992) and probably several other crops. This situation is likely to endure so long as insecticide applications are kept to a minimum. Dust presumably impairs parasitoid searching behaviour by masking semiochemicals used in host finding (see Chapter 2.5). Indirect evidence for this is provided by the fact that citrus mealybug (*Planococcus citri*, Pseudococcidae) and the coffee leaf miner (*Leucoptera coffeella*, Lyonetiidae) populations rose dramatically following the last eruption of Volcán Irazú, which coated the surrounding countryside with fine ash (Wille & Fuentes, 1975). The effectiveness of existing parasitoids can sometimes be enhanced by providing nectar sources for adult parasitoids (Mexzón & Chinchilla, 1991; Delvare & Genty, 1992) and intercropping (Altieri *et al.*, 1978; Altieri, 1991*a*).

The use of existing parasitoids against stored grain pests (Table 3.2) and insect pests of medical importance (Table 3.3) requires slightly different tactics.

Crop	Pest insect	Parasitoid
Banana	*Antichloris viridis* (Arctiidae) (Harrison, 1963)	*Brachymeria mnestor* (CHA) *Carinodes* sp. (ICH) *Elachertus ceramidiae* (EUL) *Macrocentrus* sp. (BRA) *Meteorus laphygmae* (BRA) Microgastrinae (BRA) *Miotropus* sp. (EUL) *Stenomesius ceramidiae* (EUL) *Trichogramma* sp. (TRI)
	Oiketicus kirbyi (Psychidae) (Stephens, 1962)	*Conura* sp. (CHA) *Digonogastra diversus* (BRA) *Parasierola* sp. (BET) Various ichneumonids
Beans	*Acanthoscelides obtectus* (Bruchidae)	*Conura fusiformis* (CHA) *Stenocorse bruchivorus* (BRA)
	Apion godmani (Curculionidae)	*Triaspis azteca* (BRA)
	Empoasca kraemeri (Cicadellidae)	*Anagrus empoascae* (MYM)
Cassava	*Erinnyis ello* (Sphingidae)	*Cotesia congregata* (BRA) *Cryptophion* sp. (ICH) *Telenomus dilophonotae* (SCE) *Trichogramma* sp. (TRI)
Citrus	*Aleurocanthus woglumi* (Aleyrodidae)	*Encarsia opulenta* (APH) *Eretmocerus serius* (APH)
	Conuraspis beckii (Diaspididae)	*Aphytis lepidosaphes* (APH)
Coffee	*Leucoptera coffeella* (Lyonetiidae)	*Cirrospilus* sp. (EUL) *Closterocerus* sp. (EUL) *Stiropius* sp. (BRA) *Zagramosoma americanum* (EUL)
	Planococcus citri (Pseudococcidae)	*Leptomastidea abnormis* (ENC) *Leptomastix dactylopii* (ENC)
Cole crops (Crucifers)	*Plutella xylostella* (Plutellidae) (Cordero & Cave, 1992)	*Diadegma insulare* (ICH) *Pimpla punicipes* (ICH)
Cotton	*Anthonomus grandis* (Curculionidae)	*Catolaccus grandis* (PTE) *Urosigalphus schwarzi* (BRA)
Curcurbits (melon, etc.)	*Diaphania hyalineata* (Pyralidae)	*Cardiochiles diaphaniae* (BRA) *Conura acragae* (CHA) *Glyptapanteles* (BRA)
Macadamia	*Antiteuchus costaricensis* (Pentatomidae)	*Phanuropsis semiflaviventris* (SCE) *Telenomus* sp. (SCE) *Trissolcus radix* (SCE)
	Ecdytolophatorticornis (Tortricidae)	*Ascogaster* sp. (BRA) Miscogastrinae (BRA) *Pristomerus* (ICH)

Table 3.1. Continued opposite.

Crop	Pest insect	Parasitoid
Maize	*Dalbulus maidis* (Cicadellidae)	*Gonatopus bartletti* (DRY)
	Spodoptera frugiperda (Noctuidae) (Marenco & Saunders, 1993)	*Aleoides laphygmae* (BRA)
		Campoletis sonorensis (ICH)
		Chelonus insularis (BRA)
		Cotesia marginiventris (BRA)
		Eiphosoma vitticolae (ICH)
		Glyptapanteles militaris (BRA)
		Homolobus truncator (BRA)
		Ophion flavidus (ICH)
		Pristomerus spinator (ICH)
		Temelucha sp. (ICH)
		Trichogramma sp. (TRI)
Mango	*Anastrepha obliqua* (Tephritidae) (Jirón & Mexzon, 1989)	*Aceratoneuromyia indica* (EUL)
		Doryctobracon areolatus (BRA)
		Trichopria sp. (DIA)
Peppers	*Anthonomus eugenii* (Curculionidae)	*Urosigalphus mexicanus* (BRA)
Potato	*Myzus persicae* (Aphididae)	*Diaretiella* sp. (BRA)
	Phthorimaea operculella and *Scobipalposis solanivora* (Gelechiidae) (Marsh, 1979)	*Agathis* sp. (BRA)
		Chelonus kellieae (BRA)
		Orgilus jennieae (BRA)
		Pristomerus spinator (ICH)
Rice	*Rupella albinella* (Pyralidae)	*Telenomus* sp. (SCE)
	Mocis latipes (Noctuidae) (Cave, 1992)	*Microplitis* sp. (BRA)
		Rogas nigristemmaticum (BRA)
		Scambus albitibia (ICH)
		Tricholabus lepidus (ICH)
	Tagosodes oryzicola (Delphacidae)	*Anagrus* sp. (MYM)
		Haplogonatopus hernandezae (DRY)
Sugar-cane	*Diatraea saccharallis* (Pyralidae)	*Alabagrus stigma* (BRA)
		Apanteles diatraeae (BRA)
		Digonogastra grenadensis (BRA)
		Trichogramma sp. (TRI)
Tomato	*Pseudoplusia* spp (Noctuidae)	*Copidosoma floridanum* (ENC)
	Spodoptera latifascia (Noctuidae)	*Glyptapanteles* sp. (BRA)
	S. sunia	*Euplectrus* sp. (EUL)
		Aleiodes vaughani (BRA)
		Chelonus sp. (BRA)
Various	*Liriomyza huibroidensis* (Agromyzidae)	*Chrysocharis* sp. (EUL)
		Diglyphus isaea (EUL)
		Oenonogaster sp. (BRA)
		Opius dissitus (BRA)

Table 3.1. Parasitoids associated with agricultural pests in Central America. The host records come from personal observations, Castro (1993), King and Saunders (1984), and from sources that are listed under the pest species. Abbreviations of parasitoid families are: BET = Bethylidae, BRA = Braconidae, CHA = Chalcididae, DIA = Diapriidae, DRY = Dryinidae, ENC = Encyrtidae, EUL = Eulophidae, ICH = Ichneumonidae, MYM = Mymaridae, PTE = Pteromalidae, SCE = Scelionidae, TRI = Trichogrammatidae.

Pest insect	Parasitoid
Ephestia spp. (Pyralidae)	*Bracon hebetor* (Braconidae) *Hypsicera femoralis* (Ichneumonidae) *Venturia canescens* (Ichneumonidae)
Sitophilus spp. (Curculionidae)	*Anisopteromalus calandrae* (Pteromalidae) *Holepyris sylvanidis* (Bethylidae) *Lariophagus distinguendus* (Pteromalidae) *Thecolax elegans* (Pteromalidae)
various Anobiidae Cucujidae and Tenebrionidae	*Cephalonomia gallicola* (Bethylidae) *Holepyris sylvanidis* (Bethylidae) *Plastanoxus westwoodi* (Bethylidae)

Table 3.2. Parasitoids associated with stored grain pests in Central America (from Haines, 1984, and pers. obs.).

Pest arthropod	Parasitoid
Latrodectus (Theridiidae)	*Eurytoma arachnovora* (Eurytomidae)
Musca domestica (Muscidae)	*Muscidifurax* (Pteromalidae) *Spalangia* (Pteromalidae)
Periplaneta americana (Blattidae)	*Aprostocetus hagenowi* (Eulophidae) *Evania appendigaster* (Evaniidae)
Triatoma (Reduviidae)	*Telenomus fariai* (Scelionidae)

Table 3.3. Parasitoids associated with medically important pests in Central America (LeBeck, 1991; Vargas & Fallas, 1974; Zeledón, 1957).

Hymenopterous predators

Biological control of pest insects can also be achieved by using predators such as ants, vespids, and sphecids. Because predators are usually more polyphagous than parasitoids it is generally not advisable to import exotic predators. Indeed the consequences of introducing a foreign ant species may be devastating as illustrated by the introduction of *Paratrechina fulva* from Brazil to Colombia. This ant was brought to Colombia to control leaf-cutting ants and snakes, but it now protects homopteran pests, kills chickens and displaces beneficial ants from coffee plantations! Its sheer numbers sometimes cause farms to be abandoned (Zenner-Polania, 1990). Only in the case of more monophagous predators should importations be considered. For example, the sphecid *Larra americana* (which is actually a parasitoid), was introduced from South America to Puerto Rico to control mole crickets in sugar cane fields, where it has been quite successful (*L. bicolor*: Wolcott, 1941; Castner, 1984).

In utilizing ants in biological control two questions must be addressed: which is the most desirable species of local ant (Table 3.4), and how can the population of this species be enhanced? Different ants may be required for different crop habitats. Thus *Solenopsis geminata* would probably be better in annual crops (Perfecto, 1991*b*), whereas *Azteca* species may be more appropriate in orchard crops. Examples of how populations of desirable ant species can be enhanced are provided by the Chinese, who have been moving ant nests from surrounding forests into citrus orchards since at least AD 304 (Groff & Howard, 1924; Liu, 1939; Huang & Yang, 1987). Since at least AD 1600 the Chinese have connected orange trees with bamboo bridges to facilitate the movement of ants between trees (Huang & Yang, 1987). In Cuba pieces of banana stalks are inoculated with colonies of the ants, *Pheidole megacephala*, which are fed with table scraps until all life stages are present in the colonies. These are then distributed in sweet potato fields, resulting in dramatic reduction of the sweet potato weevil (*Cylas formicarius*) populations (Levins, 1989). Other possible means of enhancing the population of a desirable ant species include suppressing competing ant species, maintaining strips of bare soil to favour open-habitat species, providing a diversity of plants to favour more closed-habitat species, and providing natural or artificial nectar sources in order to control spatial and temporal foraging patterns. Further suggestions and considerations on the manipulation of ants are provided by Lee *et al.* (1990), and Way and Khoo (1992).

There has also been some discussion of managing vespid wasps for purposes of biological control (Gillaspy, 1979, 1982, 1986). The most important predators of the cassava hornworm, *Erinnyis ello* (Sphingidae), in Colombia are said to be *Polistes canadensis* and *P. erythrocephalis* (Bellotti & Schoonhoven, 1978). *Polistes peruviana* and *P. versicolor* are thought to be important predators of the coffee leaf miner (*Leucoptera coffeella*, Lyonetiidae) in Peru (Enriquez *et al.*, 1976). In Honduras some farmers move polistine nests (at night and with much care) from the surrounding mountains into agricultural areas in order to control various pests (R. Cave, pers. comm.). There have been several attempts to manage *Polistes* wasps in order to control caterpillars on tobacco, both in Colombia (Gallego, 1950) and in the United States (Rabb, 1960). Nonetheless, observations in Colombia suggest that *Polistes* populations may be vulnerable to parasitoids and/or pathogens, and initially high populations tend to crash after a few years (W. Eberhard, pers. comm.).

Unlike ants and vespids, sphecid wasps have received little attention with respect to their potential use in biological control, despite the availability of literature on how to construct artificial nest sites (Krombein, 1967). As a family the sphecids prey on a wide variety of insects but individual species tend to be quite specialized predators, which suggests that there are many possibilities for using them in biological control.

Pollination

A very significant proportion of the human diet is derived directly or indirectly from plants that are pollinated by bees (Free, 1970). No figures are available for Central America but in the United States the total annual value of crops grown with the aid of bee pollination is estimated to be 19 billion dollars (Levin, 1983). The best known and most widespread pollinator of crop plants is the honey-bee, whose value as a pollinator far surpasses its value as a honey producer. Honey-bee hives are often moved into crops for the purpose of increasing yields, especially in large fields where flowering occurs for only a short period (Jay, 1986). Although this practice of moving honey-bee hives into crops is less common in neotropical coun-

Aculeate species	Pest attacked	Reference
FORMICIDAE		
Ectatomma ruidum	*Spodoptera frugiperda* (Lepidoptera)	Perfecto, 1990, 1991*a*
Ectatomma tuberculatum	*Anthonomus grandis* (Curculionidae)	Cook, 1904
Monomorium ebeninum	*Heteropsylla cubana* on *Leucaena*	D. Hollis (pers. comm.)
Pheidole radowszkoskii	*Spodoptera frugiperda* (Lepidoptera)	Perfecto, 1990, 1991*a*
Pheidole sp.	Eggs of *Diabrotica* (Chrysomelidae)	Risch, 1981
Solenopsis geminata	Eggs and small larvae of *Castnia licus* (Castniidae) in sugarcane	Esquival, 1983
	Eggs of *Diabrotica* (Chrysomelidae)	Risch, 1981
	Various insects in maize and squash	Risch & Caroll, 1982
Wasmannia auropunctata	*Heteropsylla cubana* on *Leucaena*	D. Hollis (pers. comm.)
VESPIDAE		
Agelaia areata	Larvae of *Estigmene acrea, S. frugiperda* (Noctuidae)	King & Saunders, 1984
Polistes spp.	Larvae of *Pseudoplusia includens, Spodoptera exigua S. frugiperda* (Noctuidae); *Diaphania nitidalis* (Pyralidae); *Erinnyis ello, Manduca sexta* (Sphingidae)	King & Saunders, 1984
Polybia spp.	Larvae of *Estigmene acrea, S. frugiperda* (Noctuidae); *Erinnyis ello* (Sphingidae)	King & Saunders, 1984

Table 3.4. Records of social aculeates preying on pest insects in Central America.

tries than in North America, many important neotropical crops are pollinated by honey-bees and presumably yields could be increased by their presence. Examples include avocado, citrus, coconut, coffee, cotton, cucumber, macadamia, mango, melon, papaya, squash and strawberry (Crane, 1990). Virtually all honey-bees in the Neotropics have recently become Africanized (transformed into 'killer bees'; see above), thus complicating the management of honey-bees for controlled pollination. They can still be used to pollinate crops but are probably best suited to situations where the hives do not have to be moved frequently, and where they can be relatively isolated from people and livestock. Feral colonies are of course probably providing a certain amount of unmanaged pollination.

Despite the preponderance of attention devoted to honey-bees, their effectiveness as pollinators has probably been overrated (Westerkamp, 1991), and this has tended to mask the importance of native bees (O'Toole, 1993). Among neotropical bees, the highly eusocial stingless bees are probably one of the more important groups of native pollinators. They are known to be important pollinators of many native tree species (Ramalho *et al.*, 1989, 1990) as well as crops such as avocado, cassava, chayote, coconut, papaya and several others (Wille *et al.*, 1983). It is quite likely that if stingless bees are to be used as pollinators of tree crops, special nest sites will have to be provided (Johnson & Hubbell, 1984). Among solitary bees, the squash bees (*Peponapis* and *Xenoglossa*) are specialized pollinators of squash, pumpkins, gourds and other Cucurbitaceae (Michelbacker *et al.*, 1971). In Costa Rica several native bees are important pollinators of guava (Hedström, 1988). In general, however, the role of native bees as pollinators of neotropical crops has received very little attention. In the north temperate zone several species have been managed for crop pollination but in the Neotropics this has been less common, one of the few examples being *Xylocopa* in the Caribbean (Corbet & Willmer, 1980).

Commercial products

Indigenous peoples of the Neotropics have utilized several hymenopterans directly as food, especially some of the eusocial species where a concentrated source of protein can be harvested from the nest. Among the Yukpa Indians of Venezuela and Colombia more hymenopterans are consumed than any other order of insects; hymenopteran dishes include the abdomens from the queens of leaf-cutting ants, the larvae of certain vespids, and the larvae of stingless bees (Ruddler, 1973). Indigenous cultures also used hymenopterans in other ways. The venom of the ant, *Paraponera clavata*, was used in preparing curare (Martinez & Vela-Diaz, 1953), and army ant mandibles were used as sutures for wounds, by letting a live ant bite with one mandible on each side of the incision and then removing the body (Gudger, 1925; Gotwald, 1982).

Since neither honey-bees nor sugarcane were present in the New World until they were introduced by Europeans, honey obtained from native hymenopterans represented virtually the only source of sugar. Native honey-producing hymenopterans include stingless bees and certain vespids (*Brachygastra* spp. and *Polybia scutellaris*). The honey of stingless bees is generally more acidic and contains more water than that of the honey-bee; in some species the honey is toxic and therefore indigenous cultures acquired detailed knowledge of the local species. The Mayan culture, which influenced the region extending from southern Mexico to northwestern Costa Rica, developed beekeeping with the stingless bee, *Melipona beecheii*. The bees were kept in hives, which consisted of a horizontal, hollow log, closed at each end with a disc of wood or soft stone (Weaver & Weaver, 1981; Kent, 1984; Crane & Graham, 1985). Inhabitants of the Azuero Peninsula of northern Panama employ dried gourds for culturing six species of stingless bees (Bennett, 1965). In various neotropical countries, including Costa Rica and especially Brazil, new approaches to beekeeping with stingless bees are being utilized in order to facilitate the harvest and minimize destruction of the nest (Nogueira-Neto, 1970*b*). Besides providing a source of sugar in the diet the honey of stingless bees has been used as a medicinal for various ailments (Schwarz, 1948).

The single most commercially important hymenopteran in Central America, and worldwide, is the honey-bee. Honey-bees belong to the genus *Apis*, which comprises three species in tropical Asia, and *A. mellifera*, which is native to Africa and Europe. It is the latter species that is commonly managed in apiaries, and more specifically, it is the Italian race that has been transported throughout the world. The total recorded world honey production is nearly a million tonnes a year, the largest producers being China, the United States, and Mexico; the latter is the largest exporter of honey. Among the seven Central American countries, El Salvador and Guatemala are the largest producers (Crane, 1990).

Honey consists mostly of fructose (about 40 percent), glucose (35 percent), and water (15 percent). The low water content allows honey to be stored for a long time without spoilage. Besides providing a source of sugar honey has many other uses, some of which are probably under utilized—for example as a dressing agent for wounds (Efem, 1988). There is an enormous literature available on the history, uses, extraction, and constituents of honey (see Crane, 1975). Beeswax is the second most important product obtained from honey-bees and is used mostly in ointments, skin creams, lipstick, candles, etc. (Coggshall & Morse, 1984; Hepburn, 1986). Lesser known commercial products obtained from honey-bees include pollen (Stanley & Linskens, 1974), propolis (Walker & Crane, 1987; Asis, 1988), royal jelly (Smith, 1959) and bee venom (Dotimus & Hider, 1987). Pollen and ROYAL JELLY, which is a secretion from the hypopharyngeal gland of the worker bee given to the developing queen, are used as dietary supplements and in cosmetics. PROPOLIS, the resin collected from plants and employed by the bees in nest construction, is used in healing creams, as a treatment for digestive disorders and for 'curing' raw distilled alcohol (W. Eberhard, pers. comm.). Honey-bee venom is currently being used as a tool in molecular biology and pharmaceutical research, and is also being tested for its anti-arthritic properties. Hymenoptera venom in general is a rich and relatively unexplored source of natural products (Schmidt, 1986); the possible utilization of parasitoid venoms, some of which have antimicrobial properties (Chapter 2.4), is almost totally unexplored.

Apiculture ('bee-keeping') in tropical countries requires certain special considerations (Espina Perez & Ordetx Ros, 1983; Crane, 1990). Nectar and pollen sources are often assumed to be readily available throughout the year but this is not always true, and supplemental food may be necessary. As in other parts of the world a number of honey-bee pests afflict the apiculturist in Central America: American foulbrood, European foulbrood, sac brood, chalk brood, nosema protozoans, mites, and wax moth (Crane, 1990; Morse & Nowogrodzki, 1990); bee parasitic mites are of particular concern (Needham *et al.*, 1988). Presently, the most important factor affecting apiculture throughout the Neotropics is the arrival of the Africanized honey-bee (see above). Although the Africanized honey-bee tends to collect more honey under conditions of poor, dispersed honeyflows, its greater aggressiveness and propensity to abscond make it more difficult to manage. However, by modifying apicultural practices the Africanized honey-bee can be utilized as a honey-producer (Gomez Rodriguez, 1986; Winston, 1992*b*).

4 The structure of Hymenoptera

Ian D. Gauld and Paul E. Hanson

INTRODUCTION

A great deal has been written about the comparative morphology of the Hymenoptera (see below) and the account included here is only intended to be a brief introduction to the subject, serving to help workers recognize homologies or determine the name of an unfamiliar structure. This synopsis has been condensed largely from Richards' (1956) excellent introduction to Hymenopteran morphology, with a few amendments that have resulted from recent research. The following conventions have been applied. When a new term is introduced and explained it is printed in upper case. Acceptable common synonyms are preceded by an equality sign (=) and follow in parentheses. Homonymous terms that are used for different structures (and that are best avoided) are enclosed in quotation marks (').

For detailed study the general morphological and anatomical texts by Snodgrass (1935), Matsuda (1965, 1970, 1976) and Scudder (1961) are recommended. Certain particularly specialized or controversial aspects of hymenopteran morphology are discussed in the following papers: Brothers, 1975, 1976 (comparative morphology of aculeates); Compere and Rosen, 1970 (prescutal morphology); Daly, 1964 (skeleto-muscular morphogenesis); Danforth & Michener, 1988 (wing folding); D'Rozario, 1942 (development of the reproductive system); Duporte & Bigelow, 1953 (clypeus); Gibson, 1985 (comparative morphology and anatomy of pro- and mesothorax); K.G.A. Hamilton, 1972*a* & *b* (wing venation); Michener, 1944 (comparative morphology of bees); Michener and Brooks, 1984 (glossae of bees); Oeser, 1961 (comparative morphology of ovipositor); Reid, 1941 (thoracic morphology of apterous and brachypterous species); Robertson, 1968 (venom apparatus); Scherbakov, 1980, 1981 and Gibson, 1993 (morphology of pterothoracic pleura); E.L. Smith, 1970 (genitalia); Snodgrass, 1941 (comparative morphology of male genitalia); Togashi, 1970 (internal reproductive organs of Symphyta); Tonapi, 1958 (comparative study of spiracles); Whitfield *et al.*, 1989 (metapostnotum in non-aculeates); Wootton, 1978 (wing venation).

4.1 GENERAL ADULT MORPHOLOGY

THE HEAD AND ITS APPENDAGES

The head is characteristically hypognathous (= orthognathous *sensu* Schröder, 1928) with a ventrally directed mouth (Figs 4.01–4.03). Three OCELLI are usually present on the VERTEX (top of the head) and laterally are a pair of well-developed compound EYES. Anteriorly, and generally between the compound eyes, is a pair of sockets, the TORULI, from which project the multisegmented ANTENNAE. The area above the antennal sockets and below the median ocellus is called the FRONS, and the lateral margins of this, adjacent to the eyes, are termed the FRONTAL ORBITS. Below the toruli are a pair of small pits, the ANTERIOR TENTORIAL PITS, which mark the point of invagination of the H-shaped internal skeleton of the head, the TENTORIUM. An EPISTOMAL SUTURE (= clypeal suture) is usually visible as a groove running from near one mandibular base through the anterior tentorial pits to meet its counterpart centrally. The area between this suture and the labrum (see below) or mouth is the CLYPEUS; in ichneumonoids the part between the clypeal margin and the toruli (which includes the clypeus and part of the frons) is called the LOWER FACE; the FACIAL ORBITS are the parts of the lower face adjacent to the eyes. In some braconids the clypeus is semicircularly emarginate and the labrum is strongly concave so they appear to have a round mouth; this concavity is termed the HYPOCLYPEAL DEPRESSION (= cyclostome condition). The shortest distance between the base of the mandible and the ventral margin of the eye is called the MALAR SPACE (which is part of the gena), and in some groups this is marked by an

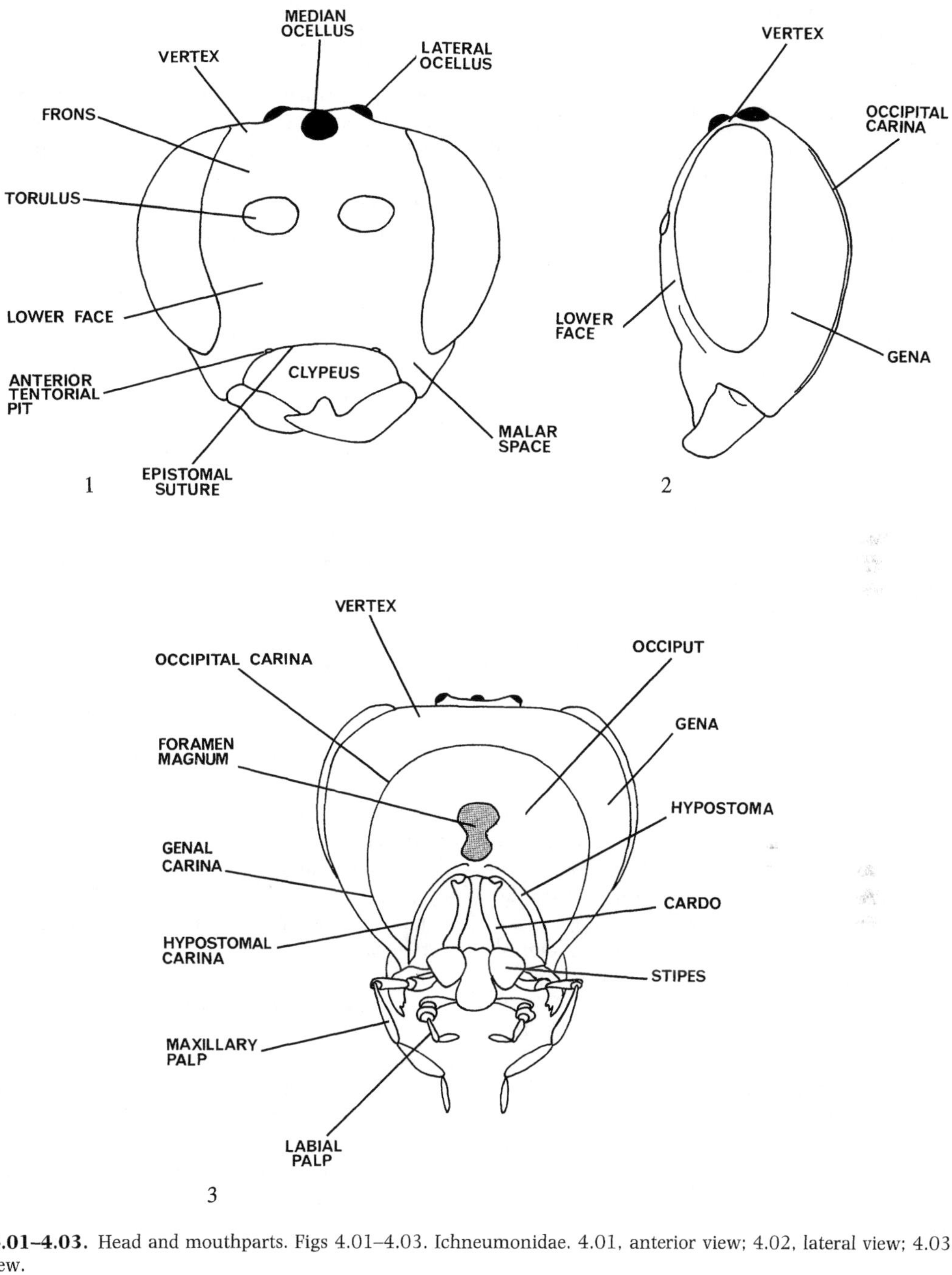

Figs 4.01–4.03. Head and mouthparts. Figs 4.01–4.03. Ichneumonidae. 4.01, anterior view; 4.02, lateral view; 4.03, posterior view.

impressed line, the MALAR SULCUS (= subocular or genal sulcus). In some formicids a pair of cuticular outgrowths, the FRONTAL LOBES, project over the toruli, partially or entirely shielding the antennal articulations. The frontal lobes may extend posteriorly onto the frons as a pair of ridges, the FRONTAL CARINAE.

In back view the head (Fig. 4.03) has large and approximately circular hole, the FORAMEN MAGNUM, in its centre. Ventrally this foramen may be contiguous with the ORAL FOSSA (= proboscidial fossa; buccal cavity), the cavity containing the mouthparts, or the two may be separated by a bridge, formed in several different ways. In some taxa it is a fusion of parts of the hypostoma (see below) and is bordered laterally by the hypostomal carinae (in which case it is called the HYPOSTOMAL BRIDGE), or it can be formed by median fusion of the postgenae (POSTGENAL BRIDGE), or genae (GENAL BRIDGE), or by a combination of these and perhaps other parts. Close to the foramen magnum are the posterior points of invagination of the tentorium, the POSTERIOR TENTORIAL PITS, and the groove extending around the foramen magnum and linking the posterior tentorial pits is the POSTOCCIPITAL SUTURE. On either side of the oral fossa is a narrow sclerotized part of the HYPOSTOMA that is delineated laterally by a ridge, the HYPOSTOMAL CARINA (= oral carina), which may extend from the postoccipital suture to near the bases of the mandibles. In taxa with a genal or postgenal bridge the hypostomal carina meets its counterpart without reaching the postoccipital suture. The area lateral to the postoccipital suture and above the hypostomal carina is the POSTGENA and this is confluent with the OCCIPUT, the area above the foramen magnum. The postgena and occiput generally are delineated peripherally by a carina that is almost concentric with the margin of the foramen magnum. The ventral part of this carina is called the GENAL CARINA, and the dorsal part is known as the OCCIPITAL CARINA. The genal carina generally meets the hypostomal carina at or near the base of the mandible, or meets its counterpart centrally. The area between the genal carina and the compound eye is called the GENA and the part adjacent to the eye is known as the GENAL ORBIT (= posterior orbit; temporal orbit). The upper part of the gena is frequently called the TEMPLE.

Various specializations of the head and its appendages occur. In many bethylids and ants the head is more or less PROGNATHOUS with the mouth directed forwards. In these groups the eyes are frequently reduced or even occasionally absent, and in worker ants the ocelli are generally lacking. In these prognathous taxa the extensive ventral surface of the head is formed by expansion of the genal bridge. In many other taxa the antennae are inserted very low down, almost adjacent to the epistomal suture so that the lower face is obliterated and in the orussids the antennae are inserted below the apparent clypeus. The genal and/or occipital carinae may be effaced so that the gena and postgena are confluent and the occiput is not separable from the vertex. In some Hymenoptera the frons has a pair of concave areas excavated to receive the basal part of the antennae; each excavation is called an ANTENNAL SCROBIS (= antennal scrobe). In many chalcids these are united into a single concavity, the SCROBES, and the area lateral to the scrobes is often raised and called the PARASCROBAL AREA.

The form of the antenna is particularly variable. Typically the first visible segment, the SCAPE, is thicker and sometimes longer than any of the remaining segments. Articulating with the scape is a somewhat smaller, subconical segment, the PEDICEL. Distal to the pedicel is the FLAGELLUM, which is nearly always divided into a number of units (except in Argidae), but is actually a single segment. Thus in using the term FLAGELLAR SEGMENTS (= flagellomeres) it is understood that these are not true segments. These flagellar segments are usually more or less cylindrical and often structurally rather similar to each other, though the distal ones may be smaller than the proximal ones, or narrower and more elongate. The most conspicuous modifications to the antenna are found in smaller Hymenoptera. In ants, bethylids and many 'microhymenoptera', the scape is half or more as long as the pedicel and flagellum; the articulation between scape and pedicel is very mobile in comparison with the degree of flexibility existing between any two other segments, so the antenna appears to be elbowed (geniculate). In many chalcidoids one or more of the basal flagellar segments is very small and ring-like and usually called an ANELLUS (plural: anelli). In Chalcidoidea, Scelionidae, Platygastridae, and a few other groups, the apical flagellar segments may be enlarged, forming a club or CLAVA. In these groups the flagellar segments situated between the anelli and the clava are collectively called the FUNICLE (= 'funiculus' of many authors).

There is considerable variation in the number of antennal segments between hymenopteran taxa (ranging from 3 to 90 or more), but within taxa there is often a fixed number of antennal segments. For example, most male aculeates have 13 antennal segments while the females usually have 12. However, some siricoids and most ichneumonoids have large numbers of flagellar segments, and in such taxa it is not uncommon to find 10 to 20 percent variation in flagellar numbers within a given species. The form of the antenna also varies between taxa. In most groups it is thread-like, or filiform, but in a few groups (e.g. male diprionids, some male eucharitids) some of the flagellar segments possess ventral prolongations so that the flagellum is plumose. In some argids the apparently single flagellar segment is bifid, and shaped like a tuning fork.

In most hymenopterans the flagellar segments bear a number of sensilla and the distribution and form of these can be of great taxonomic value (e.g. Bin, 1981). Many larger apocritans have raised structures, such as ridges or flattened areas, on some of the flagellar segments, and these are collectively termed TYLOIDS. In Chalcidoidea long conspicuous placoid sensilla, the MULTIPOROUS PLATE SENSILLA (= 'rhinaria', longitudinal sensilla; sensilla linearia), are visible (Barlin & Vinson, 1981; Wibel *et al.*, 1984). Exocrine glands also occur on the antennae of some male chalcids and proctotrupoids (Dahms, 1984*a*; Bin & Vinson, 1986). Examples of other taxa in which antennal sensilla have been studied include Aulacidae (Schmidt & Kuhbandner, 1983), Braconidae (Norton & Vinson, 1974; Borden *et al.*, 1978*b*), Chalcididae (Simser & Coppel, 1991), Ichneumonidae (Borden *et al.*, 1978*a*; Rojas-Rousse & Palevody, 1983), Scelionidae (Bin, 1981) and the aculeates (Walther, 1979).

The mouthparts of most hymenopterans are of the simple mandibulate type. The triangular to semicircular LABRUM is generally strongly sclerotized and exposed, or partially to more or less completely concealed beneath the clypeus. In formicids and bethylids the labrum folds ventrally to cover the anterior portion of the buccal cavity and is not visible in dorsal aspect. The MANDIBLES are generally well-developed and bear one to many distal teeth. In mymarommatids, vanhorniids, alysiine braconids, and in a very few others (chalcidoids and ichneumonids) the mandibles are spatulate and outcurved. This is called the EXODONT condition. The MAXILLARY and LABIAL PALPS are generally conspicuous. Primitively the former are 6-segmented and latter 4-segmented, though this 6 : 4 PALP FORMULA is often reduced. The MAXILLA and LABIUM are generally rather unspecialized, though in many higher aculeates, a few braconids and desert-living species of other taxa, they are conspicuously lengthened. In vespoids and some sphecids the elongate GLOSSA is deeply bilobed distally, while in most bees the glossa is extended as a single narrow lobe (Fig. 4.04) that terminates in a plate, the FLABELLUM. In the 'short-tongued bees' the lengthening of the proboscis is due to an elongation of the prementum, and the glossa remains short.

THE THORAX

The thorax consists of three segments, the pro-, meso-, and metathorax (Figs 4.05–4.06), with each segment being divided into a number of plates or SCLERITES. In the Apocrita the metathorax is reduced and the first abdominal segment, the PROPODEUM, is intimately associated with it (Figs 4.07–4.08). A constriction exists between this and the second abdominal segment. Thus the apparent thorax of apocritans consists of four segments; to avoid confusion with a true thorax this structure is often called the MESOSOMA (= alitrunk). In many older works the homology was not recognized and the propodeum was mistakenly referred to as the metathorax or epinotum.

The PRONOTUM is usually rather short and closely associated with the mesonotum, while the PROPLEURON is loosely attached to its notum. The PROSTERNUM is usually visible ventrally and is generally relatively small. The pronotum is often traversed dorsally by a transverse furrow and laterally this may be crossed or margined by an oblique keel, the EPOMIA. In many aculeates with a large propodeum these keels (= the pronotal carina) are continuous dorsally, and separate a more or less vertical anterior part from a longitudinal or horizontal posterior part; the anterolateral corners where they are prominent are called the HUMERI. The upper hind corner of the pronotum may partially surround or cover the mesothoracic spiracle (which primitively lies between the pro- and mesothorax), though in some taxa (e.g. Proctotrupidae, Megalyridae) the spiracle opens onto the surface of the apparent pronotum. In many aculeates and evanioids the pronotal corner is enlarged to form a PRONOTAL LOBE that completely

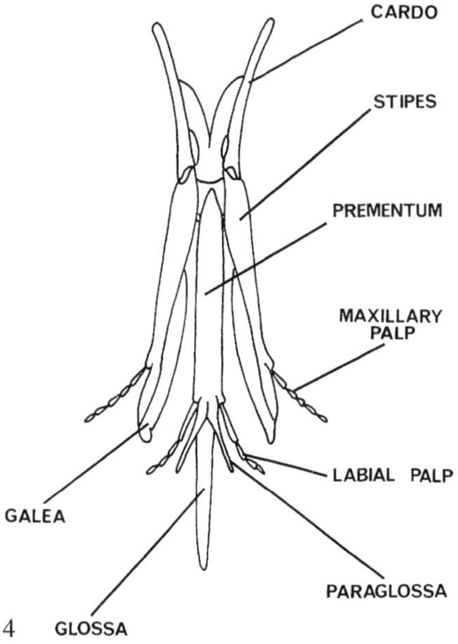

4

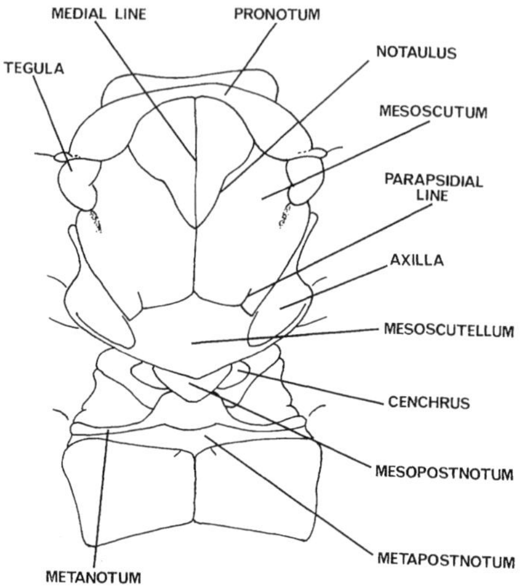

5

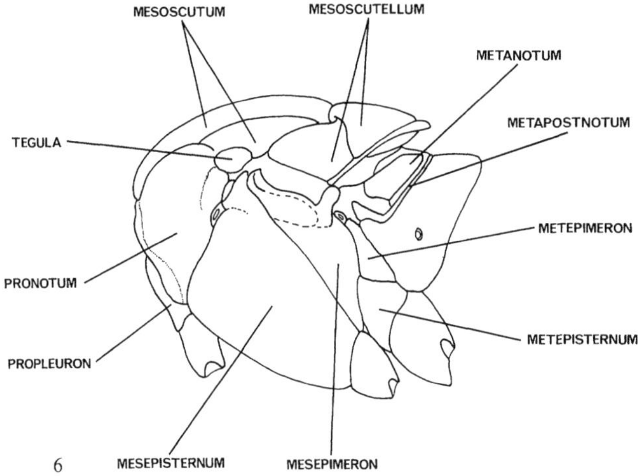

6

Figs 4.04–4.06. 4.04. Mouthparts, Apidae. Figs 4.05–4.06. Tenthredinidae (Symphyta), mesosoma. 4.05, dorsal; 4.06, lateral.

conceals the spiracle, and in some aculeates a vertical ridge, the PRETEGULAR CARINA, is present on the pronotum in front of the upper hind corner. In most apocritans the hind corner of the pronotum reaches almost to the TEGULA, a convex plate at the base of the fore wing. In the Sphecidae, Apidae, most Chalcidoidea, and some Chrysididae the hind corner of the pronotum is rather widely separated from the tegula. In the primitive condition the ventral corners of the pronotum are widely rounded and posteriorly overlap the mesepisternum, while in more derived forms they are acutely pointed and posteriorly closely co-adapted to the mesepisternum. In a number of families (e.g. Gasteruptiidae, Proctotrupidae, Scelionidae) the apparent pronotum extends round to the mid ventral line and is there fused so it forms a continuous ring behind the fore coxae.

The MESOTHORAX is the largest thoracic segment since it contains the principal flight muscles which

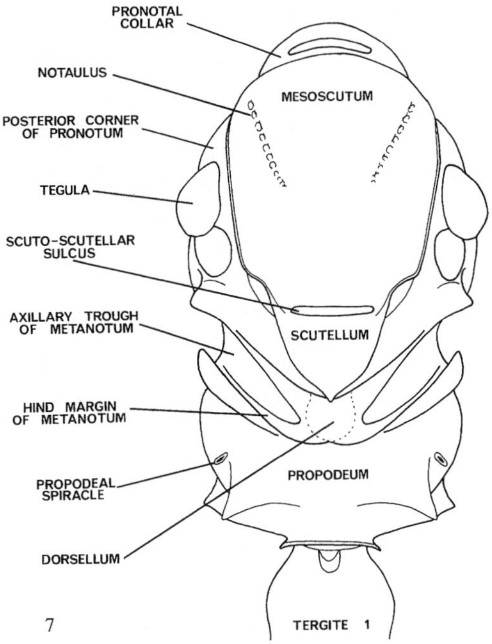

7

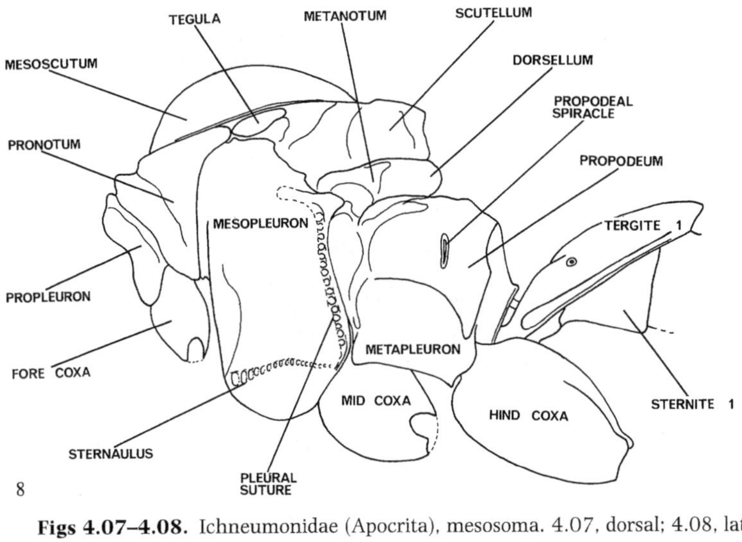

8

Figs 4.07–4.08. Ichneumonidae (Apocrita), mesosoma. 4.07, dorsal; 4.08, lateral.

power the fore wings. Dorsally it is covered by the MESONOTUM which is divided into two main parts, an anterior MESOSCUTUM, and a posterior SCUTELLUM (= mesoscutellum). The two are typically separated by a transverse groove, the SCUTO-SCUTELLAR SULCUS. In some symphytans and chalcids a transverse line of flexibility, the TRANSSCUTAL ARTICULATION is present across the mesonotum at about the level of the base of the fore wings. This separates a posterior portion of the mesoscutum, the AXILLA which extends laterally to the fore wing base. The transscutal articulation is present in most apocritans though in many groups (e.g. ichneumonids) it is effaced centrally, and is visible only laterally. In many apocritans the transscutal articulation is centrally close to the scuto-scutellar sulcus and the axillae are really only discernible laterally. In many Hymenoptera a pair of posteriorly convergent impressed lines, the NOTAULI (singular notaulus) (= 'prescutal sutures' *sensu* Richards, 1956), are present on the mesoscutum. In many sawflies the notauli meet posteriorly and define a triangular area, the MEDIAN MESOSCUTAL LOBE (= 'prescutum' *sensu* Richards, 1956), and in some of these symphytans a median longitudinal groove, the MEDIAN MESOSCUTAL SULCUS (= median scutal line), extends posteriorly from the anterior margin of the mesoscutum, bisecting the median mesoscutal lobe. This appears to be an archaic feature of Hymenoptera and while it has been widely observed in fossil apocritans (Rasnitsyn, 1980), it is uncommon in extant Apocrita, which generally lack any trace of the median mesoscutal sulcus and have the notauli shorter, or posteriorly less strongly convergent, so they do not meet. In many taxa the notauli may be vestigial or absent. Numerous apocritans have a second pair of mesoscutal marks, the PARAPSIDAL LINES, lateral to the notauli. If present, these typically extend anteriorly from the posterior region of the mesoscutum. In some scelionids the anteromedian part of the mesoscutum is smooth and concave, and margined by a carina; this is called the SKAPHION. In eumenine vespids the posterolateral corner of the mesoscutum is extended backwards into a lobe, the PARATEGULA, which overhangs the axilla.

The scutellum may be oval to somewhat shield-shaped. Its shape, sculpture and convexity are often important features for separating species, as is the degree of development of the lateral longitudinal carinae. In sawflies a MESOPOSTNOTUM is present as a transverse sclerite behind the scutellum and this is probably homologous with the FRENAL AREA (= frenum) of chalcid workers (Fig. 4.09). In sawflies the central part of this may be called the POSTSCUTELLUM, but this is not homologous with the 'postscutellum' of Apocrita.

The MESOPLEURON is primarily divided into an anteroventral MESEPISTERNUM and a posterodorsal MESEPIMERON by the PLEURAL SUTURE, an oblique groove that runs forwards and upwards from near the mid coxa to near the base of the fore wing. The mesepimeron is quite large in sawflies and most chalcidoids (except encyrtids, tanaostigmatids and many eupelmids), but in most Apocrita it is reduced to a narrow posterior strip, often delineated only by a row

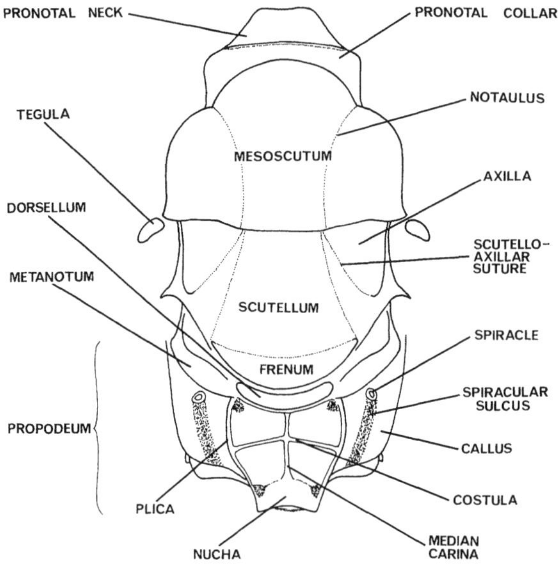

Fig. 4.09. Pteromalidae, mesosoma dorsal.

of punctures. It is not discernible at all in many aculeates. The posterodorsal corner of the mesepimeron is nearly always produced into a rounded lobe which more or less conceals the second thoracic spiracle. In most Hymenoptera a medioventral longitudinal sulcus, the MEDIAN STERNAL GROOVE is distinct. Often a PRECOXAL SUTURE runs from near the base of the mid coxa sinuously forwards to near the front end of the mesepisternum; in some ichneumonoids this may be broad and quite deeply impressed and is known as the STERNAULUS. The area between the precoxal suture and the median sternal groove is commonly called the MESOSTERNUM, though it is not homologous with the definitive mesosternum, which is probably reduced or invaginated in most hymenopterans. In xyelids, and apparently also in eupelmids, the true mesosternum is distinct.

The morphology of the anterior part of the mesopleural region is particularly complex because of the existence of intersegmental sclerites and their subsequent fusion with other parts. In most sawflies, a large subtriangular to rectangular sclerite, the PREPECTUS, is partly inserted into the excised anterior margin of the mesepisternum below the first thoracic spiracle. Morphologically the prepectus can be recognized as the origin of the spiracular occlusor muscle; in some groups in which it is extremely reduced its position can be deduced by examination of the muscles. Behind the spiracle a second sclerite, the POSTSPIRACULAR SCLERITE, is discernible in most tenthredinoids, cephids and in anaxyelids. In other Siricoidea the prepectus is elongate and slender and hidden beneath the hind margin of the pronotum, but the postspiracular sclerite is usually discernible. In orussids and all Apocrita the postspiracular sclerite is fused with the mesopleuron. In most apocritans the prepectus is usually either concealed beneath the pronotal margin, or it is fused with other parts. In the Sphecidae, Formicidae and some other vespoids it is apparently fused with the mesepisternum, while in the Vespidae and most parasitoids it is often fused with the hind margin of the pronotum. The only parasitoid group with a conspicuous independent prepectus is the Chalcidoidea, where it forms an intervening sclerite between the pronotal corner, the tegula and the mesopleuron. In many chalcids the two prepecti are confluent ventrally to form a single semi-annular sclerite. Recently an independent concealed prepectus has been found in the Stephanidae, Dryinidae, Monomachidae, Roproniidae and Austroniidae (Gibson, 1985).

The lateral mesothorax generally shows a number of other taxonomically important features. In many Apocrita a ridge, the EPICNEMIAL CARINA (= omaulus) more or less parallels the anterior margin of the mesepisternum delineating anteriorly an EPICNEMIUM (omaular area). Often the epicnemial carinae meet medioventrally so the epicnemium is continuous ventrally behind the fore coxae. Frequently the front of the subpleural area is more or less depressed to receive the backwardly projecting fore coxae. The posterior margin of this depression is sometimes defined by an ACETABULAR CARINA, which may be difficult to distinguish from the bottom of the epicnemial carina. The epicnemium is not homologous with the prepectus, and the term prepectal carina (for the epicnemial carina), though widely used, should be avoided. The part of the mesepisternum below the wing insertion is often convexly raised and called the SUBALAR PROMINENCE (= subalar area). Behind and slightly below this is an impression, the SUBALAR PIT, and in many aculeates a groove, the ANTERIOR OBLIQUE SULCUS (= 'episternal sulcus'), extends from this pit to the epicnemial carina. On the mesepisternum of virtually all apocritans, slightly before the mesepimeron and about one third of its length down, is a small pit, the EPISTERNAL SCROBE; in many aculeates a SCROBAL SULCUS extends forwards from this to the anterior oblique sulcus. The area above this suture is the HYPO-EPIMERAL AREA, which seems to be homologous with the 'speculum' of ichneumonoids. In Pompilidae and many Dryinidae the anterior oblique and scrobal sulci are not indicated, but in pompilids a groove, the POSTERIOR OBLIQUE SULCUS extends from the episternal scrobe to the lower end of the epicnemial carina.

The METATHORAX which bears the relatively small hind wings is always smaller and less differentiated than the mesothorax. A metathoracic spiracle (which is concealed behind the posterodorsal margin of the mesepimeron) is apparently functional in the Symphyta, Ichneumonoidea, Evanioidea and most aculeates, but is vestigial or lost in the Cynipoidea, Chalcidoidea, Ceraphronoidea and Proctotrupoidea. The METANOTUM is present primitively as a strap-like transverse sclerite. In apocritans its central part (the DORSELLUM) is often raised and laterally bounded by depressions; this raised portion has been called the 'postscutellum', but it is not homologous with the symphytan postscutellum. In all Symphyta, except Cephidae, a pair of rounded prominences, the CENCHRI, are present on the metanotum. These have a roughened surface to engage the wings when they are folded over the body. Cenchri are not present in any apocritans. The METAPOSTNOTUM is usually visible in symphytans, though it may be reduced. In apocritans it is variously modified by being reduced, invaginated or fused with other parts. In the Apoidea the 'propodeal triangle' is actually a posteriorly expanded metapostnotum (Brothers, 1976). The METAPLEURON is divided by the METAPLEURAL SUTURE into a METEPISTERNUM and METEPIMERON in some sawflies, while in many apocritans it is apparently undivided and dorsally confluent with the propodeum. In Sphecidae the metapleuron is primitively divided into upper and lower areas by the transmetapleural line. In many formicids the lower posterolateral quadrant of the metapleuron contains a swollen METAPLEURAL GLAND that is evident externally by the presence of a bulla.

The PROPODEUM is the first abdominal segment of apocritans that has become fused to the thorax. This segment is modified even in sawflies in which it is commonly longitudinally divided mediodorsally. The corresponding sternite is either very reduced or absent. In apocritans the propodeum is generally convex, and bears laterally a pair of conspicuous PROPODEAL SPIRACLES. The posterodorsal surface of the propodeum is variously modified in different families, and in ichneumonoids often bears several carinae enclosing AREAE. These carinae and areae have various names, which are usually explained in relevant specialist works (e.g. Gauld, 1991), as the shapes of these areae and conformations of the carinae are particularly important taxonomic features. In chalcidoids a different arrangement of carinae is present (see Fig. 4.09) and the most posterior part, which is sometimes lengthened, is called the NUCHA. The apocritan metasoma articulates via a socket, the PROPODEAL FORAMEN, that is usually at the posterior end of the propodeum, though in some taxa (e.g. the Evanioidea) it is displaced dorsally. The form of the propodeal/metasomal articulation is characteristic in several major apocritan groups, and has been detailed by Mason (1983*a*) who considered that the different forms characterize different evolutionary lineages. In ichneumonoids and most aculeates an external ligament or muscle can be seen issuing from the propodeum, immediately dorsal to the metasomal insertion, and inserting into the first metasomal segment via a dorsal orifice close to the anterior end of the tergite. In some vespids the posterolateral part of the propodeum, either side of the foramen is expanded into a flange-like lobe, the PROPODEAL VALVULA.

The APTEROUS (= wingless) mesosoma seen in some Ceraphronoidea, Proctotrupoidea, Chalcidoidea, Ichneumonoidea, Chrysidoidea, Tiphiidae, Bradynobaenidae, Mutillidae and Formicidae is frequently characterized by a reduction in the size and complexity of the mesothoracic and metathoracic sclerites. In general the subdivisions of the mesonotum (mesoscutum, scutellum, axillae) are frequently fused, and characteristic features such as the notauli, tend to be lost. The mesonotum is usually much reduced in size and no longer dominates the dorsal mesosoma as in the alate forms. A flexible suture is sometimes visible between the pronotum and mesonotum, but often the two sclerites are fused (to form a PROMESONOTUM), leaving no trace of, or a faint groove where the suture was. The metanotum is sometimes discernible as a distinct sclerite separated by sutures or impressions from the mesonotum in front and the propodeum behind, but often it is represented only by a transverse line, the METANOTAL GROOVE. In the most specialized apterous hymenopterans (female mutillids) the fusion of all mesosomal sclerites is more or less complete. In BRACHYPTEROUS species wing vestiges, or extremely reduced non-functional wings, are present. In such forms the mesosoma either retains its full complement of sclerites or shows some of the reductions characteristic of apterous species.

The wings

Hymenoptera characteristically have two pairs of wings, a large anterior pair and a smaller posterior pair. The fore and hind wings on each side are held together by a series of hooks, the HAMULI, which arise from the front edge of the hind wing and engage one or more folds (RETINACULUM) on the posterior margin of the fore wing. Characteristically, two groups of hamuli are present, a proximal group on the costal vein and a distal group on the radial vein. Large species usually have a large number of hamuli but

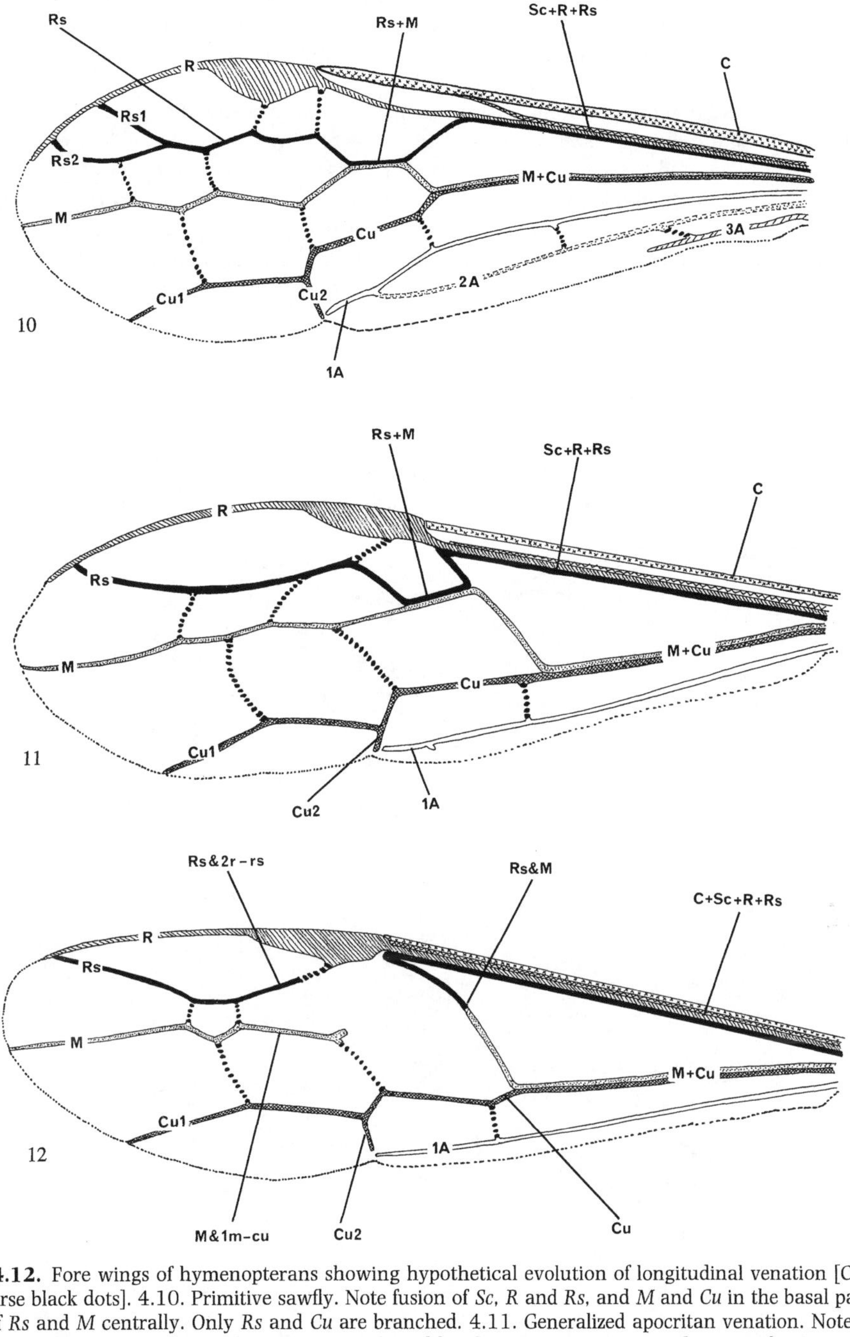

Figs 4.10–4.12. Fore wings of hymenopterans showing hypothetical evolution of longitudinal venation [Cross veins are indicated by coarse black dots]. 4.10. Primitive sawfly. Note fusion of *Sc*, *R* and *Rs*, and *M* and *Cu* in the basal part of the wing, and the fusion of *Rs* and *M* centrally. Only *Rs* and *Cu* are branched. 4.11. Generalized apocritan venation. Note the loss of *2A* and *3A*, and the branch of *Rs*. There is a characteristic Z-shaped bend in *Rs* just prior to its fusion with *M* centrally. 4.12. Typical ichneumonid venation. Note loss of *Rs+M* and the following abscissae of *Rs* and *M*, and the resultant presence of the composite veins *Rs&M*, *Rs&2r-rs* and *M&1m-cu*. [For conciseness *M&1m-cu* is often abbreviated to just 1*m-cu*.]

smaller hymenopterans have fewer and only three distal hamuli are present in most Chalcidoidea. In many Hymenoptera a large, pigmented PTEROSTIGMA (= stigma) is present about two-thirds of the way along the anterior margin of the fore wing.

The wing membrane is supported by a series of longitudinal and transverse VEINS, and is traversed by a number of FOLDS. The wing veins divide the membrane into a number of discrete areas, wing CELLS. When a cell is bordered on all sides by veins (including the pterostigma) or is proximally contiguous with the wing base, the cell is called an ENCLOSED CELL (see p. 115). The pterostigma and cells confluent with the wing margin are not counted as enclosed cells. The names and corresponding homologies of the veins, both within the Hymenoptera and throughout the Insecta, have been the subject of a great deal of controversy and in many groups of Hymenoptera idiosyncratic venational nomenclature is widely used.

The principal longitudinal veins of the hymenopterous wing are best shown in the most primitive members of the order, such as the Xyelidae (Figs 4.10,

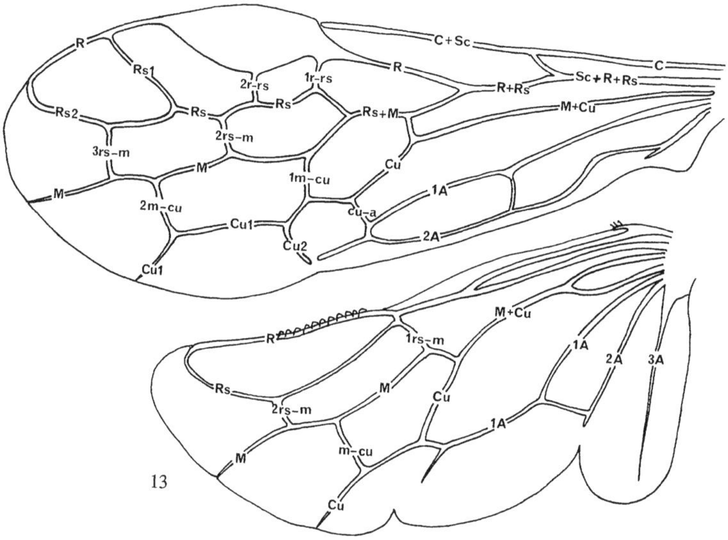

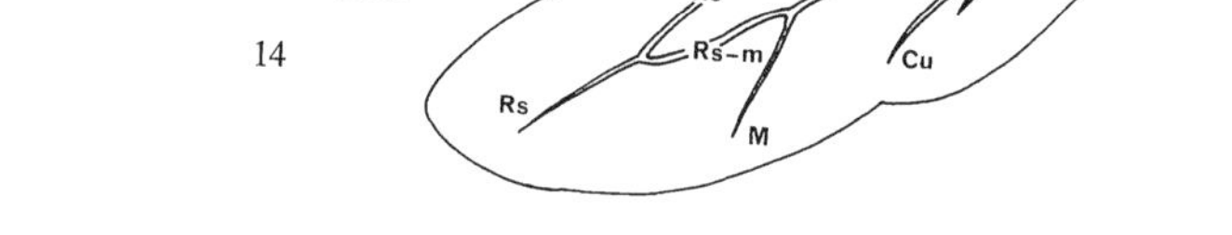

Figs 4.13–4.14. Wing venation. 4.13, Xyelidae; 4.14, Apidae.

4.13). The following are recognizable: the costa (*C*), subcosta (*Sc*), radius (*R*), radial sector (*Rs*), media (*M*), cubitus (*Cu*) and three anals (*1A*, *2A* and *3A*). A remnant of jugal vein (*J*) is often discernible near the proximal hind margin of the wings. Except in sawflies, where *Sc* is concave, all hymenopteran longitudinal veins are convex. The following cross-veins (denoted by lower case letters) have been identified: 1*r-rs*, 2*r-rs*, 1*rs-m*, 2*rs-m*, 3*rs-m*, 1*m-cu*, 2*m-cu*, 1*cu-a* and 2*cu-a*. At least one anal–anal cross-vein is present in some sawflies. The hind wing (Figs 4.13, 4.14) is rather similar except that *Rs* and *Cu* are not branched, and *Rs* and *M* are not fused centrally. Only cross-veins 1*rs-m*, 3*rs-m*, *m-cu*, *cu-a* and an anal–anal cross-vein are recognizable in the hind wing.

In the fore wing vein *R* is apparently not branched, but two branches of *Sc* (*Sc*1 and *Sc*2) are observable in some megalodontoids and two branches of *Rs* (*Rs*1 and *Rs*2) are discernible in some xyelids. Distally *Rs* or its branches generally anastomose with *R*. In some braconids, bethylids and vespids either *Rs* or *Sc*+*R* at or distal to the point of divergence of *Rs* is broadened into a PRESTIGMA immediately proximal to the pterostigma. A single unbranched *M* is present, and this is almost always fused with *Rs* (thus obliterating 1*rs-m*) for a short distance near the centre of the wing (see Figs 4.10–4.12). The single *Cu* is branched distally with *Cu*1 extending towards the hind corner of the wing, and *Cu*2 (= *Cu*1b *sensu* Richards) being short and extending to the claval notch where it anastomoses with the first anal vein. In all sawflies, except cephids, a scaly patch is present on the wing margin between the third anal vein and the jugal vein remnant.

Throughout the evolutionary history of the order there has been a reduction in venation (Figs 4.10–4.12). In the fore wings of all Hymenoptera, except for a few xyelids, 1*rs-m* has been obliterated by fusion of *Rs* and *M* centrally, while the branches of *Sc* and *Rs* were lost early in the evolutionary history of the Symphyta. Although they are now treated as separate veins (see K.G.A. Hamilton, 1972*a*, *b*) *R* and *Rs* are basally united in most insects, but in most Hymenoptera *Sc* is also fused with them, at least basally. *M* and *Cu* are also united for a considerable distance basally. In Apocrita further venational reduction has occurred. Veins *2A* and *3A* are lost or reduced to stubs so that no enclosed anal cells are present in either the fore or hind wing. In the fore wing *Sc* is completely obliterated by fusion (except in a few sphecids), and in many groups that are active fliers (e.g. Ichneumonidae) *C*, *Sc*, *R* and *Rs* are united for much of their length (Fig. 4.15). However, there is a peculiar weakness, a hinge-like mechanism, present at the base of the pterostigma, which allows deformation of the outer part of the fore wing. Associated with this is the loss of vein 1*r-rs*. The primitive number of enclosed fore wing cells for the Apocrita is probably 11, but generally only 10 are discernible (Fig. 4.18). In most lineages of apocritans there has been extensive reduction in venation and remaining veins are often composite structures. Veins comprising two or more veins fused for their length are indicated by the abbreviations separated by a plus-sign (e.g. *M*+*Cu*) while apparent single veins formed from lengths of different veins are indicated by the abbreviations separated by an ampersand (e.g. *Rs&M*). In many works concerning members of one family or subfamily composite veins may simply be referred to, for convenience, by the name of their longest component. Thus in works on the Ichneumonidae *M*&1*m-cu* is generally called 1*m-cu*.

Usually a maximum of only two enclosed cells are present in the hind wing of apocritans, but in many large vespoids *C* extends along the anterior margin of the wing enclosing a third cell, the costal. In many small Hymenoptera no discernible cells are present in the hind wings, and frequently almost no veins are visible, although hamuli are always in evidence.

The nomenclature of the wing cells is both confused and confusing. As there has been extensive fusion and obliteration of cells, any attempt to establish a series of homologous standard names is likely to be both cumbersome and controversial. We recommend the adoption of the systems shown in Figs 4.17, 4.18. In this system homologous cells in the symphytan and apocritan wings have the same names, except that the first submarginal cell of apocritans corresponds to the first and second parastigmal cells of the sawfly.

Some flexion-lines (Wootton, 1978) and fold-lines may be seen in the wings of virtually all medium to large sized Hymenoptera and where these folds cross a vein, the vein is generally transparent, and this clear area is called a FENESTRA (= bulla). The CLAVAL FURROW in fore and hind wings is a fold that is just anterior to, and parallel with, *1A*. It originates at the wing base and its distal end is marked by a notch (the PREAXILLARY EXCISION) in the wing membrane.

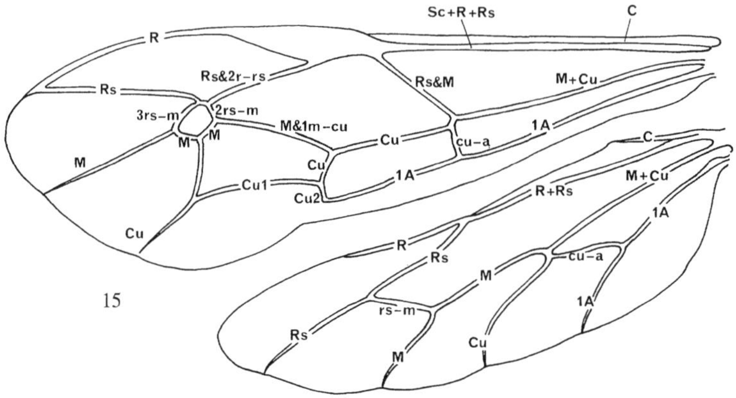

15

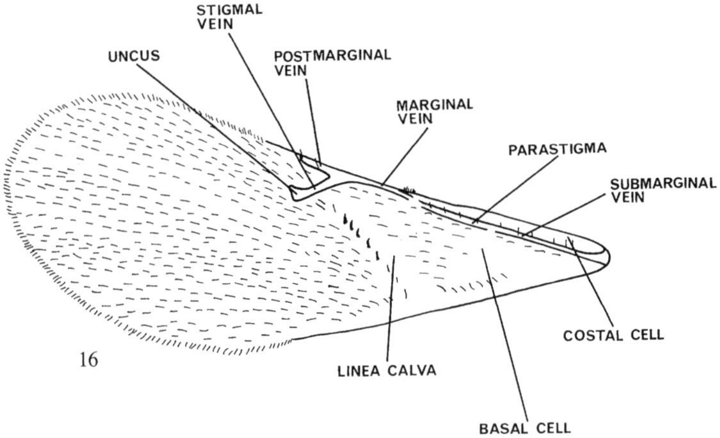

16

Figs 4.15–4.16. Wing venation. 4.15, Ichneumonidae; 4.16, Encyrtidae.

This has frequently been called the 'vannal notch', while the fold often has been referred to as the 'vannal fold', but the term 'vannal' is best avoided. Characteristically, in the hind wing a JUGAL FOLD may also be present behind the claval furrow. This fold also originates at the wing base and extends to the wing margin where it may be marked by a notch, and the area posterior to the jugal fold, the JUGUM, may be extended distally into a lobe, the JUGAL LOBE (= anal lobe). That part of the wing between the jugum and claval furrow is called the CLAVUS, and it too may occasionally be marginally expanded into a lobe, the CLAVAL LOBE (= 'vannal' or plical lobe).

In the smaller apocritans the pterostigma is often absent and the venation can be extremely reduced. In many chrysidoids only the proximal cells of the wing are delineated, while in most proctotrupids and diaprids only the anterior cells remain; in cynipoids generally only a triangular marginal (= radial) cell is delineated. In the Chalcidoidea and Scelionidae only a single, bifurcate, composite vein is developed so no cells are enclosed. The homology of the parts of this

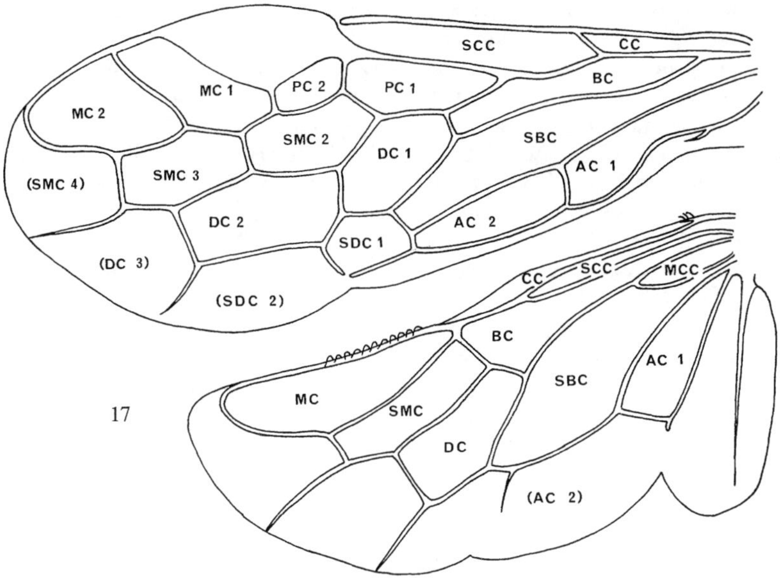

17

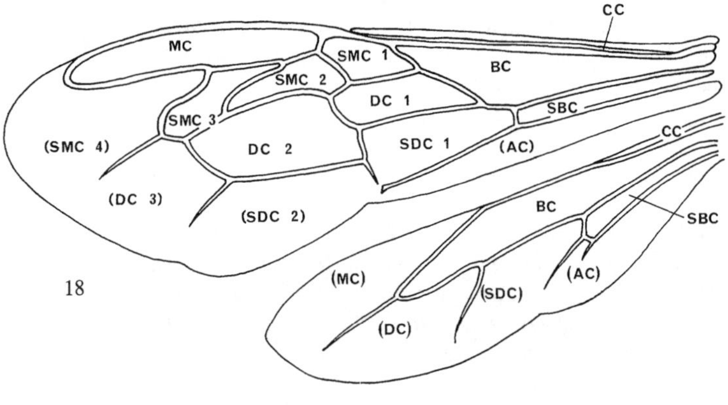

18

Figs 4.17–4.18. Wing cells. 4.17, Xyelidae; 4.18, Apidae. AC = anal cell; BC = basal cell; CC = costal cell; DC = discal cell; MC = marginal cell; PC = parastigmal cell; SBC = subbasal cell; SCC = subcostal cell; SDC = subdiscal cell; SMC = submarginal cell. The names of ***enclosed cells*** are not in parentheses, those not fully enclosed are in parentheses.

vein are difficult to establish so it is most convenient to use the very widely accepted (non-homologous) terminology: SUBMARGINAL VEIN for the proximal part that is parallel to, but not on, the wing margin; MARGINAL VEIN for the part on the wing margin; STIGMAL VEIN for the branch of the marginal projecting into the wing membrane and POSTMARGINAL VEIN for the part on the wing margin distal to the junction of the marginal and stigmal veins. At the junction of the submarginal and marginal veins a short broadened part, the PARASTIGMA, can sometimes be seen. The apex of the stigmal vein of chalcids is usually shaped like a bird's beak and is referred to as the UNCUS. The term 'stigmal vein' is also widely employed in the chrysidoid families where it refers to a free curved vein arching outwards from the pterostigma and consisting of *2r-rs* and the apical portion of *Rs*. In some platygastrids only the submarginal vein is discernible, while in others no venation is present. Chalcidoidea frequently have one or more hairless areas present on the surface of the fore wing. One area, running obliquely from the parastigma to the hind margin is termed the SPECULUM. Many aphelinids, encyrtids and eupelmids have a second oblique hairless strip, the LINEA CALVA, running from the stigmal vein to the hind margin; the setae bordering the linea calva appear to function in wing coupling while at rest (Hennessey, 1981). The wings of the very smallest apocritans are frequently fringed with long hairs.

The legs

The fore, middle and hind pairs of legs articulate with the pro-, meso-, and metathorax respectively. Each leg consists of a basal COXA, a short TROCHANTER, an elongate FEMUR, a slender TIBIA and a multisegmented TARSUS. In some Apocrita the proximal end of the femur is differentiated as a more or less distinct segment, the TROCHANTELLUS. The tibiae are provided with articulated SPURS on their distal ends. Typically the three legs have 2 : 2 : 2 tibia spur formula (most Symphyta and Ceraphronoidea) or 1 : 2 : 2 spurs (most Apocrita). This number is reduced in many lineages. Probably primitively these spurs are cylindrical, but in many taxa one or more is flattened, fringed and variously curved; such specialized spurs are generally termed CALCARIA (though occasionally this term has been used as a synonym of spur). A few symphytans (e.g. Pamphiliidae) have additional articulating PREAPICAL SPURS present on the medioventral surface of the middle and hind tibiae. In Symphyta one of the apical spurs of the fore tibia is usually enlarged and specialized to form an antennal cleaner. A similar modification (the STRIGIL) is found in Apocritans, and the first tarsal segment is also somewhat modified to function as part of this antennal cleaner. The tarsus typically has five TARSAL SEGMENTS (= tarsomeres), though the number is reduced to four or three in some chalcidoids. In many Symphyta, Trigonalyidae, Pompilidae, Sphecidae and a few Ichneumonidae a small membranous pad, the PLANTAR LOBE (= plantulae), projects from the distal apex of each of the first to fourth tarsal segments. The distal tarsal segment bears a pair of CLAWS and associated structures, the AROLIUM (= pulvillus), ORBICULA, PLANTA and CAMERA.

The legs of a number of Hymenoptera are conspicuously modified for certain purposes such as jumping (e.g. many Chalcidoidea), burrowing (e.g. some Aculeata and Ichneumonidae), prey grasping (e.g. most female Dryinidae), mate seizure and/or courtship (e.g. some male Sphecidae) or pollen carrying (most Apidae).

The abdomen

The ABDOMEN normally consists of ten segments but this number is often not visible without careful dissection. Each segment normally consists of a dorsal TERGITE and a ventral STERNITE, which may be free or secondarily fused. In many apocritans a lateral piece of each tergite is separated by a crease, or even detached from the main dorsal part of the tergite to form a LATEROTERGITE. Analogous LATEROSTERNITES are present in some Ichneumonidae. In general the last visible tergite or syntergite (whatever its number) is called the PYGIDIUM (= epipygium) and the last visible sternite the HYPOPYGIUM (= subgenital plate; hypandrium); these are specialized in diverse ways in various groups.

In the Symphyta the first seven segments are normally developed except that the first sternite is membranous; tergite VIII is also normal and it bears lateral spiracles as do tergites I to VII. The more posterior parts are modified in connection with the genitalia (see below). The tenth tergite bears a pair of piliferous processes, the CERCI (= pygostyles *sensu* Richards, 1956). In some taxa that have tergite X reduced and fused with tergite IX the cerci appear to be attached to the latter, while in a number of apocritan groups the cerci are absent.

In the Apocrita the first abdominal segment (the propodeum) is closely associated with the thorax and the apparent 'abdomen' consists of the second and following true abdominal segments. This is called the METASOMA to avoid confusion with the definitive abdomen. In ants the term GASTER has been used to refer to that part of the metasoma posterior to the petiolar, or petiolar and postpetiolar segments. In Chalcidoidea the term gaster has been used for that part of the metasoma posterior to the petiole, and thus gastral segment 1 = metasomal segment II = abdominal segment III. In this book 'gaster' is avoided wherever possible (it is here only used for the Formicidae, Chapter 16) since different workers may or may not regard it as synonymous with the metasoma.

In many Apocrita metasomal segment I is anteriorly slender and somewhat tubular, often with the tergite and sternite immovably attached to each other, though in the Ichneumonoidea sternite I is divided by a transverse suture into an anterior and posterior part and only the anterior part is closely associated with its tergite. Often the posterior margins of metasomal segment I do not overlap the anterior edges of the following segment; in many aculeate families segment I is often separated from segment II in some way, such as by reduction, grooves, constrictions or indentations, or by the development of STRIDULATORY SURFACES between the two segments. Sometimes metasomal segment II is relatively reduced or is separated behind by deep constrictions (e.g. in some ants). For these modified segments we propose that the term PETIOLAR SEGMENTS should be used. We avoid the terms 'petiole' and 'postpetiole' as these traditionally refer to different things in different groups. For example, in Formicidae the 'petiole' is metasomal segment I and the 'postpetiole' is segment II, when separated; in ichneumonoids the 'petiole' is the slender, anterior part of metasomal segment I, while the 'postpetiole' is the posterior, broader part of the same segment. In some groups of Chalcidoidea, Proctotrupoidea and Cynipoidea metasomal segment I is very small and annular; frequently it is almost obscured by or sometimes fused with the following segment. In some ichneumonids there are impressed areas, GLYMMAE, in metasomal tergite I; these may be large and impressed deeply enough to almost meet their lateral counterpart.

Typically the apocritan metasoma has spiracles on metasomal segments I to VII, but in many groups the number is reduced to I to VI or I to V and even greater reduction occurs in some groups. The Chalcidoidea, Cynipoidea, Evanioidea and some Proctotrupoidea, for example, only have a functional spiracle on metasomal tergite VII, while according to Naumann and Masner (1985) the Vanhorniidae, Scelionidae and Platygastridae lack functional metasomal spiracles entirely. In the Trigonalyidae spiracles are present only on metasomal tergite VII, and these are generally covered by the hind margin of metasomal tergite VI.

Metasomal segments II to VI are usually relatively unspecialized and somewhat resemble each other. The hind margin of each tergite generally broadly overlaps the fore margin of the following tergite. In some groups (e.g. the Ichneumonidae and some aculeates) metasomal tergite II, and sometimes also tergite III, has an elliptical or more or less circular, differently sculptured area, the THYRIDIUM, present anteriorly, though in some slender taxa the thyridia have apparently been displaced posteriorly and their course is indicated by a groove. In many Mutillidae and Bradynobaenidae a groove, the FELT LINE, is present close to and parallel with the lateral margin of metasomal tergite II. In many evolutionary lines of the Apocrita there is a tendency for some of the metasomal tergites to fuse together to form what is called a SYNTERGITE. In all Braconidae and many Chalcidoidea metasomal tergites II+III are fused and in many braconid subfamilies (especially the Cheloninae) further fusion takes place. In the genus *Chelonus* metasomal tergites I+II+III are united to form a CARAPACE, that completely covers the reduced posterior segments. The most extreme fusion occurs in the Vanhorniidae where metasomal tergites I to IV are united in a single syntergite, and metasomal sternites I to IV (in females) or I to V (in males) are fused to form a SYNSTERNITE. In the Chrysididae the posterior tergites are reduced, telescopic and retracted beneath tergite III or IV (Kimsey 1992*b*). The sternites of Hymenoptera are often relatively weakly sclerotized, and almost membranous (e.g. some Ichneumonidae), but they can be as strongly sclerotized as the tergites (e.g. Proctotrupidae, most aculeates). In some female bees they are covered with elongate pubescence and are used for pollen transport; such a hirsute area is called a VENTRAL SCOPA. The posterior segments of the metasoma are modified to accommodate the genitalia in a variety of ways in the two sexes of various groups, so it is convenient to discuss the sexes separately.

The female genitalia and associated segments

In many groups the terminal segments of the female are highly modified, reduced and invaginated. Extensive comparative anatomical study is still needed to elucidate the oviposition mechanisms of the smaller Apocrita. The following account emphasizes the

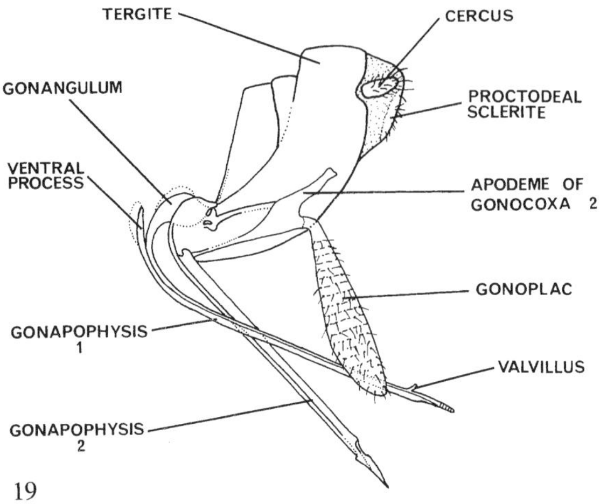

19

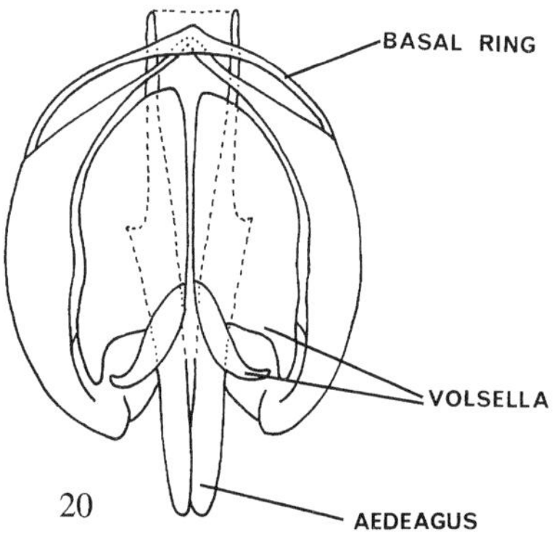

20

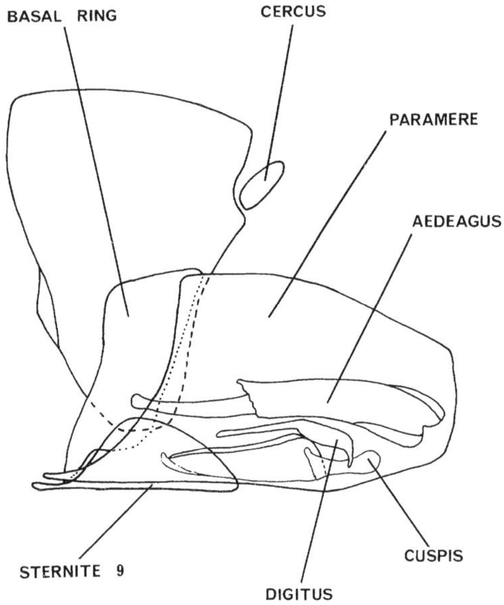

21

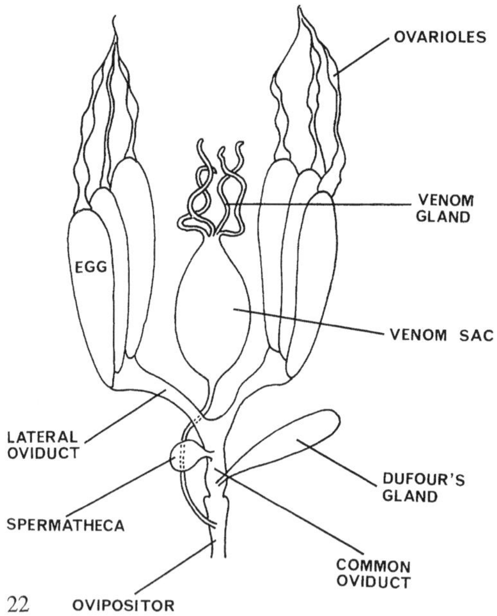

22

Figs 4.19–22. Genitalia and reproductive organs. Figs 4.19–4.21. Genitalia. 4.19, Ichneumonidae, female, lateral; 4.20, Ichneumonidae, male, dorsal; 4.21, Ichneumonidae, male, lateral. Fig. 4.22. Ichneumonidae; female genital tract.

structure, and to some extent function, of the generalized hymenopterous condition.

The last visible sternite of the female abdomen is usually sternite VII (= metasomal sternite VI), and this is often somewhat enlarged and projects backwards to cover the underside of the ovipositor. Abdominal sternites VIII and IX are reduced and membranous, and their place is taken by the parts of the ovipositor which may have been derived from coxal homologues on abdominal segments VIII and IX (Scudder, 1961). These homologues, the FIRST and SECOND GONOCOXAE, both possess ventral elongations, respectively the FIRST and SECOND GONAPOPHYSES. In Hymenoptera the first gonocoxa is apparently lost (though the VENTRAL PROCESS may represent a vestige of it), and the internal end of the first gonapophysis joins to the GONANGULUM (= first valvifer), which is probably a detached part of the second gonapophysis. The gonangulum is approximately triangular, with the first gonapophysis (= first valvula, lower valve of ovipositor or lancet) arising from one angle (Fig. 4.19), the second corner articulating with a mediodorsal process on the second gonocoxa and with the upper corner articulating with a ventral apodeme of abdominal tergite IX. The second gonocoxa (= second valvifer) is larger than the gonangulum and somewhat oblong in profile. The second gonapophysis (= second valvula, upper valve of ovipositor or stylet) arises from the anterior end of this gonocoxa, but it is generally rigidly fused with its fellow on the other side to form the inverted U-shaped upper valve of the ovipositor. Along the lower side of these fused gonapophyses is a pair of ridges which fit into grooves on the dorsal edge of the first gonapophyses. These three elements form a hollow tube, the OVIPOSITOR, through which the egg is usually passed. The second gonapophyses may bear ridges or teeth distally or may be smooth, but the first gonapophyses almost invariably are toothed distally. Internally the first gonapophysis is usually covered with microscopic, backwardly-directed ridges or denticles (see Austin & Browning, 1981), but in the Ichneumonoidea and aculeates one or more small articulating flaps, the VALVILLI, are present (Quicke, Fitton & Ingram, 1992). The posterodorsal corner of the second gonocoxa is usually produced into an apodeme and below this articulates a GONOPLAC (= third valvula or ovipositor sheath) which is normally hirsute externally, and concave internally (Fig. 4.19). The two gonoplacs project backwards and form a sheath or case for the ovipositor. Many of these structures bear sensilla (Hermann & Douglas, 1976; LeRalec & Wajnberg, 1990). In honey bees the proximal part of the ovipositor sheath bears exocrine glands, the secretions of which induce defensive behaviour in guard bees outside the hive (Cassier *et al.*, 1994).

The ovipositor is worked by two sets of antagonistic muscles, an often massive protractor that extends from the anterodorsal end of the second gonocoxa to the anterior part of abdominal tergite IX, and a retractor set extending from the posterodorsal apodeme on the second gonocoxa to an internal ridge on the side of tergite IX. As the protractor muscle contracts, tergite IX is pulled forwards and down exerting pressure on the gonangulum via the tergal/gonangular articulation. The gonangulum rocks on the gonangular/gonocoxal articulation so its front end tips downwards pushing the first gonapophysis so that it slides and its end projects beyond that of the second. Contraction of the retractor muscle reverses this process, but as the teeth on the tip of the first gonapophysis serve as barbs preventing backward movement, the net result is usually a forward movement of the second gonapophysis.

Primitively in the Hymenoptera, the female genital orifice lies at the base of the first gonapophysis and the eggs pass down the lumen of the ovipositor, but in the aculeate lineage the genital orifice is displaced forwards and the ovipositor is generally not used for egg laying.

In both the aculeates and many smaller parasitoids the entire ovipositor mechanism is retracted into the metasoma. In its simplest form this is achieved by having a short ovipositor and having the gonocoxa displaced forwards. During oviposition (or stinging) the gonocoxa is rotated so the anterior end moves downwards and backwards thus projecting the ovipositor. In many taxa adopting this strategy the ovipositor seems to be inserted near to the anterior end of the metasoma (e.g. some chalcidoids). In many cynipoids the oviposition system is modified still further and the IXth abdominal tergite is no longer directly involved. In the most modified of such taxa (the Ibaliidae) the gonangulum/gonocoxal complex is displaced to the anterior end of the metasoma and rotated about 270°, so at rest the ovipositor is coiled within the metasoma (Fergusson, 1988). In many scelionids and platygastrids the ovipositor, though retracted and longer than the metasoma, is not coiled but projects internally into a horn on abdominal tergite II. Viewed externally, this horn projects forwards over the propodeum and metathorax. In the ant subfamily Formicinae the hypopygium is modified into a NOZZLE and ACIDOPORE out of which are squirted secretions of the various venom glands associated with the ovipositor. In formicines, and also dolichoderines, the sting is vestigial and non-functional.

THE FEMALE REPRODUCTIVE ORGANS AND ASSOCIATED GLANDS

The majority of studies of the internal anatomy of Hymenoptera are restricted to few social aculeates and common pest symphytans, so it is presently not possible to review the anatomy of the order as a whole. However, the structure of the internal female reproductive organs and the associated glands has attracted widespread attention from workers studying parasitoid physiology (e.g. Davies & Vinson, 1986), symphytan/host plant interaction (e.g. Spradbery, 1973*a*), ovipositional strategy (e.g. Price, 1974), venom chemistry and defensive behaviour (e.g. Hermann & Blum, 1981). The following is merely a brief introduction to the most commonly employed names for the major structures.

The reproductive organs of the female (Fig. 4.22) consist of a pair of OVARIES positioned in the abdomen above or lateral to the gut. Each ovary consists of a number of OVARIOLES, distally tapered tube-like structures within which maturation of the oöcytes takes place (King & Richards, 1969); the number of ovarioles is very variable. Within each ovary, the ovarioles all open at more or less the same point into a LATERAL OVIDUCT. The length of the lateral oviduct varies considerably, but plesiomorphically it is probably relatively short. Many idiobionts have very short lateral oviducts and their reproductive organs contain only a few mature eggs, while koinobionts frequently have long lateral oviducts containing numerous mature eggs (Pampel, 1914; Iwata, 1960). In some species the most distal part of the lateral oviduct (near the bases of the ovarioles) is swollen and referred to as the CALYX. The two lateral oviducts unite medially to form the COMMON OVIDUCT, which extends to the base of the ovipositor. The part of the common oviduct nearest to the ovipositor is sometimes referred to as the VAGINA. A small sac-like diverticulum, the SPERMATHECA opens into the vagina dorsally. The spermatheca acts as a reservoir for sperm. It has an associated gland which secretes nutrients that enable the sperm to survive in a viable state for a considerable period of time (several years in some social aculeates).

Associated with the reproductive tract are two major accessory glands, the most conspicuous of which is the POISON GLAND (= acid gland; = mucus gland *sensu* Spradbery, 1973*a*) and its associated reservoir, the POISON SAC. In Apocrita, the often large, and subspherical poison sac has a muscular wall, while its associated gland either consists of a mass of fine convoluted tubules, as in some Symphyta and Ichneumonidae (Robertson, 1968; Gauld, 1991), or is rather simple and consists of a few thickened filaments (Robertson, 1968). Ancestrally the poison gland probably produced mucilage that coated and protected the egg (Robertson, 1968), but in many hymenopterans it produces substances which affect the ovipositional substrate or larval food source. In siricoid symphytans this gland produces a chemical that renders the hostplant more suitable for the growth of the fungal symbiont, while in Apocrita this gland produces complex venoms which primitively are used to subdue or kill the host/prey.

A second gland, which generally has the appearance of a simple, elongate sac, opens into the vagina. This structure, DUFOUR'S GLAND (= alkaline gland), is thought to have originally secreted lubricants to facilitate the passage of the egg down the ovipositor (Robertson, 1968). In the Apocrita this gland is known to secrete a range of hydrocarbons and other chemicals which have a variety of functions in different hymenopterous groups. In some parasitoids the secretion of Dufour's gland is known to act as a host-marking semiochemical (Guillot & Vinson, 1972), while in ants secretions from this gland serve in recruitment and defence (Hermann & Blum, 1981). The Dufour's gland of many bees is greatly enlarged and secretes a substance that is used to line the nest (Duffield *et al.*, 1984; Hefetz, 1987).

THE MALE GENITALIA AND ASSOCIATED SEGMENTS

In the male, abdominal segments III to IX (metasomal segments II to VIII) are usually normally developed, though fusion of some parts may occur, and in some taxa the terminal segments are retracted. In many Symphyta and possibly some Ichneumonoidea abdominal tergite X is distinct and separate, but in most apocritans tergites IX and X are fused to form a syntergite. In many aculeates, evaniids, proctotrupoids and cynipoids this syntergite may be concealed or very reduced and often the cerci are lost. As with the female, more studies of the smaller Apocrita are needed. The external genitalia of the male are attached to the hind margin of the ninth segment. Those of all Hymenoptera are quite similar, though in small species fusion of some elements has occurred. The more delicate parts of the genitalia are contained within a roughly conical capsule formed from two PARAMERES (= gonosquamae) that are surrounded proximally by a sclerotized BASAL RING (= gonocardo) (Fig. 4.20). Ventrobasally this structure is usually protected by a small IXth sternite (= subgenital plate). Projecting internally from the paramere are the VOLSELLAE. In the less specialized lineages the distal apex of each volsella is modified into a type of pincer comprising a DIGITUS (= gonolacinia) and CUSPIS (= distivolsella) (Fig. 4.21) which can be brought together to grasp the female during copulation. Positioned centrally is the bilobate intromittent organ, the AEDEAGUS (= penis).

In many primitive symphytans membranous 'cupping discs' are present on the parameres. The male genitalia are twisted in some xyelids and all tenthredinoids. In some aculeates (e.g. Pompilidae) the volsella is reduced, but a well-developed distal lobe, the PARAPENIAL LOBE, is present. The shape of structures of the male genitalia are widely used as specific characters in certain groups (e.g. Apidae, Pompilidae), but on the whole the form of the male genitalia is less useful for species recognition in the Hymenoptera than it is in the other major orders. In several large groups the male genitalia of most species are exceedingly similar, which presumably results from females only mating once and the consequent lack of sexual selection (Eberhard, 1985).

4.2 POLYMORPHISM

Polymorphism is the existence within a species of two or more distinct forms of the same developmental stage. Purely primary sexual (genitalic) differences are generally excluded from consideration, but secondary sexual differences are included as polymorphisms, as are environmentally induced phenotypic differences. Other forms of polymorphism include male dimorphism and two specialized forms of female polymorphism—alternation of generations and caste differentiation.

ENVIRONMENTALLY INDUCED POLYMORPHISM

It is well known that changes in the developmental environment can cause continuous variation in a population of adults (e.g. variation in size may result from availability of food), but differences in certain physical conditions may result in polymorphism in a population. For example, variations in temperature are known to produce different coloured morphs of many species. Harris (1987) reported that a pompilid species reared at 5°C produced black-bodied adults, while red-bodied adults emerged from pupae kept at 20°C regardless of the coloration of the parents. Distinct colour morphs of an *Aphidius* species (Braconidae) have been produced by rearing the progeny of one female under different temperature regimes (Liu & Carver, 1982), and similar environmental differences also have been shown to induce polymorphism in the cuticular sculpture patterns of other *Aphidius* species (Pungerl, 1983). In some multivoltine chalcidoids the overwintering generation is darker than summer generations and in *Trichogramma minutum* this has been shown to be the result of exposing the pupa to periods of low temperature (Flanders, 1931).

A more striking form of seasonal dimorphism occurs in several Chalcidoidea (e.g. Howard, 1937; Murakami, 1960), which have morphologically quite distinctive early and late season morphs. For example, a species of *Torymus*, which exploits more deeply concealed hosts in the spring than it does later in the year, has an autumn generation with short ovipositors and a spring generation with a high proportion of long ovipositors (Askew, 1971).

Environmentally induced wing polymorphism is quite common in many groups of parasitoids, including ichneumonoids, chalcidoids and bethylids, although little is known about its developmental basis. In some cases it seems to be the result of utilization of different species of hosts, as is the case in *Trichogramma sembidlis*, where males reared from moth eggs are winged whilst those reared from neuropteran eggs are wingless (Salt, 1937). The size of a host may determine whether or not a parasitoid is apterous, as is the case of a species of *Gelis*, where males reared on large hosts are fully winged, but those reared on small hosts are wingless (Salt, 1952). Wing polymorphism within a species may also result from local environmental selection. For example, brachyptery is often associated with populations of small species, such as orthocentrine ichneumonids and diapriids, living on windswept off-shore islands, while on the mainland such species are generally fully winged (Gauld, unpubl.).

Sexual dimorphism

It is customary to exclude the primary (reproductive) sexual differences from this category, but to include all the secondary sexual differences. Many Hymenoptera show pronounced sexual dimorphism in the extent of the development of their wings, in colour pattern and in many aspects of their general body structure. This is the result of very different selection pressures acting on the two sexes. Males are generally short-lived flower-feeders whose main function is to mate, while the females have a more protracted life, much of which is spent searching for and securing a food source for their offspring. In thynnine Tiphiidae, Mutillidae and many other groups, the female spends much of her time searching in crevices and thus does not need to be able to fly, whereas the male needs to visit flowers and to locate widely dispersed females. Consequently it is a widespread feature in the Apocrita for males to be fully winged while the females are apterous (Table 4.1). However, this pattern is often reversed in situations where the males mate with the females immediately upon emergence then generally die, while females need to fly to another site to oviposit, as is the case of the inhabitants of *Ficus* syconia (see Chapter 11.1).

Colour differences are most pronounced in groups where the males and females spend much of their time in different microhabitats. This is particularly well-illustrated in Costa Rica by the Ichneumoninae. In this subfamily genera such as *Ichneumon* and *Eutanyacra* show marked sexual dichromatism; males are black and yellow aposematically colour patterned, while females are cryptically brown and black coloured. In these genera the males are commonly encountered on flowers, but females are rarely encountered because they spend most of their time searching for noctuid pupae in the leaf litter or soil. Other genera in the same subfamily, such as *Macrojoppa* and *Tricyphus*, show almost no sexual dichromatism. In these taxa females attack hosts (such as sphingid larvae or swallowtail butterfly pupae) that occur in comparatively exposed situations, and like males they are often encountered flying around along the edges of paths and in similar habitats. In some cases, such as in certain Pompilidae, males and females may belong to separate chromatic mimicry complexes. For example, in South America females of *Chirodamus* are large blackish wasps with yellow wings and belong to a mimicry complex that includes females of several other pompilid genera, such as *Priocnemoides* and *Priocnemella*. Males, however, are orange-brown with dark margined hyaline wings, and they are mimics of species of *Apoica* and *Mischocyttarus* (Evans, 1968*b*).

Many sexually dimorphic features represent female specializations for interaction with the potential offspring's food. Some of the most obvious concern host detection, gaining access to the host and host handling. Host detection/examination generally involves using antennae, and in Costa Rica in many microhymenopterous groups (e.g. Chalcidoidea, Scelionidae, Eucoilini) there are striking sexual differences in the form of these organs. Usually those of the male are long and relatively unspecialized, whereas those of the female have the terminal segments clavate and richly endowed with sensilla. In larger apocritans the differences are less pronounced but males usually possess more flagellar segments than the female, while females sometimes have specialized sensory bristles (e.g. Xoridinae) or sensilla-rich flattened areas (many Phygadeuontinae). The metasoma of many female apocritans is highly modified for gaining access to the host/prey (see 4.3 below). Often it is longer and more compressed than that of the male, thus allowing the female to insert her rear end into narrow crevices. The most striking modification of this type is found in the family Pelecinidae (Mason, 1984) where females have an exceptionally long slender mobile metasoma, whereas that of the male is shorter and club-shaped. Many female aculeates have modifications to enable them to handle the host/prey. For example, most female dryinids have the fore tarsus modified into a large chela that is used to grasp the prey, while in

Family	Taxonomic distribution	Geographical distribution
Ceraphronidae	Rare: females of several genera	Mainly montane, above 2000 m
Megaspilidae	Rare: females of several genera	Mainly montane, above 2000 m
Platygastridae	Very rare: females of one genus	Lowlands, up to 1500 m
Scelionidae	Quite common: females of several genera	Widespread
Diapriidae	Uncommon: females of few genera	Widespread
Figitidae	Very rare: isolated females.	Montane species, above 2500 m
Agaonidae	Very common: all males of Agaoninae and many males in other subfamilies in *Ficus* syconia	Lowlands to mid altitude, up to 2000 m
Aphelinidae	Very rare: female of one species of *Aphelinus*	Montane species, above 2500 m
Encyrtidae	Rare: females of few taxa	Widespread
Eulophidae	Rare: males and females of few taxa	Widespread
Eupelmidae	Quite common: females of several genera	Widespread
Pteromalidae	Rare: females of some Diparinae	Lowland to mid altitude, up to 1500 m
Torymidae	Very rare: males of isolated *Physothorax* spp. in *Ficus* syconia	Lowlands to mid altitude, up to 1500 m
Ichneumonidae	Rare: females of *Gelis* spp.	Mid altitude to montane, above 1200 m
Braconidae	Rare: few doryctines (males and females); male of *Psenolobus* in *Ficus* syconia	Low to mid elevations, up to about 2000 m
Bethylidae	Very common: males and females of many genera	Widespread; commoner at lower altitudes
Sclerogibbidae	Common: all females	Lowlands up to 3000 m
Embolemidae	Common: females of *Embolemus*	Lowlands to mid altitude, 2000 m
Dryinidae	Common: females of many genera	Widespread
Tiphiidae	Uncommon: females of Thynninae, Methochinae	Lowlands to mid altitude, 1600 m
Mutillidae	Very common: all females	Lowlands to mid altitude, 1600 m
Rhopalosomatidae	Uncommon: both sexes of *Olixon*	Lower elevation sites up to 1300 m
Bradynobaenidae	Very common: all females	Lowlands up to 600 m
Formicidae	Very common: mainly worker females	Widespread to about 2400 m

Table 4.1. The taxonomic and geographical distribution of aptery and brachyptery in Costa Rican Hymenoptera.

rhopalosomatids the tarsi of females are very broad and concave beneath, presumably also an adaptation for maintaining a grip on prospective hosts. Most female bees are furnished with a scopa for pollen carriage whereas the males are not so modified.

Another class of sexual dimorphism involves male specializations for locating, courting or mating with the female. Diprionid and, in Costa Rica, many eucharitid males have plumose antennae which are probably used (like those of male saturniid moths) for detection of female pheromones. Other modifications of the male antenna play an important role in courtship behaviour. For example, the enlarged scapes found in males of *Melittobia* (Eulophidae) are used to cradle the female antenna in confinement during mating, and they contain glands which produce non-volatile sexual stimulants (Assem *et al.*, 1982*a*). The enlarged scape of some scelionid males, on the other hand, probably serves to protect the base of the antenna during combat with other males for the attention of a female (W.G. Eberhard, pers. comm.). The enlarged maxillary palps of male *Cyrtogaster* (Pteromalidae) probably also have a courtship function, while the enlarged hind tarsi of some male ichneumonids (Gauld, 1976*b*) possibly serve as visual stimuli to females. A number of species of Apoidea, perhaps most notably crabronine sphecids, have parts of the fore limbs enlarged (Low & Wcislo, 1992). The exact function of these structures is not known but the male covers the compound eyes of the female with them prior to copulation.

Male dimorphism: differing sexual strategies

A small number of species of Hymenoptera have large and small males (though in some cases a continuum exists between the extremes). In some species of *Centris* bees one morph is often female-like, while the other is characteristically larger and more robust (called METANDERS). Frequently the large males establish territories and fight with other males in order to have exclusive copulatory access to a group of females. Small males do not establish territories, but occasionally manage to 'sneak' into the territories of large males and copulate with females surreptitiously (Alcock *et al.*, 1978). In a few soil-nesting halictine bees the large males may have reduced wings, and a disproportionately large head and mandibles; these large, flightless morphs fight with one another for exclusive occupancy of a communal nest (Kukuk & Schwarz, 1988). A similar phenomenon occurs in several Chalcidoidea associated with figs—some males are winged and presumably mate outside the syconium, while others are wingless, fight with one another, and mate inside the fig (W.D. Hamilton, 1979). Day (1984) observed sexually dimorphic males in species of *Cryptocheilus* (Pompilidae), Masner (Chapter 9) notes that one species of *Telenomus* (Scelionidae) has normal winged males and brachypterous males with enlarged heads, and Gauld (1991) observed that a Costa Rican species of *Neotheronia* (Ichneumonidae) has a normal and an elongate male morph. The functions of the two morphs are not known in any of these cases.

Alternation of generations: female dimorphism

Most gall wasps (Cynipidae) associated with oaks (Fagaceae: *Quercus*) show a remarkable type of female dimorphism that is the result of the alternation of a sexual and an asexual generation (HETEROGONY). The sexual generation consists of both males and females. After mating, these females lay eggs that produce only females (the asexual generation), which in turn lay eggs that produce another sexual generation. The females of the two generations are often so different that they have been described as different species (sometimes in different genera!). Typically the female of the asexual generation is much larger and more robust than the relatively fragile, sexual generation female. Usually each generation produces a distinct type of gall, often on a different part of the host plant (Askew, 1984).

Caste differentiation

As well as the genetically determined differences between males and females, eusocial Hymenoptera show a differentiation of the female sex into reproductive and non-reproductive (worker) castes. This reproductive division of labour is one of the defining characteristics of eusociality. Beyond this however, males, reproductive females, or workers may themselves be differentiated into a series of castes (or subcastes) based either upon morphological differences, or upon duties on behalf of the colony which vary with the age of the individual. Wilson (1979) has defined a caste as 'an ensemble of colony members that specialize on particular tasks for prolonged periods of time'. Temporal castes based upon the changing role of the individual as it ages are not normally distinguished morphologically, but physical castes which are specialized to varying degrees for specific tasks are frequently very obvious in morphological terms. Wilson (1971) presented an excellent synopsis of earlier work on all aspects of eusocial hymenopteran caste differentiation, determination, evolution and function, and has expanded on many of these ideas in a series of more recent works (Wilson, 1976, 1979, 1985, and included references; Oster & Wilson, 1978).

4.3 FUNCTIONAL MORPHOLOGY

In this section two activities of adult Hymenoptera, gaining access to deeply concealed hosts and building of a nest, are examined in some detail. Various morphological specializations associated with these activities are outlined, showing how different lineages of Hymenoptera have often developed, in parallel, quite different adaptations to execute similar tasks.

ADAPTATIONS FOR REACHING CONCEALED HOSTS

There are at least six methods by which female parasitoids obtain access to concealed hosts, and the various taxa using each method show characteristic morphological specializations. In some cases, subtle but quite fundamental differences in, for example, how the ovipositor is manipulated suggest a particular habit has evolved independently in several fairly close lineages. The six methods of obtaining access to concealed hosts outlined below fall into two categories—remote (1 to 4) and proximate (5 to 6). In the former the parasitoid uses its ovipositor to penetrate

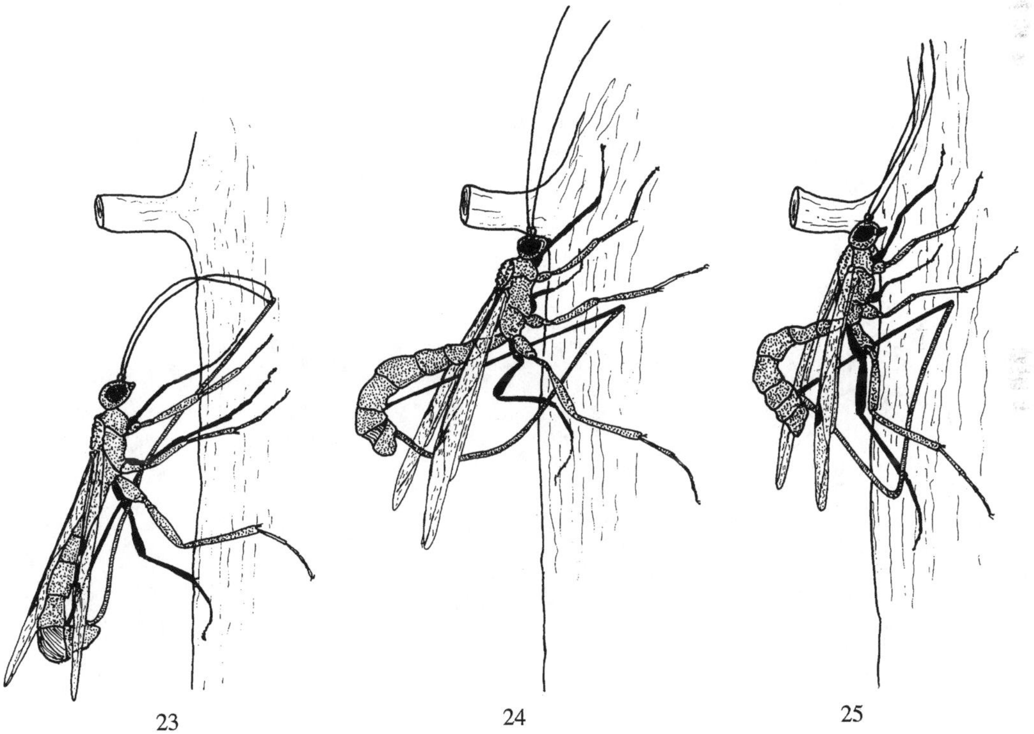

Figs 4.23–4.25. Oviposition sequence of a 'driller' (*Epirhyssa*: Ichneumonidae). 4.23, ovipositor reflexed forwards and end positioned more or less beneath head; 4.24, insect has adopted drilling posture, drilling site is now beneath mesosoma; 4.25, same with ovipositor further in.

the substrate to search for hosts, but in the latter part or all of the parasitoid's body is inserted into the substrate to reach a host.

DRILLERS. Several groups of large ichneumonoids (and perhaps stephanids, aulacids and some chalcidoids) are capable of drilling through several centimetres of tough plant tissue, or even wood, in order to reach a concealed host within. Such insects have a rigid, long and generally slightly compressed ovipositor. Boring is accomplished by the ichneumonoid positioning its body more or less perpendicularly to the substrate and bringing the tip of the ovipositor forwards, almost beneath the head (Figs 4.23–4.25). Throughout the long and apparently arduous process of drilling the ovipositor must be supported, and this has been achieved in different ways in various phylogenetic lineages. In Costa Rica rhyssine ichneumonids support the ovipositor with paired processes on the metasomal sternites, while in labenines grooves in the hind coxae perform a similar function. Some mesostenine ichneumonids apparently support the ovipositor with grooved femora, while in the Acaenitinae support seems to be provided by the greatly enlarged, ploughshare-like subgenital plate.

Other constraints of this lifeway may be expected to result in a range of various and similar adaptations being manifested by these insects. Drilling through sound timber is a time-consuming process and the ichneumonoid is potentially very vulnerable to a predator during this period. Perhaps this is why most borers are aposematically coloured. Other features that drillers have in common, such as the spherical head, short chisel-like mandibles and abundance of cuticular denticles on the head, legs or mesosoma, are probably adaptations to allow the adult to escape from its pupation chamber. The denticles or spicules possi-

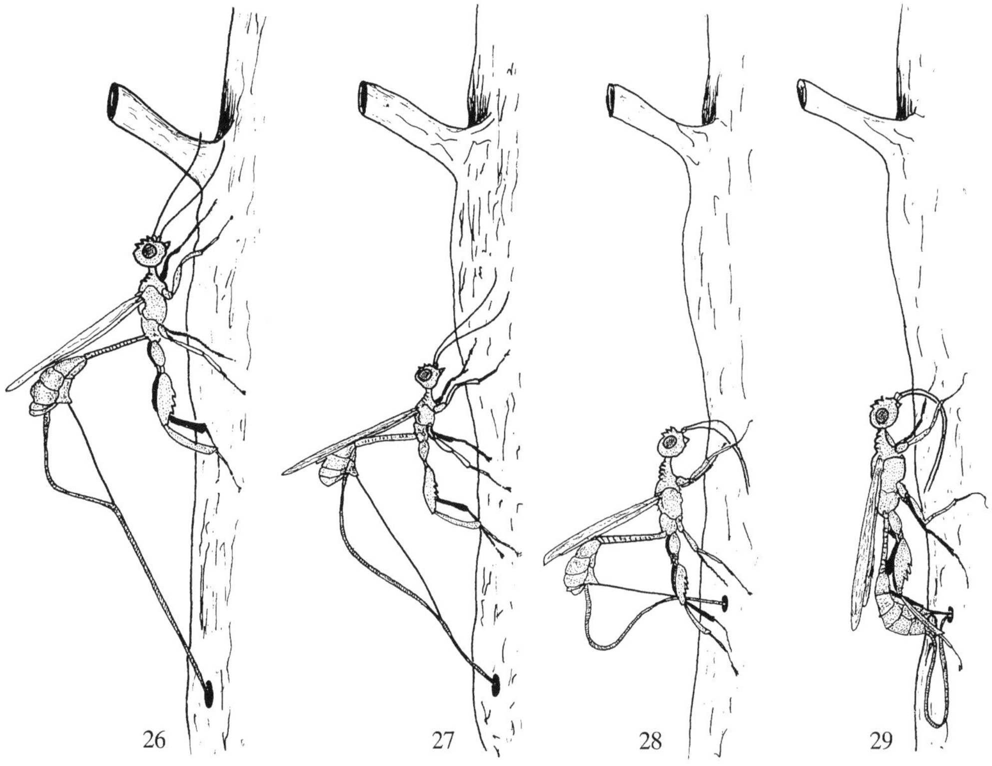

Figs 4.26–4.29. Oviposition sequence of a 'prober' (*Megischus*: Stephanidae). 4.26, apex of ovipositor is positioned into hole; 4.27–4.28, insect backs up working ovipositor into hole as described in text; 4.29, ovipositor fully inserted.

bly serve to hold the insect rigid while it twists its head back and forth, using its mandibles to bite an exit tunnel.

OVIPOSITOR PROBERS. Some groups of ichneumonoids, stephanids (Figs 4.26–4.29), and probably also other apocrite groups with long ovipositors, probe with their ovipositors through a series of cracks, tunnels and fissures in the substrate to reach a concealed host. A feature of such groups is an ability to flex the tip of the ovipositor downwards, a feat that is achieved by altering the relative position of the unequally sclerotized upper and lower valves (see Quicke, Fitton & Ingram, 1992). Although oviposition behaviour has rarely been studied in such species at least one Costa Rican species, *Dolichomitus billorum*, is known to turn during probing, causing its ovipositor to rotate (Wcislo, quoted by Gauld, 1991). This combination of rotation and flexing permits the exploration of angled side tunnels by parasitoids exploring borings. Probers do not generally have the specialized ovipositor guide structures that are found in the drillers, but they often have similar specializations to allow them to escape from the pupation site.

THREADERS. Some parasitoids, such as orussids and, in the north temperate region, ibaliids have very long ovipositors which are generally concealed coiled within the metasoma. In the ibaliids the tergal–gonocoxal muscles are degenerate and the ovipositor seems to be extruded by contraction of a muscle that swings the hypopygium down, followed by contraction of the metasoma lengthwise which has the effect of compressing the coil of the ovipositor (see Chrystal, 1930). By this means the coiled ovipositor is extruded and probably threaded through pre-existing cavities to reach a host (Figs 4.30–4.32). Threaders generally have a very slender and 'springy' ovipositor and guide it with the modified subgenital plate.

INJECTORS. Like the threaders, these are species in which the ovipositor is concealed within the metasoma,

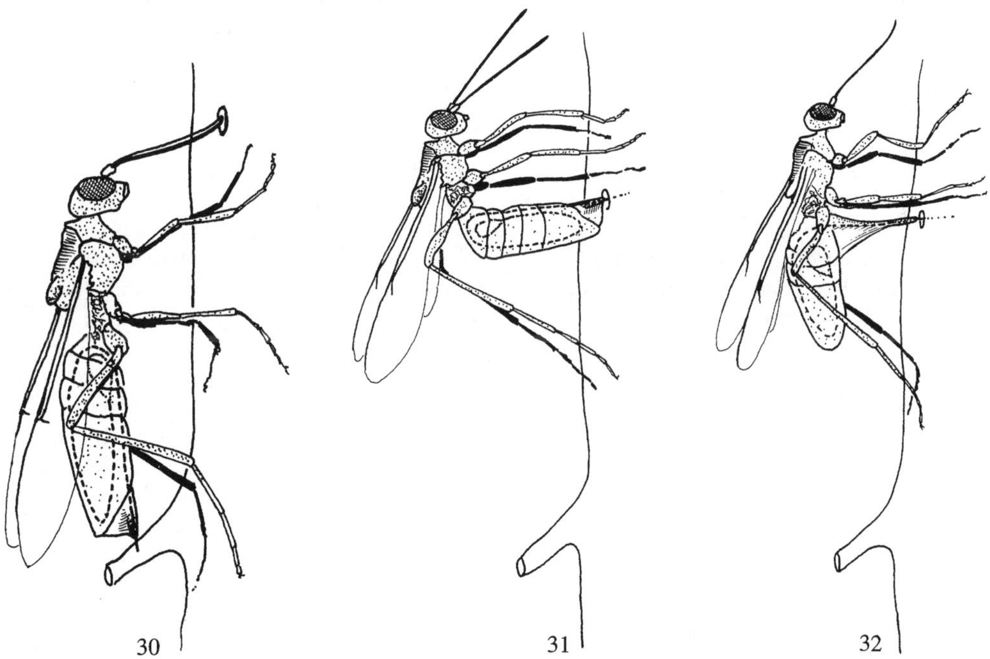

Figs 4.30–4.32. Oviposition sequence of a 'threader' (*Ibalia*: Ibaliidae). 4.30, investigating hole with antennae; 4.31, apex of metasoma positioned to hole; 4.32, subgenital plate extended down guiding ovipositor (shown as a dotted line) into hole. The metasomal segments telescope inwards extruding the ovipositor. (Redrawn and modified after Chrystal, 1930.)

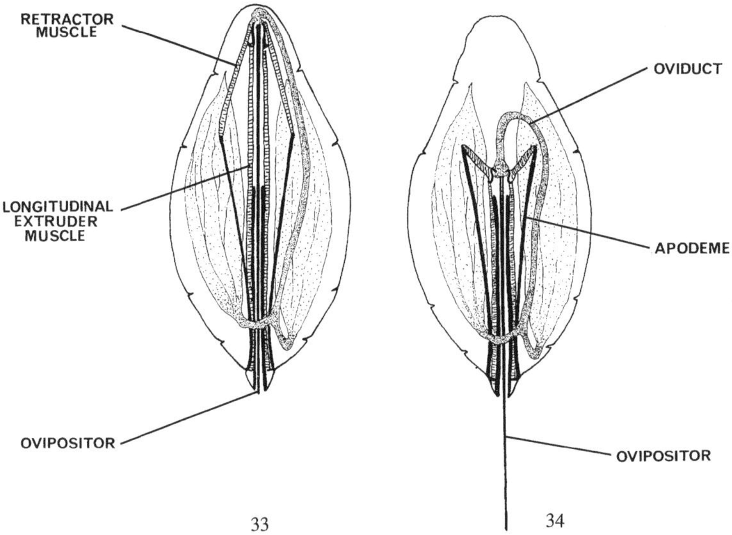

Figs 4.33–4.34. Diagram of metasomal movement of an 'injector' (*Ceratobaeus*: Scelionidae). 4.33, at rest with ovipositor retracted; 4.34, ovipositor partially extruded. (Redrawn and modified after Austin, 1983.)

but rather than being coiled and sprung, the ovipositor is straight and at rest is housed internally in a membranous tube. The invaginated ovipositor system is detached from the terminal segments of the metasoma, and is connected to the posterior metasoma only by elongated muscles. Long slender apodemes arise from the inner surface of the last visible segment and project forwards into the body cavity either side of the ovipositor (Fig. 4.33). Longitudinal muscles extending from the gonocoxa to the last tergite (abdominal tergite VIII) effect collapse of the membranous tube and eversion of the ovipositor (Fig. 4.34), while muscles extending from the gonocoxa to the apices of the slender apodemes cause retraction of the ovipositor (Austin, 1983). This type of system is found in a variety of scelionids in the *Ceratobaeus* species complex.

METASOMAL PROBERS. Several hymenopteran groups include species with highly modified, very elongate metasomas that are basically used to probe into the substrate in search of a host. Unlike the groups mentioned above that probe with their ovipositors, these species have only very short ovipositors. In Costa Rica the most extreme forms, where the entire metasoma is modified, include the Monomachidae and Pelecinidae (Mason, 1984), but rather similar though less extreme modifications are found in some ichneumonids that probe for dipterous larvae in crevices, such as species of *Neurateles* (Townes, 1971). All of these taxa have the metasoma very elongate and slender, often somewhat tapered posteriorly, and either depressed or compressed.

Other species of Hymenoptera gain access to their hosts in a similar way, but they have only the terminal metasomal segments modified. Often this modification takes the form of having the anterior or central metasomal tergites large and well-sclerotized, while the posterior ones are weakly sclerotized and telescopic (Fig. 4.35). This is found in Diapriidae where the apex of the metasoma is extruded into holes in fungi (Huggert, 1979), but such modifications are perhaps best developed in some chrysidids. Chrysidids, like other aculeates have lost the ability to use the ovipositor for laying eggs, thus those that parasitize concealed hosts must use other means of reaching the host. In Chrysididae the terminal abdominal segments are telescopic and modified into an egg-laying tube (Kimsey, 1992*b*). This tube is membranous and cannot be used to penetrate a hard substrate. Therefore, the mandibles are used to chew a hole through which the egg-laying tube can be inserted.

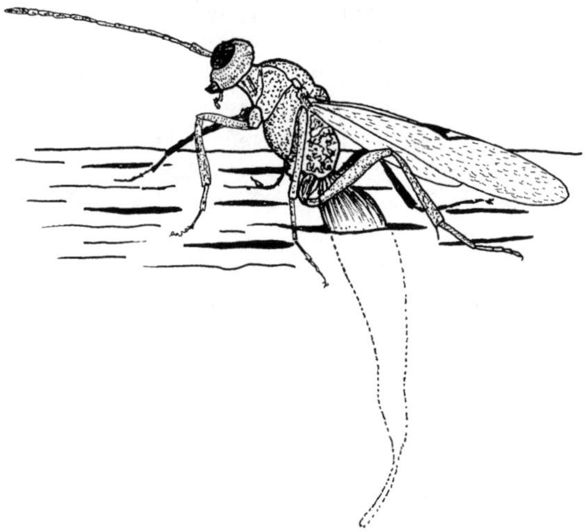

Fig. 4.35. A 'metasomal prober', a species of Diapriidae inserting its metasoma into a hole in a bracket fungus.

DIRECT ENCOUNTER. In all of the above groups the female parasitoid enters the host substrate only with her ovipositor or some portion of the metasoma. However, some parasitoids such as pristocerine bethylids, completely enter tunnels in the substrate which they then explore in order to find a host on which to oviposit. In such insects, the morphological modifications enabling them to gain ovipositional access to their host are mostly in the general form of the body. Bethylids typically have a prognathous head, a smooth flattened body, and stout (often spinose) legs. Most have the genae rather swollen to accommodate the enlarged mandibular muscles needed to bite through the substrate. Similar morphological adaptations can be found in the Scolebythidae, thynnine Tiphiidae and in members of the pteromalid subfamily Spalangiinae, which likewise enter the substrate of their hosts. Although they are not carnivorous, agaonine fig wasps must squeeze through the ostiole of a fig in order to reach the ovaries, and these wasps also have a prognathous head and flattened bodies with stout legs.

Other groups which enter different substrates, such as the soil, are frequently similarly but less strongly modified. Scoliids, tiphiine tiphiids and some ichneumonines, for example, have similar stout spinose legs.

A few lineages of Hymenoptera include species that seek their hosts in freshwater. Such species tend to have few obvious morphological specializations in common, though most have much longer tarsal claws than their non-aquatic relatives.

Adaptations for nest building

Although aculeate Hymenoptera as a group build diverse nests, from the point of view of functional morphology four general types of nesting behaviour can be distinguished: burrowers that construct tunnels in soil; 'renters', that make use of pre-existing cavities; mud daubers that construct nest pots; and resin or wax builders.

BURROWERS. The most common type of nest constructed by aculeates is a burrow that is excavated in the soil. In fossorial pompilids and sphecids the female is often equipped with rows of stout, often blade-like 'rake' setae on the front tarsi, and the tarsi themselves are commonly flattened (Bohart & Menke, 1976). Among ants, some species of *Centromyrmex* and *Pachycondyla* have stout bristles on the mid tibiae (J. Longino, pers. comm.). Other morphological adaptations of ground-nesting hymenopterans include broad mandibles for loosening soil, apically thickened hind femora (the apex often forming a flattened plate) that are used in pushing soil (scapheutine and cercerine sphecids), and a well-developed pygidial plate on the last metasomal tergite that is used in pushing and compacting soil (many sphecids, pompilids and bees).

Many bees (both burrowers and renters) have basitibial plates (scale-like structures on the outer surface of the base of the hind tibia) and spines at the apices of the tibiae (also on the outer surface) to provide traction on the burrow walls (Michener, 1974*a*). A few pompilids, ants (extratropical species), and sphecids (various Sphecini) that nest in sand have a psammophore, or 'sand basket', consisting of long curved bristles on the genal area and the ventral margin of the mandibles, which is used in conjunction with fore leg psammophores to carry sand out of the nest.

'RENTERS'. Another common type of nest are those constructed in pre-existing cavities, such as beetle burrows in wood or in hollow stems. Hymenopterans nesting in hollow stems often have elongate slender bodies, and examples of this can be found in *Pseudomyrmex* ants, pemphredonine sphecids and xeromelissine bees. Hymenopterans nesting in pre-existing holes generally close the nest with materials that are carried in from elsewhere, and some morphological features presumably facilitate this process. For example, pompilids of the genus *Dipogon* have a bifurcate brush of bristles on the maxillary cardo that is used to transport closure materials (Richards & Hamm, 1939). A rather extreme morphological modification for closing the nest entrance is found in some species of ants belonging to the genus *Zacryptocerus*, where the large, flattened head is used as a closure. Several crabronine sphecids use their large quadrate heads in a similar manner to temporarily close nests at night.

MUD DAUBERS. Many masarine and eumenine vespids burrow in hard soils and carry water to the nest site in order to soften the soil, and some 'renters' likewise employ mud in making cell partitions and closures. However, some pompilids, eumenines (potter wasps), sphecids (Sceliphrini and Trypoxylini), and a few bees ('mason bees') construct free-standing nests entirely of mud, and some of these hymenopterans show morphological modifications for manipulating mud. For example, species of *Auplopus* (Pompilidae) have a single brush of bristles on the mentum that is used in manipulating balls of mud while the last metasomal tergite is used as trowel to smooth the interior of the cell (Day, 1988). In the eumenine vespids the whole body is used to smooth the outside of the mud nest while the mandibles and glossa are used as a trowel on the inner surface (Iwata, 1976). Among mason bees, *Chalicodoma* has enlarged thoracic labial glands, which presumably provide copious amounts of saliva to moisten soil. Polistine and vespine vespids construct nests from masticated plant fibres, but show several of the modifications mud daubers possess for manipulating the viscid construction material.

RESIN AND WAX NESTS. Bees belonging to the subfamilies Euglossinae, Bombinae, Meliponinae and Apinae are characterized by the development of a corbicula on the hind tibia, which probably evolved as a basket for carrying resin to the nest (Winston & Michener, 1977; Michener *et al.*, 1978). These taxa are also characterized by metasomal wax glands and the wax secreted by these glands is used, either in conjunction with resin or separately, in nest construction (see Chapter 18). In *Bombus*, *Melipona* and *Apis* the mandibles are smooth and somewhat rounded, perhaps an adaptation that makes them more suitable for shaping wax (see Roubik, 1987).

4.4 LARVAL MORPHOLOGY

Hymenopterous larvae are basically of two types. Those of the Symphyta are more or less ERUCIFORM or caterpillar-like, with a well-developed head capsule and more or less discernible thoracic legs. Those of the Apocrita, at least in later instars are HYMENOPTERIFORM, that is rather featureless and sac-like, generally with a weakly defined head capsule and without legs (though thoracic appendages may be present). The most comprehensive overview of the larvae of the entire order is by Evans (1987*a*).

The most extensively studied hymenopterous larvae are those of the sawflies and woodwasps, particularly the exophytic forms in temperate regions. Notable studies include Yuasa (1923), Bird (1927), Parker (1935), Lorenz and Krauss (1957) and Wong (1963). There are very few morphological studies of neotropical sawfly larvae. Although in the family-level treatment we exclude the Orussidae from the sawflies and woodwasps, it is more convenient to discuss their larval morphology with that of the woodwasps as there are marked similarities between the two groups. In general the morphology of apocritan larvae is less well-known than that of sawflies, but a number of good works exist covering certain groups, especially the aculeates. Michener (1953), Grandi (1961) and Evans (1987*a*) gave overviews, and the first and last authors provided keys to families. Larvae of the parasitic Hymenoptera are reviewed by Clausen (1940*b*), Hagen (*in* DeBach, 1964), and Finlayson and Hagen (1977), the latter providing an excellent bibliography of the scattered larval descriptions. Cutler (1955) and Roskam (1982) have attempted to apply a uniform terminology to the final instar larvae of some Chalcidoidea. Other important publications on apocritan larvae include: Brothers (1972)—Mutillidae; Capek (1970, 1973)—Braconidae; Evans (1964*b*, 1964*c*)—Sphecidae; Jackson (1961)—Mymaridae; Maneval (1936)—Pompilidae, Sphecidae and Chrysididae; Maple (1947)—Encyrtidae; McGinley (1981)—Apidae; Parker (1924), Parker & Thompson (1925)—Chalcidoidea; Reid (1942)—Vespoidea; Short (1952, 1978)—Ichneumonoidea; Wharton (1989*b*)—Embolemidae; Wheeler & Wheeler, 1976—Formicidae. Despite the existence of this considerable body of literature, and a much larger number of small papers, many taxa amongst the smaller parasitoids need to be studied in more detail, and the larvae of several families, including the Sierolomorphidae and Scolobythidae, have yet to be described.

For convenience the following account of morphology is divided into three sections, the symphytan larva, the hymenopteriform apocritan larva, and the early-instar parasitoid larvae.

The symphytan larva

Symphytan larvae resemble caterpillars, though those that live in concealment generally have less well-developed features, such as legs and antennae, and a few are legless and somewhat grub-like.

The symphytan larval head is well-sclerotized, hemispherical and usually hypognathous (Fig. 4.36), though it is more or less prognathous in some leaf-mining species. It generally bears setae, which are most numerous ventrally. Viewed anteriorly, the most conspicuous feature of the head is an inverted Y-shaped suture, the EPICRANIAL SUTURE. The median part of this is the MEDIAN ECDYSIAL LINE (= coronal suture) and laterally are the FRONTAL SUTURES. The dorsal part of the head and the area lateral to these sutures is the FRONTO-PARIETAL region, and this is confluent ventrally with the GENA. Between the frontal sutures is the FRONS which is demarcated ventrally from the CLYPEUS by the EPISTOMAL SUTURE (= frontoclypeal suture). Ventrolaterally on either side of the head is an ANTENNA, and close to it a STEMMATUM (= 'ocellus'). In the Tenthredinoidea and the Macroxyelinae the stemmatum is dorsal to the antenna, but in other Xyelidae, the Megalodontoidea and the Cephoidea it is positioned more posteroventrally. In siricoids and orussoids the stemmata are absent. Antennae are present in all symphytans. Generally they are inserted on a large convex basal ANTACORIA and may have from one to seven segments. These segments are usually strongly sclerotized, more or less ring-like; the more distal segments are successively smaller in diameter, and the most distal segment is more or less conical. In most tenthredinoids the antenna is 3- to 5-segmented,

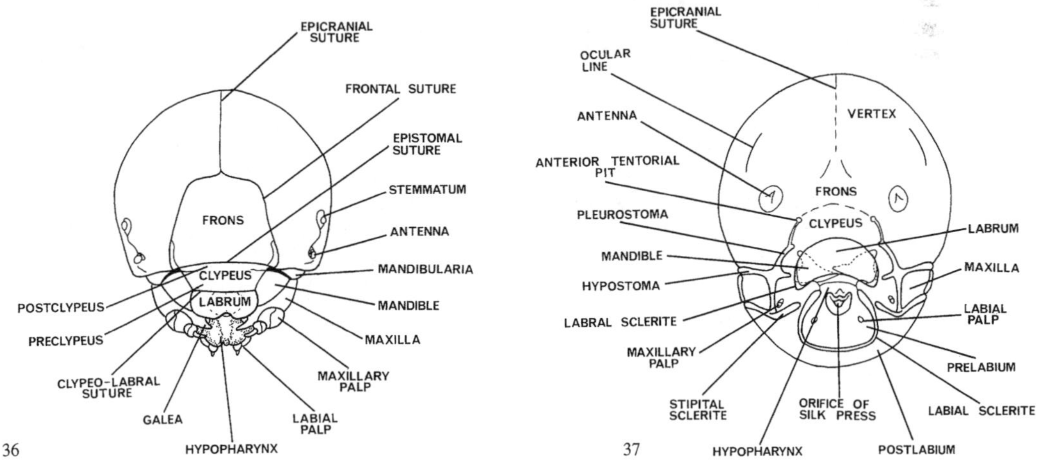

Figs 4.36–4.37. Heads of larval Hymenoptera. 4.36, Tenthredinidae; 4.37, Ichneumonidae.

though argids have one and cimbicids two segments. The more primitive sawflies tend to have more segments (six or seven in xyelids, pamphiliids and blasticotomids) while the endophytic taxa generally have a reduced number (e.g. one in siricids).

The CLYPEUS is more or less distinctly delineated. Ventrally it is separated from the labrum by the CLYPEO-LABRAL SUTURE and its lateral margins are free. Some authors recognize it as comprising two parts, the proximal POSTCLYPEUS and a distal PRECLYPEUS. It bears between 2 and 10 setae; their number and positions are of taxonomic importance. The LABRUM is usually transverse and often has a small median emargination; in many taxa the two lobes are of different size so the labrum is asymmetrical. The labrum also bears a number of taxonomically important setae. The MANDIBLES are characteristically short and stout with several short, rather blunt, apical teeth; usually they are asymmetrical. The MAXILLAE are always present and typically each comprises a basal CARDO and a distal STIPES. The PALPIFER is a more or less membranous, mound-like lobe attached to the distal end of the stipes; distally it bears one or more setae and the MAXILLARY PALP. Typically this palp is 4-segmented, but in endophytic species it is reduced; for example, it is 2-segmented in the Siricoidea. The GALEA and LACINIA are present in all taxa except orussids. The LABIUM consists of a POSTMENTUM (= prementum *sensu* Lorenz & Krauss, 1957), PREMENTUM and a MEDIAN TERMINAL LOBE (= totaglossa) which is formed from fusion of the glossae, paraglossae and part of the hypopharynx. The labial palp is normally 3-segmented, but in many endophytic taxa the number of segments is reduced. Near the distal end of the median terminal lobe is the opening for the duct of the silk gland, the SPINNERET (= sericos).

The segments of the THORAX generally bear dorsal wrinkles, the ANNULETS. A large mesothoracic spiracle is present laterally on the prothorax near its posterolateral margin (Yuasa, 1923) and a small metathoracic spiracle is usually visible above and behind the middle leg. In the Xyeloidea, Megalodontoidea and most Tenthredinoidea there are three pairs of well-developed legs and the coxa, trochanter, femur, tibia and a fused tarsus + tarsal claw are recognizable. In endophytic species of all sawfly groups the parts may be fused and the legs are rather fleshy; in orussids legs are not discernible (Middlekauff, 1983).

The ABDOMEN typically is composed of ten segments which, with the exception of the posterior two, are more or less similar in structure in that they bear spiracles laterally and are dorsally subdivided into four to seven annulets. Typically PROLEGS are present on the second to the seventh or eighth and the tenth segment. Only in the Xyelidae are prolegs found on the first and ninth segments. In many endophytic tenthredinids the prolegs are vestigial and they are entirely absent in the Blasticotomidae, Siricoidea, Cephoidea, Orussoidea and Megalodontoidea, though in the last mentioned group weak fleshy lobes may be discernible ventrally on some segments. The ninth segment does not bear a spiracle and the tenth segment is highly modified and does not have annulets. Posteriorly the tenth segment bears two membranous lobes, a SURANAL LOBE above the anus, and a SUBANAL LOBE below it. The tenth segment may possess various processes. In the Siricoidea and Cephoidea a median posterior SURANAL PROCESS (= postcornu) is present on the suranal lobe, while in Nematinae a pair of conical CAUDAL PROTUBERANCES (= pseudocerci or 'cerci') are present on the posterior margin of the tergite above the suranal lobe. In the Megalodontoidea paired, segmented SUBANAL APPENDAGES (arthrostyli or 'cerci') are present ventral to the lateral ends of the anal slit.

THE HYMENOPTERIFORM APOCRITAN LARVA

Amongst the Apocrita the generalized larval morphology is best illustrated by the HYMENOPTERIFORM (eucephalous) larva. The later instars of all taxa (except Eucharitidae) are of this general form. The hymenopteriform larva is spindle-shaped and maggot-like, with whitish translucent skin that is generally rather smooth, though it may bear small spines or bristles. The head is usually hypognathous, hemispherical, and weakly to moderately strongly sclerotized. In many mature larvae the head capsule is partially retracted into the thorax. The remainder of the body is rather featureless, and usually has 12 to 13 visible segments. The number of pairs of open spiracles in the mature larva varies from ten (most Aculeata), to nine (many Chalcidoidea and most

Ichneumonoidea), or fewer (many Proctotrupoidea); larvae of Mymaridae and Trichogrammatidae apparently never develop tracheal systems. Among ectoparasitic species the number of spiracles usually remains constant in Ichneumonoidea whereas first instar Chalcidoidea often have only four or five pairs of spiracles (Hagen *in* DeBach, 1964). Many endoparasitic first instar larvae respire cutaneously and are apneustic (without open spiracles). The form of the tracheal trunks and commissures are poorly studied, but of potential taxonomic value.

The head of the hymenopteriform larva resembles that of the Symphyta in general appearance, though it differs in that the sensory organs and appendages are generally reduced (Fig. 4.37). The antennae, if present, are 1-segmented and either nipple-like or disc-like; no stemmata are present. The head typically bears a pair of mandibles, which may be more or less elongately conical, slightly curved and terminate in a single tooth (e.g. most Ichneumonoidea; many Formicidae and Apidae), or they may be stouter with two (some polistine Vespidae), three (Stephanidae, Trigonalyidae, Gasteruptiidae, Scoliidae, eumenine Vespidae), or four teeth (Tiphiidae, Mutillidae). In some families the number of mandibular teeth in the mature larva is quite variable (e.g. 1 to 4 in Formicidae and 2 to 5 in Sphecidae), and within a species the number of teeth may increase in progressive instars (Edwards, 1980). In many taxa auxiliary denticles are present on the distal dorsal surface (e.g. many Apidae—McGinley, 1981; some Ichneumonidae—Short, 1978), or on the ventral surface (e.g. some Braconidae—Finlayson & Hagen, 1977). The maxillae are generally rounded and fleshy and protrude from each side of the head. In most aculeates two nipple-like projections (the 1-segmented palp and galea) are discernible distally, but in apids and many parasitoids the maxilla is very reduced with only a small vestigial palp discernible distally. The labium is divided by a weak transverse furrow into a postmentum and prementum; the latter bears a pair of apical tubercles which seem to be vestigial 1-segmented palps. In most Apocrita a median, slit-like orifice of the silk gland is present on the labium (the spinneret), although most Sphecidae (except Ampulicinae and Sphecinae) have a pair of openings.

In Ichneumonidea the maxillae and labium barely project and the most conspicuous features of the head are the sclerotized rods that are often present at the junction between the weak sclerites. Michener (1953) recognized six main sclerotized rods, for which we have adopted the terminology of Short (1952, 1978). The EPISTOMA follows the course of the epistomal suture and surrounds each anterior tentorial pit; it is often weakly sclerotized mediodorsally and not apparent, and in such cases the epistomal arch is said to be incomplete. Lateral to the mandible is the PLEUROSTOMA, which bears processes on which the mandible articulates. The dorsal end of the pleurostoma is confluent with the epistoma and its ventral end is confluent with the HYPOSTOMA, a rod that extends from near the ventral articulation of the mandible to the posterior tentorial pit. A HYPOSTOMAL SPUR projects ventrally from the hypostoma across the stipes and fits in a depression on the sclerotic rod (the STIPITAL SCLERITE) that supports the maxilla ventrally. The medial end of each stipital sclerite fits into a socket in the LABIAL SCLERITE, a more or less U-shaped sclerite that borders the labium laterally and ventrally. In many taxa a more or less Y-shaped PRELABIAL SCLERITE is also present; the arms of this sclerite support the opening of the silk-press and consequently this sclerite is best developed in taxa which spin a thick cocoon.

The Proctotrupoidea, Cynipoidea, and Chalcidoidea are characterized by extreme reduction of all of the mouthparts except the mandibles. The labium and maxillae are often indistinguishably fused with the hypopharynx (Roskam, 1982) and the palpi are distinguishable only as minute papillae, or are absent. In some ectoparasitic Chalcidoidea well-developed setae are present on the head and body and the position and number of these is important in discriminating species (e.g. in the Eurytomidae; Roskam, 1982).

The early instar parasitoid larvae

Although the early instars of many aculeate apocritan larvae are more or less hymenopteriform and successive larval instars resemble each other, HYPERMETAMORPHOSIS (= heteromorphosis) is relatively common amongst parasitic species. In these species the first, and less commonly the second, instar is highly modified and differs markedly from successive larval instars. Some of these specialized first instars are typical of a particular mode of life (e.g. planidia and microtype), while others are restricted to

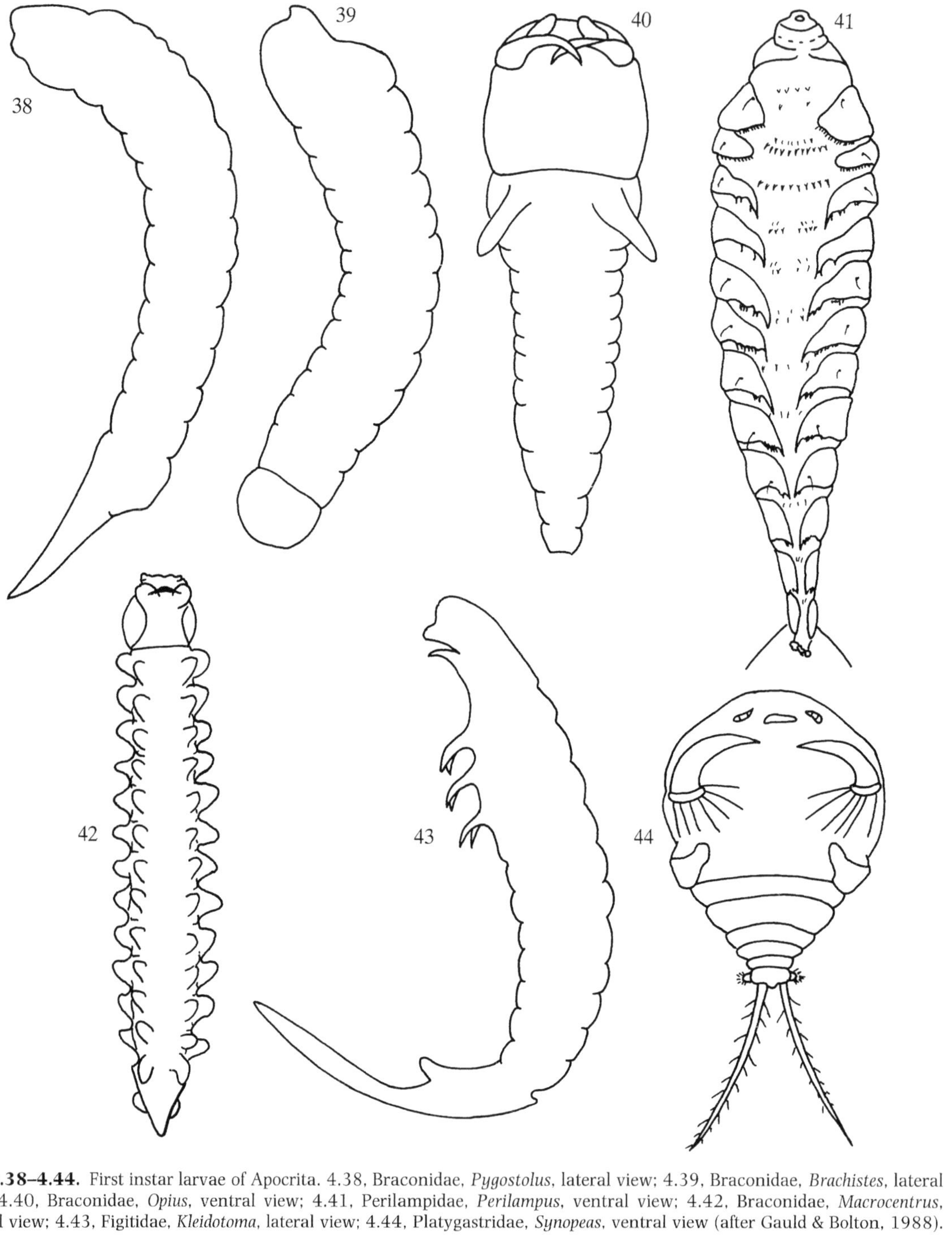

Figs 4.38–4.44. First instar larvae of Apocrita. 4.38, Braconidae, *Pygostolus*, lateral view; 4.39, Braconidae, *Brachistes*, lateral view; 4.40, Braconidae, *Opius*, ventral view; 4.41, Perilampidae, *Perilampus*, ventral view; 4.42, Braconidae, *Macrocentrus*, ventral view; 4.43, Figitidae, *Kleidotoma*, lateral view; 4.44, Platygastridae, *Synopeas*, ventral view (after Gauld & Bolton, 1988).

the members of a particular taxonomic group. The following 'classification' of larval types (first outlined by Clausen, 1940*b*) is largely an artificial one based on superficial resemblance, but it is useful because parasitoid species of many groups tend to have larvae of one or other form. Only a few taxonomically isolated species (*Agriotypus* and *Collyria* in the Ichneumonidae) with an unusual biology are known to have distinctive larvae that cannot be satisfactorily placed in this classification. However, the early stages of a very large number of parasitoids are, as yet, unknown.

PLANIDIUM (Fig. 4.41). A minute larva with a fusiform body that is usually ventrally flattened, segmented, with the segments moderately sclerotized and bearing stout setae and/or spines. This type of first instar larva is found in a variety of parasitic species that deposit their eggs apart from the host (e.g. Eucharitidae, Perilampidae, *Euceros* in the Ichneumonidae). The larva is initially free-living, during which time it has to locate its host. The sclerotized body prevents desiccation and the spines facilitate forward movement. The detailed anatomy of chalcidoid planidia was investigated by Heraty and Darling (1984).

MICROTYPE. An extremely small, maggot-like larva that hatches from a minute egg. The egg hatches only if it is ingested by a caterpillar. Upon hatching the microtype larva then makes its way from the gut to the body cavity of the caterpillar. Among Hymenoptera this type of first instar occurs only in Trigonalyidae, but outside Hymenoptera it occurs in several tachinid dipterans.

CAUDATE (Fig. 4.38). A rather unspecialized, distinctly segmented larva with a long fleshy caudal appendage arising from the last abdominal segment. There has been considerable discussion about the function of this appendage (Thorpe, 1932; Finlayson, 1964) and it may have a respiratory function. If present, a caudal appendage is usually well-developed in the first larval instar and is progressively lost in later stages; frequently it is not discernible in the final larval instar. The caudate larva is of common occurrence in many endoparasitic ichneumonoids, but is rare in other groups.

VESICULATE (Fig. 4.39). Rather similar to a hymenopteriform larva, but the proctodeum is evaginated to form a more or less spherical external vesicle (Thorpe, 1932). The vesiculate larva is common among endoparasitic Braconidae. Some ichneumonid larvae are intermediate between caudate and vesiculate in that the larva has a short caudal appendage with a small vesicle near its base (Bledowski & Krainska, 1926).

MANDIBULATE (Fig. 4.40). The thorax and abdomen of this larva taper posteriorly and are distinctly segmented; the head is large (frequently up to one third of the body length), more or less prognathous, quite heavily sclerotized and bears a pair of large sickle-shaped mandibles. In parasitoids where the early instars fight this form of larva is particularly common. It occurs in both endoparasitoids (Clausen, 1940*b*) and ectoparasitoids (Spradbery, 1969, 1970*b*).

CYCLOPIFORM (Fig. 4.44). An endophagous larva with a large unsegmented cephalothorax bearing conspicuous, well-developed falcate mandibles. The abdomen is slender, has few apparent segments, and generally terminates in a bifurcate caudal appendage. This larva is only found in some Platygastridae.

EUCOILIFORM (Fig. 4.43). A weakly segmented larva with a moderately large head, a barrel-shaped body and a long caudal appendage. The thorax bears three pairs of long fleshy ventral processes. This type of larva is characteristic of certain Cynipoidea (e.g. Figitinae). All of these endophagous larvae appear to have a polypodeiform second instar.

TELEAFORM (Fig. 4.46). An unsegmented larva with a weak constriction between the cephalothorax and abdomen. The former generally bears a pair of large ventrally directed fleshy protuberances (? mandibles). The abdomen is subspherical and has a partial ring or one or more rings of long posteriorly directed bristles. The abdomen generally terminates in a slender fleshy caudal appendage. This endophagous larva is of wide occurrence in the Scelionidae.

MYMARIFORM (Fig. 4.47). Quite similar to the teleaform larva, but without an obvious constriction between the cephalothorax and abdomen. The head bears a conical process anteriorly. This type of endophagous larva is found in many Mymaridae and Trichogrammatidae (Jackson, 1961). In at least one

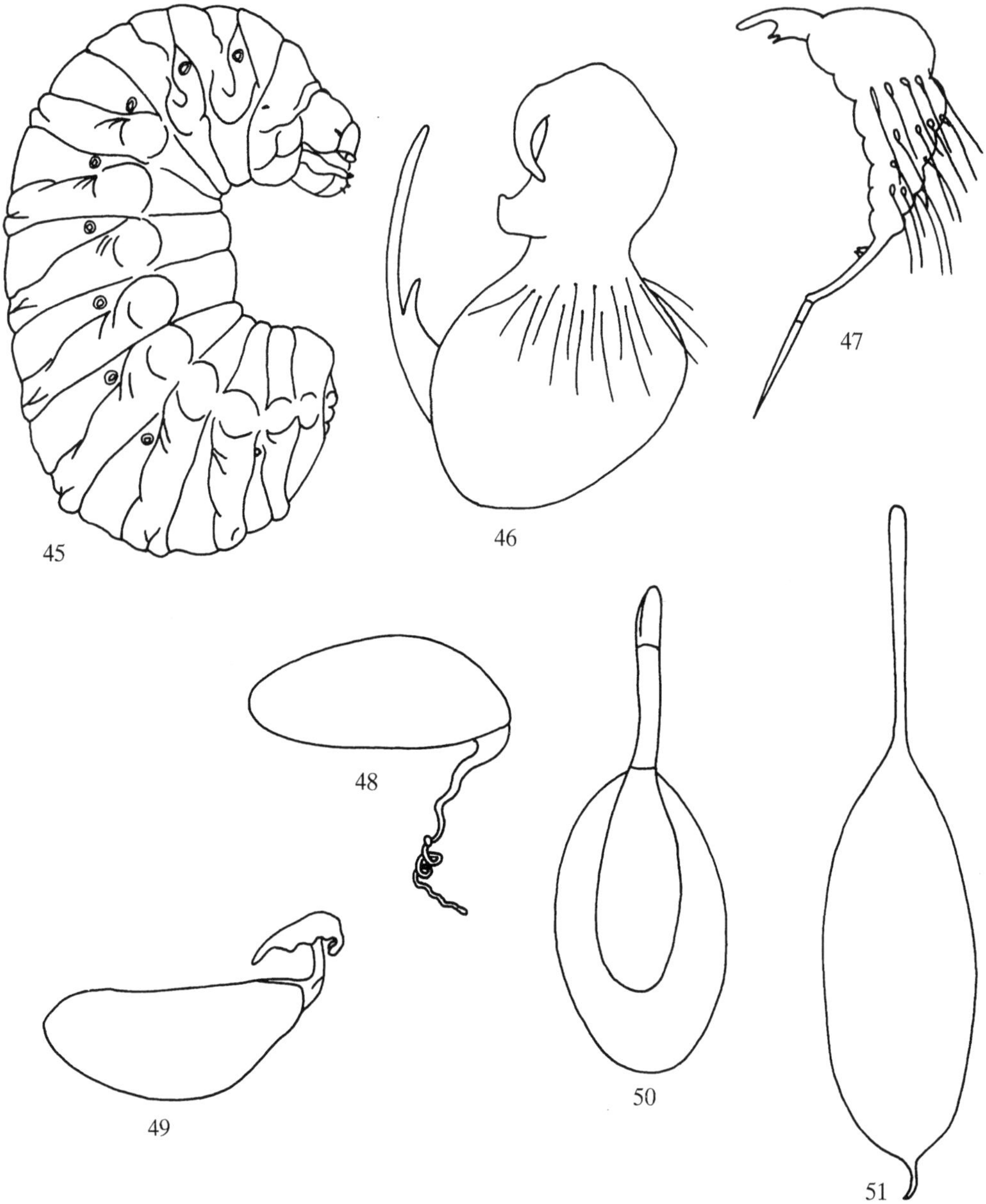

Figs **4.45–4.51.** Larvae and eggs of Apocrita. Figs 4.45–4.47. Larvae; 4.45, final instar larva of *Megachile* (Apidae), lateral view; 4.46, first instar larva of *Telenomus* (Scelionidae); 4.47, first instar larva of *Polynema* (Mymaridae). Figs 4.48–4.51. Eggs; 4.48, *Netelia* (Ichneumonidae); 4.49, *Tryphon* (Ichneumonidae); 4.50, *Blastothrix* (Encyrtidae); 4.51, *Eupelmus* (Eupelmidae) (after Gauld & Bolton, 1988).

species of trichogrammatid this larva has a segmented abdomen and is somewhat cyclopiform, though it lacks the characteristic mandibles and bifurcate tail of cyclopiform larvae.

SACCIFORM. An ovoid featureless larva that lacks visible segmentation and setae. This type of first instar is found in some species of Mymaridae, Trichogrammatidae and Dryinidae.

ENCYRTIFORM. These are among the few metapneustic larvae that occur in Hymenoptera—i.e. there is only a single pair of spiracles present and these occur on one of the posterior segments. The larva is oriented with its spiracles inside the egg stalk and the latter protrudes through the body wall of the host (Maple, 1947). This type of larva occurs in many Encyrtidae. Intermediate instars maintain contact with the egg but the final instar usually has the normal nine pairs of spiracles (peripneustic).

The intermediate larval instars usually resemble the final instar (i.e. they are hymenopteriform), but a few groups have unique intermediate stages. Certain figitids with eucoiliform first instar larvae, helorids and some proctotrupids possess a POLYPODEIFORM second instar larva—a cylindrical, segmented larva that bears paired fleshy processes ventrally. Similar larvae are also found in some braconids (Fig. 4.42). The final instar larva of some eucharitids is of a rather similar form (see Michener, 1969). Mymaridae with sacciform first instar larvae moult into a HISTRIOBDELLID second instar (Jackson, 1961)—a cylindrical larva with six body constrictions, a pair of fleshy lateroventral appendages on the head (in addition to extruded, slender mandibles) and a caudal segment bearing a pair of ear-like appendages. Trigonalyidae, which have microtype first instar larvae, moult into a hymenopteriform second instar, then into a mandibulate third instar, which is followed by two final hymenopteriform instars.

Sexual dimorphism in larvae may occur with no obvious differences in the biology between the two sexes as in some Mymaridae (Tooke, 1955), but the most striking cases of sexual dimorphism occur in Aphelinidae where female and male larvae differ biologically (see Chapter 11.2). A different type of larval dimorphism occurs in some polyembryonic Encyrtidae, where one type of larva is a 'guard' morph that fails to mature while the other is a normal feeding morph that gives rise to an adult encyrtid (Cruz, 1981, 1986*a*, *b*).

The parasitoid egg

The great majority of parasitoids have a more or less ellipsoidal egg with a thin chorion with weak or more or less no surface sculpturing. A few taxa, however, have structurally highly modified eggs. Those of many wood-boring Hymenoptera, such as rhyssine ichneumonids, are exceedingly elongate, up to 14 mm long and only 0.4 mm wide at maximum (Iwata, 1958). The slender part of these eggs passes down the lumen of the ovipositor until it protrudes apically, then the egg contents are squeezed from the innermost swelling through the ovipositor to the distal part, before the remainder of the egg passes down the ovipositor. The eggs of tryphonine ichneumonids have a thickened, incompressible chorion and are furnished with an "anchor" (Figs 4.48–4.49) which is inserted into the host's cuticle to anchor the egg externally. Eggs of anomalonine ichneumonids are shaped rather like an old-fashioned oil-lamp, and have a mushroom-like protuberance which is used to attach the egg to an organ within the host's haemocoel (Gauld, 1976*b*). Amongst the Chalcidoidea many Encyrtidae have eggs with a more or less funnel-like appendage (Fig. 4.50), the end of which protrudes from the host after oviposition. It apparently functions as a respiratory cornus (Clausen, 1940*b*). Some eurytomids, which oviposit in plant tissue, have pedicellate eggs (Fig. 4.51), and these show species-specific patterns of chorionic microsculpture (Claridge & Askew, 1960).

5 The evolution, classification and identification of the Hymenoptera

Ian D. Gauld and Paul E. Hanson

5.1 PHYLOGENETIC RELATIONSHIPS OF THE ORDER

The Hymenoptera is a very distinctive group of holometabolous insects whose monophyly is attested by several synapomorphies including: the possession of two pairs of membranous wings, the smaller hind pair of which are coupled to the larger fore pair by one or more small hooks (the hamuli), and possession of a haplo-diploid sex determination mechanism whereby diploid fertilized eggs normally give rise to females and haploid unfertilized eggs normally give rise to males. Females primitively have a lepismatid form of ovipositor (Scudder, 1961), which may be a secondarily derived (apomorphic) feature, or may represent a retention of the ancestral condition for insects.

The origin and relationships of the Hymenoptera are poorly understood and at various times it has been suggested that the group is closely related to almost every other order of holometabolous insects (see review in Hennig, 1981). In more modern works there is a marked dichotomy of opinion about the ancestry of the order. Königsmann (1976) suggested that they are probably the sister-group of the Mecopteroidea, a group that includes the orders Mecoptera, Trichoptera, Lepidoptera, Diptera and Siphonaptera, and this view is supported by Kristensen (1975). Rasnitsyn (1980) challenged this hypothesis and suggested that the Hymenoptera represents one evolutionary lineage from the extinct order Miomoptera, while all the other extant holometabolous orders constitute a separate evolutionary lineage that has arisen independently from miomopteran ancestors.

5.2 EVOLUTIONARY HISTORY OF THE ORDER

The Hymenoptera has existed for more than 220 million years. Their earliest representation in the fossil record is as xyeloids in Middle Triassic strata from Central Asia (Rasnitsyn, 1969) and the Upper Triassic from Australia (Riek, 1955). By the early Jurassic, some 180 million years ago (mya), the group had radiated extensively and twenty one families were represented (Rohdendorf, 1968; Rasnitsyn, 1969, 1980). Many of these are now extinct, but the modern families present in this fauna include Anaxyelidae, Siricidae, Xyelidae and, among the apocritans, Heloridae and Megalyridae. The Jurassic (135–200 mya) hymenopterous fauna was dominated by species of the superfamilies Xyeloidea, Siricoidea and Megalodontoidea, phytophagous groups that were probably associated with pteridophytes, gymnosperms and other non-flowering vascular plants that were the dominant vegetation at the time. Many modern representatives of these groups are now mostly confined to northern temperate regions where they are still associated with non-flowering plants, especially species of Coniferae.

Although the earliest tenthredinoids appeared in the Upper Jurassic (ca 140 mya) most of the modern tenthredinoid families appeared towards the end of the Cretaceous (95 mya). It is possible that these, and most of the apocritan families, arose with the radiation of the angiosperms in the late Cretaceous. By the middle of the Palaeocene (ca 55 mya) all extant hymenopteran superfamilies, and most of the families were represented.

THE EVOLUTIONARY HISTORY AND RELATIONSHIPS OF THE SYMPHYTA

The sawflies and woodwasps, which are traditionally classified as the suborder Symphyta, include the structurally most primitive Hymenoptera as well as those that appeared first in the fossil record. The very earliest symphytan fossils are the xyeloids (Rasnitsyn, 1980), and some members of this group show archaic features that appear to have been lost in other hymenopterans. For example, xyeloids have the most extensive wing venation of any Hymenoptera; some have veins *1rs-m* and *Rs2* present in the fore wing. The xyeloid thorax is also very primitive for, like some neuropteroids, the anepisternum is confluent (as a small anterodorsal lobe) with the preepisternum. All Hymenoptera, other than xyelids, have the anepisternum separated from the preepisternum and detached as a separate sclerite, the postspiracular sclerite (Gibson, 1993). This observation supports the hypothesis of Rasnitsyn (1969, 1980, 1988) that the Xyelidae is the ancestral group from which all other Hymenoptera are derived. However, it does not support Rasnitsyn's other postulate, a diphyletic origin for the order from a xyelid ancestor, with one lineage, the tenthredinoid line, arising from the Macroxyelinae, and all other Hymenoptera arising from the Xyelinae. Other attempts at reconstructing the phylogeny of Hymenoptera using extant material (e.g. Königsmann, 1977, 1978*a*, *b*) or by study of particular character suites (e.g. N.F. Johnson, 1988; Whitfield *et al.*, 1989) also have failed to support this diphyletic arrangement, although the monophyly of the Tenthredinoidea is supported by a number of features of both adults and larvae (Königsmann, 1977; Gibson, 1985). A synthesis of current opinion regarding the phylogenetic relationships of the higher taxa in the order is shown in Fig. 5.01. Here the five lineages of non-xyelid hymenopterans (Tenthredinoidea; Megalodontoidea; Cephoidea; Anaxyelidae; remaining siricoids + orussoids + apocritans) are shown as an unresolved polychotomy.

All extant members of the Tenthredinoidea, except the Blasticotomidae, are strophandrous, i.e. have the male genitalia rotated 180° prior to eclosion. This lineage first appeared in the fossil record in the Middle Jurassic (ca 160 mya) where it is represented by the very primitive extinct family Xyelotomidae. Rasnitsyn (1980, 1988) hypothesized that this family is

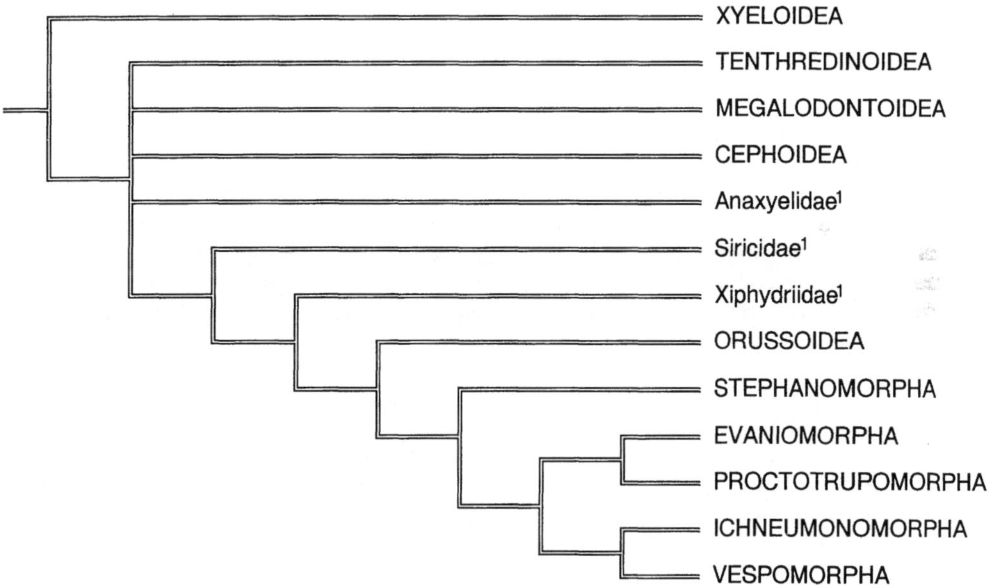

Fig. 5.01. The phylogenetic inter-relationships of the major hymenopterous groups, mainly after Rasnitsyn (1988), but leaving stephanids as Rasnitsyn (1980) and replacing Rasnitsyn's concept of a diphyletic origin for the group from the Xyelidae with Gibson's (1993) interpretation of the Xyelidae as the most primitive group of Hymenoptera. [1]= 'Siricoidea'. The relationship of the Xiphydriidae and Siricidae is problematic as there is some evidence for the relationship shown here, and some suggesting they comprise a monophyletic group (see text).

ancestral to all other tenthredinoids and may have radiated at the end of the Jurassic (135 mya) giving rise to three lineages, the Blasticotomidae, the Argidae + Pergidae + Electrotomidae, and the Tenthredinidae + Cimbicidae + Diprionidae. The Tenthredinoidea is quite well-represented in early Cenozoic strata, and during the Palaeocene and Eocene (40–65 mya) the group may well have been the dominant exophytic leaf-eating group of insects. Their major potential competitors, the macrolepidoptera, are not conspicuous until the Oligocene (ca 35 mya), but by the Pliocene (ca 9 mya) macrolepidopterans appear to have been more common than tenthredinoids (Rasnitsyn, 1980).

The remaining traditional symphytans are generally classified in three or four superfamilies, the Megalodontoidea, Cephoidea, Siricoidea and Orussoidea. The orussids are sometimes included in the Siricoidea (e.g. Krombein *et al.*, 1979), but more recently they also have been included in the Apocrita (Rasnitsyn, 1980; see below). Of these four superfamilies, the Megalodontoidea appeared first in the fossil record where it is represented in the lower Jurassic (ca 185 mya) by the extinct family Xyelydidae. The Siricoidea also appeared at the beginning of the Jurassic but attained its greatest diversity by the late Jurassic (140 mya) when it was represented by five families, Siricidae, Anaxyelidae, Gigasiricidae, Sepulcidae and Pararchexyelidae (Rasnitsyn, 1980). The last three are now extinct, and the Anaxyelidae is represented in the modern fauna by a single western North American species. Orussids date from the Upper Jurassic (ca 140 mya), while the Cephoidea did not appear until the Lower Cretaceous (ca 110 mya), possibly from a sepulcid ancestor (Rasnitsyn, 1988).

The origin of the Apocrita

It is universally accepted that the apocritan Hymenoptera arose from a symphytan ancestor, but to which group it is in fact most closely related has been the subject of some disagreement. Both the Orussoidea and Cephoidea have been postulated as being candidates (Königsmann, 1977; Rasnitsyn, 1980), but the overwhelming majority of morphological evidence now favours the former (Gibson, 1985; N.F. Johnson, 1988; Rasnitsyn, 1988; Whitfield *et al.*, 1989). Indeed none of these authors found any morphological evidence to support a sister-group relationship between the cephids and apocritans. Evidence, based on study of the pro- and mesothorax (Gibson, 1985), the mid coxal articulation (N.F. Johnson, 1988) and the metapostnotal structure (Whitfield *et al.*, 1989) suggests that the Apocrita and Orussidae are sister-groups (see also Chapter 7). The Xiphydriidae may be a sister-group to these and the Siricidae the sister-group of all three (Gibson, 1985; Rasnitsyn, 1988), although Whitfield *et al.* (1989) found evidence suggesting the Xiphydriidae and Siricidae comprise a monophyletic lineage.

The evolutionary history and relationships of the Apocrita

The oldest indisputable apocritan fossils, proctotrupoids, occur in middle Jurassic sediments (ca 160 mya) in Central Asia, and by the late Jurassic (ca 150 mya) eight families, including representatives of the Evanioidea, Stephanoidea, Megalyroidea and Proctotrupoidea are evident. The major radiation of the Apocrita apparently occurred during the Cretaceous (65–135 mya), and in late Cretaceous deposits all superfamilies, and many of the extant families, are present.

The evolutionary relationships of the major taxa of the Apocrita are far from clear, and conflicting patterns of character state distribution have yet to be resolved. Relationships between superfamilies are obscure and indeed, the monophyly of several superfamilies is not clearly demonstrable. For example, the Proctotrupoidea is generally regarded as a polyphyletic convenience group, and the position of the Evaniidae with the other evanioid families is ambiguous (cf. Gibson, 1985; N.F. Johnson, 1988). On the basis of extensive palaeontological data and morphological examination Rasnitsyn (1975, 1980, 1988) proposed a revolutionary classification. In his 1980 work Rasnitsyn divided the traditional apocritans into four infraorders—the Stephanomorpha, Evaniomorpha, Ichneumonomorpha and Vespomorpha. The first of these comprises the single extant family Stephanidae, while the Vespomorpha is equivalent to the Aculeata. The distinctness and monophyly of both is widely accepted by most workers. Rasnitsyn's (1980) concept of the Evaniomorpha comprised the evanioids, ceraphronoids, trigonalyids and megalyrids, while his Ichneumonomorpha consisted of the ichneumonoids, chalcidoids, proctotrupoids and cynipoids. In 1988

Rasnitsyn modified this arrangement somewhat, placing the proctotrupoids, chalcidoids and cynipoids in a separate group, the Proctotrupomorpha.

Gibson (1985) found synapomorphies supporting a clade Megalyridae + [Ceraphronidae + Megaspilidae] but failed to find evidence to link these groups to other evaniomorph taxa. Norman Johnson (1988) characterized Rasnitsyn's 1980 grouping of Evaniomorpha as 'a particularly bold concept' and, although he criticized Rasnitsyn's characterization of the group, he nevertheless found new evidence to support it in the form of the mid coxal articulation. Although Rasnitsyn (1988) subsequently included the stephanids in the evaniomorphs on the basis of mid coxal articulation, this was not substantiated by Johnson (1988), who continued to treat them as a distinct group.

The monophyly of the Ichneumonomorpha (*sensu* Rasnitsyn, 1980) has not been widely accepted, although some evidence for it was presented by Norman Johnson (1988). However, evidence is mounting in favour of Rasnitsyn's later arrangement for a sister-group relationship between the aculeates and the Ichneumonoidea. Aculeates and ichneumonoids both have a similar metasomal attachment to the propodeum, and both groups possess valvilli, specialized flap-like structures present on the first gonapophysis (see Gauld & Bolton, 1988; Quicke, Fitton & Ingram, 1992).

The relationships of the other parasitoid superfamilies remain largely unknown, although the Chalcidoidea is clearly monophyletic and is probably the sister-group to the Mymarommatidae (Gibson, 1986*a*). Gibson (1985) found morphological evidence supporting a clade Proctotrupidae + Pelecinidae + Vanhorniidae + Scelionidae + Platygastridae. The position of these and the other major proctotrupoid families, Monomachidae and Diapriidae, remains the subject of disagreement. Rasnitsyn's (1988) arrangement of the proctotrupomorphs is depicted in Fig. 5.02.

The aculeate superfamilies, Chrysidoidea, Vespoidea and Apoidea form a monophyletic clade characterized by having the ovipositor modified and by the loss of cerci in the females (Brothers, 1975). The evolutionary relationships of the aculeate superfamilies is more clearly understood than that of the other groups and the hypothesis proposed by Brothers (1975), that the monophyletic clade Chrysidoidea is the sister lineage to the clade comprising Vespoidea + Apoidea, is now generally accepted (see Chapters 13 and 14).

5.3 THE CLASSIFICATION AND STATUS OF MAJOR GROUPS OF HYMENOPTERA

Traditionally the Hymenoptera has been divided into two suborders, the Symphyta and the Apocrita; the latter was further generally subdivided into Parasitica and Aculeata. The superfamilies included in

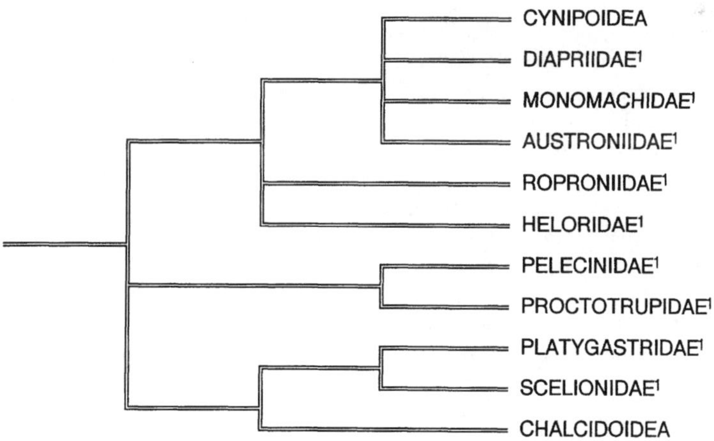

Fig. 5.02. Cladogram showing the phylogenetic relationships of the Proctotrupomorpha *sensu* Rasnitsyn (1988).
[1] = Proctotrupoidea *sensu* many authors.

these divisions generally were: Symphyta (Xyeloidea, Megalodontoidea, Cephoidea, Tenthredinoidea, Siricoidea and Orussoidea); Parasitica (Stephanoidea, Trigonalyoidea, Megalyroidea, Ceraphronoidea, Evanioidea, Ichneumonoidea, Proctotrupoidea, Cynipoidea and Chalcidoidea); Aculeata (Chrysidoidea, Vespoidea, Apoidea). Of these divisions the 'Symphyta' is undoubtedly a paraphyletic assemblage, although the Apocrita is probably monophyletic; the 'Parasitica' is paraphyletic with respect to the monophyletic Aculeata. Although both 'Symphyta' and 'Parasitica' can sometimes be useful group-names (just as can 'Reptilia' in its traditional sense) we suggest formally abandoning these subdivisions. The division between Aculeata and 'Parasitica' has indeed been gradually abandoned in some modern works (Rasnitsyn, 1980; Gauld & Bolton, 1988; Borror *et al.*, 1989), though all retained a formal symphytan-apocritan division (Rasnitsyn, however, placed the orussids in the latter). In the classification used here we recognize only superfamilies within the order; we recommend this scheme until the phylogenetic relationships of the higher taxa are more clearly understood, when perhaps stable sub- and infraorder classificatory ranks can be introduced.

Here we recognize 79 families classified in the 18 superfamilies referred to above. This is a utilitarian scheme, and inevitably will not be accepted by all. We believe it offers a compromise incorporating almost all of the more substantiated modern views, but retaining a traditional arrangement for taxa about which there is widespread disagreement. The status of the various groups and the areas of disagreement are outlined below, although classifications that persist in adhering to such indefensible concepts as placing the ceraphronoids in the Proctotrupoidea, or the stephanids in the Ichneumonoidea, are not discussed further here.

Fourteen of the superfamilies recognized here, the Megalodontoidea, Cephoidea, Tenthredinoidea, Orussoidea, Stephanoidea, Trigonalyoidea, Megalyroidea, Ceraphronoidea, Ichneumonoidea, Cynipoidea, Chalcidoidea, Chrysidoidea, Vespoidea and Apoidea, are all demonstrably monophyletic (Brothers, 1975; Königsmann, 1977, 1978*a* & *b*; Rasnitsyn, 1980; Gibson, 1985, 1986*a*; Carpenter, 1986*a*; Gauld & Bolton, 1988; N.F. Johnson, 1988; Whitfield *et al.*, 1989; Sharkey & Wahl, 1992). The position of the Mymarommatidae as sister-group to other chalcidoids is generally accepted, so their inclusion within the Chalcidoidea (e.g. by Gauld & Bolton, 1988; Borror *et al.*, 1989), or their treatment as a distinct superfamily, the Mymarommatoidea (e.g. by Goulet & Huber, 1993) is purely a matter of personal preference.

The status of the other four superfamilies, Xyeloidea, Siricoidea, Evanioidea and Proctotrupoidea, remains to be resolved. The peculiarly modified antenna of the Xyeloidea (a long composite basal flagellar segment followed by a line of small segments) suggests this group may also be monophyletic although, as with all archaic taxa, one cannot rule out the possibility that this is a ground plan feature of the order that has been modified in higher taxa. Problems with the monophyly of the Evanioidea were referred to above, but there is some evidence (e.g. in the position of the metasomal insertion into the propodeum) that the group is indeed monophyletic.

The Siricoidea and Proctotrupoidea, are almost certainly not monophyletic (cf. Figs 5.01, 5.02). The former is probably paraphyletic with respect to the Orussoidea + apocritan superfamilies (Gibson, 1985), so arguably each family (Anaxyelidae, Siricidae, Xiphydriidae) should be accorded superfamily rank. However, Whitfield *et al.* (1989) found evidence (loss of 2ph–3ph muscles and form of the attachment of t_3–3ph muscle) suggesting that the Xiphydriidae + Siricidae comprises a monophyletic lineage. Thus the Siricoidea should perhaps be restricted to just two families, and the Anaxyelidae should be placed in a separate superfamily. Until this incongruity is resolved we recommend leaving all three families in a single superfamily.

The Proctotrupoidea is an artificial aggregation of 12 families—Austroniidae, Diapriidae, Heloridae, Monomachidae, Pelecinidae, Peradeniidae, Platygastridae, Proctotrupidae, Renyxidae, Roproniidae, Scelionidae, Vanhorniidae. Arguments have been proposed for grouping most of these in several ways (see Table 5.1). Rasnitsyn (1988) advocated yet another arrangement (Fig. 5.02), suggesting that the superfamily is paraphyletic with respect to both the Cynipoidea and Chalcidoidea. In view of the widespread disagreement about the classification of these families, and bearing in mind Naumann and Masner's (1985) suggestion that it is 'premature to speculate on the phylogeny of the proctotrupoid families', we have opted to follow other workers (e.g. Gauld & Bolton, 1988; Borror *et al.*, 1989) in treating all in a single superfamily, until such time as a more robust classification is proposed. Masner (in Goulet & Huber, 1993; see also Chapter 9) now places the Platygastridae and Scelionidae in a separate superfamily from other proctotrupoids. Although there is uniformity of opinion that the two are very closely related, separating them from groups such as the Vanhorniidae and Proctotrupidae (with which they share some features) while leaving the Monomachidae and Diapriidae in the Proctotrupoidea seems artificial.

Here (following Brothers, 1975) we recognize three aculeate superfamilies, although some authors (e.g. Richards, 1956; Krombein *et al.*, 1979; Borror *et al.*, 1989; Naumann, 1991) recognize four, six, seven, or even eight. These classificatory systems are contrasted below (Table 5.2). The obvious difference in names used by Brothers (1975) and Gauld and Bolton (1988), which is used in this work, is based on the principles of zoological nomenclature. As Day (1977) has pointed out the name Chrysidoidea has temporal priority over Bethyloidea and hence is the valid name. Similarly the family-group name Apidae predates Sphecidae and is therefore the correct formative stem for the superfamily name.

In our opinion the use of six, seven or eight superfamily names is unnecessary, given the diversity of species placed in other hymenopterous superfamilies, and moreover, it poses new problems. It is certainly not indisputable, for example, that the Pompilidae should be included in a superfamily with the Rhopalosomatidae distinct from the other vespoid s.l. families (see Day, 1988 and Chapter 14), and 'Sphecoidea' is almost certainly paraphyletic with respect to 'Apoidea' in its narrower sense (Lomholdt, 1982; Alexander, 1992).

FAMILIES	**Königsmann, 1978**			**Rasnitsyn, 1980**		***Naumann & Masner, 1985**		**Gibson, 1985**	**Naumann, 1991**	
Peradeniidae	+	+	+	+	+	NM1	–	+	N1	–
Heloridae	K1	–	–	R1	–	NM1	–	+	N1	–
Proctotrupidae	K1	–	–	R1	–	+	+	G1	N1	–
Vanhorniidae	K1	–	–	R1	–	+	+	G1	N1	–
Roproniidae	K1	–	–	R1	–	+	+	+	N1	–
Pelecinidae	–	K2	–	R1	–	+	+	G1	N1	–
Monomachidae	–	K2	–	–	R2	+	+	+	N1	–
Austroniidae	+	+	+	–	R2	+	+	+	N1	–
Diapriidae	–	–	K3	–	R2	+	+	+	N1	–
Platygastridae	–	–	K3	–	R2	–	NM2	G1	–	N2
Scelionidae	–	–	K3	–	R2	–	NM2	G1	–	N2

Table 5.1. Comparison of the relationships of the proctotrupoid families as advocated by recent workers. The number refers simply to a group of taxa, thus K1, K2, K3 refer to Königsmann's first, second and third groups of taxa; + denotes that a taxon was not treated or not assigned to a group. * It is noteworthy that although Naumann and Masner stated only two groupings were obvious, they treated all eleven families in a single superfamily.

Borror *et al.*, 1989	**Krombein *et al.*, 1979**	**Richards, 1977**	**Naumann, 1991**	**Brothers, 1975**	**Gauld & Bolton, 1988**
Chrysidoidea	Bethyloidea	Chrysidoidea	Chrysidoidea	Bethyloidea	Chrysidoidea
Scolioidea	Scolioidea	Scolioidea	Vespoidea	Vespoidea	Vespoidea
Tiphioidea					
Formicoidea	Formicoidea				
Pompiloidea	Pompiloidea	Pompiloidea			
Vespoidea	Vespoidea	Vespoidea			
Sphecoidea	Sphecoidea	Sphecoidea	Sphecoidea	Sphecoidea	Apoidea
Apoidea	Apoidea	Apoidea	Apoidea		

Table 5.2. Comparison of superfamily classification adopted by recent authors.

In general there is more consensus regarding the family level classification. Disregarding the work of the obviously ill-informed or idiosyncratic, there are only five areas of confusion—the ichneumonoids, the cynipoids, the chalcidoids, the vespids and the apoids. The two extant families comprising the Ichneumonoidea have at times been increased by workers (e.g. Mackauer & Stary, 1967; Mason, 1971, 1981*a*) according various subordinate taxa family-level status. Most common of these is treating the aphidiine braconids as the Aphidiidae, or calling the paxylommatine and agriotypine ichneumonids separate families. Sound phylogenetic arguments refuting such ordinations have been presented by various authors (Gauld, 1984; Achterberg, 1984*a*; Gauld & Bolton, 1988; Sharkey & Wahl, 1992), and these are not repeated here.

The classification of the Cynipoidea has fluctuated a great deal. In the past the superfamily was often subdivided into six families, the Cynipidae, Ibaliidae, Liopteridae, Figitidae, Charipidae and Eucoilidae (e.g. Quinlan, 1979; Goulet & Huber, 1993), but Fergusson (1990) has recently undertaken a phylogenetic reconstruction, which provides a very different classification that is robustly supported by morphological, palaeontological, biogeographical and biological evidence. Following Fergusson, the Cynipoidea is here divided into four families, Cynipidae, Ibaliidae (including the 'Liopteridae'), Figitidae (including 'Charipidae' and 'Eucoilidae') and Himalocynipidae.

Although a consensus classification of the Chalcidoidea into 21 families (as used here) has recently been adopted in influential publications (e.g. Bouček, 1988; Grissell & Schauff, 1990), it is generally agreed that several of the families are not monophyletic. The Pteromalidae in particular is a 'catch-all' group for chalcidoids that cannot be placed in one of the other families and the Aphelinidae may be paraphyletic with respect to the Signiphoridae (see Chapter 11.15). The Eucharitidae and Perilampidae are undoubtedly very closely related, and from a phylogenetic perspective probably better classified in a single family (e.g. see Darling, 1992). The recent expansion of Agaonidae to include certain subfamilies previously placed in Torymidae and Pteromalidae (Bouček, 1988; see Chapter 11.1) has reopened the question of the relationship between Agaonidae and Torymidae. In general, there are undoubtedly too many families of Chalcidoidea, and eventually this number will have to be reduced. However, until there are convincing arguments based on rigorous phylogenetic analyses, the need for stability speaks louder than the need for change.

The vespids have often been classified in three families, Vespidae, Masaridae and Eumenidae (e.g. Richards, 1962, 1978; Krombein, *in* Krombein *et al.*, 1979; Gauld & Bolton, 1988). However, Carpenter (1982) produced a definitive cladistic analysis of these groups and proposed treating all as a single family. This classification has gained wide acceptance (e.g. Evans, 1987*a*; Borror *et al.*, 1989; Naumann, 1991) and is accepted in the present work.

The Sphecidae is frequently accorded superfamily status, and subdivided into eight or so families (e.g. Krombein *in* Krombein *et al.*, 1979; Goulet & Huber, 1993). However, the classification adopted by Bohart and Menke (1976) in their scholarly and now classic overview of the sphecids, that is the treatment of all groups as a single family, is gaining wide acceptance (Evans, 1987*a*; Gauld & Bolton, 1988; Borror *et al.*, 1989; Naumann, 1991). However, one outstanding classificatory problem remains—the Sphecidae as thus defined is paraphyletic with respect to the Apidae. Alexander (1992), in his exploratory analysis of the relationships between bees and sphecids, clearly demonstrated the paraphyletic nature of the Sphecidae with respect to the bees. Most frequently the apids were found to have a more or less sister-group relationship to philanthine sphecids (e.g. Alexander's figures 1, 2, 3, 5, 6, 9, 10); sphecine sphecids almost invariably emerged as the sister-group to all remaining sphecids + bees (see Fig. 5.03). When the phylogenetic interrelationships of the sphecids and bees is more clearly resolved it will undoubtedly be necessary to recognize two or more sphecid families, one of which will be the sister-group of the bees.

Since Michener's (1944) seminal classification of the bees it has been standard practice to treat them as belonging to a number of separate families classified together in a distinct superfamily, the Apoidea (e.g. Hurd *in* Krombein *et al.*, 1979; Brown, 1982; Evans, 1987*a*; Borror *et al.*, 1989). The most radical departure from the consensus view was adopted by Gauld and Bolton (1988) who treated all as a single family, Apidae, not because they disputed Michener's hierarchical classification, but because they believed his ranking of these taxa as families was disproportionately high in comparison with the other Hymenoptera.

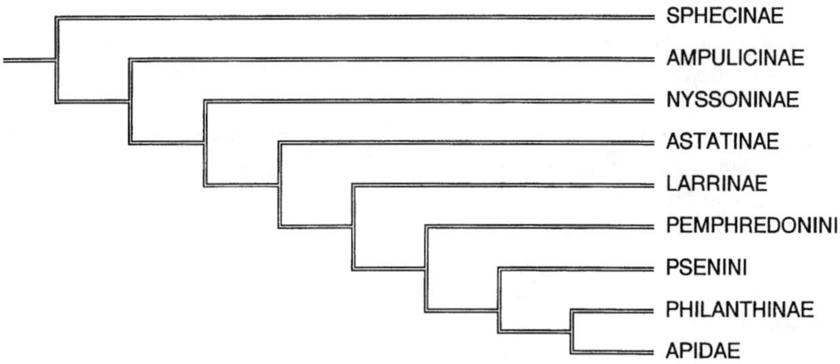

Fig. 5.03. Cladogram showing possible phylogenetic relationship between some sphecid tribes and subfamilies, and the Apidae. Modified after Alexander (1992). All taxa shown are demonstrably monophyletic, with the exception of the Nyssoninae.

Xyeloidea	Xyelidae
Megalodontoidea	Megalodontidae, Pamphiliidae
Tenthredinoidea	Argidae, Blasticotomidae, Cimbicidae, Diprionidae, Pergidae, Tenthredinidae
Cephoidea	Cephidae
Siricoidea	Anaxyelidae, Siricidae, Xiphydriidae
Orussoidea	Orussidae
Stephanoidea	Stephanidae
Trigonalyoidea	Trigonalyidae
Megalyroidea	Megalyridae
Evanioidea	Aulacidae, Evaniidae, Gasteruptiidae
Ceraphronoidea	Ceraphronidae, Megaspilidae
Proctotrupoidea	Austroniidae, Diapriidae, Heloridae, Monomachidae, Peradeniidae, Pelecinidae, Platygastridae, Proctotrupidae, Renyxidae, Roproniidae, Scelionidae, Vanhorniidae
Cynipoidea	Cynipidae,Figitidae, Ibaliidae
Chalcidoidea	Agaonidae, Aphelinidae, Chalcididae, Elasmidae, Encyrtidae, Eucharitidae, Eulophidae, Eupelmidae, Eurytomidae, Leucospidae, Mymaridae, Mymarommatidae, Ormyridae, Perilampidae, Pteromalidae, Rotoitidae, Signiphoridae, Tanaostigmatidae, Tetracampidae, Torymidae, Trichogrammatidae
Ichneumonoidea	Braconidae, Ichneumonidae
Chrysidoidea	Bethylidae, Chrysididae, Dryinidae, Embolemidae, Plumariidae, Sclerogibbidae, Scolebythidae
Vespoidea	Bradynobaenidae, Formicidae, Mutillidae, Pompilidae, Rhopalosomatidae, Sapygidae, Scoliidae, Sierolomorphidae, Tiphiidae, Vespidae
Apoidea	Apidae, Sphecidae

Table 5.3. The classification of extant families of Hymenoptera adopted in this book.

We continue to support Gauld and Bolton's interpretation, and note that Michener, who maintains usage of two superfamilies, has recently stated 'Bees are basically sphecoid wasps . . . ' (Michener & Houston *in* Naumann, 1991: p. 933). To classify the apoids (bees and sphecids) in about twenty families is, in our opinion, a distortion of the higher classification of the order. Such 'families' are certainly no more distinct than are the subfamilies of most other groups, including the pergids, tenthredinids, mutillids, ichneumonids, braconids, diapriids, bradynobaenids, tiphiids and chrysidids. Although the status of the multiplicity of apoid groups as families may be more nearly equivalent to the chalcidoid families, the chaotic nature of chalcidoid higher classification is a particularly poor model to base any justification upon.

Lastly two words about any form of compromise classification, uniting the Michener (1944) and the Bohart and Menke (1976) classifications—**avoid it**. To classify the bees in eight or so families, while treating the sphecids as a single family, as do Evans (1987*a*), Borror *et al.* (1989) and Naumann (1991), presents a quite distorted classificatory system, given that few would argue that the sphecids are the diverse paraphyletic stem-group, from within which the apids arose! The classification we adopt is shown in Table 5.3.

5.4 KEYS TO HIGHER TAXA OCCURRING IN COSTA RICA

The following keys are intended to facilitate the initial identification of the vast majority of hymenopterans encountered in Central America and northern South America. By proceeding to the pages indicated in parentheses, one can continue the identification to family (and in many cases to subfamily or genus). The characters used in the following keys are, for the most part, readily observable with an average dissecting microscope (40 × magnification). Moreover, the key has been reduced to a minimal length in order to encourage fledgeling hymenopterists. Such a simplified key excludes many of the rare exceptions (some of which are mentioned in brackets) and all identifications should therefore be verified by comparing with the diagnoses given in the family chapters.

Simplified key to major groups of fully-winged Hymenoptera in Costa Rica

1 Body without marked constriction between mesosoma (thorax) and metasoma (abdomen) (Fig. 5.04); cenchri present on metathorax; enclosed anal cell/s often present in fore and/or hind wing (Fig. 15.12).
[Fore wing often with 10 or more enclosed cells, never with less than five; hind wing usually with five or more enclosed cells (Fig. 5.12).] 2

— Body with a marked constriction between mesosoma (thorax + propodeum) and metasoma, i.e. with a 'wasp waist"; cenchri absent; anal cell never enclosed in fore or hind wing (Figs 5.13–5.20).
[In some minute species the constriction is not very evident, but these have no closed cells in the wings.] 3

2 Antennae inserted on ventral side of head adjacent to mouth and below apparent clypeus (Fig. 5.05); upper part of head with blunt tooth-like tubercles .... **Orussoidea** (p. 178)

— Antennae inserted well above mouth and clypeus (Fig. 5.06); upper part of head without tubercles. ....
... **Siricoidea** and **Tenthredinoidea** (p. 157)

3 Fore wing with venation well-developed, with one to several enclosed cells (Figs 5.13–5.19) 4

— Fore wing with venation very reduced, without any enclosed cells, usually with venation restricted to anterior margin of wing (Figs 5.20–5.24) or entirely absent. 19

4 Antennae with 16 or more segments. 5

— Antennae with 15 or fewer segments. [Heloridae have 16 segments but one of the segments is very small and the antenna therefore appears to have 15.] 8

5 Pronotum extended anteriorally into an elongate 'neck' (Fig. 5.07).
[Head spherical; top of head with small teeth; first segment of metasoma very long, slender and cylindrical.] .... **Stephanoidea** (p. 181)

— Pronotum not extended into an elongate 'neck' (Fig. 5.08) 6

6 Fore wing without a distinct costal cell, veins *C* and *Sc+R+Rs* contiguous for most of their length (Figs 5.13, 5.16) ....
.... **Ichneumonoidea** (most) (p. 389)

— Fore wing with a distinct costal cell, veins *C* and *Sc+R+Rs* separated by a membranous area for much of their length (Figs 5.14, 5.15). 7

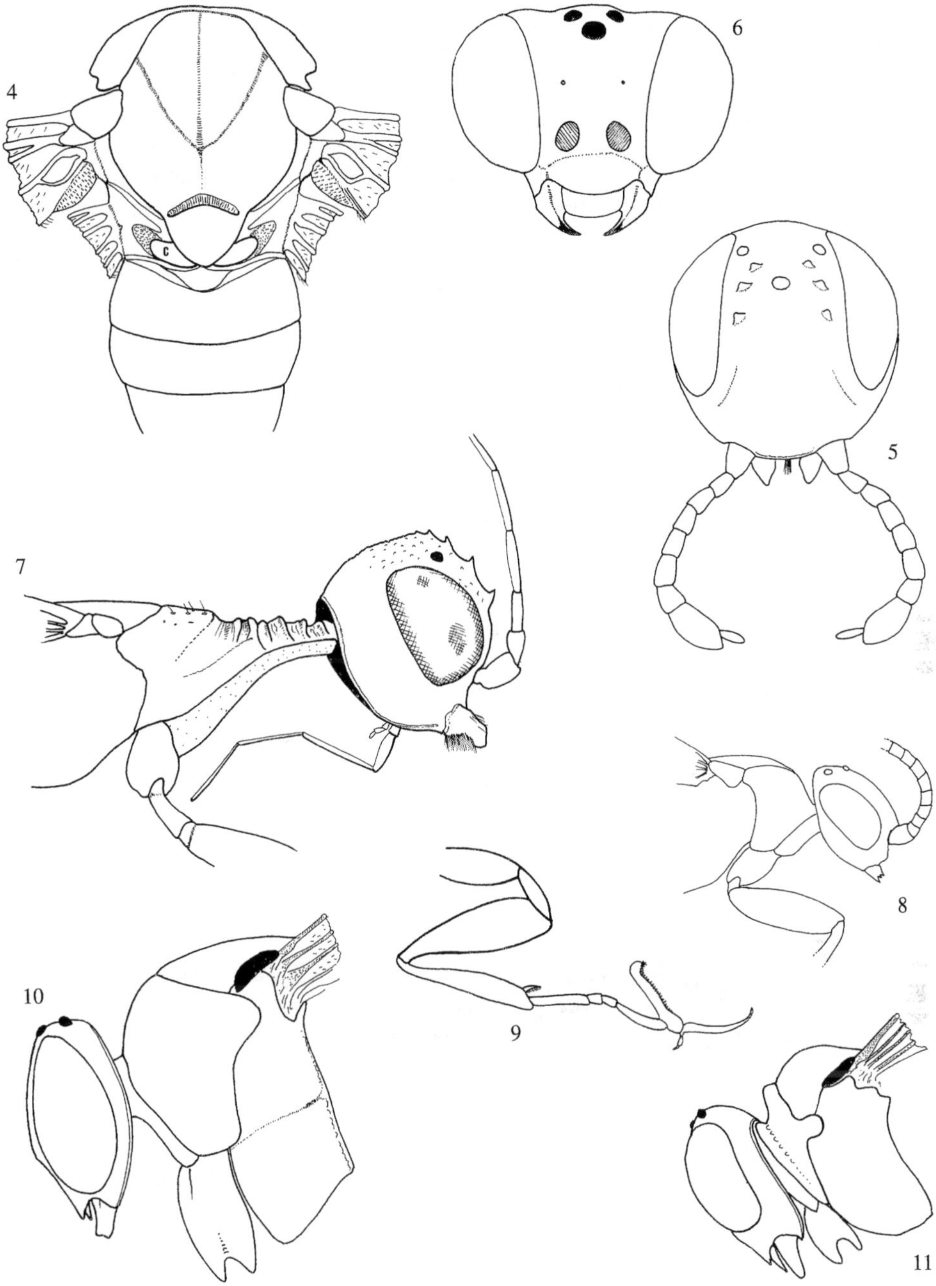

Figs 5.04–5.11. Hymenoptera, superfamilies. Fig. 5.04. Mesosoma and base of metasoma (Argidae; Tenthredinoidea); C = cenchrus. Figs 5.05–5.06. Head, anterior; 5.05, *Ophrella* sp. (Orussidae; Orussoidea); 5.06, Tenthredinidae (Tenthredinoidea). Figs 5.07–5.08. Head and pronotum, lateral; 5.07, *Megischus* sp. (Stephanidae; Stephanoidea); 5.08, *Probethylus* sp. (Sclerogibbidae: Chrysidoidea). Fig. 5.09. Fore leg, female *Trichogonatopus* sp. (Dryinidae; Chrysidoidea). Figs 5.10–5.11. Head and anterior part of mesosoma, lateral; 5.10, Pompilidae (Vespoidea); 5.11, Sphecidae (Apoidea).

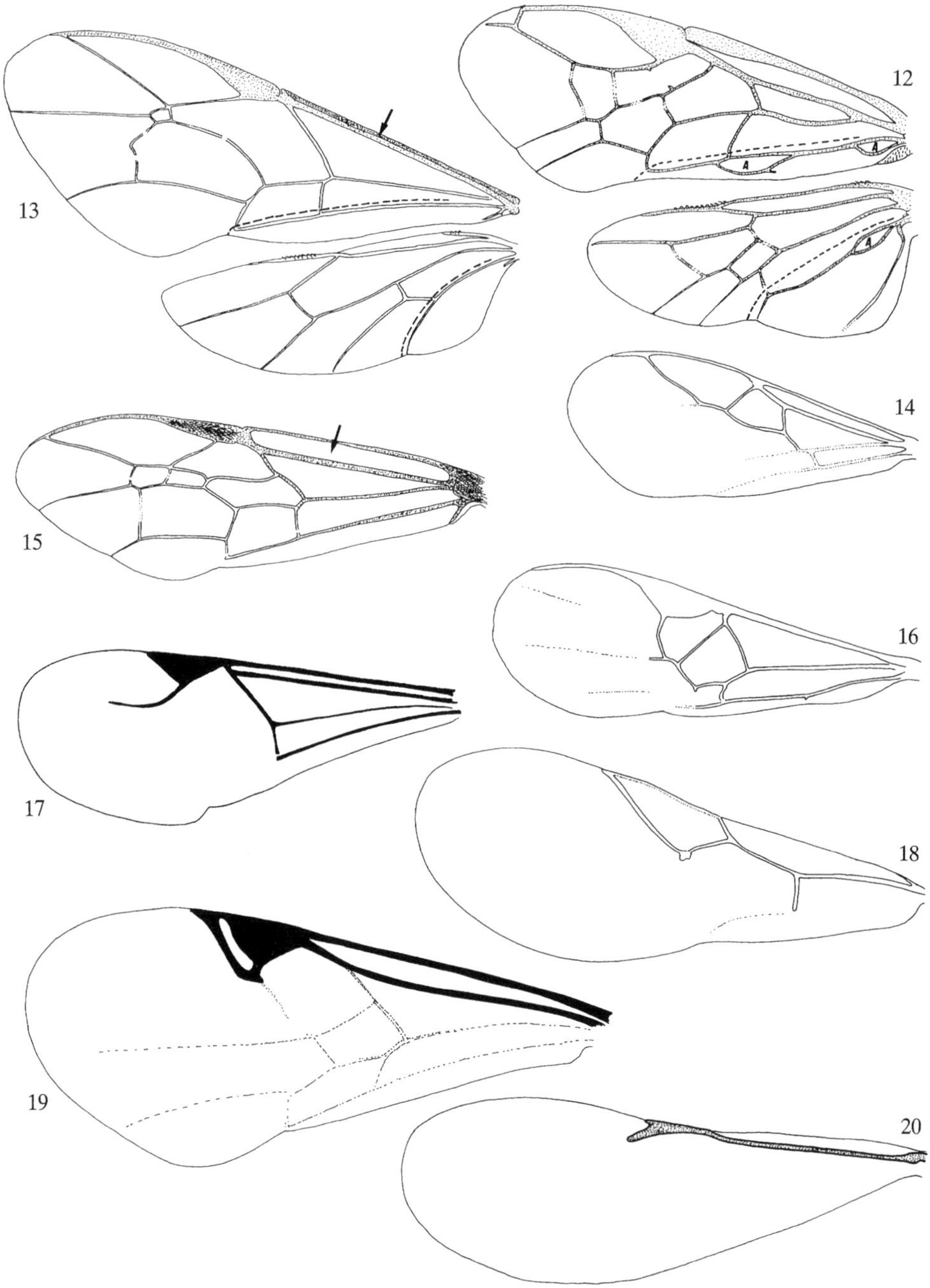

Figs 5.12–5.20. Hymenoptera, superfamilies. Figs 5.12–5.13. Fore and hind wings; 5.12, sterictiphorine Argidae (Tenthredinoidea), A = anal cell; 5.13, *Neotheronia* sp. (Ichneumonidae; Ichneumonoidea). Figs 5.14–5.20. Fore wings; 5.14, *Probethylus* sp., male (Sclerogibbidae; Chrysidoidea); 5.15, *Taeniogonalos* sp. (Trigonalyidae; Trigonalyoidea); 5.16, *Apanteles* sp. (Braconidae; Ichneumonoidea); 5.17, *Lonchodryinus* sp. (Dryinidae; Chrysidoidea); 5.18, eucoiline Figitidae (Cynipoidea); 5.19, *Sminthoserphus* sp. (Proctotrupidae; Proctotrupoidea); 5.20, Encyrtidae (Chalcidoidea).

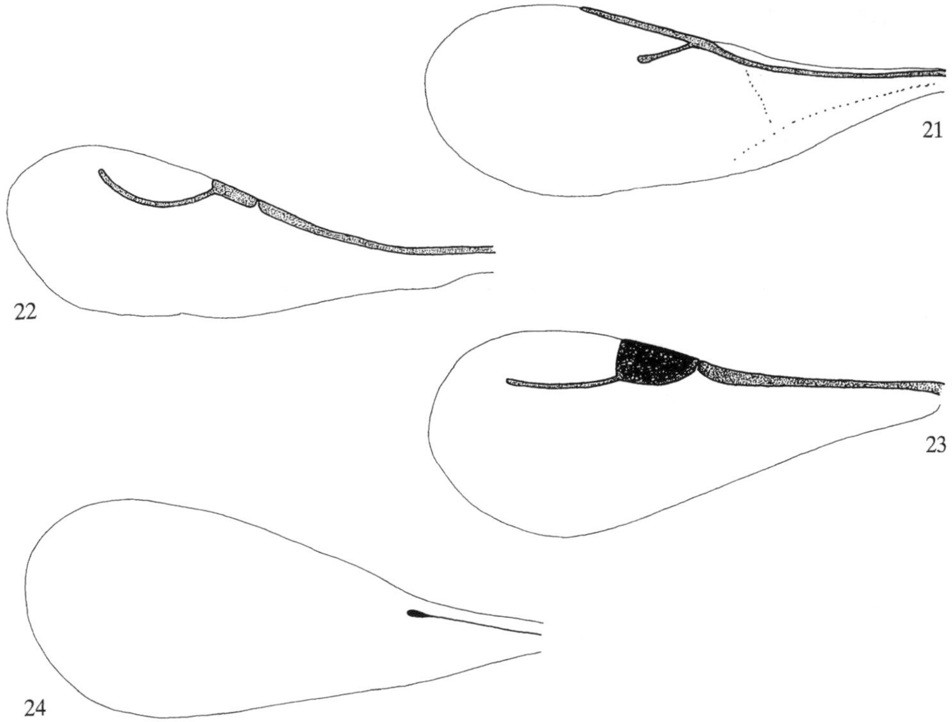

Figs 5.21–5.24. Hymenoptera, superfamilies, fore wings; 5.21, Scelionidae (Proctotrupoidea); 5.22, *Ceraphron* sp. (Ceraphronidae; Ceraphronoidea); 5.23, *Dendrocerus* sp. (Megaspilidae; Ceraphronoidea); 5.24. *Inostemma* sp. (Platygastridae; Proctotrupoidea).

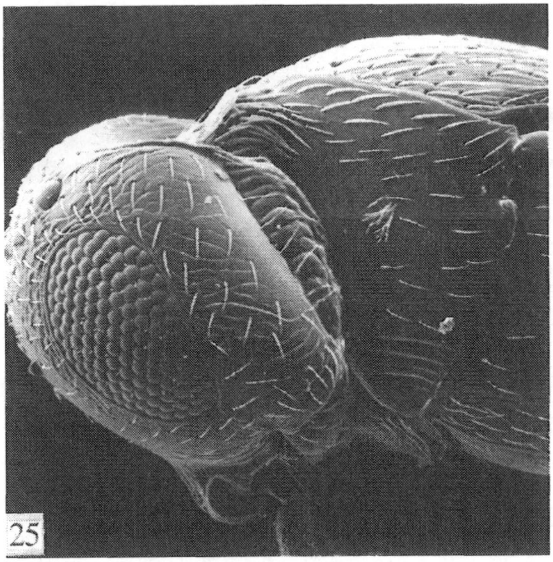

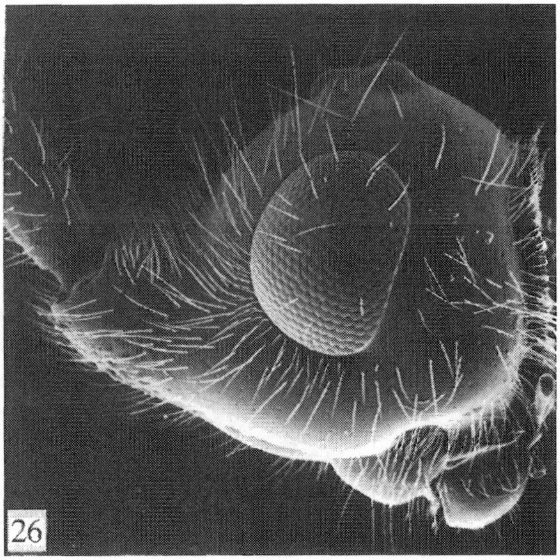

Figs 5.25–5.26. Scanning electron photomicrographs of head, lateral; 5.25, *Doddiella* (Scelionidae; Proctotrupoidea); 5.26, *Belyta* (Diapriidae; Proctotrupoidea).

7 Fore wing with nine or ten enclosed cells (Fig. 5.15); fore leg with femur unspecialized. [Mandibles often asymmetrical, the right with four, the left with three teeth; tarsi with plantar lobes; ovipositor more or less concealed in metasoma.] .... **Trigonalyoidea** (p. 187)

— Fore wing with five or fewer enclosed cells (Fig. 5.14); fore leg with femur enlarged and strongly inflated (Fig. 5.08) .... **Chrysidoidea** (Sclerogibbidae) (p. 488)

8 Metasoma inserted high up on mesosoma (Figs 8.08–8.10)............ **Evanioidea** (p. 185)

— Metasoma inserted lower down on mesosoma, close to insertion of hind coxae. 9

9 Metasoma with only three visible tergites, ***and*** body brilliantly metallic coloured, generally coarsely sculptured .... **Chrysidoidea** (some Chrysididae) (p. 479)

— Metasoma generally with more than three visible tergites, body rarely metallic coloured. 10

10 Fore wing with four or more enclosed cells ***and/or*** with hind tibia enlarged and flattened. 11

— Fore wing with one to three enclosed cells (Figs 5.17–5.19); hind tibia never flattened and enlarged. 16

11 Pronotum with upper hind corner widely separated from tegula, and lower down the side with a pronounced pronotal lobe (Fig. 5.11). .... **Apoidea** 12

— Pronotum with upper hind corner reaching to or close to tegula, usually without a pronounced pronotal lobe (Fig. 5.10)............................. 13

12 Body hairs (especially those on propodeum and near bases of wings) plumose, with short to long branches; hind legs often with tibia and tarsus flattened and broadened. .... **Apidae** (p. 650)

— Body hairs simple; hind legs not flattened. .... **Sphecidae** (p. 621)

13 Antenna with 14 or 15 segments. 14

— Antenna with 13 or fewer segments. 15

14 Fore wing with costal cell distinct (Fig. 5.19); ventral part of metasoma strongly sclerotized, rigid. .... **Proctotrupoidea** (few) (p. 209)

— Fore wing with costal cell more or less obliterated by *C* touching *Sc+R+Rs* (cf Fig. 5.16); ventral part of metasoma partially membranous. .... **Ichneumonoidea** (few) (p. 389)

15 Hind wing with one to three enclosed cells. ....**Vespoidea** (p. 504)

— Hind wing without enclosed cells. .... **Chrysidoidea** (few) (p. 464)

16 Fore wing with a fairly large triangular-shaped marginal cell, which may be open or closed on the wing margin (Fig. 5.18); mesosoma in profile with scutellum generally quite large, of similar size to propodeum (Figs 5.27, 5.28); metasoma usually somewhat flattened from side to side. .... **Cynipoidea** (p. 247)

— Fore wing without a marginal cell (Fig. 5.17), ***or if*** with one present ***then*** it is very much smaller than above (Fig. 5.19); mesosoma in profile generally with scutellum much smaller than propodeum; metasoma often somewhat flattened dorso-ventrally or extremely long. 17

17 Antenna with 14 or 15 segments. .... **Proctotrupoidea** (few) (p. 209)

— Antenna with 9 to 13 segments..................... 18

18 Either with antenna inserted high up on a slight facial protuberance or 'shelf' (best seen in lateral view; Fig. 5.26; Diapriidae), ***or*** with fore wing with a very small marginal cell, with distal abscissa of *Rs* close to and parallel with postero-distal margin of pterostigma (Fig. 5.19; Proctotrupidae). .... **Proctotrupoidea** (many) (p. 209)

— Antenna not inserted on a facial protuberance; fore wing not having a small marginal cell with distal abscissa of *Rs* close to and parallel with postero-distal margin of pterostigma; venation often as in Fig. 5.17.

[Head often large and prognathous, or females sometimes with front claw modified into a chela (Fig. 5.09).] .... **Chrysidoidea** (most) (p. 464)

19 Fore wing with stigmal vein generally curved, without a postmarginal vein, and with (Fig. 5.23) or without (Fig. 5.22) an enlarged pterostigma, but always with this region narrowly separated from the marginal vein; marginal vein always along anterior margin of wing. .... **Ceraphronoidea** (p. 199)

— Fore wing, from entirely without venation, to with some veins; **if** veins present, ***then*** either stigmal vein is not curved (Fig. 5.20, 5.21, 5.24), ***and/or*** postmarginal vein is present, ***and/or*** junction of stigmal and postmarginal veins is not separated from marginal vein, ***and/or*** proximal part of marginal vein is separated from anterior margin of wing. 20

20 Head strongly prognathous ***and*** antenna 12- or 13-segmented. .... **Chrysidoidea** (some Bethylidae) (p. 470)

— Head generally not prognathous, ***but*** if so, ***then*** antenna with a different number of segments. 21

21 Metasoma usually at least somewhat dorso-ventrally flattened and quite symmetrical, often with a lateral carina, ovipositor originating (externally) from apex of metasoma; antenna inserted either high up on a facial protuberance or 'shelf' (Fig. 5.26; Diapriidae), ***or*** inserted just above mouth (Fig. 5.25; Platygastridae and Scelionidae); coloration never metallic; pronotum, in profile, reaching back more or less to tegula. .... **Proctotrupoidea** (many) (p. 209)

— Metasoma usually not dorso-ventrally flattened, not as symmetrical, without a lateral carina, ovipositor originating from ventral surface of metasoma; antennal insertion variable; coloration sometimes metallic; pronotum, in profile, not reaching back to tegula. .... **Chalcidoidea** (p. 266)

Key to brachypterous or wingless Hymenoptera

Except for ants, brachypterous or apterous individuals of Hymenoptera are rather infrequently encountered in Costa Rica, possibly because most collecting methods have been developed to trap flying insects. Ants are ubiquitous in all trap samples (except at high altitudes). Female mutillids, because they are frequently brightly coloured, are often noticed, but other brachypterous or apterous hymenoptera are

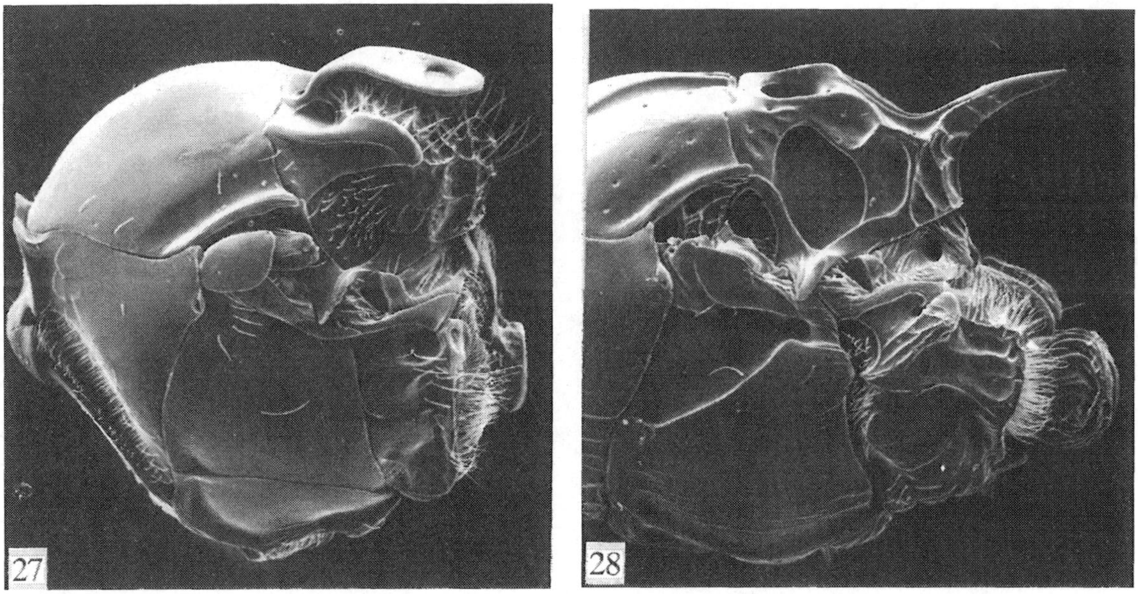

Figs 5.27–5.28. Scanning electron photomicrographs of mesosoma, lateral; 5.27, Eucoilini (Figitidae; Cynipoidea); 5.28, Figitini (Figitidae; Cynipoidea).

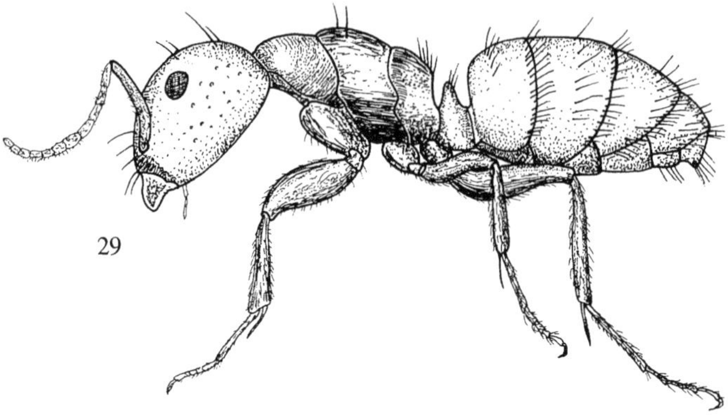

29

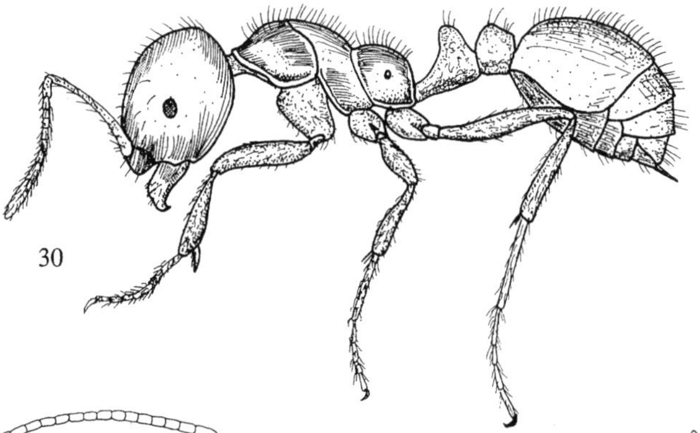

30

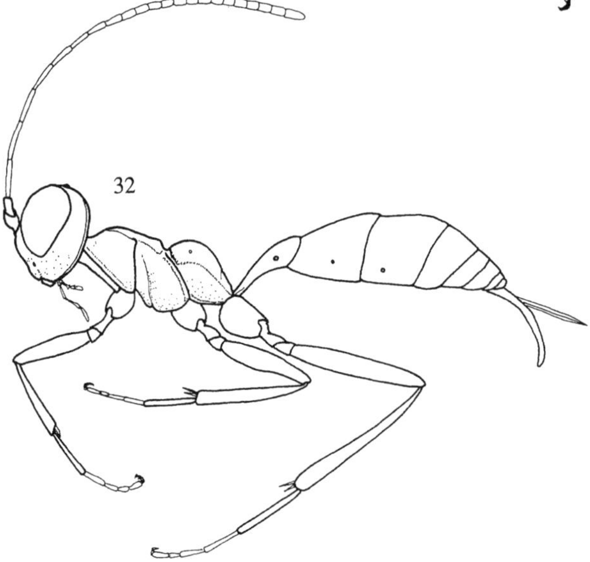

32

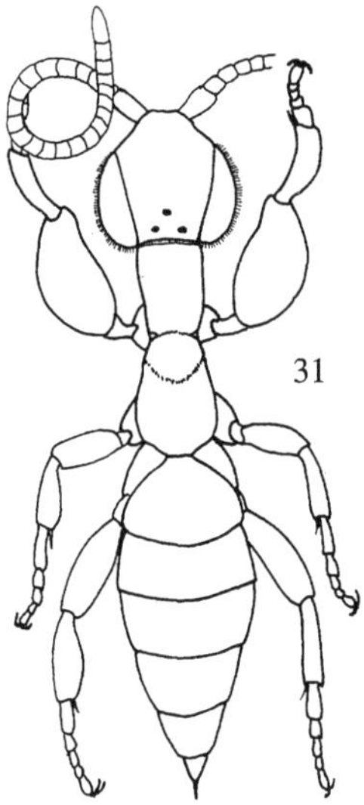

31

Figs 5.29–5.32. Wingless Hymenoptera. 5.29, *Camponotus* sp., worker (Formicidae); 5.30, *Solenopsis* sp., worker (Formicidae); 5.31, *Probethylus* sp., female (Sclerogibbidae); 5.32, *Gelis* sp., female (Ichneumonidae).

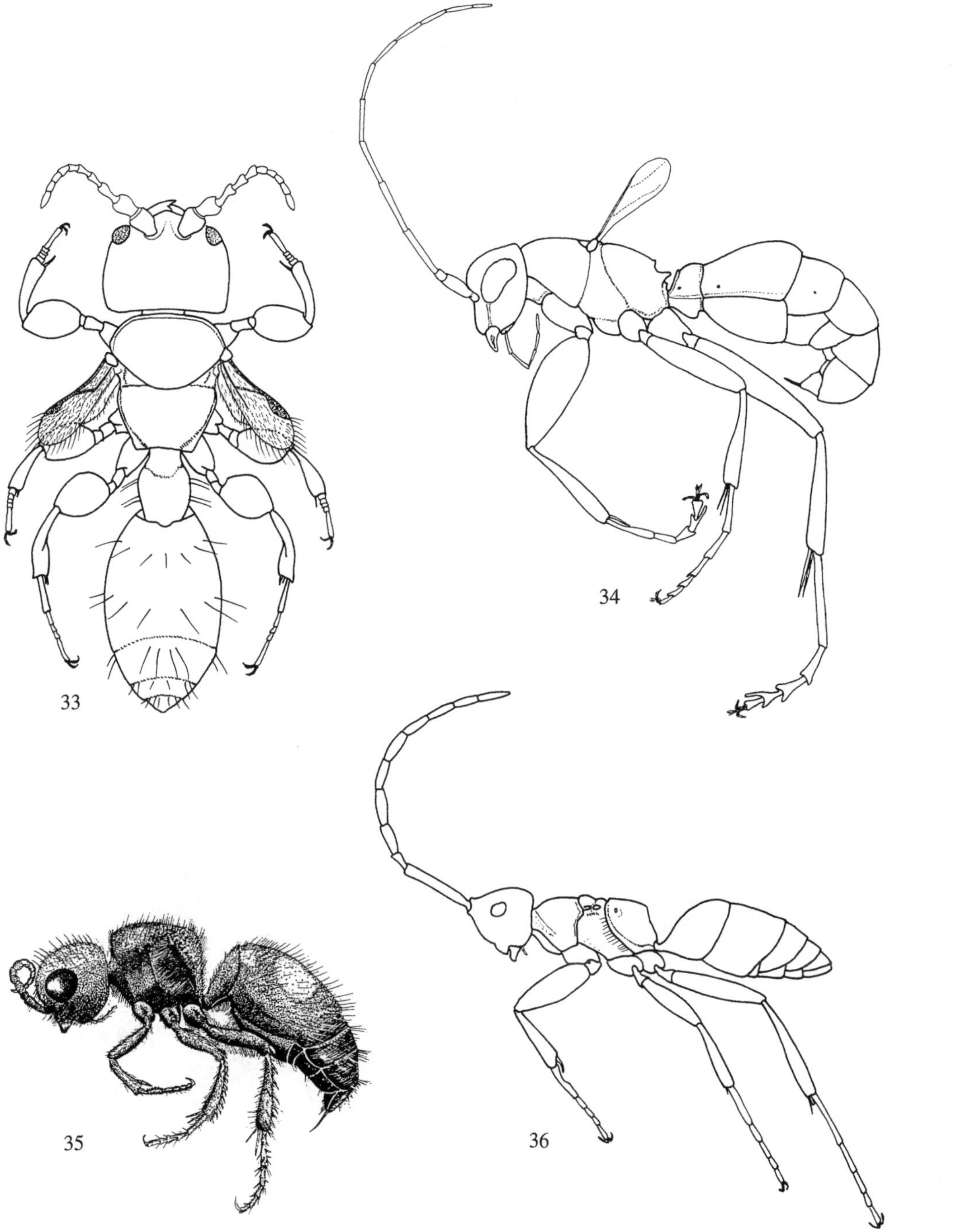

Figs 5.33–5.36. Wingless or micropterous Hymenoptera. 5.33, *Psenobolus* sp., male (Braconidae); 5.34, *Olixon* sp., female (Rhopalosomatidae); 5.35, *Pseudomethoca* sp., female (Mutillidae); 5.36, *Embolemus* sp., female (Embolemidae).

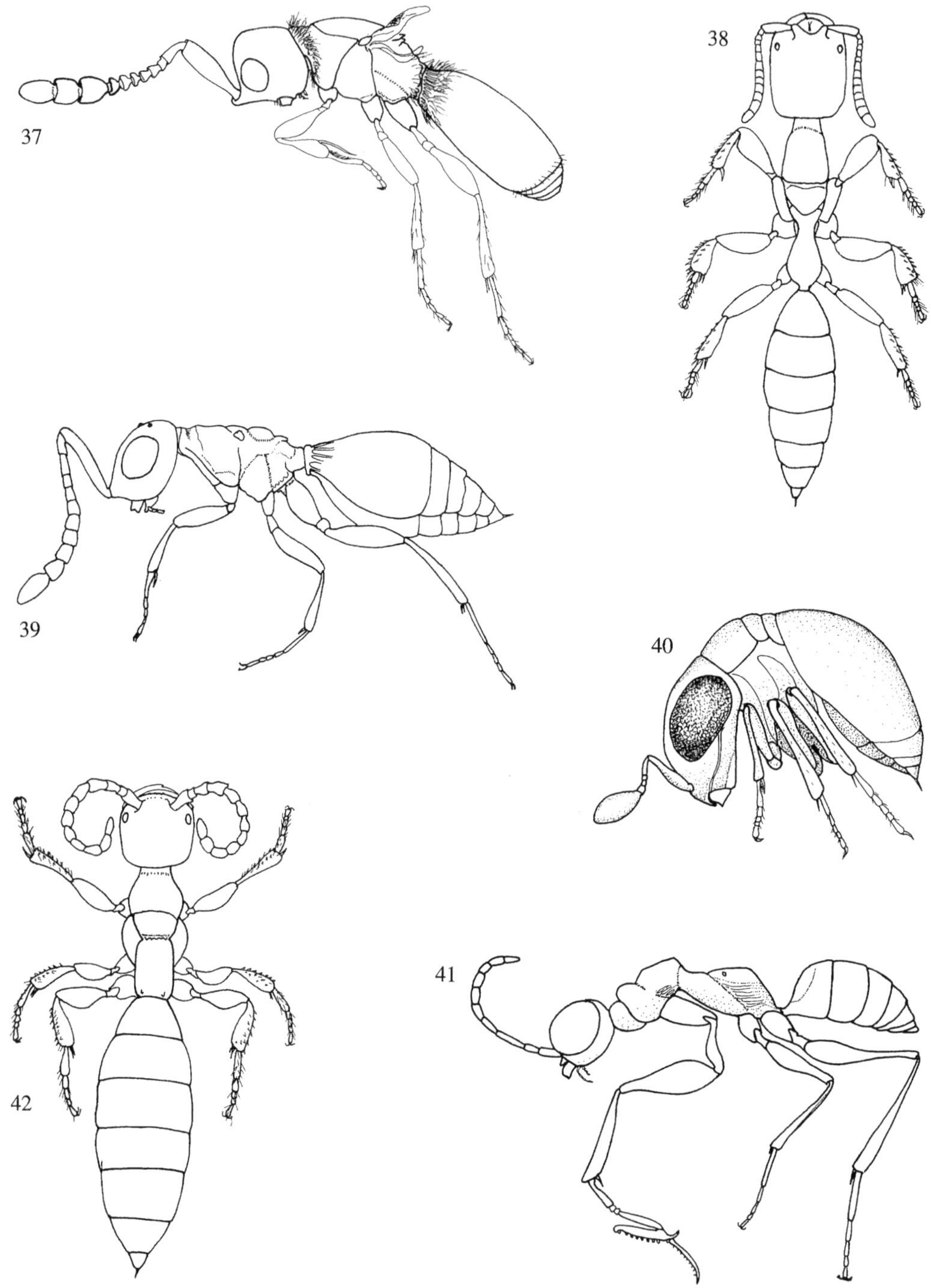

Figs 5.37–5.42. Wingless or micropterous Hymenoptera. 5.37, Diapriidae; 5.38, *Pristocera* sp. (Bethylidae); 5.39, *Conostigmus* sp. (Megaspilidae); 5.40, *Baeus* sp. (Scelionidae); 5.41, *Gonatopus* sp. (Dryinidae); 5.42, *Aleurus* sp. (Thynninae; Tiphiidae).

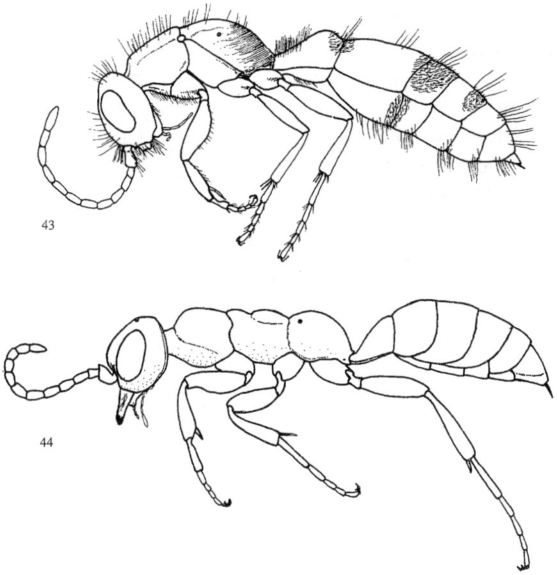

Figs 5.43–5.44. Wingless Hymenoptera. 5.43, *Typhoctes* sp. (Bradynoebaenidae); 5.44, *Methocha* sp. (Tiphiidae).

seldom noticed or collected. Many of these insects resemble their fully-winged relatives, but some are highly modified and difficult to recognize as hymenopterans.

1 Metasoma with segment I forming a node or scale (Fig. 5.29), or first two segments nodiform (Fig. 5.30). .... **Formicidae** (p. 588)
— Metasomal segment I not scale-like, if rarely slightly nodiform then segment II not deeply separated from III. 2

2 Antenna with 15 or more segments. 3
— Antenna with fewer than 15 segments. 4

3 Head prognathous, in dorsal view more or less heart-shaped, with antenna inserted at front, under a ridge, below level of lower (i.e. anterior) margin of eyes; legs short and stout, fore femur greatly thickened (Fig. 5.31); metasoma subsessile, with tergite I evenly broadened posteriorly; ovipositor sheath not projecting beyond apex of metasoma. .... **Sclerogibbidae** (p. 488)
— Head hypognathous, in dorsal view oval, with antenna inserted on plane of frons, at about the level of the centre of the eyes (Fig. 5.32); legs normal, fore femur not greatly thickened; metasoma petiolate, tergite I slender; ovipositor sheath protruding conspicuously beyond apex of metasoma. .... **Ichneumonidae** (p. 390)

4 Males. 5
— Females. 7

5 Entirely apterous species with no trace of wing vestiges.
[Only encountered inside fig fruits, possibly in the litter beneath fig trees.]. .... **Agaonidae** (p. 273)
— Brachypterous species, with at least fore wing present as a vestige. 6

6 Antenna short and 10-segmented; head massive and prognathous (Fig 5.33); metasoma with tergites II + III fused; inhabitants of the syconia of *Ficus*. .... **Braconidae** (*Psenobolus*) (p. 453)
— Antenna long and slender, 13-segmented; head hypognathous (Fig. 5.34); metasoma with tergites II and III separated; free-living insects. ... . **Rhopalosomatidae** (p. 548)

7 Mesosoma in dorsal view with pronotum, mesoscutum, scutellum, metanotum and propodeum confluent, forming a single convex surface and not separated into discrete parts.
[Body very hairy, velvet-like (Fig. 5.35), often with bright red, orange, yellow or white markings.] .... **Mutillidae** (p. 541)
— Mesosoma in dorsal view with pronotum separated, not immutably fused with mesoscutum, generally with scutellum and propodeum discernible, sometimes with all parts discrete. 8

8 Antennae inserted on shelf-like prominence, high on face (Figs 5.36, 5.37). 9
— Antennae not originating on shelf-like prominence, and/or inserted close to mouth (Figs 5.34, 5.39, 5.40). 10

9 Head, in lateral view subconical; antenna with 10 segments, the flagellum slender and relatively unspecialized (Fig. 5.36). .... **Embolemidae** (p. 490)
— Head, in lateral view subspherical, with a frontal 'shelf'; antenna usually with more than 10 (11 to 14) segments, almost always clavate, with basal flagellar segments small and the distal ones enlarged (Fig. 5.37). .... **Diapriidae** (p. 222)

10 Antennae geniculate (elbowed), with first segment (scape) markedly longer than the remaining segments, often attached quite low on face (Figs 5.38–5.40). 11
— Antennae not geniculate, with first segment usually not markedly longer than the remaining segments, attachment variable (Figs 5.41–5.44). 14

11 Metasoma with ovipositor emerging from ventral surface; metasoma usually not dorso-ventrally depressed, coloration sometimes metallic. .... **Chalcidoidea** (p. 266)
— Metasoma with ovipositor protruding (or protrusable) from the extreme apex (Fig. 5.44); metasoma centrally usually dorso-ventrally flattened; coloration never metallic. 12

12 Head very strongly prognathous, rather large and with compound eyes generally small (Fig. 5.38); legs often adapted to fossorial existence with tibiae bearing numerous short stout spines. .... **Bethylidae** (p. 470)
— Head hypognathous, not unusually large, with compound eyes quite large (Figs 5.39, 5.40); legs slender. 11

13 Metasoma with longitudinal ridges (carinae) at base (Fig. 5.39). .. **Ceraphronoidea** (p. 187)
— Metasoma without longitudinal ridges at base (Fig. 5.40) .... **Scelionidae** and **Platygastridae** (p. 229)

14 Antennae with 10 segments; fore pretarsus and claw modified as raptorial 'pincers' (Fig. 5.41). .... **Dryinidae** (p. 493)
— Antennae with 12 segments; fore tarsus not modified to form raptorial pincers. 15

15 Fore wings present as minute vestiges; legs with 4th tarsal segment asymmetrically bilobate, the longer lobe projecting almost to apex of 5th segment (Fig. 5.34).
[Bright yellow species.] .... **Rhopalosomatidae** (p. 548)
— Fore wings absent entirely (Figs 5.42–5.44); legs with 4th tarsal segment simple, not markedly asymmetrically bilobate. 16

16 Head very strongly prognathous, rather large and with compound eyes generally small (Fig. 5.42); legs often adapted to fossorial existence with tibiae bearing numerous short stout spines. .... **Tiphiidae** (Thynninae) (p. 521)
— Head hypognathous, not unusually large, with compound eyes quite large (Figs 5.43, 5.44); legs slender. 17

17 Mesosoma in dorsal view divided into two parts, an anterior pronotum and a posterior part comprising meso- and metanotum and propodeum; body hairy, hairs serrate (Fig. 5.43). .... **Bradynobaenidae** (p. 552)
— Mesosoma in dorsal view divided into three parts, an anterior pronotum, a central mesothorax (which often has a more or less discernible scutellum) and a posterior propodeum; body more sparsely hairy, hairs simple (Fig. 5.44). .... **Tiphiidae** (p. 514)

6 The sawflies and woodwasps

David R. Smith

INTRODUCTION

As is traditionally recognized (e.g. by Gauld & Bolton, 1988; Naumann, 1991; Smith, 1993), the suborder Symphyta (the sawflies and woodwasps) is a paraphyletic assemblage comprising the structurally more primitive Hymenoptera. Symphytan adults share a number of primitive (plesiomorphic) character states: their abdomen is broadly joined to the thorax and not constricted at its base, the first abdominal segment is not fused with thorax (i.e. propodeum absent), cenchri are present on the metathorax (except the non-Costa Rican family Cephidae), a trochantellus is present, and numerous enclosed cells are discernible in the wings, including one or more anal cells in the fore wing (except in most Pergidae). While the Orussidae share most of these plesiomorphic character states with the sawflies and woodwasps, they also share several apomorphic character states with the suborder Apocrita (Gibson, 1985; N.F. Johnson, 1988; Whitfield *et al.*, 1989; see Chapters 5 and 7). Moreover, the larval stages of Orussidae differ from all other symphytans in being parasitic or predatory rather than phytophagous. Therefore the orussids are treated in a separate chapter (Chapter 7).

The great majority of sawflies feed as larvae on plant material. Little is known of the feeding preferences of tropical species, but in the temperate northern regions large numbers of species are associated with relatively primitive plant families, such as anemophilous angiosperms, conifers, ferns and horsetails. Relatively few attack higher angiosperms such as the Compositae, although several feed on Gramineae and Cyperaceae. In Australia some pergids feed on dead leaves and one Costa Rican pergid has very recently been found to feed externally on jelly fungi (Auriculariaceae). A detailed account of feeding strategies of temperate sawflies was presented by Heitland and Pschorn-Walcher (1993).

Adult females of most sawflies use their saw-like ovipositor to cut into plant tissue to insert eggs, hence their common name. Oviposition is usually in the leaf, but a few oviposit in stems, petioles, or wood. Some attach their eggs to the surface of a leaf. Adult females of a few species of Argidae, Pamphiliidae and Pergidae are known to guard their eggs and young larvae (e.g. Dias, 1976; Kudô *et al.*, 1992). Larvae of most tenthredinoid and megalodontoid symphytans are external folivores, feeding either solitarily or gregariously, whilst those of other groups (Xyeloidea, Siricoidea and Cephoidea) generally feed in concealment as borers in wood or stems. A few tenthredinoid species develop as leaf-miners (e.g. some Heterarthrinae), gall-formers (e.g. some Nematinae), stem-borers (Blasticotomidae) or fruit-borers, but none of these concealed-feeders occur in Costa Rica. In Central America folivorous sawfly larvae may feed either on the upper or on the undersides of leaves. They may skeletonize the leaves or feed only on the leaf edges. When mature the last instar, a non-feeding stage called the prepupa, searches for a suitable pupation site. This site can be in the soil, in stems or rotten wood, or among the foliage or stems of plants. Some prepupae spin cocoons, and others form a cell in the soil or other substrate.

The larvae of sawflies that feed externally on plant foliage are caterpillar-like and resemble those of the Lepidoptera (e.g. Fig. 6.01), from which they can be distinguished by the possession of a single stemmatum (simple eye) on each side of the head (there are several in Lepidoptera), and the absence of crochets on the prolegs. Unlike lepidopterous larvae, the larvae of exophytic sawflies usually have prolegs on abdominal segments 2 to 7 or 2 to 8, and 10, and do not have a protruding spinneret on the labium. Larvae of some neotropical species were illustrated by Kimsey and Smith (1985). Symphytan larvae that feed in concealment are more grub-like, with reduced legs and often no stemmata (Smith & Middlekauf, *in* Evans, 1987*a*). The internal morphology of sawfly larvae has been surveyed by Maxwell (1955).

Although sawflies are often abundant in northern temperate forests, in Costa Rica they are generally rather scarce and inconspicuous, although exceptions

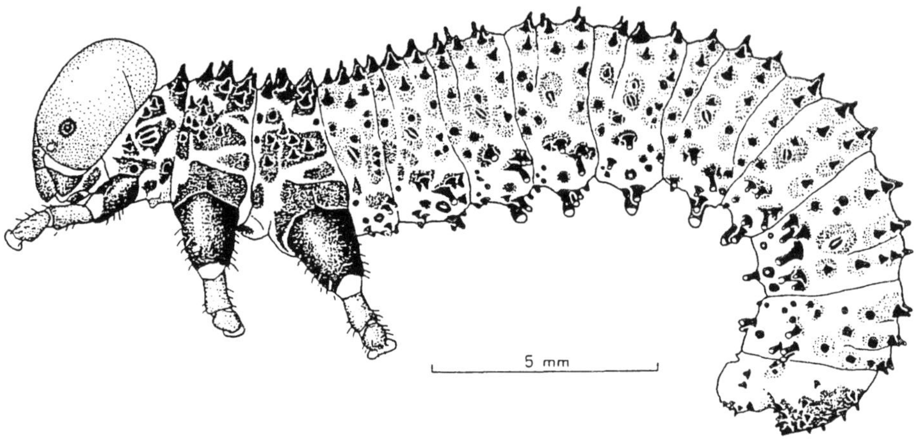

Fig. 6.01. Larva of *Sericoceros gibbus* (Argidae: Erigleninae) (after Kimsey & Smith, 1985).

do occur. For example, a tenthredinid, *Metapedias subcoerulea* is common on *Conostegia xalapensis* (Melastomataceae) at medium elevations (c. 1500 metres) in the Central Valley (Hanson, pers. comm.), and in the seasonally dry forests of northwestern Guanacaste the adults of several species of tenthredinids and argids can be common shortly after the commencement of the wet season, when trees are flushing with new growth. Little is known of the phenology of tropical species, but in temperate areas and tropical habitats with a pronounced dry season there is commonly a single generation a year. In wetter areas some sawflies, such as *M. subcoerulea*, have several generations annually. Adult sawflies usually live for only several days to a week or two, and are generally found flying about their host plants on still, sunny days.

Several sawflies are sometimes of economic importance as pests of ornamental plants, agricultural crops and timber trees in forests. Some of the most notorious pests are certain species of Diprionidae which are widespread and injurious to conifers from Canada south to northern Central America. In cooler parts of South America the tenthredinid, *Caliroa cerasi*, is on occasions a serious pest of cultivated rosaceous fruit trees. On the other hand, a few sawflies have been considered for possible use in biological control programmes targeted at eliminating invasive weeds. Two such cases are the introduction of *Ucona acaenae* into New Zealand from Chile in the 1930's to aid in control of *Acaena* (Rosaceae) (Janvier, 1933; Smith, 1973), and the potential use of *Schizocerella pilicornis* for the control of *Portulaca* (Portulacaceae) (Clement & Norris, 1982).

Classification and distribution. Sawflies and woodwasps are found on all continents except Antarctica, but are rare or absent on most oceanic islands. Throughout the world approximately 8000 species are known, and about 3000 of these occur in the western hemisphere. The extant sawflies and woodwasps, excluding Orussidae, are presently classified in five superfamilies and 13 families (Table 6.1). Although these have been traditionally classified together in the suborder Symphyta, it may eventually be more appropriate to recognize at least some of these superfamilies as separate suborders (see Chapter 5).

The sawfly fauna of Central America to the north of Lake Nicaragua differs somewhat from that of southern Central America. The Xiphydriidae is absent from northern Central America, but two additional families are present—the Diprionidae and Siricidae. The former is known to occur in northern Nicaragua (Wilkinson & Drooz, 1979; Smith, 1988) while siricine Siricidae are questionably recorded from Central America (Smith, 1988). They are associated with conifers however, so they possibly extend south to the southern limit of pines in northern Nicaragua. The cultivation of pines in Costa Rica has created an

XYELOIDEA	
Xyelidae*	Holarctic, with one genus, *Xyela*, known from pine forests in Mexico; larvae of this genus feed in developing staminate cones of pines (*Pinus* spp.).
MEGALODONTOIDEA	
Megalodontidae*	Mediterranean and Europe, to Central Asia.
Pamphiliidae*	Holarctic, with one genus, *Acantholyda*, known from pine forests in Mexico; known as web-spinners, larvae of this genus feed and develop in webs of their own making.
TENTHREDINOIDEA	
Argidae	Cosmopolitan; mostly external leaf feeders.
Blasticotomidae*	Palaearctic; endophagous in the rhachis of pteridophytes.
Cimbicidae*	Holarctic, except for the subfamily Pachylostictinae which occurs in southern Brazil and N. Argentina. All nearctic and palaearctic cimbicids are external leaf feeders, but the habits of the S. American species are unknown.
Diprionidae*	Holarctic, extending south to northern Central America; most are external leaf feeders on conifers (Pinaceae and Cupressaceae).
Pergidae	Australia and New World; most are external leaf feeders.
Tenthredinidae	Cosmopolitan; biology variable, most are external feeders but some are leaf miners, gall formers, fruit borers and others.
SIRICOIDEA ("wood wasps")	
Anaxyelidae*	Western North America; wood borers in conifers.
Siricidae*	Holarctic, probably extending south to northern Central America; wood borers in conifers and angiosperms.
Xiphydriidae	Cosmopolitan; wood borers in angiosperms.
CEPHOIDEA ('stem sawflies')	
Cephidae*	Holarctic (one genus in Mexico) and Madagascar; stem borers in woody stems and in Gramineae.

Table 6.1. The superfamilies and families of the suborder Symphyta (excluded is the Orussidae), their distribution and larval biology. [* = families not occurring in southern Central America.]

opportunity for colonization by these two families. With more intensive collecting it is possible that two other families, Pamphiliidae and Xyelidae, will eventually be found in pine forests of northern Central America. Thus, of the 13 families of sawflies and woodwasps, four occur in southern Central America, and five to possibly seven occur in northern Central America.

In Costa Rica sawflies have been collected at almost all localities sampled (except mangrove swamps), from sea-level to over 3000 metres in the mountains. Currently about 150 species are known to occur in Costa Rica and Panama. The vast majority of these species belong to the Argidae, Pergidae or the Tenthredinidae. The Costa Rican fauna represents a cross section of northern and southern elements. The northern range of a number of South American groups, such as the argid genus *Acrogymnia* and the xiphydriid genus *Derecyrta*, extends into the country and, conversely, Costa Rica represents the southernmost locality for other groups, such as the blennocampine genus *Periclista* and the allantine genus *Ametastegia*.

A synopsis of the natural history of sawflies was presented by Benson (1950). Although based on the British fauna, much of the information applies elsewhere and is a good general introduction to the Symphyta. Abe and Smith (1991) gave a list of world genera and current classification of the suborder. Taxonomic treatment of sawfly families of America south of the United States is provided by Smith (1988, 1990, 1992, in preparation).

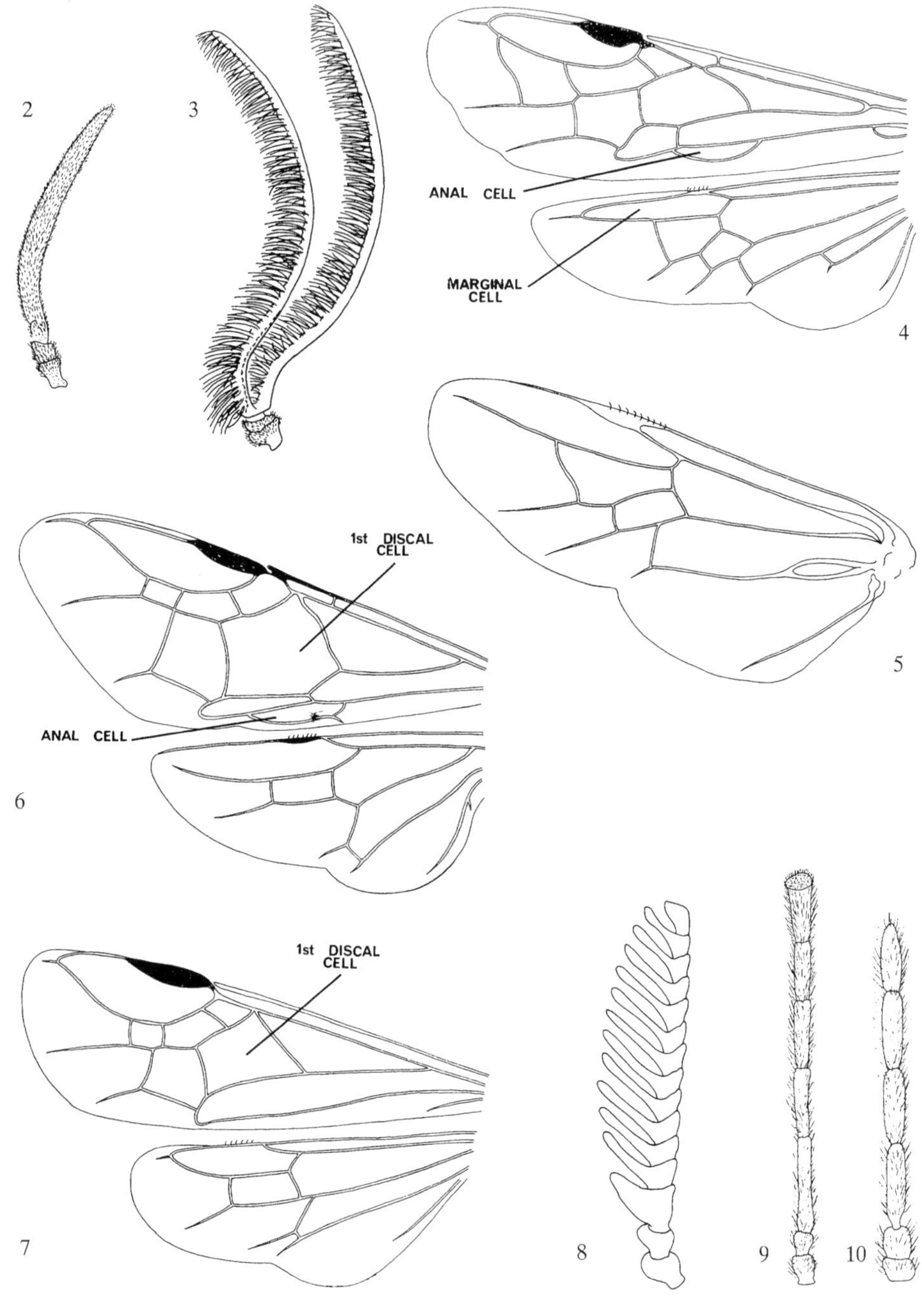

Figs 6.02–6.10. Symphyta. Figs 6.02–6.03. Antenna of *Sphacophilus* (Argidae: Sterictiphorinae); 6.02, female; 6.03, male. Fig. 6.04. Fore and hind wings of *Scobina* (Argidae: Arginae). Fig. 6.05. Hind wing, *Sphacophilus* (Argidae: Sterictiphorinae). Figs 6.06–6.07. Fore and hind wings; 6.06, *Perryia* (Pergidae: Perreyiinae); 6.07, *Aulacomerus* (Pergidae: Loboceratinae). Figs 6.08–6.10. Antennae; 6.08, *Perreyia* (Pergidae: Perreyiinae), male; 6.09, *Aulacomerus* (Pergidae: Loboceratinae); 6.10, *Acordulecera* (Pergidae: Acordulecerinae).

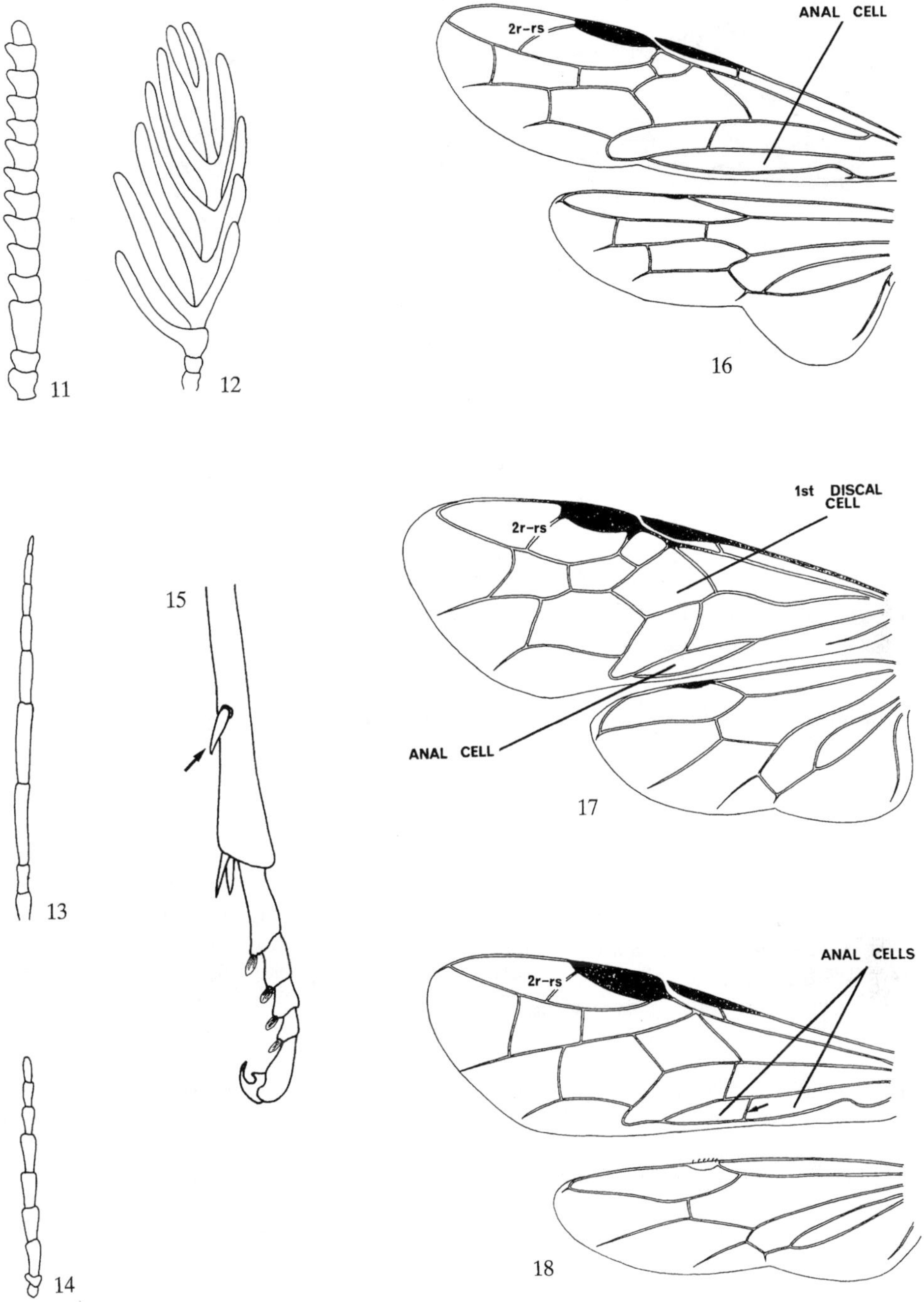

Figs 6.11–6.18. Symphyta. Figs 6.11–6.14. Antennae; 6.11, *Decameria* (Pergidae: Perreyiinae); 6.12, *Lagideus* (Pergidae: Syzygoniinae), male; 6.13, *Adiaclema* (Tenthredinidae: Selandriinae); 6.14, *Waldheimia* (Tenthredinidae: Blennocampinae). Fig. 6.15. Hind tibia of *Syzygonia* (Pergidae: Syzygoniinae). Figs 6.16–6.18. Fore and hind wings; 6.16, Selandriinae; 6.17, Blennocampinae; 6.18, Allantinae.

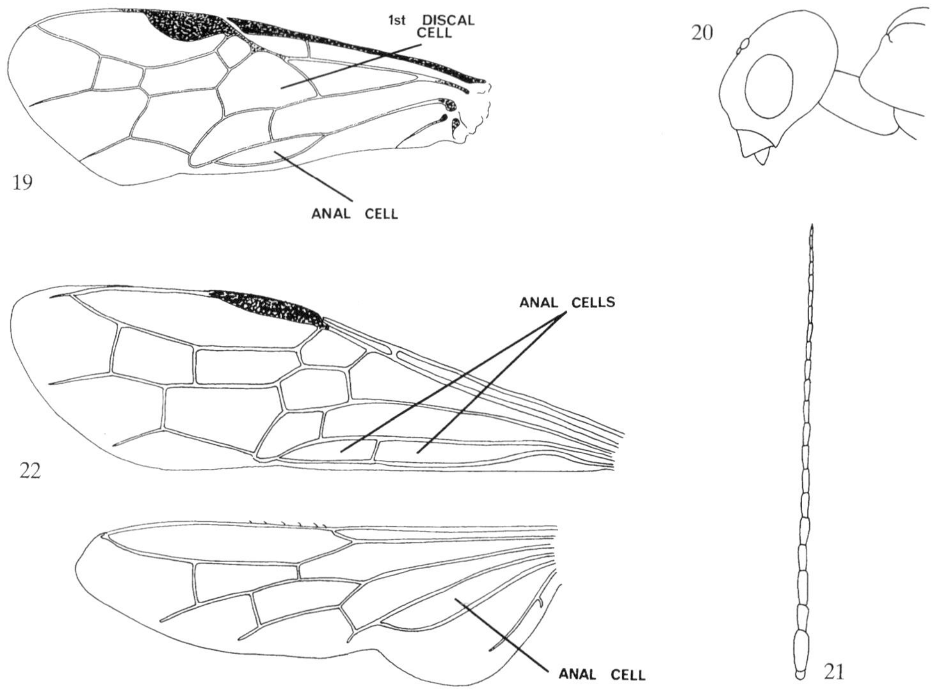

Figs 6.19–6.22. Symphyta. Fig. 6.19. Fore wing, Nematinae. Figs 6.20–6.22. *Derecyrta* (Xiphydriidae); 6.20, neck and head; 6.21, antenna; 6.22, fore and hind wings.

Key to families of sawflies and woodwasps present in Costa Rica

1 Antennal flagellum comprising a single long segment, which may be bifurcate in some males (Figs 6.02, 6.03). ...... **Argidae** (p. 163)

— Antennal flagellum with four or more segments (Figs 6.08-14, 6.21). 2

2 Antenna 9-segmented, filiform (Figs 6.13–14); fore wing usually with vein *2r-rs* present (Figs 6.16–18).
[Anal cell present in fore wing; hind wing with anal cell.] .... **Tenthredinidae** (p. 172)

— Antenna 6– to 7– or 10– to 20– or more segmented, filiform, pectinate, or serrate (Figs 6.08–12, 6.20); fore wing without vein *2r-rs* (Figs 6.06, 6.07, 6.22). 3

3 Body elongate, with a distinct neck formed from elongate cervical sclerites (Figs 6.20, 6.41); fore wing with an enclosed anal cell, crossed by anal crossvein (Fig. 6.22), but reaching to base of wing; hind wing with an enclosed anal cell (Fig. 6.22); antenna 20- or more segmented, filiform (Fig. 6.21). .... **Xiphydriidae** (p. 175)

— Body robust, not elongate, and neck not elongated (Fig. 6.39); fore wing either without an anal cell (Fig. 6.07) or with a short anal cell distally (Fig. 6.06) and *1A* fused with *2+3A* for much of its length proximally; hind wing without anal cell (Figs 6.06, 6.07); antenna 5– to 8– or 10– to 25-segmented, of various shapes, filiform, serrate, or pectinate (Figs 6.08–12). .... **Pergidae** (p. 167)

6.1 ARGIDAE

Diagnosis. Stout or elongate sawflies, ranging from 4.0 to 12.0 mm in length; coloration typically all black, black with part or all of thorax red, yellowish or yellowish with various black markings; wings hyaline, black, yellowish, or yellowish with the apices black. *Antenna 3-segmented*, with flagellum or third segment, a single segment; males of some groups with flagellum bifurcate (consisting of two long branches; Fig. 6.03). Fore wing with an enclosed anal cell, and mostly with an enclosed anal cell in hind wing (Figs 6.04, 6.05). Male genitalia strophandrous (rotated 180°).

Classification and distribution. The Argidae is cosmopolitan in distribution and comprises 51 genera and about 800 species, which are placed in six subfamilies: Arginae, Atomacerinae, Dielocerinae, Erigleninae, Sterictiphorinae and Zenarginae. All except the Zenarginae (an Australian subfamily) occur in the Neotropics and, except for the Dielocerinae, are represented in Central America. Smith (1992) has monographed the Argidae of the Neotropical region, the area where the family shows its greatest species richness. From this region (which actually included all of America south of the United States) Smith (1992) reported 32 genera and 356 species.

Biology. All Argidae are phytophagous as larvae and most feed externally on leaves, although *Schizocerella pilicornis* feeds either externally or as a leaf miner. This species is possibly native to southern South America, but it now occurs from Argentina to Canada, and its biology has been fairly well studied (references given in Smith, 1992). The food-plant associations known for species occurring in Central America are listed in Table 6.2.

Larvae of Argidae feed either singly or in groups. *Didymia unifasciata*, *Manaos kimseyae* and *Ptilia peletieri* all feed singly on the new flushes of leaves on seedlings (Kimsey & Smith, 1985). At least some species of *Sericoceros* on the other hand are known to feed gregariously. In Panama *S. gibbus* was observed laying a clump of 13 eggs in a loose cluster, which

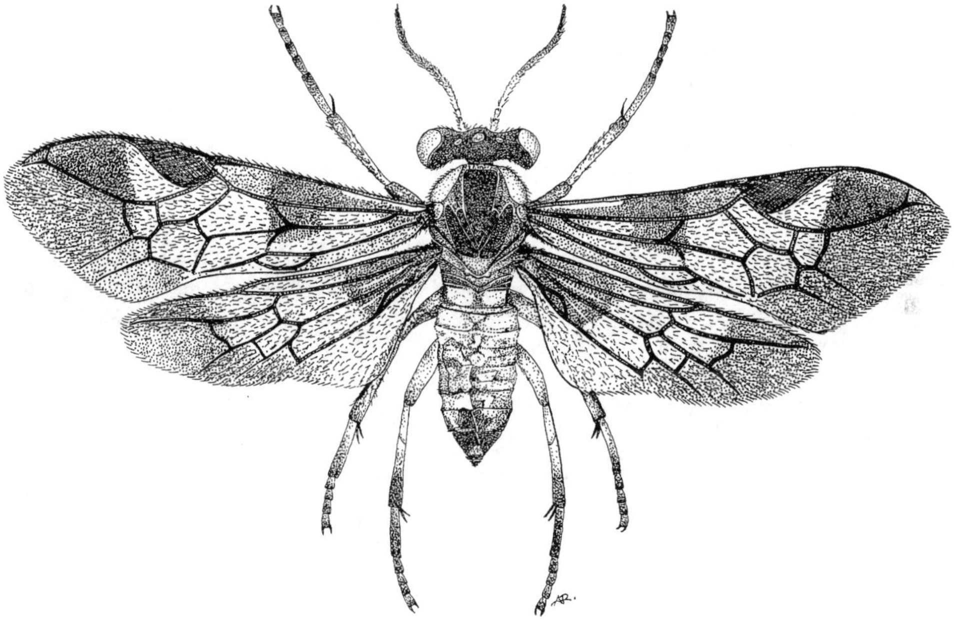

Fig. 6.23. *Ptilia versicolor* (Argidae).

Argid species	Food plant	Plant family
Didymia unifasciata	*Rourea sp.*	Connaraceae
Eriglenum crudum	*Maecherium biovulatum*	Leguminosae
Manaos kimseyae	*Inga phagifolium*	Leguminosae
Ptilia peletieri	*Cnestidium rufescens*	Connaraceae
Ptilia versicolor	*Rourea glabra*	Connaraceae
Schizocerella pilicornis	*Portulaca* sp.	Portulacaceae
Sericoceros gibbus	*Coccoloba manzanillensis*	Polygonaceae
Sericoceros vumirus	*Lonchocarpus minimiflorus*	Leguminosae
Scobina consobrina	*Sida* sp.	Malvaceae
Scobina guatemalensis	*Sida* sp.	Malvaceae
Sphacophilus janzeni	*Hymenaea coubaril*	Leguminosae
Trochophora lobata	*Rourea glabra*	Connaraceae

Table 6.2. Host plant associations of Central American argids.

was attached to the under surface of new leaves. Larval feeding took 33 days. The larvae fed gregariously along the leaf margin, holding the abdomen erect, away from the leaf margin. When disturbed they either waved their abdomen or curled it down against the leaf and secreted droplets of fluid from the coxal bases. *Sericoceros krugii*, which occurs in Puerto Rico and other Caribbean islands, shows subsocial behaviour in that the female guards the egg mass (Martorell, 1941). Similar behaviour has been observed in some South American Dielocerinae (Dias, 1975, 1976). These sawflies deposit eggs in clusters on the surface of a leaf, the female guards the eggs and young larvae, and the larvae feed in groups.

The last larval instar of Argidae, like that of other sawflies, leaves the feeding site to spin a cocoon. *Ptilia versicolor* spins a fragile, fibrous cocoon on the twigs and stems of the plant (Smith, 1992). Larvae of *Sericoceros gibbus* enter the leaf litter and construct a thick brown silk pupal case attached to twigs and dried leaves (Kimsey & Smith, 1985). In argids showing subsocial behaviour, the larvae often spin cocoons in masses. Cocoon masses of subsocial Dielocerinae are found on trunks and limbs of trees and are protected by a common covering spun co-operatively by the larvae.

Economic importance. *Schizocerella pilicornis*, which feeds on purslane (*Portulaca oleracea*), has been considered in biological control programmes of this weed (Clement & Norris, 1982).

Identification. Keys to genera and species of America south of the United States are given by Smith (1992). The following key to subfamilies is modified from this work.

Key to subfamilies of Argidae present in Central America

1 Marginal cell of hind wing closed, with accessory vein at apex (Fig. 6.04); antennae inserted above level of middle of eyes; mid- and hind tibiae each with preapical spine.
[Antennal flagellum of male simple; tarsal claws simple.] .... **Arginae** (p. 165)

— Marginal cell of hind wing open at apex (Fig. 6.05); antennae inserted at or below level of middle of eyes; tibiae without preapical spines. 2

2 Antennal flagellum of male simple; tarsal claws with a large acute basal lobe (Fig. 6.24); outer apical fore tibial spine nearly twice length of inner spine. .... **Atomacerinae** (p. 166)

— Antennal flagellum of male bifurcate (Fig. 6.03); tarsal claws usually without basal lobe (Fig. 6.25); apical fore tibial spines subequal in length. 3

3 Each mandible with two or three large teeth (Fig. 6.26); head transverse in dorsal view, broader than long. .... **Erigleninae** (p. 166)

— Left mandible simple or with two or three small teeth or with crenulations near centre, right mandible with single tooth near base (Fig. 6.27); head in dorsal view usually rounded, not notably broader than long. .... **Sterictiphorinae** (p. 166)

Synopsis of the Costa Rican fauna

Four subfamilies are represented by 15 genera and nearly 50 species in southern Central America; 13 of these genera and 31 species are recorded from Costa Rica, but more are expected to be found here.

ARGINAE

This subfamily comprises 15 genera worldwide, two of which are present in the New World. They are moderate to large sized sawflies, 8.0 to 12.0 mm in length.

ArgeEX. A holarctic and african genus comprising about 250 species. A large number occur in North America and several in Mexico, as far south as Chiapas. One, *basimacula*, occurs in Panama and Colombia. Head without sharp carinae, genal carina absent, pronotal furrow present. A key to New World species is given by Smith (1989).

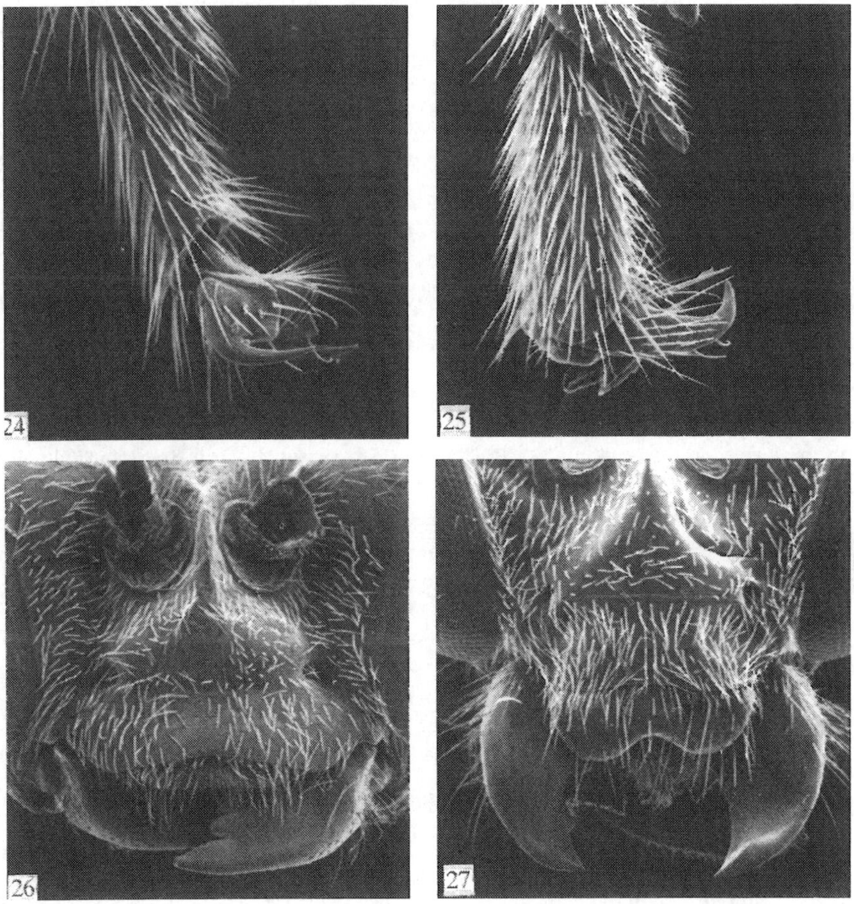

Figs 6.24–6.27. Scanning electron photomicrographs of Argidae. Figs 6.24–6.25. Tarsal claws; 6.24, *Atomacera* sp.; 6.25, *Sphacophilus* sp. Figs 6.26–6.27. Mouthparts; 6.26, *Eriglenum* sp.; 6.27, *Sphacophilus* sp.

Scobina. A neotropical genus comprising about 50 species distributed from Mexico south to northern Argentina. Eight of these occur in southern Central America and six, *consobrina, dorsalis, guatemalensis, lepida, notaticollis* and *paradorsalis*, are recorded from Costa Rica (Smith, 1992). *Scobina* is one of the most commonly collected sawfly genera in the Neotropics. Head with sharp carinae, genal carina present, pronotal furrow absent.

ATOMACERINAE

The subfamily comprises just one genus. They are small sawflies, 3.0 to 6.0 mm in length, and they are black or black with red markings on the thorax.

Atomacera. A genus comprising about 32 species, ranging from southeastern Canada to Argentina. Nine species are present in southern Central America and four, *ebena, lepidula, raza* and *tria*, have been recorded from Costa Rica (Smith, 1992).

ERIGLENINAE

The subfamily comprises just four genera, all of which are restricted to the Neotropics. Eriglenines seem to be rather uncommon insects in Malaise trap catches. They are rather plump sawflies, 8.0 to 12.0 mm in length. Social habits occur in this subfamily, but do not seem to be as well-developed as in the South American subfamily Dielocerinae. One species is known to guard its eggs and cocoons are formed in masses (Smith, 1992).

Eriglenum. A genus comprising four species. Two of these, *crudum* and *tristum* are present in Central America; both are known to occur in Costa Rica. The marginal cell in the fore wing is closed.

Sericoceros. A genus comprising 20 species, four of which are present in Central America and three, *gibbus, mexicanus* and *vumirus*, are recorded from Costa Rica. The marginal cell of the fore wing is open.

STERICTIPHORINAE

This is the largest subfamily of argids with 24 genera worldwide. Twenty of these occur in the New World, and ten are known or thought likely to occur in Costa Rica. The genera may be distinguished using the key given in Smith (1992), but for convenience an aid to generic recognition is given in Table 6.3. Sterictiphorines are medium to large sized sawflies, 6.0 to 12.0 mm in length. So far as is known, social habits are not developed, nor do the larvae form cocoon masses.

Acrogymnia. A neotropical genus extending from Costa Rica southwards to northern Argentina. It comprises 11 species, but only *A. palama* is present in Costa Rica and Panama.

Didymia[CR]. A neotropical genus ranging from Costa Rica to southern Brazil. It comprises 20 species, two of which, *D. unifasciata* and *D. nasuta*, are present in southern Costa Rica.

Hemidianeura. A New World genus ranging from the southwestern United States to northern Argentina, and containing over 30 species. Two species, *H. leucopoda* and *H. menkei* are known to occur in Costa Rica, but more species are expected to occur here.

Manaos. A neotropical genus extending from Veracruz in Mexico southwards to Argentina. It includes 12 species, two of which, *M. declivis* and *M. kimseyae*, are known to occur in Costa Rica.

Neoptilia. A New World genus with a range known to extend from the southwestern United States southwards to Venezuela and Ecuador. It includes 10 species; *biramosa* is known to occur in Costa Rica and three or four other species are present in Central America.

Ptilia. A neotropical genus ranging from Mexico to southeastern Brazil that includes seven species, two of which, *P. peletieri* and *P. versicolor*, are present in Costa Rica.

Schizocerella. A South American genus that comprises four species, one of which, *S. pilicornis*, now occurs virtually throughout the New World.

Sphacophilus. A New World genus comprising at least 60 species. Five species are known to occur in Costa Rica (*barius, holmus, janzeni, latus* and *madunus*) but others occur in Central America and may well be found here. *Sphacophilus* species are

- –Tarsal claws bifid .... *Neoptilia*
- –Tarsal claws simple
  - –Fore wing with marginal cell open at apex (Fig. 6.28)
    - –Hind wing without an enclosed anal cell (cf Fig. 6.34) .... *Schizocerella*
    - –Hind wing with an enclosed anal cell (cf Fig. 6.35)
      - –Interantennal carina Y-shaped, sharp (Fig. 6.36) .... *Acrogymnia*
      - –Interantennal carina straight, weak or absent (Fig. 6.37) .... *Sphacophilus*
  - –Fore wing with marginal cell closed at apex (Figs 6.29-6.32)
    - –Hind wing with jugal lobe grossly enlarged (Fig. 6.33) .... *Trochophora*
    - –Hind wing with jugal lobe normal (Figs 6.34-6.35)
      - –Marginal cell of fore wing with only a vestige of accessory vein present at apex (Fig. 6.29); antennal scape swollen .... *Manaos*
      - –Marginal cell of fore wing distinct accessory vein present at apex (Figs 6.30-6.32); antennal scape not swollen
        - –Fore wing with 3rd submarginal cell more or less evenly quadrate (Fig. 6.32) .... *Hemidianeura*
        - –Fore wing with 3rd submarginal cell unevenly quadrate, with anterior side longer than posterior side and/or distal side longer than proximal side (Figs 6.30, 6.31)
          - –Hind wing without enclosed anal cell (Fig. 6.34) .... *Triptenus*
          - –Hind wing with an enclosed anal cell (Fig. 6.35)
            - –Interantennal carina present (Fig. 6.38) .... *Didymia*
            - –Interantennal carina absent (cf Fig. 6.37) .... *Ptilia*

Table 6.3. Identification table for Costa Rican genera of Sterictiphorinae (modified from Smith, 1992).

amongst the most commonly collected sawflies in Malaise trap catches from the Pacific coastal plain in Costa Rica.

Triptenus[EX]. A neotropical genus occurring from Panama south to Brazil. It comprises two species one of which, *T. romani*, occurs in Panama.

Trochophora. A neotropical genus occurring from Costa Rica south to Brazil. It comprises two species, one of which, *T. lobata*, is present in Costa Rica.

6.2 PERGIDAE

Diagnosis. Diverse in structure and size, ranging from 3.0 mm (the smallest known sawflies) to 15.0 mm in length, mostly somewhat *robust* rather than elongate; black, black with red markings, or yellowish or orange with black markings; wings hyaline, black, yellowish, or yellowish with the apices blackish. *Antennae 6-, 7-, 8-, 10-, or 12- to 22-segmented, never 9-segmented*; filiform, serrate, or even pectinate (Figs 6.08–12). *Fore wing always lacking vein* 2r-rs,

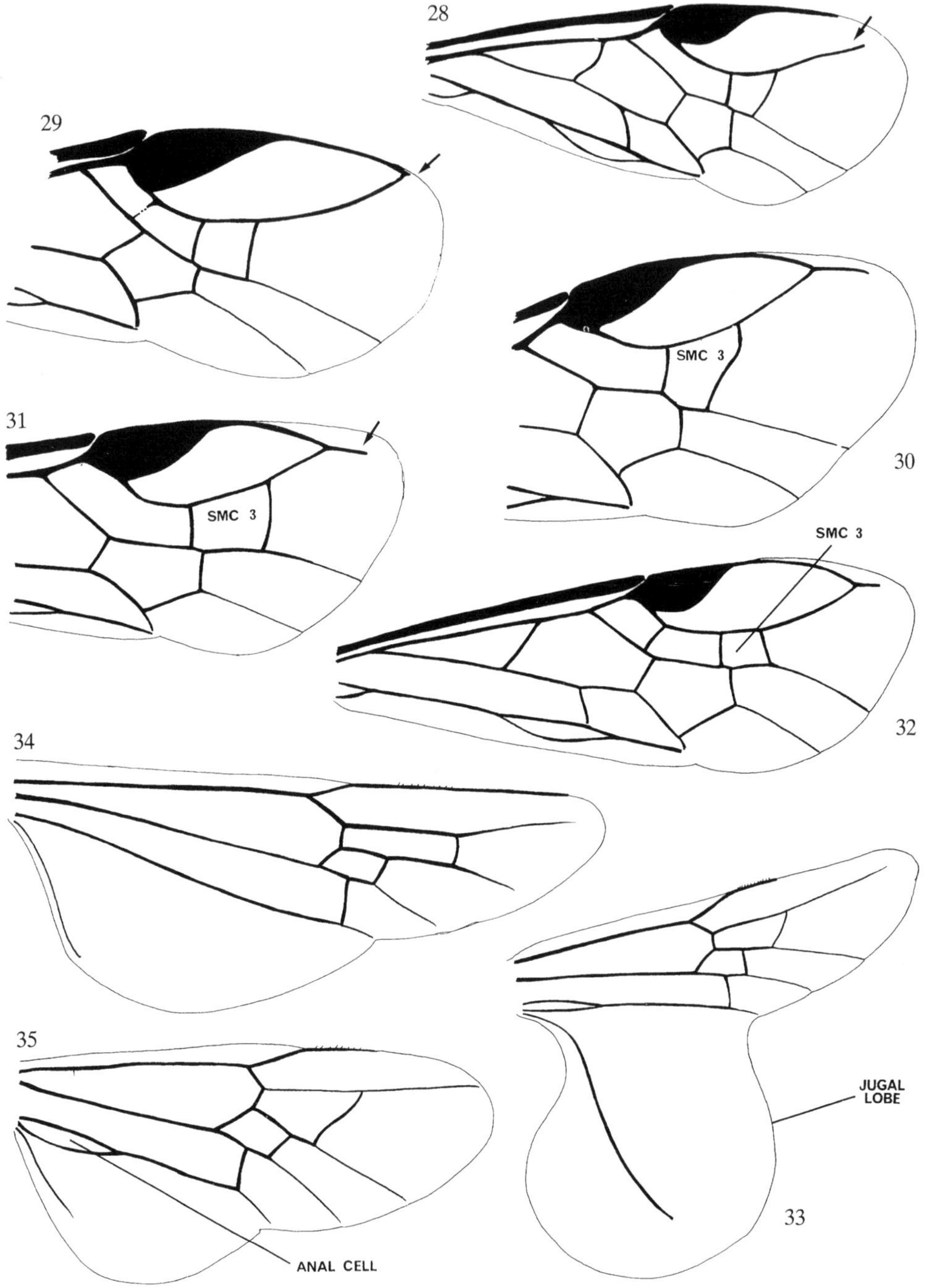

Figs 6.28–6.35. Sterictiphorine Argidae. Figs 6.28–6.32. Fore wing; 6.28, *Sphacophilus* sp.; 6.29, *Manaos* sp.; 6.30, *Didymia* sp.; 6.31, *Triptenus* sp.; 6.32, *Hemidianeura* sp. Figs 6.33–6.35. Hind wing; 6.33, *Trochophora* sp.; 6.34, *Triptenus* sp.; 6.35, *Didymia* sp.

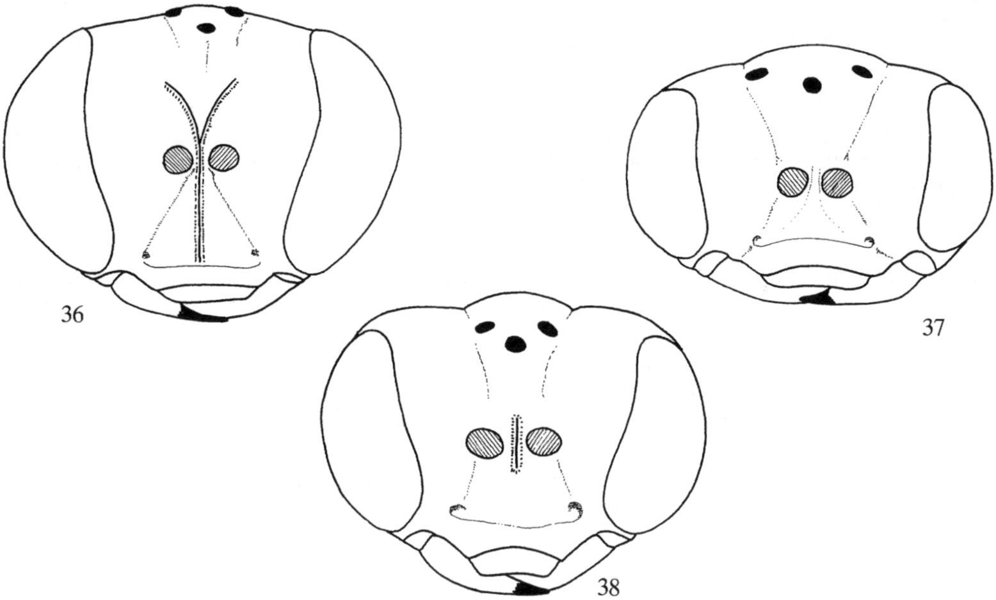

Figs 6.36–6.38. Sterictiphorine Argidae; head, anterior view; 6.36, *Acrogymnia* sp.; 6.37, *Sphacophilus* sp.; 6.38, *Didymia* sp.

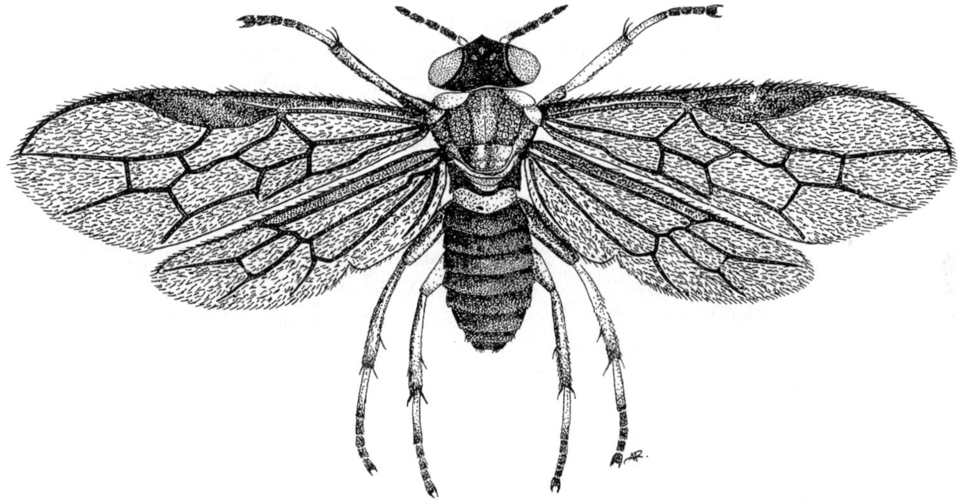

Fig. 6.39. *Acordulecera* sp. (Pergidae).

generally without an enclosed anal cell (Fig. 6.07), except Perreyiinae (Fig. 6.06); *hind wing without an enclosed anal cell* (Figs 6.06, 6.07). Male genitalia strophandrous (rotated 180°).

Classification and distribution. The Pergidae occurs in the Australian region, where it is the dominant sawfly group, and in South America, extending north to southeastern Canada. Pergids are not found in other parts of the world. The family comprises 14 subfamilies, 57 genera, and about 500 species. In the New World there are eight subfamilies (three of which are shared with the Australian region), 32 genera, and over 350 species (Smith, 1990). The Conocoxinae and Philomastiginae are restricted to southern South America, whereas the other six subfamilies are widely distributed throughout tropical America.

Biology. Like other tenthredinoids all Pergidae were thought to be phytophagous as larvae although some Australian species feed on dead and dying leaves (Tilyard, 1926). In Australia, where the habits of a number of pergids are fairly well known, most species are external feeders on foliage of woody plants, especially on species of Myrtaceae, and many apparently sequester secondary plant substances which the larvae use for defence (Macdonald & Ohmart, 1993). Unlike most other pergids phylacteophagine larvae are miners in leaves (Naumann, 1991; Macdonald & Ohmart, 1993).

Comparatively less is known about the larval foodplants and habits of New World species. A North American *Acordulecera* feeds on the leaves of Fagaceae and Juglandaceae (Smith *in* Krombein *et al.*, 1979) and a Costa Rican specimen of *Incalia americana* is labelled as 'on *Vitis*' (Vitaceae) and is associated on the same pin with part of a leaf which bears about 70 eggs in a cluster on its surface (in Paris Museum). In South America a species of *Syzygonia* is definitely known to feed on *Tibouchina* spp. (Melastomataceae), a species of *Heteroperryia* consumes *Schinus* (Anacardiaceae) (Smith, 1990) and a few other species feed on cultivated plants (see below). In Central America clues to possible hosts are found on specimen labels: *Anathulea* sp., 'from *Luehea seemanii*' (Tiliaceae); *Decameria similis*, 'on *Ficus carica*' (Moraceae) and 'on mango' (Anacardiaceae); and *Perreyia tropica*, 'reared from grass roots'. However, these 'records' must be regarded with caution as they may simply be plants on which the adults were fortuitously collected. In Brazil *Syzygonia cyanocephala* exhibits subsocial habits, with the female guarding the eggs and young larvae (Azevedo Marques, 1933).

Recently, the first record of fungivory in the Pergidae has been obtained in Costa Rica. About ten larvae of *Decameria rufiventris* were found at about 1600 metres in Monteverde feeding on a jelly fungus (*Auricularia* sp.) growing on rotten wood (Olsen & Joyce, pers. comm.). Two were reared to adulthood. Some other perreyiines possibly have similar habits, as there are a few tantalizing records of them occurring in, for sawflies, atypical situations. The wingless female of a New Guinea species of *Cladomacra*, has been found guarding its eggs under a log (Naumann, 1984). In Mexico masses of migrating larvae, possibly a *Perreyia* sp., have been observed crawling on the ground, and similar aggregations have been observed in Bolivia, Brazil and Guyana (Smith, 1990), and very recently, in Las Cruces Botanic Garden, Costa Rica (V. Thompson, pers. comm.). These aggregations of 50 or more larvae can be up to 30 cm long, 10 cm broad, and 5 cm high in the middle, and they progress as a single unit. In Costa Rica the larvae from one aggregation (*Perreyia tropica?*) were reared to adulthood on a diet of leaf litter (J. Ugalde, pers. comm).

Economic importance. South American species of *Haplostegus* and *Enjijus* are sometimes pests of guava (Pyenson, 1940), and a number of species of *Tequus* damage potato crops in Peru and Bolivia (Smith, 1980). In Brazil a species of *Perreyia* has occasionally been found to be common in pastures where they are consumed by and poison cattle (Camargo, 1956).

Identification. Keys to genera and species of America south of the United States are given by Smith (1990).

Key to subfamilies of Pergidae present in Central America

1 Anal cell present in fore wing, petiolate; first discal cell very large (Fig. 6.06).
[Mid and hind tibiae without preapical spines.] .... **Perreyiinae** (p. 171)

— Anal cell of fore wing absent, or only a basal stub of 2+3*A* present; first discal cell not unusually large (Fig. 6.07). 2

2 Fore tibia with a single apical spine; mid and hind tibiae both without preapical spines. .... **Pergulinae** (p. 171)

— Fore tibia with two apical spines; mid and sometimes hind tibia with preapical spines (Fig. 6.15)... 3

3 Hind tibia without preapical spines. ....
.... **Loboceratinae** (p. 171)
— Hind tibia with preapical spines (Fig. 6.15). 4

4 Maxillary palp 6-segmented; labial palp 4-segmented; labium trilobate. ....
.... **Acordulecerinae** (p. 171)
— Maxillary palp 3- to 5-segmented; labial palp 2- to 3-segmented; labium single lobed. ....
.... **Syzygoniinae** (p. 172)

Synopsis of the Costa Rican fauna

In Costa Rica there are probably at least ten genera and 40 species of Pergidae. Presently four subfamilies are known to occur in the country: Acordulecerinae, Loboceratinae, Perreyiinae and Syzygoniinae. A fifth subfamily, the Pergulinae, extends from Mexico to South America but it has not yet been collected in Costa Rica. Species of the Acordulecerinae and Perreyiinae are the most commonly collected pergids in Costa Rica. Most of the genera present in Costa Rica are widely distributed in the Neotropics (Mexico to northern Argentina), except for *Incalia*, which reaches the northern limit of its distribution in Costa Rica, and *Sutwanus*, which is restricted to Mexico and Central America. *Acordulecera* extends farther north than any other genus, reaching all the way to eastern North America.

ACORDULECERINAE

This subfamily only occurs in the New World where it is represented by eleven genera. Four of these are represented in Costa Rica. They are small sawflies, usually 3.0 to 5.0 mm long, with antennae short, typically filiform and 6- to 8-segmented (Fig. 6.10).

Acordulecera. A widespread genus with a range that extends from Canada to Argentina. It is one of the largest taxa in the family and probably comprises at least 200 species, 15 to 20 of which are present in Costa Rica (all undescribed). They are small sawflies with short 6-segmented antennae, with the second segment not much longer than broad.

Acordulocerídea. A neotropical genus comprising 15 species with a total range extending from Mexico to Argentina. Two or three species occur in Costa Rica. They have the antennae 7-segmented, with the third segment shorter than or subequal to the fourth, laterally flattened in males.

Anathulea. A neotropical genus comprising between 10 and 20 species, two or three of which are present in Costa Rica, including *bimaculata*. Their antennae are 6-segmented, longer than the width of the head, with second segment elongate.

Sutwanus. A mesoamerican genus containing several species, only one of which, *S. nigriceps*, is present in Costa Rica. Antennae 7-segmented with the third segment longer than the fourth.

LOBOCERATINAE

This subfamily comprises three genera, all restricted to the New World. One is known to occur in Costa Rica. They are moderately large sawflies, mostly 8.0 to 12.0 mm long, with the antennae 7-segmented (Fig. 6.09).

Aulacomerus. This widespread neotropical genus includes about 28 species, five of which are present in Costa Rica, including *hippolyte* and *lucidus*.

PERGULINAE

A small subfamily that includes two genera, one in the New World and a second in Australia. They have 7-segmented antennae.

Haplostegus^EX^. A widespread neotropical genus comprising about 15 species, known to occur in southern Mexico, Panama and throughout South America to southern Brazil. One species is known to feed on *Psidium* (Myrtaceae) in Brazil.

PERREYIINAE

This subfamily comprises seven genera, six in the New World, and one in New Guinea and Sulawesi, some species of which are unusual in having apterous females (Naumann, 1984; Shinohara, 1986). Two of the New World genera are represented in Costa Rica. Male perreyiines are more commonly collected than females; the latter are normally much larger than males, and awkward fliers that tend to be more sedentary in their habits. The antennae are 10- to 25-segmented, often serrate or pectinate in males (Figs 6.08,

6.11). Larvae of at least some species live and move in groups, and one species of *Decameria* is known to feed on a jelly fungus (*Auricularia* sp.).

Decameria. A widespread neotropical genus comprising 35 species, five of which (*championi*, *nigriventris*, *rufiventris*, *similis* and *varipes*) are present in Costa Rica. The females have the antennae 10- to 18-segmented, usually serrate whilst those of the male are filiform or serrate (not pectinate); the marginal cell of the hind wing is closed with its apex far from the wing apex.

Perreyia. A widespread neotropical genus containing 13 species. Two species, *fumipennis* and *tropica*, are known to occur in Costa Rica. The females have the antennae 16- to 24-segmented and filiform, while those of the male are pectinate; the marginal cell of the hind wing is either open or closed, its apex near the wing apex.

SYZYGONIINAE

This is an exclusively New World subfamily that includes five genera, two of which are represented in Costa Rica. All have the antennae 5- to 8-segmented, filiform or clubbed, with all or only several basal segments pectinate in some males (Fig. 6.12). The fore wing has the costa swollen.

Incalia. A neotropical genus that includes five species, one of which, *americana*, is extremely widespread and occurs from Costa Rica south to Bolivia. It is a large sawfly, more than 10.0 mm long, with an orange thorax and 7-segmented antennae.

Lagideus. A neotropical genus comprising 22 species, only one of which, *longicus*, is present in Costa Rica. It is a small sawfly, 5.0 mm in length, that is black with the fore and mid legs white, and the antennae 8-segmented.

6.3 TENTHREDINIDAE

Diagnosis. Medium sized sawflies, 4.0 to 10.0 mm in length, stout or elongate and somewhat cylindrical; commonly black, black with the thorax partly reddish or yellowish to orange with various black markings; wings hyaline, black, yellowish, or yellowish with the apices black. *Antenna 9-segmented* (at least in Costa Rican taxa), filiform (Figs 6.13, 6.14). Fore wing with

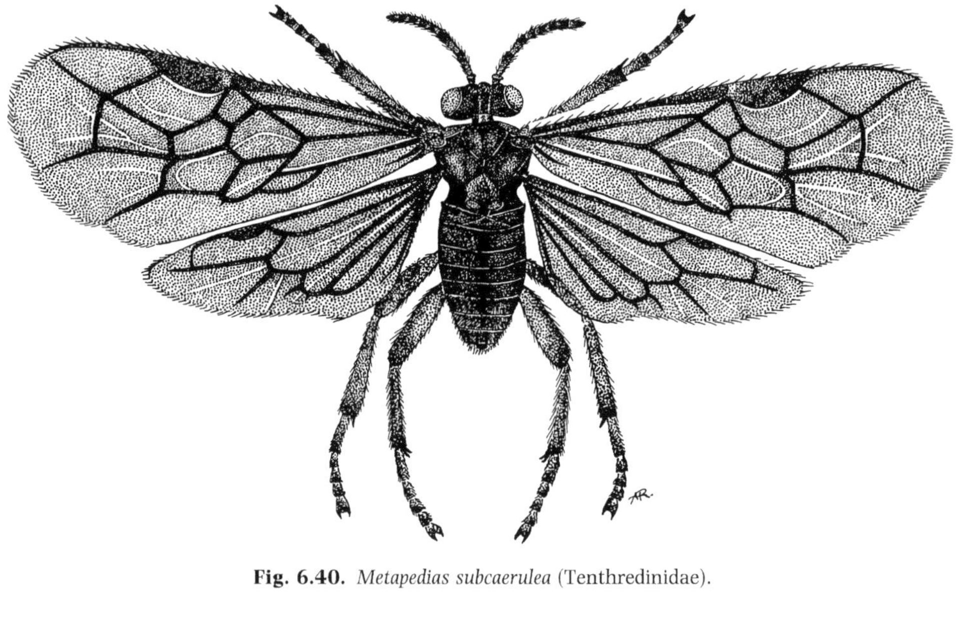

Fig. 6.40. *Metapedias subcaerulea* (Tenthredinidae).

an enclosed anal cell and usually with vein 2*r-rs* present; hind wing with an enclosed anal cell (Figs 6.16–19). Male genitalia strophandrous (rotated 180°).

Classification and distribution. Globally the Tenthredinidae is the largest family of Symphyta, comprising about 350 genera and 4000 species, but in the Neotropical region it is comparable in size to the Argidae and Pergidae. Worldwide eight subfamilies are recognized, of which four occur in southern Central America—Allantinae, Blennocampinae, Nematinae, and Selandriinae—which include about 15 genera and 40 or more species in this region. Subfamilies absent from Costa Rica include: Dolerinae (holarctic and east Africa), Heterarthrinae (holarctic and southern South America), Susaninae (just one genus in the western United States and Canada), Tenthredininae (holarctic and Asia).

Biology. In the North Temperate region the larvae of many tenthredinids are external foliage feeders, and although little is known of the habits of the Central American species, they are presumed to have a similar habit. Known host associations of Central American species are listed in Table 6.4. In Brazil some species of *Metapedias* feed on guava (Pyenson, 1940) and *Erythrina* (Leguminosae); in North America *Periclista* species are associated with *Quercus* spp. (Fagaceae), and *Ametastegia* species with *Rumex* and *Polygonum* (Polygonaceae). Most nearctic species of the Selandriinae, the largest subfamily in Costa Rica, are associated with ferns, grasses and sedges growing in humid places, and the Costa Rican species might possibly be found in similar habitats.

Kimsey and Smith (1985) described the larva and gave some life history notes on *Erythraspides interstitialis* in Panama. A female was observed inserting eggs under the epidermis on the surface of a leaf. After hatching, the larvae fed gregariously, starting at the tip of the leaf and working toward the petiole. They fed on the leaf blade but did not consume the midrib or other large veins. When not feeding, the larvae remained together in a cluster on the underside of the leaf. In the laboratory, mature larvae burrowed in the soil where they formed smooth-walled, silk-like cocoons. The period from hatching to cocoon formation was 15 days and adults emerged after ten days.

Identification. Keys to the genera and species of America south of the United States are in preparation (Smith).

Key to subfamilies of Tenthredinidae occurring in Costa Rica

1 Fore wing with anal cell extending almost to base of wing (Figs 6.16, 6.18). 2

— Fore wing with anal cell present only distally, veins 1*A* and 2+3*A* fused proximally (Figs 6.17, 6.19). 3

2 Fore wing with anal cross vein usually absent (Fig. 6.16), but present in a few species; anterior margin of mesepisternum with epicnemium. .... **Selandriinae** (p. 175)

— Fore wing with anal cross vein present (Fig. 6.18); anterior margin of mesepisternum without epicnemium. ..... **Allantinae** (p. 174)

Tenthredinid species	**Food plant**	**Plant family**
Erythraspides interstitialis	*Hamelia patens*	Rubiaceae
Metapedias intermedia	'*Erythrina rubrinervia*'	Leguminosae
	'*Spondias mombin*'	Anacardiaceae
Metapedias subcoerulea	*Conostegia xalapensis*	Melastomataceae
Waldheimia fascipennis	*Cissus sicyoides*	Vitaceae
Waldheimia suturalis	*Cissus rhombiflora*	Vitaceae

Table 6.4. Food plants of tenthredinid species in Central America (those in quotation marks are unconfirmed label data).

3 Fore wing with veins *M* and 1*m-cu*, on either side of first discal cell, parallel; vein 2*r-rs* present; vein *M* meeting *Sc+R* very close to point where *Rs+M* meets *Sc+R* so anterior corner of first subdiscal cell is quite pointed (Fig. 6.17). .... **Blennocampinae** (p. 174)

— Fore wing with veins *M* and 1*m-cu*, on either side of first discal cell, markedly divergent posteriorly; vein 2*r-rs* absent; vein *M* meeting *Sc+R* far basal to point where *Rs+M* meets *Sc+R* so anterior corner of first subdiscal cell is truncately rounded (Fig. 6.19). [Rare in Costa Rica.] .. **Nematinae** (p. 174)

Synopsis of the Costa Rican fauna

Four of the eight subfamilies of Tenthredinidae are known to occur in Costa Rica, where they are represented by about 50 species in 16 genera.

ALLANTINAE

A large and predominantly holarctic subfamily comprising about 80 genera worldwide, which is represented in the Neotropics by only about 30 species in the tribes Empriini, Acidiophorini and Allantini. Only two genera occur in Costa Rica, one in the Empriini (*Ametastegia*) and a second in the Allantini (*Probleta*). Smith (1979) revised this subfamily for the Nearctic region and included keys to all New World genera.

Ametastegia. A primarily holarctic genus for which Costa Rica is the southernmost locality. *A. championi* has been collected at Vara Blanca, at an altitude of 2100 metres. Species of this genus have both mandibles bidentate and the clypeus subtruncate or with V-shaped emargination. Keys to species are given by Smith (1979). In North America several *Ametastegia* species are known to feed as larvae on herbaceous plants, although two feed on *Salix* (Smith, 1979).

Probleta. A large neotropical allantine genus, with a range extending from Mexico south into South America. Only one unidentified species has been found in Costa Rica, at San Francisco (1100 m) in San José Province. Species of this genus have the mandibles asymmetrical and the clypeus deeply incised.

BLENNOCAMPINAE

A large and widespread subfamily with 84 genera worldwide and about 100 species occurring in the Neotropics. Four genera and at least ten species are present in southern Central America. Smith (1969*a*) has revised the nearctic fauna.

Erythraspides. A genus comprising about 15 species from eastern North America to southern Central America. *E. interstitialis* has been collected in Costa Rica. Species of this genus lack a closed discal cell in the hind wing, and have the inner tooth of the tarsal claw slightly shorter than the outer one.

Metapedias. A genus that includes about 30 species ranging from Mexico to northern Argentina; *M. biolorata, intermedia* and *subcaerulea* are known to occur in Costa Rica. *Metapedias* species lack a closed discal cell in the hind wing, and the marginal cell is short, ending some distance from the wing apex.

Periclista. A holarctic and southern South American genus, with a single species present in Costa Rica. This species, which has been collected at about 2500 metres on the Cerro de la Muerte, is closely allied to holarctic species and Costa Rica thus represents the southern limit of distribution of this species-group. It is the only genus in Costa Rica in which the basal stub of vein 2+3*A* is turned up at its apex, not straight.

Waldheimia. A primarily neotropical genus with a range extending northwards into the southwestern United States and south to Argentina. *Waldheimia* comprises about 80 species, around ten of which occur in Costa Rica, including *annulipes, fascipennis, ochra, pallens, suturalis* and *trimaculata*. These have the hind wing with the discal cell usually present, and the apex of the marginal cell close to the apical margin of the wing. The inner tooth of the hind tarsal claw is longer than the outer.

NEMATINAE

A very large and diverse subfamily comprising 42 genera worldwide. The Nematinae is numerically the dominant subfamily in the Arctic and Subarctic regions of the world, but only a few species belonging to two genera, *Nematus* and *Pristiphora*, have been recorded from America south of the United States. The former genus is represented in southern South America by a single species, *N. desantisi*, which is a pest of willows (*Salix* spp.), while the latter has scattered representatives throughout tropical America.

Pristiphora. A predominantly holarctic genus with over 100 species. It is represented by two species in Brazil, one in southern Mexico, one in Venezuela and two in Costa Rica. The Costa Rican species were collected at 1100 metres in Amistad National Park and at 1500 metres in Las Alturas biological station near the Panamanian border. They belong to the same species-group as a North American species that feeds on *Quercus*, and this could be the larval food-plant in Costa Rica.

SELANDRIINAE

This is a large subfamily that comrises 61 genera worldwide. Unlike other tenthredinids, the Selandriinae seems to have a tropical peak of species-richness. In the New World, selandriines are very diverse and common in the Neotropical region where more than 200 species are estimated to occur, whereas only about 27 species are known from North America (Smith, 1969*b*). Genera in this subfamily are rather difficult to separate and the numbers of species belonging to each has yet to be determined. Approximately 30 species have been collected in Costa Rica, belonging to nine genera.

Adiaclema. A neotropical genus with 30 described species, and probably a total of about 45 species, ranging from southern Mexico to northern Argentina and southern Brazil. Four of these, including *nobile*, occur in Costa Rica.

Dochmioglene. A neotropical genus with 19 described and perhaps a further five undescribed species ranging from Guatemala to northern Argentina and southern Brazil. Two species are known to occur in Costa Rica.

Eustromboceros. A mesoamerican genus with three described and perhaps a further seven undescribed species ranging from Mexico to Costa Rica. Only a single species, *nigromaculatus*, is known to occur in Costa Rica.

Inea. A tropical American genus with nine described and perhaps a further 16 undescribed species ranging from central Mexico to Colombia and Ecuador. Three species, *bicolorata*, *pusilla* and *pygmaeus*, are known to occur in Costa Rica.

Liliacina. A neotropical genus with five described and perhaps a further five undescribed species that range from central Mexico to northern Argentina and southern Brazil. Only one species, *diversipes*, is known to occur in Costa Rica.

Plaumanniana. A neotropical genus with 17 described and perhaps a further eight undescribed species that range from southern Mexico to northern Argentina and southern Brazil. Two species are known to occur in Costa Rica.

Proselandria. A neotropical genus with eight described and perhaps a further 17 undescribed species ranging from Guatemala to northern Argentina and southern Brazil. Two species are known to occur in Costa Rica.

Rohwerina. A small mesoamerican genus with two species, one in Mexico and a second, *ornaticornis*, occurring from southern Mexico to Costa Rica.

Stromboceridea. An American genus ranging from the southwestern United States (Arizona) south along the Andes to northwestern Argentina. It comprises 25 described species with perhaps a further 20 undescribed. Four are known from Costa Rica: *bifer*, *laeta*, *mutica* and *testaceicornis*.

6.4 XIPHYDRIIDAE

Diagnosis. Elongate, *cylindrical* sawflies, 4.0 to 25.0 mm (usually 8.0 to 15.0 mm) in length; coloration predominantly dark brown. Antennae long, filiform, 20- to 25-segmented. Head above eyes dome-shaped, usually shining. *Cervical sclerites elongated*, appearing as a short neck; posterior margin of pronotum emarginate; lateral lobes of mesoscutum usually divided by a transverse ridge. Tibiae usually without preapical spines; fore tibia with one or two apical spines. Fore wing lacking vein 2*r-rs* (in Costa Rican species), anal

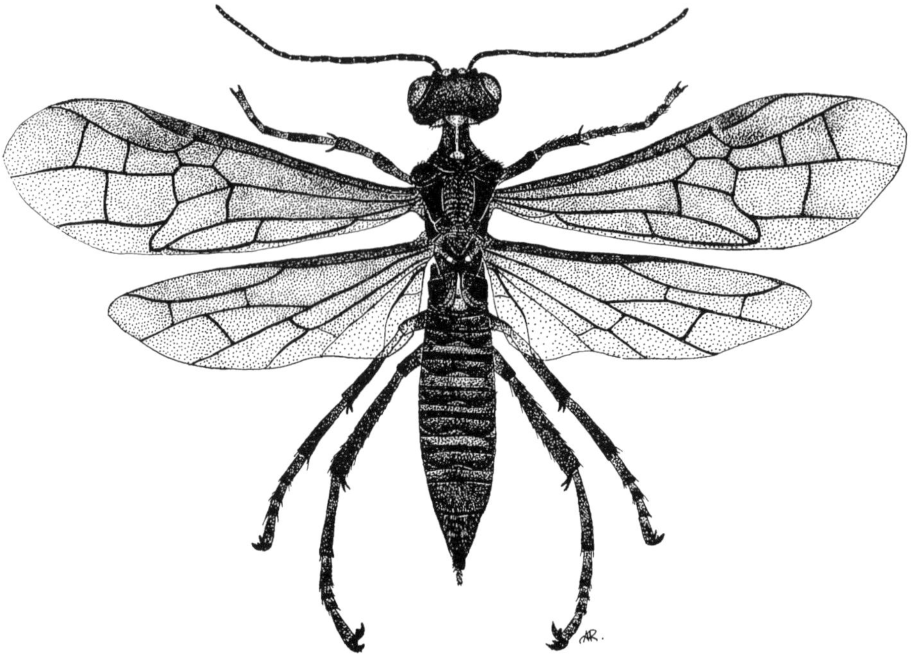

Fig. 6.41. *Derecyrta* sp. (Xiphydriidae), male.

cell complete, crossed by an anal cross vein. Abdomen with first tergite medially divided; ovipositor projecting beyond apex of abdomen; male genitalia orthandrous (not rotated 180°).

Classification and distribution. The Xiphydriidae is a small family, with 23 genera and about 100 species, which is widely distributed throughout the world except in Africa. The family is divided into two subfamilies: the Xiphydriinae (19 genera, Holarctic and Asia) and Derecyrtinae (three New World genera and one Australian genus). The Peruvian genus *Eoxiphia* is dubiously placed in the Xiphydriinae (Smith, 1988). Among New World Derecyrtinae, two genera are restricted to southern South America, whereas *Derecyrta* is widely distributed in the Neotropics. Interestingly, no xiphydriids are known from northern Central America (Honduras-Guatemala) and Mexico.

Biology. The habits and food plants of neotropical species are not known, but the larvae of other Xiphydriidae are wood-borers in angiosperms. The nearctic species, all in the genus *Xiphydria*, prefer dead or moribund small branches of trees and shrubs (Deyrup, 1984; Smith & Middlekauf, *in* Evans, 1987*a*). The adult female makes an oviposition boring in the wood, in which she deposits a number of eggs. In the Palaearctic region ovipositing females are known to inoculate the ovipositional substrate with the spores of a symbiotic fungus—*Daldinia*, an ascomycete (Francke-Grosmann, 1967)—which are stored in a pair of abdominal sacs (mycangia) that open out to the base of the ovipositor (see Chapter 2.3). The eggs hatch within two to three weeks and the larvae tunnel in the wood, but seldom penetrate deeper than 50 mm. The larvae generally complete development in a single year. The final instar larva is grub-like, with a well-sclerotized head capsule bearing 3- or 4-segmented antennae, reduced mammaform, true legs, no abdominal prolegs, and a conspicuous horn-like postcornus on the tenth abdominal segment. Pupation occurs in a chamber just below the surface of the sapwood (Chrystal & Skinner, 1932). Adult xiphydriids are most commonly collected from infested wood in rearing chambers.

Identification. A key to the species of America south of the United States is given by Smith (1988), and the nine nearctic species of *Xiphydria* are treated by Smith (1976).

Synopsis of the Costa Rican taxa

Only a single genus, *Derecyrta*, is present in Costa Rica.

Derecyrta. A neotropical genus comprising ten species occurring throughout South America north to Costa Rica. The only species present in Costa Rica is *Derecyrta rugifrons*, which is known only from the female holotype (Irazu, 2000-2300 metres elevation), and three other specimens that may represent the male of this species, collected between 700 and 1100 metres in Guanacaste and Braulio Carrillo National Parks.

7 The orussids and stephanids

INTRODUCTION

Paul E. Hanson and Ian D. Gauld

The traditional division of the order Hymenoptera into two suborders, the 'Symphyta' and Apocrita, is increasingly difficult to maintain since the characters used to unite the 'Symphyta' are mostly shared plesiomorphies. Moreover, there are several apomorphic characters that unite the Orussidae with the Apocrita, although orussids are usually classified in the 'Symphyta' (see Chapter 5). In our opinion, these features coupled with the fact that orussids develop as parasitoids, are an overwhelming argument for treating the Orussidae separately from the sawflies and woodwasps, thereby breaking with the convention of maintaining the symphytan/apocritan divide. Although the Orussidae and Stephanidae probably do not comprise a monophyletic group, they are both primitive families of parasitoids. Both comprise species that are idiobiont ectoparasitoids of wood-boring insects, which is thought to be the ancestral biology of Apocrita (Gauld, 1988*c*). Therefore, if for no other reason, it is biologically convenient to treat these two families together.

The position of the Stephanidae within the Apocrita is still unclear, since this family does not seem to show any close relationship with any other apocritan family. Rasnitsyn (1980) classified the Stephanidae as the only extant member of the infraorder Stephanomorpha, and Gibson (1985) stated that the thoracic structures he studied support Rasnitsyn's hypothesis that stephanids are a basal apocritan lineage. We agree that the Stephanidae probably represents one of the most basal clades within the Apocrita. Although stephanids are very distinctive and fully warrant inclusion in a separate superfamily (Gibson, 1985; N.F.Johnson, 1988; Whitfield *et al.*, 1989), they do have a few structural characteristics in common with orussids, which makes their inclusion together in one chapter tenable. Species in both families have a crown of teeth on the head, and they possess similar basally narrowed, scoop-like mandibles. These features could be regarded as synapomorphies, but more probably they are characteristics that evolved in the ancestor to the Orussidae + Apocrita and which were then lost in most apocritans after the Stephanidae diverged (S. Shaw, pers. comm.). Another possibility is that these features were acquired convergently as a consequence of attacking hosts living in wood, as mandibles of this form are also found in some wood-inhabiting ichneumonoids.

7.1 ORUSSIDAE

David R. Smith

Diagnosis. Elongate, cylindrical, ranging in size from 4.0 to 12.0 mm in length; usually black, sometimes with legs reddish. Female antennae 10-segmented, fusiform, last segment peg-like (Fig. 7.02); male antenna 11-segmented, setaform (Fig. 7.03); in both sexes the *antennae arise from the ventral aspect of the head, below the apparent clypeus* (Figs 7.04, 7.05). Head often punctate, sometimes with carinae on the inner margins of the eyes, and with two to three pairs of tubercles on top of the head between the eyes (Figs 7.04, 7.05); labrum tongue-like. Pronotum emarginate posteriorly. Wing venation reduced compared to that of Symphyta (Fig. 7.06); fore wing with distal ends of veins reduced to faint lines, only one radiomedial (cubital) cross vein (*2rs-m*) present, vein *2A* present. Tibiae without preapical spines; fore tibia with two apical spines; fore tarsus 5-segmented in females, 3-segmented in males. Ovipositor sheath exserted; ovipositor thread-like, longer than body length, usually coiled within body and held in median groove on eighth sternite. Male genitalia orthandrous (not rotated 180°).

Classification and distribution. The Orussidae has traditionally been classified in the suborder Symphyta (sometimes within the Siricoidea) because it shares several characters in common with the

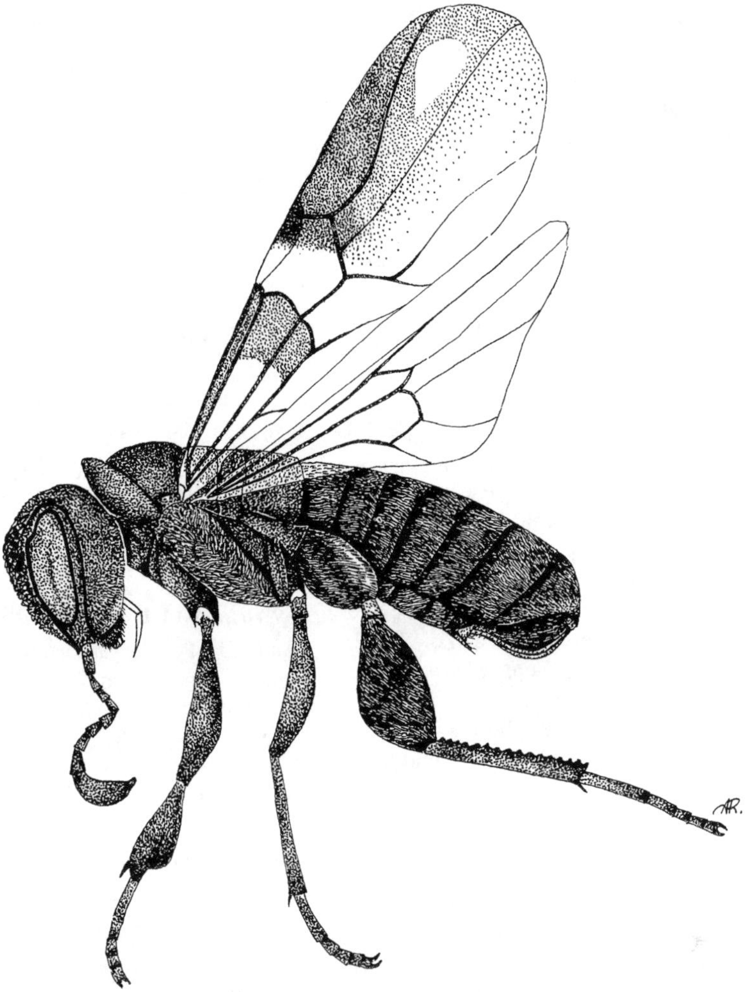

Fig. 7.01. *Ophrynopus* sp. (Orussidae).

sawflies (see Chapter 6). These characters, however, are all primitive (plesiomorphic) features and thus do not justify grouping them together. The Orussidae is a cosmopolitan family consisting of 16 genera and about 100 species, of which seven genera and about 20 species occur in the New World with three genera and about five species in Central America. The family is classified as follows (Abe & Smith, 1991):

OPHRYNOPINAE: fore wing with *2r-rs* joining pterostigma nearer its apex than its centre; male subgenital plate (= hypandrium) with three tubercles. Comprises two tribes, both of which are represented in the New World:

Guigliini—two genera, *Guiglia* (Australia and Chile) and *Kulcania* (southern Florida, and Mexico to Colombia).

Ophyrynopini—four genera, *Ophrella* (Panama and Brazil), *Ophrynopus* (southern United States to Argentina), *Ophyrynon* (southern California) and a fourth genus in southeast Asia.

ORUSSINAE: fore wing with *2r-rs* joining pterostigma near its centre; male subgenital plate simple, without tubercles. Comprises four tribes, two of which are represented in the New World, although neither occur in tropical America.

Leptorussini—four genera, *Orusella* (Chile), two others in Australia and one in the Afrotropical region.

Mocsaryini—two genera, one in the Afrotropical region and the other in Asia.

Orussini—three genera, *Orussus* (Holarctic), one Palaearctic and one Afrotropical.

Pedicristini—one Afrotropical genus.

Biology. The Orussidae are generally uncommon insects, but the females may occasionally be encountered running with an ant-like gait over dead trees and logs. They seem reluctant to fly even when disturbed. The development of few species has been studied, but some are known to develop as parasitoids on the larvae or pupae of wood-boring beetles and other holometabolous insects (Rawlings, 1957; Nuttall, 1980). Some perhaps develop, at least in their first instar, by feeding on micro-organisms in the frass-filled tunnels of other wood-borers (Cooper, 1953). Possibly such species complete development by feeding endophagously in the larvae or pupae of other insects encountered (Powell & Turner, 1975).

Nuttall (1980), working with a New Zealand species, observed the early larva developing as a solitary idiobiont ectoparasitoid on an introduced siricid. Subsequent larval instars feed endophagously in the decomposing corpse of the host. The final instar larva is hymenopteriform, without legs (although they do possess sclerotized disc-like remnants of the thoracic legs), with a 1-segmented antenna and no postcornus (Smith & Middlekauff *in* Evans, 1987*a*). The larva pupates within or by the host remains.

Identification. Middlekauff (1983) reviewed the Orussidae of North and Central America, and Smith (1988) provided a synopsis of all species from America south of the United States.

Key to genera of Orussidae present in Central America

1 Facial carinae complete, completely enclosing the frontal area (Fig. 7.05); punctures on lower part of frontal area each with several setae; maxillary palp reduced, apparently 3-segmented. .... *Kulcania*

— Facial carina absent or weak, not completely enclosing frontal area (Fig. 7.04); punctures on frontal area each with only one or no setae; maxillary palp long, 5-segmented. 2

2 Setae in punctures of frontal area distinctly broadened and laterally flattened; fore wing with *cu-a* joining *Cu*1 far distal to *M*, at centre of discal cell; 3rd antennal segment subequal in length with 4th. .... *Ophrella*

— Setae absent on frontal area, or if present slender and only expanded at apices; fore wing with *cu-a* arising nearly opposite base of *M* (Fig. 7.06); 3rd antennal segment longer than the 4th (Figs 7.02, 7.03). .... *Ophrynopus*

Synopsis of the Costa Rican fauna

All Central American orussids belong to the subfamily Ophrynopinae. Both tribes are represented, the Guigliini by *Kulcania*, and the Ophrynopini by *Ophrella* and *Ophrynopus*. In Costa Rica the majority of orussids have been collected in the seasonally dry forests in northwestern Costa Rica.

Kulcania. The genus comprises only a single species, *K. mexicana*, which occurs in Florida, Mexico, Costa Rica ('San José') and Colombia.

Ophrella. This genus includes one species in Brazil, *O. lingulata* from Panama, and at least one unidentified species in Costa Rica (P. Hanson, pers. comm.).

Ophrynopus. A genus containing seven species with a cumulative range extending from southern Arizona and Texas south to Argentina. *O. nigricans* has been recorded from the southern United States and Mexico south to Panama. Several unidentified specimens representing two or three species have been collected in Malaise traps in Santa Rosa National Park (300 m) in Guanacaste Province, and near Cuidad Colón (800 m) in San José Province.

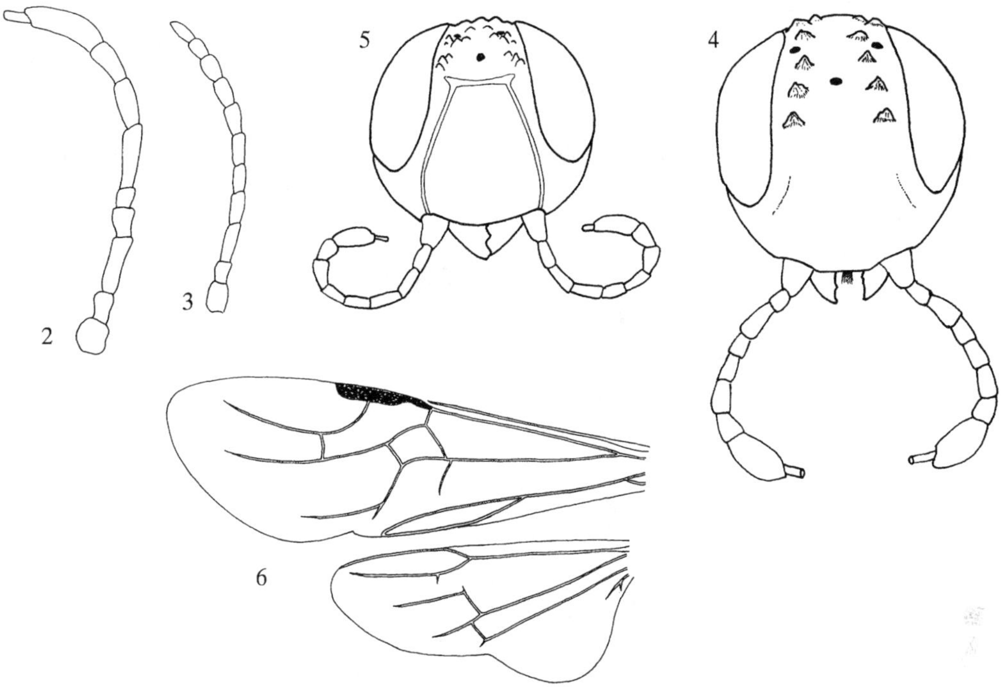

Figs 7.02–7.06. Orussidae. Figs 7.02–7.03. *Ophrynopus* sp., antenna; 7.02, female; 7.03, male. Figs 7.04–7.05. Heads; 7.04, *Ophrynopus* sp.; 7.05, *Kulcania* sp. Fig. 7.06. Fore and hind wings, *Ophrynopus* sp.

7.2 STEPHANIDAE

Ian D. Gauld

Diagnosis. Moderate to large sized insects (body 11 to over 60 mm in length, including the ovipositor; fore wing length 6.0 to 15.0 mm), elongate and cylindrical; black or dark brown in colour, ovipositor sheath with a subapical white band, a few species having a reddish head. *Antenna 30- to 42-segmented, with flagellum long and thread-like*; antennal insertion close to clypeus. *Head more or less spherical; median ocellus surrounded by a ring of five (or more) blunt to quite acute teeth* and usually with entire frons and vertex rugose to denticulate; mandible robust, tridentate, with the upper and lower teeth small and the middle tooth large and bluntly triangular. *Pronotum extended anteriorly into an elongate 'neck'*, posteriorly with upper corner more or less reaching tegula. Mesoscutum small, smaller than the scutellar region from which it is separated by a flexible transverse suture. *Fore wing with costal cell broad*; hind wing usually without enclosed cells. Fore leg with tibia centrally impressed, with a single apical spur; mid tibia without spurs; hind leg with coxa elongate and often transversely strigose, femur swollen with two or three large and several small denticles ventrally, tibia with a sub-basal constriction and two apical spurs, *tarsus of female 3-segmented* (neotropical species only), of male 5-segmented. Metasoma elongate, inserted into propodeum slightly above hind coxae; *first tergite and sternite fused into a long cylinder* (in neotropical species), that is generally annulately striate; ovipositor longer than metasoma, very slender.

Classification and distribution. The Stephanidae comprises slightly less than 200 described species classified in about six genera. The group is cosmopolitan in distribution (Townes, 1949; Tobias, 1988) but it is most species-rich in the Old World tropics (Elliott, 1922; Benoit, 1984*a*). Three genera occur in the New

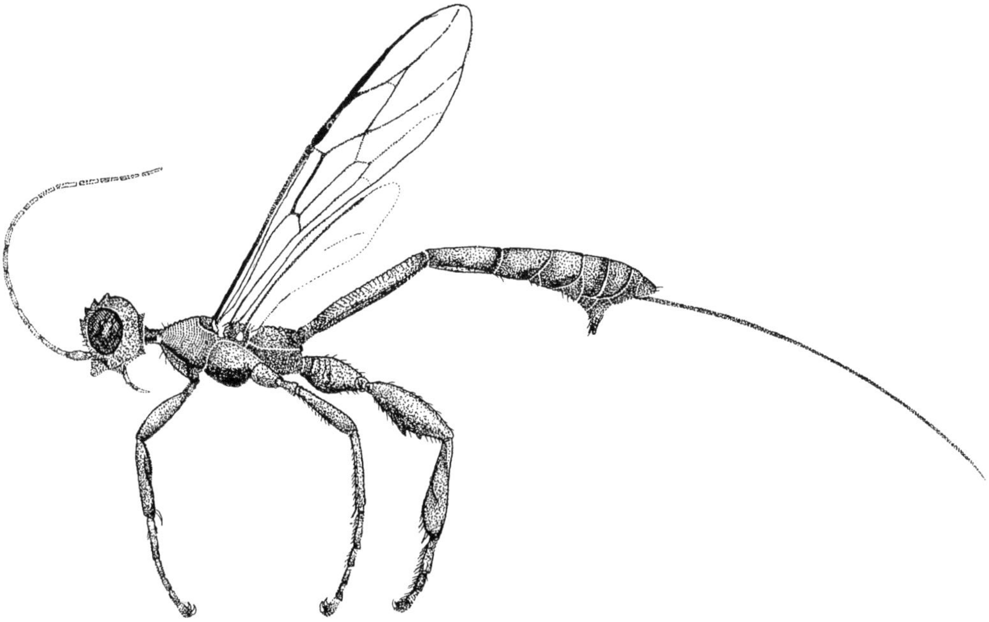

Fig. 7.07. *Megischus* sp. (Stephanidae).

World, *Foenatopus*, *Megischus* (= *Hemistephanus*) and *Schlettererius*. The last of these includes only a single species, and is restricted to pacific North America, from British Columbia south to California. In earlier works (e.g. Elliott, 1922) another genus, *Stephanus*, was also recorded from the New World. However, *Stephanus* is now restricted to include only a few Old World species, the females of which have a five-segmented tarsus. New World species previously placed within it belong to *Megischus* (Townes, 1949).

Biology. Little is known about the biology of most stephanids and the following remarks are based on very few observations of mainly extratropical species. Stephanids are solitary idiobiont ectoparasitoids of wood-boring holometabolous insects and possibly other insects using tunnels in wood, such as solitary bees (although this needs to be confirmed) (Königsmann, 1978*a*). The majority of recorded hosts in the tropics are buprestids (e.g. Townes, 1949; Benoit, 1984*b*) although in extratropical regions species are known to parasitize other wood-boring beetle larvae, such as bostrychids (Rodd, 1951*b*), and the larvae of siricids (Taylor, 1967).

Female stephanids search for hosts in dead wood, often spending much time examining, with their antennae, the undersides of fallen trees. Oviposition has been described and figured by several authors (e.g. Rodd, 1951*b*; Taylor, 1967; Jansen *et al.*, 1988), and some may actually bore through wood though others appear to be probers (see Chapter 4). The female begins by lifting her metasoma until it is at approximately right angles to the axis of the body, thus bringing the long ovipositor to an angle of about 30° with the surface of the wood. As the ovipositor is inserted into the wood, the sheaths are pushed back in a loop at the top, and the wasp moves gradually backwards whilst the metasoma is slowly lowered. The depths of penetration suggest that the ovipositor actually enters the wood at right angles to the surface. Full insertion of the ovipositor has only been observed occasionally and is apparently not essential for successful parasitism (Taylor, 1967).

The female stephanid apparently stings and immobilizes the host larva prior to laying an egg on it (Taylor, 1967). This egg is elongate and has an extremely long, thin stalk posteriorly, which presumably allows it to distort as it passes down the lumen of the ovipositor. The egg hatches 10 to 14 days after oviposition and the small active grub-like larva feeds through an integumental lesion, sometimes in an intersegmental membrane. The first instar larva has simple, sharp pointed mandibles, but later instars have characteristic tridentate mandibles. The exact number of instars is not known although three were distinguished by Taylor (1967). The final instar larva is hymenopteriform, with rather reduced mouthparts and without, or with a vestigial, salivary orifice. It consumes the entire host, except for the hardest chitinous parts, six to seven weeks after oviposition. The final instar larva pupates in the site where it finished feeding. No traces of a cocoon have been observed, and pupal exuviae occur within the final instar larval exuviae (Taylor, 1967). Male and female pupae are figured by Rodd (1951*b*).

Stephanids are most frequently encountered on or around dead standing and fallen trees, or around wooden buildings. I have observed, in Brunei, males of a species of *Foenatopus* congregating around a fallen tree from which females were emerging. This aggregation resembled those that have been observed for some tropical rhyssine ichneumonids (Eggleton, 1990), which are also parasitoids of wood-boring insects. Mating behaviour has not been described in detail, but females of at least one species are known to copulate more than once (Taylor, 1967). Females begin to oviposit about 12 days after emergence from the log.

Identification. The North American species are keyed by Townes (1949). The most complete work on neotropical species is the now dated revision by Elliott (1922), which was based on a very limited amount of material. Elliott probably recognized too many species as he had insufficient material to be able to assess the intraspecific variation that often occurs in species of wood-borer parasitoids.

Key to genera of Stephanidae present in the New World

1 Fore wing without enclosed discal or submarginal cells, with only three cells enclosed basally (Fig. 7.08); pterostigma centrally membranous and weakly pigmented; male with small, conspicuous, sharply defined disc-like tyloids on flagellar segments 2 to 11+ (Fig. 7.10). .... *Foenatopus*

— Fore wing with an enclosed discal and submarginal cell, thus with at least five fully enclosed wing cells (Fig.7.09); pterostigma centrally sclerotized and strongly pigmented; male with large, inconspicuous and ill-defined concave tyloids on basal flagellar segments (Fig. 7.11). 2

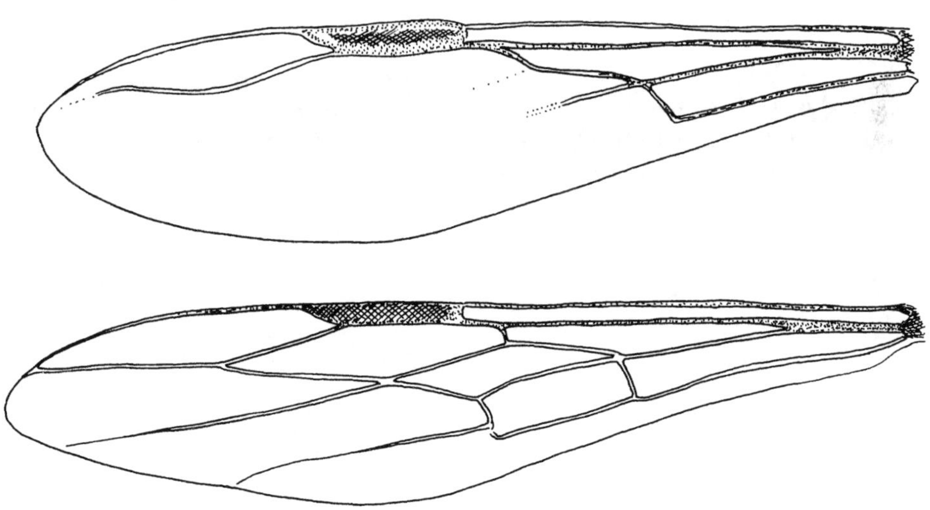

Figs 7.08–7.09. Stephanidae, fore wings. 7.08, *Foenatopus* sp.; 7.09, *Megischus* sp.

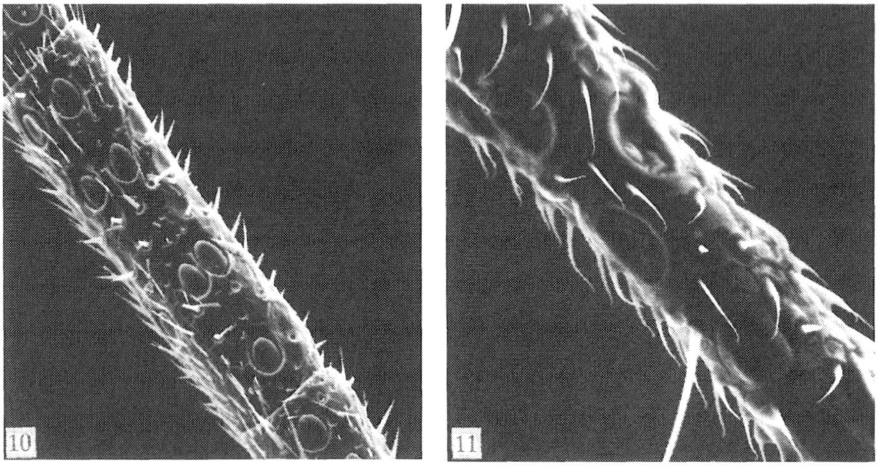

Figs 7.10–7.11. Scanning electron photomicrographs; basal segments of flagellum of male; 7.10, *Foenatopus* sp.; 7.11, *Megischus* sp.

2 First metasomal tergite at least 4.0 times as long as wide, fused with sternite; hind femur with two large teeth ventrally; female with hind tarsus 3-segmented. .... *Megischus*

— First metasomal tergite about 2.5 times as long as wide, not fused with sternite; hind femur with three large teeth ventrally; female with hind tarsus 5-segmented. .... *Schlettererius*[1]

[1] This genus is only known to occur in North America.

Synopsis of the Costa Rican fauna

In Costa Rica stephanids have only been collected at altitudes below 1400 metres, in both wet and seasonally dry forests. They are perhaps most commonly encountered below 300 metres.

Foenatopus. Principally an Old World genus, with a few isolated species in Brazil. A single individual has been collected in Costa Rica, in Corcovado National Park on the Osa Peninsula.

Megischus. This is principally a pantropical genus with about 60 described species, 25 of which occur in the Neotropics (Elliott, 1922). Five, as yet unidentified, species (which can be separated from each other by the form and sculpture of the pronotum) have been found to occur in Costa Rica. Four nominal species—*M. niger*, *M. furcatus*, *M. erythrocephalus* and *M. ruficeps*—have been recorded from Central America. Three of the five North American species—*M. texanus*, *M. arizonicus* and *M. brunneus*—have ranges that extend southwards into Mexico and the Caribbean, although they have not been recorded from Central America proper.

8 The evaniomorph parasitoid families

INTRODUCTION

Ian D. Gauld and Paul E. Hanson

Rasnitsyn (1975, 1980) proposed the Evaniomorpha, a radically new group comprising the extant families Trigonalyidae, Megalyridae, Aulacidae, Gasteruptiidae, Evaniidae, Ceraphronidae and Megaspilidae. Although the features he used to characterize the group are probably symplesiomorphic (N.F. Johnson, 1988), evidence is mounting to support Rasnitsyn's hypothesis. Johnson (*op. cit.*) found support for this group in the form of the mid coxal articulation. Gibson (1985) found evidence for the monophyly of Megaspilidae + Ceraphronidae in the form of the mesotrochanteral depressor muscle and the location of the occlusor muscle apodeme. He also found evidence for a sister group relationship between these families and the Megalyridae in the position of the mesothoracic spiracle and the form of the mesotergal-trochanteral muscle. Whitfield *et al.* (1989) produced further evidence for the monophyly of Megaspilidae + Ceraphronidae in the form of the metapostnotal region, and found enlarged phragmal lobes in both the Aulacidae and Gasteruptiidae.

The systematic position of the Evaniidae is somewhat problematical. The position of the insertion of the metasoma in the propodeum has long been used as an apomorphy for grouping the family with the Gasteruptiidae and Aulacidae, as the superfamily Evanioidea. However, Gibson (1985) found evidence suggesting that evaniids could be more closely related to some proctotrupoid families. On the other hand, the form of the mid coxal articulation observed by Norman Johnson (1988), which places the Evaniidae within the Evaniomorpha and separate from the proctotrupoids, conflicts with Gibson's evidence. We therefore suggest retaining the evaniids in their traditional systematic position, but we have included this superfamily with the trigonalyids and ceraphronoids in a single evaniomorph chapter.

The superfamilies and families comprising the Evaniomorpha are thus:

Trigonalyoidea	Trigonalyidae
Megalyroidea	Megalyridae
Evanioidea	Aulacidae, Evaniidae, Gasteruptiidae
Ceraphronoidea	Ceraphronidae, Megaspilidae

All families are represented in Mesoamerica except the Megalyridae, which is primarily a tropical Asian and Austral relict group (S.R. Shaw, 1990). Megalyrids are generally medium-sized wasps, with conspicuously exserted ovipositors; they can be distinguished from the other evaniomorphs, stephanids and ichneumonoids by the fact that the mesothoracic spiracle is apparently surrounded on all sides by the pronotum—thus appearing almost like a 'keyhole' in the sclerite. The superfamilies present in Mesoamerica can be identified using the key provided in Chapter 5. The Trigonalyoidea comprises a single family, Trigonalyidae; the families of the Evanioidea and Ceraphronoidea can be distinguished using the keys given below.

Key to families of Evanioidea

1 Metasoma small and short, in lateral view not obviously larger or longer than mesosoma, usually smaller (Fig. 8.10); hind wing with a prominent jugal lobe; propleurae unspecialized, neither elongate nor forming a neck. .... **Evaniidae** (p. 195)

— Metasoma of normal proportions, in lateral view longer than head and mesosoma combined (Fig. 8.08, 8.09); hind wing without a prominent jugal lobe; propleurae elongate, meeting dorsally and forming a neck. 2

2 Fore wing with vein *2m-cu* present (Fig. 8.01); antennae inserted low down on face, near to the lower margin of the eyes. .... **Aulacidae** (p. 192)

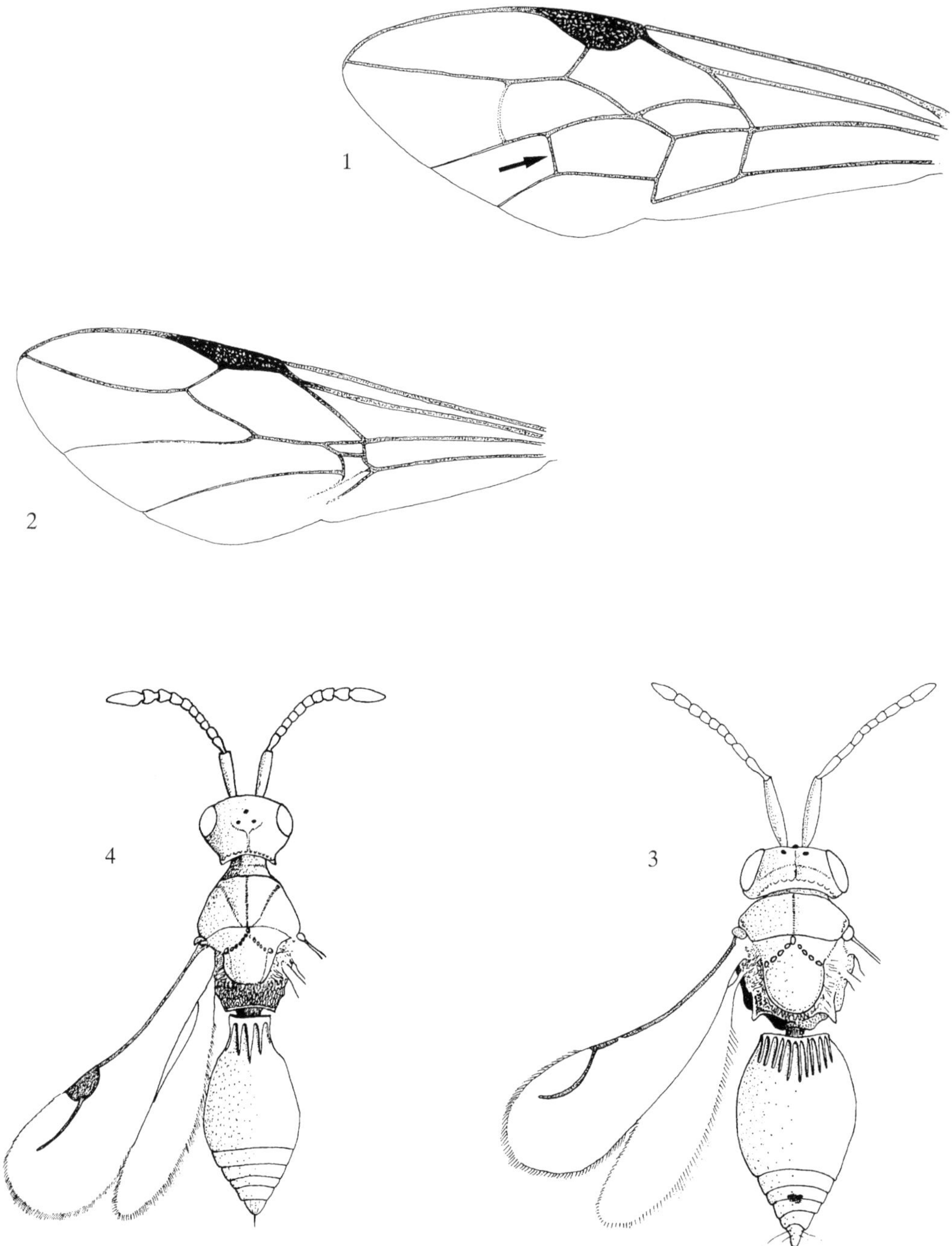

Figs 8.01–8.04. Evaniomorph families. Figs 8.01–8.02. Fore wings; 8.01, *Aulacus* (Aulacidae); 8.02, *Gasteruption* (Gasteruptiidae). Figs 8.03–8.04. Dorsal view of body and wings; 8.03, *Ceraphron* (Ceraphronidae); 8.04, *Conostigmus* (Megaspilidae).

— Fore wing without *2m-cu* (Fig. 8.02); antennae inserted higher up on face, at about mid level of eyes. .... **Gasteruptiidae** (p. 193)

Key to families of Ceraphronoidea

— Mid leg with one tibial spur; longer tibial spur of fore leg not forked apically; mesoscutum without notauli; antenna 7- to 11-segmented; metasoma with large, apparent first segment relatively broad anteriorly (Fig. 8.03); apparent fourth segment with a mediodorsal Waterston's organ present anteriorly, though this is often concealed by preceding tergite; fore wing without a pterostigma (Fig. 8.03). .. .. **Ceraphronidae** (p. 199)

— Mid leg with two tibial spurs; longer tibial spur of fore leg forked apically; mesoscutum usually with notauli well-developed; antenna 11-segmented; base of apparent first (the largest) metasomal tergite narrowed into a short 'neck' (Fig. 8.04); Waterston's organ absent; fore wing often with a large pterostigma (Fig. 8.04). .... **Megaspilidae** (p. 203)

8.1 TRIGONALYIDAE

David Carmean

Diagnosis. Moderate to large sized (body 5.0 to 13.0 mm in length, fore wing length 5.0 to 14.0 mm); generally yellow and black or brown, or blackish in colour. *Antenna 15- to 27-segmented*, males usually with tyloids (elongate to ovoid raised sensory areas) on outside of middle segments, females usually with ovoid or round sensory areas of stippling on middle segments. *Mandibles usually asymmetrical*, three teeth on left, four on right; maxillary palp usually 6-segmented. Pronotum with upper hind corner extending back to more or less reach tegula. *Fore wings usually with ten enclosed cells, including a costal cell.* Legs with trochanters 2-segmented; in most genera the second segment of the hind trochanter divided; *tarsi with plantar lobes present*, usually long and hyaline; *tarsal claws cleft.* Metasoma with sternites sclerotized; first segment conical with tergite and sternite not fused; spiracles present only on tergite VII, and then usually covered by hind margin of tergite VI.

Among Central American trigonalyids, *Nomadina* has symmetrical mandibles, rudimentary palps and antennae with as few as 13 segments. Among trigonalyids not occurring in Central America, some species of *Bareogonalos* have symmetrical mandibles, and *Orthogonalys* species have the metasomal sternites unsclerotized.

Classification and distribution. The Trigonalyidae comprises slightly more than 100 described species classified in 22 genera. The family is cosmopolitan, but its greatest species richness is in tropical regions. In Costa Rica there are eight genera and about 15 to 20 species, which is more than there are in all of America north of Mexico. The Trigonalyidae is currently divided into six subfamilies: Bareogonaloinae, Disceneinae, Lycogastrinae, Nomadininae, Seminotinae and Trigonalyinae (Weinstein & Austin, 1991). However, this classification is based on the monograph by Schulz (1907) and the subsequent catalogue by Bischoff (1938), and the group is in need of extensive revision and reclassification. Genera such as *Lycogaster* and *Seminota*, which share apparent apomorphic features (swollen antennae and plough-shaped aedeagus), are currently placed in separate subfamilies with other, less closely related taxa. Furthermore, generic placement of species is often inconsistent from continent to continent. Many species descriptions are based on too few specimens to adequately appreciate the possible intraspecific variation in size, morphology and colour patterns. The projections from the sternites appear to have been gained and lost on several occasions, and their presence or absence may be variable even within some species.

Biology. Except for some Australian species of *Taeniogonalos* which develop as primary parasitoids of pergid larvae (Raff, 1934) all trigonalyids are hyperparasitoids of endoparasitic Ichneumonoidea and Tachinidae, or parasitoids of vespid larvae. Records of lepidopterous larvae as definitive hosts (e.g. by Weinstein & Austin, 1991) are doubtful and require confirmation (Naumann, pers. comm.). The female trigonalyid lays her eggs on leaves, but these do not eclose unless they are ingested by a phytophagous insect larva (usually a caterpillar). Within this larva the eggs hatch, but no further development of the

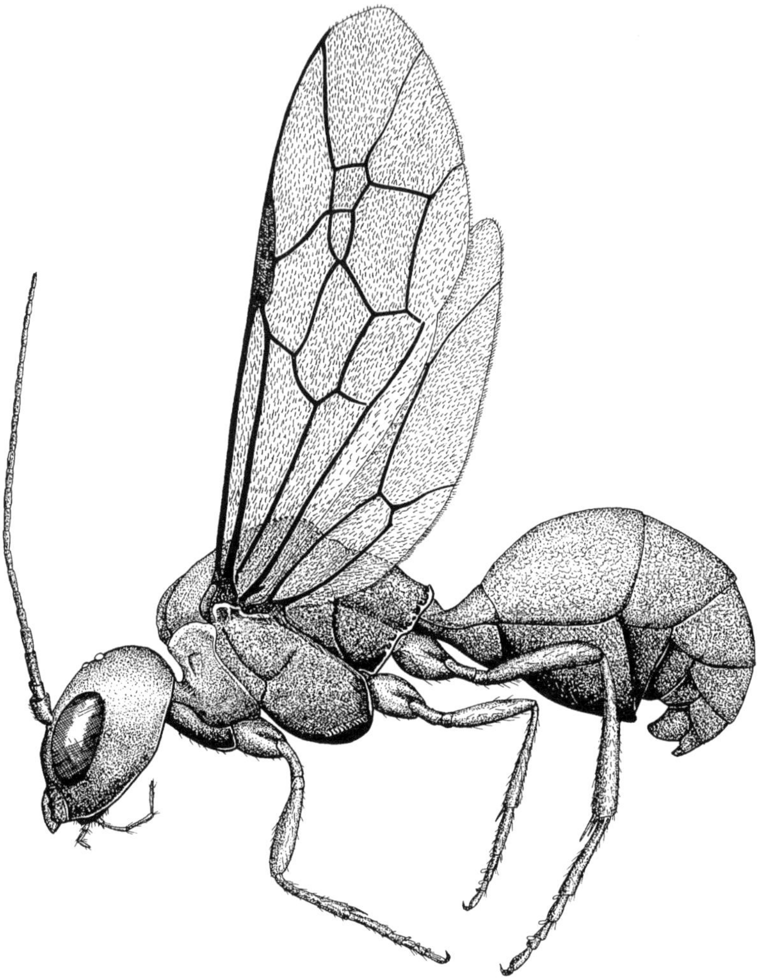

Fig. 8.05. *Taeniogonalos* sp. (Trigonalyidae).

microtype larva ensues unless the caterpillar is parasitized by another parasitoid, or is captured by a vespid and carried back to the nest. This method of encountering a host (ingestion of parasitoid eggs) is unique in Hymenoptera although it is known to occur in some Tachinidae.

Known intermediate hosts for trigonalyids worldwide include the larvae of various species of Lepidoptera, tenthredinoid sawflies, and Tipulidae. The definitive hosts are larger endoparasitoids (Ichneumonidae, Braconidae and Tachinidae) and predatory vespids (Weinstein & Austin, 1991). Although there are very few host records certain trends in host utilization are discernible. Some groups, such as *Bareogonalos*, *Seminota* and *Nomadina*, have only been reared from eusocial vespids, while many *Taeniogonalos* (= *Poecilogonalos*) species have only been reared from ichneumonoids and tachinids. However, *Lycogaster* has been reared from both parasitoids and solitary wasp predators of Lepidoptera.

Trigonalyid females oviposit minute eggs (0.1–0.15 mm long) on leaves, close to the margins. In some cases the eggs are lightly stuck to the leaf surface (Clausen, 1929), but in others the female holds the leaf in a notch formed between a projection from the second or third sternite and the last sternite, and punches the epidermis and inserts an egg into the mesophyll (Rodd, 1951*a*; Carmean, 1991). Some trigonalyids appear to be unselective in their oviposition sites and one Costa Rican species has been reared from intermediate hosts that feed on plant species in eight families. In North America *Lycogaster pullata* has been recorded as ovipositing on plants from such diverse families as Fagaceae, Rhamnaceae and Compositae (Townes, 1956). *Bareogonalos canadensis* oviposited faster (once every 2.2 seconds) and for longer periods on Douglas Fir, compared to oviposition on other foliage, which was too erratic to measure (Carmean, 1991).

Trigonalyids are extremely fecund and capable of laying several thousand eggs in their short adult life (Clausen 1929, 1931; Carmean, 1991). These eggs apparently do not develop further unless ingested by a folivorous insect larva. When in the gut of a folivorous insect the eggs hatch and it is assumed that the microtype trigonalyid larvae penetrate the gut wall and enter the haemocoel. In all New World trigonalyids there is no further development unless the folivore is preyed upon or parasitized by a suitable definitive host. In these cases further development occurs in one of two ways, either within the body of a primary parasitoid larva feeding within the host, or else within the body of a vespid larva that has consumed the folivore. Once in this definitive host the trigonalyid undergoes five larval instars (including the first). The first three instars feed endophagously, but the fourth emerges from its host and feeds ectophagously (Cooper, 1954; Yamane, 1973).

The first instar larva is minute (ca 0.1 mm long) with slender mandibles and posteriorly directed hooks and ventral spines (Clausen, 1931). The second instar larva lacks such hooks and has an unchitinized head, while the third instar larva possesses a large chitinized head and large mandibles, and is apparently adapted for eliminating supernumeraries (Yamane, 1973). The fourth instar larva is more hymenopteriform while the final instar larva is robust, with distinctive tridentate mandibles (Yamane & Kojima, 1982; Yamane & Terayama, 1983). At completion of feeding various species have been observed to construct an irregular cocoon, a silken partition or no cocoon at all (Clausen, 1929; Yamane, 1973; Carmean *et al.*, 1981; Yamane & Kojima, 1982). The size of the adult is dependent on the size of the host it fed on (Carmean *et al.*, 1981).

Key to genera of Trigonalyidae occurring in Central America

1 Tyloids not present, or female. 2

— Tyloids present (males only). 8

2 Maxillary and labial palpi rudimentary; antenna with 14 (rarely 11–16) flagellar segments. .... *Nomadina*

— Maxillary palpus extends beyond mandibles, labial palpus normal; antenna with 15–26 flagellar segments. 3

3 Antenna with 15–18 (rarely 19) flagellar segments; metasoma smooth, shiny; mesosoma strongly rugose punctate; hind trochanter two-segmented. .... *Bareogonalos*[1]

— Antenna with 19–26 flagellar segments; ***if*** metasoma smooth and shiny, ***then*** mesosoma not strongly punctate; hind trochanter apparently 3-segmented (second segment diagonally divided). 4

4 Metasomal tergites and sternites very smooth and thin (may be partially transparent), tergites folded under, intercalating with sternites; propodeum areolate-rugose, covered with network of lines; carina around propodeal foramen thick, partially double-walled, 'U' or half circle shaped.

[Ichneumonid mimics; antenna banded, when viewed dorsally head and thorax black with markings white (females and faded males) or yellowish white (males); metasoma reddish brown, may have extensive light or dark markings.] .... *Orthogonalys*[EX]

— Metasomal tergites and sternites often punctate, thick, tergites overlap sternites laterally in a straight line, without overlapping sternites ventrally; propodeum punctate, or smooth, sometimes with concentric lines around foramen, but not covered with network of lines; carina around propodeal foramen thin (though sometimes tall), and 'V' or 'U' shaped.

[Coloration various, not as above.]............ 5

5 Eye viewed from below with hind margin behind mandible; propleuron and mesopleuron yellow and entire fore wing amber; in some species vertex behind ocelli flat, at back abruptly angled towards occipital carina; propodeal foramen evenly curved dorsally.
[Very rare.] .... *Xanthogonalos*

— Eye with hind margin even with middle of mandible (less commonly at hind edge of mandible); propleuron dark, mesopleuron dark or dark with yellow markings (except *Labidogonalos* and Genus B, which also have only the leading half of the fore wing amber). 6

6 Antenna spindle-shaped; propodeum without yellow or light markings; frons between antennae broad and flat, wider than length of first flagellar segment; tyloids absent. 7

— Antenna filiform; propodeum often yellow marked; frons between antennae usually narrower and not as flat as described above; tyloids present in male. 8

7 Wings amber or hyaline, not very dark; vertex curving evenly towards occipital carina.
[Parasitoids of solitary wasps or Lepidoptera.] .... *Lycogaster*

— Wings, or part of wings, very dark or violaceous; vertex sharply angled above occipital carina.
[Parasitoids of social wasps.] .... *Seminota*

8 Gena wide, extending far beyond mandible when viewed from below (Fig. 8.07); occipital carina overlaps sharp genal angle when viewed from below; occiput sharply excavated all the way to mandible (except in Genus A); glossy between antennae and above clypeus; frons strongly angled between antennae when viewed from side; female armature, if present, on third sternite; male tyloids oval or round, less than half the flagellar segment length. 9

— Gena narrow, especially near mandible, not extending far beyond mandible when viewed from below (Fig. 8.06); occipital carina not forming top of a sharp angle with gena and occiput, occiput not sharply excavated, punctate above clypeus, flat or slightly angled between antennae in side view; female armature, if present, on second sternite; male tyloids elongate, usually more than half the flagellar segment length. 11

9 Occipital carina points toward hypostomal carina and then bends and parallels hypostomal carina to reach mandible; occiput not sharply excavated near mandible; black or very dark brown, often with white markings; most specimens with marginal cell and parts adjacent to it darkened.
[Mexico, known only from females.] .... Genus A[2]

— Occipital carina goes straight to mandible edge; occiput sharply excavated along occipital carina all the way to mandible. 10

Figs 8.06–8.07. Lower posterior region of head, Trigonalyidae; 8.06, *Labidogonalos*; 8.07, *Trigonalys*.

10 Body brown and yellow. .... Genus B[2]
— Body black and white. .... *Trigonalys*

11 Propleuron and mesopleuron yellow. ....
.... *Labidogonalos*
— Propleuron dark, mesopleuron dark or dark with yellow markings. .... *Taeniogonalos*

[1] An eastern palaearctic/North American genus whose range extends south into Mexico.
[2] Undescribed genus only known to occur in Mexico.

Synopsis of the Costa Rican fauna

Eleven genera are reported to occur in America south of the United States (Weinstein & Austin, 1991), but New World species currently assigned to *Poecilogonalos* are here placed in *Taeniogonalos*, and *Stygnogonalos* is here considered to be a synonym of *Trigonalys*. Seven of the remaining nine New World genera, and about ten species are known to be represented in Costa Rica; *Orthogonalys* may also be here. The remaining described genus present in the New World, *Bareogonalos*, is eastern palaearctic and North American in distribution, with a range that extends south only as far as Mexico. Two undescribed genera are present in Mexico: Genus A known only from females; and Genus B which contains "*Trigonalys*" *maculifrons*, a species previously placed in *Labidogonalos*, but the shape of the head and tyloids show it is more closely related to *Trigonalys*.

Labidogonalos. A small neotropical genus known from Mexico to southern Brazil. Two species probably occur in Central America. *Labidogonalos* mimics *Agelaia areata* (Vespidae) and so is more elongated than most Trigonalyidae.

Lycogaster. A medium sized genus known from southeast Asia and, in the New World, from Costa Rica northwards to Canada. A single species has been collected in Costa Rica. Species of *Lycogaster* mimic *Brachygastra* spp. (Vespidae). In Costa Rica most specimens have been collected from January to April.

Nomadina. This small genus is limited in its distribution to the Neotropical region, from Chile north to Costa Rica, but related genera are found in the Philippines and Taiwan. A single species has been collected in Costa Rica. Species have been reared from *Agelaia* and *Polybia* (Vespidae). Specimens have been collected in Costa Rica east of San José at an altitude of 1600 metres, and others have been collected in the mountains of western Panama.

Orthogonalys[EX]. This small, widespread, but rarely collected genus, is represented in both North and South America and thus likely to occur in Costa Rica, although as yet it has not been found.

Seminota. This small genus is only known to occur in Central and South America, represented in Costa Rica by a single species. *Seminota* mimics one of its host wasps, *Parachartergus apicalis*, and most collections have been made by rearing individuals from vespid nests.

Trigonalys (= *Stygnogonalos*). A small neotropical genus with a range known to extend from Brazil north into Central America. *Trigonalys melanoleuca* is common in South America, but in Central America the genus is known from only a few specimens. *T. championi*, which occurs in Central America, has a projection on the third metasomal sternite of the female and has been placed in a separate genus, *Stygnogonalos*, though in other respects it is a typical member of *Trigonalys*. In Costa Rica a specimen of *T. championi* was taken in March at Pejibaye in Cartago Province.

Taeniogonalos. It is impossible to define the genera *Poecilogonalos*, *Nanogonalos* and *Taeniogonalos* so that they are each monophyletic, and all species should probably be included in a single genus. In the New World *Taeniogonalos* is a moderate sized genus. One species, apparently the same as the only North American species, *T. costalis*, mimics *Polybia occidentalis* (Vespidae). D.H. Janzen (pers. comm.) has reared it in northwestern Costa Rica from both tachinid and ichneumonid parasitoids of various large Lepidoptera found on eight plant families. This variety indicates that *Taeniogonalos* is not specific for oviposition, intermediate, or definitive host.

Xanthogonalos. This is a small neotropical genus, but it is known from too few specimens to characterize adequately and to confidently differentiate it from the more commonly collected and similar looking *Labidogonalos*. The shape of the aedeagus does not support placement of *Xanthogonalos* by Schulz (1907) in a separate subfamily with *Seminota*.

8.2 AULACIDAE

Ian D. Gauld

Diagnosis. Body length 8.0 to 12.0 mm (excluding ovipositor), fore wing length 6.0 to 12.0 mm; reddish, black or dark brown in colour. Antenna of female 14-segmented, of male 13-segmented, with flagellum moderately long and slender; antennal insertion low down on head, close to clypeus, with concave scrobes extending ventrolaterally to malar space. Head oval, convex anteriorly; mandible robust, tridentate, with the lower tooth large and the upper two smaller and often blunt; maxillary palp 6-segmented, labial palp 4-segmented. Pronotum extending posteriorly so that upper corner more or less reaches tegula; *propleurae extended anteriorly into a conspicuous elongate "neck"* mesoscutum convex, sometimes with anterolateral low conical processes, usually coarsely sculptured, usually with distinct notauli, and posteriorly with a flexible transverse suture before scuto-scutellar groove; mesosternal region overlapping bases of mid coxae, and with posteriorly directed process extending between them; *propodeum* usually reticulately sculptured, *with point of insertion of metasoma high, on upper surface, far removed from insertion of coxae. Fore wing with eight to ten enclosed cells,* costal cell distinct and wide, pterostigma large and conspicuous; hind wing venation reduced, usually without enclosed cells, though in some extralimital large species a single more or less enclosed cell is discernible. Tibial spur formula 1 : 2 : 2. *Metasoma ellipsoidal, with first segment more or less conical,* with a conspicuous laterotergite; sternites well-sclerotized; ovipositor longer than metasoma, slender.

Classification and distribution. The Aulacidae comprises about 150 species distributed throughout the

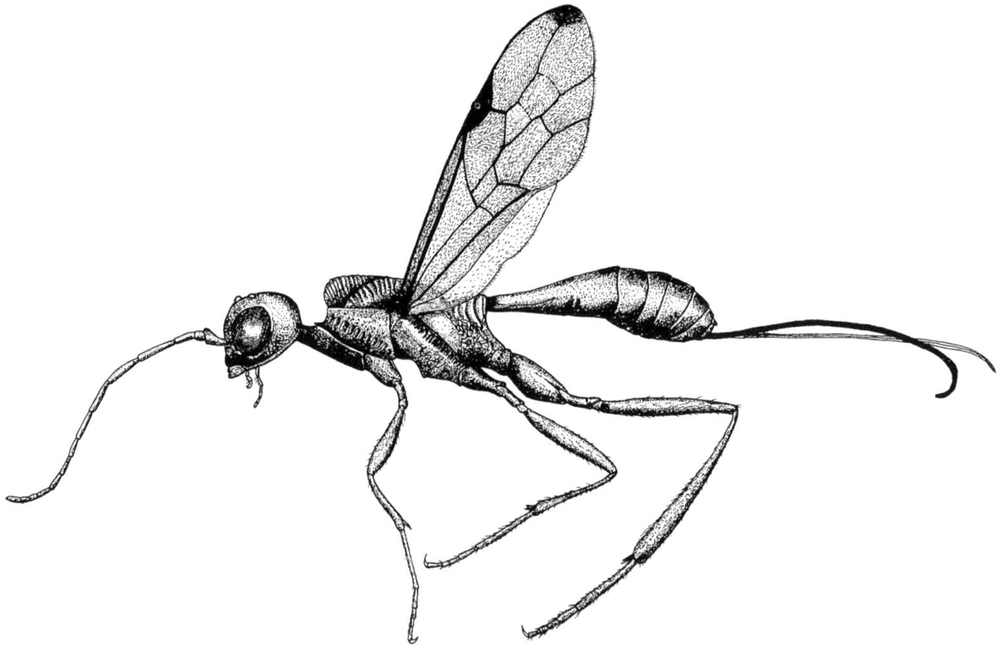

Fig. 8.08. *Aulacus* sp. (Aulacidae).

world, although the greatest species richness seems to occur in tropical South America. Three genera are currently recognized, one from the Afrotropical region (Benoit, 1984*c*; Madl, 1990) and two, *Aulacus* and *Pristaulacus* (= *Aulacostethus sensu* Townes, 1950), that are cosmopolitan (Townes, 1950; Pagliano, 1986; Konishi, 1990). Both occur widely throughout the New World, from Canada south to Chile.

Biology. Aulacids are endoparasitic koinobionts of insects living in concealment in wood or twigs. Hosts include various families of wood-boring Coleoptera (especially Cerambycidae and Buprestidae) and Hymenoptera (Xiphydriidae and possibly nematine tenthredinids that form woody stem galls) (Carlson *in* Krombein *et al.*, 1979; Barriga, 1990). The most complete study of the biology of any species (*Aulacus striatus*) was filmed by Skinner and Thompson (1960) in Britain. The female *Aulacus* locates the ovipositional boring of a woodwasp (*Xiphydria camelus*) in a crevice in the bark of a tree, inserts her ovipositor down the hole and lays a single egg in each of as many host eggs as she can reach. The aulacid larva ecloses from its egg and develops very slowly within the growing xiphydriid larva; it is characterized by having two sclerotized dorsal projections extending from the dorsolateral parts of the epistoma onto the frontal area (Short, 1978). Not until almost a year later, when the woodwasp larva is fully grown and has tunnelled up to just below the surface of the bark, does the parasitoid larva complete development and kill the host. The aulacid larva emerges from the skin of its host, spins a cocoon in which it pupates, and emerges as an adult about two weeks later. The adult escapes by gnawing a hole through the thin cap of wood left by the xiphydriid.

Identification. The nearctic species were keyed by Townes (1950) and a new nearctic species was recently described by Smith (1991); there is currently no comprehensive work on the neotropical species.

Key to genera of Aulacidae present in the New World

— Tarsal claws apparently simple or with an obscure basal tooth; occipital carina absent. ... . *Aulacus*

— Tarsal claws with distinct teeth; occipital carina present, at least on upper part of head. .... *Pristaulacus*[EX]

Synopsis of the Costa Rican fauna

Both *Aulacus* and *Pristaulacus* occur in Central America, although the latter has not yet been collected in Costa Rica.

Aulacus. A large, widespread genus present throughout the New World. Two species have been collected in Costa Rica, one in seasonally dry habitats in the northwest, the other in more uniformly wet lowland areas.

8.3 GASTERUPTIIDAE

Ian D. Gauld

Diagnosis. Body length 9.0 to 21.0 mm (excluding ovipositor), fore wing length 5.0 to 10.0 mm; black or reddish in colour. Antenna of female 14-segmented, male 13-segmented, with flagellum moderately short and fairly stout; antennal insertion about at level of centre of eyes, well above clypeus, without concave scrobes extending ventrolaterally to malar space. Head oval, convex anteriorly; mandible robust, with apical tooth sharply pointed, subapical tooth small and broadened, and with a large, laterally directed basal tooth which is obscured when mandibles are closed; maxillary palp 6-segmented, labial palp 4-segmented. Pronotum extending posteriorly so that upper corner more or less reaches tegula; *propleurae extended anteriorly into a conspicuous elongate 'neck"*; mesoscutum convex, with distinct notauli that meet centrally, and posteriorly with a flexible transverse suture before scuto-scutellar groove; mesosternal region overlapping bases of mid coxae; *propodeum usually reticulately sculptured, with point of insertion of metasoma high, on upper surface, far removed from insertion of coxae. Fore wing often folded longitudinally, with six or seven enclosed cells, costal cell distinct and wide, pterostigma narrow but distinct;* 2m-cu *absent;* hind wing without enclosed cells.

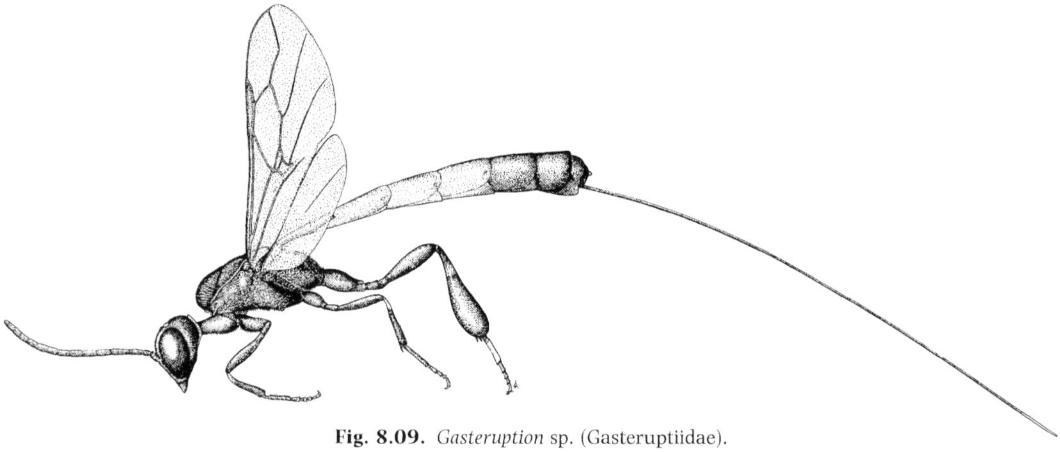

Fig. 8.09. *Gasteruption* sp. (Gasteruptiidae).

Tibial spur formula 1 : 2 : 2; *hind tibia distally strongly swollen, club-like*. Metasoma elongate and laterally compressed, with first segment elongate, subcylindrical anteriorly; ovipositor from about as long as tergite I to longer than entire metasoma, slender.

Classification and distribution. The Gasteruptiidae is a moderately small family comprising about 500 described species. It is almost worldwide in distribution, but is particularly well-represented in warm, rather dry regions such as much of Australia. The family comprises two subfamilies, the structurally rather primitive Hyptiogastrinae and the more derived Gasteruptiinae. The former group includes four genera endemic to the Australian region (*Eufoenus*, *Pseudofoenus*, *Crassifoenus* and *Hyptiogaster*), and one genus that occurs in both Australia and southern South America (*Aulacofoenus*) (Crosskey, 1953, 1962; Naumann, 1991). The Gasteruptiinae, which is cosmopolitan in distribution, has at times been divided into five or six genera (see Crosskey, 1962). Townes (1950) classified the North American species in two genera, *Gasteruption* and *Rhydinofoenus*, but Crosskey (1962) points out that all supposed genera broadly intergrade. He proposed treating all species in a single large genus, *Gasteruption*, and this interpretation is followed here.

Biology. Gasteruptiids develop, mostly as cleptoparasites, in the nests of solitary bees and wasps. Recorded hosts include a wide variety of bees that nest in the ground, hollow stems or borings in wood; species of Sphecidae and Vespidae (Eumeninae and Masarinae) nesting in similar habitats are also attacked (Crosskey, 1952; Naumann, 1991). Some species seem to have a rather broad host range. For example, the holarctic species *G. assectator* is known to attack both *Prosopis* (Apidae) and *Trypoxylon* (Sphecidae) species, whilst the European *G. jocator* is recorded as attacking species of *Prosopis*, *Osmia* (Apidae), *Pemphredon* (Sphecidae) and *Odynerus* (Vespidae) (Crosskey, 1952). In North America gasteruptiids are known to attack apid species of the genera *Ceratina*, *Dianthidium*, *Hoplitus*, *Hylaeus* and *Megachile* (Carlson *in* Krombein *et al*., 1979).

The female gasteruptiid lays an attenuated, pedunculate egg in the nest of its host. The position in which the egg is placed seems to be species-specific. For example, it can be attached to the host's egg (e.g. *G. caudatum*), on the food store near the host egg (*G. pyrenaicus*), on the cell wall (*G. thompsoni*), or even outside the cell (*G. rugulosum*) (Malyshev, 1968). In other cases it appears that the egg is laid on the bee larva (Höppner, 1904). On hatching, the gasteruptiid larva devours the host egg or larva (feeding as an idiobiont ectoparasitoid), and then generally continues feeding as a cleptoparasite on the stored food reserves, either pollen or arthropods. Where the host's nest consists of serially arranged cells, the gasteruptiid larva may go from one cell to another, devouring the contents of more than one cell (thus behaving as a predator).

There appear to be three larval instars. The final instar larva is hymenopteriform with well-developed setae on the body and head, and with spiracles present on the second thoracic and first eight abdominal segments; the mouthparts consist of tridentate mandibles, protruding maxillary palpi, and disc-like labial palpi (Malyshev, 1968; Short, 1978). The final instar larva is unusual (possibly unique amongst apocritan Hymenoptera) in having the mid and hind gut connected. It thus produces a series of pellets of excrement, rather than voiding a single faecal mass (meconium) as is generally the case in apocritan larvae. The fully-grown larva spins a brown cocoon within the cell of its host. Adults are quite commonly found feeding from flowers.

Identification. Townes (1950) provided keys to the North American species and Alayo (1972) gave a key to the two Cuban species. No keys are presently available for the Central American species.

Synopsis of the Costa Rican fauna

Only the genus, *Gasteruption*, is present in Central America.

Gasteruption. Eight species are catalogued as occurring in Mesoamerica (Hedicke, 1939), and I have seen about six species from Costa Rica. These include *G. maculicornis*, described from Guatemala, and five apparently undescribed species. In Costa Rica gasteruptiids have been most frequently collected in seasonally dry lowland habitats, but specimens have been taken up to an altitude of about 1600 metres on the edge of very wet forest.

8.4 EVANIIDAE

Mike Huben

Diagnosis. Body length 1.7 to 17.0 mm; colour predominantly black or dark brown to reddish orange, sometimes white marked, often with patches of silvery pubescence, sometimes sexually dimorphic in colour; head, mesosoma and petiole usually elaborately sculptured, the form of the sculpture varying greatly between species. Head broad; antennae 13- (or rarely 10-) segmented, sexually dimorphic, with scape elongate and often with flagellum slightly clubbed. *Mesosoma massive.* Fore wing venation from fairly complete with seven enclosed cells, to very reduced with one enclosed cell, with a tiny coriaceous lobe; *hind wing without enclosed cells, with one or two veins, with a prominent jugal lobe. Metasoma small, inserted high on propodeum, with a slender petiolate first segment and with remainder of mesosoma small and laterally compressed*; ovipositor short, barely projecting beyond apex of metasoma.

In tropical America evaniids are easily distinguished from other hymenopterans by their very small metasoma, which is attached high up on the mesosoma. A similar high metasomal insertion is found in Aulacidae and Gasteruptiidae, as well as a few Braconidae (e.g. *Coenocoelius*) and Cynipoidea (Ibaliidae: Liopterini). Most of these taxa have a much more elongate metasoma, and no other non-aculeate apocritans have a differentiated jugal lobe on the hind wing.

Classification and distribution. The Evaniidae comprises about 500 described species currently classified into 17 genera. There are large numbers of undescribed tropical species in collections of major museums. The classification of the group is still largely that outlined by Kieffer (1912) and it is in need of revision. Many of the genera are of doubtful validity. The group is more or less cosmopolitan in distribution, but the overwhelming majority of species occur in the humid tropics.

Seven genera—*Prosevania*, *Evania*, *Evaniscus*, *Evaniella*, *Semaeomyia*, *Hyptia* and Genus D—occur in the New World, but two other neotropical species-groups are candidates for generic status. Chilean specimens that key to *Semaeomyia* and resemble the Old World genus *Brachygaster*, have other features that distinguish them from both genera. Specimens that key to *Evaniella*, but have only six enclosed cells in the fore wing (vein 1*A* is shortened and encloses only one cell) are widespread. They need to be examined further to discern their placement relative to *Evaniella*.

Biology. Based on rearings in Europe and North America all evaniids are presumed to develop within

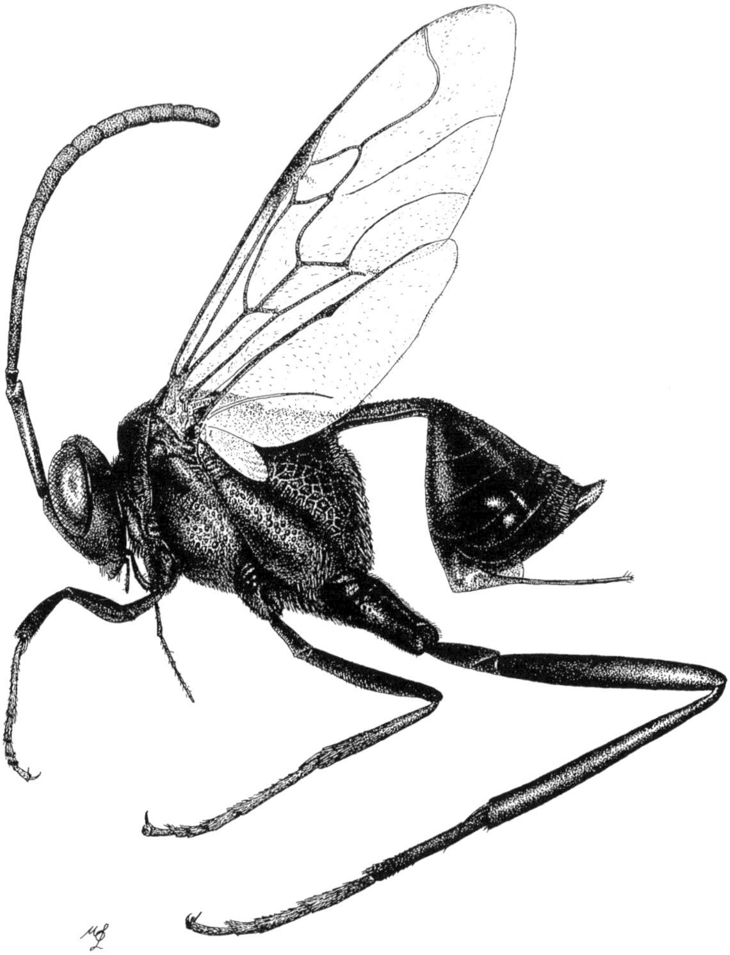

Fig. 8.10. *Evania* sp. (Evaniidae)

the oothecae of Blattaria (Lebeck, 1991), although it is worth emphasizing that virtually nothing is known about the host associations of the vast majority of tropical species. Known host records are summarized in Table 8.1.

A few species present in Central America, such as *Evania albofacialis*, are physically quite large insects, and possibly develop in the very large oothecae of cockroaches of the genera *Nyctibora* or *Megaloblatta*. Individual evaniid species do not seem to be particularly host specific. *Evania appendigaster*, for example, has been recorded as attacking five different species of synanthropic cockroach (Cameron, 1957), and *Brachygaster minuta* is known to oviposit in the egg cases of three species of *Ectobius* in Britain (Brown, 1973).

Female evaniids generally run about swiftly in the leaf litter or other places where their hosts oviposit, although they are capable of a short, jumping flight. Males fly sporadically for short distances. Individuals occasionally feed from flowers or extrafloral nectaries but they do not necessarily have to feed before oviposition. Host-feeding is unknown (Brown, 1973).

Evaniid genus	Cockroach host	References
Brachygaster	*Ectobius* spp.	Brown, 1973
Evania	*Periplaneta* spp.	Cameron, 1957
	Blatta sp.	Cameron, 1957
	Melanozosteria sp.	Swezey, 1929
	Neostylopyga sp.	Swezey, 1929
Hyptia	*Parcoblatta* spp.	Rau, 1940; Edmunds, 1952
Prosevania	*Periplaneta* sp.	Edmunds, 1954
	Blatta sp.	Edmunds, 1954
	Blattella sp.	Fahringer, 1922
Zeuxevania	*Loboptera* sp.	Genieys, 1924

Table 8.1. Recorded hosts of evaniid genera.

Female evaniids are not known to attack the ootheca while it is still being carried by the parent cockroach, but they may prefer newly deposited, relatively soft oothecae. Species that attack synanthropic cockroaches frequently enter buildings (Thoms & Robinson, 1986).

Evaniid species that attack oothecae concealed in sandy places excavate and uncover the ootheca prior to oviposition, then subsequently cover it with sand. Details of the oviposition of one species are given by Brown (1973), who observed a very precise behavioural sequence. The ovipositing female frequently turns the ootheca around to a favourable orientation, so that the long axis of the ootheca is parallel to the long axis of her body. The female often oviposits while lying on her side. Oviposition generally takes quite some time, around 30 minutes being most common (Edmunds, 1954; Cameron, 1957), but up to five hours in one instance (Brown, 1973).

Evaniid eggs are whitish and sausage-shaped, with one end slightly constricted. They are deposited singly in the ootheca, sometimes between the cockroach eggs (Edmunds, 1954) or actually within the host egg (Genieys, 1924; Brown, 1973). In some cases superparasitism is known to occur, the female wasp accepting previously parasitized hosts (Kumarasinghe & Edirisinghe, 1987), even though only a single evaniid develops to maturity. There appear to be five larval instars (Cameron, 1957; Brown, 1973), although some authors have recorded only three. The first instar is often endoparasitic within a single egg but subsequent instars behave as egg predators within the ootheca (Clausen, 1940*b*).

The larva is typically hymenopteriform; early instars lack spiracles and have simple, sharply pointed, triangular mandibles, which are often bordered with minute denticles. Later instars have nine pairs of spiracles (on the first and third thoracic segments, and the first seven abdominal segments) and the mandibles are more or less bidentate, with the lower tooth longer; larger species generally have the upper tooth blunt, broad and slightly concave so that the mandible appears tridentate (Genieys, 1924; Cameron, 1957; Short, 1978). When it has consumed the contents of the ootheca, the final instar larva rests, and has a typical prepupal form. In this state it may diapause during an unfavourable season, or it may pupate, in which case the adult emerges a few days later.

Identification. The North American species are keyed by Townes (1949). There are no satisfactory keys to the neotropical species, except for Cuba (Alayo, 1972). The older keys by Kieffer (1912) are now very outdated and not recommended. Hedicke (1939) has catalogued the World species.

Key to genera of Evaniidae present in the New World

1 Fore wing with six or seven enclosed cells (Fig. 8.11, 8.12). 2
— Fore wing with three or fewer enclosed cells (Figs 8.13, 8.14). 5

2 Hind coxa separated from the mid coxa by 2.0 or more times the length of the mid coxa (Fig. 8.15). 3
— Hind coxa closer to mid coxa, separated from it by less than 2.0 times the mid coxa length (Fig. 8.16). 4

3 Fore wing with seven enclosed cells (Fig. 8.11); hind wing with two linear veins. .... *Evania*
— Fore wing with six enclosed cells (Fig. 8.12); hind wing with a single vein. .... *Evaniscus*

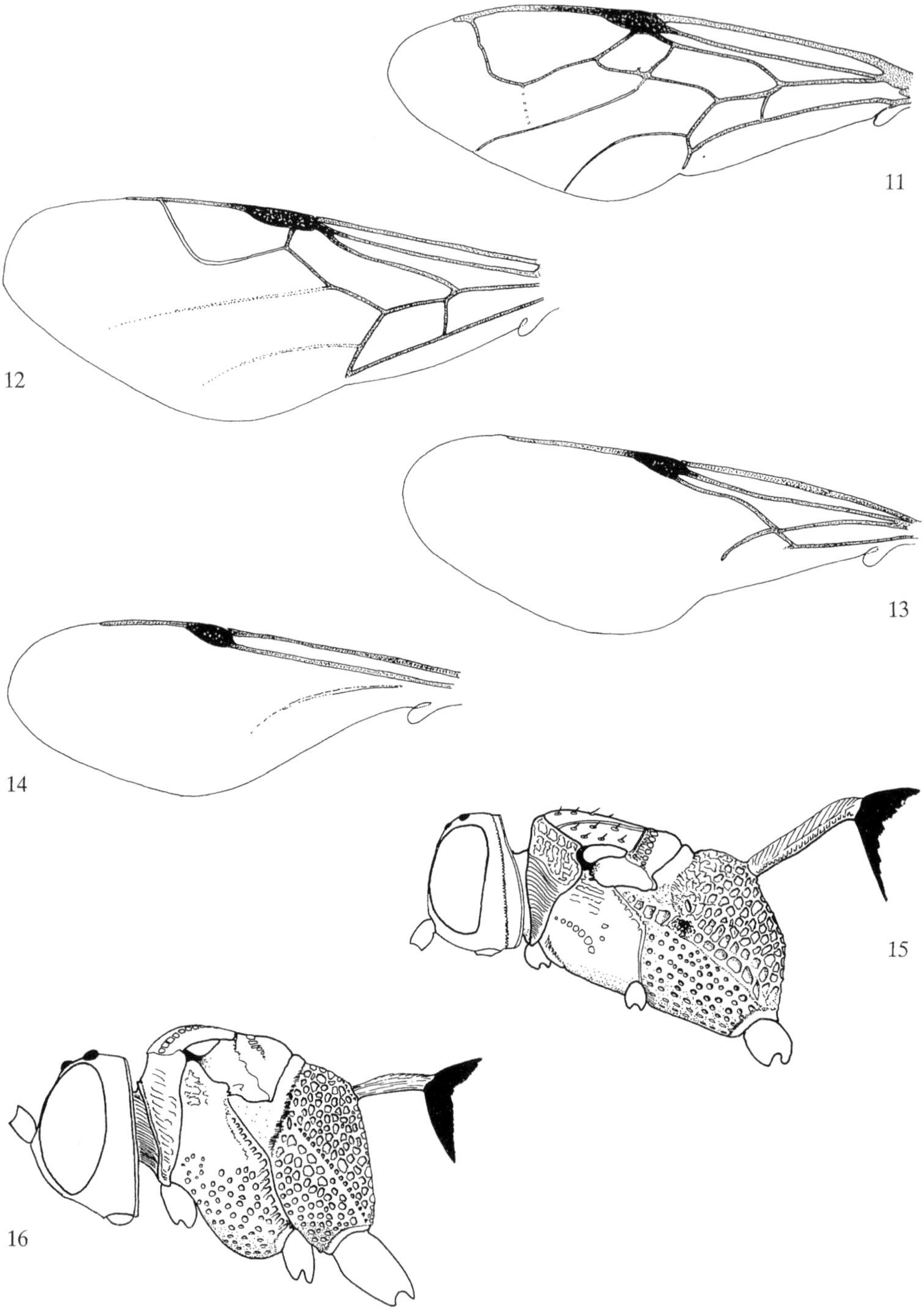

Figs 8.11–8.16. Evaniidae. Figs 8.11–8.14. Fore wings; 8.11, *Evania* sp.; 8.12, *Evaniscus* sp.; 8.13, *Semaeomyia* sp.; 8.14, *Hyptia* sp. Figs 8.15–8.16. Mesosoma and coxae; 8.15, *Evaniscus* sp.; 8.16, *Evaniella* sp.

4 Face finely striate. .... *Prosevania*[1]
— Face not striate. .... *Evaniella*

5 Fore wing with three enclosed cells (Fig. 8.13). .. .. *Semaeomyia*
— Fore wing with only a single enclosed costal cell (Fig. 8.14). 6

6 Antenna 13-segmented. .... *Hyptia*
— Antenna 10-segmented. .... Genus D

[1] A large Old World genus represented in the New World by the introduced species, *P. fuscipes*, in the eastern United States.

Synopsis of the Costa Rican fauna

All six endemic New World genera are present in Costa Rica where there are at least 30 evaniid species. The family is present at virtually all altitudes, although it is most diverse and abundant in the lowland wet forests.

Genus D. A small undescribed genus restricted to the Neotropical region, where it is most frequently collected at high altitudes, although in some areas it has been collected at much lower elevations. Three species occur in Costa Rica.

Evania. This genus (*sensu* Townes, 1949) is widespread and speciose in the Old World tropics, and is represented in Central America by perhaps five endemic species and the cosmopolitan *E. appendigaster*. This latter species is widespread in Costa Rica and common in metropolitan areas along with its synanthropic cockroach hosts. The endemic mesoamerican species, *E. albofascialis*, is the only other *Evania* species yet found in Costa Rica. This species can be common in Malaise trap samples from mid elevation forests. Many of the twenty or so described American species currently placed in this genus should be transferred to *Evaniella*.

Evaniella. This large New World genus is one of the most speciose taxa in Costa Rica, where it is represented by at least eight species.

Evaniscus. This small endemic neotropical genus is represented in Costa Rica by a single extremely uncommon species, *E. marginatus*. Isolated specimens have been collected in seasonally dry forests in northwestern Costa Rica. The Australian species included by Kieffer (1912) in this genus are improperly assigned.

Hyptia. This genus (here taken to include *Evaniellus* and *Chalcidopterella*) is a large New World taxon with seven described species in the United States, and ten additional species in Mesoamerica. Numerous other species occur in the Caribbean and South America. Eight species occur in Costa Rica.

Semaeomyia. A moderately large, neotropical genus that resembles the Old World genus *Brachygaster* (with which it has sometimes been amalgamated, although it is distinct in several features). Six species of *Semaeomyia* have been described from Central America, and about eight occur in Costa Rica.

8.5 CERAPHRONIDAE

Paul Dessart

Diagnosis. 0.5 to 3.3 mm in length; mostly black or dark brown, sometimes yellow, never metallic; fully winged or sometimes brachypterous. Antenna of female 10-segmented (very rarely 7-, 8- or 9-segmented), that of male 11-segmented (in one species 10-segmented), *attached close to mouth, geniculate with long scape*, sometimes clavate in female, anelli absent. *Pronotum extends to the tegulae*; mesoscutum usually without notauli, but with a median longitudinal furrow. *Wing venation very reduced, consisting of submarginal, marginal and stigmal (radial) veins, the latter usually long and curved; without a pterostigma*. Trochanters 1-segmented; *fore tibia with two apical spurs, longer spur not forked apically; mid tibia with one spur*, hind tibia with two. True first segment of metasoma reduced, present as a small and rather concealed petiole and forming anterior surface of the broad second metasomal segment; the large, apparent first (but in reality the second) segment of metasoma relatively broad anteriorly, with anterior transverse carina, generally followed

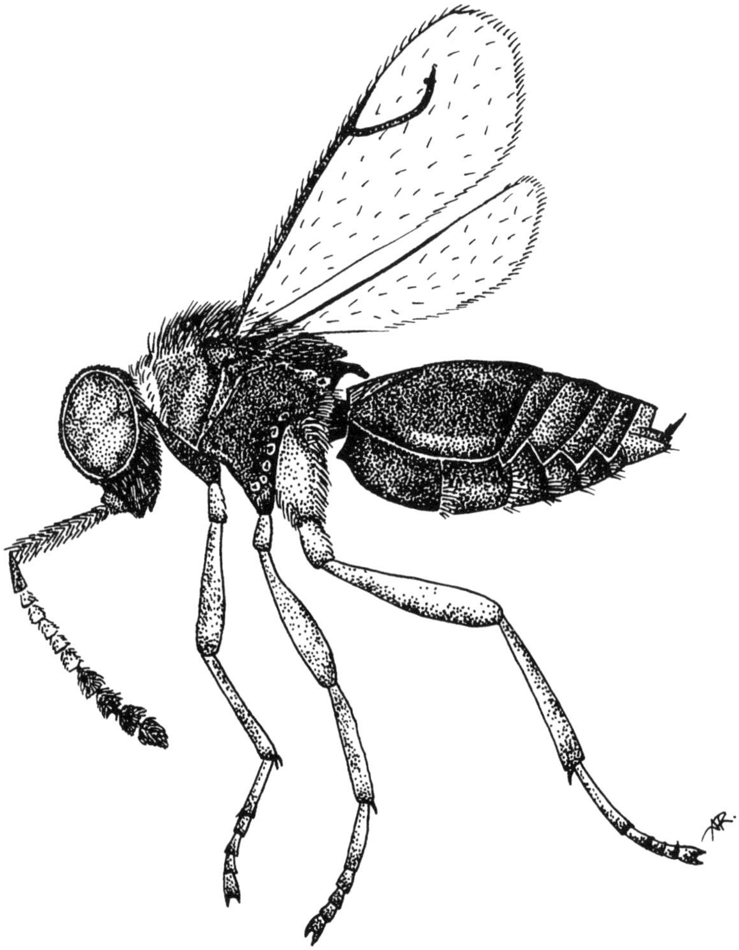

Fig. 8.17. *Ceraphron* sp. (Ceraphronidae).

by longitudinal carinae; *Waterston's organ present*, sometimes visible as a dark spot at the base of tergite V (the third short tergite after the large one) but often concealed by tergite IV (Dessart, 1992).

Brachypterous forms can be separated from other brachypterous microhymenoptera by the number of antennal segments and the form of the basal metasoma. They can be distinguished from brachypterous megaspilids by the absence of notauli. Winged forms have a very distinctive venation that is shared only with male lagynodine megaspilids. The latter however, generally have the anterior end of tergite II (the large apparent first tergite) narrowed into a short 'neck'.

Classification and distribution. The Ceraphronidae and Megaspilidae together comprise the Ceraphronoidea (Masner & Dessart, 1967). In earlier works these taxa were often included in the catch-all group 'Proctotrupoidea', but most modern workers (e.g. Königsmann, 1978*a*; Muesebeck *in* Krombein *et al.*, 1979; Gauld & Bolton, 1988) accept the separation of these taxa as a distinct superfamily.

The family Ceraphronidae comprises approximately 350 described species, which are classified in 14 genera. Eight of these genera are monotypic and the status of many is uncertain (Dessart & Cancemi, 1986). The family is cosmopolitan in distribution as are the largest genera, *Aphanogmus* and *Ceraphron*. In addition to these two cosmopolitan genera the New World fauna also includes four small genera: *Ecitonetes* and *Pteroceraphron* (both monotypic); *Synarsis* (one described North American species); and *Homaloceraphron* (three described species, all North American).

Biology. Biologically the Ceraphronidae is an enigmatic group and very little data are available about the extent of their host ranges or their interactions with hosts. What little reliable information is available concerns species of the two principal genera *Aphanogmus* and *Ceraphron*, but observations seldom extend beyond the simple recording of which species served as a host. However, these data do show that the host range of the Ceraphronidae is large, and embraces at least five orders: Diptera, Hymenoptera, Thysanoptera, Homoptera and Neuroptera. This diverse host range seems to have two biological foci:

i) small, and sometimes quite active insects living in weak concealment;
ii) the prepupae of other hymenopterous parasitoids.

Hosts in the first category include predatory cecidomyiid larvae (Parnell, 1963), phytophagous cecidomyiids (Bakke, 1955; Austin, 1984), coniopterygids (Priesner, 1936) and perhaps other small neuropterans (Chiu *et al.*, 1981), and triozid psylloids (Dessart, 1978). Muesebeck (*in* Krombein *et al.*, 1979) notes that Drosophilidae, Sciaridae, Syrphidae and Phoridae also serve as hosts for ceraphronids. Several species of *Ceraphron* and *Ecitonetes* have been collected in the nests of cricetid rodents and in association with army ants (Dessart, 1979; Muesebeck *in* Krombein *et al.*, 1979) where they might be parasitizing dipterans. Some species of *Aphanogmus* seem quite polyphagous, and have been recorded parasitizing larvae of diverse aphidophagous and acariphagous species of cecidomyiids, as well as some feeding on fungal mycelia (Dessart, 1992). Hosts in the second category include various braconids, ichneumonids (Muesebeck *in* Krombein *et al.*, 1979), bethylids (Dessart, 1988) and dryinids (Swezey, 1908).

At least some species in the first category almost certainly develop as koinobionts and are definitely endoparasitic. For example, *Aphanogmus fulmeki* from Europe and North America, develops internally in the midge larva, finally pupating within the skin of the mature host larva (Parnell, 1963; Dessart, 1992). Endoparasitism has also been observed in other species parasitizing phytophagous cecidomyiid larvae (Bakke, 1955), one having even occasionally parasitized preadult stages of thrips (Dessart & Bournier, 1971). Less is known of the biology of the hyperparasitic species of *Aphanogmus* and *Ceraphron*, but some seem to be idiobiont ectoparasitoids. For example, *Ceraphron abnormis* develops as an ectoparasitoid on Dryinidae, attacking the host at the larval/ prepupal stage after it has woven a cocoon (Swezey, 1908).

All ceraphronids, both males and females, possess a reticulate alveolar area on the metasoma mediodorsally near the anterior margin of the apparent fourth tergite. This area, the so-called Waterston's organ, is generally covered by the hind margin of the preceding tergite. Its function is unknown, but it is possibly an evaporative surface for some volatile secretion (Dessart, 1992).

Economic importance. Some species of *Aphanogmus* are sometimes considered pests since they parasitize cecidomyiid predators of aphids, mealybugs and spider mites (Oatman, 1985; Dessart, 1992; Gilkeson *et al.*, 1993). The hyperparasitic species of *Aphanogmus* and *Ceraphron* could also be considered as pests since they sometimes attack primary parasitoids of commercially important insects. For example, in Barbados, releases of *Cotesia plutellae* (Braconidae) for biological control of the diamond-back moth (*Plutella xylostella*, Plutellidae) may be negatively affected by *Aphanogmus fijiensis*, which reaches parasitism levels of 13 percent during the wet season (Cock, 1985). Parasitism of the cecidomyiid spider mite predator, *Feltiella acarivora*, by *Aphanogmus floridanus* was observed to be as high as 89.7 percent in strawberries in California (Oatman, 1985) and in commercial cultures in Canada parasitism of the cecidomyiid aphid predator *Aphidoletes aphidimyza* by *A. fulmeki* reached 50 percent (Gilkeson *et al.*, 1993).

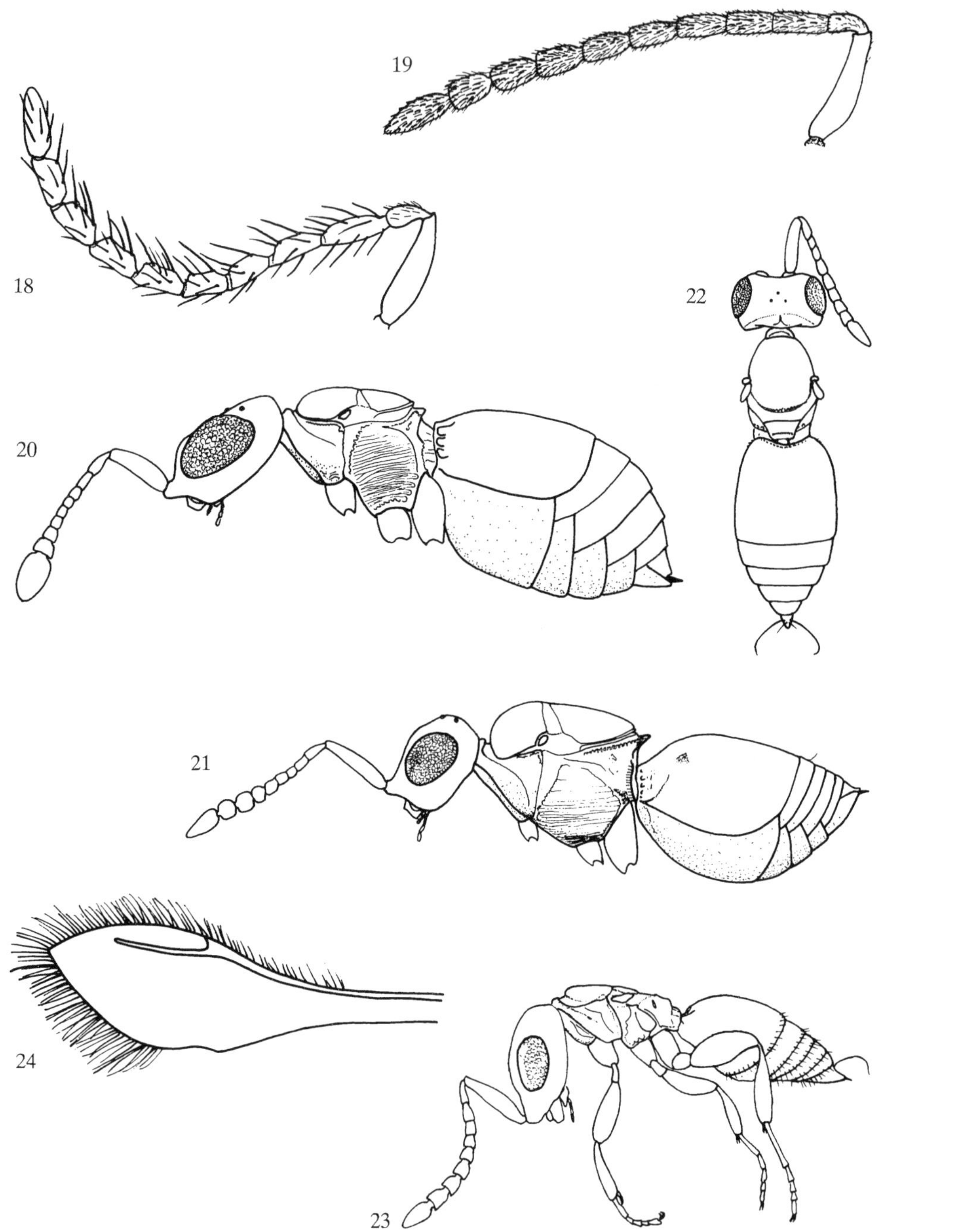

Figs 8.18–8.24. Ceraphronidae. Figs 8.18–8.19. Antenna of male; 8.18, *Aphanogmus* sp.; 8.19, *Ceraphron* sp. Figs 8.20–8.21. Females, in lateral view; 8.20, *Synarsis* sp.; 8.21, *Aphanogmus* sp. Fig. 8.22. *Ecitonetes* sp., body, dorsal view. Fig. 8.23. *Homaloceraphron* sp., whole insect, in lateral view. Fig. 8.24. Fore wing, *Pteroceraphron* sp.

Identification. Keys to the world genera are provided by Dessart and Cancemi (1986), and the six genera known from the New World can be separated by the key given below. *Ceraphron* has been divided into six subgenera, which can be separated by keys given in Dessart and Cancemi (1986).

Key to genera of Ceraphronidae known to occur in the New World

1 Mesosoma clearly compressed (higher than wide); male antenna with flagellum more or less serrate, with both short and long setae (Fig. 8.18), the longer setae longer than width of antennal segment from which they arise. 2

— Mesosoma almost always wider than high; male antenna with flagellum cylindrical (Fig. 8.19), with short setae and strong, compressed, hooked sensilla. 3

2 Mesosoma very strongly compressed and head large, in dorsal view with greatest width across head about 2.0 times width of mesosoma; head in profile about as long as mesosoma (Fig. 8.20). .... *Synarsis*[1]

— Mesosoma only moderately compressed and head of moderate size, so that in dorsal view head is only slightly broader than mesosoma; head in profile generally shorter than mesosoma (Fig. 8.21). .... *Aphanogmus*

3 Mesoscutum, axillae and scutellum fused, without furrows (Fig. 8.22).
[Carinae at base of largest metasomal segment very short; females brachypterous, males unknown.] .... *Ecitonetes*

— Mesoscutum, axillae and scutellum separated by furrows (Fig. 8.03). 4

4 Propodeum elongate; females brachypterous (Fig. 8.23).
[Carinae at base of largest metasomal segment very short.] .... *Homaloceraphron*[1]

— Propodeum not elongate; females fully winged or brachypterous. 5

5 Females (males unknown) with fore wing shiny brown, tapering apically to an angular point, with marginal setae long (Fig. 8.24). .... *Pteroceraphron*[1]

— Both sexes, fore wing not shiny brown, apically rounded, with marginal setae usually shorter (Fig. 8.03); or with wings reduced. .... *Ceraphron*

[1] Genera not yet recorded from Costa Rica.

Synopsis of the Costa Rican fauna

In Costa Rica ceraphronids have been found from the lowland coastal plains to over 3000 metres on the Cerro de la Muerte, and some species are quite common in Malaise trap samples. Brachypterous forms are fairly common at higher altitudes but, despite intensive collecting efforts, have not been found in lowland forests. As the genera are presently defined, nearly all of the ceraphronids so far encountered in Central America belong to either *Aphanogmus* or *Ceraphron*.

8.6 MEGASPILIDAE

Paul Dessart

Diagnosis. 1.0 to 4.0 mm in length; black, brown or yellow, generally never metallic (although with slight greenish or violet metallic reflections in a very few *Dendrocerus*); sometimes brachypterous. *Antenna of female and male 11-segmented, attached close to mouth, geniculate with long scape*; sometimes clavate in female; filiform, serrate, or ramose (branched) in male; anelli absent. *Pronotum extends to the tegulae*; mesoscutum usually with notauli at least anteriorly, nearly always with a median longitudinal furrow. *Wing venation very reduced, consisting of submarginal, marginal and stigmal (radial) vein, the latter long and curved; in most species with a large stigma.* Trochanters 1-segmented; *fore tibia with 2 apical spurs, longer spur forked apically*; mid tibia and hind tibia each with two spurs. Anterior end of largest metasomal tergite narrowed into a short 'neck'; Waterston's organ absent.

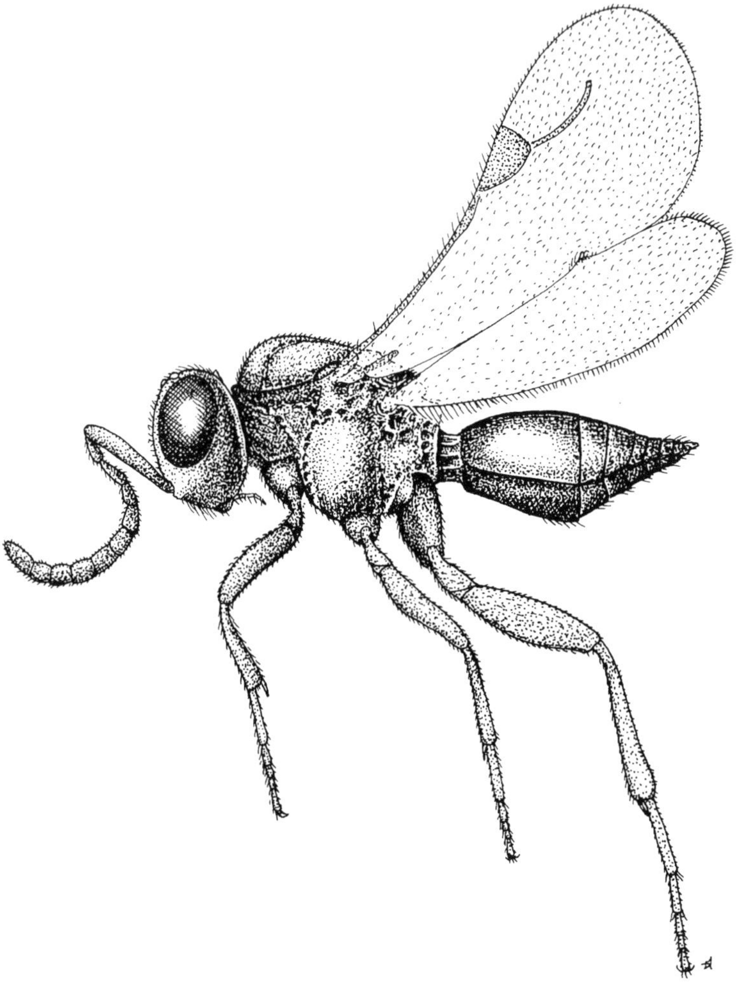

Fig. 8.25. *Dendrocerus* sp. (Megaspilidae).

Brachypterous forms can be separated from other brachypterous microhymenoptera by the number of antennal segments and the form of the large metasomal tergite; they can be distinguished from brachypterous ceraphronids by having the mesonotum very reduced or by the presence of notauli. Most winged forms have the fore wing with a very prominent pterostigma followed by a long, curved stigmal vein; male lagynodines lack the pterostigma and are thus quite similar to Ceraphronidae, but can be separated by the form of the largest metasomal segment (which is anteriorly narrowed).

Classification and distribution. The Megaspilidae is a cosmopolitan group comprising approximately 450 species in nine genera, classified in two subfamilies, Lagynodinae and Megaspilinae. The former subfamily consists of two monotypic genera (*Archisynarsis* and *Typhlolagynodes*) that are restricted to the Old World, the monotypic *Holophleps* from North America and Europe, and the cosmopolitan *Lagynodes*. The Megaspilinae, by far the larger subfamily, comprises five genera: the speciose and cosmopolitan *Conostigmus* and *Dendrocerus*, the monotypic but cosmopolitan *Trichosteresis*, and the rather poorly defined *Megaspilus* and *Platyceraphron* from North America and Europe (Dessart & Cancemi, 1986).

Biology. Biologically the Megaspilidae is somewhat better known than its sister group, the Ceraphronidae, but most of the available information seldom extends beyond the simple recording of which species served as a host. An exception, however, are various studies of some species of *Dendrocerus* that develop as aphid hyperparasitoids (e.g. Haviland, 1920; Bennett & Sullivan, 1978; Matejko & Sullivan, 1984).

Most megaspilids are ectoparasitoids of hosts concealed in cocoons or puparia (Haviland, 1920; Withycombe, 1924; Kamal, 1939; Viggiani, 1967). Many are solitary parasitoids but there are a few records of gregarious parasitism. For example, some species of *Conostigmus* (*C. triangularis*) and *Dendrocerus* (*D. rodhaini*) are gregarious ectoparasitoids of syrphid pupae within the puparium (Kamal, 1939) and, although most other species of *Dendrocerus* are generally solitary parasitoids, occasionally two undersized individuals will emerge from a single host (Bennett & Sullivan, 1978).

Considered as a whole the host range of the Megaspilidae seems to be focused around the Diptera. The only known species of *Trichosteresis*, *T. glabra*, is a parasitoid of Syrphidae, and although hosts of only three or four species of the large genus *Conostigmus* are known, two of these are from puparia of syrphids. Some species of *Conostigmus* and *Lagynodes* are myrmecophilous and possibly attack dipterous hosts in ant nests. Single species of *Megaspilus*, *Conostigmus*, and *Dendrocerus* have been found associated with xylophagous scolytids or cossids, although in the last case it was shown that they were parasitoids of saprophytic dipteran larvae (odiniids) feeding on the faeces of the cossid caterpillar. However, another species of *Conostigmus* is known to parasitize a mecopteran (Cooper & Dessart, 1975).

The remarkably diverse host range of *Dendrocerus* seems to be centred around the predators and parasitoids of sternorrhynchous Homoptera (aphids, scale-insects, and occasionally psyllids). Hosts include Hymenoptera (aphidiine braconids, and more rarely, aphelinids, encyrtids, pteromalids, and figitids), Diptera (syrphids and chamaemyiids), Neuroptera (chrysopids and coniopterygids) and Coleoptera (coccinellids) (Dessart, 1972, 1973; Takada, 1973; Fergusson, 1980). Some species attack other cyclorrhaphous Diptera and have been recorded from the puparia of chloropids and tachinids (Fergusson, 1980; Dessart, 1987, 1990). Fergusson's (1980) analysis of the host ranges of British species of *Dendrocerus* suggests that some common species are highly polyphagous, but that many show a degree of host-specialization. Some, such as *D. flavipes*, which attacks coniopterygids, and *D. pupparum*, which attacks syrphids, seem to be specialized parasitoids of predators, whereas others are secondary parasitoids, attacking various aphid primary parasitoids. At least some of these secondary parasites, such as *D. carpenteri*, are pseudohyperparasitoids since they only oviposit on a primary parasitoid larva or pupa after it has completely consumed the aphid (Haviland, 1920). While most *Dendrocerus* that attack Hymenoptera develop as secondary parasitoids, facultative tertiary and even quaternary parasitism have been observed (Haviland, 1920).

Although most megaspilids are ectoparasitoids, a north temperate species of *Conostigmus* reared from Boreidae (Mecoptera) shows a tendency towards endoparasitism. It commences development by feeding endophagously in the boreid larva, but completes development as an ectoparasitoid (Cooper & Dessart, 1975). Furthermore, incipient koinobiosis has been observed in some ectoparasitic megaspilids. For example, the venom of *D. carpenteri* does not immediately paralyze the host. The host continues to develop and remains bright yellow for up to four to six weeks before decaying (Bocchino & Sullivan, 1981).

Most megaspilids seem to pass through four larval instars. The first three are hymenopteriform and the final instar is somewhat similar in most species of *Conostigmus* (although the head is rather large), but in at least some species of *Dendrocerus* the head is smaller and a short anal cornus is present (Cooper & Dessart, 1975). Bennett and Sullivan (1978) suggested that this cornus allows the mature larva to

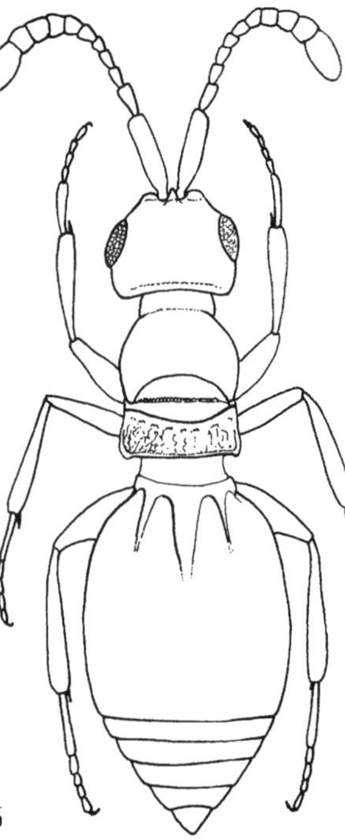

27

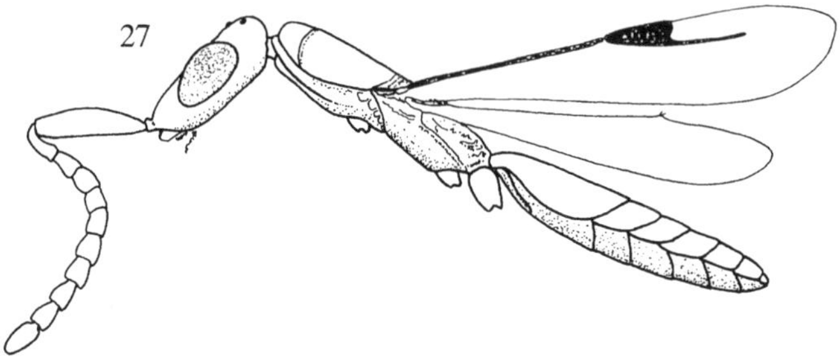

28

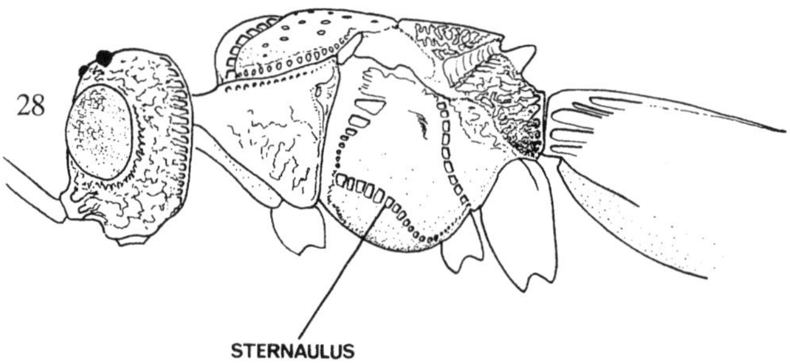

26

29

30

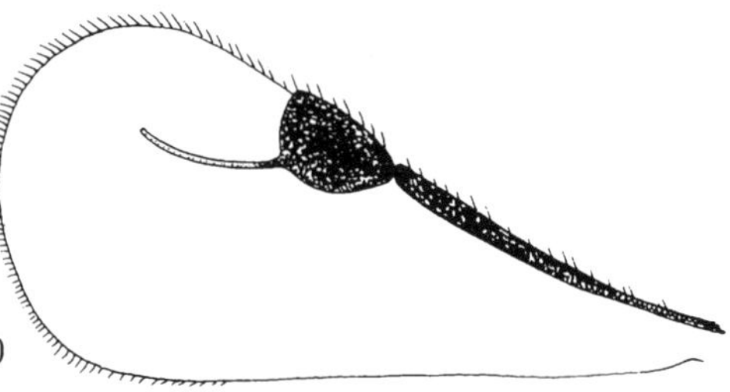

Figs 8.26–8.30. Megaspilidae. Fig. 8.26. *Lagynodes* sp., female, dorsal view. Fig. 8.27. *Platyceraphron* sp., female, lateral view. Fig. 8.28. *Megaspilus* sp., head and mesosoma, lateral view. Figs 8.29–30. Fore wings; 8.29, *Trichosteresis* sp.; 8.30, *Dendrocerus* sp.

twitch, and thus, to some extent, avoid being superparasitized.

Identification. Keys to the world genera were provided by Dessart and Cancemi (1986). *Conostigmus* has been divided into at least five subgenera, which can be separated by keys given in Dessart and Cancemi (1986).

Key to genera of Megaspilidae likely to occur in Central America

1 Females always brachypterous, mesoscutum very reduced but pronotum very large (Fig. 8.26); metasoma with the large second (but apparent first) tergite having only three longitudinal carinae; males sometimes similar, or fully winged, without a pterostigma. .... (Lagynodinae) .... *Lagynodes*[EX]

– Females, if brachypterous, with mesonotum less reduced and with large second (but apparent first) tergite generally having more than three longitudinal carinae; winged forms always with a conspicuous pterostigma. .... (Megaspilinae) 2

2 Mesosoma very depressed, much wider than high (Fig. 8.27); both sexes fully winged. .... *Platyceraphron*

– Mesosoma higher than wide or only slightly depressed; fully winged or brachypterous. ... 3

3 Ocellar triangle isosceles with narrow base, or equilaterally triangular.
[Male antenna cylindrical, never serrate in profile.] .... *Conostigmus*

– Ocellar triangle isosceles with wide base. 4

4 Propodeum with median spur dorsally grooved and apically emarginate; sternaulus present on mesopleuron (Fig. 8.28).
[Larger specimens with head foveolate, orbital and periscutellar furrows prominent; male antenna subfiliform, female antenna always slender.] .... *Megaspilus*

— Propodeum with median spur absent, or if present different from above; sternaulus absent. 5

5 Fore wing hyaline, without marginal setae and with discal setae reduced; stigmal vein shorter than pterostigma (Fig. 8.29); male antenna not serrate in profile. .... *Trichosteresis*[EX]

— Fore wing usually with normal marginal and discal setae; stigmal vein usually longer than pterostigma (Fig. 8.30); male antenna usually serrate in profile, sometimes subcylindrical or ramose. .... *Dendrocerus*

Synopsis of the Costa Rican fauna

In Costa Rica megaspilids can be found from sea level to over 3000 metres on the Cerro de la Muerte. They are rather uncommon in Malaise trap samples from dry lowland habitats, but are fairly common in most other habitats, especially at mid and high elevations (above about 1300 metres). Brachypterous forms have only been encountered at higher altitudes (above 2200 metres).

Conostigmus. A large cosmopolitan genus represented by a number of species in Costa Rica, primarily encountered at lower altitudes, between sea level and 2000 metres.

Dendrocerus. A large cosmopolitan genus represented in Costa Rica by a few species which are generally encountered only above 1000 metres. Most are perhaps associated with aphid colonies.

***Lagynodes*[EX].** A small cosmopolitan genus of tiny and rather rarely collected megaspilids. Most are probably hypogaeic.

Megaspilus. A small, primarily north temperate genus of robust megaspilids. In Costa Rica isolated specimens have been collected at about 3000 metres on the Cerro de la Muerte.

Platyceraphron. A small primarily north temperate genus of very strongly flattened megaspilids.

One species has occasionally been collected during the dry season in Santa Rosa National Park in northwestern Costa Rica.

Trichosteresis[EX]**.** A monobasic genus comprising the single widespread species, *Trichosteresis glabra*, which is known to parasitize several cosmopolitan species of aphidophagous syrphids (Dessart, 1974).

9 The proctotrupoid families

Lubomir Masner

INTRODUCTION

The superfamily Proctotrupoidea was traditionally a catch-all group for small non-chalcidoid Apocrita not readily assignable to other taxa. The Ceraphronidae and Megaspilidae have been removed to a separate superfamily (Masner & Dessart, 1967), and the Loboscelidiidae were transferred to the Chrysididae (Day, 1978), thus leaving 12 families (Naumann & Masner, 1985, Kozlov, 1994). However, these families still do not constitute a holophyletic group, and it is possibly better to recognize two superfamilies of 'proctotrupoids':

Proctotrupoidea *s. str.* (Austroniidae, Diapriidae, Heloridae, Monomachidae, Pelecinidae, Peradeniidae, Proctotrupidae, Renyxidae Roproniidae and Vanhorniidae). Antenna inserted rather high above the clypeus and without basiconic sensilla on female claval segments; fore wing generally with costal cell present and closed; metasomal segment I with tergite and sternite fused; metasomal segment VII normally developed; metasomal spiracle present; ovipositor either internal, or external but housed in tight, strongly sclerotized sheaths.

Platygastroidea (Platygastridae, Scelionidae). Antenna inserted immediately above upper margin of clypeus and with one or two basiconic sensilla on female claval segments; fore wing with costal cell open or undefined; metasomal segment I with tergite and sternite free; metasomal segment VII in females very reduced or absent; metasomal spiracle absent; ovipositor internal, often housed in tube of soft tissue.

The Proctotrupoidea *s. str.* comprises two fairly species-rich cosmopolitan families, the Diapriidae and Proctotrupidae, and seven small families, which consist mostly of one genus each and are morphologically isolated groups. The Platygastroidea comprises two speciose, cosmopolitan families. The two superfamilies differ in the focus of their host ranges. The Proctotrupoidea *s. str.* are principally parasitoids of saprophagous, mycophagous or even carnivorous insects that develop in leaf litter, fungi, soil, rotting matter and similar situations. Dipterous larvae or puparia are commonly attacked, but larval Coleoptera are also widely utilized (as are a few species of Neuroptera). All develop (more or less) as koinobionts. In the Platygastroidea the ancestral hosts are clearly arthropod eggs, in which the parasitoids develop as idiobionts. The more derived platygastrids, however, develop as koinobiont endoparasitoids of Cecidomyiidae (Diptera), although oviposition is still often into the host egg.

The formal classification of 'proctotrupoids' into two superfamilies, as described above, still leaves some unresolved problems. The Pelecinidae and Proctotrupidae are probably closely related (Gibson, 1985; Rasnitsyn, 1988) but the relationships of the other Proctotrupoidea (*s.str.*) are less certain (see chapter 5). The Monomachidae and Diapriidae do not share certain apomorphies that are present in other proctotrupoids (Gibson, 1985). Moreover, it has been suggested that the Diapriidae are possibly the sister-group of the Cynipoidea (Rasnitsyn, 1988). The relationships of the Platygastroidea are also unresolved. Gibson (1985) found that they shared certain characters with Pelecinidae + Proctotrupidae + Vanhorniidae, whereas Rasnitsyn (1988) suggested that Platygastroidea is the sister group of Chalcidoidea.

Both of the platygastroid families and five of the ten proctotrupoid families are represented in the tropical American fauna. Those absent are small, relict families—the Austroniidae and Peradeniidae, which are restricted to Australia, and the Roproniidae, Renyxidae and Vanhorniidae, which are holarctic.

Key to families of Proctotrupoidea (*s. lat.*) present in Central America

1 Antennal sockets separated from dorsal margin of clypeus by distinctly more than their own diameter. .... (Proctotrupoidea *s. str.*) 2

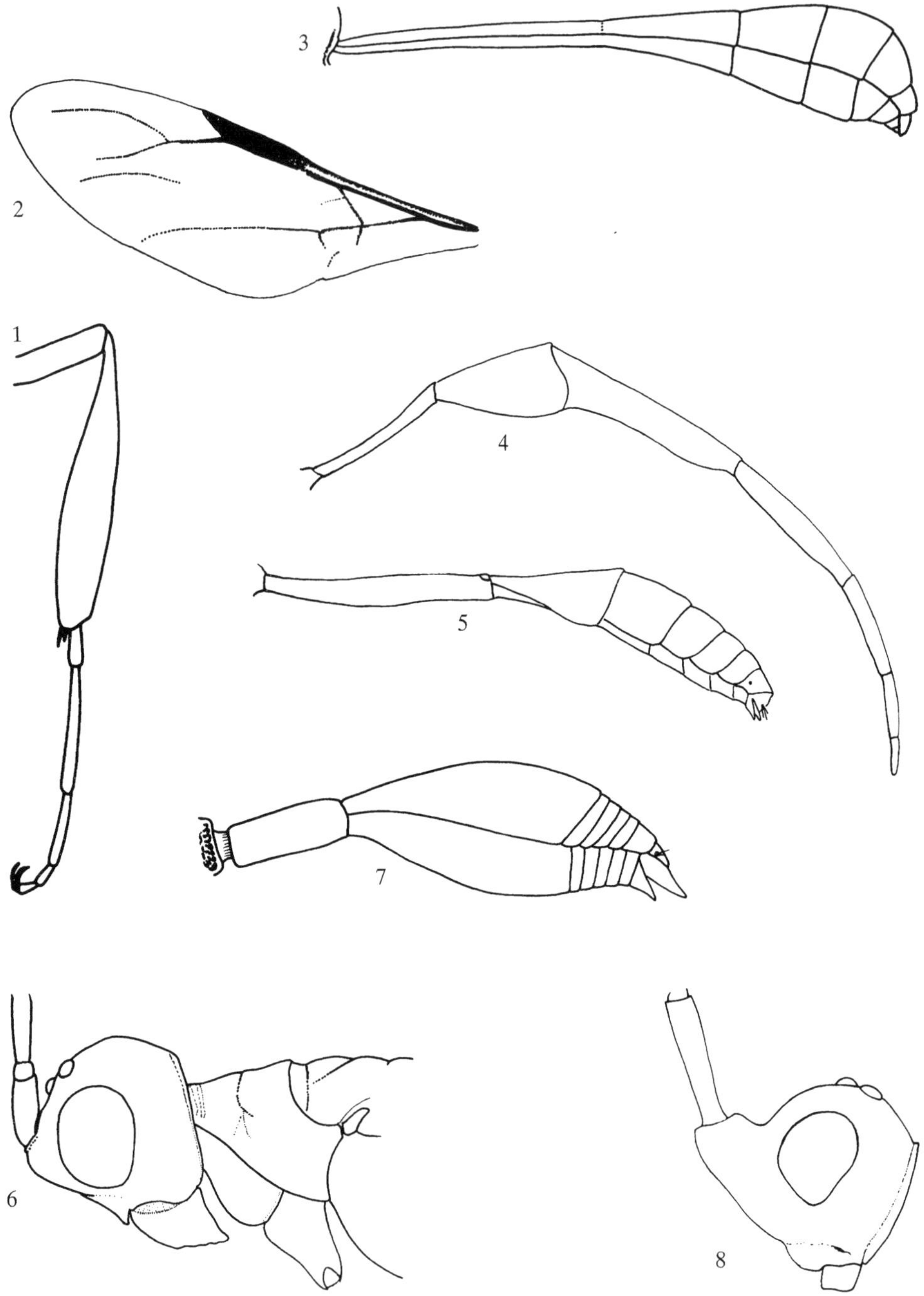

Figs 9.01–9.08. Proctotrupoidea. Figs 9.01–9.03. *Pelecinus polyturator* (Pelecinidae); 9.01, hind leg; 9.02, fore wing; 9.03, metasoma, male. Figs 9.04–9.06. *Monomachus* sp. (Monomachidae); 9.04, metasoma female; 9.05, metasoma, male; 9.06, head and prothorax, lateral. Figs 9.07–9.08. Diapriidae; 9.07, *Belyta* sp., metasoma of male, lateral; 9.08, *Belyta* sp., head, lateral.

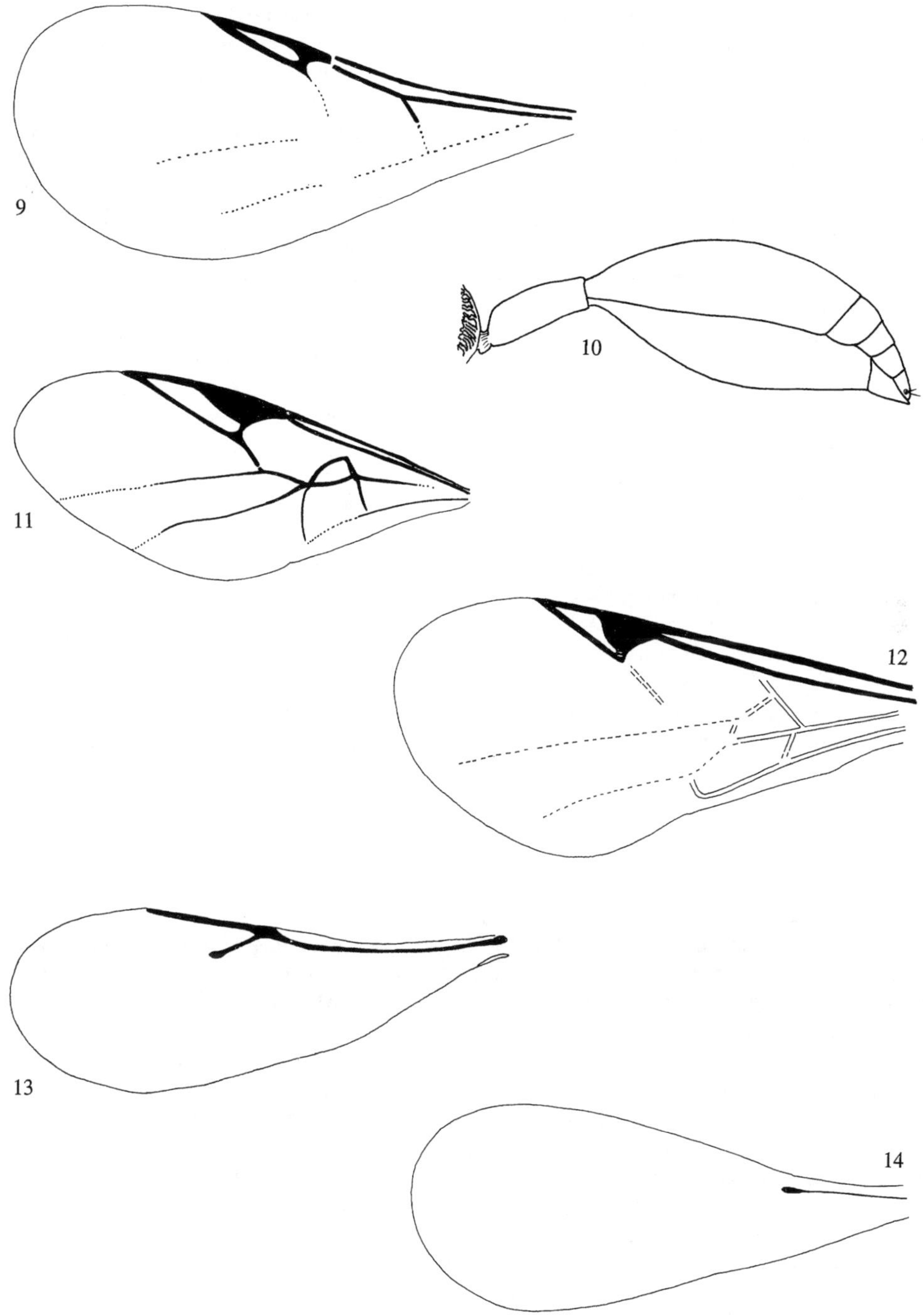

Figs 9.09–9.14. Proctotrupoidea. Fig 9.09. Diapriidae, Belytinae, fore wing. Figs 9.10–9.11. *Helorus* sp. (Heloridae); 9.10, metasoma of female, lateral; 9.11, fore wing. Figs 9.12–9.14. Fore wing; 9.12, Proctotrupidae; 9.13, Scelionidae; 9.14, Platygastridae.

— Antennal sockets contiguous with dorsal margin of clypeus, or separated from it by less than their own diameter. .... (Platygastroidea) 6

2 Hind leg with basitarsus distinctly shorter than the following tarsal segment (Fig. 9.01); fore wing with *Rs* distally bifurcate (Fig. 9.02). [Metasoma of female extremely elongate (Fig. 9.17); of male, clavate (Fig. 9.03).] .... **Pelecinidae** (p. 216)

— Hind leg with basitarsus distinctly longer than the second or following tarsal segments; fore wing with *Rs* not forked, or absent. 3

3 Metasoma of female with tergite III subequal in length to tergite IV; metasoma of both sexes elongate (Figs 9.04, 9.05); posterior margin of pronotum overlapping anterior part of mesoscutum. [Mandibles extremely large (Fig. 9.06).] ....**Monomachidae** (p. 213)

— Metasoma with tergite III considerably longer than tergite IV; metasoma not exceptionally elongate (Figs 9.07, 9.10); posterior margin of pronotum not overlapping mesoscutum, separated from latter by a deep suture. 4

4 Antenna with scape very elongate, at least 2.5 times as long as wide; fore wing without a pterostigma (Fig. 9.09); antennae usually inserted on a facial 'shelf' (Fig. 9.08). ....**Diapriidae** (p. 222)

— Antenna with scape short, at most 2.2 times as long as wide; fore wing with pterostigma present; antennae not inserted on a facial 'shelf'. 5

5 Antenna with 15 segments and an anellus; fore wing with an enclosed triangular discal cell (Fig. 9.11); mesosoma with a transscutal suture; tarsal claws pectinate. ....**Heloridae** (p. 214)

— Antenna with 13 segments; fore wing without an enclosed triangular discal cell (Fig. 9.12); mesosoma without a transscutal suture; tarsal claws not pectinate. ....**Proctotrupidae** (p. 216)

6 Metasomal tergite II not distinctly the largest of all tergites, at most subequal in length to tergite III. .... **Scelionidae** (in part) (p. 228)

— Metasomal tergite II distinctly the largest of all tergites, clearly much longer than tergite III. 7

7 Fore wing with submarginal vein (*Rs*) reaching anterior margin of wing near centre, then diverging abruptly from wing margin (as stigmal vein) (Fig. 9.13); antenna usually with 11 or 12 segments, rarely with 10; male with antennal segment 5 specialized. .... **Scelionidae** (in part) (p. 228)

— Fore wing either veinless, ***or if*** submarginal vein is present, ***then*** it does not reach the anterior margin of the wing, and a sligmal vein is not present (Fig. 9.14); antenna with 10 or fewer segments; male with antennal segment 4, or rarely 3, modified. .... **Platygastridae** (p. 241)

9.1 MONOMACHIDAE

Diagnosis. Body 10.0 to 18.0 mm long, slender; colour usually brown, yellowish brown, or in a few species even greenish, never black; smooth and glabrous. *Antenna in female 15- and in male 14-segmented*, thread-like, not geniculate, scape only slightly elongate (about three times as long as wide); *mandibles massive. Pronotum almost neck-like, capable of sliding over the anterior margin of the mesoscutum*; transscutal suture on mesoscutum (between tegulae) present; propodeum distinctly cone-shaped, without median carina, with hind coxa inserted relatively far from propodeal orifice. Fore wing with at least five more or less enclosed cells and relatively narrow pterostigma. *Metasoma* showing pronounced sexual dimorphism; in *female elongate, sickle-shaped in lateral view*, very slender in dorsal view; in *male slender but slightly broadened apically*; segment I distinctly petiolate (at least three times as long as wide in female, longer in male); *tergite III in both sexes subequal in length to tergite IV*; female with six or seven clearly discernible tergites, male with seven or eight; ovipositor extremely short, concealed inside segment VIII.

Fig. 9.15. *Monomachus* sp. (Monomachidae).

Classification and distribution. The Monomachidae is a small family that is restricted to Australia, and Central and South America. It comprises two genera, *Monomachus*, with three described species in Australia and nine in South America, and *Tetraconus*, with a single species in South America. However, the latter may not be distinct from *Monomachus* (Naumann, 1985).

Biology. Monomachids generally prefer relatively cool, humid situations (Naumann, 1985). One Australian species of *Monomachus* has been reared from mature larvae and puparia of a stratiomyiid dipteran, a species of *Boreoides* belonging to the subfamily Chiromyzinae (Riek, 1970), a group that also shows a gondwanic distribution. Only one wasp emerges from male hosts, but several wasps successfully develop in the larger female hosts. Adult monomachids are sometimes attracted to light, but they are rarely taken by diurnal sweeping (Naumann, 1985). Occasional individuals are captured in Malaise traps. In August 1992 I collected numerous males in Venezuelan cloud forests at 1100 metres (Rancho Grande) between 08.00 and 09.00 hours. By 10.00 no more individuals were found.

Synopsis of the Costa Rican fauna

Only the genus *Monomachus* has been collected in Costa Rica.

Monomachus. A single species is present in Costa Rica in forests at altitudes between 1000 and 1600 metres.

9.2 HELORIDAE

Diagnosis. About 5.0 mm in length. *Antenna* thread-like, non-geniculate with short scape, *with 15 segments plus a small ring-like anellus*; mandibles long, sickle-shaped, crossing one another in a scissor-like manner. Mesoscutum with transscutal suture (between tegulae) developed. *Fore wing with five enclosed cells including a subtriangular discal cell.* Tarsal claws pectinate. Metasoma rather short and clavate, not strongly sexually dimorphic; segment I petiolate, *tergites II+III+IV fused into a single syntergum*, and V to VIII short, generally visible posteriorly; sternites II+III+IV+V fused into a single synsternum, male with three further narrow but more or less visible sternites, female with a single triangular sternite VI; ovipositor internal.

Classification and distribution. The Heloridae is a very small, relict family with just one genus and seven described species throughout the world, all of which are uncommon. One, *Helorus brethesi*, is recorded from the neotropical region and extends from Argentina to Mexico (Townes, 1977*b*).

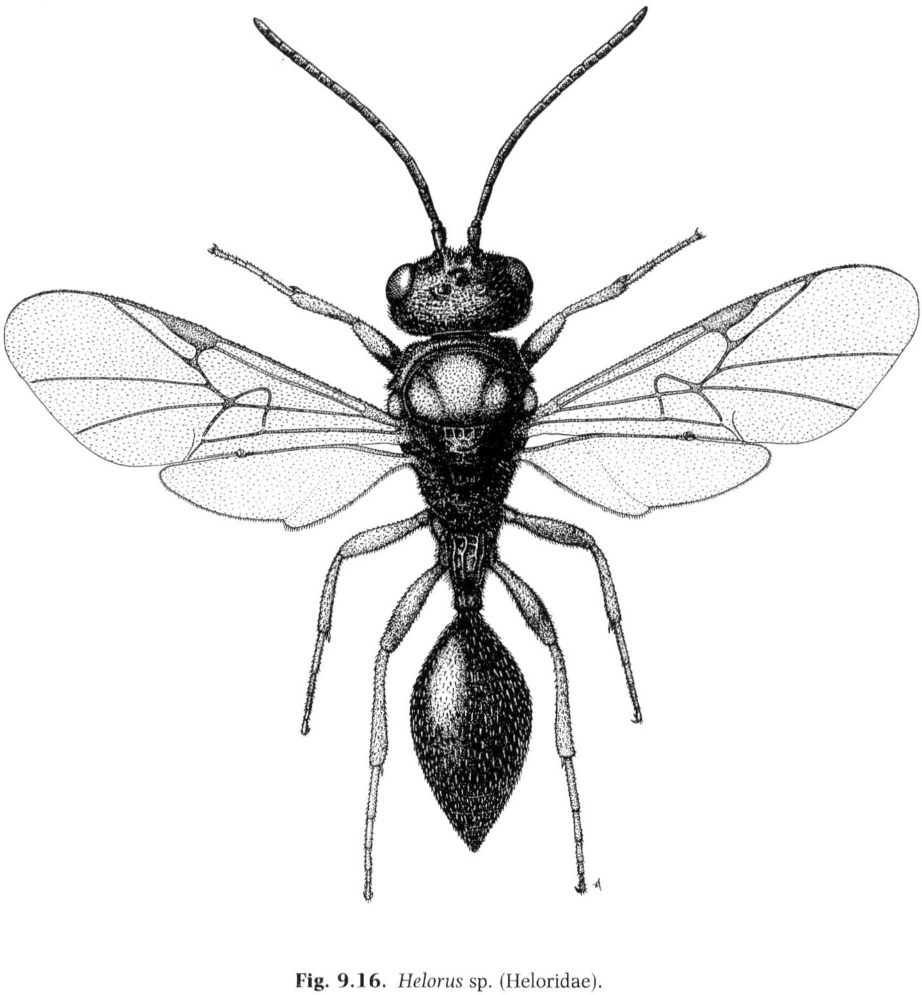

Fig. 9.16. *Helorus* sp. (Heloridae).

Biology. Helorids are solitary, endoparasitic koinobionts of Chrysopidae (Neuroptera). The female oviposits into the host larva and the egg is free floating in the haemocoel. The egg hatches in about two days and the larva remains in the first instar until the host has spun its cocoon. The first instar is polypodeiform, with paired ventral fleshy lobes on the abdominal segments, and with a strongly sclerotized head (Clancy, 1946). When the host has spun its cocoon the helorid larva completes its development rapidly, passing through two more instars. The third instar partially emerges from the skin of the host and pupates within the host cocoon. The period from oviposition to adult emergence requires about 30 days. Adults live for four to six weeks and during this period approximately 50 eggs are laid, usually at a rate of one or two per day.

Synopsis of the Costa Rican fauna

Helorus. *H. brethesi* has been collected in Costa Rica from one wet forest site at 1200 metres close to the Panama border (Las Cruces Botanic Garden). It has a bright orange mesosoma and belongs to a large aposematic complex of neotropical microhymenoptera (see discussion under Scelionidae). A

Fig. 9.17. *Pelecinus polyturator* (Pelecinidae).

second, undescribed species of *Helorus* occurs at 2600 metres on the Cerro de la Muerte.

9.3 PELECINIDAE

Diagnosis. The largest proctotrupoids, at least 20.0 mm in length and up to *70.0 mm in some females*; body black, predominantly smooth and glabrous. *Antenna* thread-like, non-geniculate, with rather short scape, *14-segmented in both sexes*. Transscutal suture on mesoscutum (between tegulae) well developed. Fore wing deeply infuscate, with broad enclosed costal cell, relatively narrow pterostigma and open marginal cell; *vein* Rs *forked distally*. Female hind tibia swollen; hind tarsus in both sexes with first segment strikingly shorter than second segment. *Metasoma showing extreme sexual dimorphism, that of female extremely elongate and tubular*, held in an arched position in flight; segments I to V very elongate and slender, II, III and IV of similar length; *male metasoma elongately club-shaped*, broadened posteriorly, with tergites II+III (perhaps more) fused to form a slender petiole. Individuals vary considerably in body length, perhaps depending on the size of the host.

Classification and distribution. The Pelecinidae is a small relict family that is thought to consist of just a single species, *Pelecinus polyturator*, which occurs from Canada to Argentina. However, this species shows great variation in size and coloration in South America and thus may be a complex of species (Muesebeck *in* Krombein *et al.*, 1979). In North America males are extremely rare while in the Neotropics both sexes are equally common, thus suggesting geographical parthenogenesis (Brues, 1928; Young, 1990).

Biology. Despite its conspicuousness, surprisingly little is known about the biology of *P. polyturator*. It has been reared from larvae of Scarabaeidae (Coleoptera), particularly species of *Phyllophaga* (Hammond, 1944; Lim *et al.*, 1980), and it appears to be a solitary endoparasitic koinobiont. The female oviposits into the host larva and in some instances the host attains the pupal stage before dying (Clausen, 1940*b*). Pupation occurs outside the body of the host, but the parasitoid pupa remains connected to the host remains (as in the family Proctotrupidae). The metasoma of female pelecinids is extremely flexible, and when handled they occasionally are capable of inflicting a sting, but it is a mild sensation like a pin-prick (Gauld, pers. comm.). Mason (1984) described the mechanics of the female metasoma.

Synopsis of the Costa Rican fauna

Pelecinus polyturator is most common in Costa Rica at intermediate altitudes (1000–2000 m) where individuals of both sexes can frequently be observed resting on the foliage of understorey plants, especially early in the rainy season.

9.4 PROCTOTRUPIDAE

Diagnosis. Body 3.0 to 10.0 mm in length; predominantly black; smooth, except on propodeum. *Antenna* thread-like, non-geniculate, with short scape, *13-segmented in both sexes*; mandible usually unidentate. *Mesoscutum with transscutal suture (between tegulae) absent*. *Fore wing* with rather broad enclosed costal cell, *large pterostigma and small enclosed marginal cell*; other veins barely visible or absent. *Metasoma spindle-shaped, in lateral view more or less curved with apex distinctly down-curved in females*; metasomal tergite I and sternite I fused to form petiole, and petiole fused posteriorly with sternite II (petiole overlapped by anterior margin of syntergum in several genera); metasomal *tergites II+III+IV fused into syntergum*; all tergites in lateral view partially or completely overlapping the sternites; ovipositor with strongly sclerotized external sheaths.

Classification and distribution. The higher classification of this group is not completely resolved and various authors have disputed the ranking of several groups. Townes and Townes (1981) in their revision of the Proctotrupidae (= Serphidae) of the world recognized 26 genera which they classified in three subfamilies: the Austroserphinae (which they called the Acanthoserphinae), the Vanhorniinae, and the Proctotrupinae. Riek (1970) treated the Austroserphinae as a separate family, but this status has not become generally accepted.

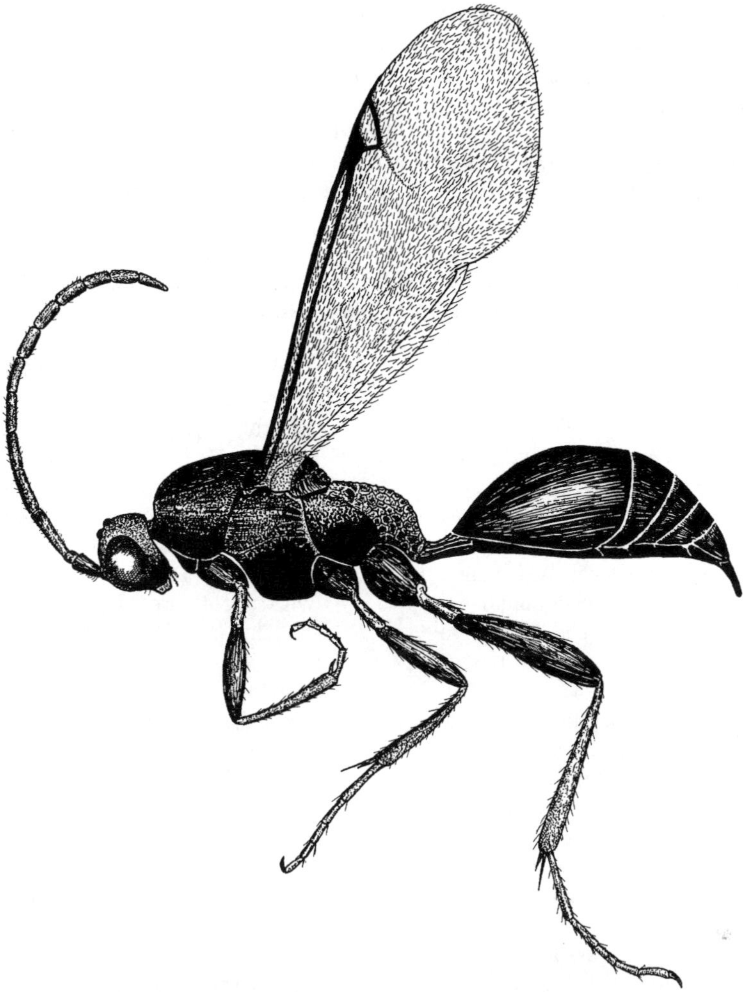

Fig. 9.18. *Exallonyx* sp. (Proctotrupidae).

The Vanhorniinae (*sensu* Townes & Townes, 1981) comprised two genera, the holarctic *Vanhornia* and the Chilean *Heloriserphus*. However, most authors (e.g. Mason, 1983*b*; Naumann & Masner, 1985; Gauld & Bolton, 1988) separate the former genus from the proctotrupids and place it in a family by itself—the Vanhorniidae. Masner (*in* Townes & Townes, 1981) suggested that *Heloriserphus* is not closely related to *Vanhornia* and should be placed in a separate new subfamily, but as yet such a taxon has not been formally proposed. The Proctotrupidae, as most generally accepted, therefore comprises two described subfamilies—Austroserphinae and Proctotrupinae—and one unassigned Chilean genus.

The Austroserphinae includes three small genera restricted to Australia and southern South America, whereas the Proctotrupinae comprises 21 genera and about 300 described species worldwide, although it has been estimated that the subfamily contains well over 1000 species (Townes & Townes, 1981). The Proctotrupinae is divided into three tribes: Disogmini, Cryptoserphini and Proctotrupini. The first of these comprises just one holarctic genus, but the other two tribes are cosmopolitan.

Biology. Proctotrupids are primary, endoparasitic koinobionts. The majority of host records are from larvae of beetles (Coleoptera), but some species parasitize larvae of Mycetophilidae (Diptera), and there is one record from a lithobiid centipede (Newman, 1867), but this has not been corroborated for over a century. Most of the host records are for the large genus, *Exallonyx*, which has been reared from five subfamilies of Staphylinidae (Townes & Townes, 1981; Frank, 1982).

Proctotrupids are solitary (e.g. Hoebeke & Wheeler, 1990) or gregarious with up to 50 individuals emerging from a single host (e.g. Crichley, 1973). The same species may be solitary or gregarious, depending upon the size of the host (e.g. Hoebeke, 1978; Hoebeke & Kovarik, 1988). Oviposition is into the host larva. Species that parasitize mycetophilids have been observed on macrofungi inserting virtually the whole of the metasoma into spore ducts in an attempt to reach mycetophilid larvae (Huggert, 1979).

The first instar larva is polypodeiform, with paired fleshy protuberances ventrally on the abdominal segments and with a strongly sclerotized head. This larva apparently remains almost quiescent until the host is about to pupate, whereupon the parasitoid larva develops rapidly. It may or may not consume the entire body contents of the host. The larvae all emerge in a characteristic fashion from the ventral surface of the host, usually through intersegmental membranes. Individuals of the gregarious species are often arranged in rows and facing the same direction. They pupate with terminal segments of the abdomen still embedded in the host (Basden, 1959) and form a thin membrane but no cocoon (Figs 9.19, 9.20). There are a number of accounts of the developmental stages of several species but one of the most complete is by Eastham (1929). Little is known of their mating behaviour, but courtship in *Phaenoserphus* involves the male entwining his antennae around those of the female (Luff, 1976).

Identification. Keys to the world genera and species are given by Townes and Townes (1981) and the key given below is modified from this source.

Key to genera of Proctotrupidae present in Costa Rica

1 Notaulus often present (Figs 9.29, 9.30), usually short and sometimes represented by an anterolateral pit; metasoma usually subsessile (Fig. 9.27); metapleuron usually with a large smooth area. .... (Cryptoserphini) 2

— Notaulus absent or represented by a shallow impression, not a distinct groove (Fig. 9.28); metasoma petiolate (Fig. 9.26); metapleuron with smooth area much smaller or absent. .... (Proctotrupini) 6

2 Fore wing with *Rs* curved obliquely or almost vertically towards *R* directly from lower part of pterostigma, without a short vertical descent from pterostigma (Fig. 9.21). .... *Brachyserphus*

— Fore wing with *Rs* descending vertically from lower corner of pterostigma, then turned at an acute angle obliquely towards *R* (Figs 9.22, 9.23). 3

3 Groove across middle of mesopleuron incomplete; propodeum reaching beyond middle of hind

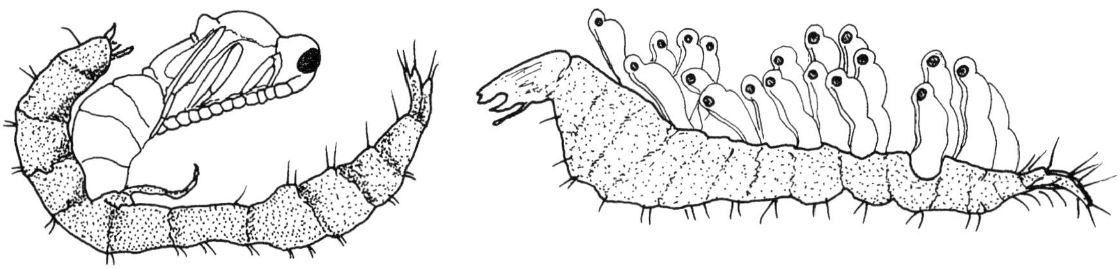

Figs 9.19–9.20. Pupae of Proctotrupidae protruding from shrivelled remains of host larva; 9.19, a solitary species of *Cryptoserphus*; 9.20, a gregarious species of *Exallonyx*.

coxa (Fig. 9.24); upper front part of smooth area of metapleuron without a ridge extending to upper lateral margin of propodeum. .... *Fustiserphus*

— Groove across middle of mesopleuron complete; propodeum never reaching beyond middle of hind coxa (Fig. 9.25); upper front part of smooth area of metapleuron with a ridge extending to upper lateral margin of metapleuron (except in *Cryptoserphus*). 4

4 Fore wing with marginal cell short, the anterior side 0.3–0.9 as long as depth of pterostigma; vertical basal part of vein *Rs* very thick and short, not longer than it is thick (Fig. 9.22). ... . *Sminthoserphus*

— Fore wing with marginal cell longer, the anterior side 0.6–2.0 as long as depth of pterostigma; vertical basal part of vein *Rs* about 2 times as long as thick (Fig. 9.23). 5

5 Longer spur of hind tibia ending between the middle and the apical 0.2 of hind basitarsus; upper front part of smooth area on metapleuron not connected by a carina to upper edge of propodeum, or sometimes connected by a weak or incomplete carina. .... *Cryptoserphus*

— Longer spur of hind tibia ending near or before middle of hind basitarsus; upper front part of smooth area on metapleuron connected by a fine short carina to upper edge of propodeum (Fig. 9.25). .... *Mischoserphus*

6 Fore and mid tarsal claws each with a long black divergent tooth near base; lateral aspect of pronotum with hairs on upper part of collar and along upper edge, usually without hairs elsewhere. .... *Exallonyx*

— Fore and mid tarsal claws simple; lateral aspect of pronotum usually with hairs generally distributed but very often with a median hairless area. .... *Phaneroserphus*

Synopsis of the Costa Rican fauna

Proctotrupids are most common in undisturbed, shaded, wet habitats, primarily at intermediate to high altitudes (above 1000 metres) although individuals have been taken in wet lowland forests and isolated specimens have been collected in seasonally dry lowland forests. The family appears to be very scarce in wet forests below 400 metres altitude. Thus on the Osa Peninsula they have been collected only from Cerro Rincón (650 m). Males are the sex more commonly caught in Malaise traps, while the females are more frequently found in yellow pan traps.

Seven genera and about 20 species occur in Costa Rica. It is interesting that genera parasitic on carabid beetle larvae have not been found in this country. As is true of the world fauna in general (except in Australia and New Zealand), about half of the Costa Rican species belong to the genus *Exallonyx*.

PROCTOTRUPINAE

PROCTOTRUPINAE: Cryptoserphini. This cosmopolitan tribe comprises 13 genera and approximately 100 described species. Five genera are known to occur in the New World tropics and all have been found to be represented in Costa Rica.

***Brachyserphus*[CR]**. A primarily holarctic genus with 11 described species; one has been collected in Costa Rica. Biology: hosts are larvae of the coleopterous families Erotylidae, Melandryidae, Mycetophagidae, Nitidulidae and Phalacridae.

***Cryptoserphus*[CR]**. A cosmopolitan genus of 13 species; two occur in Costa Rica. Biology: hosts are species of the dipterous family Mycetophilidae.

***Fustiserphus*.** An Australian and South American genus with six described species. Two species occur in Costa Rica above 2000 metres. Biology: unknown.

***Mischoserphus*[CR]**. A cosmopolitan genus with 20 described species; three occur in Costa Rica. Biology: hosts are species of the dipterous family Mycetophilidae.

***Sminthoserphus*[CR]**. A neotropical genus with five described species; one occurs in Costa Rica. Biology: unknown.

PROCTOTRUPINAE: Proctotrupini. This tribe comprises about 200 species classified in seven genera. Most genera are holarctic in distribution, but several extend into the higher areas in the oriental tropics and one,

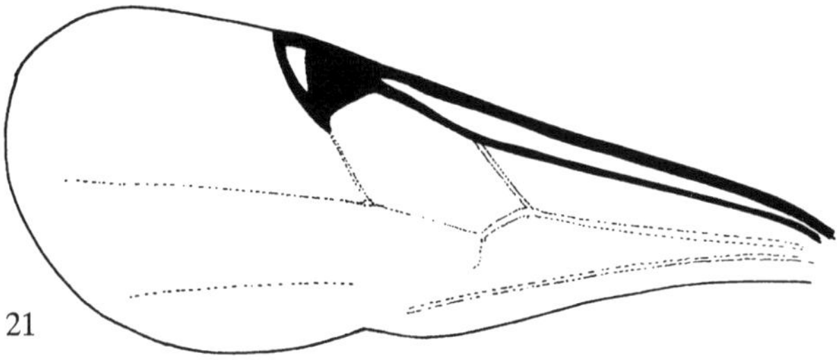
21

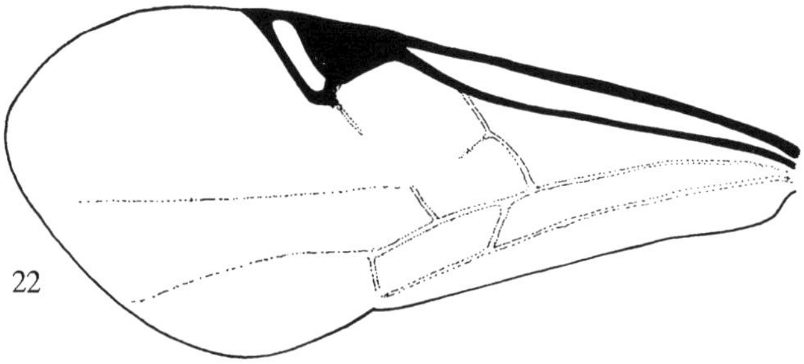
22

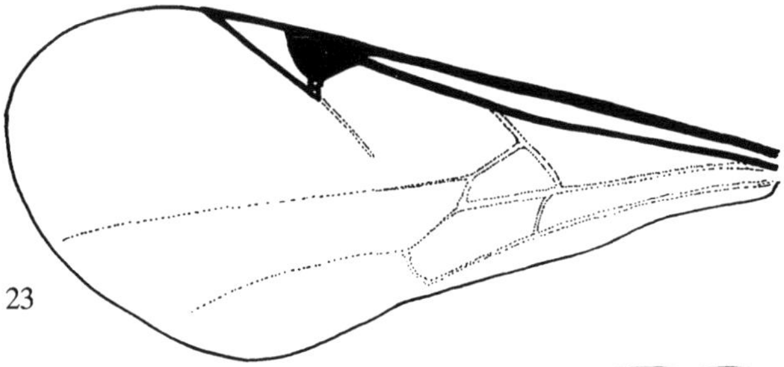
23

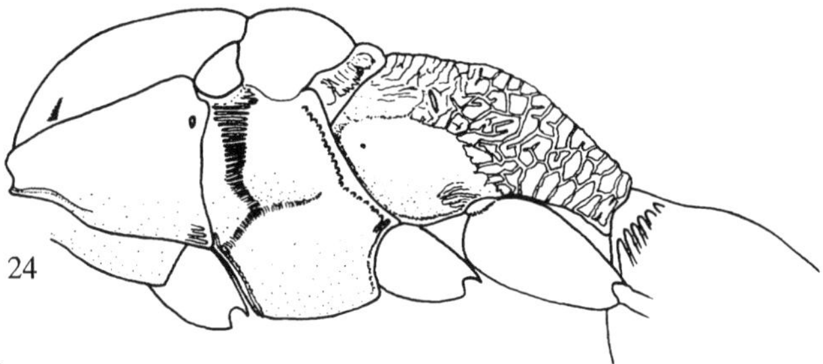
24

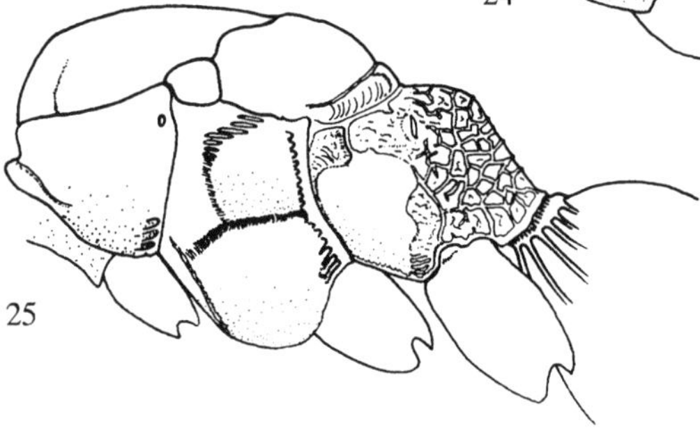
25

Figs 9.21–9.25. Proctotrupidae. Figs 9.21–9.23. Fore wing; 9.21, *Brachyserphus* sp.; 9.22, *Sminthoserphus* sp.; 9.23, *Cryptoserphus* sp. Figs 9.24–9.25. Mesosoma, lateral; 9.24, *Fustiserphus* sp.; 9.25, *Mischoserphus* sp.

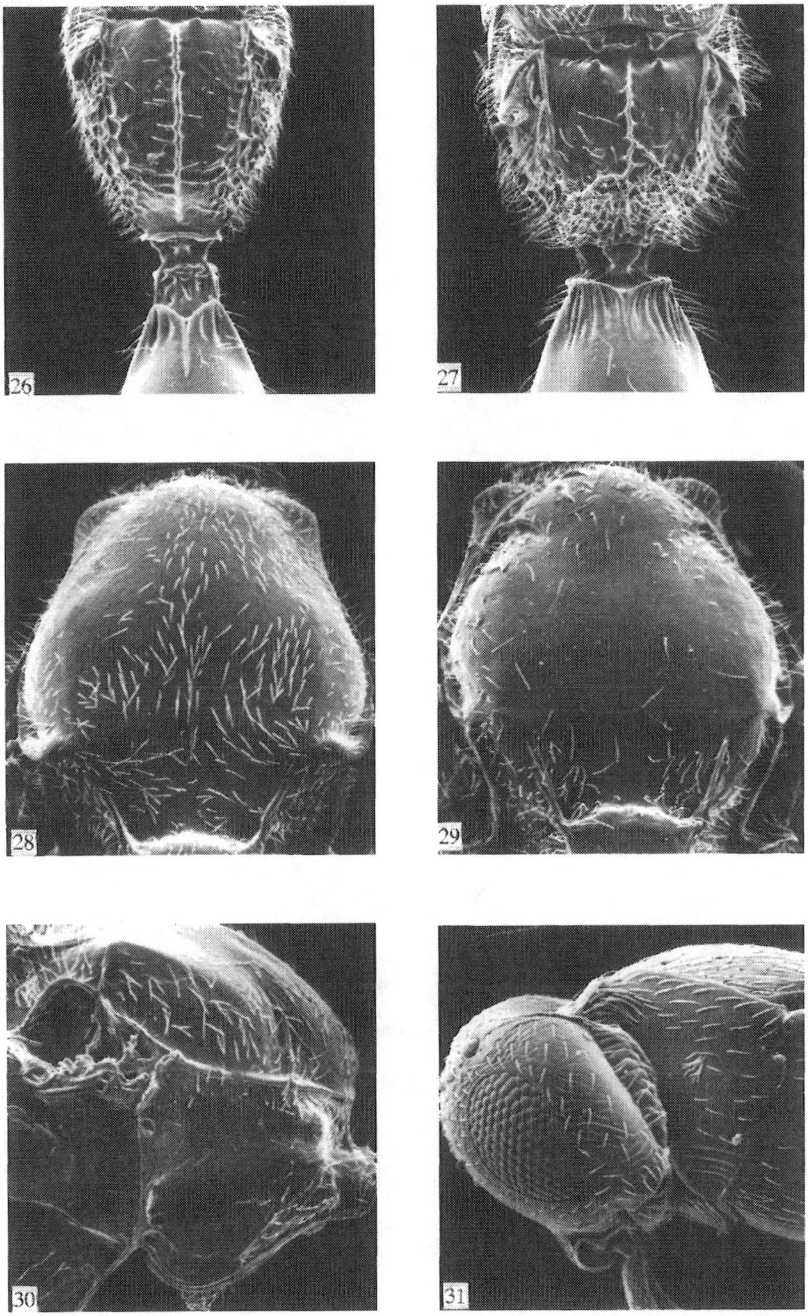

Figs 9.26–9.31. Scanning electron photomicrographs of Proctotrupoidea. Figs 9.26–9.27. Prototrupidae, propodeum and anterior end of metasoma, dorsal; 9.26, *Exallonyx* sp.; 9.27, *Cryptoserphus* sp. Figs 9.28–9.29. Proctotrupidae, mesoscutum, dorsal; 9.28, *Exallonyx* sp.; 9.29, *Cryptoserphus* sp. Fig. 9.30. Mesoscutum and pronotum, lateral, *Cryptoserphus* sp. (Prototrupidae). Fig. 9.31. Head and pronotum, lateral, *Doddiella* sp. (Scelionidae).

Phaneroserphus, similarly extends to tropical America. The largest genus, *Exallonyx*, is cosmopolitan and contains more species than all other genera in the family. Both *Phaneroserphus* and *Exallonyx* have been found to occur in Costa Rica.

***Exallonyx*.** A very large cosmopolitan genus comprising more than 150 described species, with probably very many more undescribed. In the literature from 1950–1980 many species of this genus were incorrectly placed in *Codrus*, a much smaller genus restricted to the Old World (Townes & Townes, 1981). Eight, or perhaps more species of *Exallonyx* occur in Costa Rica. Biology: hosts are the larvae of Staphylinidae.

***Phaneroserphus*^{CR}.** A primarily holarctic genus comprising five described species, and one very distinctive undescribed species which has been collected in Costa Rica. Biology: hosts are the larvae of Staphylinidae and, possibly, centipedes of the family Lithobiidae.

9.5 DIAPRIIDAE

Diagnosis. Body length 1.0 to 8.0 mm, with most species between 2.0 and 4.0 mm; usually dark brown to blackish coloured with lighter appendages, sometimes reddish or yellowish, rarely whitish; surface mostly smooth and shining. Most are fully winged but some have wings shortened to absent, especially in females. *Antennal sockets positioned high above clypeus, usually on prominent transverse ledge* (indistinct in Ismarinae), opening upwards; antenna more or less geniculate, *at most with 15 segments, usually 9 to 14*; antenna clavate in females, thread-like in males, with male antennal segments 3 or 4 modified. *Fore wing*

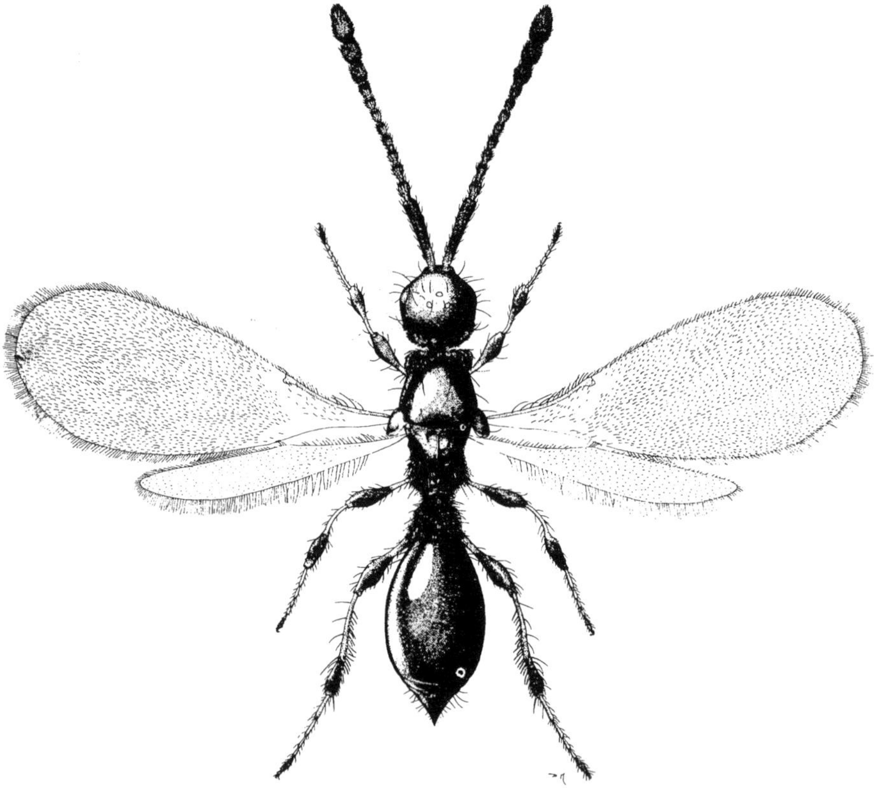

Fig. 9.32. *Trichopria* sp. (Diapriidae).

without pterostigma but sometimes with slightly thickened marginal vein, *with one to three enclosed cells, in some species sometimes virtually veinless*; hind wing often with submarginal vein reaching to hamuli, at most with one enclosed cell, less frequently hind wing without veins. Metasoma distinctly petiolate, petiole composed of fused tergite I and sternite I; tergite II, or at least apparent tergite II, the longest; functional spiracles and cerci on penultimate and apical tergites respectively; ovipositor almost entirely retracted.

Classification and distribution. The very large family Diapriidae is cosmopolitan and comprises about 150 genera classified in four subfamilies, Ambositrinae, Belytinae, Dipariinae and Ismarinae (Masner, 1976*d*; Naumann & Masner, 1985). The true size of the family is difficult to gauge as in most tropical areas very large numbers of species await description; worldwide it contains several thousand described species.

Biology. Diapriids are most commonly encountered in humid, shaded areas and in such situations the range of taxa encountered may be very diverse, and several species can be extremely common. Despite their abundance very few species have been reared and thus our knowledge of diapriid biology is extremely fragmentary. The more primitive diapriids (Belytinae and Ambositrinae) are probably parasitoids of mycetophilid and sciarid dipterous larvae or puparia found in macrofungi, leaf litter, holes in the ground and similar situations (Chambers, 1971; Naumann, 1982, 1988). Belytines can often be found ovipositing into the undersides of macrofungi, with their terminal metasomal segments extruded to form a semihyaline tube (Huggert, 1979).

Species of the small specialized subfamily Ismarinae develop as hyperparasitoids of Cicadellidae (Homoptera) via the larvae of Dryinidae (Chambers, 1955; Waloff, 1975; Jervis, 1979), but most of the more derived diapriids, the Diapriinae, are parasitoids of higher Diptera (Brachycera and Cyclorrhapha). Some have apparently switched to non-dipterous hosts, such as the larvae of staphylinid beetles found in similar habitats.

A large group of neotropical Diapriini are associated with ant nests, where they are parasitoids of ant-associated fly larvae (Paulson & Akre, 1991) or of ant larvae (Loiácono, 1987). Several of these species are symphilic myrmecophiles, mimicking the host ants in both morphology and behaviour (Huggert & Masner, 1983). Ecitonine army ants, in particular, seem to host a variety of highly specialized diapriines. After the dispersal flight the female wasp casts off her wings and becomes integrated into the ant colony.

What little detailed information that is available about diapriid biology pertains mostly to species of Diapriinae. Members of this subfamily are koinobiont endoparasitoids, with oviposition and development usually occurring within dipterous pupae in the puparium (Sanders, 1911; Pemberton & Willard, 1918*b*; Roberts, 1935; Simmonds, 1952; Knutson & Berg, 1963). Most probably oviposit into young pupae and some diapriines are known to select a particular site on the puparium in which to oviposit. For example, *Diapria conica*, a parasitoid of *Eristalsis tenax* (Syrphidae), always oviposits directly behind the respiratory horns in the suture forming the cap of the puparium (Sanders, 1911). The afrotropical species, *Trichopria stratiomyiae*, oviposits into stratiomyiid larvae and the host is killed before reaching the pupal stage (Cros, 1935).

Many diapriids are gregarious parasitoids, with from about 15 to over 300 individuals developing in a single host (Cros, 1935; Roberts, 1935; Paulson & Akre, 1991). In *Diapria conica* the female's full complement of mature eggs is deposited in one host, with oviposition requiring from one to three hours (Sanders, 1911). Although most diapriines are primary parasitoids some species of *Lepidopria* and *Trichopria* are hyperparasitoids developing in the puparia of tachinids inside the primary hosts (Huggert & Masner, 1983). *Coptera silvestrii* is a primary parasitoid of tephritid puparia, but if an opiine braconid or tetrastichine eulophid is already present, the female diapriid places her egg within the body of these parasitoids, thus developing as a facultative hyperparasitoid (Pemberton & Willard, 1918*b*).

The first instar larva of *C. silvestrii* has a large, heavily sclerotized head with very large curved mandibles (Pemberton & Willard, 1918*b*). Later instars are of more normal form without the large head. The tracheal system, with three pairs of thoracic spiracles, does not develop until the third instar. Emergence of the adult is accomplished by pushing off the cap of the tephritid puparium, and thus parasitized puparia are not easily distinguished from unparasitized puparia.

The period from oviposition to adult emergence has been reported to require from 18 days (Knutson & Berg, 1963) to nearly 40 days (Sanders, 1911). In at least some species the adults are said to be relatively long-lived (Sanders, 1911), but there is very little information available on adult biology. In *Diapria conica* courtship consists of the male placing his antennal sensory organs, located on the fourth segment, over the club of the female antenna (Sanders, 1911).

Economic importance. Diapriids have occasionally been used in classical biological control programmes against dipterous pests (Clausen, 1978). *Basalys tritoma* was imported from Europe to Canada to control the carrot rust fly (*Psila rosae*, Psilidae) but failed to become established. *Coptera silvestrii* was imported from Africa to Hawaii as part of the major campaign to control the Mediterranean fruit fly and a species of *Trichopria* was imported from Jamaica to California to control eye gnats (*Hippelates* spp., Chloropidae); neither was very successful. In Costa Rica a native species of *Trichopria* has been found attacking fruit flies (Tephritidae: *Anastrepha* spp.) in mango and guava (Jirón & Mexzon, 1989).

Identification. Currently there are no keys to the genera of neotropical Diapriidae.

Key to subfamilies of Diapriidae present in Central America

1 Metasomal tergites with sharp lateral edges, with laterotergites abruptly flexed underneath (Fig. 9.33); metasoma generally depressed dorsoventrally, distinctly wider than high; metasoma of female with only five, of male with only six visible tergites; sternite II much shorter than sternite III. .... **Ambositrinae** (p. 227)

— Metasomal tergites rounded laterally, laterotergites absent (Fig. 9.34); metasoma generally cylindrical, at most slightly wider than high; metasoma of female with more than five, of male with more than six visible tergites; sternite II of metasoma much longer than sternite III. 2

2 Antenna of female with 13 or fewer segments, antenna of male with 13 or 14 segments; fore wing with no marginal cell and vein *Rs* at most rudimentary (Fig. 9.39). .... **Diapriinae** (p. 227)

— Antenna of female with 15, of male with 14 segments; fore wing with marginal cell developed (open or closed), *Rs* present as true vein (Fig. 9.40). 3

3 Notauli well-developed and complete (Fig. 9.35); antennal sockets protruding, positioned on distinct ledge (Fig. 9.37); male antennal segment 3 modified; hind tibia not thickened. .... **Belytinae** (p. 226)

— Notauli absent, reduced to anterior pits (Fig. 9.36); antennal sockets not protruding, ledge not developed (Fig. 9.38); male antennal segment 4 (rarely also 3) modified; hind tibia thickened. .... **Ismarinae** (p. 227)

Synopsis of the Costa Rican fauna

There are over 30 genera present in Costa Rica, several of which are yet undescribed. Nine of these genera are very speciose, with more than 50 species each in the country, ten genera comprise between 12 and 50 species each, while the remaining genera are small. In total there appear to be approximately one thousand species in the country.

Diapriidae occur in all major habitats in Costa Rica, but they are especially abundant in shaded areas with high humidity. They are less common in deforested areas. Representatives of the family occur at all altitudes and are remarkably well represented at very high elevations (above 2500 m), where most of the brachypterous forms occur. Most diapriids inhabit the forest floor, where they are probably searching for hosts in leaf-litter, decaying vegetation and on fungi. Some species actually occur within the soil and several are highly modified associates of ant colonies (Masner, 1976*b*; Huggert & Masner, 1983). A few species occur near stagnant or running water and enter the water.

Where rainfall is fairly constant throughout the year diapriids occur year round, but in areas with a pronounced dry season adults of some species emerge in large numbers at the onset of the rainy season. In

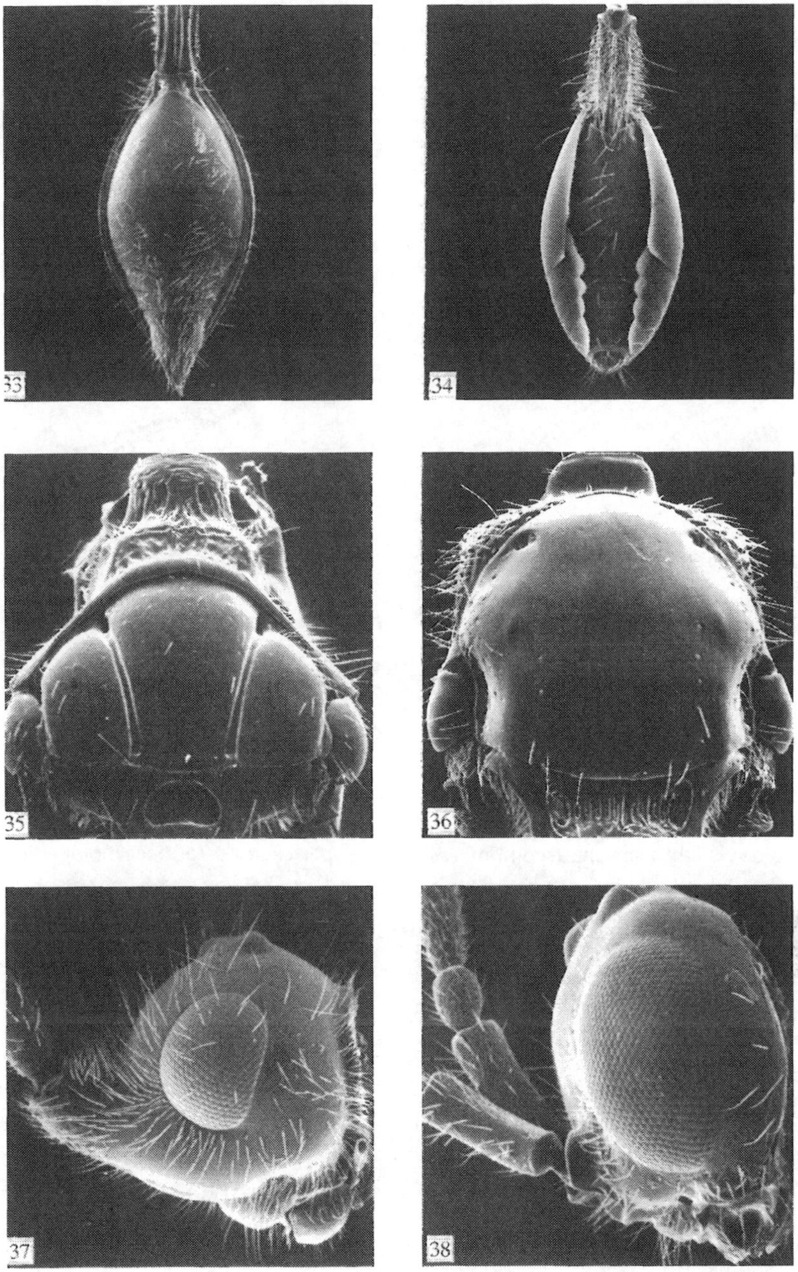

Figs 9.33–9.38. Scanning electron photomicrographs of Diapriidae. Figs 9.33–9.34. Metasoma, ventral; 9.33, *Dissoxylabis* sp. (Ambositrinae); 9.34, *Trichopria* sp. (Diapriinae). Figs 9.35–9.36. Mesoscutum; 9.35, *Belyta* sp. (Belytinae); 9.36, *Ismarus* sp. (Ismarinae). Figs 9.37–9.38. Head, showing antennal insertion; 9.37, *Belyta* sp. (Belytinae); 9.38, *Ismarus* sp. (Ismarinae).

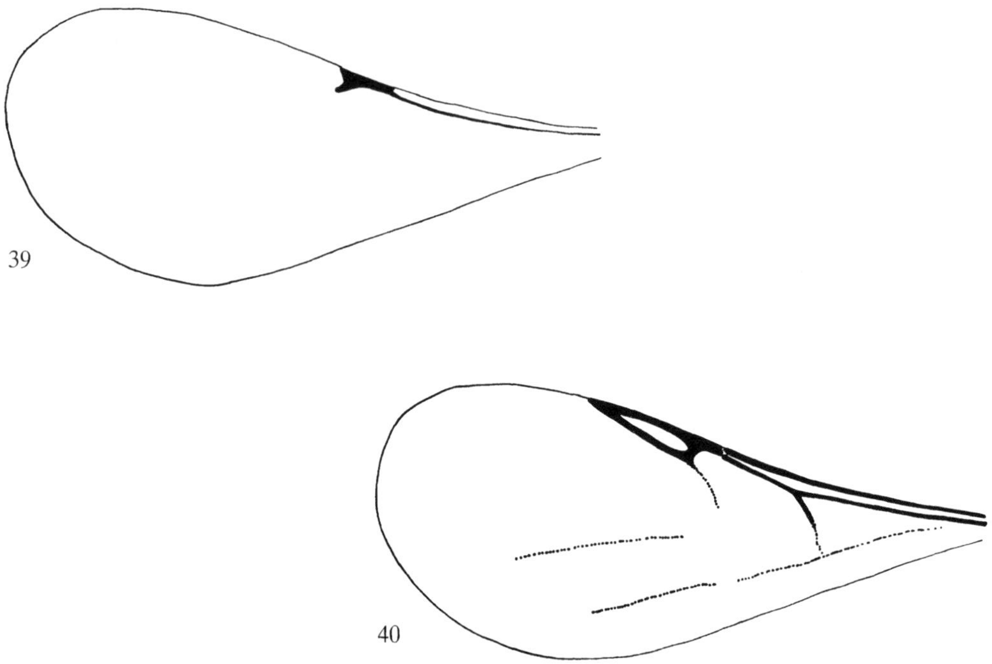

Figs 9.39–9.40. Diapriidae, fore wing; 9.39, Diapriinae; 9.40, Belytinae.

general the larger species are strong fliers and are readily captured in Malaise traps, especially males. Smaller species and brachypterous species are more effectively collected by yellow pan traps, while the ant-associates often come to light traps in large numbers.

Belytinae

The Belytinae is a very large, and morphologically a very homogeneous group. Little is known of the biology of any species, although a few have been reared from mycetophilid puparia (Muesebeck *in* Krombein *et al.*, 1979; Gauld & Bolton, 1988). It is the least derived subfamily of Diapriidae as species have plesiomorphic wing venation and relatively unspecialized antennae and metasomas. Brachypterous forms are very rare. Generic concepts of neotropical Belytinae are presently quite arbitrary. In addition to the following genera there are several undescribed genera of Belytinae in Costa Rica that are related to South American taxa.

Aclista. A moderately large genus with an estimated 20 species in Costa Rica. Biology: one North American species reared from dipterous larvae in Boletaceae (Muesebeck *in* Krombein *et al.*, 1979).

Belyta. A large, cosmopolitan genus, with about 60 species in Costa Rica. Biology: unknown.

Cinetus. A large genus with about 60 species in Costa Rica. Biology: a North American species is thought to parasitize dipterous larvae in bark beetle burrows (Muesebeck *in* Krombein *et al.*, 1979).

Miotella. A primarily neotropical genus represented by numerous species in South America and a few in the Nearctic region. At least 15 species occur in Costa Rica. Biology: unknown.

Pantoclis. A large genus with about 60 species occurring in Costa Rica. Biology: unknown.

Ismarinae

A small subfamily comprising a single widespread genus, *Ismarus* (Masner, 1976*d*). Both morphologically and biologically this is the most aberrant diapriid subfamily. The Ismarinae are structurally unique amongst Diapriidae in that the antennae are positioned rather low on the face, with the antennal ledge not developed, and the scape very short; internally they lack the tergal component of the mesotrochanteral depressor muscle that is present in the other subfamilies (Gibson, 1985). The mesosoma is relatively short and highly convex, and the metasoma is distinctly campanulate, with the tergites fused to varying degrees into a carapace.

Ismarus. Three species of this genus, *I. dux*, *I. gracilis* and *I. helavai*, have been reported from Costa Rica and/or Panama (Masner, 1976*d*), but the genus is probably represented by up to ten species in Central America. Keys to described species are given by Masner (1976*d*). Biology: parasitoids of Dryinidae (Chambers, 1955; Waloff, 1975; Jervis, 1979).

AMBOSITRINAE

The Ambositrinae is primarily a southern hemisphere group of diapriids (Masner, 1969), with about 20 genera present in Australia, New Zealand, Africa and America (Naumann, 1982, 1988). Members of this subfamily share some ancestral character states with the Belytinae (e.g. structure of the antennae), but they differ from all diapriids in the unique structure of the metasoma, especially the small sternite II and the fusion of three tergites to form the large syntergite II. Only two genera of Ambositrinae are represented in Costa Rica; one species of each genus extends into the Nearctic region.

Virtually nothing is known of the biology of any ambositrine, except that one New Zealand species has been reared from a mycetophilid puparium. Masner (1969) and Naumann (1982) both suggest that these and similar dipterans probably serve as hosts for all species.

Dissoxylabis. A New World genus ranging from Chile to the southern United States. It is represented by about 20 species in Costa Rica. Biology: unknown.

Propsilomma. A New World genus ranging from Chile and Patagonia northward to Canada. It is represented by about ten species in Costa Rica. Biology: unknown.

DIAPRIINAE

This is structurally the most derived and most diversified subfamily of the Diapriidae. The higher level (tribal) classification of the Diapriinae is not yet stabilized, but four tribes are recognizable: Psilini, Spilomicrini, Basalyini and Diapriini. The Psilini have the primitive condition of the large metasomal tergite composed of a single sclerite, whereas in all other members of the subfamily tergites II and III are fused into a single, large syntergite.

Four tribes and 21 genera of Diapriinae occur in Costa Rica. Not listed below is an enigmatic, undescribed genus that may constitute a separate subfamily; it occurs in lowland tropical rainforests of South and Central America, West Africa, and the Philippines.

DIAPRIINAE: Basalyini. Three genera of this tribe are present in Central America.

Basalys. A cosmopolitan genus with about 30 species in Costa Rica. Biology: hosts are species of Chloropidae and Anthomyiidae.

Caecopria. A neotropical genus with about 25 species occurring in Costa Rica. Biology: some species are associated with ant nests (Loiácono, 1981).

Doliopria. A New World genus with approximately 60 species occurring in Costa Rica. Biology: some species of *Doliopria* are evidently associated with ant nests, probably as parasitoids of Diptera.

DIAPRIINAE: Diapriini. In addition to the following eight genera, there are also three undescribed genera (comprising 20 species in total) associated with ants.

Acanthopria. A neotropical genus with approximately 60 species present in Costa Rica. Biology: some species are possibly associated with ants.

Corynopria. A cosmopolitan genus with approximately ten species present in Costa Rica. Biology: parasitoids of brachycerous Diptera.

Gymnopria. A neotropical (Argentina, Brazil) genus with one species collected in Costa Rica. Biology: myrmecophiles; one Argentine species has been reared from the larva of a species of the leaf-cutting ant genus *Acromyrmex* (Loiácono, 1987).

Labidopria. A neotropical genus with a single species in North America. Approximately 15 species are present in Costa Rica. Biology: associated with ants, and two species are recorded from the nest of the ecitonine ant *Labidus praedator* (Huggert & Masner, 1983).

Megaplastopria. A neotropical genus represented in Costa Rica by approximately ten species. Biology: possibly parasitoids of brachycerous Diptera.

Mimopria. A neotropical genus with about ten species represented in Costa Rica. Biology: found in association with ecitonine army ants (Masner, 1976*b*), but the exact nature of the association is not known.

Trichopria. A cosmopolitan genus represented by about 80 species in Costa Rica. Biology: parasitoids of a wide range of Diptera (Muesebeck *in* Krombein *et al.*, 1979): Stratiomyidae, Tabanidae, Syrphidae, Agromyzidae, Chloropidae, Drosophilidae, Ephydridae, Sciomyzidae, Tephritidae, Calliphoridae, Sarcophagidae and Tachinidae. Several are known to develop gregariously (e.g. Cros, 1935). A North American species has been reared from *Microdon* (Syrphidae) in ant nests (Paulson & Akre, 1991), and another occurs in the nests of the ant, *Labidus coecus*, but its host is unknown.

Xanthopria. A neotropical genus represented by about five species in Costa Rica. Biology: associated with ants, but the exact nature of the association is not known.

DIAPRIINAE: Psilini. A single genus of this tribe is present in Central America.

Coptera. A large cosmopolitan genus with approximately 100 species found in Costa Rica so far. Biology: hosts are species of Tephritidae (Muesebeck *in* Krombein *et al.*, 1979). One African species, *C. silvestrii*, develops as an endoparasitoid of tephritid puparia, or facultatively as an endophagous hyperparasitoid.

DIAPRIINAE: Spilomicrini. In addition to the five genera listed below there is an undescribed genus (represented by 15 species), some members of which exhibit unusually coarse sculpture on the mesosoma.

Entomacis. A cosmopolitan genus with about 60 species known to occur in Costa Rica. Biology: a North American species has been reared as a parasitoid of *Forcipomyia* (Ceratopogonidae) in an old ant nest. Another North American species has been taken in nests of *Aphaenogaster* (Formicidae) (Muesebeck *in* Krombein *et al.*, 1979).

Idiotypa. A cosmopolitan genus with about 40 species known to occur in Costa Rica. Biology: unknown.

Pentapria. A predominantly New World genus with isolated species in Japan and Australia. Approximately 100 species have been seen from Costa Rica. Biology: hosts are species of Stratiomyidae (Fouts, 1939).

Paramesius. A cosmopolitan genus with about 15 species known to occur in Costa Rica. Biology: unknown.

Spilomicrus. A large cosmopolitan genus with about 100 species found in Costa Rica. The North American species were revised by Masner (1991*b*). Biology: hosts are species of Syrphidae and Muscidae, and possibly Staphylinidae.

9.6 SCELIONIDAE

Diagnosis. Body 0.5 to 10.0 mm long, usually 1.0 to 2.5 mm; body shape extremely variable, varying from very thin and elongate to very short and robust, depending on the shape of the host egg; colour most commonly black, but sometimes yellow or bicoloured with head black, mesosoma reddish orange, metasoma black; usually with distinctive surface sculpture, but sometimes smooth in smaller species. Some species are brachypterous or apterous, especially females. *Antenna attached immediately above dorsal margin of clypeus, distinctly geniculate,* usually with 12 segments (occasionally with 11, 10, 9 or 7), *scape elongate,* female antenna clavate, male antenna thread-like with segment 5 modified (rarely with additional segments modified). Fore wing with submarginal vein usually reaching anterior margin of wing to continue as marginal vein (sometimes thickened into dark spot), *stigmal and postmarginal veins usually present;* hind wing of most genera with complete submarginal vein that reaches the hamuli; wings very rarely without veins or veins indistinct. *Metasoma usually moderately to strongly depressed dorso-ventrally;* metasomal segments mostly subequal in length but one segment sometimes much larger than the others; apparent segment VII in female either external or internal, with minute cerci or sensory plates, extruding with ovipositor or attached to apparent tergite VI and not extruding with ovipositor; males with eight tergites.

Classification and distribution. The family Scelionidae occurs worldwide and comprises several thousand species in roughly 150 genera, which are classified in three subfamilies, the Scelioninae, Teleasinae and Telenominae (Masner, 1976*a*). The first two subfamilies are apparently closely related whereas the third is quite different.

Biology. All scelionids are idiobiont endoparasitoids in the eggs of insects and spiders, with development always being completed within a single host egg. Most scelionids are solitary parasitoids, but gregarious development, with five to ten individuals completing development in a single host egg, is encountered in some Telenominae. Members of the subfamily Scelioninae attack the eggs of a wide range of insects (including those of the Hemiptera, Orthoptera, Embiidina, Mantodea) and spiders, whereas the Teleasinae appear to be restricted to coleopteran eggs, and Telenominae predominantly to

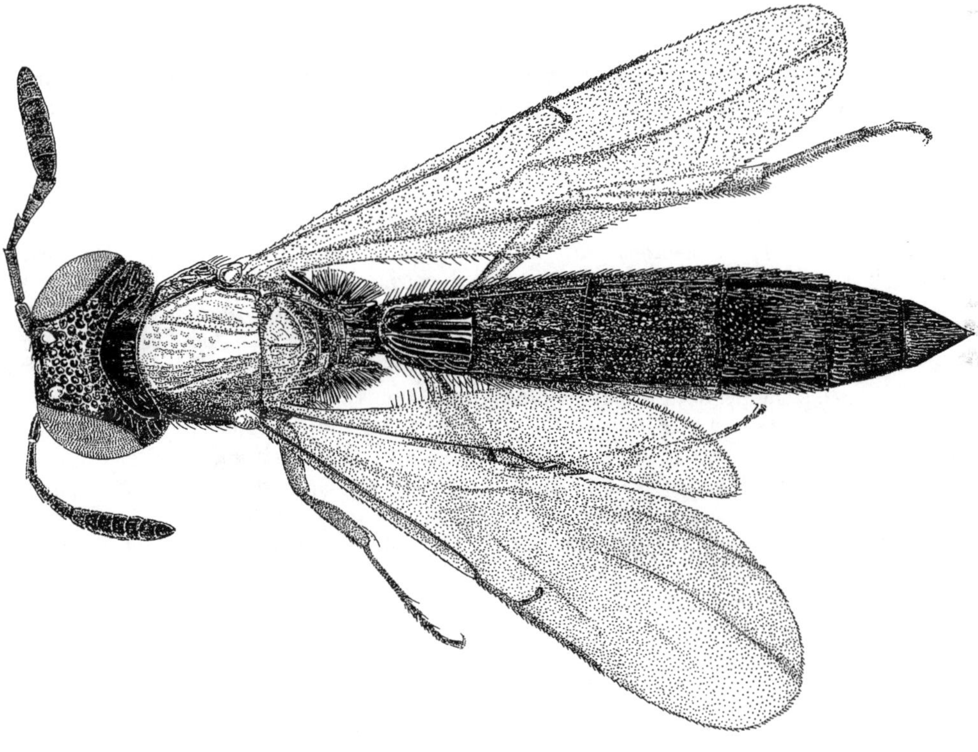

Fig. 9.41. *Chromoteleia* sp. (Scelionidae).

hemipteran and lepidopteran eggs. Most species show considerable host-specificity; many are restricted to a single host species and most are restricted to a single host family. However, such apparent host-specificity may simply reflect a lack of study (see, for example, Fedde, 1977). Only occasional records exist of scelionids ovipositing into eggs of species from several families, but in these cases the hosts always belong to the same order (N.F. Johnson, 1984).

Generally the female scelionid searches for host eggs in only one particular segment of the environment. For example, species of *Leptoteleia* seem to be restricted to oecanthine gryllids that oviposit into stems and crevices in the bark of plants (Masner, 1978), while species of *Duta* attack the eggs of nemobiine gryllids in the soil (Masner, 1991*a*). Species of *Tiphodytes* parasitize the eggs of gerrids on floating or submerged vegetation (Spence, 1986). Even within a narrow habitat, individual species often show a preference for a particular microhabitat. For example, *Telenomus sphingis* appears to spend more time searching for eggs of *Erinnyis ello* (Sphingidae) on the upper sides of leaves than on the under sides (Gold *et al.*, 1989). Scelionids attack both host eggs that are clumped (e.g. Eberhard, 1975), and those that are deposited solitarily (e.g. D.C. Allen, 1972). Some scelionids search for either eggs defended by the female insect that laid them, or gravid female hosts about to oviposit. For example, in Colombia *Phanuropsis semiflaviventris* and *Trissolcus bodkini* are attracted to immobile individuals of *Antiteuchus tripterus* (Pentatomidae), and attack the egg masses being guarded by the female pentatomid (Eberhard, 1975).

Other scelionid females are phoretic, attaching themselves to the body of the adult host and disembarking when the host is ovipositing. For example, North American species of *Synoditella* use their mandibles to attach to the intersegmental membranes of grasshoppers (Lanham & Evans, 1960), *Epigryon* attach to the ventral surface of the abdomen of Phymatidae (Masner, 1980*c*) and *Telenomus calvus* attaches with its tarsi to various locations on a pentatomid (Orr *et al.*, 1986). This behaviour allows the parasitoid access to freshly laid eggs (Clausen, 1976). Both *Synoditella* and *Epigryon* are represented in Costa Rica.

Virtually all scelionids are primary parasitoids although at least one European species has been reported to develop as a facultative hyperparasitoid (Viktorov, 1966). Most species are solitary and many strongly avoid superparasitism. Discrimination of previously parasitized hosts occurs via detection of an external marking pheromone or via an internal marker. After oviposition the female wasp applies an external marker, originating from her accessory gland (Strand, 1985), by wiping the extruded ovipositor across the surface of the host egg. The external marker is active immediately following deposition, but lasts only 12 hours to three days (Bosque & Rabinovich, 1979; Klomp *et al.*, 1980). Internal marking appears to be related to necrosis of the host egg (see below) and differs from the external marker in that it does not take effect immediately, persists throughout development, and is recognizable between taxa (Strand, 1986). When superparasitism does occur, it is always while the external marker is still detectable, but before the internal marker has taken effect.

In *Telenomus heliothidis* the ovipositing female is known to introduce a venom into the host egg which stops further host development (Strand *et al.*, 1986). Teratocytes, which are liberated by the scelionid egg membrane at the time of hatching, produce secretions that result in necrosis of the host egg (Strand, 1986). These secretions also destroy eggs of other parasitoids that happen to be present.

The first instar larva is very characteristic, without visible segmentation, but with the body constricted centrally so that two parts, a cephalothorax and abdomen, are discernible. This so-called teleaform larva has very large mandibles, and is frequently furnished with well developed processes on the cephalothorax and a long caudal appendage. The mandibles are apparently used in intraspecific combat between two scelionid larvae that hatch at roughly the same time (Strand, 1986). There are normally three larval instars and most of the feeding occurs during the last two. The second instar, at least in some species, is sacciform, and the mature larva is typically hymenopteriform (Schell, 1943; Hidaka, 1958; Gerling, 1972; Gomez de Picho, 1987). Pupation takes place in the host egg.

In those species that attack clusters of eggs most individuals that emerge are females (Waage, 1982). The males emerge first and often fight for supremacy on the egg mass (F. Wilson, 1961; Eberhard, 1975), where they then mate with their sisters as they emerge (N.F. Johnson, 1984). Courtship and mating behaviour has been described for several species, including *Phanuropsis semiflaviventris* and *Trissolcus bodkini* (Eberhard, 1975).

In the Temperate region many species of Scelionidae have adult emergence patterns that are well synchronized with the oviposition period of their host, though this is not universal (see review by Orr, 1988). A few temperate species are very long-lived as adults (Grijpma, 1984), surviving over the winter to attack host egg masses laid the following year (Torgersen & Ryan, 1981; Grijpma *et al.*, 1991). Such prolonged adult longevity also seems to occur in the seasonal parts of the tropics. For example, in the seasonally dry parts of Costa Rica a telenomine has been observed to attack egg masses of a pentatomid species, develop rapidly and emerge. It apparently passes the long dry season as an adult (Gauld, pers. comm.). Another telenomine was observed forming small aggregations on the same leaf nightly during the dry season and promptly disappeared when the rains began (W. Eberhard, pers. comm.).

In contrast with the generally uniformly black or brownish coloured scelionids found in the Old World tropics, many neotropical Scelionidae (in Costa Rica 11 genera, see below) show very striking colour patterns, with a bright reddish orange mesosoma. Most of these brightly patterned species are 3–10 mm long and occur at altitudes below 2000 metres in somewhat open habitats above the ground (e.g. on shrubs). The bright reddish orange mesosoma may be an aposematic coloration that provides protection against predators. The evidence for this coloration being aposematic is still only circumstantial, but it interesting that a similar colour pattern occurs in some Braconidae, Evaniidae, Symphyta, Mutillidae, Ichneumonidae and in the sphecid genus *Stigmus*.

Economic importance. Scelionids, particularly telenomines, have been used quite successfully in classical biological control programmes, directed principally against pest Hemiptera and Lepidoptera. Twelve scelionid species have become established following releases against 11 target pests; on six of the 23 separate occasions in which establishment occurred, the target species was effectively controlled (Greathead, 1986). *Trissolcus basalis* and *Telenomus remus* have been introduced into many different countries to control *Nezara viridula* (Pentatomidae) and noctuids, respectively (Clausen, 1978), and several species of *Telenomus* and *Trissolcus* are important factors in integrated pest management systems to control soybean stink bugs in Brazil (Kobayashi & Cosenza, 1987). Information on the biology of *T. basalis* can be found in Buschmann and Whitcomb (1980) and Powell *et al.* (1981); information on *T. remus* can be found in Gerling (1972) and Gomez de Picho (1987). One of the most successful biological control programmes involving scelionids was the introduction of *Telenomus alsophilae* from North America to Colombia to control *Oxydia trychiata*, a geometrid forest pest (Bustillo & Drooz, 1977; Drooz *et al.*, 1977).

Several telenomines that occur in Central America could possibly be used in non-classical biological control, via mass rearing or environmental modifications. For example, *Telenomus solitus* attacks eggs of *Trichoplusia ni* (Noctuidae) in Guatemala (Johnson, 1983). In Costa Rica eggs of the cassava hornworm (*Erinnyis ello*, Sphingidae) are attacked by three species of *Telenomus* (Johnson, 1990)—*T. connectans, T. dilophonotae* and *T. monilicornis*—which possibly search for hosts in different manners (Gold *et al.*, 1989). Several Central American pentatomid pests are also attacked by scelionids (Table 9.1). In Central America there has been some study of the feasibility of using *Telenomus fariai* and other telenomines against reduviid vectors of Chagas disease, but with mixed results (Zeledón, 1957; Montesinos & Rabinovich, 1979).

Orr (1988) has recently summarized the biological and ecological attributes of scelionids with respect to their potential use in biological control. In host searching ability *Telenomus* species have been reported to be generally superior to *Trichogramma* species. Two species with wide host ranges, *Telenomus alsophilae* (17 host genera) and *Trissolcus basalis* (14 host genera), have been utilized for the two most spectacular classical biological control successes involving scelionids (see above). With respect to temporal synchronization, dispersal capacity, and density responsiveness, scelionids also appear to be very suitable for biological control. Some components of the sex pheromone of *Spodoptera frugiperda* stimulate host-seeking activity by *Telenomus remus*. Thus applications of sex pheromones to cause mating disruption in the pest, would also serve to retain scelionid parasitoids in the crop (Nordlund *et al.*, 1983).

Pentatomid	Parasitoid
Antiteuchus costaricensis	*Phanuropsis semiflaviventris*
	Telenomus sp.
	Trissolcus radix
Edessa rufomarginata	*Trissolcus pustulans*
	Trissolcus urichi
Nezara viridula	*Trissolcus brochymenae*
	Trissolcus hullensis
	Trissolcus urichi
Piezodorus guildinii	*Trissolcus brochymenae*
	Trissolcus urichi

Table 9.1. Central American pentatomid pests and their telenomine parasitoids. Data from Johnson (1985*b*, 1987); records from *A. costaricensis* on macadamia, P. Hanson (pers. comm.).

Identification. Keys to the world genera of Scelionidae are given by Masner (1976*a*) and to the holarctic genera by Masner (1980*c*), and it is from these that the following key is modified.

Key to subfamilies of Scelionidae and to tribes of Scelioninae occurring in Costa Rica

1 Fore wing with marginal vein 2.5 to 4.0 times as long as stigmal vein (Fig. 9.42); metasomal tergite III by far the longest and widest tergite, as long as or longer than the following tergites combined (Fig. 9.44); lateral ocellus remote from eye margin by at least 2.0 times its own diameter, often closer to anterior ocellus than to eye margin (Fig. 9.47). .... **Teleasinae** (p. 239)

— Fore wing with marginal vein at most twice as long as stigmal vein, usually much shorter (Fig. 9.43); tergite III variable but always shorter than the combined length of the following tergites (Figs 9.45, 9.46); lateral ocellus usually closer to eye margin (Fig. 9.48). 2

2 Metasoma with wide laterotergites (Fig. 9.49) loosely attached to sternites, hence no impressed submarginal ridge; tergite II by far the largest of all the metasomal tergites (Fig. 9.45); female antenna usually 11-segmented (rarely 10-segmented), male antenna 12-segmented. .... **Telenominae** (p. 239)

— Metasoma with narrow laterotergites (Fig. 9.50) sharply impressed into sternites, with distinct submarginal ridge; tergite II seldom the largest tergite, metasoma usually with segments of more or less equal size (Fig. 9.46); female antenna usually 12-segmented, rarely with fewer segments (6–11), male antenna 12-segmented, rarely with fewer segments (8–11). ... . **(Scelioninae)** 3

3 Metasoma with tergite II longer than tergite III. [Body often short and plump, less than 3.0 mm in length.] 4

— Metasoma with tergite II at most as long as tergite III..7

4 Body smooth, with scattered long bristles; fore wing with postmarginal vein almost reaching wing apex, marginal cilia at least as long as width of wing. .... **Embidobiini** (*Embioctonus*) (p. 237)

— Body sculptured, with short bristles; fore wing with postmarginal vein and marginal cilia shorter than width of wing. 5

5 Scutellum rectangular, barely longer than metanotum (Fig. 9.51). [Females yellow, males dark brown; eyes often iridescent green; body flattened.] .... **Aradophagini** (p. 236)

— Scutellum semicircular, much longer than metanotum (Fig. 9.52). 6

6 Fore wing twice as long as metasoma, with trace of basal vein visible (Fig. 9.53). .... **Baeini** (*Baeus*, males) (p. 236)

— Fore wing at most 1.5 times as long as metasoma, with no trace of basal vein. .... **Gryonini** (p. 237)

7 Female antenna 7-segmented, club unsegmented (Fig. 9.54); male antenna with 11 or fewer segments. [Body less than 3.0 mm.] 8

— Female antenna with segmented club and thus more than 7 (usually 12) segments; male antenna 12-segmented. 9

8 Skaphion present (cf Fig. 9.52). ....**Thoronini** (*Thoronidea*) (p. 239)
— Skaphion absent. .... **Baeini** (p. 236)

9 Metasoma 'stepped' in lateral view, i.e. at least with tergites II and III markedly depressed anteriorly (Fig. 9.55). .... **Cremastobaeini** (p. 237)
— Metasoma smooth in lateral view (cf Fig. 9.50)... 10

10 Hind wing with submarginal vein incomplete, not reaching fore margin of wing (Fig. 9.56). 11
— Hind wing with submarginal vein complete, reaching fore margin of wing, or apterous (Fig. 9.57). 13

11 Mid and hind tibiae with two apical spurs. [Eyes with long scattered hairs; fore wing with the postmarginal vein absent.] .... **Sparasionini** (*Sceliomorpha*) (p. 239)
— Mid and hind tibiae with one apical spur. 12

12 Back of head with dense tufts of white setae (Fig. 9.31); metasoma mostly smooth..................... .... **Doddiellini** (p. 237)
— Back of head without dense tufts of setae; metasoma mostly sculptured. .... **Scelionini** (p. 238)

13 Frons with depression (Fig. 9.58), which has transverse striae, or is laterally carinate (Fig. 9.59). 14
— Frons not distinctly concave and laterally carinate. 16

14 Fore wing with marginal vein longer than stigmal vein (Fig. 9.60).................................... .... **Psilanteridini** (*Leptoteleia*) (p. 238)
— Fore wing with marginal vein shorter than stigmal vein (Fig. 9.61). 15

15 Body about 1 mm long, yellowish........................ .... **Embidobiini** (*Palaeogryon*) (p. 237)
— Body 2.5 mm or more in length, dark. .... **Baryconini** (*Baryconus*) (p. 236)

16 Skaphion present (cf Fig. 9.52). 17
— Skaphion absent. 19

17 Smooth and shining, without microsculpture. .... **Thoronini** (*Neothoron*) (p. 239)
— Body with microsculpture. 18

18 Metanotum extending posteriorly as semicircular or triangular plate (Fig. 9.62); notauli complete; fore wing with marginal vein shorter than stigmal vein (Fig. 9.61)........................... .... **Baryconini** (*Chromoteleia*) (p. 236)
— Without the above combination.......................... .... most **Psilanteridini** (p. 238)

19 Female antenna with less than 12 segments. [Fore wing with postmarginal and stigmal veins longer than marginal vein.]............................. .... **Embidobiini** (*Embidobia*) (p. 237)
— Female antenna with 12 segments. 20

20 Scutellum with spine on posterior margin........... .... **Psilanteridini** (*Spiniteleia*) (p. 238)
— Scutellum without spine. 21

21 Body not elongate, metasoma pedunculate (Fig. 9.63). .... **Psilanteridini** (*Anteris, Psilanteris*) (p. 238)
— Body elongate, metasoma not pedunculate (Fig. 9.64). .... **Calliscelionini** (p. 236)

Synopsis of the Costa Rican fauna

Scelionids occur at all elevations and in all major habitats in Costa Rica and they are especially diverse in undisturbed lowland rainforests and mid-elevation cloud forests. Elsewhere in the world some members of the family are known from such habitats as caves, deep soil, litter and fresh water.

All three subfamilies are represented in Costa Rica, by at least 47 genera in total. Seven of these genera each contain over 50 species in Costa Rica, and altogether there probably are between 1200 and 2500 species present in the country. The subfamilies and scelionine tribes are discussed further below, for simplicity, in alphabetical order. Certain taxa expected to occur in Costa Rica, but as yet not collected here, are also included, but denoted by the suffix ^{EX}.

SCELIONINAE

This is the largest and most variable subfamily, containing more than 90 percent of the genera in the family. It is currently subdivided into sixteen tribes

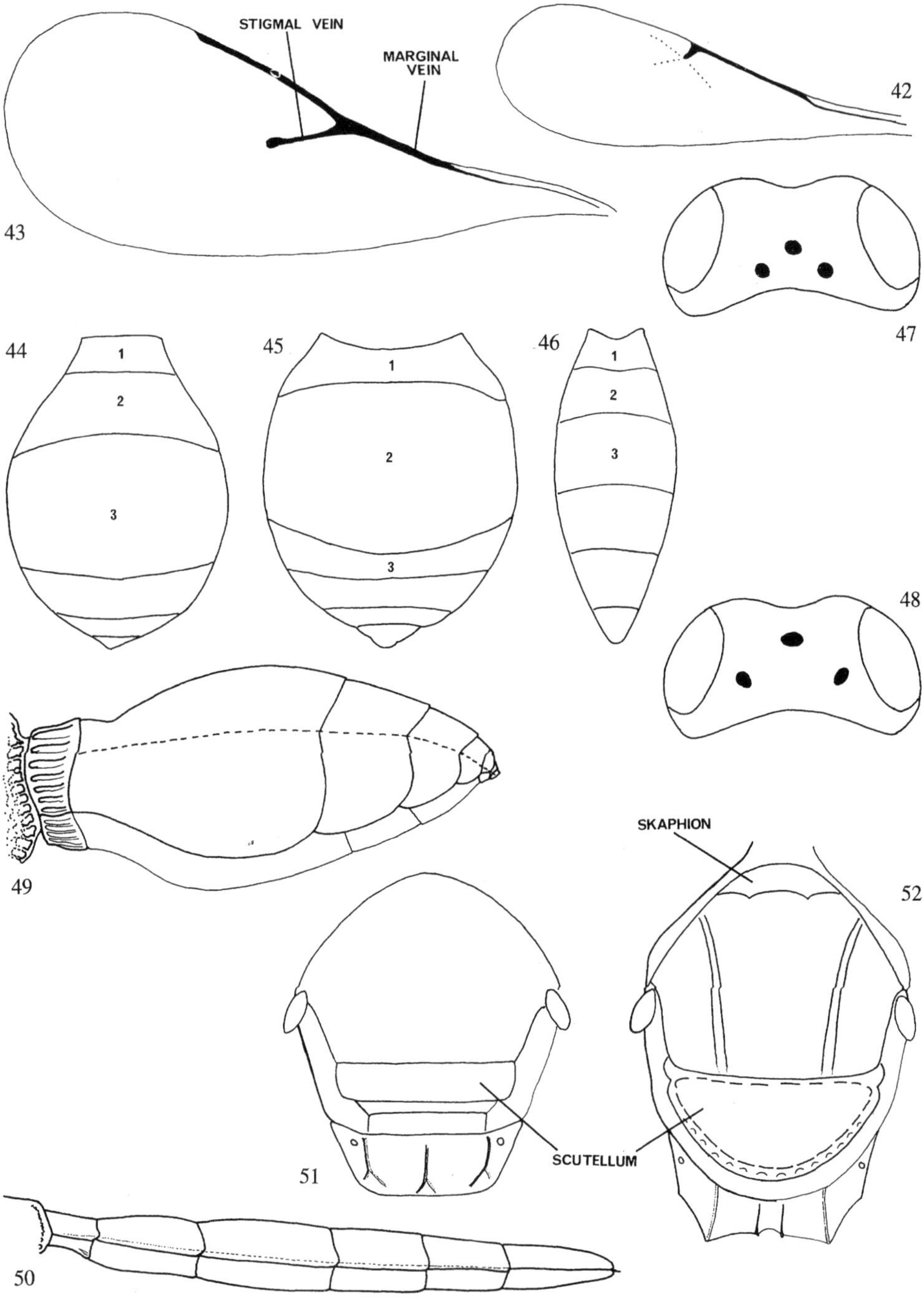

Figs 9.42- 9.52. Scelionidae. Figs 9.42–9.43. Fore wing; 9.42, Teleasinae; 9.43, *Telenomus* sp. (Telenominae). Figs 9.44–9.46. Metasoma, dorsal; 9.44, Teleasinae; 9.45, *Telenomus* sp. (Telenominae); 9.46, Scelioninae. Figs 9.47–9.48. Head, dorsal; 9.47, Teleasinae; 9.48, Scelioninae. Figs 9.49–9.50. Metasoma, lateral; 9.49, *Trissolcus* sp. (Telenominae); 9.50, *Macroteleia* sp. (Scelioninae). Figs 9.51–9.52. Mesosoma, dorsal; 9.51, *Aradophagus* sp. (Aradophagini); 9.52, Psilanteridini.

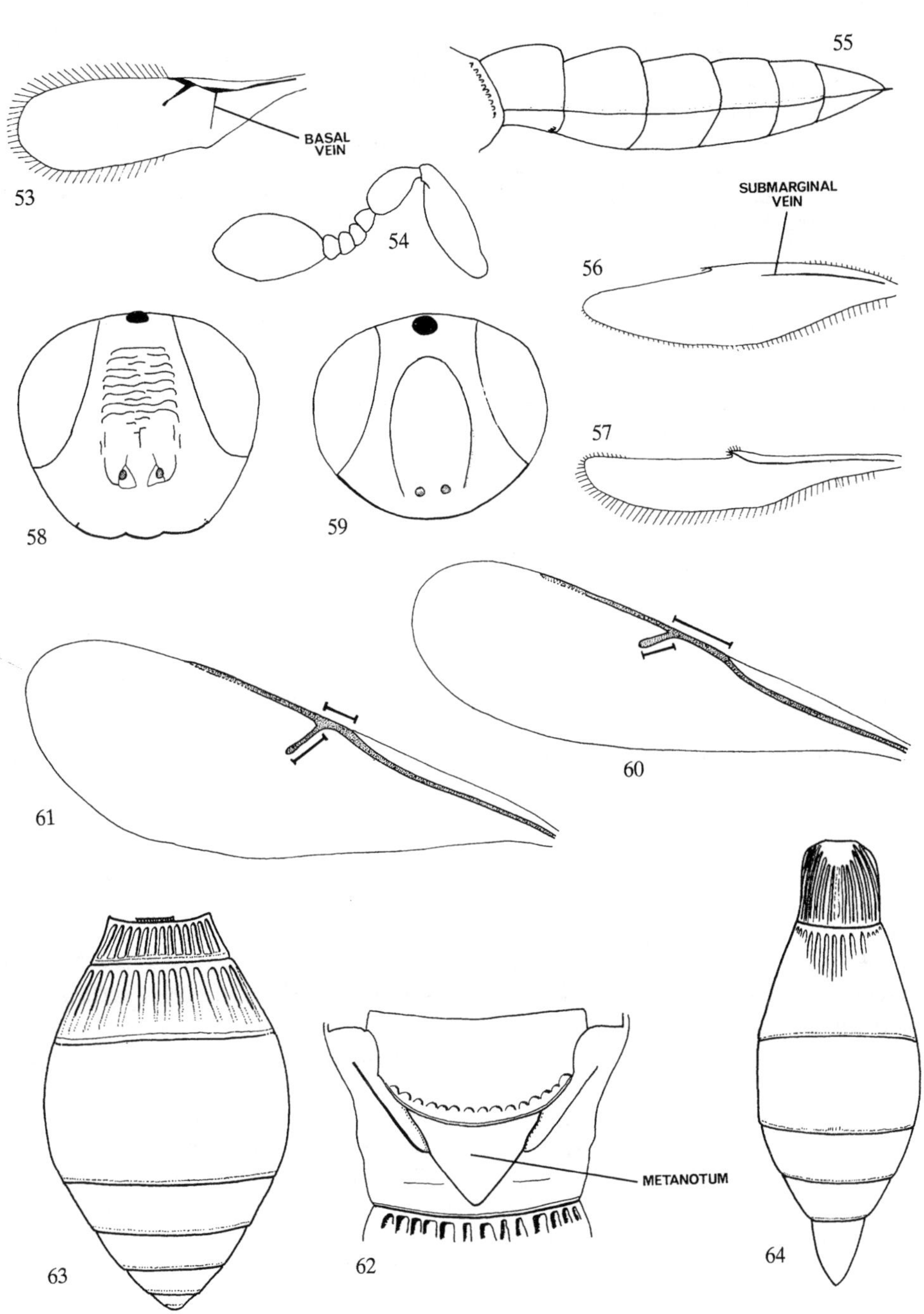

Figs 9.53–9.64. Scelionidae. Figs 9.53–9.54. *Baeus* sp.(Baeini); 9.53, fore wing of male; 9.54, antenna of female. Fig. 9.55. Metasoma, lateral, *Cremastobaeus* sp. (Cremastobaeini). Figs 9.56–9.57. Hind wing; 9.56, Sparasionini; 9.57, *Macroteleia* sp. (Calliscelionini). Figs 9.58–9.59. Head, anterior; 9.58, *Leptoteleia* sp. (Psilanteridini); 9.59, *Baryconus* sp. (Baryconini). Figs 9.60–9.61. Fore wing; 9.60, *Leptoteleia* sp. (Psilanteridini); 9.61, *Baryconus* sp. (Baryconini). Fig. 9.62. Posterior part of mesosoma, *Chromoteleia* sp. Figs 9.63–9.64. Metasoma, dorsal; 9.63, *Psilanteris* sp. (Psilanteridini); 9.64, *Calliscelio* sp. (Calliscelionini).

(Masner, 1976*a*). The Scelioninae of Costa Rica belong to 12 tribes and 39 genera. Following is a synopsis of the genera present in Costa Rica, with information on presence/absence of aposematic colour pattern ([1] = genera with some Costa Rican species having a bright reddish orange mesosoma, [2] = genera with all Costa Rican species having this colour pattern).

SCELIONINAE: Aradophagini. This small tribe includes three genera which occur widely in the Holarctic region, West Africa, southern India and Australia (Masner & Huggert, 1979*b*). Most species seem to occur in areas with a pronounced dry season. There is one very unusual undescribed genus from seasonally dry savanna habitat in northwestern Costa Rica. The hosts of this group are not known, but they might be the eggs of aradid bugs living under bark.

Aradophagus[EX]. This is primarily a holarctic genus, which extends southwards into dry tropical areas in India. Biology: possible hosts are the eggs of Aradidae (Hemiptera), which occur under bark on trees, although this suggestion is contradicted by the observation that numerous individuals occur indoors in places (G. Byers, pers. comm.).

SCELIONINAE: Baeini. This large, cosmopolitan tribe includes ten genera, three of which are present in Costa Rica. As far as is known all baeines parasitize the eggs of spiders (Austin, 1985).

Baeus. A cosmopolitan genus represented in Costa Rica by about 20 species. The females are apterous. Biology: one widespread species is known to develop in the eggs of *Latrodectus* (Theridiidae) (Pemberton & Ross, 1940), and in Costa Rica another species is known to attack eggs of *Archaearanea tepidariorum* (Theridiidae) (Valerio, 1971).

Idris (= *Ceratobaeus*). A cosmopolitan genus with at least 100 species present in Costa Rica. Females of some species are short-winged or apterous; most species are brightly coloured (yellow, orange, etc.). Biology: a North American species has been reared from the egg sacs of Salticidae (Muesebeck, *in* Krombein *et al.*, 1979).

Odontacolus. A pantropical genus that in the New World ranges as far north as Texas and Florida. It is represented by ten species in Costa Rica. Only fully-winged forms are known. Biology: unknown.

SCELIONINAE: Baryconini. Worldwide this tribe includes five genera, two of which have been found in Costa Rica. Known hosts of species of this tribe are the eggs of Tettigoniidae (Phaneropterinae) on or in the leaves of plants.

Baryconus[1]. A cosmopolitan genus with about 30 species in Costa Rica. The nearctic fauna was monographed by Ritchie and Masner (1983).

Chromoteleia[2]. A primarily neotropical genus occurring throughout South and Central America north to Mexico and the Caribbean, and in Africa. Approximately 20 species have been found in Costa Rica.

SCELIONINAE: Calliscelionini. This is the largest tribe in Scelioninae with 18 described genera worldwide, seven of which have been found in Costa Rica. Most calliscelionines are small insects and only *Macroteleia* and *Triteleia* are over 3.0 mm in length. Known hosts of this tribe are the eggs of various Orthoptera.

Anteromorpha. A cosmopolitan genus with ten species so far found in Costa Rica.

Calliscelio. A large cosmopolitan genus with about 70 species found so far in Costa Rica, some of them brightly coloured. Biology: reared from eggs of Gryllidae in North America (Muesebeck, *in* Krombein *et al.*, 1979).

Macroteleia[1]. A cosmopolitan genus with approximately 20 species present in Costa Rica. The New World species were recently revised by Muesebeck (1977). Biology: two species parasitize the eggs of species of Tettigoniidae.

Oethecoctonus[1]. A pantropical genus with a few species penetrating into some temperate areas. This genus is particularly species-rich in tropical America and approximately 20 species have been found so far in Costa Rica. The North American species were monographed by Masner (1983*a*). Biology: hosts are the eggs of tree-crickets (Gryllidae: Oecanthinae), deposited in stems or in cracks in bark.

Paridris. A cosmopolitan genus with approximately 40 species found in Costa Rica. Biology: a North American species has been reared from the eggs of *Gryllus* (Gryllidae) (Fouts, 1920).

Probaryconus[1]. A cosmopolitan genus with approximately 70 species present in Costa Rica.

Triteleia[2]. A widespread genus that is represented in Costa Rica by at least 50 species. Species of *Triteleia*, together with some *Sceliomopha*, are the largest scelionids in the country.

SCELIONINAE: Cremastobaeini. A rather small tribe comprising just two genera, both of which are represented in the Costa Rican fauna.

Cremastobaeus. A cosmopolitan genus represented in Costa Rica by about 20 species. Biology: an Indian species has been reared from the eggs of a cricket (Austin, pers. comm.).

Parascelio. A neotropical genus with species extending northwards to Mexico and in the Caribbean. Five species have been found in Costa Rica.

SCELIONINAE: Doddiellini[EX]. A small tribe comprising a single genus that occurs throughout the world in the tropics and subtropics. Its hosts are unknown, but adults are generally collected by sifting soil.

Doddiella[EX]. In the New World species of this genus have been collected throughout tropical South America, north to Panama.

SCELIONINAE: Embidobiini. This tribe comprises seven genera that occur widely in the warmer tropical and subtropical parts of the world. Three are known to occur in Costa Rica. Embidobiines are minute scelionids that are seldom caught in traps. Most are believed to parasitize the eggs of Embiidina, but one species of an ethiopian genus is known to attack the egg sac of a littoral spider (Masner & Dessart, 1972; Masner, 1976*a*).

Embidobia. This widespread genus is represented in Costa Rica by about ten species. Biology: hosts are the eggs of various Embiidina (Masner, 1976*a*).

Embioctonus. A neotropical genus, with some species extending northwards into Mexico and southern Florida. Five have been found to occur in Costa Rica. Biology: hosts are the eggs of various Embiidina (Masner, 1980*c*). Females are sometimes apterous.

Palaeogryon. This neotropical genus extends north to Mexico and the southern U.S.A. Three species are known to occur in Costa Rica. Biology: in Ecuador a species of this genus was reared from the eggs of either *Chelicerca* or *Oligambia* sp. (Embiidina) (Masner, 1976*a*).

SCELIONINAE: Gryonini. This very large tribe comprises, worldwide, ten genera, three of which occur in Costa Rica. A key to these is given by Masner (1983*b*). Except for one palaearctic species known to attack the eggs of mantids, members of this tribe seem to be confined to parasitizing the eggs of Hemiptera.

Dyscritobaeus. A pantropical genus that, in the New World, extends northwards as far as Kentucky and Maryland. It is represented in Costa Rica by approximately 25 species.

Epigryon. This neotropical genus extends northwards to southern Florida, and is represented in Costa Rica by a single species. Biology: one species of this genus, which occurs in the southern United States, is phoretic on a species of *Phymata* (Reduviidae: Phymatinae) (Masner, 1980*c*).

Gryon. An extremely large cosmopolitan genus represented in Costa Rica by at least 50 species. Masner (1979) gives a key to neotropical species in the *variicornis* species-group, and Masner (1983*b*) has revised the nearctic species. Biology: parasitoids in eggs of Hemiptera, mainly in species of Coreidae and perhaps less commonly Pentatomidae, Scutelleridae and Lygaeidae. Schell (1943) and Kenaga (1944) describe the biology of two widespread nearctic species.

SCELIONINAE: Psilanteridini. A large cosmopolitan tribe that includes 16 described genera, nine of which occur in Costa Rica. Structurally species in this tribe are very similar to members of the Calliscelionini and the two tribes should perhaps be amalgamated. The hosts of these insects are mostly unknown, but what few records are available are generally from Gryllidae although one species may attack Odonata.

Anteris. A holarctic and neotropical genus whose range extends southwards as far as Peru. It is represented in Costa Rica by about ten species.

Calotelea. A cosmopolitan genus, with at least 50 species present in Costa Rica. The nearctic species were revised by Masner (1980*a* & *b*). Biology: although there are no definite host records, it is suspected that *Calotelea* species attack the eggs of small crickets deposited in the ground. In Peru, circumstantial evidence suggests a species of *Cycloptilum* (Gryllidae, Mogoplistini) might be a host (Masner, 1980*b*).

Duta. A cosmopolitan genus with approximately 15 species occurring in Costa Rica. The nearctic species have been revised by Masner (1991*a*). Biology: hosts are eggs of the Nemobiinae (Gryllidae).

Lapitha[1]. A neotropical genus with one species extending northwards into the United States. Ten species have been found to occur in Costa Rica.

Leptoteleia[1]. A more or less cosmopolitan genus that is perhaps most species-rich in the Neotropics. It is represented in Costa Rica by at least ten species. Masner (1978) revised the world species. Biology: the hosts of two species, from the United States and Haiti, are crickets of the genera *Oecanthus* and *Chremon* (Gryllidae), groups which usually oviposit into stems or crevices in bark (Masner, 1978).

Opisthacantha. A cosmopolitan genus represented in Costa Rica by 36 species.

Psilanteris. A cosmopolitan genus with at least 50 species present in Costa Rica.

Spiniteleia. A small nearctic genus that extends south to Costa Rica where it is represented by a single species. It is also known to occur on the Galapagos Islands. Specimens of this genus are generally obtained by sifting leaf litter, or in pit-fall traps. Biology: hosts are possibly the eggs of nemobiine crickets (Masner, 1980*c*).

Thoronella. A neotropical, nearctic and ethiopian genus represented in Costa Rica by two species. Biology: a nearctic species has been observed to be phoretic on the thorax of an aeschnid dragonfly (Carlow, 1992).

SCELIONINAE: Scelionini. This cosmopolitan tribe comprises ten genera worldwide. Four of these have been found to occur in Costa Rica. The known hosts of Scelionini are eggs of the Acridoidea.

Acanthoscelio[1]. A neotropical genus that ranges north to Costa Rica, where it is represented by five species.

Pseudoheptascelio[1]. A neotropical genus that occurs as far north as Costa Rica, where it is represented by two species. Biology: one species was recently reared in Trinidad from the eggs of a species of the acridid genus *Cornops*, which is associated with water hyacinths.

Scelio[1]. A large cosmopolitan genus with at least 30 species present in Costa Rica. Biology: several North American species attack the eggs of *Melanoplus* (Acrididae) and one species, *S. ernsti*, which occurs in Costa Rica, attacks the eggs of

Schistocerca spp. (Acrididae) (Muesebeck *in* Krombein *et al.*, 1979).

Synoditella. This genus occurs in North and Central America, throughout the Caribbean, and south into Venezuela and Ecuador. One species has been collected in Costa Rica. Biology: females of *Synoditella* species are phoretic on short-horned grasshoppers (Acrididae). The biology of one North American species was described by Lanham and Evans (1958, 1960).

SCELIONINAE: Sparasionini. This small Cosmopolitan tribe comprises four genera, one of which occurs in Costa Rica. Known hosts of these scelionids are the eggs of species of Decticinae (Tettigoniidae) and Stenopelmatinae (Gryllacrididae), hosts that lay eggs individually in or on plant tissue.

Sceliomorpha[1]. A neotropical genus with one species occurring as far north as Texas. Approximately 15 species have been collected in Costa Rica. Some are amongst the largest scelionids in the country.

SCELIONINAE: Thoronini. This widespread tribe comprises seven genera, one or two of which occur in Costa Rica. Thoronines are parasitoids in the eggs of aquatic Hemiptera (Masner, 1972; Masner & Huggert, 1979*a*). Some species enter fresh water to gain access to host eggs, but they generally do so by walking on vegetation from the surface, and not by swimming (Spence, 1986; Henriquez & Spence, 1993).

Neothoron[EX]. A neotropical genus. Biology: presumed to parasitize the eggs of Nepidae.

Thoronidea. A New World genus with species occurring in Central America, the Caribbean and Maryland, U.S.A; two have been found in Costa Rica. Biology: presumed to attack eggs of Gelastocoridae (Hemiptera).

TELEASINAE

The Teleasinae is a very homogeneous cosmopolitan subfamily, containing about ten genera worldwide. Two described and one undescribed genera occur in Costa Rica. Although the Teleasinae appear to be more diverse in temperate regions than in tropical areas, it is possible that in the tropics relatively more species are found in the canopy, and the size of the group has thus been underestimated. Members of this subfamily are presumed to parasitize eggs of Carabidae (Coleoptera), many of which occur in tropical forest canopies.

Gryonoides. Primarily a neotropical genus with a range extending north to Mexico. This may be an example of a canopy dwelling group. Ten species are known from Costa Rica.

Trimorus. An extremely large cosmopolitan genus, species of which seem to prefer ground habitats, especially at higher altitudes. This genus is represented in Costa Rica by at least 100 species. The North American species were revised by Fouts (1948).

Undescribed genus (near *Xenomerus* of the Old World). About ten species of this genus have been collected at lower altitudes in Costa Rica. Males are characterized by having the antenna with some segments bottle-shaped and with very long bristles. The body of both sexes is partly smooth, the metasoma of the female is circular, and the mesoscutum has one to three longitudinal keels.

TELENOMINAE

The Telenominae is an extremely large cosmopolitan subfamily which rivals in species-richness almost any other group of parasitic Hymenoptera. However, only a small fraction of the species have been described. Morphologically it is a very homogeneous group, and the numerous species are classified in about eight genera, four of which are known to occur in Costa Rica, and one of which might spread to the country. These may be separated using Table 9.2. The Telenominae attack principally the eggs of two host groups, Hemiptera and Lepidoptera, but a few species are known to parasitize the eggs of such diverse hosts as Tabanidae (Diptera), Chrysopidae (Neuroptera) and Fulgoroidea (Homoptera).

Eumicrosoma. A cosmopolitan genus comprising relatively few species, three of which are known to

- –Marginal vein distinctly longer than stigmal vein (Fig. 9.66); body flattened dorsoventrally .... *Eumicrosoma*
- –Marginal vein usually short (Fig. 9.65); body usually not flattened
  - –Malar space with striae radiating from mandible base (Fig. 9.67) ... *Psix*
  - –Malar space without striae radiating from base of mandible
    - –At least posterior end of metasoma orange .... *Phanuropsis*
    - –Entire metasoma black
      - –Frons smooth; notauli absent .... *Telenomus*
      - –Frons sculptured; notauli often partially present .... *Trissolcus*

Table 9.2. Identification table for Costa Rican genera of Telenominae.

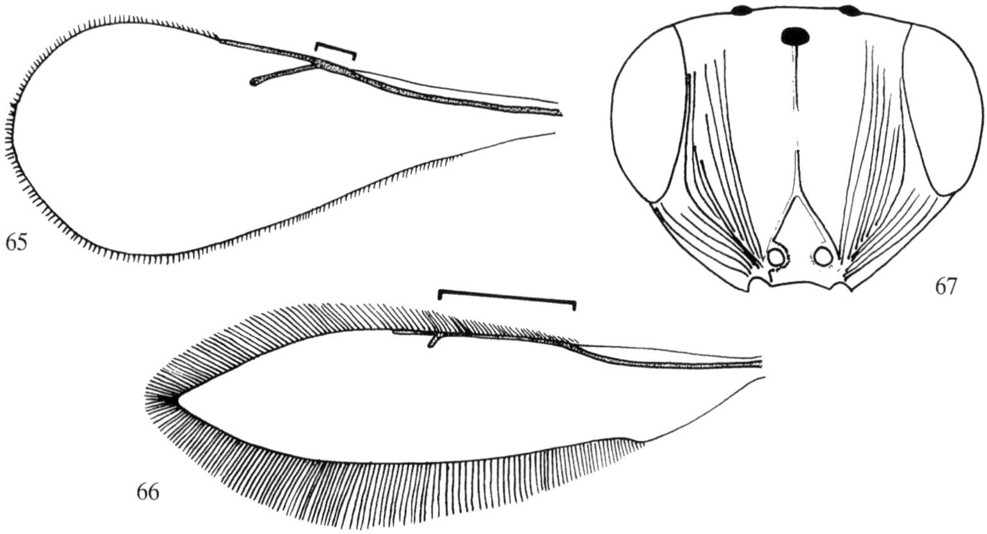

Figs 9.65–9.67. Telenominae. Figs 9.65–9.66. Fore wing; 9.65, *Telenomus* sp.; 9.66, *Eumicrosoma* sp. Fig. 9.67. Face, *Psix tunetanus*.

occur in Costa Rica. Biology: parasitoids in the eggs of species of Blissinae (Lygaeidae) (McCulloch, 1915).

Phanuropsis. A small mesoamerican and northern South American genus with a known range from Honduras to Colombia. One species is known to occur in Costa Rica. Johnson (1987) has revised the species. Biology: parasitoids in the eggs of Pentatomidae (Eberhard, 1975).

Psix^{EX}. This is an Old World genus with one species, *P. tunetanus*, apparently introduced into the southwestern United States and Venezuela, and likely to spread to Costa Rica (Johnson & Masner, 1985). Biology: parasitoids in the eggs of Pentatomidae.

Telenomus. A cosmopolitan genus comprising several hundred described species worldwide and many more undescribed ones. At least 300 species are present in Costa Rica. Johnson (1981, 1984) and Masner and Johnson (1979) have monographed some nearctic species. Biology: most species of *Telenomus* parasitize the eggs of Lepidoptera, while a few attack the eggs of Hemiptera. *T.sulculus*, reared from eggs of *Zelus* sp. (Reduviidae) in Mexico and Guyana, has dimorphic males: normal winged males and brachypterous males with enlarged heads.

Trissolcus. A large cosmopolitan genus with probably about 30 species in Costa Rica, although fewer have been recorded. Johnson (1985*a* & *b*) has revised some species groups. Biology: known hosts of species of this genus are the eggs of Pentatomidae.

9.7 PLATYGASTRIDAE

Diagnosis. Body usually 1.0 to 2.0 mm long, rarely as small as 0.7 mm or as large as 4.0 mm; predominantly black or dark brown, rarely light yellow (only in *Leptacis* and *Parabaeus*, which occur at altitudes below 800 m), never metallic; often distinctly sculptured, especially on head and dorsum of mesosoma. Brachypterous or wingless forms exceedingly rare. *Antenna attached immediately above upper margin of clypeus, usually 10-segmented* (rarely 9, 8, or 7) distinctly geniculate, *with long scape, and typical second bend between funicle and clava*; female antenna always clavate, male mostly thread-like (less commonly clavate) with segment 4 (rarely 3) modified; palpi strongly reduced, maxillary 2- or 1-segmented, labial

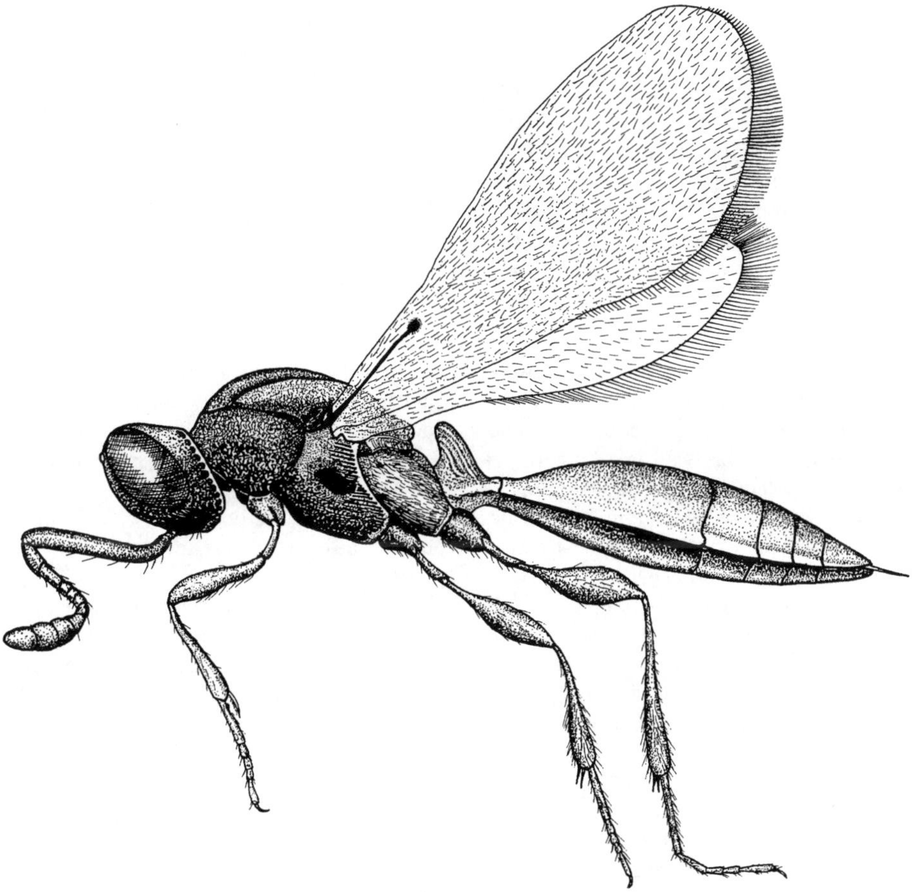

Fig. 9.68. *Inostemma* sp. (Platygastridae).

always 1-segmented. *Fore wing usually veinless, or if submarginal vein present then not reaching anterior margin of wing; marginal, postmarginal, and stigmal veins absent*; hind wing at most with short stub of submarginal vein. Metasoma beyond segment I often depressed dorsoventrally; however, in some members moderately to strongly compressed laterally, in females often tail-like; *in all members with segment II distinctly the longest and widest*; females almost always with only six apparent tergites, metasomal tergite VII (apical tergite) internal, considerably unpigmented and reduced, hidden under tergite VI and not extruded with ovipositor, without cerci or sensory plates with bristles; males with eight visible tergites.

Classification and distribution. Although platygastrids occur worldwide and are very common, the group remains one of the most poorly studied families of Hymenoptera. It is difficult to even estimate the number of species in the family, but it is certainly a large group. There are approximately 60 genera, which are now classified in two subfamilies, Sceliotrachelinae and Platygastrinae (Masner & Huggert, 1989). The previously recognized subfamily Inostemmatinae (e.g. Krombein *et al.*, 1979; Gauld & Bolton, 1988), which was distinguished primarily by the presence of a submarginal vein, has been shown to be a heterogeneous assemblage.

Biology. The biology of Platygastridae differs between the two subfamilies. Many species of Sceliotrachelinae develop similarly to Scelionidae, as endoparasitic idiobionts in the eggs of beetles or Fulgoroidea (Homoptera). Others are endoparasitic koinobionts in immobile ('egg-like') hosts, such as the immature stages of Coccoidea or Aleyrodidae (Homoptera). Most members of this subfamily are probably solitary, but a few are known to be gregarious. For example, 1 to 22 individuals of the holarctic species, *Allotropa burrelli*, develop in a single mealybug (Clancy, 1944). This species oviposits into the haemocoel of the mealybug nymph. The developing parasitoid embryo, which is surrounded by a syncytial trophamnion (see below), becomes associated with fat bodies from which it apparently absorbs nutrients prior to eclosion. The larva is more or less hymenopteriform with a single pair of spiracles on segment I and fine pointed mandibles. There is a single larval instar. When fully grown the larva forms a parchment-like cocoon within the host (Clancy, 1944).

The majority of species of the second subfamily, the Platygastrinae, are endoparasitic koinobionts of Cecidomyiidae, particularly the gall-forming cecidomyiids. In this subfamily oviposition is into the egg or early instar larva, but development is not completed until the host larva has become fully grown. Within the host egg, the platygastrid egg is located in a particular region of the embryo. For example, eggs of *Inostemma boscii* are found in the brain and those of *Trichacis remulus* in the posterior portion of the host nerve cord (Marchal, 1906; Myers, 1927). *Platygaster zosine* oviposits into the mid intestine, and its eggs fail to develop elsewhere. Most other species of *Platygaster* place the egg anywhere in the host body cavity (Clausen, 1940*b*).

The development of the European *Platygaster virgo* and the North American *P. herrickii* have been studied in some detail (M.C. Day, unpublished; Hill & Emery, 1937). Both are solitary parasitoids although in the latter, two or three individuals occasionally develop in a single host. The parasitoid egg lies free in the host haemocoel, and embryological development takes place concurrently with that of the host. The platygastrine egg develops by a series of nuclear divisions; cell walls develop between the central nuclei, but the outer layer of cytoplasm and its nuclei become separated from the underlying cells and form a syncytial trophamnion. This structure, as its name implies, surrounds the contained embryo, and its outer surface is covered with microvilli; this form of embryological development seems to be a characteristic of all platygastrids. The trophamnion is essential to the growth of the embryo, providing it with food materials absorbed from the host (Askew, 1971). It disintegrates into a number of 'pseudogerms' on eclosion of the larva and serves a teratocyte function. Other 'pseudogerms', contemporaneous with the eclosed larva, may be failed embryos or trophamnionic buds. At the time of hatching these 'pseudogerms' are released into the body cavity of the host where they undergo further growth prior to disintegrating and serving as food for the developing parasitoid larva. The number of 'pseudogerms' in a host containing a single first instar larva averages about 25 (Hill & Emery, 1937).

It is often stated that several species of Platygastridae are polyembryonic. Although this may eventually prove to be correct, at the present time polyembryony is definitely known to occur only in *Platygaster zosine*, a parasitoid of the hessian fly. The confusion perhaps stems from the fact that this species has also been treated under the names *Polygonotus minutus* (Marchal, 1904) and *Platygaster vernalis* (Hill, 1923; Leiby & Hill, 1924). In *P. zosine* the trophamnion forms a sheath around the egg, and it grows as the egg divides. Eventually each embryo becomes enveloped by a piece of the trophamnion. A simple type of polyembryony occurs in *P. hiemalis*, where development may be either monoembryonic or by 'twinning' (Leiby & Hill, 1923; Hill, 1926).

Whilst the first instar larvae of some platygastrids are more or less hymenopteriform, those of many species are quite extraordinary in appearance and superficially resemble cyclopoid copepods, with an inflated 'cephalothorax' carrying large curved mandibles, and a slender abdomen that often terminates in a bifurcate 'tail'. When fully fed these larvae are extremely bloated and ovoid in shape. The exact number of larval instars appears to vary from one to three, depending upon the species. Most of the developmental time is passed as an embryo or larva within the trophamnion. Once the larva emerges from the trophamnion it rapidly consumes its host. The platygastrid generally pupates within the larval skin of the cecidomyiid (Parnell, 1963) but some gregarious species are known to spin cocoons within the host puparium.

There is relatively little information about the biology of the adult stage of Platygastridae. They appear to be very good fliers, as evidenced by their prevalence in Malaise traps.

Economic importance. A few species of Platygastridae have occasionally been used in classical biological control. *Allotropa burrelli* was introduced from Japan to the United States to control the Comstock mealybug. The combined activity of this species and three encyrtid species resulted in excellent reductions of mealybug infestations (Haeussler & Clancy, 1944; Clausen, 1978). In Central America an undetermined species of *Amitus* attacks *Bemisia tabaci* (Aleyrodidae), but its effectiveness in reducing whitefly populations is unknown.

Identification. Masner and Huggert (1989) provided keys to the world genera previously classified in the Inostemmatinae, i.e. Sceliotrachelinae plus those Platygastrinae having a submarginal vein. Fouts' (1924) key to the genera of Platygastrinae of North America is of limited use due to subsequent changes in generic concepts and nomenclature. D. Jackson (1969) discussed some characters for separating genera of the *Synopeas-Leptacis* complex.

Key to the subfamilies of Platygastridae

— Antennal clava of female abrupt, usually 3-segmented (Fig. 9.69), or subcompact, 1-segmented; antenna of male either subclavate, 3-segmented, subcompact 1-segmented, or thread-like; metasoma with wide laterotergites (Fig. 9.71); body often rather flattened; apterous forms rare. ....**Sceliotrachelinae** (p. 243)

— Antennal clava of female at most moderately abrupt, 4- to 5-segmented (Fig. 9.70), antenna of male usually thread-like, rarely clavate; metasoma with narrow laterotergites (Figs 9.72, 9.73); body usually elongate; no apterous forms. .... **Platygastrinae** (p. 245)

Synopsis of the Costa Rican fauna

Platygastrids are present in all major habitats, including the highest altitudes, but they are relatively less common in dry, open grasslands of the lowlands. Both subfamilies occur in Costa Rica where they are represented by a total of 23 genera. Three of these genera contain over 50 species each, and six more comprise between 12 and 50 species. Based on material presently available the total number of species present in the country appears to be between 450 and 800.

SCELIOTRACHELINAE

This is the smaller of the two subfamilies but its members are much more diverse, both morphologically and biologically, than the Platygastrinae. Presently five genera and 26 species are known from Costa Rica, and two additional genera are expected to occur in the country. All except *Amitus* have the apex of the submarginal vein formed into a knob (cf Fig. 9.74).

Allotropa. A cosmopolitan genus with five species in Costa Rica. Biology: parasitoids of mealybugs (Pseudococcidae) (Clancy, 1944).

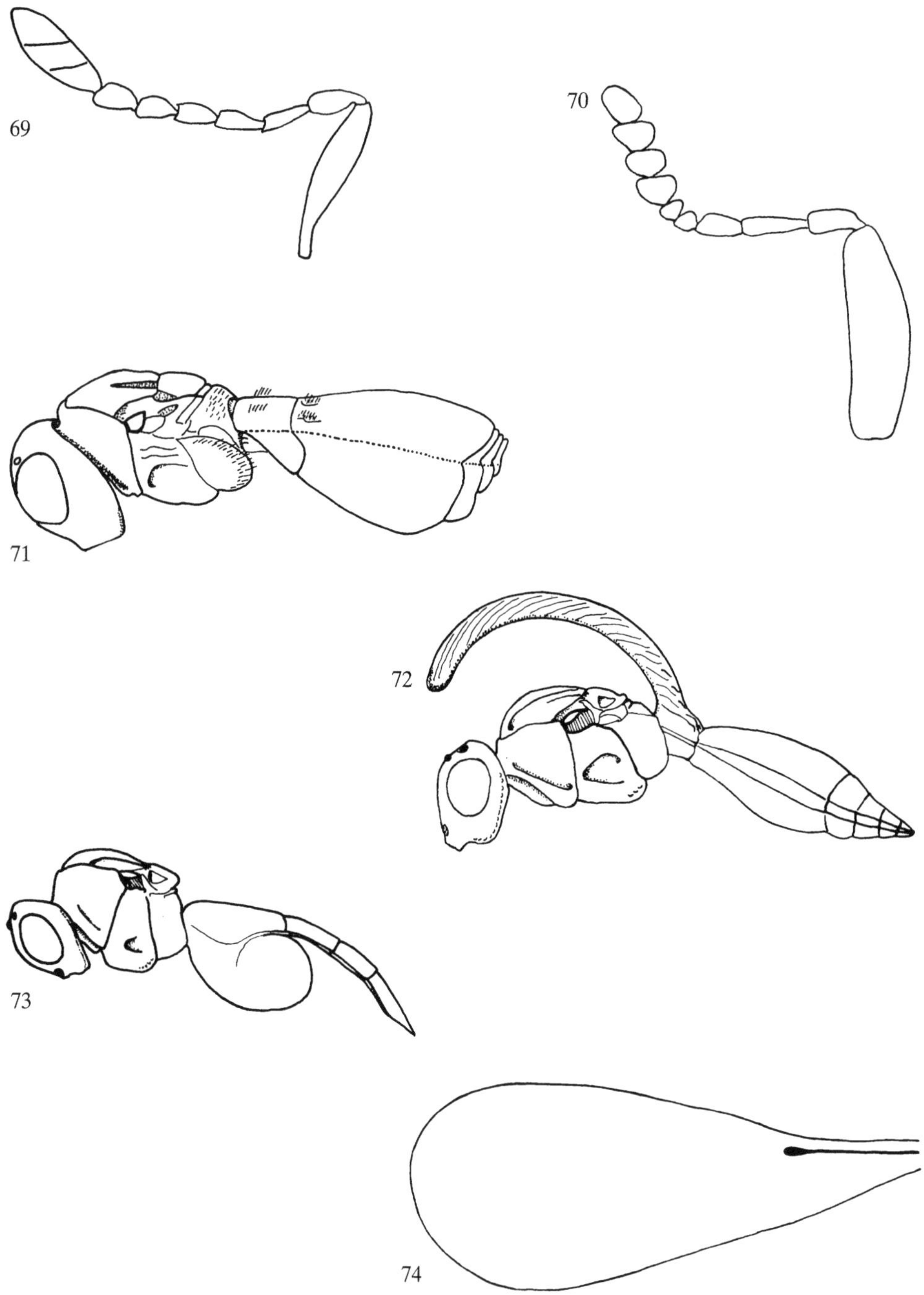

Figs 9.69–9.74. Platygastridae. Figs 9.69–9.70. Antenna of female; 9.69, *Amitus* sp. (Sceliotrachelinae); 9.70, *Inostemma* sp. (Platygastrinae). Figs 9.71–9.73. Head and body, lateral; 9.71, *Fidobia* sp. (Sceliotrachelinae); 9.72, *Inostemma* sp. (Platygastrinae); 9.73, *Synopeas* sp. (Platygastrinae). Fig. 9.74. Fore wing, *Inostemma* sp. (Platygastrinae).

Amitus. A cosmopolitan genus with five species in Costa Rica. Taxonomy: New World species, MacGown & Nebeker (1978). Biology: parasitoids of whiteflies (Aleyrodidae) (Dowell, 1979; Viggiani & Battaglia, 1983*b*).

Calomerella[EX]. A New World genus. Biology unknown.

Fidiobia. A cosmopolitan genus represented by five species in Costa Rica. Biology: *Fidobia* species are parasitoids in the eggs of Curculionidae and Chrysomelidae (Muesebeck *in* Krombein *et al*., 1979). In Jamaica *F. citri* has been reared from the eggs of the citrus weevil (*Exophthalmus* spp.).

Neobia. A New World genus with a single species in Costa Rica. Biology unknown.

Parabaeus. A widespread genus in Africa, Australia and the New World, throughout South America north to Mexico and Florida. Ten species occur in Costa Rica. They are often bright yellow-orange and are usually apterous or sometimes fully-winged. Biology unknown.

Platystasius[EX]. A widespread genus known to occur in Australia, Europe and the New World. Biology: parasitoids in the eggs of Cerambycidae.

PLATYGASTRINAE

Although this subfamily is very speciose it is very homogeneous in both morphology and biology. All species are, as far as known, koinobiont endoparasitoids of gall-midges (Cecidomyiidae). Different species seem to be associated with particular host plants, or parts of the host plant, where the host gall is located. Currently 16 genera of Platygastrinae are known to be present in Costa Rica. These contain the vast majority of the species in the family (an estimated 430 species). Genera that have a submarginal vein in the fore wing (Fig. 9.74) and are thus included in Masner and Huggert (1989) include: *Acerotella*, *Aceroteta*, *Inostemma*, *Iphitrachelus*, *Isostasius*, *Metaclisis* and *Orseta*.

Acerotella. A cosmopolitan genus with three species in Costa Rica. The nearctic species were revised by Masner (1980*d*).

Aceroteta. A cosmopolitan genus represented in Costa Rica by two species.

Amblyaspis. A large cosmopolitan genus with about 25 species in Costa Rica.

Euxestonotus. A large cosmopolitan genus with five species in Costa Rica.

Inostemma. A cosmopolitan genus with about 15 species in Costa Rica. Tergite I of female has a hump or horn (which sometimes extends over the mesosoma) and houses the ovipositor (Fig. 9.72). Some species are associated with crop plants as primary parasitoids of gall-midges.

Iphitrachelus. A cosmopolitan genus with five rarely collected species in Costa Rica. Several species have extremely wide distributions; *I. lar*, for example, is recorded from Europe, North America, Mexico, Brazil and Argentina. The nearctic species were monographed by Masner (1976*c*), and a key given to the described species of the world.

Isostasius. A cosmopolitan genus with ten species in Costa Rica.

Leptacis. A very large cosmopolitan genus with about 100 species in Costa Rica.

Metaclisis. A large holarctic and neotropical genus with ten species in Costa Rica. In tropical areas *Metaclisis* species seem to be restricted to higher altitude sites, from cloud forest up to the paramo. The nearctic species were revised by Masner (1981). Biology: Roskam (1986).

Metanopedias. A small cosmopolitan genus with five species in Costa Rica. *M. brunneipes*, which occurs from southeastern Canada to Costa Rica, has been reared from galls on orchid (*Huntleya* and *Zygopetalum*) roots in Costa Rica (Jackson & Masner, 1966).

Orseta. A small Australian and New World genus with two species in Costa Rica.

Piestopleura. A medium sized cosmopolitan genus with ten species in Costa Rica.

Platygaster. A very large cosmopolitan genus with at least 100 species in Costa Rica. These occur in all habitats sampled but form a particularly conspicuous component of the biota from high altitude sites. MacGown (1979) has revised some nearctic species.

Pyrgaspis. A small holarctic genus with two species in Costa Rica.

Synopeas. A very large polytypic cosmopolitan genus with approximately 100 species in Costa Rica. These occur in all samples examined, from all sites, but they are most numerous in samples from higher altitudes. The females of many species have the second metasomal sternite enlarged and pouch-like (Fig. 9.73), serving as a receptacle into which the elongate ovipositor can be withdrawn.

Trichacis. A medium-sized cosmopolitan genus with about 25 species in Costa Rica. Masner (1983*c*) revised the nearctic species.

10 The cynipoid families

INTRODUCTION

Nigel D. M. Fergusson

Diagnosis. Most species 1.0 to 5.0 mm in length, or up to 30 mm in the relatively rarely collected Ibaliidae; usually brown or black in colour; both sexes fully winged except in a few Cynipini and Eucoilini. Female with 13 antennal segments, male with 14 or 15 (rarely with 12 to 20 in females and 13 to 16 in males). Pronotum reaching the tegulae, frequently developed into an anterior plate which may be raised off the surrounding pronotal surface; *mesosoma in profile with scutellum almost as large as propodeum*, often with scutellum with elaborate sculpture, sometimes modified into a spine or a cup. Wing venation reduced (Fig. 10.01), *fore wing without a pterostigma* (although two very rare genera, from Australia and Africa, have a pseudopterostigma), *with a distinctive triangular marginal cell* (Figs 10.01–10.04). Tarsi 5-segmented. *Metasoma usually laterally compressed.*

Classification and distribution. The Cynipoidea share many similarities (parallelisms—especially reduction states) with the other 'microhymenoptera' (Chalcidoidea and Proctotrupoidea *s.l.*). However, the structure of the cynipoid mesosoma is most similar to that of the Proctotrupoidea *s.s.* (Gibson, 1985, 1986*a*). Likewise the cynipoid wing is similar to that of certain proctotrupoid families. This similarity in venation has been emphasized by the recent discovery of fossil cynipoids, the Archaeocynipidae, from very early Cretaceous strata (Rasnitsyn & Kovalev, 1988). The small and linear pterostigma of these fossils is similar to that of *Austroserphus* and *Acanthoserphus* (Proctotrupoidea *s.s.*) and suggests that the cynipoids lost their (possibly linear) pterostigma by gradual reduction. The closed discal cell of the Archaeocynipidae is particularly reminiscent of the venation of the proctotrupoid families Proctotrupidae, Heloridae, Roproniidae and Vanhorniidae. Thus it is suggested that the Cynipoidea plus at least some of the less derived proctotrupoid families had a common ancestor.

The Cynipoidea consists of approximately 2700 species in 250 genera. In the past the superfamily was often subdivided into six families, the Cynipidae, Ibaliidae, Liopteridae, Figitidae, Charipidae and Eucoilidae (e.g. Quinlan, 1979), but previous classifications have tended to upgrade all taxa without adequate justification. Fergusson (1990) has recently undertaken a phylogenetic reconstruction, which provides a very different classification that is robustly supported by morphological, palaeontological, biogeo- graphical and biological evidence. The Cynipoidea is now divided into four families (Table 10.1). All except the Asian family Himalocynipidae, which comprises a single species *Himalocynips vigintilis* (Yoshimoto, 1970), are represented in Costa Rica.

Family	Subfamily	Tribe
Cynipidae	Cynipinae	Aulacideini[1]
		Rhoditini[1]
		Synergini
		Cynipini
Ibaliidae	Austrocynipinae[1]	–
	Ibaliinae	Ibaliini[1]
		Liopterini
Figitidae	Anacharitinae	–
	Aspicerinae	–
	Figitinae	Charipini
		Eucoilini
		Figitini
Himalocynipidae[1]	–	–

Table 10.1. The classification of the Cynipoidea adopted in the present work. [1] = Taxa not represented in Costa Rica.

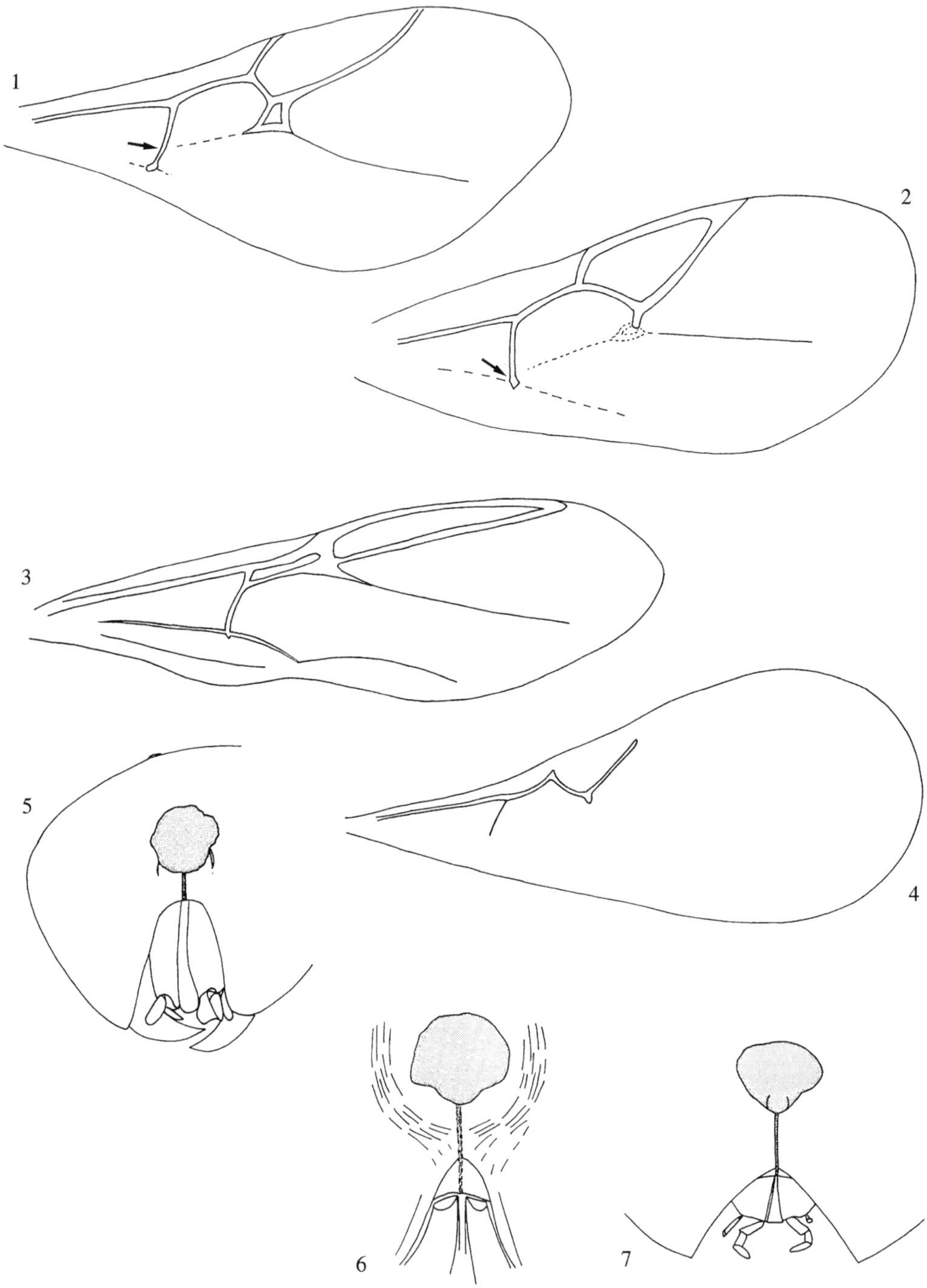

Figs 10.01–10.07. Cynipoidea. Figs 10.01–10.04. Fore wings; 10.01, Cynipidae; 10.02, Figitidae; 10.03, *Ibalia* sp. (Ibaliidae: Ibaliini); 10.04, *Dilyta* sp. (Figitidae: Charipini). Figs 10.05–10.07. Posterior part of ventral side of head; 10.05, *Melanips* sp. (Figitidae: Anacharitinae); 10.06, *Paramblynotus* sp. (Ibaliidae: Liopterini); 10.07, *Synergus* sp. (Cynipidae).

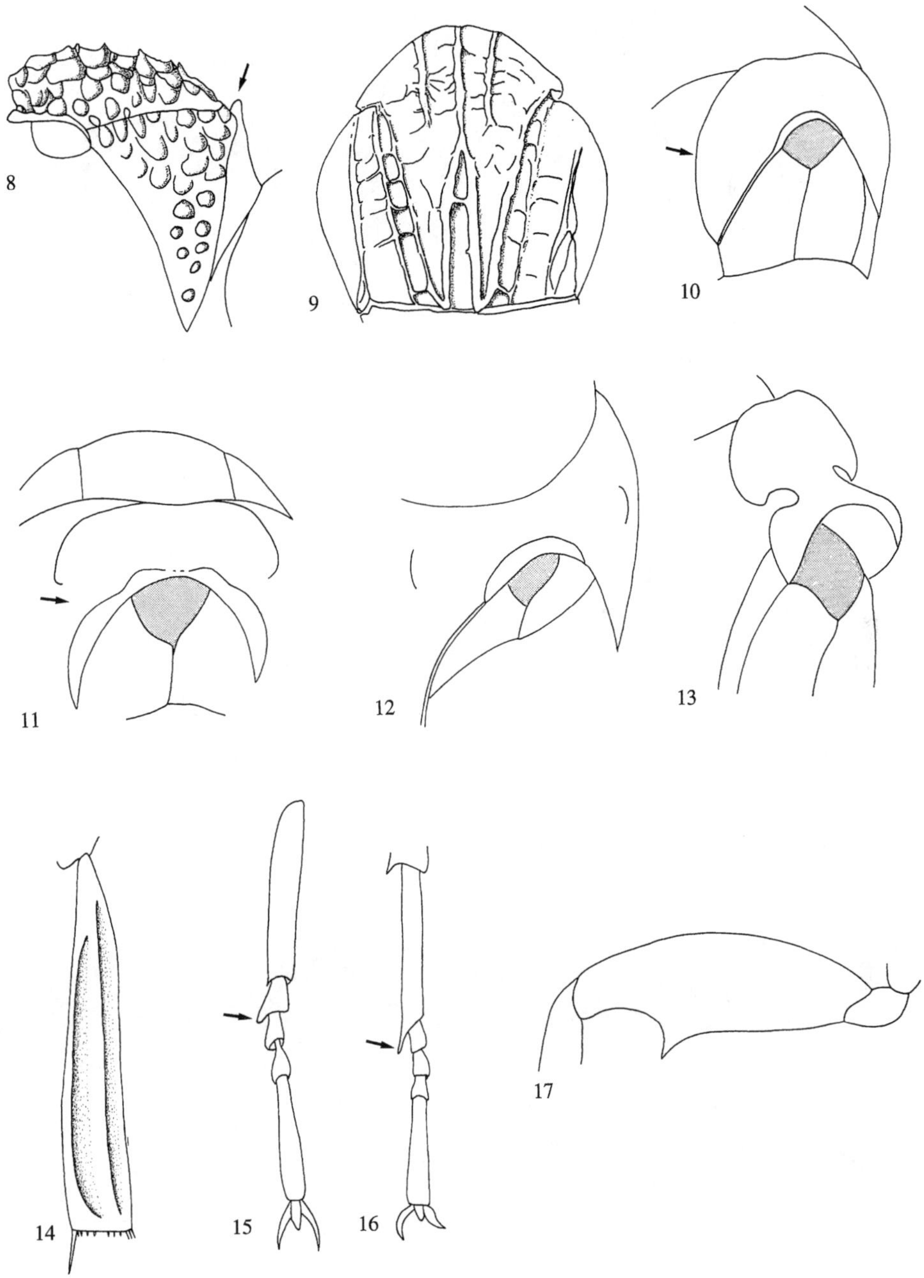

Figs 10.08–10.17. Cynipoidea. Fig. 10.08. Pronotum and mesoscutum, lateral, *Liopteron* sp. (Ibaliidae: Liopterini). Fig. 10.09. Mesoscutum, dorsal, *Aspicera* sp. (Figitidae: Aspicerinae). Figs 10.10–10.13. Mediodorsal part of pronotum, anterior view with head removed; 10.10, *Anacharis* sp. (Figitidae: Anacharitinae); 10.11, *Melanips* sp. (Figitidae: Figitini); 10.12, *Alloxysta* sp. (Figitidae: Charipini); 10.13, *Trybliographa* sp. (Figitidae: Eucoilini). Fig. 10.14. Hind tibia, *Callaspidia* sp. (Figitidae: Aspicerinae). Figs 10.15–10.16. Hind leg; 10.15, *Plastibalia* sp. (Ibaliidae: Liopterini); 10.16, *Pseudibalia* sp. (Ibaliidae: Liopterini). Fig. 10.17. Hind femur, *Oberthuerella* sp. (Ibaliidae: Liopterini).

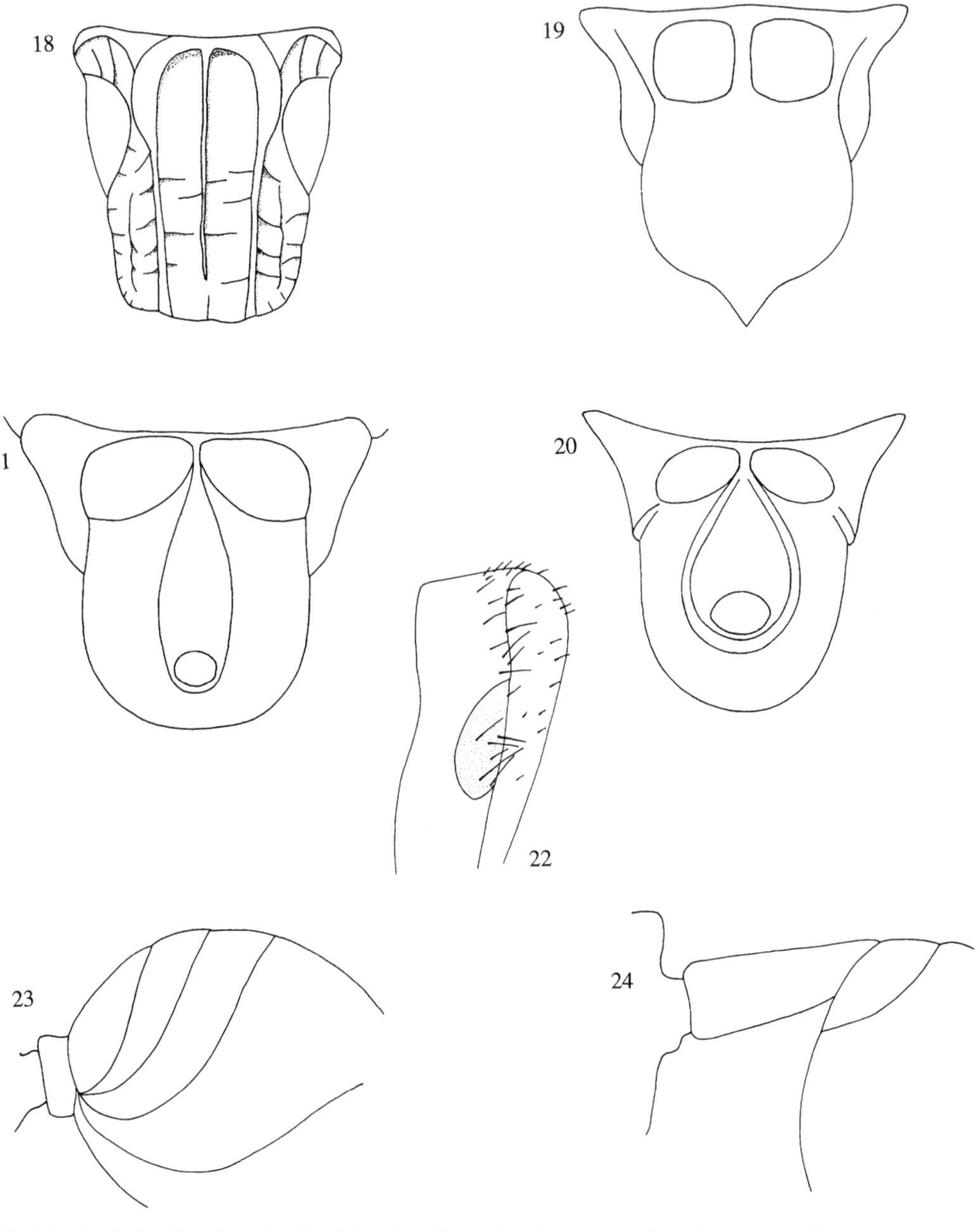

Figs 10.18–10.24. Cynipoidea. Figs 10.18–10.21. Scutellum, dorsal; 10.18, *Callaspidia* sp. (Figitidae: Aspicerinae); 10.19, *Xyalophora* sp. (Figitidae: Figitini); 10.20, *Trybliographa* sp. (Figitidae: Eucoilini); 10.21, *Kleidotoma* sp. (Figitidae: Eucoilini). Fig. 10.22. Depressed area on tergite IX, *Biorhiza* sp. (Cynipidae). Figs 10.23–10.24. Anterior part of metasoma, lateral; 10.23, *Kiefferiella* sp. (Ibaliidae: Liopterini); 10.24, *Liopteron* sp. (Ibaliidae: Liopterini).

Biology. The Cynipoidea is perhaps best known for the gall-wasps that induce galls, especially on oaks. However, most cynipoid species are parasitoids of Diptera while a few are parasitoids of Neuroptera and Homoptera (Aphidoidea and Psylloidea). The parasitic cynipoids are all koinobiont endoparasitoids, with a final instar larva that emerges from the host to finish feeding. It pupates, without forming a cocoon, still within the host's puparium (or cocoon or aphid mummy), from which the adult chews its way out. Nearly all cynipoids are solitary parasitoids; most are primary parasitoids except for those attacking Homoptera (the Charipini), which are hyperparasitoids through hymenopterous primary parasitoids.

The typical cynipoid egg has an elongate body and is pedunculate (stalked), with a peduncle that is often longer than the body of the egg. The larval instars of the gall-wasps are hymenopteriform but the larvae of the parasitoid forms undergo hypermetamorphosis,

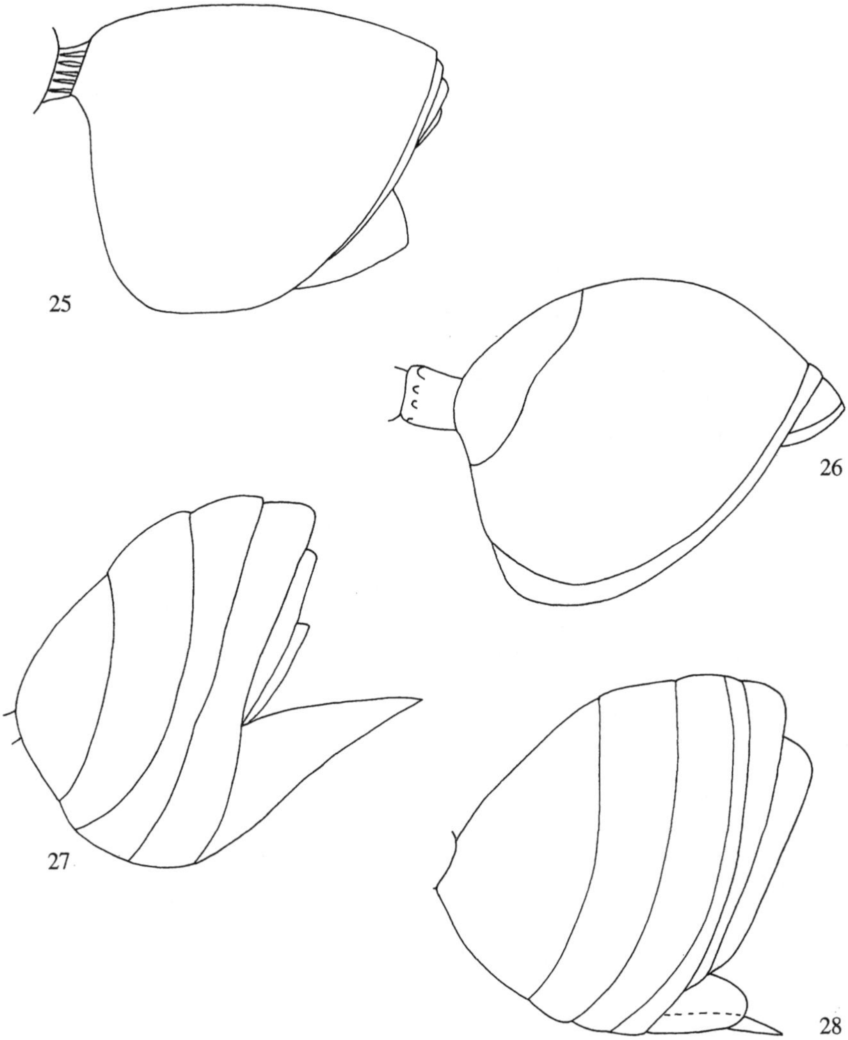

Figs 10.25–10.28. Cynipoidea, metasoma; 10.25, *Synergus* sp. (Cynipidae: Synergini); 10.26, *Aspicera* sp. (Figitidae: Aspicerinae); 10.27, *Diplolepis* (Cynipidae: Rhoditini); 10.28, *Neuroterus* (Cynipidae: Cynipini).

the first instar larva often being strikingly different from the hymenopteriform larva of later instars. For example, the first instar may have thoracic processes and a long tail, but in later larval stages the appendages and tail first shorten and then disappear. The pupa is exarate.

Identification. The literature on the Cynipoidea is mostly in the form of small papers, the only comprehensive works being those of Dalla Torre and Kieffer (1910), which is very out of date, and Weld (1952). The most recent attempts to classify and key major groups are based on the European or Afrotropical

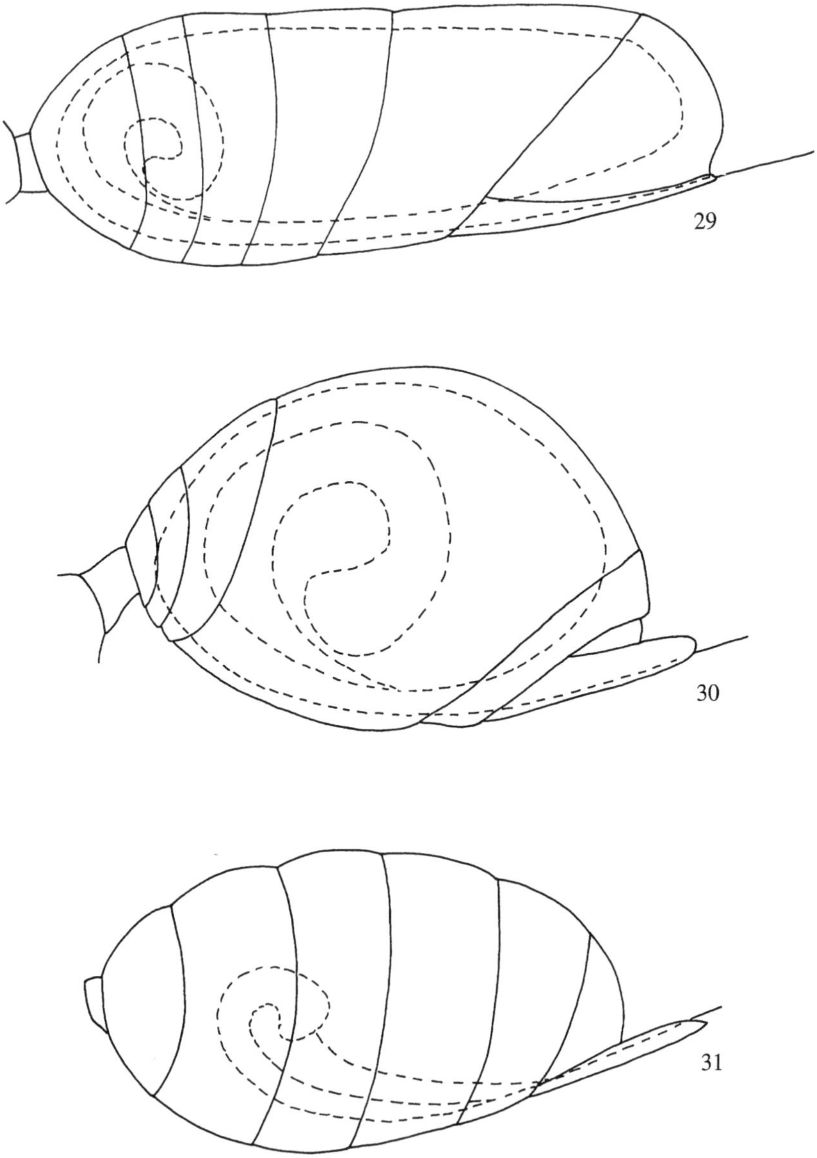

Figs 10.29–10.31. Metasoma showing position of ovipositor (dotted line); 10.29, *Ibalia* sp. (Ibaliidae: Ibaliini); 10.30, Liopterini (Ibaliidae); 10.31, hypothesized ancestral state found in Cynipidae and Figitidae.

fauna (Eady & Quinlan, 1963; Quinlan, 1978, 1979, 1986, 1988; Fergusson, 1986, 1988, 1990). The literature on neotropical species is limited and mainly concerns species from south of the equator (e.g. Houard, 1933; Diaz, 1973, 1978, 1980, 1981, 1984, 1988; Loiácono & Diaz, 1977; De Santis, 1980). Until now little was known of the Costa Rican fauna.

The families of Cynipoidea occurring in the New World can be separated with the following key. It should be noted that the venation character in the second couplet is of great phylogenetic significance, and thus is frequently used in keys to the Cynipoidea, despite the fact that it is a subtle and difficult character.

Key to the New World families of the Cynipoidea

1 Mesoscutum with rough sculpture (rugose, foveolate, or striate) (Fig. 10.08), and with strong transverse ridges; ovipositor looped in a complete circle within the metasoma (Figs 10.29, 10.30); large wasps, 5 to 30 mm, usually over 10 mm; largest segment of the metasoma, in lateral view, the fourth, fifth, or sixth (Figs 10.29, 10.30).
[Very rarely collected.] .... **Ibaliidae** (p. 256)

— Mesoscutum smooth or with light sculpture, without strong transverse ridges; ovipositor not forming a complete loop (Fig. 10.31); small wasps, 1 to 6 mm, usually under 4 mm; largest segment of the metasoma, in lateral view, the second or third (Figs 10.27, 10.28).. 2

2 Fore wing with innermost (proximal) trace of vein *Rs+M* 'pointing' to a point midway down distal part of *Rs+M* (Fig. 10.01); tergite IX of metasoma with depressed area on posterior margin (Fig. 10.22); female with the largest tergite either the second (Fig. 10.28), or the second and third fused (Fig. 10.25); mesosoma usually sculptured.
[In Central America collected only where oak trees are present.]...... **Cynipidae** (p. 253)

— Fore wing with innermost (proximal) trace of vein *Rs+M* 'pointing' to the junction of *Rs+M* with *M+Cu* (Fig. 10.02); tergite IX of metasoma without depressed area; female frequently with the third metasomal tergite the largest, sometimes the second; mesosoma usually smooth and shiny.
[The most commonly collected family.] .... **Figitidae** (p. 259)

10.1 CYNIPIDAE

Nigel D. M. Fergusson and Paul E. Hanson

Diagnosis. Except for the inquilines (see below), the hypostomes are not fused into a bridge under the foramen magnum and the lateral pronotal carinae (part of the pronotal plate) are absent. Mesosoma usually sculptured (except in the genus *Neuroterus*). *Fore wing with proximal trace of Rs+M 'pointing' to a point midway down the distal part of Rs+M* (Fig. 10.01). Mid and hind tibiae with two spurs. Metasoma only rarely with a ring of pubescence on tergite II; *largest tergite (at least in females) either tergite II or tergites II+III fused*; tergite IX with a small depressed area; hypopygium usually developed into a spine (Figs 10.27, 10.28).

Classification and distribution. Worldwide the Cynipidae comprises nearly 100 genera and over 1000 species. The majority of these taxa occur in the Holarctic region, although about ten genera have been recorded from South America. The single cynipid subfamily (Cynipinae) is subdivided into four tribes: Aulacideini, Rhoditini, Synergini and Cynipini. It is possible that all four tribes occur in northern Mesoamerica, but only the last two have been found in Costa Rica.

The Aulacideini and Rhoditini are small groups, present primarily in the Old World and North America. The least derived group, the tribe Aulacideini, have the pronotum long (in dorsal view), the metasoma with all tergites more or less the same size, and hypopygium simple. They form simple, often stem-galls, on a wide range of plants, although each species is specific to a particular genus or species of plant. In North America they form galls on Compositae and *Rubus* (Rosaceae). Members of the tribe Rhoditini have a short pronotum and a specialized ploughshare-like hypopygium (Fig. 10.27). This tribe is represented in the Holarctic region by *Diplolepis*, which forms galls on *Rosa* (Rosaceae) (e.g.

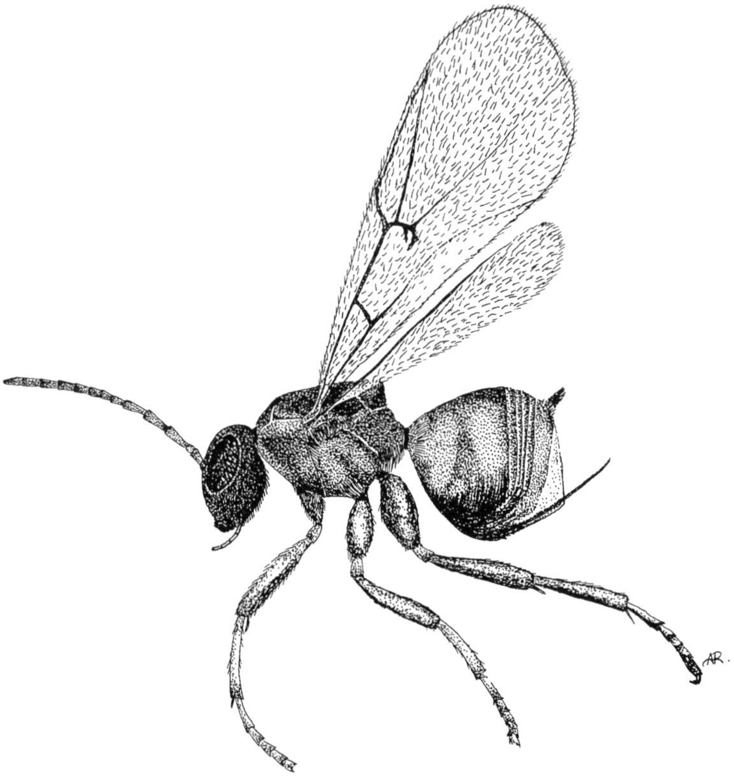

Fig. 10.32. *Callirhytis* sp. (Cynipidae: Cynipini), asexual generation, reared from acorn of *Quercus costaricensis*.

Shorthouse, 1982; Bronner, 1985), and in South America by *Eschatocerus*, which forms galls on *Acacia* and *Prosopis* (Leguminosae) (Diaz, 1981).

The Cynipini is by far the largest tribe in the family and includes about 75 to 80 percent of the known species. Like the Rhoditini, members of the Cynipini have a very short pronotum in dorsal view (median length less than 1/7 the lateral height), but unlike the Rhoditini the hypopygium is modified into a long spine with ventral hairs (Fig. 10.28). Many species form complex galls, and most have complex life cycles (see below). Nearly all species are restricted to attacking the family Fagaceae, especially oaks (*Quercus*), although some undescribed species in southern South America occur in galls on *Nothofagus* species (I. Gauld, pers. comm.).

The Synergini are morphologically derived. Their fused hypostomes (Fig. 10.07) and the presence of pronotal carinae (part of the pronotal plate) are similar to those of the Figitidae (an example of convergence). The maxillary palps are reduced to four segments and the metasoma usually appears to consist of a single syntergite (tergites II and III fused, with the remaining tergites partially hidden—Fig. 10.25). The Synergini is a very specialized tribe consisting entirely of gall inquilines (see Chapter 2.3); the majority are associated with galls formed by Cynipini.

Biology. The female gall-wasp oviposits into the host plant, and the plant reacts to larval secretions, producing a gall. Each species of gall-wasp places its egg in a precise location and each induces a specific type of gall. Thus species of cynipids can often be identified by their gall. The most actively growing tissues are usually selected as oviposition sites, and galls may be formed on buds, leaves, flowers, fruits, young stems and roots. An individual gall may be unilocular (monothalamous), i.e. contain a single larval chamber or, where several eggs are laid together, the galls may unite into a plurilocular (polythalamous) gall consisting of several chambers, with each chamber containing a larva. The grub-like larva of the gall-wasp feeds on the tissue surrounding the larval chamber, the 'nutritive zone', which is rich in nitrogenous compounds (Roth, 1949; Hough, 1953*a*-*c*). It is possible that the larva manipulates nitrogen levels to its own advantage (Hartley & Lawton, 1992).

Gall structure peripheral to the nutritive zone varies considerably between species. In addition to the distinction between unilocular versus plurilocular, it is convenient to distinguish galls that are readily separable ('detachable') from the plant tissue versus those that are not ('integral' galls) (e.g. Weld, 1957, 1959, 1960). Bud galls are often hidden within the bud scales and not visible without dissection. Stem galls may be minute, unilocular cells buried in the twig (usually not noticeable until the adult wasp chews its way out), large plurilocular swellings of branches, or detachable spheres (usually about 1 cm in diameter). Leaf galls range from unilocular cells buried in the leaf tissue with little swelling, to plurilocular swellings, to detachable, spherical or disc-shaped ('spangle') galls. In some cases the gall is naturally dehiscent—i.e. it naturally detaches from the plant and falls to the ground. The outer surface of leaf galls varies from smooth to hairy or spiny, depending on the species. In some cases there is a fairly extensive fibrous layer located between the outer surface and the larval chamber. Some of these features are presumed to be adaptations to hinder attack by parasitoids (Askew, 1984). A few cynipid galls, especially detachable stem galls, secrete 'honeydew' that attracts ants, and at least some of the ant species attracted to the gall may show aggressive behaviour toward parasitoids (Weld, 1925; Washburn, 1984; Abe, 1988).

Most Cynipini show an alternation of generations (heterogony), a sexual generation of males and females alternating with a female-only generation (the agamic or asexual generation) (Folliot, 1964; Askew, 1984). This life cycle involves two types of females which differ considerably in their morphology and in the structure of the gall they cause. The two generations of the same species are often described as different species, sometimes in different genera. There is usually one sexual generation and one asexual generation per year. The sexual generation (often in bud galls) usually develop quickly, generally in just a few weeks, while the asexual generation develops more slowly, the cynipids remaining in the gall for several months to more than a year. The eggs laid by females of the asexual generation, which are always unfertilized, give rise to males and females of the sexual generation. Some species have two types of asexual females, 'androphores' which lay mostly male eggs, and 'gynephores' which lay mostly female eggs (Askew, 1984). Because of this complexity, the complete life cycles are known for very few species, even in the north Temperate zone where they have been better studied. The life cycles of Central American gall wasps remain totally unstudied.

Species of Cynipini are quite host specific. Although a particular species may form galls on more than one species of oak, they are usually restricted to just one subgenus of oaks. Local differences in cynipine species-richness between oak species are probably due to differences in regional richness—i.e. the number of cynipine species that the oak hosts over its entire range (Cornell, 1986). Individual plants belonging to the same species often vary from having virtually no galls present, to having extremely high densities of galls. This probably reflects chemical, structural and phenological differences between individual trees (Askew, 1962; Rosenthal & Koehler, 1971; Ejlersen, 1978). Secondary plant compounds probably have little detrimental effect since the larvae feed on nutritive tissue. In fact the reverse may be true—cynipids apparently prefer plants with high levels of tannins (Taper & Case, 1987). In the only manipulative study done to date, one species of gall wasp showed a negative correlation between survivorship and gall nitrogen content (Hartley & Lawton, 1992).

Species of Synergini do not form galls of their own, but are inquilines in the developing galls of other species (mostly in galls of Cynipini). These inquilines exploit the larval food supply of a gall-forming species and this competition may or may not result in the death of the original gall-former. Synergini were probably once gall-forming species, but they have become dependent on other gall-formers. A given species of inquiline is predominantly associated with a particular type of gall and is often quite specific with respect to host plant (Askew, 1984), although the reason for this remains unknown. Although they are incapable of initiating their own gall, some produce modifications in the structure of the host gall. Several inquilines may develop in one gall, and sometimes of more than one species. Interestingly, some bivoltine species of *Synergus* show generational dimorphism (individuals of the two generations differ morphologically), although they do not necessarily attack both generations of the same host species (Askew, 1984).

Finally, it should be noted that not all inquiline cynipids belong to the Synergini: *Myrtopsen mimosae* (Cynipini) is an inquiline in galls of *Tanaostigmodes albiclavis* (Tanaostigmatidae) on *Mimosa biuncifera* (Leguminosae) in the arid areas of the southwestern United States and adjoining Mexico. Four species occur in southern South America where they have been reared from galls on Myrtaceae (*Blepharocalyx* and *Eugenia* (Diaz, 1975). Species of the endemic Australian genus, *Thrasorus*, are inquilines in pteromalid galls on *Acacia* (Leguminosae) and *Eucalyptus* (Myrtaceae) (Weld, 1952).

Identification. Alpha-level taxonomy of Cynipidae is in a very undeveloped state, particularly in the Cynipini, where the alternation of generations poses an impediment not encountered in other microhymenopteran groups. Careful field studies and/or experiments with caged plants are required to match the two generations of one species. Identification of North American taxa is more or less possible using Weld (1957, 1959, 1960), although the alternate generation remains unknown in most species. Many species of Synergini (i.e. *Synergus*) are also difficult to identify, even in the Holarctic region where taxonomic studies are available (Wiebes-Rijks, 1979). Although Kinsey's (1930) treatise included many Mexican species, the cynipid fauna of Central and South America has received very little attention, except for the publications mentioned above at the start of this chapter. The following couplet separates the two tribes occurring in Central America.

Key to tribes of Cynipidae present in Costa Rica

— Pronotum long, in dorsal view median dorsal length greater than 15 percent of its lateral length.
[Nearly all Central American specimens are *Synergus*, a genus characterized by having metasomal tergite I (petiole) in the form of a short ring that is longitudinally ridged (Fig. 10.25).] .... Synergini

— Pronotum short, in dorsal view median dorsal length less than 15 percent of its greatest lateral length; petiole not longitudinally ridged. .... Cynipini

Synopsis of the Costa Rican taxa

In Costa Rica there are twelve species of oaks (Burger, 1977), six in the subgenus *Leucobalanus* ('white oaks') and six in the subgenus *Erythrobalanus* ('red oaks'). On Cerro de la Muerte at an altitude between 2500 and 3000 metres, there are two common species, one white oak (*Q. copeyensis*) and one red oak (*Q. costaricensis*), and these two oak species do not have any gall–wasps in common. The former appears to have, by North American standards, a 'normal number' of gall wasps — perhaps as many as 15 species (= 30 types of galls), whereas *Q. costaricensis* appears to have slightly fewer (perhaps 5 to 10 species). In Costa Rica most oaks occur at mid to high altitudes (above 1200 m) in cool, wet forests, but *Q. oleoides* is exceptional in that it grows in the dry lowlands of northwestern Costa Rica. This latter oak species appears to be very depauperate in gall wasps and considerable searching efforts have revealed only two species of Cynipini, which contradicts the hypothesis that species-richness of gall-formers increases in hotter, drier sites (Fernandes & Price, 1988).

About half of all cynipid specimens collected in Costa Rica are inquilines (Synergini), and all of these belong to the genus *Synergus*. Genera of Cynipini collected thus far include *Andricus*, *Callirhytis*, *Disholcaspis*, *Dryocosmus* and *Neuroterus*.

10.2 IBALIIDAE

Nigel D. M. Fergusson

Diagnosis. *Large (4.5 to 30.0 mm), coarsely sculptured,* uncommon cynipoids. Face normally with a weak central ridge; hypostomal region set in a cavity (Fig. 10.06). Pronotum often with a dorsal tooth (Fig. 10.08). Last tergite of metasoma with a slightly upturned end; ovipositor very long, about 1.5 times the body length, coiled in a full circle inside the metasoma (Figs 10.29, 10.30). The presence of the coiled ovipositor tends to cause an expansion of the posterior

Fig. 10.33. *Paramblynotus* sp. (Ibaliidae: Liopterini).

tergites such that, *in lateral view, the apical tergites are the longest.*

Classification and distribution. The Ibaliidae is cosmopolitan, but very infrequently collected. It comprises approximately 100 species and 13 genera placed in two subfamilies: Austrocynipinae (with a single aberrant species from Australia; Riek, 1971) and Ibaliinae (cosmopolitan). The subfamily Ibaliinae is divided into two tribes, Ibaliini and Liopterini, but only the latter is known to occur in Costa Rica.

The Ibaliini comprises two genera, of which one, *Ibalia*, occurs in the New World. This is a primarily holarctic taxon and most species are associated with coniferous forests. In the New World one species, *I. ruficollis*, has a range that extends southwards into Mexico and it is possible that it could occur as far south as the natural distribution of Pinaceae, i.e. northern Nicaragua. Such a species may eventually naturally spread south onto introduced pines in Costa Rica. *Ibalia* is easily recognized by its laterally compressed and blade-like metasoma, and the very elongate marginal cell in the fore wing (Fig. 10.03).

The remaining genera of the family are placed in the tribe Liopterini. Traditionally this group has been treated as a family ('Liopteridae') and the genera were placed in three subfamilies—'Oberthuerellinae', 'Mesocynipinae' and 'Liopterinae'. However, these are not retained because only the genera related to *Oberthuerella* form a holophyletic group, the other two are paraphyletic assemblages (Fergusson, 1990). The Liopterini are principally a circumtropical group with isolated species in Australia, eastern Asia and North America. The *Oberthuerella* genus-group is limited to the Afrotropical region while five genera occur in the New World: *Kiefferiella, Liopteron, Paramblynotus, Plastibalia,* and *Pseudibalia*. Species of all are extraordinarily rare in collections.

Biology. The single species of Austrocynipinae has been reared from conifer (Araucariaceae) seeds in Australia, but its biology is unknown (Riek, 1971). Except for *Ibalia*, the biology of Ibaliinae is not well known either. The available data indicate that Ibaliinae are koinobiont endoparasitoids of wood-boring insects. *Ibalia* species oviposit into the developing eggs or young larvae of Siricoidea (Siricidae and Anaxyelidae), and emerge from a more fully grown larva (Cameron, 1965; Burks *in* Krombein *et al.*, 1979). In Argentina *Paramblynotus zonatus* has been reported attacking a larva of the cerambycid genus *Oncideres* boring in *Nectandra* sp. (Lauraceae) (Diaz, 1973). In Costa Rica, a species of *Liopteron* was observed examining a fallen branch of *Pentaclethra macroloba* (Leguminosae) (H. Hespenheide, pers. comm.).

In *Ibalia* oviposition occurs via the oviposition shaft of the siricoid host and into the haemocoel of the host embryo or larva. Hosts parasitized by *Ibalia* burrow closer to the surface than unparasitized hosts. These wood-probing parasitoids tend to have a low population density and to develop slowly, with life cycles of up to four years.

Identification. Identification of the genera is possible using Weld (1952). Kerrich (1973) provided a key to some *Ibalia* species and a useful bibliography of both systematic and biological studies. Fergusson (1988) described the morphology of *Ibalia*. Hedicke and Kerrich (1940) revised the species of *Liopteron, Plastibalia* and *Pseudibalia*.

Key to the New World genera of Liopterini

1 Petiole not or scarcely longer than broad, attached to remainder of metasoma horizontally (Fig. 10.23). 2

— Petiole at least twice as long as medially broad and attached to remainder of metasoma tangentially to its dorsal curvature (Fig. 10.24).... 3

2 Petiole smooth, without obvious sculpture; metasoma, in dorsal view with tergites IV and V of similar length. .... *Kiefferiella*[1]

— Petiole with coarse parallel longitudinal rugae; metasoma, in dorsal view, with tergite IV about 0.5 times as long as tergite V. .... *Paramblynotus*

3 Female hind tarsus without blunt spur (apical prolongation) on outer side of segment 1 or 2. ....*Liopteron*

— Female hind tarsus with a blunt spur (apical prolongation) on outer side of segment 1 or 2 (Figs 10.15, 10.16)....................................... 4

4 Spur present on second segment of hind tarsus (Fig. 10.15). .... *Plastibalia*[1]

— Spur present on first segment of hind tarsus (Fig. 10.16). .... *Pseudibalia*[EX]

[1] Genera neither recorded nor thought likely to occur in Costa Rica.

Synopsis of the Costa Rican fauna

Only two genera of Liopterini have been found to occur in Costa Rica. *Pseudibalia* could occur in the country, *Kiefferiella* occurs in the United States and *Plastibalia* is known from the Amazon Basin.

Liopteron. This genus comprises about 30 described species, most of which are known from isolated specimens collected by H.W. Bates during his trips along the Amazon. A few species occur as far north as Mexico. One species, that appears to be *L. brasiliensis*, has been collected on the Atlantic lowland plain of Costa Rica at 100 metres, and *L. westwoodii* has been collected in humid forest at an altitude of about 1600 metres. *L. tarsale* is reported to occur in Mexico and Brazil, and therefore may also be present in Costa Rica.

Paramblynotus. This genus comprises about ten species, most of which occur in the lowland south-east Asian forests, but isolated species are present in Africa and the New World. One species, *Paramblynotus zonatus*, is widely distributed throughout the Americas, from Texas south to southern

Brazil and northern Argentina. Very occasionally individuals of this have been collected in the seasonally dry lowland forests of northwestern Costa Rica.

Pseudibalia[EX]. A genus comprising two species, one from Mexico and a second from the Amazon basin of Brazil (Hedicke & Kerrich, 1940). Both have been collected only extremely infrequently.

10.3 FIGITIDAE

Nigel D. M. Fergusson

Diagnosis. Back of head with a flat, completely fused hypostomal bridge (Fig. 10.05), the line of fusion may or may not be visible. Pronotum with at least a trace of lateral carinae (Fig. 10.12), but in most species these carinae form part of a raised anterior pronotal plate (Figs 10.10, 10.11, 10.13); mesosoma usually smooth and shiny. *Fore wing with vein Rs+M 'pointing' at junction of Rs+M with M+Cu* (Fig. 10.02). *Metasomal tergite II in lateral view normally shorter than III.*

Classification and distribution. The cosmopolitan Figitidae, comprising approximately 110 genera and 1500 species, is the most species-rich and abundant cynipoid family. However, relatively few taxonomic studies have been undertaken on the diverse tropical fauna and very many species await description.

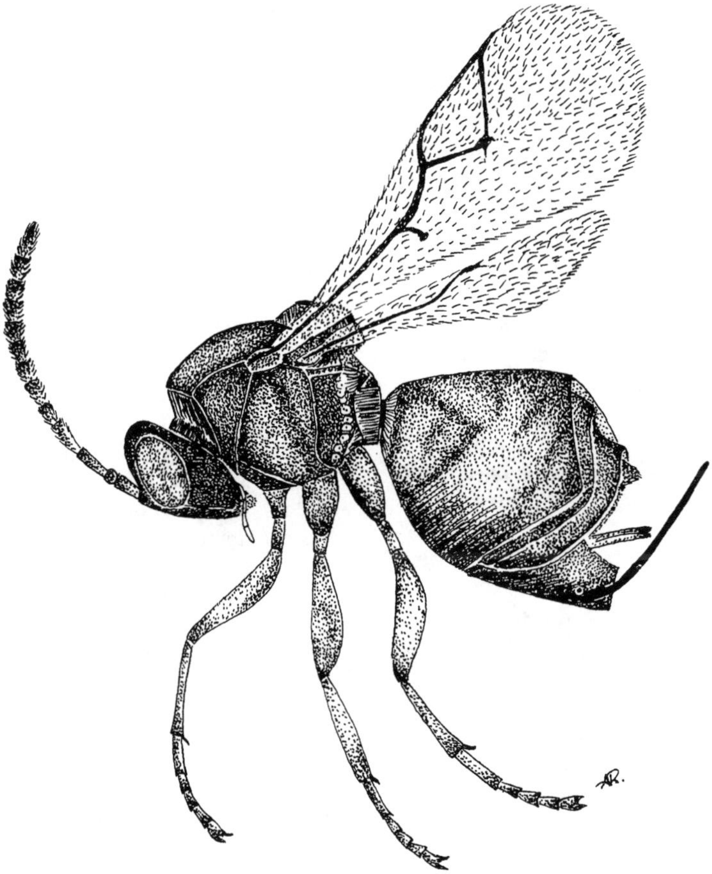

Fig. 10.34. *Ganaspis* sp. (Figitidae: Eucoilini).

Previously the taxa comprising this family were often treated as three separate families—'Figitidae', 'Eucoilidae', and 'Charipidae'—but a detailed morphological study and a cladistic analysis of the results indicate that such taxonomic ranking can no longer be justified (Fergusson, 1990). The family Figitidae is now divided into three subfamilies: Anacharitinae, Aspicerinae, and Figitinae. The last of these subfamilies comprises three tribes: Figitini, Charipini and Eucoilini.

Key to subfamilies and tribes of Figitidae

1 Pronotal carina complete (Fig. 10.10), continuously distinct from ventral region on one side across dorsum to ventral region on other side; claws with a fine basal spine; petiole with a collar ventrally and laterally but not dorsally. .... **Anacharitinae** (p. 260)

— Pronotal carina indistinct (Fig. 10.12), or with a lateral gap (Fig. 10.11), or present only as a developed dorsal plate (Fig. 10.13); claws without a fine basal spine; if petiole with a collar then it has a dorsal element. 2

2 Metasoma with tergite II saddle-shaped, with postero-lateral margin concave and central part almost tongue-like (Fig. 10.26); scutellum with three strong carinae (Fig. 10.18); hind tibia usually with a strong longitudinal ridge. . ... **Aspicerinae** (p. 261)

— Metasoma with tergite II not saddle-shaped, its postero-lateral margin slightly to strongly convex; scutellum without three strong carinae (Figs. 10.19-10.21); hind tibia without a longitudinal ridge, or with a weak ridge. .... [**Figitinae**]..... 3

3 Scutellum with an oval or tear-drop shaped, elevated plate (Figs 10.20, 10.21, 10.37)............ .... **Eucoilini** (p. 262)

— Scutellum without an elevated plate. 4

4 Pronotal carinae indistinct (Fig. 10.12); mesoscutum and scutellum smooth and shiny; veins restricted to the antero-proximal quarter of wing (Fig. 10.04); very small insects, 1.0 to 2.0 mm long. .... **Charipini** (p. 261)

— Pronotal carinae distinct (Fig. 10.10, 10.11); mesoscutum and scutellum never completely smooth; veins not restricted to the antero-proximal quarter of wing (Fig. 10.02); usually larger than 2.0 mm. .... **Figitini** (p. 264)

Anacharitinae

A cosmopolitan subfamily with at least twelve currently recognized widespread genera. Five occur in the Neotropical region and three of these are known to occur in Costa Rica; the other genera not represented in Costa Rica, are *Aegilips* and *Calofigites* which occur in southern South America. Anacharitines are koinobiont endoparasitoids of the larvae of Hemerobiidae (Neuroptera). In at least one species the female wasp oviposits in late second or early third instar host larvae (Miller & Lambdin, 1985; Cave & Miller, 1987). When the parasitoid reaches the third instar and the host has spun its cocoon, the anacharitine larva emerges from the host and feeds externally. During this time it increases notably in size, and then thrusts the apex of its abdomen through the host's cocoon to void the gut contents. The final instar larva of *Anacharis* has well-developed, tridentate mandibles, and the second to ninth body segments support paired conical, fleshy projections. The anacharitine pupates within the host cocoon, and after emerging from its pupa the adult remains within its host's cocoon for a few days.

Key to genera of Anacharitinae present in Costa Rica

1 Mesoscutum coarsely reticulate; apex of scutellum with very long, sharp spine; fore wing with marginal cell anteriorly open, not bounded by pigmented *R*1. .... *Acanthaegilips*

— Mesoscutum not coarsely reticulate; apex of scutellum with a blunt spine or with no spine; fore wing with marginal cell bounded by pigmented vein *R*1 anteriorly. 2

2 Apex of scutellum with a blunt spine. ... *Xylaspis*

— Scutellum without a spine. .... *Anacharis*

Synopsis of Costa Rican genera of Anacharitinae

Acanthaegilips. A neotropical genus with one or two Costa Rican species occurring in moist sites at altitudes between 1000 and 2100 metres.

Anacharis. A cosmopolitan genus with two or three Costa Rican species occurring in humid sites between 1000 and 3000 metres.

Xylaspis. A holarctic genus that, in the New World, apparently reaches the southern limit of its distribution in Costa Rica where one or two species have been found in various sites from near sea-level to over 3000 metres. One species is occasionally collected in seasonally dry sites.

Balna. A neotropical genus that in Costa Rica is known only from two specimens collected in somewhat dry, disturbed vegetation west of San José, at 1000 metres.

Prosapicera. A New World genus that in Costa Rica is known from a few specimens collected in wet lowland to mid altitude forests between 200 and 2200 metres. Keys to species of *Prosapicera* were provided by Diaz (1984).

ASPICERINAE

This subfamily consists of eight genera two of which occur in Costa Rica. As far as is known aspicerines are parasitoids of syrphid dipteran larvae. Species of the holarctic genera *Callaspidea* and *Omalaspis* have been reared from larvae of aphidophagous syrphids (see Fergusson, 1986), and a species of the neotropical genus *Balna* is here reported for the first time from an unidentified syrphid larva, which was preying on scale insects on citrus (P. Hanson, pers. comm.). Although detailed information is available only for *Callaspidea*, presumably all aspicerines are koinobiont endoparasitoids. The syrphid host is located via olfactory cues given off by the aphid colonies on which they prey. The female wasp temporarily paralyses a second or third instar host larva by stinging it, and then places an egg in the cerebral ganglion (Rotheray, 1979), thus avoiding any encapsulation response by the host. The adult parasitoid emerges from the syrphid puparium.

Key to genera of Aspicerinae present in Costa Rica

— Mesoscutum with a vertical protuberance; fore wing with veins normally pigmented; apex of scutellum pointed but without a long spine. ....*Balna*

— Mesoscutum without a vertical protuberance; fore wing with veins very pale; scutellum with a very long apical spine. .... *Prosapicera*

Synopsis of Costa Rican genera of Aspicerinae

Five genera are present in the New World, one possibly as an inadvertent introduction (Burks *in* Krombein *et al.*, 1979), and two of these occur in Costa Rica.

FIGITINAE

This cosmopolitan subfamily comprises three tribes, the Charipini, Figitini, and Eucoilini, all of which occur in the Neotropical region and all of which occur in Costa Rica. Biologically these are rather diverse groups, so they are discussed below separately.

FIGITINAE: Charipini. This tribe, although highly derived, is defined by size related loss of characters (e.g. loss of sculpture). It was previously divided into two subgroups ('Alloxystinae' and 'Charipinae') but recent analysis (Fergusson, 1990) shows that this cannot be substantiated. Worldwide there are about 200 species of charipines in five genera (Menke & Evenhuis, 1991): *Alloxysta* (cosmopolitan), *Apocharips* (Fergusson, 1986; Old World and Costa Rica), *Dilyta* (Holarctic and Africa), *Lytoxysta* (North America), and *Phaenoglyphus* (Holarctic). Most species belong to the genus *Alloxysta* and some, such as *A. victrix*, are very widespread (Andrews, 1978; Diaz, 1980).

The Charipini are hyperparasitoids of Homoptera. *Dilyta* and *Apocharips* species are hyperparasitoids of Psylloidea via encyrtid primary parasitoids (Menke & Evenhuis, 1991) and *Alloxysta*, *Lytoxysta* and *Phaenoglyphus* are hyperparasitoids of Aphididae via aphidiine braconid or aphelinid chalcid primary parasitoids. In general little is known about the biology of most species, although the cosmopolitan species, *A. victrix* which is a common hyperparasitoid associated with aphid crop pests, has been intensively studied. For example, its host selection behaviour has been thoroughly investigated (Gutierrez & Bosch, 1970*a* & *b*; Gutierrez, 1970*a-d*). Host specificity is not a simple association between hyperparasitoid and primary parasitoid, but rather is influenced by preferences for particular aphid/primary parasitoid combinations, and possibly certain host plants (Hafez, 1961; Andrews, 1978; Sullivan, 1987). The female charipine oviposits into the haemocoel of the primary parasitoid larva which in turn lies in the haemocoel of the aphid. More than one egg may be laid in a host, but only one will complete development. Aphids lacking a primary parasitoid are probed, but no eggs are laid. The egg apparently does not hatch until after the primary parasitoid has completely devoured the aphid host (Sullivan, 1972).

The first instar larva of *Alloxysta* is conspicuously sclerotized, and has well-developed mandibles (Haviland, 1921). Early instars feed internally, but later instars feed externally, and pupation occurs within the mummified skin of the aphid. The adult chews an irregular-shaped hole through the dorsal surface of the aphid mummy and emerges. Adults feed on sugary solutions, but host-feeding has not been observed (Matejko & Sullivan, 1980).

Alloxysta species may be parasitised by other aphid hyperparasitoids, but the endoparasitic *Alloxysta* is not invariably killed when the host is later attacked by an ectoparasitoid such as the pteromalid *Asaphes* (Sullivan, 1972). The ectoparasitic *Dendrocerus* (Megaspilidae) can successfully parasitize *Alloxysta* for only a brief period of time, shortly after the latter begins to feed externally on the primary host (Matejko & Sullivan, 1984).

Key to genera of Charipini likely to occur in Costa Rica

1 All antennal segments separated by constrictions; visible metasoma consisting of two large, subequal tergites. .... *Alloxysta*

— Last two antennal segments broadly joined; visible metasoma not as above. 2

2 Visible metasoma consisting of one large tergite. .... *Dilyta*[EX]

— Visible metasoma consisting of a small basal tergite followed by a much larger tergite. .... *Apocharips*

Synopsis of Costa Rican Charipini

Alloxysta. A cosmopolitan genus. One species, *A. victrix*, is very widely distributed and has probably been transported inadvertently to most parts of the world where it is a hyperparasitoid of aphid crop pests. In Costa Rica *Alloxysta* species have only been collected from mid to high altitude sites (above 1500 m).

Apocharips. A predominantly Old World genus that was previously unknown from the New World. *A. hansoni* has been reared from *Trioza* sp. (Psylloidea), which forms leaf galls on *Cinnamomum hammelianum* (Lauraceae) at higher elevation sites (2000–2500 m) (Menke, 1993).

Dilyta[EX]. A widespread but rarely collected North American genus (Menke & Evenhuis, 1991).

FIGITINAE: Eucoilini. Eucoilines are easily distinguished from all other microhymenoptera by the raised plate or 'cup' on the disc of the scutellum. The tribe is thus very distinctive and in previous classifications (e.g. Quinlan, 1979), it has often been accorded a higher rank than it warrants (Fergusson, 1990). Most species are fully winged but brachypterous individuals are known. It is the most speciose group of Cynipoidea and includes the majority of species encountered in most tropical situations. Indeed probably more than 95 percent of all the cynipoid wasps collected in Malaise traps in Costa Rica belong to this tribe.

Worldwide there are about 1000 described species in nearly 80 genera. In Costa Rica there are at least 20 genera present, which include perhaps between 200 and 500 species. Most species are undescribed.

Eucoilines are endoparasitoids of the larvae of a wide range of Diptera, including those in living plant tissue, such as Tephritidae, Chloropidae and Agromyzidae, but especially saprophagous species of Sepsidae, Sphaeroceridae, Drosophilidae, Ephydridae, Phoridae, Muscidae, Calliphoridae and Sarcophagidae found feeding in animal dung, carrion, rotting fruit and vegetation, birds' nests, and similar sites. Several eucoiline species occur in rotting vegetation on watersides, in swamps and along high tide lines. One European species, for example, is known to attack only the larvae of Coelopidae in rotting seaweed (Quinlan, 1978) and in other parts of the world several eucoilines attack semiaquatic ephydrids (Williams, 1938; Beardsley, 1989; Meyers & Deonier, 1993). One *Kleidotoma* species is known to enter freshwater to attack aquatic dipteran larvae of the families Ephydridae and Canaceidae (Beardsley, 1990).

Association with animal dung (e.g. by *Cothonaspis* which parasitizes Sepsidae) appears to have resulted in the loss of pubescence on the propodeum and at the base of the second metasomal tergite (Nordlander, 1978*a*). Some species of *Eucoila*, *Kleidotoma* and *Trybliographa* associated with synanthropic flies have become very widespread. A few species of Eucoilini have been fairly intensively studied, especially those that parasitize *Drosophila melanogaster* (Drosophilidae), such as *Leptopilina boulardi* and *L. heterotoma* (previously known as *Pseudeucoila bochei*; see Nordlander, 1980). Studies of these two species, many of which are concerned with odour learning by adult females, include Bakker *et al.* (1967), Meyer-Grassmann (1967), Carton and Kitano (1981), Carton *et al.* (1987), Vet and Schoonman (1988), Papaj and Vet (1990), Vet and Groenewold (1990), and DeJong and Kaiser (1991).

The female oviposits into the haemocoel of early instar dipterous larvae, often selecting the anterior or posterior end as an oviposition site. The egg is ovoid with a long tubular pedicel and following oviposition it rapidly increases in size. The first instar larva is highly specialized (known as the 'eucoiliform' type larva), having reduced mandibles and the mouth adapted for fluid feeding; it has a long tail and three pairs of long, fleshy thoracic processes, which are lost in the later instars (Keilin & Baume-Pluvinel, 1913). It also has a large anus surrounded by a ring of inwardly directed long chitinous spines (Huzimatu, 1940). As the first instar larva grows its body becomes proportionately larger with respect to its caudal appendage (Jenni, 1951). The second instar larva resembles the first in outline but possesses a shorter caudal appendage and lacks the paired thoracic appendages and modified anus (Wishart & Monteith, 1954; Eskafi & Legner, 1974*a* & *b*). There are three to five larval instars, and in species with five, the third instar resembles the second (Huzimatu, 1940) while the fourth instar is stouter and has a tracheal system with one to three pairs of spiracles. During this instar the parasitoid emerges from the host pupa (but remains within the puparium) and consumes the host from the outside (Meyers & Deonier, 1993). The final instar larva is typically hymenopteriform and usually has eight or nine pairs of spiracles. Pupation occurs within the host puparium.

Eucoilines apparently have overcome the immune defence system of their dipterous hosts by means of a venom secreted by the accessory gland of the adult female, and injected during oviposition. This venom contains virus-like particles that selectively destroy the host's lamellocytes, the dipteran blood cells involved in recognition and encapsulation of large foreign objects. The venom causes the lamellocytes to lose their adhesiveness which in turn hinders the process of encapsulation (Rizki & Rizki, 1990).

Male eucoilines are readily distinguished from females by their longer antenna. In addition the male antenna has either the third or fourth segment modified (the latter being apomorphic according to Nordlander, 1982*b*), which is used to stroke the club of the female antenna during courtship (Assem, 1969). Males mate repeatedly, but a newly mated female is not receptive to further sexual overtures from males.

Several members of the tribe are economically important as parasitoids of dipterous pests, such as fruit flies and agromyzid leaf miners. *Trybliographa daci* was introduced into Costa Rica from Hawaii in the 1950's as a biological control agent of the Mediterranean fruit fly (Tephritidae) but the outcome of this introduction is unclear (Wharton *et al.*, 1981; Jirón & Mexzon, 1989). Two native species, *Ganaspis carvalhoi* (= *G. pelleranoi*) and *Odontosema anastephae*, are known to parasitize fruit flies in Costa Rica (Wharton *et al.*, 1981). In Argentina *Agrostocynips clavatus* is reported as a parasitoid of the leaf-miner *Liriomyza huidobrensis* (Agromyzidae) (Diaz & Valladares, 1979; Valladares *et al.*, 1982). *Ganaspidium utilis*, which attacks *Liriomyza sativa* and *L. trifolii*, was introduced from Texas to Hawaii (Beardsley, 1988; Petcharat & Johnson, 1988). In Honduras two species of *Cothonaspis* have been reared from *Liriomyza sativae* (R. Cave, pers. comm.).

The identification of genera of Eucoilini is very difficult since the taxonomy of this tribe has received so little attention. Keys are available for Britain (Quinlan, 1978), Hawaii (Beardsley, 1988, 1989), and Taiwan (Lin, 1988). The only key to genera that includes the neotropical fauna is that of Weld (1952), but subsequent to Weld's work several genera have been redefined (Nordlander, 1976, 1978*a* & *b*, 1980, 1981, 1982*a*) and a few new genera have been described (Quinlan, 1976, 1984, 1986). Genera that have been identified from Costa Rica include: *Acantheucoela*,

Diceratаspis, *Ganaspis*, *Gronotoma*, *Hexacola*, *Kleidotoma*, *Leptopilina*, *Odontosema*, *Trybliographa* and *Tropideucoila* (P. Hanson & I. Gauld, pers. comm.).

FIGITINAE: Figitini. The Figitini is an unresolved paraphyletic group. The genera related to *Figites*, which have a band of hairs on the second metasomal segment, form the sister-group of the Eucoilini. The remaining Figitini (genera near *Melanips*) lack the band of pubescence and are without a synapomorphy. The tribe comprises about 125 described species in 13 genera. Thus far only two genera, *Figites* and *Neralsia*, have been found in Costa Rica.

The Figitini are endoparasitic koinobionts of a wide range of Diptera larvae. The biology of the genera *Figites* is similar to that of Eucoilini and the species are often associated with flies in dung, carrion, or rotting fruit. Species of *Melanips* often parasitize dipterous larvae that prey on aphids, i.e. Syrphidae and Chamaemyiidae (Evenhuis, 1968). Only one species, *Figites anthomyiarum*, has been extensively studied (James, 1928). The female of this species oviposits into first or second instar larvae of a variety of saprophagous cyclorrhaphans, after first temporarily paralysing the host. The elongate first instar larva has a large head and a short caudal appendage; the thoracic

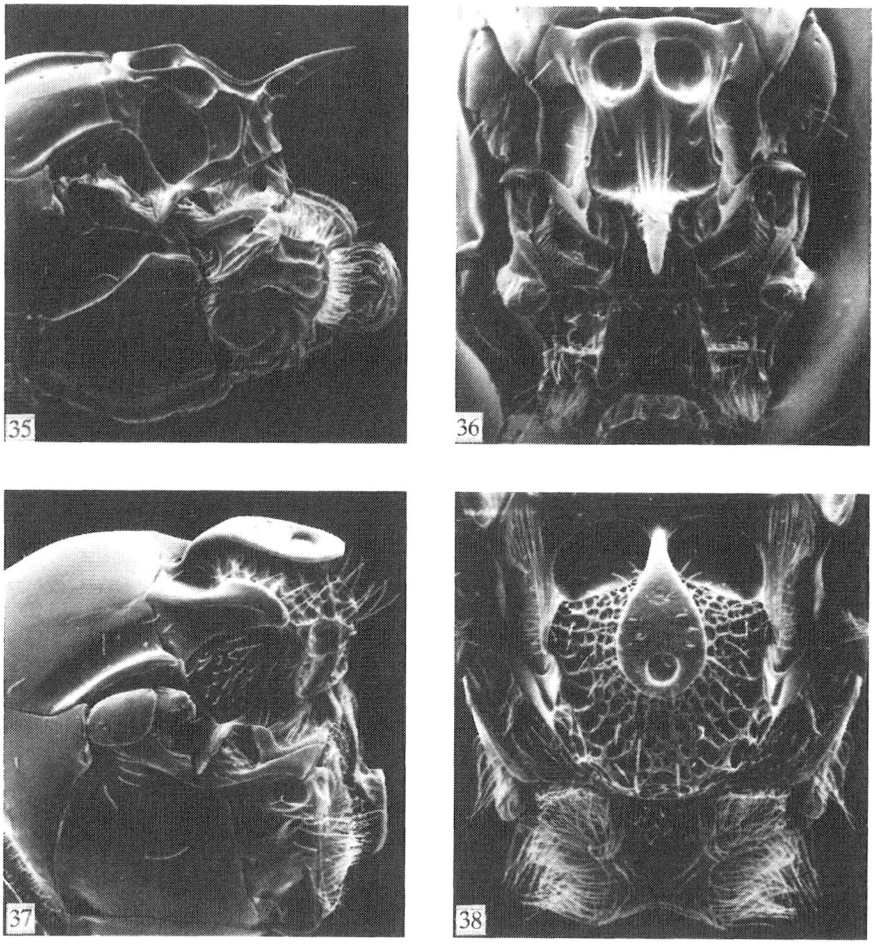

Figs 10.35–10.38. Scanning electron photomicrographs of mesosoma, Figitidae. Figs 10.35–10.36. *Neralsia* sp. (Figitini); 10.35, lateral; 10.36, dorsal. Figs 10.37–10.38. *Kleidotoma* sp. (Eucoilini); 10.37, lateral; 10.38, dorsal.

segments bear paired ventral fleshy processes. The second instar larva is polypodeiform with rudimentary paired appendages on the first ten body segments. The adult wasp emerges from the host puparium.

Key to genera of Figitini likely to occur in Costa Rica

1 Apex of scutellum with a long spine or a short tooth. 2
— Scutellum without a spine or tooth. 3

2 Scutellum with a long spine (Figs 10.36, 10.37). .... *Neralsia*
— Scutellum with a short tooth. .... *Xyalophora*[EX]

3 Eye with long hairs; mesoscutum shiny. .. *Figites*
— Eye glabrous; mesoscutum granulate and dull. .... *Melanips*[EX]

Synopsis of Costa Rican Figitini

The majority of genera in this tribe are only known to occur in temperate regions. Ten genera occur in the New World, but only two have been found in Costa Rica. Two others may occur here.

Figites. A cosmopolitan genus that in Costa Rica is known only from one site, at 1600 metres in altitude. In Europe and North America species of this genus are known to attack a range of cyclorrhaphous Diptera, especially Anthomyiidae, Calliphoridae, Muscidae and Sarcophagidae.

Melanips[EX]. A widespread genus known from all regions except the Afrotropical. Hosts are often dipterous predators of aphids (Fergusson, 1986).

Neralsia. A New World genus that in Costa Rica is widely distributed from near sea-level to 1600 metres in altitude. A North American species is known to attack Sarcophagidae (Roberts, 1935).

Xyalophora[EX]. A widespread genus occurring in Europe and Africa, and throughout the New World. A North American species is known to attack Sarcophagidae (Burks *in* Krombein *et al.*, 1979).

11 The chalcidoid families

INTRODUCTION

Paul E. Hanson and John LaSalle

The Chalcidoidea is the second most species-rich superfamily of Hymenoptera, exceeded in size only by the Ichneumonoidea. Presently the superfamily contains about 2000 genera and 19,000 described species, but the actual number of species could be as high as 100,000 (Noyes, 1990). In terms of body size, Chalcidoidea include some of the smallest of all insects; *Megaphragma*, a trichogrammatid parasitoid of thrips' eggs, is only 0.17 mm in length. The largest members of the superfamily are species of the pteromalid subfamily Leptofoeninae, which reach 25 mm in length, excluding the ovipositor. Unlike other microhymenoptera (Ceraphronoidea, Proctotrupoidea and Cynipoidea), many chalcidoids are bright metallic green or blue in coloration.

Biologically the Chalcidoidea is one of the most diverse superfamilies of Hymenoptera, with larval feeding habits ranging from carnivory to phytophagy (Bendel-Janssen, 1977). Parasitoid chalcidoids attack hosts belonging to 15 orders of Insecta (Odonata, Orthoptera, Blattaria, Mantodea, Thysanoptera, Hemiptera, Homoptera, Psocoptera, Neuroptera, Coleoptera, Strepsiptera, Lepidoptera, Diptera, Siphonaptera and Hymenoptera) and two orders of Arachnida (Araneae and Acari). A few species are predators of mites and a very few attack nematodes. Chalcidoids have been especially successful in exploiting small, concealed hosts (e.g. leaf-miners, gall-formers, etc.) and the Homoptera-Sternorrhyncha (scale-insects, etc.). Koinobiosis appears to be relatively uncommon and few Chalcidoidea attack free-living exophytic lepidopterous larvae. When they do exploit such hosts they often do so in unusual ways—for example, via egg-larval polyembryony (*Copidosoma*, Encyrtidae), or via ectoparasitic koinobiosis (*Euplectrus*, Eulophidae). In addition to parasitoids, the superfamily also includes several phytophagous species (gall-formers and seed-feeders), including the obligate pollinators of fig trees.

At present most authors include 21 families in the Chalcidoidea (Bouček, 1988; Grissell & Schauff, 1990). In comparison to other superfamilies of Hymenoptera, this number is probably too high, but our understanding of the phylogenetic relationships (and character states) within the superfamily is so meagre that any changes would be premature. Moreover, the family level classification of Chalcidoidea has already undergone so many changes that many chalcidologists appear disposed to temporarily accept the present system, despite its many problems. What is most needed to resolve the problems in the family-level classification of Chalcidoidea is more analysis of specific characters across taxa, such as those provided by Gibson (1985) on the thorax, and by Darling (1988) on the labrum.

One of the few points of agreement is that the Chalcidoidea is a monophyletic group, although the question of whether to include the enigmatic Mymarommatidae is debatable. Gibson (1986*a*) proposed three synapomorphies for Mymarommatidae + Chalcidoidea:

- the loss of two of the three ancestral sites of origin of the mesotrochanteral depressor muscle (the mesothoracic muscle that inserts into the basomedial edge of the mesotrochanter);
- the axillar phragma as the site of origin for the mesotergal-trochanteral depressor;
- the absence of the independent basal ring in the male genitalia.

In the same paper Gibson discussed three additional synapomophies for the Chalcidoidea, excluding Mymarommatidae:

- prepectus externally visible;
- mesothoracic spiracle positioned at lateral edge of mesoscutum;
- multiporous plate sensilla on the antenna.

He concluded that Mymarommatidae may possibly be the sister group to Chalcidoidea, and that Mymaridae is probably the sister group to all other Chalcidoidea.

Recently, Gibson (*in* Goulet & Huber, 1993) placed Mymarommatidae in a superfamily of its own. Although we agree that Mymarommatidae is quite distinct, we include this family under Chalcidoidea for convenience.

In the absence of a satisfactory classification we have chosen to present the families in alphabetical order. Of the 21 families of Chalcidoidea, 19 are known from Costa Rica. Rotoitidae is known only from New Zealand (Bouček & Noyes, 1987), and Mymarommatidae is known from Australia, Europe and North America. It is possible that the latter will eventually be found in Costa Rica and it is therefore included in the following key to families. The most important work on the superfamily as a whole is Bouček's (1988) revision of the Australasian genera, although it excludes Aphelinidae, Elasmidae, Encyrtidae, Mymaridae, Signiphoridae and Trichogrammatidae. Important publications treating the New World fauna include catalogues of the nearctic species (in Krombein *et al.*, 1979) and neotropical species (De Santis, 1979, 1980, 1981, 1989). Grissell & Schauff (1990) provided illustrated keys to the families. Keys to the nearctic genera are currently in preparation.

Key to the families of Chalcidoidea

Wingless male fig wasps (Agaonidae) are not included in the following key since they are only encountered inside fig fruits and in the leaf litter under a fig tree (see Chapter 11.1 for keys). Other wingless or brachypterous chalcidoids (in Costa Rica primarily Aphelinidae, Encyrtidae, Eulophidae, Eupelmidae and Pteromalidae) should run through the key. This key is designed for ease of usage and therefore does not include rare exceptions. For a more thorough key to the world families of Chalcidoidea see Gibson (*in* Goulet & Huber, 1993).

1 Hind femur enlarged and with teeth along ventral margin, hind tibia noticeably curved (Fig. 11.01). 2

— Hind femur not enlarged and usually without teeth, hind tibia not noticeably curved. 5

2 Wings folded longitudinally; tegula elongate (Fig. 11.02); ovipositor often curved over the metasoma; males with terminal metasomal tergites fused into a carapace (Fig. 11.03); relatively uncommon. ..... **Leucospidae** (p. 342)

— Wings not folded longitudinally; tegula not elongate (Fig. 11.04); ovipositor not curved over metasoma; male metasoma not as above; common. 3

3 Prepectus small (Fig. 11.04); generally non-metallic.

[Eyes not diverging ventrally; ovipositor generally not exserted; if exserted, then straight.] .... **Chalcididae** (p. 289)

— Prepectus large; generally metallic coloured. ... 4

4 Inner margins of eyes diverging ventrally; antennae inserted low on face (Fig. 11.19). .... **Pteromalidae** (p. 355)

— Inner margins of eyes more or less parallel, antennae inserted near middle of face. .... **Torymidae** (p. 377)

5 Hind wing with a basal stalk (Fig. 11.20); antennal torulus (insertion) closer to eye than to other torulus (except in Mymarommatidae); head usually with H-like pattern of lines (Fig. 11.05).

[Small (usually 1.5 mm or less in length), slender, non-metallic; fore wing with a single vein restricted to basal third of wing.] 6

— Hind wing without a basal stalk (membrane extending to base of wing); antennal torulus usually closer to other torulus than to eye; head without H-shaped pattern of lines (sometimes with V- or X-shaped pattern). 7

6 Metasoma with petiole 2-segmented; fore wing reticulate; extremely rare (not yet collected in Costa Rica). .... **Mymarommatidae***

— Metasoma with petiole 1-segmented; fore wing surface smooth; very common. .... **Mymaridae** (p. 344)

7 Antenna usually with 5 to 7 funicular segments (excluding anelli) between pedicel and club (Fig. 11.21); tarsi usually 5-segmented (4- in males of Tetracampidae). 8

— Antenna with 0 to 4 funicular segments between pedicel and club (Fig. 11.22); tarsi 3-, 4-, or 5-segmented. 19

8 Mesopleuron convex, usually longer than high (Figs 11.06, 11.09); middle leg with apical spur of tibia usually somewhat enlarged, and first tarsal segment usually with one or more rows of pegs and/or a dense pad-like cushion of setae (Fig. 11.06). 9

— Mesopleuron not convex, usually higher than long (Figs 11.15, 11.16); middle leg with apical spur of tibia not enlarged and first tarsal segment without pegs or cushion of setae. 11

9 Fore and mid coxae relatively close together, nearly contiguous (Fig. 11.06); fore wing with marginal vein usually shorter than stigmal vein (Fig. 11.23); cerci located in front of apex of metasoma (Fig. 11.07); axillae usually transverse and meeting medially (Fig. 11.08).. .... **Encyrtidae** (p. 300)

— Fore and mid coxae usually well separated (Figs 11.25, 11.09); marginal vein longer than stigmal vein; cerci located at apex of metasoma; axillae usually not transverse and meeting medially. 10

10 Notauli usually meeting, or at least strongly converging, posteriorly (Fig. 11.24); prepectus inflated and projecting forward (Fig. 11.25); rare. .... **Tanaostigmatidae** (p. 373)

— Notauli usually not visible (if so not meeting); prepectus flat (Fig. 11.09); fairly common.
[Male eupelmids are very difficult to distinguish from Pteromalidae.] .... **Eupelmidae** (p. 329)

11 Head flattened dorso-ventrally, mandible with elongate appendage extending posteriorly (Fig. 11.26); front and hind legs with tibia shorter than femur. ....**Agaonidae** (Agaoninae) (p. 273)

— Head not as above; legs with tibia as long as or longer than femur. 12

12 Pronotum not visible in dorsal view, covered by mesoscutum (Fig. 11.10); mandibles sickle-like.
[Metasoma always petiolate; sometimes with long branches on antenna (males) and/or with two long spines projecting from apex of scutellum (both sexes).] ...... **Eucharitidae** (p. 309)

— Pronotum always visible in dorsal view; mandibles not sickle-like. 13

13 Metasoma short, triangular in lateral view, with only 1 or 2 visible segments (Fig. 11.11).
[Body very short and stocky; mesosoma often with coarse punctures.] .... **Perilampidae** (p. 351)

— Metasoma not as above, with more than 2 visible segments. 14

14 Fore wing with stigmal vein very short (Fig. 11.27), rarely enlarged as a circular knob; hind coxa usually much larger than fore coxa; back of head usually with occipital carina (Fig. 11.12). 15

— Fore wing usually with stigmal vein longer; hind coxa usually not enlarged; occipital carina variable, usually absent. 16

15 Dorsal surface of metasoma with distinct pits (Fig. 11.13); ovipositor not extending beyond apex of metasoma; rare. .... **Ormyridae** (p. 350)

— Metasoma without pits; ovipositor usually extending beyond apex of metasoma; fairly common. .... **Torymidae** (p. 377)

16 Pronotum rectangular in dorsal view, about as wide as mesoscutum (Fig. 11.14); usually black or yellow (rarely metallic), dorsal surface of mesosoma often with coarse punctures.
[Metasoma often laterally flattened (Fig. 11.15), but if not, then prepectus small (Fig. 11.16).] .... **Eurytomidae** (p. 336)

— Pronotum usually shorter or else narrower than mesoscutum; coloration variable, usually without punctate sculpture. 17

17 Propodeum setose medially (Fig. 11.28); base of wing very setose; pronotum bell-shaped and at least half as long as mesoscutum; rare. .... **Tetracampidae** (p. 376)

— Propodeum with hairs restricted to lateral margins; base of wing usually less setose; pronotum variable; common. 18

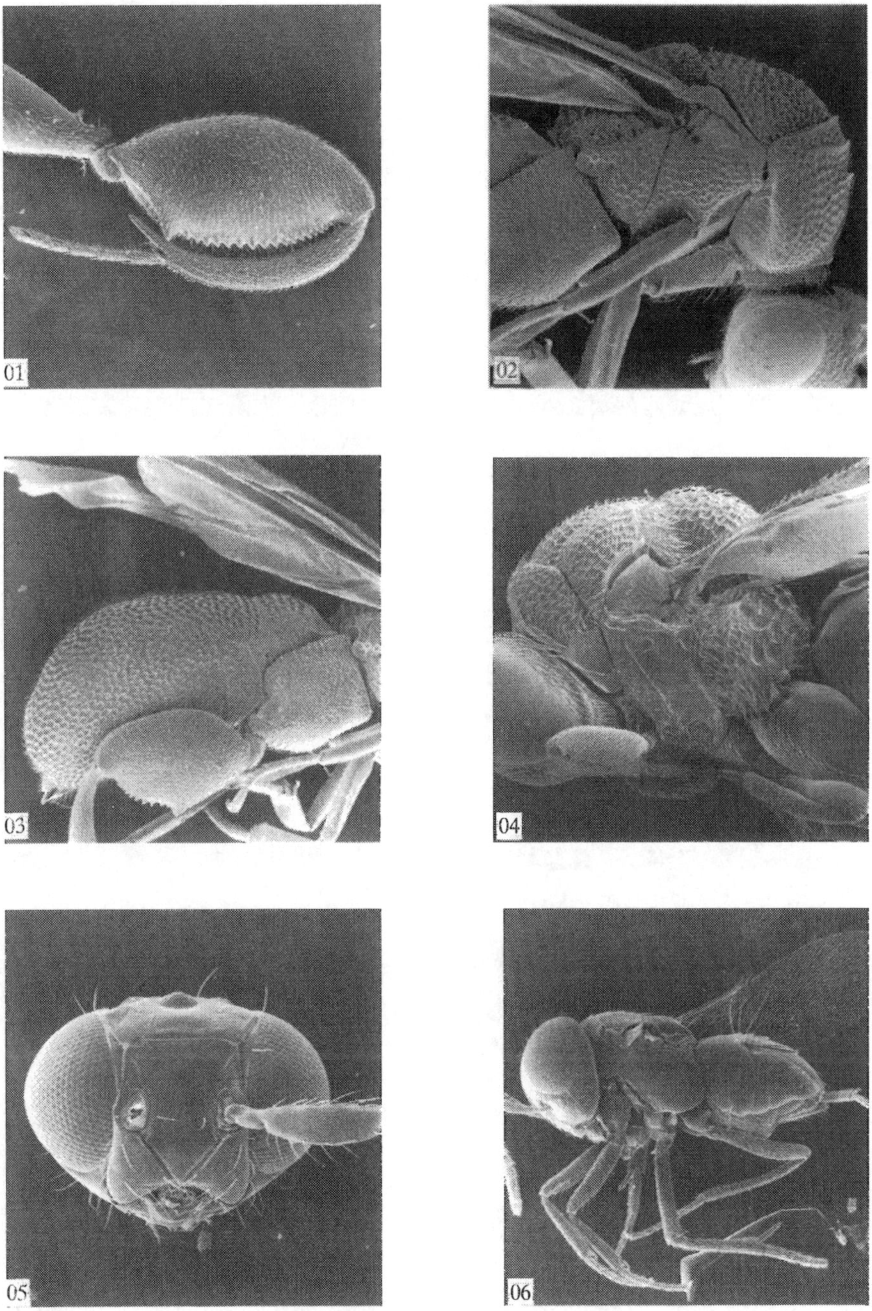

Figs 11.01–11.06. Chalcidoidea. Fig. 11.01. Chalcididae, hind leg. Fig. 11.02. Leucospidae, mesosoma, lateral view. Fig. 11.03. Leucospidae, male metasoma, lateral view. Fig. 11.04. Chalcididae, mesosoma, lateral view. Fig. 11.05. Mymaridae, head. Fig. 11.06. Encyrtidae, mesosoma and legs, lateral view.

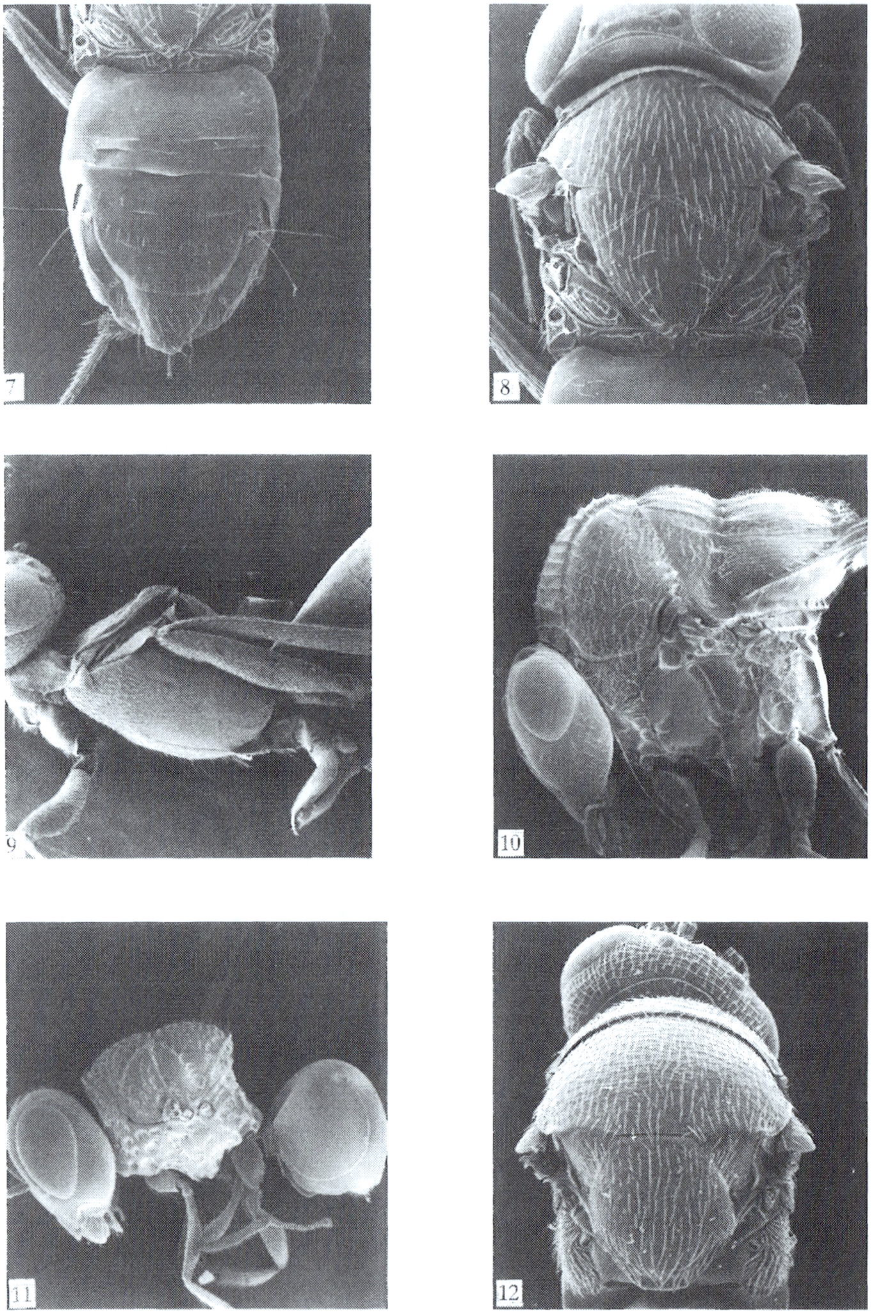

Figs 11.07–11.12. Chalcidoidea. Figs 11.07–11.08. Encyrtidae; 11.07, metasoma, dorsal view; 11.08, mesosoma, dorsal view. Figs 11.09–11.10. Mesosoma, lateral view; 11.09, Eupelmidae; 11.10, Eucharitidae. Fig. 11.11. Perilampidae. Fig. 11.12. Ormyridae, occipital carina on back of head.

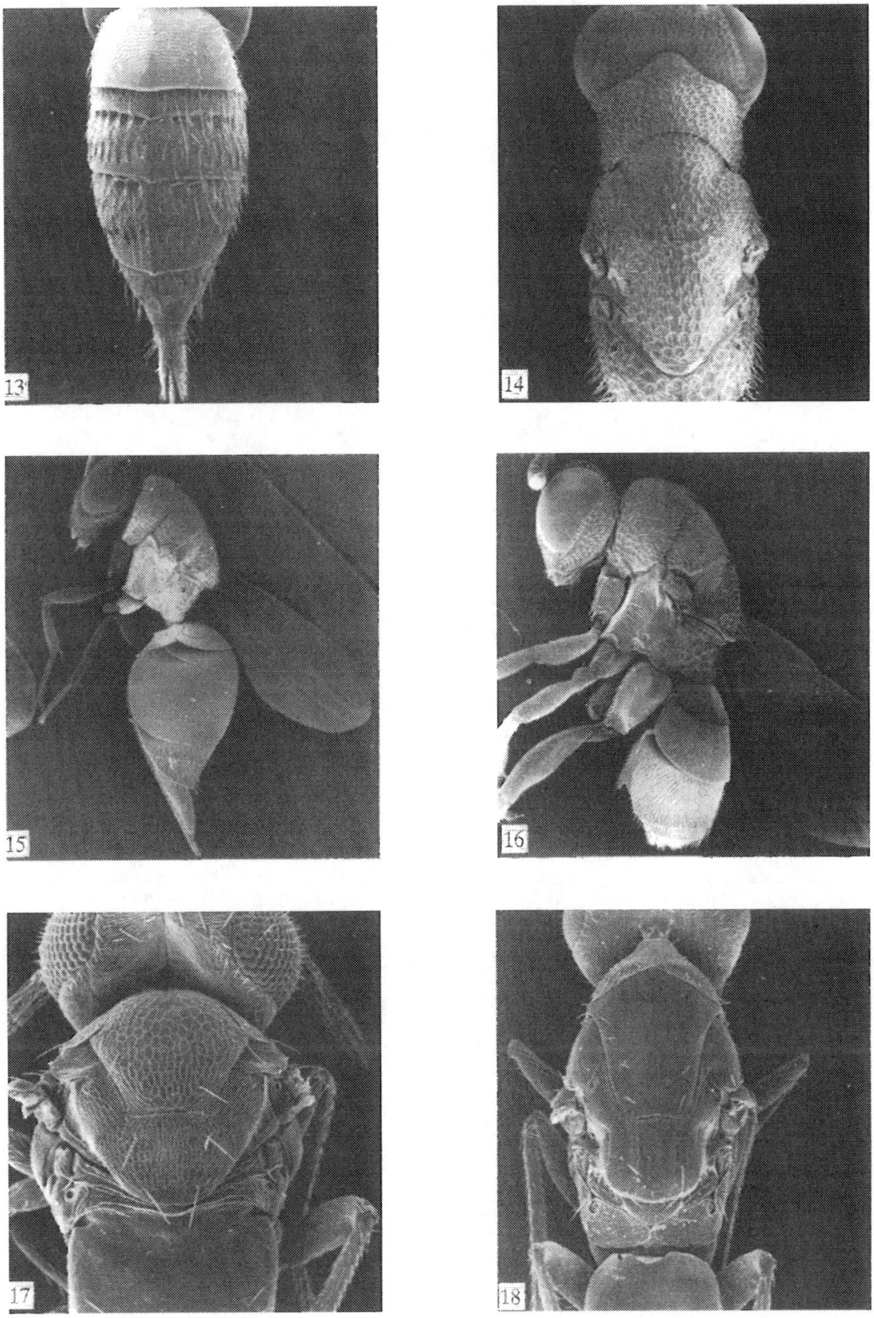

Figs 11.13–11.18. Chalcidoidea. Fig. 11.13. Ormyridae, metasoma, dorsal view. Fig. 11.14. Eurytomidae, mesosoma, dorsal view. Fig. 11.15. Eurytomidae (Eurytominae). Fig. 11.16. Eurytomidae (Rileyinae). Figs 11.17–11.18. Mesosoma, dorsal view; 11.17. Aphelinidae; 11.18, Eulophidae.

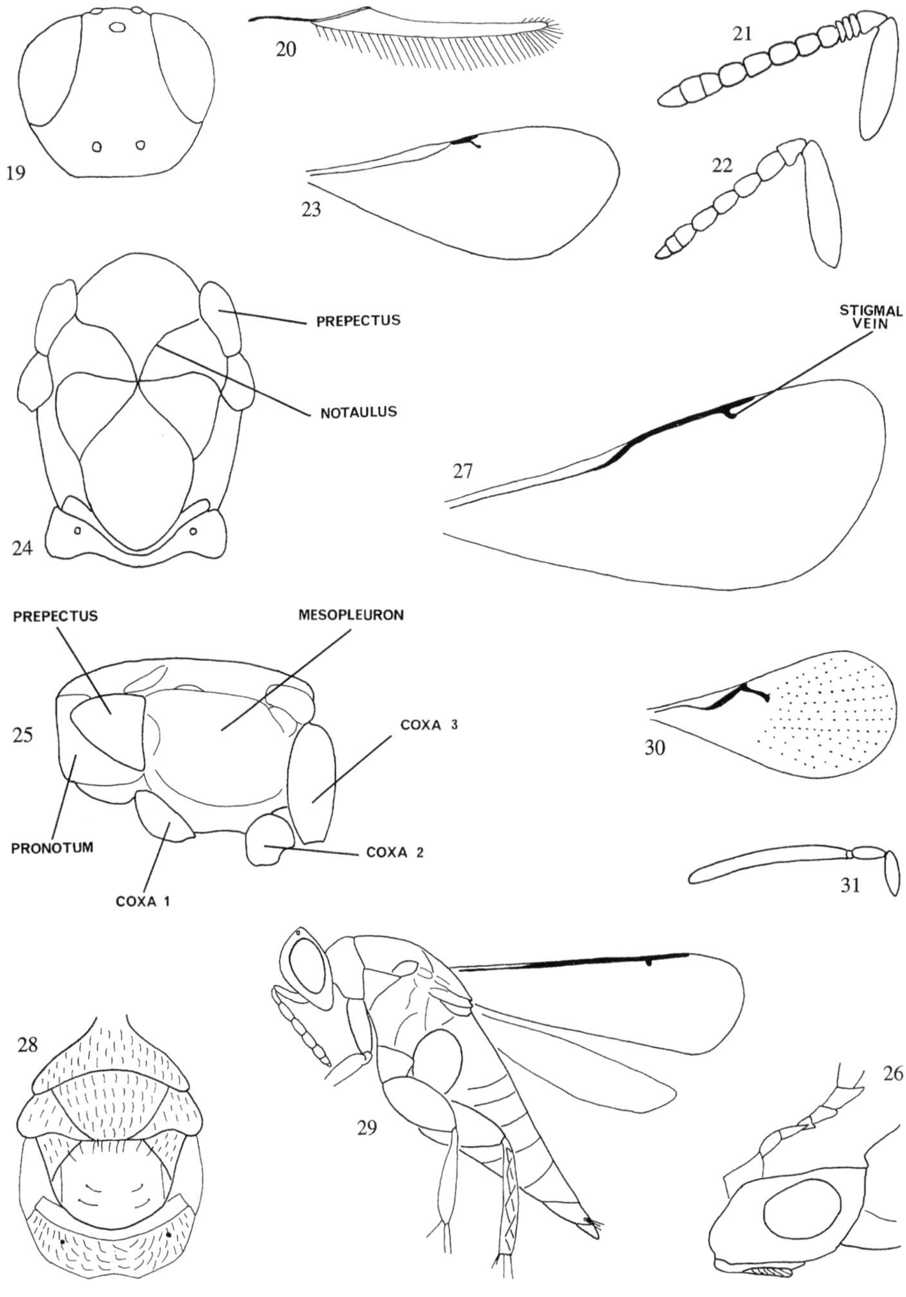

Figs 11.19–11.31. Chalcidoidea. Fig. 11.19. Pteromalidae (Cleonyminae), head. Fig. 11.20. Mymaridae, hind wing. Figs 11.21–11.22. Chalcidoid antenna; 11.21, with five funicular segments between pedicel and club; 11.22, with four funicular segments between pedicel and club. Fig. 11.23. Encyrtidae, fore wing. Figs 11.24–11.25. Tanaostigmatidae, mesosoma; 11.24, dorsal view; 11.25, lateral view. Fig. 11.26. Agaonidae (Agaoninae), head. Fig. 11.27. Torymidae, fore wing. Fig. 11.28. Tetracampidae, mesosoma, dorsal view. Fig. 11.29. Elasmidae, whole insect, lateral view. Fig. 11.30. Trichogrammatidae, fore wing. Fig. 11.31. Signiphoridae, antenna.

18 Ovipositor much longer than body, at least 2 to 3 times body length; or, apex of metasoma and slightly extended ovipositor sheaths curved downward.
[All species are associated with figs.] ....**Agaonidae** (part) (p. 273)
— Ovipositor never longer than body; apex of metasoma and ovipositor usually not curved downward. .... **Pteromalidae** (p. 355)

19 Hind coxa large, flattened and plate-like; hind tibia with dark spines arranged in a distinctive pattern (like a vertical column of X's or diamonds; Fig. 11.29).
[Fore wing long and narrow.] ....**Elasmidae** (p. 298)
— Hind coxa not greatly enlarged; hind tibia without above pattern. 20

20 Tarsi 3-segmented; fore wing without postmarginal vein, surface of wing often with rows of setae extending toward apex (Fig. 11.30).
[Very small, usually less than 1 mm in length.] .... **Trichogrammatidae** (p. 383)
— Tarsi 4- or 5-segmented; fore wing with or without a postmarginal vein, without rows of setae extending toward apex. 21

21 Antennal club very elongate and unsegmented (Fig. 11.31); mesosoma smooth, with scutellum and axillae fused and forming a narrow transverse sclerite; fore wing with membrane virtually bare, but with a fringe of long marginal setae. .... **Signiphoridae** (p. 369)
— Antennal club not so elongate; mesosoma usually not so smooth, scutellum and axillae not fused; fore wing usually with setae on membrane. 22

22 Metasoma broadly joined to mesosoma (Fig. 11.17); fore tibia with apical spur curved and bifid; very small (usually 1 mm or less), rarely metallic coloured.................................... .... **Aphelinidae** (p. 282)
— Metasoma at least somewhat constricted at base (Fig. 11.18); fore tibia with apical spur short and straight; size and coloration variable........ .... **Eulophidae** (p. 315)

11.1 AGAONIDAE

Paul E. Hanson and William Ramirez

Diagnosis. 0.8 to 3.5 mm in length (excluding the ovipositor); colour varying from non-metallic to metallic; sexual dimorphism often extreme, with apterous males sometimes having highly modified mandibles and antennae. Female antenna with 10–13 segments; occipital carina usually absent; back of head below foramen magnum closed by postgenal bridge (gula). Fore wing with stigmal vein usually elongate, often at nearly a right angle to the marginal vein. Tarsi 5-segmented. Metasoma with petiole always reduced; *ovipositor either much longer than body, or slightly exserted and curved downward.*

Classification and distribution. This family previously included only the subfamily Agaoninae, which are pollinators of figs (Moraceae, *Ficus*). However, the definition of the family has recently been expanded to include other subfamilies (formerly placed in the Torymidae and Pteromalidae) that are also exclusively associated with figs, generally as gall-formers or inquilines (Bouček, 1988). As now defined the family includes six subfamilies, four of which are pantropical—Agaoninae, Otitesellinae, Sycophaginae and Sycoryctinae—and two that are restricted to the Palaeotropics—Epichrysomallinae (one introduced genus occurs in Florida; Bouček, 1993*a*) and Sycoecinae. No genera are shared between the Neotropics and Palaeotropics, except for those that have been transported by human activity.

Agaonidae occur in tropical and subtropical areas around the world, wherever native figs are present. The greatest diversity of figs, and hence fig wasps, occurs in the region from Australasia to Madagascar (Berg, 1989). In the New World, fig species are particularly well represented in Central America and the northern Andean region. In Costa Rica figs and fig wasps occur from sea-level up to an altitude of about 2000 metres.

Biology. All species of Agaonidae are exclusively associated with figs and, where the biology is known, they are phytophagous within the ovaries of the fruit. Most of the literature available on the family pertains to the subfamily Agaoninae, which are the obligate pollinators of figs.

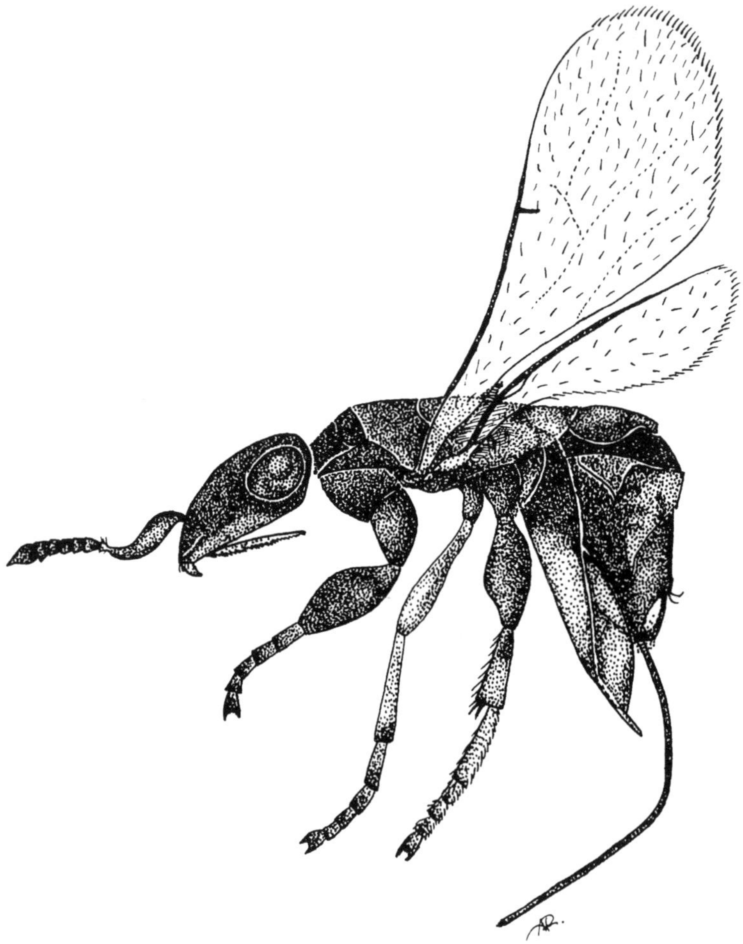

Fig. 11.32. *Tetrapus* sp. (Agaonidae: Agaoninae), female.

The pollinating fig wasps. There are about 750 species of figs worldwide (Berg, 1989), and each of these fig species usually has its own species of fig wasp on which it is entirely dependent for pollination (Ramirez, 1970*a*; Wiebes, 1979). The fig wasp, in turn, is totally dependent on a particular species of fig for its own reproduction and development. The evolution of the mutualism between figs and fig wasps has evidently occurred through a very close coevolution of the two partners (Ramirez, 1974; Herre, 1989), and is one of the best examples of coevolution between plants and insects.

There are two main groups of figs (Berg, 1989): the strictly monoecious species (female and male flowers on the same tree), and the predominantly gynodioecious species (female flowers on one tree, both male and female on another). Only the former are present in the New World and therefore the following discussion is restricted to the monoecious species.

The fig plant never produces exposed flowers; instead they are concealed within an unusual, globular hollow inflorescence resembling an immature fruit, called a SYCONIUM. The only entrance into the cavity of the syconium is the OSTIOLE, which is lined with a series of closely packed, overlapping scales at the apex of the syconium (Fig. 11.33). The development of the syconium can be divided into five phases, usually referred to as phases A to E (Galil & Eisikowitch, 1968*a*):

A. Prefemale stage. This is the period before pollination. Some abscission of inflorescences may occur during this period (Bronstein, 1988*a*).

B. Female stage. The tree becomes attractive to pollen-bearing, female fig wasps, and the scales around the ostiole loosen slightly. To enter the syconium the wasp must squeeze through the overlapping scales of the ostiole, using her serrate or lamellate mandibular appendages to pull herself in. In so doing some species lose their wings and apical antennal segments (e.g. *Pegoscapus* species). After entering the syconium she pollinates all the flowers (only female flowers are present at this time) and lays eggs in the most accessible ones—those with shorter styles. The female wasp inserts her ovipositor down the style of the flower to the floral ovule, where she places the egg (Fig. 11.33b). The egg is deposited within the ovule between the nucellus and the raphe, thus leaving the embryo sac undisturbed and not interfering with the normal fertilization process (Verkerke, 1989). Egg deposition is followed by the secretion of a drop of fluid (into the ovule), which appears to cause the subsequent swelling of the ovary. Because ovaries that are occupied by wasp larvae show abnormal morphology, they are sometimes termed 'gall-flowers' (galled flowers may be a better term since they are not intrinsically different). The female wasp never leaves and eventually dies within the syconial cavity.

Thus long-styled flowers usually produce seeds and many of the short-styled flowers produce fig wasps. The wasp's proximate aim in pollinating is to nourish her own progeny; pollination and fertilization of other flowers that produce seeds is a secondary result. From the fig's point of view, ovaries that produce female Agaoninae contribute to pollen dispersal whereas ovaries producing males or non-pollinating Agaonidae represent seed predation (since these do not disperse pollen). Fig species vary from those with physically small fruits that produce relatively few viable seeds but many pollinators, to those with physically large fruits that produce relatively more seeds but fewer pollinators (Herre, 1989).

C. Interfloral stage. After the fig embryo and endosperm have formed, the wasp egg hatches and the larva enters the enlarged nucellus in search of the embryo sac. The embryo disappears and the endosperm now serves the same function for the wasp larva as it would have done for the plant embryo—nourishing its development (Verkerke, 1989). Thus, during this stage, both seeds and wasp larvae develope.

D. Male stage. The male flowers mature and the adult fig wasps emerge from the galled ovaries into the central cavity of the syconium. The highly modified, wingless males emerge first and utilize their telescopic metasoma to mate with the females while the latter are still enclosed in the gall-flowers (usually less than 15 percent of the emerging fig wasps are males). After mating the males chew a tunnel to the outside of the syconium through which the winged females will eventually depart (males may drop to the ground but presumably serve no further function there). In some fig species, the syconium initially has an inner atmosphere with high levels of carbon dioxide, which inhibits activity by female wasps, but after a tunnel to the outside has been opened (and the inner atmosphere becomes equilibrated with the external atmosphere), the female wasps become active (Galil *et al.*, 1973*b*). Before leaving the syconium, the female wasps load up with pollen from the male flowers (see below). They then exit the syconium and fly off to search for another tree of the same species that is in the receptive stage (female stage).

E. Postfloral stage. The fruit matures, which in some fig species may be stimulated by the change in the internal atmosphere resulting from the emergence hole made by the male fig wasps (Galil *et al.*, 1973*b*). Many of the ripe fruits are eventually consumed by dispersal agents such as birds and bats (Janzen, 1979).

A fig tree generally produces fruit two or three times each year. Within a particular tree, floral development is more or less synchronized, but within a given area, different trees of the same species are in different stages of floral development, thus assuring a continuity of hosts for the fig wasps and hence pollination.

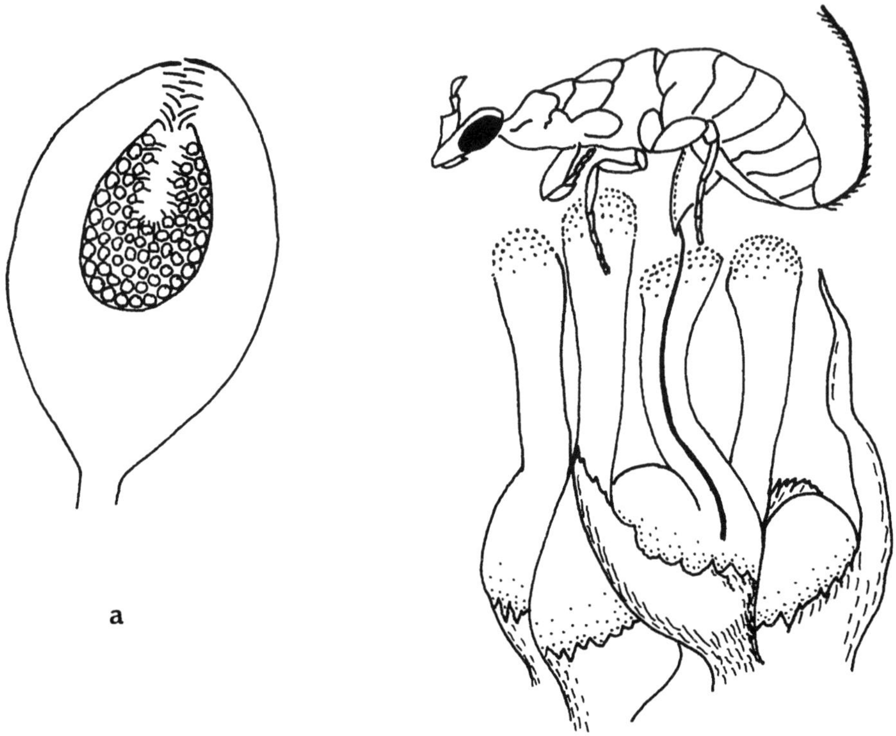

a

Fig. 11.33. Syconium, female stage: **a**) cross section of syconium; **b**) female agaonid inserting ovipositor down style of female flower (modified from Galil & Eisikowitch, 1969).

There are approximately 140 species of figs present in the Neotropics (Berg, 1989) and about 40 of these occur in Costa Rica (Burger, 1977*b*). Neotropical figs belong to two subgenera of *Ficus*—*Urostigma* (section *Americana*) and *Pharmacosycea* (section *Pharmacosycea*)—which differ in several ways (Table 11.1).

Not surprisingly, the first fig wasp to be extensively studied (e.g. by Grandi, 1929) was *Blastophaga psenes*, the pollinator of the cultivated fig. For many years it was thought that this species, and hence other fig wasps, become passively dusted with pollen on their way out of the syconium, and that pollination occurred by pollen-coated wasps crawling about over the female flowers in a syconium on another tree. It was not until 1969, after detailed studies were carried out on other species, that this view changed. Ramirez (1969), working in Costa Rica, and Galil and Eisikowitch (1969), working in East Africa, independently discovered morphological modifications for pollen transport by fig wasps. Females of *Pegoscapus estherae* and *P. tonduzi* were observed actively removing pollen from the anthers, loading it into specialized coxal corbiculae and mesosternal pockets and, upon entering a syconium on another tree, actively unloading pollen from these corbiculae (Ramirez, 1969; Galil *et al.*, 1973*a*). Subsequently, similar pollination mechanisms have been described in other fig wasp genera (e.g. Galil & Snitzer-Pasternak, 1970; Joseph & Abdurahiman, 1981; Okamoto & Tashiro, 1981), and in other species of *Pegoscapus* (Frank, 1984). *P. carlosi* and *P. mariae* are unusual in that they occur in the same species of fig and seem to be passive pollinators (Ramirez, 1970*b*).

The other genus of Agaoninae occurring in Costa Rica, *Tetrapus*, lacks specialized structures for carrying pollen; instead pollen is apparently carried in the digestive tract and among the mandibular serrations (Ramirez, 1969, 1970*b*). The figs pollinated by these

***Ficus* subgenus**	***Urostigma*: *Americana***	***Pharmacosycea*: *Pharmacosycea***
Total species	120	20
Species in Costa Rica	31	8
Tree form	mostly stranglers	mostly free-standing
Form of ostiole	helicoidal	tubular
Pollinators	*Pegoscapus* spp.	*Tetrapus* spp.
Pollen-loading mechanism	active	passive
Form of ripe fruit	red-yellow, sweet, juicy, without odour	green, not sweet, juicy, odoriferous
Seed dispersal agent	birds	bats

Table 11.1. Comparison of the two subgenera of *Ficus* present in Costa Rica.

wasps (*Pharmacosycea*) have anthers that shed pollen, rather than being actively opened by the fig wasp as in *Urostigma* figs. Re-examination of the commercial fig has shown that, while pollen transport by the wasp is indeed passive, the pollen is actually carried in the intersegmental membranes (Galil & Neeman, 1977). It seems unlikely that effective pollen transport can occur on the external surface of any wasp that has to squeeze through the ostiole of the syconium. In general, species of Agaoninae appear to vary in pollen transport efficiency, which is reflected in differences in the numbers of stamens present in different fig species (Galil & Meiri, 1981). Specialized structures for carrying pollen have apparently evolved independently and convergently in different lineages (Ramírez, 1978; Wiebes, 1982). The special structures to carry pollen probably evolved to save pollen from desiccation.

The number of female fig wasps entering the syconium, which varies from one to six or more, has important consequences for the mutualism between figs and fig wasps. Increasing numbers of foundresses (females entering a syconium) results in the production of relatively more male offspring, which may be due to local mate competition (Herre, 1987). Increasing numbers of male fig wasps in turn reduces pollen dispersal and may also lower an individual female wasp's reproduction (Herre, 1989). Moreover, fig wasp species characterized by higher numbers of foundresses tend to have more virulent nematodes (Herre, 1993). These nematodes (Diplogasteridae, *Parasitodiplogaster* species), which are probably host-specific, have immature stages that crawl onto newly emerged female fig wasps and are thus carried to a syconium on another tree (Poinar & Herre, 1991). The nematodes eventually enter the body of the fig wasp, consume it, emerge, mate, and lay eggs within the new syconium; the eggs hatch at the time the next generation of fig wasps emerge, thus completing the cycle.

The non-pollinating fig wasps. The biology of the other subfamilies of Agaonidae is less well known. Although they do not pollinate the fig, all are probably phytophagous within the ovary, either as gall-formers or as inquilines in the galled ovaries produced by Agaoninae. Females of these other subfamilies (at least among neotropical species) do not enter the syconium as do Agaoninae, but rather they oviposit through the syconial wall (as in many palaeotropical species; see Ulenberg & Nübel, 1982). Like Agaoninae, species of *Idarnes* (Sycophaginae) appear to be restricted to just one species of fig (Gordh, 1975), but the Otitesellinae and Sycoryctinae do not seem to show such specificity (Bouček, 1993*a*).

An Old World species of Sycophaginae, *Sycophaga sycomori* (which enters the syconium via the ostiole), is able to develop in figs in regions where the pollinator is absent and, although unpollinated syconia are usually aborted, oviposition by this species stimulates fruit development (Galil & Eisikowitch, 1968*b*). The female deposits an egg directly in the embryo sac and thus causes a proliferation of the nucellar cell that serves as the food source for the larva (Galil *et al.*, 1970). Three other non-pollinating agaonids occur in the same fig, one Sycophaginae and two Sycoryctinae, but all of these require the presence of the pollinator since they develop as inquilines in galled flowers (Galil & Eisikowitch, 1968*a*).

In Costa Rica, there is indirect evidence that *Idarnes* and *Aepocerus* are gall-formers (in short-styled ovaries not occupied by pollinator larvae), but unlike *Sycophaga* mentioned above, require the presence of the pollinator to prevent the syconia from aborting, and later, to chew an exit hole (Bronstein, 1991). However, more study is required, not only to confirm this suggestion, but to determine the biology of other non-pollinating species.

Many species of these other subfamilies resemble Agaoninae in that the males are wingless. However, they differ from Agaoninae in that wingless males mate with females after they have emerged from the galled ovaries into the syconial cavity. In some species all males are wingless (most *Idarnes*), in others there are both wingless and winged males, while still others have mostly winged males. In species having both wingless and winged males, fighting often occurs between the wingless males (Hamilton, 1979; but see Godfray, 1988). Winged males mate outside the fig. In Costa Rica males of *Aepocerus emarginatus* are uniformly winged, but small males have clear wings whereas larger males have black spots on the wings. The latter vie aggressively for possession of a syconium from which females are emerging, repeatedly spreading their spotted wings in apparent aggressive displays, as in some tephritid flies (W. Eberhard, pers. comm.). The clear-winged males are non-aggressive and wait for females on nearby foliage (Bronstein, 1991).

Economic importance. The very specialized pollination system of figs, which is totally dependent upon the presence of fig wasps, requires a staggered production of fruits throughout the year (Kjellberg & Maurice, 1989; Windsor *et al.*, 1989). As a consequence, figs are 'keystone' resources that can sustain numerous tropical frugivores (birds, bats, etc.) during periods when other fruits are scarce (Foster, 1982). Removal of figs would probably result in local extinction of several frugivores, which would in turn have a negative impact on other plants that depend on these frugivores for seed dispersal (Terborgh, 1986). Figs are therefore vital for the maintenance of tropical biodiversity (but see Gautier-Hion & Michaloud, 1989). Because of their unique pollination system, the conservation of figs requires a critical minimum number of trees in order to sustain the fig wasp population (McKey, 1989*b*).

In addition to fruits, the almost astronomical numbers of fig wasps emerging from a fig tree provide an important food resource for diverse predators, ranging from staphylinid beetles to birds (Bronstein, 1988*b*). Our observations suggest that fig wasps represent the predominant component of aerial plankton at low- to mid-altitudes in Costa Rica, but unfortunately the ecological significance of this aerial plankton has been poorly studied. Another role of fig trees in tropical ecosystems that is often forgotten is the physical structure of the tree itself. Because of their peculiar mode of development, mature strangler figs usually have hollow trunks, which provide shelter and nesting sites for numerous animals, ranging from bats to stingless bees.

Figs and fig wasps thus have a tremendous indirect economic importance. Of more direct economic importance is the cultivated fig, *Ficus carica*, which is native to Asia Minor and was grown in Egypt as early as 4000 B.C. Although the Adriatic variety does not require pollination for fruit development, the Smyrna variety does. Pollination is brought about by suspending inflorescences of the wild caprifig, in which fig wasps (*Blastophaga psenes*) are about to emerge, on the branches of the flowering Smyrna fig, a process known as 'caprification' (Condit, 1947; Condit & Enderud, 1956). The common commercial fig, however, does not grow well in the lowland tropics, and is mostly grown in subtropical, Mediterranean climates.

Identification. Keys to the Chalcidoidea (mostly Agaonidae) associated with neotropical figs have been given by Bouček (1993*a*), from which the following key has been adapted. We have not included introduced agaonid taxa, the pteromalid genus *Podivna*, which is known from Trinidad and Venezuela, the braconid genus *Psenobolus*, which has brachypterous males and has only been seen from *Ficus velutina*, and parasitoids associated with other inhabitants of fig syconia, such as weevil larvae (*Ceratopus* spp.) and pyralid larvae (*Boetarcha stigmosalis*) (Janzen, 1979).

Key to Chalcidoidea associated with Central American figs

1 With wings normally developed. 2
— Without wings or with wings very reduced. [Males only.] 12

2 Head dorso-ventrally flattened; mandible with a conspicuous flat appendage adhering to the underside of the head; third antennal segment with sharp projection, not ring-like (Figs 11.32, 11.34). ....(Agaoninae)........ 3
— Head not flattened; mandible normal; third antennal segment ring-like (Figs 11.38, 11.39). 4

3 Head slightly longer than wide; scape flat and subtriangular (Fig. 11.34); mandibular appendage with lamellae. .... *Pegoscapus*
— Head usually much longer than wide; scape subcylindrical and several times longer than wide (Fig. 11.32); mandibular appendage with rows of teeth. .... *Tetrapus*

4 Fore wing with marginal vein long *and* stigmal vein very short, essentially sessile (Figs. 11.171, 11.172). ....**Torymidae** (see Chapter 11.18)
— Fore wing with ***either*** marginal vein quite short, ***or if*** long, ***then*** stigmal vein of moderate length (Figs 11.35–11.39). 5

5 Ovipositor much longer than body. 6
— Ovipositor, if present, not longer than body. 7

6 Fore wing with marginal vein short, much less than 2.0× the length of the stigmal vein (Fig. 11.35); scutellum with longitudinal lines on lateral margins; last metasomal tergite normal. .... *Idarnes*
— Fore wing with marginal vein longer than 2.0× length of the stigmal vein (Fig. 11.36); scutellum without lateral lines; last metasomal tergite very elongate, forming a tubular sheath around the ovipositor. .... *Critogaster*

7 Body yellow to brown, never metallic. 8
— Body usually at least somewhat metallic. 9

8 Body smooth, without any punctures; fore wing with postmarginal vein virtually absent (Fig. 11.37). .... *Anidarnes*
— Body at least somewhat sculptured, usually with punctures on mesoscutum; postmarginal vein present. .... **Eurytomidae** (see Chapter 11.9)

9 Body with dense punctures on scutellum or raised reticulate sculpture on head and thorax; mesoscutum with extremely shallow notauli. .. ... **Pteromalidae**: *Ficicola* (see Chapter 11.14)
— Body without dense punctures or raised reticulation; notauli often deeper. 10

10 Fore and hind femora ovoid, strongly enlarged; marginal vein much shorter than stigmal vein. .... **Pteromalidae**: *Hansonita*
— Femora not unusually widened; marginal vein always longer than stigmal vein. ....(Otitesellinae)..................... 11

11 Antennae in female inserted well above lower margin of eye (Fig. 11.38); propodeum with carinae; ovipositor sheath extending slightly beyond apex of metasoma. .... *Aepocerus*
— Antennae at or below lower margin of eye (Fig. 11.39); propodeum smooth, without carinae; ovipositor sheath extending far beyond apex of metasoma. .... *Heterandrium*

12 Soft posterior half of metasoma turned forwards in U-shape under body (Fig. 11.40). ....**Agaoninae**
— Apex of metasoma never turned forwards under body. 13

13 Wing rudiments flat, reaching propodeum. ....*Aepocerus*
— Wings present as linear rudiments, or absent. 14

14 Mesoscutum with distinct notauli, scutellum and axillae (*Physothorax*); ***or***, antenna with only two segments (*Torymus*, subgenus *Nannocerus*). .... **Torymidae** (see Chapter 11.18)

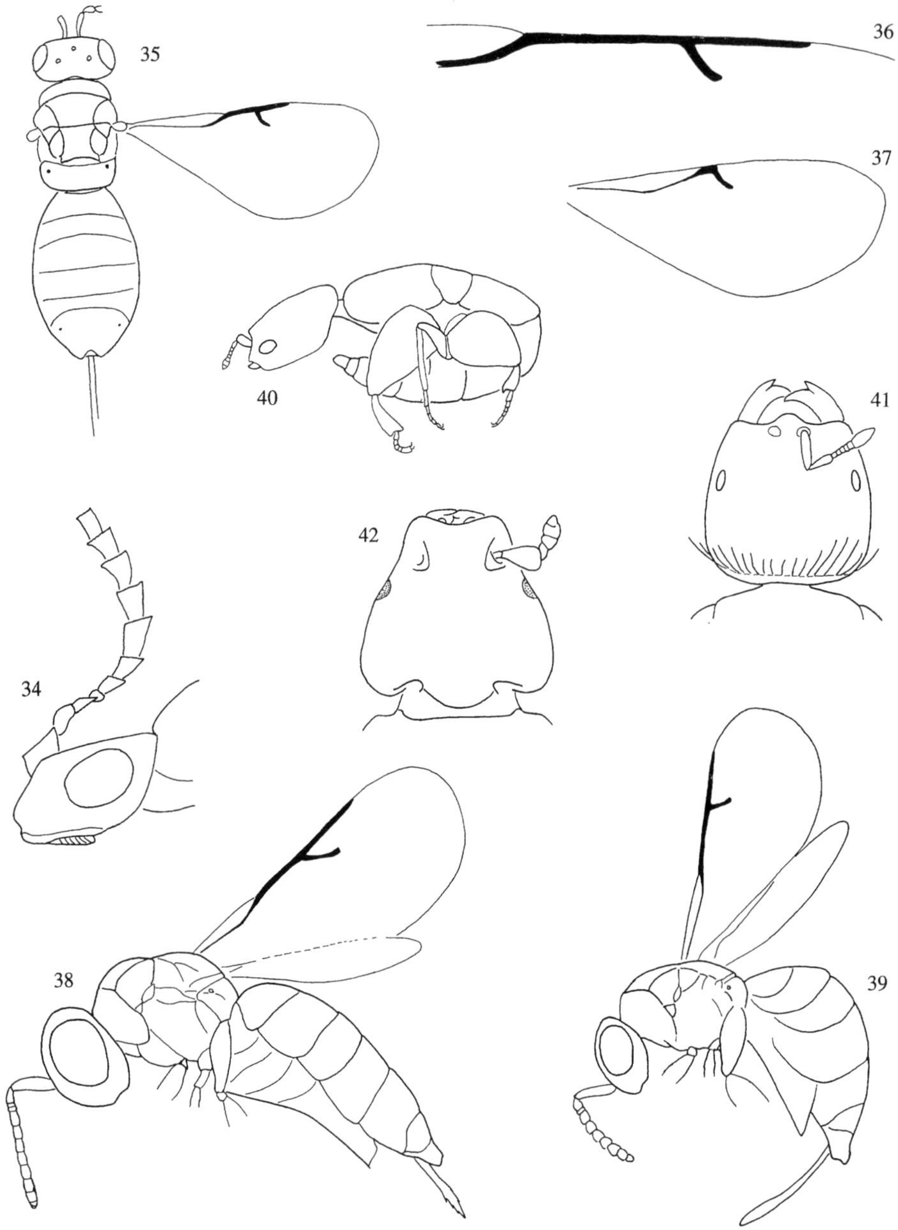

Figs 11.34–11.42. Agaonidae. Fig. 11.34. *Pegoscapus* sp. (Agaoninae), female head. Fig. 11.35. *Idarnes* sp. (Sycophaginae), female, dorsal view. Figs 11.36–11.37. Fore wing. 11.36, *Critogaster* sp. (Sycoryctinae); 11.37, *Anidarnes* sp. (Sycophaginae). Figs 11.38–11.39. Whole insect, female, lateral view; 11.38, *Aepocerus* sp. (Otitesellinae); 11.39, *Heterandrium* sp. (Otitesellinae). Fig. 11.40. Agaoninae, male, whole insect lateral view. Figs 11.41–11.42. Head, wingless male; 11.41, *Heterandrium* sp.; 11.42, *Critogaster* sp.

— Parts of mesoscutum more or less fused ***and*** antenna with more than 2 segments. 15

15 Tibiae without stout spines; posterior margin of head with row of setae (Fig. 11.41). .... *Heterandrium*
— Tibia short and broad, with numerous stout spines on dorsal side; posterior margin of head without row of setae. 16

16 Head nearly parallel-sided, quadrate. ...... *Idarnes*
— Head widening posteriorly (Fig. 11.42). .... *Critogaster*

Synopsis of the Costa Rican fauna

There are four subfamilies, seven genera, and perhaps nearly 100 species (many undescribed) of Agaonidae that occur naturally in Costa Rica. In addition to these native species it is possible that a few introduced taxa are present (or will arrive) on introduced species of figs, as has occurred in Florida (Bouček, 1993*a*). Only the native fauna is included here.

AGAONINAE

These are the obligate pollinators of figs and are divided into two tribes, Agaonini and Blastophagini. Both tribes occur in the Neotropics where they are represented by one genus each: *Pegoscapus* in the Blastophagini and *Tetrapus* in the Agaonini (see Table 11.1). The phylogeny of the subfamily has been discussed by Wiebes (1982) and Ramírez (1991).

Pegoscapus. About 30 species in Costa Rica (Table 11.2). Keys to species are given in Ramírez (1970*b*) and Wiebes (1983).

Tetrapus. Probably eight species in Costa Rica (Table 11.2).

OTITESELLINAE

Winged forms resemble some Pteromalidae. Males may be winged or wingless, or both. The placement of the two native neotropical genera in this subfamily is still open to question (Bouček, 1993*a*). The introduced genus *Walkerella* occurs on *Ficus nitida* (= *microcarpa* ?) in Honduras (specimen in USNM).

Aepocerus. Probably 20-30 species in the Neotropics (Bouček, 1993*a*), and perhaps five to ten in Costa Rica. Males often have black spots on the wings.

Heterandrium. Eight species known from the Neotropics but many more remain to be described. Five described and at least one undescribed species are known from Costa Rica. A key to the species is given by Bouček (1993*a*).

SYCOPHAGINAE

This subfamily is represented in the Neotropics by two genera and perhaps over 100 species (assuming host specificity in *Idarnes*). They are restricted to species of *Ficus* in the subgenus *Urostigma*.

Anidarnes. Recently described by Bouček (1993*a*), with three described species. Only *A. brevicauda* (reared from *Ficus costaricana*) is known from Costa Rica. Males are winged.

Idarnes. Probably about 30 species in Costa Rica (see Table 11.2). A key to the 11 described species known from Costa Rica is given in Gordh (1975). Most species have wingless males but one species recently described from Costa Rica (*I. hansoni*) has winged males (Bouček, 1993*a*).

SYCORYCTINAE

This subfamily is divided into two tribes, Apocryptini and Sycoryctini, but only the latter occurs naturally in the Neotropics, where it is represented by just one genus.

Critogaster. Five species recognized from the Neotropics, four of which are known from Panama (Bouček, 1993*a*). They occur only in figs belonging to the subgenus *Pharmacosycea*. In some species there are both apterous and winged males, whereas other species appear to have only one or the other. A preliminary key to the species is given in Bouček (1993*a*).

Ficus subgenus	*Ficus* species	*Tetrapus* or *Pegoscapus* species	*Idarnes* species
PHARMACOSYCEA	*crassiuscula*		
	crassivenosa		
	insipida (*glabrata*)	*costaricanus*	
	macbridei (*torresiana*)		
	maxima (*radula*)	*americanus*	
	tonduzii		
	werckleana		
	yoponnensis (*multinervis*)	*ecuadoranus*	
UROSTIGMA	*brevibracteata*		
	bullenei		
	caldasiana		
	cervantesiana		*bucatoma*
	citrifolia (*hemsleyana*)	*tonduzi*	*barbigera*
	colubrinae	*orozcoi*	*galbina*
	costaricana	*estherae*	
	cotinifolia	*kraussi*	
	cuatrecasana		
	davidsoniae		
	donell-smithii		
	dugandii (*turbinata*)	*baschierii*	
	goldmanii		*oscrocata*
	hartwegii (*brenesii*)		*hansoni*
	isophlebia	*urbanae*	*micheneri*
	jimenezii	*jimenezi*	*jimenezi*
	laterisyce		
	morazaniana (*lapathifolia*)	*aguilari*	*simus*
	nymphaeifolia (*duquei*)	*amabilis?*	
	obtusifolia	*hoffmeyeri*	*obtusifoliae*
	ovalis		
	paraensis (*panamensis*)		
	perforata (*oerstediana*)	*standleyi*	*camini*
	pertusa (*padifolia*)	*silvestrii*	
	popenoei (*tolimensis*)		
	schippii	*leanae*	
	trachelosyce	*aemula?*	
	trigonata	*danorum*	
	turrialbana		
	tuerckheimii	*mariae*	*ashlocki*
		carlosi	
	velutina	*torresi*	*flavicollis*

Table 11.2. Described species of Agaoninae and Sycophaginae occurring in Costa Rica.

11.2 APHELINIDAE

Paul E. Hanson

Diagnosis. Very small, about 0.5 to 1.5 mm in length, usually squat and robust or flattened, rarely elongate; yellowish to black, metallic coloration rare; usually fully winged but sometimes brachypterous. Antenna 3- to 9-segmented, flagellum of both sexes 2- to 7-segmented, that of male occasionally with one segment more. Mesoscutum with deep, straight notauli, widely separated posteriorly (generally by more than the length of the scutellum); *axillae widely separated*; mesopleuron usually obliquely divided although sometimes large, undivided, more or less convex and shield-shaped. *Fore wing with marginal vein elongate, stigmal vein very short, postmarginal vein usually absent or very short* (in the Eriaporinae long to

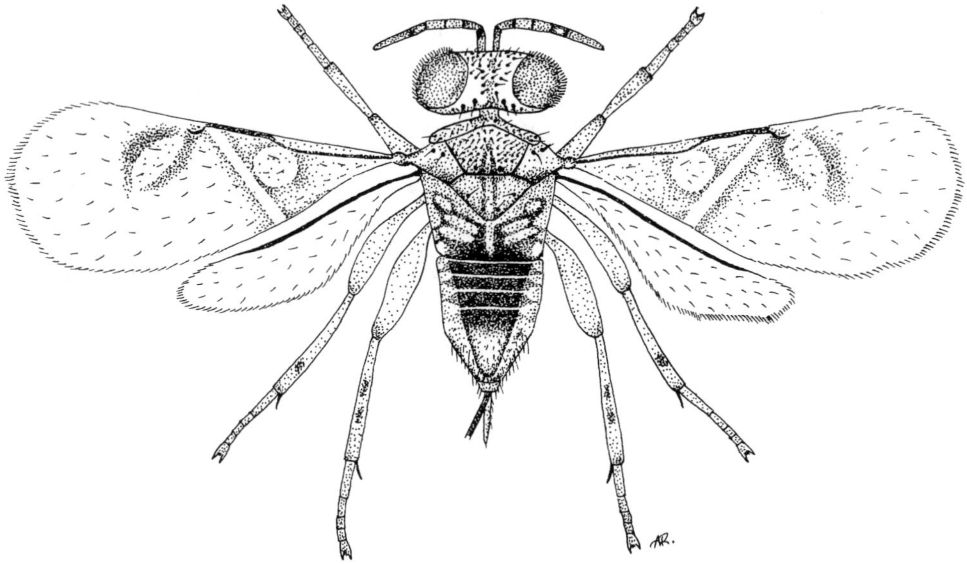

Fig. 11.43. *Aphytis* sp. (Aphelinidae) female.

very long); often with a naked, oblique streak (linea calva) from stigmal vein to posterior margin. Mid coxae well behind middle of mesopleuron; *fore tibial spur curved, bifid*; middle tibial spur fairly elongate; tarsi 4- or 5-segmented. Metasoma apparently sessile (not petiolate), broadly joining the mesosoma; ovipositor usually not or hardly exserted.

Classification and distribution. The Aphelinidae is cosmopolitan in distribution and comprises nearly 1000 described species in nearly 40 genera. The majority of species occur in tropical regions, and in Central America there are at least 15 genera, representing perhaps about 100 species.

The Aphelinidae has occasionally been treated as a subfamily of the Encyrtidae (e.g. by Gordh *in* Krombein *et al.*, 1979) but most recent authors have rejected this classification (LaSalle, 1987; Woolley, 1988; Gibson, 1989). Aphelinids may be paraphyletic with respect to Signiphoridae (Woolley, 1988) and these families together may be closely related to the Eulophidae, although exact relationships remain unknown.

The subfamilial classification is in dispute. Yasnosh (1976) recognized seven subfamilies, whereas Rosen and DeBach (1979) following De Santis (1946, 1948, 1967) recognized only three. In his seminal treatment of the family, Hayat (1983) did not follow any subfamilial classification. Here five subfamilies are recognized: Aphelininae, Azotinae, Calesinae, Coccophaginae and Eriaporinae. All but the last mentioned occur in Central America. The majority of aphelinid species belong to the Aphelininae and Coccophaginae. Azotinae may be situated between these two large subfamilies (Polaszek & Hayat, 1992) or it may be the sister-group to the Signiphoridae (Woolley, 1988). Studies by Gibson (1989) indicate that the Eriaporinae may not belong in the Aphelinidae. The subfamily Calesinae, which includes the single aberrant genus, *Cales*, probably does not belong in the Aphelinidae, and may be more correctly placed in the Eulophidae (Polaszek, 1991).

Biology. The majority of aphelinids are parasitoids of nymphal stages of sternorrhynchous Homoptera, especially of scale insects (Coccoidea), whiteflies (Aleyrodidae) and aphids (Aphidoidea). Indeed all genera of Aphelinidae have been recorded from these homopteran hosts (Yasnosh, 1980; Viggiani, 1984). However, some species of *Ablerus*, *Azotus*, and especially *Centrodora* show diverse host associations, both inter- and intraspecifically (Polaszek, 1991). Non-

sternorrhynchous host associations in these genera include mostly eggs of other insects (Orthoptera, Hemiptera, Homoptera-Auchenorrhyncha and Lepidoptera) but also larvae or pupae of Diptera, Braconidae, and Dryinidae (Viggiani, 1984; Polaszek, 1991).

The majority of Aphelinidae are primary parasitoids, but species of *Ablerus*, *Azotus* and *Centrodora* are often hyperparasitic; *Marietta* is always hyperparasitic. Most of the primary parasitoids (except the genus *Aphytis*) are endoparasitic, whereas many hyperparasitic aphelinids are ectoparasitoids. *Eretmocerus* species are ectoparasitic in the early instars and endoparasitic in the final instar (Gerling, 1966). Aphelinids are koinobionts in the sense that host development is not terminated immediately upon parasitization, although parasitoid development from egg to pupa may occur in just one host instar.

The more common host associations of aphelinid genera occurring in Central America, excluding the genera that contain highly polyphagous and/or hyperparasitic species mentioned above, are detailed in Table 11.3.

Female host selection and oviposition behaviour varies between aphelinid taxa. Female *Aphytis* tend to run about rapidly on the substrate whereas *Encarsia* move more slowly (Viggiani, 1984). *Aphytis* species are ectoparasitoids and females usually oviposit only on those hosts where there is a free space between the scale covering and the body of the host (Rosen & DeBach, 1979). The behaviour of endoparasitic aphelinids is much more varied, and in some cases may involve oviposition into particular organs; for example some species of *Coccophagus* place their eggs within the suboesophageal ganglion of the host (Flanders *et al.*, 1961). The female of *C. scutellaris* is able to locate larvae of her own species when they are at least two days old and lays unfertilized (male) eggs in them (Jarraya, 1975; see below). The fecundity of different species has been summarized by Viggiani (1984).

Host group	**Aphelinid parasitoid**
Aphidoidea	*Aphelinus*
Aleyrodidae	*Cales, Dirphys, Encarsia, Encarsiella Eretmocerus*
Coccidae	*Coccophagus*
Diaspididae	*Aphytis, Archenomus, Coccobius, Coccophagoides, Encarsia, Pteroptrix*
Pseudococcidae	*Coccophagus*

Table 11.3. The more common host associations of aphelinid genera occurring in Central America, excluding the genera that contain highly polyphagous and/or hyperparasitic species.

Several aphelinids are unique among Hymenoptera in that the males and females of the same species have different ontogenies. The females of such species always develop as primary endoparasitoids of Homoptera (usually scale insects), while conspecific males develop in one of the following ways (Walter, 1983):

i) As primary ectoparasitoids on the same host species as the female. Thus both sexes utilize the same host, but feed in different ways. This type of development ('diphagous') occurs, for example, in *Coccophagus longifasciatus*, *C. ochraceus*, and *C. scutatus*.

ii) As hyperparasitoids on their own species. Examples of this form of development ('obligate autoparasitoids') occur in *Coccophagoides utilis*, where the male is an ectoparasitoid, and in *Coccobius fulvus*, where the male is an endoparasitoid.

iii) As facultative hyperparasitoids on their own or other species. Examples include *Coccophagus lycimnia*, where males develop as ectoparasitoids, and *C. capensis*, where the males develop endoparasitically.

iv) As obligate hyperparasitoids on other species. Examples of such 'alloparasitoids' include *Coccophagus ceroplastae*, where the males develop as endoparasitoids, and *C. pulvinariae*, where males develop as ectoparasitoids.

This phenomenon, where females and males develop differently, has been termed HETERONOMY by Walter (1983); the last three categories mentioned above are examples of HETERONOMOUS HYPERPARASITISM (= Adelphoparasitism). A variety of classifications have been proposed for different forms of heteronomy (e.g. Flanders, 1959*b*, 1967; Zinna, 1961) and that presented above represents a simplification of Walter's (1983) classification. Excluded from the above list is the category known as 'heterotrophic parasitoids', where males develop on different hosts from those utilized by females. It is based on records of *Encarsia lutea* emerging from lepidopterous eggs (Stoner & Butler, 1965). However, *E. lutea* males are often found emerging from the same species of Homoptera as those in which females develop, and the single record from lepidopterous eggs may simply be the result of unfertilized females utilizing unusual hosts facultatively (Polaszek, 1991). However, recent preliminary observations suggest that males of the neotropical species, *Encarsia porteri*, may be genuinely obligate parasitoids in lepidopteran eggs (Polaszek, pers. comm.). In the case of the heteronomous hyperparasitoid, *Encarsia pergandiella*, some males develop as primary parasitoids, which appears to be caused by a paternally inherited factor that leads to the condensation and loss of the paternal chromosomes (Hunter *et al.*, 1993).

The origin of heteronomous hyperparasitism is still a mystery. However, an example of a possible starting point is provided by non-heteronomous aphelinids that lay female and male eggs in slightly different sites. For example, *Aphytis melinus* lays female eggs on the dorsum of the host while male eggs are laid under the scale-insect's body (Luck *et al.*, 1982).

Eggs of Aphelinidae are often stalked. The first instar larva of ectoparasitic species (e.g. *Aphytis* spp.) is hymenopteriform with four pairs of spiracles and a functioning tracheal system (Rosen & DeBach, 1979). In most endoparasitic species the first instar is caudate, without spiracles, and without a functioning tracheal system (Cendana, 1937). In some species where the sexes have different ontogenies, the immature stages are sexually dimorphic. Several species (e.g. *Coccophagus lycimnia*) have sexually dimorphic eggs (Flanders, 1937; Viggiani & Mazzone, 1978; Mazzone & Viggiani, 1984) and the larvae of many taxa are unusually sexually dimorphic. For example, the male first instar larva of *C. scutellaris* is clothed in long setae and has a spine-like tail, while the female larva has no long setae and is of the simple caudate type (Flanders, 1937). In some cases even the pupa (e.g. *Encarsia* spp.) may exhibit sexual dimorphism (Viggiani & Mazzone, 1978).

Pupation takes place either inside or outside the host, depending on the species. Some species pupate inside the living host, within a pupation chamber which becomes filled with air. There is some evidence that air inside this chamber is derived from the host's tracheal system as occurs in the Encyrtidae (Clausen, 1940*b*). Parasitoids of scale-insects and aphids emerge by cutting a hole through the integument of the host mummy, but if the scale has a delicate covering they push their way out from beneath it. The adults of some such species (e.g. some *Aphytis*) lack functional mandibles (Viggiani, 1984).

Adults feed on sugary solutions, and one species of *Coccophagus* has even been observed to stroke its coccid host to induce it to secrete honeydew (Cendaña, 1937). Aphelinids are synovigenic and therefore the female usually host-feeds, by macerating the host with the ovipositor or by forming a feeding tube from secretions of the accessory gland (Rosen & DeBach, 1979). Reproduction via thelytokous parthenogenesis (i.e. without males) appears to be fairly common in the Aphelinidae; for example, in *Aphytis*, about one fourth of all species whose sexuality is known are uniparental (Rosen & DeBach, 1979). In bisexual species, courtship and mating behaviour often varies between species and these differences may provide useful taxonomic characters (Viggiani & Battaglia, 1983*a*).

Economic importance. The family Aphelinidae has been more successful in classical biological control (importations) than any other group of entomophagous insects (Greathead, 1986). In Costa Rica the first successful cases of classical biological control involved aphelinids: in 1932 *Eretmocerus serius* was introduced to control the citrus blackfly, *Aleurocanthus woglumi* (Aleyrodidae), and in 1933 *Aphelinus mali* was introduced to control the woolly apple aphid, *Eriosoma lanigerum* (Aphididae). Both programmes resulted in complete control of the pest (DeBach, 1974; Clausen, 1978). Worldwide, species of *Aphytis* have proved to be the most effective natural enemies of Diaspididae, which has been attributed to the fact that they develop only as ectoparasitoids and hence are not vulnerable to encapsulation (Rosen & DeBach, 1979).

Presently one of the most serious insect pests in Central America is the tobacco whitefly (*Bemisia tabaci*, Aleyrodidae). Most of the parasitoids associated with this pest are aphelinids, i.e. species of *Encarsia* (Polaszek *et al.*, 1992) and *Eretmocerus*. The success of *Encarsia formosa* in controlling the greenhouse whitefly (*Trialeurodes vaporariorum*) would suggest that similar periodic liberations of an appropriate aphelinid species against *B. tabaci* is worth investigating, as is the possibility of importing an exotic aphelinid species (Gerling, 1992).

Several aphelinids occurring in Central America are undoubtedly very important in controlling potential scale-insect pests. *Coccophagus*, for example, is commonly reared from coccid pests such as *Coccus viridis*. In Honduras *Encarsia citrina* has been reared from *Unaspis citri* (Diaspididae) on citrus (R. Cave, pers. comm.).

Identification. Keys to the world genera of Aphelinidae are given in Hayat (1983) and Yasnosh (1983). The aphelinids of Argentina were revised by De Santis (1948). The genera present in Central America can be identified using the following key (to females), which is modified from that of Hayat. Eriaporinae, members of which have a long stigmal and postmarginal vein, is not included (see below).

Key to Central American genera of Aphelinidae

1 Antenna with 6 or fewer segments (Figs 11.44–11.46); fore wing with linea calva usually (but not always) present (Fig. 11.50). ....(Aphelininae and Calesinae)........... 2

— Antenna with 7 to 9 segments (Figs 11.47–11.49); fore wing with linea calva generally absent......(Coccophaginae and Azotinae)..... 7

2 Tarsi 4-segmented. 3

— Tarsi 5-segmented, at least on front and hind legs. 4

3 Antenna 5-segmented, clava tapered at apex (Fig. 11.44); fore wing usually (but not always) with a linea calva (Fig. 11.50). .... *Eretmocerus*

— Antenna 6-segmented, clava truncate at apex (Fig. 11.45); fore wing without a linea calva... .... *Cales*

4 Fore wing with marginal vein at most as long as costal cell (Fig. 11.51); metasoma longer than mesosoma + head (Fig. 11.53) ovipositor distinctly longer than middle tibia and exserted; antennal clava often curved at apex (Fig. 11.46). .... *Centrodora*

— Fore wing with marginal vein longer than costal cell (cf Fig. 11.52); metasoma shorter than mesosoma + head; ovipositor usually not longer than middle tibia and not exserted; clava not curved at apex. 5

5 Hypopygium prominent, extending to apex of metasoma or beyond (Fig. 11.58)..... *Aphelinus*

— Hypopygium not extending to apex of metasoma (Fig. 11.59). 6

6 Propodeum, along midline, at most as long as metanotum (Fig. 11.55); pronotum composed of single plate; fore wing strongly patterned, with infuscate markings on membrane; rare. .. .. *Marietta*

— Propodeum, along midline, longer than metanotum (Fig. 11.56); pronotum membranous in middle; fore wing less commonly patterned, but if so, usually weaker and consisting of setal patterns (Fig. 11.43); common. ....*Aphytis*

7 Female antenna 7-segmented (Figs 11.47–11.48). 8

— Female antenna 8- or 9-segmented (Fig. 11.49). 10

8 Antenna with 4 funicular segments, third funicular segment usually shorter than others (Fig. 11.47); body often slightly metallic coloured. .... *Ablerus*

— Antenna with no more than 3 funicular segments, third flagellar segment not shorter than preceding flagellar segments (Fig. 11.48); never metallic coloured. 9

9 Antenna with 3 funicular segments and 2 claval segments (Fig. 11.48); submarginal vein of fore wing with 4 or more setae; tarsi 5-segmented. .... *Coccobius*

— Antenna with 2 funicular segments and 3 claval segments; submarginal vein of fore wing with less than 4 setae (as in Fig. 11.51); tarsi 4-segmented. .... *Pteroptrix*

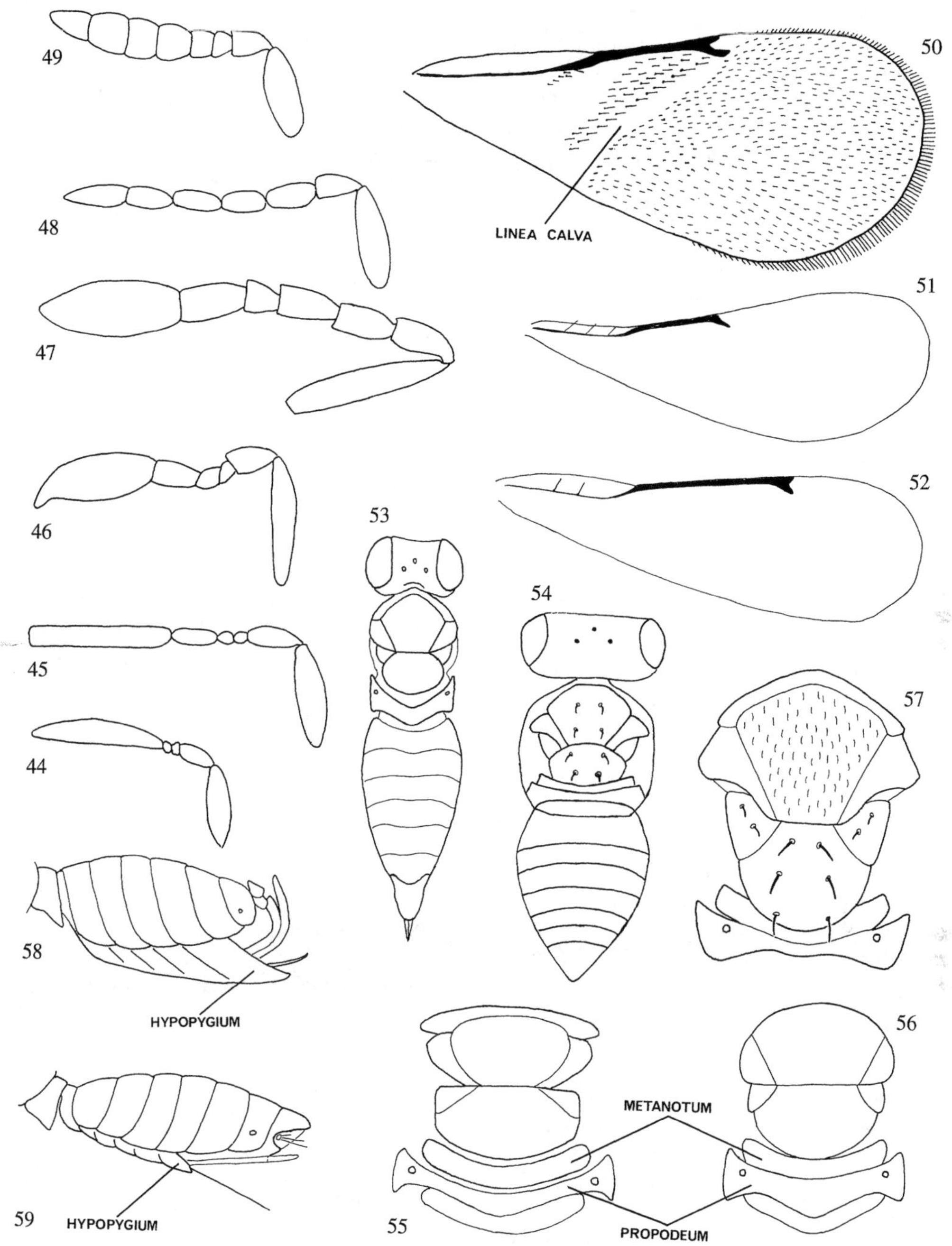

Figs 11.44–11.59. Aphelinidae. Figs 11.44–11.49. Antenna of female; 11.44, *Eretmocerus* sp.; 11.45, *Cales* sp.; 11.46, *Centrodora* sp.; 11.47, *Ablerus* sp.; 11.48, *Coccobius* sp.; 11.49, *Coccophagoides* sp. Figs 11.50–11.52. Fore wing; 11.50, showing calva linea; 11.51, *Centrodora* sp.; 11.52, *Encarsia* sp. Figs 11.53–11.54. Dorsal view of body; 11.53, *Centrodora* sp.; 11.54, *Encarsia* sp. Figs 11.55–11.57. Dorsal view of mesosoma; 11.55, *Marietta* sp.; 11.56, *Aphytis* sp.; 11.57, *Coccophagus* sp. Figs 11.58–11.59. Lateral view of metasoma; 11.58, *Aphelinus* sp.; 11.59, *Aphytis* sp.

10 Axillae small, longer than wide, projecting forwards and separated mesally by more than the maximum length of an axilla (Fig. 11.54); mesoscutum with reduced number of setae, setae bilaterally symmetrical. 11
— Axillae large, subtriangular, separated mesally by about the maximum length of an axilla (Fig. 11.57); mesoscutum with numerous setae. 13

11 Antennal flagellum spindle-shaped, often flattened, apical segment conical with pointed apex (Fig. 11.49); submarginal vein with 3 or more setae. .... *Coccophagoides*
— Antenna not as above; submarginal vein with 1 or 2 setae (Fig. 11.52). 12

12 Hind tarsi 4-segmented; rare. .... *Archenomus*
— Hind tarsi 5-segmented; common. .... *Encarsia*

13 Scutellum with at least 6 setae and each axilla with at least 2 setae (Fig. 11.57); submarginal vein usually with more than 5 setae. [Parasitoids of scale-insects.] .... *Coccophagus*
— Scutellum with 4 setae and each axilla with at most 1 seta; submarginal vein with 2 setae (as in Fig. 11.52). [Parasitoids of whiteflies.] 14

14 Sculpture of mesoscutum aciculate (with fine lines). .... *Dirphys*
— Sculpture of mesoscutum imbricate/reticulate (net-like). .... *Encarsiella*

Synopsis of the Costa Rican fauna

At least 15 genera of Aphelinidae occur in Central America, although there has been very little study of the regional fauna so this total is expected to increase. In Costa Rica aphelinids occur at all altitudes, but they are apparently less diverse at the highest elevations. Many aphelinids have been moved around the world as part of biological control programmes (which may or may not be reported), and they are also easily dispersed unintentionally to new areas through the accidental introduction of their hosts. Thus several of the aphelinid species present in Central America are not indigenous to the region. *Aphytis lingnanensis*, for example, had become unintentionally established in Mexico even before its deliberate introduction from southern China to California (DeBach, 1971). How this species arrived in Costa Rica is unknown.

Following are the genera of Aphelinidae known or thought likely to occur in Costa Rica. In addition to these there is one new genus that cannot be placed in any of the subfamilies (J. Noyes, pers. comm.). The lists of species are based on those reported from Central America by De Santis (1979, 1981, 1989), those present in Honduras (specimens in Escuela Agrícola Panamericana) and those reported to be cosmopolitan by Gordh (*in* Krombein *et al.*, 1979).

Aphelininae

Species in this subfamily may be recognized by the 3- to 6-segmented antennae, and in the possession of a distinct linea calva.

Aphelinus (= *Mesidia*, *Mesidiopsis*). About 60 species worldwide. At least four species have been found to occur in Costa Rica, including *mali* and an unidentified brachypterous species at high altitudes, at or above 2800 metres.

Aphytis. Over 100 species worldwide with perhaps at least 20 in Costa Rica. The following species are known to occur in Central America: *chilensis*, *chrysomphali*, *comperei*, *hispanicus*, *holoxanthus*, *lepidosaphes*, *lingnanensis*, *mytilaspidis*, *proclia*, *salvadorensis*. A key to the world species is given by Rosen and DeBach (1979).

Centrodora^{EX} (= *Debachiella*, *Tumidiscapus*). About 40 species worldwide.

Eretmocerus. About 30 species worldwide with *clauseni* and *serius* known from Central America. Biology: Gerling *et al.* (1990, 1991).

Marietta. About 20 species worldwide with *pulchella* reported from Central America. At least one species has been collected in northwestern Costa Rica.

Coccophaginae

Species of this subfamily may be recognized by the 7- to 9-segmented antennae, and by the absence of a linea calva.

***Archemonus*^EX** (= *Hispaniella*). About 20 species worldwide. This genus may be synonymous with *Pteroptrix* (Viggiani & Garonna, 1993).

Coccobius (= *Physcus*; see LaSalle & Bouček, 1989). About 60 species worldwide, at least two of which are common in Malaise traps in Costa Rica (one at 3000 m and the other at 1300 m).

Coccophagoides (= *Primaprospaltella*). Thirteen species worldwide. One reddish coloured species is common in Costa Rica around 2800 metres.

Coccophagus (= *Aneristus*, *Heptacritus*, *Prococcophagus*). Nearly 200 species worldwide. Among the most commonly collected genera of Aphelinidae in Costa Rica, where there are probably at least 30 species. The following species are recorded from Central America: *albicoxa*, *ceroplastae*, *lycimnia*, *mexicanus*, *occulatipennis*, *quaestor*, *saissetiae*.

***Dirphys*^EX**. A neotropical genus with four described species. *D. encantadora* has been reared from *Ceraleurodicus ingae* (Aleyrodidae) on avocado in Honduras (R. Cave, pers. comm.). *D. mexicana* is known from Costa Rica and Mexico (where it is a gregarious endoparasitoid of *C. altissimus*). Records and a key to species are given in Polaszek and Hayat (1992).

Encarsia (= *Aleurodiphilus*, *Aspidiotiphagus*, *Prospaltella*). About 200 species worldwide with perhaps at least 30 in Costa Rica. In Costa Rica *desantisi*, *nigricephala* and *pergandiella* have been reared from *Bemisia tabaci*. Species attacking diaspidids in Central America include: *aurantii*, *citrina*, *lounsburyi* and *perniciosi*. Biology: T. Williams (1991).

Encarsiella. Six species worldwide. *E. aleurodici* has been recorded from Honduras (R. Cave, pers. comm.) and Ecuador, *E. noyesi* from Honduras (R. Cave, pers. comm.) and Peru, and *E. magniclava* from Panama. Records and a key to species is given in Polaszek and Hayat (1992). Species are commonly reared from aleyrodids belonging to the genus *Aleurodicus*.

***Pteroptrix*^EX**. Twelve species worldwide.

AZOTINAE

Similar to Coccophaginae but tarsi always 5-segmented and antenna always 7-segmented (formula 1-1-4-1). A synopsis of the nearctic species was given by Darling and Johnson (1984).

Ablerus (= *Azotus*; Hayat, in press). Nearly 100 species worldwide. Biology: Pedata and Viggiani (1991).

CALESINAE

In Central America species of this subfamily may be recognized by the combination of 6-segmented antennae (females), four-segmented tarsi, discal wing hairs arranged in lines and a straight fore tibial spur.

***Cales*^EX**. Only two described species, one from Australia and a second from the Neotropics and the Palaearctic.

11.3 CHALCIDIDAE

Gérard Delvare

Diagnosis. 1.5 to 15.0 mm in length, with a hard body, surface sculpture consisting of umbilicate punctures; *colour predominantly black, yellow, or a combination of these*, sometimes with reddish markings, rarely with a metallic tinge. Sexual dimorphism minimal, being most evident in the Haltichellinae, where the flagellum of the male is usually thicker and the scape may be excavated ventrally at the apex and have a ventral projection. Antenna with 1 anellus and 7 funicular segments; *carina bordering posterior of gena usually well developed*. Pronotum separated from tegula by side lobe of scutum; notauli generally complete; *prepectus always very narrow, hardly visible*; mesopleuron with epicnemium completely delimited and horizontal subpleural area present, at least primitively; propodeum mostly areolate, less frequently rugose and rarely smooth. Hind coxa rather long; *hind femur swollen*, ventral edge with small to large teeth; hind tibia distinctly curved; tarsi 5-segmented. Metasoma subsessile to petiolate; last tergite and/or ovipositor sometimes very long.

Classification and distribution. The Chalcididae is cosmopolitan in distribution, but is particularly

Fig. 11.60. *Conura* sp. (Chalcididae).

diverse in lowland tropical areas. The family presently comprises about 1500 species in nearly 90 genera, which are classified in five subfamilies: Chalcidinae, Dirhininae, Haltichellinae, Epitraninae and Smicromorphinae. The first three subfamilies are cosmopolitan, whereas the Epitraninae and Smicromorphinae, which contain one genus each, are restricted to Africa, southern Asia, and Australia.

The Chalcidinae is divided into four tribes: Brachymeriini (two genera worldwide), Chalcidini (six genera), Cratocentrini (eight genera) and Phasgonophorini (eight genera), but the palaeotropical genus *Hovachalcis* represents another tribe. The Dirhininae contains just four genera. The Haltichellinae is divided into four tribes: Haltichellini (32 genera), Hybothoracini (25 genera), Tropimeridini (just one palaeotropical genus) and Zavoyini (one genus; Bouček, 1992).

Biology. All chalcidids are parasitoids of larvae or pupae of other insects, mostly Lepidoptera and Diptera, but also Coleoptera, Neuroptera and Hymenoptera. One European species attacks free-living Strepsiptera (Silvestri, 1943). A brief synopsis of the known hosts of taxa present in Costa Rica is given in Table 11.4. Chalcidids may be ecto- or endoparasitic. Most appear to be idiobionts although some (e.g. *Chalcis* species, see below) are koinobionts. Many Chalcididae are strictly primary parasitoids, but others are facultative or obligatory hyperparasitoids (see below). Most chalcidids are solitary parasitoids but a few are gregarious, such as species of the *Conura pygmaea* group, which attack butterfly pupae (Papilionoidea). A few chalcidids have been the subject of extended biological studies (Roberts, 1933; Hanna, 1934; Dowden, 1935; Magistretti, 1950; Arthur, 1958; Steffan, 1958 and 1959*c*; Hansen,

1980). Although much of the following information pertains to temperate species the biology of neotropical species is probably comparable.

Most chalcidids are not very host specific, although a particular species is nearly always restricted to hosts in one insect order. In *Brachymeria*, for example, many species are primary parasitoids of various families of Lepidoptera, while others are restricted to carrion infesting Diptera or to chrysomelid beetles of the subfamily Cassidinae. Species of *Conura* in the *maculata* group attack a diversity of lepidopterous hosts, some as solitary parasitoids, others as gregarious parasitoids. Such generalists may eventually prove to be habitat specific rather than host specific.

In other cases the spectrum of hosts attacked by a particular chalcidid species may be restricted to one family or even one species of host. Species of *Conura* in the *nigricornis* group are restricted to the lepidopterous families Limacodidae and Megalopygidae. The mandibles of these species are very stout, presumably for cutting their way out of the hard cocoon of the host. Species in the *oiketicusi* group attack only, as far as is known, Psychidae of the genus *Oiketicus*. Members of the *dimidiata*, *transitiva* and *pygmaea* groups are mostly restricted to butterflies of the family Nymphalidae. Those in the first two groups are solitary while those in the last group are gregarious parasitoids. Species of the *apaiis*, *maculipennis*, *melana* and *odontotae* groups are parasites of Chrysomelidae, generally leaf-miners (Delvare, 1992).

In searching for hosts, females look first for their host's habitat. Hyperparasitic chalcidids search for the primary host. Thus *Brachymeria moerens*, which attacks parasitic flies in grasshoppers, must first find the primary host and then determine whether there is a parasitic fly within (Léonide & Léonide, 1969). *B. compsilurae*, which is a secondary parasitoid of Lepidoptera, is not attracted by a free tachinid pupa, but is attracted by a pupa within the lepidopteran cocoon (Dowden, 1935). Hosts of Chalcididae are often relatively dispersed. Not surprisingly, chalcidids are good flyers and *Conura* spp. have been captured in flight 900 metres above the ground (Glick, 1939). Moreover, adult chalcidid wasps are quite long lived, varying from four to ten weeks (up to five months in *Conura side*; Arthur, 1958), with female longevity being increased by mating (Hansen, 1980). They show maximum activity under intense illumination (Desmier de Chesnon, 1973). All species that have been studied are synovigenic and total egg production by one female ranges from 150 to 200 eggs. Females have 6 to 12 ovarioles, with only one egg formed per ovariole per day (Clausen, 1940*b*). As in many synovigenic parasitoids, host-feeding has often been reported in the family, but a feeding tube has never been observed.

Oviposition behaviour varies between groups of Chalcididae. Members of the tribes Phasgonophorini and Cratocentrini attack xylophagous beetles and are morphologically adapted to ovipositing in hard wood. The petiole is very short and several metasomal tergites are reduced or fused together. These transformations evidently occurred in order to ensure rigidity of the metasoma during oviposition. The valvulae, whether they are free or enclosed in the sheath formed by the elongate last tergite, are very long in order to penetrate deeply within wood.

The female of *Brachymeria moerens*, after encountering a grasshopper with a parasitic fly within, grips it along its ventral side and oviposits across a membranous part of the body. This species does not appear to be capable of discriminating fly larvae that are already parasitized. When more than one chalcidid larva occurs within a host, a fight usually ensues, since only one parasitoid can develop in the same host

Host groups	Chalcidid parasitoid
COLEOPTERA	
Chrysomelidae	*Brachymeria*, *Conura*
Lagariidae	*Haltichella*
wood-boring beetles	Phasgonophorini (Haack *et al.*, 1981) *Acanthochalcis*, ?*Notaspidium*
STREPSIPTERA	
Mengenillidae	*Hockeria*
DIPTERA	
Muscoidea	*Brachymeria*, *Dirhinus*
Tephritidae	*Dirhinus*
LEPIDOPTERA	
various families	*Brachymeria*, *Conura*, *Haltichella*, *Hockeria*, *Psilochalcis*
HYMENOPTERA	
Ichneumonoidea	*Brachymeria*, *Conura*, *Notaspidium*

Table 11.4. The known hosts of chalcidid taxa occurring in Costa Rica.

(Léonide & Léonide, 1969). In *Chalcis canadensis*, which oviposits into stratiomyid eggs, the female rests on her hind legs while picking up a host egg with her front legs. She then oviposits into the egg, places it back in the egg mass, and repeats the process with the next egg. Females may defend host egg masses until they are ready for oviposition, fighting other females by grappling with their enlarged hind femora (Cowan, 1979). Species of *Chalcis* are koinobionts, ovipositing in eggs or larvae of Stratiomyidae, with the chalcid larva remaining in the first instar until the host pupates (Schremmer, 1960).

Perhaps the most impressive act of oviposition is that of *Lasiochalcidia* species, which parasitize ant-lions (Steffan, 1959*b*). The female, as soon as she discovers her victim, fights to prevent it from withdrawing into its sand pit, then grips the ant-lion larva with her mid and hind legs and promptly embeds her ovipositor into its prothorax while lying on her side. The host is paralysed for a short time, but can afterwards move and feed. In *Hockeria eriensis*, another parasitoid of ant-lions, the host is partly but permanently paralysed after oviposition (Steffan, 1959*b*). Thus *Lasiochalcidia* is a koinobiont whereas *Hockeria* is an idiobiont.

The first instar larvae of Chalcididae are of two types. Those of lepidopterous parasitoids are hymenopteriform and possess a functional respiratory system, whilst larvae of dipterous parasitoids are of the caudate type and apneustic. Mature larvae are always hymenopteriform. Developmental time at 25°C is two days for embryonic development, the same time for each of the first four instars, three days for the fifth instar, two days for the prepupa, and 12 days for the pupa (Dowden, 1935). The female generally needs another 24 to 48 hours after emerging to achieve her complete development. Under optimal conditions, but depending upon the size of species, development from egg to adult requires 16 to 30 days. Of course, it is prolonged in koinobiotic species, where oviposition occurs within a young larva and parasitoid larval development is delayed until the host completes its own larval development. Parasitoids of Diptera consume the entire contents of the pupa, while solitary parasitoids of Lepidoptera do not, the unconsumed part becoming dried inside the chrysalis.

Although some species of Chalcididae may be thelytokous, most species consist of both males and females. Males generally emerge before females and mating occurs very soon after the latter emerges from the host. In *Conura albifrons* the male approaches the female from a right angle, swaying his body from side to side and vibrating his antennae without touching the female (Hansen, 1980). In *C. side* the male follows the female and strokes her wings lightly with his antennae (Arthur, 1958). The time of copulation is brief (two to seven minutes) and it appears that females mate only once, although older females have been observed mating (Arthur, 1958). Two to eight days are generally required before oviposition can begin.

Adults of both sexes feed on nectar and possibly other plant secretions. In the Neotropics chalcidids are strongly attracted to plants possessing extra-floral nectaries and/or glandular hairs. Among parasitic Hymenoptera collected on such plants, Chalcididae (especially Chalcidini) often predominate. Certain plants are highly attractive to Chalcididae (Hespenheide, 1985; Mexzón & Chinchilla, 1991; Delvare & Genty, 1992): *Croton* spp. (Euphorbiaceae), *Urena* spp. (Malvaceae), *Solanum* spp. (Solanaceae), *Byttneria aculeata* (Sterculiaceae), and *Bidens cynapiifolia* (Compositae). Each plant attracts several tens of species which do not belong to the complex of natural enemies of the phytophagous insects feeding on the plant.

Large chalcidids, such as species of the *nigricornis* group of *Conura*, generally have yellow and black banded metasomas and seem to mimic vespid wasps. *Caenobrachymeria polybiaeraptor*, which parasitizes a polistine wasp, is similar to species of *Brachymeria* except that it lacks the normal punctures and has dense but short and appressed pilosity, similar to that of its host (Steffan, 1974). It is thought that within the dark nest of the host, where tactile stimuli are more important than visual stimuli, such sculpture and pilosity can mislead the vespid host.

Economic importance. Although a number of species have been reared from pest insects, various records show that their control is rather limited, since levels of parasitism are generally only between 5 to 20 percent. Some hyperparasitic species may interfere with biological control by killing primary parasitoids such as tachinids, braconids, and ichneumonids. For example, *Conura biannulata* and *immaculata* attack pupae of the ichneumonid genus *Casinaria* and may limit the latter's effectiveness in oil palm plantations and other tropical agroecosystems. The latter has also been reared from *Apechthis zapotecus* (Ichneumonidae) on *Leptophobia aripa* (Pieridae) on cruciferous crops (R. Cave, pers. comm.). *C. petiolventris* and *C. pseudofulvovariegata* are facultative hyperparasitoids of *Diadegma insulare* (Cordero & Cave, 1992).

Nonetheless, some species may be locally effective in controlling certain pest insects, especially Lepidoptera. At least 15 species of *Conura* have been reared from injurious insects in oil palm plantations in Colombia and Ecuador (Delvare, 1993), and several of these species also occur in Costa Rican oil palm plantations (Mexzón & Chinchilla, 1991). *C. maculata* for example attacks *Opsiphanes cassina* (Nymphalidae), and *C. magdalensis* and *C. miniata* attack *Tiquadra circumdata* (Tineidae). Because certain weeds growing in the understorey are highly attractive to *Conura* species, it may be possible to enhance biological control by maintaining these plants (Delvare & Genty, 1992).

Identification. A key to the 26 genera of New World Chalcididae is given by Bouček (1992).

Key to subfamilies and tribes of New World Chalcididae

1 Apex of hind tibia truncate (rarely sinuate) and mostly with two apical spurs (Fig. 11.76) ....(Haltichellinae) 2

— Apex of hind tibia tapering into a spine and with at most one apical spur (Figs 11.63, 11.70, 11.74, 11.83). 4

2 Hind tibia with one apical spur; marginal vein long, almost half as long as costal cell, thin and often discoloured; head with very deep scrobal cavity delimited by carina. .... **Zavoyini** (p. 298)

— Hind tibia with two spurs; marginal vein much shorter than half of costal cell; head rarely with very deep scrobal cavity. 3

3 Postmarginal vein of fore wing always developed, even if sometimes short; marginal vein at the margin of the wing (Fig. 11.75)....................... .... **Haltichellini** (p. 297)

— Postmarginal vein of fore wing completely absent; marginal vein somewhat removed from margin of wing (Fig. 11.78). .... **Hybothoracini** (p. 297)

4 Frons between antennal scrobes and eyes protruding into horns (Figs 11.80, 11.81); petiole present, although sometimes short, and with longitudinal carinae (Figs 11.84, 11.85); fore wing with marginal vein very long, stigmal and postmarginal extremely short (Fig. 11.82).. ....(Dirhininae) ....**Dirhinini** (p. 296)

— Frons never protruding into horns; petiole and wings different. ....(Chalcidinae) 5

5 Metasoma petiolate, or at least subsessile; petiole with a basal flange and not sculptured, mostly smooth (Fig. 11.72); propodeum with spiracle nearly in vertical position (Fig. 11.73); fore coxa with a line of bristles on posterior side.
[Antennal scrobes rarely carinate laterally.] .... **Chalcidini** (p. 296)

— Metasoma at most subsessile, petiole very short and concealed; propodeum with spiracle in oblique direction (Fig. 11.69); fore coxa without line of bristles on posterior side. 6

6 Postmarginal vein much longer than the marginal (Fig. 11.67); metasomal tergites II to IV strongly reduced and hidden for the most part under the preceding tergite (Fig. 11.66); female valvulae long but last tergite not prolonged into a long sheath.
[Body very distinctly cylindrical and with areas of dense, silvery appressed hairs.] .... **Cratocentrini** (p. 296)

— Postmarginal vein shorter than marginal; metasoma different. 7

7 Malar sulcus present, at least as carina (Fig. 11.68); postmarginal vein distinctly longer than stigmal vein, the latter very short (Fig. 11.71).
[Antennal scrobes usually carinate laterally; mostly black.] .... **Brachymeriini** (p. 296)

— Malar sulcus absent, malar space entirely punctured (Fig. 11.61); postmarginal vein very short, at most as long as the stigmal vein (Fig. 11.62); interspaces of umbilic punctures elevated into crests on mesosoma. .... **Phasgonophorini** (p. 296)

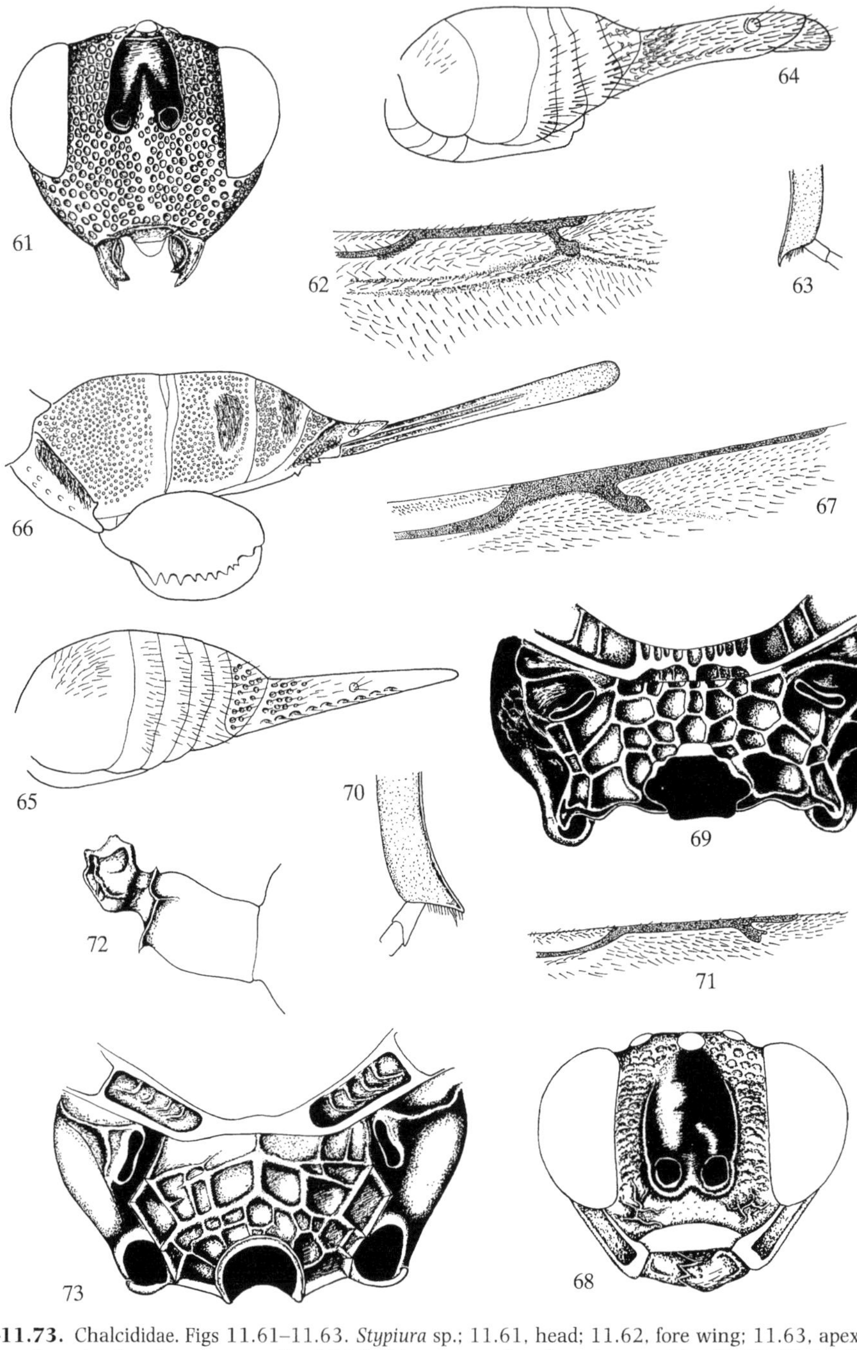

Figs 11.61–11.73. Chalcididae. Figs 11.61–11.63. *Stypiura* sp.; 11.61, head; 11.62, fore wing; 11.63, apex of hind tibia. Fig. 11.64. *Stypiura thoracica*, female metasoma. Fig. 11.65. *Trigonura* sp., female metasoma. Figs 11.66–11.67. *Acanthochalcis nigricans*; 11.66, female hind leg and metasoma; 11.67, fore wing. Figs 11.68–11.71. *Brachymeria* sp.; 11.68, head; 11.69, propodeum; 11.70, apex of hind tibia; 11.71, fore wing. Fig. 11.72. *Conura miniata*, female petiole. Fig. 11.73. *Conura femorata*, propodeum.

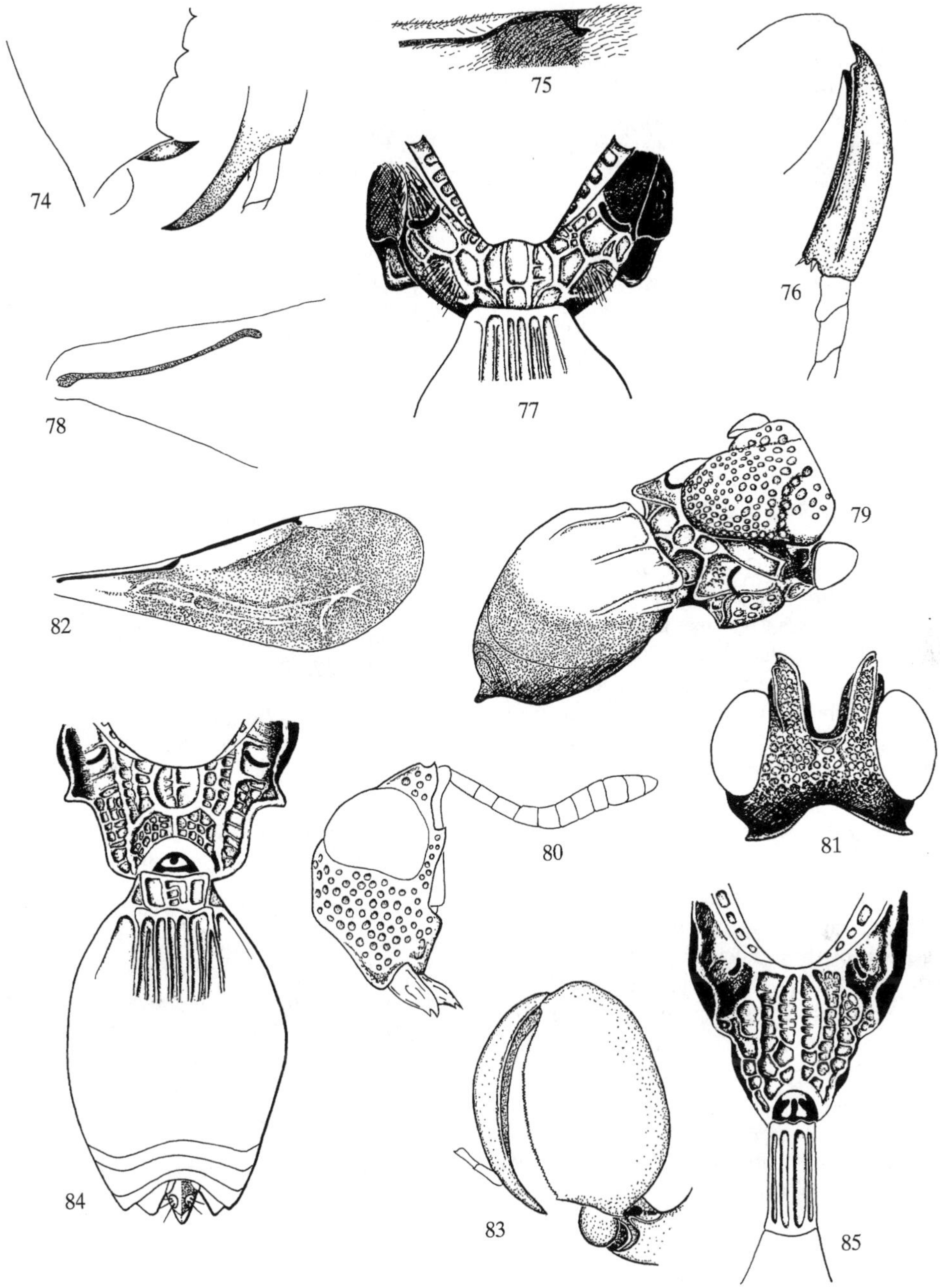

Figs 11.74–11.85. Chalcididae. Fig. 11.74, *Melanosmicra variventris*, female, apices of hind femur and tibia. Figs 11.75–11.77. *Haltichella* sp.; 11.75, fore wing; 11.76, hind tibia; 11.77, propodeum and base of metasoma. Figs 11.78–11.79. *Notaspidium* sp.; 11.78, fore wing; 11.79, posterior part of mesosoma and metasoma. Figs 11.80–11.84. *Dirhinus* sp.; 11.80, head, lateral view; 11.81, head, dorsal view; 11.82, fore wing; 11.83, hind femur and tibia; 11.84, female propodeum and metasoma. Fig. 11.85. *Dirhinus* (*Hontalia*) sp., female propodeum and metasoma.

Synopsis of the Costa Rican fauna

There are probably between 400 and 500 species of Chalcididae in Costa Rica. They are most common at low altitudes, becoming uncommon above about 1300 metres and apparently absent above 1600 metres.

CHALCIDINAE

All four tribes occur in Costa Rica.

CHALCIDINAE: Brachymeriini. This tribe comprises two genera, Brachymeria and Caenobrachymeria, the latter known only from northeastern South America (Bouček, 1992).

Brachymeria. About 200 species worldwide, with 42 described from the Neotropics, although the actual number of species is much greater. Perhaps 100 species occur in southern Central America. Most species are black although a few also have reddish markings, and I have seen one species mostly yellow and another with a metallic tinge; *B. mexicana*, with its yellowish wings and metasoma, resembles certain polistine vespids. The North American species were revised by Burks (1960). Biology: develop as parasitoids in the pupae of a wide range of lepidopterous, coleopterous and dipterous hosts, and as hyperparasitoids (Steffan, 1959*a*; Bouček, 1988).

CHALCIDINAE: Chalcidini. This tribe is most diverse in the Neotropics, where it comprises the most speciose and diverse group of the Chalcididae. The tribe was reclassified by Delvare (1992) who recognized six genera in the New World: *Chalcis* (cosmopolitan but mostly absent from the Neotropics), *Conura*, *Corumbichalcis* (Brazil), *Melanosmicra*, *Pilismicra* (Colombia) and *Stenosmicra*. Three of these are known to occur in Costa Rica. The metasoma is very mobile (unlike other Chalcidinae) and the last tergite of the female is short to very long.

Conura (= *Spilochalcis*). A very large and extremely diverse genus that recently has been divided into three subgenera, *Ceratosmicra*, *Conura s. str.* and *Spilochalcis*. These, in turn, are divided into more than 60 species-groups, keys to which are given by Delvare (1992). *Conura* is primarily a New World genus, with probably over 1000 species in the Neotropics and perhaps 250 to 300 in Costa Rica. Most species are yellow with black markings; *C. dimidiata* is unusual in having the basal two thirds of its wings black. Parasitoids of pupae of Lepidoptera, Chrysomelidae and their primary parasitoids.

Melanosmicra. This genus is restricted to the Neotropics where there are about 30 species, at least 12 of which occur in Costa Rica. They have the hypopygium sclerotized in the middle and thickly pilose at its apex. *M. flavicollis* is easily distinguished from other members of the genus by its flat body and the very sparse punctation of the mesoscutum, the interspaces being smooth and shining. Biology unknown.

Stenosmicra. At least five species are present in the Neotropics, with two, *S. exilis* and *S. tenuis*, described from Costa Rica (Bouček, 1992). They have an elongate body, the prepectus is enlarged laterally, and the notauli are faint or absent posteriorly. Biology unknown.

CHALCIDINAE: Cratocentrini. This is primarily a palaeotropical group. The world species were revised by Steffan (1959*c*). Only one of the nine described genera is known to occur in the New World, and it has been found to occur in Costa Rica.

Acanthochalcis. Three species: one in Madagascar and two in North America. *A. nigricans* (Figs 11.66 and 11.67) appears to reach the southern limit of its distribution in northwestern Costa Rica (Halstead, 1987). Up to 24 mm in length; resembles *Stypiura* species, but the propodeum lacks a hairy tooth and the mesosoma is stouter. Biology: known hosts are various wood-boring beetles.

CHALCIDINAE: Phasgonophorini. A pantropical group represented in the New World by four genera: *Parastypiura* (Brazil), *Phasgonophora* (North America), *Stypiura* (Neotropics) and *Trigonura* (North America; Fig. 11.65). Keys to neotropical genera and species are given in Steffan (1973).

Stypiura. Southern Mexico to Brazil, about ten species (several of which are undescribed). *S. dentipes* and *S. serripes* are the most commonly collected species in Costa Rica. First metasomal tergite not or hardly longer than the following tergites, the female valvulae slightly bent downwards at the apex (Fig. 11.64), and the propodeum with a hairy tooth laterally in front of the spiracle. Biology: probably parasitoids of wood boring beetles.

DIRHININAE

This subfamily is represented in the New World by a single genus. *Hontalia*, previously treated as a distinct genus, has recently been downgraded to a subgenus of *Dirhinus* by Bouček (1992).

Dirhinus. Warm regions of the world. About 15 species in the Neotropics. The subgenus *Hontalia* can be separated from *Dirhinus s. str.* by the elongate body, the longer petiole which bears only one median carina, the metasoma not dorsally flat and without longitudinal carinae at its base. Biology: parasitoids in puparia of various Diptera, especially Calliphoridae, Muscidae and Sarcophagidae, and also Tephritidae (Bouček, 1988).

HALTICHELLINAE

Three of the four tribes occur in the New World. All are represented in Costa Rica

HALTICHELLINAE: Haltichellini. This tribe is represented in the New World by six genera, four of which occur in Costa Rica. The other two genera are *Belaspidia* (California, Mediterranean, Central Asia) and *Antrocephalus* (Palaeotropics; one species associated with stored product Lepidoptera has been introduced to the New World).

Aspirrhina. Costa Rica south to Brazil. At least seven species, with *bifurca* (Halstead, 1991*a*) and *dubitator*[CR] (determined by J. Halstead) known from Costa Rica; *alvarengai* is recorded from Panama. Structurally *Aspirrhina* is very similar to *Haltichella*, but it has conspicuous preorbital carinae turning dorsally between the middle and lateral ocelli, with an additional curved carina on the lower face joining the preorbital carinae to the toruli, and the scutellum is posteriorly prolonged. Biology unknown.

Ecuada. Neotropical, comprising a single species, *E. producta* (Bouček, 1992). The head has distinct preorbital carinae, terminating dorsally in short teeth in front of the lateral ocelli; the pronotal carina forms tubercles in the middle and the scutellum has two sharp teeth. The first metasomal tergite has a transverse crest and longitudinal carinae; the sixth tergite is clothed thickly with black setae, and is dorsally produced in the female but in the male forms an acuminate spine. Biology unknown.

Haltichella. Cosmopolitan, represented in the Neotropics by five described species (some of which do not belong in the genus); there are probably actually about 20. The neotropical species all look very similar. This genus and *Notaspidium* are the most common genera of Haltichellinae in the Neotropics. Species have an additional carina present on the external side of the hind tibia (Fig. 11.76), a transverse carina present at the base of the metasoma followed posteriorly by 2 to 15 longitudinal carinae (Fig. 11.77). The nearctic species were revised by Halstead (1990*a*). Species of this genus parasitize lepidopterous pupae and one Old World species attacks Lagriidae. In Honduras *H. ornaticeps* has been reared from pupae of the ichneumonid *Diadegma insulare* (Cordero & Cave, 1992).

Hockeria. Old World, North and Central America; three rare species are recorded from the Neotropics, but the placement of one of these needs confirmation. *H. bicolor* occurs in Costa Rica. *Hockeria* species lack preorbital carinae on head, have the hind tibia without additional carina, and the metasoma regularly convex at its base, without carinae. The nearctic species have recently been revised (Halstead, 1990*b*). Several species are known to parasitize pupae of small to medium sized Lepidoptera, such as Tortricidae, Pyralidae and Psychidae, and one species has been reared from a free-living strepsipteran (Bouček, 1988).

HALTICHELLINAE: Hybothoracini. This tribe is represented in the New World by four genera, three of which are present in Costa Rica. The other genus, *Schwarzella*, comprises just one species in the southwestern United States.

Halsteadium. An exclusively neotropical genus, comprising two species, both of which occur in Costa Rica (Bouček, 1992): *alterum* and *petiolatum*. The latter has been collected from the lowland

Atlantic plain up to over 3000 metres on the Cerro de la Muerte (Bouček, 1992). *Halsteadium* species resemble *Notaspidium*, but the metasoma has a distinct petiole. Biology unknown.

Notaspidium. This genus, which also occurs in southern Asia, is more diverse in the Neotropical region, where it is represented by at least 34 species, 13 of which have been described (Halstead, 1991*b*). *N. giganteum* and *lineatum* are known from Costa Rica; *truncatum* probably occurs here as well since it is reported both from Honduras and Brazil. Species are very small (1.5 to 2.0 mm in length) and with an evident metallic tinge (unlike other Chalcididae). The scutellum often protrudes backwards and the propodeum is produced laterally backwards into sharp corners. The first metasomal tergite is very large, with a conspicuous transverse carina basally, followed by longitudinal carinae or striae (Fig. 11.79). The antennae of both sexes are similar. Biology unknown, but possibly associated with beetles in wood (Hespenheide, pers. comm.).

Psilochalcis (= *Invreia*). Primarily an Old World genus that is also represented in North America, and with one species present in Central America as far south as Costa Rica. It has the body black, the scutellum rounded or very slightly bilobed and never protruding backwards into a spine. The metasoma is without basal carinae. Three of the four North American species have been reared from *Elasmopalpus lignosellus* (Pyralidae) (Grissell & Schauff, 1981).

HALTICHELLINAE: Zavoyini. A small tribe comprising only a single genus.

Zavoya. Restricted to the Neotropics where three species have recently been described (Bouček, 1992). Two of these, *Z. cooperi* and *Z. parvula*, are known from Costa Rica. The eyes have sparse long hairs, the scutellum is produced into a long horizontal horn and the first metasomal tergite is very large, dorsally flat and striate. Biology unknown.

Acknowledgments

I thank Dr Z. Bouček and J.A. Halstead for their useful comments and correction of my English.

11.4 ELASMIDAE

Lonny D. Coote

Diagnosis. Elongate wasps, about 0.6 to 3.5 mm in length; body colour varying from black or brownish with various metallic lustres and with a few small, yellow areas, to mostly or entirely yellow. Antenna 10-segmented with a 3-segmented funicle in female and a 4-segmented funicle in male, the three basalmost segments each bearing a dorsal ramus (branch) in the male. *Dorsellum with hyaline, triangular-shaped lamella. Fore wing wedge-shaped; marginal vein long, postmarginal and stigmal veins reduced. Hind coxa expanded and flattened*; mid and hind femora flattened; fore tibial spur short, straight; *hind tibia with conspicuous, flattened spines forming elongate or diamond-shaped patterns*; tarsi 4-segmented. Metasoma subsessile, central part triangular in cross section.

Classification and distribution. The family Elasmidae contains only the genus *Elasmus*, with 203 currently recognized species worldwide (Noyes, 1990). Although the genus is cosmopolitan in distribution, most species occur in the Old World tropics. Some classifications (Peck *et al.*, 1964; Trjapitzin *in* Medvedev, 1978) have included at least one genus from what is now regarded by many systematists as the subfamily Eriaporinae of the family Aphelinidae (Ghesquiére, 1955). Structural similarities between *Elasmus* and the eriaporines are most likely due to convergence (Compere, 1947; Ferriére, 1947). The genus *Elasmus* has also been classified in the subfamily Elasminae of the family Eulophidae (Riek, 1967, 1970; Burks *in* Krombein *et al.*, 1979; De Santis, 1979, 1980), but an adequate phylogenetic analysis has not been presented to justify such a classification.

Biology. The larvae of most species of *Elasmus* are gregarious, primary ectoparasitoids of larvae or prepupae of leaf-mining, leaf-rolling, web-spinning, and casemaking Lepidoptera. Many species are facultative, some possibly obligatory, hyperparasitoids of these lepidopterans, primarily via cocooned prepupae of Braconidae and Ichneumonidae. A few species are primary ectoparasitoids of *Polistes* wasps (Vespidae) (Reed & Vinson, 1979; De Santis, 1981), and of leaf-mining beetles (Chrysomelidae: Hispinae) (Taylor,

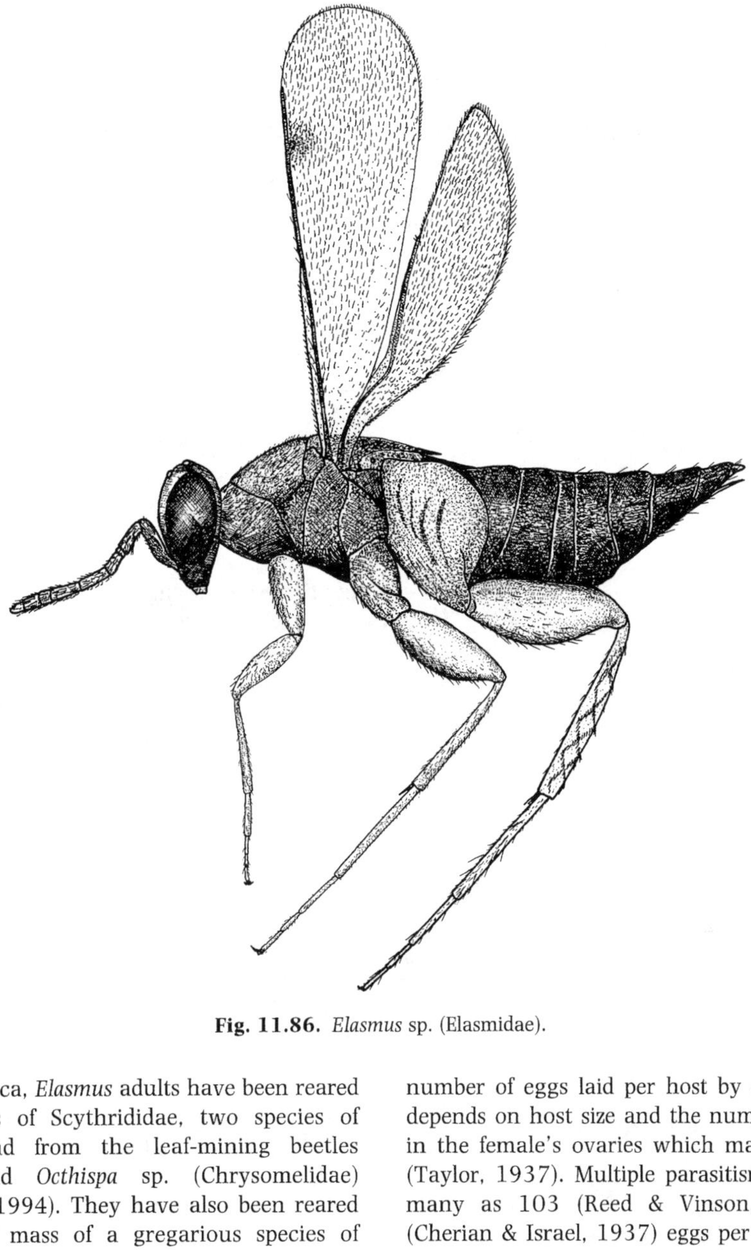

Fig. 11.86. *Elasmus* sp. (Elasmidae).

1937). In Costa Rica, *Elasmus* adults have been reared from two species of Scythrididae, two species of Gracillariidae, and from the leaf-mining beetles *Chalepus* sp. and *Octhispa* sp. (Chrysomelidae) (Memmott *et al.*, 1994). They have also been reared from the cocoon mass of a gregarious species of *Cotesia* (Braconidae) on the sphingid, *Erinnyis ello* (P. Hanson, pers. comm.).

Species of *Elasmus* employ an idiobiont strategy, whereby hosts are paralysed in a protective enclosure (e.g. leaf mine, rolled leaf, cocoon, paper nest) and the parasitoid larva consumes this defenceless host. The number of eggs laid per host by a female parasitoid depends on host size and the number of mature eggs in the female's ovaries which may be less than ten (Taylor, 1937). Multiple parasitism may result in as many as 103 (Reed & Vinson, 1979) and 170 (Cherian & Israel, 1937) eggs per host. The eggs are elongate-oval (Parker, 1924) or elongate curved with one end wider than the other (Clausen, 1940*b*; Reed & Vinson, 1979; Peter & David, 1990). The larva is hymenopteriform, distinctly segmented (Clausen, 1940*b*), and may have median pseudopodia situated intersegmentally, from the posterior margin of the second thoracic segment back to the penultimate segment. These pseudopodia are thought to aid locomotion (Taylor, 1937). There are normally four pairs of spiracles in the first instar, and nine (Clausen, 1940*b*) or ten (Reed & Vinson, 1979) pairs in later instars.

There are three or four larval instars and the life cycle from egg to adult takes around two to three weeks, although pupae may pass a prolonged dormant period inside the host enclosure. Adult sex ratio is usually female biased, but this ratio may vary depending on host stage and size, and whether or not *Elasmus* larvae feed solitarily or gregariously (Clausen, 1940*b*). Generally speaking, younger or smaller hosts, and fewer parasitoid larvae per host, will result in more male parasitoids. Although most species are arrhenotokous, some species occasionally tend towards thelytoky, with males becoming extremely rare (Miller *et al.*, 1987).

Economic importance. Species of *Elasmus* attack many economically important pests, although none have yet been found to be effective control agents. This may be due to a propensity towards hyperparasitism and gregariousness (Naumann & Sands, 1984; Miller *et al.*, 1987). There is, however, the possibility that species of *Elasmus* may be useful when combined with other species in biological control programmes (Ramachandra Rao & Cherian, 1927).

Identification. Keys to the North American *Elasmus* species were provided by Burks (1965). Keys to the New World species will soon be available (Coote, in prep.).

Synopsis of the Costa Rican fauna.

There are 31 recognized species of *Elasmus* in the New World (Burks, 1965; De Santis, 1979, 1980, 1981) and these are currently being revised (Coote, in prep.). Although *E. maculatus* is the only species that has been recorded previously from Costa Rica (De Santis, 1979), there are at least 20 species of *Elasmus* present in the country. They have been collected from sea-level up to 2600 metres altitude.

11.5 ENCYRTIDAE

John S. Noyes

Diagnosis. About 0.5 to 3.0 mm long, robust, usually fully winged, rarely brachypterous. Antennal flagellum in female cylindrical to very broadly flattened, 4- to 9-segmented, in male cylindrical to slightly flattened or with branched segments, 3- to 8-segmented; anellus usually absent in both sexes; apical segment larger than preceding segments, or apical two to four segments forming a distinct clava. Pronotum strongly transverse; *mesoscutum normally without notaular lines*, but if present extremely shallow and meeting near posterior margin of mesoscutum; anterior margins of axillae forming a straight line; scutellum shield-shaped; prepectus moderately large and free, not forming shoulders in dorsal view; *mesopleuron (or acropleuron) convex, undivided, occupying more than half the mesosoma in side view*, often touching the second segment of metasoma. *Fore wing with a bare line (linea calva) extending obliquely across wing from stigmal vein; marginal vein generally very short. Mid coxa, in profile, level with middle of mesopleuron*; mid tibial spur relatively long and stout, usually subequal in length to mid basitarsus. Metasoma most often broadly attached to mesosoma; tergites VIII and IX fused; cercal plates usually well advanced (situated anteriorly); ovipositor hidden or well exserted; outer plates of ovipositor separated from tergite VIII, at most connected by a weak membrane or an elongate sclerotized strip.

Encyrtids, like eupelmids and tanaostigmatids, have the mesopleuron greatly enlarged, but in Tanaostigmatidae and Eupelmidae the mid coxal insertion is level with the posterior margin of the mesopleuron or nearly so, whereas in all Encyrtidae the mid coxa is inserted level with the mid line of the mesopleuron. Furthermore, neither Tanaostigmatidae nor Eupelmidae have a linea calva on the fore wing. The small robust body of Encyrtidae resembles Aphelinidae, but the latter have well-marked notaular grooves and a relatively long marginal vein in the fore wing.

Classification and distribution. The Encyrtidae and the Tanaostigmatidae are probably sister-groups that arose from within a lineage that also includes the Eupelmidae (LaSalle & Noyes, 1985; LaSalle, 1987; Gibson, 1989). Synapomorphies of the Encyrtidae pos-

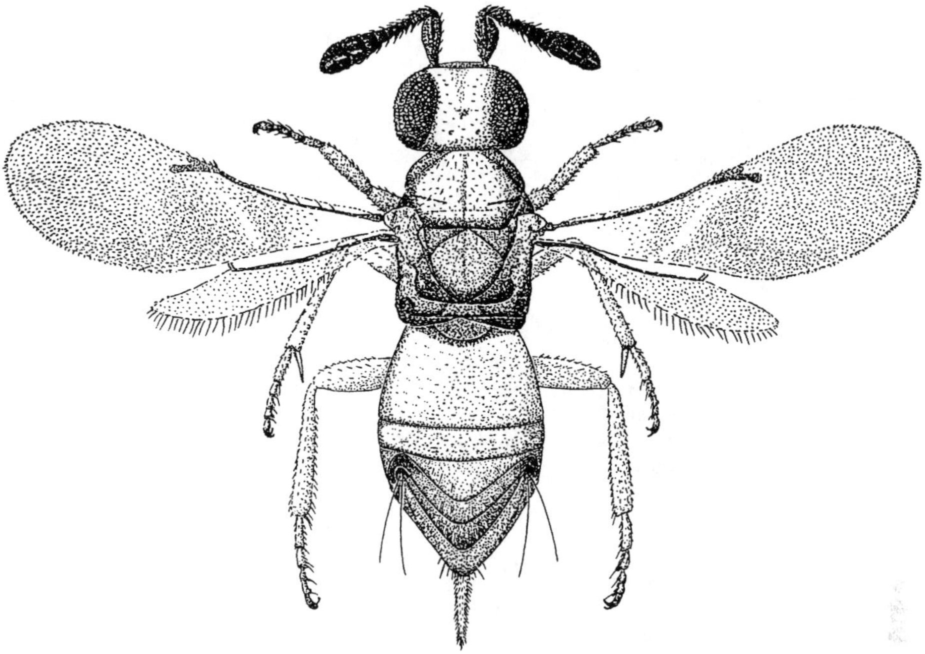

Fig. 11.87. *Mariola* sp. (Encyrtidae).

sibly include the presence of a linea calva in the fore wing, advanced mid coxae, and advanced cercal plates with separation of the outer plates of the ovipositor from the ninth abdominal tergite.

Trjapitzin (1973*a* & *b*, 1989) discussed the different systems of classification of Encyrtidae and proposed dividing the family into two subfamilies, Encyrtinae and Tetracneminae (Table 11.5). While this is perhaps the most practical division presently available, the proposed subfamilies are not necessarily monophyletic groups. Trjapitzin (1973*a*) recognized 12 tribes within the Tetracneminae and 36 tribes within the Encyrtinae, but later (Trjapitzin, 1989) reduced these to 11 and 34 tribes, respectively. Many of these tribes are based on very minor differences, and for this reason a tribal classification is not presented here.

The family Encyrtidae contains about 3000 described species worldwide and is most diverse in tropical and subtropical areas (Noyes, 1989, 1990). Currently about 450 valid genera are recognized, although some authors recognize as many as 500 to 600 genera. Thus far about 700 species and 130

Character	Tetracneminae	Encyrtinae
Fore wing linea calva	marginal setae undifferentiated	setae longer and thicker basally
Filum spinosum	usually not developed	usually present
Outer plates of ovipositor	connected to tergite IX via paratergites (narrow filaments)	not connected to tergite IX via paratergites
Syntergite	variable in shape	mostly V- or U-shaped
Hypopygium	extending slightly beyond apex of syntergites	variable
Third valvulae (gonostyli)	usually fused to second valvulae	usually articulated with second valvifers

Table 11.5. Differences between the two subfamilies of Encyrtidae.

Host group	Encyrtid parasitoid
ACARINA	
Ixodidae	*Ixodiphagus* (N)
ORTHOPTERA	
Tettigoniidae	*Ooencyrtus* (E)
various families	*Meromyzobia* (E)
HEMIPTERA	
Coreidae	*Hexacladia* (N)
Pentatomidae	*Hexacladia* (N)
Scutelleridae	*Hexacladia* (N)
various families	*Ooencyrtus* (E)
HOMOPTERA	
(Auchenorrhyncha)	
Cercopidae	*Carabunia* (N)
Cicadellidae	*Hexacladia* (N), *Meniscocephalus* (N)
Delphacidae	*Ooencyrtus* (H)
Membracidae	*Prionomastix* (N)
Various families	*Agarwalencyrtus, Cheiloneurus, Helegonatopus* (all H)
(Sternorrhyncha)	
Aclerdidae	*Bennettisca, Cheiloneurus* (H), *Meromyzobia* (H), *Microterys* (H), *Mucrencyrtus*
Aleyrodidae	*Metaphycus*
Aphididae	*Ooencyrtus, Syrphophagus* (H)
Asterolecaniidae	*Metaphycus*
Coccidae	*Ammonoencyrtus, Anicetus, Blastothrix, Cheiloneurus* (H),*Coccidaphycus, Coccidoctonus* (H), *Eusemion, Gahaniella* (FH), *Metaphycus, Microterys, Prochiloneurus* (H), *Trichomasthus*
Diaspididae	*Adelencyrtus, Ameromyzobia, Arrhenophagus, Cheiloneurus* (H), *Coccidencyrtus, Homalopoda, Epitetracnemus, Metaphycus, Neococcidencyrtus, Plagiomerus, Pseudhomalopoda, Thomsonisca, Trichomasthus, Zaomma* (H)
Dactylopiidae	*Anagyrus, Tetracnemoidea*
Eriococcidae	*Acerophagoides, Metaphycus, Microterys, Trichomasthus, Zaomma* (H)
Kermesidae	*Blastothrix,*
Kerriidae	*Metaphycus, Microterys*
Lecanodiaspididae	*Microterys*
Margarodidae	*Brethesiella*
Ortheziidae	*Cheiloneurus* (H)
Psylloidea	*Cercobelus* (N), *Metaphycus* (N), *Prionomitus* (N), *Psyllaephagus* (N), *Sectiliclava* (A), *Syrphophagus* (H), *Trechnites* (N)
Pseudococcidae	*Acerophagus, Acroaspidia, Aenasius, Anagyrus, Aphycus, Apoanagyrus, Blepyrus, Bothriocraera, Cheiloneurus* (H), *Coccidoctonus* (H), *Coccidoxenoides, Ericydnus, Euryrhopalus, Gahaniella* (H), *Gyranusoidea, Hambletonia, Leptomastix, Leptomastidea, Microterys, Neodusmetia, Prochiloneurus* (H), *Pseudaphycus, Pseudectroma, Pseudleptomastix, Rhopus, Tetracnemus, Tetracnemoidea, Zaplatycerus, Zarhopalus*
NEUROPTERA	
Ascalaphidae	*Ooencyrtus* (E)
Chrysopidae	*Isodromus* (L)
Coniopterygidae	*Trjapitzinellus* (L)
Hemerobiidae	*Isodromus* (L)
COLEOPTERA	
Anobiidae	*Cerchysius* (L)
Bruchidae	*Lamennaisia* (L)
Buprestidae	*Avetianella* (E), *Szelenyiola* (E)

Table 11.6. Continued opposite.

Cerambycidae	*Avetianella* (E), *Cerchysius*, *Ooencyrtus* (E)
Cleridae	*Cerchysius* (L)
Chrysomelidae	*Ooencyrtus* (E)
Coccinellidae	*Anagyrus* (L), *Cheiloneurus* (L), *Coccidoctonus* (L), *Homalotylus* (L), *Ooencyrtus* (L), *Prochiloneurus*
Cryptophagidae	*Cerchysiella* (L)
Erotylidae	*Cerchysiella* (L)
Lathridiidae	*Lamennaisia* (L)
Nitidulidae	*Cerchysiella* (L)
Scolytidae	*Avetianella* (E), *Szelenyiola* (E)
Silvanidae	*Cerchysiella* (L)
LEPIDOPTERA	
Gelechiidae	*Paralitomastix* (E, L)
Gracillariidae	*Paraleurocerus* (L)
Lyonetiidae	*Parablastothrix* (L)
Nepticulidae	*Parablastothrix* (L)
Psychidae	*Ageniaspis* (L), *Apsilophrys* (L)
Pterophoridae	*Ageniaspis* (L)
Pyralidae	*Copidosomopsis* (see text), *Paralitomastix* (see text)
Tineidae	*Ageniaspis* (L)
Tortricidae	*Copidosomopsis* (see text)
Yponomeutidae	*Ageniaspis* (L)
Various	*Copidosoma* (E, L, PP), *Epiencyrtus* (H), *Ooencyrtus* (E, L-H), *Tyndarichus* (L-H)
DIPTERA	
Cecidomyiidae	*Coccidoctonus* (L)
Ceratopogonidae	*Forcipestricis* (L, P)
Chamaemyiidae	*Cerchysius*, *Cheiloneurus*, *Meromyzobia* (all L)
Chloropidae	*Ooencyrtus* (P)
Muscidae	*Cerchysiella* (L), *Cerchysius* (L)
Phoridae	*Exoristobia* (P)
Pipunculidae	*Agarwalencyrtus*
Sarcophagidae	*Exoristobia* (P)
Syrphidae	*Bothriothorax* (L, P), *Exoristobia* (P), *Ooencyrtus* (L), *Syrphophagus* (L)
Tachinidae	*Exoristobia* (PP, P, FH)
Tephritidae	*Cerchysiella*
HYMENOPTERA	
Aphelinidae	*Cheiloneurus*, *Syrphophagus*
Apidae	*Coelopencyrtus*
Braconidae	*Syrphophagus*
Dryinidae	*Cheiloneurus*, *Helegonatopus*
Encyrtidae	*Cheiloneurus*, *Coccidoctonus*, *Epiencyrtus*, *Gahaniella*, *Procheiloneurus*, *Tyndarichus*, *Zaomma*
Eurytomidae	?*Lamennaisia*
Ichneumonidae	*Epiencyrtus*
Pteromalidae	*Coccidoctonus*
Various families	*Ooencyrtus*

Table 11.6. The likely host associations of the encyrtid genera occurring in Costa Rica. A = parasitoid of adult; E = egg parasitoids; FH = facultative hyperparasitoids; H = hyperparasitoids; L = larval parasitoids; N = nymphal parasitoids; P = pupal parasitoids; PP = prepupal parasitoids. Where the host stage attacked is not known or embraces many stages no letter is given.

genera of Encyrtidae have been found in Costa Rica, but most of the species (at least 90 percent) and about 20 genera are undescribed.

Biology. Most species of Encyrtidae are primary endoparasitoids of other arthropods, or hyperparasitoids via other hymenopterous parasitoids. A few species of *Microterys*, however, have been recorded as predators of the eggs of scale insects (Coccidae) (Silvestri, 1919; DeBach, 1939; Sugonjaev, 1984; Voynovich & Sugonjaev, 1989). Most encyrtids parasitize Homoptera, with about one third of the species being recorded from mealybugs (Pseudococcidae) and another third from soft scales (Coccidae).

Host range. The host-parasitoid relations of Encyrtidae in general have been discussed in some detail by Trjapitzin (1972, 1979) and summarized by Tachikawa (1978, 1981). The likely host associations of the encyrtid genera occurring in Costa Rica are given in Table 11.6.

The degree of host specificity is not very well known, but where adequate information is available it appears that specificity varies between encyrtid species. *Leptomastix dactylopii* has been recorded reliably from over 20 species of mealybugs (Noyes & Hayat, 1994) and *Ooencyrtus submetallicus* attacks not only eggs of Hemiptera and Lepidoptera, but also pupae of a chloropid fly (Noyes, 1985). *Copidosoma floridanum* attacks several species of Noctuidae, but nearly all belong to the subfamily Plusiinae (Noyes, 1989). On the other hand the Indian species *Gyranusoidea tebygi* appears to be host-species specific (Narasimham & Chacko, 1988; Willink & Moore, 1988) as does the Afrotropical species *Comperia merceti* (Gordh, 1973).

Oviposition. Host-feeding, which may be responsible for up to 25 percent of the deaths caused by an encyrtid, has been observed in *Apaoanagyrus* (Umeh, 1988), *Gyranusoidea* (Willink & Moore, 1988), *Comperiella* (Flanders, 1944), *Habrolepis* (Hafez *et al.*, 1971) and *Microterys* (Kfir & Rosen, 1980). No host-feeding was observed in *Comperia* (Gordh, 1973), nor in *Cheiloneurus paralia*, a hyperparasitoid of encyrtid primary parasitoids attacking soft scales (Kfir & Rosen, 1981). *Encyrtus infelix*, a common parasitoid of soft scale on coffee and other plants, emerges from the host with most of its eggs already developed (Embleton, 1904) and therefore probably does not need to host-feed. The same appears to be true of *Carabunia myersi*, a parasitoid of the cercopid, *Clastoptera* (Myers, 1930).

Most encyrtids oviposit into the larval/nymphal stage of their host. Among encyrtids attacking Coccoidea, some, such as *Encyrtus infelix*, *Anagyrus kamali* and *Gyranusoidea tebygi*, oviposit into the younger instars (Embleton, 1904; Moursi, 1948*a*; Willink & Moore, 1988), while others such as *Leptomastix dactylopii* and *Leptomastidea abnormis* oviposit into third instars or adults (Kirkpatrick, 1953; Jong & Alphen, 1989). All encyrtids that parasitize Orthoptera and several of those that parasitize Coleoptera or Lepidoptera are egg parasitoids. Pupal parasitoids are restricted to certain species associated with Diptera, while parasitoids of the adult stage include *Sectiliclava* from adult Psylloidea (Robinson, 1961), *Hexacladia* from adult Hemiptera (Burks, 1972) and, as stated above, many species that attack Coccoidea.

All egg parasitoids and probably most pupal parasitoids are idiobionts, whereas many of those attacking the larval stage are koinobionts. An example of a koinobiont species occurring in Costa Rica is *Carabunia myersi*, a parasitoid of cercopids of the genus *Clastoptera* (Bennett & Hughes, 1963). Some of the encyrtids parasitizing Lepidoptera, such as *Copidosoma* and *Paralitomastix*, oviposit into the host egg but develop only when the host is a mature larva or prepupa. *Bothriothorax* and *Forcipestricis*, which attack Diptera, oviposit in the host larva and complete development in the pupa.

Hyperparasitoids oviposit into the second to fifth instar larvae of the primary parasitoid (Kfir & Rosen, 1981). The hyperparasitoid is unable to discriminate between a parasitized and an unparasitized host without drilling into it with her ovipositor (Weseloh & Bartlett, 1971). The larva of the hyperparasitoid feeds within the body of its parasitoid host, but just before completing its development it emerges, feeds ectoparasitically for a short time, and then pupates near the remains of the primary parasitoid inside the mummified remains of the host (e.g. Kanuck & Sullivan, 1992).

The majority of encyrtids are solitary parasitoids, laying only one egg per host, but some species, e.g. *Carabunia myersi*, may adjust the number of eggs laid, and thus be either solitary or gregarious (Myers,

1930). A few encyrtids are polyembryonic, with multiple individuals developing from a single egg. In these species the number of individuals developing in a host varies from fewer than 20 (e.g. *Ageniaspis* spp., parasitoids of Yponomeutidae) to more than 1000 (e.g. *Copidosoma* spp., parasitoids of Noctuidae). Species of *Epiencyrtus* are polyembryonic hyperparasitoids, and in Costa Rica a species of this genus was reared from *Enicospilus* sp. (Ichneumonidae) parasitizing a saturniid caterpillar. More than 1000 individuals emerged from the prepupa of the ichneumonid (Janzen, pers. comm.). One of the most commonly encountered polyembryonic species in Costa Rica is *Copidosoma floridanum*, a cosmopolitan egg/larval parasitoid of plusiine noctuids. Only one sex develops from an individual parasitoid egg (Strand, 1989), yet both sexes typically emerge from a single parasitized caterpillar. Thus the female must be laying both unfertilized and fertilized eggs in an individual host egg (Hardy *et al.*, 1993). The embryogenesis of *C. floridanum* is described in detail in Baehrecke and Strand (1990).

Immature stages. The ovarian egg is described as being 'encyrtiform' with the egg proper connected to a bulb by a narrow stalk (Maple, 1947). There is frequently an aeroscopic plate extending the length of the egg proper and most of the stalk, and in some species there is a small cluster of cells just below the bulb. When the egg is inserted into the host it often remains attached to the host's integument by the egg stalk, with the collapsed bulb protruding to the outside. In species where the larva is apneustic (see below), the egg is easily dislodged and floats free. Generally the bulb and neck of the egg collapse shortly after oviposition and are visible only as a small peduncle (Maple, 1947; Viggiani & Maresca, 1973). In at least one species (*Ooencyrtus nezarae*) the protruding part of the egg stalk is used as a marker to enable females to discriminate between parasitized and unparasitized eggs (Takasu & Hirose, 1988).

The first instar larva is 11-segmented (Moursi, 1948*a–c*; Chumakova, 1961; Kfir & Rosen, 1980) or 13-segmented (Domenichini, 1951; Zinna, 1960*a*; Rosen & Rössler, 1966; Viggiani & Maresca, 1973; Delvare *et al.*, 1981; Boussienguet, 1988), varies in shape from spherical to elongate, and is caudate, vesiculate, or hymenopteriform in form. Caudate larvae sometimes have a bifurcate 'tail', and such an appendage is frequently found in species that parasitize Coccidae. Vesiculate larvae have a caudal vesicle and sometimes a ring of fleshy protuberances around each of the first twelve segments. In some species (e.g. of *Apoanagyrus*, *Diversinervus* and *Leptomastidea*) the early instar larva is apneustic, i.e. lacks functioning spiracles (Maple, 1947; Viggiani & Maresca, 1973; Rosen & Alon, 1983). The so-called encyrtiform larvae (such as are found in *Anagyrus*, *Microterys*, *Leptomastix* and *Pseudleptomastix*) are more or less hymenopteriform, but are metapneustic, i.e. with only a terminal pair of spiracles (Maple, 1947; Moursi, 1948*a–c*; Zinna, 1960*a*; Kfir & Rosen, 1980). Metapneustic larvae develop with their spiracles in contact with the aeroscopic plate of the egg, which protrudes from the integument of the host, thus allowing the larva access to atmospheric oxygen (Moursi, 1948*a–c*; Domenichini, 1951; Zinna, 1960*a*; Rosen and Rössler, 1966; Nenon & Biassangama, 1985). The tracheal system becomes well developed by the end of the second or third instar (Moursi, 1948*a* & *b*; Viggiani & Maresca, 1973; Delvare *et al.*, 1981) and the connection with the egg is severed during the final larval instar.

The number of larval instars varies from three to six. For example there are three in *Carabunia myersi* (Myers, 1930) and *O. venatorius* (Gerling *et al.*, 1976), four in *Apoanagyrus lopezi* (Löhr *et al.*, 1989) and *Encyrtus infidus* (Kon, 1974), five in *Ooencyrtus kuvanae* (Parker, 1933) and six in *Leptomastix phenacocci* (Moursi, 1948*a–c*) and *Anagyrus pseudococci* (Rosen & Rössler, 1966). However, the number of larval instars is difficult to ascertain accurately, as exemplified by the fact that *Ooencyrtus kuvanae* has been credited with both three (Crossman, 1925) and five (Parker, 1933), and *Encyrtus infelix* with four (Embleton, 1904) and five (Thorpe, 1936).

The encyrtid egg or larva may be encapsulated by the host (Driesche *et al.*, 1986; Blumberg, 1988), and in some cases the rate of encapsulation can be quite high. Up to 25 percent of the eggs laid by *Apoanagyrus lopezi* are encapsulated in its normal host, *Phenacoccus manihoti*, although the larvae can frequently escape from the capsule and survive (Giordanengo & Nenon, 1990). In *Leptomastix dactylopii* the egg is coated with minute proteinaceous spherulae which probably prevent encapsulation of the egg by the host (Barbier & Rambault, 1985; Barbier *et al.*, 1988). This may explain the very wide range of hosts that have been reported for this species. LeBeck (1989) observed that the female of *Comperia merceti* injected yeast-like micro-organisms into the eggs of its cockroach host. These yeast-like cells are found in the haemolymph and gut of all stages of the parasitoid, and in the poison gland reservoir of the adult female. Their function is not known, but the relationship appears to be symbiotic.

In the polyembryonic species, *Copidosoma floridanum*, there are usually more than 1000 parasitoid larvae present in the parasitized caterpillar, and of these, five percent are destined to never reach adulthood. These 'precocious' larvae differentiate sooner than normal larvae and they are larger, with well developed mandibles. When the same host is parasitized by the braconid, *Microplitis demolitor*, the latter usually dies, perhaps via physiological suppression resulting from a secretion from the labial glands of the precocious larva (Strand *et al.*, 1990). On the other hand, in the closely related *Copidosomopsis tanytmemus*, the precocious larva is thought to physically attack competitors (Cruz, 1981, 1986*b*).

During the prepupal stage of encyrtids eight pairs of spiracles become functional and the caudal pair ceases to function. The pupa is sometimes formed inside a 'cocoon' or sheath (Moursi, 1947*a*; Zinna, 1960*a* & *b*; Rosen and Rossler, 1966; Werner and Williams, 1985), which is connected to the tracheal system of the host, as in *Encyrtus* (Embleton, 1904; Clausen, 1932; Thorpe, 1936; Kon, 1974), *Carabunia* (Bennett & Hughes, 1963), and *Metaphycus* (Alam, 1957, 1959). The 'cocoon' is probably formed from a viscous secretion produced by both the salivary and ileac glands (Flanders, 1938). In some species the male pupa differs from the female pupa in being noticeably darker (Domenichini, 1951; Rosen & Alon, 1983). Although most encyrtids kill their host prior to pupation in a few species with large hosts, such as *Carabunia* spp. (in Cercopidae) and *Hexacladia* spp. (hemipteran nymphs), the host may not die until the adult parasitoids have emerged (Costa Lima, 1930; Bennett & Hughes, 1963; Rasplus *et al.*, 1990).

Development from egg to adult normally takes between 14 and 45 days, depending on the species and ambient temperature (Lloyd *in* Zinna, 1960*a*; Viggiani & Maresca, 1973; Willink & Moore, 1988). The rate of development is sometimes related to the age and size of the host, with development being more rapid in larger hosts (Moursi, 1948*c*; Löhr *et al.*, 1989; Umeh, 1988). In some species, e.g. *Comperiella* and *Zaomma*, the adults remain within the host mummy for some time after emerging from the pupa (Compere & Smith, 1927; Taylor, 1935).

Adult biology. The majority of encyrtids are arrhenotokous, although many thelytokous species are known. Species that are normally thelytokous may produce males if the female wasp is exposed to high temperatures during her development (Gordh & Lacey, 1976), or if she is exposed to a combination of high temperatures and an unfavourable host (Flanders, 1942). In *Ooencyrtus submetallicus* sexual mosaics can be produced by allowing the parasitoids to develop at a certain critical temperature (Wilson & Woolcock, 1960*a* & *b*; F. Wilson, 1962). Arrhenotokous species may produce unusually high proportions of males if reared on an abnormal host (Teran & DeBach, 1963) and the food plant of the host may also affect the sex ratio of the parasitoid (Flanders, 1942).

Mating has been observed in *Anagyrus* (Avidov *et al.*, 1967; Boussienguet, 1988), *Plagiomerus* (Gordh & Lacey, 1976), *Habrolepis* (Hafez *et al.*, 1971), *Cerchysiella* (Werner & Williams, 1985), *Tachinaephagus* (Olton & Legner, 1974) and *Comperia* (Gordh, 1973). In some species females are ready to mate immediately after emerging from the host, but in *Cerchysiella* the female may not be receptive for at least 90 minutes after emergence, and during this period will reject advances by males (Werner & Williams, 1985). Males often mate with more than one female, while females usually mate only once. In *Comperia* the male follows a female until she stops, whereupon he begins to bite at the apex of her metasoma. If he is not rejected the male places his fore legs on the female's metasoma and thrusts his metasoma forward until contact is made. Courtship and copulation may last less than 15 seconds. In *Anagyrus* the male approaches the female from the side or from the front, and if not rejected, he climbs onto her mesosoma. He places his head just above that of the female and after several seconds moves backwards to copulate for several more seconds. The male then resumes the precopulatory position at the anterior end of female's mesosoma, moving his metasoma rhythmically up and down. After about two minutes the female starts moving around and the male rides her for another two minutes before dismounting.

Under favourable conditions females of some species begin to lay eggs immediately after emergence (Smith, 1917), while in others oviposition does not occur for 48 hours (Kfir & Rosen, 1981; Willink & Moore, 1988;). Some lay between five and ten eggs per day, up to a total of 45 under laboratory conditions (Rivnay & Perzelan, 1943; Kirkpatrick, 1953).

Economic importance. The Encyrtidae is second only to Aphelinidae in terms of numbers of successes in classical biological programmes. The first recorded case of an encyrtid species being introduced purposefully for control purposes was when Albert Koebele introduced *Copidosoma floridanum* (*C. truncatellum* misident.) into Hawaii from eastern North America in 1898, in an attempt to control the noctuid *Chrysodeixis chalcites*, a pest of ornamental and garden plants. It may have become established and partially suppressed populations of the moth (Swezey, 1931). Since then there have been many successful biological control programmes utilizing Encyrtidae (e.g. Clausen, 1978; Bosch *et al.*, 1982; Carver *et al.*, 1987; Neuenschwander, 1989). Noyes & Hayat (1994) have summarized all biological control programmes, up to the end of 1993, that use Encyrtidae.

One of the most recent successful biological control programmes using an encyrtid involved the cassava mealybug, *Phenacoccus manihoti*. This pest was first reported in Zaire in 1973 and within a few years it had infested nearly the whole of tropical Africa, devastating the primary source of nutrition for 200 million people (Herren & Neuenschwander, 1991). It was thought that the mealybug had originated in Central America or northern South America, but parasitoids from this region failed to reproduce on the mealybug in Africa. It was then realized that two closely related mealybugs were involved, one from Central America and the other, the true *P. manihoti*, from central South America. In 1980 the search for natural enemies of the cassava mealybug switched to Paraguay, Bolivia and southern Brazil and very soon a number of parasitoids and predators had been recovered. One encyrtid, *Apoanagyrus lopezi*, immediately gave spectacular control in field trials in Nigeria. By 1990 this parasitoid had been successfully established in 24 African countries and had spread over more than 12.7 million square kilometres (Neuenschwander, 1990). The mealybug is now under complete control throughout the whole of its range in Africa (Neuenschwander, pers. comm.) and it is estimated that the overall benefit:cost ratio is at least $149:1, or perhaps more than $250 million per annum (Norgaard, 1988).

Pest	Encyrtid parasitoid
HOMOPTERA	
PSEUDOCOCCIDAE	
Antonina graminis	*Neodusmetia sangwani**
Dysmicoccus boninsis	*Anagyrus saccharicola**
Nipaecoccus nipae	*Pseudaphycus utilis*
Planococcus citri	*Leptomastidea abnormis**, *Leptomastix dactylopii**
Pseudococcus longispinus	*Anagyrus fusciventris**
Saccharicoccus sacchari	*A. saccharicola**
COCCIDAE	
Saissetia spp.	*Metaphycus helvolus**
various coccids	*Encyrtus infelix*, *E. lecaniorum*
PSYLLIDAE	
Heteropsylla cubana	*Psyllaephagus yaseeni*
HEMIPTERA	
Nezara viridula and others	*Ooencyrtus submetallicus*
LEPIDOPTERA	
plusiine Noctuidae	*Copidosoma floridanum*
phycitine Pyralidae	*Copidosomopsis plethoricus*
various	*Ooencyrtus submetallicus*

Table 11.7. Encyrtids and the pests they are known to attack in Central America. Species believed to be exotics are indicated with an asterisk.

There have been very few deliberate introductions of exotic encyrtid species into Central America, but many indigenous species are undoubtedly important natural enemies of some insect pests (see Table 11.7). In addition to these, some exotic species have established themselves in the region where they are playing a major role in controlling pests. These adventitious encyrtids presumably either arrived together with their host or they dispersed throughout the region from areas where deliberate importations occurred. One example is the Old World encyrtid *Neodusmetia sangwani*. This brachypterous insect, which was introduced into the southern United States to control the rhodesgrass mealybug (*Antonina graminis*), is now widespread in Central America where it is benefiting ranchers by controlling populations of the rhodesgrass mealybug. Although a monetary estimate of this benefit is not available for

Central America, in the United States it has been estimated to be worth at least $200 million per year (Dean *et al.*, 1979).

Identification. Useful references for the identification of Encyrtidae include: De Santis, 1964 (revision of Argentinian species); Trjapitzin & Gordh, 1978*a* & *b* (key to nearctic genera); Noyes, 1980 (review of neotropical genera).

Synopsis of the Costa Rican fauna

Presently 700 species of Encyrtidae are known from Costa Rica, but the true number may be closer to 1000. These species belong to 109 described genera, and possibly as many as 20 undescribed genera. Some of these genera are represented in Costa Rica by a surprisingly large number of species. For example there are at least 40 species of *Ooencyrtus*, although previously only 17 species were recognized for the entire Neotropical region (Noyes, 1985; De Santis, 1988). Other large genera in Costa Rica are *Rhytidothorax* with at least 45 species, *Copidosoma* with at least 35 species, *Anagyrus* with at least 30 species, *Forcipestricis* with at least 25 species, and *Metaphycus* with at least 25 species.

The family is present throughout Costa Rica, but the vast majority of species seem to be restricted to lower altitudes, below 1500 metres. Recent extensive collecting indicates that the encyrtid fauna of lowland areas can be divided into two distinct types. The first type is found in the northwest of the country, where there is a pronounced dry season. Here the encyrtid fauna is largely composed of species that parasitize mealybugs and scale insects. The second faunistic type is found east of the central mountain range and in the wetter regions to the west (such as the Osa Peninsula). This association primarily comprises species that belong to groups parasitizing the immature stages of holometabolous insects. The fauna above 1500 metres is relatively depauperate, and even more so above 2500 metres. At such altitudes coccoid parasitoids are very scarce and, in general, species occurring at these higher altitudes are conspicuously larger and darker than congeneric lowland species, e.g. species of *Copidosoma*.

The extensive Malaise trapping programme undertaken throughout Costa Rica has shown that coccoid parasitoids are most abundant during the later part of the dry season, which is not surprising since their hosts are more abundant during the dry season. On the other hand, groups that attack immature stages of Diptera, Lepidoptera and Coleoptera do not seem to show any distinct seasonality.

Tetracneminae

The following genera have been found in Costa Rica and can be identified with keys provided by Noyes (1980). Generic synonyms or misidentifications are given in parentheses.

> *Acroaspidia*, *Aenasius*, *Anagyrus*, *Apoanagyrus*, *Blepyrus*, *Coccidoxenoides* (= *Pauridia*), *Ericydnus*, *Euryrhopalus*, *Gyranusoidea*, *Hambletonia*, *Leptomastidea*, *Leptomastix*, *Neodusmetia*, *Rhopus*, *Tetracnemus*, *Zaplatycerus* and *Zarhopalus*.

Pseudleptomastix, a new record for the Neotropical region, runs to couplet 157 in Noyes (1980) and differs from all included genera in having the postmarginal vein of the fore wing longer than the stigmal vein, and in the tegulae being brilliant white. Males run to couplet 95, differing from all included genera by the same characters as for the female (although the tegula is white with a dark brown apex).

Encyrtinae

The following genera have been found in Costa Rica and can be identified with keys provided by Noyes (1980). Generic synonyms or misidentifications are given in parentheses.

> *Acerophagoides*, *Acerophagus*, *Adelencyrtus*, *Agarwalencyrtus* (*Epiencyrtus* misident.), *Ageniaspis*, *Ameromyzobia*, *Ammonoencyrtus*, *Anicetus*, *Aphycus*, *Apsilophrys*, *Archinus*, *Arhopoidiella*, *Arrhenophagus*, *Avetianella*, *Bennettisca*, *Blanchardiscus*, *Bothriocraera*, *Brachyplatycerus*, *Brethesiella*, *Carabunia*, *Cerapterocerus*, *Cerchysiella* (= *Zeteticontus*), *Cerchysius*, *Cheiloneurus*, *Coccidaphycus*, *Coccidencyrtus*, *Coccidoctonus*, *Coelopencyrtus*, *Coenocercus*, *Copidosoma* (= *Litomastix*), *Copidosomopsis* (= *Pentalitomastix*), *Cyderius*, *Echthrogonatopus*, *Encyrtus*, *Exoristobia*, *Forcipestricis*, *Gahaniella*, *Helegonatopus*, *Hexacladia*, *Hexencyrtus* (= *Calliencyrtus*), *Homalopoda*, *Homalotylus*, *Isodromus*, *Ixodiphagus* (= *Hunterellus*), *Lamennaisia* (= *Mercetencyrtus*), *Lirencyrtus*, *Lochitoencyrtus*, *Mahencyrtus* (*Parechthrodryinus* misident.), *Meniscocephalus* (= *Helmecephala*), *Meromyzobia* (=

Desantisella), *Metaphycus*, *Microterys*, *Mucrencyrtus*, *Neococcidencyrtus* (= *Platylyca*), *Ooencyrtus*, *Parablastothrix*, *Parablatticida* (= *Amaurilyma*, *Desobius*), *Paraleurocerus*, *Paralitomastix*, *Pentelicus* (= *Hemaenasius*; *Bothriothorax* misident.), *Plagiomerus*, *Prionomastix*, *Prochiloneurus*, *Protyndarichoides*, *Pseudaphycus*, *Pseudectroma* (= *Timberlakia*), *Pseudhomalopoda*, *Psyllaephagus*, *Rhytidothorax*, *Stemmatosteres*, *Syrphophagus* (= *Aphidencyrtus*), *Szelenyiola*, *Tetarticlava*, *Trechnites*, *Trichomasthus*, and *Zaomma*.

The following genera, except for *Paratetracnemoidea*, are new records for the Neotropical region; the previous key (Noyes, 1980) should therefore be amended as follows (in most cases female specimens only):

Blastothrix—runs with difficulty to couplet 85 but differs from both included genera (*Rhytidothorax* and *Exoristobia*) in having distinctive punctate sculpture on the frontovertex and dorsum of mesosoma.

Bothriothorax—runs to *Pentelicus* (*Bothriothorax* misident.) (couplet 184), but differs in that there is no hairless line connecting the apex of the postmarginal vein with the apex of the stigmal, while in *Pentelicus* this is present. The ovipositor is not exserted (see *Coenocercus* below).

Cercobelus—runs to *Homalopoda* (couplet 4) but differs in having hyaline fore wings, whereas in *Homalopoda* the fore wings have a distinct infuscate pattern. Males run to couplet 9, but differ from all included genera in having only four funicle segments, these all longer than broad.

Coenocercus—runs to *Pentelicus* (*Bothriothorax* misident.) (couplet 184), but differs in having the ovipositor clearly exserted by about one-third of the length of the metasoma, while in *Pentelicus* the ovipositor is more or less hidden (see also *Bothriothorax* above).

Epiencyrtus—runs to couplet 241, but differing from all other included genera except *Agarwalencyrtus* (misident.), in having the clava strongly obliquely truncate, and from *Agarwalencyrtus* in the body being completely dark with no orange areas and the thoracic dorsum quite shiny with relatively shallow sculpture. In *Agarwalencyrtus* the mesosoma is normally at least partly orange and thoracic dorsum always matt with very characteristic striate-reticulate sculpture. The funicle segments are all transverse (see *Trjapitzinellus* below).

Eusemion—runs with difficulty to couplet 63 (*Aenasius* and *Pentelicus*) and differs from these genera in the combination of the marginal vein of the fore wing being long and the stigmal and postmarginal veins relatively short, the flagellum being clavate with strongly transverse funicle segments, the hypopygium not reaching more than two-thirds along metasoma, and the body being relatively elongate.

Paratetracnemoidea—runs to couplet 156, but differs from all included species in having the interantennal prominence characteristically produced above the clypeus and between the antennal sockets in the form of a short ridge-like or horn-like structure. Males run with difficulty to couplet 69 and differ from all included genera by the same character as the female.

Philosindia—runs to couplet 67, but the antennae are inserted very nearly half way between the anterior ocellus and mouth margin and the first funicle segment is at least twice as long as the pedicel. The other genera of the group generally have the antenna inserted conspicuously nearer the mouth margin than the anterior ocellus, and the first funicle segment not longer than the pedicel.

Prionomitus—runs best to *Psyllaephagus* (couplet 198), differing in having the hypopygium reaching the apex of the metasoma.

Thomsonisca—runs to couplet 154, but differs from included genera in having the speculum interrupted by two or three lines of setae.

Trjapitzinellus—runs to couplet 241 and would fit *Agarwalencyrtus* (*Epiencyrtus* misident.), except that the clava is not twice as wide as the funicle and the marginal vein is less than three times as long as broad. The funicle segments are all quadrate or longer than broad (see *Epiencyrtus* above).

Tyndarichus—runs to couplet 89, but differs from both included genera in the submarginal vein having a strong, distinct subapical triangular expansion.

11.6 EUCHARITIDAE

John Heraty

Diagnosis. Mostly 2.0 to 5.0 mm in length; colour varying from metallic to black, or with striking yellow and black patterns. Antennae with 13 or fewer

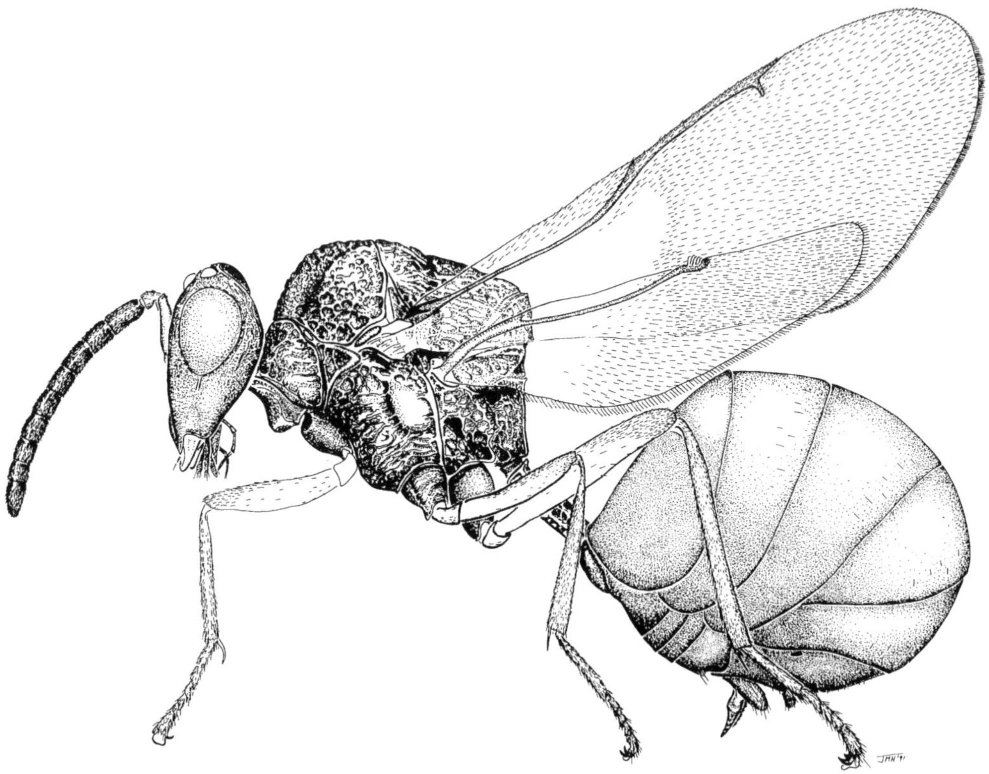

Fig. 11.88. *Orasema* sp. (Eucharitidae).

segments (up to 26 in some Old World members of the family), anellus present or absent; mandibles usually sickle-shaped with elongate apical tooth; labrum planar with 4 to 16 elongate digits; genal sulcus absent. *Pronotum reduced and ventral to mesoscutum, hidden (not visible) in dorsal view*; notauli deep and conspicuous; axillae enlarged and usually broadly fused in middle; scutellum sometimes with a pair of elongate spines projecting from apex. Fore wing with marginal vein relatively long. Tarsi 5-segmented. *Metasoma with an elongate petiole, usually longer than hind coxae*; female metasoma large, ovipositor not exceeding apex of metasoma; male metasoma small.

Only two subfamilies of Eucharitidae occur in the New World and all species can be readily recognized by these diagnostic features. Exceptions to the diagnosis are found in three additional palaeotropical subfamilies (see below) that were included in Eucharitidae by Bouček (1988).

Classification and distribution. Eucharitids are widespread and occur in almost every zoogeographical region of the world (except New Zealand), but are most abundant and diverse in the tropical regions (Heraty, 1985). Worldwide the Eucharitidae comprises 47 genera and 394 species (Heraty, unpublished), which are classified in five subfamilies by Bouček (1988):

Akapalinae	Australia, 1 genus
Echthrodapinae	Africa and New Guinea, 1 genus
Eucharitinae	cosmopolitan, 39 genera
Oraseminae	cosmopolitan, 4 genera
Philomidinae	Palaeotropics and Mediterranean, 2 genera

The higher classification of the family has not been resolved. Some studies place Oraseminae + Eucharitinae as the sister group to the Perilampidae, excluding the other three subfamilies from either family (Heraty & Darling, 1984; Darling, 1988). Based on larval morphology Darling (1992) placed Philomidinae as sister group to the Perilampinae + Eucharitinae + Oraseminae and excluded the Chrysolampinae (Fig. 11.89). This suggests that the Eucharitidae *sensu* Bouček (1988) is paraphyletic with respect to the Perilampidae. Gibson (*in* Goulet & Huber, 1993) classified Akapalinae and Philomidinae as subfamilies of Pteromalidae, and echthrodapines in the subfamily Monodontomerinae of Torymidae. Ecthrodapinae and Torymidae share an occipital carina, a setose epipygium and peg-like cerci articulated with metasomal tergite VIII.

Only two subfamilies occur in the New World, where 18 genera and 114 species are presently recognized. In Costa Rica the subfamily Oraseminae is represented by nine species of *Orasema*, while the Eucharitinae is represented by 13 species in seven genera.

Biology. Philomidinae are ectoparasitoids of ground nesting bees of the genus *Halictus* (Halictinae) (Darling, 1992), whereas Echthrodapinae are parasitoids of twig nesting bees of the genus *Braunsapis* (Allodapinae) (Michener, 1969). The host of Akapalinae is unknown. The following description of life histories refers only to the ant-parasitic subfamilies, Oraseminae and Eucharitinae, which are similar in many respects but show differences in ovipositional strategy and host choice.

Eggs are deposited away from the host into plant tissue, and the active first instar larva, termed a PLANIDIUM, is responsible for gaining access to the ant colony. The planidium is less than 0.15 mm in length with 12 well-sclerotized tergites, a caudal sucker, and an extensible labial region (Heraty & Darling, 1984). Larvae move by looping or jumping and can be extremely mobile. Access to the ant nest is achieved by some form of phoretic attachment to the host or another insect (Clausen, 1940*a* & *c*; Johnson *et al.*, 1986).

Once in the ant colony, the planidium attaches to the ant larva but does not develop until the host begins to pupate. When the ant larva pupates the eucharitid planidium relocates to the ventral thoracic region of the host pupa, where it completes three instars. In most Eucharitidae all stages are ectoparasitic. In *Orasema* and *Pseudometagea* the planidium burrows just under the cuticle of the host larva where partial feeding takes place, but further development of the first and later instars is external on the host pupa (Ayre, 1962). Thus eucharitid larvae are ectoparasitic koinobionts. They show hypermetamorphosis with later instars being more typically hymenopteriform. Usually only one parasitoid develops per host, but as many as four *Pseudochalcura* can develop on a single pupa of *Camponotus* (Wheeler, 1907; Heraty & Barber, 1990).

After pupation, the eucharitid pupae are mixed with ant pupae and cared for in the same manner as ant pupae (Williams, 1980). Eclosion of adult eucharitids is assisted by worker ants. They are completely integrated into the ant colony, being fed and groomed by the host ants. If the nest is disturbed, pupal and adult eucharitids are rescued by the ants in preference to their own brood (Wheeler, 1907; Mann, 1914).

Oraseminae. Species of Oraseminae deposit single eggs into punctures made in plant tissue using their specialized ovipositor. Eggs are deposited into involucral bracts of flowers, flower stems, undersides of oak and mango leaves, or in the skin of young banana fruits (Heraty & Darling, 1984; Johnson *et al.*, 1986). In South America *Orasema* species have earned the common name 'bicho costuriero' (seamstress insect) for their characteristic oviposition marks on leaves of various plants (Tocchetto, 1942). Planidia are often

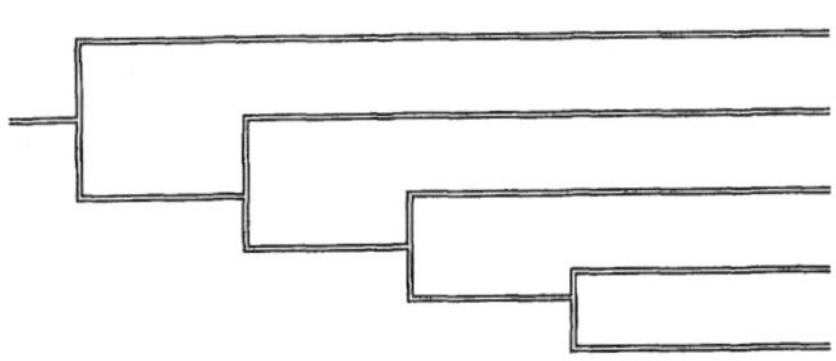

Fig. 11.89. Proposed phylogenetic relationships of the eucharitid/perilampid groups (modified after Darling, 1992).

phoretic on immature stages of thrips. Although the planidia may consume fluid while on the thrips, they apparently cannot complete their development on thrips (Wilson & Cooley, 1972; Johnson *et al.*, 1986). Ants may be collecting thrips (and planidia) as prey items, thus providing direct transport of planidia to the host larva (Heraty, 1994). Almost all Oraseminae are parasitoids of myrmicine ants, rarely of Ecitoninae and Formicinae (Table 11.8). These ants generally do not form cocoons. Feeding by the larva of *Orasema*, which is partially endoparasitic, results in a deformed pupa (Wheeler, 1907; Pelt, 1950).

Eucharitinae. Species of Eucharitinae usually deposit large egg masses into preformed cavities in plant tissue such as flower buds, and as many as 10,000 eggs may be deposited in a single oviposition (Clausen, 1940*a*). Some *Kapala* deposit their eggs into undeveloped flower buds of *Cordia* or *Hamelia*, and emergence of planidia is timed to coincide with opening of the flower. Eggs may also be scattered over the leaf surface or laid into the outer skin of small fruits (Clausen, 1940*a*; Heraty & Barber, 1990). Planidia do not require an intermediate host, but rather attach directly to adult ants for transport back to the ant nest. All New World Eucharitinae are parasitoids of Ponerinae or Formicinae (Table 11.8). Unlike hosts of Oraseminae, these ants generally form cocoons, and the eucharitid larva develops within the host cocoon.

Ant host	Eucharitid genus
PONERINAE	
Ectatomma	*Isomerala*
Odontomachus	*Kapala*
Pachycondyla	*Kapala*
MYRMICINAE	
Pheidole	*Orasema*
Solenopsis	*Orasema*
Leptothorax	*Orasema*
Wasmannia	*Orasema*
FORMICINAE	
Camponotus	*Obeza, Pseudochalcura*
Lasius	*Pseudometagea*

Table 11.8. Recorded hosts of eucharitid genera occurring in Central America (from Heraty, 1994).

Eucharitidae are generally a rare group of Hymenoptera but can be locally abundant. *Kapala* is probably the most common genus of Eucharitidae in the Neotropics, the males in particular being quite abundant in some Malaise trap samples. Males are more active in mate location and may form swarms over an ant nest with emerging females. Females mate soon after emergence from the ant nest and deposit eggs on suitable host plants in the vicinity of the ant host. Eucharitid females are usually attracted to certain types of plant structures for oviposition, and are less specific to plant taxa. If one can locate the oviposition sites it is relatively easy to collect eggs and planidia, and to gather other biological information.

Economic importance. *Orasema costaricensis* has been regarded as a pest of banana in Central America where it causes a brown spotting of fruit during oviposition (Roberts, 1958). It is possible that the females are actually ovipositing in association with thrips, which are usually more serious pests of banana (Hanson, pers. comm.). On the other hand, some eucharitids parasitize pestiferous species of ants (e.g. *Solenopsis*) and may offer possibilities for use in biological control of such pest species (D.W. Johnson, 1988).

Identification. Revisionary studies of New World groups relevant to the Central American fauna have been made for the nearctic Eucharitinae (Heraty, 1985).

Key to subfamilies and genera of Eucharitidae of Central America

1 Metallic coloured; anellus present, flagellar segments cylindrical in both sexes; prepectus not fused to pronotum (Fig. 11.88); scutellum unadorned; second metasomal (first gastral) sternite with transverse furrow; ovipositor sub-apically expanded, with strong apical teeth. ....(Oraseminae) ....*Orasema*

— Variously coloured; anellus absent, flagellar segments often serrate to branched; prepectus fused to pronotum (Figs 11.90, 11.93); scutellum often with elongate apical spines (Figs 11.90, 11.92); second metasomal sternite smooth; ovipositor needle-like, at most with minute teeth. ....(Eucharitinae) 2

2 Scutellar spines very long, reaching to apex of metasoma (Fig. 11.90). 3
— Scutellar spines absent (Fig. 11.93), or if present, much shorter than above, not reaching to apex of metasoma (Fig. 11.92). 5

3 Scuto-scutellar sulcus deeply invaginated and filled with dense, golden hairs; scutellar spines of female dorso-ventrally flattened, those of male cylindrical. .... *Dilocantha*
— Scuto-scutellar sulcus not deeply invaginated, bare; scutellar spines cylindrical. 4

4 Eyes conical (pointed above), frons strongly bulging (Fig. 11.91). .... *Isomerala*
— Eyes normal, not conical, frons not strongly bulging (Fig. 11.90). .... *Kapala*

5 Scutellum posteriorly with two blunt spines (Fig. 11.92). 6
— Scutellum rounded posteriorly (Fig. 11.93). 7

6 Mesosoma having at least some yellow to brown coloration; propodeum weakly developed; antenna cylindrical in both sexes. .... *Obeza*
— Mesosoma entirely black or metallic; propodeum with lateral margins strongly developed; male antenna branched. .... *Lophyrocera*

7 Prepectus reaching tegula; occiput broadly concave; axillae constricted medially, antennae cylindrical in both sexes. *Pseudometagea*[EX]
— Prepectus not reaching tegula (Fig. 11.93); occiput flat; axillae transverse, not constricted medially; male antennae with branches on basal flagellar segments. .... *Pseudochalcura*

Synopsis of the Costa Rican fauna

In Costa Rica eucharitids are most abundant at lower altitudes (below 1000 m) and they are entirely absent above 2200 metres, where their hosts also become scarce. Most genera of Eucharitidae found in Costa Rica are northern extensions of a more widespread South American fauna, with the exception of *Pseudometagea*, in which Costa Rica may represent the most southern record of this nearctic genus. Presently 21 to 22 species of eucharitids are known to occur in Costa Rica.

ORASEMINAE

Only a single genus of this subfamily is present in Central America.

Orasema. Africa, Asia, Australia, New World; 53 known species. Nine species occur in Costa Rica. Biology: Wheeler and Wheeler (1937); Heraty *et al.* (1993).

EUCHARITINAE

Seven genera of this subfamily are known to occur in Central America. The phylogeny of the subfamily is discussed in Heraty (1989).

Dilocantha. A neotropical genus with two or three described species whose cumulative range extends north to Mexico. Only one species is present in Costa Rica. Biology unknown.

Isomerala. A neotropical genus whose range extends north to Mexico. It comprises two species, one of which is present in Costa Rica.

Kapala. A predominantly neotropical genus whose range extends north to the southern part of the United States, and with one species recorded from Africa and Madagascar. A total of 17 species are known, five of which occur in Costa Rica. Taxonomy: Heraty and Woolley (1993).

Lophyrocera. A predominantly neotropical genus comprising three species whose cumulative range extends north to the western United States. Two species occur in Costa Rica. Biology unknown.

Obeza. A predominantly neotropical genus whose range extends north to the southern part of the United States. Eight species are known, but only one seems to be present in Costa Rica.

Pseudochalcura. A New World genus comprising 13 species, two of which occur in Costa Rica. Taxonomy: Heraty (1986).

***Pseudometagea*[EX].** A nearctic genus whose range extends south to Mexico and possibly to Costa Rica.

90

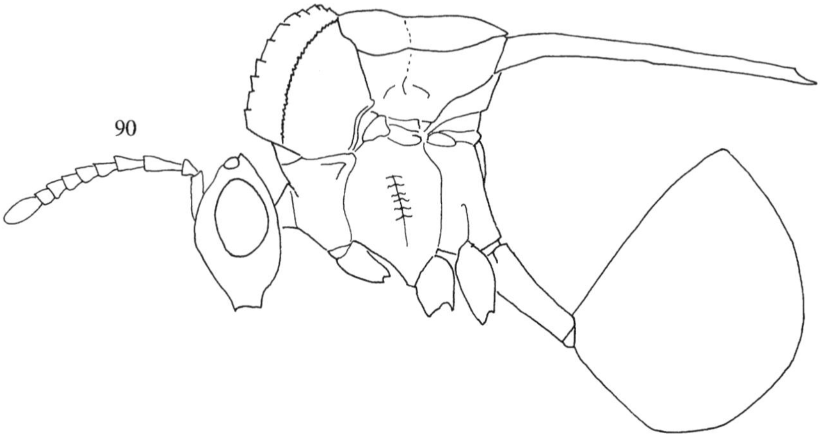

91

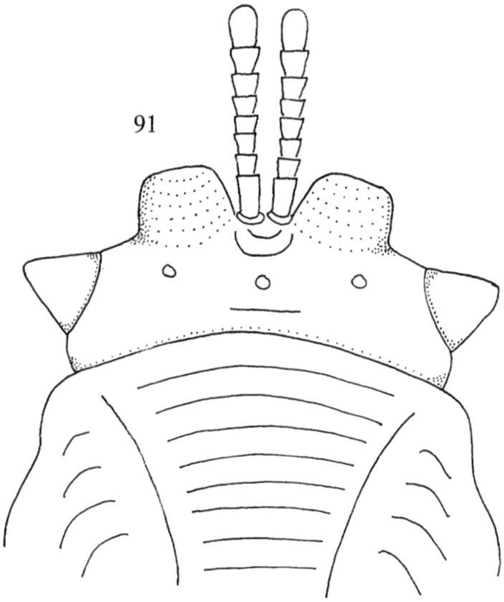

92

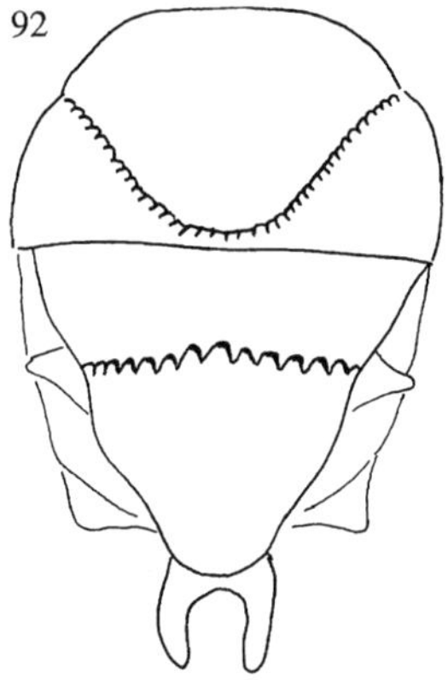

93

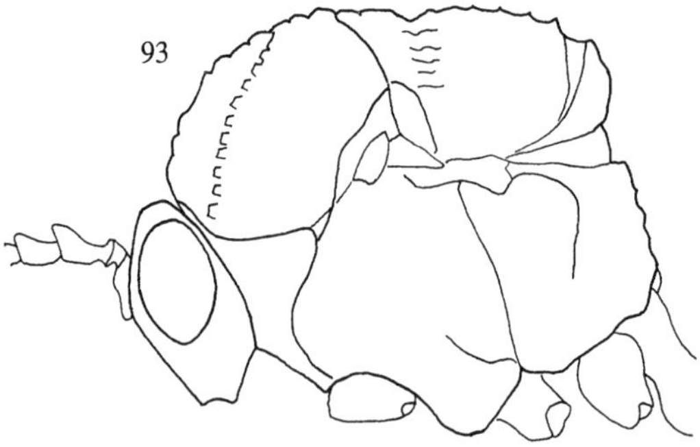

Figs 11.90–11.93. Genera of Eucharitidae. Fig. 11.90. *Kapala* sp., whole insect, lateral view. Fig. 11.91. Head of *Isomerala* sp., dorsal view. Fig. 11.92. Mesosoma of *Lophyrocera* sp., dorsal view. Fig. 11.93. Head and mesosoma of *Pseudochalcura* sp., lateral view.

Eight species are known and one of these possibly occurs in Costa Rica. Taxonomy: Heraty (1985). Biology: Ayre (1962).

11.7 EULOPHIDAE

John LaSalle and Michael E. Schauff

Diagnosis. 0.4 to 6.0 mm in length (though rarely exceeding 3.0 mm), varying from squat to elongate, robust to dorso-ventrally flattened; often soft-bodied and collapsing after death; colour varying from metallic to black to yellow. *Antenna* usually inserted at about level of lower eye margin or below, *usually with 2 to 4 funicular* segments (rarely 1 or 5), funicle of male sometimes branched. Mesoscutum with or without well marked notauli; axillae often produced forward of the scuto-scutellar suture. Fore wing with marginal vein long, postmarginal and stigmal veins often quite short. *Fore tibial spur short and straight; tarsi always 4-segmented. Metasoma constricted basally, not broadly attached to mesosoma*; ovipositor hidden or exserted.

Only one other family, the Elasmidae, has the combination of 4-segmented tarsi and a short, straight fore tibial spur. Elasmids, however, are sufficiently distinct (see Chapter 11.4) so as to not be easily confused with eulophids.

Classification and distribution. The Eulophidae are cosmopolitan in distribution, with about 3400 described species in 280 genera (Grissell & Schauff, 1990). It is the largest family of Chalcidoidea, and

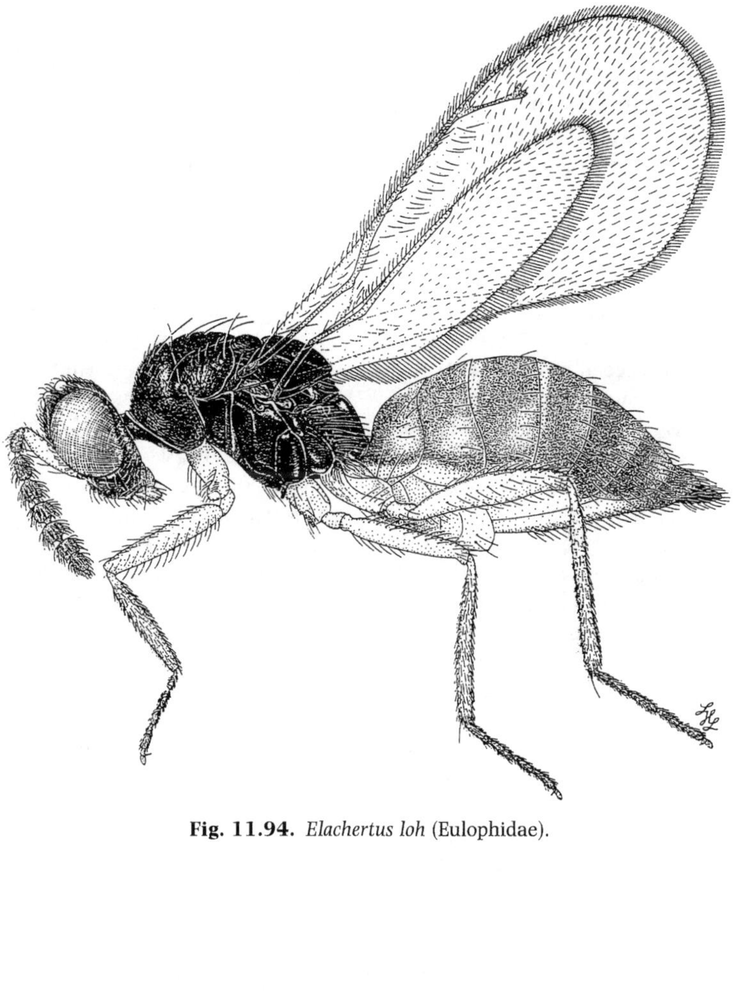

Fig. 11.94. *Elachertus loh* (Eulophidae).

they are probably the most commonly collected chalcids in all geographic realms. In Central America there are approximately 100 genera and, to the nearest order of magnitude, probably about 1000 species, most of which are undescribed.

Eulophidae appear to be closely related to the Elasmidae and Trichogrammatidae. In these three families, the number of tarsal segments is reduced (4 in eulophids and elasmids, 3 in trichogrammatids) and they have a short, straight fore tibial spur. It is possible that elasmids and trichogrammatids evolved from the Eulophidae, and the Elasmidae have often been treated as a subfamily of Eulophidae. Other families which may be related to the Eulophidae are the Aphelinidae, Signiphoridae and Tetracampidae. Further study is required to resolve the relationships of these groups.

The Eulophidae is currently divided into four subfamilies—Entedoninae, Euderinae, Eulophinae and Tetrastichinae—all of which occur in Central America. The Tetrastichinae and Entedoninae are the largest subfamilies. Another subfamily name, Elachertinae, has often been used in the literature, but this has since been included with the Eulophinae (Graham, 1987; Boucek, 1988; Grissell & Schauff, 1990). The subfamilies of Eulophidae appear to represent natural units, although they are difficult to define rigorously.

Biology. Eulophids display a remarkable diversity in their biologies. They are predominantly parasitoids of the immature stages of Lepidoptera, Coleoptera, Diptera and Hymenoptera, especially those concealed in plant tissue, such as leaf-miners, leaf-rollers, gall-formers, and stem-borers. The Eulophinae and Euderinae are fairly conservative in their host ranges while the Entedoninae and Tetrastichinae attack a wide variety of hosts. A few eulophids, especially species of Tetrastichinae, are phytophagous—mostly as inquilines in host galls, but a few are gall-formers or seed parasitoids. More detailed information on the host range of genera occurring in Central America is given below in the synopsis of the Costa Rican taxa.

At the family level it is difficult to generalize about the type of biological relationship between eulophids and their hosts. There are solitary and gregarious species, ectoparasitoids and endoparasitoids, idiobionts and koinobionts, and there appear to be roughly equal numbers of species in each category, at least among those where the biology is known. Primary parasitoids are more common than hyperparasitoids, although the latter are certainly not rare. Species that develop on small hosts such as eggs or cecidomyiid larvae are often solitary parasitoids or even predators, while those parasitizing large hosts are frequently gregarious. Predatory species are mostly members of the subfamily Tetrastichinae and their prey are always aggregated—larvae of Cecidomyiidae in seed pods (Parnell, 1963), eggs of Pyralidae in felt-covered masses (Rothschild, 1973), eggs of spiders in silken sacs (LaSalle, 1990), eriophyid mites in galls (Vereshchagina, 1961), and nematodes in galls (Berg *et al.*, 1990).

Most eulophids search for hosts associated with terrestrial plants, but some species are known to be aquatic. In Europe several species of *Aprostocetus* (subgenus *Ootetrastichus*) are known to attack the eggs of Odonata or dytiscid beetles in submerged plant stems (Fursov & Kostyukov, 1987—as *Tetrastichus*), and adult females have been seen swimming under water to reach their hosts (V. Kostyukov, V. Fursov, pers. comm.). In Mexico *Horismenus mexicanus* attacks the prepupae and pupae of psephenid beetles in water (Brown, 1968—as *Psephenivorus*). Once a host has been located the female wasp must decide whether it is acceptable for oviposition, and in gregarious species, choose how many eggs to lay and decide whether to lay fertilized (female) or unfertilized (male) eggs (Dijkstra, 1986).

Host-feeding by adult females is common in Eulophidae, although the construction of a feeding tube appears to be rare (Jervis & Kidd, 1986). When host-feeding the female may sting the host more frequently and for longer duration than when she is ovipositing. Host-feeding therefore often results in the death of the host (prey) and consequently the female must use another, often older, host individual in which to oviposit. For example, the European species, *Tetrastichus coeruleus*, utilizes newly deposited chrysomelid eggs for host-feeding and oviposits in eggs containing more advanced embryos (Capinera & Lilly, 1975; Alphen, 1980—as *asparagi*). Other examples of females choosing younger hosts for feeding and older hosts for oviposition include *Diglyphus* on agromyzid larvae (Heinz & Parrella, 1989; Kato, 1989), *Euplectrus* on lepidopterous larvae (Neser, 1973), and *Tamarixia* on psyllid nymphs (Moran *et al.*, 1969—as *Tetrastichus*). A few eulophids are known to use the same host individual for host-feeding and oviposition, for example *Melittobia* (Dahms, 1984*b*) and *Baryscapus chrysopae* (Clancy, 1946—as *Tetrastichus*).

As is true in other Hymenoptera, koinobionts usually inject a venom that causes temporary paralysis in the host whereas idiobionts commonly cause permanent paralysis or even death. Many idiobiont eulophids attack the larval stage of their host (Askew, 1968; Takada & Kamijo, 1979), either as ectoparasitoids—as in most Eulophinae (e.g. Mazanec, 1990)—or as endoparasitoids—e.g. *Chrysocharis* (some species) and *Closterocerus*. *Quadrastichus* includes both ecto- and endoparasitic idiobionts (Askew & Ruse, 1974—as *Tetrastichus*). Some Entedoninae and Tetrastichinae are endoparasitic idiobionts of eggs, including for example species of *Emersonella* and *Edovum* in chrysomelid eggs, and the cosmopolitan *Aprostocetus hagenowii* in cockroach eggs (Vargas & Fallas, 1974; Vinson & Piper, 1986). Endoparasitic, gregarious idiobionts of pupae include *Tachinobia* (Tetrastichinae) and *Trichospilus* (Eulophinae), in Diptera and Lepidoptera respectively.

Euplectrus and *Tamarixia* are ectoparasitic koinobionts of exposed hosts. The venom injected by the parasitoid not only causes temporary paralysis, to allow oviposition, but also prevents the host larva from moulting to the next instar and thus sloughing off the parasitoid larvae. Upon completion of feeding, species of *Euplectrus*, which are gregarious ectoparasitoids of leaf-feeding lepidopterous larvae, spin silken cocoons attached to the outside of their host (Neser, 1973; Wall & Berberet, 1974; Puttler *et al.*, 1980). They are one of the few chalcidoids known to spin cocoons, and the silk is secreted by the malpighian tubules, not by the labial glands as is the case in most other Hymenoptera. The South African *Tamarixia flavigaster* is a solitary ectoparasitoid on nymphs of Psylloidea (Homoptera) (Moran *et al.*, 1969). Although the host is exposed, the parasitoid larva remains hidden beneath it. When fully grown, the parasitoid larva fastens the psyllid nymph to the substrate with a silk-like substance (presumably secreted by the malpighian tubules) and then kills it.

Endoparasitic koinobiosis appears to be restricted to certain members of the subfamilies Entedoninae and Tetrastichinae. In the European species, *Entedon leucogramma*, the female wasp oviposits in scolytid eggs, but larval development occurs within the host larva (Beaver, 1966). It is possible that koinobiosis occurs in most, if not all, species of *Entedon* (List, 1932; Fisher, 1970) and *Achrysocharoides* (Bryan, 1980). Species of *Chrysocharis* exhibit a range of biologies: many are idiobionts, but others, such as *C. parksi*, which attacks *Liriomyza* (Agromyzidae) (Parkman *et al.*, 1989), are larval-pupal koinobionts (Askew, 1968; Takada & Kamijo, 1979). In the Tetrastichinae one of the better studied examples of koinobiosis is the European *Tetrastichus coeruleus*. The female of this species lays about five eggs in the egg of its chrysomelid host, but the parasitoid larva does not kill the host until it enters the soil to pupate (Johnston, 1915; Capinera & Lilly, 1975—as *asparagi*). Other examples of gregarious koinobionts include *Baryscapus chrysopae* in chrysopid larvae (Clancy, 1946—as *Tetrastichus*) and *Tetrastichus julis* in chrysomelid larvae (Dysart *et al.*, 1973).

In general very few hymenopteran parasitoids attack adult stages of holometabolous insects. In West Africa, the tetrastichine, *Phymastichus coffea*, oviposits into adult coffee berry borers (*Hypothenemus hampei*, Scolytidae), inserting the ovipositor through a pore in the elytra or thorax (Feldhege, 1992). Larvae develop as endoparasitic koinobionts, generally two per host—a male in the thorax and a female in the abdomen. Recently another species in this genus has been found in Hawaii attacking scolytids in macadamia nuts, but in this species the dagger-like ovipositor suggests that oviposition need not occur through a pore. The same species occurs in Costa Rica.

Most hyperparasitic eulophids belong to the Entedoninae and Tetrastichinae, and many are facultative rather than obligatory secondary parasitoids. Some species of *Horismenus* are hyperparasitic, mostly on Lepidoptera via braconids (e.g. Bennett, 1950), but two North American species are hyperparasitoids in spider egg sacs (Burks, 1971). Other entedonine hyperparasitoids include some species of *Pediobius* and *Teleopterus*. In the Tetrastichinae hyperparasitoids are found mostly in *Aprostocetus*, *Baryscapus*, *Tachinobia*, *Tetrastichomyia* and *Tetrastichus*. Most hyperparasitoids are ectoparasitic idiobionts but the European species, *Baryscapus endemus* (Parnell, 1964*a*—as *Aprostocetus tibialis*) and *B. nigroviolaceus* (Askew, 1968—as *Tetrastichus amethystinus*), appear to be endoparasitic koinobionts. *Oomyzus sokolowskii* is a facultative hyperparasitoid of the diamond-back moth.

Perhaps the most unusual biology shown by Eulophidae is that of *Melittobia* species, which primarily attack nest-building Hymenoptera, but are potentially some of the most polyphagous of all hymenopteran parasitoids (Dahms, 1974*b*). The adult female may enter a host cell before it is closed, in which case she delays oviposition until the host reaches the prepupal or pupal stage. If the host cell is already closed the female may gnaw her way into a cell, sometimes with several females taking turns chewing at the same hole. If the host is large she stays with this one host for the remainder of her life and is assisted in its utilization by specialized 'second-form' progeny, which are paler in colour and have reduced wings. Unlike normal females, these second-form females make no effort to disperse and can begin laying eggs almost immediately.

The immature stages of several eulophid species have been described (e.g. see Bledsoe *et al.*, 1983). The eggs are normally elongately oval or kidney-shaped (Cameron, 1939; Clancy, 1946; Askew & Ruse, 1974), or occasionally with a long anterior filament (Viggiani, 1971), which probably serves to anchor the egg to the integument of the host (Silvestri, 1911). There are from three to five larval instars. The first instar eulophid larva is hymenopteriform, 13-segmented, and occasionally with fleshy tubercles or rows of spines on the body. The mature larva is generally not hairy (Askew, 1968). The larvae of some species that attack leaf-miners (e.g. *Diglyphus*, *Chrysocharis*) construct a circle of little faecal pillars about themselves (Viggiani, 1964; Hendrickson & Barth, 1978). The larva pupates within this circle and the pillars harden to serve as supports preventing the collapse of the host mine as the plant tissue dries out.

The mating behaviour of Eulophidae has been described for several species. Details of the complex behaviour patterns involved in mating can be used to separate closely related species and may help in understanding phylogenetic relationships (Assem, 1975; Assem *et al.*, 1982*a* & *b*; Bosch & Assem, 1986). Perhaps the most unusual mating behaviour is that shown by species of *Melittobia* (Assem *et al.*, 1982*a*; Dahms, 1984*a–c*). The males have short, non-functional wings and vestigial eyes. They emerge before the females, and wait by the exit hole for the females to emerge. Males of some species are quite aggressive in their efforts to monopolize females, and will decapitate unemerged male pupae or freshly emerged males. An unmated female may remain with her developing male progeny in order to mate with an emerging son. In the closely related genus, *Tachinobia*, the males also have reduced wings and eyes, and mating behaviour may be similar to that of *Melittobia*, although fewer studies exist.

Economic importance. Eulophidae is the third most important family of Chalcidoidea used in biological control, after Aphelinidae and Encyrtidae. Some of the more important genera used in biological control projects are *Pediobius*, *Chrysocharis*, *Diglyphus*, *Closterocerus* and *Neochrysocharis*—the first against a variety of Lepidoptera and Coleoptera, the last four against leaf-miners.

There are several examples of the use of eulophids in biological control in the Neotropical region. *Aceuratoneuromyia indica* was imported into Central America for the control of various species of fruit flies (Tephritidae: *Anastrepha* spp., *Ceratitus capitata*) (Clausen, 1978; Cock, 1985). *Aprostocetus gala* has been used against citrus weevils (Curculionidae: *Diaprepes* spp., *Exophthalmus* spp.) (Cock, 1985; Schauff, 1987—both as *Tetrastichus*) and *Oomyzus sokolowskii* has been used against the diamond-back moth (Plutellidae: *Plutella xylostella*) (Cock, 1985—as *Tetrastichus*). *Tamarixia leucaenae* was recently described as a potential biological control agent of the *Leucaena* psyllid (*Heteropsylla cubana*) (Bouček, 1988).

The gregarious pupal parasitoid *Trichospilus diatreae*, was imported from Asia into the Caribbean region in 1972 as a natural enemy of lepidopterous pests, but was not released because of its polyphagy—i.e. it attacked tachinid pupae in the laboratory (Bennett *et al.*, 1987). Nonetheless it is now widespread in the Caribbean region and in Florida, although it is not clear whether it was released on another occasion and not reported, whether it arrived as an accidental immigrant, or whether it was originally more widespread than previously believed. This species is now fairly common in certain areas of Costa Rica (e.g. the Osa Peninsula), most commonly in disturbed areas but also in relatively undisturbed forest.

Several endemic (or introduced and established) species of eulophids probably play important roles in controlling populations of certain pests. Some of these are listed in Table 11.9.

Identification. Keys to the subfamilies and discussions of classification within the family are provided

Pest species	Eulophid
Anastrepha obliqua (Tephritidae)	*Aceratoneuromyia indica*
Antichloris viridis (Arctiidae)	*Elachertus ceramidiae* (Burks, 1962; Harrison, 1963) *Miotropis* sp. *Stenomesius ceramidiae* (Bouček, 1962)
Contarinia sorghicola (Cecidomyiidae)	*Aprostocetus diplosidis* (Graham, 1987)
Leucoptera coffeella (Lyonetiidae)	*Closterocerus* sp. *Zagrammosoma* sp.
Liriomyza huidobrensis (Agromyzidae)	*Diglyphus isaea*
Periplaneta americana (Blattidae)	*Aprostocetus hagenowii* (LeBeck, 1991)
Spodoptera frugiperda (Noctuidae)	*Euplectrus comstockii* *E. plathypenae*

Table 11.9. Pest species present in the northern Neotropics and eulophids known to attack them.

by Graham (1987), Bouček (1988), and Grissell and Schauff (1990). One character that has been traditionally used in identifying subfamilies is whether the submarginal vein smoothly joins the parastigma (Fig. 16), as in the Eulophinae, or whether the submarginal vein is narrowed apically and joins the parastigma slightly above the end of the vein (Figs 11.107, 11.114, 11.116). This character, however, is not always easy to assess, and displays some variation within subfamilies (Grissell & Schauff, 1990). The most important key to all eulophid genera is that for Australasia (Bouček, 1988). There are no modern keys available to all eulophid genera for any part of the New World, although keys to the nearctic eulophid genera are currently in preparation (Schauff, LaSalle & Coote, in prep.).

Key to the subfamilies of Eulophidae

1 Scutellum with one pair of setae (Fig. 11.95); submarginal vein with two setae on dorsal surface (Fig. 11.107); notauli incomplete or absent (rarely present); fronto-facial sutures, when present, situated near middle of face and distinctly removed from median ocellus (Fig. 11.96). .... **Entedoninae** (p. 323)

— Scutellum with more than one pair of setae: usually with two pairs (Figs 11.99–11.103, 11.105), occasionally with more than two pairs or (rarely) with numerous irregularly scattered setae (Fig. 11.97); fore wing usually with more than two setae on dorsal surface of submarginal vein (Figs 11.110, 11.114, 11.116); if with one or two (some Tetrastichinae), then notauli present and complete; fronto-facial sutures, when present, situated high on face and very near to median ocellus (Fig. 11.103). 2

2 Notauli absent, or incomplete posteriorly and not approaching the posterior margin of the mesoscutum or the axilla (Figs 11.97–11.100). 3

— Notauli present and either reaching the posterior margin of the mesoscutum or curving to meet the axilla (Figs 11.101–11.103, 11.105). ... 4

3 Funicle with one or two segments (Fig. 11.109) ***and*** pronotum not visible in dorsal view (Figs 11.97–11.98); ***either*** scutellum with many scattered setae and dorsum of mesosoma reticulate (Fig. 11.97), ***or*** scutellum with two pairs of setae (the second quite small and on the posterior margin) and dorsum of mesosoma relatively smooth (Fig. 11.98); scutellum never with paired, submedian lines. Parasitoids of whiteflies..................................... ....**Entedoninae** (p. 323)

— ***Either*** funicle with more than two segments (Figs 11.111, 11.113, 11.115), ***or*** pronotum clearly visible in dorsal view (Figs 11.99–11.100), ***or*** scutellum with paired, submedian lines (Fig. 11.100). Never parasitoids of whiteflies. .... **Eulophinae** (p. 323)

4 Wings absent or reduced (extending less than half the length of the metasoma). 5

— Wings present and of normal size (extending at least half the length of the metasoma). 6

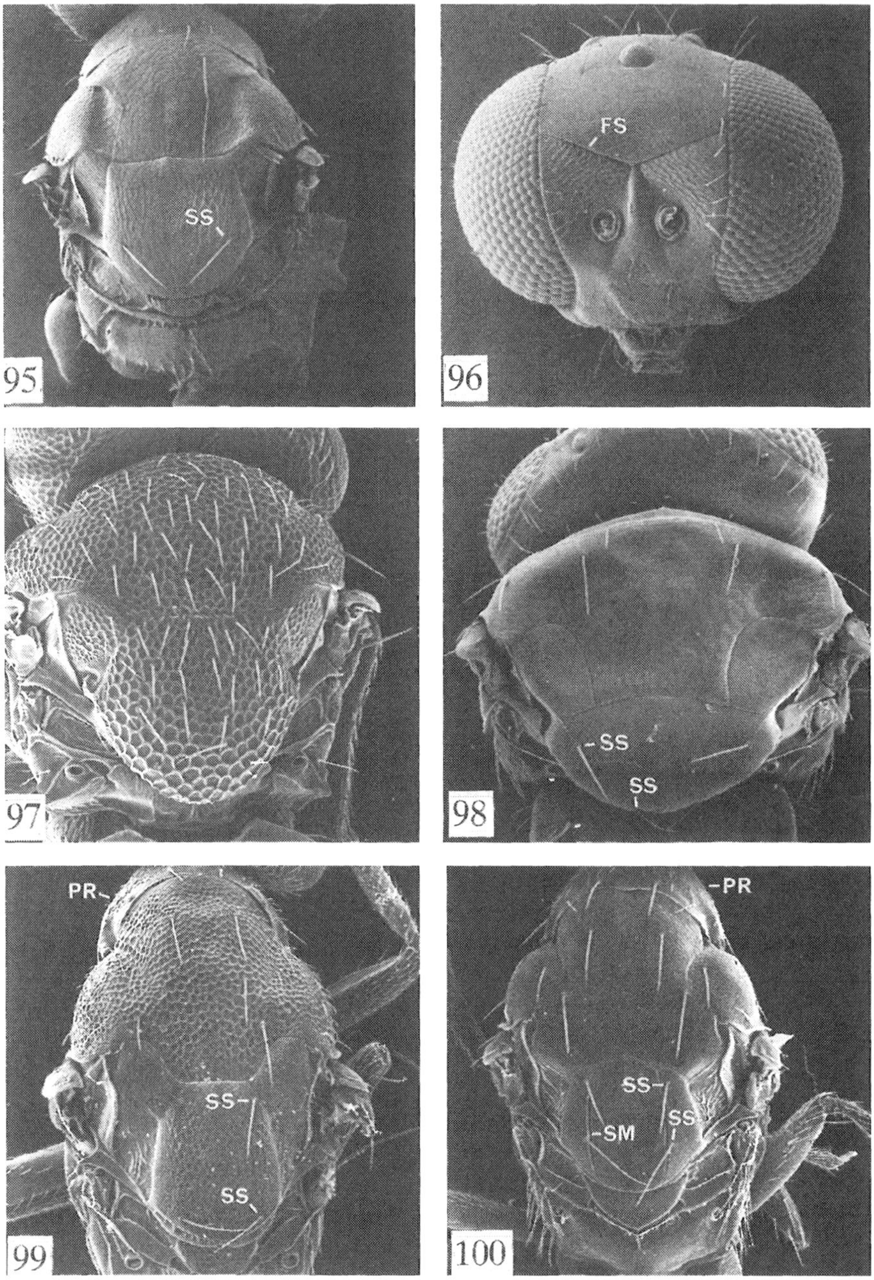

Figs 11.95–11.100. Eulophidae (all females). Figs 11.95–11.98. Entedoninae; 11.95, *Chrysocharis* sp., mesosoma; 11.96, *Chrysocharis* sp., face; 11.97, *Entedononecremnus* sp., mesosoma; 11.98, *Euderomphale* sp., mesosoma. Figs 11.99–11.100. Eulophinae; 11.99, *Sympiesis* sp., mesosoma; 11.100, *Diglyphus* sp., mesosoma.
fs = fronto-facial suture; pr = pronotum; sm = submedian line; ss = scutellar seta.

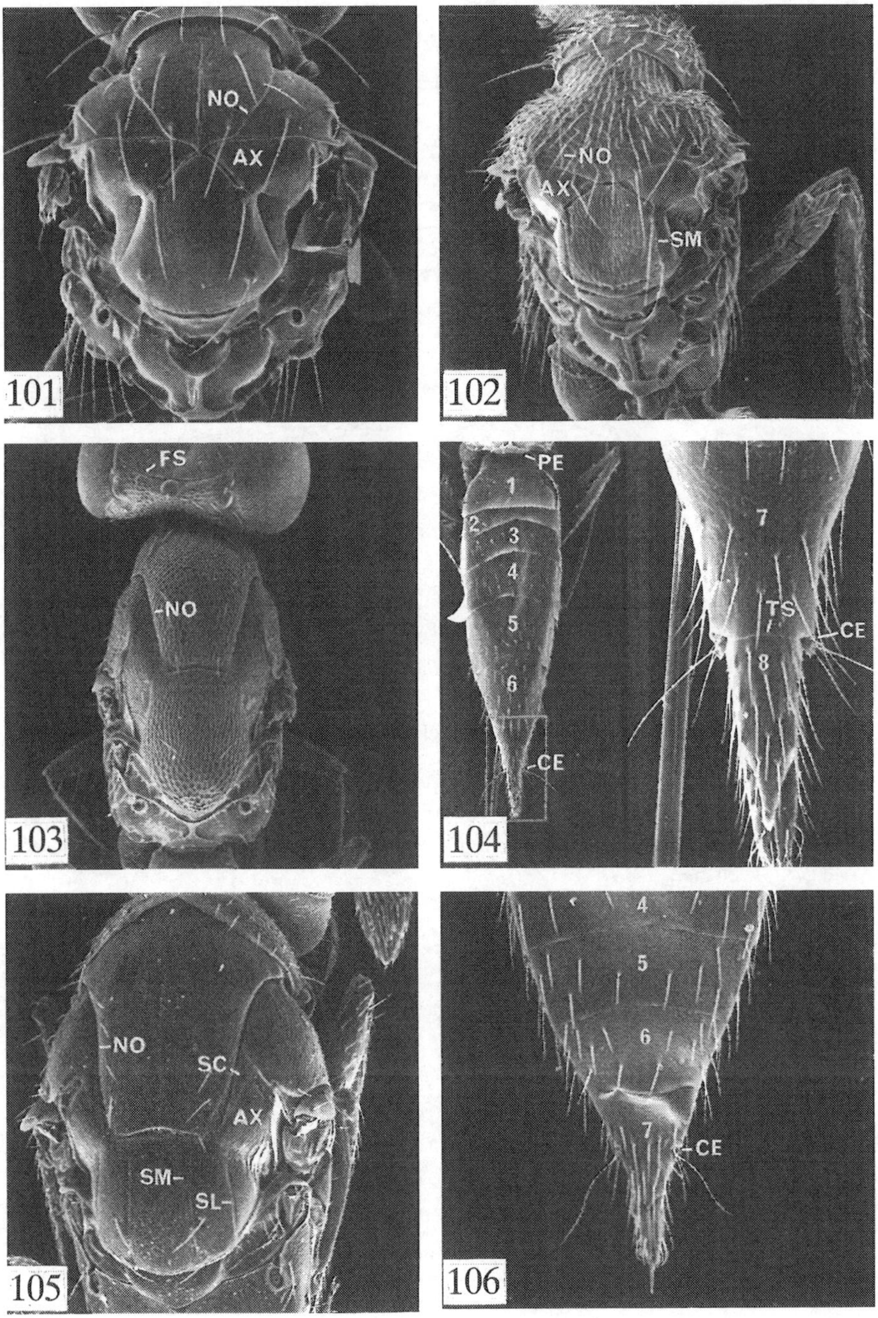

Figs 11.101–11.106. Eulophidae (all females). Figs 11.101–11.102. Eulophinae; 11.101, *Euplectrus* sp., mesosoma; 11.102, *Elachertus* sp., mesosoma. Figs 11.103–11.104. *Euderus* sp. (Euderinae); 11.103, mesosoma; 11.104, apex of metasoma. Figs 11.105–11.106. *Aprostocetus* sp. (Tetrastichinae); 11.105, mesosoma; 11.106, apex of metasoma.
ax = axilla; ce = cercus; fs = fronto-facial suture; no = notaulus; pe = petiole; sc = scapula; sl = sublateral line; sm = submedian line; ts = transverse suture.

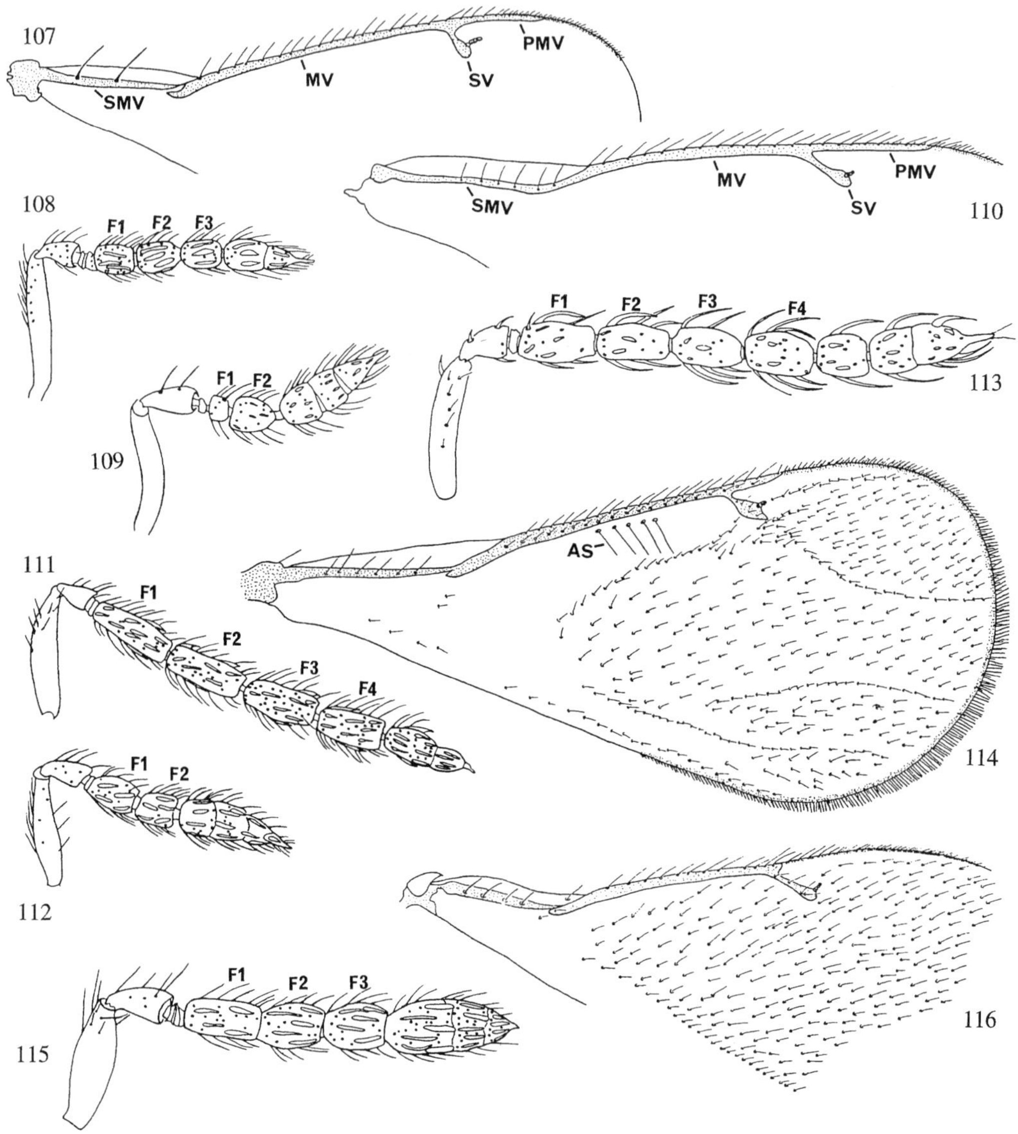

Figs 11.107–11.116. Eulophidae. Figs 11.107–11.108. *Chrysocharis* sp. (Entedoninae); 11.107, fore wing venation; 11.108, antenna. Fig. 11.109. *Entedononecremnus* (Entedoninae), antenna. Figs 11.110–11.111, *Pnigalio* sp. (Eulophinae); 11.110, fore wing venation; 11.111, antenna. Fig. 11.112. *Diglyphus* sp. (Eulophinae), antenna. Figs 11.113–11.114. *Euderus* sp. (Euderinae); 11.113, antenna; 11.114, fore wing. Figs 11.115–11.116. *Aprostocetus* sp. (Tetrastichinae); 11.115, antenna; 11.116, fore wing.
as = admarginal setae; f1-f4 = funicular segments 1-4; mv = marginal vein; pmv = postmarginal vein; smv = submarginal vein; sv = stigmal vein.

5 *Either*: female with three funicular segments (Fig. 11.115), *or* male....... **Tetrastichinae** (p. 327)

— Females only, funicle with four segments (Figs 11.111–11.112) (males fully winged)..... .... **Eulophinae** (p. 326)

6 Female with ninth metasomal tergite separated from eighth by a transverse suture which is always at the level of the cerci (Fig. 11.104), so that the metasoma has eight visible segments past the petiole; fore wing usually with area just below marginal vein bare as far as the stigmal vein except for a distinct row of admarginal setae on the underside of wing (Fig. 11.114); most forms with two to three rows of setae radiating from the stigmal vein; postmarginal vein usually distinctly longer than stigmal vein (which is often relatively short); funicle in both sexes with four (rarely five) segments (Fig. 11.113); scutellum never with submedian lines (Fig. 11.103)................. .... **Euderinae** (p. 325)

— Female with eighth and ninth metasomal tergites fused, without a transverse suture at level of cerci (Fig. 11.106), so that the metasoma has seven visible segments past the petiole; fore wing (Fig. 11.116) not as above, if with bare area beneath the marginal vein, then without rows of setae radiating from stigmal vein; postmarginal vein variable, but often rudimentary or absent; funicle in female with two to four segments; scutellum often with paired lines (Figs 11.102, 11.105)......................... 7

7 Postmarginal vein absent or rudimentary, at most half the length of the stigmal (Fig. 11.116); axillae usually strongly advanced, defining a linear or almost linear scapula (Fig. 11.105); submarginal vein usually broken or narrowed apically and not smoothly joining parastigma; scutellum usually with paired submedian lines, and often with sublateral lines (Fig. 11.105); female with three funicular segments (Fig. 11.115), male with four.

[Note: the few genera of tetrastichines which have the postmarginal vein over half the length of the stigmal vein always have the axillae strongly advanced and the submarginal vein not smoothly joining the parastigma.] .... **Tetrastichinae** (p. 327)

— Postmarginal vein developed, at least as long as stigmal and usually longer (Fig. 11.110); axillae not so strongly advanced; scapula not linear (Figs 11.101–11.102); submarginal vein smoothly joining parastigma; scutellum with or without paired submedian lines, but never with two pairs (submedian and sublateral lines) (Figs 11.101–11.102); funicle with two to four segments (Figs 11.111-11.112). ... . **Eulophinae** (p. 326)

Synopsis of the Costa Rican fauna

De Santis (1979), in the first comprehensive catalogue of neotropical Eulophidae, recorded only nine genera from Central America. Subsequent papers (De Santis, 1983*a*, 1989; Hansson, 1987; Maes, 1989; Schauff, 1991; Delvare & LaSalle, 1993; Hansson & Cave, 1993; LaSalle, 1994, 1995; La Salle & Schauff, 1994) added another twelve, but following the present work 67 genera are now known to occur in the region. A further 12 genera are likely to be present as they occur in neighbouring regions or are wide ranging in the New World. We are also aware of many Central American taxa that cannot confidently be identified to genus, as well as many that represent undescribed genera. Thus at least 100 genera of eulophids probably occur in Central America.

The fact that the only comprehensive catalogue for the region recorded less than one tenth of the genera present indicates how poorly studied the Central American eulophid fauna is. At the species level the situation is even worse. Although some preliminary attempts have been made at clarifying the status and identity of neotropical eulophid species (Bouček, 1977; LaSalle & Schauff, 1992), the rich fauna of Central America remains virtually unknown.

Entedoninae

The body in certain genera (e.g. *Entedon, Pediobius, Horismenus*) is quite strongly sclerotized, while in other genera (e.g. *Omphale, Chrysocharis, Neochrysocharis, Closterocerus*) it is weaker and specimens often shrivel after death. The most commonly collected genera are *Horismenus* and *Omphale*, but other commonly collected groups include *Chrysocharis*, *Closterocerus*, *Emersonella*, *Neochrysocharis* and *Pediobius*. Keys to Australasian (Bouček, 1988) and holarctic (Schauff, 1991) genera will allow identification of most of the Central American genera, although *Entedononecremnus*, *Microdonophagus* and *Proacrias* are not included in these keys.

The majority of entedonines are endoparasitoids and many are presumably idiobionts. Koinobiosis occurs in perhaps all species of *Entedon* and in a few species of *Chrysocharis*. Species of *Horismenus* and *Pediobius* attack a wide range of hosts (see below) and seem to be ecological homologues, with *Pediobius* being a large and important genus in the Old World, but largely replaced by *Horismenus* in the New World. Another genus with a wide host range is *Teleopterus*. The host ranges of other entedonine genera occurring in Central America are more easily defined (see Table 11.10).

Achrysocharoides[CR]. Widespread, possibly cosmopolitan. Species are endoparasitoids of small leaf-mining Lepidoptera on trees and shrubs. Both solitary and gregarious forms are known.

Alachua. Known only from a single species, *A. floridensis*, found in Florida and Costa Rica. It is a gregarious endoparasitoid in the pupae of *Camponotus abdominalis* (Formicidae), and is one of only two eulophid genera known to attack ants (Schauff & Bouček, 1987).

Host category	Entedontine parasitoid genus
Thysanoptera	*Ceranisus, Goetheana*
Aleyrodidae	*Aleuroctonus, Entedononecremnus, Euderomphale, Neopomphale*
Chrysomelidae	*Asecodes, Emersonella, Edovum*
Curculionoidea	*Entedon, Paracrias*
Syrphidae	*Microdonophagus*
Formicidae (pupae)	*Alachua*
Leaf-miners	*Achrysocharoides, Apleurotropis, Chrysocharis, Closterocerus, Kratoysma, Neochrysocharis*
Gall-formers	*Eprhopalotus, Holcopelte, Omphale*

Table 11.10. Hosts of entedontine eulophid genera, excluding *Horismenus, Pediobius* and *Teleopterus* (see text), occurring in Central America.

Aleuroctonus. Neotropics. Endoparasitoids of white flies (Aleyrodidae: Aleurodicinae) (LaSalle & Schauff, 1994).

Apleurotropis[CR]. Previously known only from the Old World. Parasitoids of leaf-miners.

Asecodes[CR]. Primarily a holarctic and southeast Asia genus; not yet known from South America. Gregarious parasitoids of immature stages of Coleoptera, particularly Chrysomelidae.

Ceranisus. Cosmopolitan; closely related to *Goetheana, Thripobius and Entedonastichus*. Idiobiont endoparasitoids in thrips larvae, particularly those of Thripidae and Phlaeothripidae (Loomans *et al.*, 1992).

Chrysocharis. A mainly holarctic genus, with some species present in tropical regions. Solitary endoparasitoids in larvae or pupae of leaf-miners (Hansson, 1985, 1987).

Chrysonotomyia[EX]. As recently re-defined (Hansson, 1994*b*) this genus contains a single species, *C. auripunctata*, from Florida and Brazil. It is closely related to both *Closterocerus* and *Neochrysocharis*. It is a parasitoid of Cecidomyiidae.

Closterocerus. Cosmopolitan. Very closely related to *Chrysonotomyia*. Endoparasitoids in eggs or young larvae of a variety of phytophagous insects, mostly of leaf-miners, but also of gall-formers; occasionally hyperparasitic.

Edovum[EX]. Known from Mexico and Colombia. This genus contains a single described species, *E. puttleri*, which is an endoparasitoid in eggs of *Leptinotarsa undecemlineata* (Chrysomelidae) (Grissell, 1981; Puttler & Long, 1983).

Emersonella. New World. Endoparasitoids in eggs of Chrysomelidae (Bouček, 1977; De Santis, 1983*b*).

Entedon[EX]. Primarily holarctic. Mainly parasitoids of phytophagous Coleoptera, predominantly Curculionidae and Bruchidae (Schauff, 1988).

Entedononecremnus. Neotropics. Endoparasitoids of whiteflies (Aleyrodidae) in the subfamily Aleurodicinae. One Costa Rican species commonly reared from *Ceraleurodicus altissimus* (LaSalle & Schauff, 1994).

Eprhopalotus. Known from Texas and Costa Rica. The only two records for this genus are as parasitoids in galls (Cynipidae and Cecidomyiidae) associated with oak (*Quercus*) (Schauff, 1991—as *Aabacharis*; Schauff & LaSalle, 1993).

Euderomphale. Cosmopolitan. Endoparasitoids of whiteflies (Aleyrodidae) in the subfamily Aleyrodinae (LaSalle & Schauff, 1994).

Goetheana^{EX}. Contains a single widely distributed species that is closely related to *Ceranisus*. Much of its large range may be due to introduction. Solitary endoparasitoid in immature thrips.

Holcopelte^{CR}. Mostly holarctic; also known from Australasia. Parasitoids of gall-forming Cecidomyiidae (Hansson, 1988).

Horismenus. Predominantly New World, with a single European species. Parasitoids or hyperparasitoids of a variety of hosts, most commonly larvae of Coleoptera (e.g. bruchids, curculionids, buprestids, cerambycids), Diptera and Lepidoptera. Two species are hyperparasitoids in spider egg cases (Burks, 1971). An undetermined species in Costa Rica has been reared from pupae of *Cladochaeta* (Drosophilidae) associated with cercopid spittle masses (P. Hanson, pers. comm.).

Kratoysma. Mainly an Old World genus, with two species recently recorded from the New World (Hansson & Cave, 1993). Parasitoids of larvae of lepidopterous leaf-miners (Gracillariidae).

Microdonophagus^{EX}. Known only from Panama. Includes a single species, *M. woodleyi*, which is a gregarious endoparasitoid in the larvae of *Microdon* (Syrphidae) (Schauff, 1986).

Neochrysocharis. Cosmopolitan. Very closely related to *Closterocerus* and *Chrysonotomyia* (see Hansson, 1990, 1994*a*,*b*, 1995). Biology similar to *Closterocerus*.

Neopomphale. Neotropics. Endoparasitoids of whiteflies (Aleyrodidae) in the subfamily Aleyrodinae. Recently described by LaSalle & Schauff (1994).

Omphale^{CR}. Cosmopolitan. Endoparasitoids of larvae of gall-forming Cecidomyiidae and a few other gall-forming insects.

Paracrias. New World, mainly neotropical. Closely related to *Horismenus*. Poorly known biologically, but what records there are indicate that species are parasitoids of weevils (Schauff, 1985*a*). *P. guatemalensis* has been reared from *Conotrachelus perseae* (Schauff, 1985*a*) and *P. anthonomi* from *Anthonomus hunteri* (Woolley & Schauff, 1987).

Pediobius. Cosmopolitan, mainly Old World and holarctic. Primary or secondary parasitoids, mostly of Lepidoptera, but also of Coleoptera, Diptera, Hymenoptera, Mantodea (eggs), Hemiptera, Thysanoptera and spider egg sacs (Burks, 1966; Kerrich, 1973; Peck, 1985).

Proacrias^{CR}. Neotropics. Probably parasitoids of leaf-miners (Bouček, 1977).

Teleopterus^{CR}. Probably cosmopolitan. Primary or secondary parasitoids of eggs or small larvae of chrysomelids, lepidopterous leaf-miners, tenthredinids, tephritids, and agromyzids.

EUDERINAE

The genera of Euderinae that occur in Central America, with the exception of *Hubbardiella*, are included in Bouček's (1988) key to the Australasian genera.

Acrias^{CR}. A circumtropical genus with several species in Central America, all of which are undescribed. Biology unknown, but may be similar to *Astichus*.

Aoridus^{CR}. Neotropics and southeast Asia. Biology unknown, possibly parasitoids of beetles in dead wood (Yoshimoto, 1971*b*).

Astichus^{CR}. Probably cosmopolitan. Several species are known to be parasitoids of small beetles (Ciidae) that develop in bracket fungi.

***Euderus*CR**. A large cosmopolitan genus. Parasitoids of larvae of Lepidoptera and Buprestidae in wood, and of eggs of Cerambycidae and Curculionidae. Sometimes develop as hyperparasitoids (Yoshimoto,1971*a*).

***Hubbardiella*CR**. Previously known from two specimens, one from Arizona and one from Trinidad. In Costa Rica specimens were collected from highly disturbed forest at an altitude of 2100 metres. Biology unknown.

EULOPHINAE

The Eulophinae appear to be the most primitive subfamily of eulophids since they show less specialized characters: submarginal vein smoothly joining parastigma (Fig. 11.110), postmarginal vein well developed, submarginal vein with three or more setae. This subfamily has previously been treated as two subfamilies: the Eulophinae (with incomplete notauli and male antenna usually with two to three branches), and Elachertinae (with complete notauli and male antenna never branched). However, these characters are not always reliable. For example, the neotropical genus *Hoplocrepis* has complete notauli and the males of many species have branched antennae.

Many of the genera can be identified using Bouček (1988), although his keys do not include *Diglyphomorpha, Grotiusomyia, Hoplocrepis, Miotropis, Necremnus* and *Paraolinx*. The most important genera in Central America are *Elachertus* and *Euplectrus*, and to a lesser extent, *Deutereulophus, Pnigalio, Hoplocrepis* and *Cirrospilus*. *Elachertus* in particular is well represented and diverse, and detailed study will be required to solve the generic limits of this and related forms.

Members of the subfamily Eulophinae are mostly ectoparasitic idiobionts of insect larvae concealed in plant tissue, although *Euplectrus* and related genera (*Metaplectrus* and *Platyplectrus*) are ectoparasitic koinobionts of exposed hosts. Species of *Cirrospilus* attack a wide range of hosts whereas the host ranges of other genera occurring in Central America are more easily defined (Table 11.11).

***Aulogymnus*EX**. Holarctic and Australia. Mainly parasitoids in galls.

***Cirrospilus*CR**. Cosmopolitan. Parasitoids (sometimes hyperparasitoids) of larvae or pupae, or sometimes eggs, of leaf-miners, leaf-rollers and gall-formers.

Host category	Eulophine parasitoid genus
Curculionidae	*Necremnus*
Lepidoptera	*Diaulomorpha, Elachertus, Euplectrus, Grotiusomyia, Hyssopus, Miotropis, Paraolinx, Stenomesius, Sympiesis, Trichospilus*
Agromyzidae	*Diaulinopsis, Diglyphus*
Leaf-miners	*Hemiptarsenus, Pnigalio, Zagrammosoma*
Gall-formers	*Aulogymnus*

Table 11.11. Hosts of eulophine eulophid genera, excluding *Cirrospilus* and the *Euplectrus* group of genera (see text), occurring in Central America.

***Deutereulophus*CR**. Australasia (Bouček, 1988—as *Entedonomorpha*), Neotropics (LaSalle & Schauff, 1992) and North America (Schauff & LaSalle, 1993). Biology unknown.

***Diaulinopsis*EX**. Mainly holarctic. Closely related to *Cirrospilus*. Parasitoids of dipterous leaf-miners (Gordh & Hendrickson, 1979).

***Diaulomorpha*CR**. Mainly Australasia, with a single described South American species. Mostly parasitoids of small caterpillars. One species in Santa Rosa National Park has been reared from leaf-miners belonging to seven families of Lepidoptera, three families of Coleoptera and one family of Diptera on a wide array of host plants (Memmott *et al.*, 1994).

***Diglyphomorpha*EX**. Caribbean (Bouček, 1977) and North America (Schauff & LaSalle, 1993). Comprises a single species, *D. aurea*, which is closely related to *Elachertus*. Biology unknown.

***Diglyphus*CR**. Predominantly holarctic, but species have been widely introduced into the Neotropics in biological control programmes. Parasitoids of dipterous leaf-miners (mainly Agromyzidae) on herbaceous plants (Gordh & Hendrickson, 1979).

***Elachertus*CR**. Cosmopolitan. Ectoparasitoids or endoparasitoids of small Lepidoptera (Schauff, 1985*b*).

Euplectrus. Cosmopolitan. Gregarious ectoparasitoids of lepidopterous larvae. They are unusual for Chalcidoidea in that they spin silken cocoons.

Grotiusomyia[CR]. New World. Parasitoids of small Lepidoptera, including leaf-miners and leaf-rollers.

Hemiptarsenus[CR] (= *Notanisomorpha*; Schauff & LaSalle, 1993). Cosmopolitan. Parasitoids of leaf-miners, mostly Diptera, but also other orders.

Hoplocrepis[CR]. Mainly neotropical, extending north to Virginia. Biology unknown.

Hyssopus[EX]. Closely related to *Elachertus*. Ectoparasitoids of small lepidopterous larvae burrowing in plant tissue (Schauff, 1985*c*).

Metaplectrus[CR]. Related to *Euplectrus*. Solitary ectoparasitoids of lepidopterous larvae.

Miotropis[CR]. Holarctic and northern Neotropics. Closely related to *Elachertus*. Primary or secondary ectoparasitoids of the larvae of small Lepidoptera.

Necremnus[CR]. Holarctic and northern Neotropics. Ectoparasitoids of larvae of Coleoptera and Lepidoptera.

Paraolinx[CR]. New World. Parasitoids of larvae of small Lepidoptera (Miller, 1964).

Platyplectrus[CR]. Related to *Euplectrus*. Solitary or gregarious ectoparasitoids of lepidopterous larvae.

Pnigalio. Cosmopolitan. Parasitoids of the larvae of various leaf-miners, mainly Lepidoptera; less commonly they attack gall-formers (Miller, 1970; Yoshimoto, 1983).

Stenomesius. Probably cosmopolitan. Ectoparasitoids of a variety of small Lepidoptera larvae (Bouček, 1962).

Sympiesis. Cosmopolitan. Solitary or gregarious ectoparasitoids of leaf-mining Lepidoptera larvae (Miller, 1970).

Trichospilus[CR]. Mainly an Old World tropical genus, with one species, *T. diatraeae*, present in New World, perhaps as an introduction (Bennett *et al.*, 1987). Gregarious parasitoids of Lepidoptera pupae (Bouček, 1976).

Host category	Tetrastichine parasitoid genus
Spider eggs	*Aranobroter, Comastichus*
Thysanoptera	*Thripastichus*
Psylloidea	*Tamarixia*
Scolytidae (adults)	*Phymastichus*
Diptera	*Aceratoneuromyia, Neotrichoporoides, Tachinobia*
Lepidoptera	*Palmistichus, Tetrastichomyia*
Galls	*Ceratoneura, Cirrospilopsis, Galeopsomyia, Paragaleopsomyia, Pentastichus*
Nest-building Hymenoptera	*Melittobia*

Table 11.12. Hosts of selected tetrastichine eulophid genera occurring in Central America.

Zagrammosoma. Cosmopolitan. Closely related to *Cirrospilus*. Parasitoids of leaf-miners, mainly Lepidoptera and Diptera (Gordh, 1978; LaSalle, 1989).

TETRASTICHINAE

This is the largest subfamily of Eulophidae. Males have a sensory plaque on the ventral surface of the scape, and often very long whorls of setae on the funicular segments. Most species have traditionally been placed in the genus *Tetrastichus*, but recently this genus has been separated into several more natural groups (Graham, 1987, 1991; Bouček, 1988). The largest genus in all regions is *Aprostocetus*, with many hundreds of species, followed by *Baryscapus* (mainly holarctic), *Tetrastichus* and *Quadrastichus*; other important genera in the Neotropics include *Galeopsomyia* and *Pentastichus*. The only modern keys to genera are for Europe (Graham, 1987, 1991), Australasia (Bouček, 1988) and North America (LaSalle, 1994). The latter includes all described genera currently known from Central America except *Cirrospilopsis*, *Palmistichus* and *Phymastichus*.

Biologically the Tetrastichinae is one of the most interesting groups of Chalcidoidea. Hosts of this subfamily are found in 100 families of insects in ten different orders; some are predators of spider eggs, gall-forming mites, and even nematodes, while a few species are partially or completely phytophagous. Species of *Aprostocetus*, *Baryscapus*, *Oomyzus*, *Quadrastichus* and *Tetrastichus* attack a wide range of hosts. The host range of other tetrastichine genera occurring in Central America are given in Table 11.12.

Most Tetrastichinae are endoparasitoids (Graham, 1987), and in several species the greater part of larval development takes place in a host stage following that in which the egg was laid (Clausen, 1940*b*). There are, however, many exceptions. *Melittobia*, *Tamarixia*, and several of the species associated with galls are ectoparasitoids, and many of these are also idiobionts. Species of *Aceratoneuromyia* and *Tachinobia* are gregarious idiobiont endoparasitoids of dipterous pupae.

Tetrastichine species associated with galls may be parasitoids of other insects in the gall, inquilines (feeding partially or entirely on gall tissue), or occasionally gall-formers, as is the case with species in the Australian genus *Quadrastichodella*. These oviposit into young flower buds of *Eucalyptus* species (Myrtaceae), transforming them into seed-like galls (Flock, 1957). Although very few tetrastichines are known to independently elicit gall formation (e.g. Ishii, 1931; Teitelbaum & Black, 1954) many probably feed partially, or even entirely, on gall tissue provided by another insect (e.g. Parnell, 1964*b*). One North American species forms its own gall within that of a cecidomyiid, but is never found forming galls on its own (Hawkins & Goeden, 1982). In the Neotropics, *Lisseurytomella* is a seed parasitoid.

Aceratoneuromyia. Mainly Old World, but one species is known from North America, and *A. indica* has been imported into the New World in biological control programmes. Parasitoids of Diptera, mainly Tephritidae, Calliphoridae, Muscidae and Sarcophagidae.

Aprostocetus. Cosmopolitan. Parasitoids of cecidomyiids or other gall insects; others attack eggs, larvae or pupae of various insects, and many are known to be hyperparasitic.

Aranobroter^{EX}. New World, mostly neotropical. Predators in spider egg sacs (LaSalle, 1990).

Baryscapus^{CR}. Mainly holarctic, with some species in more southern regions. Parasitoids (or sometimes hyperparasitoids) of a wide range of hosts including Diptera, Coleoptera, Lepidoptera, and a few others.

Ceratoneura. Mainly circumtropical, with a few species in the Nearctic. Parasitoids/inquilines in cecidomyiid galls.

Cirrospilopsis^{CR}. Known from one described species from Brazil and an undescribed species from Costa Rica. Inquilines in galls of psyllids of the genus *Neolithus* (Triozidae).

Comastichus. The single known species, *C. zopheros*, has been collected in Florida and Costa Rica. Associated with spider egg sacs.

Eriastichus. Texas to Brazil. Biology unknown.

Galeopsomyia. New World, mainly neotropical. Parasitoids/inquilines in galls formed by Cecidomyiidae and Cynipidae.

Henryana^{CR}. Mainly neotropical, with a few species known from the southeastern United States, and one undescribed species each from Africa and southeast Asia. Biology unknown.

Lisseurytomella^{EX}. Florida Texas, the Caribbean and Guyana. The single known species, *L. flava*, is phytophagous, infesting the seeds of *Cardiospermum* spp. (Sapindaceae).

Melittobia^{CR}. Cosmopolitan. Gregarious parasitoids (or sometimes hyperparasitoids) of mainly aculeate Hymenoptera, but with an extremely wide host range, particularly under laboratory conditions (Dahms, 1984*a*).

Neotrichoporoides^{CR}. Mainly palaeotropical, with a single cosmopolitan species, *N. viridimaculatus*, known from Central America. Parasitoids of Diptera in stems of grasses.

Oomyzus^{EX}. Known from all regions except the Neotropics. Parasitoids of Coleoptera, less often Neuroptera, Diptera, or Lepidoptera, attacking larvae or pupae, or sometimes eggs (of Chrysomelidae or epilachnine Coccinellidae). Several species have been used in biological control programmes, particularly *O. sokolowskii* against diamond-back moth, *Plutella xylostella*.

Palmistichus. Neotropical, with two described species. Parasitoids of Lepidoptera (e.g. Nymphalidae, Limacodidae), mostly pupae but also larvae or eggs; sometimes hyperparasitoids; also, eggs of hispine Chrysomelidae (Delvare & LaSalle, 1993).

Paragaleopsomyia[CR]. A New World tropical and subtropical genus. Species are associated with galls of Cecidomyiidae and one has been shown to form its own gall within the cecidomyiid gall.

Pentastichus[CR]. New World, mainly neotropical. Some species appear to be associated with galls, although most species are undescribed and relatively little biological information is available.

Phymastichus. The single described species of this genus occurs in Africa, but a second species is known from Hawaii, the southeastern United States and Costa Rica (LaSalle, 1995). Parasitoids of adult scolytid beetles (Feldhege, 1992).

Quadrastichus[CR]. Cosmopolitan. Parasitoids of Cecidomyiidae or various Coleoptera, although other hosts are known.

Tachinobia[CR]. Cosmopolitan, but rarely encountered. Gregarious endoparasitoids, mostly in pupae of Tachinidae associated with Lepidoptera, but also in various other Diptera.

Tamarixia[CR]. Cosmopolitan. Parasitoids of Psylloidea.

Tetrastichomyia. Mainly holarctic, with species known from Africa and Central America. Biological information is only available for one species, the holarctic *T. clisiocampae*, which attacks a variety of lepidopterous hosts as a primary, or sometimes secondary, parasitoid.

Tetrastichus. Cosmopolitan. Primary (or sometimes secondary) parasitoids of a variety of hosts, including Diptera, Coleoptera, Lepidoptera, Hymenoptera, and more rarely, Odonata and Blattaria.

Thripastichus[EX]. Widely distributed throughout the world, although much of this is probably due to introductions. The single described species, *T. gentilei*, is a parasitoid of Phlaeothripidae.

11.8 EUPELMIDAE

Gary A. P. Gibson

Diagnosis. 1.3 to 20.0 mm in length (excluding ovipositor), fully winged to brachypterous; dead individuals often contorted into U-shape, with head and metasoma reflexed upwards. *Eyes with inner margins converging above*. Prepectus not inflated anteriorly beside pronotum; *mesopleuron with at least anterior two-thirds plate-like, enlarged and almost always convexly inflated*; mesosternum usually with small semicircular membranous region anterior to each mid coxa. Fore wing with marginal vein conspicuously longer than stigmal vein; postmarginal and stigmal veins relatively long, or if very short then wing usually very sparsely setose or glabrous. Legs with *mid coxa attached distinctly behind midline of mesopleuron, adjacent to hind coxa and far from fore coxa*; fore tibia with one or more tiny, curved spicules along dorsoapical margin; mid tibial spur long, usually conspicuously robust and microsetose; tarsi 5-segmented; mid tarsus usually slightly thickened toward base and either densely pad-like setose or with peg-like spines ventrally. Metasoma without a distinct petiole; cerci not conspicuously advanced forward.

Males of the subfamily Eupelminae lack the diagnostic features listed above, except for the dorsally convergent eyes, relatively long though slender mid tibial spur, and dorsoapical fore tibial spicules. Male eupelmines are very similar to many male pteromalids, particularly cleonymine males, which also have dorsally convergent eyes and often the two described tibial features. With experience, male eupelmines can be distinguished from male pteromalids by a combination of external features. Cleonymine males often have distinctly larger hind coxae and a more coarsely reticulate femoral depression than do male eupelmines, but males of the two groups are distinguished by only a single absolute character that requires removal of the mid coxae to be seen, i.e. cleonymine males have the two mid coxal fossae completely separated by cuticle whereas male eupelmines have the fossae partly joined by a transverse membranous region.

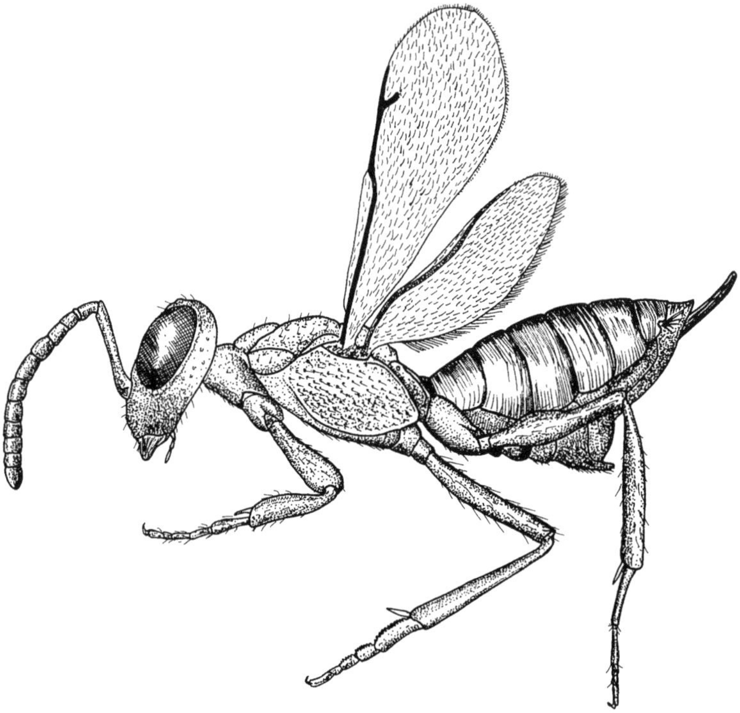

Fig. 11.117. Eupelmidae: Eupelminae.

Classification and distribution. Eupelmidae are cosmopolitan, but the family is most diverse in warmer climates. Approximately 850 species are classified in three subfamilies: Eupelminae, Calosotinae and Neanastatinae (= Metapelmatinae). The genera of Calosotinae and Neanastatinae were revised by Gibson (1989), who treated the latter subfamily as the Metapelmatinae (Bouček, 1988) because the name Neanastatinae (Kalina, 1984) had been overlooked. The Calosotinae is composed of eight genera, of which *Calosota* and *Eusandalum* are cosmopolitan, *Archaeopelma* and *Licrooides* are restricted to the New World, *Paraeusandalum* is restricted to Chile, and three genera are restricted to the Old World. The Neanastatinae consists of the cosmopolitan genus *Metapelma*, the neotropical genus *Lambdobregma*, and two Old World genera. The genera of the Eupelminae are being revised by Gibson (1995). Thirty-three genera will be recognized for the world, of which 25 are known from the Neotropical region and 19 occur in Costa Rica. Ultimately, the eupelmine species diversity in Costa Rica is likely to prove greater than that of all of America north of Mexico, from which only 14 eupelmine genera are known.

Males and females of the Calosotinae and Neanastatinae are similar in structure, but extreme sexual dimorphism characterizes the Eupelminae. Male eupelmines are largely unmodified and pteromalid-like, whereas female eupelmines are highly modified. This 'either-or' structure for Eupelminae contrasts with Calosotinae, in which the sexes are structurally similar but individuals of some genera are relatively primitive and pteromalid-like, whereas individuals of other genera are highly modified and very similar to female eupelmines. Relationships among the subfamilies and with other chalcidoids are unresolved. Because the mesosomal structures that characterize Eupelmidae are all modifications related to how they jump (Gibson, 1986*b*), it is possible that the features were derived independently in each of the three subfamilies and Eupelmidae is a grade level taxon that is not monophyletic (Gibson, 1989).

Biology. Generalizations about eupelmid biology are based almost entirely on species from the Northern Hemisphere and very little is known about the biology of the much more diverse neotropical fauna. However, most eupelmids appear to be idiobiont ectoparasitoids or predators of the immature stages of other arthropods concealed in plant tissue or silken sacs. Some species, especially those that attack exposed insect eggs, develop as endoparasitic idiobionts. Koinobiosis has never been reported in the family.

Like many other groups of idiobiont ectoparasitoids the range of hosts utilized by eupelmids is extremely diverse, including species of Orthoptera, Blattaria, Mantodea, Hemiptera (Heteroptera and Homoptera), Neuroptera, Coleoptera, Diptera, Lepidoptera and Hymenoptera, as well as the eggs of spiders. Hosts attacked occur in a wide variety of situations, including borings in wood and plant stems, aculeate nests, galls and various forms of cocoons. In some ways this parallels the diversification of host utilization observed in other groups of idiobiont ectoparasitoids such as pimpline ichneumonids (e.g. see Gauld, 1988*c* and Chapter 12.1).

Most Calosotinae and some Neanastatinae (e.g. *Metapelma*) are ectoparasitoids of beetles living in wood. Species of the extralimital genus *Neanastatus* are ectoparasitoids or predators of gall-midges (Diptera: Cecidomyiidae) and their primary parasitoids. Parasitism of wood-inhabiting beetles is much rarer in the Eupelminae, but eupelmines, like all eupelmids, possess small spicules on the dorsoapical margin of each fore tibia. Similar spicules are characteristic of a wide variety of parasitoids of wood-inhabiting beetles and apparently spicules evolved several times to assist adult parasitoids in exiting the host tunnel. It is possible that the presence of fore tibial spicules on male and female eupelmines represents a relict feature and that wood-inhabiting beetles represent the ancestral hosts for members of all three eupelmid subfamilies.

The host spectrum of Eupelminae is very wide. Species of the common genera *Brasema* and *Eupelmus* are mostly ectoparasitoids of larvae and pupae concealed in plant tissue or fibrous material. Many species are polyphagous, apparently accepting diverse hosts of similar physical appearance in a particular ecological niche. The larvae are voracious, acting as parasitoids if a single host is available, but usually as predators if more than one host is available. Hyperparasitism is common and invariably the eupelmine is the successful competitor if a primary host is parasitized by more than one parasitoid species. Superparasitism also occurs in some species, but usually only a single adult emerges. Larvae of *Eupelmus australiensis* (most literature for this species is recorded under the name *E. popa*), may even complete their development on plant tissue if they have consumed at least one larva or pupa of the sorghum midge (Cecidomyiidae) (Woodruff, 1929).

Species of *Anastatus* are mostly endoparasitoids of a wide range of insect eggs, including those of Orthoptera, Blattaria, Hemiptera and Lepidoptera. However, some species that normally attack eggs of Lepidoptera also develop as secondary ectoparasitoids of cocooned microgastrine braconids (Clausen, 1940*b*). One species of *Oozetetes* was reared as an endoparasitoid of cockroach eggs, and females of this genus, like those of *Anastatus*, have bidentate mandibles, unlike the tridentate mandibles of most eupelmids. Further investigation is required to determine whether bidentate mandibles is a common feature among eupelmid endoparasitoids of insect eggs.

There is little information on the specific cues involved in host-searching behaviour in Eupelmidae. Parasitoids of concealed hosts possibly use chemical cues, because *Macroneura vesicularis* has been shown to have increased preference for a braconid host proportional to the state of the development of the host, neglecting eggs, but having maximum preference for pupae (Morris, 1938). Chemical cues are also indicated for a species of *Eupelmus* that was shown to be attracted to larval and pupal skins removed from the mines of its leaf-mining beetle host (Taylor, 1937). *Anastatus floridanus* may be attracted by some substance in the saliva of female cockroaches (used to cement a covering for the oothecae), but may also be attracted to the adults themselves because females are known to attack oothecae while they are still attached to the cockroach (Roth & Willis, 1954).

Female eupelmines host-feed from all host stages (Clausen, 1940*b*), although a feeding tube has only been observed when feeding on hosts other than eggs (Delanoue & Arambourg, 1965). Females of some ectoparasitoid species also paralyse the larval or pupal host prior to oviposition. Usually the egg is deposited in the vicinity of the host rather than being placed directly on it. Even egg parasitoids appear to avoid ovipositing directly in the embryo, but rather place their egg in the fluid surrounding the embryo (Clausen, 1940*b*). Species of *Brasema* and *Eupelmus* usually attach their eggs to the inner wall of the host cocoon or mine, and several species use a network of fibrous material to cover the eggs (Phillips & Poos, 1921; Delanoue & Arambourg, 1965). Askew (1961*a*) suggested that the covering protected the eggs from being crushed by movements of the host larva, but it is unknown why often only some of the eggs are covered by the female.

The eupelmid egg is ellipsoidal, has an unsculptured chorion, and usually has a filament at one end and a longer stalk at the other end, which is used by some species to attach the egg to a substrate (Clausen, 1940*b*). There are normally five larval instars. The first instar larva is 13-segmented; first instar larvae of ectoparasitoid species have sparse longitudinal rows of long setae that assist in movement during feeding. First instar larvae of species of *Anastatus* and *Mesocomys*, which are both endoparasitoids, have ventral spines and a bifurcate tail, which are used for locomotion within the liquid medium of the egg. Intermediate and final instars are more similar in form, the larvae usually being cylindrical but slightly curved, lacking caudal processes, and being covered by rows of short setae or spines. The duration of the life cycle is variable, but the period devoted to feeding is always relatively short, ranging from one to two weeks (Clausen, 1940*b*); species overwinter as mature larvae.

Female eupelmines, and to a lesser extent males and females of the other two subfamilies, often die in a contorted U-shape, a structural peculiarity resulting from how they jump (Gibson, 1986*b*). Except for male eupelmines, eupelmids are very powerful and incredibly quick jumpers, and in the early literature one species was called the 'back-rolling wonder' (Clausen, 1927) because individuals often tumble upon landing. Jumping does not appear to be used for moving around and probably evolved as an escape mechanism from predators, possibly ants and spiders. Male eupelmines fly readily and female eupelmines walk very rapidly in addition to jumping, but even fully winged females either can not fly or are very poor fliers. Reduced ability to fly appears to be a consequence of mesosomal modifications to increase jumping ability, which interfere with functioning of the wing axillary sclerites. Brachyptery is not known for Calosotinae or Neanastatinae, but female brachyptery is common in Eupelminae and possibly also is a consequence of selection for increased jumping ability at the expense of flight ability. Presumably, a further consequence is a lesser potential for dispersal than for most fully winged parasitoids. In many eupelmine species reduced flight ability is associated with polyphagy, including hyperparasitism.

Economic importance. The sorghum midge (*Contarinia sorghicola*) is attacked by *Eupelmus australiensis*, and the brown-banded roach (*Supella supellectilium*) is attacked by *Anastatus tenuipes*. Both eupelmid species are cosmopolitan and probably are present in Central America. *Lecaniobius utilis* has been used for biological control of the black scale (*Saissetia oleae*) in North and South America, and *Anastatus japonicus* (most literature for this species is recorded under the name *A. disparis*) was introduced into North America from Europe for biological control of the gypsy moth (*Lymantria dispar*). Other eupelmids are parasitoids of pest insects, but few species have been utilized for biological control, probably because of the ability of many species to develop readily as secondary parasitoids. One consequence of their polyphagous potential is that eupelmines can become pests of phytophagous insects introduced for biological control of weeds. *Macroneura chrysosinamora* was shown to have parasitism rates of up to 25 percent on a case-bearer that was introduced into California for biological control of Russian thistle. It was suggested that adaptation of such native parasitoids as *M. chrysosinamora* to the case-bearer was the reason for its poor establishment and ineffectiveness as a biological control agent (Goeden *et al.*, 1987).

Identification. Keys to the genera of the Calosotinae and Neanastatinae were given by Gibson (1989), and separate keys based on sex to the genera of the Eupelminae will be given by Gibson (1995). Gibson (1989) also included a key to distinguish members of the three subfamilies from each other and from other chalcidoids, from which the following key is adapted.

Key to subfamilies of Eupelminae present in New World

1 Mesosternum with posterior margin transverse along (abutting) bases of mid coxae (Fig.

11.118); pronotum large, subtriangular and evenly sclerotized (Fig. 11.120); mid tarsus with single row of pegs along anteroventral margin of tarsal segments................................. .... **Neanastatinae** (p. 334)

— Mesosternum with a small membranous area in front of each mid coxa (Fig. 11.119), or coxa rotated anteriorly out of fossa and concealing area; pronotum more or less pentagonal (Fig. 11.121), or if subtriangular then medially divided or with a white line and often depressed; mid tarsus without pegs or with a different arrangement of pegs. 2

2 Mesoscutum with anterolateral corners conspicuously shoulder-like on either side of more or less pentagonal and evenly sclerotized pronotum (Fig. 11.121); fully winged and notauli either as V-shaped or subparallel lines .... **Calosotinae** (p. 333)

— Mesoscutum not conspicuously shoulder-like on either side of pronotum, which often is divided or has a medial white line; brachypterous, or if fully winged then notauli as V-shaped, furrow-like depressions (females only). .... **Eupelminae** (p. 334)

Synopsis of the Costa Rican fauna

CALOSOTINAE

This subfamily comprises eight genera (Gibson, 1989), five of which occur in the New World: *Paraeusandalum*, known only from Chile, and the following four genera that occur in Costa Rica.

***Archaeopelma*[CR].** Southwestern United States to Brazil; one described and at least two undescribed species, one of the latter occurring in Costa Rica. Scutellum pointed basally (anteromedial angles of axillae contiguous); notauli V-like, convergent. Biology unknown, but possibly parasitoids of xylophagous beetles.

***Calosota*.** Cosmopolitan; about 70 described species. Scutellum truncate basally (axillae not contiguous; Fig. 11.121), notauli subparallel. The nearctic species were revised by Burks (1973). Biology: parasitoids of beetles associated with dead wood (Anobiidae, Cerambycidae, Cleridae, Scolytidae), solitary wasps (Eumeninae, Sphecidae) and solitary bees. Some species are primary or secondary parasitoids of various insects in grass stems such as Cecidomyiidae and Eurytomidae.

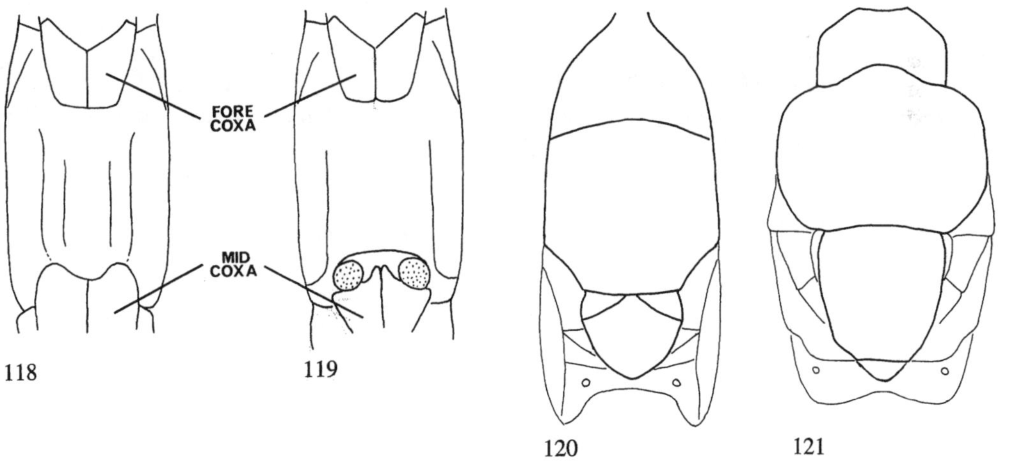

Figs 11.118–11.121. Eupelmidae. Figs 11.118–11.119. Mesosternum; 11.118, *Lambdobregma* sp. (Neanastatinae); 11.119, *Calosota* sp. (Calosotinae). Figs 11.120–11.121. Dorsal view of mesosoma; 11.120, *Lambdobregma* sp.; 11.121, *Calosota* sp.

Eusandalum. Cosmopolitan; about 60 described species. Scutellum truncate, appearing more or less petiolate basally because inner margins of axillae are incurved; notauli convergent posteriorly; mesopleuron convexly enlarged to the metapleuron. The nearctic species were revised by Girault (1917). Biology: parasitoids of xylophagous beetles—Anobiidae, Bostrichidae, Buprestidae, Cerambycidae, Curculionidae, Lyctidae and Scolytidae.

Licrooides. Southwestern United States to Argentina; one described and at least six undescribed species, one of the latter occurring in Costa Rica. Scutellum truncate basally, but with the obliquely angled inner margins of the axillae straight; notauli convergent posteriorly; mesopleuron not convexly enlarged to the metapleuron. Biology unknown, but possibly parasitoids of xylophagous beetles.

NEANASTATINAE (= Metapelmatinae)

This subfamily comprises four genera (Gibson, 1989), two of which occur in the New World.

Lambdobregma. Florida to Brazil and Chile; one described and four undescribed species. *L. schwarzi* and one undescribed species occur in Costa Rica. Hind tibia and tarsal segments not compressed; scrobes present as convergent furrows above antennal bases. Biology: apparently endoparasitoids of the eggs of crickets (Gryllidae).

Metapelma. Cosmopolitan; about 40 described species. Hind tibia and tarsal segments laminately compressed; scrobes present as short vertical depression above each antennal base. The world species were revised by Ferriére (1938). Biology: parasitoids of xylophagous beetles (Bostrichidae, Buprestidae, Cerambycidae and Curculionidae).

EUPELMINAE

This is by far the largest subfamily of Eupelmidae. As currently defined, 15 genera are known to occur in Costa Rica. Three of these, however, are likely to be reduced to subgeneric rank. On the other hand, seven undescribed genera occur in the country, thus bringing the total number of genera to 19. Keys to the world genera are not yet available (Gibson, in prep.). Males are difficult to identify to genus, but the genera, as currently recognized, can be separated into two groups based on females.

Group A. Metasoma with posterior margin of last tergite deeply emarginate (shaped like the Greek letter omega, Ω) and inner apical margin of mid tibia without groove between base of tibial spur and base of tarsus.

***Australoodera*[CR].** Circumtropical; at least 12 species, mostly undescribed (Bouček, 1988); two undescribed species occur in Costa Rica. Metasoma with penultimate tergite (tergite VII) bearing a spiracle, entire and not concealed under tergite VI; head with scrobal margin carinate and paralleling inner margin of eye; scape with a longitudinal white band and body mostly yellowish to brown, though sometimes partly white. Biology: reared from larvae of Lepidoptera and Scolytidae (Bouček, 1988).

Eupelmus. Cosmopolitan; at least 200 species (Bouček, 1988); uncommon and depauperate in the Neotropics in comparison with the Nearctic. Metasoma with penultimate tergite (tergite VII bearing a spiracle) medially divided and largely concealed under tergite VI, which extends posteriorly to the spiracle; females fully winged and almost always with a linea calva; usually uniformly metallic green. Biology: ectoparasitoids of larvae and pupae concealed in plant tissue.

***Hirticauda*[CR].** Australia and Neotropics. Similar to *Reikosiella*, except the mid tibia has apical pegs and the antennal flagellum has one or more segments white. Biology unknown, but the slim body suggests development in narrow stems (Bouček, 1988).

***Macroneura*[CR].** Cosmopolitan; about 20 species. Similar to *Eupelmus* species, except they are brachypterous. Biology: ectoparasitoids of larvae and pupae concealed in plant tissue.

Phlebopenes. New World; 14 described species; *pilipes* and *viridis* are reported from Costa Rica, though these may be conspecific, and *consors* is known from Guatemala and Colombia (Callan, 1976). Large, 5 to 20 mm long, excluding the ovipositor, which is at least twice the body length; metasoma like that of *Australoodera*; partly or entirely metallic. Biology: the long ovipositor suggests that they may also be parasitoids of xylophagous beetles; moreover, specimens have been taken from logs (H. Hespenheide, pers. comm.). One species has been reared from *Trypoxylon nitidium* (Sphecidae).

***Reikosiella*[CR].** Cosmopolitan. Metasoma like that of *Australoodera*; head with scrobal margin directed toward the inner margin of the eye so that the scrobal depression occupies virtually the entire region between the eyes; marginal vein distinctly longer than the costal cell; body mostly yellowish to dark brown, with the scape uniformly coloured. Biology: the Hawaiian type-species was reared from a pyralid larva in fruit (Bouček, 1988).

Group B. Metasoma with posterior margin of last tergite only very rarely conspicuously emarginate; if so, then inner apical margin of mid tibia with oblique groove between base of tibial spur and base of the tarsus. Last tergite almost always truncate apically, or reflexed into linear rim or fingernail-like flange.

Anastatus. Cosmopolitan; at least 200 species worldwide, many of them undescribed (Bouček, 1988). This is the most speciose genus of eupelmids in Costa Rica, where there are perhaps 30 species. Mandible bidentate; metasoma with a basal whitish region at least ventrally and usually also dorsally; fully winged individuals with fore wing infuscate, with hyaline spots or a cross-band behind the marginal vein. The nearctic species were revised by Burks (1967). Biology: parasitoids of insect eggs, mostly Hemiptera and Lepidoptera, but also Orthoptera, Blattaria and Mantodea; occasionally other hosts in similar situations.

Arachnophaga. Mostly New World, but also Africa and southwestern Europe. Mandible tridentate; head relatively flat in lateral view, antennae attached below eyes, with a rounded or carinate ridge between the torulus and lower margin of the eye; fully winged individuals with fore wing evenly infuscate behind marginal vein. The New World species were revised by Gahan (1943). Biology: ectoparasitoids or predators of arthropods enclosed in silk, i.e. lepidopterous pupae, chrysopid cocoons (Clancy,1946) and spider egg sacs. *A. opaca* was described from spider egg sacs in Panama (Gahan, 1943).

Brasema. Widespread in warm regions; about 50 species. Likely to be the most speciose genus in Costa Rica after *Anastatus*; most of the described species are, however, currently classified as *Eupelmus*. Mandible tridentate; posterior margin of last tergite truncate or slightly incurved, but not constricted or reflexed into a rim or flange; fully winged individuals usually with fore wing hyaline; posterior margins of most tergites usually distinctly emarginate. Biology: ectoparasitoids of xylophagus beetle larvae and other hosts concealed in plant tissue.

Encyrtaspis. New World; six described species. Similar to *Arachnophaga*, except that posterior margin of the conspicuously compressed hind tibia is white, and scutellum with a tuft of erect black setae. Biology: primary and secondary parasitoids of Lepidoptera.

Lecaniobius. New World. Mandible tridentate; scrobes are separate, carinately margined channels converging to median ocellus; hind tibia conspicuously compressed; fore wing evenly infuscate behind marginal vein. The nearctic species were revised by Compere (1939). Biology: predators of eggs of scale insects or hyperparasitoids of scale insects (Smith & Compere, 1928), especially Coccidae (e.g. *Saisettia*)

Macreupelmus. Central and South America, with one species known from Costa Rica. Mandible bidentate; prothorax and posterior surface of fore femur with long conspicuous black setae; last sternite with dense patch of silvery white setae along its anterolateral margin; fore wing infuscate with a tapered hyaline streak behind the marginal vein. Biology unknown.

***Ooderella*[CR].** Neotropics; moderately speciose, although most species are undescribed. Brachypterous; pronotal collar quadrangular, medially divided, and usually with right-angled to cone-like anterolateral corners; mesopleuron uniformly coloured and usually with a conical tuft of setae near the prepectus. Biology unknown.

Oozetetes[CR]**.** Florida to South America. Mandible bidentate; metasoma uniformly dark in colour; posterior margins of most tergites straight; fore wing hyaline to partly infuscate, but without well defined hyaline spots or a cross-band. Biology: parasitoids of cockroach eggs.

Zaischnopsis[CR]**.** Cosmopolitan, speciose in Costa Rica; the described species are currently classified in *Anastatus*. Mandible tridentate; fore wing usually infuscate with either a hyaline cross-band, symmetrical spots, or a tapered streak behind the marginal vein; metasoma uniformly dark; posterior margin of last tergite slightly constricted and/or reflexed as a rim or a fingernail-like flange. Biology: one species has been reared from eggs of *Orchelimum* sp. (Tettigoniidae).

11.9 EURYTOMIDAE

Paul E. Hanson

Diagnosis. Body length about 1.5 to 7.5 mm, robust to slender; *usually black and/or yellow* (only one genus is metallic coloured in the New World), strongly sclerotized. *Antenna with 9-13 segments, inserted about midway between the mouth and anterior ocellus*; male antenna often with whorls of long trichoid sensilla (setae). *Pronotum (viewed from above) quite large and broadly rectangular, often with broad punctures*; mesoscutum with notauli complete, often deep. Fore wing with marginal vein generally longer than stigmal vein, postmarginal vein always present but sometimes quite short. Tarsi 5-segmented. Metasoma very rarely broadly joined to the propodeum, petiole elongate to inconspicuous; often somewhat laterally compressed

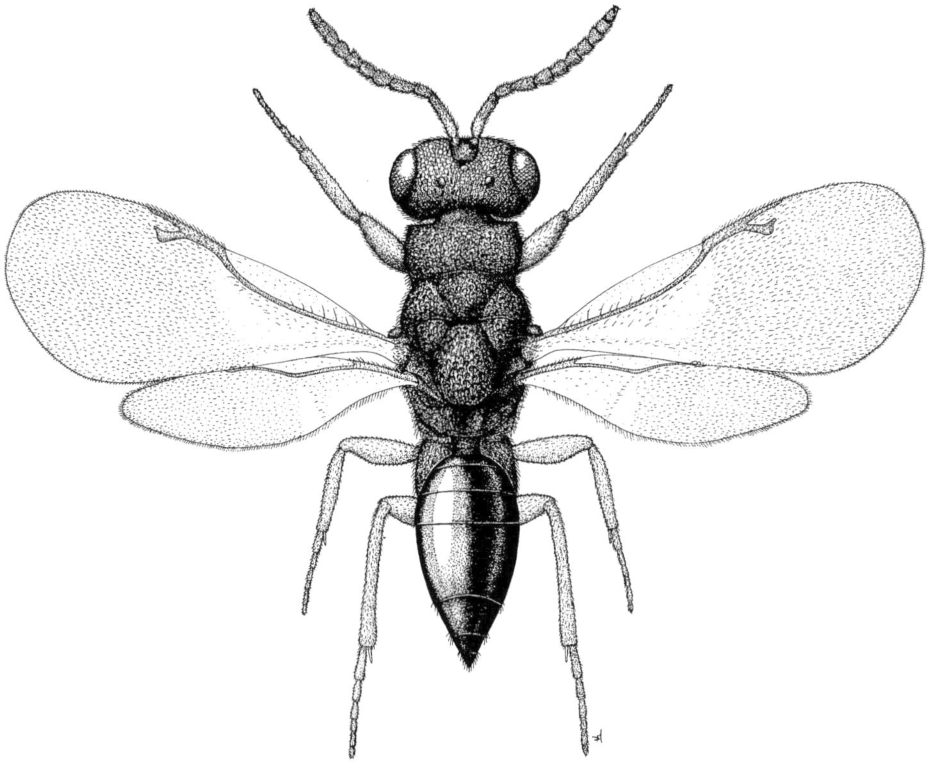

Fig. 11.122. *Eurytoma* sp. (Eurytomidae: Eurytominae).

and circular to oval in lateral view, sometimes dorsoventrally flattened; usually smooth, shining, and non-collapsing; ovipositor usually hidden, not prominent.

Classification and distribution. The Eurytomidae has a cosmopolitan distribution and contains about 1200 described species in 79 genera. Although the family has often been divided into as many as seven subfamilies (Burks, 1971; Zerova, 1989), or into several tribes (Subba Rao, 1978), there is little evidence that these groups are monophyletic. There are, however, three fairly well defined subfamilies (Stage & Snelling, 1986; Bouček, 1988): Heimbrinae (New World), Rileyinae (cosmopolitan) and Eurytominae (cosmopolitan).

Biology. Eurytomids include both carnivorous and phytophagous species. Most of the carnivorous species are ectoparasitic idiobionts of hosts concealed in plant tissue, although species of *Neorileya* are endoparasitic idiobionts in the exposed eggs of Pentatomidae. A few eurytomid species are koinobionts. For example, in the Holarctic region *Eurytoma obtusiventris, E. seratulae* and *E. tibialis (= curta)* oviposit in young gall-forming tephritid larvae; the host and parasitoid develop concurrently, and eventually the parasitoid larva stimulates early pupation of the host (Varley, 1937; Uhler, 1951; Claridge, 1961). Larvae of phytophagous eurytomids generally feed in seeds but some are gall-formers on other parts of the plant. The biological diversity outlined for the family can be seen in the largest and most common genus, *Eurytoma*. Most species in this genus are parasitoids or inquilines in galls formed by other insects, but several are parasitoids of beetle larvae in seeds or stems (primarily Bruchidae, Curculionidae and Scolytidae), or other holometabolous larvae in plant stems. In North America, phytophagous species of *Eurytoma* occur in seeds of *Juniperus* (Cupressaceae), *Ceanothus* (Rhamnaceae) and *Rhus* (Anacardiaceae) (Burks *in* Krombein *et al.*, 1979).

Ectoparasitic idiobiotic eurytomids usually attack the host larva or pupa (e.g. Taylor, 1929). Although observations are scarce, this primitive type of parasitoid behaviour presumably occurs in *Axima* (which attack bees in stems), *Bephratoides* (wood-boring beetles), *Chryseida* (bruchids in seeds), and possibly in *Rileya* (cecidomyiid gall midges). Sometimes, however, other stages of the host may be attacked. *Conoaxima*, for example, parasitizes the adult queens of *Azteca* ants living in stems of *Cecropia* (Brues, 1922) and other eurytomids attack concealed eggs. Many of these are technically predators since the parasitoid larva consumes several host eggs—for example, the North American *Macrorileya* on tettigoniid eggs (Smith, 1930), the European *Archirileya* and *Eurytoma oophaga* also on tettigoniid eggs (Silvestri, 1920), and *E. (Desantisca) arachnovora* and *E. (D.) latrodecti* on spider eggs (McMurty, 1978).

Eurytoma species attacking spider eggs are morphologically similar to species of the *E. braconidis* group (Bouček, 1988)—such as the European *E. appendigaster* and the North American *E. verticillata* (Grissell, 1985)—which are pseudohyperparasitoids in cocoons of Braconidae and Ichneumonidae. Like spider eggs, these hosts are enclosed in silk. The female wasp paralyses the host, constructs a feeding tube through the cocoon wall in order to host-feed, and then oviposits (Faure, 1926; Rosenberg, 1934). An undetermined species in Honduras was reared from a microgastrine braconid attacking a pierid (possibly *Melete*) on *Guazuma ulmifolia* (Sterculiaceae) although it is not known whether it is a facultative or obligatory hyperparasitoid (specimens in Zamorano, Honduras).

The Asian *Eurytoma monemae* has a very unusual biology. This species is a gregarious ectoparasitoid of mature lepidopteran larvae in their cocoons (Piel, 1933). The female eurytomid is dependent upon another parasitoid to gain access to the host, i.e. she follows a chrysidid and uses the puncture of the latter in which to oviposit. The larvae develop, pupate within the host cocoon, adults emerge through the same puncture hole, and sometimes another generation of parasitoids develops on the same host. Dependency upon another parasitoid to gain access to the host is a form of cleptoparasitism (see Chapter 2.5) and is known to occur in a few other eurytomids, such as the North American *E. pini*, a parasitoid of the pine shoot moth (Arthur, 1961).

The manner in which phytophagy evolved in the Eurytomidae can perhaps be seen in *Eurytoma parva*, which is initially an ectoparasitoid of young *Tetramesa tritici* larvae in grass stems. The first instar larva of *E. parva* consumes the host but the remaining instars feed on plant sap. The parasitoid is capable of developing entirely on plant tissue but apparently the host insect is required as a stimulus for oviposition (Phillips, 1927).

Phytophagy appears to be restricted to certain members of the subfamily Eurytominae. Phytophagous genera in Central America include: *Bephratelloides* (in seeds of *Annona*, Annonaceae); *Bruchophagus* (in seeds of Leguminosae); *Systole* (in seeds of Umbelliferae); and *Tetramesa* (in galls in stems or seeds of Gramineae). Some Central American species of *Eurytoma* appear to be phytophagous in seeds of *Croton* (Euphorbiaceae), orchids (including vanilla), and other plants. Various neotropical species of *Prodecatoma* are apparently phytophagous in seeds, especially those of Myrtaceae and Rubiaceae.

The eggs of Eurytomidae are usually oblong in shape with a collapsed stalk at the anterior end and a flagellum of varying length at the opposite end. *Eurytoma* species that deposit the egg external to the host generally have eggs with a heavily sculptured chorion, with pubescence or spines, giving it a darkened appearance (Claridge & Askew, 1960). The first instar larva is oval to elongate, with 13 body segments and a large, hemispherical or conical shaped head which bears sensory setae; there are four pairs of sensory setae on each thoracic segment and three on each abdominal segment. In most *Eurytoma* larvae the integument is covered in setae, either uniformly distributed or occurring in bands on each segment. There are usually four pairs of spiracles, one on the mesothorax and three on the abdomen, but some have five, eight, or ten pairs of spiracles. In the second instar larva the sensory setae are reduced. The mature larva (there are normally five instars) is more robust, with nine pairs of spiracles (second and third thoracic, and first seven abdominal segments). In *Axima* and *Conoaxima* there are median dorsal welts on the thoracic and abdominal segments.

Economic importance. The Eurytomidae contains both pest and beneficial species. In the North Temperate zone some *Tetramesa* species are pests of cereal crops (Phillips, 1936; Claridge, 1961), certain *Bruchophagus* species damage seeds of Leguminosae (Carillo & Dickason, 1963), and *Eurytoma amygdali* is a significant pest of almonds (Zerova & Fursov, 1991). In Central America *Bephratelloides* is one of the most serious pests of guanabana (Annonaceae) (Korytkowski & Peña, 1966; Grissell & Schauff, 1990; Nadel & Peña, 1991). Larvae feed on the seeds, and damage fruit production by interfering with normal ripening. *Systole coriandri* infests seeds of coriander (Umbelliferae) worldwide, and an unidentified species was taken from seeds of *Curcuma domestica* (Zingiberaceae) in quarantine in Trinidad (NHM label). The widespread tropical American species *Bruchophagus orchideatum* has been found feeding in the buds and pseudobulbs of various species of *Cattleya* (Orchidaceae), and it can be a damaging pest of cultivated orchids (Swezey, 1945).

On the other hand, some phytophagous eurytomids have been used in biological control of weeds. For example, *Eurytoma attiva*, whose larvae feed on seeds, was introduced from Trinidad to Mauritius as a biological control agent of black sage, *Cordia macrostachya* (Burks, 1958; Williams, 1960). A few carnivorous species of eurytomids have also been used in biological control programmes, but with few notable successes. One example is the attempt to use *Eurytoma latrodecti* against egg sacs of the black widow spider, *Latrodectus mactans* (McMurty, 1978).

Identification. Keys to the world genera of Eurytomidae are given by Burks (1971), but subsequently, several new genera have been described (Subba Rao, 1978; Bouček, 1988). Certain problems in Burks' keys, notably with respect to the genus *Bruchophagus*, have been discussed by Bouček (1988). Several neotropical genera have not been critically re-examined since they were originally described, and thus generic concepts will probably change with further study.

In the following key a few genera that do not seem to be very distinct have not been separated (couplet 8). It should be noted that in Burks' key *Bephrata* is included with the genera having a non-carinate gena, but I have found the gena to be more or less carinate. The character used in couplet 14 is not a convincing one for separating genera, but is repeated here from Burks' key pending further study. The majority of specimens encountered almost anywhere in the world will key out to *Eurytoma*.

Preliminary key to genera of Eurytomidae present in Costa Rica

1 Scutellum prolonged posteriorly into a blunt spine-like process; tergites II and III fused and covering most of metasoma (Fig. 11.123).
[Antenna 13-segmented.] ....
.....................(Heimbrinae) .... *Heimbra*
— Scutellum not posteriorly prolonged; tergites II and III not fused and covering most of metasoma. 2

2 Antenna with 13 segments, including two or three anelli, never sexually dimorphic; prepectus small (Fig. 11.124). .....(Rileyinae) 3
— Antenna with 9–11 segments, including one anellus, often sexually dimorphic; prepectus normal (Fig. 11.125). ...(Eurytominae) 5

3 Antenna with two anelli and six funicular segments; face narrow, scarcely broader than length of scape, and with a median, longitudinal carina on lower face. .... *Neorileya*
— Antenna with three anelli and five funicular segments; face broader than length of scape, and without median, longitudinal carina............ 4

4 Fore wing with stigmal vein apically enlarged and rounded, in form of knob. .... *Calorileya*
— Stigmal vein normal, not enlarged. .... *Rileya*

5 Fore wing with marginal vein short and broad (Fig. 11.126), usually with a dark spot below.
....6
— Fore wing not unusually stigmated. 7

6 Vertex and anterior third of pronotum flattened and on the same plane, this plane lying at nearly a right angle to the mesoscutum (Fig. 11.126); posterior end of metasoma flattened and with a dorsal concavity containing a short projection. .... *Bruchodape*
— Vertex and posterior metasoma normal. ....
.... *Sycophila*

7 Head with gena posteriorly rounded, or only vaguely carinate (Fig. 11.127). 8
— Head with gena posteriorly strongly carinate (Fig. 11.128). 9

8 Metasoma elongate, in lateral view usually over twice as long as high. ... *Isosomodes, Tetramesa*
— Metasoma circular to oval in lateral view, less than twice as long as high...............................
....*Bruchophagus, Systole*

9 Body with vivid metallic coloration. ..... *Chryseida*
— Body without metallic coloration. 10

10 Head almost twice as wide as long (dorsally), usually wider than high (Fig. 11.129), much wider than mesosoma. 11
— Head semi-quadrate (Fig. 11.130). 12

11 Fore wing with marginal vein nearly 4.0 times as long as stigmal vein; scutellum rounded; petiole nearly as long as hind femur...... *Axima*
— Fore wing with marginal vein 2.0 times as long as stigmal vein; scutellum with apical cone-shaped projection (Fig. 11.131); petiole half as long as hind femur. .... *Conoaxima*

12 Metasoma elongate, in lateral view usually over twice as long as high. .. *Aximogastra, Bephrata*
— Metasoma circular to oval in lateral view, less than twice as long as high. 13

13 Propodeum posteriorly with a distinct 'neck' which extends beyond bases of coxae.
[Apical part of submarginal vein thickened.]
.... *Aranedra*
— Propodeum without a 'neck'. 14

14 Anterior ocellus located in scrobe cavity (Fig. 11.130). 15
— Anterior ocellus located above scrobe cavity.... 16

15 Occiput (back of head) deeply concave; pronotum elongate and constricted posteriorly. ....
.... *Bephratoides*
— Occiput not deeply concave; pronotum more transverse, not sharply constricted posteriorly.
.... *Prodecatoma*

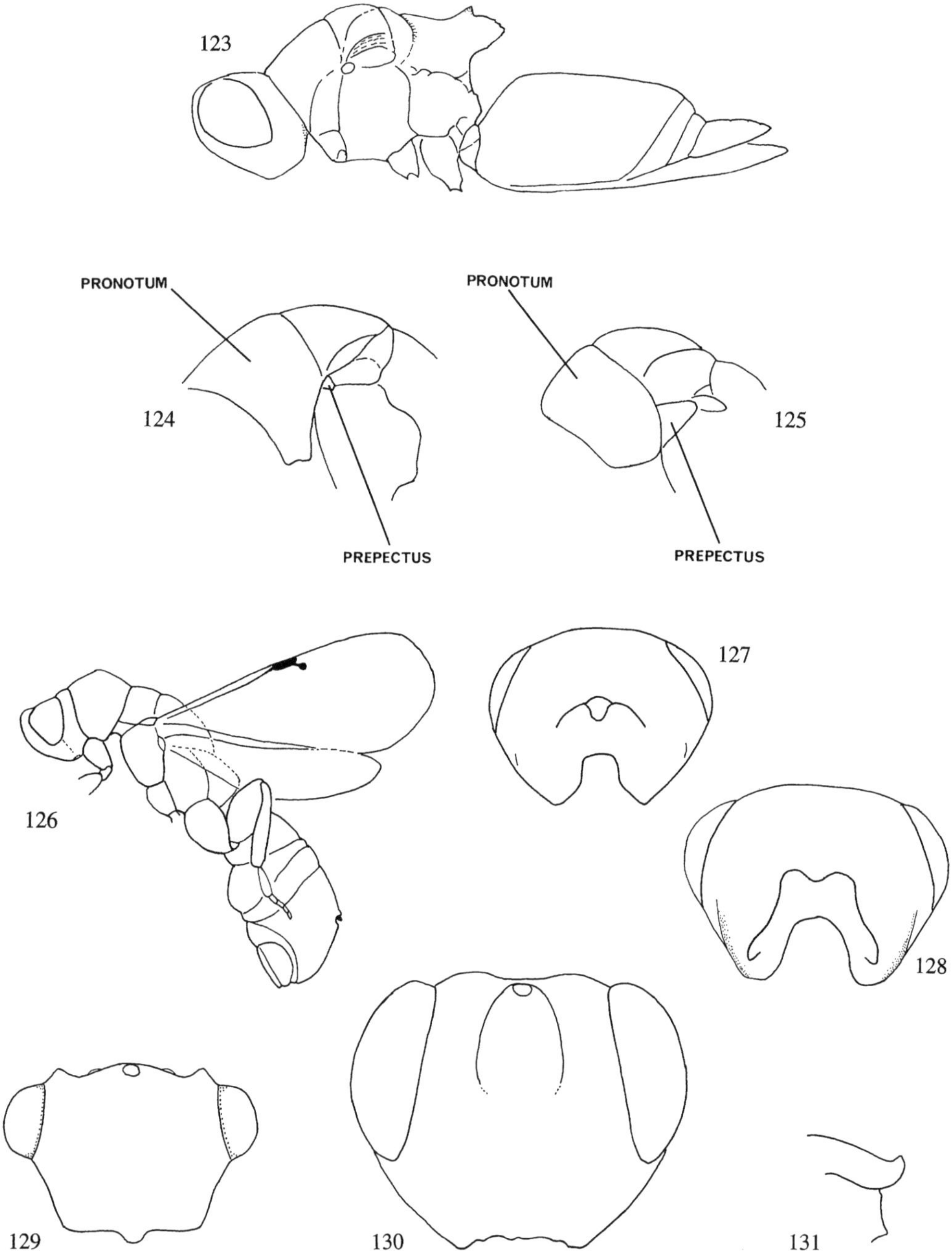

Figs 11.123–11.131. Eurytomidae. Fig. 11.123. *Heimbra* sp. (Heimbrinae), head and body, lateral view. Figs 11.124–11.125. Lateral view of front part of mesosoma; 11.124, *Neorileya* sp. (Rileyinae); 11.125, *Sycophila* sp. (Eurytominae). Fig. 11.126. *Bruchodape* sp., whole insect, lateral view. Figs 11.127–11.128. Head, posterior view; 11.127, *Bruchophagus* sp.; 11.128, *Eurytoma* sp. Figs 11.129–11.130. Head, anterior view; 11.129, *Axima*; sp. 11.130, *Bephratoides* sp. Fig. 11.131. Scutellum of *Conaxima* sp., lateral view.

16 Antenna with six long funicular segments, club solid or 2-segmented and no wider than funicle. .... *Bephratelloides*
— Antenna with five funicular segments and with 3-segmented club always differentiated from funicle. .... *Eurytoma*

Synopsis of the Costa Rican fauna

All three subfamilies of Eurytomidae are present in Costa Rica, and are represented by a total of 20 genera and perhaps between 100 and 200 species.

HEIMBRINAE[CR]

This subfamily consists of two genera—the monotypic *Symbra* from Argentina and *Heimbra* from North and South America (Stage & Snelling. 1986).

***Heimbra*[CR].** Six species, four neotropical, one in Western North America and *H. bicolor* from Mexico to Brazil. This species was recently collected on extrafloral nectaries in Atlantic forest in Costa Rica. Biology: unknown.

RILEYINAE

This subfamily is cosmopolitan in distribution but the majority of the species occur in the Neotropics.

***Calorileya*[CR].** Neotropics; two described species. *C. megastigma* is known from the Caribbean and Mexico (Subba Rao, 1978), and appears to be the species that is present in Costa Rica. Biology: unknown, but I have reared a species from leaf galls (gall-former unknown) on *Pithecellobium longifolium* (Leguminosae).

***Neorileya*.** Neotropics; four described species. Biology: endoparasitoids in eggs of Hemiptera, especially Pentatomidae, but also Coreidae and Reduviidae.

***Rileya*.** Primarily New World, with about 25 described species from the Neotropics, but with one species known from central Asia and another from Australia (Bouček, 1988). Keys to species are given in Gahan (1918) and Subba Rao (1978). Biology: in Costa Rica I have reared *Rileya* from leaf galls formed by Cecidomyiidae on a shrubby species of *Cordia* (Boraginaceae) at high altitudes (2000 m) and from cecidomyiid leaf galls on at least two different species of *Piper* (Piperaceae). In Honduras a species has been reared from *Asphondylia websteri* (Cecidomyiidae) on beans (specimens in Escuela Agrícola Panamericana).

EURYTOMINAE

This cosmopolitan subfamily is by far the largest in the family.

***Aranedra*.** Only two species are known, both of which were described from Costa Rica. Biology: reared from root galls on Araceae—*A. millsi* from *Philodendron* and *Dieffenbachia*, and *A. arae* from *Philodendron* (Burks, 1971). I have collected specimens from sea-level to 1000 metres, and have reared one specimen from a stem gall on *Piper* (presumably caused by Cecidomyiidae).

***Axima*.** New World. Four described species. Biology: the North American species is a parasitoid of bees of the genus *Ceratina* (Burks, 1966).

***Aximogastra*[CR].** The type-species was described from Brazil and subsequently another species was described from the Caribbean (De Santis, 1979). The genus seems doubtfully distinct from *Bephrata*. Several undescribed species from Costa Rica (0-1200 m) seem to belong here. Biology: The type-species was 'reared from orchids' in Brazil, and in Costa Rica one species has been reared from eggs of Tettigoniidae in twigs of citrus.

***Bephrata*.** Known only from the type species, *B. ruficollis*, which was described from Panama; it is present in Costa Rica, where it is quite common below 800 metres. Biology unknown.

***Bephratelloides*.** Neotropics; four recognized species, three of which occur in southern Central America. A key to species is given by Grissell and Schauff (1990). Biology: all are phytophagous in seeds of *Annona*.

***Bephratoides*[CR].** Known from the New World and Asia. *B. maculatus* is reported from Mexico (Burks *in* Krombein *et al.*, 1979) and Brazil (Subba Rao, 1978). A key to the five described New World species is given by Subba Rao (1978), although the specific name of one has recently been changed (Grissell & Schauff, 1990). Biology: species are parasitoids of wood-boring beetles (Buprestidae).

***Bruchodape*^CR^.** Known only from the type species, *B. ignota*, which was described from Brazil (Burks, 1971). One species (possibly the same) occurs in Costa Rica, in lowland wet forest. Biology: Burks suggested that the morphology indicated an association with Bruchidae, but evidence based on rearing is apparently absent.

***Bruchophagus*.** Cosmopolitan; undetermined number of species (because the limits of the genus are still in process of being defined; Bouček, 1988). Biology: species are phytophagous, especially in seeds of Leguminosae (Strong, 1962; Carrillo & Dickason, 1963; Batiste, 1967) but some Australian species are gall-formers on citrus (Noble, 1936—as *Eurytoma*).

***Chryseida*.** Neotropics, with three species known from Costa Rica. A key to species was provided by Burks (1956) and a new species from Costa Rica was subsequently described by Zerova (1981). Biology: parasitoids of Bruchidae.

***Conoaxima*.** South America to Costa Rica (and possibly into southern Mexico); two described species. Biology: ectoparasitoids of adult queen *Azteca* ants in stems of *Cecropia* (Brues, 1922; Wheeler, 1942) and *Cordia* (J. Longino, pers. comm.).

***Eurytoma*.** Cosmopolitan; the largest genus of Eurytomidae, with numerous undescribed species. Species-groups are discussed in Bouček (1988). Biology: extremely varied, including both phytophages and carnivores.

***Isosomodes*^CR^.** New World; three described species, two from Amazonian Brazil, and a third, *I. gigantea*, extending from the eastern United States into the Caribbean. Biology unknown.

***Prodecatoma*.** Some species are recorded from Asia and Africa, but the genus (if at all valid) is probably confined to the New World. The relationship of this genus with respect to *Eurytoma* requires further study. Biology: species have been reared from seeds and plant galls. I have seen neotropical species reared from fruits or seeds of Araceae (*Anthurium*), Myrtaceae (*Eugenia*, *Psidium*) and Rubiaceae (*Cephaelis*, *Palicourea*, *Psychotria*, *Randia*), and it is likely that many of these are phytophagous. On the other hand, a Costa Rican species occurring in seed galls of *Pithecellobium macradenia* (Leguminosae) caused by *Allorhogas* (Braconidae) is probably parasitic and/or an inquiline.

***Sycophila*.** Cosmopolitan. Biology: parasitoids/inquilines in galls, including galled fig ovaries. In Costa Rica I have reared *Sycophila* from figs, seeds of *Conostegia xalapensis* (Melastomataceae), a stem gall on *Psidium* sp. (Myrtaceae), and from seed galls on *Pithecellobium macradenia* caused by *Allorhogas* (Braconidae). I have reared several specimens from figs in Costa Rica which will key to the Old World genus, *Syceurytoma*, but Z. Bouček (pers. comm.) suggests these are probably better placed in *Sycophila*, at least for now.

***Systole*.** Holarctic and drier parts of the tropics; about 15 species (Bouček, 1988). Biology: species are phytophagous in seeds of Umbelliferae.

***Tetramesa*.** Cosmopolitan, with the majority of species known from North Temperate regions. Keys to the North American species are given in Phillips and Emery (1919) and Phillips (1936). Biology: species are gall-formers in stems and seeds of grasses (Phillips, 1920, 1936; Claridge, 1961).

11.10 LEUCOSPIDAE

Paul E. Hanson

Diagnosis. Among the largest Chalcidoidea, body length usually 8.0 to 12.0 mm; colour varying from black to black with yellow markings to orange-yellow. Antenna 13-segmented, *first flagellar segment large* (i.e. without anellus); antennal scrobe deep, carinate laterally; *posterior edge of gena rounded* (rather than carinate as in Chalcididae). Notauli reduced, tegula elongate. *Fore wing usually folded lengthwise* (as in

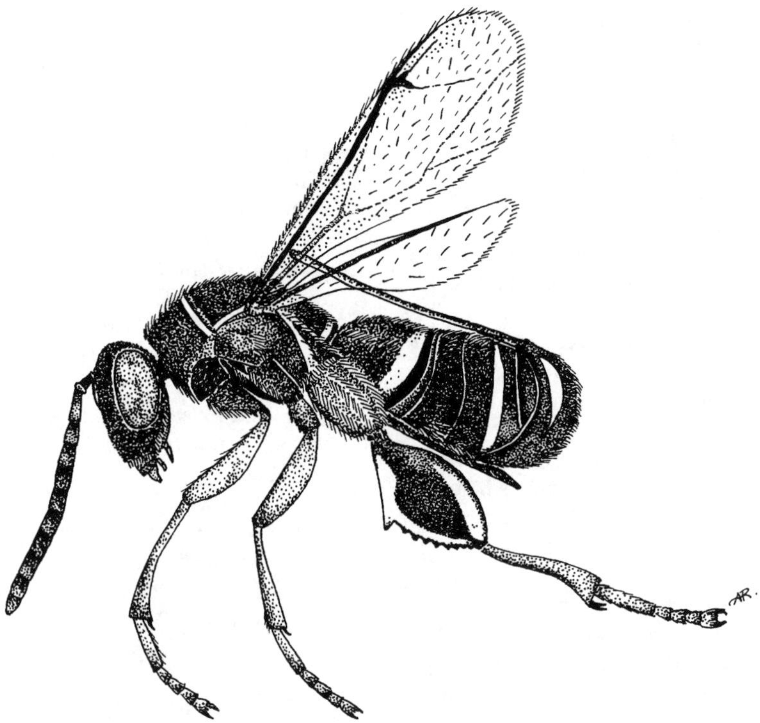

Fig. 11.132. *Leucospis* sp. (Leucospidae).

vespids), densely setose, with postmarginal vein many times longer than stigmal vein, which in turn is slightly longer than the marginal vein. *Hind femur enlarged and toothed; hind tibia strongly curved* (as in Chalcididae); tarsi 5-segmented. Metasoma of male with second tergite shorter than first, remaining tergites more or less fused into a carapace; females of the most common genus have the ovipositor curved upward and forward along the dorsal surface of the metasoma, where it often rests in a groove on the first tergite.

Classification and distribution. The Leucospidae, which is generally considered to be closely related to Chalcididae, is cosmopolitan in distribution but most species occur in warmer climates. The family comprises about 130 to 140 known species in four genera. Two genera are restricted to Africa, *Polistomorpha* is restricted to the Neotropics, while *Leucospis* is cosmopolitan (Bouček, 1974*a*).

Biology. Leucospids are ectoparasitoids (presumably idiobionts) of larvae or pupae of aculeate Hymenoptera—mostly solitary bees but also Sphecidae and eumenine Vespidae. However, host associations have been recorded for only about a quarter of the species worldwide. The only host records for species occurring in Central America are for *Leucospis affinis*, from megachiline bees (Bouček, 1974*a*), *L. klugii*, from *Xylocopa gualanensis* (Apidae: Anthophorinae) (Daly, 1976; Hurd, 1978*a*), and *Polistomorpha fasciata*, from the orchid bee, *Euglossa ignita* (Bouček, 1974*a*).

Oviposition occurs through the wall of the host nest and involves a rather elaborate process whereby the first and second metasomal tergites become widely separated dorsally. The internal part of the ovipositor is coiled back into the intersegmental membrane while the metasoma is bent downward behind the first tergite, thus bringing the distal part of the ovipositor into a perpendicular position (Clausen, 1940*b*). If the wall of the nest is very hard, drilling may require as much as three hours. Not until the host cavity has been penetrated, can the leucospid detect whether a suitable larva is present.

The egg is attached to the cell wall of the host cell, or to the host cocoon if one is present. Upon hatching the first instar larva does not begin feeding immediately, but rather searches out and destroys any competitors that may be present in the same cell. There are apparently three larval instars, and the last instar larva does not move about as much as the first. Developmental time from oviposition to emergence of the adult wasp varies from three to five weeks under optimal conditions (Clausen, 1940*b*).

Adult *Polistomorpha* species have an elongate face, with a well-developed, protractile glossa, which presumably allows them to take nectar from a greater range of flowers than is possible in most other Chalcidoidea. Adult leucospids are also noteworthy for their colour patterns, which appear to mimic polistine Vespidae. In Costa Rica *Leucospis egaia* and *L. latifrons* are black with yellow markings and resemble *Polybia occidentalis*, *L. klugii* has a dark body and reddish yellow wings with black apices similar to *Polybia rejecta*, *L. leucoteles* has a dark body and black wings with white apices similar to *Parachartergus apicalis* and *P. fraternus*, and *Polistomorpha fasciata* is orange-yellow and resembles *Agelaia myrmecophila*. In species with a wide geographic distribution mimicry can involve different models in different areas, resulting in considerable intraspecific variation in colour (Bouček, 1974*a*).

Identification. Keys to the world genera and species of Leucospidae are given in Bouček (1974*a*).

Key to New World genera of Leucospidae

— Maxillary palp 4-segmented, labial palp 3-segmented; apex of hind tibia often produced into a spine; ovipositor curved up and over the metasoma .... *Leucospis*

— Maxillary and labial palpi very reduced; apex of hind tibia perpendicular to longitudinal axis of tibia; ovipositor normal (confined to ventral surface of metasoma) .... *Polistomorpha*

Synopsis of the Costa Rican fauna

Two genera and about a dozen species of Leucospidae have been reported from Central America (Boucek, 1974*a*). In Costa Rica the family appears to be restricted to lower altitudes, and despite intensive collecting none have been found above 1500 metres.

Leucospis. There are over 100 species of *Leucospis* world-wide, at least 40 of which are restricted to the New World, and nine of these occur in Central America. The species known to occur in Costa Rica are *L. cayenensis*, *L. egaia*, *L. ignota*, *L. klugii*, *L. latifrons* and *L. leucotellus*. *L. affinis* is recorded from El Salvador and *L. rileyi* occurs in Honduras and El Salvador.

Polistomorpha. Of the seven recognized species of this neotropical genus, two or possibly three occur in Central America. Only *P. fasciata* is known from Costa Rica, but *P. atrata* is recorded from Panama.

11.11 MYMARIDAE

John T. Huber

Diagnosis. Mostly delicate, slender-bodied chalcidoids, 0.2 to 2.5 mm in length (average 0.5 to 1.0 mm); without metallic coloration; usually fully winged, but sometimes brachypterous or apterous. *Antennae usually as long as body or longer*, 8- to 13-segmented, that of female with distinct club (solid or 3-segmented), that of male filiform, *attached closer to vertex than mouth margin or about midway between mouth margin and vertex, toruli almost always closer to eyes than to each other. Head with dark bars of cuticle* (trabeculae) and associated sutures *on the vertex and frons arranged in an H-like pattern*. Scutellum divided into distinct anterior and posterior portions, marked by difference in sculpture. Fore wing with *marginal vein usually short* (rarely extending more than half way along wing), with stigmal vein very short, with postmarginal vein absent, usually with a distinct backwards projecting seta (hypochaeta) on the ventral surface of the wing blade under the marginal vein; *hind wing almost always petiolate* (stalked), wing membrane not extending to base of wing. Tarsi 4- or 5-

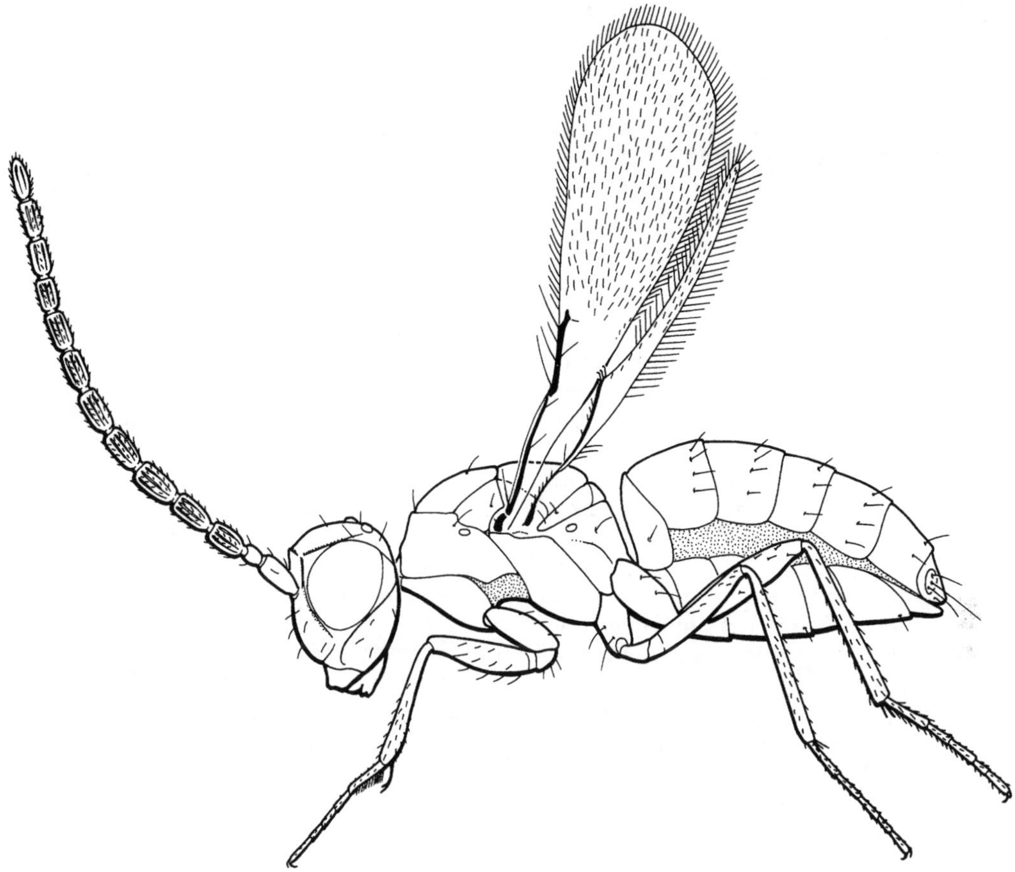

Fig. 11.133. Mymaridae.

segmented. Metasoma petiolate to sessile; ovipositor hidden to strongly exserted.

Classification and distribution. The family Mymaridae is cosmopolitan and contains about 1400 described species placed in almost 100 genera. The greatest diversity of genera occurs in Australia, New Zealand and South America, and the greatest number of species is probably in humid tropical forests. It is likely that most of the 54 genera described from the Western Hemisphere will eventually be found in Central America—at least 34 genera are known to be present.

The Mymaridae appears to be a monophyletic group based on three synapomorphies (Gibson, 1986*a*):

- fore wing with hypochaeta;
- head with frontal, median and supraorbital sutures and trabeculae;
- toruli distinctly closer to inner margin of eye than to each other.

In the same study, Gibson presented evidence for the monophyly of the Mymaridae + other Chalcidoidea (see introduction to Chapter 11). The family has been divided in different ways by different authors. Debauche (1948) used the number of tarsal segments to recognize Gonatocerinae (tarsi 5-segmented) and Mymarinae (tarsi 4-segmented), whereas Annecke and Doutt (1961) considered the structure of the metasoma to be more important and recognized Mymarinae (metasoma petiolate) and Alaptinae (metasoma sessile). In a phylogenetic analysis of the holarctic genera Schauff (1984) found greater support for the classification of Debauche, although the subfamily Gonatocerinae is paraphyletic. Previous classifications, however, need to be critically reviewed in the light of recent work (Noyes & Valentine, 1989) on the fauna from the Australian region, which appears to be the centre of diversity for the family.

Biology. The family Mymaridae contains among the smallest of all insects, some species of *Alaptus* being less than 0.2 mm long, which is smaller than some protozoans. Although they are abundant, the small size of most mymarids makes it difficult to study them alive. As a result there is a tremendous amount of new information to be discovered about any aspect of their biology. For example, there are few or no reliable host records for even some of the common genera such as *Camptoptera*. Life histories are almost completely unknown except for a few species of importance in biological control.

All species are presumed to be endoparasitic idiobionts of insect eggs, especially those laid in concealed situations such as within plant tissue or in soil (Huber, 1986). Species of the holarctic genus *Caraphractus* are known to live in ponds, where the adults use their wings as paddles to swim under water when searching for eggs of water beetles (Dytiscidae). Species of Mymaridae may be solitary or gregarious. *Caraphractus cinctus*, for example, lays two to three eggs per host with a single insertion of the ovipositor (Jackson, 1958), whereas almost 200 individuals of *Litus cynipseus* have been reared from a single egg of Staphylinidae (Viggiani, 1973). From the relatively few reliable host records available it appears that mymarids are not species specific but at most host genus specific, although some species are known to parasitize hosts in several families of a particular order. This lack of specificity can sometimes cause host induced morphological variation which may complicate species definition (Huber & Rajakulendran, 1988). In the case of *Alaptus*, species may be habitat specific, for example eggs on bark versus eggs on leaves (New, 1969). Microhabitat specificity probably occurs in other mymarid genera as well.

Host records exist for members of about one quarter of the world genera. About three quarters of the records are for the four largest and economically most important genera: *Polynema*, *Gonatocerus*, *Anaphes* and *Anagrus*. Almost half of the host records are auchenorrhynchous Homoptera. Hemiptera and Coleoptera are the next most important hosts. Other than Coleoptera, few holometabolous insects are parasitized. Some of the more frequently recorded hosts for genera occurring in Central America are given in Table 11.13. Huber (1986) gives exhaustive host lists, but cautions against the uncritical acceptance of old records.

Oviposition normally takes place before the host egg has undergone much development, and after oviposition further development of the host usually ceases. However, *Caraphractus cinctus* prefers newly laid eggs, but it is capable of developing in more advanced host embryos, in which case the female oviposits into the mid-gut of the embryo (Jackson, 1958). Another species capable of attacking well-

Host group	Mymarid parasitoid
ODONATA	
Coenagrionidae	*Anagrus*
ORTHOPTERA	
Acrididae	*Cleruchus*
Gryllidae	*Acmopolynema*
PSOCOPTERA	
various families	*Alaptus*
HEMIPTERA	
Miridae	*Anagrus, Erythmelus, Polynema*
Nabidae	*Polynema*
Tingidae	*Erythmelus*
HOMOPTERA	
Cercopidae	*Acmopolynema, ?Ooctonus*
Delphacidae	*Anagrus, Mymar*
Cicadellidae	*?Arescon, Anagrus, Erythmelus, Gonatocerus, ?Ooctonus, Polynema, ?Schizophragma, Stethynium*
Membracidae	*?Arescon, Gonatocerus, Polynema, Schizophragma*
COLEOPTERA	
Buprestidae	*Campoptera*
Chrysomelidae	*Anaphes*
Curculionidae	*Anaphes, Cleruchus*
Staphylinidae	*Litus*

Table 11.13. Probable hosts of mymarid genera present in Central America.

developed embryos is *Polynema striaticorne*, which is reported to produce three successive generations in a single brood of its membracid host (Balduf, 1928) and has been successfully introduced from North America into Europe where it controls its host, *Stictocephala bisonia* (Alma *et al.*, 1988).

The number of larval instars is often difficult to ascertain, but from two to four have been recorded (Balduf, 1928; Bakkendorf, 1934; Jackson, 1961; Sahad, 1982). The few species whose immature stages have been observed within their host egg have peculiar, mymariform first instar larvae. These larvae have the head drawn out into a conical process; the thorax + abdomen is six to eight segmented, bearing transverse rings of hairs and terminating in a long, curved process at the posterior end (see Chapter 4.4). The following instars are sac-like, without segments or appendages, but with large protruding mandibles (Clausen, 1940*b*). In a few cases (e.g. *Anagrus* spp.) the first instar is sac-like whereas the second instar (known as a histriobdellid) is cylindrical and divided into six segments, the first and last segments being the largest, and often bearing paired fleshy processes (Sahad, 1984). There does not appear to be a functional tracheal system or spiracles in any larval instar.

First instar larvae sometimes lack mandibles and appear to be incapable of attacking one another, even when there are more larvae present than can survive in one host egg (Jackson, 1958). In *Anagrus* the sac-like first instar remains relatively inactive within the host egg, whereas the mymariform first instar larva of other mymarids is quite active, using its spines and long tail to move about within the egg (Clausen, 1940*b*). In most mymarids later instars show gyrating lateral motion (e.g. Balduf, 1928), which is thought to prevent the heavier semi-solid contents of the host egg from settling to the bottom (Clausen, 1940*b*), though this requires confirmation. Pupation occurs within the host egg.

Adult mymarids are usually short-lived, pro-ovigenic, and have a relatively low reproductive capacity, total egg production probably being less than one hundred (Clausen, 1940*b*). Adult females apparently do not host-feed.

Economic importance. About ten species of mymarids, all but one in *Anaphes* or *Anagrus*, have been used in classical biological control attempts against insect pests such as Curculionidae, Chrysomelidae, Cicadellidae and Delphacidae (Huber, 1986). The best known and most successful example is that of *Anaphes nitens* against the eucalyptus snout beetle, *Gonipterus scutellatus* in southern Africa (DeBach, 1974). Other native Australian species of *Anaphes* are also being considered for release in southern Africa against *Gonipterus* (Huber & Prinsloo, 1990). *Anagrus osborni* was introduced from the Philippines to Hawaii to control the corn planthopper (*Peregrinus maidis*, Delphacidae), but information is lacking regarding the results. *Anagrus frequens*, which had been introduced for control of the sugarcane leafhopper (*Perkinsiella saccharicida*, Delphacidae), readily attacked *P. maidis* (Clausen, 1978). Because *P. maidis* is a pest in Central America these mymarids are potential candidates for importation.

In addition to classical biological control, probably much potential exists for using mymarids in inundative release programmes (e.g. Jones & Jackson, 1990) and in manipulating endemic species (see Chapter 3.2). For example, in Brazil *Acmopolynema hervali* has been used against cercopid pests of sugarcane and pasture grasses (Guagliumi, 1971). Endemic mymarids that attack pest species in the northern Neotropics are listed in Table 11.14.

Pest species	Mymarid	References
Antianthe expansa (Membracidae)	*Schizophragma latipennis*	Chaverri, 1954; Huber, 1987.
Empoasca kraemeri (Cicadellidae)	*Anagrus empoascae*	Wilde *et al.*, 1976; Gomez & Schoonhoven, 1977; Caballero & Andrews, 1985
Pycnoderes incurvus (Miridae)	*Anagrus yawi*	King & Saunders, 1984
Sogatodes oryzicola (Delphacidae)	*Anagrus* sp.	Goméz Sousa *et al.*, 1979

Table 11.14. Pest species present in the northern Neotropics and the endemic mymarids known to attack them.

Identification. Keys to genera are given by Annecke and Doutt (1961) for the world (very outdated), Yoshimoto (1990) for the Western Hemisphere, and Schauff (1984) for the Holarctic Region. No generic keys exist for any specific country in the Neotropical Region except for Cuba (Alayo & Hernández, 1978). De Santis (1979, 1981, 1989) catalogued the chalcidoids of America south of Mexico, including the Mymaridae. Huber (1986) reviewed the higher classification within the family, tabulated the genera with numbers of known species by region, listed published host records and use in biological control and provided a bibliography of much of the biological literature for the family.

Synopsis of the Costa Rican fauna

Mymarids are found in all terrestrial habitats and in Costa Rica they occur at all altitudes, although they appear to be most diverse in lowland and mid altitude rainforests. Although no aquatic mymarids have been found yet in Central America, it is possible that they occur here since at least one species of *Anagrus* is known to attack eggs of damselflies (Odonata: Zygoptera).

Following are the 34 genera of Mymaridae known or presumed to occur in southern Central America. Knowledge of the numbers of species within each genus, both at the world level and for Central America, is still very incomplete, although the largest genera, *Anaphes*, *Gonatocerus* and *Polynema*, each comprise over 200 nominal species world-wide. It is difficult to estimate the number of mymarid species present in Costa Rica, but a rough approximation is 250 species (mostly undescribed).

GONATOCERINAE

This subfamily includes eight genera that are known to occur in Costa Rica. It is a paraphyletic group that is characterized by the possession of 5-segmented tarsi.

Alaptus. Cosmopolitan; about 50 described species. Biology: parasitoids in eggs of Psocoptera. Older records of species of this genus attacking coccoids are probably incorrect (Huber, 1986).

Arescon[CR]. Cosmopolitan (except North America); about 20 described species. Biology: isolated records from eggs of Cicadellidae and Membracidae (Huber, 1986).

Camptoptera. Cosmopolitan; about 60 described species. Biology: parasitoids in eggs of Buprestidae (Schauff, 1984). Species have also been recorded from eggs of Thysanoptera, Homoptera and Lepidoptera (Noyes & Valentine, 1989), but many of these records need critical confirmation.

Dicopus[CR]. Cosmopolitan; about ten described species. Biology: hosts unknown for certain, although there are old records of a species parasitizing Diaspididae (see Huber, 1986).

Gonatocerus. Cosmopolitan; about 250 described species. Some of the most commonly collected mymarids in the Western Hemisphere. Many are quite large (average about 1.0 to 1.5 mm in length). Biology: most probably parasitize eggs of Cicadellidae, but members of one species-group apparently attack eggs of Membracidae (Huber, 1988).

Hadromymar. Costa Rica south to Ecuador. Biology unknown.

Litus[CR]. Cosmopolitan; about 15 described species. Biology: parasitoids in eggs of Staphylinidae.

Myrmecomymar[CR]. Canada south to Argentina. Females are apterous. Biology unknown.

MYMARINAE

This subfamily includes 26 genera that are known or thought very likely to occur in Costa Rica. Species of the Mymarinae are characterized by possession of 4-segmented tarsi.

Acmopolynema. Cosmopolitan; about 20 described species. Schauff (1981) reviewed the nearctic species and Fidalgo (1989) revised the neotropical species. Biology: parasitoids in eggs of Gryllidae and Cercopidae.

Anagroidea[CR]. Cosmopolitan; very few described species. Biology unknown.

Anagrus**.** Cosmopolitan; about 80 described species. Biology: parasitoids in eggs of Delphacidae, Cicadellidae, Miridae and, less commonly, Odonata.

Anaphes**^{CR}.** Cosmopolitan; over 200 described species. Huber (1992) reviewed the genus. Biology: parasitoids in eggs of Curculionidae and Chrysomelidae.

Australomymar**^{CR}.** Primarily Southern Hemisphere, especially Australia and South America, but also southeast Asia (Noyes & Valentine, 1989). In the New World this genus extends from Chile north to Costa Rica. Biology unknown.

Bruchomymar**^{CR}.** Central America south to Argentina. Biology unknown.

Caenomymar**.** Costa Rica south to Ecuador. Biology unknown.

Cleruchus**^{CR}.** Cosmopolitan; about 14 described species. Biology: one species has been reared from the eggs of Acrididae, and one from Curculionidae (Schauff, 1987). In New Zealand species of this genus seem to be associated with leaf litter and bracket fungi (Noyes and Valentine, 1989).

Cnecomymar**^{CR}.** New World. About six species have been described from the Neotropical Region. Biology unknown.

Entrichopteris**.** Costa Rica to Venezuela. Biology unknown.

Erdosiella**^{EX}.** Panama south to Argentina. Biology unknown.

Erythmelus**.** Cosmopolitan; about 50 described species. Biology: parasitoids in eggs of Tingidae, Miridae and Cicadellidae.

Krokella**.** Florida south to Brazil. Two described species, both of which occur in Costa Rica (Huber, 1993). Biology unknown.

Mymar**^{CR}.** Cosmopolitan; six described species. Biology: parasitoids in eggs of Delphacidae.

Neomymar**.** Canada south to Argentina. Biology unknown.

Omyomymar**^{CR}.** Cosmopolitan (except for Europe and Australia); four described species (Schauff, 1983). Biology unknown.

Ooctonus**^{CR}.** Primarily North Temperate zone; about 60 described species. One species extends south to Costa Rica. Biology: confirmation is required, but there are records from the eggs of cercopids and cicadellids.

Platyfrons**^{CR}.** Costa Rica south to Brazil. In Colombia it is only known from sites above 2000 metres. Biology: unknown.

Polynema**.** Cosmopolitan; over 250 described species. Biology: mostly parasitoids in eggs of Cicadellidae, Membracidae and Miridae.

Ptilomymar**^{CR}.** Cosmopolitan; about four described species. Biology unknown.

Restisoma**.** Costa Rica to Colombia. Biology unknown.

Schizophragma**.** New World, mostly in tropical latitudes; seven described species (Huber, 1987). Biology: parasitoids in eggs of Membracidae and possibly Cicadellidae.

Stephanodes**^{CR}.** Cosmopolitan; seven described species. Biology: unknown.

Stethynium**^{CR}.** Primarily Australia, but a few species occur in other parts of the world. This is the first record of the genus from tropical America. Huber (1987) revised some of the non-Australian species. Biology: parasitoids in eggs of Cicadellidae.

Tanyostethium**^{EX}.** Panama south to Brazil. Biology unknown.

Tetrapolynema**^{CR}.** Mexico south to Colombia. Biology unknown.

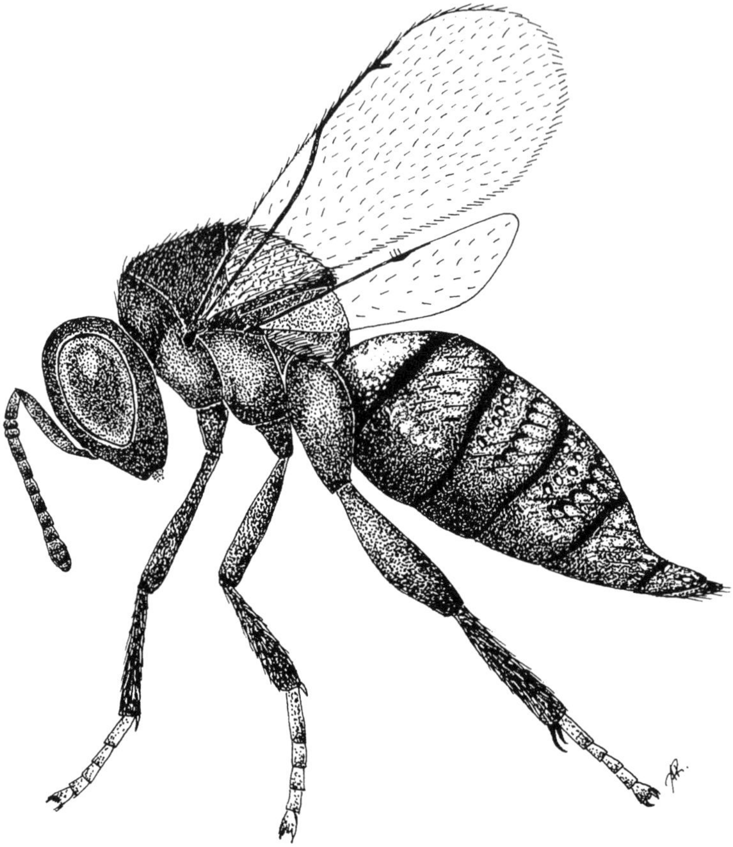

Fig. 11.134. *Ormyrus venustus* (Ormyridae).

11.12 ORMYRIDAE

Paul E. Hanson

Diagnosis. Body length 1.0 to 7.0 mm; mostly metallic blue or green. Antenna 13-segmented, ratio of anelli to funicular segments varying from 1:7, 2:6 (most common), or 3:5 (Old World only); back of head with occipital carina. Mesosoma and propodeum weakly sculptured. Fore wing with stigmal and postmarginal veins very short. Hind coxa greatly enlarged, subtriangular in cross section; *hind tibia with two stout, curved spurs at apex*, longest spur greater than half the length of first tarsal segment; tarsi 5-segmented. *Metasoma very sclerotized and heavily sculptured, with transverse rows of large foveae (pits) on the basal parts of the middle tergites*, the foveate basal part often separated from apical part of tergite by a crenulate border.

Classification and distribution. The Ormyridae share some characteristics (e.g. occipital carina, wing venation, enlarged hind coxa) with the Torymidae, but others (e.g. large hind tibial spurs) with an unplaced genus of Pteromalidae from South America. As a result the Ormyridae has been placed in each of these families in the past, but until its true relationships are better understood it seems best to maintain it as a separate family. As such the family Ormyridae comprises only two genera: the monotypic *Ormyrulus* from India (Bouček, 1986) and the cosmopolitan *Ormyrus*. However, there appear to be several undescribed genera present in southeast Asia. The genus *Ormyrus* comprises about 60 described species worldwide; 16 species are known from North America (Hanson, 1992), two of which extend south to Costa Rica.

Biology. Species of Ormyridae are parasitoids, presumably ectoparasitic idiobionts, of insects occurring in plant galls. Although some Old World species are associated with galls formed by Diptera, New World species appear to be restricted to galls formed by Hymenoptera, mostly those of Cynipidae, but also Pteromalidae (Hanson, 1992). In Costa Rica Ormyridae are known only from cynipid galls on oaks (Fagaceae: *Quercus*). In the Palaeotropics ormyrids parasitize Chalcidoidea occurring in figs (Bouček *et al.*, 1981), but thus far the family has not been found in figs of the Neotropics. Within a plant gall the ormyrid larva appears to feed either as a primary parasitoid or as a facultative hyperparasitoid.

Identification. The recent revision of the nearctic species of *Ormyrus* (Hanson, 1992) includes the two species known from Costa Rica, *O. unifasciatipennis* and *O. venustus*. The former usually has a black spot on the fore wing below the stigmal vein (though this may sometimes be very weakly pigmented), whereas the latter species entirely lacks this black spot.

Synopsis of the Costa Rican fauna

Two species of *Ormyrus* have been collected in Costa Rica (see above), only at higher altitudes (above 1500 m) where oaks are present.

11.13 PERILAMPIDAE

D. Christopher Darling

Diagnosis. 1.0 to 12.0 mm in length, metallic to black coloured. Antenna with 13 segments, including a single anellus; antenna sexually dimorphic, with the scape of the male either expanded and/or with punctures or pores on the anterior surface. Clypeus differentiated from face by sutures, anterior margin entire, without teeth or tubercles; mandibles robust, with two or three large teeth. *Mesosoma compact, usually short and high in lateral view, with pronotum visible in dorsal view, not reduced and ventral to mesoscutum* (cf. Eucharitidae); notauli complete. Tarsi 5-segmented. *Metasoma usually high and triangular in profile, with tergites II and III similar in size and fused dorsad and enclosing following segments*, tergite II with large laterotergites.

Classification and distribution. The Perilampidae is a small family comprising, worldwide, about 200 species in 12 genera and two subfamilies: Chrysolampinae and Perilampinae (Bouček, 1988). Both subfamilies have at times been placed in the Pteromalidae (e.g. Burks *in* Krombein *et al.*, 1979). There is considerable confusion over the limits of the family Perilampidae and the relationships of the included taxa with Eucharitidae (see Chapter 11.6). Most of this uncertainty revolves around the affinities of the Chrysolampinae (Heraty & Darling, 1984).

Biology. All perilampids are parasitoids of other insects and the life history is similar in many respects to that of Eucharitidae (Heraty & Darling, 1984; see Chapter 11.6). Eggs are laid away from the host, on leaves or flowers or in preformed cavities in plant tissue. The first instar larva is a sclerotized, active PLANIDIUM, which locates and attaches itself to the host. The existence of phoretic transport (cf Eucharitidae) has not been documented, but may occur in species that parasitize aculeate Hymenoptera. Once on a host a planidium searches out and kills conspecific competitors. Only a single individual develops to maturity on one host. Development is hypermetamorphic and the third instar larva is hymenopteriform.

In contrast to the Eucharitidae, the host associations of perilampids are extremely diverse. At least eight orders of insects serve as hosts, and both primary and secondary parasitism (hyperparasitism) occurs. As far as is known, the Chrysolampinae are primary parasitoids of Coleoptera that feed in enclosed situations on flowering plants. Hosts include Curculionidae in seed pods of legumes (*Chrysolampus*, *Chrysomalla*) and Nitidulidae in flower heads of mints (*Chrysolampus*). Eggs of *Chrysolampus* spp. in western North America are deposited in the seed pods of legumes and the planidia attack tychiine weevils (Darling & Miller, 1991).

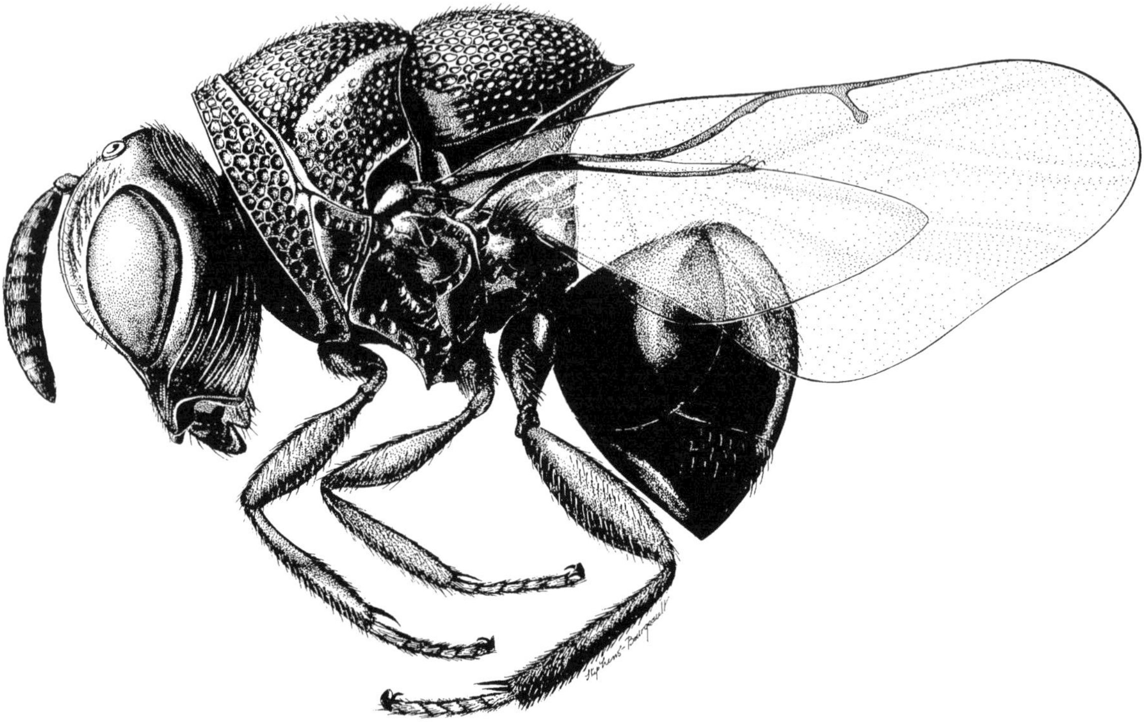

Fig. 11.135. *Perilampus* regalis (Perilampidae).

In the Perilampinae details of the life history and immature stages are known only for the genus *Perilampus* (e.g. Smith, 1912). Host associations are extremely diverse in *Perilampus* but are more restricted within species groups. Some are primary parasitoids of xylophagous Coleoptera (Anobiidae), stem-nesting wasps (Sphecidae), lacewings (Chrysopidae), or sawflies (Diprionidae, Tenthredinidae), while others are secondary parasitoids of Lepidoptera, Rhaphidioptera, Coleoptera or Orthoptera, via ichneumonids, braconids or tachinids (Diptera). There is also a species in Peru that is a hyperparasitoid in the silken webs of Embiidina, developing as a parasitoid of Sclerogibbidae (Burks, 1969). In the '*Perilampus hyalinus* complex', morphologically indistinguishable specimens can be reared as either primary parasitoids of sawflies (e.g. *Neodiprion excitans* in Belize), or as hyperparasites of Lepidoptera. This species-group is well represented in Costa Rica, as it is throughout the New World.

Among other Perilampinae, *Steffanolampus* and *Burksilampus* are primary parasitoids of xylophagous Anobiidae (Bouček, 1978), whereas *Euperilampus* species are hyperparasitoids via Ichneumonidae (Bouček, 1978*a*; Darling, 1983). In the Old World tropics *Monacon* species attack ambrosia beetles (Platypodidae) (Bouček, 1980) and *Krombeinius* species parasitize aculeate Hymenoptera (Darling, 1988).

Identification. Although only two genera of Perilampidae are definitely known to occur in Costa Rica, two other genera (*Burksilampus* and *Euperilampus*) occur in the region and may possibly be found in Costa Rica. These genera are therefore included in the following key.

Key to the genera of Perilampidae of Central America

1 Prepectus closely associated and usually fused to lateral panel of pronotum (Figs 11.136,

11.137); fore wing with marginal vein shorter, not more than 3.0 times as long as stigmal vein (Figs 11.138, 11.139); mandibles asymmetric, right mandible with three teeth, left with only two teeth. ....(Perilampinae).........................2

— Prepectus not closely associated and on the same plane as the lateral panel of the pronotum (Fig. 11.140), similar to Pteromalidae; fore wing with long marginal vein, 3.5 to 6.0 times as long as stigmal vein (Fig. 11.140); mandibles bidentate symmetrical. ....(Chrysolampinae).......... *Chrysolampus*

2 Prepectus broad and triangular, about as wide as adjacent pronotal panel (Fig. 11.136); marginal vein as long as or longer than postmarginal vein (Fig. 11.138). 3

— Prepectus narrow, less than half as wide as adjacent pronotal panel (Fig. 11.137); marginal vein shorter than postmarginal vein (Fig. 11.139). .... *Euperilampus*[EX]

3 Petiole elongate, length greater than width, sculpture uniformly alveolate or coriaceous (Fig. 11.142); malar sulcus absent (Fig. 11.144). .... *Burksilampus*[EX]

— Petiole transverse (Fig. 11.141), or if elongate with distinct rugae on dorsal surface (Fig. 11.141); malar sulcus present (Fig. 11.143) or malar region with oblique costae (Fig. 11.145)... *Perilampus*

Synopsis of the Costa Rican fauna

Both subfamilies of Perilampidae occur in Costa Rica. There are two, possibly three or four, genera present, representing 20 to 30 species. The vast majority of perilampids encountered in Costa Rica (and elsewhere) are species of *Perilampus*.

CHRYSOLAMPINAE

There are two genera of Chrysolampinae recorded in the New World, but only one of these occurs in Costa Rica. The other genus, *Chrysomalla*, appears to be restricted to the arid steppe and desert regions of western North America and central Europe and Asia (Darling, 1986), and is not expected to occur in Costa Rica.

Chrysolampus. Cosmopolitan. Biology: primary parasitoids of Coleoptera (e.g. Curculionidae, Nitidulidae) concealed in plant tissue.

PERILAMPINAE

There are four genera of Perilampinae recorded from the New World, but only *Perilampus* is known with certainty to occur in Costa Rica. *Steffanolampus* is a monotypic genus known from the Holarctic region and is not expected to occur in Costa Rica.

***Burksilampus*[EX].** This genus was described to accommodate a petiolate species of perilampine from North America (Bouček, 1978). Closely related species have been collected in Mexico (Jalisco, Veracruz) and on the island of Dominica in the West Indies. Somewhat similar species of petiolate perilampines are known from Colombia and a single species that may be referable to this genus occurs in Costa Rica. Biology: primary parasitoids of Anobiidae in wood.

***Euperilampus*[EX].** Widely distributed, though infrequently collected, in North and South America, but not yet known from Central America (Darling, 1983); species are recorded from southern Mexico and from Colombia. Biology: hyperparasitoids, attacking Ichneumonidae.

Perilampus. Cosmopolitan. Based on the material available for study, there are at least 20 species present in Costa Rica, which range in length from 1.0 to 12.0 mm and vary from brilliant metallic to black. Several species-groups have been recognized in the genus (Smulyan, 1936; Bouček, 1972*a*; Darling, 1983) and all of the New World species-groups (Smulyan, 1936) are represented in Costa Rica, as well as a striking new species that cannot be referred to any recognized species-group. There are no keys that will allow the identification of Costa Rican species. Biology: primary parasitoids or hyperparasitoids of a wide array of insects (see above); there are no host records from Central America.

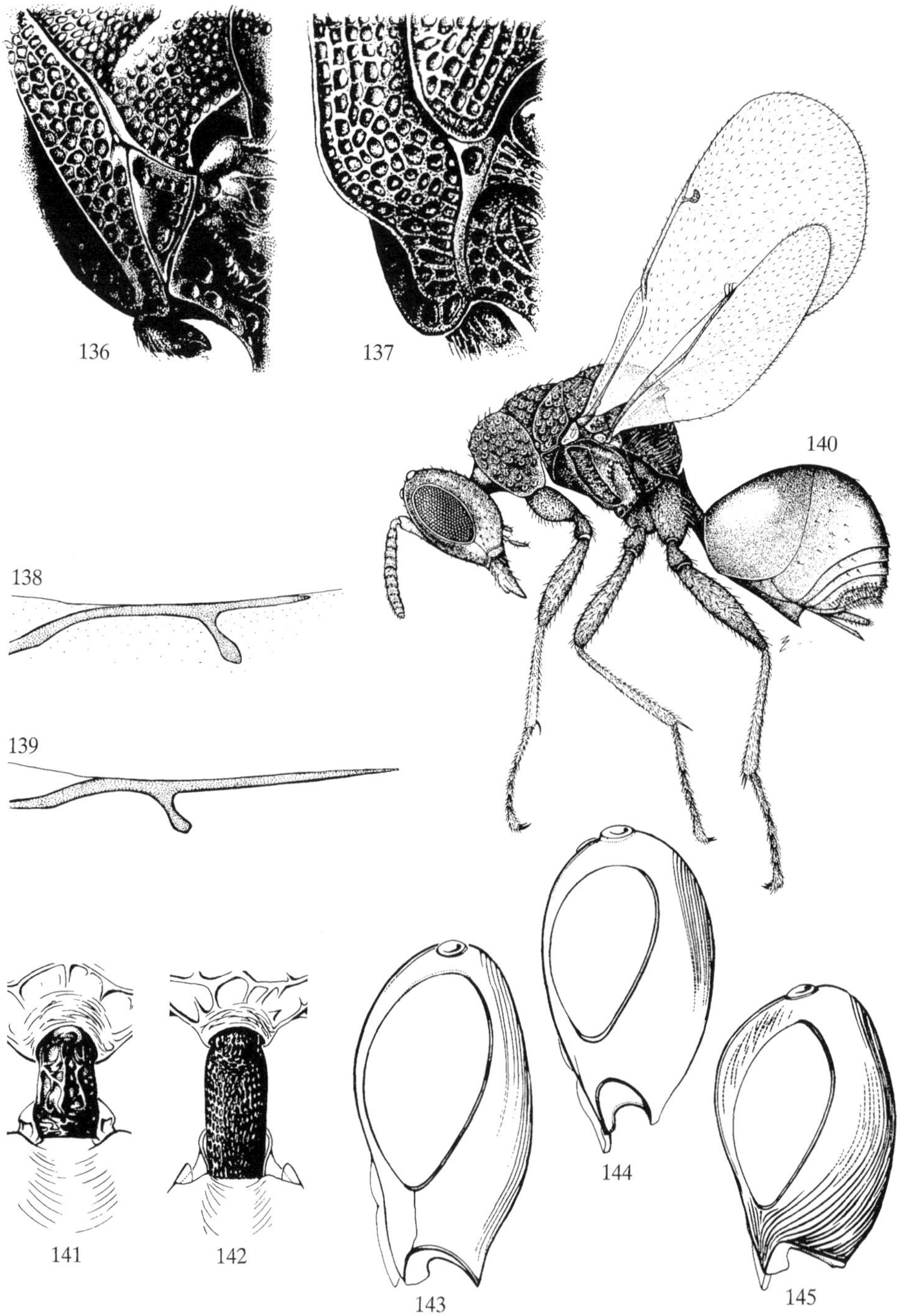

Figs 11.136–11.145. Perilampidae. Figs 11.136–11.137. Lateral view of pronotum and prepectus; 11.136, *Perilampus regalis*; 11.137, *Euperilampus triangularis*. Figs 11.138–11.139. Fore wing venation; 11.138, *Perilampus regalis*; 11.139, *Euperilampus triangularis*. Fig. 11.140. *Chrysolampus schwarzi*, Female. Figs 11.141–11.142. Petiole; 11.141, *Perilampus stygius*; 11.142, *Burksilampus anobii*. Figs 11.143–11.145. Head, lateral; 11.143, *Perilampus stygius*; 11.144, *Burksilampus* anobii; 11.145, *Perilampus regalis*.

11.14 PTEROMALIDAE

Paul E. Hanson

Diagnosis. Usually 1.0 to 7.0 mm in length, but in one genus up to 30.0 mm; colour varying from bright green or blue to black or yellow. Antenna 8- to 13-segmented (including 1–3 anelli and 5–7 *funicle segments*), antennal attachment varies from mouth margin to more than mid-way between mouth and anterior ocellus. Pronotum varies from very short to subrectangular, mesoscutum with or without notauli. Wings usually fully developed (in Central America brachypterous forms are known only in Diparinae), fore wing usually with marginal vein several times longer than broad, with postmarginal and stigmal veins usually well-developed. Tarsi 5-segmented. Metasoma subpetiolate to distinctly petiolate; ovipositor varying from completely hidden to well exserted. In general the family is *difficult to characterize except by absence of features defining other chalcidoid families*.

Classification and distribution. The Pteromalidae is cosmopolitan in distribution and is one of the largest families of Chalcidoidea, containing over 3000 recognized species in nearly 600 genera. It is generally agreed that this family is an artificial assemblage of groups that do not conveniently fit elsewhere in the Chalcidoidea. The number of subfamilies of Pteromalidae and their definitions have changed frequently. Graham (1969) recognized 15 subfamilies in his treatment of the northwestern European fauna while Bouček (1988) recognized 28 subfamilies in the Australasian region, about ten of which were newly recognized. Here I am following the classification of Bouček (1988), except that the Miscogastrini is placed as a tribe in the Pteromalinae rather than as a separate subfamily. Accordingly, 15 subfamilies are currently known from Costa Rica.

Biology. The family Pteromalidae is biologically very diverse. In North America species of the genus *Hemadas* cause galls on *Vaccinium* (Ericaceae) (McAlister & Anderson, 1932) and species of at least three genera of Ormocerinae are gall-formers on *Acacia* (Leguminosae), *Eucalyptus* (Myrtaceae) and other plants in Australia (Noble, 1938, 1940, 1941; Bouček, 1988). The vast majority of pteromalids, however, are carnivores. The modes of carnivory encountered in the family are very diverse, encompassing idiobiosis and koinobiosis, ectoparasitism and endoparasitism, solitary and gregarious parasitoids, and primary and secondary parasitoids. The hosts of Pteromalidae include spiders and eleven orders of insects (Burks *in* Krombein *et al.*, 1979; Dzhanokmen, 1991): Blattaria, Dermaptera, Hemiptera, Homoptera, Neuroptera, Coleoptera, Strepsiptera, Lepidoptera, Diptera, Siphonaptera and Hymenoptera. They attack eggs, larvae, pupae or, rarely, adults (e.g. the North American *Tomicobia* on adult Scolytidae and Curculionidae).

Idiobionts. The majority of pteromalids are ectoparasitic idiobionts of larvae or pupae of holometabolous insects, especially Coleoptera and Diptera living in concealed situations (in plant tissue, excrement, etc.). Species associated with galls may develop as inquilines, feeding on gall tissue, and/or as ectoparasitoids of larvae, pupae, or even adults of the gall-forming insect, or of other insects living in the gall (Askew, 1961*a*). Many pteromalid genera (e.g. *Spalangia, Cyrtogaster, Callitula, Muscidifurax, Nasonia, Pachycrepoideus, Sphegigaster*) comprise species that are parasitoids of dipterous pupae and develop concealed within the puparium. In Hymenoptera in general, ectoparasitoids of exposed pupae are rare, but in Pteromalidae, *Schizonotus* parasitizes exposed chrysomelid pupae (Dowden, 1939) and *Metastenus* (= *Scymnophagus*) attacks exposed coccinellid pupae (Clausen, 1940*b*). In both cases, however, the parasitoid larva is concealed beneath the pupa. One of the few truly endoparasitic idiobionts is *Pteromalus puparum* which develops in lepidopterous pupae that occur in exposed situations.

Because they consume more than a single host individual, some essentially idiobiotic pteromalids may be said to develop as predators. For example, the larvae of some species of *Systasis* have been observed to feed on a succession of small cecidomyiid larvae (Parnell, 1963). Eunotinae feed as predators on the eggs of Coccoidea and the same species can sometimes develop as an ectoparasitoid or hyperparasitoid on the adult female coccid if eggs are not available (Smith & Compere, 1928). *Arachnopteromalus* develops as a predator of eggs in uloborid spider egg sacs (Gordh, 1976*b*, 1983). True egg parasitoids appear to be rare in Pteromalidae, one of the few examples being *Erixestus* in chrysomelid eggs (Daviault, 1941; Grissell & De Santis, 1987).

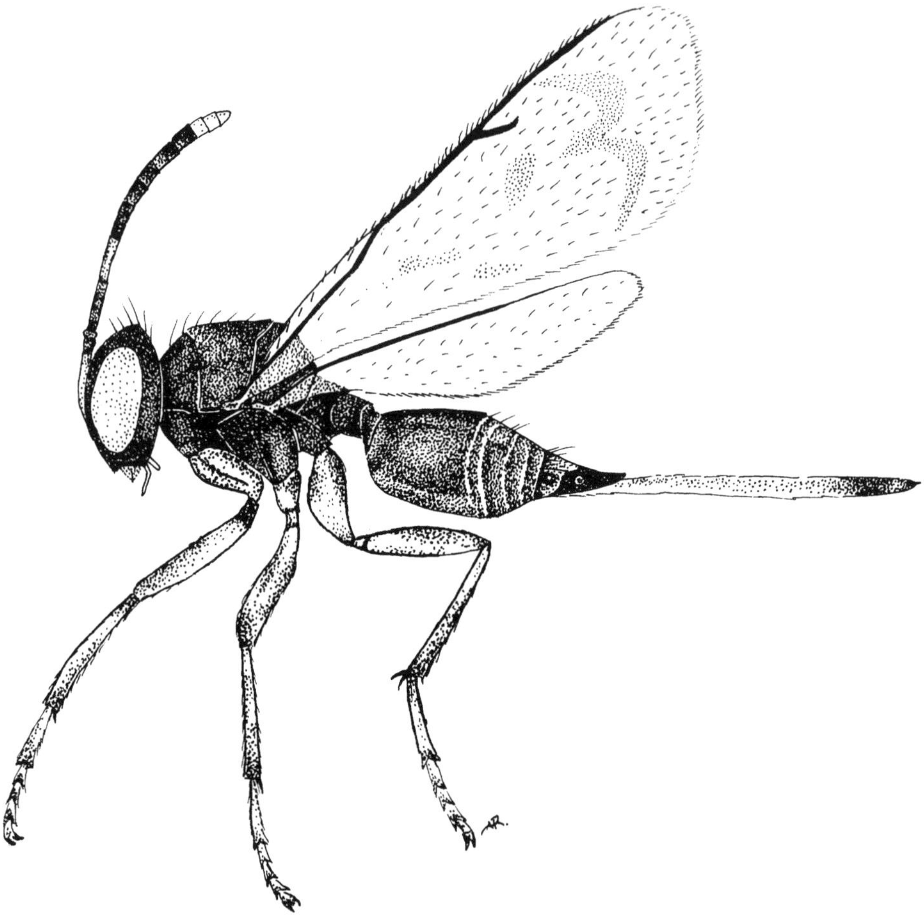

Fig. 11.146. *Lalaps* sp. (Pteromalidae).

Most species of Pteromalidae are probably solitary parasitoids, although several are known to be gregarious. Most of the gregarious species are pupal parasitoids, for example, *Pteromalus puparum*, *Metastenus*, *Nasonia* and *Schizonotus*. While several pteromalids are facultative hyperparasitoids a few are obligate hyperparasitoids. Species of *Asaphes* are, for example, obligate hyperparasitoids of aphids, attacking aphidiine braconids (Sullivan, 1972; Levine & Sullivan, 1983). *Catolaccus aeneoviridis* is another widespread obligate hyperparasitoid, attacking ichneumonoids and bethylids (Burks *in* Krombein *et al.*, 1979). *Toxeumella albipes* is a hyperparasitoid of lepidopterous larvae via microgastrine braconids (pers. obs.), and may be an obligate hyperparasitoid.

Koinobionts. Presently koinobiosis is known from only two subfamilies of Pteromalidae, the Pireninae and Pteromalinae, and all are endoparasitoids of Diptera. In Pireninae, two species of *Macroglenes* are known to be koinobionts of Cecidomyiidae, ovipositing in either the host egg or larva, and emerging from the larva (Kutter, 1934; Doane *et al.*, 1989). Other genera of Pireninae that parasitize Cecidomyiidae, such as *Gastrancistrus*, may also be koinobionts (e.g. Marchal, 1907; Askew & Ruse, 1974). The first instar larvae of these species have large mandibles, which are presumably used in fighting, as in endoparasitic koinobionts in other families (see Chapter 2.5).

In the subfamily Pteromalinae koinobionts occur in both the Miscogastrini and the Pteromalini. The best documented cases of koinobiosis are in the miscogastrine genera *Miscogaster* (Parker & Thompson, 1925) and *Halticoptera* (e.g. Takada & Kamijo, 1979). One European species of *Halticoptera* oviposits into half-grown agromyzid larvae, but the egg does not hatch until the host forms its puparium (Askew, 1968). In the tribe Pteromalini a species of *Stenomalina* is known to develop as a koinobiont in the larvae of chloropids (Kearns, 1931).

Host-feeding. Most pteromalids are synovigenic idiobionts, so it is perhaps not surprising that host-feeding has been recorded in numerous species (Jervis & Kidd, 1986). Many construct a feeding tube, and in most cases the female appears to feed and oviposit in the same host individual. For example, species of *Asaphes*, which attack aphidiine braconid larvae inside aphid mummies, inject a venom that causes immediate paralysis. Before laying an egg the female generally feeds on the host via a feeding tube from the aphidiine body to the surface of the aphid mummy (Keller & Sullivan, 1976). She feeds for about five minutes, then re-inserts her ovipositor, thus breaking the feeding tube, and lays an egg. On the other hand a female *Metastenus* may spend hours feeding on a chrysomelid pupa, after which she thrusts her ovipositor completely through the pupa and deposits an egg on the ventral surface. The original puncture may be used for repeated feeding and oviposition, resulting in a cluster of four to six eggs (Clausen, 1940*b*; Domenichini, 1957; Ricci & De Santis, 1983). In a few cases (*Muscidifurax* and *Spalangia*) host-feeding may be destructive, thus excluding the possibility of using the same host for oviposition (Legner & Gerling, 1967; Podoler & Mendel, 1977).

Immature stages. The majority of pteromalid eggs are ovate or ellipsoidal in form, often slightly curved, and without an anterior stalk or pedicel, except in *Scutellista*. Most pteromalids oviposit externally on the host, and the eggs of most of these species have a sculptured surface (with minute spicules, papillae, or granules). Eggs that are deposited internally are often smooth (Clausen, 1940*b*).

The first instar larvae of most pteromalids are hymenopteriform, with 13 body segments. The integument is often quite bare although a variable number of setae may be present (Clausen, 1940*b*). The first instar larva of the endoparasitic, koinobiotic species of Pireninae are unusual in having a large mandibulate head, and in lacking distinct body segmentation and setae (Marchal, 1907; Kutter, 1934; Doane *et al.*, 1989). There are often four pairs of spiracles, situated on the second thoracic and on the first three abdominal segments, although an open tracheal system is absent in endoparasitic species—Pireninae, *Pteromalus puparum* and *Stenomalina* (Kearns, 1931).

In most species of Pteromalidae that have been examined there appear to be five larval instars, although there are only four in *Dibrachys cavus* and *Pteromalus cerealellae*, and only three in *Macroglenes* (two feeding instars followed by a third non-feeding instar; Doane *et al.*, 1989). Later instars of most ectoparasitic pteromalids have nine pairs of spiracles, on the last two thoracic segments and on the first seven abdominal segments. In the endoparasitic koinobiont, *Stenomalina*, the first and second instars have an unusual tracheal system that has an additional pair of longitudinal trunks. Spiracles are not present until the fourth instar, a stage that is also characterized by sclerotized 'horns' on the head, which may possibly be used in breaking through the host's cuticle (Kearns, 1931).

Mating behaviour. Assem (1986) has noted a possible evolutionary sequence in courtship positions in Pteromalidae. In subfamilies that are presumed to be primitive (e.g. Spalangiinae and Cerocephalinae) the male courts the female from the posterior, that is, from the copulatory position. In more advanced subfamilies (e.g. Pteromalinae) the male is usually located more anteriorly on the female, stroking her head with his front tarsi and possibly liberating pheromones from his mandibular gland, and producing auditory stimuli via thoracic vibrations.

Economic importance. Among hymenopterous parasitoids used in classical biological control Pteromalidae rank sixth in importance (Greathead,

1986). The principal contribution of pteromalids is in the control of muscid flies, especially the house fly (*Musca domestica*) and the stable fly (*Stomoxys calcitrans*) (e.g. Cabrales *et al.*, 1985). Species of *Spalangia*, *Muscidifurax*, and *Pachycrepoideus* have been widely used and various combinations have been effective in the United States (Morgan *et al.*, 1981; Rutz & Patterson, 1990; Petersen *et al.*, 1992). Both species of flies are major pests in Central America and the use of these pteromalids offers a potential means of controlling them. Presently in Costa Rica *Pachycrepoideus vindemmiae* is being tested for its potential to control the Mediterranean fruit fly (*Ceratitis capitata*, Tephritidae). Other parasitoids of pest flies in the region include indigenous species of *Halticoptera* that attack *Liriomyza* (Agromyzidae).

Certain pteromalid parasitoids of non-dipterous hosts may also be important in biological control. For example, three cosmopolitan species are frequently found attacking beetle pests of stored grain: *Theocolax elegans* (Williams & Floyd, 1971), *Anisopteromalus calandrae* (Chatterji, 1955), and *Lariophagus distinguendus* (Kashef, 1956). Other important beetle parasitoids are some species of *Catolaccus* which attack the boll weevil, *Anthonomus grandis*. *Catolaccus grandis*, for example, has been recorded parasitizing boll weevils throughout Mesoamerica, including Costa Rica (Cross & Mitchell, 1969) and this species has been introduced into the United States where it is showing great possibilities as a biological control agent (Morales-Ramos & Cate, 1992). Some Eunotinae have been widely introduced for biological control of scale insects but with only limited success (Clausen, 1978).

A few pteromalid species may be adversarial to human interests, notably those that attack beneficial insects. For example, in Central America *Toxeumella albipes* parasitizes microgastrine braconids that help control such lepidopterous pests as brassoline nymphalids (*Caligo* and *Opsiphanes*) and limacodids (*Sibine*) in banana and oil palm plantations. Species of *Asaphes* are hyperparasitoids of aphids, and species of *Pachyneuron* attack not only aphid parasitoids but also syrphid predators of aphids. Species of *Eurydinoteloides* likewise attack syrphids. *Catolaccus aeneoviridis* is a widespread obligate hyperparasitoid that attacks economically important ichneumonoids in the New World (Burks *in* Krombein *et al.*, 1979), and finally, *Pteromalus puparum* is occasionally a pest on butterfly farms in Costa Rica.

Identification. Keys to the genera of Pteromalidae are available for northwestern Europe (Graham, 1969), the western Palaearctic (Bouček & Rasplus, 1991), Australasia (Bouček, 1988), and will soon be available for North America north of Mexico (Bouček & Heydon, in prep.). The latter should be useful for identifying many of the Central American pteromalids although genera that occur in Central America but not in North America are excluded. A major obstacle in identifying neotropical Pteromalidae is the large number of undescribed genera present in the region. The fact that 31 new genera were recently described from North America (Bouček, 1993*b*) indicates how poorly known this family is.

Key to subfamilies of Pteromalidae present in Costa Rica

1 Body very elongate, usually more than 20 mm in length; head with prominent ridge between inner eye margin and scrobal cavity; pronotum elongate, neck-like (Fig. 11.153).
[Ovipositor protruding.] .... **Leptofoeninae** (p. 365)

— Generally smaller than above, less than 20 mm in length; head usually without ridge on inner eye margin; pronotum rarely so elongate. 2

2 Head and body dorso-ventrally flattened, shiny black, metasoma petiolate (Fig. 11.148); antennae attached on lower margin of head, at sides of mouth (Fig. 11.149); antenna with no anellus, seven funicular segments and an undivided clava. .... **Spalangiinae** (p. 369)

— Without the above combination of characters. 3

3 Head with ridge or tooth between antennal bases; head usually quite globose (Fig. 11.150); body shiny yellow or brown.
[Fore wing usually without conspicuous setae except on wing margin; sometimes with a tuft of setae at apex of submarginal vein; postmarginal and stigmal veins usually short.] .... **Cerocephalinae** (p. 362)

— Head usually without tooth between antennae; if present, then head and body not as above.... 4

4 Axillae greatly advanced in front of scutellum (Figs 11.155–11.156). 5
— Axillae not advanced in front of scutellum (Fig. 11.159). 6

5 Scutellum with submedian, longitudinal grooves; axillae widely separated posteriorly (Fig. 11.155). .... **Colotrechninae** (p. 363)
— Scutellum without submedian, longitudinal grooves; axillae almost meeting posteriorly (Fig. 11.156). .. **Eutrichosomatinae** (p. 364)

6 Antenna with seven funicular segments (except in females of *Trimicrops*, which are brachypterous); ***either*** inner margins of eyes diverging on lower face, ***or*** head and mesosoma with erect dark bristles. 7
— Antenna usually with fewer than seven funicular segments; inner margins of eyes parallel; head and mesosoma without erect dark bristles. .. 8

7 Inner margins of eyes diverging ventrally in frontal view (Fig. 11.154); eyes usually setose; dorsal part of head and mesosoma without long, semi-erect, dark bristles.
[Males are very similar to male Eupelmidae.] .... **Cleonyminae** (p. 362)
— Inner margins of eyes parallel; eyes not setose; dorsal part of head and mesosoma usually with long, erect, dark bristles (Fig. 11.147).
[First tergite enlarged and covering at least 0.3 of metasoma.] .... **Diparinae** (p. 363)

8 Antenna with fewer than 13 segments. 9
— Antenna with 13 segments. 13

9 First segment of gaster enlarged; metasoma sessile (with petiole inconspicuous). 10
— First segment of gaster usually not enlarged, but if enlarged, then metasoma conspicuously petiolate. 12

10 Posterior part of head broadly concave, usually sharply carinate dorsally, with ocelli nearly touching edge (Fig. 11.151); antenna with four or five evident segments between pedicel and clava. .... **Eunotinae** (p. 364)
— Posterior part of head usually not broadly concave and not carinate; antenna with one anellus and six funicular segments. 11

11 First metasomal tergite with longitudinal striae (Fig. 11.158); head with scrobal cavity delimited by a carina; body and wings not densely setose. .... **Erotolepsiinae** (p. 364)
— First metasomal tergite without longitudinal striae; scrobal cavity not delimited by a carina; head (including eyes), dorsum of mesosoma (Fig. 11.159) and wings densely covered with short setae. .... **Herbertiinae** (p. 365)

12 Antenna with 12 segments (usually with two anelli and five funicular segments), inserted about in centre of face. ....**Ormocerinae** (p. 365)
— Antenna with less than 12 segments, inserted below centre of face. .... **Pireninae** (p. 365)

13 Propodeal spiracles situated about 0.5 of way along propodeum (Fig. 11.157); extremely small (1.0 mm or less in length). .... **Ceinae** (p. 362)
— Spiracles of propodeum situated near anterior margin; size variable, but usually larger than above. 14

14 Posterior margin of gena carinate (Fig. 11.152).
[Marginal vein short, about as long as stigmal and postmarginal veins; notauli complete; metasoma petiolate.] ....**Asaphinae** (p. 361)
— Posterior margin of gena rounded.. **Pteromalinae** (p. 366)

Synopsis of the Costa Rican fauna

Pteromalidae is one of the most poorly studied families of Hymenoptera and therefore our present knowledge of the Costa Rican fauna is very limited. I have seen more than 70 genera from the country, but the total present is likely to be at least 100. Estimating the number of species present is difficult, given that most are probably undescribed, but within an order of magnitude, there are probably at least 500 species in

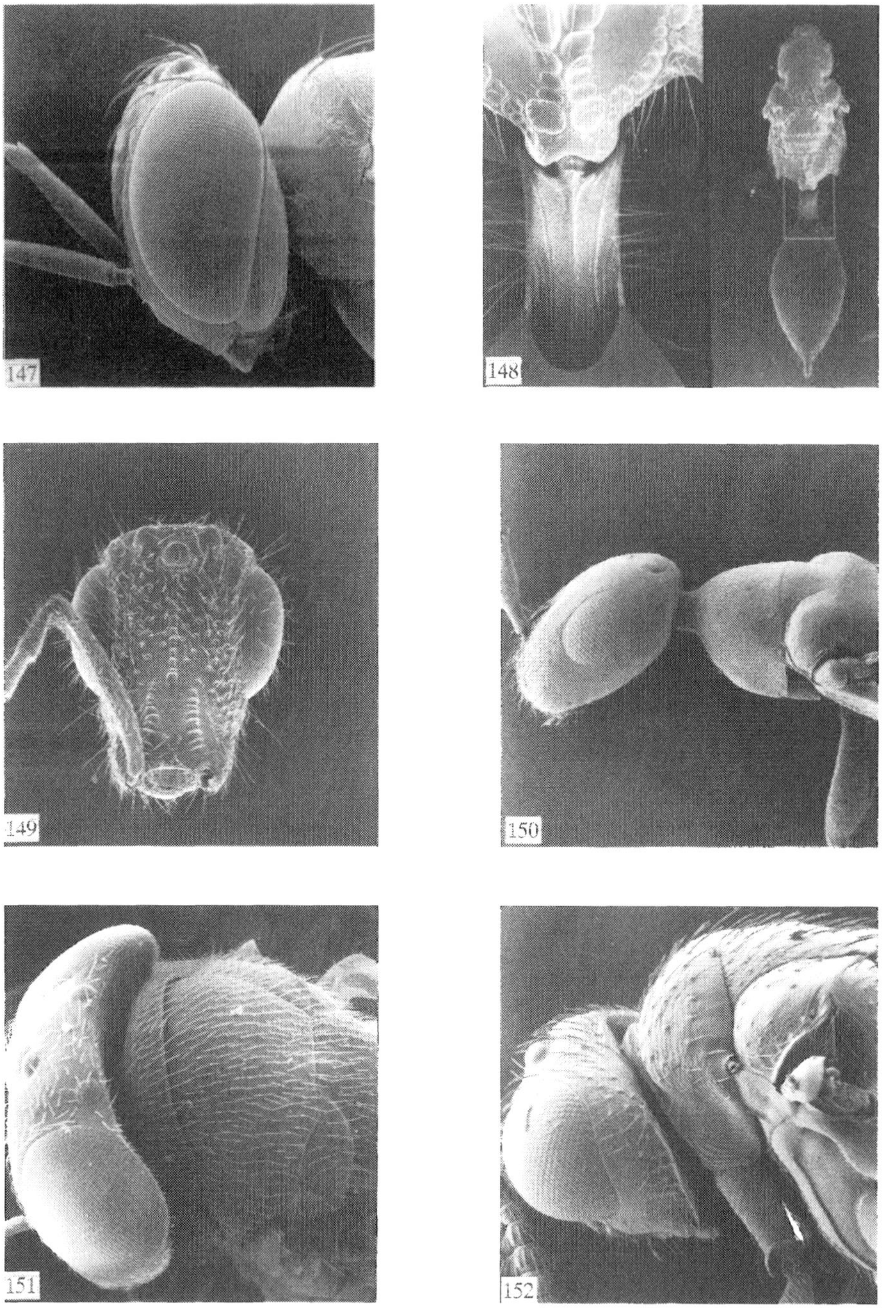

Figs 11.147–11.152. Pteromalidae. Fig. 11.147. Diparinae, head. Figs 11.148–11.149. Spalangiinae; 11.148, dorsal view of mesosoma ; 11.149, head. Figs 11.150–11.152. Head; 11.150, Cerocephalinae; 11.151, Eunotinae; 11.152, Asaphinae.

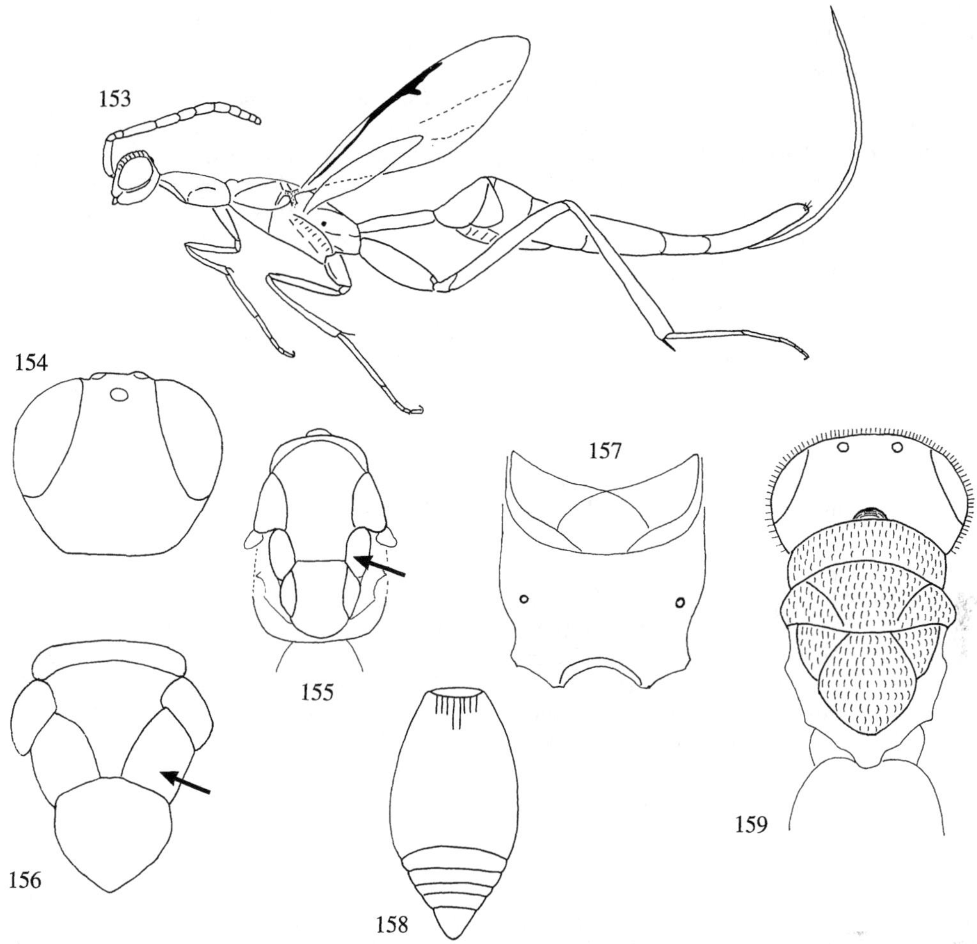

Figs 11.153–11.159. Pteromalidae. Fig. 11.153. *Leptofoenus* sp. (Leptofoeninae), whole insect, lateral view. Fig. 11.154. Cleonyminae, anterior view of head. Figs 11.155–11.156. Dorsal view of mesosoma; 11.155, Colotrechninae; 11.156, Eutrichosomatinae. Fig. 11.157. Ceinae, dorsal view of propodeum. Fig. 11.158. Erotolepsiinae, dorsal view of metasoma. Fig. 11.159. Herbertiinae, dorsal view of head and mesosoma.

Costa Rica. The family is represented at all altitudes although there are notable differences between low and high altitudes. At low altitudes (below about 1500 m) groups such as Diparinae constitute a conspicuous part of the fauna, whereas at high altitudes genera of Pteromalini are predominant.

The following is a very preliminary tabulation of the Pteromalidae presently known from Costa Rica. Because even the North American genera are still being revised (Heydon & Bouček, 1992; Bouček, 1993*b*; Bouček & Heydon, in prep.), many of the pteromalids collected in preparation for this book remain unidentified. Thus further study of the material already at hand will necessitate additions to this list in the near future.

ASAPHINAE

This subfamily comprises four genera, two of which are Australian (Bouček, 1988) and one holarctic.

Asaphes. Cosmopolitan; about 25 described species; in Costa Rica mostly at higher altitudes (above

1500 m). Biology: ectoparasitic idiobionts of aphidiine braconids in aphids (Haviland, 1922; Evenhuis, 1964). The female *Asaphes* selects an aphid that has been killed by the aphidiine larva (Dunn, 1949), injects a venom that causes immediate paralysis of the aphidiine host, feeds on the host, and then oviposits (Keller & Sullivan, 1976). The aphidiine host does not void a meconium and deteriorates into a soft, blackened mass (cf Megaspilidae; Bocchino & Sullivan, 1981). Occasionally *Asaphes* may be a tertiary parasitoid, attacking either another species of hyperparasitoid (e.g. Cynipoidea-Charipini; Sullivan, 1972) or its own species (Levine & Sullivan, 1983).

CEINAE

This small subfamily comprises three genera, only one of which has been found in Central America (Darling, 1991).

Spalangiopelta. Seldom collected; some species are more readily obtained through litter sampling (e.g. Berlese funnel) than through traditional collecting techniques. This, combined with the fact that one North American species shows continuous wing length variation (varying from fully winged to brachypterous), suggests that they attack hosts in leaf litter, probably Diptera (Darling & Hanson, 1986).

CEROCEPHALINAE

This subfamily comprises about a dozen genera and 35 species worldwide. Four genera have been collected in Central America at altitudes between seal-level and about 1600 m. *Choetospilisca* has not yet been found in the region, but may occur here as it occurs in Brazil and North America, where it parasitizes larvae of Curculionidae (Burks *in* Krombein *et al.*, 1979). A key to cerocephaline genera is given in Hedqvist (1969); a new genus, *Laesthiola*, was recently described from southern Florida (Bouček, 1993*b*).

Cerocephala. Probably Old World in origin, although two species are holarctic; *C. aquila* has become nearly cosmopolitan, possibly through human activity (hosts occur in bamboo). Sides of head rounded; fore wing with tuft of setae at apex of submarginal vein. Biology: attack Scolytidae, Anobiidae, Bostrychidae and Lyctidae, either as primary parasitoids, or as hyperparasitoids via doryctine braconids (Szczepanski, 1960; Grissell, 1981).

Neocalosoter (= *Thecolaxia*). Mostly southern hemisphere species. Sides of head rounded; without tuft of setae at apex of submarginal vein. Biology: parasitoids of scolytids in wood. In Costa Rica *N. pityopthori* has been reared from *Styphlosoma granulatum* (Scolytidae) in *Spondias mombin* logs (Matthews, 1969).

Paralaesthia. Just one species, *P. mandibularis*, which is known only from Central America. Sides of head parallel, mandibles large and widely separated; with tuft of setae at apex of submarginal cell. Biology unknown.

Theocolax (= *Choetospila*). Three species, two of which are cosmopolitan. Sides of head parallel, mandibles widely separated; with tuft of setae at apex of submarginal cell. Biology: *T. elegans* is a parasitoid of *Stegobium pamniceum* (Anobiidae) in stored grain (Williams & Floyd, 1971), and *T. formiciformis* is a parasitoid of anobiid beetle larvae in dead wood.

CLEONYMINAE

Worldwide the Cleonyminae comprises about 40 genera and 170 species (Bouček, 1958). This subfamily may be more closely related to certain Eupelmidae than to other Pteromalidae. Cleonymines are probably all ectoparasitic idiobionts, mostly of beetle larvae in dead wood or woody stems (e.g. see Grissell, 1991), but a few species may specialize on other hosts in similar situations, such as stem-nesting aculeates (Krombein, 1959). The subfamily includes five tribes: Chalcedectini, Cleonymini, Heydeniini, Lyciscini and Ooderini. No ooderines or heydeniines have been found in Central America although they occur in North America. In Costa Rica species of this subfamily occur at lower altitudes, from sea-level up to about 1600 metres.

CLEONYMINAE: Chalcedectini. This tribe is most species-rich in the New World where three genera are present: *Amotura* (pantropical), *Chalcedectus* (Brazil), *Dryadochalcis* (Brazil and Texas). Members of the Chalcedectini can be readily recognized by their greatly enlarged hind femora.

Amotura. Pantropical, north into the United States. Grissell (1991) revised the nearctic taxa. Biology: parasitoids of beetles of the families Bostrychidae and Buprestidae.

CLEONYMINAE: Cleonymini. This primarily Old World group is characterized by the possession of a lateral process on the preclaval segment of the antenna of females. Two genera occur in the New World: *Cleonymus* (cosmopolitan) and *Notanisus* (= *Pannoniella*; apparently introduced into North America from Europe). The former is known to occur in Costa Rica.

Cleonymus (= *Ptinobius*). More than 20 described species. Biology: parasitoids of beetles in wood.

CLEONYMINAE: Lyciscini. This primarily southern hemisphere group comprises approximately 25 genera (Hedqvist, 1961; Boucek, 1988). It is represented in Central America by at least three genera, possibly more as the neotropical genera need revision. Members of the tribe are readily recognizable by their densely punctate, dorso-laterally carinate metasoma.

Epistenia. New World; about 20 described species. Pronotum transverse; posterior edge of gena carinate. Biology: parasitoids of aculeate Hymenoptera nesting in stems (Krombein, 1959).

Lycisca. Neotropical; nine described species. Pronotum at least as long as wide; scutellum often quite flat; body usually very elongate. A key to species is given in Hedqvist (1961). Biology unknown.

Parepistenia. South America, Africa, southern Asia and Australia; the genus comprises at least 50 species worldwide (Bouček, 1988). Similar to *Epistenia* but with posterior margin of gena rounded, not carinate. Biology: reared from logs and stems containing beetles.

COLOTRECHINAE: (= Dvaliniinae; Grissell, 1985)
Bouček (1988) recognizes five tribes of Colotrechninae, but only one of these, the Colotrechnini, is known from the New World. This tribe comprises three Australian and six New World genera, and the widely distributed genus *Colotrechnus*. Of the seven genera present in the New World four have not been seen in Costa Rica: *Baridobius* (North America), *Colotrechnus* (North America), *Dvalinia* (Brazil) and *Elachertodomyia* (North America). Host records are from Curculionidae and Scolytidae, except *Colotrechnus*, which has been reared from Agromyzidae in flower heads.

Bofuria. Neotropics; two described species from Colombia and St. Vincent (Hedqvist, 1978). This is the most commonly collected colotrechnine genus in Costa Rica where at least two species are present. Individuals are most commonly taken between 700 and 1600 metres, although isolated specimens have been found at sea-level, and up to 2100 metres. Antenna with six funicular segments; pronotum carinate. Biology unknown.

Bomburia. Neotropics; two described species from Brazil. Antenna with five funicular segments; antennae attached above lower eye margin. Biology: a specimen from Costa Rica was reared from the scolytid *Phloeotribus armatus* (specimen label in USNM).

Dipachystigma. Transferred from Pireninae to Colotrechninae by Heydon and Bouček (1992). New World; two described species from North America. Antenna with four funicular segments; apex of submarginal vein enlarged. Biology: parasitoids of scolytid larvae (Burks *in* Krombein *et al.*, 1979).

DIPARINAE
Worldwide there are fewer than 100 described species of Diparinae, but the actual number is certainly much higher. Bouček (1988) divided the subfamily into two tribes: Lieparini (one Australian genus) and the cosmopolitan Diparini comprising about 20 genera, most species of which occur in tropical and subtropical regions. Some species in this subfamily have brachypterous females, while males are always fully winged. The hosts are unknown except for one species in India, which was reared from the larva of a weevil (Curculionidae) feeding on the roots of *Cyperus* (Cyperaceae) (Bouček, 1988). In Costa Rica the subfamily is represented by four described genera (and possibly one or two undescribed genera), and is

common from sea-level to about 1600 metres in altitude. Keys to genera are given in Hedqvist (1969) and Yoshimoto (1977).

Dipara (= *Apterolaelaps*). Cosmopolitan; less than 20 described species. At least two species are present in Costa Rica (0–1600 m), one of which has brachypterous females. Petiole of female about as long as broad; male antenna with flagellar segments constricted at the ends and with whorls of long setae.

Lelaps. Neotropics; about 50 described species; at least 15 species are present in Costa Rica. Female antenna with first funicular segment much longer than pedicel; male antenna with long setae.

Netomocera. Cosmopolitan; less than 10 described species. Only one (apparently undescribed) species has been seen in Costa Rica. Small (2 mm), robust; petiole of female subtriangular (broader than long); male antenna not noticeably setose, petiole at most 1.5 times as long as broad.

Trimicrops. Three described species; at least two species are present in Costa Rica. Females brachypterous, antenna with two or three anelli; male antenna not noticeably setose, petiole very elongate.

EROTOLEPSIINAE

This small subfamily comprises two Australian genera and the neotropical *Erotolepsia*; it is possible that *Balrogia* from Brazil belongs here as well (Bouček, 1988). In Costa Rica this subfamily has been collected at lower altitudes, from sea-level up to 1200 metres.

Erotolepsia. Texas to Brazil, including the Caribbean; one described species. Biology: apparently reared from Stratiomyidae (Bouček & Heydon, in prep.).

EUNOTINAE

Members of this subfamily often oviposit under (as opposed to through) the scale covering of mature Coccoidea, and the larvae, which are gregarious, develop as predators on the eggs contained within. The mature larva spins a light web of silk between the host body and the plant surface (Smith & Compere, 1928; Flanders, 1952, 1959*a*). Eunotinae is divided into three tribes (Bouček, 1988), two of which are known from Costa Rica, and the third is expected to occur here.

EUNOTINAE: Eunotini. This tribe contains five genera, two of which are restricted to the Northern Hemisphere. The other three are all known to occur in Costa Rica.

Cephaleta. Neotropics and tropical Asia; about 10 described species. This is the most commonly collected member of the subfamily in Costa Rica (0–2100 m). Biology: associated with Coccidae.

Mesopeltita. Neotropics and Africa; represented in Costa Rica by *M. truncatipennis*. Biology: associated with Coccidae.

Scutellista. Probably native to the Palaeotropics. *S. caerulea* (= *cyanea*), which has been widely used in classical biological control (Clausen, 1978), is present in Central America although there are no reports of its introduction into the region. Biology: parasitoids of a variety of Coccidae, especially *Ceroplastes*, *Lecanium* and *Saissetia*.

EUNOTINAE: Moranilini. Members of the Moranilini are characterized by reduced pilosity and presence of paired bristles on the mesosoma.

***Moranila*[EX].** *M. californica* has been widely introduced throughout the warmer parts of the world. Biology: species are associated with Coccidae.

EUNOTINAE: Tomocerodini. Members of this tribe differ from all other Eunotinae in having metasomal tergite I shorter than tergite II. There is only one described genus.

Tomocerodes. Arizona to Argentina; one described species. Biology: parasitoids of *Lecanodiaspis* (Lecanodiaspididae) (Burks *in* Krombein *et al.*, 1979).

EUTRICHOSOMATINAE

This small subfamily contains just two New World genera, both of which are found in Central America, and possibly one Australian genus (Bouček, 1974*b*, 1988).

***Eutrichosoma*.** Two described species. Marginal vein barely longer than stigmal vein. *E. flabellatum* has been collected in low elevation dry forest in northwestern Costa Rica and in mid elevation (1600 m) wet forest near San José. Biology: the host of *E. flabellatum* is unknown but the only other species in the genus has been reared from weevil larvae (Curculionidae) in North America and Mexico (Bouček, 1974*b*).

***Peckianus*.** One described species, *P. laevis*. Marginal vein much longer than stigmal vein. Biology: reared from bean pods containing weevils of the genus *Apion*.

HERBERTIINAE

This subfamily contains only one genus. In Costa Rica it has been collected from sea-level to 1600 metre altitude.

***Herbertia*.** Cosmopolitan; small genus, most species-rich in tropical and subtropical regions. A key to described species is given by Burks (1959). Biology: parasitoids of leaf-mining Agromyzidae (Diptera) (Burks, 1959). An unidentified species was reared from an agromyzid leaf miner in *Smilex spinosa* (Liliaceae) in northwestern Costa Rica (Memmott *et al.*, 1994).

LEPTOFOENINAE

This subfamily comprises one Australian and one neotropical genus.

***Leptofoenus*.** Neotropical with four species present in Central America, all from lowland sites. *Leptofoenus* species are exceptionally large, elongate pteromalids, often more than 12.0 mm in length. A key to species is given by LaSalle and Stage (1985). Biology unknown, but their form and size suggest they may be parasitoids of beetles in wood.

ORMOCERINAE

As defined by Bouček (1988) this subfamily includes three tribes, Ormocerini, Melanosomellini and Systasini. The first of these is holarctic and includes a single genus *Ormocerus*. The other two tribes are known to be present in Central America.

ORMOCERINAE: Melanosomellini. This tribe, previously known as Brachyscelidiphaginae, contains over 30 genera, most of which are restricted to Australia, but some of which occur in southern Asia, Africa and South America (Yoshimoto, 1972). It is uncertain whether the North American *Hemadas*, which causes galls on *Vaccinium* (Ericaceae) (McAlister & Anderson, 1932), belongs here. The neotropical taxa are mostly unstudied. Most Melanosomellini are associated with plant galls, as gall-formers (e.g. Noble, 1938, 1940, 1941) or inquilines.

ORMOCERINAE: Systasini. This tribe comprises two cosmopolitan genera.

***Semiotellus*.** Cosmopolitan; about a dozen described species. Mesosoma punctate. Biology unknown.

***Systasis*[EX].** Cosmopolitan; about 30 described species. Fore wing with a distinct row of admarginal setae on the ventral side. Biology: reared from the galls of Cecidomyiidae, and from the seeds of grasses (*Panicum* spp.).

PIRENINAE

This subfamily contains about 12 genera worldwide, at least three of which are presently known from Costa Rica. Many are parasitoids of larvae of Cecidomyiidae (Diptera).

***Bugacia*.** Primarily holarctic; in Costa Rica it has been collected only at 1600 metres altitude. Similar to *Gastrancistrus* but pronotum with transverse carina. Biology: parasitoids of Cecidomyiidae.

***Erixestus*.** New World; two described species, one from the United States and a second from Argentina. At least one species is present in Costa Rica. Body robust, black; sides of propodeum densely covered with white setae. Biology: solitary endoparasitoids of eggs of *Calligrapha* species (Coleoptera: Chrysomelidae) (Grissell & De Santis, 1987). A new species is being described from eggs of *Zygogramma magica* (Chrysomelidae) on *Tithonia* *tubaeiformis* (Compositae) in Honduras (Grissell & Cave, in press).

Gastrancistrus (= *Parasyntomocera*, *Stigmatocrepis*, *Parecrizotes* and *Mesecrizotes*). More than 100 species worldwide; the most commonly collected genus of Pireninae in Costa Rica (all altitudes). Marginal vein less than 3.5 times as long as stigmal vein. Biology: endoparasitic koinobionts of Diptera, ovipositing in the larva and emerging from the pupa (e.g. Askew & Ruse, 1974).

Macroglenes (= *Pirene*; see Bouček, 1988). Cosmopolitan; about 20 described species. Marginal vein at least 3.5 times as long as stigmal vein. Biology: endoparasitic koinobionts of Cecidomyiidae, ovipositing in the egg and emerging from the larva (Doane *et al.*, 1989).

PTEROMALINAE

This is by far the largest subfamily of Pteromalidae, containing over half of all the genera and species in the family. Considerable taxonomic difficulties still remain to be resolved about the classification of this group, which is largely an aggregation of genera that do not fit elsewhere. The three currently recognized tribes—Trigonoderini, Miscogastrini and Pteromalini —are all represented in Costa Rica.

PTEROMALINAE: Trigonoderini. This tribe, which was placed in the Miscogastrinae by Graham (1969), is characterized by having a fairly elongate pronotum and a long postmarginal vein, which usually exceeds the marginal vein in length. Most species are parasitoids of larvae of wood-boring beetles, especially Anobiidae and Bostrychidae. A key to the genera is given in Hedqvist (1968).

Gastracanthus. Primarily holarctic; about 12 described species. One species is known to occur in Costa Rica at altitudes between 900 and 1600 metres. Clypeus edentate; prepectus normal.

Ogloblinisca. United States south to Argentina; three described species, but five species occur in Costa Rica below about 2200 metres. Clypeus edentate; prepectus with a delimited triangular region posteriorly.

Trigonoderus. Holarctic and South America; about 30 described species. Four species have been found in Costa Rica, all above 1000 metres. Clypeus with a median tooth.

PTEROMALINAE: Miscogastrini. Members of this primarily holarctic tribe are characterized by having three asymmetrically placed teeth on the anterior margin of the clypeus. Most if not all are parasitoids of Diptera. At least one genus appears to be common in Costa Rica although others are undoubtedly present, especially at higher altitudes (above 1500 m).

Halticoptera. Widespread; about 30 described species. Biology: endoparasitic koinobionts of Agromyzidae and Tephritidae, ovipositing in the larva and emerging from the puparium (Askew, 1968). In Peru *H. arduine* (holotype collected by Charles Darwin) and *H. peruviana* are parasitoids of *Liriomyza huidobrensis* (De Santis, 1987), and I have seen an unidentified species from this host in Panama. *H. circulus* has been reared from *L. sativae* in Honduras (R. Cave, pers. comm.).

PTEROMALINAE: Pteromalini. Most of the neotropical species belonging to this tribe are probably undescribed and several new genera await description. Most species are probably ectoparasitic idiobionts, although at least *Stenomalina* is known to be an endoparasitic koinobiont. Following are the genera that have been identified thus far from Costa Rica; many more are certainly present.

Acaenacis (= *Megatrydymus*). Mainly holarctic; about 5 described species. Anterior tentorial pits evident. Biology: a North American species has been reared from cynipid twig galls and circumstantial evidence suggests that a Costa Rican species has the same biology. Another species (in the USNM) has been reared from *Inga* (Leguminosae) in Panama.

Alticornis. New World; two described species (Bouček, 1993*b*). Antennal scape extending well above vertex; male with thorn-like projections on face. Biology unknown.

Anisopteromalus. Cosmopolitan; about 20 described species, mainly in the Old World. Head and dorsum of mesosoma covered with dense decumbent hairs. Biology: the widespread *A. calandrae* parasitizes beetle larvae associated with stored grain. An African species develops as an ectoparasitoid of a curculionid in legume pods (Rasplus, 1988).

Arachnopteromalus. New World; one described species, *A. dasys*, which occurs from the United States south to Brazil. In Costa Rica it has been collected mostly from seasonally dry forests. Eyes, mesosoma and fore wing very setose. Biology: oviposits in egg sacs of uloborid spiders, including species of *Uloborus*, *Octonoba* and *Philoponella*, where the larvae develop as predators (Gordh, 1976*b*, 1983).

Callitula. Cosmopolitan; about 30 described species. Antennae inserted quite high on face, club pointed. Biology: parasitoids of Diptera, especially Agromyzidae and Chloropidae (e.g. Simmonds, 1952).

Catolaccus. Widespread. Antenna with six funicular segments; gena concave. Biology: *C. hunteri*, a widespread parasitoid of Bruchidae and Curculionidae, has been found to occur in Guatemala; *C. grandis* parasitizes the boll weevil, *Anthonomus grandis*.

Chrysoglyphe. New World. Biology: one species (which is only provisionally assigned to this genus and may not belong here) has been reared in Nicaragua from a drosophilid predator of tree-frog eggs (Gordh, 1977). The fly larva feeds on the eggs of *Centrolenella fleishmanni* which are stuck to the undersides of leaves. Oviposition by the parasitoid has not been observed but it may be into the drosophilid larva; emergence is from the puparium.

Cryptoprymna. Widespread; less than 10 described species. In Costa Rica it apparently only occurs above about 1600 metres. Clypeus edentate; antennal club with large area of micropilosity; pronotum carinate; petiole elongate, reticulate; tergite I large, hypopygium extending to apex of metasoma. A key to nearctic species was given by Heydon (1988). Biology: parasitoids of Syrphidae associated with Homoptera (Heydon, 1988).

Cyrtogaster (= *Polycystus*; Heydon, 1989*a*). Primarily holarctic; about 40 described species. In Costa Rica species of this genus have only been found above 1500 metres. Clypeus with 3 symmetrical teeth; pronotum carinate; petiole elongate with longitudinal carinae. A key to species is given in Heydon (1989*a*). Biology: solitary ectoparasitoids of dipterous pupae, especially those of leaf-miners and stem-borers (e.g. see Cameron, 1939).

Dibrachys. Cosmopolitan, apparently most species-rich in the Holarctic; about a dozen described species. Postmarginal vein very short; occipital carina present. *D. cavus* is one of the most polyphagous of all Chalcidoidea, developing as a primary or secondary parasitoid on a range of insect orders and spider families (Burks *in* Krombein *et al.*, 1979).

Euneura. Cosmopolitan; about 10 described species. In Costa Rica it occurs at all elevations. Similar to *Pachyneuron* but petiole is shorter. Biology: hyperparasitoids of aphids (Kamijo & Takada, 1983).

Eurydinoteloides (= *Aeronea*, *Protolaccus*). Neotropics; less than 10 described species. Similar to *Lyrcus* but gena more concave. Biology: parasitoids of syrphid pupae.

Hansonita. Presumably neotropical; one described species (Bouček, 1993*a*). Although tentatively placed in the Pteromalinae, the relationships of this genus require further study (Z. Bouček, pers. comm.). Biology: the type-species was reared from the figs of *Ficus pertusa* (Moraceae) collected at about 1300 metres.

Heteroschema. New World; four described species, but many more undescribed. This is one of the most commonly collected pteromalids in Costa Rica. Antenna with five funicular segments; propodeum with a transverse carina, reticulate, extended posteriorly. Biology: in northwestern Costa Rica, one species has been reared from a buprestid leaf-miner (*Brachys* sp.) on *Gouania polygama* (Rhamnaceae), and another from an unidentified agromyzid leaf-miner on *Lantana camara* (Verbenaceae) (Memmott *et al.*, 1994).

Lariophagus. Cosmopolitan; about 15 described species. Six funicular segments; postmarginal vein often short; dorsum of mesosoma flattened, black and densely reticulate. Biology: *L. distinguendus* is a cosmopolitan parasitoid of beetles associated with stored grain while *L. texanus* is a parasitoid of bruchid larvae and is commonly reared from seeds of *Acacia farnesiana* in Costa Rica. Details of the biology of an Old World species are given by Kashef (1956, 1959).

Lyrcus (= *Zatropis*). New World; very speciose and one of the most commonly collected pteromalids in Costa Rica. Antenna usually with five, sometimes six, funicular segments; fore wing with a distinct row of admarginal setae on ventral side; head and mesosoma often with flattened white setae (Heydon & Bouček, 1992). Biology: mostly parasitoids of weevils and bruchids.

Metastenus (= *Scymnophagus*). Cosmopolitan; less than 10 described species. Marginal vein enlarged; five funicular segments. Biology: ectoparasitoids of coccinellid pupae (Clausen, 1940*b*; Domenichini, 1957; Ricci & De Santis, 1983). *M. cryptognathae* has been reared from *Cryptognatha nodiceps* in Trinidad and *M. townsendi* has been reared from *Pentilia castanea* in Colombia (specimen labels in USNM).

Miristhma. New World; two described species (Bouček, 1993*b*). Propodeum extremely prolonged posteriorly. Biology unknown.

Muscidifurax. Predominantly New World; one species cosmopolitan. One species, *M. raptoroides*, has been described from Costa Rica (Kogan & Legner, 1970). Marginal vein enlarged basally. Biology: solitary ectoparasitoids in puparia of calypterate Diptera.

Nasonia. (= *Mormoniella*). Cosmopolitan; previously thought to comprise just one species but now known to consist of three species, at least in North America (Darling & Werren, 1990). Postmarginal vein short; body sculpture very delicate; occipital carina present; pronotum carinate. Biology: gregarious ectoparasitoids in the puparia of calypterate Diptera; one of the best studied of all Hymenoptera (see review in Whiting, 1967 and Godfray, 1994).

Neocatolaccus. Most species-rich in the New World and Australia. Antenna with five funicular segments; propodeum with transverse carina, not extended posteriorly; hind tibia with two apical spurs. Biology: mostly parasitoids of Curculionidae, although one North American species attacks Cerambycidae (Burks *in* Krombein *et al.*, 1979).

Norbanus. Cosmopolitan; about 20 described species. Body very robust; six funicular segments, club pointed. Biology: parasitoids of Curculionidae.

Notoglyptus. Pantropical; about 10 described species, two of which, *luteicrus* and *tzeltales*, are known from Central America (Heydon, 1989*b*). Gena concave; notauli complete; petiole about as long as wide; tergite I large; hypopygium extending to the tip of the metasoma. Biology unknown.

Pachycrepoideus. One cosmopolitan species (*vindemmiae*) and two African species (Bouček, 1988). Marginal vein enlarged; antennae inserted low on face. Biology: ectoparasitoids in puparia of Diptera (Crandell, 1939).

Pachyneuron. Cosmopolitan; about a dozen described species. In Costa Rica it occurs only above 2500 metres. Marginal vein enlarged; antenna with six funicular segments. Biology: primary parasitoids of Syrphidae and secondary parasitoids in aphids.

Polstonia. New World; two described species in the United States. In Costa Rica this genus seems only to occur at higher elevations. Gena concave, antennal club with terminal spine, petiole elongate and reticulate; tergites I and II subequal in length. Keys to the North American species are given by Heydon (1988). Biology unknown.

Psilocera (= *Lophocomodia*). Cosmopolitan; about 20 described species, with at least 15 species (mostly undescribed) in Costa Rica (very common at low altitudes but also present at mid altitudes). Antennal club enlarged and asymmetrical; gena concave. Biology: there is one record of a North American species from a case-building chrysomelid

(Chlamisinae) larva (Burks *in* Krombein *et al.*, 1979).

Pteromalus (= *Habrocytus*). Cosmopolitan; over 100 species (Bouček, 1988). Biology: species of this genus have a diverse range of host associations. Some are gregarious endoparasitoids in the pupae of various large Lepidoptera (especially those of the Nymphaloidea), but other hosts include stem-nesting aculeates, stem-boring microlepidopterous, coleopterous, hymenopterous and dipterous larvae and various galls, especially those of Cynipini (Burks *in* Krombein *et al.*, 1979). *P. semotus* has been reared from *Sitotroga cerealella* on sorghum in Honduras (specimens in Escuela Agrícola Panamericana).

Schizonotus. Predominantly holarctic; a small genus represented by a single species in Costa Rica. Postmarginal vein short; frenal groove distinct. Biology: gregarious ectoparasitoids of chrysomelid pupae.

Sphegigaster. Cosmopolitan; about 20 described species. Clypeus bidentate; notauli incomplete; petiole elongate; tergite I emarginate posteriorly. Biology: parasitoids of Agromyzidae (Cameron, 1939).

Syntomopus. Cosmopolitan; about a dozen described species. Mesosoma flattened dorso-ventrally; pronotum rectangular; petiole elongate. Biology: parasitoids of stem-mining Agromyzidae, but stem-mining Lepidoptera and possibly also cynipids are used as hosts (Lange, 1944; Heydon, 1993).

Toxeumella. A mainly neotropical genus with one described species, *T. albipes*. Head with occipital carina, gena concave; notauli complete. Biology: hyperparasitoid of lepidopterous larvae via microgastrine braconids.

Toxeumelloides. New World; two described species (Bouček, 1993*b*). Petiole elongate, narrower posteriorly; gena concave. Biology unknown.

Trichokaleva. Costa Rica to South America; one described species, *T. microstigmi*. In Costa Rica it occurs below 1000 metres. Eyes hairy and gena concave. Biology: reared from the nests of *Microstigmus* (Sphecidae) in Colombia (Bouček, 1972*b*).

Spalangiinae

This subfamily comprises only two genera, the Sonoran Mexican genus *Paraspalangia* that was collected from intertidal algae (Yoshimoto, 1976*b*), and the cosmopolitan *Spalangia*. In Costa Rica members of this subfamily have been collected from sea-level up to 2100 metres.

Spalangia. Between 25 and 50 described species. The world species of *Spalangia* were revised by Bouček (1963). Biology: solitary, idiobiont ectoparasitoids of fly pupae (within puparia), especially those of Muscidae, Calliphoridae and Sarcophagidae, which occur in excrement, carrion, etc. Some species are also known to attack the pupae of Chloropidae and Drosophilidae (Richardson, 1913; Gerling & Legner, 1968). Because of their economic importance, the biologies of several species have been fairly intensively studied. In searching for hosts in excrement female *Spalangia* are capable of penetrating deeper than other parasitoids associated with this habitat. The female uses her ovipositor to differentiate between parasitized and unparasitized hosts (Wylie, 1972). In feeding, the larva makes over 100 incisions in surface of the host pupa, especially on the dorsal surface of the abdomen.

11.15 SIGNIPHORIDAE

Paul E. Hanson

Diagnosis. About 0.5 to 2.0 mm in length, robust; usually fully-winged, rarely brachypterous; colour varying from completely black to almost completely yellowish orange. Antenna inserted at or a little above mouth margin; 4- to 7-segmented, *flagellum composed of 1–4 anelli and a long unsegmented club*. Mesoscutum without notauli; scutellum strongly transverse with anterior and posterior margins subparallel; *axillae not distinctly marked off from scutellum, the two together forming a transverse band*; *propodeum with a characteristic large triangular median area*, delimited by two lateral sulci. Fore and hind wings with well-developed mar-

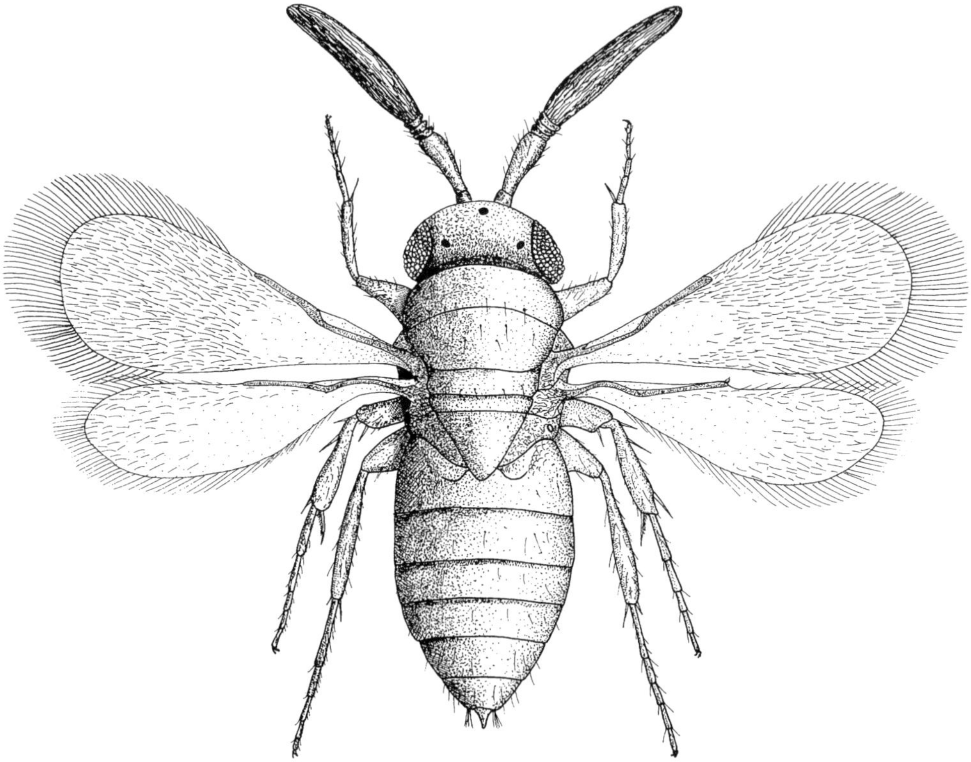

Fig. 11.160. *Signiphora* sp. (Signiphoridae).

ginal setae, but *without microtrichiae* (except for very few large setae that occur in characteristic locations); fore wing with marginal vein quite long (about as long as submarginal vein), *postmarginal and stigmal veins not developed or extremely short*. Middle femur with at least one long spine on antero-ventral margin; middle tibia often with long spines on dorsal margin; tarsi 5-segmented. Metasoma sessile; ovipositor usually hidden.

Classification and distribution. The Signiphoridae (previously called the Thysanidae) is a small family containing about 100 species in four genera (Woolley, 1988). It is a cosmopolitan group that is most species-rich in the Neotropical Region. The Signiphoridae appears to be related to the Aphelinidae and a sister-group relationship between it and azotine aphelinids is suggested by at least one synapomorphy—the presence of anterior apodemes on metasomal sternites III - VI. These appear to be absent in all other Chalcidoidea (Woolley, 1988). Such a relationship would make the Aphelinidae paraphyletic with respect to Signiphoridae, and is an argument for including the latter within the former family.

Biology. Signiphorids are most commonly parasitoids of sternorrhynchous Homoptera, especially scale-insects (Coccoidea) and white-flies (Aleyrodidae), but also Aphidoidea and Psylloidea. Other signiphorids are known to parasitize larvae or pupae of Diptera, especially those that are in some way associated with these homopterans (e.g. Chamaemyiidae and Drosophilidae). Most signiphorids are thought to be solitary endoparasitoids (apparently idiobionts), but at least some species of *Signiphora* at least complete development as ectoparasitoids (Matta, 1979) and

Chartocerus elongatus develops as a gregarious ectoparasitoid (Clausen, 1924). Although some signiphorids, such as *S. borinquensis* and *S. flavella* which attack Diaspididae (DeBach *et al.*, 1958), are primary parasitoids, many that attack homopterans are hyperparasitic via other chalcidoids or platygastrids. These attack mature larvae, pupae, or even newly transformed adult wasps. Among species occurring in Central America *Signiphora aleyrodis* and *S. townsendi* are obligate hyperparasitoids in Aleyrodidae (Woolley & Vet, 1981), *S. flavopalliata* is an obligate hyperparasitoid in Diaspididae (DeBach, 1953) and *Chartocerus elongatus* is an obligate hyperparasitoid in Pseudococcidae (Clausen, 1924).

One of the best studied species is *Signiphora borinquensis*, which is a primary parasitoid of diaspidids, having been originally reared from the coconut scale (*Aspidiotus destructor*) on banana (Quezada *et al.*, 1973). This species is thelytokous, the normal production of males being less than one percent (except at very high temperatures). Although newly emerged females have about six eggs ready for oviposition, host-feeding generally occurs first, and is accomplished via a feeding tube. This feeding, if done by a single parasitoid, does not kill the host. Oviposition follows and, under optimal conditions, about three quarters of the total number of eggs (average of 25) are laid in the first week of the parasitoid's life. Oosorption can occur under suboptimal conditions. The first instar larva is hymenopteriform with four pairs of spiracles, one on the mesothoracic segment and one on each of the first three abdominal segments. The first two larval instars of *S. borinquensis* are endoparasitic, feeding mostly on the host's eggs, whereas the last two instars are ectoparasitic. Pupation takes place outside the body of the host but under the scale covering.

The female of *S. coquilletti*, after ovipositing in a whitefly, spins a web over the host (Woolley & Vet, 1981), the significance of which requires further study.

Economic importance. Signiphorids have rarely been used in classical biological control, probably because many of the species are obligate hyperparasitoids. In Central America *S. aleyrodidis* is occasionally reared as a hyperparasitoid of *Bemisia tabaci* (Aleyrodidae), attacking aphelinid primary parasitoids. However, it is doubtful whether this signiphorid has any significant effect on the populations of the primary parasitoids.

Identification. Keys to the genera of Signiphoridae and to the species-groups of *Signiphora* are given in Woolley (1988), from which the following key is adapted.

Key to genera of Signiphoridae

1 Fore leg with tibial calcar bearing a comb of fine setae (Fig. 11.161); propodeum with lamelliform process posteriorly (Fig. 11.162); antenna of female usually with three or fewer anelli. .... *Signiphora*

— Fore leg with tibial calcar without comb of fine setae; propodeum without lamelliform process posteriorly; antenna of female with four anelli. 2

2 Fore wing with more than five setae present on marginal vein (Fig. 11.163); mid leg with femur bearing three or four long spines. ....*Chartocerus*

— Fore wing with five or fewer setae present on marginal vein; mid leg with femur usually bearing 1 long spine (rarely more). 3

3 Head prognathous, subrectangular in dorsal view (Fig. 11.164); antenna of female with four anelli, of male with three; mid tibia subcylindrical and without long spines on dorsal surface. .... *Clytina*

— Head hypognathous and generally circular in dorsal or frontal view; antennae of both sexes with four anelli; mid tibia widening apically and bearing long spines on dorsal surface....... .... *Thysanus*

Synopsis of the Costa Rican fauna

All four genera of Signiphoridae are present in Costa Rica where the family is represented by at least 10 species, occurring from sea-level up to 2100 metres.

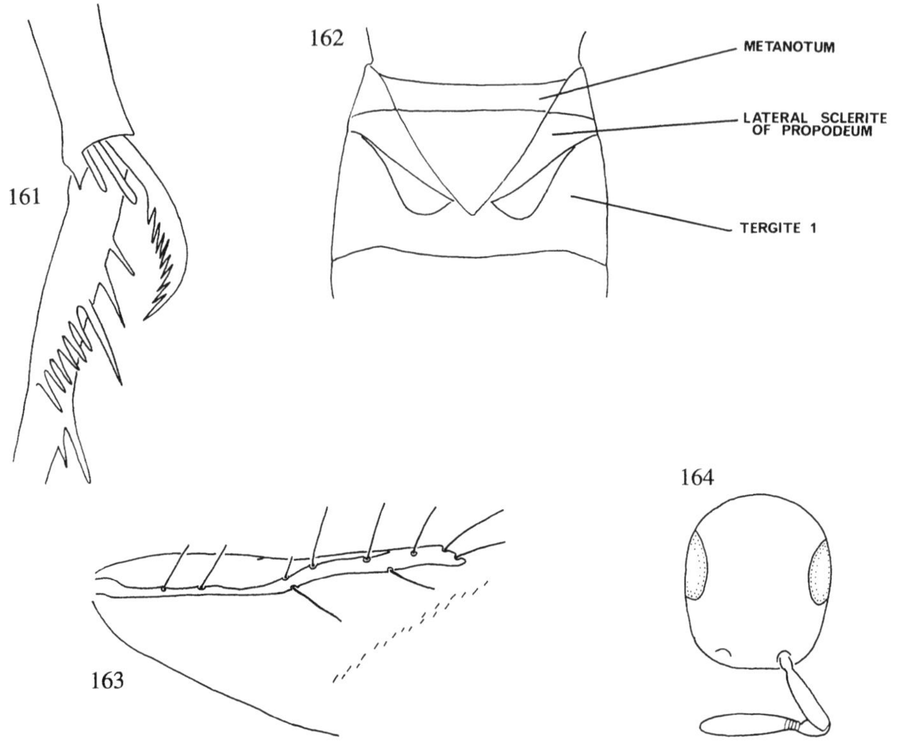

Figs 11.161–11.164. Signiphoridae. Figs 11.161–11.162. *Signiphora* sp.; 11.161, calcar of fore leg; 11.162, propodeal region. Fig. 11.163. *Chartocerus* sp., basal part of fore wing. Fig. 11.164. *Clytina* sp., head.

***Chartocerus*.** Cosmopolitan, about 30 species. Biology: secondary parasitoids of Pseudococcidae, Coccidae, Psylloidea and Aphidoidea though some are primary parasitoids of larvae or pupae of chamaemyiids that prey on mealybugs.

***Thysanus*.** This genus comprises about a dozen species, one cosmopolitan, one from South Africa and the rest in the New World. Biology: *Thysanus* species have been reared from Diaspididae, possibly as secondary parasitoids.

***Clytina*.** This genus comprises two species, one in the Palaearctic and one undescribed species from Costa Rica. Biology: The palaearctic species is a pupal parasitoid of Chloropidae (Diptera) (Erdös, 1957).

***Signiphora*.** This genus comprises about 50 species, classified in four species-groups (Woolley, 1988):

The *flavopalliata* species-group. Two species in this group are cosmopolitan, three Australian and the rest occur in the New World. Biology: primary or secondary parasitoids of Diaspididae, or secondary parasitoids of Aleyrodidae. All of the species of *Signiphora* discussed above, under 'Biology', belong to this group.

The *bifasciata* species-group. A predominantly neotropical group. Biology: reared mostly from Coccidae, Pseudococcidae and Psylloidea.

The *dipterophaga* species-group. A predominantly neotropical group. Biology: reared from Diaspididae, pupae of drosophilid predators of Pseudococcidae and from Tachinidae.

The *coleoptratus* species-group[EX]. This group comprises two rare neotropical species in the

Caribbean and northern South America. Biology: reared from mealybugs, possibly as secondary parasitoids.

11.16 TANAOSTIGMATIDAE

John LaSalle

Diagnosis. Body length 0.7 to 3.8 mm long, robust; usually black to brown or yellow, rarely metallic. Head almost always wider than high, clypeus bilobed; antenna with two anelli and six funicular segments (rarely three anelli and five funicular segments), male funicle sometimes branched. *Prepectus large, distinctly swollen and rounded anteriorly* (Fig. 11.166), *easily visible in dorsal view as prominent 'shoulders' at the side of the mesosoma* (Fig. 11.167); notauli present, sinuate, usually converging or meeting at the midline of the mesoscutum (Fig. 11.167); mesoscutum and scutellum joined by a flexible membrane; mesopleuron large, convex, longer than high (Fig. 11.166). Mid coxa large, subspherical, attached near posterior margin of mesopleuron; mid tibia with large spur

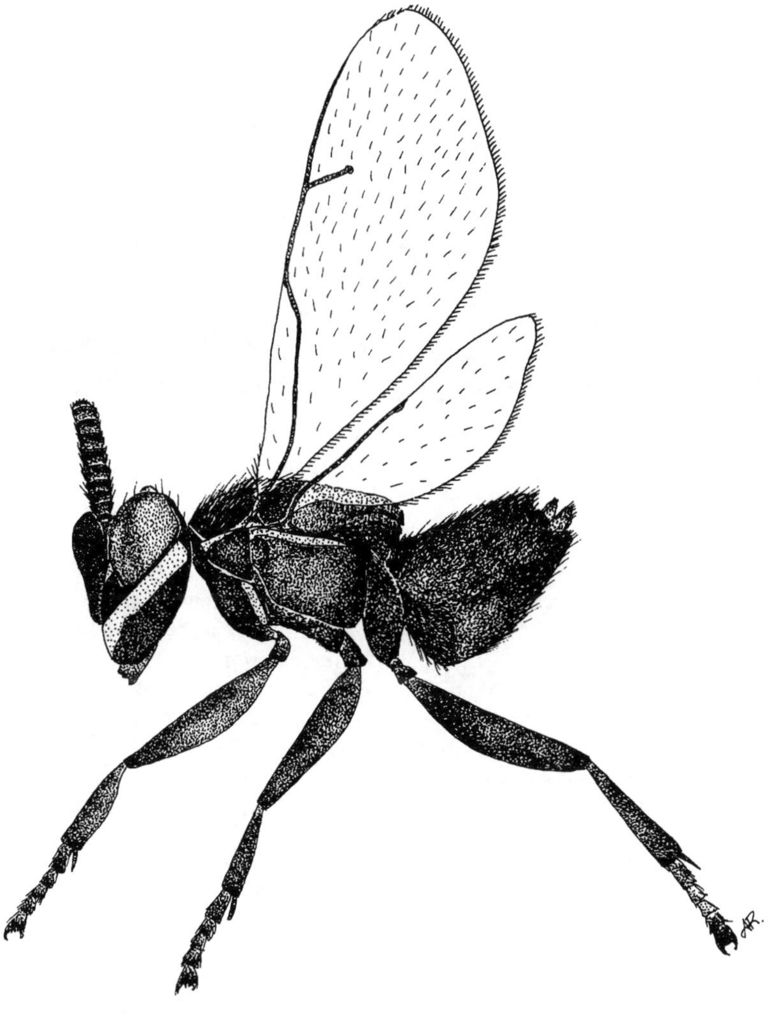

Fig. 11.165. *Tanaostigma coursetiae* (Tanaostigmatidae).

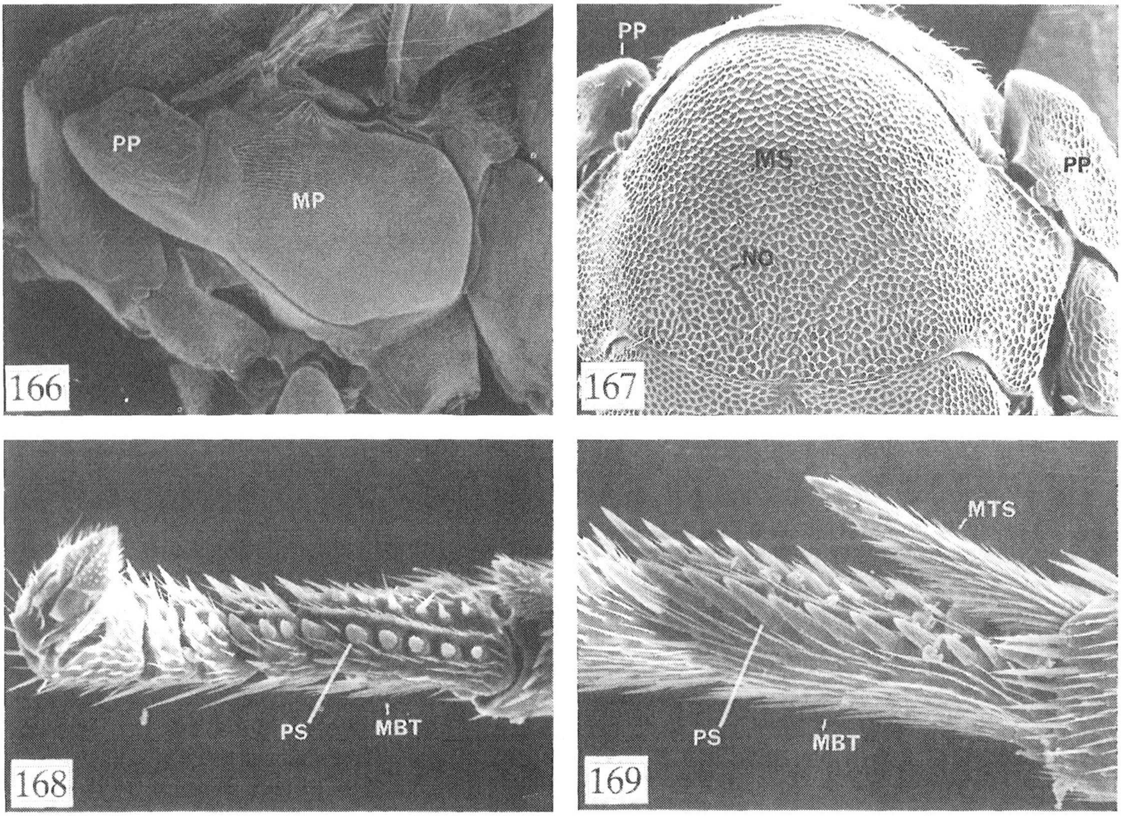

Figs 11.166–11.169. *Tanaostigmodes* sp.; 11.166, mesosoma, lateral view; 11.167, mesosoma, dorsal view; 11.168, middle tarsus; 11.169, middle basitarsus.
mbt = basitarsus; mp = mesopleuron; ms = mesoscutum; mts = tibial spur; no = notaulus; pp = prepectus; ps = peg-like spine.

(Fig. 11.169). Fore wing with marginal vein longer than postmarginal and stigmal veins.

Classification and distribution. Tanaostigmatids are a small group, and species are not often encountered by the general collector. They are found throughout the tropical, and in some cases subtropical, regions of the world, but are most abundant in the Neotropics. There are seven genera: four restricted to the New World, two to the Old World, and *Tanaostigmodes* which occurs in the New World and in Australasia (LaSalle, 1987; Bouček, 1988). There are 88 described species, of which 74 are from the New World, where they range from the southern United States to northern Argentina (LaSalle, 1987).

Tanaostigmatids are related to Encyrtidae and Eupelmidae, all of which have a swollen, convex mesopleuron (Fig. 11.166), and the mid leg with a large tibial spur and a row of peg-like spines on the basitarsus (Figs 11.168, 11.169). Tanaostigmatids have been traditionally treated as more closely related to Eupelmidae than Encyrtidae, and have occasionally been placed as a subfamily of the Eupelmidae. However, they are now considered as more closely related to the Encyrtidae than Eupelmidae (LaSalle, 1987; Gibson, 1989). Differences between encyrtids and tanaostigmatids are given by LaSalle and Noyes (1985), and LaSalle (1987).

The Tanaostigmatidae is a well-defined family, with one exception. The Japanese genus *Cynipencyrtus* was transferred to the Tanaostigmatidae by LaSalle and

Noyes (1985), but not all authors are convinced it belongs here (Gibson, 1989).

Biology. Most tanaostigmatids are apparently phytophagous, usually either gall-formers or as inquilines in the galls of other insects. Most are associated with woody shrubs and trees in the family Leguminosae, although other host plants are known (e.g. Polygonaceae). Some species of *Tanaostigmodes* are apparently phytophagous in the seeds of leguminous plants. For example, the Indian species *Tanaostigmodes cajaninae* infests seeds of pigeon pea (*Cajanus cajan*) as well as several other legumes (Lateef *et al.*, 1985), and in the New World *T. haematoxyli* attacks the seeds of *Haematoxylon*. The Japanese *Cynipencyrtus* is a parasitoid in galls formed by Cynipidae.

It has generally been assumed that tanaostigmatids initiate gall formation, but it now appears that inquilinism may be more common than previously suspected. In Brazil *Tanaostigmodes fernandesi* has been reported as an inquiline in an *Anadiplosis* (Cecidomyiidae) gall on *Machaerium aculeatum* (Leguminosae) (Fernandes *et al.*, 1987; LaSalle, 1987). Galls that contained the tanaostigmatid were larger, and of a different shape and consistency, than the normal cecidomyiid gall. The cecidomyiid larvae were never found in galls with *T. fernandesi*, and it was presumed that they were killed by the inquiline larva which then took over the gall. In Costa Rica, *Tanaostigma coursetiae* has been reared as an inquiline from galls of *Euphalerus certus* (Psyllidae) on *Lonchocarpus atropurpureus* (Leguminosae) (P. Hanson, pers. comm.).

Economic importance. Only one case of an economically important tanaostigmatid is known from the New World. Dozier (1932) reported on the biology of *Tanaostigmodes haematoxyli*, which was at that time a severe pest infesting seeds of logwood, *Haematoxylon campechianum* (Leguminosae) in Haiti. Logwood, or 'campeche', is used in the dye industry, and was also the chief honey plant on the island at that time. *T. haematoxyli* caused such extensive damage to seed pods that in some areas it prevented natural reseeding of this tree.

Identification. The New World species of Tanaostigmatidae were revised by LaSalle (1987), who included keys to all genera and species. The key below is based on this work but excludes *Minapis*, which is only known from Argentina.

Key to genera of Tanaostigmatidae occurring in Central America

1 Large flattened white setae present on head and/or mesosoma ***or*** if such setae are absent ***then*** all funicular segments distinctly wider than long ***and*** stigmal vein is slender, straight or only slightly curved and more or less perpendicular to postmarginal vein. . *Tanaostigma*

— Without large flattened white setae; funicular segments not all distinctly wider than long; stigmal vein various, if straight and perpendicular to postmarginal vein then stigma distinctly swollen. 2

2 Face and frons without large setiferous punctures. .... *Tanaostigmodes* (most)

— Face and frons with large setiferous punctures. 3

3 Propodeum with strong median carina and very strong plicae which converge and meet medially before posterior margin of propodeum....... .... *Microprobolos*

— Propodeum not as above, if plicae present then they are parallel and do not meet medially... 4

4 Interantennal projection present and prominent; antenna with funicle segments 1 to 5, and usually also 6, distinctly longer than wide....... .... *Tanaoneura*

— ***Either*** with interantennal projection very small or absent, ***or*** with at least several funicular segments not distinctly longer than wide........ ....*Tanaostigmodes* (few)

Synopsis of the Costa Rican fauna

LaSalle (1987) recorded ten species in the four genera present in southern Central America (Nicaragua to Panama), but the recent collecting programme in Costa Rica has yielded several new species and there are perhaps between 25 and 50 species in the region.

They occur primarily at low to mid altitude sites, from sea-level up to about 1500 metres.

Microprobolos. This genus comprises a single described species, *M. titan*, which occurs from Mexico south to Brazil, and has been collected in lowland forest on the Pacific coast of Costa Rica. Biology: host-plant is unknown.

Tanaoneura. A genus comprising 11 described species, four of which are present in Central America: *T. darwini*, *T. hirticoxa*, *T. portoricensis* and *T. smicropleura*. Recent extensive sampling has yielded several undescribed species. Biology: *T. hirticoxa* and *T. smicropleura* in galls on *Inga*; *T. portoricensis* in seed galls on *Inga* (Leguminosae).

Tanaostigma. A genus comprising 11 described species, two of which are present in Central America: *T. coursetiae* and *T. impilum*. Biology: *T. coursetiae* has been found in psyllid galls on *Lonchocarpus* (Leguminosae).

Tanaostigmodes. A genus with about 50 described species, 45 of which occur in the New World. Three described species are present in Central America: *T. aulafrons*, *T. basilaris* and *T. gracilis*, but recent sampling has yielded several undescribed species. Biology: *T. basilaris* has been reared from seeds of *Triplaris* (Polygonaceae).

11.17 TETRACAMPIDAE

Paul E. Hanson

Diagnosis. Body length 0.8 to 3.3 mm; mostly metallic green. *Antenna 12-segmented* in most species, often including *one anellus and usually six funicular segments* (rarely five or seven); clava sometimes undi-

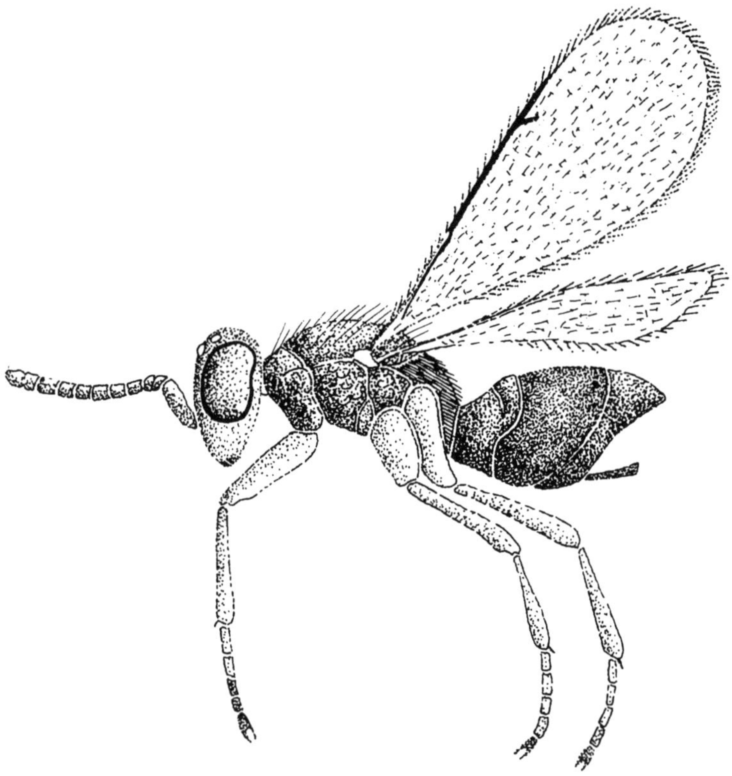

Fig. 11.170. *Epiclerus* sp. (Tetracampidae).

vided. Pronotum long, from three-quarters as long as to longer than mesoscutum, rounded without prominent anterior corner; notauli complete; scutellum often with two pairs of long bristles; *propodeum usually setose medially or submedially. Fore wing usually very setose* (at least in New World species); marginal vein relatively long. *Tarsi 5-segmented in females, 4-segmented in males* (male tarsi 5-segmented in one holarctic genus); *fore tibial spur small and nearly straight.* Metasoma with a short petiole; hind edge of basal tergite strongly invaginated.

Classification and distribution. The Tetracampidae is a small family whose relationships are poorly understood, but in some features it is intermediate between the Eulophidae and Pteromalidae. Worldwide there are about 50 species and 14 genera, which have been classified in three subfamilies. The subfamily Mongolocampinae, comprising three genera from the Eastern Palaearctic Region, is placed in Tetracampidae by Sugonjaev (1971), but Bouček (1988) suggested that this subfamily is closer to Aphelinidae. Species of Platynocheilinae, which contains just one holarctic genus, have the base of the fore wing bare, the marginal vein in males very swollen and five tarsal segments in both sexes. Tetracampinae, which is cosmopolitan and contains the remaining ten genera, have the base of the fore wing densely setose and the males with 4-segmented tarsi. In the Tetracampinae only *Epiclerus* and *Dipriocampe* have been reported from the New World, and the latter appears to be restricted to North America.

Biology. The biologies of most species of Tetracampidae are unknown, but the few host records that are available suggest they attack leaf-miners and eggs. Species of *Epiclerus* are recorded from Agromyzidae. Other tetracampids are endoparasitoids in insect eggs: *Dipriocampe* in eggs of Diprionidae, and *Foersterella* and *Cassidocida* from eggs of cassidine Chrysomelidae (Nakamura & Abbas, 1987).

Identification. Bouček and Askew (1968) provided a list of the world species, but the group was known only from the Old World at that time. Yoshimoto (1978) subsequently described two species of *Epiclerus* from the New World (one from North America and one from Cuba), and Bouček (1993*b*) has recently described two species, one *Epiclerus* and one *Platynocheilus*, from Canada.

Synopsis of the Costa Rican fauna

Only *Epiclerus* is known from Costa Rica (400–2100 m altitude), but *Dipriocampe* could possibly occur in northern Central America, in pine forests where its diprionid hosts are present. *Epiclerus* is distinguished from other tetracampids by its elongate petiole and very setose propodeum.

Epiclerus. This genus appears to be nearly cosmopolitan in distribution although it has not yet been reported from South America. A single species is known from Costa Rica.

11.18 TORYMIDAE

Paul E. Hanson

Diagnosis. Body length usually 1.0 to 7.5 mm (excluding ovipositor); mostly metallic green or blue, but sometimes entirely yellow; surface sculpture quite weak; always fully winged except in males of some species in figs. Antenna 13-segmented, usually with *one anellus and seven funicular segments*, rarely with two or three anelli; posterior margin of gena not carinate; back of head usually with horseshoe-like occipital carina. Mesoscutum with notauli complete, but sometimes shallow. Fore wing usually with *marginal vein quite long*, and with *postmarginal and stigmal veins very short*, the stigmal vein often sessile, except in Megastigminae where it is enlarged and circular. *Hind coxa often much larger than other coxae*; tarsi 5-segmented. Metasoma rarely broadly attached to the propodeum, first tergite (petiole) usually concealed but occasionally distinct; tergite IX very reduced, cerci elongate, almost peg-like; *female with ovipositor always extending beyond apex of metasoma* and sometimes longer than the body.

Classification and distribution. The Torymidae is a moderate sized family, containing 84 genera and about 1000 species world-wide. Previously, groups such as Ormyridae, Sycophaginae (= Idarninae) and even Agaoninae were included within the family, but

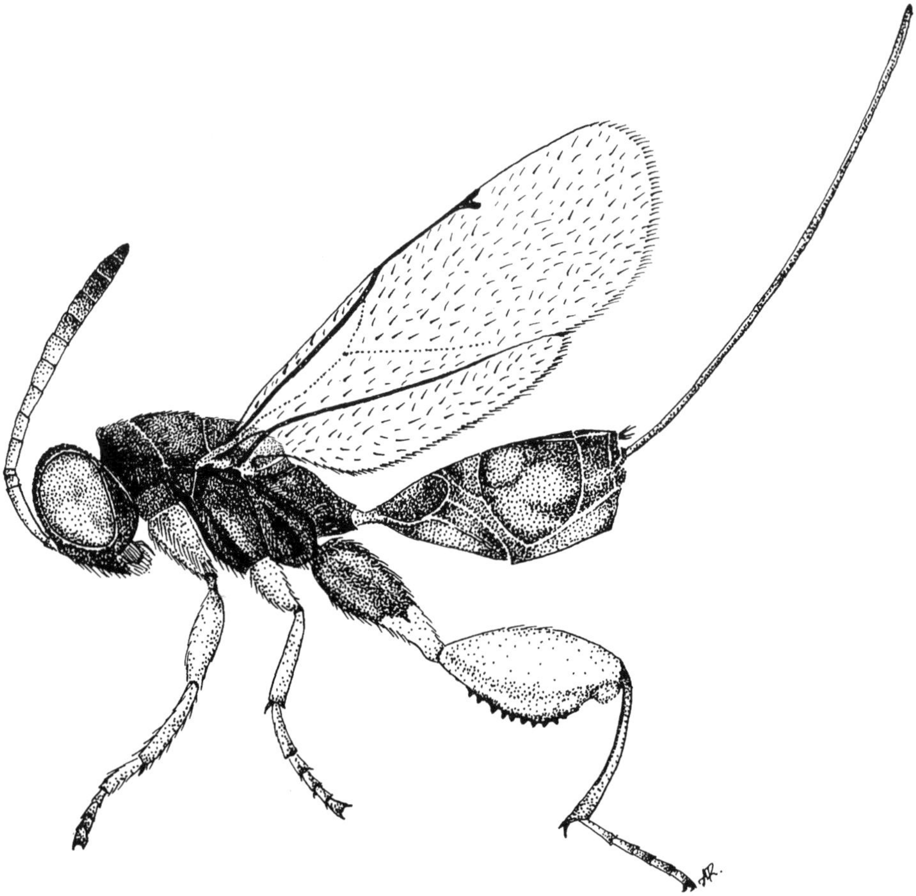

Fig. 11.171. *Podagrion* sp. (Torymidae).

recently (Bouček, 1988) redefined the Torymidae in a more restricted way. As currently defined the Torymidae (*sensu* Bouček) comprises four subfamilies—Thaumatoryminae, Toryminae, Megastigminae and Monodontomerinae. The first of these is only known only to occur in the Mediterranean region whilst the other three subfamilies are more or less cosmopolitan in distribution.

The Megastigminae is most diverse in the Australasian region (Bouček, 1988), and only the cosmopolitan genus *Megastigmus* is present in the New World. The Monodontomerinae is divided into five tribes (Bouček, 1978*b*, 1988)—Erimerini (North America), Chrysochalcissini (Palaeotropical), Monodontomerini (cosmopolitan), Palachiini (mostly palaeotropical, but also known from the Dominican Republic and Tobago in the Neotropics) and Podagrionini (cosmopolitan). The Toryminae is divided into two tribes, the Odopoiini comprising just one palaeotropical genus, and the cosmopolitan Torymini (Bouček, 1988).

Biology. The family Torymidae comprises species with a very diverse spectrum of biological habits, ranging from phytophages to carnivores. The former are mostly seed feeders, while the majority of carnivorous species are probably solitary, ectoparasitic idiobionts, several of which are facultatively hyperparasitic.

A brief summary of the host relationships of genera that occur in Central America is given in Table 11.15.

Most phytophagous torymids belong to the Megastigminae. Although a few Old World species of *Megastigmus* are parasitoids in plant galls (Noble, 1938), the larvae of most species (including all New World species, as far as is known) feed on the highly nutritious tissue in developing seeds (Milliron, 1949; Hussey, 1955; see Chapter 2.3). Most are restricted to a single plant genus, with the known host-plants including various conifers, Anacardiaceae, Aquifoliaceae (Grissell, 1989), arborescent Rosaceae and a few others (Lessmann, 1962). In Costa Rica *Megastigmus* has recently been found developing in seeds of *Symplocos* (Symplocaceae) (Hanson, in prep.). As in the North Temperate zone (e.g. Annila, 1982), fully developed larvae appear to diapause within the seeds for variable periods of time. Outside the Megastigminae, a few species of Old World *Torymus* are known to be phytophagous in seeds of various arborescent Rosaceae.

Among the carnivorous torymids most species are presumably solitary idiobiont ectoparasitoids. The biology of Podagrionini is very uniform in that all known species attack mantid eggs in their oothecae. The podagrionine larva feeds inside the egg case and both ectoparasitic and endoparasitic habits have been reported. The female wasps are sometimes phoretic on the adult female mantid, thus ensuring that they have access to freshly deposited eggs, which they must reach before the froth of the ootheca hardens (Bordage, 1913). The mating behaviour of one, *Podagrion mantis*, was described by Grissell and Goodpasture (1981).

Within the Monodontomerini species of *Monodontomerus* parasitize various lepidopterous and hymenopterous hosts. Specific host records from the Neotropics include *M. cubensis* from *Leucophobetron argentiflua* (Limacodidae), *M. mexicanus* from *Trypoxylon mexicanus* (Sphecidae), and *M. schrottkyi* from eumenine Vespidae (De Santis, 1979). Monodontomerines attacking sphecid and eumenine wasps probably have biologies similar to species that parasitize solitary bees (see Eves, 1970). The holarctic species, *M. aereus*, is somewhat unusual in that it is a gregarious endoparasitoid of lymantriid pupae, but it can also develop as a hyperparasitoid. In the latter instance it feeds externally on the ichneumonoid or tachinid primary parasitoid, unless the tachinid puparium happens to be very fresh, in which case *M. aereus* will develop endophagously (Muesebeck, 1931). Species of the South American genus *Perissocentrus* also attack lepidopterous pupae and can be facultatively hyperparasitic (Grissell, 1992).

In the Toryminae all species of the genus *Physothorax* and the subgenus *Nannocerus* (genus

Torymid taxon	**Host range**
MEGASTIGMINAE	
Megastigmus	seed-feeders
MONODONTOMERINAE	
Liodontomerus	parasitoids of *Bruchophagus* (Eurytomidae)
Monodontomerus	parasitoids of Lepidoptera or Hymenoptera
Pachytomoides	parasitoids of mantid eggs
Podagrion	parasitoids of mantid eggs
Torymoides	parasitoids in cecidomyiid galls
Zaglyptonotus	parasitoids of Bruchidae
TORYMINAE	
Diomorus	parasitoids of sphecids in twigs
Physothorax	associated with figs
Torymus	parasitoids in galls
T. (*Nannocerus*)	associated with figs

Table 11.15. The host relationships of torymid genera occurring in Central America. Data from De Santis, 1979; Grissell *in* Krombein *et al.*, 1979; Grissell, 1992.

Torymus) are strictly associated with figs, but very little is known about the nature of this association. In Costa Rica I have reared two different species of *Physothorax* from the same species of *Ficus*, and on the other hand, the same *Physothorax* species from more than one fig species. The wasps appear to emerge from enlarged ovary galls that are presumably caused by non-pollinating Agaonidae. In Florida a species of *Physothorax* has been reported as a parasitoid of cecidomyiid larvae that form galls on the inner wall of the syconium (Bouček, 1993*a*). The much smaller size of *Nannocerus* suggests that their biology differs from that of *Physothorax*. In both groups of fig-inhabiting torymids the males may be winged or wingless, as in many non-pollinating Agaonidae (see Chapter 11.1).

The vast majority of *Torymus* species are associated with various types of plant galls (Grissell, 1976). In Costa Rica they have been reared from galls induced by Cecidomyiidae, Cynipidae and Lepidoptera, but they are rarely encountered in galls formed by Psylloidea (Homoptera). Most *Torymus* species are solitary, ectophagous idiobionts and several may be facultatively hyperparasitic. Some may feed partially or even entirely upon the gall tissue as inquilines (Askew, 1961*a*, 1965). A few species of *Torymus* are associated with hosts other than gall-formers. For example, in Europe one species has been observed to be a hyperparasitoid developing ectophagously on a pteromalid parasitoid of beetle larvae in seed pods. This seemed to be obligate hyperparasitism as the torymid was never observed as a primary parasitoid of the beetle larva (Parnell, 1964*a*).

Torymids lay kidney-shaped to very elongate ovoid eggs in, on, or near the food source (Askew, 1966; Skrzypczynska, 1978). The eggs of *Monodontomerus* bear numerous minute recurved spines (Muesebeck, 1931). The larvae are hymenopteriform, and their cuticle bears setae, which are most conspicuous in parasitic species (Varley, 1937; Danks, 1971*a* & *b*; Askew & Ruse, 1974) but less so in phytophagous species (Hussey, 1955; Askew, 1966). The female pupa has the ovipositor externally visible and bent over its dorsum (Skrzypczynska, 1978).

Economic importance. Some of the northern temperate species of *Megastigmus* that attack conifer seeds are minor pests (Skrzypczynska, 1978). Among the parasitic torymids very few have been employed in biological control, probably because their hosts are seldom economically important pests.

Identification. Keys to torymid genera are available for Australasia (Bouček, 1988) and North America (Grissell, in prep.). The following key should permit identification of genera occurring in Central America (excluding Mexico). *Neopalachia* (Monodontomerinae: Palachiini) is known from the Caribbean and in the following key it would key out to Podagrionini, but can be separated by the apex of the hind tibia being truncate (not prolonged into a spur). *Perissocentrus*, which is known from Colombia, has long hind tibial spurs situated distinctly before the tibial apex, whereas in *Zaglyptonotus* the spurs are inserted on the apex of the tibia (Grissell, 1992). Also not included in the key is the tribe Erimerini, which can be distinguished from other torymids by having the hind tibia with only one apical spur.

Key to genera of Torymidae present in Central America

(For wingless males of species associated with figs see key in Chapter 11.1)

1 Macropterous, fore wing with stigmal vein enlarged and knob-like (Fig. 11.172); body entirely yellow, without any metallic sheen..... ....(Megastigminae).... *Megastigmus*

— Macropterous (rarely apterous), fore wing with stigmal vein not enlarged; body usually with at least some metallic sheen. 2

2 Hind femur distinctly enlarged and bearing at least three large teeth on ventral edge; hind tibia distinctly curved. (Fig. 11.171)................ ....(Podagrionini)........................ 3

— Hind femur not greatly enlarged, not distinctly toothed, or at most with two teeth, or sometimes finely serrate beneath. 4

3 Antenna with first flagellar segment as long as wide (i.e. not ring-like), clava very elongate (Fig. 11.173); metasternum (between hind coxae) with two submedian carinae, and at narrowest point narrower than diameter of coxal cavity (Fig. 11.175). .... *Pachytomoides*

— Antenna with first flagellar segment wider than long (i.e. ring-like), clava usually less elongate (Fig. 11.174); metasternum with a single median carina, and at narrowest point about equal to diameter of coxal cavity (Fig. 11.176). .... *Podagrion*

4 Groove between mesopleuron and metapleuron strongly sinuate, the metapleuron forming a distinct lobe (Fig. 11.177). ...(Toryminae).... 5

— Groove between mesopleuron and metapleuron virtually straight (Fig. 11.178). ....(Monodontomerini) 8

5 Hind femur without teeth. .... *Torymus*

— Hind femur with one or more teeth (Fig. 11.179). 6

6 Hind femur with at least two teeth (Fig. 11.179); mesoscutum often with coarse punctures. .... *Physothorax*

— Hind femur with only one tooth; mesoscutum usually without coarse punctures. 7

7 Propodeum smooth; ovipositor usually longer than body. .... *Torymus* (*Nannocerus*)

— Propodeum roughly sculptured; ovipositor usually shorter than body................................ .... *Diomorus* (+ a few *Torymus*)

8 Hind femur with a broad conspicuous tooth; scutellum with a distinct transverse (frenal) groove (Fig. 11.180). .... *Monodontomerus*

— Hind femur without a tooth; scutellum without a transverse groove. 9

9 Antenna with one anellus and seven funicular segments; apex of hind tibia with longest spur nearly equal to length of first tarsal segment (Fig. 11.181). .... *Zaglyptonotus*

— Antenna with two anelli and six funicular segments; apex of hind tibia with longest spur much shorter than length of first tarsal segment. 10

10 Head with occipital carina (as in Fig. 11.180)..... .... *Torymoides*

— Head without evident occipital carina. ....*Liodontomerus*

Synopsis of the Costa Rican fauna

Three subfamilies of Torymidae are present in Central America. Two tribes of Monodontomerinae are presently known from the region, but it is possible that one or two other tribes may eventually be found—Erimerini is present in North America and Palachiini is known from the Dominican Republic and Tobago. Within Costa Rica species of Torymidae have been collected at all altitudes. Ten genera are known or expected to occur in Costa Rica and these appear to be represented by between 50 and 100 species, the majority of which belong to *Torymus* or *Podagrion*.

MEGASTIGMINAE

Only one genus is known to occur in the New World. As far as is known all the New World species are phytophagous.

Megastigmus [CR]. A cosmopolitan genus previously unreported from the Neotropics. At least two or three species have been found at all altitudes in Costa Rica. A key to the nearctic species of *Megastigmus* was provided by Milliron (1949). Biology: Costa Rican species have been reared from seeds of *Ilex* (Aquifoliaceae) and *Symplocos* (Symplocaceae), and in Mexico *M. albifrons* has been found attacking seeds of *Pinus* (Pinaceae) (Yoshimoto, 1979; Grissell, 1985). An introduced species has been reported on *Schinus terebinthifolius* (Anacardiaceae) in Florida (Habeck *et al*., 1989).

MONODONTOMERINAE

MONODONTOMERINAE: Monodontomerini. This tribe is infrequently collected in Costa Rica.

Liodontomerus [EX]. Primarily a holarctic genus, where it is represented by about 15 species, with a few isolated records from the Neotropics. Biology: North American *Liodontomerus* species are parasitoids of *Bruchophagus* species (Eurytomidae) in leguminous seeds (Grissell *in* Krombein *et al*., 1979).

Monodontomerus. Primarily a holarctic genus comprising 20 to 30 species. Three species are

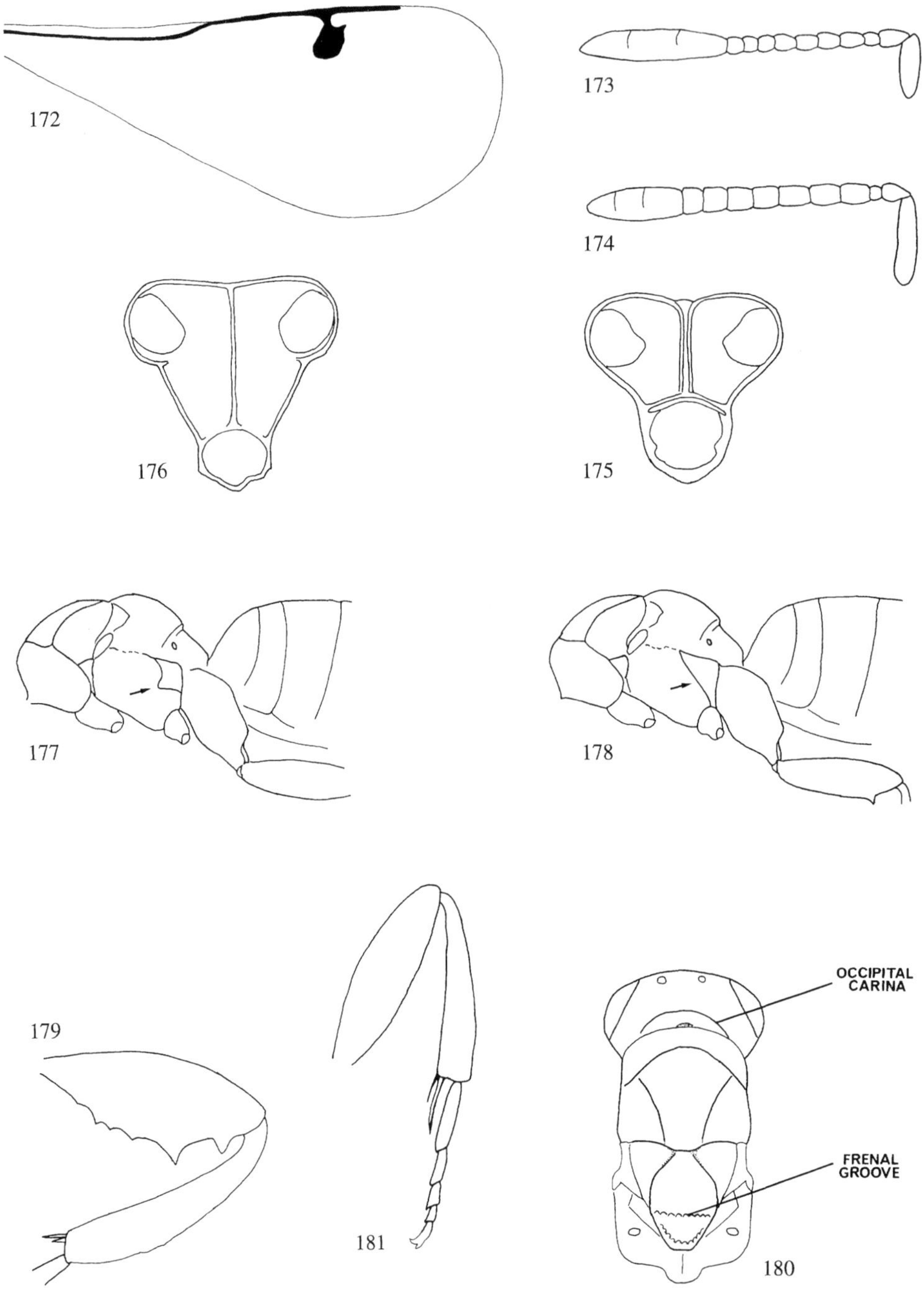

Figs 11.172–11.181. Torymidae. Fig. 11.172. *Megastigmus*, fore wing. Figs 11.173–11.176. Podagrionini; 11.173, *Pachytomoides* sp., antenna; 11.174, *Podagrion* sp., antenna; 11.175, *Pachytomoides* sp., metasternum; 11.176, *Podagrion* sp., metasternum. Figs 11.177–11.178. Mesosoma, lateral; 11.177, *Torymus* sp.; 11.178, *Monodontomerus* sp. Fig. 11.179. *Physothorax* sp., hind leg. Fig. 11.180. *Monodontomerus* sp., dorsal view of head and mesosoma. Fig. 11.181. *Zaglyptonotus* sp., hind leg.

known from South America (De Santis, 1979) and at least two species occur in mid altitude sites (800-2000 m) in Costa Rica. Biology: in Honduras *M. viridiscapus* has been reared from *Spodoptera* (Noctuidae) (specimen label data, Zamorano, Honduras) while in North America *Monodontomerus* species are recorded as primary or secondary parasitoids of macrolepidopteran and symphytan larvae, and as parasitoids in the nests of sphecids and bees (Grissell *in* Krombein *et al.*, 1979).

Torymoides (= *Dimeromicrus*). A cosmopolitan genus comprising 30 to 40 species. It is represented in Costa Rica by *T. cecidomyiae* (yellow body, green head). Biology: *T. cecidomyiae* has been reared from a stem gall on *Psidium guineense* (Myrtaceae) and from a cecidomyiid leaf gall on mangrove (Rhizophoraceae).

Zaglyptonotus[CR]. A genus known from just two North American species (Grissell *in* Krombein *et al.*, 1979). In Costa Rica the genus has only been taken from the seasonally dry forests in the northwestern part of the country, which may represent the southern limit of its distribution. Biology: North American species have been reared from the immature stages of various holometabolous insects living in concealment—Curculionidae, Gelechiidae and Tephritidae (Grissell *in* Krombein *et al.*, 1979).

MONODONTOMERINAE: Podagrionini. In contrast to the Monodontomerini, species of this tribe are commonly collected, although they appear to be absent at altitudes above about 1500 metres. All known species of Podagrionini develop on mantid eggs in an ootheca. Keys to the nearctic species are given by Grissell and Goodpasture (1981).

Pachytomoides. This genus occurs in warmer regions of the world and comprises about 20 species.

Podagrion. This genus occurs in warmer regions of the world and is represented by about 60 species.

TORYMINAE

Three genera of this subfamily are known to occur in Central America, although one has not yet been collected in Costa Rica.

Diomorus[EX]. A cosmopolitan genus comprising fewer than ten species. One is reported from Guatemala (De Santis, 1979). Biology: parasitoids of stem-nesting sphecids and bees (Krombein, 1960; Parker & Bohart, 1966; Danks, 1971*b*).

Physothorax. A neotropical genus with about ten described species. Bouček (1993*a*) provided a key to seven of the species, but the two or three Costa Rican species appear to be undescribed. Biology: associated with the syconia of *Ficus* species (Moraceae) but the larval diet is unknown.

Torymus. With about 200 described species worldwide, and many more undescribed, this is the largest genus in the family. In Costa Rica this genus occurs at all altitudes and may comprise as many as 30 to 50 species. Grissell (1976) provided keys to the species of *Torymus* of western North America. Biology: species of *T.* (*Nannocerus*) are restricted to figs (Bouček, 1993*a*). A few European species are phytophagous in seeds, one or two are inquilines in galls, but the majority are parasitic (often facultatively hyperparasitic) in galls (Grissell *in* Krombein *et al.*, 1979).

11.19 TRICHOGRAMMATIDAE

John D. Pinto

Diagnosis. Among the smallest of insects, *0.2 to 1.5 mm* in length, mostly 1.0 mm or less. Body shape compact or elongate, but always lacking a distinct constriction at junction of meso- and metasoma; wings usually present. Colour light yellow to dark brown, often a combination of both; rarely orange, red or slightly metallic; cuticle smooth, rarely noticeably sculptured. Antenna with 2 to 9 (usually 3 to 7) flagellar segments, including 1-3 anelli, 0-2 funicular segments, and 1-5 claval segments; segment number, as well as characteristics of the setae and sensilla vary between the sexes in several genera (e.g. *Trichogramma*, *Trichogrammatoidea*, *Paratrichogramma* and *Ufens*). Mesosoma with a short pronotum and complete notauli. Fore wings varying in shape from extremely narrow and strap-like, to very broad and slightly rounded apically; *postmarginal vein entirely absent or extremely short*; *discal setae of both fore and*

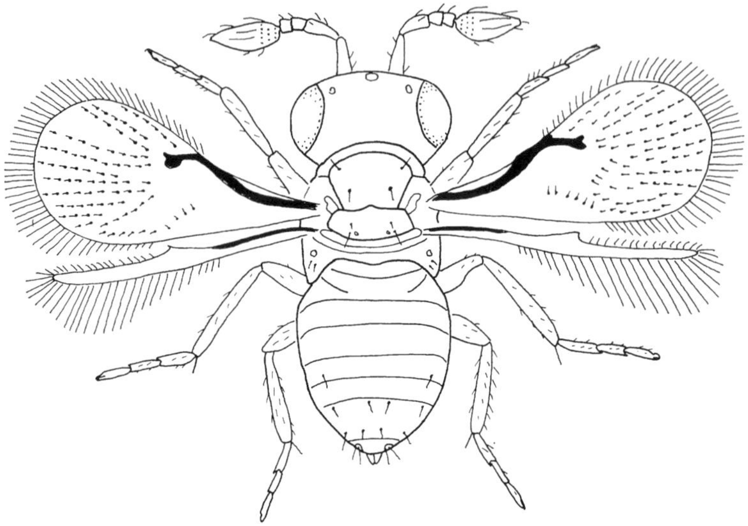

Fig. 11.182. *Trichogrammatoidea* sp. (Trichogrammatidae).

hind wings often linearly arranged. Tarsi 3-segmented. Ovipositor of varying length, strongly exserted or not.

Classification and distribution. The Trichogrammatidae is a cosmopolitan family containing over 500 described species in 79 genera. Several classifications of the family have been proposed. The most widely followed is that of Viggiani (1971) which divides the group into two subfamilies, the Trichogrammatinae and the Oligositinae. The Trichogrammatinae consists of the tribes Trichogrammatini and Paracentrobiini; the Oligositinae includes the Oligositini and Chaetostrichini. This classification is based primarily on male genitalia, and requires testing by additional characters. All four tribes are cosmopolitan and all occur in Costa Rica. Presently, those genera where the males are unknown cannot be confidently placed within this classification.

Distributions of genera remain poorly known but several common genera such as *Trichogramma*, *Oligosita* and *Paracentrobia* have representatives throughout the world and in a wide variety of habitats. The family remains poorly sampled everywhere but indications are that diversity is greatest in warm, mesic climates.

Biology. Trichogrammatids are solitary or gregarious, endophagous idiobionts in the eggs of other insects. The vast majority are primary parasitoids but one species of *Trichogramma* is known to occasionally develop as a facultative hyperparasitoid on *Telenomus* (Scelionidae) in lepidopteran eggs (Strand & Vinson, 1984). The few recorded exceptions of egg parasitism (e.g. from cecidomyid puparia; Viggiani, 1981) should be corroborated. In general, the biology of this family is poorly known, and most of the available information concerns *Trichogramma*, a relatively well studied group because of its importance in biological control of lepidopterous pests. Although most of our knowledge of the family is based on this genus, it should be kept in mind that *Trichogramma* may not be particularly typical of the family as a whole, and that the genera parasitizing eggs of Lepidoptera actually are in a minority in the family.

Trichogrammatids attack a wide variety of insect eggs, but those of Hemiptera, Homoptera, Lepidoptera and Coleoptera are the most common hosts. Some genera appear to have a restricted host range while others are notoriously polyphagous. For example, *Megaphragma* (unrecorded but probably in Costa Rica) has been associated only with eggs of Thysanoptera, and *Poropoea* is known only from the beetle family Attelabidae. In contrast, species of *Trichogramma* are notoriously non-specific, with certain species known to successfully develop not only in eggs of numerous families of Lepidoptera, but in those of five other orders of insects as well (Salt, 1935; Pinto *et al.*, 1986). Yet, *Trichogramma* does show host preferences, and certain potential hosts are avoided. Flanders (1937) noted that the number of natural hosts of *Trichogramma* was fewer than the known suitable hosts based on laboratory studies. He attributed this to habitat preferences, and indeed, it appears that habitat may be of primary importance in determining the host array of any particular species (Pinto & Oatman, 1988).

The process of host location and host recognition has been fairly well studied in *Trichogramma*. Stimuli that affect searching behaviour include: host sex pheromones (Noldus, 1989), accessory gland secretions deposited by a female moth around her eggs (Nordlund *et al.*, 1987), moth scales (Zaborski *et al.*, 1987), and physical characteristics of the host egg such as its size, colour, and texture (Renou *et al.*, 1989). Foraging behaviour and host recognition may vary between strains of the same species (Pak & DeJong, 1987; Chassain *et al.*, 1988).

Insects ovipositing on plants sometimes place their eggs on the plant surface or insert them at varying depths within plant tissue. Several genera, such as *Trichogramma* and *Trichogrammatoidea*, parasitize eggs placed on the surface of plants, although at least one species, *Trichogramma semblidis*, is known to enter the galleries of bark beetles (Scolytidae) and parasitize their eggs (Michalski & Seniczak, 1974). Others, such as *Oligosita* and *Paracentrobia* commonly parasitize Homoptera and Hemiptera eggs placed beneath the plant epidermis. At least two genera, *Hydrophylita* and *Prestwichia*, attack eggs of aquatic insects (e.g. Odonata, Notonectidae, Dytiscidae), and are known to swim beneath the surface of the water to search for hosts (Lubbock, 1863; Matheson & Crosby, 1912; Henriksen, 1922).

Adult females of certain Trichogrammatidae are phoretic on their hosts, which presumably facilitates access to host eggs. Phoresy has been recorded in a species of *Oligosita* associated with a tettigoniid (*Conocephalus longipenne*) in Java (Ferriére, 1926), and in *Pseudoxenufens forsythi* associated with the brassolid butterflies, *Opisphanes cassina* and *Caligo eurilochus*, in Ecuador (Malo, 1961; Yoshimoto, 1976*a*). Recently, females of a new species of *Brachista*, which occurs from the southwestern United States to Costa Rica, have been found commonly attached to adult robber flies (Asilidae) of the genus *Efferia* (Pinto, 1994).

The female wasp punctures the host egg chorion with her ovipositor and places one or more eggs within. *Trichogramma pretiosum* females, ovipositing in eggs of *Helicoverpa virescens*, inject a venom that stops host development. The host is digested before the eggs hatch, allowing the larva to completely consume the contents of the eggs within ten hours of hatching (Strand, 1986). Successful development of the trichogrammatid larva depends on various factors including the structure of the host egg chorion, i.e. the ability of the chorion to resist desiccation (Pak *et al.*, 1990).

Two general forms of first instar larvae have been described in the family, mymariform (e.g. in *Ophioneurus* and *Poropoea*) and sacciform (e.g. in *Oligosita*, *Trichogramma* and *Trichogrammatoidea*). Mymariform larvae bear long setae and have the posterior end drawn out into a long 'tail'. Sacciform larvae on the other hand are globular to cylindrical in shape and lack setae. The number of larval instars reported in the family varies. One has been reported in *Oligosita* (Bakkendorf, 1934), two in *Trichogrammatoidea* and *Ophioneurus* (Bakkendorf, 1934; Hutchinson *et al.*, 1990), three in *Trichogramma* (Pak & Oatman, 1982; Manweiler, 1986), and five in *Poropoea* (Silvestri, 1916). Whether this variation is due to actual differences or errors in observation remains to be determined. The mature larva is generally robust, distinctly segmented and without integumental spines or setae; the tracheal system appears to be lacking. Pupation takes place within the remains of the host egg and the adult parasitoid emerges by biting a hole in the chorion of the egg.

Time needed to complete development depends largely on temperature. In *Trichogramma evanescens* and *Trichogrammatoidea lutea* the duration of development varies from about six days at 30°C to four to five weeks at 15°C (Bourarach & Hawlitzky, 1989). In temperate climates species of *Trichogramma* are known to overwinter within the host egg as diapausing pupae (Parker & Pinnell, 1971; Curl & Burbutis, 1977). This probably occurs in other genera as well (e.g. see Bakkendorf, 1934).

The number of trichogrammatids developing within a single egg depends on egg size. In *Trichogramma* it can vary from one on small hosts, to 50 or more on large hosts (Flanders, 1935). An average of 34 individuals of *Pseudoxenufens forsythi* (as *Xenufens ruskini*) have been reported emerging from single eggs of the brassolid butterfly, *Caligo eurilochus*, in Ecuador (Malo, 1961). As in other egg parasitoids female trichogrammatids apply an external marking pheromone by wiping the extruded ovipositor across the host egg after oviposition. This external marker persists for 12 to 18 hours and subsequently host necrosis serves as an internal marker (Strand, 1986). Both superparasitism and multiparasitism are known to occur in *Trichogramma*, but they usually occur before the onset of host necrosis (Klomp *et al.*, 1980; Waage & Godfray, 1985; Thomson & Stinner, 1989).

Polymorphism of adults has been observed in a few trichogrammatid species and results from development occurring in different hosts (Salt, 1937) or from development occurring in hosts exposed to different temperatures (Flanders, 1931). Sex ratios vary considerably within species, and probably among species and genera as well. Males are unknown or extremely rare in some species of *Trichogramma*. It is likely that several species of this genus occur as both uniparental (= thelytokous) and biparental (= arrhenotokous) forms (Stouthamer *et al.*, 1990*b*). *Xenufens*, a relatively common genus in Costa Rica, is known only from females and is probably thelytokous. Adult trichogrammatids appear to be relatively short-lived (about a week) and reproductive capacity varies from about 30 to 200 eggs per female (Clausen, 1940*b*).

Economic importance. Although *Trichogramma* species have not been notably successful in classical biological control, they are the most commonly and widely used entomophagous arthropods in programmes of mass rearing for periodic liberation in the field, i.e. augmentative releases (Ridgway *et al.*, 1981; King *et al.*, 1985; Ridgway & Morrison, 1985; Hassan, 1988). In Nicaragua *T. pretiosum* is being released against various noctuid pests: *Alabama argillacea* and *Trichoplusia ni* in cotton, *Pseudoplusia includens* in beans, and *Helicoverpa zea* in various crops. Although at present *Trichogramma* is not being used in Costa Rica, it was used in the past and there is talk of doing so again in the future. The only other genera currently used to any extent in biological control are *Trichogrammatoidea*, which has a more limited distribution than *Trichogramma*, but is also associated with Lepidoptera eggs (Nagaraja, 1978, 1987; Hutchinson *et al.*, 1990), and *Doirania*, used against tettigoniid (Orthoptera) pests in oil palm in Papua New Guinea (Takis Solulu, *in litt.*).

Species of *Trichogramma* are also found occurring naturally in field collected eggs of Central American pest insects. In Costa Rica *T. pretiosum* has been reared from *Helicoverpa* eggs on maize, and from noctuid and sphingid eggs on potato (Pinto, unpubl.). Three species—*T. beckeri*, *T. pretiosum*, and *T.* sp. (as *T. semifumatum*)—have been reared from the mahogany shoot borer, *Hypsiphila grandella* (Pyralidae) (Nagarkatti *in* Whitmore, 1976); *Trichogrammatoidea hypsipylae* has also been reported from the same host (Nagaraja, 1978). Other records from Central American pests include *Trichogramma pretiosum* from *Spodoptera* sp. on maize and from *Alabama argillacea*, *T. exiguum* from *Helicoverpa zea* on maize and from *Spodoptera frugiperda*, and *T. atopovirilia* from *Spodoptera* sp. on maize and from *Spodoptera frugiperda* (Pinto, unpubl.).

Identification. A key to the world genera of Trichogrammatidae was published by Doutt and Viggiani (1968). This extremely useful work includes, in addition to the key, complete generic synonymies, diagnoses and illustrations based on type species, and complete species listings. A key, in Spanish, to the genera of Trichogrammatidae of Peru was published by Ruiz and Korytkowski (1979). A useful summary of the geographic distribution of all trichogrammatid genera occurs in Yousuf and Shafee (1987). These authors also recently published a catalogue of genus-group names and a species checklist (Yousuf & Shafee, 1986*a* & *b*).

The Trichogrammatidae remains very poorly known taxonomically and only a small fraction of its diversity is appreciated. Consequently, generic concepts will remain tentative and in a state of flux in the foreseeable future. This should be kept in mind when using the available keys. For example, several of the species collected in Costa Rica cannot be unambiguously assigned to genus at this time.

Synopsis of the Costa Rican fauna

Only three genera of Trichogrammatidae had been reported from Central America prior to collections preparatory to this volume, and only two, *Trichogramma* and *Trichogrammatoidea*, had been reported from Costa Rica (Nagaraja, 1978; De Santis 1979, 1981, 1989). Recent collections have increased the genera known for Costa Rica to 22. Eight of these (*Brachista, Ceratogramma, Hydrophylita, Lathrogramma, Pintoa, Trichogrammatella, Trichogrammatomyia* and *Xenufens*) are known only from the New World. One genus, *Haeckeliania*, previously unknown in the New World, turns out to be quite common in collections from North and Central America. Additional genera not yet recorded from Costa Rica but likely to occur there include *Doirania, Epoligosita, Megaphragma, Mirufens, Tumidiclava* and *Ufens*.

Within Costa Rica Trichogrammatidae have been collected at altitudes up to 3500 metres, but most come from forested areas at relatively low elevations (<1000 m). The only genus commonly found in the few available collections from high altitude, primarily from the Cerro de la Muerte (2500–3500 m), is *Trichogramma*. Generic diversity can be quite high at any one site. For example, 17 of the 22 genera known to occur in the country have been collected in Santa Rosa and Guanacaste National Parks. Except for one species of *Trichogrammatoidea, Trichogrammatella tristis*, and a few *Trichogramma* species, the hosts of the Costa Rican trichogrammatids are unknown. The host records given below are primarily from the North American literature.

Aphelinoidea^CR^. A cosmopolitan genus characterized by having an antenna with no funicle and a 2- to 3-segmented club; fore wing with a reduced stigmal vein, densely setose, without distinct setal lines. Biology: parasites in eggs of Homoptera.

Brachista^CR^. *Brachista* is the correct name for a group of New World species formerly assigned to *Giraultiola* (Pinto, 1993). They have an antenna with no funicle and a 4-segmented club; metasoma elongate. Biology: parasites in eggs of Coleoptera and Homoptera.

Ceratogramma^CR^. A neotropical genus ranging from Costa Rica south to Argentina. Species have an antenna with three anelli, two funicle segments and a 3-segmented club. Biology: parasites in eggs of Coleoptera.

Chaetogramma^CR^. A North and Central American genus characterized by having an antenna with one funicle segment (may be incompletely divided) and an unsegmented club. Biology unknown.

Chaetostricha^CR^. A cosmopolitan genus, characterized by having an antenna with a 2-segmented funicle, the first segment of which is short and appressed to second, and a 3-segmented club; ovipositor usually exserted. Biology: parasites in eggs of Hemiptera (Heteroptera).

Haeckeliania^CR^. A widespread genus represented in the New World, Australia and Asia, and characterized by having an antenna with no funicle and a complex 5-segmented club, with elongate placoid sensilla extending well beyond the surface. Biology unknown.

Hydrophylita^CR^. A New World genus, species of which have very narrow, pointed fore wings. Biology: parasites in eggs of Odonata in water.

Ittys^CR^. A palaearctic and New World genus characterized by having an antenna with two funicle segments, the second shorter than the first and about as long as wide, and a 3-segmented club. Biology: parasites in eggs of Hemiptera (Heteroptera) and Homoptera.

Lathrogramma^CR^. A New World genus characterized by having an antenna with no funicle and a 5-segmented club, the first segment of which is short, inconspicuous and appressed to second. Biology: parasites in eggs of Homoptera.

Lathromeroidea^CR^. A New World, Australian and palaearctic genus characterized by having an antenna with no funicle and a 5-segmented club; fore wing with setal lines conspicuously converging near apex of the retinaculum. Biology: parasites in eggs of Odonata, Homoptera and aquatic Hemiptera.

Oligosita^CR^. A cosmopolitan genus, believed to be the most species-rich group in the family. The genus is characterized by having an antenna with a single anellus, one funicle segment and a 3-segmented club; fore wing relatively narrow, the disc often nearly devoid of setae. Biology: parasites in eggs of Coleoptera, Hemiptera (Heteroptera), Homoptera and Orthoptera.

***Paracentrobia*^CR^**. A cosmopolitan genus, similar to *Ittys*, except the second funicle segment is wider than long. Biology: parasites in eggs of Homoptera.

***Paratrichogramma*^CR^**. A New World, Australian and South African genus. The antenna of the female has one funicle segment and an unsegmented club; fore wing with a short marginal vein that does not reach the margin, stigmal vein is obsolescent. Biology: parasites in eggs of Lepidoptera.

***Pintoa*^CR^**. A New World genus, similar to *Oligosita* except that the funicle segment is closely associated with the club and bears a long curved placoid sensillum. Biology unknown.

***Poropoea*^CR^**. Primarily a holarctic genus, characterized by having an antenna with two funicle segments and a 3-segmented club; venation sigmoid, with the marginal vein not reaching wing margin. Biology: parasites in eggs of Coleoptera.

Trichogramma. A cosmopolitan genus comprising about 140 described species; eight species are known from Costa Rica but several more undoubtedly occur. The female antenna consists of two funicle segments and an unsegmented club; fore wing venation is sigmoid. Keys are available for the world species (Voegele & Pintureau, 1982; Sorokina, 1993), the New World species (Nagaraja & Nagarkatti, 1973) and the Peruvian species (Ruiz & Korytkowski, 1979). Two morphologically aberrant species were recently described from Costa Rica (Pinto, 1992). Because of the widespread misapplication of names (Pinto *et al.*, 1978), unforeseen intraspecific variation (Pinto *et al.*, 1989), and what remains an extremely poorly known fauna with numerous undescribed species, current keys to *Trichogramma* are unfortunately of little value. Identification is complicated by the existence of morphologically homogeneous species complexes. Investigators are attempting to use reproductive compatibility and allozymic variation to clarify species limits in these forms (e.g. Nagaraja, 1987; Pinto *et al.*, 1991, 1992). Biology: parasites in the eggs of a wide range of insects, including Coleoptera, Diptera, Lepidoptera, Hymenoptera, Hemiptera and Neuroptera.

***Trichogrammatella*^CR^**. A monotypic genus previously known only from Trinidad and Panama. It has an antenna with an elongate 5-segmented club; fore wing with an elongate stigmal vein that curves gradually from the marginal vein. Biology: Costa Rican specimens have been reared from eggs of Membracidae (Homoptera).

Trichogrammatoidea. A widespread genus, except in the Holarctic region. Females are characterized by having an antenna with two funicle segments and an unsegmented club; fore wing venation sigmoid. It may be separated from *Trichogramma* by the absence of the RS-1 setal line beneath the stigmal vein, and by the form of the male genitalia. Biology: parasites in eggs of Lepidoptera.

***Trichogrammatomyia*^CR^**. North and Central America. Females have antennae similar to *Trichogramma* and *Trichogrammatoidea*, but their wing venation is normal, not sigmoid. Biology: parasites in eggs of Lepidoptera.

***Uscana*^CR^**. A cosmopolitan genus characterized by having an antenna with no funicle and a 4-segmented club. Biology: parasites in eggs of Coleoptera.

***Xenufens*^CR^**. A New World genus, characterized by having an antenna with two short funicle segments and a short 2-segmented club, with an incomplete third segment; fore wing venation sigmoid. Biology: parasites in eggs of Lepidoptera.

***Zagella*^CR^**. A New World genus, characterized by having an antenna with two funicle segments, the first of which is short and appressed to the second, and a 3-segmented club. *Zagella* species are similar to *Chaetostricha* except that the ovipositor is not exserted. Biology unknown.

Genus questionable. Several undescribed species in the Costa Rican fauna cannot be placed to genus. Most key unsatisfactorily to *Lathromeroidea*, *Pterygogramma* or *Uscana* in Doutt and Viggiani (1969).

12 The ichneumonoid families

INTRODUCTION

Ian D. Gauld and Scott Richard Shaw

The superfamily Ichneumonoidea is a monophyletic group that is characterized by at least six apomorphic features:

- Adult mandible with two teeth. The presumed plesiomorphic condition is possession of three or four teeth. The derived condition occurs convergently in some other lineages of the Microhymenoptera. In some ichneumonoids the number is further reduced to one. In the Alysiinae, some opiine braconids, the Diplazontinae and a few banchine ichneumonids it is secondarily increased to three or four (Mason, 1987*a*; Sharkey & Wahl, 1992).
- Prepectus fused to the posterior lateral (vertical) margin of the pronotum, mesothoracic spiracle positioned directly above the prepectus, and external pit indicating origin of spiracular occlusor muscle lying near posterior pronotal margin (Gibson, 1985).
- Fore wing with veins *C* and *Sc+R+Rs* contiguous for almost their entire basal length, with the virtual obliteration of the costal cell. This condition has been independently derived in some Rhopalosomatidae and Sphecidae.
- Metasoma with sternite I divided into a heavily sclerotized anterior part and a comparatively weakly sclerotized posterior part (Mason, 1981*a*, 1987*a*).
- Metasomal tergites I and II articulating by means of a pair of dorsolateral condyles on the hind margin of tergite I and on the anterior margin of tergite II (Mason, 1987*a*). The plesiomorphic condition is that these tergites do not articulate on condyles and can telescope.
- Cephalic capsule of final instar larva with a hypostomal spur (Sharkey & Wahl, 1992). This structure has been secondarily lost in some ichneumonoids, usually taxa that do not spin a cocoon (Short, 1978).

Another possible apomorphic feature of the group is the presence of longitudinal and transverse carinae delineating a number of discrete areas on the propodeum, but these structures have been lost in many groups (Sharkey & Wahl, 1992).

The number of families placed in the Ichneumonoidea has fluctuated, but it has always included the Braconidae and Ichneumonidae. Some workers have included the Stephanidae, Megalyridae and even the evanioid families (e.g. Pagliano, 1986; Pagliano & Scaramozzino, 1990), but these families lack the apomorphic features characterizing the superfamily and are more logically treated in separate superfamilies. Various groups of ichneumonids and braconids, such as the Paxylommatinae, Agriotypinae, Apozyginae and Aphidiinae, have from time to time been accorded family-group status. Despite this the first two clearly belong in the Ichneumonidae and the last two are part of the Braconidae (Gauld, 1984; Gauld & Bolton, 1988; Sharkey & Wahl, 1992). The Paxylommatinae, Agriotypinae and Apozyginae are all small and infrequently collected groups and arguments about their status have involved only a few specialists (e.g. Mason, 1971, 1978*a*,1981*a*; Achterberg, 1976) The Aphidiinae, however, is a large, widespread, and economically important subfamily, and many workers (e.g. Stary, 1960, 1966, 1970; Mackauer & Stary, 1967; Mackauer, 1968; Finlayson *in* Evans, 1987*a*) continue to insist, despite the overwhelming phylogenetic evidence (e.g. Achterberg 1984*a*; Quicke & Achterberg, 1990), on treating it as a distinct family.

The two families treated here, the Ichneumonidae and Braconidae, comprise one of the most species-rich superfamilies of organisms known. Worldwide they probably include at least 150,000 species. In Costa Rica they are amongst the most abundant and commonly collected of all Hymenoptera. They may be separated by the following key:

Key to the families of Ichneumonoidea present in Costa Rica

1 Fully-winged species.. 2
— Brachypterous or apterous species. 4

2 Fore wing with *2m-cu* distinct (Fig. 12.01).
[1st discal and submarginal cells confluent, forming a single large discosubmarginal cell; hind wing with *rs-m* joining *Rs* after divergence of *Rs* from *Sc+R+Rs* (Fig. 12.01).] ....
.... **Ichneumonidae** (most) (p. 390)
— Fore wing with *2m-cu* absent (Figs 12.02–12.04). 3

3 Hind wing with *rs-m* joining *Rs* distinctly after divergence of *Rs* from *Sc+R+Rs* (Figs 12.01, 12.02); metasoma with tergites II and III separated by a membranous articulation, the hind margin of the former overlapping the anterior margin of the latter.
[Fore wing with marginal cell enclosed (Fig. 12.02).] ....
.... **Ichneumonidae** (*Ophionellus*) (p. 409)
— Hind wing with *rs-m* joining *Sc+R+Rs* before, or more or less at point of divergence of *Rs* (Fig. 12.03) or with veins indistinct (Fig 12.04); metasoma with tergites II and III confluent and fused (Fig. 12.06).
[Fore wing with 1st discal and submarginal cells separated or confluent; marginal cell enclosed or open.] ....
.... **Braconidae** (most) (p. 431)

4 Metasoma with tergites II and III separated by a membranous articulation, the hind margin of the former overlapping the anterior margin of the latter (Fig. 12.05); entirely apterous, female, ant-like insects. ....
.... **Ichneumonidae** (*Gelis*) (p. 416)
— Metasoma with tergites II and III confluent and fused (Fig. 12.06); generally brachypterous with at least a vestige of the fore wing discernible; males or females. ....
.... **Braconidae** (few) (p. 431)

12.1 ICHNEUMONIDAE

Ian D. Gauld

Diagnosis. Body 2.0 to 61.0 mm in length (excluding ovipositor and antennae), fore wing length 2.0 to 35.0 mm, rarely brachypterous or apterous; sexual dimorphism very slight to moderate. *Antenna usually more than 16-segmented*, generally with 30+ segments, not geniculate, and usually not apically clavate; mandible usually bidentate, rarely unidentate or apparently tridentate; maxillary palp 5-segmented, labial palp 4-segmented, sometimes with both reduced. Upper hind corner of pronotum extending backwards to more or less touch tegula. Fore wing with four or more enclosed cells; *first discal and submarginal cells confluent forming a single discosubmarginal cell; costal and subcostal cells obliterated by virtual fusion of C+Sc+R+Rs; 2m-cu almost always present*; pterostigma present and usually well differentiated; hind wing with *rs-m* arising from *Rs* after *Rs* and *Sc+R* have diverged. Metasoma with five or more visible tergites, usually (in Costa Rica always) with tergites II and III separate.

Some species of the morphologically aberrant genus *Ophionellus* (Anomaloninae) have only a single enclosed cell in the fore wing. Females of a few species of *Gelis* (Cryptinae) are apterous and superficially ant-like, but with a protruding ovipositor.

Classification and distribution. The family Ichneumonidae as here recognized comprises 36 subfamilies. It includes the Paxylommatinae and Agriotypinae (for reasons see Gauld, 1984; Gauld & Bolton, 1988), although these groups have sometimes been treated as separate families (e.g. by Mason, 1981*a*). The subfamilial classification of the Ichneumonidae is still far from stable as subfamilies are gradually being split up into more natural, holophyletic units (see Eggleton, 1989; Wahl, 1990). The major differences between the classification used here (which incorporates most of the findings of Eggleton and Wahl) and the most widely used system proposed by Townes (1969), is that the Pimplinae and the Oxytorinae have been subdivided into more natural groups, the Helictinae has been included in the Orthocentrinae and, following Wahl (1993*b*), the Brachycyrtinae are recognized as a distinct subfamily separate from the Labeninae.

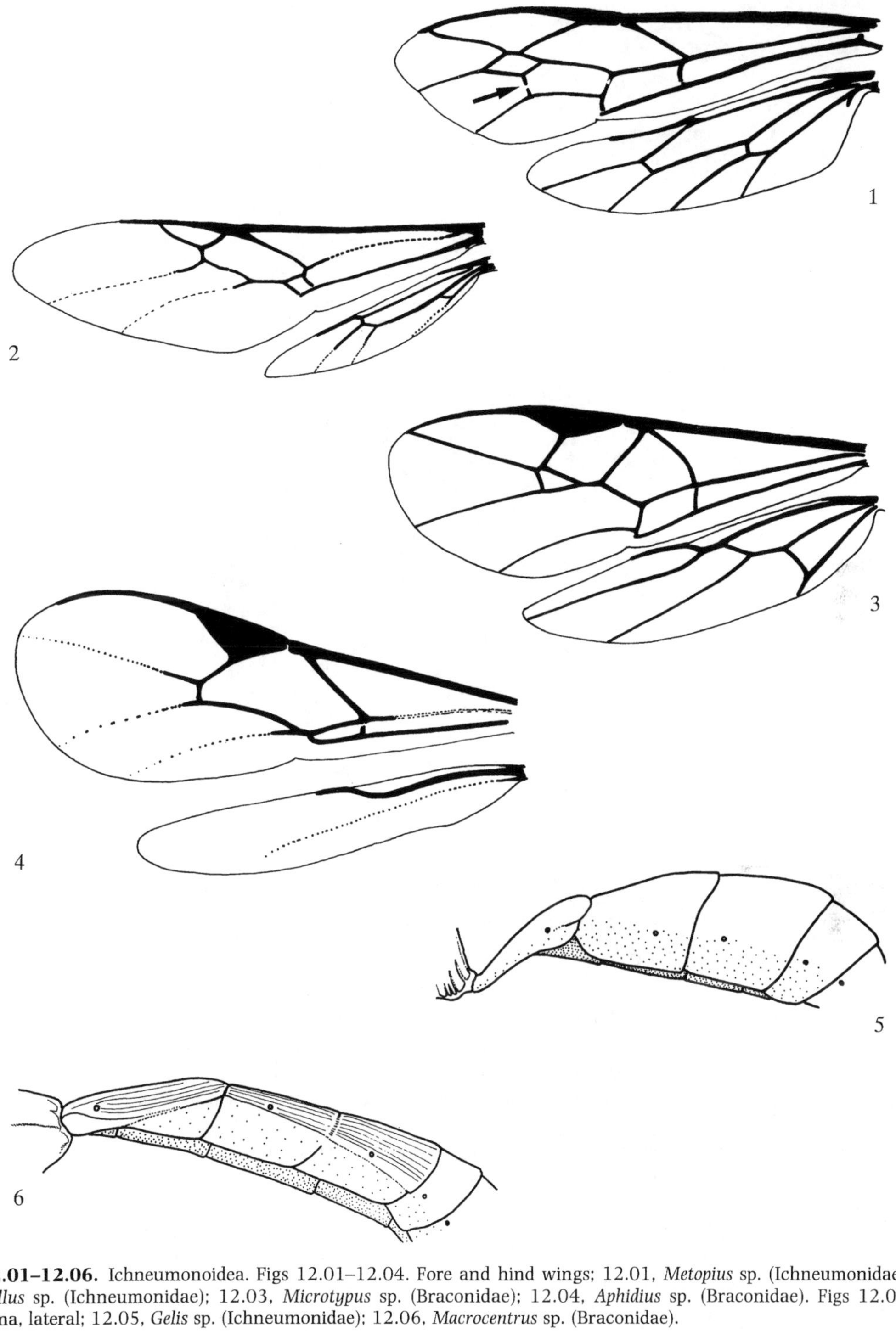

Figs 12.01–12.06. Ichneumonoidea. Figs 12.01–12.04. Fore and hind wings; 12.01, *Metopius* sp. (Ichneumonidae); 12.02, *Ophionellus* sp. (Ichneumonidae); 12.03, *Microtypus* sp. (Braconidae); 12.04, *Aphidius* sp. (Braconidae). Figs 12.05–12.06. Metasoma, lateral; 12.05, *Gelis* sp. (Ichneumonidae); 12.06, *Macrocentrus* sp. (Braconidae).

Fig. 12.07. *Cestrus* sp. (Cryptinae).

Twenty five of the 36 ichneumonid subfamilies are known to occur in Costa Rica (Gauld, 1991). The subfamilies not found here—Adelognathinae, Agriotypinae, Collyriinae, Diacritinae, Eucerotinae, Microleptinae, Neorhacodinae, Orthopelmatinae, Paxylommatinae, Stilbopinae and Tatogastrinae—are very small subfamilies each comprising about 40 or fewer species. All, except the Eucerotinae and Tatogastrinae, are primarily restricted to the North Temperate region. The Tatogastrinae is restricted to temperate southern South America, but the Eucerotinae is widely distributed (Barron, 1976, 1978) and may possibly be found in Costa Rica.

Biology. The vast majority of ichneumonids are parasitoids of holometabolous insects and spiders. Those associated with insects normally attack the larval or pupal stage, although a few oviposit into the egg of their host, but complete development in the host larva or pupa. Those that attack spiders may, depending upon species, parasitize any stage, including the adults. Egg sacs of both arachnids and insects are attacked by a few ichneumonid species. The larvae of these species feed on a succession of eggs, and thus can be considered to be predators. Many ichneumonid subfamilies are restricted to a particular order of insects although some subfamilies (e.g. Pimplinae) attack a fairly wide range of hosts. Compared with Braconidae, hyperparasitism is more common and gregarious parasitism is less common. Gregarious parasitism is mostly restricted to a few ectoparasitic species, for example pimplines such as *Iseropus* and related genera (Gauld, 1991), tryphonines such as *Netelia* and some generations of *Adelognathus* species (Kopelke, 1987). Numerous ichneumonids are ectoparasitic idiobionts, including many species of Labeninae, Pimplinae, Cryptinae, Brachycyrtinae, Rhyssinae and Xoridinae. Endoparasitic idiobiosis, usually in more or less exposed lepidopterous pupae with thick cuticles, occurs in a few Cryptinae, some Pimplinae and many Ichneumoninae. Ectoparasitic koinobiosis seems to have evolved separately in Adelognathinae (not present in Costa Rica), polysphinctine Pimplinae and Tryphoninae. Most of the remaining subfamilies are composed primarily of endoparasitic koinobiont species, many of which have complex physiological interactions with their host (see Chapter 2.5).

Oviposition. Ichneumonid females may have as few as ten (in some Ichneumoninae and Cryptinae) to several thousand (in Eucerotinae) potential eggs (Iwata, 1958, 1960). There is a close correlation between the size of the egg and the stage of the host attacked; koinobionts that emerge from the host's pupa but oviposit in early instar larvae tend to have large numbers of small, microlecithal eggs, whilst idiobionts attacking pupae usually have few large eggs (Price, 1973, 1974). The eggs of most species are ovoid, pale and without apparent chorionic sculpture (Iwata, 1958). Many species with very long ovipositors have an extremely elongate egg that is compressible to facilitate its passage down the lumen of the ovipositor. Tryphonines (which are ectoparasitic koinobionts) have eggs furnished with a variety of different types of hooks or anchors that are embedded in the host cuticle and firmly attach the egg to the outside of the host (Mason, 1967; Kasparayan, 1981). The eggs of some Anomaloninae have a mushroom-like protuberance which is used to anchor the egg internally within the host (Gauld, 1976*b*).

Larval and pupal stages. In ichneumonids there may be five larval instars, but the number may be reduced, for example to four in some Cremastinae (Giraldo-Vanegas & García, 1992) and three in Anomaloninae (Tothill, 1922). The exact number is often difficult to ascertain (Rojas-Rousse & Benoit, 1977). The first larval instar of many species is spindle-shaped with a well-developed head capsule, conspicuous antennae and strongly sclerotized, sharply pointed mandibles. In ectoparasitoids the cuticle may be quite strongly sclerotized and furnished with robust setae (Pschorn-Walcher & Zinnert, 1971; Kasparayan, 1981). The spiracles of at least some species are open on the prothorax and first eight abdominal segments (Spradbery, 1970*a*).

The more specialized endoparasitoids (e.g. most of the ophioniformes group of subfamilies) have a caudate first instar larva, which often has a weakly sclerotized head capsule, vestigial antennae and no spiracles. The mandibles may be well-developed (Tothill, 1922; Pschorn-Walcher, 1967) or vestigial (Gauld, 1976*a*). The tail-like caudal appendage may be long or short, but in general its relative length decreases as the larva grows (Bradley & Burgess, 1934). Successive larval instars have shorter appendages (Fisher, 1959; Frilli, 1965), and in all known cases the final instar larva is hymenopteriform (Clausen, 1940*b*), with the head capsule small and partially retracted into the thorax. Because the final instar larval skin is generally retained in the cocoon, and consequently is relatively easily collected, considerable study has been made of the morphology of this stage (e.g. Beirne, 1941; Finlayson, 1975; Short, 1978; Wahl, 1990, 1993*a*).

Most koinobionts kill their host as a final instar larva within the host's pupation chamber and thus they are afforded some degree of protection. Even so, many ichneumonids spin an ovoid, dense silken cocoon (e.g. see Slovák, 1984). Some species that spend a very short period in their cocoon spin a quite frail cocoon (Danthanarayana *et al.*, 1977), and in multivoltine species there may be cocoon dimorphism—diapausing larvae spin a thicker cocoon than their non-diapausing siblings. The cryptine vespid parasitoid, *Sphecophaga*, has three forms of cocoons, a white form that gives rise to a brachypterous female, a thin yellow cocoon giving rise to winged females in the same summer, and a thick yellow cocoon that gives rise to winged males and females the following summer (Donovan, 1991). The cocoons of some ichneumonids that undergo protracted diapause (e.g. *Enicospilus lebophagus*) are very dense and have a nacreous inner wall that is probably resistant to dessication and penetration by micro-organisms. Ichneumonids that pupate within their host's pupa (such as Anomaloninae, Metopiinae and Ichneumoninae) generally spin only a rudimentary cocoon, as do many idiobionts (such as Rhyssinae) that attack hosts deeply concealed in woody tissue. Many species seem to diapause as a prepupa, although a few north temperate species diapause as pharate adults within the cocoon (Morley, 1915).

Some campoplegines (e.g. *Cryptophion* spp.) are unusual in that they kill their host caterpillar on the food-plant before the caterpillar has become fully grown (Gauld & Janzen, 1994). In such an exposed situation the ichneumonid cocoon is extremely vulnerable to scavengers or hyperparasitoids and campoplegines have adopted a variety of strategies for concealing their cocoons. Many species have black and white mottled cocoons that resemble bird-droppings. Those of *Cryptophion* are concealed under the host larval remains, whilst some northern temperate species of *Hyposoter* construct a false cocoon (Finlayson, 1966). Other species ensure that their cocoons are removed from the leaf surface. Those of *Charops*, for example, are suspended from the host food-plant by a long thread, and species of *Bathyplectes* and some *Phobocampe* have 'jumping cocoons' that break free of the substrate so that cocoon and pupa fall amongst the leaf litter (Gauld & Bolton, 1988).

Adult biology. The great majority of ichneumonids are bisexual, although a few cosmopolitan, synanthropic species (e.g. *Diplazon laetatorius*, *Venturia canescens*) are thelytokous throughout much of their range. *Sphecophaga* appears to have a deuterotokous form which exploits locally abundant resources, as well as normal bisexual forms that overwinter (Donovan, 1991). Sexual dimorphism is generally not as pronounced amongst the Ichneumonidae as it is in many other groups of apocrite Hymenoptera, although in a few genera it can be very striking. For example, some species of Cryptinae (e.g. many *Gelis* spp.) have apterous, ant-like females and normal alate males. Species of the pimpline genus *Dolichomitus* have females of relatively normal proportions, but the males may be very elongate and slender (Gauld, 1991). Sexual dimorphism is most apparent in idiobiont groups, the males of which are relatively short-lived flower-feeders who spend most of their lives flying low through, or resting on, vegetation. The females, on the other hand, are long-lived and spend most of their lives on the ground searching for hosts in leaf litter or in concealment under bark etc. Thus in some Ichneumoninae males are often aculeate mimics and aposematically coloured, whilst the females are cryptic, black or brownish insects. Many idiobionts (e.g. many Cryptinae, Pimplinae and Labeninae) show considerable sexual dimorphism in size, which is a result of facultative arrhenotoky (Flanders, 1956; Aubert, 1959)—i.e. females lay a disproportionately large number of female eggs on large hosts and male eggs on small hosts (Arthur & Wylie, 1959; Kishi, 1970). In general, koinobionts show less pronounced sexual dimorphism; males and females frequently resemble each other closely in shape, colour pattern and even size (Gauld & Fitton, 1987) and both males and females can often be seen flying amongst or resting on the vegetation. The nocturnally active koinobionts are the least sexually dimorphic of all the ichneumonids, females and males generally differing only in the density of pectination of the tarsal claws (Gauld, 1988*b*).

In many ichneumonids the males emerge before the females. In the Rhyssinae (and probably also other wood-borer parasitoids) males congregate around tree trunks from which conspecific females are in the process of emerging, and some of these males establish territories which they defend against other males (Eggleton, 1990). Other species show scramble competition for mates and in these cases males are sometimes specialized for mating with the female prior to her full emergence (Nuttall, 1973). In temperate regions males of some species of Orthocentrinae, Tersilochinae and Diplazontinae form conspicuous swarms, although this has not been observed in Costa Rica. However, in Santa Rosa National Park I have seen males of *Joppidium* species (Cryptinae) forming groups flying above and around newly emerged females. Such behaviour suggests that the female may be releasing a pheromone, as has also been suggested for some Campopleginae (Vinson, 1972; Gordh & Hendrickson, 1976). Further evidence for sexual pheromones in Ichneumonidae is the observation that males of a species of *Exetastes* (Banchinae) are attracted to 8-dodecenyl and 11-tetradecenyl acetates (Hrdy & Sedivy, 1979).

Mating is rather brief (usually from one to ten minutes in duration) and has rarely been observed. The females of many species are believed to mate only once (Gordh & Hendrickson, 1976; Slovák, 1986*b*), but some mate repeatedly (Mathur, 1967*a*). The male often approaches the female from the rear and repeatedly strokes her with his antennae (Slobodchikoff, 1973; Barrows, 1976), but some species approach face to face with the male vibrating his wings (Veen, 1982). During copulation the male may be situated behind, above, or rarely to the side of the female, and he usually continues to antennate her throughout copulation (Barrows, 1976; Gordh & Hendrickson, 1976; Danthanarayana *et al.*, 1977).

Little is known about the phenology of ichneumonids in Costa Rica or indeed any other tropical area. In regions that have a pronounced dry season, such as northwestern Costa Rica, the majority of koinobiont species are active shortly after the start of the rains, when they exploit the first flush of larvae that appear when plants put on new growth. Most of these ichneumonids will have diapaused as prepupae in their host's pupal retreat. Many show a high degree of synchrony with their hosts, and emerge as adults for a very short period. Sometimes, as in the case of *Enicospilus lebophagus*, only some of a year's population emerges the following year; a small proportion may remain in diapause for another twelve months (D.H. Janzen, pers. comm.). The idiobionts, which mostly attack prepupae or pupae, tend to appear slightly later in the season. Many continue to fly about (and thus be collected in flight interception traps) through the wet season and well into the following dry season. Thus in the seasonal tropics, unlike the northern temperate regions, idiobiont parasitoids may be active throughout the entire inclement season; potential hosts that are in diapause are thus susceptible to attack throughout their diapause (Gauld, 1987). A few species of genera such as *Camera* (Cryptinae) and *Clistopyga* (Pimplinae) have only been collected in the dry season, suggesting they may diapause in the wetter periods. In the northern temperate regions some species, particularly of the Ichneumoninae, diapause as adults, concealing themselves during the winter beneath the bark of trees and in grass tussocks.

Defences against predators. Ichneumonids have a variety of defensive strategies against predators. Some are furnished with cuticular processes, which may offer physical protection. For example, some species of *Certonotus* (Labeninae) have spines on the pronotum, propodeum and metanotum, and *Acrotaphus* species (Pimplinae) have a pronotal shield and occipital flange protecting the cervical membrane against attacks by asilids. Many medium to large-sized species of mesostenines are black and white striped, an adaptation that renders them very difficult to focus on (at least for the human eye) as they fly through patches of light and shade in the forest understorey.

Many ichneumonids, especially those with relatively short ovipositors such as species of *Netelia*, *Ophion*, *Carinodes* and *Joppa* are capable of stinging—a sensation not unlike a pin-prick. Some of these species are aposematically coloured. Species of the pimpline genera *Neotheronia* and *Xanthopimpla* have large tarsal claws that have an internal poison-sac, and these claws function as a type of poison fang (Townes, 1940). Most of these species are also aposematically coloured and, like those capable of stinging, many are Müllerian mimics of aggressive aculeates (Gauld, 1991). Many other species are also brightly coloured and are apparently Batesian mimics of aculeates (Evans, 1966*e*, 1968*b*). Some (e.g. *Dolichomitus* spp.) are yellow and black banded, or brown and yellow, and thus appear to mimic polistine vespids (see Table 15.2). Others (e.g. *Rhynchophion* sp.) are metallic blue with violet and orange wings, and in flight closely resemble species of *Pepsis* (Pompilidae); some mesostenines twitch their wings while running and can easily be mistaken for pompilids. A few pimplines are orange with black-banded wings and in flight they resemble noxious braconids (see Chapter 12.2). A few ichneumonids, most notably some *Pimpla*, *Apechthis* and *Exochus* species, emit a pungent odour when handled and this may serve to deter predators. Many small ichneumonids are black with a bright red mesosoma, as in species of Metopiinae, Cryptinae, Pimplinae, Tersilochinae and Orthocentrinae. This colour pattern occurs in several other families of Hymenoptera (see Chapter 9.6) but its significance is not understood.

Although striking colour patterns are very obvious features of diurnally active lowland tropical ichneumonids, at high altitudes (above about 2600 m) a large proportion of diurnal species are black or brown, and generally inconspicuous. There are, however, a few exceptions. Several species of Ichneumoninae and Banchinae are particularly conspicuous; they have the head and most of the mesothorax black, the prothorax, epicnemium, metathorax and propodeum bright lemon yellow, and the metasoma reddish. I have not seen this colour pattern in any other region except at high altitudes in Central America. Several other montane species (e.g. Pimplinae, Tryphoninae and Ichneumoninae) are black with slightly yellowish wings and with the anterior margin of the fore wing infumate, thus closely resembling a common high altitude species of *Polybia* (see Table 15.2).

A large proportion of the lowland tropical koinobiont ichneumonids are nocturnally active, and this may help them avoid diurnally active predators (Gauld, 1987). Virtually all nocturnal species are rather uniformly orange or brownish. In comparison to the lowlands (1600 m or less), at higher altitudes nocturnally active species apparently comprise a much smaller proportion of the number of ichneumonid koinobionts present.

Economic importance. In other parts of the world ichneumonids have been used in classical biological control programmes, especially against symphytan pests (e.g. Taylor, 1976). There are no documented cases of deliberate introductions of ichneumonid species into Central America for biological control purposes, although there is certainly a potential for doing so. Several endemic species undoubtedly play important roles in lowering the populations of various pest species in Central America.

As can be seen from Table 12.1, the most important ichneumonid subfamilies in biological control are the Campopleginae (e.g. Clausen, 1956; Morey, 1971; Rao, 1971; Ashley, 1983; Cock *et al.*, 1987; García, 1991) and the Cremastinae (Patel & Habib, 1984). *Microcharops anticarsiae* has recently been introduced from Costa Rica into the United States to control *Anticarsia gemmatalis* (Noctuidae) (Gupta, 1988), and attempts have been made to introduce *Eiphosoma vitticolle* into the United States to control the fall arm worm, *Spodoptera frugiperda* (Noctuidae) (Ashley *et al.*, 1982).

Pest species	Common name	Ichneumonid parasitoid
Anticarsia gemmatalis (Noctuidae)	Velvetbean caterpillar	*Microcharops anticarsiae* (Campopleginae)
Antichloris viridis (Arctiidae)	Banana moth	*Carinodes* sp. (Ichneumoninae)
Ecdytolopha torticornis (Tortricidae)	Macadamia nut borer	*Pristomerus* sp. (Cremastinae)
Erinnyis ello (Sphingidae)	Cassava hornworm	*Cryptophion* sp. (Campopleginae)
Mocis latipes (Noctuidae)	Guinea grass moth	*Scambus albitibia* (Pimplinae) *Tricholabus lepidus* (Ichneumoninae)
Opsiphanes cassina (Nymphalidae)		*Casinaria* sp. (Campopleginae)
Phthorimaea operculella (Gelechiidae)	Potato tuber moth	*Pristomerus spinator* (Cremastinae)
Plutella xylostella (Plutellidae)	Diamondback moth	*Diadegma* spp. (Campopleginae) *Pimpla punicipes* (Pimplinae)
Polygrammodes elevata (Pyralidae)	Sweetpotato borer	*Eiphosoma* sp. (Cremastinae)
Rothschildia orizaba (Saturniidae)		*Enicospilus* sp. (Ophioninae)
Sibene spp. (Limacodidae)	Oil palm slug moths	*Casinaria* sp. (Campopleginae)
Spodoptera frugiperda (Noctuidae)	Fall armyworm	*Campopletis grioti* (Campopleginae) *Eiphosoma vitticolle* (Cremastinae) *Ophion flavidus* (Ophioninae) *Pristomerus spinator* (Cremastinae) *Temelucha* sp. (Cremastinae)

Table 12.1. Major pests in Central America and the native ichneumonid species that attack them.

Identification. Townes (1969, 1970*a* & *b*, 1971) gives keys to the world genera of all subfamilies except the Ichneumoninae. A key to the neotropical Ichneumoninae was included as an appendix to Townes and Townes' (1966) catalogue of neotropical species. Unfortunately Townes adopted an idiosyncratic system of nomenclature and did not follow the principle of priority for family-group names, but formed names from the oldest included genus. Fitton and Gauld (1976, 1978) have attempted to rectify this situation, but there are still disputed areas.

The most recent key to the subfamilies of Central America was published by Gauld (1991) and is not repeated here, but a brief identification guide to the common subfamilies is given below. This is intended to be simple and ignores the rather uncommon exceptions; others run out to the last couplet.

A guide to females of the ichneumonid subfamilies commonly encountered in Central America

1 Fore wing lacking *2rs-m*, with *3rs-m* joining *M* distal to *2m-cu* (Figs 12.08, 12.09). 2

— Fore wing not as above, ***either*** with an areolet ***or*** with *2rs-m* but lacking *3rs-m* (Figs 12.10–12.23). 4

2 Subgenital plate very large, projecting far beyond apex of metasoma (Fig. 12.42); fore and mid tarsal claws with a small auxiliary tooth (Fig. 12.60).

[Large black and white patterned insects.] .. .. **Acaenitinae** (p. 409)

— Subgenital plate not projecting beyond apex of metasoma; tarsal claws simple or pectinate. 3

3 Fore wing with a spurious vein extending from claval notch to outer corner (Fig. 12.08); tarsal claws always closely pectinate (Fig. 12.61); hind wing with distal abscissa of *Cu*1 present (Fig. 12.24). .... **Ophioninae** (p. 422)

— Fore wing without a spurious vein (Fig. 12.09); tarsal claws sparsely and inconspicuously pectinate; hind wing with distal abscissa of *Cu*1 entirely absent (Fig. 12.25). .... **Anomaloninae: Anomalonini** (p. 409)

4 Tergite I of metasoma with spiracle behind centre, obviously closer to hind margin than to anterior end (Figs 12.28, 12.29). 5

— Tergite I of metasoma with spiracle near or in front of the centre (Figs 12.30, 12.31). 14

5 Fore wing with areolet quadrate to pentagonal (Figs 12.11–12.13), sometimes with outer vein (3*rs-m*) absent (Figs 12.10, 12.14) ***and*** with ovipositor without a dorsal subapical notch (cf Fig. 12.45).

[Metasoma with tergites II-IV usually somewhat depressed; antenna often with central part flattened ventrally; tergite I never with deep glymmae (pits or troughs) laterally (cf Fig. 12.29).] 6

— Not entirely as above, ***either*** with fore wing with areolet rhombic and somewhat petiolate above (Figs 12.15, 12.18), oblique to triangular (Fig. 12.19) or entirely absent (Figs 12.16, 12.17), ***or if*** very rarely pentagonal ***then*** ovipositor with a distinct dorsal subapical notch (Fig. 12.44); ovipositor otherwise with or rarely without a dorsal subapical notch.

[Metasoma sometimes with tergites II-IV strongly laterally compressed; antenna almost never ventrally flattened.] 8

6 Fore wing with *cu-a* distal to base of *Rs&M* by more than 0.8 times its own length (Fig. 12.10). .... **Brachycyrtinae** (p. 410)

— Fore wing with *cu-a* closer to base of *Rs&M* (Figs 12.11–12.14). 7

7 Clypeus flat and rather broad (Fig. 12.35); ovipositor generally not protruding beyond apex of metasoma, or if doing so then with sheaths rigid. .... **Ichneumoninae** (p. 417)

— Clypeus somewhat convex (12.34); ovipositor generally protruding well beyond apex of metasoma, its sheaths flexible. .... **Cryptinae** (p. 414)

8 Fore wing with two distinct and widely separated bullae (clear areas resembling short breaks) in vein 2*m-cu* (cf Figs 12.20–12.23). 9

— Fore wing with a single bulla in vein 2*m-cu* (Figs 12.16–12.19). 10

9 Head with a distinct impressed subocular sulcus (Fig. 12.41); mandibles very slender and delicate; mesoscutum very convex in profile......... .... **Orthocentrinae** (part) (p. 426)

— Head without an impressed subocular sulcus; mandibles often long but quite robust; mesoscutum not exceptionally convex.

[Tergite I often with deep glymmae; ovipositor sheath often slightly broadened centrally.] .... **Tryphoninae** (part) (p. 430)

10 Occipital carina closer to lateral ocelli than ocellar diameter; propodeum reticulate (Fig. 12.46).

[Areolet never present; metasoma slender, strongly laterally compressed.] .... **Anomaloninae: Gravenhorstiini** (p. 409)

— Occipital carina separated from lateral ocelli by far more than ocellar diameter; propodeum various, sometimes with discernible areae (Figs 12.47, 12.48), never reticulate. 11

11 Subgenital plate large; ovipositor very slender enclosed in rigid sheath (Fig. 12.43).

[Areolet usually very large, rhombic (Fig. 12.15); males with gonosquamae posteriorly elongate, rod-like.] .... **Mesochorinae** (part) (p. 420)

— Subgenital plate smaller; ovipositor moderately stout, enclosed in flexible sheaths. 12

12 Clypeus very broad, fringed with long parallel setae (Fig. 12.32).

[Small insects, fore wing length generally <8 mm; fore wing with marginal cell deep, pterostigma large (Fig. 12.16.] .... **Tersilochinae** (p. 430)

— Clypeus neither unusually broad nor fringed with long parallel setae (Fig. 12.33). 13

13 Tibial spurs inserted into separate area from that of tarsus (Fig. 12.62); hind wing with first abscissa of *Rs* much shorter than *rs-m* (Fig. 12.26).

[Face usually pale; tergite II often polished or striate; ovipositor sometimes sinuous; hind tibia sometimes with a ventral tooth; propodeum often with clearly defined area, with a coffin-shaped area superomedia (Fig. 12.48).] .... **Cremastinae** (p. 412)

— Tibial spurs inserted into common area with that of tarsus (Fig. 12.63); hind wing with first abscissa of *Rs* as long as or longer than *rs-m* (Fig. 12.27).

[Face often black; tergite II often polished but slightly granulate; ovipositor never sinuous; hind tibia never with a ventral tooth; propodeum often not regularly areated (Fig. 12.47), almost never with a completely enclosed area superomedia.] .... **Campopleginae** (p. 411)

14 Areolet very large, regularly rhombic (Fig. 12.15) ***and*** subgenital plate large ***and*** the ovipositor very slender and enclosed in rigid sheath (Fig. 12.43).

[Males with gonosquamae posteriorly elongate, rod-like.] .... **Mesochorinae** (part) (p. 420)

— Not exactly as above. 15

15 Face and clypeus weakly to very strongly convex, with the clypeus confluent and indistinguishable from the face (Fig. 12.37) ***or*** face with a shield-shaped raised flat area (Fig. 12.36). 16

— Face flat to weakly convex, with clypeus delineated dorsally or laterally by weak impressions (Figs 12.40, 12.41). 17

16 Antenna with scape cylindrical, much longer than broad (Fig. 12.38); subocular sulcus present and malar space very broad (Fig. 12.37).

[Small species.] ... **Orthocentrinae** (p. 426)

— Antenna with scape goblet-shaped, not much longer than broad (Fig. 12.39); subocular sulcus absent and malar space very short. .... **Metopiinae** (p. 420)

17 Mandible tridentate (Fig. 12.40). .... **Diplazontinae** (p. 417)

— Mandible bidentate or rarely unidentate. 18

18 Metasoma with impressed triangular areas on tergites II-IV (Fig. 12.50); metanotum produced laterally into a small hook (or "catch") that engages a process on metapostnotum (Fig. 12.53). ..... **Lycorininae** (p. 419)

— Metasoma without impressed triangular areas of tergites II–IV (Fig. 12.52), although sometimes V-shaped impressions are present (Fig. 12.51); metanotum not produced into a hook laterally (Fig. 12.54). 19

19 Scuto-scutellar groove with a median carina; mesoscutum with a transverse flexible suture before scuto-scutellar groove (Fig. 12.55); antenna with angulation near apex, the segments proximal to the angulation bearing one or two long stout bristles (Fig 12.70). .... **Xoridinae** (p. 431)

— Scuto-scutellar groove without a median carina; mesoscutum without a transverse flexible suture before scuto-scutellar groove; antenna without an angulation near apex and without long stout bristles. 20

20 Head with a distinct impressed subocular sulcus (Fig. 12.41); mandibles very small, slender and delicate; apex of hind tibia with a fringe of fine close hairs internally (Fig. 12.57); mesoscutum very convex in profile. ....**Orthocentrinae** (most) (p. 426)

— Head without an impressed subocular sulcus; mandibles generally quite large, neither unusually slender nor delicate; apex of hind tibia without a fringe of fine close hairs internally (Fig. 12.56); mesoscutum variously convex in profile. 21

21 Ovipositor with a distinct dorsal subapical notch (cf Fig. 12.44). 22

— Ovipositor without a dorsal subapical notch (Fig. 12.45). 23

22 Ovipositor shorter than apical depth of metasoma (Fig. 12.69) **and** propodeum variously sculptured but never smooth with only the posterior transverse carina present.

[First flagellar segment sometimes with a tyloid on outer side; mandibles robust often with lower tooth the longer.] .... **Ctenopelmatinae** (p. 416)

Ovipositor generally conspicuously longer than apical metasomal depth, ***or if*** shorter, ***then*** propodeum with only posterior transverse carina present (Fig. 12.54).

[Submetapleural carina often expanded laterally into a lobe (Fig. 12.66).] .... **Banchinae** (most) (p. 409)

23 Fore wing with areolet large and pentagonal (Fig. 12.20); metasoma inserted in propodeum quite far above insertion of hind coxa (cf Fig. 12.64). .... **Labeninae** (p. 419)

— Fore wing without an areolet, or if present then areolet is small and obliquely triangular to rhombic; metasoma inserted in propodeum very close to insertion of hind coxa (cf Figs 12.65, 12.66). 24

24 Mesoscutum with coarse, file-like rugae (Fig. 12.49). .... **Rhyssinae** (p. 429)

— Mesoscutum smoother, without coarse rugae. 25

25 Mid and hind tarsal claw with an acute basal lobe (Fig. 12.58), ***and/or*** simple, very large and generally with a spatulate bristle tending to point to apex (Fig. 12.59); metasoma very strongly sclerotized and usually coarsely punctate. .... **Pimplinae** (p. 429)

— Mid and hind tarsal claw without an acute basal lobe, often small, sometimes pectinate, never with a spatulate bristle; metasoma various, not coarsely punctate. 26

26 Tarsal claw densely pectinate to apex (cf Fig. 12.61); ocelli very large, the lateral ones separated from eye by less than their own diameter.

[Generally large, yellow or orange coloured insects.] .... **Tryphoninae: *Netelia*** (p. 430)

— Tarsal claw not densely pectinate to apex; ocelli generally smaller, the lateral ones separated from eye by more than their own diameter. .... VARIOUS SUBFAMILIES[1]

[[1] A variety of relatively infrequently encountered taxa will run here, including Poemeniinae (p. 429), Oxytorinae (p. 428), Phrudinae (p. 428), Cylloceriinae (p. 417), some Tryphoninae (p. 430) and a very few species of Banchinae, Cryptinae and Ctenopelmatinae. Many can be recognized using the keys provided by Gauld, 1991.]

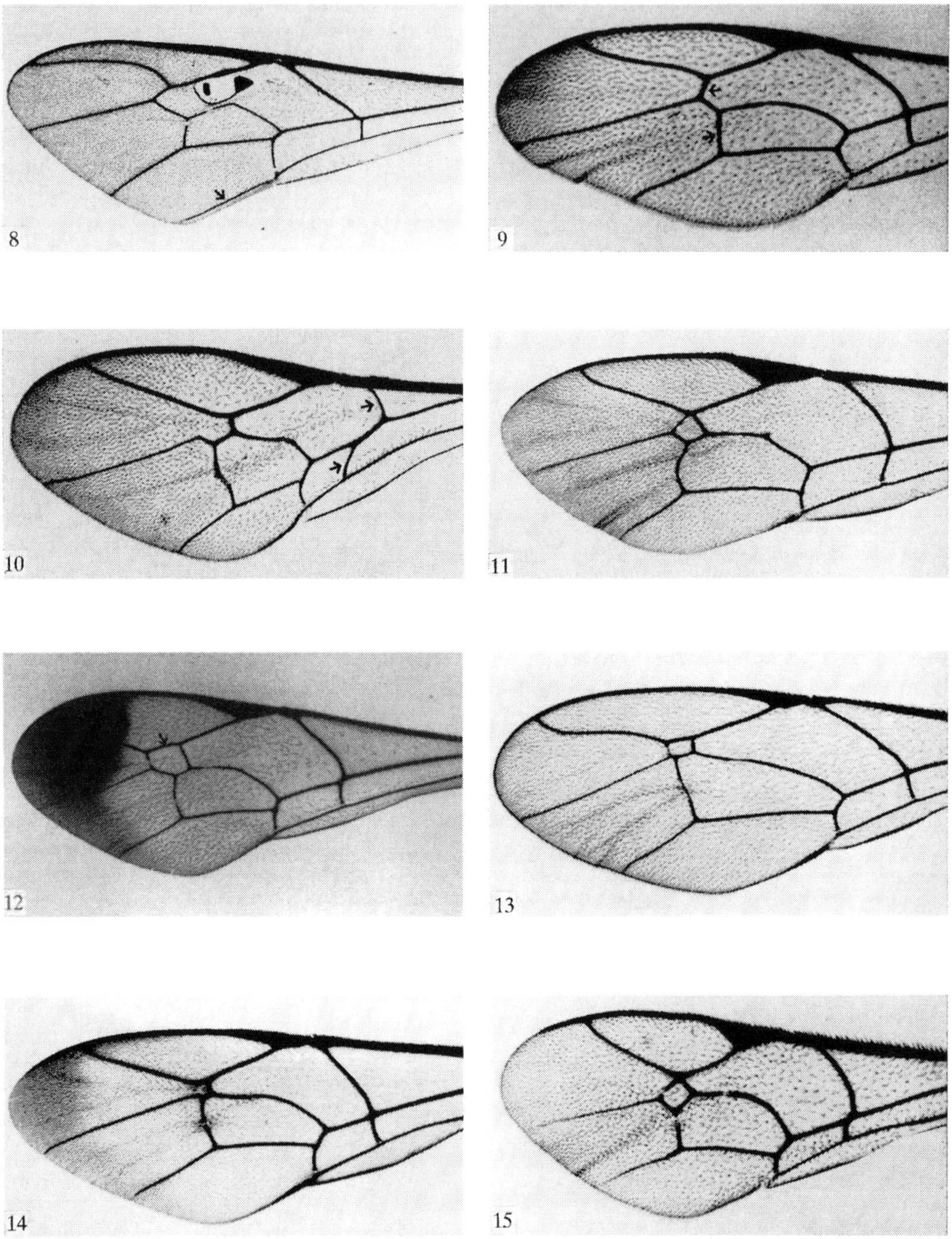

Figs 12.08–12.15. Ichneumonidae, fore wing; 12.08, *Enicospilus* sp. (Ophioninae); 12.09, *Anomalon* sp. (Anomaloninae); 12.10, *Brachycyrtus* sp. (Brachycyrtinae); 12.11, *Carinodes* sp. (Ichneumoninae); 12.12, *Joppa* sp. (Ichneumoninae); 12.13, *Cryptanura* sp. (Cryptinae); 12.14, *Lymeon* sp. (Cryptinae); 12.15, *Mesochorus* sp. (Mesochorinae).

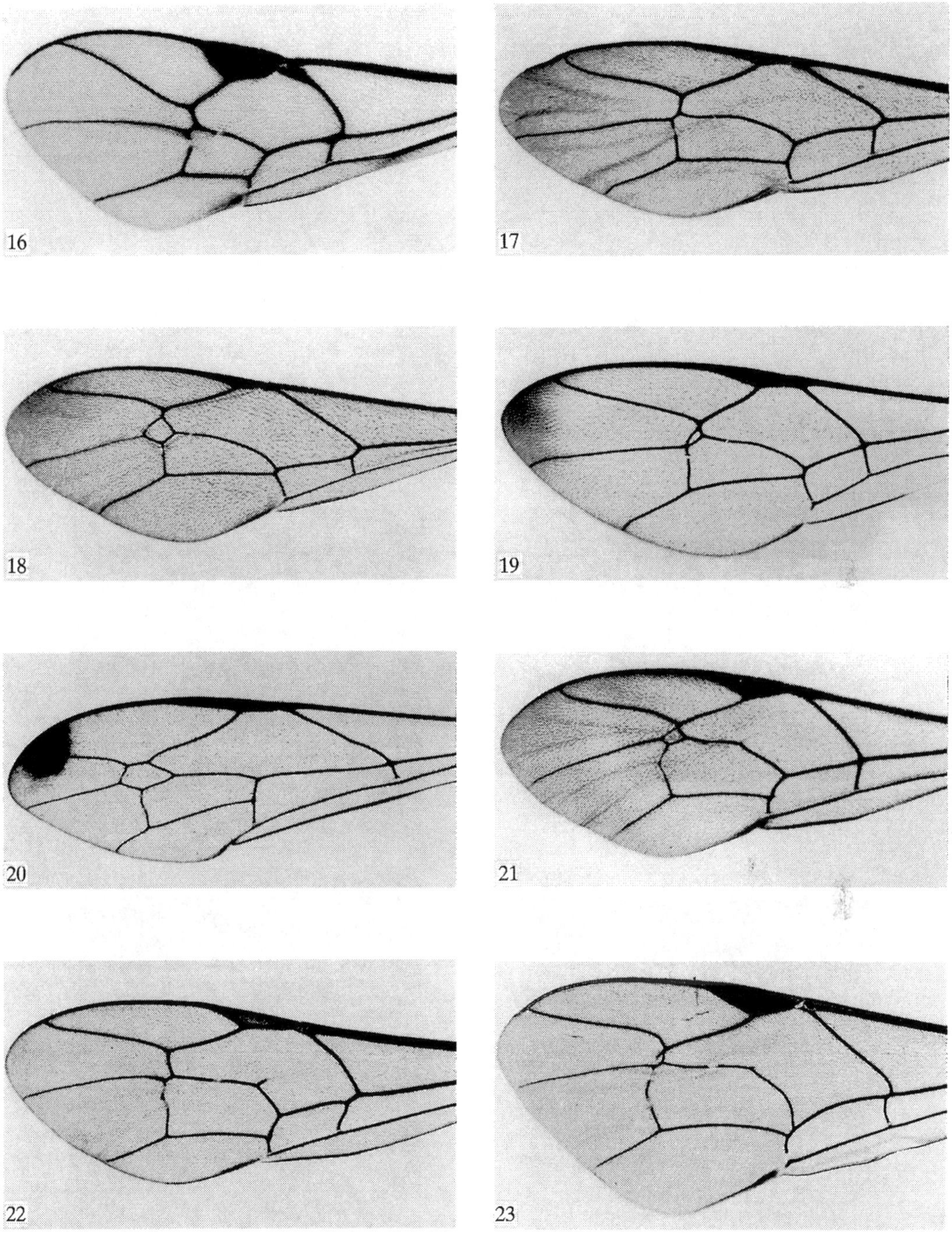

Figs 12.16–12.23. Ichneumonidae, fore wing; 12.16, *Stethantyx* sp. (Tersilochinae); 12.17, *Aphanistes* sp. (Anomaloninae); 12.18, *Delopia* sp. (Campopleginae); 12.19, *Eiphosoma* sp. (Cremastinae); 12.20, *Labena* sp. (Labeninae); 12.21, *Pimpla* sp. (Pimplinae); 12.22, *Mnioes* sp. (Banchinae); 12.23, *Netelia* sp. (Tryphoninae).

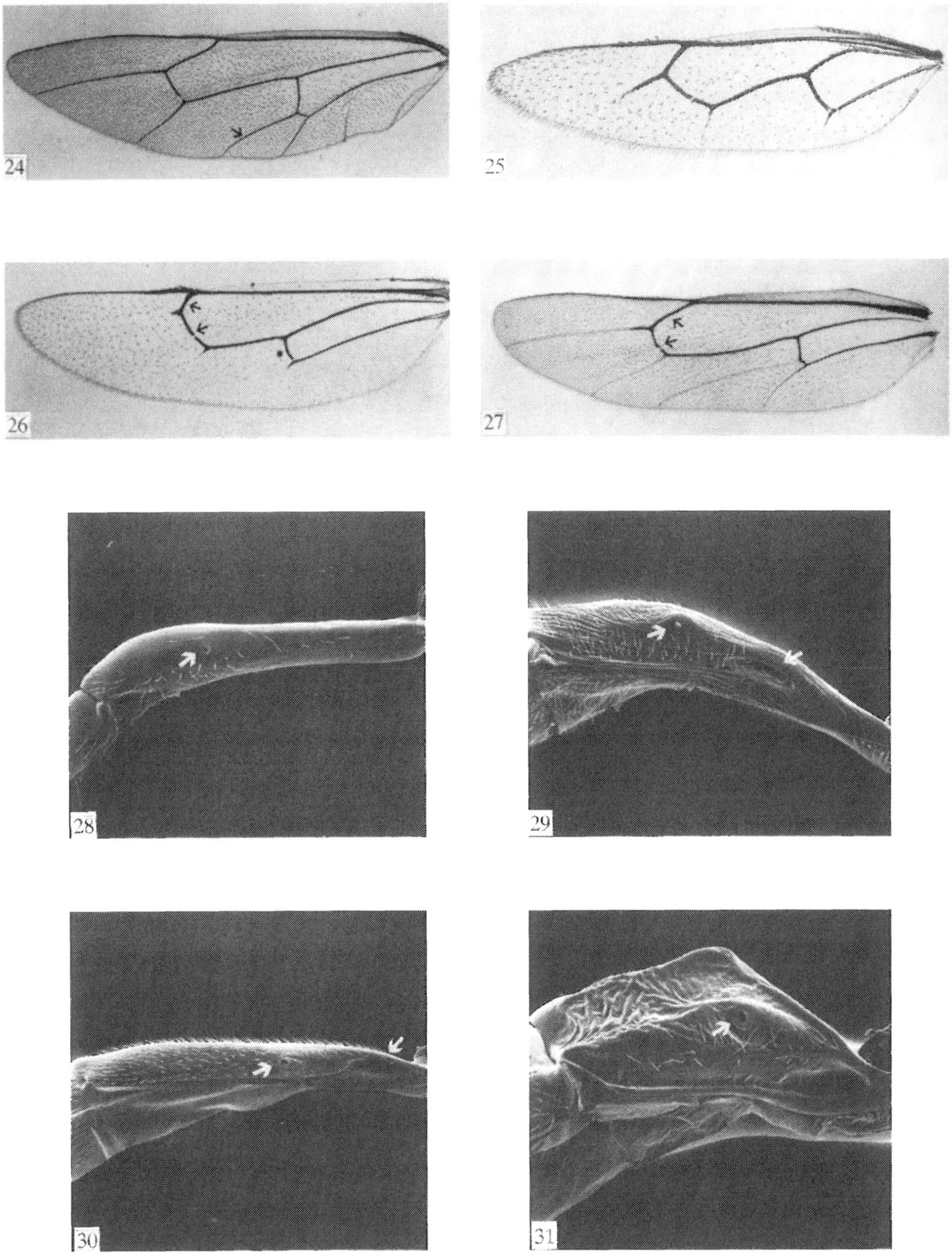

Figs 12.24–12.31. Ichneumonidae. Figs 12.24–12.27. Hind wing; 12.24, *Enicospilus* sp. (Ophioninae); 12.25, *Anomalon* sp. (Anomaloninae); 12.26, *Eiphosoma* sp. (Cremastinae); 12.27, *Delopia* sp. (Campopleginae). Figs 12.28–12.31. Scanning electron photomicrographs of first segment of metasoma; 12.28, *Cryptanura* sp. (Cryptinae); 12.29, *Zagryphus* sp. (Tryphoninae); 12.30, *Netelia* sp. (Tryphoninae); 12.31, *Pimpla* sp. (Pimplinae).

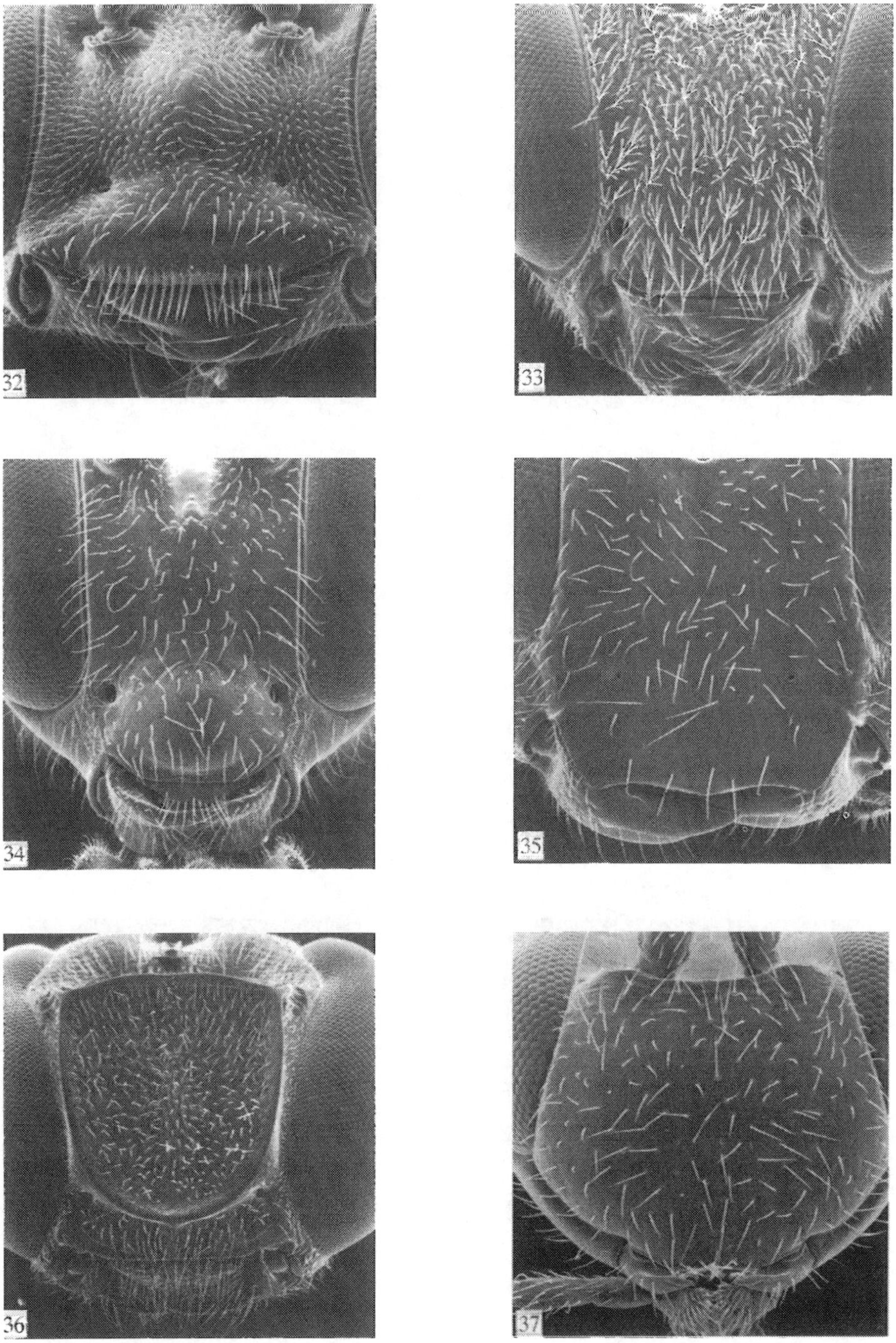

Figs 12.32–12.37. Ichneumonidae, scanning electron photomicrographs, clypeus and lower face; 12.32, *Stethantyx* sp. (Tersilochinae); 12.33, *Delopia* sp. (Campopleginae); 12.34, *Cryptanura* sp. (Cryptinae); 12.35, *Joppa* sp. (Ichneumoninae); 12.36, *Metopius* sp. (Metopiinae); 12.37, *Orthocentrus* sp. (Orthocentrinae).

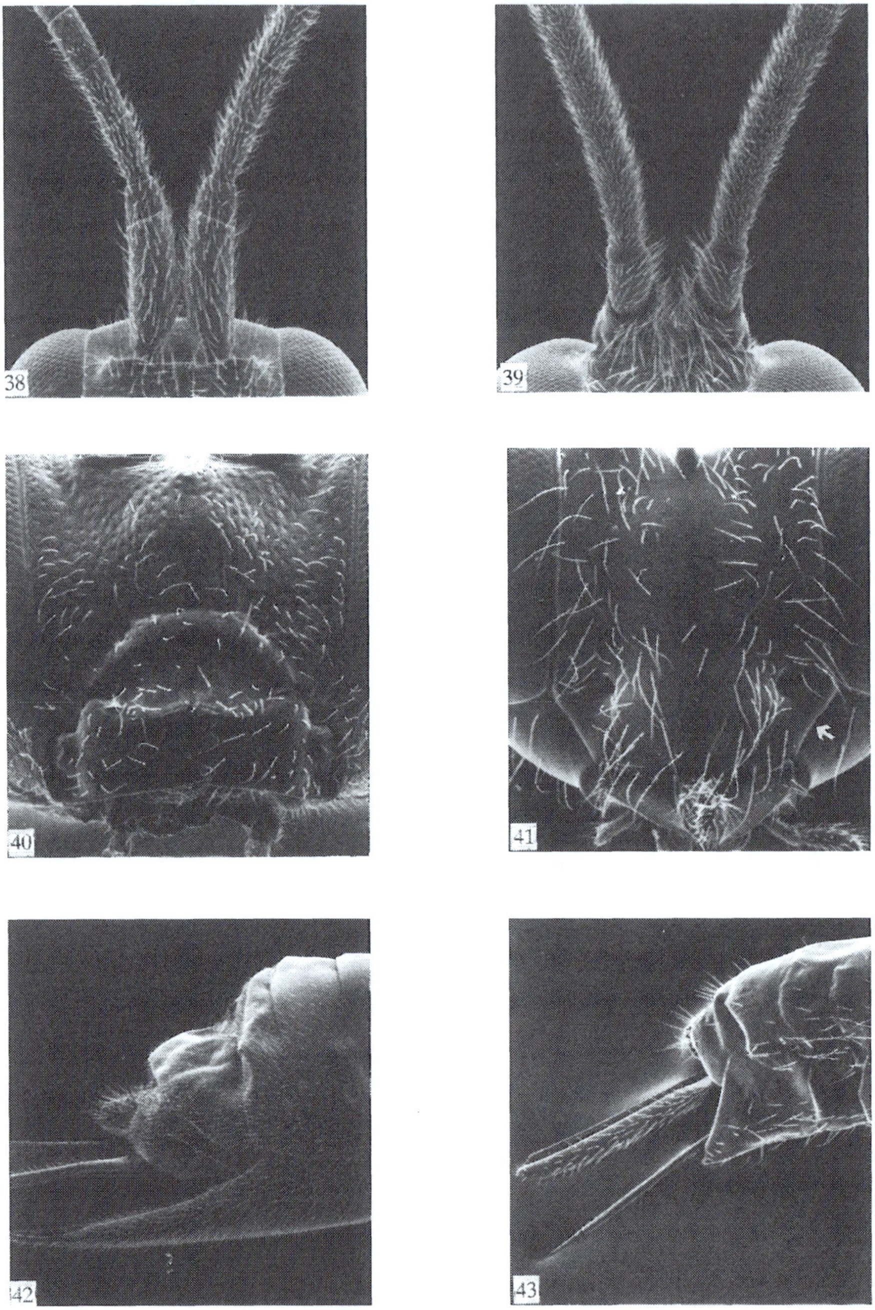

Figs 12.38–12.43. Ichneumonidae, scanning electron photomicrographs. Figs 12.38–12.39, base of antenna; 12.38, *Orthocentrus* sp. (Orthocentrinae); 12.39, *Colpotrochia* sp. (Metopiinae). Figs 12.40–12.41. Clypeus and lower face; 12.40, *Diplazon* sp. (Diplazontinae); 12.41, *Eusterinx* sp. (Orthocentrinae). Figs 12.42–12.43. Apex of metasoma, lateral; 12.42, *Arotes* sp. (Acaenitinae); 12.43, *Mesochorus* sp. (Mesochorinae).

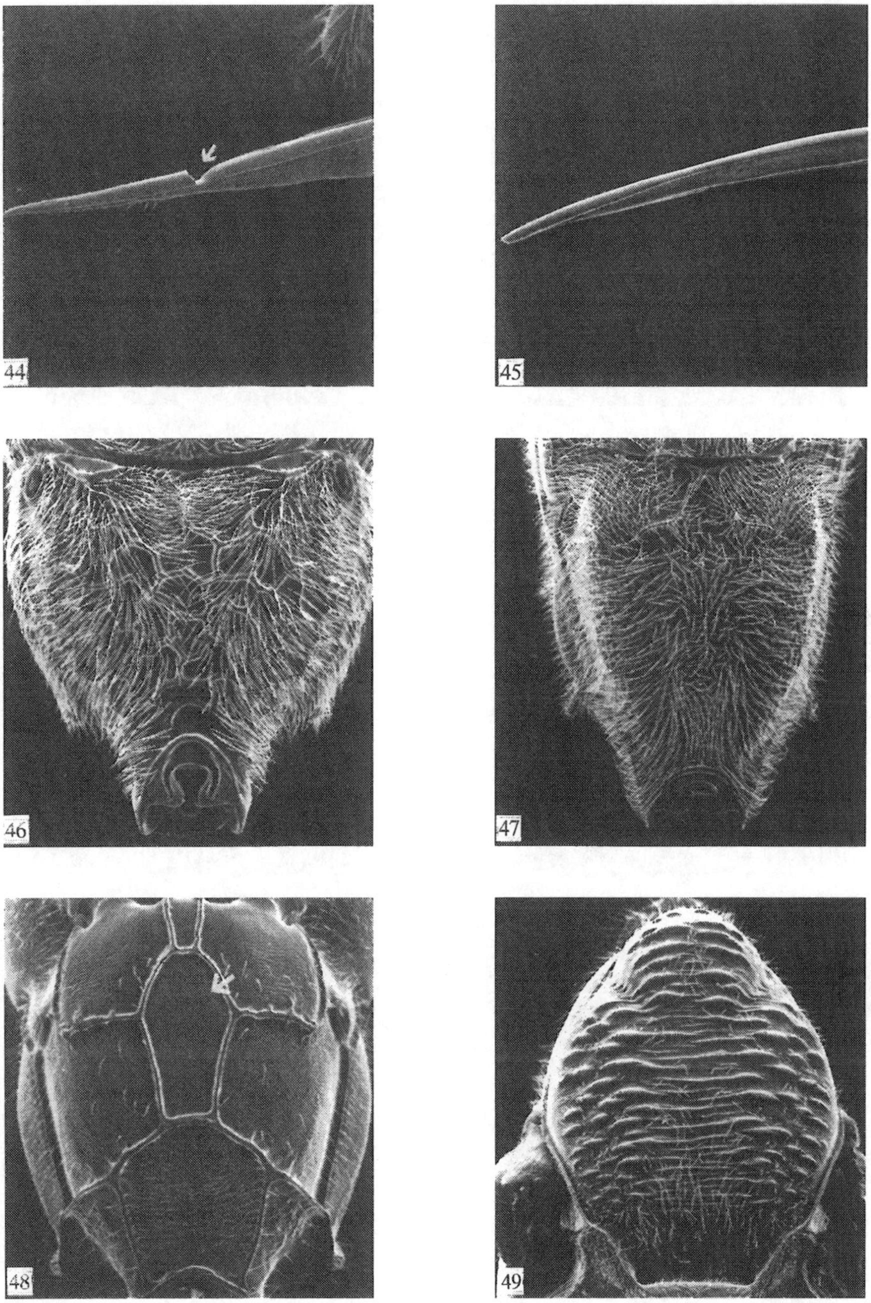

Figs 12.44–12.49. Ichneumonidae, scanning electron photomicrographs. Figs 12.44–12.45. Apex of ovipositor; 12.44, *Delopia* sp. (Campopleginae); 12.45, *Zagryphus* sp. (Tryphoninae). Figs 12.46–12.48. Propodeum, dorsal; 12.46, *Corsonchus* sp. (Anomaloninae); 12.47, *Delopia* sp. (Campopleginae); 12.48, *Eiphosoma* sp. (Cremastinae). Fig. 12.49, mesoscutum, dorsal, *Epirhyssa* sp. (Rhyssinae).

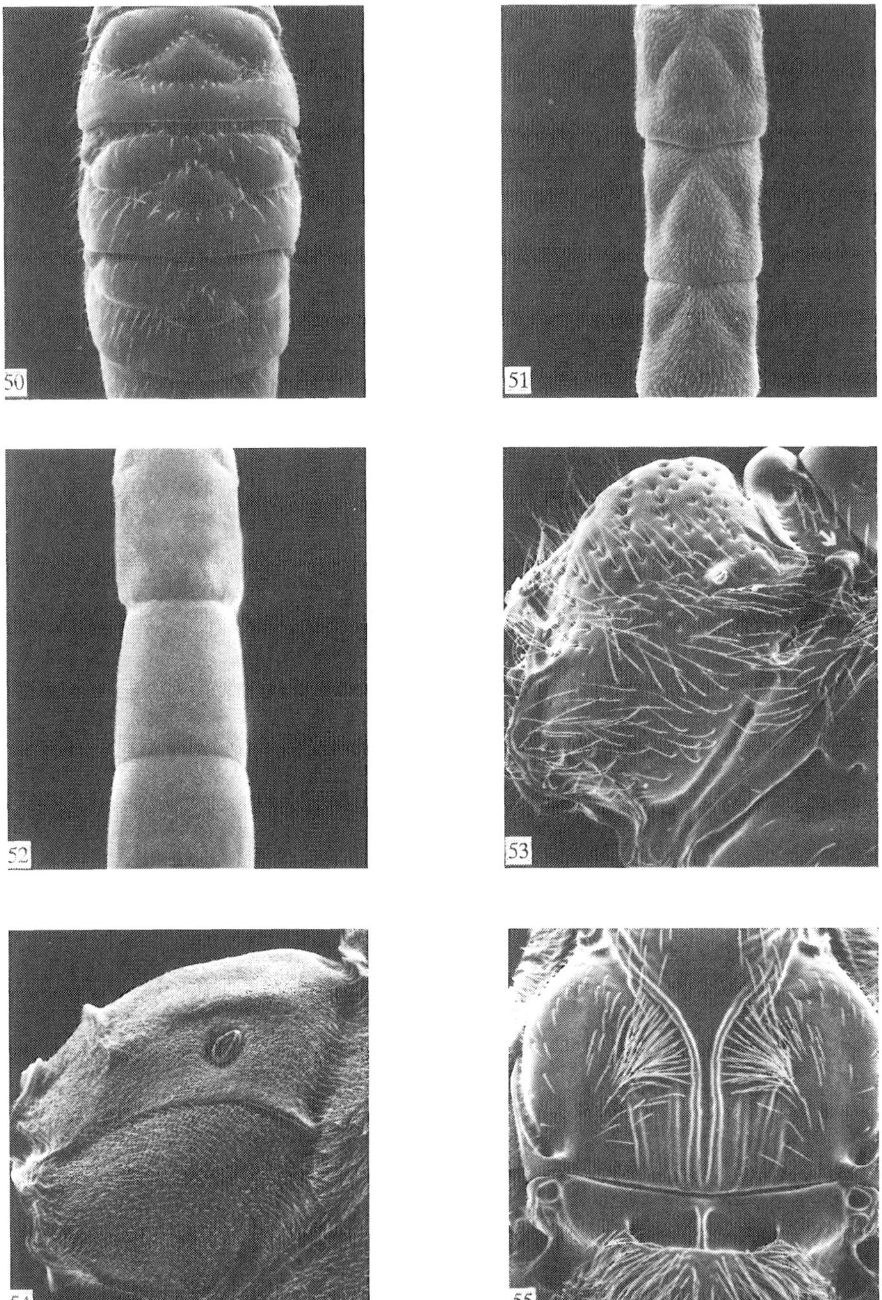

Figs 12.50–12.55. Ichneumonidae, scanning electron photomicrographs. Figs 12.50–12.52. Metasomal tergites II–IV; 12.50, *Lycorina* sp. (Lycorininae); 12.51, *Glypta* sp. (Banchinae); 12.52, *Mnioes* sp. (Banchinae). Figs 12.53–12.54. Propodeum, lateral; 12.53, *Lycorina* sp. (Lycorininae); 12.54, *Mnioes* sp. (Banchinae). Fig. 12.55. Scuto-scutellar suture of *Xorides* sp. (Xoridinae).

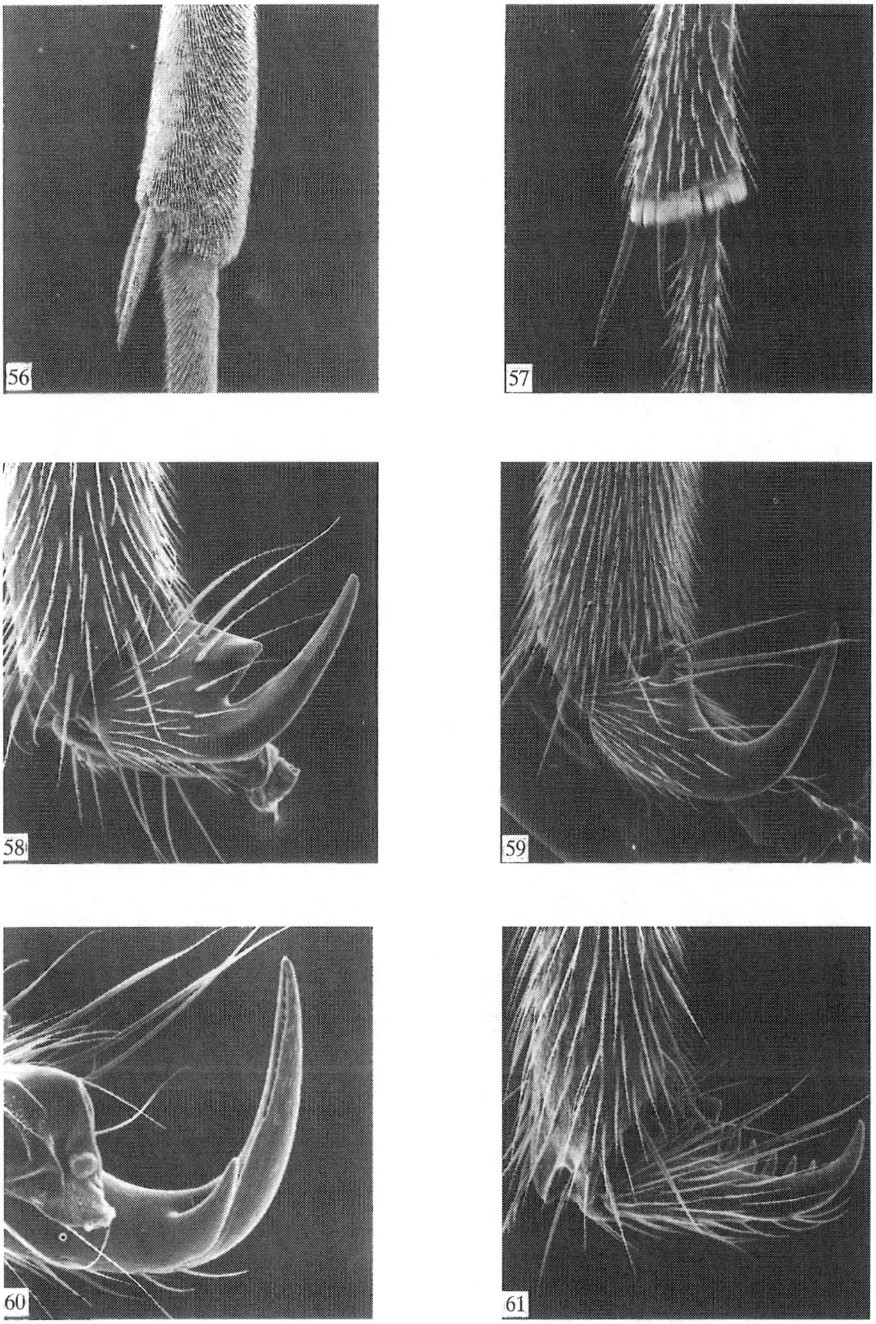

Figs 12.56–12.61. Ichneumonidae, scanning electron photomicrographs. Figs 12.56–12.57. Apex of hind tibia, internal; 12.56, *Pimpla* sp. (Pimplinae); 12.57, *Megastylus* sp. (Orthocentrinae). Figs 12.58–12.61. Mid tarsal claw; 12.58, *Anastelgis* sp. (Pimplinae); 12.59, *Xanthopimpla* (Pimplinae); 12.60, *Arotes* sp. (Acaenitinae); 12.61, *Enicospilus* sp. (Ophioninae).

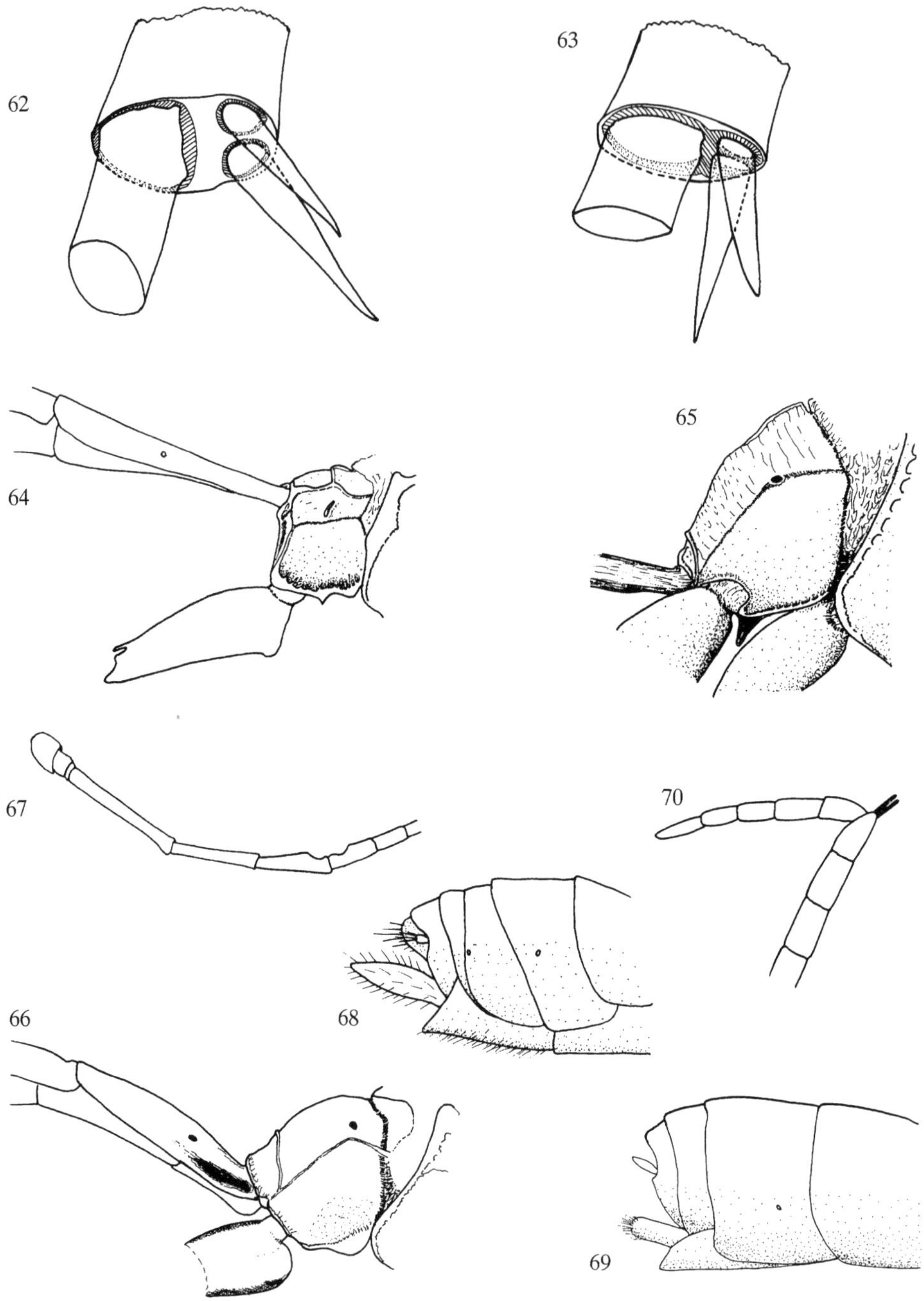

Figs 12.62–12.70. Ichneumonidae. Figs 12.62–12.63. Insertion of hind tibial spurs; 12.62, *Eiphosoma* sp. (Cremastinae); 12.63, *Delopia* sp. (Campopleginae). Figs 12.64–12.66. Insertion of metasoma into propodeum; 12.64, *Certonotus* sp. (Labeninae); 12.65, *Lissocaulus* sp. (Banchinae); 12.66, *Lissonota* sp. (Banchinae). Fig. 12.67. Base of antenna of male *Cylloceria* (Cylloceriinae). Figs 12.68–12.69. Apex of metasoma of female; 12.68, *Oxytorus* sp. (Oxytorinae); 12.69, Ctenopelmatinae. Fig. 12.70. Apex of female antenna, *Xorides* sp. (Xoridinae).

A synopsis of the Costa Rican fauna

Ichneumonids occur in virtually all terrestrial habitats in Costa Rica, from the mangrove coastal forests to paramo-like areas above 3300 metres. Species are generally most numerous in forests, whilst relatively few seem to occur in more open, dry habitats such as pastures or pineapple plantations. Although the greatest species-richness seems to occur in pristine lower montane forests (Gauld, 1991), large numbers of species are often present in relatively disturbed habitats such as suburban gardens.

The 25 subfamilies of Ichneumonidae known from Costa Rica, although arranged alphabetically here for convenience, can be placed in informal groups that share a number of common features (Gauld, 1991): pimpliform subfamilies (Rhyssinae, Pimplinae, Poemeniinae, Acaenitinae, Cylloceriinae, Diplazontinae and Orthocentrinae), tryphoniform subfamilies (Tryphoninae), ichneumoniform subfamilies (Labeninae, Brachycyrtinae, Cryptinae, Ichneumoninae and possibly Xoridinae), ophioniform subfamilies (Ctenopelmatinae, Banchinae, Mesochorinae, Campopleginae, Ophioninae, Cremastinae, Tersilochinae and Anomaloninae) and the remaining subfamilies, which are presently unplaced.

Acaenitinae

The Acaenitinae is a moderately small, mainly holarctic and Old World tropical taxon with one genus, *Arotes*, extending south into Central America. Little is known about their biology, but one European acaenitine is a koinobiont endoparasitoid of the larvae of curculionids boring in plant tissue (Shaw & Wahl, 1989; Zwakhals, 1989). Female acaenitines are easily recognized by the elongate 'plough-share' shaped subgenital plate (Fig. 12.42).

Arotes. A small, primarily holarctic genus represented by two species in Central America. One has been collected in lower montane forest in northwestern Costa Rica. The Central American species are keyed by Gauld (1991).

Anomaloninae

(Whole insect, Fig. 12.71). This is a moderately sized, cosmopolitan subfamily. Anomalonines are koinobiont endoparasitoids of, mostly, lepidopterous larvae, although some species of one genus, *Anomalon*, attack tenebrionid beetle larvae (Townes, 1971). Most anomalonines parasitize geometroid, tortricoid or pyraloid Lepidoptera (Gauld & Mitchell, 1977), although some larger species attack Noctuidae and Saturniidae, and a few species are known to parasitize Papilionoidea. Oviposition is into the host larva and the adult ichneumonid always emerges from the host pupa (Tothill, 1922).

Anomalonines are easily recognized by their elongate slender appearance (Fig. 12.71), lack of any areolet (Figs 12.09, 12.17) and reticulate propodeum (Fig. 12.46). The subfamily is represented in Costa Rica by about 30 species. Species of *Anomalon* are common throughout the country, from over 2500 metres on the Cerro de la Muerte, to the coastal plain in Corcovado National Park. Many (formerly placed in *Neogreeneia*) are restricted to very humid sites, whilst typical *Anomalon* species occur in very dry areas. Species of *Corsoncus* may also be found in the seasonally dry habitats, whilst several species of *Philodrymus* and *Podogaster* occur in humid lowland forests, as do some species of the morphologically aberrant genus *Ophionellus*, a taxon with very reduced wing venation and a very elongate metasoma. Other species of this genus are to be found in very dry open scrubby areas. At higher altitudes species of *Aphanistes* and *Habronyx* occur. Other genera occurring in Costa Rica include *Agrypon*, *Gravenhorstia*, *Ophiopterus*, *Parania* *Trichomma* and *Phaenolabrorychus*. Gauld (1976*a*) gives a key to world genera, and a monograph covering the Costa Rican species is currently in preparation.

Banchinae

The Banchinae is a large and diverse subfamily that is well-represented in most regions. Banchines are koinobiont endoparasitoids of lepidopterous larvae; many of the species with long ovipositors attack ditrysian microlepidopterous hosts, especially Pyralidae, concealed in leaf rolls and soft plant tissue, while species with short ovipositors attack more exposed, macrolepidopterous larvae, especially Noctuidae (Fitton, 1985). Oviposition is often into the first, second or third instar host larva, although sometimes all instars except the last are acceptable (Veen, 1982; Slovák, 1986*a*). The host is killed as a

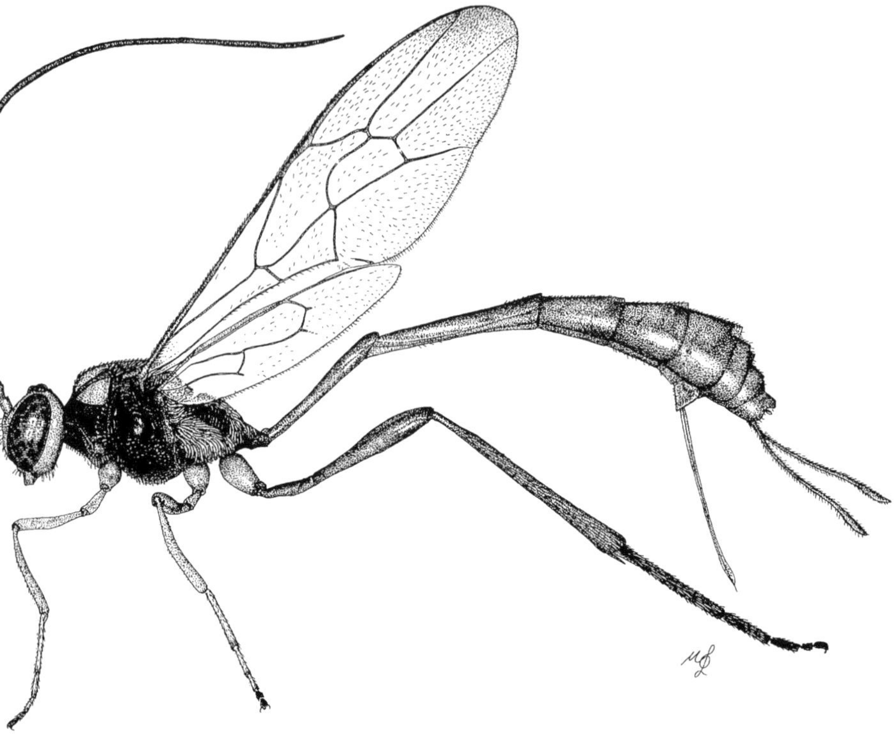

Fig. 12.71. *Philodrymus* sp. (Anomaloninae).

prepupa (Mathur, 1967*b*; Danthanarayana *et al.*, 1977).

As a group, the Banchinae is rather difficult to characterize. Females of most species have a dorsal subapical notch on the ovipositor though this is lacking in one undescribed Costa Rican genus. Species of the tribe Glyptini have impressed V-shaped grooves on metasomal tergites II to IV (Fig. 12.51). Species in the other two tribes, Banchini and Atrophini, often have the propodeum smooth with only the posterior transverse carina present (Fig. 12.54) and many have the submetapleural carina anteriorly expanded into a lobe.

Banchines are abundant in Costa Rica, where the fauna comprises about 150 species (Gauld, 1991), and specimens can be encountered in most habitats. In dry forests some species of *Mnioes* have been the most frequently collected of all ichneumonids for much of the wetter part of the year, whilst in wetter areas species of *Deleboea*, *Eudeleboea*, *Hapsinotus* and *Syzeuctus* are often common. *Diradops* is species-rich in open drier habitats, and species of *Exetastes* are common in similar habitats, especially above 2000 metres. At high altitude, above 2500 metres, species of *Glypta*, *Lissonota* and *Zaglyptomorpha* can be found throughout most of the year. Other genera known to be represented in Costa Rica include *Hadrostethus*, *Isomeris*, *Lissocaulus*, *Loxodocus*, *Meniscomorpha*, *Occia* and *Procestus*.

Currently there are no usable keys to species in Central America, but many genera may be recognized using Townes (1970*b*). However, a number of quite distinctive species-groups do not fit well into any of the described genera. A study of the Costa Rican fauna is currently being undertaken by J. Ugalde.

Brachycyrtinae

The Brachycyrtinae is a small group of taxa that were previously included in the subfamily Labeninae (Townes, 1969; Gauld, 1991). Wahl (1993*b*) presented evidence for treating them as a separate subfamily and his interpretation is followed here.

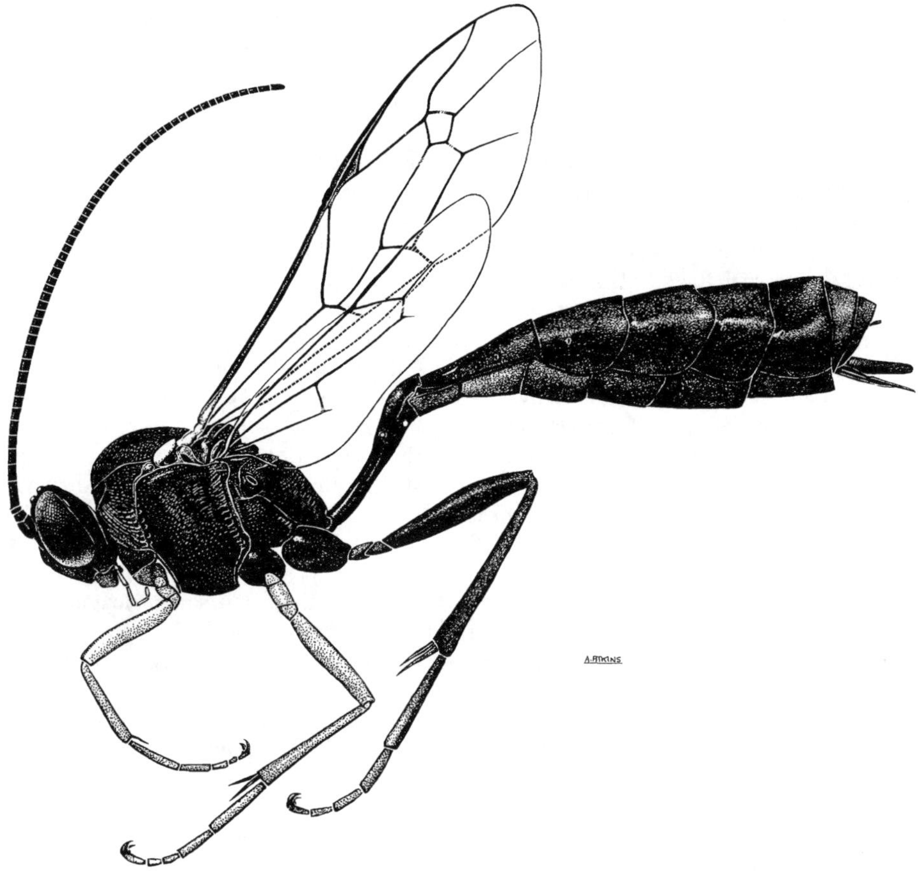

Fig. 12.72. *Delopia* sp. (Campopleginae).

The Brachycyrtinae is most species-rich in southern South America and Australia (Gauld, 1983), although one genus, *Brachycyrtus*, is cosmopolitan. Brachycyrtines are idiobiont ectoparasitoids of a variety of cocooned hosts: species of some Australian genera attack spider egg sacs (Gauld, 1984) and species of *Brachycyrtus* attack cocooned Chrysopidae (Neuroptera) (Cushman, 1936).

Only the nominate genus is present in Costa Rica. It is easily recognized by its characteristic venation (Fig. 12.10).

Brachycyrtus. A small cosmopolitan genus of Brachycyrtinae represented in Costa Rica by about five species, mostly in lowland forests. Walkley (1956) gave a key to species.

CAMPOPLEGINAE (= Porizontinae) (Whole insect, Fig. 12.72)

The Campopleginae is a very large, cosmopolitan subfamily. Species are koinobiont endoparasitoids; some attack the larvae of tenthredinoid sawflies, a small number parasitize the larvae of phytophagous beetles (Chrysomelidae, Curculionidae and Cerambycidae) or Rhaphidiidae (Neuroptera), but the great majority attack larvae of Lepidoptera. A wide variety of lepidopterans are used as hosts including members of most macrolepidopteran families and many ditrysian microlepidoptera. Species of several genera, such as *Microcharops* and *Casinaria*, attack the larvae of butterflies (Gupta, 1988; Jerman & Gauld, 1988).

Although the host range of the subfamily as a whole is extremely wide, some genera have very narrow host ranges. For example, in Costa Rica, members of the genus *Cryptophion* have only been reared from sphingid or ceratocampine saturniid larvae (Gauld & Janzen, 1994). Like other koinobiont ichneumonids, most campoplegines delay killing the host until that host is a prepupa and in some way concealed in a pupation retreat. However, in some genera, such as *Charops*, *Microcharops* and *Cryptophion*, the host is killed when it is only partly grown and exposed on the plant.

The Campopleginae are a fairly uniform group. All have a dorsal subapical notch in the apex of the ovipositor, and often the ovipositor is evenly upcurved. Most are black or mainly black with the metasoma strongly laterally compressed, and a slender first tergite with the spiracles far behind the centre (Fig. 12.71). The sculpture of the face and mesoscutum is often more or less granulate and area superomedia of the propodeum is generally open posteriorly.

The Campopleginae is represented in Costa Rica by about 150 species. The large cosmopolitan genera *Delopia* (= *Dusona*), *Diadegma* (= *Angitia*), *Casinaria*, *Venturia* and *Hyposoter* are all quite species-rich, as are some of the more or less endemic neotropical genera such as *Microcharops* and *Cryptophion*. Many of the cosmopolitan genera are widely distributed throughout the country, but *Microcharops* and *Cryptophion* have only been collected below about 1600 metres. Another cosmopolitan genus, *Enytus*, is well-represented at higher altitude sites, and several species of the morphologically aberrant genus *Nonnus* occur in mid-altitude humid forests. Other widespread genera represented in Costa Rica include *Campoctonus*, *Charops*, *Cymodusa* and *Prochas*. A number of species do not fit well into any described genus and several groups of Central American species probably warrant separate generic status.

The species of *Microcharops* have recently been revised by Gupta (1988), and a revision of the species of *Cryptophion* has been completed by Gauld and Janzen (1994), but there are no keys to the majority of campoplegines present in Central America. The generic key presented by Townes (1970*b*) is the best available work on the group, but it is far from satisfactory.

CREMASTINAE

The Cremastinae is a moderately large, mainly tropical and subtropical subfamily which, unlike many other ichneumonid groups, can be quite species-rich in dry, sunny areas. Cremastines are koinobiont endoparasitoids of the larvae of Lepidoptera and less commonly of phytophagous Coleoptera. Many attack hosts feeding in weakly concealed situations, such as leaf rolls, tunnels in shoots, and soft fruits, although a number of common species of *Eiphosoma* attack more exposed hosts such as noctuid larvae. The host is usually killed as a prepupa, or less commonly as a pupa.

Cremastines are often mistaken for campoplegines which they superficially resemble. However, they differ in the position of the insertion of the hind tibial spurs (see Figs 12.62, 12.63). Most also have the first abscissa of *Rs* in the hind wing much shorter than *rs-m* (Fig. 12.26). Many also have tergite II striate and the propodeum with a clearly defined coffin-shaped area superomedia (Fig 12.48). Unlike campoplegines some cremastines have the ovipositor apex decurved or sinuous, and/or the hind tibia with a ventral tooth; no campoplegine is known to have these features.

The Cremastinae is represented in Costa Rica by about 50 species belonging to eight genera. These may be distinguished using Table 12.2. A comprehensive generic key is given by Townes (1971).

***Creagrura*[CR].** A small neotropical genus with a single species, *C. nigripes*, present widely throughout Costa Rica from sea-level up to about 1500 metres.

***Eiphosoma*.** A very large New World genus with about ten species in Costa Rica. Species of *Eiphosoma* are often encountered along the edges of forests and in overgrown pastures. These slender insects are particularly conspicuous as they have a rather slow up-and-down flight, with their legs and antennae outstretched. Costa Lima (1953) gave a key to some of the neotropical species.

***Eutanygaster*[CR].** A small neotropical genus with a single species in mid altitude forests in Costa Rica.

***Pristomerus*.** A very large, cosmopolitan genus with at least 15 species present in Costa Rica. *Pristomerus* species occur in a variety of habitats, from sea-level up to over 2500 metres. There are no keys to species.

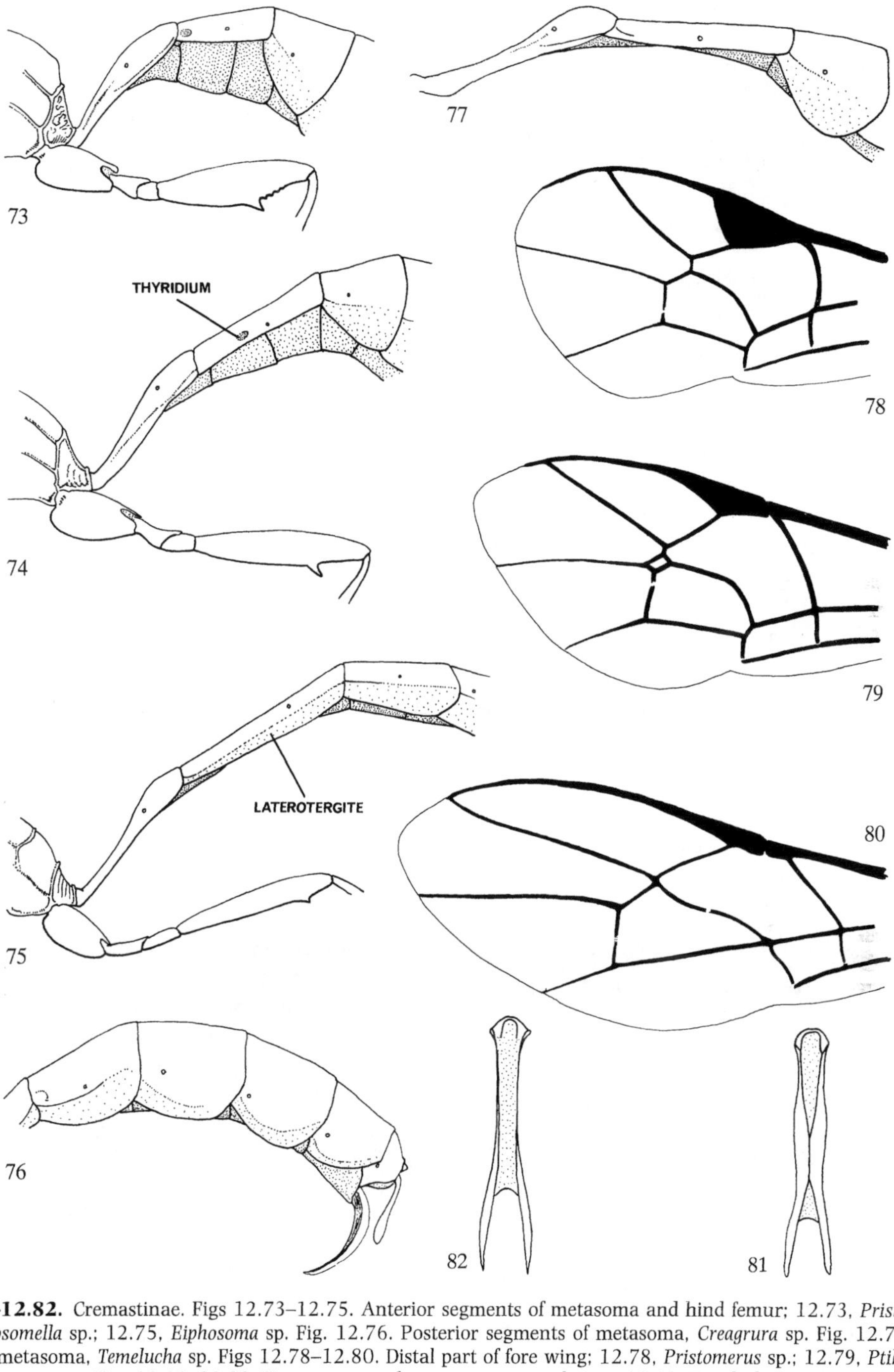

Figs 12.73–12.82. Cremastinae. Figs 12.73–12.75. Anterior segments of metasoma and hind femur; 12.73, *Pristomerus* sp.; 12.74, *Xiphosomella* sp.; 12.75, *Eiphosoma* sp. Fig. 12.76. Posterior segments of metasoma, *Creagrura* sp. Fig. 12.77. Anterior segments of metasoma, *Temelucha* sp. Figs 12.78–12.80. Distal part of fore wing; 12.78, *Pristomerus* sp.; 12.79, *Ptilobaptus* sp.; 12.80, *Eutanygaster* sp. Figs 12.81–12.82. First segment of metasoma, ventral view; 12.81, *Temelucha* sp.; 12.82, *Trathala* sp.

- –Tergite II with thyridium (Figs 12.73–74)
  - –Areolet present **and** hind femur without tooth below; ovipositor short, very strongly curved down (Fig. 12.76) .... *Creagrura*
  - –Areolet absent **or** hind femur with tooth below (Fig. 12.73); ovipositor long, proximally straight
    - –Thyridium close to anterior end of tergite II (Fig. 12.73); areolet absent (Fig 12.78); ovipositor tip often sinuate .... *Pristomerus*
    - –Thyridium close to centre of tergite II (Fig. 12.74); areolet usually present, ovipositor tip not sinuate .... *Xiphosomella*
- –Tergite II without thyridium (Figs 12.75, 12.77)
  - –Laterotergite II folded up under tergite (Fig 12.77)
    - –Areolet present (Fig. 12.79) .... *Ptilobaptus*
    - –Areolet absent (*cf* Fig. 12.78)
      - –Margins of tergite I ± touching ventrally (Fig. 12.81) .... *Temelucha*
      - –Margins of tergite I parallel, not touching below (Fig. 12.82) .... *Trathala*
  - –Laterotergite II pendant, exposed (Fig. 12.75)
    - –Hind femur with ventral tooth (Fig. 12.75); fore wing with 2*rs-m* distinct, often with areolet complete (*cf* Fig. 12.79) .... *Eiphosoma*
    - –Hind femur without ventral tooth; fore wing with 2*rs-m* virtually obliterated as *M* is contiguous with *Rs* (Fig. 12.80) .... *Eutanygaster*

Table 12.2. Identification table for Costa Rican genera of Cremastinae.

***Ptilobaptus*[CR].** A small neotropical genus with a single species present in Costa Rica in lowland Atlantic forests between 300 and 600 metres.

***Temelucha*.** A very large cosmopolitan genus with about ten species present in Costa Rica. Species have been encountered in most habitats from sea-level to over 2500 metres.

***Trathala*.** A very large cosmopolitan genus with about ten species present in Costa Rica. Most species have been collected in mid altitude forest habitats between 1000 and 2200 metres. Around the upper limits of this range several unusually large species, with a fore wing length about 15 mm, occur.

***Xiphosomella*.** A large and primarily neotropical genus with two species in the southern United States. It is represented by about ten species in Costa Rica, mostly below 2000 metres. *Xiphosomella* species are often encountered along the edges of forests. They are very slender insects with a slow up-and-down flight. There are no keys to species.

CRYPTINAE (= Gelinae or Phygadeuontinae). (Whole insect Fig. 12.07).

The Cryptinae is the largest subfamily of Ichneumonidae and species of this cosmopolitan group may be encountered in virtually all terrestrial habitats. The subfamily includes some of the most conspicuous and abundant of all Costa Rican ichneumonids.

Nomenclaturally this group has had a particularly complex history (see Fitton & Gauld, 1976, 1978), but for the previous 15 years the oldest valid name based on rigorous interpretation of the *International Code of Zoological Nomenclature*, Phygadeuontinae, had gained a wide measure of acceptance. However, using their plenary powers, the International Commission on Zoological Nomenclature recently overturned good scholarship and the strict application of their own rules, preferring instead to validate names used by a few workers who had earlier chosen to ignore established ICZN rules of nomenclature. The Commission validated the use of *Cryptus* Fabricius, 1804 (for the genus *Itamoplex*), and the family-group name Cryptinae, by suppressing the heretofore perfectly valid senior name *Cryptus* Panzer, 1804 (*Opinion, 1757*). Unfortunately, there is no appeal against such arbitrary abuse of plenary power by those charged with supposedly conserving nomenclatural stability, so I am here reluctantly adopting the use of Cryptinae for the group I have, in previous publications, referred to as the Phygadeuontinae. Certain changes in tribal and subtribal nomenclature must also follow, and these are summarized in Table 12.3.

Townes (1970*a*) divided the subfamily into three tribes. Two of these, Cryptini (= Mesostenini) and Phygadeuontini, are quite distinctive, but the limits of the third (and consequently its name) have been subject to some disagreement. Townes (1970*a*) included two dissimilar groups of genera, the *Hemigaster* and *Aptesis* groups, in this tribe. The former do not share any derived characteristics with the *Aptesis* group of genera, so I prefer instead to place the hemigastrines in the Cryptini, leaving a distinctive third tribe, the Aptesini, comprising predominantly north temperate genera.

Most tropical species of Cryptinae belong to the Cryptini, and the Aptesini and Phygadeuontini are apparently more species rich in temperate regions. Almost all studied cryptines are idiobiont ectoparasitoids, although a very few north temperate species are more or less endophagous (e.g. *Gelis vicinus* in butterfly pupae), and a few others (the phygadeuontine subtribe Stilpnina) are koinobiont endoparasitoids

Townes, 1970*a*	**Fitton & Gauld, 1978**	**Present work, following Opinion1757**
GELINAE	PHYGADEUONTINAE	CRYPTINAE
Gelini (14 subtribes)	Phygadeuontini (same 14 subtribes)	Phygadeuontini (same 14 subtribes)
Echthrini (no subtribes)	Hemigasterini (no subtribes)	Aptesini[1] (no subtribes)
Mesostenini	Mesostenini	Cryptini
Agrothereutina	Agrothereutina	Agrothereutina
Ischnina	Hedycryptina	Cryptina
Baryceratina	Baryceratina	Baryceratina
Mesostenina	Mesostenina	Mesostenina
Glodianina	Glodianina	Glodianina
Goryphina	Goryphina	Goryphina
Lymeonina	Lymeonina	Lymeonina
Coesulina	Coesulina	Coesulina
Vagenathina	Vagenathina	Vagenathina
Ateleutina	Ateleutina	Ateleutina
Melanocryptina	Melanocryptina	Melanocryptina
Ceratocryptina	Ceratocryptina	Ceratocryptina
Nematopodiina	Ospry nchotina	Osprynchotina
Gabuniina	Gabuniina	Gabuniina
Sphecophagina	Sphecophagina	Sphecophagina

Table 12.3. A summary of principal usages of family-group names in the Cryptinae from 1970–1994. [1] Note that the usage of Aptesini for this group results not from nomenclatural change, but from the fact that the *Hemigaster* group of genera are probably best placed in the Cryptini, not with the large *Aptesis* group of genera.

of Diptera (Gauld, 1988*c*). Some tropical cryptines, particularly those in the *Polycyrtus* genus-group, develop as endoparasitic koinobionts of Lepidoptera (D.H. Janzen, pers. comm.). The commonest hosts of Cryptinae are endopterygote pupae or prepupae enclosed in cocoons or plant tissue. Several are known to attack lepidopterans, including a species of *Mallochia*, which in Mexico attacks a pyralid (Smith *et al.*, 1990), *Chirotica* spp. which parasitize bagworms (Psychidae) throughout the World, and *Baryceros* spp. which attack limacodids. Many smaller cryptines, especially species of *Isdromas* and related genera, develop as pseudohyperparasitoids in the cocoons of other ichneumonids. Other cocoon-like structures, such as spider egg sacs are also attacked (Fitton *et al.*, 1987), and an African species attacks the egg masses of a curculionid (Kerrich, 1969). Some Cryptini, including those of the large and predominantly tropical subtribe Gabuniina, attack hosts concealed in wood (Gupta & Gupta, 1983) and members of the subtribes Osprynchotina and Sphecophagina attack the immature stages of aculeate Hymenoptera in their nests (Janvier, 1967; Donovan, 1991). Some members of the subtribe Lymeonina, such as species of *Toechorychus* and *Pachysomoides*, are common parasitoids of social vespids, whilst a species of *Lymeon* has been reared in Brazil from the pupa of an argid sawfly (Dias, 1976). Although a considerable amount of information is available about the host relationships of some Cryptinae, virtually nothing is known about the biology of many of the large tropical genera.

Structurally this is a very diverse subfamily. Most species have the metasoma depressed and tergite I with the spiracles at or beyond the centre, the fore wing with a pentagonal or quadrate areolet (that may lack 3*rs-m* externally) and possess a long ovipositor (Fig. 12.73). The ovipositor has no dorsal notch (but it may have a ridge or nodus) and the tarsal claws are simple. Smaller species resemble some ichneumonines but they often have the clypeus more convex, possess a deep sternaulus and have flexible ovipositor sheaths.

The Cryptinae is very species-rich in Costa Rica, and probably includes more than 500 species. A considerable number belong to just four genera, *Cryptanura*, *Polycyrtus*, *Lymeon* and *Diapetimorpha*, but other cryptine genera represented in Costa Rica include *Acerastes*, *Agonocryptus*, *Baryceros*, *Biconus*, *Bicristella*, *Bicryptella*, *Camera*, *Cestrus*, *Cryptohelcostizus*, *Cryptopteryx*, *Digonocryptus*, *Joppidium*, *Lamprocryptidea*, *Latosculum*, *Mallochia*, *Messatoporus*, *Pachysomoides*, *Photocryptus*, *Polycyrtidea*, *Priotomis*, *Prosthoporus*, *Toechorychus*, *Trachysphyrus* and *Whymperia*. Most of these genera include medium sized to large, black-and-white striped cryptines that are common in forests from sea-level up to 1500 metres. In drier habitats brownish species with yellowish wings are abundant. In almost all areas small, gaudy black-white-red patterned species of *Lymeon* and *Diapetimorpha* abound. In the cooler regions above 2500 metres the Cryptinae is represented by numerous drab small blackish, reddish or brown species. At these higher altitude sites species of the other large tribe, the Phygadeuontini seem to be more common than they are in the lowlands. Genera represented include *Lissaspis*, *Isdromas*, *Distathma*, *Mastrus*, *Amphibulus* and *Stilpnus*, and one or two apterous, antlike species of *Gelis* occur on the Cerro de la Muerte.

Townes (1970*a*) provides a key to genera, but this can be difficult to use, at least at the subtribal level. Although many of the species present in Costa Rica can be clearly assigned to a genus many cannot and considerable effort will be necessary to establish generic limits in tropical America. Modern species-level treatments include Porter (1967) on *Trachysphyrus*, Porter and O'Neill (1985) on *Chromocryptus*, S. Gupta (1982) on *Agonocryptus* and S. Gupta (1983) on *Prosthoporus*. Cushman (1931, 1945) provides useful keys to neotropical species of *Polycyrtus* and *Cryptanura* although many additional and undescribed species are now known.

CTENOPELMATINAE (= Scolobatinae).

This is a large, mainly north temperate subfamily. In sub-boreal parts of northern Europe ctenopelmatines comprise about 18 percent of the species (Jussila, 1965, 1984) and farther south, in Britain, the group still accounts for more than 10 percent of the species-richness (Gauld, 1984). However, the subfamily is generally very poorly represented in tropical habitats, and only about ten species have been described from the Neotropical region (Townes & Townes, 1966). The vast majority of ctenopelmatines are koinobiont endoparasitoids of larval tenthredinoid and megalodontoid sawflies, although a very few species are known to attack Lepidoptera (Heath, 1961; Gauld, 1984). The female generally oviposits into a middle or late instar larva, but many species that have slender ovipositors oviposit into the egg or young larva

(McConnell, 1938). Usually the host is not killed until after it has spun a cocoon (Graham, 1953; Pschorn-Walcher, 1967).

The Ctenopelmatinae is a very difficult group to characterize. Species are generally rather stout, and most have short ovipositors that generally have a long subapical notch, although a few species with very slender ovipositor apices lack a notch; the ovipositor sheath frequently is only hirsute near its apex (Fig. 12.69). The metasoma is generally cylindrical with tergite I with spiracles at or before the centre. Many have rather stout, long and weakly tapered mandibles and often the upper tooth is slightly shorter than the lower one.

Ctenopelmatines are generally quite scarce in Costa Rica, although there are at least three times as many species in the country than have been described for all of America south of the United States. I have seen about 30 species, mostly belonging to the genera *Coelorhachis*, *Physotarsus* and *Phobetes*. Other genera include *Asthenara*, *Lathrolestes Nanium*, *Onarion*, *Perilissus Rhorus*, and several apparently undescribed taxa. The generic keys given by Townes (1970*b*) are extremely difficult to use. A monograph on the Costa Rican species is currently nearing completion.

CYLLOCERIINAE (= Microleptinae in part)

The Cylloceriinae is a small subfamily containing only two genera, *Allomacrus* and *Cylloceria* (Wahl, 1990). The former is restricted to the Holarctic region whilst the latter extends from the Holarctic region into the Neotropics. Males may easily be recognized by the presence of deeply concave areas on the third and fourth flagellar segments (Fig. 12.67), although females resemble banchines, but lack a dorsal subapical notch in the ovipositor.

Cylloceria. A small genus represented in Costa Rica by four species; three occur at altitudes above 2000 metres, while the fourth is widespread in lower montane wet forests. A key to species is provided by Gauld (1991). A North American species of *Cylloceria* is a koinobiont endoparasitoid of the larvae of Tipulidae (Diptera) (Wahl, 1986).

DIPLAZONTINAE

The Diplazontinae is a relatively small cosmopolitan subfamily that is most diverse in the North Temperate region. Diplazontines are koinobiont endoparasitoids, mostly of aphidophagous Syrphidae (Diptera). Females oviposit into the host larva, or in a few species, into the egg (Rotheray, 1981*b*), and adult progeny emerge from the host puparium. In temperate regions diplazontines can be very common around aphid colonies and the males of some species form swarms.

In Costa Rica Diplazontines are easily recognized by the apparently tridentate mandible (Fig 12.40). The neotropical species have been monographed by Dasch (1964), but this work is based on very little material and no species were recorded from Costa Rica. In Costa Rica the approximately 13 species of diplazontines are generally scarce and most species have been encountered above 2000 metres. Most belong to the genus *Woldstedtius* (= *Syrphoctonus sensu* Townes, 1971), but other genera present include *Diplazon*, *Enizemum*, *Promethes*, *Schachticraspedon*, *Sussaba*, *Syrphidepulo* and *Syrphoctonus* (= *Homotropus sensu* Townes,1971). A revision of the Costa Rican species by Paul Hanson and Ian Gauld is currently nearing completion.

ICHNEUMONINAE

(Whole insect, Fig. 12.83). The Ichneumoninae is an extremely large, cosmopolitan subfamily that rivals the Cryptinae in size. Ichneumonines are specialist endoparasitoids of Lepidoptera. Many seem to be extremely host specific and some groups of species have very narrow host ranges. For example, the large colourful species of the subtribe Trogina attack Papilionidae, the subtribe Callojoppina attack Sphingidae (Ward & Gauld, 1987), the Listrodromini attack Lycaenidae and the majority of Platylabini attack Geometridae (Heinrich, 1960). All ichneumonines emerge from the pupa of their host; some actually oviposit into the pupa, but many species oviposit into the final instar larva (Hinz, 1983). Many of these koinobionts place their eggs in sites where they are not in direct contact with the host's haemolymph, such as the gut wall (Veen, 1981) or a salivary gland (Strickland, 1923), a behaviour which perhaps protects them against the immuno-defensive system of their host (Salt, 1968).

Ichneumonines abound in all terrestrial habitats in Costa Rica and they have even been collected in the centre of a banana plantation. I have seen between 300 and 400 species, but the total number in Costa Rica could be considerably higher as many of the larger species, collected by hand-netting, are often not present in Malaise trap samples. Two of the largest

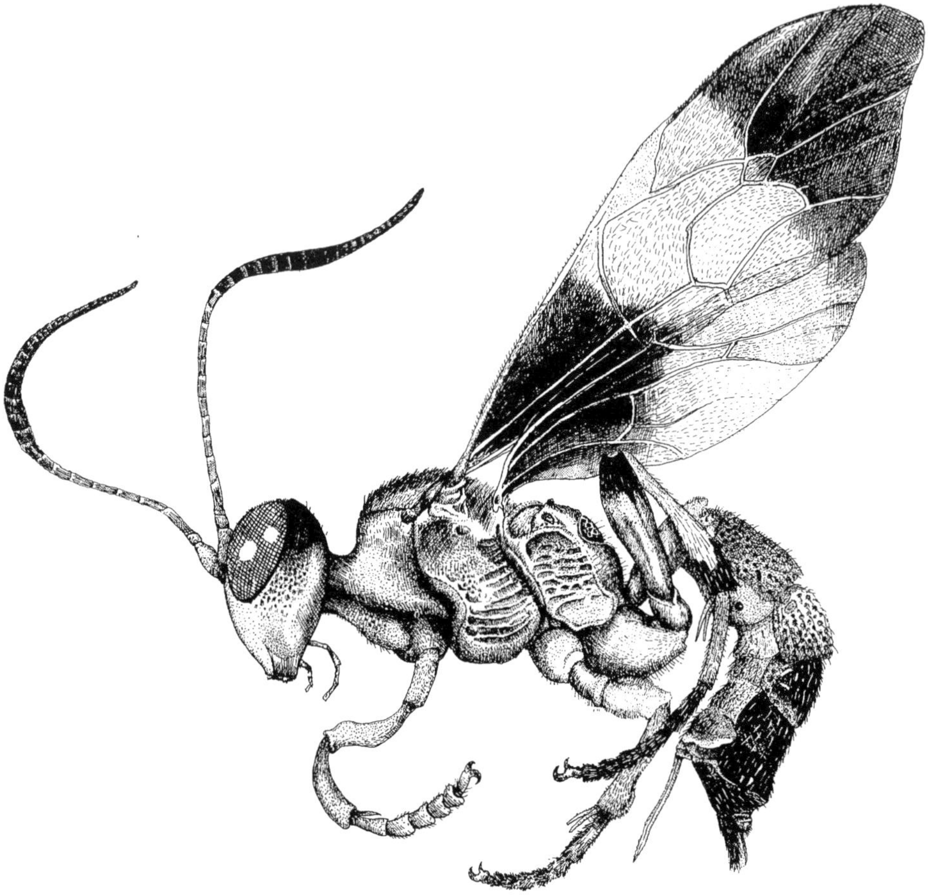

Fig. 12.83. *Joppa* sp. (Ichneumoninae).

ichneumonine genera in Costa Rica are *Carinodes* and *Joppa*. Species of the former are common in seasonally dry habitats and agroecosystems, while the characteristically yellow and black patterned *Joppa* species can be common in wet forests. For neither genus is a single host known. At high altitudes species of Platylabini are frequently encountered and the Phaeogenini is very species-rich. Other genera known to occur in Costa Rica include *Abzaria*, *Areoscelis*, *Coelichneumon*, *Conopyge*, *Cryptojoppa*, *Diacantharius*, *Dilopharius*, *Diphyus*, *Eutanyacra*, *Ichneumon*, *Joppocryptus*, *Limonethe*, *Lobaegis*, *Lophojoppa*, *Lusius*, *Macrojoppa*, *Matara*, *Microsage*, *Notacma*, *Oedicephalus*, *Ortezia*, *Pedinopelte*, *Plagiotrypes*, *Platylabus*, *Protopelmus*, *Rhabdotus*, *Setanta*, *Tetragonochora*, *Tricholabus*, *Tricyphus* and *Trogomorpha*.

The Ichneumoninae is a distinctive group. Most species have the metasoma dorsoventrally depressed, tergite I with spiracles far behind the centre, and a short ovipositor that is slender, lacks a dorsal subapical notch and is unusual in being encased in a very rigid sheath (Fig. 12.72). All have a complete areolet that is either pentagonal or virtually rhombic, but sessile (Figs 12.11, 12.12), and most have the clypeus/lower face flat (Fig. 12.35). Many are large and brightly coloured insects.

There are no keys to the Central American genera except for the synoptic appendix provided by Townes and Townes (1966). Heinrich (1977) monographed the species of Florida and provided illustrated keys to the nearctic genera of Ichneumoninae; this publication may be of some use in Central America. Ward and Gauld (1987) provide a key to the Mesoamerican genera of the Callajoppina. Like the Cryptinae, considerable problems remain with defining generic limits for tropical American ichneumonines and many species cannot currently be assigned to any genus.

Labeninae

The Labeninae is a moderately small subfamily which is most species-rich in South America and Australia (Gauld, 1983; Wahl, 1993*b*); a few species occur northwards into North America and Indonesia, but this subfamily is otherwise not present elsewhere. It is divided into three rather distinctive tribes, Labenini, Groteini and Poecilocryptini. The first two occur in the New World and the Australian region, whilst the Poecilocryptini is endemic to Australia. Labenines are idiobiont ectoparasitoids; species of Labenini parasitize the pupae or prepupae of wood-boring Coleoptera and Siricidae (Townes & Townes, 1960; Hocking, 1967) and Groteini are parasitoids of aculeate Hymenoptera, especially of bees (Graenicher, 1905; Janvier, 1967; Slobodchikoff, 1967). Some complete their larval development by consuming the host's pollen store.

Labenines are easily recognized by their rather large complete areolets in the fore wing (Fig 12.20), and by having the metasoma inserted into the propodeum above and slightly away from the insertion of the hind coxae (Fig. 12.64). About 30 species of Labeninae in three genera occur in Costa Rica. These genera may be distinguished using Table 12.4. Comprehensive keys are given by Townes (1969).

Certonotus (= *Apechoneura*). This large genus of Labenini is widespread in Australia, and in the New World from southern South America north to Mexico. About 10 species occur in Costa Rica. *Certonotus* species are, on account of the extremely long ovipositor and bright colour pattern, amongst the most spectacular of Costa Rican ichneumonids. Most are generally only encountered in lowland or premontane wet forest habitats below 1000 metres. There is no key to species.

Grotea. A small genus of Groteini widely distributed in the New World, and with about five species in Costa Rica in low to mid altitude sites. Slobodchikoff (1970) gave a key to North American species of *Grotea*.

Labena. A very large genus of Labenini distributed throughout Australia and the New World north to southern Canada. About 15 species occur in Costa Rica. *Labena* species are common in seasonally dry forests during the dry season and I have seen individuals investigating holes in fence posts in open countryside in Guanacaste. There is no key to the New World tropical species.

Lycorininae

This subfamily contains a single cosmopolitan genus, *Lycorina*. Little is known about the biology of the group, but some species are known to parasitize weakly concealed lepidopterous larvae such as pyralids (Finlayson, 1976; Short, 1978).

Lycorina. About eight species have been found in Costa Rica. One is common in lowland Pacific forests, but the rest are seldom collected and seem to be restricted to humid forests from sea-level up to about 2000 metres. The dry forest species has been reared from a species of *Lampronia* (Prodoxidae). *Lycorina* species are easily recognized by the

- –Mesoscutum with transverse sharp ridges .... *Certonotus*
- –Mesoscutum smooth to punctate
  - –Hind wing with 1–3 hamuli in basal group .... *Labena*
  - –Hind wing with about 6 hamuli in basal group .... *Grotea*

Table 12.4. Identification table for Costa Rican genera of Labeninae.

impressed triangular areas on metasomal tergites II–IV (Fig. 12.50) and the possession on the metanotum of a small hook (or 'catch') that engages a process on metapostnotum (Fig. 12.53). A key to species is currently nearing completion (Gauld, in prep.).

MESOCHORINAE

The Mesochorinae is a moderately large, cosmopolitan subfamily. Most species are small and easily overlooked, but a few nocturnal species (of *Cidaphus*) are large and resemble ophionines. Mesochorines are endoparasitic koinobionts; they are obligatory hyperparasitoids that develop within the larva of another parasitoid living on (Gauld & Bolton, 1988), or most commonly within (Wahl, 1993*a*), a primary host. Oviposition is into the primary parasitoid larva even when it is within the primary host (Yeargan & Braman, 1989); the parasitoid host is generally killed after it has killed the primary host and has constructed a cocoon. Although it has been claimed (Dasch, 1971) that some mesochorines are primary parasitoids, there is little evidence to support such a supposition (Haeussler, 1940; Wahl, 1993*a*).

The primary host of parasitoids attacked by mesochorines is frequently a lepidopteran or symphytan larva, but curculionid larvae (Coseglia *et al.*, 1977) and even mirids or psocopterans may serve (Waloff, 1967; King, 1971; Carlson *in* Krombein *et al.*, 1979). The smaller mesochorine species are often reared from microgastrine braconids, but a variety of ichneumonoids including meteorine, euphorine and macrocentrine braconids, and anomalonine, banchine, campoplegine, metopiine and tryphonine ichneumonids commonly serve as hosts (Carlson *in* Krombein *et al.*, 1979; Wahl, 1993*a*). Species have also been reared from tachinids (Short, 1978).

With practice mesochorines are one of the easiest ichneumonid groups to recognize. Males have the gonosquama elongated and rod-like while females have a large subgenital plate and the very slender ovipositor encased in a stout and rigid sheath (Fig. 12.43). Most species are small and have a large rhombic areolet in the fore wing (Fig. 12.15).

The Mesochorinae is well-represented in Costa Rica and I have seen about 40 species from a variety of habitats belonging to two genera. Dasch (1974) gave keys to the genera and species of the Neotropical Region. Wahl (1993*a*) presents a phylogenetic classification of the genera of the group. The two genera in Costa Rica are easily distinguishable; *Cidaphus* species are large with a distinct distal abscissa of *Cu*1 in the hind wing while *Mesochorus* species are smaller and have no trace of this vein.

Cidaphus. A small cosmopolitan genus of large species with an ophionoid facies. Two species occur in Costa Rica, *C. rostratus*, which is common in low and mid altitude sites and an undescribed species from high altitude on the Cerro de la Muerte. The former has been reared by D.H. Janzen in Santa Rosa National Park from a *Thyreodon* species (Ichneumonidae: Ophioninae) attacking a sphingid larva.

Mesochorus (including *Stictopisthus*). An exceedingly large cosmopolitan genus of small to moderately large species, at least 40 of which have been collected in Costa Rica where they are widely distributed and present in all habitats sampled. Several species have been reared from microgastrine braconid parasitoids of sphingids.

METOPIINAE

The Metopiinae is a medium-sized, cosmopolitan subfamily. Metopiines are koinobiont endoparasitoids of Lepidoptera; oviposition is into the host larva and the parasitoid emerges from the host's pupa. Some metopiines oviposit into a specific site within the host, such as the cephalic ganglion (Gerig, 1960), or in the wall of the hind gut by inserting the ovipositor into the host's anus (Aeschlimann, 1974). In the North Temperate region species of several genera have been reared as parasitoids of lepidopteran forest pests (Carlson *in* Krombein *et al.*, 1979); since numerous species of these genera occur in tropical forests, many will no doubt be found to attack Lepidoptera that could assume economically important status in forest regeneration programmes.

Metopiines, except for *Metopius* species, have a convex face with no impression separating the clypeus dorsally or laterally; *Metopius* species have a shield-shaped area on the face bounded by carinae (Fig. 12.36). All metopiines have a dorsoventrally depressed metasoma and a very short ovipositor. Most also have stout legs with swollen femora which tend to 'fit into' mesosomal depressions.

The Metopiinae is represented in Costa Rica by between 60 and 70 species belonging to 11 genera.

These may be discriminated using Table 12.5. Townes (1971) provides a comprehensive key to genera, but there are no keys to species for Central America.

Aplolophus^{EX}. A small New World genus with species present in the Andes of South America and in the Holarctic region, so it is expected to occur, probably at high altitude, in Costa Rica.

Chorinaeus^{CR}. A large and predominantly holarctic genus with a single species present in Costa Rica.

Colpotrochia. A large cosmopolitan genus that is best represented in the Oriental and American tropics. Three or four species have been found in Costa Rica. Species of *Colpotrochia* are quite commonly encountered in lowland disturbed habitats. In flight they resemble vespids.

Cubus^{CR}. A small neotropical genus with a single, rather uncommonly collected species present in the seasonally dry forests of northwestern Costa Rica.

Exochus^{CR}. A very large genus comprising the majority of species in the subfamily. Most are rather small and mainly black insects. In Costa Rica about 40 species have been collected, mainly in lowland and mid altitude forests, although species have been collected up to 3100 metres.

Hypsicera. A large Old World tropical and subtropical genus with one or two introduced species in tropical America, *H. femoralis* and *H. curvator* (Townes & Townes, 1966). The former has been found in Costa Rica. These insects are parasitoids of several synanthropic microlepidopterans that attack stored products and it is believed that they have been spread by human commerce (Fitton *in* Gauld, 1984).

Leurus. A small New World genus that is most species-rich in South America. Two or three species, including *L. caeruliventris*, have been found to occur in Costa Rica at altitudes between sea-level and 1600 metres.

Metopius. A large cosmopolitan genus represented in Costa Rica by about six species. They are quite commonly encountered in lowland disturbed habitats and none have been colleced above 1500 metres. In flight, and in their buzzing when netted, they resemble vespids.

Seticornuta^{CR}. A predominantly Old World genus with one doubtfully placed nearctic species. This, or a very similar species, occurs at high altitude in Costa Rica.

Synosis^{CR}. A small genus present in the Holarctic region and South America. A single species has been collected at 2700 metres in Costa Rica.

Triclistus^{CR}. A cosmopolitan genus represented in Costa Rica by about five species. These have been found to occur in a variety of mid to high altitude sites, between 1000 and 3000 metres.

Trieces^{CR}. A large cosmopolitan genus represented by about five species in Costa Rica. Most occur in relatively disturbed mid and low altitude sites and one is very common in lowland dry forest.

OPHIONINAE

The Ophioninae is a moderately large, cosmopolitan subfamily that is rather poorly represented in temperate regions, but is extremely species rich in the wet tropics (Gauld, 1985). Ophionines are koinobiont endoparasitoids of, predominantly, the larvae of macrolepidoptera that feed in exposed situations. The host is generally killed after it has constructed a pupation retreat but before pupating. Some species pupate in the host pupa and one is known to pupate within the skin of the final-instar host larva, partially mummifying it. The great majority of ophionines are nocturnally active and the females search for host larvae that feed at night; a few lowland species, belonging to the genera *Thyreodon* and *Rhynchophion*, are diurnally active. The greatest species-richness of ophionines, mainly *Enicospilus* species, occurs in lowland or mid-altitude forest sites. At high altitude, above 2500 metres, species of *Ophion* predominate. Several species, such as *Enicospilus trilineatus*, *E. glabratus* and *Ophion flavidus*, can be extremely common in disturbed habitats and agroecosystems, where they attack a variety of noctuid pests.

Ophionines are one of the easiest ichneumonid subfamilies to recognize. All lack 2*rs-m* and have

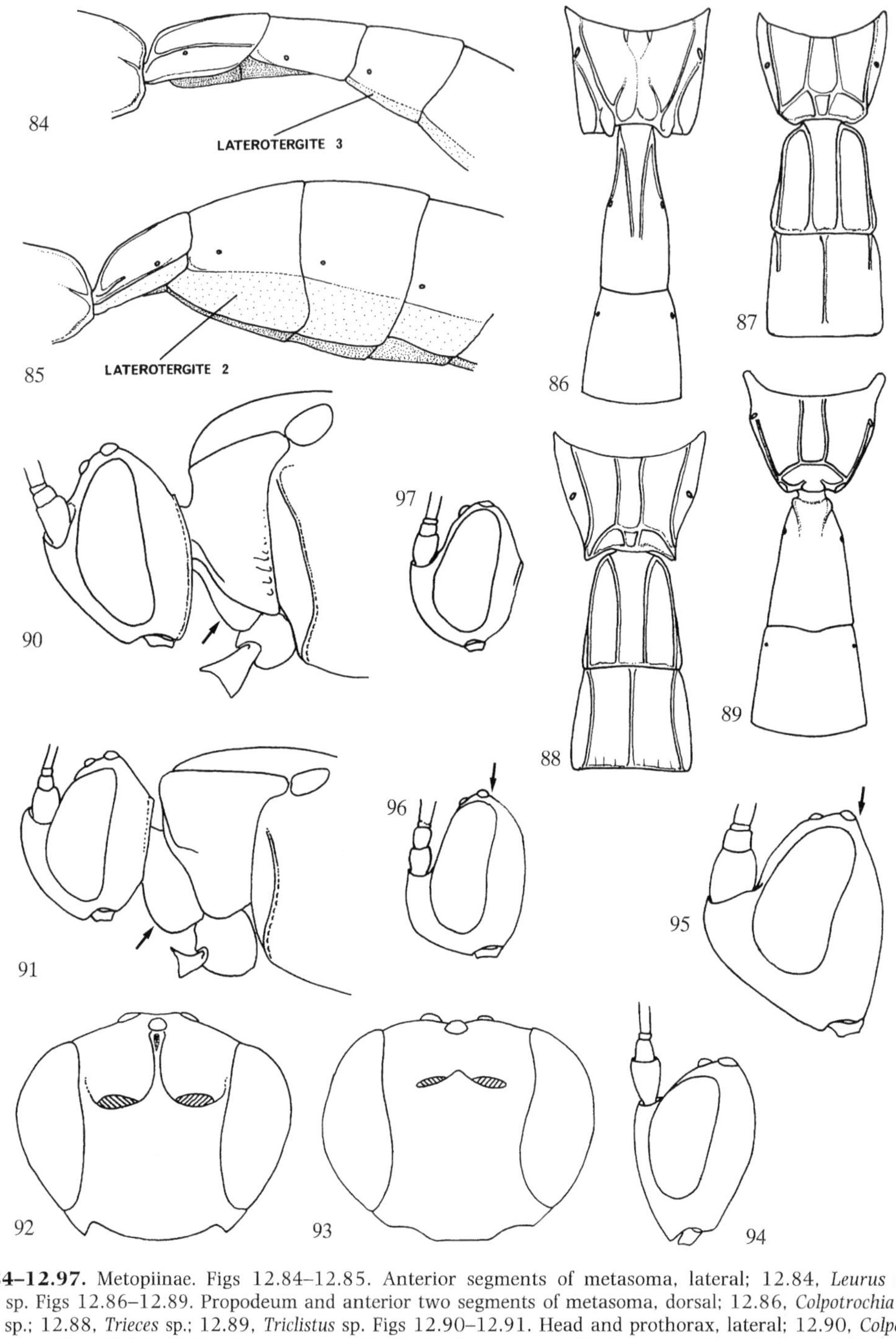

Figs 12.84–12.97. Metopiinae. Figs 12.84–12.85. Anterior segments of metasoma, lateral; 12.84, *Leurus* sp.; 12.85, *Seticornuta* sp. Figs 12.86–12.89. Propodeum and anterior two segments of metasoma, dorsal; 12.86, *Colpotrochia* sp.; 12.87, *Chorinaeus* sp.; 12.88, *Trieces* sp.; 12.89, *Triclistus* sp. Figs 12.90–12.91. Head and prothorax, lateral; 12.90, *Colpotrochia* sp.; 12.91, *Cubus* sp. Figs 12.92–12.93. Face, anterior; 12.92, *Triclistus* sp.; 12.93, *Exochus* sp. Figs 12.94–12.97. Head, lateral; 12.94, *Apolophus* sp.; 12.95, *Hypsicera* sp.; 12.96, *Exochus* sp.; 12.97, *Leurus* sp.

- –Face flat, with carina enclosing shield-shaped area (Fig. 12.36) .... *Metopius*
- –Face weakly to strongly convex
  - –Laterotergites III–V absent
    - –Tergite II sublaterally carinate to apex (Fig. 12.88) .... *Trieces*
    - –Tergite II sublaterally not carinate to centre (Fig. 12.87) .... *Chorinaeus*
  - –Laterotergites III–V distinct (Figs 12.84–12.85)
    - –Antennal sockets separated by high lamella with a deep median groove dorsally (Fig. 12.92)
      - –Propleuron inflated, coxa-like (Fig. 12.91) .... *Cubus*
      - –Propleuron weakly convex (Fig. 12.90)
        - –Tergite I slender (Fig. 12.86), sternite reaching ±0.4 of its length .... *Colpotrochia*
        - –Tergite I stouter (Fig. 12.89), sternite reaching 0.2 of its length .... *Triclistus*
    - –Antennal sockets not separated by high lamella (Fig. 12.93), **or** if one present it lacks a deep median groove dorsally
      - –Areolet present
        - –Propodeal spiracle circular; face weakly convex (Fig. 12.94) .... *Apolophus*[EX]
        - –Propodeal spiracle elliptical; face strongly convex (Fig. 12.97)
          - –Laterotergite II distinct (Fig. 12.85) .... *Seticornuta*
          - –Laterotergite II ± absent (Fig. 12.84) .... *Leurus*
      - –Areolet absent
        - –Back of head vertical behind posterior ocelli (Fig. 12.95) .... *Hypsicera*
        - –Back of head sloping behind posterior ocelli (Fig. 12.96)
          - –Front spur of mid tibia longer than hind spur .... *Synosis*
          - –Hind spur of mid tibia longer than front spur .... *Exochus*

Table 12.5. Identification table for Costa Rican genera of Metopiinae.

3rs-m distal to *2m-cu* (Fig. 12.08) and differ from the only other taxa with such fore wing venation by possessing a spurious vein that extends from the vannal notch to the hind corner of the wing. All have the metasoma compressed, tergite I slender with its spiracle near the hind end, and possess a short, notched ovipositor.

The Ophioninae is represented in Costa Rica by about 110 species, belonging to ten genera. These may be discriminated using Table 12.6. More exhaustive keys to Central American genera and species are given by Gauld (1988*b*), and a key to neotropical genera in Spanish is furnished by Gauld and Lanfranco (1987).

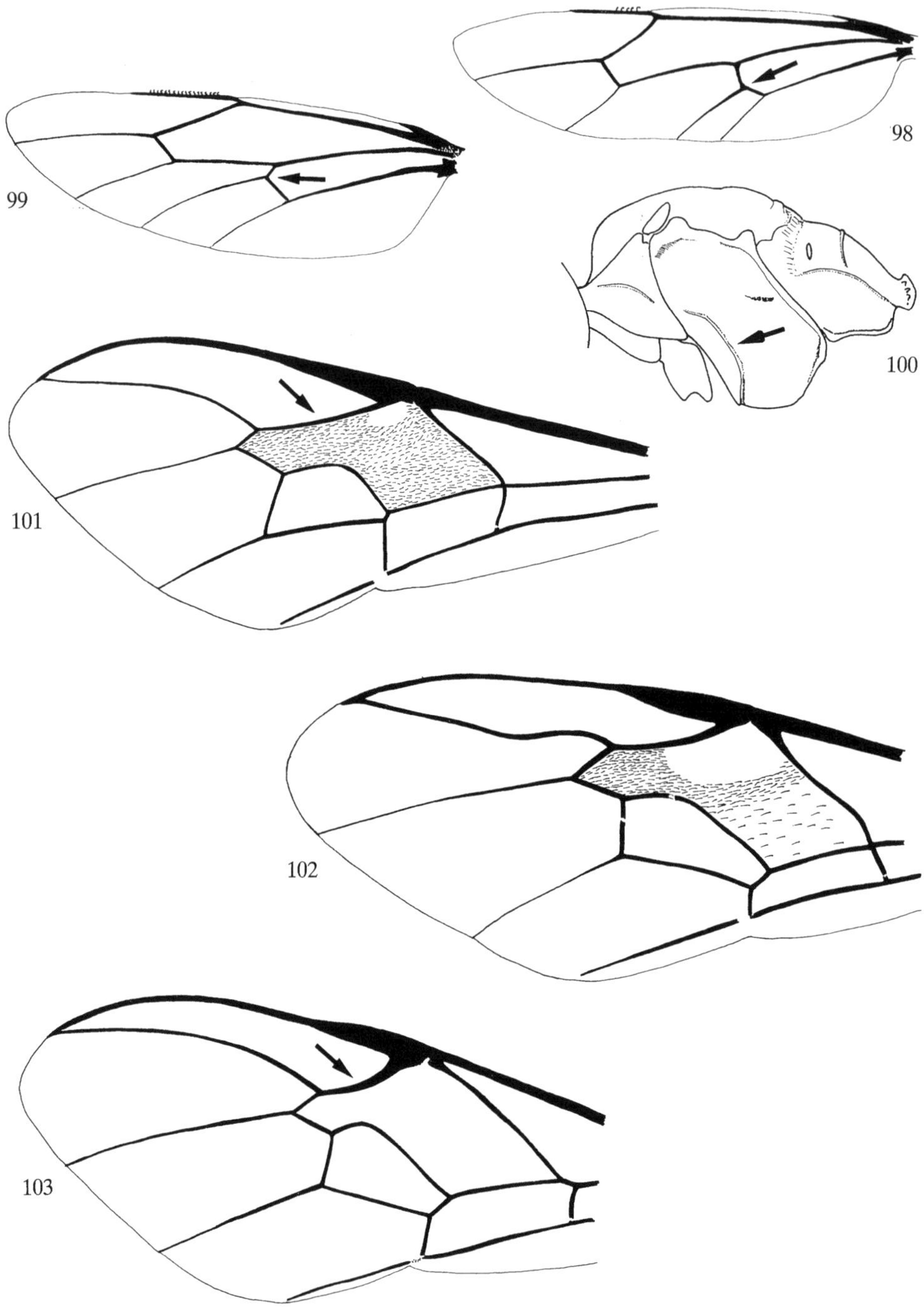

Figs 12.98–12.103. 105 Ophioninae. Figs 12.98–12.99. Hind wing; 12.98, *Ophiogastrella* sp.; 12.99, *Thyreodon* sp. Fig. 12.100. Mesosoma lateral, *Ophiogastrella* sp. Figs 12.101–12.103. Fore wing, distal part; 12.101, *Ophiogastrella* sp.; 12.102, *Sicophion* sp.; 12.103, *Janzophion* sp.

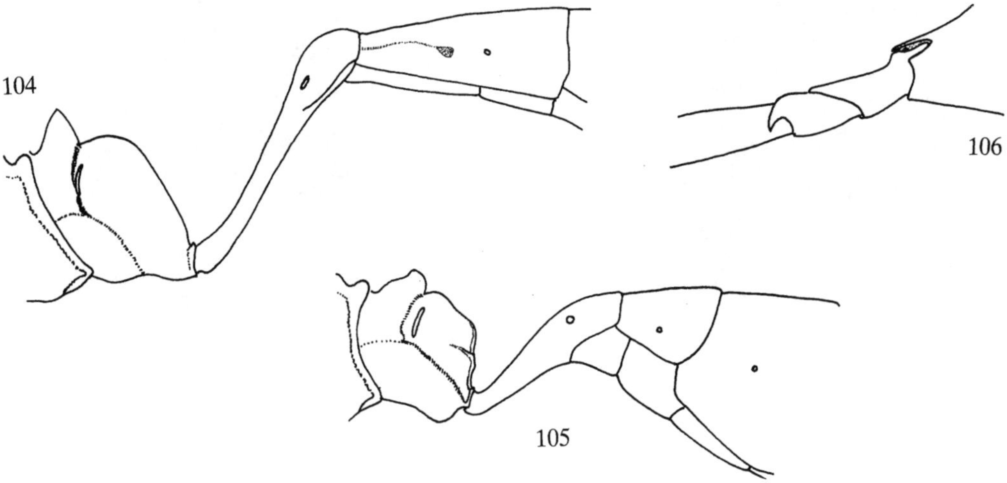

Figs 12.104–12.106. Ophioninae. Figs 12.104–12.105. Propodeum and anterior segments of metasoma; 12.104, *Thyreodon* sp.; 12.105, *Rhynchophion* sp. Fig. 12.106, Hind trochantellar segments, *Stauropoctonus* sp.

Enicospilus. An extremely large cosmopolitan genus whose greatest species-richness is found in lowland to mid-altitude forests throughout the tropics. About 80 species occur in Costa Rica in all habitats from sea-level up to about 1600 metres. Above this very few species occur, and only *E. devriesi* and *E. georginae* have been taken above 2200 metres in Costa Rica. Known hosts of *Enicospilus* species include a wide variety of macrolepidopterous caterpillars (principally Noctuidae, Geometridae, Notodontidae, Lymantriidae, Lasiocampidae, Arctiidae and Saturniidae) that feed exposed on leaves (Gauld, 1988*b*).

Janzophion. A small mesoamerican genus comprising two species, one in Mexico and *J. nebosus* in Costa Rica and Panama (Gauld, 1988*b*). In Costa Rica this species seems to be restricted to rather high altitude wet forests between 1500 and 2500 metres. Its host is unknown.

Ophiogastrella. A small neotropical genus that is most species-rich in lowland dry forest habitats. Four species have been found to occur in Costa Rica, all in the dry northwest. One has been reared as a parasitoid of a geometrid (D.H. Janzen, pers. comm.).

Ophion. A very large cosmopolitan genus that is most species-rich in temperate northern areas. It is represented in Costa Rica by about 12 species, eight of which only occur above 2000 metres (Gauld, 1988*b*). The only species that occurs on the coastal plain, *O. flavidus*, is common in agricultural areas where it attacks noctuid pests.

Prethophion. A neotropical genus comprising a single species, *P. latus*, that ranges from Bolivia north to Costa Rica. It is rarely encountered and a few individuals have been collected at altitudes between 600 and 1300 metres. Its host is unknown.

Rhynchophion. A small New World genus with a single widespread species, *R. flammipennis* that occurs from Arizona south to Ecuador. In Costa Rica individuals have only been collected at the beginning of the wet season, in the seasonally dry lowlands in the northwest of the country. Its host is a sphingid (D.H. Janzen, pers. comm.).

Sicophion. A small neotropical genus known only from montane wet forest sites, from Bolivia north to Costa Rica. A single species, *S. fenestralis*, occurs in Costa Rica where it has only been collected at

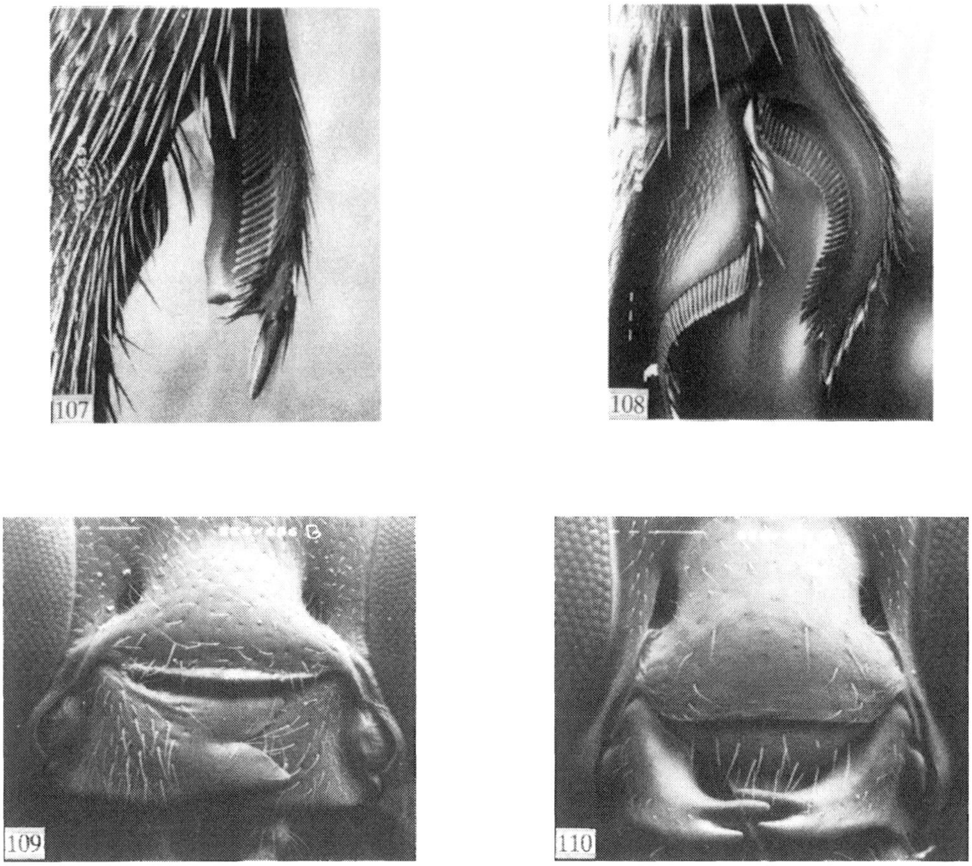

Figs 12.107–12.110. Ophioninae, scanning electron photomicrographs. Figs 12.107–12.108. Fore tibial spur; 12.107, *Ophion* sp.; 12.108, *Enicospilus* sp. Figs 12.109–12.110. Mandibles. 12.109, *Ophiogastrella* sp.; 12.110, *Enicospilus* sp.

altitudes between 2000 and 2500 metres (Gauld, 1988*b*). Its host is unknown.

Simophion[EX]. A small genus widespread in the drier parts of the world, including the southwestern United States and northern Mexico. One species is known to occur in Panama (Gauld, 1988*b*) and may be present in Costa Rica. Its host is unknown.

Stauropoctonus. A small genus with isolated species in the Old World and tropical America. A single species, *S. bicarinatus*, has occasionally been collected in premontane and lower montane forests in Costa Rica.

Thyreodon. A moderately large American genus, that is represented in Costa Rica by eight or nine species. They occur from sea-level up to about 1000 metres though there are occasional records from higher altitudes. Many species inhabit rather open, dry country and seasonally dry forests whilst others are confined to humid forests. All are parasitoids of sphingids or of ceratocampine saturniids (Gauld, 1988*b*).

ORTHCENTRINAE (includes Helictinae *sensu* Gauld, 1991; = Microleptinae in part).

This is a moderately large, cosmopolitan subfamily that is most species-rich in fairly cool, humid

- Occipital carina absent
  - Hind trochantellus with acute tooth (Fig. 12.106); mandibles twisted 90° .... *Stauropoctonus*
  - Hind trochantellus simple; mandibles barely twisted
    - Fore wing with *Rs+2r* slender (cf Fig. 12.101); epicnemial carina absent laterally .... *Prethophion*
    - Fore wing with *Rs+2r* basally broadened (Fig. 12.103); epicnemial carina present laterally .... *Janzophion*
- Occipital carina ± complete
  - Hind wing with distal abscissa of *Cu*1 joining *cu-a* close to *M* (Fig. 12.99)
    - Discosubmarginal cell broadly glabrous anteriorly (Fig. 12.102) .... *Sicophion*
    - Discosubmarginal cell without a large glabrous area anteriorly
      - Propodeum enlarged; tergite II elongate (Fig. 12.104) .... *Thyreodon*
      - Propodeum not enlarged; tergite II quadrate (Fig. 12.105) .... *Rhynchophion*
  - Hind wing with distal abscissa of *Cu*1 joining *cu-a* ± midway between *M* and 1*A*, or closer to 1*A* (Fig. 12.98)
    - Fore tibial spur with membranous flange (Fig. 12.107) .... *Ophion*
    - Fore tibial spur without membranous flange (Fig. 12.108)
      - Fore wing with *Rs+2r* straight (Fig. 12.101) **and** mandibles not twisted (Fig. 12.109)
        - Epicnemial carina present laterally (Fig. 12.100) .... *Ophiogastrella*
        - Epicnemial carina absent laterally .... *Simophion*[EX]
      - Fore wing with *Rs+2r* sinuous (often with free sclerite in discosubmarginal cell) **and/or** the mandibles twisted (Fig. 12.110) .... *Enicospilus*

Table 12.6. Identification table for Costa Rican genera of Ophioninae.

shaded areas. In such habitats these small, and often rather delicate ichneumonids can be extremely numerous and comprise more than 30 percent of the ichneumonids in a Malaise trap sample. Very little is known about the biology of the group, but some species are koinobiont endoparasitoids of the larvae of Mycetophilidae (Diptera) (Wahl, 1986) and related families of Diptera. A species of *Chilocyrtus* was recently reared in Costa Rica from a sciarid larva feeding on fungi on frass in an abandoned stem mine (P. Hanson, pers. comm.) and a species of *Megastylus* known to occur in Costa Rica has been reared in Colombia (in NHM collections) from the larva of the mycetophilid *Neoditomyia colombiana*.

Species of the Orthocentrinae may be recognized by the possession of a fine fringe of close hairs on the apex of the hind tibia, by their slender mandibles and generally by the possession of an impressed subocular sulcus. Many are delicate small insects, and often have a rather humped appearance to the mesosoma, and a 'pinched' appearance to the face. Others have a convex face, rather like that of some metopiines, but they may be distinguished from metopiines by their long cylindrical antennal scapes.

The Orthocentrinae is one of the least studied subfamilies of Ichneumonidae and only the European and North American species have received close taxonomic attention. The fauna of the tropics is virtually unworked and despite the fact that orthocentrines are common in montane forests, only three species have been described from the entire Neotropical region (Townes & Townes, 1966). The Orthocentrinae is represented in Costa Rica by about 75 species, many of which occur in the wet forests above 1200 metres. I have seen species belonging to *Chilocyrtus*, *Eusterinx*, *Laepserus*, *Megastylus*, *Orthocentrus*, *Plectiscidea*, *Proclitus* and *Helictes*, as well as some that cannot be placed to genus. Dasch (1992) has revised the nearctic species of the entire subfamily.

OXYTORINAE (= Microleptinae in part)

The Oxytorinae is a small subfamily comprising a single genus, *Oxytorus* (Wahl, 1990).

Oxytorus. A small and primarily holarctic genus that is represented in Costa Rica by two or three species that occur between sea-level and 1600 metres. *Oxytorus* is one of the most difficult of ichneumonid taxa to recognize, but females may be discerned by the simple "leaf-shaped" ovipositor sheath that conceals a very short ovipositor (Fig. 12.68). Nothing is known about the biology of any member of the group.

PHRUDINAE

The Phrudinae is a small cosmopolitan subfamily that is represented in Costa Rica by a species of *Erythrodolius*. Virtually nothing is known about the biology of phrudines, except that one European species has been reared as a koinobiont endoparasitoid of a species of Derodontidae (Coleoptera) (Franz, 1958). As a group phrudines are rather difficult to characterize, although species of most genera have the genal carina

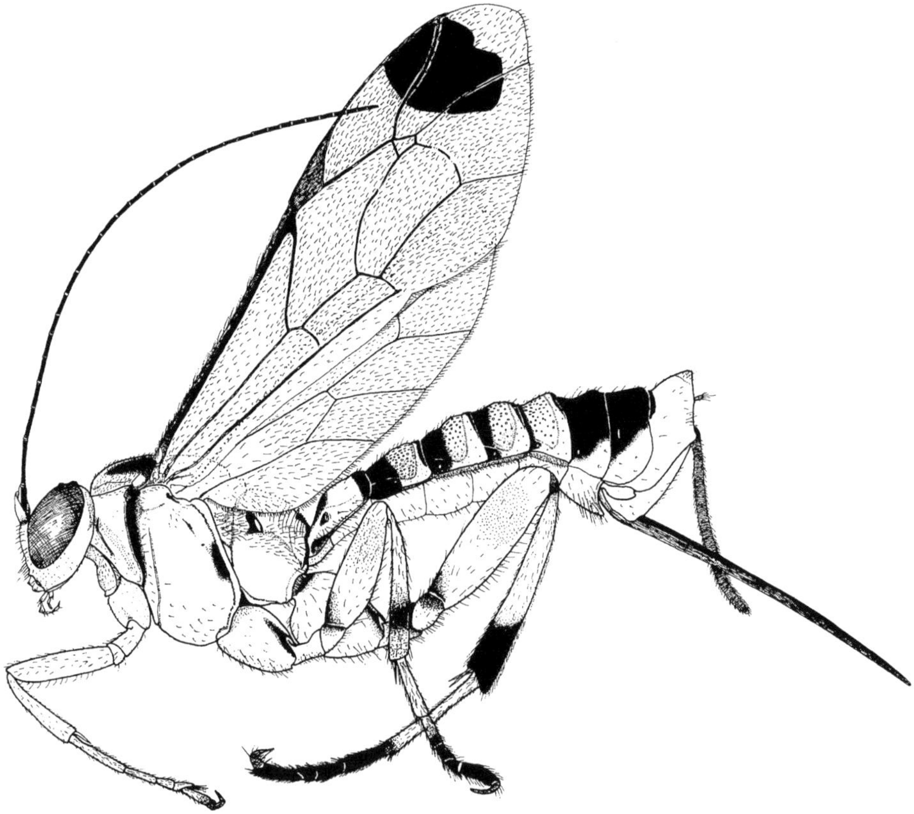

Fig. 12.111. *Pimpla sumichrasti* (Pimplinae).

joining the hypostomal carina away from and internal to the base of the mandible; in most other ichneumonids in which the occipital carina joins the hypostomal carina distant from the mandible this junction is directly above, not internal to the mandibular articulation.

***Erythrodolius*[CR]**. A small genus previously only known to occur in the Afrotropical region. A single undescribed species is widespread in mid-altitude forest throughout Costa Rica.

PIMPLINAE (= Ephialtinae in part) (Whole insect, Fig. 12.111).

The Pimplinae is a moderately large, cosmopolitan subfamily that is divided into three tribes—Ephialtini, Pimplini and Polysphinctini. The group exhibits a wider range of biologies and host associations than any other subfamily (Fitton *et al.*, 1988). Primitively pimplines are ectoparasitic idiobionts that attack hosts in concealment and many species show marked specializations for gaining access to such hosts. From this basic strategy a number of evolutionary lineages have arisen which specialize in different ways. *Tromatobia* and *Zaglyptus* in the Ephialtini, have larvae that devour spider eggs, whilst the Polysphinctini develop as koinobiont ectoparasitoids of spiders; the Pimplini are idiobiont endoparasitoids of the prepupae and pupae of a wide variety of Lepidoptera. Facultative hyperparasitism is a common feature of many members of the group, and some species of *Neotheronia* may be obligate hyperparasitoids. Some pimplines are common parasitoids of a variety of economically important insects, but their general lack of host-specificity means that few have been introduced in biological control programmes.

Morphologically pimplines are an extraordinarily diverse group. Most have large claws, an irregular rhombic areolet, two bullae in *2m-cu*, a dorsoventrally depressed metasoma, and no notch on the ovipositor. Tergite I is generally stout with the spiracles at or in front of the centre, and tergites II - IV are generally strongly sclerotized and frequently biconvex, or with impressions, and usually with at least some coarse punctures.

Pimplines occur in a wide variety of habitats in Costa Rica, from the coastal lowlands to the tops of the highest mountains, and the group includes some of the commonest ichneumonids encountered in any locality (e.g. species of *Neotheronia* and *Pimpla*). Species richness is greatest in premontane/lower montane forests where, in some sites, more than 40 percent of the entire Costa Rican fauna has been collected, but endemicity seems to be highest in montane situations. All three tribes of Pimplinae occur in Costa Rica where the group is represented by 27 genera and about 150 species. Genera of Ephialtini present are: *Anastelgis*, *Calliephialtes*, *Dolichomitus*, *Clydonium*, *Clistopyga*, *Iseropus*, *Leptopimpla*, *Odontopimpla*, *Scambus*, *Tromatobia*, *Umanella*, *Zaglyptus* and *Zonopimpla*; genera of Pimplini present are: *Apechthis*, *Itoplectis*, *Neotheronia*, *Nomosphecia*, *Pimpla* and *Xanthopimpla*. Genera of Polysphinctini present are: *Acrotaphus*, *Dreisbachia*, *Eruga*, *Flacopimpla*, *Hymenoepimecis*, *Polysphincta*, *Ticapimpla* and *Zatypota*. The Costa Rican fauna has been monographed by Gauld (1991) and a supplement is being prepared.

POEMENIINAE

This is a small, widespread subfamily that is represented in all regions except Africa and Australia. Previously poemeniine were included within the Pimplinae, but this classification was based on shared plesiomorphic features (Eggleton, 1989). Poemeniines are idiobiont ectoparasitoids of hosts concealed in wood; hosts include wood-boring beetles of the families Cerambycidae, Curculionidae and Melandryidae, as well as Sphecidae nesting in holes in timber (Townes & Townes, 1960; Fitton *et al.*, 1988). Species of one north temperate genus, *Pseudorhyssa*, are obligate multiparasitoids of siricids, and oviposit via the oviposition hole made by a rhyssine (Spradbery, 1969).

Poemeniines are elongate insects with long ovipositors. The subfamily is represented in the Neotropical region by two genera, *Ganodes* and *Rodrigama*. Each genus includes a single Costa Rican species (Gauld, 1991) and they have only been collected in lower montane forests.

RHYSSINAE

The Rhyssinae is a moderately small cosmopolitan subfamily that is most species-rich in the lowland forests of south-east Asia (Kamath & Gupta, 1972). Previously this subfamily was included as a tribe within the Pimplinae, but such a classification is based on shared plesiomorphic features (Eggleton, 1989). Rhyssines are idiobiont ectoparasitoids of concealed holometabolous insects. The adult females are highly specialized for oviposition into hosts tunnelling in wood and are capable of penetrating several centimetres of sound timber with their ovipositors. The north temperate species parasitize the larvae of Siricoidea (Chrystal & Skinner, 1932; Hanson, 1939), but wood-boring beetles probably serve as hosts for the tropical species (Porter, 1978).

In Costa Rica rhyssines may easily be recognized by the possession of file-like rugae on the mesoscutum (Fig. 12.49) and the absence of an areolet in the fore wing. A single genus occurs in Costa Rica.

Epirhyssa. A large genus occurring throughout tropical America and south-east Asia. In Costa Rica *Epirhyssa* is represented by ten described species in the lowland and premontane forests between sea-level and about 1300 metres. An undescribed species has recently been collected between 1600 and 2000 metres, around the Central Valley. Gauld (1991) provides keys to species.

TERSILOCHINAE

The Tersilochinae is a medium-sized cosmopolitan subfamily that comprises mostly small-sized ichneumonids with a fore wing length of 5 mm or less. Tersilochines are koinobiont endoparasitoids that oviposit into the host larva and frequently kill the host in its pupation chamber. The majority of species parasitize the larvae of phytophagous beetles, especially the families Curculionidae, Nitidulidae and Chrysomelidae (Cushman, 1916; Parker *et al.*, 1950), but species of one northern temperate genus attack the larvae of xyelid sawflies (Gauld & Bolton, 1988). Some species of one South American genus, *Stethantyx*, are common parasitoids of vegetable weevils (*Listroderes* spp.) and several have been introduced into Australia in an attempt to control these pests (Wilson & Wearne, 1962; Gauld, 1984).

Tersilochines are small to medium sized insects that resemble cremastines in form. They are easily recognized by the broad clypeus that has a fringe of stout bristles (Fig. 12.32). Many have a rather characteristic venation consisting of no areolet, a deep marginal cell, a large pterostigma and often with the veins around 2*rs-m* rather swollen.

The Tersilochinae is represented in Costa Rica by about 35 species. In the lowlands and lower montane regions many species of *Stethantyx* are common, and in mid-altitude humid forests *Meggoleus* species have been taken. A considerable diversity of genera and species occur at higher elevations. Keys to world genera are given by Townes (1971), but the work of Horstmann (1971) is also useful for generic recognition. There are no keys to the majority of neotropical species, and some are difficult to assign to any described genus.

TRYPHONINAE

The Tryphoninae is a large, cosmopolitan subfamily, but the greatest generic diversity of the group occurs in the North Temperate regions. The subfamily is subdivided into eight tribes. The Ankylophonini is endemic to Australia, whilst the Eclytini, Exenterini, Idiogrammatini, Sphinctini and Tryphonini are predominantly north temperate insects. The remaining two tribes, Oedemopsini and Phytodietini, are cosmopolitan.

Tryphonines are koinobiont ectoparasitoids of the larvae of Symphyta and Lepidoptera. They are unique amongst ichneumonids in having a highly modified egg that is large and bears a stalk and anchor. This anchor passes down the lumen of the ovipositor while the actual egg travels externally (Kasparayan, 1981). Using the ovipositor, the female embeds the anchor into the host's cuticle, thus attaching the egg externally to the host larva (Vance, 1927; Simmonds, 1947). Development is usually completed after the host larva has constructed a pupation retreat.

The Tryphoninae is almost impossible to characterize morphologically except by the fact that females often have an egg hanging beneath their ovipositor. All tryphonines have no subapical notch on the ovipositor, many have tergite I with the spiracles at or in front of the centre, they often possess deep glymmae and many have pectinate claws. In some the ovipositor sheath is slightly broadened centrally, and species of one very common and species-rich genus, *Netelia*, are ophionoid in appearance though they generally have an areolet in the fore wing.

The primarily north temperate tribes Sphinctini, Exenterini and Tryphonini are represented in Costa Rica by isolated, and apparently uncommon, species in the genera *Boethus*, *Chiloplatys*, *Eridolius*, *Polyblastus* and *Sphinctus*. The Oedemopsini is represented, mostly in humid montane forests, by *Atopotrophos*, *Cladeutes*, *Neliopisthus*, *Oedemopsis*,

Thymaris and *Zagryphus*. The Phytodietini comprises a few species of *Phytodietus* and more than 50 species of *Netelia*, a nocturnally active genus that is encountered in virtually all habitats. A key to tryphonine genera is given by Townes (1969). A revision of the Costa Rican species of all genera, except *Netelia*, is currently nearing completion (Gauld, in prep.).

XORIDINAE

The Xoridinae is a small, widely distributed subfamily that is most species-rich in the central European and eastern North American deciduous forests. Xoridines are idiobiont ectoparasitoids of the larvae, pupae and even cocooned adults of various families of wood-boring beetles, and possibly also of other wood-boring Holometabola such as siricoids (Chrystal & Skinner, 1931; Townes & Townes, 1960). Both solitary and gregarious species are known (Gauld & Bolton, 1988). A single genus is present in Costa Rica.

Xorides. A very large cosmopolitan genus that is apparently very species-rich throughout the tropics, although species are generally rarely collected. Approximately eight species occur in Costa Rica, mostly in the lower montane forests. *Xorides* is a very distinctive group with a characteristic geniculation in the antenna of the female (Fig. 12.70), and a uniquely modified mesoscutum (Fig. 12.55). A revision of the Costa Rican species is currently nearing completion (Gauld, in prep.).

12.2 BRACONIDAE

Scott Richard Shaw

Diagnosis. Body 1.0 to 14.0 mm in length (excluding the antennae and ovipositor); almost always fully-winged, with the fore wing length 1.0 to 15.0 mm, rarely brachypterous or apterous; sexual dimorphism negligible to moderate. *Antenna ordinarily with 14 or more flagellar segments* (up to 98), very rarely with fewer (eight to 13); flagellum usually filiform, rarely geniculate, serrate, or clavate. Mandible typically bidentate, sometimes exodont with three to seven teeth, or rarely unidentate. Upper hind corner of pronotum nearly reaching to tegula. Fore wing venation quite variable throughout the family but mostly stable within genera, sometimes even within subfamilies, often reduced apically; first discal and submarginal cells sometimes separated by a distinct vein (*M* or *Rs+M*), sometimes confluent; *vein 2m-cu absent*; pterostigma present and usually well-differentiated, sometimes elongate or linear. Hind wing with cross vein *rs-m* joining *Sc+R+Rs* before the separation of *Rs*. *Metasoma with tergites II and III usually rigidly fused*, occasionally flexibly fused; junction of fused tergites II and III often indicated by a sculptured or smooth suture, but if obliterated then apparent second 'segment' bears two sets of spiracles; metasoma sometimes with additional fused tergites.

Classification and distribution. The Braconidae is cosmopolitan and the second largest family of Hymenoptera (Achterberg, 1988*c*), surpassed only by the Ichneumonidae in species-richness. More than 10,000 braconid species have already been named, and estimates of the total number of species range from 40,000 (Marsh & Carlson *in* Krombein *et al.*, 1979) to 50,000 (Mason *in* Danks, 1979). There may be around 2,000 species present in Costa Rica.

Currently about 40 subfamilies of Braconidae are recognized (Quicke & Achterberg, 1990), but the subfamily classification has been in a state of flux for many years and the number of subfamilies has varied depending upon the interpretation of individual authors. Most of the disagreements concern the ranking or alignment of relatively minor lineages; in general the limits of the speciose subfamilies are stable. The system adopted here is a consensus classification developed from Achterberg (1984*a*), Gauld and Bolton (1988), Quicke and Achterberg (1990) and Sharkey (*in* Goulet & Huber, 1993). The most notable changes are the restriction of the definition of the Rogadinae and the consequent recognition of subfamilies Rhyssalinae, Pambolinae and Hormiinae (see Whitfield, 1989, 1992; Quicke, Ficken & Fitton, 1992; Quicke, Fitton & Ingram, 1992; Wharton, 1993*b*), and the acceptance of the Meteorinae as a subfamily distinct from the Euphorinae (S.R. Shaw, 1985, 1988*a*; Maetô, 1990).

Twenty-eight braconid subfamilies are known to occur in Costa Rica, and the Ypsistocerinae and Adeliinae are likely to occur in the country. Members of certain other small subfamilies occurring in North America (e.g. Meteorideinae and Neoneurinae) may also eventually be found in Costa Rica.

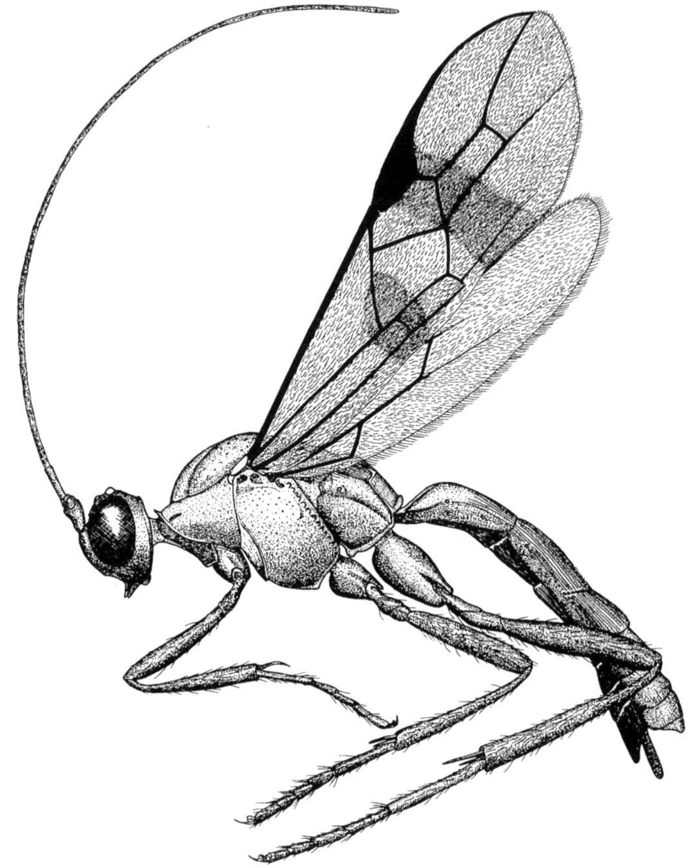

Fig. 12.112. *Rogas* sp. (Rogadinae).

Biology. The vast majority of species of Braconidae are primary parasitoids of other insects, although phytophagy is now known to occur rarely in the group. A Brazilian and a Costa Rican species of *Allorhogas* develop in seeds (Macêdo & Monteiro, 1989; Hanson, pers. comm.), and other possible instances of phytophagy occur in *Monitoriella* and *Psenobolus* (see below). The hosts of most entomophagous braconids belong to the orders Coleoptera, Lepidoptera, Diptera and, to a much lesser extent, the Hemiptera, Homoptera and rarely Hymenoptera (Matthews, 1974; Gauld & Bolton, 1988; M.R. Shaw & Huddleston, 1991; Wharton, 1993*a*). It is common for whole braconid subfamilies to utilize only a single host order (Matthews, 1974), although the Euphorinae attack hosts from at least six different orders (S.R. Shaw, 1985, 1988*a*).

Most idiobiont braconids (e.g. Doryctinae, Braconinae) are ectoparasitoids of concealed subcorticolous or endophytic hosts, but the Rogadinae includes ectoparasitic idiobionts, and both ecto- and endoparasitic koinobionts (M.R. Shaw, 1983). Most other major groups of Braconidae are koinobiont endoparasitoids, but a final obligatory ectoparasitic stage is known to occur in several groups. Although many braconids are solitary parasitoids, gregariousness has evolved independently in many different lineages, including ectoparasitoid idiobiont genera (e.g. *Bracon*, *Hormius*, *Oncophanes*, *Spathius*) and several koinobiont groups (e.g. some Rogadinae,

Macrocentrinae, Euphorinae and Microgastrinae). In general brood size corresponds to the number of eggs deposited, but polyembryony is known to occur in the Macrocentrinae (Parker, 1931*a*; Daniel, 1932). Gregariousness is more common in the Braconidae than it is in the Ichneumonidae.

Braconids are primarily larval parasitoids, ovipositing in and emerging from the host larva. Species of Alysiinae, Opiinae and Meteorideinae are larval-pupal parasitoids, and the Cheloninae, some Ichneutinae (Nageli, 1936), some Alysiinae (Wharton, 1984), and a few species of Helconinae (Alauzet, 1987) are egg-larval parasitoids. There are no true egg parasitoids in the Braconidae. Although most braconids parasitize the larval stage of their host, euphorines commonly utilize adult insects (S.R. Shaw, 1985, 1988*a*) as do members of the Aphidiinae, the *Sericobracon* lineage of Doryctinae and possibly the Ypsistocerinae. Although the Neoneurinae have not been definitively reared, observations of ovipositional strikes at worker ants, and their short sickle-like ovipositor, strongly suggest that they are koinobiont endoparasitoids of adult worker ants (Huddleston, 1980; S.R. Shaw, 1992, 1993). Some euphorines that parasitize adult beetles will also oviposit into pupae when these occur in the same microhabitat (Obrycki *et al.*, 1985) but no other braconid (except possibly some aspidobraconine braconines) is known to oviposit directly into a host pupa.

Hyperparasitism is rare in the Braconidae and, unlike the Ichneumonidae, none are obligate. A European *Syntretus* species develops as a pseudohyperparasitoid of adult ichneumonids (Cole, 1959*b*), a holarctic *Aphaereta* species is a hyperparasitoid of a sarcophagid (Kulman, 1965) and Wharton (1984) cites several other alysiines that are hyperparasitoids of tachinids or phorids.

Host searching and oviposition. Many studies indicate that braconids are attracted to a particular habitat, regardless of the presence of hosts (Nishida, 1956) and in such cases attractants have been found to include plant volatiles (Read *et al.*, 1970; Achterberg, 1983*a*; Ayal, 1987) or odours emanating from decaying substrata, such as carrion, dung, fungi, and rotting fruit (Wharton, 1984). Once the host habitat has been located, actual host location may be accomplished by a combination of tactile, chemical, and visual cues. Tactile cues, such as vibrations from host feeding, may aid in locating hosts concealed in rotting wood (DeLeon, 1935; Ryan & Rudinsky, 1962), in leaf mines (Sugimoto *et al.*, 1988*a* & *b*), in fruit (Lawrence, 1981), or in carrion (Evans, 1933). Many koinobiont braconids that attack more exposed hosts, locate them by specific semiochemical cues (Eller *et al.*, 1988), such as kairomones released from host feeding, host mandibular secretions (Vinson, 1968), host cuticular proteins, lepidopteran wing scales, silk (Weseloh, 1976), frass (Hendry *et al.*, 1973; Fukushima *et al.*, 1989) or honeydew (Ayal, 1987). At close range, some leaf-miner parasitoids may visually recognize the characteristic shape of the host leaf-mine (Sugimoto *et al.*, 1988*a*). Visual cues are particularly important to euphorines, which are stimulated by host motion (Bryden & Bishop, 1945; O.J. Smith, 1952; M.F. Walker, 1961) and may be induced to oviposit on artificial host models (Richerson & DeLoach, 1972).

Host acceptance by braconids commonly involves four behavioural steps: antennation of the host, ovipositor probing, insertion of the ovipositor (drilling), and egg deposition (Arthur, 1981). In *Alysia manducator*, a host may be rejected at any of these stages (Chernoguz & Reznik, 1987). Idiobiont species, which stop host development at oviposition, are most likely to be affected by the host biomass at time of oviposition, and in gregarious species such as *Habrobracon hebetor*, the size of emerging adults is density-dependent and the carrying capacity of the host is directly related to host mass (Taylor, 1988*a*). Survival is lower on smaller hosts and competition more intense, and consequently ovipositing females may exhibit a preference for larger hosts (Taylor, 1988*b*). Although it may be more difficult for koinobiont braconids to evaluate host suitability at the time of oviposition, it is nevertheless common for koinobiont species to exhibit a decided preference for particular host stages. For example, *Cotesia glomerata* prefers first instar host larvae, although the first through third instars may be parasitized (Dreische, 1988) and *Dinocampus* prefers adult coccinellid hosts, although late instar larvae or pupae may sometimes be attacked (Obrycki *et al.*, 1985). In koinobionts host suitability may also be affected by the plant on which the host is feeding (Mueller, 1983; Vinson, 1984; Gauld & Gaston, 1994).

Two types of venom apparatus have been distinguished in Braconidae (Edson & Vinson, 1979) and the venoms injected into the host during oviposition serve a variety of functions (Beard, 1978). Among idiobiont ectoparasitoids, such as *Habrobracon* species, the venom typically causes a permanent paralysis (Beard, 1952; Laing & Caltagirone, 1969) whilst koinobionts either do not paralyse the host, or induce only a temporary paralysis (M.R. Shaw, 1983). However, the venom may induce a delayed inhibition of host development (M.R. Shaw, 1981). Many koinobionts also inject symbiotic viruses, which are maintained within the calyx lining at the base of the ovaries and coat the egg at the time of oviposition (Stoltz, 1986; Styer *et al.*, 1987; Tanaka, 1987*b*; Stoltz & Whitfield, 1992). These polydnaviruses, along with venoms and calyx fluids, play an important role in the suppression of the host immune response (Guillot & Vinson, 1972; Kitano, 1986; Tanaka, 1987*a*; Stoltz *et al.*, 1988; Wago & Tanaka, 1989), and may also induce elevated host haemolymph protein levels (Beckage *et al.*, 1989), delay or prolong host development, induce extra host instars and delay or inhibit host diapause (Moore, 1989). It has been suggested that these viruses are fundamental to the diversification of endoparasitoid braconids (Whitfield, 1990*b* & *c*).

Idiobiont ectoparasitoid species typically require a long period of time to oviposit because the host usually must be reached by drilling through a substrate, and must be paralysed. Field studies of *Ecphylus costaricensis*, a bark beetle parasitoid, indicated that drilling took 25 to 42 minutes and an additional 11 to 19 minutes were required for egg deposition and ovipositor withdrawal (Matthews, 1969). Koinobiont endoparasitoid species usually complete the oviposition sequence much more rapidly, especially if the host is more or less exposed. For example, the egg-larval parasitoid *Chelonus curvimaculatus* completes its entire oviposition sequence in 17 to 21 seconds (Leluk & Jones, 1989) and the microgastrine, *Glyptapanteles thompsoni*, deposits 20 to 25 eggs in a single thrust of the ovipositor lasting only a few seconds (Vance, 1931).

Larval development. In braconids there are probably ancestrally five larval instars, although in many groups that number is reduced, for example to four in some Opiinae (Willard, 1920), and to three in some Microgastrinae (Vance, 1931), Cheloninae (Vance, 1932*a*; Boyce, 1936) and Meteorinae (West & Miller, 1989; Fernández & Terán, 1990*a*). Although first instar larvae have been described for many species, O'Donnell's (1982) study of the Aphidiinae is the only comparative work on this stage. First instar larvae are usually hymenopteriform in idiobiont ectoparasitoid braconids and often have transverse bands of setae on each segment (Finlayson, 1964). In many koinobionts the first instar larvae have a disproportionately large head with long mandibles (e.g. Cheloninae—Vance, 1932*a*; Boyce, 1936), and/or are caudate, as in Aphidiinae (Couchman & King, 1977) and Euphorinae (Loan & Holdaway, 1961*a* & *b*). In the Cheloninae, Microgastrinae and some other subfamilies the larva is initially caudate but subsequently develops a bladder-like vesicle posteriorly (Cardona & Oatman, 1975). The Agathidinae have polypodeiform first instar larvae, which resemble caudate larvae, but have paired fleshy ventral protuberances on some segments (Parker, 1931*a*; Simmonds, 1947). Hypermetamorphosis occurs commonly in koinobiont endoparasitoid braconids (Finlayson & Hagen, 1977), with the later instars differing dramatically from the first instar larva (Clausen, 1940*b*).

The first instar larva of most koinobiont endoparasitoid braconids is enveloped within a trophamnion after eclosion (Jackson, 1928; Brindley, 1939; Polaszek, 1986). The trophamnion is composed of oversized cells, which ultimately separate and often survive independently within the host haemocoel where they enlarge rapidly, increasing their size ten to twenty times. These separate cells, called teratocytes, are sometimes consumed by the growing parasitoid larva (Hinton, 1954; Polaszek, 1986), although they may have other functions (see Chapter 2.5).

In many koinobionts intermediate larval instars have very reduced mandibles and principally feed on haemolymph and fat globules. Final instar braconid larvae are usually hymenopteriform, with few differentiating morphological structures other than the sclerotized parts of the head capsule and the usually rather well-developed mandibles. They have been extensively studied because the exuviae can be recovered from cocoons (Finlayson & Hagen, 1977). Short (1952) and Capek (1973) studied the morphology of the head of final instar braconid larvae and provided keys for their identification. Capek (1970) presented a revised classification of the Braconidae at subfamily and tribal level based upon an analysis of larval morphology and host relationships.

The mature larvae of most braconid species pupate within an ovoid, dense silken cocoon. Ectoparasitoid idiobionts usually spin flimsy cocoons in the concealed sites where they have fed. These are often pale brown and elongate oval in appearance. However, *Atanycolus* species that spin their cocoons under loose bark form a coffin-like structure that is flat on both top and bottom (Shenefelt, 1943). Among koinobionts, some do not kill the host until the latter has concealed itself within its pupal retreat. Thus chelonines allow the host caterpillar to first form a cocoon, before killing the host and spinning its own cocoon within (Harbo & Kraft, 1969). Species of Opiinae and Alysiinae may spin thin cocoons, or they may pupate naked, inside the puparia of their dipterous hosts (Wharton, 1984). On the other hand, many more braconids than ichneumonids kill the host on vegetation rather than within its pupal retreat, and this is reflected in a wider range of cocoon-forming habits in the family (Gauld & Bolton, 1988). The larva generally emerges from the body of the host before producing its cocoon, but most aphidiine genera pupate within the mummified body of the host aphid, and the Rogadini pupate within the mummified remains of caterpillars. In such cases, the parasitoid larva often glues the host mummy to the substrate through a ventral opening, and the actual silken cocoon within is reduced. The cocoon of *Dinocampus coccinellae* is spun beneath its coccinellid host, with the dead or dying beetle forming a protective dome over the parasitoid pupa (Balduf, 1926). Similarly, cocoons of several solitary species of *Microgaster* (= *Microplitis sensu* Nixon, 1970) are formed between the prolegs of aposematic or cryptic caterpillars, and adults often emerge before the host dies. In this way the parasitoid gains continued protection from the host's appearance or behaviour throughout its vulnerable cocoon stage (Gauld & Bolton, 1988). The cocoons of most solitary *Meteorus* species are spun suspended from long threads (Huddleston, 1980) whilst those of gregarious meteorines may be formed in loose heaps or organized into spherical masses (Huddleston, 1983; S.R. Shaw, 1985). In most of these braconids that pupate on vegetation the time spent in the pupal stage is shortened, with emergence generally occurring within a week or two.

Adult biology. As with the ichneumonids, sexual dimorphism is most apparent in the idiobionts, especially in the Doryctinae, Braconinae and Ypsistocerinae. Although rather unusual in the Hymenoptera, male brachyptery (flightlessness) occurs commonly in several doryctine genera (Marsh, 1965). In Costa Rica apterous males are found in the fig-inhabiting *Psenobolus* (W. Ramirez, P. Hanson, pers. comm.). W.D. Hamilton (1978) has hypothesized that the flightless-male condition is a trait associated with insects occupying ephemeral niches, such as under bark or in rotting wood, that inbreed and multiply rapidly, then disperse after several generations when the quality of the microhabitat deteriorates.

In many braconid species the males often emerge before the females. Some may form aggregations at female emergence sites and the males of *Coeloides dendroctoni* are able to locate pre-emergent females still in their exit tunnels under bark (DeLeon, 1935). Among koinobionts the emerging females may be more dispersed because the host may move some distance from the feeding site to a cryptic pupation retreat prior to being killed, or the parasitoid may exit the host and pupate some distance from the host remains. Consequently, many koinobiont male braconids locate females using species-specific sex pheromones (e.g. Pemberton & Willard, 1918*a*; Fink, 1926; Parker, 1931*b*; Hagen, 1953; Bousch & Baerwald, 1967; Cole, 1970; Read *et al*., 1970; Vinson, 1972; Obara & Kitano, 1974; Kamano *et al*., 1989). In a few koinobiont groups, such as the Blacini (Konig, 1967; Achterberg, 1975, 1977, 1988*c*; Syrjamaki, 1976) and to some extent the Microgastrinae (Whitfield, 1987) and Cheloninae (S.R. Shaw, 1991), male-dominated mating aggregations are known to occur.

A short premating period may be required in some braconids (Hagen, 1953), but the majority of species are ready to mate upon emergence from the cocoon (Matthews, 1974). Courtship and mating of braconids typically involves the following behavioural steps: attraction, antennal searching and recognition, orientation and pursuit, wing fanning and vibration, approach and mounting, antennation, copulation, and post-copulatory grooming (Matthews, 1974; Askari & Coppel, 1978). Rapid wing vibration by courting males is an almost universal component of braconid mating behaviour (e.g. Kitano, 1975), and may serve to orient the male to the female's odour (Vinson, 1972) or vice versa, or to produce acoustic courtship signals (Sivinski & Webb, 1989). Actual copulation is brief, normally one minute or less (Oatman *et al*., 1969; Lewis, 1970; Cardona & Oatman, 1971; Kainoh, 1986). Generally the male mounts the female dorsally, but some chelonines assume an end-to-end position (Harbo & Kraft, 1969), and some euphorines assume an end-to-end position with the male upside down (Seitner & Notzl, 1925; Jackson, 1928). Multiple mating is common for males, but females of many species mate only once (Matthews, 1974), although some, such as *Cotesia flavipes*, may mate up to eight times (Gifford & Mann, 1967).

Relatively little is known about the feeding habits of adult braconids in Costa Rica although many species have been observed apparently feeding from flowers, and many species (such as agathidines) are attracted to and consume baits containing hydrolysed tryptophan and sugar (Hanson & Gauld, pers. comm.). Numerous species of Agathidinae (Sharkey, 1985; Sharkey & Mason, 1986), Cardiochilinae (Mao, 1949; Sharkey & Mason, 1986; Huddleston & Walker, 1988), and Cheloninae have the maxillo-labial complex modified into a proboscis for flower feeding. *Pholetesor ornigis* may acquire nutrients from flower nectar, oozing tree sap, ripe fruit, or honeydew from homopterous insects (Ridgway & Mahr, 1990). Host-feeding (see Chapter 2.5) is not widespread in the Braconidae, but has been observed in several Braconinae (Genieys, 1925; Doner, 1934; Beard, 1952; Clark, 1963; Angalet, 1964) and some Rogadinae (Ahmad, 1943; Chatterjee, 1943; M.R. Shaw, 1983). Some aphidiines that parasitize ant-tended root aphids have been observed to solicit regurgitated food from tending ants (Hincks, 1958; Takada & Hashimoto, 1985; Gross, 1993).

A preoviposition period may be required for egg maturation in many braconids, especially the synovigenic idiobionts. For example, *Spathius vulnificus* undergoes a preoviposition period averaging 13.5 days (Matthews, 1974). Potential egg production varies enormously in braconids, both between species and within species. In *Habrobracon* species the number of eggs per female can vary as much as from 32 to 954 (Laing & Caltagirone, 1969). If hosts are not available at the time of egg maturation, oocytes may be reabsorbed—a strategy that may enable some braconids to achieve increased searching ability in times when hosts are scarce (Ramadan *et al.*, 1989).

Like most other Hymenoptera the majority of braconids are arrhenotokous, although some species of Euphorinae, Microgastrinae, and Agathidinae are thelytokous (Balduf, 1926; Vance, 1931; Clausen, 1940*b*; Hummelen, 1974; S.R. Shaw, 1988*b*).

Little is known about the phenology of braconids in Costa Rica or in most other tropical areas, but the studies of Beeson and Chatterjee (1935) in India indicate that multivoltinism is common among tropical braconids, often with two to four generations occurring per year. Malaise trap samples suggest that similar multivoltinism may occur in humid Costa Rican forests. Virtually no information is available on how tropical braconids may survive the pronounced dry season in northwestern Costa Rica. However, it seems likely that many diapause, probably as prepupae or pupae within their cocoons, or as endoparasitoid larvae within the parasitized hosts. Among koinobiont braconids in the Temperate region it is common for the first instar larva to diapause within the parasitized host until the host larva is fully fed and is ready to pupate (Balduf, 1963).

Defence against predators. Tropical braconids exhibit a wide range of defensive strategies. Some (e.g. *Gronaulax*) are known to feign death (thanatosis) for several minutes when disturbed (Achterberg, 1986). Many of the larger, diurnally active species are aposematically coloured and may be Batesian or Müllerian mimics of aggressive aculeates or distasteful Lepidoptera, Coleoptera or Diptera (Quicke, 1984, 1986*a* & *b*). Other braconines, such as *Digonogastra* and *Ipobracon*, are orange with patterned wings and are themselves probably distasteful models for mimicry complexes involving other ichneumonoids (Gauld, 1991). In Costa Rica aposematic coloration occurs most frequently in the larger species of the Doryctinae, Braconinae, Cenocoelini and Agathidinae, and to a lesser extent in the Rogadinae, Alysiinae, Orgilinae, Cardiochilinae, Microgastrinae, and Cheloninae.

Ant mimicry probably also occurs commonly in tropical braconids, although little information is available on the subject. Many small braconids are superficially ant-like in appearance, and this resemblance is enhanced by the host searching behaviour of most braconids, which typically spend a considerable amount of time walking rapidly about on foliage or fallen trees, flying mostly when disturbed or abandoning a site. Field studies of *Ecphylus costaricensis* by Matthews (1969) found that ants were also common on the same fallen tree, and encounters with ants were surprisingly frequent. *Paradelius rubra*, from California, run about rapidly on foliage, resembling ants while in motion (Whitfield, 1988*c*). When disturbed these wasps emit 'a choking, formic acid-like chemical into the air'.

Many of the most common and abundant Costa Rican braconid species (Microgastrinae, Alysiinae,

Opiinae) are only a few millimetres or less in length and are inconspicuously coloured. Small body size is a convergent trend that has evolved many times within the Braconidae (Tobias, 1975), possibly as a result of selection pressure from vertebrate predators. The adaptive value of small body size may have been a factor in the diversification of leaf-miner parasitoids in the tropics, as groups such as Hormiinae, Gnamptodontinae, Opiinae and Miracinae are more species-rich and abundant in Costa Rica than in temperate regions. While small braconids may escape the attention of vertebrate predators, they are still vulnerable to invertebrate predators such as ants and spiders. Like the large, aposematically coloured species, these small, inconspicuous braconids are endowed with abdominal exocrine glands that may function in defence (Buckingham & Sharkey, 1988).

A large proportion of the lowland tropical koinobiont ichneumonids are nocturnally active (and attracted to lights), which may be an adaptation to escape from diurnally active, visually oriented predators (Gauld, 1987). Gauld hypothesized that nocturnal activity should be especially favourable in the tropics, especially among koinobionts that parasitize nocturnal, leaf-feeding lepidopterous larvae. In the Costa Rican fauna, the most abundant and diverse groups of nocturnally active Braconidae are the Rogadinae (*Rogas, Aleiodes, Cystomastax*), Meteorinae (*Meteorus*), Macrocentrinae (*Macrocentrus*), and Homolobinae (*Homolobus, Exasticolus*). These taxa tend to be pale yellowish brown in colour (resembling ophionine ichneumonids) and they are all koinobiont endoparasitoids of lepidopterous larvae, corroborating Gauld's hypothesis. Although they do not show the typical pale coloration of nocturnal ichneumonoids, many Blacinae (*Blacus*) are also nocturnally active (Huddleston & Gauld, 1988), and these are also well diversified in the Costa Rican fauna.

Economic importance. Because many braconids favour warm arid environments (Juillet, 1964) and are common in agroecosystems, the group has been used extensively in biological control programmes. They are amongst the most frequently established hymenopterans employed in classical biological control (Greathead, 1986). Although many subfamilies have been investigated for bio-control applications, the most frequently employed species belong to the Microgastrinae, Braconinae and Opiinae. Several braconids occurring in Central America (Table 12.7) are important in the natural suppression of insect pests, especially pyralid stem-borers of maize and sugar-cane (Hummelen, 1974; Austin & Dangerfield, 1989; Quicke & Wharton, 1989; Rodriguez-del-Bosque & Smith, 1989; Wharton *et al.*, 1989), pyralid pests of Cucurbitaceae (Marsh, 1986) and pigeon pea (Bennett, 1950), noctuids (Marsh, 1978), the potato tuberworm (Marsh, 1979), and tephritids (Wharton & Marsh, 1978) (see Table 12.7).

In 1955 the Mediterranean fruit fly (*Ceratitis capitata*) appeared in Costa Rica and shortly thereafter several parasitoids were introduced from Hawaii, most of which were opiine braconids (Wharton *et al.*, 1981; Gilstrap & Hart, 1987; Jiron & Mexon, 1989). Of the five opiines introduced only *Diachasmimorpha longicaudata* (= *Biosteres*; see Wharton, 1987) and *Biosteres arisanus* (= *B. oophilus*) became established. The results of later introductions from Africa (Steck *et al.*, 1986) are unknown. Because the two established species do not appear to be very effective further biological control programmes are probably necessary.

The most serious and widespread sugar-cane pests in the Neotropics are the pyralid stem-borers belonging to the genus *Diatraea*. Because insecticides are neither very effective nor economically feasible against these pests considerable work has been done on biological control (e.g. Bennett, 1971; Alam, 1980). In Costa Rica the most important braconid parasitoids are two native species (see Table 12.7), and the introduced species *Cotesia flavipes*, which is being mass reared in the laboratory for periodic, inundative releases in the field.

Identification. Shenefelt (1969, 1970*a* & *b*, 1972, 1973*a* & *b*, 1974, 1975, 1978, 1980), Shenefelt and Marsh (1976), Mackauer and Stary (1967) and Mackauer (1968) have produced catalogues of the world species of Braconidae, and Marsh (*in* Krombein *et al.*, 1979) catalogued the North American species. The majority of Costa Rican braconid species are apparently undescribed. In all cases, except for the Cheloninae, no generic identification aids are given below as the Costa Rican Braconidae fauna has been less extensively studied than the ichneumonid fauna, and such simplified diagnostic aids could result in taxa

Pest species	Common name	Braconid parasitoid
Acanthoscelides obtectus (Bruchidae)	Bean weevil	*Stenocorse bruchivorus*
Anastrepha obliqua (Tephritidae)	Mango fruit fly	*Doryctobracon areolatus* *Opius anastrephae* *Opius bellus*
Anastrepha striata (Tephritidae)	Guava fruit fly	*Doryctobracon crawfordi* *Doryctobracon zeteki* *Opius vierecki*
Antichloris viridis (Arctiidae)	Banana moth	*Macrocentrus* sp. *Meteorus laphygmae*
Anthonomus grandis (Curculionidae)	Cotton boll weevil	*Urosigalphus schwarzi*
Apion godmani (Curculioninae)	Bean-pod weevil	*Triaspis azteca*
Diaphania spp. (Pyralidae)	Melonworm, pickleworm	*Cardiochiles diaphaniae* *Hypomicrogaster diaphaniae*
Diatraea saccharalis (Pyralidae)	Sugar-cane borer	*Alabagrus stigma* *Apanteles diatraeae* *Cotesia flavipes* *Digonogastra grenadensis*
Erinnyis ello (Sphingidae)	Cassava hornworm	*Cotesia* spp.
Heliothis and *Spodoptera* spp. (Noctuidae)	Corn earworm, armyworms	*Aleiodes laphygmae* *Aleiodes vaughani* *Cardiochiles nigriceps* *Chelonus insularis* *Cotesia marginiventris* *Homolobus truncator* *Meteorus arizonensis* *Meteorus laphygmae*
Leucoptera coffeella (Lyonetiidae)	Coffee leaf-miner	*Stiropius* sp.
Liriomyza huibroidensis (Agromyzidae)	Serpentine leaf-miner	*Oenonogaster* sp. *Opius dissitus*
Mocis latipes (Noctuidae)	Guinea grass moth	*Microplitis* sp. *Aleiodes nigristemmaticum*
Myzus persicae (Aphididae)	Peach-potato aphid	*Diaretiella* spp.
Oiketicus kirbyi (Psychidae)	Bagworm	*Digonogastra diversus*
Phthorimaea operculella (Gelechiidae)	Potato tuber moth	*Agathis* sp. *Chelonus kellieae* *Orgilus jennieae*
Polygrammodes elevata (Pyralidae)	Sweetpotato borer	*Apanteles thurberiae*
Toxotrypana curvicauda (Tephritidae)	Papaya fruit fly	*Doryctobracon toxotrypanae*
Trichoplusia ni (Noctuidae)	Cabbage looper moth	*Chelonus* sp. *Glyptapanteles caffreyi*

Table 12.7. Major insect pests in Central America that are partially suppressed by Braconidae, and the braconids responsible.

being wrongly determined. Although there are no keys for the neotropical fauna, North American ones (e.g. Marsh, 1963; Marsh *et al.*, 1987) are useful for generic identification. The key given below will facilitate the identification of subfamilies known or expected to occur in Costa Rica.

Key to the subfamilies of Braconidae known or likely to occur in Costa Rica

1 Mandibles exodont, teeth directed away from longitudinal axis of body and tips usually not

meeting when mandibles are closed (Fig. 12.147). .... **Alysiinae** (p. 448)

— Mandibles normal, teeth directed towards each other and tips meeting or overlapping when mandibles are closed (Figs 12.148–12.150). 2

2 Metasomal tergite II+III basally with a polished transverse or semi-circular area, posteriorly usually with a crenulate sulcus (Fig. 12.134); minute species less than 2.0 mm in body length. .... **Gnamptodontinae** (p. 455)

— Metasomal tergite II+III not so sculptured, without a polished transverse or semi-circular area basally, or polished area triangular and body size larger than 2.0 mm. 3

3 Face strongly protruding anteriorly; antennae arising from a distinct shelf (Fig. 12.128).
[Very rare.] .... **Betylobraconinae** (p. 449)

— Face not strongly protruding anteriorly; antennae not arising from a distinct shelf (Fig. 12.129). 4

4 Mandibles, when closed, close to lower margin of clypeus and concealing labrum, without a conspicuous oral cavity (Figs 12.149, 12.150). 5

— Mandibles, when closed, remote from lower margin of clypeus, thus a conspicuous oral cavity (Fig. 12.148) is present between the mandibles and clypeus; labrum normally visible and often concave.
[Some Opiinae are intermediate and can be keyed via either couplet.] 23

5 First metasomal spiracle situated within a membranous or weakly sclerotized area of the laterotergite (Figs 12.129, 12.135, 12.136), ***or*** spiracle not visible in air-dried specimens due to shrinking. 6

— First metasomal spiracle situated within a well sclerotized area of tergite I (Figs 12.130–12.133). 8

6 Antennal flagellum 18- to 49-segmented; fore wing vein *Rs* strongly arched towards anterior margin of wing (Fig. 12.115); maxillary palp 6-segmented. .... **Cardiochilinae** (p. 450)

— Antennal flagellum 12- to 16-segmented; fore wing vein *Rs* straight to slightly curved, usually reduced apically (Fig. 12.116); maxillary palp 5-segmented.
[Microgastrinae have placodes arranged evenly in two ranks per segment, thus the 16-segmented flagellum may appear to be 32-segmented.] 7

7 Antennal flagellum 16-segmented; first metasomal tergite of various shapes, but not extremely narrow and linear medially (Fig. 12.136). .... **Microgastrinae** (p. 459)

— Antennal flagellum 12-segmented; first metasomal tergite extremely narrow and linear medially (Fig. 12.135).
[Minute species.] .... **Miracinae** (p. 459)

8 Hind trochantellus laterally with a cluster of stout spines (Fig. 12.141); occipital carina absent.
[Yellowish brown species.] ....**Macrocentrinae** (p. 456)

— Hind trochantellus laterally without a cluster of stout spines; occipital carina usually present, at least laterally, but sometimes absent. 9

9 Fore wing with vein *M* separated from prestigma by a wing fold and usually an associated bulla; 2nd submarginal cell very small, marginal cell narrow and vein *Rs* tubular to wing margin (Fig. 12.117), rarely with this apical venation absent.
[Often brightly coloured with conspicuous colour patterning on wings.] .... **Agathidinae** (p. 447)

— Fore wing venation not as above: vein *M* joining prestigma; 2nd submarginal cell larger or absent; marginal cell of various shapes, but usually larger and not so narrow, more rarely open or absent (Figs 12.118–12.127). 10

10 First three metasomal tergites forming a rigid carapace, with at most one flexible joint; fore wing with vein 2*rs-m* present, thus a closed 2nd submarginal cell present (Fig. 12.118). ... .11

— Metasoma tergites not forming a carapace, ***or if*** carapace-like ***then*** metasoma with more than three visible tergites, ***or*** fore wing with vein *2rs-m* absent (Figs 12.119, 12.120). 12

11 Metasomal carapace without flexible joints, at most with two transverse grooves and often without any apparent segmentation (Fig. 12.151).
[Common and sometimes locally abundant species.] .... **Cheloninae** (p. 450)

— Metasomal carapace with a flexible joint separating tergite I from tergites II+III.
[Rarely collected species.] .... **Sigalphinae** (p. 462)

12 Fore wing with distal part of *Rs* arising from pterostigma, not reaching to wing margin (Fig. 12.119); metasomal tergites I to III fused into a single flat surface comprising more than 0.5 of dorsal surface of metasoma.
[Minute.] .... **Adeliinae^{EX}** (p. 447)

— Fore wing with distal part of *Rs* reaching wing margin (Figs 12.122, 12.123), ***or if*** reduced, ***then*** arising from *2r-rs*, not directly from pterostigma (Figs 12.120, 12.121), ***and*** metasomal tergites I to III not fused into a single flat surface comprising more than 0.5 of dorsal surface of metasoma. 13

13 Fore wing with *M* abruptly curved just before joining *Rs* (Fig. 12.120).
[Minute.] .... **Ichneutinae** (p. 456)

— Fore wing with *M* straight or gradually curved to join *Rs* (Figs 12.121–12.127). 14

14 Fore wing with vein *Rs* not extending to wing margin; 2nd submarginal cell large, enclosed and of characteristic shape (Fig. 12.121). .... **Mendesellinae** (p. 456)

— Fore wing with vein *Rs* usually extending to wing margin (Figs 12.122–12.127), ***or if*** reduced or absent ***then*** 2nd submarginal cell is also absent (Fig. 12.113); 2nd submarginal cell, when present, of various angular shapes. 15

15 Hind wing with vein *Cu&cu-a* absent (Fig. 12.113); metasoma with tergites weakly sclerotized, tergites II+III flexible and capable of bending at line of fusion; scutellar sulcus smooth.
[Fore wing venation generally reduced; parasitoids of aphids; most commonly collected above 1500 m.] .... **Aphidiinae** (p. 448)

— Hind wing with vein *Cu&cu-a* present (Fig. 12.114); metasoma with tergites well-sclerotized, tergite II+III solidly fused and not capable of bending at line of fusion; scutellar sulcus usually foveate, only rarely smooth. 16

16 Occipital carina entirely absent, dorsally effaced ***or if*** weakly present, ***then*** wing venation as in Fig. 12.114, with long 2nd submarginal cell. 17

— Occipital carina entire, well-developed dorsally. 18

17 Fore wing with 2nd submarginal cell absent (Fig. 12.122), ***or if*** present ***then*** small, triangular and pointed above (Fig. 12.123); marginal cell bounded below by *2r-rs* and a single abscissa of *Rs*.................... **Orgilinae** (p. 461)

— Fore wing with 2nd submarginal cell present long, so that marginal cell is bounded posteriorly by *2r-rs* and two abscissae of *Rs* (Fig. 12.114). .... **Opiinae** (in part) (p. 459)

18 Fore wing with *2rs-m* present, thus a closed 2nd submarginal cell present (Fig. 12.124); tergite I usually not long and slender, if so, widening abruptly beyond middle. 19

— Fore wing with *2rs-m* absent, thus no closed 2nd submarginal cell (Figs 12.126, 12.127) ***or*** rarely ***if*** *2rs-m* present (Fig. 12.125), ***then*** tergite I is long and slender, not widening abruptly beyond middle. 21

19 Metasoma with tergite I broadened abruptly behind the middle (Fig. 12.140). ....**Meteorinae** (p. 457)

— Metasoma with tergite I more or less parallel-sided, not broadened abruptly behind middle (Fig. 12.139). 20

20 Hind tibia with inner spur half as long as basitarsus (Fig. 12.142). ..... **Homolobinae** (p. 456)

— Hind tibia with inner spur much less than half as long as basitarsus (Fig. 12.143). .... **Helconinae** (most tribes) (p. 455)

21 First three tergites carapace-like, covering most of the metasoma (Fig. 12.131). .... **Helconinae** (Brachistini) (p. 455)
— First three tergites not carapace-like (Fig. 12.130). 22

22 Fore wing with distal abscissa of *Rs* moderately to very strongly curved, reaching wing margin well before wing apex (Fig. 12.127), ***or*** rarely absent; metasomal tergite I often petiolate. .... **Euphorinae** (p. 454)
— Fore wing with distal abscissa of *Rs* always present and relatively straight, reaching wing margin at or near wing apex (Fig. 12.126); metasomal tergite I usually not petiolate. .... **Blacinae** (p. 449)

23 Fore tibia with a longitudinal row of stout peg-like setae along anterior side (Fig. 12.144) (most easily seen in profile at high magnifications), ***or*** (if difficult to observe in small specimens) fore wing with *2rs-m* sometimes partially or completely absent, ***or*** males with pterostigma present in hind wing (Fig. 12.145). .... **Doryctinae** (most) (p. 452)
— Fore tibia without a longitudinal row of stout peg-like setae along anterior side; fore wing with *2rs-m* normally present; males never with a pterostigma present in hind wing. 24

24 Occipital carina at least dorsally absent ***and/or*** epicnemial carina absent. 25
— Occipital carina and epicnemial carina both distinctly present (Fig. 12.133). 28

25 Occipital carina completely absent. 26
— Occipital carina partially present, at least laterally. 27

26 Mandible bidentate; compound eye of normal size (Fig. 12.138); body not densely setose. .... **Braconinae** (p. 449)
— Mandible unidentate; compound eye abnormally small (Fig. 12.137), sometimes rudimentary; body densely setose.
[Small, rarely encountered termitophilous species.] .... **Ypsistocerinae**^{EX} (p. 463)

27 Fore wing with *2rs-m* unusually short, much less than 0.5 length of *Rs* between *2r-rs* and *2rs-m* (Fig. 12.114), ***or*** (rarely) ***if*** *2r-rs* is longer, ***then*** pterostigma is greatly elongated and *2r-rs* arises near its extreme base. .... **Opiinae** (in part) (p. 459)
— Fore wing with *2rs-m* not unusually short, nearly equal to 0.5 length of *Rs* between *2r-rs* and *2rs-m* (Fig. 12.125); pterostigma never greatly elongated, *2r-rs* arising 0.3–0.5 of way along pterostigma length, never near its extreme base. .... **Exothecinae** (p. 454)

28 Spiracles of metasomal tergites II+III situated on epipleura, below lateral margins of dorsal surface (Fig. 12.132); pterostigma normal (Rhyssalini) ***or*** narrow and linear, extending to wing apex (Hydrangeocolini). .... **Rhyssalinae** (p. 462)
— Spiracles of metasomal tergites II+III situated on dorsal surface near, but above, lateral margins (Fig. 12.133) ***and*** pterostigma normal, never narrow and linear. 29

29 Propodeum with tubercles or spines at lateral corners (Fig. 12.133); tergites II+III fully sclerotized, entirely smooth and polished ***and*** without a sclerotized dividing line between tergite II and tergite III. .... **Pambolinae** (p. 461)
— Propodeum without tubercles or spines at lateral corners (Fig. 12.112); tergites II+III ***either*** with distinct surface sculpture, not entirely smooth and polished, with a distinct dividing line between tergite II and tergite III, ***or*** extensively desclerotized. 30

30 Propodeum with a large area superomedia (Fig. 12.146); dorsal carinae of metasomal tergite I widely separated posteriorly; tergites II+III extensively membranous. .... **Hormiinae** (p. 456)

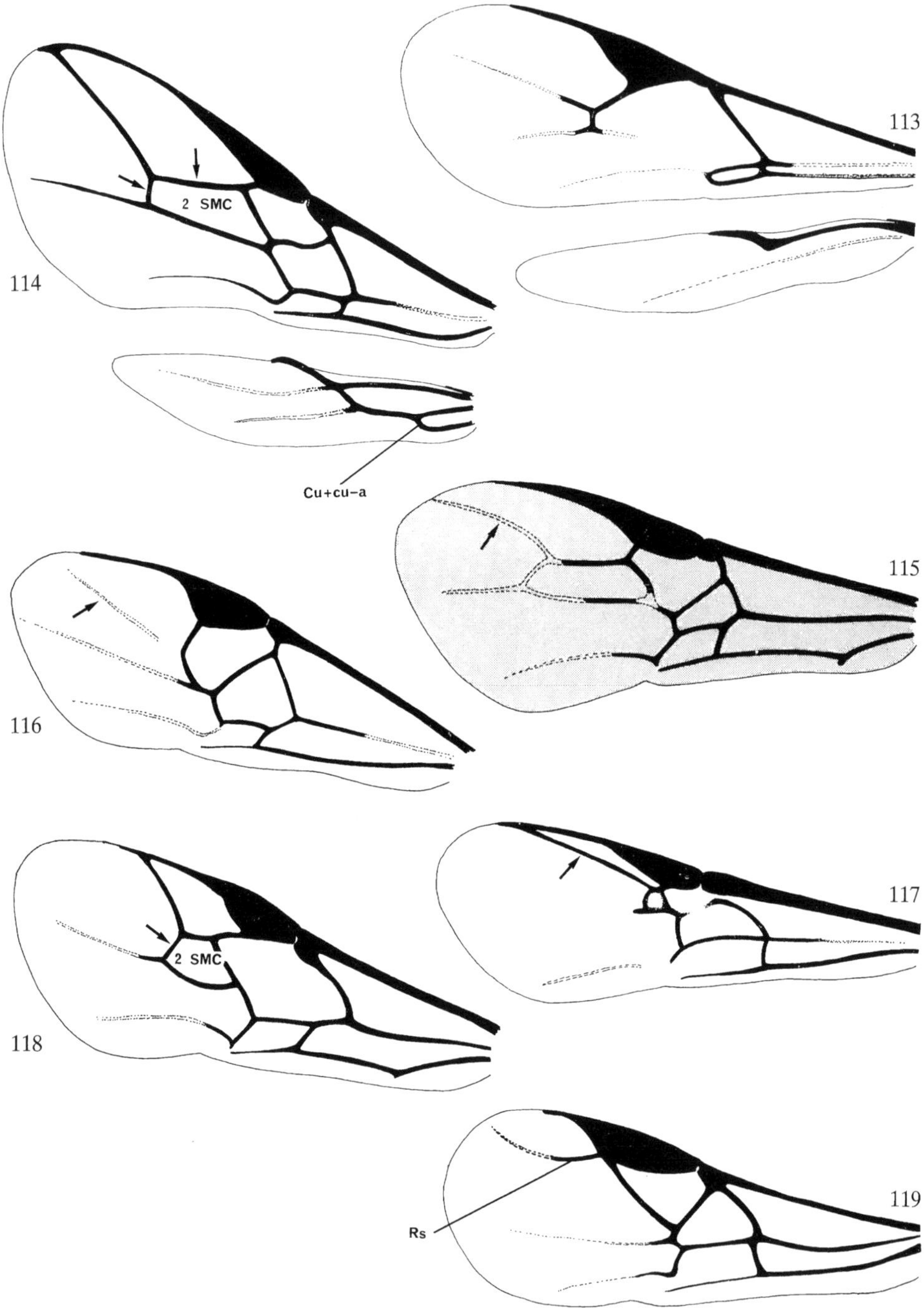

Figs 12.113–12.119. Braconidae. Figs 12.113–12.114. Fore and hind wings; 12.113, *Lysiphlebus* sp. (Aphidiinae); 12.114, *Opius* sp. (Opiinae). Figs 12.115–12.119. Fore wing; 12.115, *Cardiochiles* sp. (Cardiochilinae); 12.116, *Cotesia* sp. (Microgastrinae); 12.117, *Coccygidium* sp. (Agathidinae); 12.118, *Chelonus* sp. (Cheloninae); 12.119, *Adelius* sp. (Adeliinae).

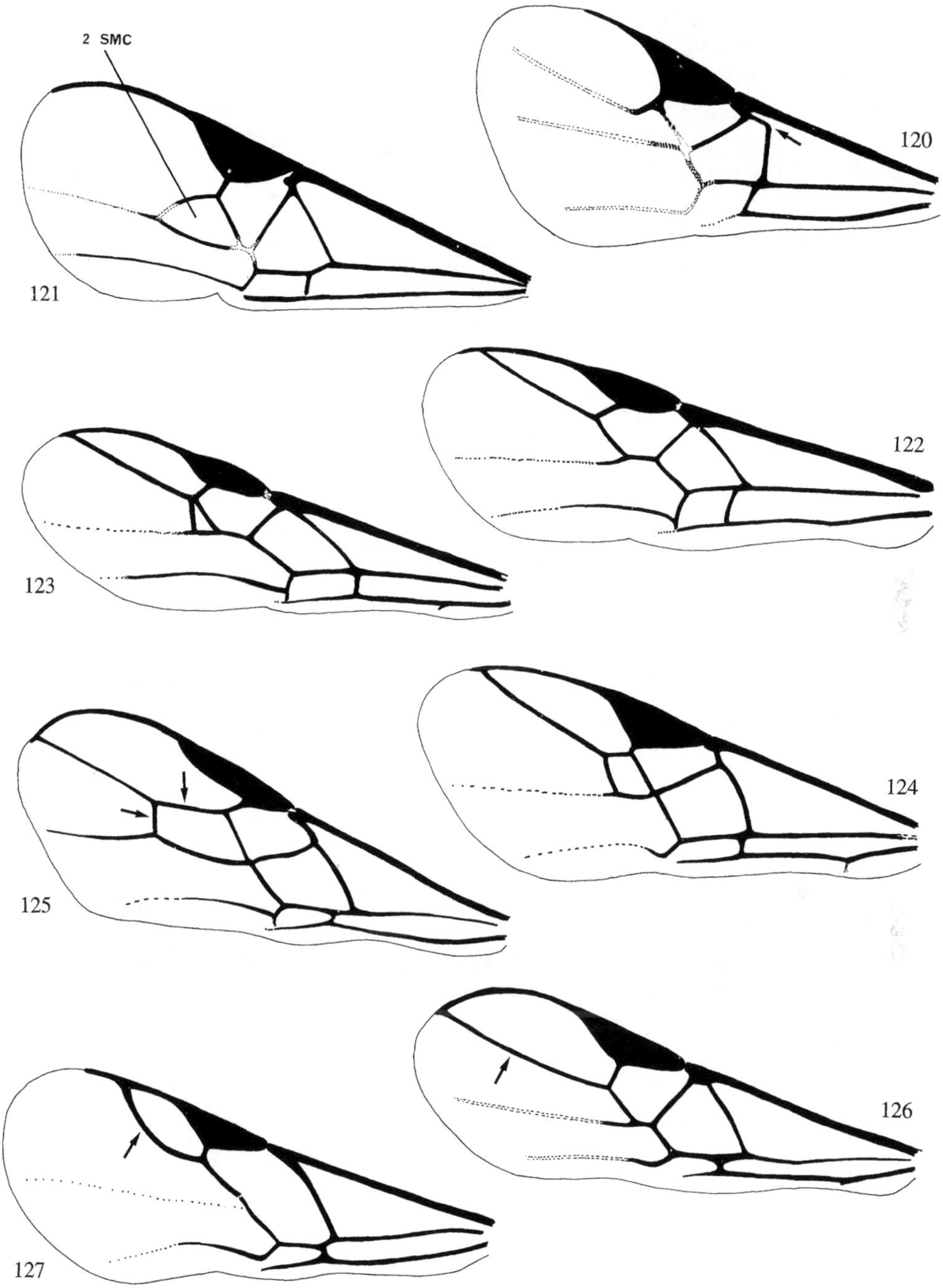

Figs 12.120–12.127. Braconidae, fore wing; 12.120, *Muesebeckia* sp. (Ichneutinae); 12.121, *Epsilogaster* sp. (Mendesellinae); 12.122, *Orgilus* sp. (Orgilinae); 12.123, *Stantonia* sp. (Orgilinae); 12.124, *Meteorus* sp. (Meteorinae); 12.125, *Colastes* sp (Exothecinae); 12.126, *Blacus* sp. (Blacinae); 12.127, *Microctonus* sp. (Euphorinae).

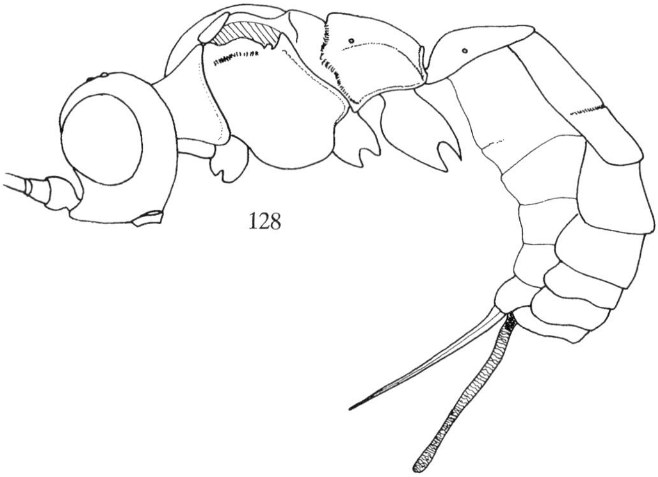

128

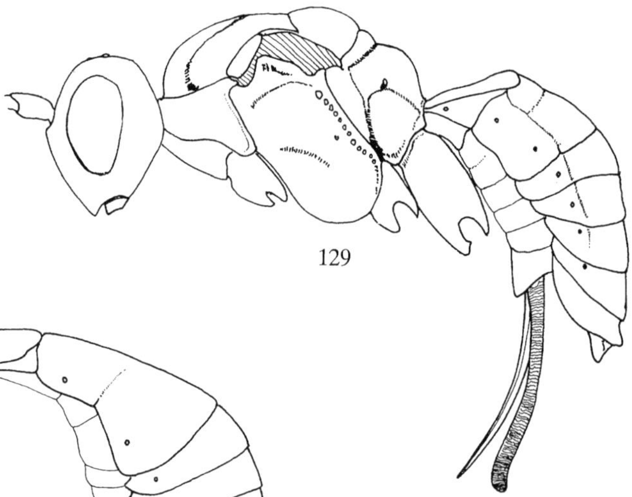

129

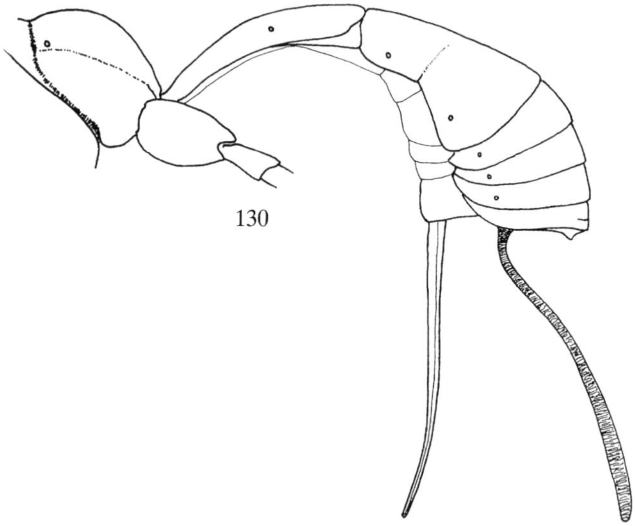

130

Figs 12.128–12.130. Braconidae. Figs 12.128–12.129. Head and body, lateral; 12.128, Betylobraconinae; 12.129, *Cardiochiles* sp. (Cardiochilinae); Fig. 12.130, Propodeum and metasoma, lateral; *Microctonus* sp. (Euphorinae).

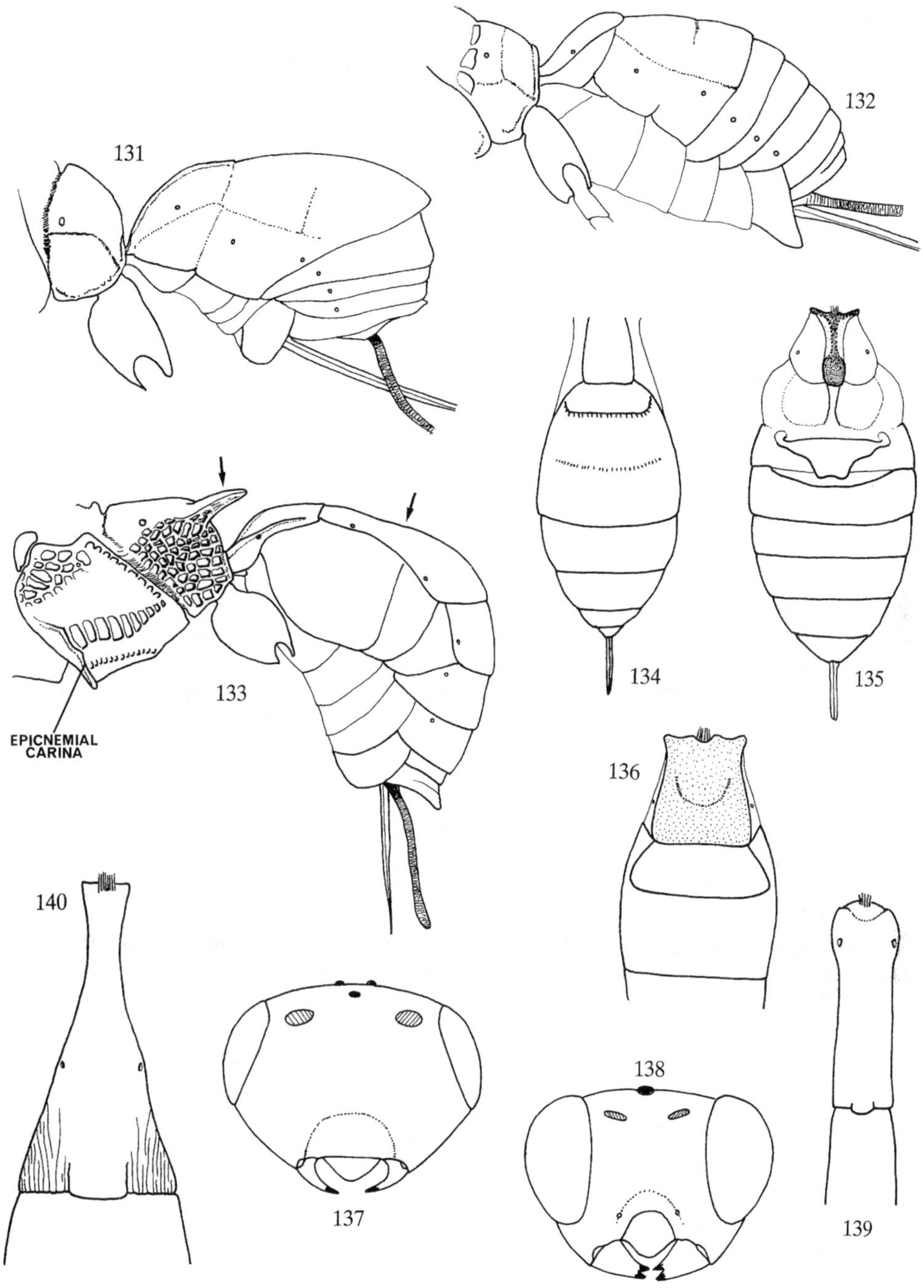

Figs 12.131–12.140. Braconidae. Figs 12.131–12.133. Meso- and metasoma lateral; 12.131, Brachistini; 12.132, *Hydrangeocola* sp. (Rhyssalinae); 12.133, *Pambolus* sp. (Pambolinae). Figs 12.134–12.136. Metasoma, dorsal; 12.134, Gnamptodontinae; 12.135, *Mirax* sp. (Miracinae); 12.136, *Cotesia* sp. (Microgastrinae). Figs 12.137–12.138. Head, anterior; 12.137, Ypsistocerinae; 12.138, *Bracon* sp. (Braconinae). Figs 12.139–12.140. Tergite I of metasoma, dorsal; 12.139, *Homolobus* sp. (Homolobinae); 12.140, *Meteorus* sp. (Meteorinae).

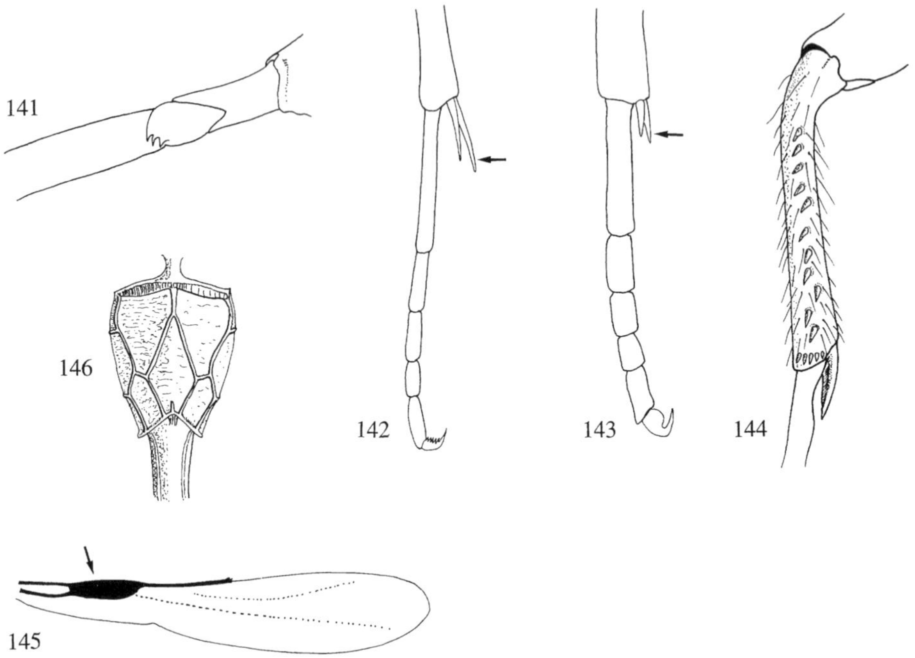

Figs 12.141–12.146. Braconidae. Fig. 12.141, Hind trochantellar segments, *Macrocentrus* sp. (Macrocentrinae). Figs 12.142–12.143. Hind tibial spurs; 12.142, *Homolobus* sp. (Homolobinae); 12.143, *Diospilus* sp. (Helconinae). Figs 12.144–12.145. Doryctinae. 12.144, fore tibia, *Heterospilus* sp.; 12.145, hind wing, *Heterospilus* sp. male. Fig. 12.146. Propodeum, dorsal, *Hormius* sp. (Hormiinae).

— Propodeum usually without a distinct area superomedia, ***or if*** present ***then*** it is relatively small; dorsal carinae of tergite I converging posteriorly (often meeting near the front of the tergite); tergites II+III always well sclerotized ***and*** with extensive and distinct surface sculpturing (Fig. 12.112). .... **Rogadinae** (p. 461)

Synopsis of the Costa Rican taxa

The Braconidae occur in virtually all terrestrial habitats in Costa Rica, but although many are common in open dry areas, the greatest number of species occurs in moist forested areas. The Opiinae and Alysiinae, whose dipteran hosts are most frequently encountered in moist situations (Wharton, 1984), are significantly more abundant and species-rich in humid areas. The braconid fauna of freshwater habitats has not been investigated in Costa Rica, but in temperate regions some dacnusine Alysiinae are associated with such habitats, where they parasitize agromyzid and ephydrid larvae mining in leaves and stems of emergent aquatic vegetation (Riegel, 1950; Griffiths, 1964). Some species may even enter the water to oviposit (Hagen, 1984). Some Opiinae and Braconinae are also semi-aquatic. For example, the palaearctic braconine genus *Baryproctus* has enlarged distal tarsal segments enabling it to run on the surface of still water (Quicke & Sharkey, 1989).

The subfamilies of Braconidae known or expected to occur in Costa Rica, although arranged alphabetically here for convenience, can be placed in informal groups that share a number of common features (Wharton *et al.*, 1992). The most distinctive of these are: the microgastroid complex (including Adeliinae, Cardiochilinae, Cheloninae, Dirrhopinae, Ichneutinae, Mendesellinae, Microgastrinae, Miracinae) and the helconoid complex (including Agathidinae, Blacinae, Euphorinae, Helconinae, Homolobinae, Macrocentrinae, Meteorinae, Orgilinae, Sigalphinae). There is continu-

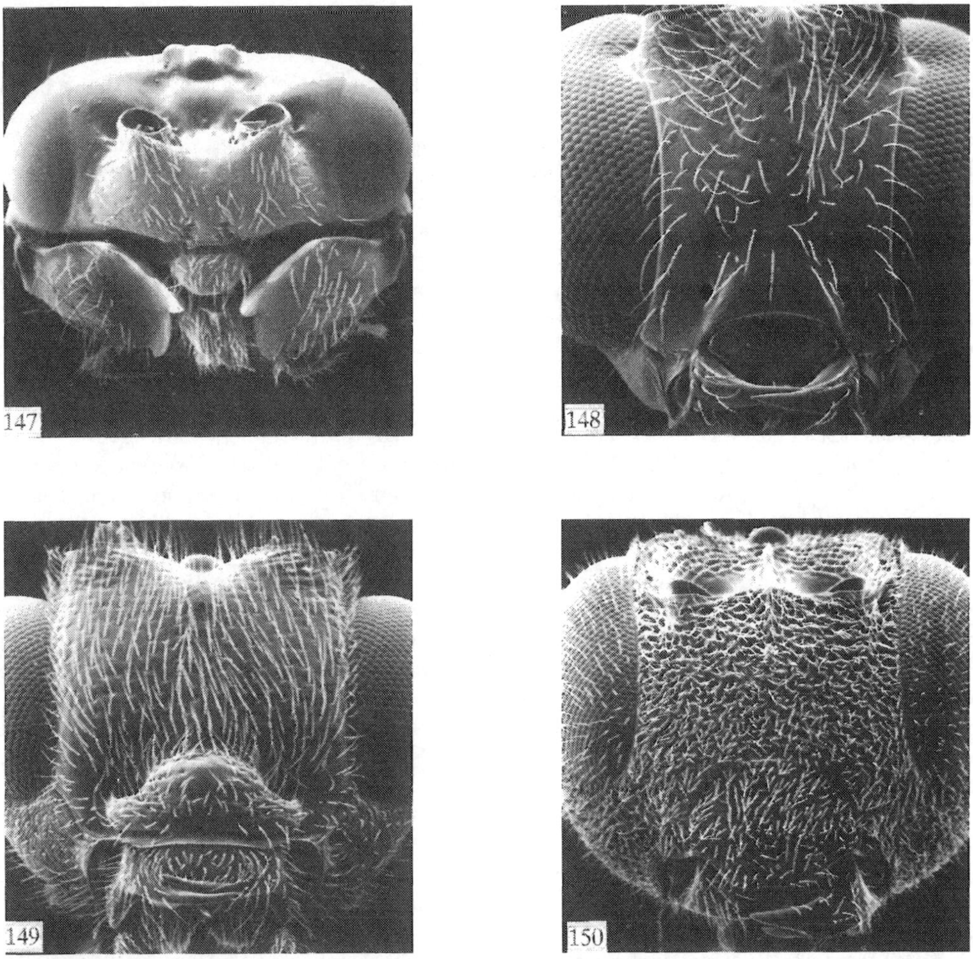

Figs 12.147–12.150. Scanning electron photomicrographs of faces of Braconidae; 12.147, *Gnathopleura* sp. (Alysiinae); 12.148, *Aleiodes* sp. (Rogadinae); 12.149, *Orgilus* sp. (Orgilinae); 12.150, *Chelonus* sp. (Cheloninae).

ing debate as to whether the so-called "cyclostome" subfamilies (including the Betylobraconinae, Braconinae, Doryctinae, Exothecinae, Gnamptodontinae, Hormiinae, Pambolinae, Ypsistocerinae, Rogadinae, Rhyssalinae and the Alysiinae + Opiinae) comprise a paraphyletic grade or a monophyletic lineage. The Aphidiinae appear to comprise a solitary clade intermediate between these other groups.

ADELIINAE^{EX}

This small subfamily comprises about 20 species worldwide, and is represented in North America by only five described species in two genera, *Adelius* and *Paradelius* (Marsh *in* Krombein *et al.*, 1979; Whitfield, 1988*c*), the latter of which also occurs in the Afrotropics. Adeliines are inconspicuous and are rarely encountered. They develop as koinobiont endoparasitoids of leaf-mining larvae of Nepticulidae. Oviposition is probably into the egg of the host (M.R. Shaw & Huddleston, 1991) and pupation is within the cocoon of the host.

AGATHIDINAE

The Agathidinae is a large and homogeneous subfamily of more than 48 genera and 840 species worldwide (Chou & Sharkey, 1989). They have a distinctive fore wing venation and many tropical species are brightly coloured, often with conspicuous colour patterns on the wings. Monophyly of the subfamily is indicated by two synapomorphies: vein *M* of the fore wing is separated from the prestigma by a wing fold and usually an associated bulla (Sharkey, 1986*a*), and males have setal brushes associated with glands on metasomal terga VII and VIII (Buckingham & Sharkey, 1988).

Agathidines are koinobiont endoparasitoids of endophytic lepidopterous larvae. Most species attack hosts living in concealment in leaf rolls, silk webs, flower heads or plant stems, especially species of Coleophoridae, Pyralidae, Tortricidae, Gelechiidae, Cochylidae and Incurvariidae (Muesebeck, 1927; Marsh, 1961; Shenefelt, 1970*b*; Sharkey, 1985, 1988; Chou & Sharkey, 1989), while a common species of *Bassus* in Santa Rosa National Park is known to attack a range of hesperiids (D.H. Janzen, pers. comm.). Some neotropical agathidines with short ovipositors may parasitize exposed larvae, possibly those of catocaline Noctuidae (Sharkey, 1983) and definitely some Arctiidae (D.H. Janzen, pers. comm.). Agathidines use their long ovipositors to probe for hosts in frass-ridden tunnels and retreats. Most species oviposit into an early instar host larva, often into a nerve ganglion or the protocerebrum (Quednau, 1970; Odebiyi & Oatman, 1977), but in some species into the haemocoel (Simmonds, 1947). The first instar larva is polypodeiform (Hagen, 1964) and development is delayed until after the host is in its pupation retreat. The parasitoid emerges and spins its cocoon within the host's pupation chamber (Thorpe, 1933; Simmonds, 1947; Dondale, 1954; Hummelen, 1974).

Genera recorded from Costa Rica are: *Agathis*, *Alabagrus* (18 spp. recorded from Costa Rica; Sharkey, 1988), *Bassus* (Sharkey, 1985), *Coccygidium* (= *Zelomorpha*), *Crassomicrodus*, *Cremnops*, *Earinus*, *Mesocoelus*, *Pharpa* (Sharkey, 1986*b*), *Plesiocoelus* and *Zacremnops* (Sharkey & Wharton, 1985; Sharkey, 1990).

Alysiinae

Alysiinae is a very large cosmopolitan subfamily that comprises more than 1100 species worldwide, currently placed in about 60 genera (Shenefelt, 1974; Fischer, 1975; Wharton, 1977*b*) and two tribes, Dacnusini and Alysiini. A conspicuous and distinctive characteristic of the group is their exodont (outturned) mandibles; this trait has evolved independently in some Hormiinae and Opiinae (Wharton, 1977*a*) but these other taxa are extremely rare. Alysiines are koinobiont endoparasitoids of the larvae of cyclorrhaphous Diptera. Most are solitary, but a few *Aphaereta* species are gregarious (Salkeld, 1959; Hughes & Woolcock, 1976). Dacnusini are almost exclusively parasitoids of Agromyzidae, and generally species are quite host-specific (Griffiths, 1964). Alysiini have a much broader host range, utilizing 25 different cyclorrhaphan families (Wharton, 1984), although there is some degree of host preference at the generic level. For example, *Aspilota* species usually parasitize phorids and many *Phaenocarpa* species utilize anthomyiid or scathophagid hosts. *Alloea* species are restricted to the Lonchopteridae and *Gnathopleura* species are only recorded from calypterate muscoids (Wharton, 1984). Some species are hyperparasitoids, developing in phorid, tachinid and sarcophagid primary parasitoids (Kulman, 1965; Wharton, 1984). Oviposition by alysiines is usually into a larval fly, or rarely into a dipteran egg (Evans, 1933). Pupation is internal, within the host puparium, from which the adult alysiine escapes by using its exodont mandibles to pry apart an exit.

Both tribes occur in Costa Rica but the Alysiini are much more diversified and frequently encountered. The nearctic genera of the tribe Alysiini were reviewed by Wharton (1980), and a key to the New World genera was included. Genera of Alysiini present in Costa Rica include: *Alysia* (Wharton, 1986, 1988*b*), *Alysiata*, *Aphaereta* (Wharton, 1977*b*), *Asobara*, *Aspilota* (especially diverse in Costa Rica and abundant at high elevations; Wharton, 1985), *Gnathopleura*, *Oenonogastra* and *Phaenocarpa*. Thus far the only genera of Dacnusini (Riegel, 1952, 1982) encountered in Costa Rica are *Chorebus* and *Coelinius*.

Aphidiinae

Although quite species-rich in the Holarctic region, this subfamily is sparsely represented in Costa Rica, occurring primarily at altitudes above 1500 metres and up to 3000 metres on the Cerro de la Muerte, or in crops. Aphidiines are solitary koinobiont endoparasitoids of nymphs and adults of ovoviviparous Aphidoidea; even aphid embryos may be parasitized while still inside their mother (Mackauer & Kambhampati, 1988). The exclusively oviparous groups, such as Phylloxeridae and Adelgidae, are apparently not attacked. Aphidiines have a variety of attack strategies including grasping the host with the legs or metasomal process, rapid thrusting with the ovipositor and laying an egg in any site or, less commonly stinging and inducing temporary paralysis before ovipositing into the fused thoracico-abdominal ganglia (Griffiths, 1960; Schlinger & Hall, 1960, 1961; Calvert & van den Bosch, 1972). It is possible that the aphidiine leaves a pheromone-like external marker on the host after oviposition (Chow & Mackauer, 1986). Host suitability is determined by the age of the host and the parasitoid (Tardieux & Rabasse, 1988) as well as the number and depth of sting (Tardieux & Rabasse, 1990). There are three larval instars (O'Donnell, 1982, 1987). Most aphidiines pupate within the mummified host cuticle, although the larvae of some genera (e.g. *Praon*) emerge from the host and spin a tent-like cocoon directly beneath the aphid, attaching the remains to the substrate (Beirne, 1942; Schlinger & Hall, 1960; Johnson, 1987). Stary (1970) provided a review of the biology of the subfamily and Mackauer (1968) listed host records for the world species. The Aphidiinae represent one of the most important groups of biological control agents of aphid pests (Stary, 1967*a* & *b*, 1968*a–c*).

Keys to species are available for Mexico (Stary & Remaudière, 1982, 1983; Stary, 1983) and Cuba (Stary, 1981). However, caution is required in using them since some characters traditionally used to separate species are more variable than previously supposed (Pungerl, 1983). Genera present in Costa Rica include: *Aphidius*, *Binodoxys*, *Diaretiella* (reared from *Myzus persicae*), *Lysiphlebus* (reared from *Aphis nerii* and *Toxoptera aurantii*), *Monoctonus* (collected only from very high altitudes, above 2500 m), *Praon* and *Trioxys*.

Betylobraconinae

A very small subfamily occurring in the Australian (Achterberg, 1991), nearctic (Quicke & Achterberg, 1990) and neotropical (Achterberg, pers. comm.) regions. Their phylogenetic placement is by no means resolved (Quicke & Achterberg, 1990; Wharton *et al.*, 1992). Although they are often treated as a distinct subfamily, some workers consider them to be an aberrant lineage of the Rogadinae (Wharton, pers. comm.). Betylobraconines are represented in Costa Rica by two minute species (body length 1.5 to 2.5 mm) which superficially resemble some diapriids. One has been collected in the Golfo Dulce Forest Reserve (c. 100 m) whilst the second is from Zurquí de Moravia (1600 m). Their hosts are not known.

Blacinae

A relatively small, cosmopolitan subfamily comprising about 165 described species (Achterberg, 1988*c*). The Blacinae, as it was traditionally recognized (e.g. Marsh *in* Krombein *et al.*, 1979), was a "catch-all" subfamily. Several genera formerly placed in it have been transferred to the Helconinae (Mason, 1974), Euphorinae (S.R. Shaw, 1985) or Orgilinae (Muesebeck, 1970). The monophyly of the Blacinae, even in its more restricted sense, has yet to be convincingly demonstrated (Quicke & Achterberg, 1990).

Very little is known about the biology of Blacinae, but *Blacus* species are thought to be koinobiont endoparasitoids of a variety of coleopteran larvae in damp decaying vegetation, including those of Staphylinidae, Nitidulidae, Cryptophagidae, Scolytidae, Anobiidae, Curculionidae and Cerambycidae. Records of them as parasitoids of dipterous larvae in similar situations require verification, and are believed by some authors (e.g. M.R. Shaw & Huddleston, 1991) to be erroneous. Blacine species have been revised by Achterberg (1975, 1988*d*), but relatively little is known about the neotropical fauna. Twenty-two blacine species are described from the Neotropics, but none has previously been recorded from Costa Rica. Several *Blacus* species are now known to occur in the country, including an undescribed species of the subgenus *Artocrus*, previously known only from Brazil and Argentina.

Braconinae

A very large, cosmopolitan subfamily comprising more than 150 genera and 5000 species worldwide (Quicke, 1988*b*, 1989*a*) and one genus, *Bracon*, is estimated to include more than 2000 species worldwide (Achterberg, 1988*b*). Although braconines are very diverse in the tropics their centre of diversity seems to be the Old World, and the group is rather less species-rich in the New World (Quicke, pers. comm.). The vast majority of braconines are solitary or gregarious idiobiont ectoparasitoids of subcorticolous or endophytic holometabolous insect larvae, although Afrotropical species of the Aspidobraconina are unusual in being endoparasitoids of exposed butterfly pupae (Achterberg, 1984*c*). The hosts of most braconines are the larvae of Coleoptera or Lepidoptera (Muesebeck, 1925; Shenefelt, 1943, 1978; Quicke, 1988*d*, 1989*a*), although occasionally Diptera or Hymenoptera (Munro, 1917) may be attacked. Several species of *Bracon* were reared in a survey of leaf-miners in northwestern Costa Rica (Memmott *et al.*, 1994) including two that attack a wide range of coleopterous and lepidopterous hosts mining in several plant families. These host records are summarized in Table 12.8, to serve as an example of the range of hosts attacked by a species in a given area.

Many braconine species parasitize stem-boring hosts, so they are one of the parasitoid groups most frequently reared from lepidopterous borers of gramineous crops (e.g. Wharton *et al.*, 1989). Several species have been investigated for biological control of endophytic crop pests.

Recent studies of the neotropical fauna are Quicke (1988*a*, *c* & *d*; 1989*a–c*); work on the nearctic fauna (Mason, 1978*b*; Quicke, 1988*b*; Quicke & Sharkey, 1989) will also be useful in helping to identify tropical American genera. Those present in Costa Rica include: *Atanycolus*, *Bracon* (Muesebeck, 1925, as *Microbracon*), *Calobracon*, *Compsobraconoides*, *Compsobracon*, *Cyanopterus*, *Digonogastra* (Quicke, 1988*c*), *Habrobracon*, *Lasiophorus*, *Megabracon*, *Myosoma*, *Palabracon*, *Pheloura*, *Sacirema* and *Vipio* (= *Isomecus*).

Cardiochilinae

This is a moderately diverse cosmopolitan subfamily, species of which are solitary koinobiont endoparasitoids of endophytic lepidopteran larvae, especially Pyralidae and Noctuidae (Shenefelt, 1973*a*; Huddleston & Walker, 1988). The group is represented in Costa Rica by several species of *Cardiochiles*, a large genus comprising more than about 150 species worldwide, as well as by some species of *Toxoneuron* and *Tenthredoides*. The Mexican (Mao, 1945) and North American (Mao, 1949) *Cardiochiles* species have been revised, and to some extent these studies are useful for identifying species present in Costa Rica.

Cheloninae

The Cheloninae is a large cosmopolitan subfamily with more than 500 known species worldwide (Walker & Huddleston, 1987). Chelonines have a distinctive metasomal carapace, formed by the fusion the first three metasomal terga (Dudarenko, 1974). Although members of several other braconid subfamilies (e.g. Rogadinae, Helconinae) have convergently

Parasitoid	Host	Host food-plant
Bracon sp 1	*Chalepus* spp. (Chrysomelidae)	*Lasiacis* sp. (Gramineae)
	Cosmopterix sp (Cosmopterigidae)	*Lasiacis* sp. (Gramineae)
	Elachista sp. (Elachistidae)	*Olyra latifolia* (Gramineae)
	Gracillariidae indet.	*Luehea* sp. (Tiliaceae)
	Oxychalepus auticornis (Chrysomelidae)	*Centrosema pubescens* (Leguminosae)
	Oxychalepus postdatus (Chrysomelidae)	*Cassia hayesiana* (Leguminosae)
	Scythridae indet.	*Lasiacis* sp. (Gramineae)
	Uroplata fusca (Chrysomelidae)	*Arrabidaea mollisima* (Bignoniaceae)
Bracon sp 2	*Brachys* sp. (Buprestidae)	*Gouania polygama* (Rhamnaceae)
	Chalepus spp. (Chrysomelidae)	*Lasiacis* sp. (Gramineae)
	Cosmopterix sp (Cosmopterigidae)	*Juncus* sp. (Juncaceae)
	Baliosus sp. (Chrysomelidae)	*Helicteres guazumaefolia* (Sterculiaceae)
	Oxychalepus auticornis (Chrysomelidae)	*Centrosema pubescens* (Leguminosae)
	Oxychalepus auticornis (Chrysomelidae)	Leguminosae indet.
	Oxychalepus postdatus (Chrysomelidae)	*Cassia hayesiana* (Leguminosae)
	Uroplata fusca (Chrysomelidae)	*Arrabidaea mollisima* (Bignoniaceae)
	Uroplata sp. (Chrysomelidae)	*Lantana camara* (Verbenaceae)

Table 12.8. Host relationships of two species of *Bracon* parasitizing leaf miners in Santa Rosa National Park. This table is abstracted from data based on extensive sampling of a range of leaf-miners in dry forest over a two year period (Memmott *et al.*, 1994). Note none were found to attack any of the common agromyzids present.

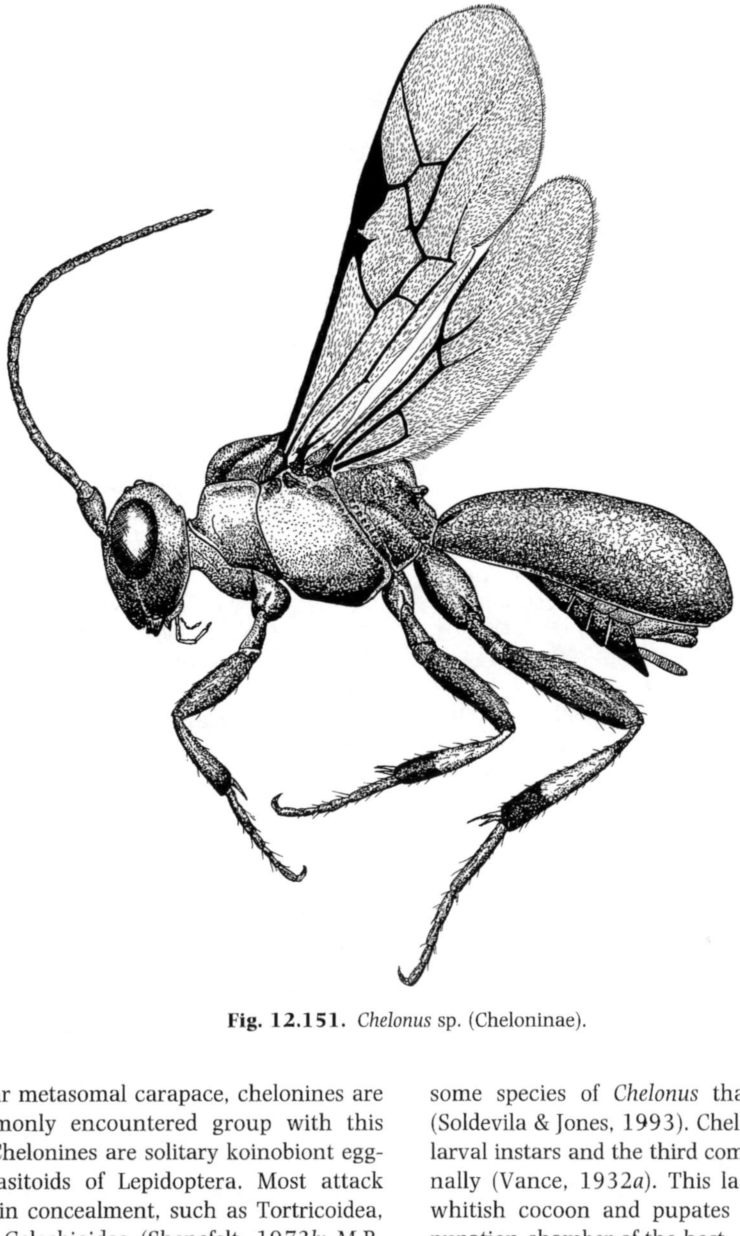

Fig. 12.151. *Chelonus* sp. (Cheloninae).

evolved a similar metasomal carapace, chelonines are the most commonly encountered group with this characteristic. Chelonines are solitary koinobiont egg-larval endoparasitoids of Lepidoptera. Most attack hosts that feed in concealment, such as Tortricoidea, Pyraloidea and Gelechioidea (Shenefelt, 1973*b*; M.R. Shaw & Huddleston, 1991), although some *Chelonus* species attack noctuids (Marsh, 1978). Some have a fairly broad host range (Broodryk, 1969*a*). The host larva is not killed until after it has spun its cocoon (Vance, 1932*a*), but cocoon-spinning behaviour is initiated prematurely in a pre-ultimate instar host by some species of *Chelonus* that attack larger hosts (Soldevila & Jones, 1993). Cheloninae have only three larval instars and the third completes its feeding externally (Vance, 1932*a*). This last instar spins a small whitish cocoon and pupates within the cocoon or pupation chamber of the host.

Genera present in Costa Rica are *Ascogaster*, *Chelonus*, *Leptodrepana* and *Phanerotoma*; they may be distinguished using Table 12.9. Although species of these genera are common in Central America, none have been revised in the region. The nearctic species of *Ascogaster* were reviewed by S.R. Shaw (1983) and

- –Metasomal carapace with two transverse grooves .... *Phanerotoma*
- –Metasomal carapace without transverse grooves
  - –1st discal and submarginal cell confluent .... *Chelonus*
  - –1st discal and 1st submarginal cells separated
    - –Eyes more or less hairless .... *Ascogaster*
    - –Eyes bearing even short hairs .... *Leptodrepana*

Table 12.9. Identification table for Costa Rican genera of Cheloninae.

the nearctic species of *Chelonus* (subgenus *Microchelonus*) by McComb (1967).

Doryctinae

The Doryctinae is a large, cosmopolitan subfamily which is perhaps most species-rich in the New World tropics, where its diversity rivals that of the Braconinae in the Old World tropics. The subfamily is demonstrably monophyletic based on two synapomorphies: the possession of a double node at the apex of the ovipositor (Quicke *et al.*, 1992*b*); and presence of a row or cluster of stout spines along the anterior edge of the fore tibia (Marsh, 1965; Achterberg, 1984*a*). These spines may assist the adult parasitoid to escape from the endophytic habitat of its host in which it pupates. The spines are easily observable in larger doryctines, but they may be quite difficult to see in minute species (e.g. many *Heterospilus*). They are apparently secondarily absent in *Sericobracon* (S.R. Shaw & Edgerly, 1985).

Doryctines are mostly idiobiont ectoparasitoids of subcorticolous or endophytic Coleoptera larvae, especially those of the families Bostrichidae, Buprestidae, Cerambycidae, Lyctidae and Scolytidae (Marsh, 1965; Shenefelt & Marsh, 1976). Consequently in the field individuals are frequently found in association with dead or damaged trees, firewood and brush piles. Species of *Stenocorse* parasitize bruchids in leguminous tree seeds (Marsh, 1968) and one species, *S. bruchivora*, has been observed to be the most abundant braconid reared from bruchid-infested fruits in Costa Rica (Whitehead, 1975). Some *Allorhogas* species have also been reared from legume seeds (Whitehead, 1975) and others are associated with plant galls (Guimaraes, 1957; Marsh, *in* Krombein *et al.*, 1979). The exact nature of most of these associations of *Allorhogas* remains to be clarified, but a Brazilian species is phytophagous, forming galls in the seeds of *Pithecellobium* (Macêdo & Monteiro, 1989). A related Costa Rican species appears also to be phytophagous (P. Hanson, pers. comm.). A very different host association for a doryctine is shown by *Sericobracon*, which in Trinidad has been reared from webspinners (Embiidina) (S.R. Shaw & Edgerly, 1985). Field observations by J. Edgerly indicate that *Sericobracon* is an idiobiont endoparasitoid.

The most frequently encountered doryctines in Costa Rica belong to *Heterospilus*, a genus represented in the country by at least 39 species. These small sized doryctines are easily distinguished from other doryctine genera by the reduction of vein *2rs-m* in the fore wing, which is always desclerotized posteriorly and often completely absent. Males have a prestigma in the hind wing, an unusual characteristic that is seen in a few other doryctine genera, but otherwise is unique in the Hymenoptera. *Heterospilus* species parasitize a wide range of endophytic, mostly stem-boring, hosts (Marsh, 1982*a*), including, in the Coleoptera: Anobiidae, Bostrichidae, Bruchidae, Buprestidae, Cerambycidae, Curculionidae, Languriidae, Mordellidae and Scolytidae, and in the Lepidoptera: Gelechiidae, Incurvariidae, Pyralidae and Tortricidae. One Costa Rican species, *H. microstigmi* attacks the larvae of *Microstigmus comes* (Sphecidae) (Matthews, 1991). Two Mexican species have been reared from the cotton boll weevil (Marsh, 1982*b*).

Members of the tribe Spathiini are distinguished from other doryctines by their petiolate first metasomal segment. *Spathius* is a cosmopolitan genus with over 300 described forms, which are mostly ectoparasitoids of wood-boring coleopterous larvae, predominantly Scolytidae (Matthews, 1970*a*). Spathiines have radiated extensively in the Neotropics, the diversity being perhaps comparable to that found by Nixon (1943) in the Old World Malesian archipelago, which he considered to be the centre of distribution. Many of the neotropical species comprise a distinct lineage (Matthews & Marsh, 1969; Matthews, 1970*a*) that has been treated as a separate genus, *Notiospathius* (Matthews & Marsh, 1973; Shenefelt & Marsh, 1976). There are at least 14 species of *Notiospathius* in Costa Rica from sea-level up to 1600 metres, but only one species of *Spathius* has been collected.

Other doryctine genera present in Costa Rica include: *Acanthorhogas, Acrophasmus* (Marsh, 1968), *Aivalykus, Curtisella, Doryctes* (Marsh, 1969), *Ecphylus* (at least eight species, one described; Matthews, 1969), *Heerz* (one species, *H. lukenatcha*, present in Atlantic lowland forest; Marsh, 1993), *Janzenia* (one species, *J. gauldi*, has been collected in lowland dry forest between April and June; Marsh, 1993), *Johnsonius* (two species in Pacific lowland wet forest, Marsh, 1993), *Liobracon* (Marsh, 1970*a*), *Masonius* (one species, *M. fasciatus* collected in Santa Rosa National Park; Marsh, 1993), *Megaloproctus* (Marsh, 1983), *Monolexis* (Marsh, 1965), *Nervellius* (Marsh, 1988), *Odontobracon* (with a large, curved dorsal tooth on the hind coxa; Marsh, 1970*b*, 1988), *Pedinotus* (Matthews, 1969; Marsh, 1970*a*), *Pioscelus, Psenobolus* (found in fruits of certain fig species), *Rhaconotus* (Marsh, 1976), *Rhoptrocentroides* (single widespread species, *R. platyfemur;* Marsh, 1993) and *Semirhytus*. Other genera expected to occur are *Coiba, Glyptocolastes, Hecabolus, Leluthia, Parallorhogas, Whartonius* and *Whitfieldia*. A key to doryctine genera is given by Marsh (1993).

A few doryctine species are among the largest and most strikingly coloured Braconidae in Costa Rica. For example *Megaloproctus xanthus*, from the Caribbean lowlands up to Turrialba, is 17.0 mm long, including the ovipositor, and has a bright yellowish orange mesosoma and metasoma.

MONITORIELLA (genus incertae sedis). The enigmatic genus *Monitoriella* is here provisionally placed with the Doryctinae as the anterior surface of its fore tibia does have modified stout setae, especially at the apex, although these are neither so peg-like nor so apparent as those of many doryctines. Although Hedqvist (1963) included *Monitoriella* in the Hormiinae, he stated that it is 'related to some genera of Doryctinae'. However, *Monitoriella* does not appear to have the

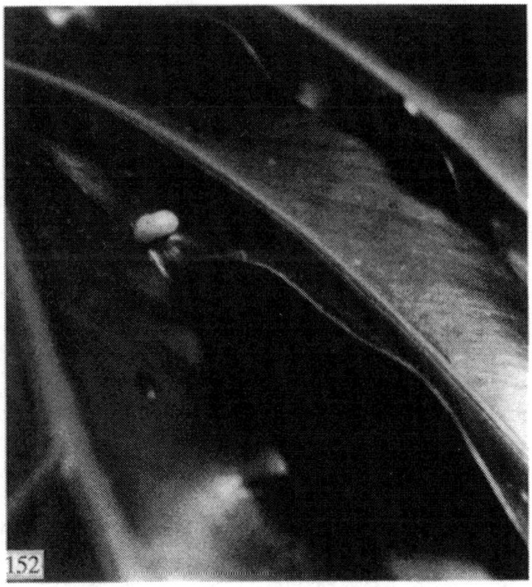

Fig. 12.152. Gall on leaf of *Philodendron* believed to be produced by *Monitoriella* sp. (Braconidae).

double node of the ovipositor and Paul Marsh (pers. comm.) has not found any synapomorphy uniting it with any group of Doryctinae. Clearly further study is needed to resolve its placement. It has been placed in the Hormiini (Hedqvist, 1963; Marsh *in* Krombein *et al.*, 1979) but studies of this group have demonstrated that it should be excluded (Whitfield, 1989, 1992; Wharton, 1993*b*). The latter author concluded that 'the relationships of *Monitoriella* are unresolved' and suggested placement in the Pambolini based on 'head shape and weak propodeal tubercles'. However, it lacks the smooth and polished tergites II+III (an important synapomorphy for the Pambolini), whilst the head shape varies in both groups and the propodeal tubercles of *Monitoriella* are never so developed as in the Pambolini.

Monitoriella occurs from southern Texas south through Central America to Peru (Hedqvist, 1963; Marsh *in* Krombein *et al.*, 1979; Wharton, 1993*b*). At least three species occur in Costa Rica at sites between 200 and 1600 metres. Species have been reared from galls on the leaves of *Philodendron* (Araceae) (Fig. 12.152). It is probable that they actually cause this gall, and that their larvae are phytophagous (P. Hanson, pers. comm.).

Euphorinae

A large and diverse subfamily, with 32 genera worldwide, which are classified in three sections and nine tribes (S.R. Shaw, 1985). Euphorines are solitary or gregarious koinobiont endoparasitoids of adult Coleoptera (*Centistes*, *Dinocampus*, *Microctonus*, *Perilitus*; Balduf, 1926; Jackson, 1928; Loan & Holdaway, 1961*a* & *b*; Waloff, 1961), adult Hymenoptera (*Syntretus*; Cole, 1959*b*; Alford, 1968), adult Neuroptera (*Chrysopopthorus*; Séméria, 1976), nymphal and adult Hemiptera (*Aridelus*, *Leiophron*, *Wesmaelia*), and nymphal and adult Psocoptera (*Euphoriella*; Sommerman, 1956). In those species attacking the adult stage of the host, the latter may remain alive for a day or so after the emergence of the final instar euphorine larva. Parasitism of female hosts usually results in reduced egg production but parasitism may not affect fertility of male hosts (Weiss & Williams, 1980). Euphorines that parasitize Heteroptera usually oviposit into early nymphal instars and emerge from later nymphal instars or the adult (Waloff, 1967; Loan, 1980; Debolt, 1981).

At the subfamily level, the Euphorinae have a host range that is substantially broader than that of any other braconid subfamily, presumably because of the evolution of parasitism of adult insects (Muesebeck, 1936; Tobias, 1966; S.R. Shaw, 1985, 1988*a*). It is hypothesized that adult-parasitism evolved through parasitism of Chrysomelidae, whose adults and larvae occur on the same plants and have similar feeding habits (Tobias, 1966; S.R. Shaw, 1988*a*). Subsequent evolution mostly involved diversification among phytophagous coleopteran host groups (especially Chrysomelidae, Curculionidae and Scolytidae), as well as exploitation of some predaceous beetles (e.g. Carabidae, Coccinellidae). Adoption of other host orders probably evolved initially through the parasitism of other common host species with similar feeding habits in the same microhabitat. Parasitism of adult hymenopterans by the Syntretini probably involves attacking hosts while they are feeding at flowers (S.R. Shaw, 1988*a*; Walker *et al.*, 1990*a*).

The euphorine genera of the world were reviewed by S.R. Shaw (1985), and their phylogeny and biology was reviewed by S.R. Shaw (1988*a*). Although euphorines are rather diverse in the tropics, only a few studies are pertinent to neotropical species, and most of the Costa Rican species are undescribed. Seven tribes and 15 genera have been found in Costa Rica: *Aridelus* (Papp, 1965), *Centistes*, *Chrysopophthorus* (Mason, 1964), *Cryptoxilos*, *Dinocampus*, *Euphoriella*, *Leiophron*, *Microctonus*, *Orionis* (S.R. Shaw, 1987), *Perilitus*, *Peristenus*, *Streblocera*, *Syntretus*, *Townesilitus* (Haeselbarth & Loan, 1983), and *Wesmaelia*. The genus *Betelgeuse* was recently described from Mexico but has not yet been seen from Costa Rica (S.R. Shaw, 1988*b*). In Costa Rica a species of *Cryptoxilos* has been reared from adults of the Cecropia petiole-borer (Scolytidae: *Scolytodes* sp.) (P. Hanson, pers. comm.).

Exothecinae

A small subfamily comprising the predominantly holarctic tribe Exothecini (*sensu* Achterberg, 1983*b*; Whitfield, 1989), represented in Costa Rica by at least one species each of the genera *Colastes* and *Xenosternum*. Exothecines are mostly solitary ectoparasitoids of leaf-mining Lepidoptera, Hymenoptera, Diptera, or Coleoptera larvae (Muesebeck, 1932*b*; Shenefelt, 1975; Achterberg, 1983*b*; M.R. Shaw, 1983). Some species, such as *Colastes braconius*, are broadly polyphagous, attacking a wide range of leaf-mining hosts (M.R. Shaw, 1983).

GNAMPTODONTINAE

A small subfamily comprising the single tribe Gnamptodontini (*sensu* Whitfield, 1989), represented in Costa Rica by several species of *Gnamptodon* and *Pseudognamptodon*. Gnamptodontines are thought to be solitary endoparasitoids of leaf-mining Nepticulidae (Achterberg, 1988*a*; M.R. Shaw & Huddleston, 1991). The host larva regains mobility after parasitization (Whitfield, 1989).

HELCONINAE

A moderately large subfamily, divided into four tribes: Brachistini, Cenocoelini, Diospilini and Helconini. All except the last are known to be present in Costa Rica. Helconines are koinobiont endoparasitoids, most usually utilizing subcorticolous or endophytic Coleoptera larvae, and less commonly Lepidoptera larvae, as hosts. Where known species are solitary parasitoids and pupate within the host's pupation site (Pechuman, 1939; Obrtel, 1960; Haeselbarth, 1962), but details of the biology of most species are unknown.

HELCONINAE: Brachistini. This is the largest tribe (*sensu* Mason, 1974) in the subfamily and species parasitize a wide array of hosts, most frequently species of Bruchidae and Curculionidae, but less commonly also members of the coleopteran families Anobiidae, Cerambycidae and Mordellidae, and the lepidopteran family Coleophoridae (Martin, 1956; Mason, 1974; Gauld & Bolton, 1988). Some brachistines (*Eubazus*) are egg-larval parasitoids (Martin, 1956; Alauzet, 1987), and the habit may be typical of all members of the tribe (M.R. Shaw & Huddleston, 1991). There are three larval instars and the final instar larva emerges from the host in its pupation chamber to complete feeding externally. Genera present in Costa Rica include: *Aliolus, Allodorus, Eubazus, Nealiolus, Triaspis* and *Urosigalphus*. Species of *Urosigalphus* are mostly endoparasitoids of seed-infesting bruchid and curculionid larvae, although a few species parasitize twig and gall infesting weevils (L. Gibson, 1972*a* & *b*). Gibson (1972*b*) treated 21 species of *Urosigalphus* occurring in Central America, including the Costa Rican species *U. avocadoae* and *U. flavens*, associated with avocado and cacao, respectively.

HELCONINAE: Cenocoelini.. Species of this tribe are mainly tropical in distribution, and are sometimes treated as a separate subfamily (Achterberg, 1984*a*; Gauld & Bolton, 1988; Quicke & Achterberg, 1990). Cenocoelines are quite distinctive in having the metasoma inserted very high upon the propodeum, much as in an evaniid. Although the Cenocoelini is undoubtedly monophyletic, it is more difficult to establish the phylogenetic position of the tribe relative to other lineages (Wharton *et al.*, 1992). Since Don Quicke and Kees van Achterberg were unable to establish the phylogenetic position of the Cenocoelini, and they did not establish the monophyly of the Helconinae in the absence of the cenocoeline lineage, a more conservative and traditional classification is followed here (e.g. Marsh *in* Krombein *et al.*, 1979; Saffer, 1982). Most *Cenocoelius* species are solitary endoparasitoids of wood-boring beetle larvae, especially Cerambycidae, but less commonly Scolytidae or Buprestidae (Saffer, 1982). There are at least ten undescribed species of *Cenocoelius* in Costa Rica. The only named species, *C. andirae*, is associated with *Cleogonus* (Curculionidae) in seeds of *Andira inermis* (Leguminosae) (Saffer, 1977).

HELCONINAE: Diospilini. Species of this tribe are parasitoids of immature Coleoptera (Mason, 1987*b*), and have been reared from Anobiidae, Cerambycidae, Chrysomelidae, Curculionidae, Melandryidae and Nitidulidae (Marsh *in* Krombein *et al.*, 1979; Gauld & Bolton, 1988). Little is known about their biology except for that of a species of *Diospilus* that is a parasitoid of *Meligithes* (Osborne, 1960). Osborne recorded that the braconid's egg has a pedicel which is sometimes embedded in the host's gut wall, that a trophamnion is present around the embryo, that the first instar larva has falcate mandibles and develops an anal vesicle, and that the final instar larva emerges from the host in its pupation chamber and concludes feeding externally. Some species may be egg-larval parasitoids (M.R. Shaw & Huddleston, 1991). Genera present in Costa Rica include *Baeacis* and several species of *Diospilus*.

HELCONINAE: Helconini^{EX}. Members of this tribe are commonly associated with wood-boring or bark-mining beetles, especially Cerambycidae, but also Buprestidae, Curculionidae, and Melandryidae

(Shenefelt, 1970*a*; Watanabe, 1972; Marsh *in* Krombein *et al.*, 1979; Achterberg, 1987*a*). Little is known of their biology.

HOMOLOBINAE.

A small subfamily represented in Costa Rica by two genera, *Homolobus* and *Exasticolus*. Homolobines are solitary koinobiont endoparasitoids of macrolepidopteran larvae, especially Noctuidae and Geometridae. *Homolobus* completes its feeding as an ectoparasitoid (Allen, 1982), after the host has formed its pupation chamber. Homolobine species were revised by Achterberg (1979, 1984*b*).

HORMIINAE.

A small sufamily comprising two tribes, the Hormiini and Lysitermini, both of which are represented in the Costa Rican fauna. As far as is known, all are facultative gregarious idiobiont ectoparasitoids, usually of concealed microlepidopterous larvae (Beeson & Chatterjee, 1935; Nixon, 1940). Hormiini are recorded from Gelechiidae, Pyralidae and Tortricidae (Marsh *in* Krombein *et al.*, 1979). The Old World lysitermine genus *Cedria* is unusual in being the only known braconid where maternal brood care occurs; after ovipositing 12 to 26 eggs on a pyralid larva, the female remains with the host caterpillar, tenaciously guarding her brood against chalcidoid hyperparasitoids (Beeson & Chatterjee, 1935). Genera of Hormiinae present in Costa Rica include *Allobracon*, *Hormius* and *Lysitermes*.

ICHNEUTINAE

The Ichneutinae is a small, mostly holarctic subfamily, comprising two tribes, the Ichneutini and the Muesebeckiini. Both are represented in Costa Rica.

ICHNEUTINAE: Ichneutini. Species of this tribe are solitary koinobiont endoparasitoids, mainly of endophytic or exposed-feeding tenthredinoid sawfly larvae. Many are egg-larval parasitoids (Nageli, 1936), but some that attack exposed hosts only oviposit into early instar host larvae. The host is finally killed after it has constructed a cocoon (Zinnert, 1969). Ichneutines are most diversified in cool temperate areas of the Holarctic region, but are comparatively rare in the tropics. In Costa Rica there are two species of *Ichneutidea*: one collected in Pacific dry forest, on Cerro el Hacha in Guanacaste National Park, and the other collected in humid lowland Atlantic forest near Bribri in the southeast.

ICHNEUTINAE: Muesebeckiini. Species of this tribe are endoparasitoids of leaf-mining lepidopteran larvae, and are moderately diversified in the Neotropics (Mason, 1969). Some Muesebeckiini are moderately common in Santa Rosa National Park (Whitfield, pers. comm.) and at least four species of *Oligoneurus* have been collected in a variety of sites from 400 metres up to 3000 metres on the Cerro de la Muerte. Species of *Muesebeckia* and *Lispixys* occur in Mexico, Ecuador and Peru so may eventually be found in Costa Rica.

MACROCENTRINAE

A cosmopolitan subfamily, with seven genera worldwide, and represented in Costa Rica by several species of the genus *Macrocentrus*. Adult macrocentrines are pale yellowish brown and are nocturnally active, as in ophionine ichneumonids (Huddleston & Gauld, 1988). Macrocentrines are solitary or gregarious koinobiont endoparasitoids mostly of endophytic lepidopteran larvae, especially Tortricidae and Noctuidae (Shenefelt, 1969). The macrocentrine larva completes its growth by feeding externally on the host (Daniel, 1932; Dittrick & Chiang, 1982). The parasitoid larva spins its cocoon in the host's pupation site. Some gregarious species are known to be polyembryonic (Parker, 1931*a*; Daniel, 1932), with eight to ten individuals developing from a single egg. Since the individuals emerging from a parasitized host are often of the same sex, polyembryonic development presumably occurs in most of the gregarious species (Muesebeck, 1932*a*). The North American species of *Macrocentrus* were revised by Muesebeck (1932*a*).

MENDESELLINAE

This is the most recently discovered braconid subfamily (Whitfield & Mason, 1994) and it comprises two rarely collected genera, *Mendesella*, which is exclusively South American, and *Epsilogaster*, which occurs from the southern United States south to Brazil and Ecuador. The latter genus is represented in Costa Rica by two species which resemble miracines, except that the spiracle of the first metasomal tergite is located on the margin of the tergite, and not in the lateral membrane, and the fore wing has a large and distinctive, apically ovoid

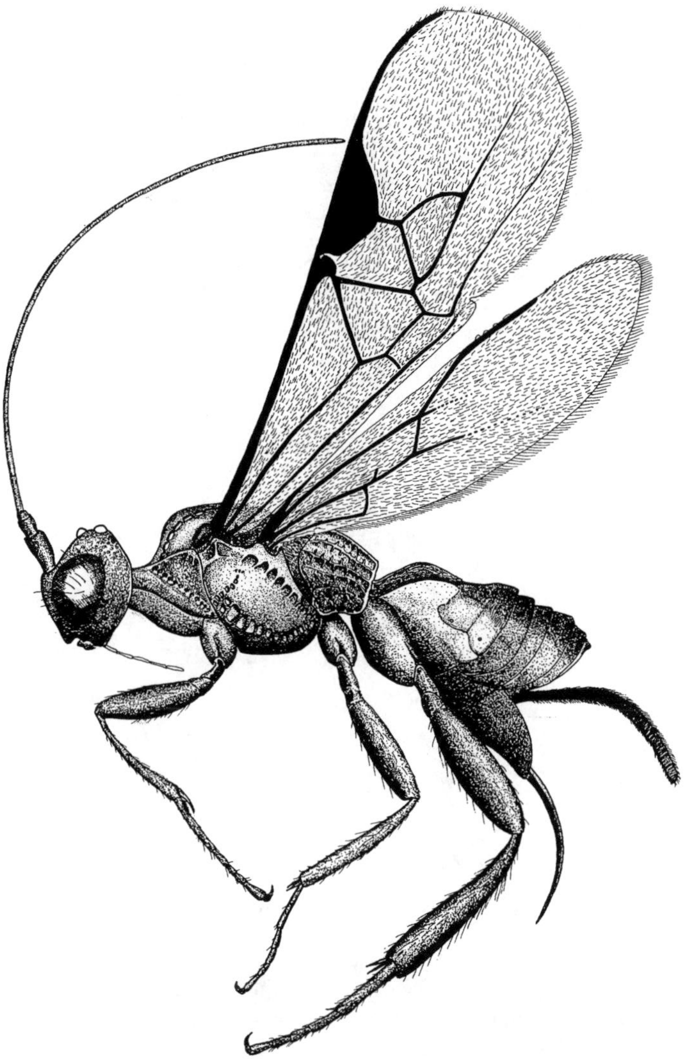

Fig. 12.153. *Epsilogaster* sp. (Mendesellinae).

areolet (second submarginal cell). One of the Costa Rican species, *E. bicolor*, occurs in the Pacific lowlands, whilst the other, *E. tico*, occurs at higher elevations (700-1300 m) on the wetter Atlantic watershed. Details of the biology of mendesellines is not known, but *E. bicolor* has been reared from a momphid (Lepidoptera) stem-miner in *Cephalanthus* (Rubiaceae).

Meteorinae

The Meteorinae (*sensu* S.R. Shaw, 1985) is a subfamily that comprises the cosmopolitan genus *Meteorus* with about 174 species worldwide (S.R. Shaw, 1988*a*). Meteorines are solitary or gregarious koinobiont endoparasitoids of larval Lepidoptera or Coleoptera, and many species are polyphagous (West & Miller, 1989; Maetô, 1990). The vast majority of *Meteorus* species are solitary parasitoids attacking exophytic lepidopteran larvae, and many of these are nocturnally active. Others utilize hosts that are only weakly concealed (e.g. in leaf-rolls or under webbing). The most frequently utilized hosts are Noctuidae, Geometridae and Tortricidae, but many other lepidopterans including Hepialidae,

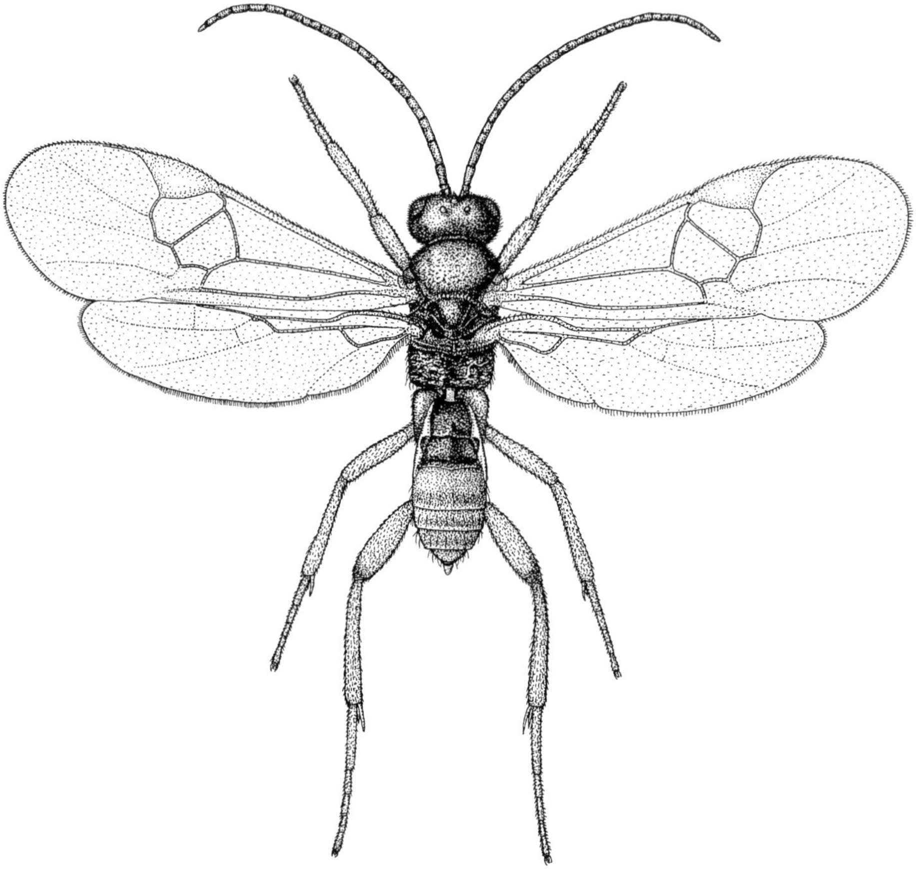

Fig. 12.154. *Cotesia* sp. (Microgastrinae).

Zygaenidae, Tineidae, Gelechiidae, Pyralidae, Lycaenidae, Nymphalidae, Lasiocampidae, Notodontidae and Arctiidae are also parasitized (Muesebeck, 1923; Huddleston, 1980). Many of these hosts are considered pest insects, consequently several *Meteorus* species have been studied for purposes of biological control (Huddleston, 1980, 1983; Fernández & Terán, 1990*a* & *b*). The hosts of those *Meteorus* species that attack beetles are primarily species living in wood, tree bark and bracket fungi, especially Cerambycidae, Tenebrionidae, Scolytidae, Cryptophagidae, Melandryidae and Ciidae (Huddleston, 1983), although one is known to attack a leaf-mining chrysomelid (M.R. Shaw & Huddleston, 1991).

The solitary parasitoids of arboreal lepidopterous larvae are unusual in that typically they emerge from the host larva while it is still feeding, and the host larva sometimes continues feeding after parasitization though death ensues. The *Meteorus* larva drops from the vegetation by a thread, from which it hangs and spins an ovoid cocoon in which it pupates (Madel, 1963). The gregarious *Meteorus* species usually form their cocoons together in a loose, irregular heap (Huddleston, 1983), although the terminal threads may still be present, but intertwined (S.R. Shaw, 1985). A few highly specialized gregarious species in the Old World form a very organized spherical cocoon mass, with all the exit holes facing outwards (Huddleston, 1983). Those species attacking coleopterous larvae form stalkless cocoons within the beetle gallery (DeLeon, 1933; Mason, 1973). As far as is known there are three larval instars. The first is typically caudate and of long duration, especially in species that spin a pendant cocoon. An ectophagous final feeding stage is unknown, and those species that emerge from a feeding host larva may have a non-feeding final instar (M.R. Shaw & Huddleston, 1991).

MICROGASTRINAE

This extremely large, cosmopolitan subfamily is the largest group in the Braconidae and comprises over 2500 described species worldwide (Shenefelt, 1972, 1973*a*). Mason (1981*b*) estimated that the actual world total is between 5000 and 10,000 species. The majority of species were placed in *Apanteles* (*sensu* Muesebeck, 1920; Shenefelt, 1972), but Mason (1981*b*) reclassified many species in a large number of separate genera. A recent re-analysis of Mason's data by Walker *et al.* (1990*b*) suggests that the limits of some groups need critical reassessment, but Mason's generic concepts have been widely adopted.

Microgastrines are common solitary or gregarious koinobiont endoparasitoids of lepidopteran larvae. In Britain, where the fauna is comparatively well known, almost every exophytic lepidopteran species is attacked by one or more microgastrines and many hosts in weak concealment are also parasitized (M.R. Shaw & Huddleston, 1991). Many microgastrines are fairly host specific, but some *Pholetesor* species attack several different families of leaf-mining microlepidopterans on a range of plant species (Whitfield & Wagner, 1988). As many microgastrines are very common parasitoids of lepidopterous larvae, including pest species (e.g. Nixon, 1965, 1968, 1970; Mason, 1981*b*; Austin & Dangerfield, 1989). They are among the most successfully used parasitoid species in biological control programmes. Oviposition is usually into an early instar host larva although a few species are known to oviposit into the host egg (Zorin, 1930; Wilbert, 1960). In the gregarious species (e.g. *Apanteles*, *Cotesia*, *Glyptapanteles*), all of the eggs (up to several hundred) are injected during a single quick insertion of the ovipositor. In a few genera a final ectoparasitic feeding period is known (Vance, 1932*b*). Microgastrine larvae emerge from the host larva and pupate on, or very near the host remains, and in gregarious species the form and placement of the cocoon mass is often diagnostic at the species level (Mason, 1981*b*). For example, *Cotesia americana*, a common Central American parasitoid of sphingid larvae, forms a large cottony mass, up to three inches in length, whilst others pupate in neatly arranged rows resembling a honey-comb. Although most microgastrines emerge and pupate before the host has formed its cocoon, the larva of *Deuterixys* pupates inside the cocoon of its lyonetiid host (Mason, 1981*b*; Whitfield, 1985).

The microgastrine genera occurring in Costa Rica can be identified by using Mason's (1981*b*) keys. Species identifications are extremely difficult since most of the genera have not been revised recently, although a few widespread species can be identified by using Muesebeck's (1920, 1922) keys to North American species. In Costa Rica, as in most areas, microgastrine species normally comprise the most conspicuous element of the braconid fauna. About 200 species belonging to 29 described genera occur in the country and specimens potentially representing three undescribed genera have been seen (Whitfield, pers. comm.). Five additional genera are expected to occur here (Table 12.10).

MIRACINAE

A small subfamily comprising solitary koinobiont endoparasitoids of leaf-mining lepidopteran larvae, especially Nepticulidae (Papp, 1989) and Lyonetiidae (Achterberg, 1976). Pupation occurs within the host cocoon (Gauld & Bolton, 1988). Only three neotropical species have been described (Marsh, 1979), two of which are of some economic importance. *Mirax malcolmi* from Colombia parasitizes *Scrobipalpa* on *Solanum saponaceum* and *M. insularis* is a common parasitoid of *Leucoptera coffeella*, the coffee leaf-miner, in Guadeloupe and the Dominican Republic (Muesebeck, 1937). An unidentified species of *Mirax* has been reared in northwestern Costa Rica from a *Phyllonorycter* sp. (Gracillariidae) mining the leaves of *Lonchocarpus rugosus* (Leguminosae) (Memmott *et al.*, 1994). At least 30 other undescribed *Mirax* species occur in Costa Rica in a wide variety of low to mid altitude sites, sea-level to 1600 metres, though one specimen is known from 2600 metres on the Cerro de la Muerte (G. Wright, pers. comm.).

OPIINAE

A very large, cosmopolitan subfamily comprising more than 1300 described species (Wharton, 1988*a*), the vast majority of which are placed in the genus *Opius*. The subfamilies Opiinae and Alysiinae presumably comprise a monophyletic group within the

Genus	Species in Costa Rica	Host relationships
Alphomelon	3	Hesperiidae
Apanteles	40–50	Microlepidoptera –S, g
Choeras	5–6	Microlepidoptera, esp Pyraloidea –S
Clarkinella	1	unknown
Cotesia	10–15	Macrolepidoptera –s, G
Dasylagon[EX]	–	Sesiidae, Thyrididae –S?
Deuterixys	1	Lyonetiidae –S
Diolcogaster	25–30	Noctuidae, Geometridae –s, g
Distatrix	1	Macrolepidoptera –s, G
Dolichogenidea	8–10	various Lepidoptera –S, g
Exix	2–3	Noctuidae –S
Fornicia	1	Limacodidae
Glyptapanteles	25–30	Macrolepidoptera –s, G
Hypomicrogaster	8–10	Microlepidoptera esp miners –S
Illidops	1	Psychidae, Scythrididae –S
Larissimus[EX]	–	unknown
Lathrapanteles[EX]	–	(see Williams, 1985),
Microplitis	8–10	usually Noctuidae–s, g
Papanteles	2–3	unknown
Parapanteles	2–3	Noctuidae, Notodontidae –G
Pholetesor	4–5	leafmining microlepidoptera –S
Prasmodon	1	unknown
Promicrogaster	8–10	mycetophagous Lepidoptera
Protapanteles	1	usually Geometridae –S, g
Protomicroplitis	1	Noctuidae
Pseudapanteles	5–6	unknown
Rasivalva[EX]	–	Macrolepidoptera –S
Rhygoplitis	3	esp. Pyraloidea –S
Sathon[EX]	–	Noctuidae –G
Sendaphne	2	unknown
Snellenius	1	Sphingidae –G?
Venanides	1	esp Pyraloidea, Tortricoidea –G, s
Veranus	2–3	leafmining microlepidoptera –S
Xanthomicrogaster	3	unknown

Table 12.10. Genera of Microgastrinae known or expected to occur in Costa Rica, together with their recorded hosts, and host relationships (S = solitary; G = gregarious; lower case letter indicates relatively minor frequency of the habit in a genus). The numbers of species known from Costa Rica were determined by J. Whitfield after examination of over 2500 specimens.

Braconidae, based on the apomorphic feature of endoparasitism of cyclorrhaphous Diptera (Wharton, 1988*a*). However, the monophyly of the Opiinae has not yet been firmly established (Quicke & Achterberg, 1990), and it seems likely that the Opiinae, as presently defined, may be a paraphyletic group. Opiines are solitary koinobiont endoparasitoids of the larvae of cyclorrhaphous Diptera (especially Agromyzidae), particularly those that mine in fruit and leaves (Cameron, 1941); many are economically important as parasitoids of fruit flies (Wharton & Marsh, 1978). A few opiines are known to oviposit into the host egg (Prokopy & Webster, 1978) but in general oviposition is into the host larva, sometimes while it is young, but frequently often only late in its development. Some species lay female eggs preferentially in larger hosts encountered (Avilla & Albajes, 1984), a habit commonly seen in idiobionts but much less frequently observed in koinobionts. This, together with the obligatory pre-ovipositional period manifested by at least some species (Willard, 1920) and the temporary paralysis induced in the host at the time of oviposition (Glas & Vet, 1983), suggest that opiines may be quite closely related to idiobiont groups (M.R. Shaw & Huddleston, 1991). The egg is deposited just below the host's cuticle, sometimes attached to it by one end (Willard, 1920). The mandibulate first instar larva persists until about the time the host pupates, when the accompanying hormonal changes trigger the parasitoid's moult (Lawrence, 1986*a* & *b*). Two further instars feed on fluids within the host, while the fourth and final instar consumes the entire contents of the host puparium. Pupation is internal, within the cleaned-out host puparium.

The New World opiine species have been extensively studied by Fischer (1964, 1965, 1967*a* & *b*, 1973, 1977). However, the tribes, subtribes and several genera were recently redefined and reclassified by Wharton (1987, 1988*a*). Genera present in Costa Rica include: *Biosteres, Diachasmimorpha, Doryctobracon, Fopius, Opius* and *Utetes* (*Bracanastrepha* is provisionally treated as a subgenus of this genus).

ORGILINAE

The Orgilinae is a small subfamily with about 200 described species, most of which are slender and medium-sized (4.0 to 5.0 mm long). The subfamily is difficult to define, but all known species have a synapomorphic subapical dorsal notch on the ovipositor (Sharkey, 1986*a*). It is divided into two tribes: Microtypini and Orgilini. Species of Microtypini (e.g. *Stantonia*) are sometimes classified as a separate subfamily (Quicke & Achterberg, 1990). Since these authors were unable to establish the phylogenetic position of the Microtypini, and they did not establish the monophyly of the Orgilinae in the absence of the microtypine lineage, a more conservative and traditional classification is followed here (e.g. Muesebeck, 1970; Marsh *in* Krombein *et al.*, 1979).

Orgilinae are koinobiont endoparasitoids of caterpillars feeding in concealed sites, especially Pyralidae, Tortricidae, Coleophoridae, Gelechiidae, Oecophoridae and Gracillariidae (Muesebeck, 1938, 1970; Achterberg, 1987*b*). The presence of frass may be an important cue for the parasitoid as individuals have been observed probing frass-laden tunnels with their ovipositors (M.R. Shaw & Huddleston, 1991). Oviposition is into the larval host's haemocoel and the caudate first instar larva gradually develops an anal vesicle. There are apparently three larval instars and the final instar larva emerges from the host to conclude feeding. The host is killed as a prepupa and the orgiline cocoon is spun within the host's cocoon (Broodryk, 1969*b*; Oatman *et al.*, 1969).

In Costa Rica most of the commonly collected species belong to the widespread genus *Orgilus* (Muesebeck, 1970), but a species of *Stantonia* (Muesebeck, 1938) has been collected on the Atlantic lowands and the Osa Peninsula. A species of *Orgilus* has been reared in Santa Rosa National Park from the larvae of leaf-mining lepidopterans of the families Gracillariidae and Scythrididae in Acanthaceae and Malvaceae (Memmott *et al.*, 1994).

PAMBOLINAE

A small subfamily that has most often been included (often as a tribe) within the Hormiinae (Hedqvist, 1963) or Rogadinae (Marsh, 1979; M.R. Shaw & Huddleston, 1991). Recent studies justify classifying pambolines as a separate subfamily (Quicke, Ficken & Fitton, 1992; Quicke, Fitton & Ingram, 1992; Quicke *et al.*, 1992). Pambolines are thought to be facultative gregarious idiobiont ectoparasitoids of Lepidoptera or Coleoptera larvae in seeds, wood or stored products (Whitfield, 1989; M.R. Shaw & Huddleston, 1991). Although relatively rare in temperate braconid faunas (Marsh, 1979; M.R. Shaw & Huddleston, 1991), pambolines are moderately well represented in Costa Rica where there are at least ten species of *Pambolus* occurring throughout the country at elevations ranging from sea-level to 2600 metres. Most *Pambolus* species have very distinctive spines at the lateral corners of the propodeum, although these are reduced to small tubercles in some minute species. Recently a single species of *Acrisis* was collected in Costa Rica at Las Alturas Biological Station (1500 m), close to the Panamanian border. This is the most southern record for the genus (see Muesebeck, 1935*b*).

ROGADINAE

A very large, cosmopolitan subfamily, represented in Costa Rica by two tribes, the Rhysipolini and Rogadini, of which the latter is vastly more species-rich.

The Rhysipolini (*sensu* Whitfield, 1989, 1992) is rather poorly represented in Costa Rica, but *Rhysipolis* is known to occur in Santa Rosa National Park (300 m) (Whitfield, pers. comm.) and *Cantharoctonus* has been collected above San Antonio de Escazú (1300 m). Rhysipolines are solitary ectoparasitoids of concealed microlepidoptera and *Rhysipolis* species usually parasitize gracillariid leaf-miners (Whitfield & Achterberg, 1987).

The Rogadini are koinobiont endoparasitoids, mostly of exposed-feeding macrolepidopteran larvae (M.R. Shaw, 1983). However, species of the *Stiropius* group (*Stiropius, Choreborogas, Polystenidea* and *Viridipyge*) parasitize leaf-mining lyonetiid and gracillariid larvae (Whitfield, 1988*a*). For example, a species of *Stiropius* has been reared from the coffee leaf-miner (*Leucoptera coffeella*, Lyonetiidae) in Nicaragua (P. Hanson, pers. comm.). Most rogadines are solitary parasitoids, although a few gregarious species are known. Pupation is internal, within the shrunken and mummified remains of the host caterpillar. Species of *Stiropius, Choreborogas* and *Viridipyge* are numerous in Costa Rica, but may be easily overlooked because of their minute size and nocturnal activity (Whitfield, 1988*a*, 1990*a*).

Adults of *Aleiodes, Cystomastax* and *Rogas* are nocturnally active and may be commonly collected at lights. Species of *Aleiodes* are common in Costa Rica but they are relatively less diverse than in the Holarctic region. Species of this genus parasitize a wide array of macrolepidopteran hosts with the Noctuoidea, Geometroidea and Sphingoidea being most commonly utilized (Muesebeck, 1960; M.R. Shaw, 1983). The host mummy is commonly glued down by the parasitoid larva through a hole chewed in the bottom of the host's prothorax. Consequently the host mummies are commonly found stuck to a leaf or twig. Emerging *Aleiodes* adults normally cut a neat, circular emergence hole at the posterior end of the mummified caterpillar. The larger species of *Cystomastax* and *Rogas* are particularly diverse in tropical America, and form a very conspicuous element of the Costa Rican braconid fauna. As far as is known, most *Rogas* species parasitize Limacodidae, Zygaenidae, Lycaenidae and Riodinidae. The host mummy is normally not glued down by the parasitoid larva, and the emergence hole is ragged and irregular (O.J. Smith *et al.*, 1955—as *Pelecystoma*). Other genera of Rogadinae present in Costa Rica include: *Clinocentrus, Polystenidea* and *Yelicones*.

RHYSSALINAE

A very small subfamily comprising two tribes, the Rhyssalini (*sensu* Whitfield, 1988*b*) and Hydrangeocolini (*sensu* Whitfield, 1989). The Rhyssalinae have traditionally been included in the Rogadinae, but recent phylogenetic studies (Whitfield, 1989, 1992; Quicke & Achterberg, 1990; Quicke, Ficken & Fitton, 1992; Quicke *et al.*, 1992; Wharton *et al.*, 1992; Wharton, 1993*a*) demonstrate that the rhyssalines should be regarded as a separate, relatively archaic subfamily. The placement of the Hydrangeocolini is a more difficult problem. Although the Hydrangeocolini have traditionally been included in the Opiinae, Hormiinae or Rogadinae, recent phylogenetic studies demonstrate that they should not be included in any of these groups (Wharton, 1988*a*, 1993*a*; Whitfield, 1989, 1992). The position of the spiracles of metasomal tergites II+III laterally on the laterotergites may be a synapomorphy for Rhyssalini + Hydrangeocolini, but there are two interpretations of the phylogenetic significance of this character (Quicke & Achterberg, 1990; Whitfield, 1992). Since this character is equivocal a conservative approach is followed here and the Hydrangeocolini are provisionally assigned to the Rhyssalinae.

RHYSSALINAE: Hydrangeocolini. This tribe is represented in Costa Rica by at least three species of *Hydrangeocola* occurring at elevations from 400 metres on the Atlantic slopes, to 3000 metres on the Cerro de la Muerte. Although they are rare, the elongate pterostigma and greatly enlarged second submarginal cell make them quite easy to recognize. The biology of the Hydrangeocolini is not known in detail, but they are thought to be ectoparasitoids of gall-forming Lepidoptera (Whitfield, 1989).

RHYSSALINAE: Rhyssalini. Facultatively gregarious ectoparasitoids, usually of concealed microlepidoptera or Coleoptera larvae (Muesebeck, 1935*a*; Whitfield, 1989). One undescribed genus is known from Santa Rosa National Park (Whitfield, pers. comm.).

SIGALPHINAE

A small cosmopolitan subfamily comprising about ten described species. Some species are known to be solitary or gregarious koinobiont endoparasitoids of the larvae of Noctuidae (Shaw & Huddleston, 1991). An undescribed species of *Sigalphus* has been reared by D.H. Janzen and W. Hallwachs in Santa Rosa National Park, in northwestern Costa Rica (M. Sharkey, pers.

comm.). It develops as a solitary endoparasitoid of two species of agaristine noctuids, destroying the final instar host larva after that larva has constructed a pupal chamber within rotten wood (D.H. Janzen, pers. comm.).

YPSISTOCERINAE[EX]

A very small and rarely encountered neotropical subfamily that have only been collected in termite nests (Brues, 1923; Cushman, 1923). Quicke, Ficken & Fitton, (1992); Quicke *et al.* 1992) show that ypsistocerines share two synapomorphies with the Doryctinae and treat them as a tribe in that subfamily.

Comprising two genera, *Termitobracon* and *Ypsistocerus*, the Ypsistocerinae is morphologically very distinctive, having the antennae inserted high on the head, unidentate mandibles, reduced palpi, small to rudimentary compound eyes, and densely setose body and wings. In *Termitobracon*, males are always brachypterous, whilst females occur as fully winged and brachypterous morphs (Brues, 1923). Reduction of the eyes and brachyptery are presumably adaptations to subterranean existence. Behavioural observations suggest that ypsistocerines are true termitophiles since the host termites display no hostility towards them (Brues, 1923). The immature stages are unknown.

Termitobracon has been collected from Panama, and since the presumed host termites, *Nasutitermes*, occur as far north as Mexico (Emerson, 1955) it is possible that ypsistocerines will eventually be discovered in Costa Rica.

13 The chrysidoid families

INTRODUCTION

Ian D. Gauld and Paul E. Hanson

The superfamily Chrysidoidea is a monophyletic clade that is the sister-group to the other two aculeate superfamilies (Brothers, 1975). It comprises seven extant monophyletic families, the Bethylidae, Chrysididae, Dryinidae, Embolemidae, Plumariidae, Sclerogibbidae and Scolebythidae. The most recent and comprehensive investigation of the phylogeny of the group is the work of Carpenter (1986*a*) who postulated the phylogenetic relationships shown in Fig. 13.01.

All chrysidoids develop either as parasitoids of other insects, or as cleptoparasites in the nests of aculeate Hymenoptera. Rasnitsyn (1980) suggested that host-searching within a friable substrate is the primitive condition of the aculeate lineage, and indeed it is widespread in diverse groups (Carpenter, 1986*a*) including the Chrysidoidea, where the ancestral condition could have been searching for beetle larvae in soil or rotten wood. Little is known about the biology of most primitive chrysidoids. Plumariid females have been found in the soil (Evans, 1966*f*) and scolebythids have been found in association with wood boring beetles (Brothers, 1981), though the exact nature of this association is not known. More is known of the biology of the more derived groups. Bethylids are ectoparasitic idiobionts, and develop by feeding through an integumentary lesion of their concealed host. Presumably this is a plesiomorphic feature of the superfamily (Fig. 13.02). The Bethylidae are primarily parasitoids of cryptic coleopterous larvae, although one lineage exploits microlepidopterous larvae, especially those found in similar cryptic situations to beetles. One species also attacks hymenopterans nesting in wood. Like some bethylids, the most primitive chrysidids, the Cleptinae, search for hosts in soil or leaf litter, but unlike other members of the superfamily, they do not subdue active hosts; instead they attack relatively immobile prepupal sawflies enclosed in cocoons. The higher chrysidids continue to exploit such "packaged" hosts but have radiated in two ways; amesigines and loboscelidiines attack large insect eggs, while most chrysidines attack prepupae of solitary nest-building hymenopterans, such as eumenine vespids, sphecids and bees (Kimsey & Bohart, 1990). Some members of the most plesiomorphic tribe in the Chrysidinae, the Elampini, attack hymenopterous nests in the ground, but other chrysidines attack aculeates nesting in stems and cavities. Several members of the most derived lineage, the Chrysidini, are cleptoparasites of aculeates that construct mud nests.

Koinobiosis seems to be a feature of the sclerogibbid + embolemid + dryinid lineage (Fig. 13.02). Sclerogibbids utilize embiid hosts and develop, at least ini-

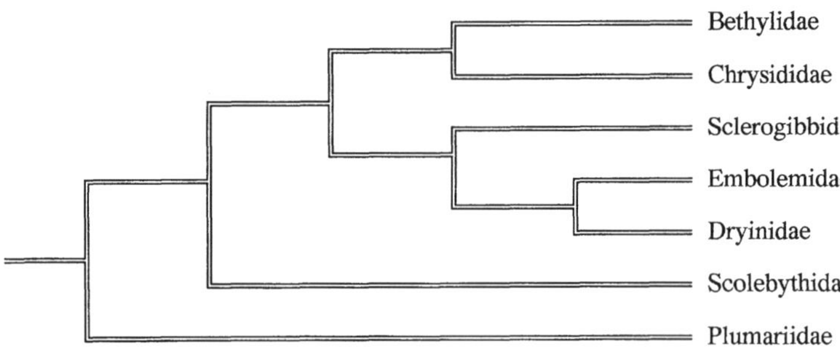

Fig. 13.01. Postulated phylogenetic relationships of the Chrysidoidea. For details of characters supporting this cladogram see Carpenter (1986).

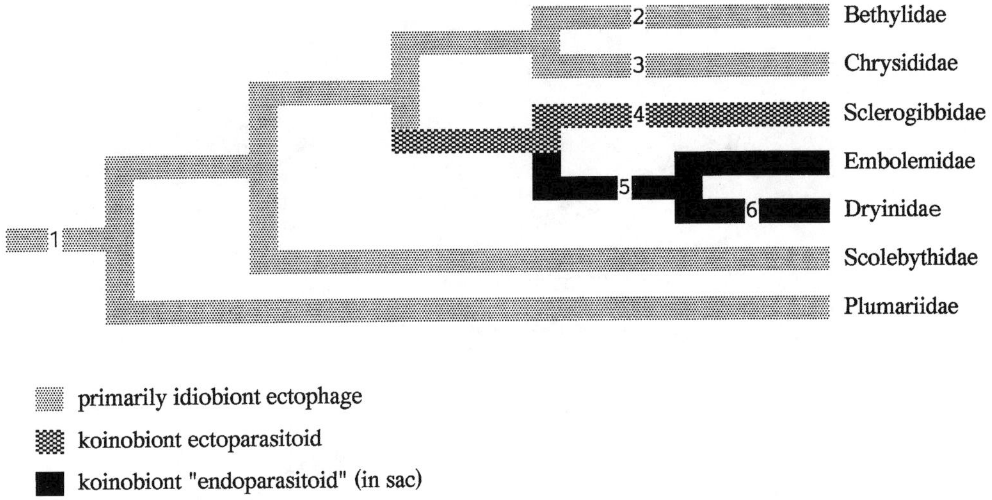

Fig. 13.02. Suggested biological evolution within the Chrysidoidea. **1**: exploiting beetle hosts in soil, leaf litter and rotten wood; **2**: primarily exploiting beetles in the ancestral habitat, but with some groups attacking lepidopterous larvae in similar situations; **3**: primitively attacking 'encased' hosts in ancestral situation; **4**: attacking embiids in ancestral situation; **5**: attacking Auchenorrhyncha in ancestral situation; **6**: attacking Auchenorrhyncha in exposed situation. As with many other groups of parasitoids the evolution of koinobiosis has apparently facilitated the exploitation of free-living hosts.

tially, as ectoparasitic koinobionts. Embolemids and dryinids attack Auchenorrhyncha (Homoptera), but develop in an essentially similar manner as koinobionts within protruding sacs on their hosts. While embolemids retain the plesiomorphic habit of attacking concealed hosts associated with rotten wood (Bridwell, 1958), dryinids have radiated to exploit primarily free-living fulgoroids and cicadellids (Olmi, 1984).

Six of the seven chrysidoid families are present in Central America. Only the Plumariidae, a small group confined to the southern part of the southern hemisphere, is absent from Costa Rica. The six Central American families can be separated by the following key.

Key to families of Chrysidoidea occurring in Central America

1 Antenna with 20 to 24 segments; coxae with a dorsal furrow bounded on both sides by raised keels (Fig. 13.15).
[Head in profile oval, with antenna inserted under an overhang (Fig. 13.14); fore femur strongly swollen.] ... **Sclerogibbidae** (p. 488)

— Antenna with 13 or fewer segments; coxae without a furrow bounded by keels. 2

2 Antenna with 12 or 13 segments. 3

— Antenna with 10 segments. 5

3 Propleuron extended into a neck-like prolongation, exposed dorsally (Fig. 13.16); fore wing with six enclosed cells, including the marginal and 1st submarginal (Fig. 13.03); fore coxa with point of insertion of trochanter close to articulation, the coxa extended posteriorly (Fig. 13.11). ....**Scolebythidae** (p. 468)

— Propleuron not extended into a neck, dorsally concealed by pronotum; fore wing with five or fewer enclosed cells (Figs 13.04–13.05), never with marginal and 1st submarginal cells enclosed. 4

4 Metasoma with five or fewer visible tergites; head more or less hypognathous, almost oval in profile (Fig. 13.06), or flattened dorsally, if slightly elongate then never extended far above and behind eyes, the eyes always quite large (Fig. 13.08); fully-winged species, with

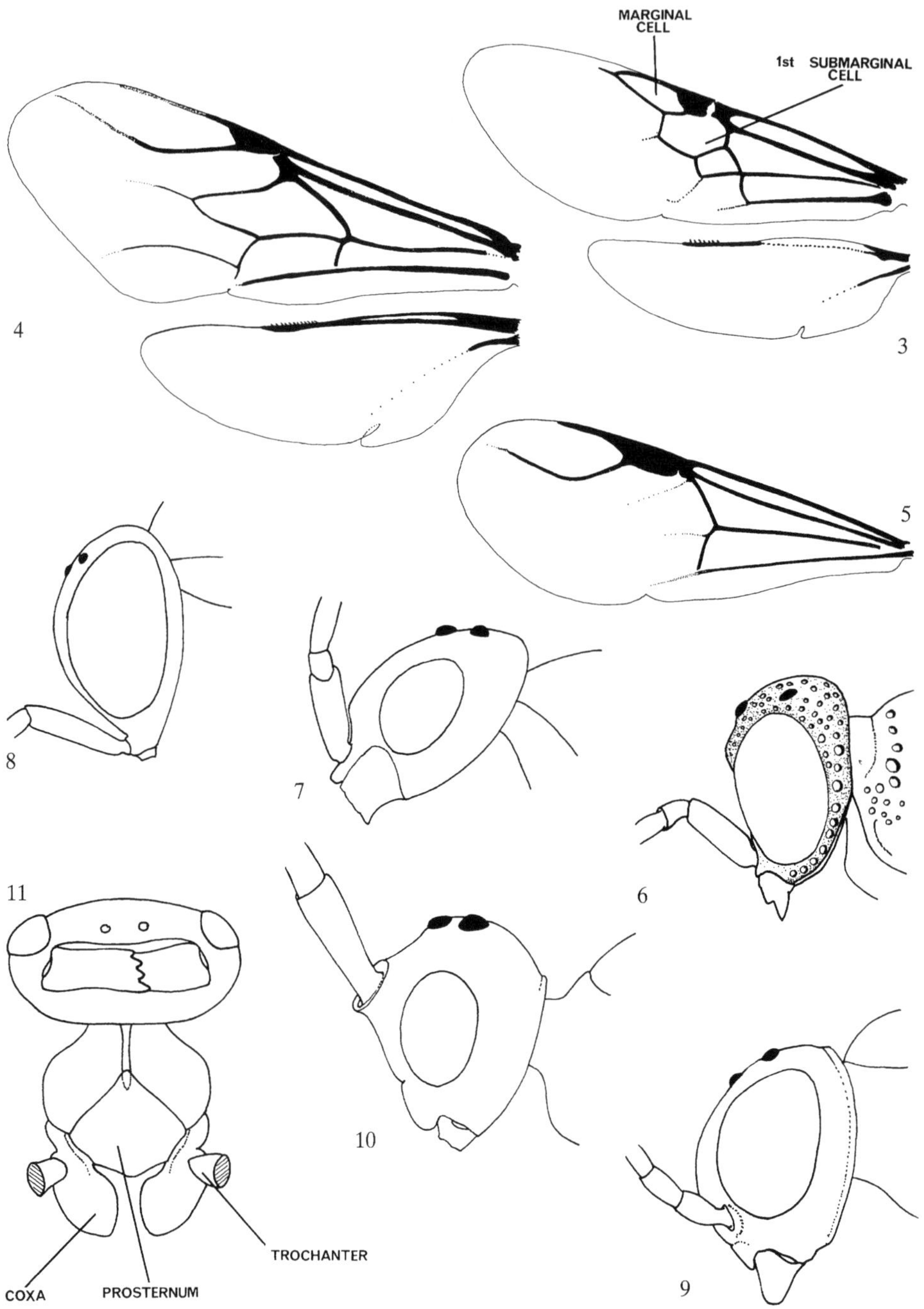

Figs 13.03–13.11. Chrysidoidea. Figs 13.03–13.04. Fore and hind wings; 13.03, *Clystopsenella* sp. (Scolebythidae); 13.04, *Chrysis* sp. (Chrysididae). Fig. 13.05. Fore wing, *Thaumatodryinus* sp. (Dryinidae). Figs 13.06–13.10. Heads, lateral; 13.06, *Chrysis* sp. (Chrysididae); 13.07, *Pristocera* sp. (Bethylidae); 13.08, *Amisega* sp. (Chrysididae); 13.09, *Aphelopus* sp. (Dryinidae); 13.10, *Embolemus* sp. (Embolemidae). Fig. 13.11. Prothorax, ventral, *Clystopsenella* sp. (Scolebythidae).

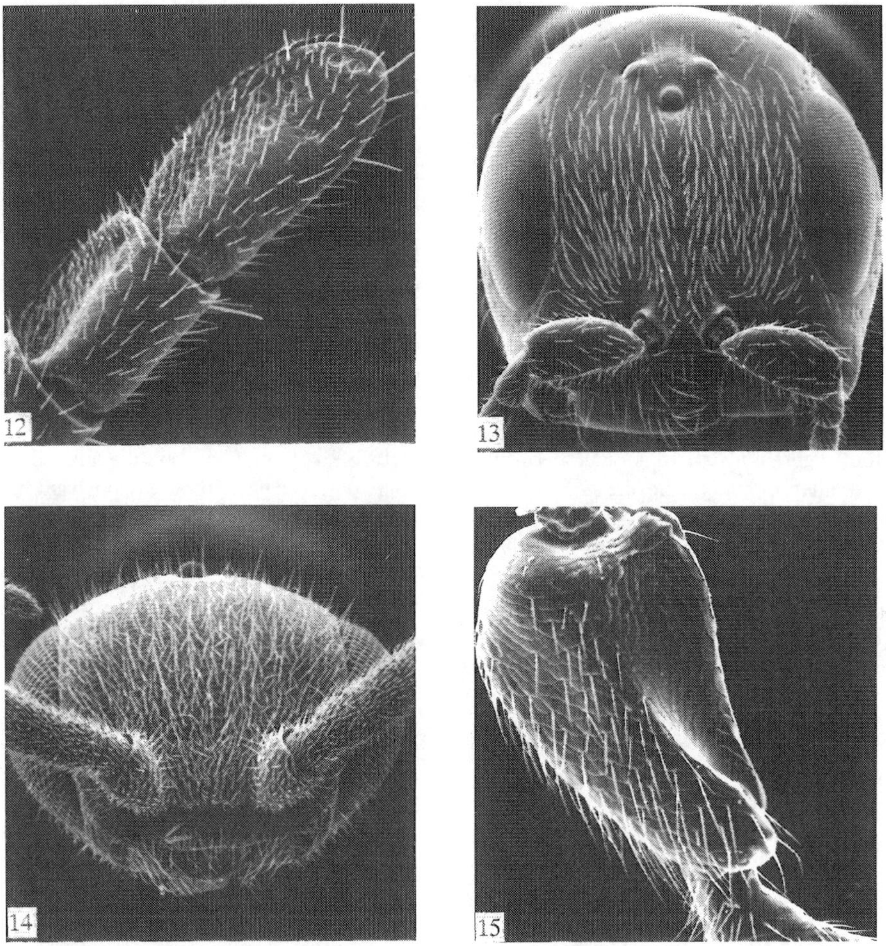

Figs 13.12–13.15. Chrysidoidea; scanning electron photomicrographs. Figs 13.12–13.13. *Clystopsenella* sp. (Scolebythidae); 13.12, apex of female antenna; 13.13, head. Figs 13.14–13.15. *Probethylus* sp. (Sclerogibbidae); 13.14, head, female; 13.15, hind coxa, male.

hind corner of pronotum generally not more or less touching the tegula. .... **Chrysididae** (p. 479)

— Metasoma with six or seven visible tergites; head large and in profile somewhat elongated, often somewhat "egg-shaped", extended above and behind the rather small eyes, and sometimes prognathous (Fig. 13.07); apterous, micropterous or macropterous species, if fully-winged then with hind corner of pronotum more or less contiguous with tegula. .... **Bethylidae** (p. 470)

5 Head with antennal insertion close to clypeus, below the imaginary centre line of the head, not arising from an anteriorly directed prominence (Fig. 13.09); fore leg often with tarsus modified into a grasping chela. .... **Dryinidae** (p. 493)

— Head with antennal insertion far from clypeus, above imaginary centre line of the head, arising from an anteriorly directed prominence (Fig. 13.10); fore leg never with tarsus chelate. .... **Embolemidae** (p. 490)

13.1 SCOLEBYTHIDAE

Ian D. Gauld

Diagnosis. Body length 6.0 to 10.0 mm; predominantly shiny black; males and females fully winged (fore wing length 3.0 to 5.0 mm). *Antenna short, 13-segmented in both sexes, inserted very low down on head,* just above mouth, thus obscuring clypeus and lower face; apex of antenna of female slightly flattened with large subcircular sensilla (Fig. 13.12). Head quite large, hemispherical in shape (Fig. 13.13), with mouth usually directed downwards, occasionally with head tilted slightly forwards; eyes moderately large, oval; *mouth very wide*; mandibles with four teeth, the ventral one the largest and the remainder becoming progressively smaller dorsally, the dorsal one very small; maxillary palp 6-segmented, labial palp 4-segmented. *Propleuron very long, projecting far in front of pronotum; prosternum large and rhombic, slightly rounded behind* (Fig. 13.11); pronotum reaching back to tegula, its hind margin transverse; mesoscutum with parapsidal furrows discernible; mesopleuron without a transverse sulcus; mesosternal region not projecting backwards over bases of hind coxae. *Fore wing with six enclosed cells—the costal, basal, subbasal, discal, submarginal and marginal cells*; the wing distad of these cells without venation (Fig. 13.03); pterostigma distinct; hind wing without enclosed cells, with a sharply differentiated claval lobe. Fore leg with *coxa flattened, extended posteriorly, with insertion of trochanter close to articulation*; fore femur swollen; tibial spur formula 1: 2: 2, mid and hind legs with outer tibial spur short, slender and bristle-like; tarsal claws simple. Metasoma somewhat spindle-shaped, tapered to a point posteriorly; without a constriction between first and second segments; *metasomal segments III-VI bearing fields of long conspicuous setae.*

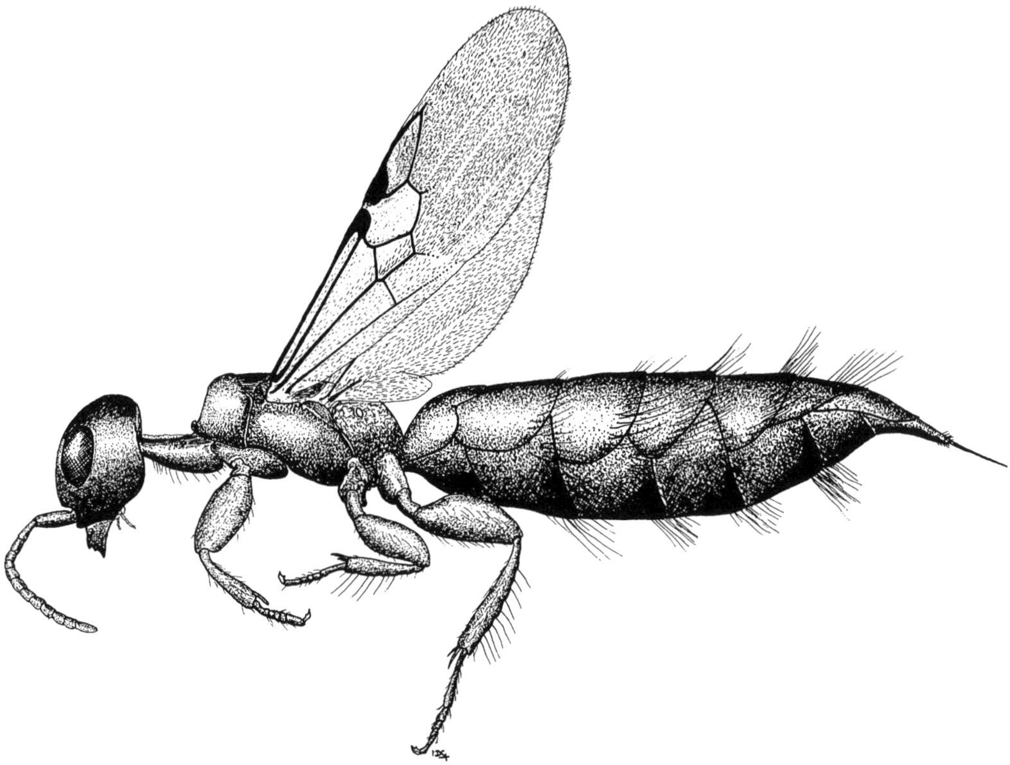

Fig. 13.16. *Clystopsenella* sp. (Scolebythidae).

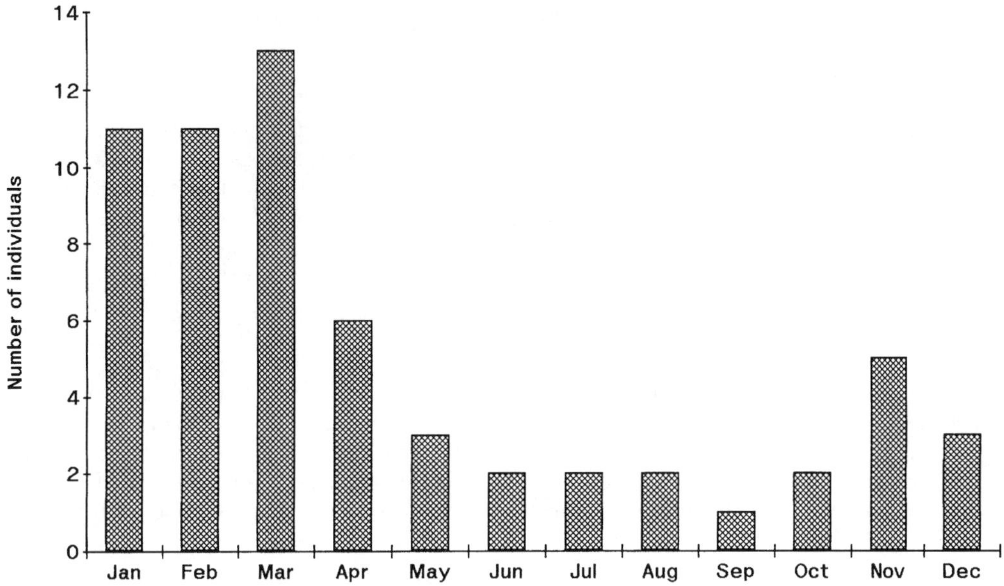

Fig. 13.17. Monthly abundance of *Clystopsenella* in Malaise traps in forest in Santa Rosa National Park, averaged from data for 1985–7.

Classification and distribution. The Scolebythidae is a small family comprising less than ten species, in three genera. *Clystopsenella* occurs in the Neotropics and Australia, *Scolebythus* is restricted to Madagascar, and *Ycaploca* occurs in South Africa and Australia (Naumann, 1991). Only a single species has been described in each genus, but a few undescribed species of *Clystopsenella* are present in museum collections.

Biology. Although published information suggests that scolebythids possibly develop as gregarious ectoparasitoids of the larvae of wood-boring beetles (Evans, 1963*b*; Nagy, 1975; Day, 1977; Evans *et al.*, 1979), all evidence is circumstantial. Nagy's (1975) record of the African scolebythid, *Ycaploca evansi*, attacking the larvae of *Hylotrupes bajulus* (Cerambycidae) is questionable since no detailed biological information is preserved with the specimens (see Day, 1977; Brothers, 1981). Another series of *Y. evansi* was collected from a burrow in wood along with the remains of a large cerambycid larva, and in this instance the beetle larva would have been sufficiently large to permit the development of several scolebythids (Brothers, 1981). However, no cocoons were found. Specimens of the Madagascan species, *Scolebythus madecassus*, were collected as a group (five females, one male) within a hollow stick (Evans *et al.*, 1979), but again, there is no definitive host association. Nonetheless, all these pieces of circumstantial evidence, coupled with the fact that scolebythids are morphologically modified apparently for life in tunnels, strongly suggest a wood-boring host.

Although most species of Scolebythidae are known from both sexes, in six years of Malaise trapping in Costa Rica all specimens collected (more than 100) were female. This is unusual because males of small aculeates are generally more likely to be taken in traps than females. Perhaps this species is thelytokous, or alternatively, perhaps males do not leave the tunnels in which they presumably develop.

Synopsis of the Costa Rican fauna

Scolebythidae is represented in Costa Rica by one, or possibly two species of *Clystopsenella*. The uncertainty about the number of species present stems from the fact that one individual (collected in Santa Rosa

National Park) has a rather different pattern of setae on the metasoma from all the other specimens, and may be either a second species or a deformed specimen of the first. The commoner Costa Rican species is undescribed and differs in several features from the only described member of the genus, the Brazilian species, *C. longiventris*.

All specimens found in Costa Rica were collected in Malaise traps operated at low altitudes, many from rather disturbed habitats or seasonally dry forests, but a few from undisturbed wet forests (La Selva Biological Station). In Santa Rosa National Park individuals of *Clystopsenella* have been taken in all months of the year, and almost equal numbers have been collected in three forests of different ages. However, data from continuous comparable sampling showed a clear seasonal peak in abundance in the early to mid dry season, between January and March (Fig. 13.17).

13.2 BETHYLIDAE

Albert T. Finnamore and Ian D. Gauld

Diagnosis. Body length 1.0 to 15.0 mm; usually dark brown to black, sometimes faintly metallic, apterous females yellowish; males usually fully winged, females fully winged to apterous. *Antenna inserted close to clypeus, 12- or 13-segmented,* with same number of segments in both sexes. *Head usually elongate and depressed,* prognathous in females. Pronotum, in macropterous species, with upper hind corner more or less reaching tegula, always with anterior flange well developed, concealing propleura in dorsal view; prosternum usually small, more or less transverse, often concealed in ventral view. *Fore wing with reduced venation, with four or fewer enclosed cells discernible* (up to six in some extralimital taxa); marginal cell, if discernible, not closed distally; one or no discal cells

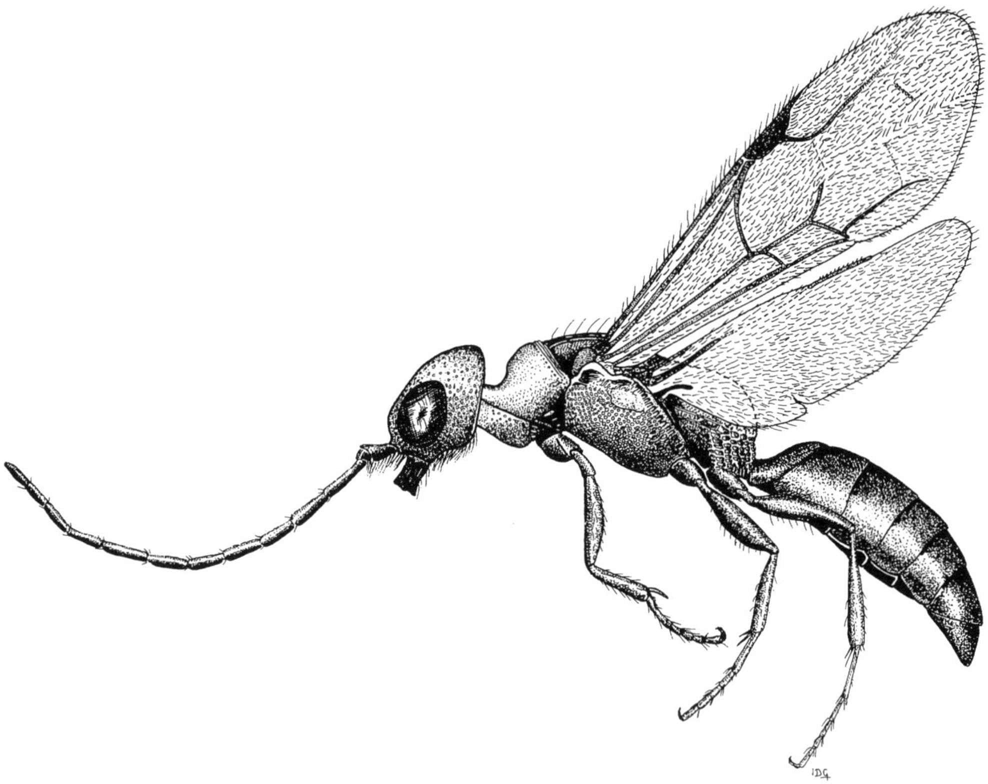

Fig. 13.18. *Pristocera* sp. male (Bethylidae).

present; hind wing without enclosed cells; claval lobe present, jugal lobe absent. *Metasoma with six or seven exposed terga.*

Classification and distribution. The family Bethylidae is cosmopolitan in distribution, but the greatest species-richness occurs in tropical regions. It is probably the largest family in the Chrysidoidea and, although there are about 2200 described species worldwide, this figure may represent as little as 30 percent of the true size of the group. The family comprises four subfamilies: Bethylinae, Epyrinae, Mesitiinae, and Pristocerinae. The Mesitiinae is confined to the Old World whilst the other three are cosmopolitan in distribution and are represented in Costa Rica.

Biology. Comparatively little is known about the biology and host relationships of the vast majority of tropical bethylids, but all species that have been investigated have been found to be idiobiont or incipient koinobiont ectoparasitoids of the larvae of Coleoptera (Pristocerinae and many Epyrinae), microlepidoptera (most Bethylinae and some Epyrinae) and Sphecidae (one species of Bethylinae). In Costa Rica one bethylid is known to attack adult scolytids (Kirkendahl, pers. comm.). Bethylids usually attack species living in cryptic situations such as in the soil, leaf litter, rotten wood, seeds and so forth. Several neotropical species (e.g. of *Sclerodermus*) have been collected in association with ants, and in the laboratory they can develop on ant larvae (Evans, 1964*a*), but whether such insects are used as hosts in natural situations has yet to be shown. Hawkins and Gordh (1986) provide a bibliography of the world literature on Bethylidae.

Little is known about how females locate hosts and what determines their host range. A few species are capable (at least under laboratory conditions) of developing on a range of hosts of different orders (see Evans, 1964*a*). Genera more usually seem to be associated with a few families of one order. The Pristocerinae and most Epyrinae attack a range of beetle families, including Scolytidae, Cerambycidae, Ciidae, Buprestidae, Anobiidae, Dermestidae, Tenebrionidae, Elateridae, Bruchidae, Cucujidae, Ptinidae, Bostrichidae, Lyctidae, Cleridae and some Curculionidae. Species of Bethylinae attack a range of microlepidopterous groups, including species of the superfamilies Tineoidea, Gelechioidea, Tortricoidea and Pyraloidea.

Many bethylids search for hosts in concealed situations: some species of *Pristocera* search for elaterid larvae in the soil, other species in this genus search for xylophagous weevils in wood borings, and many Bethylinae search for microlepidopteran larvae in seed pods or leaf rolls (Baker, 1976; Gordh, 1976*a*; Gordh & Hawkins, 1981). Adult female bethylids display a number of adaptations for seeking hosts in cryptic habitats. These include one or more of the following: small size, flattened body, wedge-shaped prognathous head, powerful enlarged fore femora, and aptery.

Laelius species generally encounter and attack dermestid larvae in concealment, but sometimes they will attack and conceal dermestid larvae found in the open (Evans, 1964*a*). In some other genera moving the host to a place of concealment seems to be a common behaviour. For example, *Parascleroderma* drags its clerid host into a crack after stinging it (Maneval, 1930) and species of *Cephalonomia* also conceal their parasitized beetle hosts (Powell, 1938; Finlayson, 1950). In *Epyris* the female stings a tenebrionid larva that is more than twice her size, then drags it over her back to a suitable nesting site. The host is temporarily concealed between lumps of soil whilst the wasp excavates a simple burrow; the host is then placed in the burrow, and an egg is laid on it (Williams, 1919*a*).

In some cases the host larva is antennated prior to attack and may be rejected as unsuitable at this stage (Abraham *et al.*, 1990). A rather typical form of attack was described by Gordh (1976*a*) who observed that a *Goniozus* species embeds her mandibles in the host's cuticle immediately behind the head capsule. She then thrusts her metasoma forward, stinging the caterpillar in the ventral nerve cord. Gordh noted that females failing to attack in this manner and hold the host with their mandibles were often decapitated by the host. However, in other species the female bethylid stings the host without holding it (Baker, 1976). Most accounts of attacks note that the bethylid attempts to sting the host in the gular region between the head capsule and the thorax.

Many bethylids attack and subdue hosts considerably larger than themselves and so are equipped with a very powerful sting. Their sting is sometimes painful to humans, and can cause swelling and dermatological reactions (Dauphin & Castro, 1991; see also review by Guiglia, 1958), although these tiny wasps are often not noticed as the cause of these symptoms. Bethylid venom sometimes paralyses the host almost immediately (Peter & David, 1991*a*), or requires an hour or so to take effect (Remadevi *et al.*, 1978), during which time the parasitoid host-feeds, grooms herself, or even moves away, sometimes for several days, and subsequently returns. Generally a single female attacks and subdues a host larva, but *Sclerodermus* females sometimes co-operate in subduing a large wood-boring beetle larva (Bridwell, 1920; Wheeler, 1928). The effect of the venom varies from species to species, but may induce permanent or temporary paralysis. Some envenomated final instar larvae pupate or spin a cocoon (Bridwell, 1920; and see Gordh, 1976*a*).

Oviposition sites vary from group to group. *Goniozus* and *Prosierola* species generally deposit the egg(s) dorsally, whilst species of *Cephalonomia* deposit an egg dorsally or in the intersegmental folds ventrally. *Laelius* species remove the specialized sticky setae from their dermestid hosts, disposing of them by grooming, and then bite off the short setae on the venter of the host abdomen before laying an egg (Yamada, 1955; Mertins, 1980). Many authors have observed that eggs are most commonly deposited on the central body segments. The number of eggs deposited per host varies greatly. Some *Pristocera* and *Epyris* species only deposit a single egg, whereas others in these genera lay several. Most species of *Goniozus* and *Prosierola* lay several eggs per host, and they can estimate the host size and deposit a predictable number of eggs on a given host (Kishitani, 1961; Gordh, 1976*a*). Gregarious species generally produce strongly female biased broods, and Griffiths and Godfray (1988) demonstrated a significant negative relationship between clutch size and (male: female) sex ratio; larger brooded gregarious species tend to produce a more female biased brood than smaller brooded species. In some bethylid species the female remains with the host, guarding her brood until they pupate (Baker, 1976; Remadevi *et al.*, 1981; Hardy & Blackburn, 1991).

The actual number of larval instars has not been determined accurately but most species probably have three to five (Gordh, 1976*a*). The first instar larva is sacciform and develops externally on the host, feeding via an integumentary lesion. Larval development is often quite rapid. In the last instar the anterior third of the body is sometimes inserted through a hole in the host's body so that the parasitoid's body is perpendicular to that of the host. The last larval instar is somewhat hymenopteriform but has a prognathous head, with the labium tending to project well beyond the other mouthparts and with strongly tapered denticulate mandibles (Evans, 1987*a*). The fully-fed larva generally spins a whitish, silken cocoon in close proximity to the dead host. Sometimes *Laelius* larvae pupate within the hollowed out dermestid skin (Spencer, 1942). Gregarious species tend to spin cocoons close together.

Most species are believed to be bisexual and reproduce arrhenotokously in the normal hymenopteran manner, although there are a few reports of thelytoky (e.g. Busck, 1917). Little is known about mate-finding and mating in solitary species, but males of some Pristocerinae apparently carry females in flight during copulation (Evans, 1963*a*). In gregarious species sib-mating is common with males sometimes biting open and entering the cocoon of the female and copulating with her prior to her emergence (Gordh *et al.*, 1983; Gordh & Medved, 1986). In other cases females bite a hole in their own cocoon and wait within for a male (Abraham *et al.*, 1990), or mate after emergence (Peter & David, 1991*b*). Courtship, when observed was very perfunctory, and copulation brief. In *Laelius pedatus* the male begins on top of the female, but after the genitalia are engaged, he flips over onto his back behind the female and intromission occurs in this position (Mertins, 1980). In some cases males and females copulate repeatedly (Yamada, 1955), but in others females were receptive to males only once (Gordh *et al.*, 1983). However, females are often very long-lived and apparently once the sperm supply has been depleted a female will mate again, sometimes with her son (Gordh, 1976*a*). Females generally require a pre-ovipositional period lasting up to thirteen days (Peter & David, 1991*a*; Abraham *et al.*, 1990), during which time they sometimes attack and feed from potential hosts. Both males and females have been observed feeding on honeydew secretions (Evans, 1964*a*).

Economic importance. Bethylids are known to attack beetle larvae in stems, seeds and stored produce, as well as a range of microlepidopterous pests. In the Central American region species of *Goniozus* attack the pink bollworm (*Pectinophora* *gossypiella*, Gelechiidae) on cotton (Cock, 1985), *Oiketikus kirbyi* (Psychidae) on banana (Stephens, 1962), and *Calibitia picata* (Pyralidae) on avocado (specimens in Zamarano). Not surprisingly, several bethylids have been investigated for biological control purposes (Clausen, 1978; Cock, 1985).

Probably the most economically important bethylids in tropical America are two exotic species that have been introduced to control the coffee berry borer (*Hypothenemus hampei*, Scolytidae). This beetle was accidentally introduced from Africa into the New World, where it is now one of the most serious pests of coffee (Quezada & Urbina, 1987). Attempts to control it through classical biological control have centred around the use of two bethylid species from Africa, *Cephalonomia stephanoderis* and *Prorops nasuta*. As a result there is an enormous amount of literature treating the biology and utilization of these two bethylid species (e.g. see Klein Koch *et al.*, 1988; Murphy & Moore, 1990).

Identification. Evans (1957*b*, 1961*c*, 1963*a* & *c*, 1964*a*, 1978) provides keys to genera and many Central American species. The key to the genera of Bethylidae occurring in Costa Rica given below is modified from Evans (1964*a*, 1978). A few additional genera have been included that almost certainly occur in, but have not yet been recognized from Costa Rica.

Key to genera of Bethylidae of Costa Rica

1 Fore wing with vein *Rs&M* simple, straight or evenly arched, not angled or giving rise to any vein or stub (Figs 13.19–13.22), sometimes without veins *Rs&M* (Fig. 13.23); frons usually without a median longitudinal carina or polished streak extending from clypeus; macropterous, brachypterous or apterous. 2

— Fore wing with veins *Rs&M* obtusely angled or giving rise to a vein (*Rs+M*) or vein stub (Fig. 13.24), sometimes with an enclosed discal cell (Fig. 13.25); frons with a median longitudinal carina or polished streak extending a short distance from clypeus (Fig. 13.33); macropterous or brachypterous, never completely apterous ....(Bethylinae)............20

2 Male: macropterous; scutellum not in contact with propodeum, metanotum well developed with an emargination or fovea opposite posterior apex of scutellum (Figs 13.26, 13.27). Female: apterous, without tegula or ocelli; eye small or absent, eye height up to 0.25 of head width (Figs 13.34, 13.35); propodeum often constricted anteriorly ....(Pristocerinae)......3

— Both sexes macropterous, brachypterous or apterous, if apterous then usually with tegula and ocelli; scutellum in contact with propodeum or nearly so, metanotum occasionally narrowly transverse but without fovea opposite posterior apex of scutellum (Figs 13.28–13.32); female with eye height more than 0.25 of head width (Figs 13.36, 13.37); female propodeum not constricted anteriorly. ....(Epyrinae)...................8

3 Male: macropterous. 4

— Female: apterous. 6

4 Metasomal tergite II usually with one or two pairs of tubercles, pits, depressions or pale spots, sometimes unmodified; clypeus expanded beneath antennal socket, without indentation. ....*Dissomphalus*

— Metasomal tergite II without modifications; clypeus indented beneath antennal socket, sometimes only slightly, most often abruptly and deeply. 5

5 Median lobe of clypeus strongly projecting, truncate or triangular, often with a small median tooth; lateral indentation deep, often abrupt, forming a right angle or defining basal angles of a triangular median lobe; eyes hairy; pronotal disc rounded anteriorly, without transverse carina (Fig. 13.26) ....*Pseudisobrachium*

— Median lobe of clypeus variable, if similar to above then lateral indentation slight, a shallow notch; eyes usually glabrous; pronotum often with an anterior transverse carina (Fig. 13.27). ....*Pristocera*

6 Propodeum, in dorsal view, more or less parallel-sided, without a strong constriction anteriorly, only slightly narrowed at spiracles

(Fig. 13.34); maxillary palpi short, at most 2-segmented. ....*Dissomphalus*

— Propodeum, in dorsal view, strongly constricted anteriorly or at spiracles (Fig. 13.35); maxillary palpi with 3 to 5 segments. 7

7 Propodeum constricted at extreme anterior end; eye absent or consisting of a single facet. ....*Pseudisobrachium*

— Propodeum constricted at spiracles (Fig. 13.35); eye present, consisting of more than one facet. .... *Pristocera*

8 Antenna with 10 flagellar segments; if fully winged then fore wing with subbasal and often basal cells open, at most enclosed by weak veins (Fig. 13.23)..................................... ....(Cephalonomiini)......................9

— Antenna with 11 flagellar segments (first flagellar segment short in some species); fore wing with basal and usually subbasal cells enclosed (Figs 13.19–13.21). 10

9 Apterous, micropterous, ***or if*** fully winged, ***then*** fore wing with vein *Rs* completely absent. ....*Cephalonomia*[EX]

— Fully winged, fore wing with vein *Rs* present, at least in part (Fig. 13.23). .... *Plastanoxus*

10 Clypeus short, truncate, broadly rounded or emarginate, barely projecting beyond antennal sockets. ....(Sclerodermini)................11

— Clypeus with a projecting median lobe which is angular or narrowly rounded. ....(Epyrini)...12

11 Propodeal disc margined posteriorly by a transverse carina (Fig. 13.36); occipital carina present. ....*Glenosema*

— Propodeal disc not margined posteriorly by a transverse carina (Fig. 13.37); occipital carina absent. ....*Sclerodermus*[EX]

12 Scutellum with an anterior uninterrupted, transverse groove, simple or with posterior deflections laterally or broadened to form lateral pits, but still connected by a deep groove. (Figs 13.28–13.30). 13

— Scutellum with a pair of anterolateral pits that are usually separated, at least by a thin, shallow, septum (Figs 13.31, 13.32); occasionally pits are connected by a thin superficial line. 17

13 Fully winged, fore wing with vein *Rs* short, subequal to length of *Rs&M* (Fig. 13.21). ....*Laelius*[EX]

— If fully winged, then fore wing with *Rs* very much longer than *Rs&M* (Figs 13.19, 13.20), rarely brachypterous or apterous. 14

14 Clypeus with three prominent lobes, median lobe slightly exceeding rounded lateral lobes; mesoscutum usually without notauli or with grooves weakly developed. ....*Holepyris*

— Clypeus with only the median lobe developed, lateral lobes absent or undeveloped; mesoscutum usually with notauli developed (Figs 13.29, 13.30). 15

15 Pronotal disc rounded anteriorly and laterally, without carinae (Fig. 13.29); male antenna with first flagellar segment distinct from second and slightly shorter than pedicel. ....*Rhabdepyris*

— Pronotal disc carinate at least anteriorly and often laterally (Fig. 13.30); male antenna with first flagellar segment often not distinct from second and usually much shorter than pedicel. 16

16 Eyes strongly hairy; antenna of male simple. ....*Anisepyris*

— Eyes glabrous; antenna of male (the only sex known) strongly pectinate (Fig. 13.38). ....*Procalyoza*[EX]

17 Pronotum with its posterior part elevated and prolonged backwards so as to overlie the anterior end of the mesoscutum (Fig. 13.32); mesopleuron with subalar fovea large, deep; female unknown. .... *Aspidepyris*

— Pronotum with posterior margin simple, not overlaying mesoscutum (Fig. 13.31) mesopleuron with subalar fovea normal, not unusually deep. 18

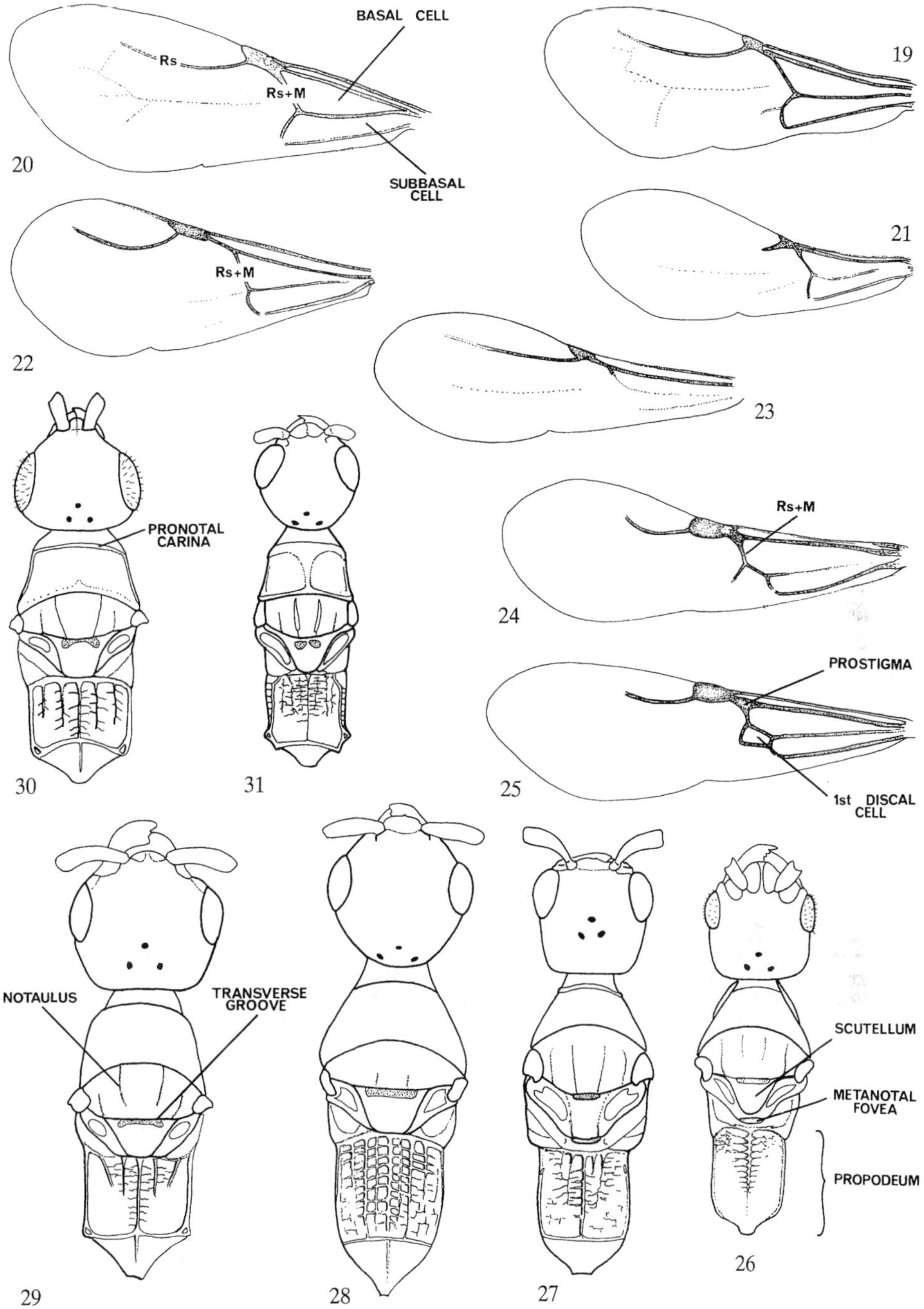

Figs 13.19–13.31. Bethylidae. Figs 13.19–13.25. Fore wings; 13.19, *Anisepyris* sp.; 13.20, *Rhabdepyris* sp.; 13.21, *Laelius* sp.; 13.22, *Dissomphalus* sp.; 13.23, *Plastanoxus* sp.; 13.24, *Goniozus* sp.; 13.25, *Prosierola* sp. Figs 13.26–13.31. Head and mesosoma of males, dorsal; 13.26, *Pseudisobrachium* sp.; 13.27, *Pristocera* sp.; 13.28, *Laelius* sp.; 13.29, *Rhabdepyris* sp.; 13.30, *Anisepyris* sp.; 13.31, *Bakeriella* sp.

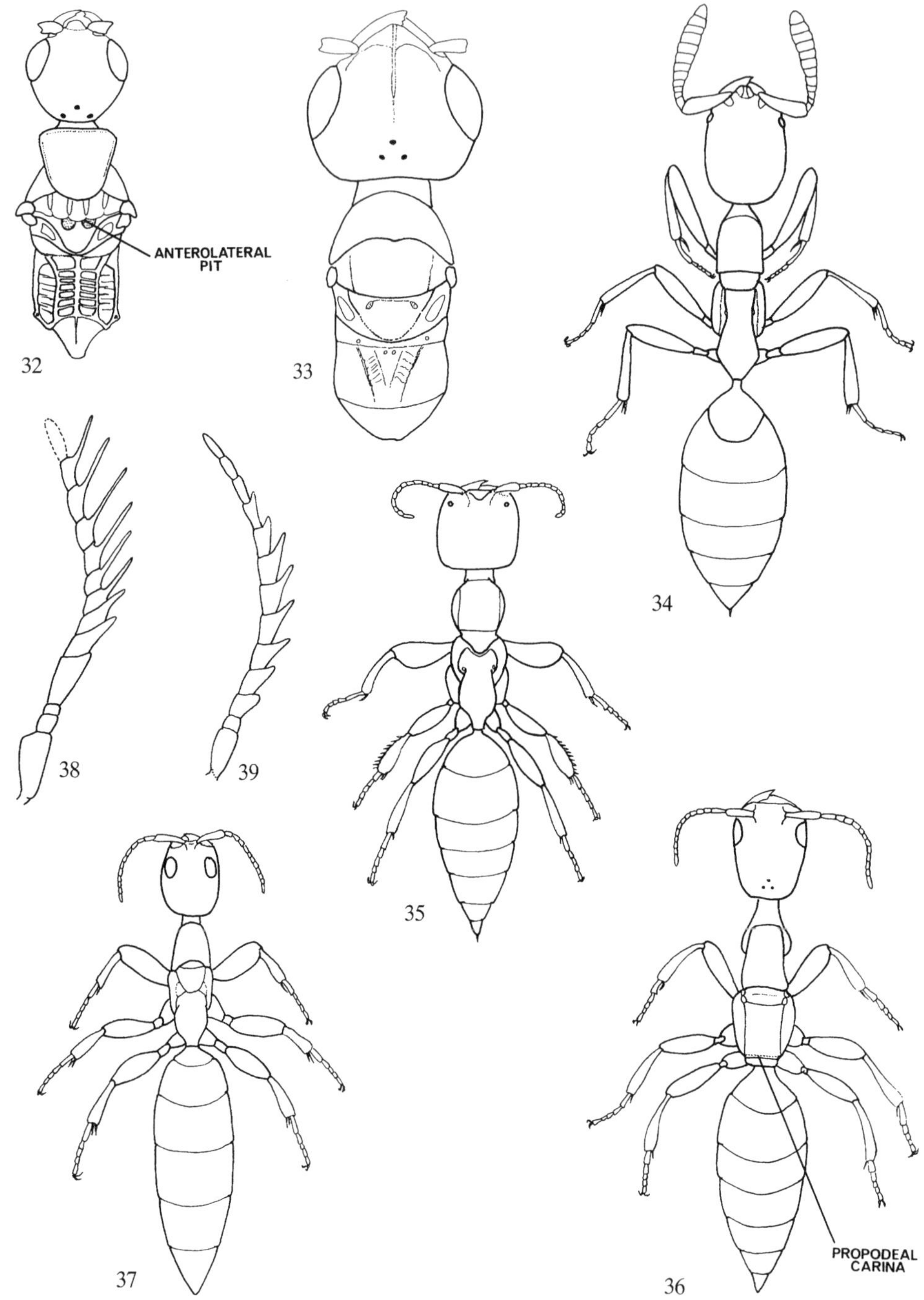

Figs 13.32–13.39. Bethylidae. Figs 13.32–13.33. Head and mesosoma of males, dorsal; 13.32, *Aspidepyris* sp.; 13.33, *Prosierola* sp. Figs 13.34–13.37. Females; 13.34, *Dissomphalus* sp.; 13.35, *Pristocera* sp.; 13.36, *Glenosema* sp.; 13.37, *Sclerodermus* sp. Figs 13.38–13.39. Antenna of male; 13.38, *Procalyoza westwoodi*; 13.39, *Calyozina* sp.

18 Pronotal disc carinate anteriorly, often laterally and medially as well (Fig. 13.31). ....*Bakeriella*
— Pronotal disc rounded, without carinate margins. 19

19 Antenna simple; scutellar disc with rounded edges. .... *Epyris*
— Antenna pectinate (Fig. 13.39); scutellar disc flat, sharp-edged; female unknown. ....*Calyozina*

20 Propodeum without well-developed lateral carinae; scutellum with scuto-scutellar groove deeply impressed, with small deep grooves extending backwards posteriorly; fore wing with, or more usually without (Fig. 13.24) an enclosed 1st discal cell.........................*Goniozus*
— Propodeum with well-developed lateral carinae; scutellum with scuto-scutellar groove broad and shallow, with large oval or round foveae at either end; fore wing with an enclosed 1st discal cell (Fig. 13.25)........................*Prosierola*

Synopsis of the Costa Rican fauna

The Costa Rican bethylid fauna is very species-rich. I have seen over 140 species (based on examination of 1700 specimens) which is almost as many species as are currently known for all of America north of Mexico. About 60 percent of the Costa Rican species are pristocerines, but this is possibly an artifact due to bias in the collecting technique (Malaise traps). Many more species of Epyrinae and Bethylinae are likely to be collected in Costa Rica.

PRISTOCERINAE

There are over 25 nominal genera in this subfamily, four of which, *Dissomphalus*, *Parascleroderma*, *Pristocera* and *Pseudisobrachium*, are present in the New World. The subfamily is represented in Costa Rica by three cosmopolitan genera. The fourth genus with American representatives, *Parascleroderma*, is only known in the New World from species in the southern United States and northern Mexico (Evans, 1964*a*).

Dissomphalus. Large cosmopolitan genus with at least 30 species in Costa Rica. Biology: little-known, but they possibly attack beetle larvae. Evans (1964*a*) notes that some neotropical species are found in association with army ants, possibly as parasitoids of myrmecophilous beetles.

Pristocera (includes *Apenesia* **syn. n.**). A large cosmopolitan genus with at least 25 species in Costa Rica. Biology: nearctic and Asian species attack the larvae of Elateridae (Yasumatsu, 1955) and some Old World species have been reared from curculionid larvae in roots and stems (Evans, 1964*a*; Baker, 1976).

Pseudisobrachium. A large genus recorded from all regions of the world except Australia. At least 33 species occur in Costa Rica. Biology: many species have been found in ants' nests, where they possibly attack myrmecophilous beetles (Evans, 1964*a*).

EPYRINAE

This subfamily has been divided into three tribes, Epyrini, Sclerodermini and Cephalonomiini. These are discussed separately below.

EPYRINAE: Epyrini. This tribe includes nine genera in the Western Hemisphere, several of which are cosmopolitan. Seven of these are known to occur in Costa Rica and two others are expected to occur here.

Anisepyris. A widespread American genus, centred in the tropics, that probably includes over 100 species. Eight species have been found in Costa Rica. Biology unknown.

Aspidepyris. A small genus that is only recorded from Mesoamerica. Three species have been taken in Costa Rica. Biology unknown.

Bakeriella. A moderate sized genus confined to the New World, with most species restricted to tropical areas. Three species have been found so far in Costa Rica. Biology unknown.

Calyozina. A small genus occurring in tropical America, Botswana and Taiwan (Krombein, 1992*b*). A single species has been found in Costa Rica. Biology unknown.

Epyris. A very large cosmopolitan genus with at least ten species present in Costa Rica. Biology: *Epyris* species attack tenebrionid larvae and transport them to a suitable site where a shallow nest burrow is excavated to receive them (Williams, 1919*a*; Evans, 1964*a*). In temperate regions several species overwinter gregariously under bark (Krombein *in* Krombein *et al.*, 1979).

Holepyris. A large cosmopolitan genus with at least six species in Costa Rica. Biology: *Holepyris* species can often be found under bark, where they might be attacking larvae of subcorticolous beetles. A few species of *Holepyris* attack larvae of Tenebrionidae and possibly microlepidopterous pests of stored food, and have been transported throughout the world by human activity.

Laelius^{EX}. This is a moderately large widespread genus. Biology: several species are encountered in human domiciles, where they attack dermestid larvae (Back, 1940; Krombein *in* Krombein *et al.*, 1979).

Procalyoza^{EX}. This genus includes a single species, *P. westwoodi*, known only from the type specimen collected in the mid nineteenth century in Panama. It is easily distinguished by its pectinate antennae. Biology unknown.

Rhabdepyris. A large cosmopolitan genus represented in Costa Rica by at least ten species. Biology unknown.

EPYRINAE: Cephalonomiini. This tribe is represented in the New World by two genera, *Cephalonomia* and *Plastanoxus*. Species of a third genus, *Prorops*, have been introduced from Africa into Brazil in an attempt to control the coffee berry-borer. Only *Plastanoxus* has been found in Costa Rica, but several cosmopolitan species of *Cephalonomia* commonly occur in warehouses and similar situations and are likely to be present.

Cephalonomia^{EX}. A relatively small genus. Biology: parasitoids of a variety of beetles, in bark, fungi and in stored products (Evans, 1964*a*).

Plastanoxus. A small genus with a single species collected in Costa Rica. Biology: widespread tramp species, such as *P. westwoodi*, are parasitoids of beetles in stored products.

EPYRINAE:. Sclerodermini. This tribe is represented in the New World by five genera, *Nesepyris*, *Chilepyris*, *Lepidosternopsis*, *Sclerodermus* and *Glenosema*. Only the last mentioned genus has been recorded from Costa Rica, but species of *Sclerodermus* occur widely throughout the Caribbean and South America (Evans, 1964*a*; Krombein *in* Krombein *et al.*, 1979).

Glenosema. A small holarctic genus with a single species found in Costa Rica. Biology unknown.

Sclerodermus^{EX}. A moderately large cosmopolitan genus. Biology: several species attack wood-boring beetle larvae, including species found in human habitations. Some South American species have been found in association with ants' nests and in Hawaii one species is known to attack tineoid moth larvae boring in wood (Evans, 1964*a*).

BETHYLINAE

This subfamily was recently cladistically analysed by Polaszek and Krombein (1994) who recognized seven genera. Four of these occur in the Western Hemisphere. *Lytopsenella* (which has six enclosed cells in the fore wing) is restricted to Chile, while *Bethylus* (which has 12-segmented antennae) is apparently strictly holarctic (Evans, 1964*a*). The other two genera, both of which have 13 segmented antennae, are represented in Costa Rica.

Goniozus. An extremely large cosmopolitan genus with at least ten species in Costa Rica. In many earlier works (e.g. Evans, 1964a) *Parasierola* was recognized as distinct, but it is now generally treated as a synonym of *Goniozus*. Biology: *Goniozus* species are gregarious parasitoids of a range of microlepidopteran larvae, although one Brazilian species is known to attack *Microstigmus* nests (Sphecidae) in abandoned beetle borings (Melo & Evans, 1993). *Goniozus* species are amongst the most widely studied of all bethylids, as several species attack commercially important lepidopteran larvae, including such notorious pests as the Pink Bollworm in cotton (Busck, 1917)

Prosierola. A large neotropical genus with representatives in the extreme southern parts of the United States. A single species has been collected so far in Costa Rica. Biology: *Prosierola* species have been recorded attacking microlepidoptera (Evans, 1964*a*).

13.3 CHRYSIDIDAE

Lynn S. Kimsey

Diagnosis. Body 2.0 to 15.0 mm in length. *Species of the most common subfamily are usually heavily sclerotized and metallic blue or green in colour*. Sexual dimorphism minimal, both sexes having 13-segmented antennae; in some genera the sexes cannot be distinguished without dissecting the genitalia. Most species with tooth-like posterolateral angles on the propodeum. *Wing venation reduced, with 5 or fewer closed cells in the fore wing* (at most marginal, basal, subbasal, 1st discal and 1st subdiscal), and one or no closed cells in the hind wing. *Metasoma with reduction in the number of visible segments, 3 or 5 in males, 3 or 4 in females*, depending on the subfamily; the remaining segments are telescoped within the abdomen, forming an ovipositor tube in the female or a genital tube in the male; the sting is reduced and functions more as an egg guide than as a defensive structure, a feature unique amongst the aculeate families.

Classification and distribution. Worldwide the Chrysididae comprises about 3000 described species in 84 genera. These are currently classified in four subfamilies, the Cleptinae, Amiseginae, Loboscelidiinae and Chrysidinae. The last of these taxa, which includes about 80 percent of the species in the family, is subdivided into four tribes, Elampini, Allocoeliini, Parnopini and Chrysidini. The hypothesized phylogenetic relationships of these taxa is shown in Fig. 13.41.

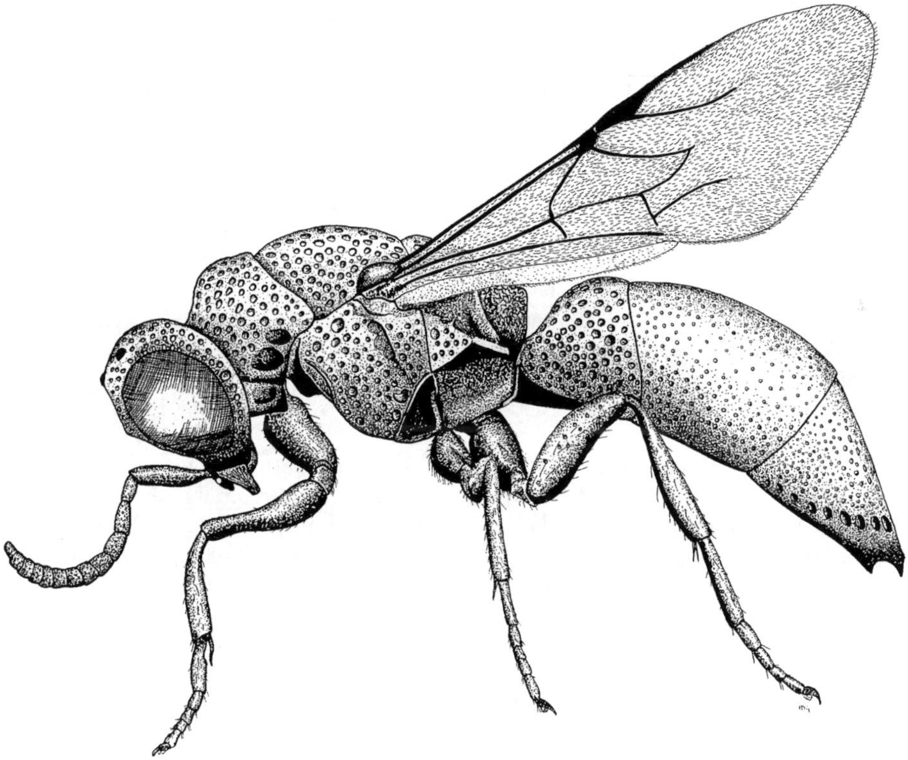

Fig. 13.40. *Pleurochrysis* sp. (Chrysididae).

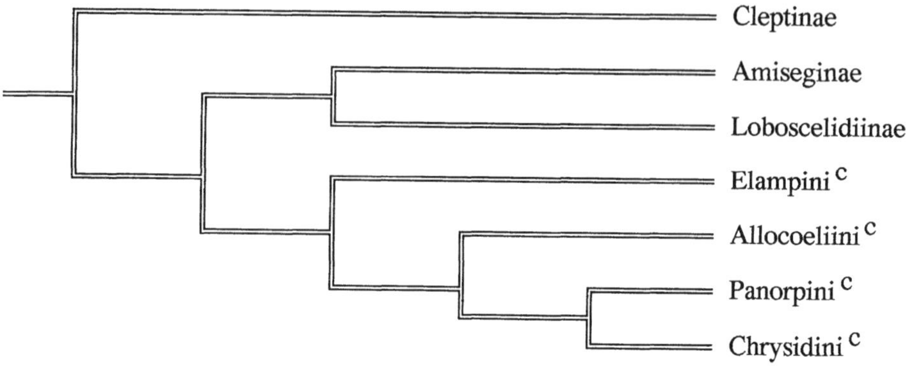

Fig. 13.41. Postulated phylogenetic relationships of the Chrysididae. Taxa labelled c are included in the Chrysidinae. For details of characters supporting this cladogram see Kimsey and Bohart (1991).

The family has a cosmopolitan distribution, but it is relatively depauperate in Australia. In the Neotropical region there are three subfamilies—Cleptinae, Amiseginae and Chrysidinae—and 20 genera; 16 of these genera are represented in Costa Rica and neighbouring countries by a total of at least 70 species. The most primitive subfamily, the Cleptinae, occurs in the Holarctic and Neotropical regions, and consists of only two genera. The Amiseginae occurs primarily in tropical areas of the world and contains about 30 genera, while the Loboscelidiinae is restricted to the tropics of Asia and Australia and includes only two genera. The Chrysidinae, which includes 50 genera, is divided into four tribes, two of which, the Elampini and Chrysidini, occur in the Neotropics. The other two tribes are rather small and restricted to the Old World; the Allocoeliini includes only one African genus, and Parnopini includes three genera that occur in Africa and the Holarctic region (Kimsey & Bohart, 1991).

Biology. Chrysidid wasps include ectoparasitoids of sawfly prepupae, endoparasitoids of phasmid eggs, and ectoparasitoids and/or cleptoparasites of nest-building Hymenoptera. All are solitary and most appear to be idiobionts, although some members of the subfamily Chrysidinae allow the host to develop in the manner of koinobionts. The biology of most tropical species of chrysidids is unknown, and even when information is available it usually consists of nothing more than host records.

Female Cleptinae search for sawfly (Tenthredinidae and Diprionidae) cocoons on the ground. Once a cocoon is located, the female chews a hole in it with her heavy mandibles and inserts a single egg with her long, robust ovipositor tube, laying the egg ventrally on the abdomen of the host. Finally she closes the hole with a mucilaginous material. There are no host records for Central American species although *Cleptidea panamensis* has been observed flying around bushes heavily infested with the sawfly *Erythraspides interstitialis* (Tenthredinidae) (Kimsey, 1986*a*).

Amiseginae and the extralimital subfamily Loboscelidiinae parasitize the eggs of Phasmatidae (Costa Lima, 1936; Kimsey & Bohart, 1991). In tropical America male amisegines are more often collected than females, perhaps because the latter sex spend most of their time searching in the leaf litter and on plants for their hosts. The female wasp bites a small hole in the host egg chorion with her slender mandibles and then uses her needle-like ovipositor tube to place an egg inside the host egg (Krombein, 1983). Virtually nothing is known about the immature stages of Amiseginae. Adult wasps emerge through the operculum of the host egg.

Chrysidinae are parasitoids and/or cleptoparasites of solitary, nest-building Hymenoptera—eumenine vespids, sphecids, and bees—although species of one Old World genus, *Praestochrysis*, attacks lepidopterous larvae enclosed within their hardened cocoons (Yamada, 1987*a* & *b*). Table 13.1 is a summary of host records for genera occurring in Central America.

The female chrysidine sometimes uses her mandibles, or rarely her ovipositor, to penetrate the host nest, or gain access to the host by entering an open cell that is

Chrysidinae	Host
Elampini	
Elampus	Ground-nesting Sphecidae
Hedychridium	Cavity and ground-nesting Sphecidae and Apidae
Hedychrum	Ground-nesting philanthine Sphecidae
Holopyga	Ground-nesting Sphecidae
Chrysidini	
Caenochrysis	Mud-nesting trypoxylonine Sphecidae
Chrysis	Mud-nesting eumenine Vespidae and Sphecidae; various Apidae
Ipsiura	Mud-nesting Eumeninae and Sphecidae
Exochrysis	Cavity-nesting Sphecidae
Neochrysis	Mud-nesting Sphecidae
Pleurochrysis	Mud-nesting Sphecidae and Eumeninae

Table 13.1. Host records of genera of Chrysidinae occurring in Central America (data from Kimsey & Bohart, 1991).

in the process of being provisioned (Clausen, 1940*b*). Nest-builders attacked by these wasps are quite capable of defending their nests by stinging or biting, but the chrysidine's body is heavily armoured, and the concave metasoma allows the cleptoparasite to roll up into an impenetrable ball if she is attacked by the host. If detected penetrating the nest, the female chrysidid will be evicted by the host, but she is often very persistent in her attempts to re-enter the nest.

In some species the newly hatched larva attaches to the host, or the cell wall, and only begins to feed after the host has devoured the nest provisions and moulted to a prepupa. In other species the chrysidid larva begins feeding immediately, consuming the host larva and the nest provisions. If more than one chrysidid egg is laid in a host cell the first larva to hatch usually destroys the other eggs. Chrysidines are predominantly ectoparasitoids. One species is reported to be an endoparasitoid (Clausen, 1940*b*), but this needs to be confirmed.

An interesting variation occurs in *Pseudolopyga carrilloi* in California. This species oviposits in free-living nymphal bugs (Lygaeidae), but the chrysidid larva will not develop beyond the first instar unless the bug in which it resides is captured by a sphecid wasp (*Solierella*). Once in the sphecid nest, the larva moults to the second instar, emerges from the bug, destroys the sphecid egg or larva, and then consumes the bugs that were provisioned by the female sphecid (Carrillo & Caltagirone, 1970).

Chrysidid larvae are reported to pass through five instars. The first instar has 12 (Cleptinae) or 13 (Chrysidinae) indistinct body segments and a large, non-sclerotized, quadrangular head with small, sickle-shaped mandibles (Clausen, 1940*b*). First instar Cleptinae have spiracles on the first thoracic segment and on the first seven abdominal segments, whilst Chrysidinae have spiracles on the second thoracic segment and the first eight abdominal segments. The mature larvae are relatively generalized Aculeata, with 10 pairs of spiracles and broad mandibles with two to four teeth (Evans, 1987*a*). A parchment-like cocoon is spun within that of the host.

Identification. Keys to the subfamilies, tribes, and genera of the world have been provided by Kimsey and Bohart (1991). The following key is modified from this work.

Key to genera of Chrysididae of Central America

1 Metasoma with five visible external segments in males, four in females (Fig. 13.42); sternites strongly convex, visible in profile; propodeum generally box-like in profile, usually with some horizontal dorsal surface; colour usually dark brown to slightly metallic, not heavily sculptured. 2

— Metasoma with three visible external segments in both sexes (Fig. 13.43); sternites strongly concave or flat, only the first generally visible in profile; propodeum sloping abruptly, without a horizontal dorsal surface; colour usually metallic blue-green, usually heavily sculptured. ....(Chrysidinae)................5

2 Face above antennal sockets convex, without indication of scapal basin; clypeus deeply emarginate below each antennal socket and with a protruding medial truncation (Fig. 13.44). ....(Cleptinae)...........*Cleptidea*

— Face above antennal sockets flat or concave with some indication of scapal basin; clypeus not deeply emarginate below each antennal socket, without protruding medial truncation (Fig. 13.45). ....(Amiseginae)..............3

3 Propodeum with posterolateral angle or tooth; occipital carina well developed at least dorsally; pronotum with transverse anterior sulcus (Fig. 13.46). .... *Adelphe*

— Propodeum posterolaterally rounded, not angulate or dentate; occipital carina absent; pronotum without transverse anterior sulcus (Figs 13.47, 13.48). 4

4 Malar space short, much less than half as long as eye height (Fig. 13.47); gena without carina; pronotum without a posteromedial groove. ....*Amisega*

— Malar space long, at least half as long as eye height (Fig. 13.48); gena with blade-like carina; pronotum with a short, deep posteromedial groove. .... *Duckeia*

5 Tarsal claws dentate, with one to several subsidiary teeth (Figs 13.57, 15.60); metasomal tergite III never with a row of pits; mesopleuron with scrobal sulcus oblique. ......(Elampini).......6

— Tarsal claws simple, without subsidiary teeth; metasomal tergite III generally with subapical row of pits (Fig. 13.43), though sometimes these are faint or absent; mesopleuron with scrobal sulcus more or less horizontal (Fig. 13.61), or absent (Figs 13.62, 13.63)..... ....(Chrysidini).........................11

6 Scutellum anterolaterally with tubercle on edge of wing fossa (Fig. 13.49); male with metasomal tergite III with medial area depressed and differentially sculptured with setose stripe along midline. ....*Exallopyga*

— Scutellum anterolaterally evenly curved, without tubercle on edge of wing fossa (Figs 13.50, 13.51); male with tergite III unmodified. 7

7 Tarsal claw with single, perpendicular, submedial tooth (Fig. 13.57); face flat or slightly concave with at least narrow area of fine transverse striation in scapal basin. ....*Hedychridium*

— Tarsal claw with one or more subparallel subsidiary teeth (Figs 13.58-13.60); face usually concave, without an area of fine transverse striation in scapal basin. 8

8 Fore wing with vein *Rs&M* straight or nearly so (Fig. 13.52); hind tarsal claw with single subsidiary tooth (Fig. 13.58); female with metasomal sternite III with transverse sulcus near base; male with hind femur dilated and non-metallic black or brown. .... *Hedychrum*

— Fore wing with vein *Rs&M* strongly arched medially (Figs 13.53, 13.54); tarsal claw with two or more subsidiary teeth (Figs 13.59, 13.60); female with sternite III without transverse sulcus; male with hind femur not dilated and usually metallic coloured. 9

9 Fore wing with basal cell setose (Fig. 13.53). ....*Holopyga*

— Fore wing without setae in basal cell (Fig. 13.54). 10

10 Metanotum mediodorsally extended into a more or less acute, posteriorly directed process, flat dorsally; mesoscutum coarsely and regularly punctate (Fig 13.50); metasomal tergite III with membrane-filled snout-like apex; scapal basin flat or shallowly concave, usually sculptured; female with gena with row of erect short setae. .... *Elampus*

— Metanotum rounded or slightly projecting; mesoscutum sparsely punctate (Fig. 13.51); tergite III without apical snout-like structure; scapal basin usually deeply concave and smooth; female with gena without row of setae. ....*Holophris*

11 Mesopleuron divided horizontally by scrobal sulcus (Fig. 13.61); fore wing with vein *Rs* long and ending close to anterior wing margin (Fig. 13.55), or forming a short stub from pterostigma. 12

— Mesopleuron not divided horizontally by scrobal sulcus (Figs 13.62, 13.63); fore wing with vein *Rs* long and ending at least 1.5 median ocellus diameters away from anterior wing margin (Fig. 13.56). 13

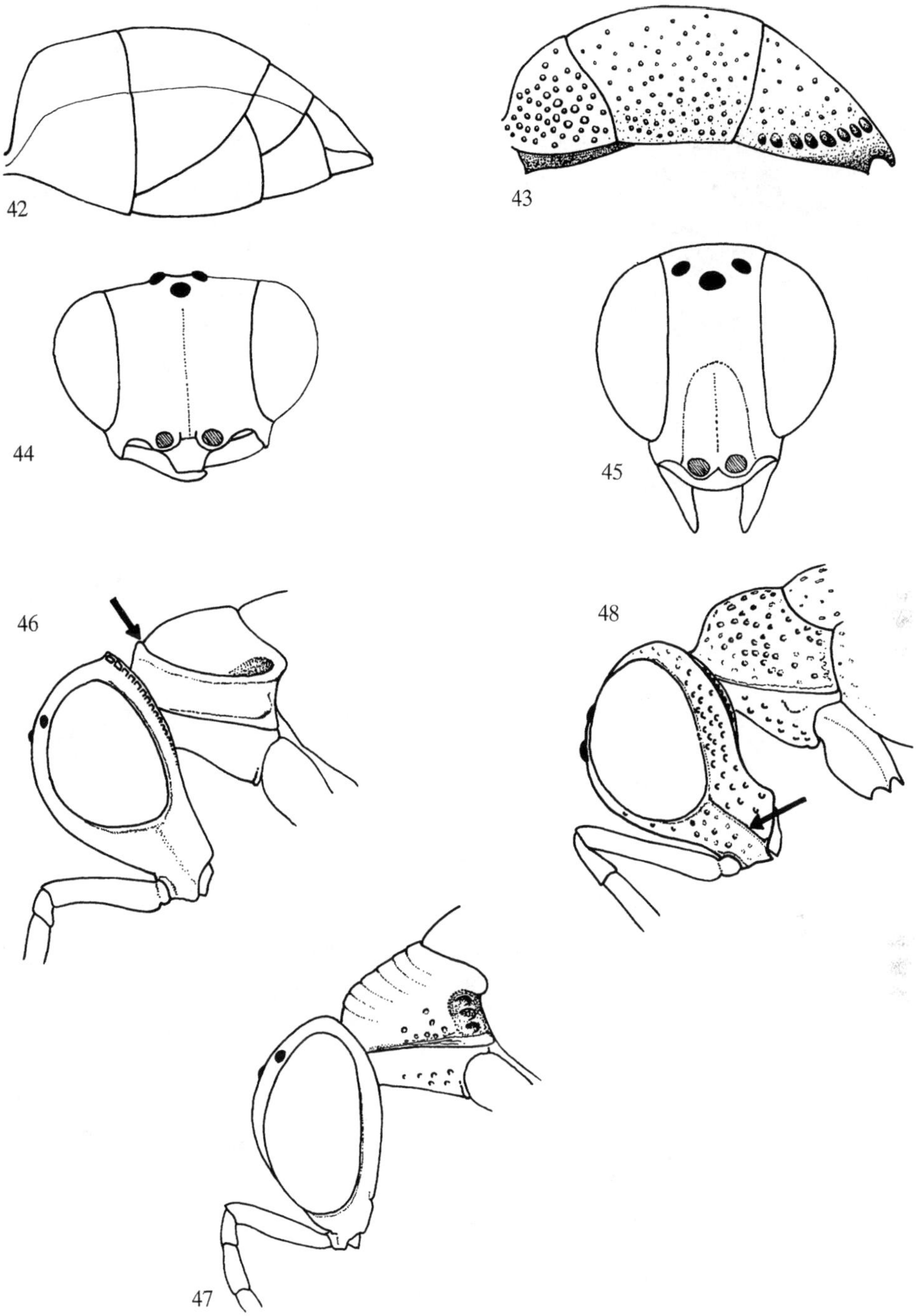

Figs 13.42–13.48. Chrysididae. Figs 13.42–13.43. Metasomas, female; 13.42, *Duckeia* sp. 13.43, *Pleurochrysis* sp. Figs 13.44–13.45. Face; 13.44, *Cleptidea* sp.; 13.45, *Amisega* sp. Figs 13.46–13.48. Head and pronotum; 13.46, *Adelphe* sp.; 13.47, *Amisega* sp.; 13.48, *Duckeia* sp.

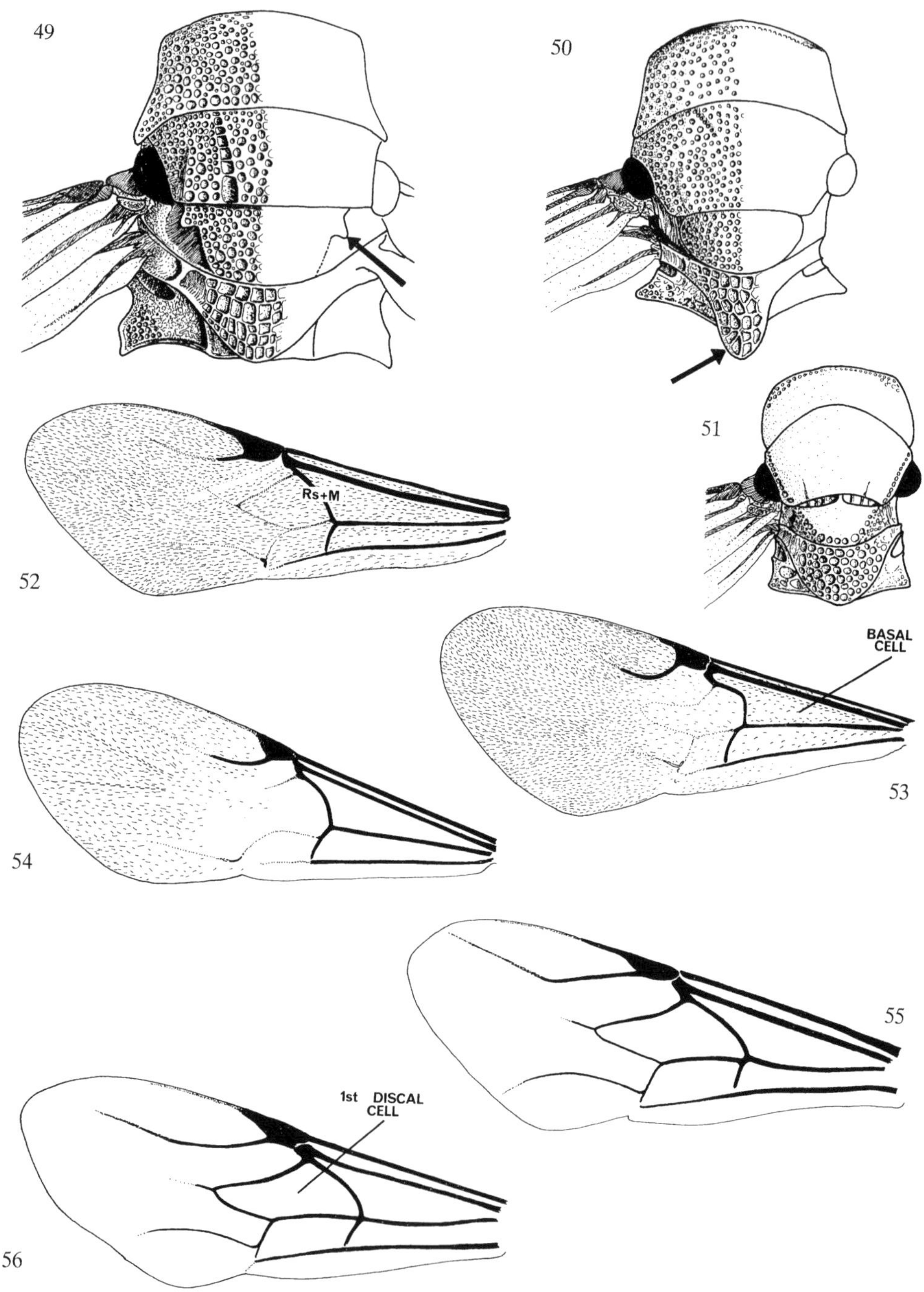

Figs 13.49–13.56. Chrysididae. Figs 13.49–13.51. Mesosoma, dorsal; 13.49, *Exallopyga* sp.; 13.50, *Elampus* sp.; 13.51, *Holophris* sp. Figs 13.52–13.56. Fore wing; 13.52, *Hedychrum* sp.; 13.53, *Holopyga* sp.; 13.54, *Holophris* sp.; 13.55, *Chrysis* sp.; 13.56, *Ipsiura* sp.

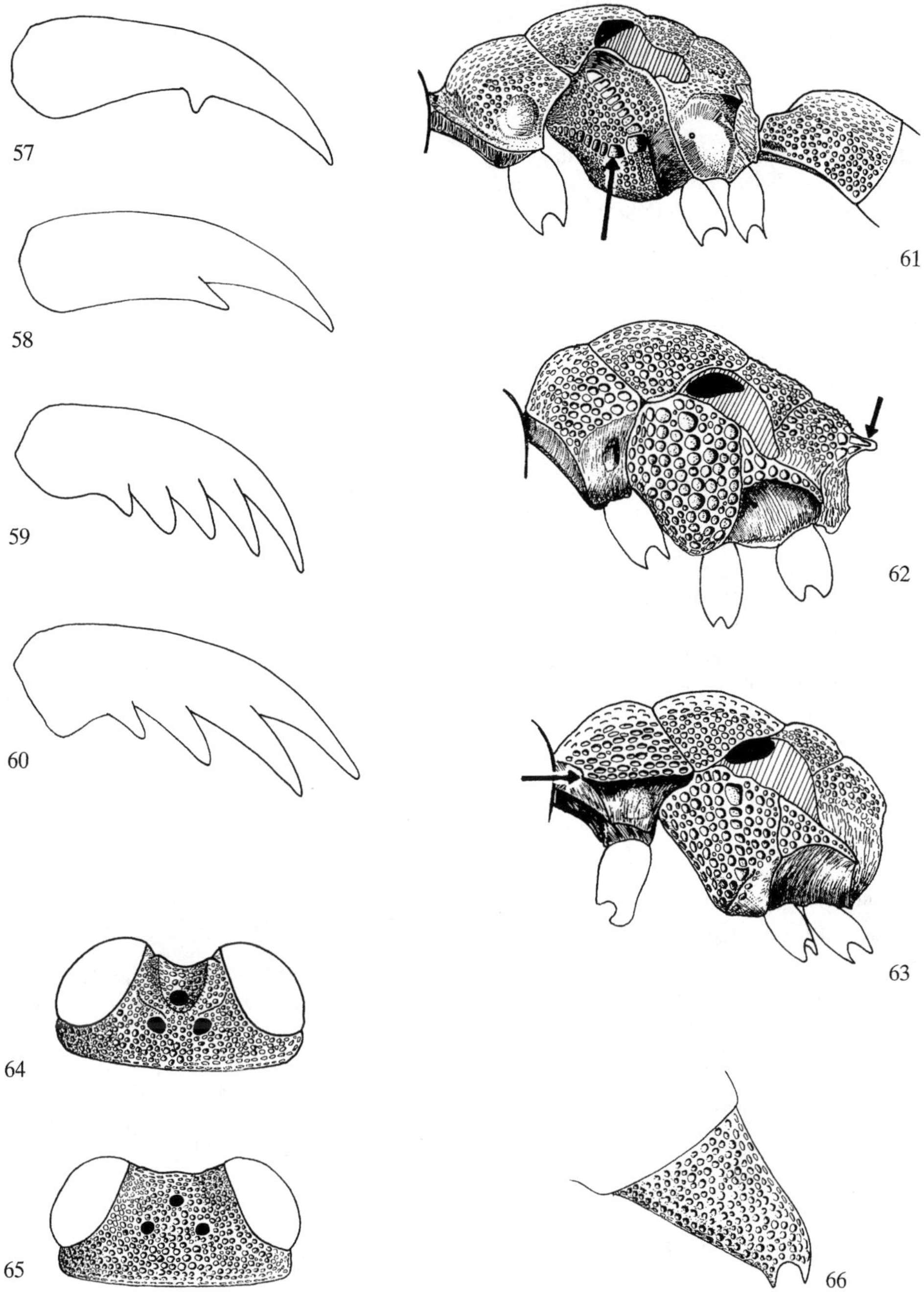

Figs 13.57–13.66. Chrysididae. Figs 13.57–13.60. Hind tarsal claw; 13.57, *Hedychridium* sp.; 13.58, *Hedychrum* sp.; 13.59, *Elampus* sp.; 13.60, *Holophris* sp. Figs 13.61–13.63. Mesosoma, lateral; 13.61, *Chrysis* sp.; 13.62, *Exochrysis* sp.; 13.63, *Ipsiura* sp. Figs 13.64–13.65. Head, dorsal; 13.64, *Ipsiura* sp.; 13.65, *Neochrysis* sp. Fig. 13.66, *Neochrysis* sp., tergite III of metasoma.

12 Rim of metasomal tergite III with four to six teeth or angles. .... *Chrysis*
— Rim of metasomal tergite III with zero to three teeth or angles. .... *Caenochrysis*

13 Pronotum with sharp, straight, protruding sublateral carina (Fig. 13.63); face with transverse frontal carina extending back to enclose median ocellus Fig. 13.64). .... *Ipsiura*
— Pronotum without sublateral carina or with carina irregular and punctate, not projecting (Fig. 13.62); frontal carina usually not enclosing median ocellus, incomplete dorsally, sometimes more or less entirely absent (Fig. 13.65). 14

14 Propodeum with dorsomedial tooth (Fig. 13.62); subgenal area discrete and bounded by carinae; fore wing with 1st discal cell complete (cf Fig. 13.56). .... *Exochrysis*
— Propodeum without dorsomedial tooth (cf Fig. 13.63), ***or if*** one present ***then*** fore wing with 1st discal cell incomplete ***and*** subgenal area not defined; fore wing otherwise generally with 1st subdiscal cell enclosed; subgenal area defined or not. 15

15 Metasomal tergite III with pit row absent (Fig. 13.66) or only indicated by very small pits, with little or no prepit swelling; pronotal anterior declivity usually without submedial pits, or these indistinguishable from nearby punctures. .... *Neochrysis*
— Metasomal tergite III with pit row plainly indicated by a depression and at least some pits large and deep (Fig. 13.43), with prepit swelling usually present; pronotal anterior declivity with two deep submedial pits. .... *Pleurochrysis*

Synopsis of the Costa Rican fauna

All three of the subfamilies present in tropical America—Cleptinae, Amiseginae and Chrysidinae—are present in Costa Rica. The Costa Rican fauna comprises 16 genera and at least 70 species.

Cleptinae

This subfamily consists of two genera: *Cleptes* (predominantly holarctic, but with one species in Argentina) and *Cleptidea* (neotropical).

Cleptidea. This genus includes 18 species, seven of which occur in Mesoamerica: *balboana* (Panama), *fasciata* (Guatemala, Panama), *janzeni* (Costa Rica), *nigrocincta* (southern Mexico), *panamensis* (Costa Rica, Panama), *scutellaris* (southern Mexico) and *viridiceps* (southern Mexico). In Costa Rica isolated individuals have been collected in lowland dry forests in northwestern Guanacaste. Keys to species are given by Kimsey (1981, 1986*a*).

Amiseginae

This primarily tropical subfamily comprises 28 genera, five of which occur in the New World: *Anadelphe* (Brazil, Ecuador; Kimsey, 1987*a*), *Nesogyne* (Dominican Republic; Krombein, 1957) and the following three genera.

Adelphe. This neotropical genus comprises 37 described species, one of which occurs in America north of Mexico. At least seven species are present in Costa Rica: *confusa, hansoni, limonae, nitida, paradoxa, paralaevis* and *robusta*, where they have been collected from sea-level up to an elevation of about 3000 metres. Taxonomy: Kimsey (1986*b*, 1993*b*). Biology: the hosts of the tropical species are unknown, but the North American species has been reared from the eggs of *Anisomorpha ferruginea* (Phasmatidae) (Krombein, 1957).

Amisega. This genus comprises 25 described species and occurs from southern Ontario to central Chile, with the majority of the species occurring in the Neotropics. Six species occur in Central America, with *cooperi, geminata* and *striata* known from Costa Rica (below 500 metres). Taxonomy: Kimsey (1987*a*, 1993*b*).

Duckeia. A small neotropical genus comprising three species, one of which, *D. vagabunda*, occurs in Mexico and Central America. It has been collected in Alajuela Province of Costa Rica in the Upper Valley of the Rio Peñas Blancas at about 700 metres. Keys to species are given by Kimsey (1987*a*). Biology: the Brazilian species *D. cyanea*, has been reared from the eggs of a species of *Prisopus* (Phasmatidae) attached in rows on a twig (Costa Lima, 1936).

CHRYSIDINAE

Two of the four tribes occur in the Neotropics. Both are represented in Costa Rica.

CHRYSIDINAE: Elampini. Six genera of this tribe are known to occur in the Central American region.

***Elampus*^EX^**. A large genus represented throughout the world, except Australasia, by about 60 species. Only three of these occur in the Neotropics, one in Brazil, and two from Mexico to El Salvador. Biology: in North America one species has been reared from the nests of the ground nesting sphecids *Hoplisoides* and *Mimumesa* (Krombein, 1967).

Exallopyga. A small neotropical genus comprising three species, one of which occurs in Central America. This species, *E. guatemalensis*, has been collected in Costa Rica in the Central Valley at about 1300 metres.

Hedychridium. A very large genus of over 200 species distributed nearly worldwide except Australia and South America. A single species, *centrale*, is known to occur in Central America (Bohart & Kimsey, 1978). Biology: in North America one species of *Hedychridium* has been reared from the nests of *Solierella* species (Parker & Bohart, 1968; Carrillo & Caltigirone, 1970).

Hedychrum. A very large genus comprising about 150 species, mostly in the Old World. A single species, *aztecum*, occurs in Central America. Biology: a North American species has been reared from the nest of a *Cerceris* species (Sphecidae) (Byers, 1978).

Holophris. A medium sized genus represented in the Old World and Neotropics by 23 species, one of which, *albolimbatus*, occurs in Central America.

Holopyga. A large genus with 91 described species, mostly in the Old World. Three species, *boutheryi*, *luzulina* and *ventralis*, are recorded as occurring in Central America. Biology: a North American species has been reared from the nests of *Bicyrtes* species (Sphecidae) (Evans, 1966*d*).

CHRYSIDINAE: Chrysidini. Only six genera of this tribe occur in the Central American region.

Caenochrysis. A large neotropical genus with 51 described species, ten of which occur in Central America (Kimsey & Bohart, 1981) and at least seven of which have been found in Costa Rica: *areolata, azteca, doriae, mucronata, parvula, saussurei* and *tridens*. Biology: the widespread species *C. doriae*, *C. mucronata* and *C. tridens* have been reared from the mud nests of *Trypoxylon* species (Sphecidae) (Krombein, 1967; Parker & Bohart, 1967; Parker, 1967; Johnson, 1974).

Chrysis. An extremely large cosmopolitan genus comprising about 1000 species, at least 15 of which are known to occur in Costa Rica including: *antennalis, brasiliensis, conica, inflata, intricans, intricata, nisseri, nitidula, oreadis, rastellum, smaragdula* and *venusta*. Several other species, such as *Chrysis angolensis*, have become widely distributed through human activity and are expected to occur in Costa Rica. Biology: hosts of *Chrysis* species include a wide variety of Sphecidae, Apidae and Vespidae (for summary see Kimsey & Bohart, 1991).

Exochrysis. A rather small, neotropical genus comprising twelve species, six of which are known to occur in Central America; *panamensis, silvanus, spinigera* and *tolteca* are known to occur in Costa Rica between sea-level and 800 metres. Taxonomy: Kimsey (1985). Biology: *E. tolteca* has been reared from the nests of species of *Podium* (Sphecidae) (Krombein, 1967).

Ipsiura. A medium sized neotropical genus with 37 described species, 13 of which occur in Central America (Bohart, 1985), and six of which have been collected in Costa Rica: *cooperi, covillei, irwini, pilifrons, tropicalis* and *venezuelana*, all at altitudes below 500 metres. Biology: hosts include species of *Eumenes* and *Pachodynerus* (Vespidae), and *Trypoxylon* and *Sceliphron* (Sphecidae) (Kimsey & Bohart, 1991).

Neochrysis. A medium sized genus with 20 described species. It is primarily neotropical, but two species occur in the southern United States. Seven species occur in Central America (Kimsey, 1985) and two are recorded from Costa Rica: *lecointei* and *montezuma* (the latter is not uncommon in Costa Rica below about 1300 metres). Taxonomy: Kimsey (1985). Biology: Species of this genus are commonly reared from *Trypoxylon* (Sphecidae) nests (Coville, 1981*a*).

Pleurochrysis. A medium sized neotropical genus with 36 described species, three of which occur in Central America. All three, *alfkeni*, *morosa* and *postica*, have been been recorded from Costa Rica (Kimsey & Bohart, 1981; Kimsey, 1985). Taxonomy: Kimsey (1985). Biology: *P. postica* has been taken in the drier northwest of Costa Rica and *alfkeni* has been collected in more humid situations between sea-level and 800 metres. *P. postica* has been reared from the nest of *Sceliphron* (Sphecidae) (Kimsey & Bohart, 1991).

13.4 SCLEROGIBBIDAE

Ian D. Gauld

Diagnosis. Body length 2.3 to 6.0 mm. Usually yellowish brown. Females apterous, males fully winged. Both sexes with *antenna inserted close to clypeus beneath a frontal shelf* (Fig. 13.14); *antenna (of New World species) 20 to 24-segmented.* Female with head prognathous (Fig. 13.67); mesosoma highly modified, somewhat hour-glass shaped; *fore femur grossly swollen; coxae with a dorsal furrow bounded on both sides by raised keels.* Males with head subtriangular (Fig. 13.68); pronotum extending posteriorly so that upper corner more or less reaches tegula; five (or rarely six) enclosed cells in fore wing—marginal, submarginal, costal, basal and subbasal; hind wing without cells, but with a well-developed claval lobe (which is basally positioned and easily mistaken for a jugal lobe); legs slightly modified in the manner of those of the female, but less conspicuously so (Fig. 13.15). Metasoma spindle-shaped with six or seven visible tergites.

Classification and distribution. The Sclerogibbidae is a very small family comprising less than 20 described species, which are currently classified in three genera—*Caenosclerogibba*, *Probethylus* and *Sclerogibba* (Richards, 1939; Yasumatsu, 1958). The first of these is known from a single species in Japan (Yasumatsu, 1958), whilst the other two are primarily subtropical and tropical in distribution. *Sclerogibba* is Afrotropical with isolated species in southeast Asia and southern Europe, whilst *Probethylus* occurs in Australia and the New World.

Biology. Sclerogibbids develop as solitary or gregarious ectoparasitic koinobionts of nymphal or adult

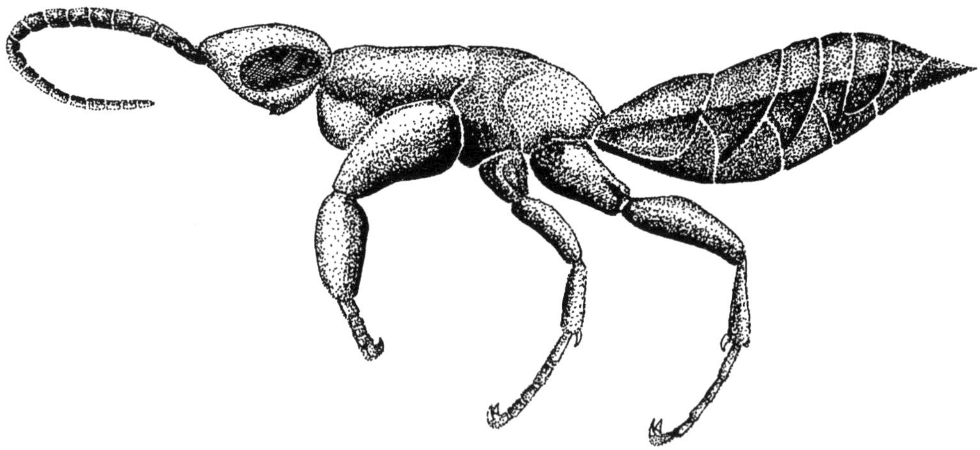

Fig. 13.67. *Probethylus* sp. female (Sclerogibbidae).

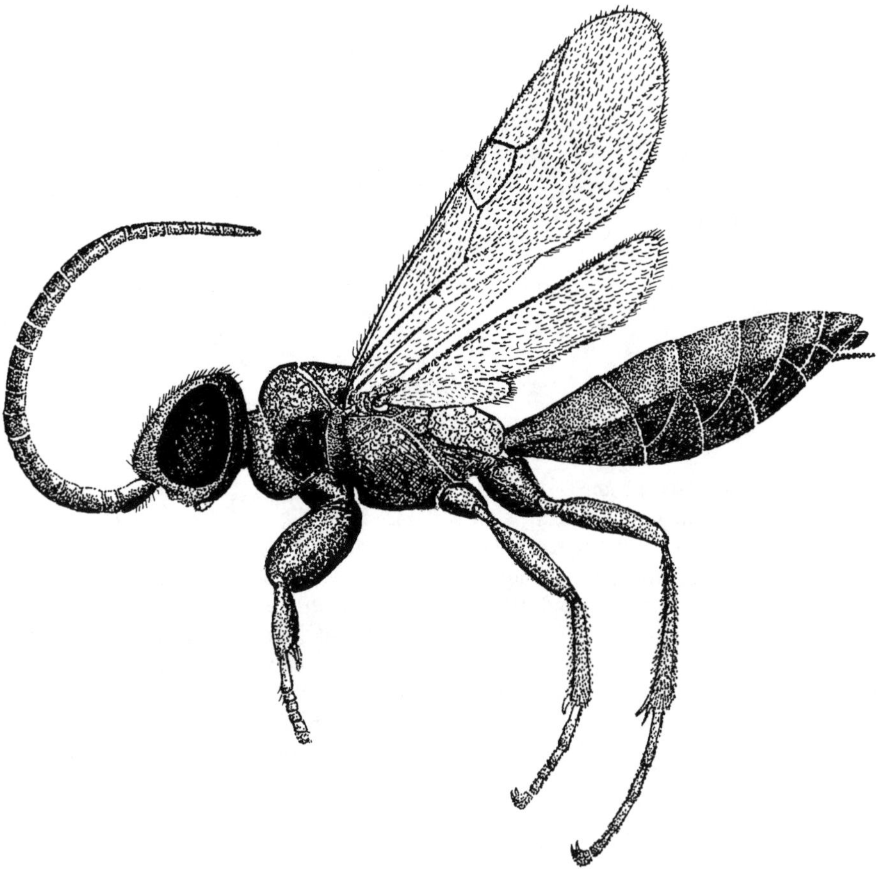

Fig. 13.68. *Probethylus* sp. male (Sclerogibbidae).

Embiidina. Callan (1939), working in Trinidad, observed the larva of *Probethylus callani* attached transversely, dorsally on its nymphal embiid host, to an intersegmental membrane, either between the head and prothorax, or between the pro- and mesothoracic segments. Shetlar (1973), working in the United States, observed one or two larvae of *Probethylus schwarzi* feeding on the nymph of *Anisembia texana* through the first and second abdominal intersegmental membranes. Yokoyama and Tsuneyoshi (1958) also observed that a Japanese sclerogibbid, *Caenosclerogibba japonica*, developed gregariously, with 8 to 13 eggs being placed on various abdominal sternites of its mature female embiid host. In all studies it was observed that up to the time that the larvae were half grown (or later), they appeared to have little effect on the parasitized embiids except that these hosts were slightly smaller than normal and sometimes had the abdomen slightly deflected laterally. They remained as active as their unparasitized siblings. Later, however, the embiids became sluggish and were eventually killed by the parasitoid.

Few observations have been made on the larval development. Yokoyama and Tsuneyoshi (1958) observed that the 'early' larva of *C. japonica* was spindle-shaped and pale in colour, whereas the 'late' larva was slightly flattened dorsoventrally. The latter had an enlarged first abdominal segment, and the second to fourth segments were interrupted mediodorsally, and tuberculate lateromedially. The sclerogibbid larva continued to feed on its dead host until the body contents were consumed, and then pupated near the

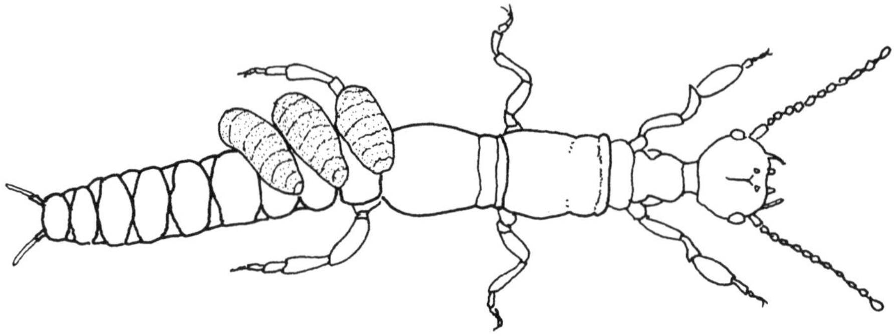

Fig. 13.69. Larvae of *Probethylus* sp. feeding on an embiid.

host remains in a tough whitish cocoon. These were rendered somewhat inconspicuous in the silken embiid tunnels due to adhering embiid faecal pellets. In Peru a species of sclerogibbid attacking Embiidina is known to be parasitized by a perilampid (Burks, 1969).

Identification. A key to the described New World species is given by Shetlar (1973).

Synopsis of the Costa Rican fauna

Only *Probethylus* is known to occur in the New World. It includes three described species: *P. callani* from the Caribbean and Mexico, *P. mexicanus* from Mexico, and *P. schwarzi* from the United States south of about 35°N. *P. callani* has been found in Costa Rica where numerous males and isolated females were collected in Malaise traps in seasonally dry forest in Santa Rosa National Park. Most were taken between April and June in the late dry season and early wet (Fig. 13.70).

A single female, possibly of a different species, has been collected in oak forest on the Cerro de la Muerte, at about 2800 metres.

13.5 EMBOLEMIDAE

Ian D. Gauld

Diagnosis. Body length 2.0 to 4.0 mm; fore wing length 1.5 to 3.5 mm; brown or black; females apterous or fully winged; males fully winged. *Antenna 10-segmented in both sexes, arising from a prominence (which is most strongly developed in females) well above clypeus.* Eyes small and oval; lower face usually with a pair of impressed lines that originate near anterior tentorial pits and converge to antennal insertion; clypeus apically convex; mandibles with four teeth, the ventral one the largest and the remainder becoming progressively smaller dorsally; *maxillary palp 4-segmented (apparently 5 in male), labial palp 2-segmented.* Pronotum reaching back to tegula, hind margin (in winged forms) strongly arcuate to almost transverse; mesoscutum with notauli impressed near anterior margin, parapsidal furrows usually discernible; mesopleuron of winged forms with a transverse sulcus. *Fore wing with proximal three cells enclosed,* first discal cell sometimes closed but the veins delineating it posteriorly are often weak, pterostigma distinct, marginal cell open at extreme apex, first subdiscal cell open distally; hind wing without enclosed cells, with a sharply differentiated claval lobe. Fore leg simple, or in females with femur swollen; tibial spur formula 1: 2: 2, sometimes with mid tibia with only a single spur; *fore tibial spur strongly curved and truncate*; outer hind tibial spur short and broad; tarsal claws simple. Metasoma spindle-shaped without constriction between first and second segments.

Classification and distribution. The Embolemidae is a small family comprising about 20 species classified in two genera, *Embolemus* and *Ampulicomorpha*. The former is cosmopolitan in distribution, whilst the latter

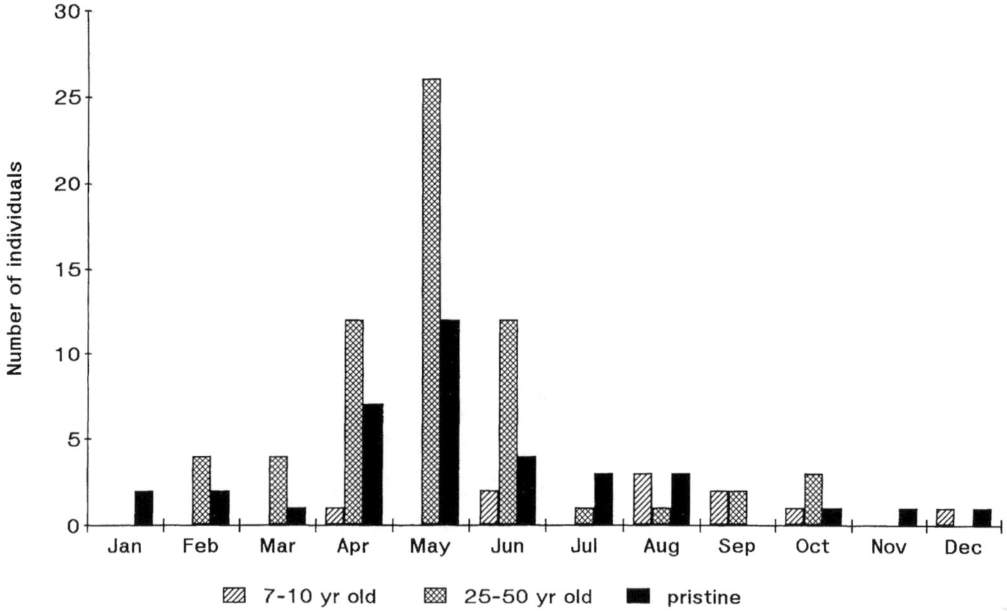

Fig. 13.70. Monthly occurrence of *Probethylus callani* (Sclerogibbidae) in Malaise trap catches in different aged forest habitats in Santa Rosa National Park, averaged from data for 1985–7.

is predominantly holarctic. Both have been found in Costa Rica.

Biology. *Ampulicomorpha confusa* has been reared in North America as a koinobiont ectoparasitoid of the nymphs of Achilidae (Homoptera: Fulgoroidea) which live under loose bark (Bridwell, 1958). The female wasp stings its host and feeds on the exuding body fluids; after the attack the achilid seems to show no ill effects. The egg has not been observed, but the embolemid larva develops enclosed in a sac bulging from beneath the wing pad (Bridwell, 1958) or between abdominal tergites II and III of its host (Wharton, 1989*b*). The sac is of parasitoid origin and is formed from larval exuviae. The final instar larva has a distinct, more or less prognathous head capsule with tridentate mandibles (Wharton, 1989*b*). After destroying its host the embolemid larva constructs a cocoon beneath the bark.

The biology of *Embolemus* is unknown, but Hirashima and Yamagishi (1975) noted that one species has been collected in an ant's nest, and they suggest that it perhaps attacks ants. I consider that unlikely since *Ampulicomorpha confusa*, and all members of the sister group to the Embolemidae, the Dryinidae, attack homopterous hosts. Various Homoptera, including fulgorids, are known to be symbionts of ants (Hölldobler & Wilson, 1990*a*), and it is perhaps one of these that is the host of *Embolemus*.

Key to genera of Embolemidae present in the New World

1 Apterous. ....*Embolemus*
— Wings fully developed. 2

2 Hind margin of pronotum strongly concave (Fig. 13.73), the pronotum in dorsal view short and not strongly impressed medially; wings hyaline. .... *Embolemus*
— Hind margin of pronotum more or less transverse (Fig. 13.74), the pronotum in dorsal view elongate, with a fairly strong median longitudinal impression; wings infumate. ....
....*Ampulicomorpha*

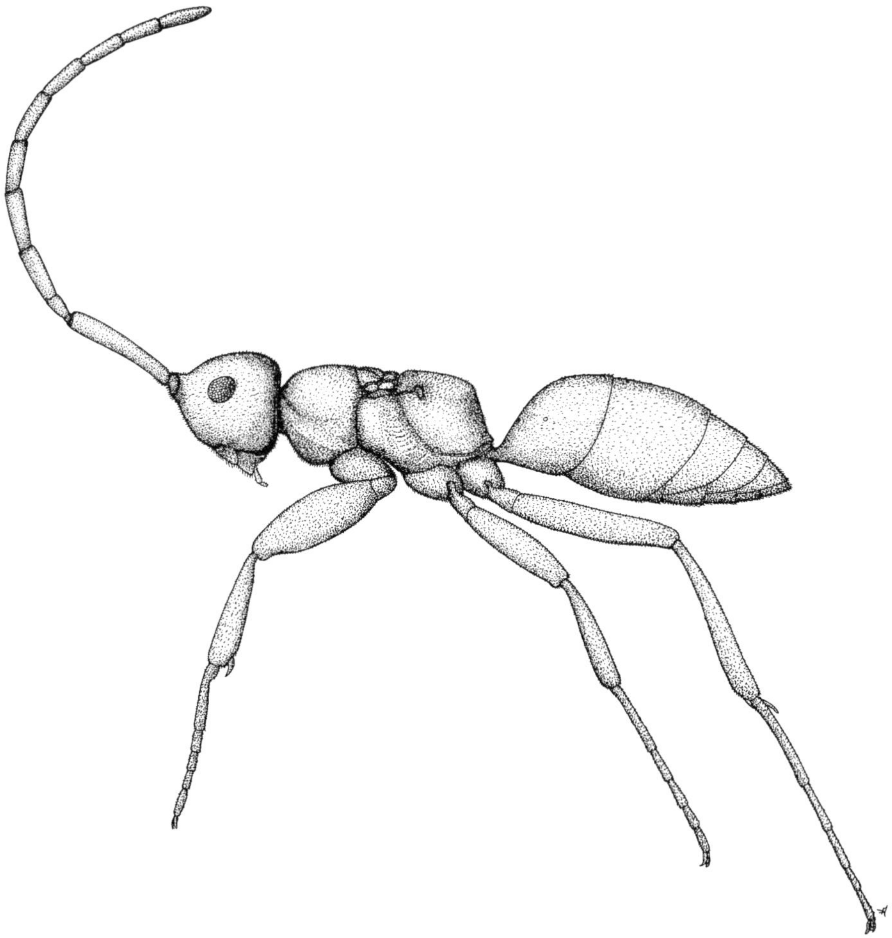

Fig. 13.71. *Embolemus* sp. female (Embolemidae).

Synopsis of the Costa Rican fauna

Embolemidae are rather infrequently collected in Costa Rica, but occasional individuals have been taken in Malaise trap samples from altitudes between about sea-level and 2800 metres.

Ampulicomorpha. This genus is known from two described species, one in the Nearctic and a second in the Palaearctic region (Rasnitsyn & Matveev, 1989). Two specimens, representing an undescribed species of *Ampulicomorpha*, have been collected in Costa Rica, one from north east of San José (Zurquí de Moravia, 1600 m), the other from the southern part of the country (Las Alturas, 1500 m). Both localities are in humid forest.

Embolemus. This is a cosmopolitan genus that contains most of the species in the family (Hirashima & Yamagishi, 1975; Gauld & Bolton, 1988). A moderate number of males representing three putative species have been collected in forested localities from sea level up to about 2800 metres. We have not collected a single female of this genus.

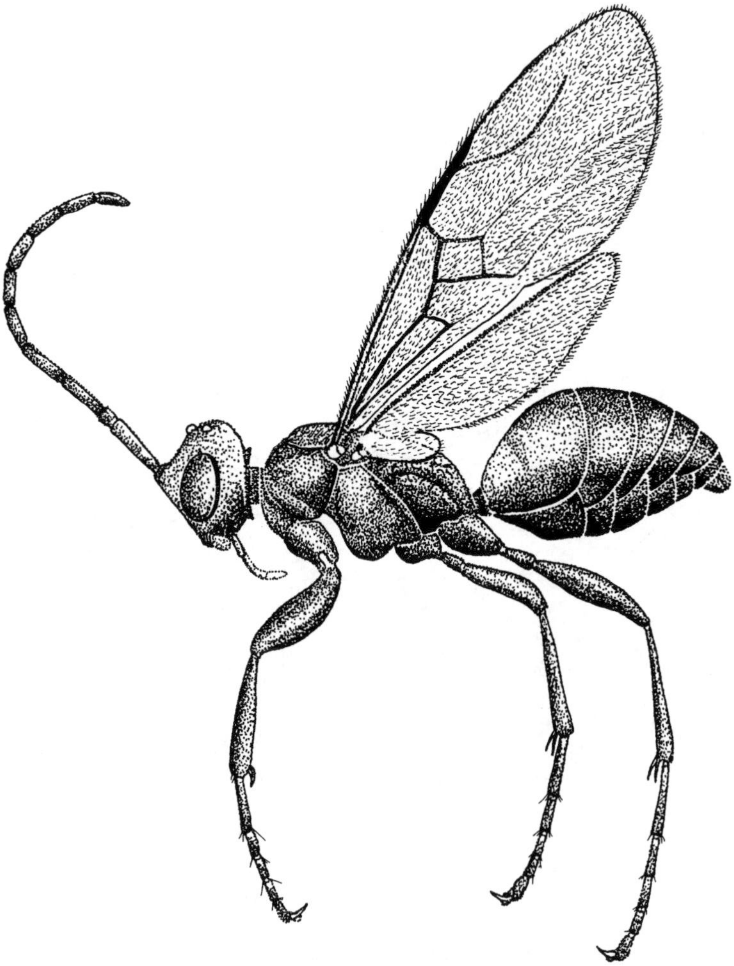

Fig. 13.72. *Embolemus* sp. male (Embolemidae).

13.6 DRYINIDAE

Massimo Olmi

Diagnosis. Body length 0.9 to 11.6 mm; black (many males) to brightly coloured (many females); *females apterous, brachypterous, or fully winged*; males fully winged. Sexual dimorphism slight (in Aphelopinae) to very pronounced (making it impossible to associate the sexes without appropriate rearings). *Antenna 10-segmented* in both sexes, filiform or slightly clavate, sometimes geniculate, rarely pectinate; flagellar segments of many females with multiporous plate sensilla (especially in species parasitizing Fulgoroidea); antennal insertion close to clypeus. *Fore wing usually with three enclosed cells* (costal, basal and subbasal), rarely with a fourth (first subdiscal in a few *Thaumatodryinus* spp), or only one (costal cell in Aphelopinae) or two (costal and basal cells in Biaphelopinae and Conganteoninae); stigmal vein usually present, arising from the pterostigma and consisting of *2r-rs* plus the apical portion of *Rs*, or rarely absent; when present this vein is usually free, or rarely intersected by a faint to vestigial track of the proximal portion of *Rs*. Hind wing with marginal vein present on leading edge, otherwise without veins or

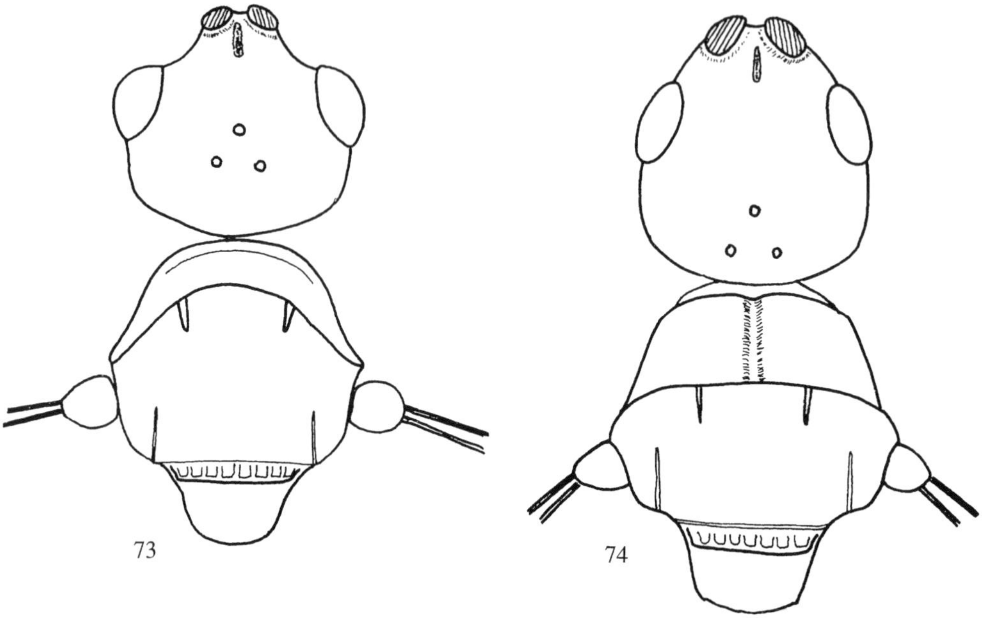

Figs 13.73–74. Embolemidae; pronotum in dorsal view; 13.73, *Embolemus* sp., male; 13.74, *Ampulicomorpha* sp., male.

enclosed cells; claval lobe conspicuous; jugal lobe absent. *Fore tarsi of females (except Aphelopinae and Biaphelopinae) chelate*, the chela composed of a strongly projecting fifth tarsal segment, opposed by an enlarged claw; inner edge of enlarged claw and tarsal segment 5 commonly bearing teeth, bristles or lamellae. More specialised dryinid females have fore legs with elongate coxae, trochanters and tarsal segments (to improve their raptorial ability).

Classification and distribution. The Dryinidae comprises about 850 species worldwide classified in 10 subfamilies: the Anteoninae, Apodryininae, Aphelopinae, Biaphelopinae, Bocchinae, Conganteoninae, Dryininae (including *Thaumatodryinus*), Gonatopodinae, Plesiodryininae and Transdryininae (Olmi, 1984, 1986, 1987*b*, 1989, 1993*b*). Seven of these, the Anteoninae, Aphelopinae, Apodryininae, Biaphelopinae, Bocchinae, Dryininae and Gonatopodinae, occur in the New World, and all except the Apodryininae (only known from Australia and southern Chile) are represented in Costa Rica (Olmi, 1993*a*). Subfamilies not represented in Costa Rica are generally small groups comprising less than ten species.

Biology. Dryinids develop as koinobiont parasitoids of auchenorrhynchan Homoptera belonging to the superfamilies Cicadelloidea and Fulgoroidea (leafhoppers, treehoppers and planthoppers). They are more or less endoparasitic during their early instars, but are ectoparasitic in later instars, except for species of *Crovettia*, which are fully endophagous. This genus differs from all other dryinids in two other features of its biology: it is the only dryinid genus known to attack membracids, and it is the only aculeate hymenopteran to show polyembryony (Kornhauser, 1919). The hosts of the subfamilies found in Costa Rica are listed in Table 13.2. It should be noted that Bocchinae and Gonatopodinae attack non-typhlocybine cicadellids (Olmi, 1984). Most dryinids attack either the nymphal or adult stage of their host, but *Crovettia* always attacks nymphs.

Little is known about how dryinids actually locate hosts, but possibly they are attracted by the secretions of their hosts. Many dryinids, especially the apterous species, are ant-like in general appearance. This resemblance possibly allows the female parasitoid to approach the host more readily since honeydew-secreting Homoptera are accustomed to ants

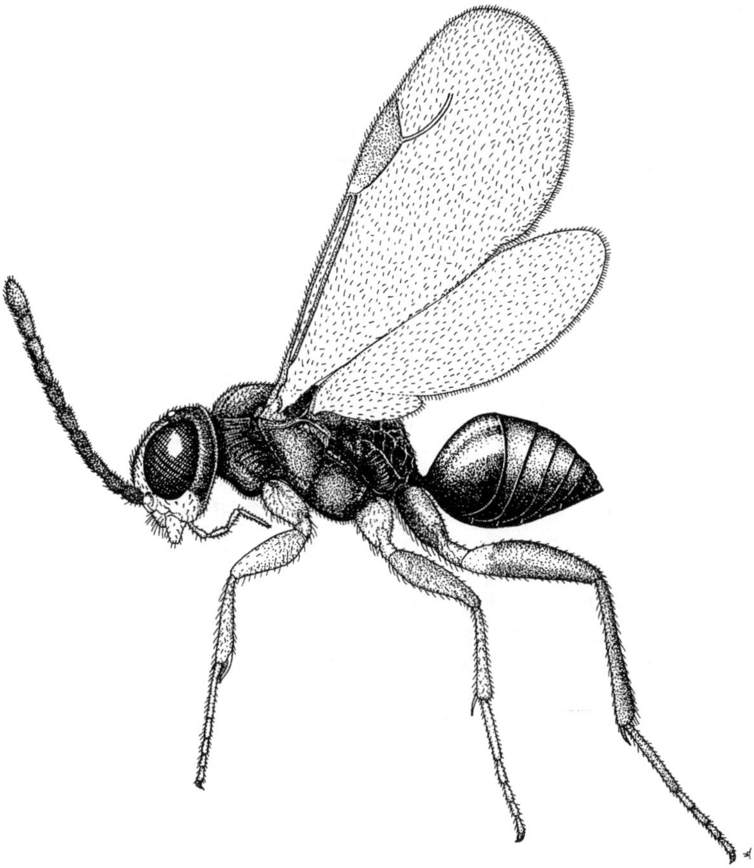

Fig. 13.75. *Aphelopus* sp. (Dryinidae: Aphelopinae).

attending them. The Australian species *Anteon myrmecophilum* (= *Paranteon myrmecophilum*) imitates typical postures of ants, sub-erect on their mid and hind legs, face to face, stroking one another, licking each other's mouth, and soliciting food (Perkins, 1905).

Dryinid subfamily	Host group
Anteoninae	Cicadellidae
Aphelopinae: *Aphelopus*	Cicadellidae: Typhlocybinae
Aphelopinae: *Crovettia*	Membracidae
Bocchinae	Fulgoroidea and Cicadellidae
Dryininae	Fulgoroidea
Gonatopodinae	Fulgoroidea and Cicadellidae

Table 13.2. The hosts of the dryinid subfamilies found in Costa Rica.

Females consume honeydew and other sugary solutions, but in most groups, except the Aphelopinae, they also feed on the same homopteran species that serve as hosts. The female dryinid grasps the host insect with its chelate fore tarsi and feeds on the haemolymph and body tissues of the host (Jervis *et al.*, 1987). In the more specialized subfamilies (Gonatopodinae, Dryininae) host-feeding frequently kills the host, whereas in the more primitive subfamily Anteoninae it commonly does not. In those groups where feeding results in the death of the host, mortality due to host-feeding may equal mortality due to parasitization.

The manner in which the host is attacked differs in different groups, depending upon whether or not the females have chelate fore tarsi. In the Aphelopinae the females, which have no chelae, grasp the body of the host between the anterior two pairs of legs, and some species use their mandibles as well. In other subfamilies the females capture the hosts using the chelae. The host is stung through an intersegmental membrane, the venom causing temporary paralysis, and an egg is laid with the ovipositor (unlike virtually all other Aculeata). In most groups the egg is laid into the haemocoel of the host.

Larval development. Development of the immature stages of Dryinidae has been best studied in species of *Aphelopus*. The small, alecithal egg develops into an embryo which is surrounded by a trophamnion. Instead of typical eclosion, the chorion seems to dissolve, leaving the first instar larva within the trophamnion, through which respiration and food exchange occurs. The second instar larva, still within the trophamnion, at first remains motionless near the wound inflicted by the dryinid ovipositor, but eventually it is compressed against the host integument and pushed out of the host through the oviposition wound. Its body thus becomes visible externally, although it is to some extent hidden by the overlapping sclerites. There are five larval instars, unless the first instar is considered to be an embryo, in which case there would be four larval instars.

Eventually only the head of the dryinid larva remains within the host. As the overlapping sclerites of the leafhopper move, the exposed posterior part of the dryinid larva runs the risk of being damaged. To avoid this damage, a cuticular ring, which has been formed around the head, is shed during the second moult, and this together with the old cuticle, plugs the intersegmental gap. The posterior extremity of the larva is attached to the host's integument by an adhesive substance which it secretes. The exuviae discarded during the second and subsequent moults adhere to the host by the same adhesive substance. These exuviae form a characteristic cyst or sac (the thylacium) that projects from the host body. At first it is small, but after each moult it grows, and ranges in colour from black or brown to green or yellow. The thylacium of aphelopines is elongate and subrectangular, quite different from those of the other dryinids which are spherical.

When the posterior end of the larval body is pushed out of the host, the epidermis of the host starts a scar reaction around the parasitoid larva. This scar tissue is anomalous in that the constituent cells are larger than normal. These cells penetrate between the trophamnion and the larval cuticle, and form a layer surrounding the dryinid larva. The trophamnion around the posterior part of the larva dissolves, and thus in the thylacium the larva is surrounded only by scar tissue. The trophamnion remains only around the head in the haemocoel, where it continues its nutritional and respiratory functions. The scar tissue is interrupted in front of the larva's face and through this hole the dryinid sucks in the nutrients absorbed through the trophamnion. The hole is kept open by the larva feeding.

The trophamnion is in contact with tracheae of the host, thus allowing gaseous exchange to the parasitoid larva. Respiration is, however, constantly cutaneous, because the tracheae of the parasitoid are filled with liquid (Buyckx, 1948). During the fourth instar the larval cuticle becomes very thick, and in the subsequent moult this thick larval skin is not cast, but it forms a sheath around the final instar larva. Eventually the penultimate instar larval skin breaks and the final instar larva emerges. The typhlocybine host ceases activity and remains attached to the food plant by its rostrum. The final instar dryinid larva is structurally very different from the other instars as it is hymenopteriform, instead of sacciform. It also has well-developed mouthparts, whereas the previous instars had rudimentary mouthparts, and it has functional tracheae (Buyckx, 1948). The mature dryinid larva then kills its host by consuming all the host's internal tissue. On completion of feeding, the dryinid larva splits open the thylacium, crawls out, and pupates in the soil (Keilin & Thompson, 1915; Buyckx, 1948; Jervis, 1980*a* & *b*). The pupae are exarate and enclosed in a silk cocoon, the silk being produced by labial glands of the mature larvae.

Development of *Crovettia* is similar to *Aphelopus*, but within the trophamnion the egg divides to form up to 70 morulae, each of which forms an embryo. Eventually each embryo becomes enclosed in a separate trophamnion, which is permeable to some substances but chemically transforms others before passing them to the embryo (Ivanova-Kasas, 1972). The trophamnion also protects the embryos by separating them from the haemolymph. The larvae develop within the host haemocoel, and finally consume all the tissues of the membracid host after the latter has reached either its final nymphal instar or adulthood. After moulting the dryinid larvae emerge from the abdominal sternal regions of the membracid corpse.

In Gonatopodinae a small egg is glued externally (Fenton, 1918) or internally (Giri & Freytag, 1989) on the intersegmental membrane. If the egg is positioned externally, the first instar larva penetrates the host through the ovipositor wound immediately after eclosion. In the first instar the posterior part of the parasitoid body protrudes from the host between two overlapping sclerites. A trophamnion and scar tissue are absent. The spiracles of the dryinid larva are open, the tracheae are not fluid-filled, and respiration is tracheal, using atmospheric air taken in via the posterior part of the body (Ponomarenko, 1975). The mouthparts are rudimentary and the larva is incapable of feeding directly on the host haemolymph. In front of the larval face two oval vesicles are visible; their function is not clear, but they are probably the equivalent of the *Aphelopus* trophamnion and act as mediators for obtaining food (Ponomarenko, 1975). During the first moult the larval cuticle is cast outside the host body, remaining around the posterior end of the dryinid larva. During the succeeding moults other cast larval skins are added to this thylacium. In some species the penultimate instar larva (third or fourth) begins to consume all the tissue of its host, killing it during the nymphal or adult stage (Hernandez, 1984; Hernandez & Belloti, 1984; Giri & Freytag, 1989). In other species it is the last instar larva that devours all the body contents of the host (Fenton, 1918). Pupation occurs in a silken cocoon on leaves or stems of the food-plant of the host, or occasionally in the soil.

Development of Dryininae is essentially similar to that of Gonatopodinae (Haupt, 1932; Ramirez, 1992). Development of Anteoninae and Bocchinae is poorly known, but seems to be intermediate between that of *Aphelopus* and that of the Gonatopodinae and Dryininae (Ponomarenko, 1975). The larva is surrounded by a tissue-like trophamnion, but it is much thinner than in other subfamilies. On the sides of the mouth two conical epithelial processes are visible, which seem to be homologous with the oval vesicles of the Gonatopodinae and Dryininae. The mature larva pupates in a silken cocoon in the soil.

Effects of parasitization on the host. In some species of Gonatopodinae and in the few species of Dryininae that have been studied (Pillault, 1951; Subba Rao, 1957), parasitization of host nymphs may prevent the transformation of the host to the adult stage. Pillault (1951) suggested that this developmental arrestment is caused by the dryinid larva and not by the venom injected by the adult female during oviposition. Host development is halted so long as the dryinid larva remains alive, but if the parasitoid larva is killed host development resumes. In one gonatopodine species inhibition of development may or may not occur, probably depending upon the individual host's reaction to the dryinid larva. Dryinid larvae may also inhibit the development of competing parasitoids. For example, the first eclosed dryinid larva may stop the development of other dryinids, pipunculids and strepsipterans developing in the same host (Ponomarenko, 1971, 1975). In some cases, however, development of other dryinid larvae is not affected and two mature larvae may emerge from one host. In such cases the adult dryinids are small and possibly display anomalous colours.

In other dryinids parasitization of the host nymph does not prevent its transformation to the adult stage, and in these cases the adult host can show pronounced anatomical and physiological changes resulting from parasitization. Hosts attacked by aphelopines, for example, often show external colour changes, and internally there may be a degeneration of the ovarioles. Adult males are sometimes parasitically castrated and show many of the sexual characteristics of the females. In such cases male stridulatory organs are lost, and the external genital appendages show considerable modifications. The rate of development of parasitized nymphs may be retarded as the parasitoid extracts nourishment from the host. The parasitized host may eat more, and a part of this extra food, together with substances coming from the degeneration of internal organs, is transformed into fat tissue (Buyckx, 1948; Fenton, 1918; Jervis, 1980*a*; Kornhauser, 1919). Gonatopodinae induce anatomical and physiological changes equivalent to those produced by the Aphelopinae (Lindberg, 1950; Raatikainen, 1967; Ponomarenko, 1971, 1975; Hernandez, 1984; Hernandez & Belloti, 1984; Giri & Freytag, 1989; Kitamura, 1983, 1988). When adult hosts are attacked by dryinids, the host does not show such anatomical changes because its organs are fully developed prior to parasitization. However, its reproductive capacity is often reduced.

Life cycle. In the Neotropics a few life cycles of dryinids have been studied. In Colombia the time taken for development of *Haplogonatopus hernandezae*, a gonatopodine parasitoid of *Tagosodes orizicolus* (= *Sogatodes oryzicola*), takes 21 to 39 days to develop from the egg to adult at 28°C and 75.5 percent relative humidity; 2 to 5 days are required for embryonic development, 14 to 23 days for larval development, 2 days for the prepupal stage and 3 to 9 days for the pupal stage. The adult female lives 5 to 17 days (Hernandez, 1984; Hernandez & Belloti, 1984). Other species that have been studied include *Gonatopus bartletti* (Gonatopodinae) in El Salvador (Quezada, 1979) and *Dryinus poecilopterae* (Dryininae) in Colombia (Ramirez, 1992). In some dryinids the pupae remain in their cocoons for many months until the temperature and humidity conditions permit the emergence of the adults (Perkins, 1905).

Adult male dryinids are short-lived and either they do not eat, or they feed only on sugary solutions such as honeydew. The females are apparently more active, and move constantly and rapidly in search of hosts, although they are poor flyers. For dispersal dryinids probably rely on their parasitized hosts since macropterous Auchenorrhyncha are known to undertake long migrations, even across hundreds of kilometres of open sea, and the flying ability of parasitized individuals is apparently not impaired (Kitamura & Nishikata, 1987).

Dryinid reproduction is either arrhenotokous (the normal hymenopteran form), or less commonly, in some Gonatopodinae, Anteoninae, Dryininae and *Crovettia* (Aphelopinae) thelytokous.

Economic importance. Because many leafhoppers and planthoppers are important agricultural pests, the dryinid species that attack them are potentially beneficial pest control organisms. Table 13.3 lists leafhopper and planthopper pests known to be attacked by dryinids in the Central American region. In some places up to 78 percent of the pest populations sampled are parasitized (LeQuesne, 1972) although parasitization rates vary seasonally, and may be higher in the wet season than at other times of the year (Peña & Shepard, 1986). Many species of Dryinidae inflict far higher mortality on the host than simple assessment of parasitism rates suggest because additional host individuals are killed and used for host-feeding. For example, in Colombia *Haplogonatopus hernandezae* kills an average of five rice planthoppers per day by parasitization and an approximately equal number by host-feeding. Since the adult female dryinid lives 5 to 17 days, it can kill nearly 100 additional planthoppers (Hernandez, 1984; Hernandez & Belloti, 1984). The efficiency of *Gonatopus bartletti* attacking *Dalbulus maidis* has been studied in Puerto Rico (Bartlett, 1939) and Mexico (Moya Raygoza, 1990).

Some dryinids have been used in classical biological control programmes. For example, in 1906 and 1907 *Gonatopus* (= *Pseudogonatopus*) *hospes* and *Haplogonatopus vitiensis* were introduced into Hawaii from China and Fiji respectively, in order to control the sugar-cane planthopper, *Perkinsiella saccharicida* (Swezey, 1928), but these introductions failed apparently because the dryinids were attacked by hyperparasitic Encyrtidae (Williams, 1931). Although dryinids have not been extensively employed in biological control programmes in Central America, there are several species that potentially could be utilized. For example, species of *Gonatopus*

Pest species	Dryinid parasitoid	References
Dalbulus maidis	*Gonatopus bartletti*	Quezada, 1979; Vega & Barbosa, 1990; Moya Raygoza, 1990
Perkinsiella saccharicida	*Gonatopus morenoi*	Olmi, 1984
Poekilloptera phalaenoides	*Dryinus poecilopterae*	Ramirez, 1992
Saccharosydne saccharivora	*Gonatopus variistriatus*	Olmi, 1984
Tagosodes orizicolus	*Haplogonatopus hernandezae*	Hernandez, 1984
	Gonatopus flavus	Olmi, 1984, 1993*b*

Table 13.3. Major planthopper and leafhopper pests in Central America and the dryinids that attack them.

that are present in Colombia could be introduced against *Perkinsiella saccharicida* and *Saccharosydne saccharivora* if they are not already present (Quezada, 1986).

As mentioned above in the Hawaiian example, the effectiveness of dryinids as biological control agents may be limited by hyperparasitoids. These include Diapriidae on Anteoninae and Aphelopinae (Waloff, 1975; Masner, 1976*d*; Jervis, 1979), Ceraphronidae on Gonatopodinae (Swezey, 1908; Pagden, 1934), Encyrtidae on Gonatopodinae and Dryininae (Perkins, 1906; Pagden, 1934; Subba Rao, 1957; Tachikawa, 1981) and Chalcididae on Dryininae. In their turn the hyperparasitic Encyrtidae may be parasitized by Eulophidae (Perkins, 1906).

Identification. Keys to the world genera and species of Dryinidae are given by Olmi (1984, 1993*b*). More recent treatments of genera or species-groups are mentioned below under the faunal synopsis.

Key to subfamilies of Dryinidae present in Central America

(Note: males of the subfamily Biaphelopinae are unknown.)

1 Females. 2
— Males. 8

2 Front tarsus not chelate (Fig. 13.75). 3
— Front tarsus chelate (Fig. 13.76). 4

3 Fore wing with only the costal cell enclosed by pigmented veins (Fig. 13.77). ....**Aphelopinae** (p. 502)
— Fore wing with costal and basal cells enclosed by pigmented veins (Fig. 13.78). ....**Biaphelopinae** (p. 502)

4 Chela without rudimentary claw (Fig. 13.85). [Tibial spur formula 1: 1: 2; occipital carina complete; macropterous or brachypterous, at least with scale-like vestiges of fore wings discernible.] .... **Anteoninae** (p. 502)
— Chela with rudimentary claw (Fig. 13.86). 5

5 Antenna with tufts of long hairs on segments 5 to 10 (Fig. 13.88). ....*Thaumatodryinus*) (p. 503)
— Antenna without tufts of long hairs (Fig. 13.72). 6

6 Mid leg without a tibial spur, the tibial spur formula 1: 0: 1 or 1: 0: 2. ....**Gonatopodinae** (p. 503)
— Mid leg with a tibial spur, the tibial spur formula 1: 1: 1 or 1: 1: 2. 7

7 Mandible with four teeth, the anterior one the largest and the others progressively smaller (Fig. 13.81); 5th segment of fore tarsus always with more than one lamella (Fig. 13.86). ....**Dryininae** (most genera) (p. 503)
— Mandible with one to four teeth (Figs 13.82–84); quadridentate mandibles commonly with three large teeth and a rudimentary fourth tooth between the two posterior teeth (Fig. 13.83), ***or*** rarely with four teeth, the anterior one the largest and the others progressively smaller, ***but then*** with 5th segment of fore tarsus with only one preapical lamella (Fig. 13.87). ....**Bocchinae** (p. 502)

8 Fore wing with only the costal cell enclosed by pigmented veins (Fig. 13.77); occipital carina complete. ....**Aphelopinae** (p. 502)
— Fore wing with costal, basal and subbasal cells enclosed by pigmented veins (Figs 13.79, 13.80), or occasionally with only the costal cell enclosed, but then occipital carina is absent. 9

9 Mandible with four teeth, the anterior one the largest and the others progressively smaller (Fig. 13.81), or occasionally teeth similar to each other, never with a rudimentary tooth. 10
— Mandible with one to four teeth (Fig. 13.82–13.84); quadridentate mandibles with three large teeth and a rudimentary tooth between the two posterior teeth (Fig. 13.83). 11

10 Fore wing with distal abscissa of *R* about as long as or longer than the pterostigma (Fig. 13.80). .... *Thaumatodryinus* (p. 503)
— Fore wing with distal abscissa of *R* shorter than the pterostigma (Fig. 13.79). .... **Anteoninae** (p. 502)

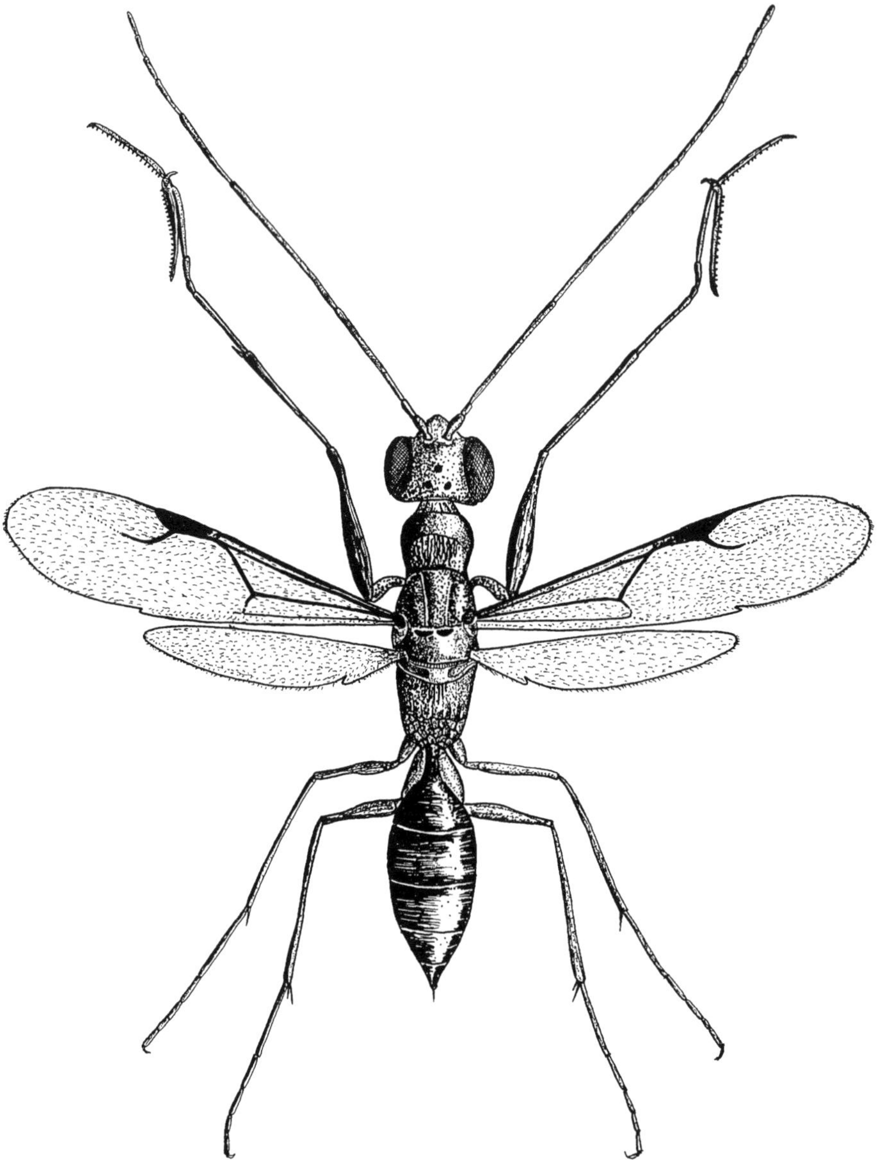

Fig. 13.76. *Gonadryinus hansoni* (Dryinidae: Dryininae).

11 Mandible with one to four teeth; tridentate mandibles with teeth not progressively larger from the anterior one to the posterior one (Fig. 13.82). .... **Bocchinae** (p. 502)

— Mandible with three teeth progressively larger from the anterior one to the posterior one (Fig. 13.84). 12

12 Occipital carina complete or incomplete; male genitalia simple, without dorsal process on gonoforceps (Fig. 13.89). **Dryininae** (p. 503)

— Occipital carina absent; occasionally complete, but then with male genitalia with dorsal process on gonoforceps (Fig. 13.90). .... **Gonatopodinae** (p. 503)

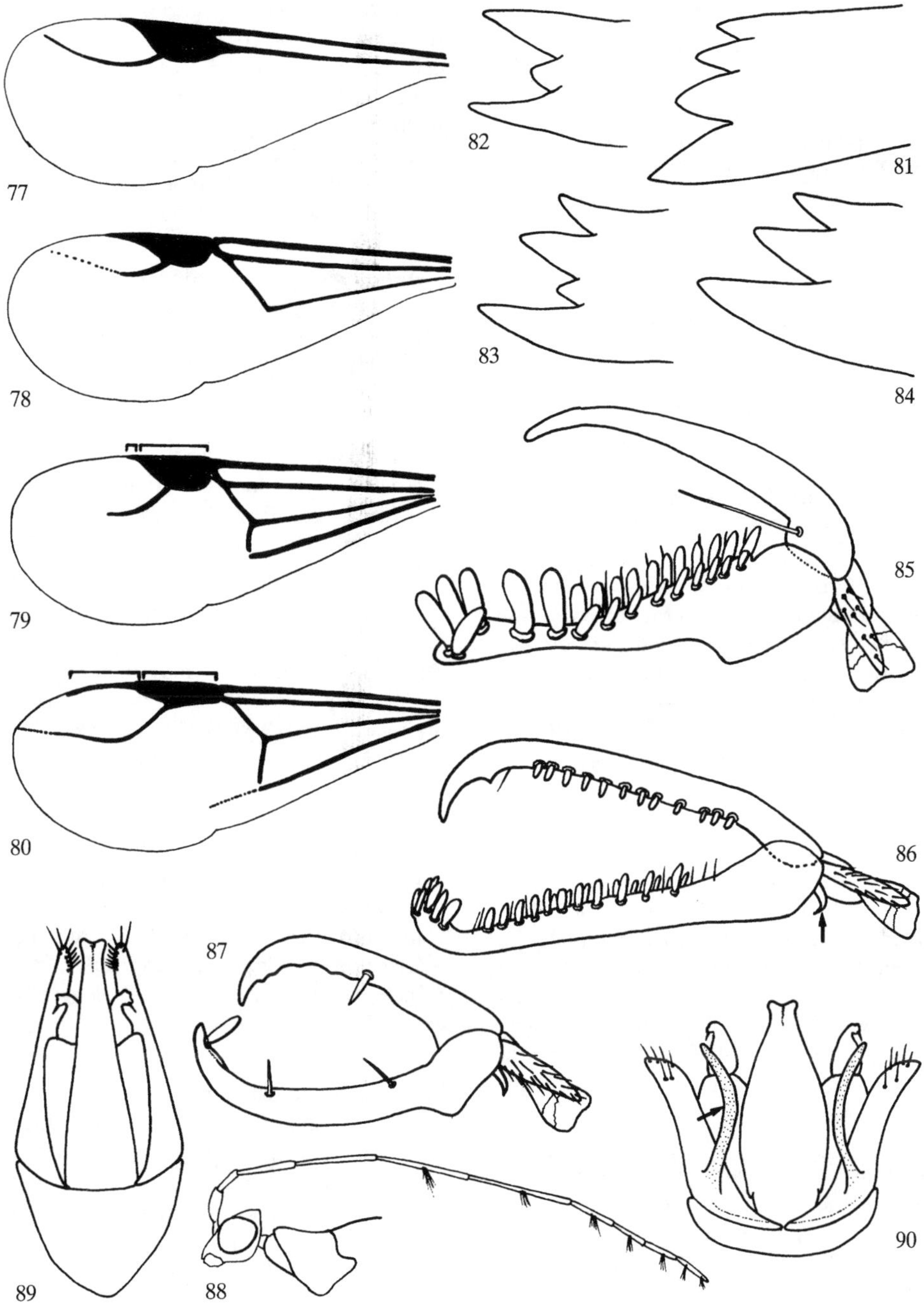

Figs 13.77–13.90. Dryinidae. Figs 13.77–13.80. Fore wing; 13.77, *Aphelopus* sp. (Aphelopinae); 13.78, *Biaphelopus* sp. (Biaphelopinae); 13.79, *Anteon* sp. (Anteoninae); 13.80, *Thaumatodryinus* sp. (Dryininae). Figs 13.81–13.84. Mandible; 13.81, Dryininae (female); 13.82, Bocchinae; 13.83, Bocchinae; 13.84, Dryininae (male). Figs 13.85–13.87. Fore tarsal chela; 13.85, Anteoninae; 13.86, Dryininae; 13.87, Bocchinae. Fig. 13.88. Antenna, *Thaumatodryinus* sp. (Dryininae). Figs 13.89–13.90. Male genitalia; 13.89, *Dryinus* sp. (Dryininae); 13.90. *Gonatopus* sp. (Gonatopodinae).

Synopsis of the Costa Rican fauna

Seven of the eleven subfamilies of Dryinidae are represented in Costa Rica by a total of 89 species in 16 genera (Olmi, 1993*a* & *b*). Dryinids occur in virtually all terrestrial habitats, from the coast to paramo-like areas above 2800 metres on the Cerro de la Muerte. They are present in both natural forests and agroecosystems.

ANTEONINAE

The Anteoninae is a large, cosmopolitan subfamily comprising five genera, three of which are represented in Costa Rica by a total of 41 species; they have been collected from sea-level to 3000 metres. Anteonines are small to moderately large dryinids (body length of 0.9 to 7.5 mm) and the females are black or brightly coloured, while the males are commonly black. Both sexes are usually fully-winged, but rarely the females are brachypterous, but they always have at least the fore wing discernible as a scale-like vestige. In fully-winged individuals the fore wing has three enclosed cells, the costal, basal and subbasal (Fig. 13.79). Although the females have chelate fore tarsi, the chela is primitive. The inner edge of the enlarged claw lacks lamellae or peg-like teeth (Fig. 13.85), which are characteristic of the more specialized dryinids.

Anteon. A large cosmopolitan genus represented in Costa Rica by 24 species. Keys to the neotropical species are given by Olmi (1989). Biology: species of *Anteon* attack various Cicadellidae, but none are recorded from Typhlocybinae.

Deinodryinus. A large, widespread genus represented in Costa Rica by 16 species. Keys to the neotropical species are given by Olmi (1989). Biology: unknown.

Lonchodryinus. A small cosmopolitan genus with a single species present in Costa Rica. Keys to the neotropical species are given by Olmi (1989). Biology: *Lonchodryinus* species attack various Cicadellidae.

APHELOPINAE

The Aphelopinae is a small cosmopolitan subfamily comprising just two genera, both of which are present in Costa Rica, where species can be found between sea-level and 3000 metres. Aphelopines are small dryinids (body length 1.1 to 3.1 mm) which are commonly black coloured, but with the head often partly whitish. Fore wing venation is very reduced, with a single enclosed cell (Fig. 13.77). Species of this subfamily exhibit very little sexual dimorphism and are believed to be very primitive.

Aphelopus. A moderate sized, cosmopolitan genus with six species known to occur in Costa Rica. Keys to the neotropical species are given by Olmi (1984). Biology: *Aphelopus* species parasitize typhlocybine cicadellids.

Crovettia. A primarily neotropical genus with two species present in North America. Three species are recorded from Costa Rica (Olmi, 1993*a*). Keys to the neotropical species are given by Olmi (1984). Biology: *Crovettia* species parasitize Membracidae.

BIAPHELOPINAE

This is a very small subfamily comprising only a single genus, *Biaphelopus* (Olmi, 1984). Males and biology are unknown. These are small insects (body length 2.4 to 2.6 mm) which are black in colour, and with two enclosed cells in the fore wing (Fig. 13.78).

Biaphelopus. A very small genus comprising only two species worldwide. One species is present in Nepal, whilst the other, *Biaphelopus hansoni*, occurs in Costa Rica (Olmi, 1989) where it has been collected on the edge of dense lowland wet forest at 400 metres on the Atlantic slope. Biology: unknown.

BOCCHINAE

The Bocchinae is a small subfamily comprising six genera, one of which occurs in Costa Rica. Bocchines are small to large dryinids (body length 1.4 to 10.6 mm) with very pronounced sexual dimorphism. The males commonly are black whilst the females vary from black to brightly coloured. Both sexes are usually winged and females are only rarely brachypterous.

Bocchus. A moderately large, cosmopolitan genus represented in Costa Rica by two species. Keys to the neotropical species of *Bocchus* are given by Olmi (1989). Biology: hosts are Issidae (Olmi, 1984).

DRYININAE

A large subfamily comprising, worldwide, five extant genera. Three of these have been found in Costa Rica, at low and mid altitudes. A key to genera is given by Olmi (1993*b*) and keys to the neotropical species are given in Olmi (1984, 1986, 1987*a*, *b* & *c*, 1989, 1990). Dryinines are small to large dryinids (body length 1.6 to 11.6 mm) with both sexes fully winged, but sexual dimorphism very pronounced. The males are commonly black and the females brightly coloured. The fore legs of the female are very elongate and chelate.

Dryinus (= *Alphadryinus, Mesodryinus, Perodryinus, Tridryinus*, see Olmi, 1993*b*). A large cosmopolitan genus that, worldwide, includes almost all of the species in the subfamily. Thirteen species are known to be present in Costa Rica. Biology: parasitoids of species of Dictyopharidae, Cixiidae, Flatidae, Acanaloniidae, Issidae, Ricaniidae, Delphacidae and Fulgoridae (Olmi, 1984). The biology of *D. poecilopterae* has been studied in Colombia (Ramirez, 1992).

Gonadryinus. A small neotropical genus represented in Costa Rica by a single species, *G. hansoni*. Biology: unknown.

Thaumatodryinus. A small tropical and subtropical genus with one species, *T. clarus*, recorded from Costa Rica. Keys to the neotropical species are given by Olmi (1984, 1987*a*). Biology: *Thaumatodryinus* species parasitize Flatidae.

GONATOPODINAE

This is the largest subfamily of dryinids, comprising more than half of the described species. Fourteen genera are recognized and six of these are represented in Costa Rica where individuals have been collected from sea-level to 2100 metres. A key to genera is given by Olmi (1993*b*) and keys to the neotropical species are given by Olmi (1984, 1986, 1987*a*). Gonatopodines are small to moderately large dryinids (body length 1.2 to 7.8 mm) that exhibit pronounced sexual dimorphism—the females commonly being apterous (winged in a few genera, rarely brachypterous) and black or brightly coloured, while the winged males are usually black. The females have very elongate and chelate fore legs and the male genitalia have a more or less developed dorsal process of gonoforceps.

Eucamptonyx. A small neotropical genus represented in Costa Rica by two species. Biology: unknown.

Gonatopus (= *Acrodontochelys, Apterodryinus, Dicondylus, Donisthorpina, Pseudogonatopus* see Olmi, 1993*b*). A very large, cosmopolitan genus represented in Costa Rica by 14 species. Biology: *Gonatopus* species parasitize a variety of Cicadellidae (but not Typhlocybinae, Idiocerinae or Macropsinae), Delphacidae, Acanaloniidae, Flatidae, Tropiduchidae and Lophopidae.

Haplogonatopus. A small but widespread genus that is represented in Costa Rica by a single species, *H. hernandezae*. Biology: *Haplogonatopus* species parasitize Delphacidae.

Neodryinus. A moderate sized, pantropical genus with a few species also occurring in North America. It is represented in Costa Rica by two species. Biology: *Neodryinus* species parasitize Flatidae, Ricaniidae and Nogodinidae (Olmi, 1984).

Pareucamptonyx. A small neotropical genus represented in Costa Rica by a single species, *P. costaricanus*. Biology: unknown.

Trichogonatopus. A small predominantly neotropical genus with one species occurring northwards into the southern United States. It is represented in Costa Rica by a single species, *T. albomarginatus*. Biology: unknown.

14 The vespoid families (except vespids and ants)

INTRODUCTION

Denis J. Brothers

The superfamily Vespoidea is almost certainly a monophyletic clade which is the sister group of the Apoidea (Brothers, 1975; Brothers & Carpenter, 1993). Unlike the Chrysidoidea and the Apoidea, however, the Vespoidea does not appear to possess any unique synapomorphies, mainly being characterized by the reduced prepectus. It comprises ten extant monophyletic families (Fig. 14.01). The most recent and comprehensive study of the phylogeny of the families (and some subfamilies) of Vespoidea is that of Brothers and Carpenter (1993) who re-evaluated the work of Brothers (1975) and subsequent authors. The phylogeny shown in Fig. 14.01 represents our present best estimate of their relationships.

The major differences from the previous schemes (e.g. Brothers, 1975) involve consideration of the Sierolomorphidae as the most 'primitive' family, and the Pompilidae as more closely related to the Mutillidae, Tiphiidae and Sapygidae than to the Rhopalosomatidae. The Pompilidae is unequivocally a member of the Vespoidea contrary to Rasnitsyn's (1980, 1988) contention that it should be considered part of the Apoidea.

In attempting to trace the changes in biology reflected in the above phylogeny (Fig. 14.02), difficulties are encountered, since the biology of some of the taxa is unknown, and others show a considerable range of behaviour patterns. Species of Sierolomorphidae are thought to be idiobiont ectophagous parasitoids of other arthropods found within a friable solid substrate, since this is hypothesized to be the primitive condition in the aculeates (Rasnitsyn, 1980) and their morphology does not contradict it (Brothers & Carpenter, 1993), but nothing is actually known of their habits. Members of the Tiphiidae are primarily idiobiont ectophagous parasitoids of solitary subterranean beetle larvae, so presumably retaining the primitive condition. The Pompilidae are specialized in attacking spiders and in closing the shelter in which the host (prey) is placed, but they are otherwise primitive in utilizing single host/prey individuals and paralysing them. Mutillidae

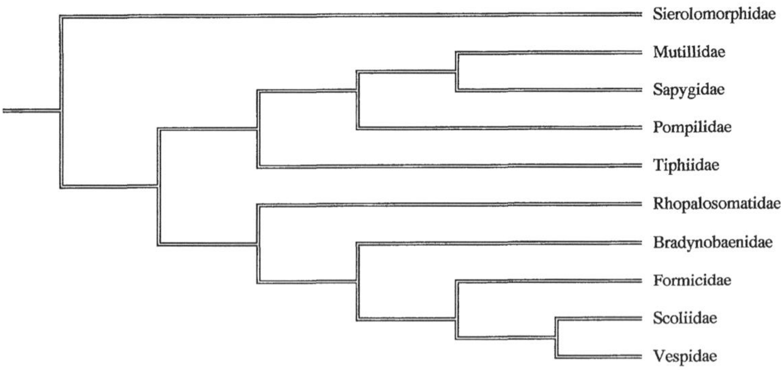

Fig. 14.01. Postulated phylogenetic relationships of the Vespoidea. For details of characters supporting this cladogram see Brothers and Carpenter (1993).

and Sapygidae are essentially similar, but utilize the immatures of other aculeate Hymenoptera as hosts and do not transport them.

The other major lineage of Vespoidea shows a greater range of behaviour patterns. Rhopalosomatidae are koinobiont ectoparasitoids of crickets, developing in a manner superficially similar to that of Embolemidae and Dryinidae (Chrysidoidea), and thus very different from other vespoids. The biology of the Bradynobaenidae is almost completely unknown (Chapter 14.1); one species of Typhoctini has been recorded as a koinobiont ectoparasitoid of immature Solifugae, but this may not be typical of other members of the family. The entirely social Formicidae have a range of behaviour patterns, but the most primitive ants are predators on other arthropods, utilizing numerous prey individuals which are transported to the nest. The Vespidae show various degrees of sociality, but primitively are predators of other arthropods which are captured and transported to a previously constructed cell already containing an egg. The Scoliidae are somewhat anomalous in this series, because they seem to have retained an essentially primitive pattern, being idiobiont ectoparasitoids of beetle larvae within a substrate, although they appear to transport the host individual a short distance and construct a rudimentary cell in which it is placed.

All of the vespoid families are represented in Central America, although some (such as the

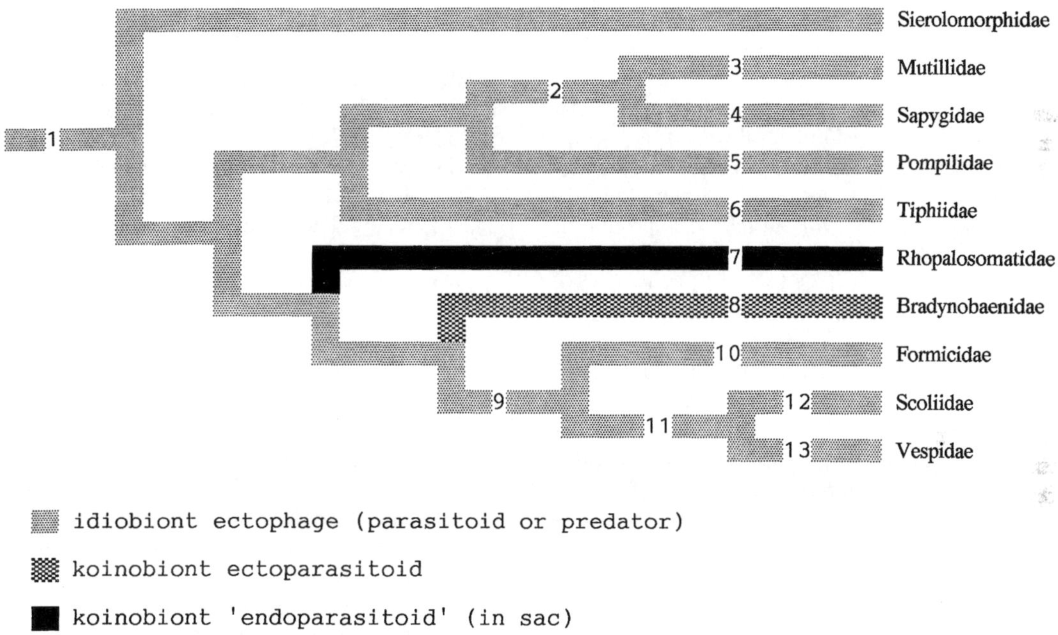

Fig. 14.02. Suggested biological evolution within the Vespoidea. **1**: exploiting single arthropod (probably immature beetle) hosts in soil (the ancestral habitat); **2**: exploiting single immature aculeate Hymenoptera in host cells in ancestral habitat; **3**: some species exploiting hosts in other situations or attacking other immature stages in enclosed 'packages'; **4**: some species exploiting hosts in other situations; **5**: exploiting single spiders in ancestral habitat, but many are relocated and concealed; many species construct cells and a few are koinobiont ectoparasitoids of active spiders; **6**: one species attacking and re-locating single gryllotalpids in ancestral habitat and a few attacking ancestral hosts in burrows in wood; **7**: attacking single gryllids on ancestral habitat; **8**: attacking single immature Solifugae in ancestral habitat (based on one species); **9**: relocating host (prey); **10**: eusocial, attacking many and varied arthropods in ancestral habitat, but many species in different habitats, and/or utilizing fungal or plant material as food; **11**: constructing and closing cell containing host (prey); **12**: exploiting single ancestral hosts in ancestral habitat; **13**: exploiting several insect hosts placed in ancestral habitat after egg laying, but many species utilizing different habitats, and some social to various degrees. Complex nest construction has evolved independently in Pompilidae, Vespidae and Formicidae.

Rhopalosomatidae) are very much better represented than others (such as the Scoliidae). The remainder of this chapter covers the eight families without eusocial representatives. Summarizing the vast body of literature that has accrued about ants and vespids requires treating each as a separate chapter.

The winged members of the Vespoidea may be identified to family using the key given below; apterous individuals can be recognized using the keys provided in Chapter 5.4.

Key to alates of families of Vespoidea present in Central America

Ian D. Gauld

[Note: Masarine vespids and chyphotine bradynobaenids are accommodated in this key. Neither have been collected in Costa Rica; the former are expected, but the latter (indicated by an asterisk) are unlikely to be present.]

1 Metasoma with segment I or segments I and II, node-like, humped in lateral view, sharply separated from following segments (Figs 14.03–14.06); eyes with inner margin convex; tergite II without felt lines; pronotum in dorsal view with posterior margin very strongly concave (Figs 14.07, 14.08). ....**Formicidae** (p. 588)

— Metasoma with segment I not node-like (Figs 14.10, 14.11), ***if*** slightly humped in lateral view and somewhat separated from following segments (Fig. 14.09) ***then either*** eyes with inner margin notched (Figs 14.12–14.15) ***and/or*** tergite II with felt lines (Fig. 14.9) ***and/or*** pronotum in dorsal view with posterior margin weakly concave (Figs 14.19–14.20). 2

2 Distal part of wings finely corrugated (Fig. 14.36); mesosoma ventrally with metepisternum extending posteriorly to cover bases of hind coxae (Fig. 14.27); male with three stout spines projecting beyond apex of last metasomal sternite. .... **Scoliidae** (p. 555)

— Distal part of wings not corrugated (Figs 14.30–14.34); mesosoma ventrally with metepisternum not covering bases of hind coxae (Fig. 14.28); male with apex of metasoma various—simple, with hooked last tergite or with two projecting spines, never with three long spines. 3

3 Mesopleuron with a straight diagonal sulcus which traverses the mesopleural scrobe (Fig. 14.21); hind legs relatively long and generally with tibia somewhat spinose.

[Eye generally without a deep notch on inner side opposite antennal socket; longest apical spur of hind tibia with 'brush' on inner side basally, this 'brush' coinciding with a 'brush' on the inner surface of the basitarsus.] .... **Pompilidae** (p. 522)

— Mesopleuron without a straight diagonal sulcus which traverses the mesopleural scrobe (Figs 14.10–14.11); hind legs various. 4

4 Eye with a deep notch on inner side nearly opposite antennal socket, this notch U- or V-shaped (Figs 14.13–14.15) 5

— Eye with inner margin simply convex or straight (Figs 14.16, 14.18), or with a very broad, shallow notch (Fig. 14.17). 10

5 Antenna with one or two stout setae on the dorsal apical margin of first to fifth flagellar segments; fore wing with costal cell obliterated, *C* and *Sc+R+Rs* contiguous for most of their length (Fig. 14.30); females with distal four tarsal segments flattened and broadened, ventrally concave. ....**Rhopalosomatidae** (p. 548)

— Antenna without stout setae on the dorsal apical margin of the basal flagellar segments; fore wing with costal cell discernible, though sometimes narrow, *C* not contiguous with *Sc+R+Rs* for most of its length (Figs 14.31–14.34); females with distal four tarsal segments unspecialized. 6

6 Antenna with flagellum short and very strongly swollen apically, club-like. ....**Vespidae** (Masarinae^{EX}) (p.564)

— Antenna with flagellum short to moderately long, not swollen apically, not club-like. 7

7 Fore wing with 1st discal cell very elongate, at least 0.35 of length of wing, often more (Fig. 14.31); antennal sockets more or less flush with front of head, without overhanging dorsal lobe or ridge (Fig. 14.14); fore wing always folded longitudinally; upper hind corner of pronotum right-angled or acutely pointed (Fig. 14.11)..**Vespidae** (p. 561)

— Fore wing with 1st discal cell short to moderately elongate (Figs 14.32–14.34); antennal sockets facing downwards or outwards, with an overhanging dorsal lobe or ridge (Fig. 14.15); fore wing seldom folded longitudinally; upper hind corner of pronotum truncated, rounded or obtuse (Fig. 14.10). 8

8 Male only: apex of sternite VIII produced into a long, up-curved hook; mesosternal region produced posteriorly into a pair of large triangular lobes that partially cover the bases of the mid coxae (Fig. 14.28). .... **Tiphiidae** (male Myzininae) (p. 514)

— Males or females: male with apex of sternite VIII not produced into a long up-curved hook; mesosternal region not posteriorly produced into a pair of large triangular lobes. 9

9 Metasoma petiolate, tergite I rather slender, much narrower and shorter than tergite II (Fig. 14.19); mid and hind tarsal claws simple; hind wing without a distinct jugal lobe (Fig. 14.32). .... **Mutillidae** (part) (p. 541)

— Metasoma not petiolate, tergite I gradually narrowed anteriorly, posteriorly as broad as tergite II, and longer than tergite II (Fig. 14.20); mid and hind tarsal claws with a median tooth; hind wing with a distinct jugal lobe (Fig. 14.37). .... **Sapygidae** (p. 539)

10 Metasoma with anterior end of first segment slender, formed from sternite only (Fig. 14.29).

[Ocelli enlarged; tergite II with 'felt-lines' laterally; last sternite posteriorly produced into a stout up-curved hook.] ....**Bradynobaenidae** (Chyphotinae*)

— Metasoma with anterior end of first segment broad to slender, formed from tergite as well as sternite. 11

11 Males only (antenna 13-segmented): posterior end of metasoma with a pair of slender, often hooked projections (the apices of the parameres) (Fig. 14.22), the subgenital plate apically truncate or with a short median point; tergite II often with felt-lines (an elongate area bearing fine close short pubescence) laterally (Fig. 14.9); metasoma in dorsal view generally petiolate anteriorly, with tergite I either much narrower than tergite II and differentiated from it, and/or abruptly narrowed into a 'neck' anteriorly (Figs 14.23–14.25).

[Very hairy wasps, upper part of head bearing pubescence that is longer than interocellar distance.] 12

— Males or females: male with posterior end of metasoma without paired projecting parameres, sometimes with the subgenital plate posteriorly elongately pointed and up-curved to form a hook, or sometimes rounded to acute; male and female with tergite II without felt-lines laterally; metasoma, in dorsal view, spindle-shaped, with tergite I evenly narrowed, posteriorly as broad as tergite II, and more or less confluent with it. 13

12 Metasoma with tergite I abruptly constricted anteriorly into a narrow 'neck' (Fig. 14.25); head without large lobe overhanging antennal sockets; eye quite elongately oval, with a very shallow indentation opposite antennal insertion, weakly convex (Fig. 14.17); fore wing with ten enclosed cells (Fig. 14.35). ....**Bradynobaenidae** (Typhoctinae)(p. 552)

— Metasoma with tergite I evenly constricted anteriorly, or not constricted, not forming a narrow 'neck' (Figs 14.23–14.24); head with lobe or ridge overhanging the antennal sockets; eye subcircular, without a shallow indentation opposite antennal insertion, and very strongly convex (Fig. 14.16); fore wing with six to ten enclosed cells (e.g. Fig. 14.32). ....**Mutillidae** (p. 541)

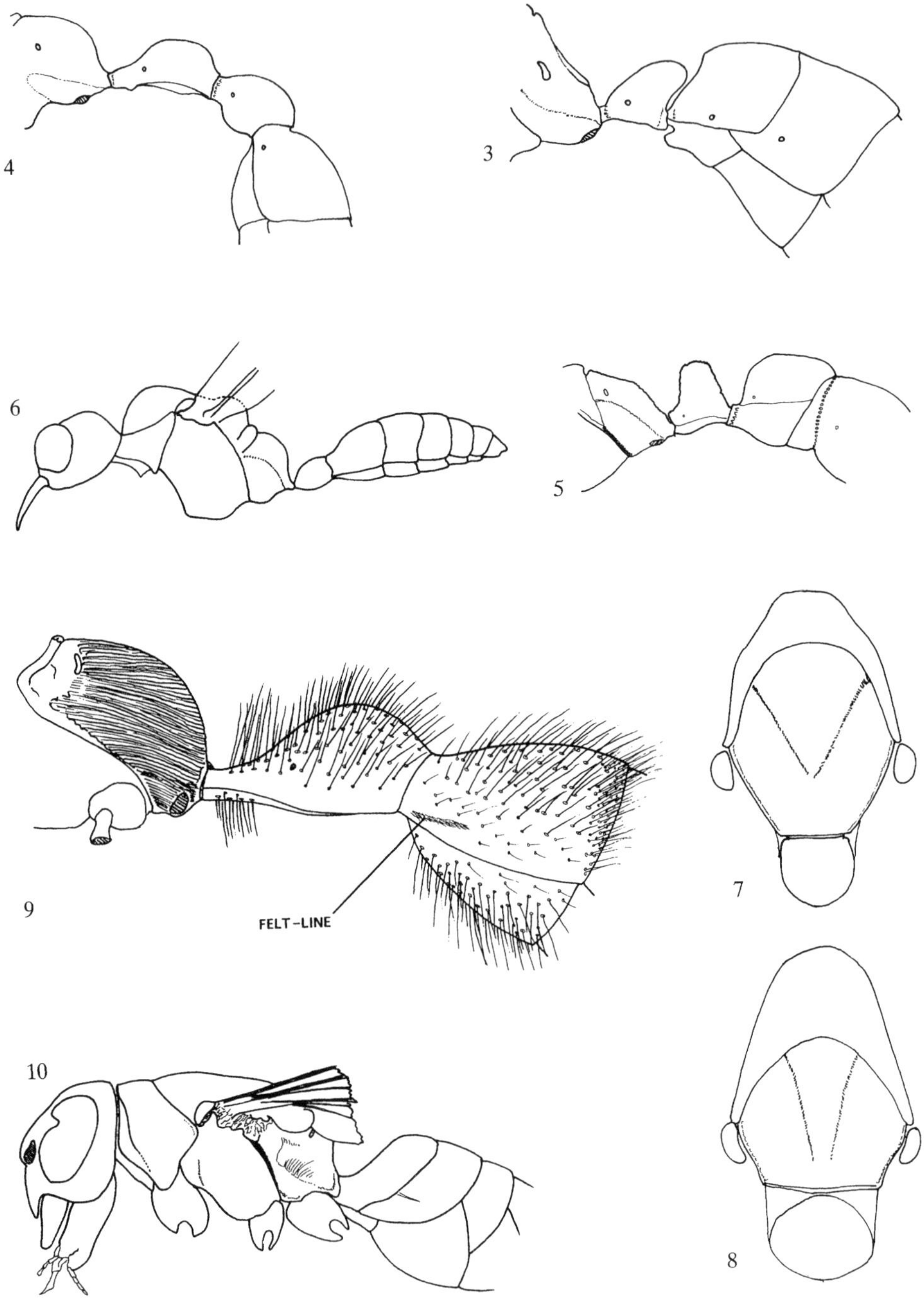

Figs 14.03–14.10. Vespoidea. Figs 14.03–14.06. Formicidae, lateral view petiolar region; 14.03, *Labidus* sp.; 14.04, *Pseudomyrmex* sp.; 14.05, *Leptogenys* sp.; 14.06, male formicine. Figs 14.07–14.08. Formicidae, males, anterior part of mesosoma, dorsal; 14.07, *Leptogenys* sp.; 14.08, *Pseudomyrmex* sp. Fig. 14.09. *Typhoctes* sp. (Bradynobaenidae), base of metasoma, lateral. Fig. 14.10. Mesosoma and base of metasoma, lateral, *Huarpa* sp. (Sapygidae).

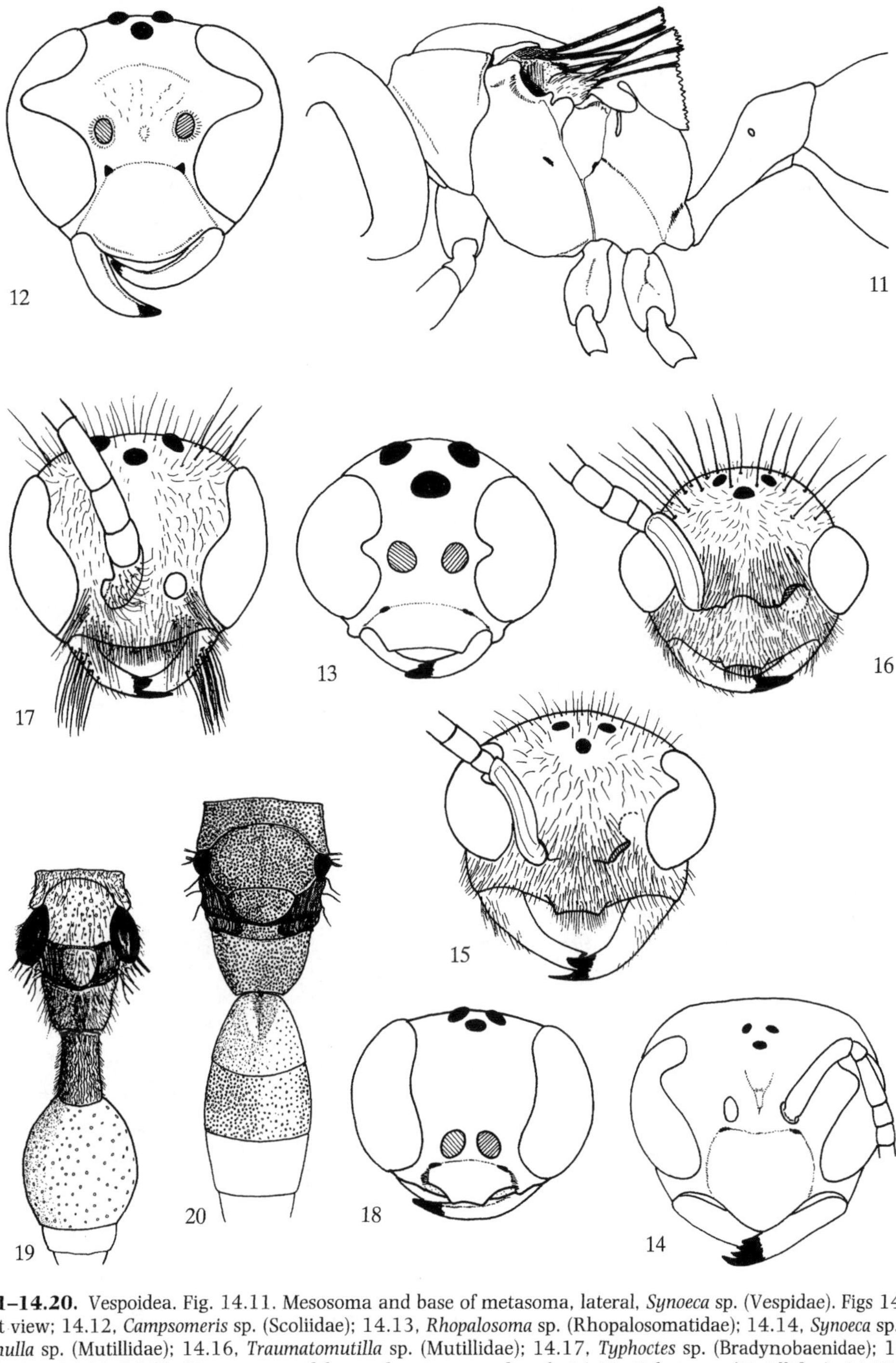

Figs 14.11–14.20. Vespoidea. Fig. 14.11. Mesosoma and base of metasoma, lateral, *Synoeca* sp. (Vespidae). Figs 14.12–14.18. Head, front view; 14.12, *Campsomeris* sp. (Scoliidae); 14.13, *Rhopalosoma* sp. (Rhopalosomatidae); 14.14, *Synoeca* sp. (Vespidae); 14.15, *Timulla* sp. (Mutillidae); 14.16, *Traumatomutilla* sp. (Mutillidae); 14.17, *Typhoctes* sp. (Bradynobaenidae); 14.18, *Tiphia* (Tiphiidae). Figs 14.19–14.20. Mesosoma and base of metasoma, dorsal; 14.19, *Ephuta* sp. (Mutillidae); 14.20, *Sapyga* sp. (Sapygidae).

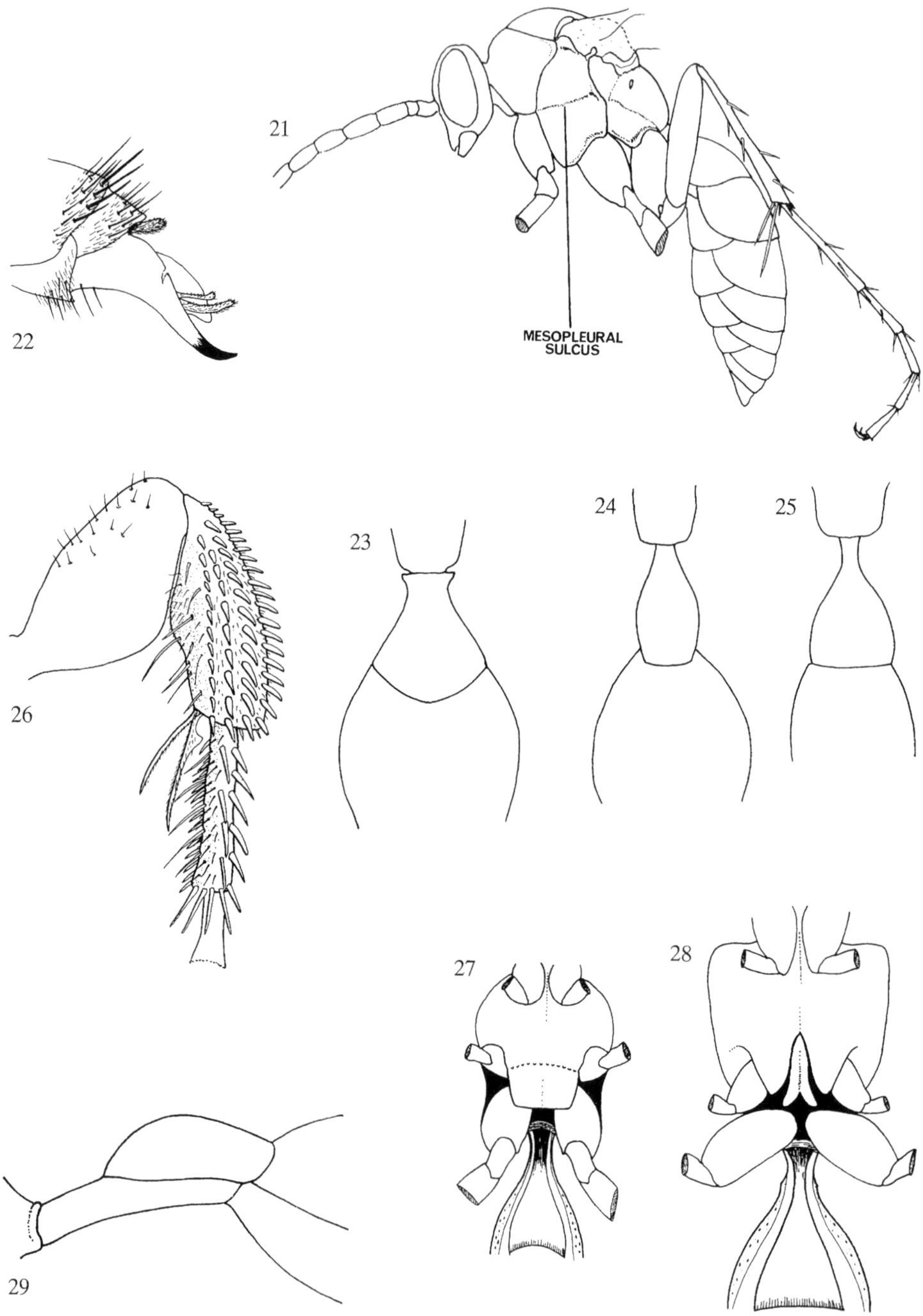

Figs 14.21–14.29. Vespoidea. Fig. 14.21. Pompilidae, body, lateral. Fig. 14.22. Male genitalia, *Traumatomutilla* sp. (Mutillidae). Figs 14.23–14.25. Anterior part of metasoma, dorsal; 14.23, 14.24, Mutillidae; 14.25, *Typhoctes* sp. (Bradynobaenidae). Fig. 14.26. Hind leg, female, *Myzinum* sp. (Tiphiidae). Figs 14.27–14.28. Mesosoma, ventral; 14.27, *Campsomeris* sp. (Scoliidae); 14.28, *Tiphia* sp. (Tiphiidae). Fig. 14.29. Base of metasoma, lateral, *Chyphotes* sp. (Bradynobaenidae).

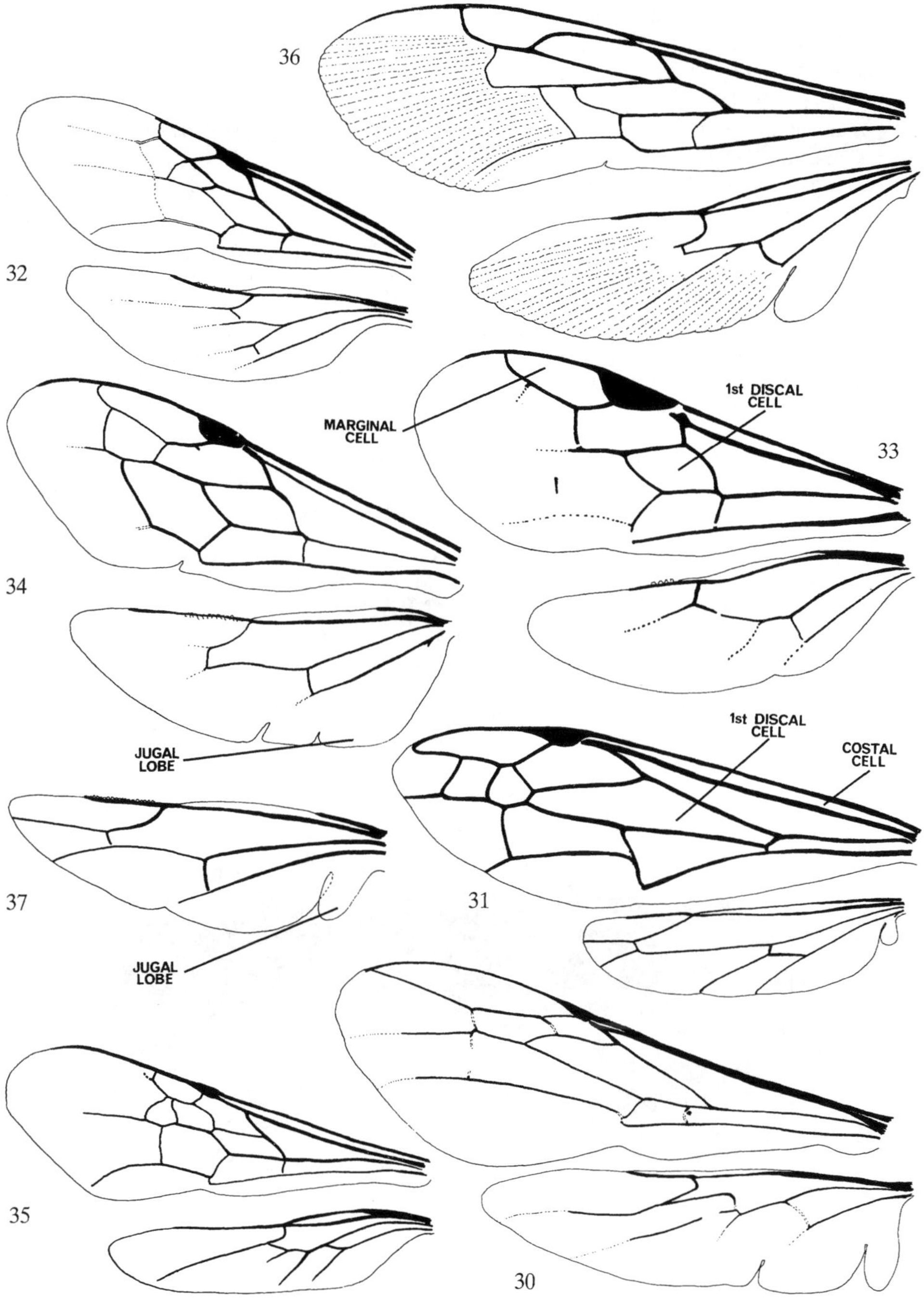

Figs 14.30–14.37. Vespoidea. Figs 14.30–14.36. Fore and hind wings; 14.30, *Rhopalosoma* sp. (Rhopalosomatidae); 14.31, *Synoeca* sp. (Vespidae); 14.32, Mutillidae; 14.33, *Sierolomorpha* sp. (Sierolomorphidae); 14.34, *Tiphia* sp. (Tiphiidae); 14.35, *Typhoctes* sp. (Bradynobaenidae); 14.36, *Campsomeris* sp. (Scoliidae). Fig. 14.37. Hind wing, *Sapyga* sp. (Sapygidae).

13 Hind wing without a discernible jugal lobe; fore wing with seven enclosed cells, the marginal cell always longer than the pterostigma (Fig. 14.33); male with subgenital plate unspecialized; male antenna often with slender tyloids on central flagellar segments; mesosternal region not posteriorly produced into a pair of large triangular lobes. .... **Sierolomorphidae** (p. 512)

— Hind wing with a jugal lobe; fore wing usually with eight, nine or ten enclosed cells (Fig. 14.34), ***if*** fewer (six or seven) ***then*** marginal cell shorter than pterostigma (Fig. 14.46); male often with apex of sternite VIII produced into a long, up-curved hook, or ***if*** simple, ***then*** with fringed posteriorly with a row of short stout spines; mesosternal region sometimes produced posteriorly into a pair of large triangular lobes that partially cover the bases of the mid coxae (Fig. 14.28). .... **Tiphiidae** (p. 514)

14.1 SIEROLOMORPHIDAE

Paul E. Hanson

Diagnosis. Body length 3.5 to 6.0 mm; *shiny black*, sometimes with yellowish legs; with barely visible short,

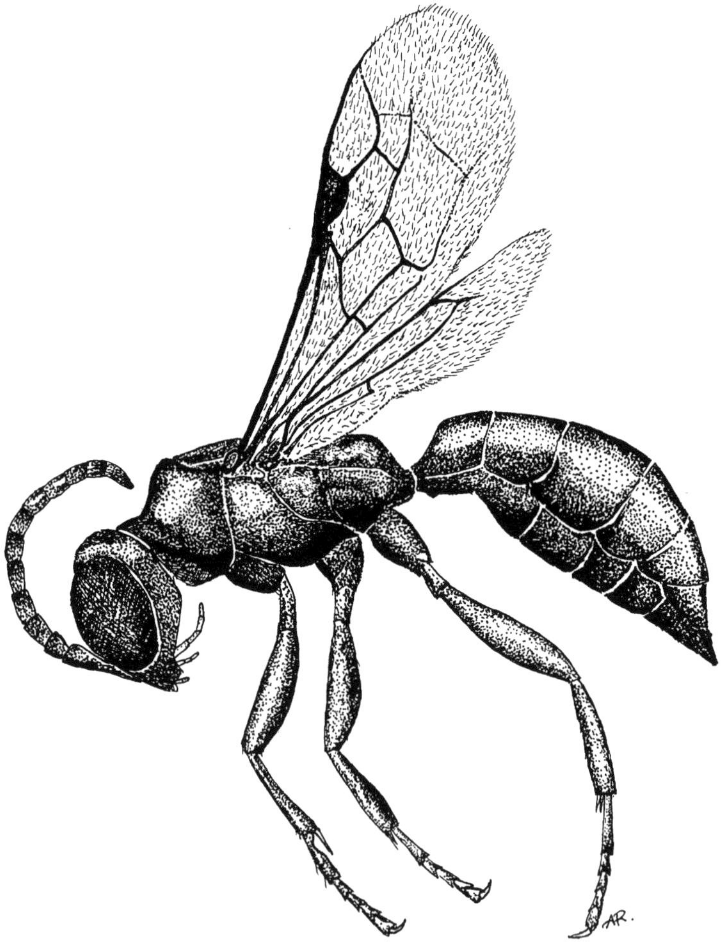

Fig. 14.38. *Sierolomorpha* sp. (Sierolomorphidae).

decumbent hairs; sexual dimorphism minimal, both sexes usually fully winged. Antennae inserted low on face, in female 12-segmented, in male 13-segmented, male antenna often with tyloids; eye oval with inner margin very shallowly sinuate, not protuberant. Pronotum large, freely articulating with mesothorax, hind margin shallowly arcuate, posterolateral angle rounded and reaching tegula. *Fore wing with seven enclosed cells, pterostigma large and sclerotized,* 2rs-m *and* 2m-cu *absent*; hind wing with two enclosed cells, without distinct claval and jugal lobes. *Fore and hind coxae contiguous*; tibial spur formula 1: 2: 2, spurs simple; tarsal claws with submedian tooth. *Metasoma with first segment broad* (not petiolate), tergite I with lateral margin overlying sternite I and not fused to it; female with tergite VII hidden; male sternite VIII simple.

Classification and distribution. This small enigmatic family is known from the single genus, *Sierolomorpha*, which comprises seven described species in the New World (Canada to Panama), one in Hawaii, and two in the Old World (northern China and Tadzhikistan). The relationship of the family is not well understood. Brothers (1975) placed the Sierolomorphidae in the Vespoidea, just above the basal Tiphiidae – Sapygidae – Mutillidae branch, but in their re-analysis of vespoid phylogeny Brothers and Carpenter (1993) now consider the sierolomorphids as the sister-lineage to all remaining vespoids.

Biology. Nothing is known about the biology of this family, although given what is understood about the phylogenetic position of the Sierolomorphidae and the habitats of its relatives, it is possibly attacking coleopterous larvae that are concealed in soil or another substrate. However, this is pure speculation at present.

Identification. Evans (1961*b*) provided a key to the nearctic species and an additional species was described by Miller (1986).

Synopsis of Costa Rican taxa

Two species of *Sierolomorpha* have been collected in Costa Rica: *similis* in the seasonally dry, lowland (below 1000 m) forests of northwestern Costa Rica and west of San José (Ciudad Colón), and an undescribed species from mid-altitude (1500 m) forests in southern Costa Rica.

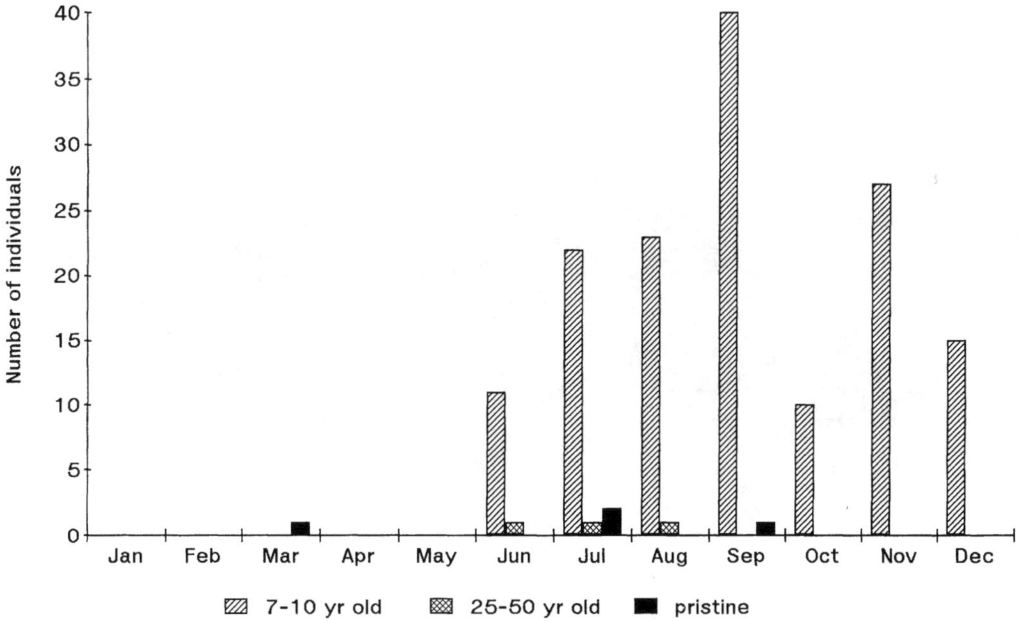

Fig. 14.39. Monthly occurrence of *Sierolomorpha similis* (Sierolomorphidae) in Malaise trap catches in different aged forest habitats in Santa Rosa National Park, averaged from data for 1985–7.

S. similis has been collected most commonly in scrubby regenerating wooded areas in Santa Rosa National Park west of the *Area Administrativa*. Despite intensive collecting few individuals have been taken in more mature areas of forest such as the Bosque San Emilio or the Bosque Humedo. Virtually all specimens were collected between June and December (the wet season) (Fig. 14.39), and only a single individual was taken during the dry season from January to May (Gauld, pers. comm.).

14.2 TIPHIIDAE

Lynn S. Kimsey

Diagnosis. Body length 2.8 to 20.0 mm; usually black, sometimes black with yellow markings. Sexual dimorphism slight to extreme; males always winged, females either winged or apterous, the latter slender and ant-like in general appearance. Eye not emarginate internally (except in males of some Myzininae). Antenna of females 12-, of males 13-segmented. Pronotum separated from mesoscutum by a strong dorsal suture, the two not fused; dorsal mesosoma with other transverse sutures present behind the promesonotum; in winged forms the *pronotum produced back to, or almost to, the tegulae*; mesosoma in ventral view with the *mid coxae widely separated, distinctly not contiguous, their bases preceded by a pair of teeth or laminate lobes which may overlap the coxae*. Venation nearly complete, the fore wing with six or (usually) more enclosed cells; hind wing almost always with two (very rarely three) enclosed cells, and usually with both claval and jugal lobes present,

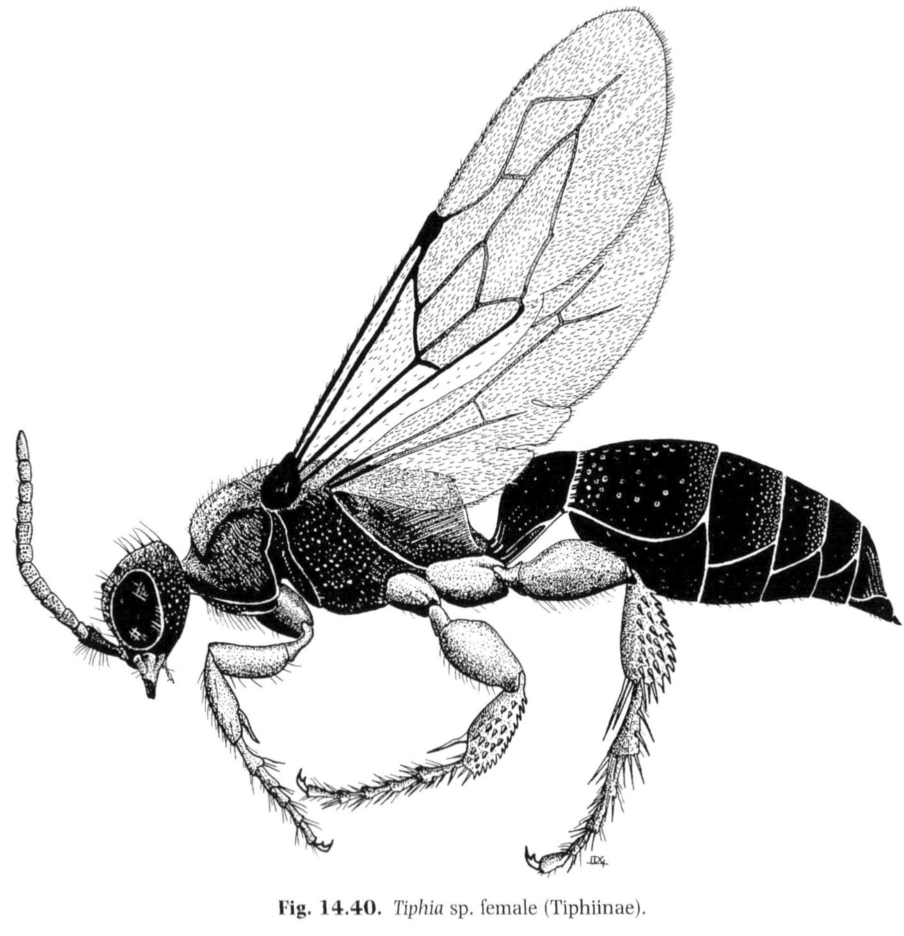

Fig. 14.40. *Tiphia* sp. female (Tiphiinae).

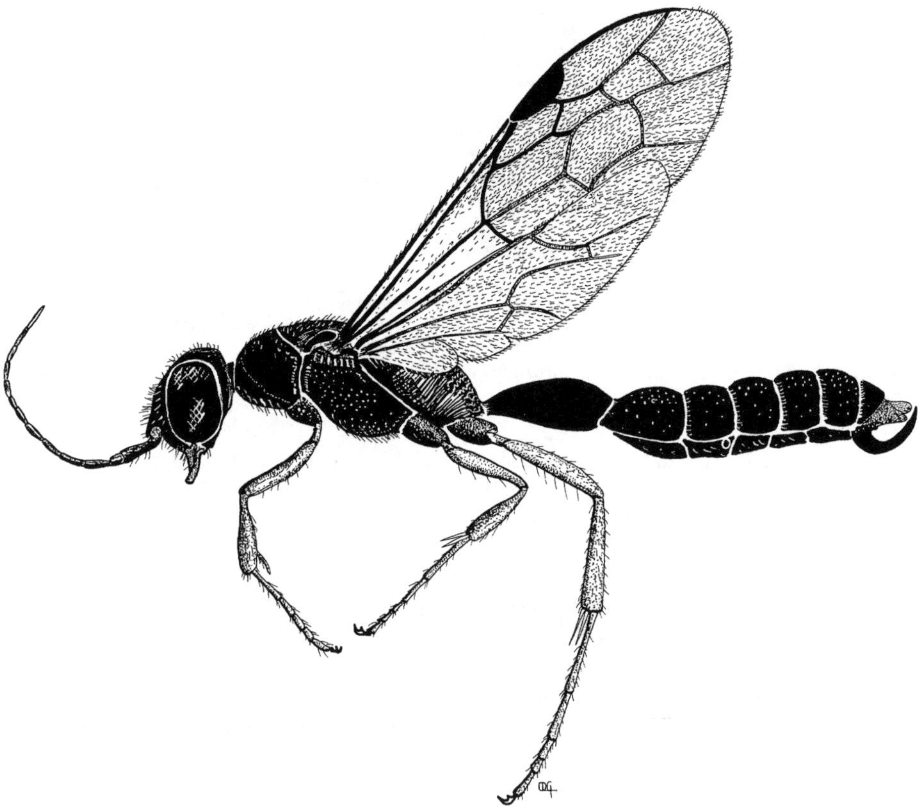

Fig. 14.41. *Pterombrus* sp. male (Myzininae).

though the former is more or less indistinguishable in the Thynninae. First metasomal tergite usually separated from the second by a constriction; *terminal metasomal sternite of males with a distinct up-curved spiniform process* (except in the subfamily Thynninae, and in the extralimital Diamminae and Anthoboscinae).

Classification and distribution. The family Tiphiidae is a relatively small group of aculeate wasps, comprising approximately 2000 described species in 90 genera worldwide (Kimsey, 1991*b*). These are classified into seven subfamilies, the Anthoboscinae, Diamminae, Tiphiinae, Brachycistidinae, Methochinae, Thynninae and Myzininae. The putative phylogenetic relationships (Kimsey, 1991*b*) between these subfamilies are shown in Fig. 14.42. Brothers and Carpenter, in their re-analysis of aculeate phylogeny (1993), presented a slightly different phylogenetic arrangement; they suggest the Thynninae is the sister-group to the (Tiphiidae + Brachycistidinae) + (Myzininae + Methochinae).

All subfamilies except for Diamminae, which comprises only one monotypic genus endemic to southern Australia, occur in the New World (Kimsey, 1991*b*). The Anthoboscinae includes four genera restricted to South America, one genus restricted to North America, and the genus *Anthobosca*, which is cosmopolitan and may possibly be found in Mesoamerica (although it has not been seen yet). During the course of intensive sampling in preparation for this book individuals of Brachycistidinae, a subfamily previously only known to occur in western North America, were collected in the dry part of northwestern Costa Rica, thus recording this subfamily from the Neotropics for the first time. The Methochinae, Myzininae and Tiphiinac are widely distributed throughout the world while the Thynninae has a primarily transantarctic distribution.

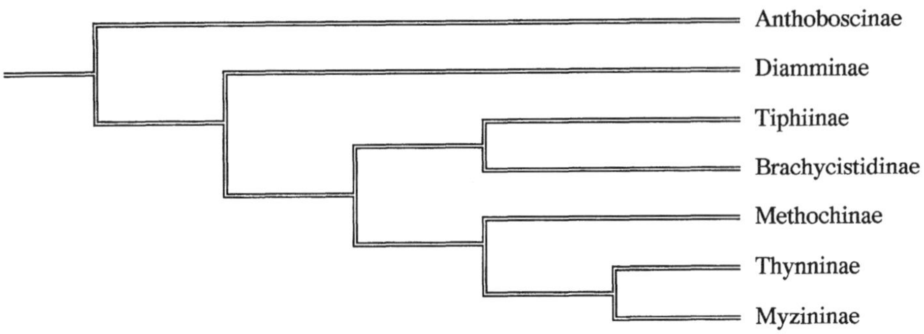

Fig. 14.42. Postulated phylogenetic relationships of the Tiphiidae. For details of characters supporting this cladogram see Kimsey (1991*b*).

Biology. The biology of species of the Anthoboscinae and Brachycistidinae is unknown, although many species of the latter are nocturnally active and probably fossorial (Kimsey, 1991*b*). Species of the other tiphiid subfamilies are solitary, ectoparasitic idiobionts of large soil-dwelling insects. The majority attack beetle larvae, and many show a preference for more or less fully-grown individuals. Species of Tiphiinae and Thynninae attack scarabaeids (Clausen, 1940*b*), the Methochinae attack cicindelids (Burdick & Wasbauer, 1959) and the Myzininae attack both scarabs and cicindelids (Kimsey, 1991*b*). In Argentina one thynnine has been reported as attacking a curculionid (Lloyd, 1952), and in the Old World tiphiids are also recorded from larvae of Tenebrionidae. The single species in the subfamily Diamminae, *D. bicolor*, attacks mole crickets (Gryllotalpidae) (Clausen, 1940*b*). It is worth emphasizing here that the hosts of most tiphiid species, especially the tropical species, are unknown.

In the most extensively studied genus, *Tiphia* (Tiphiinae), the female wasp burrows into the soil in order to gain access to a scarabaeid larva. Beginning in the venter of the thorax, the female wasp repeatedly stings the host, injecting a venom that causes temporary paralysis. She then often commences an extensive 'kneading' of the ventral surface of the abdomen with her mandibles, after which she curls her body around the host, moves her ovipositor backward and forward in a groove between two abdominal segments, and finally attaches an egg in this groove (Clausen, 1940*b*). The exact intersegmental groove chosen and the precise position in which the egg is placed on the body of the host appears to be constant for any given species (e.g. see Clausen *et al.*, 1932). Kneading or 'malaxation' of host cuticle in preparation for oviposition appears to be lacking in *Tiphia* species that oviposit on the dorsum of the host, but has been observed in Methochinae (Williams, 1919*b*) and Thynninae (Lloyd, 1952). Methochinae, Myzininae and Thynninae are reported to cause permanent paralysis in their hosts and to place the egg longitudinally rather than transversely on the body of the host (Clausen, 1940*b*). The majority of tiphiids leave the parasitized host in the burrow in which it was found and fill it with soil. The Australian Diamminae on the other hand is reported to drag the temporarily paralysed host to a previously prepared burrow (Clausen, 1940*b*).

Tiphiid larvae are quite generalized and uniform, with quadridentate mandibles (Evans, 1965*a*). The second pair of thoracic spiracles is reduced as in Mutillidae and Pompilidae (Evans, 1987*a*). Larvae of some Thynninae are covered in liquid excreted from a transverse slit in the last abdominal segment (Lloyd, 1952). There are five larval instars and each successive instar may make a new feeding puncture a short distance in front of the previous one (Clausen, 1940*b*). In the latter part of the fifth instar suctorial feeding is abandoned and the entire host body is consumed with the exception of the head and legs. It is in this stage that the host is finally killed. Cocoons of Tiphiidae are more slender than those of Scoliidae, being broader at the anterior end, and the emergence hole is more irregular (Clausen, 1940*b*).

Adult tiphiids feed on honeydew and nectar, often during the morning hours. Mating may take place on the vegetation during this period, except in Thynninae (see below). An additional source of nutrition for the adult female is host-feeding, which usually entails biting off one of the legs of a host on which she has just oviposited and imbibing the fluids that exude from the wound (Clausen, 1940*b*). Thynninae, on the other hand, are reported to use different hosts for oviposition and for host-feeding (Janvier, 1933). In Costa Rica males of a species of *Myzinum* have been observed to form sleeping aggregations on small trees (Gauld, pers. comm.).

Economic importance. Some species of *Tiphia* have been used in classical biological control programmes against scarab pests such as the sugar-cane grub (*Phyllophaga smithi*) and the Japanese beetle (*Popillia japonica*) (Clausen, 1978). Although in several cases the tiphiids became established, and often exerted some control, their effectiveness in lowering the pest populations appears to have been limited by availability of food sources for the adults.

Identification. Keys to the subfamilies of the world are provided by Kimsey (1991*b*). The subfamilies and genera occurring in Costa Rica can be separated with the following key. The Anthoboscinae (denoted by an asterisk) are not known to occur in Costa Rica, but are included in the key because they occur in both North and South America.

Key to subfamilies and genera of Tiphiidae present in Costa Rica

1 Wingless (some females). 2
— Winged (all males and most females). 4

2 Mid and hind tibiae with one spur, hind tibial spur with row of comb-like spines; mesopleural lamella small and digitate. ....(Methochinae) .... *Methocha*
— Mid and hind tibiae with two spurs, hind tibial spur without comb; mesopleural lamellae broad and flat. 3

3 Frons not expanded and overhanging antennal socket in dorsal view. .... Brachycistidinae
— Frons bulging above, and somewhat overhanging antennal socket. ....(Thynninae).... *Aelurus*

4 Male: sternite VIII with a flattened subtriangular apical platform, margined with short, stout spines; gonostylus with coiled ribbon-like setae usually showing externally; hind wing with claval lobe barely indicated, without an incision (Fig. 14.43).
[Fore wing, of Costa Rican species, hyaline with a large infumate patch towards distal end.] ....(Thynninae) .... *Aelurus*
— Males or females: male with sternite VIII forming a long up-curved spine apically (Fig. 14.41) (NB in one extralimital group, simple, but this does not have an infumate patch in the fore wing); gonostylus without ribbon-like apical setae; hind wing with claval lobe generally clearly indicated by a fairly deep incision (Figs 14.44–14.45), or if lobe weakly defined then marginal cell shorter than pterostigma (Fig. 14.46); marginal cell otherwise long, sometimes open distally. 5

5 Fore wing with marginal cell extremely short (sometimes obliterated), distal abscissa of *Rs* more or less joining *R* at distal end of pterostigma; fore wing of Costa Rican species with 6 or 7 enclosed cells (Fig. 14.46). ....(Brachycistidinae) .... *Quemaya*
— Fore wing with marginal cell always very long, sometimes open distally (Fig. 14.40), with *Rs* approaching or joining anterior margin of wing far distad of distal end of pterostigma; fore wing with 8 to 10 enclosed cells (Figs 14.44–14.45). 6

6 Frons with projections or lobes overhanging antennal sockets, the sockets thus being directed slightly downwards or outwards from the plane of the front of the head (Fig. 14.47); females with marginal cell of fore wing closed distally (Fig. 14.44); metasternum without subtriangular plate between mesopleural lamellae. 7
— Frons flat, without projections or lobes overhanging antennal socket, the socket thus being more or less flush with front of head (Fig.

14.48); females with marginal cell of fore wing open distally (Fig. 14.40); metasternum forming a subtriangular plate between mesopleural lamellae. 11

7 Males: sternite VIII forming a long up-curved spine apically (Fig. 14.41). 8
— Females. ....(Myzininae) 10

8 Metasoma with sternites III to V with a deeply invaginate transverse furrow delineating a closely punctate anterior area and a weakly biconvex, sparsely punctate posterior area (Fig. 14.49); hind coxa with a triangular dorsal flange (Fig. 14.51); fore wing with two submarginal cells (Fig. 14.45). ....(Methochinae) .... *Methocha*
— Metasoma with sternites III to V simply convex, without a deeply invaginate transverse furrow; hind coxa without a triangular dorsal flange; fore wing with three submarginal cells (Fig. 14.44). ....(Myzininae) 9

9 Mesoscutum without notauli; antenna 13-segmented; eyes not emarginate; usually black or black and red coloured. .... *Pterombrus*
— Mesoscutum with notauli; antenna appearing 12-segmented as pedicel is hidden in apex of scape; eyes strongly emarginate (notched) near antennal insertion; usually black with extensive yellow or orange markings. .... *Myzinum*

10 Slender, gracile insects, with legs slender, unspecialized, the longer hind tibial spur without a conspicuous basal notch; fore wing with pterostigma distinct and distal part of *R* adjacent to fore margin of wing (Fig. 14.41); mesoscutum without notauli; antenna 12-segmented; usually black or black and red coloured. .... *Pterombrus*
— Robust insects, with legs stout, spinose and fossorial (cf Fig. 14.40), the longer hind tibial spur with a conspicuous basal notch; fore wing with pterostigma indistinct and distal part of *R* behind fore margin of wing (Fig. 14.44); mesoscutum with notauli; antenna appearing 11-segmented as pedicel is hidden in apex of scape; usually black with extensive yellow or orange markings. .... *Myzinum*

11 Mid tibia with two spurs; male with sternite VIII apically rounded, unspecialized. ....Anthoboscinae*
— Mid tibia with one spur; male with sternite VIII forming a long up-curved spine apically. ....(Tiphiinae) 12

12 Body black with yellow markings; tegulae very large and elongate, extending beyond metanotum and covering bases of hind wings. .... *Epomidopteron*
— Body entirely black or with reddish areas, or off-white markings on face; tegulae large, but never reaching back to metanotum (Fig. 14.40). 13

13 First metasomal tergite without transverse medial carina or well-defined lateral depressions. .... *Tiphia*
— First metasomal tergite with transverse medial carina and often with well-defined lateral depressions (Fig. 14.52). .... *Paratiphia*

Synopsis of the Costa Rican fauna

Currently about 30 species in eight genera and five subfamilies are known to occur in Costa Rica. A sixth subfamily, Anthoboscinae, which is represented in both North and South America, may also occur here. Tiphiids are commonly encountered in all lowland and mid altitude sites throughout Costa Rica, but they appear to be much rarer at high altitude sites above 2500 metres.

Brachycistidinae

This subfamily consists of 13 nominal genera, virtually all of which are restricted to western North America. There is probably much need for generic synonymy since all genera, except *Brachycistis*, are only known from one sex. Most of the described taxa comprise rather large to small, brownish species with large prominent ocelli—a typical adaptation to a nocturnally active existence. However, the males of *Quemaya* are small and brown to black. A key to the North American genera of Brachycistidinae is given by Wasbauer (1968). One genus of Brachycistidinae, *Quemaya*, has recently been found in Costa Rica, a southern extension of the known range of the subfamily of more than 2000 kilometres.

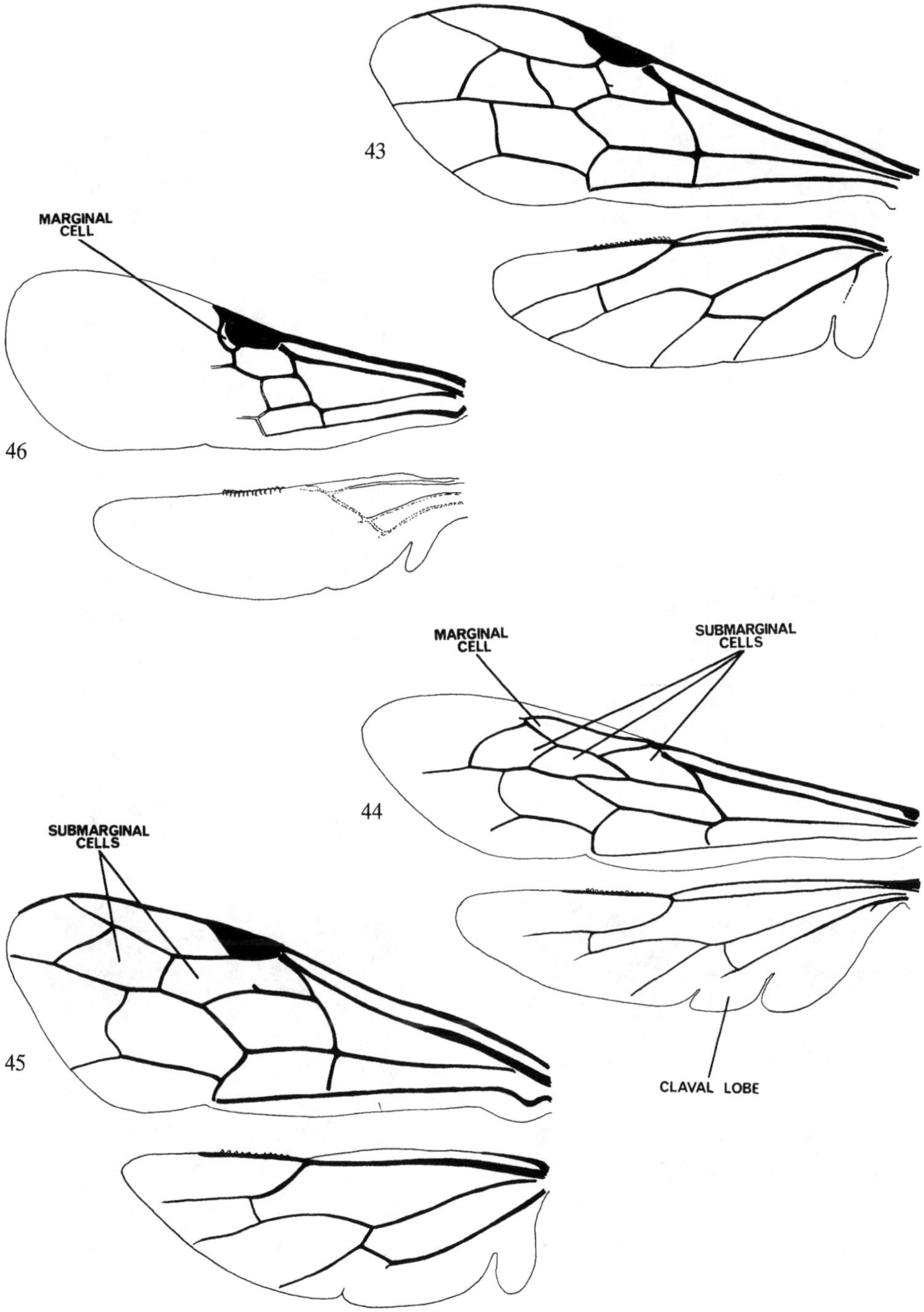

Figs 14.43–14.46. Tiphiidae, fore and hind wings; 14.43, *Aelurus* sp., male (Thynninae); 14.44, *Myzinum* sp., female (Myzininae); 14.45, *Methocha* sp., male (Methochinae); 14.46, *Quemaya* sp., male (Brachycistidinae).

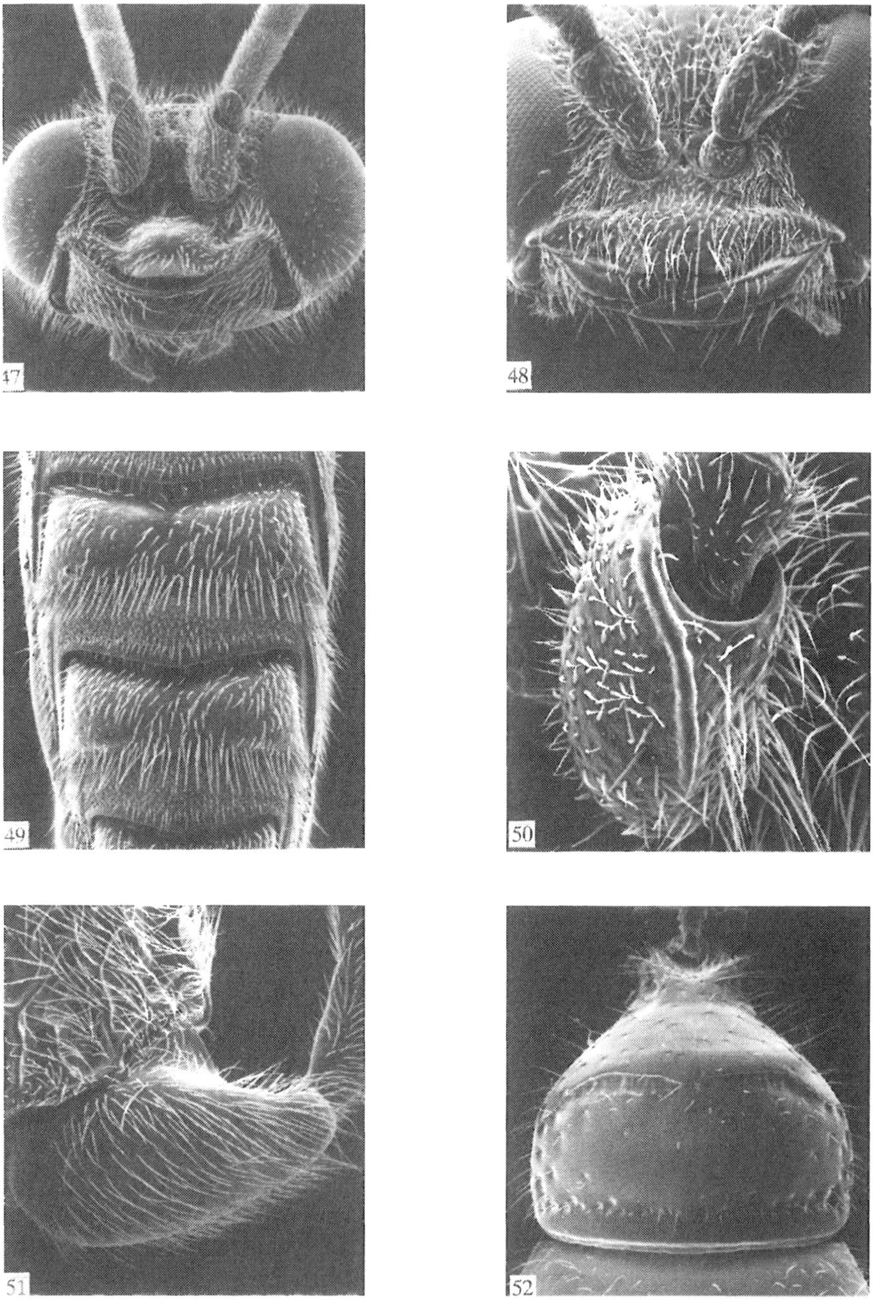

Figs 14.47–14.52. Scanning electron photomicrographs of Tiphiidae. Figs 14.47–14.48. Faces; 14.47, *Methocha* sp. (Methochinae); 14.48, *Paratiphia* sp. (Tiphiinae). Figs 14.49–14.51. *Methocha* sp. (Methochinae); 14.49, male metasoma, ventral view; 14.50, scape; 14.51, male hind coxa. Fig. 14.52. Tergite I of metasoma, *Paratiphia* (Tiphiinae).

***Quemaya*^CR^**. Males of two or more species of *Quemaya* are sometimes very common in the seasonally dry parts of northwestern Costa Rica. Species of this genus are very tiny wasps (body length 2.8 to 3.5 mm), black in colour, and not specialized for nocturnal activity. The presumably wingless females have not been found yet. Large numbers of one species have been collected in a Malaise trap operating in open woodland adjacent to the administrative area in Santa Rosa National Park during February and March—well into the dry season.

METHOCHINAE

This subfamily consists of only two genera, one of which, *Karlissa*, is restricted to Sri Lanka, whilst the other, *Methocha*, occurs throughout the world except for Australia (Kimsey, 1991*b*).

Methocha. This genus is represented in Costa Rica by at least two species. It occurs from sea level to 1300 m, and is quite common in banana and sugar-cane plantations. Males are winged and average 10.0 mm in length; females are wingless, somewhat smaller, ant-like, and are rarely collected. Biology: species of *Methocha* are parasitoids of cicindelid beetle larvae and may attack arboreal species as well as terrestrial ones. In some instances the female allows herself to be seized by the cicindelid larva, whereupon she quickly stings the host in the thorax (Williams, 1919*b*). If the host escapes from the burrow it is stung again and dragged back into the burrow (Iwata, 1936). Some reports suggest that paralysis is permanent (Williams, 1919*b*), while others have observed that it is temporary (Iwata, 1936).

MYZININAE

This subfamily consists of 12 genera worldwide, ten in the Old World and two in the New World. Both sexes are large wasps (8.0 to 20.0 mm in length) and, in the New World, fully winged. Males tend to be slender and elongate with straight antennae; females of *Myzinum* are heavy-bodied, with coiled antennae, whereas females of *Pterombrus* are quite similar to males.

Myzinum. A New World genus with four described species in Costa Rica. Species of *Myzinum* have been taken at sites from sea-level up to about 1500 metres; they are apparently most commonly encountered in rather open, second growth areas (e.g. coffee plantations in the Central Valley). Biology: *Myzinum* species are parasitoids of scarabaeid larvae.

Pterombrus. A New World genus with at least four species in Costa Rica (sea-level to 1600 metres). Two species were described by Krombein and Evans (1976), who also described the larva of one. Biology: *Pterombrus* species attack cicindelid larvae. In Costa Rica *P. piceus* has been observed attacking larvae of *Pseudoxychila tarsalis*, which burrow in vertical banks at intermediate elevations around 1200 metres (Palmer, 1976; *in* Janzen, 1983). Although they are strong fliers the female wasps search for hosts while walking over the ground. When a larval burrow of a tiger beetle is located the wasp usually waits for the larva to retreat into the burrow before entering it. Once inside she stings the cicindelid, positions it ventral side up, oviposits on the venter of the abdomen, exits from the burrow and finally closes the burrow with soil. The host has three larval instars, but the tiphiid prefers the second. Interestingly, the wasps appear to show no interest in cicindelid larvae removed from their burrows. The egg hatches in five days, larval development requires 8 to 13 days, another 25 days are passed in the prepupal stage and 20 days in the pupal stage.

THYNNINAE

This subfamily shows a typical transantarctic distribution, with 36 genera in Australia, and 23 in South America—most of which are confined to the more temperate southern part of the continent (Genise & Kimsey, 1991; Kimsey 1991*c*). The South American fauna consists of three tribes; the genus *Aelurus*, which is the only New World genus belonging to the tribe Rhagigasterini, is the only genus that extends as far north as Central America (Kimsey, 1992*c*).

Aelurus. This New World genus comprises 13 species, of which only *A. septentrionalis* occurs in Costa Rica, achieving the northernmost distribution of the subfamily (Kimsey, 1991*a*). This species belongs to a species-group that occurs along the Andes, from northern Argentina to Colombia, and in the highlands of Costa Rica and Panama. In

Costa Rica *A. septentrionalis* has been collected at several humid forest sites (Monteverde, Pitilla, Volcán Cacao, Zurquí de Moravia and Las Alturas) at altitudes between 700 and 1600 metres and between March and July. Biology: although nothing is known about the biology of any *Aelurus* species, it is probably similar to other genera in the subfamily, which attack scarab beetle larvae. Females are wingless whereas males are strong fliers and carry females in phoretic copulation (Lloyd, 1952; Alcock, 1981; Toro & Carvajal, 1989). Receptive females crawl up on vegetation and call the patrolling males by releasing a pheromone; the male removes the female with very little preliminary courtship but often carries her to flowers where she can feed while they remain coupled. After mating the female is dropped to the ground. Males of a species in Chile have been reported to form 'sleeping' aggregations in the soil or on foliage (Janvier, 1933).

TIPHIINAE

This subfamily comprises ten genera, three of which are restricted to the Old World, two (*Krombeinia* and *Neotiphia*) to the Nearctic and two (*Mallochia* and *Megatiphia*) to South America (Kimsey, 1993*c*). Three genera have been found to occur in Costa Rica, the neotropical *Epomidopteron*, the holarctic *Paratiphia*, and the cosmopolitan *Tiphia*. All tiphiines are presumed to be parasitoids of soil-dwelling scarab beetle larvae.

Epomidopteron. This is a rarely collected, monotypic genus, represented by *E. julii*, which is known to occur from the southwestern United States to Brazil (H.W. Allen, 1972). In Costa Rica this species has been collected in forests between about 600 and 700 metres around Volcán Orosi, in Guanacaste Province.

Paratiphia. A holarctic genus represented in Costa Rica by one or two species.

Tiphia. A cosmopolitan genus represented in Costa Rica by at least five species. Several may be very common, and in the late wet season—September/October—species of this genus are one of the most abundant non-social aculeates collected in Malaise traps in Santa Rosa National Park. In Costa Rica, *Tiphia* species have been collected from sea-level up to 2800 metres on Cerro de la Muerte. The South American species of this genus were monographed by Harry W. Allen (1972). This author has also reviewed the Panamanian (Allen, 1964), West Indian (Allen, 1970) and North American species (Allen, 1966, 1971). Biology: *Tiphia* species are ectoparasitoids of the larvae of soil-dwelling scarabaeids and the biology of a number of species have been investigated by Clausen and co-workers (e.g. Clausen *et al.*, 1932; see Clausen, 1940*b*).

14.3 POMPILIDAE

Marius S. Wasbauer

Diagnosis. Size: small to very large wasps (fore wing length 2.0 to 50.0 mm in Costa Rican species); black, bluish black, reddish brown, black with yellow spots or yellow with black and/or red markings. Antennae filiform or occasionally somewhat serrate, 12-segmented (often curled in preserved specimens) in females, 13-segmented (rarely twelve) in males; compound eyes with inner margins straight or somewhat emarginate, never notched; mandible with one or two teeth in addition to the apical cusp; maxillary palpi 6-segmented, labial palpi 4-segmented. Mesosoma usually unmodified, occasionally laterally compressed or elongate; *pronotum hinged to mesothorax, its sides extending posteriorly to touch the tegulae, mesepisternum with a transverse suture* (Figs 14.62–14.65); metapleuron transverse, often with anteroventral pit. *Legs long*, hind femora usually reaching or exceeding tip of metasoma; fore tibia with single calcar, mid and hind tibia with two apical spurs, those of hind tibia modified as calcaria; interior surface of hind tibia with a longitudinal brush of appressed hairs, tarsi with plantulae. Fully winged, except females brachypterous or apterous in a few extralimital forms. *Fore wing with 10 or rarely 9 enclosed cells; hind wing with 3.* Metasoma rarely petiolate, with six (females) or seven (males) visible sternites, female sternite VI envelops ovipositor above, male sternite VI with a pair of sclerotized hooks posteriorly, sternite VII reduced, internal, bearing cerci. Females with a potent sting.

Classification and distribution. The Pompilidae are worldwide in distribution and form a rich component of the aculeate wasp fauna of tropical countries.

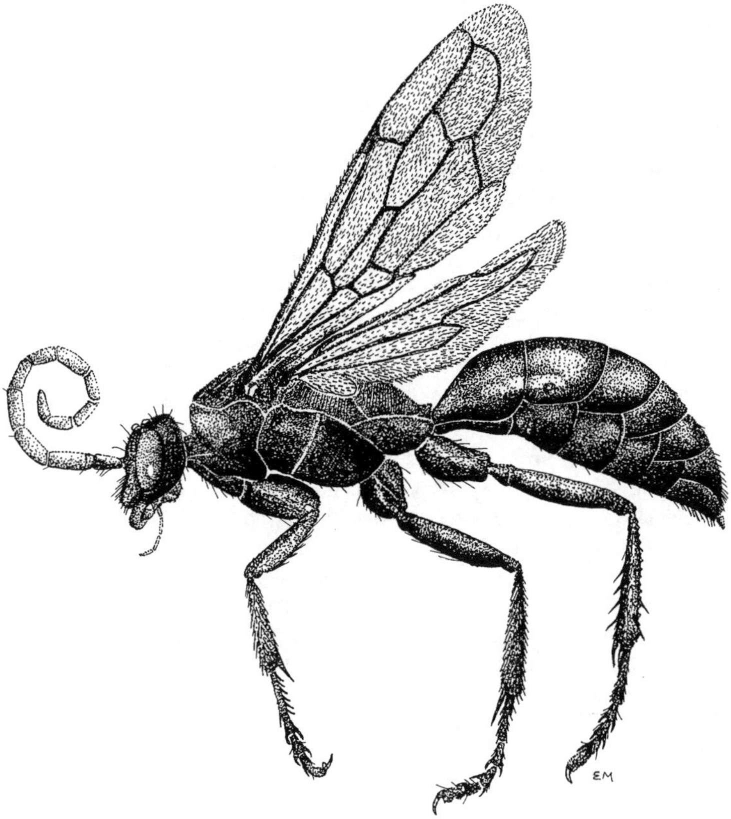

Fig. 14.53. *Pepsis* sp. (Pompilidae).

Subtropical areas, including arid zones, also have a rich pompilid fauna, but the group is less species-rich in the cooler temperate regions. Costa Rica, with about 230 species, has almost 80 per cent as many taxa as are known to occur in America north of Mexico (289 spp.) and approximately six times the number of pompilid species that occur in Britain (Day, 1988), a cool temperate country with nearly five times the surface area.

Worldwide the Pompilidae consists of about 5000 species classified in approximately 120 genera. However, there has been no inclusive work treating the classification of the Pompilidae on a world basis and much synonymy exists in the family both at the generic and infrageneric levels. The family is divided unequally into four subfamilies, the Pompilinae, Pepsinae, Ceropalinae and Notocyphinae. The first two of these, the Pepsinae and Pompilinae, are cosmopolitan and comprise the great majority of species. The Ceropalinae consists of two closely related genera of cleptoparasitic species which occur in all zoogeographic regions. The Notocyphinae, also containing two genera, consists of species occurring in austral North, Central and South America and the Oriental region. All four subfamilies are represented in Costa Rica.

Biology. Pompilids are commonly collected in most habitats throughout Costa Rica, although relatively few individuals have been taken in the high altitude sites on the Cerro de la Muerte. They are rapid flyers when in the air and fast runners on the ground, relying on long legs to carry them through the vegetation. Some are quite fugitive, refusing to fly up into a net when it is placed over them, taking refuge in a crevice or crack or hiding under leaves or debris and then flying away when the net is removed. When searching for spiders, females may fly for short distances, but they usually spend most of their time running quickly over the ground, constantly flicking their wings (a field diagnostic feature for the family).

Biologically the Pompilidae, or spider wasps as they are colloquially known, are rather uniform in that:

a) all species utilize spiders exclusively as prey;
b) all species provision a single spider per cell, in contrast with most spider-hunting sphecid wasps (see Chapter 17);
c) all species lay a single egg on or in the abdomen of the spider.

These three biological canons impose an unusual degree of uniformity on the life histories of spider wasps, but within this framework, considerable diversity exists. Some pompilids develop in much the same way as idiobiont parasitoids. The females hunt for spiders, such as species of Ctenizidae, that live in burrows. The wasp simply stings and paralyses the spider, oviposits, then closes the burrow and departs. However, most pompilids search for free living spiders and, after locating, stinging and paralysing one the female spider wasp carries it to some secluded place before interring it and ovipositing. The place of concealment may be little more than a pre-existing crevice, but many species excavate burrows in the ground and some, such as *Auplopus* species, construct elaborate mud nests. A few pompilids, such as species of *Notocyphus* and Old World *Homonotus*, develop as koinobiont ectoparasitoids; they temporarily paralyse the host and oviposit on its abdomen. The spider then recovers and lives normally until the wasp larva eventually kills it. This way of life is very similar to that in *Polysphincta* and related genera in the pimpline Ichneumonidae (Kaston, 1959; Gauld, 1991). Finally, as is the case with most groups of nest provisioning aculeates, a few spider wasps such as *Evagetes* and *Ceropales* are cleptoparasites of other pompilids.

Worldwide, the range of spiders attacked by pompilids is very broad encompassing all the mygalomorph families and about half of the families of the Labidognatha (see Table 14.1 for New World records). Two striking absences from this host-list are the Linyphiidae, a large family of physically small spiders and the Theridiidae—spiders that spin complex three-dimensional webs. Similar patterns of spider exploitation have been observed in Britain (Day, 1988). At the generic level pompilids seem either to specialize in attacking spiders of the mygalomorph or errant labidognath families. Although there are a few records of more generalist genera, such as *Sericopompilus*,

Suborder Orthognatha (The mygalomorphs)	
Ctenizidae	*Aporus*; *Pepsis*; *Psorthaspis*.
Theraphosidae	*Hemipepsis*; *Notocyphus*; *Pepsis*.
Suborder Labidognatha	
Acanthoctenidae	*Priocnemella*
Agelenidae	*Ageniella*; *Anoplius*; *Cryptocheilus*; *Priocnessus*.
Amaurobiidae	*Anoplius*.
Anyphaenidae	*Auplopus*; *Sericopompilus*.
Araneidae	*Agenioideus*; *Anoplius*; *Caliadurgus*; *Episyron*; *Poecilopompilus*; *Sericopompilus*.
Clubionidae	*Ageniella*; *Anoplius*; *Auplopus*; *Dipogon*; *Priocnemis*.
Ctenidae	*Auplopus*
Dictynidae	*Cryptocheilus*
Gnaphosidae	*Ageniella*; *Anoplius*; *Dipogon*; *Cryptocheilus*; *Priocnemis*.
Heteropodidae	*Auplopus*
Lycosidae	*Ageniella*; *Anoplius*; *Cryptocheilus*; *Entypus*; *Minagenia*; *Paracyphononyx*; *Priocnemis*; *Tachypompilus*.
Oxyopidae	*Ageniella*; *Aporinellus*; *Priocnemis*; *Sericopompilus*.
Philodromidae	*Anoplius*; *Auplopus*; *Sericopompilus*.
Pisauridae	*Auplopus*; *Tachypompilus*.
Salticidae	*Ageniella*; *Anoplius*; *Aporinellus*; *Auplopus*; *Priocnemis*; *Sericopompilus*.
Thomisidae	*Ageniella*; *Aporinellus*; *Auplopus*; *Dipogon*; *Sericopompilus*.

Table 14.1. Families of spiders and the genera of pompilids (present in Costa Rica) known to attack them in the New World.

attacking orb-web spiders, these araneids are particularly exploited by species of one or two specialist genera such as *Episyron* and *Caliadurgus*.

At the species level some pompilid species are rather generalized hunters, utilizing a number of genera or even families of spiders, while others are much more restrictive in their prey selection. For example, in the eastern United States *Anoplius* (*Arachnophroctonus*) *marginatus* is known to provision with at least 22 species of spiders belonging to seven different families (Evans & Yoshimoto, 1962). On the other hand, *Anoplius* (*Anoplius*) *imbellis*, a species widespread from Alaska to Costa Rica is known to provision with spiders in only three genera of Lycosidae, perhaps because it hunts only in areas of heavy soil next to bodies of still or slow moving water (Wasbauer, 1957, 1983). More details of provisioning and nesting behaviour are given below in the subfamily, tribal and generic accounts.

The larvae of Pompilidae are rather typical apodous mandibulate aculeate larvae. They are little studied, but the available information indicates that the vestigial second thoracic spiracles, distinct antennal papillae and somewhat trilobate clypeus together will separate most larvae from those of other families. More details of larval morphology are given by Evans (1987*a*). The pupal stage is passed within a brownish silken cocoon, which internally has a parchment-like quality.

Very little information is available on the mating behaviour of Pompilidae. The smaller males usually emerge before the females and spend much of their time flying low over the ground or vegetation looking for mates. They spend a much greater proportion of their time in flight than do the females and probably for this reason are much more frequently taken in flight interception traps. In many ground nesting Pompilidae, courtship and mating occur on the nesting site (Wasbauer, 1955, 1957; Shimizu, 1989). Alcock (1978, 1984) observed that males of *Hemipepsis ustulata* exhibit territoriality by defending their perch sites in the tops of palo verde trees on ridge tops in the Sonoran Desert of Arizona. Each male selects an area and vigorously defends it against other conspecific males while waiting for a receptive female to enter the territory. It may be that in certain species with polymorphic males, more than one mating strategy is in effect. Day (1984) suggests that in certain *Cryptocheilus* species, males of one morph may enter the ground to mate with newly emerging females and those of another position themselves above ground and wait for emerging females to reach the surface.

Adult feeding habits of pompilids are not documented for many species, but studies suggest that nectar from flowers is an important source of food (Krombein *in* Krombein *et al.*, 1979; Wasbauer & Kimsey, 1985). Most appear to visit a wide variety of flowers but species of *Pepsis* appear to prefer milkweed flowers (*Asclepias*). Some species of *Pepsis* have elongate mouthparts which might aid in securing nectar from the milkweed and other plants. Other genera of pompilids with elongate mouthparts such as *Notocyphus*, have a number of flower records for some of the species (Townes, 1957) but a definite correlation with nectar feeding from tubular flowers has not been established. There are also a few records of pompilids feeding from extrafloral nectaries and on plant and insect produced honeydew. The author has seen adults of a species of *Aporinellus* feeding on a peanut brittle candy in open stalls in Mexico. In addition to sugary substances, the females of some species also imbibe haemolymph oozing from captured spiders. For example, *Anoplius* (*Notiochares*) *lepidus* is a frequent flower visitor and also captures spiders specifically to feed on them (Evans & Yoshimoto, 1962). On the other hand, *A.* (*Arachnophroctonus*) *apiculatus* feeds not at flowers at all but only on haemolymph of captured spiders. In Costa Rica, *Auplopus semialatus* also imbibes haemolymph from captured spiders (W. Eberhard, pers. comm.). The same may be true of the tribe Ageniellini in which the females often amputate the spider's legs before prey transport. There do not appear to have been any attempts to synthesize the literature on adult feeding and to relate it to other aspects of biology (e.g. a possible correlation between host-feeding and synovigenesis; see Chapter 2.5).

As is true of most of the groups of Hymenoptera in the tropics, little is known about the biology of Costa Rican Pompilidae (but see Wcislo *et al.*, 1988). The reader is referred to the summaries provided by Olberg (1959) for the European species and Evans and Yoshimoto (1962) for the northeastern United States species. Evans (1953) and Iwata (1976) have provided evolutionary perspectives on the ethology of spider wasps.

Identification. The Mexican and Central American species of the subfamily Pompilinae were revised by

Evans (1966*e*, 1968*a*) and the Pompilinae of California were reviewed by Wasbauer and Kimsey (1985). The nearctic species of Pepsinae and Ceropalinae were revised by Townes (1957), but other than a key to the genera of neotropical Ageniellini (Evans, 1973*b*), no keys are available for Central American representatives of these two subfamilies.

Key to subfamilies and tribes of Pompilidae present in Costa Rica

1 Apex of hind tibia with auxiliary spines of unequal length, irregularly spaced and somewhat radiating or splayed out (Fig. 14.54); fore wing with 2nd discal cell usually produced into a pocket at its posterobasal corner (Fig.14.57); sternite II without a well defined transverse groove; female labrum at least partially hidden beneath clypeus (Fig. 14.59). .... (Pompilinae) 2

— Apex of hind tibia with spines (if present) of rather uniform length and spacing, more or less parallel and not splayed out (Fig. 14.55); fore wing with 2nd discal cell not produced into a pocket at its posterobasal corner (Fig. 14.58); ***either*** with sternite II with a well defined transverse groove (Fig. 14.71) ***or*** female labrum fully exposed (Figs 14.60, 14.61). 5

2 Females. 3

— Males. 4

3 Pronotum with collar not well differentiated from disc, streptaulus absent medially and collar on nearly same plane as disc, or if on lower plane, streptaulus absent altogether (Fig. 14.63); pronotum longer than mesoscutum in most genera; eyes sometimes covered with short hairs. .... **Aporini**

— Pronotum with collar separated from disc by a complete streptaulus, disc sloping upward strongly from collar (Fig. 14.62); pronotum not longer than mesoscutum along the midline; eyes never hairy. .... **Pompilini**

4 Fore wing with two submarginal cells, or if with three then the pronotum is nearly as long as mesoscutum; propodeum without sharp conical processes posterolaterally. ..... **Aporini**

— Pronotum considerably shorter than mesoscutum; fore wing with three submarginal cells (Fig. 14.47), or if with two, propodeum with sharp, conical processes posterolaterally (Fig. 14.62). .... **Pompilini**

5 Female labrum at least partially hidden beneath clypeus; sternite II usually with a well developed transverse groove (Fig. 14.71) (absent in some males); hind tibia often with raised teeth (Fig. 14.56); hypopygium of female usually more or less circular in cross section (laterally compressed in *Minagenia*); fore wing with 3*rs-m* usually straight or evenly arcuate (Fig. 14.58). ....(Pepsinae) 6

— Female labrum fully exposed (Fig. 14.60, 14.61); sternite II without a transverse groove; hind tibia smooth (Fig. 14.55); hypopygium of female laterally compressed (Figs 14.66–67); fore wing with 3*rs-m* angled at middle or somewhat sinuate. 7

6 Tergite I, in dorsal view, with the sides evenly convergent anteriorly or slightly convex (Fig. 14.69), in lateral view with a suture delimiting a laterotergite; male genitalia with parapenial lobe not decurved at apex (Fig. 14.68). ....**Pepsini**

— Tergite I, in dorsal view, with the sides somewhat concave, giving it a somewhat petiolate appearance (Fig. 14.70), in lateral view without a suture delimiting a laterotergite; male genitalia with parapenial lobe slender, decurved at apex.......................... **Ageniellini**

7 Fore wing with pterostigma very small, about 2.5 times as long as wide; hind wing with *cu-a* meeting *Cu*1 at or distal to point of divergence of *M* and *Cu*1 (Fig. 14.72); pronotum with definite dorsal surface, separated from anterior surface by distinct angulation (Fig. 15.64); inner margins of compound eyes nearly parallel (Fig. 14.61), at most slightly sinuate; sting downcurved. .... **Notocyphinae**

— Fore wing with pterostigma at least 3.5 times as long as wide; hind wing with *cu-a* meeting *M*+*Cu*1 proximal to point of divergence of *M* and *Cu*1 (Fig. 14.73); pronotum strap-shaped, dorsal surface very short, not usually sepa-

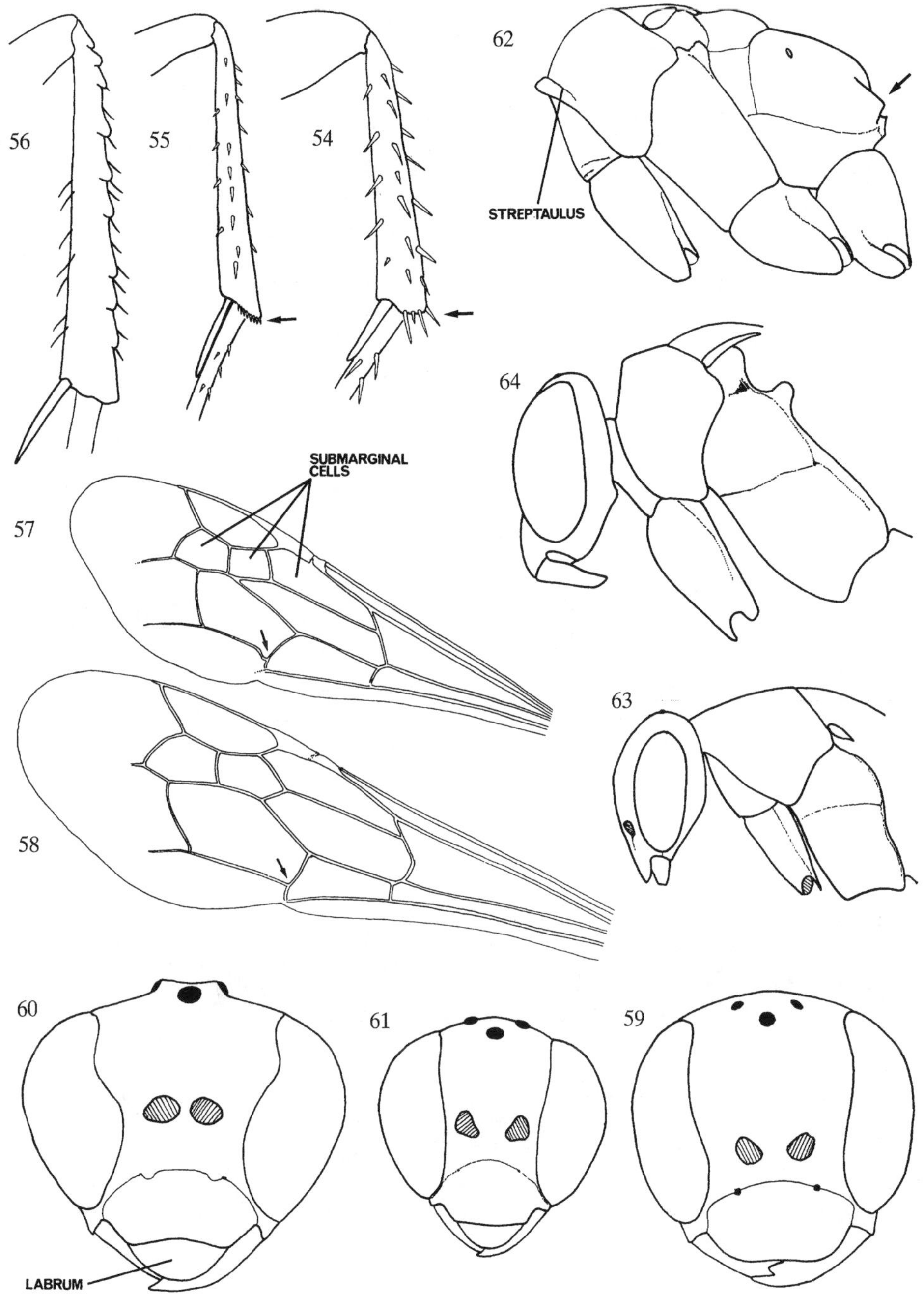

Figs 14.54–14.64. Pompilidae. Fig. 14.54–14.56. Hind tibia; 14.54, *Psorthaspis* sp. (Pompilinae); 14.55, *Notocyphus* sp. (Notocyphinae); 14.56, *Priocnemis* sp. (Pepsinae). Figs 14.57–14.58. Fore wings; 14.57, *Pompilus* sp. (Pompilinae); 14.58, *Ageniella* sp. (Pepsinae). Figs 14.59–14.61. Face. 14.59, *Pompilus* sp. (Pompilinae); 14.60, *Ceropales* sp. (Ceropalinae); 14.61, *Notocyphus* sp. (Notocyphinae). Figs 14.62–14.64. Mesosoma, lateral; 14.62, *Aporinellus* sp. (Pompilinae); 14.63, *Psorthaspis* sp. (Pompilinae); 14.64, *Notocyphus* sp. (Notocyphinae).

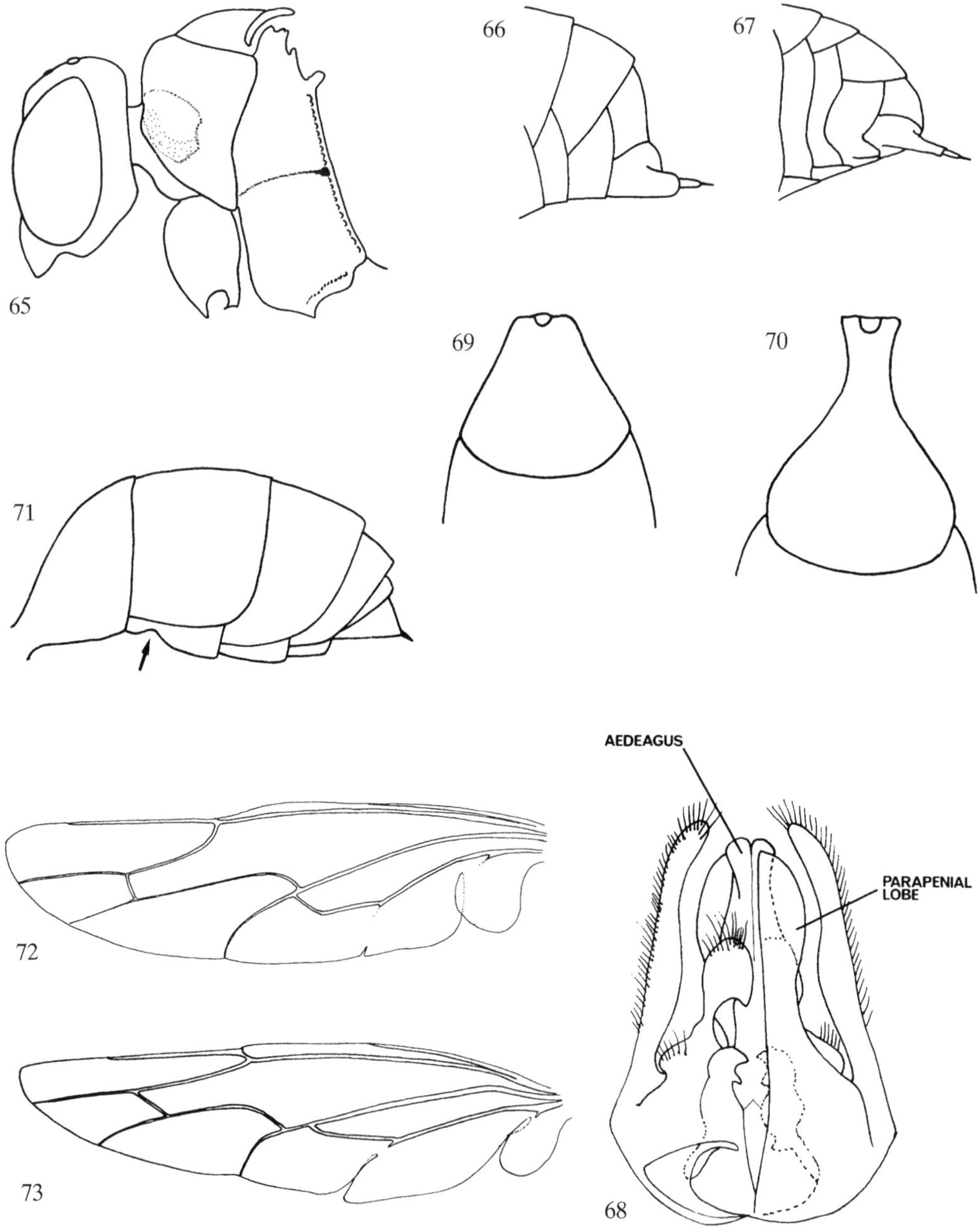

Figs 14.65–14.73. Pompilidae. Fig. 14.65. Mesosoma, lateral, *Ceropales* sp. (Ceropalinae). Figs 14.66–14.67. Subgenital plate of female *Ceropales* sp. (Ceropalinae). Fig. 14.68. Male genitalia, Pepsini. Figs 14.69–14.70. Tergite I, dorsal; 14.69, *Calliadurgus* (Pepsinae); 14.70, *Ageniella* sp. (Pepsinae). Fig. 14.71. Metasoma, lateral, *Ageniella* sp. (Pepsinae). Figs 14.72–14.73. Hind wings; 14.72, *Notocyphus* sp. (Notocyphinae); 14.73, *Ceropales* sp. (Ceropalinae).

rated from anterior surface by an angulation (Fig. 14.65); inner margins of compound eyes strongly diverging dorsally (Fig. 14.60), emarginate or strongly sinuate; sting straight. .... **Ceropalinae**

Synopsis of the Costa Rican fauna

The classification presented here differs somewhat from that of Townes (1957), who included in the Ceropalinae groups as diverse as *Notocyphus* and *Minagenia*. Perhaps when good cladistic studies are available, a sound higher classification will emerge, but for the present, I have accorded *Notocyphus* and *Minotocyphus* subfamily status and placed *Minagenia* in the Pepsinae on the basis of its similarity to *Caliadurgus* and *Priocnemis*. This arrangement results in four subfamilies and will accommodate the New World fauna, but may not be satisfactory for other zoogeographic regions.

In Costa Rica, all four subfamilies are present and the family is represented by a total of 34 genera and about 250 species.

PEPSINAE

This subfamily is represented in Costa Rica by the two tribes Pepsini and Ageniellini.

PEPSINAE: Pepsini. This cosmopolitan tribe includes genera of both temperate and tropical affinities. Nine genera are known to occur in Costa Rica. With respect to biology, Pepsini includes more generalized species than any other group of pompilids. Females of some genera attack mygalomorph spiders in the families Theraphosidae and Ctenizidae and inter them in their own burrows. What little is known of *Minagenia* indicates that the female wasp searches out an errant (non-sedentary) spider, paralyses it, deposits an egg on the side of the abdomen and flies off without attempting to bury it. The effects of the sting are temporary and the spider recovers, continuing to feed until it is eventually killed by the developing wasp larva. In the genus *Priocnemis*, some species construct multicellular nests in the ground (Yoshimoto, 1954) which are provisioned with a range of labidognath spiders. Members of one genus, *Dipogon*, nest in hollow plant stems or in holes in wood and make a linear series of cells.

Caliadurgus (= *Calicurgus*). A neotropical and oriental genus, with a single holarctic species, that is most species-rich in tropical America where it comprises a rather homogeneous group of small to medium sized spider wasps. Ten species, all unidentified, have been found in Costa Rica. Females are uniformly black or black with a red or partially red metasoma, and males have a black head and body that is sometimes variously marked with yellow or white; wings banded in both sexes, less strongly so in the males. The females may be distinguished from those of the related *Priocnemis* by the dorsally strongly convergent compound eyes and the presence of a stout curved spine at the apex of the fore tibia; the males by the very short pronotum with a vertical or frontward sloping anterior face. The argentine species were revised by Roig Alsina (1982). Biology: the South American species, *C. gayi*, constructs a short burrow with a single terminal cell, and provisions it with an orb-web spider (Claude-Joseph, 1930). Similar observations have been made on the holarctic species, which also provisions with Araneidae (Day, 1988).

Cryptocheilus. A large cosmopolitan genus that is most species-rich in the Old World, where many species are handsomely patterned with integumental spots of contrasting colours. The six nearctic species are uniformly black with either black or reddish wings, but the two un-named species found in Costa Rica resemble the Old World forms in that they exhibit similar integumental colour patterns and males show structural polymorphism like that in several european species. Biology: european species have been recorded provisioning with species of Agelenidae, Dictynidae and, less commonly, Gnaphosidae and Lycosidae (Day, 1988), but the prey of North American species appears to be principally Lycosidae (Krombein *in* Krombein *et al.*, 1979).

Dipogon. A cosmopolitan genus, occurring in every zoogeographic region except possibly Australia. There are two subgenera and five species in Costa Rica: *Dipogon* (*Deuteragenia*) *aztecus*, *D.* (*Deuteragenia*) *calipterus nubifer*, an undescribed species of *Deuteragenia* and two undetermined species in the nominate subgenus. *Dipogon* are small to moderate-sized species that are easily recognized by the three-toothed mandible, which is rare in Pompilidae, and by the fascicle of long curved hairs that arises from each maxillary cardo, nearly attaining the base of the mandible of the females. The males are distinctive in their short, crenulate antennae and their prolonged subgenital plate (viewed in profile). Females and usually males of all species have banded wings. Evans (1974) has reviewed the Mexican and Central American species. Biology: In California *D.* (*Deuteragenia*) *calipterus nubifer* was found nesting in the hollow stems of herbaceous plants, where the females apparently spend the night. Gnaphosid, thomisid and clubionid spiders were transported by the wasp walking backwards and dragging the spider by the spinnerets. The number of nest cells varied from three to five (Williams, 1966). Many temperate species of this genus are commonly found constructing nests in pre-existing tunnels in wood or even holes in walls.

Entypus (= *Priocnemioides*). A widespread New World genus with only one species found in Costa Rica, but another widespread South American species, *E. urichi*, has been reported to occur as far north as Chiriquí in Panama. The body colour of all species is black and the wings and antennae may be orange or black. Biology: host records are known for about half of the seven nearctic species and all are species of *Lycosa* (Lycosidae) (Krombein *in* Krombein *et al.*, 1979).

Hemipepsis. Large, brightly coloured wasps found in both the Asian and African tropics, and in the New World. The Old World fauna is rich in species, whereas the New World has only a few species in the southwestern United States and Central America. Three or four species of *Hemipepsis* occur in Costa Rica, all of which are black with red or brown wings. One, *H. mexicana*, is reported to occur from the Big Bend country of Texas south to Panama (Townes, 1957). Biology: *Hemipepsis* species attack mygalomorph spiders (Theraphosidae).

Minagenia. Relatively small, dark wasps, that are one of the most abundant groups of Pompilidae in the tropical rainforests. There are 12 species in Costa Rica, none of which have been determined. Until recently *Minagenia* was placed in the Ceropalinae, mainly on the basis of the laterally compressed pygidial area of the female. However, most morphological characteristics align it with *Priocnemis* and *Caliadurgus*. The flattened female pygidium and the lack of hair and spines is probably associated with the search for and capture of spiders in cryptic situations. Biology: the nearctic species, *Minagenia osoria*, has been reared from immature Lycosidae, with eggs and larvae found on the base of the abdomen of several other lycosids (Kaston, 1959). It is assumed from this that the female seeks out an errant spider in its retreat (although it should be noted most lycosids have no retreat) and paralyses it, oviposits and simply leaves to repeat the process without building a nest or transporting the prey. The paralysis is temporary and the spider recovers to carry on its normal activities until it is killed by the developing pompilid larva, which spins its cocoon under bark or wherever the spider succumbs.

Pepsis. The 'tarantula hawks' are restricted to the New World and are most diverse in tropical habitats. Eighteen species are presently known from Costa Rica (C. Vardy, pers. comm.): *aquila, atalanta, basifusca, cassiope, festiva, grossa, ianthina, lepida, menechma, mexicana, mildei, montezuma, optimus, sommeri, terminata, tricuspidata, vitripennis* and *xanthocera*. Keys to species are currently being prepared by C. Vardy. *Pepsis* species are among the largest wasps known and are brightly hued with black bodies often rendered brilliant blue, green, purple or golden by the appressed pubescence and wings which can be fiery red, black, metallic green, purple or sometimes clear. Biology: species of *Pepsis* prey on mygalomorph spiders, particularly those of the genera *Aphonopelma* and *Cyrtopholis* (Theraphosidae) and less commonly on trapdoor spiders (Ctenizidae). *Pepsis thisbe*, a species which occurs in Costa Rica, has been studied in California (Williams, 1956). In artificial study chambers, this wasp attacked a species of *Aphonopelma*, dug a burrow with a terminal cell in friable soil and dragged the spider inside. The wasp abrades the side of the spider's abdomen with the fore legs presumably to remove the hair before depositing an elongate, slightly curved egg. Some species of *Pepsis* inter the paralysed spider in its own burrow (Passmore, 1936).

Priocnemis. Small to moderate sized wasps characterized by the presence of teeth on the hind tibia of the female and, in certain forms (subgenus

Umbripennis), of the males as well. Nearly all the species are black bodied sometimes with the metasoma and rarely the mesosoma red. The wings are most often lightly infuscate, darker at the tips, often banded, especially in the females. With the removal of the neotropical subgenus *Sphictostethus* (Roig Alsina, 1987), the genus is mostly holarctic and oriental with a few Australian and African species. The species occurring in the Neotropics are largely extensions of northern elements. There are 14 species in North America, and one of the most common of these, *P.* (*Priocnemis*) *cornica*, extends south to Costa Rica where it is rarely encountered. The female of this species often has a red metasoma in Mexican individuals and in the three specimens found to date in Costa Rica. Biology: *P. cornica* hunts among stones and clods on open soil as well as among debris and low vegetation. When she catches her prey she stings it several times anywhere on the body and when it is immobile, she stings it more slowly on the ventral side of the cephalothorax (Evans and Yoshimoto, 1962). The host range encompasses 18 species of spiders in five families—Clubionidae, Gnaphosidae, Lycosidae, Oxyopidae and Salticidae. Females transporting their spider commonly hide it temporarily while searching for a place to build a nest. The nest is usually excavated in a pre-existing cavity, such as a rodent burrow. From the side of this cavity a short burrow is constructed, terminating in a single cell. The complete life cycle requires three weeks and there are several generations per year.

Priocnessus. A New World genus, consisting of medium sized to moderately large species with black, reddish brown or yellow bodies, often with spots or stripes of contrasting integumental colour and/or silver or gold pubescent markings. The majority of *Priocnessus* species occur in Central America, although six are present in the United States. Dreisbach (1960, 1961) described some Central American species and provided keys, but many species remain unrecognizable. Seven species have been recorded in Costa Rica: *flavidus*, *hurdi*, *ornamentatus* and four undetermined species. Biology: one North American species hunts *Agelenopsis* species (Agelenidae) (Krombein *in* Krombein *et al.*, 1979).

PEPSINAE: Ageniellini. This tribe is cosmopolitan in distribution. The greatest species diversity is in the tropics. Four genera are known to occur in Costa Rica, but a fifth is present in Panama and parts of tropical South America, and may occur north into Costa Rica. Female ageniellines characteristically transport their spider prey by walking forward carrying the prey under their body. Associated with this manner of transport is the tendency of the female to amputate the legs of the spider at the coxae. One group of genera including the North American *Phanagenia* and the widespread *Auplopus* contain species which construct simple to elaborate, single to multicelled mud nests. In some cases, nesting is gregarious and females guard the entrance to the nest. Females of *naumanni*, the single known species of *Dimorphagenia*, apparently construct mud cells inside the nests of a vespid, *Agelaia* sp. Males of *naumanni* are macrocephalic (Evans, 1973*b*). For some speciose genera, surprisingly little is known about the biology, and no biological information exists for some more recently described neotropical genera, such as *Mystacagenia* and *Atopagenia*.

Ageniella. A large New World genus divided into seven subgenera (Evans, 1973*b*). It is the largest genus of Pompilidae present Costa Rica where it is represented by about 50 species belonging to five subgenera. In addition, there is another segregate of subgeneric rank that apparently is unnamed. *Ageniella* females are recognizable by the lack of a stout curved spine at the apex of the front tibia, the presence on the mentum of not more than a few slender setae, the mandibles without basal hair tufts, labrum completely hidden and the apical tergite covered with bristles. Males can be separated from other ageniellines by the antennae usually being unicolorous (or if not, there is no contrasting colour band on the middle flagellar segments), the metapleural suture absent or at most faintly indicated and usually without yellow facial markings. Biology: in North America one or two species have been observed constructing burrows or nesting in pre-existing cavities (Krombein *in* Krombein *et al.*, 1979). Females seek out pre-formed depressions in the soil and begin their nests from these (Evans & Yoshimoto, 1962; Wasbauer & Leech, 1973; Kurczewski & Kurczewski, 1987). Prey consists of mainly Lycosidae but several other families of errant spiders are also used, including Clubionidae, Gnaphosidae, Oxyopidae, Salticidae and Thomisidae. *A. anconis* has been observed in Panama carrying a clubionid by the spinnerette, without amputating the spider's legs (Kimsey, 1980*a*). Evans (1973*b*) provided a key to the subgenera in the Neotropics; the Costa Rican subgenera are:

***Ageniella* (*Ageniella*).** A mainly nearctic subgenus with a few neotropical species. There are eight species in Costa Rica including the widespread *A. utilis*.

***Ageniella* (*Alasagenia*).** Large, robust pompilids found only in the neotropical region. The two Costa Rican species have black bodies; the wings of one species are black, those of the other hyaline with a dark stripe along the costal margin.

***Ageniella* (*Ameragenia*).** This neotropical subgenus occurs from Florida to Southern Chile. About 30 species have been recognized in Costa Rica.

***Ageniella* (*Nemagenia*).** A predominantly neotropical subgenus comprising moderate sized species that are entirely black or black with a red metasoma. The one described species, *longula*, is widely distributed, occurring from Kansas and Missouri south at least as far as Bolivia. It is known to provision its nests with lycosid spiders. The Costa Rican fauna consists of *longula* and one other possibly undescribed species.

***Ageniella* (*Priophanes*).** A nearctic and neotropical subgenus with seven species in Costa Rica including *arcuata*, which extends as far north as 50° in Canada. Aspects of the biology of this species have been documented by Evans and Yoshimoto (1962).

Atopagenia. These are among the largest of all the agenielline wasps, rivalling the Asian *Macromeris* and approaching the size of a small *Pepsis*. There is a single species, *menkei*, occurring in southern Costa Rica and Panama. The body is black and the wings black with blue metallic reflection. The genus is unique in many morphological features and probably represents an early evolutionary offshoot from the agenielline line. Biology unknown.

Auplopus. A large, cosmopolitan genus that is particularly species-rich in the tropics. Dreisbach (1963) recognized 81 species from America south of the United States, 56 of which were from Central America. Approximately 35 species occur in Costa Rica. Those treated by Dreisbach, together with those described from Panama (denoted by an asterisk) but not yet confirmed as occurring in Costa Rica, include: *abnormalis**, *aeruginosus**, *albifrons**, *anthracinus**, *argentinus**, *callainus**, *dietzi**, *earinus**, *exilis**, *femurrubrus**, *fuscus*, *gertschi**, *grossus**, *incognitus*, *lineatus**, *magnus*, *marginalis*, *minus**, *minusculus**, *nebulosus*, *niger*, *nigriculus**, *olivarus**, *opacus**, *panamensis**, *purpureus**, *quartus*, *sapphirus*, *semialatus*, *shannoni**, *splendens**, *venetus**, *villosus*, *violaceus** and *zeteki**. Biology: *Auplopus* species construct mud nests. Sometimes several individuals build nests in close proximity to one another and the two published observations of neotropical species both indicate communal nesting. Kimsey (1980*a*) reported that in Panama a communal nest of *A. esmeralda* consisted of 95 mud cells in a palm frond base (mud was collected from a termite nest). Wcislo *et al.* (1988) described the nesting behaviour of the primitively social species, *A. semialatus* in Costa Rica. Two to eight females co-operatively built and maintained mud nests. Cohabiting females were usually tolerant of one another and defended the nest against natural enemies, but they often became intensively competitive when a spider was brought to the nest. *Auplopus* species are known to attack a very wide range of spiders including, in Costa Rica, species of Anyphaenidae, Clubionidae, Ctenidae, Heteropodidae and Salticidae (Wcislo *et al.*, 1988). In North America species of Philodromidae, Pisauridae and Thomisidae are also utilized (Krombein *in* Krombein *et al.*, 1979).

Mystacagenia[EX]. A neotropical genus comprising four species (Evans, 1973*b*, 1980), one of which, *elegantula*, occurs in Panama. *Mystacagenia* is known only from the females. It is characterized by the tufts of long, pale, apically projecting setae at the base of the mandibles. The wings are banded in all known species and the body has a complex pattern with light and dark areas which is distinctive for each species. Because of the disruptive colour pattern, it is assumed that all species are forest dwellers. Biology: unknown, but Kimsey (1980*a*) reported *M. elegantula* collecting mud from a termite (*Nasutitermes*) nest.

Priocnemella. A small predominantly neotropical genus one species of which has a range extending northwards into the southern United States. Seven species have been found in Costa Rica, *rufothorax*, *tabascoensis* and five undetermined species. They are small to moderate sized slender wasps that resemble *Ageniella*, but the females differ in having an outwardly curved spine at the apex of the fore tibia. Males are often difficult to separate from those of *Ageniella*, but usually have the middle segments of the flagellum orange or yellow and the rest of the antenna black. Species vary in colour from black to red and black to black marked with yellow to orange-brown; some have areas of appressed golden pubescence. Wing colour can be orange, slightly yellowish or clear. The majority of species have banded wings. Biology: unknown, but *P. rufothorax* has been observed carrying acanthoctenid spiders, without amputating its legs (Kimsey, 1980*a*).

POMPILINAE

This subfamily is represented in Costa Rica by two tribes, Aporini and Pompilini.

POMPILINAE: Aporini. A holarctic, neotropical and Australian tribe that includes both temperate and tropical species. Five genera occur in Costa Rica. Species of Aporini attack spiders, mostly Ctenizidae, found in concealed situations such as underground burrows. The spider is stung and left in its burrow after oviposition.

Allaporus. A primarily nearctic genus consisting of small species that are black or black and red in colour with hyaline, banded or infuscate wings. One species, *smithianus*, extends as far south as Costa Rica.

Aporus. A palaearctic, nearctic and neotropical genus with five subgenera in the New World; three are represented in Costa Rica by five taxa—(*A.*) *concolor*, (*A.*) *idris comptus*, (*A.*) *notabilis notabilis*, (*Neoplaniceps*) *chiapanus*, and (*Notoplaniceps*) *innotatus*. They are small to medium sized wasps with dark wings and black integument, sometimes with yellow spots. One species, *idris*, is variegated with banded wings. Evans (1973*a*) provides keys to the South American taxa. Biology: the nesting behaviour of a nearctic species, (*Plectraporus*) *hirsutus*, was described by Williams (1928).

Epipompilus. A predominantly gondwanic genus that, in the New World, extends north to Florida. Eleven New World species were recognized by Evans (1961*a*, 1967), and four occur in Costa Rica: *aztecus*, *delicatus*, *nigribasis* and an undetermined species. *Epipompilus* is generally considered one of the most primitive groups of Pompilidae, but its exact placement is uncertain. Evans (1966*e*) included it in the Aporini but, because it has so many unique features, Bradley (1944) and later, Harris (1987) placed it in a distinct tribe. Biology: unknown for neotropical species, but Harris (1987) provided information about *E. insularis* in New Zealand. This species attacks retreat-building spiders in their retreats or occasionally in the open. The effects of the initial paralysis are very transitory and the spider recovers to resume its normal activities. Eclosion and larval development take place on the abdomen of the active spider.

Euplaniceps. A rarely encountered and primarily South American group with one or two species extending north into Central America. *Euplaniceps* contains a number of described species and Bradley (1944) revised some. Other descriptions are scattered throughout the literature and the genus is in need of revision. *E. ceres*, known only from the female, occurs in Panama (Evans, 1966*e*), and a single specimen from Costa Rica may be the male of this species. *Euplaniceps*, like *Allaporus*, has the second submarginal cell receiving both 1*m-cu* and 2*m-cu*. This feature distinguishes these two genera from the other Central American Aporini with two submarginal cells.

Psorthaspis. A primarily mesoamerican genus with 13 species in Central America, nine further species in the southern United States (Bradley, 1944) and one species confined to Colombia. Five species occur in Costa Rica: *coelestis*, *connexa connexa*, *formosa*, *laevifrons* and *variegata variegata*.

Psorthaspis resembles *Aporus*, but the former has three submarginal cells and the female possesses a specialised clypeus. The males are characterized by short antennae and the low, excavated propodeum. Several Costa Rican species have areas of dense golden and white pubscence on the meso- and metasoma, thus resembling certain mutillids. Biology: a North American species, *P. planata*, attacks the ctenizid trapdoor spider, *Bothriocyrtum californicum* (Davidson, 1905). The female wasp attacks the spider in its burrow after prying open the trap door and lays an egg on the dorsal surface of the abdomen (Jenks, 1938). The wasp larva completes its development and pupates within the burrow.

POMPILINAE: Pompilini. A large cosmopolitan tribe represented in Costa Rica by 13 genera. Worldwide, members of this tribe are primarily ground nesting wasps which excavate a burrow with a single enlarged terminal cell. Some species, including some *Agenioideus* and *Priochilus*, nest in pre-formed cavities. Females in the latter genus line their nests with large pieces of leaf or blades of grass (Williams, 1928). The majority of pompilines capture prey before building a nest. Species of *Evagetes* are cleptoparasites of other Pompilini.

Agenioideus. A primarily holarctic genus with a few representatives in the oriental and neotropical regions. Three species occur in Costa Rica, the primarily nearctic species (*Agenoideus*) *humilis* and (*Gymnochares*) *birkmanni*, together with a species of the mainly South American subgenus *Enbanksia*, (*E.*) *accoleus* (Evans, 1965*b*). A fourth species, (*E.*) *fumosus*, occurs in Panama. Biology: *humilis* is frequently seen around buildings where it preys exclusively on orb-web spiders (Araneidae). Female wasps, which possess a tarsal rake, excavate short burrows, often in cracks and fissures in walls of buildings. After interment and oviposition, the female fills the burrow by raking loose material into it with her fore legs. Another widespread North and Central American species, *birkmanni*, has been recorded transporting gnaphosid spiders and sometimes flying with its prey. This species and those belonging to the subgenus *Enbanksia* probably do not dig burrows since they lack a tarsal rake.

Anoplius. A very large and diverse cosmopolitan genus. Females are easily recognized by the presence of stout bristles on the pygidium and the well developed pulvillar comb; males are characterized by the cleft tarsal claws. *Anoplius* is divided into a number of subgenera, four of which occur in Costa Rica. They are considered separately below:

***Anoplius* (*Anopliodes*)**. A primarily neotropical subgenus with one species extending north in the eastern United States. Five species occur in Costa Rica: *chiriqui* (also known from Panama), *parsonsi* (Panama to Florida), *varius* (Peru to Costa Rica) and two unidentified species, one or both of which may be undescribed. *Anopliodes* are moderate sized wasps, black with dark wings; the males and sometimes the females have erect white hair on the propodeum. The marginal cell of the fore wing is long; males have a yellow stripe on the posterior border of the pronotum and are extensively patterned with whitish appressed pubescence. The distal fore tarsal segment of the male is slender and unmodified.

***Anoplius* (*Anoplius*)**. A mainly holarctic subgenus, but a few species are present in both the Old and New World tropics. The Costa Rican fauna includes three named species, *fulgidus* (Brazil and Peru through Central America and the Antilles to the southern United States), *imbellis* (Alaska to Costa Rica) and *papago* (southern United States to Costa Rica). In addition there are three unidentified species, some of which may be undescribed. *Anoplius* (*Anoplius*) species are moderate sized wasps, black or rarely blue in integumental colour. The wings vary from nearly hyaline to deeply infuscate, violaceous. Females lack a tarsal rake and the males have the distal segment of the fore tarsus asymmetrical, slightly to strongly lobed on the inner margin. Males often have dense hair brushes on some of the apical sterna. Biology: in Texas females of *fulgidus* hunt for *Pirata* species (Lycosidae) along slow-moving streams and during prey transport alight on the surface film of the water (Wasbauer, 1955). In California, *imbellis* females hunt along the margins of still water and sometimes pursue their wolf spider prey over the surface film of the water and subdue them a considerable distance from shore, (Wasbauer, 1957). They use their wings when towing the paralysed spider over the water surface. Prey transport over land is in the usual manner, the wasp walking backward, grasping the spider by a posterior coxa and dragging it sideways. The prey is deposited near a twig or tuft of grass, or sometimes in the open, while the wasp searches for a suitable spot for burrow construction. Burrows are usually started from an existing depression, but sometimes from a flat surface. They are short, 1.0 to 3.5 cm in length, and terminate in an enlarged cell. Paralysis of the spider is temporary and the egg is placed vertically on the side of the abdomen. The wasp fills the burrow by scraping soil into it from around the entrance with the fore legs. The end of the metasoma is used to pound down the soil. Several genera of Lycosidae are used as prey (Wasbauer & Kimsey, 1985).

***Anoplius* (*Arachnophroctonus*).** A large and diverse subgenus found throughout the warmer regions of the world. Several species occur in South America and five have been found in Costa Rica: *americanus ambiguus* (widely distributed in North, Central and northern South America), *chiapanus* (western Mexico to Costa Rica), *cuatemoc* (eastern Mexico to Costa Rica) and *echinatus* (Brazil to Costa Rica). A sixth species, *apiculatus apiculatus*, occurs in both Nicaragua and Panama, but has not yet been found in Costa Rica. *Arachnophroctonus* includes small to fairly large black wasps, some of which have the metasoma partly to entirely reddish. The wings are most often dark, but may be nearly hyaline. The distal fore tarsal segment of the male is asymmetrical with the inner margin slightly produced in most species. Biology: Evans and Yoshimoto (1962) have summarized biological information available on some subspecies of *apiculatus* present in the United States. Adults rarely if ever visit flowers or honeydew, but apparently derive their food by host-feeding. Females of this species hunt for immatures of several genera of Lycosidae. The spider is chased from its shallow burrow in the sand, pursued and then stung to immobility with several quick thrusts followed by a slower sting on the venter of the cephalothorax. The paralysed spider is grasped by a hind leg and usually dragged backward, although smaller spiders are sometimes carried forward by the wasp. The spider is left on the sand while the wasp commences digging a burrow and as digging continues, is brought close to the entrance where it often becomes covered with sand. When the terminal cell is completed, the wasp enters the burrow and drags the spider in by the spinnerets. The nests vary in depth from 5 to 17 cm. To fill the nest, the wasp scrapes sand from inside the burrow and later from the outside. The sand is tamped down using the tip of the metasoma. The female wasp obliterates evidence of the nest by scraping sand in various directions over the location of the entrance. Other North American species have been recorded visiting flowers. Many construct short burrows in sandy places which they provision principally with species of Lycosidae, though Agelenidae, Amaurobiidae, Araneidae, Clubionidae, Gnaphosidae, Philodromidae and Salticidae have also been recorded as prey (Krombein *in* Krombein *et al.*, 1979).

***Anoplius* (*Notiochares*).** A predominantly neotropical subgenus found in northern South America and Central America with two species in the southern United States. Two species are known to occur in Costa Rica: *amethystinus amethystinus* and *lepidus lepidus*. *Notiochares* species are large, relatively robust wasps, black but sometimes with the appressed pubescence imparting bright bluish or greenish tones. The wings are dark. Females are characterized by having the anterior margin of the clypeus emarginate, and the males by a velvety semilunar patch of specialized pubescence on sternite IV. Biology: the nearctic subspecies, *amethystinus atramentarius*, occurs in open fields near water, and adults are frequently found on flowers although females also feed on the body fluids of spiders, which are often captured specifically for feeding (Evans & Yoshimoto, 1962). These are different from the species of the wolf spider genus *Lycosa* (Lycosidae) that are taken for larval provisions. The spider prey is deposited on the ground while the nest is being constructed, but is often moved during this period. Burrows are located in bare areas surrounded by tall vegetation. They, are initiated from the flat surface of the soil and vary in depth to 10 cm. The spider is oriented in the terminal cell with the head facing outward, the legs of the spider forming a plug against which the fill is packed. The egg is placed diagonally on the lateroventral portion of the base of the abdomen. The burrow is filled and sometimes lumps of sand, leaves and twigs are placed over the top.

Aporinellus. This widespread genus is found in nearly all the warmer regions of the world except Australia. Four species are known to occur in Costa Rica, *basalis*, *medianus*, *taeniatus* and *yucatanensis*. They are mostly small sized wasps, black, usually with extensive greyish pubescence that is often coalesced into bands on the metasoma. Both sexes have conical teeth posterolaterally on the propodeum and the metapostnotum is absent dorsally. Biology: adults of *medianus* feed extensively at flowers, extrafloral nectaries and honeydew. Hunting is done over vegetation in sandy locations. Once the prey (thomisid, salticid and oxyopid spiders) has been paralysed, it is grasped by the mouthparts and dragged over the ground by the wasp walking backwards. The spider is often left on a low plant while the nest is being excavated. After a short burrow (3 to 6 cm) with an enlarged terminal cell is constructed in the sand, the wasp enters and drags the spider inside, grasping it by the spinnerets (Evans & Yoshimoto, 1962; Wasbauer & Kimsey, 1985). The other Costa Rican species also have been recorded attacking species of Oxyopidae, Thomisidae and Salticidae in the United States (Krombein *in* Krombein *et al.*, 1979).

Aplochares. A small tropical genus comprising two species: *adrastes* (Santa Catarina, Brazil) and *A. imitator* (southern Brazil to Honduras). They are medium sized, rather stout bodied wasps, characterized by a large pulvillar pad and strongly emarginate clypeus of the female and the short maxillary palpi and presence of a short malar space in the male. The wings of both sexes have a tendency to fold longitudinally. As Evans (1966*e*) has pointed out, *imitator* is aptly named, for with its black body and dark, apically whitish wings, it resembles the vespid *Parachartergus apicalis*. Males are now known from several localities in Costa Rica and are patterned somewhat differently. The legs are black, fading to dull reddish apically as in the female, but the middle and hind tibiae are striped dorsally with whitish and the posterior margin of the pronotum has a narrow white stripe. The wings are hyaline with a dark stripe longitudinally along the costal margin. Biology: unknown.

Balboana. A neotropical genus ranging from Argentina, Brazil and Paraguay north to Mexico. There are six described South American species, but probably at least as many species are undescribed. Six have been collected in Costa Rica: *cameroni*, *pulchella*, *tarsalis* and three unidentified species, at least one of which is undescribed. In addition, the widespread species, *auripennis*, recorded from Brazil to Panama, probably occurs in Costa Rica. The genus is not taxonomically well-understood because specimens are not commonly collected. *Balboana* is closely related to *Priochilus* from which it differs principally in the form of the pronotum in the female, which is angulate in lateral view, the anterior surface forming a sharp angle with the dorsal surface and in some species appearing toothed in dorsal view. Additionally, the metasoma is somewhat compressed apically and the antennae (in Central American forms) are somewhat swollen medially, thus appearing slightly clavate. Some of the outer antennal segments of the male are crenulate and bear distinct sensory areas beneath. The basic ground colour of both sexes is black but in one undetermined species, the mesosoma of the male is red and the metasoma black. The wings may be clear or suffused with yellow, and are banded in both sexes. The banding pattern may be specifically distinct. Biology: unknown.

Episyron. A cosmopolitan genus with a single widespread neotropical species, *conterminus*, ranging from Brazil to New York. *Episyron* is characterized by wings that fold longitudinally and by the presence of scale-like pubescence on the propodeum and tergite I, and often on other parts of the mesosoma. The New World species are medium sized wasps, mostly black with limited yellow markings; occasionally the legs are red. The wings vary from extensively infuscate to hyaline, usually with the apex darker. Biology: females of *conterminus* excavate burrows (2.5 to 7.5 cms deep) in sandy soil and provision the terminal cell with araneid spiders;

they often fly short distances carrying their prey (Krombein, 1953).

Evagetes. A moderate sized, cosmopolitan genus present in all regions except Australia. *Evagetes* is represented by five species in Central America, two of which, *parvus* and an unidentified species, occur in Costa Rica. *Evagetes* species are similar to *Arachnospila*, but the antennae are relatively short and, in the female, stout and flattened on one side. Biology: cleptoparasites of nest-building Pompilidae (Evans & Yoshimoto, 1962). Females use their thickened antennae to search out the newly provisioned nest of another pompilid. They then dig down and expose the provisioned spider in the cell. They eat the egg of the original provisioner and lay one of their own, filling the burrow again prior to leaving.

Paracyphononyx. A fairly large genus that is widespread throughout the warmer parts of the world. Several species are present in South America and one, *unicolor*, occurs as far north as Guatemala. It is the only species found so far in Costa Rica, but another species, *semiplumbeus*, has been taken in Panama. Morphological variation between the species is great and the group may well be polyphyletic. *Paracyphononyx* species are medium sized wasps, the New World species black or with the metasoma red, the males frequently and the females occasionally with areas of grayish pubescence on the mesosoma and metasoma, often forming bands on the latter. The wings of the females are usually dark but those of the males subhyaline to hyaline with darker apices. Males have a malar space, which may or may not be present in the females. The head is attached low on the thorax sometimes giving a somewhat humpbacked appearance. The tarsal claws are cleft in both sexes and the labrum is fully exposed. The males are further characterized by crenulate antennae and dense erect pubescence on the posterior slope of the propodeum. Biology: a North American species has been recorded attacking lycosid spiders (Hurd & Wasbauer, 1956).

Poecilopompilus. This new world genus ranges from southern Canada to southern South America. Five species occur in Costa Rica, two of which, *algidus marcidus* and *interruptus dubitatus*, are subspecies of northern taxa. The other three are neotropical: *badius* (Ecuador to Costa Rica), *flavopictus flavopictus* (Colombia to southern Texas) and *polistoides polistoides* (northern South America becoming scarcer in Central America) (Evans, 1966*e*). *P. polistoides* has only been collected once in Costa Rica, but *flavopictus* is common throughout the country. They are mostly yellowish vespid mimics. Biology: in the United States *interruptus interruptus* hunts in tall vegetation for species of Araneidae. The spider is dragged over the ground, being grasped in the wasp's mandibles by a hind coxa. It is often placed on a plant before nest construction begins. Burrow length averages 7 cm. After oviposition, the wasp fills the burrow by raking in loose material and tamping it down with the tip of the metasoma. The last two segments are turned under so that the ramming of the soil is done with the dorsal surfaces of these segments (Evans & Yoshimoto, 1962).

Priochilus. A neotropical genus characteristic of rain forest situations. It occurs from northern South America to southern Mexico. There are about 20 species, nine of which occur in Costa Rica: *admirationis admirationis, captivum, fustiferum, gloriosum gloriosum, gracile, scrupulum, sericeifrons, splendidulum splendidulum* and *veraepacis*. In addition, there is one unidentified species, which is possibly undescribed. These are medium to large wasps that are all black, or (in one species) black with a red metasoma. The males frequently have yellow or cream coloured integumental markings. Females are sometimes elegantly marked with patches of silvery or gold pubescence. The wings are black or hyaline with strong dark banding. As in some Pepsinae, the hind tibia of the females of certain species is serrate. Biology: *Priochilus* species do not excavate burrows and none possess a tarsal rake. In Ecuador *gloriosum* nests under loose bark and another species constructs a nest in a rolled up palm frond base (Williams, 1928). In both cases the nests were lined with bits of leaves and other plant material.

Sericopompilus. This small genus is characteristic of the southern Nearctic region. There are three species, one of which, *neotropicalis*, ranges from the western and central United States, south to Costa Rica. *Sericopompilus* species are rather slender wasps with long wings. Females are black with a red metasoma. In the northern portion of the range the wings of the females are uniformly dark, but farther south, as in Costa Rica, the wings show considerable variation and can be entirely dark or nearly hyaline with darker markings. In Costa Rica the posterior margin of the pronotum is always marked with yellow. The males have a red and black metasoma and the wings are hyaline with a broad apical dark spot. Biology: the North American *apicalis* provisions with spiders of the families Anyphaenidae, Araneidae, Oxyopidae, Philodromidae, Salticidae and Thomisidae (Krombein *in* Krombein *et al.*, 1979).

Tachypompilus. A large pantropical genus that, in the New World, ranges north into the southern United States. There are ten described species in South America (Banks, 1946, 1947). Costa Rican species include *unicolor cerinus* (southwestern United States to Costa Rica) and *ferrugineus affinis*. This latter species is widespread throughout South America north to Costa Rica where there is a zone of intergradation between it and the northern subspecies *ferrugineus ferrugineus*. A third species, *T. mendozae*, is widespread from Argentina to Panama and might occur in Costa Rica. *Tachypompilus* species are moderate to large sized red or red and black pompilids. Biology: in North America often common around walls or piles of stones where they nest in crevices, provisioning the nest with lycosid and pisaurid spiders.

Xenopompilus[CR]. *Xenopompilus* was originally proposed as a subgenus of *Pompilus*. Day (1981) redefined *Pompilus* and restricted it to *cinereus* and a few other Old World species, leaving *Xenopompilus* and several other New World subgenera in an uncertain position. Until a careful study of the relationships of these groups is made, the name *Xenopompilus* can be used in a generic sense for this small, uncommonly encountered group. All three species occur in Mexico, with one extending as far north as southern Arizona, and one, *tarascanus*, now recorded from Costa Rica. Individuals of all species are black, the females without stout spines at the apex of the metasoma, the ultimate tarsal segments without spines beneath, the tarsal rake absent and the frons abruptly declivous between the antennal sockets; the males have the distal tarsal segment slender and symmetrical, the tarsal claws toothed and the metasoma uniformly silvery pubescent, not banded. Biology: unknown.

CEROPALINAE

This small subfamily, occurring in every zoogeographic region of the world, is very distinct from the Pepsinae and Pompilinae and has been considered a separate family by some authors (e.g. Móczár, 1986*a*). It comprises two genera, *Ceropales* and *Irenangelus*, both of which are highly specialized cleptoparasites of other Pompilidae.

Ceropales. A large, cosmopolitan genus containing about 135 species worldwide, *Ceropales* was revised by Móczár in a series of papers (1986*a*, *b*, 1987, 1988, 1989, 1990, 1991). The genus includes small to moderate sized wasps, of rather compact form, but with very long, slender legs. The basic integumental colour is black although a number of species have the metasoma red. Yellow or white markings on the head and body are the rule in this genus. The legs are often red or yellow or are marked with these colours. The form of the male genitalia and the shape of the subgenital plate as well as the kidney-shaped eyes and right-angled hind tarsal claws are very diagnostic for *Ceropales*, and the genus is thus easy to recognize. There are four subgenera, but species of only one, *Bifidoceropales*, have been recorded in Costa Rica: *cubensis albopicta*, *cubensis menkei*, *femoralis*, *splendida* and *rugata*. Biology: of New World species not well-known, but the european *maculata maculata* attacks species of the genera *Pompilus*, *Agenioideus*, *Anoplius*, *Episyron*, *Priocnemis* and *Auplopus*. The *Ceropales* female either rests on vegetation in the nesting area of the host wasp, or actually follows females about whether or not they are carrying prey. When the spider is exposed during transport, or when left while the host wasp explores the area for a suitable nesting spot, the female *Ceropales* runs to the spider and inserts an egg into the book lung. The host wasp may attack the *Ceropales* and drive her off repeatedly, but *Ceropales* is usually successful in depositing an egg. The host pompilid then continues her nesting and inters the spider. The egg of *Ceropales* hatches first and the first instar larva devours the host larva when it ecloses (Richards & Hamm, 1939; Grandi, 1954).

Irenangelus. A neotropical, oriental and Madagascan genus of rather slender wasps. Specimens are seldom collected and are probably rare or very localized. Evans (1969*b*) revised the neotropical species and described several new ones, including subsequently one from Costa Rica (Evans 1987*b*). *Irenangelus* species resemble *Ceropales*, but the genitalia and subgenital plate are not as strongly modified and the hind tarsal claws are evenly curved rather than right-angled. Nearly all the species are uniformly reddish in colour, sometimes with limited yellow markings on the head or mesosoma. The wings are clear, often with a strongly contrasting black spot at the tip. There are two species-groups separated by abundant characters, such as the shape of the metapostnotum, presence or absence of parapsidal furrows and presence or absence of hair brushes on the tarsi of the males. The Costa Rican fauna includes three described species, *eberhardi*, *ichneumonoides*, *lucidus* and two undescribed. Biology: in Asia *Irenangelus* has been reared as a cleptoparasite of a nest-building *Tachypompilus* species (Williams, 1919*b*), and in Costa Rica, females of *I. eberhardi* have been observed to wait motionless near the uncapped, prey-containing mud nest of the gregariously nesting *Auplopus semialatus*. The instant the nest is unattended by the provisioning *semialatus* female, the *Irenangelus* female quickly approaches the nest and deposits an egg on an unattended spider in an open cell (Wcislo *et al.*, 1988).

NOTOCYPHINAE

This subfamily is almost exclusively tropical in distribution. It comprises two genera, the Oriental *Minotocyphus* and the New World *Notocyphus*.

***Notocyphus*^CR^**. A fairly large and diverse genus occurring throughout the American tropics. Banks (1947) recognized 29 species in South America but additional species occur in Central America and in the Antilles. The Costa Rican fauna consists of eight species, all of them un-named except for the polytypic *dorsalis*, which is the only species occurring in the United States. Townes (1957) recorded *dorsalis* from Texas to northeastern and north-central Mexico, *dorsalis arizonicus* from Arizona and California and *dorsalis restrictus* from Central Mexico to Guatemala. The last subspecies also occurs in Costa Rica. Species of *Notocyphus* vary from medium to quite large in size. The basic colour of the mesosoma is black but in several species the mesoscutum is reddish and in one Costa Rican species the entire insect is orange. The metasoma varies from entirely black to black with yellow spots, black with terminal segments red, entirely yellow or yellow with various combinations of red or orange. The wings are most frequently dark, but in some species the males have hyaline wings with darker apices. The genus is easily recognized by the cleft tarsal claws in both sexes, the large, fully exposed labrum, laterally compressed terminal metasomal segments of the female and tent-like subgenital plate of the male. Biology: in Ecuador *tyrannicus* was reared from a theraphosid spider in a retreat in a folded leaf about two metres above the ground. The spider was active, having partially recovered from the effects of the sting and bore on its abdomen a small pompilid larva which eventually killed it (Williams, 1928). In the nearctic region, Simons (1989) reared an adult male of *dorsalis* from a very small larva he discovered on the side of an immature tarantula, *Aphonopelma chalcodes* (Theraphosidae). These accounts both indicate that the spider regains normal activity after the attack and maintains it until shortly before it is killed by the wasp larva.

14.4 SAPYGIDAE

Paul E. Hanson

Diagnosis. Body length usually about 7.0 to 10.0 mm; mostly yellow and black; both sexes fully winged. Antenna 12-segmented in females and 13-segmented in males, dorsal rim of torulus with small tubercle or transverse ridge; *eyes large and, in Neotropical taxa, with the inner borders emarginate*; mandible tridentate. *Pronotum abruptly truncated anteriorly and reaching back to the tegulae*; mesosternum simple, without plates or denticles in front of or overlapping the middle coxae. *Fore wing venation complete*; hind wing with two enclosed cells, claval lobe reduced, jugal lobe present. Mid coxae situated close together or contiguous, never widely separated. *First and second metasomal tergites not separated by a constriction; sternite VI of female expanded laterally*, with sides overlapping each other dorsally and projecting posteriorly beyond tergite VI;

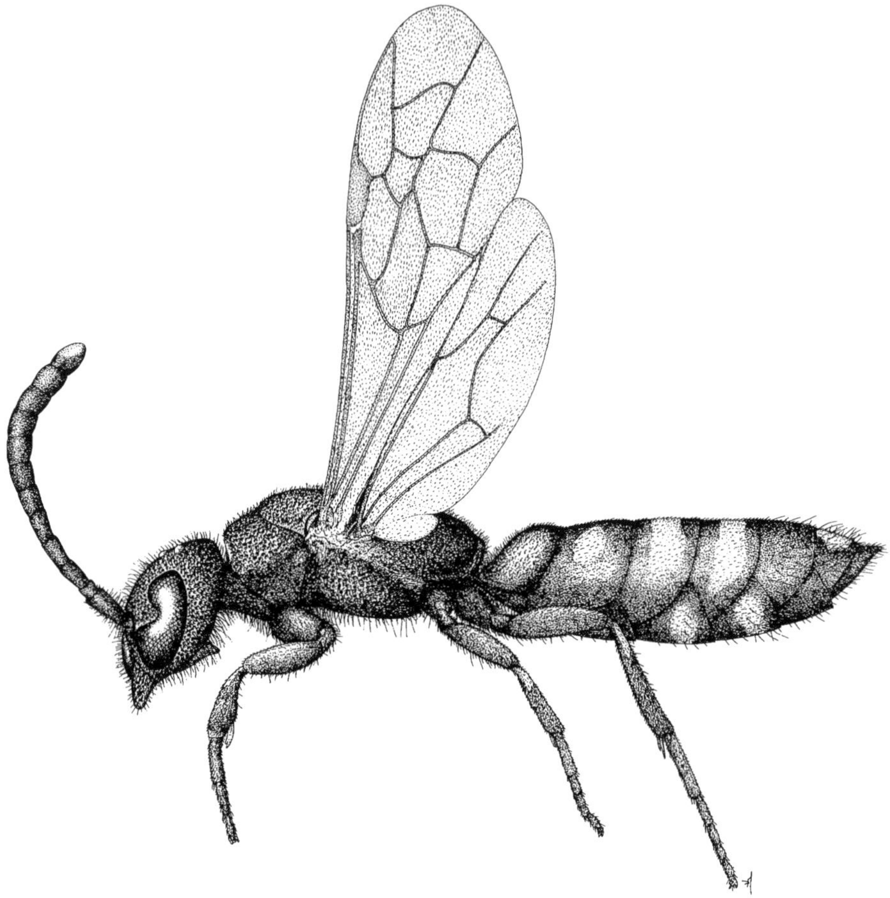

Fig. 14.74. *Sapyga* sp. (Sapygidae).

sting sheath with barbs posterodorsally; sternite VIII (hypopygium) of male simple, entirely exposed.

Classification and distribution. The Sapygidae is a small family with a nearly cosmopolitan distribution (absent from Australia), comprising about 80 to 90 species. The family is divided into two subfamilies: Fedtschenkiinae and Sapyginae. The former contains just one holarctic genus (mostly in arid areas), which is distinguished from all other members of the family by its non-emarginate inner eye margins. The Sapyginae consists of eight genera, four of which are restricted to the Old World. In the New World there are four genera: the Chilean *Araucania*, the nearctic *Eusapyga*, the neotropical *Huarapea* and the holarctic *Sapyga* (Pate, 1947). Only the latter two have been found in Costa Rica.

Biology. Larvae of Fedtschenkiinae are ectoparasitoids on the larvae of soil-nesting Eumeninae. As far as is known, all members of the Sapyginae are cleptoparasites in the nests of bees, the group of bees attacked apparently being characteristic of the sapygid genus (Pate, 1947). Species of *Huarapea* attack nests of *Xylocopa* whereas species of *Sapyga* attack nests of Megachilinae. There has been very little study of the biology of *Huarapea* (Hurd & Moure, 1961), although it is probably very similar to that of

Polochrum, an Old World genus that also attacks nests of *Xylocopa*.

The most detailed biological studies are those of the North American species, *Sapyga pumila*, which attacks the alfalfa leafcutter bee, *Megachile rotundata* (Torchio, 1972). The female of *S. pumila* deposits one or more eggs in freshly completed cells of *Megachile* by inserting her ovipositor through the cell cap. If the bee returns while the sapygid is in the nest, the bee will often bite off a portion of the wasp's antenna; toward the end of the season many sapygids have parts of their antennae missing. The active, first instar larva crawls about within the cell and punctures other wasp eggs or larvae with its long sickle shaped mandibles, until there is only one sapygid per cell. The surviving larva eventually destroys the host egg (if one is present), a process that requires considerable contortion of the larva's mouthparts, since the egg is twice the size of the larva. The larval sapygid then feeds on the nectar and pollen provisions in the nest, passing through four instars before spinning a cocoon. Adult *S. pumila* visit flowers and alfalfa pollen can be found in the midgut of dissected individuals. Mating apparently occurs in mating stations, or leks; adults pass the night in host nests.

Economic importance. In North America *Sapyga pumila* is a serious pest of *Megachile rotundata*, the introduced alfalfa leafcutter bee. This bee is a very important pollinator of alfalfa in many of the western states in the U.S., and a successful business has been founded propagating it. *Sapyga pumila* is native to North America and originally probably attacked widely dispersed native bees. However, with the mass propagation of the alfalfa leafcutter bee, the sapygid began to attack this bee. By the 1970's the sapygid became so prevalent that it threatened commercial rearing of the bee (Torchio, 1979).

Identification. Keys to the New World genera of Sapygidae are given in Pate (1947). The two genera present in Costa Rica can be separated by the following couplet:

— Head with the occipital carina distinct. ....*Huarapea*

— Head without a distinct occipital carina. ....*Sapyga*

Synopsis of the Costa Rican fauna

Only two genera occur in Costa Rica, *Huarapea* and *Sapyga*, each of which appears to be represented by just one or two rare species. Thus far they have been collected only in the drier parts of the country (including the San José area), below 1500 m altitude.

14.5 MUTILLIDAE

Denis J. Brothers

Diagnosis. Body length 1.5 to 25.0 mm; predominantly black, brown, or black and reddish, usually with brightly coloured patches or fringes of silver, gold, yellow, or red, usually formed by pubescence but often integumental in Neotropical species; cuticle usually very hard and heavily sculptured. Most mutillids are strongly pubescent and there is extreme sexual dimorphism in body form and colour, the males being fully winged (although rarely brachypterous or apterous) and the females entirely apterous (hence the common name: 'velvet ants'). *Head with dorsal rim of the antennal socket expanded to form a strong antennal tubercle; mid coxae very close to each other*; first metasomal segment usually more or less sessile, but sometimes forming a short petiole (especially in males); with a notch between first and second metasomal sterna; second tergite, and sometimes sternite, usually with a longitudinal felt line on each side. *Males* with inner margin of eye convex, weakly sinuate, or deeply notched; pronotum more or less transverse and usually strongly concave posteriorly; wings often infuscated, with strong venation on basal three-quarters; hind wing generally without claval and jugal lobes. *Females* with eye oval to circular; mesosoma with pronotum + meso- + metathorax + propodeum fused into a solid box-like structure, many of the sutures not visible (pronotum articulating with mesothorax in a few non-neotropical species); sting often very long and powerful.

Classification and distribution. The family is cosmopolitan but predominantly tropical and contains about 8000 species in seven subfamilies: Myrmosinae (including Kudakrumiini), Pseudophotopsidinae, Ticoplinae, Rhopalomutillinae, Sphaeropthalminae, Myrmillinae and Mutillinae (Brothers, 1975). Only two

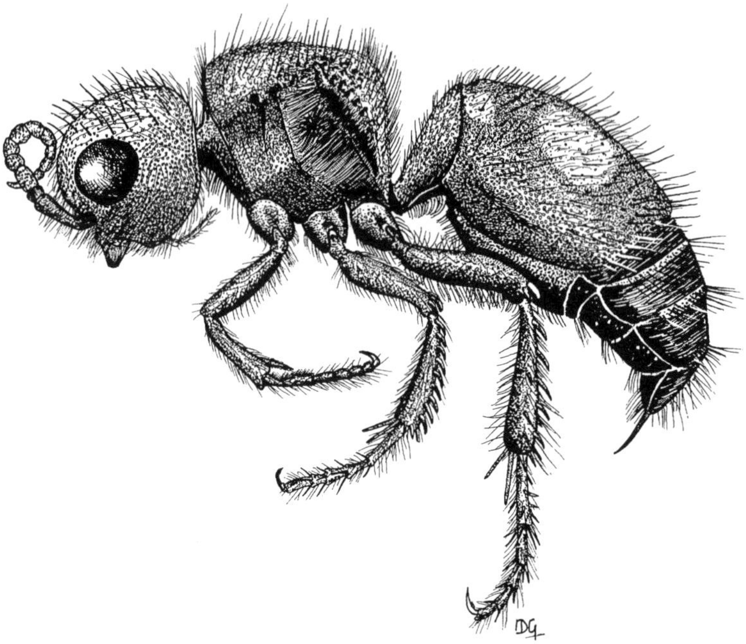

Fig. 14.75. *Pseudomethoca* sp. (Mutillidae) female.

of these subfamilies, the Sphaeropthalminae and Mutillinae, occur in Central America. The former comprises two tribes, Dasylabrini and Sphaeropthalmini, with only the latter occurring in the New World. The tribe Sphaeropthalmini in turn comprises two subtribes, both of which are represented in Central America. The subfamily Mutillinae comprises two tribes, Mutillini and Ephutini, both of which occur in Central America.

Biology. On a worldwide basis, mutillids have been most commonly collected in relatively dry habitats, such as savannah and semi-desert, areas where their putatively commonest hosts (other aculeate Hymenoptera) also tend to occur in greatest numbers. This may be a function of ease of collection, however, since the wingless females are most readily seen when running over bare soil or sand. They are very difficult to see and almost impossible to catch in dense undergrowth. Widespread use of Malaise traps in a variety of habitats in Costa Rica (and elsewhere) has shown that males may be quite common in areas previously considered unsuitable for mutillids, including lowland rainforests.

Despite the size of the group, our knowledge of the basic biology of mutillids is very limited. Worldwide fewer than ten species have been studied in any detail, although there is some scanty information on another 140 species or so. General accounts of the biology of Mutillidae can be found in Mickel (1928), Clausen (1940*b*), Ferguson (1962) and Brothers (1972, 1978, 1989).

The larvae are solitary ectoparasitoids of the resting terrestrial immatures of other insects. The recorded range of hosts (in order of decreasing frequency) includes the fully fed larvae or pupae of a wide variety of bees and aculeate wasps (especially Sphecidae) within cells and/or cocoons; the puparia of flies, specially Glossinidae (tsetse), but also other Cyclorrhapha; the pupae of moths (Limacodidae) within hard cocoons; the pupae of clythrine and cryptocephaline beetles (Chrysomelidae) in hard cases; and the oothecae of cockroaches (Polyphagidae). Since this list is based on knowledge of the hosts of only about two percent of the species, it must be viewed with caution, and the range of potential hosts is certainly greater. A common requirement, as indicated by the host records,

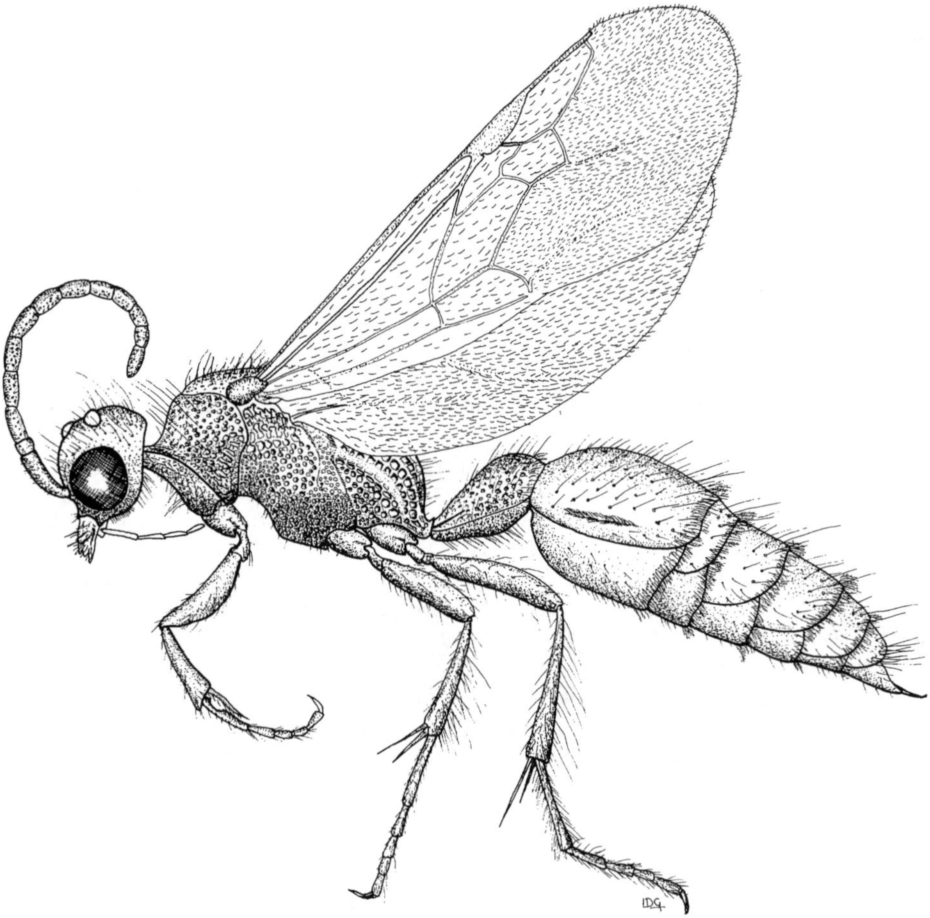

Fig. 14.76. *Sphaeropthalma* s.l. sp. (Mutillidae) male.

seems to be that the stage attacked be enclosed in some sort of 'package' with a rigid wall, and that it be of such a size as to permit the full development of the larva.

Host records of mutillid genera occurring in Central America are given in Table 14.2. These were gleaned from the following sources: Callan, 1942, 1977 & pers. comm. [*Traumatomutilla*]; Mickel, 1973; Krombein *in* Krombein *et al.*, 1979; Kimsey, 1980*a*; Roubik, 1989, 1990; Cambra & Quintero, 1992; Krombein, 1992*a*; Yanega,1994. Specific host records from Central America include *Ephuta* sp. on *Auplopus esmeralda* (Kimsey 1980*a*), *Paramutilla halicta* on *Augochlorella edentata* (Mickel, 1973), *Pseudomethoca hesperus* and doubtfully *P. transversa* on *Halictus hesperus* (Brothers, 1982; Brooks & Roubik, 1983), *P. willei* on *Lasioglossum umbripennis* (Mickel, 1969), *Hoplomutilla xanthocerata* on *Eulaema meriana* (Roubik, 1990) and *Pappognatha myrmiciformis* on *Euglossa dodsoni* (Yanega, 1994). We know little about the degree of host specificity, however. At least some species appear to be situation-specific rather than host-specific, and this may be a general characteristic. For example, in North America *Pseudomethoca frigida* attacks at least eight species of halictine bees which all nest in similar situations in the soil or rotting logs, and *Sphaeropthalma orestes* is reported from the underground nests of bees, sphecids and eumenine vespids, and *Sphaeropthalma unicolor* from the underground

Mutillid genus	Host group
Dasymutilla	Anthidiini, Anthophorinae, Bombini, Halictinae, Megachilinae, Polistinae, Sphecidae, Tiphiidae
Ephuta	Pompilidae
Hoplognathoca	Anthophorinae
Hoplomutilla	Euglossini
Paramutilla	Halictinae
Pappognatha	Euglossini
Pseudomethoca	Andreninae, Anthophorinae, Halictinae
Sphaeropthalma s.l.	Anthidiini, Anthophorinae, Eumeninae, Megachilinae, Pompilidae, Sapygidae, Sphecidae
Timulla	Anthophorinae, Eumeninae, Sphecidae
Traumatomutilla	Sphecidae
Xystromutilla	Sphecidae

Table 14.2. Host records of mutillid genera known to occur in Central America.

cells of six genera of bees, but also from cells constructed in hollow plant stems by bees and sphecids (Ferguson, 1962; Krombein *in* Krombein *et al.*, 1979).

The females search the ground or vegetation (depending on the species) for host nests or cocoons. Species that attack multicellular nests of Hymenoptera may spend several days within the nest moving from cell to cell (Krombein, 1992*a*). The female oviposits directly into the cell or cocoon of the host after making an opening with her mandibles. The potential host inside is first investigated with the antennae and/or the sting, and the egg is laid only if the host is in a resting state (has finished feeding or defaecating in the case of a larva). The host may or may not be stung before oviposition takes place, and even if stung it may or may not be paralysed. After oviposition the female reseals the host container, either by plugging the opening in the cell with soil or by applying a salivary secretion to the host cocoon.

Upon hatching, the larva punctures the cuticle of the host with its sharp mandibles and feeds on the haemolymph. Development is rapid and there are about five larval instars. As in Tiphiidae, the larvae have the second pair of thoracic spiracles greatly reduced and the mandibles quadridentate (Evans *in* Evans, 1987*a*). Pupation occurs within a papery cocoon spun in the host container, although the larva may sometimes merely add some silk to the inside of the host cocoon and not form a separate cocoon. The pupa has numerous processes on the body, and the female usually has the sting exserted and curved above the metasomal apex. Generally only a single mutillid develops per host individual (or ootheca) and in cases of superparasitism one larva kills the others. A few Australian species are exceptional, however, in producing up to about four mutillids from a single host cell (Brothers, 1984).

Depending on the species, mutillids may be diurnal or nocturnal (many Sphaeropthalmini). Diurnal species are usually not active during the hottest time of the day (Nonveiller, 1963; Brothers, pers. obs.). Adult males generally fly and feed on nectar. Females, on the other hand, run and may feed on adults or immatures of the host, or on moist pollen-nectar masses in the cells of bee hosts, in addition to or instead of honeydew from Homoptera or sweet exudates from plants. Females are notorious for their painful sting (hence the name 'cow-killers' for some large North American species), which is used mainly in defence against predators rather than in attacking the host. Both sexes stridulate under stress (e.g. when grasped) by rubbing a finely striated medial area at the base of the third metasomal tergite against a transverse ridge on the underside of the apex of the second tergite. Oscillographic analyses of the sounds produced may be useful taxonomically and in associating sexes (Genise & Straneck, 1991).

Males are often larger than their conspecific females, although the difference may be insignificant, or sometimes even reversed. Where the sexes are not very different in size, mating generally takes place on the substrate and lasts only a few seconds. Size differentiation (e.g. in species of *Timulla*), where a male only 1.3 times the length of the female has about twice her mass, permits transportation of the female by the male during or before copulation. Such a male typically clasps the female in the neck region using his mandibles, holds her mesosoma with his legs and flies for some distance before settling. Mating generally seems to take place after the flight but true phoretic copulation is found in some species (none known from Costa Rica). In these the male is relatively even larger, usually at least two times and up to about four times the length of the female (about 8 to 64 times her mass) and mating may persist in flight for many minutes. During this process the male may visit flowers for nectar and the female may then also be able to feed. Such phoresy may be an important means of overcoming the reduced vagility of the females because of their lack of wings, although there is no evidence to suggest that non-phoretic species are any less widely dispersed than phoretic ones.

Identification. Because the Mutillidae are poorly studied taxonomically, even generic limits are sometimes unclear. For example, various Central American species combine the characteristics of the well-known genus *Dasymutilla* (very common in North America and with one or two species as far south as Chile) and *Traumatomutilla* (common in South America with one species as far north as Panama). These two genera should thus probably either be synonymized or split into more genera. It is also usually impossible to associate male and female of a particular species in the absence of direct biological information, such as observation of mating or rearing of specimens from the same host at the same time and place. Furthermore, within a single species, and even within the same population and sex, specimens may vary considerably in size as a result of differences in the amount of food available to the larvae. Colour may also vary, with particular body regions grading from reddish-brown to black or pubescent markings from silver to gold, for example. Recognition of all specimens of one sex as representing a particular species may thus also be difficult. Unlike the situation in some other groups of insects where sexual dimorphism is extreme but where only one sex is used as the basis for species description, both sexes are used in the Mutillidae, with the result that there are many species, and even genera (including about nine from Central America), known from one sex only.

The subfamilies, tribes, and subtribes of Mutillidae occurring in Costa Rica can be identified by the key below. Cambra and Quintero (1992) provided keys to the genera of Panamanian mutillids, but they did not include eight genera known to occur in Costa Rica (*Darditilla*, *Haplognathoca*, *Lophomutilla*, *Paramutilla*, *Sphaeropthalma* and three undescribed genera). On the other hand, their keys include three genera not yet known from Costa Rica (*Hoplocrates*, *Horcomutilla* and *Nanotopsis*).

Key to suprageneric groups of Mutillidae in Central America

1 Females; apterous individuals. 2

— Males; fully-winged individuals. 5

2 Eye almost circular, usually strongly projecting, and often with the surface smooth (facets virtually indistinguishable); mesosoma in dorsal view distinctly widest across mesothorax or pronotum, and narrowest posteriorly (Figs 14.77, 14.78), lateral margins abruptly convex or strongly sinuate; mesopleural carina ending dorsally some distance posterior to pronotal spiracle. (Sphaeropthalminae: Sphaeropthalmini) .. 3

— Eye oval, only weakly protruding and with surface distinctly faceted; mesosoma in dorsal view parallel-sided, or widest posteriorly (Figs 14.79, 14.80),lateral margins more or less straight or concave, or oval (evenly narrowed anteriorly and posteriorly); mesopleural carina ending dorsally at or immediately posterior to the pronotal spiracle, or obliterated. .. ..(Mutillinae) 4

3 First metasomal segment generally short, sessile and not constricted posteriorly; head usually more or less transversely quadrate and as large as or larger than mesosoma in dorsal view (Fig. 14.77); dorsolateral margin of mesosoma usually abrupt or even carinate, often with teeth. .... **Pseudomethocina**

— First metasomal segment generally distinctly petiolate and slightly constricted posteriorly; head usually more or less oval and smaller than mesosoma in dorsal view (Fig. 14.78); dorsolateral margin of mesosoma usually rounded, sometimes carinate or toothed. ....**Sphaeropthalmina**

4 First metasomal segment sessile to weakly campanulate and more or less evenly broadened from front to rear, almost as broad posteriorly as second metasomal segment (Fig. 14.80); tergite II of metasoma with longitudinal felt line on each side; body usually entirely black or black with reddish mesosoma, with paired longitudinal and transverse silver or gold pubescent markings on metasoma; mesosoma about as broad as head and metasoma, almost rectangular from above and with pleural regions smooth. ....**Mutillini**

— First metasomal segment predominantly cylindrical, very much narrower posteriorly than second metasomal segment (Fig. 14.79); tergite II of metasoma without longitudinal felt lines; body usually entirely black or brownish, sometimes with a pair of pale spots on tergite II; mesosoma narrower than head and metasoma, narrowly oval and evenly rounded anteriorly and posteriorly, with pleural regions densely and heavily punctured. .... **Ephutini**

5 Eye almost circular (Fig. 14.81), usually strongly projecting, and often with surface smooth (facets virtually indistinguishable), with inner margin of the eye convex or weakly sinuate; tegula almost circular, evenly convex and ending at or anterior to level of transscutal articulation; pterostigma well sclerotized and distinct (Fig. 14.83). .... (Sphaeropthalminae: Sphaeropthalmini) 6

— Eye oval, only weakly protruding and with surface distinctly faceted, with inner margin of eye deeply and abruptly notched (Fig. 14.82); tegula elongate, ending posterior to level of transscutal articulation, distinctly reflexed posteriorly or with a distinct longitudinal angle anteriorly; pterostigma unsclerotized and cell-like or absent (Fig. 14.84). .....(Mutillinae)....7

6 First metasomal segment generally short but sometimes somewhat elongate, sessile and not constricted posteriorly. .... **Pseudomethocina**

— First metasomal segment generally distinctly petiolate and constricted posteriorly. .... **Sphaeropthalmina**

7 First metasomal segment sessile to weakly campanulate and more or less evenly broadened from front to rear, almost as broad posteriorly as second metasomal segment; tergite II with longitudinal felt line on each side; body generally black with metasoma yellowish and some pale pubescence on mesosoma; tegula evenly convex anteriorly and distinctly reflexed posteriorly. .... **Mutillini**

— First metasomal segment predominantly cylindrical, very much narrower posteriorly than second metasomal segment; tergite II without longitudinal felt line; body generally entirely black with transverse silvery pubescent bands on metasoma; tegula longitudinally angulate anteriorly and strongly longitudinally convex posteriorly. .... **Ephutini**

Synopsis of the Costa Rican fauna

Unfortunately, the very rich neotropical fauna is poorly studied, so that estimates of the actual numbers of species present are difficult. According to a recent catalogue of the neotropical Mutillidae (Nonveiller, 1990), nearly 1500 species occur in the neotropical region (including Mexico) and 180 species in Central America. Cambra and Quintero (1992) recorded 67 species from Panama, based mostly on females ('the sex most commonly encountered', i.e. taken by hand collecting). Only 37 species have been recorded from Costa Rica itself.

These figures are likely to be considerable underestimates of the true size of the Central American mutillid fauna, however, since I have seen almost 200 species from Costa Rica alone. These taxa are represented almost exclusively by male specimens collected in Malaise traps over the last few years. It is likely that at least 300 species occur in Costa Rica. The genera that have been collected, with very rough estimates of

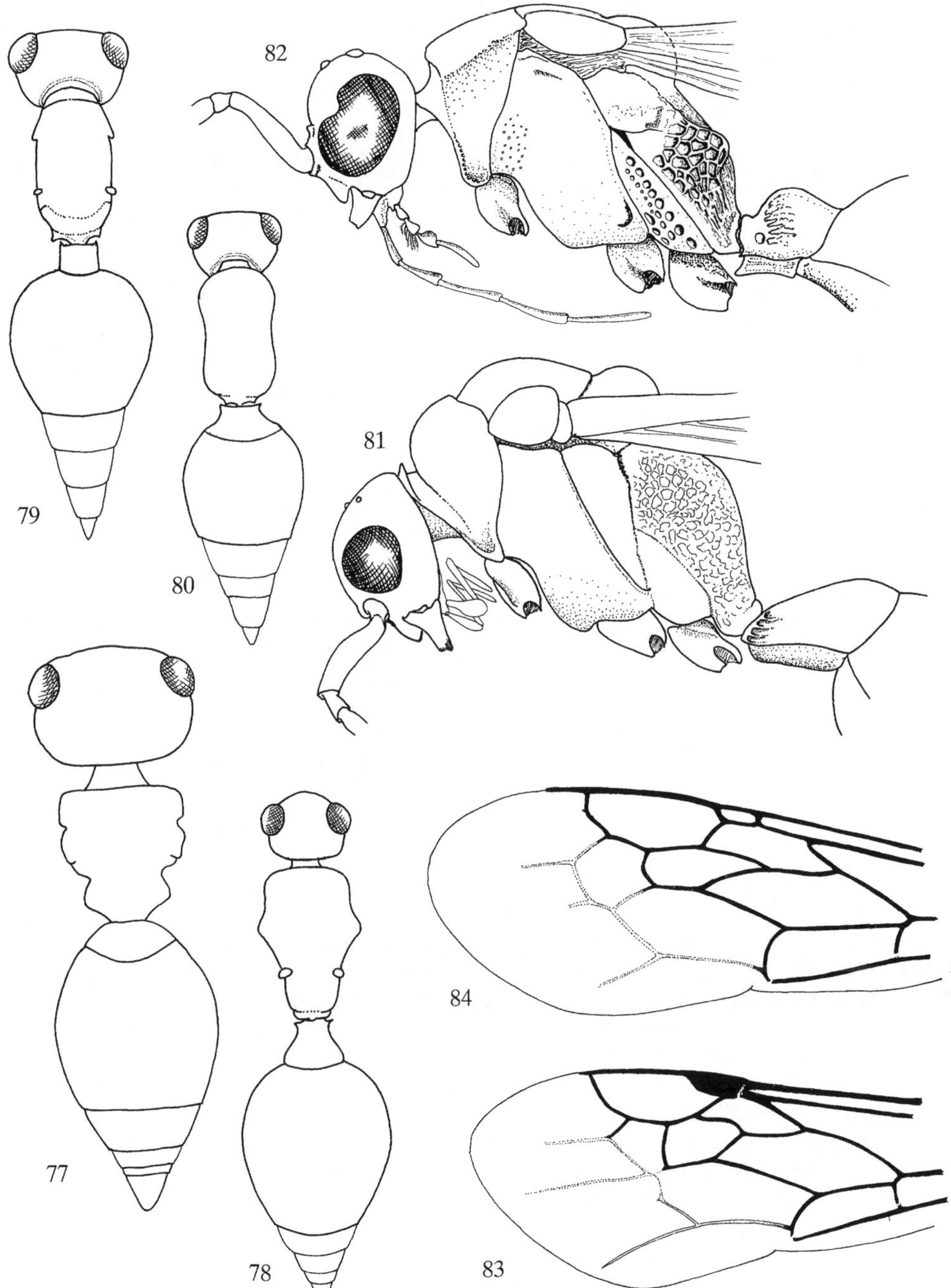

Figs 14.77–14.84. Mutillidae. Figs 14.77–14.80. Female, head and body; 14.77, *Pseudomethoca* sp. (Pseudomethocina); 14.78, *Dasymutilla/Traumatomutilla* sp. (Sphacropthalmina); 14.79, *Ephuta* sp. (Ephutini); 14.80, *Timulla* sp. (Mutillini). Figs 14.81–14.82. Head and mesosoma; 14.81. *Hoplomutilla* sp. (Pseudomethocina); 14.82. *Ephuta* sp. (Ephutini). Figs 14.83–14.84. Fore wing; 14.83, *Dasymutilla/Traumatomutilla* sp. (Sphaeropthalmina); 14.84, *Timulla* sp. (Mutillini).

- SPHAEROPTHALMINAE:
  - Sphaeropthalmini, Pseudomethocina (61 spp.)
    - *Darditilla* (7, m f)
    - *Hoplognathoca* (3, f)
    - *Hoplomutilla* (3, m f)
    - *Pappognatha* (1, f)
    - *Pertyella* (2, f)
    - *Pseudomethoca* (*s. l.*) (43, m f)
    - New genus (2, f)
  - Sphaeropthalmini, Sphaeropthalmina (31 spp.)
    - *Dasymutilla*/*Traumatomutilla* (14, m f)
    - *Lophomutilla* (2, f)
    - *Lophostigma* (2, f)
    - *Paramutilla* (1, m)
    - *Sphaeropthalma* (including *Photopsis*) (3, m f)
    - *Xystromutilla* (2, m f)
    - Two new genera (5, f; 2, m)
- MUTILLINAE:
  - Mutillini
    - *Timulla* (30, m f)
  - Ephutini
    - *Ephuta* (77, m f)

Table 14.3. The genera of Mutillidae known to occur in Costa Rica together with an estimate of the numbers of species and sexes.

numbers of species and sexes for each, are given in Table 14.3.

14.6 RHOPALOSOMATIDAE

Ian D. Gauld

Diagnosis. Body length 3.0 to 20.0 mm; fully winged or both sexes brachypterous, fore wing length of fully winged forms 6.0 to 17.0 mm; colour varying from yellowish to black. Antenna 12-segmented in female, 13-segmented in male, often with one or two conspicuous setae near distal end of proximal flagellar segments; antenna inserted at or slightly below centre of eyes, lower in brachypterous forms, but in these the eyes are reduced in size. *Eyes of fully winged forms strongly emarginate on inner border opposite antennal base*, of brachypterous forms unspecialized; clypeus apically concave, exposing labrum; mandible tridentate, with a large apical and two small subsidiary teeth, sometimes with subsidiary teeth weak; maxillary palp 6-segmented, labial palp 4-segmented. Pronotum reaching back to tegula, in fully winged forms with its upper hind margin strongly arcuate, thin; mesoscutum, in fully winged forms with parapsidal furrows discernible; mesopleuron without a transverse sulcus; mesosternal region with a pair of conspicuous backwardly directed lamellate lobes partially covering bases of mid coxae. Fully winged forms with fore wing having six or seven enclosed cells, veins *C* and *Sc+R+Rs* contiguous so *costal cell is more or less obliterated*; hind wing with one or two enclosed cells, with claval and jugal lobes present; brachypterous forms with fore wing about as long as propodeum, without enclosed cells, hind wing reduced to a short, thread-like vestige. Fore leg with femur swollen, especially in female; tibial spur formula 1: 2: 2; *all legs of females with tarsal segments 2 or 3 to 4 or 5 broadened* (Fig 14.90), *concave beneath, and clothed with fine hairs*; tarsal claws bifid or with a subapical tooth. Metasoma with first segment goblet-shaped to elongate and club-shaped, with a very slight constriction separating it from the second segment; female with six, male with seven visible tergites; male with parameres apically produced into elongate, upwardly curved, needle-like processes which resemble a sting when the genitalia are retracted.

Classification and distribution

Rhopalosomatids have been hypothesized as being rather closely related to the Pompilidae (Brothers, 1975), but Day (1988) questioned this supposed relationship and concluded that the rhopalosomatids are probably an early and distinctive divergence of the vespoid lineage, having no special affinity with the Pompilidae. In their recent re-analysis of vespoid phylogeny, Brothers and Carpenter (1993) placed the rhopalosomatids as the sister-lineage to the Bradynobaenidae + Formicidae + Scoliidae + Vespidae (Fig 14.01), and quite separate from the Pompilidae.

The Rhopalosomatidae is a small family of aculeate Hymenoptera comprising about 40 species worldwide, which are classified in four genera (Townes, 1977*a*): *Liosphex* (one New World species and one in the Philippines), *Olixon* (New World, Africa and Australia), *Paniscomima* (Old World) and *Rhopalosoma* (New World). The family is primarily tropical in distribution although species of *Rhopalosoma* and *Olixon* occur in the United States. In Costa Rica we have collected all three genera

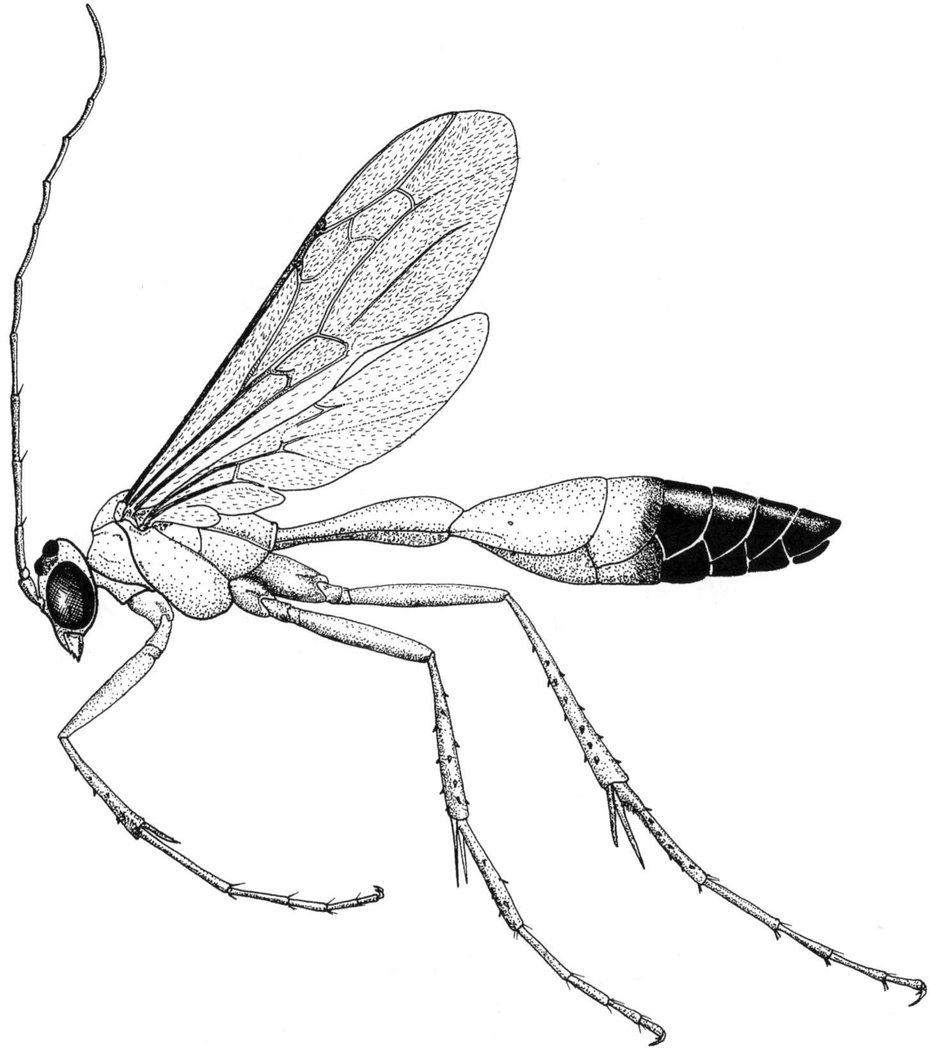

Fig. 14.85. *Rhopalosoma* sp. (Rhopalosomatidae).

known from the New World, representing about eight species.

Biology. As far as is known rhopalosomatids are parasitoids of crickets (Gryllidae). The common *Rhopalosoma nearcticum*, which occurs from the southeastern United States south to Panama and is common throughout lowland Costa Rica, has been reared in the United States as a koinobiont ectoparasitoid of *Hapithus* and *Orocharis* species (Hood, 1914; Gurney, 1953). *Olixon testaceum*, an even more widely distributed species occurring from Ohio south to Brazil, has been reared from species of *Hapathis* and *Nemobius* in North America (Carlson *in* Krombein *et al.*, 1979); there is circumstantial evidence that *O. testaceum* attacks *Cycloptilum* spp. in Honduras (Townes, 1977*a*). The host of *Liosphex* is unknown.

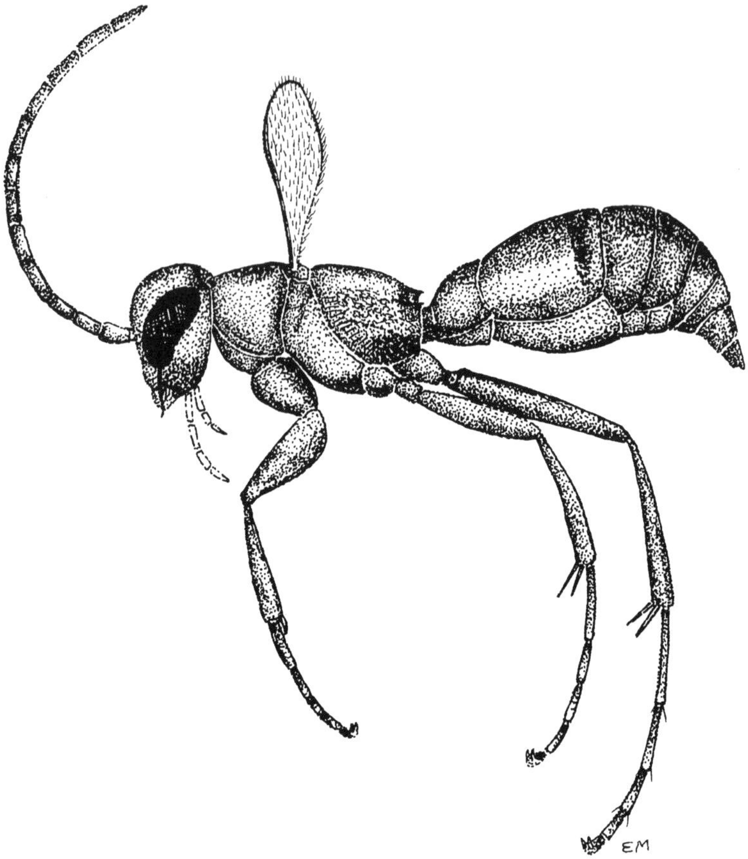

Fig. 14.86. *Olixon testaceum* (Rhopalosomatidae).

In the few species that have been studied the female rhopalosomatid attaches an egg externally to the cricket immediately behind its hind coxa, and the larva develops by feeding through lesions made by its mandibles. At least in *R. nearcticum*, each successive instar shifts its feeding site posteriorly along the side of the host's abdomen, in the membrane between the abdominal tergites and sternites, while maintaining its anal attachment site behind the host's hind coxae. The developing larva causes the cricket's hind leg to be held out at an unnatural angle (Gurney, 1953). The eggshell and larval exuviae remain attached to the gryllid even after the fifth instar parasitoid larva has concluded feeding and abandoned the host. The fifth instar, which has ten pairs of spiracles (two thoracic and eight abdominal) and a pair of projecting spinnerets on the labium, tunnels into soft soil and constructs an ovoid brownish cocoon (Guerney, 1953).

Identification. Keys to the world genera and species are given in Townes (1977*a*) and a modified form of this key is given below.

Key to Central American genera of Rhopalosomatidae

1 Wings very reduced, the anterior one about as long as propodeum (Fig. 14.86), and without enclosed cells; inner margin of eye not notched opposite antennal insertion. ... *Olixon*

— Wings fully developed, the anterior one longer than mesosoma and with six or seven enclosed cells (Fig. 14.85); inner margin of eye notched opposite antennal insertion. 2

2 Ocelli large, the lateral one separated from eye by less than its own maximum diameter; occipital carina present on upper part of head; tergite I of metasoma more than 2.5 times as long as wide (Fig. 14.85); predominantly yellowish brown insects. .... *Rhopalosoma*

— Ocelli small, the lateral one separated from eye by more than twice its own maximum diameter; occipital carina absent; tergite I of metasoma less than 2.0 times as long as wide; black or black and yellowishinsects....*Liosphex*

Synopsis of the Costa Rican fauna

Three genera and about eight species of Rhopalosomatidae occur in Costa Rica, although the exact number of species is uncertain since there seems to be considerable intraspecific variation in both *Rhopalosoma* and *Liosphex*. Rhopalosomatids have often been considered to be rare insects, but specimens of *Rhopalosoma* are quite commonly collected in the lowlands of Costa Rica. Species of the other two American genera, *Olixon* and *Liosphex*, also occur in lowland Malaise trap samples. No species of the family have been found above 2000 metres altitude.

Rhopalosoma. Two species of *Rhopalosoma*, *R. nearcticum* (=*simile*) and an undescribed slightly infuscate species, are regularly collected at light and in Malaise traps. The former is common in a variety of lowland habitats below 1500 metres, but it is most usually collected in dry forest or disturbed areas below 500 metres. In Santa Rosa National Park data from continuous Malaise trap sampling in forests of three different ages (Fig. 14.87) suggest *R. nearcticum* has a preference for open young regenerating woodlands. In such areas individuals are present most months of the year, although there is a marked peak in abundance around the start of the wet season. In progressively more mature woodland it is both less common and more seasonal in its occurrence.

The undescribed species of *Rhopalosoma* seems to be restricted to lowland wet forest habitats. A third species, *R. alvarengai* has been recorded from Costa Rica, but we have failed to collect it, and it is only

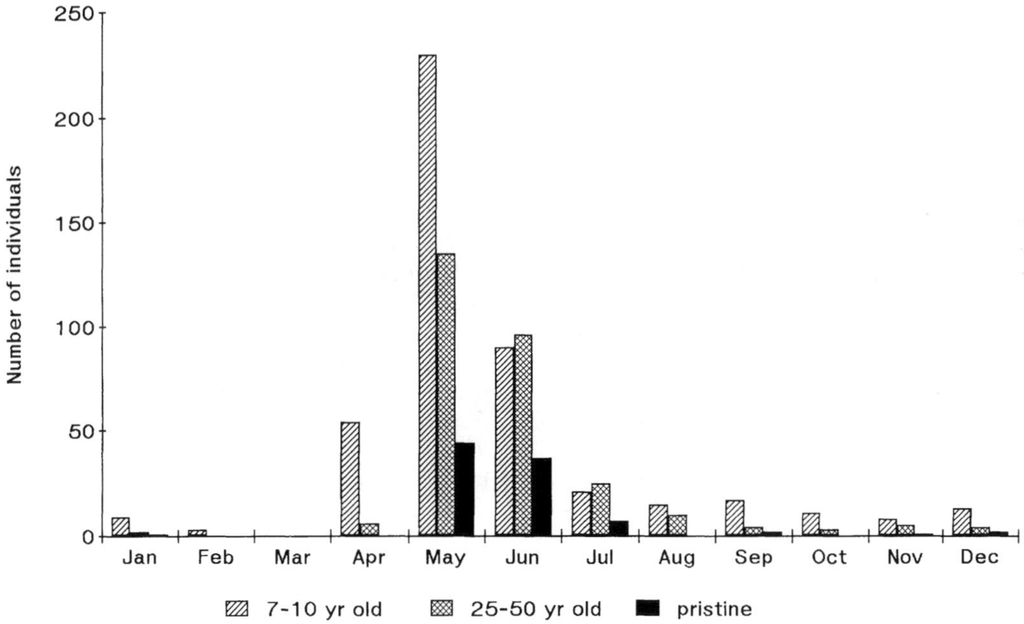

Fig. 14.87. Monthly occurrence of *Rhopalosoma nearcticum* (Rhopalosomatidae) in Malaise trap catches in different aged forest habitats in Santa Rosa National Park, averaged from data for 1985–7.

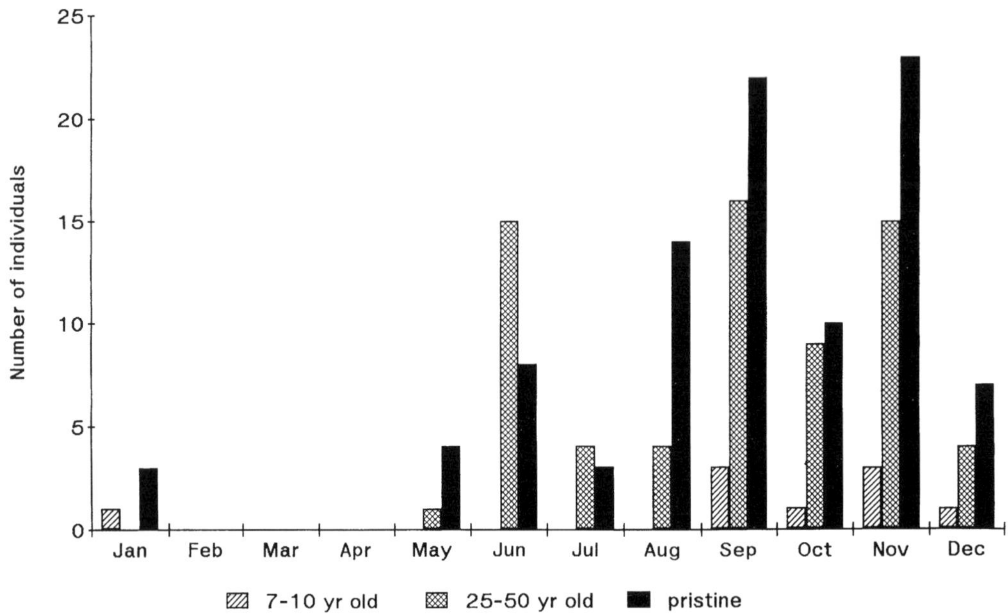

Fig. 14.88. Monthly occurrence of *Olixon testaceum* (Rhopalosomatidae) in Malaise trap catches in different aged forest habitats in Santa Rosa National Park, averaged from data for 1985–7.

known from a single specimen from an unknown locality (Townes, 1977*a*).

Olixon. *O. testaceum*, a small, ant-like and predominantly yellowish species is the only Mesoamerican species of the genus. It has been collected in Costa Rica in most habitats sampled on the Pacific and Atlantic coasts below 1000 metres. In Santa Rosa National Park data from continuous comparable sampling in forests of three different ages suggest that *O. testaceum* is most common in denser forests (Fig. 14.88). Individuals have been collected from the beginning of the wet season in May to the early dry season in January, but they appear to be most abundant from August to November.

Liosphex. The one described New World species of this genus, *L. varius*, and perhaps three undescribed species of the same genus have occasionally been collected in a variety of lowland habitats throughout Costa Rica. No species has been found above 1000 metres. Specimens are most frequently taken in lowland humid forest sites, although one species has been collected in Pacific dry forest during the wet season.

14.7 BRADYNOBAENIDAE

Denis J. Brothers

Diagnosis. As in the Mutillidae, with which most members of this family have been confused, there is extreme sexual dimorphism, the males being fully winged (Fig. 14.89) and the females entirely apterous (Fig. 5.43); the body forms and usually the colours of the two sexes are also different. Because of the marked morphological diversity within the family, it is difficult to provide a simple diagnosis for all members. Since Central American species belong only to the subfamily Typhoctinae the following diagnosis pertains to that subfamily.

Body length 4.0 to 8.0 mm. Head with *only slight elaboration of the dorsal rim of the antennal socket* (Fig. 14.91); *mid coxae slightly separated by metasternum*; first metasomal segment almost triangular, with

Fig. 14.89. *Typhoctes* sp. (Bradynobaenidae).

sternum flat; deep notch between first and second metasomal sterna; second tergite with a short longitudinal felt line anteriorly on each side. Males slender, black with some white pubescence; *eye oval, with inner margin sinuate or shallowly notched*; pronotum more or less transverse and shallowly concave posteriorly; wings fairly narrow, with strong venation on the basal two-thirds or half; fore wing with small pterostigma and ten enclosed cells; hind wing without claval and jugal lobes; first metasomal tergite evenly swollen, *with a narrow 'neck' anteriorly*. Females less slender, black, but sometimes partly reddish, with some white or gold pubescence; eye oval; mesosoma elongate; pronotum trapezoidal from above, slightly narrower posteriorly, loosely articulated with the more or less rectangular and fused meso- + metathorax + propodeum (figs. 14.93, 14.94); first metasomal tergite with short transverse horizontal posterior face.

The other New World subfamilies can be distinguished as follows:

Chyphotinae—dull brownish; metasoma petiolate, with petiole apparently composed only of sternite I; tergite II with felt lines; male with fairly complete wing venation and a strong up-curved spine at apex

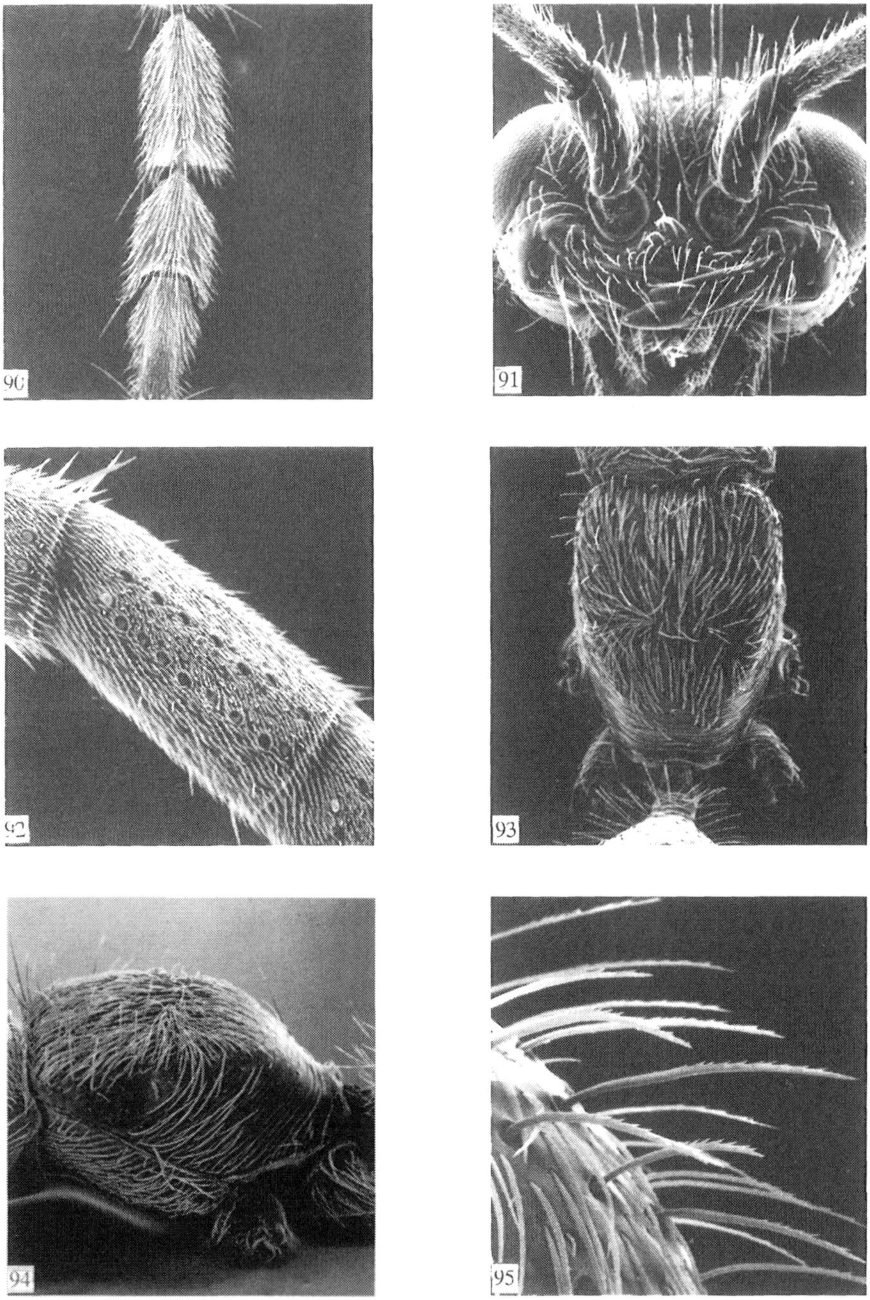

Figs 14.90–14.95. Scanning electron photomicrographs of Rhopalosomatidae and Bradynobaenidae. Fig. 14.90. *Rhopalosoma* sp., female, hind tarsus. Figs 14.91–14.95. *Typhoctes* spp.; 14.91, face of female, anterior view; 14.92, sensilla on antenna of male; 14.93, propodeum of female, dorsal view; 14.94, propodeum of female, lateral view; 14.95, barbed hairs on head of female.

of metasoma; female with pronotum rectangular, rest of mesosoma somewhat globular.

Bradynobaeninae—black, black and reddish, or brownish; metasoma sessile, with cuticular grooves in place of felt lines; male with reduced wing venation and three short spines at apex of metasoma; female with entire mesosoma somewhat globular.

Classification and distribution. The family comprises about 200 species in four subfamilies: Apterogyninae (Afrotropical and Palaearctic), Bradynobaeninae (Peru to Argentina), Chyphotinae (Nearctic), and Typhoctinae (including Typhoctini and Eotillini; New World). Its limits were established by Brothers (1975) but the subfamilies have been considered as separate families by Genise (1986), although not by Nonveiller (1990) in a catalogue of the neotropical species.

Biology. Almost nothing is known of the biology of the family. Central American species (all Typhoctinae) are diurnal and rarely found. The females run rapidly and the males have been collected in Malaise traps. Males apparently visit flowers for nectar since specimens have been collected with pollen on the head, but what the females feed on is unknown. Both males and females bear stout erect slightly branched bristles (Fig. 14.95) as do some Mutillidae. The larvae presumably develop as parasitoids of other arthropods. The only host record is for a nearctic species of *Typhoctes*, larvae of which were found on immature Solifugae (Arachnida) (Brothers, MS).

As far as the other subfamilies are concerned, Chyphotinae are nocturnal, Bradynobaeninae diurnal, and the Apterogyninae includes both nocturnal and diurnal species. Hosts for these three subfamilies are unknown.

Synopsis of the Costa Rican fauna

Only the Typhoctinae has been found in Central America. One genus, *Typhoctes*, is known to occur in Central America; the other genera of Typhoctinae (*Typhoctoides*, *Eotilla* and *Prototilla*) occur farther south. In the key to families provision has been made for recognition of the Chyphotinae. Although species of this subfamily have not been found south of the dry Sonoran-southwestern United States, the recent discovery of brachycistidines (another Sonoran-southwestern United States group) in northwestern Costa Rica suggests that some of these rare dry habitat taxa could possibly occur southwards into Central America.

Typhoctes. Males of at least three species and females of two have been collected in Costa Rica at altitudes between 300 and 800 metres. One species of *Typhoctes* has been described from Mexico, one from Guatemala and another, *T. evansi*, recently from Panama (Quintero & Rodríguez, 1990). A few undescribed species have been collected in Colombia and Jamaica. *T. evansi* is quite commonly collected in Malaise traps in Santa Rosa National Park. Individuals have been taken between January and May, during the dry season.

14.8 SCOLIIDAE

Albert T. Finnamore and Paul E. Hanson

Diagnosis. *Mostly 20.0 to 30.0 mm in length; robust and hairy*; mostly black with yellow or orange markings or all black; both sexes fully winged. Sexual dimorphism slight; males smaller, more slender and with more elongate antennal segments. Antenna of female 12- and male 13-segmented. *Members of Scoliinae (all New World taxa) with inner borders of eyes deeply emarginate.* Pronotum with posterior dorsal margin U-shaped, posterior lateral margin truncate and extending back to tegula; propodeum with two widely separated dorsal longitudinal sutures; fore coxae contiguous, middle and hind coxae widely separated; meso- and metasterna flattened and with small plate-like projections over the coxal bases; *metasternum quadrangular* (not triangular as in many Tiphiidae). Fore wings with ten enclosed cells, hind wings with three; hind wing without a claval lobe, but with a long jugal lobe; *distal portions of wings finely, longitudinally striate*, with a densely corrugated appearance. Middle and hind tibiae of female stout, with oblique conical teeth or setae; tarsal claws simple. First metasomal sternite separated from second by a deep constriction. *Terminal metasomal sternite of male armed with three spines.*

Classification and distribution. The Scoliidae is a small family containing about 300 species worldwide, with the majority occurring in tropical and subtropical regions. They are absent from high latitudes such

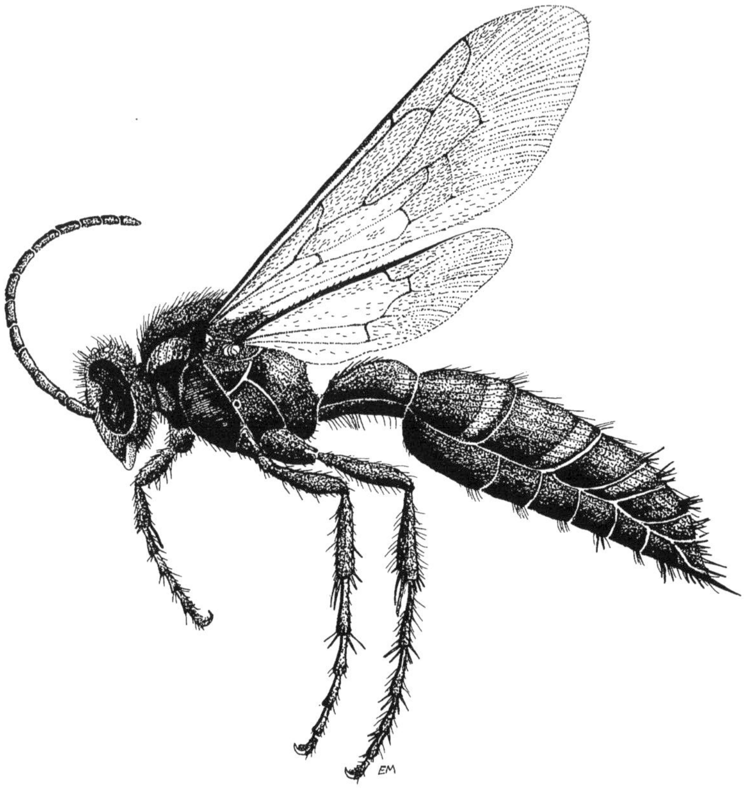

Fig. 14.96. *Campsomeris* sp. (Scoliidae).

as Canada and southern South America. The family is divided into two subfamilies, the Proscoliinae and the Scoliinae, the former comprising just one genus with two primitive species known only from the eastern Mediterranean region (Day *et al.*, 1981). The cosmopolitan subfamily Scoliinae is divided into two tribes, the Scoliini, with six genera, and the Campsomerini, with about 15 genera. Although this is a rather small family, the nomenclature is rather confusing as there is a tradition of using both subgeneric and subspecific names.

Biology. Despite their large size, and the fact that where they occur they can be quite conspicuous, there is surprisingly little information available about the biology of Scoliidae. Females are powerfully built and have spiny legs adapted for burrowing into the soil where they seek out large coleopterous larvae, usually Scarabaeidae. Upon detecting the host the female stings and paralyses it. In common with Vespidae, the venom of Scoliidae contains kinins, which apparently block synaptic transmission in the nervous system (Piek, 1986). After the host is paralysed the female lays an egg on it, the precise position varying between species. In at least some cases the female drags the paralysed host more deeply into the soil and prepares a crude cell before ovipositing. Females can lay a maximum of two eggs per day over a period of about two months (Clausen, 1940*b*).

The larval scoliid feeds on the host as an ectophagous idiobiont parasitoid, at first with only its head embedded in the feeding puncture but eventually more and more of its body entering the host. Everything except the host integument is devoured (Clausen, 1940*b*). The first instar larva has simple mandibles whereas the following instars have tridentate mandibles. The mature larva has ten pairs of well developed spiracles, and a unique feature of the family is the presence of five or six sensilla in each antennal orbit instead of the usual three (Evans *in* Evans, 1987*a*). At the end of larval development a cocoon is spun, and eventually the adult emerges through a circular cap at the anterior end (unlike tiphiids where the hole is more irregular and cut to one side). The larval period, consisting of four instars, lasts about one week, and roughly three weeks are passed in the cocoon.

In the lower Rio Grande area of Texas, males are active for a maximum of four hours per day, during which time they can be found visiting flowers (Porter, 1981). 'Courtship' in *Scolia dubia* apparently consists of both sexes flying horizontally low over the ground. Eventually a female alights on the ground and shows her willingness to copulate by spreading her wings, thereby exposing her orange abdomen. In this species pre-copulatory females apparently "sleep" with males on vegetation whereas after copulation they sleep in the soil (Rau, 1932).

Economic importance. In other parts of the world there are at least two cases of classical biological control programmes utilizing scoliids against scarab pests (Clausen, 1978). In Central America *Campsomeris dorsata* has been reported attacking *Ligyrus nasutus* and *Phyllophaga* spp. (King & Saunders, 1984).

Identification. The following key to genera and subgenera is based on Bradley (1945, 1964) and examination of over 300 specimens.

Key to genera and subgenera of Scoliidae of Costa Rica

1 Fore wing with vein 2*m-cu* absent (Fig. 14.97), or present and arching to vein 1*m-cu*. .... (Scoliini) .... *Scolia*
— Fore wing with vein 2*m-cu* present, extending from *Cu*1 to *M* (Fig. 14.98). .... (Campsomerini).... *Campsomeris* 2

2 Mesosoma with lateral part of propodeum coarsely and obviously striate. ....subgenus *Rhabdotomeris*
— Mesosoma with lateral part of propodeum not obviously striate. 3

3 Antenna at least three times as long as the width of the head; males. 4
— Antennal about equal in length to the width of the head; females. 11

4 Hind tibia with apical spurs white. 5
— Hind tibia with apical spurs black. 7

5 Fore wing with apex setose only along anterior margin. .... subgenus *Dielis*
— Fore wing with apex uniformly setose. 6

6 Metasoma with tergite IV yellow marked. ....subgenus *Xanthocampsomeris*
— Metasomal with tergite IV entirely black. ....*C.* (*Lissocampsomeris*) *wesmaeli*

7 Fore wing with apex uniformly setose. ....subgenus *Stygocampsomeris*
— Fore wing with apex setose only along anterior margin 8

8 Propodeum and clypeus yellow marked. .... subgenus *Aelocampsomeris*
— Propodeum and clypeus entirely black. 9

9 Metasoma with orange markings. .... subgenus *Pygodasis*
— Metasoma entirely black. 10

10 All setae black. .... subgenus *Campsomeris*
— Some setae, at least on ventral surface of metasoma, white. ..... *C.* (*Lissocampsomeris*) *columba*

11 Mesosoma with posterior surface of propodeum coarsely punctate, the punctures bearing setae; hind tibial spurs pointed (cf Fig. 14.99). ....subgenus *Pygodasis*
— Mesosoma with posterior surface of propodeum impunctate, ***or if*** punctate ***then*** longer hind tibial spur spatulate (cf Fig. 14.100). 12

12 Hind tibia with apical spurs white. 13
— Hind tibia with apical spurs black. 14

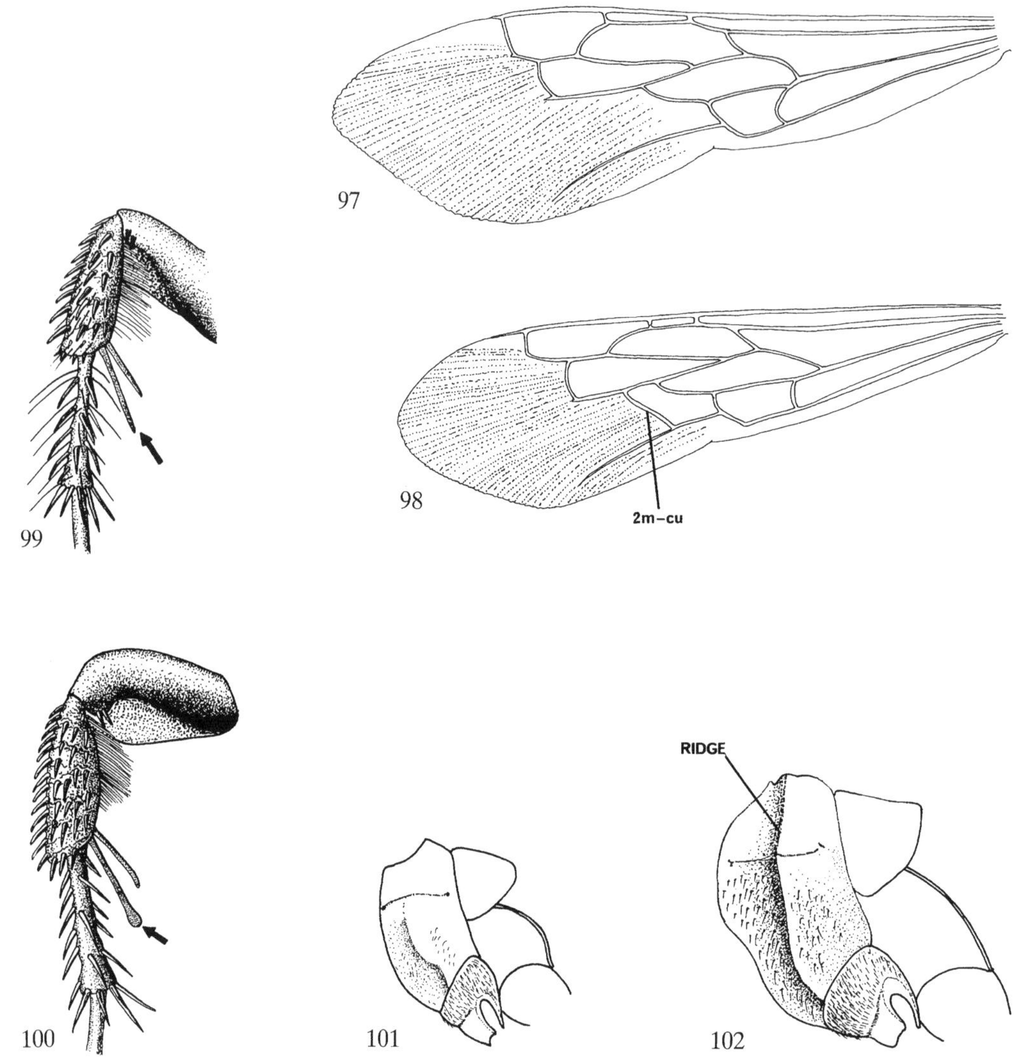

Figs 14.97–14.102. Scoliidae. Figs 14.97–14.98. Fore wings; 14.97, *Scolia* sp.; 14.98, *Campsomeris* sp. Figs 14.99–14.100. Hind tibial spur; 14.99, *C.* (*Xanthocampsomeris*) sp.; 14.100, *C.* (*Dielis*) sp. Figs 14.101–14.102. Mesopleuron; 14.101, *C.* (*Lissocampsomeris*) sp.; 14.102, *C.* (*Aelocampsomeris*) sp.

13 Hind tibia with longer spur pointed (Fig. 14.99). .... subgenus *Xanthocampsomeris*

— Hind tibia with longer spur spatulate (Fig. 14.100). .... *C.* (*Dielis*) *plumipes*

14 Metasoma with tergites II and III (and sometimes also IV) predominantly orange. .... *C.* (*Dielis*) *dorsata*

— Metasoma with tergites II and III black, ***or*** I-III (and sometimes also IV) black with yellow markings. 15

15 Mesopleural ridge weakly indicated just above mid coxa (Fig. 14.101); vertical upper part of mesopleuron impunctate. .... subgenus *Lissocampsomeris*

— Mesopleuron with a distinct ridge (Fig. 14.102), ***or*** vertical portion of upper mesopleuron with numerous punctures.....................................16

16 Hind tibia with longer spur pointed, less than half length of basal tarsal segment. .... subgenus *Aelocampsomeris*

— Hind tibia with longer spur spatulate, or if pointed, then over half length of basal tarsal segment. 17

17 Hind tibia with longer spur pointed. .... subgenus *Stygocampsomeris*

— Hind tibia with longer spur spatulate. .... subgenus *Campsomeris*

Synopsis of the Costa Rican fauna

Scoliids are widespread, and at times common, throughout Costa Rica, although they are notably absent at altitudes above 2000 metres. There appear to be about 15 species present.

SCOLIINAE: Scoliini. This tribe comprises about six genera, two of which, *Scolia* and *Triscolia*, are represented in the New World. Only the former occurs in the Neotropics.

Scolia. Two species of this genus are known to occur in Costa Rica.

SCOLIINAE: Campsomerini. This tribe contains about 15 genera, but only three of these—*Campsomeris*, *Colpa* (= *Trielis*) and *Crioscolia*—occur in the New World. Only *Campsomeris*, which is exclusively New World in distribution, occurs in the Neotropics (Bradley, 1957; Betrem & Bradley, 1972).

Campsomeris. This genus is represented in Costa Rica by about 12 species in eight subgenera.

Campsomeris (Aelocampsomeris). A single species, *variegata*, is recorded as occurring in Costa Rica. It has been collected in mid altitude sites.

Campsomeris (Campsomeris). We have only found one species of this subgenus, *regifica*, in Costa Rica. It has been collected from sea-level up to 1200 metres.

Campsomeris (Dielis). This subgenus is represented in Costa Rica by two species, *dorsata* and *plumipes*. The former has the yellow pronotal mark extending to the tegula whereas the yellow mark does not reach the tegula in the latter species. *C. plumipes* has only been collected from sites between 900 and 1600 metres, while *dorsata* appears to occur only below 1100 metres. In Santa Rosa National Park, *dorsata* is common in Malaise trap collections towards the end of the wet season (October to December).

Campsomeris (Lissocampsomeris). Represented by two species, *columba* and *wesmaeli*. The former is a lowland species (below 1300 metres), but it is rarely collected on the Osa Peninsula. On the other hand *wesmaeli* has most frequently been collected from the lowlands in the southwestern part of the country (Osa Peninsula to Carara Reserve) although isolated individuals have been taken as far north as Guanacaste National Park.

Campsomeris (Pygodasis). A single species of this subgenus, *ephippium* has been found in Costa Rica, at altitudes between 800 and 2000 metres. This species is easily recognized by its large size (female up to 40 mm) and by the fact that both sexes have tergites II and III predominantly orange. *C. vittata* (male tergites I and II with yellow) has been recorded from Panama and Mexico, and *ianthina* (male all black) from Colombia and Mexico; both are thus expected to occur here.

Campsomeris (Rhabdotomeris). A single species of this subgenus, *rokitanskyi*, occurs in Costa Rica. It seems to occur generally in mid altitude sites but very little material has been collected.

Campsomeris (Stygocampsomeris). A single species of this subgenus, *servillei*, occurs in Costa Rica. It has primarily been found between 600 and 1500 metres although isolated individuals have been collected from lower altitudes. An intensive Malaise trap survey of the Osa

Peninsula has found that this species is common on Cerro Rincon (750 m) but only one individual has been collected from any of the other sites.

***Campsomeris* (*Xanthocampsomeris*).** This subgenus is represented in Costa Rica by a single species, *hesterae*, which appears to be restricted to altitudes between 600 and 1200 metres.

15 The vespid wasps (Vespidae)

Mary Jane West-Eberhard,
James M. Carpenter and Paul E. Hanson

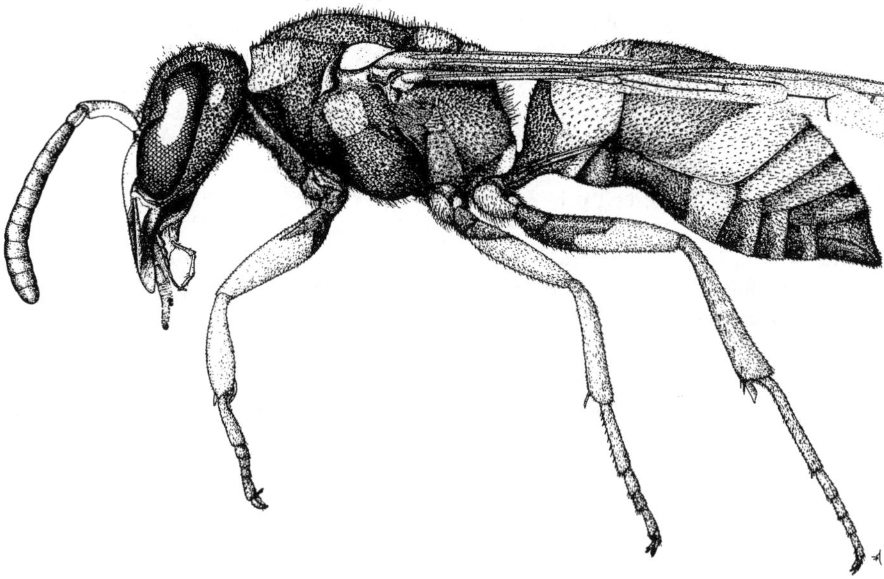

Fig. 15.01. *Ancistrocerus* sp. (Eumeninae).

Diagnosis. Medium to large sized wasps (4.5 to 25.0 mm in length); colour variable, often yellow with black markings to all black; both sexes fully winged. Antenna 12-segmented in females, 13-segmented in males, male antenna often more curved at apex; *inner margin of eye emarginate medially*. Pronotum reaching back to tegulae. *Wings folded longitudinally at rest* and fore wing with an elongate first discal (discoidal) cell (except in Masarinae); hind wing with closed cells. Mid coxae contiguous. First metasomal tergite and sternite partially fused, the former overlapping the latter; a distinct constriction present between the first and second sternites.

Classification and distribution. Traditionally (e.g. Spradbery, 1973*b*; Krombein *in* Krombein *et al.*, 1979; Gauld & Bolton, 1988) this group has been divided into three families, the Masaridae, Eumenidae and Vespidae. However, following Carpenter's (1982) phylogenetic analysis, it has become increasingly accepted that all should be united in a single family, the Vespidae. This family is now classified in six apparently monophyletic subfamilies (Carpenter, 1982, 1991); their postulated phylogenetic relationships are shown in Fig. 15.02.

The most primitive subfamily, the Euparagiinae, comprises only nine extant species classified in a single genus that is presently restricted to the xeric areas of the southwestern United States and Mexico. The subfamily was widespread in the Cretaceous Period (Carpenter & Rasnitsyn, 1990; Brothers, 1992), occurring in central Asia and southern Africa. The

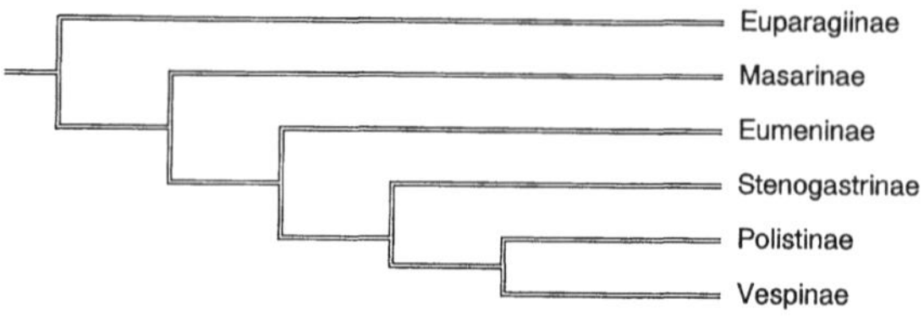

Fig. 15.02. Proposed phylogenetic relationships of the subfamilies of Vespidae. Details of the characters supporting this reconstruction are given in Carpenter (1982).

Masarinae is a larger group comprising about 250 species in 18 genera. It is cosmopolitan and primarily restricted to dry areas in Australia, southern Africa, the Mediterranean Basin and the New World, where it has a disjunct distribution with species present in the western United States, Mexico, Guatemala and Panama, and in the western and southern parts of South America (see below). The Eumeninae is by far the largest subfamily comprising about 3000 species throughout the world, classified in 203 genera. The sister-group of the two best known vespid subfamilies is the Stenogastrinae, a rather small, tropical Asian group comprising about 50 species in six genera. The remaining two subfamilies, the Polistinae and Vespinae, include some of the most commonly encountered of all wasps in most parts of the World. The former group is cosmopolitan and includes about 800 species classified in 28 genera, while the latter is both smaller and has a more restricted distribution. The Vespinae is primarily Holarctic and tropical Asian, though species have been inadvertently introduced into South Africa, Chile, Hawaii, Australia and New Zealand, and includes about 60 species placed in four genera.

Only the Eumeninae and Polistinae have been found in Costa Rica, but the Masarinae may occur here since they are present in countries to the north and south of Costa Rica. Among the eusocial groups of Vespidae, the Central American fauna differs significantly from that of North America in that the subfamily Polistinae predominates (17 genera in Costa Rica versus three genera in the United States). The subfamily Vespinae extends from North America (where there are three genera) south to Guatemala, but is absent from the rest of Central America.

Biology. Vespids, especially the eusocial species (Polistinae), form a common and conspicuous part of the fauna of Central America, and individuals can be encountered in almost all situations from the coastal plains to high altitude on the Cerro de la Muerte. Several species frequently nest in and around human habitations and are particularly attracted to sweet substances and kitchen refuse. Because of their painful sting the polistines are feared by the general public and are usually regarded as a nuisance. Although some species have a predilection for sugary food, most vespids are predators on other arthropods. Masarines are a notable exception as they provision their nests with pollen and nectar.

All vespids construct nests, usurp nests, or occupy pre-existing cavities in which to nest. Usually the nest is made of mud or masticated plant fibres, and some may be quite large and complex structures with a form characteristic of the particular species building it (Figs 15.03-15.08). The Euparagiinae, Masarinae and Eumeninae are solitary to primitively social (nest-sharing without a worker caste); Stenogastrinae, Polistinae and Vespinae are eusocial (group-living with sterile workers) (Spradbery, 1973*b*). Cleptoparasitism has evidently not evolved in the family, although in temperate regions there are socially parasitic species (inquilines) of *Polistes* and *Vespula*. Oviposition is into an empty cell and Carpenter (1982) considers this habit to be an autapomorphy for the family.

The mature larvae differ considerably between subfamilies and genera, but in general are 7 to 35 mm long, with the head rather large, lacking antennal papillae (present in Stenogastrinae), with ten pairs of spiracles, and middle body segments often divided into two dorsal annulets by a transverse crease (Evans, 1987*a*). The pleural lobes may be weakly developed (Eumeninae) or well developed (Polistinae). Eumenines have a smooth integument with minute setae whereas polistines have the integument covered with spinules and with areas of fairly long setae on the venter of the more anterior segments. Larvae of *Mischocyttarus*

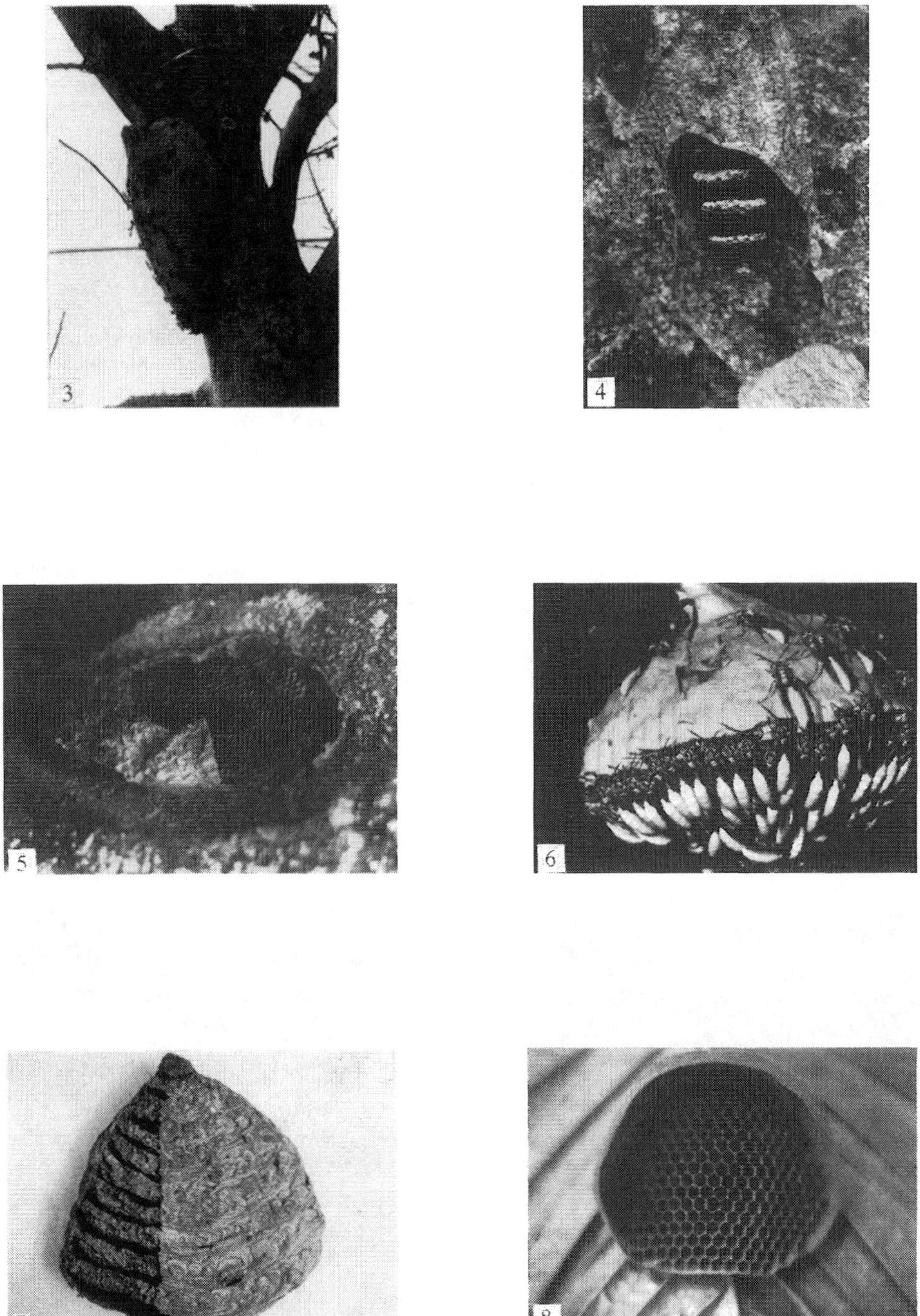

Figs 15.03–15.08. Vespid nests; 15.03, *Synoeca* sp.; 15.04, *Parachartergus* sp.; 15.05, *Metapolybia* sp.; 15.06, *Apoica* sp.; 15.07, *Polybia* sp.; 15.08, *Charterginus* sp.

species have conspicuous abdominal lobes which, by covering the retracted head, evidently serve to protect them from adults aggressively soliciting regurgitated liquids (Hunt, 1988). Full-grown eumenine larvae spin a delicate silken cocoon (although a few species do not spin cocoons) whereas polistines spin only a thick silken cap over the mouth of the cell. Information on masarine larvae is very incomplete (Carpenter, 1982).

Key to the subfamilies of Vespidae of Central America

1 Fore wings not longitudinally folded at rest; 1st discal cell shorter than subbasal cell.
[Very rare.] .... **Masarinae** (p. 564)
— Fore wings longitudinally folded at rest; 1st discal cell longer than subbasal cell. 2

2 Mesoscutum with a parategula, a flattened lobe projecting from posterolateral corner (Fig. 15.09); claws with an auxiliary sub-basal tooth.
[Mid tibia usually with a single spur.] .... **Eumeninae** (p. 565)
— Mesoscutum without parategula (Fig. 15.10), but often with vertical lamella along postero-lateral edge; claws simple.
[Mid tibia usually with two apical spurs.] ... 3

3 Hind coxa with dorsal carina on posterior surface; hind wing without a jugal lobe.
[Absent south of Guatemala.] .... **Vespinae***
— Hind coxa without dorsal carina; hind wing usually with a jugal lobe. .... **Polistinae** (p. 574)

15.1 THE SUBFAMILY MASARINAE ('POLLEN WASPS')

Classification and distribution. Species of Masarinae occur mostly in warm dry regions of the world. The subfamily is classified in two tribes (Carpenter, 1982): the Gayellini, which comprises ten species restricted to the Neotropics, and the Masarini, with over 200 species that are widely distributed throughout warm dry parts of the Old World, western North America and South America (Richards, 1962).

The Gayellini includes only two genera — the Patagonian *Gayella* and the tropical American *Paramasaris* (Carpenter, 1989). *Paramasaris fuscipennis* has been collected in Mexico, Guatemala, Panama and Colombia (Carpenter, 1989), but not yet in Costa Rica.

The Masarini comprises 16 genera, three of which occur in the New World (Richards, 1962; Vecht & Carpenter, 1990): *Ceramiopsis* and *Trimeria* from South America, and *Pseudomasaris* from western North America and Mexico.

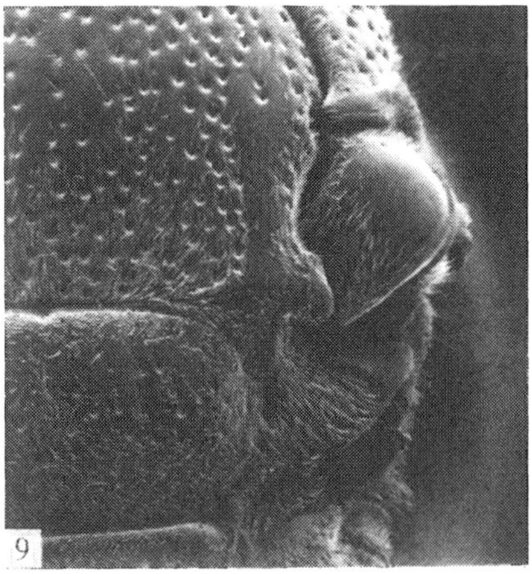

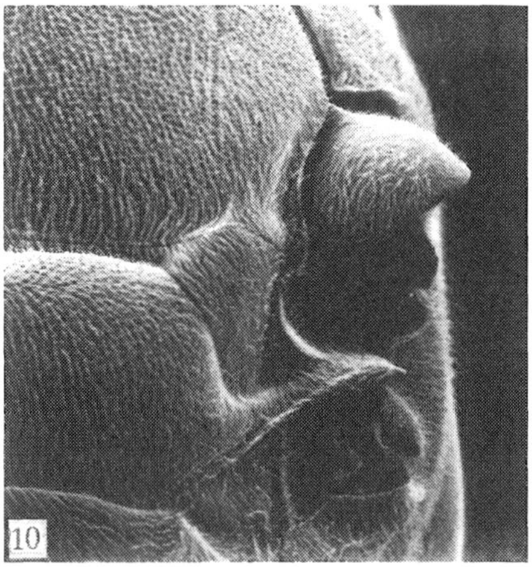

Figs 15.09–15.10. Scanning electron photomicrographs of Vespidae, posterolateral part of mesoscutum; 15.09, *Eumenes* sp. (Eumeninae); 15.10, *Polistes* sp. (Polistinae).

Biology. The masarines are the only Hymenoptera other than bees to provision their nest entirely with pollen and nectar. The only information available on the biology of the tribe Gayellini is for the South American species *Gayella eumenoides* (Claude-Joseph, 1930), whereas much more information is available for the tribe Masarini (e.g. Richards, 1962; Torchio, 1970; Zucchi *et al.*, 1976; Gess & Gess, 1980; Naumann & Cardale, 1987). Within the subfamily as a whole, the excavation of burrows via the use of regurgitated water is probably the ancestral condition, whilst the construction of cylindrical mud cells attached to rocks or plant stems (as in *Gayella eumenoides*) is probably derived. A further evolutionary sequence can be seen in the substitution of nectar, or even silk (probably produced by the mandibular glands of the adult female), for water as a bonding agent (Gess & Gess, 1992). The nest is provisioned with nectar and pollen, which the female carries in her crop, and individual species are often oligolectic (i.e. specialists on a few species of flowers). The thin-walled, translucent cocoon consists of salivary secretion and silk. Mating behaviour has not been observed in Gayellini but a few studies exist for Masarini (e.g. Longair, 1987).

Identification. Richards (1962) provided keys to the species of Masarinae of the world and Carpenter (1989) provided revised keys to the Gayellini.

15.2 THE SUBFAMILY EUMENINAE ('POTTER WASPS')

Classification and distribution. The Eumeninae was previously given family status and divided into three subfamilies (Richards, 1962): Eumeninae, Raphioglossinae and Zethinae. However, subsequent

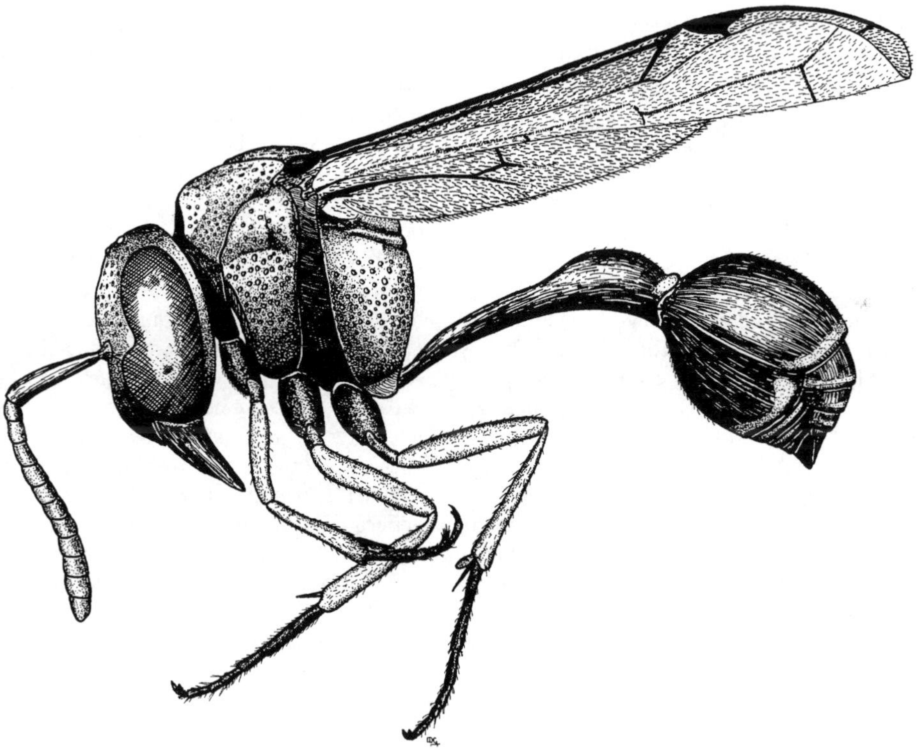

Fig. 15.11. *Omicron* sp. (Eumeninae).

phylogenetic analysis has shown that the Zethinae is a paraphyletic group, and that a group consisting of Eumeninae + Zethinae, without Raphioglossinae, is also paraphyletic (Carpenter, 1982). Thus the Eumeninae is now considered to include these other two subfamilies. Worldwide the Eumeninae includes about 3000 species in 203 currently recognized genera. About 80 species in 22 genera are currently known in Costa Rica. However, because eumenines have not been intensively collected (and are infrequently taken in Malaise traps), the actual number of species present in the country is probably well over 100.

Biology. Eumenines are much less commonly seen than polistines. All species are predators, mostly of Lepidoptera larvae, and all construct nests, which usually involves the use of mud in one way or another. Some species build free-standing, jug-like nests, hence the common name, 'potter wasps'. Most eumenines are solitary, but a few are primitively social.

Nesting behaviour. Some species excavate burrows in horizontal or vertical earth surfaces, especially in clay soils, by regurgitating water and removing pellets of mud (Iwata, 1976; Cowan, 1991). *Euodynerus crypticus*, for example, nests in vertical ground burrows in small colonies, the nest consisting of about three cells one above the other (Krombein *in* Krombein *et al.*, 1979). Other species build single cells at the end of branching tunnels (Evans, 1956). Partitions between cells and final closures are made with mud. Several burrowing species construct a chimney-like cylinder of mud ('turret') over the nest entrance although in some species this turret may be eventually levelled and used to seal the entrance of the burrow. There is still no agreement on the function of these chimneys (Cowan, 1991).

Non-burrowing species generally employ mud in nest construction, and this mud is usually obtained by drinking water and flying to another site where the water is regurgitated and mixed with soil. Most eumenines nest in pre-existing cavities, such as hollow plant stems, old beetle borings, abandoned nests of other Hymenoptera, and even empty snail shells. Mud and sometimes resinous material is often used in lining, partitioning, and closing the nest. *Pachodynerus nasidens*, for example, occupies old nests of *Sceliphron* and *Zeta*, or other cavities, and constructs a unique crescent-shaped temporary plug (Jayasingh & Taffe, 1982). *Pseudodynerus quadrisectus* is reported to use cavities in wood (Krombein *in* Krombein *et al.*, 1979). The yellow and black *Zethus histrionicus* is a common occupant of holes around houses in Costa Rica.

Several eumenines build exposed nests of mud or plant material. Some of these mud-daubing species, for example *Montezumia cortesioides* (Evans, 1973*c*, see Willink, 1982), use a rock crevice or other protected site as a support for a nest of mud cells placed upon each other and surrounded by additional mud, forming an amorphous ball of mud. These mud nests are usually distinguishable from those of sphecid mud-daubers by their more granular texture and shorter tubes that are only one cell long. Other mud-daubing eumenines (*Eumenes* and *Pachymenes*) make elegant mud cells in the shape of pots or jugs, which are widest in the middle and narrow to a small opening at the top. These mud pots may occur singly or in groups of two to four and are frequently attached to twigs. It has been suggested that the very restricted aperture in these nests limits the size of prey that can be provisioned. Methods of mud manipulation vary. Some apply the mortar from the inside with their mandibles, while others (e.g. *Eumenes*) construct from the outside with their mandibles and fore tarsi (Iwata, 1976; A.P. Smith, 1978).

A few eumenines use plant material in addition to or instead of mud in nest construction. *Ancistroceroides bogotanus* covers its cylindrical mud cells with kneaded plant material (Iwata, 1976). Species of *Zethus* belonging to the subgenera *Zethus* and *Zethusculus* nest in abandoned beetle burrows, closing the cell with cut pieces of leaves, while species of the subgenus *Zethoides* (e.g. *Z. aztecus*, *Z. clypearis*, *Z. matzicatzin*, *Z. miniatus*, *Z. olmecus*, *Z. toltecus*) construct nests entirely of masticated vegetable material and resin (Bohart & Stange, 1965). The nest types built by genera of Eumeninae that occur in Costa Rica are listed in Table 15.1. Nests are so far undescribed for *Hypalastoroides*, *Parazumia*, *Pararhaphidoglossa* and *Pseudacaromenes*.

Some species make more than one type of nest. For example *Parancistrocerus fulvipes* uses borings in wood, old mud dauber nests, burrows of ground nesting bees, or it constructs its own burrows in the ground (Evans, 1956; Krombein, 1967). *Pachodynerus nasidens* may use pre-existing cavities or make free mud nests

	Burrow in soil	Cavities	Exposed
Ancistroceroides	–	–	X
Ancistrocerus	X	X	x
Eumenes	–	–	X
Euodynerus	x	X	x
Leptochilus	–	–	X
Minixi	–	–	X
Monobia	–	X	–
Montezumia	X	–	X
Omicron	–	X	?
Pachodynerus	–	X	x
Pachymenes	–	–	X
Parancistrocerus	x	X	–
Pirhosigma	–	–	X
Pseudodynerus	–	X	–
Santamenes	–	–	X
Stenodynerus	x	X	–
Zeta	–	–	?
Zethus	–	X	X

Table 15.1. Nest sites used by genera of Eumeninae occurring in Costa Rica. When more than one type occurs within a genus, X = many species and x = a few species. Information is from Evans & Matthews (1974), Iwata (1976), Soika (1978, 1990) and Krombein (*in* Krombein *et al.*, 1979).

(Jayasingh & Taffe, 1982; Carpenter, 1986*b*). This plasticity at both the species and generic level makes it difficult to deduce phylogenetic patterns in nesting behaviour (Carpenter & Cumming, 1985). Burrowing into the soil is generally assumed to be primitive and according to one hypothesis the building of exposed mud nests may have evolved from the habit of removing moistened pellets of earth to construct a turret over the burrow (Iwata, 1976; Evans & West-Eberhard, 1970). On the other hand the stem-group of Eumeninae (Zethinae and Raphiglossinae of authors) are all cavity nesters ('renters') or builders of exposed nests, and cavity nesting occurs in the majority of other genera (Carpenter & Cumming, 1985).

Provisioning behaviour. When the cell is completed, but before provisioning begins, a single egg is laid within it, usually attached to the upper surface of the cell by a short thread. Prey consists mostly of Lepidoptera larvae but may also include phytophagous Coleoptera larvae (Chrysomelidae and Curculionidae). Most eumenines search for concealed prey such as leaf-folders (Isley, 1913) and the female wasps are attracted by caterpillar silk and frass (Steiner, 1984). The prey is stung on the underside of the head and thorax and the venom usually causes permanent paralysis (Cooper, 1953). In addition, the prey may be kneaded (i.e. gently chewed without breaking the skin), which may help incapacitate the prey (Isley, 1913) and/or provide a means for detecting endoparasitoids (Lashomb & Steinhauer, 1975; Cowan, 1981; Bohart *et al.*, 1982). The prey is carried to the nest, supported by the mandibles and legs, and placed in the cell.

Most eumenines are mass provisioners, i.e. the cell is fully provisioned and sealed before the larva begins to develop. The exact number of prey individuals provided depends on their size, with more being amassed when they are small (Brooke, 1981). The developing larvae of mass provisioning eumenines take several days to devour all the prey individuals furnished. To avoid contamination of the cell by putrefying flesh, the prey are generally only lightly paralysed. Progressive provisioners, on the other hand, which sequentially provide new prey individuals for their developing larva, sometimes kill the prey. Apparently very few masticate the prey as do polistines.

The vast majority of eumenines are solitary or subsocial but a few demonstrate parasocial (communal) behaviour. Progressive provisioning combined with communal behaviour has been observed in the South American species *Montezumia cortesioides* (Evans, 1973*c*, as *M. dimidiata*) and *Zethus (Zethoides) miniatus* (West-Eberhard, 1987). The latter species, in which 15 or more females may occupy a common nest made from masticated leaves, demonstrates many traits that are more characteristic of the Polistinae (Ducke, 1914; West-Eberhard, 1987, 1988, unpublished): construction of a nest that is attached at only one point to the substrate; nest sharing among adults; cooperative brood care with temporary reproductive division of labour; partial progressive provisioning; brood care (defence of nest) extended until adult eclosion; and cell re-use.

Mating behaviour. Although most species have not been studied, mating behaviour in eumenines appears to be quite diverse. In some twig nesting species males emerge first and compete among themselves for access to their emerging sisters (Cowan, 1986). In ground-nesting species males may patrol nesting aggregations and in arid areas males may defend or patrol water pools (Cowan, 1991). In some species females mate only once whereas in others the female will mate repeatedly. In some species the female's wings remain free after being mounted by the male and the pair may fly off to mate in dense vegetation. Most male eumenines have hooked antennae that are used during courtship, and many females have secretory pits on the top of the head (Cumming & Leggett, 1985) which may also play a role in sexual behaviour. Post-intromission antennal stroking has been observed in *Eumenes consobrinus* (Eberhard, 1990). After intromission the male of some species releases his grip on the female's thorax, and with genitalia still linked he dangles from her held only by the genitalia. Finally, postcopulatory displays are reported from some eumenines.

Biotic associations. Several different parts of the body of adult eumenines are modified to accommodate the dispersing deutonymphs of mites, which live in the nest, feeding on fungi (Cooper, 1955). The most frequently used body parts include axillary fossae, propodeal cavity, and cavities in metasomal segments (Krombein, 1967). All species of *Parancistrocerus* have a complex symbiotic relationship with winterschmidtiid mites, each wasp having a host-specific mite. The hypopial stage of the mite congregates in large numbers in the acarinarium, a chamber at the base of the second metasomal tergite which is covered by the apex of the first tergite (Krombein, 1967).

Economic importance. As predators of lepidopteran larvae eumenines are potential biological control agents (Bohart *et al.*, 1982), although their capacity to control pest populations is probably rather small.

Identification. Carpenter and Cumming (1985) provided a key to the nearctic genera, on which the following key is based in part. Soika (1978, 1982, 1990) provided keys to neotropical species of *Eumenes*, *Hypalastoroides*, *Minixi*, *Omicron*, *Pararhaphidoglossa*, *Pirhosigma*, *Pachymenes*, *Santamenes* and *Pseudacaromenes*. Other keys to species that include Central American taxa are mentioned below in the generic synopses.

Key to the genera of Eumeninae occurring in Costa Rica

1 Metasoma petiolate (Fig. 15.11); segment I in dorsal view with width half or less that of II, and at least twice as long as wide, usually longer (Fig. 15.12). 2
— Metasoma not petiolate (Fig. 15.1); segment I with width more than half that of II, much less than twice as long as wide (Fig. 15.13). 11

2 Propodeal orifice narrowly acute dorsally; propodeal valvula elongate, rectangular (Fig. 15.14). .... *Zethus*
— Propodeal orifice broadly rounded dorsally; propodeal valvula short, rounded (Fig. 15.15). 3

3 Pronotum with oblique humeral carina (Fig. 15.16). .... *Zeta*
— Pronotum without oblique humeral carina (Figs. 15.17–15.19). 4

4 Lateral margins of tergite I not meeting ventrally, sternite I visible along entire petiole length. .... *Pseudacaromenes*
— Lateral margins of tergite I fused ventrally, sternite I reduced to posterior crescentic sclerite. 5

5 Pronotal carina absent laterally below humeri (Fig. 15.18). 6
— Pronotal carina well-developed along entire length, reaching down to or beyond level of middle of pronotum in profile (Figs 15.17, 15.19). 7

6 Tergite II without translucent apical lamella. ....*Pachymenes*
— Tergite II with translucent apical lamella. ....*Santamenes*

7 Pronotum without pretegular carina (Fig. 15.17), or with carina present only posterior to spiracular operculum.......................8
— Pronotum with complete pretegular carina (Fig. 15.19). 10

8 Tergite I with apical margin not thickened into blunt ridge and without transverse preapical furrow. .... *Pirhosigma*
— Tergite I with apical margin thickened into blunt ridge and with a transverse furrow anterior to this (Fig. 15.17). 9

9 Epicnemial carina present (Fig. 15.17). . *Omicron*
— Epicnemial carina absent. .... *Eumenes*

10 Free apical part of clypeus very short (Fig. 15.20); fore wing with *2m-cu* joining *M* on posterior margin of 3rd submarginal cell (Fig. 15.22) distal to, or opposite *2rs-m*. ....*Pararhaphidoglossa*
— Free apical part of clypeus almost as long as basal width (Fig. 15.21); fore wing with *2m-cu* joining *M* on posterior margin of 2nd submarginal cell, proximal to *2rs-m* (Fig. 15.23). ....*Minixi*

11 Fore wing with *2rs-m* fused for part of its length with *2r-sm*, before *2rs-m* joins *2r-rs*, so 2nd submarginal cell is petiolate above (Fig. 15.24). .... *Hypalastoroides*
— Fore wing with *2rs-m* joining *Rs* after *Rs* has joined *2r-rs*, so 2nd submarginal cell is sessile above (Figs 15.32, 15.33). 12

12 Anterior face of pronotum with two small, close set, deeply impressed medial foveae (Fig. 15.25), which may be contiguous; tegula abruptly expanded and posterolaterally broadly rounded. 13
— Anterior face of pronotum without two close set, deeply impressed foveae (Fig. 15.26), sometimes with one; tegula usually more evenly convex. 14

13 Tergite II smooth basally, forming acarinarium beneath apex of tergite I that is often full of mites (may be concealed by preceding tergite). .... *Parancistrocerus*
— Tergite II ridged basally, not forming acarinarium. .... *Stenodynerus*

14 Submarginal carina extended posteriorly as pointed process above valvula; propodeal valvula enlarged, free posteriorly from submarginal carina, sometimes extending as a rectangular lamella (Fig. 15.27). 15
— Submarginal carina usually not projecting, if forming pointed process then fused to valvula; propodeal valvula usually not enlarged and free posteriorly, never rectangular (Fig. 15.28). 16

15 Tergite I with transverse carina at crest of anterior declivity (Fig. 15.27). ...... *Ancistroceroides*
— Tergite I without carina. .... *Leptochilus*

16 Dorsal surface of propodeum nearly horizontal, more or less on the same level as the metanotum (Fig. 15.29). .... *Pseudodynerus*
— Dorsal surface of propodeum below plane of metanotum, sloping posteroventrally (Fig. 15.31). 17

17 Pronotum with complete oblique humeral carina (Fig. 15.30); metasomal sternite II not strongly sulcate (cf Fig. 15.29). . *Pachodynerus*
— Pronotum without oblique humeral carina; ***or if*** partial carina is present, **then** metasomal sternite II is sulcate (Fig. 15.31). 18

18 Tergite I with transverse carina at crest of anterior declivity (cf Fig. 15.27). ...... *Ancistrocerus*
— Tergite I without carina (Fig. 15.31). 19

19 Prestigma at most little more than half the length of the pterostigma (measured along posterior border) (Fig. 15.32). ...... *Euodynerus*
— Prestigma almost as long as or longer than the pterostigma (Fig. 15.33). 20

20 Mesepisternum without epicnemial carina; maxillary palpus 6-segmented, labial palpus 4-segmented. .... *Parazumia*
— Mesepisternum with epicnemial carina (Fig. 15.31); maxillary palpus 5-segmented, labial palpus 3-segmented. 21

21 Propodeum with lateral angles rounded (Fig. 15.34); submarginal carina continuous above propodeal orifice. .... *Montezumia*

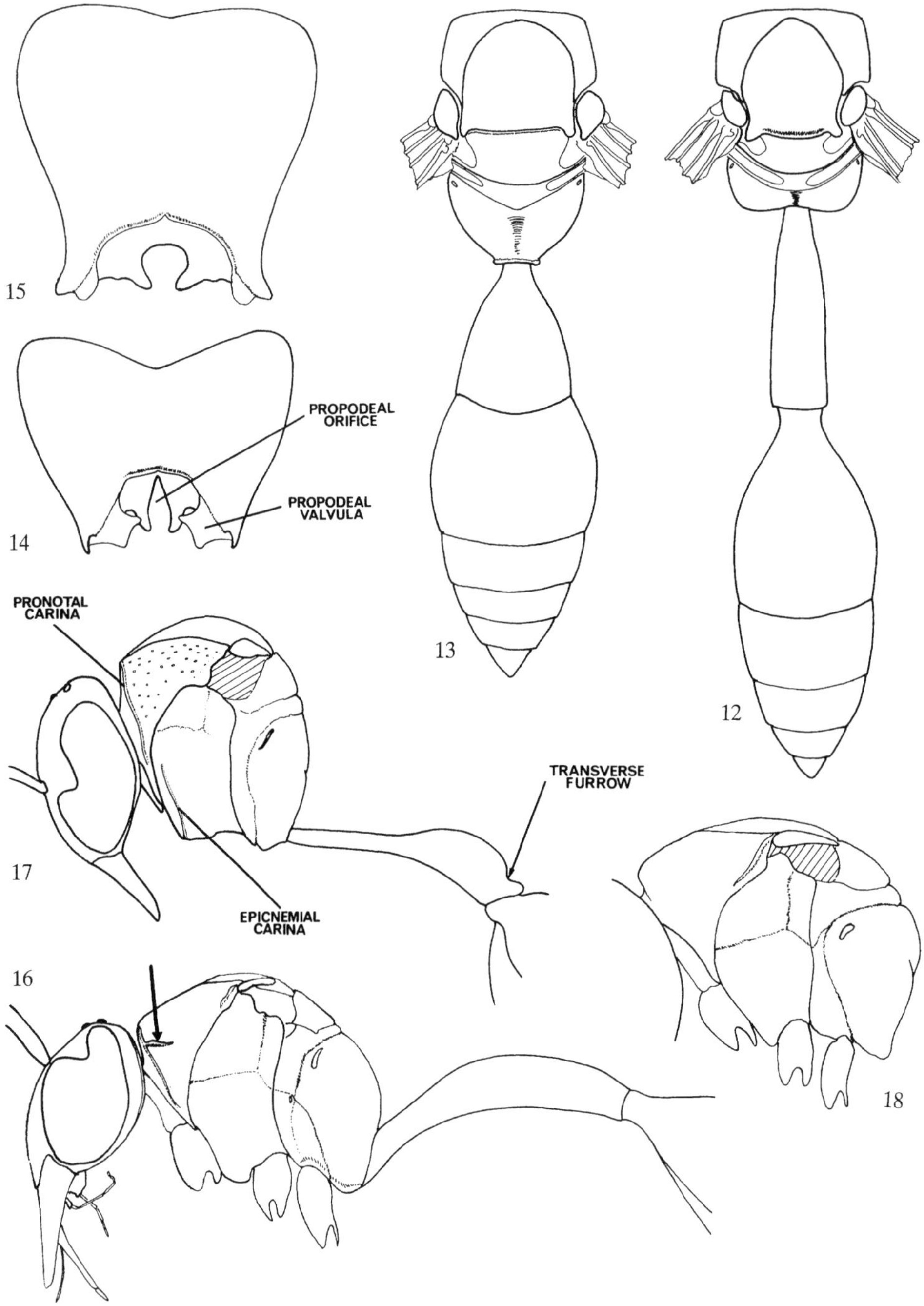

Figs 15.12–15.18. Eumeninae. Figs 15.12–15.13. Meso- and metasoma, dorsal; 15.12, *Zeta* sp.; 15.13, *Montezumia* sp. Figs 15.14–15.15. Propodeum, posterior; 15.14, *Zethus* sp.; 15.15, *Pachymenes* sp. Figs 15.16–15.17. Head, mesosoma and tergite I, lateral; 15.16, *Zeta* sp.; 15.17, *Omicron* sp. Fig. 15.18. Mesosoma, lateral, *Pachymenes* sp.

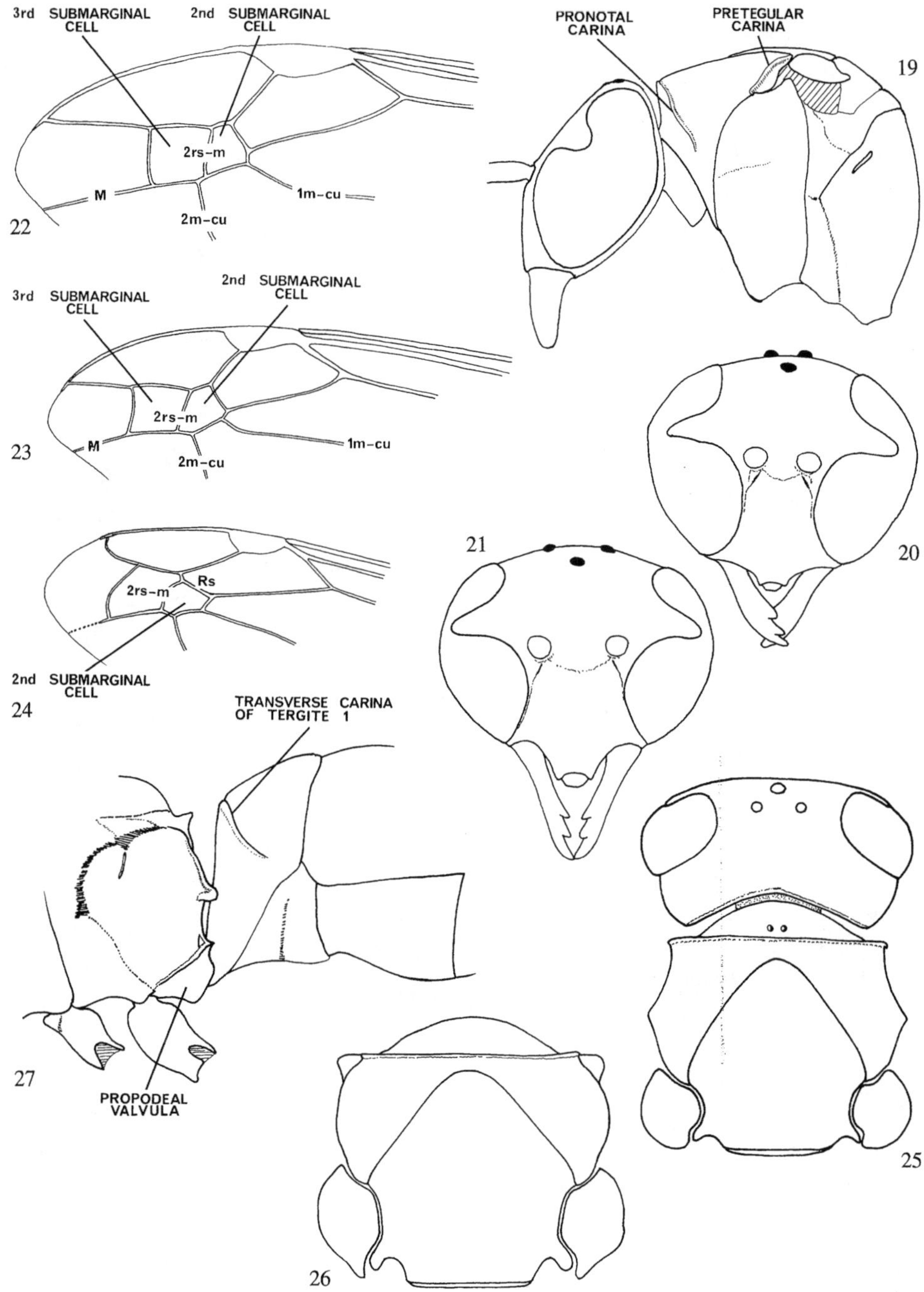

Figs 15.19–15.27. Eumeninae. Fig. 15.19. Mesosoma, lateral, *Pararhaphidoglossa* sp. Figs 15.20–15.21. Face, frontal; 15.20, *Pararhaphidoglossa* sp.; 15.21, *Minixi* sp. Figs 15.22–15.24. Anterodistal part of fore wing; 15.22, *Pararhaphidoglossa* sp.; 15.23, *Minixi* sp.; 15.24, *Hypalastoroides* sp. Figs 15.25–15.26. Anterior part of mesosoma, dorsal; 15.25, *Stenodynerus* sp.; 15.26, *Ancistroceroides* sp. Fig. 15.27. Propodeum and anterior segments of metasoma, *Ancistroceroides* sp.

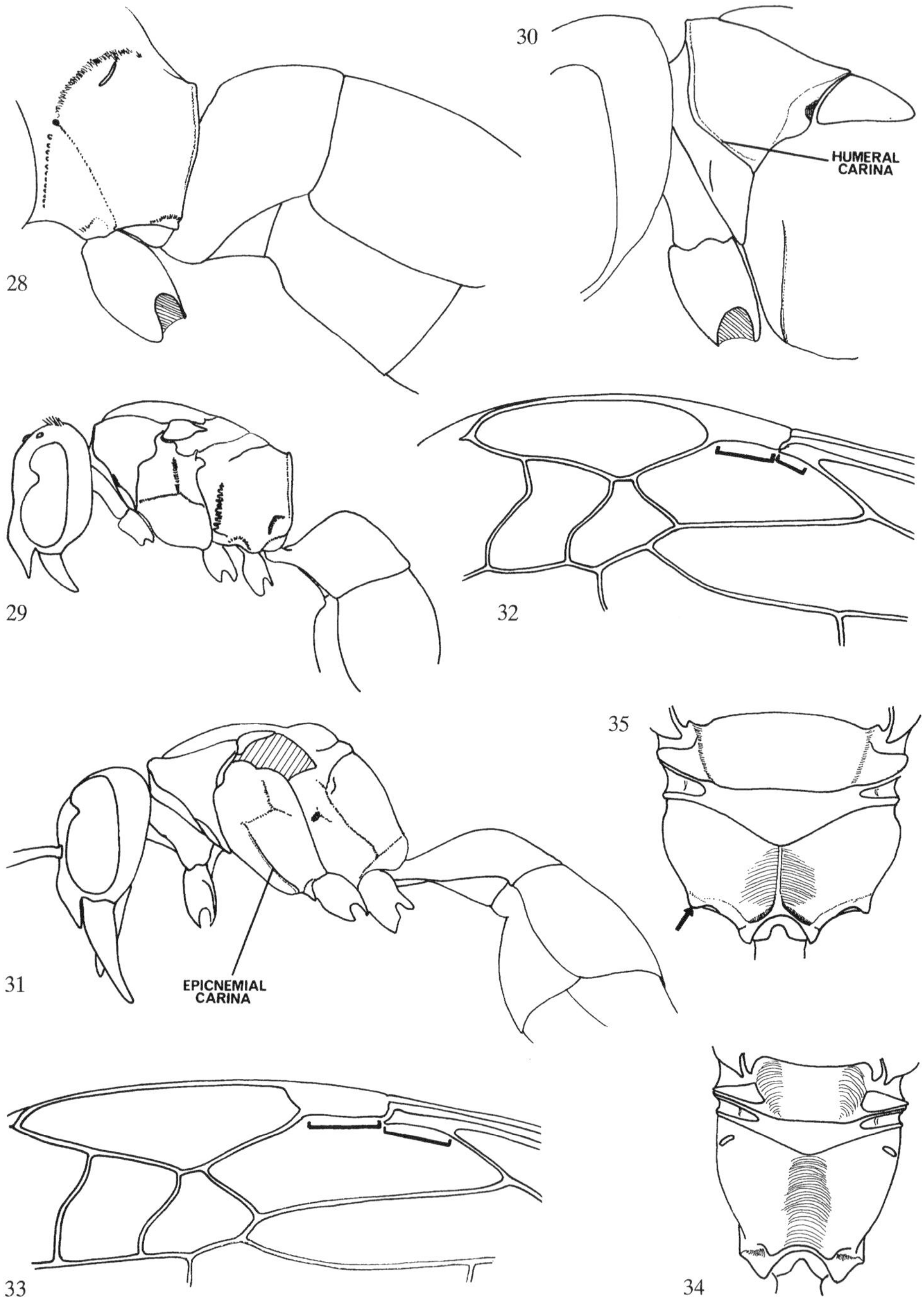

Figs 15.28–15.35. Eumeninae. Fig. 15.28. Propodeum and anterior segments of metasoma, *Pachodynerus* sp. Fig. 15.29. Mesosoma, lateral, *Pseudodynerus* sp. Fig. 15.30. Prothorax, lateral, *Pachodynerus* sp. Fig. 15.31. Mesosoma and anterior segments of metasoma, lateral, *Montezumia* sp. Figs 15.32–15.33. Anterocentral part of fore wing; 15.32, *Euodynerus* sp.; 15.33, *Montezumia* sp. Figs 15.34–15.35. Propodeum, posterodorsal view; 15.34, *Montezumia* sp.; 15.35, *Monobia* sp.

— Propodeum with lateral angles usually pointed (Fig. 15.35); submarginal carina interrupted at propodeal orifice. .... *Monobia*

Synopsis of the Costa Rican fauna

Currently 22 genera and nearly 80 species of Eumeninae are known to occur in Costa Rica. Many of these genera are primarily neotropical taxa (e.g. *Monobia, Montezumia, Pseudodynerus, Zethus*), whilst a few are principally north temperate genera with isolated representation in tropical America (e.g. *Ancistrocerus, Leptochilus*). Within Costa Rica most species of eumenines appear to be restricted to low elevation sites (< 1500 m) but one species of *Ancistrocerus* occurs at 3000 metres.

The following are the genera and species presently known to occur (or presumed to occur) in Costa Rica (^{CR} = new record for Costa Rica; ^{EX} = not yet seen but probably present based on its presence to the north and south of Costa Rica):

Ancistroceroides. The taxonomic status of this genus has only recently been clarified, but the name is applicable to most of the South American species described as *Ancistrocerus* (Carpenter & Vecht, 1991). The twenty or so described species have a range extending from Mexico to Argentina. One species is known to occur in Costa Rica: *Ancistroceroides bogotanus* (Zavattari) **n. comb.**^{CR}.

Ancistrocerus. A moderately large, primarily holarctic genus with some six species endemic to the Neotropics. One species, *similis*, has been collected in Costa Rica; its range extends north to Mexico. Biology: only lepidopteran larvae are recorded as prey in the New World, but a european species is reported to prey on coleopterous larvae (Krombein *in* Krombein *et al.*, 1979).

Eumenes. A fairly large, cosmopolitan genus, with three species occurring in Costa Rica: *americanus, consobrinus* and *filiformis*. Most of the species making nests in the shape of mud pots were originally described in this genus, which has been gradually split during this century into several dozen genera. Biology: only known to provision with larvae of Lepidoptera.

Euodynerus. A moderately large genus, most diverse in the north temperate regions, but with a few species found in the tropics. The species so far found in Costa Rica, *crypticus*^{CR}, is primarily a nearctic species, occurring as far north as Canada. Biology: only known to provision with larvae of Lepidoptera.

Hypalastoroides. This is a neotropical genus, with 27 species ranging from Argentina to the southern United States. Three species have been recorded from Costa Rica: *brasiliensis, costaricensis* and *mexicanus*.

Leptochilus. A moderately large, holarctic genus, with a few species extending into the Neotropics. The species found in Costa Rica, *tropicanus*^{CR}, occurs as far south as Colombia. Keys to species: Parker (1966). Biology: species are known to provision with both the larvae of Lepidoptera and Coleoptera.

Minixi. A small neotropical genus, with one species in Mesoamerica (*mexicanum*) that is distributed from Colombia to Texas.

Monobia. A moderately sized, neotropical genus, with more than 30 species. Some species are found throughout Central America, and one species is widespread in the United States. Three species have so far been collected in Costa Rica: *angulosa, nigripennis* and *texana*. Keys to species: Willink (1982). Biology: only known to provision with larvae of Lepidoptera.

Montezumia. A moderately large genus, with almost 50 species. It is neotropical, with some species extending north to the southwestern United States. Five species have been collected in Costa Rica: *azurescens, cortesia, dimidiata, huasteca* and *infernalis*. Keys to species: Willink (1982). Biology: one individual of *M. cortesia* (in NHM) has been reared from an old *Sceliphron*-like nest found under an overhang. Species are only known to provision with larvae of Lepidoptera.

Omicron. A fairly large genus of small wasps, endemic to the Neotropics. Numerous species occur in Mesoamerica, and the following is doubtless a partial list for Costa Rica: *acapulcense*^{EX}, *aequale*^{EX}, *aggressor*[CR], *aviculum*, *criticum*[EX], *deminutum*, *flavonigrum*, *foxi*[CR], *globicolle*[CR], *nanum*, *nymphale*, *regulum regulum*, *reguloide*, *thoracicum* and *totonacum*.

Pachodynerus. A moderately sized genus, endemic to the Neotropics, but with some species occuring as far north as the United States, and one (*nasidens*) that has been introduced widely onto islands in the Pacific. Two species occur in Costa Rica: *nasidens* and *praecox*. Keys to species: Carpenter (1986*b*). Biology: only known to provision with larvae of Lepidoptera.

Pachymenes. A small genus of medium-sized wasps, endemic to the Neotropics. Of the 11 species, three occur in Mesoamerica and Costa Rica: *aztecus*[CR], *ghilianii* and *obscurus*.

Parancistrocerus. A moderately large genus with most included species found in the New World. The neotropical species are diverse but little described, and the species recorded from Costa Rica, *fulvipes*[CR], is common in the United States. Biology: only known to provision with larvae of Lepidoptera.

Pararhaphidoglossa. This is a fairly small (20 or so species), neotropical genus with a few species distributed in Mesoamerica. Three have been found in Costa Rica: *fulvior*, *invenusta* and *nigrofulva*.

Parazumia. A small genus with one South American species that has also been collected in Costa Rica: *carinulata*[CR]. There are also a few species occurring in Mexico and the southern United States.

Pirhosigma. A small neotropical genus, with one species widespread in Central America: *aenigmaticum*.

Pseudacaromenes. A neotropical genus, with one large species, *alfkeni*. The few specimens of this genus have been collected in Brazil, Costa Rica and Belize.

Pseudodynerus. A small neotropical genus, ranging from Paraguay north to the eastern coast of the United States. Two species have been found in Costa Rica: *maxillaris* and *quadrisectus*[CR]. Keys to species: Bequaert (1941). Biology: only known to provision with larvae of Lepidoptera.

Santamenes. A small neotropical genus, recently separated from *Pachymenes*. The four species occur from Argentina to Mexico, and one has been collected in Costa Rica: *novarae*[CR].

Stenodynerus. A fairly large genus of small wasps, most diverse in the Holarctic but with numerous neotropical representatives. So far five species have been collected in Costa Rica: *farias*, *iolans*, *otomitus*, *temoris* and *undiformis*[CR]. Keys to species: Bohart (1980). Biology: most species provision with lepidopterous larvae, but in the United States two ground nesting species also use coleopterous larvae (Krombein *in* Krombein *et al.*, 1979).

Zeta. A small genus (four species) of large wasps, widespread in the Neotropics, including the Greater Antilles. One species is found in Central America, and occurs in Costa Rica: *argillaceum*.

Zethus. A large genus, with about 200 species, of which a handful are African and a few nearctic, with the remainder endemic to the Neotropics. Several dozen species have been recorded from Central America, and the following list of Costa Rican species is doubtless incomplete: *apicalipenns*, *aztecus*[CR], *cerceroides*, *clypearis*, *clypeolaris*, *costarricensis*, *dodgei*, *dreisbachi*[CR], *fuscus*, *histrionicus*, *isthmicus*, *lopezi*, *matzicatzin*, *miscogaster*[EX], *montezuma*[EX], *olmecus*, *spinosus*[CR], *strigosus*, *toltecus* and *westwoodi*. Keys to species: Bohart and Stange (1965). Biology: known to progressively provision with larvae of Lepidoptera.

15.3 THE SUBFAMILY POLISTINAE ('PAPER WASPS')

Classification and distribution. Polistines occur throughout the world but are most diverse in tropical regions, especially the Neotropics. Worldwide there are 28 genera and 800 species, and in Costa Rica 17 genera and nearly 100 species have been recorded. The Costa Rican fauna belong to three tribes: Polistini (*Polistes*), Mischocyttarini (*Mischocyttarus*) and Epiponini (the remaining genera).

Biology. Most social vespids (Stenogastrinae, Polistinae and Vespinae) demonstrate the following

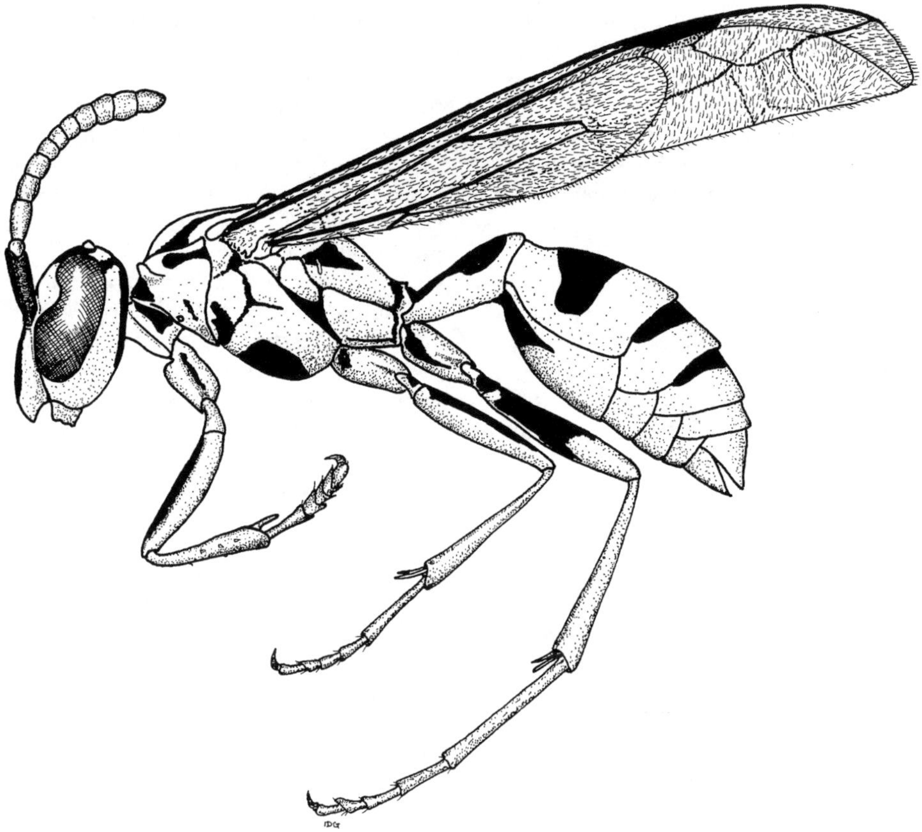

Fig. 15.36. *Agelaia areata* (Polistinae).

behavioural traits: construction of nests of vegetable material; nest sharing among adults; co-operative brood care; reproductive division of labour (sterile workers); simultaneous progressive provisioning with masticated prey (mostly lepidopterous larvae); brood care extended to adult eclosion; adult–adult trophallaxis; and cell re-use.

Colony founding. Polistines initiate new colonies in one of two ways (West-Eberhard, 1969; Jeanne, 1980): independent founding and swarm founding. In the former type, practiced by *Polistes* and *Mischocyttarus* species, one or several queens initiate a new colony by themselves, without workers. Often a colony is founded by a single female ('foundress') who is soon joined by one or more additional females. However, most of the egg laying is done by a single female, usually the one that initiated the nest. Although the other females are fertile they are prevented from laying eggs by the highly territorial foundress, whose physical dominance (attacks and chasing), threat behaviour, and (in some species) egg-eating, keeps the other females from reproducing.

In contrast, members of the tribe Epiponini initiate colonies via a swarm of workers, accompanied by a smaller number of queens, leaving an established nest. The workers select the site of a new nest and movement to this site is co-ordinated by means of pheromone trails (Naumann, 1971; Forsyth, 1981*a* & *b*; West-Eberhard *in* Jaisson, 1982). Individual 'scout' workers move back and forth between the old and new nest sites, rubbing an abdominal scent gland on prominent leaves. Then many wasps take flight at once and individually follow the scent trail.

Swarm founders are generally polygynous (i.e. with more than one queen) although in some (e.g. *Metapolybia aztecoides*) there are cycles between polygyny and monogyny (West-Eberhard, 1978*b*), a pattern that may prove widespread in the group (Strassmann *et al.*, 1991, 1992). Reproductive dominance and queen recognition in at least one species appears to occur by means of pheromones (West-Eberhard, 1977).

Swarm founding has two potential advantages over independent founding. First, socialization of dispersal and colony founding reduces the risk of queen mortality. Second, the initially larger colony allows greater specialization among colony members, more effective defence against predators, and (combined with the usually enveloped nest) greater homeostatic control over physical conditions within the nest (Jeanne, 1991).

Nest construction and defence. Paper wasps, as their name implies, typically build a nest of chewed vegetable fibres. Depending on the species, construction material may include a combination of long woody fibres, plant hairs, short vegetable chips, mud, and glandular secretions (Wenzel, 1991). Independent founding polistines (*Polistes* and *Mischocyttarus*) construct a single unenveloped comb, which is attached to the substrate by a tough pedicel. All swarm founding genera (tribe Epiponini), except *Apoica*, include species that build an envelope over the nest; such enveloped nests often comprise several combs. The nest may be located on the undersurface of a leaf (e.g. species of *Polistes, Mischocyttarus, Leipomeles*), flat on a tree trunk (*Metapolybia, Synoeca*), hanging from branches (*Brachygastra, Polybia*), or hidden in cavities above or below ground, such as hollow tree trunks (some species of *Agelaia*).

Nest architecture has been greatly influenced by ant predation, as several kinds of special architectural defences against ants have been demonstrated. In *Polistes* and *Mischocyttarus* an ant repellent substance secreted from a gland on the sixth metasomal sternite is applied to the pedicel (Post & Jeanne, 1981; Post *et al.*, 1984; Espelie & Hermann, 1990). Enveloped nests probably evolved primarily as a defence against ants (Jeanne, 1975), although attacks by parasitoids may also have played a role (see Gadagkar, 1991; Jeanne, 1991). *Leipomeles dorsata* and *Nectarinella championi* have an additional defence in the form of sticky stalks produced from oral secretions, which are densest in the area near the nest entrance.

In addition to defensive features incorporated into the nest when it is built, adult wasps actively guard the nest against ants. Such behavioural defences include 'pecking' with the mandibles accompanied by wing fanning (West-Eberhard, 1969), physical removal of ants (Jeanne, 1991), and removal of pheromone trails left by ants (West-Eberhard, 1989).

Army ants pose a particularly serious problem (Chadab, 1979) and possible defences include: nesting very high in the canopy (e.g. *Agelaia areata* and *Epipona* spp.), workers blockading the nest entrance with their bodies (e.g. *Synoeca septentrionalis*), blockading combined with a tough nest covering (e.g. *Polybia emaciata*), or nesting in close association with *Azteca* or *Pseudomyrmex* ants (e.g. *Agelaia myrmecophila* and *Polybia rejecta*). The latter of course possibly entails a need for some defence against these ants, and *Parachartergus aztecus*, which lives with *Pseudomyrmex*, chemically camouflages the nest from the ants by coating the nest envelope with lipids similar to those in the ant's cuticle (Espelie & Hermann, 1988). Most polistines however, when confronted by an army ant raid, have few options except to rapidly evacuate the nest (Chadab, 1979).

In the seasonally dry, open forests of northwestern Costa Rica ants appear to be less important enemies of social wasps than vertebrate predators. In this region bird predation, especially by the grey-headed kite (*Leptodon cayanensis*: Accipitridae) on nests of *Polybia occidentalis* and *P. barbouri*, has been reported to reach as high as fifty per cent towards the end of the dry season (Windsor, 1976). Other threats to polistine nests, especially unenveloped nests, include brood parasites such as pyralid moth larvae (Nelson, 1968, 1971; Jeanne, 1979*b*) and hymenopteran parasitoids (see Chapter 2.6).

Castes. In the few tropical species that have been studied in this resepct—for example, *Polistes* spp. (West-Eberhard, 1969, 1986), *Metapolybia aztecoides* (West-Eberhard, 1978*b*), and *Synoeca surinama* (West-Eberhard, 1981)—caste (queen vs. worker) determination depends upon dominance relations within the colony when a female is young. If a queen or group of queens is present newly emerged females become workers, but in the absence of a queen they become queens. Male production in these species does not seem to correlate with queen production. For example, a colony of *Metapolybia* observed for 18 months produced numerous reproductive swarms and queens without producing a single male (West-Eberhard, 1978*b*). Reproductive swarming (swarm production not accompanied by nest abandonment) occurred during periods of strong aggressive competition among queens.

Queens generally appear larger than workers due to their enlarged abdomen containing developed oocytes, and old queens of some species (e.g. *Polybia, Metapolybia, Agelaia*) are darker or browner in colour. Morphological differences between queens and workers are most pronounced in *Agelaia* (Jeanne & Fagen, 1974) and, at the other extreme, virtually non-existent in genera such as *Nectarinella* and *Protopolybia* (West-Eberhard, unpublished). Living queens can often be identified by their behaviour. In some species of *Polistes* queens aggressively attack and dominate other females, while in other species queens perform threat displays and other females avoid them or approach slowly. In *Metapolybia* and *Synoeca* (West-Eberhard, 1978*b*, 1981) and *Leipomeles* (West-Eberhard, unpublished) queens walk slowly and curve the abdomen and antennae toward other approaching queens in characteristic 'bending' behaviour.

Diet. Adult polistines feed partially on nectar and honeydew, which is sometimes stored as viscous drops in egg-containing cells in the nest. Some species tend honeydew-secreting Homoptera; for example, *Parachartergus fraternus* has been observed tending Aetalionidae (Letourneau & Choe, 1987) and another *Parachartergus* species has been seen to tend membracid nymphs in San Antonio de Escazú (W. Eberhard, pers. comm.). Adult females hunt caterpillars and often extensively masticate the prey at the capture site, so that when returning to the nest their abdomens are heavy with ingested fluids, and the solid pellet they carry in their mouth is relatively dry. The ingested fluids are regurgitated to adults and larvae at the nest, and the solid food is also divided among adults and larvae. Where the prey organisms are brought into the nest more or less intact (e.g. *Polybia occidentalis*), the prey has been found to be dominated by larval Lepidoptera, although many other arthropods are also used (Gobbi *et al.*, 1984). For example, early in the wet season in northwestern Costa Rica *P. occidentalis* captures large numbers of alate termites, which may be cached in the nest (Forsyth, 1978). In addition to arthropod prey, polistine larvae may also receive some nectar (Hunt *et al.*, 1987).

Nesting cycle. In tropical polistines, nests of both independent- and swarm-founding species are often perennial. In swarm-founding polistines brood production and nest expansion (addition of cells) may occur in pulses, with nest expansion often coinciding with periods when there are relatively few larvae to feed (West-Eberhard, 1978*a*). Lack of nest expansion in at least some tropical *Polistes* coincides with periods of dominance conflict when there is no established egg-laying queen (West-Eberhard, 1969). However, the duration of the 'nesting cycle' (occupancy of a single nest – Jeanne, 1991) is independent of offspring emigration (swarm production) and the production of sexuals (males and/or queens).

The nest may be abandoned if the brood is destroyed, by army ants or birds for example. In the areas with a long dry season, such as northwestern Costa Rica, some species migrate to higher (more humid) elevations, seasonally abandoning breeding nests to occupy dry season nests that are not used for brood rearing (reviewed in West-Eberhard, 1982 and Jeanne, 1991). Barring nest destruction or strong seasonality, nothing is known about what additional factors (if any) regulate the nesting cycle.

Mating behaviour. Although copulation in nature has apparently been reported in only one genus of tropical paper wasp (*Polistes*—see West-Eberhard, 1969; Polak, 1992), several different patterns of male behaviour have been observed that likely function in mate acquisition. The most common is 'patrolling'—flight along a repeated path, often by several males, which hover near or briefly land on prominent leaves or other objects in their path. Upon landing, patrolling males sometimes rub the substrate with a sinuous or rapid movement of the metasoma, presumably applying a pheromone (brief review in West-Eberhard, 1981). This patrolling and marking behaviour is easily confused with the trail-marking behaviour of swarming females. In Costa Rica it occurs in at least some populations of *Synoeca, Polistes, Epipona, Parachartergus* and *Mischocyttarus* males (West-Eberhard, 1981 and unpublished; Polak MS.). Male behaviour, however, may vary within species due to the presence of alternative tactics (see Polak, 1992) and geographic variation. In males of some swarm-founding species there is a very conspicuous glandular area on the ventral metasoma, presumably the source of the marking pheromone (Downing *et al.*, 1985). An unusual marking behaviour was observed by Wenzel (1987) in males of Costa Rican *Polistes major*, which use a secretion produced from a mandibular gland.

Males of Costa Rican *Agelaia* species (probably *A. multipicta*), rather than patrolling and hovering, have been observed flying very rapidly about the crown of a low tree on a ridge, but no scent marking was seen. Males of *Parachartergus apicalis* 'perched' and flew at passing objects. Although most such behaviour is not closely associated with nest sites, and some large patrolling areas (e.g. of *Mischocyttarus mexicanus*) persist for years with no obvious connection with nest locations, sites of male activity do sometimes appear to be associated with particular nests (e.g. *Synoeca septentrionalis*, *Epipona* sp.; West-Eberhard, unpublished). In a population of *Polistes canadensis* in which unmated nest foundresses sometimes occurred, males visited sites of newly founded nests (West-Eberhard, 1982). Polak (1992) observed territorial males of *Polistes canadensis* and *P. carnifex* along ridges in Guanacaste, but never saw them attempt copulation with females on or near nests. Territorial males scent marked on perches within their territories, and in *P. canadensis* territorial males sometimes switched to an alternative 'cruising' behaviour.

Biotic associations and mimicry. Because social wasps often defend their nests by inflicting a painful sting, many other animals, ranging from katydids (Downhower & Wilson, 1973) to oropendolas (Smith *in* Janzen, 1983), associate with them and thereby probably gain some measure of protection from predators. In an experimental study of bird/wasp associations in the seasonally dry forests of northwestern Costa Rica, Joyce (in press) demonstrated that birds commonly found nesting in association with polistine wasps built significantly more nests in trees containing artificial *Polybia*-like nests than in trees lacking these artificial nests. As mentioned previously, some polistines nest in association with certain ants. In addition, some polistines nest together, the less pugnacious species tending to associate with more pugnacious species. For example, in northwestern Costa Rica *Polistes carnifex* often nests with *Polybia occidentalis* (Corn, 1972), and *Mischocyttarus immarginatus* tends to nest in association with *Polybia occidentalis* and *P. diguetana* (Windsor, 1972; Gorton, 1978); interestingly *M. angulatus* was never found nesting with *Polybia* (Starr, 1988).

A variety of polistines (Richards, 1978) probably gain protection from predators by resembling other common aggressive polistines. In many cases *Agelaia* species may serve as models for less aggressive or numerous species such as *Mischocyttarus* species. The former are much more aggressive and have more painful stings (Starr, 1985). The fact that *Agelaia* species have large colonies and thereby provide large numbers of stinging models to 'train' predators may also play a role. Many other insects, including various hymenopterans, Mantispidae (Opler, 1981) and arctiid moths, perhaps also gain protection from predators in a similar way. Table 15.2 lists possible examples of mimicry complexes in Costa Rica. This table was assembled by inspection of specimens in the Hymenoptera collection at the University of Costa Rica, looking for insects resembling aggressive wasps in size, colour and shape. But we emphasize that there is so far no experimental verification that any of these wasp 'mimicry complexes' actually function to deter predators.

In Table 15.2 only colour patterns are described, but many of the species have been observed to show behavioural similarities as well, reinforcing the impression that they have evolved under selection for adaptive resemblance (Starrett, 1993). It should be noted, however, that recurrent colour patterns are common in insects of different genera and even families, including as developmental anomalies. Moreover, even visually complex colour patterns are sometimes genetically simple and evolutionarily labile (Nijhout, 1991). Such variation may lead to convergences (or parallelisms) without selection for mimicry, although it may also facilitate the evolution of mimicry.

Economic importance. Because of their painful stings the polistine wasps, like the Vespinae of the north temperate regions, are probably amongst the most feared of all insects. In a small fraction of the human population (no more than one percent) a sting from these wasps can represent a life-threatening danger due to an extreme allergic reaction; anyone can develop an allergy to stings (see Chapter 3.1). These wasps are most problematic when they nest near human habitations, and in a world that is

Colour pattern	Vespine models	'Mimics'
Blue-black body, dark wings	*Synoeca septentrionalis,* *Polistes bicolor*	*Montezumia* sp. (Eumeninae); *Pimpla caeruleata* (Ichneumonidae); *Pepsis* spp. (Pompilidae).
Dark brown body, dark wings	*Agelaia panamaensis,* *Mischocyttarus labiatus,* *M. melanarius*	*Neogorytes* sp. (Sphecidae); *Ochloptera* sp. (Sphecidae).
Black body, black wings with white tips	*Parachartergus apicalis* *P. fraternalis,* *Mischocyttarus atramentarius,* *M. atrocyaneus*	*Monobia* sp. (Eumeninae); *Hoplisoides* sp. (Sphecidae); *Sagenista* sp. (Sphecidae); indet males (Mutillidae); *Leucospis leucoteles* (Leucospidae); *Clydonium cabrerai* (Ichneumonidae); *Seminota* sp. (Trigonalyidae); *Aplochares imitator* (Pompilidae); *Odyneropsis* sp. (Apidae).
Black body, wings yellowish-red with black tips	*Polybia rejecta*	*Pachodynerus nasidens* (Eumeninae); *Hoplisoides* sp. (Sphecidae); *Leucospis klugii* (Leucospidae); *Priocnemis* sp. (Pompilidae); *Brachymeria mexicana* (Chalcididae).
Blackish/dark brown body, wings yellowish brown, anteriorly black	*Polybia aequatorialis tristis*	*Phytodietus* sp. (Ichneumonidae); *Pimpla ichneumoniformis* (Ichneumonidae); some indet Ichneumoninae (Ichneumonidae).
Black body with slight yellow markings (margins of pronotum, metasomal tergites, etc.)	*Polybia occidentalis* sp. gp. *Mischocyttarus immarginatus,* *M. pallidipectus,* *M. tolensis,* *Polistes pacificus*	*Leucospis latifrons* (Leucospidae); *Taeniogonalos costalis* (Trigonalyidae); *Phytodietus* sp. (Ichneumonidae); *Sphinctus* sp. (Ichneumonidae).
Body yellow with black markings	*Agelaia areata,* *A. cajennensis,* *A. multipicta,* *A. pallipes,* *Mischocyttarrus alienus,* *M. basimacula,* *M. cleomenes,* *M. collarellus,* *M. costaricensis,* *M. fraudulentus,* *Polybia emaciata*	*Zethus histrionicus* (Eumeninae); *Cerceris* sp. (Sphecidae); *Megistommum* sp. (Sphecidae); *Psammaletes* sp. (Sphecidae); *Neotheronia* spp. (Ichneumonidae); *Dolichomitus zonatus* (Ichneumonidae); *Hyposoter* sp. (Ichneumonidae); *Carinodes* spp. (Ichneumonidae); *Labidogonalos* (Trigonalyidae); some *Conura* spp. (Chalcididae); *Neoplisus* sp. (Sphecidae); *Pseudoplisus* sp. (Sphecidae).
Body yellowish brown with black markings, metasoma orange-brown; wings yellowish	*Agelaia myrmecophila* *Mischocyttarus angulatus*	*Polistomorpha fasciata* (Leucospidae); *Zeta argillaceum* (Eumeninae); *Neotheronia* spp. (Ichneumonidae); *Dolichomitus annulicornis* (Ichneumonidae).
Body as above; wings dark brown	*Polistes canadensis*	*Joppidium* sp. (Ichneumonidae).
Body mainly black, end of metasoma bright yellow	*Brachygastra* spp. *Clypearia* sp.	*Lycogaster* (Trigonalyidae); *Cryptopteryx* sp. (Ichneumonidae).

Table 15.2. Recurrent colour patterns including social vespids in Costa Rica: material for the occurrence of protective mimicry (see text).

becoming increasingly urbanized their importance as urban pests will probably continue to grow (Davis, 1977).

The most frequent and problematic urban wasp pests in Costa Rica are *Polistes* species, which often nest on houses and become aggressive when disturbed by movements near the nest. *Polybia* species commonly build their familiar light grey globe-like nests, high up beneath the eaves of buildings or on large leaves of cultivated plants. Since these colonies are innocuous if undisturbed many families leave them intact due to the beauty of the nests. Occasionally the cavity nesting *Agelaia* species take advantage of wall-spaces or discarded containers to build their unenveloped nests. Fortunately the most aggressive, large-colony species such as *Agelaia areata* and *Brachygastra* species, usually build nests either high up in trees, or in rural areas, where they seldom harm people.

If a wasp nest must be removed, the most effective way to do it is at night, when most wasps are inside the nest and relatively calm. It is important not to shine a light directly on the nest, or jar the surrounding area, as this will alert the wasps. By careful planning it is usually possible to place a jar over the nest or surround it completely with a plastic bag (without disturbing the wasps), and then quickly scrape it from the substrate. If the nest is to be sprayed this is also best done at night, by a fleet-footed person wearing protective clothing.

On the other hand, as predators, polistines are useful biological control agents of agricultural pests, and entomophobic tendencies to exterminate them wantonly ought to be controlled. *Polistes canadensis* and *P. erythrocephalus* for example, are important predators of the cassava hornworm, *Erinnyis ello* (Sphingidae) in Colombia (Belotti & Schoonhoven, 1978). There has been some discussion of the possibility of managing vespid wasps for purposes of biological control (Gallego, 1950; Rabb, 1960; Gillaspy, 1979, *in* Breed *et al.*, 1982, *in* Vinson, 1986). Although attempts to manipulate social wasps in pest management programmes have been unsuccessful, natural populations are probably common predators of lepidopterous pests (see also Chapter 3).

Identification. Richards (1978) provided keys to the genera and species of Polistinae of the New World. Subsequent nomenclatural changes and descriptions of new Central American species have been published by Snelling (1983*a*), Carpenter and Day (1988), Carpenter and Wenzel (1988, 1989), and Carpenter (1993). Keys to the Costa Rican species of Polistinae are being prepared by Starr and Snelling.

Key to the genera of Polistinae occurring in Costa Rica

1 Metasomal segment I subsessile, evenly conical in dorsal view (Fig. 15.37); propodeum with orifice dorsally acute; pronotum with a carina posterior to the lateral fovea only (Fig. 15.43). [Large wasps, fore wing length 10 mm+.] .. .. *Polistes*

— Metasomal segment I differently shaped, in dorsal view basally petiolate to long-petiolate or sessile with width greater than length (Figs 15.38–15.42); propodeum with orifice often broadly rounded dorsally; pronotum with a carina anterior to the lateral fovea, if the latter is present (Fig. 15.44). 2

2 Mid and hind tarsi with third and fourth segments asymmetrical, inner lobe longer than outer lobe (Fig. 15.45); fore coxa not dorsolaterally produced; metasomal segment I petiolate, in dorsal view with width half or less that of II, and at least twice as long as wide. .... *Mischocyttarus*

— Mid and hind tarsi with third and fourth segments symmetrical; fore coxa usually dorsolaterally produced; metasomal segment I variously shaped. 3

3 Pronotum without lateral fovea (Figs 15.46, 15.47). 4

— Pronotum with lateral fovea, which is sometimes shallow (Figs 15.48–15.51). 8

4 Scutellum angled in profile; metanotum compressed, vertical (Fig. 15.46); head with occipital carina present dorsolaterally; pronotum without pretegular carina. [Body stout.] .... *Brachygastra*

— Scutellum rounded in profile (Fig. 15.47); metanotum not compressed or vertical; head without occipital carina; pronotum with pretegular carina. 5

5 Fore coxa dorsolaterally produced (Fig. 15.52). 6
— Fore coxa not dorsolaterally produced (Fig. 15.53). 7

6 Clypeus apically emarginate (Fig. 15.54); propleuron without a lateral, crenate groove; cuticle black, bare and shining. .... *Epipona*
— Clypeus apically bluntly pointed (Fig. 15.55); propleuron with lateral, crenate groove (Fig. 15.47); cuticle not shining. . *Metapolybia*

7 Head, in lateral view, with gena narrower than eye at ocular sinus (Fig. 15.56); clypeus longer than wide; propodeum posteromedially concave. .... *Clypearia*
— Head, in lateral view, with gena as wide or wider than eye at ocular sinus (Fig. 15.57); clypeus at most as long as wide; propodeum posteromedially convex.
[Large, bluish-black wasps.] .... *Synoeca*

8 Pronotum with dorsal carina sinuous in lateral view (Fig. 15.48); fore wing with prestigma about as long as pterostigma.
[Most species have dark wings with pale tips.] .... *Parachartergus*
— Pronotum without sinuous carina; fore wing with prestigma shorter than pterostigma. 9

9 Mesoscutum with posterolateral lamella absent anteriorly, not adjoining tegula (Fig. 15.58). 10
— Mesoscutum with posterolateral lamella present adjoining tegula (Fig. 15.59). 12

10 Metanotum produced posteromedially into elongate lobe overlapping the propodeum (Fig. 15.40) .... *Protopolybia*
— Metanotum without posterior lobe. 11

11 Clypeus apically truncate, longer than wide (Fig. 15.60); head, in lateral view, with gena narrower than eyes just below ocular sinus; metasomal segment I, in dorsal view, with short petiole, posteriorly about as broad as tergite II (Fig. 15.41) .... *Charterginus*
— Clypeus apically bluntly pointed or rounded, at most as long as wide (Fig. 15.61); head, in lateral view, with gena usually wider than eyes just below ocular sinus; metasomal segment I, in dorsal view, posteriorly much narrower than tergite II (Fig. 15.42) ... *Polybia*

12 Ocelli enlarged, separated from eyes by less than an ocellar diameter (Fig. 15.62); hind wing with jugal lobe reduced. .... *Apoica*
— Ocelli normal, separated from eyes by more than an ocellar diameter (Fig. 15.63); hind wing with jugal lobe normal, not reduced. 13

13 Head, in dorsal view, with occipital carina present (cf Fig. 15.62). 14
— Head, in dorsal view, with occipital carina absent (Fig. 15.63). 15

14 Mesepisternum with a dorsal subhorizontal groove (cf Fig. 15.50). .... *Agelaia*
— Mesepisternum without a dorsal subhorizontal groove (cf Figs 15.49, 15.51).
[Small pale yellow wasps.] .... *Leipomeles*

15 Metanotum vertical; scutellum more or less angled in profile (Fig. 15.49); pronotal keel extending down behind lateral fovea.
[Small yellow wasps.] .... *Nectarinella*
— Metanotum not vertical; scutellum rounded in profile (Figs. 15.50, 15.51); pronotal keel not extending down behind lateral fovea. 16

16 Mesepisternum without dorsal subhorizontal groove (Fig. 15.51); palp formula 5: 3.
[Very stout, black with red face.] ....*Chartergellus*
— Mesepisternum with dorsal subhorizontal groove present, at least as anterior trace (Fig. 15.50); palp formula 6: 4. .... *Pseudopolybia*

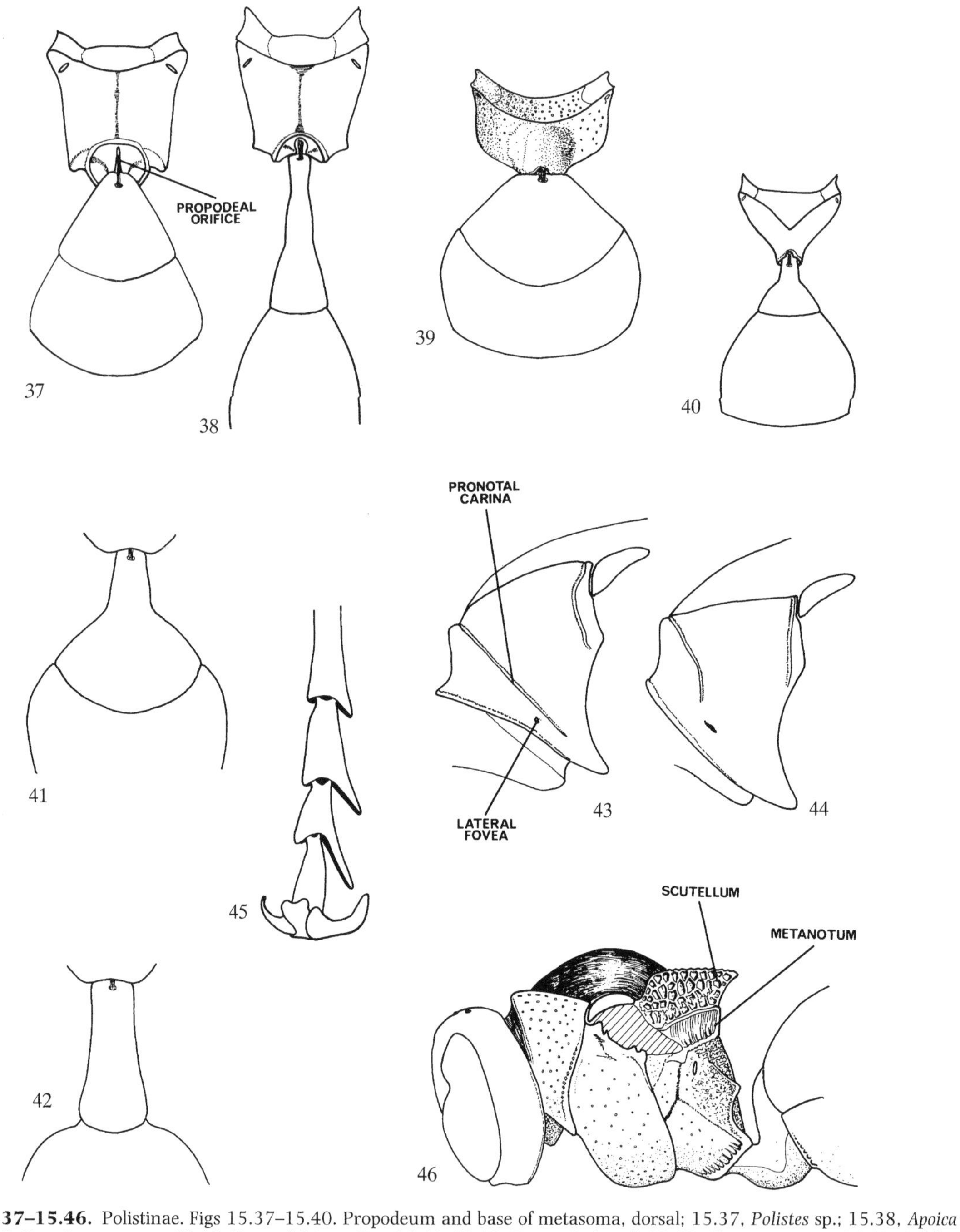

Figs 15.37–15.46. Polistinae. Figs 15.37–15.40. Propodeum and base of metasoma, dorsal; 15.37, *Polistes* sp.; 15.38, *Apoica* sp.; 15.39, *Parachartergus* sp.; 15.40, *Protopolybia* sp. Figs 15.41–15.42. Tergite I, dorsal; 15.41, *Charterginus* sp.; 15.42, *Polybia* sp. Figs 15.43–15.44. Pronotum, lateral; 15.43, *Polistes* sp.; 15.44, *Apoica* sp. Fig. 15.45. Distal part of hind tarsus, *Mischocyttarus* sp. Fig. 15.46. Mesosoma and base of metasoma, lateral, *Brachygastra* sp.

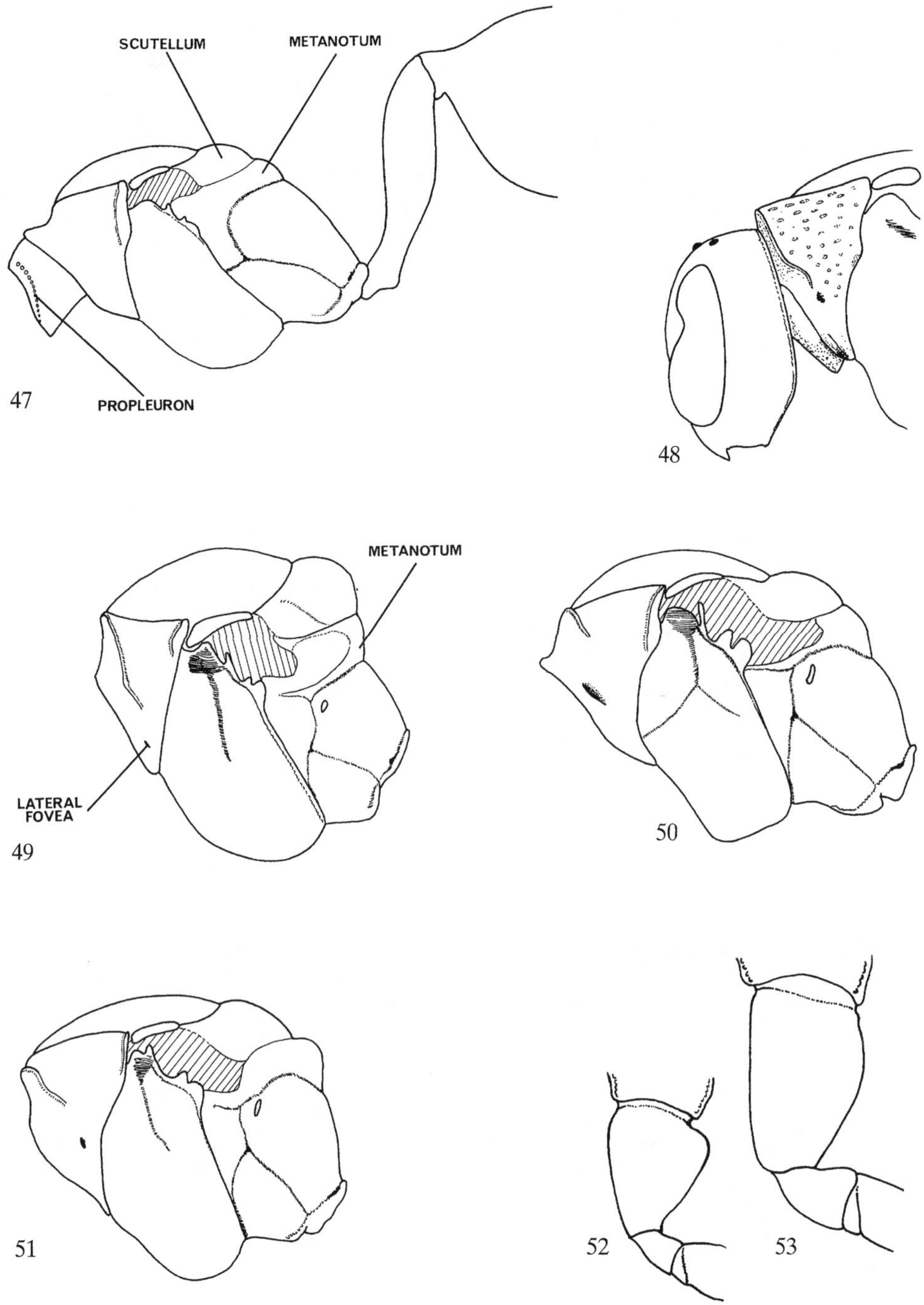

Figs 15.47–15.53. Polistinae. Fig. 15.47. Mesosoma and base of metasoma, lateral, *Metapolybia* sp. Fig. 15.48. Head and pronotum, lateral; *Parachartergus* sp. Figs 15.49–15.51. Mesosoma, lateral; 15.49, *Nectarinella* sp.; 15.50, *Pseudopolybia* sp.; 15.51, *Chartergellus* sp. Figs 15.52–15.53. Fore coxa, anterior; 15.52, *Metapolybia* sp.; 15.53, *Synoeca* sp.

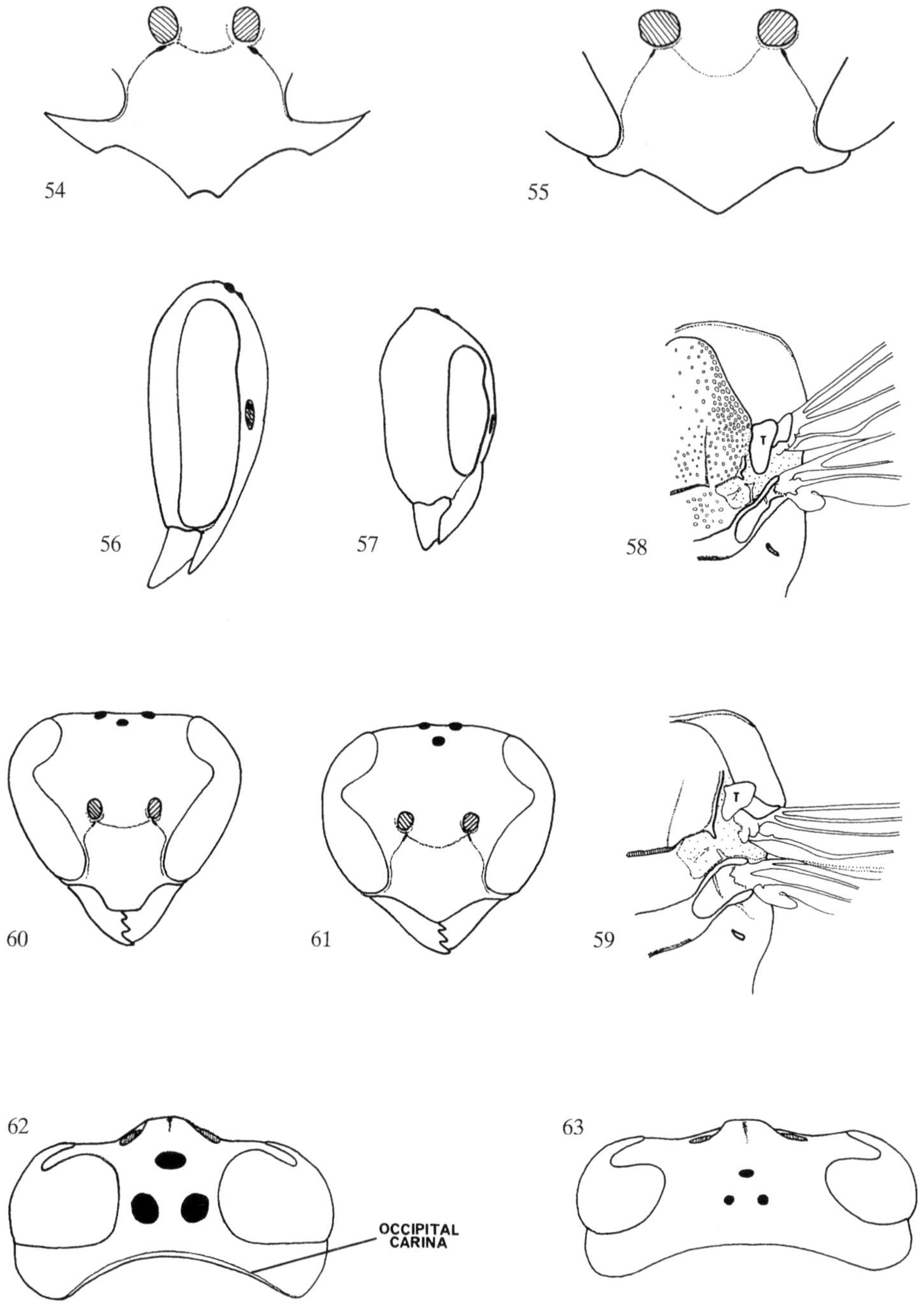

Figs 15.54–15.63. Polistinae. Figs 15.54–15.55. Clypeus; 15.54, *Epipona* sp.; 15.55, *Metapolybia* sp. Figs 15.56–15.57. Head, lateral; 15.56, *Clypearia* sp.; 15.57, *Synoeca* sp. Figs 15.58–15.59. Wing base; 15.58, *Charterginus* sp.; 15.59, *Agelaia* sp. Figs 15.60–15.61. Head, frontal; 15.60, *Charterginus* sp.; 15.61, *Polybia* sp. Figs 15.62–15.63. Head, dorsal; 15.62, *Apoica* sp.; 15.63, *Pseudopolybia* sp.

Synopsis of the Costa Rican fauna.

In Costa Rica there are 17 genera and about 90 species of Polistinae. The Costa Rican polistines are predominantly South American in their affinities with several species reaching the northern limit of their distribution here, especially on the Osa Peninsula (Starr & Snelling, in prep.). A few species reach the southern limit of their distribution in Costa Rica—*Agelaia xanthopus, Brachygastra mellifica, Metapolybia azteca, Mischocyttarus immarginatus, M. melanarius, Polistes dorsalis, P. instabilis,* and *P. oculatus* (Starr & Snelling, in prep.). The majority of species occur in the lowlands (below about 1200 m) but a few are characteristic of higher altitudes (above 1500 m) and these tend to be dark coloured (e.g. *Agelaia xanthopus, Polistes aterrimus, Polybia aequatorialis tristis*).

Following are the genera and species of Polistinae present in Costa Rica, with brief descriptions of nests [CR = new record for Costa Rica identified by Carpenter; CR (S) = identified by Starr and Snelling].

POLISTINI

This tribe comprises a single genus *Polistes* which is cosmopolitan.

Polistes. A genus comprising more than 200 species, that is most species-rich in the tropics. Sixteen species are known to occur in Costa Rica: *apicalis* CR (S), *aterrimus, bicolor* CR (S), *canadensis, carnifex, dorsalis, erythrocephalus, goeldii* CR (S), *infuscatus* CR (S), *instabilis, major, myersi, oculatus* CR (S), *pacificus, testaceicolor* and *versicolor*. Nest: simple, exposed, stalked comb.

MISCHOCYTTARINI

An exclusively New World tribe comprising a single genus.

Mischocyttarus. A large, New World genus comprising over 200 described species with a cumulative range extending from southwestern Canada south to northern Argentina. Nineteen species occur in Costa Rica: *alienus, angulatus, atramentarius, atrocyaneus, basimacula, cleomenes, collarellus* CR (S), *costaricensis, fraudulentus, immarginatus, labiatus* CR (S), *mastigophorus, melanarius, mexicanus* CR, *nigroclavatus, pallidipectus, pelor, tolensis* and *vaqueroi* CR. Nest: simple exposed, stalked comb; more variable than in *Polistes*.

EPIPONINI

This large tribe (which is more or less equivalent to the Polybiini *sensu* (Richards, 1978) excluding *Mischocyttarus* and three Old World genera) comprises 22 genera in the New World, 15 of which are represented in Costa Rica. A preliminary phylogeny of the group was presented by Carpenter (1991). The New World genera not represented in Costa Rica are small groups principally confined to the equatorial parts of South America; only one, *Angiopolybia*, is known to have a range extending into Central America, where it is represented in Panama (Richards, 1978).

Agelaia (= *Stelopolybia*). A rather small tropical American genus comprising 22 described species with a cumulative range from Mexico to northern Argentina. Eight species are known to occur in Costa Rica: *areata, cajennensis, multipicta, myrmecophila, pallipes, panamaensis, xanthopus* and *yepocapa*. Nest: usually in cavities, several concentric curving combs joined by pedicels and suspended from above; *A. myrmecophila* may be an obligate associate with *Azteca* ants (Richards, 1978); *A. areata* builds an enveloped nest in exposed sites (Jeanne, 1973).

Apoica. A small tropical genus with a cumulative range extending from Mexico southwards to northern Argentina. Eight species have been described, and three are known to occur in Costa Rica: *pallens, strigata* CR (S) and *thoracica*. *Apoica* species are nocturnal, and individuals can be frequently collected at lights at night. Nest: flat comb in form of inverted saucer or cone, attached (sessile) to small twig or stem (Fig. 15.06).

Brachygastra. A moderately small genus comprising 16 species with a cumulative range that extends from the southwestern United States to Argentina. Five species are known to occur in Costa Rica: *augusti, lecheguana, mellifica, scutellaris* and *smithii*. Nest: each comb covered and fused with envelope. Some species store greater quantities of nectar in the nest than in other polistines hence their common name, 'honey wasps'.

Chartergellus. A small genus containing eight described species that range from southeastern Brazil north to Costa Rica where one species is known to occur: *atectus*. Nest: several combs hang from a leaf or branch by lateral pedicels, parallel to each other, not attached to envelope.

Charterginus. A small genus comprising six species with a cumulative range extending from Honduras south to equatorial South America. Only one species, *C. nevermanni*, is known to occur in Costa Rica, though a second species occurs in Panama (Richards, 1978). Nest: single widely stalked comb surrounded by an envelope (often star-shaped), exit hole beneath (between comb and substrate) (Fig. 16.08).

Clypearia. A small tropical genus comprising seven species with a cumulative range extending from Southern Mexico to Bolivia. Only one species, *C. naumanni*, occurs in Mesoamerica, and it has been found to occur in Costa Rica. Nest: unstalked comb, flat on substrate, covered by envelope.

Epipona. A small genus comprising three widespread species that have a range extending from Mexico south to Bolivia. Two species, *guerini* and *niger*, occur in Costa Rica. Nest: each comb covered and fused with envelope. Envelope of strong cardboard-like material tapering to eccentric entrance at bottom.

Leipomeles. A small genus with two species that cumulatively range from Costa Rica to Brazil. Only one, *dorsata*, has been taken in Costa Rica. Nest: several horizontal stalked combs, which may become fused, on the undersurface of leaves, with a very delicate flattened envelope; exit hole near leaf tip.

Metapolybia. A moderately small genus with 11 described species that have a cumulative range extending from Mexico to Paraguay. Three species occur in Costa Rica: *azteca*, *aztecoides* and *docilis*. Nest: unstalked comb, flat on substrate, covered by envelope, often containing many small transparent 'windows' (Fig. 15.05).

Nectarinella. A small genus that includes a single species, *N. championi* that ranges from Costa Rica south to Colombia. Nest: unstalked comb, flat on substrate, covered by envelope bearing minute hair-like sticky stalks.

Parachartergus. A moderate sized genus comprising 16 described species that have a cumulative range extending from Mexico south to Argentina. Three species occur in Costa Rica: *apicalis*, *aztecus* and *fraternus*. Nest: several combs hang from a leaf or branch by lateral pedicels, parallel to each other, not attached to envelope. Envelope of delicate paper, tapering downward to entrance which is often produced as a tube (Fig. 15.04).

Polybia. A moderately large genus comprising more than 50 described species with a cumulative range extending from the southwestern United States to Argentina. Fifteen described species are known to occur in Costa Rica: *aequatorialis*, *barbouri*, *belemensis*, *diguetana*, *emaciata*, *flavitincta*, *flavifrons*, *nidulatrix*, *occidentalis*, *parvulina*, *raui*, *rejecta*, *scrobalis*, *simillima* and *tinctipennis*. Nest: each comb covered and fused with envelope (e.g. Fig. 15.07); in *P. emaciata* the envelope is made of mud; the large nests (up to 60–70 cm long) of *P. rejecta* are often associated with *Azteca* or *Pseudomyrmex* ants (Richards, 1978; Windsor *in* Janzen, 1983).

Protopolybia (= *Pseudochartergus*). A moderate sized genus containing 28 species with a cumulative range from Guatemala south to Argentina. Seven species are known to occur in Costa Rica: *acutiscutis*, *chartergoides*, *exigua*, *panamensis*, *perfulvula*, *picteti* and *scutellaris*. Nest: one, two or (in *P. acutiscutis*) many combs, each in its own envelope and with its own exit hole.

Pseudopolybia. A small genus with four species that cumulatively range from Nicaragua to Bolivia. Only one species, *P. compressa*, occurs in Costa Rica. Nest: several combs, one below the other, surrounded by an independent envelope.

Synoeca. A small genus with fivc dcscribcd species that cumulatively range from Mexico to Argentina. Two species occur in Costa Rica: *chalibea* CR (S) and *septentrionalis*. Nest: unstalked comb, covered by envelope, flat on substrate, especially on smooth barked trees (Fig. 15.03) such as *Bursera simaruba* (Burseraceae) (Stevens *in* Janzen, 1983). Envelope highest at centre, with marked corrugations, entrance above, slightly produced as short lip.

16 The ants (Formicidae)

John T. Longino and Paul E. Hanson

Fig. 16.01. *Paraponera clavata.*

Diagnosis. Completely winged (males, and at least for a short time after eclosion, queens) or apterous (workers). Head prognathous in female castes (worker and queen); *antennae elbowed, with an elongate basal segment* (scape), female castes with 4 to 12 segments, males with 9 to 13 segments. When wings present, the hind wing generally lacking lobes (except in some Ponerinae) and with 0 to 2 enclosed cells present. Metapleural gland present on mesosoma, opening above hind coxa, sometimes the gland secondarily reduced or lost. *Metasomal segment I always reduced and isolated from the remaining metasomal segments to form a distinct node or scale* (petiole); frequently metasomal segment II also reduced and isolated (postpetiole). Males often bear little resemblance to queens and workers, being smaller, more fragile, and with smaller heads and relatively larger eyes and ocelli.

Classification and distribution. The phylogenetic relationship of the Formicidae is uncertain and authors disagree about where to place the family within the Aculeata. Wilson *et al.* (1967*a* & *b*) placed the Formicidae closest to the methochine Tiphiidae whereas Brothers (1975) and Brothers and Carpenter (1993) place the ants with the Rhopalosomatidae, Bradynobaenidae, Scoliidae and Vespidae, apart from the Tiphiidae, Pompilidae, Mutillidae and Sapygidae (see Chapter 14).

Worldwide the Formicidae comprises 16 subfamilies (Baroni Urbani *et al.*, 1992), about 300 genera, and roughly 8800 species (Hölldobler & Wilson, 1990*a*). In Costa Rica there are seven subfamilies and 81 genera. At the time of this writing the number of Costa Rican species known to the authors is approximately 620. Most future increases in known Costa Rican species-richness will probably come from teasing apart complexes of sibling species, rather than discovering conspicuously distinct species. The four largest subfamilies, in terms of both species-richness and abundance, are the Ponerinae, Myrmicinae, Dolichoderinae and Formicinae. Three smaller subfamilies, the Cerapachyinae, Ecitoninae and Pseudomyrmecinae are also present in Costa Rica. All of these groups are cosmopolitan, except Ecitoninae, which is restricted to the New World.

Local species-richness of ants in Costa Rica, as elsewhere in the tropics, can be quite high. Crowns of large canopy trees can host up to 50 species. The soil and leaf litter on the forest floor is similarly species-rich, and there is almost no overlap in species between the canopy and the forest floor (Longino unpubl.). In a study of Peruvian rainforest canopy, Wilson (1987) reported high within-tree richness (43 species in one crown), but relatively low between-tree heterogeneity. The same is true in lowland rainforest in Costa Rica. Most canopy ant species are widespread and can be repeatedly sampled in trees of many different species. The marked degree of host-plant specificity that contributes to the high diversity of phytophagous insects does not occur in ants. Ant diversity remains high from sea-level up to between 800 and 1000 metres, but drops precipitously above this altitude. At 2500 metres very few species of ants occur and above this elevation they are essentially absent, other than as occasional alates that have flown up from lower elevations.

Ongoing research suggests that many ant species have narrow altitudinal ranges. Collections made along elevational transects reveal numerous cases of altitudinal zonation, particularly in genera such as *Cyphomyrmex, Stenamma, Neostruma, Strumigenys, Hypoponera* and *Pachycondyla*. What previously were thought to be single ant species are being resolved into sets of several cryptic species, segregated by elevation, and with largely parapatric distributions (Longino, unpubl.). This seems to be particularly true of leaf litter ants. Broad biogeographic patterns of ants are similar to those for plants (e.g. Gentry, 1982). For example, many ant species in Guanacaste are at the southern limits of their ranges, which extend northward through Central America to southern Mexico. Some species of the Atlantic lowlands (e.g. *Azteca xanthochroa*) appear endemic to Central America, but have their closest relatives in South America (Longino, 1991). The wet forest of the southern Pacific coast seems even more South American in affinities than the forests of the Atlantic lowlands. The high density and diversity of *Azteca* species, in particular, is more similar to South American forests than to other Costa Rican forests.

A number of ant species are referred to as 'tramp ants'. These are closely associated with human activity and have been inadvertently transported by commerce, so that they now are widely distributed throughout the tropics (and in some cases in centrally-heated buildings in temperate regions). *Tapinoma melanocephalum*, the 'crazy ant', can be found on kitchen tables almost anywhere in the lowland tropics. Other tramp species found in Costa Rica include *Paratrechina longicornis, Monomorium pharaonis, Tetramorium bicarinatum* and *Wasmannia auropunctata* (perhaps the only native tramp). Tramp ants are remarkably quick at colonizing human habitations, even isolated households deep in the forest, yet they are very rarely found in surrounding forested habitats.

Biology. There is a vast amount of literature available on the biology of ants, much of it being summarized by Wheeler (1910), Sudd and Franks (1987), and especially Hölldobler and Wilson (1990*a*). It is only possible to present a brief synopsis of this literature here.

Sociality. Ants are the only family of Hymenoptera where all species are eusocial—i.e. live in colonies of related individuals, have overlapping generations, and exhibit reproductive division of labour—or have secondarily lost sociality by becoming workerless social parasites (see chapter 2.6). Female ants occur as two or more distinct castes, the principal division being between queens and workers. In general, queens have a functional reproductive system, ocelli, and for a short period following eclosion, wings. Workers are generally either sterile or capable of producing only unfertilized, male-producing eggs, often lack ocelli, and never have wings. The degree of differentiation between queens and workers varies considerably. In army ants and some myrmicines the queen is many times larger than workers, whereas in some ponerines, workers and queens are barely distinguishable and differ only in the presence or absence of ocelli. Typically, an enlarged mesosoma with black wing scars on the sides also distinguishes queens from workers.

In many species of ants the workers in a colony are all similar in size and shape, and such species are termed MONOMORPHIC. In other species there are comparatively large and substantially smaller worker individuals (castes) in the same colony. Such caste differentiation is determined by pheromonal control and/or food supply. CONTINUOUSLY POLYMORPHIC species such as *Eciton, Atta, Azteca* and most *Camponotus* exhibit continuous variation in worker size from the smallest (minor) workers to the largest (major) workers. In contrast, DIMORPHIC species, such as *Pheidole*, some *Camponotus* and most *Zacryptocerus*, have two discrete size classes of workers, the minor and major workers, with no intermediates. Size differences between workers is often correlated with division of labour, and in these cases the size variation is also distinctly allometric (Hölldobler & Wilson, 1990*a*). However, there are usually more tasks than there are castes. For example, in leaf-cutting ants (*Atta*) four physical castes perform a total of 20 to 30 tasks.

Worker polymorphisms are found in some species of only 15 percent of the ant genera (Hölldobler & Wilson, 1990*a*). Nearly all Ponerinae are monomorphic, the major exception being the specialized termite predators (Wheeler, 1936), such as *Pachycondyla laevigata* (D. Feener, pers. comm.). Although physically distinct worker castes are relatively uncommon in ants, nearly all species have temporal castes, i.e. the adult workers change roles as they grow older, typically passing from nurses within the nest to foragers outside the nest. Even in polymorphic ants each physical caste often passes through changes in behaviour with ageing. In the case of leaf-cutting ants, for example, three of the four physical castes pass through two temporal castes; thus the total number of castes, physical and temporal, is seven (Wilson, 1980*a*).

A typical colony life cycle begins with a newly mated queen who initiates a colony by herself. Depending on the species, she may periodically forage, or remain permanently sealed in a nest chamber, rearing a first generation of workers on her own body stores of fat and protein. Once workers are produced they take over all duties of brood care, nest construction, and foraging. The colony undergoes a growth phase, during which only workers are added to the population. This growth phase is followed by a reproductive phase, when males and new winged queens are produced. These males and new queens (called alates) fly from the nest and mate during a NUPTIAL FLIGHT, which usually occurs at a particular hour, depending on the species (McCluskey, 1965, 1987). Males die very soon after leaving the nest. The newly mated females generally search for new nest sites, dropping their wings at some point during the process. The season of reproduction varies between species. Some, such as the *Azteca* species that inhabit *Cecropia* trees, produce alates continuously over long time periods, while others, such as *Atta* species, have periodic synchronized flights that are presumably triggered by environmental cues such as rainfall. Within species, there often appears to be great variation in the proportion of males to new queens produced; some colonies will produce almost all males, while others produce almost all females (a phenomenon discussed by Helms & Rissing, 1990).

Most ant species are MONOGYNOUS, with a single, egg-laying queen. A few species are POLYGYNOUS, with more than one dealate, mated queen (not to be confused with multiple unmated alate queens residing in the parent colony prior to the nuptial flight). In general polygyny appears to be correlated with small populations, specialized nest sites, or variable food supply (Hölldobler & Wilson, 1990*a*). In Costa Rica, several species of arboreal, polydomous (see below under nests) ants are polygynous, with queens spread throughout the colony in many nests. UNICOLONIAL species represent an extreme form of polygyny. They have enormous, sprawling populations with no territorial boundaries or obvious limits to the colony. A possible example in Costa Rica is a species in the *Crematogaster parabiotica* species-complex, which in certain sites in the Atlantic lowlands, appears to occur in nearly every tree canopy, with clusters of queens, workers and brood dispersed under epiphyte mats and in dead wood.

An exception to the standard colony cycle is SOCIAL PARASITISM, where one ant species is parasitic on other ant species. In some cases, a parasitic queen enters the colony of a host species, kills the colony queen, and uses the host workers to rear her own brood. In other cases, parasitic species have workers that raid colonies of host species, bringing workers back as slaves. The slaves then work as foragers for the parasite colony, helping to rear the parasites' brood. In general, parasitic species tend to be locally distributed and to exist in small populations, and thus new cases continue to be discovered. In Switzerland, which probably has been more thoroughly studied than anywhere else, one third of the species are parasitic. However, social parasitism is virtually unknown in the tropics, and it is unclear whether this absence is real or simply due to lack of study.

In contrast to obvious social parasitism, cohabitation of the same nest (termed PARABIOSIS by Forel, 1898; see also Swain, 1980; Wheeler, 1986*a*) and use of the same foraging trails (*Camponotus beebei* and *Azteca chartifex*—Wilson, 1965; *Zacryptocerus maculatus* and *Azteca*—Adams, 1990) are common in the tropics. The exact nature of these relationships, whether mutualistic, parasitic or commensalistic, is unknown.

Nests. An ant nest is not the same as an ant colony, and the distinction is particularly important for tropical ant faunas. A colony is a genetically defined unit composed of the ants themselves: one or more queens and their descendants, functioning as a family unit, and maintaining their identity with respect to other such units (except in unicolonial species). In contrast, a nest is a physical structure that ants use for shelter. This distinction is important because of POLYDOMOUS species—those where a colony is distributed in more than one nest. A nest of a polydomous colony contains only a portion of the colony, sometimes only workers, sometimes workers and brood. Often in large polydomous colonies the outlying nests are relatively conspicuous and contain only workers and brood, while the nest with the colony queen is very well hidden and/or highly defended. In Costa Rica polydomous colonies occur in some species of *Crematogaster, Azteca, Dolichoderus, Camponotus* and *Myrmelachista*, at least one species of *Solenopsis* and in *Wasmannia auropunctata*.

Nesting behaviour is extremely varied in ants, and is an understudied aspect of ant biology. Differences in nesting habits may be far more striking than the morphological traits that distinguish species and genera. Species that are difficult to distinguish by standard morphological traits may be easy to identify in the field, based on nesting habits. For example, *Azteca* workers are notoriously difficult to distinguish by morphology, yet some are *Cecropia* specialists, some nest in live stems of other plants, some nest in dead stems, and some build external carton nests.

The simplest nests are pre-existing cavities in soil, leaf litter, or dead wood. Some *Cyphomyrmex* species nest between two dead leaves in the leaf litter. Species of *Hypoponera, Gnamptogenys, Pheidole, Strumigenys* and many others, occupy cavities left by beetles and other burrowing insects in rotten logs. Highly mobile species of *Paratrechina, Tapinoma* and *Camponotus* will readily move into a closed umbrella or a box of clothes left on a porch overnight. Other ants excavate their own nests in the soil. For many *Pheidole*, the nest is a single gallery leading to a simple chamber a few centimetres below the ground. A nest of *Atta cephalotes* or *Megalomyrmex modestus* consists of many galleries leading to many different chambers. The excavated soil may be distributed so as to leave no sign of the entrance hole, or it may be piled in variously shaped cones, craters, or mounds many metres across, as produced by *Atta* species (Fig. 16.02).

Arboreal ants either occupy pre-existing cavities excavated by stem-boring insects or they excavate their own cavities. Species that nest in live stems have a range of specificity, some using a number of different plant species and others only a single plant species—the most specific ants often forming ant–plant associations (see below). Species of *Azteca, Camponotus, Crematogaster* and *Pheidole* use construction materials to form walls, galleries and nests (Longino, 1986). In wet forest understorey, some species of *Pheidole, Crematogaster* and *Cyphomyrmex* form very fragile

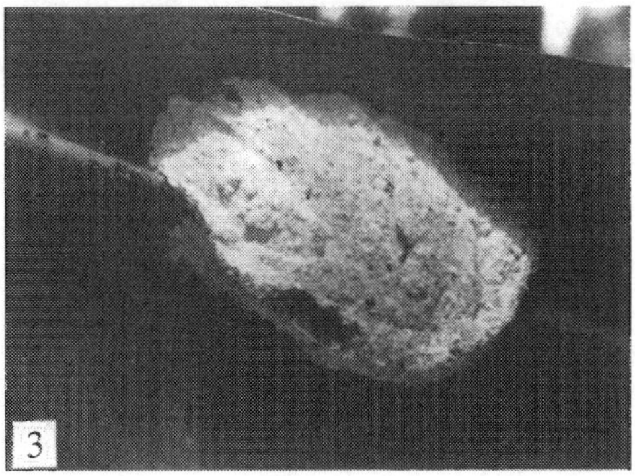

Fig. 16.02–16.03. Nests of the Attini. Fig. 16.02. Part of the site occupied by the entrance to a nest of *Atta cephalotes* in La Selva. Note the metre rule in the foreground. Fig. 16.03. A nest of *Apterostigma* sp. on the underside of a leaf in La Selva. The entire nest, made from fungus, is around 5cm in length.

nests from accreted soil. Several species of *Apterostigma* nest in trees, their fungal gardens hanging under leaves (Fig. 16.03) and branches, in cavities on tree trunks, or in the stilt roots of palms. In more exposed areas, higher in the canopy and at forest edges some ants, such as species of *Azteca* and *Crematogaster*, manufacture a more resistant carton of chewed plant fibres. The carton varies between ant species in coarseness, thickness, texture, colour and strength. *Dendromyrmex* species and *Camponotus senex* reinforce their carton with larval silk, making a very strong material (Schremmer, 1979; Wilson, 1981; Hölldobler & Wilson, 1983).

Ants also vary greatly in how they use carton. Some, such as *Crematogaster torosa*, use it only for partitions inside plant stems, and often for baffles at the nest entrance. Others, such as *C. stolli* and *Azteca foreli*, build extensive gallery systems over trunks and branches, connecting dispersed nest cavities and/or sheltering Homoptera. Some species build exposed nests entirely out of carton. In Costa Rica, *Azteca schimperi* and a species in the *A. chartifex* species-complex build massive nests suspended from branches of canopy trees. Species of *Tapinoma*, *Paratrechina* and *Pheidole* build tiny carton nests clinging to the undersides of individual leaves. Some ant species tolerate and/or foster epiphytes growing on their carton nests, and in some cases the epiphyte species are specialized inhabitants of these 'ant gardens' (Wheeler, 1921; Weber, 1943; Davidson, 1988; Davidson & Epstein, 1989). Highly developed ant gardens in Costa Rica are formed by:

1) an *Azteca* species near *uli*, possibly *gnava*;
2) an undescribed *Crematogaster*;
3) a parabiotic association between a *Crematogaster* species in the *parabiotica* complex and *Odontomachus panamensis*;
4) a recently discovered *Pheidole* in Monteverde.

In other cases, especially in the wet forest understorey, ant carton is colonized by a diverse assemblage of unspecialized epiphytes (Longino, 1986). This type of unspecialized or diffuse use of ant carton is exemplified by the nests of *Crematogaster longispina* (Kleinfeldt, 1978).

Foraging. Although a foraging function is general to all ant workers, there is great variety in when and how they forage, what foods they take, and how specific they are. Each ant species has a daily foraging schedule, some being primarily diurnal, others nocturnal. *Paraponera clavata* for example tends to be mostly a nocturnal forager, although there is some activity during the day (Harrison & Breed, 1987). Two different daily patterns have been observed in *Azteca alfari* (Barnwell, 1967). In the leaf-cutting ant, *Atta cephalotes*, diel foraging pattern is determined by the activity of phorid flies—the ants forage less when the phorid parasitoids are most active (Wetterer, 1990*a*; Orr, 1992; Feener & Brown, 1992).

Ant species use a variety of procedures for discovering (hunting) and retrieving food (Traniello, 1989). These procedures may be understandable in terms of optimal foraging theory (specifically, central place foraging) which assumes that foraging behaviour is shaped by the need to maximize net energy yield. Foraging ants also need to avoid predators, especially visually hunting vertebrates. The diverse foraging strategies of ants can be classified by the particular combination of hunting, retrieving and defence that is employed (Hölldobler & Wilson, 1990*a*):

Hunting: 1) by solitary workers, 2) by individual workers following trunk trails, 3) by groups of workers searching in concert.

Retrieving: 1) by solitary workers who return to the nest on their own, 2) by individual workers who return along persistent trunk trails, 3) by individual workers recruited to the food sites by scouts that lay down chemical trails (see Chapter 2.6), 4) by groups of workers who gang-carry the food items.

Defence: 1) by guard workers during hunting, 2) by guard workers during harvesting and retrieval of food (e.g. in *Atta*, the small 'hitchhiking' caste that rides on leaf fragments being carried back to the nest in order to defend against phorid parasitoids—Feener & Moss, 1990; see Chapter 2.7).

The kinds of foods ants use vary from species to species. Many are generalized scavengers, feeding mainly on dead insects and various sugar sources. The sugar sources include extrafloral nectaries on plants, and exudates (honeydew) from Homoptera (in the tropics, primarily Coccoidea, Membracidae and Aetalionidae). Omnivorous scavenger species are the ants most often encountered at baits of tuna or honey. More specialized diets are sometimes reflected in modifications of the mandibles. A common modification in predatory species is for the mandibles to be greatly elongate with fork-like teeth at the apex, and with the ability to snap shut like a trap (Moffett, 1986). Costa Rican ants with snapping mandibles include *Acanthognathus*, *Anochetus*, *Odontomachus* and *Strumigenys* (Brown & Wilson, 1959; Carlin & Gladstein, 1989). Another modification is seen in some seed-harvesting ants, in which the major workers have short, thick mandibles that they use to crack seeds. In Costa Rica, *Solenopsis geminata* is a facultative seed harvester, and the major workers have seed-milling mandibles.

Carnivorous ants that capture live prey include generalized predators, which take a wide range of prey species and sizes, and highly specialized predators, which attack only a narrow range of prey (Hölldobler & Wilson, 1990*a*). Costa Rica hosts a rich assortment of prey specialists (Table 16.1), and undoubtedly many more await discovery and documentation. A good example is provided by *Leptogenys*, which includes at least three Costa Rican species that specialize on isopods: *ambigua*, *punctaticeps* and cf. *imperatrix*. The last species is particularly conspicuous since it commonly nests in vertical banks along road cuts, and the ejected isopod shells form a grey streak below the nest entrance.

Many specialized predators forage solitarily, but others attack *en masse*. Groups of up to 100 workers of *Gnamptogenys bispinosa* simultaneously attack large millipedes many times their size, each grasping a leg of the millipede and thus immobilizing it. *Pachycondyla laevigata* and *Simopelta* species are convergent with army ants (see below), attacking their prey (other social insects) in large, co-ordinated raids (Wheeler, 1936; Gotwald & Brown, 1967).

Army ants. Army ants have been called 'the Huns and Tartars of the insect world' (Wheeler, 1910), 'comparable to a wolf pack, but with fifty thousand miniature wolves' (Rettenmeyer *in* Janzen 1983). Also known as legionary ants, the army ants include three subfamilies: the Dorylinae and Aenictinae of the Old World, and the Ecitoninae of the New World. They are characterized by Hölldobler and Wilson (1990*a*) thus:

1) frequent and orderly colony migration;
2) group foraging and raiding (i.e. individual workers move exclusively in groups);
3) permanently wingless queens that have reduced eyes, strong legs, and a massive abdomen.

The Ecitoninae, which occurs from the southern United States to northern Argentina, comprises about

Ant species	Prey	Source
PONERINAE		
Belonopelta deletrix	Campodeidae, Japygidae, small Chilopoda	Wilson, 1955
Discothyrea spp.	arthropod eggs	Brown, 1979
Gnamptogenys bispinosa	Myriapoda	pers. obs.
Gnamptogenys hartmani (= *turmalis*; J. Lattke, pers. comm.)	nests of Attini	pers. obs.; Brown, 1961; Mackay & Vinson, 1989
Gnamptogenys regularis	other ants (?)	pers. obs.
Leptogenys spp.	terrestrial Isopoda	pers. obs.
Pachycondyla laevigata	Isoptera	Wheeler, 1936; pers. obs.
Prionopelta amabili	Campodeidae, Geophilomorphidae	Wilson & Hölldobler, 1986
Proceratium spp.	arthropod eggs	Brown, 1979
Simopelta spp.	*Pheidole* colonies	Gotwald & Brown, 1967; pers. obs.
Thaumatomyrmex spp.	Polyxenidae	Brandao *et al.*, 1991
CERAPACHYINAE		
Cerapachys spp.	Ants, esp. *Pheidole*	Hölldobler, 1982
Cylindromyrmex spp.	Isoptera	Overal & Bandeira, 1985
ECITONINAE		
Neivamyrmex spp.	other ants	Schneirla, 1971; pers. obs.
MYRMICINAE		
Smithistruma spp.	Collembola	Wilson, 1953
Strumigenys spp.	Collembola	Wilson, 1950, 1953
Trichoscapa membranifera	Collembola	Wilson, 1953

Table 16.1. Ants that are actual or likely specialized predators in Costa Rica.

150 species in five genera (Kempf, 1972; Watkins, 1976). Army ant colonies are large, ranging from about 30,000 to more than one million workers, but with only one queen (Rettenmeyer & Watkins, 1978). Workers are weakly to strongly polymorphic and in large species, such as *Eciton burchelli*, there are soldiers with long, hook-shaped mandibles. Smaller species (less than 5 mm long) usually lack a distinct soldier caste. Instead of a permanent nest, army ants form a massive cluster of ants—a BIVOUAC—which reassembles daily (in nocturnal species) or nightly (in diurnal species), and is located either above ground (epigaeic species) or below ground (hypogaeic species). Epigaeic species form hanging chains of ant bodies, linked to one another by their strong tarsal claws, inside hollow trees, between buttresses, under logs, etc. It should be noted, however, that most species of Ecitoninae are cryptic and subterranean (e.g. *Neivamyrmex*, the largest genus).

Colonies of *Eciton* and at least some other army ant species follow an endogenous cycle whereby the behavioural and reproductive cycles are closely synchronized (Schneirla, 1971). In the best studied species, *E. burchelli*, the statary phase, when the colony remains in one bivouac site, coincides with transformation of larvae into pupae, and egg-laying by the queen. The migratory phase, when the colony changes bivouac site at the end of each day's foraging, coincides with the emergence of new adult workers and the hatching of new larvae. Emigration to a new bivouac site occurs around dusk, when the workers stop bringing back food, and begin carrying larvae and accumulated food to a new site (usually 50 to 70 metres distant). After most of the larvae have been transported (by about 20:00–22:00 hours) the queen makes the journey, surrounded by a large retinue of soldiers and unladen smaller workers (Rettenmeyer *et al.*, 1978).

Army ant species differ in their temporal foraging patterns (i.e. diurnal versus nocturnal foragers) and spatial foraging patterns. *E. burchelli* is a swarm raider, that is workers spread out into a fan-shaped swarm with a broad, carpet-like front some 10–15 metres wide. *Labidus praedator* is also a swarm raider but the front is usually less than four metres wide. Most other army ant species are column raiders, pushing outward along narrow dendritic odour trails. Swarm raiders prey on a wide variety of arthropods, including other social Hymenoptera, and particularly large-bodied prey such as scorpions, spiders, cockroaches and orthopterans. Column raiders tend to specialize on other ants, but they sometimes raid social wasp or termite nests.

Army ants have the fastest recruiting system known for any social insect and can recruit a hundred or more workers within one minute (Chadab & Rettenmeyer, 1975). The workers carry the prey under their bodies while running rapidly on their long legs. Much of the captured prey is kept in temporary caches along raid columns and is carried to the bivouac later. In *E. burchelli* teams of large workers carry large prey items very efficiently, since carrying an intact prey item by a group requires fewer workers than would dismembering the same prey item and carrying the pieces individually (Franks, 1986). During an average raid by this species 30,000 prey items are carried back to the bivouac, from an average distance of about 100 metres (Franks, 1982).

A raid by *E. burchelli* can often be heard—the rustling of thousands of ants, the sound of frantic insects trying to escape, and the twittering of antbirds. Numerous birds (Willis & Oniki, 1978) and parasitoid flies (Conopidae and Tachinidae—Rettenmeyer, 1961; Phoridae—B.V. Brown, pers. comm.) follow the swarm, attacking insects flushed out by the advancing ants. Ithomiine butterflies may also follow the swarm to feed on the droppings of antbirds (Ray & Andrews, 1980). In rural houses army ant raids can be a welcome intrusion, since they flush out vermin such as cockroaches and scorpions. In the wake of an army ant raid the forest floor may be left depleted of leaf-litter arthropods (Otis *et al.*, 1986). It has been suggested that army ants selectively remove other ant colonies, thereby maintaining various stages of ant succession and hence preserving species diversity (Franks & Bossert, 1983). Finally, there is an incredible diversity of obligate and facultative arthropod symbionts associated with army ant colonies (see Chapter 2.7). Thus, because of their numerous ecological interactions, army ants are often considered to be 'keystone' species in tropical ecosystems.

Colony multiplication in *Eciton* species occurs annually, during the dry season or early wet season. A brood is produced that consists of larvae destined to become new queens (a few) and up to 4,000 males. By the time these larvae are mature the bivouac is divided into a brood-free zone, containing the old queen and her affiliated workers, and another zone that contains the sexual brood and their affiliated workers. The new queens emerge first, followed by the males. The latter fly off in search of another colony. The colony then splits, with the old queen and her workers moving off in one direction, and the new queens with their workers in another direction. Only one of the new queens is allowed to enter the new bivouac. Queen longevity appears to be at least four years, but in some cases the original queen is replaced by a new queen, probably when she becomes senescent. In either case, the end result is the division of the original colony into two colonies. The new queens mate with males from other colonies within a few days of emergence and it is possible that the old queens mate annually (Rettenmeyer, 1963*a*). Workers may be involved in sexual selection, favouring males that are superficially similar in size and shape to their conspecific queen (Franks & Hölldobler, 1987). Sexual selection by workers may also explain why the males are so well endowed with metasomal glands (Hölldobler & Engel-Siegel, 1982) and why males lose their wings upon entering an alien colony (Schneirla, 1971).

Fungus-growing ants. The cultivation of fungus is a specialized feeding behaviour exhibited by ants of the tribe Attini (subfamily Myrmicinae). This group of 11 genera and about 200 species is restricted to the New World, primarily in tropical and subtropical regions (Weber, 1972, 1982). Rather than foraging directly for food attine ants forage for substrate on which to grow fungus. They carry the substrate to the nest, where other workers incorporate it into the fungus gardens, which serve as the food supply for the colony. Fungus-growing ants differ greatly in colony size and in the types of substrate they use. A *Cyphomyrmex* nest may be confined between two leaves in the leaf litter with a fungus garden growing on two or three caterpillar droppings and some beetle wing covers. In contrast, a mature nest of the leaf-cutting ant, *Atta cephalotes*, is evident as a large bare area (250 m^2 or more in surface area) on the forest floor, and may contain 500 to 1000 underground chambers and five million workers. Leaf-cutting ants are capable of harvesting up to 17 percent of leaf production in a forest and thus have a significant impact on the vegetation, nutrient cycling, etc. (Cherrett, 1989).

Fungus-growing ants live in an obligate mutualistic association with a fungus. Worker ants harvest the specialized swollen hyphae of the fungus (GONGLYDIA or STAPHYLAE) and feed these to the larvae. The fungus in turn depends upon the ants for dispersal. A new queen starts her fungus garden from a fragment of fungus, carried from her natal nest in her infrabuccal chamber (behind the base of the labium). Because the fungus does not produce spore-bearing structures, there is still some question as to its identity, but it appears to belong to the basidiomycete family Agaricaceae and is variously placed in *Agaricus*, *Lepiota*, *Leucoagaricus* and *Leucocoprinus* (for a review see Wetterer *in* Hunt & Nalepa, 1994). The evolutionary origins of fungus cultivation by Attini is still unresolved, but two possibilities have been suggested. Garling (1979) proposed that the ants originally fed on mycorrhizal fungi, whereas Hölldobler and Wilson (1990*a*) suggest it originated by fungi colonizing refuse or stored food in ant nests.

Ecologically and economically, the most important fungus-growing ants are the leaf-cutting ants, which belong to the genera *Acromyrmex* and *Atta* (Lofgren & Vander Meer, 1986; Vander Meer *et al.*, 1990). Like several other genera of Attini leafcutters recruit nestmates to foraging sites, but unlike other Attini they show extreme worker size-polymorphism and extensive division of labour (Wilson, 1980*a* & *b*; Wetterer, 1990*b*, 1991*a*). Among the species occurring in Costa Rica, *Atta cephalotes* cuts leaves from large trees in prolonged mass attacks lasting up to several days, and foraging ants vary greatly in size. On the other hand, *Acromyrmex octospinosus* cuts leaves from small herbs and scavenges fallen leaves and other plant parts. Compared with *A. cephalotes*, its foragers tend to be larger and have a narrower range of body mass (Wetterer, 1991*b*). *Acromyrmex volcanus* appears to be quite similar to *A. octospinosus*, except that foragers are even larger (Wetterer, 1993).

In Costa Rica colonies of *Atta* use up to 77 percent of the plant species growing in the area (Rockwood, 1976; Blanton & Ewel, 1985; Rockwood & Hubbell, 1987). Of those plants that are used, not all are equally exploited, and often a particular plant species may be attacked only during a certain season (Fowler & Robinson, 1979). Furthermore, leafcutters often utilize only certain individuals of a plant species, or only specific leaves of an individual plant (Cherrett, 1968; Rockwood, 1976; Nichols-Orians & Schultz, 1990). Leaf-cutting ants appear to be capable of making subtle distinctions in leaf quality, and when preferred leaves decline in abundance the colony may switch to less-preferred leaves (Shepherd, 1985). Colonies are more selective of plants the more distant they are from the nest, although near the nest they sometimes do not harvest leaves from certain, apparently palatable species (Rockwood & Hubbell, 1987).

Studies on the effects of leaf water content have produced conflicting results, varying from a positive correlation between water content and palatability (Cherrett, 1972; Stradling, 1978; Fowler & Robinson, 1979; Bowers & Porter, 1981), to no correlation (Howard, 1987, 1988), to a negative correlation (Blanton & Ewel, 1985). Leaf water content probably covaries with other plant chemicals and colonies probably vary seasonally in their water requirements. In general *Atta cephalotes* seems to prefer young leaves over old leaves (Cherrett, 1972; Rockwood, 1975; Littledyke & Cherrett, 1978), soft leaves over tough leaves (Cherrett, 1972; Nichols-Orians & Schultz, 1990; but see Howard, 1988), leaves without epiphylls (liverworts, lichens, etc.) over leaves with epiphylls (Mueller & Wolf-Mueller, 1991), and sun leaves over shade leaves (Hubbell & Wiemer *in* Jaisson, 1983). Thus leafcutters usually do more damage in early successional habitats (Blanton & Ewel, 1985) and are generally located close to gaps in the forest (Jaffe & Vilela, 1989). In areas with pronounced seasonality maximum foraging intensity often coincides with peak production of new leaves and flowers (Rockwood, 1975).

The fungal symbionts of leaf-cutting ants secrete enzymes capable of breaking down various phenolic compounds present in leaves, such as hydrolysable tannins (Powell & Stradling *in* Huxley & Cutler, 1991), but leaves with very high levels of condensed tannins are avoided by the ants. The unequal exploitation of individuals of a particular plant species may result from intraspecific differences in condensed tannin concentrations. This, in turn, may be mediated by variation in resource availability—i.e. leaves that develop when growth is nutrient-limited tend to have very high tannin levels (Nichols-Orians, 1991). Other secondary plant compounds such as terpenoids may also be detrimental to the fungus and repellant to the ants (Vander Meer *et al.*, 1990). In addition, leaf-cutters appear to prefer leaves without latex over those with latex (Stradling, 1978; Hubbell *et al.*, 1984).

In several studies leaf-cutting ants have been shown to harvest substrates that are toxic to their fungus (Mullenax, 1979; Bueno *et al. in* Vander Meer *et al.*, 1990). Since adult ants imbibe sap while cutting leaves (Littledyke & Cherrett, 1976), and use the sugars thus obtained to satisfy their own energy requirements (Quinlan & Cherrett, 1979), host plant selection may include gustatory responses by the ants. Knapp, Howse and Kermarrec (*in* Vander Meer *et al.*, 1990) have clarified the complexities involved in the host selection process by distinguishing four categories of rejection response in *Acromyrmex octospinosus* colonies:

1) Plants such as *Peperomia* are inspected but not cut, suggesting odour-mediated rejection.
2) On plants such as *Annona squamosa* and *Cucurbita moschata* cutting begins but is not completed, which suggests rejection mediated by gustatory stimuli.
3) With other plants there is a delayed rejection, suggesting that the rejected plant is not repellent to the ant, but rather is unsuitable for the fungus. In banana, for example, recruitment ceases with time, but the rejection is short-lived and foraging resumes each time the plant is encountered in the future.
4) In other plants, such as sweet potato, the rejection is delayed but persistent, suggesting avoidance conditioning.

These results help explain why different authors may reach different conclusions about the palatability of a given plant species.

To reach favoured plants leaf-cutting ants follow trunk trails and branch trails, which encompass a foraging radius of up to 250 metres in *A. cephalotes* (Lewis *et al.*, 1974). Nestmates are recruited with a pheromone, 4-methylpyrole-2-carboxylate (Moser & Blum, 1963; Vilela *et al.*, 1987), and unladen workers clear conspicuous trunk trails ('highways') along which the ants can move unhindered (Fowler, 1978; Shepherd, 1982). Foraging ants sometimes drop leaf pieces from the tree to the ground, or at the junctions of new branch trails and established trunk trails, where they are picked up by other workers (Hubbell *et al.*, 1980). The colony itself appears to have an auxiliary labour force that is effectively on a stand-by basis as active foragers (Wilson, 1983). A given colony may forage primarily at night or during the day, but daily rhythms often vary between colonies, and colonies can change rhythms over time (Lewis *et al.*, 1974). Such foraging patterns are often affected by the presence of phorid parasitoids (Feener & Moss, 1990; Wetterer, 1990*a*; Orr, 1992; Feener & Brown, 1992).

Once the pieces of leaves have been passed to smaller worker castes within the nest, they undergo a rather elaborate process of preparation before being incorporated into the fungal gardens. This process consists of cutting the leaves into smaller pieces, removal of the superficial wax layer, and then maceration of the leaf tissue (Quinlan & Cherrett, 1977). Once prepared in this manner the worker ant applies some of its own faecal material to the leaf tissue and incorporates it into the fungal garden. The faeces contain protease enzymes, which the ant acquires by consuming fungus and which are not broken down in the ant's digestive system. The ant thus distributes the enzyme from where it is produced to where it is most needed (Boyd & Martin, 1975; Martin *et al.*, 1975; Martin, 1987).

During the course of maintaining the fungal gardens, workers of *Atta sexdens* add secretions from their metapleural glands to the gardens (Schildknecht *et al.*, 1973). These secretions contain several compounds, one of which (indoleacetic acid) is believed to serve as a stimulant to mycelial growth (Schildknecht & Koob, 1970), another (phenylacetic acid) is an inhibitor of bacterial growth (Maschwitz *et al.*, 1970), and a third (D-3-hydroxydecanoic acid) is an inhibitor of germination of alien fungal spores (Schildknecht & Koob, 1971).

The life cycle of a colony of leaf-cutters begins with the departure of winged males and virgin queens from the parent nest. In *Atta cephalotes* this occurs just before dawn, usually at the beginning of the rainy season. During their nuptial flight the corpulent queens ascend to a great height where they mate with perhaps three to eight males (Kerr, 1962*b*). Following this flight the queen casts off her wings and begins digging a nest in the soil. In a small chamber she spits out a wad of fungal mycelium and 'fertilizes' it with her own excrement. The first brood of larvae are fed special eggs produced by the queen. Mortality during the nuptial flight and immediately afterwards, as the queens attempt to found new colonies, is very high, about 90 percent in *A. cephalotes*. Much of this seems to be due to pathogenic fungi (Stevens *in* Janzen, 1983).

Once adult workers begin foraging for fungal substrate, the young queen ceases attending the brood and the fungal gardens. In these early stages, the nest of *A. cephalotes* is characterized by having an entrance in the form of a turret built of mud globules. In three or four years the colony attains the size of a mature colony although the production of reproductives may be less (Stevens *in* Janzen, 1983). As the colony matures the production of males and virgin queens begins and colony growth rate tapers off. Maximal life span of a colony appears to vary from seven to twenty years, depending on the site.

Ant–plant associations. A variety of phenomena traditionally have been grouped under the subject of ant–plant associations (Huxley & Cutler, 1991). These associations are typically potential or known mutualisms, and display apparent special adaptation by the plants, by the ants, or both. The ants benefit by receiving food and/or a nest site while the plants benefit by receiving protection or seed dispersal. However, in some cases only the ants benefit from the association.

Seed harvesting (seed predation) by ants is more common in deserts and dry grasslands than it is in tropical rainforests (Hölldobler & Wilson, 1990*a*). In Costa Rica *Solenopsis geminata* and some species of *Pheidole* occasionally carry seeds back to their nests, and if some are lost before being fed to the larvae, seed dispersal may result. A less fortuitous type of seed dispersal occurs in plants having seeds that bear fat bodies (ELAIOSOMES) or other ant-attractant properties. Ants harvest the seeds, chew off the fat body, and eventually discard the intact seed. Such specialized seed dispersal by ants is known as MYRMECOCHORY (Beattie, 1985; Davidson & Epstein, 1989). In southern Mexico, seeds of *Calathea* (Marantaceae) are dispersed by the primarily predatory ants, *Odontomachus* and *Pachycondyla* (Horvitz & Beattie, 1980) and in Panama, seeds of *Chrysothemis friedrichsthaliana* (Gesneriaceae) are removed from the plant by *Azteca* spp., *Paratrechina* sp. and *Pheidole* sp. (Lu & Mesler, 1981). Some neotropical legumes may also be specialized for dispersal by ants (McKey, 1989*a*).

Many plants have extrafloral nectaries on leaf blades, petioles, flower bases, or around fruits (Bentley, 1977), and/or produce lipid-containing 'pearl bodies' from leaf blades or stems (O'Dowd, 1982). These traits are thought to be special adaptations that reduce herbivory by attracting ants. At least in the case of aggressive species, the ants act as 'pugnacious bodyguards' around nectaries, driving off or preying upon herbivores (e.g. Smiley, 1986; but see Hespenheide, 1985 and Koptur, 1985). In tropical dry forests in Mexico ants use extrafloral nectaries primarily in the dry season (Rico-Gray, 1993).

Cases in which ants show apparent special adaptation for particular plants usually involve nest sites. Ants that nest in live stems (or other live structures) of plants show a range of behavioural specializations for plants. Although a continuum of specialization probably exists, the most specialized species are the most conspicuous, and have traditionally received most attention by naturalists. PLANT-ANTS are those ant species that inhabit only one or a few species of plants, and rarely or never forage off their host plant. MYRMECOPHYTES are those plant species that have evolved specialized structures that house ants, or, in the absence of known specialized structures, are nevertheless regularly inhabited by one or more plant-ants. The diversity of both plant-ants and myrmecophytes appears to be greater in the Neotropics than in other tropical regions, and the most important force driving the evolution of ant-plant symbioses appears to be competition among arboricolous ants (see Davidson & McKey, 1993). In Costa Rica, these ant-plant associations include eight sets of myrmecophytes:

1. *Cecropia*. In Costa Rica three species of *Cecropia* (Urticaceae), common trees in forest gaps and second growth forest, are specialized myrmecophytes (Burger, 1977; Longino, 1989*a*, *in* Huxley & Cutler, 1991). They have hollow stems that house ants, and they produce food bodies (MULLERIAN BODIES) from specialized patches of trichomes (TRICHILIA) at the base of each petiole. There are five species of *Azteca* that are obligate inhabitants of *Cecropia* in Costa Rica, and they vary in nest structure and behaviour, some being more aggressive toward intruders than others (Longino, *in* Huxley & Cutler, 1991). In addition to the obligate *Azteca* ants, Costa Rica also has an undescribed species of *Pachycondyla* that is a specialized inhabitant of *Cecropia* saplings in forested areas throughout the country. The association between ants and *Cecropia* has apparently been lost on most Caribbean islands and in cloud forest *Cecropia*, such as *C. polyphlebia* in Costa Rica (Janzen, 1973).

Earlier studies have shown that in some cases *Cecropia* ants clip tendrils of vines that touch the plant (Janzen, 1969), reduce damage to the leaves by herbivores (Downhower, 1975), and enhance growth and survivorship of saplings (Schupp, 1986). Other studies have failed to show protection of *Cecropia* by ants (e.g. Andrade & Carauta, 1982), and there is considerable debate in the older literature on whether the association is a mutualism or not (Bequaert, 1922; Wheeler, 1942). This confusion is probably due to a failure to appreciate differences in aggressiveness between ant species.

2. *Acacia*. In Central America several species of *Acacia* (Leguminosae) are specialized myrmecophytes (Janzen, 1974) supporting aggressive species of *Pseudomyrmex* as obligate inhabitants (Ward *in* Huxley & Cutler, 1991). These acacia trees are characteristic of the Pacific lowlands, both the dry habitats in the northwest and the wet forest of the Osa Peninsula. They have large thorns, which ants hollow out and inhabit. They produce two food sources for ants, extrafloral nectar from nectaries situated on the leaf rachis and specialized food bodies (BELTIAN BODIES) from the tips of leaflets. Brown (1960) made a strong case based on comparative data that the swollen thorn acacias evolved in the context of a mutualism with ants, and Janzen (1966*a*) experimentally demonstrated mutualism between *Pseudomyrmex ferrugineus* and an *Acacia* species at a site in southern Mexico. All species of the *P. ferrugineus* group are obligate inhabitants of swollen-thorn acacias. This group occurs from Mexico to northern Colombia and comprises ten species, four of which are reported from Costa Rica (Ward, 1993):

P. flavicornis—northwestern Costa Rica; *A. collinsii*, less frequently *A. cornigera* and *A. hindsii*.

P. nigrocinctus—northwestern Costa Rica; *A. collinsii*, *A. cornigera* and *A. hinsii*.

P. particeps—southwestern Costa Rica; *A. allenii*.

P. spinicola—western Costa Rica; *A. collinsii*, less frequently *A. allenii* and *A. cornigera*; biology described by Janzen (*in* Janzen, 1983, as *ferruginea*).

Two nonaggressive species belonging to other species groups of *Pseudomyrmex* also regularly occur in swollen-thorn acacias in Costa Rica. *P. nigropilosus* is a parasite of the mutualism in that it occupies uninhabited acacias without protecting the plant (Janzen, 1975). *P. subtilissimus* likewise does not protect the plant and appears to be an inquiline on acacias occupied by *P. flavicornis* (Ward, 1989).

3. *Piper*. Three species of myrmecophytic *Piper* (Piperaceae)— *P. cenocladum*, *P. fimbriulatum* and *P. sagittifolium*—occur in the understorey of lowland wet forest throughout Costa Rica (Burger, 1971; Risch *et al.*, 1977; Letourneau, 1983). They have clasping petioles which form chambers, and ants inhabit these swollen petioles and cavities which the ants excavate in the stem. The plants produce food bodies on the inside walls of the petiolar chambers. Only one ant species, *Pheidole bicornis*, is known to be a specialized plant-ant in *Piper*, and the occupancy rate in these myrmecophytic *Piper* is nearly 100 percent. A clerid beetle may be a parasite on this mutualism (Letourneau, 1990).

4. *Cordia alliodora*. The laurel, *Cordia alliodora* (Boraginaceae), is a myrmecophytic tree that occurs throughout lowland Costa Rica, in both dry and wet forest. The tips of stem internodes, where whorls of branches emerge, are swollen and hollow, and ants inhabit these chambers. There are no known food bodies. Specialized plant-ants of *Cordia* in Costa Rica include an unresolved species-complex of small, black *Azteca*, and *Zacryptocerus setulifer*. Colonies of *Azteca* and *Zacryptocerus* often coexist on the same tree.

5. *Triplaris melaenodendron*. The myrmecophytic tree *Triplaris melaenodendron* (Polygonaceae) is widespread throughout Pacific slope dry forest (Brandbyge, 1986). Its stems are hollow and harbour ant colonies, but there are no known food bodies or nectaries. Common plant-ant inhabitants of this tree in the lowlands are *Pseudomyrmex viduus* and *Azteca beltii*, both of which are generalists in myrmecophytes (see below). Mid-elevation *Triplaris* are frequently inhabited by *Azteca longiceps*, which may be a specialist in *Triplaris*. *Triplaris* trees are also commonly inhabited by generalized arboreal ants.

6. Melastomataceae. Various myrmecophytic species of Melastomataceae occur in wet forests throughout Costa Rica. All are small understorey plants which have myrmecophytic structures, such as leaf or petiole pouches, but do not appear to provide any food bodies or nectaries. They lack specialized plant-ants, and are instead inhabited by a range of opportunistic species. One melastome species, *Conostegia setosa*, is most often occupied by generalist *Pheidole* spp. (including *P. bilimeki*), several small species of *Solenopsis*, and the generalist *Wasmannia auropunctata* (L. Tennant, pers. comm.).

7. Lauraceae and Meliaceae. Several species of understorey and subcanopy Lauraceae (*Licaria*, *Ocotea*) and Meliaceae (*Guarea*) are inhabited by a diverse radiation of specialized *Myrmelachista* (Stout, 1979; Hammel, 1986; Burger and Werff, 1990; see synopsis of genera below), and in some locations by *Azteca*. Occupation of particular *Ocotea* species by *Myrmelachista* is so dependable that presence or absence of ants is used in keys to *Ocotea* species (Hammel, 1986; Burger and Werff, 1990). This is one of the most common ant-plant associations in Costa Rican primary wet forest, and yet has been largely overlooked because the plants lack obvious myrmecophytic traits, and the ants are timid and generally remain inside the stems. This may be a case where ants show special adaptations for particular plants, but not vice versa.

8. *Tetrathylacium costaricense*. The tree, *T. costaricense* (Flacourtiaceae), is a small, weakly myrmecophytic plant that occurs in the rainforests of southwestern Costa Rica (L. Tennant, pers. comm.). The stems are hollow and split when mature, allowing a range of opportunistic arboreal species to nest inside, including several *Azteca* species that build carton trails along the branches.

Only *Acacia*, *Cecropia* and *Piper* directly provide their ants with a conspicuous food source. Chambers of ant-occupied *Cordia* and *Triplaris* are always packed with coccoid Homoptera, presumably the ants' main food source since workers do not forage off the tree. The ants in melastomes, being non-specific, are generalist foragers on leaf surfaces and around the plant. The ants of *Tetrathylacium* are likewise generalist foragers, although the *Azteca* presumably also maintain Homoptera in the stems. The food of *Myrmelachista* in *Ocotea* and *Guarea* is a mystery. The stems of these plants are packed with a very high density of workers and brood. Workers rarely venture outside of the stems, and only once have workers been seen foraging off a plant (a short column leading from the base of the tree to a dead moth). The stems harbour mealybugs, but their abundance seems far too low to support the population of workers. The plant must be directly supporting the ants in some way, but the mechanism has not been investigated.

The degree to which a myrmecophyte benefits by the presence of ants depends on the aggressiveness of the ant species inhabiting the plant. All these myrmecophytes, including the highly specialized *Acacia*, *Cecropia* and *Piper*, may occasionally and in some circumstances frequently be inhabited by generalized, opportunistic species of arboreal ants. Two ant species appear to be specialized plant-ants, but generalized in their use of several myrmecophytes. *Pseudomyrmex viduus* has been collected from *Triplaris*, *Ocotea* (Lauraceae) and *Cordia alliodora*. *Azteca beltii* has been collected from *Triplaris melaenodendron*, *Cordia alliodora* and *Cecropia peltata*.

In addition to the examples given above, another important group of myrmecophytes include various epiphytic plants (Davidson & Epstein, 1989). Ant-epiphyte associations are based on: seed dispersal of epiphytes by ants, the use of carton nests constructed by ants as a substrate for epiphyte growth, extrafloral nectaries, and/or hollow cavities in epiphytes that house ants. Although a number of epiphytes exhibit some ant-associated traits, they often have no specific relation to a particular ant species. For example, the Costa Rican fern *Solanopteris brunei* (Polypodiaceae) has swollen secondary rhizomes that may be inhabited by ants, although six different species have been recorded in these rhizomes and some individual plants did not host ants (Gómez, 1974). In tropical Asia, in similar ant/fern associations, ant-derived nitrogenous substances have been shown to be incorporated into the fern tissues (Gay, 1993). In Panama the orchid, *Caularthron bilamellatum*, produces extrafloral nectaries and has pseudobulbs occupied (in 85 percent of the plants examined) by ants, mostly *Azteca velox* (Fisher, 1992).

Economic importance (see also Chapter 3). Ants, through their omnipresence and their diverse ecological roles, as predators, scavengers, herbivores and mutualists, are arguably the dominant group of animals in neotropical lowland ecosystems. It is likely that most plants and animals are affected, directly or indirectly, by their presence. Thus it is not surprising that ants are a very significant factor in tropical agriculture, though the magnitude of their impact, both beneficial and detrimental, is probably not yet fully appreciated. Leaf-cutting ants (*Atta* and *Acromyrmex*) are the most obvious damaging species in agroecosystems, but other ant species that protect and tend homopterous pests can also promote economic damage. For example, *Solenopsis geminata* and *Pheidole* species protect a mealybug that is the vector of a serious virus disease in pineapple.

A few ants, most notably the 'bullet ant' (*Paraponera clavata*) and the 'fire ant' (*Solenopsis geminata*), are capable of inflicting very painful stings, although these are usually not life threatening. Several ant species, including the pantropical tramp species (see above), are common in buildings. In houses these ants are merely annoying, but in hospitals they may spread contaminants. In a Costa Rican candy factory the presence of *Solenopsis geminata* resulted in a shipment of products being rejected in quarantine, before reaching a foreign market, which was extremely costly for the factory (P. Hanson, unpub.). Similar problems have occurred with alate *Camponotus* on ships being loaded with bananas. Another species of *Camponotus* (subgenus *Tanymyrmex*) sometimes utilizes electronic equipment as a nest site, and by so doing, recently damaged hundreds of facsimile machines in Costa Rica.

Identification. Keys to the subfamilies and genera of neotropical ants, and references to taxonomic revisions are given in Hölldobler and Wilson (1990*a*) and Bolton (1994), and it is from these works that the keys given in this chapter are derived.

Key to the subfamilies of Formicidae present in Costa Rica

1 Second metasomal segment much smaller than and/or highly differentiated from third (i.e. with a postpetiole) (Figs 16.04, 16.05); never with a row of short, peg-like teeth on each side of the last visible metasomal tergite (pygidium). 2

— Second and third metasomal segments of similar size and shape, and often no constriction between them (i.e. without postpetiole) (Figs 16.06–16.09) ***or if*** second segment somewhat reduced (i.e. difficult to determine if it is a postpetiole) ***then*** pygidium with a row of peg-like teeth (Fig. 16.10). 4

2 Eyes absent or represented by a single ommatidium; antennal sockets not at all concealed by frontal lobes (Fig. 16.11). .... **Ecitoninae** (p. 609)

— Eyes usually present, with many ommatidia; if eyes reduced then antennal sockets at least partially covered by frontal lobes (Fig. 16.12). 3

3 Promesonotal suture absent; eye length along longest axis usually much less than 40 percent of head length, measured from vertex to anterior margin of clypeus (Fig. 16.05); posterior margin of clypeus usually projecting back between antennal sockets (Fig. 16.12); antenna with 12 or fewer segments. ....**Myrmicinae** (p. 609)

— Promesonotal suture present; eye length 40 percent or more of head length (Fig. 16.04); posterior margin of clypeus not projecting back between antennal sockets; antenna always with 12 segments. ....**Pseudomyrmecinae** (p. 609)

4 Sting present; gaster fairly elongate and cylindrical, often with its first segment slightly constricted posteriorly so that it is somewhat separated from remaining segments (Fig. 16.08). 5

— Sting absent; gaster short and generally subspherical, never with its first segment separated from remaining segments (Figs 16.06, 16.09). 6

5 Pygidium with a row of erect peg-like teeth on each side, the two rows converging posteriorly toward the sting (Fig. 16.10).

[Antennal sockets nearly touching; postpetiole fairly distinct; rare.] .... **Cerapachyinae** (p. 607)

— Pygidium lacking peg-like teeth (sometimes with stiff, short setae). .... **Ponerinae** (p. 601)

6 Apex of metasoma with a semicircular to circular acidopore formed from last sternite (hypopygium), often projecting as a small cone or nozzle and fringed with a ring of diverging setae (Fig. 16.09); ***if*** difficult to observe, ***then*** antennal sockets separated from clypeus (Fig. 16.13); sour taste of formic acid in live specimens. .... **Formicinae** (p. 618)

— Apex of metasoma without acidopore; antennal sockets always touching clypeus (Fig. 16.14); live specimens, when crushed, often with characteristic odour of rotten cheese. ....**Dolichoderinae** (p. 616)

Synopsis of Costa Rican genera

The ant genera known to occur in Costa Rica are listed below. Genera that have not yet been seen in Costa Rica but may possibly occur in the country include *Acanthostichus* (South America, Mexico), *Cheliomyrmex* (South America, northern Central America), and *Probolomyrmex* (Panama and South America).

PONERINAE

A large subfamily comprising, worldwide, about 50 genera classified in six tribes (Bolton, 1994). All six tribes and 21 genera are known to occur in Costa Rica. Ponerines are mostly predaceous, solitary hunters (no recruitment), with small colonies. It is a species-rich group with many small cryptic species in leaf litter and large conspicuous species on vegetation.

PONERINAE: Amblyoponini. Comprises seven extant genera, two of which, *Amblyopone* and *Prionopelta*, occur in Costa Rica.

Amblyopone. A cosmopolitan genus, three species of which occur in Costa Rica, *A.* cf. *degenerata*, *A. mystriops* and *A. orizabana*. The first two are known only from single sites in the Atlantic lowlands, but the third is more widespread, though rare. Taxonomy: Lattke (1991). Biology: these ants live in small colonies in leaf litter and are predaceous.

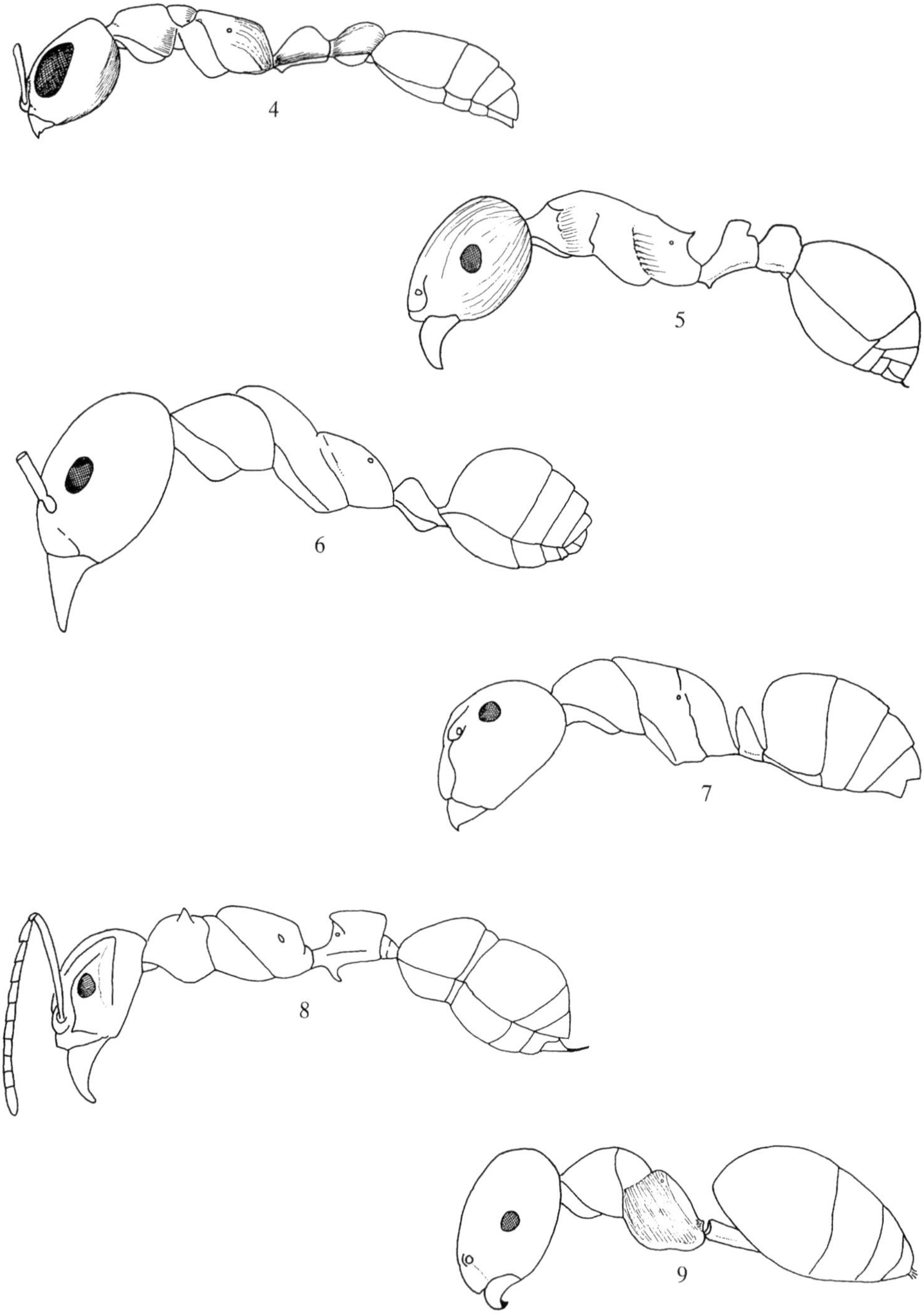

Figs 16.04–16.09. Formicidae. Figs 16.04–16.09. Head and body, lateral; 16.04, *Pseudomyrmex* sp. (Pseudomyrmicinae); 16.05, *Tetramorium* sp. (Myrmicinae); 16.06, *Azteca* sp. (Dolichoderinae); 16.07, *Camponotus* sp. (Formicinae); 16.08, *Paraponera* sp. (Ponerinae); 16.09, *Brachymyrmex* sp. (Formicinae).

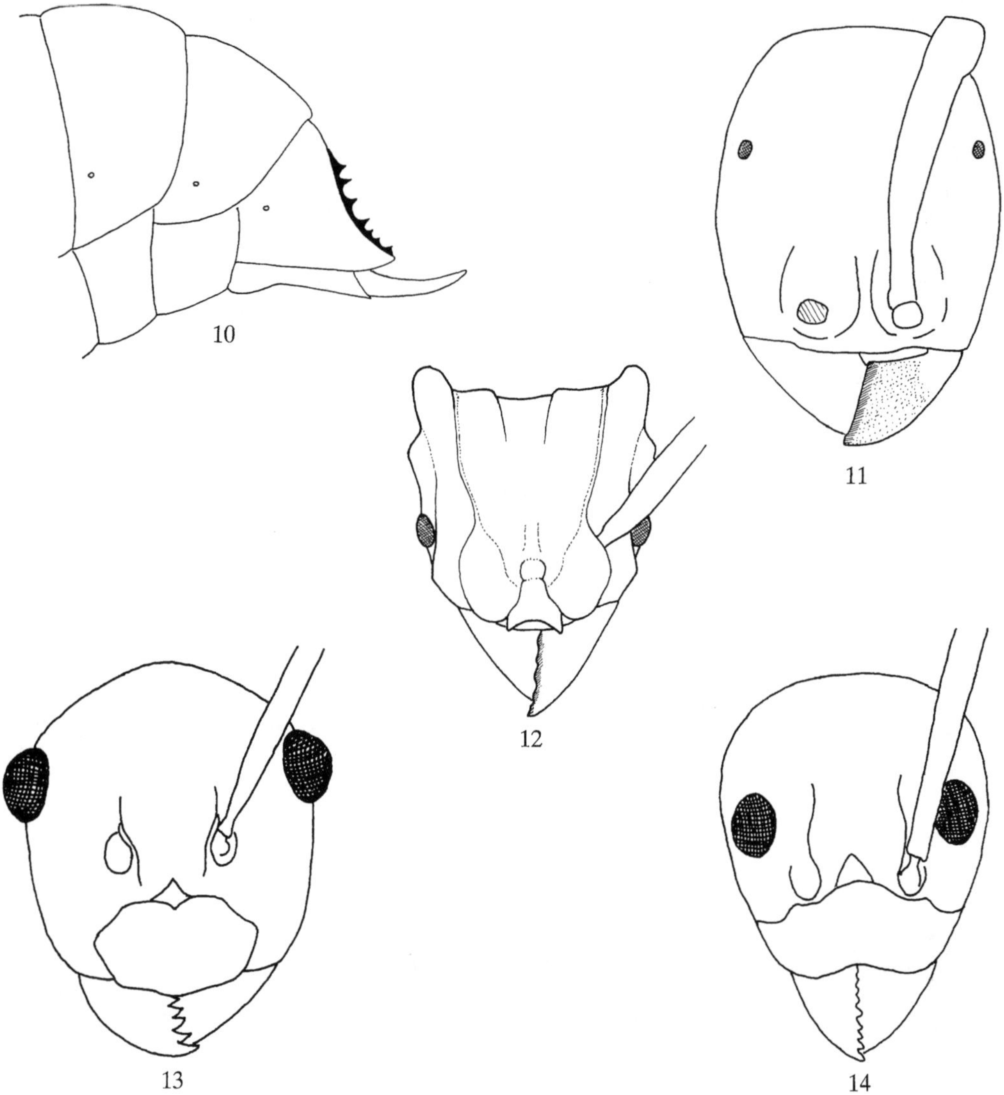

Figs 16.10–16.14. Formicidae. Fig. 16.10. Apex of metasoma of Cerapachyinae. Figs 16.11–16.14. Head, anterior; 16.11, Ecitoninae; 16.12, *Cyphomyrmex* sp. (Myrmicinae); 16.13 *Dendromyrmex* sp. (Formicinae); 16.14, *Dolichoderus* sp. (Dolichoderinae).

Prionopelta. Pantropical; in the New World, extending north to the southern United States; two common and widespread species occur in Costa Rica. They are yellowish and characterized by the denticulate median lobe of the clypeus, mandibles with three teeth and the small eyes (Fig. 16.16). Biology: predaceous ants living in large colonies in leaf litter. The pantropical *P. amabilis* nests in rotting logs, preys on campodeid diplurans, and shows temporal division of labour (Hölldobler & Wilson, 1986).

PONERINAE: Ectatommini. Comprises nine extant genera, seven of which occur in Costa Rica.

Acanthoponera. A neotropical genus represented in Costa Rica by apparently two species. *A. minor* is

Figs 16.15–16.20. Ponerinae. Figs 16.15–16.17. Head, anterior; 16.15, *Anochetus* sp.; 16.16, *Prionopelta* sp.; 16.17, *Thaumatomyrmex* sp. Figs 16.18–16.20. Head and body lateral; 16.18, *Ectatomma* sp.; 16.19, *Hypoponera* sp.; 16.20, *Pachycondyla* sp.

present on the Pacific side and *A. crassa* (which is currently treated as a junior synonym of *minor* but is probably distinct) is present on the Atlantic side; nowhere are these sympatric. Biology: common, widespread ants in wet forest habitats. The nests are in the ground, and workers forage nocturnally on low vegetation. Their feeding habits are unknown.

Discothyrea. A primarily tropical and southern temperate genus, with at least three species present in Costa Rica. They appear to be rare, but are widespread. Biology: small colonies live in leaf litter. The single North American species is a predator of the eggs of other arthropods (Brown, 1958).

Ectatomma. A neotropical genus represented in Costa Rica by four species (e.g. Fig. 16.18). Two, *E. ruidum* and *E. tuberculatum*, are exceedingly common; *E. gibbum* is less common, being present in primary forest; *E. edentatum* is at the northern limit of its range in the highlands of the Osa Peninsula (and perhaps also in Talamancas). Biology: *Ectatomma* species live in small colonies, nest in the ground, and forage on the ground and on vegetation. They are predaceous but also visit extrafloral nectaries. *E. tuberculatum* often preys on individually foraging social Hymenoptera. References to biology: *ruidum*—Breed *et al.* (1990), Lachaud (1990); *tuberculatum*—Wheeler (1986*a*), Dejean & Lachaud (1992).

Gnamptogenys. Asia, Australia and the Neotropics (one species reaches the southern United States); at least 21 species occur in Costa Rica. Biology: common, widespread ants that live in small colonies in the ground, leaf litter and live or dead stems. They are predaceous, with some species specializing in millipedes, and some group raiding.

Heteroponera. An Australian/neotropical genus. Only one species, *H. panamensis*, has been found in Costa Rica. Biology: common in mid-upper elevation wet forest (1000–2000 m) where it lives in small colonies, most often in the live stems of understorey plants.

Paraponera. A neotropical genus with only one species, the notorious *P. clavata* ('giant tropical ant', 'bala') (see Fig. 16.01). Biology: very common in Atlantic slope wet forest, but it is conspicuously absent from similar habitats on the Osa Peninsula. These ants live in large colonies, often in the ground at bases of trees having buttresses and extrafloral nectaries, especially *Pentaclethra macroloba* (Leguminosae) (Bennett & Breed, 1985; Breed & Harrison, 1989; Hölldobler & Wilson, 1990*b*). Workers forage arboreally, mostly nocturnally (McCluskey & Brown, 1972; Hermann, 1975; Young & Hermann, 1980). They are predaceous, but also visit extrafloral nectaries. Trail pheromones are used mostly for orientation of individual foragers, but recruitment is possible if a trail-laying ant enters the colony (Barrett *et al.*, 1985; Breed & Bennett, 1985; Breed *et al.*, 1987; Breed & Harrison, 1989); temporal learning has been demonstrated (Harrison & Breed, 1987). Alarmed workers release an odour that attracts phorid parasitoids (Brown & Feener, 1991*a* & *b*). More details of biology: Jorgensen *et al.* (1984); Hermann & Young (1980); Breed & Harrison (1988); Breed *et al.* (1991); Kannowski (1991); Nelson *et al.* (1991).

Proceratium. Cosmopolitan; perhaps most species-rich in warmer parts of the North Temperate region; only two species have been found in Costa Rica. Taxonomy: Ward (1988). Biology: small colonies in leaf litter, and are reputedly specialist predators on spider eggs.

PONERINAE: Platythreini. A small tribe comprising two genera of which one, *Platythyrea*, occurs in Costa Rica.

Platythyrea. Pantropical; in the New World, extending north to the southern United States; four species are known from Costa Rica. Body without erect hairs. Biology: common and frequently widespread predaceous ants that live in small colonies. They are usually arboreal and most often encountered running on tree trunks, although *P. punctata* often nests in dead wood on the ground (R. Snelling, pers. comm.). One species (*P. prizo*) described by Kugler (1976), is still known only from La Selva Biological Station.

PONERINAE: Ponerini. A large tribe with 23 genera worldwide, nine of which occur in Costa Rica.

Anochetus. A pantropical genus with at least five species in Costa Rica (Snelling, pers. comm.) of which three are nominate (one of which is a complex of species). These ants have long, slender mandibles, abruptly bent inward near the apex, like *Odontomachus* spp. (Fig. 16.15). Biology: *A. emarginatus* occurs in understorey ant gardens on trunks; the *A. mayri* complex are moderately common ants living in small colonies in leaf litter or under epiphyte mats; *A. orchidicola* is a rare species found under epiphyte mats. They are all widespread in wet forest, and assumed to be predaceous.

Belonopelta. A neotropical and oriental genus with a single very rare species (cf. *deletrix*) in Costa Rica. Biology: individuals have been found in leaf litter.

Centromyrmex. A pantropical genus with only one species, *C. alfaroi* known to occur in Costa Rica. It is very rarely collected. Biology: it has been found in soil.

Cryptopone. A widespread genus that is most species-rich in southeast Asia. In the New World a single species, *C. gilva*, is known to occur in North America, and it has been found to occur as far south as Costa Rica. Biology: common in cloud forest (absent in lowlands?) where small colonies live in rotten wood.

Hypoponera. A large, cosmopolitan genus with at least ten, and perhaps many more, species occurring in Costa Rica (e.g. Fig. 16.21). Biology: very common, widespread predaceous ants living in small colonies in leaf litter and under epiphyte mats.

Leptogenys. Primarily pantropical; in the New World extending north to the southern margin of the United States; at least five species are present in Costa Rica. Tarsal claws finely pectinate. These are apparently rare, but widespread ants. Biology: live in small colonies in leaf litter and in dead wood. They are predaceous, with some specialized predators of isopods (Wheeler, 1904).

Odontomachus. Mainly pantropical; in the New World, extending north to the southern United States; ten species have been found in Costa Rica. They may be very common, and are widespread. Biology: live in small to large colonies in leaf litter, rotten wood, suspended arboreal debris and epiphyte clumps. They forage on the ground and/or arboreally. They are predaceous, using their long snapping mandibles for defence and for subduing prey (Garlin & Gladstein, 1989). Individuals are occasionally found at extrafloral nectaries.

Pachycondyla. A large, predominantly pantropical genus; in the New World extending north to the southern United States; at least 20 species occur in Costa Rica (e.g. Fig. 16.20). Biology: very common and widespread ants living in small to large colonies in leaf litter, dead wood, dead or live stems or under epiphyte mats. They are predaceous, with one species, *P. laevigata*, a specialized group raider of termites. Occasional individuals visit extrafloral nectaries.

Simopelta. Neotropical; at least five species present in Costa Rica in mid-upper elevation wet forest (1000–2000 m), including cloud forest where they can be locally common. Biology: form large colonies, and are nomadic group raiders on *Pheidole* colonies.

PONERINAE: Thaumatomyrmecini. A small neotoropical tribe comprising one genus, *Thaumatomyrmex*. It is known to occur in Costa Rica.

Thaumatomyrmex. Neotropical; two very rare species in Costa Rica. Mandibles with bizarre, extremely long, slender teeth (Fig. 16.17). Biology: widespread wet forest ants that live in small colonies in the leaf litter. They are specialized predators of polyxenid millipedes (Brandao *et al.*, 1991).

PONERINAE: Typhlomyrmecini. A small neotropical tribe comprising a single genus, *Typhlomyrmex*. It is known to occur in Costa Rica.

Typhlomyrmex. Neotropical; three species in Costa Rica. Biology: *T. rogenhoferi* is common and widespread, living in small colonies in rotten wood. Species of this genus are predaceous.

- –Tarsal claws simple, lacking a median tooth .... *Neivamyrmex*
- –Tarsal claws with a single median tooth
  - –Width of scape near apex more than 1/3 its entire length (Fig. 16.21)........ *Nomamyrmex*
  - –Width of scape near apex less than 1/3 its length (Fig. 16.22).
    - –Propodeum with teeth or lamellae (Fig. 16.23)........ *Eciton*
    - –Propodeum without teeth or lamellae........ *Labidus*

Table 16.2. Simple diagnostic characteristics of workers of the genera of Ecitoninae occurring in Costa Rica (adapted from Hölldobler & Wilson, 1990*a*).

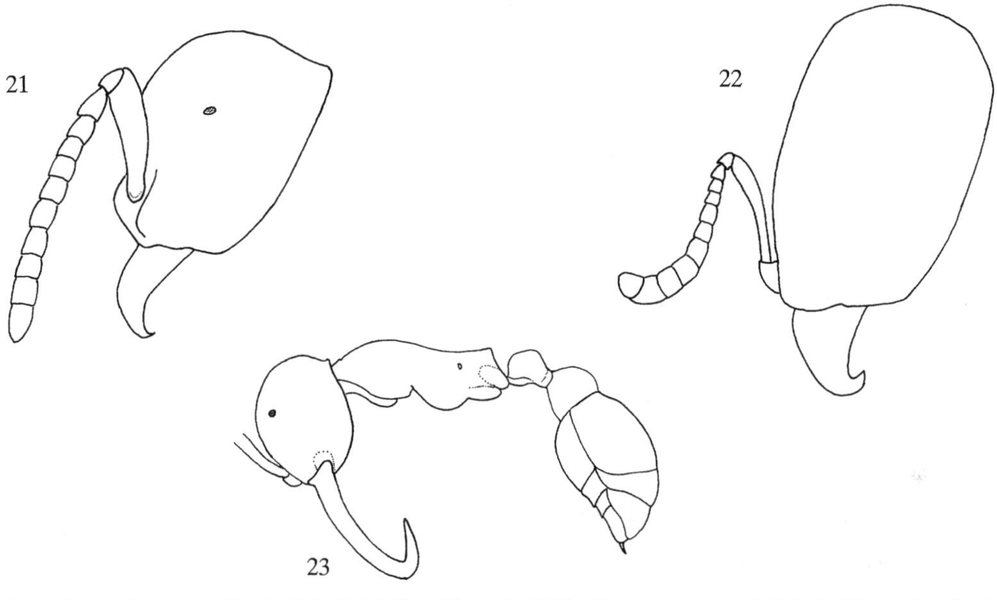

Figs 16.21–16.23. Ecitoninae. Figs 16.21–16.22. Head, lateral; 16.21, *Nomamyrmex* sp.; 16.22, *Labidus* sp. Fig. 16.23. Head and body, lateral, *Eciton* sp.

CERAPACHYINAE (BOLTON, 1990*a*)

A small, predominantly pantropical subfamily with seven genera worldwide. It is represented in Costa Rica by two genera classified in separate tribes the Cerapachyini and Cylindromyrmecini respectively (Bolton, 1994).

Cerapachys. Predominantly pantropical; in the New World extending as far north as the southern United States. At least two species occur in Costa Rica. Gena margined ventrally by a carina. Biology: rarely encountered in wet forest habitats in leaf litter.

Cylindromyrmex. A neotropical genus with at least three species present in Costa Rica. Gena without carina. Biology: apparently very rare, live in dead wood.

- Polymorphic, with discrete minor and major castes; antennae with 10 (minute arboreal species), 11 (one uncommon species), or more commonly 12 segments, apical three segments always forming a differentiated club; mandible triangular with 3rd tooth from apex disproportionately small relate to first two; minor workers with short frontal carinae and no antennal scrobes (majors of a few species have scrobes); eyes usually well-developed; promesonotum lacking spines; propodeum usually with pair of spines, but may be reduced to tubercles or ridges, rarely totally absent; propodeal suture always distinct; dorsal profile of mesosoma not flat nor evenly arched; petiole with distinct dorsal node and anterior peduncle ..... *Pheidole*
- Not fitting above description
  - Antennal scrobes present, running longitudinally below eyes (Fig. 6.24) ..... BASICEROTINI
  - Antennal scrobes absent, or present and running longitudinally above eyes, or present with eye situated at posterior apex of scrobe
    - Antenna 6-segmented ..... DACETONINI
    - Antenna 7-segmented ..... *Tatuidris*
    - Antenna 9-segmented ..... *Oligomyrmex* (part)
    - Antenna 10-segmented (with 2-segmented club) ..... *Solenopsis, Carebarella, Xenomyrmex* (part)
    - Antenna 11-segmented
      - Mandible elongate. sublinear with inwardly directed teeth at apex (cf Fig. 16.25) ..... *Acanthognathus*
      - Mandible subtriangular
        - Postpetiole articulated on dorsal surface of first gastral segment; gaster in dorsal view heart-shaped and capable of reflection over mesosoma (Fig. 16.28) ..... *Crematogaster*
        - Antennal scrobes distinct; sculpture of head and mesosoma usually very coarse; frontal carina often greatly expanded, covering gena in front view; entirely arboreal in stems Fig. 16.33) ..... CEPHALOTINI
        - Surface texture usually dull and opaque; promesonotum with spines or tubercles (Figs 16.29, 16.30), or body covered with dense long hairs (Fig. 16.34); nests in litter, soil, or under epiphytes ..... ATTINI
        - Not fitting above descriptions
          - Propodeum with teeth
            - With antennal scrobes (Fig. 16.31) ..... *Wasmannia Lachnomyrmex*
            - Without antennal scrobes ..... *Oligomyrmex* (part), *Leptothorax* (part)
          - Propodeum without teeth ..... *Tranopelta, Xenomyrmex* (part)
    - Antenna 12-segmented
      - Antenna with 3-segmented club; sculpture finely granular and opaque; lacking erect setae; postpetiole grossly swollen ..... *Cardiocondyla*
      - Not fitting above description
        - Frontal lobes not projecting over base of mandible (Fig. 16.27)
          - Petiole as in Fig. 16.35; hind tibial spur pectinate ..... *Hylomyrma*
          - Petiole shaped differently; hind tibia spur simple or lacking
            - Propodeal suture not impressed (Fig. 16.32)
              - Clypeus forming thin projecting ridge anterior to antennal socket ..... *Tetramorium*
              - Clypeus not as above ..... *Rogeria, Leptothorax* (part)
            - Propodeal suture impressed
              - Antennal club 2-segmented (Fig. 16.27) ..... *Adelomyrmex*
              - Antennal club 3-segmented ..... *Monomorium, Megalomyrmex*
              - Antennal club 4-segmented ..... *Stenamma*
              - Antennal filiform, without club ..... *Aphaenogaster*
        - Frontal lobes projecting over base of mandible (Fig. 16.26) ..... *Stegomyrmex*

Table 16.3. Simple diagnostic characteristics of workers of the genera of Myrmicinae occurring in Costa Rica (adapted from Hölldobler & Wilson, 1990*a* and Bolton, 1994).

ECITONINAE (BOLTON, 1990*b*)

This subfamily comprises the New World army ants (see above). It includes five genera, four of which are known to occur in Costa Rica. These may be distinguished using Table 16.2.

Eciton. An exclusively neotropical genus with at least seven species present in Costa Rica. They are very common throughout the country. Biology: *Eciton* species forage on the ground surface and up into vegetation. Workers of the *E. burchelli* complex are largely black, and very common diurnal swarm raiders. *E. hamatum* workers are bright yellow-orange, and also very common, diurnal column raiders. The remaining species forage mainly at night, and are apparently less common.

Labidus. A small mainly neotropical genus. One widespread species, *L. coecus*, occurs as far north as the southern United States. Three or four species occur in Costa Rica, and they can be very common throughout the country. Biology: *L. praedator* has workers that are all black. It is the most common swarm raider and is active both day and night. A second, less common species, *L. spininodis*, differs only in having a ventrally projecting spine beneath the petiole. *L. coecus* has bright red workers. It is very common and usually reported as nocturnal, although an underground raid on other ant nests has been described from a coffee plantation during daylight hours (Perfecto, 1992). It is more omnivorous than other army ants, and often forages in houses for kitchen scraps.

Neivamyrmex. Argentina to Iowa; over 100 described species, with more than 20 in Costa Rica, a few of which are quite common. Biology: *N. pilosus* workers are black, diurnally active ants that are common in lowland (0–500 m) habitats. *N. sumichrasti* workers are dark brown, diurnally active ants that are common in cloud forests. The remaining species (workers often reddish) forage mainly at night and are rarely encountered. They are column raiders.

Nomamyrmex. A very small mainly neotropical genus, with one species (*esenbeckii*) reaching the extreme south of Texas. Two species occur in Costa Rica. Biology: *N. esenbeckii* is common throughout the country. It is a diurnally active column raider. It is the only army ant that has been observed attacking large *Atta* colonies, and all observed *N. esenbeckii* prey has been *Atta* brood (Longino, unpubl.).

PSEUDOMYRMECINAE (WARD, 1989, 1990)

A rather small subfamily comprising three genera, one in the Old World and two in the New World. *Myrcidris* contains a single species known only from Brazil, while *Pseudomyrmex* contains many species and is widespread.

Pseudomyrmex. Primarily neotropical (north to the extreme south of the United States); nearly 200 described species; at least 50 species are present in Costa Rica and many are common. Biology: typically timid ants with small colonies living in dead or live stems, foraging solitarily for live prey and visiting extrafloral nectaries. One species is known to tear open the leaf mines of a hispine beetle and predate the beetle larvae (Memmott *et al.*, 1993). A few species of *Pseudomyrmex* are obligate plant-ants in *Acacia*, forming large colonies and aggressively defending their host plant.

MYRMICINAE

A very large and diverse subfamily comprising about 150 extant genera that have been divided into about 20 tribes, though not all genera can easily be assigned to a tribe. Indeed, Hölldobler and Wilson (1990*a*) remark that 'the tribal classification of the Myrmicinae is still in a state of severe disarray'. However, Bolton (1994) tentatively places all extant myrmicine genera into tribes, and for ease of reference this classification is used below. Currently 42 myrmicine genera are known to occur in Costa Rica.

MYRMICINAE: Agroecomyrmecini. This small tribe comprises *Tatuidris* plus three extinct genera.

Tatuidris. A neotropical genus with *T. tatusia* and at least one undescribed species in Costa Rica. Biology: very rarely encountered in forest leaf litter.

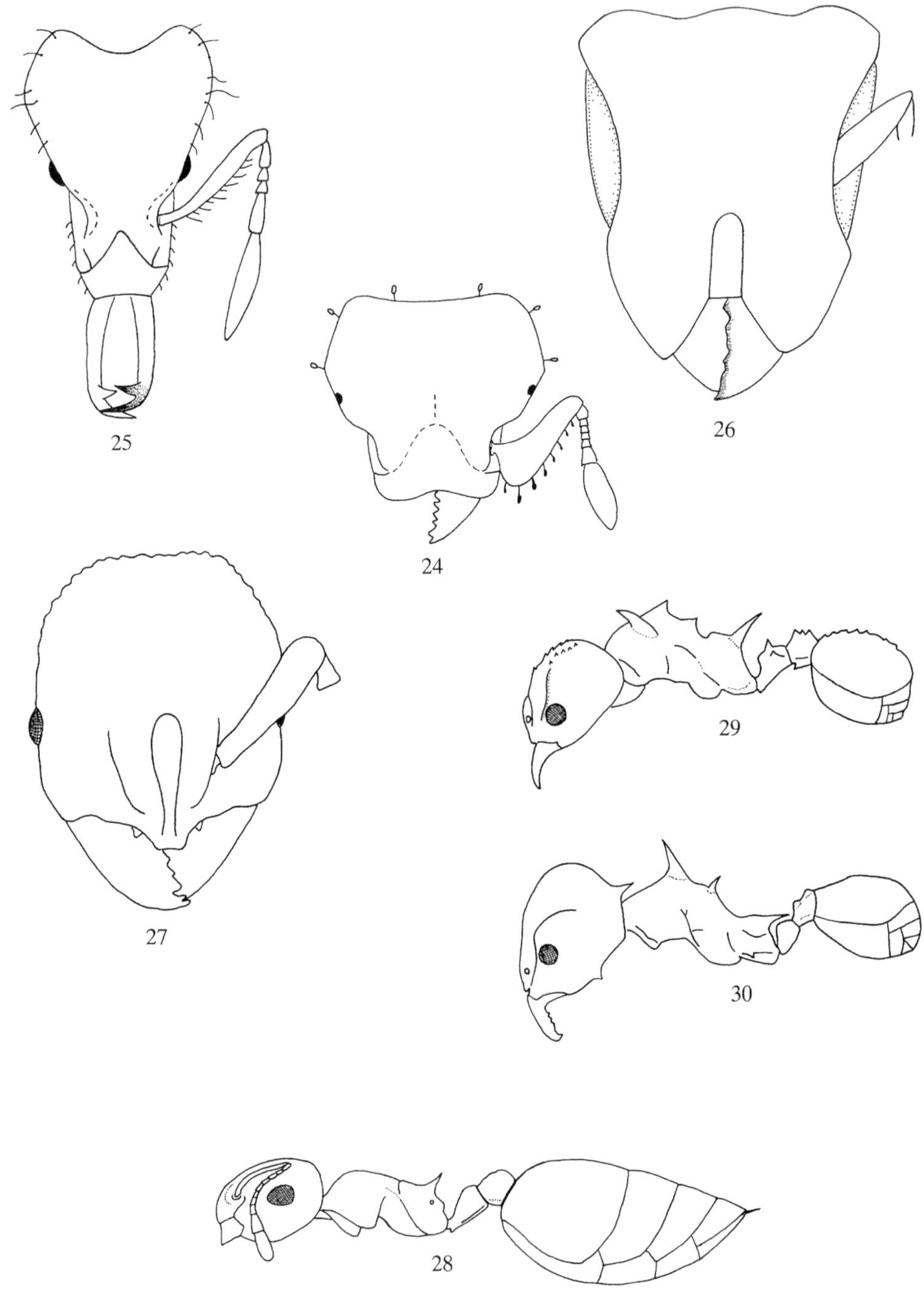

Figs 16.24–16.30. Myrmicinae. Figs 16.24–16.27. Head, anterior; 16.24, *Octostruma* sp. (Basicerotini); 16.25, *Strumigenys* sp. (Dacetini); 16.26, *Stegomyrmex* sp.; 16.27, *Adelomyrmex* sp. Figs 16.28–16.30. Head and body, lateral; 16.28, *Crematogaster* sp. (Crematogastrini); 16.29, *Acromyrmex* sp. (Attini); 16.30, *Atta* sp. (Attini).

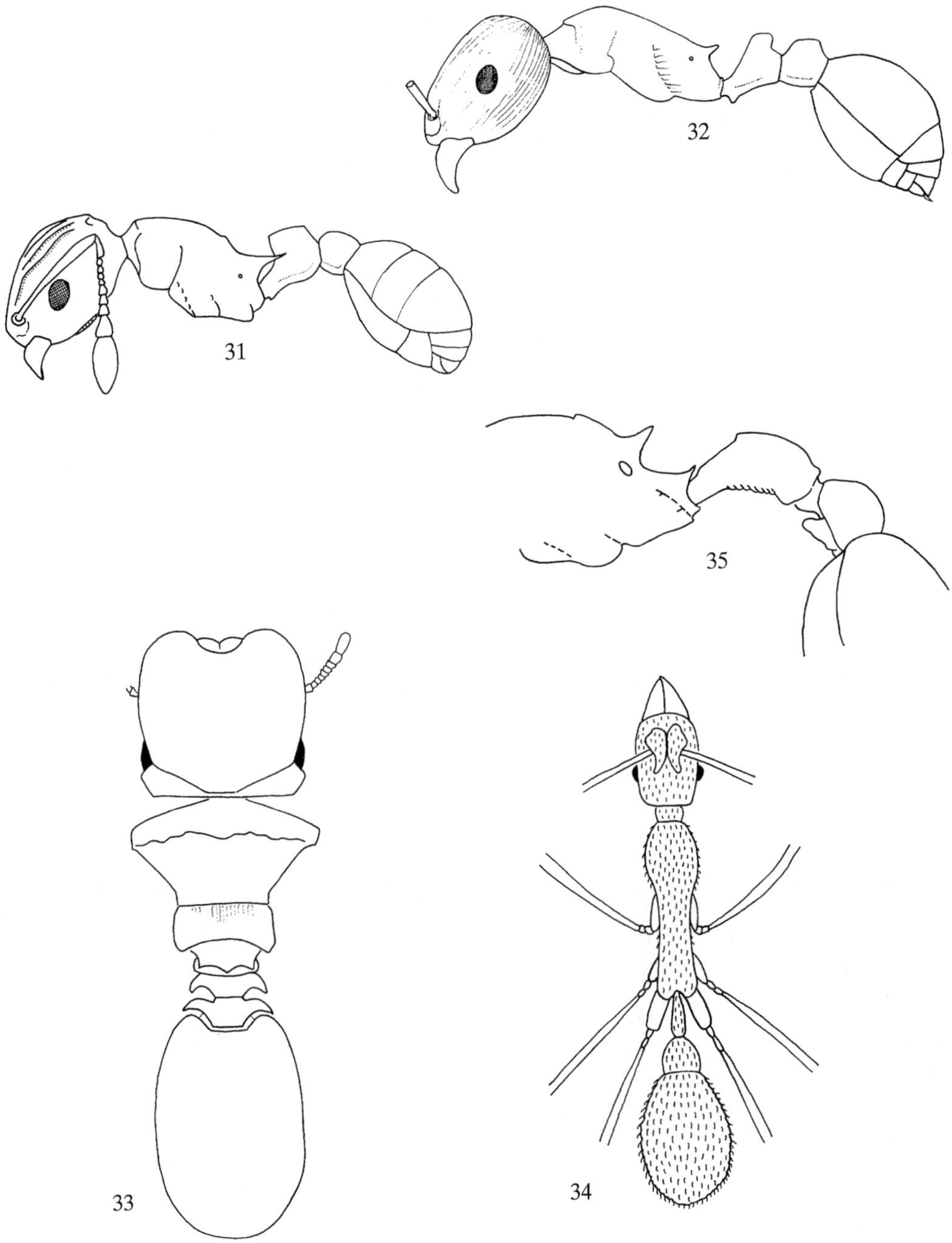

Figs 16.31–16.35. Myrmicinae. Figs 16.31–16.32. Head and body, lateral; 16.31, *Wasmannia* sp.; 16.32, *Tetramorium* sp. (Tetramoriini). Figs 16.33–16.34. Head and body, dorsal view; 16.33, *Zacryptocerus* sp. (Cephalotini); 16.34, *Apterostigma* sp. (Attini). Fig. 16.35. Petiolar segments, lateral, *Hylomyrma* sp.

MYRMICINAE: Attini. This exclusively New World tribe of fungus-growing ants (see above) comprises 12 genera, eight of which are represented in Costa Rica.

Acromyrmex. Subtropical South America north to southwestern United States. Workers are moderately polymorphic; dorsum of mesosoma bears four pairs of spines (Fig. 16.29). Many species of this genus occur in grasslands and semidesert areas. Three species occur in Costa Rica. Biology: *A. coronatus* is common in montane areas, *A. octospinosus* is a widespread species whose range includes urban areas and *A. volcanus* is present in primary rainforests on Atlantic slopes and lowlands. *A. volcanus* primarily has arboreal nests. *Acromyrmex* species use cut leaves, flowers and fruit to grow fungus. (Additional biology: Wetterer, 1991*b*).

Apterostigma. Neotropical; at least seven species present in Costa Rica. Upper body covered with dense, appressed hairs; head and mesosoma elongate in dorsal view (Fig. 16.34). Biology: very common in forests where they live in small colonies. Most nest in dead wood, but a few construct nests low on tree trunks or under leaves of low vegetation (Fig. 16.03). The walls of such exposed nests are composed of the fungus itself (the white flocculent masses are a common sight in lowland rainforests). Seeds and anthers are used as a substrate for fungus gardens. (Additional biology: Forsyth, 1981; Black, 1987—but species identifications unreliable).

Atta. Predominantly neotropical, but extending into the southern United States. Workers are extremely polymorphic; dorsum of mesosoma bears three pairs of spines (Fig. 16.30). Three species are recorded from Costa Rica. Biology: *A. cephalotes* (orange-brown with shiny patches on sides of mesosoma) is both the most common and widespread species in the country (nest, Fig. 16.02). It deposits nest refuse in abandoned subterranean chambers. *A. colombica* (red-brown, without shiny patches on sides of mesosoma) is more common on the Osa Peninsula, but occurs sporadically in Atlantic slope forests. It deposits nest refuse above-ground. A third species, *A. sexdens*, has been reported from Costa Rica (Kempf, 1972), but its occurrence has not been verified.

Cyphomyrmex. Mostly neotropical, but extending into the southern United States. Body nearly hairless, without spines; antennal scrobes present. At least ten species are present in Costa Rica, some of which are very common and widespread. Taxonomy: Snelling and Longino (1992). Biology: live in small colonies and there are both ground-nesting and arboreal species. *Cyphomyrmex* species use caterpillar droppings and dead insect parts on which to grow fungus.

Mycocepurus. Neotropics. Pronotum and anterior of mesonotum bearing a circle of eight upwardly directed spines. Only one species, *M. smithi*, is known from Costa Rica. Biology: found on beach margins on the Atlantic coast, where it may be locally abundant and from leaf litter in forested sites throughout the country.

Myrmicocrypta. Neotropics. Head and mesosoma moderately tuberculate, with erect, clavate hairs. Only a single rare species (cf. *ednaella*) is known from Costa Rica. Biology: nests in the soil.

Sericomyrmex. Neotropics. Body covered with long hairs; head heart-shaped. The taxonomy is poorly understood. Biology: one species (*amabilis?*) is both common and widespread in forested lowland habitats in Costa Rica. These ants live in small colonies in the soil. The fungus substrate they use includes leaf, flower, and fruit fragments (J. Wetterer, pers. comm.).

Trachymyrmex. A large genus, extending from Argentina to New York. Mesosoma frequently with spines; metasoma tuberculate. At least four species are present in Costa Rica. Biology: common and widespread in lowland habitats. They live in small colonies in soil and never use trunk trails in foraging. Their fungus substrate includes leaf fragments (J. Wetterer, pers. comm.), and in the United States insect excrement is also incorporated into the substrate.

MYRMICINAE: Basicerotini. Primarily neotropical, but also present in the Australian and Oriental regions.

Seven genera are currently recognized, all of which occur in the Neotropics; four of these are present in Costa Rica.

Basiceros. Neotropics. Antenna 12-segmented. Represented in Costa Rica by *B. manni*. Biology: a rarely collected nocturnal ant that forages on the ground in wet forests. It forms small colonies in the leaf litter. Although they are predaceous, these ants are very cryptic and slow moving (Wilson & Hölldobler, 1986).

Eurhopalothrix. An Indo-Australian/neotropical genus with a single species occurring in southern Florida. Antenna 7-segmented. At least five species are present in Costa Rica. Biology: common but inconspicuous predaceous ants living in small colonies in forest leaf litter.

Octostruma. Neotropics. Antenna 8-segmented. At least three species occur in Costa Rica. Biology: common but inconspicuous predaceous ants living in small colonies in forest leaf litter.

Rhopalothrix. Neotropics and Australia. Antenna 7-segmented. At least one species in Costa Rica, known from fewer than ten specimens in the country. Biology: collected in wet forest leaf litter.

MYRMICINAE: Blepharidattini. Contains two genera, one of which, *Wasmannia*, is present in Costa Rica.

Wasmannia. A neotropical genus with one tramp species, *W. auropunctata*, present in other areas of the world. At least three species occur in Costa Rica. Biology: *W. auropunctata* is very common in disturbed lowland habitats (< 500 m) in Costa Rica and workers can be very abundant in second growth, where they are dominant predators and scavengers, and are often common at extrafloral nectaries. They have strong recruitment capabilities. Their large polydomous colonies have no discrete colony centre and the nests may be found almost anywhere: under stones, in dead wood, in leaf litter, in stems or under bark. Bibliography: Ulloa-Chacon *et al.* (1991). The two other *Wasmannia* species occur in primary forest and are rare.

MYRMICINAE: Cephalotini. Primarily neotropical, with a few species of one genus extending northwards into the southern United States. Three genera are currently recognized, all of which occur in Costa Rica.

Cephalotes. Neotropics. Head with spines. Represented in Costa Rica by two species, *C. serraticeps* and *C. atratus* (R. Snelling, pers. comm.). Biology: *C. serraticeps* is a rare, lowland primary wet forest ant that lives in large colonies in boles of live canopy trees. The workers are continuously polymorphic. Their feeding habits are poorly known.

Procryptocerus. Neotropics. Head circular in dorsal view. The 12 Costa Rican species are generally rare but locally common. Biology: arboreal with small nests in live or dead stems. They are generally sparsely distributed and restricted to high canopy in lowland forest, but they are more common on low vegetation in mid to upper elevation forests. *Procryptocerus* species are solitary foragers, but their feeding habits are poorly known (Wheeler, 1984).

Zacryptocerus. Mainly neotropical, but extending into the southern United States. Head quadrate, especially in soldier caste (Fig. 16.33); body dorsoventrally flattened. At least 13 species are present in Costa Rica. Biology: common ants in lowland (0–500 m) habitats throughout the country. All are arboreal with small to large nests in live or dead stems. A few species are characteristic of second growth and *Z. setulifer* is an obligate plant-ant in *Cordia alliodora*. They are solitary foragers, but multiple workers may accumulate at extrafloral nectaries. *Z. maculatus* may share foraging trails with *Azteca* (Adams, 1990).

MYRMICINAE: Crematogastrini. This tribe includes only the cosmopolitan genus *Crematogaster*.

Crematogaster. A very large, cosmopolitan genus with at least 33 species present in Costa Rica. Biology: can be very common ants throughout much of the country. Most species are arboreal, but a few nest in dead sticks in leaf litter. Some species are canopy dominants, forming large polydomous colonies that span one or more canopy trees. They are generalized omnivores or scavengers and are common on extrafloral nectaries.

MYRMICINAE: Dacetonini. A large primarily pantropical and Australian tribe comprising about 25 genera, of which 11 are native to the New World and one, *Trichoscapa*, is introduced to the Americas from the Old World. This genus and five of the native neotropical genera occur in Costa Rica.

Acanthognathus. A neotropical genus represented in Costa Rica by two species: *A. ocellatus* (with pits on face) and *A. teledectus* (lacking pits on face). Biology: rare predaceous species, present in wet forest leaf litter or under epiphytes. They live in small colonies.

Glamyromyrmex. Neotropics. Mandibles triangular. At least two species occur in Costa Rica. Biology: very rare predaceous ants that live in small colonies in wet forest leaf litter.

Neostruma. Neotropics. Mandibles elongate. At least six species in Costa Rica. Biology: common but inconspicuous ants in forest leaf litter or under rotten wood. They live in small colonies and are predaceous.

Smithistruma. Cosmopolitan. Mandibles triangular; body often with scale-like hairs. Six species occur in Costa Rica. Biology: generally rare, living in small colonies in forest leaf litter or under rotten wood. They are predaceous, and at least some species are specialist predators on Collembola (Smith *in* Krombein *et al.*, 1979).

Strumigenys. Predominantly pantropical. Mandibles elongate and apically forked (Fig. 16.25). More than 50 species occur in the New World, with at least 20 species in Costa Rica. Biology: very common, but inconspicuous ants living in small colonies in forest leaf litter or under rotten wood. They are predaceous on small arthropods.

Trichoscapa. An Old World, mainly tropical genus, with one species, *T. membranifera*, spread throughout the tropical and warm subtropical parts of the world although its presence in Costa Rica has not been confirmed by us. Mandibles triangular. Biology: a rare ant, living in small colonies in leaf litter of disturbed areas. They are predaceous on a wide variety of small, soft bodied arthropods.

MYRMICINAE: Formicoxenini. Comprises about 25 genera of which two, *Cardiocondyla* and *Leptothorax* occur in Costa Rica.

Cardiocondyla. A widespread Old World genus with at least two tramp species occurring in Costa Rica. Head and mesosoma without hairs. Biology: common only in disturbed habitats, and occasionally along the edges of beaches. They are tiny ants with small colonies in soil.

Leptothorax. A large cosmopolitan genus, with at least seven inconspicuous but common species throughout Costa Rica. Biology: small-colonied, arboreal ants with scattered, solitary foragers. In North America several species are closely associated with other ants, and either lestobiotic or dulotic (see Chapter 2.7). In Costa Rica *L. echinatinodis* is common in second growth and disturbed habitats, with individual workers being common at extrafloral nectaries. The remaining species generally occur high in the canopy or are otherwise rarely encountered.

MYRMICINAE: Metaponini. Comprises four genera one of which, *Xenomyrmex*, occurs in Costa Rica. It has only tentatively been associated with the other genera in this tribe (Bolton, 1994).

Xenomyrmex. A small neotropical genus with at least two species present in Costa Rica. Biology: very rarely encountered, tiny ants that live in high canopy bark cavities.

MYRMICINAE: Myrmicini. Comprises six extant genera one of which, *Hylomyrma*, occurs in Costa Rica.

Hylomyrma. A neotropical genus with at least two species present in Costa Rica. Biology: common, but inconspicuous ants in wet forest leaf litter.

MYRMICINAE: Ochetomyremecini. Comprises two genera one of which, *Tranopelta*, occurs in Costa Rica.

Tranopelta. A neotropical genus with one species, *T. gilva*, in Costa Rica. Biology: widely distributed but apparently rare, tiny yellow ants found under rocks in primary forest or second growth areas. The large queens are occasionally found at lights.

MYRMICINAE: Pheidolini. Thisa tribe comprises eight extant genera two of which, *Aphaenogaster* and *Pheidole*, occur in Costa Rica.

Aphaenogaster. A large cosmopolitan genus represented in Costa Rica by at least three species. Similar to *Pheidole* but larger and with an undifferentiated antennal club. Biology: live in small colonies with a nest in the ground. They are solitary foragers and scavengers.

Pheidole. A very large cosmopolitan genus with more than 100 species in Costa Rica. Antennal club 3-segmented; propodeum toothed. Biology: very common and species are extremely variable in their habits. They construct nests in the ground, leaf litter, live or dead stems, or make carton nests. Most are generalized scavengers and predators, and many frequent extrafloral nectaries, but some harvest seeds, and at least one forms ant gardens. *P. bicornis* is an obligate plant-ant in *Piper* (Piperaceae). Further biological information: *cephalica* (Wilson, 1986*b*), *fiorii* (Black, 1987).

MYRMICINAE: Pheidologetonini. Comprises ten extant genera one of which, *Oligomyrmex*, occurs in Costa Rica.

Oligomyrmex. Cosmopolitan. Eyes very reduced; workers dimorphic. There are at least three species in Costa Rica, two of which belong to a group of species accorded distinction as a separate genus, *Erebomyrma*, by some authors (e.g. Wilson, 1986*a*; Hölldobler & Wilson, 1990*a*), and a third which is an extremely tiny species and probably belongs in *Oligomyrmex sensu stricto*. Biology: moderately common, leaf litter and soil inhabitants.

MYRMICINAE: Solenopsidini. This tribe comprises 13 genera. Four of these, *Carebarella*, *Megalomyrmex*, *Monomorium* and *Solenopsis*, occur in Costa Rica.

Carebarella. A neotropical genus with only a single, very rare species, *C. bicolor*, known to occur in Costa Rica. Biology: only found in soil samples.

Megalomyrmex. A neotropical genus with at least ten species present in Costa Rica. Taxonomy: Brandao (1990). Biology: most are known only from occasional workers in leaf litter, but a few are locally common in primary wet forest. These latter species nest in soil and may be found foraging on low vegetation and tree trunks, tending Homoptera or at extrafloral nectaries.

Monomorium. A widespread, predominantly Old World genus with many synanthropic species that are widely distributed. At least three of these tramp species occur in Costa Rica. Biology: common only in human domiciles and other disturbed habitats. They are generalized omnivores and scavengers and very similar to *Solenopsis* in the field. *M. pharaonis* (Pharaoh's ant) is all yellow and a common pest in dwellings. *M. ebeninum*, which is all black, is moderately common in pastures in northwestern Costa Rica and *M. floricola* (bicoloured, with lighter mesosoma) is widespread in brushy areas.

Solenopsis. A cosmopolitan genus with a potentially large number of species present in Costa Rica. *S. geminata* is the only 'fire ant' (named for its painful, alkaloid-containing venom) present. Taxonomy: Trager (1991). Biology: *S. geminata* are large, continuously polymorphic ants with well-developed compound eyes. They are very common in all lowland second growth and agricultural habitats, and common to rare in primary forests. They construct large conspicuous nests in the soil, and are generalized omnivores and scavengers, foraging on low vegetation, or more commonly on the ground. The remaining species of *Solenopsis* are generally smaller, monomorphic (though at least one rare leaf litter species is dimorphic, with a discrete soldier caste), and with eyes reduced to a few facets. They are often dominant ants in leaf litter and under epiphyte mats in the canopy. Forest clearings on the wet Atlantic slope are often dominated by a polydomous shiny black species of *Solenopsis*. Many *Solenopsis* species have strong recruitment capabilities.

MYRMICINAE: Stegomyrmecini. This tribe comprises a single genus, *Stegomyrmex*. It has been found to occur in Costa Rica.

Stegomyrmex. A neotropical genus with one species, *S. manni*, present in Costa Rica. Taxonomy: Diniz (1990). Biology: rare, wet forest ants that inhabit leaf litter.

MYRMICINAE: Stenammini. This tribe comprises 18 genera of which four, *Adelomyrmex*, *Lachnomyrmex*, *Rogeria* and *Stenamma*, occur in Costa Rica. The first of these is only very tentatively associated with the other genera in the tribe.

Adelomyrmex. A neotropical/Indo-australian genus that is represented in Costa Rica by at least five species. Biology: usually rare ants that can be common on upper Atlantic slopes and in cloud forest. They occur in leaf litter.

Lachnomyrmex. A neotropical genus with at least two species present in Costa Rica. Biology: generally rare, but can be common on upper Atlantic slopes (700–1500 m) and in cloud forest. They are found in leaf litter, often together with *Adelomyrmex*.

Rogeria. A New World and Australian genus with at least five species in Costa Rica. Taxonomy: Kugler, 1994. Biology: rare, but widely distributed, inconspicuous small ants living in leaf litter, or occasionally arboreal.

Stenamma. A widespread genus with at least 15 species present in Costa Rica. Biology: can be very common in montane cloud forest, but they decrease in abundance at lower elevations, below about 700 metres. These inconspicuous ants live in small colonies, in leaf litter and under epiphyte mats.

MYRMICINAE: Tetramoriini. A primarily Old World tribe comprising seven genera, two of which occur in North America (possibly as introductions). One of these genera, *Tetramorium*, is represented in Costa Rica.

Tetramorium. A widespread Old World genus that includes a number of cosmopolitan tramp species, two of which occur in Costa Rica. These are *T. bicarinatum* (bicoloured, large) and *T. simillimum* (uniform orange, small, superficially similar to *Wasmannia auropunctata*). Biology: very common in and around lowland dwellings, but generally very rare beyond house clearings. However, *T. bicarinatum* is common in the *Typha* swamp at Palo Verde.

DOLICHODERINAE (Shattuck, 1992*a* & *b*)

This cosmopolitan subfamily comprises about 23 genera, of which 11 occur in the Neotropics and five are present in Costa Rica. These may be distinguished by Table 16.4. In the New World, dolichoderines are a

- –Integument thickened, sometimes conspicuously sculptured; mesosoma often with spines or sharp angles .... *Dolichoderus*
- –Integument thin and flexible, rarely sculptured; thoracic spines generally absent
  - –Petiole scale-like (Fig. 16.36)
    - –Propodeum with upward projecting tooth (Fig. 16.36) .... *Dorymyrmex*
    - –Propodeum without tooth .... *Linepithema*
  - –Petiole not scale-like (Figs 16.37, 16.38)
    - –Petiolar node readily visible; mesonotum raised slightly above pronotum (Fig. 16.37); head often 'heart-shaped' .... *Azteca*
    - –Petiolar node tubular and often hidden; mesonotum not raised above pronotum (Fig. 16.38) .... *Tapinoma*

Table 16.4. Simple diagnostic characteristics of workers of the genera of Dolichoderinae occurring in Costa Rica (adapted from Hölldobler & Wilson, 1990*a*).

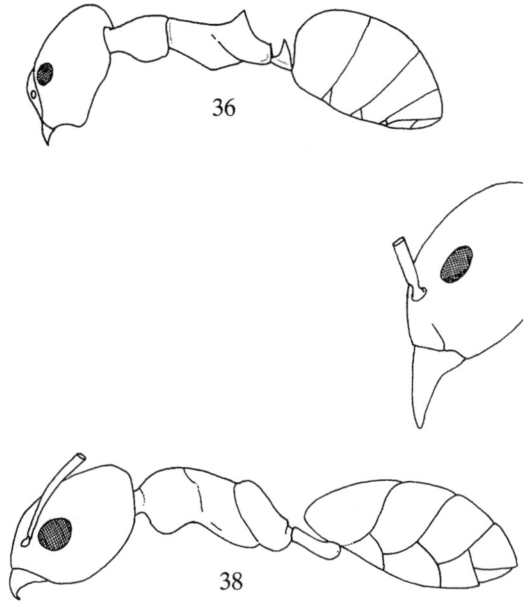

Figs 16.36–16.38. Dolichoderinae, lateral view of head and body; 16.36, *Dorymyrmex* sp.; 16.37, *Azteca* sp.; 16.38, *Tapinoma* sp.

dominant component of the arboreal ant community, but they are nearly absent from the leaf litter. They rely heavily on carbohydrates for sustenance. Instead of a sting they are furnished with repugnatorial glands.

***Azteca*.** A neotropical genus with at least 25 species in Costa Rica. Biology: very common arboreal ants in lowland habitats (< 500 m) where they are generalized scavengers. Many are very common at extrafloral nectaries and some tend Homoptera. Some species form large polydomous colonies that dominate the canopy; all tend to have large colonies. They nest in carton nests, ant gardens, living tree cavities and living or dead stems. Five species are obligate inhabitants of *Cecropia* (Longino, 1989*a* & *b*, *in* Huxley & Cutler, 1991).

Dolichoderus (= *Hypoclinea, Monacis*). A cosmopolitan genus with at least seven species in Costa Rica. Taxonomy: MacKay (1993). Biology: arboreal ants that can be very common in lowland habitats (< 500 m) where they are generalized scavengers. Many are very common at extrafloral nectaries and some tend Homoptera. Some species have large polydomous colonies. They nest in abandoned termite nests, loose-weave carton nests, or in dead wood.

Dorymyrmex (= *Conomyrma*). A small neotropical genus with at least one species (cf. *pyramicus*) in Costa Rica. Biology: moderately common in lowland habitats (< 500 m), more common in open areas and secondary growth. They are generalized scavengers and often frequent visitors to extrafloral nectaries. The medium sized colonies nest in the ground.

Linepithema (formerly placed in *Iridomyrmex*, which as now defined is absent from the New World). This genus is represented by at least one species (cf. *iniquus*) in Costa Rica. Biology: a rare, primary forest species. Workers have been collected at extrafloral nectaries and in leaf litter.

***Tapinoma*.** A widespread mainly pantropical genus, several species of which are common tramps. At least four species occur in Costa Rica. Biology:

Characteristics	Genus
Antenna 8- to 11-segmented; eyes greatly reduced (Fig. 16.39); pale yellow colour; strictly subterranean	*Acropyga*
Antenna 9- to 10-segmented; eyes of normal size (Fig. 16.40); colour variable; generally arboreal	
— Antenna without club; petiole hidden under gaster (Fig. 16.40)	*Brachymyrmex*
— Antenna with club; petiolar scale-like and prominent (Fig. 16.41)	*Myrmelachista*
Antenna 12-segmented.	
— Antennae contiguous with posterior border of clypeus (Fig. 16.43); dorsum of mesosoma with stout, stiff hairs	*Paratrechina*
— Antennae separated from posterior border of clypeus (Fig. 16.42); dorsum of mesosoma with thin, flexible hairs	
— — Polymorphic; eyes not usually bulging as strongly as Fig. 16.42; common	*Camponotus*
— — Monomorphic; eyes strongly bulging (Fig. 16.42); rare	*Dendromyrmex*

Table 16.5. Simple diagnostic characteristics of workers of the genera of Formicinae occurring in Costa Rica (adapted from Hölldobler & Wilson, 1990*a*).

very common generalized scavengers and frequent visitors to extrafloral nectaries. Most are arboreal, with small or large colonies that often have very ephemeral nest sites in dead stems or in debris. *T. melanocephalum* is a pantropical tramp species that is common in houses in Costa Rica. Another species, cf. *ramulorum*, makes small carton nests under leaves (Black, 1987).

FORMICINAE

A large subfamily with about 50 genera classified in 13 tribes throughout the world. Only nine genera occur in the Neotropics, six in Costa Rica. Species of this subfamily are often very abundant, and many are largely arboreal, but not dominant in Central America (unlike parts of South America where *Camponotus femoratus* may be a canopy dominant). Formicines rely heavily on carbohydrates for sustenance. Instead of a sting, formic acid is sprayed from the tip of the abdomen.

FORMICINAE: Brachymyrmecini. This tribe comprises five genera, one of which is present in Costa Rica.

Brachymyrmex. Mostly neotropical, but extending northwards into the southern United States. At least eight species occur in Costa Rica. Biology: very common in all habitats. Most are arboreal, nesting most often in dead stems or in various plant cavities. Workers are common at extrafloral nectaries.

FORMICINAE: Camponotini. Worldwide, a very species-rich tribe comprising ten extant genera. Two, *Camponotus* and *Dendromyrmex* occur in Costa Rica.

Camponotus. An extremely large cosmopolitan genus with at least 40 species present in Costa Rica where they are sometimes known as 'hormigas agrias', or sour ants. Biology: very common in most habitats, where they are generalized scavengers; many are very common at extrafloral nectaries. Most are arboreal, with small or large, frequently polydomous colonies; nests are most often in dead stems, but also in live stems and in tree cavities. One species, *C. sericeiventris*, is a particularly common, diurnal ant in Costa Rica.

Dendromyrmex. A neotropical genus with at least two rare species in Costa Rica. Biology: arboreal ants in primary wet forest where their small colonies construct nests of leaves bound together with carton and silk.

FORMICINAE: Lasiini. This tribe comprises eight extant genera, one of which is present in Costa Rica.

Paratrechina. Cosmopolitan. At least nine species occur in Costa Rica. Biology: *P. longicornis* is a pantropical tramp species, common in Costa Rican houses. The other species are native and may be very common in most habitats. They include both arboreal and leaf litter inhabitating forms. Many are generalized scavengers, and workers are common at extrafloral nectaries. Their nests are usually ephemeral, constructed in dead stems, various plant cavities, or in suspended debris, or they may construct small flimsy carton nests. *P. caeciliae*, for example, forms carton nests under leaves (Black, 1987).

FORMICINAE: Myrmelachistini. A small New World tribe comprising a single genus, *Myrmelachista*. It has been found to occur in Costa Rica.

Myrmelachista. A neotropical genus with a single species present in Florida, where it might have been introduced. At least 14 species occur in Costa Rica.

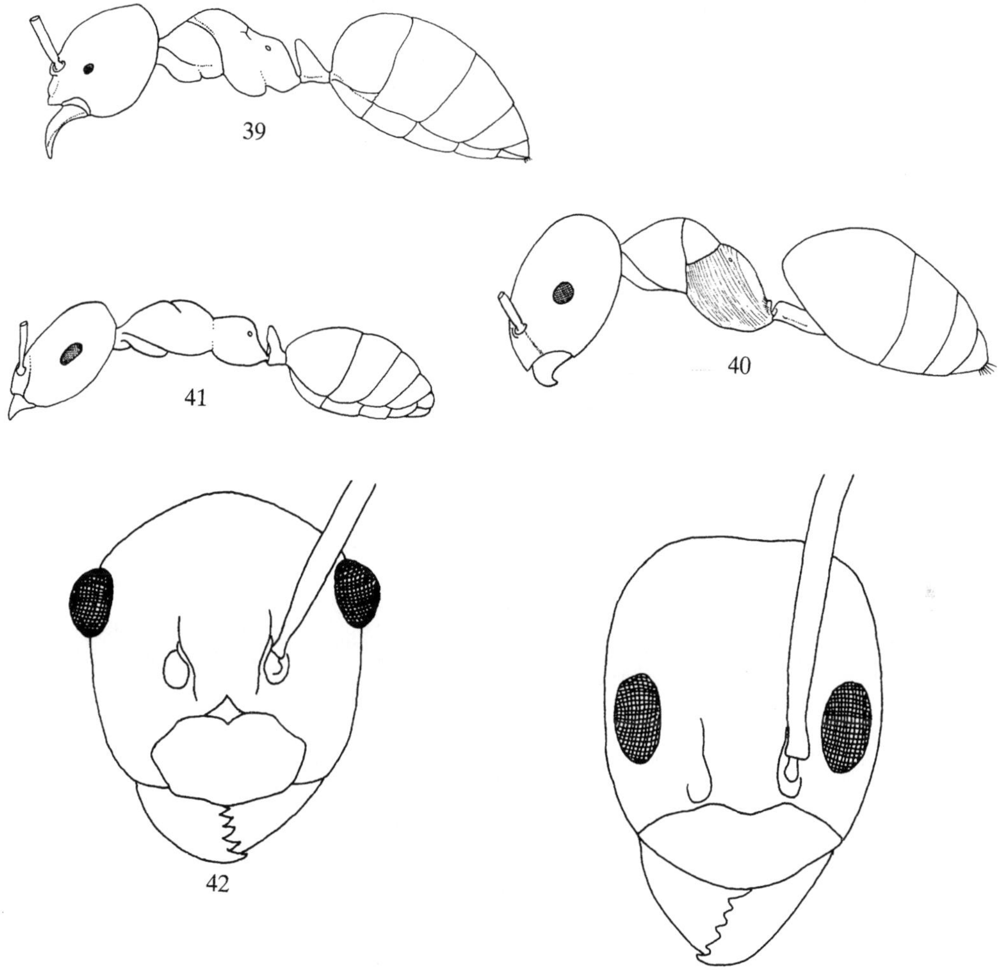

Figs 16.39–16.43. Formicinae. Figs 16.39–16.41. Lateral view of head and body; 16.39, *Acropyga* sp.; 16.40, *Brachymyrmex* sp.; 16.41, *Myrmelachista* sp. Figs 16.42–16.43. Head, front view; 16.42, *Dendromyrmex* sp.; 16.43, *Paratrechina* sp.

Biology: one set of species is common in mid-elevation (*ca.* 1000 m) forests. These ants forage externally on vegetation and can be common at extrafloral nectaries. They have large polydomous colonies, nesting in live or dead stems. Another set of species is very common, but cryptic, in low-elevation (< 500 m) wet forest. They are specialized plant-ants inhabiting live stems of understorey Lauraceae (Stout, 1979; Hammel, 1986) and Meliaceae.

FORMICINAE: Plagiolepidini. This tribe comprises six extant genera, of which one, *Acropyga*, is known to occur in Costa Rica.

Acropyga. Mostly pantropical; in the New World, extending northwards into the southern United States. At least four species occur in Costa Rica. Biology: probably common in soil (though infrequently collected), but are most often encountered under stones or occasionally in leaf litter samples. According to Weber (1944) many species tend scale-insects, and some may be of economic importance by dispersing mealybugs on the roots of coffee plants.

Acknowledgements

This work was supported by the National Science Foundation and by the Office of Forestry, Environment and Natural Resources, Bureau of Science and Technology, of the United States Agency for International Development under NSF Grant No. BSR-9025024, and by Apple Computer.

17 The sphecid wasps (Sphecidae)

Paul Hanson and Arnold S. Menke

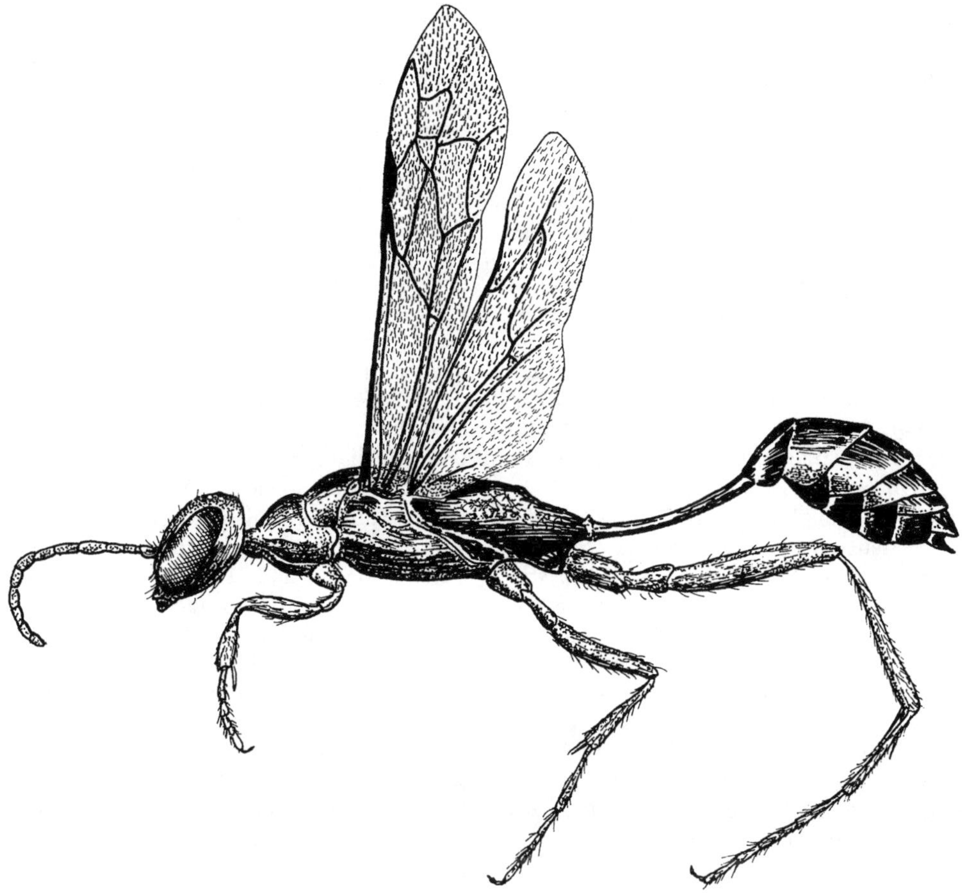

Fig. 17.01. *Chalybion* sp. (Sceliphrini).

Diagnosis. Body length 2.0 to 50.0 mm; both sexes fully winged (except in the Old World subfamily Heterogynainae); sexual dimorphism slight, although heads, and to some extent legs, vary between the sexes, often strikingly so; colour variable; *body vestiture simple, setae not branched or plumose*. Antenna 12-segmented in females, 13 in males (infrequently 12); inner eye margins straight, arcuate, sinuate or notched. Pronotum joined to mesothorax immovably, its hind margin straight and typically separated from mesoscutum by a constriction that contributes to the formation of a characteristic raised collar; *posterolateral margin of pronotum produced into a rounded lobe that covers mesothoracic spiracle*; pronotal lobe typically not attaining tegula, the mesoscutum and mesopleuron in broad contact there. Fore wing usually with numerous veins and enclosed cells, with a pterostigma; hind wing with enclosed cells, claval lobe present, jugal lobe usually present. *Hind leg with cleaning pecten on inner side of basitarsus* (absent in bees).

Classification and distribution. The Sphecidae is a large family comprising, worldwide, over 8000 species in 245 genera classified into 11 subfamilies: Ampulicinae, Sphecinae, Pemphredoninae, Astatinae, Laphyragoginae, Crabroninae (= Larrinae of authors), Entomosericinae, Xenosphecinae, Nyssoninae, Philanthinae and Heterogynainae (Bohart & Menke, 1976; Day, 1984). The Larrinae has traditionally been separated from the Crabroninae, but they are treated here as a single subfamily following the opinions of Evans (1964*c*), Lomholdt (1985) and Menke (1988); Crabroninae is the older family-group name and is thus the correct name for the unified group. Although all eleven sphecid subfamilies are usually treated as comprising a single family, as is done in this chapter, Finnamore (*in* Goulet & Huber, 1993) treated the sphecids as nine distinct families. Recent phylogenetic analyses do suggest that some changes in the classificatory status of the various groups may need to be made eventually as the Sphecidae, as a single family, is almost certainly paraphyletic with respect to the Apidae. Lomholdt (1982) suggested that Ampulicinae + Sphecinae is the sister group to the bees + the rest of the sphecids, but Alexander's (1992) more comprehensive cladistic analysis failed to corroborate this. The status of the Heterogynainae remains subject to dispute. Alexander (1992) placed *Heterogyna* + most Sphecidae and the Apidae *s.l.* as the sister-group to the Ampulicinae + Sphecinae, thereby confirming Day's (1984) placement of the group as a subfamily within the Sphecidae. Brothers and Carpenter (1993), however, placed *Heterogyna* as a separate family, the sister-lineage to the Sphecidae + Apidae. As Sphecidae, with or without *Heterogyna*, seems to be paraphyletic with repect to the Apidae, we here continue to follow Day's interpretation, preferring to avoid recognizing a separate family until monophyletic families within the Apoidea are defined.

Seven of the 11 sphecid subfamilies occur in Central America. Not present are the Laphyragoginae, Entomosericinae and Heterogynainae, which have distributions centred in the dry Mediterranean region of the Old World and variously extend into the Central Asia, northern India or the Afrotropical region, and the Xenosphecinae which is confined to the south-western United States. These four subfamilies are very small groups, comprising in total about 20 species. The seven subfamilies present in Central America are represented in Costa Rica by 82 genera and about 290 species. The actual number of sphecid genera in the Costa Rican fauna is likely to be 87 or more, and the number of species will probably exceed 400.

Biology. Sphecids are primarily solitary wasps, although a few species exhibit presocial behaviour and one species of *Microstigmus* is eusocial (Matthews, 1991). The larvae develop by feeding externally on arthropods provided for them by their female parent. In the majority of species the female provisions each cell in the nest with multiple prey, but in some of the more primitive sphecids (Ampulicinae and some Sphecinae) the female provides each larva with only a single prey item. In most instances the adult female catches live prey, which is paralysed by being stung, but species of *Microbembex* provide their young with dead insects scavenged from the surface of sandy areas. The fact that they still go through the motions of stinging suggests that they are derived from predatory ancestors (Evans, 1966*a*; Matthews & Evans, 1974). The venoms injected by stinging generally cause permanent paralysis or rarely even death of the host organism, and sphecids that do this conceal the immobilized arthropod, generally in a specially constructed hideaway, the nest. However, a few species of *Larra* (Steiner, 1984; Menke, 1992*a*) and *Chlorion* that attack hosts living in burrows have venoms that only temporarily paralyse the host. These wasps lay an egg on the immobilized host, then leave. The host revives and continues to live in its burrow until killed by the developing wasp larva. Thus these wasps can be considered to be koinobiont ectoparasitoids. Different species of *Chlorion* show different stages in the transition from parasitoid to nest-building.

The females of most sphecid species that do construct nests build them prior to hunting for prey. Some species of *Prionyx* and *Podalonia*, however, capture their prey before excavating the nest, sometimes temporarily caching the prey on a plant before digging (Evans, 1958; Kurczewski *et al.*, 1992). As in other groups of solitary nest-provisioning aculeates, cleptoparasitism—ovipositing in nests constructed by other species—has evolved amongst the sphecids. Members of the tribe Nyssonini exhibit this habit. In Central America, for example, *Epinysson* species oviposit in nests of *Hoplisoides* (Reinhard, 1929—as *Nysson* and *Hoplisus*) and *Zanysson* species are cleptoparasites of *Tachytes* (Evans, 1966*b*).

In most sphecids the lifetime fecundity is low, and an individual female seldom lays more than 15 eggs. The oocytes mature relatively slowly and the resultant eggs are quite large (O'Neill, 1985). Species of *Larra*, however, which do not construct nests, lay smaller eggs and are more fecund (Bohart & Menke, 1976). Sex ratios are usually 1:1 except where first-generation males overlap and mate with females of the second generation, in which case the second generation is often female-biased (Brockmann & Grafen, 1992). An example of a Costa Rican species in which the demography has been intensively studied is *Sceliphron assimile* (Hunt, 1993).

Nesting behaviour. Nests may be built below or above ground, but the former is probably the ancestral condition. The females of species that excavate burrows in soil are usually equipped with rows of specialized 'rake' setae on the tarsus of the front leg. These fossorial species also tend to have broad mandibles for loosening soil and a well-developed pygidial plate on the last metasomal tergite, which is used for pushing soil. In Costa Rica ground-nesting sphecids include a few Sceliphrini (some *Chlorion* and *Penepodium*), some Sphecini (*Prionyx* and *Sphex*), Ammophilini, most Psenini, Astatinae, Larrini (often called 'digger wasps'), Bothynostethini, Oxybelini, many Crabronini, Nyssoninae ('sand wasps') and Philanthinae.

Four modes of digging behaviour have been recognized (Olberg, 1959; Evans, 1966*b*). 'Rakers' (e.g. Bembicini) excavate oblique burrows by vigorously moving the front legs while holding the abdomen high, thus throwing a stream of soil particles beneath and behind the body. 'Pullers' (e.g. *Mellinus*) also dig oblique burrows, but instead of throwing the soil backward they gather it in a mass between the head and front legs and then back out of the burrow to deposit it at the entrance. 'Carriers' (e.g. *Ammophila*) proceed in a similar manner, but rather than dump the soil at the burrow entrance they walk or fly with it, dropping the soil some distance from the entrance. Rakers often dig in sandy soil whereas pullers and carriers require soil particles that adhere in clumps. Several species use a combination of raking and pulling depending on the type of soil. Finally, 'pushers' (e.g. *Cerceris, Crabro*) dig vertical burrows by using the pygidial plate to push the soil as they back out of the burrow.

Nests containing just one cell (e.g. Ampulicinae, *Prionyx*, Ammophilini) probably represent the ancestral state in Sphecidae (Evans & West-Eberhard, 1970; Iwata, 1976), although in some cases unicellular nests may have evolved secondarily from multicellular nests (e.g. *Microbembex, Steniolia, Stictia*). Most sphecids construct multicellular nests, and among ground-nesters the pattern in which individual cells are arranged varies between species and often within species (e.g. in different soil types). Cell arrangement may be irregular (diffuse) or serial, and the first cells may begin at the far end of the burrow or near the entrance. Nests of Astatinae, Philanthinae and some Crabroninae often include a temporary storage area just before the area containing the brood cells (Evans, 1957*a*, 1971; Evans & Matthews, 1973).

Many sphecids, including several primitive groups (Ampulicinae, some Sphecinae, many *Liris*) are not diggers, but instead use existing cavities as nest sites. *Dynatus*, for example, has been observed using burrows in wood that were previously occupied by the bee, *Centris labrosa* (Kimsey, 1978). More advanced groups of sphecids often modify the substrate to a greater extent by, for example, using their mandibles to excavate in old wood or plant stems (*Psenulus*, Pemphredonini, many Miscophini, *Ectemnius, Lestica* and *Rhopalum*). Most use tubular cavities such as beetle borings or plant stems and these sphecids will often adopt 'trap nests' placed in the field (Krombein, 1967). Species that nest in tubular cavities generally construct a series of cells separated by partitions, which may consist of compacted wood fragments, mud (*Chalybion*, Trypoxylini), or even compacted fibres from blades of grass (*Isodontia* spp.). Some species of *Isodontia* have abandoned partitions, so that the larvae develop in a single large brood cell, which is unique in Sphecidae. Not all cavity-nesters use tubular cavities. *Pison cressoni*, for example, constructs mud cells in hummingbird nests (Oniki, 1970), abandoned vespid nests, and larval tents of lasiocampid moths (Menke, 1988).

A few sphecids do not require cavities, but just a surface on which to build a nest entirely of material gathered by the female. Preferred substrates are in sheltered places and include rock faces, tree trunks, undersides of leaves, buildings, etc. The most common construction material is mud and within the Sphecidae 'mud-daubers' have evolved in the Sceliphrini and in the Trypoxylini. *Sceliphron* uses its mandibles to gather a ball of mud from the margins of puddles, flies with it to the nest site, and applies it with its mouthparts and legs; 30-40 loads of mud are required to build one cell (Shafer, 1949; Evans & West-Eberhard, 1970). The completed nest consists of four to ten parallel cells and the entire mass is plastered over with more mud. *Trigonopsis* differs in that it, at least sometimes, quarries dry soil by adding water

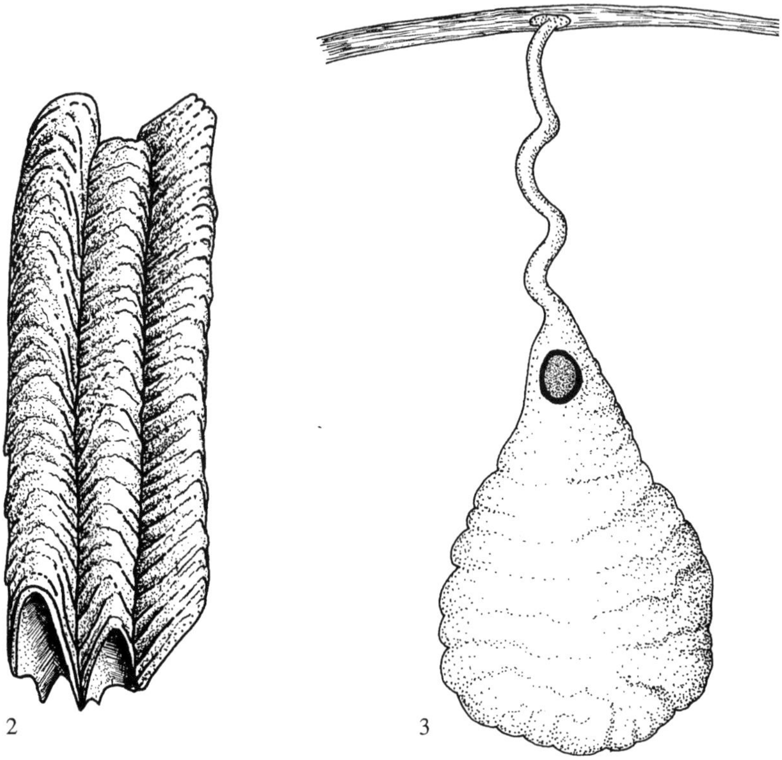

2 3

Figs 17.02–17.03. Sphecidae, nests; 17.02, *Trypoxylon* sp.; 17.03, *Microstigmus* sp.

and it does not use its legs in construction, its long neck apparently serving to add mud from within the nest (Eberhard, 1974). Like *Sceliphron* some mud-daubing species of *Trypoxylon* gather mud with which they construct a series of parallel tubes (Fig. 17.02), but unlike *Sceliphron* these tubes open at the bottom and each consists of a linear series of cells. In some cases the nest is not covered with a final application of mud, and inverted V-shaped ridges may be evident (Cross *et al.*, 1975). Old mud-dauber nests are often re-used by other species; *Chalybion* for example commonly uses old *Sceliphron* nests.

Species of *Microstigmus* (Pemphredonini) also construct exposed nests, but instead of mud they use plant material, combined with silk secreted by glands in the tip of the abdomen of the female. Silk production by adults is rare in Hymenoptera (occurring in a few other Pemphredonini), and the use of silk with plant material to form a pedicellate bag is unique to *Microstigmus* (Matthews, 1991). The plant material used in nest construction varies from species to species but in Costa Rica *Microstigmus comes* utilizes the waxy bloom on the undersides of leaves of the palm, *Cryosophila guagara* (Matthews, 1968*b*, *in* Janzen, 1983), *M. adelphus* uses plant hairs (Richards, 1972) and *M. thripoctenus* uses wood chips (Matthews, 1970*b*). The small, bag-like nests of *Microstigmus* species are suspended from the undersides of leaves, rootlets, sheltered rock faces or even overhanging roofs of buildings (see Fig. 17.03). Unlike other social insects, nest size depends on the amount of raw material gathered at the outset and there is no opportunity for subsequent expansion (Matthews & Starr, 1984). The larvae are reared in cells in the lower part of the nest, whilst the upper chamber, near the entrance, is occupied by adults. Nests of *Microstigmus* may be usurped by *Trypoxylon latro*, which line abandoned nests with mud (Matthews, 1983).

Social behaviour. Although the vast majority of sphecid species are strictly solitary in their nesting habits, presocial behaviour appears to have evolved independently in several lineages, and eusocial behaviour is now well documented in one species. The most common type of presocial behaviour is communal nesting, whereby several females co-operate in building a common nest with each female provisioning her own brood cells. Several ground-nesting sphecids such as *Bembecinus quinquespinosus* (Evans *et al.*, 1986), *Stictia heros* (Sheehan, 1984) and *S. maculata* (Matthews *et al.*, 1981) nest in aggregations, but each female builds her own nest and these species are therefore solitary. Co-operative nest building on the other hand is known to occur in certain species of Sphecinae, Pemphredoninae, Crabroninae, Nyssoninae and Philanthinae (Matthews, 1991). Examples of nest sharing in species occurring in Costa Rica include *Sphex ichneumoneus* (Brockmann & Dawkins, 1979; Brockmann *et al.*, 1979), *Trigonopsis cameronii* (Eberhard, 1972, 1974), *Microstigmus* spp. (see below), *Tachytes distinctus* (Lin & Michener, 1972), *Trypoxylon* (*Trypoxylon*) *fabricator* (Sakagami *et al.*, 1990), *Trachypus mexicanus* (Menke, 1980), *T. petiolatus* (Evans & Matthews, 1973) and *Cerceris intricata* (Alcock, 1975). Different populations of the same species may vary in the extent to which nest sharing occurs.

Communal nesting, where several females of the same generation co-operate in constructing and defending the nest, is generally considered to be an intermediate stage in the evolution of eusocial behaviour via the parasocial route (see chapter 2.6). The other route to eusociality, the subsocial route, can occur when adult offspring remain in their natal nest and live in association with their mother. Although many sphecids tend to nest close to their natal nest sites, and some even re-use the same nest in which they developed (e.g. many *Cerceris*; Hook, 1987), the mother often does not live long enough to associate with her adult offspring. Thus the relatively short female life span, such that mothers rarely overlap with daughters, may be an important factor explaining the lack of significant social evolution in Sphecidae (Evans & West-Eberhard, 1970).

Eusocial behaviour occurs in *Microstigmus comes* (Matthews 1968*a* & *b*; Ross & Matthews, 1989*a* & *b*, 1991). Several females (usually three to eight) co-operate in constructing a nest, and then in the provisioning of the cells within, one at a time, with collembolans. Nests can be constructed entirely by a single foundress, but groups of co-operating females establish a nest more rapidly. Genetic relatedness of female nestmates suggests that colonies consist of mother-daughter groups. The putative mother of a colony (the 'queen') weighs more than other females in the nest and appears to be the exclusive egg-layer. The achievement of eusociality in *M. comes* is perhaps best explained by a combination of attributes: unique mode of nest construction, sufficiently long adult life spans to allow continuously overlapping generations and use of tiny but abundant prey (Matthews, 1991).

Hunting and provisioning behaviour. The type of food organisms provided for the larva by the adult female is often characteristic of the genus or tribe (Table 17.1). In the case of holometabolous prey the adult stage is usually utilized, although Ammophilini use lepidopterous larvae to provision their nests. When the prey is a hemimetabolous insect, however, either the immature or adult stage may be taken indiscriminately.

Species of sphecids show varying degrees of prey specificity. At La Selva Biological Station in northeastern Costa Rica *Trypoxylon (Trypargilum) xanthandrum* preys mostly on spiders of the family Senoculidae, *T. superbum* mostly on Salticidae (including ant-mimicking species), and *T. agamemnon* and *T. vagum* mostly on Araneidae (Coville, 1981*b*; Coville & Griswold, 1983, 1984). On the other hand *T. spinosum* and *T. tenoctitlan* take a wide range of spiders. Different species groups of *Cerceris* appear to specialize on Tenebrionidae or Buprestidae, Chrysomelidae or Curculionidae (Scullen & Wold, 1969; Callan, 1990), and those that attack weevils often show a great deal of specificity (Evans, 1971). Among the most polyphagous sphecids are some species of *Microbembex*, which scavenge a wide variety of dead Diptera and other insects (Bohart & Menke, 1976).

Individual wasps sometimes specialize on particular prey items, suggesting that they learn to search certain types of areas more intensively. Within these areas there are probably certain stimuli that elicit more intensive local searching behaviour. *Mellinus* species are reported to hunt flies at dung, the female wasp stalking an individual fly in a 'cat-like' manner (Hamm & Richards, 1930; Evans, 1989). At a close distance (about 10–15 cm) odour appears to be the predominant stimulus, which may explain why sphecids are rarely deceived by visually mimetic prey.

Some sphecids possess morphological adaptations for capturing particular types of prey. The long mandibles of *Podium* and *Trigonopsis*, for example, may serve in capturing cockroaches (Eberhard, 1974). Although definite evidence is lacking, it is possible

Prey group	Sphecid taxa
Araneae	*Chalybion*, *Sceliphron*, *Miscophus*, Trypoxylini
Blattaria	Ampulicinae, *Dynatus*, *Penepodium*, *Podium*, *Trigonopsis*
Coleoptera	*Bothynostethus*, *Entomognathus*, Cercerini
Collembola	*Microstigmus*
Diptera	Oxybelini, *Crabro*, *Ectemnius*, *Podagritus*, *Rhopalum*, *Mellinus*, *Bembix*, *Rubrica*, *Steniolia*, *Stictia*
Hemiptera	Astatinae, *Solierella*, *Anacrabro*, *Bicyrtes*
Homoptera	Psenini, *Polemistus*, *Stigmus*, *Entomocrabro*, *Quexua*, Gorytini, Stizini
Hymenoptera	Philanthini
Lepidoptera	Ammophilini, *Lestica*
Orthoptera	Sphecini, Larrini, Scapheutini
Psocoptera	*Nitela*
Thysanoptera	*Microstigmus*, *Spilomena*, *Xysma*

Table 17.1. Principal prey of sphecid taxa occurring in Central America. Prey records mostly from Bohart & Menke (1976); records for Scapheutini from Cooper (1988, 1993*a*) and Vardy (1987); record for *Quexua* from Cooper (1986*b*).

that the prominent clypeal protuberances in female *Cerceris* are related to the capture of adult beetles.

Upon capturing the prey the female sphecid stings it and injects a venom. Generally envenomation induces permanent paralysis, although in some Sphecinae and Larrini only temporary paralysis results (Steiner *in* Piek, 1986). The venom affects the neuromuscular system and does not appear to be specific to a particular type of prey. Nonetheless, anatomical differences between prey seem to necessitate different stinging procedures since Hemiptera and adult holometabolous insects are generally stung just once in the venter of the thorax, whereas orthopteroid insects and caterpillars are stung several times in different regions of the body. Paralysing (rather than outright killing) the prey prevents it from rotting, which is especially important for mass-provisioners (see below). Among sphecids that practice progressive provisioning there is less need to ensure that the prey remains fresh for long periods, and many bembicines that progressively provision have a venom that kills their prey (Evans, 1966*a* & *b*). Many Pemphredonini that prey on small delicate insects—such as aphids, thrips and collembolans—often kill and masticate the prey before carrying it to the nest.

The vast majority of sphecids utilize mass provisioning, whereby the cell is fully provisioned and sealed before the larva begins to develop. Many Stizini and Bembicini demonstrate a stage intermediate between mass and progressive provisioning. Here the egg is usually laid on the first prey item brought into the nest (in the case of *Bembix* it is laid in the empty cell) and then after the larva begins to feed, several more prey are rapidly added and the cell is sealed. A North American species of *Ammophila* simultaneously tends several nests containing developing larvae and the female wasp is in contact with her offspring throughout their larval development (Evans, 1965*c*).

Defence against natural enemies. Sphecid prey and larvae are vulnerable to attack by a range of cleptoparasites, parasitoids and predators (see Chapter 2.6). Many aspects of sphecid behaviour appear to be adaptations for minimizing mortality inflicted by these natural enemies (Spofford *et al.*, 1986; Evans & O'Neill, 1988). Such behavioural responses can be grouped in four categories:

1) Returning to the nest. Female sphecids return to the nest using visual landmarks that they learned during 'orientation flights' while building the nest. Frequently the prey-laden female is followed by cleptoparasitic 'satellite flies' (Sarcophagidae: Miltogramminae), and to thwart these pursuers many sphecids employ diversionary tactics in their flight to the nest (McCorquodale, 1986). Another means of avoiding pursuing cleptoparasites is to forage during a time of the day when the latter are less active. Some species of *Crabro*, for example, provision very early in the morning and at dusk, when miltogrammine flies are less active (Evans *et al.*, 1980).

The manner in which prey is carried by the female when returning to the nest has evidently evolved independently of prey type, and different types of prey carriage show phylogenetic patterns (Evans, 1962). The putative ancestral form can be seen in Ampulicinae, where the female grasps the cockroach prey by the antennal bases with her mandibles and walks backward, dragging it to the nest. This form of prey carriage is unique in Sphecidae, but common in Pompilidae where females frequently walk backwards dragging a paralysed spider (Evans, 1953). Some Sphecinae and Larrini also grasp the prey with their mandibles and drag it to the nest, but they walk forwards. If smaller prey are utilized the prey can be carried with the mandibles in flight (Sphecinae, Larrini, Pemphredonini, Astatinae, Trypoxylini, Mellinini, Cercerini). However, carrying prey in the mandibles entails risks upon arrival at the nest, since either the nest must be left open, and hence vulnerable to intruders, or the prey must be temporarily abandoned while the nest is opened, and thus vulnerable to waiting cleptoparasites. Presumably these risks selected for prey carriage utilizing the middle legs. In such cases the prey is always carried in flight and is not abandoned while the nest is opened (Crabronini, Gorytini, Stizini, Bembicini, Philanthini). The most advanced stage of this evolutionary sequence is thought to be prey carriage utilizing the hind legs, or even the sting (Oxybelini).

2) Nesting behaviour. Many sphecids attempt to conceal their nests, which is thought to reduce the likelihood of their discovery by natural enemies. This may be done in a variety of ways depending upon the nesting behaviour of the species involved. Mounds of freshly excavated soil can serve as an indication of the presence of a nest, and many ground-nesting sphecids, such as *Bembix multipicta* and *Stictia maculata*, level the mounds of soil that accumulate around their burrow entrance (Evans, 1966*a*; Matthews *et al.*, 1981). Other species, such as *Bicyrtes* and *Microbembex*, throw the soil while excavating and thus there is no need for levelling.

Many sphecids, such as *Bembix multipicta* and *Stictia heros*, block the nest entrance with a temporary closure, in order to impede entry by intruders while the female is away hunting (Cane & Miyamoto, 1979; Sheehan, 1984). The disadvantage of such temporary closure is that when the female arrives carrying prey, more time is required at the entrance, allowing certain miltogrammine flies an opportunity to larviposit on the prey before the female enters the nest. Perhaps for this reason certain sphecids, such as *Cerceris binodis*, leave the entrance open, thus permitting the returning female to drop directly into the entrance and close it from inside (Evans, 1971).

Final nest closure in ground-nesting groups is related to digging behaviour. Thus 'rakers' (e.g. Bembicini) face away from the burrow and scrape soil into the hole, whereas 'carriers' (e.g. *Ammophila*) collect pebbles or other particles that they carry back to the nest in order to fill the burrow. Most sphecids use the tip of the abdomen for packing soil in the burrow but Sphecinae pound the soil with their heads, or in the case of some *Ammophila*, with a pebble held in the mandibles. In stem-nesters the final plug is usually preceded by an empty vestibular cell and the plug itself is often similar to the material used in making nest partitions.

Several ground-nesting sphecids excavate accessory burrows, which may serve to divert the attention of parasitoids and cleptoparasites (Evans, 1966*c*), for example in *Stictia heros* (Sheehan, 1984). Other aspects of nesting behaviour that may serve to reduce mortality by natural enemies include nesting in dense aggregations (Rosenheim, 1990), construction of nests with few cells, and construction of intricate tunnels in multicellular nests.

3) Nest guarding by males. In species of *Trypoxylon* (*Trypargilum*) the male remains in the nest while his mate is out hunting (Richards 1934; Coville & Coville, 1980; Brockmann & Grafen, 1992). These male nest guards sometimes drive off solitary foraging ants, but are probably less effective against bombyliid and miltogrammine flies (Krombein, 1967). They also prevent other females of the same or different species from taking over the nest. Males of *Trypoxylon nitidum*, for example, were usually able to discourage females of *T. tenoctitlan* from entering the nest (Coville, 1981*a*). *T. superbum* does not make a final thick nest closure as do other members of the subgenus, probably because the males of this species guard the nest until all larvae have formed cocoons (Coville & Griswold, 1984). Similar nest guarding behaviour by males occurs in *Dynatus* (Kimsey, 1978), *Tachytes distinctus* (Lin & Michener, 1972), *Trypoxylon (Trypoxylon) fabricator* (Sakagami *et al.*, 1990), some *Pison* (Antropov, 1990) and *Oxybelus* (Hook & Matthews, 1980).

4) Direct removal of natural enemies. Some *Oxybelus* chase intruders away from the nest site (Bohart & Menke, 1976), and both *Bembix multipicta* and *Stictia maculata* have been observed physically removing ants from the vicinity of the nest (Cane & Miyamoto, 1979; Matthews *et al.*, 1981). Within the nest itself, progressive provisioners are in contact with the larva in its cell for a longer period of time, and thus presumably have more opportunities for cleaning the nest of intruders than do mass provisioners.

Immature stages. Sphecid larvae are relatively immobile and perhaps for this reason the adult female places the egg in a specific location on the prey, the exact position being more or less constant within sphecid taxa. In the case of hard-bodied prey the egg is usually placed on some weak point in the integument such as the front coxa (Sceliphrini, Sphecini), mid coxa (Ampulicinae), hind coxa (Psenini, some Gorytini), prosternum (Astatinae, some Crabronini), the beak of Hemiptera (*Bicyrtes*), or the wing base of flies (*Rubrica*). Others attach the egg by the posterior end only, usually on the underside of the prey, so that the emerging larva can move its head about (e.g. Cercerini). Those utilizing spiders usually place the egg on the ventral part of the abdomen (Trypoxylini). Some progressive provisioners lay the egg in the empty cell (Stizini, *Bembix*).

Sphecid larvae have ten pairs of well developed spiracles, mandibles with one or more lateral setae or sensilla, labrum usually somewhat bilobed, and maxilla with both palpus and galea well developed. In Ampulicinae and Sphecinae the spinneret is a long transverse slit, whereas in the other subfamilies the spinnerets are paired (Evans, 1987*a*). The cocoon spun by the fifth instar larva varies between groups. Many Nyssoninae incorporate sand grains into the walls of the cocoon and construct an elaborate series of pores, which are presumably adaptations allowing the pupa to survive should the nest become exposed (Evans *et al.*, 1974). In contrast the cocoons of Philanthinae, which are situated deeper in the soil, are delicate and thin-walled (Evans & O'Neill, 1988). Twig nesters likewise have fragile cocoons and *Microstigmus* species form none at all.

Adult feeding and 'sleeping'. As in other Hymenoptera, adult sphecids feed on sugary solutions from flowers, extrafloral nectaries and honeydew. Species of Sphecinae and Philanthinae are common on flowers whereas Pemphredoninae and Crabronini tend to prefer honeydew. Most sphecids have short tongues and are thus more common on flowers with shallow corollas such as Euphorbiaceae and Umbelliferae. However, *Ammophila*, *Bembix* and especially *Steniolia* have elongate mouthparts and exploit a variety of flowers (Evans, 1966*b*). *Trigonopsis cameronii* has been observed feeding on leaves and green berries of *Solanum*, where they show intraspecific aggression (Eberhard, 1974). In addition, many sphecids (e.g. Ampulicinae, Crabroninae, Mellinini) feed on body fluids exuding from prey that will eventually be given to the larva, or even completely devour a prey item (e.g. Evans, 1989).

Like many other insects sphecids pass the nights and periods of inclement weather in a non-active state that has been called 'sleeping' (Evans, 1966*b*). Individual females sometimes spend the night in the nest (*Bembix*), in a short burrow apart from the nest (*Microbembex*), or on plants, where they often grasp the plant in a characteristic pose. Males sometimes dig special burrows of their own in which to spend the night (e.g. *Astata, Bembix, Microbembex*), in which case they often possess rake setae on the fore basitarsus. In other species male and female wasps form large sleeping aggregations on vegetation (*Ammophila, Bembecinus, Steniolia*) or on the undersides of rock ledges (*Bembecinus quinquespinosus*—Evans *et al.*, 1986). Some loose aggregations consist of different species, but compact aggregations usually consist of a single species (both sexes or only males). In Costa Rica aggregations of *Chalybion* are often seen on porches, the adults returning each night to the same site (F. Parker, pers. comm.; Prentice & Walker, 1991).

Mating behaviour. Most female sphecids are thought to mate only once in their lives, the notable exceptions being those species of *Trypoxylon* and *Oxybelus* where males guard the nest (Coville & Coville, 1980; Brockmann & Grafen, 1992; Hook & Matthews, 1980), and at least one species of *Mellinus* where males are also territorial. In some sphecids the males patrol for receptive females in areas where the latter are emerging and/or nesting. For example, among species occurring in Central America, this type of mate searching is reported in *Bembecinus quinquespinosus* (O'Neill & Evans, 1983), *Microbembex monodonta, Bembix multipicta* (Evans, 1966*a*) and *Cerceris truncata* (Werner, 1961). Unlike bees, very few sphecids mate in areas where the females are foraging. Among those that do are *Ammophila procera* (Bohart & Knowlton, 1953) and a North American species of *Mellinus*, the latter defending territories on dung where females come to hunt flies (Evans, 1989). In other species males defend territories near areas where females are nesting/emerging, as in *Tachysphex terminatus* (Kurczewski, 1966) and many Philanthinae (Evans & O'Neill, 1988). In all species of Philanthinae where territoriality has been observed, the males mark their territories with pheromones produced by the mandibular glands, applying the chemicals to stems of plants by means of hair brushes on the clypeus (Evans & O'Neill, 1988). Although territoriality is widespread in the Sphecidae, scent-marking of territories appears to be restricted to this subfamily.

In many species of *Crabro*, including *C. costaricensis*, the male forelegs are greatly modified, and these modifications appear to be used in courtship by the male (Low & Wcislo, 1992). Another Costa Rican crabronine (apparently a species of *Crossocerus*) has extremely modified hind legs. In many Bembicini males fly about over the nesting area, occasionally pausing to assume a characteristic pose with the body close to the ground and the legs extended rather rigidly; males and females typically come together in the air, but fly off to mate on the ground or on a plant (Evans, 1966*a*).

Mimicry. A few Costa Rican sphecids, especially in the Gorytini, appear to mimic polistine vespids (see Chapter 15). At mid-altitudes (1500–2000 m) *Neogorytes* and one species of *Ochleroptera* have slender, dark bodies with darkened fore wings as in *Agelaia panamensis*. At lower altitudes a species of *Sagenista* has a robust, dark body with the fore wings dark basally but transparent distally, as in *Parachartergus*. A species of *Hoplisoides* is similar but the wings are yellowish with darkened apices, as in *Polybia rejecta*. Species of *Megistommum*, *Neoplisus*, *Psammaletes*, *Pseudoplisus* and many *Cerceris* are predominantly yellow with black markings, as in *Agelaia areata*. At least one species of *Stigmus* has the black—orange—black pattern that is so common in Scelionidae (see Chapter 9), although in this case it is not clear what the model is.

Economic importance. Certain species of Philanthini, especially in the Old World, are serious predators of honey bees. Many sphecids, however, are beneficial since they prey on groups of insects that are potentially pests, and a few have been used in classical biological control programmes (Clausen, 1978). *Larra bicolor* was introduced from Brazil into Puerto Rico to control the mole cricket, *Scapteriscus vicinus* (Gryllotalpidae). Although it became established, its effectiveness was apparently dependent upon the presence of certain plants whose flowers provide food for the adult wasps (Menke, 1992*a*). *Ampulex compressa* was introduced from New Caledonia into Hawaii to control the american cockroach, *Periplaneta americana*. It became established and evidently plays at least some role in lowering the population of the cockroach (Williams, 1942). *A. compressa* is now present in Venezuela as well (Menke & Yustiz, 1983).

Although most biological control efforts have emphasized introductions of exotic species, native sphecids could potentially be manipulated through construction of trap nests, providing nesting sites in preferred soil types (e.g. *Ammophila* and *Tachysphex*), sand piles (e.g. Bembicini), a dripping source of water (for mud) and flowers (for food) (F. Parker, pers. comm.).

Identification. The subfamilies and tribes of Sphecidae occurring in Costa Rica can be identified with the following key. Keys to the world genera of Sphecidae are provided in Bohart & Menke (1976); keys (in Spanish) to the neotropical genera will soon be available (Fernandez & Menke, in prep.).

Key to the subfamilies and tribes of Costa Rican Sphecidae

1 Metasoma petiolate, petiole composed of sternite I only; jugal lobe of hind wing large (Fig. 17.04).
[Usually medium to large wasps.] .... (Sphecinae) 2

— Metasoma variable, ***if*** petiolate, ***then*** jugal lobe small (Figs 17.05, 17.06); jugal lobe otherwise various. 4

2 Fore wing with 2*m-cu* joining *M* distal to 2*rs-m*, on third submarginal cell (Fig. 17.07); tarsal claws with two teeth near base (Fig. 17.09). .. .. **Sphecini** (p. 637)

— Fore wing with 2*m-cu* joining *M* proximal to 2*rs-m*, on second submarginal cell (Fig. 17.04); tarsal claws with one tooth near base (Fig. 17.10), or toothless. 3

3 Tarsi with plantar lobes ventrally (Fig. 17.08) and/or claws with mesal tooth (Fig. 17.10). ... . **Sceliphrini** (p. 637)
— Tarsi without plantar lobes, and claws simple, without tooth. .... **Ammophilini** (p. 638)

4 Mid tibia with two apical spurs. 5
— Mid tibia with one apical spur. 11

5 Tarsal claws with tooth (cf Fig. 17.10); notauli usually present. .... (Ampulicinae) 6
— Tarsal claws without tooth; notauli present or absent. 7

6 Metasoma petiolate; fore wing with distal apex of marginal cell removed from wing margin (Fig. 17.11). .... **Ampulicini** (p. 637)
— Metasoma sessile; fore wing with distal apex of marginal cell ending on wing margin (Fig. 17.12). .... **Dolichurini** (p. 637)

7 Hind wing with jugal lobe large (cf Fig. 17.04); fore wing with distal apex of marginal cell truncate (Fig. 17.13).
[Body stout, usually black.] .... **Astatinae** (p. 639)
— Hind wing with jugal lobe small (cf Figs 17.05, 17.06); fore wing with marginal cell variable. 8

8 Fore wing with abscissa of *Sc+R* between *Rs* and base of pterostigma very long, much more than half length of first submarginal cell (Fig. 17.14). .... **Nyssoninae: Stizini** (p. 648)
— Fore wing with abscissa of *Sc+R* between *Rs* and base of pterostigma shorter, about half or less of the length of first submarginal cell. 9

9 Fore wing with second submarginal cell not receiving either 1*m-cu* or 2*m-cu* (Fig. 17.15); metasoma petiolate. .... **Nyssoninae: Mellinini** (p. 644)
— Fore wing with second submarginal cell receiving either 1*m-cu* or 2*m-cu* or both (Figs 17.16, 17.17); metasoma sessile. 10

10 Fore wing with second submarginal cell triangular, petiolate above so that 2*rs-m* joins *Rs* before *Rs* meets 2*r-rs* (Fig. 17.16), or with two enclosed submarginal cells; propodeum with posterolateral projection or spinose process (Fig. 17.26). .... **Nyssoninae: Nyssonini** (p. 644)
— Fore wing with three submarginal cells, the second neither triangular nor petiolate above, thus with 2*rs-m* joining *Rs* after *Rs* meets 2*r-rs* (Fig. 17.17); posterolateral corners of propodeum rounded. .... **Nyssoninae: Gorytini** (p. 645)

11 Inner margins of eyes notched, sometimes weakly so (Fig. 17.21). 12
— Inner margins of eyes straight, sinuate or arcuate, but not notched. 13

12 Hind wing with jugal lobe large (cf Fig. 17.04); clypeus always broad, at least twice as broad as long, and with antennal sockets not contiguous with clypeofacial suture (cf Fig. 17.24). Infrequently collected. .... **Philanthinae: Philanthini** (p. 649)
— Hind wing with jugal lobe small (cf Fig. 17.06); clypeus various, usually narrow, less than 1.5 times as broad as long (Fig. 17.21) ***or*** if broad, ***then*** with antennal sockets contiguous with clypeofacial suture. Very common in collections. .... **Crabroninae: Trypoxylini** (p. 639)

13 Fore wing with two or fewer submarginal cells (Figs 17.18-17.20). 14
— Fore wing with three submarginal cells (Figs 17.05, 17.06) 17

14 Metanotum posterolaterally with pair of foliaceous squamae (Fig. 17.27); propodeum with basal projection (mucro); fore wing with first submarginal and first discal cells confluent, forming single cell (Fig. 17.18). ....**Crabroninae: Oxybelini** (p. 640)
— Metanotum not foliaceous; propodeum without basal process; fore wing with first submarginal and first discal cells separate (Figs 17.19, 17.20). 15

15 Fore wing with two submarginal cells, ***or if*** only one or none, ***then*** pterostigma is large, its hind margin strongly arcuate (Fig. 17.19). .....
Pemphredoninae: Pemphredonini (p. 638)

— Fore wing with a single submarginal cell ***and*** pterostigma narrow (Fig. 17.20). 16

16 Antennal scape longer than combined length of pedicel and first two flagellar segments, usually equal to pedicel plus at least first four flagellar segments (Fig. 17.22). ....**Crabroninae: Crabronini** (p. 640)

— Antenna with scape not longer than combined length of pedicel and first two flagellar segments. .... **Crabroninae: some Miscophini** (p. 639)

17 Metasoma with slender petiole composed of sternite only, tergite I displaced to apex of petiole. ....**Pemphredoninae: Psenini** (p. 638)

— Metasoma sessile, or with stout petiole composed of tergite and sternite. 18

18 Ocelli specialized, ***either*** with the front and/or hind ones deformed, flattened (Figs 17.28, 17.29) ***or*** all ocelli large and sunken into vertex. 19

— Ocelli unspecialized, circular with convex lens and not sunken into vertex. 20

19 Labrum elongate, noticeably longer than wide (Fig. 17.23); hind wing with jugal lobe at most a little more than half as long as claval area (Fig. 17.30); robust wasps, often with yellow and black patterns. .... **Nyssoninae: Bembicini** (p. 648)

— Labrum about as long as wide, or shorter; hind wing with jugal lobe subequal to length of claval area (Fig. 17.31); body typically black, or metasoma red. .... **Crabroninae: Larrini** (p. 639)

20 Hind femur thickest towards its base, its apex simple. .... **Crabroninae: some Miscophini** (p. 639)

— Hind femur thickest distally, outer apex with flattened plate that projects downwards, plate teardrop-shaped, its surface often setose, often margined by carina (Fig. 17.32). 21

21 Inner margins of eyes parallel or nearly so; antennal sockets not contiguous with frontoclypeal suture (Fig. 17.24). .... **Philanthinae: Cercerini** (p. 649)

— Inner margins of eyes strongly converging below; antennal sockets contiguous with frontoclypeal suture (Fig. 17.25). 22

22 Hind wing with *M* and *Cu* diverging proximal to cross vein *cu-a* (Fig. 17.05); fore wing with 3*rs-m* joining *Rs* near apex of marginal cell. ... . **Crabroninae: Scapheutini** (p. 640)

— Hind wing with *M* diverging from *Cu* distal to cross vein *cu-a* (Fig. 17.06); fore wing with 3*rs-m* joining *Rs* near middle of marginal cell. .... **Crabroninae: Bothynostethini** (p. 640)

Synopsis of the Costa Rican taxa

Seven of the eight subfamilies known in the New World are represented in Costa Rica; absent is the Xenosphecinae, which is restricted to southwestern North America. We have seen 82 genera from Costa Rica, but five additional genera are expected to occur in the country (Table 17.2). About 290 species are known from Costa Rica but the actual number is probably 400 or more.

Costa Rica (and Panama) is a zoogeographic crossroads as far as Sphecidae is concerned. A number of South American genera (e.g. *Microstigmus*, *Quexua*, *Neogorytes*, *Parataruma*, *Scapheutes*, and *Bohartella*) reach this area, but seemingly go no farther north. Several North American genera (e.g. *Podalonia*, *Chalybion*, *Mimumesa*, *Psenulus*, *Crabro*, *Lestiphorus*, *Psammaletes*, *Pseudoplisus*, *Tanyoprymnus* and *Eucerceris*) extend southward as far as Costa Rica or Panama. The genus *Pisonopsis* has a disjunct distribution: North and Central America and southern South America; the northern element extends southward into northwestern Costa Rica. South American species that appear to reach the northern limit of their distribution in Costa Rica include *Chlorion viridicoeruleum*, *Dynatus nigripes*, *Pison abathes*, *P. arachniraptor*, *P. maculipenne*, *P. cooperi* and *P. gnythos*. The genera to which these species belong extend northward into Mexico or North America. Likewise, certain northern species reach south only as far as the seasonally dry northwest of Costa Rica: *Ammophila procera*, *Eremnophila aureonotata*, *Prionyx parkeri*, *Tachysphex*

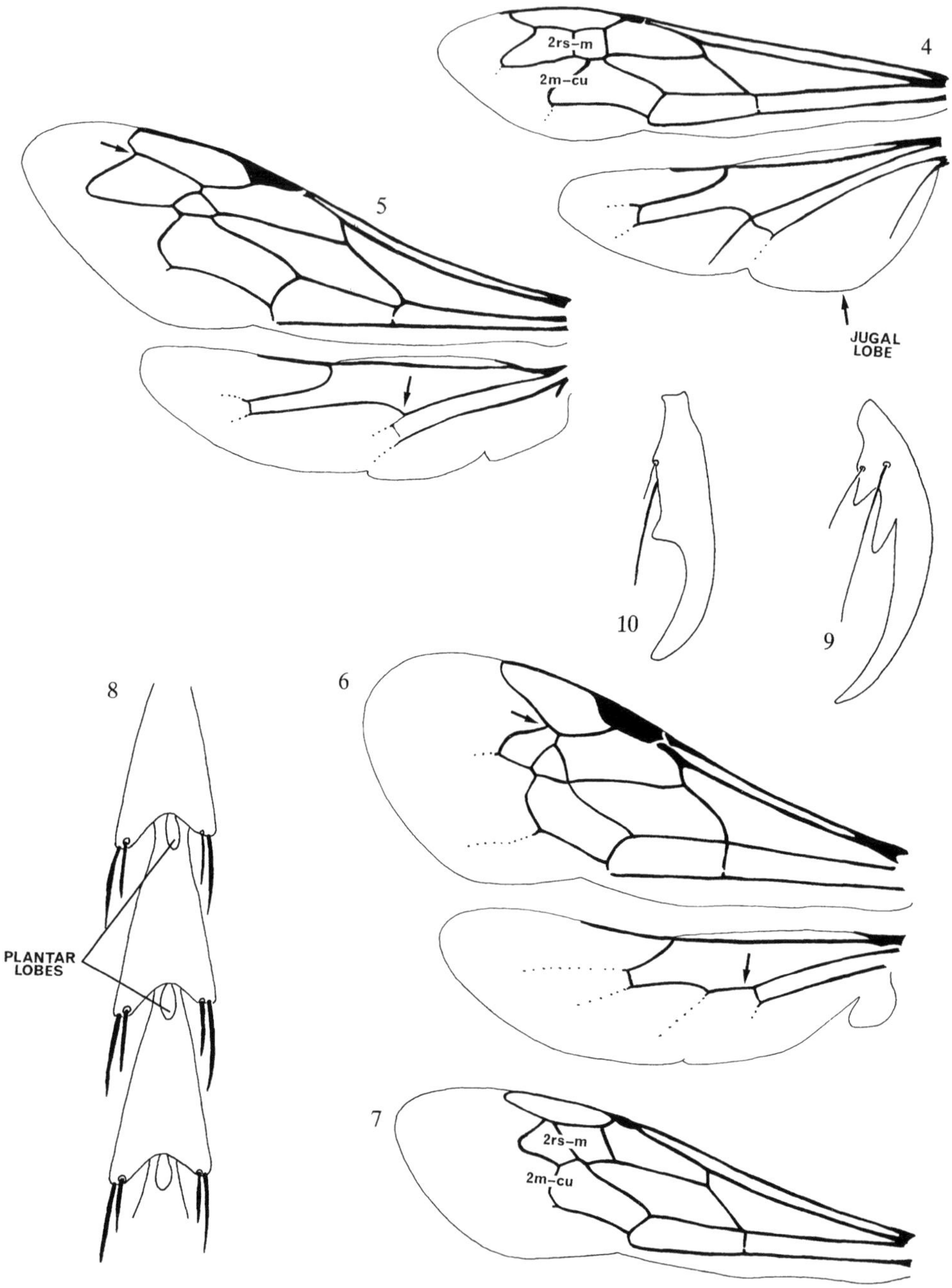

Figs 17.04–17.10. Sphecidae. Figs 17.04–17.06. Fore and hind wing; 17.04, *Penepodium* sp. (Sceliphrini); 17.05, *Scapheutes* sp. (Scapheutini); 17.06, *Bothynostethus* sp. (Bothynostethini). Fig. 17.07. Fore wing, *Sphex* sp. (Sphecini). Fig. 17.08. Hind tarsal segments 2–4, *Sceliphron* sp. (Sceliphrini). Figs 17.09–17.10. Tarsal claws; 17.09, *Isodontia* sp. (Sphecini); 17.10, *Chlorion* sp. (Sceliphrini).

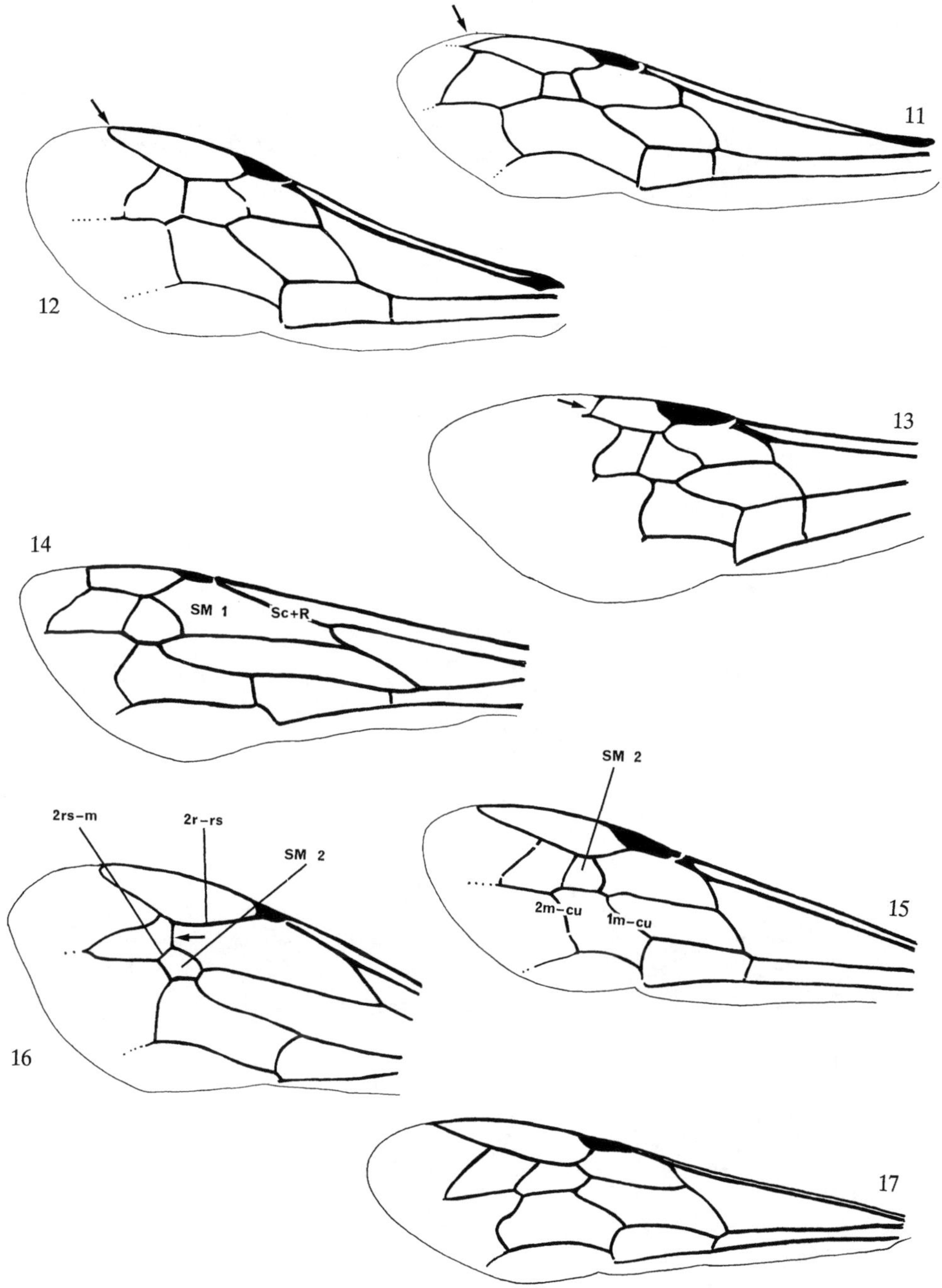

Figs 17.11–17.17. Fore wing of Sphecidae; 17.11, *Ampulex* sp. (Ampulicini); 17.12, *Dolichurus* sp. (Dolichurini); 17.13, *Astata* sp. (Astatini); 17.14, *Bembecinus* sp. (Stizini); 17.15, *Mellinus* sp. (Mellinini); 17.16, *Zanysson* sp. (Nyssonini); 17.17 *Neogorytes* sp. (Gorytini).

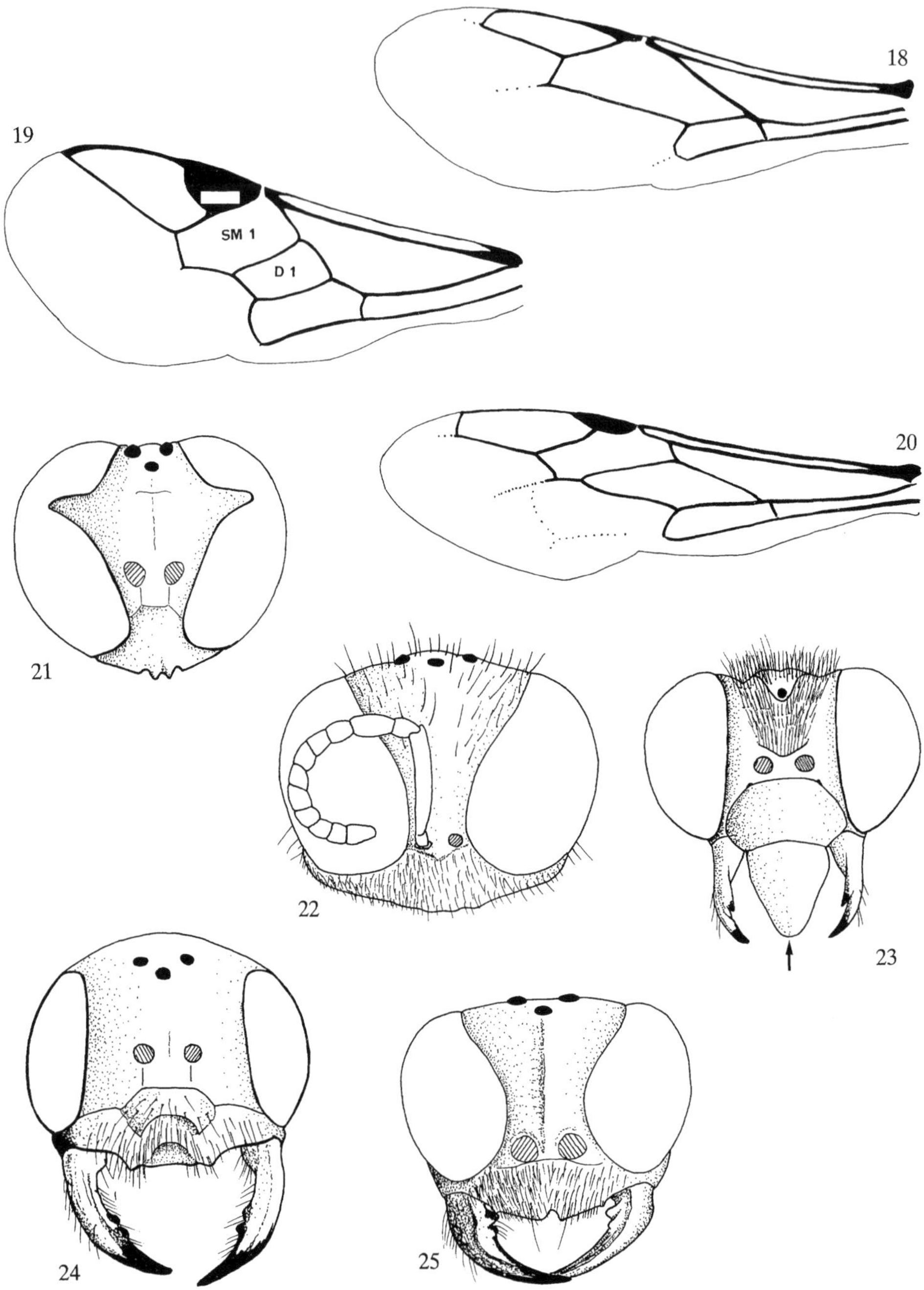

Figs 17.18–17.25. Sphecidae. Figs 17.18–17.20. Fore wing; 17.18, *Oxybelus* sp. (Oxybelini); 17.19, *Microstigmus* sp. (Pemphredonini); 17.20, *Ectemnius* sp. (Crabronini). Figs 17.21–17.25. Head, anterior; 17.21, *Trypoxylon* sp. (Trypoxylonini); 17.22, *Crabro* sp. (Crabronini); 17.23, *Steniolia* sp. (Bembicini); 17.24, *Cerceris* sp. (Cercerini); 17.25, *Scapheutes* sp. (Scapheutini).

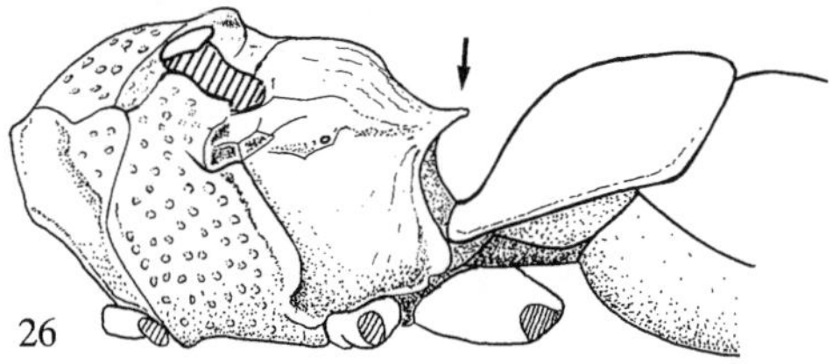

26

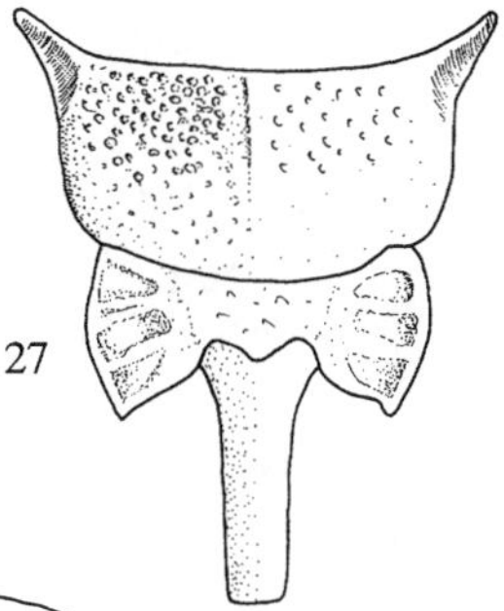

27

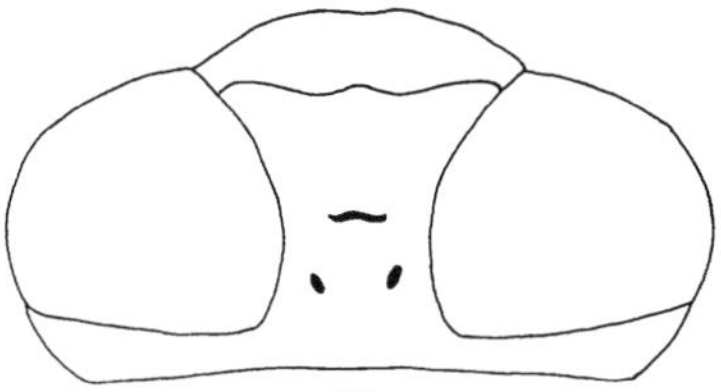

28

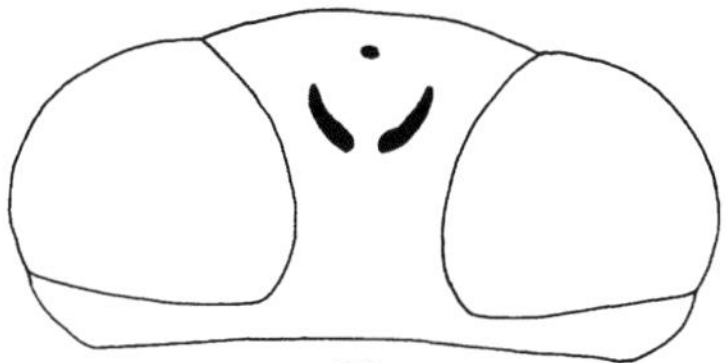

29

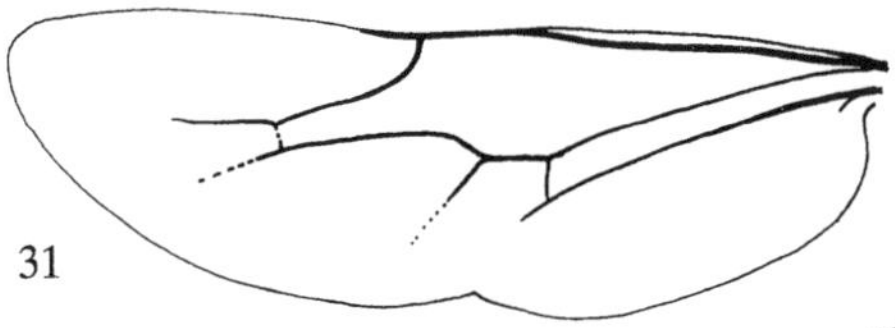

31

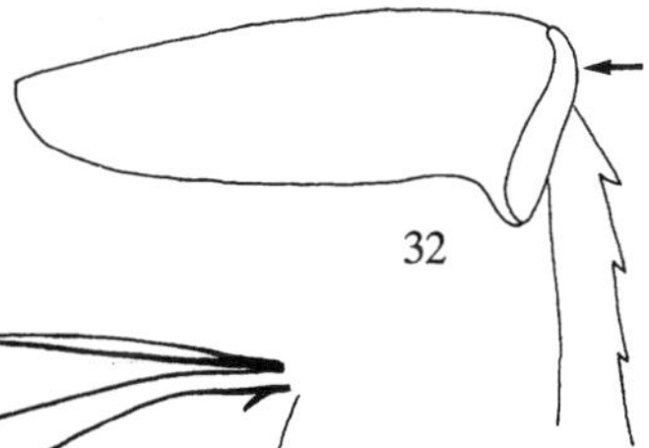

32

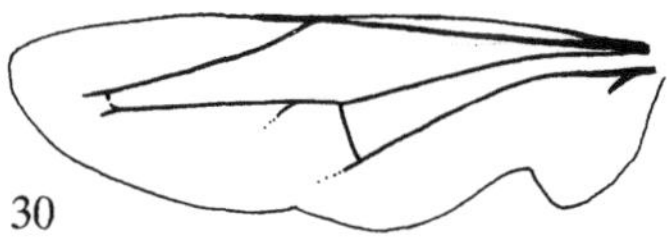

30

Figs 17.26–17.32. 17.25, *Sphecidae*. Fig. 17.26. Mesoma and base of metasoma, lateral, *Zanysson* sp. (Nyssonini). Fig. 17.27. Scutellum, metanotum and base of propodeum, *Oxybelus* sp. (Oxybelini). Figs 17.28–17.29. Head, dorsal; 17.28, *Bembix* sp. (Bembicini); 17.29, *Tachytes* sp. (Larrini). Figs 17.30–17.31. Hind wing; 17.30, *Microbembix* sp. (Bembicini); 17.31, *Larra* sp. (Larrini). Fig. 17.32. Hind femur, *Cerceris* sp. (Cercerini).

maya, *T. ashmeadii* and *T. maurus*, but the genera to which these species belong are found throughout the Neotropical region.

Within Costa Rica the majority of sphecid species occur at lower altitudes, below 1500 metres. In the lowlands many species occur in wet forests, with species of *Trypoxylon* being particularly common in Malaise trap samples. Bembicini (Nyssoninae) generally prefer more open sunny areas with sandy soil, such as beaches, although *Stictia maculata* occurs in shaded habitats in wet forests on the Osa Peninsula in southwestern Costa Rica (Matthews *et al.*, 1981). Many species occurring at mid altitudes (1500–2200 m) tend to be dark coloured, for example *Mellinus pygmaeus*. At high altitudes, above 2500 metres, sphecids are rather scarce though *Podalonia communis*, a common species in North America, occurs up to 3000 metres on the Cerro de la Muerte in Costa Rica.

In the following list the total numbers of genera and species in the world have been updated from those given in Bohart and Menke (1976). The numbers of species occurring in Costa Rica are based on both literature and personal observation. These latter numbers should be viewed as minimum estimates. In a few cases mention is made of species reported from Nicaragua, Panama or 'Central America' when it is thought likely that these will occur in Costa Rica. Only those taxonomic references published since Bohart and Menke (1976), or not listed by them, are given. Because so few of the Central American species have been studied, notes on biology generally refer to non-Central American species (references given in Bohart & Menke, 1976), except where noted.

Sphecid group	Number genera	Number species
AMPULICINAE		
Dolichurini	2	2
Ampulicini	1	4
SPHECINAE		
Sceliphrini	7	16
Sphecini	3	14
Ammophilini	3	9
PEMPHREDONINAE		
Psenini	4	11
Pemphredonini	4	12
ASTATINAE		
Astatini	1	6
CRABRONINAE		
Larrini	4	28
Miscophini	3	7
Trypoxylini	4	51
Bothynostethini	1	1
Scapheutini	2	3
Oxybelini	1	9
Crabronini	17	44
XENOSPHECINAE	–	–
NYSSONINAE		
Mellinini	1	3
Heliocausini	–	–
Alyssonini	–	–
Nyssonini	3	4
Gorytini	11	20
Stizini	1	4
Bembicini	6	14
PHILANTHINAE		
Philanthini	1	4
Aphilanthopini	–	–
Odontosphecini	–	–
Cercerini	2	22
Totals	82	288

Table 17.2. New World subfamilies and tribes of Sphecidae with numbers of genera and species known from Costa Rica. These numbers do not include five genera that are expected to occur in Costa Rica.

Additional information about sphecid biology in general can be found in Evans (1963*d*, 1966*a-d* & *g*), Evans and West-Eberhard (1970), Bohart and Menke (1976) and Iwata (1976).

AMPULICINAE

Two tribes are present in the New World. Both occur in Costa Rica.

AMPULICINAE: Dolichurini. Five genera, two of which are restricted to Australia and one to Asia.

***Dolichurus*[CR].** Cosmopolitan, primarily palaeotropical; 45 species, one in Costa Rica; 0-1400 m. Notauli well developed; hind wing jugal lobe present. Prey: Blattaria.

***Paradolichurus*.** Southern California to Brazil; four species, in Costa Rica only *boharti* (Kimsey, 1993*a*). Notauli undeveloped; hind wing jugal lobe absent. Prey: unknown.

AMPULICINAE: Ampulicini. Only one genus.

Ampulex. Cosmopolitan, primarily tropical; about 130 species, four in Costa Rica—*angusticollis, collator, maculicornis* and *neotropica*; 0–1400 m. Some species are dark coloured with banded wings (presumably ant mimics) while others are brilliant metallic green. Prey: Blattaria.

SPHECINAE

The three New World tribes are all represented in Costa Rica.

SPHECINAE: Sceliphrini. Eight genera; all but the monotypic *Stangeella* (South America) occur in Costa Rica.

Chalybion. (Fig. 17.01). Cosmopolitan but absent in South America; 40 species, only two in the New World, in Costa Rica only *zimmermanni* (Menke, 1991); 200–1200 m. Dark metallic blue; propodeum without dorsal U-shaped enclosure. Taxonomy: Hensen (1988). Prey: Araneae.

Chlorion. Cosmopolitan but absent in Australia; 19 species, in Costa Rica only *viridicoeruleum*. Metallic blue or green; propodeum with dorsal U-shaped enclosure. Prey: Gryllidae.

Dynatus. Central Mexico to Argentina; three species in Costa Rica, including *crassipes*; (*nigripes* occurs from Panama to Brazil). Very large (30 to 50 mm long); all black, wings often orange. Prey: Blattaria. Biology: *nigripes* (Kimsey, 1978; Cooper, 1980).

Penepodium. Central Mexico to Argentina; 22 species, three or more in Costa Rica—*foeniforme, goryanum* and one or more undetermined species; (*albovillosum* and *luteipenne* occur in Panama). Dark coloured, legs all black. Prey: Blattaria. Biology: *goryanum* nests in abandoned beetle burrows; *luteipenne* excavates burrows in soil, often using water to loosen the soil.

Podium. New World, primarily neotropical; 20 species, four in Costa Rica—*denticulatum, friesei, fumigatum* and *rufipes*; 0–1800 m. Head and mesosoma black, legs often partially red; wings infuscate or spotted. Prey: Blattaria.

Sceliphron. Cosmopolitan, primarily Old World; 30 species, three in Costa Rica—*assimile, caementarium* and *fistularium*; 0–1200 m. Body with yellow markings; propodeum with dorsal U-shaped enclosure. Prey: Araneae. Biology: *assimile* (Hunt, 1993); *fistularium* (Freeman, 1982).

Trigonopsis. Southern Mexico to southern Brazil; 16 species, in Central America only *cameronii*; 0–800 m. Head narrowed posteriorly, triangular; pronotum elongate, neck-like. Taxonomy: Vardy (1978). Prey: Blattaria. Biology: Eberhard (1974).

SPHECINAE: Sphecini. Five genera; all except *Chilosphex* (Mediterranean) and *Palmodes* (holarctic) occur in Costa Rica.

Isodontia. Cosmopolitan; 60 species, three in Costa Rica—*cyanipennis, fuscipennis* and *mexicana*; 0–700 m. Blue, black or black with red metasoma; side of propodeum with spiracular groove absent or incomplete. Prey: Gryllidae, Tettigoniidae.

Prionyx. Cosmopolitan, mostly in the Mediterranean area; 54 species, three in Costa Rica—*fervens, thomae* and *parkeri*; 0–1200 m. Second submarginal cell narrow above (anteriorly). Prey: Acrididae.

Sphex. Cosmopolitan; 116 species, eight in Costa Rica—*caliginosus, dorsalis, ichneumoneus, melanopus, nitidiventris, opacus, servillei* and *tinctipennis*; 0–1600 m. Side of propodeum with spiracular groove complete. Prey: mostly Tettigoniidae, but also Gryllidae and Gryllacrididae. Biology: *S. ichneumoneus* (Brockmann, 1979, 1985; Brockmann & Dawkins, 1979).

SPHECINAE: Ammophilini. Six genera, three of which are restricted to the Old World.

Ammophila. Cosmopolitan; 202 species, four in Costa Rica—*centralis, gaumeri, picipes* and *procera* (Menke, 1991); 0–1300 m. Mesepisternal sulcus extending straight down from subalar fossa. Prey: lepidopterous and symphytan larvae. Biology: *procera* (Evans, 1959).

Eremnophila. New World, primarily neotropical; nine species, three in Costa Rica—*aureonotata* (Menke, 1991), *melanaria* and *opulenta*; 0–1100 m. Mesepisternal sulcus curving posteriorly, then downward. Prey: *E. aureonotata*, notodontid larvae.

Podalonia. Cosmopolitan but absent in South America; 65 species, two in Costa Rica—*communis* and *robusta*; 220–3000 m. Metasoma abruptly widening beyond petiole (unlike the other two genera). Prey: *P. robusta*, larval Noctuidae; unearths cutworms (Kurczewski *et al.*, 1992).

PEMPHREDONINAE

The two tribes Psenini and Pemphredonini are both represented in Costa Rica.

PEMPHREDONINAE: Psenini. Ten genera; one is restricted to Japan, two to Hawaii, one (*Ammopsen*) to the southwestern United States and one (*Mimesa*) is essentially holarctic. In *Mimumesa* and *Pseneo* vein *cu-a* in the hind wing joins *Cu* after *Cu* and *M* have diverged, whereas in *Pluto* and *Psenulus cu-a* joins *M+Cu* at or before the point where the two veins diverge.

Mimumesa. Primarily holarctic; 31 species, two are restricted to the Caribbean while *longicornis* occurs in North America, Cuba and the Caribbean coast of Central America (0-400 m). Petiole with longitudinal carinae. Prey: Cicadellidae and Delphacidae.

Pluto. New World; 59 species, two in Costa Rica—*annulipes*, *punctatellus*; two additional species are known from Nicaragua and two others from Panama; 0–400 m. Hind coxa with a down-curved bristle. Taxonomy: Lith (1979). Prey: Cicadellidae.

***Psen*[EX].** Cosmopolitan but not yet collected from countries between Honduras and Peru. About 100 species. Prey: Homoptera-Auchenorrhyncha.

Pseneo. Considered a subgenus of *Psen* by Lith (1975*a*). Primarily New World but also southeast Asia; 25 species, about five in Costa Rica—*longiventris* (= *montezuma*) and four undetermined species; 0–1600 m. Petiole without longitudinal carinae. Taxonomy: Lith (1975*a* & *b*). Prey: Cicadellidae.

***Psenulus*[CR].** Cosmopolitan but absent in South America; about 165 species, three in Costa Rica—*mayorum* and two undetermined species; 0–1300 m. Hind coxa without a down-curved bristle. Prey: Aphidoidea, Psylloidea, or Cicadellidae and Delphacidae.

PEMPHREDONINAE: Pemphredonini. Seventeen genera, 13 in the New World; six of the latter do not occur south of Mexico. *Parastigmus* is known only from Argentina. *Polemistus* has three discoidal cells whereas the other genera have only two.

Microstigmus. Costa Rica to Paraguay; 22 species, four in Costa Rica—including *comes* and *thripoctenus*; (*miconiae* and *adelphus* occur in Panama); 0–1300 m. Petiole short; with one closed submarginal cell. Prey: Collembola (*comes*) or Thysanoptera (*thripoctenus*).

***Polemistus*[EX].** Cosmopolitan; 35 species, *yoda* is known from Nicaragua. Taxonomy: Menke and Vincent (1983). Prey: Aphidoidea.

Spilomena. Cosmopolitan; 73 species, three in Costa Rica—*foxii* and two undetermined species; 0-1600 m. Petiole short; with two closed submarginal cells. Prey: mostly immature Thysanoptera, but other records include immature Psylloidea, Coccoidea and Aphidoidea.

Stigmus. Cosmopolitan but absent in Africa and Australia; 31 species, about four in Costa Rica—*temporalis* and three undetermined species; 0–1600 m. Petiole elongate (longer than wide). Prey: Aphidoidea.

***Xysma*[CR].** Only two described species, one from the eastern United States and one from South Africa; one undetermined species in Costa Rica; also known from Ecuador (Cooper, 1993*b*) and Brazil. Petiole short; with no closed submarginal cells. Prey: Thysanoptera.

ASTATINAE

Four genera: *Astata* (see below), *Diploplectron* (holarctic and South Africa, primarily western United States and Mexico), *Dryudella* (primarily holarctic, south to Mexico) and *Uniplectron* (monotypic, Mexico).

Astata. Cosmopolitan but absent in Australia; 81 species, six in Costa Rica; 300–1500 m. Prey: nymphal and adult Hemiptera, especially Pentatomidae.

CRABRONINAE

All seven New World tribes of Crabroninae are present in Costa Rica.

CRABRONINAE: Larrini. Fifteen genera, eight in the New World. *Larropsis* occurs in northern Mexico, and *Ancistromma* and *Prosopigastra* probably occur there also. *Parapiagetia* is primarily an Old World genus but two species occur in southern South America. In *Larra* and *Liris* the frons has a transverse swelling just below the mid ocellus; this is absent in the other genera.

Larra. Cosmopolitan, primarily tropical; about 60 species, two in Costa Rica—*godmani* and *praedatrix*; 0–1200 m. Metasoma with at least some red. Taxonomy and biology: Menke (1992*a*). Parasitoids of Gryllotalpidae.

Liris. Cosmopolitan, primarily tropical; probably over 350 species, nine in Costa Rica—including *apicipennis*, *bellus*, *lutusator*, *sternalis* and *truncatulus*; (*panamensis*, *tenebrosus*, *tinctipennis* and *violaceipennis* occur in Panama); 0–1600 m. Metasoma all black. Taxonomy: North American species (Krombein & Gingras, 1984). Prey: Gryllidae.

Tachysphex. Cosmopolitan; over 410 species, twelve in Costa Rica; 0–1500 m. Ocellar scars oval. Taxonomy: Pulawski (1988). Prey: Acrididae, Tettigoniidae, Mantidae or Blattidae.

Tachytes. Cosmopolitan; about 300 species, five in Costa Rica—*amazonus*, *chrysopyga*, *distinctus*,*fraternus* and *ornatipes*; 0–1300 m. Ocellar scars very elongate, comma-like. Taxonomy: Bohart (1978). Prey: Acrididae, Tettigoniidae, Tetrigidae or Tridactylidae.

CRABRONINAE: Miscophini. Sixteen genera, five of which occur in the New World. *Lyroda* is known from North America and Brazil while *Plenoculus* extends as far south as southern Mexico in the New World.

Miscophus[CR]. Cosmopolitan but absent in Australasia; about 173 species, one in Costa Rica. Fore wing with two submarginal cells, the second triangular and petiolate above. Prey: small, usually immature Araneae.

Nitela. Cosmopolitan; about 50 species, five in Costa Rica—*bifida*, *carinifrons*, *costaricensis*, *guiana*[CR] and *schmidti*; 0–2000 m. Fore wing with one submarginal cell. Prey: Psocoptera.

Solieriella. Cosmopolitan but absent in Australia; 110 species, in Costa Rica only *costaricae*; 0–1300 m. Fore wing with two or three submarginal cells, if two then second not petiolate above. Taxonomy: Bohart (1990). Prey: mostly Hemiptera.

CRABRONINAE: Trypoxylini. Five genera; the monotypic *Pisoxylon* is known only from Peru and Brazil.

Aulacophilus[CR]. Mexico to Brazil; three species, in Central America only *jansoni*. Fore wing with two submarginal cells; metasoma petiolate. Prey: Araneae. Biology: Cooper (1986*a*).

Pison. Cosmopolitan, primarily Southern Hemisphere; about 190 species, thirteen in Costa Rica—*abathes*, *arachniraptor*, *cameronii*, *chrysops*, *conforme*, *cooperi*, *cressoni*, *eu*, *gynthos*, *krombeini*, *longicorne*, *maculipenne* and *pilosum* (Menke, 1988, 1990, 1992*b*); 0–1600 m. Fore wing with three (sometimes two) submarginal cells; marginal cell acute distally; metasoma not petiolate. Taxonomy: Menke (1988). Prey: Araneae.

Pisonopsis. New World; five species, two from southern South America and three from western North America; one of the latter, *birkmanni*, has recently been reported in Costa Rica (Menke, 1990). Fore wing with three submarginal cells; marginal cell rounded distally. Prey: Araneae.

Trypoxylon. Cosmopolitan; over 660 species, 36 in Costa Rica—20 in the subgenus *Trypargilum* and 16 in the subgenus *Trypoxylon*; 0–2000 m. Fore wing with one submarginal cell. Taxonomy: *T. Trypargilum* (Coville, 1981*b*). Prey: Araneae.

Biology: *fabricator* (Sakagami *et al.*, 1990); *monteverdeae* (Brockmann & Grafen, 1992); *nitidum* (Coville, 1981*a*); *superbum* (Coville & Griswold, 1984); *tenoctitlan* (Coville & Coville, 1980); *xanthandrum* (Coville & Griswold, 1983).

CRABRONINAE: Bothynostethini. Three genera; *Sanaviron* (Vardy, 1987) and *Willinkiella* are restricted to South America.

***Bothynostethus*.** Mexico to Brazil; ten species, one in Costa Rica; 0-400 m. Prey: Chrysomelidae.

CRABRONINAE: Scapheutini. Two genera. Taxonomy: Menke and Vardy (1980).

***Bohartella*[CR].** Costa Rica to Brazil; only two described species, with one undetermined species in Costa Rica. Propodeum coarsely areolated posteriorly. Prey: nymphs of Acrididae. Biology: apparently nest in soil (Vardy, 1987).

***Scapheutes*[CR].** Costa Rica to Brazil; three species, two from Costa Rica; 0–400 m. Propodeum smooth posteriorly. Prey: nymphs of Tettigoniidae (Cooper, 1988, 1993*a*).

CRABRONINAE: Oxybelini. Five genera, three in the New World. *Belomicrus* occurs in North America (south to Mexico), the Palaearctic and Africa, while the monotypic *Enchemicrum* is known only from the southern United States.

***Oxybelus*.** Cosmopolitan except Australia; about 250 species, nine in Costa Rica—*andinus, argentiopilosus, aztecus, irwini, marginatus, mexicanus* (= *bugabensis*), *peruvicus, sericeus* and *sparideus* (Bohart 1993); (*analis* occurs in Nicaragua); 0–1300 m. Prey: Diptera, especially Muscoidea.

CRABRONINAE: Crabronini. Forty seven genera, 23 of which occur in the New World. New World genera absent from Costa Rica include: *Alinia* (Venezuela to Argentina; Leclercq, in press), *Echucoides* (Peru and Bolivia), *Encopognathus* (mostly Old World but also California), *Huavea* (southwestern United States and Mexico), *Lindenius* (holarctic), *Moniaecera* (North America to southern Mexico), and *Tracheliodes* (holarctic and Brazil). The genera known to occur in Costa Rica may be separated using Table 17.3. It should be noted that *Podagritus* and *Rhopalum* are difficult to separate.

***Anacrabro*.** New World; 12 species, in Costa Rica only *mocanus*. Prey: Lygaeidae, Miridae.

***Chimila*[CR].** Mexico (Leclercq, pers. comm.) to Brazil; four species, one in Costa Rica. Taxonomy: Leclercq (1980*a*). Biology: unknown.

***Crabro*.** Mainly holarctic, in the New World south to Costa Rica; about 90 species, with *costaricensis* known from Costa Rica and *peltista* from Nicaragua. Prey: Diptera.

***Crossocerus*.** Cosmopolitan; about 230 species, three in Costa Rica, including *gemblacensis* and *viennensis*; 0–2000 m. Prey: mostly on small Diptera but also other insects. Biology: Costa Rican species have a broad pygidial plate and are therefore probably fossorial.

***Ectemnius*.** Cosmopolitan; about 180 species, 12 in Costa Rica—including *aztecus, basiflavus, burgdorfi, carinatus* (= *saxatilis*), *centralis, chagrinatus, excavatus, hector, pacuarus, schwarzi, semipunctatus* and *sonorensis* (*recuparatus* occurs in Panama). Mostly low altitudes, but *semipunctatus* (the most common species) occurs from sea-level to 1600 m. *E. azteca* and *E. centralis* have yellowish orange wings with darkened apices. Taxonomy: Leclercq (1991). Prey: various Diptera, especially Syrphidae, Stratiomyidae and Muscoidea. Biology: *E. basiflavus* nests in pith of *Cedrella odorata* (Meliaceae) seedlings (Leclercq, 1991).

***Enoplolindenius*.** New World; 18 species, four in Costa Rica—*chibcha, orotina, pugnans* and *serrei*; 0–2000 m. Biology: unknown.

***Entomocrabro*.** Neotropics; ten species, in Central America only *bequaerti* and *narinensis*. Taxonomy: Leclercq, (1981*a*). Prey: Cicadellidae.

- –Tergites I–IV or I–VI laterally carinate; venter of metasoma flat or concave .... *Anacrabro*
- –Only tergite I, or I-II, laterally carinate; venter of metasoma convex
  - –Mandible notched ventrally (cf Fig. 17.34)
    - –Ocellar triangle equilateral (cf Fig. 17.39); eyes glabrous, or at most with a few scattered hairs .... *Entomocrabro*
    - –Ocellar triangle posteriorly wider than long (cf Fig. 17.38); eyes setose .... *Entomognathus*
  - –Mandible not notched ventrally (Fig. 17.33)
    - –Vertex with tubercle above and behind eye (Fig. 17.33); apex of mandible simple
      - –Gena carinate from postocular tubercle to mandible (Fig. 17.33); acetabular carina present .... *Quexua*
      - –Gena without carina; acetabular carina absent .... *Holcorhopalum*
    - –Vertex without tubercle above and behind eye; apex of mandible variable
      - –Scapal basin (scrobe) carinate laterally (and often dorsally) (Fig. 17.35)
        - –Scapal basin carinate laterally but not dorsally
          - –Female pygidium with sharp medial ridge (Fig. 17.40), deep submedial notch and terminating in a sharp process surrounded by stout setae; gena with a vertical carina closely parallel to outer orbit .... *Parataruma*
          - –Female pygidial plate triangular; gena without carina .... *Taruma*
        - –Scapal basin carinate laterally and dorsally
          - –Mesoscutum with transverse anterolateral carinae (Fig. 17.47); female pygidial plate wide and flat. .... *Enoplolindenius*
          - –Mesoscutum simple; female pygidial plate narrow and concave .... *Foxita*
      - –Scapal basin not carinate
        - –First metasomal segment petiolate anteriorly, enlarged distally (Fig. 17.37); hind tibia often swollen apically
          - –Epicnemial and postspiracular carina present (Fig. 17.46), if not 1*m-cu* joining *Rs+M* well distal to middle of submarginal cell (cf Fig. 17.45) .... *Podagritus*
          - –Epicnemial and postspiracular carina absent, or if epicnemium is angulate, 1*m-cu* joining *Rs+M* at middle of submarginal cell .... *Rhopalum*
        - –Metasoma sessile (Fig. 17.36); hind tibia usually not swollen apically
          - –Ocellar triangle equilateral (Fig. 17.39) .... *Crossocerus*
          - –Ocellar triangle wider than long (Fig. 17.38)
            - –Upper inner margin of eye with distinct adjoining depression (fovea) which lacks setae (Fig. 17.38) .... *Lestica*
            - –Upper inner margin of eye without adjoining depression (fovea); if shallow depression present it is setose
              - –Female pygidium with a median longitudinal carina (cf Fig. 17.40) .... *Lecrenierus*
              - –Female pygidium flat (Fig. 17.41) .... *Crabro*
              - –Female pygidium narrow and concave (Fig. 17.42)
                - –Occipital carina flat dorsally (Fig. 17.43) .... *Pae*
                - –Occipital carina convex dorsally
                  - –Fore wing with first abscissa of *Rs+M* no more than twice length of second abscissa of *Rs+M* (Fig. 17.44) .... *Chimila*
                  - –Fore wing with first abscissa of *Rs+M* more than twice length of second abscissa of *Rs+M* (Fig. 17.45) .... *Ectemnius*

Table 17.3. Identification table for Costa Rican genera of Crabronini (modified from Bohart & Menke, 1976, and Fernandez & Menke, in prep.)

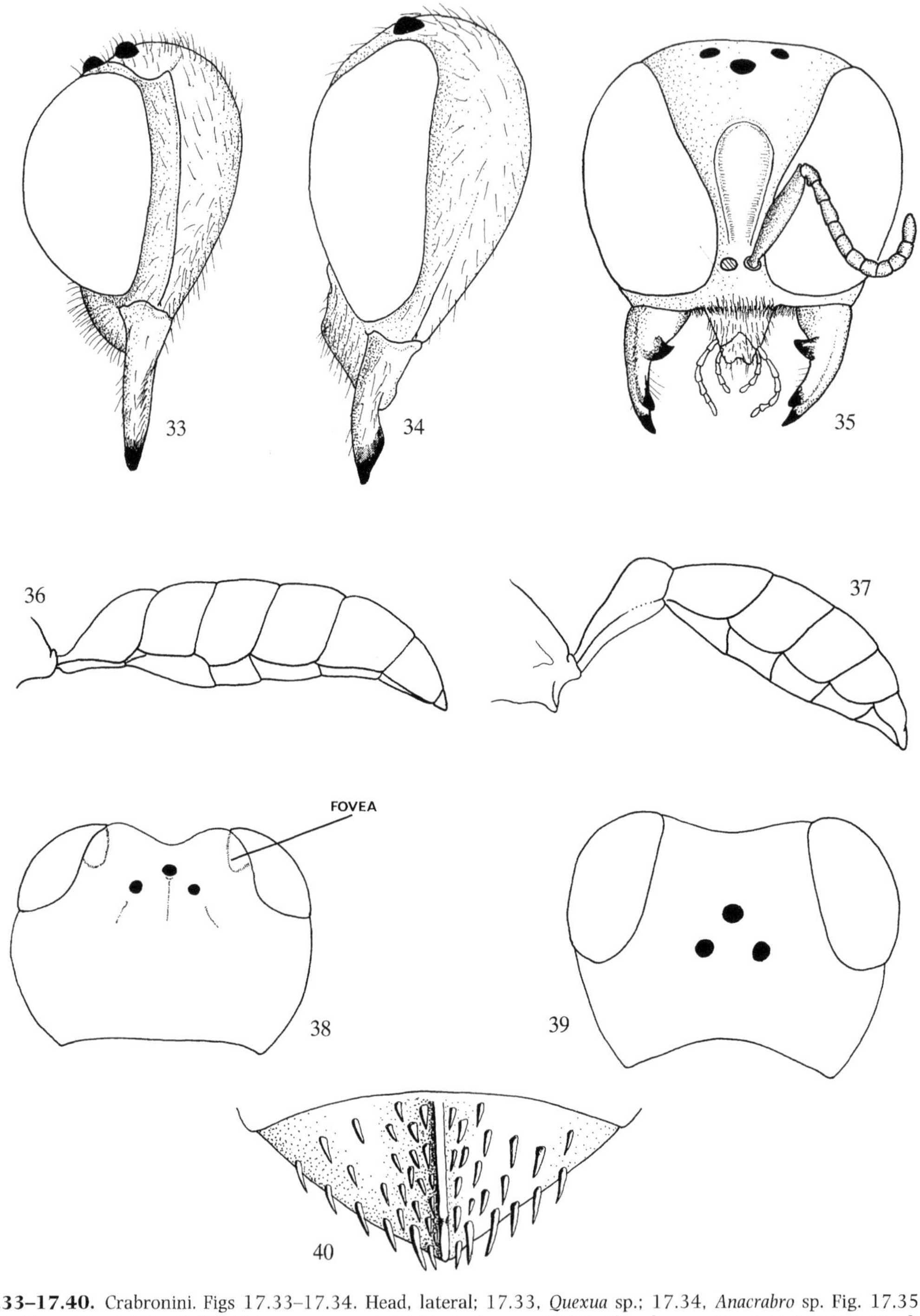

Figs 17.33–17.40. Crabronini. Figs 17.33–17.34. Head, lateral; 17.33, *Quexua* sp.; 17.34, *Anacrabro* sp. Fig. 17.35. Head, frontal, *Foxita* sp. Figs 17.36–17.37. Metasoma, lateral; 17.36, *Crabro* sp.; 17.37, *Rhopalum* sp. Figs 17.38–17.39. Head, dorsal; 17.38, *Lestica* sp.; 17.39, *Crossocerus* sp. Fig 17.40. Last visible tergite (pygidium) of female, *Parataruma* sp.

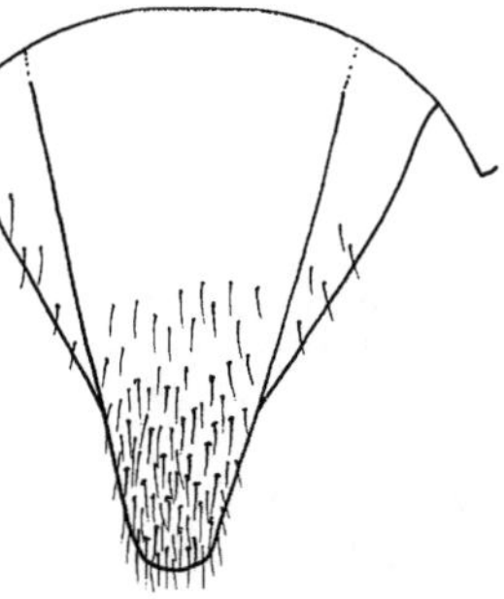

41

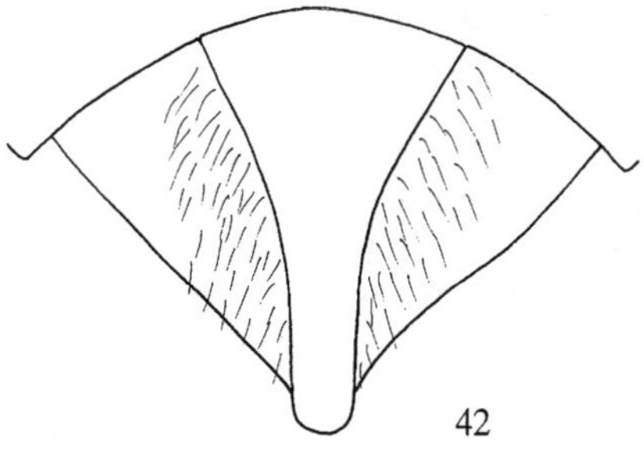

42

43

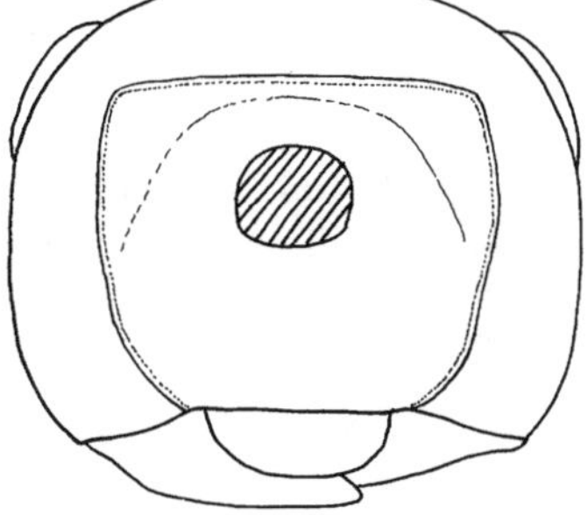

44

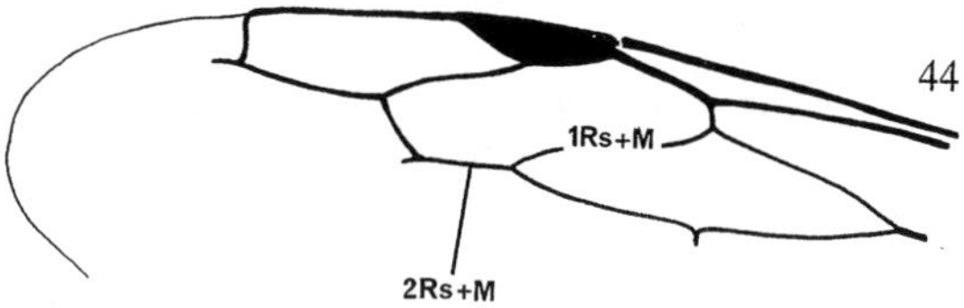

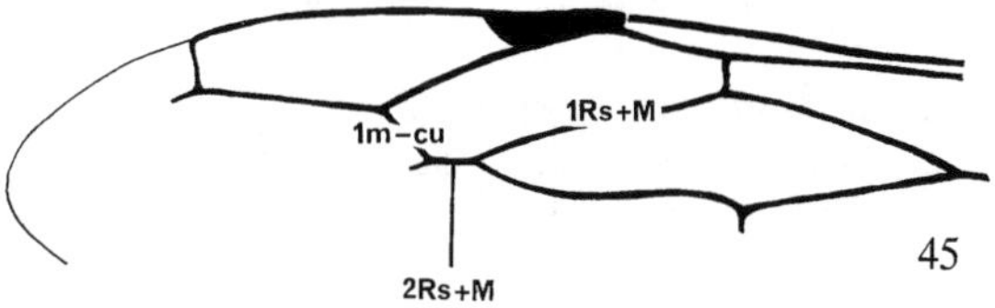

45

46

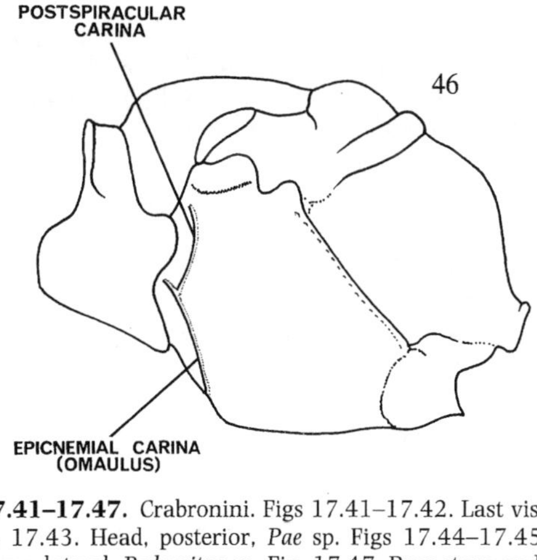

47

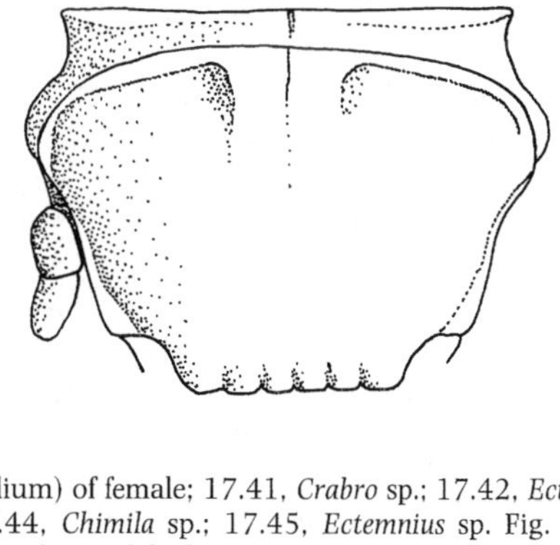

Figs 17.41–17.47. Crabronini. Figs 17.41–17.42. Last visible tergite (pygidium) of female; 17.41, *Crabro* sp.; 17.42, *Ectemnius* sp. Fig. 17.43. Head, posterior, *Pae* sp. Figs 17.44–17.45. Fore wing; 17.44, *Chimila* sp.; 17.45, *Ectemnius* sp. Fig. 17.46. Mesosoma, lateral, *Podagritus* sp. Fig. 17.47. Pronotum and mesoscutum, dorsal, *Enoplolindenius* sp.

Entomognathus. Cosmopolitan, absent from Australia; 45 species, in Costa Rica only *geometricus*; 300–1200 m. Prey: Chrysomelidae.

Foxita. Costa Rica to Paraguay; 17 species, five in Costa Rica—*galibi*, *woyowai* (common on the Osa Peninsula) and three undetermined species; 0–600 m. Taxonomy: Leclercq, (1980*a*). Biology: unknown.

Holcorhopalum. Mexico to Guyana; only one species—*foveatum*; 0–400 m. Biology: unknown.

Lecrenierus^{CR}**.** Costa Rica to Argentina; 11 species, one in Costa Rica. Taxonomy: Leclercq (1977). Biology: unknown.

Lestica. Cosmopolitan; 45 species, two in Costa Rica—*constanceae* and *florkini*; 400–600 m. Prey: adult Lepidoptera (e.g. small Noctuidae, Pyralidae, Tortricidae).

Pae (= *Lamocrabro*). Mexico to Brazil; three species, in Costa Rica only *paniquita*. Biology: unknown.

Parataruma. Southern Mexico to Brazil; two species, in Central America only *leclercqi*; 0–400 m. Taxonomy: Kimsey (1982*b*). Biology: unknown.

Podagritus. Mexico to Argentina and Chile, Australia and New Zealand; about 90 species, with three species in Costa Rica—*fulvohirtus* and two undetermined; 1000–2000 m. Taxonomy: Leclercq (1981*b*, 1982). Prey: various Diptera, but in New Zealand some species locally prey on Ephemeroptera, Trichoptera and some Coleoptera (Cantharidae, Chrysomelidae, Scirtidae) (Harris, 1990).

Quexua. Costa Rica to Bolivia and Brazil; twelve species, two in Costa Rica—*nericata* (tentative identification) and *ricata*; 0–800 m. Taxonomy: Leclercq (1980*b*). Prey: Cicadellidae (Cooper, 1986*b*).

Rhopalum. Cosmopolitan; 127 species, three in Costa Rica—*calverti*, *nicaraguense* and *tristani*; 0–2600 m. Prey: small flies (e.g. Chironomidae), Psocoptera, Aphidoidea and other insects.

Taruma^{EX}**.** Mexico, Peru and Brazil; just one species—*bara*; 0–800 m. May be a synonym of *Foxita*. Taxonomy: Leclercq (1980*a*). Biology: unknown.

NYSSONINAE

Five of the seven New World tribes are represented in Costa Rica.

NYSSONINAE: Mellinini. One genus.

Mellinus. Primarily holarctic, south in the New World to Venezuela, possibly Brazil; ten species, possibly three in Costa Rica—*alpestris* and perhaps two undescribed species (Menke, in press). Taxonomy: Siri and Bohart (1974). Prey: Diptera, especially Muscoidea.

NYSSONINAE: Nyssonini. Eighteen genera, 13 of which occur in the New World. *Antomartinezius*, *Cresson*, *Neonysson* and *Perisson* are restricted to southern South America. *Idionysson* and *Losada* occur further north in South America (the latter in Venezuela) but have not been found in Central America. *Hyponysson* is nearctic while *Synnevrus* is holarctic, but neither has been reported from Central America. *Nysson* is cosmopolitan but apparently absent in Australasia and the Neotropics.

Epinysson. New World; 22 species, two in Costa Rica including *guatemalensis*; (*sigua* occurs in Panama). Hind tibia without teeth along posterior margin; metasoma with sternites normal. Biology: *E. guatemalensis* has been reported as a cleptoparasite of *Hoplisoides costalis*.

Foxia^{EX}**.** Southwestern United States to Argentina; ten species, with *cuna* reported from Panama. Hind tibia without teeth along posterior margin; metasoma with sternites toothed laterally. Biology: unknown.

Metanysson^{CR}**.** Southwestern United States to Argentina; 16 species, one species in Costa Rica. Hind tibia with teeth along posterior margin; fore wing with two submarginal cells. Taxonomy: Fritz (1973). Biology: apparently cleptoparasites of *Cerceris* species.

Zanysson^{CR}**.** New World; 17 species, in Costa Rica only *changuina*. Hind tibia with teeth along posterior margin; fore wing with three submarginal cells. Biology: cleptoparasites of *Tachytes*.

NYSSONINAE: Gorytini. Over 400 described species in 31 genera. Ten genera occur in both the Old and New World, nine are restricted to the Old World and 12 to the New World. Of the latter, five (*Arigorytes, Hapalomellinus, Harpactus, Trichogorytes* and *Xerogorytes*) are restricted to North America (primarily southwestern United States and Mexico) while two (*Liogorytes* and *Pterygorytes*) are restricted to South America. *Clitemnestra* is restricted to Chile and Australia. *Gorytes* occurs in the Holarctic and Africa. *Oryttus* occurs in the Holarctic, Africa and Chile. Care is needed to separate genera, but Table 17.4 will facilitate their recognition. It should be noted that *Megistommum* and *Pseudoplisus* are difficult to separate.

Argogorytes. Cosmopolitan but absent from Africa; 27 species, two in Costa Rica; 0–1400 m. Prey: *Aphrophora* species, Cercopidae (spittlebug nymphs).

Hoplisoides. Cosmopolitan but absent from Australia; 70 species, four in Costa Rica—*costalis, denticulatus, iridipennis* and *vespoides*; 0–1400 m.

- –Hind wing with *cu-a* joining *M*+*Cu* almost its own length or more before *M* and *Cu* diverge (Figs 17.50, 17.51)
    - –Frons broader at level of mid ocellus than below it (Fig. 17.61) .... *Ochleroptera*
    - –Frons narrower at level of mid ocellus than below it (Fig. 17.60)
        - –Hind wing with *M* diverging from *Cu* at more or less 90° (Fig. 17.50) .... *Neogorytes*
        - –Hind wing with *M* diverging from *Cu* at an acute angle (Fig. 17.51) .... *Argogorytes*
- –Hind wing with *cu-a* joining *Cu* distal to divergence of M, or joining *M*+*Cu* close to or slightly before point of divergence (Figs 17.48, 17.49)
    - –Female with two rake setae on fore basitarsus before apex (Fig. 17.52) ***and*** spiracular groove present on side of propodeum (Fig. 17.53)
        - –Tergite I club-like, distal part of tergites convexly enlarged (Fig. 17.58). .... *Megistommum*
        - –Tergite I variable, ***if*** club-like ***then*** distal part of tergites not convexly enlarged (Fig. 17.59) .... *Pseudoplisus*
    - –Female usually with three or more rake setae on fore basitarsus before apex; spiracular groove present or absent on side of propodeum.
        - –Acetabular carina complete (Fig. 17.57)
            - –Propodeum coarsely areolate dorsolaterally .... *Sagenista*
            - –Propodeum punctate or smooth dorsolaterally .... *Hoplisoides*
        - –Acetabular carina absent or incomplete
            - –Metasomal segment I not elongate, not noticeably narrower than II (cf Fig. 17.55); mesopleuron without any trace of sternaulus
                - –Clypeus laterally almost a right angle or obtuse; large wasps, at least 25 mm in length .... *Sphecius*
                - –Clypeus laterally acute; body 13 mm or less in length .... *Tanyoprymnus*
            - –Metasomal segment I elongate, often noticeably narrower than II (Fig. 17.56); mesopleuron usually with sternaulus (Fig. 17.54)
                - –Sternaulus complete, continuous with epicnemial carina; anterior median part of propodeum smooth .... *Neoplisus*
                - –Sternaulus incomplete, not as well defined as epicnemial carina; anterior median part of propodeum longitudinally striate
                    - –Hind wing with *cu-a* evenly sinuous (Fig. 17.48); anterior median part of propdoeum striate only at the base .... *Lestiphorus*
                    - –Hind wing with *cu-a* strongly curved anteriorly (Fig. 17.49); propodeal inclusion completely covered with longitudinal striae. .... *Psammaletes*

Table 17.4. Identification table for genera of Gorytini present in Costa Rica (modified from Bohart & Menke, 1976, and Fernandez & Menke, in prep.)

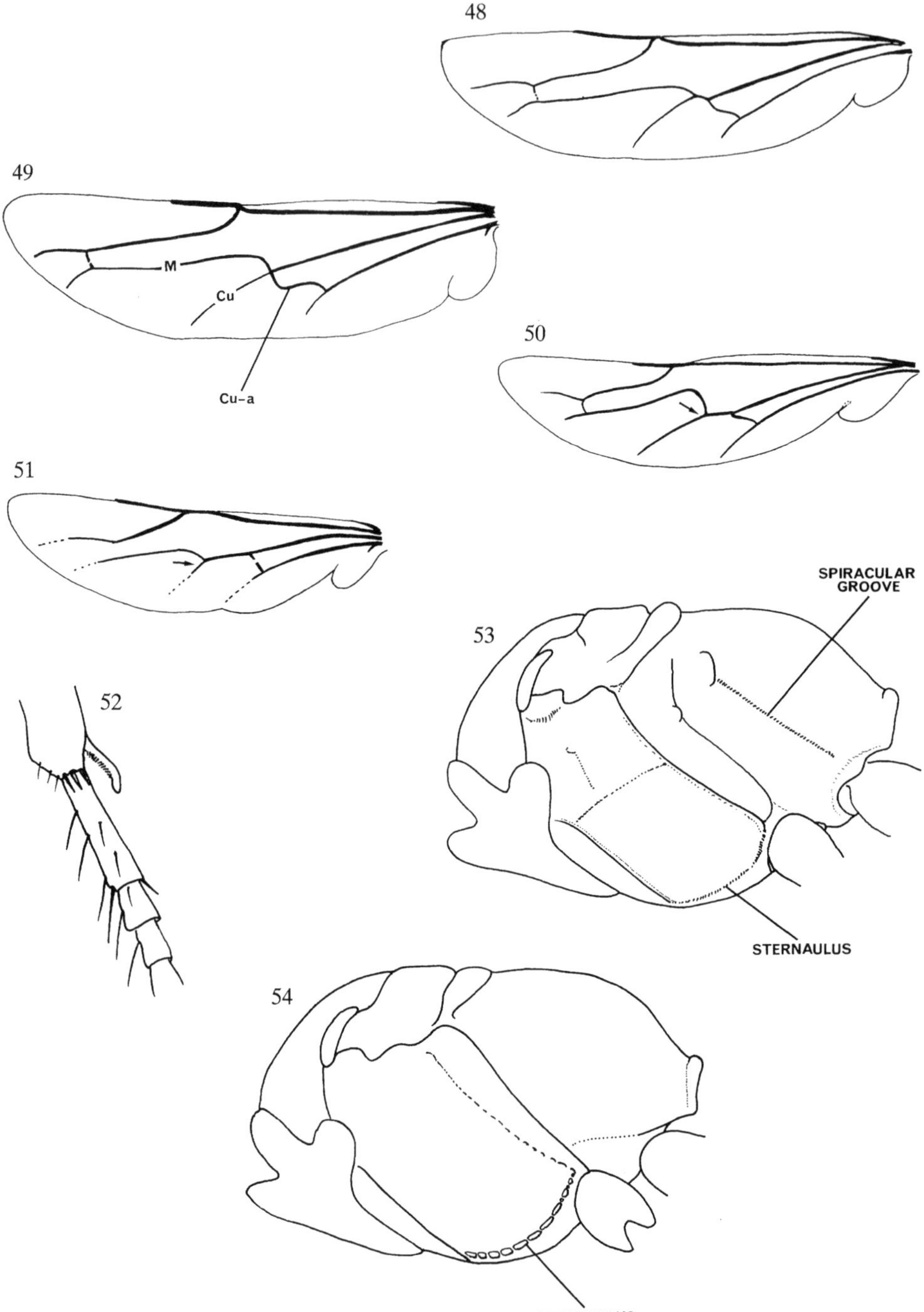

Figs 17.48–17.54. Gorytini. Figs 17.48–17.51. Hind wing; 17.48, *Lestiphorus* sp.; 17.49, *Psammaletes* sp.; 17.50, *Neogorytes* sp.; 17.51, *Argogorytes* sp. Fig. 17.52. Fore basitarsus *Megistommum* sp. Fig. 17.53. Propodeum lateral, *Megistommum* sp. Fig. 17.54. Mesosoma, lateral, *Psammaletes* sp.

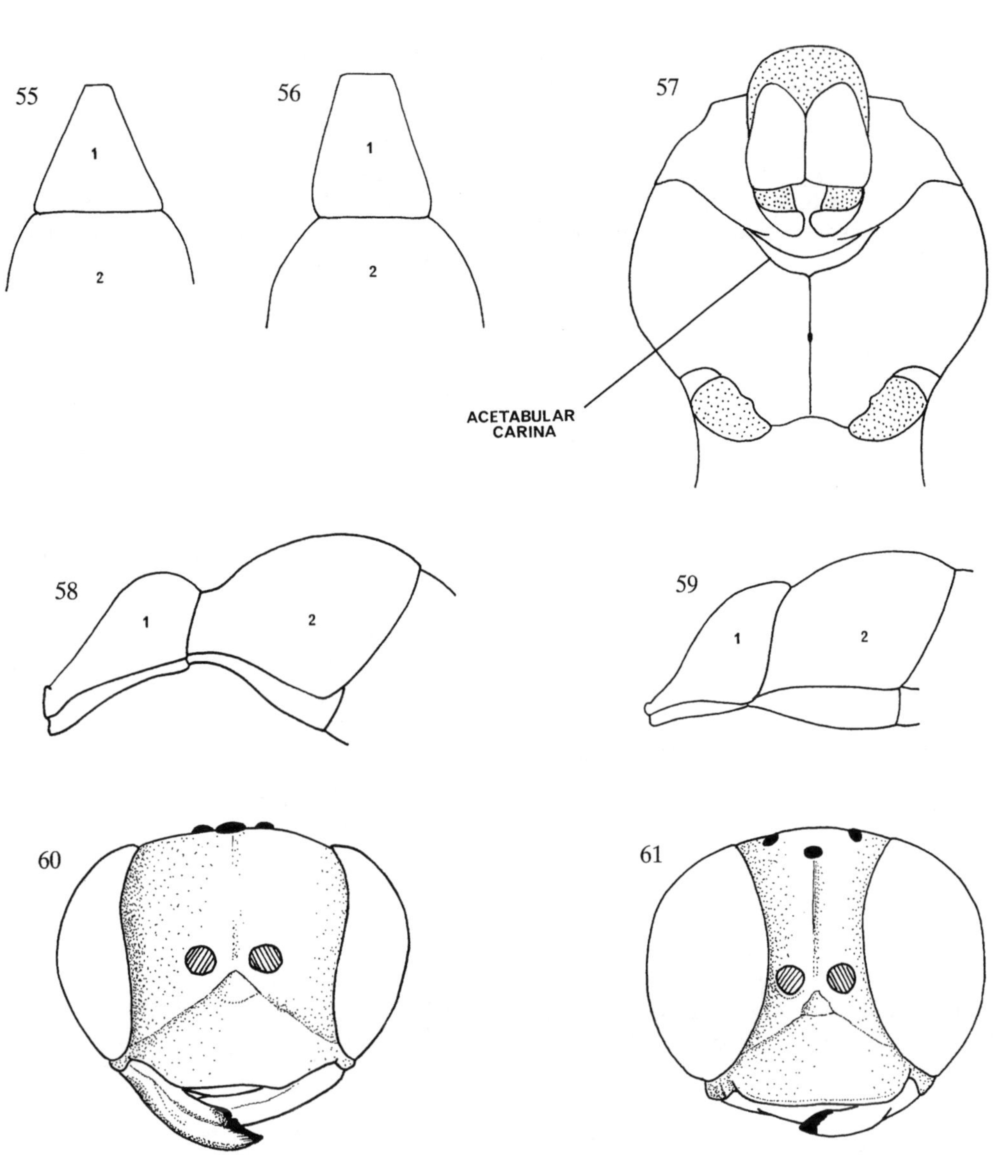

Figs 17.55–17.61. Gorytini. Figs 17.55–17.56. Tergites I and II, dorsal; 17.55, *Sagenista* sp.; 17.56, *Psammaletes* sp. Fig. 17.57. Pro- and mesosternal region, *Sagenista* sp. Figs 17.58–17.59. Tergite I, lateral; 17.58, *Megistommum* sp.; 17.59, *Pseudoplisus* sp. Figs 17.60–17.61. Head, anterior; 17.60, *Argogorytes* sp.; 17.61, *Ochleroptera* sp.

Prey: Homoptera-Auchenorrhyncha; *H. costalis*, Membracidae; *H. denticulatus*, Cicadellidae (Evans, 1966*a*).

Lestiphorus[CR]**.** Cosmopolitan but absent in Australasia; 17 species, one in Costa Rica. Prey: Cercopidae.

***Megistommum*.** Mexico to Brazil; eight species, in Costa Rica only *splendidum*; 0–600 m. Biology: unknown.

Neogorytes[CR]**.** Highlands of Peru and Ecuador to Costa Rica; one described species; 1600 m. Biology: unknown.

Neoplisus. Mexico to Argentina; nine species, in Costa Rica *notabilis* and possibly another species; 0–400 m. Biology: unknown.

Ochleroptera. New World and New Guinea; 12 species, four in Costa Rica; 0–2000 m. Prey: Cicadellidae, Cercopidae, Fulgoroidea, Membracidae and Psylloidea.

Psammaletes**[CR].** North America south to Costa Rica; three species, two in Costa Rica; 0–600 m. Prey: Fulgoroidea.

Pseudoplisus. North America (primarily in the southwest) south to Costa Rica, also Europe and Africa; 33 species, in Costa Rica only *tritospilus*; 200 m. Prey: apparently Cercopidae. Biology: unknown.

Sagenista. Mexico to Brazil; seven species, two in Costa Rica—*cayennensis* and an undetermined species; 0–400 m. Prey: nymphal and adult Fulgoroidea.

Sphecius**[EX].** Cosmopolitan; 21 species in the Neotropics, known from the Caribbean, Honduras, Brazil and Argentina. Prey: Cicadidae.

Tanyoprymnus. North America south to Costa Rica; only one described species—*moneduloides*. Prey: Dictyopharidae (Fulgoroidea).

NYSSONINAE: Stizini. Three genera; *Stizoides* and *Stizus* are widely distributed but absent from the Neotropics, southeast Asia and Australia.

Bembecinus. Cosmopolitan; 162 species, four in Costa Rica—*agilis, bolivari, mexicanus* (tentative identification) and *quinquespinosus*. Prey: Cicadellidae and other Homoptera-Auchenorrhyncha. Biology: *agilis* (Callan, 1991); *bolivari* (Callan, 1991); *quinquespinosus* (Evans *et al.*, 1986).

NYSSONINAE: Bembicini. Seventeen genera; except for the cosmopolitan *Bembix*, all are restricted to the New World. Two genera (*Chilostictia* and *Xerostictia*) are restricted to North America (including Baja California), while three others (*Glenostictia, Microstictia* and *Stictiella*) are primarily North American but have species that reach into central and southern Mexico. Six genera (*Carlobembix, Editha, Hemidula, Selman, Trichostictia* and *Zyzzyx*) are restricted to South America. In Costa Rica bembicines are most frequently found along sandy beaches. The genera of Bembicini are rather difficult to discriminate, but a key to them is given by Bohart and Menke (1976).

Bembix. Cosmopolitan; about 350 species, in Costa Rica only *multipicta*. Prey: a variety of Diptera (but usually not Nematocera). Biology: Cane and Miyamoto (1979).

Bicyrtes. New World; 23 species, four in Costa Rica—including *dicisa* and *variegata*; (*spinosa* occurs in Panama and *fodiens* is reported from 'Central America'). Prey: Hemiptera (usually nymphs), especially Coreidae, Pentatomidae and Reduviidae. Biology: *B. dicisa* (Callan, 1991), *B. variegata* (Martin & Martin, 1990; Callan, 1991).

Microbembex. New World; 28 species, in Costa Rica only *monodonta*. Prey: nest is provisioned with a wide variety of dead or disabled insects.

Rubrica. Southern Mexico to Argentina; four species, two in Costa Rica—*denticornis* and *nasuta*. Prey: flies, especially Syrphidae and Tabanidae. Biology of *nasuta* (= *surinamensis*): Evans *et al.* (1974), Callan (1991).

Steniolia. North America to northwestern South America; 15 species, two in Costa Rica—*guatemalensis* and *longirostra*. Prey: flies, especially Bombyliidae. Biology: *S.longirostra* (Larsson, 1990).

Stictia. New World, primarily South America; 28 species, four in Costa Rica—*heros, maculata, punctata* and *signata*. Biology: *S. heros* (Sheehan, 1984; Larsson & Larsson, 1989); *S. maculata* (Matthews *et al.*, 1981); *S. signata* (Post, 1981; Philippi & Eberhard, 1986; Martin & Martin, 1990; Callan, 1991).

PHILANTHINAE

Two of the four New World tribes of Philanthinaae are represented in Costa Rica.

PHILANTHINAE: Philanthini. Two genera. *Philanthus* occurs in the Holarctic and Africa; the North American species *P. gibbosus* extends south to El Salvador. Biology: Evans and O'Neill (1988).

Trachypus. Southern Texas to Argentina; 31 species, four in Costa Rica—*gracilis, mexicanus, petiolatus* and *spinosus*; 0–3000 m. Taxonomy: Rubio Espina (1976). Prey: *T. petiolatus* (= *mendozae*) has been observed preying on halictine bees in Costa Rica (Eickwort & Eickwort, 1972*a*). Biology: *mexicanus* (Menke, 1980), *petiolatus* (Evans & Matthews, 1973; Menke, 1980).

PHILANTHINAE: Cercerini. Two genera.

Cerceris. Cosmopolitan; the largest genus of Sphecidae with over 930 described species, 21 in Costa Rica. Fore wing with 3*rs-m* joining *Rs* about two-thirds of way along marginal cell. Prey: North American species prey exclusively on Coleoptera (primarily Curculionidae, Buprestidae, Chrysomelidae, Tenebrionidae and Bruchidae) but a few Old World species prey on bees or even other sphecids. Biology: Scullen and Wold (1969), Evans (1971), Callan (1990).

***Eucerceris*[CR].** North America (primarily the southwestern United States and Mexico) south to Panama; 41 species, in Costa Rica only *violaceipennis*. Fore wing with 3*rs-m* joining *Rs* close to distal end of marginal cell. Prey: Curculionidae.

18 The bees (Apidae)

Terry Griswold, Frank D. Parker and Paul E. Hanson

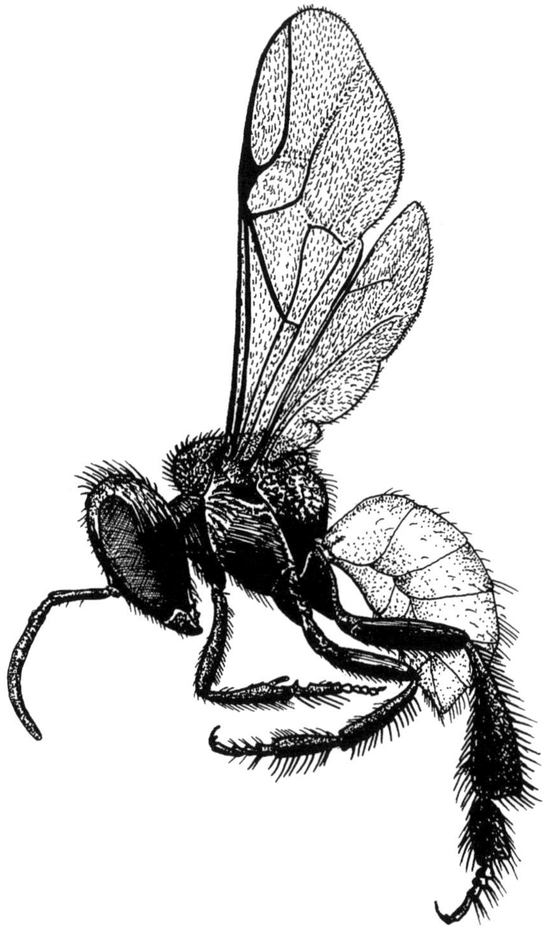

Fig. 18.01. *Trigona* sp. (Meliponinae).

Diagnosis Body length 1.5 to 46.0 mm, in Costa Rica ranging from the minute stingless bee *Trigonisca discolor* (less than 3 mm) to the robust carpenter bee *Xylocopa nautlana* (more than 30 mm). Black, brown, reddish, yellow, sub-metallic, or brilliantly metallic blue, green, or coppery; a few are marked with yellow or white. *Pubescence ranges from nearly absent to very long and dense*, varying in colour but *always having at least some plumose hairs*, especially in the vicinity of the mesothoracic spiracle, the upper part of the propodeum and the lateral portions of the metanotum. Both sexes always fully winged; sexual dimorphism slight. Antenna of female 12- and male 13-segmented; mouthparts frequently with glossa very long and pointed. *Pronotum short, anterolaterally with a lobe that extends posteriorly, almost reaching tegula (as in Sphecidae)*; admedian lines of mesoscutum fused into a single median line; metapostnotum extensive. *Hind basitarsus flattened, wider than other tarsal segments*, especially in females; strigil lacking. Fore and hind wings with many well defined cells (except in some Meliponinae); hind wing with claval and usually jugal lobe present. Female metasoma with six exposed tergites and six sternites; male usually with seven exposed tergites and six sternites, but apical sclerites concealed in a few taxa.

Classification and distribution. Traditionally the bees have been treated as 11 separate families comprising the superfamily Apoidea (Michener, 1965; McGinley, 1980; Michener & Greenberg, 1980), but here (following Gauld & Bolton, 1988) they are placed in a single family Apidae, which together with the Sphecidae constitute the superfamily Apoidea. Alexander (1992), in his exploratory analysis of the relationships between bees and sphecids, clearly demonstrated the paraphyletic nature of the Sphecidae with respect to the bees. Several synapomorphies support the monophyly of the bees, and most frequently the bees were found to have a more or less sister-group relationship to philanthine sphecids. While the classification adopted here leaves the paraphyly of the Sphecidae unresolved, the eventual classification of the Apoidea will probably recognize just one family of bees.

COLLETIDAE Euryglossinae* Hylaeinae Xeromelissinae Colletinae Diphaglossinae	MELITTIDAE Meganomiinae* Melittinae* Dasypodinae*
STENOTRITIDAE [Stenotritinae*]	CTENOPLECTRIDAE [Ctenoplectrinae*]
OXAEIDAE [Oxaeinae*]	FIDELIIDAE [Fideliinae*]
HALICTIDAE Rophitinae* Halictinae Nomiinae*	MEGACHILIDAE Lithurginae Megachilinae
ANDRENIDAE Andreninae Panurginae	ANTHOPHORIDAE Nomadinae Anthophorinae Xylocopinae
	APIDAE Euglossinae Bombinae Meliponinae Apinae

Table 18.1. The 'traditional' classification of bees into 11 families. In this work all bees are placed in the single family Apidae, but the previously recognized subfamilies are retained. Names in square brackets have not been used in the traditional classification as all genera in that particular group have been included in one 'family'. Subfamilies followed by an asterisk are not present in Costa Rica.

The subfamilies used here (see table 18.1) are not the 11 'traditional' families reduced in rank (with the exception of Oxaeidae, Ctenoplectridae and Stenotritidae, which have never been divided into subfamilies) but rather the subfamilies commonly recognized within these families. Accordingly 26 such subfamilies are currently recognized (as listed by Roubik, 1989, plus the Euglossinae) and of these, 16 are known to occur in Costa Rica. The Oxaeinae probably also occurs here as well. Of the remaining nine subfamilies Euryglossinae and Stenotritinae occur in Australia, Meganomiinae in Africa, Ctenoplectrinae in the Palaeotropics, Fideliinae in Chile and Africa, Rophitinae primarily in the Holarctic (except *Penapis* in Chile and Peru), Nomiinae in the Old World and the Nearctic, and Melittinae and Dasypodinae in the Holarctic and Africa (Michener, 1979).

Worldwide the family Apidae comprises over 4000 genera and 25,000–30,000 species. Ninety-five genera and over 650 species of bees have been identified during the course of this study of the Costa Rican fauna. Although the Costa Rican bee fauna is significantly smaller than that of Mexico or the continental United States (see Table 18.2), the diversity is remarkable when the great disparity in land mass is taken into account. On a per area basis Costa Rica has 16 times the number of species as does Mexico and 27 times as many as the continental United States (see below, under 'Synopsis').

Biology. Unlike most other Aculeata, virtually all bee larvae are phytophagous, developing exclusively on a diet of pollen and nectar (or plant oils), which are collected and carried to the nest by the adult female. The only known exceptions are three species

	Costa Rica		Mexico		U.S.A.	
	Gen	spp	Gen	spp	Gen	spp
Colletinae	1	17	2	48	2	115
Diphaglossinae	3	6	4	13	2	6
Hylaeinae	1	19	1	20	1	52
Xeromelissinae	1	1	1	2	–	–
Oxaeinae	–	–	1	10	1	10
Rophitinae	–	–	5	20	7	104
Nomiinae	–	–	2	16	2	26
Halictinae	21	195	19	160	11	400
Andreninae	1	6	2	89	3	448
Panurginae	4	10	8	382	9	855
Melittinae	–	–	1	1	2	12
Dasypodinae	–	–	1	9	1	21
Lithurginae	1	1	1	5	1	6
Megachilinae	12	96	23	278	18	642
Nomadinae	5	20	12	60	14	542
Anthophorinae	26	143	42	337	33	419
Xylocopinae	2	37	2	58	2	42
Meliponinae	11	48	11	29	–	–
Euglossinae	4	58	4	30	1	1
Bombinae	1	6	2	21	2	49
Apinae	1	1	1	1	1	1
Totals	**95**	**664**	**145**	**1589**	**113**	**3745**
Spp per 10^4 sq km		129.9		8.1		4.8

Table 18.2. Subfamilies of bees present in Central and North America with numbers of genera and species present in Costa Rica, Mexico and the continental United States (U.S.A.) together with the numbers of bee species per 10^4 square kilometres. Data for Mexico and U.S.A. from Ayala *et al.*, 1993; Costa Rica based on over 24,000 records.

belonging to the *hypogea* group of *Trigona* (Meliponinae), which are obligate necrophages, i.e. provision the nest with carrion (Camargo & Roubik, 1991). Roughly 15 to 20 percent of all bee species are parasites (mostly cleptoparasites) that utilize the provisions collected by another species of bee (Bohart, 1970; Wcislo, 1987). The majority of bees are solitary but several groups have evolved various levels of sociality. The predominant group of bees in tropical areas are the stingless bees (Meliponinae), all of which are highly eusocial. An extensive review of the biology of social bees is given by Michener (1974a) and of tropical bees in general by Roubik (1989).

Nests. Detailed classifications have been proposed to categorize the different types of nests made by bees (Malyshev, 1936; Sakagami & Michener, 1962; Stephen *et al.*, 1969). In general most bees can be divided into three groups, based on their nesting habits:

i) burrowers, which actively excavate nests, usually in soil ('digger bees') but sometimes in plants;
ii) 'renters' or 'lodgers', which utilize pre-existing holes (beetle burrows in wood, hollow plant stems, pre-existing holes in the ground, etc.);
iii) bees that construct most of the nest from collected materials (mud, resin, etc.) and/or secreted wax. These often utilize pre-existing cavities or semi-sheltered locations, although a few construct freely exposed nests.

Several species in the first and third groups construct a chimney-like structure (TURRET) over the nest entrance, which may be made of mud (ground–nesting bees) or resin (many Meliponinae). Regardless of the nest type, the brood cell is the basic unit of all bee nests, except for a few bees such as *Lithurge*, Allodapini (Africa) and Bombinae, which lay eggs in a communal chamber. Cells may or may not be separable from the substrate, and their orientation, shape and composition differ (often characteristically) between taxa.

Bees that burrow in soil utilize vertical embankments or relatively flat ground. In the latter, freshly excavated soil often accumulates around the burrow entrance (the TUMULUS). The depth of the burrow varies from a couple of centimetres to almost three metres (e.g. *Andrena helianthi*). The burrow may or may not have lateral branches, and the brood cells are arranged singly, in a linear series, or in clusters. Cell walls of burrowing bees are usually constructed of reworked substrate material, and the inner surface of the cell is often lined with a wax–like or cellophane-like substance secreted by the Dufour's and/or salivary glands of the adult female. This cell lining probably serves to prevent the cell from becoming too wet or too dry, and to defend against microbial infection. Colletinae, Diphaglossinae, Oxaeinae and Halictinae secrete macrocyclic lactones (Cane, 1983b). In *Calliopsis* (Panurginae) the Dufour's gland secretion is dominated by a diversity of aliphatic hydrocarbons, or paraffins, while in other Panurginae and in Andreninae the secretion contains a variety of terpenes such as geranyl and farnesyl esters (Cane, 1983*a*). In soil-nesting Anthophorinae a water-resistant lining may or may not be present; in the case of *Anthophora* (Anthophorinae) the Dufour's gland produces unusual triglycerides, which are hydrolysed to diglycerides in the cell lining (Norden *et al.*, 1980). Many of the bees mentioned above probably combine Dufour's gland secretions with those from the labial glands (Duffield *et al.*, 1984), applying the mixture with the mouthparts. For example, in Colletinae (and presumably Diphaglossinae) labial gland secretions may hasten the polymerization of the lactones to form the cellophane-like cell lining. Besides providing a hydrophobic cell lining, the chemical complexity of Dufour's gland secretions suggests that they may also serve a communicative function, for example in nest recognition (especially when the secretion is applied to the nest entrance) and possibly in nest-mate recognition in social species (Hefetz, 1987).

Most Xylocopinae ('carpenter bees') burrow in plant materials, mostly in rotting wood (*Xylocopa*) or in pithy plant stems (*Ceratina*). Some species of *Xylocopa* utilize hollow stems of bamboo, but in order to reach this cavity, they must chew their way through the hard outer part of the stem. Although it has often been stated that carpenter bees lack a brood-cell lining, in fact a hydrophobic lining is present (Gerling *et al.*, 1989).

Bees that nest in existing holes often make their cells from materials brought from outside the nest. In Hylaeinae, and presumably Xeromelissinae, the Dufour's gland secretes lactones (as in Colletinae and Diphaglossinae), although secretions from the hypertrophied labial glands may be the primary source of the silk-like cell lining (Batra, 1972). Nests of most other bees utilizing existing holes apparently lack a secreted cell lining, although it is probable that an inconspicuous one may actually be present (F.D. Parker, unpublished). In Megachilinae cell partitions are constructed of resin (mixed with tiny stones or leaf bits), mud, plant hairs, or cut or masticated pieces of leaves (Iwata, 1976). A few Anthophorinae also use pre-existing holes (e.g. some *Centris*).

Some bees build exposed nests. A few Megachilinae ('mason bees') produce copious amounts of saliva from their enlarged thoracic labial glands, and this saliva is mixed with dry soil to produce a mortar for constructing free-standing nests (Duffield *et al.*, 1984). *Anthidiellum* builds exposed nests of resin. All Euglossinae construct nests of resin (sometimes combined with mud or bark fragments), usually in more or less sheltered locations. The eusocial subfamilies utilize wax in nest construction, which is secreted by integumental glands on the anterior portions of metasomal tergites and sternites III–VI in Bombinae, tergites III–VI in Meliponinae, and sternites III–VI in Apinae. The wax, which accumulates as scales, is groomed from the metasoma with the hind legs, stored within the nest and then used to make free-standing cells or brood chambers (Bombinae). In Meliponinae, most of which nest in tree cavities, large amounts of resin (PROPOLIS) are mixed with the wax (the mixture is termed CERUMEN), whereas in Apinae cerumen is used only in the periphery of the nest.

Sociality. Eusocial behaviour has arisen many times in the bees (Michener, 1974*a*). Eusociality occurs in some Old World Xylocopinae (Allodapini), some Halictinae, all Bombinae (except for the social parasites), all Meliponinae and all Apinae. Allodapine bees appear to have evolved eusociality via subsocial behaviour (female shares the nest with her offspring), whereas the other groups appear to have evolved eusociality via parasocial behaviour (females of the same generation share a nest; see Chapter 2.6). Mutual tolerance among females of the same generation that share a nest is probably favoured by the need for defence (Michener, 1974*a*).

Three types of parasocial colonies are commonly distinguished: communal colonies (each female makes, provisions and oviposits in her own cells), quasisocial colonies (females co-operate in building and provisioning cells, but all females lay eggs), and semisocial colonies (as in quasisocial, but some females lay eggs while others do not). Several Costa Rican bees typically live in communal colonies, for example *Agapostemon nasutus* (Eickwort & Eickwort, 1969). Examples of quasisocial colonies in Costa Rica include some orchid bees such as *Euglossa imperialis* and *E. championi* (Roberts & Dodson, 1967). Semisocial colony foundation and co-operative behaviour appear to be regular ways of life in some Costa Rican Halictinae, for example *Caenaugochlora costaricensis*, *Pseudaugochloropsis graminea* and *P. sordicutis* (Michener & Kerfoot, 1967). Limited observations suggest that *Augochlorella edentata* may be semisocial (Eickwort & Eickwort, 1973*a*; Michener, 1974*a*).

Primitive eusocial behaviour is known to occur in several species of Costa Rican Halictinae. Colonies of *Augochlora nominata*, at least in some localities and perhaps seasonally, consist of large-headed queens, and a few smaller workers with normal heads and empty spermathecae (Eickwort & Eickwort, 1972a). Other primitively eusocial Halictinae include *Halictus ligatus* (Michener & Bennett, 1977), *H. hesperus* (Brooks & Roubik, 1983) and *H. lutescens*, which may have up to 600 females in a single nest (Sakagami & Okazawa, 1985). In the case of *H. ligatus* eusociality apparently occurs only in young colonies, at least in the tropics (in Canada the colonies remain small and eusocial). Eusocial behaviour in *Lasioglossum (Dialictus) umbripenne* differs between two populations located in different parts of Costa Rica (Michener, 1974*a*). In a wet forest in the eastern part of the country queens may be replaced by daughters, older colonies often have several queens, and males and queens are produced at the same time as workers (Eickwort & Eickwort, 1971). In a seasonally dry forest in the western part of the country the foundress queen lives through the year (if she dies she is not replaced), colonies usually have only one queen, and males and queens are produced after worker production has stopped (Wille & Orozco, 1970). In the latter population queens are much larger than workers, with no overlap in size, which is also the case in *H. hesperus*.

Among the primitively eusocial bees, the bumblebees (Bombinae) approach most closely the highly eusocial bees in that food is stored in the nest for adult consumption. A colony is started by an individual queen who must do her own foraging initially. Her first batches of eggs produce workers, which are usually smaller but morphologically similar. The queen suppresses oogenesis in workers by means of pheromones. Eventually males and young queens are produced, and mating occurs. Fertilized queens are able to pass a period of inclement weather by themselves before initiating a new colony. In tropical regions the establishment of new colonies does not follow the synchronized annual pattern of temperate areas. Colonies sometimes last only six months before they are destroyed by natural enemies (e.g. stratiomyid fly larvae; Dias, 1958) or they may last several years (Sakagami & Zucchi, 1965). Large colonies contain over 100 queens, many of them fertilized. Some new queens continue to use the same nest, each defending her own territory (by buzzing and ejecting liquid faeces), but eventually the number of queens is reduced to one.

The highly eusocial bees, namely the stingless bees and honey-bees, are characterized by: large perennial colonies (often consisting of many thousands of individuals) that are established by swarming, pronounced caste differentiation, inability of queens to forage, elaborate nest architecture (including effective thermoregulation) and highly developed communication among members of the colony (including the use of alarm pheromones for mounting a massive defence of the nest) (Michener, 1974*a*). As queens are responsible for laying very large numbers of eggs they have modified internal reproductive tracts, but the increased capacity for egg-laying is achieved in different ways in the two highly eusocial groups. Honey-bee queens have 100–150 ovarioles per ovary whereas stingless bee queens usually have four greatly lengthened ovarioles, although some species with large colonies have as many as 10 to 15 ovarioles per ovary (Sakagami, 1982). In honey-bees worker ovarioles are usually not developed, but in stingless bees workers often lay eggs (trophic eggs and male-producing eggs). Honey-bees oviposit in an empty cell, feed the larva progressively for five days and then close the cell, whereas stingless bees mass provision the cell, oviposit and close the cell. In honey-bees there is no structural difference between cells used for food storage and those used for brood rearing, and the cells are re-used, whereas in stingless bees, brood cells differ from storage cells and the cells are rebuilt rather than re-used. In honey-bees new colonies are formed by fission, the old queen and a swarm of workers leaving to find a new nest site. In stingless bees swarming is a gradual process—workers begin by going back and forth from an existing colony, carrying building materials and food, and eventually a swarm of workers and one to several young queens fly to the new nest.

Foraging. All non-parasitic bees forage for pollen and nectar (or floral oils) with which to feed their larvae. In addition, some groups (e.g. Megachilinae, Centridini, Euglossinae, Meliponinae, Apinae) forage for nesting materials. Many of the morphological characteristics used to distinguish females of non-parasitic taxa are structures for resource gathering and the manipulation of provisions in the nest.

Although Hylaeinae carry pollen internally in the crop, most non-parasitic bees carry pollen externally in a specialized hairy structure, the SCOPA, which is most often located on the hind legs, but in Lithurginae and Megachilinae is located on the venter of the metasoma (Roberts & Vallespir, 1978). Some other bees such as Xeromelissinae and Halictinae possess a rather well-developed scopal apparatus on the metasoma in addition to that on the hind legs. During grooming the pollen is scraped off the body by the fore, or sometimes fore and mid legs, and is passed to the hind legs which pack the pollen into the scopa (Jander, 1976; Thorp, 1979). Euglossinae, Bombinae, Meliponinae and Apinae first wet the pollen with regurgitated nectar and then pack it into the hairless CORBICULAE on the hind legs (Michener *et al.*, 1978; Wille, 1979*a*; Kimsey, 1984*b*). Some Panurginae also wet pack the pollen in a sparsely-haired scopa. Corbiculae have an equally important function in transporting resins for use in nest-building (Megachilinae also collect resins but carry it in their mandibles).

The most common morphological adaptation for nectar collecting is the possession of modified mouthparts (labio-maxillary complex). Bees are often divided into the 'short-tongued' groups (Colletinae, Hylaeinae, Xeromelissinae, Andreninae, Panurginae, Oxaeinae, Halictinae) and the 'long-tongued' groups (the remaining subfamilies). The latter can collect nectar from flowers with longer corolla tubes not available to many short-tongued bees, although they differ not so much in length of the mouthparts as they do in functional morphology (Harder, 1983; Michener & Brooks, 1984; Winston, 1979). Some short-tongued bees are able to gain access to nectar in deep-throated flowers due to their small body size. Nectar is transported in the crop and, upon returning to the nest, the contents are regurgitated into a cell or into a honey pot for storage.

Some bees, principally Anthophorinae (Centridini, Exomalopsini and Tetrapediini) and Ctenoplectrinae, collect floral oils as a substitute for nectar. Plants having oil flowers (e.g. many Malpighiaceae and some Melastomataceae) are most abundant in the Neotropics (Buchmann, 1987). The oil is secreted by special glands (elaiophores) located within the flower, and bees that utilize this oil often have blade-like setae on the fore and mid tarsi with which to rupture the oil-containing glands (Neff & Simpson, 1981). The oil is mopped up with the front and mid tarsi, and then transferred to the scopa on the hind legs. Nectar for the adult bee's own energy requirements must be obtained from other flowers since oil-producing flowers typically do not produce nectar.

Bees vary in the degree of fidelity to particular floral resources (e.g. Heithaus, 1979*b* & *c*). Most show more restricted use in pollen collection than in nectar gathering. Some bees collect pollen from just a few closely related plant species (OLIGOLECTY), while others utilize a wide variety of plants (POLYLECTY). The few examples of oligolectic bees in Costa Rica belong mostly to the Anthophorinae: *Ancyloscelis* and *Melitoma* on *Ipomoea* (Convolvulaceae) (Schlising, 1970; Real, 1981), and *Peponapsis* and *Xenoglossa* on *Cucurbita* (Cucurbitaceae). In addition *Lithurge* (Lithurginae) is oligolectic on *Opuntia* (Cactaceae). Bees may also specialize in foraging at a particular height (e.g. Frankie & Coville, 1979), or at a certain time of day. In general, tropical bees are most active in the morning, although several Diphaglossinae and a few Anthophorinae (*Peponapis* and *Xenoglossa*) are matinal (forage at dawn), and the halictine *Megalopta* is nocturnal (Wolda & Roubik, 1986). Many tropical bees forage only during a certain part of the year when resources are most abundant, although most have a longer flight season than temperate zone bees, and stingless bees and many sweat bees are essentially aseasonal (Heithaus, 1979*a*).

Flowers differ in the rewards that they provide for the bee and how these rewards are presented. Besides the conventional rewards of pollen, nectar and oils, some plants provide more unusual products. For example, certain species of *Dalechampia* (Euphorbiaceae) and *Clusia* (Clusiaceae) provide floral resins that some Megachilinae, Euglossinae and Meliponinae utilize in nest construction (Armbruster & Herzig, 1984). Many orchids provide male Euglossinae with fragrances (see below). Even conventional rewards differ in how they are offered. In plants such as Cochlospermaceae, Melastomataceae, Solanaceae and certain Leguminosae (e.g. *Cassia*) the pollen is enclosed in tubular anthers having an apical pore, which certain bees vibrate vigorously in order to shake out the pollen ('buzz pollination'; Michener, 1962*a*; Wille, 1963; Buchmann & Hurley, 1978). Other bees may collect pollen from these flowers by biting open the anther, gleaning pollen left by other bees (Wille, 1963), or reaching into the pore with their front tarsi (e.g. *Chilicola ashmeadi* on Solanaceae; Eickwort, 1967). Other plants possess a specialized, 'explosive' pollination mechanism, whereby only certain pollinators such as large bees are capable of 'triggering' the flower (e.g. Davis, 1987).

The interaction between plant and bee is not always mutualistic. In some cases the flower provides no reward and attracts a pollinator through deception or mimicry (e.g. Dafni, 1984; Ackerman, 1986; Agren & Schemske, 1991). On the other hand, regular visitation of a flower by a particular species of bee does not necessarily mean that it is the pollinator. Non-beneficial visits result when bees 'rob' flowers by piercing the corolla from the outside to access the nectar, by harvesting pollen while avoiding contact with the stigma, by interfering with the true pollinator, etc.

Larval development. After preparing a cell the female bee provisions it with some combination of pollen, nectar, honey (processed nectar), or floral oils. In *Augochlora*, *Megachile*, and *Anthophora* secretions from Dufour's gland are present in the larval provisions (Cane & Carlson, 1984), but their role is still a matter of speculation (Duffield *et al.*, 1984). In honey-bees the young larvae are progressively fed primarily with secretions from the hypopharyngeal and mandibular glands of nurse bees.

In Bombinae and Apinae the female oviposits into an empty cell (or chamber in the case of Bombinae) and larval food is provided progressively. In most other Central American bees the larvae are fed by mass provisioning—the female provisions, oviposits, closes the cell, and after finishing with all the cells she plugs the nest. Usually each cell is constructed, provisioned and sealed before the next cell is begun although exceptions occur in some Halictinae (Sakagami & Michener, 1962), some Meliponinae, and in all Bombinae and Apinae.

The nature of the provisions varies from a fermenting liquid (*Ptiloglossa*) to a dry mass of pollen, and in some cases a thin layer of nectar is added to the top of the pollen mass. Some bees mould the provisions into loaves (e.g. Xylocopinae) or balls (e.g. Panurginae, Halictinae), which may serve to prevent water loss. The female may attach the egg to the cell wall (Colletinae), or she may place the egg on top of the food mass (most bees), under it (Emphorini) or in it (some *Osmia*, *Lithurge*). In *Chilicola ashmeadi* and *Ptiloglossa* the egg floats on fluid or semi–fluid provisions.

Upon eclosion the larva begins to feed on the provisions, often remaining rather stationary in the first instar. All bees appear to pass through five larval instars although the first may remain within the egg chorion (Torchio, pers. comm.). In most bees the larva does not defaecate until it has finished feeding, and the faeces are often pressed against the cell wall (or the cell cap in Eucerini). In Megachilinae the larva usually defaecates earlier, but apparently avoids contaminating its food by spinning silk strands that entangle the faecal pellets and keep them away from the food (Michener, 1953). Early defaecation also occurs in Xylocopinae, but in this case the faecal pellets merely drop and accumulate in the bottom of the cell (Michener, 1962*b*).

Cocoon formation by larvae is often associated with bees that do not secrete a cell lining (Michener, 1964), but exceptions occur (*Colletes*, *Cadeguala*, some parasitic bees). Thus cocoons appear to be universally present in Megachilinae, Euglossinae, Bombinae, Meliponinae and Apinae, whereas they are completely lacking in Hylaeinae, Andreninae, Panurginae, Halictinae, Nomadinae and Xylocopinae (Stephen *et al.*, 1969). Except for the tribe Anthophorini, most

Anthophorinae appear to spin cocoons; those of Emphorini and Exomalopsini consist of just one layer whereas those of Eucerini consist of two or more layers (Torchio, 1974). In the Diphaglossinae, cocoons of *Ptiloglossa guinnae* have perforations in the top (Roberts, 1971), and those of *Crawfordapis luctuosa* are peculiar, apparently isolating liquid and faeces in a lower cell (Roubik & Michener, 1984).

Parasitic bees. Parasitic bees do not forage for pollen but instead utilize the pollen stores in the nests of other bees. Non-parasitic bees may occasionally parasitize members of the same species in a facultative manner, but the obligate parasites nearly always attack other species, usually members of the same subfamily (except Nomadinae).There are three categories of parasitic bees (see Chapter 2.6):

i) CLEPTOPARASITES—solitary bees that lay their egg in the nest of another bee;
ii) CLEPTOBIONTS—eusocial bees that rob provisions and often nesting materials from the nests of another eusocial species;
iii) SOCIAL PARASITES—eusocial bees that usurp the nest of another eusocial species and never produce their own worker caste.

In parasitic bees the structures associated with pollen gathering (plumose hairs, the scopa, etc.) and nest building (e.g. the pygidial plate) are often reduced (Arduser & Michener, 1987). However, cleptobionts, which usually build their own nest, generally retain structures associated with nest construction. Cleptoparasites and social parasites generally have more ovarioles per ovary and more mature oocytes per ovariole, which presumably allows them to oviposit in rapid succession if they find a number of host cells in a short period of time (Alexander & Rozen, 1987).

Most species of parasitic bees are cleptoparasites, which appears to be true in tropical and temperate regions alike. In Costa Rica there are 20 genera of cleptoparasites (Table 18.3). All members of the subfamily Nomadinae (Rozen, 1966), and all members of the anthophorine tribes Ericrocini (Snelling & Brooks, 1986), Osirini and Rhathymini are presumed to be cleptoparasites, accounting for over half (12) of the cleptoparasitic genera in the region. Three parasitic genera of North America whose hosts reach Costa Rica are not known to be present here. These include (with their hosts): *Stelis* (Osmiini, Anthidiini), *Holcopasites* (*Calliopsis*, *Pseudopanurgus*) and *Paranomada* (*Exomalopsis*); none of these are known south of Mexico.

Cleptoparasitic genus	**Host bees**
Halictinae	
Microsphecodes	Halictinae
Ptilocleptis	unknown
Sphecodes	various, primarily
Temnosoma	unknown
Megachilinae	
Coelioxys	Megachilinae: *Megachile*, Anthophorinae: *Centris*
Dolichostelis	Megachilinae: *Megachile*
Hoplostelis	Euglossinae
Nomadinae	
Epeolus	Colletinae:
Nomada	various, primarily *Andrena*
Odyneropsis	Diphaglossinae: *Ptiloglossa*
Thalestria	Oxaeinae
Triepeolus	various, primarily Eucerini
Anthophorinae	
Aglaomelissa	Anthophorinae: Centris
Coelioxoides	unknown
Ctenioschelus	Anthophorinae: *Centris*
Mesocheira	Anthophorinae: *Centris*
Mesoplia	Anthophorinae: *Centridini*
Osiris	unknown
Protosiris	unknown
Rhathymus	Anthophorinae: *Epicharis*
Euglossinae	
Exaerete	Euglossinae

Table 18.3. Genera of cleptoparasitic bees occurring in southern Central America. Data from Roubik (1989) and also from below under 'Synopsis of the Costa Rican fauna'.

Cleptoparasitic bees often have a thickened cuticle, which presumably provides them with protection against the sting of the host bee. The reduced pilosity of cleptoparasites often gives them a wasp-like appearance and in a few cases the colour patterns resemble those of polistine vespids; *Odyneropsis gertschi* for example appears to mimic *Polybia flavitincta* (Roberts, 1971), and other species of the same genus are mimics of *Parachartergus* species. Some female cleptoparasites open a cell that has recently been closed by the host, oviposit, and then reseal the cell. In cleptoparasitic halictines the host egg or larva is often destroyed by the adult female, who sometimes takes the nest by force, occasionally even killing the adult host (Michener, 1978). Females of *Exaerete* likewise destroy the host egg (Bennett, 1972). In other cleptoparasites the female simply conceals her egg in the host provisions or cell wall, and the first or second instar cleptoparasite larva kills the host egg or larva (and any other parasitic larvae) with its elongate mandibles. It then proceeds to feed on the provisions. Although some cleptoparasites wait until the host cell is sealed before ovipositing, females of most species, including *Nomada* (Rozen, 1977), *Stelis* (F.D. Parker, unpublished) and *Epeolus* (Torchio & Burdick, 1988) enter cells that are still being provisioned in order to oviposit, thus leaving the host to seal the cell.

Unlike cleptoparasites, cleptobionts are eusocial and generally build their own nests. Obligate cleptobionts are restricted to tropical regions and in the Neotropics include all species of *Lestrimelitta* (Meliponinae) (Nogueira-Neto, 1970*a*). These bees conduct mass raids on the nests of other stingless bees, carrying off food in their crops and nest material on their hind tibiae. Although they generally build their own nest, they may occasionally establish themselves in the host nest. *Trigona angustula* usually forages and builds its own nests but opportunistically usurps the nests of other stingless bees.

In contrast to cleptobionts, social parasites appear to be more common in temperate regions than in the tropics. The social parasite *Psithyrus* (Bombinae), which parasitizes other bumblebees, extends only as far south as Guatemala. In the rest of Central America the only social parasites appear to be facultative, e.g. *Halictus ligatus* (Packer, 1986). Although all Meliponinae are eusocial, and many rob food and nesting materials, this appears not to have led to the evolution of permanent social parasitism. This may be a result of having permanent colonies, since seasonality is thought to be a central factor in the evolution of parasitism (Wcislo, 1987).

Mating. Females of most bees are assumed to mate only once, usually soon after emergence, but in Panurginae, Anthidiini and honey-bees the female mates several times (Alcock *et al.*, 1977, 1978). The site where mating occurs is often correlated with the nesting habits. In solitary species that nest in aggregations mating often occurs at the nesting site. When nests are scattered mating may occur at flowers visited by females. However, in other bees mating takes place neither at the nest site nor at the floral resource, but rather at landmarks such as a hilltops, tops of canopies, open areas in the forest, light coloured tree trunks and so forth.

Male bees exhibit one of two basic behavioural patterns at an encounter site (Eickwort & Ginsberg, 1980): one to many males may patrol a site non-aggressively, or each male may defend a territory against other males. Among Costa Rican species, males of *Melissodes persimilis* patrol the nesting area (Buchmann & Jones, 1980) while *Chilicola ashmeadi* males patrol flowers that are frequented by females (Eickwort, 1967). Territorial males show two types of flight behaviour within their territories. The male may move more or less constantly within his territory, as occurs in *Calliopsis* (Shinn, 1967). Alternatively, the male may hover about, occasionally perching or darting after passing insects. This type of behaviour occurs in males of *Ptiloglossa guinnae*, which defend territories at the nesting site (Roberts, 1971), and in *Centris*, *Xylocopa*, Euglossinae and some Bombinae, which often defend landmarks.

In Bombinae (bumblebees) one of four types of mate searching strategies may be employed by males, depending on the density and dispersion of females (P. Williams, 1991):

i) patrolling (a series of objects are marked with a secretion from the labial gland and the sites are patrolled)—low density, even dispersion;
ii) cruising (like patrolling but male hovers in one small area)—high density, clumped dispersion;
iii) racing (potential mates pursued from a perch)—high density, even dispersion;
iv) territorial (like racing but males compete for perches)—low density, clumped dispersion.

Sexual dimorphism in relative eye size, antennal length, wing shape, and mesosoma shape is least developed in patrolling species and most developed in racing species. Males of the Costa Rican *Bombus pullatus* appear to utilize patrolling as do those of the nearctic *B. mixtus*, but male flight paths of the former are fewer, more widely dispersed, and used by more males. This difference possibly results from the larger colonies and longer season in the tropics, which in turn make it is easier to find existing paths and more likely that a path will contain siblings (Stiles, 1976).

In the seasonally dry forests of northwestern Costa Rica 14 species of *Centris* have been observed to defend and display in territories, which are usually marked with glandular secretions (Frankie *et al.*, 1980, 1989; Vinson *et al.*, 1982, 1984; Williams *et al.*, 1984; Coville *et al.*, 1986). Each species is unique in: territory size, plants associated with the territory, chemical nature of the marking pheromones, frequency of marking, and daily territorial period. Species of the subgenera *Hemisiella*, *Heterocentris* and *Trachina* have scent glands in the hind legs whereas species in the subgenus *Centris* have mandibular scent glands. Several species maintain territories in the crowns of trees (although differing in precise location within the crown), whereas *C. adanae* defends territories in openings in the grass. *C. aethyctera* is unique in lacking scent glands and in defending a larger territory, which may include several flowering trees.

Another group of bees that appear to use scent secretions in conjunction with territoriality is the genus *Xylocopa*. In males of *X. fimbriata* and *X. gualanensis* tubules are evident beneath the scutellum, and the scent gland opens between the metanotum and the propodeum, the secretions perhaps being pumped onto the body surface while hovering (Vinson *et al.*, 1986; Williams *et al.*, 1987). Male orchid bees often defend territories that include prominent tree trunks on which they display, apparently releasing a pheromone. Prior to establishing territories, males visit certain flowers in order to obtain 'perfume', although the exact role of these fragrances is still unclear (Dressler, 1982; Williams & Whitten, 1983; Zimmerman & Madriñán, 1988) (see also below under Euglossinae).

Economic importance. Both the introduced honeybee (Crane, 1990) and native stingless bees (Schwarz, 1948; Nogueira-Neto, 1970*b*), provide honey, wax and other valuable products (see Chapter 3.2). Among stingless bees, species of *Melipona* are the most highly esteemed honey-producers. Bees are, however, even more important as pollinators. They are the single most important group of pollinators of tropical plants (Bawa *et al.*, 1985), including even some plants that are adapted to other pollinators (Haber & Frankie, 1982). A very significant proportion of the human diet is derived directly or indirectly from plants that are pollinated by bees (Free, 1970). Among native neotropical species, stingless bees are undoubtedly one of the most important groups of pollinators of both native and crop plants. It has been estimated that the stingless bees inhabiting a square kilometre of lowland forest harvest about one billion kilojoules and five litres of nitrogen during the year (Roubik, 1989, pp. 353–355). The role of stingless bees in the pollination of Central American crops is relatively unstudied, although achiote (*Bixa orellana*, Bixaceae) and vanilla (*Vanilla planifolia*, Orchidaceae) are known to be pollinated by species of *Melipona* (Wille, 1976), and chayote (*Sechium edule*, Cucurbitaceae) by *Partamona cupira* and *Trigona corvina* (Wille et al., 1983). Among solitary species, the 'squash bees' (*Peponapis* and *Xenoglossa*) are the most important pollinators of several native cucurbitaceous crops such as squash, pumpkins and gourds (Michelbacher *et al.*, 1971). *Xylocopa* greatly increases fruit setting in passionfruit (*Passiflora edulis*, Passifloraceae) (Corbet & Willmer, 1980), and various native bees are important pollinators of guava (*Psidium guajava*) (Hedström, 1988). Because of their importance in crop pollination (O'Toole, 1993), as well as in maintaining native plant diversity (Neff & Simpson, 1993), more attention needs to be directed toward conservation of native bees (Vinson *et al.*, 1993).

Although bees are generally beneficial, a few species can become pests under certain circumstances. Certain stingless bees (especially *Trigona corvina*, *T. fuscipennis*, *T. silvestriana* and *T. nigerrima*) occasionally cause damage to crops such as citrus, macadamia, mango, palms and rambutan by biting off flower buds and leaves, or scraping young leaves with their mandibles, in order to obtain resins and chewed leaf material that they use in nest construction (Wille, 1965; Camacho, 1966; Haueisen Freire & Gara, 1970). A few stingless bees may occasionally transmit plant pathogens while visiting flowers (Buddenhagen & Elsasser, 1962). In many crops, controlling stingless bees is complicated by the fact that they are needed for pollination.

Certain stingless bees aggressively defend their nests by biting, and they have the very annoying habit of crawling into one's hair, nose, eyes, etc. Although less aggressive, species of *Oxytrigona* (known as 'fire bees') have very caustic bites that can cause long-lasting blisters (Bian *et al.*, 1984; Roubik *et al.*, 1987). Honeybees can pose serious medical risks to persons who are hypersensitive to their stings. The probability of being stung by a honey-bee has increased in recent years with the spread of the more aggressive, Africanized

- –Fore wing with 0–1 submarginal cells (Fig. 18.02) .... MELIPONINAE
- –Fore wing with 2 submarginal cells (Fig. 18.03)
    - –Small, nearly hairless bees; glossa truncate
        - –Black; face elongate with extensive yellow .... HYLAEINAE
        - –Brown; face less elongate, less yellow .... XEROMELISSINAE
    - –Small to large, hairy bees; glossa pointed
        - –With 2 subantennal sutures (cf Fig. 18.10) .... most PANURGINAE
        - –With 1 subantennal suture (cf Fig. 18.11)
            - –Pygidial plate present; jugal lobe extending to level of *cu-a* (cf Fig. 18.08) .... LITHURGINAE
            - –Pygidial plate absent; jugal lobe not reaching to level of *cu-a* (cf Fig. 18.09) .... MEGACHILINAE
- –Fore wing with 3 submarginal cells (Figs 18.04–18.09)
    - –Hind wing with jugal lobe extending to level of *cu-a* (Fig. 18.08)
        - –With 2 subantennal sutures (Fig. 18.10) .... ANDRENINAE/*Protandrena*
        - –With 1 subantennal suture (Fig. 18.11)
            - –Fore wing with *M* between *Cu* and *Rs* + *M* bowed (Fig. 18.08); usually metallic .... HALICTINAE
            - –Fore wing with *M* between *Cu* and *Rs* + *M* almost straight (Fig. 18.04); non-metallic
                - –Small to medium sized; *2m-cu* in fore wing bowed out posteriorly (Fig. 18.04) .... COLLETINAE
                - –Large; *2m-cu* straight (cf Fig. 18.09) .... DIPHAGLOSSINAE
    - –Hind wing with jugal lobe far short of level of *cu-a* (Fig. 18.09)
        - –2nd submarginal cell < 1st along posterior margin (Fig. 18.09)
            - –*M* between Cu and *Rs* + *M* bowed (Fig. 18.09), usually dark blue .... *Ceratina*
            - –*M* between Cu and *Rs* + *M* not bowed (cf Fig. 18.04) .... ANTHOPHORINAE/NOMADINAE
        - –2nd submarginal cell > 1st along posterior margin (Figs 18.05–18.07)
            - –Eyes hairy; third submarginal cell oblique (Fig. 18.07) .... APINAE
            - –Eyes not hairy; third submarginal cell subrectangular (Figs 18.05, 18.06)
                - –2nd submarginal cell almost triangular (Fig. 18.06) .... *Xylocopa*
                - –2nd submarginal cell anteriorly broad (Fig. 18.05)
                    - –Scutellum overhanging propodeum .... EUGLOSSINAE
                    - –Scutellum not overhanging propodeum .... BOMBINAE

Table 18.4. Simple diagnostic characteristics of the subfamilies of bees occurring in Costa Rica.

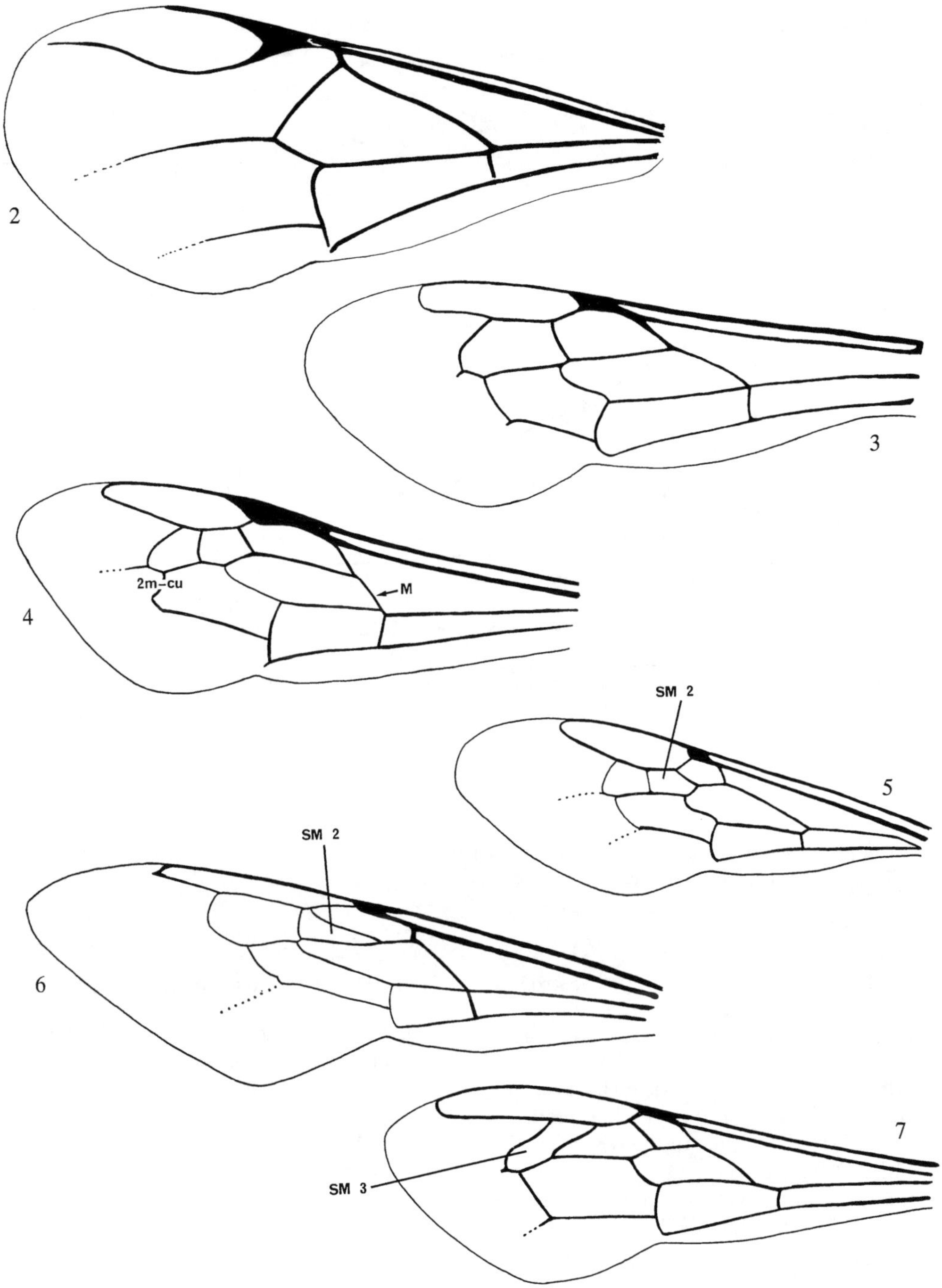

Figs 18.02–18.07. Apidae. Fore wing; 18.02, Meliponinae; 18.03, Megachilinae; 18.04, Colletinae; 18.05, *Bombus* sp. (Bombinae); 18.06, *Xylocopa* sp. (Xylocopinae); 18.07, *Apis mellifera*. (Apinae).

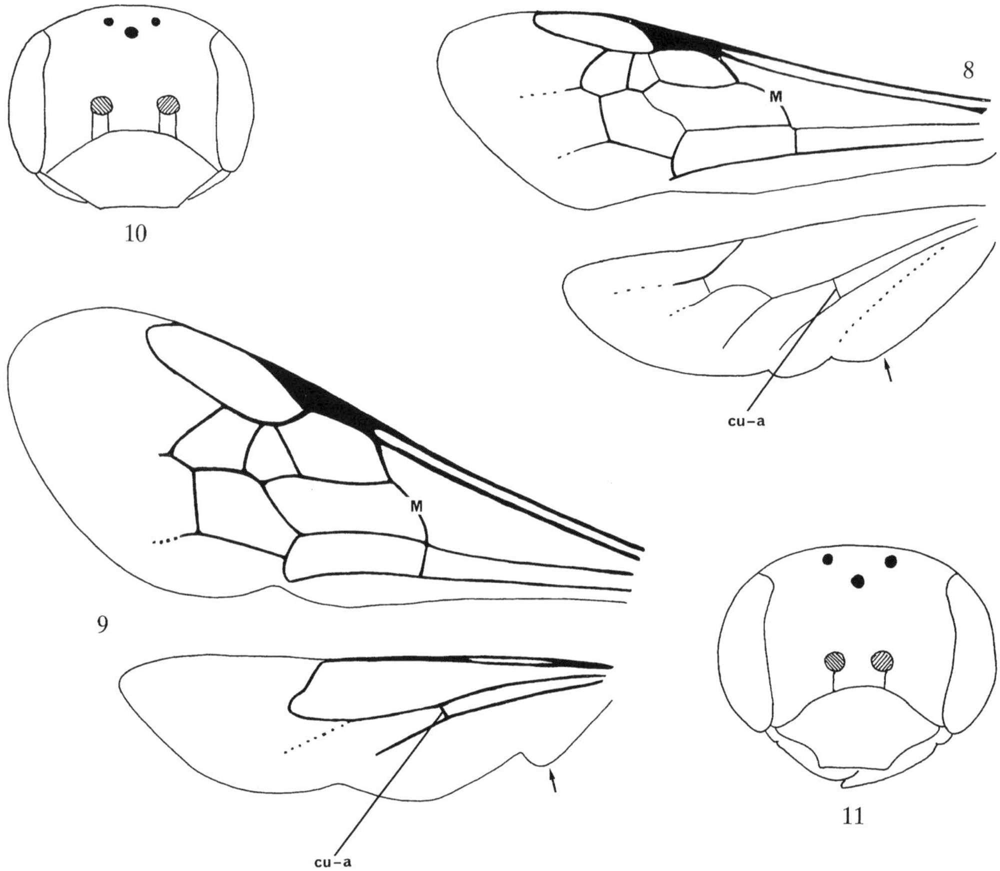

Figs 18.08–18.11. Apidae. Figs 18.08–18.09. Fore and hind wings; 18.08, Halictinae; 18.09, *Ceratina* sp. (Xylocopinae). Figs 18.10–18.11. Head, anterior; 18.10, Andreninae; 18.11, Halictinae.

honey-bee ('killer bee') throughout the Neotropics, although the problem is not as grave as the media would have one believe (see Chapter 3.1).

Identification Many of the Costa Rican bees can be identified to subfamily using the characteristics presented in Table 18.4, although the serious reader should consult Michener *et al.* (1994) which gives keys to the genera of North and Central America. The bee fauna of Costa Rica is still not well known taxonomically; only 27 per cent of the 95 genera have been revised, compared to 58 percent of the Mexican genera and 81 per cent of the bee genera of the United States (Ayala *et al.*, 1993).

Synopsis of the Costa Rican fauna

Presently, 95 genera and over 650 species of bees have been identified from Costa Rica. On a per area basis, Costa Rica has 16 times the number of species as does Mexico (Table 18.2), which challenges the long-held belief that there is a decrease in bee diversity along a latitudinal gradient from the speciose temperate regions (especially the arid regions) to the relatively depauperate tropical rainforests. Competition for floral resources by social bees is often suggested as a cause for the perceived decline in species diversity in tropical regions. This hypothesis is challenged by results from three Costa Rican sites that have been intensively collected by Frank Parker: Cañas, Escazú and Bijagua. More than 200 species are recorded from each of the sites with Bijagua, the wettest site and the site with the greatest number of highly eusocial bees (stingless bees), having the greatest overall bee diversity. Although it is difficult to compare local faunas due to differences in size, habitat diversity, and collecting intensity, the bee diversity at sites in Costa Rica appears to compare favourably with a Mexican faunal study, where 228 species are recorded from a dry deciduous forest in Jalisco (Ayala, 1988), and with nearctic studies, which range from 59 to 187 species (Rust *et al.*, 1983, 1985; Tepedino, 1982; Bohart & Knowlton, 1973).

COLLETINAE

The Colletinae is a cosmopolitan subfamily that is most diverse in Australia and South America (Michener, 1979). Of the 14 genera only *Colletes* is known from Costa Rica, although *Eulonchopria* possibly occurs here as well. Colletines are moderate-sized black bees with distinct fovea on frons along inner margin of eye. These and Diphaglossinae are the only subfamilies with the glossa bifid. Nests are constructed in the ground and all are solitary. The American genera have been revised by Michener (1989).

Colletes. Worldwide, except in Australia; in Costa Rica there are at least 18 species, which are best represented at intermediate elevations. Small to moderate-sized, hairy bees, lacking a pygidial plate; scopa present on hind femur and tibia as well as the second sternite. In one species-group that includes *C. isthmicus* and *C. mexicanus*, there is also a mesosternal scopa consisting of hairs that are apically hooked. Among Costa Rican bees only *Manoeca* shares this condition. *C. punctipennis* is the only Costa Rican bee with pictured wings. Nests are constructed in flat ground or cliff faces, and may be scattered or aggregated.

***Eulonchopria*^EX^.** This genus has been reported from Nicaragua and northern South America (Ayala *et al.*, 1993). They are coarsely, contiguously punctate bees with white apical maculations on some of the tergites.

DIPHAGLOSSINAE

A subfamily that is restricted to the New World. It includes three tribes, Caupolicanini, Diphaglossini and Dissoglotini. The Diphaglossini occurs mostly in southern South America (Michener, 1966, 1986*b*) and is not expected to occur in Costa Rica. Diphaglossinae are large bees, with bifid glossa (as in Colletinae). Many are most active at dawn or dusk and have enlarged ocelli. Keys to tribes and genera are provided by Michener (1966, 1986b).

DIPHAGLOSSINAE: Caupolicani. This group comprises three genera, all of which are present in Costa Rica. Their antennae have the first flagellar segment elongate, subequal to or longer than scape.

***Caupolicana*^CR^.** New World; not previously recorded from Central America; an undescribed species has been collected from wet lowland forest (400 m). Large, non-metallic, hairy bees; the undescribed species has greatly enlarged ocelli. Biology unknown.

Crawfordapis. Mexico to Panama; one species, *luctuosa*. Large, robust black bee with dark wings; covered with dense black pubescence, except for reddish hairs laterally near the posterior end of the metasoma. In Costa Rica *C. luctuosa* is found in forests above 1500 metres and is active throughout the year. Nests are constructed in exposed sites on flat ground or nearly vertical banks in aggregations, and cells are provisioned mostly with pollen from buzz-pollinated plants (Otis et al., 1982; Roubik & Michener, 1984).

Ptiloglossa. Four species are known from Costa Rica: *guinnae*, *mayarum*, *tenuimarginata* and an undetermined species. Large, robust hairy bees with enlarged ocelli and, in Costa Rican species, with a submetallic metasoma. In a cloud forest east of Volcán Poas *P. guinnae* was found nesting in an aggregation of over 1300 nests in nearly flat ground. Females forage at dawn, utilizing primarily buzz-pollinated plants as a source of pollen. The cell provisions are quite soupy and may harbour fermenting yeasts (Roberts, 1971).

DIPHAGLOSSINAE: Dissoglotini. This group comprises three genera which cumulatively range from Mexico to Argentina (Michener, 1986*b*). Their antennae have the first flagellar segment much shorter than scape.

***Mydrosoma*.** Nine species; one, *brooksi*, in Costa Rica in southern Puntarenas Province. Taxonomy: Michener, 1986*b*. These bees fly in the late afternoon or just after sunset (Ayala, pers. comm.).

HYLAEINAE

A cosmopolitan subfamily with its greatest generic diversity in Australia. Only one genus, *Hylaeus*, occurs in the New World (Michener, 1979). These are small, mostly black bees, which often have contrasting light markings.

***Hylaeus*.** Cosmopolitan; in Costa Rica there are 20 species in two subgenera, *Hylaena* and *Hylaeopsis* (Snelling, 1982), mostly in the highlands. Nearly hairless, although some species have dense velvety hairs on the propodeum and the apical margins of the tergites. Pollen is carried in the crop (there is no scopa). There are no studies on Costa Rican species but elsewhere *Hylaeus* species are solitary or occasionally parasocial, nesting in hollow stems and preexisting cavities in wood or soil.

XEROMELISSINAE

A New World subfamily that is most diverse in the temperate forests of South America. Only one genus is present in Mesoamerica. These are mostly small dark bees. Their nest cells are lined with a cellophane-like material (Michener, 1979).

Fig. 18.12. *Hylaeus* sp. (Hylaeinae).

Chilicola. Five species in Mesoamerica of which (*Anoediscelis*) *ashmeadi* and (*Hylaeosoma*) *polita* are known from Costa Rica, where they occur at intermediate altitudes. Minute, slender, hairless, dark brown bees (except for a yellow clypeal mark in males); females differ from *Hylaeus* in the presence of a ventral metasomal scopa composed of long loosely plumose hairs. Taxonomy: Toro and Michener (1975) and Michener (1994). *C. ashmeadi* was found nesting in hollow dry twigs of *Erythrina poeppigiana* (Leguminosae) and provisioning with pollen from Solanaceae (oligolectic) (Eickwort, 1967).

OXAEINAE^{EX}

These moderate sized, fast-flying bees are restricted to the New World, with most species occurring in the tropics. There are three genera: one occurs from the southern United States to Mexico, one monotypic genus is known only from Brazil—Argentina, and the genus *Oxaea* is centred in northern Brazil but one species, *O. fuscescens*, is known from Guatemala.

Oxaea^{EX}**.** Species are narrowly polylectic, especially on *Solanum* and *Cassia*, and utilize buzz pollination; all species appear to be solitary (Hurd & Linsley, 1976). In Colombia *Oxaea fuscescens* nests were found in grassy pastures in red sandy soil devoid of rocks, and appeared to be perennial. Females butt intruders at the nest site; males are territorial, aggressive near flowers, and can hover in flight (Roberts, 1973).

HALICTINAE ('sweat bees')

This cosmopolitan subfamily is divided into three tribes: Augochlorini (New World), Halictini (cosmopolitan) and Nomioidini (Old World) (Michener, 1978). The New World Halictini can be further divided into two informal groups (see Table 18.5). A catalogue of the New World species is given by Moure and Hurd (1987).

Halictines are among the most numerous and diverse groups of bees in Costa Rica, and are rivalled in numbers only by the Meliponinae. They are small to medium sized, often metallic green bees; non-parasitic species with scopa composed of dense plumose hair on hind femur and outer surface of hind tibia; females commonly with loose, mostly simple hairs on lateral surface of propodeum, and a distinct brush of hair on venter of metasoma, which appear to aid in pollen collection; fore basitarsus usually covered with stiff suberect to erect hair in form of bottlebrush and with longitudinal comb composed of a single row of dense bristles. Nests of Halictinae consist of burrows in the ground or in rotting wood.

	Augochlorini	**Agapostemonines**	***Halictus* and *Lasioglossum***
Distribution	New World	New World	cosmopolitan
Diversity centre	warm Neotropics	cool Neotropics	Holarctic
Sociality	solitary/semisocial	solitary/communal	solitary/eusocial
Nest architecture	cells next to main burrow	cells far from main burrow	cells next to main burrow
Head/mesosoma	shiny metallic	shiny metallic	often black
Male metasoma	without yellow markings	often with yellow markings	without yellow markings
Male hind tarsomeres	basal pair articulated	basal pair usually fused	basal pair articulated
Metasomal scopal pubescence....	simple	plumose	plumose
Female tergite V	deeply cleft	entire	entire
Male pygidial plate	absent	present	present

Table 18.5. Characteristics of the New World groups of Halictinae (modified from Roberts & Brooks, 1987).

HALICTINAE: Augochlorini. This tribe comprises 25 genera and nearly 500 species, all of which are restricted to the New World, with 14 genera and about 75 percent of the species exclusively South American. In Costa Rica there are ten genera, a key to which is given by Eickwort (1969*a*). Although the majority are diurnal pollen-collectors, some are crepuscular and a few are parasitic. Most nest in soil (often in banks), but *Augochlora* (*Augochlora*), *Megalopta* and probably *Augochlora* (*Mycterochlora*) nest in dead wood. The cells are either excavated in the substrate and concentrated to one side of the main burrow (*Augochloropsis*), or constructed in a cluster in a previously excavated cavity (Eickwort & Sakagami, 1979; see Fig. 18.14). Cells may be arranged horizontally, vertically, or irregularly (versus primarily horizontal in Halictini). Re-use of cells is common (rare in Halictini).

***Augochlora*.** Canada to Argentina; over 130 described species, with at least 27 in Costa Rica. The species are placed in four subgenera, three of which occur in Costa Rica.

***A.* (*Augochlora*).** Female mandible bidentate, teeth subequal. Species nest in dead wood; nests of *hallinani*, *sidaefoliae*, and *smaragdina* have been described by Eickwort & Eickwort (1973*b*).

***A.* (*Mycterochlora*).** Female mandible bidentate, teeth subequal. Two apparently undescribed species occur in Costa Rica. Species are presumed to nest in wood.

***A.* (*Oxystoglossella*).** Female mandible bearing a small subapical tooth. Species nest in the soil. The nests of *cordiaefloris* and *nominata* are constructed rather like those of Halictini (Eickwort & Eickwort, 1972a); *nominata* is primitively eusocial.

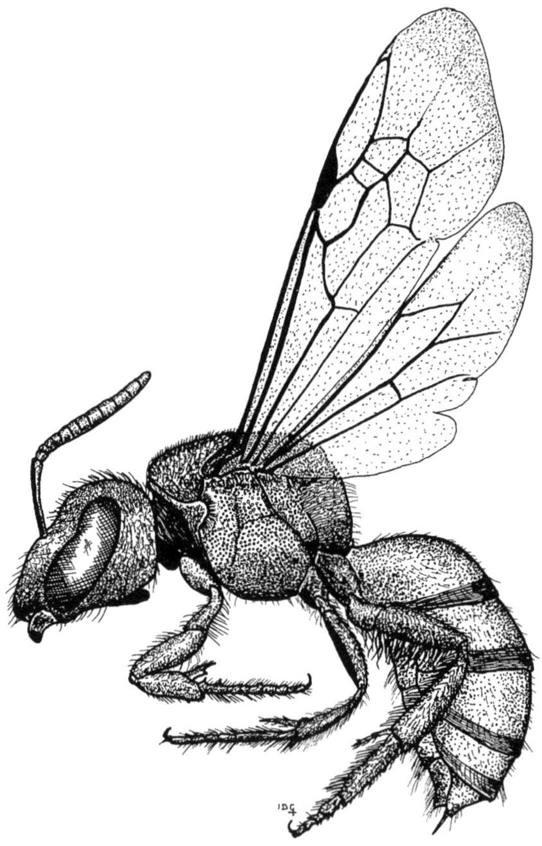

Fig. 18.13. *Augochlorella* sp. (Halictinae).

Augochlorella. (Fig. 18.13). Canada to Argentina; 16 species, three in Costa Rica (low to mid elevations sites): *edentata*, *neglectula* and *pomoniella*. Taxonomy: Ordway (1966). Nests of *edentata* are in vertical, vegetation-covered moist banks, with the entrances produced into long turrets (Eickwort & Eickwort, 1973*a*); it may be primitively eusocial (Michener, 1974*a*).

Augochloropsis. Primarily neotropical, with species extending into the eastern and central United States; the largest genus of Augochlorini, with 138 described species, 14 in Costa Rica (widespread and abundant). These are about equally divided between the subgenera *Augochloropsis* and *Paraugochloropsis*. Brilliant green, gold, coppery, purple, or red bees, with head and mesosoma sometimes dark; inner posterior margin of tegula produced; first two tergites often with dense apical fringe of stout hairs. Nests aggregated or not, in flat ground or vertical banks; mostly solitary or facultatively communal, though *ignita* may be semisocial or eusocial (Michener & Lange, 1959; Michener, 1974*a*).

Caenaugochlora. Mexico to Ecuador; 14 described species, but 18 species occur in Costa Rica, where they are widespread, but absent from the drier Pacific parts. Many species have an elongate head; several species have clouded or bicoloured wings, and some have brilliant red metasomas; males with tufts of specialized hair on sternite IV (as in *Pseudaugochloropsis*); hind tibial spur of the female loosely pectinate (subgenus *Caenaugochlora*) or finely and densely pectinate (subgenus *Ctenaugochlora*). The nest of *costaricensis*, which is semisocial, was described by Michener and Kerfoot (1967; as *Pseudaugochloropsis*).

Megalopta. Mexico to Argentina; 28 species, three in Costa Rica, including *centralis* and *genalis*. Medium-sized bees with metallic green head and mesosoma and contrasting yellowish brown metasoma; ocelli greatly enlarged; ventral hairs of hind tibial scopa loosely plumose (as opposed to densely plumose in most Augochlorini). *M. centralis* has been observed nesting in a vertical, 22 mm diameter branch in the upper part of a dead tree (Janzen, 1968*b*). Species are nocturnal and buzz-pollinate Solanaceae.

Megommation[CR]. This genus, previously known only from South America, comprises four described species, in three subgenera, and two apparently undescribed species of *M. (Megaloptina)* in Costa Rica. Dull metallic with reddish metasoma or mostly dark brown; mouthparts very long, slender and set in a narrow fossa; clypeus strongly protruding; ventral hind tibial hairs of female scopa have reduced plumosity (as in *Megalopta*). They are crepuscular bees that construct nests in flat ground. The entrance has an erect turret (Michener & Lange, 1958).

Neocorynura. Mexico to Argentina; about 65 species in two subgenera; all 25 of the Costa Rican species belong to the nominate subgenus. They are absent from the drier parts of northwestern Costa Rica. Variable in form, but often elongate with a clavate metasoma; remarkably similar in shape and markings to polistine wasps; wings transparent, yellowish, black or bicoloured. Their nests are in moist dirt banks and are sometimes aggregated. The nest of *pubescens* was described by Michener *et al.* (1966). The nesting behaviour of *N. fumipennis* differs from most other halictines in that each cell is provisioned and sealed before the next cell is begun; the female defends the nest with her jaws instead of blocking the entrance with the metasoma as do most other halictines (Michener *et al.*, 1966).

Pereirapis. This genus, comprising perhaps just a single species, *semiaurata*, occurs from Mexico to Brazil. It is widespread and common in Costa Rica. Small bees, very similar to *Augochlorella*, but with a non-metallic brown metasoma that contrasts with a metallic head and mesosoma. The nest was described by Eickwort and Sakagami (1979).

Pseudaugochloropsis. Texas to Argentina; seven species, three in Costa Rica (absent from drier areas): *graminea*, *sordicutis* and an undetermined species. Males have tufts of specialized hairs on sternite IV (as in *Caenaugochlora*), and the terminal antennal segment hooked (unique among Halictinae). The vertex of the female has a ridge above the ocelli. *P. sordicutis* is dichromatic, with both black and green forms from the same nest. Nests of *graminea* and *sordicutis* (as *nigerrima*) are in vertical banks and in the latter species are aggregated (Michener & Kerfoot, 1967). About two thirds of the nests are small and inhabited by lone females that remain in the nest until about the time of emergence of the offspring (the mother probably dies at this time). Some nests are subsequently enlarged by semisocial colonies that may consist of up to seven females.

Temnosoma. Arizona to Argentina; seven species in two subgenera. Two species belonging to the nominate subgenus occur in Costa Rica: *smaragdinum* and an undescribed species. Small, bright metallic, coarsely punctured bees, resembling chrysidids. They are cleptoparasites.

HALICTINAE: Halictini. This tribe comprises close to 30 genera worldwide with over 700 currently recognized species in the New World. In the Americas there are two groups of non-parasitic Halictini (see Table 18.5): 'agapostemonines' (most of the genera listed below) and other Halictini (*Halictus* and *Lasioglossum*). Keys to the genera of agapostemonines of Mesoamerica and to species (except *Agapostemon*, *Caenohalictus* and *Habralictus*) are given in Roberts and Brooks (1987). Nests of agapostemonines have long lateral tunnels leading to cells (Fig. 18.14), which are arranged singly or in series. Nests of other Halictini have long to short, obsolescent laterals; nests having cells in clusters differ from those of Augochlorini in that the cells are typically excavated in the soil and the cavity then dug around them (Eickwort, 1969*b*; Eickwort & Sakagami, 1979).

Agapostemon. Canada to Paraguay; 43 species, six in Costa Rica: *atrocaeruleus*, *cockerelli*, *intermedius*, *leunculus*, *nasutus* and *texanus*. Most *Agapostemon* species occur below 1500 metres, except *leunculus* which is found between 1000 and 3000 metres. Posterior surface of propodeum surrounded by a carina (as in *Agapostemonoides*). Taxonomy: Roberts (1972). *A. nasutus* nests in vertical, vegetated banks, and is colonial (Eickwort & Eickwort, 1969).

Agapostemonoides. Costa Rica to Bolivia; only one species, *hurdi*, is known. It has only been collected below 1500 metres in Costa Rica. It is like *Agapostemon*, but darker and with hairy eyes. There are two colour phases in females: metasoma black with yellow maculae, or amber with yellow maculae. *A. hurdi* may be oligolectic on *Piper* (Roberts & Brooks, 1987). Its nest is unknown.

Caenohalictus. Mexico to southern Argentina; about 50 species, most of which occur in South

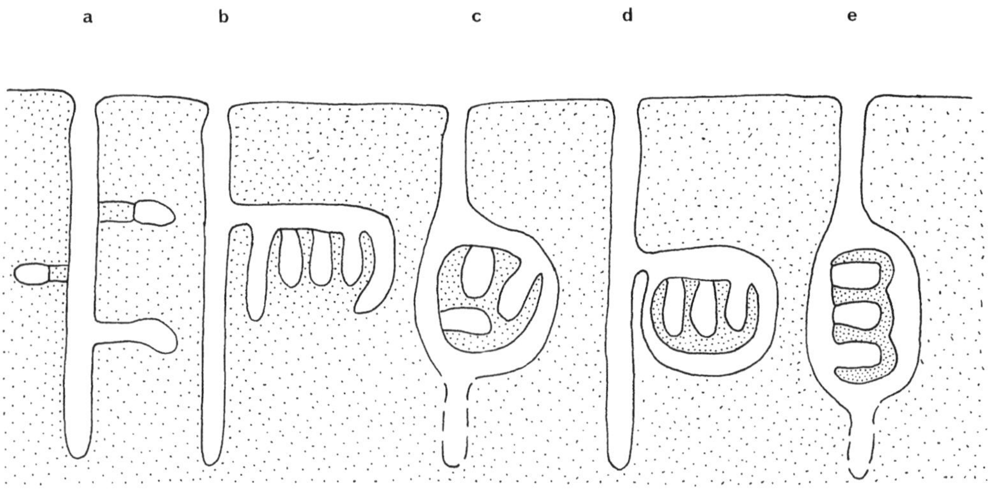

Fig. 18.14. Schematic diagram of nests of Costa Rican Halictinae (modified from Eickwort & Sakagami, 1979): **a**) Halictini (agapostemonines); **b**) *Augochloropsis*; **c**) *Caenaugochlora*, *Megalopta*, *Neocorynura*, *Pseudaugochloropsis*; **d**) *Megommation*; **e**) *Augochlora* (*Augochlora*), *Augochlorella*, *Pereirapis*.

America; apparently only one undescribed species occurs in Costa Rica. Eyes hairy; mesoscutum shiny. The nest of a Colombian species was described by Michener *et al.* (1979); cells are clustered or single at the ends of lateral burrows.

Dinagapostemon. Rare bees occurring in Mesoamerica and northern South America. Only two species (*costaricensis* and *orestes*) are known from Costa Rica (both from only one sex); they occur in moist forests at elevations between 1000 and 2400 metres. Their eyes have long hairs. Nests are unknown.

Habralictus. Mexico to southern Brazil; 21 described species in South America; in Costa Rica there are 14 species (most are probably undescribed). Small slender bees with glabrous eyes; the male metasoma is very slender. They nest communally in dirt banks; those of a Brazilian (Michener & Lange, 1958) and a Colombian species (Michener *et al.*, 1979) have been described.

Halictus. Nearly cosmopolitan (absent in Australia and southern South America); over 200 species, most of which occur in the Palaearctic region. Small bees distinguished from other Halictini by the presence of distinct apical hair-bands on the metasomal tergites. The genus is divided into three subgenera, two of which occur in the New World.

H. (Halictus). A primarily Old World genus represented by four species in the Nearctic region, one of which, *ligatus*, extends south to Colombia and Trinidad. Black; females are unique among Costa Rican bees in having a greatly enlarged gena with a distinct ventral angle. Nests of this species were studied by Michener and Bennett (1977) in Colombia; young colonies are eusocial.

H. (Seladonia). In the New World this group extends south to Brazil; seven species, two in Costa Rica: *hesperus* (metasoma black) and *lutescens* (metasoma orange). Head and mesosoma metallic. Nests of *hesperus* were studied by Brooks and Roubik (1983) and Brooks and Cane (1984) in Panama, and those of *lutescens* by Sakagami and Okazawa (1985) in Guatemala. Nests of both species were studied by Wille and Michener (1971) in northwestern Costa Rica, where they apparently pass the rainy season in a state of ovarian diapause. Both species are eusocial.

Lasioglossum. Cosmopolitan; approximately 420 species in the New World. It is probably the largest genus of bees in Costa Rica. Small bees, which may have pubescent bands on the metasoma, but these are never apical as in *Halictus*. There are three subgenera which are sometimes accorded generic rank.

L. (Dialictus). About 40 species in Costa Rica. Small, submetallic bees with reddish metasoma. They vary from solitary to primitively eusocial. *L. umbripenne* is primitively eusocial; nests have multiple entrances with branching burrows, and guard bees block the narrowed entrances with their heads (Eickwort & Eickwort, 1971).

L. (Evylaeus). At least ten species occur in Costa Rica. Black bees.

L. (Lasioglossum). Absent from South America, but there are four species in Costa Rica: *costale*, *eickworti*, *katyae* and *uyacicola*. Black bees. Taxonomy: McGinley (1986).

Microsphecodes. Neotropical; seven species, with perhaps two in Costa Rica. Small, shiny, black and orange bees, with a propodeum similar to *Sphecodes*. *M. kathleenae* has been reared as a cleptoparasite of *Lasioglossum umbripenne* (Eickwort & Eickwort, 1972*b*; Eickwort & Stage, 1972).

Ptilocleptis. A small, rarely collected neotropical genus with three described species. Two species occur in Costa Rica: *tomentosa* and one undescribed species. Rather small bees with rugose sculpturing on the mesosoma; surface of tergites II—VI obscured by dense appressed plumose hair. Cleptoparasites, host unknown.

Rhinetula. This genus contains just a single species, *denticrus*, which occurs in lowland wet forests from Honduras to Bolivia. Moderate-sized, metasoma dark, eyes hairy; unlike other agapostemonines,the male antennae are short, as in the female. *R. denticrus* appears to be crepuscular but its nests and floral preferences are unknown.

Sphecodes. Cosmopolitan; two subgenera: *S. (Sphecodes)*, which is cosmopolitan except for South America, and *S. (Austrosphecodes)*, which has 32 species in South America. The former subgenus comprises 86 species in North America and at least four species in Costa Rica, including *aeneiceps*, *clypeatus* and *costaricensis*. Small, sparsely hirsute, rather coarsely punctate bees; males often black; females usually with a red metasoma; basal zone of propodeum well delineated and coarsely areolate. *Sphecodes* species are cleptoparasites.

ANDRENINAE

This subfamily comprises six genera—two in southwestern North America, one in Peru, two in Chile, and the primarily holarctic genus *Andrena* (Michener, 1986*a*). It consists of small to moderate sized bees which can be recognized by the presence in females of a hairy facial fovea, and the absence on the inner surface of the hind tibia of short, apically modified hairs (keirotrichia).

Andrena. Primarily holarctic, south to Panama; over a thousand species. The New World species are classified in 49 subgenera, of which only the subgenus *Callandrena* (*discreta*, *vidalesi*, and two undetermined species) has been reported from Costa Rica, although there are also two species (apparently undescribed) belonging to an unknown subgenus. They occur uncommonly in the highlands where they have been collected from December to March, and in the dry forests of northwestern Costa Rica in August. Black, sometimes with faint metallic reflections, occasionally with metasoma red; lower face with light markings (especially in males). The scopa on the hind legs is augmented by a more or less developed propodeal corbicula on the lateral surface. Taxonomy: LaBerge (1967, 1986). Many are oligolectic; all nest in soil, and most are solitary, although a few may be quasisocial (communal). The biology of the Costa Rican species is unknown.

PANURGINAE

Panurgines occur in the Palaearctic and drier parts of Africa but are most abundant and diverse in the New World. There are 38 genera/subgenera in the Nearctic and more than 20 genera/subgenera in the Neotropics (Michener, 1979). Four genera have been found in Costa Rica. Panurgines are most diverse in dry areas and are poorly represented in the wet lowland and montane tropical forests. They are usually small, sparsely pubescent bees with facial maculations. Females have well developed but hairless facial foveae. The mouthparts, fore coxae, and mesosternum often have modified hairs, apparently related to pollen collection; they lack the long scopal hairs on the hind trochanter found in *Andrena*. All panurgines are ground nesters and most are solitary; many are oligolectic. Rozen (1989) has reviewed the biology of the less derived genera, *Protandrena*, *Pseudopanurgus*, *Pterosarus* and *Heterosarus*.

Calliopsis. Primarily southwestern United States and adjacent Mexico; eight subgenera and 38 species; two subgenera and two species in Costa Rica: (*Calliopsima*) *hurdi* and (*Calliopsis*) *hondurasica*. They are rare and occur at intermediate elevations and in the dry forests of the Pacific coast. In the latter area they are active during the dry season. Moderately hairy bees with yellow or white markings on the face, pronotum, and legs (better developed in males). Species nest in aggregations and the burrow entrance remains closed throughout provisioning (Shinn, 1967).

Heterosarus. The genus comprises about 120 species, which are grouped in two subgenera, *Heterosarus* and *Pterosarus*. Five species are known from Costa Rica: (*P.*) *bidentis*, (*H.*) *parvulus*, (*H.*) *setiger*, and two undescribed species. They occur in the highlands and on the dry Pacific slope, and have been collected from July to January. Shiny to dull, black bees, with light facial markings in the males; males may also have marks on the pronotal lobe and legs. Their nests are scattered, the entrance lacks a tumulus, and the main burrow remains open during provisioning (Rozen, 1989). *H. setiger* has been collected on *Senecio benestedianus* (Compositae) on the Cerro de la Muerte (Timberlake, 1977).

Protandrena. One undescribed species of this genus is known from northwestern Costa Rica. It is the only Costa Rican panurgine with three submarginal cells. Nest structure is similar to *Heterosarus* (Rozen, 1989), but nothing is known about the biology of the Costa Rican species.

Pseudopanurgus. This genus comprises 32 species, two of which occur in Costa Rica: *costaricensis* and *crenulatus*. Rather shiny black bees with dark wings; the face and legs of males have yellow markings; females are remarkable for their greatly elongated fore coxal spine, which is clothed with long curved hairs; scopal hairs plumose. The nests of Costa Rican species are unknown, but those of other species are similar to *Heterosarus* (Rozen, 1989).

LITHURGINAE

This subfamily comprises three genera, *Lithurge* (nearly cosmopolitan), *Microthurge* (southern Brazil, Argentina and Bolivia) and the relatively primitive *Trichothurgus* (Argentina, Chile and Peru) (Michener, 1983). All have mouthparts reaching the metasoma (usually shorter in Megachilinae). The outer surfaces of tibiae bear coarse hairless spicules, the hind basitarus is nearly cylindrical (it is flattened in females of most bees) and the first metasomal tergite is flattened.

Lithurge. This genus is divided into two subgenera: *Lithurge* (Old World, but adventive in New World) and *Lithurgopsis* (New World, possibly absent from wet tropics). There are about 12 New World species, of which only *planifrons* occurs in Costa Rica. Moderate-sized, black bees with narrow but distinct apical hair bands on the tergites; female subantennal area with a pair of projections. Taxonomy: Snelling (1983b, 1986a). Species of *Lithurgopsis* are oligolectic on *Opuntia* (Cactaceae). Nests are excavated in dead wood.

MEGACHILINAE ('leaf-cutter bees' and 'mason bees')

This subfamily is divided into four tribes, of which three occur in Costa Rica. Females of the nonparasitic species carry pollen on the ventral surface of the metasoma. They nest mostly in pre-existing cavities, although a few species excavate their own burrows in the ground or build exposed nests. Cell walls or partitions are made of materials brought into the nest from the outside and include pieces of leaves, chewed leaf material, resin or mud. Most species are solitary but some species of *Megachile* are communal or perhaps quasisocial.

MEGACHILINAE: Osmiini. A primarily holarctic and afrotropical tribe, but it is present in much of Asia and in the New World south to Panama. Of the eight North American genera only two, *Osmia* and *Heriades*, reach Central America. They have a large pterostigma and arolia are present. Most species nest in existing holes in stems, wood and even in discarded snail shells. A few Old World groups excavate their own nests or build exposed nests on twigs.

Heriades. Holarctic region, tropical Africa and Asia; in the New World there are approximately 28 species, half of which are undescribed; three species in Costa Rica: *bruneri*, *currani* and an undescribed species. Small, slender, cylindrical, black bees with a coarsely sculptured integument and a sharp carina bordering the anterior dorsal face of tergite I. In Costa Rica they show strong seasonality, flying from December to April. Their nesting biology is unknown in Costa Rica, but elsewhere they construct a linear series of cells in holes in wood, and use resin to cap cells and to plug the nest entrance.

Osmia. Holarctic, but most diverse in western North America where there are about 115 species in 12 subgenera. Only the subgenus *Diceratosmia* reaches Costa Rica, where it is represented by two species in the highlands: *azteca* and an undescribed species. Moderate-sized, black bees, sometimes clothed in rather dense reddish hair. Nests consist of a linear series of cells in stems or wood, and cell partitions are made of masticated leaves. *O. azteca* specializes on Compositae.

MEGACHILINAE: Anthidiini. This cosmopolitan tribe consists of 80 genera and subgenera in the Old World, and 35 genera and subgenera in the New World. Eight genera, two of which are parasitic, are recorded from Costa Rica. Anthidiines are small to moderate sized bees which are usually black with yellow markings; the pterostigma is small and the posterior margin of the scutellum is at a distinct angle to the dorsal surface.

Anthidiellum. Holarctic, in the New World as far south as Costa Rica. There are about eight species in North America, one of which, *apicale*, reaches Costa Rica. Yellow wings with dark tips; preoccipital carina strong on vertex but not extending ventrally onto the gena; omaular carina strong; scutellum short, truncate, carinate posteriorly and overhanging propodeum. Species construct exposed, resin nests on twigs.

Anthidium. Holarctic, Afrotropical and Neotropical regions; about 60 species in the New World, two in Costa Rica: *hallinani* and *maculifrons*. Hind tibia with longitudinal carina; arolia absent; mandible of female with six or seven teeth; tergite VII of male trilobed or spined. Species nest in existing cavities (*hallinani* uses trap-nests) or excavate nests in soft soil; they use cottony plant fibres in nest construction.

Anthodioctes. Neotropical; 21 species in three subgenera; two subgenera in Costa Rica: *Anthodioctes* (wavy hair on the fore basitarsus—*calcaratum* and an undescribed species) and *Nanathidium* (scopal hair clubbed—*gualanense*). These bees have a pair of carinae present between the antennal sockets; preoccipital carina strong, extending ventrally almost to hypostomal carina; omaular carina distinct, incomplete; basal zone of propodeum horizontal, pitted. Species nest in existing holes and utilize trap nests.

Dolichostelis. North America to Costa Rica; six species, in Costa Rica only *costaricensis*. Propodeum with distinct pitted basal zone; fore and mid tibiae with two outer apical spines. Taxonomy: Parker and Bohart (1979). *D. costaricensis* has been reared as a cleptoparasite of *Megachile (Chelostomoides) otomita* in northwestern Costa Rica (Parker *et al.*, 1987).

Hoplostelis. Neotropical; about 12 species. Two subgenera have been recognized: larger robust species (*Hoplostelis*) and smaller slender species (*Austrostelis*) (Michener & Griswold, 1994). The former is represented in Costa Rica by *bivittata* (wings yellow with dark tips) while the latter group is known both from South America and Mexico, and has now been found to be present in Costa Rica. A third group of species is represented by an undescribed species in northwestern Costa Rica (Rio Naranjo). Propodeum with a distinct but narrow pitted basal zone; parantennal carinae present; fore and mid tibiae with one outer apical spine. *H. bivittata* is a cleptoparasite of orchid bees.

Hypanthidioides. Neotropical; seven species, in Costa Rica only (*Saranthidium*) *panamense*. Mostly red and yellow, small bees with some black on head and mesosoma; parantennal and omaular carinae present; a semicircular carina on the propodeum forms a dorsal anterior shelf and passes behind the spiracle, but a large postspiracular pit is absent; arolia absent; males with a truncation on the apical margin of tergite VI, tergite VII bilobed, sternite V with apical comb of red teeth; scopal hairs are blunt, appearing singed. Biology unknown.

Hypanthidium. Neotropical; more than eight species, two apparently undescribed species occur in Costa Rica. Omaular carina present but incomplete; propodeum with a large quadrate pit behind the spiracle; arolia absent; scopal hairs blunt, appearing singed; tergite VII of male large, flat, semicircular. Biology unknown.

Paranthidium. North America to Costa Rica; about six species in two subgenera; in Costa Rica only the rarely collected (*Rapanthidium*) *vespoides*. Preoccipital, omaular and propodeal carinae absent; fore and mid tibiae without apical spines; mandible of female slender medially, greatly expanded apically, blade-like and without teeth. Biology unknown.

MEGACHILINAE: Megachilini. This tribe consists of two cosmopolitan genera, both with numerous subgenera. They are small to moderate sized bees with a large pterostigma (longer than wide) and without arolia.

Coelioxys. A large cosmopolitan genus which, in the New World, comprises 133 species in 14 subgenera. In Costa Rica there are nearly 30 species in 11 subgenera: *Acrocoelioxys* (6 species), *Boreocoelioxys* (1), *Cyrtocoelioxys* (9), *Dasycoelioxys* (2), *Glyptocoelioxys* (2), *Haplocoelioxys* (1), *Melanocoelioxys* (2), *Neocoelioxys* (2), *Platycoelioxys* (1), *Rhinocoelioxys* (2) and *Synocoelioxys* (1). Taxonomy: Mitchell (1973). Mainly black bees, sometimes with red on basal tergites; eyes hairy; lower gena often with an excavated area; axilla with a spine; scutellum often modified; metasoma of female elongate and pointed apically; male with multiple spines on tergite VI. *Coelioxys* species are cleptoparasites of *Megachile*, and in the Neotropics also of *Centris*.

Megachile. In the New World this genus comprises about 500 species in approximately 40 subgenera. Nearly 50 species in 15 subgenera are recorded from Costa Rica: *Acentron* (2 species), *Argyropile* (1), *Austromegachile* (8), *Chelostomoides* (previously placed in *Chalicodoma*; 5), *Chrysosarus* (2), *Cressoniella* (1), *Holcomegachile* (1), *Leptorachis* (5), *Melanosarus* (1), *Moureana* (2), *Neomegachile* (3), *Pseudocentron* (7), *Ptilosarus* (3), *Sayapis* (3) and *Tylomegachile* (1). Taxonomy: Mitchell (1980). Black bees, often appearing yellowish because of dense light hair; mandible of female usually with one or more cutting edges between the teeth; fore leg of male, especially the basitarsus, often highly modified. Most species utilize leaf pieces to line the cells, but some use soil and/or resin.

NOMADINAE

This subfamily comprises over 30 genera worldwide; 15 are present in North America, mostly in desert regions; five occur in Costa Rica. They are small to moderately large, sparsely pubescent bees, and typically have distinct markings due to integumental maculations or contrasting hair bands. All species are cleptoparasites ('cuckoo bees').

Epeolus. Cosmopolitan; more than 100 species. Both subgenera, *Epeolus* and *Trophocleptria*, occur in Costa Rica, with a total of six species. Moderate sized bees with clouded wings; body sometimes all black, more often with off-white transverse pubescent bands composed of dense squamose hairs, and with pronotum, axilla and scutellum red. The biologies of the Costa Rican species are unknown, but elsewhere species are cleptoparasites of *Colletes*.

Nomada. Primarily holarctic, uncommon in the tropics. Of the eleven North American subgenera only three reach Costa Rica: *Hypochrotaenia* (three species), *Micronomada* (two species, mostly from Pacific lowlands), and *Nomada* (single species in drier, northwestern part of country). Small, wasp-like bees with distinct yellow and red-brown markings, including transverse bands on two or more of the tergites; fore wing with a strong dark stripe along anterior margin. Taxonomy: Snelling (1986b). The biologies of the Costa Rican species are unknown, but elsewhere species are cleptoparasites of *Andrena* (primarily), *Agapostemon* and *Exomalopsis* (Snelling, 1986*b*).

Odyneropsis. Southern Arizona south throughout the tropics; about ten species, three of which are rare, but apparently widely distributed in Costa Rica. Moderately large, wasp-like bees, all black with the fore wing dark, but sometimes with a white apex; axilla spined. *O. gertschi* is a cleptoparasite of *Ptiloglossa guinnae* (Roberts, 1971).

Thalestria. South America; monotypic, with one unconfirmed record from Costa Rica. Moderately large black bee that appears blue to green because of dense reflective scale-like hairs; axilla spined; scutellum biangulate. It is a cleptoparasite of *Oxaea*.

Triepeolus. New World; over 130 species, eight in Costa Rica (highlands and Pacific side). Moderate-sized black bees with dense white or yellow squamose pubescent bands; axilla often spined. Species are cleptoparasites of Eucerini, *Protoxaea* and *Ptiloglossa*.

ANTHOPHORINAE

This large, cosmopolitan subfamily is represented in Costa Rica by nine tribes. Anthophorine bees usually nest in the ground (sometimes in wood). Nest cell shape and cocoon characteristics of the tribes of Anthophorinae were briefly compared by Torchio (1974).

ANTHOPHORINAE: Anthophorini. This tribe comprises four genera with greatest diversity in xeric and Mediterranean climates. Only one genus occurs in Costa Rica.

Deltoptila. Endemic to the highlands of Mesoamerica (Chihuahua, Mexico to Costa Rica; Ayala, pers. comm.); about eight species, two in Costa Rica: *costaricensis* and an undescribed species. It has often been collected from *Cuphea* (Lythraceae) (LaBerge & Michener, 1963).

ANTHOPHORINAE: Centridini. This New World tribe is most diverse in the Neotropics and comprises three genera, *Centris*, *Epicharis* and *Ptilotopus*. The last of these is so far known only from South America and Panama, but may be present in Costa Rica. The other two genera are both well-represented in Costa Rica. Centridine females have longitudinal combs on the fore and mid basitarsi, and a large basal hind tibial plate; the scopa is very large, composed of long, highly branched hair on the hind tibia and the enlarged basitarsus. Keys to genera and species are given in Snelling (1984).

Centris. Deserts of North America south throughout the Neotropics; over 130 species in 13 subgenera, of which 34 species in eight subgenera occur in Costa Rica. Moderate-sized to large black bees; metasoma often red, yellow or metallic green, and covered with dense, short, black to yellow pubescence; clypeus only slightly protuberant, and usually marked with yellow in both sexes; hind femur of male often enlarged. The mating behaviour (see above, under Biology) and nesting behaviour of several species occurring in northwestern Costa Rica have been intensively studied (e.g. by Frankie *et al.*, 1988, 1989, 1990; Vinson *et al.*, 1993). *Centris* species either dig burrows in soil or use pre-existing holes in wood. Nests contain one to several cells, which are provisioned with oil, collected from oil-bearing plants (e.g. *Byrsonima crassifolia*, Malpighiaceae), and pollen.

C. (Centris). Metasoma metallic blue. Seven species in Costa Rica: *adani* (Frankie *et al.*, 1980; Vinson *et al.*, 1982), *aethiocesta* (Vinson & Frankie, 1988), *aethyctera* (Vinson & Frankie, 1977, 1991), *flavifrons* (Vinson & Frankie, 1988), *flavofasciata* (Vinson *et al.*, 1987), *inermis* (= *segregata*) (Coville *et al.*, 1983) and *obscurior*. Many species have 'metanders' (unusually large, robust males). Species nest in the soil.

C. (Hemisiella). Four species in Costa Rica: *dichrootricha*, *nitida* (Williams *et al.*, 1984), *trigonoides* (Williams *et al.*, 1984) and *vittata*. They nest in holes in wood, lined with soil.

C. (Heterocentris). Four species in Costa Rica: *analis*, *bicornuta*, *difformis* and *labrosa*. They nest in holes in wood, lined with wood chips.

C. (Melanocentris). Black, or with the metasoma dusky ferruginous. Six species in Costa Rica: *agilis*, *agiloides*, *flavilabris*, *fusciventris*, *obsoleta* and *plumipes*.

C. (Paracentris). Two species in Costa Rica: *atripes* and *nigrocaerulea*.

C. (Ptilocentris). Only *festiva* occurs in Costa Rica.

C. (Trachina). Six species in Costa Rica: *fuscata*, *heithausi* (Coville *et al.*, 1986), *labiata*, *longimana*, *similis* and *vidua*. They nest in soil, or in the sides of inhabited arboreal termite nests.

C. (Xanthemisia). Three species in Costa Rica: *carolae*, *lutea* and *rubella*. They nest in holes in wood, lined with wood chips.

Epicharis. Primarily neotropical but extending into North America; nine subgenera, four in Costa Rica: *Epicharana* (4 species), *Eicharoides* (2), *Hoplepicharis* (1) and *Parepicharis* (1). A fifth subgenus, *Epicharitides*, is known from Panama and may occur in Costa Rica. Large, black bees, usually with yellow markings on the metasoma; vertex with a plume of long setae behind each eye. Species collect oil. One species has been found nesting at the back of a small cave (G. Frankie, pers. comm.).

ANTHOPHORINAE: Emphorini. Emphorines are restricted to the New World. The tribe comprises three genera, only one of which, *Melitoma*, is known from Costa Rica; *Diadasia* may also occur here since it is known from temperate North America and from Panama south (Michener, 1954). Emphorines are relatively large bees with a convex vertex and a protruding clypeus. The burrow has a turret over the entrance.

Melitoma. Southern United States south throughout the Neotropics; about ten species, four in Costa Rica: *marginella*, *monozonula* and two undescribed species. Rather broad and flat black bees with narrow transverse bands of white hair on the tergites; tongue very long. Species are oligolectic on *Ipomoea* (Convolvulaceae) and fly mostly from December through to March. Nests of M. *segmentaria* (= *euglossoides*) have been described in Colombia (Michener, 1974*a*) and Mexico (Torchio, 1974). This species nests near water, which it collects and uses to soften the hard clay in which it burrows. Because it nests in exposed vertical banks it is rare in lowland tropical areas, since embankments quickly become covered with vegetation, and adobe is not used as a construction material by humans in these areas (Michener, 1974a; Torchio, 1974; Linsley *et al.*, 1980).

ANTHOPHORINAE: Ericrocini. This tribe is represented by four genera in Costa Rica. They are moderate-sized to large bees that often have metallic scales. The labrum is distinctive in having a pair of apical hair tufts. The mid tibial spur and scutellum are also modified. A key to the genera is given in Snelling and Brooks (1985). All ericrocines are cleptoparasites and, where known, their hosts are Centridini.

Aglaomelissa. Costa Rica to Venezuela; monotypic. *A. duckei* has a metallic blue to green body and the fore wing with a dark apex. It is cleptoparasitic, apparently on *Centris*.

Ctenioschelus. Mexico (Ayala *et al.*, 1993) to Brazil; monotypic. *C. goryi* has green scales on its body; the antenna of the male is longer than the body. Host unknown.

Mesocheira. Mexico to Brazil; monotypic. In Costa Rica *M. bicolor* is known to occur only in the northwest. Black and reddish bee with the dorsum of the metasoma covered with blue-green scale-like hairs and wings with dark apical marks; scutellum with a distinctive bilobed plate extending posteriorly. It is a cleptoparasite of *Centris* species (Parker, 1977).

Mesoplia. Southwestern United States to Argentina; about a dozen species in two subgenera, *Eumelissa* and *Mesoplia*. Both subgenera, and seven species occur in Costa Rica. Body covered with metallic blue to green scales, with the legs and tegula sometimes red; fore wing sometimes with a dark apical spot. They fly from December to May, and are cleptoparasites of *Centris* and *Epicharis* (Snelling & Brooks, 1985).

ANTHOPHORINAE: Eucerini. The Eucerini is a large tribe found on all continents except Australia. It is represented by 30 genera in the New World. Male eucerines characteristically have very long antennae extending beyond the mesosoma. The North and Central American genera were revised by LaBerge (1957).

Florilegus. New World; 11 species in three subgenera, two species in Costa Rica: *(F.) condignus* and *(Floriraptor) melectoides*. The former occurs from the United States to southern South America, but in Costa Rica it is known only from the northwestern part of the country. *F. melectoides* occurs from Costa Rica to South America. Black bees with a bluish or greenish sheen; head and mesosoma densely clothed in light (male) or black and white (female) hair; metasoma with distinct transverse hair bands. Taxonomy: Urban (1970). They are polylectic, but with an apparent preference for the Leguminosae. Nests of *condignus* were described from the United States (LaBerge & Ribble, 1966), where they were constructed in flat ground and consisted of shallow (less than 20 mm deep), vertical burrows with single vertical cells at the ends of short lateral tunnels.

Gaesischia. Southern Arizona south through the Neotropics; 19 species in three subgenera, in Costa Rica only *(Gaesischiana) exul*. Black with a largely tan metasoma, densely covered with light hair; clypeus with a broad white stripe; scopa composed of simple hairs. *G. exul* is active from January to March in drier parts of the highlands and the Pacific slope. Nest unknown.

Melissodes. New World; more than 120 species in eight subgenera, in Costa Rica there are seven species in three subgenera: (*Ecplectica*) *raphaelis*, (*Eumelissodes*) *persimilis*, (*Melissodes*) *panamensis*, (*M.*) *tepaneca*, (*M.*) *thelypodii* and two undetermined species. Black, with abundant reddish hair; tegula with the outer anterior margin incurved rather than convex; scopal hairs highly plumose. Taxonomy: LaBerge (1956, 1961). In southern Costa Rica at mid altitudes (1100 m) *persimilis* is oligolectic on Compositae (Buchmann & Jones, 1980).

Melissoptila. Southern Texas south through the Neotropics; 16 species in three subgenera, in Costa Rica two species in one subgenus (*Ptilomelissa*): *joseana* and *pinguis*. Small, black bees, usually with dense appressed golden pubescence concealing the surface of metasoma; scopal hairs long and loosely plumose. Taxonomy: Urban (1968). They fly from November to March and nest gregariously in soil.

Peponapis. New World; 17 species, seven in Costa Rica, all in the subgenus (*Peponapis*). Black bees, with or without light hair on mesosoma and light hair bands on tergites; scopal hairs coarse and sparse, especially on inner surface of tibia and basitarsus; male antenna not greatly elongate. They are active early in the morning, are oligolectic on Cucurbitaceae and have species-specific pollen collecting devices, which apparently influences their pollen collecting ability on particular *Cucurbita* species (Hurd & Linsley, 1966, 1970; Hurd *et al.*, 1971).

Svastra**[CR]**. One subgenus, *S.* (*Svastra*), is present in Argentina and Chile, while *S.* (*Brachymelissodes*) and *S.* (*Epimelissodes*) occur in temperate North America. Although not previously recorded south of Guatemala, two species of *Epimelissodes* (*nitida* and aff. *albocollaris*) occur in the drier Pacific side of Costa Rica. Black; female with particularly dense plumose hair on venter of metasoma in addition to the long, dense, plumose scopal hairs on the hind leg; male antenna not greatly elongate. Taxonomy: LaBerge (1956, 1958). One species flies in July and August, and the other has been collected in November. The biologies of the Costa Rican species are unknown.

Tetraloniella**[CR]**. Primarily holarctic; approximately 40 species in North America, six in Costa Rica. Black with pale tergal bands and a moderately plumose scopa. They fly from October to January. Nests are unknown.

***Thygater*.** Neotropical; 25 species in two subgenera; in Costa Rica there are six species in the nominate subgenus: *aethiops*, *analis*, *cockerelli*, *colombiana*, *crawfordi* and *montezuma*. Moderate-sized to large, black bees; clypeus strongly protuberant; malar space wide; scopal hairs sparsely and loosely plumose. Taxonomy: Urban (1967). Species tend to nest communally in embankments.

***Xenoglossa*.** North America south to northwestern Costa Rica; seven species, in Costa Rica only *gabbii*. Large black bee with the tergites (except the first) clothed in yellow-brown hair; ocelli enlarged. Like *Peponapis*, species are active in the early morning and are oligolectic on *Cucurbita* (Cucurbitaceae).

ANTHOPHORINAE: Exomalopsini. This tribe is restricted to the New World with its greatest diversity in the tropics. Four of the eleven genera occur in Costa Rica. They are small to moderate-sized, rather robust bees.

***Ancyloscelis*.** Southern United States south throughout the Neotropics; at least 15 species (half of them undescribed), three in Costa Rica: *apiformis* (= *armatus*) and two undescribed. These are recorded from the highlands and the Pacific side. Rather broad, flat, black bees, with narrow white hair bands on the tergites; head projecting forward, tongue rather long; female scopa on hind tibia and basitarsus large, densely plumose, black dorsally, white ventrally; male hind femur greatly swollen. Species are oligolectic on *Ipomoea* (Convolvulaceae) and fly from December through March. Nests of several species were described by Michener (1954, 1974b) and Torchio (1974); nests are gregarious, in sloping ground or in vertical embankments (including adobe walls).

***Exomalopsis*.** Southwestern United States south through the Neotropics; seven subgenera, two in Costa Rica: *Exomalopsis* (9 species) and *Megomalopsis* (*mellipes* and an undetermined species). Black, sometimes appearing brown to yellow because of relatively dense hair; head very narrow in lateral view, mouthparts rather short; scopa large, dense, highly plumose, usually bicoloured. Taxonomy: Timberlake (1980). Biologies of the Costa Rican species are unknown but elsewhere species nest in soil.

***Monoeca*.** Neotropical; approximately ten species, two in Costa Rica. Black bees, with contrasting reddish metasoma in females; fore basitarsus with a comb on the inner side; females with stout, posteriorly bent hairs on mesosternal region. Species collect oil from flowers of Malpighiaceae (Ayala, pers. comm.). Nests unknown.

***Paratetrapedia*.** Neotropical; over 50 species in seven subgenera; in Costa Rica there are 18 species in four subgenera: *Lophopedia* (8 species), *Paratetrapedia* (7), *Tropidopedia* (1) and *Xanthopedia* (2). Generally black but sometimes with a red metasoma and occasionally with the body tan; fore basitarsus with a comb on the outer side. Taxonomy: Michener and Moure (1957). These bees collect oil from flowers.

ANTHOPHORINAE: Osirini. This tribe is found in the Holarctic and Neotropical regions, two of the six genera are known from Costa Rica. Osirines are rare, slender, wasp-like bees. The females have very elongate stings and all are cleptoparasites. The genera were revised by Roig-Alsina (1989).

Osiris. Neotropical; 20 species, 11 in Costa Rica. They are rarely encountered but they are apparently widespread throughout the country, except in montane areas. Predominantly yellow-green or brown, usually with dark scutal markings and black apical bands on the tergites. Taxonomy: Shanks (1986). Their hosts are unknown but suspected to be *Paratetrapedia*.

Protosiris CR. Costa Rica south through tropical South America; five species, of which only one rare, undescribed species occurs in Costa Rica at altitudes between 400 and 700 metres. Similar in appearance to larger *Osiris*. Taxonomy: Shanks (1986; as part of *Osiris*).

ANTHOPHORINAE: Rhathymini. This tribe comprises a single genus, Rhathymus. They have paired tufts of apical hair on the labrum (as do Ericrocini) but the body is more elongate and not metallic, the tongue is very long, and the mid tibial spur is not modified. They are cleptoparasites.

Rhathymus. Neotropical; at least five species, in Costa Rica two undescribed species. Both are rarely encountered. Moderate-sized to large bees, greenish yellow with clouded wings, or black with yellow wings. They are cleptoparasites of *Epicharis*.

ANTHOPHORINAE: Tetrapediini. This tribe consists of two genera. *Tetrapedia* and *Coelioxoides* which has recently been transferred to the tribe (Roig-Alsina, 1990).

Coelioxoides. Neotropical; three species, in Costa Rica only the rare *punctipennis*. Elongate, slender, wasp-like bees with dark wings. Taxonomy: Roig-Alsina, 1990. Cleptoparasites, with hosts unknown.

Tetrapedia. Neotropical. This genus has been divided into two subgenera. One species, belonging to *T.* (*Tetrapedia*), occurs in Costa Rica. Rather elongate, shiny, black bee with dark wings; mid and hind tibial spurs pectinate; scopa composed of long, dense, simple, apically-directed hairs. Species collect oil from flowers of Malpighiaceae and nest in holes in wood (Ayala, pers. comm.).

XYLOCOPINAE

This large subfamily is divided into three tribes: Allodapini (Old World), Ceratini (cosmopolitan) and Xylocopini (cosmopolitan). A phylogenetic analysis of the tribes is given by Sakagami and Michener (1987). Most xylocopine species nest in wood or stems.

XYLOCOPINAE: Ceratini ('small carpenter bees'). This tribe comprises four genera: *Ceratina* (cosmopolitan), *Manuelia* (Argentina and Chile), *Megaceratina* (Afrotropical) and *Pithitis* (Palaeotropical). The genus Manuelia appears to be the most primitive group in the subfamily, sharing many features with the Xylocopini (Daly *et al.*, 1987).

Ceratina. Cosmopolitan. In Costa Rica there are 24 species in four subgenera. *C.* (*Calloceratina*) (6 species) occurs mostly on the Pacific side of the country while *C.* (*Crewella*) is represented by one species in the Atlantic lowlands. *C.* (*Ceratina*) with 12 species and *C.* (*Zadontomerus*) with five, are both widespread. Small to moderate-sized, dorso-ventrally flattened, weakly pubescent, dull to brilliant metallic; often with white or yellow facial marks; scopa consisting of dense plumose hairs on tibia but weak on basitarsus. These bees usually nest in cavities in stems or soft wood. The females sometimes progressively feed offspring and there are varying degrees of co-operation among adults within the nest (Roubik, 1989). *C.* aff. *rectangulife* and *C. ignara* have been studied near San José (Michener & Eickwort, 1966). The latter was found provisioning its nest with pollen from *Gomphrena dispersa* (Amaranthaceae).

XYLOCOPINAE: Xylocopini ('large carpenter bees'). This tribe comprises three genera: *Lestis* (eastern Australia), *Proxylocopa* (arid parts of the southwestern Palaearctic) and *Xylocopa* (nearly cosmopolitan). In the Galapagos Islands *X. darwini* is the only bee present and may play a role in the establishment of immigrant plant species (Linsley *et al.*, 1966). The common name of the group derives from the fact that most species nest in solid wood (except in *Proxylocopa*, which nest in the ground), sometimes causing structural damage.

***Xylocopa*.** (Fig. 18.15). Cosmopolitan; about 700 species, the majority in Africa and the Neotropics (Hurd, 1978*a*). In the New World there are about 100 species in 17 subgenera, in Costa Rica 13 species in six subgenera. Large to very large, generally black or submetallic although the males of some species are yellow-brown; scopal hairs very dense and simple; dorsal surface of metasoma mostly bare. Species are often active very early in the morning. Some species have special abdominal pouches for carrying phoretic mites. Biology: Gerling *et al.* (1989).

***X.* (*Calloxylocopa*).** Only *tenuata* occurs in Costa Rica, where it is restricted to the highlands.

***X.* (*Megaxylocopa*).** Three species occur in Costa Rica: *fimbriata* (biology in Janzen, 1966b), *frontalis* and *nautlana*. Males are yellowish brown.

***X.* (*Neoxylocopa*).** Two species occur in Costa Rica: *gualanensis* (Sage, 1968; Frankie & Daly, 1983) and an undetermined species. Males are yellowish brown.

***X.* (*Notoxylocopa*).** Only *tabaniformis* occurs in Costa Rica, where it is restricted to the highlands. Taxonomy: O'Brien and Hurd (1965).

***X.* (*Schoenherria*).** Five species occur in Costa Rica: *barbatella* (Wille, 1963), *lateralis*, *muscaria*, *subvirescens* and *viridis*.

***X.* (*Stenoxylocopa*).** Only *strandi* occurs in Costa Rica. Taxonomy: Hurd (1978*b*). Species excavate nests in dead culms of bamboo and related plants, and the females have features on the mandibles that appear to be adapted for this habit (Hurd & Moure, 1960).

EUGLOSSINAE ('orchid bees').

This subfamily is confined to the Neotropics, with a cumulative range extending from Mexico to Argentina. The common name of these bees derives from the fact that males are pollinators of orchids from which they gather fragrances attractive to females. Euglossines are moderate-sized (8.5 mm long) to very large (29 mm long) bees; some are bril-

Fig. 18.15. *Xylocopa* sp. (Xylocopinae).

Fig. 18.16. *Euglossa* sp. (Euglossinae).

liant, smooth metallic while others are very hairy and dark coloured with yellow hairs at the apex of the metasoma. The mouthparts (proboscis) are very long (reaching at least to the base of the metasoma). The relationships of the genera are discussed in Kimsey (1987*b*) and a list of the 166 recognized species is given in Kimsey and Dressler (1986).

The nests are either exposed (some species of *Euglossa*) or in pre-existing cavities in banks and tree trunks. They consist principally of brood cells since there are no storage cells or pots as in bumblebees and stingless bees. The nest is constructed of resin (*Euglossa*), resin mixed with bark fragments (*Eufriesea*), or mud with a thin layer of resin on the inner surface (*Eulaema*). Many are solitary, although some species of *Eufriesea* nest in aggregations (Myers & Loveless, 1976), and some species of *Euglossa* and perhaps all *Eulaema* regularly have several females per nest (Roberts & Dodson, 1967; Young, 1985; Eberhard, 1989). The nature of the interactions between these females is poorly understood, but thus far no examples of eusociality have been found in the subfamily.

Male orchid bees often defend territories centered around a prominent tree trunk situated in a canopy gap (Kimsey, 1980b; Schemske & Lande, 1984). They alternate between perching on the tree trunk and making periodic forays to patrol the territory. During perching the male bee performs a series of buzzes with the wings, which appears to serve a thermoregulatory function (Stern & Dudley, 1991). Prior to establishing territories, males visit certain flowers in order to obtain 'perfume', although the exact role of these fragrances is still unclear (Dressler, 1982; Williams & Whitten, 1983; Zimmerman & Madriñán, 1988). At least 55 genera and 625 species of neotropical orchids provide fragrances as their primary attractant and the flowers are often designed so as to place pollinia on specific locations on the body of the male orchid bees (Dodson, 1975; Dressler, 1981; Williams, 1982; Fig. 18.17). The same pollination syndrome is shown by a few Gesneriaceae, Araceae, and a few other plants (Buchmann, 1980; Williams & Dressler, 1976).

About half of the 60 compounds contained in orchid bee fragrances have been identified (most are essential oils widely distributed in nature), and it appears that specificity is achieved not by species-specific compounds but rather by species-specific blends. After landing on the flower the male mops up fragrances with his front tarsi, and then while hovering transfers them to the mid basitarsi and then to the hind tibiae. The hind tibia has an elongate slit leading to a glandular pouch within, where the chemicals appear to be modified and passed to other parts of the body (Roberts *et al.*, 1982). The fragrances could provide a means for a female to indirectly assess the quality of a potential mate, or the males might pass the chemicals to females during copulation (Roubik, 1989).

With the discovery of artificial chemical baits for attracting males, these rapid and wary bees have become easier to study. Such baits have been used to study seasonality (Janzen *et al.*, 1982; Ackerman, 1983) and daily foraging patterns (Armbruster & McCormick, 1990) although future studies need to consider between-site differences in species composition (Armbruster, 1993). Other aspects of orchid bee biology that have been studied include pollinator fidelity (Roubik & Ackerman, 1987) and grooming behaviour (Kimsey, 1984*b*).

Eufriesea. About 52 species, 13 in Costa Rica. Many of these reach the northern limit of their

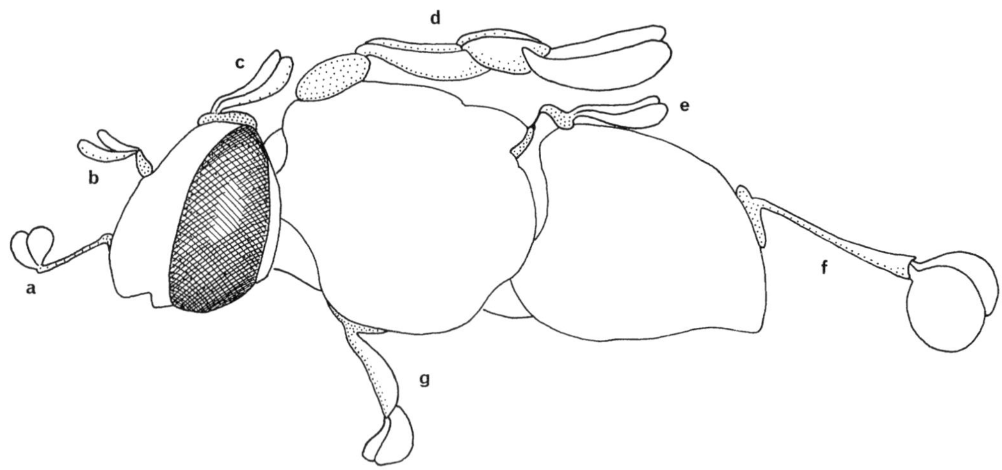

Fig. 18.17. Euglossine bee showing the sites of attachment of the pollinia of various Costa Rican orchids (modified after Dressler, 1981); **a**) *Notylia*; **b**) *Kefersteinia*; **c**) *Peristeria elata*; **d**) *Catasetum*; **e**) *Acineta*; **f**) *Cycnoches*; **g**) *Dressleria*.

***Eufriesea* sp.**	**Orchid pollinia**
E. chrysopyga	*Polycycnis gratiosa, Stanhopea wardii, Acineta chrysantha, A. superba*
E. concava	*Kegeliella atropilosa*
E. duckei	*Sievekingia fimbriata*
E. mexicana	*Sievekingia fimbriata, Trichopilia subulata*
E. mussitans	*Dressleria suavis*
E. purpurata	*Cynoches aureum, Coryanthes* sp., *Dressleria helleri, Gongora* sp.
E. schmidtiana	*Stanhopea costaricensis, Coryanthes rodriguezii*
E. superba	*Cynoches pentadactylon, Sievekingia* sp.
E. surinamensis	*Cypholoron* sp.

Table 18.6. Records of orchid pollinia attached to males of *Eufriesea* species known from Costa Rica (from Kimsey, 1982).

distribution in Costa Rica, although *caerulescens* and *mexicana* are at the southern limit of their distribution here. Mostly metallic, though some have yellow and black hairs resembling *Eulaema*, but *Eufriesea* species have a metallic face without white markings; males with hind tibial slit long (reaching the apical margin); females lack a black scutellar tuft. Taxonomy: Kimsey (1977, 1982). Unlike other genera, many species appear to fly only during the rainy season (Ackerman, 1983). Males use orchids in the groups Catosetinae, Stanhopeinae and Zygopetalinae (see Table 18.6).

Euglossa. Slightly over 100 species, of which 32 (in the subgenera *Euglossa, Euglossella, Glossura* and *Glossurella*) occur in Costa Rica. Brilliant metallic, not densely hairy; labrum whitish with two dark oval spots (labrum totally dark in other euglossines); males with hind tibial slit short (not reaching the apex of the tibia); females with a black scutellar tuft.

Eulaema. About 13 species in two subgenera (*Apeulaema* and *Eulaema*), ten species in Costa Rica. Black, hairy, with conspicuous patterns of yellow or orange hairs, and sometimes with limited metallic tints on the metasoma; face dark coloured, often with white markings; males with hind tibial slit long; females with black scutellar tufts. *E. cingulata* and *E. polychroma* have been found together in a mixed colony in Panama (Roubik, 1990).

Exaerete. Six species, three in Costa Rica: *dentata* (rare), *frontalis* and *smaragdina*. Brilliant green; hind tibiae narrow. Taxonomy: Kimsey (1979). Cleptoparasites of *Eufriesea* and *Eulaema*.

BOMBINAE ('bumblebees')

The subfamily comprises about 250 species and is usually divided into two genera, *Bombus* and *Psithyrus*, although the former may be paraphyletic with respect to the latter (P.H. Williams, 1985). For biological reasons it is convenient to maintain these two genera (Michener, 1990), since all species of *Psithyrus* are social parasites. Bombines are medium-sized (9 mm long) to large (22 mm) hairy bees; the proboscis is shorter than in orchid bees (not reaching beyond the middle coxae).

Bumblebees occur primarily in cool temperate parts of the Holarctic region, extending south to Tierra del Fuego, northern Africa, the Himalayas, the Philippines and the mountains of Java. In the New World *Psithyrus* reaches only as far south as Guatemala. In tropical regions in general, bumblebees are more common at mid to high elevations, except for *B. atratus* in Brazil, which occurs in lowland rainforests (Moure & Sakagami, 1962). In the Old World tropics bumblebees are totally absent from the lowlands, whereas in the Neotropics they are present, though less common than at higher altitudes (Michener, 1990).

Except for the social parasites, all bumblebees are primitively eusocial. Individual queens establish nests in rodent burrows, cavities under or on vegetation, and similar situations. Recently established nests are highly vulnerable to usurpation by other bumblebees, usually of the same species, and in these conflicts the larger queen tends to prevail. The nest may or may not be covered by a thin layer of wax and pollen. The 'disten-

Bumblebee	Status and distribution
Bombus (*Fervidobombus*)	
digressus	Rare; 1500–3300 m
mexicanus	Fairly common; mostly 1000–1500 m
pullatus	Fairly common; below 1500 m
weisi	Rare; one Costa Rican record from 1500 m
Bombus (*Pyrobombus*) ephippiatus	Abundant; above 1500 m
Bombus (*Robustobombus*) *volucelloides*	Locally common (in Amistad N.P.[1]); 500–2000 m

Table 18.7. Abundance and distribution of Costa Rican bumblebees (Bombinae).
[1]G. Chavarría, pers. comm.

sible cells' are unique among aculeates in that they are closed but grow with the growing larvae, commonly contain several eggs and later larvae, develop a separate bulge for each larva, and may become divided into several cells as the larvae mature (Katayama, 1989). The acquisition of distensible cells has enabled bumblebees to adopt batched oviposition, resulting in the production of offspring in distinct cohorts. Larvae are fed progressively, either by food introduced through the tops of the cells or by food pressed in through pockets at the bases of the cells. After the larva spins its pupal cocoon the adult bees remove the pollen-wax cell material covering the cocoon and use it for new egg cells. Both honey and pollen are stored in pots separate from brood cells and these pots are often made of old cocoons, but may be entirely constructed by the bees.

Bumblebees are frequently employed in studies of optimal foraging (e.g. Hodges, 1985; Harder, 1986), resource partitioning (e.g. Johnson, 1986) and mimicry (Plowright & Owen, 1980). More general information on the biology of bumblebees is given in Alford (1975), Heinrich (1976), Morse (1982) and Plowright and Laverty (1984).

Bombus. This genus has been divided into various subgenera (e.g. 35 by Richards, 1968), although these subgenera are less distinct than those in other bee genera (Michener, 1990). The New World species were monographed by Milliron (1973) and a key to the species of Mexico and Central America is given in Labougle (1990). Six species are known to occur in Costa Rica (Table 18.7).

MELIPONINAE ('stingless bees')

A primarily circumtropical subfamily which, in the New World, ranges from Mexico to southern Brazil and northern Argentina. They have achieved a level of eusociality comparable to that of honey-bees and are the dominant group of bees in lowland tropical rain forests, where they may occur at densities of up to 600 colonies per square kilometre (Roubik, 1989). The common name of the group derives from the fact that the sting is reduced and not exsertable. Despite their inability to sting, they are by no means defenceless. Some species swarm out of the nest to attack intruders—crawling into hair, ears, nose, eyes, etc.—and biting.

Nests of most Costa Rican species have been described by Wille and Michener (1973) and by subsequent authors (Roubik, 1983; Camargo & Roubik, 1991). These are summarized in Table 18.8; nests are unknown for *Melipona yucatanica* and *Trigona (Geotrigona) acapulconis.* The most common nesting site is in cavities of tree trunks or large branches, although *Plebeia tica* utilizes smaller diameter (2–3 cm), hollow branches and lianas. Among those nesting in cavities in the ground, *P. mirandula* and *Trigona buchwaldi* sometimes occupy abandoned mammal burrows (Wille, 1966). Others construct completely or partly exposed nests, and these species tend to defend their nests more aggressively than do cavity-nesters. Partly exposed nests (e.g. those of *Partamona cupira*) are often situated in the bases of palm leaves, underneath clumps of epiphytes, etc. Although many species occasionally utilize abandoned termite or ant nests (e.g. *T. fuscipennis* sometimes uses cavities made by birds in termite nests), a few regularly excavate their own cavities in active nests: *Plebeia latitarsis* and *Trigona ferricauda* in exposed termite nests, *Paratrigona peltata* in aerial nests of *Camponotus senex*, and *Nannotrigona mellaria* in tree cavities containing aggressive ants (Wille & Michener, 1973; Roubik, 1983).

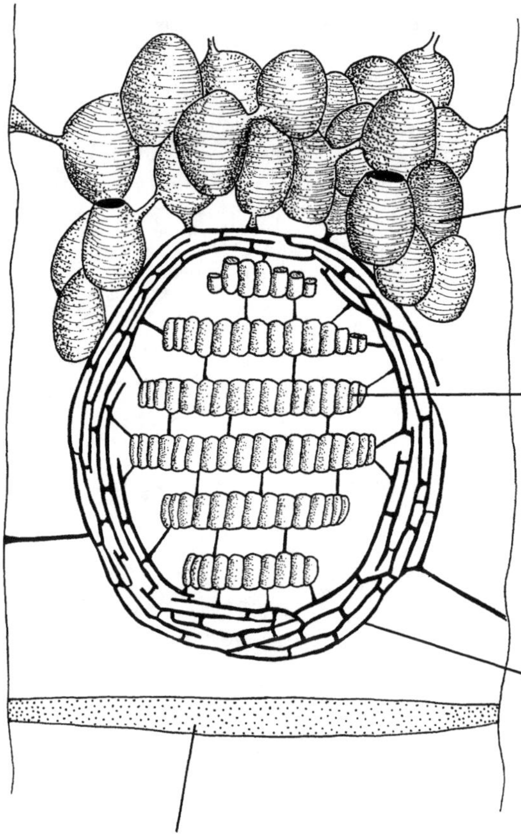

Fig. 18.18. Schematic diagram of cross section of stingless bee nest inside hollow tree showing brood cells (arranged in combs) within the involucrum with food pots located above.

The nests are constructed primarily of CERUMEN — a mixture of wax, which is secreted from the metasomal tergites, and resins, which must be collected (Haueisen Freire & Gara, 1970). Resins are secreted by injured or diseased plants, although resin secretion occurs naturally in several tropical trees (e.g. Leguminosae), and a few plants produce floral resins. Some stingless bees, especially those building exposed nests, mix the resin with chewed leaf material, wood pulp (*T. nigerrima*), faeces or mud. Since most nests are concealed in cavities, usually only the entrance is readily observable. This nest entrance consists of a resinous tube, except in *Plebeia tica*, which utilizes such tiny holes that there is no need to narrow it with resin (however, an internal tube is present). In many species the entrance is so narrow that only one to four bees can occupy it at the same time, but in others it is larger (e.g. *Lestrimellita, Nannotrigona, Partamona, Scaptotrigona* and *Trigona-Trigona*). In some species the entrance tube projects from the substrate and this projecting tube varies not only in length and shape (Table 18.8), but also in rigidity, ornamentation, and thickness of the lip. In *Nannotrigona, Plebeia mirandula, Trigona buchwaldi* and *T. dorsalis* the tube is perforated.

The nest covering is termed BATUMEN and usually consists of a single layer of resin and wax. Several layers are present in nests of *Partamona* and *Trigona (Trigona)*, while batumen is absent in nests of *Trigonisca*. In exposed nests the batumen is thick all the way around (up to 24 cm thick in *Trigona corvina*, which incorporates mud). In cavity-nesters the

Genus species	Nest site	Entrance tube	Involucrum	Pots	Broodcells
Cephalotrigona					
capitata	ground, tree	0	++	A	Co, ICo, a
Lestrimelitta					
limao	tree	++ C, F	+	A	Co, a
Melipona					
beecheii	tree	0	++		Co, s
fasciata	tree	0	++	A	Co, a
fuliginosa	tree	0	++		Co
marginata	tree	0	++	U	SCo
Nannotrigona					
mellaria	tree	++ C	++		Co, a
perilampoides	tree	+, ++ C	++	A,U	Co
Oxytrigona					
daemoniaca	tree				
mellicolor	tree	0	++	R	SCo, a
Paratrigona					
opaca	exp	+ F	++	S/U	Co, s
ornaticeps	tree	+ I	++	R	Co, s
peltata	p-exp, exp	++ C	++	U	Co, a
Partamona					
cupira	p-exp	0, + F	+	A, R, S, U	Co, a
peckolti	exp	+ F	+	U	Co, a
Plebeia (Nogueirapis)					
mirandula	ground	+ C	+	U	Co
Plebeia (Plebeia)					
frontalis	tree	0	+,++	A, U	Co, s
jatiformis	tree	0	0,++		Co, s
tica	tree	0	0	D	C
Plebeia (Scaura)					
latitarsis	tree	++ C	0	A, U	Co, a
Scaptotrigona					
luteipennis	tree	++ C			
mexicana	tree	+ C, F, I	++	A, S, U	Co
pectoralis	tree	++ C, I			
postica	tree				ISCo
wheeleri					
Trigona (Frieseomelitta)					
nigra	tree	0	0	D	C
Trigona (Geotrigona)					
leucogastra	ground				
Trigona (Tetragona)					
dorsalis	roots, tree	++ C	++	A, U	SCo, s
lurida	roots	0, + C		A, U	
perangulata	tree	0 F	++	A, S, U	SCo
Trigona (Tetragonisca)					
angustula	tree	+, ++ C	++	A, S	Co, s
buchwaldi	ground	+ C	++	S, U	Co, a
Trigona (Trigona)					
cilipes	tree	++, C	0	R	ISCo
corvina	exposed	0	+	S	ISCo, a

Table 18.8. Continued opposite.

Genus species	Nest site	Entrance tube	Involucrum	Pots	Brood cells
ferricauda	termite nests				
fulviventris	tree	0, +, ++ F	0	R	Co, ICo
fuscipennis	tree	0, + F	0, +	A, R, U	ISCo, s
necrophaga	tree	+ F	0	U	SCo, a
nigerrima	exposed	++ C	+, ++	U	Co, a
pallens	tree	+ C			Co
silvestriana	exposed	+, ++ C, F	0, +	S, U	Co, ISCo
Trigonisca					
atomaria	tree	0	0	D	C
buyssoni	tree	0	0	D	C, a

Table 18.8. Nests of Costa Rican stingless bees (modified from Wille & Michener, 1973). Nest sites: exp = exposed, p-exp = partly exposed, tree = in tree cavity. Entrance tube (projecting from substrate): 0 = absent, + = as long as or shorter than wide, ++ = longer than wide, C = cylindrical, F = funnel-shaped, I = irregular. Involucrum: 0 = absent, + = discontinuous, ++ = conspicuous and continuous. Storage pots (position in relation to brood chamber, ignoring discrepancies due to unusual nest sites): A = above, D = dispersed (lacking distinct brood chamber), S = at the side, R = all around, U = under (below). Brood cell arrangement: C = clusters, Co = combs, ICo = irregular combs, ISCo = irregular spiral combs, SCo = spiral combs. Brood cell production: a = asynchronously, s = synchronously.

batumen usually consists of thick plates sealing off the top and bottom of the cavity (in *Melipona* these are made largely of mud), and a thin layer against the inner surface of the wood (Fig. 18.18). Within the batumen are storage pots and brood cells, which open upward and are supported by short pillars and connectives (these are very long in *Partamona, Trigona fulviventris* and *T. fuscipennis*). The storage pots contain either honey or pollen, except in species of *Lestrimelitta*. These bees rob pollen and honey from other species, and transport a mixture of these nutrients back to their own nest in their crop, and then store it as a mixture. In most species the honey pots may be completely separated from the pollen pots (*T. buchwaldi*), partially separated (most species) or fully intermixed (*Oxytrigona mellicolor, T. corvina, T. nigerrima* and *T. silvestriana*). Storage pots are usually oval or subspherical in shape but are elongate in *Trigona buchwaldi* and *Trigonisca*. The heart of the nest is the brood chamber, which is commonly surrounded by an INVOLUCRUM. Within this chamber are the brood cells, which are smaller than the storage pots and are arranged in combs or clusters (Table 18.8).

Brood cells are constructed asynchronously (i.e. successively—new cells in various stages of construction) or synchronously (new cells in the same stage of construction), depending on the species. Provisioning can also be asynchronous or synchronous, or a combination of the two (Sakagami, 1982). Cells are built by several 'nursing' workers (as opposed to one individual in Bombinae). These same workers regurgitate food, obtained from the storage pots, into the completed cell (unlike Apinae where the tasks of cell construction and provisioning are performed by different aged workers). Although the queen is present during cell construction her interaction with workers increases after the cell has been completed. Once it is provisioned the queen lays an egg and a worker then closes the cell. After the cell has been provisioned workers frequently lay unviable trophic eggs, which are eaten by the queen. In *Lestrimelitta* and *Plebeia* (*Plebeia*) worker egg-laying is not in conjunction with queen oviposition and such eggs are eaten by workers or the queen. Male-producing eggs are sometimes laid after oviposition by the queen but before the cell is capped. In most genera castes (queen versus worker) are determined trophically, and only a few queens are produced in special, large cells. *Melipona* is unique in that there appears to be a strong genetic component to caste determination, and numerous small queens (most of which are killed by the workers) are reared in cells that are identical to worker cells. After the larvae spin cocoons workers remove the surrounding cell

- –Large (8–14 mm in length); thorax and upper part of head densely hairy; wings relatively short, barely reaching apex of abdomen .... *Melipona*
- –Smaller (2-8 mm in length); thorax and upper head less hairy; wings reaching well beyond apex of abdomen.
  - –Inner surface of hind tibia without depressed, smooth area along posterior edge, i.e. inner surface uniformly hairy (Fig. 18.19).
    - –Hind tibia lacking corbicula (on outer surface); head shiny, without hair .... *Lestrimelitta*
    - –Hind tibia with corbicula on outer surface; head not shiny, hirsute
      - –Scutellum extending posteriorly over propodeum; [< 5 mm long].
        - –Posterior margin of scutellum emarginate, anterior margin with median, V-shaped depression (Fig. 18.22); thorax coarsely sculptured .... *Nannotrigona*
        - –Posterior margin of scutellum rounded, anterior margin without median depression (Fig. 18.23); thorax finely punctate .... *Paratrigona*
      - –Scutellum not extending beyond propodeum.
        - –Anterior margin of scutellum with median depression (Fig. 18.24) .... *Scaptotrigona*
        - –Anterior margin of scutellum without median depression (cf Fig. 18.23)
          - –Base of propodeum hairy; hind tibia spoon-shaped; > 5 mm long .... *Partamona*
          - –Base of propodeum glabrous; hind tibia not spoon-shaped; < 5mm long .... *Plebeia (Nogueirapis)*
  - –Inner surface of hind tibia with a very narrow, depressed smooth area along posterior margin (Fig. 18.20).
    - –Outer surface of hind basitarsus very convex, bulbous (Fig. 18.25); face without yellow markings .... *Plebeia (Scaura)*
    - –Outer surface of hind basitarsus not markedly swollen; face with yellow markings .... *Plebeia (Plebeia)*
  - –Inner surface of hind tibia with a broad, depressed smooth area along posterior margin, distinct from raised hairy area (Fig. 18.21)
    - –Posterior margin of hind tibia without branched hairs
      - –Fore wing with vein *Rs* strongly curved basally (Fig. 18.28); 2–4 mm in length .... *Trigonisca*
      - –Fore wing with vein *Rs* not strongly curved basally (Fig. 18.29); 4–9 mm in length
        - –Base of propodeum hairy; head with occipital carina (Fig. 18.30); mesosoma usually black .... *Cephalotrigona*
        - –Base of propodeum glabrous; head without occipital carina; mesosoma usually orange .... *Oxytrigona*
    - –Posterior margin of hind tibia with simple and branched hairs

Table 18.9. Continued opposite.

–Mandible with 4–5 pointed teeth (Fig. 18.31)	*Trigona (Trigona)*
–Mandible with only two distinct pointed teeth (Fig. 18.32)	
–Body with at least some orange color	
–Basal inner surface of hind basitarsus with velvety hairless area (Fig. 18.26)	*Trigona (Tetragonisca)*
–Basal inner surface of hind basitarsus hairy (Fig. 18.27)	*Trigona (Tetragona)*
–Body entirely black	
–Upper face with sparse, long pale hairs; metasoma with white hairs basally	*Trigona (Geotrigona)*
–Upper face without long pale hairs; metasoma without white hairs basally	*Trigona (Frieseomelitta)*

Table 18.9. Simple diagnostic characteristics of the genera and subgenera of stingless bees (Meliponinae) occurring in Costa Rica.

wall so that the cocoons are exposed, and when adult bees emerge, space becomes available for construction of new brood cells. Thus, there are often two advancing fronts of cells—one on the top comb and the other at the emptied space at the bottom.

Each adult worker bee passes through four stages as it ages: callow, 'nursing' (discussed above), 'household' (receiving nectar, removing wastes, etc.) and forager. The secretion and manipulation of wax are not confined to the household stage (as in honey-bees), but begins very early. Flight and foraging begins later than in honey-bees. Stingless bees are generalized flower visitors and different species appear to partition food resources via different foraging strategies based on: presence or absence of recruitment, solitary versus group foraging, and aggressiveness (Johnson & Hubbell, 1974, 1975; Hubbell & Johnson, 1977, 1978; L.K. Johnson, 1980, 1981, 1983; Johnson *et al.*, 1987). Small meliponines (with the exception of *Plebeia mirandula*) usually show little capacity for recruitment and tend to forage solitarily on scattered resources, whereas many of the larger species are capable of following a trail of odour droplets deposited on the vegetation by another member of the colony—e.g. in *Cephalotrigona*, *Lestrimelitta*, *Oxytrigona*, *Scaptotrigona*, and *Trigona* (subgenera *Tetragona* and *Trigona*). These recruitment pheromones (which may serve as alarm pheromones at higher concentrations) are secreted by the mandibular glands (Roubik, 1989). For example, *T. fulviventris* normally forages solitarily, but can quickly recruit hundreds of workers to a concentrated resource. *T. silvestriana* is another facultative group forager but shows a marauding behaviour apparently adapted to localized resources that are quickly depleted. Persistent group foragers such as *Scaptotrigona pectoralis* and *T. fuscipennis* are slower in discovering a resource, but once discovered they recruit rapidly, and aggressively defend it against other bees. However, these foraging patterns may vary somewhat, depending on the distance of the food source from the nest. In addition to flowers, species of *Oxytrigona* and *Trigona* sometimes collect honeydew from Homoptera. Species of the *Trigona hypogea* group, which includes *T. necrophaga*, are obligate necrophages that utilize pheromone trails to recruit foragers to carrion (Camargo & Roubik, 1991).

As in honey-bees, colony multiplication occurs via swarming. Unlike honey-bees the process is gradual, being initiated by a group of workers that searches for a new nest site (Wille & Orozco, 1974). Once a site has been selected nest construction begins and eventually honey is brought from the mother colony (pollen is usually freshly collected). After the new nest has been provisioned, a small group of workers departs from the mother colony with a virgin queen. At the new nest site (or at the mother nest in the case of supersedure), the virgin queen makes a nuptial flight, mates (apparently with just one male in most cases, unlike honey-bees), loses her flight capability and begins ovipositing.

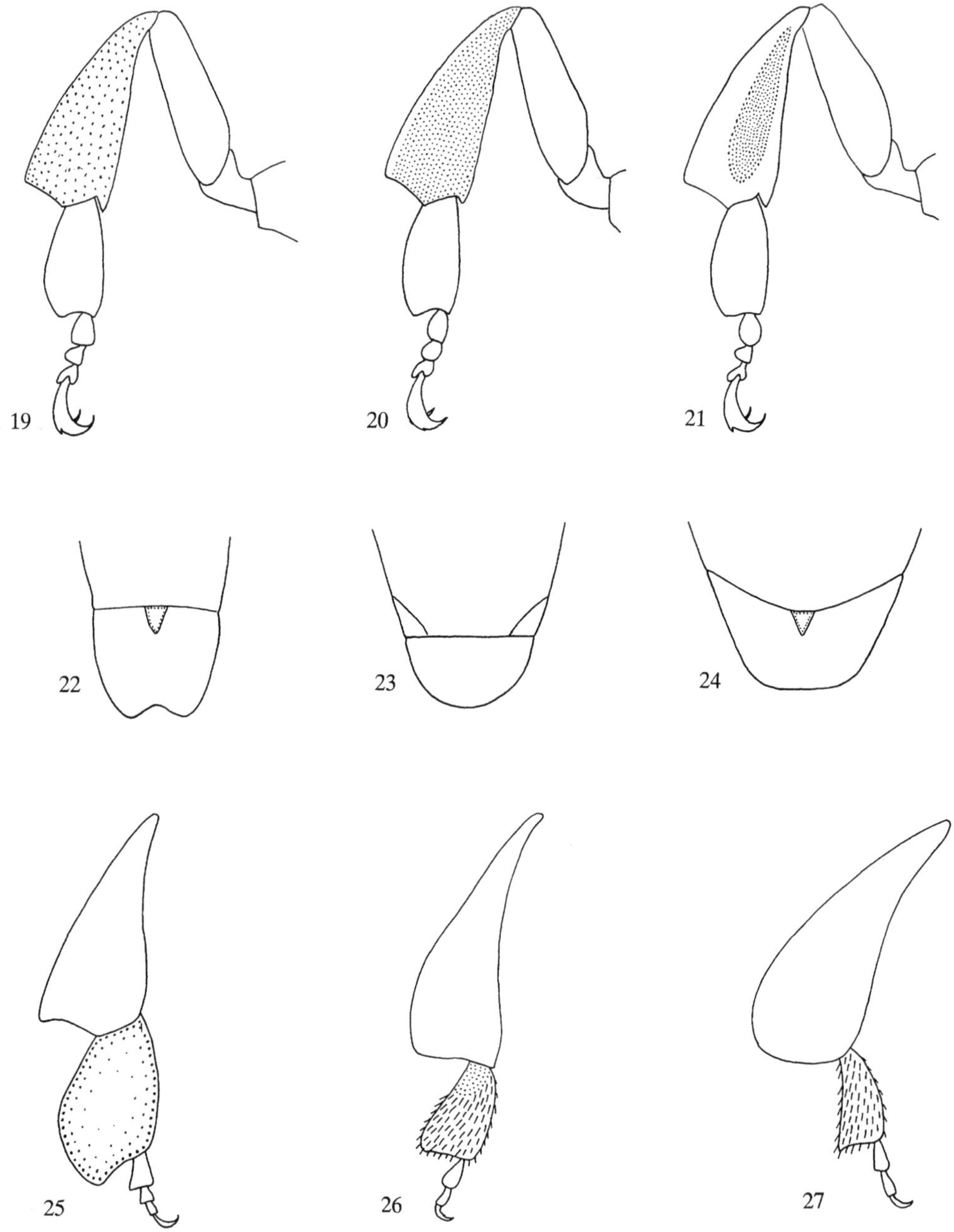

Figs 18.19–18.27. Meliponinae. Figs 18.19–18.21. Hind left leg posterior view; 18.19, *Scaptotrigona* sp.; 18.20, *Plebeia* (*Plebeia*) sp.; 18.21, *Trigona* sp. Figs 18.22–18.24. Scutellum, dorsal; 18.22, *Nannotrigona* sp.; 18.23, *Paratrigona* sp.; 18.24, *Scaptotrigona* sp. Figs 18.25–18.27. Hind left leg, posterior view; 18.25, *Plebeia* (*Scaura*) sp.; 18.26, *Trigona* (*Tetragonisca*) sp.; 18.27, *Trigona* (*Tetragona*) sp.

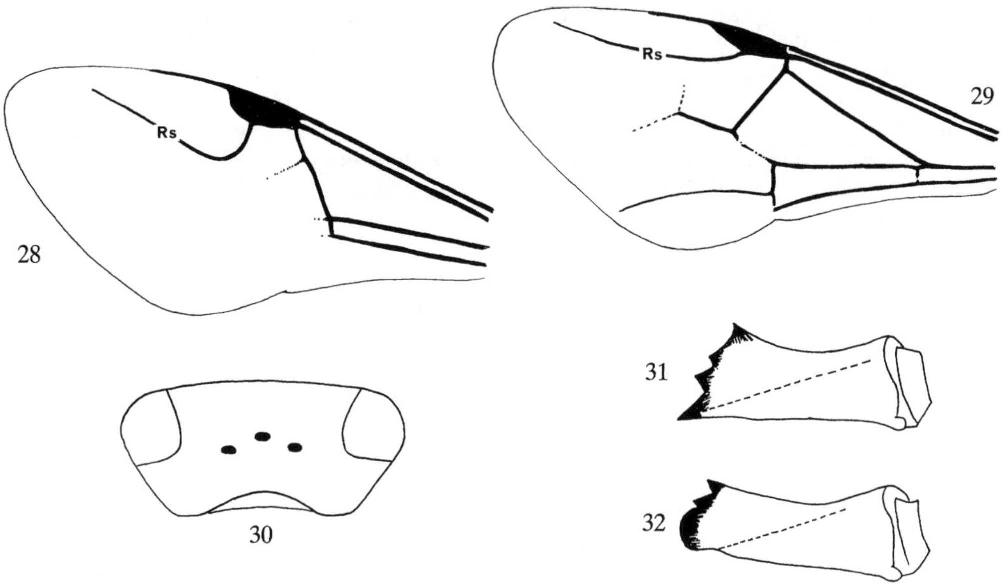

Figs 18.28–18.32. Meliponinae. Figs 18.28–18.29. Fore wing; 18.28, *Trigonisca* sp.; 18.29, *Cephalotrigona* sp. Fig. 18.30. Head, dorsal, *Cephalotrigona* sp. Figs 18.31–18.32. Mandible; 18.31, *Trigona* (*Trigona*) sp.; 18.32, *Trigona* (*Tetragona*) sp.

More information on the biology of stingless bees is provided by Schwarz (1948), Wille (1961, 1983), Camargo (1970), Wille and Michener (1973), Sakagami (1982) and Roubik (1989).

The number of recognized genera of Meliponinae has varied considerably. Wille (1979*b*) recognized 8 genera, all of the Central American species being placed in *Lestrimelitta*, *Melipona* or *Trigona*. On the other hand Camargo (1989), following earlier work of Moure, recognized about 50 genera. Here, we follow the classification of Michener (1990), who recognized 21 genera—the pantropical *Trigona*, 9 palaeotropical genera and 11 neotropical genera. Of the 12 genera occurring in the Neotropics, only *Trichotrigona* is absent from Central America. The genera and subgenera occurring in Costa Rica can be identified by using the key provided in Table 18.9. In addition to these 11 genera, a new genus is being described by Roubik, Camargo and Lobo for a new species that occurs at mid altitude sites (1500–2000 m) in Costa Rica and Panama. A key to the species of stingless bees of Panama is given by Roubik (1992).

Cephalotrigona. Three species, in Costa Rica only *capitata* (subspecies *zexmeniae*).

Lestrimelitta. Possibly ten species, most of which are undescribed (Michener, 1990); two species occur in Costa Rica. Although the name *limao* has traditionally been used, this species is probably restricted to Brazil (Ayala, pers. comm.). All are 'robber' bees, that is, workers do not forage except in nests of other stingless bees (especially species of *Nannotrigona*, *Paratrigona*, *Plebeia* and *Scaptotrigona*).

Melipona. About 40 species, five in Costa Rica: *beecheii*, *fasciata* (the true *fasciata* may be restricted to Mexico; Ayala, pers. comm.), *fuliginosa* (= *flavipennis*), *marginata*, and *yucatanica*. Two pairs of sister species show mutually exclusive distributions in Panama and Costa Rica: *beecheii* occurs in Costa Rica but not in Panama while the closely related *compressipes* occurs in Panama but not in Costa Rica, and the same pattern is shown by *yucatanica* and *favosa* (Camargo *et al.*, 1988). Taxonomy and biology of the Costa Rican species: Wille (1976; what he calls *fulvipes* is actually *yucatanica*, see Camargo *et al.*, 1988). Nests of *fuliginosa* are made of seeds and resin of *Vismia* (Clusiaceae) (Roubik, 1983).

Nannotrigona. Nine species, two in Costa Rica: *mellaria* and *perilampoides* (both have been classified as subspecies of *testaceicornis*).

Oxytrigona. About eight species, two in Costa Rica: *daemoniaca* (southeastern part of the country) and *mellicolor* (both have been classified as subspecies of *tataira*). These species are sometimes called 'fire bees', for their very caustic bites, which can cause long-lasting blisters (Bian *et al*., 1984; Roubik *et al*., 1987).

Paratrigona (= *Aparatrigona*). About a dozen species, three in Costa Rica: *opaca* (including subspecies *guatemalensis* and *opaca*), *ornaticeps* and *peltata*.

Partamona. About 16 species, three in Costa Rica: *cupira*, *grandipennis* and *peckolti* (= *testacea musarum*).

Plebeia. This genus is divided into four subgenera, three of which occur in Costa Rica. The monotypic subgenus *P*. (*Schwarziana*) is restricted to Brazil, Paraguay, and northern Argentina.

P. (Nogueirapis). Costa Rica to Bolivia; three species, in Costa Rica only *mirandula*. Taxonomy: Wille (1962). Biology: Wille (1964, 1966).

P. (Plebeia). About 30 species, six in Costa Rica: *frontalis*, *jatiformis*, *tica* (Wille, 1969) and three undescribed species.

P. (Scaura). Four species, in Costa Rica only *latitarsis*.

Scaptotrigona. About two dozen species, five in Costa Rica: *luteipennis*, *mexicana* (= *pachysoma*; Ayala, in press), *pectoralis*, *postica* and *wheeleri*.

Trigona. (Fig. 18.01). This is the largest and most widely distributed meliponine genus. It occurs both in the Neotropics and in the Palaeotropical region from India to Australia. It is divided into ten subgenera, of which four are palaeotropical and six neotropical. Five of the latter occur in Central America; only *T. (Duckeola)* is absent. It should be noted that the first four subgenera listed below have often been combined in the subgenus *Tetragona* (e.g. Wille & Michener, 1973).

T. (Frieseomelitta). About 10 species, in Costa Rica only *nigra*.

T. (Geotrigona). About ten species, two in Costa Rica: *acapulconis* and *leucogastra*.

T. (Tetragona). About 13 species, three in Costa Rica: *dorsalis*, *lurida* (placed in a separate subgenus, *Ptilotrigona*, by Roubik, 1992) and *perangulata*.

T. (Tetragonisca). About four species, two in Costa Rica: *angustula* (= *jaty*) and *buchwaldi*. Biology: Wille (1966).

T. (Trigona). About 30 species, nine in Costa Rica: *cilipes*, *corvina*, *ferricauda*, *fulviventris*, *fuscipennis*, *necrophaga*, *nigerrima*, *pallens* and *silvestriana*. Taxonomy and biology: Wille (1965), Camargo and Roubik (1991). Biology: Johnson and Hubbell (1975), Wille and Orozco (1974), L.K. Johnson (1981).

Trigonisca. Sometimes treated as a subgenus of the Old World genus *Hypotrigona*; about 17 species, five in Costa Rica: *atomaria*, *buyssoni*, *discolor*, *rhodopta* and *schulthessi*. *T. buyssoni* is actually a Peruvian species, but the name has generally been used in a broader sense, thus including various undescribed species (Ayala, in press).

APINAE ('honey-bees')

This subfamily comprises one genus, *Apis*, which has been traditionally regarded as containing just four species (Ruttner, 1988), although there may be at least nine species (Michener, 1990). The group is endemic to the Old World tropics, but one species, *A. mellifera*, has been domesticated and moved throughout the world. It was brought to the New World in the 16th century by the Spanish and Portuguese colonists from Europe. Recently, a new race was introduced, resulting in the 'Africanization' of virtually all honey-bees in the Neotropics (see Chapter 3.1, 'killer bees'). These Africanized honey-bees can readily survive in feral colonies.

Like stingless bees, honey-bees are highly eusocial. Nests are exposed or in cavities, and are made primarily of wax secreted by the sternal wax glands of workers. The cells are more or less horizontal, forming vertical combs of two layers of cells that open in opposite directions. The larvae are fed progressively and the cells are not closed until the larval development is completed. Brood cells and storage cells are similar; queen-producing cells tend to hang from brood combs. Recruitment to a food source occurs via the well known WAGGLE DANCE (Frisch, 1967; Lindauer, 1967; Kirchner & Sommer, 1992; Michelsen *et al.*, 1992; Seeley & Towne, 1992; see also Chapter 2.6).

The honey-bee, *Apis mellifera*, is one of the most intensively investigated of all insects. Detailed studies have been made of adult and larval morphology (Snodgrass, 1956; Torchio & Torchio, 1975), foraging behaviour (Calderone & Page, 1992; Oldroyd *et al.*, 1992; Greggers & Menzel, 1993) and other aspects of its biology (summaries are provided by Seeley, 1985 and Winston, 1987).

NOTE ADDED IN PROOF

Costa Rica has been the focus of considerable attention by mellitologists, yet much remains to be learned about its bee fauna, as evidenced by the very recent discovery of two additional genera, *Chlerogella* and *Diadasia*. These additions bring the generic count for Costa Rica to 97.

HALICTINAE: Augochlorini

Chlerogella. This rare genus, previously only known from Panama and Peru, comprises three described species. One occurs in Costa Rica. Small slender bees with a red metasoma; no other Costa Rican bee such an elongate head — the malar space fully three times as long as broad. Males of this genus remain unknown and nothing is known of its nesting habits.

ANTHOPHORINAE: Emphorini

Diadasia. New World, predominantly in temperate regions and approximately 30 species in North America. *Diadasia* (*Dasiapis*) *olivacea* has been found near the coast in the extreme north west of Guanacaste Province. Black bees with cream-coloured clypeus (males) and mandibles (both sexes); metasoma obscured by dense, appressed ochre pile; fore tibia of female with dense ventral brush of long plumose hairs. Known species nest gregariously in flat or gently sloping ground; a turret is formed over the nest entrance.

and maintaining their identity with respect to other such units.

Deuterotoky A form of parthenogenesis whereby unmated females may produce both male and female offspring.

Eclosion —of an egg. Hatching.

Ectoparasitoid A parasitoid (*qv*) that feeds externally on its host.

Ectophagous Feeding on the outside, of, for example, a host larva.

Edentate Without teeth.

Emarginate With a distinct notch.

Endoparasitoid A parasitoid (*qv*) that feeds internally within a host organism.

Endophagous —of a parasitoid. Feeding within a host larva.

Endophytic Feeding in concealment within plant tissue.

Exophytic Feeding in a more or less exposed situation externally on plant tissue.

Folivore An organism that consumes leaves.

Fossorial Adapted for digging in the earth

Gondwanic A distribution pattern believed to have been centred on the ancient southern supercontinent of Gondwanaland, and to have become disjunct with the breakup of this land mass. Typically gondwanic groups (such as the thynnine tiphiids) are currently more or less restricted to South America and Australia.

Gongylidia The specialized swollen hyphae of a mutualistic fungus used as food by fungus-growing ants.

Gynephores Asexual female cynipines which lay only female eggs.

Hemizygote A normal male hymenopteran, i.e. one with a haploid chromosome compliment.

Heterogyny Alternation of generations (in the Cynipini) where a bisexual generation comprising males and females gives rise to a female-only generation.

Heteronomous hyperparasitism A form of hyperparasitism, found in some members of the family Aphelinidae, where the males develop as parasitoids of the larva of primary parasitic females of their own species.

Hyperparasitoid A parasitoid that develops by feeding on or within another parasitoid; strictly one that develops whilst the primary parasitoid is still developing on the primary host (see also pseudohyperparasitoid).

Hypogaeic Living underground.

Idiobiont A parasitoid (*qv*) that feeds as a larva on a host that has been immobilized and developmentally arrested by the female parent prior to oviposition.

Immunodefensive system Of a host, the cellular and/or physiological systems responsible for encapsulating and destroying foreign bodies within the haemocoel.

Intermediate host The host used by the first instar of some parasitoids (e.g. Trigonalyidae) that do not oviposit directly on the definitive hosts.

Kairomone A chemical substance produced by one organism that evokes a response in another species of organism, and is adaptively favourable to the receiver but not the transmitter.

Koinobiont A parasitoid (*qv*) that feeds as a larva on or within a host which continues development after parasitization.

Malaxation The act of 'kneading' a host or prey organism by applying repeated pressure with the mandibles, but not piercing the cuticle. Malaxation of paralysed arthropod hosts is practicized particularly by many solitary aculeates, such as Tiphiidae.

Monobasic A taxon that includes only a single species.

Monogynous —of an ant colony. With a single egg-laying female or queen

Monomorphic —literally, of one form. Used in ant literature to describe a situation where workers are all of virtually the same size and shape.

Müllerian bodies Small structures in the hollow stems of certain myrmecophytes (*qv*), derived from specialized patches of trichomes (trichilia) at the base of each petiole, which serve as food for associated plant-ants.

Multilocular A gall containing several chambers, each of which houses a cynipid larva.

Multiparasitism A form of superparasitism (*qv*) involving more than one species of parasitoid.

Myrmecophytes = ant plants. Plants such as *Cecropia* spp., that are regularly inhabited by one or more species of ants (the plant-ants *qv*).

Nuptial flight The mating flight of ants, where alates (*qv*) leave the nest and unmated females fly high pursued by males that attempt to copulate with them.

Orthandrous —of sawflies and woodwasps. With the male genitalia in the normal orientation, not rotated (cf. strophandrous).

Parasitoid An organism which develops on or in another, single (host) organism, extracts nourishment from it, and kills it as a direct or indirect result of that development.

Planidium A minute, fusiform, ventrally flattened first instar larva, that is usually quite well sclerotized and

	active. It generally seeks out and attaches itself to a host or an organism likely to contact the host
Plant-ants	Species of ants that live in myrmecophytes (*qv*).
Polydomous	—of ant species. With a single colony (*qv*) occupying two of more discrete nests.
Polyembryony	The development of multiple embryos and thus larvae, from a single egg.
Polygynous	—of an ant colony. With more than one egg-laying female or queen.
Polymorphic	—literally, with many forms. Used in ant literature to describe the situation where different sized workers (such as majors and minors) occur in a single colony.
Polyphagous	Using a wide range of items as food; of a parasitoid having a host range that usually embraces several orders or many families
Pro-ovigenic	Emerging from the pupa with a full complement of more or less mature eggs in the oviducts, and failing to develop more eggs to maturity subsequently.
Pseudohyperparasitoid	A parasitoid that attacks another parasitoid after the primary parasitoid has completed the destruction of its own (primary) hosts.
Pseudoparasitized	The condition of a host when it has been stung and envenomated by certain parasitoids (esp. chelonine braconids), yet does not contain a developing parasitoid (Jones, 1985). Pseudoparasitized hosts may fail to develop normally.
'Pseudoparasitoid'	A hymenopteran which is evolutionarily derived from typical parasitoids, but feeds on several hosts in close proximity, such as a series of eggs in a cocooned egg-mass or several cecidomyiid larvae in a seed pod. This term is best avoided (e.g. by use of term predator) as confusion can arise with pseudoparasitized (*qv*).
Reniform	Kidney-shaped
Spiniform	Shaped like a sharp spine.
Strophandrous	—of sawflies. With the male genitalia rotated 180° prior to eclosion (*cf* orthandrous).
Superparasitism	The allocation of one or more eggs to a previously parasitized host.
Synovigenic	Developing a number of eggs successively to maturity throughout life. At any one time synovigenic females will only have a few mature eggs in the oviduct.
Teratocyte	A dissociated cell, usually of the trophamnion (*qv*), that lives within the host haemocoel.
Thelytoky	A form of parthenogenesis whereby diploid females are produced from unfertilized eggs.

Thylacium A sac composed of the cast exuviae of an ectoparasitic hymenopteran larva; these are characteristic of dryinids and can often be seen projecting from between the abdominal segments of homopteran hosts. Thylacia are also produced by larvae of a few other hymenopteran families.

Trophamnion The layer of cells surrounding the developing embryo of certain endoparasitoids which takes in nutrients from the host's haemolymph to nourish the developing parasitoid.

Unilocular A cynipid gall that contains a single larval chamber.

Venoms Fluid secretions from glands associated with the female reproductive tract; these substances are injected into hosts or potential enemies via the ovipositor.

References

Compiled and checked by P.A. Mitchell

The references included here have, where possible, been checked to original source. In less than 0.5 percent of the cases this was not possible and in these instances the reference was checked against a secondary source. In some cases dates of publication given on the journal or book differ from the actual publication date, which here follows the printed date in parentheses.

I thank the staff of the Department of Library Services of The Natural History Museum, London, particularly Ms C. Gokce and Ms L. Mitchell, for their invaluable help.

Abe, M. & Smith, D.R. 1991. The genus-group names of Symphyta (Hymenoptera) and their type-species. *Esakia*, **31**: 1–115.

Abe, T. 1982. Ecological role of termites in a tropical rain forest, pp. 71–75. *In* : Breed, M.D., Michener, C.D. & Evans, H.E. (eds) *The Biology of Social Insects*. 419pp. Westview Press, Boulder.

Abe, Y. 1988. Trophobiosis between the gall wasp, *Andricus symbioticus*, and the gall-attending ant, *Lasius niger*. *Applied Entomology and Zoology*, **23**: 41–44.

Abe, Y. 1991. Notices on courtship behaviour and copulation of *Synergus japonicus* Walker (Hym., Cynipidae). *Journal of Applied Entomology*, **111**: 478–483.

Abraham, Y.J., Moore, D. & Godwin, G. 1990. Rearing and aspects of biology of *Cephalonomia stephanoderis* and *Prorops nasuta* (Hymenoptera: Bethylidae) parasitoids of the coffee berry borer, *Hypothenemus hampei* (Coleoptera: Scolytidae). *Bulletin of Entomological Research*, **80**: 121–128.

Achterberg, C. van, 1975. A revision of the tribus Blacini (Hymenoptera, Braconidae, Helconinae). *Tijdschrift voor Entomologie*, **118**: 159–322.

Achterberg, C. van, 1976. A preliminary key to the subfamilies of the Braconidae (Hymenoptera). *Tijdschrift voor Entomologie*, **119**: 33–78.

Achterberg, C. van, 1977. The function of swarming in *Blacus* species (Hymenoptera, Braconidae, Helconinae). *Entomologische Berichten*, **37**: 151–152.

Achterberg, C. van, 1979. A revision of the subfamily Zelinae auct. (Hymenoptera, Braconidae). *Tijdschrift voor Entomologie*, **122**: 241–479.

Achterberg, C. van, 1983*a*. The preference of zoophagous Hymenoptera for certain types of plants as shown by the subfamily Gnaptodontinae (Braconidae). *Verhandelingen SIEEC*, **10**: 97–98.

Achterberg, C. van, 1983*b*. Revisionary notes on the Palaearctic genera and species of the tribe Exothecini Foerster (Hymenoptera, Braconidae). *Zoologische Mededelingen*, **57**: 339–355.

Achterberg, C. van, 1984*a*. Essay on the phylogeny of Braconidae (Hymenoptera: Ichneumonoidea). *Entomologisk Tidskrift*, **105**: 41–58.

Achterberg, C. van, 1984*b*. Addition to the revision of the genus *Zele* Curtis (Hymenoptera: Braconidae). *Entomologische Berichten*, **44**: 110–112.

Achterberg, C. van, 1984*c*. Revision of the genera of Braconini with the first and second metasomal tergites immovably joined (Hymenoptera, Braconidae, Braconinae). *Tijdschrift voor Entomologie*, **127**: 137–164.

Achterberg, C. van, 1986. The oviposition behaviour of parasitic Hymenoptera with very long ovipositors (Ichneumonoidea: Braconidae). *Entomologische Berichten*, **46**: 113–115.

Achterberg, C. van, 1987*a*. Revision of the European Helconini (Hymenoptera: Braconidae: Helconinae). *Zoologische Mededelingen*, **61**: 263–285.

Achterberg, C. van, 1987*b*. Revisionary notes on the subfamily Orgilinae (Hymenoptera: Braconidae). *Zoologische Verhandelingen*, **242**: 1–111.

Achterberg, C. van, 1988*a*. A new species of the genus *Gnamptodon* from Italy (Hymenoptera: Braconidae). *Entomologische Berichten*, **48**: 159–161.

Achterberg, C. van, 1988*b*. *Bracon lineifer* spec. nov., a peculiar new species from the Netherlands (Hymenoptera: Braconidae). *Entomologische Berichten*, **49**: 191–194.

Achterberg, C. van, 1988*c*. Parallelisms in the Braconidae (Hymenoptera) with special reference to the biology. *Advances in Parasitic Hymenoptera Research*, **1988**: 85–115.

Achterberg, C. van, 1988*d*. Revision of the subfamily Blacinae Foerster (Hymenoptera, Braconidae). *Zoologische Verhandelingen*, **249**: 1–324.

Achterberg, C. van, 1991. Key to the subfamilies of Braconidae known in Australia, pp. 947–948. *In* : *The Insects of Australia*, II. 1137pp. Melbourne University Press, Melbourne.

Ackerman, J.D. 1983. Diversity and seasonality of male euglossine bees (Hymenoptera: Apidae) in central Panamá. *Ecology*, **64**: 274–283.

Ackerman, J.D. 1986. Mechanisms and evolution of food-deceptive pollination systems in orchids. *Lindleyana*, **1**: 108–113.

Adams, E.S. 1990. Interaction between the ants *Zacryptocerus maculatus* and *Azteca trigona* : interspecific parasitization of information. *Biotropica*, **22**: 200–206.

Aeschlimann, J.-P. 1974. Biologie et comportement de *Chorinaeus funebris* Gravenhorst (Hymenoptera: Ichneumonidae). *Annales de Zoologie-Écologie Animale*, **6**: 529–538.

Aeschlimann, J.-P. 1990. Simultaneous occurrence of thelytoky and bisexuality in hymenopteran species, and its implications for the biological control of pests. *Entomophaga*, **35**: 3–5.

Ågren, J. & Schemske, D.W. 1991. Pollination by deceit in a Neotropical monoecious herb, *Begonia involucrata*. *Biotropica*, **23**: 235–241.

Ahmad, M. 1943. The morphology and biology of *Rhogas testaceus* (Spin.) a braconid internal parasite of spotted bollworms of cotton. *Indian Journal of Entomology*, **5**: 189–205.

Akre, R.D. 1968. The behavior of *Euxenister* and *Pulvinister*, histerid beetles associated with army ants. *Pan-Pacific Entomologist*, **44**: 87–101.

Akre, R.D. & Rettenmeyer, C.W. 1966. Behavior of Staphylinidae associated with army ants (Formicidae: Ecitonini). *Journal of the Kansas Entomological Society*, **39**: 745–782.

Akre, R.D. & Rettenmeyer, C.W. 1968. Trail-following by guests of army ants (Hymenoptera: Formicidae: Ecitonini). *Journal of the Kansas Entomological Society*, **41**: 165–174.

Alam, M.M. 1980. Biological and ecological factors affecting populations of sugarcane moth-borer, *Diatraea saccharalis* (Lep.: Pyralidae) Barbados, W.I. *Entomophaga*, **25**: 401–414.

Alam, S.M. 1957. The biology of *Metaphycus taxi* Alam (Encyrtidae: Hymenoptera) in the constant temperature room, with notes on the anatomy of its pre-imaginal stages. *Indian Journal of Entomology*, **19**: 231–240.

Alam, S.M. 1959. The life-history and the larval anatomy of *Euaphycus variolosus* Alam. (Hymenoptera, Encyrtidae)—an endoparasite of *Asterolecanium variolosum* (Ratzh) (Hemiptera, Coccidae). *Proceedings of the Zoological Society, Calcutta*, **12**: 35–40.

Alauzet, C. 1987. Bioecologie de *Eubazus semirugosus, Coeloïdes abdominalis* et *C. sordidator* (Hymenoptera: Braconidae) parsites de *Pissodes notatus* (Col.: Curculionidae) dans le sud de la France. *Entomophaga*, **32**: 39–47.

Alayo D., Pastor 1972. Estudios sobre los himenópteros de Cuba. VI- Familias Evaniidae y Gasteruptiidae. *Poeyana*, **95**: 1–15.

Alayo D., Pastor & Hernández, L.R. 1978. *Introducción al estudio de los Himenópteros de Cuba. Superfamilia Chalcidoidea.* 105pp. Academia de Ciencias de Cuba, La Habana.

Alcock, J. 1975. Social interactions in the solitary wasp *Cerceris simplex* (Hymenoptera: Sphecidae). *Behaviour*, **54**: 142–152.

Alcock, J. 1978. Notes on male mate-locating behavior in some bees and wasps of Arizona. *Pan-Pacific Entomologist*, **54**: 215–225.

Alcock, J. 1981. Notes on the reproductive behavior of some Australian thynnine wasps (Hymenoptera: Tiphiidae). *Journal of the Kansas Entomological Society*, **54**: 681–693.

Alcock, J. 1984. Convergent evolution in perching and patrolling site preferences of some hilltopping insects of the Sonoran Desert. *Southwestern Naturalist*, **29**: 475–480.

Alcock, J., Barrows, E.M., Gordh, G., Hubbard, L.J., Kirkendall, L., Pyle, D.W., Ponder, T.L. & Zalom, F.G. 1978. The ecology and evolution of male reproductive behaviour in the bees and wasps. *Zoological Journal of the Linnean Society*, **64**: 293–326.

Alcock, J., Eickwort, G.C. & Eickwort, K.R. 1977. The reproductive behavior of *Anthidium maculosum* (Hymenoptera: Megachilidae) and the evolutionary significance of multiple copulations by females. *Behavioral Ecology and Sociobiology*, **2**: 385–396.

Alexander, B. 1986. Alternative methods of nest provisioning in the digger wasp *Clypeadon laticinctus* (Hymenoptera: Sphecidae). *Journal of the Kansas Entomological Society*, **59**: 59–63.

Alexander, B.A. 1992. An exploratory analysis of cladistic relationships within the superfamily Apoidea, with special reference to sphecid wasps. (Hymenoptera). *Journal of Hymenoptera Research*, **1**: 25–61.

Alexander, B. & Rozen, J.G. Jr. 1987. Ovaries, ovarioles, and oocytes in parasitic bees (Hymenoptera: Apoidea). *Pan-Pacific Entomologist*, **63**: 155–164.

Alexander, R.D. 1974. The evolution of social behavior. *Annual Review of Ecology and Systematics*, **5**: 325–383.

Alexander, R.D., Noonan, K.M. & Crespi, B.J. 1991. The evolution of eusociality. pp. 3–44. *In* : Sherman, P.W., Jarvis, J.U.M. & Alexander, R.D. *The Biology of the Naked Mole-rat*. 518pp. Princeton University Press, Princeton, New Jersey.

Alford, D.V. 1968. The biology and immature stages of *Syntretus splendidus* (Marshall) (Hymenoptera: Braconidae, Euphorinae), a parasite of adult bumblebees. *Transactions of the Royal Entomological Society of London*, **120**: 375–393.

Alford, D.V. 1975. *Bumblebees*, 352pp. Davis-Poynter, London.

Allen, A.A. 1982. Observations on the ectoparasitic stage of two *Homobbus* [sic!] spp. (Hym.: Braconidae). *Proceedings and Transactions of the British Entomological and Natural History Society*, **15**: 11.

Allen, D.C. 1972. Insect parasites of the saddled prominent, *Heterocampa guttivitta* (Lepidoptera: Notodontidae) in the northeastern United States. *Canadian Entomologist*, **104**: 1609–1622.

Allen, H.W. 1964. New species of *Tiphia* (Hymenoptera: Tiphiidae) from Panama. *Notulae Naturae*, **371**: 1–6.

Allen, H.W. 1966. A revision of the Tiphiinae (Hymenoptera: Tiphiidae) of Eastern North America. *Transactions of the American Entomological Society*, **92**: 231–356.

Allen, H.W. 1970. *Tiphia* of the West Indies (Hymenoptera: Tiphiidae). *Proceedings of the Entomological Society of Washington*, **72**: 252–269.

Allen, H.W. 1971. A monographic study of the genus *Tiphia* (Hymenoptera: Tiphiidae) of Western North America. *Transactions of the American Entomological Society*, **97**: 201–359.

Allen, H.W. 1972. A monographic study of the subfamily Tiphiinae (Hymenoptera: Tiphiidae) of South America. *Smithsonian Contributions to Zoology*, **113**: 1–76.

Alma, A., Arno, C. & Vidano, C. 1988. Particularities of *Polynema striaticorne* as egg parasite of *Stictocephala bisonia* (Rhynchota Auchenorrhyncha), pp. 597–603. *In* : Vidano, C. & Arzone, A. (eds) *6th Auchenorrhyncha Meeting, Turin, Italy, September 7–11 1987. Proceedings*. 652pp. Consiglio Nazionale delle Ricerche, Italy.

Al-Maliky, S.K. & Al-Izzi, M.A.J. 1990. The progeny production of a hymenopterous parasitoid, *Apanteles* sp. group *ultor*, as affected by temperature. *Entomophaga*, **35**: 217–222.

Alphen, J.J.M. van 1980. Aspects of the foraging behaviour of *Tetrastichus asparagi* Crawford and *Tetrastichus* spec. (Eulophidae), gregarious egg parasitoids of the asparagus beetles *Crioceris asparagi* L. and *C. duodecimpunctata* L. (Chrysomelidae). I. Host-species selection, host-stage selection and host discrimination. *Netherlands Journal of Zoology*, **30**: 307–325.

Alphen, J.J.M. van, Dijken, M.J. van & Waage, J.K. 1987 (1988). A functional approach to superparasitism: host discrimination needs not be learnt. *Netherlands Journal of Zoology*, **37**: 167–179.

Alphen, J.J.M. van & Nell, H.W. 1982. Superparasitism and host discrimination by *Asobara tabida* Nees (Braconidae: Alysiinae), a larval parasitoid of Drosophilidae. *Netherlands Journal of Zoology*, **32**: 232–260.

Alphen, J.J.M. van & Thunnissen, I. 1983 (1984). Host selection and sex allocation by *Pachycrepoideus vindemiae* Rondani (Pteromalidae) as a facultative hyperparasitoid of *Asobara tabida* Nees (Braconidae: Alysiinae) and *Leptopilina heterotoma* (Cynipoidea; Eucoilidae). *Netherlands Journal of Zoology*, **33**: 497–514.

Alphen, J.J.M. van & Visser, M.E. 1990. Superparasitism as an adaptive strategy for insect parasitoids. *Annual Review of Entomology*, **35**: 59–79.

Altieri, M.A. 1991 *a*. How best can we use biodiversity in Agroecosystems? *Outlook on Agriculture*, **20**: 15–23.

Altieri, M.A. 1991 *b*. Classical biological control and social equity. *Bulletin of Entomological Research*, **81**: 365–369.

Altieri, M.A., Francis, C.A., Schoonhoven, A. van & Doll, J.D. 1978. A review of insect prevalence in maize (*Zea mays* L.) and bean (*Phaseolus vulgaris* L.) polycultural systems. *Field Crops Research*, **1**: 33–49.

Altieri, M.A., Trujillo, J., Campos, L., Klein-Koch, C., Gold, C.S. & Quezada, J.R. 1989. El control biológico clásico en América Latina en su contexto histórico. *Manejo Integrado de Plagas* (CATIE, Costa Rica), **12**: 82–107.

Andersson, M. 1984. The evolution of eusociality. *Annual Review of Ecology and Systematics*, **15**: 165–189.

Andrade, J.C. de & Carauta, J.P.P. 1982. The *Cecropia-Azteca* association: a case of mutualism? *Biotropica*, **14**: 15.

Andrews, F.G. 1978. Taxonomy and host specificity of Nearctic Alloxystinae with a catalog of the world species (Hymenoptera: Cynipidae). *California Department of Food and Agriculture, Laboratory Services/Entomology, Occasional Papers*, **25**: 1–128.

Angalet, G.W. 1964. *Bracon greeni* (Hymenoptera: Braconidae), a potential parasite of the boll weevil. *Indian Journal of Entomology*, **26**: 447–452.

Annecke, D.P. & Doutt, R.L. 1961. The genera of the Mymaridae. Hymenoptera: Chalcidoidea. *Republic of South Africa Department of Agricultural Technical Services Entomology Memoirs*, **5**: 1–71.

Annila, E. 1982. Diapause and population fluctuations in *Megastigmus specularis* Walley and *Megastigmus spermotrophus* Wachtl. (Hymenoptera, Torymidae). *Annales Entomologici Fennici*, **48**: 33–36.

Antropov, A.V. 1990. Another nest-guarding male of *Pison*. *Sphecos*, **20**: 20.

Arduser, M.S. & Michener, C.D. 1987. An African genus of cleptoparasitic halictid bees (Hymenoptera: Halictidae). *Journal of the Kansas Entomological Society*, **60**: 324–329.

Armbruster, W.S. 1993. Within-habitat heterogeneity in baiting samples of male euglossine bees: possible causes and implications. *Biotropica*, **25**: 122–128.

Armbruster, W.S. & Herzig, A.L. 1984. Partitioning and sharing of pollinators by four sympatric species of *Dalechampia* (Euphorbiaceae) in Panama. *Annals of the Missouri Botanical Garden*, **71**: 1–16.

Armbruster, W.S. & McCormick, K.D. 1990. Diel foraging patterns of male euglossine bees: ecological causes and evolutionary response by plants. *Biotropica*, **22**: 160–171.

Arthur, A.P. 1958. Development, behaviour, and descriptions of immature stages of *Spilochalcis side* (Walk.) (Hymenoptera: Chalcididae). *Canadian Entomologist*, **90**: 590–595.

Arthur, A.P. 1961. The cleptoparasitic habits and the immature stages of *Eurytoma pini* Bugbee (Hymenoptera: Chalcidae), a parasite of the European pine shoot moth, *Rhyacionia buoliana* (Schiff.) (Lepidoptera: Olethreutidae). *Canadian Entomologist*, **93**: 655–660.

Arthur, A.P. 1981. Host acceptance by parasitoids, pp. 97–120. *In* : Nordlund, D.A., Jones, R.L. & Lewis, W.J. (eds) *Semiochemicals, Their Role in Pest Management*. 306pp. Wiley & Sons, New York.

Arthur, A.P. & Ewen, A.B. 1975. Cuticular encystment: a unique and effective defense reaction by cabbage looper larvae against parasitism by *Banchus flavescens* (Hymenoptera: Ichneumonidae). *Annals of the Entomological Society of America*, **68**: 1091–1094.

Arthur, A.P., Hegdekar, B.M. & Batsch, W.W. 1972. A chemically defined, synthetic medium that induces oviposition in the parasite *Itoplectis conquisitor* (Hymenoptera: Ichneumonidae). *Canadian Entomologist*, **104**: 1251–1258.

Arthur, A.P., Hegdekar, B.M. & Rollins, L. 1969. Component of the host haemolymph that induces oviposition in a parasitic insect. *Nature*, **223**: 966–967.

Arthur, A.P., Stainer, J.E.R., & Turnbull, A.L. 1964. The interaction between *Orgilus obscurator* (Nees) (Hymenoptera: Braconidae) and *Temelucha interruptor* (Grav.) (Hymenoptera: Ichneumonidae), parasites of the pine shoot moth, *Rhyacionia buoliana* (Schiff.) (Lepidoptera: Olethreutidae). *Canadian Entomologist*, **96**: 1030–1034.

Arthur, A.P. & Wiley, H.G. 1959. Effects of host size on sex ratio, development time and size of *Pimpla turionellae* (L.) (Hymenoptera: Ichneumonidae). *Entomophaga*, **4**: 297–301.

Artigues, M., Avilla, J., Sarasúa, M.J. & Albajes, R. 1992. Primary host discrimination and elimination of supernumeraries in the facultative autoparasitoid *Encarsia tricolor* (Hymenoptera: Aphelinidae). *Entomologia Generalis*, **17**: 131–138.

Ashley, T.R. 1983. Growth pattern alterations in fall armyworm, *Spodoptera frugiperda*, larvae after parasitization by *Apanteles marginiventris, Campoletis grioti, Chelonus insularis and Eiphosoma vitticole* [sic]. *Florida Entomologist*, **66**: 260–266.

Ashley, T.R., Waddill, V.H., Mitchell, E.R. & Rye, J. 1982. Impact of native parasites on the fall armyworm, *Spodoptera frugiperda* (Lepidoptera: Noctuidae), in south Florida and release of the exotic parasite, *Eiphosoma vitticolle* (Hymenoptera: Ichneumonidae). *Environmental Entomology*, **11**: 833–837.

Asis, M. 1989. *Propoleo: el Oro purpura de las Abejas*. 255pp. Centro de Información y Documentación Agropecuario, La Habana.

Askari, A. & Coppel, H.C. 1978. Observations on courtship and mating behavior of *Meteorus pulchricornis*, a gypsy moth parasitoid. *Annals of the Entomological Society of America*, **71**: 364–366.

Askew, R.R. 1961*a*. *Eupelmus urozonus* Dalman (Hym., Chalcidoidea) as a parasite in cynipid oak galls. *The Entomologist*, **94**: 196–201.

Askew, R.R. 1961*b*. A study of the biology of species of the genus *Mesopolobus* Westwood (Hymenoptera: Pteromalidae) associated with cynipid galls on oak. *Transactions of the Royal Entomological Society of London*, **113**: 155–173.

Askew, R.R. 1962. The distribution of galls of *Neuroterus* (Hym: Cynipidae) on oak. *Journal of Animal Ecology*, **31**: 439–455.

Askew, R.R. 1965. The biology of the British species of the genus *Torymus* Dalman (Hymenoptera: Torymidae) associated with the galls of Cynipidae (Hymenoptera) on oak, with special reference to alternation of forms. *Transactions for the Society for British Entomology*, **16:** 217–232.

Askew, R.R. 1966. Observations on the British species of *Megastigmus* Dalman (Hym., Torymidae) which inhabit oak galls. *The Entomologist*, **99**: 124–128.

Askew, R.R. 1968. Hymenoptera 2. Chalcidoidea. Section (b). *Handbooks for the Identification of British Insects* **8** (2)b: 1–39.

Askew, R.R. 1971. *Parasitic Insects*. xvii+316pp. Heinemann, London.

Askew, R.R. 1984. The biology of gall wasps. pp. 223–271. *In* : Ananthakrishnan, T.N. (ed.) *The Biology of Gall Insects*. 362pp. Arnold, London.

Askew, R.R. 1990. Species diversities of hymenopteran taxa in Sulawesi, pp. 255–260. *In* : Knight, W.J. & Holloway, J.D. (eds) *Insects and Rain Forests of South East Asia (Wallacea)*. 343pp. Royal Entomological Society, London.

Askew, R.R. & Ruse, J.M. 1974. Biology and taxonomy of species of the genus *Enaysma* Delucchi (Hym., Eulophidae, Entedontinae) with special reference to the British fauna. *Transactions of the Royal Entomological Society of London*, **125**: 257–294.

Askew, R.R. & Shaw, M.R. 1979. Mortality factors affecting the leaf-mining stages of *Phyllonorycter* (Lepidoptera: Gracillariidae) on oak and birch. 2. Biology of the parasite species. *Zoological Journal of the Linnean Society*, **67**: 51–64.

Askew, R.R. & Shaw, M.R. 1986. Parasitoid communities: their size, structure and development, pp. 225–264. *In* : Waage, J. & Greathead, D. *Insect Parasitoids*. 389pp. Academic Press, London.

Assem, J, van den, 1969. Reproductive behaviour of *Pseudeucoila bochei* (Hymenoptera: Cynipidae). I. A description of courtship behaviour. *Netherlands Journal of Zoology*, **19**: 641–648.

Assem, J. van den 1975. Temporal patterning of courtship behaviour in some parasitic Hymenoptera, with special reference to *Melittobia acasta*. *Journal of Entomology* (A), **50**: 137–146.

Assem, J. van den 1986. Mating behaviour in parasitic wasps, pp. 137–167. *In* : Waage, J. & Greathead, D. (eds) *Insect Parasitoids*. 389pp. Academic Press, London.

Assem, J. van den, Bosch, H.A.J. in den & Prooy, E. 1982*a*. *Melittobia* courtship behaviour: a comparative study of the evolution of display. *Netherlands Journal of Zoology*, **32**: 427–471.

Assem, J. van den, Gijswijt, M.J. & Nübel, B.K. 1982*b*. Characteristics of courtship and mating behaviour used as classificatory criteria in Eulophidae-Tetrastichinae (Hymenoptera), with special reference to the genus *Tetrastichus* s.l. *Tijdschrift voor Entomologie*, **125**: 205–220.

Assem, J. van den & Kuenen, D.J. 1958. Host finding of *Choetospila elegans* Westw. (Hym. Chalcid.) a parasite of *Sitophilus granarius* L. (Coleopt. Curcul.). *Entomologia Experimentalis et Applicata*, **1**: 174–180.

Assem, J. van den & Putters, F.A. 1980. Patterns of sound produced by courting chalcidoid males and its biological significance. *Entomologia Experimentalis et Applicata*, **27**: 293–302.

Assem, J. van den, Putters, F.A. & Prins, T.C. 1984. Host quality effects on sex ratio of the parasitic wasp *Anisopteromalus calandrae* (Chalcidoidea, Pteromalidae). *Netherlands Journal of Zoology*, **34**: 33–62.

Aubert, J.-F. 1959. Biologie de quelques Ichneumonidae Pimplinae et examen critique de la thèorie de Dzierzon. *Entomophaga*, **4:** 75–188.

Auger, J., LeCompte, C., Paris, J. & Thibout, E. 1989. Identification of leek-moth and diamondback-moth frass volatiles that stimulate the parasitoid, *Diadromus pulchellus*. *Journal of Chemical Ecology*, **15**: 1391–1398.

Austin, A.D. 1983. Morphology and mechanics of the ovipositor system of *Ceratobaeus* Ashmead (Hymenoptera: Scelionidae) and related genera. *International Journal of Insect Morphology and Embryology*, **12**: 139–155.

Austin, A.D. 1984. The fecundity, development and host relationships of *Ceratobaeus* spp. (Hymenoptera: Scelionidae), parasites of spider eggs. *Ecological Entomology*, **9**: 125–138.

Austin, A.D. 1985. The function of spider egg sacs in relation to parasitoids and predators, with special reference to the Australian fauna. *Journal of Natural History*, **19**: 359–376.

Austin, A.D. & Browning, T.O. 1981. A mechanism for movement of eggs along insect ovipositors. *International Journal of Insect Morphology and Embryology*, **10**: 93–108.

Austin, A.D. & Dangerfield, P.C. 1989. The taxonomy of New World microgastrine braconids (Hymenoptera) parasitic on *Diatraea* spp. (Lepidoptera: Pyralidae). *Bulletin of Entomological Research*, **79**: 131–144.

Avidov, Z., Rössler, Y. & Rosen, D. 1967. Studies on an Israel strain of *Anagyrus pseudococci* (Girault) (Hym., Encyrtidae). II. Some biological aspects. *Entomophaga*, **12**: 111–118.

Avilla, J & Albajes, R. 1984. The influence of female age and host size on the sex ratio of the parasitoid *Opius concolor*. *Entomologia Experimentalis et Applicata*, **35**: 43–47.

Ayal, Y. 1987. The foraging strategy of *Diaeretiella rapae* I. The concept of the elementary unit of foraging. *Journal of Animal Ecology*, **56**: 1057–1068.

Ayala, R. 1988. Abejas silvestres (Hymenoptera: Apoidea) de Chamela, Jalisco, Mexico. *Folia Entomologica Mexicana*, **77**: 395–493.

Ayala, R., Griswold, T.L. & Bullock, S.H. 1993. The Native bees of Mexico, pp. 179–227. *In* : Ramamoorthy, T.P., Bye, R., Lot, A. & Fa, J. (eds). *Biological Diversity of Mexico: Origins and Distribution*. 812pp. Oxford University Press, New York.

Ayre, G.L. 1962. *Pseudometagea schwarzii* (Ashm.) (Eucharitidae: Hymenoptera), a parasite of *Lasius neoniger* Emery (Formicidae: Hymenoptera). *Canadian Journal of Zoology*, **40**: 157–164.

Azevedo Marques, L.A. de. 1933. Tenthredinidae con hecida por 'mosca de serra' cuja larva, ou 'falsa lagarta' é nociva a varias espécies do genero *Tibouchina* (Biologia de *Bergiana cyanocephala* (Klug, 1824) Konow, 1899). *Ministerio da Agricultura, Instituto Biologico de Defesa Agricola, Rio de Janeiro*, **12**: 1–11.

Back, E.A. 1940. A new parasite of *Anthrenus vorax* Waterhouse. *Proceedings of the Entomological Society of Washington*, **42**: 110–113.

Baehrecke, E.H. & Strand, M.R. 1990. Embryonic morphology and growth of the polyembryonic parasitoid *Copidosoma floridanum* (Ashmead) (Hymenoptera: Encyrtidae). *International Journal of Insect Morphology and Embryology*, **19**: 165–175.

Bai, B. & Mackauer, M. 1990. Host discrimination by the aphid parasitoid *Aphelinus asychis* (Hymenoptera: Aphelinidae): when superparasitism is not adaptive. *Canadian Entomologist*, **122**: 363–372.

Bai, B. & Mackauer, M. 1991. Recognition of heterospecific parasitism: competition between aphidiid (*Aphidius ervi*) and aphelinid (*Aphelinus asychis*) parasitoids of aphids (Hymenoptera: Aphidiidae; Aphelinidae). *Journal of Insect Behavior*, **4**: 333–345.

Baker, G.L. 1976. The biology of *Pristocera rufa* Kieffer (Hymenoptera: Bethylidae), a parasite of *Pantorhytes szentivanyi* Marshall (Coleoptera: Curculionidae) in Papua New Guinea. *Journal of the Australian Entomological Society*, **15**: 153–160.

Baker, H.G. & Baker, I. 1983. floral nectar sugar constituents in relation to pollinator type. pp. 117–141. *In*: Jones, C.E. & Little R.J. (eds) *Handbook of Experimental Pollination Biology*. 558pp. Van Nostrand Reinhold, New York.

Baker, W.L. 1972. *Eastern Forest Insects*. 642pp. USDA Forest Service Miscellaneous Publication, 1175. Washington D.C.

Bakke, A. 1955. Insects reared from spruce cones in Northern Norway 1951. Norsk *Entomologisk Tidsskrift*, **9**: 152–212.

Bakkendorf, O. 1934. Biological investigations on some Danish hymenopterous egg-parasites, especially in homopterous and heteropterous eggs, with taxonomic remarks and descriptions of new species. *Entomologiske Meddelelser*, **19**: 1–134.

Bakker, K., Bagchee, S.N., Zwet, W.R. van & Meelis, E. 1967. Host discrimination in *Pseudeucoila bochei* (Hymenoptera: Cynipidae). *Entomologia Experimentalis et Applicata*, **10**: 295–311.

Balduf, W.V. 1926. The bionomics of *Dinocampus coccinellae* Schrank. *Annals of the Entomological Society of America*, **19**: 465–498.

Balduf, W.V. 1928. Observations on the buffalo tree hopper *Ceresa bubalus* Fabr. (Membracidae, Homoptera), and the bionomics of an egg parasite, *Polynema striaticorne* Girault (Mymaridae, Hymenoptera). *Annals of the Entomological Society of America*, **21**: 419–435.

Balduf, W.V. 1963. A distinct type of host-parasite relation among insects. *Annals of the Entomological Society of America*, **56**: 386–391.

Ball, J.C. & Dahlsten, D.L. 1973. Hymenopterous parasites of *Ips paraconfusus* (Coleoptera: Scolytidae) larvae and their contribution to mortality. I. Influence of host tree and tree diameter on parasitization. *Canadian Entomologist*, **105**: 1453–1464.

Ballou, H.A. 1934. Notes on some insect pests in the Lesser Antilles. *Tropical Agriculture*, **11**: 210–212.

Baltensweiler, W. von & Moreau, J.P. 1957. Ein Beitrag biologische-systematischer Art zur Kenntnis der Gattung *Phytodietus* (Hymenoptera). *Zeitschrift für Angewandte Entomologie*, **41**: 272–276.

Banks, N. 1946. Studies of South American Psammocharidae. Part I. *Bulletin of the Museum of Comparative Zoology*, **96**: 311–525.

Banks, N. 1947. Studies of South American Psammocharidae. Part II. *Bulletin of the Museum of Comparative Zoology*, **99**: 371–486.

Barbier, R. & Raimbault, J.-P. 1985 (1986). Formation et mise en place de spherules autour du chorion de l'oeuf, chez *Leptomastix dactylopii* (Hymenoptera, Encyrtidae). *Bulletin de la Société Zoologique de France*, **110**: 339–346.

Barbier, R., Raimbault, J.P. & Nenon, J.P. 1988. The sphaerulae of the egg chorion of *Leptomastix dactylopii* H. (Hymenoptera, Encyrtidae). *In* : Bouletreau, M. & Bonnot, G. (eds) *Parasitoid Insects: European Workshop, Lyon, September 7–10, 1987*. *Les Colloques de l'INRA*, **48**: 97–98.

Barlin, M.R. & Vinson, S.B. 1981. Multiporous plate sensilla in antennae of the Chalcidoidea (Hymenoptera). *International Journal of Insect Morphology and Embryology*, **10**: 29–42.

Barnwell, F.H. 1967. Daily patterns in the activity of the arboreal ant *Azteca alfari*. *Ecology*, **48**: 991–993.

Baroni Urbani, C., Bolton, B. & Ward, P.S. 1992. The internal phylogeny of ants (Hymenoptera: Formicidae). *Systematic Entomology*, **17**: 301–329.

Barr, S.E. 1971. Allergy to Hymenoptera stings. Review of the world literature: 1953–1970. *Annals of Allergy*, **29**: 49–66.

Barras, R. 1965. Variations in the receptivity of the female *Mormoniella vitripennis* (Walker) (Hymenoptera, Pteromalidae). *Proceedings of the XII International Congress of Entomology* : 299.

Barrett, B.A., Jorgensen, C.D. & Looman, S.J. 1985. Foraging recruitment by the giant tropical ant, *Paraponera clavata* (Hymenoptera, Formicidae). *Pan-Pacific Entomologist*, **61**: 334–338.

Barriga T., J. E. 1990. Parásitos y depredadores de larvas de Cerambycidae y Buprestidae (Coleoptera) de Chile. *Revista Chilena Entomologia*, **18**: 57–59.

Barron, J.R. 1976. Systematics of *Nearctic Euceros* (Hymenoptera: Ichneumonidae: Eucerotinae). *Naturaliste Canadien*, **103**: 285–375.

Barron, J.R. 1978. Systematics of the world Eucerotinae (Hymenoptera, Ichneumonidae). Part II: Non-Nearctic species. *Naturaliste Canadien*, **105**: 327–374.

Barrows, E.M. 1976. Sexual behavior in *Hyposoter fugitivus* (Hymenoptera: Ichneumonidae). *Entomological News*, **87**: 101–102.

Bartlett, B.R. 1964. Patterns in the host-feeding habit of adult parasitic Hymenoptera. *Annals of the Entomological Society of America*, **57**: 344–350.

Bartlett, B.R. 1978. Pseudococcidae, pp. 137–170. *In* : Clausen, C.P. (ed.), *Introduced Parasites and Predators of Arthropod Pests and Weeds: A World Review*. 545pp. USDA Agricultural Handbook No. 480.

Bartlett, K.A. 1939. A dryinid parasite attacking *Baldulus maidis* in Puerto Rico. *Journal of Agriculture of the University of Puerto Rico*, **22**: 497–498.

Basden, E.B. 1959. *Phaenoserphus viator* (Hal.)(Hym., Proctotrupidae), a parasite of carabid larvae, in Scotland. *Entomologist's Monthly Magazine*, **95**: 35–36.

Batiste, W.C. 1967. Biology of the trefoil seed chalcid *Bruchofagus kolobovae* Fedoseeva (Hymenoptera: Eurytomidae). *Hilgardia* **38**: 427–469.

Batra, S.W.T. 1972. Some properties of the nest-building secretions of *Nomia, Anthophora, Hylaeus* and other bees. *Journal of the Kansas Entomological Society*, **45**: 208–218.

Batra, S.W.T. & Schuster, J.C. 1977. Nests of *Centris, Melissodes*, and *Colletes* in Guatemala (Hymenoptera: Apoidea). *Biotropica*, **9**: 135–138.

Bawa, K.S., Bullock, S.H, Perry, D.R., Coville, R.E. & Grayum, M.H. 1985. Reproductive biology of tropical lowland rain forest trees. II. Pollination systems. *American Journal of Botany*, **72**: 346–356.

Beard, R.L. 1952. The toxicology of *Habrobracon* venom: a study of a natural insecticide. *Connecticut Agricultural Experiment Station Bulletin*, **562**: 3–27.

Beard, R.L. 1978. Venoms of Braconidae, pp. 773–800. *In* : Bettini, S. (ed.) *Arthropod Venoms*, 977pp. Springer-Verlag, Berlin.

Beardsley, J.W. Jr 1988. Eucoilid parasites of agromyzid leafminers in Hawaii (Hymenoptera: Cynipoidea). *Proceedings of the Hawaiian Entomological Society*, **28**: 33–47.

Beardsley, J.W. Jr 1989. Hawaiian Eucoilidae (Hymenoptera: Cynipoidea), key to genera and taxonomic notes on apparently non-endemic species. *Proceedings of the Hawaiian Entomological Society*, **29**: 165–193.

Beardsley, J.W. 1990. The genus *Kleidotoma* Westwood in Hawaii, with descriptions of three new species (Hymenoptera: Cynipoidea: Eucoilidae). *Proceedings of the Hawaiian Entomological Society*, **30**: 131–146.

Beattie, A.J. 1985. *The Evolutionary Ecology of Ant-Plant Mutualisms*. 182pp. Cambridge University Press, Cambridge.

Beaver, R.A. 1966. The biology and immature stages of *Entedon leucogramma* (Ratzburg). (Hymenoptera: Eulophidae), a parasite of bark beetles. *Proceedings of the Royal Entomological Society, London* (A), **41**: 37–41.

Beck, L. 1971. Bodenzoologische Gliederung und charakterisierung des amazonischen regenwaldes. *Amazoniana*, **3**: 69–132.

Beckage, N.E. 1985. Endocrine interactions between endoparasitic insects and their hosts. *Annual Review of Entomology*, **30**: 371–413.

Beckage, N.E., Metcalf, J.S., Nesbit, D.J., Schleifer, K.W., Zetlan, S.R. & de Buron, I. 1990. Host hemolymph monophenoloxidase activity in parasitized *Manduca sexta* larvae and evidence for inhibition by wasp polydnavirus. *Insect Biochemistry*, **20**: 285–294.

Beckage, N.E., Nesbit, D.J., Nielsen, B.D., Spence, K.D. & Barman, M.A.E. 1989. Alteration of hemolymph polypeptides in *Manduca sexta* larvae parastized by *Cotesia congregata* : a two-dimensional electrophoretic analysis and comparison with major bacteria-induced proteins. *Archives of Insect Biochemistry and Physiology*, **10**: 29–45.

Beckage, N.E. & Riddiford, L.M. 1978. Developmental interactions between the tobacco hornworm *Manduca sexta* and its braconid parasite *Apanteles congregatus*. *Entomolgia Experimentalis et Applicata*, **23**: 139–151.

Beckage, N.E. & Riddiford, L.M. 1982. Effects of parasitism by *Apanteles congregatus* on the endocrine physiology of the tobacco hornworm *Manduca sexta*. *General and Comparative Endocrinology*, **47**: 308–322.

Beckage, N.E. & Riddiford, L.M. 1983. Growth and development of the endoparasitic wasp *Apanteles congregatus* : dependence on host nutritional status and parasite load. *Physiological Entomology*, **8**: 231–241.

Beckage, N.E. & Templeton, T.J. 1985. Temporal synchronization of emergence of *Hyposoter exiguae* and *H. fugitivus* (Hymenoptera: Ichneumonidae) with apolysis preceding larval molting in *Manduca sexta* (Lepidoptera: Sphingidae). *Annals of the Entomological Society of America*, **78**: 775–782.

Beeson, C.F.C. & Chatterjee, S.N. 1935. On the biology of the Braconidae (Hymenopt.). *Indian Forest Records* (N.S.), **1**: 105–138.

Beevers, M., Lewis, W.J., Gross, H.R. Jr & Nordlund, D.A. 1981. Kairomones and their use for management of entomophagous insects: X. Laboratory studies on manipulation of host-finding behavior of *Trichogramma pretiosum* Riley with a kairomone extracted from *Heliothis zea* (Boddie) moth scales. *Journal of Chemical Ecology*, **7**: 635–648.

Beirne, B.P. 1941. A consideration of the cephalic structures and spiracles of the final instar larvae of the Ichneumonidae (Hym.). *Transactions of the Society for British Entomology*, **7**: 123–190.

Beirne, B.P. 1942. Observations on the life-history of *Praon volucre* Haliday (Hym.: Braconidae), a parasite of the mealy plum aphis (*Hyalopterus arundinis* Fab.). *Proceedings of the Royal Entomological Society of London* (A), **17**: 42–47.

Bell, W.J. 1991. *Searching Behaviour. The Behavioural Ecology of Finding Resources.* 358pp. Chapman & Hall, London.

Belloti, A. & Schoonhoven, A. van 1978. Mite and insect pests of cassava. *Annual Review of Entomology*, **23**: 39–67.

Belt, T. 1874. *The Naturalist in Nicaragua.* 403pp. University of Chicago Press, Chicago. [1985 edition.]

Bendel-Janssen, M. 1977. Zur Biologie, Ökologie und Ethologie der Chalcidoidea (Hym.). *Mitteilungen aus der Biologischen Bundesanstalt für Land-und Forstwirtschaft*, **176**: 1–163.

Bennett, A.W. & Sullivan, D.J. 1978. Defensive behavior against tertiary parasitism by the larva of *Dendrocerus carpenteri* an aphid hyperparasitoid. *Journal of the New York Entomological Society*, **86**: 153–160.

Bennett, B. & Breed, M.D. 1985. On the association between *Pentaclethra macroloba* (Mimosaceae) and *Paraponera clavata* (Hymenoptera: Formicidae) colonies. *Biotropica*, **17**: 253–255.

Bennett, C.F. Jr. 1965. Beekeeping with stingless bees in western Panama. *Bee World*, **46**: 23–24.

Bennett, F.D. 1950. Parasites of *Ancylostomia stercorea* (Zell.), (Pyralidae, Lepidoptera) a pod borer attacking pigeon pea in Trinidad. *Bulletin of Entomological Research*, **50**: 737–757.

Bennett, F.D. 1971. Current status of biological control of the small moth borers of sugar cane *Diatraea* spp. (Lep. Pyralididae). *Entomophaga*, **16**: 111–124.

Bennett, F.D. 1972. Observations on *Exaerete* spp. and their hosts *Eulaema terminata* and *Euplusia surinamensis* (Hymen., Apidae, Euglossinae) in Trinidad. *Journal of the New York Entomological Society*, **80**: 118–124.

Bennett, F.D. 1981. Hyperparasitism in the practice of biological control, pp. 43–49. *In*: Rosen, D. (ed.) *The Role of Hyperparasitism in Biological Control—A Symposium* (Number 4103). 52pp. University of California, Berkeley.

Bennett, F.D. 1993. Do introduced parasitoids displace native ones? *Florida Entomologist*, **76**: 54–63.

Bennett, F.D., Glenn, H., Yaseen, M. & Baranowski, R.M. 1987. Records of *Trichospilus diatraeae*, an Asian parasite (Hymenoptera: Eulophidae) from the Caribbean and Florida. *Florida Entomologist*, **70**: 184–186.

Bennett, F.D. & Hughes, I.W. 1963. Studies on the life history and biological control of the spittlebug, *Clastoptera undulata* Uhler (Cercopidae: Hemiptera). *Entomophaga*, **8**: 49–66.

Benoit, P.L.G. 1984*a*. Stephanidae de l'Afrique occidentale et centrale. (Hymenoptera). *Revue de Zoologie Africaine*, **98**: 215–228.

Benoit, P.L.G. 1984*b*. Stephanidae du Sahara. (Hymenoptera). *Revue de Zoologie Africaine*, **98**: 434–439.

Benoit, P.L.G. 1984*c*. Aulacidae, famille nouvelle pour la faune de L'Afrique tropicale (Hymenoptera). *Revue de Zoologie Africaine*, **98**: 799–803.

Benson, R.B. 1950. An introduction to the natural history of British sawflies. *Transactions of the Society for British Entomology*, **10**: 45–142.

Bentley, B.L. 1976. Plants bearing extrafloral nectaries and the associated ant community: interhabitat differences in the reduction of herbivore damage. *Ecology*, **57**: 815–820.

Bentley, B.L. 1977. Extrafloral nectaries and protection by pugnacious bodyguards. *Annual Review of Ecology and Systematics*, **8**: 407–427.

Bentley, B.L. 1983. Nectaries in agriculture, with an emphasis on the tropics. pp. 202–222. *In* : Bentley, B. & Elias, T. (eds) *The Biology of Nectaries*. 259pp. Columbia University Press, New York.

Bequaert, J. 1922. Ants of the American Museum Congo expedition. IV. Ants in their diverse relations to the plant world. *Bulletin of the American Museum of Natural History*, **45**: 333–583.

Bequaert, J. 1926. Medical and economic entomology. *Contribution from the Harvard Institute for Tropical Biology and Medicine*, **4**: 160–257.

Bequaert, J. 1941. *Pseudodynerus*, a neotropical complex of eumenine wasps (Hymenoptera, Vespidae). *American Museum Novitates*, **1106**: 1–10.

Berg, C.C. 1989. Classification and distribution of *Ficus*. *Experientia*, **45**: 605–611.

Berg, E. van den, Prinsloo, G.L. & Neser, S. 1990. An unusual host association: *Aprostocetus* sp. (Eulophidae), a hymenopterous predator of the nematode *Subanguina mobilis* (Chit & Fisher, 1975) Brzeski, 1981 (Anguinidae). *Phytophylactica*, **22**: 125–127.

Betrem, J.G. & Bradley, J.C. 1971 (1972). The African Campsomerinae (Hymenoptera, Scoliidae). *Monografieën van de Nederlandse Entomologische Vereniging*, **6**: 1–326.

Bian, Z., Fales, H.M., Blum, M.S., Jones, T.H., Rinderer, T.E. & Howard, D.F. 1984. Chemistry of cephalic secretion of fire bee *Trigona (Oxytrigona) tataira*. *Journal of Chemical Ecology*, **10**: 451–461.

Bin, F. 1981. Definition of female antennal clava based on its plate sensilla in Hymenoptera Scelionidae Telenominae. *Redia*, **64**: 245–261.

Bin, F. & Vinson, S.B. 1986. Morphology of the antennal sex-gland in male *Trissolcus basalis* (Woll.) (Hymenoptera: Scelionidae), an egg parasitoid of the green stink bug, *Nezara viridula* (Hemiptera: Pentatomidae). *International Journal of Insect Morphology and Embryology*, **15**: 129–138.

Bird, R.D. 1927. The external anatomy of the larva of *Hoplocampa halcyon* Nort. with a key to the instars and to those of related species (Tenthredinidae, Hymenoptera). *Annals of the Entomological Society of America*, **20**: 481–487.

Bischoff, H. 1938. *Hymenopterorum Catalogus*, Pars 5: Trigonaloidae. 18pp. Junk, The Hague.

Black, R.W. II. 1987. The biology of leaf nesting ants in a tropical wet forest. *Biotropica*, **19**: 319–325.

Blanton, C.M. & Ewel, J.J. 1985. Leaf-cutting ant herbivory in successional and agricultural tropical ecosystems. *Ecology*, **66**: 861–869.

Bledowski, R. & Krainska, M.K. 1926. Die Entwicklung von *Banchus femoralis* Thoms. (Hymenoptera, Ichneumonidae). *Bibliotheca Universitatis Liberae Polonae* (A), **16:** 1–50.

Bledsoe, L.W., Flanders, R.V. & Edwards, C.R. 1983. Morphology and development of the immature stages of *Pediobius foveolatus* (Hymenoptera: Eulophidae). *Annals of the Entomological Society of America*, **76**: 953–957.

Blumberg, D. 1988. Encapsulation of eggs of the encyrtid wasp, *Metaphycus swirskii*, by the hemispherical scale, *Saissetia coffeae*: effects of host age and rearing temperature. *Entomologia Experimentalis et Applicata*, **47:** 95–99.

Blumberg, D. & Luck, R.F. 1990. Differences in the rates of superparasitism between two strains of *Comperiella bifasciata* (Howard) (Hymenoptera: Encyrtidae) parasitizing California Red Scale (Homoptera: Diaspididae): an adaptation to circumvent encapsulation? *Annals of the Entomological Society of America*, **83**: 591–597.

Blunk, H. & Kerrich, G.J. 1956. Polymorphismus bei *Haplaspis nanus* (Grav.) (= *Hemiteles fulvipes* Grav.) (Hym. Ichneumonidae) und die Beschreibung einer neuen *Haplaspis* Art aus Ceylon. *Bolletino del Laboratoria di Zoologia Generale e Agraria della Facolta Agraria in Portici*, **33**: 546–563.

Bocchino, F.J. & Sullivan, D.J. 1981. Effects of venoms from two aphid hyperparasitoids, *Asaphes lucens* and *Dendrocerus carpenteri* (Hymenoptera: Pteromalidae and Megaspilidae), on larvae of *Aphidius smithi* (Hymenoptera: Aphidiidae). *Canadian Entomologist*, **113**: 887–889.

Boethel, D.J. & Eikenbary, R.D. 1986. *Interactions of Plant Resistance and Parasitoids and Predators of Insects*. 224pp. Ellis Horwood, Chichester.

Bohart, G.E. 1970. Commercial production and management of wild bees—a new entomological industry. *Bulletin of the Entomological Society of America*, **16**: 8–9.

Bohart, G.E. & Knowlton, G.F. 1953. Notes on mating, prey provisioning, and nesting of *Sphex procerus* (Dahlbom). *Proceedings of the Entomological Society of Washington*, **55**: 100–101.

Bohart, G.E. & Knowlton, G.F. 1973. The bees of Curlew Valley. *Proceedings of the Utah Academy of Sciences, Arts, and Letters*, **50**: 1–9.

Bohart, G.E., Parker, F.D. & Tepedino, V.J. 1982. Notes on the biology of *Odynerus dilectus* (Hym.: Eumenidae), a predator of the alfalfa weevil, *Hypera postica* (Col.: Curculionidae). *Entomophaga*, **27**: 23–31.

Bohart, R.M. 1978. *Tachytes* of South America (Hymenoptera, Sphecidae, Larrinae). *Transactions of the American Entomological Society*, **104**: 435–505.

Bohart, R.M. 1980. The Middle American species of *Stenodynerus* (Hymenoptera, Eumenidae). *Polskie Pismo Entomologiczne*, **50**: 71–108.

Bohart, R.M. 1985. New *Ipsiura* and a key to known species of the genus (Hymenoptera: Chrysididae). *Journal of the Kansas Entomological Society*, **58**: 708–720.

Bohart, R.M. 1990 (1991). New species and a key to North American *Solierella* in the *inermis* group (Hymenoptera: Sphecidae: Larrinae: Miscophini). *Psyche*, **97**: 229–240.

Bohart, R.M. 1993. A synopsis of Central American and Caribbean *Oxybelus* (Hymenoptera, Sphecidae). *Insecta Mundi*, **7**: 159–168.

Bohart, R.M. & Kimsey, L.S. 1978. A revision of the New World species of *Hedychridium* (Hymenoptera, Chrysididae). *Proceedings of the Biological Society of Washington*, **91**: 590–635.

Bohart, R.M. & Menke, A.S. 1976. *Sphecid Wasps of the World. A Generic Revision.* 695pp. University of California Press, Berkeley.

Bohart, R.M. & Stange, L.A. 1965. A revision of the genus *Zethus* Fabricius in the Western Hemisphere (Hymenoptera: Eumenidae). *University of California Publications in Entomology*, **40**: 1–208.

Boinski, S. & Fowler, N.L. 1989. Seasonal patterns in a tropical lowland forest. *Biotropica*, **21**: 223–233.

Bolton, B. 1990*a*. Abodominal characters and status of the cerapachyine ants (Hymenoptera, Formicidae). *Journal of Natural History*, **24**: 53–68.

Bolton, B. 1990*b*. Army ants reassessed: the phylogeny and classification of the doryline section (Hymenoptera, Formicidae). *Journal of Natural History*, **24**: 1339–1364.

Bordage, E. 1913 (1914). Notes biologiques recueillies á l'Ile de la Réunion. II. Sur la biologie et l'éthologie de divers hyménoptères. *Bulletin Scientifique de la France et de la Belgique*, **47**: 377–412.

Borden, J.H., Chong, L. & Rose, A. 1978*a*. Morphology of the elongate placoid sensillum on the antennae of *Itoplectis conquisitor*. *Annals of the Entomological Society of America*, **71**: 223–227.

Borden, J.H., Rose, A. & Chorney, R.J. 1978*b*. Morphology of the elongate sensillum placodeum on the antennae of *Aphidius smithi* (Hymenoptera: Aphidiidae). *Canadian Journal of Zoology*, **56**: 519–525.

Borror, D.J., Triplehorn, C.A. & Johnson, N.F. 1989. *An Introduction to the Study of Insects*. 875pp. Saunders College Publishing, Philadelphia.

Bosch, H.A.J. in den & Assem, J. van den 1986. The taxonomic position of *Aceratoneuromyia granularis* Domenichini (Hymenoptera: Eulophidae) as judged by characteristics of its courtship behaviour. *Systematic Entomology*, **11**: 19–23.

Bosch, R. van den, 1964. Encapsulation of the eggs of *Bathyplectes curculionis* (Thomson) (Hymenoptera: Ichneumonidae) in larvae of *Hypera brunneipennis* (Boheman) and *Hypera postica* (Gyllenhal) (Coleoptera: Curculionidae). *Journal of Insect Pathology*, **6**: 343–367.

Bosch, R. van den, Messenger, P.S. & Gutierrez, A.P. 1982. *An Introduction to Biological Control*, 247pp. Plenum Press, New York & London.

Bosque, C. & Rabinovich, J.E. 1979. Population dynamics of *Telenomus fariai* (Hymenoptera: Scelionidae), a parasite of Chagas' disease vectors VII. Oviposition behavior and host discrimination. *Canadian Entomologist*, **111**: 171–180.

Bouček, Z. 1958. Eine Cleonyminen-Studie; Bestimmungsstabelle der Gattungen mit Beschreibungen und Notizen, eingeschlossen einige Eupelmidae (Hym. Chalcidoidea). *Acta Entomologica Musei Nationalis Pragae*, **32**: 353–386.

Bouček, Z. 1962. A new *Stenomesius* (Hym., Eulophidae) reared from *Ceramidia viridis* in Ecuador. *Entomophaga*, **7**: 189–192.

Bouček, Z. 1963. A taxonomic study in *Spalangia* Latr. (Hymenoptera, Chalcidoidea). *Acta Entomologica Musei Nationalis Pragae*, **35**: 429–512.

Bouček, Z. 1972*a* (1971). Mediterranean Perilampinae: *Euperilampus* and genera allied to *Chrysomalla* (Hym., Chalcidoidea). *Mitteilungen Münchner Entomologischen Gesellschaft*, **61**: 90–107.

Bouček, Z. 1972*b*. A new genus and species of Pteromalidae (Hym.) parasitic on sphecids in South America. *Mitteilungen der Schweizerischen Entomologischen Gesellschaft*, **45**: 113–116.

Bouček, Z. 1974*a*. A revision of the Leucospidae (Hymenoptera: Chalcidoidea) of the World. *Bulletin of the British Museum (Natural History)* Entomology, Supplement **23**: 1–241.

Bouček, Z. 1974*b*. The pteromalid subfamily Eutrichosomatinae (Hymenoptera: Chalcidoidea). *Journal of Entomology* (B), **43**: 129–138.

Bouček, Z. 1976. The African and Asiatic species of *Trichospilus* and Cotterellia (Hymenoptera, Eulophidae). *Bulletin of Entomological Research*, **65**: 669–681.

Bouček, Z. 1977. Descriptions of two new species of neotropical Eulophidae (Hymenoptera) of economic interest, with taxonomic notes on related species and genera. *Bulletin of Entomological Research*, **67**: 1–15.

Bouček, Z. 1978*a*. A generic key to Perilampinae (Hymenoptera, Chalcidoidea), with a revision of *Krombeinius* n.gen. and *Euperilampus* Walker. *Entomologica Scandinavica*, **9**: 299–307.

Bouček, Z. 1978*b*. A study of the non-podagrionine Torymidae with enlarged hind femora, with a key to the African genera (Hymenoptera). *Journal of the Entomological Society of Southern Africa*, **41**: 91–134.

Bouček, Z. 1980. A revision of the genus *Monacon* Waterston (Hymenoptera: Chalcidoidea: Perilampinae), parasites of ambrosia beetles (Coleoptera: Platypodidae). *Bulletin of Entomological Research*, **70**: 73–96.

Bouček, Z. 1986. Taxonomic study of chalcidoid wasps (Hymenoptera) associated with gall midges (Diptera: Cecidomyiidae) on mango trees. *Bulletin of Entomological Research*, **76**: 393–407.

Bouček, Z. 1988. *Australasian Chalcidoidea* (Hymenoptera). *A Biosystematic Revision of Genera of Fourteen Families, with a Reclassification of Species*. 832pp. C.A.B. International, Wallingford.

Bouček, Z. 1992. The New World genera of Chalcididae, pp. 49–117. *In*: Delvare, G. & Boucek, Z. On the New World Chalcididae (Hymenoptera). *Memoirs of the American Entomological Institute*, **53**: 1–466.

Bouček, Z. 1993*a*. The genera of chalcidoid wasps from *Ficus* fruit in the New World. *Journal of Natural History*, **27**: 173–217.

Bouček, Z. 1993*b*. New taxa of North American Pteromalidae and Tetracampidae (Hymenoptera), with notes. *Journal of Natural History*, **27**: 1239–1313.

Bouček, Z. & Askew, R.R. 1968. Index of world Tetracampidae, pp. 3–19. *In*: Delucchi, V. & Remaudière, G. (eds) *Index of Entomophagous Insects 4*, 19pp. Le Francois, Paris.

Bouček, Z. & Noyes, J.S. 1987. Rotoitidae, a curious new family of Chalcidoidea (Hymenoptera) from New Zealand. *Systematic Entomology*, **12**: 407–412.

Bouček, Z. & Rasplus, J.-Y. 1991. *Illustrated Key to West-Palearctic Genera of Pteromalidae* (Hymenoptera: Chalcidoidea), 140pp. INRA, Paris.

Bouček, Z., Watsham, A. & Wiebes, J.T. 1981. The fig wasp fauna of the receptacles of *Ficus thonningii* (Hymenoptera, Chalcidoidea). *Tijdschrift voor Entomologie*, **124**: 149–233.

Bourarach, K. & Hawlitzky, N. 1989. Étude comparative des potentialités biologiques de deux Trichogrammes: *Trichogramma evanescens* et *Trichogrammatoidea lutea* (Hym.: Trichogrammatidae). *Entomophaga*, **34**: 95–104.

Boush, G.M. & Baerwald, R.J. 1967. Courtship behavior and evidence for a sex pheromone in the apple maggot parasite, *Opius alloeus*. *Annals of the Entomological Society of America*, **60**: 865–866.

Boussienguet, J. 1988. Morphologie et biologie d'*Anagyrus nyombae*, n.sp., parasite de *Phaenacoccus manihoti*, au Gabon (Hymenoptera, Encyrtidae; Homoptera, Pseudococcidae). *Revue Française d'Entomologie* (N.S.), **10**: 277–283.

Bowers, M.A. & Porter, S.D. 1981. Effect of foraging distance on water content of substrates harvested by *Atta columbica* (Guerin). *Ecology*, **62**: 273–275.

Boyce, H.R. 1936. Laboratory breeding of *Ascogaster carpocapsae* Vier. with notes on biology and larval morphology. *Canadian Entomologist*, **68**: 241–246.

Boyd, N.D. & Martin, M.M. 1975. Faecal proteinases of the fungus-growing ant, *Atta texana*: their fungal origin and ecological significance. *Journal of Insect Physiology*, **21**: 1815–1820.

Boza, M.A. & Mendoza, R. 1981. *The National Parks of Costa Rica*. 310pp. Gráficas Alvi, Madrid.

Bradley, J.C. 1944. A preliminary revision of the Pompilinae (exclusive of the tribe Pompilini) of the Americas (Hymenoptera: Pompilidae). *Transactions of the American Entomological Society*, **70**: 23–157.

Bradley, J.C. 1945. The Scoliidae (Hymenoptera) of northern South America, with especial reference to Venezuela.—I. The genus *Campsomeris*. *Boletín de Entomología Venezolana*, **4**: 1–36.

Bradley, J.C. 1957. The taxa of *Campsomeris* (Hymenoptera: Scoliidae) occurring in the New World. *Transactions of the American Entomological Society*, **83**: 65–77.

Bradley, J.C. 1964. Further notes on the American taxa of *Campsomeris* (Hymenoptera: Scoliidae). *Entomological News*, **75**: 101–108.

Bradley, W.G. & Burgess, E.D. 1934. The biology of *Cremastus flavoorbitalis* (Cameron), an ichneumonid parasite of the European corn borer. *Technical Bulletin of the United States Department of Agriculture*, **441**: 1–15.

Brandao, C.R.F. 1990. Systematic revision of the neotropical ant genus *Megalomyrmex* Forel (Hymenoptera: Formicidae: Myrmicinae), with the description of thirteen new species. *Arquivos de Zoologia, Sao Paulo*, **31**: 411–481.

Brandao, C.R.F., Diniz, J.L.M. & Tomotake, E.M. 1991. *Thaumatomyrmex* strips millipedes for prey: a novel predatory behaviour in ants and the first case of sympatry in the genus. *Insectes Sociaux*, **38**: 335–344.

Brandbyge, J. 1986. A revision of the genus *Triplaris* (Polygonaceae). *Nordic Journal of Botany*, **6**: 545–570.

Breed, M.D., Abel, P., Bleuze, T.J. & Denton, S.E. 1990. Thievery, home ranges, and nestmate recognition in *Ectatomma ruidum*. *Oecologia* **84**: 117–121.

Breed, M.D. & Bennett, B. 1985. Mass recruitment to nectar sources in *Paraponera clavata*: a field study. *Insectes Sociaux*, **32**: 198–208.

Breed, M.D., Fewell, J.H., Moore, A.J. & Williams, K.R. 1987. Graded recruitment in a ponerine ant. *Behavioral Ecology and Sociobiology*, **20**: 407–411.

Breed, M.D. & Harrison, J.M. 1988. Worker size, ovary development and division of labor in the giant tropical ant, *Paraponera clavata* (Hymenoptera: Formicidae). *Journal of the Kansas Entomological Society*, **61**: 285–291.

Breed, M.D. & Harrison, J. 1989. Arboreal nesting in the giant tropical ant *Paraponera clavata*, (Hymenoptera: Formicidae). *Journal of the Kansas Entomological Society*, **62**: 133–135.

Breed, M.D, Michener, C.D. & Evans, H.E. (eds.) 1982. *The Biology of Social Insects*. 419pp. Westview Press, Boulder.

Breed, M.D. & Stiller, T.M. 1992. Honey bee, *Apis mellifera*, nestmate discrimination: hydrocarbon effects and the evolutionary implications of comb choice. *Animal Behaviour*, **43**: 875–883.

Breed, M.D., Stiller, T.M., Fewell, J.H. & Harrison, J.M. 1991. Intercolonial interactions and nestmate discrimination in the giant tropical ant, *Paraponera clavata*. *Biotropica*, **23**: 301–306.

Breeuwer, J.A.J. & Werren, J.H. 1990. Microorganisms associated with chromosome destruction and reproductive isolation between two insect species. *Nature*, **346**: 558–560.

Brehélin, M. (ed.) 1986. *Immunity in Invertebrates. Cells, Molecules, and Defense Reactions*, 233pp. Springer-Verlag, Berlin.

Brian, M.V. 1980. Social control over sex and caste in bees, wasps and ants. *Biological Reviews (Cambridge)*, **55**: 379–415.

Bridwell, J.C. 1920. Some notes on Hawaiian and other Bethylidae (Hymenoptera) with description of a new genus and species. 2nd paper. *Proceedings of the Hawaiian Entomological Society*, **4**: 291–314.

Bridwell, J.C. 1958. Biological notes on *Ampulicomorpha confusa* Ashmead and its fulgoroid host. *Proceedings of the Entomological Society of Washington*, **60**: 23–26.

Brindley, M.D. 1939. Observations on the life-history of *Euphorus pallipes* (Curtis) (Hym.: Braconidae), a parasite of Hemiptera-Heteroptera. *Proceedings of the Royal Entomological Society of London* (A), **14**: 51–56.

Brockmann, H.J. 1979. Nest-site selection in the great golden digger wasp, *Sphex ichneumoneus* L. (Sphecidae). *Ecological Entomology*, **4**: 211–224.

Brockmann, H.J. 1980. Diversity in the nesting behavior of mud-daubers (*Trypoxylon politum* Say; Sphecidae). *Florida Entomologist*, **63**: 53–64.

Brockmann, H.J. 1984. The evolution of social behaviour in insects. pp. 340–361. *In*: Krebs, J.R. & Davies, N.B. (eds), *Behavioural Ecology: An Evolutionary Approach*, 2nd ed. 493pp. Blackwells, Oxford.

Brockmann, H.J. 1985. Provisioning behavior of the great golden digger wasp, *Sphex ichneumoneus* (L.) (Sphecidae). *Journal of the Kansas Entomological Society*, **58**: 631–655.

Brockmann, H.J. & Dawkins, R. 1979. Joint nesting in a digger wasp as an evolutionarily stable preadaptation to social life. *Behaviour*, **71**: 203–245.

Brockmann, H.J. & Grafen, A. 1992. Sex ratios and life-history patterns of a solitary wasp, *Trypoxylon (Trypargilum) politum* (Hymenoptera: Sphecidae). *Behavioral Ecology and Sociobiology*, **30**: 7–27.

Brockmann, H.J., Grafen, A. & Dawkins, R. 1979. Evolutionarily stable nesting strategy in a digger wasp. *Journal of Theoretical Biology*, **77**: 473–496.

Brodeur, J. 1992. Caterpillars are more than a nutrient reservoir for parasitic wasps. *Proceedings in Experimental and Applied Entomology*, **3**: 57–61.

Brodeur, J. & McNeil, J.N. 1992. Host behaviour modification by the endoparasitoid *Aphidius nigripes*: a strategy to reduce hyperparasitism. *Ecological Entomology*, **17**: 97–104.

Bronner, R. 1977. Contribution à l'étude histochimique des tissus nourriciers des zoocecidies. *Marcellia*, **40**: 1–134.

Bronner, R. 1985. Anatomy of the ovipositor and oviposition behavior of the gall wasp *Diplolepis rosae* (Hymenoptera: Cynipidae). *Canadian Entomologist*, **117**: 849–858.

Bronskill, J.F. 1960. The capsule and its relation to the embryogenesis of the ichneumonid parasitoid *Mesoleius tenthredinis* Morl. in the larch sawfly, *Pristiphora erichsonii* (Htg.)(Hymenoptera: Tenthredinidae). *Canadian Journal of Zoology*, **38**: 769–775.

Bronstein, J.L. 1988*a*. Limits to fruit production in a monoecious fig: consequences of an obligate mutualism. *Ecology*, **69**: 207–214.

Bronstein, J.L. 1988*b*. Predators of fig wasps. *Biotropica*, **20**: 215–219.

Bronstein, J.L. 1991. The nonpollinating wasp fauna of *Ficus pertusa*: exploitation of a mutualism? *Oikos*, **61**: 175–186.

Broodryk, S.W. 1969*a*. The biology of *Chelonus (Microchelonus) curvimaculatus* Cameron (Hymenoptera: Braconidae). *Journal of the Entomological Society of Southern Africa*, **32**: 169–189.

Broodryk, S.W. 1969*b*. The biology of *Orgilus parcus* Turner (Hymenoptera: Braconidae). *Journal of the Entomological Society of Southern Africa*, **32**: 243–257.

Brooke, M. de L. 1981. The nesting biology and population dynamics of the Seychelles potter wasp *Eumenes alluaudi* Perez. *Ecological Entomology*, **6**: 365–377.

Brooks, R.W. & Cane, J.H. 1984. Origin and chemistry of the secreted nest entrance lining of *Halictus hesperus* (Hymenoptera: Apoidea). *Journal of the Kansas Entomological Society*, **57**: 161–165.

Brooks, R.W. & Roubik, D.W. 1982 (1983). A halictine bee with distinct castes: *Halictus hesperus* (Hymenoptera: Halictidae) and its bionomics in central Panamá. *Sociobiology*, **7**: 263–282.

Brooks, R.W. & Wahl, D.B. 1987. Biology and mature larva of *Hemipimpla pulchripennis* (Saussure), a parasite of *Ropalidia* (Hymenoptera: Ichneumonidae, Vespidae). *Journal of the New York Entomological Society*, **95**: 547–552.

Brothers, D.J. 1972. Biology and immature stages of *Pseudomethoca f. frigida*, with notes on other species (Hymenoptera: Mutillidae). *University of Kansas Science Bulletin*, **50**: 1–38.

Brothers, D.J. 1975. Phylogeny and classification of the aculeate Hymenoptera, with special reference to Mutillidae. *University of Kansas Science Bulletin*, **50**: 483–648.

Brothers, D.J. 1976. Modifications of the metapostnotum and origin of the 'propodeal triangle' in Hymenoptera Aculeata. *Systematic Entomology*, **1**: 177–182.

Brothers, D.J. 1978. Biology and immature stages of *Myrmosula parvula* (Hymenoptera: Mutillidae). *Journal of the Kansas Entomological Society*, **51**: 698–710.

Brothers, D.J. 1981. Note on the biology of *Ycaploca evansi* (Hymenoptera: Scolebythidae). *Journal of the Entomological Society of Southern Africa*, **44**: 107–108.

Brothers, D.J. 1982. Two new species of Mutillidae associated with *Halictus hesperus* (Halictidae) in Panama (Hymenoptera). *Sociobiology*, **7**: 205–212.

Brothers, D.J. 1984. Gregarious parasitoidism in Australian Mutillidae (Hymenoptera). *Australian Entomological Magazine*, **11**: 8–10.

Brothers, D.J. 1989. Alternative life-history styles of mutillid wasps (Insecta, Hymenoptera), pp. 279–291. *In*: Bruton, M.N. (ed.) *Alternative Life-History Styles of Animals*. 616pp. Kluwer Academic Publishers, Dordrecht.

Brothers, D.J. 1992. The first Mesozoic Vespidae (Hymenoptera) from the Southern Hemisphere, Botswana. *Journal of Hymenoptera Research*, **1**: 119–139.

Brothers, D.J. & Carpenter, J.M. 1993. Phylogeny of Aculeata: Chrysidoidea and Vespoidea. *Journal of Hymenoptera Research*, **2**: 227–302.

Brown, B.V. 1993. Convergent adaptations in Phoridae (Diptera) living in the nests of social insects: a review of the New World Aenigmatiinae. *Memoirs of the Entomological Society of Canada*, **165**: 115–137.

Brown, B.V. & Feener, D.H. Jr. 1991*a*. Behavior and host location cues of *Apocephalus paraponerae* (Diptera: Phoridae), a parasitoid of the giant tropical ant, *Paraponera clavata* (Hymenoptera: Formicidae). *Biotropica*, **23**: 182–187.

Brown, B.V. & Feener, D.H. Jr. 1991*b*. Life history parameters and description of the larva of *Apocephalus paraponerae* (Diptera: Phoridae), a parasitoid of the giant tropical ant *Paraponera clavata* (Hymenoptera: Formicidae). *Journal of Natural History*, **25**: 221–231.

Brown, B.V. & Feener, D.H. Jr. 1993. Life history and immature stages of *Rhyncophoromyia maculineura*, an ant-parasitizing phorid fly (Diptera: Phoridae) from Peru. *Journal of Natural History*, **27**: 429–434.

Brown, B.V., Francoeur, A. & Gibson, R.L. 1991. Review of the genus *Styletta* (Diptera: Phoridae), with description of a new genus. *Entomologica Scandinavica*, **22**: 241–250.

Brown, H.P. 1968. *Psephenus* (Coleoptera: Psephenidae) parasitized by a new chalcidoid (Hymenoptera: Eulophidae). II. Biology of the parasite. *Annals of the Entomological Society of America*, **61**: 452–456.

Brown, J.J. & Kainoh, Y. 1992. Host castration by *Ascogaster* spp. (Hymenoptera: Braconidae). *Annals of the Entomological Society of America*, **85**: 67–71.

Brown, J.R. & Phillips, J.R. 1991. Survival of *Microplitis croceipes* in *Heliothis* NPV-infected *Heliothis virescens* larvae. *Southwestern Entomologist*, **16**: 25–29.

Brown, V.K. 1973. The biology and development of *Brachygaster minutus* Olivier (Hymenoptera: Evaniidae), a parasite of the oothecae of *Ectobius* spp. (Dictyoptera: Blattidae). *Journal of Natural History*, **7**: 665–674.

Brown, W.L. Jr. 1958. Contributions toward a reclassification of the Formicidae, II. Tribe Ectatommini (Hymenoptera). *Bulletin of the Museum of Comparative Zoology*, **118**: 173–362.

Brown, W.L. Jr. 1960. Ants, acacias and browsing mammals. *Ecology*, **41**: 587–592.

Brown, W.L. 1961. A note on the ant *Gnamptogenys hartmani* Wheeler. *Psyche*, **68**: 69.

Brown, W.L. Jr. 1979 (1980). A remarkable new species of *Proceratium*, with dietary and other notes on the genus (Hymenoptera: Formicidae). *Psyche*, **86**: 337–346.

Brown, W.L. Jr. 1982. Hymenoptera, pp. 652–680. *In*: Parker, S.P. (ed.) *Synopsis and Classification of Living Organisms 2*. 1232pp. McGraw-Hill, New York.

Brown, W.L. Jr. & Wilson, E.O. 1959 (1960). The evolution of the dacetine ants. *Quarterly Review of Biology*, **34**: 278–294.

Brues, C.T. 1922. *Conoaxima*, a new genus of the hymenopterous family Eurytomidae, with a description of its larva and pupa. *Psyche*, **29**: 153–158.

Brues, C.T. 1923. *Termitobracon*, a termitophilous braconid from British Guiana. *Zoologica: Scientific Contributions of the New York Zoological Society from the Tropical Research Station in British Guiana*, **3**: 427–432.

Brues, C.T. 1928. A note on the genus *Pelecinus*. *Psyche*, **35**: 205–209.

Bryan, G. 1980. Courtship behaviour, size differences between the sexes and oviposition in some *Achrysocharoides* species (Hym., Eulophidae). *Netherlands Journal of Zoology*, **30**: 611–621.

Bryden, J.W. & Bishop, M.W.H. 1945. *Perilitus coccinellae* (Schrank) (Hym., Braconidae) in Cambridgeshire. *Entomologist's Monthly Magazine*, **81**: 51–52.

Buchmann, S.L. 1980. Preliminary anthecological observations on *Xiphidium caeruleum* Aubl. (Monocotyledoneae: Haemodoraceae) in Panama. *Journal of the Kansas Entomological Society*, **53**: 685–699

Buchmann, S.L. 1983. Buzz pollination in angiosperms, pp. 73–113. *In*: Jones, C.E. & Little R.J. (eds) *Handbook of Experimental Pollination Biology*. 558pp. Van Nostrand Reinhold, New York.

Buchmann, S.L. 1987. The ecology of oil flowers and their bees. *Annual Review of Ecology and Systematics*, **18**: 343–369.

Buchmann, S.L. & Hurley, J.P. 1978. A biophysical model for buzz pollination in angiosperms. *Journal of Theoretical Biology*, **72**: 639–657.

Buchmann, S.L. & Jones, C.E. 1980. Observations on the nesting biology of *Melissodes persimilis* Ckll. (Hymenoptera: Anthophoridae). *Pan-Pacific Entomologist*, **56**: 200–206.

Buckingham, G.R. & Sharkey, M.J. 1988. Abdominal exocrine glands in Braconidae (Hymenoptera), pp. 199–242. *In*: Gupta, V.K. (ed.) *Advances in Parasitic Hymenoptera Research*. 546pp. Brill, Leiden.

Buckley, R.C. 1987. Interactions involving plants, Homoptera, and ants. *Annual Review of Ecology and Systematics*, **18**: 111–135.

Buddenhagen, I.W. & Elsasser, T.A. 1962. An insect-spread bacterial wilt epiphytotic of bluggoe banana. *Nature*, **194**: 164–165.

Budenberg, W.J. 1990. Honeydew as a contact kairomone for aphid parasitoids. *Entomologia Experimentalis et Applicata*, **55**: 139–147.

Bühler, A., Hanzlik, T.N. & Hammock, B.D. 1985. Effects of parasitization on *Trichoplusia ni* by *Chelonus* sp. *Physiological Entomology*, **10**: 383–394.

Bull, J.J. 1981. Sex ratio evolution when fitness varies. *Heredity*, **46**: 9–26.

Bull, J.J. 1983. *Evolution of Sex Determining Mechanisms*. 316pp. Benjamin/Cummings Publishing Co., Inc., Menlo Park.

Burdick, D.J. 1961. A taxonomic and biological study of the genus *Xyela* Dalman in North America. *University of California Publications in Entomology*, **17**: 285–356.

Burdick, D.J. & Wasbauer, M.S. 1959. Biology of *Methoca californica* Westwood (Hymenoptera: Tiphiidae). *Wasmann Journal of Biology*, **17**: 75–88.

Burger, W. 1971. Flora Costaricensis: Piperaceae. *Fieldiana* Botany, **35**: 5–227.

Burger, W. 1977*a*. Flora Costaricensis: Fagaceae. *Fieldiana* Botany, **40**: 59–82.

Burger, W. 1977*b*. Flora Costaricensis: Moraceae. *Fieldiana* Botany, **40**: 94–215.

Burger, W. & Werff, H. van der 1990. Flora Costaricensis: Lauraceae. *Fieldiana* Botany, New Series, **23**: 1–129.

Burks, B.D. 1956. The species *Chryseida* (Hymenoptera, Eurytomidae). *Bulletin of the Brooklyn Entomological Society*, **51**: 109–116.

Burks, B.D. 1958. Three species of *Eurytoma* important in biological control of weeds (Hymenoptera, Eurytomidae). *Entomological News*, **69**: 177–185.

Burks, B.D. 1959. The species of the genus *Herbertia* Howard (Hymenoptera, Pteromalidae). *Proceedings of the Entomological Society of Washington*, **61**: 249–255.

Burks, B.D. 1960. A revision of the genus *Brachymeria* Westwood in America north of Mexico (Hymenoptera: Chalcididae). *Transactions of the American Entomological Society*, **86**: 225–273.

Burks, B.D. 1962. A new *Elachertus* parasitic on a pest of bananas (Hymenoptera, Eulophidae). *Entomological News*, **73**: 233–235.

Burks, B.D. 1965. The North American species of *Elasmus* Westwood (Hymenoptera, Eulophidae). *Proceedings of the Biological Society of Washington*, **78**: 201–207.

Burks, B.D. 1966. The immature stages of *Axima zabriskiei* Howard (Hymenoptera, Eurytomidae). *Entomological News*, **77**: 199–202.

Burks, B.D. 1967. The North American species of *Anastatus* Motschulsky (Hymenoptera, Eupelmidae). *Transactions of the American Entomological Society*, **93**: 423–431.

Burks, B.D. 1969. New Perilampidae (Hymenoptera: Chalcidoidea). *Proceedings of the Entomological Society of Washington*, **71**: 73–81.

Burks, B.D. 1971. A synopsis of the genera of the family Eurytomidae (Hymenoptera: Chalcidoidea). *Transactions of the American Entomological Society*, **97**: 1–89.

Burks, B.D. 1972. The genus *Hexacladia* Ashmead (Hymenoptera: Encyrtidae). *Proceedings of the Entomological Society of Washington*, **74**: 363–371.

Burks, B.D. 1973. North American species of *Calosota* Curtis (Hymenoptera: Eupelmidae). *Journal of the Washington Academy of Sciences*, **63**: 26–31.

Buschman, L.L. & Whitcomb, W.H. 1980. Parasites of *Nezara viridula* (Hemiptera: Pentatomidae) and other Hemiptera in Florida. *Florida Entomologist*, **63**: 154–162.

Busck, A. 1917. Notes on *Perisierola emigrata* Rohwer, a parasite of the pink bollworm (Hymenoptera, Bethyliidae). *Insecutor Inscitiae Menstruus*, **5**: 3–5

Bustillo, A.E. & Drooz, A.T. 1977. Cooperative establishment of a Virginia (USA) strain of *Telenomus alsophilae* on *Oxydia trychiata* in Colombia. *Journal of Economic Entomology*, **70**: 767–770.

Buyckx, J.E. 1948. Recherches sur un Dryinide, *Aphelopus indivisus*, parasite de Cicadines. *La Cellule*, **52**: 63–155.

Byers, G.W. 1978. Nests, prey, behavior and development of *Cerceris halone* (Hymenoptera: Sphecidae). *Journal of the Kansas Entomological Society*, **51**: 818–831.

Caballero, R. & Andrews, K.L. 1985. *Anagrus* sp. y otros enemigos naturales del lorito verde (*Empoasca* spp.) en el cultivo del frijol en Honduras. *Ceiba*, **26**: 149–152.

Cabrales, G., Figueroa, R., Uribe, F. & Trochez, C.I. 1985. Evaluacion del efecto del parasitismo por *Spalangia endius* W. (Hym. Pteromalidae) sobre la dinamica de poblacion de *Musca domestica* L. en galpones para aves (La Florida-Risaralda). *Acta Agronomica*, **35**: 93–105.

Calderone, N.W. & Page, R.E. Jr 1992. Effects of interactions among genotypically diverse nestmates on task specialization by foraging honey bees (*Apis mellifera*). *Behavioral Ecology and Sociobiology*, **30**: 219–226.

Callan, E.McC. 1939. A note on the breeding of *Probethylus callani* Richards (Hymenopt., Bethylidae), an embiopteran parasite. *Proceedings of the Royal Entomological Society of London* (B), **8**: 223–224.

Callan, E.McC. 1942. A note on *Timulla (Timulla) eriphyla* Mickel (Hym., Mutillidae), a parasite of *Tachysphex blatticidus* F.X. Williams (Hym., Larridae), from Trinidad, B.W.I. *Proceedings of the Royal Entomological Society of London* (A), **17**: 18.

Callan, E. McC. 1976. Observations on the nesting behavior and prey of gorytine wasps in Trinidad (Hymenoptera, Sphecidae). *Psyche*, **83**: 324–335.

Callan, E. McC. 1977. Observations on *Centris rufosuffusa* Cockerell (Hymenoptera: Anthophoridae) and its parasites. *Journal of Natural History*, **11**: 127–135.

Callan, E. McC. 1990. Nesting behaviour and prey of *Cerceris* wasps in Trinidad (Hymenoptera: Sphecidae). *The Entomologist*, **109**: 194–199.

Callan, E. McC. 1991. Nesting behaviour and prey of sand wasps in Trinidad (Hymenoptera: Sphecidae: Nyssoninae). *The Entomologist*, **110**: 134–138.

Calvert, D.J. 1973. Experimental host preferences of *Monoctonus paulensis* (Hymenoptera: Braconidae), including a hypothetical scheme of host selection. *Annals of the Entomological Society of America*, **66**: 28–33.

Calvert, D.J. & Bosch, R. van den 1972. Behavior and biology of *Monoctonus paulensis* (Hymenoptera: Braconidae), a parasite of dactynotine aphids. *Annals of the Entomological Society of America*, **65**: 773–779.

Camacho, E. 1966. Daño que las abejas jicotes del género *Trigona* causan a los árboles de Macadamia. *Turrialba*, **16**: 193–194.

Camargo, J.M.F. de 1968 (1970). Ninhos e biologia de algumas espécies de Meliponideos (Hymenoptera: Apidae) de região de Pôrto Velho, Território de Rondônia, Brasil. *Revista de Biología Tropical*, **16**: 207–239.

Camargo, J.M.F. de 1989. Comentários sobre a sistemática de Meliponinae (Hymen. Apidae). *Anais do XIV Simposia anuale da ACIESP*, (Supplement), **68**: 41–61.

Camargo, J.M.F. de, Moure, J.S. & Roubik, D.W. 1988. *Melipona yucatanica* new species (Hymenoptera: Apidae: Meliponinae); stingless bee dispersal across the Caribbean arc and post-Eocene vicariance. *Pan-Pacific Entomologist*, **64**: 147–158.

Camargo, J.M.F. & Roubik, D.W. 1991. Systematics and bionomics of the apoid obligate necrophages: the *Trigona hypogea* group (Hymenoptera: Apidae; Meliponinae). *Biological Journal of the Linnean Society*, **44**: 13–39.

Camargo, O.R. 1956. As larvas 'mata-porco' no Rio Grande do Sul. *Boletim da Directoria da Producção Animal*, Rio Grande do Sul, Porto Alegre, **13**: 23–29.

Cambra T., R. & Quintero Arias, D. 1992. Velvet ants of Panama: distribution and systematics (Hymenoptera: Mutillidae), pp. 459–478. *In*: Quintero, D. & Aiello, A. (eds) *Insects of Panama and Mesoamerica*. 692pp. Oxford University Press, Oxford.

Cameron, E. 1941 (1939). The holly leaf-miner (*Phytomyza ilicis* Curt.) and its parasites. *Bulletin of Entomological Research*, **30**: 173–208.

Cameron, E. 1957. On the parasites and predators of the cockroach. II.—*Evania appendigaster* (L.). *Bulletin of Entomological Research*, **48**: 199–209.

Cameron, E.A. 1965. The Siricinae (Hymenoptera: Siricidae) and their parasites. *Technical Bulletin. Commonwealth Institute of Biological Control*, **5**: 1–31.

Campbell, R.W. 1963. Some ichneumonid-sarcophagid interactions in the gypsy moth *Porthetria dispar* (L.) (Lepidoptera: Lymantriidae). *Canadian Entomologist*, **95**: 337–345.

Cane, J.H. 1983*a*. Preliminary chemosystematics of the Andrenidae and exocrine lipid evolution of the short-tongued bees (Hymenoptera: Apoidea). *Systematic Zoology*, **32**: 417–430.

Cane, J.H. 1983*b*. Chemical evolution and chemosystematics of the Dufour's gland secretions of the lactone-producing bees (Hymenoptera: Colletidae, Halictidae, and Oxaeidae). *Evolution*, **37**: 657–674.

Cane, J.H. & Carlson, R.G. 1984. Dufour's gland triglycerides from *Anthophora*, *Emphoropsis* (Anthophoridae) and *Megachile* (Megachilidae) bees (Hymenoptera: Apoidea). *Comparative Biochemistry and Physiology* (B), **78**: 769–772.

Cane, J.H. & Miyamoto, M.M. 1979. Nest defense and foraging ethology of a Neotropical sand wasp, *Bembix multipicta* (Hymenoptera: Sphecidae). *Journal of the Kansas Entomological Society*, **52**: 667–672.

Capek, M. 1970. A new classification of the Braconidae (Hymenoptera) based on the cephalic structures of the final instar larva and biological evidence. *Canadian Entomologist*, **102**: 846–875.

Capek, M. 1973. Key to the final instar larvae of the Braconidae (Hymenoptera). *Acta Instituti Forestalis Zvolenensis*, **1973**: 259–268.

Capinera, J.L. & Lilly, J.H. 1975. *Tetrastichus asparagi*, parasitoid of the asparagus beetle: some aspects of host-parasitoid interaction. *Annals of the Entomological Society of America*, **68**: 595–596.

Cardé, R.T. & Baker, T.C. 1984. Sexual communication with pheromones, pp. 355–383. *In*: Bell, W.J. & Cardé, R.T. (eds) *Chemical Ecology of Insects*. 524pp. Chapman & Hall, London.

Cardona, C. & Oatman, E.R. 1971. Biology of *Apanteles dignus* (Hymenoptera: Braconidae), a primary parasite of the tomato pinworm. *Annals of the Entomological Society of America*, **64**: 996–1007.

Cardona, C. & Oatman, E.R. 1975. Biology and physical ecology of *Apanteles subandinus* Blanchard (Hymenoptera: Braconidae), with notes on temperature responses of *Apanteles scutellaris* Muesebeck and its host, the potato tuberworm. *Hilgardia*, **43**: 1–51.

Carillo S., J.L. & Dickason, E.A. 1963. Biology and economic importance of seed chalcids infesting red clover and alfalfa in Oregon. *Oregon Agricultural Experiment Station Technical Bulletin*, **68**: 1–35.

Carlin, N.F. & Gladstein, D.S. 1989. The 'bouncer' defense of *Odontomachus ruginodis* and other odontomachine ants (Hymenoptera: Formicidae). *Psyche*, **96**: 1–20.

Carlow, T. 1992. *Thoronella* sp. (Hymenoptera: Scolionidae [*sic*]) discovered on the thorax of an Aeshnidae (Anisoptera). *Notulae Odonatologicae*, **3**: 149–150.

Carmean, D. 1991. Biology of the Trigonalyidae (Hymenoptera), with notes on the vespine parasitoid *Bareogonalos canadensis*. *New Zealand Journal of Zoology*, **18**: 209–214.

Carmean, D., Akre, R.D., Zack, R.S. & Reed, H.C. 1981. Notes on the yellowjacket parasite *Bareogonalis* (sic) *canadensis* (Hymenoptera: Trigonalidae). *Entomological News*, **92**: 23–26.

Carpenter, J.M. 1982. The phylogenetic relationships and natural classification of the Vespoidea (Hymenoptera). *Systematic Entomology*, **7**: 11–38.

Carpenter, J.M. 1986*a*. Cladistics of the Chrysidoidea (Hymenoptera). *Journal of the New York Entomological Society*, **94**: 303–330.

Carpenter, J.M. 1986*b*. The genus *Pachodynerus* in North America (Hymenoptera: Vespidae: Eumeninae). *Proceedings of the Entomological Society of Washington*, **88**: 572–577.

Carpenter, J.M. 1989. Additions to social wasps of Costa Rica. *Sphecos*, **18**: 10–11.

Carpenter, J.M. 1991. Phylogenetic relationships and the origin of social behavior in the Vespidae, pp. 7–32. *In*: Ross, K.G. & Matthews, R.W. (eds) *The Social Biology of Wasps*. 678pp. Cornell University Press, Ithaca.

Carpenter, J.M. 1993. Biogeographic patterns in the Vespidae (Hymenoptera): two views of Africa and South America, pp. 139–155. *In*: Goldblatt, P. (ed.) *Biological Relationships Between Africa and South America*. 630pp. Yale University Press, New Haven.

Carpenter, J.M. & Cumming, J.M. 1985. A character analysis of the North American potter wasps (Hymenoptera: Vespidae; Eumeninae). *Journal of Natural History*, **19**: 877–916.

Carpenter, J.M. & Day, M.C. 1988. Nomenclatural notes on Polistinae (Hymenoptera: Vespidae). *Proceedings of the Entomological Society of Washington*, **90**: 323–328.

Carpenter, J.M. & Rasnitsyn, A.P. 1990. Mesozoic Vespidae. *Psyche*, **97**: 1–20.

Carpenter, J.M. & Vecht, J. van der, 1991. A study of the Vespidae described by William J. Fox (Insecta: Hymenoptera), with assessment of taxonomic implications. *Annals of Carnegie Museum* **60**: 211–241.

Carpenter, J.M. & Wenzel, J.W. 1988. A new species and nest type of *Mischocyttarus* from Costa Rica (Hymenoptera: Vespidae; Polistinae), with descriptions of nests of three related species. *Psyche*, **95**: 89–99.

Carpenter, J M. & Wenzel, J.W. 1989. Synonymy of the genera *Protopolybia* and *Pseudochartergus* (Hymenoptera: Vespidae; Polistinae). *Psyche* **96**: 177–186.

Carrillo S., J.L. & Caltagirone, L.E. 1970. Observations on the biology of *Solierella peckhami*, *S. blaisdelli* (Sphecidae), and two species of Chrysididae (Hymenoptera). *Annals of the Entomological Society of America*, **63**: 672–681.

Carton, Y., Chibani, F., Haouas, S. & Marrakchi, M. 1987. Egg-laying strategy under natural conditions of *Leptopilina boulardi*, a hymenopteran parasitoid of *Drosophila* spp. *Entomologia Experimentalis et Applicata*, **43**: 193–201.

Carton, Y. & Kitano, H. 1981. Evolutionary relationships to parasitism by seven species of the *Drosophila melanogaster* subgroup. *Biological Journal of the Linnean Society*, **16**: 227–241.

Carver, M., Inkerman, P.A. & Ashbolt, N.J. 1987. *Anagyrus saccharicola* Timberlake (Hymenoptera: Encyrtidae) and other biota associated with *Saccharicoccus sacchari* (Cockerell) (Homoptera: Pseudococcidae) in Australia. *Journal of the Australian Entomological Society*, **26**: 367–368.

Cassier, P., Tel-Zur, D. & Lensky, Y. 1994. The sting sheaths of honey bee workers (*Apis mellifera* L.): structure and alarm pheromone secretion. *Journal of Insect Physiology*, **40**: 23–32.

Castner, J.L. 1984. Suitability of *Scapteriscus* spp. mole crickets (Ort.: Gryllotalpidae) as hosts of *Larra bicolor* (Hym.: Sphecidae). *Entomophaga*, **29**: 323–332.
Castro, B (ed.) 1993. *Inventario de Enemigos Naturales de Plagas Insectiles en Varios Cultivos en Honduras*. 352pp. Departmento de Protección Vegetal, Escuela Agricola Panamericana, 488. Zamorano.
Cave, R.D. 1992. Inventory of parasitic organisms of the striped grass looper, *Mocis latipes* (Lepidoptera: Noctuidae), in Honduras. *Forida Entomologist*, **75**: 592–598.
Cave, R.D., Gaylor, M.J. & Bradley, J.T. 1987. Host handling and recognition by *Telenomus reynoldsi* (Hymenoptera: Scelionidae), an egg parasitoid of *Geocoris* spp. (Heteroptera: Lygaeidae). *Annals of the Entomological Society of America*, **80**: 217–223.
Cave, R.D. & Miller, G.L. 1987. Notes on *Anacharis melanoneura* (Hymenoptera: Figitidae) and *Charitopes mellicornis* (Hymenoptera: Ichneumonidae) parasitizing *Micromus posticus* (Neuroptera: Hemerobiidae). *Entomological News*, **98**: 211–216.
Cendaña, S.M. 1937. Studies on the biology of *Coccophagus* (Hymenoptera), a genus parasitic on nondiaspidine Coccidae. *University of California Publications in Entomology*, **6**: 337–399.
Chadab, R. 1979 (1980). Early warning cues for social wasps attacked by army ants. *Psyche*, **86**: 115–123.
Chadab, R. & Rettenmeyer, C.W. 1975. Mass recruitment by army ants. *Science*, **188**: 1124–1125.
Chambers, V.H. 1955. Some hosts of *Anteon* spp. (Hym., Dryinidae) and a hyperparasite *Ismarus* (Hym., Belytidae). *Entomologist's Monthly Magazine*, **91**: 114–115.
Chambers, V.H. 1970 (1971). Large populations of Belytinae (Hym., Diapriidae). *Entomologist's Monthly Magazine*, **106**: 149–154.
Chapman, R.F. 1969. *The Insects, Structure and Function*. 819pp. English Universities Press, London.
Charnov, E.L. 1982. *The Theory of Sex Allocation*. 255pp. Princeton University Press, Princeton.
Charnov, E.L. & Skinner, S.W. 1984. Evolution of host selection and clutch size in parasitoid wasps. *Florida Entomologist*, **67**: 5–21.
Chassain, C., Boulétreau, M. & Fouillet, P. 1988. Host exploitation by parasitoids: local variations in foraging behaviour of females among populations of *Trichogramma* species. *Entomologia Experimentalis et Applicata*, **48**: 195–202.
Chatterjee, P.N. 1943 (1944). Description of *Rhogas plecopterae* sp. nov. from India with notes on its biology and morphology. *Indian Journal of Entomology*, **5**: 213–221.
Chatterji, S. 1955. Studies on the biology of *Aplastomorpha calandrae* Howard [Insecta: Hymenoptera: Chalcidae] parasitic on some storage pests. *Proceedings of the Zoological Society of Calcutta*, **8**: 11–23.
Chaud-Netto, J. 1980. Biological studies on *Pseudohypocera kerteszi* (Phoridae, Diptera). *Experientia*, **36**: 61–62.
Chaverri, E. 1954. Anotaciones sobre la biología del *Antianthe expansa* Germar, plaga del pimiento en Costa Rica. *Revista Biologia Tropical*, **2**: 269–282.
Cherian, M.C. & Israel, P. 1937. Studies on *Elasmus zehntneri* Ferr., a parasite of the sugarcane white moth borer (Scirpophaga). *Madras Agricultural Journal*, **25**: 273–279.
Chernoguz, D.G. & Reznik, S.Y. 1987. Etological and physiological components of host specificity of the braconid *Alysia manducator* Panz. (Hymenoptera, Braconidae). *Entomologicheskoe Obozrenie*, **66**: 499–510. [In Russian with English summary.]

Cherrett, J.M. 1968. The foraging behaviour of *Atta cephalotes* L. (Hymenoptera, Formicidae), I. Foraging pattern and plant species attacked in tropical rain forest. *Journal of Animal Ecology*, **37**: 387–403.

Cherrett, J.M. 1972. Some factors involved in the selection of vegetable substrate by *Atta cephalotes* (L.) (Hymenoptera: Formicidae) in tropical rain forest. *Journal of Animal Ecology*, **41**: 647–660.

Cherrett, J.M. 1986. History of the leaf-cutting ant problem, pp. 10–17. *In*: Lofgren C.S. & Vander Meer, R.K. (eds) *Fire Ants and Leaf-cutting Ants: Biology and Management*. 435pp. Westview Press, Boulder.

Cherrett, J.M. 1989. Leaf-cutting ants, pp. 473–488. *In*: Lieth, H. & Werger, M.J.A. (eds), *Ecosystems of the World 14B. Tropical Rain Forest Ecosystems. Biogeographical and Ecological Studies*. Elsevier, Amsterdam.

Cherrett, J.M. & Jutsum, A.R. 1983. The effects of some ant species, especially *Atta cephalotes* (L.), *Acromyrmex octospinosus* (Reich.) and *Azteca* sp. (Hym. Form.) on citrus growing in Trinidad, pp. 155–163. *In*: Jaisson, P. (ed.) *Social Insects in the Tropics 2*. 252pp. Université Paris Nord, Paris.

Chittampalli, P. & Mulcahy, F.D. 1990. Honey and sugar in the treatment of wounds and ulcers in biomedicine and in the Ayurveda, pp. 155–164. *In*: Posey, D.A. & Overal, W.L. (eds), *Ethnobiology: Implications and Applications. 2*. 256pp. Museu Paraense Emílio Goeldi, Belém.

Chiu, S.-C., Chou, L.Y. & Chou, K.C. 1981. A preliminary survey on the natural enemies of *Kerria lacca* (Kerr) in Taiwan. *Journal of Agricultural Research of China*, **30**: 420–425. [In Chinese: English summary.]

Chou, L.Y. & Sharkey, M.J. 1989. The Braconidae (Hymenoptera) of Taiwan. I. Agathidinae. *Journal of Taiwan Museum*, **42**: 147–223.

Chow, F.J. & Mackauer, M. 1986. Host discrimination and larval competition in the aphid parasite *Ephedrus californicus*. *Entomologia Experimentalis et Applicata*, **41**: 243–254.

Chrystal, R.N. 1930. Studies of the *Sirex* parasites. *Oxford Forestry Memoirs*, **11**: 1–63.

Chrystal, R.N. & Skinner, E.R. 1931. Studies in the biology of *Xylonomus brachylabris* Kr., and *X. irrigator* F., parasites of the larch longhorn beetle, *Tetropium gabrieli* Weise. *Forestry*, **5**: 21–331.

Chrystal, R.N. & Skinner, E.R. 1932. Studies in the biology of the woodwasp *Xiphydria prolongata* Geoffr. (*dromedarius* F.) and its parasite *Thalessa curvipes* Grav. *Scottish Forestry Journal*, **46**: 36–57.

Chumakova, B.M. 1961. Parasites of injurious scale-insects from Kabardino-Balkaria (Hymenoptera, Chalcidoidea). *Entomologicheskoe Obozrenie*, **40**: 315–338. [In Russian.]

Clancy, D.W. 1944. Biology of *Allotropa burrelli*, a gregarious parasite of *Pseudococcus comstocki*. *Journal of Agricultural Research, Washington D.C.* **69**: 159–167.

Clancy, D.W. 1946. The insect parasites of the Chrysopidae (Neuroptera). *University of California Publications in Entomology*, **7**: 403–496.

Claret, J., Porcheron, P. & Dray, F. 1978. La teneur en ecdysones circulantes au cours du dernier stade larvaire de l'Hyménoptère endoparasite *Pimpla instigator*, et l'entrée en diapause. *Comptes Rendus Hebdomadaires des Séances de l'Académie des Sciences* (Série D), **286**: 639–641.

Claridge, M.F. 1961*a*. A contribution to the biology and taxonomy of some Palaearctic species of *Tetramesa* Walker (= *Isosoma* Walk.; = *Harmolita* Motsch.) (Hymenoptera: Eurytomidae), with particular reference to the British fauna. *Transactions of the Royal Entomological Society of London*, **113**: 175–216.

Claridge, M.F. 1961*b*. Biological observations on some eurytomid (Hymenoptera: Chalcidoidea) parasites associated with Compositae, and some taxonomic implications. *Proceedings of the Royal Entomological Society of London* (A), **36**: 153–158.

Claridge, M.F. & Askew, R.R. 1960. Sibling species in the *Eurytoma rosae* group (Hym., Eurytomidae). *Entomophaga*, **5**: 141–153.

Clark, A.M. 1963. The influence of diet upon the adult life span of two species of *Bracon*. *Annals of the Entomological Society of America*, **56**: 616–619.

Clark, D.B., Guayasamín, C., Pazmiño, O., Donoso, C. & Páez de Villacís, Y. 1982. The tramp ant Wasmannia auropunctata: autecology and effects on ant diversity and distribution on Santa Cruz Island, Galapagos. *Biotropica*, **14**: 196–207.

Claude-Joseph, F. 1930. Recherches biologiques sur les prédateurs du Chili. *Annales des Sciences Naturales. Zoologie* (10) **13**: 235–354.

Clausen, C.P. 1924. The parasites of *Pseudococcus maritimus* (Ehrhorn) in California (Hymenoptera, Chalcidoidea). Part II. Biological studies and life histories. *University of California Publications, Technical Bulletin, Entomology*, **3**: 253–292.

Clausen, C.P. 1927. The bionomics of *Anastatus albitarsis* Ashm., parasitic in the eggs of *Dictyoploca japonica* Moore (Hymen.). *Annals of the Entomological Society of America*, **20**: 461–473.

Clausen, C.P. 1929. Biological studies on *Poecilogonalos thwaitesii* (Westw.), parasitic in the cocoons of *Henicospilus* (Hymen: Trigonalidae). *Proceedings of the Entomological Society of Washington*, **31**: 67–79.

Clausen, C.P. 1931. Biological notes on the Trigonalidae (Hymenoptera). *Proceedings of the Entomological Society of Washington*, **33**: 72–81.

Clausen, C.P. 1932. The biology of *Encyrtus infidus* Rossi, a parasite of *Lecanium kunoensis* Kuw. (Hymen.). *Annals of the Entomological Society of America*, **25**: 670–687.

Clausen, C.P. 1940*a*. The oviposition habits of the Eucharidae (Hymenoptera). *Journal of the Washington Academy of Sciences*, **30**: 504–516.

Clausen, C.P. 1940*b*. *Entomophagous Insects*. 688pp. McGraw Hill, New York.

Clausen, C.P. 1940c. The immature stages of the Eucharidae. *Proceedings of the Entomological Society of Washington*, **42**: 161–170.

Clausen, C.P. 1956. Biological control of insect pests in the continental United States. *United States Department of Agriculture Technical Bulletin*, **1139**: 1–151.

Clausen, C.P. 1976. Phoresy among entomophagous insects. *Annual Review of Entomology*, **21**: 343–368.

Clausen, C.P. (ed.) 1978. *Introduced Parasites and Predators of Arthropod Pests and Weeds: a World Review*. 545pp. United States Department of Agriculture, (Handbook 480). Washington D.C.

Clausen, C.P., Gardner, T.R. & Sato, K. 1932. Biology of some Japanese and Chosenese grub parasites (Scoliidae). *United States Department of Agriculture Technical Bulletin*, **308**: 1–26.

Clausen, C.P., Jaynes, H.A. & Gardner, T.R. 1933. Further investigations of the parasites of *Popillia japonica* in the Far East. *United States Department of Agriculture Technical Bulletin*, **366**: 1–58.

Clement, S.L. & Norris, R.F. 1982. Two insects offer potential biological control of common purslane. *California Agriculture*, **36**(2): 16–18.

Cloutier, C. & Baudin, F. 1990. Searching behavior of the aphid parasitoid *Aphidius nigripes* (Hymenoptera: Aphidiidae) foraging on potato plants. *Environmental Entomology*, **19**: 222–228.

Cloutier, C., Lévesque, C.A., Eaves, D.M. & Mackauer, M. 1991. Material adjustment of sex ratio in response to host size in the aphid parasitoid *Ephedrus californicus. Canadian Journal of Zoology*, **69**: 1489–1495.

Cock, M.J.W. (ed.) 1985. *A Review of Biological Control of Pests in the Commonwealth Caribbean and Bermuda up to 1982*. Commonwealth Institute of Biological Control, Technical Communication, 9. 218pp. Commonwealth Agricultural Bureaux, Slough.

Cock, M.J.W., Godfray, H.C.J. & Holloway, J.D. 1987. *Slug and Nettle Caterpillars. The Biology, Taxonomy and Control of the Limacodidae of Economic Importance on Palms in South-east Asia.* 270pp. C.A.B. International, Wallingford.

Coddington, J.A., Griswold, C.E., Dávila, D.S., Peñaranda, E. & Larcher, S.F. 1991. Designing and testing sampling protocols to estimate biodiversity in tropical ecosystems, pp. 44–60. *In*: Dudley, E.C. (ed.), *The Unity of Evolutionary Biology. Proceedings of the Fourth International Congress of Systematic and Evolutionary Biology, 1*. 588pp. Dioscorides Press, Portland, Oregon.

Coggshall, W.L. & Morse, R.A. 1984. *Beeswax: Production, Harvesting, Processing and Products*. 192pp. Wicwas, Ithaca.

Cole, L.R. 1959*a*. On the defences of lepidopterous pupae in relation to the oviposition behaviour of certain Ichneumonidae. *Journal of the Lepidopterists' Society*, **13**: 1–10.

Cole, L.R. 1959*b*. On a new species of *Syntretus* Foerster (Hym., Braconidae) parasitic on an adult ichneumonid, with a description of the larva and notes on its life history and that of its host, *Phaeogenes invisor* (Thunberg). *Entomologist's Monthly Magazine*, **95**: 18–21.

Cole, L.R. 1967. A study of the life-cycles and hosts of some Ichneumonidae attacking pupae of the green oak leaf roller moth, *Tortrix viridana* (L.) (Lepidoptera: Tortricidae). *Transactions of the Royal Entomological Society of London*, **119**: 267–281.

Cole, L.R. 1970. Observations on the finding of mates by male *Phaeogenes invisor* and *Apanteles medicaginis* (Hymenoptera: Ichneumonoidea). *Animal Behaviour*, **18**: 184–189.

Common, I.F.B. 1954. A study of the ecology of the adult bogong moth, *Agrotis infusa* (Boisd.) (Lepidoptera: Noctuidae), with special reference to its behaviour during migration and aestivation. *Australian Journal of Zoology*, **2**: 223–263.

Compere, H. 1939. The insect enemies of the black scale, *Saissetia oleae* (Bern.) in South America. *University of California Publications in Entomology*, **7**: 75–90.

Compere, H. 1947. A new genus and species, *Eurymyiocnema aphelinoides* (Hymenoptera, Aphelinidae), and a history of the genera *Euryischia* Riley and *Myiocnema* Ashmead. *Bulletin of Entomological Research*, **38**: 381–388.

Compere, H. & Rosen, D. 1970. The prescutum in Hymenoptera. *Proceedings of the Royal Entomological Society of London* (A), **45**: 91–97.

Compere, H. & Smith, H.S. 1927. Notes on the life-history of two oriental chalcidoid parasites of *Chrysomphalus. University of California Publications in Entomology*, **4**: 63–73.

Condit, I.J. 1947. *The Fig*. 222pp. Chronica Botanica, Waltham, Mass.

Condit, I.J. & Enderud, J. 1956. A bibliography of the fig. *Hilgardia*, **25**: 1–663.

Cook, O.F. 1904. Report on the habits of the kelep, or Guatemalan cotton-boll-weevil ant. *United States Department of Agriculture, Bureau of Entomology Bulletin*, **49**: 1–15.

Cooper, K.W. 1953. Egg gigantism, oviposition, and genital anatomy: their bearing on the biology and phylogenetic position of *Orussus* (Hymenoptera: Siricoidea). *Proceedings of the Rochester Academy of Science*, **10**: 38–68.

Cooper, K.W. 1954. Biology of eumenine wasps, IV.—a trigonalid wasp parasitic on *Rygchium rugosum* (Saussure) (Hymenoptera, Trignalidae). *Proceedings of the Entomological Society of Washington*, **56**: 280–288.

Cooper, K.W. 1955. II. Venereal transmission of mites by wasps, and some evolutionary problems arising from the remarkable association of *Ensliniella trisetosa* with the wasp *Ancistrocerus antilope*. Biology of eumenine wasps. *Transactions of the American Entomological Society*, **80**: 119–174.

Cooper, K.W. 1957. Biology of eumenine wasps. V. Digital communication in wasps. *Journal of Experimental Zoology*, **134**: 469–514.

Cooper K.W. & Dessart, P. 1975. Adult, larva and biology of *Conostigmus quadratogenalis* Dessart & Cooper, sp. n., (Hym. Ceraphronoidea), parasite of *Boreus* (Mecoptera) in California. *Bulletin & Annales de la Sociétè Royale Belge d'Entomologie*, **111**: 37–53.

Cooper, M. 1980. Notes on the biology of *Dynatus nigripes* (Westw.) and *Penepodium albovillosum* (Cam.) (Hym., Sphecidae). *Entomologist's Monthly Magazine*, **116**: 87–89.

Cooper, M. 1986*a*. A note on the biology of *Aulacophilus eumenoides* Ducke (Sphecidae). *Sphecos*, **11**: 16.

Cooper, M. 1986*b*. Nest of *Quexua verticalis* (F.Smith) (Sphecidae—Crabronini). *Sphecos*, **11**: 17.

Cooper, M. 1988. Prey of *Scapheutes brasiliensis* Handlirsch (Sphecidae). *Sphecos*, **17**: 13.

Cooper, M. 1993*a*. A second prey record for *Scapheutes* (Sphecidae). *Sphecos*, **24**: 17.

Cooper, M. 1993*b*. *Xysma* (Sphecidae) in Ecuador. *Sphecos*, **24**: 17.

Corbet, S.A. 1971. Mandibular gland secretion of larvae of the flour moth, *Anagasta kuehniella* contains an epideictic pheromone and elicits oviposition movements in a hymenopteran parasite. *Nature*, **232**: 481–484.

Corbet, S.A. 1987. More bees make better crops. *New Scientist*, **1570**: 40–43.

Corbet, S.A. & Willmer, P.G. 1980. Pollination of the yellow passionfruit: nectar, pollen and carpenter bees. *Journal of Agricultural Science*, Cambridge, **95**: 655–666.

Cordero, J. & Cave, R.D. 1992. Natural enemies of *Plutella xylostella* (Lep.: Plutellidae) on crucifers in Honduras. *Entomophaga*, **37**: 397–407.

Corn, M.L. 1972 (1973). Notes on the biology of *Polistes carnifex* (Hymenoptera, Vespidae) in Costa Rica and Colombia. *Psyche*, **79**: 150–157.

Cornell, H.V. 1983. The secondary chemistry and complex morphology of galls formed by the Cynipinae (Hymenoptera): why and how? *American Midland Naturalist*, **110**: 225–234.

Cornell, H.V. 1986. Oak species attributes and host size influence cynipine wasp species richness. *Ecology*, **67**: 1582–1592.

Coseglia, A.F., Simpson, R.G. & Eklund, L.R. 1977. Biology of *Mesochorus nigripes*: a hyperparasite of *Bathyplectes* spp. *Annals of the Entomological Society of America*, **70**: 695–698.

Costa Lima, A. da, 1930. Sobre insectos que vivem em maracujás (*Passiflora* spp.). *Memorias do Instituto Oswaldo Cruz*, **23**: 159–162.

Costa Lima, A. da 1936. Sur un nouveau chryside: *Duckeia cyanea*, parasite des oeufs de phasmide, pp. 173–175. *Livre Jubilaire de M. Eugène-Louis Bouvier*. 379pp. Fermin-Didot, Paris.

Costa Lima, A. da 1953. Espécies de *Eiphosoma* Cresson (Hymenopt. Ichneum. Ophion.). *Arquivos do Museu Nacional, Rio de Janeiro*, **42**: 175–189.

Couchman, J.R. & King, P.E. 1977. Morphology of the larval stages of *Diaeretiella rapae* (M'Intosh) (Hymenoptera: Aphidiidae). *International Journal of Insect Morphology & Embryology*, **6**: 127–136.

Coudron, T.A., Kelly, T.J. & Puttler, B. 1990. Developmental responses of *Trichoplusia ni* (Lepidoptera: Noctuidae) to parasitism by the ectoparasite *Euplectrus plathypenae* (Hymenoptera: Eulophidae). *Archives of Insect Biochemistry and Physiology*, **13**: 83–94.

Coville, R.E. 1981*a* (1982). Biological observations on three *Trypoxylon* wasps in the subgenus *Trypargilum* from Costa Rica: *T. nitidum schulthessi, T. saussurei*, and *T. lactitarse* (Hymenoptera: Sphecidae). *Pan-Pacific Entomologist*, **57**: 332–340.

Coville, R.E. 1981*b* (1982). Wasps of the genus *Trypoxylon* subgenus *Trypargilum* in North America. *University of California Publications in Entomology*, **97**: 1–147.

Coville, R.E. & Coville, P.L. 1980. Nesting biology and male behavior of *Trypoxylon (Trypargilum) tenoctitlan* in Costa Rica (Hymenoptera: Sphecidae). *Annals of the Entomological Society of America*, **73**: 110–119.

Coville, R.E., Frankie, G.W., Buchmann, S.L., Vinson, S.B. & Williams, H.J. 1986. Nesting and male behavior of *Centris heithausi* (Hymenoptera: Anthophoridae) in Costa Rica with chemical analysis of the hindleg glands of males. *Journal of the Kansas Entomological Society*, **59**: 325–336.

Coville, R.E., Frankie, G.W. & Vinson, S.B. 1983. Nests of *Centris segregata* (Hymenoptera: Anthophoridae) with a review of the nesting habits of the genus. *Journal of the Kansas Entomological Society*, **56**: 109–122.

Coville, R.E. & Griswold, C. 1983. Nesting biology of *Trypoxylon xanthandrum* in Costa Rica with observations on its spider prey (Hymenoptera: Sphecidae; Araneae: Senoculidae). *Journal of the Kansas Entomological Society*, **56**: 205–216.

Coville, R.E. & Griswold, C. 1984. Biology of *Trypoxylon (Trypargilum) superbum* (Hymenoptera: Sphecidae), a spider-hunting wasp with extended guarding of the brood by males. *Journal of the Kansas Entomological Society*, **57**: 365–376.

Cowan, D.P. 1979. The function of enlarged hind legs in oviposition and aggression by *Chalcis canadensis* (Hymenoptera: Chalcididae). *Great Lakes Entomologist*, **12**: 133–136.

Cowan, D.P. 1981. Parental investment in two solitary wasps *Ancistrocerus adiabatus* and *Euodynerus foraminatus* (Eumenidae: Hymenoptera). *Behavioral, Ecological and Sociobiology*, **9**: 95–102.

Cowan, D.P. 1986. Sexual behavior of eumenid wasps (Hymenoptera: Eumenidae). *Proceedings of the Entomological Society of Washington*, **88**: 531–541.

Cowan, D.P. 1991. The solitary and presocial Vespidae, pp. 33–73. *In*: Ross, K.G. & Matthews, R.W. (eds) *The Social Biology of Wasps*. 678pp. Cornell University Press, Ithaca.

Crandell, H.A. 1939. The biology of *Pachycrepoideus dubius* Ashmead (Hymenoptera), a pteromalid parasite of *Piophila casei* Linne (Diptera). *Annals of the Entomological Society of America*, **32**: 632–654.

Crane, E. 1975. *Honey: a Comprehensive Survey*. 608pp. Heinemann, London.

Crane, E. 1990. *Bees and Beekeeping: Science, Practice and World Resources*. 614pp. Heinemann Newnes, Oxford.

Crane, E. 1992. The past and present status of beekeeping with stingless bees. *Bee World*, **73**: 29–42.

Crane, E. & Graham, A.J. 1985. Bee hives in the ancient world. 2. *Bee World*, **66**: 148–170.

Critchley, B.R. 1973. Parasitism of the larvae of some Carabidae (Coleoptera). *Journal of Entomology* (A) **48**: 37–42.

Cronin, J.T. & Gill, D.E. 1989. The influence of host distribution, sex, and size on the level of parasitism by *Itoplectis conquisitor* (Hymenoptera: Ichnemonidae). *Ecological Entomology*, **14**: 163–173.

Cros, A. 1935. Biologie du *Trichopria stratiomyiae* Kieffer (Hymén. Proctotrypidae). *Bulletin de la Société d'Histoire Naturelle de l'Afrique du Nord*, **26**: 131–136.

Cross, E.A., Stith, M.G. & Bauman, T.R. 1975. Bionomics of the organ-pipe mud-dauber, *Trypoxylon politum* (Hymenoptera: Sphecoidea). *Annals of the Entomological Society of America*, **68**: 901–916.

Cross, W.H. & Mitchell, H.C. 1969. Distribution and importance of *Heterolaccus grandis* as a parasite of the boll weevil. *Annals of the Entomological Society of America*, **62**: 235–236.

Crosskey, R.W. 1953. A revision of the genus *Hyptiogaster* Kieffer (Hymenoptera: Gasteruptionidae), with descriptions of two new genera and three new species. *Transactions of the Royal Entomological Society London*, **104**: 347–384.

Crosskey, R.W. 1962. The classification of the Gasteruptiidae (Hymenoptera). *Transactions of the Royal Entomological Society London*, **114**: 377–402.

Crossman, S.S. 1925. Two imported egg parasites of the gipsy moth *Anastatus bifasciatus* Fonsc. and *Schedius kuvanae* Howard. *Journal of Agricultural Research*, **30**: 643–675.

Crozier, R.H. 1975. Hymenoptera, pp. 1–95. *In*: Bernard, J. (ed.) *Animal Cytogenetics 3. Insecta, 7.* 95pp. Gebrüder Bornträger, Berlin.

Crozier, R.H. 1977. Evolutionary genetics of the Hymenoptera. *Annual Review of Entomology*, **22**: 263–288.

Cruz, Y.P. 1981. A sterile defender morph in a polyembryonic hymenopterous parasite. *Nature*, **294**: 446–447.

Cruz, Y.P. 1986*a*. Development of the polyembryonic parasite *Copidosomopsis tanytmemus* (Hymenoptera: Encyrtidae). *Annals of the Entomological Society of America*, **79**: 121–127.

Cruz, Y.P. 1986*b*. The defender role of the precocious larvae of *Copidosomopsis tanytmemus* Caltagirone (Encyrtidae, Hymenoptera). *Journal of Experimental Zoology*, **237**: 309–318.

Cumming, J.M. & Leggett, F.L. 1985. Cephalic foveae of eumenine wasps (Hymenoptera: Vespidae). *Journal of Natural History*, **19**: 1197–1207.

Curl, G.D. & Burbutis, P.P. 1977. The mode of overwintering of *Trichogramma nubilale* Ertle and Davis. *Environmental Entomology*, **6**: 629–632.

Cushman, R.A. 1916. *Thersilochus conotracheli* a parasite of the plum curculio. *Journal of Agricultural Research*, **6**: 847–855.

Cushman, R.A. 1923. A new subfamily of Braconidae (Hym.) from termite nests. *Proceedings of the Entomological Society of Washington*, **25**: 54–56.

Cushman, R.A. 1931. Notes on ichneumon-flies of the genus *Polycyrtus* with descriptions of new species. *Proceedings of the United States National Museum*, **78**: 1–62.

Cushman, R.A. 1936. The ichneumon-flies of the genus *Brachycyrtus* Kriechbaumer. *Proceedings of the United States National Museum*, **84**: 17–24.

Cushman, R.A. 1945. The ichneumon-flies of the genus *Cyrptanura* Brullé, mainly tropical American. *Proceedings of the United States National Museum*, **96**: 139–176.

Cutler, J.R. 1955. The morphology of the head of the final instar larva of *Nasonia vitripennis* Walker (Hymenoptera: Chalcidoidea). *Proceedings of the Royal Entomological Society of London* (A), **30**: 73–81.

Dafni, A. 1984. Mimicry and deception in pollination. *Annual Review of Ecology and Systematics*, **15**: 259–278.

Dahlman, D.L. 1990. Evaluation of teratocyte functions: an overview. *Archives of Insect Biochemistry & Physiology*, **13**: 159–166.

Dahlman, D.L. 1991. Teratocytes and host/parasitoid interactions. *Biological Control*, **1**: 118–126.

Dahms, E.C. 1984*a*. Revision of the genus *Melittobia* (Chalcidoidea: Eulophidae) with the description of seven new species. *Memoirs of the Queensland Museum*, **21**: 271–336.

Dahms, E.C. 1984*b*. An interpretation of the structure and function of the antennal sense organs of *Melittobia australica* (Hymenoptera: Eulophidae) with the discovery of a large dermal gland in the male scape. *Memoirs of the Queensland Museum*, **21**: 361–385.

Dahms, E.C. 1984*c*. A review of the biology of species in the genus *Melittobia* (Hymenoptera: Eulophidae) with interpretations and additions using observations on *Melittobia australica*. *Memoirs of the Queensland Museum*, **21**: 337–360.

Dalla Torre, K.W. & Kieffer, J.J. 1910. Cynipidae. *Tierreich*, **24**: 1–891.

Daly, H.V. 1964. Skeleto-muscular morphogenesis of the thorax and wings of the honey bee, *Apis mellifera* (Hymenoptera: Apidae). *University of California Publications in Entomology*, **39**: 1–77.

Daly, H.V. 1976. *Leucopsis klugii* (Hymenoptera, Chalcidoidea) reared from *Xylocopa brasilianorum* (Hymenoptera, Apoidea) in Costa Rica. *Pan-Pacific Entomologist*, **52**: 271.

Daly, H.V. 1983. Taxonomy and ecology of Ceratinini of North Africa and the Iberian Peninsula (Hymenoptera: Apoidea). *Systematic Entomology*, **8**: 29–62.

Daly, H.V., Michener, C.D., Moure, J.S. & Sakagami, S.F. 1987. The relictual bee genus *Manuelia* and its relation to other Xylocopinae (Hymenoptera: Apoidea). *Pan-Pacific Entomologist*, **63**: 102–124.

Daly, H.V., Stage, G.I. & Brown, T. 1967. Natural enemies of bees of the genus *Ceratina* (Hymenoptera: Apoidea). *Annals of the Entomological Society of America*, **60**: 1273–1282.

Damman, H. 1986. The osmaterial glands of the swallowtail butterfly *Eurytides marcellus* as a defence against natural enemies. *Ecological Entomology*, **11**: 261–265.

Damman, H. & Cappuccino, N. 1991. Two forms of egg defence in a chrysomelid beetle: egg clumping and excrement cover. *Ecological Entomology*, **16**: 163–167.

Danforth, B.N. & Michener, C.D. 1988. Wing folding in the Hymenoptera. *Annals of the Entomological Society of America*, **81**: 342–349.

Daniel, D.M. 1932. *Macrocentrus ancylivorus* Rohwer, a polyembryonic braconid parasite of the oriental fruit moth. *Technical Bulletin of New York State Agricultural Experimental Station*, **187**: 5–101.

Danks, H.V. 1971*a*. Nest mortality factors in stem-nesting aculeate Hymenoptera. *Journal of Animal Ecology*, **40**: 79–82.

Danks, H.V. 1971*b*. Biology of some stem-nesting aculeate Hymenoptera. *Transactions of the Royal Entomological Society of London*, **122**: 323–399.

Danks, H.V. 1979. Summary of the diversity of Canadian terrestrial arthropods, pp. 240–244. *In*: Danks, H.V. (ed.) Canada and its insect fauna. *Memoirs of the Entomological Society of Canada*, **108**: 1–573.

Danthanarayana, W., Farrugia, D. & Gauld, I.D. 1977. Studies on the biology and systematic position of a new species of ichneumonid parasitising the light brown apple moth, *Epiphyas postvittana* (Walker) (Lepidoptera: Tortricidae), in Australia. *Bulletin of Entomological Research*, **67**: 607–617.

Darling, D.C. 1983. A review of the New World species of *Euperilampus* (Hymenoptera: Chalcidoidea), with notes about host associations and phylogenetic relationships. *Quaestiones Entomologicae*, **19**: 1–40.

Darling, D.C. 1986. Revision of the New World Chrysolampinae (Hymenoptera: Chalcidoidea). *Canadian Entomologist*, **118**: 913–940.

Darling, D.C. 1988. Comparative morphology of the labrum in Hymenoptera: the digitate labrum of Perilampidae and Eucharitidae (Chalcidoidea). *Canadian Journal of Zoology*, **66**: 2811–2835.

Darling, D.C. 1991. Revision of the world species of *Spalangiopelta* (Hymenoptera: Chalcidoidea: Pteromalidae: Ceinae). *Royal Ontario Museum Life Sciences Contributions*, **155**: 1–43.

Darling, D.C. 1992. The life history and larval morphology of *Aperilampus* (Hymenoptera: Chalcidoidea: Philomidinae), with a discussion of the phylogenetic affinities of the Philomidinae. *Systematic Entomology*, **17**: 331–339.

Darling, D.C. & Hanson, P.E. 1986. Two new species of *Spalangiopelta* from Oregon (Hymenoptera: Chalcidoidea), with a discussion of wing length variation. *Pan-Pacific Entomologist*, **62**: 153–164.

Darling, D.C. & Johnson, N.F. 1984. Synopsis of Nearctic Azotinae (Hymenoptera: Aphelinidae). *Proceedings of the Entomological Society of Washington*, **86**: 555–562.

Darling, D.C. & Miller, T.D. 1991. Life history and larval morphology of *Chrysolampus* (Hymenoptera: Chalcidoidea: Chrysolampinae) in western North America. *Canadian Journal of Zoology*, **69**: 2168–2177.

Darling, D.C. & Packer, L. 1988. Effectiveness of Malaise traps in collecting Hymenoptera: the influence of trap design, mesh size, and location. *Canadian Entomologist*, **120**: 787–796.

Darling, D.C. & Werren, J.H. 1990. Biosystematics of *Nasonia* (Hymenoptera: Pteromalidae): two new species reared from birds' nests in North America. *Annals of the Entomological Society of America*, **83**: 352–370.

Dasch, C.E. 1964. The neotropic Diplazontinae (Hymenoptera, Ichneumonidae). *Contributions of the American Entomological Institute*, **1** (1): 1–77.

Dasch, C.E. 1971. Ichneumon-flies of America north of Mexico: 6. Subfamily Mesochorinae. *Memoirs of the American Entomological Institute*, **16**: 1–376.

Dasch, C.E. 1974. Neotropic Mesochorinae (Hymenoptera: Ichneumonidae). *Memoirs of the American Entomological Institute*, **22**: 1–509.

Dasch, C.E. 1992. The Ichneumon-flies of America north of Mexico: Part 12. Subfamilies Microleptinae, Helictinae, Cylloceriinae and Oxytorinae (Hymenoptera: Ichneumonidae). *Memoirs of the American Entomological Institute*, **52**: 1–470.

Dauphin, P. & Castro, J. 1991. *Scleroderma abdominalis* West, responsable de dermatoses dans le Sud-Ouest de la France (Hym. Bethylidae). *L'Entomologiste*, **47**: 327–328.

Daviault, L. 1941. La Chrysomele du Saule. *Naturaliste Canadien*, **68**: 89–111.

Davidson, A. 1905. An enemy of the trap door spider. *Entomological News*, **16**: 233–234.

Davidson, D.W. 1988. Ecological studies of neotropical ant gardens. *Ecology*, **69**: 1138–1152.

Davidson, D.W. & Epstein, W.W. 1989. Epiphytic associations with ants, pp. 200–233. *In*: Lüttge, U. (ed.), *Vascular Plants as Epiphytes. Evolution and Ecophysiology*. 270pp. Springer-Verlag, New York.

Davidson, D.W. & McKey, D. 1993. The evolutionary ecology of symbiotic ant-plant relationships. *Journal of Hymenoptera Research*, **2**: 13–83

Davies, D.H., Strand, M.R. & Vinson, S.B. 1987. Changes in differential haemocyte count and *in vitro* behaviour of plasmatocytes from host *Heliothis virescens* caused by *Campoletis sonorensis* polydnavirus. *Journal of Insect Physiology*, **33**: 143–153.

Davies, D.H. & Vinson, S.B. 1986. Passive evasion by eggs of braconid parasitoid *Cardiochiles nigriceps* of encapsulation *in vitro* by haemocytes of host *Heliothis virescens*. Possible role for fibrous layer in immunity. *Journal of Insect Physiology*, **32**: 1003–1010.

Davies, M. & Kathirithamby, J. 1986. *Greek Insects*. 211pp. Duckworth, London.

Davis, H.G. 1978. Yellowjacket wasps in urban environments, pp. 163–185. *In*: Frankie, G.W. & Koehler, C.S.(eds), *Perspectives in Urban Entomology*. 417pp. Academic Press, New York.

Davis, M.A. 1987. The role of flower visitors in the explosive pollination of *Thalia geniculata* (Marantaceae), a Costa Rican marsh plant. *Bulletin of the Torrey Botanical Club*, **114**: 134–138.

Day, M.C. 1977. A new genus of Plumariidae from Southern Africa, with notes on Scolebythidae (Hymenoptera: Chrysidoidea). *Cimbebasia* (A), **4**: 171–177.

Day, M.C. 1978 (1979). The affinities of *Loboscelidia* Westwood (Hymenoptera: Chrysididae, Loboscelidiinae). *Systematic Entomology*, **4**: 21–30.

Day, M.C. 1981. A revision of *Pompilus* Fabricius (Hymenoptera: Pompilidae), with further nomenclatural and biological considerations. *Bulletin of the British Museum (Natural History)* Entomology, **42**: 1–42.

Day, M.C. 1984. Male polymorphism in some Old World species of *Cryptocheilus* Panzer (Hymenoptera: Pompilidae). *Zoological Journal of the Linnean Society*, **80**: 83–101.

Day, M.C. 1988. Spider wasps Hymenoptera: Pompilidae. *Handbooks for the Identification of British Insects*, **6**(4): 1–60.

Day, M.C., Else, G.R. & Morgan, D. 1981. The most primitive Scoliidae (Hymenoptera). *Journal of Natural History*, **15**: 671–684.

Dean, H.A., Schuster, M.F., Boling, J.C. & Riherd, P.T. 1979. Complete biological control of *Antonina graminis* in Texas with *Neodusmetia sangwani* (a classic example). *Bulletin of the Entomological Society of America*, **25**: 262–267.

DeBach, P. 1939. *Microterys titiani* Gir., an egg predator of *Leucanium corni* Bouché *Journal of Economic Entomology*, **32**: 728–729.

DeBach, P. 1953. *Thysanus flavopalliatus* (Ashm.) parasitic on *Comperiella bifasciata* How. in California red scale. *Journal of Economic Entomology*, **46**: 1112.

DeBach, P. (ed.) 1964. *Biological Control of Insect Pests and Weeds*. 844pp. Chapman & Hall, New York.

DeBach, P. 1971. Principios y posibilidades del control biologico de las plagas. *Boletin de la Sociedad Entomológica del Perú*, **6**: 39–47.

DeBach, P. 1974. *Biological Control by Natural Enemies*. 323pp. Cambridge University Press, Cambridge.

DeBach, P., Kennett, C.E. & Pence, R.J. 1958. Species of *Thysanus* as primary parasites. *Journal of Economic Entomology*, **51**: 114–115.

Debauche, H.R. 1948. Étude sur les Mymarommidae et les Mymaridae de la Belgique (Hymenoptera Chalcidoidea). *Mémoires du Musée Royal d'Histoire Naturelle de Belgique*, **108**: 1–248.

Debolt, J.W. 1981. Laboratory biology and rearing of *Leiophron uniformis* (Gahan) (Hymenoptera: Braconidae), a parasite of *Lygus* spp. (Hemiptera: Miridae). *Annals of the Entomological Society of America*, **74**: 334–337.

Debolt, J.W. 1991. Behavioral avoidance of encapsulation by *Leiophron uniformis* (Hymenoptera: Braconidae), a parasitoid of *Lygus* spp. (Hemiptera: Miridae): relationship between host age, encapsulating ability, and host acceptance. *Annals of the Entomological Society of America*, **84**: 444–446.

Dejean, A. & Lachaud, J.-P. 1992. Growth-related changes in predation behavior in incipient colonies of the ponerine ant *Ectatomma tuberculatum* (Olivier). *Insectes Sociaux*, **39**: 129–143.

DeJong, D., Morse, R.A. & Eickwort, G.C. 1982. Mite pests of honeybees. *Annual Review of Entomology*, **27**: 229–252.

DeJong, R. & Kaiser, L. 1991. Odor learning by *Leptopilina boulardi*, a specialist parasitoid (Hymenoptera: Eucoilidae). *Journal of Insect Behavior*, **4**: 743–750.

Delanoue, P. & Arambourg, Y. 1965. Contribution a l'étude en laboratoire d'*Eupelmus urozonus* Dalm. (Hym. Chalcidoidea Eupelmidae). *Annals de la Société Entomologique de France* (N.S.), **1**: 817–842.

Delage-Darchen, B. & Darchen, R. 1982. Les enzymes digestives des glandes salivaires et de l'intestin moyen d'une abeille sociale du Mexique, *Melipona beecheii* (B.). *Annales des Sciences Naturales, Zoologie*, (13) **4**: 91–96.

DeLeon, D. 1933. Notes on the biology of *Meteorus hypophloei* Cushman (Hymenoptera—Braconidae). *Bulletin of the Brooklyn Entomological Society*, **28**: 32–36.

DeLeon, D. 1935. The biology of *Coeloides dendroctoni* Cushman (Hymenoptera—Braconidae) an important parasite of the mountain pine beetle (*Dendroctonus monticolae* Hopk.). *Annals of the Entomological Society of America*, **28**: 411–424.

Delvare, G. 1992. A reclassification of the Chalcidini with a checklist of the New World species, pp. 119–441. *In*: Delvare, G. & Bouček, Z. On the New World Chalcididae (Hymenoptera). *Memoirs of the American Entomological Institute*, **53**: 1–466.

Delvare, G. 1993. Les Chalcididae d'importance économique dans les palmeraies d'Amérique tropicale (Hymenoptera). *Bulletin de la Société Entomologique de France*, **97**: 349–372.

Delvare, G. & Genty, P. 1992. Intérêt des plantes attractives pour la faune auxiliaire dans les palmeraies d'Amérique tropicale. *Oléagineux*, **47**: 551–559.

Delvare, G. & LaSalle, J. 1993. A new genus of Tetrastichinae (Hymenoptera: Eulophidae) from the Neotropical region, with the description of a new species parasitic on key pests of oil palm. *Journal of Natural History*, **27**: 435–444.

Delvare, G., Mirkarimi, A.A. & Nguyen, T.X. 1981. Redescription de *Prionomitus mitratus* Dalman (Hym. Encyrtidae) endoparasite de psylles (Hom. Psyllidae). *Annales de la Société Entomologique de France* (N.S.), **17**: 171–178.

Denlinger, D.L. 1986. Dormancy in tropical insects. *Annual Review of Entomology*, **31**: 239–264.

Dennill, G.B. 1988. Why a gall former can be a good biocontrol agent: the gall wasp *Trichilogaster acaciaelongifoliae* and the weed *Acacia longifolia*. *Ecological Entomology*, **13**: 1–9.

De Santis, L. 1946. Taxonomía de la familia Aphelinidae (Hymenoptera, Chalcidoidea). *Revista del Museo de La Plata* (N.S.) Zoologia, **31**: 1–21.

De Santis, L. 1948. Estudio monográfico de los Afelínidos de la República Argentina (Hymenoptera, Chalcidoidea). *Revista del Museo de La Plata* (N.S.) Zoologia, pp. 23–280.

De Santis, L. 1964. Encirtidos de la Republica Argentina (Hymenoptera: Chalcidoidea). *Anales de la Comisión de Investigación Cientifica. Provincia de Buenos Aires Gobernación*, **4**: 9–422.

De Santis, L. 1967. *Catálogo de los Himenópteros Argentinos de la Serie Parasitica, Incluyendo Bethyloidea*, 337 pp. Provincia de Buenos Aires Gobernacion, Comision de Investigacion Cientifica, La Plata.

De Santis, L. 1979. *Catálogo de los Himenópteros Calcidoideos de América al Sur de los Estados Unidos*. 488pp. Comisión de Investigaciones Cientificas de la Provincia de Buenos Aires. La Plata.

De Santis, L. 1980. *Catálogo de los Himenópteros Brasileños de la Serie Parasitica Incluyendo Bethyloidea*. 395pp. Editora da Universidade Federal do Paraná, Curitiba.

De Santis, L. 1981 (1983). Catálogo do los himenópteros calcidoideos de America al sur de los Estados Unidos—Primer Suplemento. *Revista Peruana de Entomologia*, **24**: 1–38.

De Santis, L. 1983. Las espéces Argentinas, Uruguayas y Brasileñas del género *Emersonella* Girault, 1916 (Insecta, Hymenoptera, Eulophidae). *Anais da Sociedade Entomológica do Brasil*, **12**: 249–259.

De Santis, L. 1987 (1985). Las especies peruanas de *Halticoptera* (Insecta, Hymenoptera, Pteromalidae). *Revista Peruana de Entomología*, **28**: 1–3.

De Santis, L. 1988. Tres calcidoideos (Hymenoptera) Brasileños parasitoides de los huevos de *Dirphia araucariae* Jones, 1908 (Lepidoptera, Attacidae). *Anais de Sociedade Entomólogica do Brasil*, **17**: 165–171.

De Santis, L. 1989. Catálogo de los himenópteros calcidoideos (Hymenoptera) al sur de los Estados Unidos. Segundo Suplemento. *Acta Entomología Chilena*, **15**: 9–90.

Desmier de Chenon, R. 1973. Etude d'*Hispoleptis* et de ses parasites. *Rapport de Mission Entomologique en Colombie et Equateur* (6.09.1972–11.02.1973). 66pp. I.N.R.A., Versailles.

Dessart, P. 1972 Révision des espèces européennes du genre *Dendrocerus* Ratzeburg, 1852 (Hymenoptera Ceraphronoidea). *Mémoires de la Société Royale Belge d'Entomologie*, **32**: 1–310.

Dessart, P. 1973. *Dendrocerus propodealis* sp. n. (Hym. Ceraphronoidea) parasite de *Chrysopa madestes* Banks (Neur. Chrysopidae) en Inde. *Bulletin et Annals de la Société Royale d'Entomologie Belgique*, **109**: 269–276.

Dessart, P. 1974. Les Mégaspilides européens (Hym. Ceraphronoidea) parasites des Diptères Syrphides avec une révision du genre *Trichosteresis*. *Annals de la Société Entomologique de France* (N.S.), **10**: 395–448.

Dessart, P. 1978. Four new species of African Ceraphronidae (Hymenoptera). *Journal of the Entomological Society of Southern Africa*, **41**: 275–284.

Dessart, P. 1979. Ceraphronoidea nord-americains nouveaux ou peu connus (Hymenoptera). *Bulletin & Annals de la Société Royale Belge d'Entomologique*, **115**: 147–159.

Dessart, P. 1987. *Dendrocerus pallocrates* sp. n., d'Amérique du Sud (Hymenoptera Ceraphronoidea Megaspilidae). *Bulletin & Annales de la Société Royale Belge d'Entomologie*, **123**: 137–140.

Dessart, P. 1988. *Aphanogmus goniozi* sp.n., hyperparasite d'un Béthylidé au Sri Lanka (Hymenoptera Ceraphronoidea Ceraphronidae). *Bulletin & Annals de la Société Royale Belge d'Entomologie*, **124**: 99–104.

Dessart, P. 1990. *Dendrocerus chloropidarum* n.sp., nouvelle espèce européenne à notaulices incomplètes (Hymenoptera Ceraphronoidea Megaspilidae). *Bulletin de l'Institut Royal des Sciences Naturelles de Belgique*, Entomologie, **60**: 69–70.

Dessart, P. 1992. L'organe de Waterston des Ceraphronidae (Hymenoptera Ceraphronoidea). *Bulletin et Annals de la Société Royale d'Entomologie Belgique*, **128**: 203–212.

Dessart, P. & Bournier, A. 1971. *Thrips tabaci* Lindman (Thysanoptera) hôte inattendu d'*Aphanogmus fumipennis* (Thomson) (Hym. Ceraphronidae). *Bulletin & Annales de la Société Royale d'Entomologique de Belgique*, **107**: 116–118.

Dessart, P. & Cancemi, P. 1986 (1984–5). Tableau dichotomique des genres de Ceraphronoidea (Hymenoptera) avec commentaires et nouvelles espèces. *Frustula Entomologica* (N.S.), **7–8**: 307–372.

DeVries, P.J. 1977. *Eumaeus minyas*, an aposematic lycaenid butterfly. *Brenesia*, **12**: 269–270.

DeVries, P.J. 1987. *The Butterflies of Costa Rica and Their Natural History*. 327 pp. Princeton University Press, Princeton.

DeVries, P.J. 1990. Enhancement of symbioses between butterfly caterpillars and ants by vibrational communication. *Science*, **248**: 1104–1106.

DeVries, P.J. 1991*a*. Mutualism between *Thisbe irenea* butterflies and ants, and the role of ant ecology in the evolution of larval-ant associations. *Biological Journal of the Linnean Society*, **43**: 179–195.

DeVries, P.J. 1991*b*. Evolutionary and ecological patterns in myrmecophilous riodinid butterflies, pp. 143–156. *In*: Huxley, C.R. & Cutler, D.F. (eds) *Ant-Plant Interactions*. 601pp. Oxford University Press, Oxford.

DeVries, P.J., Cocroft, R.B. & Thomas, J. 1993. Comparison of acoustical signals in *Maculinea* butterfly caterpillars and their obligate host *Myrmica* ants. *Biological Journal of the Linnean Society*, **49**: 229–238.

Deyrup, M.A. 1984. A maple wood wasp, *Xiphydria maculata*, and its insect enemies (Hymenoptera: Xiphydriidae). *Great Lakes Entomologist*, **17**: 17–28.

Dias, B.F. de Souza, 1975. Comportamento pré-social de Sínfitas do Brasil Central. I. *Themos olfersii* (Klug) (Hymenoptera, Argidae). *Studia Entomologica* (N.S.), **18**: 401–432.

Dias, B.F. de Souza, 1976. Comportamento pré-social de Sínfitas do Brasil Central. II. *Dielocerus diasi* Smith, 1975 (Hymenoptera, Argidae). *Studia Entomologica* (N.S.), **19**: 461–501.

Dias, D. 1958. Contribuiçao para o conhecimento da bionomia de *Bombus incarum* Franklin da Amazônia (Hymenóptera: Bombidae). *Revista Brasileira de Entomologia*, **8**: 1–20.

Diaz, N.B. 1973. Una familia de Cynipoidea nueva para la Argentina. *Neotropica*, **19**: 141–144.

Diaz, N.B. 1975. Anotaciones sobre cinipoideos Argentinos III (Hymenoptera). *Neotropica*, **21**: 32–36.

Diaz, N.B. 1978. Estudio ecologico y sistematico de cinipoideos neotropicales IV (Hymenoptera Cynipidae). *Neotropica*, **24**: 123–125.

Diaz, N.B. 1980. Nota sobre los Alloxystinae de la república Argentina (Hym. Cynipoidea). *Revista de la Sociedad Entomológica Argentina*, **39**: 15–18.

Diaz, N.B. 1981 (1980). Cinipoideos galigenos e inquilinos de la republica Argentina. *Revista de la Sociedad Entomológica Argentina*, **39**: 221–226.

Diaz, N.B. 1984 (1985). Revisión del género *Prosaspicera* Kieffer, 1907 (Hymenoptera: Cynipoidea). *Revista de la Sociedad Entomológica Argentina*, **43**: 221–238.

Diaz, N.B. 1987 (1988). Presencia de *Leptopilina boulardi* en la Republica Argentina (Cyn. Eucoilidae). *Neotropica*, **33**: 36.

Díaz, N.B. & Valladares, G. 1979. Nota sobre *Agrostocynips clavatus* y los agromicidos hospedantes (Hymenoptera Cynipoidea). *Neotropica*, **25**: 23–25.

Dicke, M., Sabelis, M.W., Takabayashi, J., Bruin, J. & Posthumus, M.A. 1990. Plant strategies of manipulating predator-prey interactions through allelochemicals: prospects for application in pest control. *Journal of Chemical Ecology*, **16**: 3091–3118.

Dijken, M.J. van & Waage, J.K. 1987. Self and conspecific superparasitism by the egg parasitoid *Trichogramma evanescens*. *Entomologia Experimentalis et Applicata*, **43**: 183–192.

Dijkstra, L.J. 1986 (1987). Optimal selection and exploitation of hosts in the parasitic wasp *Colpoclypeus florus* (Hym., Eulophidae). *Netherlands Journal of Zoology*, **36**: 177–301.

Diniz, J.L.M. 1990. Systematic revision of the tribe Stegomyrmicini, with description of a new species (Hymenoptera, Formicidae). *Revista Entomologica do Brasil*, **34**: 277–295.

Disney, R.H.L. 1991. The fire-ant parasitoids of the *Pseudacteon spatulatus* complex (Diptera, Phoridae; Hymenoptera, Formicidae). *Sociobiology*, **18**: 283–298.

Dittrick, L.E. & Chiang, H.C. 1982. Developmental characteristics of *Macrocentrus grandii* as influenced by temperature and instar of its host, the European corn borer. *Journal of Insect Physiology*, **28**: 47–52.

Dixon, A.F.G., Kindlmann, P., Leps, J. & Holman, J. 1987. Why there are so few species of aphids, especially in the tropics. *American Naturalist*, **129**: 580–592.

Dmoch, J., Lewis, W.J., Martin, P.B. & Nordlund, D.A. 1985. Role of host-produced stimuli and learning in host selection behavior of *Cotesia* (= *Apanteles*) *marginiventris* (Cresson). *Journal of Chemical Ecology*, **11**: 453–463.

Doane, J.F., DeClerck-Floate, R., Arthur, A.P. & Affolter, F. 1989. Description of the life cycle stages of *Macroglenes penetrans* (Kirby) (Hymenoptera: Chalcidoidea, Pteromalidae), a parasitoid of the wheat midge, *Sitodiplosis mosellana* (Géhin) (Diptera: Cecidomyiidae). *Canadian Entomologist*, **121**: 1041–1048.

Dodson, C.H. 1966. Ethology of some bees of the tribe Euglossini (Hymenoptera: Apidae). *Journal of the Kansas Entomological Society*, **39**: 607–629.

Dodson, C.H. 1975. Coevolution of orchids and bees, pp. 91–99. *In*: Gilbert, L.E. & Raven, P.H. (eds) *Coevolution of Animals and Plants*. 246pp. University of Texas Press, Austin.

Domenichini, G. 1951. Parassiti e iperparassiti di *Pseudococcus citri* Risso in Italia e nel Perù. *Bollettino di Zoologia Agraria e Bachicoltura*, **17**: 157–180.

Domenichini, G. 1956 (1957). Contributo alla conoscenza dei parassiti e iperparassiti dei Coleoptera Coccinellidae. *Bollettino di Zoologia Agraria e Bachicoltura*, **22**: 215–246.

Dondale, C.D. 1954. Biology of *Agathis laticinctus* (Cress.) (Hymenoptera: Braconidae), a parasite of the eye-spotted bud moth, in Nova Scotia. *Canadian Entomologist*, **86**: 40–44.

Doner, M.H. 1934. Observations on the biology of *Microbracon pygmaeus* (Prov.), an important parasite of *Coleophora pruniella* Cl. *Annals of the Entomological Society of America*, **27**: 435–442.

Donnelly, T.W. 1992. Geological setting and tectonic history of Mesoamerica, pp. 1–13. *In*: Quintero, D. & Aiello, A. (eds), *Insects of Panama and Mesoamerica. Selected Studies*. 692pp. Oxford University Press, Oxford.

Donovan, B.J. 1991. Life cycle of *Sphecophaga vesparum* (Curtis) (Hymenoptera: Ichneumonidae), a parasitoid of some vespid wasps. *New Zealand Journal of Zoology*, **18**: 181–192.

Dotimas, E.M. & Hider, R.C. 1987. Honeybee venom. *Bee World*, **68**: 51–70.

Doutt, R.L. 1952. The teratoid larva of polyembryonic Encyrtidae (Hymenoptera). *Canadian Entomologist*, **84**: 247–250.

Doutt, R.L. 1973. Maternal care of immature progeny by parasitoids. *Annals of the Entomological Society of America*, **66**: 486–487.

Doutt, R.L., Annecke, D.P. & Tremblay, E. 1976. Biology and host relationships of parasitoids, pp. 143–168. *In*: Huffaker, C.B. & Messenger, P.S. (eds) *Theory and Practice of Biological Control*. 788pp. Academic Press, New York.

Doutt, R. L. & Viggiani, G. 1968. The classification of the Trichogrammatidae (Hymenoptera: Chalcidoidea). *Proceedings of the California Academy of Sciences*. 4th Series, **35**: 477–586.

Dover, B.A., Davies, D.H., Strand, M.R., Gray, R.S., Keeley, L.L. & Vinson, S.B. 1987. Ecdysteroid-titre reduction and developmental arrest of last-instar *Heliothis virescens* larvae by calyx fluid from the parasitoid *Campoletis sonorensis*. *Journal of Insect Physiology*, **33**: 333–338.

Dover, B.A., Strand, M.R., Davies, D.H. & Vinson, S.B. 1989. Ultrastructure of host tissues exposed to the calyx fluid of the parasitoid, *Campoletis sonorensis* (Cameron) (Hymenoptera: Ichneumonidae). *International Journal of Insect Morphology and Embryology* **18**: 47–57.

Dover, B.A. & Vinson, S.B. 1990. Stage-specific effects of *Campoletis sonorensis* parasitism on *Heliothis virescens* development and prothoracic glands. *Physiological Entomology*, **15**: 405–414.

Dowden, P.B. 1935. *Brachymeria intermedia* (Nees), a primary parasite, and *B. compsilurae* (Cwfd.), a secondary parasite, of the gypsy moth. *Journal of Agricultural Research*, **50**: 495–523.

Dowden, P.B. 1939. *Schizonotus sieboldi*, an important parasite of the imported willow leaf beetle (*Plagiodera versicolora*). *Journal of Agricultural Research*, **58**: 581–592.

Dowell, R.V. 1979. Synchrony and impact of *Amitus hesperidum* (Hym.: Platygasteridae) on its host, *Aleurocanthus woglumi* (Hom.: Aleyrodidae) in southern Florida. *Entomophaga*, **24**: 221–227.

Downhower, J.F. 1975. The distribution of ants on *Cecropia* leaves. *Biotropica*, **7**: 59–62.

Downhower, J.F. & Wilson, D.E. 1973. Wasps as a defense mechanism of katydids. *American Midland Naturalist*, **89**: 451–455.

Downing, H.A. 1991. The function and evolution of exocrine glands, pp. 540–569. *In*: Ross, K.G. & Matthews, R.W. (eds) *The Social Biology of Wasps*. 678pp. Cornell University Press, Ithaca.

Downing, H.A., Post, D.C. & Jeanne, R.L. 1985. Morphology of sternal glands in male polistine wasps (Hymenoptera: Vespidae). *Insectes Sociaux*, **32**: 186–197.

Dozier, H.L. 1932. Two important West Indian seed-infesting Chalcid wasps. *Journal of the Department of Agriculture of Puerto Rico*, **16**: 103–112.

Dreisbach, R.R. 1960. Seventeen new species in the genus *Priocnessus* (Hymenoptera: Psammocharidae) with keys to males and females of all known Neotropical species. *American Midland Naturalist*, **63**: 335–364.

Dreisbach, R.R. 1961. Additional new species in the genus *Priocnessus* Banks (Hymenoptera: Psammocharide) with photomicrographs of genitalia of all the new males. *American Midland Naturalist*, **65**: 215–235.

Dreisbach, R.R. 1963. New species of spider wasps, genus *Auplopus*, from the Americas south of the United States. *Proceedings of the United States National Museum*, **114**: 137–211.

Dressler, R.L. 1981. *The Orchids Natural History and Classification*, 332pp. Harvard University Press, Cambridge, Mass.

Dressler, R.L. 1982. Biology of the orchid bees (Euglossini). *Annual Review of Ecology and Systematics*, **13**: 373–394.

Driesche, R.G., van, 1988. Field measurement of population recruitment of *Apanteles glomeratus* (L.) (Hymenoptera: Braconidae), a parasitoid of *Pieris rapae* (L.) (Lepidoptera: Pieridae), and factors influencing adult parasitoid foraging success in kale. *Bulletin of Entomological Research*, **78**: 199–208.

Driesche, R.G. van, Bellotti, A., Herrera, C.J. & Castillo, J.A. 1986. Encapsulation rates of two encyrtid parasitoids by two *Phenacoccus* spp. of cassava mealybugs in Colombia. *Entomologia Experimentalis et Applicata*, **42**: 79–82.

Drooz, A.T., Bustillo, A.E., Fedde, G.F. & Fedde, V.H. 1977. North American egg parasite successfully controls a different host genus in South America. *Science*, **197**: 390–391.

D'Rozario, A.M. 1942. On the development and homologies of the genitalia and their ducts in Hymenoptera. *Transactions of the Royal Entomological Society, London*, **92**: 363–415.

DuBois, M.B. 1986. A revision of the native New World species of the ant genus *Monomorium* (*minimum* group) (Hymenoptera: Formicidae). *University of Kansas Science Bulletin*, **53**: 65–119.

Ducke, A. 1914. Über Phylogenie und Klassifikation der sozialen Vespiden. *Zoologische Jahrbücher Abteilung für Systematik, Geographie und Biologie der Tiere*, **36**: 303–330.

Dudarenko, G.P. 1974. Formation of the abdominal carapace in braconids (Hymenoptera, Braconidae) and some aspects of the classification of the family. *Entomological Review*, **53**: 80–90.

Duffey, S.S., Bloem, K.A. & Campbell, B.C. 1986. Consequences of sequestration of plant natural products in plant-insect-parasitoid interactions, pp. 31–60. *In*: Boethel, D.J. & Eikenbary, R.D. (eds) *Interactions of Plant Resistance and Parasitoids and Predators of Insects*. 224pp. Ellis Horwood, Chichester.

Duffield, R.M., Wheeler, J.W. & Eickwort, G.C. 1984. Sociochemicals of bees, pp. 387–428. *In*: Bell, W.J. & Cardé, R.T. (eds) *Chemical Ecology of Insects*. 524pp. Chapman & Hall, London.

Dunn, J.A. 1949. The parasites and predators of potato aphids. *Bulletin of Entomological Research*, **40**: 97–122.

Duodu, Y.A. & Davis, D.W. 1974. Selection of alfalfa weevil instars by, and mortality due to, the parasite *Bathyplectes curculionis* (Thomson). *Environmental Entomology*, **3**: 549–552.

DuPorte, E.M. & Bigelow, R.S. 1953. The clypeus and the epistomal suture in Hymenoptera. *Canadian Journal of Zoology*, **31**: 20–29.

Dysart, R.J., Maltby, H.L. & Brunson, M.H. 1973. Larval parasites of *Oulema melanopus* in Europe and their colonization in the United States. *Entomophaga*, **18**: 133–167.

Dzhanokmen, K.A. 1990. Trophic links of parasitic Hymenoptera of the family Pteromalidae (Chalcidoidea). *Entomologicheskoe Obozrenie*, **69**: 764–781. [In Russian: English translation, *Entomological Review*, **70**(5): 45–66.]

Eady, R.D. & Quinlan, J. 1963. Hymenoptera, Cynipoidea. Key to families and subfamilies and Cynipinae (including galls). *Handbooks for the Identification of British Insects*, **8**, 1a: 1–81.

Eastham, L.E.S. 1929. The post-embryonic development of *Phaenoserphus viator* Hal. (Proctotrypoidea), a parasite of the larva of *Pterostichus niger* (Carabidae), with notes on the anatomy of the larva. *Parasitology*, **21**: 1–21.

Eberhard, W.G. 1972. Altruistic behavior in a sphecid wasp: support for kin-selection theory. *Science*, **175**: 1390–1391.

Eberhard, W.G. 1974. The natural history and behaviour of the wasp *Trigonopsis cameronii* Kohl (Sphecidae). *Transactions of the Royal Entomological Society of London*, **125**: 295–328.

Eberhard, W.G. 1975. The ecology and behavior of a subsocial pentatomid bug and two scelionid wasps: strategy and counterstrategy in a host and its parasites. *Smithsonian Contributions to Zoology*, **205**: 1–39.

Eberhard, W.G. 1978. Mating swarms of a South American *Acropygia* (Hymenoptera: Formicidae). *Entomological News*, **89**: 14–16.

Eberhard, W.G. 1985. *Sexual Selection and Animal Genitalia*. 244pp. Harvard University Press, Cambridge Massachusetts.

Eberhard, W.G. 1989 (1988). Group nesting in two species of *Euglossa* bees (Hymenoptera: Apidae). *Journal of the Kansas Entomological Society*, **61**: 406–411.

Eberhard, W.G. 1990. Copulatory courtship in the wasp *Eumenes consobrinus* (Hymenoptera, Vespidae). *Journal of the Kansas Entomological Society*, **63**: 342–343.

Edgerly, J.S. 1987. Maternal behaviour of a webspinner (order Embiidina). *Ecological Entomology*, **12**: 1–11.

Edmunds, L.R. 1952. Some notes on the habits and parasites of native wood-roaches in Ohio (Orthoptera: Blattidae). *Entomological News*, **63**: 141–145.

Edmunds, L.R. 1954. A study of the biology and life history of *Prosevania punctata* (Brullé) with notes on additional species (Hymenoptera: Evaniidae). *Annals of the Entomological Society of America*, **47**: 575–592.

Edson, K.M. & Vinson, S.B. 1977. Nutrient absorption by the anal vesicle of the braconid wasp, *Microplitis croceipes*. *Journal of Insect Physiology*, **23**: 5–8.

Edson, K.M. & Vinson, S.B. 1979. A comparative morphology of the venom apparatus of female braconids (Hymenoptera: Braconidae). *Canadian Entomologist*, **111**: 1013–1024.

Edson, K.M., Vinson, S.B., Stoltz, D.B. & Summers, M.D. 1981. Virus in a parasitoid wasp: suppression of the cellular immune response in the parasitoid's host. *Science*, **211**: 582–583.

Edwards, R. 1980. *Social Wasps. Their Biology and Control*. 398pp. Rentokil Limited, East Grinstead.

Efem, S.E.E. 1988. Clinical observations on the wound healing properties of honey. *British Journal of Surgery*, **75**: 679–681.

Eggleton, P. 1989. The Phylogeny and Evolutionary Biology of the Pimplinae (Hymenoptera: Ichneumonidae). 295pp. Unpublished PhD thesis, University of London.

Eggleton, P. 1990. Male reproductive behaviour of the parasitoid wasp *Lytarmes maculipennis* (Hymenoptera: Ichneumonidae). *Ecological Entomology*, **15**: 357–360.

Eggleton, P. & Belshaw, R. 1992. Insect parasitoids: an evolutionary overview. *Philosophical Transactions of the Royal Society, London* (B), **337**: 1–20.

Eggleton, P. & Belshaw, R. 1993. Comparisons of dipteran, hymenopteran and coleopteran parasitoids: provisional phylogenetic explanations. *Biological Journal of the Linnean Society*, **48**: 213–226.

Eggleton, P. & Gaston, K.J. 1990. 'Parasitoid' species and assemblages: convenient definitions or misleading compromises? *Oikos*, **59**: 417–421.

Ehler, L.E. 1989. Observations on *Scutellista cyanea* Motsch. (Hymenoptera: Pteromalidae). *Pan-Pacific Entomologist*, **65**: 151–155.

Eickwort, G.C. 1967. Aspects of the biology of *Chilicola ashmeadi* in Costa Rica (Hymenoptera: Colletidae). *Journal of the Kansas Entomological Society*, **40**: 42–73.

Eickwort, G.C. 1969*a*. A comparative morphological study and generic revision of the augochlorine bees (Hymenoptera: Halictidae). *University of Kansas Science Bulletin*, **48**: 325–524.

Eickwort, G.C. 1969*b*. Tribal positions of Western Hemisphere green sweat bees, with comments on their nest architecture (Hymenoptera: Halictidae). *Annals of the Entomological Society of America*, **62**: 652–660.

Eickwort, G.C. 1979. Mites associated with sweat bees (Halictidae), pp. 575–581. *In*: Rodriguez, J.G. (ed.) *Recent Advances in Acarology*, I. 631pp. Academic Press, New York.

Eickwort, G.C. 1981. Presocial insects, pp. 199–280. *In*: Hermann, H.R. (ed.) *Social Insects*, 2. 491pp. Academic Press, New York.

Eickwort, G.C. & Eickwort, K.R. 1969. Aspects of the biology of Costa Rican halictine bees, I. *Agapostemon nasutus* (Hymenoptera: Halictidae). *Journal of the Kansas Entomological Society*, **42**: 421–452.

Eickwort, G.C. & Eickwort, K.R. 1971. Aspects of the biology of Costa Rican halictine bees, II. *Dialictus umbripennis* and adaptations of its caste structure to different climates. *Journal of the Kansas Entomological Society*, **44**: 343–373.

Eickwort, G.C. & Eickwort, K.R. 1972*a*. Aspects of the biology of Costa Rican halictine bees, IV. *Augochlora* (*Oxystoglossella*) (Hymenoptera: Halictidae). *Journal of the Kansas Entomological Society*, **45**: 18–45.

Eickwort, G.C. & Eickwort, K.R. 1972*b*. Aspects of the biology of Costa Rican halictine bees, III. *Sphecodes kathleenae*, a social cleptoparasite of *Dialictus umbripennis*. *Journal of the Kansas Entomological Society*, **45**: 529–541.

Eickwort, G.C. & Eickwort, K.R. 1973*a*. Aspects of the biology of Costa Rican halictine bees, V. *Augochlorella edentata* (Hymenoptera: Halictidae). *Journal of the Kansas Entomological Society*, **46**: 3–16.

Eickwort, G.C. & Eickwort, K.R. 1973*b*. Notes on the nests of three wood-dwelling species of *Augochlora* from Costa Rica (Hymenoptera: Halictidae). *Journal of the Kansas Entomological Society*, **46**: 17–22.

Eickwort, G.C. & Ginsberg, H.S. 1980. Foraging and mating behavior in Apoidea. *Annual Review of Entomology*, **25**: 421–446.

Eickwort, G.C. & Sakagami, S.F. 1979. A classification of nest architecture of bees in the tribe Augochlorini (Hymenoptera: Halictidae; Halictinae), with description of a Brazilian nest of *Rhinocorynura inflaticeps*. *Biotropica*, **11**: 28–37.

Eickwort, G.C. & Stage, G.I. 1972. A new subgenus of Neotropical *Sphecodes* cleptoparasitic upon *Dialictus* (Hymenoptera: Halictidae, Halictinae). *Journal of the Kansas Entomological Society*, **45**: 500–515.

Eijsackers, H.J.P. & Bakker, K. 1971. Elimination by physical attack of supernumerary larvae of *Pseudeucoila bochei* Weld (Cynipidae) in their hosts, larvae of *Drosophila*. *Netherlands Journal of Zoology*, **21**: 205–207.

Ejlersen, A. 1978. The spatial distribution of spangle galls (*Neuroterus* spp.) on Oak (Hymenoptera, Cynipidae). *Entomologiske Meddelelser*, **46**: 19–25.

Eliescu, G. 1932. Beiträge zur Kenntnis der Morphologie, Anatomie und Biologie von *Lophyrus pini* L. *Zeitschrift für angewandte Entomologie*, **19**: 22–67, 188–206.

Eller, F.J., Bartlett, R.J., Jones, R.L. & Kulman, H.M. 1984. Ethyl (Z)-9-hexadecenoate: a sex pheromone of *Syndipnus rubiginosus*, a sawfly parasitoid. *Journal of Chemical Ecology*, **10**: 291–300.

Eller, F.J., Tumlinson, J.H. & Lewis, W.J. 1988. Beneficial arthropod behavior mediated by airborne semiochemicals. II. Olfactometric studies of host location by the parasitoid *Microplitis croceipes* (Cresson) (Hymenoptera: Braconidae). *Journal of Chemical Ecology*, **14**: 425–434.

Elliott, E.A. 1922. Monograph of the hymenopterous family Stephanidae. *Proceedings of the Zoological Society of London*, **1922**: 705–831.

Elliott, N.B. & Elliott, W.M. 1987. Nest usurpation by females of *Cerceris cribrosa* (Hymenoptera: Sphecidae). *Journal of the Kansas Entomological Society*, **60**: 397–402.

Ellis, W.N. 1967. Studies on Neotropical Collembola, I: Some Collembola from Guatemala. *Beaufortia*, **14**: 93–107.

Embleton, A.L. 1904. V. On the anatomy and development of *Comys infelix*, Embleton, a hymenopterous parasite of *Lecanium hemisphaericum*. *Transactions of the Linnean Society of London*, (2) Zoology, **9**: 231–254.

Emerson, A.E. 1955. Geographical origins and dispersions of termite genera. *Fieldiana*: Zoology, **37**: 465–521.

Enriquez, E., Bejarano, S. & Vila, V. 1976 (1975). Observaciones sobre avispas predatoras de *Leucoptera coffeella* Guer-Men, en el centro y sur del Peru. *Revista Peruana de Entomologia*, **18**: 82–83.

Erdös, J. 1957. Recentiores observationes entomocoenologicae in *Pragmites communis* Trin. *Allattani Közlemények*, **46**: 60–65.

Erwin, T.L. 1983. Beetles and other insects of tropical forest canopies at Manaus, Brazil, sampled by insecticidal fogging, pp. 59–75. *In*: Sutton, S.L., Whitmore, T.C. & Chadwick, A.C. (eds) *Tropical Rain Forest: Ecology and Management*. 498pp. Blackwells, Oxford.

Escalante, G. & Rabinovich, J.E. 1979. Population dynamics of *Telenomus fariai* (Hymenoptera: Scelionidae), a parasite of Chagas' disease vectors IX. Larval competition and population size regulation under laboratory conditions. *Researches on Population Ecology*,, **20**: 235–246.

Escoubas, P. & Blum, M.S. 1990. The biological activities of ant-derived alkaloids, pp. 482–489. *In*: Vander Meer, R.K., Jaffe, K. & Cedeno, A. (eds). *Applied Myrmecology, A World Perspective*, 741pp. Westview Press, Boulder.

Eskafi, F.M. & Legner, E.F. 1974*a*. Fecundity, development, and diapause in *Hexacola* sp. near *websteri*, a parasite of *Hippelates* eye gnats. *Annals of the Entomological Society of America*, **67**: 769–771.

Eskafi, F.M. & Legner, E.F. 1974*b*. Descriptions of immature stages of the cynipid *Hexacola* sp. near *websteri* (Eucoilinae: Hymenoptera), a larval-pupal parasite of *Hippelates* eye gnats (Diptera: Chloropidae). *Canadian Entomologist*, **106**: 1043–1048.

Espelie, K.E. & Hermann, H.R. 1988. Congruent cuticular hydrocarbons: biochemical convergence of a social wasp, an ant and a host plant. *Biochemical Systematics and Ecology*, **16**: 505–508.

Espelie, K.E. & Hermann, H.R. 1990. Surface lipids of the social wasp *Polistes annularis* (L.) and its nest and nest pedicel. *Journal of Chemical Ecology*, **16**: 1841–1852.

Espina, D. 1986. *Beekeeping of the Assassin Bees*. 170pp. Editorial Technológica de Costa Rica, Cartago.

Espina Perez, D. & Ordetx Ros, G.S. 1983. *Flora Apicola Tropical*. 406pp. Editorial Technológica de Costa Rica, Cartago.

Esquival R., E.A. 1983. Effective control of the giant mothborer *Castnia licus* (Drury) in Panama, utilizing biological-cultural methods. *Entomology Newsletter, International Society of Sugarcane Technologists*, **14**: 6–7.

Evans, A.C. 1933. Comparative observations on the morphology and biology of some hymenopterous parasites of carrion-infesting Diptera. *Bulletin of Entomological Research*, **24**: 385–405.

Evans, H.E. 1953. Comparative ethology and the systematics of spider wasps. *Systematic Zoology*, **2**: 155–172

Evans, H.E. 1956. Notes on the biology of four species of ground-nesting Vespidae (Hymenoptera). *Proceedings of the Entomological Society of Washington*, **58**: 265–270.

Evans, H.E. 1957*a*. Studies on the comparative ethology of digger wasps of the genus *Bembix*. 248pp. Comstock, Ithaca.

Evans, H.E. 1957*b*. The North and Central American species of *Propristocera* (Hymenoptera: Bethylidae). *Proceedings of the Entomological Society of Washington*, **59**: 289–296.

Evans, H.E. 1958. Studies on the nesting behavior of digger wasps of the tribe Sphecini. Part I. Genus *Prionony x* Dahlbom. *Annals of the Entomological Society of America*, **51**: 177–186.

Evans, H.E. 1959. Observations on the nesting behavior of digger wasps of the genus *Ammophila*. *American Midland Naturalist*, **62**: 449–473.

Evans, H.E. 1961*a*. A reconsideration of the genus *Epipompilus* (Hymenoptera: Pompilidae). *Psyche*, **68**: 25–37.

Evans, H.E. 1961*b*. A preliminary review of the Nearctic species of *Sierolomorpha* (Hymenoptera). *Breviora*, **140**: 1–12.

Evans, H.E. 1961*c*. A revision of the genus *Pseudisobrachium* in North and Central America (Hymenoptera, Bethylidae). *Bulletin of the Museum of Comparative Zoology*, **126**: 211–318.

Evans, H.E. 1962. The evolution of prey-carrying mechanisms in wasps. *Evolution*, **16**: 468–483.

Evans, H.E. 1963*a*. A revision of the genus *Pristocera* in the Americas (Hymenoptera, Bethylidae). *Bulletin of the Museum of Comparative Zoology*, **129**: 241–290.

Evans, H.E. 1963*b*. A new family of wasps. *Psyche*, **70**: 7–16.

Evans, H.E. 1963*c*. A revision of the genus *Apenesia* in the Americas (Hymenoptera, Bethylidae). *Bulletin of the Museum of Comparative Zoology*, **130**: 249–359.

Evans, H.E. 1963*d*. Predatory wasps. *Scientific American*, **208**: 145–154.

Evans, H.E. 1964*a*. A synopsis of the American Bethylidae (Hymenoptera, Aculeata). *Bulletin of the Museum of Comparative Zoology*, **132**: 1–222.

Evans, H.E. 1964*b*. Further studies on the larvae of digger wasps (Hymenoptera: Sphecidae). *Transactions of the American Entomological Society*, **90**: 235–299.

Evans, H.E. 1964*c*. The classification and evolution of digger wasps as suggested by larval characters (Hymenoptera: Sphecoidea). *Entomological News*, **75**: 225–237.

Evans, H.E. 1965*a*. A description of the larva of *Methocha stygia* (Say), with notes on other Tiphiidae (Hymenoptera). *Proceedings of the Entomological Society of Washington*, **67**: 88–95.

Evans, H.E. 1965*b*. Studies on Neotropical Pompilidae (Hymenoptera) I. The genus *Agenioideus* Ashmead in South America. *Breviora*, **234**: 1–7.

Evans, H.E. 1965*c*. Simultaneous care of more than one nest by *Ammophila azteca* Cameron (Hymenoptera, Sphecidae). *Psyche*, **72**: 8–23.

Evans, H.E. 1966*a*. The behavior patterns of solitary wasps. *Annual Review of Entomology*, **11**: 123–154.

Evans, H.E. 1966*b*. *The Comparative Ethology and Evolution of the Sand Wasps*. 526pp. Harvard University Press, Cambridge, Mass.

Evans, H.E. 1966*c*. *Life on a Little Known Planet*. 318pp. Dell, New York.

Evans, H.E. 1966*d*. Nests and prey of two species of *Philanthus* in Jackson Hole, Wyoming (Hymenoptera, Sphecidae). *Great Basin Naturalist*, **26**: 35–40.

Evans, H.E. 1966*e*. A revision of the Mexican and Central American spider wasps of the subfamily Pompilinae (Hymenoptera: Pompilidae). *Memoirs of the American Entomological Society*, **20**: 1–442.

Evans, H.E. 1966*f* (1967). Discovery of the female *Plumarius* (Hymenoptera: Plumariidae). *Psyche*, **73**: 229–237.

Evans, H.E. 1966*g*. The accessory burrows of digger wasps. *Science*, **152**: 465–471.

Evans, H.E. 1967. Studies on neotropical Pompilidae (Hymenoptera) III. Additional notes on *Epipompilus* Kohl. *Breviora, Museum of Comparative Zoology*, **273**: 1–15.

Evans, H.E. 1968*a*. Mexican and Central American Pompilinae (Hymenoptera, Pompilidae): supplementary notes, I. *Entomological News*, **79**: 158–167.

Evans, H.E. 1968*b*. Studies on Neotropical Pompilidae (Hymenoptera) IV. Examples of dual sex-limited mimicry in *Chirodamus*. *Psyche*, **75**: 1–22.

Evans, H.E. 1969*a*. Phoretic copulation in Hymenoptera. *Entomological News*, **80**: 113–124.

Evans, H.E. 1969*b*. Studies on neotropical Pompilidae (Hymenoptera) VII. *Irenangelus* Schulz. *Studia Entomologica*, **12**: 417–431.

Evans, H.E. 1971. Observations on the nesting behavior of wasps of the tribe Cercerini. *Journal of the Kansas Entomological Society*, **44**: 500–523.

Evans, H.E. 1973*a*. VIII. The genus *Aporus* Spinola in South America. Studies on neotropical Pompilidae (Hymenoptera). *Studia Entomologica* (N.S.), **16**: 353–370.

Evans, H.E. 1973*b*. Studies on neotropical Pompilidae (Hymenoptera). IX. The genera of Auplopodini. *Psyche*, **80**: 212–226.

Evans, H.E. 1973*c*. Notes on the nests of *Montezumia* (Hymenoptera, Eumenidae). *Entomological News*, **84**: 285–290.

Evans, H.E. 1974. A review of the species of *Dipogon* occurring in Central America, Mexico and extreme southwestern United States (Hymenoptera: Pompilidae). *Transactions of the American Entomological Society*, **100**: 29–51.

Evans, H.E. 1977. Extrinsic versus intrinsic factors in the evolution of insect sociality. *Bioscience*, **27**: 613–617.

Evans, H.E. 1978. The Bethylidae of America north of Mexico. *Memoirs of the American Entomological Institute*, **27**: 1–332.

Evans, H.E. 1980. A new species of *Mystacagenia* from Panama (Hymenoptera: Pompilidae). *Pan-Pacific Entomologist*, **56**: 185–186.

Evans, H.E. 1987*a*. (Coordinator) Order Hymenoptera, pp. 597–710. *In*: Stehr, F.W. (ed.) *Immature Insects*. 754pp. Kendall Hunt, Dubuque.

Evans, H.E. 1987*b*. A new species of *Irenangelus* from Costa Rica (Hymenoptera: Pompilidae: Ceropalinae). *Proceedings of the Entomological Society of Washington*, **89**: 559–561.

Evans, H.E. 1989. The mating and predatory behavior of *Mellinus rufinodus* Cresson (Hymenoptera: Sphecidae). *Pan-Pacific Entomologist*, **65**: 414–417.

Evans, H.E., Kugler, C. & Brown, W.L. Jr. 1979. Rediscovery of *Scolebythus madecassus*, with a description of the male and of the female sting apparatus (Hymenoptera: Scolebythidae). *Psyche*, **86**: 45–51.

Evans, H.E., Kurczewski, F.E. & Alcock, J. 1980. Observations on the nesting behaviour of seven species of *Crabro* (Hymenoptera, Sphecidae). *Journal of Natural History*, **14**: 865–882.

Evans, H.E. & Matthews, R.W. 1973. Observations on the nesting behavior of *Trachypus petiolatus* (Spinola) in Colombia and Argentina (Hymenoptera: Sphecidae: Philanthini). *Journal of the Kansas Entomological Society*, **46**: 165–175.

Evans, H.E. & Matthews, R.W. 1974. Notes on nests and prey of two species of ground-nesting Eumenidae from So. America (Hymenoptera). *Entomological News*, **85**: 149–153.

Evans, H.E., Matthews, R.W. & Callan, E.Mc.C. 1974. Observations of the nesting behavior of *Rubrica surinamensis* (DeGeer) (Hymenoptera, Sphecidae). *Psyche*, **81**: 334–352.

Evans, H.E. & O'Neill, K.M. 1985. Male territorial behavior in four species of the tribe Cercerini (Sphecidae: Philanthinae). *Journal of the New York Entomological Society*, **93**: 1033–1040.

Evans, H.E. & O'Neill, K.M. 1988. *The Natural History and Behavior of North American Beewolves*. 278pp. Cornell University Press, Ithaca, New York.

Evans, H.E., O'Neill, K.M. & O'Neill R.P. 1986. Nesting site changes and nocturnal clustering in the sand wasp *Bembecinus quinquespinosus* (Hymenoptera: Sphecidae). *Journal of the Kansas Entomological Society*, **59**: 280–286.

Evans, H.E. & West Eberhard, M.J. 1970. *The Wasps*. 265pp. University of Michigan, Ann Arbor.

Evans, H.E. & Yoshimoto, C.M. 1962. The ecology and nesting behavior of the Pompilidae (Hymenoptera) of the northeastern United States. *Miscellaneous Publications of the Entomological Society of America*, **3**: 66–119.

Evenhuis, H.H. 1964. The interrelations between apple aphids and their parasites and hyperparasites. *Entomophaga*, **9**: 227–231.

Evenhuis, H.H. 1968. The natural control of the apple-grass aphid, *Rhopalosiphum insertum*, with remarks on the control of apple aphids in the Netherlands in general. *Netherlands Journal of Plant Pathology*, **74**: 106–117.

Evers, C.A. & Seeley, T.D. 1986. Kin discrimination and aggression in honey bee colonies with laying workers. *Animal Behaviour*, **34**: 924–925.

Eves, J.D. 1970. Biology of *Monodontomerus obscurus* Westwood, a parasite of the alfalfa leafcutting bee, *Megachile rotundata* (Fabricius) (Hymenoptrera: Torymidae; Megachilidae). *Melanderia*, **4**: 1–18.

Ewen, A.B. & Arthur, A.P. 1976. Cuticular encystment in three noctuid species (Lepidoptera): induction by acid gland secretion from an ichneumonid parasite (*Banchus flavescens*). *Annals of the Entomological Society of America*, **69**: 1087–1090.

Fabres, G. & Reymonet, C. 1991. L'induction maternelle de la diapause larvaire chez *Dinarmus acutus* (Hym.: Pteromalidae). *Entomophaga*, **36**: 121–129.

Fahringer, J. 1922. Beiträge zur Kenntnis der Lebensweise einiger Schmarotzerwespen unter besonderer Berücksichtigung ihrer Bedeutung für biolog. Bekämpfung von Schädlingen. *Zeitschrift für Angewandte Entomologie*, **8**: 325–388.

Faure, J.C. 1926. Contribution à l'étude d'un complexe biologique: la Piéride du chou (*Pieris brassicae* L.) et ses parasites hyménoptères. Thesis du Faculté de Sciences de la Université Lyon. 223pp.

Fedde, G.F. 1977. A laboratory study of egg parasitization capabilities of *Telenomus alsophilae*. *Environmental Entomology*, **6**: 773–776.

Feddersen, I., Sander, K. & Schmidt, O. 1986. Virus-like particles with host protein-like antigenic determinants protect an insect parasitoid from encapsulation. *Experientia*, **42**: 1278–1281.

Feener, D.H. Jr & Brown, B.V. 1992. Reduced foraging of *Solenopsis geminata* (Hymenoptera: Formicidae) in the presence of parasitic *Pseudacteon* spp. (Diptera: Phoridae). *Annals of the Entomological Society of America*, **85**: 80–84.

Feener, D.H. Jr & Brown, B.V. 1993. Oviposition behavior of an ant-parasitizing fly, *Neodohrniphora curvinervis* (Diptera: Phoridae), and defense behavior by its leaf-cutting ant host *Atta cephalotes* (Hymenoptera: Formicidae). *Journal of Insect Behavior*, **6**: 675–688.

Feener, D.H. Jr & Moss, K.A.G. 1990. Defense against parasites by hitchikers in leaf-cutting ants: a quantitative assessment. *Behavioral Ecology and Sociobiology*, **26**: 17–29.

Feldhege, M.R. 1992. Rearing techniques and aspects of biology of *Phymastichus coffea* (Hymenonoptera: Eulophidae), a recently described endoparasitoid of the coffee berry borer, *Hypothenemus hampei* (Coleoptera: Scolytidae). *Café Cacao Thé*, **36**: 45–54.

Fenton, F.A. 1918. The parasites of Leaf-Hoppers. With special reference to the Biology of the Anteoninae. Parts I–III. *Ohio Journal of Science*, **18**: 177–212, 243–278, 285–291.

Ferguson, W.E. 1962. Biological characteristics of the mutillid subgenus *Photopsis* Blake, and their systematic values (Hymenoptera). *University of California Publications in Entomology*, **27**: 1–91.

Fergusson, N.D.M. 1980. A revision of the British species of *Dendrocerus* Ratzeburg (Hymenoptera: Ceraphronoidea) with a review of their biology as aphid hyperparasites. *Bulletin of the British Museum (Natural History)* (Entomology) **41**: 255–314.

Fergusson, N.D.M. 1986. Charipidae, Ibaliidae & Figitidae. Hymenoptera: Cynipoidea. *Handbooks for the Identification of British Insects*, **8**(1c): 1–55.

Fergusson, N.D.M. 1988. A comparative study of the structures of phylogenetic importance of female genitalia of the Cynipoidea (Hymenoptera). *Systematic Entomology*, **13**: 13–30.

Fergusson, N.D.M. 1990. A Phylogenetic Study of the Cynipoidea (Hymenoptera). 425pp. Unpublished PhD thesis, City of London Polytechnic.

Fernandes, G.W. & Price, P.W. 1988. Biogeographical gradients in galling species richness. Tests of hypotheses. *Oecologia*, **76**: 161–167.

Fernandes, G.W., Martins, R.P. & Tameirao Neto, E. 1987. Food web relationships involving *Anadiplosis* sp. galls (Diptera: Cecidomyiidae) on *Machaerium aculeatum* (Leguminosae). *Revista Brasileira de Botanica*, **10**: 117–123.

Fernández B., R.I. & Terán B., J.B. 1990*a*. Biología de *Meteorus laphygmae* Viereck (Hymenoptera: Braconidae). Parte I: Fases de huevo, larva y pupa. *Revista de Facultad de Agronomía, Maracay*, **16**: 177–198.

Fernández B., R.I. & Terán B., J.B. 1990*b*. Biología de *Meteorus laphygmae* Viereck (Hymenoptera: Braconidae). Parte II: adulto, partenogenesis e interrelaciones parásito-hospedero. *Revista de Facultad de Agronomía, Maracay*, **16**: 199–206.

Ferrière, C. 1926. Un nouveau cas de phorésie: trichogrammides sur sauterelles. *Treubia*, **8**: 274–278.

Ferrière, C. 1938. Eupelmides exotiques (Hymenopt. Chalcididae). I. Les genres *Metapelma* Westw., *Anastatoidea* Gahan et *Neanastatus* Girault. *Annales de la Société Entomologique de France*, **107**: 25–72.

Ferrière, C. 1947. Les espèces européennes du genre *Elasmus* Westw. (Hym. Chalc.). *Mitteilungen der Schweizerischen Entomologischen Gesellschaft*, **20**: 565–580.

Fidalgo, P. 1989. Revisión de las especies neotropicales del género *Acmopolynema* Ogl. (Hymenoptera: Mymaridae). *Revista de la Sociedad Entomologica Argentina*, **46**: 3–67.

Fiedler, K. 1991. Systematic, evolutionary, and ecological implications of mymecophily within the Lycaenidae (Insecta: Lepidoptera: Papilionoidea). *Bonner Zoologische Monographien*, **31**: 1–210.

Fincke, O.M., Higgins, L. & Rojas, E. 1990. Parasitism of *Nephila clavipes* (Araneae, Tetragnathidae) by an ichneumonid (Hymenoptera, Polysphinctini) in Panama. *Journal of Arachnology*, **18**: 321–329.

Fink, D.E. 1926. The biology of *Macrocentrus ancylivora* Rohwer, an important parasite of the strawberry leaf roller (*Ancylis comptana* Froehl.). *Journal of Agricultural Research*, **32**: 1121–1134.

Finlayson, L.H. 1950. The biology of *Cephalonomia waterstoni* Gahan (Hym., Bethylidae), a parasite of *Laemophloeus* (Col., Cucujidae). *Bulletin of Entomological Research*, **41**: 79–97.

Finlayson, T.F. 1964. The final-instar larva of *Bracon nuperus* Cress. (Hymenoptera: Braconidae). *Canadian Entomologist*, **96**: 1470–1471.

Finlayson, T. 1966. The false cocoon of *Hyposoter parorgyiae* (Vier.) (Hymenoptera: Ichneumonidae). *Canadian Entomologist*, **98**: 139.

Finlayson, T. 1975. The cephalic structures and spiracles of final-instar larvae of the subfamily Campopleginae, tribe Campoplegini (Hymenoptera: Ichneumonidae). *Memoirs of the Entomological Society of Canada*, **94**: 1–137.

Finlayson, T. 1976. Cephalic structures and spiracles of final-instar larvae of the genus *Toxophoroides* (Hymenoptera: Ichneumonidae: Lycorinae). *Canadian Entomologist*, **108**: 981–984.

Finlayson, T. & Hagen, K.S. 1977. Final-instar larvae of parasitic Hymenoptera. *Simon Fraser University Pest Management Papers*, **10**: 1–111.

Fischer, M. 1964. Die Opiinae der nearktischen region (Hymenoptera, Braconidae), I Teil. *Polskie Pismo Entomologiczne*, **34**: 197–530.

Fischer, M. 1965. Die Opiinae der nearktischen region (Hymenoptera, Braconidae), II Teil. *Polskie Pismo Entomologiczne*, **35**: 3–212.

Fischer, M. 1967*a*. Die amerikanischen Arten der Gattungen *Euopius*, *Gnaptodon* und *Pseudognaptodon* (Hymenoptera: Braconidae). *Beiträge zur Entomologie*, **17**: 959–976.

Fischer, M. 1967*b*. Zusammenfassung der neotropischen Opiinae mit Ausschluss der Gattung *Opius* Wesm. (Hymenoptera, Braconidae). *Beiträge zur Neotropischen Fauna*, **5**: 1–21.

Fischer, M. 1973 (1972). Hymenoptera Braconidae (Opiinae I). *Das Tierreich*, **91**: 1–620.

Fischer, M. 1975. Taxonomische untersuchungen an neotropischen Alysiini, besonders solchen aus dem British Museum (Natural History) in London (Hym. Braconidae: Alysiinae). *Studia Entomologica*, **18**: 127–152.

Fischer, M. 1977. Hymenoptera Braconidae (Opiinae II-Amerika). *Das Tierreich*, **96**: 1–1001.

Fisher, B.L. 1992. Facultative ant association benefits a Neotropical orchid. *Journal of Tropical Ecology*, **8**: 109–114.

Fisher, R.C. 1959. Life history and ecology of *Horogenes chrysostictos* Gmelin (Hymenoptera, Ichneumonidae), a parasite of *Ephestia sericarium* Scott (Lepidoptera, Phycitidae). *Canadian Journal of Zoology*, **37**: 429–446.
Fisher, R.C. 1963. Oxygen requirements and the physiological supression of supernumerary insect parasitoids. *Journal of Experimental Biology*, **40**: 531–540.
Fisher, R.C. 1971. Aspects of the physiology of endoparasitic Hymenoptera. *Biological Reviews of the Cambridge Philosophical Society*, **46**: 243–278.
Fisher, R.C. & Ganesalingam, V.K. 1970. Changes in the composition of host haemolymph after attack by an insect parasitoid. *Nature*, **227**: 191–192.
Fittkau, E.J. & Klinge, H. 1973. On biomass and trophic structure of the Central Amazonian rainforest ecosystem. *Biotropica*, **5**: 2–14.
Fitton, M.G. 1985. The ichneumon-fly genus *Banchus* (Hymenoptera) in the Old World. *Bulletin of the British Museum (Natural History)* Entomology, **51**: 1–60.
Fitton, M.G. & Gauld, I.D. 1976. The family-group names of the Ichneumonidae (excluding Ichneumoninae) (Hymenoptera). *Systematic Entomology*, **1**: 247–258.
Fitton, M.G. & Gauld, I.D. 1978. Further notes on family-group names of the Ichneumonidae (Hymenoptera). *Systematic Entomology*, **3**: 245–247.
Fitton, M.G., Shaw, M.R. & Austin, A.D. 1987. The Hymenoptera associated with spiders in Europe. *Zoological Journal of the Linnean Society*, **90**: 65–93.
Fitton, M.G., Shaw, M.R. & Gauld, I.D. 1988. Pimpline ichneumon-flies. *Handbooks for the Identification of British Insects*, **7**(1): 1–110.
Flanders, S.E. 1931. The temperature relationships of *Trichogramma minutum* as a basis for racial segregation. *Hilgardia*, **5**: 395–406.
Flanders, S.E. 1935. Host influence on the prolificacy and size of *Trichogramma*. *Pan-Pacific Entomologist*, **11**: 175–177.
Flanders, S.E. 1937. Habitat selection by *Trichogramma*. *Annals of the Entomological Society of America*, **30**: 208–210.
Flanders, S.E. 1938. Cocoon formation in endoparasitic chalcidoids. *Annals of the Entomological Society of America*, **31**: 167–180.
Flanders, S.E. 1942. Abortive development in parasitic Hymenoptera, induced by the food-plant of the insect host. *Journal of Economic Entomology*, **35**: 834–835.
Flanders, S.E. 1944. Observations on *Comperiella bifasciata*, an endoparasite of diaspine coccids. *Annals of the Entomological Society of America*, **37**: 365–371.
Flanders, S.E. 1952 (1953). Biological observations on parasites of the black scale. *Annals of the Entomological Society of America* **45**: 543–549.
Flanders, S.E. 1956. The mechanisms of sex-ratio regulation in the (Parasitic) Hymenoptera. *Insectes Sociaux*, **3**: 325–334.
Flanders, S.E. 1959*a*. Biological control of *Saissetia nigra* (Nietn.) in California. *Journal of Economic Entomology*, **52**: 596–600.
Flanders, S.E. 1959*b*. Differential host relations of the sexes in parasitic Hymenoptera. *Entomologia Experimentalis et Applicata*, **2**: 125–142.
Flanders, S.E. 1967. Deviate-ontogenies in the aphelinid male (Hymenoptera) associated with the ovipositional behavior of the parental female. *Entomophaga*, **12**: 415–427.
Flanders, S.E. 1971. Multiple parasitism of armored coccids (Homoptera) by host-regulative aphelinids (Hymenoptera); ectoparasites versus endoparasites. *Canadian Entomologist*, **103**: 857–872.
Flanders, S.E., Bartlett, B.R. & Fisher, T.W. 1961. *Coccophagus basalis* (Hymenoptera: Aphelinidae): its introduction into California with studies of its biology. *Annals of the Entomological Society of America*, **54**: 227–236.

Flechtmann, C.H.W. & Camargo, C.A. de 1979. Acari associated with stingless bees (Meliponidae [sic], Hymenoptera) from Brazil, pp. 315–319. *In*: Piffl, E. (ed.) *Proceedings of the 4th International Congress of Acarology*. 752pp. Akadémiai Kiadó, Budapest.

Fleming, J.G.W. 1992. Polydnaviruses: mutualists and pathogens. *Annual Review of Entomology* **37**: 401–425.

Fletcher, D.J.C. & Ross, K.G. 1985. Regulation of reproduction in eusocial Hymenoptera. *Annual Review of Entomology*, **30**: 319–343.

Flock, R.A. 1957. Biological notes on a new chalcid-fly from seed-like Eucalyptus galls in California. *Pan-Pacific Entomologist*, **33**: 153–155.

Folliot, R. 1964. Contribution a l'étude de la biologie des cynipides gallicoles (Hyménoptères, Cynipoidea). *Annales des Sciences Naturelles, Zoologie*. (12) **6**: 407–564.

Force, D.C. 1975. Succession of *r* and *K* strategists in parasitoids, pp. 112–129. *In*: Price, P.W. (ed.) *Evolutionary Strategies of Parasitic Insects and Mites*. 224pp. Plenum Press, New York.

Forel, A. 1898. La parabiose chez les fourmis. *Bulletin de la Société Vaudoise des Sciences Naturelles*, **34**: 380–384.

Forsyth, A. 1981*a*. Sex ratio and parental investment in an ant population. *Evolution*, **35**: 1252–1253.

Forsyth, A. 1981*b*. Swarming activity of polybiine social wasps (Hymenoptera: Vespidae: Polybiini). *Biotropica*, **13**: 93–99.

Forsyth, A.B. 1978. *Studies on the Behavioral Ecology of Polygynous Social Wasps*. Ph.D. dissertation, Harvard University, Cambridge, Mass.

Foster, R.B. 1982. Famine on Barro Colorado Island, pp. 201–212. *In*: Leigh, E.G. Jr, Rand, A.S. & Windsor, D.M. (eds) *The Ecology of a Tropical Forest. Seasonal Rhythms and Long-term Changes*. 468pp. Smithsonian Institution Press, Washington, D.C.

Fouts, R.M. 1920. Some new parasites, with remarks on the genus *Platygaster* (Hymenoptera). *Proceedings of the Entomological Society of Washington*, **22**: 61–72.

Fouts, R.M. 1924. Revision of the North American wasps of the subfamily Platygasterinae. *Proceedings of the United States National Museum*, **2484**: 1–145.

Fouts, R.M. 1939. Descriptions of one new genus and three new species of Diapriidae (Hymenoptera). *Proceedings of the Entomological Society of Washington*, **41**: 260–264.

Fouts, R.M. 1948. Parasitic wasps of the genus *Trimorus* in North America. *Proceedings of the United States National Museum*, **3225**: 91–148.

Fowler, H.G. 1978. Foraging trails of leaf-cutting ants. *Journal of the New York Entomological Society*, **86**: 132–136.

Fowler, H.G. & Robinson, S.W. 1979. Foraging by *Atta sexdens* (Formicidae: Attini): seasonal patterns, caste and efficiency. *Ecological Entomology*, **4**: 239–247.

Francke-Grosmann, H. 1967. Ectosymbiosis in wood-inhabiting insects, pp. 142–205. *In*: Henry, S.M. (ed.) *Symbiosis*, II. 443pp. Academic Press, New York.

Frank, J.H. 1982. The parasites of the Staphylinidae (Coleoptera) *Bulletin University of Florida Agricultural Experimental Station* (Technical), **824**: 1–118.

Frank, S.A. 1984 (1985). The behavior and morphology of the fig wasps *Pegoscapus assuetus* and *P. jimenezi*: descriptions and suggested behavioral characters for phylogenetic studies. *Psyche*, **91**: 289–308.

Frankie, G.W. & Coville, R. 1979. An experimental study on the foraging behavior of selected solitary bee species in the Costa Rican dry forest (Hymenoptera: Apoidea). *Journal of the Kansas Entomological Society*, **52**: 591–602.

Frankie, G.W. & Daly, H.V. 1983. *Xylocopa gualanensis* (Xicote, Avispa Carpintera, Carpenter Bee), pp. 777–779. *In*: Janzen, D.H. (ed.), *Costa Rican Natural History*, 816pp. University Chicago Press, Chicago.

Frankie, G.W., Vinson, S.B. & Coville, R.E. 1980. Territorial behavior of *Centris adani* and its reproductive function in the Costa Rican dry forest (Hymenoptera: Anthophoridae). *Journal of the Kansas Entomological Society*, **53**: 837–857.

Frankie, G.W., Vinson, S.B., Newstrom, L.E. & Barthell, J.F. 1988. Nest site and habitat preferences of *Centris* bees in the Costa Rican dry forest. *Biotropica*, **20**: 301–310.

Frankie, G.W., Vinson, S.B., Newstrom, L.E., Barthell, J.F., Haber, W.A. & Frankie, J.K. 1990. Plant phenology, pollination ecology, pollinator behavior and conservation of pollinators in Neotropical dry forest, pp. 37–47. *In*: Bawa, K.S. & Handley, M. (eds), *Reproductive Ecology of Tropical Forest Plants*, 421pp. Cambridge University Press, Cambridge.

Frankie, G.W., Vinson, S.B. & Williams, H. 1989. Ecological and evolutionary sorting of 12 sympatric species of *Centris* bees in Costa Rican dry forest, pp. 535–549. *In*: Bock, J.H. & Linhart, Y.B. (eds), *The Evolutionary Ecology of Plants*. 600pp. Westview Press, Boulder, Colorado.

Franks, N.R. 1982. Ecology and population regulation in the army ant *Eciton burchelli*, pp. 389–395. *In*: Leigh, E.G. Jr, Rand, A.S. & Windsor, D.M. (eds), *The Ecology of a Tropical Forest. Seasonal Rhythms and Long-term Changes*. 468pp. Smithsonian Institute Press, Washington, D.C.

Franks, N.R. 1986. Teams in social insects: group retrieval of prey by army ants (*Eciton burchelli*, Hymenoptera: Formicidae). *Behavioral Ecology and Sociobiology*, **18**: 425–429.

Franks, N.R. & Bossert, W.H. 1983. The influence of swarm raiding army ants on the patchiness and diversity of a tropical leaf litter ant community, pp. 151–163. *In*: Sutton, S.L., Whitmore, T.C. & Chadwick, A.C. (eds) *Tropical Rain Forest: Ecology and Management*. 498pp. Blackwell, Oxford.

Franks, N.R. & Hölldobler, B. 1987. Sexual competition during colony reproduction in army ants. *Biological Journal of the Linnean Society*, **30**: 229–243.

Franz, J.M. 1958. Studies on *Laricobius erichsonii* Rosenh. (Coleoptera: Derodontidae) a predator on chermesids. Part I, Distribution, life-history and ecology. *Entomophaga*, **3**: 109–164.

Free, J.B. 1970. *Insect Pollination of Crops*. 544pp. Academic Press, London.

Free, J.B. 1987. *Pheromones of Social Bees*. 218pp. Chapman & Hall, London.

Freeman, B.E. 1982. The comparative distribution and population dynamics in Trinidad of *Sceliphron fistularium* (Dahlbom) and *S. asiaticum* (L.) (Hymenoptera: Sphecidae). *Biological Journal of the Linnean Society*, **17**: 343–360.

Freeman, B.E. & Ittyeipe, K. 1982. Morph determination in *Mellitobia*, a eulophid wasp. *Ecological Entomology*, **7**: 355–363.

Frilli, F. 1965. Studi sugli imenotteri icneumonidi. 1. *Devorgilla canescens* (Grav.). *Annali dell'Istituto di Entomologia Agraria della Universitá di Bari*, **1**: 119–207.

Frisch, K. von. 1967. *The Dance Language and Orientation of Bees*. 566pp. Belknap Press, Cambridge, Mass.

Fritz, M.A. 1973. Nyssonini Neotropicales VI (Hym. Sphecidae: Nyssoninae). *Anales del Museo de Historia Natural de Valparaiso*, **6**: 191–202.

Fuester, R.W., Taylor, P.B. & Eisenberg, A. 1989. Suitability of prepupae and pupae of the gypsy moth (Lepidoptera: Lymantriidae) as hosts of *Coccygomimus disparis* (Hymenoptera: Ichneumonidae). *Environmental Entomology*, **18**: 338–341.

Führer, E. & Elsufty, R. 1979. Produktion fungistatischer metabolite durch teratocyten von *Apanteles glomeratus* L. (Hym., Braconidae). *Zeitschrift für Parasitenkunde*, **59**: 21–25.

Führer, E. & Keja, T.D. 1976. Physiologische Wechselbeziehungen zwischen *Pieris brassicae* und dem Endoparasiten *Apanteles glomeratus*. Der Einflüss der Parasitierung auf Wachstum und Körpergewicht des Wirtes. *Entomologia Experimentalis et Applicata*, **19**: 287–300.

Führer, E. & Kilincer, N. 1972. Die motorische aktivität der endoparasitischen larven von *Pimpla turionellae* L. und *Pimpla flavicoxis* Ths. (Hym., Ichneum.) in der Wirtspuppe. *Entomophaga*, **17**: 149–163.

Führer, E. & Willers, D. 1986. The anal secretion of the endoparasitic larva *Pimpla turionellae*: sites of production and effects. *Journal of Insect Physiology*, **32**: 361–367.

Fukushima, J.I., Kuwahara, Y. & Suzuki, T. 1989. Isolation and identification of a kairomone responsible for the stinging behavior of *Bracon hebetor* Say (Hymenoptera: Braconidae) from frass of the almond moth *Cadra cautella* Walker. *Agricultural and Biological Chemistry*, **53**: 3057–3059.

Fursov, V.N. & Kostyukov, V.V. 1987. New species of the genus *Tetrastichus* (Hymenoptera, Eulophidae), egg parasites of damselflies and dragonflies and of predaceous diving beetles. *Zoologicheskii Zhurnal*, **66**: 217–228. [In Russian: English summary.]

Gadagkar, R. 1991. *Belonogaster, Mischocyttarus, Parapolybia*, and independent-founding *Ropalidia*, pp. 149–190. *In*: Ross, K.G. & Matthews, R.W. (eds) *The Social Biology of Wasps*. 678pp. Cornell University Press, Ithaca.

Gahan, A.B. 1918. A synopsis of the species belonging to the chalcidoid genus *Rileya* Ashmead (Hym.). *Proceedings of the Entomological Society of Washington*, **20**: 136–150.

Gahan, A.B. 1943. Revisions of two genera of chalcid-flies belonging to the family Eupelmidae from North and South America. *Proceedings of the United States National Museum*, **3173**: 339–369.

Galil, J., Dulberger, R. & Rosen, D. 1970. The effects of *Sycophaga sycomori* L. on the structure and development of the synconia in *Ficus sycomorus* L. *New Phytologist*, **69**: 103–111.

Galil, J. & Eisikowitch, D. 1968*a*. On the pollination ecology of *Ficus sycomorus* in East Africa. *Ecology*, **49**: 259–269.

Galil, J. & Eisikowitch, D. 1968*b*. Flowering cycles and fruit types of *Ficus sycomorus* in Israel. *New Phytologist*, **67**: 745–758.

Galil, J. & Eisikowitch, D. 1969. Further studies on the pollination ecology of *Ficus sycomorus* L. (Hymenoptera, Chalcidoidea, Agaonidae). *Tijdschrift voor Entomologie*, **112**: 1–13.

Galil, J. & Meiri, L. 1981. Number and structure of anthers in fig syconia in relation to behaviour of the pollen vectors. *New Phytologist*, **88**: 83–87.

Galil, J. & Neeman, G. 1977. Pollen transfer and pollination in the common fig (*Ficus carica* L.). *New Phytologist*, **79**: 163–171.

Galil, J., Ramirez B., W. & Eisikowitch, D. 1973*a*. Pollination of *Ficus costaricana* and *F. hemsleyana* by *Blastophaga esterae* and *B. tonduzi* in Costa Rica (Hymenoptera: Chalcidoidea, Agaonidae). *Tijdschrift voor Entomologie*, **116**: 175–183.

Galil, J. & Snitzer-Pasternak, Y. 1970. Pollination in *Ficus religiosa* L. as connected with the structure and mode of action of the pollen pockets of *Blastophaga quadraticeps* Mayr. *New Phytologist*, **69**: 775–784.

Galil, J., Zeroni, M. & Shalom, D.B. 1973*b*. Carbon dioxide and ethylene effects in the co-ordination between the pollinator *Blastophaga quadraticeps* and the syconium in *Ficus religiosa*. *New Phytolologist*, **72**: 1113–1127.

Gallego M., F.L. 1950. Estudios entomológicos; el gusano de las hojas de la yuca. *Revista Facultad Nacional de Agronomía* (Medellín, Colombia), **11**: 84–110.

Gamboa, G.J., Reeve, H.K. & Pfennig, D.W. 1986. The evolution and ontogeny of nestmate recognition in social wasps. *Annual Review of Entomology*, **31**: 431–454.

García, J.L. 1991. Variaciones estacionales de la 'Polilla del Repollo' *Plutella xylostella* (L.) (Lep.: Yponomeutidae) y del parasito *Diadegma insulare* Cresson (Hym.: Ichneumonidae), en La Estacion Experimental Cataurito, Edo Aragua. *Boletín de Entomología Venezolana* (N.S.), **6**: 27–35.

Gardner, S.M., Ward, S.A. & Dixon, A.F.G. 1984. Limitation of superparasitism by *Aphidius rhopalosiphi*: a consequence of aphid defensive behaviour. *Ecological Entomology*, **9**: 149–155.

Garling, L. 1979. Origin of ant-fungus mutualism: a new hypothesis. *Biotropica*, **11**: 284–291.

Gaston, K.J. 1991*a*. Body size and probability of description: the beetle fauna of Britain. *Ecological Entomology*, **16**: 505–508.

Gaston, K.J. 1991*b*. The magnitude of global insect species richness. *Conservation Biology*, **5**: 283–296.

Gastreich, K.R., Strassmann, J.E. & Queller, D.C. 1993. Determinants of high genetic relatedness in the swarm-founding wasp *Protopolybia exigua*. *Ethology Ecology and Evolution*, **5**: 529–539.

Gauld, I.D. 1976*a*. The classification of the Anomaloninae (Hymenoptera: Ichneumonidae). *Bulletin of the British Museum (Natural History)* Entomology, **33**: 1–135.

Gauld, I.D. 1976*b*. The taxonomy of the genus *Heteropelma* Wesmael (Hymenoptera: Ichneumonidae). *Bulletin of the British Museum (Natural History)* Entomology, **34**: 155–219.

Gauld, I.D. 1983. The classification, evolution and distribution of the Labeninae, an ancient southern group of Ichneumonidae (Hymenoptera). *Systematic Entomology*, **8**: 167–178.

Gauld, I.D. 1984. *An Introduction to the Ichneumonidae of Australia*. 413pp. British Museum (Natural History), London

Gauld, I.D. 1985. The phylogeny, classification and evolution of parasitic wasps of the subfamily Ophioninae (Ichneumonidae). *Bulletin of the British Museum (Natural History)* Entomology, **51**: 61–185.

Gauld, I.D. 1986. Latitudinal gradients in ichneumonid species-richness in Australia. *Ecological Entomology*, **11**: 155–161.

Gauld, I.D. 1987. Some factors affecting the composition of tropical ichneumonid faunas. *Biological Journal of the Linnean Society*, **30**: 299–312.

Gauld, I.D. 1988*a*. The species of the *Enicospilus americanus* complex (Hymenoptera: Ichneumonidae) in eastern North America. *Systematic Entomology*, **13**: 31–53.

Gauld, I.D. 1988*b*. A survey of the Ophioninae (Hymenoptera: Ichneumonidae) of tropical Mesoamerica with special reference to the fauna of Costa Rica. *Bulletin of the British Museum (Natural History)* Entomology, **57**: 1–309.

Gauld, I.D. 1988*c*. Evolutionary patterns of host utilization by ichneumonoid parasitoids (Hymenoptera: Ichneumonidae and Braconidae). *Biological Journal of the Linnean Society*, **35**: 351–377.

Gauld, I.D. 1991. The Ichneumonidae of Costa Rica, 1. *Memoirs of the American Entomological Institute*, **47**: 1–589.

Gauld, I.D. & Bolton, B. 1988. *The Hymenoptera*. 332pp. Oxford University Press and British Museum (Natural History), Oxford.

Gauld, I.D., Collins, N.M. & Fitton, M.G. 1990. *The Biological Significance and Conservation of Hymenoptera in Europe*. 47pp. Council of Europe, Strasbourg.

Gauld, I.D. & Fitton, M.G. 1987. Sexual dimorphism in Ichneumonidae: a response to Hurlbutt. *Biological Journal of the Linnean Society*, **31**: 291–300.

Gauld, I.D. & Gaston, K.J. 1994. The taste of enemy-free space: parasitoids and nasty hosts, pp 279–299. *In*: Hawkins, B.A. & Sheehan, W. (eds) *Parasitoid Community Ecology*. Oxford University Press, Oxford.

Gauld, I.D., Gaston, K.J. & Janzen, D.H. 1992. Plant allelochemicals, tritrophic interactions and the anomalous diversity of tropical parasitoids: the 'nasty' host hypothesis. *Oikos*, **65**: 353–357.

Gauld, I.D. & Janzen, D.H. 1994. The classification, evolution and biology of the Costa Rican species of *Cryptophion* (Hymenoptera: Ichneumonidae). *Zoological Journal of the Linnean Society*, **110**: 297–324.

Gauld, I.D. & Lanfranco L., D. 1987. Los géneros de Ophioninae de Centro y Sudamérica. *Revista Biología Tropical*, **35**: 257–267.

Gauld, I.D. & Mitchell, P.A. 1977. Ichneumonidae, Orthopelmatinae & Anomaloninae. *Handbooks for the Identification of British Insects*, **7**(2b): 1–32.

Gauld, I.D. & Mitchell, P.A. 1981. *The Taxonomy, Distribution and Host Preferences of Indo-Papuan Parasitic Wasps of the Subfamily Ophioninae (Hymenoptera: Ichneumonidae)*. 611pp. Commonwealth Institute of Entomology, C.A.B., Slough.

Gautier-Hion, A. & Michaloud, G. 1989. Are figs always keystone resources for tropical frugivorous vertebrates? A test in Gabon. *Ecology*, **70**: 1826–1833.

Gay, H. 1993. Animal-fed plants: an investigation into the uptake of ant-derived nutrients by the far-eastern epiphytic fern *Lecanopteris* Reinw. (Polypodiaceae). *Biological Journal of the Linnean Society*, **50**: 221–233.

Genieys, P. 1924. Contribution a l'étude des Evaniidae: *Zeuxevania splendidula* Costa. *Bulletin Biologique de la France et de la Belgique*, **58**: 482–494.

Genieys, P. 1925. *Habrobracon brevicornis* Wesm. *Annals of the Entomological Society of America*, **18**: 143–202.

Genise, J.F. 1986. Las Bradynobaenidae y algunas modificaciones a la clasificación general de Hymenoptera Aculeata. *Physis (Buenos Aires)*, C, **44**: 39–53.

Genise, J.F. & Kimsey, L.S. 1991. New genera of South American Thynninae (Hymenoptera: Tiphiidae). *Psyche*, **98**: 57–69.

Genise, J.F. & Straneck, R. 1991. Preliminary studies on the taxonomic value of stridulation and the stridulatory organ in the Mutillidae. *Sphecos*, **21**: 7–11.

Gentry, A.H. 1982. Neotropical floristic diversity: phytogeographical connections between Central and South America, Pleistocene climatic fluctuations, or an accident of the Andean orogeny? *Annals of the Missouri Botanical Garden*, **69**: 557–593.

Gerig, L. 1960. Zur Morphologie der Larvenstadien einiger parasitischer Hymenopteren des Grauen Lärchenwicklers (*Zeiraphera griseana* Hübner). *Zeitschrift für Angewandte Entomologie*, **46**: 121–177.

Gerling, D. 1966. Studies with whitefly parasites of southern California. II. *Eretmocerus californicus* Howard (Hymenoptera: Aphelinidae). *Canadian Entomologist*, **98**: 1316–1329.

Gerling, D. 1972. The developmental biology of *Telenomus remus* Nixon (Hym., Scelionidae). *Bulletin of Entomological Research*, **61**: 385–388.

Gerling, D. 1992. Approaches to the biological control of whiteflies. *Florida Entomologist*, **75**: 446–456.

Gerling, D., Condé, J.E. & Rabinovich, J.E. 1976. The comparative development of two egg parasites of *Rhodnius prolixus* (Hemiptera: Reduviidae), vector of Chagas' disease in Venezuela. *Canadian Entomologist*, **108**: 427–432.

Gerling, D. & Legner, E.F. 1968. Developmental history and reproduction of *Spalangia cameroni*, parasite of synanthropic flies. *Annals of the Entomological Society of America*, **61**: 1436–1443.

Gerling, D. & Orion, T. 1973. The giant cells produced by *Telenomus remus* (Hymenoptera: Scelionidae). *Journal of Invertebrate Pathology*, **21**: 164–171.

Gerling, D., Orion, T. & Delarea, Y. 1990. *Eretmocerus* penetration and immature development: A novel approach to overcome host immunity. *Archives of Insect Biochemistry and Physiology*, **13**: 247–253.

Gerling, D., Tremblay, E. & Orion, T. 1991. Initial stages of the vital capsule formation in the *Eretmocerus-Bemisia tabaci* association. *Redia*, **74** (3) Appendice: 411–415.

Gerling, D., Velthuis, H.H.W. & Hefetz, A. 1989. Bionomics of the large carpenter bees of the genus *Xylocopa*. *Annual Review of Entomology*, **34**: 163–190.

Gerritsen, J. 1980. Sex and parthenogenesis in sparse populations. *American Naturalist*, **99**: 489–494.

Gess, F.W. & Gess, S.K. 1980. Ethological studies of *Jugurtia confusa* Richards, *Ceramius capicola* Brauns, *C. linearis* Klug and *C. lichtensteinii* (Klug) (Hymenoptera: Masaridae) in the Eastern Cape Province of South Africa. *Annals of the Cape Province Museum of Natural History*, **13**: 63–83.

Gess, F.W. & Gess, S.K. 1992. Ethology of three southern African ground nesting Masarinae, two *Celonites* species and a silk-spinning *Quartinia* species, with a discussion of nesting by the subfamily as a whole (Hymenoptera: Vespidae). *Journal of Hymenoptera Research*, **1**: 145–155.

Gherna, R.L., Werren, J.H., Weisburg, W., Cote, R., Woese C.R., Mandelco, L. & Brenner, D.J. 1991. *Arsenophonus nasoniae* new genus, new species, the causative agent of the son-killer trait in the parasitic wasp *Nasonia vitripennis*. *International Journal of Systematic Bacteriology*, **41**: 563–565.

Ghesquière, J. 1955. Contribution à l'étude du genre *Eriaporus* Waterston et genres affins (Hym. Chalcidoidea Aphelinidae). *Mémoires de la Société Royale d'Entomologie de Belgique*, **27**: 216–238.

Gibson, G.A.P. 1985. Some pro- and mesothoracic structures important for phylogenetic analysis of Hymenoptera, with a review of terms used for the structures. *Canadian Entomologist*, **117**: 1395–1443.

Gibson, G.A.P. 1986*a*. Evidence for monophyly and relationships of Chalcidoidea, Mymaridae, and Mymarommatidae (Hymenoptera: Terebrantes). *Canadian Entomologist*, **118**: 205–240.

Gibson, G.A.P. 1986*b*. Mesothoracic skeletomusculature and mechanics of flight and jumping in Eupelminae (Hymenoptera, Chalcidoidea: Eupelmidae). *Canadian Entomologist*, **118**: 691–728.

Gibson, G.A.P. 1989. Phylogeny and classification of Eupelmidae, with a revision of the world genera of Calosotinae and Metapelmatinae (Hymenoptera: Chalcidoidea). *Memoirs of the Entomological Society of Canada*, **149**: 1–121.

Gibson, G.A.P. 1993. Groundplan structure and homology of the pleuron in Hymenoptera based on a comparison of the skeletomusculature of Xyelidae (Hymenoptera) and Raphidiidae (Neuroptera). *Memoirs of the Entomological Society of Canada*, **165**: 165–187.

Gibson, L.P. 1972*a*. Revision of the genus *Urosigalphus* of the United States and Canada (Hymenoptera: Braconidae). *Miscellaneous Publications of the Entomological Society of America*, **8**: 83–134.

Gibson, L.P. 1972*b*. *Urosigalphus* of Mexico and Central America (Hymenoptera: Braconidae). *Miscellaneous Publications of the Entomological Society of America*, **8**: 135–157.

Gifford, J.R. & Mann, G.A. 1967. Biology, rearing, and a trial release of *Apanteles flavipes* in the Florida Everglades to control the sugarcane borer. *Journal of Economic Entomology*, **60**: 44–47.

Gilbert, W.M. 1973. Foraging behavior of *Trigona fulviventris* in Costa Rica (Hymenoptera: Apidae). *Pan-Pacific Entomologist*, **49**: 21–25.

Gilkeson, L.A., McLean, J.P. & Dessart, P. 1993. *Aphanogmus fulmeki* Ashmead (Hymenoptera: Ceraphronidae), a parasitoid of *Aphidoletes aphidimyza* Rondani (Diptera: Cecidomyiidae). *Canadian Entomologist*, **125**: 161–162.

Gillaspy, J.E. 1979. Management of *Polistes* wasps for caterpillar predation. *Southwestern Entomologist*, **4**: 334–352.

Gillaspy, J.E. 1982. The impact on man of *Polistes* wasps, with special reference to caterpillar suppression, pp. 129–133. *In*: Breed, M.D., Michener, C.D. & Evans, H.E. (eds). *The Biology of Social Insects*, 419pp. Westview Press, Boulder.

Gillaspy, J.E. 1986. *Polistes* wasps: biology and impact on man, pp. 332–352. *In*: Vinson, S.B. (ed.). *Economic Impact and Control of Social Insects*, 421pp. Praeger Publishers, New York.

Gilstrap, F.E. & Hart, W.G. 1987. Biological control of the Mediterranean fruit fly in the United States and Central America. *United States Department of Agriculture, Agricultural Research Service*, **56**: 1–64.

Giordanengo, P. & Nenon, J.P. 1990. Melanization and encapsulation of eggs and larvae of *Epidinocarsis lopezi* by its host *Phenacoccus manihoti*, effects of superparasitism and egg laying patterns. *Entomologia Experimentalis et Applicata*, **56**: 155–163.

Giraldo-Vanegas, H. & García, J.L. 1992. Determinación del número de instares de *Eiphosoma vitticolle* Cresson (Hymenoptera: Ichneumonidae). *Boletín de Entomología Venezolana* (N.S.), **7**: 133–137.

Girault, A.A. 1917. Notes on chalcid flies, chiefly from California. *Journal of Entomology and Zoology*, **9**: 8–12.

Giri, M.K. & Freytag, P.H. 1989 (1986). Development of *Dicondylus americanus* (Hymenoptera: Dryinidae). *Frustula Entomologica* (N.S.), **9**: 215–222.

Glas, P.C.G. & Vet, L.E.M. 1983. Host-habitat location and host location by *Diachasma alloeum* Muesebeck (Hym.; Braconidae), a parasitoid of *Rhagoletis pomonella* (Dipt.; Tephritidae). *Netherlands Journal of Zoology*, **33**: 41–54.

Glick, P.A. 1939. The distribution of insects, spiders and mites in the air. *United States Department of Agriculture Technical Bulletin*, **673**: 1–150.

Gobbi, N., Machado, V.L.L. & Tavares Filho, J.A. 1984. Sazonalidade das presas utilizadas na alimentacaçao de *Polybia occidentalis occidentalis* (Olivier, 1791) (Hym., Vespidae). *Anais da Sociedade Entomológico do Brasil*, **13**: 63–69.

Godfray, H.C.J. 1987. The evolution of clutch size in parasitic wasps. *American Naturalist*, **129**: 221–233.

Godfray, H.C.J. 1988. Virginity in haplodiploid populations: a study on fig wasps. *Ecological Entomology*, **13**: 283–291.

Godfray, H.C.J. 1994. *Parasitoids. Behavioral and Evolutionary Ecology*. 473pp. Princeton University Press, Princeton N.J.

Goeden, R.D., Ricker, D.W. & Müller H. 1987. Introduction, recovery, and limited establishment of *Coleophora klimeschiella* (Lepidoptera: Coleophoridae) on Russian thistles, *Salsola australis*, in Southern California. *Environmental Entomology*, **16**: 1027–1029.

Gold, C.S., Altieri, M.A. & Bellotti, A.C. 1989. Relative oviposition rates of the cassava hornworm, *Erinnyis ello* (Lep.: Sphingidae), and accompanying parasitism by *Telenomus sphingis* (Hym.: Scelionidae), on upper and lower leaf surfaces of cassava. *Entomophaga*, **34**: 73–76.

Gomez L., L.A. & Schoonhoven, A. v. 1977. Oviposición del *Empoasca kraemeri* en frijol y evalucación del parasitismo por *Anagrus* sp. *Revista Colombiana de Entomología*, **3**: 29–38.

Gómez, L.D. 1974. Biology of the potato-fern *Solanopteris brunei*. *Brenesia*, **4**: 37–61.

Gómez, L.D. 1986. Vegetación de Costa Rica, pp. 1–327. *In*: Gómez, L.D. (ed.) *Vegetación y Clima de Costa Rica*, 1. 327pp. EUED, San José.

Gómez-Núñez, J.C. 1971. *Tapinoma melanocephalum* as an inhibitor of *Rhodnius prolixus* populations. *Journal of Medical Entomology*, **8**: 735–737.

Gómez Rodriguez, R. 1986. *Manejo de la Abeja Africanizada*. 280pp. Apicultura Venezolana, Caracas.

Gómez Sousa, J., Meneses Carbonell, R. & Grillo Ravelo, H., 1979. Principales enemigos naturales de *Sogatodes orizicola* (Muir) (Homoptera, Delphacidae) en la zona arrocera de Sancti Spiritus, Cuba. *Centro Agricola*, **6**: 3–13.

Gordh, G. 1973. Biological investigations on *Comperia merceti* (Compere), an encyrtid parasite of the cockroach *Supella longipalpa* (Serville). *Journal of Entomology* (A), **47**: 115–123.

Gordh, G. 1975. The comparative external morphology and systematics of the Neotropical parasitic fig wasp genus *Idarnes* (Hymenoptera: Torymidae). *University of Kansas Science Bulletin*, **50**: 389–455.

Gordh, G. 1976*a*. *Goniozus gallicola* Fouts, a parasite of moth larvae, with notes on other bethylids (Hymenoptera: Bethylidae; Lepidoptera: Gelechiidae). *Technical Bulletin of the United States Department of Agriculture*, **1524**: 1–27.

Gordh, G. 1976*b*. A new genus of Pteromalidae from Missouri, the type-species of which parasitizes *Uloborus octonarius* Muma (Hymenoptera: Chalcidoidea; Araneida: Uloboridae). *Journal of the Kansas Entomological Society*, **49**: 100–104.

Gordh, G. 1977. A new neotropical species of Pteromalidae (Hymenoptera: Chalcidoidea) that parasitizes Diptera. *Brenesia*, **10/11**: 85–94.

Gordh, G. 1978. Taxonomic notes on *Zagrammosoma*, a key to the Nearctic species and descriptions of new species from California (Hymenoptera: Eulophidae). *Proceedings of the Entomological Society of Washington*, **80**: 344–359.

Gordh, G. 1981. The phenomenon of insect hyperparasitism and its taxonomic occurrence in the Insecta, pp. 10–18. *In*: Rosen, D. (ed.) *The Role of Hyperparasitism in Biological Control: a Symposium* (Number 4103). 52pp. University of California, Berkeley.

Gordh, G. 1983. New distributional and host records for *Arachnopteromalus dasys* Gordh (Hymenoptera: Pteromalidae), an egg-sac parasite of uloborid spiders (Araneida). *Proceedings of the Entomological Society of Washington*, **85**: 181.

Gordh, G. & DeBach, P. 1978. Courtship behavior in the *Aphytis lingnanensis* group, its potential usefulness in taxonomy, and a review of sexual behavior in the parasitic Hymenoptera (Chalcidoidea: Aphelinidae). *Hilgardia*, **46**: 37–75.

Gordh, G. & Hall, J.C. 1979. A critical point drier used as a method of mounting insects from alcohol. *Entomological News*, **90**: 57–59.

Gordh, G. & Hawkins, B. 1981. *Goniozus emigratus* (Rohwer), a primary external parasite of *Paramyelois transitella* (Walker), and comments on bethylids attacking Lepidoptera (Hymenoptera: Bethylidae; Lepidoptera: Pyralidae). *Journal of the Kansas Entomological Society*, **54**: 787–803.

Gordh, G. & Hendrickson, R. 1976. Courtship behavior in *Bathyplectes anurus* (Thomson) (Hymenoptera: Ichneumonidae). *Entomological News*, **87**: 271–274.

Gordh, G. & Hendrickson, R. Jr. 1979. New species of *Diglyphus*, a world list of the species, taxonomic notes and a key to New World species of *Diglyphus* and *Diaulinopsis* (Hymenoptera: Eulophidae). *Proceedings of the Entomological Society of Washington*, **81**: 666–684.

Gordh, G. & Lacey, L. 1976. Biological studies of *Plagiomerus diaspidis* Crawford, a primary internal parasite of diaspidid scale insects (Hymenoptera: Encyrtidae; Homoptera: Diaspididae). *Proceedings of the Entomological Society of Washington*, **78**: 132–144.

Gordh, G. & Medved, R.E. 1986. Biological notes on *Goniozus pakmanus* Gordh (Hymenoptera: Bethylidae), a parasite of pink bollworm, *Pectinophora gossypiella* (Saunders) (Lepidoptera: Gelechiidae). *Journal of the Kansas Entomological Society*, **59**: 723–734.

Gordh, G., Woolley, J.B. & Medved, R.E. 1983. Biological studies on *Goniozus legneri* Gordh (Hymenoptera: Bethylidae) a primary external parasite of the navel orangeworm *Amyelois transitella* and pink bollworm *Pectinophora gossypiella* (Lepidoptera: Pyralidae, Gelechiidae). *Contributions of the American Entomological Institute*, **20**: 433–468.

Gorton, R.E. Jr. 1978. Observation on the nesting behavior of *Mischocyttarus immarginatus* (Rich.) (Vespidae: Hymenoptera) in a dry forest in Costa Rica. *Insectes Sociaux*, **25**: 197–204.

Gotwald, W.H. Jr 1982. Army ants, pp. 157–254. *In*: Hermann, H.R. (ed.), *Social Insects IV*, 385pp. Academic Press, New York.

Gotwald, W.H. Jr & Brown, W.L. Jr 1967 (1966). The ant genus *Simopelta* (Hymenoptera: Formicidae). *Psyche*, **73**: 261–277.

Goulet, H. & Huber, J.T. (eds). 1993. *Hymenoptera of the World: An Identification Guide to Families*. 668pp. Agriculture Canada Publications, Ottawa.

Graenicher, P. 1905. On the habits of two ichneumonid parasites of the bee *Ceratina dupla* Say. *Entomological News*, **16**: 43–49.

Graham, A.R. 1953. Biology and establishment in Canada of *Mesoleius tenthredinis* Morley (Hymenoptera: Ichneumonidae), a parasite of the larch sawfly, *Pristiphora erichsonii* (Hartig) (Hymenoptera: Tenthredinidae). *Report of the Quebec Society for the Protection of Plants*, **35**: 61–75.

Graham, M.W.R. de V. 1969. The Pteromalidae of north-western Europe (Hymenoptera: Chalcidoidea). *Bulletin of the British Museum (Natural History)* Entomology Supplement, **16**: 1–908.

Graham, M.W.R. de V. 1987. A reclassification of the European Tetrastichinae (Hymenoptera: Eulophidae), with a revision of certain genera. *Bulletin of the British Museum (Natural History)* Entomology, **55**: 1–392.

Graham, M.W.R. de V. 1991. A reclassification of European Tetrastichinae (Hymenoptera: Eulophidae): revision of the remaining genera. *Memoirs of the American Entomological Institute*, **49**: 1–322.

Grandi, G. 1929. Studio morfologico e biologico della *Blastophaga psenes* (L.). *Bollettino del Laboratorio di Entomologia del R. Istituto Superiore Agraria di Bologna*, **2**: 1–147.

Grandi, G. 1954. Contributi alla conoscenza degli Imenotteri aculeati XXVI. *Bollettino dell'Istituto di Entomologia dell'Università degli studi di Bologna*, **20**: 81–255.

Grandi, G. 1961. Studi di un entomologo sugli Imenotteri superiori. *Bollettino dell'Istituto di Entomologia dell'Università degli studi di Bologna*, **25**: 1–659.

Grant, B., Burton, S., Contoreggi, C., & Rothstein, M. 1980. Outbreeding via frequency-dependent mate selection in the parasitoid wasp, *Nasonia (= Mormoniella) vitripennis* Walker. *Evolution*, **34**: 983–992.

Graur, D. 1985. Gene diversity in Hymenoptera. *Evolution*, **39**: 190–199.

Greathead, D.J. 1986. Parasitoids in classical biological control, pp. 289–318. *In*: Waage, J. & Greathead, D. (eds) *Insect Parasitoids*. 389pp. Academic Press, London.

Greggers, U. & Menzel, R. 1993. Memory dynamics and foraging strategies of honeybees. *Behavioral Ecology and Sociobiology*, **32**: 17–29.

Grijpma, P. 1984. Host specificity of *Telenomus nitidulus* (Thomson) (Hymenoptera: Scelionidae), egg parasite of the satin moth, *Leucoma salicis* L. *Nederlands Bosbouwtijdschrift*, **56**: 201–207.

Grijpma, P., Lenteren, J.C. van & Sonderen, L.M. van 1991. Host specificity and oviposition behaviour of *Telenomus nitidulus*, egg parasite of the satin moth, *Leucoma salicis*. *Proceedings in Experimental and Applied Entomology*, **2**: 169–170.

Griffiths, G.C.D. 1960. The behaviour and specificity of *Monoctonus paludum* Marshall (Hym., Braconidae), a parasite of *Nasonovia ribes-nigri* (Mosley) on lettuce. *Bulletin of Entomological Research*, **51**: 303–319.

Griffiths, G.C.D. 1964. The Alysiinae (Hym. Braconidae) parasites of the Agromyzidae (Diptera). I. General questions of taxonomy, biology and evolution. *Beiträge zur Entomologie*, **14**: 823–914.

Griffiths, N.T. & Godfray, H.C.J. 1988. Local mate competition, sex ratio and clutch size in bethylid wasps. *Behavioural Ecology and Sociobiology*, **22**: 211–217.

Grissell, E.E. 1974. A new *Dibrachys* with a key to the Nearctic species (Hymenoptera: Pteromalidae). *Florida Entomologist*, **57**: 313–320.

Grissell, E.E. 1976. A revision of Western Nearctic species of *Torymus* Dalman (Hymenoptera: Torymidae). *University of California Publications in Entomology*, **79**: 1–120.

Grissell, E.E. 1981. *Edovum puttleri*, n.g., n.sp. (Hymenoptera: Eulophidae), an egg parasite of the Colorado potato beetle (Chrysomelidae). *Proceedings of the Entomological Society of Washington*, **83**: 790–796.

Grissell, E.E. 1985. Some nomenclatural changes in the Chalcidoidea (Hymenoptera). *Proceedings of the Entomological Society of Washington*, **87**: 350–355.

Grissell, E.E. 1989. *Megastigmus floridanus* (Hymenoptera: Torymidae), newly discovered in Ilex seed (Aquifoliaceae). *Proceedings of the Entomological Society of Washington*, **91**: 604–610.

Grissell, E.E. 1991. A revision of Nearctic Chalcedectini (Chalcidoidea: Pteromalidae) with a New World checklist. *Proceedings of the Entomological Society of Washington*, **93**: 1–15.

Grissell, E.E. 1992. A revision of *Perissocentrus* Crawford (Hymenoptera: Torymidae). *Journal of Hymenoptera Research*, **1**: 91–102.

Grissell, E.E. & De Santis, L. 1987. A new species of *Erixestus* (Hymenoptera: Pteromalidae), an egg parasitoid of *Calligrapha polyspila* (Coleoptera: Chrysomelidae) in Argentina. *Proceedings of the Entomological Society of Washington*, **89**: 264–268.

Grissell, E.E. & Goodpasture, C.E. 1981. A review of Nearctic Podagrionini, with description of sexual behavior of *Podagrion mantis* (Hymenoptera: Torymidae). *Annals of the Entomological Society of America*, **74**: 226–241.

Grissell, E.E. & Schauff, M.E. 1981. New Nearctic *Invreia* (Hymenoptera: Chalcididae) from lepidopterous pests of peanut. *Proceedings of the Entomological Society of Washington*, **83**: 1–12.

Grissell, E.E. & Schauff, M.E. 1990. *A Handbook of the Families of Nearctic Chalcidoidea (Hymenoptera)*. 85pp. The Entomological Society of Washington, Washington, D.C.

Griswold, T.L. & Michener, C.D. 1988. Taxonomic observations on Anthidiini of the Western Hemisphere. *Journal of the Kansas Entomological Society*, **61**: 22–45.

Groff, G.W. & Howard, C.W. 1924 (1925). The cultured citrus ant of South China. *Lingnaam Agricultural Review*, **2**: 108–114.

Groombridge, B. (ed.) 1992. *Global Biodiversity. Status of the Earth's Living Resources*. 585pp. Chapman & Hall, London.

Gross, P. 1991. Influence of target pest feeding niche on success rates in classical biological control. *Environmental Entomology*, **20**: 1217–1227.

Gross, P. 1993. Insect behavioral and morphological defenses against parasitoids. *Annual Review of Entomology*, **38**: 251–273.

Gross, P. & Price, P.W. 1988. Plant influences on parasitism of two leafminers: a test of enemy free space. *Ecology*, **69**: 1506–1516.

Guagliumi, P. 1971. Entomofauna della canna da zucchero nel nord-est del Brasile. *Rivista di Agricoltura Subtropicale e Tropicale*, **65**: 49–66; 95–127.

Gudger, E.W. 1925. Stitching wounds with the mandibles of ants and beetles. A minor contribution to the history of surgery. *Journal of the American Medical Association*, **84**: 1861–1864.

Guerrant, E.O. Jr & Fiedler, P.L. 1981. Flower defenses against nectar-pilferage by ants. *Biotropica* (Supplement), **13**(2): 25–33.

Guiglia, D. 1958. Les sclerodermines par rapport à l'homme. *Proceedings of the Tenth International Congress of Entomology, Montreal*, **3**: 883–887.

Guillot, F.S. & Vinson, S.B. 1972. The role of the calyx and poison gland of *Cardiochiles nigriceps* in the host-parasitoid relationship. *Journal of Insect Physiology*, **18**: 1315–1321.

Guimaraes, J.A. 1957. *Contribuicao ao estudio da Cecidologia Brasiliana*. [Unpublished Thesis, UFRRJ, Brasil.]

Gupta, A.P. (ed.) 1986. *Hemocytic and Humoral Immunity in Arthropods*. 535pp. Wiley-Interscience, New York.

Gupta, S. 1982. A revision of the genus *Agonocryptus* (Hymenoptera: Ichneumonidae). *Contributions of the American Entomological Institute*, **19**(2): 1–45.

Gupta, S. 1983. Studies on the genus *Prosthoporus* (Hymenoptera: Ichneumonidae). *Contributions of the American Entomological Institute*, **20**: 245–253.

Gupta, S. & Gupta, V. 1983. Ichneumonologia Orientalis, 9. The tribe Gabuniini (Hymenoptera: Ichneumonidae). *Oriental Insects Monograph*, **10**: 1–313.

Gupta, V.K. 1962. Taxonomy, zoogeography and evolution of Indo-Australian *Theronia* (Hymenoptera: Ichneumonidae). *Pacific Insects Monograph*, **4**: 1–142.

Gupta, V.K. 1988 (1987). A revision of the genus *Microcharops* (Hymenoptera: Ichneumonidae). *Contributions of the American Entomological Institute*, **23**(2): 1–42.

Gurney, A.B. 1953. Notes on the biology and immature stages of a cricket parasite of the genus *Rhopalosoma*. *Proceedings of the United States National Museum*, **103**: 19–34.

Gutierrez, A.P. 1970*a*. Studies on host selection and host specificity of the aphid hyperparasite *Charips victrix* (Hymenoptera: Cynipidae) 3. Host suitability studies. *Annals of the Entomological Society of America*, **63**: 1485–1491.

Gutierrez, A.P. 1970*b*. Studies on host selection and host specificity of the aphid hyperparasite *Charips victrix* (Hymenoptera: Cynipidae) 4. The effect of age of host on host selection. *Annals of the Entomological Society of America*, **63**: 1491–1494.

Gutierrez, A.P. 1970*c*. Studies on host selection and host specificity of the aphid hyperparasite *Charips victrix* (Hymenoptera: Cynipidae) 5. Host selection. *Annals of the Entomological Society of America*, **63**: 1495–1498.

Gutierrez, A.P. 1970*d*. Studies on host selection and host specificity of the aphid hyperparasite *Charips victrix* (Hymenoptera: Cynipidae) 6. Description of sensory structures and a synopsis of host selection and host specificity. *Annals of the Entomological Society of America*, **63**: 1705–1709.

Gutierrez, A.P. & Bosch, R. van den, 1970*a*. Studies on host selection and host specificity of the aphid hyperparasite *Charips victrix* (Hymenoptera: Cynipidae). 1. Review of hyperparasitism and the field ecology of *Charips victrix*. *Annals of the Entomological Society of America*, **63**: 1345–1354.

Gutierrez, A.P. & Bosch, R. van den, 1970*b*. Studies on host selection and host specificity of the aphid hyperparasite *Charips victrix* (Hymenoptera: Cynipidae). 2. The bionomics of *Charips victrix*. *Annals of the Entomological Society of America*, **63**: 1355–1360.

Guzo, D. & Stoltz, D.B. 1985. Obligatory multiparasitism in the tussock moth, *Orgyia leucostigma*. *Parasitology*, **90**: 1–10.

Guzo, D. & Stoltz, D.B. 1987. Observation on the cellular immunity and parasitism in the tussock moth. *Journal of Insect Physiology*, **33**: 19–31.

Habeck, D.H., Bennett, F.D. & Grissell, E.E. 1989. First record of a phytophagous seed chalcid from Brazilian peppertree in Florida. *Florida Entomologist*, **72**: 378–379.

Haber, W.A. & Frankie, G.W. 1982. Pollination of *Luehea* (Tiliaceae) in Costa Rican deciduous forest. *Ecology*, **63**: 1740–1750.

Haeselbarth, E. 1962. Zur Biologie, Entwicklungsgeschichte und Okologie von *Brachistes atricornis* Ratz. (Hym., Brac.) als eines Parasiten von *Pissodes piceae* (Ill.) (Col., Curc.). *Zeitschrift für Angewandte Entomologie*, **49**: 233–289.

Haeselbarth, E. & Loan, C. 1983. *Townesilitus*, a new genus for a species group in *Microctonus* (Hymenoptera: Braconidae, Euphorinae). *Contributions of the American Entomological Institute*, **20**: 384–387.

Haeussler, G.J. 1940. Parasites of the Oriental fruit moth in Japan and Chosen and their introduction into the United States. *Technical Bulletin of the United States Department of Agriculture*, **728**: 1–62.

Haeussler, G.J. & Clancy, D.W. 1944. Natural enemies of Comstock mealybug in the eastern States. *Journal of Economic Entomology*, **37**: 503–509.

Hafez, M. 1961. Seasonal fluctuations of population density of the cabbage aphid, *Brevicoryne brassicae* (L.), in the Netherlands, and the role of its parasite *Aphidius (Diaretiella) rapae* (Curtis). *Tijdschrift voor Plantenziekten*, **67**: 445–548.

Hafez, M., Tawfik, M.F.S. & Raouf, A. 1970 (1971). On the population dynamics of the black scale *Chrysomphalus ficus* Ashm., in U.A.R. Part III. On the bionomics of *Habrolepis pascuorum* Mercet (Hymenopte [sic] Encyrtidae), a parasite of the black scale, *Chrysomphalus ficus* Ashm. *Technical Bulletin, United Arab Republic, Ministry of Agriculture and Agrarian Reform, Plant Protection Department*, **2**: 35–89.

Hagen, K.S. 1953. A premating period in certain species of the genus *Opius* (Hymenoptera: Braconidae). *Proceedings of the Hawaiian Entomological Society*, **15**: 115–116.

Hagen, K.S. 1964. Developmental stages of parasites, pp. 168–246. *In*: DeBach, P. (ed.) *Biological Control of Insect Pests and Weeds*. 844pp. Chapman & Hall, New York.

Hagen, K.S. 1984. Aquatic Hymenoptera, pp. 438–447. *In*: Merritt, R.W. & Cummings, K.W. (eds) *An Introduction to the Aquatic Insects of North America*, second edition. 722pp. Kendall-Hunt, Dubuque.

Hagen, K.S. 1986. Ecosystem analysis: plant cultivars (HPR), entomophagous species and food supplements, pp. 151–197. *In*: Boethel, D.J. & Eikenbary, R.D. (eds) *Interactions of Plant Resistance and Parasitoids and Predators of Insects*. 224pp. Ellis Horwood, Chichester.

Hågvar, E.B. 1988. Multiparasitism of the green peach aphid, *Myzus persicae*: competition in the egg stage between *Aphidius matricariae* and *Ephedrus cerasicola*. *Entomologia Experimentalis et Applicata*, **47**: 275–282.

Hågvar, E.B. & Hofsvang, T. 1989. Effect of honeydew and hosts on plant colonization by the aphid parasitoid *Ephedrus cerasicola*. *Entomophaga*, **34**: 495–501.

Haines, C.P. 1984. Biological methods for integrated control of insects and mites in tropical stored products III: the use of predators and parasites. *Tropical Stored Products Information*, **48**: 17–25.

Halffter, G. 1987. Biogeography of the montane entomofauna of Mexico and Central America. *Annual Review of Entomology*, **32**: 95–114.

Hall, R.W., Ehler, L.E. & Bisabri-Ershadi, B. 1980. Rate of success in classical biological control of arthropods. *Bulletin of the Entomological Society of America*, **26**: 111–114.

Halstead, J.A. 1987. *Acanthochalcis nigricans* Cameron — New distribution information, including Central America (Hymenoptera: Chalcididae). *Pan-Pacific Entomologist*, **63**: 236.

Halstead, J.A. 1990*a*. Review of *Haltichella* Spinola in the Nearctic region (Hymenoptera: Chalcididae). *Proceedings of the Entomological Society of Washington*, **92**: 153–159.

Halstead, J.A. 1990*b*. Revision of *Hockeria* Walker in the Nearctic Region with descriptions of males and five new species (Hymenoptera: Chalcididae). *Proceedings of the Entomological Society of Washington*, **92**: 619–640.

Halstead, J.A. 1991*a*. New species of *Aspirhina* Kirby from the Neotropical Region (Hymenoptera: Chalcididae). *Pan-Pacific Entomologist*, **67**: 65–71.

Halstead, J.A. 1991*b*. New species of *Notaspidium* Dalla Torre from the Nearctic and Neotropical Regions (Hymenoptera: Chalcididae). *Pan-Pacific Entomologist*, **67**: 229–242.

Hamilton, K.G.A. 1972*a*. The insect wing, part III. Venation of the orders. *Journal of the Kansas Entomological Society*, **45**: 145–162.

Hamilton, K.G.A. 1972*b*. The insect wing, part IV. Venational trends and the phylogeny of the winged orders. *Journal of the Kansas Entomological Society*, **45**: 295–308.

Hamilton, W.D. 1964. The genetical evolution of social behaviour. I and II. *Journal of Theoretical Biology*, **7**: 1–52.

Hamilton, W.D. 1967. Extraordinary sex ratios. *Science*, **156**: 477–488.

Hamilton, W.D. 1972. Altruism and related phenomena, mainly in social insects. *Annual Review of Ecology and Systematics*, **3**: 193–232.

Hamilton, W.D. 1974. Evolution sozialer Verhaltensweisen bei sozialen Insekten, pp. 60–93. *In*: Schmidt, G.H. (ed.) *Sozialpolymorphismus bei Insekten Probleme der Kastenbildung im Tierreich*. 974pp. Wissenschaftliche Verlagsgesellschaft MBH, Stuttgart.

Hamilton, W.D. 1978. Evolution and diversity under bark, pp. 154–175. *In*: Mound, L.A. & Waloff, N. (eds) *Diversity of Insect Faunas*. 204pp. Blackwell Scientific Publications, London.

Hamilton, W.D. 1979. Wingless and fighting males in fig wasps and other insects, pp. 167–220. *In*: Blum, M.S. & Blum, N.A. (eds), *Sexual Selection and Reproductive Competition in Insects*. 463pp. Academic Press, New York.

Hamm, A.H. & Richards, O.W. 1930. The biology of the British fossorial wasps of the families Mellinidae, Gorytidae, Philanthidae, Oxybelidae, and Trypoxylidae. *Transactions of the Entomological Society of London*, **78**: 95–131.

Hammel, B. 1986. New species and notes on Lauraceae from the Caribbean lowlands of Costa Rica. *Journal of the Arnold Arboretum*, **67**: 123–136.

Hammond, G.H. 1944. Economic importance and host relationship of *Pelecinus polyturator* Drury. *Canadian Entomologist*, **76**: 130.

Hanna, A.D. 1934. The male and female genitalia and the biology of *Euchalcidia caryobori* Hanna (Hymenoptera: Chalcidinae). *Transactions of the Royal Entomological Society of London*, **82**: 107–136.

Hansen, J.D. 1980. The life history and behavior of *Spilochalcis albifrons* (Hymenoptera: Chalcididae), a parasite of the larch casebearer, *Coleophora laricella* (Lepidoptera: Coleophoridae). *Journal of the Kansas Entomological Society*, **53**: 553–566.

Hanson, H.S. 1939. Ecological notes on the *Sirex* wood wasps and their parasites. *Bulletin of Entomological Research*, **30**: 27–65.

Hanson, P.E. 1991. Los parásitos associados al cafeto en Costa Rica. *Manejo Integrado de Plagas* (Costa Rica), **20/1**: 8–10.

Hanson, P. 1992. The Nearctic species of *Ormyrus* Westwood (Hymenoptera: Chalcidoidea: Ormyridae). *Journal of Natural History*, **26**: 1333–1365.

Hansson, C. 1985. Taxonomy and biology of the Palearctic species of *Chrysocharis* Förster, 1856 (Hymenoptera: Eulophidae). *Entomologica Scandinavica*, **26**: 1–130.

Hansson, C. 1987. Revision of the New World species of *Chrysocharis* Förster (Hymenoptera: Eulophidae). *Entomologica Scandinavica*, **31**: 1–86.

Hansson, C. 1988. A revision of the genus *Mestocharis* and a review of the genus *Grahamia* (Hymenoptera, Eulophidae). *Proceedings of the Entomological Society of Washington*, **90**: 28–36.

Hansson, C. 1990. A taxonomic study on the palearctic species of *Chrysonotomyia* Ashmead and *Neochrysocharis* Kurdjumov (Hymenoptera: Eulophidae). *Entomologica Scandinavica*, **21**: 29–52.

Hansson, C. 1994*a*. Re-evaluation of the genus *Closterocerus* Westwood (Hymenoptera: Eulophidae), with a revision of the Nearctic species. *Entomologica Scandinavica*, **25**: 1–25.

Hansson, C. 1994*b*. The classification of *Chrysonotomyia* Ashmead and *Teleopterus* Silvestri (Hymenoptera: Eulophidae), with a review of the species in the Nearctic region. *Proceedings of the Entomological Society of Washington*, **96**: 665–673.

Hansson, C. 1995. Revision of the Nearctic species of *Neochrysocharis* Kurdjumov (Hymenoptera: Eulophidae). *Entomologica Scandinavica*, **26**: 27–46.

Hansson, C. & Cave, R.D. 1993. First record of *Kratoysma* (Hymenoptera: Eulophidae) from the New World, including the description of two new species. *Proceedings of the Entomological Society of Washington*, **95**: 253–257.

Harbo, J.R. & Kraft, K.J. 1969. A study of *Phanerotoma toreutae*, a parasite of the pine cone moth *Laspeyresia toreuta*. *Annals of the Entomological Society of America*, **62**: 214–220.

Harder, L.D. 1983. Functional differences of the proboscides of short- and long-tongued bees (Hymenoptera, Apoidea), *Canadian Journal of Zoology*, **61**: 1580–1586.

Harder, L.D. 1986. Effects of nectar concentration and flower depth on flower handling efficiency of bumble bees. *Oecologia*, **69**: 309–315.

Hardy, I.C.W. & Blackburn, T.M. 1991. Brood guarding in a bethylid wasp. *Ecological Entomology*, **16**: 55–62.

Hardy, I.C.W., Ode, P.J. & Strand, M.R. 1993. Factors influencing brood sex ratios in polyembryonic Hymenoptera. *Oecologia*, **93**: 343–348.

Hare, J.D. & Luck, R.F. 1991. Indirect effects of citrus cultivars on life history parameters of a parasitic wasp. *Ecology*, **72**: 1576–1585.

Harris, A.C. 1987. Pompilidae (Insecta: Hymenoptera). *Fauna of New Zealand*, **12**: 1–154.

Harrison, E.G., Fisher, R.C. & Ross, K.M. 1985. The temporal effects of Dufour's gland secretion in host discrimination by *Nemeritis canescens*. *Entomologia Experimentalis et Applicata*, **38**: 215–220.

Harrison, J.M. & Breed, M.D. 1987. Temporal learning in the giant tropical ant, *Paraponera clavata*. *Physiological Entomology*, **12**: 317–320.

Harrison, J.O. 1963. The natural enemies of some banana insect pests in Costa Rica. *Journal of Economic Entomology*, **56**: 282–285.

Hartley, S.E. & Lawton J.H. 1992. Host-plant manipulation by gall-insects: a test of the nutrition hypothesis. *Journal of Animal Ecology*, **61**: 113–119.

Haskins, C.P. 1978 (1979). Sexual calling behavior in highly primitive ants. *Psyche*, **85**: 407–415.

Hassan, S. A. 1988. The commercial use of *Trichogramma* and other egg parasites, information from 20 countries. *Trichogramma News*, **4**: 12–14.

Haueisen Freire, J.A. & Gara, R.I. 1970. Algumas observaçoes sôbre o comportamento de algumas especies do genero *Trigona* (Apidae—Meliponini). *Turrialba*, **20**: 351–356.

Haupt, H. 1932. Die Mundteile der Dryinidae (Hym.). *Zoologischer Anzeiger*, **99**: 1–18.

Haviland, M.D. 1920. On the bionomics and development of *Lygocerus testaceimanus*, Kieffer, and *Lygocerus cameroni*, Kieffer, (Proctotrypoidea-Ceraphronidae), parasites of *Aphidius* (Braconidae). *Quarterly Journal of Microscopical Science*, **65**: 101–127.

Haviland, M.D. 1921. On the bionomics and post-embryonic development of certain cynipid hyperparasites of aphides. *Quarterly Journal of Microscopical Science*, **65**: 451–478.

Haviland, M.D. 1922. On the post-embryonic development of certain chalcids, hyperparasites of aphides, with remarks on the bionomics of hymenopterous parasites in general. *Quarterly Journal of Microscopical Science*, **66**: 321–338.

Havron, A., Rosen, D., Rössler, Y. & Hillel, J. 1987. Selection of the male hemizygous genotype in arrhenotokous insects and mites. *Entomophaga*, **32**: 261–268.

Hawkins, B.A. 1990. Global patterns of parasitoid assemblage size. *Journal of Animal Ecology*, **59**: 57–72.

Hawkins, B.A. & Goeden, R.D. 1982. Biology of a gall-forming *Tetrastichus* (Hymenoptera: Eulophidae) associated with gall midges on saltbush in southern California. *Annals of the Entomological Society of America*, **75**: 444–447.

Hawkins, B.A. & Gordh, G. 1986. Bibliography of the world literature of the Bethylidae (Hymenoptera: Bethyloidea). *Insecta Mundi*, **1**: 261–283.

Hawkins, B.A. & Lawton, J.H. 1987. Species richness for parasitoids of British phytophagous insects. *Nature*, **326**: 788–790.

Hayat, M. 1983. The genera of Aphelinidae (Hymenoptera) of the world. *Systematic Entomology*, **8**: 63–102.

Heath, J. 1961. Some parasites of Eriocraniidae (Lep.). *Entomologist's Monthly Magazine*, **97**: 163.

Hedicke, H. 1939. *Hymenopterorum Catalogus, Pars 11: Gasteruptiidae*. 54pp. Junk, s'Gravenhage.

Hedicke, H. & Kerrich, G.J. 1940. A revision of the family Liopteridae (Hymenopt., Cynipoidea). *Transactions of the Royal Entomological Society of London*, **90**: 177–225.

Hedlin, A.F. 1956. Notes on the life history and habits of a chalcid, *Bruchophagus caraganae* (Nik.), (Hymenoptera: Chalcididae), infesting seeds of caragana. *Canadian Entomologist*, **88**: 622–625.

Hedqvist, K.-J. 1961. Notes on Cleonymidae (Hym. Chalcidoidea). I. *Entomologisk Tidskrift*, **82**: 91–110.

Hedqvist, K.-J. 1963. Notes on Hormiinae with description of new genera and species (Hym., Ichneumonoidea, Braconidae). *Entomologisk Tidskrift*, **84**: 30–61.

Hedqvist, K.-J. 1969. New genera and species of *Diparini* with notes on the tribe (Hym., Chalcidoidea). *Entomologisk Tidskrift*, **90**: 174–202.

Hedqvist, K.-J. 1978. A new subfamily and two new genera and species from the New World (Hymenoptera, Chalcidoidea: Pteromalidae). *Entomologica Scandinavica*, **9**: 135–139.

Hedström, I. 1988. Pollen carriers and fruit development of *Psidium guajava* L. (Myrtaceae) in the Neotropical region. *Revista de Biología Tropical*, **36**: 551–553.

Hefetz, A. 1987. The role of DuFour's gland secretions in bees. *Physiological Entomology*, **12**: 243–253.

Hegazi, E.M. & Führer, E. 1985. Instars of *Microplitis rufiventris* (Hym.: Braconidae) and their relative developmental speed under different photoperiods. *Entomophaga*, **30**: 231–243.

Hegdekar, B.M. & Arthur, A.P. 1973. Host hemolymph chemicals that induce oviposition in the parasite *Itoplectis conquisitor* (Hymenoptera: Ichneumonidae). *Canadian Entomologist*, **105**: 787–793.

Heinrich, B. 1976. Resource partitioning among some eusocial insects: bumblebees. *Ecology*, **57**: 874–889.

Heinrich, B. 1979. *Bumble Bee Economics*. 245pp. Harvard University Press, Cambridge, Mass.

Heinrich, B. 1983. Do bumblebees forage optimally, and does it matter? *American Zoologist*, **23**: 273–281.

Heinrich, G.H. 1960 (1961). Synopsis of Nearctic Ichneumoninae Stenopneusticae with particular reference to the Northeastern region (Hymenoptera). Part I. *Canadian Entomologist* (Supplement), **15**: 1–87.

Heinrich, G.H. 1977. Ichneumoninae of Florida and neighboring States. *Arthropods of Florida and Neighboring Land Areas*, **9**: 1–350.

Heinz, K.M. & Parrella, M.P. 1989. Attack behavior and host size selection by *Diglyphus begini* on *Liriomyza trifolii* in chrysanthemum. *Entomologia Experimentalis et Applicata*, **53**: 147–156.

Heithaus, E.R. 1979*a*. Community structure of Neotropical flower visiting bees and wasps: diversity and phenology. *Ecology*, **60**: 190–202.

Heithaus, E.R. 1979*b*. Flower-feeding specialization in wild bee and wasp communities in seasonal neotropical habitats. *Oecologia*, **42**: 179–194.

Heithaus, E.R. 1979*c*. Flower visitation records and resource overlap of bees and wasps in northwest Costa Rica. *Brenesia*, **16**: 9–52.

Heitland, W. & Pschorn-Walcher, H. 1993. Feeding strategies of sawflies, pp 93–118. *In*: Wagner, M.R. & Raffa, K.F. (eds) *Sawfly Life History Adaptations to Woody Plants*. 564pp. Academic Press, San Diego.

Helms, K.R. & Rissing, S.W. 1990 (1991). Single sex alate production by colonies of *Pheidole desertorum* and *Pheidole xerophila tucsonica* (Hymenoptera: Formicidae). *Psyche*, **97**: 213–216.

Henaut, A. 1990. Study of the sound produced by *Pimpla instigator* (Hymenoptera, Ichneumonidae) during host selection. *Entomophaga*, **35**: 127–139.

Henaut, A. & Guerdoux, J. 1982. Location of a lure by the drumming insect *Pimpla instigator* (Hymenoptera, Ichneumonidae). *Experientia*, **38**: 346–347.

Hendrickson, R.M. Jr. & Barth, S.E. 1978. Notes on the biology of *Diglyphus intermedius* (Hymenoptera: Eulophidae), a parasite of the alfalfa blotch leafminer, *Agromyza frontella* (Diptera: Agromyzidae). *Proceedings of the Entomological Society of Washington*, **80**: 210–215.

Hendry, L.B., Greany, P.D. & Gill, R.J. 1973. Kairomone mediated host-finding behavior in the parasitic wasp *Orgilus lepidus*. *Entomologia Experimentalis et Applicata*, **16**: 471–477.

Hennessey, R.D. 1981. At-rest setal wing coupling and restraining mechanisms in the Encyrtidae and Aphelindae (Hymenoptera: Chalcidoidea). *Annals of the Entomological Society of America*, **74**: 172–176.

Hennig, W. 1981. *Insect Phylogeny*. 514pp. Wiley, Chicester. [English translation by A.C. Pont.]

Henriksen, K.L. 1922. Notes upon some aquatic Hymenoptera (*Anagrus Brocheri* Schulz *Prestwichia aquatica* Lubb., *Agriotypus armatus* Walk.). *Annales de Biologie Lacustre*, **11**: 19–37.

Henriquez, N.P. & Spence, J.R. 1993. Host location by the gerrid egg parasitoid *Tiphodytes gerriphagus* (Marchal) (Hymenoptera: Scelionidae). *Journal of Insect Behavior*, **6**: 455–466.

Hensen, R.V. 1988. Revision of the nominate subgenus *Chalybion* Dahlbom (Hymenoptera, Sphecidae). *Tijdschrift voor Entomologie*, **131**: 13–64.

Hepburn, H.R. 1986. *Honeybees and Wax. An Experimental Natural History*. 205pp. Springer-Verlag, Berlin.

Hérard, F., Keller, M.A., Lewis, W.J. & Tumlinson, J.H. 1988. Beneficial arthropod behavior mediated by airborne semiochemicals. IV. Influence of host diet on host-oriented flight chamber responses of *Microplitis demolitor* Wilkinson. *Journal of Chemical Ecology*, **14**: 1597–1606.

Heraty, J.M. 1985. A revision of the Nearctic Eucharitinae (Hymenoptera: Chalcidoidea: Eucharitidae). *Proceedings of the Entomological Society of Ontario*, **116**: 61–103.

Heraty, J.M. 1986. *Pseudochalcura* (Hymenoptera: Eucharitidae), a New World genus of chalcidoids parasitic on ants. *Systematic Entomology*, **11**: 183–212.

Heraty, J.M. 1989. Morphology of the mesosoma of *Kapala* (Hymenoptera: Eucharitidae) with emphasis on its phylogenetic implications. *Canadian Journal of Zoology*, **67**: 115–125.

Heraty, J.M. 1994. Classification and evolution of the Oraseminae in the Old World, including revisions of two closely related genera of Eucharitinae (Hymenoptera: Eucharitidae). *Royal Ontario Museum Life Sciences Contributions*, **157**: 1–174.

Heraty, J.M. & Barber, K.N. 1990. Biology of *Obeza floridana* (Ashmead) and *Pseudochalcura gibbosa* (Provancher) (Hymenoptera: Eucharitidae). *Proceedings of the Entomological Society of Washington*, **92**: 248–258.

Heraty, J.M. & Darling, D.C. 1984. Comparative morphology of the planidial larvae of Eucharitidae and Perilampidae (Hymenoptera: Chalcidoidea). *Systematic Entomology*, **9**: 309–328.

Heraty, J.M., Wojcik, D.P. & Jouvenaz, D.P. 1993. Species of *Orasema* parasitic on the *Solenopsis saevissima*-complex in South America (Hymenoptera: Eucharitidae, Formicidae). *Journal of Hymenoptera Research*, **2**: 169–182.

Heraty, J.M. & Woolley, J.B. 1993. Separate species or polymorphism: a recurring problem in *Kapala* (Hymenoptera: Eucharitidae). *Annals of the Entomological Society of America*, **86**: 517–531.

Hermann, H.R. 1975. Crepuscular and nocturnal activities of *Paraponera clavata* (Hymenoptera: Formicidae: Ponerinae). *Entomological News*, **86**: 94–98.

Hermann, H.R. Jr. & Blum, M.S. 1966. The morphology and histology of the hymenopterous poison apparatus. I. *Paraponera clavata* (Formicidae). *Annals of the Entomological Society of America*, **59**: 397–409.

Hermann, H.R. & Blum, M.S. 1981. Defensive mechanisms in the social Hymenoptera, pp. 77–197. *In*: Hermann, H.R., *Social Insects*, 2. 491pp. Academic Press, New York.

Hermann, H.R. & Douglas, M.E. 1976. Comparative survey of the sensory structures on the sting and ovipositor of hymenopterous insects. *Journal of the Georgia Entomological Society*, **11**: 223–239.

Hermann, H.R. & Young, A.M. 1980. Artificially elicited defensive behavior and reciprocal aggression in *Paraponera clavata* (Hymenoptera: Formicidae: Ponerinae). *Journal of the Georgia Entomological Society*, **15**: 8–10.

Hernandez, M.P. 1984. *Ciclo de vida y hábitos de* Haplogonatopus hernandezae *Olmi (Hymenoptera: Dryinidae) controlador natural del saltahojas del arroz* Sogatodes orizicola *(Muir)*. Trabajo de Grado presentado como requisito para optar al título de Biologo (Entomologo). 67pp. Universidad del Valle, Departamento de Biología, Cali, Colombia.

Hernandez, M.P. & Belloti, A. 1984. Ciclos de vida y hábitos de *Haplogonatopus hernandezae* Olmi (Hymenoptera: Dryinidae) controlador natural del saltahojas del arroz *Sogatodes orizicola* (Muir). *Revista Colombiana de Entomologia*, **10**(3+4): 3–8.

Herre, E.A. 1987. Optimality, plasticity and selective regime in fig wasp sex ratios. *Nature*, **329**: 627–629.

Herre, E.A. 1989. Coevolution of reproductive characteristics in 12 species of New World figs and their pollinator wasps. *Experientia*, **45**: 637–647.

Herre, E.A. 1993. Population structure and the evolution of virulence in nematode parasites of fig wasps. *Science*, **259**: 1442–1445.

Herren, H.R.. & Neuenschwander, P. 1991. Biological control of cassava pests in Africa. *Annual Review of Entomology*, **36**: 257–283.

Herrera, W. 1986. Clima de Costa Rica, pp. 1–118. *In*: Gómez, L.D. (ed.) *Vegetación y Clima de Costa Rica* 2. 118pp. EUNED, San José.

Hespenheide, H.A. 1979. Are there fewer parasitoids in the tropics? *American Naturalist*, **113**: 766–769.

Hespenheide, H.A. 1985. Insect visitors to extrafloral nectaries of *Byttneria aculeata* (Sterculiaceae): relative importance and roles. *Ecological Entomology* **10**: 191–204.

Heydon, S.L. 1988. A review of the Nearctic species of *Cryptoprymna* Förster, with the description of a new genus, *Polstonia* (Hymenoptera: Pteromalidae). *Proceedings of the Entomological Society of Washington*, **90**: 1–11.

Heydon, S.L. 1989*a*. Relationships among Holarctic genera in the *Cyrtogaster*-group with a review of the species of North America north of Mexico (Hymenoptera: Pteromalidae). *Journal of the New York Entomological Society*, **97**: 192–217.

Heydon, S.L. 1989*b* A review of the world species of *Notoglyptus* Masi (Hymenoptera: Pteromalidae). *Proceedings of the Entomological Society of Washington*, **91**: 112–123.

Heydon, S.L. 1993. *Syntomopus* Walker: the nearctic species with a review of known host associations (Hymenoptera: Pteromalidae). *Journal of Hymenoptera Research*, **2**: 107–116.

Heydon, S.L. & Bouček, Z. 1992. Taxonomic changes in Nearctic Pteromalidae, with the description of some new taxa (Hymenoptera: Chalcidoidea). *Proceedings of the Entomological Society of Washington*, **94**: 471–489.

Heywood, V.H. 1985. *Flowering Plants of the World*. 336pp. Croom Helm, London.

Hidaka, T. 1958. Biological investigations on *Telenomus gifuensis* Ashmead (Hym.: Scelionidae), an egg-parasite of *Scotinophara lurida* Burmeister (Hem.: Pentatomidae) in Japan. *Acta Hymenopterologica*, **1**: 75–93.

Hill, C.C. 1923. *Platygaster vernalis* Myers, an important parasite of the hessian fly. *Journal of Agricultural Research, Washington D.C.*, **25**: 31–42.

Hill, C.C. 1926. *Platygaster hiemalis* Forbes, a parasite of the hessian fly. *Journal of Agricultural Research, Washington D.C.*, **32**: 261–275.

Hill, C.C. & Emery, W.T. 1937. The biology of *Platygaster herrickii*, a parasite of the hessian fly. *Journal of Agricultural Research, Washington D.C.*, **55**: 119–213.

Hincks, W.D. 1958. *Myrmecobosca mandibularis* Maneval (Hym., Braconidae), a myrmecophilous aphid parasite in Britain. *Entomologist's Monthly Magazine*, **94**: 20–21.

Hinton, H.E. 1951. Myrmecophilous Lycaenidae and other Lepidoptera — a summary. *Proceedings and Transactions of the South London Entomological and Natural History Society*, **1949–50**: 111–175.

Hinton, H.E. 1954. Entomology. Insect blood. *Science Progress*, **52**: 684–696.

Hinton, H.E. 1955. Protective devices of endopterygote pupae. *Transactions of the Society for British Entomology*, **12**: 49–92.

Hinz, R., 1983. The biology of the European species of the genus *Ichneumon* and related species (Hymenoptera: Ichneumonidae). *Contributions of the American Entomological Institute*, **20**: 151–152.

Hirashima, Y. & Yamagishi, K. 1975. Embolemidae of Japan, with description of a new species of *Embolemus* from Hachijo Island (Hymenoptera, Bethyloidea). *Esakia*, **9**: 25–30.

Hobby, B.M. 1932. The prey of sawflies (Hym., Tenthredinidae). *Proceedings of the Entomological Society of London*, **7**: 14–15.

Hocking, H. 1967. A native ichneumonid, *Certonotus tasmaniensis* Turn., parasitising *Sirex noctilio* F. (Siricidae) in Tasmania. *Journal of the Australian Entomological Society*, **6**: 57–60.

Hodges, C.M. 1985. Bumble bee foraging: the threshold departure rule. *Ecology*, **66**: 179–187.

Hoebeke, E.R. 1978. Notes on the biology of *Codrus carolinensis* (Hymenoptera: Proctotrupidae), a parasite of *Platydracus violaceus* (Coleoptera: Staphylinidae). *Journal of the Kansas Entomological Society*, **51**: 507–511.

Hoebeke, E.R. & Kovarik, P.W. 1988. *Exallonyx obsoletus* (Hymenoptera: Serphidae), a larval parasite of the rove beetle *Quedius laevigatus* (Coleoptera: Staphylinidae). *Entomological News*, **99**: 217–220.

Hoebeke, E.R. & Wheeler, Q.D. 1990. Notes on the biology of *Brachyserphus barberi* Townes (Hymenoptera; Serphidae) a parasitoid of the fungus beetle *Mycetophagus melsheimeri* LeConte (Coleoptera: Mycetophagidae). *Journal of the New York Entomological Society*, **98**: 376–378.

Hofsvang, T. 1990. Discrimination between unparasitized and parasitized hosts in hymenopterous parasitoids. *Acta Entomologica Bohemoslovakia*, **87**: 161–175.

Hofsvang, T. & Hågvar, E.B. 1986. Oviposition behaviour of *Ephedrus cerasicola* (Hym.: Aphidiidae) parasitizing different instars of its aphid host. *Entomophaga*, **31**: 261–267.

Holdridge, L.R. 1967. *Life Zone Ecology. Revised edition*. 206pp. Tropical Science Center, San José, Costa Rica.

Holdridge, L.R., Grenke, W.C., Hatheway, W.H., Liang, T. & Tosi, J.A. Jr, 1971. *Forest Environments in Tropical Life Zones. A pilot Study*. 747pp. Pergamon Press, Oxford.

Hölldobler, B. 1982. Communication, raiding behavior and prey storage in *Cerapachys* (Hymenoptera: Formicidae). *Psyche*, **89**: 3–23.

Hölldobler, B. & Engel-Siegel, H. 1982. Tergal and sternal glands in male ants. *Psyche*, **89**: 113–132.

Hölldobler, B. & Wilson, E.O. 1983. The evolution of communal nest-weaving in ants. *American Scientist*, **71**: 490–499.

Hölldobler, B. & Wilson, E.O. 1986. Ecology and behavior of the primitive cryptobiotic ant *Prionopelta amabilis* (Hymenoptera: Formicidae). *Insectes Sociaux*, **33**: 45–58.

Hölldobler, B. & Wilson, E.O. 1990*a*. *The Ants*. 732pp. Belknap, Harvard University Press, Cambridge, Mass.

Hölldobler, B. & Wilson, E.O. 1990*b*. Host tree selection by the neotropical ant *Paraponera clavata* (Hymenoptera: Formicidae). *Biotropica*, **22**: 213–214.

Höller, C., Bargen, H., Vinson, S.B. & Braune, H.J. 1993. Sources of the marking pheromones used for host discrimination in the hyperparasitoid *Dendrocerus carpenteri*. *Journal of Insect Physiology*, **39**: 649–656.

Höller, C., Williams, H.J. & Vinson, S.B. 1991. Evidence for a two-component external marking pheromone system in an aphid hyperparasitoid. *Journal of Chemical Ecology*, **17**: 1021–1035.

Holmes, H.B. 1972. Genetic evidence for fewer progeny and a higher percent males when *Nasonia vitripennis* oviposits in previously parasitized hosts. *Entomophaga*, **17**: 79–88.

Hood, J.D. 1914 (1913). Notes on the life history of *Rhopalosoma poeyi* Cresson. *Proceedings of the Entomological Society of Washington*, **15**: 145–148.

Hood, W.G. & Tschinkel, W.R. 1990. Desiccation resistance in arboreal and terrestrial ants. *Physiological Entomology*, **15**: 23–35.

Hook, A.W. 1987. Nesting behavior of Texas *Cerceris* digger wasps with emphasis on nest reutilization and nest sharing (Hymenoptera: Sphecidae). *Sociobiology*, **13**: 93–118.

Hook, A.W. & Matthews, R.W. 1980 (1981). Nesting biology of *Oxybelus sericeus* with a discussion of nest guarding by male sphecid wasps (Hymenoptera). *Psyche*, **87**: 21–37.

Hooker, M.E. & Barrows, E.M. 1992. Clutch size reduction and host discrimination in the superparasitizing gregarious endoparasitic wasp *Pediobius foveolatus* (Hymenoptera: Eulophidae). *Annals of the Entomological Society of America*, **85**: 207–213.

Höppner, H. 1904. Zur Biologie der Rubus-Bewohner I–III. *Allgemeine Zeitschrift für Entomologie. Berlin*, **9**: 97–103, 129–134, 161–171.

Horstmann, K. 1971. Revision der europäischen Tersilochinen (Hymenoptera: Ichneumonidae). Teil I. *Veröffentlichungen der Zoologischen Staatssammlung München*, **15**: 45–138.

Horvitz, C.C. & Beattie, A.J. 1980. Ant dispersal of *Calathea* (Marantaceae) seeds by carnivorous ponerines (Formicidae) in a tropical rain forest. *American Journal of Botany*, **67**: 321–326.

Houard, C. 1933. *Les Zoocécidies des Plantes de l'Amérique du Sud et de l'Amérique Centrale*, 519pp. Hermann et Cie, Paris.

Hough, J.S. 1953*a*. Studies on the common spangle gall of oak I. The developmental history. *New Phytologist*, **52**: 149–177.

Hough, J.S. 1953*b*. Studies on the common spangle gall of oak II. A general consideration of past work on gall induction. *New Phytologist*, **52**: 218–228.

Hough, J.S. 1953*c*. Studies on the common spangle gall of oak III. The importance of the stage in laminar extension of the host leaf. *New Phytologist*, **52**: 229–237.

Howard, J.J. 1985. Observations on resin collecting by six interacting species of stingless bees (Apidae: Melliponinae). *Journal of the Kansas Entomological Society*, **58**: 337–345.

Howard, J.J. 1987. Leafcutting ant diet selection: the role of nutrients, water, and secondary chemistry. *Ecology*, **68**: 503–515.

Howard, J.J. 1988. Leafcutting ant diet selection: relative influence of leaf chemistry and physical features. *Ecology*, **69**: 250–260.

Howard, L.O. 1937. Resumé and conclusions to Paul Marchal's extended paper on the Trichogrammas. *Annals of the Entomological Society of America*, **30**: 551–557.

Howarth, F.G. 1991. Environmental impacts of classical biological control. *Annual Review of Entomology*, **36**: 485–509.

Hrdy, I. & Sedivy, J. 1979. Males of *Exetastes cinctipes* (Hymenoptera, Ichneumonidae) attracted to 8-dodecenyl and 11-tetradecenyl acetates. *Acta Entomologica Bohemoslovaca*, **76**: 59–61.

Huang, H.T. & Yang, P. 1987. The ancient cultured citrus ant. *Bioscience*, **37**: 665–671.

Hubbard, S.F., Marris, G., Reynolds, A. & Rowe, G.W. 1987. Adaptive patterns in the avoidance of superparasitism by solitary parasitic wasps. *Journal of Animal Ecology*, **56**: 387–401.

Hubbell, S.P., Howard, J.J. & Wiemer, D.F. 1984. Chemical leaf repellency to an attine ant: seasonal distribution among potential host plant species. *Ecology*, **65**: 1067–1076.

Hubbell, S.P. & Johnson, L.K. 1977. Competition and nest spacing in a tropical stingless bee community. *Ecology*, **58**: 949–963.

Hubbell, S.P. & Johnson, L.K. 1978. Comparative foraging behavior of six stingless bee species exploiting a standardized resource. *Ecology*, **59**: 1123–1136.

Hubbell, S.P., Johnson, L.K., Stanislav, E., Wilson, B. & Fowler, H. 1980. Foraging by bucket-brigade in leaf-cutter ants. *Biotropica*, **12**: 210–213.

Huber, J.T. 1986. Systematics, biology, and hosts of the Mymaridae and Mymarommatidae (Insecta: Hymenoptera): 1758–1984. *Entomography*, **4**: 185–243.

Huber, J.T. 1987. Review of *Schizophragma* Ogloblin and the non-Australian species of *Stethynium* Enock (Hymenoptera: Mymaridae). *Canadian Entomologist*, **119**: 823–855.

Huber, J.T. 1988. The species groups of *Gonatocerus* Nees in North America with a revision of the *sulphuripes* and *ater* groups (Hymenoptera: Mymaridac). *Memoirs of the Entomological Society of Canada*, **141**: 1–109.

Huber, J.T. 1992. The subgenera, species groups, and synonyms of *Anaphes* (Hymenoptera: Mymaridae) with a review of the described Nearctic species of the *fuscipennis* group of *Anaphes s.s.* and the described species of *Anaphes (Yungaburra). Proceedings of the Entomological Society of Ontario*, **123**: 23–110.

Huber, J.T. 1993. New genus and two new species of Mymaridae (Hymenoptera) from Florida and tropical America. *Florida Entomologist*, **76**: 348–358.

Huber J.T. & Prinsloo, G.L. 1990. Redescription of *Anaphes nitens* (Girault) and description of two new species of *Anaphes* Haliday (Hymenoptera: Mymaridae), parasites of *Gonipterus scutellatus* Gyllenhal (Coleoptera: Curculionidae) in Tasmania. *Journal of the Australian Entomological Society*, **29**: 333–341.

Huber, J.T. & Rajakulendran, V.K. 1988. Redescription of and host-induced antennal variation in *Anaphes iole* Girault (Hymenoptera: Mymaridae), an egg parasite of Miridae (Hemiptera) in North America. *Canadian Entomologist*, **120**: 893–901.

Huddleston, T. 1980. A revision of the western palearctic species of the genus *Meteorus* (Hymenoptera: Braconidae). *Bulletin of the British Museum (Natural History)* (Entomology), **41**: 1–58.

Huddleston, T. 1983. *Meteorus* (Hymenoptera: Braconidae) of Australia and New Guinea. *Systematic Entomology*, **8**: 393–420.

Huddleston, T. & Gauld, I. 1988. Parasitic wasps (Ichneumonoidea) in British light-traps. *The Entomologist*, **107**: 134–154.

Huddleston, T. & Walker, A.K. 1988. *Cardiochiles* (Hymenoptera: Braconidae), a parasitoid of lepidopterous larvae, in the Sahel of Africa, with a review of the biology and host relationships of the genus. *Bulletin of Entomological Research*, **78**: 435–461.

Huffaker, C.B. & Matsumoto, B.M. 1982. Differences in egg wastage by superparasitism, contrasting *Venturia (= Nemeritis) canescens* searching singly versus searching in groups. *Researches on Population Ecology*, **24**: 270–275.

Huffaker, C.B. & Messenger, P.S. 1976. *Theory and Practice of Biological Control*. 788pp. Academic Press, New York.

Huger, A.M., Skinner, S.W. & Werren, J.H. 1985. Bacterial infections associated with the son-killer trait in the parasitoid wasp *Nasonia (=Mormoniella) vitripennis* (Hymenoptera: Pteromalidae). *Journal of Invertebrate Pathology*, **46**: 272–280.

Huggert, L. 1979. *Cryptoserphus* and Belytinae wasps (Hymenoptera, Proctotrupoidea) parasitizing fungus- and soil-inhabiting Diptera. *Notulae Entomologicae*, **59**: 139–144.

Huggert, L. & Masner, L. 1983. A review of myrmecophilic-symphilic diapriid wasps in the holarctic realm, with descriptions of new taxa and a key to genera (Hymenoptera: Proctotrupoidea: Diapriidae). *Contributions of the American Entomological Institute*, **20**: 63–89.

Hughes, R.D. & Woolcock, L.T. 1976. *Aphaereta aotea* sp. n. (Hymenoptera: Braconidae), an alysiine parasite of dung breeding flies. *Journal of the Australian Entomological Society*, **15**: 191–196.

Hummelen, P.J. 1974. Relations between two rice borers in Surinam, *Rupela albinella* (Cr.) and *Diatraea saccharalis* (F.), and their hymenopterous larval parasites. *Mededelingen ŸLandbouwhogeschool, Wageningen*, **74** (1): 1–88.

Hunt, J.H. 1988. Lobe erection behavior and its possible social role in larvae of *Mischocyttarus* paper wasps. *Journal of Insect Behavior*, **1**: 379–386.

Hunt, J.H. 1991. Nourishment and the evolution of the social Vespidae, pp. 426–450. *In*: Ross K.G. & Matthews, R.W. (eds) *The Social Biology of Wasps*. 678pp. Cornell University Press, Ithaca.

Hunt, J.H. 1993. Survivorship, fecundity, and recruitment in a mud dauber wasp, *Sceliphron assimile* (Hymenoptera: Sphecidae). *Annals of the Entomological Society of America*, **86**: 51–59.

Hunt, J.H., Brown, P.A., Sago, K.M. & Kerker, J.A. 1991. Vespid wasps eat pollen (Hymenoptera: Vespidae). *Journal of the Kansas Entomological Society*, **64**: 127–130.

Hunt, J.H., Jeanne, R.L., Baker, I. & Grogan, D.E. 1987. Nutrient dynamics of the swarm-founding social wasp species, *Polybia occidentalis* (Hymenoptera: Vespidae). *Ethology*, **75**: 291–305.

Hunt, J.H. & Nalepa, C.A. (eds) 1994. *Nourishment & Evolution in Insect Societies*. 448pp. Westview Press, Boulder, Colorado.

Hunter, K.W. Jr. & Stoner, A. 1975. *Copidosoma truncatellum*: effect of parasitization on food consumption of larval *Trichoplusia ni*. *Environmental Entomology*, **4**: 381–382.

Hunter, M.S., Nur, U. & Werren, J.H. 1993. Origin of males by genome loss in an autoparasitoid wasp. *Heredity*, **70**: 162–171.

Hurd, P.D. Jr. 1978*a*. *An Annotated Catalog of the Carpenter Bees (Genus* Xylocopa *Latreille) of the Western Hemisphere* (Hymenoptera: Anthophoridae). 106pp. Smithsonian Institution Press, Washington, D.C.

Hurd, P.D. 1978*b*. Bamboo-nesting carpenter bees (genus *Xylocopa* Latreille) of the subgenus *Stenoxylocopa* Hurd and Moure (Hymenoptera: Anthophoridae). *Journal of the Kansas Entomological Society*, **51**: 746–764.

Hurd, P.D. Jr. & Linsley, E.G. 1966. The Mexican squash and gourd bees of the genus *Peponapis* (Hymenoptera: Apoidea). *Annals of the Entomological Society of America*, **59**: 835–851.

Hurd, P.D. Jr. & Linsley, E.G. 1970. A classification of the squash and gourd bees *Peponapis and Xenoglossa*. *University of California Publications in Entomology*, **62**: 1–39.

Hurd, P.D. Jr. & Linsley, E.G. 1976. The bee family Oxaeidae with a revision of the North American species (Hymenoptera: Apoidea). *Smithsonian Contributions to Zoology*, **220**: 1–75.

Hurd, P.D. Jr. Linsley, E.G. & Whitaker, T.W. 1971. Squash and gourd bees (*Peponapis, Xenoglossa*) (Hymenoptera: Apoidea) and the origin of the cultivated Cucurbita. *Evolution*, **25**: 218–234.

Hurd, P.D. Jr. & Moure, J.S. 1960. A New-World subgenus of bamboo-nesting carpenter bees belonging to the genus *Xylocopa* Latreille (Hymenoptera: Apoidea). *Annals of the Entomological Society of America*, **53**: 809–821.

Hurd, P.D. Jr. & Moure, J.S. 1961. Some notes on sapygid parasitism in the nests of carpenter bees belonging to the genus *Xylocopa* Latreille (Hymenoptera: Aculeata). *Journal of the Kansas Entomological Society*, **34**: 19–22.

Hurd, P.D. Jr. & Wasbauer, M.S. 1956. New host records for North American spider-wasps (Hymenoptera: Pompilidae). *Journal of the Kansas Entomological Society*, **29**: 168–176.

Hussey, N.W. 1955. The life-histories of *Megastigmus spermotrophus* Wachtl (Hymenoptera: Chalcidoidea) and its principal parasite, with descriptions of the developmental stages. *Transactions of the Royal Entomological Society of London*, **106**: 133–151.

Hutchison, W. D., Moratorio, M. & Martin, J. M. 1990. Morphology and biology of *Trichogrammatoidea bactrae* (Hymenoptera: Trichogrammatidae), imported from Australia as a parasitoid of Pink Bollworm (Lepidoptera: Gelechiidae) eggs. *Annals of the Entomological Society of America*, **83**: 46–54.

Huxley, C.R. & Cutler, D.F. (eds) 1991. *Ant-Plant Interactions*. 601pp. Oxford University Press, Oxford.

Huzimatu, K. 1940. The life history of a new cynipid fly *Kleidotoma japonica* n.sp. *The Science Reports of the Tôhoku Imperial University* (Series 4) **15**: 457–480.

Ikan, R., Gottlieb, R., Bergmann, E.D. & Ishay, J. 1968. Ubiquinones of the queens of the oriental hornet, *Vespa orientalis*. *Journal of Insect Physiology*, **14**: 1215–1220.

Inouye, C.W. 1978. Resource partitioning in bumblebees: experimental studies of foraging behavior. *Ecology*, **59**: 672–678.

Isely, D. 1913 (1914). The biology of some Kansas Eumenidae. *University of Kansas Science Bulletin*, **8**: 235–309.

Ishii, T. 1931. Notes on the phytophagous habit of some chalcidoids, with descriptions of new species. *Kontyû*, **5**: 132–138.

Ivanova-Kasas, O.M. 1972. Polyembryony in insects, pp. 243–271. *In*: Counce, S.J. & Waddington, C.H. (eds) *Developmental Systems: Insects*, 1. 304pp. Academic Press, London.

Iwata, K. 1936. Biology of two Japanese species of *Methoca* with the description of a new species (Hymenoptera, Thynnidae). *Kontyû*, **10**: 57–89.

Iwata, K. 1958. Ovarian eggs of 233 species of the Japanese Ichneumonidae (Hymenoptera). *Acta Hymenopterologica*, **1**: 63–74.

Iwata, K. 1960. The comparative anatomy of the ovary in Hymenoptera. Part V. Ichneumonidae. *Acta Hymenopterologica*, **1**: 115–169.

Iwata, K. 1976. *Evolution of Instinct. Comparative Ethology of Hymenoptera*. 535pp. Amerind, New Delhi.

Jackson, D. 1969. New characters for generic separation in the *Synopeas-Leptacis* complex (Hymenoptera: Platygasteridae). *Proceedings of the Entomological Society of Washington*, **71**: 400–404.

Jackson, D. & Masner, L. 1966. A redescription of *Metanopedias brunneipes* (Ashmead) (Hymenoptera: Platygasteridae) with notes on its geographical distribution and host relationships. *Canadian Entomologist*, **98**: 1214–1219.

Jackson, D.J. 1928. The biology of *Dinocampus (Perilitus) rutilus* Nees, a braconid parasite of *Sitona lineata* L. Part 1. *Proceedings of the Zoological Society of London*, **1928**: 597–630.

Jackson, D.J. 1958. Observations on the biology of Caraphractus cinctus Walker (Hymenoptera: Mymaridae), a parasitoid of the eggs of Dytiscidae. I. Methods of rearing and numbers bred on different host eggs. *Transactions of the Royal Entomological Society of London*, **110**: 533–554.

Jackson, D.J. 1961. Observations on the biology of *Caraphractus cinctus* Walker (Hymenoptera: Mymaridae), a parasitoid of the eggs of Dytiscidae. II. Immature stages and seasonal history with a review of mymarid larvae. *Parasitology*, **51**: 269–294.

Jackson, D.J. 1969. Observations on the female reproductive organs and the poison apparatus of *Caraphractus cinctus* Walker (Hymenoptera: Mymaridae). *Zoological Journal of the Linnean Society*, **48**: 59–81.

Jaffe, K. & Vilela, E. 1989. On nest densities of the leaf-cutting ant *Atta cephalotes* in tropical primary forest. *Biotropica*, **21**: 234–236.

Jaisson, P. (ed.) 1983. *Social Insects in the Tropics*. Vol. 1, 280pp. Vol. 2, 252pp. Université Paris-Nord, Paris.

Jalali, S.K., Singh, S.P., Ballal, C.R. & Kumar, P. 1988. Competitive interaction between *Cotesia kazak* and *Hyposoter didymator*, exotic parasitoids of *Heliothis armigera*. *Entomologia Experimentalis et Applicata*, **46**: 221–225.

James, H.C. 1928. On the life-histories and economic status of certain cynipid parasites of dipterous larvae, with descriptions of some new larval forms. *Annals of Applied Biology*, **15**: 287–316.

Jander, R. 1976. Grooming and pollen manipulation in bees (Apoidea): The nature and evolution of movements involving the foreleg. *Physiological Entomology*, **1**: 179–194.

Jansen, E., Bense, U. & Schrameyer, K. 1988. *Stephanus serrator* (Fabricius, 1798) in der Bundesrepublik Deutschland (Hymenoptera, Stephanidae). *Entomofauna*, **9**: 421–428.

Janssen, A., Alphen, J.M. van, Sabelis, M.W. & Bakker, K. 1991 (1992). Microhabitat selection behaviour of *Leptopilina heterotoma* changes when odour of competitor is present. *Redia*, Appendice, **74**(3): 203–210.

Janvier, H. 1933. Étude biologique de quelques Hyménoptères du Chili. *Annales des Sciences Naturelles* (Zoologie) (10), **16**: 209–356.

Janvier, H. 1967. Estudio biológico de algunos Himenópteros de Chile. *Publicaciones del Centro de Estudios Entomologicos, Facultad de Filosophia y Educacion, Universidad de Chile*, **9**: 1–67. [A collection and Spanish translation of Janvier's small papers published in French in Chilean journals between 1919 and 1935.]

Janzen, D.H. 1966*a*. Coevolution of mutualism between ants and acacias in Central America. *Evolution*, **20**: 249–275.

Janzen, D.H. 1966*b*. Notes on the behavior of the carpenter bee *Xylocopa fimbriata* in Mexico (Hymenoptera: Apoidea). *Journal of the Kansas Entomological Society*, **39**: 633–641.

Janzen, D.H. 1968*a*. Reproductive behavior in the Passifloraceae and some of its pollinators in Central America. *Behavior*, **32**: 33–48.

Janzen, D.H. 1968*b*. Notes on nesting and foraging behavior of *Megalopta* (Hymenoptera: Halictidae) in Costa Rica. *Journal of the Kansas Entomological Society*, **41**: 342–350.

Janzen, D.H. 1969. Allelopathy by myrmecophytes: the ant *Azteca* as an allelopathic agent of *Cecropia*. *Ecology*, **50**: 147–153.

Janzen, D.H. 1971. The ecological significance of an arboreal nest of *Bombus pullatus* in Costa Rica. *Journal of the Kansas Entomological Society*, **44**: 210–216.

Janzen, D.H. 1973. Dissolution of mutualism between *Cecropia* and its *Azteca* ants. *Biotropica*, **5**: 15–28.

Janzen, D.H. 1974. Epiphytic myrmecophytes in Sarawak: mutualism through the feeding of plants by ants. *Biotropica*, **6**: 237–259.

Janzen, D.H. 1975. *Pseudomyrmex nigropilosa*: a parasite of a mutualism. *Science*, **188**: 936–937.

Janzen, D.H. 1979. How to be a fig. *Annual Review of Ecology and Systematics*, **10**: 13–51.

Janzen, D.H. 1981. The peak in North American ichneumonid species richness lies between 38° and 42°N. *Ecology*, **62**: 532–537.

Janzen, D.H. 1983. *Costa Rican Natural History*. 816pp. University of Chicago, Chicago.

Janzen, D.H. 1984. Two ways to be a tropical big moth: Santa Rosa saturniids and sphingids. *Oxford Surveys in Evolutionary Biology*, **1**: 85–140.

Janzen, D.H. 1987. How moths pass the dry season in a Costa Rican dry forest. *Insect Science and its Application*, **8**: 489–500.

Janzen, D.H., DeVries, P.J., Higgins, M.L. & Kimsey, L.S. 1982. Seasonal and site variation in Costa Rican euglossine bees at chemical baits in lowland deciduous and evergreen forests. *Ecology*, **63**: 66–74.

Janzen, D.H. & Pond, C.M. 1975. A comparison, by sweep sampling, of the arthropod fauna of secondary vegetation in Michigan, England and Costa Rica. *Transactions of the Royal Entomological Society of London*, **127**: 33–50.

Jay, S.C. 1986. Spatial management of honey bees on crops. *Annual Review of Entomology*, **31**: 49–65.

Jarraya, A. 1975. Contribution á l'étude des interactions hôte-parasite chez *Coccus hesperidum* L. (Hom. Coccidae) et son parasite *Coccophagus scutellaris* Dalman (Hym. Aphelinidae). I - Etude expérimentale du comportement de ponte du parasite. *Archives de l'Institute Pasteur de Tunis*, **52**: 415–456.

Jayasingh, D.B. & Taffe, C.A. 1982. The biology of the eumenid mud-wasp *Pachodynerus nasidens* in trapnests. *Ecological Entomology*, **7**: 283–289.

Jeanne, R.L. 1973. Aspects of the biology of *Stelopolybia areata* (Say) (Hymenoptera: Vespidae). *Biotropica*, **5**: 183–198.

Jeanne, R.L. 1974. Nest site selection by *Metapolybia* in Costa Rica (Hymenoptera: Vespidae). *Journal of the Kansas Entomological Society*, **48**: 285–290.

Jeanne, R.L. 1975. The adaptiveness of social wasp nest architecture. *Quarterly Review of Biology*, **50**: 267–287.

Jeanne, R.L. 1979*a*. A latitudinal gradient in rates of ant predation. *Ecology*, **60**: 1211–1224.

Jeanne, R.L. 1979*b*. Construction and utilization of multiple combs in *Polistes canadensis* in relation to the biology of a predaceous moth. *Behavioral Ecology and Sociobiology*, **4**: 293–310.

Jeanne, R.L. 1980. Evolution of social behavior in the Vespidae. *Annual Review of Entomology*, **25**: 371–396.

Jeanne, R.L. 1981. Alarm recruitment, attack behavior, and the role of the alarm pheromone in *Polybia occidentalis* (Hymenoptera: Vespidae). *Behavioral Ecology and Sociobiology*, **9**: 143–148.

Jeanne, R.L. 1982. Evidence for an alarm substance in *Polistes canadensis*. *Experientia*, **38**: 329–330.

Jeanne, R.L. (ed.) 1988. *Interindividual Behavioral Variability in Social Insects*. 456pp. Westview Press, London.

Jeanne, R.L. 1991. The swarm-founding Polistinae, pp. 191–231. *In*: Ross K.G. & Matthews, R.W. (eds) *The Social Biology of Wasps*. 678pp. Cornell University Press, Ithaca.

Jeanne, R.L. & Fagen, R. 1974. Polymorphism in *Stelopolybia areata* (Hymenoptera, Vespidae). *Psyche*, **81**: 155–166.

Jenks, G.E. 1938. Marvels of metamorphosis. *National Geographic Magazine*, **74**: 807–828.

Jenni, W. 1951. Beitrag zur Morphologie und Biologie der Cynipide *Pseudeucoila bochei* Weld, eines larvenparasiten von *Drosophila melanogaster* Meig. *Acta Zoologica*, **32**: 177–254.

Jerman, E.J. & Gauld, I. 1988. *Casinaria*, a paraphyletic ichneumonid genus (Hymenoptera), and a revision of the Australian species. *Journal of Natural History*, **22**: 589–609.

Jervis, M.A. 1979. Parasitism of *Aphelopus* species (Hymenoptera: Dryinidae) by *Ismarus dorsiger* (Curtis)(Hymenoptera: Diapriidae). *Entomologist's Gazette*, **30**: 127–129.

Jervis, M.A. 1980*a*. Life history studies on *Aphelopus* species (Hymenoptera, Dryinidae) and *Chalarus* species (Diptera, Pipunculidae), primary parasites of typhlocybine leafhoppers (Homoptera, Cicadellidae). *Journal of Natural History*, **14**: 769–780.

Jervis, M.A. 1980*b*. Ecological studies on the parasite complex associated with typhlocybine leafhoppers (Homoptera, Cicadellidae). *Ecological Entomology*, **5**: 123–136.

Jervis, M.A. & Kidd, N.A.C. 1986. Host-feeding strategies in hymenopteran parasitoids. *Biological Reviews, Cambridge*, **61**: 395–434.

Jervis, M.A., Kidd, N.A.C., Fitton, M.G., Huddleston, T. & Dawah, H.A. 1993. Flower-visiting by hymenopteran parasitoids. *Journal of Natural History*, **27**: 67–105.

Jervis, M.A., Kidd, N.A.C. & Sahragard, A. 1987. Host-feeding in Dryinidae: its adaptative significance and its consequences for parasitoid-host population dynamics, pp. 591–596. *In*: Vidano, C. & Arzone, A. (eds) *6th Auchenorrhyncha Meeting, Turin, Italy, September 7–11 1987. Proceedings*. 652pp. Consiglio Nazionale delle Ricerche, Italy.

Jirón, L.F. & Mexzon, R.G. 1989. Parasitoid hymenopterans of Costa Rica: geographical distribution of the species associated with fruit flies (Diptera: Tephritidae). *Entomophaga*, **34**: 53–60.

Johansson, A.S. 1951. Studies on the relation between *Apanteles glomeratus* L. (Hym., Braconidae) and *Pieris brassicae* L. (Lepid., Pieridae). *Norsk Entomologisk Tidsskrift*, **8**: 145–186.

Johnson, B. 1959. Effect of parasitization by *Aphidius platensis* Brèthes on the developmental physiology of its host, *Aphis craccivora* Koch. *Entomologia Experimentalis et Applicata*, **2**: 82–99.

Johnson, D.W. 1988. Eucharitidae (Hymenoptera: Chalcidoidea): biology and potential for biological control. *Florida Entomologist*, **71**: 528–537.

Johnson, J.B., Miller, T.D., Heraty, J.M. & Merickel, F.W. 1986. Observations on the biology of two species of *Orasema* (Hymenoptera: Eucharitidae). *Proceedings of the Entomological Society of Washington*, **88**: 542–549.

Johnson, J.W. 1987. A revision of the species of *Praon* Haliday in North America north of Mexico (Hymenoptera: Aphidiidae). *Canadian Entomologist*, **119**: 999–1025.

Johnson, L.K. 1980. Alarm response of foraging *Trigona fulviventris* (Hymenoptera: Apidae) to mandibular gland components of competing bee species. *Journal of the Kansas Entomological Society*, **53**: 357–362.

Johnson, L.K. 1981. Effect of flower clumping on defense of artificial flowers by aggressive stingless bees. *Biotropica*, **13**: 151–157.

Johnson, L.K. 1983*a*. Foraging strategies and the structure of stingless bee communities in Costa Rica, pp. 31–58. *In*: Jaisson, P. (ed.), *Social Insects in the Tropics*, 2. 252pp. Université Paris-Nord, Paris.

Johnson, L.K., Haynes, L.W., Carlson, M.A., Fortnum, H.A. & Gorgas, D.L. 1985. Alarm substances of the stingless bee, *Trigona silvestriana*. *Journal of Chemical Ecology*, **11**: 409–416.

Johnson, L.K. & Howard, J.J. 1987. Olfactory disc number in bees of different sizes and ways of life (Apidae: Meliponinae). *Journal of the Kansas Entomological Society*, **60**: 380–388.

Johnson, L.K. & Hubbell, S.P. 1974. Aggression and competition among stingless bees: field studies. *Ecology*, **55**: 120–127.

Johnson, L.K. & Hubbell, S.P. 1975. Contrasting foraging strategies and coexistence of two bee species on a single resource. *Ecology*, **56**: 1398–1406.

Johnson, L.K. & Hubbell, S.B. 1984. Nest tree selectivity and density of stingless bee colonies in a Panamanian forest, pp. 147–154. *In*: Chadwick, A.C. & Sutton, S.L. (eds) *Tropical Rain-Forest. The Leeds Symposium*. 335pp. Leeds Philosophical & Literary Society, Leeds.

Johnson, L.K., Hubbell, S.P. & Feener, D.H. Jr 1987. Defense of food supply by eusocial colonies. *American Zoologist*, **27**: 347–358.

Johnson, M.D. 1974. *Trypargilum politum* (Say) as a host for *Trichrysis tridens* (Lepeletier) (Hymenoptera: Sphecidae; Chrysididae). *Proceedings of the Entomological Society of Washington*, **76**: 448–449.

Johnson, N.F. 1981. The New World species of the *Telenomus nigricornis* group (Hymenoptera: Scelionidae). *Annals of the Entomological Society of America*, **74**: 73–78.

Johnson, N.F. 1983. Description of *Telenomus solitus*, n. sp. (Hymenoptera: Scelionidae), a noctuid egg parasitoid. *Proceedings of the Entomological Society of Washington*, **85**: 226–228.

Johnson, N.F. 1984. Systematics of Nearctic *Telenomus*: Classification and revisions of the *podisi* and *phymatae* species groups (Hymenoptera: Scelionidae). *Bulletin of the Ohio Biological Survey* (New Series), **6**(3): 1–113.

Johnson, N.F. 1985*a*. Revision of the New World species of the *thyantae* group of *Trissolcus* (Hymenoptera: Scelionidae). *Canadian Entomologist*, **117**: 107–112.

Johnson, N.F. 1985*b*. Systematics of New World *Trissolcus* (Hymenoptera: Scelionidae): species related to *T. basalis*. *Canadian Entomologist*, **117**: 431–445.

Johnson, N.F. 1987. The neotropical telenomine genus *Phanuropsis* Girault (Hymenoptera: Scelionidae). *Annals of the Entomological Society of America*, **80**: 660–663.

Johnson, N.F. 1988. Midcoxal articulations and the phylogeny of the order Hymenoptera. *Annals of the Entomological Society of America*, **81**: 870–881.

Johnson, N.F. 1990. *Telenomus* (Hymenoptera: Scelionidae) egg parasites of *Erinnyis ello* (Lepidoptera: Sphingidae). *Proceedings of the Entomological Society of Washington*, **92**: 306–309.

Johnson, N.F. & Masner, L. 1985. Revision of the genus *Psix* Kozlov & Lê (Hymenoptera: Scelionidae). *Systematic Entomology*, **10**: 33–58.

Johnson, R.A. 1986. Intraspecific resource partitioning in the bumble bees *Bombus ternarius* and *B. pennsylvanicus*. *Ecology*, **67**: 133–138.

Johnston, F.A. 1915. Asparagus-beetle egg parasite. *Journal of Agricultural Research*, **4**: 303–314.

Joiner, R.L., Vinson, S.B. & Benskin, J.B. 1973. Teratocytes as a source of juvenile hormone activity in a parasitoid-host relationship. *Nature*, **246**: 120–121.

Jones, D. 1985. Parasite regulation of host insect metamorphosis: a new form of regulation in pseudoparasitized larvae of *Trichoplusia ni*. *Journal of Comparative Physiology* (B), **155**: 583–590.

Jones, D. 1987. Material from adult female *Chelonus* sp. directs expression of altered developmental programme of host Lepidoptera. *Journal of Insect Physiology*, **33**: 129–134.

Jones, D., Jones, G., Rudnicka, M., Click, A., Reck-Malleczewen, V. & Iwaya, M. 1986. Pseudoparasitism of host *Trichoplusia ni* by *Chelonus* spp. as a new model system for parasite regulation of host physiology. *Journal of Insect Physiology*, **32**: 315–328.

Jones, W.A. & Jackson, C.G. 1990. Mass production of *Anaphes iole* for augmentation against *Lygus hesperus*: effects of food on fecundity and longevity. *Southwestern Entomologist*, **15**: 463–468.

Jones, W.T. 1982. Sex ratio and host size in a parasitoid wasp. *Behavioral Ecology and Sociobiology*, **10**: 207–210.

Jong, P.W. de & Alphen, J.J.M. van 1989. Host size selection and sex allocation in *Leptomastix dactylopii*, a parasitoid of *Planococcus citri*. *Entomologia Experimentalis et Applicata*, **50**: 161–169.

Jorgensen, C.D., Black, H.L. & Hermann, H.R. 1984. Territorial disputes between colonies of the giant tropical ant *Paraponera clavata* (Hymenoptera: Formicidae: Ponerinae). *Journal of the Georgia Entomological Society*, **19**: 156–158.

Joseph, M. & Abdurahiman, V.C. 1981. Oviposition behaviour of *Ceratosolen fusciceps* Mayr (Agaonidae: Hymenoptera) and the mechanism of pollination of *Ficus racemosa* L. *Journal of the Bombay Natural History Society*, **78**: 287–291.

Juillet, J.A. 1964. Influence of weather on flight activity of parasitic Hymenoptera. *Canadian Journal of Zoology*, **42**: 1133–1141.

Jussila, R. 1965. Reports from the Kevo subarctic research station. 2. The Ichneumonidae of the Kevojoki area in Inari Lapland (Finland). *Annales Universitatis Turkuensis*, (A,II) **34**: 1–186.

Jussila, R. 1984. Ichneumonidae (Hymenoptera) of Inari Lapland. *Kevo Notes*, **7**: 83–99.

Jussila, R. & Käpylä, M. 1975. Observations on *Townesia tenuiventris* (Hlmgr) (Hym., Ichneumonidae) and its hosts *Chelostoma maxillosum* (L.) (Hym., Megachilidae) and *Trypoxylon figulus* (L.) (Hym., Sphecidae). *Annales Entomologici Fennici*, **41**: 81–86.

Jutsum, A.R. 1988. Commercial application of biological control: status and prospects. *Philosophical Transactions of the Royal Society of London, Biological Science*, **318**: 357–373.

Jutsum, A.R., Cherrett, J.M. & Fisher, M. 1981. Interactions between the fauna of citrus trees in Trinidad and the ants *Atta cephalotes* and *Azteca* sp. *Journal of Applied Ecology*, **18**: 187–195.

Kaas, J.P., Elzen, G.W. & Ramaswamy, S.B. 1990. Learning in *Microplitis croceipes* Cresson (Hym., Braconidae). *Journal of Applied Entomology*, **109**: 268–273.

Kaas, J.P., Elzen, G.W. & Ramaswamy, S.B. 1992. Flight responses in a wind tunnel of the parasitoid *Microplitis croceipes* to three spring and three summer host plants of *Heliothis* spp. *Entomologia Experimentalis et Applicata*, **63**: 207–212.

Kainoh, Y. 1986. Mating behavior of *Ascogaster reticulatus* Watanabe (Hymenoptera: Braconidae), an egg-larval parasitoid of the smaller tea tortrix moth, *Adoxophyes* sp. (Lepidoptera: Tortricidae) I. Diel patterns of emergence and mating, and some conditions for mating. *Applied Entomology and Zoology*, **21**: 1–7.

Kainoh, Y. 1988. Some factors influencing sex ratio in *Ascogaster reticulatus* Watanabe (Hymenoptera: Braconidae). *Applied Entomology and Zoology*, **23**: 35–40.

Kainoh, Y., Nemoto, T., Shimizu, K., Tatsuki, S., Kusano, T. & Kuwahara, Y. 1991. Mating behavior of *Ascogaster reticulatus* Watanabe (Hymenoptera: Braconidae), an egg-larval parasitoid of the smaller tea tortrix, *Adoxophyes* sp. (Lepidoptera: Tortricidae). III. Identification of a sex pheromone. *Applied Entomology and Zoology*, **26**: 543–549.

Kalina, V. 1984. New genera and species of Palaearctic Eupelmidae (Hymenoptera, Chalcidoidea). *Silvaecultura Tropica et Subtropica*, **10**: 1–29.

Kamal M. 1939. Biological studies on some hymenopterous parasites of aphidophagous Syrphidae. *Technical Science Bulletin, Ministry of Agriculture, Egypt*, **207**: 1–111.

Kamano, Y., Shimizu, K., Kainoh, Y. & Tatsuki, S. 1989. Mating behavior of *Ascogaster reticulatus* Watanabe (Hymenoptera: Braconidae), an egg-larval parasitoid of the smaller tea tortrix, *Adoxophyes* sp. (Lepidoptera: Tortricidae). II. Behavioral sequence and a role of sex pheromone. *Applied Entomology and Zoology*, **24**: 372–378.

Kamath, M.K. & Gupta, V.K. 1972. Ichneumonologia Orientalis, 2. The tribe Rhyssini (Hymenoptera: Ichneumonidae). *Oriental Insects Monograph*, **2**: 1–300.

Kannowski, P.B. 1991. Occurrence of alates of the neotropical ant, *Paraponera clavata* (Hymenoptera, Formicidae), at lights during the dry-wet seasons interface in Panama. *Journal of Entomological Science*, **26**: 375–377.

Kanuck, M.J. & Sullivan, D.J. 1992. Ovipositional behavior and larval development of *Aphidencyrtus aphidivorus* (Hymenoptera: Encyrtidae), an aphid hyperparasitoid. *Journal of the New York Entomological Society*, **100**: 527–532.

Kashef, A. 1955 (1956). Étude biologique de *Stegobium paniceum* L. (Col. Anobiidae) et de son parasite: *Lariophagus distinguendus* Först. (Hym. Pteromalidae). *Annales de la Société Entomologique de France*, **124**: 1–88.

Kas[c]hef, A.H. 1959. *Lariophagus distinguendus* Först. (Hymenoptera: Pteromalidae) ectoparasite on *Rhizopertha domminica* Fab. (Coleoptera: Bostrychidae). *Bulletin de la Société Entomologique d'Égypte*, **43**: 165–184.

Kasparyan, D.R. 1981. *Fauna of the USSR. Hymenoptera 3(1). Ichneumonidae (Subfamily Tryphoninae) Tribe Tryphonini.* 414pp. Oxonian Press, New Delhi. [English translation of a book first published in Russian in 1973.]

Kaston, B.J. 1959. Notes on pompilid wasps that do not dig burrows to bury their spider prey. *Bulletin of the Brooklyn Entomological Society*, **54**: 103–113.

Katayama, E. 1989. Comparative studies on the egg-laying habits of some Japanese species of bumblebees (Hymenoptera, Apidae). *Entomological Society of Japan, Occasional Publications*, **2**: 1–161.

Kathirithamby. J. 1989. Review of the order Strepsiptera. *Systematic Entomology*, **14**: 41–92.

Kathirithamby, J. 1992. *Strepsiptera* of Panama and Mesoamerica, pp. 421–431. *In*: Quintero, D. & Aiello, A. (eds) *Insects of Panama and Mesoamerica*. 692pp. Oxford University Press, Oxford.

Katô, M. 1989. Host-handling behavior of the parasitoid, *Diglyphus minoeus* (Hym.: Eulophidae), parasitizing the honeysuckle leaf-miner, *Phytomyza lonicerae* (Dipt.: Agromyzidae). *Entomophaga*, **34**: 503–509.

Kearns, H.G.H. 1931. The larval and pupal anatomy of *Stenomalus micans* Ol. (Pteromalidae), a Chalcid endoparasite of the gout-fly of barley (*Chlorops taeniopus* Meig.), with some details of the life history of the summer generation. *Parasitology*, **23**: 380–395.

Keeler, K.H. 1978. Insects feeding at extrafloral nectaries of *Ipomoea carnea* (Convolvulaceae). *Entomological News*, **89**: 163–168.

Keeping, M.G. & Crewe R.M. 1983. Parasitoids, commensals and colony size in nests of *Belonogaster* (Hymenoptera: Vespidae). *Journal of the Entomological Society of Southern Africa*, **46**: 309–323.

Keilin, D. & Baume Pluvinel, G. de la. 1913. Formes larvaires et biologie d'un cynipide entomophage. *Eucoila keilini* Kieffer. *Bulletin de Science de France et Belgique*, **47**: 88–104.

Keilin, D. & Thompson, W.R. 1915. Sur le cycle évolutif des Dryinidae, Hyménoptères parasites des Hémiptères Homoptères. *Comptes Rendus Hebdomadaires de Séances de la Société de Biologie*, **78**: 83–87.

Keller, L.J. & Sullivan, D.J. 1976. Oviposition behavior and host feeding of *Asaphes lucens* an aphid hyperparasitoid. *Journal of the New York Entomological Society*, **84**: 206–211.

Kempf, W.W. 1972. Catálogo abreviado das formigas de Ragiao Neotropical (Hymenoptera: Formicidae). *Studia Entomologica*, **15**: 3–344.

Kenaga, E.E. 1944. Notes on *Hadronotus parkeri* Fouts (Hymenoptera: Scelionidae). *Entomological News*, **55**: 173–174.

Kennedy, B.H. 1979. The effect of multilure on parasites of the European elm bark beetle, *Scolytus multistriatus*. *Bulletin of the Entomological Society of America*, **25**: 116–118.

Kent, R.B. 1984. Mesoamerican stingless beekeeping. *Journal of Cultural Geography*, **4**: 14–28.

Kerfoot, W.B. 1967. Correlation between ocellar size and the foraging activities of bees (Hymenoptera; Apoidea). *American Naturalist*, **101**: 65–70.

Kerr, W.E. 1957. Introdução de abelhas africanas no Brasil. *Brasil Apicola*, **3**: 211–213.

Kerr, W.E. 1962*a*. Genetics of sex determination. *Annual Review of Entomology*, **7**: 157–176.

Kerr, W.E. 1962*b*. Tendências evolutivas na reprodução dos himenópteros sociais. *Arquivos do Museu Nacional, Rio de Janeiro*, **52**: 115–116.

Kerr, W.E. & Nielsen, R.A. 1967. Sex determination in bees (Apinae). *Journal of Apicultural Research*, **6**: 3–9.

Kerrich, G.J. 1969. Description of an ichneumonid (Hym.) that preys on egg-masses of weevils harmful to tea culture in Kenya. *Bulletin of Entomological Research*, **59**: 469–472.

Kerrich G.J. 1973. On the taxonomy of some forms of *Ibalia* Latreille (Hymenoptera: Cynipoidea) associated with conifers. *Zoological Journal of the Linnean Society*, **53**: 65–79.

Kfir, R. & Rosen, D. 1980. Biological studies of *Microterys flavus* (Howard) (Hymenoptera: Encyrtidae), a primary parasite of soft scales. *Journal of the Entomological Society of Southern Africa*, **43**: 223–237.

Kfir, R. & Rosen, D. 1981. Biology of the hyperparasite *Cheiloneurus paralia* (Walker) (Hymenoptera: Encyrtidae) reared on *Microterys flavus* (Howard) in brown soft scale. *Journal of the Entomological Society of Southern Africa*, **44**: 131–139.

Kidd, N.A.C. & Jervis, M.A. 1991. Host-feeding and oviposition strategies of parasitoids in relation to host stage. *Researches on Population Ecology*, **33**: 13–28.

Kieffer, J.J. 1912. *Das Tierreich, Evaniidae*, 431pp. Friedländer und Sohn, Berlin

Kimsey, L.S. 1977. New species of bees in the genera *Euplusia* and *Eufriesia* (Hymenoptera: Apidae: Euglossini). *Pan-Pacific Entomologist*, **53**: 8–18.

Kimsey, L.S. 1978. Nesting and male behavior in *Dynatus nigripes spinolae* (Lepeletier) (Hymenoptera, Sphecidae). *Pan-Pacific Entomologist*, **54**: 65–68.

Kimsey, L.S. 1979. An illustrated key to the genus *Exaerete* with descriptions of male genitalia and biology (Hymenoptera: Euglossini, Apidae). *Journal of the Kansas Entomological Society*, **52**: 735–746.

Kimsey, L.S. 1980*a*. Notes on the biology of some Panamanian Pompilidae, with a description of a communal nest (Hymenoptera). *Pan-Pacific Entomologist*, **56**: 98–100.

Kimsey, L.S. 1980*b*. The behaviour of male orchid bees (Apidae, Hymenoptera, Insecta) and the question of leks. *Animal Behaviour*, **28**: 996–1004.

Kimsey, L.S. 1981. The Cleptinae of the Western Hemisphere (Chrysididae: Hymenoptera). *Proceedings of the Biological Society of Washington*, **94**: 801–818.

Kimsey, L.S. 1982*a*. Systematics of bees of the genus *Eufriesea* (Hymenoptera, Apidae). *University of California Publications in Entomology*, **95**: 1–125.

Kimsey, L.S. 1982*b*. *Parataruma*, a new genus of neotropical Crabronini (Hymenoptera, Sphecidae). *Psyche*, **89**: 169–173.

Kimsey, L.S. 1984. The behavioural and structural aspects of grooming and related activities in euglossine bees. *Journal of Zoology*, London, **204**: 541–550.

Kimsey, L.S. 1985. Distinction of the '*Neochrysis*' genera and description of new species (Chrysididae, Hymenoptera). *Psyche*, **92**: 269–286.

Kimsey, L.S. 1986*a*. *Cleptidea* revisited (Hymenoptera, Chrysididae). *Journal of the Kansas Entomological Society*, **59**: 314–324.

King, P.E. & Richards, J.G. 1968. Oösorption in *Nasonia vitripennis* (Hymcnoptera: Pteromalidae). *Journal of Zoology, London*, **154**: 495–516.

King, P.E. & Richards, J.G. 1969. Oögenesis in *Nasonia vitripennis* (Walker) (Hymenoptera: Pteromalidae). *Proceedings of the Royal Entomological Society of London* (A), **44**: 143–157.

Kinsey, A.C. 1930. The gall wasp genus *Cynips*. *Indiana University Studies*, **16**: 1–577.

Kirchner, W.H. & Sommer, K. 1992. The dance language of the honeybee mutant *diminutive wings*. *Behavioral Ecology and Sociobiology*, **30**: 181–184.

Kirkpatrick, T.W. 1953. Insect pests of cacao and insect vectors of cacao virus disease. *Report of Cacao Research in Trinidad*, **1945–1951**: 122–125.

Kishi, Y. 1970. Difference in the sex ratio of the pine bark weevil parasite, *Dolichomitus* sp. (Hymenoptera: Ichneumonidae), emerging from different host species. *Applied Entomology and Zoology*, **5**: 126–132.

Kishitani, Y. 1961. Observations on the egg-laying habit of *Goniozus japonicus* Ashmead (Hymenoptera, Bethylidae). *Kontyû*, **29**: 175–179.

Kistner, D.H. 1979. Social and evolutionary significance of social insect symbionts, pp. 339–413. *In*: Hermann, H.R. (ed.), *Social Insects*, Volume I. 437pp. Academic Press, New York.

Kistner, D.H. 1982. The social insects' bestiary, pp. 1–244. *In*: Hermann, H.R. (ed.), *Social Insects*, Volume III. 459pp. Academic Press, New York.

Kitamura, K. 1983. Comparative studies on the biology of Dryinid wasps in Japan. (2) Relationship between temperature and the developmental velocity of *Haplogonatopus atratus* Esaki & Hashimoto (Hymenoptera: Dryinidae). *Bulletin of the Faculty of Agriculture, Shimane University*, **17**: 147–151.

Kitamura, K. 1988. Comparative studies on the biology of Dryinid wasps in Japan. 5) Development and reproductive capacity of hosts attacked by *Haplogonatopus apicalis* (Hymenoptera, Dryinidae) and the development of progenies of the parasites in their hosts. *Kontyû*, **56**: 659–666.

Kitamura, K. & Nishikata, Y. 1987. A monitor-trap survey of parasitoids of the leaf- and planthoppers supposedly migrated from the mainland China (Homoptera: Auchenorrhyncha). *Bulletin of the Faculty of Agriculture, Shimane University*, **21**: 171–177.

Kitano, H. 1975. Studies on the courtship behavior of *Apanteles glomeratus* L. 2. Role of the male wing during courtship and the releaser of mounting and copulatory behavior in the males. *Kontyû*, **43**: 513–521.

Kitano, H. 1986. The role of *Apanteles glomeratus* venom in the defensive response of its host, *Pieris rapae crucivora*. *Journal of Insect Physiology*, **32**: 369–375.

Kitano, H., Wago, H. & Arakawa, T. 1990. Possible role of teratocytes of the gregarious parasitoid, *Cotesia* (= *Apanteles*) *glomerata* in the suppression of phenoloxidase activity in the larval host, *Pieris rapae crucivora*. *Archives of Insect Biochemistry & Physiology*, **13**: 177–185.

Kjellberg, F. & Maurice, S. 1989. Seasonality in the reproductive phenology of *Ficus*: its evolution and consequences. *Experientia*, **45**: 653–660.

Kleinfeldt, S.E. 1978. Ant-gardens: the interaction of *Codonanthe crassifolia* (Gesneriaceae) and *Crematogaster longispina* (Formicidae). *Ecology*, **59**: 449–456.

Klein Koch, C., Espinosa, O., Tanadazo, A., Cisneros, P. & Delgado, D. 1988. Factores naturalcs dc regulacion y control biologico de la broca del café (*Hypothenemus hampei* Ferr.). *Sanidad Vegetal*, **3**: 5–30.

Klomp, H. & Teerink, B.J. 1962. Host selection and number of eggs per oviposition in the egg-parasite *Trichogramma embryophagum* Htg. *Nature*, **195**: 1020–1021.

Klomp, H., Teerink, B.J. and Wei Chun Ma 1980. Discrimination between parasitized and unparasitized hosts in the egg parasite *Trichogramma embryophagum* (Hym.: Trichogrammatidae): a matter of learning and forgetting. *Netherlands Journal of Zoology*, **30**: 254–277.

Knight, A.L. & Croft, B.A. 1989. Host discrimination by the gregarious parasitoid *Oncophanes americanus* (Hymenoptera: Braconidae). *Pan-Pacific Entomologist*, **65**: 146–150.

Knutson, L. 1978. Insects: Diptera (flies), pp. 128–137. *In*: Morse R.A. (ed.) *Honey Bee Pests, Predators, and Diseases*. 430pp. Cornell Univesity Press, Ithaca.

Knutson, L.V. & Berg, C.O. 1963. *Phaenopria popei* (Hymenoptera: Diapriidae) reared from puparia of sciomyzid flies. *Canadian Entomologist*, **95**: 724–726.

Kobayashi T. & Cosenza, G.W. 1987. Integrated control of soybean stink bugs in the Cerrados. *Japan Agricultural Research Quarterly*, **20**: 229–236.

Kogan, M. & Legner, E.F. 1970. A biosystematic revision of the genus *Muscidifurax* (Hymenoptera: Pteromalidae) with descriptions of four new species. *Canadian Entomologist*, **102**: 1268–1290.

Kon, V.Q. 1974. Morphological and biological peculiarities of the preimaginal stages of *Encyrtus infidus* Rossi (Hymenoptera, Chalcidoidea), the internal parasite of *Eulecanium caraganae* Borchs. *Entomologicheskoe Obozrênie*, **53**: 732–751. [In Russian; English translation: *Entomological Review, Washington*, **53**: 8–21.]

König, R. 1967. Ein Beitrag zur Kenntnis einiger Braconidenarten in Schleswig-Holstein (Hymenoptera: Braconidae, Blacinae). *Faunistisch-Ökologische Mitteilungen*, **3**: 112–121.

Königsmann, E. 1976. Das phylogenetische system der Hymenoptera Teil 1: Einführung, Grundplanmerkmale, Schwestergruppe und Fossilfunde. *Deutsche Entomologische Zeitschrift* (N.F.), **23**: 253–279.

Königsmann, E. 1977. Das phylogenetische System der Hymenoptera Teil 2: 'Symphyta'. *Deutsche Entomologische Zeitschrift* (N.F.), **24**: 1–40.

Königsmann, E. 1978*a*. Das phylogenetische system der Hymenoptera. Teil 3: 'Terebrantes' (Unterordnung Apocrita). *Deutsche Entomologische Zeitschrift* (N.F.), **25**: 1–55.

Königsmann, E. 1978*b*. Das phylogenetische system der Hymenoptera. Teil 4: Aculeata (Unterordnung Apocrita). *Deutsche Entomologische Zeitschrift* (N.F.), **25**: 365–435.

Konishi, K. 1990. A revision of the Aulacidae of Japan (Hymenoptera, Evanioidea). *Japanese Journal of Entomology*, **58**: 637–655.

Kopelke, J.-P. 1987. *Adelognathus cubiceps* Roman 1924. (Ichneumonidae: Adelognathinae)—ein ungewöhnlicher Parasitoid der gallenbildenden *Pontania*-Arten (Tenthredinidae: Nematinae) (Insecta: Hymenoptera). *Senckenbergiana Biologia*, **67**: 253–259.

Koptur, S. 1985. Alternative defenses against herbivores in *Inga* (Fabaceae: Mimosoideae) over an elevational gradient. *Ecology*, **66**: 1639–1650.

Koptur, S. 1992. Extrafloral nectary-mediated interactions between insects and plants, pp 81–129. *In*: Bernays, E. (ed.) *Insect Plant Interactions 4*. CRC Press, Boca Raton.

Kornhauser, S.I. 1919. The sexual characteristics of the membracid, *Thelia bimaculata* (Fabr.). I. External changes induced by *Aphelopus theliae* (Gahan). *Journal of Morphology*, **32**: 531–636.

Korytkowski, G. & Peña, D.O. 1966 (1968). *Bephrata cubensis* Ashmead (Hym.: Eurytomidae), una nueva especie dañina a las Anonáceas en el Peru. *Revista Peruana de Entomología*, **9**: 56–60.

Kouamé, K.L. & Mackauer, M. 1992. Influence of aphid size, age and behaviour on host choice by the parasitoid wasp *Ephedrus californicus*: a test of host-size models. *Oecologia*, **88**: 197–203.

Kozlov, M.A. 1994. Renyxidae fam. n. a remarkable family of parasitic Hymenoptera from the Russian Far East. *Far Eastern Entomologist*, **1**: 1–7.

Krell, P.J. 1991. Polydnaviridae, pp. 321–338. *In*: Adams, J.R. & Bonami, J.R. (eds), *Atlas of Invertebrate Viruses*, 684pp. C.R.C. Press, Boca Raton, Florida.

Kristensen, N.P. 1975. The phylogeny of hexapod 'orders'. A critical review of recent accounts. *Zeitschrift für Zoologische Systematik und Evolutionsforschung*, **13**: 1–44.

Krombein, K.V. 1952. Biological and taxonomic observations on the wasps in a coastal area of North Carolina (Hymenoptera: Aculeata). *Wasmann Journal of Biology*, **10**: 257–341.

Krombein, K.V. 1953. Kill Devil Hills wasps, 1952 (Hymenoptera: Aculeata). *Proceedings of the Entomological Society of Washington*, **55**: 113–135.

Krombein, K.V. 1955. Miscellaneous prey records of solitary wasps. I. (Hymenoptera: Aculeata). *Bulletin of the Brooklyn Entomological Society*, **50**: 13–17.

Krombein, K.V. 1957. A generic review of the Amiseginae, a group of phasmatid egg parasites, and notes on the Adelphinae (Hymenoptera, Bethyloidea, Chrysididae). *Transactions of the American Entomological Society*, **82**: 147–215.

Krombein, K.V. 1959. Biological notes on *Chrysis (Ceratochrysis) enhuycki* Cooper, and its host, *Leptochilus republicanus zendaloides* (Robertson) (Hymenoptera: Chrysididae, Vespidae). *Entomological News*, **70**: 17–23.

Krombein, K.V. 1960. Biological notes on some Hymenoptera that nest in Sumach pith. *Entomological News*, **71**: 29–36, 63–69.

Krombein, K.V. 1967. *Trap-Nesting Wasps and Bees: Life Histories, Nests and Associates*. 570pp. Smithsonian Press, Washington D.C.

Krombein, K.V. 1983. Biosystematic studies of Ceylonese wasps, XI: A monograph of the Amiseginae and Loboscelidiinae (Hymenoptera: Chrysididae). *Smithsonian Contributions to Zoology*, **376**: 1–79.

Krombein, K.V. 1992*a*. Host relationships, ethology and systematics of *Pseudomethoca* Ashmead (Hymenoptera: Mutillidae, Andrenidae, Halictidae and Anthophoridae). *Proceedings of the Entomological Society of Washington*, **94**: 91–106.

Krombein, K.V. 1992*b*. Systematics of the genera of Epyrinae with ramose male antennae (Hymenoptera: Bethylidae). *Proceedings of the Entomological Society of Washington*, **94**: 345–360.

Krombein, K.V. & Evans, H.E. 1976. Three new neotropical *Pterombrus* with description of the diapausing larva (Hymenoptera: Tiphiidae). *Proceedings of the Entomological Society of Washington*, **78**: 361–368.

Krombein, K.V. & Gingras, S.S. 1984. Revision of North American *Liris* Fabricius (Hymenoptera: Sphecoidea: Larridae). *Smithsonian Contributions to Zoology*, **404**: 1–96.

Krombein, K.V., Hurd, P.D. Jr, Smith, D.R. & Burks, B.D. (eds) 1979. *Catalog of Hymenoptera in America North of Mexico*. 3 vols, 2735pp. Smithsonian Institution Press, Washington D.C.

Kudô, S., Maetô, K. & Ozaki, K. 1992. Maternal care in the red-headed spruce web-spinning sawfly, *Cephalcia isshikii* (Hymenoptera: Pamphiliidae). *Journal of Insect Behavior*, **5**: 783–795.

Kugler, C. 1976. A new species of *Platythyrea* (Hymenoptera; Formicidae) from Costa Rica. *Psyche*, **83**: 216–221.

Kugler, C. 1994. Revision of the ant genus *Rogeria* (Hymenoptera: Formicidae) with descriptions of the sting apparatus. *Journal of Hymenoptera Research*, **3**: 17–89.

Kukor, J.J. & Martin, M.M. 1986. The transformation of *Saperda calcarata* (Coleoptera: Cerambycidae) into a cellulose digester through the inclusion of fungal enzymes in its diet. *Oecologia*, **71**: 138–141.

Kukuk, P.F., Eickwort, G.C., Raveret-Richter, M., Alexander, B., Gibson, R., Morse, R.A. & Ratnieks, F. 1989. Importance of the sting in the evolution of sociality in the Hymenoptera. *Annals of the Entomological Society of America*, **82**: 1–5.

Kukuk, P.F. & Schwarz, M. 1988. Macrocephalic male bees as functional reproductives and probable guards. *Pan-Pacific Entomologist*, **64**: 131–137.

Kulman, H.M. 1965. Natural control of the eastern tent caterpillar and notes on its status as a forest pest. *Journal of Economic Entomology*, **58**: 66–70.

Kumar, P. & Ballal, C.R. 1992. The effect of parasitism by *Hyposoter didymator* (Hym.: Ichneumonidae) on food consumption and utilization by *Spodoptera litura* (Lep.: Noctuidae). *Entomophaga*, **37**: 197–203.

Kumarasinghe, N.C. & Edirisinghe, J.P. 1987. Oothecal parasites of *Periplaneta americana*: parasitization and development in relation to host age. *Insect Science and its Application*, **8**: 225–228.

Kurczewski, F.E. 1966. *Tachysphex terminatus* preying on Tettigoniidae- an unusual record (Hymenoptera: Sphecidae, Larrinae). *Journal of the Kansas Entomological Society*, **39**: 317–322.

Kurczewski, F.E. 1990. Additional observations on *Tachypompilus ferrugineus* with emphasis on male behavior (Hymenoptera: Pompilidae). *Great Lakes Entomologist*, **23**: 159–163.

Kurczewski, F.E. & Kurczewski, E.J. 1987. Nest and prey of *Ageniella (Leucophrus) fulgifrons* (Hymenoptera: Pompilidae). *Great Lakes Entomologist*, **20**: 75–80.

Kurczewski, F.E., O'Brien, M.F. & Spofford, M.G. 1992. Nesting behavior of *Podalonia robusta* (Cresson) (Hymenoptera: Sphecidae). *Journal of Hymenoptera Research*, **1**: 235–239.

Kuris, A.M. 1974. Trophic interactions: similarity of parasitic castrators to parasitoids. *Quarterly Review of Biology*, **49**: 129–148.

Kutter, H. 1934. Weitere Untersuchungen über *Kakothrips robustus* Uzel und *Contarinia pisi* Winn., sowie deren Parasiten, insbesondere *Pirene graminae* Hal. *Mitteilungen der Schweizerischen Entomologischen Gesellschaft*, **16**: 1–82.

LaBerge, W.E. 1956. A revision of the bees of the genus *Melissodes* in North and Central America. Part I (Hymenoptera, Apidae). *University of Kansas Science Bulletin*, **37**: 911–1194.

LaBerge, W.E. 1957. The genera of bees of the tribe Eucerini in North and Central America (Hymenoptera, Apoidea). *American Museum Novitates*, **1837**: 1–44.

LaBerge, W.E. 1958. Notes on the North and Central American bees of the genus *Svastra* Holmberg. *Journal of the Kansas Entomological Society*, **31**: 266–273.

LaBerge, W.E. 1961. A revision of the bees of the genus *Melissodes* in North and Central America. Part III (Hymenoptera, Apidae). *University of Kansas Science Bulletin*, **42**: 283–663.

LaBerge, W.E. 1967. A revision of the bees of the genus *Andrena* of the Western Hemisphere. Part I. *Callandrena*. (Hymenoptera: Andrenidae). *Bulletin of the University of Nebraska State Museum*, **7**: 1–316.

LaBerge, W.E. 1986 (1985). A revision of the bees of the genus *Andrena* of the Western Hemisphere. Part XI. Minor subgenera and subgeneric key. *Transactions of the American Entomological Society*, **111**: 441–567.

LaBerge, W.E. & Michener, C.D. 1963. *Deltoptila*, a middle American genus of anthophorine bees (Hymenoptera, Apoidea). *Bulletin of the University of Nebraska State Museum*, **4**: 211–225.

LaBerge, W.E. & Ribble, D.W. 1966. Biology of *Florilegus condignus* (Hymenoptera: Anthophoridae), with a description of its larva, and remarks on its importance in alfalfa pollination. *Annals of the Entomological Society of America*, **59**: 944–950.

Labougle, J.M. 1990. *Bombus* of México and Central America (Hymenoptera, Apidae). *University of Kansas Science Bulletin*, **54**: 35–73.

Lachaud, J.P. 1990. Foraging activity and diet in some neotropical ponerine ants. I. *Ectatomma ruidum* Roger (Hymenoptera, Formicidae). *Folia Entomológica Mexicana*, **78**: 241–256.

Laing, D.R. & Caltagirone, L.E. 1969. Biology of *Habrobracon lineatellae* (Hymenoptera: Braconidae). *Canadian Entomologist*, **101**: 135–142.

Landis, D.A. & Haas, M.J. 1992. Influence of landscape structure on abundance and within-field distribution of European corn borer (Lepidoptera: Pyralidae) larval parasitoids in Michigan. *Environmental Entomology*, **21**: 409–416.

Lange, W.H. Jr 1944. Insects affecting Guayule with special reference to those associated with nursery plantings in California. *Journal of Economic Entomology*, **37**: 392–399.

Lanham, U.N. & Evans, F.C. 1958. Phoretic scelionids on grasshoppers of the genus *Melanoplus*. *Pan-Pacific Entomologist*, **34**: 213–214.

Lanham, U.N. & Evans, F.C. 1960. Observations on the scelionid component of a grassland insect fauna. *Bulletin of the Brooklyn Entomological Society*, **55**: 84–87.

Larsson, F.K. 1990. Thermoregulation and activity patterns of the sand wasp *Steniolia longirostra* (Say) (Hymenoptera: Sphecidae) in Costa Rica. *Biotropica*, **22**: 65–68.

Larsson, F.K. & Larsson, K. 1989. Body size and nesting behaviour of the sand wasp *Stictia heros* (Hymenoptera: Sphecidae) in Costa Rica. *Journal of Tropical Ecology*, **5**: 117–120.

LaSalle, J. 1987. New World Tanaostigmatidae (Hymenoptera, Chalcidoidea). *Contributions of the American Entomological Institute*, **23**(1): 1–181.

LaSalle, J. 1989. Notes on the genus *Zagrammosoma* (Hymenoptera: Eulophidae) with description of a new species. *Proceedings of the Entomological Society of Washington*, **91**: 230–236.

LaSalle, J. 1990. Tetrastichinae (Hymenoptera: Eulophidae) associated with spider egg sacs. *Journal of Natural History*, **24**: 1377–1389.

LaSalle, J. 1994. North American genera of Tetrastichinae (Hymenoptera: Eulophidae). *Journal of Natural History*, **28**: 109–236.

LaSalle, J. 1995. A new species of *Phymasticus* (Hymenoptera : Eulophidae) parasitic on adult *Xyleborus perforans* (Coleoptera: Scolytidae) on macadamia trees in Hawai'i. *Proceedings of the Hawaiian Entomological Society*, **32**: 95–101.

LaSalle, J. & Gauld, I.D. 1991. Parasitic Hymenoptera and the biodiversity crisis. *Redia*, Appendice, **74**(3): 315–334.

LaSalle, J. & Gauld, I.D. (eds) 1993. *Hymenoptera and Biodiversity*. 348 pp. C.A.B. International, Wallingford.

LaSalle, J. & Noyes, J.S. 1985. New family placement for the genus *Cynipencyrtus* (Hymenoptera; Chalcidoidea; Tanaostigmatidae). *Journal of the New York Entomological Society*, **93**: 1261–1264.

LaSalle, J. & Schauff, M.E. 1992. Preliminary studies on neotropical Eulophidae (Hymenoptera: Chalcidoidea): Ashmead, Cameron, Howard and Walker species. *Contributions of the American Entomological Institute*, **27**(1): 1–47.

LaSalle, J. & Schauff, M.E. 1994. Systematics of the tribe Euderomphalini (Hymenoptera: Eulophidae): parasitoids of whiteflies (Homoptera: Aleyrodidae). *Systematic Entomology*, **19**: 235–258.

LaSalle, J. & Stage, G.I. 1985. The chalcidoid genus *Leptofoenus* (Hymenoptera: Pteromalidae). *Systematic Entomology*, **10**: 285–298.

Lashomb, J.H. & Steinhauer, A.L. 1975. Observations of *Zethus spinipes* Say (Hymenoptera: Eumenidae). *Proceedings of the Entomological Society of Washington*, **77**: 164.

Lateef, S.S., Reed, W. & Lasalle, J. 1985. *Tanaostigmodes cajaninae* LaSalle sp.n. (Hymenoptera: Tanaostigmatidae), a potential pest of pigeon pea in India. *Bulletin of Entomological Research*, **75**: 305–313.

Lattke, J.E. 1991. Studies of neotropical *Amblyopone* Erichson (Hymenoptera: Formicidae). *Contributions in Science. Natural History Museum of Los Angeles County*, **428**: 1–7.

Lawrence, P.O. 1981. Host vibration — a cue to host location by the parasite, *Biosteres longicaudatus*. *Oecologia*, **48**: 249–251.

Lawrence, P.O. 1986*a*. Host-parasite hormonal interactions: an overview. *Journal of Insect Physiology*, **32**: 295–298.

Lawrence, P.O. 1986*b*. The role of 20-hydroxyecdysone in the moulting of *Biosteres longicaudatus*, a parasite of the Caribbean fruit fly, *Anastrepha suspensa*. *Journal of Insect Physiology*, **32**: 329–337.

Lawrence, P.O. 1990. The biochemical and physiological effects of insect hosts on the development and ecology of their insect parasites: an overview. *Archives of Insect Biochemistry and Physiology*, **13**: 217–228.

LeBeck, L.M. 1989. Extracellular symbiosis of a yeast-like microorganism within *Comperia merceti* (Hymenoptera: Encyrtidae). *Symbiosis*, **7**: 51–66.

LeBeck, L.M. 1991. A review of the hymenopterous natural enemies of cockroaches with emphasis on biological control. *Entomophaga*, **36**: 335–352.

Leclercq, J. 1977. *Lecrenierus* genre nouveau de Crabroniens sud-américains (Hymenoptera, Sphecidae). *Bulletin des Recherches Agronomiques de Gembloux*, **12**: 55–70.

Leclercq, J. 1980*a*. Crabroniens d'Amérique Latine appartenant aux genres que Vernon S. L. Pate nomma *Chimila, Foxita et Taruma. Bulletin de la Société Royale des Sciences de Liège*, **49**: 70–83.

Leclercq, J. 1980*b*. Crabroniens du genre *Quexua* Pate trouvés en Amérique latine (Hymenoptera, Sphecidae). *Bulletin des Recherches Agronomiques de Gembloux*, **15**: 61–70.

Leclercq, J. 1981*a*. Craboniens d'Amérique Latine appartenant aux genres *Entomocrabro* Kohl et *Tracheliodes* Morawitz (Hymenoptera Sphecidae). *Bulletin et Annales de la Société Royale Belge D'Entomologie*, **116**: 233–245.

Leclercq, J. 1981*b*. *Podagritus (Chilichuca) brieni* n. sp. et 27 autres espèces de *Podagritus* de l'Amérique du Sud (Hymenoptera Sphecidae Crabroninae). *Annales de la Société Royale Zoologique de Belgique*, **111**: 65–88.

Leclercq, J. 1982. Identification de 39 espèces de *Podagritus* Spinola sous-genre *Parechuca* Leclercq trouvées en Amérique Latine (Hymenoptera Sphecidae Crabroninae). *Annales de la Société Royale Zoologique de Belgique*, **112**: 251–279.

Leclercq, J. 1991. Hymenopteres sphecides crabroniens du genre *Ectemnius* Dahlbom trouves en Amérique Latine. *Bulletin de la Société Royale des Sciences de Liège*, **60**: 3–29.

Lee, J.-H., Johnson, S.J. & Wright, V.L. 1990. Quantitative survivorship analysis of the velvetbean caterpillar (Lepidoptera: Noctuidae) pupae in soybean fields in Louisiana. *Environmental Entomology*, **19**: 978–986.

Legner, E.F. 1987. Pattern of thelytoky acquisition in *Muscidifurax raptor* Girault and Sanders (Hymenoptera: Pteromalidae). *Bulletin of the Society for Vector Ecology*, **12**: 517–527.

Legner, E.F. 1991. Estimations of number of active loci, dominance and heritability in polygenic inheritance of gregarious behavior in *Muscidifurax raptorellus* (Hymenoptera: Pteromalidae). *Entomophaga*, **36**: 1–18.

Legner, E.F. & Gerling, D. 1967. Host-feeding and oviposition on *Musca domestica* by *Spalangia cameroni*, *Nasonia vitripennis*, and *Muscidifurax raptor* (Hymenoptera: Pteromalidae) influences their longevity and fecundity. *Annals of the Entomological Society of America*, **60**: 678–691.

Leiby, R.W. 1922. The polyembryonic development of *Copidosoma gelechiae* with notes on its biology. *Journal of Morphology*, **37**: 195–285.

Leiby, R.W. & Hill, C.C. 1923. The twinning and monembryonic development of *Platygaster hiemalis*, a parasite of the hessian fly. *Journal of Agricultural Research*, **25**: 337–350.

Leiby, R.W. & Hill, C.C. 1924. The polyembryonic development of *Platygaster vernalis*. *Journal of Agricultural Research*, **28**: 829–840.

Leius, K. 1960. Attractiveness of different food and flowers to the adults of some hymenopterous parasites. *Canadian Entomologist*, **92**: 369–376

Leius, K. 1961*a*. Influence of food on fecundity and longevity of adults of *Itoplectis conquisitor* (Say) (Hymenoptera: Ichneumonidae). *Canadian Entomologist*, **93**: 771–780.

Leius, K. 1961*b*. Influence of various foods on fecundity and longevity of adults of *Scambus buolianae* (Htg.) (Hymenoptera: Ichneumonidae) *Canadian Entomologist*, **93**: 1079–1084.

Leluk, J. & Jones, D. 1989. *Chelonus* sp. near *curvimaculatus* venom proteins: analysis of their potential role and processing during development of host *Trichoplusia ni*. *Archives of Insect Biochemistry and Physiology*, **10**: 1–12.

LeMasurier, A.D. 1987. A comparative study of the relationship between host size and brood size in *Apanteles* spp. (Hymenoptera: Braconidae). *Ecological Entomology*, **12**: 383–393.

Lenteren, J.C. van, 1976. The development of host discrimination and the prevention of superparasitism in the parasite *Pseudeucoila bochei* Weld (Hym.: Cynipidae). *Netherlands Journal of Zoology*, **26**: 1–83.

Lenteren, J.C. van, 1981. Host discrimination by insect parasitoids, pp. 153–179. *In*: Nordlund, D.A., Jones, R.L. & Lewis, W.J. (eds) *Semiochemicals. Their Role in Pest Control*. 306pp. Wiley & Sons, New York.

Lenteren, J.C. van & Bakker, K. 1975. Discrimination betweeen parasitised and unparasitised hosts in the parasitic wasp *Pseudeucoila bochei*: a matter of learning. *Nature*, **254**: 417–419.

Léonide, J. & Léonide, J.-C. 1969. Insectes parasites et prédateurs des Diptères acridiphages; parasites occasionnels des Orthoptères en Provence. *Bulletin de la Société Entomologique de France*, **74**: 21–32.
Le Quesne, W.J. 1972. Studies on the coexistence of three species of *Eupteryx* (Hemiptera: Cicadellidae) on nettle. *Journal of Entomology* (Series A), **47**: 37–44.
Le Ralec, A. & Wajnberg, E. 1990. Sensory receptors of the ovipositor of *Trichogramma maidis* (Hym.: Trichogrammatidae). *Entomophaga*, **35**: 293–299.
Lessman, D. von 1962. Übersicht der bisher bekannten *Megastigmus*-Arten und ihrer Wirtspflanzen. *Zeitschrift für Angewandte Entomologie*, **50**: 233–238.
Letourneau, D.K. 1983. Passive aggression: an alternative hypothesis for the *Piper-Pheidole* association. *Oecologia*, **60**: 122–126.
Letourneau, D.K. 1990. Code of ant-plant mutualism broken by parasite. *Science*, **248**: 215–217.
Letourneau, D.K. & Choe, J.C. 1987. Homopteran attendance by wasps and ants: the stochastic nature of interactions. *Psyche*, **94**: 81–91.
Levin, D.A. 1976. Alkaloid-bearing plants: an ecogeographic perspective. *American Naturalist*, **110**: 261–284.
Levin, D.A. & York, B.M. Jr 1978. The toxicity of plant alkaloids: an ecogeographical perspective. *Biochemical Systematics and Ecology*, **6**: 61–76.
Levin, M.D. 1983. Value of bee pollination to U.S. agriculture. *Bulletin of the Entomological Society of America*, **29**(4): 50–51.
Levine, L. & Sullivan, D.J. 1983. Intraspecific tertiary parasitoidism in *Asaphes lucens* (Hymenoptera: Pteromalidae), an aphid hyperparasitoid. *Canadian Entomologist*, **115**: 1653–1658.
Levins, R. 1989. Science, Technology, and Agriculture in Cuba. Address to New World Agriculture Group, Boston, Massachusetts.
Lewis, T., Pollard, G.V. & Dibley, G.C. 1974. Micro-environmental factors affecting diel patterns of foraging in the leaf-cutting ant *Atta cephalotes* (L.) (Formicidae: Attini). *Journal of Animal Ecology*, **43**: 143–153.
Lewis, W.J. 1970. Life history and anatomy of *Microplitis croceipes* (Hymenoptera: Braconidae), a parasite of *Heliothis* spp. (Lepidoptera: Noctuidae). *Annals of the Entomological Society of America*, **63**: 67–70.
Lewis, W.J. & Jones, R.L. 1971. Substance that stimulates host-seeking by *Microplitis croceipes* (Hymenoptera: Braconidae), a parasite of *Heliothis* species. *Annals of the Entomological Society of America*, **64**: 471–473.
Lewis, W.J. & Tumlinson, J.H. 1988. Host detection by chemically mediated associative learning in a parasitic wasp. *Nature*, **331**: 257–259.
Lewis, W.J., Tumlinson, J.H. & Krasnoff, S. 1991. Chemically mediated associative learning: an important function in the foraging behavior of *Microplitis croceipes* (Cresson). *Journal of Chemical Ecology*, **17**: 1309–1325.
Lewis, W.J., Vet, L.E.M., Tumlinson, J.H., Lenteren, J.C. van & Papaj, D.R. 1990. Variations in parasitoid foraging behavior: essential element of a sound biological control theory. *Environmental Entomology*, **19**: 1183–1193
Lim, K.P., Yule, W.N. & Stewart, R.K. 1980. A note on *Pelecinus polyturator* (Hymenoptera: Pelecinidae), a parasite of *Phyllophaga anxia* (Coleoptera: Scarabaeidae). *Canadian Entomologist*, **112**: 219–220.
Lin, K.S. 1988. The Eucoilidae from Taiwan, I. (Hymenoptera: Cynipoidea). *Journal of Taiwan Museum*, **41**(2): 1–66.

Lin, N. & Michener, C.D. 1972. Evolution of sociality in insects. *Quarterly Review of Biology*, **47**: 131–159.

Lindauer, M. 1967. Recent advances in bee communication and orientation. *Annual Review of Entomology*, **12**: 439–470.

Lindberg, H. 1950. Notes on the biology of dryinids. *Societas Scientiarum Fennica Commentationes Biologicae*, **10**(15): 1–19.

Linsley, E.G., MacSwain, J.W. & Michener, C.D. 1980. Nesting biology and associates of *Melitoma* (Hymenoptera, Anthophoridae). *University of California Publications in Entomology*, **90**: 1–45.

Linsley, E.G. Rick, C.M. & Stephens, S.G. 1966. Observations on the floral relationships of the Galápagos carpenter bee. *Pan-Pacific Entomologist*, **42**: 1–18.

List, G.M. 1932. A cherry pest in Colorado. *Bulletin of the Colorado Agricultural Experimental Station*, **385**: 1–106.

Lith, J.P. van 1975*a*. Neotropical species of *Psen* and *Pseneo* (Hymenoptera, Sphecidae, Psenini). *Tijdschrift voor Entomologie*, **118**: 1–41.

Lith, J.P. van 1975*b*. New neotropical species of *Pseneo* (Hymenoptera, Sphecidae, Psenini). *Entomologische Berichten*, **35**: 170–173.

Lith, J.P. van 1979. The New World genus *Pluto* (Hymenoptera, Sphecidae, Psenini). *Tijdschrift voor Entomologie*, **122**: 127–239.

Litte, M. 1981. Social biology of the polistine wasp *Mischocyttarus labiatus*: survival in a Colombian rain forest. *Smithsonian Contributions to Zoology*, **327**: 1–27.

Littledyke, M. & Cherrett, J.M. 1976. Direct ingestion of plant sap from cut leaves by the leaf-cutting ants *Atta cephalotes* (L.) and *Acromyrmex octospinosus* (Reich) (Formicidae, Attini). *Bulletin of Entomological Research*, **66**: 205–217.

Littledyke, M. & Cherrett, J.M. 1978. Defence mechanisms in young and old leaves against cutting by the leaf-cutting ants *Atta cephalotes* (L.) and *Acromyrmex octospinosus* (Reich) (Hymenoptera: Formicidae). *Bulletin of Entomological Research*, **68**: 263–271.

Liu, G. 1939. Some extracts from the history of entomology in China. *Psyche*, **46**: 23–28.

Liu, S.-S. & Carver, M. 1982. The effect of temperature on the adult integumental coloration of *Aphidius smithi*. *Entomologia Experimentalis et Applicata*, **32**: 54–60.

Lloyd, D.C. 1940. Host selection by hymenopterous parasites of the moth *Plutella maculipennis* Curtis. *Proceedings of the Royal Society of London* (B), **128**: 451–484.

Lloyd, D.C. 1952. Biological observations on some thynnids of western Patagonia. *Bulletin of Entomological Research*, **42**: 707–719.

Loan, C.C. 1980. Plant bug hosts (Heteroptera: Miridae) of some euphorine parasites (Hymenoptera: Braconidae) near Belleville, Ontario, Canada. *Le Naturaliste Canadien*, **107**: 87–93.

Loan, C. & Holdaway, F.G. 1961*a*. *Pygostolus falcatus* (Nees) (Hymenoptera, Braconidae), a parasite of *Sitona* species (Coleoptera, Curculionidae). *Bulletin of Entomological Research*, **52**: 473–488.

Loan, C. & Holdaway, F.G. 1961*b*. *Microctonus aethiops* (Nees) auctt. and *Perilitus rutilus* (Nees) (Hymenoptera: Braconidae), European parasites of *Sitona* weevils (Coleoptera: Curculionidae). *Canadian Entomologist*, **93**: 1057–1079.

Lofgren, C.S. & Vander Meer, R.K. 1986. *Fire Ants and Leaf-cutting Ants: Biology and Management*. 435pp. Westview Press, Boulder.

Löhr, B., Santos, B. & Varela, A.M. 1989. Larval development and morphometry of *Epidinocarsis lopezi* (DeSantis) (Hym., Encyrtidae), parasitoid of the cassava mealybug,

Phenacoccus manihoti Mat.-Ferr. (Hom., Pseudococcidae). *Journal of Applied Entomology*, **107**: 334–343.

Loiácono, M.S. 1981. Diapridos ecitofilos sinfilos de la region neotropical (Hymenoptera — Proctotrupoidea — Diapriidae). *Revista de la Sociedad Entomologica Argentina*, **40**: 297–310.

Loiácono, M.S. 1987 (1985). Un nuevo diaprido (Hymenoptera) parasitoide de larvas de *Acromyrmex ambiguus* (Emery) (Hymenoptera, Formicidae) en el Uruguay. *Revista de la Sociedad Entomologica Argentina*, **44**: 129–136.

Loiácono, M.S. & Díaz, N.B. 1977. Anotaciones sobre himenopteros proctotrupoideos y cinipoideos argentinos (Hymenoptera: Proctotrupoidea, Cynipoidea). *Neotropica*, **23**: 95–102

Lomholdt, O. 1982. On the origin of the bees (Hymenoptera: Apidae, Sphecidae). *Entomologica Scandinavica*, **13**: 185–190.

Lomholdt, O.C. 1985. A reclassification of the larrine tribes with a revision of the Miscophini of southern Africa and Madagascar (Hymenoptera: Sphecidae). *Entomologica Scandinavica*, (Suppl.) **24**: 1–183.

Longair, R.W. 1987. Mating behavior at floral resources in two species of *Pseudomasaris* (Hymenoptera: Vespidae: Masarinae). *Proceedings of the Entomological Society of Washington*, **89**: 759–769.

Longino, J.T. 1986. Ants provide substrate for epiphytes. *Selbyana*, **9**: 100–103.

Longino, J.T. 1989*a*. Geographic variation and community structure in an ant-plant mutualism: *Azteca* and Cecropia in Costa Rica. *Biotropica*, **21**: 126–132.

Longino, J.T. 1989*b*. Taxonomy of the *Cecropia*-inhabiting ants in the *Azteca alfari* species group (Hymenoptera: Formicidae): evidence for two broadly sympatric species. *Contributions in Science. Natural History Museum of Los Angeles County*, **412**: 1–16.

Longino, J.T. 1991. Taxonomy of the *Cecropia*-inhabiting *Azteca* ants. *Journal of Natural History*, **25**: 1571–1602.

Longino, J.T. & Nadkarni, N.M. 1990. A comparison of ground and canopy leaf litter ants (Hymenoptera: Formicidae) in a neotropical montane forest. *Psyche*, **97**: 81–93.

Loomans, A.J.M., Silva, I. & Lenteren, J.C. van 1992. *Ceranisus menes* (Hymenoptera: Eulophidae), a potential biological control agent of *Frankliniella occidentalis* (Thysanoptera: Thripidae)? *Proceedings in Experimental and Applied Entomology*, **3**: 40–45.

Lorenz, H. & Kraus, M. 1957. Die larvalsystematik der Blattwespen (Tenthredinoidea und Megalodontoidea). *Abhandlungen zur Larvalsystematik der Insekten*, **1**: 1–339.

Low, B.S. & Wcislo, W.T. 1992. Male foretibial plates and mating in *Crabro cribrellifer* (Packard) (Hymenoptera: Sphecidae), with a survey of expanded male forelegs in Apoidea. *Annals of the Entomological Society of America*, **85**: 219–223.

Lu, K.L. & Mesler, M.R. 1981. Ant dispersal of a Neotropical forest floor gesneriad. *Biotropica*, **13**: 159–160.

Lubbock, J. 1863. On two aquatic Hymenoptera, one of which uses its wings in swimming. *Transactions of the Linnaean Society of London*, **24**: 135–143.

Lubin, Y.D. 1984. Changes in the native fauna of the Galápagos Islands following invasion by the little red fire ant, *Wasmannia auropunctata*. *Biological Journal of the Linnean Society*, **21**: 229–242.

Luck, R.F. 1981. Parasitic insects introduced as biological control agents for arthropod pests, pp. 125–284. *In*: Pimentel, D. (ed.), *CRC Handbook of Pest Management in Agriculture, II*. 501pp. CRC Press, Boca Raton, Florida.

Luck, R.F., Messenger, P.S. & Barbieri, J.F. 1981. The influence of hyperparasitism on the performance of biological control agents, pp. 34–42. *In*: Rosen, D. (ed.) *The Role of Hyperparasitism in Biological Control —A Symposium*, Number 4103. 52pp. University of California, Berkeley.

Luck, R.F. & Podoler, H. 1985. Competitive exclusion of *Aphytis lingnanensis* by *A. melinus*: potential role of host size. *Ecology*, **66**: 904–913.

Luck, R.F., Podoler, H. & Kfir, R. 1982. Host selection and egg allocation behaviour by *Aphytis melinus* and *A. lignanensis*: comparison of two facultatively gregarious parasitoids. *Ecological Entomology*, **7**: 397–408.

Luck, R.F., Stouthamer, R. & Nunney, L.P. 1993. Sex determination and sex ratio patterns in parasitic Hymenoptera, pp. 442–476. *In*: Wrensch, D.L. & Ebbert, M.A. (eds) *Evolution and Diversity of Sex Ratio in Insects and Mites*. 630pp. Chapman & Hall, New York.

Luff, M.L. 1975 (1976). Notes on the biology of the developmental stages of *Nebria brevicollis* (F.) (Col., Carabidae) and on their parasites, *Phaenoserphus* spp. (Hym., Proctotrupidae). *Entomologist's Monthly Magazine*, **111**: 249–255.

Lum, P.T.M. 1977. Effect of glucose on autogenous reproduction of *Bracon hebetor* Say. *Journal of the Georgia Entomological Society*, **12**: 150–153.

MacArthur, R.H. & Wilson, E.O. 1967. *The Theory of Island Biogeography*. 203pp. Princeton University Press, Princeton, N.J.

Macdonald, J. & Ohmart, C.P. 1993. Life history strategies of Australian pergid sawflies and their interactions with host plants, pp. 485–502. *In*: Wagner, M.R. & Raffa, K.F. (eds) *Sawfly Life History Adaptations to Woody Plants*. 564pp. Academic Press, San Diego.

MacDonald, J.F. & Matthews, R.W. 1984. Nesting biology of the southern yellowjacket, *Vespula squamosa* (Hymenoptera: Vespidae): social parasitism and independent founding. *Journal of the Kansas Entomological Society*, **57**: 134–151.

Macêdo, M.V. de & Monteiro, R.F. 1989. Seed predation by a braconid wasp, *Allorhogas* sp. (Hymenoptera). *Journal of the New York Entomological Society*, **97**: 358–362.

MacGown, M.W. 1979. The Platygastridae (Hymenoptera: Proctotrupoidea) parasitic on midges (Cecidomyiidae) found on conifers in Canada and the United States. *Mississippi Agricultural & Forestry Experiment Station, Information Bulletin*, **9**: 1–131.

MacGown, M.W. & Nebeker, T.E. 1978. Taxonomic review of *Amitus* (Hymenoptera: Proctotrupoidea, Platygastridae) of the Western Hemisphere. *Canadian Entomologist*, **110**, 275–283.

Mackauer, M. 1968. *Hymenopterorum Catalogus* (nova editio) Pars 3. Aphidiidae. 103pp. Junk, 's-Gravenhage.

Mackauer, M. 1990. Host descrimination and larval competition in solitary endoparasitoids, pp. 41–62. *In*: Mackauer, M., Ehler, L.E. & Roland, J. (eds) *Critical Issues in Biological Control*. 330pp. Intercept, Andover.

Mackauer, M. & Kambhampati, S. 1988. Parasitism of aphid embryos by *Aphidius smithi*: some effects of extremely small host size. *Entomologia Experimentalis et Applicata*, **49**: 167–173.

Mackauer, M. & Stary, P. 1967. *Hym. Ichneumonoidea World Aphidiidae. Index of Entomophagous Insects*, 195pp. Le Francois, Paris.

Mackay, W.P. 1993. A review of the new world ants of the genus *Dolichoderus* (Hymenoptera: Formicidae). *Sociobiology*, **22**: 1–148.

MacKay, W.P. & Vinson, S.B. 1989. Rediscovery of the ant *Gnamptogenys hartmani* (Hymenoptera: Formicidae) in eastern Texas. *Proceedings of the Entomological Society of Washington*, **91**: 127.

Madden, J.L. & Coutts, M.P. 1979. The role of fungi in the biology and ecology of woodwasps (Hymenoptera: Siricidae), pp. 165–174. *In*: Batra, L.R. (ed.) *Insect Fungus Symbiosis. Nutrition, Mutualism, and Commensalism*. 276pp. Allanheld, Osmun, Montclair, New Jersey.

Madel, G. 1963. Beiträge zur Morphologie und Biologie von *Meteorus fragilis* Wesm. (Hym. Brac.) — ein Endoparasit des Mondvogels *Phalera bucephala* L. (Lep. Notodontidae). *Zeitschrift für Angewandte Entomologie*, **53**: 1–47.

Madl, M. 1990. Eine neue *Panaulix*-Art Benoit aus Kenya (Hymenoptera, Evanioidea, Aulacidae). *Mitteilungen der Münchner Entomologischen Gesellschaft*, **80**: 85–88.

Maes, J.-M. 1989. Catalogo de los insectos controladores biologicos en Nicaragua. Volumen III. Insectos parasitoides. *Revista Nicaraguense de Entomologia*, **10**: 1–138.

Maetô, K. 1990. Phylogenetic relationships and host associations of the subfamily Meteorinae Cresson (Hymenoptera, Braconidae). *Japanese Journal of Entomology*, **58**: 383–396.

Magistretti, G. 1950. Biología de la *Psychidosmicra brethesi* Blanchard, enemigo natural del 'bicho de cesto' común, *Oiketicus kirbyi* Guild. *Revista de la Facultad de Ciencias Agrarias* (Mendoza), **2**(2): 1–15.

Magrini, E.A. & Botelho, P.S.M. 1991. Influência do alimento do hospedeiro *Diatraea sacharalis* (Lepidoptera, Pyralidae), sobre *Trichogramma galloi* (Hymenoptera, Trichogrammatidae). *Anais da Sociedade Entomológica do Brasil*, **20**: 99–108.

Malo, F. 1961. Phoresy in *Xenufens* (Hymenoptera: Trichogrammatidae), a parasite of *Caligo eurilochus* (Lepidoptera: Nymphalidae). *Journal of Economic Entomology*, **54**: 465–466.

Malyshev, S.I. 1936. The nesting habits of solitary bees. A comparative study. *Eos*, **11**: 201–309.

Malyshev, S.I. 1968. *Genesis of the Hymenoptera and the Phases of Their Evolution*. 319pp. Methuen, London.

Maneval, H. 1930. Description et moeurs de l'adulte et de la larve d'une espèce nouvelle du genre *Parascleroderma* (Hym. Bethylidae). *Bulletin de la Société Entomologique de France*, **1930**: 53–61.

Maneval, H. 1936. Nouvelles notes sur divers Hyménoptères et leurs larves. *Revue Française d'Entomologie*, **3**: 18–32.

Mani, M.S. 1964. *The Ecology of Plant Galls*. Monographs in Biology, XII. 434pp. Junk, The Hague.

Mann, W.M. 1914. Some myrmecophilous insects from Mexico. *Psyche*, **21**: 171–184.

Mann, W.M. 1915. Some myrmecophilous insects from Hayti. *Psyche*, **22**: 161–166.

Manweiler, S. A. 1986. Developmental and ecological comparisons of *Trichogramma minutum* and *Trichogramma platneri* (Hymenoptera: Trichogrammatidae). *Pan-Pacific Entomologist*, **62**: 128–139.

Mao, Y.-T. 1945. Synopsis of the Mexican species of *Cardiochiles* Nees (Hymenoptera, Braconidae). *Pan-Pacific Entomologist*, **21**: 125–134.

Mao, Y.-T. 1949. The species of ichneumon-flies of the genus *Cardiochiles* occurring in America north of Mexico. *Proceedings of the United States National Museum*, **99**: 229–266.

Maple, J.D. 1947. The eggs and first instar larvae of Encyrtidae and their morphological adaptations for respiration. *University of California Publications in Entomology*, **8**: 25–122.

Marchal, P. 1904. Recherches sur la biologie et le développement des Hyménoptères parasites. I. La polyembryonie spécifique ou germinogonie. *Archives de Zoologie Expérimentale et Générale*, (4) **2**: 257–335.

Marchal, P. 1906. Recherches sur la biologie et le développement des Hyménoptères parasites. II. Les platygasters. *Archives de Zoologie Expérimentale et Générale*, **4**: 485–640.

Marchal, P. 1907. Utilisation des insectes auxiliaires entomophages dans la lutte contre les insectes nuisibles à l'agriculture. *Annales de l'Institute d'Agronomie, Paris*, (Sér. 2), **6**: 281–354.

Marenco, R.A. & Saunders, J.L. 1993. Parasitoides del gusano cogollero, *Spodoptera frugiperda* (Lepidoptera: Noctuidae) en maiz, en Turrialba Costa Rica. *Manejo Integrado de Plagas*, **27**: 18–23.

Mariconi, F.A.M. 1970. As *Suavas*, 167 pp. Editora Agronomica 'Ceres', São Paulo.

Mari Mutt, J.A. 1977. New species of melittophilous *Pseudocyphoderus* from Costa Rica and remarks on taxonomy (Collembola: Cyphoderidae). *Entomological News*, **88**: 229–234.

Markl, H. 1985. Manipulation, modulation, information, cognition: some of the riddles of communication, pp. 163–194. *In*: Hölldobler, B. & Lindauer, M. (eds), *Experimental Behavioral Ecology and Sociobiology in Memoriam Karl von Frisch 1886–1982* (Fortschritte der Zoologie, Band 31), 488pp. Gustav Fischer Verlag, Stuttgart.

Marsh, P.M. 1961. A taxonomic study of the genus *Cremnops* Foerster in America north of Mexico (Hymenoptera, Braconidae). *Annals of the Entomological Society of America*, **54**: 851–861.

Marsh, P.M. 1963. A key to the Nearctic subfamilies of the family Braconidae (Hymenoptera). *Annals of the Entomological Society of America*, **56**: 522–527.

Marsh, P.M. 1965. The Nearctic Doryctinae. I. A review of the subfamily with a taxonomic revision of the tribe Hecabolini (Hymenoptera: Braconidae). *Annals of the Entomological Society of America*, **58**: 668–699.

Marsh, P.M. 1968. The Nearctic Doryctinae, VI. The genera *Acrophasmus, Glyptocolastes, Doryctinus*, and a new genus, *Stenocorse* (Hymenoptera: Braconidae). *Proceedings of the Entomological Society of Washington*, **70**: 101–113.

Marsh, P.M. 1969 (1968). The Nearctic Doryctinae, VII. The genus *Doryctes* Haliday (Hymenoptera: Braconidae). *Transactions of the American Entomological Society*, **94**: 379–405.

Marsh, P.M. 1970*a*. The Nearctic Doryctinae, VIII. The genera *Liobracon* and *Pedinotus*, with notes on the definition of the subfamily (Hymenoptera: Braconidae). *Proceedings of the Entomological Society of Washington*, **72**: 313–317.

Marsh, P.M. 1970*b*. The Nearctic Doryctinae, IX. The genus *Odontobracon* and notes on related genera (Hymenoptera: Braconidae). *Pan-Pacific Entomologist*, **46**: 275–283.

Marsh, P.M. 1976. The Nearctic Doryctinae, X. The genus *Rhaconotus* Ruthe (Hymenoptera: Braconidae). *Proceedings of the Entomological Society of Washington*, **78**: 389–403.

Marsh, P.M. 1978. The braconid parasites (Hymenoptera) of *Heliothis* species (Lepidoptera: Noctuidae). *Proceedings of the Entomological Society of Washington*, **80**: 15–36.

Marsh, P.M. 1979. Descriptions of new Braconidae (Hymenoptera) parasitic on the potato tuberworm and on related Lepidoptera from Central and South America. *Journal of the Washington Academy of Sciences*, **69**: 12–17.

Marsh, P.M. 1982*a*. Descriptions of two new species of *Heterospilus* (Hymenoptera: Braconidae) from South America. *Boletín de Entomología Venezolana* N.S., **2**: 57–60.

Marsh, P.M. 1982*b*. Two new species of *Heterospilus* (Hymenoptera: Braconidae) from Mexico being introduced against the cotton boll weevil, *Anthonomus grandis* (Coleoptera: Curculionidae). *Proceedings of the Entomological Society of Washington*, **84**: 849–854.

Marsh, P.M. 1983. A taxonomic study of the South American genus *Megaloproctus* Schulz (Hymenoptera: Braconidae: Doryctinae). *Contributions of the American Entomological Institute*, **20**: 363–380.

Marsh, P.M. 1986. A new species of *Cardiochiles* (Hymenoptera: Braconidae) introduced into Florida to control *Diaphania* spp. (Lepidoptera: Pyralidae). *Proceedings of the Entomological Society of Washington*, **88**: 131–133.

Marsh, P.M. 1988. Revision of the tribe Odontobraconini in the Western Hemisphere (Hymenoptera: Braconidae: Doryctinae). *Systematic Entomology*, **13**: 443–464.

Marsh, P. 1991. Description of a phytophagous doryctine braconid from Brasil (Hymenoptera: Braconidae). *Proceedings of the Entomological Society of Washington*, **93**: 92–95.

Marsh, P. 1993. Descriptions of new Western Hemisphere genera of the subfamily Doryctinae (Hymenoptera: Braconidae). *Contributions of the American Entomological Institute*, **28**(1): 1–58.

Marsh, P.M., Shaw, S.R., & Wharton, R.A. 1987. An identification manual for the North American genera of the family Braconidae (Hymenoptera). *Memoirs of the Entomological Society of Washington*, **13**: 1–98.

Martin, J.C. 1956. A taxonomic revision of the triaspidine braconid wasps of Nearctic America (Hymenoptera). *Canada Department of Agriculture Publication*, **965**: 1–157.

Martin, M.M. 1979. Biochemical implications of insect mycophagy. *Biological Reviews*, **54**: 1–21.

Martin, M.M. 1987. *Invertebrate-Microbial Interactions. Ingested Fungal Enzymes in Arthropod Biology*. 148pp. Cornell University, Ithaca.

Martin, M.M., Boyd, N.D., Gieselmann, M.J. & Silver, R.G. 1975. Activity of faecal fluid of a leaf-cutting ant toward plant cell wall polysaccharides. *Journal of Insect Physiology*, **21**: 1887–1892.

Martin, W.F. & Martin, R.F. 1990. Reproduction of the sand wasps *Stictia signata* (L.) and *Bicyrtes variegata* (Olivier) (Hymenoptera: Sphecidae) on the Caribbean coast of Quintana Roo, Mexico. *Pan-Pacific Entomologist*, **66**: 71–78.

Martinez, J.M. & Vela-Diaz, R. 1953. Contribución española a la historia del curare. *Hypnos*, **1**: 7–64.

Martorell, L.F. 1941. Biological notes on the sea-grape sawfly, *Schizocera krugii* Cresson, in Puerto Rico. *Caribbean Forester*, **2**: 141–144.

Maschwitz, U., Koob, K. & Schildknecht, H. 1970. Ein Beitrag zur Funktion der Metathoracaldrüse der Ameisen. *Journal of Insect Physiology*, **16**: 387–404.

Masner, L. 1969. The geographic distribution of recent and fossil Ambositrinae (Hymenoptera, Proctotrupoidea: Diapriidae) *Tagungsberichte Deutsche Akademie der Landwirtschaftswissenschaften zu Berlin*, **80**: 105–109.

Masner, L. 1972. The classification and interrelationships of Thoronini (Hymenoptera: Proctotrupoidea, Scelionidae). *Canadian Entomologist*, **104**: 833–849.

Masner, L. 1976*a*. Revisionary notes and keys to world genera of Scelionidae (Hymenoptera: Proctotrupoidea). *Memoirs of the Entomological Society of Canada*, **97**: 1–87.

Masner, L. 1976*b*. Notes on the ecitophilous diapriid genus *Mimopria* Holmgen (Hymenoptera: Proctotrupoidea, Diapriidae). *Canadian Entomologist*, **108**: 123–126.

Masner, L. 1976*c*. The Nearctic species of *Iphitrachelus* Walker (Hymenoptera, Proctotrupoidea, Platygastridae), with a key to world species. *Canadian Entomologist*, **108**: 1065–1068.

Masner, L. 1976*d*. A revision of the Ismarinae of the New World (Hymenoptera: Proctotrupoidea, Diapriidae). *Canadian Entomologist*, **108**: 1243–1266.

Masner, L. 1978. A revision of the New World species of *Leptoteleia* Kieffer (Hymenoptera: Scelionidae), egg parasites of crickets. *Canadian Entomologist*, **110**: 353–380.

Masner, L. 1979. The *variicornis*-group of *Gryon* Haliday (Hymenoptera: Scelionidae). *Canadian Entomologist*, **111**: 791–805.

Masner, L. 1980*a*. The identity of *Calotelea ocularis* Ashmead, 1894 (Hymenoptera, Proctotrupoidea, Scelionidae). *Canadian Entomologist*, **112**: 393–396.

Masner, L. 1980*b*. A revision of the Nearctic species of *Calotelea* Westwood (Hymenoptera, Proctotrupoidea, Scelionidae). *Canadian Entomologist*, **112**: 397–408.

Masner, L. 1980*c*. Key to genera of Scelionidae of the Holarctic region, with descriptions of new genera and species (Hymenoptera: Proctotrupoidea). *Memoirs of the Entomological Society of Canada*, **113**: 1–54.

Masner, L. 1980*d*. The Nearctic species of *Acerotella* Masner (Hymenoptera, Proctotrupoidea, Platygastridae). *Canadian Entomologist*, **112**: 1291–1303.

Masner, L. 1981. Revision of the Nearctic species of *Metaclisis* Foerster (Hymenoptera, Platygastridae, Inostemmatinae). *Canadian Entomologist*, **113**: 1069–1091.

Masner, L. 1983*a*. The genus *Oethecoctonus* Ashmead in North America (Hymenoptera: Proctotrupoidea: Scelionidae). *Canadian Entomologist*, **115**: 17–24.

Masner, L. 1983*b*. A revision of *Gryon* Haliday in North America (Hymenoptera: Proctotrupoidea: Scelionidae). *Canadian Entomologist*, **115**: 123–174.

Masner, L. 1983*c*. Revision of the Nearctic species of *Trichacis* Foerster (Hymenoptera: Proctotrupoidea: Platygastridae). *Canadian Entomologist*, **115**: 1071–1093.

Masner, L. 1990. Status report on taxonomy of Hymenoptera in North America, pp. 231–240. *In*: Kosztarab, M. & Schaefer, C.W. (eds) *Systematics of the North American Insects and Arachnids: Status and Needs*. 247pp. Virginia Agricultural Experiment Station Information Series 90–1. Virginia Polytechnic Institute and State University.

Masner, L. 1991*a*. The Nearctic species of *Duta* Nixon (Hymenoptera: Scelionidae), egg parasitoids of ground crickets (Orthoptera: Gryllidae). *Canadian Entomologist*, **123**: 777–793.

Masner, L. 1991*b*. Revision of *Spilomicrus* Westwood in America north of Mexico (Hymenoptera: Proctotrupoidea, Diapriidae). *Canadian Entomologist*, **123**: 107–177.

Masner, L. & Dessart, P. 1967. La reclassification des catégories taxonomiques supérieures des Ceraphronoidea (Hymenoptera). *Bulletin d'Institut Royal des Sciences Naturelles de Belgique*, **43**(22): 1–33.

Masner, L. & Dessart, P. 1972. Notes on Embidobiini (Scelionidae: Hymenoptera) with description of a new genus. *Canadian Entomologist*, **104**: 505–510.

Masner, L. & Goulet, H. 1981. A new model of flight-interception trap for some hymenopterous insects. *Entomological News*, **92**: 199–202.

Masner, L. & Huggert, L. 1979*a*. Descriptions of new taxa in the Thoronini (Hymenoptera, Proctotrupoidea, Scelionidae). *Canadian Entomologist*, **111**: 911–917.

Masner, L. & Huggert, L. 1979*b*. Revision of the world species of Aradophagini (Hymenoptera: Scelionidae). *Canadian Entomologist*, **111**: 1089–1100.

Masner, L. & Huggert, L. 1989. World review and keys to genera of the subfamily Inostemmatinae with reassignment of the taxa to the Platygastrinae and

Sceliotrachelinae (Hymenoptera: Platygastridae). *Memoirs of the Entomological Society of Canada*, **147**: 1–214.

Masner, L. & Johnson, N.F. 1979. A new species of *Telenomus* (Hymenoptera: Scelionidae), an egg parasite of the ambush bug, *Phymata* sp. (Heteroptera: Phymatidae). *Canadian Entomologist*, **111**: 1115–1119.

Mason, W.R.M. 1964. The genus *Chrysopophthorus* Goidanich (Hymenoptera: Braconidae). *Canadian Entomologist*, **96**: 1005–1017.

Mason, W.R.M. 1967. Specialization in the egg structure of *Exenterus* (Hymenoptera: Ichneumonidae) in relation to distribution and abundance. *Canadian Entomologist*, **99**: 375–384.

Mason, W.R.M. 1969. Muesebeckiini, a new tribe of Braconidae (Hymenoptera). *Proceedings of the Entomological Society of Washington*, **71**: 263–278.

Mason, W.R.M. 1971. An Indian *Agriotypus* (Hymenoptera: Agriotypidae). *Canadian Entomologist*, **103**: 1521–1524.

Mason, W.R.M. 1973. Recognition of *Zemiotes* (Hymenoptera: Braconidae). *Proceedings of the Entomological Society of Washington*, **75**: 213–215.

Mason, W.R.M. 1974. A generic synopsis of Brachistini (Hymenoptera: Braconidae) and recognition of the name *Charmon* Haliday. *Proceedings of the Entomological Society of Washington*, **76**: 235–246.

Mason, W.R.M. 1978*a*. A new genus, species and family of Hymenoptera (Ichneumonoidea) from Chile. *Proceedings of the Entomological Society of Washington*, **80**: 606–610.

Mason, W.R.M. 1978*b*. A synopsis of the Nearctic Braconini, with revisions of Nearctic species of *Coeloides* and *Myosoma* (Hymenoptera: Braconidae). *Canadian Entomologist*, **110**: 721–768.

Mason, W.R.M. 1981*a*. Paxylommatidae: the correct family-group name for *Hybrizon* Fallén (Hymenoptera: Ichneumonoidea), with figures of unusual antennal sensilla. *Canadian Entomologist*, **113**: 433–439.

Mason, W.R.M. 1981*b*. The polyphyletic nature of *Apanteles* Foerster (Hymenoptera: Braconidae): A phylogeny and reclassification of Microgastrinae. *Memoirs of the Entomological Society of Canada*, **115**: 1–147.

Mason, W.R.M. 1983*a*. *The phylogeny of the Apocrita* (Unpublished lecture notes handed out at workshop on the taxonomy and biology of the parasitic Hymenoptera, Gainesville Fla.)

Mason, W.R.M. 1983*b*. The abdomen of *Vanhornia eucnemidarum* (Hymenoptera: Proctotrupoidea). *Canadian Entomologist*, **115**: 1483–1488.

Mason, W.R.M. 1984. Structure and movement of the abdomen of female *Pelecinus polyturator* (Hymenoptera: Pelecinidae). *Canadian Entomologist*, **116**: 419–426.

Mason, W.R.M. 1987*a*. Discovery of female *Apozyx* (Hymenoptera: Apozygidae) and comments on its taxonomic position. *Proceedings of the Entomological Society of Washington*, **89**: 226–229.

Mason, W.R.M. 1987*b*. *Vadum*, a new genus of nearctic Braconidae (Hymenoptera). *Proceedings of the Entomological Society of Washington*, **89**: 325–328.

Matejko, I. & Sullivan, D.J. 1984. Interspecific tertiary parasitoidism between two aphid hyperparasitoids: *Dendrocerus carpenteri* and *Alloxysta megourae* (Hymenoptera: Megaspilidae and Cynipidae). *Journal of the Washington Academy of Sciences* **74**: 31–38.

Matheson, R. & Crosby, C.R. 1912. Aquatic Hymenoptera in America. *Annals of the Entomological Society of America*, **5**: 65–71.

Mathur, K.C. 1967*a*. Notes on *Apistephialtes* sp., an ichneumonid larval parasite of *Hypsipyla robusta* Moore in India. *Technical Bulletin of the Commonwealth Institute of Biological Control*, **9**: 133–135.

Mathur, K.C. 1967*b*. Notes on *Banchopsis ruficornis* Cameron (Hym.: Ichneumonidae) an internal larval parasite of *Heliothis armigera* (Hübner) in India. *Current Science*, **36**: 356–357.

Matsuda, R. 1965. Morphology and evolution of the insect head. *Memoirs of the American Entomological Institute*, **4**: 1–334.

Matsuda, R. 1970. Morphology and evolution of the insect thorax. *Memoirs of the Entomological Society of Canada*, **76**: 1–431.

Matsuda, R. 1976. Morphology and evolution of the insect abdomen. *International Series in Pure and Applied Biology, Zoology Division*, **56**: 1–534.

Matsuda, R. 1987. *Animal Evolution in Changing Environments with Special Reference to Abnormal Metamorphosis*. 355pp. John Wiley & Sons, New York.

Matta V., A. 1979. Enemigos naturales de las conchuelas blancas del olivo en el valle de Azapa, Arica-Chile. *Idesia*, **5**: 231–242.

Matthews, R.W. 1968*a*. Nesting biology of the social wasp *Microstigmus comes* (Hymenoptera: Sphecidae, Pemphredoninae). *Psyche*, **75**: 23–45.

Matthews, R.W. 1968*b*. *Microstigmus comes*: sociality in a sphecid wasp. *Science*, **160**: 787–788.

Matthews, R.W. 1969. The behavior of three wasp parasites of a Costa Rican bark beetle, with description of a new species of *Ecphylus* (Hymenoptera, Braconidae). *Journal of the Georgia Entomological Society*, **4**: 111–118.

Matthews, R.W. 1970*a*. A revision of the genus *Spathius* in America north of Mexico (Hymenoptera, Braconidae). *Contributions of the American Entomological Institute*, **4**(5): 1–86.

Matthews, R.W. 1970*b*. A new thrips-hunting *Microstigmus* from Costa Rica (Hymenoptera: Sphecidae, Pemphredoninae). *Psyche*, **77**: 120–126.

Matthews, R.W. 1974. Biology of Braconidae. *Annual Review of Entomology*, **19**: 15–32.

Matthews, R.W. 1975. Courtship in parasitic wasps, pp. 66–86. *In*: Price, P.W. (ed.) *Evolutionary Strategies of Parasitic Insects and Mites*. 224pp. Plenum Press, New York.

Matthews, R.W. 1983. Biology of a new *Trypoxylon* that utilizes nests of *Microstigmus* in Costa Rica (Hymenoptera: Sphecidae). *Pan-Pacific Entomologist*, **59**: 152–162.

Matthews, R.W. 1991. Evolution of social behavior in sphecid wasps. pp. 570–602. *In*: Ross, K.G. & Matthews, R.W. *The Social Biology of Wasps*. 678pp. Cornell University Press, Ithaca.

Matthews, R.W. & Evans, H.E. 1974. Notes on the behavior of three species of *Microbembex* (Hymenoptera: Sphecidae) in South America. *Journal of the Georgia Entomological Society*, **9**: 79–85.

Matthews, R.W. & Marsh, P.M. 1969. On the status of Cresson's *Stenophasmus* species from Cuba (Hymenoptera: Braconidae: Doryctinae). *Proceedings of the Entomological Society of Washington*, **71**: 94–98.

Matthews, R.W. & Marsh, P.M. 1973. *Notiospathius*, a new Neotropical genus (Hymenoptera: Braconidae). *Journal of the Washington Academy of Sciences*, **63**: 73–75.

Matthews, R.W., Saunders, R.A. & Matthews, J.R. 1981. Nesting behavior of the sand wasp *Stictia maculata* (Hymenoptera: Sphecidae) in Costa Rica. *Journal of the Kansas Entomological Society*, **54**: 249–254.

Matthews, R.W. & Starr, C.K. 1984. *Microstigmus comes* wasps have a method of nest construction unique among social insects. *Biotropica*, **16**: 55–58.

Maxwell, D.E. 1955. The comparative internal larval anatomy of sawflies (Hymenoptera: Symphyta). *Canadian Entomologist*, supplement 1, **87**: 1–132.

Maynard Smith, J. 1982. *Evolution and the Theory of Games*. 220pp. Cambridge University Press, Cambridge.

Mazanec, Z. 1990. The immature stages and life history of *Diaulomorpha* sp. (Hymenoptera: Eulophidae), a parasitoid of *Perthida glyphopa* Common (Lepidoptera: Incurvariidae). *Journal of the Australian Entomological Society*, **29**: 147–159.

Mazzone, P. & Viggiani, G. 1984. Osservazioni morfo-biologiche sugli stadi preimmaginali di *Prococcophagus varius* Silv. e *P. saissetiae* Ann. e Myn. (Hym. Aphelinidae), parassiti di *Saissetia oleae* Oliv. (Hom. Coccoidea). *Bollettino del Laboratorio di Entomologia Agraria 'Filippo Silvestri'*, **41**: 143–148.

McAlister, L.C. Jr & Anderson, W.H. 1932. The blueberry stem-gall in Maine. *Journal of Economic Entomology*, **25**: 1164–1169.

McAuslane, H.J., Vinson, S.B. & Williams, H.J. 1991. Stimuli influencing host microhabitat location in the parasitoid *Campoletis sonorensis*. *Entomologia Experimentalis et Applicata*, **58**: 267–277.

McCalla, D.R., Genthe, M.K. & Hovanitz, W. 1962. Chemical nature of an insect gall growth factor. *Plant Physiology*, **37**: 98–103.

McCluskey, E.S. 1965. Circadian rhythms in male ants of five diverse species. *Science*, **150**: 1037–1039.

McCluskey, E.S. 1987. Circadian rhythm in the tropical ant *Ectatomma* (Hymenoptera: Formicidae). *Psyche*, **94**: 245–251.

McCluskey, E.S. & Brown, W.L. Jr 1972. Rhythms and other biology of the giant tropical ant *Paraponera*. *Psyche*, **79**: 335–347.

McComb, C.W. 1967. A revision of the *Chelonus* subgenus *Microchelonus* in North America north of Mexico (Hymenoptera: Braconidae). *University of Maryland Agricultural Experiment Station Bulletin* A-**149**: 1–148.

McConnell, H.S. 1938. Additional notes on *Oocenteter tomostethae*. *Proceedings of the Entomological Society of Washington*, **40**: 23–24.

McCorquodale, D.B. 1986. Digger wasp (Hymenoptera: Sphecidae) provisioning flights as a defence against a nest parasite, *Senotainia trilineata* (Diptera: Sarcophagidae). *Canadian Journal of Zoology*, **64**: 1622–1627.

McCulloch, J.W. & Yuasa, H. 1915. Further data on the life economy of the chinch bug egg parasite. *Journal of Economic Entomology*, **8**: 248–260.

McGinley, R.J. 1980. Glossal morphology of the Colletidae and recognition of the Stenotritidae at the family level (Hymenoptera: Apoidea). *Journal of the Kansas Entomological Society*, **53**: 539–552.

McGinley, R.J. 1981. Systematics of the Colletidae based on mature larvae with phenetic analysis of apoid larvae. *University of California Publications in Entomology*, **91**: 1–307.

McGinley, R.J. 1986. Studies of Halictinae (Apoideae: Halictidae), I: Revision of New World *Lasioglossum* Curtis. *Smithsonian Contributions to Zoology*, **429**: 1–294.

McIver, J.D. & Stonedahl, G. 1993. Myrmecomorphy: morphological and behavioral mimicry of ants. *Annual Review of Entomology*, **38**: 351–379.

McKey, D. 1989*a*. Interactions between ants and leguminous plants. pp. 673–718. *In*: Stirton, C.H. & Zarucchi, J.L. (eds.), *Advances in Legume Biology. Monographs in Systematic Botany*, **29**. 842pp. Missouri Botanical Garden.

McKey, D. 1989*b*. Population biology of figs: applications for conservation. *Experientia*, **45**: 661–673.

McMurtry, J.A. 1978. Theridiidae, p.1. *In*: Clausen, C.P. *Introduced Parasites and Predators of Arthropod Pests and Weeds: A World Review*, 545pp. USDA Agriculture Handbook No.480.

Medvedev, G.S. (ed.) 1978. *Keys to Insects of the European Part of the USSR*, **3**. 756pp. Nauka Publishers, Leningrad. [Translation: 1987. 1341 pp. Amerind Publ. Co., New Delhi.]

Melo, G.A.R. de & Evans, H.E. 1993. Two new *Microstigmus* species (Hymenoptera, Sphecidae), with the description of their parasite, *Goniozus microstigmi* sp. n. (Hymenoptera, Bethylidae). *Proceedings of the Entomological Society of Washington*, **95**: 258–263.

Meloche, F. & Guppy, J.C. 1990. Encapsulation of eggs and larvae of *Dacnusa dryas* (Hymenoptera: Braconidae) by the alfalfa blotch leafminer (Diptera: Agromyzidae) and its importance in host-parasite coexistence. *Environmental Entomology*, **19**: 423–427.

Memmott, J., Godfray, H.C.J. & Bolton, B. 1993. Predation and parasitism in a tropical herbivore community. *Ecological Entomology*, **18**: 348–352.

Memmott, J., Godfray, H.C.J. & Gauld, I.D. 1994. The structure of a tropical host-parasitoid community. *Journal of Animal Ecology*, **63**: 521–540.

Menke, A.S. 1980. Biological notes on *Trachypus mexicanus* Saussure and *T. petiolatus* (Spinola) (Hymenoptera: Sphecidae). *Journal of the Kansas Entomological Society*, **53**: 235–236.

Menke, A.S. 1988. *Pison* in the New World: A revision (Hymenoptera: Sphecidae: Trypoxylini). *Contributions of the American Entomological Institute*, **24**(3): 1–171.

Menke, A.S. 1990. *Pisonopsis* in Costa Rica (Sphecidae). *Sphecos*, **19**: 26.

Menke, A.S. 1991. Wasping in Costa Rica — 1991. *Sphecos*, **22**: 10–12.

Menke, A.S. 1992*a*. *Pison* in Costa Rica (Sphecidae). *Sphecos*, **23**: 8–10.

Menke, A.S. 1992*b*. Mole cricket hunters of the genus *Larra* in the New World (Hymenoptera: Sphecidae, Larrinae). *Journal of Hymenoptera Research*, **1**: 175–234.

Menke, A.S. 1993. A new species of *Apocharips* from Costa Rica (Hymenoptera: Cynipoidea, Charipidae). *Journal of Hymenoptera Research*, **2**: 97–100.

Menke, A.S. & Evenhuis, H.H. 1991. North American Charipidae: Key to genera, nomenclature, species checklists, and a new species of *Dilyta* Förster (Hymenoptera: Cynipoidea). *Proceedings of the Entomological Society of Washington*, **93**: 136–158.

Menke, A.S. & Vardy, C. 1980. A synopsis of the tribe Scapheutini (Hymenoptera: Sphecidae). *Papéis Avulsos de Zoologia*, **34**: 73–85.

Menke, A.S. & Vincent, D.L., 1983. A review of the genus *Polemistus* in the New World (Hymenoptera: Sphecidae). *Pan-Pacific Entomologist*, **59**: 163–175.

Menke, A.S. & Yustiz, E. 1983. *Ampulex compressa* (F.) in Venezuela (Hymenoptera: Sphecidae). *Proceedings of the Entomological Society of Washington*, **85**: 180.

Mertins, J.W. 1980. Life history and behavior of *Laelius pedatus*, a gregarious bethylid ectoparasitoid of *Anthrenus verbasci*. *Annals of the Entomological Society of America*, **73**: 686–693.

Mexzón, R.G. & Chinchilla, C.M. 1991. Entomofauna perjudicial, enemigos naturales y malezas útiles en palma aceitera (*Elaeis guineensis*) J. en América Central. *Manejo Integrado de Plagas*, **20–21**: 1–7.

Meyer-Grassmann, A. 1967. *Drosophila* und *Pseudeucoila* V: Beiträge zur parasitierungsbiologie von *Pseudeucoila bochei* Weld (Cynipidae, Hymenoptera) und Bericht über zwei neue mutanten. *Revue Suisse de Zoologie*, **74**: 409–437.

Meyers, D.M. & Deonier, D.L. 1993. A behavioral-ecological study of *Kleidotoma parydrae* Beardsley (Hymenoptera: Eucoilidae), with notes on *Anaphes* sp. (Hymenoptera: Mymaridae) parasites of *Parydra* spp. (Diptera: Ephydridae). *Contributions of the American Entomological Institute*, **27**: 7–29.

Michalski, J. & Seniczak, S. 1974. *Trichogramma semblidis* [Chalcidoidea : Trichogrammatidae] as a parasite of the bark beetle eggs [Coleoptera : Scolytidae]. *Entomophaga*, **19**: 237–242.

Michelbacher, A.E., Hurd, P.D. Jr & Linsley, E.G. 1971. Experimental introduction of squash bees (*Peponapis*) to improve yields of squashes, gourds and pumpkins. *Bee World*, **52**: 156–166.

Michelsen, A., Andersen, B.B., Storm, J., Kirchner, W.H. & Lindauer, M. 1992. How honeybees perceive communication dances, studied by means of a mechanical model. *Behavioral Ecology and Sociobiology*, **30**: 143–150.

Michener, C.D. 1944. Comparative external morphology, phylogeny, and a classification of the bees (Hymenoptera). *Bulletin of the American Museum of Natural History*, **82**: 151–326.

Michener, C.D. 1953. Comparative morphological and systematic studies of bee larvae with a key to the families of hymenopterous larvae. *The University of Kansas Science Bulletin*, **35**: 987–1102.

Michener, C.D. 1954. Bees of Panamá. *Bulletin of the American Museum of Natural History*, **104**: 1–175.

Michener, C.D. 1962. An interesting method of pollen collecting by bees from flowers with tubular anthers. *Revista de Biologia Tropical*, **10**: 167–175.

Michener, C.D. 1964. Evolution of the nests of bees. *American Zoologist*, **4**: 227–239.

Michener, C.D. 1965. A classification of the bees of the Australian and South Pacific regions. *Bulletin of the American Museum of Natural History*, **130**: 1–362.

Michener, C.D. 1966. The classification of the Diphaglossinae and North American species of the genus *Caupolicana* (Hymenoptera, Colletidae). *University of Kansas Science Bulletin*, **46**: 717–751.

Michener, C.D. 1969. Immature stages of a chalcidoid parasite tended by allodapine bees (Hymenoptera: Perilampidae and Anthophoridae). *Journal of the Kansas Entomological Society*, **42**: 247–250.

Michener, C.D. 1974*a*. *The Social Behavior of the Bees. A Comparative Study*. 404pp. Belknap Press, Cambridge, Mass.

Michener, C.D. 1974*b*. Further notes on nests of *Ancyloscelis* (Hymenoptera: Anthophoridae). *Journal of the Kansas Entomological Society*, **47**: 19–22.

Michener, C.D. 1975. The Brazilian bee problem. *Annual Review of Entomology*, **20**: 399–416.

Michener, C.D. 1977. Nests and seasonal cycle of *Neocorynura pubescens* in Colombia (Hymenoptera: Halictidae). *Revista de Biologia Tropical*, **25**: 39–41.

Michener, C.D. 1978. The parasitic groups of Halictidae (Hymenoptera, Apoidea). *University of Kansas Science Bulletin*, **51**: 291–339.

Michener, C.D. 1979 (1980). Biogeography of the bees. *Annals of the Missouri Botanical Garden*, **66**: 277–347.

Michener, C.D. 1983. The classification of the Lithurginae (Hymenoptera: Megachilidae). *Pan-Pacific Entomologist*, **59**: 176–187.

Michener, C.D. 1986*a*. New Peruvian genus and a generic review of Andreninae (Hymenoptera: Apoidea: Andrenidae). *Annals of the Entomological Society of America*, **79**: 62–72.

Michener, C.D. 1986*b*. A review of the tribes Diphaglossini and Dissoglottini (Hymenoptera, Colletidae). *University of Kansas Science Bulletin*, **53**: 183–214.

Michener, C.D. 1989. Classification of American Colletinae (Hymenoptera, Apoidea). *University of Kansas Science Bulletin*, **53**: 622–703.

Michener, C.D. 1990. Classification of the Apidae (Hymenoptera). *University of Kansas Science Bulletin*, **54**: 75–163.

Michener, C.D. & Bennett, F.D. 1977. Geographical variation in nesting biology and social organization of *Halictus ligatus*. *University of Kansas Science Bulletin*, **51**: 233–260.

Michener, C.D., Breed, M.D. & Bell, W.J. 1979. Seasonal cycles, nests, and social behavior of some Colombian halictine bees (Hymenoptera: Apoidea). *Revista de Biologia Tropical*, **27**: 13–34.

Michener, C.D. & Brooks, R.W. 1984. A comparative study of the glossae of bees (Apoidea). *Contributions of the American Entomological Institute*, **22**(1): 1–73.

Michener, C.D. & Brothers, D.J. 1974. Were workers of eusocial Hymenoptera initially altruistic or oppressed? *Proceedings of the National Academy of Sciences of the United States of America*, **71**: 671–674.

Michener, C.D. & Eickwort, K.R. 1966. Observations on nests of *Ceratina* in Costa Rica (Hymenoptera, Apoidea). *Revista Biologia Tropical*, **14**: 279–286.

Michener, C.D. & Greenberg, L. 1980. Ctenoplectridae and the origin of long-tongued bees. *Zoological Journal of the Linnean Society*, **69**: 183–203.

Michener, C.D. & Kerfoot, W.B. 1967. Nests and social behavior of three species of *Pseudaugochloropsis* (Hymenoptera: Halictidae). *Journal of the Kansas Entomological Society*, **40**: 214–232.

Michener, C.D., Kerfoot, W.B. & Ramirez B., W. 1966. Nests of *Neocorynura* in Costa Rica. *Journal of the Kansas Entomological Society*, **39**: 245–258.

Michener, C.D. & Lange, R.B. 1958. Observations on the behavior of Brasilian halictid bees, III. *University of Kansas Science Bulletin*, **39**: 473–505.

Michener, C.D. & Lange, R.B. 1959. Observations on the behaviour of Brazilian Halictid bees (Hymenoptera, Apoidea) IV. *Augochloropsis* with notes on extralimital forms. *American Museum Novitate*, **1924**: 1–41.

Michener, C.D. & Moure, J.S. 1957. A study of the classification of the more primitive non-parasitic anthophorine bees (Hymenoptera, Apoidea). *Bulletin of the American Museum of Natural History*, **112**: 395–452.

Michener, C.D., Winston, M.L. & Jander, R. 1978. Pollen manipulation and related activities and structures in bees of the family Apidae. *University of Kansas Science Bulletin*, **51**: 575–601.

Mickel, C.E. 1928. Biological and taxonomic investigations on the mutillid wasps. *Smithsonian Institution United States National Museum Bulletin*, **143**: 1–351.

Mickel, C.E. 1969. *Pseudomethoca willei* n. sp. reared from cells of the bee *Lasioglossum* (*Dialictus*) *umbripenne* (Ellis) (Hymenoptera: Mutillidae; Apoidea). *Journal of the Kansas Entomological Society*, **42**: 524–526.

Mickel, C.E. 1973. *Paramutilla halicta* n. genus, n. species, a parasite of the halictine bee *Augochlorella edentata* (Hymenoptera: Mutillidae). *Journal of the Kansas Entomological Society*, **46**: 1–3.

Middlekauf, W.W. 1983. A revision of the sawfly family Orussidae for North and Central America (Hymenoptera: Symphyta, Orussidae). *University of California Publications in Entomology*, **101**: 1–46.

Miller, C.D. 1964. Some species of the New World genus *Paraolinx* Ashmead (Hymenoptera: Eulophidae). *Canadian Entomologist*, **96**: 1352–1362.

Miller, C.D. 1970. The Nearctic species of *Pnigalio* and *Sympiesis* (Hymenoptera: Eulophidae). *Memoirs of the Entomological Society of Canada*, **68**: 1–121.

Miller, F.D. Jr, Cheetham, T., Bastian, R.A. & Hart, E.R. 1987. Parasites recovered from overwintering mimosa webworm *Homadaula anisocentra* (Lepidoptera: Plutellidae). *Great Lakes Entomologist*, **20**: 143–146.

Miller, G.L. & Lambdin, P.L. 1985. Observations on *Anacharis melanoneura* (Hymenoptera: Figitidae), a parasite of *Hemerobius stigma* (Neuroptera: Hemerobiidae). *Entomological News*, **96**: 93–97.

Miller, J.C. 1982. Life history of insect parasitoids involved in successful multiparasitism. *Oecologia*, **54**: 8–9.

Miller, T.D. 1986. New species of the genus *Sierolomorpha* from New Mexico (Hymenoptera: Sierolomorphidae). *Pan-Pacific Entomologist*, **62**: 121–123.

Milliron, H.E. 1949. Taxonomic and biological investigations in the genus *Megastigmus* with particular reference to the taxonomy of the Nearctic species (Hymenoptera: Chalcidoidea: Callimomidae). *American Midland Naturalist*, **41**: 257–420.

Milliron, H.E. 1970. *Pyrobombus (P.) cascadensis*, an undescribed species of bumblebee from the Pacific Northwest, USA (Hymenoptera: Bombinae). *Canadian Entomologist*, **102**: 382–383.

Mills, N.J. 1992. Parasitoid guilds, life-styles, and host ranges in the parasitoid complexes of tortricoid hosts (Lepidoptera: Tortricoidea). *Environmental Entomology*, **21**: 230–239.

Mitchell, T.B. 1973. *A Subgeneric Revision of the Bees of the Genus Coelioxys of the Western Hemisphere (Hymenoptera: Megachilidae)*. 129pp. North Carolina State University, Raleigh.

Mitchell, T.B. 1980. *A Generic Revision of the Megachiline Bees of the Western Hemisphere (Hymenoptera: Megachilidae)*. 95pp. North Carolina State University, Raleigh.

Móczár, L. 1986*a*. Revision of the genus *Hemiceropales* Priesner, 1969 (Hymenoptera: Ceropalidae). *Acta Zoologica Hungarica*, **32**: 317–342.

Móczár, L. 1986*b*. Revision of the *fulvipes-*, *ruficornis-* and *variegata*-groups of the genus *Ceropales* Latreille (Hym., Ceropalidae). *Acta Biologica*, Szeged (N.S.), **32**: 121–136.

Móczár, L. 1987. Revision of the *maculata* and *albicincta* groups of the genus *Ceropales* Latreille (Hymenoptera: Ceropalidae). *Acta Zoologica Hungarica*, **33**: 121–156.

Móczár, L. 1988. Revision of the subgenus *Priesnerius* Móczár (Hymenoptera, Ceropalidae). *Linzer Biologische Beiträge*, **20**: 119–160.

Móczár, L. 1989. Revision of the *helvetica*-group of the genus *Ceropales* Latreille (Hym.: Ceropalidae). *Beiträge zur Entomologie*, **39**: 9–61.

Móczár, L. 1990. Revision of the subgenus *Bifidoceropales* Priesner of the genus *Ceropales* Latreille (Hymenoptera: Ceropalidae). *Acta Zoologica Hungarica*, **36**: 58–85.

Móczár, L. 1991. Supplement to the revision of the genus *Ceropales* Latreille, I. (Hymenoptera: Ceropalidae). *Acta Zoologica Hungarica*, **37**: 67–74.

Moffett, M.W. 1986. Trap-jaw predation and other observations on two species of *Myrmoteras* (Hymenoptera: Formicidae). *Insectes Sociaux*, **33**: 85–99.

Momoi, S. & Okamoto, K. 1965. Notes on an ichneumonid parasite (Hymenoptera) of *Parasa consocia* (Lepidoptera). *Proceedings of the Entomological Society of Washington*, **67**: 238–243

Montesinos, H.J. & Rabinovich, J.E. 1979. Population dynamics of *Telenomus fariai* (Hym.: Scelionidae), a parasite of Chagas' disease vectors. VIII. Morphological and ecological comparisons between two allopatric populations under laboratory conditions. *Entomophaga*, **24**: 177–183.

Moore, S.D. 1989. Regulation of host diapause by an insect parasitoid. *Ecological Entomology*, **14**: 93–98.

Morales-Ramos, J.A. & Cate, J.R. 1992. Laboratory determination of age-dependent fecundity, development, and rate of increase of *Catolaccus grandis* (Burks) (Hymenoptera: Pteromalidae). *Annals of the Entomological Society of America*, **85**: 469–476.

Moran, V.C., Brothers, D.J. & Case, J.J. 1969. Observations on the biology of *Tetrastichus flavigaster* Brothers & Moran (Hymenoptera: Eulophidae), parasitic on psyllid nymphs (Homoptera). *Transactions of the Royal Entomological Society of London*, **121**: 41–58.

Morey, C.S. 1971. Biología de *Campoletis grioti* (Blanchard) (Hymen.: Ichneumonidae) parasito de la 'lagarta cogollera del maiz' *Spodoptera frugiperda* (J.E. Smith). *Revista Peruana de Entomologia*, **14**: 263–271.

Morgan, P.B., Weidhaas, D.E. & Patterson, R.S. 1981. Host-parasite relationship: augmentative releases of *Spalangia endius* Walker used in conjunction with population modeling to suppress field populations of *Musca domestica* L. (Hymenoptera: Pteromalidae and Diptera: Muscidae). *Journal of the Kansas Entomological Society*, **54**: 496–504.

Morley, C. 1915. *Ichneumonologia Britannica* 5. 400pp. H & W Brown, London.

Morris, K.R.S. 1937. The prepupal stage in Ichneumonidae, illustrated by the life-history of *Exenterus abruptorius* Thb. *Bulletin of Entomological Research*, **28**: 525–534.

Morris, K.R.S. 1938. *Eupelmella vesicularis* Retz. (Chalcididae) as a predator of another chalcid, *Microplectron fuscipennis* Zett. *Parasitology*, **30**: 20–32.

Morrison, G., Auerbach, M. & McCoy, E.D. 1979. Anomalous diversity of tropical parasitoids: a general phenomenon? *American Naturalist*, **114**: 303–307.

Morse, D.H. 1982. Behavior and ecology of bumble bees, pp. 245–322. *In*: Hermann, H.R. (ed.) *Social Insects, Volume III*. 459pp. Academic Press, New York.

Morse, R.A. (ed.) 1978. *Honey Bee Pests, Predators, and Diseases*. 430pp. Cornell University Press, Ithaca, N.Y.

Morse, R.A. & Nowogrodzki, R. (eds) 1990. *Honey Bee Pests, Predators, and Diseases*. [2nd edn]. 474pp. Cornell University Press, Ithaca, N.Y.

Moser, J.C. 1963. Contents and structure of *Atta texana* nest in summer. *Annals of the Entomological Society of America*, **56**: 286–291.

Moure, J.S. & Hurd, P.D. Jr 1987. *An Annotated Catalog of the Halictid Bees of the Western Hemisphere (Hymenoptera: Halictidae)*. 405pp. Smithsonian Institution Press, Washington, D.C.

Moursi, A.A. 1948*a*. Contributions to the knowledge of the natural enemies of mealybugs 2. *Anagyrus kamali* Moursi, a parasite of the Hibiscus melaybug, *Phenacoccus hirsutus* Green (Hymenoptera: Encyrtidae). *Bulletin de la Société Fouad 1er D'Entomologie*, **32**: 9–16.

Moursi, A.A. 1948*b*. Contributions to the knowledge of the natural enemies of mealybugs 3. *Anagyrus aegyptiacus* Moursi, a parasite of the Lebbek mealybug, *Pseudococcus filamentosus* Ckll. (Hymenoptera: Encyrtidae). *Bulletin de la Société Fouad 1er D'Entomologie*, **32**: 17–32.

Moursi, A.A. 1948*c*. Contributions to the knowledge of the natural enemies of mealybugs 4. *Leptomastix phenacocci* Compere, a parasite of the Lebbek mealybug, *Pseudococcus filamentosus* Ckll. (Hymenoptera: Encyrtidae). *Bulletin de la Société Fouad 1er D'Entomologie*, **32**: 33–40.

Moya Raygoza, G. 1990. *Parasitoides de* Dalbulus *spp. (Homoptera: Cicadellidae) en Jalisco, Mexico.* 62 pp. Unpublished. Tesis de Maestria en Ciencias. Colegio de Postgraduados, Chapingo, Mexico.

Mueller, T.F. 1983. The effect of plants on the host relations of a specialist parasitoid of *Heliothis* larvae. *Entomologia Experimentalis et Applicata*, **34**: 78–84.

Mueller, U.G. & Wolf-Mueller, B. 1991. Epiphyll deterrence to the leafcutter ant *Atta cephalotes. Oecologia*, **86**: 36–39.

Muesebeck, C.F.W. 1920. A revision of the North American species of the ichneumon-flies belonging to the genus *Apanteles. Proceedings of the United States National Museum*, **58**: 483–576.

Muesebeck, C.F.W. 1922. A revision of the North American ichneumon-flies belonging to the subfamilies Neoneurinae and Microgastrinae. *Proceedings of the United States National Museum*, **61**: 1–76.

Muesebeck, C.F.W. 1923. A revision of the North American species of ichneumon-flies belonging to the genus *Meteorus* Haliday. *Proceedings of the United States National Museum*, **63**: 1–44.

Muesebeck, C.F.W. 1925. A revision of the parasitic wasps of the genus *Microbracon* occurring in America north of Mexico. *Proceedings of the United States National Museum*, **67**: 1–85.

Muesebeck, C.F.W. 1927. A revision of the parasitic wasps of the subfamily Braconinae occurring in America north of Mexico. *Proceedings of the United States National Museum*, **69**: 1–73.

Muesebeck, C.F.W. 1931. *Monodontomerus aereus* Walker, both a primary and a secondary parasite of the Brown-tail Moth and the Gipsy Moth. *Journal of Agricultural Research*, **43**: 445–460.

Muesebeck, C.F.W. 1932*a*. Revision of the nearctic ichneumon-flies belonging to the genus *Macrocentrus. Proceedings of the United States National Museum*, **80**: 1–55.

Muesebeck, C.F.W. 1932*b*. Two new species of *Phanomeris* Foerster (Hymenoptera, Braconidae) parasitic on leaf-mining sawflies. *Proceedings of the Entomological Society of Washington*, **34**: 81–83.

Muesebeck, C.F.W. 1935a. On the genus *Oncophanes* Foerster, with descriptions of two new related genera (Hymenoptera: Braconidae). *Annals of the Entomological Society of America*, **28**: 241–250.

Muesebeck, C.F.W. 1936. The genera of parasitic wasps of the braconid subfamily Euphorinae, with a *review of the nearctic species. United States Department of Agriculture, Miscellaneous Publication*, **241**: 1–38.

Muesebeck, C.F.W. 1937. A new West Indian species of *Mirax* Haliday parasitic on the coffee leaf-miner (Hymenoptera: Braconidae). *Proceedings of the Entomological Society of Washington*, **39**: 139–141.

Muesebeck, C.F. 1938. Two reared North American species of the genus *Stantonia* Ashmead (Hymenoptera: Braconidae). *Proceedings of the Entomological Society of Washington*, **40**: 89–91.

Muesebeck, C.F.W. 1956. Two new braconid parasites of the avocado looper. *Pan-Pacific Entomologist*, **32**: 25–28.

Muesebeck, C.F.W. 1959. New reared species of *Lygocerus* Foerster (Hymenoptera: Ceraphronidae). *Entomological News*, **70**: 91–96.

Muesebeck, C.F.W. 1960. New reared neotropical species of *Rogas* Nees (Hymenoptera: Braconidae). *Entomological News*, **71**: 257–261.

Muesebeck, C.F.W. 1970. The nearctic species of *Orgilus* Haliday (Hymenoptera: Braconidae). *Smithsonian Contributions to Zoology*, **30**: 1–104.

Muesebeck, C.F.W. 1977. The parasitic wasps of the genus *Macroteleia* Westwood of the New World (Hymenoptera, Proctotrupoidea, Scelionidae). *Technical Bulletin of the United States Department of Agriculture*, **1565**: 1–57.

Muesebeck, C.F.W. & Dohanian, S.M. 1927. A study in hyperparasitism, with particular reference to the parasites of *Apanteles melanoscalus* (Ratzeburg). *Bulletin of the United States Department of Agriculture. Washington*, **1487**: 1–35.

Mullenax, C.H. 1979. The use of jackbean (*Canavalia ensiformis*) as a biological control for leaf-cutting ants (*Atta* spp.). *Biotropica*, **11**: 313–314.

Munro, J.W. 1917. The structure and life-history of *Bracon* sp.: a study in parasitism. *Proceedings of the Royal Society of Edinburgh*, **36**: 313–333.

Murakami, Y. 1960. Seasonal dimorphism in the Encyrtidae (Hymenoptera, Chalcidoidea). *Acta Hymenopterologica*, **1**: 199–204.

Murphy, S.T. & Moore, D. 1990. Biological control of the coffee berry borer, *Hypothenemus hampei* (Ferrari) (Coleoptera, Scolytidae): previous programmes and possibilities for the future. *Biocontrol News and Information*, **11**: 107–117.

Myers, J. & Loveless, M.D. 1976. Nesting aggregations of the euglossine bee *Euplusia surninamensis* (Hymenoptera: Apidae): individual interactions and the advantage of living together. *Canadian Entomologist*, **108**: 1–6.

Myers, J.G. 1927. Natural enemies of the pear leaf-curling midge, *Perrisia pyri*, Bouché (Dipt., Cecidom.). *Bulletin of Entomological Research*, **18**: 129–138.

Myers, J.G. 1930. *Carabunia myersi*, Watrst. (Hym., Encyrtidae), a parasite of nymphal froghoppers (Hom., Cercopidae). *Bulletin of Entomological Research*, **21**: 341–351.

Myers, J.H., Higgins, C. & Kovacs, E. 1989. How many insect species are necessary for the biological control of insects? *Environmental Entomology*, **18**: 541–547.

Nadel, H. 1987. Male swarms discovered in Chalcidoidea (Hymenoptera: Encyrtidae: Pteromalidae). *Pan-Pacific Entomologist*, **63**: 242–246.

Nadel, H. & Peña, J. 1991. Hosts of *Bephratelloides cubensis* (Hymenoptera: Eurytomidae) in Florida. *Florida Entomologist*, **74**: 476–479.

Nafus, D.M. & Schreiner, I.H. 1988. Parental care in a tropical nymphalid butterfly *Hypolimnas anomala*. *Animal Behaviour*, **36**: 1425–1431.

Nagaraja, H. 1978. Studies on *Trichogrammatoidea* (Hymenoptera: Trichogrammatidae). *Oriental Insects*, **12**: 489–530.

Nagaraja, H. 1987. Recent advances in biosystematics of *Trichogramma* and *Trichogrammatoidea* (Hymenoptera, Trichogrammatidae). *Proceedings of the Indian Academy of Sciences (Animal Science)*, **96**: 469–477.

Nagaraja, H. & Nagarkatti, S. 1973. A key to some New World species of *Trichogramma* (Hymenoptera: Trichogrammatidae), with descriptions of four new species. *Proceedings of the Entomological Society of Washington*, **75**: 288–297.

Nägeli, W. 1936. Die kleine Fichtenblattwespe (*Lygaeonematus pini* Retz. = *Nematus abietinus* Christ.). *Mitteilungen der Schweizerischen Anstalt fur das forstliche Versuchswesen*, **19**: 213–381.

Nagy, C.G. 1975. A new genus of Scolebythidae (Hymenoptera) from South Africa and Australia. *Journal of the Entomological Society of Southern Africa*, **38**: 75–78.

Nakamura, K. & Abbas, I. 1987. Preliminary life table of the spotted tortoise beetle, *Aspidomorpha miliaris* (Coleoptera: Chrysomelidae) in Sumatra. *Researches on Population Ecology*, **29**: 229–236.

Narasimham, A.U. & Chacko, M.J. 1988. *Rastrococcus* spp. (Hemiptera: Pseudococcidae) and their natural enemies in India as potential biocontrol agents for *R. invadens* Williams. *Bulletin of Entomological Research*, **78**: 703–708.

Narendran, T.C. 1984. Chalcids and sawflies associated with plant galls, pp. 273–303. *In*: Ananthakrishnan, T.H. (ed.) *Biology of Gall Insects*, 362pp. Arnold, London.

Naumann, I.D. 1982. Systematics of the Australian Ambositrinae (Hymenoptera: Diapriidae), with a synopsis of non-Australian genera of the subfamily. *Australian Journal of Zoology*, Supplementary Series, **85**: 1–239.

Naumann, I.D. 1984. An apterous female sawfly (Hymenoptera: Symphyta) from Papua New Guinea. *Systematic Entomology*, **9**: 339–349.

Naumann, I.D. 1985. The Australian species of Monomachidae (Hymenoptera: Proctotrupoidea), with a revised diagnosis of the family. *Journal of the Australian Entomological Society*, **24**: 261–274.

Naumann, I.D. 1988. Ambositrinae (Insecta: Hymenoptera: Diapriidae). *Fauna of New Zealand*, **15**: 1–165.

Naumann, I.D. 1991. Hymenoptera (Wasps, Bees, Ants, Sawflies), pp 916–1000. *In*: *The Insects of Australia, Volume II*. 1137pp. Melbourne University Press, Melbourne.

Naumann, I.D. & Cardale, J.C. 1987. Notes on the behaviour and nests of an Australian masarid wasp *Paragia (Paragia) decipiens decipiens* Shuckard (Hymenoptera: Vespoidea: Masaridae). *Australian Entomological Magazine*, **13**: 59–65.

Naumann, I.D. & Masner, L. 1985. Parasitic wasps of the proctotrupoid complex: a new family from Australia and a key to world families (Hymenoptera: Proctotrupoidea *sensu lato*). *Australian Journal of Zoology*, **33**: 761–783.

Naumann, I.D. & Sands, D.P.A. 1984. Two Australian *Elasmus* spp. (Hymenoptera: Elasmidae), parasitoids of *Pectinophora gossypiella* (Saunders)(Lepidoptera: Gelechiidae): their taxonomy and biology. *Journal of the Australian Entomological Society*, **23**: 25–32.

Needham, G.R., Page, R.E. Jr, Delfinado-Baker, M. & Bowman, C.E. (eds) 1988. *Africanized Honey Bees and Bee Mites*. 572pp. Ellis Horwood, Chichester.

Neff, J.L. & Simpson, B.B. 1981. Oil-collecting structures in the Anthophoridae (Hymenoptera): morphology, function, and use in systematics. *Journal of the Kansas Entomological Society*, **54**: 95–123.

Neff, J.L. & Simpson, B.B. 1993. Bees, pollination systems and plant diversity, pp. 143–167. *In*: LaSalle, J. & Gauld, I.D. (eds), *Hymenoptera and Biodiversity*, 348pp. CAB International, Wallingford.

Nelson, C.R., Jorgensen, C.D., Black, H.L., & Whiting, J. 1991. Maintenance of foraging trails by the giant tropical ant *Paraponera clavata* (Insecta: Formicidae: Ponerinae). *Insectes Sociaux*, **38**: 221–228.

Nelson, J.M. 1966 (1967). Parasites in *Polistes* nests. *Proceedings North Central Branch Entomological Society of America*, **21**: 83–84.

Nelson, J.M. 1968. Parasites and symbionts of nests of *Polistes* wasps. *Annals of the Entomological Society of America*, **61**: 1528–1539.

Nelson, J.M. 1970 (1971). Nesting habits and nest symbionts of *Polistes erythrocephalus* Latreille (Hymenoptera Vespidae) in Costa Rica. *Revista de Biología Tropical*, **18**: 89–98.

Nenon, J.P. & Biassangama, A. 1985. Mise en evidence du siphon respiratoire de l'hyménoptère *Leptomastix dactylopii* (Encyrtidae) lors de son développement dans la cochenille *Planococcus citri* (Pseudococcidae). *Comptes Rendus Hebdomadaires des Séances de L'Académie des Sciences (III)*, **300**: 519–524.

Neser, S. 1973. Biology and behaviour of *Euplectrus* species near *laphygmae* Ferrière. *Entomology Memoir. Department of Agricultural Technical Services, Republic of South Africa*, **32**: 1–32.

Neuenschwander, P. 1989. Biocontrol of mango mealybug. *IITA Research Briefs*, **9**: 5–6.

Neuenschwander, P. 1990. Biological control of the cassava mealybug by *Epidinocarsis lopezi* in Africa: a review of impact. *IITA Research*, **1**: 1–4.

Neumann, F.G. & Minko, G. 1981. The *Sirex* woodwasp in Australian radiata pine plantations. *Australian Forestry*, **44**: 46–63.

New, T.R. 1969. The biology of some species of *Alaptus* (Mymaridae) parasitising eggs of Psocoptera. *Transactions of the Society for British Entomology*, **18**: 181–193

Newman, E. 1867. Description of the larva of *Xanthia gilvago*. *The Entomologist*, **3**: 342.

Nguyen, R. & Sailer, R.I. 1987. Facultative hyperparasitism and sex determination of *Encarsia smithi* (Silvestri) (Hymenoptra: Aphelinidae). *Annals of the Entomological Society of America*, **80**: 713–719.

Nichols-Orians, C.M. 1991. Environmentally induced differences in plant traits: consequences for susceptibility to a leaf-cutter ant. *Ecology*, **72**: 1609–1623.

Nichols-Orians, C.M. & Schulz, J.C. 1990. Interactions among leaf toughness, chemistry, and harvesting by attine ants. *Ecological Entomology*, **15**: 311–320.

Nijhout, H.F. 1991. *The Development and Evolution of Butterfly Wing Patterns*. 297pp. Smithsonian Institution Press, Washington, D.C.

Nishida, T. 1956. An experimental study of the ovipositional behavior of *Opius fletcheri* Silvestri (Hymenoptera: Braconidae), a parasite of the melon fly. *Proceedings of the Hawaiian Entomological Society*, **16**: 126–134.

Nixon, G.E.J. 1940. New genera and species of Hormiinae, with a note on *Hormiopterus* Giraud (Hym., Braconidae). *Annals and Magazine of Natural History*, (Series 11), **5**: 473–493.

Nixon, G.E.J. 1943. A revision of the Spathiinae of the Old World (Hymenoptera, Braconidae). *Transactions of the Royal Entomological Society of London*, **93**: 173–456.

Nixon, G.E.J. 1965. A reclassification of the tribe Microgasterini (Hymenoptera: Braconidae). *Bulletin of the British Museum (Natural History)* Entomology, Supplement, **2**: 1–284.

Nixon, G.E.J. 1968. A revision of the genus *Microgaster* Latreille (Hymenoptera: Braconidae). *Bulletin of the British Museum (Natural History)* Entomology, **22**: 33–72.

Nixon, G.E.J. 1970. A revision of the N.W. European species of *Microplitis* Förster (Hymenoptera: Braconidae). *Bulletin of the British Museum (Natural History)* Entomology, **25**: 3–30.

Noble, N.S. 1936. The citrus gall wasp, (*Eurytoma fellis* Girault). *Department of Agriculture, New South Wales, Science Bulletin*, **53**: 1–41.

Noble, N.S. 1938. *Epimegastigmus (Megastigmus) brevivalvus* Girault: a parasite of the citrus gall wasp (*Eurytoma fellis* Girault); with notes on several other species of hymenopterous gall inhabitants. *Department of Agriculture, New South Wales, Science Bulletin*, **65**: 1–46.

Noble, N.S. 1940. *Trichilogaster acaciae-longifoliae* (Froggatt) (Hymenopt., Chalcidoidea), a wasp causing galling of the flower-buds of *Acacia longifolia* Willd., *A. floribunda* Sieber and *A. sophorae* R.Br. *Transactions of the Royal Entomological Society of London*, **90**: 13–38.

Noble, N.S. 1941. *Trichilogaster maideni* (Froggatt) (Hymenopt., Chalcidoidca), a wasp causing galls on *Acacia implexa* Benth., and *A. maideni* F.v.M., with observations on

Australian chalcidoid galls. *Proceedings of the Linnean Society of New South Wales*, **66**: 178–200.

Noda, T., Kitamura, C. & Takahashi, S. 1982. Host selection behavior of *Anicetus beneficus* Ishii et Yasumatsu (Hymenoptera: Encyrtidae) I. Ovipositional behavior for the natural host *Ceroplastes rubens* Maskell (Hemiptera: Coccidae). *Applied Entomology and Zoology*, **17**: 350–357.

Nogueira-Neto, P. 1970*a*. Behavior problems related to the pillages made by some parasitic stingless bees (Meliponinae, Apidae), pp. 416–434. *In*: Aronson, L.R., Tobach, E., Lehrman, D.S. & Rosenblatt, J.S. (eds) *Development and Evolution of Behavior, Essays in Memory of T.C. Schneirla*. 656pp. Freeman and Co., San Francisco.

Nogueira-Neto, P. 1970*b*. *A Criaçao de Abelhas Indígenas sem Ferrao (Meliponinae)*, 2nd Edition. 365pp. Chácaras & Quintais, Sao Paulo.

Nogueira, S.B. & Martinho, M.R. 1983. Leaf-cutting ants (*Atta* sp.), damage to and distribution along Brazilian roads, pp. 181–186. *In*: Jaisson, P. (ed.), *Social Insects in the Tropics*, 2. 252pp. Université Paris-Nord, Paris.

Noldus, L.P.J.J. 1989. *Chemical Espionage by Parasitic Wasps. How Trichogramma Species Exploit Moth Sex Pheromone Systems* 252pp. Noldus, Wageningen.

Noldus, L.P.J.J., Lenteren, J.C. van & Lewis, W.J. 1991*a*. How *Trichogramma* parasitoids use moth sex pheromones as kairomones: orientation behaviour in a wind tunnel. *Physiological Entomology*, **16**: 313–327.

Noldus, L.P.J.J., Lewis, W.J. & Tumlinson, J.H. 1990. Beneficial arthropod behavior mediated by airborne semiochemicals. IX. Differential response of *Trichogramma pretiosum*, an egg parasitoid of *Heliothis zea*, to various olfactory cues. *Journal of Chemical Ecology*, **16**: 3531–3544.

Noldus, L.P.J.J., Potting, R.P.J. & Barendregt, H.E. 1991*b*. Moth sex pheromone adsorption to leaf surface: bridge in time for chemical spies. *Physiological Entomology*, **16**: 329–344.

Nonveiller, G. 1963. Quelle est la cause de la rareté des Mutillides? Resultats de l'étude de certains de leurs caractères biologiques et ecologiques (Hymenoptera Mutillidae). *Memorie della Società Entomologica Italiana*, **42**: 24–57.

Nonveiller, G. 1990. Catalogue of the Mutillidae, Myrmosidae and Bradynobaenidae of the Neotropical Region including Mexico. *Hymenopterorum Catalogus* (nova editio), **18**: 1–150.

Noonan, K.M. 1981. Individual strategies of inclusive-fitness maximizing in *Polistes fuscatus* foundresses, pp. 18–44. *In*: Alexander, R.D. & Tinkle, D.W. (eds.), *Natural Selection and Social Behavior*. 532pp. Chiron Press, New York.

Norden, B., Batra, S.W.T., Fales, H.M., Hefetz, A. & Shaw, G.J. 1980. *Anthophora* bees: unusual glycerides from maternal Dufour's glands serve as larval food and cell lining. *Science*, **207**: 1095–1097.

Nordlander, G. 1976. Studies on Eucoilidae (Hym., Cynipoidea). I A revision of the north-western European species of *Cothonaspis* Htg. with description of a new species and notes on some other genera. *Entomologisk Tidskrift*, **97**: 65–77.

Nordlander, G. 1978*a*. Revision of the genus *Rhoptromeris* Förster, 1869 with reference to north-western European species. Studies on Eucoilidae (Hym.: Cynipoidea) II. *Entomologica Scandinavica*, **9**: 47–62.

Nordlander, G. 1978*b*. Parasitoids of the frit fly, *Oscinella frit* (L.) on oats. *Norwegian Journal of Entomology*, **25**: 89–90.

Nordlander, G. 1980. Revision of the genus *Leptopilina* Förster. 1869, with notes on the status of some other genera (Hymenoptera, Cynipoidea: Eucoilidae). *Entomologica Scandinavica*, **11**: 428–453.

Nordlander, G. 1981. A review of the genus *Trybliographa* Förster, 1869 (Hymenoptera, Cynipoidea: Eucoilidae). *Entomologica Scandinavica*, **12**: 381–402.

Nordlander, G. 1982*a*. Identities and relationships of the previously confused genera *Odonteucoila*, *Coneucoela*, and *Trichoplasta* (Hymenoptera, Cynipoidea: Eucoilidae) *Entomologica Scandinavica*, **13**: 269–292.

Nordlander, G. 1982*b*. Systematics and phylogeny of an interelated group of genera within the family Eucoilidae (Insecta: Hymenoptera, Cynipoidea). 34pp. Thesis, University of Stockholm, Sweden.

Nordlund, D.A., Lewis, W.J. & Gueldner, R.C. 1983. Kairomones and their use for management of entomophagous insects XIV. Response of *Telenomus remus* to abdominal tips of *Spodoptera frugiperda*, (Z)-9-tetradecene-1-ol acetate and (Z)-9-dodecene-1-ol acetate. *Journal of Chemical Ecology*, **9**: 695–701.

Nordlund, D.A., Strand, M.R., Lewis, W.J. & Vinson, S.B. 1987. Role of kairomones from host accessory gland secretion in host recognition by *Telenomus remus* and *Trichogramma pretiosum*, with partial characterization. *Entomologia Experimentalis et Applicata*, **44**: 37–43.

Norgaard, R.B. 1988. The biological control of cassava mealybug in Africa. *American Journal of Agricultural Economics*, **70**: 366–371.

Norton, W.N. & Vinson, S.B. 1974. Antennal sensilla of three parasitic Hymenoptera. *International Journal of Insect Morphology and Embryology*, **3**: 305–316.

Noyes, J.S. 1980. A review of the genera of Neotropical Encyrtidae (Hymenoptera: Chalcidoidea). *Bulletin of the British Museum (Natural History)* Entomology, **41**: 107–253.

Noyes, J.S. 1982. Collecting and preserving chalcid wasps (Hymenoptera: Chalcidoidea). *Journal of Natural History*, **16**: 315–334.

Noyes. J.S. 1985. A review of the Neotropical species of *Ooencyrtus* Ashmead, 1900 (Hymenoptera: Encyrtidae). *Journal of Natural History*, **19**: 533–554.

Noyes, J.S. 1989. The diversity of Hymenoptera in the tropics with special reference to Parasitica in Sulawesi. *Ecological Entomology*, **14**: 197–207.

Noyes, J.S. 1990. The number of described chalcidoid taxa in the world that are currently regarded as valid. *Chalcid Forum*, **13**: 9–10.

Noyes J.S. & Hayat, M. 1994. *Oriental Mealybug Parasitoids of the Anagyrini (Hymenoptera: Encyrtidae)*. 554pp. CAB International, Wallingford.

Noyes, J.S. & Valentine, E.W. 1989. Mymaridae (Insecta: Hymenoptera) -- introduction, and review of genera. *Fauna of New Zealand*, **17**: 1–95.

Nuttall, M.J. 1973. Pre-emergence fertilisation of *Megarhyssa nortoni nortoni* (Hymenoptera: Ichneumonidae). *New Zealand Entomologist*, **5**: 112–117.

Nuttall, M.J. 1980. Insect parasites of *Sirex*. *Forest and Timber Insects in New Zealand*, **47**. [Unpaginated.]

Oatman, E.R. 1985. *Aphanogmus floridanus*, a primary parasite of *Feltiella acarivora*, a predator of spider mites on strawberry (Hymenoptera: Ceraphronidae; Diptera: Cecidomyiidae). *Proceedings of the Entomological Society of Washington*, **87**: 714–716.

Oatman, E.R. & Platner, G.R. 1989. Parasites of the potato tuberworm, tomato pinworm, and other, closely related gelechiids. *Proceedings of the Hawaiian Entomological Society*, **29**: 23–30.

Oatman, E.R., Platner, G.R., & Greany, P.D. 1969. The biology of *Orgilus lepidus* (Hymenoptera: Braconidae), a primary parasite of the potato tuberworm. *Annals of the Entomological Society of America*, **62**: 1407–1414.

Obara, M. & Kitano, H. 1974. Studies on the courtship behavior of *Apanteles glomeratus* L. I. Experimental studies on releaser of wing-vibrating behavior in the male. *Kontyû*, **42**: 208–214.

O'Brien, L.B. & Hurd, P.D. Jr 1965. Carpenter bees of the subgenus *Notoxylocopa* (Hymenoptera: Apoidea). *Annals of the Entomological Society of America*, **58**: 175–196.

Obrtel, R. 1960. Premature stages of *Triaspis caudatus* (Nees) (Hym.: Braconidae). *Folia Zoologica*, **9**: 35–38.

Obrycki, J.J., Tauber, M.J. & Tauber, C.A. 1985. *Perilitus coccinellae* (Hymenoptera: Braconidae): parasitization and development in relation to host-stage attacked. *Annals of the Entomological Society of America*, **78**: 852–854.

Odebiyi, J.A. & Oatman, E.R. 1977. Biology of *Agathis unicolor* (Schrottky) and *Agathis gibbosa* (Say) (Hymenoptera: Braconidae), primary parasites of the potato tuberworm. *Hilgardia*, **45**: 123–151.

O'Donnell, D.J. 1982. Taxonomy of the immature stages of parasitic Hymenoptera associated with aphids. Unpublished Thesis, University of London. 428pp.

O'Donnell, D.J. 1987. Larval development and the determination of the number of instars in aphid parasitoids (Hymenoptera: Aphidiidae). *International Journal of Insect Morphology and Embryology*, **16**: 3–15.

O'Donnell, S. 1995. Necrophagy by neotropical swarm-founding wasps (Hymenoptera: Vespidae, Epiponini). *Biotropica*, **27**: 133–136.

O'Dowd, D.J. 1982. Pearl bodies as ant food: an ecological role for some leaf emergences of tropical plants. *Biotropica*, **14**: 40–49.

Oeser, R. 1961. Vergleichend-morphologische untersuchungen über den ovipositor der Hymenopteren. *Mitteilungen aus dem Zoologischen Museum in Berlin*, **37**: 1–119.

Ogloblin, A.A. 1913 (1914). [Contribution à la biologie des Coccinelles.] *Russkoe Entomologicheskoe Obozrenie*, **13**: 27–43. [In Russian.]

Ohsaki, N. & Sato, Y. 1990. Avoidance mechanisms of three *Pieris* butterfly species against the parasitoid wasp *Apanteles glomeratus*. *Ecological Entomology*, **15**: 169–176.

Okamoto, M. & Tashiro, M. 1981. Mechanism of pollen transfer and pollination in *Ficus erecta* by *Blastophaga nipponica*. *Bulletin of the Osaka Museum of Natural History*, **34**: 7–16.

Olberg, G. 1959. *Das Verhalten der Solitären Wespen Mitteleuropas (Vespidae, Pompilidae, Sphecidae)*. 402pp. Veb Deutscher Verlag der Wissenschaften, Berlin.

Oldroyd, B.P., Rinderer, T.E. & Buco, S.M. 1992. Intra-colonial foraging specialism by honey bees (*Apis mellifera*) (Hymenoptera: Apidae). *Behavioral Ecology and Sociobiology*, **30**: 291–295.

Olkowski, H. & Olkowski, W. 1976. Entomophobia in the urban ecosystem, some observations and suggestions. *Bulletin of the Entomological Society of America*, **22**: 313–317.

Olmi, M. 1984. A revision of the Dryinidae (Hymenoptera). *Memoirs of the American Entomological Institute*, **37**: 1–1913.

Olmi, M. 1986 (1984-5). New species and genera of Dryinidae (Hymenoptera Chrysidoidea). *Frustula Entomologica* (N.S.), **7–8**: 63–105.

Olmi, M. 1987*a*. New species of Dryinidae (Hymenoptera, Chrysidoidea). *Fragmenta Entomologica*, **19**: 371–456.

Olmi, M. 1987*b*. New species of Dryinidae, with description of a new subfamily from Florida and a new species from Dominica amber (Hymenoptera, Chrysidoidea). *Bollettino del Museo Regionale di Scienze Naturali di Torino*, **5**: 211–238.

Olmi, M. 1987*c*. Nuove specie americane di Dryinidae (Hymenoptera Chrysidoidea). *Bollettino della Società Entomologica Italiana*, **119**: 99–116.

Olmi, M. 1989 (1991). Supplement to the revision of the world Dryinidae (Hymenoptera Chrysidoidea). *Frustula Entomologica* (N.S.), **12**: 109–395.

Olmi, M. 1990. Description de cinq nouvelles espèces de Dryinidae de la région paléarctiques (Hymenoptera, Chrysidoidea). *Revue Française d'Entomologie* (N.S.), **12**: 135–142.

Olmi, M. 1993. Dryinidae di Costa Rica: catalogo e considerazioni biogeografiche ed evolutive (Hymenoptera Chrysidoidea). *Bollettino della Società Entomologica Italiana*, **124**: 186–200.

Olton, G.S. & Legner, E.F. 1974. Biology of *Tachinaephagus zealandicus* (Hymenoptera: Encyrtidae), parasitoid of synanthropic Diptera. *Canadian Entomologist*, **106**: 785–800.

O'Neill, K.M. 1985. Egg size, prey size, and sexual size dimorphism in digger wasps (Hymenoptera: Sphecidae). *Canadian Journal of Zoology*, **63**: 2187–2193.

O'Neill, K.M. & Evans, H.E. 1983. Body size and alternative mating tactics in the beewolf *Philanthus zebratus* (Hymenoptera; Sphecidae). *Biological Journal of the Linnean Society*, **20**: 175–184.

Oniki, Y. 1970. Brazilian sphecid wasps in occupied hummingbird nests. *Journal of the Kansas Entomological Society*, **43**: 354–356.

Opinion 1757. 1994. Cryptus Fabricius, 1804 and Cryptinae Kirby, 1837 (Insecta, Hymenoptera): conserved. *Bulletin of Zoological Nomenclature*, **51**: 74–75.

Opler, P.A. 1981. Polymorphic mimicry of polistine wasps by a neotropical neuropteran. *Biotropica*, **13**: 165–176.

Ordway, E. 1966. Systematics of the genus *Augochlorella* (Hymenoptera, Halictidae) north of Mexico. *University of Kansas Science Bulletin*, **46**: 509–624.

Orr, D.B. 1988. Scelionid wasps as biological control agents: a review. *Florida Entomologist*, **71**: 506–528.

Orr, D.B., Russin, J.S. & Boethel, D.J. 1986. Reproductive biology and behavior of *Telenomus calvus* (Hymenoptera: Scelionidae), a phoretic egg parasitoid of *Podisus maculiventris* (Hemiptera: Pentatomidae). *Canadian Entomologist*, **118**: 1063–1072.

Orr, M.R. 1992. Parasitic flies (Diptera: Phoridae) influence foraging rhythms and caste division of labor in the leaf-cutter ant, *Atta cephalotes* (Hymenoptera: Formicidae). *Behavioral Ecology and Sociobiology*, **30**: 395–402.

Osborne, P. 1960. Observations on the natural enemies of *Meligethes aeneus* (F.) and *M. viridescens* (F.) (Coleoptera: Nitidulidae). *Parasitology*, **50**: 91–110.

Osman, S.E. & Führer, E. 1979. Histochemical analysis of accessory genital gland secretions in female *Pimpla turionellae* L. (Hymenoptera: Ichneumonidae). *International Journal of Invertebrate Reproduction*, **1**: 323–332.

Oster, G.F. & Wilson, E.O. 1978. *Caste and ecology in the social insects* (Monographs in Population Biology, 12). xv+352pp. Princeton University Press, Princeton, N.J.

Otis, G.W., Bath, J.B. & Ramirez B., W. 1988. Have Africanized bees brought honey-bee tracheal mites to Costa Rica? pp. 541–545. *In*: Needham, G.R., Page, R.E. Jr, Delfinado-Baker, M. & Bowman, C.E. (eds) *Africanized Honey Bees and Bee Mites*. 572pp. Ellis Horwood Ltd. Chichester.

Otis, G.W., McGinley, R.L., Garling, L. & Malaret, L. 1982 (1983). Biology and systematics of the bee genus *Crawfordapis* (Colletidae, Diphaglossinae). *Psyche*, **89**: 279–296.

Otis, G.W., Santana C.E., Crawford, D.L. & Higgins, M.L. 1986. The effect of foraging army ants on leaf-litter arthropods. *Biotropica*, **18**: 56–61.

O'Toole, C. 1993. Diversity of native bees and agroecosystems, pp. 169–196. *In*: LaSalle, J. & Gauld, I.D. (eds), *Hymenoptera and Biodiversity*, 348pp. CAB International, Wallingford.

Overal, W.L. & Bandeira, A.G. 1985. Nota sobre hábitos de *Cylindromyrmex striatus* Mayr, 1870, na Amazônia (Formicidae, Ponerinae). *Revista Brasiliera de Entomologia*, **29**: 521–522.

Overal, W.L. & Posey, D.A. 1990. Uso de formigas *Azteca* spp. para controle biológico de pragas agrícolas entre os índios Kayapó do Brasil. pp. 219–225. *In*: Posey, D.A. & Overal, W.L. (eds), *Ethnobiology: Implications and Applications. 1*. 363pp. Museu Paraense Emílio Goeldi, Belém.

Owen, D.F. & Owen, J. 1974. Species diversity in temperate and tropical Ichneumonidae. *Nature*, **249**: 583–584.

Owen, R.E. 1983. Sex ratio adjustment in *Asobara persimilis* (Hymenoptera: Braconidae), a parasitoid of *Drosophila*. *Oecologia*, **59**: 402–404.

Paarmann, W. & Stork, N.E. 1987. Canopy fogging, a method of collecting living insects for investigations of life history strategies. *Journal of Natural History*, **21**: 563–566.

Packer, L. 1986. The biology of a subtropical population of *Halictus ligatus* IV: A cuckoo-like caste. *Journal of the New York Entomological Society*, **94**: 458–466.

Pagden, H. 1934. Notes on hymenopterous parasites of padi pests in Malaya. *Scientific Series of the Department of Agriculture, Straits Settlements and Federated Malay States*, **15**: 1–13.

Page, R.E. Jr 1986. Sperm utilization in social insects. *Annual Review of Entomology*, **31**: 297–320.

Pagliano, G. 1986. Aulacidae, Stephanidae ed Evaniidae d'Italia con descrizione di un nuovo Stephanidae del Marocco (Hymenoptera, Ichneumonoidea). *Atti del Museo Civico di Storia Naturale* (Grosseto), **9–10**: 5–20.

Pagliano, G. & Scaramozzino, P. 1989 (1990). Elenco dei generi di Hymenoptera del mondo. *Memorie della Società Entomologica Italiana*, **68**: 1–210.

Pak, G.A. & de Jong, E.J. 1987 (1988). Behavioural variations among strains of *Trichogramma* spp.: host recognition. *Netherlands Journal of Zoology*, **37**: 137–166.

Pak, G.A., Dalen, A. van, Kaashoek, N. & Dijkman, H. 1990. Host egg chorion structure influencing host suitability for the egg parasitoid *Trichogramma* Westwood. *Journal of Insect Physiology*, **36**: 869–875.

Pak, G.A. & Oatman, E.R. 1982. Comparative life table, behavior and competition studies of *Trichogramma brevicapillum* and *T. pretiosum*. *Entomologia Experimentalis et Applicata*, **32**: 68–79.

Palmer, M. 1976. Notes on the biology of *Pterombrus piceus* Krombein (Hymenoptera: Tiphiidae). *Proceedings of the Entomological Society of Washington*, **78**: 369–375.

Pampel, W. 1914. Die weiblichen Geschlechtsorgane der Ichneumoniden. *Zeitschrift für Wissenschaftliche Zoologie*, **108**: 290–357.

Papaj, D.R. & Lewis, A.C. (eds) 1993. *Insect Learning. Ecological and Evolutionary Perspectives*, 398pp. Chapman & Hall, New York.

Papaj, D.R. & Vet, L.E.M. 1990. Odor learning and foraging success in the parasitoid, *Leptopilina heterotoma*. *Journal of Chemical Ecology*, **16**: 3137–3150.

Papp, J. 1965. A monograph of the genus *Aridelus* Marsh. (Hymenoptera, Braconidae). *Acta Zoologica, Hungaricae*, **11**: 181–201.

Papp, J. 1989. *Mirax heinrichi* sp. n. from the U.S.A. and taxonomic bionomic data of two European *Mirax* species (Hymenoptera: Braconidae, Adeliinae). *Folia Entomologica Hungarica*, **50**: 105–110.

Park, O. 1964. Observations upon the behavior of myrmecophilous pselaphid beetles. *Pedobiologia*, **4**: 129–137.

Parker, D.L. 1933. The interrelations of two hymenopterous egg parasites of the gypsy moth, with notes on the larval instars of each. *Journal of Agricultural Research*, **46**: 23–34.

Parker, F.D. 1966. A revision of the North American species in the genus *Leptochilus* (Hymenoptera: Eumenidae). *Miscellaneous Publications of the Entomological Society of America*, **5**: 151–229.

Parker, F.D. 1967. Notes on the nests of three species of *Pseudomasaris* Ashmead. *Pan-Pacific Entomologist*, **43**: 213–216.

Parker, F.D. 1977. Biological notes on some Mexican bees. (Hymenoptera: Megachilidae, Anthophoridae). *Pan-Pacific Entomologist*, **53**: 189–192.

Parker, F.D. & Bohart, G.E. 1979. *Dolichostelis*, a new genus of parasitic bees (Hymenoptera: Megachilidae). *Journal of the Kansas Entomological Society*, **52**: 138–153.

Parker, F.D. & Bohart, R.M. 1966. Host-parasite associations in some twig-nesting Hymenoptera from western North America. *Pan-Pacific Entomologist*, **42**: 91–98.

Parker, F.D. & Bohart, R.M. 1968. Host-parasite associations in some twig-nesting Hymenoptera from western North America, Part II. *Pan-Pacific Entomologist*, **44**: 1–6.

Parker, F.D., Cane, J.H., Frankie, G.W. & Vinson, S.B. 1987. Host records and nest entry by *Dolichostelis*, a kleptoparasitic anthidiine bee (Hymenoptera: Megachilidae). *Pan-Pacific Entomologist*, **63**: 172–177.

Parker, F.D. & Pinnell, R.E. 1971. Overwintering of some *Trichogramma* spp. in Missouri. *Journal of Economic Entomology*, **64**: 80–81.

Parker, H.-L. 1924. Recherches sur les formes post-embryonnaires des chalcidiens. *Annales de la Société Entomologique de France*, **93**: 261–379.

Parker, H.L. 1931*a*. *Macrocentrus gifuensis* Ashmead, a polyembryonic braconid parasite in the European corn borer. *United States Department of Agriculture, Washington, D.C. Technical Bulletin*, **230**: 1–62.

Parker, H.L. 1931*b*. Note on *Meteorus (Zemiotes) nigricollis* Thomson, an occasional parasite of the European corn borer. *Proceedings of the Entomological Society of Washington*, **33**: 93–103.

Parker, H.L. 1935. Notes on the anotomy of tenthredinid larvae, with special reference to the head. *Bollettino del Laboratorio di Zoologia Generale e Agraria del R. Istituto Superiore Agrario in Portici*, **28**: 159–191.

Parker, H.L., Berry, P.A. & Silveria, A. 1950. Vegetable weevils and their natural enemies in Argentina and Uruguay. *United States Department of Agriculture, Technical Bulletin*, **1016**: 1–28.

Parker, H.L. & Thompson, W.R. 1925. Notes on the larvae of the Chalcidoidea. *Annals of the Entomological Society of America*, **18**: 384–395.

Parkman, P., Dusky, J.A. & Waddill, V.H. 1989. Leafminer and leafminer parasitoid incidence on selected weeds in south Florida [USA]. *Florida Entomologist*, **72**: 559–561.

Parnell, J.R. 1963. Three gall midges (Diptera: Cecidomyidae) and their parasites found in the pods of broom (*Sarothamnus scoparius* (L.) Wimmer). *Transactions of the Royal Entomological Society of London*, **115**: 261–275.

Parnell, J.R. 1964*a*. The parasite complex of the two seed beetles *Bruchidius ater* (Marsham) (Coleoptera: Bruchidae) and *Apion fuscirostre* Fabricius (Coleoptera: Curculionidae). *Transactions of the Royal Entomological Society of London*, **116**: 73–88.

Parnell, J.R. 1964*b*. Investigations on the biology and larval morphology of the insects associated with the galls of *Asphondylia sarothamni* H. Loew (Diptera: Cecidomyidae) on broom (*Sarothamnus scoparius* (L.) Wimmer.). *Transactions of the Royal Entomological Society of London*, **116**: 255–273.

Passmore, L. 1936. Tarantula and tarantula hawk. *Nature Magazine*, **27**: 155–159.

Pate, V.S.L. 1947. Neotropical Sapygidae, with a conspectus of the family (Hymenoptera: Aculeata). *Acta Zoologica Lilloana*, **4**: 393–426.

Patel, P.N. & Habib, M.E.M. 1984. Levantamento e eficiencia de insetos parasitos de *Spodoptera frugiperda* (Abbot & Smith, 1797) (Lepidoptera, Noctuidae). *Revista de Agricultura*, Brasil, **59**: 229–237.

Patterson, J.T. 1921. The development of *Paracopidosomopsis*. *Journal of Morphology*, **36**: 1–69.

Paulson, G.S. & Akre, R.D. 1991. *Trichopria* sp. (Hymenoptera: Diapriidae) reared from *Microdon albicomatus* Novak (Diptera: Syrphidae). *Canadian Entomologist*, **123**: 719.

Pechuman, L.L. 1939. The insects found in the bark and wood of the American elm (*Ulmus americana* L.). Ph.D. Dissertation, Cornell University, Ithaca, New York. Pp. 136–138.

Peck, O. 1985. The taxonomy of the Nearctic species of *Pediobius* (Hymenoptera: Eulophidae), especially Canadian and Alaskan forms. *Canadian Entomologist*, **117**: 647–704.

Peck, O., Bouček, Z. & Hoffer, A. 1964. Keys to the Chalcidoidea of Czechoslovakia (Insecta: Hymenoptera). *Memoirs of the Entomological Society of Canada*, **34**: 1–120.

Pedata, P.A. & Viggiani, G. 1991. Preliminary morpho-biological observations on *Azotus perspeciosus* (Girault) (Hymenoptera: Aphelinidae), hyperparasitoid of *Pseudaulacaspis pentagona* (Targioni Tozzetti) (Homoptera: Diaspididae). *Redia*, Appendice **74**(3): 343–350.

Pelt, A.F. van 1950. *Orasema* in nests of *Pheidole dentata* Mayr (Hymenoptera: Formicidae). *Entomological News*, **61**: 161–163.

Pemberton, C.E. & Rosa, J.S. 1940. Notes on the life history of *Baeus californicus* Pierce, an egg parasite of the black widow spider. *The Hawaiian Planters' Record*, **44**: 73–80.

Pemberton, C.E. & Willard, H.F. 1918*a*. Interrelations of fruit-fly parasites in Hawaii. *Journal of Agricultural Research*, **12**: 285–296.

Pemberton, C.E. & Willard, H.F. 1918*b*. A contribution to the biology of fruit-fly parasites in Hawaii. *Journal of Agricultural Research*, **15**: 419–466.

Peña, N. & Shepard, M. 1986. Seasonal incidence of parasitism of brown planthoppers, *Nilaparvata lugens* (Homoptera: Delphacidae), green leafhoppers, *Nephotettix* spp., and whitebacked planthoppers, *Sogatella furcifera* (Homoptera: Cicadellidae) in Laguna Province, Philippines. *Environmental Entomology*, **15**: 263–267.

Pennacchio, F., Vinson, S.B. & Tremblay, E. 1992. Host regulation effects on *Heliothis virescens* (F.) larvae induced by teratocytes of *Cardiochiles nigriceps* Viereck (Lepidoptera, Noctuidae—Hymenoptera, Braconidae). *Archives of Insect Biochemistry and Physiology*, **19**: 177–192.

Pennacchio, F., Vinson, S.B. & Tremblay, E. 1994. Morphology and ultrastructure of the serosal cells (teratocytes) in *Cardiochiles nigriceps* (Viereck) (Hymenoptera: Braconidae) embryos. *International Journal of Insect Morphology and Embryology*, **23**: 93–104.

Perfecto, I. 1990. Indirect and direct effects in a tropical agroecosystem: the maize-pest-ant system in Nicaragua. *Ecology*, **71**: 2125–2134.

Perfecto, I. 1991*a*. Ants (Hymenoptera: Formicidae) as natural control agents of pests in irrigated maize in Nicaragua. *Journal of Economic Entomology*, **84**: 65–70.

Perfecto, I. 1991*b*. Dynamics of *Solenopsis geminata* in a tropical fallow field after ploughing. *Oikos*, **62**, 139–144.

Perfecto, I. 1992 (1993). Observations of a *Labidus coecus* (Latreille) underground raid in the central highlands of Costa Rica. *Psyche*, **99**: 214–220.

Perfecto, I. & Vandermeer, J.A. 1993. Cleptobiosis in the ant *Ectatomma ruidum* in Nicaragua. *Insectes Sociaux*, **40**: 295–299.

Perkins, R.C.L. 1905. Leaf-hoppers and their natural enemies (Pt. I Dryinidae). *Report of Hawaiian Sugar Planters Association Experiment Station*, **1**(1): 1–69.

Perkins, R.C.L. 1906. Leaf-hoppers and their natural enemies (Pt. VIII. Encyrtidae, Eulophidae, Trichogrammidae). *Report of Hawaiian Sugar Planters Association Experiment Station*, **1**(8): 241–267.

Perry, D.R. & Williams, J. 1981. The tropical rain forest canopy: a method providing total access. *Biotropica*, **13**: 283–285.

Petcharat, J. & Johnson, M.W. 1988. Biology of the leafminer parasitoid *Ganaspidium utilis* Beardsley (Hymenoptera: Eucoilidae). *Annals of the Entomological Society of America*, **81**: 477–480.

Peter, C. & David, B.V. 1990. Biology of *Elasmus brevicornis* Gahan (Hymenoptera: Elasmidae) a parasite of the pumpkin caterpillar *Diaphania indica* (Saunders) (Lepidoptera: Pyraustidae). *Entomon*, **15**: 165–169.

Peter, C. & David, B.V. 1991*a*. Observations on the oviposition behaviour of *Goniozus sensorius* (Hymenoptera: Bethylidae) a parasite of *Diaphania indica* (Lepidoptera: Pyralidae). *Entomophaga*, **36**: 403–407.

Peter, C. & David, B.V. 1991*b*. Biology of *Goniozus sensorius* Gordh (Hymenoptera: Bethylidae) a parasitoid of the pumpkin caterpillar *Diaphania indica* (Saunders) (Lepidoptera: Pyralidae). *Insect Science and its Application*, **12**: 339–345.

Petersen, J.J., Watson, D.W. & Pawson, B.M. 1992. Evaluation of field propagation of *Muscidifurax zaraptor* (Hymenoptera: Pteromalidae) for control of flies associated with confined beef cattle. *Journal of Economic Entomology*, **85**: 451–455.

Philippi, T. & Eberhard, W.G. 1986. Foraging behavior of *Stictia signata* (Hymenoptera: Sphecidae). *Journal of the Kansas Entomological Society*, **59**: 604–608.

Phillips, W.J. 1927. *Eurytoma parva* (Girault) Phillips and its biology as a parasite of the wheat jointworm, *Harmolita tritici* (Fitch). *Journal of Agricultural Research*, **34**: 743–758.

Phillips, W.J. 1936. A second revision of the chalcid flies of the genus *Harmolita* (*Isosoma*) of America, north of Mexico, with descriptions of 20 new species. *United States Department of Agriculture Technical Bulletin*, **518**: 1–25.

Phillips, W.J. & Emery, W.T. 1919. A revision of the chalcid-flies of the genus *Harmolita* of America north of Mexico. *Proceedings of the United States National Museum*, No.2281, **55**: 433–471.

Phillips, W.J. & Poos, F.W. 1921. Life-history studies of three jointworm parasites. *Journal of Agricultural Research*, **21**: 405–426.

Piek, T. (ed.) 1986. *Venoms of the Hymenoptera. Biochemical, Pharmacological and Behavioural Aspects*. 570pp. Academic Press, London.

Piek, T. 1992. A toxinological argument in favour of the close relationship of Vespidae, Scoliidae, Tiphiidae, Mutillidae and Formicidae (Hymenoptera). *Proceedings in Experimental and Applied Entomology*, **3**: 99–104.

Piel, O. 1933. *Monema flavescens* Wkr. and its parasites (Lepidoptera, Heterogeneidae). *Lingnan Journal of Science* (Supplement), **12**: 173–201.

Pierce, N.E. 1987. The evolution and biogeography of associations between lycaenid butterflies and ants, pp. 89–116. *In*: Harvey, P.H. & Partridge, L. (eds) *Oxford Surveys in Evolutionary Biology*, Volume 4, 271pp. Oxford University Press, Oxford.

Pierce, N.E., Kitching, R.L., Buckley, R.C., Taylor, M.F.J. & Benbow, K.F. 1987. The costs and benefits of cooperation between the Australian lycaenid butterfly, *Jalmenus evagoras*, and its attendant ants. *Behavioral Ecology and Sociobiology*, **21**: 237–248.

Pijls, J.W.A.M., Hofker, C.D., Staalduinen, M.J. van & Alphen, J.J.M. van 1991. Interspecific host discrimination and competition in *Epidinocarsis lopezi* and *E. diversicornis*, parasitoids used for biological control of the cassava mealybug, *Phenacoccus manihoti*. *Redia*, Appendice **74**(3): 191–194.

Pillault, R. 1951 (1952). Notes sur *Dryinus tarraconensis* (Hym. Dryinidae) prédateur d'une cicadelle. *Annales de la Société Entomologique de France*, **120**: 67–76.

Pimental, D. 1955. Relationship of ants to fly control in Puerto Rico. *Journal of Economic Entomology*, **48**: 28–30.

Pinto, J.D. 1992. Novel taxa of *Trichogramma* from the New World tropics and Australia (Hymenoptera: Trichogrammatidae). *Journal of the New York Entomological Society*, **100**: 621–633.

Pinto, J.D. 1993. Correct application of the nominal genus *Brachista* Walker (Hymenoptera: Trichogrammatidae). *Proceedings of the Entomological Society of Washington*, **95**: 298–299.

Pinto, J.D. 1994. A taxonomic study of *Brachista* (Hymenoptera: Trichogrammatidae) with a description of the two new species phoretic on robberflies of the genus *Efferia* (Diptera: Asilidae_) *Proceedings of the Entomological Society of Washington*, **96**: 120–132.

Pinto, J.D., Kazmer, D.J., Platner, G.R. & Sassaman, C.A. 1992. Taxonomy of the *Trichogramma minutum* complex (Hymenoptera: Trichogrammatidae): allozymic variation and its relationship to reproductive and geographic data. *Annals of the Entomological Society of America*, **85**: 413–422.

Pinto, J.D. & Oatman, E.R. 1988. *Trichogramma* species in a chaparral community of southern California, with a description of a new species (Hymenoptera: Trichogrammatidae). *Pan-Pacific Entomologist*, **64**: 391–402.

Pinto, J.D., Oatman, E.R. & Platner, G.R. 1986. *Trichogramma pretiosum* and a new cryptic species occurring sympatrically in southwestern North America (Hymenoptera: Trichogrammatidae). *Annals of the Entomological Society of America*, **79**: 1019–1028.

Pinto, J.D., Platner, G.R. & Oatman, E R. 1978. Clarification of the identity of several common species of North American *Trichogramma* (Hymenoptera: Trichogrammatidae). *Annals of the Entomological Society of America*, **71**: 169–180.

Pinto, J.D., Stouthamer, R., Platner, G.R. & Oatman, E.R. 1991. Variation in reproductive compatibility in *Trichogramma* and its taxonomic significance (Hymenoptera: Trichogrammatidae). *Annals of the Entomological Society of America*, **84**: 37–46.

Pinto, J.D., Velten, R.K., Platner, G.R. & Oatman, E.R. 1989. Phenotypic plasticity and taxonomic characters in *Trichogramma* (Hymenoptera: Trichogrammatidae). *Annals of the Entomological Society of America*, **82**: 414–425.

Pitcairn, M.J. & Gutierrez, A.P. 1992. Influence of adult size and age on the fecundity and longevity of *Tetrastichus incertus* (Hymenoptera: Eulophidae). *Annals of the Entomological Society of America*, **85**: 53–57.

Plowright, R.C. & Laverty, T.M. 1984. The ecology and sociobiology of bumble bees. *Annual Review of Entomology*, **29**: 175–199.

Plowright, R.C. & Owen, R.E. 1980. The evolutionary significance of bumble bee color patterns: a mimetic interpretation. *Evolution*, **34**: 622–637.

Plsek, R.W., Kroll, J.C. & Watkins II, J.F. 1969. Observations of carabid beetles, *Helluomorphoides texanus*, in columns of army ants and laboratory experiments on their behavior. *Journal of the Kansas Entomological Society*, **42**: 452–456.

Podoler, H. & Mendel, Z. 1977. Analysis of solitariness in a parasite-host system (*Muscidifurax raptor*, Hymenoptera: Pteromalidae—*Ceratitis capitata*, Diptera: Tephritidae). *Ecological Entomology*, **2**: 153–160.

Poinar, G.O. Jr & Herre, E.A. 1991. Speciation and adaptive radiation in the fig wasp nematode, *Parasitodiplogaster* (Diplogasteridae: Rhabditida) in Panama. *Revue de Nematologie*, **14**: 361–374.

Polak, M. 1992. Distribution of virgin females influences mate-searching behavior of male *Polistes canadensis* (L.) (Hymenoptera: Vespidae). *Journal of Insect Behavior*, **5**: 531–535.

Polak, M. (submitted). Landmark territoriality in *Polistes canadensis* and *P. carnifex*.

Polaszek, A. 1986. The effects of two species of hymenopterous parasitoid on the reproductive system of the pea aphid, *Acyrthosiphon pisum*. *Entomologia Experimentalis et Applicata*, **40**: 285–292.

Polaszek, A. 1991. Egg parasitism in Aphelinidae (Hymenoptera: Chalcidoidea) with special reference to *Centrodora* and *Encarsia* species. *Bulletin of Entomological Research*, **81**: 97–106.

Polaszek, A., Evans, G.A. & Bennett, F.D. 1992. *Encarsia* parasitoids of *Bemisia tabaci* (Hymenoptera: Aphelinidae, Homoptera: Aleyrodidae): a preliminary guide to identification. *Bulletin of Entomological Research*, **82**: 375–392.

Polaszek, A. & Hayat, M. 1992. A revision of the genera *Dirphys* Howard and *Encarsiella* Hayat (Hymenoptera: Aphelinidae). *Systematic Entomology*, **17**: 181–197.

Polaszek, A. & Krombein, K.V. 1994. The genera of Bethylinae (Hymenoptera: Bethylidae). *Journal of Hymenoptera Research*, **3**: 91–105.

Ponomarenko, N.G. 1971. Some peculiarities of development of Dryinidae. *Proceedings of the XIII International Congress of Entomology, Moscow*, **1**: 281–282.

Ponomarenko, N.G. 1975. Peculiarities of larval development in Dryinidae (Hymenoptera). *Entomologicheskoe Obozrenie*, **54**: 534–540. [In Russian with English Summary.]

Porter, C.C. 1967. A revision of the South American species of *Trachysphyrus*. *Memoirs of the American Entomological Institute*, **10**: 1–368.

Porter, C.C. 1978. A revision of the genus *Epirhyssa* (Hymenoptera, Ichneumonidae). *Studia Entomologia*, **20**: 297–412.

Porter, C.C. 1981. Scoliidae (Hymenoptera) of the lower Río Grande Valley. *Florida Entomologist*, **64**: 441–453.

Porter, C.C. 1986. South American and Floridian disjuncts in the Sonoran genus *Compsocryptus* (Hymenoptera: Ichneumonidae). *Psyche*, **93**: 13–33.

Porter, C.C. 1989. *Compsocryptus* of the northern Caribbean with description of a new species from Hispaniola (Hymenoptera: Ichneumonidae). *Florida Entomologist*, **72**: 665–673.

Porter, C.C. & O'Neill, T.J. 1985. A revision of the New World genus *Chromocryptus* (Hymenoptera: Ichneumonidae). *Psyche*, **92**: 407–445.

Post, D.C. 1981. Observations on female nesting and male behavior of *Stictia signata* (Hymenoptera: Sphecidae) in Brazil. *Revista de Biologia Tropical*, **29**: 105–113.

Post, D.C. & Jeanne, R.L., 1981. Colony defense against ants by *Polistes fuscatus* (Hymenoptera: Vespidae) in Wisconsin. *Journal of the Kansas Entomological Society*, **54**: 599–615.

Post, D.C., Mohamed, M.A., Coppel, H.C. & Jeanne, R.L. 1984. Identification of ant repellent allomone produced by social wasp *Polistes fuscatus* (Hymenoptera: Vespidae). *Journal of Chemical Ecology*, **10**: 1799–1807.

Powell, D. 1938. The biology of *Cephalonomia tarsalis* (Ash.), a vespoid wasp (Bethylidae: Hymenoptera) parasitic on the sawtoothed grain beetle. *Annals of the Entomological Society of America*, **31**: 44–49.

Powell, J.A. & Turner, W.J. 1975. Observations on oviposition behavior and host selection in *Orussus occidentalis* (Hymenoptera: Siricoidea). *Journal of the Kansas Entomological Society*, **48**: 299–307.

Powell, J.E., Shepard, M. & Sullivan, M.J. 1981. Use of heating degree day and physiological day equations for predicting development of the parasitoid *Trissolcus basalis*. *Environmental Entomology*, **10**: 1008–1011.

Powell, W. 1986. Enhancing parasitoid activity in crops, pp. 319–340. *In*: Waage, J. & Greathead, D. (eds) *Insect Parasitoids*. 389pp. Academic Press, London.

Prance, G.T. 1976. The pollination and androphore structure of some Amazonian Lecythidaceae. *Biotropica*, **8**: 235–241.

Prance, G.T. 1987. Biogeography of neotropical plants, pp 46–65. *In*: Whitmore, T.C. & Prance, G.T. (eds) *Biogeography and Quaternary History in Tropical America*. 214pp. Oxford University Press, Oxford.

Prebble, M.L. 1941*a*. The diapause and related phenomena in *Gilpinia polytoma* (Hartig). I. Factors influencing the inception of diapause. *Canadian Journal of Research* (D) Zoological Sciences, **19**: 295–346.

Prebble, M.L. 1941*b*. The diapause and related phenomena in *Gilpinia polytoma* (Hartig). III. Bioclimatic relations. *Canadian Journal of Research* (D) Zoological Sciences, **19**: 350–362.

Prebble, M.L. 1941*c*. The diapause and related phenomena in *Gilpinia polytoma* (Hartig). IV. Influence of food and diapause upon reproductive capacity. *Canadian Journal of Research* (D) Zoological Sciences, **19**: 417–454.

Prentice, T.R. & Walker, G.P. 1991. Nocturnal clustering of the solitary wasp, *Chalybion californicum* (Saussure) (Hymenoptera: Sphecidae). *Pan-Pacific Entomologist*, **67**: 222–224.

Prévost, G., Davies, D.H. & Vinson, S.B. 1990. Evasion of encapsulation by parasitoid correlated with the extent of host hemocyte pathology. *Entomologia Experimentalis et Applicata*, **55**: 1–10.

Price, P.W. 1972. Parasitoids utilizing the same host: adaptive nature of differences in size and form. *Ecology*, **53**: 190–195.

Price, P.W. 1973. Reproductive strategies in parasitoid wasps. *American Naturalist*, **107**: 684–693.

Price, P.W. 1974. Strategies for egg production. *Evolution*, **28**: 76–84.

Price, P.W. & Clancy, K.M. 1986. Interactions among three trophic levels: gall size and parasitoid attack. *Ecology*, **67**: 1593–1600.

Price, P.W., Fernandes, G.W. & Waring, G.L. 1987. Adaptive nature of insect galls. *Environmental Entomology*, **16**: 15–24.

Priesner, H. 1936. *Aphanogmus steinitzi* spec. nov., ein Coniopterygiden-parasit (Hymenoptera Proctotrupoidea). *Bulletin de la Société Royale Entomologique D'Egypte*, **20**: 248–251.

Prokopy, R.J. & Webster, R.P. 1978. Oviposition-deterring pheromone of *Rhagoletis pomonella* a kairomone for its parasitoid *Opius lectus*. *Journal of Chemical Ecology*, **4**: 481–494.

Pschorn-Walcher, H. 1967. Biology of the ichneumonid parasites of *Neodiprion sertifer* (Geoffroy) (Hym.: Diprionidae) in Europe. *Technical Bulletin of the Commonwealth Institute of Biological Control*, **8**: 7–51.

Pschorn-Walcher, H. & Zinnert, K.D. 1971. Investigations on the ecology and natural control of the larch sawfly (*Pristiphora erichsonii* Htg. Hym.: Tenthredinidae) in Central Europe. Part II. Natural enemies: their biology and ecology, and their role as mortality factors in *P. erichsonii*. *Technical Bulletin of the Commonwealth Institute of Biological Control*, **14**: 1–50.

Pulawski, W.J. 1988. Revision of North American *Tachysphex* wasps including Central American and Caribbean species (Hymenoptera: Sphecidae). *Memoirs of the California Academy of Sciences*, **10**: 1–211.

Pungerl, N.B. 1983. Variability in characters commonly used to distinguish *Aphidius* species (Hymenoptera: Aphidiidae). *Systematic Entomology*, **8**: 425–430.

Puttler, B. 1961. Biology of *Hyposoter exiguae* (Hymenoptera: Ichneumonidae), a parasite of lepidopterous larvae. *Annals of the Entomological Society of America*, **54**: 25–30.

Puttler, B. 1974. *Hypera postica* and *Bathyplectes curculionis*: encapsulation of parasite eggs by host larvae in Missouri and Arkansas. *Environmental Entomology*, **3**: 881–882.

Puttler, B., Gordh, G. & Long, S.H. 1980. Bionomics of *Euplectrus puttleri*, new species, an introduced parasite of the velvetbean caterpillar, *Anticarsia gemmatalis*, from South America. *Annals of the Entomological Society of America*, **73**: 28–35.

Puttler, B. & Long, S.H. 1983. Host specificity tests of an egg parasite, *Edovum puttleri* (Hymenoptera: Eulophidae), of the Colorado potato beetle, *Leptinotarsa decemlineata* (Coleoptera: Chrysomelidae). *Proceedings of the Entomological Society of Washington*, **85**: 384–387.

Pyenson, L. 1940. Notes on the biology of three tenthredinid (Hym.) pests of the guava. *Bulletin of Entomological Research*, **30**: 467–469.

Pyke, G.H. 1982. Foraging in bumblebees: rule of departure from an inflorescence. *Canadian Journal of Zoology*, **60**: 417–428.

Quednau, F.W. 1970. Notes on life-history, fecundity, longevity and attack pattern of *Agathis pumila* (Hymenoptera: Braconidae), a parasite of the larch casebearer. *Canadian Entomologist*, **102**: 736–745.

Queller, D.C., Strassman, J.E. & Hughes, C.R. 1988. Genetic relatedness in colonies of tropical wasps with multiple queens. *Science*, **242**: 1155–1157.

Quezada, J.R. 1974. Biological control of *Aleurocanthus woglumi* (Homoptera: Aleyrodidae) in El Salvador. *Entomophaga*, **19**: 243–254.

Quezada, J.R. 1979. Hallazgo de *Agonatopus* sp: (Hymenoptera: Dryinidae), parasito del *Dalbulus maidis* (Homoptera: Cicadellidae) en El Salvador. *Ceiba*, **23**: 1–12.

Quezada, J.R. 1986. Utilización del control biológico clásico. *Manejo Integrado de Plagas*, **2**: 16–31.

Quezada, J.R., DeBach, P. & Rosen, D. 1973. Biological and taxonomic studies of *Signiphora borinquensis*, new species, (Hymenoptera: Signiphoridae), a primary parasite of diaspine scales. *Hilgardia*, **41**: 543–603.

Quezada, J.R. & Urbina, N.E. 1987. La broca del fruto del cafeto, *Hypothenemus hampei*, y su control. *In*: Pinochet, J. (ed.), *Plagas y Enfermedades de Caracter Epidémico en Cultivos Frutales de la Región Centroamericana*, Centro Agronómico Tropical de Investigación y Enseñanza, Informe Técnico, **110**: 48–59.

Quicke, D.L.J. 1984. Are some rogadine (Hym.: Braconidae) puparia hoverfly mimics? *Proceedings and Transactions of the British Entomological and Natural History Society*, **17**: 60.

Quicke, D.L.J. 1986*a*. Preliminary notes on homeochromatic associations within and between the Afrotropical Braconinae (Hym., Braconidae) and Lamiinae (Col., Cerambycidae). *Entomologist's Monthly Magazine*, **122**: 97–109.

Quicke, D. 1986*b*. Warning coloration and mimicry, pp. 74–80. *In*: Betts, C. (ed.) *Hymenopterists's Handbook*. 208pp. Amateur Entomologist's Society, Middlesex.

Quicke, D.L.J. 1988*a*. A new genus and species of Braconinae (Hymenoptera: Braconidae) parasitic on *Diatraea* sp. (Lepidoptera: Pyralidae) in Ecuador. *Bulletin of Entomological Research*, **78**: 15–18.

Quicke, D.L.J. 1988*b*. *Digonogastra*: the correct name for nearctic *Iphiaulax* of authors (Hymenoptera, Braconidae). *Proceedings of the Entomological Society of Washington*, **90**: 196–200.

Quicke, D.L.J. 1988*c*. Host relationships in the Braconinae (Hymenoptera: Braconidae) — how little we know! *News Bulletin, Entomological Society of Queensland*, **16**: 85–92.

Quicke, D.L.J. 1988*d*. Reclassification of some Neotropical Braconinae (Hym., Braconidae). *Entomologist's Monthly Magazine*, **124**: 195–199.

Quicke, D.L.J. 1989*a*. Further new host records for genera and species of Braconinae (Hym., Braconidae). *Entomologist's Monthly Magazine*, **125**: 199–206.

Quicke, D.L.J. 1989*b*. Reclassification of some New World species of Braconinae (Hym., Braconidae). *Entomologist's Monthly Magazine*, **125**: 119–121.

Quicke, D.L.J. 1989*c*. A new Neotropical genus and species of Coeloidini (Hym., Braconidae, Braconinae). *Entomologist's Monthly Magazine*, **125**: 9–12.

Quicke, D.L.J. & van Achterberg, C. 1990. Phylogeny of the subfamilies of the family Braconidae (Hymenoptera: Ichneumonoidea). *Zoologische Verhandelingen*, **258**: 1–95.

Quicke, D.L.J., Ficken, L.C. & Fitton, M.G. 1992. New diagnostic ovipositor characters for doryctine wasps (Hymenoptera, Braconidae). *Journal of Natural History*, **26**: 1035–1046.

Quicke, D.L.J., Fitton, M.G. & Ingram, S. 1992. Phylogenetic implications of the structure and distribution of ovipositor valvilli in the Hymenoptera (Insecta). *Journal of Natural History*, **26**: 587–608.

Quicke, D.L.J. & Sharkey, M. 1989. A key to and notes on the genera of Braconinae (Hymenoptera: Braconidae) from American North of Mexico with descriptions of two new genera and three new species. *Canadian Entomologist*, **121**: 337–361.

Quicke, D.L.J., Tunstead, J., Falco, J.V. & Marsh, P.M. 1992. Venom gland and reservoir morphology in the Doryctinae and related braconid wasps (Insecta, Hymenoptera, Braconidae). *Zoologica Scripta*, **21**: 403–416.

Quicke, D.L.J. & Wharton, R.A. 1989. *Myosoma nyanzaensis* sp.n. (Hymenoptera: Braconidae) parasitic on the stem-borer pest *Chilo partellus* (Swinhoe) (Lepidoptera: Pyralidae) in Kenya. *Bulletin of Entomological Research*, **79**: 1–5.

Quinlan, J. 1976. Two new genera of Eucoilidae (Hymenoptera: Cynipoidea). *Systematic Entomology*, **1**: 65–69.

Quinlan, J. 1978. Hymenoptera Cynipoidea: Eucoilidae. *Handbooks for the Identification of British Insects*, **8**(1b): 1–58.

Quinlan, J. 1979. A revisionary classification of the Cynipoidea (Hymenoptera) of the Ethiopian zoogeographical region. *Bulletin of the British Museum (Natural History)* Entomology, **39**: 85–133.

Quinlan, J. 1984. *Stentorceps*, a remarkable new genus of eucoilid (Hymenoptera) from Africa. *Systematic Entomology*, **9**: 479–485.

Quinlan, J. 1986. A key to the Afrotropical genera of Eucoilidae (Hymenoptera), with a revision of certain genera. *Bulletin of the British Museum (Natural History)* Entomology, **52**: 243–366.

Quinlan, J. 1988. A revision of some Afrotropical genera of Eucoilidae (Hymentoptera). *Bulletion of the British Museum (Natural History)*, Entomology, **56**: 171–229.

Quinlan, R.J. & Cherrett, J.M. 1977. The role of substrate preparation in the symbiosis between the leaf-cutting ant *Acromyrmex octospinosus* (Reich) and its food fungus. *Ecological Entomology*, **2**: 161–170.

Quinlan, R.J. & Cherrett, J.M. 1978. Aspects of the symbiosis of the leaf-cutting ant *Acromyrmex octospinosus* (Reich) and its food fungus. *Ecological Entomology*, **3**: 221–230.

Quinlan, R.J. & Cherrett, J.M. 1979. The role of fungus in the diet of the leaf-cutting ant *Atta cephalotes* (L.). *Ecological Entomology*, **4**: 151–160.

Quintero A., D. & Rodríguez, R.R. 1990. New species of *Typhoctes* from Panama (Hymenoptera: Typhoctidae). *Annals of the Entomological Society of America*, **83**: 30–35.

Raatikainen, M. 1967. Bionomics, enemies and population dynamics of *Javesella pellucida* (F.) (Hom. Delphacidae). *Annales Agriculturae Fenniae*, **6** (Suppl. 2): 1–149.

Rabaud, E. 1922. Notes sur le comportement de *Rielia manticida* Kieff., proctotrypide parasite des oothèques de mantes. *Bulletin de la Société Zoologique de France*, **47**: 10–15.

Rabb, R.L. 1960. Biological studies of *Polistes* in North Carolina (Hymenoptera: Vespidae). *Annals of the Entomological Society of America*, **53**: 111–121.

Rabb, R.L. & Bradley J.R. 1970. Marking host eggs by *Telenomus sphingis*. *Annals of the Entomological Society of America*, **63**: 1053–1056.

Raff, J.W. 1934. Observations on saw-flies of the genus *Perga*, with notes on some reared primary parasites of the families Trigonalidae, Ichneumonidae and Tachinidae. *Proceedings of the Royal Society of Victoria*, **47**: 54–77.

Ralston, J.S. 1977. Egg guarding by male assassin bugs of the genus *Zelus* (Hemiptera: Reduviidae). *Psyche*, **84**: 103–107.

Ramachandra Rao, R.S.Y. & Cherian, M.C. 1926 (1927). Notes on the life-history and habits of *Elasmus nephantidis*, Rohw. *Madras Agricultural Department Yearbook*, **1926**: 39–50.

Ramadan, M.M., Wong, T.T.Y. & Beardsley, J.W. 1989. Survivorship, potential, and realized fecundity of *Biosteres tryoni* (Hymenoptera: Braconidae), a larval parasitoid of *Ceratitis capitata* (Diptera: Tephritidae). *Entomophaga*, **34**: 291–297.

Ramalho, M., Kleinert-Giovannini, A. & Imperatriz-Fonseca, V.L. 1989. Utilization of floral resources by species of *Melipona* (Apidae, Meliponinae): floral preferences. *Apidologie*, **20**: 185–195.

Ramalho, M., Kleinert-Giovannini, A. & Imperatriz-Fonseca, V.L. 1990. Important bee plants for stingless bees (*Melipona* and Trigonini) and Africanized honeybees (*Apis mellifera*) in neotropical habitats: a review. *Apidologie*, **21**: 469–488.

Ramirez, G.A. 1992. Observaciones sobre la biologia del *Tridryinus poecilopterae* (Rich) (Hymenoptera: Dryinidae) parasito de la seudopolilla algondonosa *Poekilloptera phalaenoides* (L.) (Homoptera: Flatidae). *Entomólogo, Boletín de Noticias (Sociedad Colombiana de Entomologia Socolen)*, **71**: 1–5.

Ramírez B., W. 1969. Fig wasps: mechanisms of pollen transfer. *Science* **163**: 580–581.

Ramírez B., W. 1970*a*. Host specificity of fig wasps (Agaonidae). *Evolution*, **24**: 680–691.

Ramírez B., W. 1970*b*. Taxonomic and biological studies of Neotropical fig wasps (Hymenoptera: Agaonidae). *University of Kansas Science Bulletin*, **49**: 1–44.

Ramírez B., W. 1974. Coevolution of *Ficus* and Agaonidae. *Annals of the Missouri Botanical Garden*, **61**: 770–780.

Ramírez B., W. 1978. Evolution of mechanisms to carry pollen in Agaonidae (Hymenoptera Chalcidoidea). *Tijdschrift voor Entomologie*, **121**: 279–293.

Ramirez, W. 1982. *Bombus mexicanus* Cresson, un hospedante de *Melaloncha*, moscas parásitas de la abeja de miel en el nuevo mundo. *Revista Biología Tropical*, **30**: 177.

Ramirez, W. 1984. Biología del género *Melaloncha* (Phoridae), moscas parasitoides de la abeja doméstica (*Apis mellifera* L.) en Costa Rica. *Revista de Biología Tropical*, **32**: 25–28.

Ramirez B., W. 1991. Evolution of the mandibular appendage in fig wasps (Hymenoptera: Agaonidae). *Revista de Biología Tropical*, **39**: 87–95.

Rao, V.P. 1971. Biological control of pests in Fiji. *Commonwealth Institute of Biological Control, Miscellaneous Publications*, **2**: 1–38.

Rasnitsyn, A.P. 1969 [The origin and evolution of the lower Hymenoptera.] *Trudy Paleontologeskogo Instituta, Akademia Nauk SSSR*, **123**: 1–195 [In Russian.]

Rasnitsyn, A.P. 1975. Early evolution of the higher Hymenoptera (Apocrita). *Zoologicheskii Zhurnal* **54**: 848–860. [In Russian; English summary.]

Rasnitsyn, A.P. 1980. [The origin and evolution of hymenopteran insects.] *Trudy Paleontologeskogo Instituta, Akademia Nauk SSSR*, **174**: 1–191 [In Russian.]

Rasnitsyn, A.P. 1988. An outline of evolution of the hymenopterous insects (Order Vespida). *Oriental Insects*, **22**: 115–145.

Rasnitsyn, A.P. & Kovalev, O.V. 1988. The oldest Cynipoidea (Hymenoptera, Archaeocynipidae fam. n.) from the Early Cretaceous Transbaikalia. *Vestnik Zoologii*, **1988(1)**: 18–21. [In Russian; English summary.]

Rasnitsyn, A.P. & Matveev, D.G. 1989. First Palaearctic representative of the genus *Ampulicomorpha* Ashmead (Hymenoptera, Embolemidae). *Entomologicheskoe Obozrenie*, **68**: 657–661. [In Russian; English Summary.]

Rasplus, J-Y. 1988. Description de deux nouvelles espèces du genre *Anisopteromalus* Rutschka. Clé des espèces afrotropicales (Hym. Pteromalidae). *Bulletin de la Société Entomologique de France*, **93**: 119–127.

Rasplus, J.-Y., Pluot-Sigwalt, D., Llosa, J.F. & Couturier, G. 1990. *Hexacladia linci*, n.sp. (Hymenoptera: Encyrtidae) endoparasite de *Lincus malevolus* Rolston (Heteroptera: Pentatomidae) au Pérou. *Annales de la Société Entomologique de France* (N.S.) **26**: 255–263.

Rau, P. 1922. Ecological and behavior notes on Missouri insects. *Transactions of the Academy of Science of St. Louis*, **24**: 1–71.

Rau, P. 1932. The courtship dance and sleeping habits of *Scolia dubia*. *Bulletin of the Brooklyn Entomological Society*, **27**: 59–62.

Rau, P. 1940. The life history of the wood-roach, *Parcoblatta pennsylvanica* DeGeer, (Orthoptera: Blattidae). *Entomological News*, **51**: 33–35.

Raw, A. 1977. The biology of two *Exomalopsis* species (Hymenoptera: Anthophoridae) with remarks on sociality in bees. *Revista Biologia Tropical*, **25**: 1–11.

Raw, A. 1984. The nesting biology of nine species of Jamaican bees (Hymenoptera). *Revista Brasileira de Entomologia*, **28**: 497–506.

Rawlings, G.B. 1957. *Guiglia schauinslandi* (Ashmead) (Hym., Orussidae) a parasite of *Sirex noctilio* (Fabricius) in New Zealand. *The Entomologist*, **90**: 35–36.

Ray, T.S. & Andrews, C.C. 1980. Antbutterflies: butterflies that follow army ants to feed on antbird droppings. *Science*, **210**: 1147–1148.

Read, D.P., Feeny, P.P. & Root, R.B. 1970. Habitat selection by the aphid parasite *Diaeretiella rapae* (Hymenoptera: Braconidae) and hyperparasite *Charips brassicae* (Hymenoptera: Cynipidae). *Canadian Entomologist*, **102**: 1567–1578.

Real, L.A. 1981. Nectar availability and bee-foraging on *Ipomoea* (Convolvulaceae). *Biotropica* (Reproductive Botany Supplement) **13**: 64–69.

Reed, H.C. & Vinson, S.B. 1979. Observations of the life history and behavior of *Elasmus polistis* Burks (Hymenoptera: Chalcidoidea: Eulophidae). *Journal of the Kansas Entomological Society*, **52**: 247–257.

Reid, J.A. 1941. The thorax of the wingless and short-winged Hymenoptera. *Transactions of the Royal Entomological Society of London*, **91**: 367–446.

Reid, J.A. 1942. On the classification of the larvae of the Vespidae (Hymenoptera). *Transactions of the Royal Entomological Society of London*, **92**: 285–331.

Reinhard, E.G. 1929. *The Witchery of Wasps*, 291pp. The Century Co., New York & London.

Remadevi, O.K., Mohamed, U.V.K., Abdurahiman, U.C. & Narendran, T.C. 1978. Oviposition behaviour of *Perisierola nephantidis* Muesbeck (Bethylidae: Hymenoptera) a larval parasite of *Nephantis serinopa* Meyrick (Xylorictidae: Lepidoptera). *Entomon*, **3**: 303–305.

Remadevi, O.K., Mohamed, U.V.K. & Abdurahiman, U.C. 1981. Some aspects of the biology of *Parasierola nephantidis* Muesebeck (Hymenoptera, Bethylidae), a larval parasitoid of *Nephantis serinopa* Meyrick (Lepidoptera, Xylorictidae). *Polskie Pismo Entomologiczne*, **51**: 597–604.

Renou, M., Hawlitzky, N., Berthier, A., Malosse, C. & Ramiandrasoa, F. 1989. Mise en évidence d'une activité kairomonale des oeufs de la pyrale du maïs sur les femelles de *Trichogramma maidis*. *Entomophaga*, **34**: 569–580.

Renzik, S.Ya. & Umarova, T.Ya. 1991. Host population density influence on host acceptance in *Trichogramma*. *Entomologia Experimentalis et Applicata*, **58**: 49–54.

Rettenmeyer, C.W. 1960. Behavior, abundance and host specificity of mites found on Neotropical army ants (Acarina: Formicidae: Dorylinae). *Proceedings of the Eleventh International Congress of Entomology*, **1**: 610–612.

Rettenmeyer, C.W. 1961. Arthropods associated with Neotropical army ants with a review of the behavior of these ants (Arthropoda: Formicidae: Dorylinae). [Unpublished Ph.D. dissertation, University of Kansas, Lawrence, xv+605pp.]

Rettenmeyer, C.W. 1962. The behavior of millipeds found with Neotropical army ants. *Journal of the Kansas Entomological Society*, **35**: 377–384.

Rettenmeyer, C.W. 1963*a*. Behavioral studies of army ants. *University of Kansas Science Bulletin*, **44**: 281–465.

Rettenmeyer, C.W. 1963*b*. The behavior of Thysanura found with army ants. *Annals of the Entomological Society of America*, **56**: 170–174.

Rettenmeyer, C.W. & Akre, R.D. 1968. Ectosymbiosis between phorid flies and army ants. *Annals of the Entomological Society of America*, **61**: 1317–1326.

Rettenmeyer, C.W., Topoff, H. & Mirenda, J. 1978. Queen retinues of army ants. *Annals of the Entomological Society of America*, **71**: 519–528.

Rettenmeyer, C.W. & Watkins, J.F. II, 1978. Polygyny and monogyny in army ants (Hymenoptera: Formicidae). *Journal of the Kansas Entomological Society*, **51**: 581–591.

Ricci, J.G. & De Santis, L. 1983. Presencia de *Metastenus concinnus* Walker (Hym., Pteromalidae) parasitoide de *Cryptognatha signata* Korsch. (Col., Coccinellidae) en la región Neotropical. CIRPON — *Revista de Investigacion*, **1**: 187–191.

Rice, R.E. 1969. Response of some predators and parasites of *Ips confusus* (LeC.) (Coleoptera: Scolytidae) to olfactory attractants. *Contributions. Boyce Thompson Institute for Plant Research*, **24**: 189–194.

Richards, O.W. 1932. Observations on the genus *Bethylus* Latr. (= *Perisemus* Foerst) (Hymenoptera, Bethylidae). *Transactions of the Entomological Society of the South of England*, **8**: 35–40.

Richards, O.W. 1934. The American species of the genus *Trypoxylon* (Hymenopt., Sphecoidea). *Transactions of the Royal Entomological Society of London*, **82**: 173–362.

Richards, O.W. 1939. The Bethylidae subfamily Sclerogibbinae (Hymenoptera). *Proceedings of the Royal Entomological Society of London* (B), **8**: 211–223.

Richards, O.W. 1956. Hymenoptera. Introduction and keys to families. *Handbooks for the Identification of British Insects*, **6**(1): 1–94.

Richards, O.W. 1962. *A Revisional Study of the Masarid Wasps (Hymenoptera, Vespoidea)*. 294pp. British Museum (Natural History), London.

Richards, O.W. 1968. The subgeneric divisions of the genus *Bombus* Latreille (Hymenoptera: Apidae). *Bulletin of the British Museum (Natural History)* Entomology, **22**: 209–276.

Richards, O.W. 1972. The species of the South American wasps of the genus *Microstigmus* Ducke (Hymenoptera: Sphecoidea, Pemphredoninae). *Transactions of the Royal Entomological Society of London*, **124**: 123–148.

Richards, O.W. 1978. *The Social Wasps of the Americas Excluding the Vespinae*, 580pp. British Museum (Natural History), London.

Richards, O.W. & Hamm, A.H. 1939. The biology of the British Pompilidae (Hymenoptera). *Transactions of the Society for British Entomology*, **6**: 51–114.

Richardson, C.H. 1913. Studies on the habits and development of a hymenopterous parasite, *Spalangia muscidarum* Richardson. *Journal of Morphology*, **24**: 513–557.

Richerson, J.V. & Borden, J.H. 1972. Host finding behavior of *Coeloides brunneri* (Hymenoptera: Braconidae). *Canadian Entomologist*, **104**: 1235–1250.

Richerson, J.V. & DeLoach, C.J. 1972. Some aspects of host selection by *Perilitus coccinellae*. *Annals of the Entomological Society of America*, **65**: 834–839.

Riches, H.R.C. 1989. Bee venom hypersensitivity update. *Bee World*, **70**: 12–18.

Rico-Gray, V. 1993. Use of plant-derived food resources by ants in the dry tropical lowlands of coastal Veracruz, Mexico. *Biotropica*, **25**: 301–315.

Ridgway, N.M. & Mahr, D.L. 1990. Reproduction, development, and longevity of *Pholetesor ornigis* (Hymenoptera: Braconidae), a parasitoid of spotted tentiform leafminer (Lepidoptera: Gracillariidae), in the laboratory. *Annals of the Entomological Society of America*, **83**: 790–794.

Ridgway, R.L., Ables, J.R., Goodpasture, C. & Hartstack, A.W. 1981. *Trichogramma* and its utilization for crop protection in the U.S.A. pp. 41–48. *In*: Coulson, J. R. (ed.) *Proceedings of the Joint American-Soviet Conference on Use of Beneficial Organisms in the Control of Crop Pests, Washington, D.C., U.S.A. August 13–14, 1979*. 62pp. Entomological Society of America Publication.

Ridgway, R.L. & Morrison, R.K. 1985. Worldwide perspective on practical utilization of *Trichogramma* with special reference to control of *Heliothis* on cotton. *Southwestern Entomologist*, Supplement **8**: 190–198.

Ridgway, R.L. & Vinson, S.B. (eds) 1977. *Biological Control by Augmentation of Natural Enemies Insect and Mite Control with Parasites and Predators*. 480pp. Plenum Press, New York.

Riegel, G.T. 1950. A new genus and species of Dacnusini (Hym.: Braconidae). *Entomological News*, **61**: 125–129.

Riegel, G.T. 1952. Neotropical exodont Braconidae (Hymenoptera). *Transactions of the Illinois Academy of Science*, **45**: 177–181.

Riegel, G.T. 1982. The American species of Dacnusinae, excluding certain Dacnusini (Hymenoptera: Braconidae). *Novitates Arthropodae*, **1**(3): 1–185.

Riek, E.F. 1955. Fossil insects from the Triassic beds at Mt. Crosby, Queensland. *Australian Journal of Zoology*, **3**: 654–691.

Riek, E.F. 1967. Australian Hymenoptera Chalcidoidea family Eulophidae, subfamily Elasminae. *Australian Journal of Zoology*, **15**: 145–199.

Riek, E.F. 1970. Hymenoptera (Wasps, bees, ants), pp. 867–959. *In*: CSIRO, *The Insects of Australia*, 1029 pp. Melbourne University Press, Carlton.

Riek, E. 1971. A new subfamily of cynipoid wasps (Hymenoptera: Cynipoidea) from Australia. *Entomological Essays to Commemorate the Retirement of Professor K. Yasumatsu*. pp. 107–112.

Risch, S. 1981. Ants as important predators of rootworm eggs in the neotropics. *Journal of Economic Entomology*, **74**: 88–90.

Risch, S.J. & Carroll, C.R. 1982. Effect of a keystone predaceous ant, *Solenopsis geminata*, on arthropods in a tropical agroecosystem. *Ecology*, **63**: 1979–1983.

Risch, S., McClure, M., Vandermeer, J. & Waltz., S. 1977. Mutualism between three species of tropical *Piper* (Piperaceae) and their ant inhabitants. *American Midland Naturalist*, **98**: 433–444.

Ritchie, A.J. & Masner, L. 1983. Revision of the Nearctic species of *Baryconus* (Hymenoptera: Scelionidae, Scelioninae). *Canadian Journal of Zoology*, **61**: 704–720.

Rivnay, E. & Perzelan, J. 1943. Insects associated with *Pseudococcus* spp. (Homoptera) in Palestine, with notes on their biology and economic status. *Journal of the Entomological Society of Southern Africa*, **6**: 9–28.

Rizki, R.M. & Rizki, T.M. 1984. Selective destruction of a host blood cell type by a parasitoid wasp. *Proceedings of the National Academy of Sciences of the United States of America*, **81**: 6154–6158.

Rizki, R.M. & Rizki, T.M. 1990. Parasitoid virus-like particles destroy *Drosophila* cellular immunity. *Proceedings of the National Academy of Sciences of the United States of America*, **87**: 8388–8392.

Roberts, D.R., Alecrim, W.D., Heller, J.M., Ehrhardt, S.R. & Lima, J.B. 1982. Male *Eufriesia purpurata*, a DDT-collecting bee in Brazil. *Nature*, **297**: 62–63.

Roberts, F.S. 1958. Insects affecting banana production in Central America. *Proceedings of the Tenth International Congress of Entomology*, **3**: 411–415.

Roberts, R.A. 1933. Biology of *Brachymeria fonscolombei* (Dufour), a hymenopterous parasite of blowfly larvae. *United States Department of Agriculture Technical Bulletin*, **365**: 1–22.

Roberts, R.A. 1935. Some North American parasites of blowflies. *Journal of Agricultural Research*, **50**: 479–494.

Roberts, R.B. 1971. Biology of the crepuscular bee *Ptiloglossa guinnae* n. sp. with notes on associated bees, mites, and yeasts. *Journal of the Kansas Entomological Society*, **44**: 283–294.

Roberts, R.B. 1972. Revision of the bee genus *Agapostemon* (Hymenoptera: Halictidae). *University of Kansas Science Bulletin*, **49**: 437–590.

Roberts, R.B. 1973. Nest architecture and immmature stages of the bee *Oxaea flavescens* and the status of Oxaeidae (Hymenoptera). *Journal of the Kansas Entomological Society*, **46**: 437–446.

Roberts, R.B. & Brooks, R.W. 1987 Agapostemonine bees of Mesoamerica (Hymenoptera: Halictidae). *University of Kansas Science Bulletin*, **53**: 357–392.

Roberts, R.B. & Dodson, C.H. 1967. Nesting biology of two communal bees, *Euglossa imperialis* and *Euglossa ignita* (Hymenoptera: Apidae), including description of larvae. *Annals of the Entomological Society of America*, **60**: 1007–1014.

Roberts, R.B. & Vallespir, S.R. 1978. Specialization of hairs bearing pollen and oil on the legs of bees (Apoidea: Hymenoptera). *Annals of the Entomological Society of America*, **71**: 619–627.

Robertson, P.L. 1968. A morphological and functional study of the venom apparatus in representatives of some major groups of Hymenoptera. *Australian Journal of Zoology*, **16**: 133–166.

Robinson, D.M. 1961. The parasites of the Psyllidae-2. *Parapsyllaephagus adulticolus* gen. et sp. nov., the first hymenopterous parasite of an adult psyllid (Homoptera). 3. Some notes on the biology and host relationships of *Parapsyllaephagus adulticolus* Robinson (Hymenoptera). *Annals and Magazine of Natural History; including Zoology, Botany and Geology*, (13)**4**: 117–121 and 155–159.

Robinson, G.E. 1992. Regulation of division of labor in insect societies. *Annual Review of Entomology*, **37**: 637–665.

Robinson, S.W. & Cherrett, J.M. 1978. The possible use of methyl 4-methylpyrrole-2-carboxylate, an ant trail pheromone, as a component of an improved bait for leaf-cutting ant (Hymenoptera: Formicidae) control. *Bulletin of Entomological Research*, **68**: 159–170.

Rockwood, L.L. 1975. The effects of seasonality on foraging in two species of leaf-cutting ants (*Atta*) in Guanacaste Province, Costa Rica. *Biotropica*, **7**: 176–193.

Rockwood, L.L. 1976. Plant selection and foraging patterns in two species of leaf-cutting ants (*Atta*). *Ecology*, **57**: 48–61.

Rockwood, L.L. & Glander, K.E. 1979. Howling monkeys and leaf-cutting ants: comparative foraging in a tropical deciduous forest. *Biotropica*, **11**: 1–10.

Rockwood, L.L. & Hubbell, S.P. 1987. Host-plant selection, diet diversity, and optimal foraging in a tropical leafcutting ant. *Oecologia*, **74**: 55–61.

Rodd, N.W. 1951*a*. Notes on the biology of an Australian trigonalid wasp (*Taeniogonalos heterodoxus* Raym.). *Australian Zoologist*, **11**: 338–340.

Rodd, N.W. 1951*b*. Some observations on the biology of Stephanidae and Megalyridae (Hymenoptera). *Australian Zoologist*, **11**: 341–346.

Rodendorf, B.B. 1968. Insect phylogeny and paleontological data. *Entomological Review, Washington*, **47**: 189–199.

Rodriguez-del-Bosque, L.A. & Smith, J.W. Jr, 1989. Parasitization of *Diatraea lineolata* pupae and diapausing larvae by several exotic parasites. *Florida Entomologist*, **72**: 703–705.

Roig Alsina, A. 1982. Revision del genero *Caliadurgus* Pate en la Argentina (Hymenoptera: Pompilidae). *Revista de la Sociedad Entomológica Argentina*, **41**: 233–252.

Roig Alsina, A. 1985 (1987). Contribucion al conocimiento de los Pepsinae Sudamericanos. IV El genero *Sphictostethus* Kohl (Hymen., Pompilidae). *Revista de la Sociedad Entomológica Argentina*, **44**: 277–315.

Roig Alsina, A. 1989. The tribe Osirini, its scope, classification, and revisions of the genera *Parepeolus* and *Osirinus* (Hymenoptera, Apoidea, Anthophoridae). *University of Kansas Science Bulletin*, **54**: 1–23.

Roig-Alsina, A. 1990. *Coelioxoides* Cresson, a parasitic genus of Tetrapediini (Hymenoptera: Apoidea). *Journal of the Kansas Entomological Society*, **63**: 279–287.

Roitberg, B.D. & Lalonde, R.G. 1991. Host marking enhances parasitism risk for a fruit-infesting fly *Rhagoletis basiola*. *Oikos*, **61**: 389–393.

Rojas-Rousse, D. & Benoit, M. 1977. Morphology and biometry of larval instars of *Pimpla instigator* (F.) (Hymenoptera: Ichneumonidae). *Bulletin of Entomological Research*, **67**: 129–141.

Rojas-Rousse, D. & Palevody, C. 1983. Organogenèse et ultrastucture des sensilles placoïdes des antennes de *Diadromus pulchellus* Wesmael (Hymenoptera: Ichneumonidae). *International Journal of Insect Morphology and Embryology*, **12**: 171–185.

Rosen, D. 1981. *The role of Hyperparasitism in Biological Control: A Symposium*. 52pp. No 4103, University of California, Berkeley.

Rosen, D. & Alon, A. 1983. Taxonomic and biological studies of *Diversinervus cervantesi* (Girault) (Hymenoptera: Encyrtidae), a primary parasite of soft scale insects. *Contributions of the American Entomological Institute*, **20**: 336–362.

Rosen, D. & DeBach, P. 1979. Species of *Aphytis* of the world (Hymenoptera: Aphelinidae). *Series Entomologica*, **17**: 1–801.

Rosen, D. & Rössler, Y. 1966. Studies on an Israel strain of *Anagyrus pseudococci* (Girault) (Hymenoptera, Encyrtidae). I. Morphology of the adults and developmental stages. *Entomophaga*, **11**: 269–277.

Rosenberg, H.T. 1934. The biology and distribution in France of the larval parasites of *Cydia pomonella* L. *Bulletin of Entomological Research*, **25**: 201–256.

Rosenheim, J.A. 1990. Density-dependent parasitism and the evolution of aggregated nesting in solitary Hymenoptera. *Annals of the Entomological Society of America*, **83**: 277–286.

Rosenheim, J.A. & Rosen, D. 1991. Foraging and oviposition decisions in the parasitoid *Aphytis lingnanensis*: distinguishing the influences of egg load and experience. *Journal of Animal Ecology*, **60**: 873–893.

Rosenheim, J.A. & Rosen, D. 1992. Influence of egg load and host size on host-feeding behaviour of the parasitoid *Aphytis lingnanensis*. *Ecological Entomology*, **17**: 263–272.

Rosenthal, S.S. & Koehler, C.S. 1971. Intertree distributions of some cynipid (Hymenoptera) galls on *Quercus lobata*. *Annals of the Entomological Society of America*, **64**: 571–574.

Roskam, J.C. 1982. Larval characters of some eurytomid species (Hymenoptera, Chalcidoidea). *Proceedings of the Koninklijke Nederlandse Akademie van Wetenschappen* (C), **85**: 293–305.

Roskam, J.C. 1986. Biosystematics of insects living in female birch catkins. IV. Egg-larval parasitoids of the genera *Platygaster* Latreille and *Metaclisis* Förster (Hymenoptera, Platygastridae). *Tijdschrift voor Entomologie*, **129**: 125–140.

Ross, K.G. & Carpenter, J.M. 1991. Phylogenetic analysis and the evolution of queen number in eusocial Hymenoptera. *Journal of Evolutionary Biology*, **4**: 117–130.

Ross, K.G. & Matthews, R.W. 1989*a*. Population genetic structure and social evolution in the sphecid wasp *Microstigmus comes*. *American Naturalist*, **134**: 574–598.

Ross, K.G. & Matthews, R.W. 1989*b*. New evidence for eusociality in the sphecid wasp *Microstigmus comes*. *Animal Behavior*, **38**: 613–619.

Ross, K.G. & Matthews R.W. (eds) 1991. *The Social Biology of Wasps.* 678pp. Cornell University Press, Ithaca.

Rössler, Y. & DeBach, P. 1973. Genetic variability in a thelytokous form of *Aphytis mytilaspidis* (Le Baron) (Hymenoptera: Aphelinidae). *Hilgardia,* **42**: 149–175.

Roth, L.M. & Willis, E.R. 1954. *Anastatus floridanus* (Hymenoptera: Eupelmidae) a new parasite on the eggs of the cockroach *Eurycotis floridana. Transactions of the American Entomological Society,* **80**: 29–41.

Roth, P. 1949. Beiträge zur Biologie der Gallwespen. *Verhandlungen der Naturforschenden Gesellshaft in Basel,* **60**: 104–178.

Rotheram, S. 1967. Immune surface of eggs of a parasitic insect. *Nature,* **214**: 700.

Rotheram, S. 1973. The surface of the egg of a parasitic insect. II. The ultrastructure of the particulate coat on the egg of *Nemeritis. Proceedings of the Royal Society, London* (B), **183**: 195–204.

Rotheray, G.E. 1979. The biology and host searching behaviour of a cynipoid parasite of aphidophagous syrphid larvae. *Ecological Entomology,* **4**: 75–82.

Rotheray, G.E. 1981*a*. Courtship, male swarms and a sex pheromone of *Diplazon pectoratorius* (Thunberg) (Hymenoptera: Ichneumonidae). *Entomologist's Gazette,* **32**: 193–196.

Rotheray, G.E. 1981*b*. Host searching and oviposition behaviour by some parasitoids of aphidophagous Syrphidae. *Ecological Entomology,* **6**: 79–87.

Rotheray, G.E. 1984. Host relations, life cycles and multiparasitism in some parasitoids of aphidophagous Syrphidae (Diptera). *Ecological Entomology,* **9**: 303–310.

Rothschild, G.H.L. 1970. Parasites of rice stemborers in Sarawak (Malaysian Borneo). *Entomophaga,* **15**: 21–51.

Roubik, D.W. 1983. Nest and colony characteristics of stingless bees from Panamá (Hymenoptera: Apidae). *Journal of the Kansas Entomological Society,* **56**: 327–355.

Roubik, D.W. 1989. *Ecology and Natural History of Tropical Bees.* 514pp. Cambridge University Press, Cambridge.

Roubik, D.W. 1990. A mixed colony of *Eulaema* (Hymenoptera: Apidae), natural enemies, and limits to sociality. *Journal of the Kansas Entomological Society,* **63**: 150–157.

Roubik, D.W. 1992. Stingless bees: a guide to Panamanian and Mesoamerican species and their nests (Hymenoptera: Apidae: Meliponinae), pp. 495–524. *In*: Quintero, D. & Aiello, A. *Insects of Panama and Mesoamerica,* 692pp. Oxford University Press, Oxford.

Roubik, D.W. & Ackerman, J.D. 1987. Long-term ecology of euglossine orchid-bees (Apidae: Euglossini) in Panama. *Oecologia,* **73**: 321–333.

Roubik, D.W. & Michener, C.D. 1984. Nesting biology of *Crawfordapis* in Panamá (Hymenoptera, Colletidae). *Journal of the Kansas Entomological Society,* **57**: 662–671.

Roubik, D.W., Moreno, J.E., Vergara, C. & Wittmann, D. 1986. Sporadic food competition with the African honey bee: projected impact on neotropical social bees. *Journal of Tropical Ecology,* **2**: 97–111.

Roubik, D.W., Smith, B.H. & Carlson, R.G. 1987. Formic acid in caustic cephalic secretions of stingless bee, *Oxytrigona* (Hymenoptera: Apidae). *Journal of Chemical Ecology,* **13**: 1079–1086.

Rozen, J.G. Jr 1966. The larvae of the Anthophoridae (Hymenoptera, Apoidea) Part 2. The Nomadinae. *American Museum Novitates,* **224**: 1–38.

Rozen, J.G. Jr 1977. Immature stages of and ethological observations on the cleptoparasitic bee tribe Nomadini (Apoidea, Anthophoridae). *American Museum Novitates.* **2638**: 1–16.

Rozen, J.G. Jr 1989. Life history studies of the 'primitive' panurgine bees (Hymenoptera: Andrenidae: Panurginae). *American Museum Novitates.* **2962**: 1–27.

Rubio Espina, E. 1976 (1974/5). Revisión del género *Trachypus* Klug (Hymenoptera: Sphecidae). *Revista de la Facultad de Agronomía*, Universidad del Zulia, **3**: 7–87.

Ruddle, K. 1973. The human use of insects: examples from the Yukpa. *Biotropica*, **5**: 94–101.

Ruiz A., E.R. & Korytkowski G., C.A. 1979. Contribucion al conocimiento de los Trichogrammatidae (Hymenoptera: Chalcidoidea) del Peru. *Revista Peruana de Entomologia*, **22**: 1–12.

Rust, R.W., Hanks, L.M. & Bechtel, R.C. 1983. Aculeata Hymenoptera of Sand Mountain and Blow Sand Mountains, Nevada. *Great Basin Naturalist*, **43**: 403–408.

Rust, R., Menke, A. & Miller, D. 1985. A biogeographic comparison of the bees, sphecid wasps, and mealybugs of the California Channel Islands (Hymenoptera, Homoptera), pp. 29–58. *In*: Menke, A.S. & Miller, D.R. (eds). *Entomology of the California Channel Islands*. 178pp. Santa Barbara Museum of Natural History, Santa Barbara.

Ruttner, F. 1988. *Biogeography and Taxonomy of Honeybees*. 284pp. Springer-Verlag, Berlin.

Rutz, D.A. & Patterson, R.S. (eds) 1990. *Biocontrol of Arthropods Affecting Livestock and Poultry*. 316pp. Westview Press, Boulder.

Ryan, R.B. & Rudinsky, J.A. 1962. Biology and habits of the douglas-fir beetle parasite, *Coeloides brunneri* Viereck (Hymenoptera: Braconidae), in western Oregon. *Canadian Entomologist*, **94**: 748–763.

Saffer, B. 1977. A new species of *Cenocoelius* from Costa Rica (Hymenoptera: Braconidae). *Proceedings of the Entomological Society of Washington*, **79**: 593–596.

Saffer, B. 1982. A systematic revision of the genus *Cenocoelius* (Hymenoptera, Braconidae) in North America including Mexico. *Polskie Pismo Entomologiczne*, **52**: 73–167.

Sage, R.D. 1968. Observations on feeding, nesting, and territorial behavior of carpenter bees genus *Xylocopa* in Costa Rica. *Annals of the Entomological Society of America*, **61**: 884–889.

Sahad, K.A. 1982. Biology and morphology of *Gonatocerus* sp. (Hymenoptera, Mymaridae), an egg parasitoid of the green rice leafhopper, *Nephotettix cincticeps* Uhler (Homoptera, Deltocephalidae). *Kontyû*, **50**: 467–476.

Sahad, K.A. 1984. Biology of *Anagrus optabilis* (Perkins) (Hymenoptera, Mymaridae), an egg parasitoid of delphacid planthoppers. *Esakia*, **22**: 129–144.

Sakagami, S.F. 1982. Stingless bees, pp. 361–423. *In*: Hermann, H.R. (ed.) *Social Insects*, Volume III. 459pp. Academic Press, New York.

Sakagami, S.F., Gobbi, N. & Zucchi, R. 1990. Nesting biology of a quasisocial sphecid wasp *Trypoxylon fabricator*. I. Nests and inhabitants. *Japanese Journal of Entomology*, **58**: 846–862.

Sakagami, S.F. & Okazawa, T. 1985. A populous nest of the halictine bee *Halictus* (*Seladonia*) *lutescens* from Guatemala (Hymenoptera, Halictidae). *Kontyû*, **53**: 645–651.

Sakagami, S.F. & Michener, C.D. 1962. *The Nest Architecture of the Sweat Bees (Halictinae), A Comparative Study of Behavior*. 135pp. University of Kansas Press, Lawrence, Kansas.

Sakagami, S.F. & Michener, C.D. 1987. Tribes of Xylocopinae and origin of the Apidae (Hymenoptera: Apoidea). *Annals of the Entomological Society of America*, **80**: 439–450.

Sakagami, S.F. & Zucchi, R. 1965. Winterverhalten einer neotropischen Hummel, *Bombus atratus*, innerhalb des Beobachtungskastens. Ein Beitrag zur Biologie der

Hummeln. *Journal of the Faculty of Science, Hokkaido University, Series VI, Zoology*, **15**: 712–762.

Salkeld, E.H. 1959. Notes on anatomy, life-history, and behaviour of *Aphaereta pallipes* (Say) (Hymenoptera: Braconidae), a parasite of the onion maggot, *Hylemya antiqua* (Meig.). *Canadian Entomologist*, **91**: 93–97.

Salt, G. 1929. A contribution to the ethology of the Meliponinae. *Transactions of the Entomological Society of London*, **77**: 431–470.

Salt, G. 1935. Experimental studies in insect parasitism III—Host selection. *Proceedings of the Royal Society of London* (B), **117**: 413–435.

Salt, G. 1937. The egg-parasite of *Sialis lutaria*: a study in the influence of a host upon a dimorphic parasite. *Parasitology*, **29**: 539–553.

Salt, G. 1938. Experimental studies in insect parasitism. VI.—Host suitability. *Bulletin of Entomological Research*, **29**: 223–246.

Salt, G. 1952. Trimorphism in the ichneumonid parasite *Gelis corruptor*. *Quarterly Journal of Microscopical Science*, **93**: 453–475.

Salt, G. 1961. Competition among insect parasitoids. *Symposia of the Society for Experimental Biology*, **15**: 96–119.

Salt, G. 1964. The ichneumonid parasite *Nemeritis canescens* (Gravenhorst) in relation to the wax moth *Galleria mellonella* (L.). *Transactions of the Royal Entomological Society of London*, **116**: 1–14.

Salt, G. 1968. The resistance of insect parasitoids to the defence reactions of their hosts. *Biological Reviews of the Cambridge Philosophical Society*, **43**: 200–232.

Salt, G. 1970. *The Cellular Defence Reactions of Insects*. Cambridge Monographs in Experimental Biology, 16. 118pp. Cambridge University Press, Cambridge.

Salt, G. 1971. Teratocytes as a means of resistance to cellular defence reactions. *Nature*, **232**: 639.

Sanders, G.E. 1911. Notes on the breeding of *Tropidopria conica* Fabr. *Canadian Entomologist*, **43**: 48–50.

Sandlan, K. 1979. Sex ratio regulation in *Coccygomimus turionella* Linnaeus (Hymenoptera: Ichneumonidae) and its ecological implications. *Ecological Entomology*, **4**: 365–378.

Sato, Y. & Ohsaki, N. 1987. Host-habitat location by *Apanteles glomeratus* and effect of food-plant exposure on host-parasitism. *Ecological Entomology*, **12**: 291–297.

Sato, Y., Tagawa, J. & Hidaka, T. 1986. Effects of the gregarious parasitoids, *Apanteles ruficrus* and *A. kariyai*, on host growth and development. *Journal of Insect Physiology*, **32**: 281–286.

Sato, Y. & Tanaka, T. 1984. Effect of the number of parasitoid (*Apanteles kariyai*) eggs (Hym.: Braconidae) on the growth of host (*Leucania separata*) larvae (Lep.: Noctuidae). *Entomophaga*, **29**: 21–28.

Schatz, G.E. 1990. Some aspects of pollination biology in Central American forests, pp. 69–84. *In*: Bawa, K.S. & Hadley, M. (eds), *Reproductive Ecology of Tropical Forest Plants*. 421pp. UNESCO, Paris, and The Parthenon Publishing Group, Carnforth (UK).

Schauff, M.E. 1981. A review of nearctic species of *Acmopolynema* Ogloblin (Hymenoptera: Mymaridae). *Proceedings of the Entomological Society of Washington*, **83**: 444–460.

Schauff, M.E. 1983. A new genus of Mymaridae (Hymenoptera: Chalcidoidea) from the New World. *Proceedings of the Entomological Society of Washington*, **85**: 543–551.

Schauff, M.E. 1984. The holarctic genera of Mymaridae (Hymenoptera: Chalcidoidea). *Memoirs of the Entomological Society of Washington*, **12**: 1–67.

Schauff, M.E. 1985*a*. The new world genus *Paracrias* Ashmead (Hymenoptera: Eulophidae). *Proceedings of the Entomological Society of Washington*, **87**: 98–109.

Schauff, M.E. 1985*b*. Taxonomic study of the nearctic species of *Elachertus* Spinola (Hymenoptera: Eulophidae). *Proceedings of the Entomological Society of Washington*, **87**: 843–858.

Schauff, M.E. 1985*c*. Revision of the nearctic species of *Hyssopus* Girault (Hymenoptera: Eulophidae). *Journal of the New York Entomological Society*, **93**: 1096–1108.

Schauff, M.E. 1986. *Microdonophagus*, a new entedontine genus (Hymenoptera: Eulophidae) from Panama. *Proceedings of the Entomological Society of Washington*, **88**: 167–173.

Schauff, M.E. 1987. Taxonomy and identification of the egg parasites (Hymenoptera: Platygastridae, Trichogrammatidae, Mymaridae, and Eulophidae) of citrus weevils (Coleoptera: Curculionidae). *Proceedings of the Entomological Society of Washington*, **89**: 31–42.

Schauff, M.E. 1988. The species of *Entedon* in America north of Mexico (Hymenoptera: Eulophidae). *Journal of the New York Entomological Society*, **96**: 30–62.

Schauff, M.E. 1991. The Holarctic genera of Entedoninae (Hymenoptera: Eulophidae). *Contributions of the American Entomological Institute*, **26**: 1–109.

Schauff, M.E. & Bouček, Z. 1987. *Alachua floridensis*, a new genus and species of Entedoninae (Hymenoptera: Eulophidae) parasitic on the Florida carpenter ant, *Camponotus abdominalis* (Formicidae). *Proceedings of the Entomological Society of Washington*, **89**: 660–664.

Schauff, M.E. & LaSalle, J. 1993. Nomenclatural notes on genera of North American Eulophidae (Hymenoptera: Chalcidoidea). *Proceedings of the Entomological Society of Washington*, **95**: 488–503.

Schedl, W. 1991. Hymenoptera, Unterordnung Symphyta. *Handbuch der Zoologie, IV. Arthropoda: Insecta*, **31**: 1–136.

Schell, S.C. 1943. The biology of *Hadronotus ajax* Girault (Hymenoptera-Scelionidae), a parasite in the eggs of squash-bug (*Anasa tristis* DeGeer). *Annals of the Entomological Society of America*, **36**: 625–635.

Schemske, D.W. & Lande, R. 1984. Fragrance collection and territorial display by male orchid bees. *Animal Behavior*, **32**: 935–937.

Scherbakov, D.E. 1980. Morphology of pterothoracal pleura in Hymenoptera. 1. Ground plan. *Zoologicheskii Zhurnal* **59**: 1644–1653. [In Russian; English summary.]

Scherbakov, D.E. 1981. Morphology of pterothoracal pleura in Hymenoptera. 2. Modifications of the ground plan. *Zoologicheskii Zhurnal* **60**: 205–213. [In Russian; English summary.]

Schildknecht, H. & Koob, K. 1970. Plant bioregulators in the metathoracic glands of myrmicine ants. *Angewandte Chemie*, International Edition in English, **9**: 173.

Schildknecht, H. & Koob, K. 1971. Myrmicacin, the first insect herbicide. *Angewandte Chemie*, International Edition in English, **10**: 124–125.

Schildknecht, H., Reed, P.B., Reed, F.D. & Koob, K. 1973. Auxin activity in the symbiosis of leaf-cutting ants and their fungus. *Insect Biochemistry*, **3**: 439–442.

Schlinger, E.I. & Hall, J.C. 1960. The biology, behavior and morphology of *Praon palitans* Muesebeck, an internal parasite of the spotted alfalfa aphid, *Therioaphis maculata* (Buckton) (Hymenoptera: Braconidae, Aphidiinae). *Annals of the Entomological Society of America*, **53**: 144–160.

Schlinger, E.I. & Hall, J.C. 1961. The biology, behavior and morphology of *Trioxys utilis*, an internal parasite of the spotted alfalfa aphid, *Therioaphis maculata* (Buckton)

(Hymenoptera: Braconidae, Aphidiinae). *Annals of the Entomological Society of America*, **54**: 34–45.

Schlising, R.A. 1970. Sequence and timing of bee foraging in flowers of *Ipomoea* and *Aniseia* (Convolvulaceae). *Ecology*, **51**: 1061–1067.

Schmidt, J.M. 1992 (1991). The role of physical factors in tritrophic interactions. *Redia*, **74**(3) Appendice: 43–93.

Schmidt, J.M. & Smith, J.J.B. 1985*a*. Host volume measurement by the parasitoid wasp *Trichogramma minutum*: the roles of curvature and surface area. *Entomologia Experimentalis et Applicata*, **39**: 213–221.

Schmidt, J.M. & Smith, J.J.B. 1985*b*. The mechanism by which the parasitoid wasp *Trichogramma minutum* responds to host clusters. *Entomologia Experimentalis et Applicata*, **39**: 287–294.

Schmidt, J.M. & Smith, J.J.B. 1987. Short interval time measurement by a parasitoid wasp. *Science*, **237**: 903–905.

Schmidt, J.O. 1986. Chemistry, pharmacology, and chemical ecology of ant venoms, pp. 425–508. *In*: Piek, T. (ed.). *Venoms of the Hymenoptera, Biochemical, Pharmacological and Behavioural Aspects*, 570pp. Academic Press, London.

Schmidt, J.O. 1990. Hymenopteran venoms: striving toward the ultimate defense against vertebrates, pp. 387–419. *In*: Evans, D.L. & Schmidt, J.O. (eds), *Insect Defenses. Adaptive Mechanisms and Strategies of Prey and Predators*. 482pp. State University of New York Press, Albany.

Schmidt, K. & Kuhbandner, B. 1983. Ontogeny of the sensilla placodea on the antennae of *Aulacus striatus* Jurine (Hymenoptera: Aulacidae). *International Journal of Insect Morphology and Embryology*, **12**: 43–57.

Schmidt, O., Andersson, K., Will, A. & Schuchmann-Feddersen, I. 1990. Viruslike particle proteins from a hymenopteran endoparasitoid are related to a protein component of the immune system in the lepidopteran host. *Archives of Insect Biochemistry and Physiology*, **13**: 107–115.

Schmidt, O. & Schuchmann-Feddersen, I. 1989. Role of virus-like particles in parasitoid-host interaction of insects, pp. 91–119. *In*: Harris, J.R. (ed.) *Subcellular Biochemistry, Volume 15 Virally Infected Cells*. 450pp. Plenum Press, New York.

Schneider, F. 1950. Die Entwicklung des Syrphidenparasiten *Diplazon fissorius* Grav. (Hym., Ichneum.) in uni-, oligo- und polyvoltinen Wirten und sein Verhalten bei parasitärer Aktivierung der Diapauselarven durch *Diplazon pectoratorius* Grav. *Mitteilungen der Schweizerischen Entomologischen Gesellschaft*, **23**: 155–194.

Schneider, F. 1951. Einige physiologische Beziehungen zwischen Syrphidenlarven und ihren Parasiten. *Zeitschrift für Angewandte Entomologie*, **33**: 150–162.

Schneirla, T.C. 1971. *Army Ants: a Study in Social Organization*. 349pp. Freeman, San Francisco.

Schoenly, K. 1990. The predators of insects. *Ecological Entomology*, **15**: 333–345.

Schremmer, F. 1960. Beitrag zur Biologie der in Stratiomyiidenlarven parasitierenden Chalcididen der gattung *Smicra* Spin. (*Chalcis* F.). *Zeitschrift der Arbeitsgemeinschaft Oesterreichischer Entomologen*, **12**: 83–89.

Schremmer, F. 1979. Das Nest der neotropischen Weberameise *Camponotus* (*Myrmobrachys*) *senex* Smith (Hymenoptera, Formicidae). *Zoologischer Anzeiger*, **203**: 273–282.

Schröder, C. 1928–29. Die psychischen Fähigkeiten der Insekten mit exkurs in die probleme der allgemeinen biologie, pp. 1059–1292. *In*: Schröder, C. (ed.) *Handbuch der Entomologie* 2. 1410pp. Gustav Fischer, Jena.

Schulz, W.A. 1907. Hymenoptera, Fam. Trigonaloidae, pp. 1–24. *In*: Wytsman, P. *Genera Insectorum*. Fasc 61. 24 pp.

Schumacher, M.J., Schmidt, J.O. & Egen, N.B. 1989. Lethality of 'killer' bee stings. *Nature*, **337**: 413.

Schupp, E.W. 1986. *Azteca* protection of *Cecropia*: ant occupation benefits juvenile trees. *Oecologia*, **70**: 379–385.

Schwarz, H.F. 1948. Stingless bees (Meliponidae) of the Western Hemisphere. *Bulletin of the American Museum of Natural History*, **90**: 1–546.

Scoble M.J. 1992. *The Lepidoptera*. 404pp. Oxford University Press, Oxford.

Scudder, G.G.E. 1961. The comparative morphology of the insect ovipositor. *Transactions of the Royal Entomological Society of London*, **113**: 25–40.

Scullen, H.A. & Wold, J.L. 1969. Biology of wasps of the tribe Cercerini, with a list of the Coleoptera used as prey. *Annals of the Entomological Society of America*, **62**: 209–214.

Seeley, T.D. 1985. *Honey Bee Ecology, A Study of Adaptation in Social Life*. 210pp, Princeton University Press, Princeton, New Jersey.

Seeley, T.D. & Towne, W.F. 1992. Tactics of dance choice in honey bees: do foragers compare dances? *Behavioral Ecology and Sociobiology*, **30**: 59–69.

Seitner, M. & Nötzl, P. 1925. *Pityophthorus henscheli* Seitner und sein Parasit *Cosmophorus henscheli* Ruschka. *Zeitschrift für Angewandte Entomologie*, **11**: 187–196.

Séméria, Y. 1976. Un braconide, *Chrysopophthorus chrysopimaginis* Goidanich (Hymenoptera, Braconidae) parasite des imagos de Chrysopinae (Planipennia, Chrysopidae). *Bulletin Mensual de la Société Linnéenne de Lyon* (N.S.), **45**: 102–127.

Sevcik, C. & Hernandez, C. J. 1990. Some findings on neurotoxins from the venom of the giant ant, *Paraponera clavata*, pp. 461–471. *In*: Vander Meer, R.K., Jaffe, K. & Cedeno, A. (eds) *Applied Myrmecology, A World Perspective*, 741pp. Westview Press, Boulder.

Shafer, G.D. 1949. *The Ways of a Mud-Dauber*. 78pp. Stanford University Press, Stanford.

Shanks, S.S. 1986 (1987). A revision of the neotropical bee genus *Osiris* (Hymenoptera: Anthophoridae). *Wasmann Journal of Biology*, **44**: 1–56.

Sharkey, M.J. 1983. *Marjoriella*, a new neotropical genus of Agathidinae (Braconidae: Hymenoptera). *Contributions of the American Entomological Institute*, **20**: 94–100.

Sharkey, M.J. 1985. Notes on the genera *Bassus* Fabricius and *Agathis* Latreille, with a description of *Bassus arthurellus* n. sp. (Hymenoptera: Braconidae). *Canadian Entomologist*, **117**: 1497–1502.

Sharkey, M.J. 1986*a*. The phylogenetic affinities of *Mesocoelus* Schulz (Agathidinae: Braconidae: Hymenoptera). *Canadian Entomologist*, **118**: 283–286.

Sharkey, M.J. 1986*b*. *Pharpa*, a new genus of neotropical Agathidinae (Hymenoptera: Braconidae) with a discussion of phylogentic relationships. *Canadian Entomologist*, **118**: 1231–1239.

Sharkey, M.J. 1988. A taxonomic revision of *Alabagrus* (Hymenoptera: Braconidae). *Bulletin of the British Museum (Natural History)* Entomology, **57**: 311–437.

Sharkey, M.J. 1990. A revision of *Zacremnops* Sharkey and Wharton (Hymenoptera: Braconidae: Agathidinae). *Proceedings of the Entomological Society of Washington*, **92**: 561–570.

Sharkey, M.J. & Mason, W.R.M. 1986. The generic validity of *Aenigmostomus* and *Asiacardiochiles* (Hymenoptera: Braconidae). *Proceedings of the Entomological Society of Washington*, **88**: 300–302.

Sharkey, M.J. & Wahl, D.B. 1992. Cladistics of the Ichneumonoidea (Hymenoptera). *Journal of Hymenoptera Research*, **1**: 15–24.

Sharkey, M.J. & Wharton, R.A. 1985. Redefinition of *Megagathis* Kriechbaumer, and reassignment of new world species to *Zacremnops* new genus (Hymenoptera: Braconidae: Agathidinae). *Canadian Entomologist*, **117**: 599–603.

Shattuck, S.O. 1992*a*. Higher classification of the ant subfamilies Aneuretinae, Dolichoderinae and Formicinae (Hymenoptera: Formicidae). *Systematic Entomology*, **17**: 199–206.

Shattuck, S.O. 1992*b*. Generic revision of the ant subfamily Dolichoderinae (Hymenoptera: Formicidae). *Sociobiology*, **21**: 1–181.

Shaw, M.R. 1981. Delayed inhibition of host development by nonparalyzing venoms of parasitic wasps. *Journal of Invertebrate Pathology*, **37**: 215–221.

Shaw, M.R. 1983. On(e) evolution of endoparasitism: the biology of some genera of Rogadinae (Braconidae). *Contributions of the American Entomological Institute*, **20**: 307–328.

Shaw, M.R. & Askew, R.R. 1979. Hymenopterous parasites of Diptera (Hymenoptera Parasitica), pp. 164-171. *In*: Stubbs, A. & Chandler, P. (eds) A Dipterist's Handbook, *The Amateur Entomologist* (1978) **15**: 1–255.

Shaw, M.R. & Huddleston, T. 1991. Classification and biology of braconid wasps (Hymenoptera: Braconidae). *Handbooks for the Identification of British Insects*, **7**(11): 1–126.

Shaw, M.R. & Wahl, D.B. 1989. The biology, egg and larvae of *Acaenitus dubitator* (Panzer) (Hymenoptera, Ichneumonidae: Acaenitinae). *Systematic Entomology*, **14**: 117–125.

Shaw, S.R. 1983. A taxonomic study of Nearctic *Ascogaster* and a description of a new genus *Leptodrepana* (Hymenoptera: Braconidae). *Entomography*, **2**: 1–54.

Shaw, S.R. 1985. A phylogenetic study of the subfamilies Meteorinae and Euphorinae (Hymenoptera: Braconidae). *Entomography*, **3**: 277–370.

Shaw, S.R. 1987. *Orionis*, a new genus from Central America, with an analysis of its phylogenetic placement in the tribe Euphorini (Hymenoptera: Braconidae). *Systematic Entomology*, **12**: 103–109.

Shaw, S.R. 1988*a*. Euphorine phylogeny: the evolution of diversity in host-utilization by parasitoid wasps (Hymenoptera: Braconidae). *Ecological Entomology*, **13**: 323–335.

Shaw, S.R. 1988*b*. A new Mexican genus and species of Dinocampini with serrate antennae (Hymenoptera: Braconidae: Euphorinae). *Psyche*, **95**: 289–297.

Shaw, S.R. 1990. Phylogeny and biogeography of the parasitoid wasp family Megalyridae (Hymenoptera). *Journal of Biogeography*, **17**: 569–581.

Shaw, S.R. 1991. An unusual manner of aggregation in the braconid *Chelonus (Microchelonus) hadrogaster* McComb (Hymenoptera). *Journal of Insect Behavior*, **4**: 537–542.

Shaw, S.R. 1992. Seven new North American species of *Neoneurus* (Hymenoptera: Braconidae). *Proceedings of the Entomological Society of Washington*, **94**: 26–47.

Shaw, S.R. 1993. Observations on the ovipositional behavior of *Neoneurus mantis*, an ant-associated parasitoid from Wyoming (Hymenoptera: Braconidae). *Journal of Insect Behavior*, **6**: 649–658.

Shaw, S.R. & Edgerly, J. 1985. A new braconid genus (Hymenoptera) parasitizing webspinners (Embiidina) in Trinidad. *Psyche*, **92**: 505–511.

Sheehan, W. 1984. Nesting biology of the sand wasp *Stictia heros* (Hymenoptera: Sphecidae: Nyssoninae) in Costa Rica. *Journal of the Kansas Entomological Society*, **57**: 377–386.

Sheehan, W. 1991. Host range patterns of hymenopteran parasitoids of exophytic lepidopteran folivores, pp. 209–249. *In*: Bernays, E. (ed.) *Insect-Plant Interactions, Volume III*. 258pp. CRC Press, Boca Raton.

Sheehan, W. & Hawkins, B.A. 1991. Attack strategy as an indicator of host range in metopiine and pimpline Ichneumonidae (Hymenoptera). *Ecological Entomology*, **16**: 129–131.

Sheehan, W. & Shelton, A.M. 1989. The role of experience in plant foraging by the aphid parasitoid *Diaeretiella rapae* (Hymenoptera: Aphidiidae). *Journal of Insect Behavior*, **2**: 743–759.

Shenefelt, R.D. 1943. The genus *Atanycolus* Foerster in America north of Mexico. *Research Studies of the State College of Washington*, **11**: 51–163.

Shenefelt, R.D. 1969. *Hymenopterorum Catalogus* Pars 4 (nova editio) Braconidae 1, Hybrizoninae, Euphorinae, Cosmophorinae, Neoneurinae, Macrocentrinae. Pp. 1–176. Junk, 's-Gravenhage.

Shenefelt, R.D. 1970*a*. *Hymenopterorum Catalogus* Pars 5 (nova editio) Braconidae 2, Heliconinae (sic), Calyptinae, Mimagathidinae, Triaspinae. Pp. 177–306. Junk, 's-Gravenhage.

Shenefelt, R.D. 1970*b*. *Hymenopterorum Catalogus* Pars 6 (nova editio) Braconidae 3, Agathidinae. Pp. 307–428. Junk, 's-Gravenhage.

Shenefelt, R.D. 1972. *Hymenopterorum Catalogus* Pars 7 (nova editio) Braconidae 4, Microgasterinae *Apanteles*. Pp. 429–668. Junk, 's-Gravenhage.

Shenefelt, R.D. 1973*a*. *Hymenopterorum Catalogus* Pars 9 (nova editio) Braconidae 5, Microgasterinae & Ichneutinae. Pp. 669–812. Junk, 's-Gravenhage.

Shenefelt, R.D. 1973*b*. *Hymenopterorum Catalogus* Pars 10 (nova editio) Braconidae 6, Cheloninae. Pp. 813–936. Junk, 's-Gravenhage.

Shenefelt, R.D. 1974. *Hymenopterorum Catalogus* Pars 11 (nova editio) Braconidae 7, Alysiinae. Pp. 937–1113. Junk, 's-Gravenhage.

Shenefelt, R.D. 1975. *Hymenopterorum Catalogus* Pars 12 (nova editio) Braconidae 8, Exothecinae Rogadinae. Pp. 1115–1262. Junk, 's-Gravenhage.

Shenefelt, R.D. 1978. *Hymenopterorum Catalogus* Pars 15 (nova editio) Braconidae 10, Braconinae, Gnathobraconinae, Mesostoinae, Pseudodicrogeniinae, Telengainae, Ypsistocerinae, plus Braconidae in general, major groups, unplaced genera and species. Pp. 1425–1872. Junk, 's-Gravenhage.

Shenefelt, R.D. 1980. *Hymenopterorum Catalogus* Pars 16 (nova editio) Braconidae 11, Introduction, guide to host names, index to braconid names. 1–384. Junk, The Hague.

Shenefelt, R.D. & Marsh, P.M. 1976. *Hymenopterorum Catalogus* Pars 13 (nova editio) Braconidae 9, Doryctinae. Pp. 1263–1424. Junk, 's-Gravenhage.

Shepherd, J.D. 1982. Trunk trails and the searching strategy of a leaf-cutter ant *Atta colombica*. *Behavioral Ecology and Sociobiology*, **11**: 77–84.

Shepherd, J.D. 1985. Adjusting foraging effort to resources in adjacent colonies of the leaf-cutter ant, *Atta colombica*. *Biotropica*, **17**: 245–252.

Sherman, P.W., Seeley, T.D. & Reeve, H.K. 1988. Parasites, pathogens, and polyandry in social Hymenoptera. *American Naturalist*, **131**: 602–610.

Shetlar, D.J. 1973. A redescription and biology of *Probethylus schwarzi* Ashmead (Hymenoptera: Sclerogibbidae) with notes on related species. *Entomological News*, **84**: 205–210.

Shiga, M. & Nakanishi, A. 1968. Variation in the sex ratio of *Gregopimpla himalayensis* Cameron (Hymenoptera: Ichneumonidae) parasitic on *Marucosoma neustria testacea*

Motschulsky (Lepidoptera: Lasiocampidae), with considerations on the mechanism. *Kontyû*, **36**: 369–376.

Shimizu, A. 1989. An ethological study of *Agenioideus ishikawai* (Hymenoptera, Pompilidae). *Japanese Journal of Entomology*, **57**: 654–662.

Shinn, A.F. 1967. A revision of the bee genus *Calliopsis* and the biology and ecology of *C. andreniformis* (Hymenoptera, Andrenidae). *University of Kansas Science Bulletin*, **46**: 753–936.

Shinohara, A. 1986. A new apterous sawfly from Sulawesi, Indonesia (Hymenoptera: Pergidae: Perreyiinae), and the pleural origin of the ventral region of the sawfly mesothorax. *Systematic Entomology*, **11**: 247–253.

Short, J.R.T. 1952. The morphology of the head of larval Hymenoptera with special reference to the head of the Ichneumonoidea, including a classification of the final instar larvae of the Braconidae. *Transactions of the Royal Entomological Society of London*, **103**: 27–84.

Short, J.R.T. 1978. The final larval instars of the Ichneumonidae. *Memoirs of the American Entomological Institute*, **25**: 1–508.

Shorthouse, J.D. 1982. Resource exploitation by gall wasps of the genus *Diplolepis*, pp. 193–198. *In*: Visser, J.H. & Minks, A.K. (eds) *Proceedings of the 5th International Symposium on Insect-Plant Relationships. Wageningen, The Netherlands. 1–4 March 1982*. 464pp. Pudoc, Wageningen.

Shorthouse, J.D. & Rohfritsch, O. (eds) 1992. *Biology of Insect-Induced Galls*. 480pp. Oxford University Press, Oxford.

Silvestri, F. 1911. Di una nova di *Aleurodes* vivent e sull' olivo. *Bollettino del Laboratorio di Zoologia Generale e Agraria della R. Scuola Superiore d'Agricoltura in Portici*, **5**: 214–225.

Silvestri, F. 1916. Contribuzione alla conoscenza del genere *Poropoea* Förster (Hymenoptera, Chalcididae). *Bollettino del Laboratorio di Zoologia Generale e Agraria in Portici*, **11**: 120–135.

Silvestri, F. 1919. Contribuzioni alla conoscenza degli insetti dannosi e dei loro simbionti. V. La cocciniglia del Nocciuolo (*Eulecanium coryli* L.). *Bollettino del Laboratorio di Zoologia Generale e Agraria della R. Scuola Superiore d'Agricoltura in Portici*, **13**: 127–192.

Silvestri, F. 1920. Contribuzione alla conoscenza de parassiti delle ova del grilletto canterino (*Oecanthus pellucens* Scop., Orthoptera, Achetidae). *Bollettino del Laboratorio di Zoologia Generale e Agraria della R. Scuola Superiore d'Agricoltura in Portici*, **14**: 219–250.

Silvestri, F. 1921. Contribuzioni alla conoscenza biologica degli imenottori parassiti. 5, Sviluppo del *Platygaster dryomyiae* Silv. *Bollettino del Laboratorio di Zoologia Generale e Agraria della R. Scuola Superiore d'Agricultura in Portici*, **11**: 299–326.

Silvestri, F. 1943. Studi sugli 'Strepsiptera' (Insecta). III. Descrizione e biologia di 6 specie italiane de *Mengenilla*. *Bollettino del Laboratorio di Zoologia Generale e Agraria della Facoltá Agraria in Portici*, **32**: 197–282.

Simmonds, F.J. 1947. The biology of the parasites of *Loxostege sticticalis* L., in North America—*Bracon vulgaris* (Cress.) (Braconidae, Agathinae). *Bulletin of Entomological Research*, **38**: 145–155.

Simmonds, F.J. 1952. Parasites of the frit-fly, *Oscinella frit* (L.), in eastern North America. *Bulletin of Entomological Research*, **43**: 503–542.

Simoes, D., Bego, L.R., Zucchi, R. & Sakagami, S.F. 1980. *Melaloncha sinistra* Borgmeier, an endoparasitic phorid fly attacking *Nannotrigona (Scaptotrigona) postica* Latreille (Hymenoptera, Meliponinae). *Revista Brasileira de Entomologia*, **24**: 137–142.

Simons, L.H. 1989. A second record of tarantula parasitism by *Notocyphus dorsalis arizonicus* Townes (Hymenoptera: Pompilidae). *Pan-Pacific Entomologist*, **65**: 34–37.

Simser, D.H. & Coppel, H.C. 1991. An array of spatulate sensilla on antennae of male *Brachymeria lasus* (Hymenoptera: Chalcididae). *The Great Lakes Entomologist*, **24**: 103–108.

Siri, M.L. & Bohart, R.M. 1974. A review of the genus *Mellinus*. *Pan-Pacific Entomologist*, **50**: 169–176.

Sivinski, J. & Webb, J.C. 1989. Acoustic signals produced during courtship in *Diachasmimorpha* (= *Biosteres*) *longicaudata* (Hymenoptera: Braconidae) and other Braconidae. *Annals of the Entomological Society of America*, **82**: 116–120.

Skinner, E.R. & Thompson, G.H. 1960. Film: *The Alder woodwasp and its Insect Enemies*.

Skinner, S.W. 1982. Maternally-inherited sex ratio in the parasitoid wasp *Nasonia vitripennis*. *Science*, **215**: 1133–1134.

Skinner, S.W. 1985. Clutch size as an optimal foraging problem for insects. *Behavioral Ecology and Sociobiology*, **17**: 231–238.

Skrzypczynska, M. 1978. *Megastigmus suspectus* Borries, 1895 (Hymenoptera, Torymidae), its morphology, biology and economic significance. *Zeitschrift für Angewandte Entomologie*, **85**: 204–215.

Slansky, F. Jr 1986. Nutritional ecology of endoparasitic insects and their hosts: an overview. *Journal of Insect Physiology*, **32**: 255–261.

Slansky, F. Jr & Scriber, J.M. 1985. Food consumption and utilization, pp. 87–163. *In*: Kerkut, G.A. & Gilbert, L.I. (eds), *Comprehensive Insect Physiology, Biochemistry and Pharmacology, Volume 4. Regulation: Digestion, Nutrition, Excretion*. 639pp. Pergamon Press, Oxford.

Slobodchikoff, C.N. 1967. Bionomics of *Grotea californica* Cresson, with a description of the larva and pupa (Hymenoptera: Ichneumonidae). *Pan-Pacific Entomologist*, **43**: 161–168.

Slobodchikoff, C.N. 1970. A revision of the genus *Grotea* (Hymenoptera: Ichneumonidae). *Pan-Pacific Entomologist*, **46**: 50–63.

Slobodchikoff, C.N. 1973. Behavioral studies of three morphotypes of *Therion circumflexum* (Hymenoptera: Ichneumonidae). *Pan-Pacific Entomologist*, **49**: 197–206.

Slovák, M. 1984. New data about developmental stages of *Exetastes cinctipes* (Hym., Ichneumonidae). *Biológia*, **39**: 611–616.

Slovák, M. 1986*a*. Longevity of adults and fecundity in *Exetastes cinctipes* (Hym., Ichneumonidae). *Biológia*, **41**: 105–114.

Slovák, M. 1986*b*. Mating behaviour in laboratory reared *Exetastes cinctipes* (Hym., Ichneumonidae). *Biológia*, **41**: 543–548.

Slüss, R. 1968. Behavioral and anatomical responses of the convergent lady beetle to parasitism by *Perilitus coccinellae* (Schrank) (Hymenoptera: Braconidae). *Journal of Invertebrate Pathology*, **10**: 9–27.

Slüss, R.R. & Leutenegger, R. 1968. The fine structure of the trophic cells of *Perilitus coccinellae* (Hymenoptera: Braconidae). *Journal of Ultrastructure Research*, **25**: 441–451.

Smiley, J. 1986. Ant constancy at *Passiflora* extrafloral nectaries: effects on caterpillar survival. *Ecology*, **67**: 516–521.

Smith, A.P. 1978. An investigation of the mechanisms underlying nest construction in the mud wasp *Paralastor* sp. (Hymenoptera: Eumenidae). *Animal Behaviour*, **26**: 232–240.

Smith, B.H. & Roubik, D.W. 1983. Mandibular glands of stingless bees (Hymenoptera: Apidae): chemical analysis of their contents and biological function in two species of *Melipona*. *Journal of Chemical Ecology*, **9**: 1465–1472.

Smith, D.R. 1969*a*. Nearctic sawflies 1. Blennocampinae: adults and larvae (Hymenoptera: Tenthredinidae). *United States Department of Agriculture, Technical Bulletin*, **1397**: 1–179.

Smith, D.R. 1969*b*. Nearctic sawflies II. Selandriinae: adults (Hymenoptera: Tenthredinidae). *United States Department of Agriculture, Technical Bulletin*, **1398**: 1–48.

Smith, D.R. 1973. Sawflies of Chile: a new genus and species and key to genera of Tenthredinidae (Hymenoptera: Symphyta). *Proceedings of the Entomological Society of Washington*, **75**: 402–408.

Smith, D.R. 1976. The xiphydriid woodwasps of North America (Hymenoptera: Xiphydriidae). *Transactions of the American Entomological Society*, **102**: 101–131.

Smith, D.R. 1979. Nearctic sawflies IV. Allantinae: adults and larvae (Hymenoptera: Tenthredinidae). *United States Department of Agriculture, Technical Bulletin*, **1595**: 1–172.

Smith, D.R. 1980. Identification of the *Acordulecera* "potato" sawflies of Peru and Bolivia, with descriptions of these and related species from South America (Hymenoptera: Pergidae). *Journal of the Washington Academy of Sciences*, **70**: 89–103.

Smith, D.R. 1988. A synopsis of the sawflies (Hymenoptera: Symphyta) of America south of the United States: introduction, Xyelidae, Pamphiliidae, Cimbicidae, Diprionidae, Xiphydriidae, Siricidae, Orussidae, Cephidae. *Systematic Entomology*, **13**: 205–261.

Smith, D.R. 1989. The sawfly genus *Arge* (Hymenoptera: Argidae) in the Western Hemisphere. *Transactions of the American Entomological Society*, **115**: 83–205.

Smith, D.R. 1990. A synopsis of the sawflies (Hymenoptera, Symphyta) of America south of the United States: Pergidae. *Revista Brasileira de Entomologia*, **34**: 7–200.

Smith, D.R. 1991. A new *Aulacus* (Hymenoptera: Gasteruptionidae: Aulacinae) from Virginia [USA]. *Entomological News*, **102**:187–191.

Smith, D.R. 1992. A synopsis of the sawflies (Hymenoptera: Symphyta) of America south of the United States: Argidae. *Memoirs of the American Entomological Society*, **39**: 1–201.

Smith, D.R. 1993. Systematics, life history, and distribution of sawflies, pp. 3–32. *In*: Wagner, M.R. & Raffa, K.F. (eds) *Sawfly Life History Adaptations to Woody Plants*. 564pp. Academic Press, San Diego.

Smith, E.L. 1970. Evolutionary morphology of the external insect genitalia. 2. Hymenoptera. *Annals of the Entomological Society of America*, **63**: 1–27.

Smith, H.D. 1932. *Phaeogenes nigridens* Wesmael, an important Ichneumonid parasite of the pupa of the European corn borer. *Technical Bulletin of the United States Department of Agriculture*, **331**: 1–45.

Smith, H.S. 1912. Technical results from the gypsy moth parasite laboratory. IV. The chalcidoid genus *Perilampus* and its relations to the problem of parasite introduction. *United States Department of Agriculture, Bureau of Entomology, Technical Series*, **19**: 33–69.

Smith, H.S. 1917. On the life-history and successful introduction into the United States of the Sicilian mealy-bug parasite. *Journal of Economic Entomology*, **10**: 262–268.

Smith, H.S. 1929. Multiple parasitism: its relation to the biological control of insect pests. *Bulletin of Entomological Research*, **20**: 141–149.

Smith, H.S. & Compere, H. 1928. A preliminary report on the insect parasites of the black scale, *Saissetia oleae* (Bernard). *University of California Publications in Entomology*, **4**: 231–334.

Smith, J.W. Jr, Rodriguez del Bosque, L.A. & Agnew, C.W. 1990. Biology of *Mallochia pyralidis* (Hymenoptera: Ichneumonidae), an ectoparasite of *Eoreuma loftini* (Lepidoptera: Pyralidae) from Mexico. *Annals of the Entomological Society of America*, **83**: 961–966.

Smith, K.G.V. 1974. Rearing the Hymenoptera Parasitica. *Leaflets of the Amateur Entomologists' Society*, **35**: 1–15.

Smith, L.M. 1930. *Macrorileya oecanthi* Ashm. a Hymenopterous egg parasite of tree crickets. *University of California Publications in Entomology*, **5**: 165–172.

Smith, M.V. 1959. The production of royal jelly, *Bee World*, **40**: 250–254.

Smith, O.J. 1952. Biology and behavior of *Microctonus vittatae* Muesebeck (Braconidae) with descriptions of its immature stages. *University of California Publications in Entomology*, **9**: 315–343.

Smith, O.J., Diboll, A.G. & Rosenberger, J.H. 1955. Laboratory studies of *Pelecystoma harrisinae* (Ashmead), an adventive braconid parasite of the western grape leaf skeletonizer. *Annals of the Entomological Society of America*, **48**: 232–237.

Smulyan, M.T. 1936. A revision of the Chalcid flies of the genus *Perilampus* Latreille occurring in America north of Mexico. *Proceedings of the United States National Museum*, **83**: 369–412.

Snelling, R.R. 1982. The taxonomy of some neotropical *Hylaeus* and descriptions of new taxa (Hymenoptera: Colletidae). *Bulletin of the Southern California Academy of Science*, **81**: 1–25.

Snelling, R.R. 1983*a*. Taxonomic and nomenclatural studies on American polistine wasps (Hymenoptera: Vespidae). *Pan-Pacific Entomologist*, **59**: 267–280.

Snelling, R.R. 1983*b*. The North American species of the bee genus *Lithurge* (Hymenoptera: Megachilidae). *Contributions in Science. Natural History Museum of Los Angeles County*, **343**: 1–11.

Snelling, R.R. 1984. Studies on the taxonomy and distribution of American Centridine Bees (Hymenoptera: Anthophoridae). *Contributions in Science. Natural History Museum of Los Angeles County*, **347**: 1–69.

Snelling, R.R. 1986*a*. The taxonomic status of two North American *Lithurge* (Hymenoptera: Megachilidae). *Bulletin of the Southern California Academy of Sciences*, **85**: 29–34.

Snelling, R.R. 1986*b*. Contributions toward a revision of the New World nomadine bees. A partitioning of the genus *Nomada* (Hymenoptera: Anthophoridae). *Contributions in Science. Natural History Museum of Los Angeles County*, **376**: 1–32.

Snelling, R.R. & Brooks, R.W. 1985. A review of the genera of cleptoparasitic bees of the tribe Ericrocini (Hymenoptera: Anthophoridae). *Contributions in Science. Natural History Museum of Los Angeles County*, **369**: 1–34.

Snelling, R.R. & Longino, J.T. 1992. Revisionary notes on the fungus-growing ants of the genus *Cyphomyrmex*, *rimosus* group (Hymenoptera: Formicidae: Attini), pp. 479–494. *In*: Quintero, D. & Aiello, A. (eds) *Insects of Panama and Mesoamerica*. 692pp. Oxford University Press, Oxford.

Snodgrass, R.E. 1935. Principles of Insect Morphology. 667pp. McGraw Hill, New York.

Snodgrass, R.E. 1941. The male genitalia of Hymenoptera. *Smithsonian Miscellaneous Collections*, **99** (14): 1–86.

Snodgrass, R.E. 1956. *Anatomy of the Honey Bee*. 334pp. Comstock Publ. Ass., New York.

Soika, A.G. 1978. Revisione degli Eumenidi neotropicali appartenenti ai generi *Eumenes* Latr., *Omicron* Sauss., *Pararaphidoglossa* Schulth. ed affini. *Bolletino del Museo Civico di Storia Naturale di Venezia*, **29**: 1–420.

Soika, A.G. 1981 (1982). Contributo alla conoscenza del genere neotropicale *Hypalastoroides* Sauss. (Hym. Vespoidea). *Bollettino del Museo Civico di Storia Naturale di Venezia*, **32**: 33–59.

Soika, A.G. 1990. Revisione degli Eumenidi neotropicali appartenenti ai generi *Pachymenes* Sauss., *Santamenes* n. gen., *Brachymenes* G. S., *Pseudacaromenes* G. S., *Stenosigma* G. S., e *Gamma* Zav. (Hymenoptera). *Bollettino del Museo Civico di Storia Naturale di Venezia*, **39**: 71–172.

Sokolowski, M.B. & Turlings, T.C.J. 1987. *Drosophila* parasitoid-host interactions: vibrotaxis and ovipositor searching from the host's perspective. *Canadian Journal of Zoology*, **65**: 461–464.

Soldevila, A.I. & Jones, D. 1993. Expression of a parasitism-specific protein in lepidopteran hosts of *Chelonus* sp. *Archives of Insect Biochemistry and Physiology*, **24**: 149–169.

Sommerman, K.M. 1956. Parasitization of nymphal and adult psocids (Psocoptera). *Proceedings of the Entomological Society of Washington*, **58**: 149–152.

Soper, R.S., Shewell, G.E. & Tyrrel, D. 1976. *Calcondamyia auditrix* nov. sp., (Diptera: Sarcophagidae) a parasite which is attracted by the mating song of its host, *Okanagana rimosa* (Homoptera: Cicadidae). *Canadian Entomologist*, **108**: 61–68.

Sorokina, A.P. 1993. *Key to the species of the genus* Trichogramma *Westw. (Hymenoptera: Trichogrammatidae) of the World Fauna*. 77pp. Kolos Publishing House, Moscow.

Speirs, D.C., Sherratt, T.N. & Hubbard, S.F. 1991. Parasitoid diets: Does superparasitism pay? *Trends in Ecology & Evolution*, **6**: 22–25.

Spence, J.R. 1986. Interactions between the scelionid egg parasitoid *Tiphodytes gerriphagus* (Hymenoptera) and its gerrid hosts (Heteroptera). *Canadian Journal of Zoology*, **64**: 2728–2738.

Spencer, G.J. 1942. A note on *Laelius* sp., a parasite of the carpet beetle *Anthrenus scrophulariae* (L.) (Hymenoptera, Bethylidae). *Proceedings of the Entomological Society of British Columbia*, **39**: 21–22.

Spivak, M., Fletcher, D.J.C. & Breed, M.D. (eds) 1991. The *'African' Honey Bee*. 435pp. Westview Press, Boulder, Colorado.

Spofford, M.G., Kurczewski, F.E. & Peckham, D.J. 1986. Cleptoparasitism of *Tachysphex terminatus* (Hymenoptera: Sphecidae) by three species of Miltogrammini (Diptera: Sarcophagidae). *Annals of the Entomological Society of America*, **79**: 350–358.

Spradbery, J.P. 1969. The biology of *Pseudorhyssa sternata* Merrill (Hym., Ichneumonidae), a cleptoparasite of siricid woodwasps. *Bulletin of Entomological Research*, **59**: 291–297.

Spradbery, J.P. 1970*a*. Host finding by *Rhyssa persuasoria* (L.) an ichneumonid parasite of siricid woodwasps. *Animal Behaviour*, **18**: 103–114.

Spradbery, J.P. 1970*b*. The immature stages of European ichneumonid parasites of siricine woodwasps. *Proceedings of the Royal Entomological Society of London* (A), **45**: 14–28.

Spradbery, J.P. 1973*a*. A comparative study of the phytotoxic effects of siricid woodwasps on conifers. *Annals of Applied Biology*, **75**: 309–320.

Spradbery, J.P. 1973*b*. *Wasps*. 408pp. Sidgwick & Jackson, London.

Spradbery, J.P. 1977. The oviposition biology of siricid woodwasps in Europe. *Ecological Entomology*, **2**: 225–230.

Spradbery, J.P. & Kirk, A.A. 1978. Aspects of the ecology of siricid woodwasps (Hymenoptera: Siricidae) in Europe, North Africa and Turkey with special reference to the biological control of *Sirex noctilio* F. in Australia. *Bulletin of Entomological Research*, **68**: 341–359.

Staden, J. van & Davey, J.E. 1978. Endogenous cytokinins in the laminae and galls of *Erythrina latissima* leaves. *Botanical Gazette*. London, **139**: 36–41.

Stage, G.I. & Snelling, R.R. 1986. The subfamilies of Eurytomidae and systematics of the subfamily Heimbrinae (Hymenoptera: Chalcidoidea). *Contributions in Science. Natural History Museum of Los Angeles County*, **375**: 1–17.

Stamp, N.E. 1981. Behavior of parasitized aposematic caterpillars: advantageous to the parasitoid or the host? *American Naturalist*, **118**: 715–725.

Stamp, N.E. 1982. Behavioral interactions of parasitoids and Baltimore checkerspot caterpillars (*Euphydryas phaeton*). *Environmental Entomology*, **11**: 100–104.

Stanley, R.G. & Linskens, H.F. 1974. *Pollen: Biology, Biochemistry, Management*. 307pp. Springer-Verlag, Berlin.

Starr, C.K. 1979. Origin and evolution of insect sociality: a review of modern theory, pp. 35–79. *In*: Hermann, H.R. (ed.), *Social Insects* Volume I. 437pp. Academic Press, New York.

Starr, C.K. 1985. Enabling mechanisms in the origin of sociality in the Hymenoptera—the sting's the thing. *Annals of the Entomological Society of America*, **78**: 836–840.

Starr, C.K. 1988. The nesting association of the social wasps *Mischocyttarus immarginatus* and *Polybia* spp. in Costa Rica. *Biotropica*, **20**: 171–173.

Starr, C.K. 1989. In reply, is the sting the thing? *Annals of the Entomological Society of America*, **82**: 6–8.

Starr, C.K. & Bozzoli de Wille, M.E. 1990. Social wasps among the Bribri of Costa Rica, pp. 187–194, *In*: Posey, D.A. & Overal, W.L. (eds), *Ethnobiology: Implications and Applications*. 1. 363pp. Museu Paraense Emílio Goeldi, Belém.

Starrett, A. 1993. Adaptive resemblance: a unifying concept for mimicry and crypsis. *Biological Journal of the Linnean Society*, **48**: 299–317.

Stary, P. 1960. The generic classification of the family Aphidiidae (Hymenoptera). *Acta Societatis Entomologicae Cechosloveniae, Prague*, **57**: 238–252.

Stary, P. 1966. *Aphid Parasites of Czechoslovakia. A Review of the Czechoslovak Aphidiidae (Hymenoptera)*. 242pp. Junk. The Hague.

Stary, P. 1967*a*. A review of hymenopterous parasites of citrus pest aphids of the world and biological control projects (Hym., Aphidiidae; Hom., Aphidoidea). *Acta Entomologica Bohemoslovaca*, **64**: 37–61.

Stary, P. 1967*b*. Control biológico de áfidos que atacan al café y al cacao, en Sur América e Indias Occidentales. *Turrialba*, **17**: 388–397.

Stary, P. 1968*a*. Biological control of aphid pests (Homoptera Aphidoidea) by parasites (Hym. Aphidiidae) in the West Indies. *Annales de la Société Entomologique de France* (N.S.), **4**: 27–43.

Stary, P. 1968*b*. Parasites and their role in limitation and control of aphids attacking Annonaceous trees in the West Indies (Hymenoptera: Aphidiidea; Homoptera: Aphidoidea). *Turrialba*, **18**: 129–132.

Stary, P. 1968*c*. Parasites and their utilization in aphid control in the tropics (Hymenoptera: Aphidiidae, Homoptera: Aphidoidea). *Turrialba*, **18**: 387–390.

Stary, P. 1970. Biology of aphid parasites (Hymenoptera: Aphidiidae) with respect to integrated control. *Series Entomologica*, **6**: 1–643.

Stary, P. 1981. Aphid parasitoids (Hymenoptera, Aphidiidae) of Cuba. *Acta Entomologica Bohemoslovaca*, **78**: 33–42.

Stary, P. 1983. New species and records of aphid parasitoids from Mexico (Hymenoptera, Aphidiidae). *Acta Entomologica Bohemoslovaca*, **80**: 35–48.

Stary, P. & Remaudière, G. 1982. New genera, species, and host records of aphid parasitoids (Hymenoptera, Aphidiidae) from Mexico. *Annales de la Société Entomologique de France* (N.S.), **18**: 107–127.

Stary, P. & Remaudière, G. 1983. Complements to the aphid parasitoid fauna of Mexico (Hymenoptera, Aphidiidae). *Annales de la Société Entomologique de France* (N.S.), **19**: 113–116.

Steck, G.J., Gilstrap, F.E., Wharton, R.A. & Hart, W.G. 1986. Braconid parasitoids of Tephritidae (Diptera) infesting coffee and other fruits in west-central Africa. *Entomophaga*, **31**: 59–67.

Steffan, J.-R. 1957. Morphologie du pétiole abdominal des Chalcididae (Hymenoptera). *Bulletin du Muséum d'Histoire naturelle, Paris* (2e série), **29**: 315–322.

Steffan, J.R. 1958. Comportement de *Lasiochalcidia igiliensis* (Ms.) et de l'espèce nouvelle *L. pugnatrix* (Hym. Chalcididae), parasites de fourmilions. *Bulletin du Muséum d'Histoire naturelle, Paris* (2e série), **30**: 506–512.

Steffan, J.R. 1959*a*. Les espèces françaises du genre *Brachymeria* (Westw.) et commentaires sur leur biologie (Hym. Chalcididae). *Cahiers des Naturalistes, Bulletin des Naturalistes Parisiens*, **15**: 35–43.

Steffan, J.-R. 1959*b*. Les chalcidiens parasites de fourmilions (I). *Vie et Milieu*, **10**: 303–317.

Steffan, J.R. 1959*c*. Révision de la tribu des Cratocentrini (Hym. Chalcididae). *Acta Entomologica Musei Nationalis Pragae*, **33**: 287–325.

Steffan, J.R. 1973. Révision des genres *Stypiura* Kirby et *Parastypiura* Steffan (Hym. Chalcididae) de la Région Néotropicale. *Annales de la Société Entomologique de France* (N.S.), **9**: 391–412.

Steffan, J.R. 1974. Trois cas d'adaptations morphologiques chez les Chalcididae (Hym.). *Annales de la Société Entomologique de France* (N.S.), **10**: 565–575.

Stehli, F.G. & Webb, S.D. (eds) 1985. *The Great American Biotic Interchange*. 532pp. Plenum Press, New York.

Steiner, A.L. 1984. Why can mole crickets stung by *Larra* wasps (Hymenoptera, Sphecidae: Larrinae) resume normal activities? The evolution of temporary paralysis and permanent deactivation of the prey. *Journal of the Kansas Entomological Society*, **57**: 152–154.

Stephen, W.P., Bohart, W.P. & Torchio, P.F. 1969. *The Biology and External Morphology of Bees*. 140pp. Agricultural Experiment Station, Oregon State University, Corvallis.

Stephens, C.S. 1962. *Oiketicus kirbyi* (Lepidoptera: Psychidae) a pest of bananas in Costa Rica. *Journal of Economic Entomology*, **55**: 381–386.

Stephens, C.S. 1984. Ecological upset and recuperation of natural control of insect pests in some Costa Rican banana plantations. *Turrialba*, **34**: 101–105.

Stern, D.L. & Duldley, R. 1991. Wing buzzing by male orchid bees, *Eulaema meriana* (Hymenoptera: Apidae). *Journal of the Kansas Entomological Society*, **64**: 88–94.

Sternlicht, M. 1973. Parasitic wasps attracted by the sex pheromone of their coccid host. *Entomophaga*, **18**: 339–342.

Stiles, E.W. 1976. Comparison of male bumblebee flight paths: temperate and tropical (Hymenoptera: Apoidea). *Journal of the Kansas Entomological Society*, **49**: 266–274.

Stiles, F.G. & Skutch, A.F. 1989. *A Guide to the Birds of Costa Rica*. 511 pp. Cornell University Press, Ithaca.

Stiling, P. 1990. Calculating the establishment rates of parasitoids in classical biological control. *American Entomologist*, **36**: 225–230.

Stillwell, M.A. 1964. The fungus associated with woodwasps occurring in beech in New Brunswick. *Canadian Journal of Botany*, **42**: 495–496.

Stillwell, M.A. 1967. The pigeon tremex, *Tremex columba* (Hymenoptera: Siricidae), in New Brunswick. *Canadian Entomologist*, **99**: 685–689.

Stinner, R.E. 1977. Efficacy of inundative releases. *Annual Review of Entomology*, **22**: 515–531.

Stoltz, D.B. 1986. Interactions between parasitoid-derived products and host insects: an overview. *Journal of Insect Physiology*, **32**: 347–350.

Stoltz, D.B. 1990. Evidence for chromosomal transmission of polydnavirus DNA. *Journal of General Virology*, **71**: 1051–1056.

Stoltz, D.B. & Guzo, D. 1986. Apparent haemocytic transformations associated with parasitoid-induced inhibition of immunity in *Malacosoma disstria* larvae. *Journal of Insect Physiology*, **32**: 377–388.

Stoltz, D.B., Guzo, D., Belland, E.R., Lucarotti, C.J. & MacKinnon, E.A. 1988. Venom promotes uncoating *in vitro* and persistence *in vivo* of DNA from a braconid polydnavirus. *Journal of Genetics and Virology*, **69**: 903–908.

Stoltz, D.B. & Vinson, S.B. 1979. Viruses and parasitism in insects. *Advances in Virus Research*, **24**: 125–171.

Stoltz, D. & Whitfield, J.B. 1992. Viruses and virus-like entities in the parasitic Hymenoptera. *Journal of Hymenoptera Research*, **1**: 125–139.

Stoner, A. & Butler, G.D. Jr 1965. *Encarsia lutea* as an egg parasite of bollworm and cabbage looper in Arizona cotton. *Journal of Economic Entomology*, **58**: 1148–1150.

Stout, J. 1979. An association of an ant, a mealy bug, and an understory tree from a Costa Rican rain forest. *Biotropica*, **11**: 309–311.

Stouthamer, R. & Luck, R.F. 1991. Transition from bisexual to unisexual cultures in *Encarsia perniciosi* (Hymenoptera: Aphelinidae): new data and a reinterpretation. *Annals of the Entomological Society of America*, **84**: 150–157.

Stouthamer, R., Luck, R.F. & Hamilton, W.D. 1990*a*. Antibiotics cause parthenogenetic *Trichogramma* (Hymenoptera/ Trichogrammatidae) to revert to sex. *Proceedings of the National Academy of Sciences of the United States of America*, **87**: 2424–2427.

Stouthamer, R., Luck, R.F. & Werren, J.H. 1992. Genetics of sex determination and the improvement of biological control using parasitoids. *Environmental Entomology*, **21**: 427–435.

Stouthamer, R., Pinto, J.D., Platner, G.R. & Luck, R.F. 1990*b*. Taxonomic status of thelytokous forms of *Trichogramma* (Hymenoptera: Trichogrammatidae). *Annals of the Entomological Society of America*, **83**: 475–481.

Stradling, D.J. 1978. The influence of size on foraging in the ant, *Atta cephalotes*, and the effect of some plant defence mechanisms. *Journal of Animal Ecology*, **47**: 173–188.

Strand, M.R. 1986. The physiological interactions of parasitoids with their hosts and their influence on reproductive strategies, pp. 97–136. *In*: Waage, J. & Greathead, D. (eds) *Insect Parasitoids*. 389pp. Academic Press, London.

Strand, M.R. 1989. Development of the polyembryonic parasitoid *Copidosoma floridanum* in *Trichoplusia ni*. *Entomologia Experimentalis et Applicata*, **50**: 37–46.

Strand, M.R., Baehrecke, E.H. & Wong, E.A. 1991. The role of host endocrine factors in the development of polyembryonic parasitoids. *Biological Control*, **1**: 144–152.

Strand, M.R., Johnson, J.A. & Culin, J.D. 1990. Intrinsic interspecific competition between the polyembryonic parasitoid *Copidosoma floridanum* and solitary endoparasitoid *Microplitis demolitor* in *Pseudoplusia includens*. *Entomologia Experimentalis et Applicata*, **55**: 275–284.

Strand, M.R., Meola, S.M. & Vinson, S.B. 1986. Correlating pathological symptoms in *Heliothis virescens* eggs with development of the parasitoid *Telenomus heliothidis*. *Journal of Insect Physiology*, **32**: 389–402.

Strand, M.R. & Vinson, S.B. 1982. Behavioral response of the parasitoid *Cardiochiles nigriceps* to a kairomone. *Entomologia Experimentalis et Applicata*, **31**: 308–315.

Strand, M.R. & Vinson, S.B. 1983. Host acceptance behavior in *Telenomus heliothidis* (Hymenoptera: Scelionidae) toward *Heliothis virescens* (Lepidoptera: Noctuidae). *Annals of the Entomological Society of America*, **76**: 781–785.

Strand, M.R. & Vinson, S.B. 1984. Facultative hyperparasitism by the egg parasitoid *Trichogramma pretiosum* (Hymenoptera: Trichogrammatidae). *Annals of the Entomological Society of America*, **77**: 679–686.

Strassmann, J.E., Gastreich, K.R., Queller, D.C. & Hughes, C.R. 1992. Demographic and genetic evidence for cyclical changes in queen number in a neotropical wasp, *Polybia emaciata*. *American Naturalist*, **140**: 363–372.

Strassmann, J.E., Queller, D.C., Solis, C.R. & Hughes, C.R. 1991. Relatedness and queen number in the neotropical wasp, *Parachartergus colobopterus*. *Animal Behaviour*, **42**: 461–470.

Streams, F.A. & Greenberg, L. 1969. Inhibition of the defense reaction of *Drosophila melanogaster* parasitized simultaneously by the wasps *Pseudeucoila bochei* and *Pseudeucoila mellipes*. *Journal of Invertebrate Pathology*, **13**: 371–377.

Strickland, E.H. 1923. Biological notes on parasites of prairie cutworms. *Bulletin of the Department of Agriculture of the Dominion of Canada* (N.S.), **26**: 1–40.

Strong, D.R., Lawton, J.H. & Southwood, R. 1984. *Insects on Plants*. 313pp. Blackwell Scientific Publications, Oxford.

Strong, F.E. 1962. Laboratory studies of the biology of the alfalfa seed chalcid, *Bruchophagus roddi* Guss. (Hymenoptera: Eurytomidae). *Hilgardia*, **32**: 229–249.

Styer, E.L., Hamm, J.J. & Nordlund, D.A. 1987. A new virus associated with the parasitoid *Cotesia marginiventris* (Hymenoptera: Braconidae): replication in noctuid host larvae. *Journal of Invertebrate Pathology*, **50**: 302–309.

Subba Rao, B.R. 1957. The biology and bionomics of *Lestodryinus pyrillae* Kieff. (Dryinidae: Hymenoptera) a nymphal parasite of *Pyrilla perpusilla* Walk., and a note on its role in the control of *Pyrilla*. *Journal of the Bombay Natural History Society*, **54**: 741–749.

Subba Rao, B.R. 1978. New genera and species of Eurytomidae (Hymenoptera: Eurytomidae). *Proceedings of the Indian Academy of Sciences* (B) (Animal Sciences), **87** (12): 293–319.

Sudd, J.H. & Franks, N.R. 1987. *The Behavioural Ecology of Ants*. 206pp. Blackie, Glasgow.

Sugimoto, T., Ichikawa, T., Mitomi, M. & Sakuratani, Y. 1988*b*. Foraging for patchily-distributed leaf-miners by the parasitoid *Dapsilarthra rufiventris* (Hymenoptera: Braconidae) IV. Analyses of sounds emitted by a feeding host. *Applied Entomology and Zoology*, **23**: 209–211.

Sugimoto, T., Imoarai, T. & Tsuji, H. 1983. Oosorption in eulophid wasp, *Chrysocharis pentheus* Walker (Hymenoptera: Eulophidae). *Applied Entomology and Zoology*, **18**: 287–289.

Sugimoto, T., Shimono, Y., Hata, Y., Nakai, A. & Yahara, M. 1988*a*. Foraging for patchily-distributed leaf-miners by the parasitoid, *Dapsilarthra rufiventris* (Hymenoptera: Braconidae) III. Visual and acoustic cues to a close range patch-location. *Applied Entomology and Zoology*, **23**: 113–121.

Sugimoto, T., Uenishi, M. & Machida, F. 1986. Foraging for patchily-distributed leaf-miners by the parasitoid, *Dapsilarthra rufiventris* (Hymenoptera: Braconidae) I. Discrimination of previously searched leaflets. *Applied Entomology and Zoology*, **21**: 500–508.

Sugonjaev, E.S. 1971. A new subfamily of chalcids, Mongolocampinae subfam.n. (Hymenoptera, Chalcidoidea, Tetracampidae) from Mongolia and Kazakhstan. *Entomologicheskoe Obozrenie*, **50**: 664–675. [In Russian.]

Sugonjaev, E.S. 1984. Chalcid-flies (Hymenoptera, Chalcidoidea) parasites of soft scales Lecaniidae fauna of the USSR. Combined research of host-parasite systems in insects (Homoptera, Coccoidea). *Trudy Zoologicheskogo Instituta Akademiya Nauk SSSR*, **117**: 1–122. [In Russian.]

Sullivan, D.J. 1972. Comparative behavior and competition between two aphid hyper-parasites: *Alloxysta victrix* and *Asaphes californicus* (Hymenoptera: Cynipidae; Pteromalidae). *Environmental Entomology*, **1**: 234–244.

Sullivan, D.J. 1987. Insect hyperparasitism. *Annual Review of Entomology*, **32**: 49–70.

Suomalainen, E., Saura, A. & Lokki, J. 1987. *Cytology and Evolution in Parthenogenesis*. 216pp. CRC Press, Boca Raton, Florida.

Swain, R.B. 1980. Trophic competition among parabiotic ants. *Insectes Sociaux*, **27**: 377–390.

Swezey, O.H. 1908. On peculiar deviations from uniformity of habit among chalcids and proctotrupids. *Proceedings of the Hawaiian Entomological Society*, **2**: 18–22.

Swezey, O.H. 1928. Present status of certain insect pests under biological control in Hawaii. *Journal of Economic Entomology*, **21**: 669–676.

Swezey, O.H. 1929. Notes on the egg parasites of insects in Hawaii. *Proceedings of the Hawaiian Entomological Society*, **7**: 282–292.

Swezey, O.H. 1931. *Litomastix floridana* (Ashm.), a recent immigrant in Hawaii. *Proceedings of the Hawaiian Entomological Society*, **7**: 369–370, 390, 419–421.

Swezey, O.H. 1945. Insects associated with orchids. *Proceedings of the Hawaiian Entomological Society*, **12**: 343–403.

Swezey, S.L. & Cano Vasquez, E. 1991. Biological control of citrus blackfly (Homoptera: Aleyrodidae) in Nicaragua. *Environmental Entomology*, **20**: 1691–1698.

Syrjämäki, J. 1976. The mystery of the missing females in connection with male swarming of *Blacus ruficornis* Nees (Hym., Braconidae). *Annales Entomologici Fennici*, **42**: 66–68.

Szczepánski, H. 1960. On chalcidoid wasps (Hymenoptera) parasitizing at the bark-beetles (Coleoptera, Scolytidae) in Borecka forests (distr. Wegorzewo, Poland). *Polskie Pismo Entomologiczne*, **30**: 405–416. [In Polish; English summary.]

Tachikawa, T. 1978. Hosts of the Encyrtidae in the world (Hymenoptera: Chalcidoidea). *Transactions of the Shikoku Entomological Society*, **14**: 43–63.

Tachikawa, T. 1981. Hosts of encyrtid genera in the world (Hymenoptera: Chalcidoidea). *Memoirs of the College of Agriculture, Ehime University*, **25**: 85–110.

Tagawa, J. & Kitano, H. 1981. Mating behaviour of the braconid wasp, *Apanteles glomeratus* L. (Hymenoptera: Braconidae) in the field. *Applied Entomology and Zoology*, **16**: 345–350.

Takabayashi, J. & Takahashi, S. 1986. Effect of kairomones in the host searching behavior of *Apanteles kariyai* Watanabe (Hymenoptera: Braconidae), a parasitoid of the common armyworm, *Pseudaletia separata* Walker (Lepidoptera: Noctuidae). II. Isolation and identification of arrestants produced by the host larvae. *Applied Entomology and Zoology*, **21**: 114–118.

Takada, H. 1973. Studies on aphid hyperparasites of Japan. I, aphid hyperparasites of the genus *Dendrocerus* Ratzeburg occurring in Japan (Hymenoptera: Ceraphronidae). *Insecta Matsumurana* (N.S.), **2**: 1–37.

Takada, H. & Hashimoto, Y. 1985. Association of the root aphid parasitoids *Aclitus sappaphis* and *Paralipsis eikoae* (Hymenoptera, Aphidiidae) with the aphid-attending ants *Pheidole fervida* and *Lasius niger* (Hymenoptera: Formicidae). *Kontyû*, **53**: 150–160.

Takada, H. & Kamijo, K. 1979. Parasite complex of the garden pea leaf-miner, *Phytomyza horticola* Gourea in Japan. *Kontyû*, **47**: 18–37.

Takagi, M. 1985. The reproductive strategy of the gregarious parasitoid, *Pteromalus puparum* (Hymenoptera: Pteromalidae) 1. Optimal number of eggs in a single host. *Oecologia*, **68**: 1–6.

Takasu, K. & Hirose, Y. 1988. Host discrimination in the parasitoid *Ooencyrtus nezarae*: the role of the egg stalk as an external marker. *Entomologia Experimentalis et Applicata*, **47**: 45–48.

Tanaka, T. 1986. Effects of the calyx and venom fluids of *Apanteles kariyai* Watanabe (Hymenoptera: Braconidae) on the fat body and hemolymph protein contents of its host *Pseudaletia separata* (Walker) (Lepidoptera: Noctuidae). *Applied Entomology and Zoology*, **21**: 220–227.

Tanaka, T. 1987*a*. Effect of the venom of the endoparasitoid, *Apanteles kariyai* Watanabe, on the cellular defence reaction of the host, *Pseudaletia separata* Walker. *Journal of Insect Physiology*, **33**: 413–420.

Tanaka, T. 1987*b*. Morphology and functions of calyx fluid filaments in the reproductive tracts of endoparasitoid, *Microplitis mediator* (Hym.: Braconidae). *Entomophaga*, **32**: 9–17.

Tanaka, T. & Vinson, S.B. 1991. Depression of prothoracic gland activity of *Heliothis virescens* by venom and calyx fluids from the parasitoid, *Cardiochiles nigriceps*. *Journal of Insect Physiology*, **37**: 139–144.

Tanaka, T. & Wago, H. 1990. Ultrastructural and functional maturation of teratocytes of *Apanteles kariyai*. *Archives of Insect Biochemistry and Physiology*, **13**: 187–197.

Taper, M.L. & Case, T.J. 1987. Interactions between oak tannins and parasite community structure: Unexpected benefits of tannins to cynipid gall-wasps. *Oecologia*, **71**: 254–261.

Tardieux, I. & Rabasse, J.M. 1988. Some aspects of host immunity and physiological suitability in aphids attacked by *Aphidius colemani*, pp. 311–315. *In*: Niemczyk, E. & Dixon, A.F.G. (eds) *Ecology and Effectiveness of Aphidophaga*. 341pp. SPB Academic Publishing, The Hague.

Tardieux, I. & Rabasse, J.M. 1990. Role of some epigenetic factors influencing the host suitability of *Myzus persicae* for the parasitoid *Aphidius colemani*. *Entomologia Experimentalis et Applicata*, **54**: 73–80.

Tauber, M.J., Tauber, C.A. & Masaki, S. 1986. *Seasonal Adaptations of Insects*. 411pp. Oxford University Press, Oxford.

Tawfik, M.F.S. 1991. The teratocytes of the braconid *Apanteles glomeratus* L. *Egyptian Journal of Biological Pest Control*, **1**: 129–136.

Taylor, A.D. 1988*a*. Host effects on larval competition in the gregarious parasitoid *Bracon hebetor*. *Journal of Animal Ecology*, **57**: 163–172.

Taylor, A.D. 1988*b*. Host effects on functional and ovipositonal responses of *Bracon hebetor*. *Journal of Animal Ecology*, **57**: 173–184.

Taylor, K.L. 1967. Parasitism of *Sirex noctilio* F. by *Schlettererius cinctipes* (Cresson) (Hymenoptera: Stephanidae). *Journal of the Australian Entomological Society*, **6**: 13–19.

Taylor, K.L. 1976. The introduction and cstablishment of insect parasitoids to control *Sirex noctilio* in Australia. *Entomophaga*, **21**: 429–440.

Taylor, R.L. 1929. The biology of the white pine weevil, *Pissodes strobi* (Peck), and a study of its insect parasites from an economic viewpoint. *Entomologica Americana*, **9**: 167–246; **10**: 1–86.

Taylor, T.H.C. 1935. The campaign against *Aspidiotus destructor* Sign. in Fiji. *Bulletin of Entomological Research*, **26**: 1–100.

Taylor, T.H.C. 1937. *The Biological Control of an Insect in Fiji: An Account of the Coconut Leaf-mining Beetle and its Parasite Complex*. 239pp. IIE, London.

Teitelbaum, S.S. & Black, L.M. 1954. The effect of a phytophagous species of *Tetrastichus*, new to the United States, on sweet clover infected with wound-tumor virus. *Phytopathology*, **44**: 548–550.

Tepedino, V.J. 1982. Flower visitation and pollen collection records for bees of high altitude shortgrass prairie in southeastern Wyoming. *Southwestern Entomologist*, **7**: 16–25.

Teran, A.L. & DeBach, P. 1963. Observaciones sobre *Comperiella bifasciata* How. (Hymen. Encyrtidae). *Revista Agronomica de Noroeste Argentina*, **4**: 5–23.

Terborgh, J. 1986. Keystone plant resources in the tropical forest, pp. 330–344. *In*: Soulé, M.E. (ed.), *Conservation Biology. The Science of Scarcity and Diversity*. 584pp. Sinauer, Sunderland, Massachussetts.

Thibout, E., LeCompte, C. & Auger, J. 1988. *Diadromus pulchellus*: search for a host and specificity, pp. 7–14. *In*: Bouletreau, M. & Bonnot, G. (eds) *Parasitoid Insects: European Workshop, Lyon, September 7–10, 1987*. 170pp. Les Colloques de l'INRA, Versailles.

Thompson, S.N. 1983. The nutritional physiology of *Trichoplusia ni* parasitized by the insect parasite, *Hyposoter exiguae*, and the effects of parallel-feeding. *Parasitology*, **87**: 15–28.

Thompson, S.N. 1986. Effect of the insect parasite *Hyposoter exiguae* (Viereck) on the carbohydrate metabolism of its host, *Trichoplusia ni* (Hübner). *Journal of Insect Physiology*, **32**: 287–293.

Thoms, E.M. & Robinson, W.H. 1986. Distribution, seasonal abundance, and pest status of the oriental cockroach (Orthoptera: Blattidae) and an evaniid wasp (Hymenoptera: Evaniidae) in urban apartments. *Journal of Economic Entomology*, **79**: 431–436.

Thomson, M.S. & Stinner, R.E. 1988. Comparative responses of feral and laboratory *Trichogramma* spp. (Hymenoptera: Trichogrammatidae) to *Heliothis* spp. (Lepidoptera: Noctuidae) moth scales and inert particles. *Journal of Entomological Science*, **23**: 245–250.

Thomson, M.S. & Stinner, R.E. 1989. *Trichogramma* spp. (Hymenoptera: Trichogrammatidae): field hosts and multiple parasitism in North Carolina. *Journal of Entomological Science*, **24**: 232–240.

Thornhill, R. & Alcock, J. 1983. *The Evolution of Insect Mating Systems*. 547 pp. Harvard University Press, Cambridge, Mass.

Thorp, R.W. 1979 (1980). Structural, behavioral, and physiological adaptations of bees (Apoidea) for collecting pollen. *Annals of Missouri Botanical Garden*, **66**: 788–812.

Thorpe, W.H. 1932. Experiments upon respiration in the larvae of certain parasitic Hymenoptera. *Proceedings of the Royal Society of London* (B), **109**: 450–471.

Thorpe, W.H. 1933. Notes on the natural control of *Coleophora laricella*, the larch case-bearer. *Bulletin of Entomological Research*, **24**: 271–291.

Thorpe, W.H. 1936. On a new type of respiratory interrelation between an insect (chalcid) parasite and its host (Coccidae). *Parasitology*, **28**: 517–540.

Thorpe, W.H. & Caudle, H.B. 1938 (1939). A study of the olfactory responses of insect parasites to the food plant of their host. *Parasitology*, **30**: 523–528.

Thorpe, W.H. & Jones, F.G.W. 1937. Olfactory conditioning in a parasitic insect and its relation to the problem of host selection. *Proceedings of the Royal Society of London* (B), **124**: 56–81.

Thurber, D.K., Belk, M.C., Black, H.L., Jorgensen, C.D., Hubbell, S.P. & Foster, R.B. 1993. Dispersion and mortality of colonies of the tropical ant *Paraponera clavata*. *Biotropica*, **25**: 215–221.

Tilden, R.L. & Ferkovich, S.M. 1988. Kairomonal stimulation of oviposition into an artificial substrate by the endoparasitoid *Microplitis croceipes* (Hymenoptera: Braconidae). *Annals of the Entomological Society of America*, **81**: 152–156.

Tillman, P.G. & Powell, J.E. 1992. Intraspecific host discrimination and larval competition in *Microplitis croceipes, Microplitis demolitor, Cotesia kazak* (Hym.: Braconidae) and *Hyposoter didymator* (Hym.: Ichneumonidae), parasitoids of *Heliothis virescens* (Lep.: Noctuidae). *Entomophaga*, **37**: 229–237.

Tillman, P.G. & Cate, J.R. 1993. Effect of host size on adult size and sex ratio of *Bracon mellitor* (Hymenoptera: Braconidae). *Environmental Entomology*, **22**: 1161–1165.

Tillyard, R.J. 1926. *The Insects of Australia and New Zealand*. 560pp. Angus & Robertson, Sydney.

Timberlake, P.H. 1977. Description of two new species of *Heterosarus* Robertson (Hymenoptera, Apoidea). *Pan-Pacific Entomologist*, **53**: 56–59.

Timberlake, P.H. 1980. Review of North American *Exomalopsis* (Hymenoptera, Anthophoridae). Parts I–IV. *University of California Publications in Entomology*, **86**: 1–158.

Tobias, V.I. 1966. Generic groupings and evolution of parasitic Hymenoptera of the subfamily Euphorinae (Hymenoptera, Braconidae) II. *Entomological Review*, **45**: 348–358.

Tobias, V.I. 1975. Body sizes in the braconids (Hymenoptera, Braconidae) and their evolution. *Entomological Review*, **54**: 98–107.

Tobias, V.I. 1988. Family Stephanidae, pp. 236–237. *In*: Zhelokhovtsev, A.N., Tobias, V.I. & Kozlov, M.A. (eds). *Key to the Identification of Insects of European USSR*, 3. Hymenoptera, 6. 267pp. Moscow.

Tocchetto, A. 1942. Bicho costuriero. *Revista Agronomia, Porto Alegre*, **6**: 587–588.

Togashi, I. 1970. The comparative morphology of the internal reproductive organs of the Symphyta (Hymenoptera). *Mushi* (Suppl.), **43**: 1–114.

Tonapi, G.T. 1958. Presence of propolis in the prothoracic spiracles of *Apis florea* Fab. *Current Science*, **27**: 133–134.

Tooke, F.G.C. 1955. The eucalyptus snout-beetle, *Gonipterus scutellatus* Gyll. A study of its ecology and control by biological means. *Entomology Memoirs. Department of Agriculture, Union of South Africa*, **3**: 1–282.

Topoff, H.R. 1969. A unique predatory association between carabid beetles of the genus *Helluomorphoides* and colonies of the army ant *Neivamyrmex nigrescens*. *Psyche*, **76**: 375–381.

Torchio, P.F. 1970. The ethology of the wasp, *Pseudomasaris edwardsii* (Cresson), and a description of its immature forms (Hymenoptera: Vespoidea, Masaridae). *Contributions in Science. Natural History Museum of Los Angeles County*, **202**: 1–32.

Torchio, P.F. 1972. *Sapyga pumila* Cresson, a parasite of *Megachile rotundata* (F.) (Hymenoptera: Sapygidae; Megachilidae). I: biology and description of immature stages. *Melanderia*, **10**: 1–22.

Torchio, P.F. 1974. Notes on the biology of *Ancyloscelis armata* Smith and comparisons with other anthophorine bees (Hymenoptera: Anthophoridae). *Journal of the Kansas Entomological Society*, **47**: 54–63.

Torchio, P.F. 1979. An eight-year field study involving control of *Sapyga pumila* Cresson (Hymenoptera: Sapygidae), a wasp parasite of the alfalfa leafcutter bee, *Megachile pacifica* Panzer. *Journal of the Kansas Entomological Society*, **52**: 412–419.

Torchio, P.F. & Burdick, D.J. 1988. Comparative notes on the biology and development of *Epeolus compactus* Cresson, a cleptoparasite of *Colletes kincaidii* Cockerell (Hymenoptera: Anthophoridae, Colletidae). *Annals of the Entomological Society of America*, **81**: 626–636.

Torchio, P.F. & Torchio, D.M. 1975. Larvae of the Apidae (Hymenoptera, Apoidea). Part I. Apini, *Apis. Agricultural Experiment Station, Utah State University Research Report*, **20**: 1–36.

Torgersen, T.R. & Ryan, R.B. 1981. Field biology of *Telenomus californicus* Ashmead, an important egg parasite of douglas-fir tussock moth. *Annals of the Entomological Society of America*, **74**: 185–186.

Torgerson, R.L. & Akre, R.D. 1969. Reproductive morphology and behavior of a thysanuran, *Trichatelura manni*, associated with army ants. *Annals of the Entomological Society of America*, **62**: 1367–1374.

Torgerson, R.L. & Akre, R.D. 1970. The persistence of army ant chemical trails and their significance in the ecitonine-ecitophile association (Formicidae: Ecitonini). *Melanderia*, **5**: 1–28.

Toro, H. & Carvajal, C. 1989. Ajuste genital en la copula de Thynninae (Hymenoptera, Tiphiidae). *Acta Entomológica Chilena*, **15**: 123–130.

Toro, H. & Michener, C.D. 1975. The subfamily Xeromelissinae and its occurrence in Mexico (Hymenoptera: Colletidae). *Journal of the Kansas Entomological Society*, **48**: 351–357.

Tosi, J.A. 1969. *Ecological Map of Costa Rica*. Tropical Science Center, San José. 1 pp.

Tothill, J.D. 1922. The natural control of the fall webworm (*Hyphantria cunea* Drury) with an account of its several parasites. *Bulletin of the Department of Agriculture of Canada, Entomological Branch*, **19**: 1–107.

Townes, H.K. Jr 1940. A revision of the Pimplini of eastern North America (Hymenoptera, Ichneumonidae). *Annals of the Entomological Society of America*, **33**: 283–323.

Townes, H. 1949. The Nearctic species of the family Stephanidae (Hymenoptera). *Proceedings of the United States National Museum*, **99**: 361–370.

Townes, H. 1950. The Nearctic species of Gasteruptiidae (Hymenoptera). *Proceedings of the United States National Museum*, **100**: 85–145.

Townes, H. 1956. The Nearctic species of trigonalid wasps. *Proceedings of the United States National Museum*, **106**: 295–304.

Townes, H. 1957. Nearctic wasps of the subfamilies Pepsinae and Ceropalinae. *Bulletin of the United States National Museum*, **209**: 1–286.

Townes, H. 1969. Genera of Ichneumonidae 1. *Memoirs of the American Entomological Institute*, **11**: 1–300.

Townes, H. 1970*a*. Genera of Ichneumonidae 2. *Memoirs of the American Entomological Institute*, **12**: 1–537.

Townes, H. 1970*b*. Genera of Ichneumonidae 3. *Memoirs of the American Entomological Institute*, **13**: 1–307.

Townes, H. 1971. Genera of Ichneumonidae 4. *Memoirs of the American Entomological Institute*, **17**: 1–372.

Townes, H. 1972. A light-weight Malaise trap. *Entomological News*, **83**: 239–247.

Townes, H. 1973. Comments on the design of insect nets. *Entomological News*, **84**: 1–7.

Townes, H. 1977*a*. A revision of the Rhopalosomatidae (Hymenoptera). *Contributions of the American Entomological Institute*, **15** (1): 1–34.

Townes, H. 1977*b*. A revision of the Heloridae (Hymenoptera). *Contributions of the American Entomological Institute*, **15** (2): 1–12.

Townes H. & Townes, M. 1960. Ichneumon-flies of America north of Mexico: 2. Subfamilies Ephialtinae, Xoridinae, Acaenitinae. *United States National Museum Bulletin*, **216** (2): 1–676.

Townes, H. & Townes, M. 1966. A catalogue and reclassification of neotropic Ichneumonidae. *Memoirs of the American Entomological Institute*, **8**: 1–367.

Townes, H. & Townes, M. 1981. A revision of the Serphidae (Hymenoptera). *Memoirs of the American Entomological Institute*, **32**: 1–541.

Trager, J.C. 1991. A revision of the fire ants, *Solenopsis geminata* group (Hymenoptera: Formicidae: Myrmicinae). *Journal of the New York Entomological Society*, **99**: 141–198.

Traniello, J.F.A. 1989. Foraging strategies of ants. *Annual Review of Entomology*, **34**: 191–210.

Tremblay, E. 1966. Ricerche sugli Imenotteri parassiti II. Osservazioni sull'origine e sul destino dell'involucro embrionale degli Afidiini (Hymenoptera: Braconidae: Aphidiinae) e considerazioni sul significato generale delle membrane embrionali. *Bollettino del Laboratorio di Entomologia Agraria 'Filippo Silvestri' Portici*, **24**: 119–166.

Tripp, H.A. 1961. The biology of a hyperparasite, *Euceros frigidus* Cress. (Ichneumonidae) and description of the planidial stage. *Canadian Entomologist*, **93**: 40–58.

Trivers, R.L. 1974. Parent-offspring conflict. *American Zoologist*, **14**: 249–264.

Trjapitzin, V.A. 1972. Host-parasite relations in the family Encyrtidae (Hymenoptera, Chalcidoidea), pp. 31–48. *In*: Zaslavsky, V.A. (ed.) *Host-Parasite Relations in Insects*, 130pp. Nauka, Leningrad. [In Russian with English summary.]

Trjapitzin, V.A. 1973*a*. Classification of parasitic Hymenoptera of the family Encyrtidae (Chalcidoidea). Part 1. Review of systems of classifications. The subfamily Tetracneminae Howard, 1892. *Entomologicheskoe Obozrênie*, **52**: 163–175. [In Russian; English translation: *Entomological Review, Washington*, **52**: 118–125.]

Trjapitzin, V.A. 1973*b*. Classification of parasitic Hymenoptera of the family Encyrtidae (Chalcidoidea). Part II. The subfamily Encyrtinae Walker, 1837. *Entomologicheskoe Obozrênie*, **52**: 416–429. [In Russian; English translation: *Entomological Review, Washington*, **52**: 287–295.]

Trjapitzin, V.A. 1979. Hypothetic trends of evolution of host-parasite relations of parasitic Hymenoptera of the family Encyrtidae (Chalcidoidea). *Trudy Zoologicheskogo Instituta. Akademiya Nauk SSSR. Leningrad*, **83**: 120–125. [In Russian.]

Trjapitzin, V.A. 1989. Parasitic Hymenoptera of the Fam. Encyrtidae of Palaearctics. *Opredeliteli po faune SSSR Izdavavaemiye Zoologiya In-Tom AN SSSR*, **158**: 1–489. [In Russian.]

Trjapitzin, V.A. & Gordh, G. 1978*a*. Review of genera of Nearctic Encyrtidae (Hymenoptera, Chalcidoidea). I. *Entomologicheskoe Obozrênie*, **57**: 364–385. [In Russian; English translation: *Entomological Review, Washington*, **57**: 257–270.]

Trjapitzin, V.A. & Gordh, G. 1978*b*. Review of genera of Nearctic Encyrtidae (Hymenoptera, Chalcidoidea). II. *Entomologicheskoe Obozrênie*, **57**: 636–653. [In Russian; English translation: *Entomological Review, Washington*, **57**: 437–448.]

Turlings, T.C.J. & Tumlinson, J.H. 1992. Systemic release of chemical signals by herbivore-injured corn. *Proceedings of the National Academy of Science of the United States of America*, **89**: 8399–8402.

Turlings, T.C.J., Tumlinson, J.H. & Lewis, W.J. 1990. Exploitation of herbivore-induced plant odors by host-seeking parasitic wasps. *Science*, **250**: 1251–1253.

Uhler, L.D. 1951. Biology and ecology of the goldenrod gall fly, *Eurosta solidaginis* (Fitch). *Memoir Cornell University Agricultural Experiment Station*, **300**: 1–51.

Ulenberg, S.A. & Nübel, B.K. 1982. The oviposition of *Apocrypta* Coquerel (fig wasp parasites; Hymenoptera, Chalcidoidea, Torymidae). *Proceedings of the Koninklijke Nederlandse Akademie van Wetenschappen* (C), **85**: 607–613.

Ulloa-Chacon, P. & Cherix, D. 1990. The little fire ant *Wasmannia auropunctata* (R.) (Hymenoptera: Formicidae), pp. 281–289. *In*: Vander Meer, R.K., Jaffe, K. & Cedeno, A. (eds). *Applied Myrmecology, A World Perspective*, 741pp. Westview Press, Boulder.

Ulloa-Chacón, P. Cherix, D. & Meier, R. 1991. Bibliografia de la hormiga colorada, *Wasmannia auropunctata* Roger (Hymenoptera: Formicidae). *Noticias de Galápagos*, **50**: 8–12.

Umeh, E.-D.N. 1988. Development, oviposition, host feeding and sex determination in *Epidinocarsis lopezi* (De Santis) (Hymenoptera: Encyrtidae). *Bulletin of Entomological Research*, **78**: 605–611.

Unruh, T.R. & Messing, R.H. 1993. Intraspecific biodiversity in Hymenoptera: implications for conservation and biological control, pp. 27–52. *In*: LaSalle, J. & Gauld, I.D. (eds), *Hymenoptera and Biodiversity*, 348pp. CAB International, Wallingford.

Urban, D. 1967. As espécies do gênero *Thygater* Holmberg, 1884. (Hymenoptera, Apoidea) *Boletim da Universidade Federal do Paraná, Zoologia*, **2**: 177–309.

Urban, D. 1968. As espécies do gênero *Melissoptila* Holmberg, 1884 (Hymenoptera—Apoidea). *Revista Brasileira de Entomologia*, **13**: 1–94.

Urban, D. 1970. As espécies do gênero *Florilegus* Robertson, 1900. (Hymenoptera—Apoidea). *Boletim da Universidade Federal do Paraná, Zoologia*, **3**: 245–280.

Valerio, C.E. 1971. Parasitismo en huevos de araña *Achaearanea tepidariorum* (Koch) (Aranea: Theridiidae) en Costa Rica. *Revista de Biología Tropical*, **18**: 99–106.

Valladares, G., Diaz, N. & DeSantis, L. 1982. Tres notas sobre dipteros agromicidos de la republica Argentina y sus himenopteros parasitoides (Insecta). *Revista de la Sociedad Entomólogica Argentina*, **41**: 319–330.

Vance, A.M. 1927. On the biology of some ichneumonids of the genus *Paniscus* Schrk. *Annals of the Entomological Society of America*, **20**: 405–417.

Vance, A.M. 1931. *Apanteles thompsoni* Lyle, a braconid parasite of the European corn borer. *United States Department of Agriculture, Technical Bulletin*, **233**: 1–28.

Vance, A.M. 1932*a*. The biology and morphology of the braconid *Chelonus annulipes* Wesm., a parasite of the European corn borer. *United States Department of Agriculture, Technical Bulletin*, **294**: 1–48.

Vance, A.M. 1932*b*. *Microgaster tibialis* Nees as a hymenopterous parasite of *Pyrausta nubilalis* Hubn. in Europe. *Annals of the Entomological Society of America*, **25**: 121–134.

Vander Meer, R.K., Jaffe, K. & Cedeno, A. 1990. *Applied Myrmecology, A World Perspective*, 741pp. Westview Press, Boulder.

Vardy, C.R. 1978. A revision of the Neotropical wasp genus *Trigonopsis* Perty (Hymenoptera: Sphecidae). *Bulletin of the British Museum (Natural History)* Entomology, **37**: 117–152.

Vardy, C.R. 1987. Three new taxa of Neotropical Larrinae (Hym., Sphecidae) and a new prey record. *Entomologists' Monthly Magazine*, **123**: 99–105.

Vargas V., M. & Fallas B., F. 1974. Notes on the biology of *Tetrastichus hagenowii* (Hymenoptera, Eulophidae) a parasite of cockroach oothecae. *Entomological News*, **85**: 23–26.

Varley, G.C. 1937. Description of the eggs and larvae of four species of chalcidoid Hymenoptera parasitic on the knapweed gall-fly. *Proceedings of the Royal Entomological Society of London* (B), **6**: 122–130.

Varley, G.C. & Butler, C.G. 1933. The acceleration of development of insects by parasitism. *Parasitology*, **25**: 263–268.

Vecht, J. van der & Carpenter, J.M. 1990. A catalogue of the genera of the Vespidae (Hymenoptera). *Zoologische Verhandelingen*, **260**: 3–62.

Veen, J.C. van, 1981. The biology of *Poecilostictus cothurnatus* (Hymenoptera, Ichneumonidae) an endoparasitoid of *Bupalus piniarius* (Lepidoptera, Geometridae). *Annales Entomologici Fennici*, **47**: 77–93.

Veen, J.C. van 1982. Notes on the biology of *Banchus femoralis* Thomson (Hym., Ichneumonidae) an endoparasitoid of *Panolis flammea* (D.& S.) (Lep., Noctuidae). *Zeitschrift für Angewandte Entomologie*, **94**: 300–311.

Vega, F.E. & Barbosa, P. 1990. *Gonatopus bartletti* Olmi (Hymenoptera: Dryinidae) in México: a previously unreported parasitoid of the corn leafhopper *Dalbulus maidis* (Delong & Wolcott) and the Mexican corn leafhopper *Dalbulus elimatus* (Ball) (Homoptera: Cicadellidae). *Proceedings of the Entomological Society of Washington*, **92**: 461–464.

Vereshchagina, V.V. 1961. *Tetrastichus (Myiomisa) sajoi* Szélnyi—a predator of the plum shoot mite—*Eriophyes phloeocoptes* Nal. *Trudy Moldavskogo Nauchno-Issledovatel'skogo Instituta Sadovodstva*, **7**: 31–33.

Verkerke, W. 1989. Structure and function of the fig. *Experientia*, **45**: 612–622.

Vet, L.E.M. 1985. Olfactory microhabitat location in some eucoilid and alysiine species (Hymenoptera), larval parasitoids of Diptera. *Netherlands Journal of Zoology*, **35**: 720–730.

Vet, L.E.M. 1988. The influence of learning on habitat location and acceptance by parasitoids, pp. 29–34. *In*: Bouletreau, M. & Bonnot, G. (eds) *Parasitoid Insects: European Workshop, Lyon, September* 7–10, 1987. 170pp. Les Colloques de l'INRA, Versailles.

Vet, L.E.M., DeJong, R., Giessen, W.A. van & Visser, J.H. 1990. A learning-related variation in electroantennogram responses of a parasitic wasp. *Physiological Entomology*, **15**: 243–247.

Vet, L.E.M. & Groenewold, A.W. 1990. Semiochemicals and learning in parasitoids. *Journal of Chemical Ecology*, **16**: 3119–3135.

Vet, L.E.M., Janse, C., Achterberg, C. van & Alphen, J.J.M. van, 1984. Microhabitat location and niche segregation in two sibling species of drosophilid parasitoids: *Asobara tabida* (Nees) and *A. rufescens* (Foerster) (Braconidae: Alysiinae). *Oecologia*, **61**: 182–188.

Vet, L.E.M. & Opzeeland, K. van, 1985 (1986). Olfactory microhabitat selection in *Leptopilina heterotoma* (Thomson) (Hym.: Eucoilidae), a parasitoid of Drosophilidae. *Netherlands Journal of Zoology*, **35**: 497–504.

Vet, L.E.M. & Schoonman, G. 1988. The influence of previous foraging experience on microhabitat acceptance in *Leptopilina heterotoma*. *Journal of Insect Behavior*, **1**: 387–392.

Viggiani, G. 1964. La specializzazione entomoparassitica in alcuni eulofidi (Hym., Chalcidoidea). *Entomophaga*, **9**: 111–118.

Viggiani, G. 1967. Richerche sugli Hymenoptera Chalcidoidea XIII. Sugli entomoparassiti della *Semidalis aleurodiformis* Steph. (Neur. Coniopterygidae), con descrizione di un nuovo genere di Encyrtidae (Prima nota). *Bollettino del Laboratorio di Entomologia Agraria 'Filippo Silvestri', Portici*, **25**: 163–175.

Viggiani, G. 1971. Ricerche sugli Hymenoptera Chalcidoidea XXVIII. Studio morfologico comparativo dell'armatura genitale esterna maschile dei Trichogrammatidae. *Bollettino del Laboratorio di Entomologia Agraria 'Filippo Silvestri', Portici*, **29**: 181–222.

Viggiani, G. 1973. Ricerche sugli Hymenoptera Chalcidoidea XXXVIII. Descrizione del mascio di *Litus* Haliday (1833) (Mymaridae). *Bollettino del Laboratorio di Entomologia Agraria 'Filippo Silvestri', Portici*, **30**: 231–234.

Viggiani, G. 1981. Note su alcune specie di *Oligosita* Walker (Hym. Trichogrammatidae) e descrizione di quattro nuove specie. *Bollettino del Laboratorio di Entomologia Agraria 'Filippo Silvestri', Portici*, **38**: 125–132.

Viggiani, G. 1984. Bionomics of the Aphelinidae. *Annual Review of Entomology*, **29**: 257–276.

Viggiani, G. & Battaglia, D. 1983*a*. Courtship and mating behaviour in a few Aphelinidae (Hym. Chalcidoidea). *Bolletino del Laboratorio di Entomologia Agraria 'Filippo Silvestri', Portici*, **40**: 89–96.

Viggiani, G. & Battaglia, D. 1983*b*. Courtship and mating behaviour of two species of *Amitus* Hald. *(Hymenoptera Platygastridae)*. *Bollettino del Laboratorio di Entomologia Agraria 'Filippo Silvestri', Portici*, **40**: 115–118.

Viggiani, G. & Garonna, A.P. 1991 (1993). Le specie italiane del complesso *Archenomus* Howard, *Archenomiscus* Nikolskaja, *Hispaniella* Mercet e *Pteroptrix* Westwood, con nuove combinazioni generiche (Hymenoptera: Aphelinidae). *Bollettino del Laboratorio di Entomologia Agraria 'Filippo Silvestri', Portici*, **48**: 57–88.

Viggiani, G. & Maresca, A. 1973. Ricerche sugli Hymenoptera Chalcidoidea XXXIV. Dati morfologici e biologici sulla *Leptomastidea abnormis* (Grlt.) (Hym. Encyrtidae), important parassita del *Planococcus citri* (Risso). *Bollettino del Laboratorio di Entomologia Agraria 'Filippo Silvestri', Portici*, **30**: 55–65.

Viggiani, G. & Mazzone, P. 1978. Morfologia, biologia e utilizzazione di *Prospaltella lahorensis* How. (Hym. Aphelinidae), parassita esotico introdotto in Italia per la lotta biologica al *Dialeurodes citri* Ashm. *Bollettino del Laboratorio di Entomologia Agraria 'Filippo Silverstri', Portici*, **35**: 99–161.

Viitasaari, M. 1984. Sahapistiäiset 3. Siricoidea, Orussoidea ja Cephoidea. *University of Helsinki, Department of Agricultural and Forest Zoology Reports*, **6**: 1–66.

Viktorov, G.A. 1966. [*Telenomus sokolovi* Mayr (Hymenoptera, Scelionidae) as a secondary parasite of the eggs of *Eurygaster integriceps* Put.]. *Doklady Akademia Nauk SSSR* (Biol.) **169**: 741–744. [In Russian.]

Vilela, E.F., Jaffé, K., & Howse, P.E. 1987. Orientation in leaf-cutting ants (Formicidae: Attini). *Animal Behaviour* **35**: 1443–1453.

Villa, J.D. & Weiss, M.R. 1990. Observations on the use of visual and olfactory cues by *Trigona* spp foragers. *Apidologie*, **21**: 541–545.

Vinson, S.B. 1968. Source of a substance in *Heliothis virescens* (Lepidoptera: Noctuidae) that elicits a searching response in its habitual parasite *Cardiochiles nigriceps* (Hymenoptera: Braconidae). *Annals of the Entomological Society of America*, **61**: 8–10.

Vinson, S.B. 1970. Development and possible functions of teratocytes in the host-parasite association. *Journal of Invertebrate Pathology*, **16**: 93–101.

Vinson, S.B. 1972. Courtship behavior and evidence for a sex pheromone in the parasitoid *Campoletis sonorensis* (Hymenoptera: Ichneumonidae). *Environmental Entomology*, **1**: 409–414.

Vinson, S.B. 1975. Biochemical coevolution between parasitoids and their hosts, pp. 14–48. *In*: Price, P.W. (ed.) *Evolutionary Strategies of Parasitic Insects and Mites*. 224pp. Plenus Press, New York.

Vinson, S.B. 1976. Host selection by insect parasitoids. *Annual Review of Entomology*, **21**: 109–133.

Vinson, S.B. 1978. Courtship behavior and source of a sexual pheromone from *Cardiochiles nigriceps*. *Annals of the Entomological Society of America*, **71**: 832–837.

Vinson, S.B. 1981. Habitat location, pp. 51–77. *In*: Nordland, D.A., Jones, R.L. & Lewis, W.J. (eds) *Semiochemicals. Their Role in Pest Control*. 306pp. Wiley & Sons, New York.

Vinson, S.B. 1984*a*. Parasitoid-host relationship, pp. 205–233. *In*: Bell, W.J. & Cardé, R.T. (eds) *Chemical Ecology of Insects*. 524pp. Chapman & Hall, London.

Vinson, S.B. 1984*b*. How parasitoids locate their hosts: a case of insect espionage, pp. 325–348. *In*: Lewis, T. (ed.) *Insect Communication*. 414pp. Academic Press, London.

Vinson, S.B. 1985. The behavior of parasitoids, pp. 417–469. *In*: Kerkut, G.A. & Gilbert, L.I. (eds) *Comprehensive Insect Physiology Biochemistry and Pharmacology, 9. Behavior*. 735pp. Pergamon Press, New York.

Vinson, S.B. (ed.) 1986. *Economic Impact and Control of Social Insects*, 421pp. Praeger, New York.

Vinson, S.B. 1990*a*. How parasitoids deal with the immune system of their host: an overview. *Archives of Insect Biochemistry and Physiology*, **13**: 3–27.

Vinson, S.B. 1990*b*. Potential impact of microbial insecticides on beneficial arthropods in the terrestrial environment, pp. 43–64. *In*: Laird, M., Lacey, L.A. & Davidson, E.W. (eds), *Safety of Microbial Insecticides*, 259pp. CRC Press, Boca Raton, Florida.

Vinson, S.B. 1991. Chemical signals used by parasitoids. *Redia*, Appendice, **74** (3): 15–42.

Vinson, S.B. & Ables, J.R. 1980. Interspecific competition among endoparasitoids of tobacco budworm larvae (Lep.: Noctuidae). *Entomophaga*, **25**: 357–362.

Vinson, S.B. & Barbosa, P. 1987. Interrelationships of nutritional ecology of parasitoids, pp. 673–695. *In*: Slansky, F. Jr & Rodriguez, J.G. (eds) *Nutritional Ecology of Insects, Mites, Spiders, and Related Invertebrates*. 1016pp. Wiley, New York.

Vinson S.B. & Barras, D.J. 1970. Effects of the parasitoid, *Cardiochiles nigriceps*, on the growth, development, and tissues of *Heliothis virescens*. *Journal of Insect Physiology*, **16**: 1329–1338.

Vinson, S.B. & Frankie, G.W. 1977. Nests of *Centris aethyctera* (Hymenoptera: Apoidea: Anthophoridae) in the dry forest of Costa Rica. *Journal of the Kansas Entomological Society*, **50**: 301–311.

Vinson, S.B. & Frankie, G.W. 1988. A comparative study of the ground nests of *Centris flavifrons* and *Centris aethiocesta* (Hymenoptera: Anthophoridae). *Entomologia Experientalis et Applicata*, **49**: 181–187.

Vinson, S.B. & Frankie, G.W. 1991. Nest variability in *Centris aethyctera* (Hymenoptera: Anthophoridae) in response to nesting site conditions. *Journal of the Kansas Entomological Society*, **64**: 156–162.

Vinson, S.B., Frankie, G.W. & Barthell, J. 1993. Threats to the diversity of solitary bees in a neotropical dry forest in Central America, pp. 53–81. *In*: LaSalle, J. & Gauld, I.D. (eds), *Hymenoptera and Biodiversity*, 348pp. C.A.B International, Wallingford.

Vinson, S.B., Frankie, G.W. & Coville, R.E. 1987. Nesting habits of *Centris flavofasciata* Friese (Hymenoptera: Apoidea: Anthophoridae) in Costa Rica. *Journal of the Kansas Entomological Society*, **60**: 249–263.

Vinson, S.B., Frankie, G.W. & Williams, H.J. 1986. Description of a new dorsal mesosomal gland in two *Xylocopa* species (Hymenoptera: Anthophoridae) from Costa Rica. *Journal of the Kansas Entomological Society*, **59**: 185–189.

Vinson, S.B. & Guillot, F.S. 1972. Host marking: a source of a substance that results in host discrimination in insect parasitoids. *Entomophaga*, **17**: 241–245.

Vinson, S.B. & Iwantsch, G.F. 1980*a*. Host suitability for insect parasitoids. *Annual Review of Entomology*, **25**: 397–419.

Vinson, S.B. & Iwantsch, G.F. 1980*b*. Host regulation by insect parasitoids. *Quarterly Review of Biology*, **55**: 143–165.

Vinson, S.B. & Lewis, W.J. 1973. Teratocytes: growth and numbers in the hemocoel of *Heliothis virescens* attacked by *Microplitis croceipes*. *Journal of Invertebrate Pathology*, **22**: 351–355.

Vinson, S.B. & MacKay, W.P. 1990. Effects of the fire ant, *Solenopsis invicta*, on electrical circuits and equipment, pp.496–503. *In*: Vander Meer, R.K., Jaffe, K. & Cedeno, A. (eds). *Applied Myrmecology, A World Perspective*, 741pp. Westview Press, Boulder.

Vinson, S.B. & Piper, G.L. 1986. Source and characterization of host recognition kairomones of *Tetrastichus hagenowii*, a parasitoid of cockroach eggs. *Physiological Entomology*, **11**: 459–468.

Vinson, S.B. & Scarborough, T.A. 1991. Interactions between *Solenopsis invicta* (Hymenoptera: Formicidae), *Rhopalosiphum maidis* (Homoptera: Aphididae), and the parasitoid *Lysiphlebus testaceipes* Cresson (Hymenoptera: Aphidiidae). *Annals of the Entomological Society of America*, **84**: 158–164.

Vinson, S.B. & Scott, J.R. 1974. Ultrastructure of teratocytes of *Cardiochiles nigriceps* Viereck (Hymenoptera: Braconidae). *International Journal of Insect Morphology and Embryology*, **3**: 293–304.

Vinson, S.B. & Sroka, P. 1978. Effects of superparasitism by a solitary endoparasitoid on the host, parasitoid and field samplings. *Southwestern Entomologist*, **3**: 299–304.

Vinson, S.B. & Stoltz, D.B. 1986. Cross-protection experiments with two parasitoid (Hymenoptera; Ichneumonidae) viruses. *Annals of the Entomological Society of America*, **79**: 216–218.

Vinson, S.B. & Williams, H.J. 1991. Host selection behavior of *Campoletis sonorensis*: a model system. *Biological Control*, **1**: 107–117.

Vinson, S.B., Williams, H.J., Frankie, G.W. & Coville, R.E. 1984. Comparative morphology and chemical contents of male mandibular glands of several *Centris* species (Hymenoptera: Anthophoridae) in Costa Rica. *Comparative Biochemistry and Physiology*. (A. Comparative Physiology), **77**: 685–688.

Vinson, S.B., Williams, H.J., Frankie, G.W., Wheeler, J.W., Blum, M.S. & Coville, R.E. 1982. Mandibular glands of male *Centris adani*, (Hymenoptera: Anthophoridae) their morphology, chemical constituents, and function in scent marking and territorial behavior. *Journal of Chemical Ecology*, **8**: 319–327.

Vinson, S.B., Williams, H.J., McAuslane, H.J. & Frankie, G.W. 1989. Chemical contents of male mandibular glands of three *Centris* species (Hymenoptera:

Anthophoridae) from Costa Rica. *Comparative Biochemistry and Physiology.* (B. Comparative Biochemistry), **93**: 73–75.

Visser, M.E., Alphen, J.J.M. van & Nell, H.W. 1992. Adaptive superparasitism and patch time allocation in solitary parasitoids: the influence of pre-patch experience. *Behavioral Ecology and Sociobiology*, **31**: 163–171.

Voegelé, J., Brun, P. & Daumal, J. 1974. Les trichogrammes. I. Modalites de la prise de possession et de l'elimination de l'hôte chez le parasite embryonnaire *Trichogramma brasiliensis. Annals Société d'Entomologie de France* (N.S.), **10**: 757–762.

Voegele, J. & Pintureau, B. 1982. Caractérisation morphologique des groupes et espèces du genre *Trichogramma* Westwood, pp. 45–75. *In*: *Les Trichogrammes*. Ier Symposium International Antibes, 20–23 Avril 1982. 307pp. Les Colloques de l'INRA, No. 9. Versailles.

Vogt, E.A. & Nechols, J.R. 1991. Diel activity patterns of the squash bug egg parasitoid *Gryon pennsylvanicum* (Hymenoptera: Scelionidae). *Annals of the Entomological Society of America*, **84**: 303–308.

Volker, K.C. & Simpson, R.G. 1975. Behavior of alfalfa weevil larvae affecting the establishment of *Tetrastichus incertus* in Colorado. *Environmental Entomology*, **4**: 742–744.

Völkl, W. & Mackauer, M. 1990. Age-specific pattern of host discrimination by the aphid parasitoid *Ephedrus californicus* Baker (Hymenoptera: Aphidiidae). *Canadian Entomologist*, **122**: 349–361.

Völkl, W. & Stadler, B. 1991. Interspecific larval competition between *Lysiphlebus testaceipes* and *Aphidius colemani* (Hym., Aphidiidae). *Journal of Applied Entomology*, **111**: 63–71.

Voynovich, N.D. & Sugonjaev, E.S. 1989. Peculiarities of parasitism of chalcids (Hymenoptera, Chalcidoidea) on soft scales (Homoptera, Coccoidea). IV. Biology and preimaginal stages of *Microterys lunatys* [sic] Dalman—parasite and predator of *Physokermes hemicryphus* Dalman. *Trudy Zoologicheskogo Instituta Akademiya Nauka SSSR*, **191**: 31–41. [In Russian.]

Waage, J.K. 1979. Foraging for patchily-distributed hosts by the parasitoid *Nemeritus canescens. Journal of Animal Ecology*, **48**: 353–371.

Waage, J.K. 1982. Sib-mating and sex ratio strategies in scelionid wasps. *Ecological Entomology*, **7**: 103–112.

Waage, J. 1986. Family planning in parasitoids: adaptive patterns of progeny and sex allocation, pp. 63–95. *In*: Waage, J. & Greathead, D. (eds) *Insect Parasitoids*. 389pp. Academic Press, London.

Waage, J.K. & Godfray, H.C.J. 1985. Reproductive strategies and population ecology of insect parasitoids, pp. 449–470. *In*: Sibley, R.M. & Smith, R.H. (eds) *Behavioural Ecology, Ecological Consequences of Adaptive Behaviour*. 620pp. Blackwells, Oxford.

Waage, J. & Greathead, D. 1986. *Insect Parasitoids*. 389pp. Academic Press, London.

Waage, J.K. & Lane, J.A. 1984. The reproductive strategy of a parasitic wasp II. Sex allocation and local mate competition in *Trichogramma evanescens. Journal of Animal Ecology*, **53**: 417–426.

Waage, J.K. & Ng, S.M. 1984. The reproductive strategy of a parasitic wasp I. Optimal progeny and sex allocation in *Trichogramma evanescens. Journal of Animal Ecology*, **53**: 401–415.

Wago, H. & Tanaka, T. 1989. Synergistic effects of calyx fluid and venom of *Apanteles kariyai* Watanabe (Hymenoptera: Braconidae) on the granular cells of

Pseudaletia separata Walker (Lepidoptera: Noctuidae). *Zoological Science*, **6**: 691–696.

Wahl, D.B. 1986. Larval structures of oxytorines and their significance for the highe-classification of some Ichneumonidae (Hymenoptera). *Systematic Entomology*, **11**: 117–127.

Wahl, D.B. 1990. A review of the mature larvae of Diplazontinae, with notes on larvae of Acaenitinae and Orthocentrinae and proposal of two new subfamilies (Insecta: Hymenoptera, Ichneumonidae). *Journal of Natural History*, **24**: 27–52.

Wahl, D.B. 1993*a*. Cladistics of the genera of Mesochorinae (Hymenoptera: Ichneumonidae). *Systematic Entomology*, **18**: 371–387.

Wahl, D. 1993*b*. Cladistics of the ichneumonid subfamily Labeninae (Hymenoptera: Ichneumonidae). *Entomologia Generalis*, **18**: 91–105.

Walker, A.K., Joshi, N.K. & Verma, S.K. 1990*a*. The biosystematics of *Syntretomorpha szaboi* Papp (Hymenoptera: Braconidae: Euphorinae) attacking the Oriental honey bee, *Apis cerana* Fabricius (Hymenoptera: Apidae), with a review of braconid parasitoids attacking bees. *Bulletin of Entomological Research*, **80**: 79–83.

Walker, A.K., Kitching, I.J. & Austin, A.D. 1990*b*. A reassessment of the phylogenetic relationships within the Microgastrinae (Hymenoptera: Braconidae). *Cladistics*, **6**: 291–306.

Walker, A.K. & Huddleston, T. 1987. New Zealand chelonine braconid wasps (Hymenoptera). *Journal of Natural History*, **21**: 339–361.

Walker, M.F. 1961. Some observations on the biology of the ladybird parasite, *Perilitus coccinellae* (Schrank) (Hym., Braconidae), with special reference to host selection and recognition. *Entomologist's Monthly Magazine*, **97**: 240–244.

Walker, P. & Crane, E. 1987. Constituents of propolis. *Apidologie*, **18**: 327–334.

Walkley, L.M. 1956. A tribal revision of the brachycyrtine wasps of the world (Cryptinae-Ichneumonidae). *Proceedings of the United States National Museum*, **106**: 315–329.

Wall, R. & Berberet, R.C. 1974. The life cycle of *Euplectrus plathypenae*, a gregarious external parasitoid of peanut foliage feeders in Oklahoma. *Environmental Entomology*, **3**: 744–746.

Waller, D.A. & Moser, J.C. 1990. Invertebrate enemies and nest associates of the leaf-cutting ant *Atta texana* (Buckley) (Formicidae, Attini), pp. 255–273. *In*: Vander Meer, R.K., Jaffe, K. & Cedeno, A. (eds). *Applied Myrmecology, A World Perspective*, 741pp. Westview Press, Boulder.

Waloff, N. 1961. Observations on the biology of *Perilitus dubius* (Wesmael) (Hymenoptera: Braconidae), a parasite of the chrysomelid beetle *Phytodecta olivacea* Forster. *Proceedings of the Royal Entomological Society of London* (A), **36**: 96–102.

Waloff, N. 1967. Biology of three species of *Leiophron* (Hymenoptera: Braconidae, Euphorinae) parasitic on Miridae on broom. *Transactions of the Royal Entomological Society of London*, **119**: 187–213.

Waloff, N. 1974. Biology and behaviour of some species of Dryinidae (Hymenoptera). *Journal of Entomology* (A), **49**: 97–109.

Waloff, N. 1975. The parasitoids of the nymphal and adult stages of leafhoppers (Auchenorrhyncha: Homoptera) of acidic grassland. *Transactions of the Royal Entomological Society of London*, **126**: 637–686.

Walter, G.H. 1983. 'Divergent male ontogenies' in Aphelinidae (Hymenoptera, Chalcidoidea): a simplified classification and a suggested evolutionary sequence. *Biological Journal of the Linnean Society of London*, **19**: 63–82.

Walter, G.H. 1988. Activity patterns and egg production in *Coccophagus bartletti*, an aphelinid parasitoid of scale insects. *Ecological Entomology*, **13**: 95–105.

Walther, J.R. 1979. Vergleichende morphologische Bertrachtung der antennalen Sensillenfelder einiger ausgewählter Aculeata (Insecta, Hymenoptera). *Zeitschrift für Zoologische Systematik und Evolutionsforschung*, **17**: 30–56.

Ward, P.S. 1988. Mesic elements in the western Nearctic ant fauna: taxonomic and biological notes on *Amblyopone*, *Proceratium*, and *Smithistruma* (Hymenoptera: Formicidae). *Journal of the Kansas Entomological Society*, **61**: 102–124.

Ward, P.S. 1989. Systematic studies on pseudomyrmicine ants: revision of the *Pseudomyrmex oculatus* and *P. subtilissimus* species groups, with taxonomic comments on other species. *Quaestiones Entomologicae*, **25**: 393–468.

Ward, P.S. 1990. The ant subfamily Pseudomyrmecinae (Hymenoptera: Formicidae): generic revision and relationship to other formicids. *Systematic Entomology*, **15**: 449–489.

Ward, P.S. 1993. Systematic studies on *Pseudomyrmex* acacia-ants (Hymenoptera: Formicidae: Pseudomyrmecinae). *Journal of Hymenoptera Research*, **2**: 117–168.

Ward, S. & Gauld, I. 1987. The callajoppine parasitoids of sphingids in Central America (Hymenoptera: Ichneumonidae). *Systematic Entomology*, **12**: 503–508.

Wasbauer, M.S. 1955. Observations on the biology of *Anoplius fulgidus* Cresson. *Pan-Pacific Entomologist*, **31**: 90–92.

Wasbauer, M.S. 1957. A biological study of *Anoplius* (*Anoplius*) *imbellis* Banks (Hymenoptera: Pompilidae). *Wasmann Journal of Biology*, **15**: 81–97.

Wasbauer, M.S. 1968. New genera of male Brachycistidinae with a redescription of *Brachycistellus* Baker and a key to North American genera. *Pan-Pacific Entomologist*, **44**: 184–197.

Wasbauer, M.S. 1982 (1983). Prey records for some North American spider wasps (Hymenoptera: Pompilidae). *Pan-Pacific Entomologist*, **58**: 223–230.

Wasbauer, M.S. 1987. A new genus of Ageniellini from Central America (Hymenoptera: Pompilidae: Pepsinae). *Psyche*, **94**: 181–187.

Wasbauer, M.S. & Kimsey, L.S. 1985. California spider wasps of the subfamily Pompilinae (Hymenoptera: Pompilidae). *Bulletin of the California Insect Survey*, **26**: 1–130.

Wasbauer, M.S. & Leech, H.B. 1973. A biological note on two species of *Ageniella* from California (Hymenoptera: Pompilidae) *Pan-Pacific Entomologist*, **49**: 182–183.

Washburn, J.O. 1984. Mutualism between a cynipid gall wasp and ants. *Ecology*, **65**: 654–656.

Watanabe, C. 1972. A revision of the Helconini of Japan and a review of helconine genera of the world (Hymenoptera, Braconidae). *Insecta Matsumurana*, **35**: 1–18.

Watkins II, J.F. 1976. *The Identification and Distribution of New World Army Ants (Dorylinae: Formicidae)*. 102pp. Baylor University Press, Waco, Texas.

Way, M.J. 1963. Mutualism between ants and honeydew-producing Homoptera. *Annual Review of Entomology*, **8**: 307–344.

Way, M.J. & Khoo, K.C. 1992. Role of ants in pest management. *Annual Review of Entomology*, **37**: 479–503.

Wcislo, W.T. 1987. The roles of seasonality, host synchrony, and behaviour in the evolutions and distributions of nest parasites in Hymenoptera (Insecta), with special reference to bees (Apoidea). *Biological Review*, **62**: 515–543.

Wcislo, W.T. 1990. A new species of *Lasioglossum* from Costa Rica (Hymenoptera: Halictidae). *Journal of the Kansas Entomological Society*, **63**: 450–453.

Wcislo, W.T., West-Eberhard, M.J. & Eberhard, W.G. 1988. Natural history and behavior of a primitively social wasp, *Auplopus semialatus*, and its parasite, *Irenangelus eberhardi* (Hymenoptera: Pompilidae). *Journal of Insect Behavior*, **1**: 247–260.

Weaver, N. & Weaver, E.C. 1981. Beekeeping with the stingless bee *Melipona beecheii*, by the Yucatecan Maya. *Bee World*, **62**: 7–19.

Weber, N.A. 1943. Parabiosis in neotropical 'ant gardens'. *Ecology*, **24**: 400–404.

Weber, N.A. 1944. The Neotropical coccid-tending ants of the genus *Acropyga* Roger. *Annals of the Entomological Society of America*, **37**: 89–122.

Weber, N.A. 1966. Fungus growing ants. *Science*, **153**: 587–604.

Weber, N.A. 1972. *Gardening Ants: the Attines*. 146pp. American Philosophical Society, Philadelphia.

Weber, N.A. 1982. Fungus ants, pp. 255–363. *In*: Hermann, H.R. (ed.) *Social Insects*, IV. 385pp. Academic Press, New York.

Weinstein, P. 1989. Cycloalexy in an Australian pergid sawfly (Hymenoptera, Pergidae). *Bulletin et Annales de la Sociètè Royale Belgique de Entomologique*, **125**: 53–60.

Weinstein, P. & Austin, A.D. 1991. The host relationships of trigonalyid wasps (Hymenoptera: Trigonalyidae), with a review of their biology and catalogue to world species. *Journal of Natural History*, **25**: 399–433.

Weiss, M.J. & Williams, R.N. 1980. Some host-parasite relationships of *Microctonus nitidulidis* and *Stelidota geminata*. *Annals of the Entomological Society of America*, **73**: 323–326.

Weld, L.H. 1925. Another gall that secretes honeydew. *Bulletin of the Brooklyn Entomological Society*, **20**: 175–179.

Weld, L.H. 1952. *Cynipoidea (Hym.) 1905–1950*. 351pp. (Privately printed) Ann Arbor, Michigan.

Weld, L.H. 1957. *Cynipid Galls of the Pacific Slope*. 64pp. (Privately printed) Ann Arbor, Michigan.

Weld, L.H. 1959. *Cynipid Galls of the Eastern United States*. 124pp. (Privately printed) Ann Arbor, Michigan.

Weld, L.H. 1960. *Cynipid Galls of the Southwest*. 35pp. (Privately printed) Ann Arbor, Michigan.

Wenzel, J.W. 1987. Male reproductive behavior and mandibular glands in *Polistes major* (Hymenoptera: Vespidae). *Insectes Sociaux*, **34**: 44–57.

Wenzel, J.W. 1991. Evolution of nest architecture, pp. 480–519. *In*: Ross K.G. & Matthews, R.W. (eds) *The Social Biology of Wasps*. 678pp. Cornell University Press, Ithaca.

Werner, F.G. 1961 (1960). A note on the prey and nesting site of *Cerceris truncata* Cameron (Hymenoptera: Sphecidae). *Psyche*, **67**: 43–44.

Werner, J.J. & Williams, R.N. 1985. Observations on the development and behavior of *Zeteticontus utilis* (Hymenoptera: Encyrtidae), a parasite of *Carpophilus* spp. (Coleoptera: Nitidulidae). *Journal of Entomological Science*, **20**: 450–453.

Werren, J.H. 1980. Sex ratio adaptations to local mate competition in a parasitic wasp. *Science*, **208**: 1157–1159.

Werren, J.H. 1983. Sex ratio evolution under local mate competition in a parasitic wasp. *Evolution*, **37**: 116–124.

Werren, J.H. 1991. The paternal-sex-ratio chromosome of *Nasonia*. *American Naturalist*, **137**: 392–402.

Werren, J.H. 1993. The evolution of inbreeding in haplodiploid organisms, pp. 42–59. *In*: Thornhill, N.W. (ed.), *The Natural History of Inbreeding and Outbreeding, Theoretical and Empirical Perspectives*. 575pp. University of Chicago Press, Chicago.

Werren, J.H. & Assem, J. van den, 1986. Experimental analysis of a paternally inherited extrachromosomal factor. *Genetics*, **114**: 217–233.

Weseloh, R.M. 1972. Spatial distribution of the gypsy moth (Lepidoptera: Lymantriidae) and some of its parasitoids within a forest environment. *Entomophaga*, **17**: 339–351.

Weseloh, R.M. 1974. Host recognition by the gypsy moth larval parasitoid, *Apanteles melanoscelus*. *Annals of the Entomological Society of America*, **67**: 583–587.

Weseloh, R.M. 1976. Behavioral responses of the parasite, *Apanteles melanoscelus*, to gypsy moth silk. *Environmental Entomology*, **5**: 1128–1132.

Weseloh, R.M. 1981. Host location by parasitoids, pp. 79–95. *In*: Nordland, D.A., Jones, R.L. & Lewis, W.J. (eds) *Semiochemicals. Their Role in Pest Control*. 306pp. Wiley & Sons, New York.

Weseloh, R.M. & Andreadis, T.G. 1982. Possible mechanism for synergism between *Bacillus thuringiensis* and the gypsy moth (Lepidoptera: Lymantriidae) parasitoid, *Apanteles melanoscelus* (Hymenoptera: Braconidae). *Annals of the Entomological Society of America*, **75**: 435–438.

Weseloh, R.M. & Bartlett, B.R. 1971, Influence of chemical characteristics of the secondary scale host on host selection behavior of the hyperparasite *Cheiloneurus noxius* (Hymenoptera: Encyrtidae). *Annals of the Entomological Society of America*, **64**: 1259–1264.

West, K.J. & Miller, J.C. 1989. Patterns of host exploitation by *Meteorus communis* (Hymenoptera: Braconidae). *Environmental Entomology*, **18**: 537–540.

West-Eberhard, M.J. 1969. The social biology of polistine wasps. *Miscellaneous Publications. Museum of Zoology, University of Michigan*, **140**: 1–101.

West Eberhard, M.J. 1973. Monogyny in 'polygynous' social wasps. *Proceedings of the VIIth International Congress of the International Union for the Study of Social Insects*, **7**: 396–403.

West Eberhard, M.J. 1975. The evolution of social behavior by kin selection. *Quarterly Review of Biology*, **50**: 1–33.

West-Eberhard, M.J. 1977. The establishment of reproductive dominance in social wasp colonies. *Proceedings of the Eighth International Congress of the International Union for the Study of Social Insects*, **8**: 223–227.

West-Eberhard, M.J. 1978*a*. Polygyny and the evolution of social behavior in wasps. *Journal of the Kansas Entomological Society*, **51**: 832–856.

West-Eberhard, M.J. 1978*b*. Temporary queens in *Metapolybia* wasps: nonreproductive helpers without altruism? *Science*, **200**: 441–443.

West-Eberhard, M.J. 1981. Intragroup selection and the evolution of insect societies. pp. 3–17. *In*: Alexander, R.D. & Tinkle, D.W. (eds). *Natural Selection and Social Behavior*. 532pp. Chiron Press, New York.

West-Eberhard, M.J. 1982. The nature and evolution of swarming in tropical social wasps (Vespidae, Polistinae, Polybiini), pp 97–128. *In*: Jaisson, P. (ed.) Social Insects in the Tropics. *Proceedings of the First International Symposium organised by the International Union for the Study of Social Insects and the Sociedad Mexicana de Entomologia*. 280pp. Université Paris Nord.

West-Eberhard, M.J. 1983. Sexual selection, social competition, and speciation. *Quarterly Review of Biology*, **58**: 155–183.

West-Eberhard, M.J. 1986. Dominance relations in *Polistes canadensis* (L.), a tropical social wasp. *Monitor Zoologico Italiano* (N.S.), **20**: 263–281.

West-Eberhard, M.J. 1987. Flexible strategy and social evolution, pp. 35–51. *In*: Ito, Y., Brown, J.L. & Kikkawa, J. (eds) *Animal Societies: Theories and Facts*, 291pp. Japan Scientific Societies Press, Tokyo.

West-Eberhard, M.J. 1989. Scent-trail diversion, a novel defense against ants by tropical social wasps. *Biotropica*, **21**: 280–281.

West-Eberhard, M.J. 1990. The genetic and social structure of polygynous social wasp colonies (Vespidae: Polistinae), pp. 254–255. *In*: Veeresh, G.K., Mallik, B. & Viraktamath, C.A. (eds). *Social Insects and the Environment. Proceedings of the 11th International Congress of IUSSI, 1990*. 765pp. Oxford & IBH, New Delhi.

Westerkamp, C. 1991. Honeybees are poor pollinators—why? *Plant Systematics and Evolution*, **177**: 71–75.

Wetterer, J.K. 1990*a*. Diel changes in forager size, activity, and load selectivity in a tropical leaf-cutting ant, *Atta cephalotes*. *Ecological Entomology*, **15**: 97–104.

Wetterer, J.K. 1990*b*. Load-size determination in the leaf-cutting ant, *Atta cephalotes*. *Behavioral Ecology*, **1**: 95–101.

Wetterer, J.K. 1991*a*. Allometry and the geometry of leaf-cutting in *Atta cephalotes*. *Behavioral Ecology and Sociobiology*, **29**: 347–351.

Wetterer, J.K. 1991*b* (1992). Foraging ecology of the leaf-cutting ant *Acromyrmex octospinosus* in a Costa Rican rain forest. *Psyche*, **98**: 361–371.

Wetterer, J.K. 1993. Foraging and nesting ecology of a Costa Rican leaf-cutting ant, *Acromyrmex volcanus*. *Psyche*, **100**: 65–76.

Wharton, R. 1977*a*. Exodontiellini, a new tribe of Opiinae with exodont mandibles (Hymenoptera: Braconidae). *Pan-Pacific Entomologist*, **53**: 297–303.

Wharton, R.A. 1977*b*. New world *Aphaereta* species (Hymenoptera: Braconidae: Alysiinae), with a discussion of terminology used in the tribe Alysiini. *Annals of the Entomological Society of America*, **70**: 782–803.

Wharton, R.A. 1980. Review of the nearctic Alysiini (Hymenoptera, Braconidae) with discussion of generic relationships within the tribe. *University of California Publications, Entomology*, **88**: 1–112.

Wharton, R.A. 1984. Biology of the Alysiini (Hymenoptera: Braconidae), parasitoids of cyclorrhaphous Diptera. *Texas Agricultural Experiment Station, Technical Monograph*, **11**: 1–39.

Wharton, R.A. 1985. Characterization of the genus *Aspilota* (Hymenoptera: Braconidae). *Systematic Entomology*, **10**: 227–237.

Wharton, R.A. 1986. The braconid genus *Alysia* (Hymenoptera): a description of the subgenera and a revision of the subgenus *Alysia*. *Systematic Entomology*, **11**: 453–504.

Wharton, R.A. 1987. Changes in nomenclature and classification of some opiine Braconidae (Hymenoptera). *Proceedings of the Entomological Society of Washington*, **89**: 61–73.

Wharton, R.A. 1988*a*. Classification of the braconid subfamily Opiinae (Hymenoptera). *Canadian Entomologist*, **120**: 333–360.

Wharton, R.A. 1988*b*. The braconid genus Alysia (Hymenoptera): A revision of the subgenus *Anarcha*. *Contributions of the American Entomological Institute*, **25** (2): 1–69.

Wharton, R.A. 1989*a*. Classical biological control of fruit-infesting Tephritidae, pp. 303–313. *In*: Robinson, A.S. & Hooper, G. (eds) *Fruit Flies, Their Biology, Natural Enemies and Control, 3B*. 447pp. Elsevier Science Publishers, Amsterdam.

Wharton, R.A. 1989*b*. Final instar larva of the embolemid wasp, *Ampulicomorpha confusa* (Hymenoptera). *Proceedings of the Entomological Society of Washington*, **91**: 509–512.

Wharton, R.A. 1993*a*. Bionomics of the Braconidae. *Annual Review of Entomology*, **38**: 121–143.

Wharton, R.A. 1993*b*. Review of the Hormiini (Hymenoptera: Braconidae) with description of new taxa. *Journal of Natural History*, **27**: 107–171.

Wharton, R.A. & Gilstrap, F.E. 1983. Key to and status of opiine braconid (Hymenoptera) parasitoids used in biological control of *Ceratitis* and *Dacus* s.l. (Diptera: Tephritidae). *Annals of the Entomological Society of America*, **76**: 721–742.

Wharton, R.A., Gilstrap, F.E., Rhode, R.H., Fischel-M., M. & Hart, W.G. 1981. Hymenopterous egg-pupal and larval-pupal parasitoids of *Ceratitis capitata* and *Anastrepha* spp. (Dip.: Tephritidae) in Costa Rica. *Entomophaga*, **26**: 285–290.

Wharton, R.A. & Marsh, P.M. 1978. New World Opiinae (Hymenoptera: Braconidae) parasitic on Tephritidae (Diptera). *Journal of the Washington Academy of Sciences*, **68**: 147–167.

Wharton, R.A., Shaw, S.R., Sharkey, M.J., Wahl, D.B., Woolley, J.B., Whitfield, J.B., Marsh, P.M. & Johnson, W. 1992. Phylogeny of the subfamilies of the family Braconidae (Hymenoptera: Ichneumonoidea): a reassessment. *Cladistics*, **8**: 199–235.

Wharton, R.A., Smith, J.R., Quicke, D.L.J., & Browning, H.W. 1989. Two new species of *Digonogastra* Viereck (Hymenoptera: Braconidae) parasitic on Neotropical pyralid borers (Lepidoptera) in maize, sorghum and sugarcane. *Bulletin of Entomological Research*, **79**: 401–410.

Wheeler, D.E. 1984. Behavior of the ant, *Procryptocerus scabriusculus* (Hymenoptera: Formicidae), with comparisons to other cephalotines. *Psyche*, **91**: 171–192.

Wheeler, D.E. 1986*a*. *Ectatomma tuberculatum*: foraging biology and association with *Crematogaster* (Hymenoptera: Formicidae). *Annals of the Entomological Society of America* **79**: 300–303.

Wheeler, D.E. 1986*b*. Polymorphism and division of labor in *Azteca chartifex* laticeps (Hymenoptera: Formicidae). *Journal of the Kansas Entomological Society*, **59**: 542–548.

Wheeler, G.C. & Wheeler, E.W. 1937. New hymenopterous parasites of ants (Chalcidoidea: Eucharidae). *Annals of the Entomological Society of America*, **30**: 163–175.

Wheeler, G.C. & Wheeler, J. 1976. Ant larvae: review and synthesis. *Memoirs of the Entomological Society of Washington*, **7**: 1–108.

Wheeler, W.M. 1904. A crustacean-eating ant (*Leptogenys elongata* Buckley). *Biological Bulletin of the Marine Biological Laboratory, Woods Hole*, **6**: 251–259.

Wheeler, W.M. 1907. The polymorphism of ants, with an account of some singular abnormalities due to parasitism. *Bulletin of the American Museum of Natural History*, **23**: 1–94.

Wheeler, W.M. 1910. *Ants. Their Structure, Development and Behavior*. 663pp. Columbia University Press, New York.

Wheeler, W.M. 1921. A new case of parabiosis and the 'ant gardens' of British Guiana. *Ecology*, **2**: 89–103.

Wheeler, W.M. 1925. A new guest-ant and other new Formicidae from Barro Colorado Island, Panama. *Biological Bulletin of the Marine Biological Laboratory, Woods Hole*, **49**: 150–181.

Wheeler, W.M. 1928. *The Social Insects Their Origin and Evolution*. 378pp. Harcourt Brace, New York.

Wheeler, W.M. 1936. Ecological relations of ponerine and other ants to termites. *Proceedings of the American Academy of Arts and Sciences*, **71**: 159–243.

Wheeler, W.M. 1942. Studies of neotropical ant-plants and their ants. *Bulletin of the Museum of Comparative Zoology*, **90**: 1–262.

White, M.J.D. 1973. *Animal Cytology and Evolution*. 3rd edition. 961pp. Cambridge University Press, Cambridge.

Whitehead, D.R. 1975. Parasitic Hymenoptera associated with bruchid-infested fruits in Costa Rica. *Journal of the Washington Academy of Science*, **65**: 108–116.

Whitfield, J.B. 1985. The nearctic species of *Deuterixys* Mason (Hymenoptera: Braconidae). *Pan-Pacific Entomologist*, **61**: 60–67.

Whitfield, J.B. 1987. Male swarming by a microgastrine braconid, *Apanteles coniferae* (Haliday) (Hymenoptera). *Proceedings and Transactions of the British Entomological and Natural History Society*, **20**: 133–135.

Whitfield, J.B. 1988*a*. Revision of the Nearctic species of the genus *Stiropius* Cameron (= *Bucculatriplex* Auct.) with the description of a new related genus (Hymenoptera: Braconidae). *Systematic Entomology*, **13**: 373–385.

Whitfield, J.B. 1988*b*. Taxonomic notes on Rhyssalini and Rhysipolini (Hymenoptera: Braconidae) with first nearctic records of three genera. *Proceedings of the Entomological Society of Washington*, **90**: 471–473.

Whitfield, J.B. 1988*c*. Two new species of *Paradelius* (Hymenoptera: Braconidae) from North America with biological notes. *Pan-Pacific Entomologist*, **64**: 313–319.

Whitfield, J.B. 1989. The limits and composition of the rogadine tribes. [Unpublished lecture outline distributed at the national meeting of the Entomological Society of America, Lexington, Kentucky.]

Whitfield, J.B. 1990*a*. Phylogenetic review of the *Stiropius* group of genera (Hymenoptera: Braconidae, Rogadinae) with description of a new Neotropical genus. *Proceedings of the Entomological Society of Washington*, **92**: 36–43.

Whitfield, J.B. 1990*b*. Symbiont-induced speciation and endoparasitoid insects. *Evolutionary Theory*, **9**: 211–213.

Whitfield, J.B. 1990*c*. Parasitoids, polydnaviruses and endosymbiosis. *Parasitology Today*, **6**: 381–384.

Whitfield, J.B. 1992. The polyphyletic origin of endoparasitism in the cyclostome lineages of Braconidae (Hymenoptera). *Systematic Entomology*, **17**: 273–286.

Whitfield, J.B. & Achterberg, C. van. 1987. Clarification of the taxonomic status of the genera *Cantharoctonus* Viereck, *Noserus* Foerster, and *Pseudavga* Tobias (Hymenoptera: Braconidae). *Systematic Entomology*, **12**: 509–518.

Whitfield, J.B., Johnson, N.F. & Hamerski, M.R. 1989. Identity and phylogenetic significance of the metapostnotum in nonaculeate Hymenoptera. *Annals of the Entomological Society of America*, **82**: 663–673.

Whitfield, J.B. & Mason, W.R.M. 1994. Mendesellinae, a new subfamily of braconid wasps (Hymenoptera: Braconidae) with a review of relationships within the microgastroid assemblage. *Systematic Entomology*, **19**: 61–76.

Whitfield, J.B. & Wagner, D.L. 1988. Patterns in host ranges within the Nearctic species of the parasitoid genus *Pholetesor* Mason (Hymenoptera: Braconidae). *Environmental Entomology*, **17**: 608–615.

Whitfield, J.B. & Wagner, D.L. 1991. Annotated key to the genera of Braconidae (Hymenoptera) attacking leafmining Lepidoptera in the Holarctic region. *Journal of Natural History*, **25**: 733–754.

Whiting, A.R. 1967. The biology of the parasitic wasp *Mormoniella vitripennis* [= *Nasonia brevicornis*] (Walker). *Quarterly Review of Biology*, **42**: 333–406.

Whitman, D.W. 1988. Plant natural products as parasitoid cuing agents, pp. 386–396. *In*: Cuttler, H.G. (ed.) *Biologically Active Natural Products Potential Use in Agriculture*. 496pp. American Chemical Society, Washington, D.C.

Whitman, D.W., Blum, M.S. & Alsop, D.W. 1990. Allomones: chemicals for defense. pp. 289–351. *In*: Evans, D.L. & Schmidt, J.O. (eds) *Insect Defenses*. 482pp. State University of New York Press, New York.

Whitman, D.W. & Eller, F.J. 1992. Orientation of *Microplitis croceipes* (Hymenoptera: Braconidae) to green leaf volatiles: dose-response curves. *Journal of Chemical Ecology*, **18**: 1743–1753.

Whitmore, J.L. (ed.) 1976. *Studies on the Shoot Borer* Hypsipyla grandella *(Zeller) Lep. Pyralidae. Vol.II.* 139pp. IICA Miscellaneous Publication No. 101, CATIE, Turrialba (Costa Rica).

Wibel, R.G, Cassidy, J.D., Buhse, H.E. Jr, Cummings, M.R., Bindokas, V.P., Charlesworth, J. & Baumgartner, D.L. 1984. Scanning electron microscopy of antennal sense organs of *Nasonia vitripennis* (Hymenoptera: Pteromalidae). *Transactions of the American Microscopical Society*, **103**: 329–340.

Wiebes, J.T. 1979. Co-evolution of figs and their insect pollinators. *Annual Review of Ecology and Systematics*, **10**: 1–12.

Wiebes, J.T. 1982. The phylogeny of the Agaonidae (Hymenoptera, Chalcidoidea). *Netherlands Journal of Zoology*, **32**: 395–411.

Wiebes, J.T. 1983. Records and descriptions of *Pegoscapus* Cameron (Hymenoptera Chalcidoidea, Agaonidae). *Proceedings of the Koninklijke Nederlandse Akademie van Wetenschappen*, Series C, **86**: 243–253.

Wiebes-Rijks, A.A. 1979. A character analysis of the species of *Synergus* Hartig, section II (Mayr, 1872) (Hymenoptera, Cynipidae). *Zoologische Mededelingen*, **53**: 297–321.

Wilbert, H. 1960. *Apanteles pieridis* (Bouché) (Hym., Braconidae), ein Parasit von *Aporia crataegi* (L.) (Lep., Pieridae). *Entomophaga*, **5**: 183–211.

Wilde, G., Schoonhoven, A. van & Gomez-Laverde, L. 1976. The biology of *Empoasca kraemeri on Phaseolus vulgaris*. *Annals of the Entomological Society of America*, **69**: 442–444.

Wilkinson, R.C. & Drooz, A.T. 1979. Oviposition, fecundity, and parasites of *Neodiprion excitans* from Belize, C.A. *Environmental Entomology*, **8**: 501–505.

Willard, H.F. 1920. *Opius fletcheri* as a parasite of the melon fly in Hawaii. *Journal of Agricultural Research*, **20**: 423–438.

Wille, A. 1961. Las abejas jicotes de Costa Rica. *Revista de la Universidad de Costa Rica*, **22**: 1–30.

Wille, A. 1962. A revision of the subgenus *Nogueirapis*; an archaic group of stingless bees (Hymenoptera: Apidae). *Journal of the New York Entomological Society*, **70**: 218–234.

Wille, A. 1963. Behavioral adaptations of bees for pollen collecting from *Cassia* flowers. *Revista de Biología Tropical*, **11**: 205–210.

Wille, A. 1964. Notes on a primitive stingless bee, *Trigona (Nogueirapis) mirandula*. *Revista de Biología Tropical*, **12**: 117–151.

Wille, A. 1965 (1966). Las abejas atarrá de la región mesoamericana del género y subgénero *Trigona* (Apidae-Meliponini). *Revista de Biología Tropical*, **13**: 271–291.

Wille, A. 1966 (1967). Notes on two species of ground nesting stingless bees (*Trigona mirandula* and *T. buchwaldi*) from the Pacific rain forest of Costa Rica. *Revista de Biología Tropical*, **14**: 251–277.

Wille, A. 1967 (1969). A new species of stingless bee *Trigona* (*Plebeia*) from Costa Rica, with descriptions of its general behavior and cluster-type nest. *Revista de Biología Tropical*, **15**: 299–313.

Wille, A. 1976. Las abejas jicotes del género *Melipona* (Apidae: Meliponini) de Costa Rica. *Revista de Biología Tropical*, **24**: 123–147.

Wille, A. 1979*a*. A comparative study of the pollen press and nearby structures in the bees of the family Apidae. *Revista de Biología Tropical*, **27**: 217–221.

Wille, A. 1979*b*. Phylogeny and relationships among the genera and subgenera of the stingless bees (Meliponinae) of the world. *Revista de Biología Tropical*, **27**: 241–277.

Wille, A. 1983. Biology of the stingless bees. *Annual Review of Entomology*, **28**: 41–64.

Wille, A. & Fuentes, G. 1975. Efecto de la ceniza del Volcán Irazú (Costa Rica) en algunos insectos. *Revista de Biología Tropical*, **23**: 165–175.

Wille, A. & Michener, C.D. 1971. Observations on the nests of Costa Rican *Halictus* with taxonomic notes on Neotropical species (Hymenoptera: Halictidae). *Revista de Biología Tropical*, **18**: 17–31.

Wille, A. & Michener, C.D. 1973. The nest architecture of stingless bees with special reference to those of Costa Rica (Hymenoptera: Apidae). *Revista de Biología Tropical*, **21**: 9–278.

Wille, A. & Orozco, E. 1969 (1970). The life cycle and behavior of the social bee *Lasioglossum (Dialictus) umbripenne*. (Hymenoptera: Halictidae). *Revista de Biología Tropical*, **17**: 199–245.

Wille, A. & Orozco, E. 1974 (1975). Observations on the founding of a new colony by *Trigona cupira* (Hymenoptera: Apidae) in Costa Rica. *Revista de Biología Tropical*, **22**: 253–287.

Wille, A., Orozco, E. & Raabe, C. 1983. Polinización del chayote *Sechium edule* (Jacq.) Swartz en Costa Rica. *Revista de Biología Tropical*, **31**: 145–154.

Williams, D.J.M. 1985. The new world genus *Lathrapanteles* n. gen.: phylogeny and placement in the Microgastrinae (Hymenoptera: Braconidae: Cotesiini). *Canadian Journal of Zoology*, **63**: 1962–1981.

Williams, F.X. 1919*a*. *Epyris extraneus* Bridwell (Bethylidae), a fossorial wasp that preys on the larva of the tenebrionid beetle, *Gonocephalum seriatum* (Boisduval). *Proceedings of the Hawaiian Entomological Society*, **4**: 55–63.

Williams, F.X. 1919*b*. Philippine wasp studies. Part II. Descriptions of new species and life history studies. *Bulletin of the Experiment Station of the Hawaiian Sugar Planters' Association, Entomological Series*. **14**: 19–186.

Williams, F.X. 1928. Studies in tropical wasps—their hosts and associates (with descriptions of new species). *Bulletin of the Experiment Station of the Hawaiian Sugar Planters' Association, Entomological Series*. **19**: 1–179.

Williams, F.X. 1931. *Handbook of the Insects and Other Invertebrates of Hawaiian Sugarcane Fields*. 400pp. Hawaiian Sugar Planters Association, Experiment Station, Honolulu.

Williams, F.X. 1938. Biological studies in Hawaiian water-loving insects Part III Diptera or Flies A, Ephydridae and Anthomyiidae. *Proceedings of the Hawaiian Entomological Society*, **10**: 85–115.

Williams, F.X. 1942. *Ampulex compressa* (Fabr.), a cockroach-hunting wasp introduced from New Caledonia into Hawaii. *Proceedings of the Hawaiian Entomological Society*, **11**: 221–233.

Williams, F.X. 1956. Life history studies of *Pepsis* and *Hemipepsis* wasps in California (Hymenoptera, Pompilidae). *Annals of the Entomological Society of America*, **49**: 447–466.

Williams, F.X. 1966. A spider-hunting wasp found nesting in hollow plant stems (Hymenoptera: Pompilidae: Pepsinae). *Wasmann Journal of Biology*, **24**: 33–47.

Williams, H.J., Vinson, S.B. & Frankie, G.W. 1987. Chemical content of the dorsal mesosomal gland of two *Xylocopa* species (Hymenoptera: Anthophoridae) from Costa Rica. *Comparative Biochemistry and Physiology*. (B. Comparative Biochemistry), **86**: 311–312.

Williams, H.J., Vinson, S.B., Frankie, G.W., Coville, R.E. & Ivie, G.W. 1984. Morphology, chemical contents and possible function of the tibial gland of males of the Costa Rican solitary bees *Centris nitida* and *Centris trigonoides subtarsata* (Hymenoptera: Anthophoridae). *Journal of the Kansas Entomological Society*, **57**: 50–54.

Williams, J.L. 1978. Insects: Lepidoptera (moths), pp. 105–127. *In*: Morse R.A. (ed.) *Honey Bee Pests, Predators, and Diseases*. 430pp. Cornell Univesity Press, Ithaca.

Williams, J.R. 1960. The control of black sage (*Cordia macrostachya*) in Mauritius: the introduction, biology and bionomics of a species of *Eurytoma* (Hymenoptera, Chalcidoidea). *Bulletin of Entomological Research*, **51**: 123–133.

Williams, N.H. 1982. The biology of orchids and euglossine bees, pp. 119–171. *In*: Arditti, J. (ed.), *Orchid Biology Reviews and Perspectives, II*. 390pp. Cornell University Press, Ithaca.

Williams, N.H. & Dressler, R.L. 1976. Euglossine pollination of *Spathiphyllum* (Araceae). *Selbyana*, **1**: 349–356.

Williams, N.H. & Whitten, W.M. 1983. Orchid floral fragrances and male euglossine bees: methods and advances in the last sesquidecade. *Biological Bulletin*, **164**: 355–395.

Williams, P.H. 1985. A preliminary cladistic investigation of relationships among the bumble bees (Hymenoptera, Apidae). *Systematic Entomology*, **10**: 239–255.

Williams, P.H. 1991. The bumble bees of the Kashmir Himalaya (Hymenoptera: Apidae, Bombini). *Bulletin of the British Museum (Natural History)* Entomology, **60**: 1–204.

Williams, R.N. 1980 (1978). Insect natural enemies of fire ants in South America with several new records. *Proceedings of the Tall Timbers Conference on Ecological Animal Control by Habitat Management*, **7**: 123–134.

Williams, R.N. & Floyd, E.H. 1971. Effect of two parasites, *Anisopteromalus calandrae* and *Choetospila elegans* upon populations of the maize weevil under laboratory and natural conditions. *Journal of Economic Entomology*, **64**: 1407–1408.

Williams, T. 1991. Host selection and sex ratio in a heteronomous hyperparasitoid. *Ecological Entomology*, **16**: 377–386.

Willink, A. 1982. Revision de los generos *Montezumia* Saussure y *Monobia* Saussure (Hymenoptera: Eumenidae). *Bolétin de la Academia Nacional de Ciencias*, **55**: 1–321.

Willink, E. & Moore, D. 1988. Aspects of the biology of *Rastrococcus invadens* Williams (Hemiptera: Pseudococcidae), a pest of fruit crops in West Africa, and one of its primary parasitoids, *Gyranusoidea tebygi* Noyes (Hymenoptera: Encyrtidae). *Bulletin of Entomological Research*, **78**: 709–715.

Willis, E.O. & Oniki, Y. 1978. Birds and army ants. *Annual Review of Ecology and Systematics*, **9**: 243–263.

Wilson, D.D., Ridgway, R.L. & Vinson, S.B. 1974. Host acceptance and oviposition behavior of the parasitoid *Campoletis sonorensis* (Hymenoptera: Ichneumonidae). *Annals of the Entomological Society of America*, **67**: 271–274.

Wilson, E.O. 1950. Notes on the food habits of *Strumigenys louisianae* Roger (Hymenoptera: Formicidae). *Bulletin of the Brooklyn Entomological Society*, **45**: 85–86.

Wilson, E.O. 1953. The ecology of some North American dacetine ants. *Annals of the Entomological Society of America*, **46**: 479–495.

Wilson, E.O. 1955. Ecology and behavior of the ant *Belonopelta deletrix* Mann (Hymenoptera: Formicidae). *Psyche*, **62**: 82–87.

Wilson, E.O. 1965. Trail sharing in ants. *Psyche*, **72**: 2–7.

Wilson, E.O. 1971. *The Insect Societies*. 548pp. Belknap Press, Harvard University Press, Cambridge, Mass.

Wilson, E.O. 1976. Behavioral discretization and the number of castes in an ant species. *Behavioral Ecology and Sociobiology*, **1**: 141–154.

Wilson, E.O. 1979. The evolution of caste systems in social insects. *Proceedings of the American Philosophical Society*, **123**: 204–210.

Wilson, E.O. 1980*a*. Caste and division of labor in leaf-cutter ants (Hymenoptera: Formicidae: *Atta*) I. The overall pattern in *A.sexdens*. *Behavioral Ecology and Sociobiology*, **7**: 143–156.

Wilson, E.O. 1980*b*. Caste and division of labor in leaf-cutter ants (Hymenoptera: Formicidae: *Atta*) II. The ergonomic optimization of leaf cutting. *Behavioral Ecology and Sociobiology*, **7**: 157–165.

Wilson, E.O. 1981. Communal silk-spinning by larvae of *Dendromyrmex* tree-ants (Hymenoptera: Formicidae). *Insectes Sociaux*, **28**: 182–190.

Wilson, E.O. 1983. Caste and division of labor in leaf-cutter ants (Hymenopterra: Formicidae: *Atta*) IV. Colony ontogeny of *A. cephalotes*. *Behavioral Ecology and Sociobiology*, **14**: 55–60.

Wilson, E.O. 1985. The priniciples of caste evolution, pp. 307–324. *In*: Hölldobler, B. & Lindauer, M. (eds) *Experimental Behavioral Ecology and Sociobiology in Memoriam Karl von Frisch 1886–1982*. (*Fortschritte der Zoologie*, Band 31) 488pp. Gustav Fischer Verlag, Stuttgart.

Wilson, E.O. 1986*a*. Caste and division of labor in *Erebomyrma*, a genus of dimorphic ants (Hymenoptera: Formicidae: Myrmicinae). *Insectes Sociaux*, **33**: 59–69.

Wilson, E.O. 1986*b*. The organization of flood evacuation in the ant genus *Pheidole* (Hymenoptera: Formicidae). *Insectes Sociaux*, **33**: 458–469.

Wilson, E.O. 1987. The arboreal ant fauna of Peruvian Amazon forests: a first assessment. *Biotropica*, **19**: 245–251.

Wilson, E.O. (ed.) 1988. *Biodiversity*. 521pp. National Academy Press, Washington D.C.

Wilson, E.O., Carpenter, F.M. & Brown, W.L. Jr 1967*a*. The first Mesozoic ants. *Science*, **157**: 1038–1040.

Wilson, E.O., Carpenter, F.M. & Brown, W.L. 1967*b*. The first Mesozoic ants, with the description of a new subfamily. *Psyche*, **74**: 1–19.

Wilson, E.O. & Hölldobler, B. 1986. Ecology and behavior of the neotropical cryptobiotic ant *Basiceros manni* (Hymenoptera: Formicidae: Basicerotini). *Insectes Sociaux*, **33**: 70–84.

Wilson, F. 1961. Adult reproductive behaviour in *Asolcus basalis* (Hymenoptera: Scelionidae). *Australian Journal of Zoology*, **9**: 739–751.

Wilson, F. 1962. Sex determination and gynandromorph production in aberrant and normal strains of *Ooencyrtus submetallicus* (Hymenoptera: Encyrtidae). *Australian Journal of Zoology*, **10**: 349–359.

Wilson, F. & Wearne, G.R. 1962. The introduction into Australia of parasites of *Listroderes obliquus* Klug. *Australian Journal of Agricultural Research*, **13**: 249–257.

Wilson, F. & Woolcock, L.T. 1960*a*. Temperature determination of sex in a parthenogenetic parasite, *Ooencyrtus submetallicus* (Howard) (Hymenoptera: Encyrtidae). *Australian Journal of Zoology*, **8**: 153–169.

Wilson, F. & Woolcock, L.T. 1960*b*. Environmental determination of sex in a parthenogenetic parasite. *Nature*, **186**: 99–100.

Wilson, T.H. & Cooley, T.A. 1972. A chalcidoid planidium and an entomophilic nematode associated with the western flower thrips. *Annals of the Entomological Society of America*, **65**: 414–418.

Windsor, D.M. 1972. Nesting association between two neotropical polybiine wasps (Hymenoptera, Vespidae). *Biotropica*, **4**: 1–3.

Windsor, D.M. 1976. Birds as predators on the brood of *Polybia* wasps (Hymenoptera: Vespidae: Polistinae) in a Costa Rican deciduous forest. *Biotropica*, **8**: 111–116.

Windsor, D.M., Morrison, D.W., Estribi, M.A. & de Leon, B. 1989. Phenology of fruit and leaf production by 'strangler' figs on Barro Colorado Island, Panamá. *Experientia*, **45**: 647–653.

Winston, M.L. 1979. The proboscis of the long-tongued bees: A comparative study. *University of Kansas Science Bulletin*, **51**: 631–667.

Winston, M.L. 1987. *The Biology of the Honey Bee*. 281pp. Harvard University Press, Cambridge, Mass.

Winston, M.L. 1992*a*. The biology and management of africanized honey bees. *Annual Review of Entomology*, **37**: 173–193.

Winston, M.L. 1992*b*. *Killer Bees The Africanized Honey Bee in the Americas*. 162pp. Harvard University Press, Cambridge, Mass.

Winston, M.L. & Michener, C.D. 1977. Dual origin of highly social behavior among bees. *Proceedings of the National Academy of Science of the United States of America*, **74**: 1135–1137.

Wishart, G. & Monteith, E. 1954. *Trybliographa rapae* (Westw.) (Hymenoptera: Cynipidae), a parasite of *Hylemya* spp. (Diptera: Anthomyiidae). *Canadian Entomologist*, **86**: 145–154.

Withycombe, C.L. 1924. Note on the economic value of the Neuroptera, with special reference to the Coniopterygidae. *Annals of Applied Biology*, **11**: 112–125.

Wolcott, G.N. 1941. The establishment in Puerto Rico of *Larra americana* Saussure. *Journal of Economic Entomology*, **34**: 53–56.

Wolda, H. & Roubik, D.W. 1986. Nocturnal bee abundance and seasonal bee activity in a Panamanian forest. *Ecology*, **67**: 426–433.

Wojcik, D.P. 1989 (1988). Behavioral interactions between ants and their parasites. *Florida Entomologist*, **72**: 43–51.

Wong, H.R. 1963. The external morphology of the adults and ultimate larval instar of the larch sawfly, *Pristiphora erichsonii* (Htg.) (Hymenoptera: Tenthredinidae). *Canadian Entomologist*, **95**: 897–921.

Woodruff, L.C. 1929. *Eupelmus popa* Girault, a parasite of the Sorghum midge, *Contarinia sorghicola* Coquillett. *Journal of Economic Entomology*, **22**: 160–167.

Woolley, J.B. 1988. Phylogeny and classification of the Signiphoridae (Hymenoptera: Chalcidoidea). *Systematic Entomology*, **13**: 465–501.

Woolley, J.B. & Schauff, M.E. 1987. A new species of *Paracrias* (Hymenoptera: Eulophidae) parasitic on *Anthonomus* spp. (Coleoptera: Curculionidae). *Proceedings of the Entomological Society of Washington*, **89**: 770–775.

Woolley, J.B. & Vet, L.E.M. 1981. Postovipositional web-spinning behavior in a hyperparasite, *Signiphora coquilletti* Ashmead (Hymenoptera: Signiphoridae). *Netherlands Journal of Zoology*, **31**: 627–633.

Wootton, R.J. 1978. Function, homology and terminology in insect wings. *Systematic Entomology*, **4**: 81–93.

Wylie, H.G. 1965. Discrimination between parasitized and unparasitized house fly pupae by females of *Nasonia vitripennis* (Walk.) (Hymenoptera: Pteromalidae). *Canadian Entomologist*, **97**: 279–286.

Wylie, H.G. 1966. Some effects of female parasite size on reproduction of *Nasonia vitripennis* (Walk.) (Hymenoptera: Pteromalidae). *Canadian Entomologist*, **98**: 196–198.

Wylie, H.G. 1970. Oviposition restraint of *Nasonia vitripennis* (Hymenoptera: Pteromalidae) on hosts parasitized by other hymenopterous species. *Canadian Entomologist*, **102**: 886–894.

Wylie, H.G. 1971*a*. Observations on intraspecific larval competition in three hymenopterous parasites of fly puparia. *Canadian Entomologist*, **103**: 137–142.

Wylie, H.G. 1971*b*. Oviposition restraint of *Muscidifurax zaraptor* (Hymenoptera: Pteromalidae) on parasitized housefly pupae. *Canadian Entomologist*, **103**: 1537–1544.

Wylie, H.G. 1972. Larval competition among three hymenopterous parasite species on multiparasitized housefly (Diptera) pupae. *Canadian Entomologist*, **104**: 1181–1190.

Wylie, H.G. 1979 Sex ratio variability of *Muscidifurax zaraptor* (Hymenoptera: Pteromalidae). *Canadian Entomologist*, **111**: 105–109.

Wylie, H.G. 1980. Factors affecting facultative diapause of *Microctonus vittatae* (Hymenoptera: Braconidae). *Canadian Entomologist*, **112**: 747–749.

Yamada, Y. 1955. Studies on the natural enemy of the woollen pest, *Anthrenus verbaci* Linné (*Allepyris microneurus* Kieffer) (Hymenoptera, Bethylidae). *Mushi*, **28**: 13–29.

Yamada, Y. 1987*a*. Characteristics of the oviposition of a parasitoid, *Chrysis shanghaiensis*. *Applied Entomology and Zoology*, **22**: 456–464.

Yamada, Y. 1987*b*. Factors determining the rate of parasitism by a parasitoid with a low fecundity, *Chrysis shanghaiensis* (Hymenoptera: Chrysididae). *Journal of Animal Ecology*, **56**: 1029–1042.

Yamane, S. 1973. Descriptions of the second to final instar larvae of *Bareogonalos jezoensis* with some notes on its biology (Hymenoptera, Trigonalidae). *Kontyû*, **41**: 194–202.

Yamane, S. & Kojima, J. 1982. *Pseudonomadina* new genus, with description of a new species from the Philippines (Hymenoptera, Trigonalidae). *Kontyû*, **50**: 183–188.

Yamane, S. & Terayama, M. 1983. Description of a new subspecies of *Bakeronymus typicus* Rohwer parasitic on the social wasp *Parapolybia varia* Fabricius in Taiwan (Hymenptera: Trigonalidae). *Memoirs of the Kagoshima University Research Center for the South Pacific*, **3**(2): 169–173.

Yanega, D. 1994. Arboreal, ant-mimicking mutillid wasps, *Pappognatha*; parasites of Neotropical *Euglossa* (Hymenoptera: Mutillidae and Apidae). *Biotropica*, **26**: 465–468.

Yasnosh, V.A. 1976. Classification of the parasitic Hymenoptera of the family Aphelinidae (Chalcidoidea). *Entomologicheskoe Obozrenie*, **55**: 159–168. [In Russian with English summary.][Translated in: *Entomological Review*, **55**: 114–120.]

Yasnosh, V.A. 1980. Peculiarities of morpho-biological evolution and classification of parasitic Hymenoptera of the family Aphelinidae. *International Congress of Entomology Abstracts*, **16**: 29.

Yasnosh, V.A. 1983. Peculiarities in the geographical distribution of Aphelinidae (Hymenoptera) in the European part of the USSR and the Caucasus, pp. 259–266. *In*: Eliava, I. Ya. (ed.) *The Fauna and Ecology of the Invertebrates of Georgia*, 268pp. Metsniereba, Tbilisi. [In Russian.]

Yasumatsu, K. 1955. Taxonomic notes on three wireworm parasites of the genus *Pristocera* from the Far East (Hymenoptera: Bethyloidea). *Journal of the Faculty of Agriculture, Kyushu University*, **10**: 233–249.

Yasumatsu, K. 1958. A new addition to the genera of the Sclerogibbidae (Hymenoptera). *Kontyû*, **26**: 20–24.

Yeargan, K.V. & Braman, S.K. 1986. Life history of the parasite *Diolcogaster facetosa* (Weed) (Hymenoptera: Braconidae) and its behavioral adaptation to the defensive response of a lepidopteran host. *Annals of the Entomological Society of America*, **79**: 1029–1033.

Yeargan, K.V. & Braman, S.K. 1989. Life history of the hyperparasitoid *Mesochorus discitergus* (Hymenoptera: Ichneumonidae) and tactics used to overcome the defensive behavior of the green cloverworm (Lepidoptera: Noctuidae). *Annals of the Entomological Society of America*, **82**: 393–398.

Yokoyama, A. & Tsuneyoshi, M. 1958. Discovery of a hymenopterous ectoparasite of *Oligotoma japonica* Okajima (Embioptera). *Kontyû*, **26**: 25–28.

Yoshida, S. 1977. Pseudo-mated female in Pteromalidae (Hymenoptera). *Japanese Journal of Applied Entomology and Zoology*, **21**: 173–174.

Yoshikawa, K. 1962. Introductory studies on the life economy of polistine wasps. VII. Comparative consideration and phylogeny. *Journal of Biology, Osaka City University*, **13**: 45–64.

Yoshimoto, C.M. 1954. A study of the biology of *Priocnemis minorata* Banks (Hymenoptera, Pompilidae). *Bulletin of the Brooklyn Entomological Society*, **49**: 130–138.

Yoshimoto, C.M. 1970. A new subfamily of Cynipoidea (Hymenoptera) from Nepal. *Canadian Entomologist*, **102**: 1583–1585.

Yoshimoto, C.M. 1971. Revision of the genus *Euderus* of America north of Mexico (Hymenoptera: Eulophidae). *Canadian Entomologist*, **103**: 541–578.

Yoshimoto, C.M. 1972. Notes on two species of Brachyscelidiphaginae (Pteromalidae: Chalcidoidea, Hymenoptera). *Canadian Entomologist*, **104**: 969–976.

Yoshimoto, C.M. 1976*a*. *Pseudoxenufens forsythi* a new genus and species of Trichogrammatidae (Hymenoptera: Chalcidoidea) from western Ecuador. *Canadian Entomologist*, **108**: 419–422.

Yoshimoto, C.M. 1976*b*. *Playaspalangia* a new genus of Spalagiinae [sic] (Hymenoptera, Chalcidoidea: Pteromalidae) from Mexico. *Canadian Entomologist*, **108**: 475–478.

Yoshimoto, C.M. 1977. Revision of the *Diparinae* (Pteromalidae: Chalcidoidea) from America north of Mexico. *Canadian Entomologist*, **109**: 1035–1056.

Yoshimoto, C.M. 1978. Two new species of *Epiclerus* from the New World (Hymenoptera: Chalcidoidea, Tetracampidae). *Canadian Entomologist*, **110**: 1207–1211.

Yoshimoto, C.M. 1979. A new species of *Megastigmus* from Mexico (Chalcidoidea: Torymidae, Megastigminae). *Canadian Entomologist*, **111**: 201–206.

Yoshimoto, C.M. 1983. Review of North American *Pnigalio* Schrank (Hymenoptera: Eulophidae). *Canadian Entomologist*, **115**: 971–1000.

Yoshimoto, C.M. 1990. A review of the genera of the New World Mymaridae (Hymenoptera: Chalcidoidea). *Flora and Fauna Handbook*, **7**: 1–166.

Young, A.M. 1985. Notes on the nest structure and emergence of *Euglossa turbinifex* Dressler (Hymenoptera: Apidae: Bombinae: Euglossini) in Costa Rica. *Journal of the Kansas Entomological Society*, **58**: 538–543.

Young, A.M. & Hermann, H.R. 1980. Notes on foraging of the giant tropical ant *Paraponera clavata* (Hymenoptera: Formicidae: Ponerinae). *Journal of the Kansas Entomological Society*, **53**: 35–55.

Young, D.K. 1990. Distribution of *Pelecinus polyturator* in Wisconsin (Hymenoptera: Pelecinidae), with speculations regarding geographical parthenogenesis. *The Great Lakes Entomologist*, **23**: 1–4.

Yousuf, M. & Shafee, S.A. 1986*a*. Catalogue of genus-group names of world Trichogrammatidae (Hymenoptera). *Indian Journal of Systematic Entomology*, **3**: 13–27.

Yousuf, M. & Shafee, S.A. 1986*b*. Checklist of species and bibliography of the world Trichogrammatidae (Hymenoptera). *Indian Journal of Systematic Entomology*, **3**: 29–82.

Yousuf, M. & Shafee, S.A. 1987 (1988). Taxonomy of Indian Trichogrammatidae (Hymenoptera: Chalcidoidea). *Indian Journal of Systematic Entomology*, **4**: 55–200.

Yuasa, H. 1922 (1923). A classification of the larvae of the Tenthredinoidea. *Illinois Biological Monographs*, **7**(4): 1–172.

Zaborski, E., Teal, P.E.A. & Laing, J.E. 1987. Kairomone-mediated host finding by spruce budworm egg parasite, *Trichogramma minutum*. *Journal of Chemical Ecology*, **13**: 113–122.

Zeledón, R. 1957. Sobre la biología del *Telenomus fariai* Lima, 1927 (Hymenoptera: Scelionidae), parásito endófago de huevos de algunos Triatominae. *Revista de Biología Tropical*, **5**: 1–17.

Zenner-Polania, I. 1990. Biological aspects of the 'Hormiga Loca,' *Paratrechina (Nylanderia) fulva* (Mayr), in Colombia, pp. 290–297. In: Vander Meer, R.K., Jaffe, K. & Cedeno, A. (eds). *Applied Myrmecology, A World Perspective*. 741pp. Westview Press, Boulder.

Zerova, M.D. 1980. *Chryseida* Spinola, a new species of the genus (Hymenoptera, Eurytomidae) from Central America. *Entomologicheskoe Obozrênie*, **59**: 176–180. [In Russian.]

Zerova, M.D. 1989. A review of the chalcid-flies the parasites of *Diplolepis mayri* in the USSR with description of a new species of the family Eurytomidae (Hymenoptera, Chalcidoidea). *Trudy Zoologicheskogo Instituta. Akademiya Nauk SSSR*, **191**: 50–55. [In Russian.]

Zerova, M.D. & Fursov, V.N. 1991. The palaearctic species of *Eurytoma* (Hymenoptera: Eurytomidae) developing in stone fruits (Rosaceae: Prunoideae). *Bulletin of Entomological Research*, **81**: 209–219.

Zhang, D., Dahlman, D.L. & Gelman, D.B. 1992. Juvenile hormone esterase activity and ecdysteroid titer in *Heliothis virescens* larvae injected with *Microplitis croceipes* teratocytes. *Archives of Insect Biochemistry and Physiology*, **20**: 231–242.

Zimmerman, J.K. & Madriñán R., S. 1988. Age structure of male *Euglossa imperialis* (Hymenoptera: Apidae: Euglossini) at nectar and chemical sources in Panama. *Journal of Tropical Ecology*, **4**: 303–306.

Zinna, G. 1960*a*. Richerche sugli insetti entomofagi I. Specializzazione entomoparassitica negli Encyrtidae: Studio morfologico, etologico e fisiologico del *Leptomastix dactylopii* Howard (con note del Dr D.C. Lloyd, Commonwealth Institute of Biological

Control, Fontana, California, U.S.A.). *Bollettino del Laboratorio di Entomologia Agraria 'Filippo Silvestri'*, **18**: 1–148.

Zinna, G. 1960*b*. Esperimenti di lotta biologica contro il cotonello degli agrumi (*Pseudococcus citri* (Rossi)) nel l'Isola di Procida mediante l'impiego di due parassiti esotici, *Pauridia peregrina* Timb. e *Leptomastix dactylopii* How. (Con note del Dr D.C. Lloyd, Commonwealth Institute of Biological Control, Fontana, California, U.S.A.). *Bollettino del Laboratorio di Entomologia Agraria 'Filippo Silvestri'*, **18**: 257–284.

Zinna, G. 1961. Richerche sugli insetti entomofagi II. Specializzazione entomoparassitica negli Aphelinidae: studio morfologico, etologico e fisiologico del *Coccophagus bivittatus* Compere, nuovo parassita del *Coccus hesperidum* L. per l'Italia. *Bollettino del Laboratorio di Entomologia Agraria 'Fillippo Silvestri'*, **19**: 301–358.

Zinnert, K.-D. 1969. Vergleichende Untersuchungen zur Morphologie und Biologie der Larvenparasiten (Hymenoptera: Ichneumonidae und Braconidae) mitteleuropäischer Blattwespen aus der Subfamilie Nematinae (Hymenoptera: Tenthredinidae). *Zeitschrift für Angewandte Entomologie*, **64**: 180–217, 277–306.

Zorin, P.V. 1930. On the biology of *Microgaster marginatus* Nees (Hymenoptera). *Russkoe Entomologicheskoe Obozrênie*, **24**: 220–224. [In Russian.]

Zucchi, R., Yamane, S. & Sakagami, S.F. 1976. Preliminary notes on the habits of *Trimeria howardi*, a Neotropical communal masarid wasp, with description of the mature larva (Hymenoptera: Vespoidea). Studies on the vespoid larvae. II. *Insecta Matsumurana* (N.S.), **8**: 47–57.

Zwakhals, C.J. 1989. De Nederlandse Acaenitinae en de gastheer van *Acaenitus dubitator* (Hymenoptera: Ichneumonidae). *Entomologische Berichten*, **49**: 21–25.

Taxonomic index

The following list contains only genera and suprageneric classificatory categories. Where reference has been made in the text to "pimplines" or pompilids or similar informal categories, these have been indexed under the formal name (i.e. as Pimplinae or Pompilidae). Throughout indexing has been done to the most precise taxonomic level given, thus references to *Ablerus* (Aphelinidae) are indexed under the generic name and not also indexed under Aphelinidae. Principal references are given in **bold**; extinct taxa are denoted by ✠.

Subject index

The manufacturer's authorised representative in the EU for product safety is Oxford University Press España S.A. of El Parque Empresarial San Fernando de Henares, Avenida de Castilla, 2 - 28830 Madrid (www.oup.es/en or product.safety@oup.com). OUP España S.A. also acts as importer into Spain of products made by the manufacturer.
Printed and bound by CPI Group (UK) Ltd, Croydon, CR0 4YY
14/09/2026
02216301-0001